LOW TEMPERATURE PHYSICS

To learn more about the AIP Conference Proceedings, including the Conference Proceedings Series, please visit the webpage
http://proceedings.aip.org/proceedings

LOW TEMPERATURE PHYSICS

24th International Conference on Low Temperature Physics

LT24

Orlando, Florida 10 – 17 August 2005

PART A

EDITORS
Y. Takano
S. P. Hershfield
S. O. Hill
P. J. Hirschfeld
University of Florida, Gainesville
A. M. Goldman
University of Minnesota, Minneapolis

All papers have been peer-reviewed

SPONSORING ORGANIZATIONS
University of Florida, Department of Physics
University of Florida, College of Liberal Arts and Sciences
University of Florida, Division of Sponsored Research
National Science Foundation
23rd International Conference on Low Temperature Physics
International Union of Pure and Applied Physics
The Abdus Salam International Centre for Theoretical Physics
National High Magnetic Field Laboratory

Melville, New York, 2006
AIP CONFERENCE PROCEEDINGS ■ VOLUME 850

PHYS

Editors:

Yasu Takano
Selman P. Hershfield
Stephen O. Hill
Peter J. Hirschfeld

Department of Physics
University of Florida
P.O. Box 118440
Gainesville, FL 32611-8440
USA

E-mail: takano@phys.ufl.edu
selman@phys.ufl.edu
hill@phys.ufl.edu
pjh@phys.ufl.edu

Allen M. Goldman
School of Physics and Astronomy
University of Minnesota
P.O. Box 790
Minneapolis, MN 55455
USA

E-mail: goldman@physics.umn.edu

L.C. Catalog Card No. 2006906631
ISBN 0-7354-0347-3
ISSN 0094-243X

CD-ROM available: ISBN 0-7354-0348-1

Printed in the United States of America

CONTENTS

LONDON PRIZE AND SIMON PRIZE LECTURES

London Prize Lectures

Simon Prize Lecture

QUANTUM GASES, FLUIDS, AND SOLIDS

Quantum Gases

Liquid ^{3}He

Liquid ^{4}He

Turbulence and Vortices

Helium in Aerogel

Low-Dimensional and Confined Helium

Solid ^{3}He and Two-Dimensional ^{3}He

Solid ^{4}He

Hydrogen

Other Quantum Liquids and Solids

SUPERCONDUCTIVITY

Cuprates and Ruthenates—Experiment

Cuprates and Ruthenates—Theory

MgB_2

Organics

Other Compounds

Magnetic Superconductors

Fulde-Ferrell-Larkin-Ovchinnikov States

Mesoscopic Superconductivity

Vortices and Field Distribution in Superconductors

Ferromagnet-Superconductor Structures

Intrinsic Josephson Junctions

Flux Qubits

Superconductor-Insulator Transitions

Superconductivity—General

MAGNETISM AND MISCELLANEOUS PROPERTIES OF SOLIDS

Low-Dimensional Magnetism—Experiment

Low-Dimensional Magnetism—Theory

Frustration: Spin and Structural Glasses

Molecular Magnets and Nanomagnetism

Heavy Fermions

Manganites, Ruthenates, Cobaltates, and Vanadates

Magnetism—General

Nonmagnetic Solids

Phonons and Thermal Transport

CONDUCTING ELECTRONS IN CONDENSED MATTER

Quantum Hall Effect and 2D Electron Gases

Kondo Effect in Quantum Dots

Aharonov-Bohm and Mesoscopic Effects

Coulomb Blockade and Single Electron Tunneling

Metal-Insulator Transition

Electrons on Helium

Hopping and Granular Materials

Nanotubes

Spin and Magnetic Transport

Optical and Time Dependent Response

Other Transport Phenomena

MATERIALS, TECHNIQUES, AND APPLICATIONS

Refrigeration and Thermometry

Cryogenic Sensors and Detectors

Cryogenic Devices and Instrumentation

Techniques and Properties of Materials

Preface to the Proceedings of the

LT-24
THE 24th INTERNATIONAL CONFERENCE ON LOW TEMPERATURE PHYSICS

August 10-17, 2005, Orlando, Florida, USA

The LT series has been held every two or three years since 1949 (Cambridge, Mass, USA), LT23 being in Hiroshima, Japan. It is run under the auspices of the Commission on Low Temperature Physics, C5, of the International Union of Pure and Applied Physics. This meeting is the most important meeting of the low temperature research community, which embraces much of the largest section of physics, condensed matter research. Because the meeting is held only every third year, important, new discoveries are always presented. It also brings the entire community together for important informal discussions. Several satellite conferences of from 100 to 300 participants are also associated with the LT meetings, helping to focus the various specialties and economizing on time and travel funds. The meeting is also important in establishing contacts between researchers in scientifically advanced nations and developing nations.

International meetings are becoming more important in the current technologically advanced era, paradoxical as this might seem. In fact, full use of technology was made in the operation of this conference and the preparation of the proceedings, from on-line registration to manuscript refereeing. But this technology does not replace face-to-face interaction, where serious discussions among scientists resolve the important questions of the day. And particularly, such meetings are the incubating oven for young scientists, who, for the first time, encounter the full scientific international community. Most meetings, by necessity, are international meetings. More than ever, science is a diverse, global endeavor, with the responsibility to provide the data and theories with which the world is run.

The proceedings of LT24 represent a comprehensive presentation of the current state of low temperature research and the directions in which it will likely go in the next three years. They have been refereed, edited, and prepared for on-line publication for full, rapid dissemination to the entire world. Hard copies have been prepared for distribution to libraries with limited inter-net resources.

The papers are organized following the pattern of their presentation at the conference, with additional subsections to promote comprehension of the scope of the works presented. The main headings are:

1) Quantum Gases, Fluids and Solids
2) Superconductivity
3) Magnetism and Miscellaneous Properties of Solids (structural, thermal)
4) Conducting Electrons in Condensed Matter (localization, electron transport, mesoscopics)
5) Materials, Techniques and Applications (refrigeration, detectors, superconducting devices, etc.).

Organizing the conference and preparing these proceedings was challenging and not without excitement. It could not have been accomplished without the skilled assistance of the various committees listed in these proceedings, and our sincere thanks are extended to all of them. We also thank the many sources of funding that made the meeting possible. In the end, for us it was a labor of love that raised our spirits and increased our appreciation of our fellow scientists in this endeavor all over the world. We look forward to the next iteration at LT25 in Amsterdam and wish the organizers well.

Gary Ihas
Mark Meisel
Yasu Takano

Gainesville, June 28, 2006

LT-24 Organizing Committee
G. G. Ihas, University of Florida (chair)
M. W. Meisel, University of Florida (secretary)
Y. Takano, University of Florida (publication)

Program Committee
A. M. Goldman, University of Minnesota (chair)
J. E. Crow, National High Magnetic Field Laboratory (chair posthumous)
J. Beamish, University of Alberta (Section 1 director)
E. Cornell, University of Colorado
T.-L. Ho, Ohio State University
I. A. Fomin, Kapitza Institute
J. E. Thomas, Duke University
M. R. Beasley, Stanford University (Section 2 director)
P. B. Littlewood, University of Cambridge
F. Steglich, Max Planck Institute, Dresden
M. Randeria, Ohio State University
I. K. Schuller, University of California, San Diego (Section 3 director)
C. Pepin, CEA-Saclay
S. Sachdev, Yale University
H. Tanaka, Tokyo Institute of Technology
D. C. Ralph, Cornell University (Section 4 director)
R. H. McKenzie, University of Queensland
V. J. Goldman, Stony Brook University
Y. Iye, University of Tokyo
G. M. Seidel, Brown University (Section 5 director)
G. Frossati, Leiden Univesity
J. R. Owers-Bradley, University of Nottingham
M. Paalanen, Helsinki University of Technology

Publication Committee
Y. Takano, University of Florida (chair)
A. M. Goldman, University of Minnesota
S. P. Hershfield, University of Florida
S. O. Hill, University of Florida
P. J. Hirschfeld, University of Florida
C. J. Stanton, University of Florida

Financial Support Committee
A. Feher, P.J. Safarik University (chair)
A. Lacerda, NHMFL, Los Alamos National Laboratory
H.-J. Lee, Pohang Institute of Science and Technology
J. J. Niemela, The Abdus Salam International Centre for Theoretical Physics
M. W. Meisel, University of Florida

LONDON PRIZE AND SIMON PRIZE LECTURES

Berkeley Experiments on Superfluid Macroscopic Quantum Effects

Richard Packard

Physics Department, University of California, Berkeley, CA 94720, USA

Abstract. This paper provides a brief history of the evolution of the Berkeley experiments on macroscopic quantum effects in superfluid helium. The narrative follows the evolution of the experiments proceeding from the detection of single vortex lines to vortex photography to quantized circulation in ^{3}He to Josephson effects and superfluid gyroscopes in both ^{4}He and ^{3}He.

Keywords: macroscopic quantum effects, superfluid helium
PACS: 67.40.-w, 67.57.z, 01.65.+g

INTRODUCTION

I am greatly honored to receive the 2005 London Memorial Prize. The body of work that is mentioned in my citation was done collaboratively with a group of graduate students and postdocs who have shared this research adventure with me since 1969. We all share the honor of this prize and thank the London Prize Committee for selecting our work. Although my many co-authors over the years have contributed greatly to our various projects, there are several people that I wish to single out.

Michael Sanders was my thesis advisor. He introduced me to the topic of superfluidity and by example taught me a great deal about how to select experimental problems. Gary Williams was my first graduate student. He taught me more than I taught him and has been a continuing source of knowledge on all things superfluid. I especially thank him for organizing my nomination for this prize. Greg Swift and Jim Eisenstein shared lab space together and were my first students to study superfluid ^{3}He. Jim showed me how little I knew about formal quantum theory and Greg taught me how far I needed to go to be a more creative inventor. Jukka Pekola encouraged me to investigate ^{3}He flow in single submicron apertures. His frequent statement of “Let's do it!” was inspiring. Stefano Vitale has been a close friend, sailing buddy, and final arbiter of everything related to SQUIDs, gravity, signals and noise. I wish he were a more frequent visitor to Berkeley. My longest collaboration has been with Seamus Davis, one of the co-recipients of this year's prize. First as a graduate student, then a postdoc and finally a faculty colleague, Seamus taught me a great deal about organization and how to keep my mouth shut. He is a fine addition to the Cornell faculty and a great loss to me and Berkeley. More recently, Kostya Penanen let me realize again how much more I have to learn about physics and how lacking the US undergraduate education is compared to the FSU. Finally I wish to mention my two present students, Emile Hoskinson and Yuki Sato. Emile reminds me often how careful analysis can tease information out of data. His careful analysis is an excellent example of experimental detective work. Yuki is still rather new in the group but he gives me great hope for the future of low temperature physics. As long as young scientists like Yuki and Emile continue to probe Nature below 2 K, I am confident that the intriguing discoveries that have persisted in low temperature physics will continue unabated.

Since all the work cited in the London Prize is well documented in the literature, I will focus here on the personal narration of how some of the work came to fruition. For those who have no interest in the human or historical side of physics I encourage them to turn to the next paper in these proceedings. For those unfamiliar with our work, please visit my group web site (http://www.physics.berkeley.edu/research/packard/) and click on the list of publications. Since this is not a research manuscript I make no attempt to cite the literature broadly. Those important references are all included in the Berkeley papers which are included as footnotes.

ENTERING LOW TEMPERATURE PHYSICS

Surprisingly this story begins with the construction of a sailboat, something that occupied much of my time while I was an undergraduate. Having this many-faceted project to occupy my time gave me a convenient excuse to be absent from many of my physics classes, resulting of course in rather mediocre grades. Consequently upon graduation my first graduate experience was not in a diversified research department. Leaving the boat behind

CP850, *Low Temperature Physics: 24th International Conference on Low Temperature Physics;*
edited by Y. Takano, S. P. Hershfield, S. O. Hill, P. J. Hirschfeld, and A. M. Goldman

and focusing on classes let me attain rather high grades in an physics masters degree program. I used that improved academic record to reapply to graduate schools and achieved considerably greater acceptance than in the first round. In 1966 I chose to attend the University of Michigan for two reasons. 1. My wife Roseanne could also get a graduate degree there in her specialty and 2. They had eliminated the requirement of a comprehensive exam for the beginning grad students, substituting instead the requirement of good grades in all graduate courses. That seemed to me a system more logical than the usual prelim exams given by many other schools. It seemed unwise to me then (and now) to ask physics students to perform under enormous pressure on an exam that might mean more than their previous four years of classroom examinations.

At Michigan, my ignorance of most research fields made all the research offerings seem uninteresting. I had the good fortune to hear Michael Sanders describe a recent visit to the Soviet Union to attend a low temperature physics conference. His description of that trip made me feel he would be a good choice as an advisor. However the graduate student grapevine said that his group was at that time so large that he was not taking on new students.

Believing this to be true, I first approached another professor to inquire about research. That professor asked me about the phase transition in my grades from mediocre to the highest in my previous graduate school. I gave the feeble (but true) excuse of my building a sailboat and missing many classes. He offered me a project that involved classified research, a sphere of endeavor I did not care to join considering the political climate of the Vietnam War.

Following the dictum, "If you don't fail once a day you aren't trying hard enough." I decided to go see Sanders and apply for a position. To my surprise Sander's first question was, "Aren't you the guy who built a boat?" When I answered yes he said, "Well if you can build a boat you can probably do the experiments I have in mind." At the time I didn't understand the basis for that judgment but upon graduation three years later I realized that my thesis was really just another boat project. All the skills it takes to do one (planning, design, materials acquisition, multitasking, fabrication, testing, and perseverance) are equally applicable to the other.

DETECTION OF SINGLE QUANTUM VORTICES

The problem Mike assigned (I note that it took three years to get up the nerve to call him Mike rather than Professor Sanders) was to use ion trapping to detect the creation of vortices in rotating superfluid helium. My first task was to build a rotating cryostat. This was enjoyable for me because it involved lots of machining and designing. I assumed that it all had to be built cheaply and looking back I don't think the entire investment was more than $500. I took special delight in showing the 2 m tall spinning apparatus to passing grad students. The rotating electrical contacts were a series of about 18 concentric troughs of mercury into which dipped small nickel blades. This worked quite well except that when the speed of the contacts grew too large the mercury was thrown out of the troughs onto the floor. I just scooped it up, cleaned off the dust, and poured it back in. Contrast that action to the reaction on a campus today of a small mercury spill!

Sander's suggestion was to detect the presence of vortices using the fact that electrons form bubble states (with diameter near 3 nm) which can become trapped on quantized vortices. This was well known from work in R.J. Donnelly's group in Chicago and Careri's group in Rome. Presumably the amount of charge trapped in a sample of rotating superfluid should be proportional to the number of vortices present. If we could measure the amount of charge trapped in the liquid we would have a measure of how many vortices were present. We hoped to see a staircase increase in the trapped charge as the number of vortices changed from 0 to 1 to 2, etc.

The primary technical challenge in the experiment was that one could only trap about 100 electrons on a centimeter length of vortex line. This was about one order of magnitude below the resolution of the best charge measuring amplifiers of the day. We hoped that the electrons could be brought out through the free surface of the rotating helium and then accelerated in the vapor to create charge multiplication due to collisions. However we knew from previous work that electron bubbles cannot be brought through the liquid surface in nonrotating helium (there is an effective work function). Would the presence of vortices change that?

To our delight and relief we found that the electrons emerged easily out the top of the vortices. We should have published that result but I was too focused on moving ahead with the experiment and didn't think anyone but we would care. However another group discovered the same phenomenon and had the good sense to publish a Physical Review Letter on the topic.

Up to that moment progress had been good but we then hit a barrier. Our plan was to accelerate the electrons emerging out the top of the vortices in order to achieve ion multiplication and collect the amplified charge on a metal plate. As I increased the accelerating voltage there was no noticeable gain until at some voltage the gas completely discharged in an electric breakdown. I assumed that the breakdown was caused by the collector electrode not being sufficiently smooth or not sufficiently planar with respect to the surface. I spent several months working on improving those electrodes but always found the same result: no amplification and then catastrophic breakdown.

One of the golden rules of science is "an afternoon in the library can save you six months in the laboratory". Unfortunately that rule was not included in the graduate course curriculum. Forced by desperation I finally did spend an afternoon in the library studying what was known about gaseous breakdown and the construction of gas proportional counters for nuclear physics. It finally dawned on me that the worst possible geometry for controlled amplification was the parallel plate system I was employing. What was needed was a small diameter wire collector that presented an electric field gradient. With parallel plate electrodes the electric field is uniform so that when there is sufficient acceleration to create amplification, the multiplication is enormous, on the order of 2^n where n is the number of mean free paths traversed. In contrast, a wire presents an electric field varying as r^{-1} so that the region of space active for multiplication is controlled and limited.

Armed with that late afternoon insight I went directly to the lab, made a wire collector, cooled down and, after working all night, by morning was able to show Sanders stable gain on the order of 100. I remember him saying, "Why didn't you think of that before?" to which I recall responding, "You're the advisor, why didn't you think of it?"

This was the final technical hurdle. Figure 1 shows some typical data[1] giving the charge trapped in the fluid as a function of rotation speed of the cryostat. The steps heralding the appearance of the individual vortices are clearly evident.

Although it was exciting to see this predicted phenomenon, one doesn't learn much new by simply proving a prediction. The more interesting aspects of this work came over the succeeding months as we learned a great deal about the nature of rotating superfluids[2]. The most significant qualitative thing we learned was that rotating superfluid helium is a very metastable system. The vortices often appear at angular velocities much greater than the equilibrium predictions (i.e. the speed where the free energy is minimized with the presence of the vortex) and once formed, tend to not leave. Even when the cryostat was stopped there remains remnant sections of vortex filament. This is presumably due to vortex pinning at microscopic pinnacles on the surrounding walls.

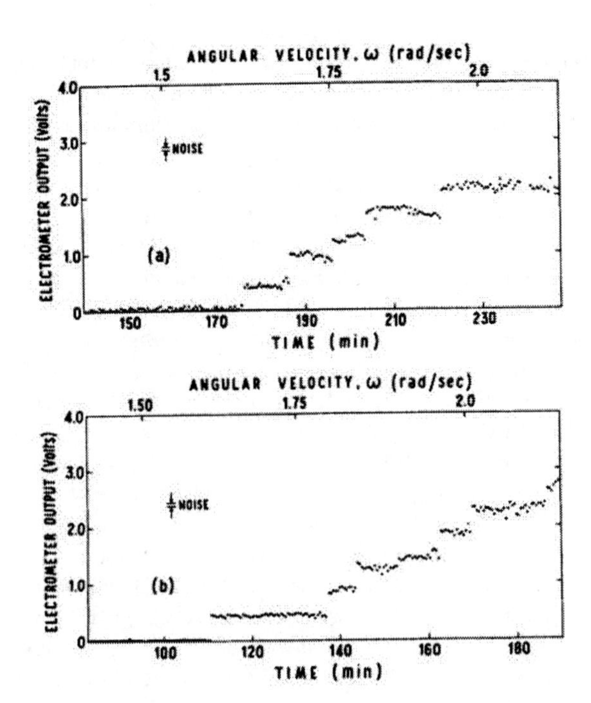

FIGURE 1. A plot of trapped charge as a function of time while the angular velocity is steadily increasing. The upward steps indicated the creation of individual vortices

PULSAR GLITCHES AND SUPERFLUID VORTICES

After graduation I moved to the University of California at Berkeley, first as a postdoctoral researcher working with Fred Reif and later as an assistant professor. Soon after arriving in Berkeley I overheard a lunch conversation describing the strange sudden acceleration events (called glitches) that were being observed in pulsars. I learned that the pulsars were believed to be rotating neutron stars with superfluid interiors. Based on my work with rotating helium it was clear to me that if a rotating container of superfluid was slowly decelerating, the stick-slip metastable relaxation to equilibrium of the vortices would cause abrupt torques on the container. If the container was freely suspended the walls would glitch upward in speed. It took only a few days to write up this idea to explain the pulsar glitching[3]. A few years later P. W. Anderson developed this idea into a more detailed theory of pinning which is now part of the standard paradigm in the pulsar community. I was delighted that the things we learned about vortices in helium could contribute to a seemingly unrelated field. Experiments that I suggested in that paper were later performed successfully in Tblissi, Georgia, where the superfluid glitches were clearly seen.

PHOTOGRAPHY OF VORTEX ARRAYS

I joined the Berkeley faculty in 1971. Although my postdoctoral work focused on studying ultra violet emission from liquid helium and liquid and solid neon, my appointment to the faculty was going to require something more substantial to be promoted to tenure. One path to follow was to extend the vortex charge-trapping experiments to see if I could accelerate the electrons emerging from the surface and image them onto a phosphor screen in order to map the vortex distribution. This had been a long term goal of my work with Mike Sanders although there was insufficient time to pursue it at Michigan. Many years later I heard that Prof. Careri in Rome had conceived of similar experiments calling it the "TV method".

This experiment faced several technical challenges. My first estimates of the image brightness were based on the assumption that the electrons would be accelerated through 10 kV before striking a phosphor. However early efforts showed that there was electric breakdown above the helium when only 500 volts were applied between electrodes. This implied that only a few photons would be available to "see" each vortex line. Furthermore, it was not known if any of the conventional phosphors would emit light at cryogenic temperatures after electron bombardment. In additon, to perform electron optics above the liquid helium surface one needs a long electron mean-free-path. Typical pumped-bath temperatures are not sufficiently low to achieve the correct limit. This technical challenge called for a ^{3}He refrigerator, and a rotating one to boot. Fortunately Mike Sanders had offered to send my old rotating Dewar set to Berkeley since it wasn't being used in Michigan. At that point I set out to make an adsorption pumped ^{3}He refrigerator along the lines described in a paper by J. Daunt. This was my first involvement with sub-Kelvin technology.

Early in this project I received the best gift any assistant professor could receive. Gary Williams, a new graduate student at Berkeley, walked into my lab and said he would like to work on superfluid helium. It became clear very quickly that Gary had already learned more about superfluids while he was an undergraduate than I knew at that time as a beginning academic. Since we were only a few years apart in age we made a good team. We were immediately equals in the project and set out on a challenging yet very exciting path. Both of us immensely enjoyed the challenge before us and our wives had a difficult time keeping us out of the lab.

Our first attempts to image the vortices were frustrating: the photographs showed only blurs of light. Gary developed a cold cathode emitter that produced a triangular pattern of emerging electrons to simulate the predicted vortex lattice. We learned from this "phantom" that the electron paths diverged due to space charge. We fixed this problem by surrounding the experiment with a home-made superconducting magnet which created electron focusing using the small cyclotron orbits characterizing the system. We again tried to image the vortices but still found only blurred patterns with no structure. However we now knew that the electron optics were not a limitation.

One feature that didn't make sense was the measured lifetime of the electrons on the vortices. Since the electrons escape by thermal activation, the lifetime, τ, should have been immeasurably long at 300 mK, our operating temperature. However we never found a τ greater than a few seconds. Gary suggested that perhaps we were not measuring the lifetime of electrons on the lines but rather the lifetime of the lines in the container. If the vortices were continually moving about they would eventually encounter a wall and be destroyed, or at least electrically discharged. At 0.3 K there is insufficient normal fluid to damp vortex motion so it seemed reasonable that the vortices might be randomly moving around searching for their equilibrium position. We decided to trick the vortices by introducing a bit of ^{3}He into the fluid. The ^{3}He would serve as a fixed amount of normal fluid even at the lowest temperatures. The vortice's motion would then be damped, thus driving them to their equilibrium positions where they are at rest in the rotating frame.

After adding the ^{3}He, the observed trapped lifetime immediately jumped to immeasurably high values thus suggesting our model was correct. Now we thought the vortices might be immobile enough to let their picture be taken. However, the addition of the ^{3}He raised the vapor pressure so that at 0.3 K sharply focused electron imaging was no longer possible above the liquid. Lower temperatures were required but that would require a dilution refrigerator, and a rotating one at that. While I was a graduate student I had observed one of the first dilution refrigerators being assembled. It was an enormous undertaking and I decided then that I would never be involved with something so complex. However since any temperature below 0.1 K would suffice to get the vapor pressure low enough, we didn't need a great dilution refrigerator

A short while earlier John Wheatley has started selling dilution refrigerators through his company SHE. Since there was no such thing as start-up funds in those days I had little money to spend on equipment. However, for about $1000 I could just afford to purchase the dilution stage of what SHE called their "minifridge", which used a simple countercurrent heat exchanger capable of cooling to ~ 40 mK.

With the tenure clock ticking we decided to undertake to build from scratch a rotating dilution refrigerator based on the SHE stage. Everything but the stage would be built by ourselves or by our department shops. Fortunately that was an era where we paid no shop charge, one of the main attractions of Berkeley. If I had to pay a substantial shop charge I never would have undertaken to build that machine. Gary and I put in a very intensive seven weeks, building as fast a possible. At the end of that period we had our rotating dilution refrigerator and even used it to perform a quick experiment to look for thinning of flowing superfluid films[4].

Soon the day came when everything was in place and we began the vortex photography run. We had run out of tricks to play and Gary was probably thinking that he had made a poor choice in research topic. Like almost all low temperature work, it wasn't until the middle of the night that we were ready for the ultimate test. We rotated the cryostat at a speed favorable for one vortex to exist in the little bucket. We charged up the lines with electrons for about 20 seconds, discharged the lines, and accelerated the charge toward a home-made phosphor screen deposited on a coherent fiberoptics light pipe that extended from 40 mK up to a home-made room temperature image-intensified camera. We then reached into the slowly rotating machine to remove the single piece of high speed film that might hold the vortex image, Gary's Ph.D. and my professional fate.

Going into the dark room together, Gary ran the film though the developer and fixer, turned on the lights and held the still dripping film up to a light. He looked at me and made the greatest understatement I have yet heard: "I think there is a spot in the center of the film."

Some of the early images we acquired were published in Phys. Rev. Letters[5]. I received tenure and Gary became a faculty member at UCLA. A few years later my postdoc Ed Yarmchuck made signal-averaged pictures[6] of the vortices which displayed the regular predicted patterns up to about N=10. Figure 2 shows some of the vortex states. Ed even made some movies showing how the vortices moved to adjust to a transition from N to N+1 vortices. This was an enormously fun experiment for me. We were all very young and did so much with our own hands. One of the nicest souvenirs of that time is a letter that Gary and I received from Richard Feynman. He told us how pleased he was "to see visions in my head at night coming out in black and white reality."

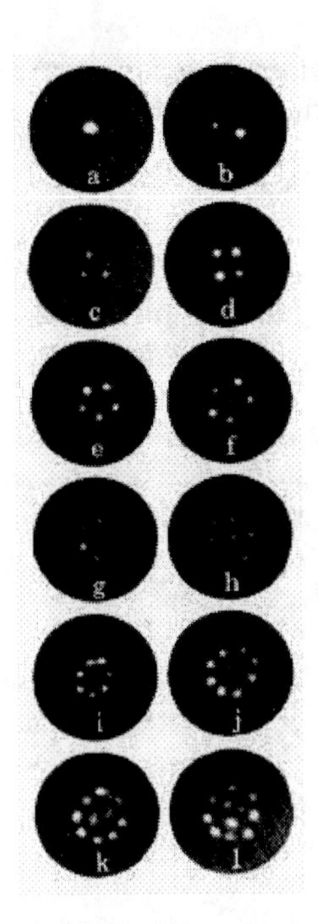

FIGURE 2. Photographs of the low lying stable vortex states in rotating HeII. The black outline is the boundary of a 2 mm diameter rotating cylinder. The angular velocity is varied from a 0.3 rad/s to 0.59 rad/s. Each image is a 60-fold multiple exposure taken at the same angular velocity. From Ref. [6].

ENTERING SUPERFLUID ^{3}HE PHYSICS

By the mid 1970's the discovery of new phases of liquid ^{3}He began to turn my attention away from vortices. Here was a new quantum liquid waiting to be explored. Unfortunately this state of matter only exists below about 2 mK and I had no experience working below 40 mK. The thought of building dilution refrigerators, etc. seemed overwhelming to me. However in the Summer of 1976 I attended a Gordon Conference in New Hampshire where all the talk was about ^{3}He. Two short conversations there had a big impact on my research direction.

The first occurred when I was wading in Lake Winnipesaukee with Matti Krusius. Matti was then a postdoc of John Wheatley. I must have mentioned to Matti my trepidation about trying to do millikelvin physics, especially concerning the small size of my group which was never more than two students. Matti made the optimistic statement: " The colder you go the easier it becomes." He meant that thermal conductivities go to zero with temperature so thermal isolation becomes ever easier. Fortunately he forgot to mention that the thermal boundary resistances go to infinity which, combined with vanishing heat capacities of ^{3}He, makes the very lowest temperatures still rather difficult to achieve.

The second conversation occurred the next evening. I was tagging along with some East Coast folks and we settled down in a pizza parlor. At that time there were three methods being used to cool into the ^{3}He superfluid phases. Wheatley at San Diego used CMN demagnetization. Cornell used Pomeranchuk cooling and Helsinki used copper nuclear adiabatic demagnetization. I asked John Reppy, "If you were starting from scratch to go into the ^{3}He business, which cooling technique would you choose?" After draining the last of his large glass of beer and wiping the suds from his ever- present beard, he replied "Absolutely nuclear cooling."

I returned to Berkeley with the sure knowledge that nuclear cooling was the way to go and that "things would get easier when we got colder". At that moment a document serendipitously passed my way. My chemistry colleague Norman Philips had just returned from a sabbatical leave in Helsinki and brought back to Berkeley a thesis by Robert Gylling entitled something like: "The design, construction and operation of a nuclear demagnetization cryostat." Armed with this instruction manual on how to be an ultra low temperature physicist I convinced my second graduate student Keith DeConde to help me build a cryostat capable of doing experiments on superfluid ^{3}He. It is significant to mention that Keith had already gathered data on an experiment on ^{4}He that might have served as a thesis. However, he had used a rather simple apparatus and we both agreed he should build something more substantial before graduation. I somehow neglected to tell him that the team in Helsinki consisted of about 6 experienced researchers and that these machines typically cost much more money than I possessed.

My thought was that I needed to buy only the dilution stage of an SHE fridge, a demagnetization magnet, and a dewar. All the rest I planned to beg, borrow, build or steal. Using again our zero charge machine shop we made some drawings and got to work. By the time that the dilution stage arrived in its little box we had the homemade cryostat built and had read enough papers to figure out how to do nuclear orientation (NO) thermometry. It was easy to borrow the nuclear counting electronics because there was a lot of this type of physics performed at the Lawrence Berkeley Laboratory. I was even able to borrow a radioactive Cobalt source to provide the thermometry signal.

Within about six months we had the DR working as determined from the NO measurements and then had to figure out how to do the nuclear demag part. The most difficult thing was measuring the temperature. It seemed that platinum NMR thermometry was the most promising technique but neither of us knew anything about NMR. Furthermore, although Berkeley had some of the world's greatest NMR experts (Hahn, Knight, Portis), they all knew too much to be of help to novices like us. Also, none of them had any experience doing NMR at the relatively low frequency of 200 kHz using op amps for the electronics. We thought we would just copy the things the Helsinki group had published and couldn't go too far wrong. They even specified the kind of platinum powder to use. It came from a company called Leico and we ordered a similar sample .

Our first attempts to determine temperature using NMR thermometry failed. In fact we couldn't even detect an NMR signal. Assuming that being novices this was par for the course, we carefully redesigned the electronics and tried again. Still nothing. Over and over this went, week after week until after a few months we lowered our sights and simply desired to see any kind of NMR signal. My colleague Walter Knight still had one viable NMR system and he helped us to see a signal with an old platinum sample from Erwin Hahn. We then tried our much better Leico powder and saw nothing! This just didn't make sense. We then sent our sample to an analysis laboratory which returned a report telling that the "very pure" Leico powder was highly contaminated with iron, a magnetic impurity that surely would wash out any NMR features. We had always wondered why the powder didn't look silvery the way we expected but was instead a deep black. We then called the manufacturer who said something to the effect, "Oh yes, we discovered we had sent a bad batch of platinum but we didn't want to upset you." These were really considerate people. Later I read a remark in a paper by George Pickett who wrote something like: "This powder looks black because it is black and any thoughts about the wavelength of light are completely irrelevant."

With hindsight the months spent chasing the elusive NMR signal were not a complete loss. By the time we finally acquired the NMR signal in powder given to us by Erwin Hahn, we had become quite expert in the NMR technology. Thermometry with this technique has never let us down yet. If the signal had appeared immediately we never would have gained the depth of understanding of the electronics or the physics which came in very useful in later experiments.

After Keith graduated two new students agreed to work with me: Jim Eisenstein and Greg Swift. They were an interesting combination. Jim was a committed Easterner and loved formal quantum theory. He considered the West Coast a primitive environment totally lacking class. Greg was thoroughly a Nebraskan and had won the state science fair by building a binary computer out of cloths pins mounted on rubber bands. Their skills and personalities complemented each other very well and we all had a good time. I enjoyed playing the Philistine for Jim's refined sensibility and I enjoyed observing the inventive genius of Greg who could find short cuts for some very difficult technical tasks. I still smile when I reread Greg's paper on how to repair a millikelvin cryostat in 24 hours turnaround.[7]

Greg and Jim's experiments were selected to be the simplest we could do to enter the field of ^{3}He superfluid physics. Greg would demonstrate a predicted anisotropy in the superfluid dielectric constant[8] and Jim would determine the depairing critical current for ^{3}He flowing in small tubes[9].

The effect that Greg sought turned out to be almost four orders of magnitude smaller than the original predictions but he eventually succeeded in seeing the effect at $\delta\varepsilon/\varepsilon \sim 10^{-10}$. I think that is still probably a record for dielectric measurements. The clever geometry of Greg's capacitor was something he learned in kindergarten in a paper-folding lesson. The capacitance bridge expertise that we gained from this experiment proved invaluable in future years.

Jim's experiment also worked very well and he was able to make several measurements that we had not planned originally. It was this experiment, measuring the flow through tubes of ~25 micron diameter that led eventually years later to our research on Josephson weak links. But that came long afterwards.

I had enormous good fortune in having as my first four students: Gary Williams, Keith DeConde, Greg Swift and Jim Eisenstein. Because of all that we accomplished I came out of that period feeling we could do anything! That was a too optimistic assessment but it served me well in the future to go in directions I would not have moved had we not had the early successes.

DETECTION OF ^{3}HE PERSISTENT CURRENTS

In 1984 I had the good fortune to spend six months in Olli Lounasmaa's laboratory in Helsinki. Olli had first spent several months in my lab helping to develop a new experiment to search for persistent currents in ^{3}He. I was focusing on this problem because Jim Eisenstein and I had difficulty in understanding the large flow dissipation in his ^{3}He depairing apparatus. We were led to speculate[10] that there might be an unsuspected intrinsic dissipation present that would prevent the ^{3}He from being truly "super". The way to rigorously test for superfluidity is through the presence of persistent currents. It turned out that our mystery dissipation was just our lack of understanding of the phenomenon called "second viscosity"[11]. Even with this explained I felt it was useful to observe persistence in flow. Our plan was to detect trapped angular momentum by detecting the reaction to imposed torques on a powder-filled torus that had been previously rotated. The apparatus was built in Berkeley and then I took a six month leave to work in Olli's lab where he had the ROTA 1 rotating cryostat.

While in Helsinki I worked closely with graduate student Jukka Pekola. We were very compatible and the experiment progressed rapidly. The apparatus showed clear evidence of persistent currents[12] and in addition confirmed a vortex phase transition[13] that had previously been observed through their NMR studies.

I took away two wonderful souvenirs from that visit. The first, my son Ben, was born in the Helsinki women's hospital. Since it never occurred to me to take a marriage license to Finland, Ben was issued a passport by the U.S. Consulate under the illegitimacy statute, which states that the mother is a U.S. citizen but the father is unknown. The second souvenir was Jukka Pekola who, after his graduation, came to Berkeley to work as a postdoc with me.

DETERMINATION OF QUANTIZED CIRCULATION AND COOPER PAIRING

After returning from Finland I decided to follow the example of the Helsinki laboratory and acquire two nuclear demagnetization cryostats. One cryostat could be used for developing new experiments while the other was used to take data on more mature projects. Working with Seamus Davis, Rena Zieve and John Close, we developed a rotating refrigerator capable of achieving temperatures in ^{3}He as low as 160 μK. Although my original motivation to have the new machine rotate was to perform NMR experiments on ^{3}He vortices, we began to formulate more ambitious plans when we saw that we were achieving really quite low temperatures.

For many years I had harbored the desire to determine if fluid circulation in superfluid ^{3}He was quantized in units of $h/2m_3$. Such an experiment would simultaneously demonstrate macroscopic quantum phase coherence (this is the origin of quantized circulation) and also provide the smoking gun for Cooper pairing by the presence of the factor 2 in the denominator. Everyone in this field accepted the paradigm that included both these points but paradigms need experimental confirmation.

The method to determine circulation in superfluids was invented by W. F. Vinen in about 1960. Using ^{4}He, he monitored the precession of the vibration plane of a vibrating wire immersed in rotating helium. The rate of precession of a round wire gives directly the fluid circulation around the wire[14]. For this measurement to be feasible requires the damping due to normal fluid to be sufficiently small for the wire's vibration to have a high Q. This criterion is easily achieved in ^{4}He where the fluid viscosity is small even above the transition temperature. By contrast, in ^{3}He the normal viscosity above T_c is similar to machine oil and a thin wire won't vibrate at all. However at temperatures below $0.2T_c$ the normal fluid density becomes exponentially small and the background damping drops dramatically. When we found that we could achieve 160 μK (~$0.2T_c$ for ^{3}He) we decided to try the Vinen technique in ^{3}He.

The experimental cell is really quite simple. A persistent magnet provides a field transverse to a vertical superconducting wire. Leads across the wire's end provide both for the input of a current pulse (which plucks the wire) and a route to convey the voltage induced by the vibration up to a room temperature preamplifier. When the wire vibrated perpendicular to the magnetic field there is maximum signal and when the wire vibrates parallel to the field there is no induced voltage. The cryostat is rotated about the vertical axis. One simply monitors the modulation envelope of the vibration and computes the circulation from a simple formula.

Figure 3 shows the measured circulation as a function

of the cryostat's rotation speed[15]. The quantization levels 1, 0, and –1 are very clear and the unit of quantization is $h/2m_3$, just as predicted from theory. So the accepted paradigm is correct but we were fortunate to discover a new phenomenon that was completely unexpected.

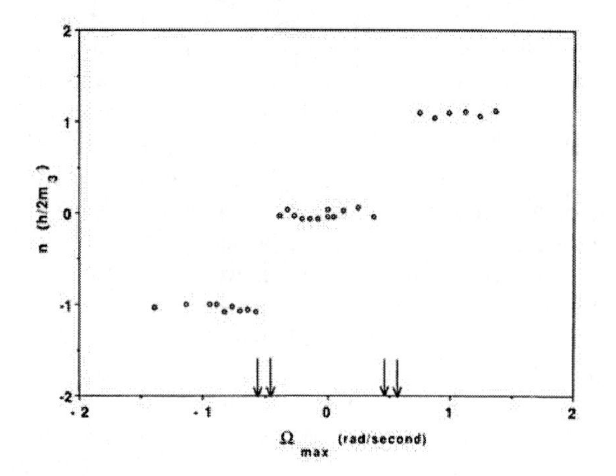

FIGURE 3. The value of circulation around a wire as a function of the maximum angular speed at which the cryostat had been rotated. The circulation level is unstable between the sets of arrows.

Figure 4 shows the circulation as a function of time after the cryostat has been brought to rest. The circulation remains trapped for many hours but then, as one end of the vortex filament becomes unpinned from the wire, the circulation slowly decreases as the filament becomes "unzipped" from the wire. The new feature is the slow oscillation in the circulation that is superimposed on the decreasing signal. Through a phenomenological force balance model we conjectured that the unpinned segment of vortex must exhibit a kind of "helicopter" motion in order for the vortex tension to be balanced by Magnus force. We derived an expression for the helicopter frequency that agreed within 1% of the observed value[16]. Later Klaus Schwartz observed this helicopter motion in numerical simulations[17] and GrishaVolovik[18] showed that the frequency was just the ac-Josephson frequency associated with cyclic 2π phase decrements which must accompany the hydrodynamic pressure head across the cylindrical container. Discovering and then explaining the helicopter motion was at least as exciting for us as observing the quantized circulation.

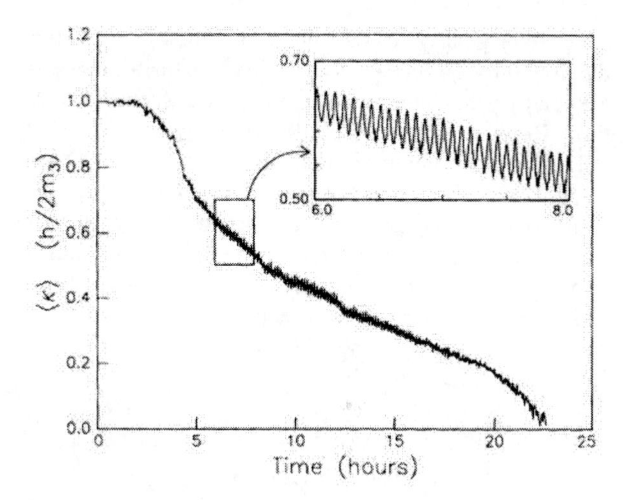

FIGURE 4. The decay of circulation trapped on a vibrating wire. The effective circulation varies periodically as the detached end of the vortex precesses around the wire at the Josephson frequency.

SMALL APERTURE EXPERIMENTS

Shortly after my stay in Helsinki, Jukka Pekola joined my group as a postdoctoral fellow. The project we pursued came out of discussions we began in Helsinki. I had recently finished studying ^{3}He flow in single long tubes of about 25 μm diameter. This dimension is much greater than the ^{3}He healing length, $\xi_3 \sim$60 nm. Jukka had previously pursued a search for Josephson signatures in Nucleopore filter, a material containing millions of long tubes of submicron dimensions. On the blackboard we had determined that it should be possible to study flow in a single submicron pore. Although the experiment looked difficult our desire to find a superfluid "weak link" was high.

In 1983, just before Jukka arrived, Seamus Davis joined my lab as a new graduate student. He and Jukka worked well together and soon they were making single submicron holes by bombarding materials with heavy ion fragments from a radioactive source. We were able to do this because Reimer Spohr was visiting the Berkeley group of Buford Price, the co-inventor of track etching technology. Reimer helped us to make a simple apparatus that permitted a single nuclear fragment to pass through our sample foil. Chemical etching later turned that track into our pore.

We used our sensitive capacitive techniques to study the deflection of a flexible diaphragm that formed one wall of a chamber connected to a larger bath through our small hole. Jukka and Seamus studied the flow of ^{3}He through the small pores under a variety of conditions. Unfortunately there was no flow signature that resembled anything that we anticipated as connected to weak link behavior. However, the flow signature of the pores also did not fit into our understanding of superfluidity. In particular there was no obvious critical velocity: dissipation was always present[19]!

While Jukka and Seamus continued the project I went off for a one semester stay in the Kyoto Laboratory of

Akira Hirai and Takao Mizusaki. While there I became even more interested in pursuing Josephson physics. My goal was to develop a superfluid analog of a SQUID, (superconducting quantum interference device). By hand written letter (this was pre-email days) I urged Seamus and Jukka to detour from their I-P measurements to seek interference effects. This suggestion of changing direction coming from so far away was not well received by my young colleagues. I have never been very successful in changing the course of a student researcher's direction, and in this case, as in several others, I just had to wait for new opportunities.

Returning to Berkeley my students Seamus, Ajay Amar, and I began a short experiment to investigate some interesting claims about the superflow properties of ^{3}He films. Our experiments observed superfluid film flow for sufficiently thick films and allowed us to map out the phase boundary (thickness vs. T_c) for ^{3}He films[20]. We found behavior consistent with prevailing theory and established some familiarity with film flow which would serve us well ten years later when we detected third sound in ^{3}He.

After the film work was concluded Ajay pursued experiments on ^{4}He flowing through small apertures. In 1985 Eric Varoquaux and Olivier Avenel detected single 2π phase slip events in superfluid ^{4}He. Using an ion-milled slit in submicron thick nickel foil, they presented clear evidence of the phase slippage process and showed how one could determine the nucleation processes for primordial vortices[21, 22]. Their apparatus used the geometry equivalent to an rf-SQUID. It seemed possible that the phase slip process in this geometry might be used to make a sensitive gyroscope. Such an instrument might be sufficiently sensitive to make contributions to geodesy or general relativity. Amar's thesis project was initially intended to demonstrate the superfluid ac-SQUID but ended up being a study of the phase slip process[23]. We had a lot to learn before we could make a rotation sensor using superfluids.

In 1992 I spent one semester in Trento, Italy as a guest of Stefano Vitale and Massimo Cerdonio. They were also pursuing development of a superfluid gyroscope using phase slippage. Their goal was to develop instruments to test general relativity. Stefano was already an expert on rf SQUIDs and understood all the subtleties contributing to the noise of these devices. For five months I shared Stefano's office and had an extraordinarily enjoyable visit. It turned out that Stefano shared my passions for sailing and physics. In fact at age 18 he had been European champion in a small dinghy class. During that sabbatical we wrote a paper describing the principles of superfluid gyroscopes, including devices that might exploit a dc-Josephson effect if that was ever achieved, as well as linear phase slip devices[24].

I returned to Berkeley fully enthused to pursue both ac and dc versions of superfluid gyros. In ^{4}He it became clear that we would have to understand the phase slip process better. The Saclay group were observing flow signatures that were elusive in Berkeley. This led us to abandon the double path geometry of the ac-SQUID in order to focus on the processes occurring solely in the small aperture. Fortunately this different geometry enabled us to study two regimes of flow: low frequency Helmholtz motion, similar to the Saclay research, in addition to dc pressure drives. Comparing the critical velocities for phase slippage in the two kinds of flow led us to test a theory that Stefano and I had developed[25]. The experiments were successful and led to discovering what seems to be a universal energy barrier for the creation of primordial vortices[26]. This provides important insight into the long standing problem of understanding how vortices enter superfluid ^{4}He.

With the knowledge and experience acquired from our phase slip experiments we returned with renewed enthusiasm to developing a superfluid analog of an ac-SQUID. Whereas quantum phase coherence throughout a sample lets the superconducting SQUID exhibit sensitivity to magnetic flux, the superfluid analog would be sensitive to rotation flux, $\Omega \cdot A$ where A is the area vector normal to the plane of the pick-up loop. Due to the dot product in the flux, the specific signature for the phase slip gyroscope would be a cosine modulation in the phase slip critical velocity as the plane of the device was reoriented with respect to the rotation axis of the Earth.

Keith Schwab and Jeffrey Steinhauer were the two students who, in 1995, felt the satisfaction of seeing the signal induced by the Earth's rotation[27]. Figure 5 shows the modulation in the phase slip critical velocity caused by reorientation of the ^{4}He phase slip gyroscope with respect ot he Earth's rotation axis. Avenel and Varoquaux had observed a similar phenomenon a few months earlier[28] but we were nevertheless thrilled to have achieved our first gyroscope in Berkeley. It had been almost 13 years since I first embarked on the path to see this phenomenon! It still seems miraculous to me that all the atoms in a toroidal sample of helium communicate in order to maintain the spatial phase coherence that underlies the superfluid gyroscope.

Much of our research overlapped the Saclay group and led to some hard feelings between the two laboratories. This was the only time in my career that we did not enjoy friendly relations with competing groups. It was a great pity and a loss for both groups. Professional scientists are too often more concerned about who is first rather than the more important questions about what is actually learned. When ego takes priority over knowledge, science is not well served.

My student Niels Bruckner later developed a multi-turn geometry for the phase slip gyroscope that increased its sensitivity by almost two orders of magnitude[29]. No

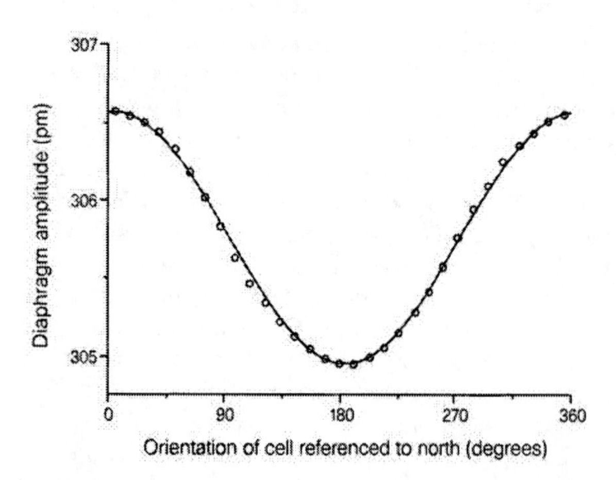

FIGURE 5. The modulation of phase slip apparent critical velocity as a function of the orientation of the superfluid gyroscope. The solid line is the predicted cosine pattern arising from the rotation flux.

new noise sources have appeared thus far so it may be possible to extend the device to useful sensitivities. Recently however our efforts in Berkeley have been focused on other kinds of rotation sensors which are analogs of dc-SQUIDs.

At the same time that some of my students were pursuing phase slip studies in ^{4}He, others were trying to observe physics related to a sine-like current-phase relation characterized by a Josephson weak link. The criterion to enter this regime is to have an aperture whose spatial dimensions are on the order of the superfluid healing length. Our program evolved from Eisenstein's 25 μm tube, progressed to Pekola and Davis's single submicron long tube, and eventually reached small single apertures machined in thin silicon nitride using e-beam technology. None of these apertures displayed flow characteristics that we could associate with Josephson-like behavior. This was a very discouraging time for this investigative avenue.

In the late 1980's the Saclay group reported flow experiments in ^{3}He using a slit aperture whose width was comparable to the superfluid healing length. The aperture was part of the inductive element in a hydro-mechanical oscillator. This so-called Helmholtz oscillator displayed nonlinear behavior which they modeled as consistent with a sine-like current phase relation, the signature of a Josephson weak link[30]. I found their findings sufficiently encouraging to continue our investigation of ^{3}He weak link physics.

The most striking signature displayed by a Josephson weak link is the quantum oscillation resulting from a chemical potential differential. For superfluid ^{3}He, a chemical potential difference is associated with an applied pressure head. Our experiments therefore often consisted of measuring the mass-current through an aperture resulting from a given pressure head. These current-pressure characteristics are the analog of I-V curves studied in superconducting systems. For several years we observed neither Josephson oscillations nor an I-P curve that resembled anything anticipated for a weak link. Our expectation was that the I-P curve should extrapolate to zero pressure at a finite critical current. In contrast all the I-P curves passed through the origin. An understanding of this mystery came almost ten years after we first observed the phenomenon[31].

As a final effort in the project I wanted to see if an array of many nanometer sized apertures might provide signatures that were more informative than our single aperture experiments. My hope was that an array might behave quantum coherently with all the apertures locked together. If there are N holes undergoing coherent Josephson oscillations, the mass current would be amplified by N and single aperture signals hidden by noise might be revealed. The basis of this hope was that gradients in quantum phase represent velocity fields. Therefore the lowest state of mechanical energy should have no phase gradient along a surface running parallel with the array. A crude estimate of the phase gradients induced by $k_B T$ fluctuations indicated that phase coherence should be quite strong at the 1 mK temperature regime characteristic of superfluid ^{3}He.

Our first attempt at an array experiment produced an I-P curve similar to those in one aperture. However Yuri Mukharsky, a postdoc in my lab, suggested that the array we were using contained apertures too closely spaced. He explained that close spacing would induce an effective linear kinetic inductance in series with the apertures that would hide the possible nonlinear Josephson effects we were seeking.

During these early years of aperture experiments the technology of e-beam nano-fabrication was not readily available to us. Our first array samples were made gratis by Perkin Elmer corporation. Later my students began to learn the technology in the steadily developing microlab in our electrical engineering department. It took many months for a student to learn how to manufacture the apertures and our existing e-beam writers would only hold one small sample at a time. We found it most efficient to send a trained student to the Cornell National Microfabrication Facility, to work with a technician there. In the Cornell facility one could write on an entire 4" wafer and generate dozens of aperture arrays with various hole sizes and grid dimensions. My student Alex Loshak brought back many samples of aperture arrays, some of which we are still using today, almost ten years later.

One challenge that has not been satisfactorily met is to accurately determine the aperture size before we do experiments. The problem is due to the difficulty of performing e-beam microscopy on insulating materials such as the silicon nitride "window" that contains our arrays. When the nitride becomes charged by the electron beam, the images get easily distorted. In addition, the act of viewing an aperture changes the microstructure of the device due to the e-beam bombardment. Some of the images are so deceiving that what appears as an aperture in a picture is in fact a closed indentation in the surface. At present it is sufficient to make optical microscope observations of the typically 50 nm apertures. Although this dimensions is only 1/10 the wavelength of the light, the observer can discern whether or not the holes are open.

The arrays that have proven most productive were those designed following Mukharsky's suggestion that the apertures be well spaced from each other. We used a 65x65 array of nominally 60 nm diameter holes arranged on a square lattice with 3 μm spacing. With the array immersed in superfluid ^{3}He below 1 mK, we applied a sudden pressure differential across it. We sensed the mass current through the array by the resultant deflection of one flexible wall of the chamber. This wall is typically a plastic membrane about 10 μm thick. The plastic is coated with a superconducting layer which, when it moves relative to an adjacent coil, changes that coil's inductance in a manner detectable using a superconducting SQUID. This type of displacement transducer was developed for gravity wave research and can detect deflection as small as 10^{-15} m.

In our first attempts with these arrays the transient signals were viewed in real time on an oscilloscope. The traces revealed no Josephson oscillation behavior. However when we put the electrical signal into an amplifier and listened to it on a set of head phones we could clearly distinguish a whistling tone that changed continuously in frequency. Sergei Pereversev, a postdoc in the group, was the first person to hear these Josephson oscillations. My student Scott Backhaus had been looking for Josephson sound in phase slip phenomena in ^{4}He. He agreed to join our ^{3}He effort and quickly developed a computer program to Fourier transform the transient signals and look for a quantum whistle signal that was correlated with the changing pressure head across the apertures. The result of that analysis was striking. The frequency of the oscillation increased linearly with the pressure head and the slope of that curve agreed within the systematic error of our pressure calibration with the Josephson frequency formula[32] $f_j = \Delta\mu/h$, where μ is the chemical potential and h is Planck's constant. For the first time we were convinced that we had Josephson weak links.

Our discovery of Josephson oscillations opened up a very rich field of physics to us. The Josephson frequency provided an absolute in-situ calibration of our membrane pressure gauge. This in turn led us to a method which permits determination of the current-phase relation for the weak link arrays. We discovered that although the arrays were characterized by the Josephson sine-like relation when the temperature dependent healing length was comparable to the aperture dimensions, at lower temperatures the $I(\phi)$ function morphed into an almost π periodic curve. This unexpected feature was later explained in microscopic theory by E. Thuneberg and his students[33]. The $I(\phi)$ curve also explained a mysterious π-state we discovered that had attracted considerable theoretical attention. We also found that the arrays had two stable states characterized by different $I(\phi)$ functions and different critical currents. We speculated that the origin of the bi-stability was due to two possible orientations of an internal vector field near the array surfaces. This was later shown to be the case by further calculations by Thuneberg's group[33].

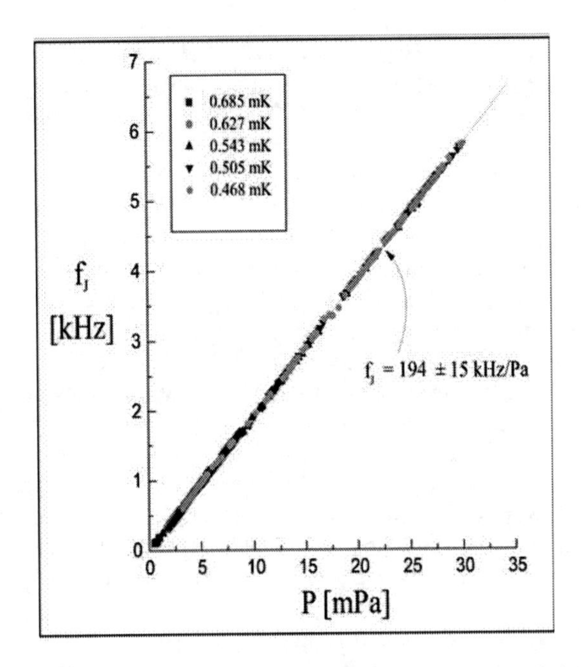

FIGURE 6. A plot of the frequency of the "quantum whistle" as a function of the pressure difference across a weak link array. The various symbols correspond to different temperatures. All the points lie on a straight line whose slope is given by the Josephson frequency formula.

We went on to observe several classic phenomena characteristic of superconducting Josephson weak links[34]. These included quantitative agreement with the Shapiro step theory, Fiske steps and plasma mode motion. In fact the superfluid array turned out to be a better example of these phenomena than the 1000's of papers on the superconducting analogs!

Much of the research on weak links had been driven by the goal of making sensitive rotation sensors, i.e. superfluid gyroscopes. Since most superfluid phenomena are of a rather esoteric nature, it would be satisfying to see some superfluid phenomenon make a contribution to applied technology. As described earlier we had already made a ^{4}He gyroscope using quantized phase slippage but we felt that a true dc-SQUID device would offer potentially lower noise. A superfluid-SQUID is a direct analog of the superconducting dc SQUID wherein rotation plays the role of magnetism.

Graduate student Ray Simmonds had the primary goal for his thesis to make the first proof-of-principle demonstration of such a device. After several discussions related to the geometry of this experiment, Ray settled on a design that minimized the superfluid complexity. He included a pair of similar weak link arrays in a flow circuit containing an enclosed area of sufficient size so that the Earth's rotation would provide slightly more than one quantum of rotation flux. The idea was to reorient the "pick-up loop" with respect to the Earth's rotation axis. Our rotating cryostat, originally used to observe quantized circulation was a perfect reorientable platform for this project.

Early trials showed the unmistakable interference signature we sought but building vibrations were limiting our signal to noise. Our cryostat already contained extensive vibration isolation and our laboratory was in a second basement which had the lowest vibration noise in our department. Nevertheless, the sensitive superfluid device was limited in its rotation resolution by motion in the local environment. Almost all vibration in buildings is generated by the occupants and the machinery that supports their occupation of the building. It was clear that the way to get the quietest conditions was to shut down all of the buildings mechanical infrastructure and evacuate all the occupants except for ourselves. Although this is a simple technical solution it is politically very difficult. After extensive negotiations we were able to have the building "shut down" over one Christmas and one New Year weekend. The building on these days was remarkably quiet. We were able to get good quality data which we published in Nature showing a respectable proof-of-principle demonstration of the superfluid ^{3}He dc-SQUID gyroscope[35]. Figure 7 shows the interference pattern induced by reorientation of the gyroscope loop with respect to the Earth's rotation axis.

This proof-of-principle experiment was in some sense the goal of much of our research for over a decade. Getting the results, which agreed so well with theory, was very satisfying. On the other hand it raised the obvious questions of how sensitive a gyroscope could be made and could such a device find utility in geodesy, seismology or navigation. On paper the sensitivity of a superfluid SQUID looks very good and might be greater than competing technologies including laser gyroscopes and atomic beam devices. Superfluid gyros, atomic beam gyros and optical gyros all utilize the same principle: a rotationally induced phase shift in a wave function that senses the absolute inertial frame. The relevant phase

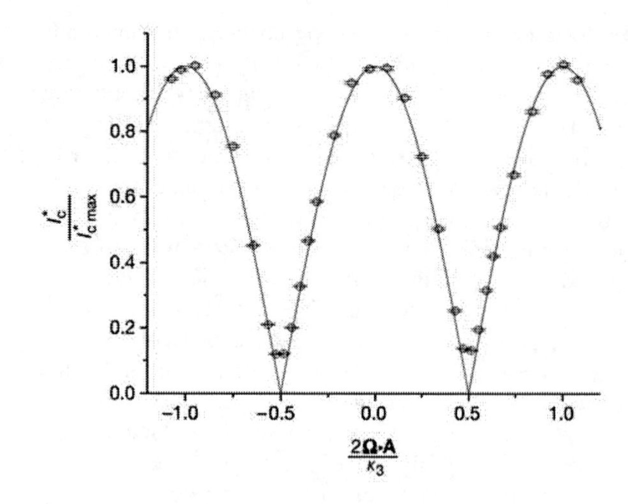

FIGURE 7. The critical current of a superfluid ^{3}He SQUID gyroscope. Two spatially separated weak link arrays behave as a single weak link whose critical current is modulated by rotation flux. The solid line is the theoretical prediction.

shift can be written

$$\delta\phi = \frac{2\pi\Omega \cdot A}{h/m}, \tag{1}$$

where Ω is the angular velocity of the interference loop, A is the area vector, and m is the mass of the relevant particle: the atomic mass or, in the case of photons, $h\nu/c^2$. Since this photon "mass" is at least 10^{10} times smaller than atomic masses, the superfluid devices and atomic devices offer great potential. On the other hand, there has been fifty years of progress in measuring small laser shifts. Atomic beam devices may be limited by shot noise in the particle number involved and also are problematic thus far regarding creation of a substantial enclosed area. Helium gyroscopes have large phase shifts relative to photons, very large particle numbers and simple methods to enclose a large area. However they inherently involve cryogenic technology which is not user friendly, particularly to the targeted user communities.

A SURPRISE IN ^{4}HE

Following our proof-of-principle ^{3}He gyroscope I hoped to extend the technology to determine its practical limitations as a rotation sensor. Postdoc Tom Haard and graduate student Emile Hoskinson joined the project and were assigned the task of developing a more sensitive ^{3}He gyro. By the time Tom left, a cell had been designed which was intended to make measurements of ^{3}He Josephson phenomena away from zero pressure which was the only regime we had explored previously. Testing ^{3}He experiments is quite expensive and time consuming since submillikelvin cryostats are inherently large, complex and consume large amounts of liquid helium and researcher time. It is therefore common for us to pretest experimental cells in superfluid ^{4}He in simple pumped-bath cryostats. Much can be learned from such tests. Emile's cell used our standard 65x65 array of 50 nm apertures. Although his test goal was to determine if the displacement transducer was operational and if the apertures behaved as expected, he made a discovery which has temporarily shifted our research direction and raised scientific questions that are presently the focus of our research program. One paragraph is needed to set the stage for Emile's discovery.

I mentioned earlier that Scott Backhaus had been looking for "Josephson phase slip sound", a phenomenon I conjectured while on sabbatical leave working with Stefano Vitale in Trento. The idea was an extension of the concept of quantized phase slippage. As P.W. Anderson envisioned the dissipation process in ^{4}He, a pressure gradient will accelerate superfluid through an aperture until at some critical velocity, v_c, a quantized vortex is stochastically nucleated near some asperity. The vortex grows in size at the expense of the flow energy in the aperture and eventually passes across the aperture removing energy from the flow, causing the phase difference across the aperture to change by 2π. If the pressure gradient is maintained, these 2π phase slips will repeat at the Josephson frequency, $f_j = \Delta\mu/h$, where $\Delta\mu$ is the chemical potential difference. I assumed that such a repetitive process would lead to the creation of an acoustic wave at the same frequency and that this sound could be detected by coupling the aperture to a resonant organ pipe tuned to the Josephson frequency. However Scott demonstrated that due to the stochastic nature of the phase slip nucleation process, there is no detectable signal at f_j unless $\Delta v_c/\delta v_{\text{slip}} < 1$, where Δv_c is the statistical width of the stochastic distribution in critical velocity and δv_{slip} is the velocity drop in the aperture due to a 2π phase slip. In all of Scott's experiments this ratio was greater than unity and the sound is similar to a shot noise source rather than a periodic excitation[36]. Thus as Emile began his test experiments using ^{4}He to determine the functionality of his cell (for future ^{3}He experiments) there was no reason to think that an array of apertures would exhibit phase slip sound.

One evening in March 2004 Emile was testing his cell by applying small differential pressure steps to an aperture array. The experimental cell was immersed in a simple dewar with no sophisticated temperature regulation. When the vacuum pump was turned off the temperature of the bath and apparatus slowly drifted toward T_λ. It had become practice in my laboratory to wear head phones connected to the output of our cell's displacement gauge. This permits us to gain insight into the various environmental acoustic noise sources that might degrade our experiments. As the temperature drifted upward, just below T_λ Emile heard a faint and very brief "chirp" sound in the head phones. Could the ^{4}He in the aperture array be exhibiting some form of Josephson oscillations several millikelvin below T_λ? Our experiments for the past year have now focused on answering this and related questions.

Emile added a temperature regulation system that stabilized the temperature of the entire helium bath to ± 50 nK. The Josephson frequency is given by

$$f_j = \frac{\Delta\mu}{h} = \frac{m_4}{h}\left(\frac{\Delta P}{\rho} - s\Delta T\right). \qquad (2)$$

In his first experiments Emile did not have a method to determine ΔT during a transient so we could not test this complete formula. However, in the first instant after applying a pressure step, there is no temperature differential and $\Delta\mu$ is given solely by the pressure term which can be measured from the deflection of a diaphragm. A plot of frequency vs. that initial pressure confirmed the Josephson frequency formula for this limited case[37]. Soon after, an internal heater was added to the cell that permitted thermal measurements, giving calibration constants needed to determine the time evolution of ΔT during the flow transient. With this in hand we were able to confirm the Josephson frequency formula containing both pressure and temperature terms[38].

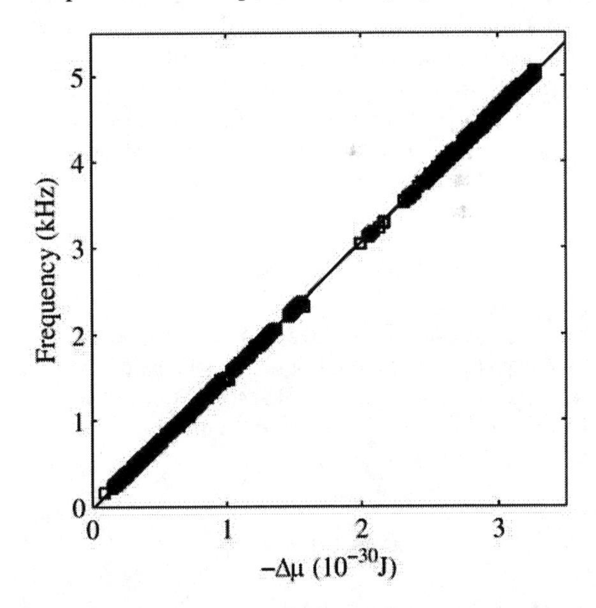

FIGURE 8. A plot of the ^{4}He quantum whistle as a function of chemical potential difference. The straight line slope is given by h^{-1} to within the experimental accuracy.

A subsequent experiment revealed the current-phase relation as the aperture array evolves from a sine-like $I(\phi)$ when ξ_4 is comparable to the aperture dimensions to a linear form when, at lower temperatures, ξ_4 is much smaller than the apertures. We have also recently investigated how the 4225 apertures can oscillate collectively while exhibiting dissipative phase slippage at the lower temperatures. At present this is the immediate focus of our research. In the near future we hope to develop a ^{4}He SQUID gyroscope. Since this device might operate near 2 K rather than the submillikelvins of the ^{3}He device, it may be possible to use a cryocooler to maintain the required temperatures. A cryo-cooler based superfluid gyroscope would not require that the user have specialized cryogenic knowledge. Perhaps we are approaching the time of creating a practical device using the macroscopic quantum properties of the superfluid state.

CONCLUSION

This paper is a personal sketch of the historical development of the Berkeley experiments on macroscopic quantum phenomena. I have focused on those experiments where the results explicitly involve Plank's constant in the macroscopic world. Since this is not a research article I have not referenced much of the work of others in this field although their contributions have been enormous and their work is referenced in the Berkeley research papers.

I hope the reader will feel some of the excitement of the discovery process that we have been privileged to experience. The superfluid states of helium have proven to be outstanding model systems to clarify principles of correlated states of matter. They can be considered the hydrogen atom of condensed matter systems because the properties displayed are often in perfect agreement with theory.

ACKNOWLEDGMENTS

The research described here has been supported continuously by the National Science Foundation. My coworkers and I are grateful for that support. In recent years we have also received support from the NASA program in Fundamental Physics.

REFERENCES

1. T. M. Sanders and R. Packard. Detection of single quantized vortex lines in rotating He II. Phys. Rev. Lett. **22**, 823 (1969).
2. R. Packard and T. M. Sanders. Observations of single vortex lines in rotating superfluid helium. Phys. Rev. A **6**, 799 (1972).
3. R. Packard. Pulsar speedups related to metastability of the superfluid neutron-star core. Phys. Rev. Lett. **28**, 1080 (1972).
4. G. A. Williams and R. Packard. Thickness of the moving superfluid film at temperatures below 1 K. Phys. Rev. Lett. **32**, 587 (1974).
5. G. A. Williams and R. Packard. Photographs of quantized vortex lines in rotating He II. Phys. Rev. Lett. **33**, 280 (1974).
6. E. J. Yarmchuk, M. J. V. Gordon and R. Packard. Observation of stationary vortex arrays in rotating superfluid helium. Phys. Rev. Lett. **43**, 214 (1979).
7. G. W. Swift and R. E. Packard. Rapid shutdown and restart of a millikelvin cryostat. Cryogenics, April, p. 241 (1981).
8. G. W. Swift, J. P. Eisenstein and R. E. Packard. A measurement of anisotropy in the dielectric constant of ^{3}He-A. Physica **107B**, 283 (1981).
9. J. Eisenstein, G. W. Swift and R. Packard. Observation of a critical current in ^{3}He-B Phys. Rev. Lett., **43**, 1676 (1979).
10. J. P. Eisenstein and R. E. Packard. Observation of flow dissipation in ^{3}He-B. Phys. Rev. Lett., **49**, 564 (1982).
11. H. Brand and M. Cross, Explanation of Flow dissipation in ^{3}He-B, Phys. Rev. Lett., **49**, 1959 (1982)
12. J. P. Pekola, J. T. Simola, K. K. Nummila, O. V. Lounasmaa, and R. E. Packard. Persistant current experiments on superfluid ^{3}He-Band ^{3}He-A. Phys. Rev. Lett. **53**, 70 (1984).
13. J. P. Pekola, J. T. Simola, P. J. Hakonen, M. Krusius, O. V. Lounasmaa, K. K. Nummila, G. Mamniashvili, G. E. Volovik and R. E. Packard. Phase diagram of the first-order vortex-core transition in superfluid ^{3}He-B. Phys. Rev. Lett. **53**, 584 (1984).
14. W. F. Vinen, Proc. Roy. Soc., A**260**, 218 (1961)
15. J.C. Davis, R.J. Zieve, J.D. Close and R.E. Packard, Observation of Quantized Circulation in Superfluid ^{3}He, Phys. Rev. Lett., **66**, 329 (1990)
16. R.J. Zieve, Yu. Mukharsky, J. D. Close, J. C. Davis and R. E. Packard, Precession of a Single Vortex Line in Superfluid ^{3}He, Phys. Rev. Lett., **68**, 1327 (1992)
17. K. W. Schwarz, Phys. Rev. B, **47**, 12030 (1993)
18. T. Sh. Misirpashaev and G. E. Volovik, JETP Lett. **56**, 41 (1992)
19. J. P. Pekola, J. C. Davis and R. E. Packard. Thermally activated dissipation in flowing superfluid ^{3}He. J. Low Temp. Phys. **71**, 141 (1988).
20. J.C. Davis, A. Amar, J.P. Pekola and R.E. Packard. On the superfluidity of ^{3}He films. Phys. Rev. Lett. **60**, 302 (1988).
21. O. Avenel, E. Varoquaux, Phys. Rev. Lett. **55**, 2704 (1985)
22. E. Varoquaux, O. Avenel, and M.W. Meisel, Can. Jour. Of Phys., **65**, 1377 (1987)
23. A. Amar, Y. Sasaki, J. C. Davis and R.E. Packard, Quantized Phase Slips in Superfluid ^{4}He, Phys. Rev. Lett. **68**, 2624, 1992
24. R. E. Packard and S. Vitale, Principles of Superfluid Gyroscopes, Phys. Rev. B, **46**, 3540 (1992)
25. R.E. Packard and S. Vitale, Some Phenomenological Theoretical Aspects of Superfluid Critical Velocities, Phys. Rev. B, **45** 2512 (1992)
26. J. Steinhauer, K. Schwab, Yu. Mukharsky, J.C. Davis and Richard Packard, Vortex Nucleation in Superfluid ^{4}He, Phys. Rev. Lett., **74**, 5056, (1995)
27. K. Schwab, N. Bruckner and R.E. Packard, Detection of the Earth's rotation using superfluid phase coherence,

Nature, **386**, 585 (1997)
28. O. Avenel and E. Varoquaux, Czech. J. Phys. (Suppl. S6) **48**, 3319 (1996)
29. N. Bruckner and R.E. Packard, Development of a multi-turn superfluid phase-slip gyroscope. J. of App. Physics **93**, 1798 (2003)
30. O. Avenel and E. Varoquaux, Josephson effect and quantum pahse slippage in superfluids, Phys. Rev. Lett., **60**, 416 (1988)
31. R. W. Simmonds, A Marchenkov, S. Vitale, J.C. Davis and R.E. Packard, New flow dissipation mechanisms in superfluid ^{3}He, Phys. Rev. Lett., **84**, 6062 (2000)
32. S. V. Pereversev, A Loshak, S. Backhaus, J. C. Davis and R. E. Packard, Quantum Oscillations in a superfluid ^{3}He-B weak link, Nature, **338**, 449-451, (1997)
33. E. Thuneberg, these proceedings
34. J.C. Davis and R.E. Packard, Superfluid Josephson weak links, Rev. of Mod. Physics, **74**, 741 (2002) This review covers all of the Berkeley ^{3}He weak link experiments up to the present.
35. R. W. Simmonds, A. Marchenkov, E. Hoskinson, J. C. Davis and R. E. Packard, Quantum interference in superfluid ^{3}He, Nature, **412**, 58 (2001)
36. S. Backhaus and R.E. Packard, Shot-noise acoustic radiation from a ^{4}He phase slip aperture, Phys. Rev. Lett., **81**, 1893 (1998).
37. E. Hoskinson, R. E. Packard, T. M. Haard, Quantum Whistling in Superfluid ^{4}He, Nature, **433**, 376 (2005)
38. E. Hoskinson and R. Packard, Thermally driven Josephson oscillations in superfluid ^{4}He, Phys. Rev. Lett., **94**, 155303 (2005)

Rotons, Superfluidity, and Helium Crystals

Sébastien Balibar

Laboratoire de Physique Statistique de l'Ecole Normale Supérieure,
associé au CNRS et aux Universités Paris 6 et 7,
24 Rue Lhomond, 75231 Paris Cedex 05, France

Abstract. Fritz London understood that quantum mechanics could show up at the macroscopic level, and, in 1938, he proposed that superfluidity was a consequence of Bose-Einstein condensation. However, Lev Landau never believed in London's ideas; instead, he introduced quasiparticles to explain the thermodynamics of superfluid ^{4}He and a possible mechanism for its critical velocity. One of these quasiparticles, a crucial one, was his famous "roton" which he considered as an elementary vortex. At the LT0 conference (Cambridge, 1946), London criticized Landau and his "theory based on the shaky grounds of imaginary rotons". Despite their rather strong disagreement, Landau was awarded the London prize in 1960, six years after London's death. Today, we know that London and Landau had both found part of the truth: BEC takes place in ^{4}He, and rotons exist.

In my early experiments on quantum evaporation, I found direct evidence for the existence of rotons and for evaporation processes in which they play the role of photons in the photoelectric effect. But rotons are now considered as particular phonons which are nearly soft, due to some local order in superfluid ^{4}He. Later we studied helium crystals which are model systems for the general study of crystal surfaces, but also exceptional systems with unique quantum properties. In our recent studies of nucleation, rotons show their importance again: by using acoustic techniques, we have extended the study of liquid ^{4}He up to very high pressures where the liquid state is metastable, and we wish to demonstrate that the vanishing of the roton gap may destroy superfluidity and trigger an instability towards the crystalline state.

Keywords: superfluid helium, solid helium, crystal surfaces, nucleation, London
PACS: 67.40.-w, 67.80.-s, 81.10.Aj, 43.35.+d

This London prize lecture is an opportunity to review some aspects of superfluid helium, a macroscopic quantum system which surprises us since London's time. Rotons will be my guide because I wish to recall the controversy which opposed Fritz London to Lev Landau (Section 1). After this historical introduction, we shall see that Landau's famous quasiparticles are more or less present in most of my own work: quantum evaporation (Section 2), the surface of helium crystals (Section 3), and the stability limits of liquid helium (Section 4).

1. FRITZ LONDON AND LEV LANDAU

Fritz London arrived in Paris in 1937. He was trying to escape from the antisemitic Germany and Paul Langevin had offered him a position at the Institut Henri Poincaré. Langevin was a professor at the Collège de France and an influential member of the "Front populaire", the coalition of political parties from the French left. In January 1938, London had already understood that the large molar volume of liquid helium was a consequence of the quantum kinetic energy of He atoms [1]. He then realized that the superfluid transition temperature, the "lambda" point T_λ = 2.17 K, was close to the temperature at which an ideal gas with the same density would undergo a Bose-Einstein condensation (BEC). He eventually noticed that the singularity in the specific heat at T_λ was similar to the one expected at a BEC transition, although not quite the same as could be expected since liquid helium is not an ideal gas. He thus proposed that superfluidity was a consequence of some kind of BEC [2].

One month later, Laszlo Tisza developed London's idea and introduced his "two fluid model" [3] to describe the non-classical properties of superfluid helium which had been discovered by J.F. Allen, in particular the fountain effect [4]. Laszlo Tisza had been in a Hungarian jail for 14 months after being accused of being a communist; after his liberation he went to Kharkov (1935-37) as a post-doc in Landau's group and finally arrived in Paris where he found help from the same group of left intellectuals around Langevin. He had a position at the Collège de France, 300 m from Fritz London. When London first heard about Tisza's two fluid model, he could not believe that, in a liquid which was pure and simple, there could be two independent velocity fields; this was indeed quite a revolutionary idea. But the joined works of London and Tisza explained most of the helium properties which were known in 1938.

At the same time in Moscow, Piotr Kapitza had invented the word "superfluidity" in analogy with superconductivity because he had the remarkable intuition that these two phenomena should have a common explanation [5]. As for Lev Landau, Stalin and Molotov

CP850, *Low Temperature Physics: 24th International Conference on Low Temperature Physics;*
edited by Y. Takano, S. P. Hershfield, S. O. Hill, P. J. Hirschfeld, and A. M. Goldman

had put him in jail because he was the co-author of an anti-Stalin leaflet [6]. After a heroic fight, Kapitza succeeded in liberating Landau [7]. In 1941, Landau published the famous article in which he invented the concept of quasiparticle in quantum fluids [8]. There had to be two kinds of quasiparticles to describe superfluid ^{4}He, phonons and some others which he called "rotons" because he thought that they were elementary vortices. Together with phonons, rotons formed the normal component in a two fluid model which was similar to the one first introduced by Tisza. However, Tisza had reasoned in the frame of gases and he thought that the normal fluid was made of the atoms left out of the condensate.

In his 1941 article, Landau was very critical about his former post-doc :
"the explanation advanced by Tisza not only has no foundations in his suggestions but is in direct contradiction with them" [8].
Furthermore, he never mentioned BEC and never referred to London. How could this be? For a long time, I thought that Landau had an objection similar to the one raised by London himself: the properties of an ideal gas could not apply to a liquid. We know how much work was necessary to extend BEC to interacting systems. More recently, a rather likely explanation came to me from a discussion with Lev Pitaevskii. The year 1941 was long before the BCS theory and the introduction of Cooper pairs. Since Kapitza and Landau thought that superfludity and superconductivity had the same origin, and since ^{4}He atoms were bosons while electrons were fermions, the quantum statistics could not be involved!

A few years later, when ^{3}He became available, it was crucial to see if liquid ^{3}He was superfluid at temperatures of order 1 K, also to verify if Landau's rotons had any reality. The first LT meeting took place at Cambridge (UK) in 1946, and the opening talk was given by Fritz London. He insisted on his explanation of superfluidity, also on the fact that Peshkov's early experiments on thermal waves (second sound) could not distinguish between Tisza's and Landau's predictions. He must have been also quite upset by Landau's attitude to comment on "*Landau's theory based on the shaky ground of rotons*" in the following way: *"The quantization of hydrodynamics [by Landau] is a very interesting attempt... however quite unconvincing as far as it is based on a representation of the states of the liquid by phonons and what he calls "rotons". There is unfortunately no indication that there exists anything like a "roton"; at least one searches in vain for a definition of this word... nor any reason given why one of these two fluids should have a zero entropy (inevitably taken by Landau from Tisza)"* [9]. Clearly, London and Landau had rather different approaches to superfluidity and had both easily recognized the weak points of their opponent's theory.

A few years later only, Osborne et al. [10] showed that ^{3}He was not superfluid down to 1 K and Peshkov found that the second sound velocity increased below 1 K, as predicted by Landau but not by Tisza [11]. Some more years later, the existence of rotons was demonstrated by neutron scattering experiments [12].

Both London and Landau probably died too early to admit that they both had part of the truth: BEC takes place in superfluid ^{4}He, also in ^{3}He and in superconductors thanks to the pairing of fermions. As for rotons, nobody doubts of their existence, it's only their physical nature which is still somewhat controversial, as we shall see below. One of my motivations in recalling this old time is the following message which I received from Laszlo Tisza himself, on the 17th of June 2005 :

Dear Sebastien,
I am delighted to read in Physics Today that you are to receive the Fritz London Prize. [...] This is wonderful! Please receive my warmest congratulations. Yesterday I was leafing through old correspondence and I found a letter in which I nominated Landau for the Prize. I am sure I was not alone. I was actually at LT-7 in Toronto when the Prize was announced. It is actually unconscionable of Landau not to have taken note of the remarkable Simon - London work on helium in Oxford 1934-5! I never heard a word of it while at UFTI. All he said was that London was not a good physicist. I am looking forward to your book to straighten out matters. With warmest regards, Laszlo

The content of this message would need a lot more comments, but, for the present lecture, let me only wish Laszlo Tisza to enjoy many more years of scientific activity after turning 100 in 2007.

2. QUANTUM EVAPORATION

When I started playing with heat pulses in superfluid ^{4}He, I was in fact trying to detect the emission of vortices by a flow through an orifice. It was shown later by the group of Varoquaux and Avenel [13], followed by the group of Packard and Davis [14] that much more sensitive techniques were needed for this. But I was surprised to observe that heat pulses could propagate through the liquid-gas interface during the filling of our cell. After observing that, at high temperature, heat propagated as a second sound wave in the liquid and could emit ordinary sound in the gas when hitting the liquid-gas interface, I realized that the low temperature regime was much more interesting. The vapor pressure vanishes exponentially as T goes to zero, so that atoms propagate ballistically over macroscopic distances. Furthermore, the mean free path of phonons and rotons also becomes large, so that a heat pulse propagates as a burst of ballistic rotons and phonons. In such a regime, I heard from Horst Meyer who had worked on this phenomenon [15] that P.W. Anderson [16] had predicted that it should be similar to the photoelectric effect. Quasiparticles incident on the liquid-vapor interface could evaporate atoms in a way similar to photons ejecting electrons from the surface of a metal. The conservation of energy would imply that the kinetic energy E_a of the evaporated atom would be equal to the difference between the quasiparticle energy and the binding energy E_b of atoms to the liquid. Since rotons have a minimum energy $\Delta = 8.65$ K and $E_b = 7.15$ K, atoms evaporated by rotons should have a minimum kinetic energy $\Delta - E_b = 1.5$ K. This corresponds to a minimum velocity $v = 79$ m/s.

By varying the liquid level in my cell and by measuring the flight time from a heater in the liquid to a bolometer in the gas, about 1 cm above, I could measure the velocity of atoms evaporated by rotons. I found the first preliminary evidence for the minimum velocity predicted by Anderson [17]. The photoelectric effect is the experimental evidence for the quantization of light. Similarly, my experiments showed that heat in a superfluid can be quantized as quasiparticles, especially Landau's famous rotons. Rotons are difficult to detect with a bolometer because their reflection probability on any solid surface is high. After being transformed into ballistic atoms, rotons were easy to detect and evidence for their minimum energy was found.

Given these preliminary results, I asked Adrian Wyatt if we could continue together on this subject and use his experimental techniques which were much more sensitive than mine, in order to be more quantitative. Adrian called the phenomenon "quantum evaporation" and we got particularly interested in the case of anomalous rotons. This is because we expected the component of the momentum parallel to the surface to be conserved, as usual, not the velocity of course. Now, rotons have two branches on each side of the minimum in their dispersion curve: R^+ have their momentum parallel to their group velocity but R^- have it antiparallel. As a consequence, the evaporation by R^- rotons traveling to the right should evaporate atoms traveling to the left. For this, I started building a cell in Adrian's laboratory at Exeter, where heaters and detectors could rotate in a vertical plane, but my postdoc time ended long before this cell could be finished and work. In fact, Adrian and his group worked on quantum evaporation for more than two decades and performed a beautiful analysis of the whole phenomenon: they not only observed evaporation by rotons but also by phonons; they could also separate evaporation from R^+ and from R^- rotons and obtain clear evidence for the anomalous evaporation we had imagined [18–20]. They also measured most evaporation probabilities by the various kinds of quasiparticles and compared them with calculations by Dalfovo et al [21]. This calculation included predictions on the reverse process, quantum condensation of atoms incident on the free surface of liquid helium, which was first observed by D.O. Edwards et al. The condensation of atoms was found to depend on their momentum [22]. I am not sure that quantitative agreement is well established between theory and experiments on quantum evaporation and condensation, and I have always been a little surprised to see that very few groups performed experiments on this. Among them is the one led by H.J. Maris and G. Seidel whose particle detection method involves quantum evaporation, and whose results on solar neutrinos might be of great importance [23].

3. THE SURFACE OF HELIUM CRYSTALS

Together with Harry Alles and Alexandr Parshin, we have written a detailed review of this whole field which appeared this year [24]. I wish to focus here on some aspects of my own work only. I started studying the surface of helium crystals after noticing that, when crystallizing superfluid helium in a cell, the crystal position was sensitive to gravity: it occupied the bottom part as if it was water in a glass. David Edwards came to visit me in Paris and he suggested that we could try to measure the surface tension α of solid helium by measuring its capillary rise in a narrow-gap capacitor. We did this together [17] and obtained the first direct measurement of α, in fact an average value of the surface stiffness $\gamma = \alpha + \partial^2\alpha/\partial\phi^2$ (ϕ is the angular orientation of the surface) which governs the surface curvature in a generalized Laplace equation [24]. We also found two surprises. The capillary rise was negative, it was a depression because, apparently, the copper walls of our capacitor were preferentially wet by liquid helium. Moreover, we found an anomaly around 1 K: helium crystals looked much stiffer below this temperature than above. At the same time in Haifa, Jud Landau and Steve Lipson measured γ in their optical cryostat and found agreement with our results above 1 K but not below where facets showed up on the shape [26]. Also at the same time but in Moscow, Konstantin Keshishev, Alexandr Parshin, and Alexeï Babkin discovered that capillary waves could propagate at the surface of helium crystals below about 0.6 K as if it was a free liquid surface. They called them "melting freezing" or "crystallization" waves and obtained a value for γ at low T in agreement with our high temperature value but not with our low temperature value. I soon realized that, when facetted, a helium crystal could not pop through a hole as if it were a liquid, so that, below 1 K, our measurement was wrong. I also realized that, in order to study these crystals, it was extremely useful to see through the cryostat walls, that is to drill holes in the stainless steel and put windows.

In 1926, when Keesom discovered that superfluid

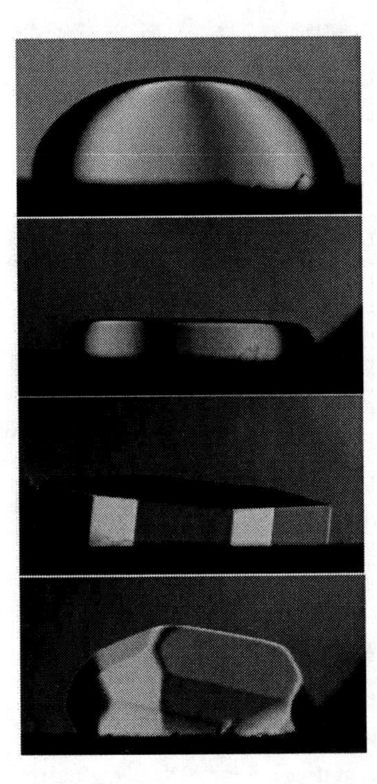

FIGURE 1. As temperature goes down, more facets appear at the surface of ^{4}He crystals.

helium solidified at 25 bar, he tried to observe the liquid-solid interface but he failed. Through the walls of his glass Dewar, "...there was nothing peculiar to be seen...". Some fifty years later, we had better techniques and the optical observation of this interface brought very interesting information (Fig. 1). Among the numerous results obtained by many groups, I wish to mention here the roughening transitions and the crystallization waves.

3.1. The roughening transitions

Most of the static properties of helium crystals are common to all other crystals. A central one is the existence of successive roughening transitions where new facets appear on the equilibrium shape as T decreases; one says that the crystal surface changes from *rough* to *smooth* in order to express that its large scale fluctuations disappear (it is *not* a change at atomic scale). Thanks to a complete study of the first transition where so called "c" facets of the hexagonal structure appear, we have found precise agreement with the set of renormalization-group theories which predict that roughening transitions belong to the Kosterlitz-Thouless universality class. In particular, we have found that the step free energy vanishes exponentially as T approaches the roughening temperature T_R from below; we also found agreement with the universal relation between T_R and the surface stiffness $\gamma_R = \gamma(T_R)$,

$$k_B T_R = \frac{2}{\pi} \gamma_R d^2 \quad (1)$$

where d is the step height. We also found agreement for the critical variation of γ as a function of orientation, the critical variation of the growth rate as a function of temperature and growth driving force, and the critical behavior of the surface stiffness as a function of orientation and temperature. For a detailed review, please see ref. [24]. Here I only wish to emphasize on the universal relation (Eq. 1) which is the best known property. Agreement was found between Nozières' theory [28] and all the results obtained by us and by the Moscow group [29]; for this we had to adjust three parameters, namely the roughening temperature T_R itself, the strength of the coupling of the interface to the underlying lattice, and a small scale cutoff where fluctuations start.

The c facet is the simplest one for such a comparison because ^{4}He crystals are easier to orient with a c facet horizontal than with any other facet horizontal, also because the c axis is a six-fold symmetry axis so that there is only one component of the surface stiffness tensor in this direction. The study of the other facets in ^{4}He is still incomplete so that the agreement with Nozières' theory is not as precise, but all measurements are compatible with its predictions (see ref. [24]). As for ^{3}He crystals which are bcc, their (110) facets could only be seen below about 100 mK [30] although the roughening temperature was predicted to occur at 260 mK. This has been a puzzle for a long time. Thanks to the recent work of Todoshchenko et al. in Helsinki [31], one now understands that the coupling of the surface is extremely weak close to the roughening temperature, so that facets have a negligible size except if the temperature is much lower than 260 mK. A surprising result by the Helsinki group is that this coupling becomes strong at low temperature since quantum fluctuations of the surface are strongly damped when the viscosity of liquid ^{3}He is high.

Thanks to the study of helium crystals, the universal relation of roughening is now well established and we have used it to explain the very large number of facets observed by P. Pieranski at the surface of some lyotropic liquid crystals [32]. These cubic crystals show up to 60 different types of facets at room temperature, which are arranged in sets of "devil's staircases" around high symmetry orientations. Together with Nozières, we showed that these crystals are soft in the sense that their typical elastic energy is much smaller than their surface energy, so that their steps are in fact embedded as edge dislocations below the crystal surface. We could calculate their stiffness components and show that the large interaction between steps is compensated by the small step energy in the calculation of the roughening temperatures, so that the existence of a large number of facets at room temperature is a simple consequence of the large value of the lattice spacing [33].

3.2. Crystallization waves

Contrary to most of their static properties which could be generalized to classical crystals, the dynamic properties of helium crystals are obviously particular to quantum systems. For example, around 100 mK, the growth rate of rough crystal surfaces is larger by 11 orders of magnitude in ^{4}He than in ^{3}He. No classical crystal would show such a difference between two isotopes. In ^{4}He, the growth rate increases as temperature goes down while, for classical crystals, everything depends on thermal activation above energy barriers, so that the growth rate always decreases when temperature goes down. The striking difference between ^{4}He and ^{3}He has two different origins. Firstly, excitations are different in superfluid ^{4}He and in a Fermi liquid such as ^{3}He. The crystallization of superfluid ^{4}He is mainly limited by collisions of the moving crystal surface with excitations (phonons and rotons); as a consequence, since thermal excitations disappear at low T, the growth rate diverges to infinity as T tends to zero [34, 35]. This is reminiscent of the mobility of electrons in metals which also increases as the density of thermal phonons vanishes. Since the excitations in liquid ^{3}He are Fermi quasiparticles which have a large momentum, the intrinsic mobility of the liquid-solid interface is much smaller than in ^{4}He (there is more momentum exchange during a collision with Fermi quasiparticles than during collisions with thermal phonons). As a consequence, even at 320 mK where the latent heat of crystallization is zero in ^{3}He, the shape of a ^{3}He crystal relaxes to equilibrium in a few seconds, which is about 6 orders of magnitude slower than in ^{4}He at the same temperature. But as soon as the latent heat is non-zero, since the thermal conductivity of liquid ^{3}He is poor, the resistance to growth involves a bulk thermal resistance which may be very large while it is negligible in ^{4}He. As a result, the growth dynamics of ^{4}He crystals looks strikingly fast while it looks as slow as for classical crystals in ^{3}He.

A famous consequence of the fast dynamics of ^{4}He crystals is the existence of crystallization waves. These waves are well defined on rough surfaces below about 0.6 K, when thermal rotons disappear and the growth rate is only limited by phonons. In these waves, the restoring forces are gravity and surface tension as for waves at a free liquid surface; the kinetic energy comes from the mass transport which is necessary to change a liquid into a solid phase whose density is larger. As a consequence, one can use a measurement of their dispersion relation to obtain precise values of the surface tension (more

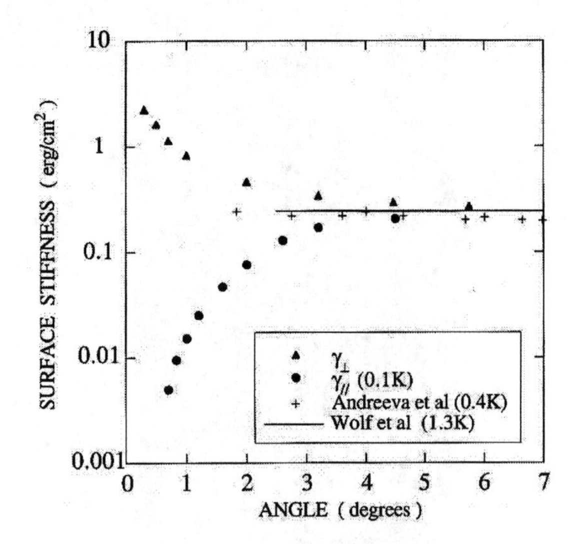

FIGURE 2. The anisotropy of the surface stiffness of ^{4}He crystals, as measured by Rolley et al [36].

precisely the surface stiffness). This proved particularly interesting to study vicinal surfaces which are tilted by a small angle with respect to facets.

A vicinal surface is called "stepped" if it is made of well separated steps. For this the tilt angle has to be small enough. At small tilt angles, the vicinal surface properties are governed by step properties while at larger angles the surface is isotropically rough. With E. Rolley [36], we studied stepped surfaces to measure the properties of steps. For this, we built a cell which could rotate by $\pm$ 6° around two perpendicular axes. In this cell we first grew a crystal with a vertical c axis; it was wide enough for gravity to force the crystal surface to be horizontal. When rotating the cell, the surface kept horizontal but its crystalline orientation rotated. This allowed us to propagate waves either perpendicular to steps or parallel to them, also to vary the step density and of course the temperature. Fig. 2 shows some of the results of this delicate experiment. One sees that the surface stiffness becomes highly anisotropic as the tilt angle ϕ tends to zero. We verified that the component $\gamma_\perp$ of the stiffness tensor diverged as $1/\phi$ while the component $\gamma_{//}$ was proportional to ϕ. A stepped surface is somewhat like a corrugated board, very soft to bending in one direction but very stiff in the other direction. From a fit with theory, we obtained the step energy and the interaction between steps which had two parts, an entropic repulsion and an elastic one. We could also estimate the step width from the crossover angle (about 2.5°) from rough to stepped behavior and we found very good agreement with what we had already learned when studying the roughening transition. We also understood the details of step-phonon collisions from the damping of crystallization waves on

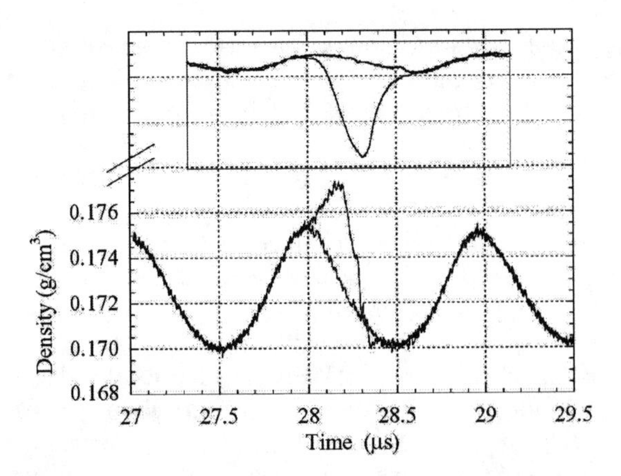

FIGURE 3. Acoustic nucleation of ^{4}He crystals occurs 4.3 bar above P_m = 25.3 bar, the liquid-solid equilibrium pressure. One records the density at the acoustic focus from the light reflected at the glass-helium interface (lower traces); the nucleation is also easily detected from the transmitted light (upper traces). At the threshold, one observes either one or the other type of signal with a probability 0.5 [41].

stepped surfaces, and we obtained evidence for the adsorption of ^{3}He impurities on these steps.

At the end of this long study, I could compare our results on He steps with those obtained by Ellen Williams and her group on Si crystals. At high enough temperature on vicinal surfaces of Si crystals, the steps fluctuate, consequently the width of terraces between them. Williams *et al.* had measured the distribution of terrace widths and found it narrower in the presence of the elastic repulsion [37]. Our description of step-step interactions was perfectly consistent with hers and this convinced me of the great interest of helium as a model system: with helium crystals one has access to some physical quantities which are not easy to measure in more classical systems but the information obtained in helium can be generalized to all others.

4. THE STABILITY LIMITS OF LIQUID HELIUM

In the recent years, we have extended the phase diagram of liquid helium to negative pressure where it is metastable with respect to the gas and to high pressure where it is metastable with respect to the solid. Thanks to acoustical techniques which I learned from Humphrey Maris, we are now able to study superfluid ^{4}He from – 9.5 bar up to 160 bar, possibly more, a pressure range much larger than the stability one (0 to 25 bar). This is because the liquid-gas and the liquid-solid transitions are first order, so that the nucleation of the stable phase occurs a certain distance away from the equilibrium line. If there are no impurities and if one eliminates the influence of walls, no "heterogeneous nucleation" takes place; the nucleation has to be "homogeneous". Our experiments use piezoelectric transducers with a hemispherical shape. They are excited at resonance in a thickness mode, so that bursts of ultrasound (typically 1 MHz) can be focused at the center of the transducer, far from any wall.

One difficulty is the calibration of the pressure amplitude at the acoustic focus. By studying the variation of the nucleation threshold as a function of the static pressure in the cell, we could measure this amplitude within about 10 percent. We found that, in the low temperature limit, bubbles nucleate, i.e. cavitation occurs near the spinodal limit which has now been calculated by several methods as -9.5 ± 0.2 bar near $T = 0$ in ^{4}He [38]. We also found a crossover from a quantum nucleation regime below about 0.2 K to a thermally activated one at higher temperature. In ^{3}He, we showed that cavitation occurs near -3 bar and that quantum cavitation is not possible in a short time there because of the existence of zero sound whose velocity does not vanish at the spinodal limit [39]. All these results and some others are discussed in a review article which I wrote on "Nucleation in quantum liquids" [40].

More recently, we developed our research in two different directions. One is the extension to liquid water of acoustic cavitation. This is another example of the model character of helium: we are extending what we learned in helium, a simple liquid whose spinodal line is well established, to THE complex liquid where the spinodal line is a matter of controversy; homogeneous cavitation studies could distinguish between two competing theories of water. The other direction is the homogeneous crystallization threshold of superfluid helium. For this we use the same type of acoustic technique but we study nucleation in the positive swings of the waves instead of the negative ones. This will be my last subject for this lecture, and, as we shall see, it will take us back to rotons.

In a first series of experiments, we focused the acoustic waves on a clean glass plate [41]. From the reflectivity at the glass-helium interface, we could measure the local density, also the local pressure from the known equation of state of liquid helium. We found crystal nucleation 4.3 bar above the liquid-solid equilibrium pressure P_m = 25.3 bar (see Fig. 3). This overpressure was 2 to 3 orders of magnitude larger than in previous experiments where favorable defects or impurities must have been present on cell walls. This experiment also showed that crystals could grow at 100 m/s, reach a 10 μm size in 100 ns and be easily observed. Eventually, we studied the statistics of nucleation and showed that the nucleation was heterogeneous, taking place on one particular defect of the glass plate.

In a second series of experiments, we removed the glass plate [42]. Apparently, we observed cavitation in the negative swings but no crystallization in the positive

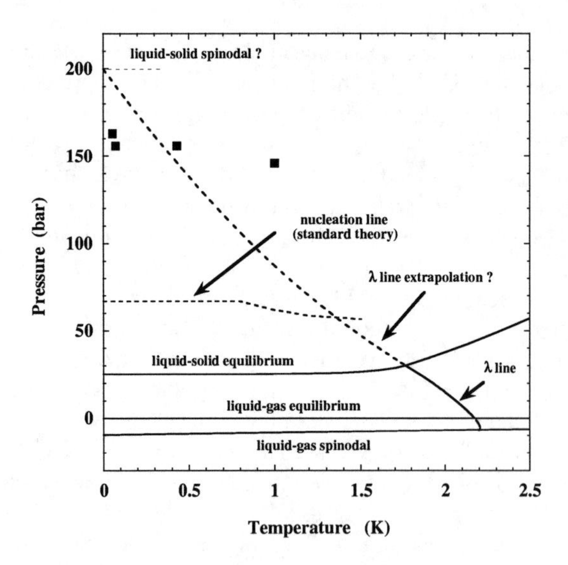

FIGURE 4. The extended phase diagram of helium. The liquid phase can be observed in a metastable state at negative pressure down to -9.5 bar and at high pressure up to 160 bar, possibly up to an instability line where the roton gap vanishes [42].

swings, even at 25 bar near P_m. This was a surprising result which showed that the standard nucleation theory could not be used for precise predictions far from equilibrium. Indeed, it uses the "thin wall approximation" to calculate the nucleation barrier with the value of the macroscopic surface tension of crystals in equilibrium with liquid helium. This elementary calculation predicts homogeneous nucleation at 65 bar while we could overpressurize liquid helium up to 160 bar without nucleation. According to Maris and Caupin [43], the liquid-solid interfacial energy increases with density so that the standard theory underestimates the nucleation barrier.

How far can one pressurize liquid helium before it crystallizes? This is in fact a rather open question. In 1971, Schneider and Enz proposed the existence of a stability limit where the roton gap Δ vanishes [44]. Δ is known to decrease with pressure from 0 to 25 bar. If Δ vanishes, rotons become a soft mode which triggers an instability: a density wave should spontaneously grow and break the translation symmetry of the liquid. Since the wavevector of rotons is the inverse of the interatomic distance, the periodic phase is likely to evolve into the stable crystalline phase. Maris estimated that $\Delta = 0$ around 200 bar from an extrapolation of Dalfovo's density functional [45]. More recently, it was found from Monte Carlo simulations that this stability limit probably occurs at even higher pressure [46].

In my opinion, this instability illustrates the nature of rotons. Landau chose this name because he thought that they could be elementary vortices. Later, Feynman modified Landau's view and tried to consider rotons as elementary vortex rings, but this new image had a difficulty [47]. Indeed, the group velocity of rotons with energy Δ is zero, and this is hardly compatible with the image of a vortex ring. Nozières recently noticed [48] that the existence of a roton minimum in the dispersion relation $\omega(q)$ of helium excitations is just a sign of local order in liquid helium. Feynman himself explains that $\omega(q)$ is proportional to the inverse of the structure factor $S(q)$, so that the roton minimum is a consequence of the existence of a large maximum in $S(q)$ (a "Bragg peak" as Nozières says). There is some local order in liquid helium, consequently a large probability to find atoms at an atomic distance from another atom, and the dispersion of phonons with a wavelength equal to the interatomic distance resembles that of a periodic crystal. As pressure increases, the local order increases and the roton minimum tends to zero. In this representation, rotons are density fluctuations signaling the proximity of a crystalline phase. When the roton minimum vanishes, the liquid becomes unstable. This is the new view of rotons which we hope to verify experimentally.

With this in mind, we have started a third series of measurements, now with two hemispherical transducers forming a sphere. The amplitude of positive peaks is higher but the calibration is more difficult because the acoustic focusing is more non-linear than in the hemispherical geometry [49]. According to our preliminary results [50], we have found homogeneous nucleation of ^{4}He crystals in the bulk of liquid ^{4}He, a long standing challenge.

In order to measure the nucleation pressure, we plan Brillouin scattering measurements inside the acoustic wave. We should obtain the local instantaneous sound velocity and relate it to the local pressure from the most recent equation of state [46]. In fact, we expect Brillouin scattering to tell us about another important issue, namely the vanishing of superfluidity as a function of pressure. Superfluidity is a long range quantum order which requires exchange between atoms. The higher the pressure, consequently the density, the more difficult is this exchange. This is why the superfluid transition temperature decreases as a function of density, contrary to the BEC transition in a weakly interacting gas. We wonder what happens to superfluidity in highly pressurized liquid ^{4}He, how the lambda line extrapolates at densities of order 0.23 g/cm^3, 30 percent more than at 25 bar. At first sight, we expected the lambda line to join the liquid-solid instability line at T=0, as drawn on Fig. 4. But Nozières argued that superfluidity could disappear before Δ becomes zero [48]. Since Brillouin scattering allows one to detect superfluidity from the existence of second sound, we plan to get some information on this issue. If we could perform Raman scattering as well, we could perhaps also measure the vanishing of the roton gap. Such experiments look difficult but worth trying in the coming years.

ACKNOWLEDGMENTS

I wish to express my deep gratitude to all my students, postdocs and visitors: none of my results, especially the best ones, could have been obtained without them. Among the collaborators who deeply influenced me, I wish to thank four great scientists. My PhD supervisor Albert Libchaber triggered my curiosity and forced me to define my research subjects by myself. David Edwards taught me the necessity of rigor in experimental physics and the power of thermodynamics. My collaboration with Philippe Nozières has been an enthusiastic fight for more than two decades; from him I learned the real beauty of theoretical physics. With Humphrey Maris, I discovered the great usefulness of numerics and shared the pleasure of scientific dreams. I am very grateful to all four of them, also to Adrian Wyatt, Steve Lipson, Kostia Keshishev, Paul Leiderer, Will Saam, Takao Mizusaki, Harry Alles and Sasha Parshin who invited me in their labs for a series of very stimulating visits.

REFERENCES

1. F. London, *Proc. Roy. Soc.* **A 153**, 576 (1936).
2. F. London, *Nature* **141**, 643 (1938).
3. L. Tisza, *Nature* **141**, 913 (1938).
4. J.F. Allen and A.D. Misener, *Nature* **141**, 75 (1938); J.F. Allen and H. Jones, *Nature* **141**, 243 (1938).
5. P.L. Kapitza, *Nature* **141**, 74 (1938).
6. G. Gorelik, *Scientific American* **277**, 72 (1997).
7. L. Pitaevskii, *J. Low Temp. Phys.* **87**, 127 (1992).
8. L.D. Landau, *Zh. Exp. Teor. Fiz.* **11**, 592 (1941) and *J. Phys. (USSR)* **5**, 71 (1941)].
9. F. London, *Report on an Internat. Conf. on Fund. particles and low Temp.*, Cambridge 22-27 July 1946, published by The Physical Society, Taylor and Francis, London 1947, p. 1.
10. D.W. Osborne, B. Weinstock, and B.M. Abraham, *Phys. Rev.* **75**, 988 (1949).
11. V.P. Peshkov, *Zh. Eksp. Teor. Fiz.* **18**, 951 (1948); **19**, 270 (1949); **23**, 687 (1952).
12. D.G. Henshaw and A.D.B. Woods, *Phys. Rev.* **121**, 1266 (1961)
13. E.Varoquaux, M.W. Meisel and O. Avenel, *Phys. Rev. Lett.* **57**, 2291 (1986).
14. J. Steinbauer, K. Schwab, Yu. Mukharski, J.C. Davis and R.E. Packard, *Phys. Rev. Lett.* **74**, 5056 (1995).
15. D.T. Meyer, H. Meyer, W. Hallidy, and C.F. Kellers, *Cryogenics* **3**, 150 (1963).
16. P.W. Anderson, *Phys. Lett.* **A 29**, 563 (1969).
17. S. Balibar, J. Buechner, B. Castaing, C. Laroche, A. Libchaber, *Phys. Rev.* **B 18**, 3096 (1978).
18. F.R. Hope, M.J. Baird, and A.F.G. Wyatt, *Phys. Rev. Lett.* **52**, 1528 (1984)
19. M. Brown and A.F.G. Wyatt, *J. Phys.: Condens. Matter* **2**, 5025 (1990).
20. M.A.H. Tucker and A.F.G. Wyatt, *Science* **283**, 1150 (1999).
21. F. Dalfovo, A. Frachetti, A. Lastri, L. Pitaevskii, and S. Stringari, *J. Low Temp. Phys.* **104**, 367 (1996).
22. D.O. Edwards, G.G. Ihas, and C.P. Tam, *Phys. Rev.* **B 16**, 3122 (1977).
23. S. R. Bandler, R. E. Lanou, H. J. Maris, T. More, F. S. Porter, G. M. Seidel, and R. H. Torii, *Phys. Rev. Lett.* **68**, 2429 (1992).
24. S. Balibar, H. Alles, and A. Ya. Parshin, *Rev. Mod. Phys.* **77**, 317 (2005).
25. S. Balibar, D.O. Edwards, and C. Laroche, *Phys. Rev. Lett.* **42**, 782 (1979).
26. J. Landau, S. G. Lipson, L. M. Määttänen, L. S. Balfour, and D. O. Edwards, *Phys. Rev. Lett.* **45**, 31 (1980).
27. K.O. Keshishev, A. Ya. Parshin, and A. V. Babkin, *Pis'ma Zh. Eksp. Teor. Fiz.* **30**, 63 (1979) [*Sov. Phys. JETP Lett.* **30**, 56 (1980)].
28. P. Nozières and F. Gallet, *J. Phys. (Paris)* **48**, 353 (1987).
29. S. Balibar, C. Guthmann, and E. Rolley, *J. Phys. I (France)* **3**, 1475 (1993).
30. E. Rolley, S. Balibar, F. Gallet, F. Graner, and C. Guthmann, *Europhys. Lett.* **8**, 523 (1989).
31. I. A. Todoshchenko, H. Alles, H. J. Junes, A. Ya. Parshin, and V. Tsepelin, *Phys. Rev. Lett.* **93**, 175301 (2004).
32. P. Pieranski, P. Sotta, D. Rohe, M. Imperor-Clerc, *Phys. Rev. Lett.* **84**, 2409 (2000).
33. P. Nozières, F. Pistolesi, and S. Balibar, *Eur. Phys. J.* **B 24**, 387-394 (2001).
34. K.O. Keshishev, A. Y. Parshin, and A. V. Babkin, *Zh. Eksp. Teor. Fiz.* **80**, 716 (1981) [Sov. Phys. JETP **53**, 362 (1981)]
35. J. Bodensohn, K. Nicolai, and P. Leiderer, *Z. Phys. B: Condens. Matter* **64**, 55 (1986).
36. E. Rolley, C. Guthmann, E. Chevalier, and S. Balibar, *J. Low Temp. Phys.* **99**, 851 (1995).
37. E.D. Williams and N. C. Bartelt, *Science* **251**, 393 (1991).
38. F. Caupin and S. Balibar *Phys. Rev.* **B 64**, 064507 (2001).
39. F. Caupin, H.J. Maris, and S. Balibar, *Phys. Rev. Lett.* **87**, 145302 (2001).
40. S. Balibar, *J. Low Temp. Phys.* **129**, 363 (2002).
41. X. Chavanne, S. Balibar, and F. Caupin, *Phys. Rev. Lett.* **86**, 5506 (2001).
42. F. Werner, G. Beaume, A. Hobeika, S. Nascimbene, C. Herrmann, F. Caupin and S. Balibar, *J. Low Temp. Phys.* **136**, 93 (2004).
43. H.J. Maris and F. Caupin, *J. Low Temp. Phys.*, **131**, 145 (2003).
44. T. Schneider and C.P. Enz, *Phys. Rev. Lett.* **27**, 1186 (1971).
45. F. Dalfovo , A. Lastri, L. Pricaupenko, S. Stringari, and J. Treiner, *Phys. Rev.* **B 52**, 1193 (1995).
46. L. Vranjes, J. Boronat, J. Casulleras and C. Cazorla, *J. Low Temp. Phys.* **138**, 43 (2005).
47. R.P. Feynman in *Progress in Low Temperature Physics*, Vol. **1**, C. G. Gorter (ed.), North Holland (1955), Chap. 2.
48. P. Nozières, *J. Low Temp. Phys.* **137**, 45 (2004).
49. C. Appert, C. Tenaud, X. Chavanne, S. Balibar, F. Caupin, and D. d'Humières, *Eur. Phys. J.* **B 35**, 531 (2003).
50. R. Ishiguro, F. Caupin, and S. Balibar, "Homogeneous Nucleation of Solid ^{4}He", this conference.

Emergent Physics
on Vacuum Energy and Cosmological Constant

G.E. Volovik

Low Temperature Laboratory, Helsinki University of Technology, P.O.Box 2200, FIN-02015 HUT, Finland
L.D. Landau Institute for Theoretical Physics, Kosygin Str. 2, 119334 Moscow, Russia

Abstract. The phenomenon of emergent physics in condensed-matter many-body systems has become the paradigm of modern physics, and can probably also be applied to high-energy physics and cosmology. This encouraging fact comes from the universal properties of the ground state (the analog of the quantum vacuum) in fermionic many-body systems, described in terms of the momentum-space topology. In one of the two generic universality classes of fermionic quantum vacua the gauge fields, chiral fermions, Lorentz invariance, gravity, relativistic spin, and other features of the Standard Model gradually emerge at low energy. The condensed-matter experience provides us with some criteria for selecting the proper theories in particle physics and gravity, and even suggests specific solutions to different fundamental problems. In particular, it provides us with a plausible mechanism for the solution of the cosmological constant problem, which I will discuss in some detail.

Keywords: physical vacuum; effective theory; cosmological constant
PACS: 71.27.+a; 98.80.Es; 12.60.-i; 04.20.-q

INTRODUCTION

In condensed matter physics we deal with many different strongly correlated and/or strongly interacting systems. There are no small parameters in such a system and we cannot treat it perturbatively. However, from our experience we know that at length scales much larger than the inter-atomic spacing, rather simple behavior emerges which is described by an effective theory. This theory is determined by the universality class to which the system belongs and does not depend on microscopic details of the system. There are several types of effective theories.

- A typical example of an effective theory is provided by the Ginzburg-Landau theory describing superconductivity in the vicinity of the transition temperature T_c. This theory, extended to multicomponent superfluids, superconductors and Bose condensates, as well as to the critical phenomena close to T_c, is determined by the symmetry of the system above T_c and describes the symmetry breaking below T_c.
- Effective theories of hydrodynamic type deal with the low-frequency collective modes away from the critical region. These are the two-fluid hydrodynamics of superfluid ^{4}He; the London theory of superconductivity; their extension to spin and orbital dynamics of superfluid ^{3}He; elasticity theory in crystals, etc. This type of effective theories also describes topologically non-trivial configurations (including the topological defects – singularities of the collective fields protected by topology, such as quantized vortices) and their dynamics (see the book [1] for review on the role of the topological quantum numbers in physics).
- In the limit $T \to 0$ an effective quantum field theory (QFT) emerges. It deals with the ground state of the system (the quantum vacuum), quasiparticle excitations above the vacuum (analog of elementary particles), and their interaction with low-energy collective modes (bosonic fields). The QFT kind of effective theories includes the Landau Fermi-liquid theory with its extension to non-Landau fermionic systems; the quantum Hall effect; the theory of superfluids and superconductors at $T \ll T_c$, etc. Here one encounters a phenomenon which is opposite to the symmetry breaking: the symmetry is enhanced in the limit $T \to 0$ [2]. An example is provided by high-temperature superconductors with gap nodes: close to the nodes quasiparticles behave as 2+1 Dirac fermions, i.e. their spectrum acquires the Lorentz invariance. In superfluid ^{3}He-A other elements of the relativistic QFT (RQFT) emerge at $T \to 0$: chiral (Weyl) fermions, gauge invariance, and even some features of effective gravity [3].

In most cases effective theories cannot be derived from first principles, i.e. from the underlying microscopic theory [4]. If we want to check that our principles of construction of effective theories are correct and also to search for other possible universality classes, we use some very simple models, which either contain a small parameter, or are exactly solvable. Example is the BCS theory of a weakly interacting Fermi gas, from which all the types of the effective theories of superconductivity – Ginzburg-Landau, London and QFT – can be derived within their regions of applicability.

CP850, *Low Temperature Physics: 24th International Conference on Low Temperature Physics;*
edited by Y. Takano, S. P. Hershfield, S. O. Hill, P. J. Hirschfeld, and A. M. Goldman

In particle physics effective theories are also major tools [5]. The Standard Model of quark and leptons and electroweak and strong interactions operating below 10^3GeV is considered as an effective low-energy RQFT emerging well below the "microscopic" Planck energy scale $E_P \sim 10^{19}$GeV. It is supplemented by the Ginzburg-Landau type theory of electroweak phase transition, and by the hydrodynamic type theory of gravity – the Einstein general relativity theory. The chiral symmetry and nuclear physics are the other examples of effective theories; they emerge in the low-energy limit of the quantum chromodynamics. In addition, the condensed matter examples (^{3}He-A in particular) suggest that not only these effective theories, but even the fundamental physical laws on which they are based (relativistic invariance, gauge invariance, general relativity, relation between spin and statistics, etc.) can be emergent. According to this view the quantum vacuum – the modern ether – can be thought of as some kind of condensed-matter medium. This may or may not be true, but in any case it is always instructive to treat the elementary particle physics with the methods and experiences of the condensed matter physics.

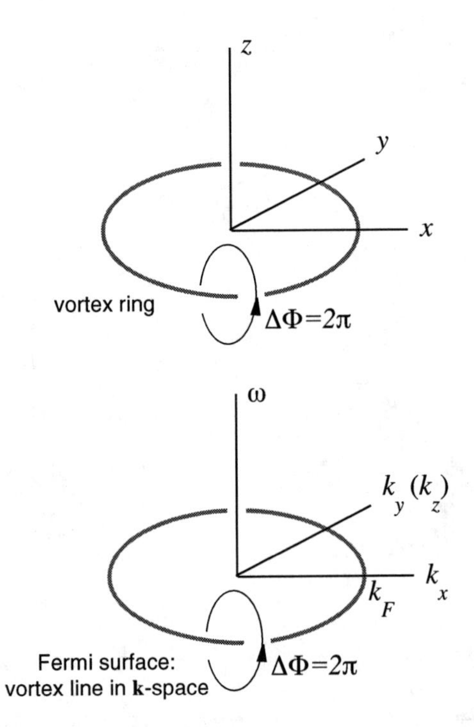

FIGURE 1. *Top*: vortex loop in superfluids and superconductors. The phase Φ of the order parameter $\Psi = |\Psi|e^{i\Phi}$ changes by 2π around the vortex line and is not determined at the line. *Bottom*: a Fermi surface is a vortex in momentum space. The Green's function near the Fermi surface is $G = (i\omega - v_F(k - k_F))^{-1}$. Let us consider the two-dimensional (2D) system, where $k^2 = k_x^2 + k_y^2$. The phase Φ of the Green's function $G = |G|e^{i\Phi}$ changes by 2π around the line situated at $\omega = 0$ and $k = k_F$ in the 3D momentum-frequency space (ω, k_x, k_y). In the 3D system, where $k^2 = k_x^2 + k_y^2 + k_z^2$, the vortex line becomes the surface in the 4D momentum-frequency space (ω, k_x, k_y, k_z) with the same winding number.

FERMI POINT AND STANDARD MODEL

The universality classes of QFT are based on the topology in momentum space. All the information is encoded in the low-energy asymptote of the Green's function for fermions $G(\mathbf{k}, i\omega)$. The singularities in the Green's function in momentum space remind the topological defects living in real space [3, 6]. Such a singularity in the $\mathbf{k}$-space as the Fermi surface is analogous to a quantized vortex in the $\mathbf{r}$-space. It is described by the same topological invariant – the winding number (Fig. 1). Protected by topology, the Fermi surface survives in spite of the interaction between fermions. On the emergence of a Fermi surface in string theory see Ref. [7].

Another generic behavior emerges in superfluid ^{3}He-A. The energy spectrum of the Bogoliubov–Nambu fermionic quasiparticles in ^{3}He-A is

$$E^2(\mathbf{k}) = v_F^2(k - k_F)^2 + \Delta^2(\mathbf{k}) \ , \ \Delta^2(\mathbf{k}) = c_\perp^2 \left(\mathbf{k} \times \hat{\mathbf{l}}\right)^2 , \quad (1)$$

where p_F is the Fermi momentum, v_F is the Fermi velocity, and $\hat{\mathbf{l}}$ is the direction of the angular momentum of the Cooper pairs.

As distinct from conventional superconductors with s-wave pairing, the gap Δ in this p-wave superfluid is anisotropic and vanishes for $\mathbf{k} \parallel \hat{\mathbf{l}}$ (Fig. 2). As a result the energy spectrum $E(\mathbf{k})$ has zeroes at two points $\mathbf{k} = \pm k_F \hat{\mathbf{l}}$. Such point nodes in the quasiparticle spectrum are equivalent to point defects in real space – the hedgehogs – and thus are protected by topology. Moreover the spectrum of elementary particles in the Standard Model has also the same kind of topologically protected zeroes (Fig. 3). The quarks and leptons above the electroweak transition are massless, and their spectrum $E^2(\mathbf{k}) = c^2k^2$ has a zero at $\mathbf{k} = 0$ described by the same topological invariant as the point nodes in ^{3}He-A. This is the reason why superfluid ^{3}He-A shares many properties of the vacuum of the Standard Model.

Close to the zeroes the spectrum (1) acquires the "relativistic" form:

$$E^2(\mathbf{k}) = c_\parallel^2(k_z \pm k_F)^2 + c_\perp^2 k_x^2 + c_\perp^2 k_y^2 \ , \ c_\parallel \equiv v_F \ , \quad (2)$$

where the z-axis is chosen along $\hat{\mathbf{l}}$. For an experimentalist working with ^{3}He-A at low temperature, quasiparticles in Eq. (1) look like one-dimensional: they move only along the direction of the nodes (along $\hat{\mathbf{l}}$); otherwise they are Andreev reflected [8]. A more accurate consid-

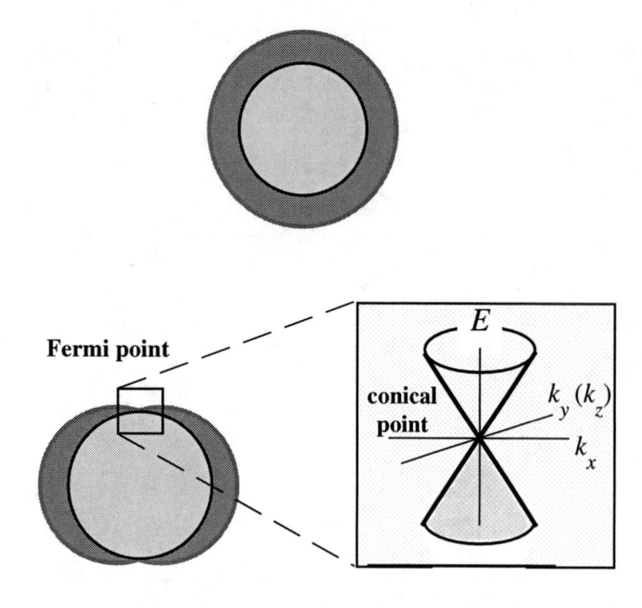

FIGURE 2. *Top*: isotropic gap in an s-wave superconductor. *Bottom left*: in p-wave superfluid ^{3}He-A the gap is anisotropic and vanishes for $\mathbf{k} \parallel \hat{\mathbf{l}}$. The energy spectrum (1) has two point nodes – Fermi points. *Bottom right*: close to the node the spectrum (2) is similar to the conical spectrum of right-handed or left-handed fermions of the Standard Model.

eration in the vicinity of the node in Eq. (2) reveals that they can move in the transverse direction too but about thousand times slower: the velocity of propagation in the transverse direction $c_\perp \sim 10^{-3} c_\parallel$.

On the other hand, low-energy inner observers living in the ^{3}He-A vacuum would not notice this huge anisotropy. They would find that their massless elementary particles move in all directions with the same speed, which is also the speed of light. The reason for this is that for their measurements of distance they would use rods made of quasiparticles: this is their matter. Such rods are not rigid and their lengths depend on the orientation. Also, the inner observers would not notice the "ether drift", i.e. the motion of the superfluid vacuum: Michelson–Morley-type measurements of the speed of "light" in moving "ether" would give a negative result. This resembles the physical Lorentz–Fitzgerald contraction of length rods and the physical Lorentz slowing down of clocks. Thus the inner observers would finally rediscover the fundamental Einstein principle of special relativity in their Universe, while we know that this Lorentz invariance is the phenomenon emerging at low energy only.

The physics emerging in the vicinity of the point nodes is remarkable. In addition to the Lorentz invariance, the other phenomena of the RQFT are reproduced. The collective motion of ^{3}He-A cannot destroy the topologically

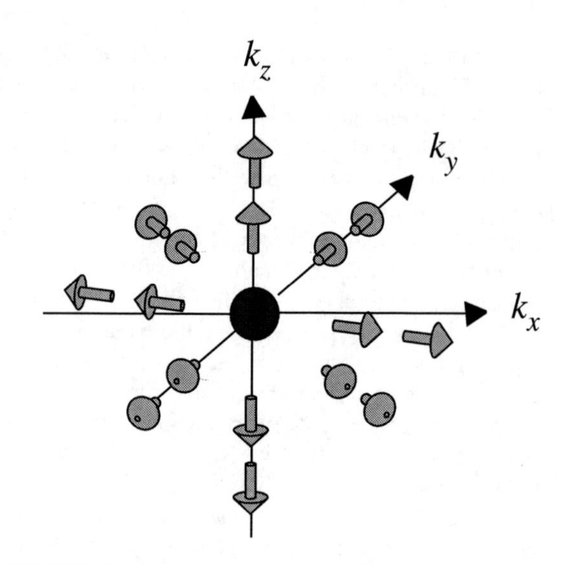

FIGURE 3. Fermi point is the hedgehog in momentum space. The Hamiltonian of the fermionic quadiparticles living close to the Fermi point is the same as either the Hamiltonian for right-handed particles $H = \hbar c\, \boldsymbol{\sigma} \cdot \mathbf{k}$ or that for the left-handed particles $H = -\hbar c\, \boldsymbol{\sigma} \cdot \mathbf{k}$. For each momentum $\mathbf{k}$ we draw the direction of the particle spin $\boldsymbol{\sigma}$, which for right-handed particles is oriented along the momentum $\mathbf{k}$. The spin distribution in momentum space looks like a hedgehog, whose spines are represented by spins. The spines point outward for the right-handed particls, while for the left-handed particles for which spin is anti-parallel to momentum the spines of the hedgehog point inward. Direction of spin is not determined at singular point $\mathbf{k} = 0$ in the momentum space. The topological stability of the hedgehog singularity under deformations provides the generic behavior of the system with Fermi points in the limit of low energy. This is the reason why the chiral particles are protected in the Standard Model and why superfluid ^{3}He-A shares many properties of the vacuum of the Standard Model.

protected nodes, it can only shift the position of the nodes and the slopes of the "light cone". The resulting general deformation of the energy spectrum near the nodes can be written in the form

$$g^{\mu\nu}(k_\mu - eA_\mu^{(a)})(k_\nu - eA_\nu^{(a)}) = 0\,. \tag{3}$$

Here the four-vector A_μ describes the degrees of freedom of the ^{3}He-A vacuum which lead to the shift of the nodes. This is the dynamical "electromagnetic" field emerging at low energy, and $e = \pm 1$ is the "electric" charge of particles living in the vicinity of north and south poles correspondingly. The elements of the matrix $g^{\mu\nu}$ come from the other collective degrees of freedom which form the effective metric and thus play the role of emerging dynamical gravity. These emergent phenomena are background independent, if the system stays within the Fermi-point universality class. Background independence is the main criterion for the correct quantum theory of gravity. [10]

One may try to construct a condensed matter system with a large number of point nodes in the spectrum which would reproduce all the elements of the Standard Model: 16 chiral fermions per generation; $U(1)$, $SU(2)$ and $SU(3)$ gauge fields; and gravity. There are many open problems on this way especially with gravity: in ^{3}He-A the equations for the "gravitational field" (i.e. for the metric $g^{\mu\nu}$) only remotely resemble Einstein's equations; while the equation for the "electromagnetic" field A_μ coincides with Maxwell's equation only in a logarithmic approximation. However, even in the absence of exact correspondence between the condensed matter system and the Standard Model, there are many common points which allow us to make conclusions concerning some unsolved problems in particle physics and gravity. One of them is the problem of the weight of the vacuum – the cosmological constant problem [11, 12].

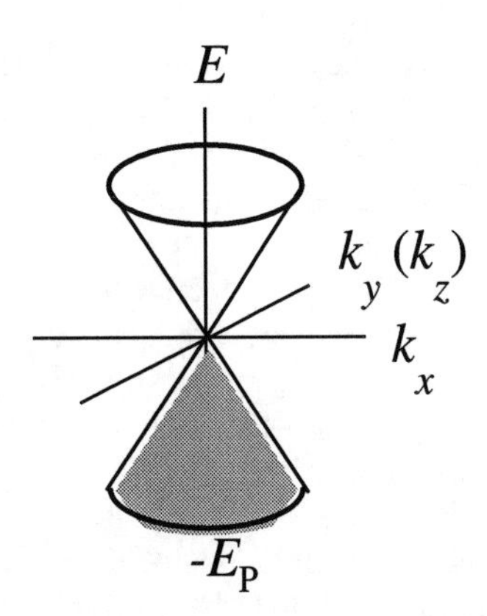

FIGURE 4. Occupied negative energy levels in the Dirac vacuum produce a huge negative contribution to the vacuum energy and thus to the cosmological constant. Summation of all negative energies $E(\mathbf{k}) = -\hbar ck$ in the interval $0 < E < -E_\text{P}$, where E_P is the Planck energy scale, gives the energy density of the Dirac vacuum: $\epsilon_{\text{Dirac vacuum}} = -\int (d^3k/(2\pi)^3)\hbar ck \sim -\hbar c/a_\text{P}^4$, where $a_\text{P} = \hbar c/E_\text{P}$ is the Planck length.

VACUUM ENERGY AND COSMOLOGICAL CONSTANT

Cosmological Term and Zero Point Energy

In 1917, Einstein proposed the model of our Universe with geometry of a three-dimensional sphere [9]. To obtain this perfect Universe, static and homogeneous, as a solution of equations of general relativity, he added the famous cosmological constant term – the λ-term. At that time the λ-term was somewhat strange, since it described the gravity of the empty space: the empty space gravitates as a medium with energy density $\epsilon = \lambda$ and pressure $p = -\lambda$, where λ is the cosmological constant. This medium has an equation of state

$$p = -\epsilon = -\lambda \,. \tag{4}$$

When it became clear that our Universe was not static, Einstein removed the λ-term from his equations.

However, later with development of quantum fields it was recognized that even in the absence of real particles the space is not empty: the vacuum is filled with zero point motion which has energy, and according to general relativity, the energy must gravitate. For example, each mode of electromagnetic field with momentum $\mathbf{k}$ contributes to the vacuum energy the amount $\frac{1}{2}\hbar\omega(\mathbf{k}) = \frac{1}{2}\hbar ck$. Summing up all the photon modes and taking into account two polarizations of photons one obtains the following contribution to the energy density of the empty space and thus to λ:

$$\lambda = \epsilon_{\text{zero point}} = \int \frac{d^3k}{(2\pi)^3}\hbar ck \,. \tag{5}$$

Now it is non-zero, but it is too big, because it diverges at large k. The natural cut-off is provided by the Planck length scale a_P, since the effective theory of gravity – the Einstein general relativity – is only applicable at $k > 1/a_\text{P}$. Then the estimate of the cosmological constant, $\lambda \sim \hbar c/a_\text{P}^4$ exceeds by 120 orders of magnitude the upper limit posed by astronomical observations.

There are also contributions to the vacuum energy from the zero point motion of other bosonic fields, and a contribution from the occupied negative energy states of fermions (Fig. 4). If there is a supersymmetry – the symmetry between fermions and bosons – the contribution of bosons would be canceled by the negative contribution of fermions. However, since the supersymmetry is not exact in our Universe, it can reduce the discrepancy between theory and experiment only by about 60 orders of magnitude. The physical vacuum remains too heavy, and this poses the main cosmological constant problem.

One may argue that there must exist some unknown but very simple principle, which leads to nullification of the cosmological constant. Indeed, in theories in which gravity emerges from the quantum matter fields, the flat space with $\lambda = 0$ appears as a classical equilibrium solution of the underlying microscopic equations [14]. But what to do with our estimation of the zero point energy of quantum fields and the energy of the Dirac vacuum, which are huge irrespective of whether the vacuum is in equilibrium or not?

Recently the experimental evidence for non-zero λ was established: it is on the order of magnitude of the energy density of matter, $\lambda \sim 2 - 3\epsilon_{\text{matter}}$ [13]. People find it easier to believe that the unknown mechanism of cancellation, if existed, would reduce λ to exactly zero rather than the observed very low value. So, why is λ non-zero? And also, why is it on the order of magnitude of the matter density? None of these questions has an answer within the effective quantum field theory, and that is why our condensed matter experience is instructive, since we know both the effective theory and the underlying microscopic physics, and are able to connect them.

Since we are looking for the general principles governing the energy of the vacuum, it should not be of importance for us whether the QFT is fundamental or emergent. Moreover, we expect that these principles should not depend on whether or not the QFT obeys all the symmetries of the RQFT: these symmetries (Lorentz and gauge invariance, supersymmetry, etc.) still did not help us to nullify the vacuum energy. That is why, to find these principles, we can look at the quantum vacua whose microscopic structure is well known at least in principle. These are the ground states of the quantum condensed-matter systems such as superfluid liquids, Bose-Einstein condensates in ultra-cold gases, superconductors, insulators, systems experiencing the quantum Hall effect, etc. These systems provide us with a broad class of Quantum Field Theories which are not restricted by Lorentz invariance. This allows us to consider the cosmological constant problems from a more general perspective.

Zero Point Energy in Condensed Matter

The principle which leads to the cancellation of zero-point energy is more general; it comes from a thermodynamic analysis which is not constrained by symmetry or a universality class. To see it, let us consider two quantum vacua: the ground states of two quantum liquids, superfluid ^{4}He and one of the two superfluid phases of ^{3}He, the A-phase. We have chosen these two liquids because the spectrum of quasiparticles playing the major role at low energy is "relativistic". This allows us to make the connection to the RQFT. In superfluid ^{4}He the relevant quasiparticles are phonons (the quanta of sound waves), and their spectrum is $E(k)=\hbar ck$, where c is the speed of sound. In superfluid ^{3}He-A the relevant quasiparticles are fermions. The corresponding "speed of light" c (the slope in the linear spectrum of these fermions in Eq. (2)) is anisotropic; it depends on the direction of their propagation.

Lets us start with superfluid ^{4}He and apply the same reasoning as we did in the case of the electromagnetic field, i.e. we assume that the energy of the ground state of the liquid comes from the zero point motion of the phonon field. Then according to Eq. (5) one has for the energy density

$$\epsilon_{\text{zero point}}=\frac{1}{2}\int\frac{d^3k}{(2\pi)^3}\hbar ck\sim\frac{\hbar c}{a_{\text{P}}^4}\sim\frac{E_{\text{P}}^4}{\hbar^3c^3}\,, \quad (6)$$

where the role of the Planck length a_{P} is played by the interatomic spacing, and the role of the Planck energy scale $E_{\text{P}}=\hbar c/a_{\text{P}}$ is provided by the Debye temperature, $E_{\text{P}}=E_{\text{Debye}}\sim 1$ K; $c\sim 10^4$ cm/s.

The same reasoning for the fermionic liquid ^{3}He-A suggests that the vacuum energy comes from the Dirac sea of "elementary particles" with spectrum (2), i.e. from the occupied levels with negative energy (see Fig. 4):

$$\epsilon_{\text{Dirac vacuum}}=-2\int\frac{d^3k}{(2\pi)^3}E(\mathbf{k})\sim-\frac{E_{\text{P}}^4}{\hbar^3c_{\parallel}c_{\perp}^2}\,. \quad (7)$$

Here the Planck energy cut-off is provided by the gap amplitude, $E_{\text{P}}=\Delta\sim c_{\perp}p_F\sim 1$ mK; $c_{\parallel}\sim 10^4$ cm/s; $c_{\perp}\sim 10$ cm/s.

The above estimates were obtained by using the effective QFT for the "relativistic" fields in the two liquids in the same manner as we did for the quantum vacuum of the Standard Model. Now let us consider what the exact microscopic theory tells us about the vacuum energy.

Real Vacuum Energy in Condensed Matter

The underlying microscopic physics of these two liquids is the physics of a system of N atoms obeying the conventional quantum mechanics and described by the N-body Schrödinger wave function $\Psi(\mathbf{r}_1,\mathbf{r}_2,\ldots,\mathbf{r}_i,\ldots,\mathbf{r}_N)$. The corresponding many-body Hamiltonian is

$$\mathcal{H}=-\frac{\hbar^2}{2m}\sum_{i=1}^{N}\frac{\partial^2}{\partial\mathbf{r}_i^2}+\sum_{i=1}^{N}\sum_{j=i+1}^{N}U(\mathbf{r}_i-\mathbf{r}_j)\,, \quad (8)$$

where m is the bare mass of the atom, and $U(\mathbf{r}_i-\mathbf{r}_j)$ is the pair interaction of the bare atoms i and j. In the thermodynamic limit where the volume of the system $V\to\infty$ and N is macroscopically large, there emerges an equivalent description of the system in terms of quantum fields, in a procedure known as second quantization. The quantum field in the ^{4}He (^{3}He) system is presented by the bosonic (fermionic) annihilation operator $\psi(\mathbf{x})$. The Schrödinger many-body Hamiltonian (8) becomes the Hamiltonian of the QFT [15]:

$$\hat{H}_{\text{QFT}}=\hat{H}-\mu\hat{N}=\int d\mathbf{x}\psi^\dagger(\mathbf{x})\left[-\frac{\nabla^2}{2m}-\mu\right]\psi(\mathbf{x})$$
$$+\frac{1}{2}\int d\mathbf{x}d\mathbf{y}U(\mathbf{x}-\mathbf{y})\psi^\dagger(\mathbf{x})\psi^\dagger(\mathbf{y})\psi(\mathbf{y})\psi(\mathbf{x}). \quad (9)$$

Here $\hat{N}=\int d^3x\ \psi^\dagger(\mathbf{x})\psi(\mathbf{x})$ is the operator of the particle number (number of atoms); μ is the chemical potential – the Lagrange multiplier introduced to take into account the conservation of the number of atoms. Putting aside the philosophical question of what is primary – quantum mechanics or quantum field theory – let us discuss the vacuum energy.

The energy density of the vacuum in the above QFT is given by the vacuum expectation value of fhe Hamiltonian $\hat{H}_{\rm QFT}$ in (9):

$$\epsilon = \frac{1}{V}\left\langle \hat{H}_{\rm QFT}\right\rangle_{\rm vac} . \qquad (10)$$

In this thermodynamic limit one can apply the Gibbs-Duhem relation, $E - \mu N - TS = -pV$, which at $T = 0$ states:

$$\left\langle \hat{H}\right\rangle_{\rm vac} - \mu\left\langle \hat{N}\right\rangle_{\rm vac} = -pV , \qquad (11)$$

where p is the pressure. Using Eqs. (9) and (10) one obtains the relation between the pressure and energy density in the vacuum state:

$$p = -\epsilon . \qquad (12)$$

It is a general property, which follows from thermodynamics, that the vacuum behaves as a medium with the above equation of state. Thus it is not surprising that the equation of state (12) is applicable also to the particular case of the vacuum of the RQFT in Eq. (4). This demonstrates that the problem of the vacuum energy can be considered from a more general perspective not constrained by the relativistic Hamiltonians. Moreover, it is not important whether gravity emerges or not in the system.

Nullification of Vacuum Energy

Let us consider a situation in which the quantum liquid is completely isolated from the environment. For example, we launch the liquid in space where it forms a droplet. The evaporation at $T = 0$ is absent in the liquid, that is why the ground state of the droplet exists. In the absence of external environment the external pressure is zero, and thus the pressure of the liquid in its vacuum state is $p = 2\sigma/R$, where σ is the surface tension and R the radius of the droplet. In the thermodynamic limit where $R \to \infty$, the pressure vanishes. Then according to the equation of state (12) for the vacuum, one has $\epsilon = -p = 0$. This nullification of the vacuum energy occurs irrespective of whether the liquid is made of fermionic or bosonic atoms.

If observers living within the droplet measure the vacuum energy (or the vacuum pressure) and compare it with their estimate, Eq. (6) or Eq. (7) depending on in which liquid they live, they will be surprised by the disparity of many orders of magnitude between the estimates and observations (see Fig. 5). But we can easily explain to these observers where their theory goes wrong. Equations (6) and (7) take into account only the

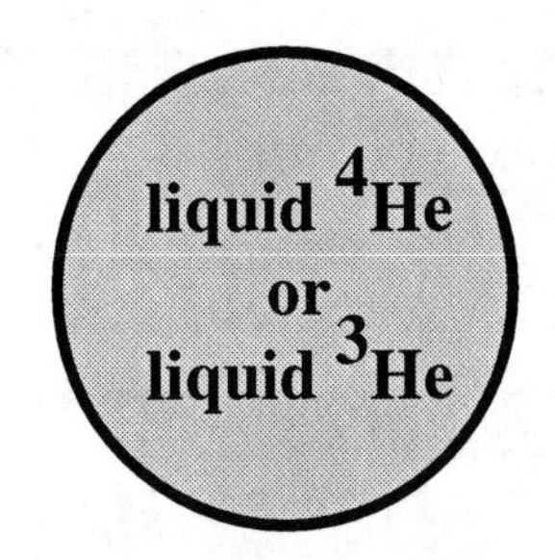

FIGURE 5. Droplet of quantum liquid. Naive estimation of the vacuum energy density in superfluid ^{4}He as the zero point energy of the phonon field gives $\epsilon_{\rm zero\ point} \sim E_{\rm P}^4/\hbar^3c^3$, where $E_{\rm P}$ is the Debye energy. Naive estimation of the vacuum energy in superfluid ^{3}He-A as the energy of the Dirac vacuum gives $\epsilon_{\rm Dirac\ vacuum} \sim -E_{\rm P}^4/\hbar^3c_{\parallel}c_{\perp}^2$, where $E_{\rm P}$ is the amplitude of the superfluid gap. But the real energy density of the vacuum in the droplets is much smaller: for both liquids it is $\epsilon_{\rm vac} = -2\sigma/R$, where σ is the surface tension and R is the radius of the droplet. It vanishes in the thermodynamic limit: $\epsilon_{\rm vac}(R \to \infty) = 0$. Inner observers living within the droplet would be surprised by the disparity of many orders of magnitude between their estimates and observations. For them it would be a great paradox, which is similar to our cosmological constant problem.

degrees of freedom below the "Planck" cut-off energy, which are described by an effective theory. At higher energies, the microscopic energy of interacting atoms in Eq. (9) must be taken into account, which the low-energy observers are unable to do. When one sums up all the contributions to the vacuum energy, sub-Planckian and trans-Planckian, one obtains the zero result. The exact nullification occurs without any special fine-tuning, due to the thermodynamic relation applied to the whole equilibrium vacuum.

This thermodynamic analysis does not depend on the microscopic structure of the vacuum and thus can be applied to any quantum vacuum (Fig. 6), including the vacuum of the RQFT. The main lesson from condensed matter, which the particle physicists may or may not accept, is this: the energy density of the homogeneous equilibrium state of the quantum vacuum is zero in the absence of an external environment. The higher-energy (trans-Planckian) degrees of freedom of the quantum vacuum, whatever they are, perfectly cancel the huge positive contribution of the zero-point motion of the quantum fields as well as the huge negative contribution of the Dirac vacuum.

This conclusion is supported by the relativistic model, in which our world represents the $(3+1)$-dimensional membrane embedded in the $(4+1)$-dimensional anti-de Sitter space. Huge contributions to the cosmological constant coming from different sources cancel each other without fine-tuning [16]. This is the consequence of the vacuum stability.

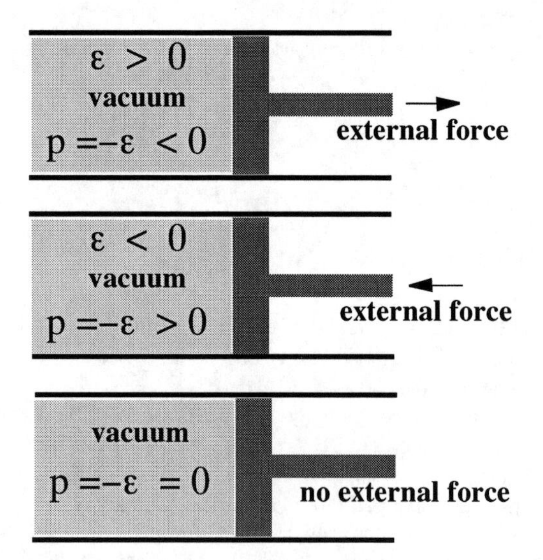

FIGURE 6. If the vacuum energy is positive, the vacuum tries to reduce its volume by moving the piston to the left. To reach an equilibrium, the external force must be apllied which pulls the piston to the right and compensates for the negative vacuum pressure. In the same manner, if the vacuum energy is negative, the applied force must push the piston to the left to compensate for the positive vacuum pressure. If there is no external force from the environment, the self-sustained vacuum must have zero energy.

Why the Vacuum Energy is Non-Zero

Let us now try to answer the question why, in the present Universe, the energy density of the quantum vacuum is on the same order of magnitude as the energy density of matter. For that let us again exploit our quantum liquids as a guide. Till now we discussed the pure vacuum state, i.e. the state without a matter. In the QFT of quantum liquids the matter is represented by excitations above the vacuum – quasiparticles. We can introduce quasiparticles to the liquid droplets by raising their temperature T a non-zero value. The quasipartcles in both liquids are "relativistic" and massless. The pressure of the dilute gas of quasiparticles as a function of T has the same form as the pressure of ultra-relativistic matter (or radiation) in the hot Universe, if one uses the determinant of the effective (acoustic) metric:

$$p_{\mathrm{matter}} = \gamma T^4 \sqrt{-g} \, . \tag{13}$$

For the quasipartcles in ^{4}He, one has $\sqrt{-g} = c^{-3}$ and $\gamma = \pi^2/90$; for the fermionic quasiparticles in ^{3}He-A, $\sqrt{-g} = c_\perp^{-2} c_\parallel^{-1}$ and $\gamma = 7\pi^2/360$. The gas of quasiparticles obeys the ultra-relativistic equation of state:

$$\epsilon_{\mathrm{matter}} = 3 p_{\mathrm{matter}} \, . \tag{14}$$

Let us consider again the droplet of a quantum liquid which is isolated from the environment, but now at a finite T. In the absence of an environment and for a sufficiently big droplet, where we can neglect the surface tension, the total pressure in the droplet must be zero. This means, that in equilibrium, the partial pressure of the matter (quasiparticles) in Eq. (13) must be necessarily compensated by the negative pressure of the quantum vacuum (superfluid condensate):

$$p_{\mathrm{matter}} + p_{\mathrm{vac}} = 0 \, . \tag{15}$$

The induced negative vacuum pressure leads to the positive vacuum energy density according the equation of state (12) for the vacuum, and one obtains the following relation between the energy density of the vacuum and that of the ultra-relativistic matter (or radiation) in thermodynamic equilibrium:

$$\epsilon_{\mathrm{vac}} = -p_{\mathrm{vac}} = p_{\mathrm{matter}} = \frac{1}{3}\epsilon_{\mathrm{matter}} \, . \tag{16}$$

This is actually what occurs in quantum liquids, but the resulting equation,

$$\epsilon_{\mathrm{vac}} = \kappa \epsilon_{\mathrm{matter}} \, , \tag{17}$$

with $\kappa = \frac{1}{3}$, does not depend on the details of the system. It is determined by the equation of state for the matter and is equally applicable to: (i) a superfluid condensate + quasiparticles with a linear "relativistic" spectrum; and (ii) the vacuum of relativistic quantum fields + an ultra-relativistic matter (but still in the absence of gravity).

What is the implication of this result to our Universe? It demonstrates that when the vacuum is disturbed, the vacuum pressure responds to the perturbation; as a result the vacuum energy density becomes non-zero. In the above quantum-liquid examples the vacuum is perturbed by a "relativistic matter". The vacuum is also perturbed by the surface tension of the curved 2D surface of the droplet which adds its own partial pressure. The corresponding response of the vacuum pressure is $2\sigma/R$.

Applying this to the general relativity, we can conclude that the homogeneous equilibrium state of the quantum vacuum without a matter is not gravitating, but the disturbed quantum vacuum has a weight. In the Einstein Universe the vacuum is perturbed by the matter and also by the gravitational field (the 3D space curvature). These perturbations induce the non-zero cosmological constant, which was first calculated by Einstein who found that $\kappa = \frac{1}{2}$ for the cold static Universe [9] (for the hot static Universe filled with ultrarelativistic matter, $\kappa = 1$). In the expanding or rotating Universe the vacuum is perturbed by expansion or rotation, etc. In all these cases, the value of the vacuum energy density is proportional to the magnitude of perturbations. Since all the perturbations of the vacuum are small in the present Universe, the present cosmological constant must be small.

CONCLUSION

What is the condensed matter experience good for? It provides us with some criteria for selecting the proper theories in particle physics and gravity, For example, some scenarios of inflation are prohibited, since according to the Gibbs-Duhem relation the metastable false vacuum also has zero energy [17]. The condensed matter experience suggests its specific solutions to different fundamental problems, such as cosmological constant problem. It demonstrates how the symmetry and physical laws emerge in different corners of parameters, including the zero energy corner. It also provides us with a variety of universality classes and corresponding effective theories, which are not restricted by Lorentz invariance and by other imposed symmetries.

The effective field theory is the major tool in condensed matter and particle physics. But it is not appropriate for the calculation of the vacuum energy in terms of the zero-point energy of effective quantum fields. Both in condensed matter and particle physics, the contribution of the zero-point energy to the vacuum energy exceeds, by many orders of magnitude, the measured vacuum energy. The condensed matter, however, gives a clue to this apparent paradox: it demonstrates that this huge contribution is cancelled by the microscopic (trans-Planckian) degrees of freedom that are beyond the effective theory. We may know nothing about the trans-Planckian physics, but the cancellation does not depend on the microscopic details, being determined by the general laws of thermodynamics. This allows us to understand, in particular, what happens after the cosmological phase transition, when the vacuum energy decreases and thus becomes negative. The microscopic degrees of freedom will dynamically readjust themselves to the new vacuum state, relaxing the vacuum energy back to zero [17]. Actually, the observed compensation of zero-point energy suggests that there exists an underlying microscopic background and the general relativity is an effective theory rather than a fundamental one.

In the disturbed vacuum, the compensation is not complete, and this gives rise to the non-zero vacuum energy proportional to disturbances. The cosmological constant is small simply because in the present Universe all the disturbances are small: the matter is very dilute, and the expansion is very slow, i.e. the vacuum of the Universe is very close to its equilibrium state. One of the disturbing factors in our Universe is the gravitating matter, this is why it is natural that the measured cosmological constant is on the order of the energy density of the matter: $\kappa \sim 3$ in Eq. (17).

Thus, from the condensed matter point of view, there are no great paradoxes related to the vacuum energy and cosmological constant. Instead we have the practical problem to be solved: how to calculate κ and its time dependence. Of course, this problem is not simple, since it requires the physics beyond the Einstein equations, and there are too many routes on the way back from the effective theory to the microscopic physics.

ACKNOWLEDGMENTS

This work is supported in part by the Russian Ministry of Education and Science, through the Leading Scientific School grant #2338.2003.2, and by the European Science Foundation COSLAB Program.

REFERENCES

1. D. J. Thouless, *Topological Quantum Numbers in Nonrelativistic Physics*, World Scientific, Singapore, 1998.
2. S. Chadha and H. B. Nielsen, *Nucl. Phys.*, **B 217**, 125–144 (1983).
3. G. E. Volovik, *The Universe in a Helium Droplet*, Clarendon Press, Oxford, 2003.
4. R. B. Laughlin and D. Pines, *Proc. Natl Acad. Sci. USA*, **97**, 28–31 (2000).
5. G. Ecker, "Effective Field Theories", hep-ph/0507056.
6. P. Horava, *Phys. Rev. Lett.*, **95**, 016405 (2005).
7. P. Horava and C. A. Keeler, "Noncritical M-Theory in 2+1 Dimensions as a Nonrelativistic Fermi Liquid", hep-th/0508024.
8. N. A. Greaves and A. J. Leggett, *J. Phys. C: Solid State Phys.*, **16**, 4383–4404 (1983).
9. A. Einstein, *Sitzungberichte der Preussischen Akademie der Wissenschaften*, **1**, 142–152 (1917); also in a translated version in *The Principle of Relativity*, Dover, 1952.
10. L. Smolin, "The Case for Background Independence," hep-th/0507235.
11. S. Weinberg, *Rev. Mod. Phys.*, **61**, 1–23 (1989).
12. T. Padmanabhan, *Phys. Rept.*, **380**, 235–320 (2003).
13. D. N. Spergel, L. Verde, H. V. Peiris, *et al.*, *Astrophys. J. Suppl.*, **148**, 175–194 (2003).
14. D. Amati and G. Veneziano, *Phys. Lett.*, **105 B**, 358–362 (1981).
15. A. A. Abrikosov, L. P. Gorkov, and I. E. Dzyaloshinskii, *Quantum Field Theoretical Methods in Statistical Physics*, Pergamon, Oxford, 1965.
16. A. A. Andrianov, V. A. Andrianov, P. Giacconi, and R. Soldati, *JHEP*, **0507**, 003 (2005).
17. G. E. Volovik, *Annalen der Physik*, **14**, 165–176 (2005).

QUANTUM GASES, FLUIDS, AND SOLIDS

Path-Integral Monte Carlo and the Squeezed Trapped Bose-Einstein Gas

Juan Pablo Fernández[1,2] and William J. Mullin[1]

[1]*Department of Physics, University of Massachusetts, Amherst, MA 01003, U.S.A.*
[2]*Lebanon College, Lebanon, NH 03766, U.S.A.*

Abstract. Bose-Einstein condensation has been experimentally found to take place in finite trapped systems when one of the confining frequencies is increased until the gas becomes effectively two-dimensional (2D). We confirm the plausibility of this result by performing path-integral Monte Carlo (PIMC) simulations of trapped Bose gases of increasing anisotropy and comparing them to the predictions of finite-temperature many-body theory. PIMC simulations provide an essentially exact description of these systems; they yield the density profile directly and provide two different estimates for the condensate fraction. For the ideal gas, we find that the PIMC column density of the squeezed gas corresponds quite accurately to that of the exact analytic solution and, moreover, is well mimicked by the density of a 2D gas at the same temperature; the two estimates for the condensate fraction bracket the exact result. For the interacting case, we find 2D Hartree-Fock solutions whose density profiles coincide quite well with the PIMC column densities and whose predictions for the condensate fraction are again bracketed by the PIMC estimates.

Keywords: Bose-Einstein condensation; Two-dimensional Bose systems; Path-integral Monte Carlo simulations.
PACS: 03.75.Hh, 05.30.Jp, 05.70.Fh, 32.80.Pj

Though it has been rigorously proved that two-dimensional (2D) Bose-Einstein condensation cannot occur in the thermodynamic limit, the fact that it has been experimentally detected [1] in finite trapped gases is not too surprising. Nothing in principle forbids taking an anisotropic "pancake"-shaped trap and increasing the largest of its confining frequencies ($\omega_z \equiv \lambda\omega$, say, where $\omega \equiv \omega_x = \omega_y$) to an arbitrary degree, and it would be reasonable to expect some sort of quasi-2D behavior beyond a certain threshold. An ideal gas of N atoms, for example, condenses at a higher temperature than its 3D counterpart ($T_{2D} = 0.829N^{1/2}T_{3D}$), and thus it should be possible to condense a system by compressing it. A zero-temperature model based on the Thomas-Fermi approximation [1] predicts that the crossover for an interacting gas occurs at a compression ratio $\lambda = (225m\omega a^2N^2 / 32\hbar)^{1/3}$, where m is the mass of each atom and a is its s-wave scattering length. Beyond this point, the system acquires an effective two-dimensional coupling constant given by $g_{2D} = (\lambda m\omega / 2\pi\hbar)^{1/2} g_{3D}$, where $g_{3D} = 4\pi\hbar^2 a / m$; this result provides a characterization of a 2D system in terms of experimentally known parameters.

The smooth lines in Figure 1 show both the exact number density—integrated over z, expressed as a function of $\xi = (m\omega/\hbar)^{1/2} (x^2 + y^2)^{1/2}$, and multiplied by $2\pi\xi$ so that $\int d\xi\, N(\xi) = N$—of an ideal gas of 100 atoms at $T = 1.3\,T_{3D}$ compressed to $\lambda = 50$ and the density profile of a 2D gas at the same actual temperature, which here is $T = 0.728T_{2D}$. The close agreement makes the curves almost indistinguishable. Both densities have been resolved into their condensate and noncondensate components.

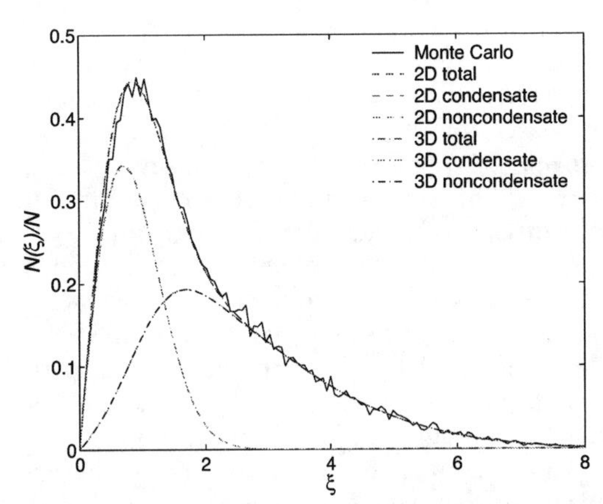

FIGURE 1. Surface number density of a compressed 3D ideal Bose gas and density of an identical 2D gas at the same temperature.

CP850, *Low Temperature Physics: 24th International Conference on Low Temperature Physics;*
edited by Y. Takano, S. P. Hershfield, S. O. Hill, P. J. Hirschfeld, and A. M. Goldman

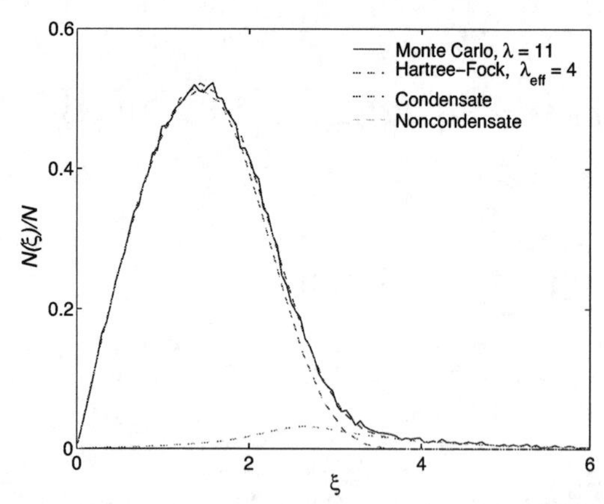

FIGURE 2. Monte Carlo number surface density and best-fit 2D profile of an interacting 3D Bose gas in a highly anisotropic trap.

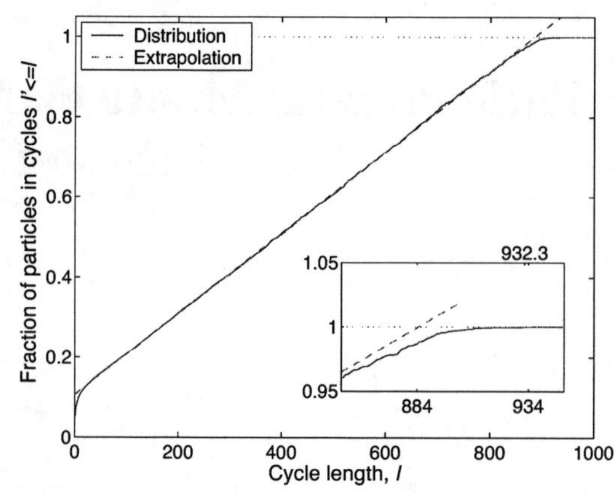

FIGURE 3. Fraction of atoms in cycles of lengths $l' \leq l$ for the gas in Fig. 2. To estimate N_0 we can take the first l for which this is unity or extrapolate the linear segment up to 1.

The jagged line, which also agrees quite well with the total density profiles, is the result of a path-integral Monte Carlo simulation of the compressed system. The method, pioneered in this context by W. Krauth [2], calculates ensemble averages using as a weighting function the N-body density matrix of the system—expanded via Feynman path integrals into slices with tractable high-temperature actions and fully symmetrized to account for Bose statistics. The resulting high-dimensional integral and sum over $N!$ possible label reshufflings is sampled by generating multiparticle, multislice moves appropriate for the ideal gas and sifting them through the Metropolis algorithm. The simulation incorporates interactions between the atoms by solving exactly the quantum two-body problem for hard spheres of radius a and using a pair-product approximation. The program outputs histograms of particles' positions, from which the number density can be inferred directly. Two ways of extracting the condensate fraction N_0 use the fact that every single configuration of the system may be characterized by a permutation that can in turn be decomposed into c_1 1-cycles, c_2 2-cycles, etc., which information the program can easily store and output; the presence of nonzero c_l's for high values of l signals the occurrence of BEC. Krauth's original method [2] reasons that the longest cycle must contain condensed atoms exclusively; a second method [3] assumes that, while the non-condensed particles occur only in small cycles, the condensate is composed in equal parts of atoms from all cycles up to a maximum size, accessible through extrapolation, that corresponds to one large permutation cycle involving the entire condensate. For example, in the ideal-gas simulation described above, the program yields $N_0 = 62$ and $N_0 = 39$ respectively, which bracket the exact value, $N_0 = 40$. (The uncertainty is large due to the small N but decreases in larger systems.)

Figure 2 depicts the Monte Carlo number surface density of an *interacting* three-dimensional Bose gas of $N = 1000$ rubidium atoms (which should become 2D at $\lambda \geq 5.07$) at $T = 0.5316\,T_{3D}$, with a compression ratio $\lambda = 11$. The Monte Carlo result is superimposed on a pure-2D mean-field profile (broken again into condensed and thermal components) obtained by solving the finite-temperature Gross-Pitaevskii equation for the condensate with a minimization-collocation method and treating the thermal cloud in the Hartree-Fock approximation [4]. The only adjustable parameter is the effective anisotropy parameter in the coupling constant; the profile that best fits the Monte Carlo density in this case has $\lambda_{eff} = 4$. In general we find this effective compression factor to be lower than the Thomas-Fermi prediction; this is presumably a finite-temperature effect.

Figure 3 shows the two methods used to find the condensate fraction of the same system. The inset zooms in on the region of interest and displays Krauth's and our estimates (934 and 884, respectively) along with the fraction predicted by the best-fit 2D Hartree-Fock calculation (932.3). The Monte Carlo values again bracket the mean-field result. We conclude that there is a phenomenon resembling a condensation into a single state in a 3D system so compressed as to exhibit quasi-2D behavior.

REFERENCES

1. A. Görlitz *et al.*, *Phys. Rev. Lett.* **87**, 130402 (2001).
2. W. Krauth, *Phys. Rev. Lett.* **77**, 3695 (1996).
3. W. J. Mullin, S. D. Heinrichs, and J. P. Fernández, *Physica B* **284–288**, 9 (2000).
4. J. P. Fernández and W. J. Mullin, *J. Low Temp. Phys.* **128**, 233 (2002); **138**, 687 (2005).

Composite Fermions and Quartets in Optical Traps and in High-T_c Superconductors

M.Yu. Kagan*, I.V. Brodsky*, A.V. Klaptsov*, D.V. Efremov†, R. Combescot** and X. Leyronas**

*P.L. Kapitza Institute for Physical Problems, Kosygin str., 2, Moscow, 119334, Russia
†Technische Universitat Dresden Institut fur Theoretische Physik, Dresden, 01062, Germany
**Ecole Normale Supérieure, 24 rue Lhomond, 75231 Paris Cedex 05, France

Abstract. We consider a possibility of the creation of composite fermions in optical traps and in high-T_c superconductors. For optical traps we study a model of Fermi-Bose mixture with resonant attraction between particles of different sorts. In this case a pairing between fermion and boson of the type bf is possible. This pairing corresponds to creation of composite fermions. At low temperatures and equal densities of fermions and bosons composite fermions are further paired in quartets. In the 2D case we exactly solve Skorniakov-Ter-Martirosian type of integral equations [1] and find the binding energies of two bosons plus one fermion fbb and two bosons plus two fermions $fbfb$. For high-T_c superconductors we consider a quartet - a bound state of two composite holes $\Delta = < hh >$, where each composite hole $h = fb$ consists of a spinon and a holon bound by the stringlike potential. Our investigations are important for recent experiments on the observation of weakly bound composite fermions and bosons in optical traps in the regime of Feshbach resonance.

Keywords: BCS-BEC crossover, composite particles, superconductivity
PACS: 32.80.Pj, 05.30.Fk, 05.30.Jp, 03.75.Mn

INTRODUCTION

The model of a Fermi-Bose mixture is very popular nowadays in connection with different problems in condensed matter physics such as high-T_c superconductivity (SC), superfluidity in ^{3}He-^{4}He mixture [2], fermionic superfluidity in magnetic traps and so on.

In high-T_c superconductivity this model was first proposed by Ranninger and co-workers [3] to describe simultaneously the high transition temperature and short coherence length of SC pairs on the one hand and the presence of a well defined Fermi surface on the other hand. Later on Anderson [4] and Laughlin [5] reformulated this model introducing bosonic degrees of freedom (holons) and fermionic degrees of freedom (spinons) bound by a string-like potential [6].

In this paper we show that, for low temperatures and equal densities of fermions and bosons $n_B = n_F = n$ the composite fermions are further paired in quartets $bfbf$.

FERMI-BOSE MIXTURE OF NEUTRAL PARTICLES

We start from the Fermi-Bose mixture of neutral particles in optical traps. Here we consider the Hamiltonian with a short-range resonant attraction between fermions and bosons. It leads to a shallow bound state of fermion and boson $|E_b| = 1/ma^2$ with a large scattering length $a \gg r_0$, where r_0 is the range of the potential, and we assume for simplicity $m_B = m_F = m$. To complete the construction of the phase diagram of this model, we consider a residual interaction between a molecule bf and unpaired particles b or f as well as between two molecules $bf_\uparrow$ and $bf_\downarrow$. This residual interaction proves to be repulsive for $bf_\uparrow$ and $f_{\uparrow,\downarrow}$. However, it is attractive [7] for bf and b. In the 2D case we exactly solve Skorniakov-Ter-Martirosian integral equations (see Fig. 1) and find $E_3 = 2.39E_b$ for the bound state of bfb [8]. We also derive exact integral equation for quartets [8] and find $E_4 = 3E_b$ for the most shallow bound state of $bfbf$.

The phase diagram of the model is dominated by unpaired fermions and bosons at high temperatures. For low temperatures $T < T_* = |E_4|/3\ln(|E_4|/4T_0)$, where $T_0 = 2\pi n/m$ is degeneracy temperature for fermions and bosons, the quartets prevail in the system. They are bose-condensed below the temperature ([9])

$$T_c = \frac{T_0}{\ln\ln(4/na^2)} \tag{1}$$

STRONGLY INTERACTING FERMI-BOSE MIXTURE OF SPINONS AND HOLONS

For high-T_c superconductors our idea is starting from strongly interacting mixture of spinons $f_{i\sigma}$ and holons b_i to derive a one-band model for weakly interacting

CP850, *Low Temperature Physics: 24th International Conference on Low Temperature Physics;*
edited by Y. Takano, S. P. Hershfield, S. O. Hill, P. J. Hirschfeld, and A. M. Goldman

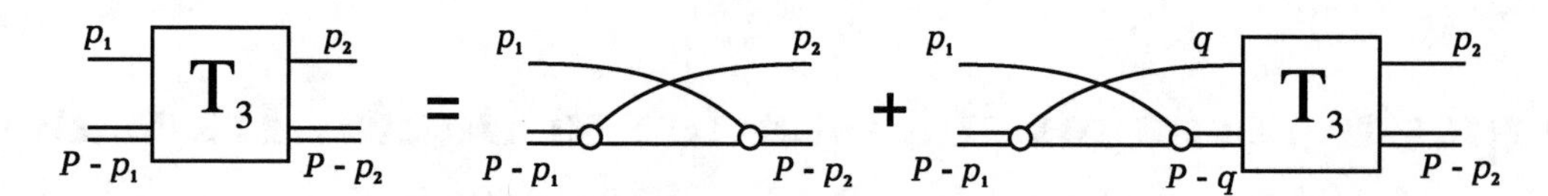

FIGURE 1. Skorniakov-Ter-Martirosian integral equation for trios.

composite holes $h_{i\sigma} = f_{i\sigma} b_i$. Here we utilize the classical solution of Ref. [6]. According to Ref. [6] the composite hole (spin polaron) is a quasiparticle with a spectrum:

$$E = E_b + JQ^2, \quad (2)$$

where J is antiferromagnetic exchange interaction and $E_b \sim J^{2/3} t^{1/3}$ is a bound state of spinon and holon in a string-like potential $V(R) = JzS^2 R/2d$ (z is the number of nearest neighbors, d is intersite distance, $S = 1/2$ is a spin value).

As it was shown in Ref. [10], the residual interaction between two composite holes has a dipole-dipole character $V(r) \sim \lambda/r^2$ for small hole concentration $x \ll 1$. This interaction can lead to a shallow bound state of two holes (to the quartet formation) in the d-wave channel [11]. It is quite appealing to consider superconductive curve for T_c versus concentration x as BSC-BEC crossover for pairing of composite holes in the d-wave channel.

CONCLUSIONS

Our investigations enrich the phase diagram for ultracold Fermi-Bose mixtures in optical traps. They are important in connection with recent experiments of Jin et al. [12] on heteronuclear Feshbach resonance. In these experiments a bound state of ^{87}Rb boson and ^{40}K fermion (composite fermion in our terminology) was observed. They are also building the bridge between the physics of ultracold Fermi-Bose gases and high-T_c superconductors.

ACKNOWLEDGMENTS

This work was supported by Russian Foundation for Basic Research (Grant No. 04-02-16050), CRDF (Grant No. RP2-2355-MO-02) and the grant of Russian Ministry of Science and Education. We are grateful to A.F. Andreev, I.A. Fomin, P. Fulde, Yu. Kagan, L.V. Keldysh, Yu. Lozovik, S.V. Maleev, B.E. Meierovich, A.Ya. Parshin, T.M. Rice, V.N. Ryzhov, G.V. Shlyapnikov, G.S. Strinati, V.B. Timofeev, D. Volhardt and P. Woelfle for fruitful discussions. M.Yu.K is grateful to the University of Pierre and Marie Curie, where this work was completed.

REFERENCES

1. G. V. Skorniakov and K. A. Ter-Martirosian, *Zh. Eksp. Teor. Phys* **31**, 775 (1956) [*Sov. Phys. - JETP* **4**, 648 (1957)].
2. J. Bardeen, G. Baym, and D. Pines, *Phys. Rev.* **156** 207 (1967).
3. J. Ranninger and S. Robaszkiewicz, *Physica B* **135**, 468 (1985); B. K. Chakraverty, J. Ranninger, and D. Feinberg, *Phys. Rev. Lett.* **81**, 433 (1998).
4. P. W. Anderson, *Science* **235**, 1196 (1987).
5. R. B. Laughlin, *Phys. Rev. Lett.* **60**, 2677 (1988); B. A. Bernevig, D. Giuliano, and R. B. Laughlin *Phys. Rev. Lett.* **87**, 177206 (2001).
6. L. N. Bulaevskii, E. L. Nagaev, and D. I. Kholmskii, *JETP* **27**, 836 (1968); W. F. Brinkman and T. M. Rice *Phys. Rev. B* **2**, 1324 (1970).
7. M. Yu. Kagan, I. V. Brodsky, D. V. Efremov, and A. V. Klaptsov, *Phys. Rev. A* **70**, 023607 (2004).
8. I. V. Brodsky, M. Yu. Kagan, A. V. Klaptsov, R. Combescot, and X. Leyronas *Phys. Rev. A*, in preparation.
9. D. S. Fisher and P. C. Hohenberg, *Phys. Rev. B* **37**, 4936 (1988); M. Yu. Kagan and D. V. Efremov *Phys. Rev. B* **65**, 195103 (2002).
10. B. I. Shraiman and E. D. Siggia, *Phys. Rev. B* **42**, 2485 (1990).
11. V. I. Belinicher, A. L. Chernyshev, and V. A. Shubin, *Phys. Rev. B* **56**, 3381 (1997).
12. S. Inouye, J. Goldwin, M. L. Olsen, C. Ticknor, J. L. Bohn, and D. S. Jin, *Phys. Rev. Lett.* **93**, 183201 (2004).

Elementary Excitations of Condensates in a Kronig-Penney Potential

I. Danshita[a], S. Kurihara[a], and S. Tsuchiya[b]

[a]*Department of Physics, Waseda University, Ohkubo, Shinjuku, Tokyo 169-8555, Japan*
[b]*Institute of Industrial Science, University of Tokyo, Komaba, Meguro, Tokyo 153-8505, Japan*

Abstract. We investigate the excitation spectrum of a Bose-Einstein condensate in a Kronig-Penney potential. We solve the Bogoliubov equations analytically and obtain the band structure of the excitation spectrum. We find that the excitation spectrum is gapless and linear at low energies. This property is found to be attributed to the anomalous tunneling behavior of low energy excitations, which has been predicted by Kagan *et al.*

Keywords: Bose-Einstein condensation, optical lattice, elementary excitation, Kronig-Penney model
PACS: 03.75.Lm 05.35.Jp 03.75.Kk

Recently, experimental observation of elementary excitations of Bose-Einstein condensates (BECs) in an optical lattice[1] has attracted much attention. In the present paper, we theoretically study elementary excitations of BECs in an optical lattice with use of a Kronig-Penney (KP) potential. The KP model is useful in understanding the problem of the periodic potential qualitatively, because it allows an analytical treatment of the problem for arbitrary values of the lattice depth. Moreover, the excitation spectrum can be related to tunneling properties of the excitations through a single potential barrier in the KP model. Kagan *et al.*[2] studied the tunneling problem of the excitations and predicted that a barrier is transparent for excitations within a limited range of low energies; they called such a behavior the anomalous tunneling. We shall show that the anomalous tunneling is crucial to the phonon-like form of the low energy excitations in the KP potential.

We consider a BEC confined in a combined potential of an axisymmetric harmonic potential and a one-dimensional periodic potential along the axial direction (the x axis). As the periodic potential, we adopt a Kronig-Penney potential

$$V(x) = V_0 \sum_{n=-\infty}^{\infty} \delta(x-na), \tag{1}$$

where a is the lattice constant, and V_0 is the strength of the delta-function potential barrier. It is assumed that the condensate is so elongated along the axial direction that the axial confinement can be neglected. We assume that the frequency of the harmonic potential of the radial direction is large enough to regard the problem as one-dimensional. Our formulation of the problem is based on the mean-field theory[3], which consists of the time-independent Gross-Pitaevskii equation and the Bogoliubov equations. They are

$$\left[-\frac{\hbar^2}{2m}\frac{d^2}{dx^2} + V(x) + g\,|\Psi_0|^2\right]\Psi_0 = \mu\Psi_0, \tag{2}$$

and

$$\begin{pmatrix} H_0 & -g\Psi_0(x)^2 \\ g\Psi_0^*(x)^2 & -H_0 \end{pmatrix}\begin{pmatrix} u(x) \\ v(x) \end{pmatrix} = \varepsilon \begin{pmatrix} u(x) \\ v(x) \end{pmatrix}, \tag{3}$$

$$H_0 = -\frac{\hbar^2}{2m}\frac{d^2}{dx^2} - \mu + V(x) + 2g\,|\Psi_0(x)|^2 \quad . \tag{4}$$

Here μ is the chemical potential and g is the coupling constant of the interparticle interaction.

At first, we shall solve Eq. (2). We assume that the condensate does not have a supercurrent. Assuming that the lattice constant is much larger than the healing length, one can approximately obtain the ground state solution of Eq. (2) as

$$\Psi_0(x) = \sqrt{\frac{\mu}{g}}\tanh\left(\frac{|x-na|+x_0}{\xi}\right), \quad (n-\tfrac{1}{2})a < x < (n+\tfrac{1}{2}), \tag{5}$$

where $\xi \equiv \hbar/\sqrt{m\mu}$ is the healing length and x_0 is determined by the boundary condition at $x=na$.

CP850, *Low Temperature Physics: 24th International Conference on Low Temperature Physics;*
edited by Y. Takano, S. P. Hershfield, S. O. Hill, P. J. Hirschfeld, and A. M. Goldman

Next, we analytically calculate the band structure of the excitation spectrum. We shall first find solutions of Eq. (3) in the region $|x|<a/2$ where only a single barrier exists. There exist two independent solutions with the excitation energy ε of the single barrier problem, corresponding to two types of scattering process. One solution $\psi^l(x)$ describes the process where a Bogoliubov excitation comes from left, and the other solution $\psi^r(x)$ describes the process where a Bogoliubov excitation comes from right. The solution $\psi^l(x)$ is written[2] as

$$\psi^l(x)=\begin{cases}\begin{pmatrix}u_1\\v_1\end{pmatrix}e^{ip_+x/\hbar}+r\begin{pmatrix}u_2\\v_2\end{pmatrix}e^{-ip_+x/\hbar} \\ +b\begin{pmatrix}u_3\\v_3\end{pmatrix}e^{p_-x/\hbar}, & x<0, \\ t\begin{pmatrix}u_1\\v_1\end{pmatrix}e^{ip_+x/\hbar}+c\begin{pmatrix}u_4\\v_4\end{pmatrix}e^{-p_-x/\hbar}, & x>0,\end{cases} \tag{6}$$

where $p_\pm=\sqrt{2m(\sqrt{\mu^2+\varepsilon^2}\mp\mu)}$ satisfies the Bogoliubov spectrum for a uniform system. The coefficients r, b, t, and c are the amplitudes of reflected, left localized, transmitted, and right localized components, respectively. All the coefficients can be determined by the boundary condition at x=0. Expanding $|t|$ and the phase δ of t around ε =0, one can analytically obtain

$$|t|\cong 1-\alpha\left(\tfrac{\varepsilon}{\mu}\right)^2, \quad \delta\cong\beta\tfrac{\varepsilon}{\mu}, \tag{7}$$

where α and β are functions of V_0. The transmission coefficient $T=|t|^2$ approaches unity and the phase approaches zero as the energy is reduced to zero. This means that the potential barrier is transparent for low energy excitations — anomalous tunneling of Kagan *et al.*[2] One can also calculate $\psi^r(x)$ in the same way.

We can write a general solution of Eq. (3) in $|x|<a/2$ as a linear combination of $\psi^l(x)$ and $\psi^r(x)$:

$$\left(u(x),v(x)\right)^{\mathbf{t}}\equiv\psi(x)=\chi\psi^l(x)+\eta\psi^r(x) , \tag{8}$$

where χ and η are arbitrary constants. Extending this solution to all regions of x by means of Bloch's theorem, one can obtain the relation between the excitation energy and the quasi-momentum q:

$$\cos(\tfrac{p_+a}{\hbar}+\delta)/|t|=\cos(\tfrac{qa}{\hbar}). \tag{9}$$

This relation is exactly the same form as that of a single particle.[4] We clearly see from Eq. (7) that the tunneling properties in the single barrier problem determine the band structure of the excitation spectrum.

Solving Eq. (9) for the excitation energy, we can obtain the band structure of the excitation spectrum as shown in Fig. 1, which shows that the first band is phonon-like in the low energy regions. One can verify that the form of the excitation spectrum is phonon-like for arbitrary values of the potential depth by substituting Eq. (7) into Eq. (9). Thus, the anomalous tunneling properties of Eq. (7), namely the perfect transmission of very low energy excitations, leads to the phonon dispersion.

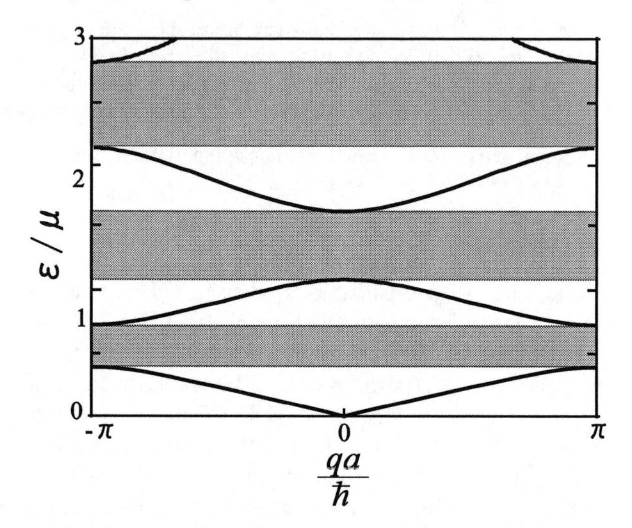

FIGURE 1. Band structure of the excitation spectrum in a KP potential with a = 5ξ and V_0 = 3$\mu\xi$ are shown. Shaded areas represent band gaps.

ACKNOWLEDGMENTS

The authors would like to thank T. Kimura, N. Yokoshi, T. Nikuni, and N. Hatano for valuable discussions. I. D. is supported by a Grant-in-Aid for JSPS fellows.

REFERENCES

1. T. Stöferle, H. Moritz, C. Schori, M. Köhl and T. Esslinger, *Phys. Rev. Lett.* **92**, 130403 (2004).
2. Yu. Kagan, D. L. Kovrizhin and L. A. Maksimov, *Phys. Rev. Lett.* **90**, 130402 (2003).
3. L. P. Pitaevskii and S. Stringari, *Bose-Einstein Condensation*, NY: Oxford University Press, 2003, pp. 177-179.
4. N. W. Ashcroft and N. D. Mermin, *Solid State Physics*, Philadelphia: Saunders College Publishing, 1976, pp. 146-149.

Elementary Excitations of Condensates in Double-Well Traps

Kyota Egawa, Ippei Danshita, Nobuhiko Yokoshi, and Susumu Kurihara

Department of Physics, Waseda University, 3-4-1 Ohkubo, Shinjuku, Tokyo 169-8555, Japan

Abstract. We investigate collective excitations of Bose-Einstein condensates in double-well traps. We numerically solve the Bogoliubov-de Gennes equations with a double-well trap and show that a crossover from the dipole mode to the Josephson plasma oscillation occurs in the lowest excitation. It is found that the anomalous tunneling property of the low energy excitations dominates the crossover.

Keywords: Bose-Einstein condensation, double-well trap, Bogoliubov excitation, Josephson plasma
PACS: 03.75.Lm 05.35.Jp 03.75.Kk

Recently Josephson plasma oscillation was observed by Albiez *et al.*[1] in Bose-Einstein condensates (BECs) of atomic gases trapped in double wells. Because of the high tunability of system parameters, such a system provides an ideal stage for observing the phase coherence and tunneling phenomena. Here it is expected that, as far as the non-linearity is small, the low energy excitations can be described by the Bogoliubov theory.[2] Salasnich *et al.*[3] solved the Bogoliubov-de Gennes (BdG) equations and found that the lowest excitation energy, which is the dipole mode energy when the barrier is switched off, reduces as the barrier height increases. One of the present authors analytically investigated the relation between the reduction of the lowest excitation energy and a crossover to the Josephson plasma mode in a box-shaped barrier potential.[4] It was found that anomalous tunneling, which was suggested by Kagan *et al.*,[5] plays a crucial role in the crossover. In this paper, we numerically solve the BdG equations and show that the mechanism of tunneling is valid also in an experimentally accessible trap.

We consider BECs in an axi-symmetric harmonic potential with a barrier; the explicit form is given by

$$V_{ext}(\mathbf{r}) = \frac{m\omega_\rho^2\rho^2}{2} + \frac{m\omega_z^2 z^2}{2} + U_0 \exp\left(-\frac{z^2}{\sigma^2}\right), \quad (1)$$

where the confinement in the radial direction $\rho=(x^2+y^2)^{1/2}$ is much stronger than in the z direction ($\omega_\rho >> \omega_z$). Then, we assume one-dimensional treatment of the problem can be justified. The barrier is created by focusing a blue-tuned laser beam, and one can control the strength U_0 and width σ of the barrier by varying the intensity and focusing of the laser. When the barrier height is zero, the lowest excitation is the dipole mode, which can be understood as the motion of the center of mass. On the other hand, when the barrier potential is high enough, the density of the condensates around z=0 is dilute and the lowest excitation mode of the system can be interpreted as the Josephson plasma oscillation.

In this system, the Gross-Pitaevskii (GP) equation can be written as

$$\left(-\frac{\hbar^2}{2m}\frac{\partial^2}{\partial z^2} + V_{ext}(z) - \mu + g|\Psi(z)|^2\right)\Psi(z) = 0, \quad (2)$$

where $g=2\hbar^2 a_s/m\alpha_\rho^2$ with a_s and α_ρ being the s-wave scattering length and harmonic oscillator length in the radial direction, respectively. μ is the chemical potential. In order to solve the GP equation, we expand $\Psi(z)$ in the basis of the trap-potential eigenfunctions.[6] Inserting this function into the GP equation and taking the integration after multiplying each basis function with the equation, one finds nonlinear simultaneous equations. These equations can be solved by the Newton-Raphson method.

One can obtain the energy of elementary excitations by solving the BdG equations, which are written as

$$\begin{pmatrix} H_0 & g\Psi(z)^2 \\ -g\Psi^*(z)^2 & -H_0 \end{pmatrix}\begin{pmatrix} u(z) \\ v(z) \end{pmatrix} = \varepsilon \begin{pmatrix} u(z) \\ v(z) \end{pmatrix}, \quad (3)$$

$$H_0 = -\frac{\hbar^2}{2m}\frac{d^2}{dz^2} - \mu + V_{ext}(z) + 2g|\Psi(z)|^2, \quad (4)$$

where $\Psi(z)$ is the solution of the GP equation.

CP850, *Low Temperature Physics: 24th International Conference on Low Temperature Physics;*
edited by Y. Takano, S. P. Hershfield, S. O. Hill, P. J. Hirschfeld, and A. M. Goldman

Figure 1 shows the lowest Bogoliubov excitation and the Josephson plasma energy. When the strength of the Gaussian barrier U_0 equals zero, the lowest Bogoliubov excitation energy approaches $\hbar\omega_z$. This mode can be regarded as the dipole mode described above. As the barrier becomes higher, the dipole mode energy decreases and approaches the Josephson plasma energy $\varepsilon_{pl}=(E_C E_J)^{1/2}$, which is estimated from the GP equation. The capacitive energy E_C is defined as $4d\mu/dN$. As for the Josephson energy E_J, one can interpret it as the difference in the mean-field energy between the even-parity ground state and the odd-parity first excited state.[7]

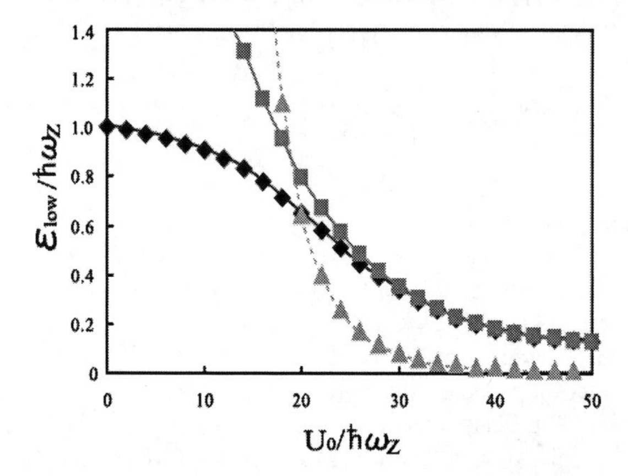

FIGURE 1. The lowest Bogoliubov excitation energy is plotted as a function of the barrier height (diamonds). The squares are the Josephson plasma energy obtained with use of GP result. The triangles are the width of the peak of the tunneling probability. We consider sodium atoms of N=3000. The parameter values are as follow: ω_p=250×2π (1/s), ω_z=10 ×2π (1/s), a_s=3 nm and σ=1.33 μm.

Let us consider the tunneling property of the excitations to elucidate the crossover between the dipole mode and the Josephson plasma mode. The transmission coefficient through the barrier has a peak at zero energy (anomalous tunneling), i.e. the low excitations can be easily transmitted even through the high barrier. This tunneling effect is restricted to the low excitations. We can estimate the peak width $\Delta\varepsilon$ of the tunneling probability within the WKB approximation as

$$\Delta\varepsilon = \frac{e^{-\kappa_0 d}}{\kappa_0}\mu, \kappa_0 = \frac{1}{d}\int_{-d/2}^{d/2}\sqrt{\frac{2m}{\hbar^2}(V_{\text{barrier}}(z)-\mu)}dz. \quad (5)$$

Here d is the classical turning point. Then, by comparing the lowest Bogoliubov excitation energy with the width $\Delta\varepsilon$, one can confirm whether the tunneling behavior is effective or not.

In Fig. 1, the width $\Delta\varepsilon$ crosses the excitation energy around the point where the crossover between the dipole mode and the Josephson plasma mode occurs. This leads to the following interpretation. When the lowest Bogoliubov excitation energy is small enough compared to the width $\Delta\varepsilon$, the excitations can pass through the barrier. Therefore the lowest excitation is not affected by the barrier potential, and its energy almost corresponds to the dipole mode energy. On the other hand, when the lowest excitation energy is larger than the width $\Delta\varepsilon$, the excitations cannot pass through the barrier. Then the lowest excitation energy approaches the Josephson plasma energy.

In summary, we solved GP equation numerically in an experimentally accessible trap and calculated the Bogoliubov excitations. Focusing on the lowest Bogoliubov excitation, we confirmed the crossover from the dipole mode to the Josephson plasma oscillation. We estimated the contribution of the anomalous tunneling to the crossover.

This work was partially supported by The 21st Century COE Program at Waseda University. I.D. is supported by a JSPS fellowship.

REFERENCES

1. M. Albiez, R. Goti, J. Folling, S. Hunsmann, M. Cristiani, and M. K. Oberthaler, *Phys. Rev. Lett.* **95**, 010402 (2005).
2. G. S. Paraoanu, S. Kohler, F. Sols, and A. J. Leggett, *J. Phys. B* **34**, 4689 (2001).
3. L. Salasnich, A. Parola, and L. Reatto, *Phys. Rev. A* **60**, 4171 (1999).
4. I. Danshita, "Tunneling Phenomena in Bose Condensed Systems", Master Thesis, Waseda University, 2005.
5. Y. Kagan, D. L. Kovrizhin, and L. A. Maksimov, *Phys. Rev. Lett.* **90**, 130402 (2003).
6. M. Edwards, R. J. Dodd, C. W. Clark, and K. Burnett, *J. Res. Natl. Inst. Stand. Technol.* **101**, 553 (1996).
7. L. P. Pitaevskii and S. Stringari, "*Bose-Einstein Condensation*", Oxford: Oxford University Press, 2002, pp. 287.

Dipolar Oscillations of Strongly Correlated Bosons on One-Dimensional Lattices

M. Rigol, V. Rousseau, R. T. Scalettar, and R. R. P. Singh

Physics Department, University of California, Davis, CA 95616, USA

Abstract. We study by means of an exact approach the dipolar oscillations of strongly correlated bosons on one-dimensional lattices. At low densities damping is always present. It increases for larger initial displacements of the trap, and produces dramatic changes in the momentum distribution function. For higher densities, when a Mott insulating core appears in the middle of the system, the center of mass almost does not move after an initial displacement. In this case the momentum distribution function of the system remains, at any stage of the evolution, very similar to the one in the initial ground state.

Keywords: damping, one-dimensional, lattice, Mott-insulator
PACS: 03.75.Kk, 03.75.Lm, 05.30.Jp

INTRODUCTION

Recent experiments studying the dipolar oscillations of trapped one-dimensional (1D) bosons [1, 2] have reported a large damping when an axial lattice is present in the system. In addition, an overdamping was observed in which the center of mass (COM) remains displaced from the middle of the trap [1, 2]. We show here that these experimental features can be well understood by means of an exactly solvable model of impenetrable bosons (HCB) on a 1D lattice [3].

The HCB Hamiltonian can be written as

$$H = -t\sum_i \left(b_i^\dagger b_{i+1} + h.c.\right) + V_2 \sum_i x_i^2\, n_i, \qquad (1)$$

with the addition of the on-site constraints $b_i^{\dagger 2} = b_i^2 = 0$, $\left\{b_i, b_i^\dagger\right\} = 1$, which avoid double or higher occupancy. In Eq. (1), t is the hopping parameter, and V_2 is the curvature of the harmonic confining potential.

Hamiltonian (1) can be exactly diagonalized by means of the Jordan-Wigner transformation, which maps HCB into noninteracting fermions. After diagonalization the remaining nontrivial problem is the calculation of the HCB off-diagonal one-particle correlations. They are needed to obtain the momentum distribution function (n_k), a quantity measured in the experiments. An exact numerical approach for solving this problem was presented in Refs. [4, 5], for the ground-state [4] and non-equilibrium [5] cases. We follow this approach to study the oscillations of a trapped system when at time $\tau = 0$ the trap is displaced a distance x_0. We show that at low densities damping grows with the initial displacement, and in the large damping regime it causes dramatic changes in n_k. On the contrary, when there is a Mott-insulator (MI) in the trap, the COM of the system barely moves from its initial position, and n_k remains similar to the one in the ground state.

DIPOLAR OSCILLATIONS

In Fig. 1(a),(b) we show the COM oscillations of a trapped HCB system that at time $\tau = 0$ has a maximum density $n = 0.48$, and a cloud radius $R_0 \sim 145a$ (a is the lattice constant). The increase of the initial trap displacement produces two effects (i) the damping increases, (ii) for large values of x_0 the COM remains displaced from the middle of the trap. Since HCB can be mapped into noninteracting fermions, one can understand that damping occurs for any initial density and trap displacement. In the fermion language, the damping is caused by a dephasing of the particles, which arises due to their dispersion in the lattice-harmonic trap system [6, 7]. For large displacements, Bragg scattering starts to occur and produces the shift of the COM oscillations in Fig. 1(b). In the latter case a very large damping can be also observed.

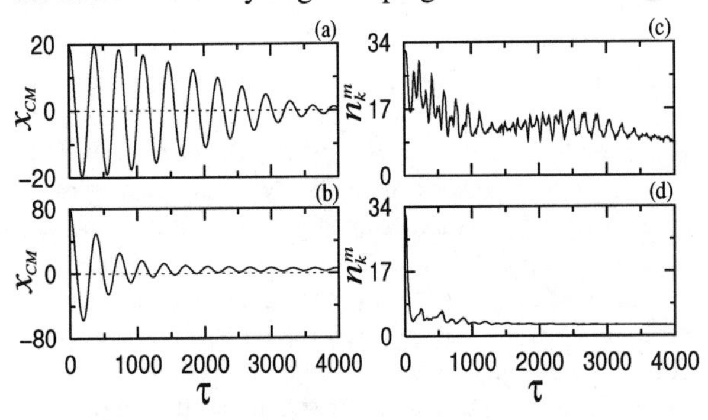

FIGURE 1. Evolution of the COM position (a),(b), and the maximum value of n_k (c),(d), vs time (τ, in units of $\hbar/t$). The initial trap displacements are $x_0 = 20a$ [$x_0/R_0 \sim 0.14$] (a),(c), and $x_0 = 80a$ [$x_0/R_0 \sim 0.55$] (b),(d). These systems have $N_b = 101$, and $V_2 a^2 = 10^{-4} t$. (a denotes the lattice constant.)

We next analyze the consequences of the damping of the COM motion in n_k. In Fig. 1(c),(d) we show how the maximum value of n_k (n_k^m) evolves as a function of time.

CP850, *Low Temperature Physics: 24th International Conference on Low Temperature Physics;* edited by Y. Takano, S. P. Hershfield, S. O. Hill, P. J. Hirschfeld, and A. M. Goldman

The changes in n_k^m become dramatic with increasing x_0. In the large damping regime [Fig. 1(d)], one can see that n_k^m reduces almost to its minimum value in the first oscillation of the COM. The reduction of n_k^m is accompanied by a large increment in the full width at half maximum (w), as shown in Fig. 2. These results are in contrast with the ones for the equivalent noninteracting fermions [6], where after the damping of the COM oscillations, n_k^m and w are almost the same as the ones at $\tau = 0$. (Contrary to the density, the evolution of n_k for HCB is very different to the one of noninteracting fermions.)

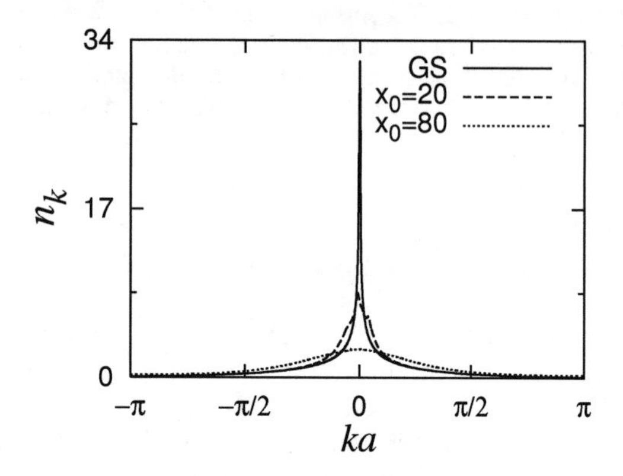

FIGURE 2. n_k of the ground state (GS) compared with the ones at $\tau = 4000\hbar/t$ for two different initial displacements of the trap [$N_b = 101$, and $V_2a^2 = 10^{-4}t$]. As shown in Fig. 1(a),(b), at $\tau = 4000\hbar/t$ the motion of the CM is totally damped.

We have also studied the consequences of approaching the MI regime on the damping of the COM oscillations. We find that in this case the final COM position (after damping) remains displaced from the center of the trap even if the ratio between the initial trap displacement x_0 and the cloud radius R_0 is kept constant. To approach the formation of the MI domain we keep the number of particles in the system fixed while increasing the curvature of the trap. Another important consequence of approaching the formation of the MI core is that the damping rate increases. When the MI domain is present, the COM barely displaces from its original position, and the oscillations of the COM are overdamped [inset in Fig. 3(a)]. The above results are in agreement with the experiments reported by Stoferle *et al.* [1], and differ from the ones in the weakly interacting regime where for small initial trap displacements no shift of the COM position is observed after damping [8].

One interesting feature that appears with the formation of the MI can be seen in Fig. 3(b). Although n_k^m at $\tau = 0$ is small compared with n_k^m when there is no MI [Fig. 1(c),(d)], its value remains almost the same during the evolution of the system. The same occurs with w, as seen in Fig. 3(b). The above result can be intuitively understood by the fact that almost no COM energy is

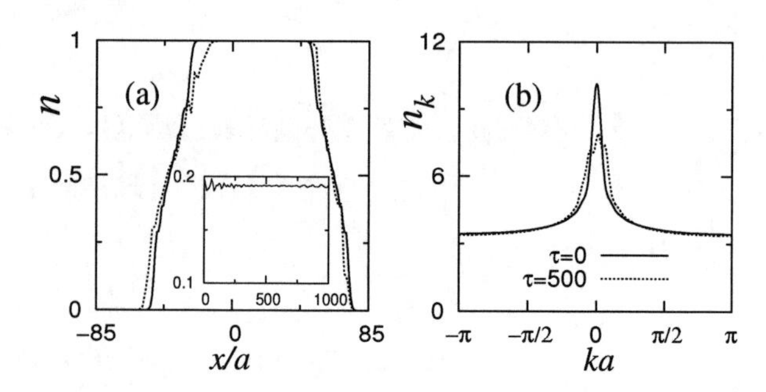

FIGURE 3. Density and n_k in a trap with a MI, $N_b = 101$ and $V_2a^2 = 1.6 \times 10^{-3}t$. The inset shows x_{CM}/R_0 vs τ.

"transfered" into excitation energy. This keeps the system in a state with n_k similar to the ones in the ground state. Something similar may occur in experiments in the overdamped regime, when the COM almost does not change from its initial position [2]. However, we should remark that we do obtain big changes of n_k^m and w (Fig. 2) as a consequence of the damping in systems with no MI. This is in contrast with the behavior of w one infers from the constant cloud widths reported experimentally in Ref. [2]. Since in experiments w is measured after time of flight, the difference between experiments and theory may be due to interparticle interactions during the expansion, which could change the n_k initially present in the trap. (An explicit example for the expansion in 1D was presented in Ref. [9].) If this is the case, other experimental techniques may be needed to determine the effects of the damping on n_k.

ACKNOWLEDGMENTS

We thank G.G. Batrouni, T. Bryant, V. Dunjko, A. Muramatsu, M. Olshanii, and D. Thouless for insightful discussions. This work was supported by NSF-DMR-0312261 and NSF-DMR-0240918.

REFERENCES

1. T. Stöferle *et al.*, *Phys. Rev. Lett.* **92**, 130403 (2004).
2. C. D. Fertig *et al.*, *Phys. Rev. Lett.* **94**, 120403 (2005).
3. M. Rigol, V. Rousseau, R. T. Scalettar, and R. R. P. Singh, *Phys. Rev. Lett.* **95**, 110402 (2005).
4. M. Rigol and A. Muramatsu, *Phys. Rev. A* **70**, 031603(R) (2004); **72**, 013604 (2005).
5. M. Rigol and A. Muramatsu, *Phys. Rev. Lett.* **93**, 230404 (2004).
6. M. Rigol and A. Muramatsu, *Phys. Rev. A* **70**, 043627 (2004).
7. L. Pezzè *et al.*, *Phys. Rev. Lett.* **93**, 120401 (2004).
8. A. Polkovnikov and D.-W. Wang, *Phys. Rev. Lett.* **93**, 070401 (2004).
9. M. Rigol and A. Muramatsu, *Phys. Rev. Lett.* **94**, 240403 (2005).

Expansion of Tonks-Girardeau Gases on a One-Dimensional Optical Lattice

Marcos Rigol* and Alejandro Muramatsu†

*Physics Department, University of California, Davis, CA 95616, USA
†Institut für Theoretische Physik III, Universität Stuttgart, Pfaffenwaldring 57, D-70550 Stuttgart, Germany.

Abstract. We study the free expansion of Tonks-Girardeau gases on one-dimensional lattices after removing the confining potential. We find that for low initial densities the momentum distribution of the expanding gas rapidly approaches the one of noninteracting fermions. In the opposite limit, i.e, when the initial system is a pure Mott insulator with one particle per lattice site, the expansion leads to the emergence of quasi-condensates at finite momentum. These traveling quasi-condensates develop quasi-long range one-particle correlations similar to the ones shown to be universal in the equilibrium case.

Keywords: one-dimensional, Tonks-Girardeau, lattice, bosons
PACS: 03.75.Kk, 03.75.Lm, 05.30.Jp

INTRODUCTION

At low temperatures and densities, bosons in one dimension (1D) are expected to enter the so called Tonks-Girardeau (TG) regime [1]. In this regime they become strongly correlated, and behave as impenetrable particles also known as hardcore bosons (HCB's) [2]. (Two recent experiments have achieved the TG regime [3, 4].) Contrary to weakly interacting bosons, HCB's share many properties with noninteracting spinless fermions to which they can be mapped [2]. Properties like energy and density profiles are identical in both systems. However, quantities like the momentum distribution function (n_k) and the so-called natural orbitals (NO's) are very different due to the different behavior of the off-diagonal one-particle correlations.

We show in this work that the nonequilibrium dynamics of TG gases on 1D lattices exhibits remarkable features. At low densities, far from the regime where a Mott insulator appears in the center of the trap, n_k of expanding HCB's rapidly approaches that of noninteracting fermions [5]. This occurs without any loss in coherence as reflected by a large occupation of the lowest NO. The NO's (ϕ^η) are defined as the eigenfunctions of the one-particle density matrix $\rho_{ij} = \langle b_i^\dagger b_j \rangle$ [6],

$$\sum_{j=1}^{N} \rho_{ij}\phi_j^\eta = \lambda_\eta \phi_i^\eta, \qquad (1)$$

and have occupations λ_η. They are effective one-particle states, like a Bose-Einstein condensate that would correspond to the highest occupied (lowest) NO in a higher dimensional dilute system. At high densities, when the system starts its free evolution from a pure Mott-insulating (Fock) state, we show that quasicondensates of HCB's emerge at finite momentum [7].

EXPANSION OF A 1D TG GAS

On a 1D lattice the HCB Hamiltonian can be written as

$$H = -t\sum_i \left(b_i^\dagger b_{i+1} + h.c.\right) + V_2 \sum_i x_i^2\, n_i, \qquad (2)$$

with the addition of the on-site constraints $b_i^{\dagger 2} = b_i^2 = 0$, $\left\{b_i, b_i^\dagger\right\} = 1$, which avoid a double or higher occupancy. In Eq. 2, the hopping parameter is denoted by t, and the curvature of the harmonic confining potential by V_2.

In Fig. 1(a) we show n_k for an expanding TG gas at four different times after turning off the confining potential. We compare it with the one of noninteracting fermions, which does not change during the expansion. (The initial density in the middle of the trap is $n = 0.33$.) One can see in Fig. 1(a) that the peak at $n_{k=0}$

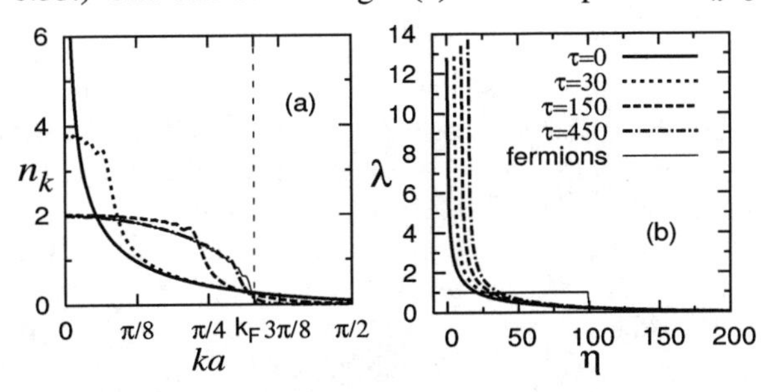

FIGURE 1. n_k (a) and λ (b) for 100 HCB's expanding from an initial state with $V_2a^2 = 2.6\times10^{-5}t$, and compared to the ones for the corresponding fermions. Times (τ) are given in units of $\hbar/t$, k_F denotes de Fermi momentum, and a the lattice constant.

disappears shortly after switching off the trap, and that n_k approaches the one of the fermions as time passes. In contrast to n_k, Fig. 1(b) shows that coherence, as reflected in the occupation of the lowest NO (λ_0), slightly

CP850, *Low Temperature Physics: 24th International Conference on Low Temperature Physics;*
edited by Y. Takano, S. P. Hershfield, S. O. Hill, P. J. Hirschfeld, and A. M. Goldman

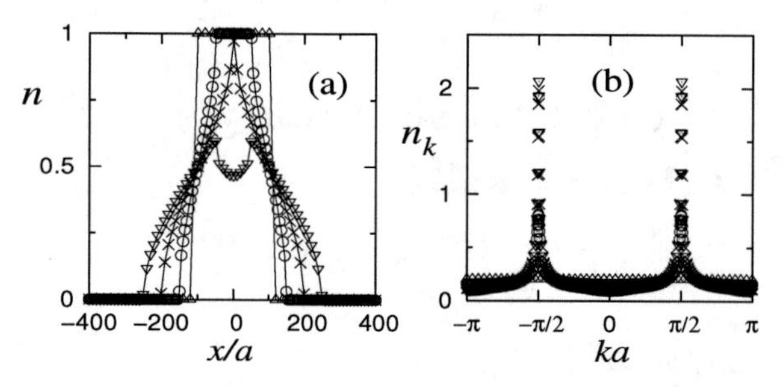

FIGURE 2. Evolution of density (a) and momentum (b) profiles of 201 HCB's in 1000 lattice sites. The times are $\tau = 0$ ($\triangle$), $25\hbar/t$ ($\bigcirc$), $50\hbar/t$ ($\times$), and $75\hbar/t$ (∇). Positions (a) and momenta (b) are normalized by the lattice constant a.

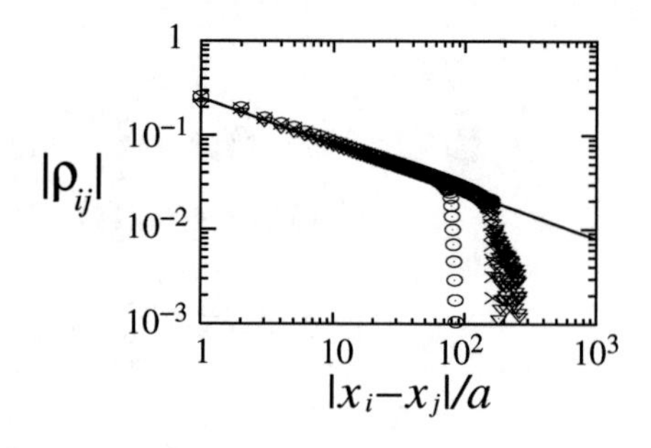

FIGURE 3. $|\rho_{ij}|$ vs $|(x_i - x_j)|$ for density profiles and n_k depicted in Fig. 2. Different evolution times are plotted as τ=$25\hbar/t$ ($\bigcirc$), $50\hbar/t$ ($\times$), and $75\hbar/t$ (∇). The straight line exhibits a power law $|(x_i - x_j)/a|^{-1/2}$.

increases as a function of time. [In Fig. 1(b) we have displaced the λ curves at different times so that the effect can be better seen.] The apparent contradiction between the fermionization of n_k, and hence the reduction of $n_{k=0}$, and the increase of λ_0 can be understood by studying the Fourier transform of the lowest NO at $\tau = 0$ and $\tau > 0$. Initially $|\phi_k^0|$ has a peak at $k = 0$ showing that quasi-condensation occurs around $k = 0$, and this is reflected in n_k [5]. For $\tau > 0$ the lowest NO extends over k-space so that the HCB's forming the lowest NO have many different momenta, following n_k in Fig. 1 [5]. Hence, there is no contradiction between the observed behavior of the NO occupations and n_k. They are just different from the equilibrium case.

The fermionization process discussed above can be understood by studying the one-particle density matrix $\rho_{il} = |\rho_{il}|e^{i\theta_{il}}$. As shown in Ref. [5], $|\rho_{il}| \sim |x_i - x_l|^{-1/2}$ so that as the system expands quasi-long correlations increase the lowest NO occupation. This is what would happen to the ground state of a confined Bose gas if the curvature of the trapping potential is reduced. Coherence increases with increasing the system size because bosons become more delocalized in space. On the other hand the phase θ_{il}, which appears only during the nonequilibrium dynamics, is the one accounting for the destruction of the $n_{k=0}$ peak, and the full fermionization of n_k.

So far we have presented results for a system that at $\tau = 0$ is distant from the regime where a MI appears in the middle of the trap. We find that, as the density in the trap increases and particles become localized over smaller distances in the center of the cloud, a different behavior of n_k is observed at short times. Figure 2 shows density profiles (a), and their corresponding n_k (b), for the time evolution of an initial MI (Fock) state. At $\tau = 0$, n_k is flat as corresponds to a pure Fock state (identical for the fermions), and during the evolution of the system sharp peaks appear at $k = \pm\pi/2a$. The formation of these peaks generates a HCB n_k that departs at short times from the one of the fermions.

The peaks appearing at $k = \pm\pi/2a$ signal the emergence of quasi-condensates of HCB's at finite momentum [7]. This can be understood as follows. When the system is released from the trap, quasi-long range correlations start to develop between initially uncorrelated particles. We find that a power-law decay can be observed in the one-particle density matrix $|\rho_{ij}| \sim |(x_i - x_j)/a|^{-1/2}$ (Fig. 3), which has the same exponent $(-1/2)$ shown to be universal in the ground state [8]. As the number of particles in the initial MI state (N_b) is increased, the size of the traveling quasi-condensate increases $\sim N_b$. This, and the behavior of $|\rho_{ij}|$, implies that the occupation of the lowest NO increases $\sim \sqrt{N_b}$ [7].

We should remark, however, that for long expansion times the high of the $k = \pm\pi/2a$ peaks reduces, and the n_k of the TG gas starts approaching again the one of the equivalent noninteracting fermions. The time scale for the fermionization process is very large in this case, and it starts affecting the low-momenta region first. This implies that the dynamically generated quasi-condensates with $k \sim \pm\pi/2$ could be used as atom lasers [7].

ACKNOWLEDGMENTS

We thank HLR-Stuttgart for allocation of computer time, and SFB 382 for financial support.

REFERENCES

1. M. Olshanii, *Phys. Rev. Lett.* **81**, 938 (1998).
2. M. Girardeau, *J. Math. Phys.* **1**, 516 (1960).
3. B. Paredes *et al.*, *Nature (London)* **429**, 277 (2004).
4. T. Kinoshita, T. Wenger, and D. S. Weiss, *Science* **305**, 1125 (2004).
5. M. Rigol and A. Muramatsu, *Phys. Rev. Lett.* **94**, 240403 (2005).
6. O. Penrose and L. Onsager, *Phys. Rev.* **104**, 576 (1956).
7. M. Rigol and A. Muramatsu, *Phys. Rev. Lett.* **93**, 230404 (2004).
8. M. Rigol and A. Muramatsu, *Phys. Rev. A* **70**, 031603(R) (2004); **72**, 013604 (2005).

Non-Conservation of Transverse Magnetization in Spin Diffusion in Trapped Boltzmann Gases

R. J. Ragan[1] and W. J. Mullin[2]

[1]Physics Department, University of Wisconsin at Lacrosse, La Crosse, WI 54601, USA
[2]Physics Department, Hasbrouck Laboratory, University of Massachusetts, Amherst, MA 01003, USA

Abstract. Experiments in a mixture of two hyperfine states of trapped Bose gases show behavior analogous to a spin-1/2 system, including transverse spin waves and other familiar Leggett-Rice-type effects. We have derived the kinetic equations applicable to these systems, including the spin dependence of interparticle interactions in the collision integral. We find that the hydrodynamic diffusive modes cease to exist because interactions with different scattering lengths for up-up, up-down, and down-down spins lead to a spin-spin relaxation that causes non-conservation of transverse magnetization. We give results for the quadrupole modes, the modes studied in experiments with equal scattering lengths. Instead of a linear dependence on relaxation time τ for the diffusive mode, we find a divergence at small τ. No such effect occurs in Fermi gases.

Keywords: Spin diffusion, quantum gases, kinetic equation, ultra-cold gases
PACS: 03.75.Mn,05.30Jp,05.60.Gg,51.10.+y,67.20.+k.

Spin waves and spin diffusion in Fermi and Bose fluids have been of interest since Leggett and Rice suggested unusual mean-field effects in 1970[1]. Recent progress has been made in sorting out experimental[2] and theoretical[3] uncertainties in anisotropic spin diffusion. A mixture of two trapped hyperfine states of a gas can be viewed as a pseudo-spin system and analogous spin experiments can be performed. Indeed experimental striations in hyperfine state densities have been interpreted in terms of spin waves[4]. We have recently begun examining anisotropic spin diffusion for these trapped gas systems[5]. In Ref. 5 we found that diffusion constants were anisotropic in Bose systems, even in the Boltzmann limit, but not in the Fermi case, if the scattering lengths for different hyperfine states are unequal. However that reference considered only dipole spin modes, while experiments looked at the quadrupole modes. Here we extend our computations to the quadrupole modes and emphasize an important difference in Bose and Fermi systems, namely, that transverse spin waves and spin diffusion are not good modes in the Bose gas, because the transverse magnetization is not conserved when the scattering lengths differ among the two hyperfine states. This feature is not present in the Fermi gas, where transverse quadrupolar diffusive modes are well-behaved.

We solve a kinetic equation for the non-equilibrium part δm_{p+} of the transverse polarization, which in the trapped gas case takes the form

$$\frac{\partial \delta m_{p+}}{\partial t}+i\frac{\eta t_{12}}{\hbar}\left(m_p{}^{(0)}\delta M_+ - M_0\delta m_{p+}\right)-i\Omega_0\delta m_{p+} + \sum_i\left[\frac{p_i}{m}\frac{\partial \delta m_{p+}}{\partial r_i}-\frac{\partial U}{\partial r_i}\frac{\partial \delta m_{p+}}{\partial p_i}\right]=2(2|\hat{L}_p|1) \tag{1}$$

where the 2x2 density has been written as $\hat{n}_p=(1/2)(f_p\hat{I}+\mathbf{m}_p\cdot\hat{\sigma})$ where $\hat{\sigma}$ is a Pauli matrix; $(1/2)(f_p \pm m_{pz})$ gives the diagonal components of the density; $n_{pi}=n_{pii}$; and $\mathbf{m}_p$ represents the polarization, which in equilibrium is along the $\hat{\mathbf{z}}$ axis. The transverse components are $m_{p\pm}(\mathbf{r})=(m_{px}\pm i m_{py})$. The total magnetization is $\mathbf{M}(\mathbf{r})=\int d\mathbf{p}/h^3\,\mathbf{m}(\mathbf{r})$ with M_0 evaluated at the equilibrium polarization $m_p{}^{(0)}$. The t's can be evaluated in terms of the measured scattering lengths $a_{\alpha\beta}$ by using $t_{\alpha\beta}=4\pi\hbar a_{\alpha\beta}/m$. These are taken to be different for the two hyperfine states 1 and 2. Experiments to date using Rb[4] have had equal t values. U is the average external potential while Ω_0 is proportional to the difference in external potentials for the two states. This latter quantity leads to an important coupling between modes, but we will

CP850, *Low Temperature Physics: 24th International Conference on Low Temperature Physics;* edited by Y. Takano, S. P. Hershfield, S. O. Hill, P. J. Hirschfeld, and A. M. Goldman

neglect it in this brief report. The collision integral has been evaluated explicitly[5] and $\hat{L}_p$ is the linearized form of it. The constant η is +1(-1) for bosons (fermions).

In our moments approach in the Boltzmann limit[5,6], we consider a trial form $\delta m_{p+} = (a_0 + a_1 z^2 + a_2 z p_z + a_3 p_z^2) m_p^{(0)}$. We integrate over position and momentum, arriving at a set of coupled equations for the parameters a_i, which we then solve. Our transverse results depend on two relaxation rates arising from our expression for the collision integral. The first, $1/\tau_\perp$, adds corrections to the leading longitudinal rate proportional to $1+\eta$ times factors depending on $t_{\sigma\sigma} - t_{12}$. Thus it shows an anisotropy between longitudinal and transverse modes for bosons, but none for fermions, even in the Boltzmann limit. The second relaxation rate, γ_T, depends in *leading* order on $1+\eta$ times factors depending on $t_{\sigma\sigma} - t_{12}$ and so vanishes completely for fermions and for equal t values. It is analogous to a $1/T_2$ relaxation rate in NMR. For bosons this relaxation rate occurs in the expression for the monopole mode, which means the transverse magnetization is not conserved; it also appears in the expressions for the quadrupole modes and drastically affects their diffusive lifetime.

In Fig. 1 we show the imaginary part ω_I of the quadrupole diffusive mode frequency ω as a function of $\omega_z \tau_\perp$ (where ω_z is the frequency in the long axis of the trap) both with and without the γ_T rate. For bosons with $\gamma_T = 0$, the moments method in the hydrodynamic limit ($\omega_z \tau_\perp \to 0$) gives the following quadrupole rate:

$$\omega = 4\omega_z^2 \tau_\perp (i - \omega_M \tau_\perp), \qquad (2)$$

whereas for bosons with $\gamma_T > 0$ one finds that $\omega_I \sim \gamma_T$ in the small $\omega_z \tau_\perp$ limit, as one can see from the figure. In Eq. 2 there is a mean field of frequency $\omega_M \sim \eta t_{12}$, which gives rise to the spin waves.

Also shown in the figure is the quadrupole rate calculated numerically from the local hydrodynamic equations for the polarization density. As in the dipole mode case[5], the spatial dependence of the relaxation time weakens the singularity in the quadrupole spectrum, $\omega_I \sim \sqrt{\log(1/\omega_z \tau_\perp)}$, as $\omega_z \tau_\perp \to 0$.

Our result suggests that for Bose systems with unequal scattering lengths spin waves and diffusive behavior will not be possible because of the short lifetime of this mode—with the magnetization not being conserved—while for Fermi systems the magnetization is conserved and spin waves and spin diffusion can exist. We estimate that ^{23}Na, for which the scattering lengths differ, could have a rate of $\gamma_T \approx 0.02/\tau_\perp$.

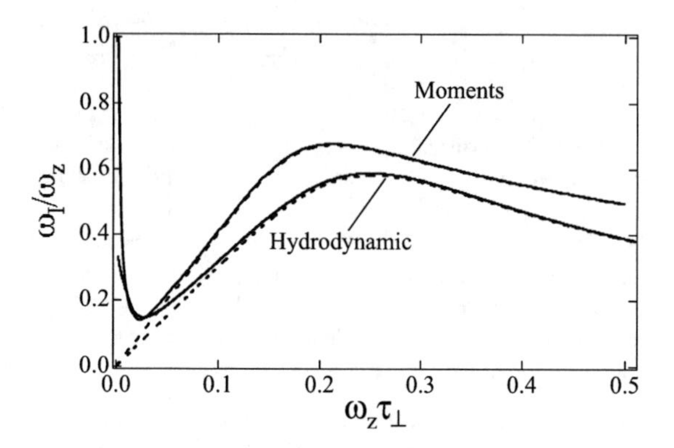

FIGURE 1. Quadrupole diffusive mode. We present calculations from both a moments approach and a numerical solution of the local hydrodynamic equations. For the dashed curves the non-conserving relaxation γ_T is neglected. The solid curves show the rapid decay when that rate is $0.002/\tau_\perp$. (The hydrodynamic results are valid only at small $\omega_z \tau_\perp$.)

REFERENCES

1. A. J. Leggett, *J. Phys. C* **3**, 448 (1970); A. J. Leggett and M. J. Rice, *Phys. Rev. Lett.* **20**, 586; **21**, 506 (1968).
2. H. Akimoto, D. Candela, J. S. Xia, W. J. Mullin, E. D. Adams, and N. S. Sullivan, *Phys. Rev. Lett.* **90**, 105301 (2003). This paper has a rather complete list of theoretical and experimental references for homogeneous real-spin Fermi fluids.
3. W. J. Mullin and R. J. Ragan, *Jour. Low Temp. Phys.* **138**, 73 (2005).
4. H. J. Lewandowski, D. M. Harber, D. L. Whitaker, and E. A. Cornell, *Phys. Rev. Lett.* **88**, 070403 (2002); J. M. McGuirk, H. L. Lewandowski, D. M. Harper, T. Nikuni, J. E. Williams, and E. A. Cornell, *Phys. Rev. Lett.* **89**, 090402 (2002); M. O. Oktel and L. S. Levitov, *Phys. Rev. Lett.* **88**, 230403 (2002); J. N. Fuchs, D. M. Gangardt, and F. Laloë, *Phys. Rev. Lett.* **88**, 230404 (2002); J. E. Williams, T. Nikuni, and C. W. Clark, *Phys. Rev. Lett.* **88**, 230405 (2002); B. DeMarco and D. S. Jin, *Phys. Rev. Lett.* **88**, 040405 (2002).
5. W. J. Mullin and R. J. Ragan, *Jour. Low Temp. Phys.*, **138**, 711 (2005).
6. T. Nikuni, J. E. Williams, and C. W. Clark, *Phys. Rev. A* **66**, 043411 (2002).

Matter-Wave Solitons In Optical Superlattices

Pearl J. Y. Louis, Elena A. Ostrovskaya, and Yuri S. Kivshar

Nonlinear Physics Centre and ARC Centre of Excellence for Quantum-Atom Optics, Research School of Physical Sciences and Engineering, The Australian National University, Canberra ACT 0200, Australia

Abstract. In this work we show that the properties of both bright and dark Bose-Einstein condensate (BEC) solitons trapped in optical *superlattices* can be controlled by changing the shape of the trapping potential whilst maintaining a constant periodicity and lattice height. Using this method we can control the properties of bright gap solitons by *dispersion management.* We can also control the interactions between dark lattice solitons. In addition we demonstrate a method for controlled generation of matter-wave gap solitons in *stationary* optical lattices by interfering two condensate wavepackets, producing a single wavepacket at a gap edge with properties similar to a gap soliton. As this wavepacket evolves, it forms a bright gap soliton.

Keywords: Bose-Einstein condensates, optical lattices, solitons
PACS: 03.75.Lm

INTRODUCTION

One of the advantages of using an optical lattice to trap BECs is the flexible and precise control we have over the lattice parameters. In this work we show that by changing the lattice parameters we can control the properties of both bright and dark BEC lattice solitons. We investigate double-periodic optical *superlattices* which allow the lattice height and periodicity to be kept constant while still allowing the properties of the lattice solitons to be controlled by manipulating the *shape* of the trapping potential. Currently, BEC gap solitons have only been successfully created in moving optical lattices [1]. Here, we show that they can also be created in *stationary* optical lattices.

MODEL

We model the system using the dimensionless 1D Gross-Pitaevskii (GP) equation for the condensate wavefunction $\psi(x,t)$,

$$i\frac{\partial}{\partial t}\psi(x,t)=\left[-\frac{1}{2}\frac{\partial^2}{\partial x^2}+V(x)+g|\psi(x,t)|^2\right]\psi(x,t). \qquad (1)$$

The strength of the two-body interactions are characterised by g. $V(x)$ is the superlattice potential,

$$V(x)=U\left[\varepsilon\sin^2\left(\frac{\pi}{d_1}x\right)+(1-\varepsilon)\sin^2\left(\frac{\pi}{d_2}x\right)\right]. \qquad (2)$$

It has two periodicities, d_1 and d_2. In this work we assume that $d_1/d_2=2$. U is proportional to the total laser intensity and $0 \le \varepsilon \le 1$ controls the shape of the superlattice (see Fig. 1 below).

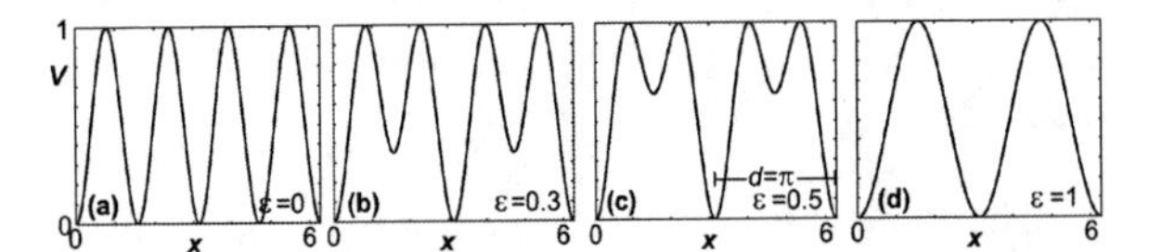

FIGURE 1. Examples of the superlattice potential given by Eq. (2) for $d_1/d_2=2$ and for varying ε (i.e. shape) but fixed lattice height $V_0=1$ and superperiod $d=\pi$.

BRIGHT GAP SOLITONS

In Ref. [1] bright gap solitons were created in a repulsive BEC by accelerating the lattice so that the condensate was adiabatically transferred to the lower gap edge where the effective dispersion, D, is negative. We show that similar bright gap solitons can be created in *stationary* optical lattices.

Our method is based on one first used to create optical gap solitons [2-4]. We assume that the initial BEC wavepacket is formed by interfering two identical counter-propagating BEC wavepackets with phases θ_1 and θ_2 and momentums $k_1=1$ and $k_2=-1$ respectively. When they interfere they produce a single wavepacket at the Bragg condition with an internal structure that mimics that of a weakly nonlinear gap soliton at either the top or the bottom of a gap depending on the choice of $\Delta\theta=\theta_1-\theta_2$. In all the

CP850, *Low Temperature Physics: 24th International Conference on Low Temperature Physics;* edited by Y. Takano, S. P. Hershfield, S. O. Hill, P. J. Hirschfeld, and A. M. Goldman

results that follow, the interatomic interactions are repulsive unless otherwise specified.

Figs. 2(a)-(c) start with the same initial wavepacket and g is identical in all three cases. However, the superlattice potential varies (for the parameter values used see Fig. 2). $\Delta\theta=0$, which places the BEC at the bottom of the first finite gap where D is negative. In each case a bright gap soliton forms. Due to the discrepancies between the initial shape of the wavepacket and the spatial structure of the gap soliton that forms, transient dynamics are observed that gradually become damped as excess atoms are radiated into the low density background.

To provide further confirmation that the localised states seen in Figs. 2(a)-(c) are gap solitons, we compare them with gap soliton solutions found by solving the time-independent Gross-Pitaevskii equation (which also gives an estimate of the chemical potentials, μ, for the gap solitons in Fig. 2). We also investigate the positive D regime, the non-interacting case, and attractive interactions. In each case the condensate behaves as expected taking into account the values of D and g, providing further evidence that the localised states seen in Fig. 2 form because of the dual effects of dispersion and nonlinearity.

The three gap solitons shown in Fig. 2 clearly have different properties despite the fact that the initial wavepacket and g are identical for all three cases. The superlattice parameters are also identical except for ε, which as seen in Fig. 1 controls the shape of the superlattice. As $\varepsilon\rightarrow1$, $|D|$ at the edges of the first gap decreases due to the increased size of the gap. If N remains constant, as $|D|$ decreases, in order to obtain a fundamental gap soliton, the width of the wavepacket decreases and its peak density increases. Hence Fig. 2 shows the properties of bright gap solitons being controlled by *dispersion management*.

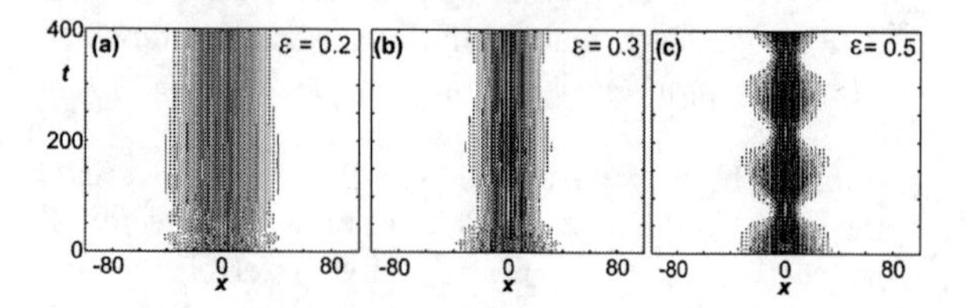

FIGURE 2. $\psi(x,t=0)$ is identical for (a)-(c): N=1000 and the BEC initially occupies ~30 superperiods. $\Delta\theta=0$ which places the BEC at the bottom of the first gap. The superlattice potentials in (a)-(c) have the same height (V_0=1) and superperiod, $d=\pi$. However the shape of the superlattice (i.e. ε) varies. As $\varepsilon\rightarrow1$, $|D|$ decreases.

DARK LATTICE SOLITONS

In Fig. 3 we show that interactions between solitons can be controlled using *shallow* optical superlattices. In each case we start with a pair of mutually out-of-phase dark solitons separated by six superperiods. μ is chosen so that the maximum density of the nonlinear Bloch wave background is the same for all the cases studied. All the superlattice parameters are fixed (V_0=1, $d=\pi$) except for the shape of the superlattice which is given by ε.

Our numerical simulations show that dark solitons, initially in the large wells of a superlattice [Fig. 3(a)], are *pinned* by the periodic potential even if the lattice is shallow. If the dark solitons are initially in the small wells of a superlattice the strength of the interaction between them can be controlled by changing the contrast between the small and large wells i.e. the shape of the superlattice [compare Figs. 3(b) and (c)]. The differences in behaviour can be explained by differences in the Peierls-Nabarro (PN) potential barrier. The PN potential barrier is the energy required to move a soliton one lattice site and is equal to the difference in energy between a soliton at a potential maximum and a soliton at a neighbouring potential minimum.

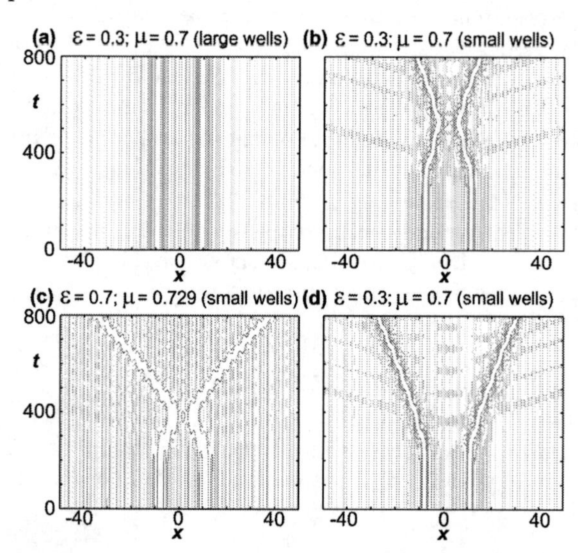

FIGURE 3. Soliton interactions. In (d) a small-amplitude periodic perturbation is added to the initial state. Based on Fig. 7 from P.J.Y. Louis *et al., J. Opt. B* **6**, S309 (2004).

REFERENCES

1. B. Eiermann *et al., Phys. Rev. Lett.* **92**, 230401 (2004).
2. J. Feng, *Opt. Lett.* **18**, 1302 (1993).
3. D. Neshev *et al., Phys. Rev. Lett.* **93**, 083905 (2004).
4. D. Mandelik *et al., Phys. Rev. Lett.* **92**, 093904 (2004).

Effect of Quantum Correction in the Bose-Hubbard Model

Hideki Matsumoto, Kiyoshi Takahashi and Yoji Ohashi

Institute of Physics, University of Tsukuba, Ibaraki 305-8571, Japan

Abstract. Effects of quantum correction in the Bose-Hubbard model at finite temperature are investigated for a homogeneous atomic Bose gas in an optical lattice near its superfluid-insulator transition. Starting from a strong coupling limit, higher order quantum corrections due to the hopping interaction is included in a local approximation (a dynamical mean field approximation) of the non-crossing approximation. When the upper or lower Hubbard band approaches zero energy, there appears a shallow band in the middle of the Hubbard gap due to a strong correlation in the system.

Keywords: Bose-Einstein condensation, Bose-Hubbard model, superfluid-insulator transition, strongly correlated particle system
PACS: 03.75.Lm, 05.30.Jp

INTRODUCTION

Experiments of Bose-Einstein condensation of dilute atomic gases have stimulated a large number of theoretical studies of quantum nature of Bose and Fermi gases. The attractive features of these systems are that one can control many physical parameters, not only temperature and particle number, but also an applied potential and even the atomic interaction strength. In particular, the superfluid-insulator transition in the optical lattice[1] is expected to provide us with a good laboratory system of correlated particles with controllable parameters.

It has been shown that a Bose system near its superfluid-insulator transition can be described by the Bose-Hubbard model[2], whose Hamiltonian is given by $H = \sum_i[-\mu\psi^\dagger(i)\psi(i) + \frac{1}{2}U\psi^\dagger(i)^2\psi(i)^2]$ $-J_0\sum_{<ij>}\psi^\dagger(i)\psi(j)$. Here $\psi(i)$ is the single component boson field at site i, μ is the chemical potential, U is the repulsive interaction strength and J_0 is the hopping energy. The notation $<ij>$ indicates the nearest neighbor hopping.

In the fermionic case, it is known that a correlated narrow band occurs at the Fermi level[3]. In this paper we investigate whether or not a similar correlated band appears near the superfluid-insulator transition even in the bosonic case.

FORMULATION

We start from the strong coupling limit and choose an unperturbed Hamiltonian as

$$H_0 = \sum_i[-\mu\psi^\dagger(i)\psi(i) + \frac{1}{2}U\psi^\dagger(i)^2\psi(i)^2]\,. \quad (1)$$

At each site, the Hamiltonian is diagonalized by a bosonic number state $|n(i)>$, which has the eigenenergy $E_n = -\mu n + \frac{1}{2}Un(n-1)$. The boson field $\psi(i)$ is expressed as $\psi(i) = \sum_n |(n-1)(i)> \sqrt{n} < n(i)|$ in this notation.

Let us put $<\psi(i)>= v(i)$ and define $\varphi(i) = \psi(i) - v(i)$, where $<\cdots>$ indicates the thermal average. Consider a doublet $\Phi(i) = \begin{pmatrix}\varphi(i)\\ \varphi^\dagger(i)\end{pmatrix}$. It satisfies the Heisenberg equation of the form

$$\tau_3\, i\hbar\frac{\partial}{\partial t}\Phi(i) = -J\Phi^\alpha(i) + \begin{pmatrix} j_\varphi(i)\\ j_\varphi^\dagger(i)\end{pmatrix}\,. \quad (2)$$

Here we used the notation $\Phi^\alpha(i) = \sum_j \alpha(i,j)\Phi(j)$ and $\tau_3 = \begin{pmatrix}1 & 0\\ 0 & -1\end{pmatrix}$. We take $J = 6J_0$ by assuming the cubic lattice. The Fourier transform of the retarded function $G(\omega,\vec{k})$=$-iF.T. < R\Phi(i)\Phi^\dagger(j) >$ is expressed as $G(\omega,\vec{k}) = 1/(\omega\tau_3 + J\alpha(\vec{k}) - \Sigma(\omega,\vec{k}))$. Here the system is assumed to be homogeneous, $\alpha(\vec{k})$ is given by $\alpha(\vec{k}) = \frac{1}{3}(\cos(k_xa) + \cos(k_ya) + \cos(k_za))$ and $\Sigma(\omega,\vec{k})$ is a 2×2 matrix, which expresses the energy and self-energy obtained from j_φ and $j_\varphi^\dagger$. Note that $v(i)$ is constant ($v(i) = v$). When we approximate $\Sigma(\omega,\vec{k})$ locally by evaluating quantum corrections at the single site, we have

$$G(\omega,\vec{k}) = \frac{1}{G_0(\omega)^{-1} + J\alpha(\vec{k})} \quad (3)$$

with $G_0(\omega)$ being an irreducible single site propagator $G_0 =< R\Phi(i)\Phi^\dagger(i) >_0$. Here the suffix '0' indicates the exclusion of direct hoppings.

Since G_0 is the single site propagator, it can be evaluated by considering a site $i = 0$ denoted by 's' and the rest by 'R', $H = H_s + H_R + H_{sR}$, where H_s is given by $H_s = -\mu\psi^\dagger(0)\psi(0) + \frac{1}{2}U\psi^\dagger(0)^2\psi(0)^2 - J(v^*\psi(0) + \psi^\dagger(0)v)$, $H_{sR} = -J(\varphi^{\dagger\alpha}\psi(0) + \psi^\dagger(0)\varphi^\alpha)$ [3] and H_R is the Hamiltonian of the rest. In order to evaluate quantum corrections to local energy

CP850, *Low Temperature Physics: 24th International Conference on Low Temperature Physics;*
edited by Y. Takano, S. P. Hershfield, S. O. Hill, P. J. Hirschfeld, and A. M. Goldman

levels, define a resolvent $R_{nn'}(t-t') = \theta(t-t')$ $Tr_R[e^{-\beta H_R} < n|e^{-iH_{st}}u(t,t')e^{iH_{st'}}|n'>]/Tr_R[e^{-\beta H_R}]$, where $u(t,t') = Texp\{-i\int_{t'}^{t} dt'' H_{sR}(t'')\}$ with T indicating the time-ordered product. The Fourier transform of $R_{nn'}$ is given in the form

$$R_{nn'}(\omega) = \left(\frac{1}{\omega - E - \Sigma_R(\omega)}\right)_{nn'} \quad (4)$$

with $E_{nn} = E_n$, $E_{n,n-1} = \sqrt{n}v$, $E_{n-1,n} = \sqrt{n}v^*$ and other $E_{nn'}$ are zero. The spectral function of $R_{nn'}(\omega)$ is denoted as $\sigma_{nn'}(\omega)$. Then v is obtained from $v = \frac{1}{z}\int_{e_0}^{\infty} d\omega \sum_n \sqrt{n}\sigma_{n,n-1}(\omega)e^{-\beta(\omega-e_0)}$ with $z = \int_{e_0}^{\infty} d\omega \sum_n \sigma_{nn}(\omega)e^{-\beta(\omega-e_0)}$ and e_0 being a threshold energy.

The self energy is evaluated by a one-loop approximation as $\Sigma_{nn'}(\omega) = J^2\int d\nu d\kappa \frac{(1+f_B(\nu))}{\omega-\kappa-\nu+i\delta}$ $[\sqrt{n}\ (\sqrt{n'}\sigma_{n-1,n'-1}(\kappa)\rho_R^{11}(\nu) + \sqrt{n'+1}\sigma_{n-1,n'+1}(\kappa)$ $\rho_R^{12}(\nu)) + \sqrt{n+1}(\sqrt{n'}\sigma_{n+1,n'-1}(\kappa)\rho_R^{21}(\nu) + \sqrt{n'+1}$ $\sigma_{n+1,n'+1}(\kappa)\ \rho_R^{22}(\nu))]$. Here δ is an infinitesimally small positive number. Spectrals $[\rho(\omega)]$ and $[\rho_R(\omega)]$, which are 2×2 matrices, are defined by

$$[\rho(\omega)] = -\frac{1}{\pi}\mathrm{Im}\frac{a^3}{(2\pi)^3}\int d^3k \frac{1}{G_0(\omega)^{-1} + J\alpha(\vec{k})} \quad (5)$$

$$[\rho_R(\omega)] = -\frac{1}{\pi}\mathrm{Im}\frac{a^3}{(2\pi)^3}\int d^3k \frac{\alpha(\vec{k})^2}{G_0(\omega)^{-1} + J\alpha(\vec{k})} \quad (6)$$

The onsite propagator $G_0(\omega)$ is evaluated by realizing $< R\Phi\Phi^\dagger > = < R\Psi\Psi^\dagger >$, where $\Psi = \begin{pmatrix}\psi \\ \psi^\dagger\end{pmatrix}$. We have, for example, in the noncrossing approximation[4]

$$G_0^{11}(\omega) = \int d\kappa_1 d\kappa_2 \sqrt{n+1}\sigma_{n+1,n'+1}(\kappa_1) \times \sqrt{n'+1}\sigma_{n',n}(\kappa_2)\frac{e^{-\beta\kappa_1} - e^{-\beta\kappa_2}}{\omega - \kappa_1 + \kappa_2 + i\delta} \quad (7)$$

RESULTS

A self-consistent numerical calculation of Σ_R, R, G_0, $[\rho]$ and $[\rho_R]$ is performed. In Figure 1, the density of states $\rho(\omega)$ $(=\rho_{11}(\omega))$ for $U/J = 10.0$ is presented. The chemical potential is changed as indicated. The dotted lines are results of the mean field theory. As the upper band approaches zero energy, a shallow band develops at $\omega < 0$. This band appears because of development of correlated levels in the resolvent due to the correlation with the surrounding particles. Studies on the properties of this band will be presented elsewhere.

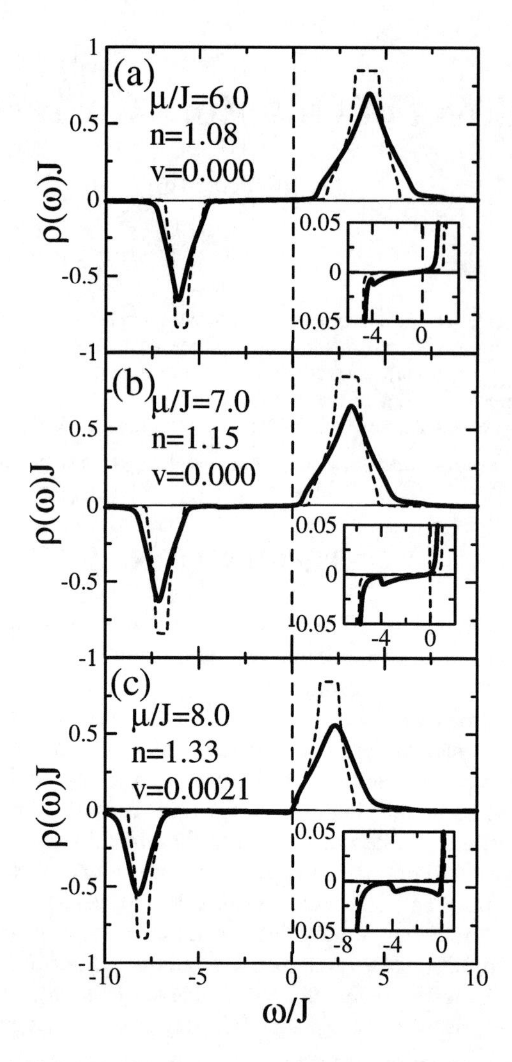

FIGURE 1. Density of states $\rho(\omega)$. The parameters are $U/J = 10.0$ and $T/J = 1.0$. Chemical potential is chosen to be $\mu/J = 6.0(a), 7.0(b), 8.0(c)$. The insets show the behavior of $\rho(\omega)$ inside the gap in detail.

ACKNOWLEDGMENTS

This work was supported by a Grant-in-Aid for Scientific Research from the Japan Society for the Promotion of Science.

REFERENCES

1. M. Greiner, et al., Nature **415**, 39 (2002).
2. D. van Oesten, et al., Phys. Rev. A**63**, 063605 (2001).
3. H. Matsumoto, *Highly Correlated Particle Systems and Composite Operator Methods* in Lectures on the Physics of Highly Correlated Electron Systems (to be published, AIP, 2005).
4. Y. Kuramoto, Z. Phys. B**53**, 37 (1983).

Continuum Theory of Tkachenko Modes in Rapidly Rotating Bose-Einstein Condensate

Edouard B. Sonin

Racah Institute of Physics, Hebrew University of Jerusalem, Givat Ram, Jerusalem 91904, Israel

Abstract. The continuum theory of Tkachenko modes in a rotating 2D Bose–Einstein condensate (BEC) is considered taking into account density inhomogeneity and compressibility of the condensate. Two regimes of rotation are discussed: (i) the Vortex Line Lattice (VLL) regime, in which the vortex array consists of thin vortex lines with the core size much less than the intervortex distance; (ii) the Lowest-Landau Level (LLL) regime, in which vortex cores overlap and the BEC wave function is a superposition of states at the lowest Landau level. The theory is in good agreement with experimental observation of Tkachenko modes in the VLL regime. But theoretical frequencies in the LLL regime are essentially higher than those observed for very rapidly rotating BEC. This provides evidence that the experiment has not yet reached the LLL regime.

Keywords: BEC, rotation, vortices, Tkachenko wave, lowest Landau level
PACS: 03.75.Kk, 67.40.Vs

INTRODUCTION

Tkachenko mode [1] is a transverse sound wave in a vortex lattice, which exists in the ground state of a rotating superfluid. It has attracted a lot of attention in vortex dynamics of laboratory and astrophysical superfluids. Observation of this extremely soft mode on the order of, or less than, a few Hz in laboratory superfluids was very difficult because even slight pinning of vortex ends shades the contribution of vortex shear rigidity, transforming the Tkachenko wave into a classical inertial wave [2]. However, the discovery of Bose–Einstein condensate (BEC) and possibility to rapidly rotate it has made an observation of Tkachenko modes more feasible, and has resulted in clear experimental detection of them [3].

The continuum theory of Tkachenko waves in BEC must take into account a number of important features absent in "old" superfluids. In a rapidly rotating BEC one cannot ignore the compressibility of the superfluid. The effect of finite compressibility was investigated theoretically in the past [2]. But in rotating He II one could observe the effect only in containers about a few hundred meters in diameter. In contrast, for Tkachenko waves in BEC the effect may be very important as was pointed out by Baym [4]. Another feature of rotating BEC, also connected with its compressibility, is that the liquid density is inhomogeneous. This was taken into account by Anglin and Crescimanno [5], but they neglected compressibility. In addition, experimentalists can rotate BEC much faster than helium in a container. As a result two limiting regimes should be considered. At moderate rotation speed the vortex lattice consists of vortex lines with a core size on the order of the coherence length ξ, which is essentially smaller than the intervortex distance $b = \sqrt{\kappa/\sqrt{3}\Omega}$. Here Ω is the angular velocity of rotation and $\kappa = h/m$ is the circulation quantum. This is the Vortex Line Lattice (VLL) regime. With increasing Ω, the vortex lattice becomes more dense and at $\Omega \gg \Omega_{c2} \sim \kappa/\xi^2$ vortex cores overlap since $\xi \gg b$. This regime, inaccessible in rotating helium, is analogous to the mixed state of a type-II superconductor close to the second critical magnetic field $H_{c2} \sim \Phi_0/\xi^2$ (Φ_0 is the magnetic flux quantum). It is called the Lowest Landau Level (LLL) regime, since the BEC wave function is a coherent superposition of single-particle states similar to the lowest Landau levels of a charged particle in a magnetic field. Below we present the summary of the theory, taking into account the aforementioned features of the BEC, and compare it with experiments.

EQUATIONS OF HYDRODYNAMICS

We consider a rotating 2D BEC in a pancake geometry with the density profile $n_0 = n_0(0)(1 - r^2/R^2)$ determined by the Thomas–Fermi approximation and dependent only on the distance r from the cloud center. Here $R = \sqrt{2}c_s(0)/\sqrt{\omega_\perp^2 - \Omega^2}$ is the radius of the BEC cloud (Thomas–Fermi radius), $c_s(0)$ is the sound velocity at the symmetry axis $r = 0$, and $\omega_\perp$ is the trap frequency, which characterizes the curvature of the parabolic potential confining the BEC cloud. Only axisymmetric Tkachenko eigenmodes are considered. In polar coordinates, the equations of motion for the monochromatic mode $\propto e^{-i\omega t}$ are [6]:

$$2\Omega i\omega v_r = -\omega^2 v_t - \frac{c_T^2}{n_0}\frac{1}{r^2}\frac{\partial}{\partial r}\left[n_0 r^3 \frac{\partial}{\partial r}\left(\frac{v_t}{r}\right)\right], \quad (1)$$

CP850, *Low Temperature Physics: 24th International Conference on Low Temperature Physics;*
edited by Y. Takano, S. P. Hershfield, S. O. Hill, P. J. Hirschfeld, and A. M. Goldman

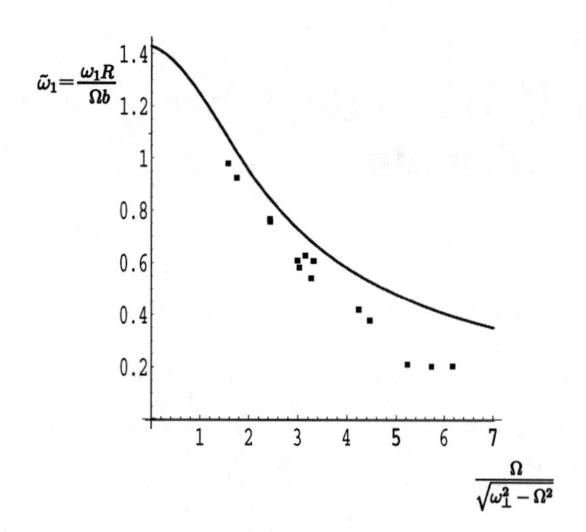

FIGURE 1. Comparison between the theory (solid line) and the experiment (black squares)

$$2\Omega i\omega v_t = \frac{c_s^2}{n_0}\frac{\partial}{\partial r}\left[\frac{1}{r}\frac{\partial(n_0 r v_r)}{\partial r}\right]. \qquad (2)$$

Here v_r is the radial liquid velocity, and v_t is the azimuthal velocity component, which is approximately the same for the liquid and for the vortices. For a weakly interacting Bose gas c_s^2 is proportional to the density n_0. Therefore the ratio c_s^2/n_0 can be replaced by its value $c_s^2(0)/n_0(0)$ at the BEC center $r=0$. The Tkachenko velocity $c_T = \sqrt{2C_2/mn_0}$ in Eq. (1) is determined by the shear elastic modulus C_2 of the vortex lattice.

These equations should be supplemented by boundary conditions at the cloud boundary $r=R$ [6]:

$$\frac{dv_t(R)}{dr} - \frac{v_t(R)}{R} = 0\,, \qquad (3)$$

$$\frac{dv_r(R)}{dr} + \frac{v_r(R)}{R} = -\frac{i\omega\Omega R}{2c_s(0)^2}v_t(R)\,. \qquad (4)$$

SOLUTION OF EQUATIONS

Solution of this boundary problem yields eigenfrequencies for all axisymmetric Tkachenko modes of the BEC cloud of radius R. The results of the numerical solution can be presented as a set of functions $\tilde{\omega}_i = f_i(s)$ of the parameter $s = 2\Omega R/c_s(0) = 2\sqrt{2}\Omega/\sqrt{\omega_\perp^2-\Omega^2}$, which characterizes the effect of compressibility. Here $i \geq 1$ are the eigenmodes, and $\tilde{\omega} = \omega R/\Omega b$ is the reduced frequency. At $s \to 0$ the functions $f_i(s)$ become constant. But at large $s \gg 1$ the functions $f_i(s)$ are inversely proportional to s: $\tilde{\omega}_i = \gamma_i/s$. The two lowest eigenmodes correspond to $\gamma_1 = 7.17$ and $\gamma_2 = 16.9$. Eventually the theory yields simple expressions for the eigenfrequencies in the high-compressibility limit $s \gg 1$: $\omega_i = \sqrt{2\pi/\sqrt{3}}\gamma_i(c_T/c_s)(\omega_\perp - \Omega)$.

The presented continuum theory can be used both in the VLL and in the LLL regimes [7]. But one should use different expressions for the shear elastic modulus $C_2 = mn_0c_T^2/2$. In the VLL regime $C_2 = mn_0\kappa\Omega/8\pi$ as directly follows from Tkachenko's work. In the LLL regime the shear modulus $C_2 = 0.1027mn_0c_s^2$ can be obtained from the old calculations for type II superconductors near H_{c2} [8]. Eventually the eigenfrequencies in the LLL regime ($s \gg 1$) are: $\omega_i \approx 0.8633\gamma_i(\omega_\perp - \Omega)$.

COMPARISON WITH EXPERIMENTS

The solid line in Fig. 1 shows the first reduced eigenfrequency $\tilde{\omega}_1$ found numerically and plotted as a function of $\Omega/\sqrt{\omega_\perp^2-\Omega^2}$. Black squares are experimental points. They were obtained for various parameters, but in dimensionless variables they collapse on the same curve, as expected from the present analysis. Quantitative agreement between the theory and the experiment looks quite good, but becomes worse at larger $\Omega/\sqrt{\omega_\perp^2-\Omega^2}$. This disagreement is connected with the growth of the ratio ξ/b and is a signal of the crossover to the LLL regime. Coddington *et al.* [3] measured also the ratio of the two first frequencies $\omega_2/\omega_1 = 1.8$ at $\Omega/\omega_\perp = 0.95$, which corresponds to $s = 8.61$, The present theory predicts the ratio $\omega_2/\omega_1 = 2.09$.

In order to reach the LLL regime Schweikhard *et al.* [9] performed experiments with more rapid rotation of the BEC cloud. Their results roughly agree with expected linear dependence of eigenfrequencies on $\omega_\perp - \Omega$, but the slope in the experiment was about 4 times less than predicted for the LLL regime. This is evidence that the experiment has not yet reached the LLL limit, although experimental values of $(\omega_\perp - \Omega)/\omega_\perp$ look small enough. Apparently the LLL regime can be realized in experiments with a smaller number of atoms.

REFERENCES

1. V.K. Tkachenko, *Zh. Eksp. Teor. Fiz.* **50**, 1573 (1966) [*Sov. Phys.-JETP* **23**, 1049 (1966)].
2. E.B. Sonin, *Rev. Mod. Phys.* **59**, 87 (1987).
3. I. Coddington *et al.*, Phys. Rev. Lett. **91**, 100402 (2003).
4. G. Baym, *Phys. Rev. Lett.* **91**, 110402 (2003).
5. J.R. Anglin and M. Crescimanno, cond-mat/0210063.
6. E.B. Sonin, *Phys. Rev. A* **71**, 011603 (2005).
7. E.B. Sonin, *Phys. Rev. A* **72**, 021606 (2005).
8. R. Labusch, *Phys. Stat. Sol.* **32**, 439 (1969); E.H. Brandt, *Phys. Stat. Sol.* **36**, 381 (1969).
9. V. Schweikhard *et al.*, *Phys. Rev. Lett.* **92**, 040404 (2004).

Vortex Structures in Rotating Two-Component Bose-Einstein Condensates in an Anharmonic Trapping Potential

Hiromitsu Takeuchi*, Kenichi Kasamatsu† and Makoto Tsubota*

*Department of Phisics, Osaka City University, Sumiyosi-Ku,Osaka 558-8585, Japan
†Department of General Education,Ishikawa National College of Technology, Tsubata, Ishikawa 929-0392, Japan

Abstract. We investigate numerically stable structures of vortex states in rotating two-component Bose-Einstein condensates (BECs) in a quadratic plus quartic potential. We discover various new vortex structures by changing the intracomponent coupling constants, the intercomponent coupling constant and the rotation frequency.

Keywords: Bose-Einstein condensate, vortex, multicomponent condensates
PACS: 03.75.Lm, 03.75.Mn

The study of quantized vortices is one of the most important fields in BEC physics. It is known that quantized vortices form a triangular lattice in a rotating one-component BEC trapped by a harmonic potential. In a quadratic plus quartic potential $V(r) = \frac{1}{2}m\omega^2 r^2 + \frac{1}{4}Kr^4$, if a BEC rotates with a frequency Ω higher than ω, the effective potential including the centrifugal term makes a "Mexican hat" structure $V_{\rm eff}(r) = \frac{1}{2}m(\omega^2 - \Omega^2)r^2 + \frac{1}{4}Kr^4$. Then, a giant vortex with a circular superflow can be created[1]. In two-component BECs trapped by a harmonic potential, we predicted that quantized vortices form various structures such as square lattices, vortex sheets *etc.* depending on the intercomponent interaction and the rotation frequency[2]. Actually, the square lattices were observed[3]. Hence we can expect that new structures appear in rotating two-component BECs in a quadratic plus quartic potential.

We solve the time-dependent coupled Gross-Pitaevskii equations in a frame rotating with a frequency Ω around the z axis. The calculation is done for the two-dimensional system perpendicular to the z axis. If the atomic mass, the trapping frequency and the particle number of one component are equal to those of the other, the coupled equations are reduced to

$$i\frac{\partial \psi_i}{\partial t} = [h_i + U_{ii}|\psi_i|^2 + U_{ij}|\psi_j|^2]\psi_i, \qquad i,j \in 1,2. \tag{1}$$

Here $h_i = -\nabla^2 + V - \mu_i - \Omega\hat{L}_z$, μ_i is the chemical potential of the i-th component and $\hat{L}_z$ is the angular momentum along the rotation axis. The wave functions are normalized as $\int |\psi_i|^2 d\mathbf{r} = 1$. The intra- and intercomponent coupling constants are reduced to $U_{ij} \propto N a_{ij}$, where N is the particle number and a_{ij} refers to the s-wave scattering length between components i and j. The effective potential is

$$V_{\rm eff}(r) = \frac{1}{4}[(1-\Omega^2)r^2 + \frac{1}{4}r^4]. \tag{2}$$

We calculate these equations by using the imaginary time propagation for finding stable states.

For a uniform system, the energy functional is

$$E = \int d\mathbf{r}[\frac{1}{2}U_{11}|\psi_1|^4 + \frac{1}{2}U_{22}|\psi_2|^4 + U_{12}|\psi_1|^2|\psi_2|^2]. \tag{3}$$

Here the integral of the third term corresponds to the intercomponent energy (E_{12}). In this system with $U_{11} > 0$ and $U_{22} > 0$, it is generally known that if $U_{11}U_{22} > U_{12}^2$, the two BECs overlap because U_{ii} is dominant. If $U_{11}U_{22} < U_{12}^2$, since U_{12} is dominant, the two BECs separate in the case $U_{12} > 0$ or collapse in the case $U_{12} < 0$. Thus U_{12} has the critical value $U_c = \sqrt{U_{11}U_{22}}$, which is applicable to our system in a potential too. Hence we study the vortex states for the following cases:

1. $U_{11}U_{22} > U_{12}^2$,
   (a) $U_{12} > 0$,
   (b) $U_{12} < 0$,
2. $U_{11}U_{22} < U_{12}^2$,
   (a) $U_{12} > 0$,
   (b) $U_{12} < 0$.

The typical stable vortex states for $\Omega = 1.5$, 2.5 and 3.5 are shown in Fig.1.

First, we investigate stable structures for $U_{11} = U_{22} = 250$; the critical value is $U_c = 250$. In case 1(a), the positions of vortices in each component are alternating and the density distribution reflects the shape of a "Mexican hat" potential with each Ω (A, B, D, I of FIG.1). For case 2(a), the two components separate and their distributions

CP850, *Low Temperature Physics: 24th International Conference on Low Temperature Physics;*
edited by Y. Takano, S. P. Hershfield, S. O. Hill, P. J. Hirschfeld, and A. M. Goldman

are interlaced with each other for smaller U_{12} (C, E, F, J) like "vortex sheets"[2] and BECs make "sliced watermelon" structures for larger U_{12} (G, H, K) ; the density distribution averaged over the two components recovers roughly the same density profile as that of the one-component BEC. These vortex structures are reflected in E_{12}; E_{12} is maximized around $U_{12} = U_c$ and decreases as U_{12} increases. It is expected that, if $U_{11} = U_{22}$, the ground states should have symmetrical patterns and the two BECs will have the same energy. All pairs of 1(a) are almost symmetric (A, B, D, I), but most of the pairs of 2(a) are asymmetric (C, E, F, G, J, K). These asymmetrical states are metastable, having long lifetimes. There are a lot of metastable states in case 2(a) especially for smaller U_{12}, so it takes very long time to get the ground (symmetrical) states; real experiments could find such metastable states. In case 1(b), the two BECs completely coincide with each other. In case 2(b), the BECs collapse.

Second, we investigate stable structures for $U_{11} = 100$ and $U_{22} = 250$; the critical value is $U_c \simeq 158$. It is known that a single-component BEC has more vortices as U_{ii} increases even if Ω is fixed[4]. In the case of non-rotating two-component systems, a BEC with larger U_{ii} surrounds the other or has a larger area than the other[5]. These features are seen in our rotating systems too, and the "Mexican hat" creates unique vortex structures. In cases 1(a) and 2(a), for $\Omega = 1.5$, component 1 has less vortices and component 2 has more vortices as U_{12} increases (L,M,N). For $\Omega = 2.5$, when we increase U_{12} the two BECs become more separated (P, Q, R), eventually breaking the axisymmetry (S). For $\Omega = 3.5$, the axisymmetry is broken more easily (U) than for smaller values of Ω. Together with states U and V, we have found similar asymmetric states for the same values of Ω and U_{12}, indicating that all these states are metastable. In case 1(b), the positions of the vortices of the two components coincide but their density distributions do not, unlike when $U_{11} = U_{22}$. In case 2(b), the BECs collapse.

Detailed calculations including the three-dimensional analyses are in progress and will be reported shortly elsewhere.

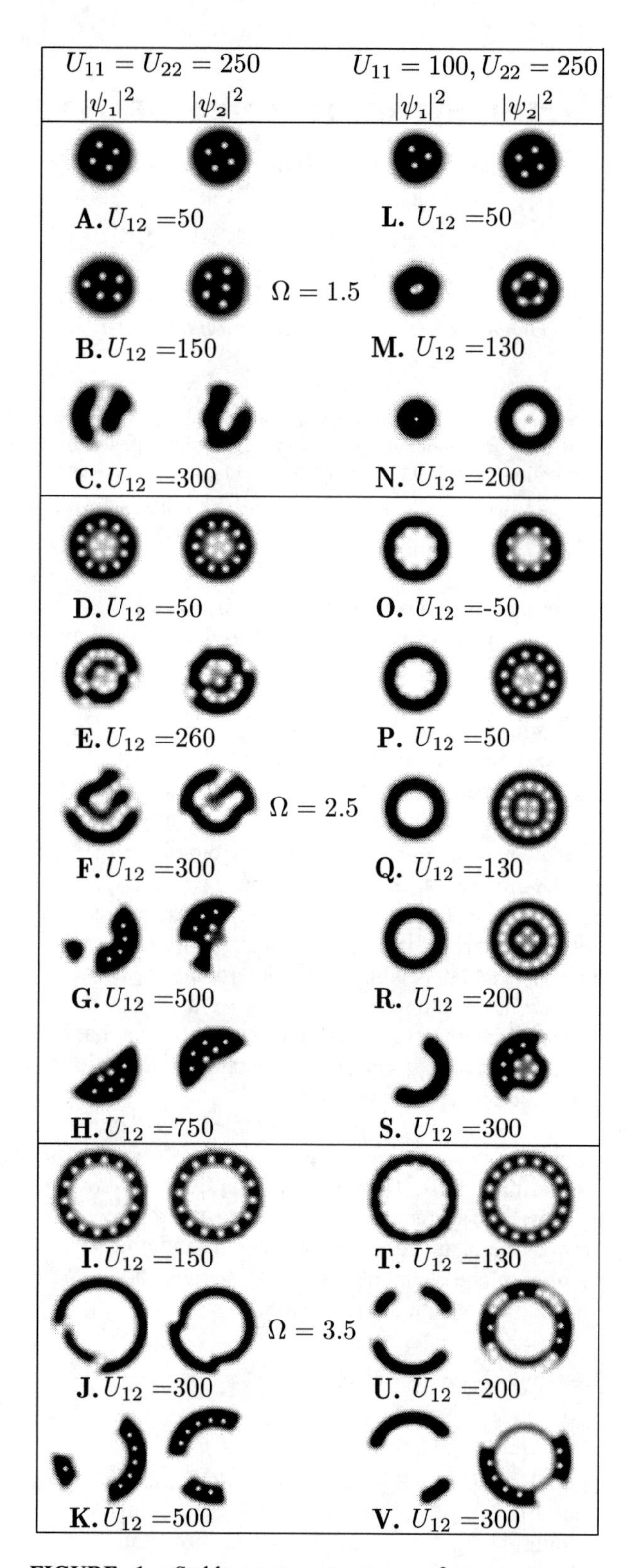

FIGURE 1. Stable vortex structures of two-component BECs in an quadratic plus quartic potential. In the case of $U_{11} = U_{22} = 250$, the critical value of U_{12} is $\sqrt{U_{11}U_{22}} = 250$, and in the case of $U_{11} = 100$ and $U_{22} = 250$, $\sqrt{U_{11}U_{22}} \simeq 158$.

REFERENCES

1. K. Kasamatsu, M. Tsubota and M.Ueda, *Phys. Rev. A*, **66**, 053606 (2002).
2. K. Kasamatsu, M. Tsubota and M.Ueda, *Phys. Rev. Lett*, **91**, 150406 (2003).
3. V. Schweikhard, I. Coddington, P. Engles, S.Tung and E. A. Cornell, *Phys. Rev. Lett*, **93**, 210403 (2004).
4. K. Kasamatsu, M. Tsubota and M.Ueda, *Phys. Rev. A*, **67**, 033610 (2003).
5. T.-L. Ho and V. B. Shenoy, *Phys. Rev. Lett*, **77**, 3276–3279 (1996).

Energy Spectrum of Fermions in a Rotating Boson-Fermion Mixture

Rina Kanamoto and Makoto Tsubota

Department of Physics, Osaka City University, Osaka 558-8585, Japan

Abstract. We investigate single-particle quantum states of degenerate fermions, which interact repulsively with a Bose-Einstein condensate in a rotating harmonic trap. The boson-fermion coupling partially lifts the degeneracies of the fermionic energy spectrum in a slow rotating regime, and causes level discontinuities of occupied states upon the formation of quantized vortices in the condensate. In a fast rotating regime, the Fermi energy is located in a highly quasidegenerate single band, and hence the lowest-Landau-level description is applicable. The probability of finding fermions on the vortex cores is significantly enhanced when the numbers of vortices and fermions are comparable. However, increasing the number of fermions expands the fermionic cloud outward, and phase separation occurs due to the Pauli pressure while the shape of the bosonic condensate remains unchanged.

Keywords: Boson-Fermion mixture, vortex lattice, energy spectrum
PACS: 03.75.Lm, 03.75.Ss

Cooling atoms with different quantum statistics into their degenerate temperatures [1] provides us with possibilities of observing various macroscopic quantum phenomena, such as Bose-Einstein condensation and superfluidity. Feshbach resonance between bosonic and fermionic atoms has been identified recently [2], and this enables us to investigate the degenerate atoms for a wide rage of interactions. In ultracold atomic gases, not only the atomic interaction but also rotating drive plays a crucial role in studies of superfluid properties. Rotating drive allows penetrations of quantized vortices into the degenerate cloud as experimentally observed in bosonic [3] and fermionic superfluids [4], and would cause another quantum phenomenon, e.g., a quantum Hall state [5] in the fast rotating regime. Details of behaviors of bosons and fermions under rotation are still elusive, and extensive analyses have been continued.

In this paper, we address the single-particle properties of degenerate (not paired) fermionic atoms which interact with a bosonic condensate. Let us consider a mixture of $N_{\rm B}$ bosons and spin-polarized $N_{\rm F}$ fermions in a rotating two-dimensional (2D) harmonic trap. The trapping frequencies and atomic masses for both components are assumed to be identical, and denoted by ω and M, respectively. Any angular momenta, angular frequencies, energies, and lengths are measured in units of $\hbar$, ω, $\hbar\omega$, and $l=\sqrt{\hbar/(M\omega)}$, throughout. The atomic interactions are characterized by the s-wave scatterings with the dimensionless 2D coupling constants g for boson-boson coupling, and h for boson-fermion coupling, whereas s-wave scattering between identical fermions is prohibited because of the Pauli principle. We suppose that bosonic atoms form a Bose-Einstein condensate described by a condensate wave function $\psi^{(\rm B)}$, and fermionic atoms degenerate into single-particle orbitals $\psi^{(\rm F)}_{j=1,\dots,N_{\rm F}}$. The mean-field energy functional $E=E^{(\rm F)}+E^{(\rm B)}+E^{(\rm BF)}$ in the rotating frame thus consists of

$$E^{(\rm B)} = N_{\rm B}\int d\mathbf{r}\psi^{(\rm B)*}(\mathbf{r})\left[\hat{K}+\frac{g}{2}n^{(\rm B)}(\mathbf{r})\right]\psi^{(\rm B)}(\mathbf{r}), \quad (1)$$

$$E^{(\rm F)} = \sum_{j=1}^{N_{\rm F}}\varepsilon_j^{(\rm F)};\quad \varepsilon_j^{(\rm F)}=\int d\mathbf{r}\psi_j^{(\rm F)*}(\mathbf{r})\hat{K}\psi_j^{(\rm F)}(\mathbf{r}), \quad (2)$$

$$E^{(\rm BF)} = h\int d\mathbf{r}n^{(\rm F)}(\mathbf{r})n^{(\rm B)}(\mathbf{r}), \quad (3)$$

where $\mathbf{r}=(x,y)$, and the densities are given by $n^{(\rm B)}(\mathbf{r})=N_{\rm B}|\psi^{(\rm B)}(\mathbf{r})|^2$, and $n^{(\rm F)}(\mathbf{r})=\sum_{j=1}^{N_{\rm F}}|\psi_j^{(\rm F)}(\mathbf{r})|^2$, respectively. The Hamiltonian for a free atom is denoted as $\hat{K}\equiv(-\nabla^2+r^2)/2-\Omega\hat{L}_z$ with the angular-momentum operator $\hat{L}_z=-i(x\partial_y-y\partial_x)$. The eigensolutions of $\hat{K}$ are given by $\varepsilon_{nm}^{(0)}=n+m+1-\Omega(m-n)$ and $u_{nm}(r)=e^{r^2/2}(\partial_x+i\partial_y)^m(\partial_x-i\partial_y)^n e^{-r^2}/\sqrt{\pi n!m!}$, where n,m are zero or positive integers, and u_{nm} is the well-defined angular momentum state, i.e., $\hat{L}_z u_{nm}(\mathbf{r})=(m-n)u_{nm}(\mathbf{r})$. In order to minimize the total energy E, we expand $\psi^{(\rm B)}$ and $\psi^{(\rm F)}$ in terms of u_{nm}, and iterate minimization of $\tilde{\varepsilon}^{(\rm B)}=(E^{(\rm B)}+E^{(\rm BF)})/N_{\rm B}$ [6] and diagonalization of $\tilde{\varepsilon}^{(\rm F)}=\varepsilon^{(\rm F)}+h\int d\mathbf{r}\psi^{(\rm F)*}n^{(\rm B)}\psi^{(\rm F)}$ until a self-consistent convergence [7].

In the absence of a boson-fermion coupling and within the lowest-Landau-level (LLL) approximation for the condensate, the total angular momentum of bosons $L_{\rm B}$ jumps from zero to $N_{\rm B}$ at $\Omega_{\rm cr}^{(h=0)}=1-gN_{\rm B}/(8\pi)$, which corresponds to the onset of vortex formation. In contrast, the total angular momen-

CP850, *Low Temperature Physics: 24th International Conference on Low Temperature Physics;*
edited by Y. Takano, S. P. Hershfield, S. O. Hill, P. J. Hirschfeld, and A. M. Goldman

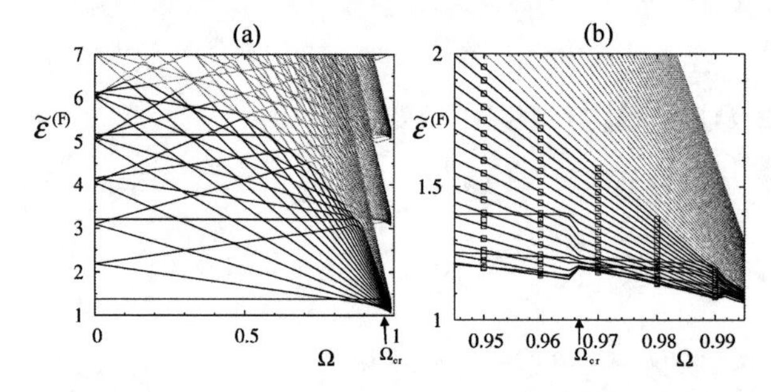

FIGURE 1. (a) Energy spectrum of fermions. Thick and thin lines denote occupied and unoccupied levels, respectively. The highest-occupied level corresponds to the Fermi energy. (b) Square dots enlarge occupied states in (a), and thick and thin lines are occupied and unoccupied states obtained within LLL.

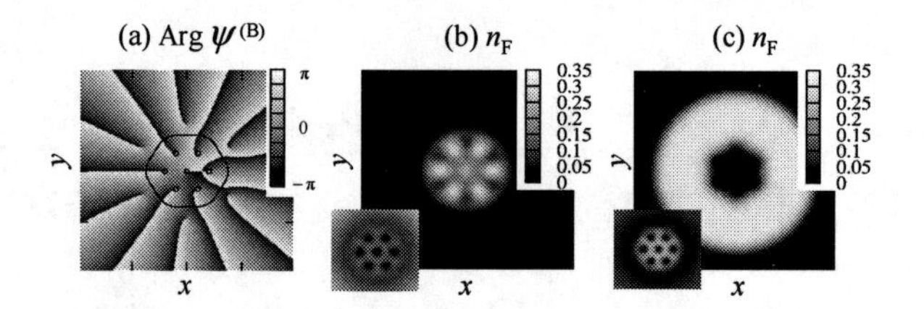

FIGURE 2. (a) Phase profiles of condensate wave function. Density profiles of fermions for (b) small ($N_F = 8$), and (c) large ($N_F = 50$) number of fermions. Insets show the effective potentials for fermions $V_F^{(\text{eff})}$.

tum of free fermions, which is the summation of the single-particle angular momentum $(m-n)$ over occupied states, increases monotonically for $\Omega > 0$. The total value is well described by the Thomas-Fermi approximation (TFA) with the rigid-body rotation, given by $L_F^{(RB)} = \Omega \int r^2 n^{(F)}(r) d^2r = (8N_F)^{3/2}\Omega/(24\sqrt{1-\Omega^2})$. The chemical potential is evaluated by the TFA as $\mu_F = \sqrt{2N_F(1-\Omega^2)}$, which takes huge value of the order of $\sqrt{N_F}$ at $\Omega = 0$, but becomes of the order of unity in the fast rotating limit $\Omega \to 1$.

Next we introduce the boson-fermion coupling, and show the energy spectrum $\tilde{\varepsilon}^{(F)}$ of fermions in Fig. 1 [$h = 30g = 1.5 \times 10^{-2}$, $N_B = 50N_F = 10^3$]. While the $(n+m)$-th level is $(n+m+1)$-fold degenerate when $h = 0$ at $\Omega = 0$, the boson-fermion coupling lifts these degeneracies by the amount $\varepsilon_{|m-n|}^{(BF)}$, and only degeneracies with respect to the same $|m-n|$ are held. For slow rotating regime where the condensate has no vortices, the deviation from $\varepsilon_{nm}^{(0)}$ is also given by $\varepsilon_{|m-n|}^{(BF)}$, and the highest occupied level gradually decreases due to the narrowing of level spacings between the occupied states. The highest occupied level and the total angular momentum well agree with μ_F and $L_F^{(RB)}$ given by the TFA for free fermions. When a single vortex enters the condensate at $\Omega_{cr} \simeq \Omega_{cr}^{(h=0)}$, the spectrum exhibits discontinuous jump, and the boson-fermion interaction causes level crossings. In the fast rotating regime, the spectrum becomes quasidegenerate and the Fermi energy is located in the band characterized by the quantum number $n = 0$. The LLL approximation is thus valid for the occupied degenerate states as shown in Fig. 1 (b), whereas the TFA fails in this regime.

We show the phase profile of the condensate wave function in Fig. 2 (a) for $h = 1.25g = 10^{-3}$, $N_B = 10^3$, and $\Omega = 0.998$. The small circles enclosing branch cuts denote the positions of the quantized vortices, and the outer large circle shows the spread of the condensate cloud. When the number of vortices N_v and the number of fermions N_F are comparable, the fermions plug up the vortex cores in the condensate [Fig. 2 (b), $N_F = 8$] in order to reduce the interaction energy. In the inset, we show the effective potential for fermions, given by $V_F^{(\text{eff})}(\mathbf{r}) = (1-\Omega^2)r^2/2 + hn_B(\mathbf{r})$ where the first parenthetic term is the sum of the trapping and the centrifugal potentials, and the last term is the potential created by the boson-fermion coupling. Since the trapping and centrifugal potentials almost cancel each other, the bosonic density $n^{(B)}$ dominates $V_F^{(\text{eff})}$ in the limit $\Omega \to 1$.

The fermions are thus subjected to a periodic potential whose depth is determined by the sign and strength of interaction h when the Abrikosov vortex lattice is formed in the condensate. As the number of fermions increases as $N_v \ll N_F$, both the centrifugal potential and the Pauli pressure help the occurrence of phase separation as shown in Fig. 2 (c), where the fermionic cloud expands outward even though the other parameters and the bosonic density remain unchanged from those in Fig. 2 (a). Such density patterns are also formed when the repulsive boson-fermion coupling or angular frequency of the rotating drive is increased.

In summary, we found that the occupied fermionic states in the rotating boson-fermion mixture drastically change from those of free fermions, depending on the angular frequency of rotating drive, boson-fermion coupling, and number of fermions.

REFERENCES

1. A.G. Truscott *et al.*, *Science* **291**, 2570 (2001); F. Schreck *et al.*, *Phys. Rev. Lett.* **87**, 080403 (2001).
2. S. Inouye *et al.*, *Phys. Rev. Lett.* **93**, 183201 (2004).
3. K.W. Madison *et al.*, *Phys. Rev. Lett.* **84**, 806 (2000); J.R. Abo-Shaeer *et al.*, *Science* **292**, 476 (2001).
4. M.W. Zwierlein *et al.*, *Nature* **435**, 1047 (2005).
5. T.L. Ho and C.V. Ciobaru, *Phys. Rev. Lett.* **85**, 4648 (2000).
6. D.A. Butts and D.S. Rokhsar, *Nature* **397**, 327 (1999).
7. T. Karpiuk *et al.*, *Phys. Rev. A* **69**, 043603 (2004).

Spin Correlation and Superfluidity of Trapped Fermi Atoms on an Optical Lattice

Masahiko Machida[1,3], Susumu Yamada[1], Yoji Ohashi[2], and Hideki Matsumoto[2]

[1]*CCSE, Japan Atomic Energy Agency, 6-9-3 Higashi-Ueno, Taito-ku, Tokyo 110-0015, Japan*
[2]*Institute of Physics, University of Tsukuba, 1-1-1 Tennodai, Tsukuba, Ibaraki 305-8571, Japan*
[3]*CREST (JST), 4-1-8 Honcho, Kawaguchi, Saitama 332-0012, Japan*

Abstract. We investigate a possibility of superfluidity in a trapped gas of Fermi atoms with repulsive interaction U in the presence of an optical lattice. Applying the exact diagonalization method to a one-dimensional fermion-Hubbard model including the trap potential, we find that, when U exceeds a critical value, the binding energy of two Fermi atoms becomes negative indicating that an attractive interaction effectively works. In this case, a "Mott core" appears in the trap center, where each site is occupied by one atom, and a large Cooper-pair amplitude develops between atoms in the left and right hand sides of this core. In order to clarify the origin of the attractive force, we compute a pseudo spin correlation between neighboring sites given by $<S_i \cdot S_{i+1}>$. As a result, we find that a creation of a singlet pair between both sides of the Mott core is strongly coupled with a pseudo-spin singlet re-formation leading to an energy gain process.

Keywords: Atomic Fermi Gas, Optical Lattice, Hubbard Model, Mott Insulator, Cooper Pair, Exact Diagonalization
PACS: 03.75Ss, 05.30.Fk, 74.20.Mn, 71.30.+h

INTRODUCTION

An ultra-cold atom gas placed in an optical lattice presents a quite intriguing experimental situation. It covers rich physics of the Hubbard model showing strongly correlated behaviors. Since the discovery of high-Tc cuprates, the fermion-Hubband model has attracted much attention as a typical model to investigate the mechanism of superconductivity originating from the on-site repulsion. However, it is still elusive whether or not the model really describes the superfluid phase. Thus, in a Fermi atom gas loaded into an optical lattice, studies of the Cooper pairing via a repulsive interaction in the presence of the optical lattice will be a next big challenge.

In the ultra-cold atomic Fermi gas in the optical lattice, one should include the effect of a harmonic trap potential. In the presence of the trap, atoms tend to form a cluster in the center of the trap, which naturally leads to a dome-like density profile [1]. On the other hand, the on-site repulsion excludes the double occupancy of atoms. As a result, a flat density profile with one atom per one site is favorable [1]. We call the flat region "Mott core" below. Since the trap potential and the repulsive interaction thus have opposite effects as described above, their interplay is an interesting problem [1]. In the previous paper [2], we proposed a novel mechanism of fermion superfluidity associated with repulsive interaction U and trap potential V [2]. The superfluidity is expected to emerge when the system spatially consists of two regions, i.e., a Mott core in the center of the trap and metallic wings growing around the edges of the Mott core [1]. In such a situation, the compressibility in the atom density is almost zero in the center of the trap, whereas it has a finite value around the surface of the core [1]. This makes it possible to vary the atom number around the Mott core and to form a singlet pair between both edges [2]. In this paper, we numerically demonstrate that the paring amplitude indeed develops between both side-edges of the Mott core [2]. We also calculate the local spin correlation in the Mott core region, and clarify the spin structures of the Mott core. The calculation result reveals that spin singlets are formed inside the Mott core in a zigzag like manner. This result indicates that the Mott core is not antiferro-magnetic but similar to the RVB state. Such a frustrated region is expected to make it possible to yield a spin-mediated attractive interaction and to lead to a Cooper pairing instability like in under-doped High-T_c superconductors.

CP850, *Low Temperature Physics: 24th International Conference on Low Temperature Physics;*
edited by Y. Takano, S. P. Hershfield, S. O. Hill, P. J. Hirschfeld, and A. M. Goldman

NUMERICAL CALCULATION

The target model is the one-dimensional Hubbard model with a harmonic trap [1,2] given by

$$H_{Hubbard} = -t\sum_{i,j,k}\left(a^{\dagger}_{j\sigma}a_{i\sigma} + H.C.\right) + U\sum_{i} n_{i\uparrow}n_{i\downarrow} + V\left(\frac{2}{N-1}\right)^2 \sum_{i,\sigma} n_{i\sigma}\left(i - \frac{N+1}{2}\right)^2, \quad (1)$$

where a two-component Fermi gas is assumed and Fermi atoms with pseudo spin "up (↑)" or "down (↓)" hop between neighboring sites. The model Hamiltonian includes the on-site repulsive interaction U, as well as the trap potential characterized by a parameter V, which can be experimentally controlled by the intensity of the laser light. In the previous paper [2], we calculated the binding energy (E_b) of two Fermi atoms based on the ground state energy obtained by exact diagonalization. If E_b is negative, it then means that an attractive interaction works between two atoms. However, we note that E_b is also sensitive to conditions except for the pairing. The number of site N is 14 in this paper.

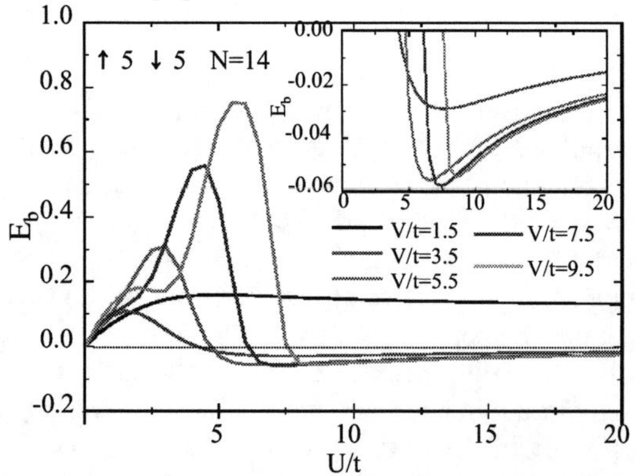

FIGURE 1. U/t dependence of E_b normalized by t for various V/t .The inset is an enlarged view of the region of the minimum E_b.

Figure 1 presents the U/t dependence of E_b for various V/t. It is found that every case except for V/t = 1.5 shows a sign change at a certain U/t which depends on V/t [2]. Before the negative E_b appears, the particle distribution shows a dome-like shape, while it changes into a flat "mesa" shape, i.e., the Mott core, after the sign change of E_b. Then, the metallic region, in which the particle density is less than unity, develops at both edges of the flat Mott core [1,2]. Only in the metallic regions, particles can move and fluctuate. In fact, when we add two particles into the system, these particles cling to the edges.

Figure 2 shows the U/t dependence of the singlet Cooper pair function [2] given by

$$\Delta_{i,j} = \frac{1}{\sqrt{2}}\langle n+1\uparrow, n+1\downarrow | C_{i\uparrow}C_{j\downarrow} - C_{i\downarrow}C_{j\uparrow} | n\uparrow, n\downarrow\rangle, \quad (2)$$

where i is fixed at a metallic edge (i=2) and the j dependence of Δ_{ij} is presented. The figure clearly indicates that the singlet pair amplitude develops between the both metallic edges when the flat Mott core emerges [2]. This result reflects a fact that the particle density can fluctuate only at the metallic edges [1,2]. When two particles with up and down pseudo-spins come to the system, the attractive interaction works between them, causing them to form a singlet Cooper pair at the edges. Figure 1 and 2 clearly indicate that such a distant pairing is possible and instability into a superfluid state can occur.

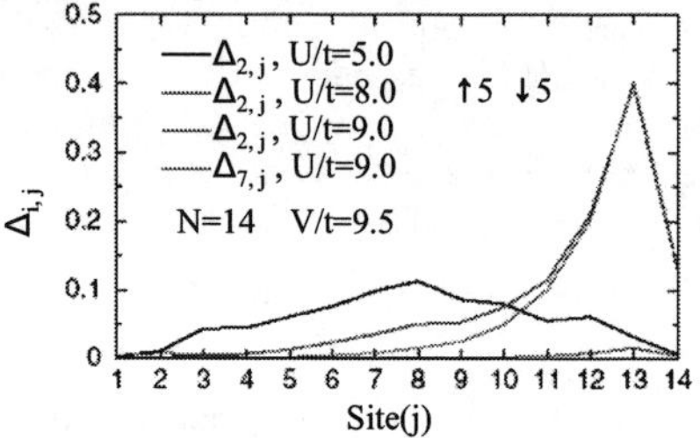

FIGURE 2. U/t dependence of the singlet pair function Δ_{ij} normalized by t for the case of V/t=9.5. The site i is fixed to be 2 and 7, and their j dependences for various U/t are shown.

Figure 3 presents site dependences of the local spin correlation $\langle S_i \cdot S_{i+1}\rangle$ for two U/t values. It is found that when E_b is negative (U/t=8.0), i.e., the Mott core emerges, the local spin correlation in $\langle S_i \cdot S_{i+1}\rangle$ shows a zigzag structure. By noticing that $\langle S_i \cdot S_{i+1}\rangle = -3/4$ for a complete spin singlet, the spin singlet is found to grow locally. Thus, it is found that the singlet Cooper pair can easily develop between the metallic wings.

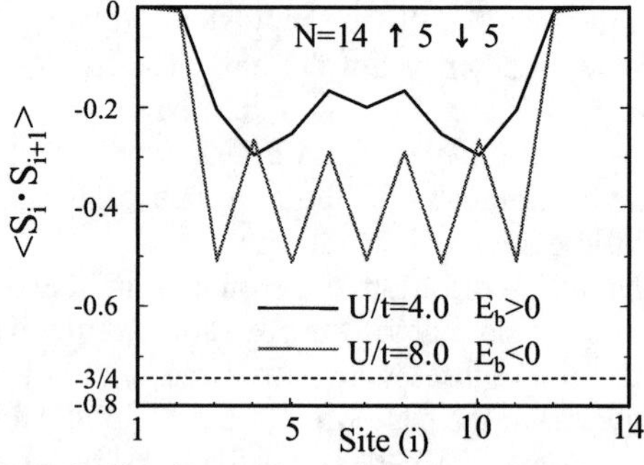

FIGURE 3. Site dependence of $\langle S_i \cdot S_{i+1}\rangle$ for two U/t values, in which E_b is positive and negative for V/t=9.5.

REFERENCES

1. M. Rigol et al., *Phys. Rev. Lett.* **91**, 130403 (2003).
2. M. Machida et al., *Phys. Rev. Lett.* **93**, 200402 (2004).

Structure of a Quantized Vortex in Fermi Atom Gas

Masahiko Machida[1,3] and Tomio Koyama[2,3]

[1]*CCSE, Japan Atomic Energy Research Institute, 6-9-3 Higashi-Ueno, Taito-ku, Tokyo 110-0015, Japan*
[2]*IMR, Tohoku University 2-1-1 Katahira Aoba-ku, Sendai 980-8577, Japan*
[3]*CREST(JST), 4-1-8 Honcho, Kawaguchi, Saitama 332-0012, Japan*

Abstract. In atomic Fermi gases, the pairing character changes from BCS-like to BEC-like when one decreases the threshold energy of the Feshbach resonance. With this crossover, the system enters the strong-coupling regime through the population enhancement of diatom molecules, and the vortex structure becomes much different from well-known core structures in BCS superfluid since the superfluid order parameter is given by a sum of BCS pairs and BEC molecular condensates. In this paper, we study the structure of a vortex by numerically solving the generalized Bogoliubov-de Gennes equation derived from the fermion-boson model and clarify how the vortex structure changes with the threshold energy of the Feshbach resonance. We find that the diatom boson condensate enhances the matter density depletion inside the vortex core and the discreteness of localized quasi-particle spectrum.

Keywords: Atomic Fermi Gas, Vortex Structure, fermion-boson model, Bogoliubov-de Gennes Equation
PACS: 03.75Ss, 03.75.Lm, 03.75.Kk, 74.25.Op

INTRODUCTION

The most clear-cut confirmation of a BCS-like pairing state in the superfluid atomic Fermi gas is the observation of a quantized vortex [1]. Thus, the study of a quantized vortex in an ultra-cold atomic Fermi gas is one of the most important issues. In this paper, we theoretically investigate the single-vortex state of a superfluid atomic Fermi gas on the basis of the generalized Bogoliubov de Gennes (BdG) equation [2,3] derived from the fermion-boson model which can describe the conversion of two fermions to a molecule and vice versa [4].

In conventional BCS superconductors, the vortex core has been extensively studied both theoretically and experimentally. The matter density depression inside the core is found to be negligible in these superconductors [5]. This is because the superconducting gap energy Δ is much less than the Fermi energy E_F. However, the situation is different in strong coupling superconductors since $\Delta \sim E_F$. As a result, the depletion of the matter density is expected to be large and detectable [1-3,5]. In fact, vortices in a neutron star and an atomic Fermi gas [1] have been very recently studied based on this idea.

In the atomic Fermi gas the pairing character in the superfluid state changes from a BCS-like one to a BEC-like one when the threshold energy of the Feshbach resonance decreases and approaches the chemical potential. In the neighborhood of this crossover the system goes into the strong-coupling limit, and moreover, it can go to BEC regime; no analogous phenomenon have been investigated in the studies of vortex cores in superconductors. In the crossover and BEC regimes, since the superfluid is a mixture of BCS pairs and BEC molecules [4], the vortex core structure is theoretically an intricate issue. Therefore, in this paper, we numerically study the structure of a vortex in the fermion-boson model [4]. This model fully describes the BCS-BEC crossover in the atomic Fermi gas as the Feshbach resonance threshold parameter is varied [4]. We numerically solve the generalized BdG equation for the single vortex depending on the threshold parameter [2]. In the calculation, both fermionic and bosonic gap functions in the single-vortex state are self-consistently calculated. Using the solutions, we clarify the core structure of the single vortex in the crossover and BEC regimes.

NUMERICAL CALCULATIONS AND THEIR RESULTS

In order to study the vortex core structure, we solve a generalized BdG equation derived from the fermion-boson model, which describes conversion processes between a boson and two fermions due to the Feshbach resonance as well as a standard BCS-like attractive

CP850, *Low Temperature Physics: 24th International Conference on Low Temperature Physics;*
edited by Y. Takano, S. P. Hershfield, S. O. Hill, P. J. Hirschfeld, and A. M. Goldman

interaction [4]. In this paper, we consider an isolated vortex in the two-dimensional s-wave case [1-3]. In this case, since the system has 2D cylindrical symmetry, two gap-functions, i.e., the BCS gap function and the molecular condensate function can be expanded by a set of eigen-functions in the radial direction. Thus, the solution of the generalized BdG equation is obtained by the matrix diagonalization method, and the two gap functions are calculated by the gap equation and the relation between the BCS gap and the molecular condensate function. Furthermore, we keep a constraint for the total number of particles, $N=N_F + 2N_B$ = const., where N_F is the number of fermions and N_B is that of the condensate molecules [2,3].

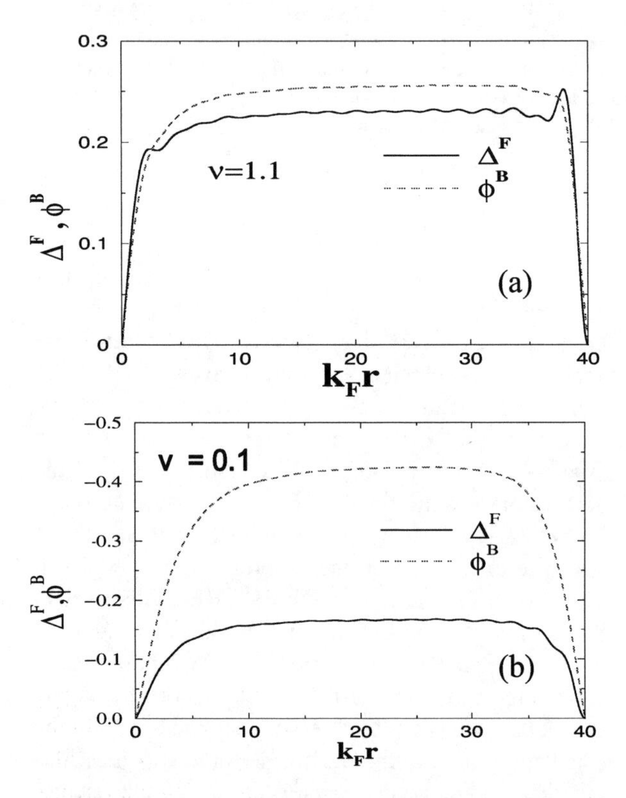

FIGURE 1. Radial distribution of the BCS gap function Δ^F and the molecular-condensate gap function φ^B at (a) ν =1.1 and at (b) ν = 0.1. In both cases, the attractive interaction of two-body fermions U=0.5 and the coupling–constant for the boson-fermion conversion g=0.6. Both functions are normalized by E_F

Let us now present the results of numerical calculations. In this paper, we concentrated on the region of ν (the Feshbach-resonance threshold parameter in the fermion-boson model) < 1 where the system goes from the crossover regime to the BEC regime by decreasing ν. In the previous paper [2], we presented how the vortex core structure changes when one approaches the crossover regime ($\nu \sim 1$) from the BCS side (ν >1). Thus, a main aim of this paper is to compare the vortex core structure between the crossover regime and the BEC limit.

Figure 1 shows the radial distribution of both the BCS gap function Δ^F and the molecular-condensate function φ^B at ν =1.1 and 0.1. As seen in Fig. 1(a), at which the system is located at the crossover regime, the molecular condensate function develops and its amplitude is comparable to that of the BCS gap function. In this case, the BCS gap function shows a Friedel-oscillation-like structure [5], whereas the molecular condensate shows no such structure. In contrast, when ν =0.1, which is located almost in the BEC regime, neither of them shows such a structure as seen in Fig. 1(b). This is because low-lying bound states inside the vortex core almost disappear as seen in Fig. 2, which is the energy spectrum of the fermionic excitation as a function of the angular momentum, and the Meissner Gap opens. Furthermore, we find that the vortex core size at ν =0.1 is clearly larger than that at ν =1.1. This indicates that the relation between the vortex core size and the threshold parameter becomes asymmetric at the crossover [1,2].

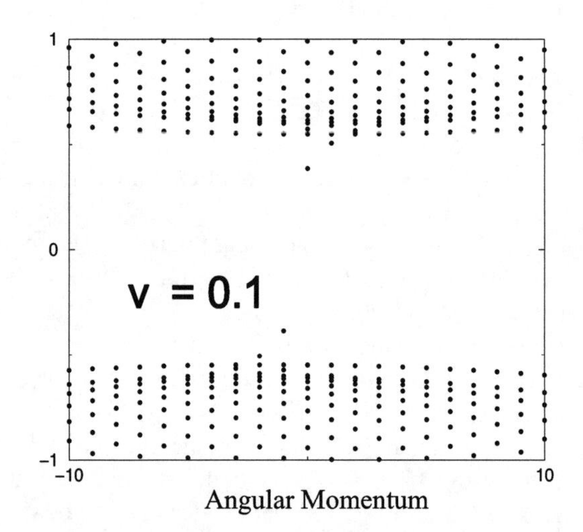

FIGURE 2. Energy spectrum of fermionic excitations in a vortex core as a function of the angular momentum at ν =0.1. U=0.5 and g=0.6.

REFERENCES

1. A. Bulgac and Y. Yu, *Phys. Rev. Lett.* **91**, 190404 (2005).
2. M. Machida and T. Koyama, *Phys. Rev. Lett.* **94**, 140401 (2005).
3. Y. Kawaguchi and T. Ohmi, cond-mat/0411018.
4. Y. Ohashi and A. Griffin, *Phys. Rev. Lett.* **89**, 130402 (2002).
5. M.Machida and T.Koyama. *Phys. Rev. Lett.* **90**, 077003 (2003).

BCS-BEC Crossover in a Gas of Fermi Atoms with a *P*-wave Feshbach Resonance

Yoji Ohashi

Institute of Physics, University of Tsukuba, Tsukuba, Ibaraki, 305, Japan

Abstract. We present a theoretical study of the possibility of p-wave superfluidity in a gas of Fermi atoms with a p-wave Feshbach resonance. This anisotropic Feshbach resonance leads to a tunable pairing interaction in the p-wave Cooper channel. The character of superfluidity is shown to continuously change from the p-wave BCS type to the Bose-Einstein condensation (BEC) of tightly-bound molecules with finite orbital angular momenta, $L_z=\pm1$ and 0, as the threshold energy 2ν of the Feshbach resonance is lowered.

Keywords: BCS-BEC crossover; Feshbach resonance; fermion superfluidity,
PACS: 03.75.Ss, 03.70.+k, 03.75.Mn

One of the hottest topics in the current research of cold atom physics is the BCS-BEC crossover, where the character of fermion superfluidity continuously changes from a weak-coupling BCS-type to the BEC of tightly bound molecules, with increasing a pairing interaction [1, 2, 3]. In a Fermi gas, an effective pairing interaction associated with a Feshbach resonance (FR) [4, 5] is tunable by adjusting the threshold energy 2ν of the Feshbach resonance. Using this interaction, the BCS-BEC crossover has been realized [6, 7]. Since the discovered superfluidity is a s-wave type, the search for p-wave superfluidity is the next big challenge.

In this paper, we consider a single-component Fermi gas with a p-wave FR, which can be described by the coupled fermion-boson model Hamiltonian [8]

$$H=\sum_{p}\varepsilon_{p}c_{p}^{\dagger}c_{p}+\sum_{q,j}E_{q}b_{q,j}^{\dagger}b_{q,j} + \frac{g_r}{\sqrt{2}}\sum_{p,q,j}p_j[b_{q,j}c^{\dagger}_{p+q/2}c^{\dagger}_{-p+q/2}+h.c.] - \frac{U}{2}\sum_{p,p',q,j}(p.p')c^{\dagger}_{p+q/2}c^{\dagger}_{-p+q/2}c_{-p'+q/2}c_{p'+q/2}. \quad (1)$$

Here c_p is the annihilation operator of a Fermi atoms with the kinetic energy $\varepsilon_p=p^2/2m$, and b_{qj} ($j=x,y,z$) describe three kinds of molecules in the p-wave channel, associated with a p-wave FR. The kinetic energy of these molecules is given by $E_q=q^2/2M+2\nu$, where $M=2m$ is the mass of a molecule, and 2ν is the threshold energy of FR. [E_q has no j-dependence due to the spherical symmetry of the system we are considering.] The second line in (1) describes FR, where two atoms form a molecule with $L_z=\pm1$or 0, and it dissociates into two atoms. (1) also includes a non-resonance p-wave interaction U, which is taken to be weakly attractive. Since one molecule consists of two Fermi atoms, the total number N of Fermi atoms must be conserved. This constraint can be absorbed into (1), by considering the grand canonical form $H=H-\mu N$. Here $N=N_F+2\Sigma_j N_{Bj}$, where N_F is the number of Fermi atoms, and N_{Bj} is the number of Feshbach molecules in the j-th channel. For simplicity, we consider a uniform gas in (1), and neglect effects of a trap potential.

We calculate the superfluid phase transition temperature T_c in the p-wave BCS-BEC crossover. In the strong-coupling regime, fluctuation effects in the p-wave Cooper channel are crucial, which we include within the Gaussian fluctuations. This is an extension of the strong-coupling s-wave superconductivity theory [3] to a p-wave superfluid. T_c is determined from the Thouless criterion that the phase transition occurs when the particle-particle scattering vertex has a pole at $q=\omega=0$. The resulting equation has the same form as the p-wave BCS gap equation at Tc, with the pairing interaction $U_{\rm eff}\equiv U+g^2/(2\nu-2\mu)$. $U_{\rm eff}$ becomes large when $2\nu\to2\mu+0$. The chemical potential μ is determined from the equation for N, which involves effects of fluctuations [2].

CP850, *Low Temperature Physics: 24th International Conference on Low Temperature Physics;*
edited by Y. Takano, S. P. Hershfield, S. O. Hill, P. J. Hirschfeld, and A. M. Goldman

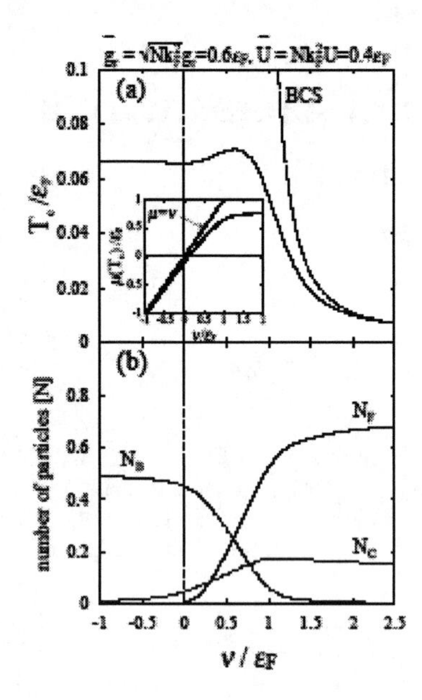

FIGURE 1. (a) T_c as a function of 2ν in a narrow FR. 'BCS' represents the mean-field BCS result. In the s-wave BCS-BEC crossover, it has been shown [9] that the peak structure disappears when higher order fluctuation effects beyond the Gaussian level are included. Therefore, the peak around ν/ε_F~0.5, as well as that in Fig. 2(a), may be an artifact of the Gaussian approximation. Inset: μ at T_c. (b) The numbers of Fermi atoms (N_F), Feshbach molecules [with L_z=±1 and 0] (N_B), and preformed Cooper pairs (N_C) at T_c. N_C also involves the contribution of scattering states.

Figure 1(a) shows the self-consistent solution of T_c in a narrow FR. As 2ν is lowered, μ deviates from the Fermi energy ε_F and approaches ν [see the inset], so that the pairing interaction U_{eff} becomes strong. Then T_c gradually deviates from the weak-coupling BCS result, and approaches a constant value in the strong-coupling regime (ν<0). In this regime, since dominant particles are Feshbach molecules [N_B in Fig. 1(b)], the phase transition can be understood as the BEC of these molecules. In the BEC limit ($\nu\to-\infty$), we obtain $Tc=0.066\varepsilon_F$, which equals T_c of a *three*-component $N/2$ ideal Bose gas [with L_z=±1 and 0]. In a s-wave superfluid, the BEC regime is described by a *one*-component $N/2$ molecular Bose gas[1-3]. This smaller number of molecular components in the s-wave case leads to higher $T_c=0.218\varepsilon_F$ [2,3] in the BEC limit.

Figure 2(a) shows T_c in a broad Feshbach resonance. Because of the large Feshbach coupling g_r, the BCS-BEC crossover occurs when $\nu\gg\varepsilon_F$. We note that, in the narrow resonance in Fig. 1, the crossover occurs when $\nu\sim\varepsilon_F$. However, when the pairing interaction is scaled by the p-wave atomic scattering length a_p, we find that the both results are very similar to each other (see the inset). In the crossover regime (ν/ε_F~25), Fig. 2(b) shows that preformed Cooper pairs N_C are dominant. Feshbach molecules (N_B) are almost absent due to the large threshold energy ν/ε_F >>1. The dominant bosons continuously change from the preformed Cooper pairs to the Feshbach molecules around ν=0. The Feshbach molecules are dominant in the BEC regime where ν<0, i.e., the molecular band $E_q=q^2/2M+2\nu$ is lower than the fermion band ε_p.

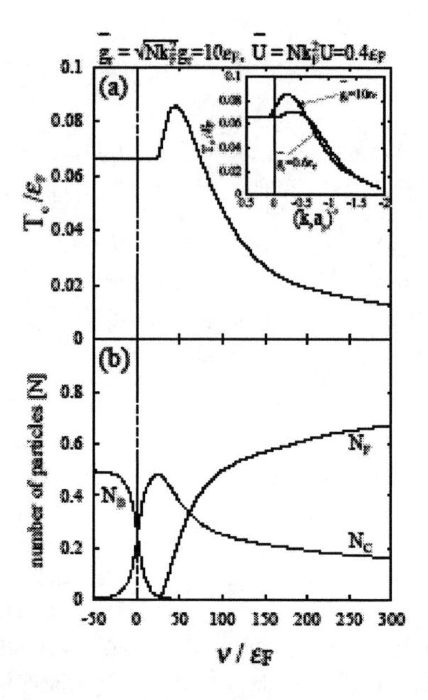

FIGURE 2. (a) T_c as a function of 2ν in a broad FR. The inset shows T_c as a function of the p-wave scattering length a_p. (b) N_F, N_B, and N_C in the BCS-BEC crossover.

In summary, we have discussed the p-wave BCS-BEC crossover in a single-component Fermi gas with a p-wave Feshbach resonance. T_c in the BEC limit is given by $T_c=0.066T_F$, which is accessible experimentally. The search for molecular BEC would be the first step to realize p-wave superfluidity.

REFERENCES

1. A. Leggett, in *Modern Trends in the Theory of Condensed Matter*, edited by A. Pekalski and J. Przystawa (Springer Verlag, Berlin, 1980) p.14.
2. P. Nozières and S. Schmitt-Rink, *J. Low Temp. Phys.* **59**, 195 (1985).
3. Y. Ohashi and A. Griffin, *Phys. Rev. Lett.* **89**, 130402 (2002); *Phys. Rev. A* **67**, 033603, 063612 (2003).
4. E. Timmermans *et al.*, *Phys. Lett. A* **285**, 228 (2001).
5. M. Holland *et al*, *Phys. Rev. Lett.* **87**, 120406 (2001).
6. C. Regal *et al.*, *Phys. Rev. Lett.* **92**, 040403 (2004).
7. M. Bartenstein *et al.*, *Phys. Rev. Lett.* **92**, 120401 (2004).
8. Y. Ohashi, *Phys. Rev. Lett.* **94**, 050403 (2005); T. Ho and R. Diener, *Phys. Rev. Lett.* **94**, 090402 (2005).
9. R. Haussmann, *Phys. Rev. B* **49**, 12975 (1994).

Instability of Superfluid across Feshbach Resonances

C.-H. Pao*, Shin-Tza Wu* and Sungkit Yip†

*Department of Physics, National Chung Cheng University, Chiayi 621, Taiwan
†Institute of Physics, Academia Sinica, Nankang, Taipei 115, Taiwan

Abstract. We consider a dilute atomic gas comprising two species of Fermions with unequal concentrations. We study at zero temperature the atom-molecule mixture with inter-species low-energy scatterings induced by a Feshbach resonance. To examine the stability of this superfluid, we calculate the chemical potential and the superfluid density for fixed individual concentrations. We find that the system can have distinct properties compared to the ordinary Bose-Fermi mixture due to the unbound fermions.

Keywords: Degenerate Fermi gases, BEC-BCS crossover, Feshbach resonances
PACS: 03.75.Ss, 05.30.Fk, 34.90+q

Laser cooling experiments in dilute fermionic atoms [1] open a new field to study the superfluid properties. Through the Feshbach resonance [2], the effective interaction between the atoms can be varied over a wide range from a weak coupling BCS regime to a strong coupling regime where they form diatomic molecules and undergo Bose Einstein condensation (BEC) at low temperature.

Most investigations in this field are focused on a system with equal concentrations of the two fermionic species. In this study, we deal with a general case involving unequal concentrations of the two species. The system can then have mismatched Fermi surfaces instead of a single Fermi surface.

A fermion system with mismatched Fermi surfaces was first studied by Fulde and Ferrell, and Larkin and Ovchinnikov (FFLO) [3] in the 1960's. They found that the system is likely to have an inhomogeneous gapless superconducting phase. The related problem in dilute ultracold atoms has revived interests [4] in this work. However, these studies are restricted to the weak-coupling regime. We present below an analysis to all coupling strengths.

We consider two fermion species, denoted as "spin" $\uparrow$ and $\downarrow$, of equal mass. Because of unequal concentrations of these two species and the possible pairing, the system includes three fields: the chemical potentials μ_σ ($\sigma = \uparrow$ and $\downarrow$) and the pairing field Δ. It is convenient to introduce the average chemical potential $\mu \equiv (\mu_\uparrow + \mu_\downarrow)/2$ and the difference $h \equiv (\mu_\uparrow - \mu_\downarrow)/2$ to describe the system. Through the generalized BCS mean field approach at zero temperature [5], the excitation spectrum for each spin is found to be

$$E_{\uparrow,\downarrow}(\mathbf{k}) = \mp h + \sqrt{\xi(\mathbf{k})^2 + \Delta^2}\,. \tag{1}$$

where $\xi(\mathbf{k}) = \hbar^2 k^2/2m - \mu$. The total density $n = n_\uparrow + n_\downarrow$ and the density difference $n_d = n_\uparrow - n_\downarrow$ are then [6]

$$n = \int \frac{d^3k}{(2\pi)^3}\left\{1 + \frac{2\xi(\mathbf{k})}{E_\uparrow + E_\downarrow}\left[f(E_\uparrow) - f(E_\downarrow)\right]\right\}, \tag{2}$$

$$n_d = \int \frac{d^3k}{(2\pi)^3}\left[f(E_\uparrow) - f(-E_\downarrow)\right], \tag{3}$$

with $f(E_\sigma)$ is the Fermi function.

The scattering between fermions is short ranged and can be modelled as an effective s-wave interaction characterized by the corresponding a scattering length a. The order parameter is then calculated by

$$\frac{m}{4\pi a}\Delta = -\Delta \int \frac{d^3k}{(2\pi)^3}\left[\frac{1 - f(E_\uparrow) - f(E_\downarrow)}{E_\uparrow + E_\downarrow} - \frac{m}{\hbar^2 k^2}\right]. \tag{4}$$

We solve Eqs. (2) to (4) self-consistently for fixed n and n_d. We take the spin up to be the majority species so that $h \geq 0$ and $E_\downarrow < 0$ always. Then the density difference reduces to $n_d = \int \frac{d^3k}{(2\pi)^3} f(E_\uparrow)$, and the integration is only determined by $E_\uparrow(\mathbf{k}) < 0$ [8].

Here $k_F = (3\pi^2 n)^{1/3}$. The dimensionless coupling constant is defined as $g \equiv 1/(\pi k_F a)$, which varies from $+\infty$ for large negative detuning to $-\infty$ for large positive detuning.

The inset of Fig. 1 shows a typical behavior of μ, Δ, and h as functions of g for a given density difference n_d/n. The behaviors of μ and Δ are rather like the case of equal concentrations [5]. However, the coupling constant g has to be larger than a minimum value g_c in order for a finite Δ to exist. For $g < g_c$, Eq. (4) requires that $\Delta = 0$ and the system is in the normal state. The main frame of Fig. 1 shows h/ε_F as function of g in the intermediate region ($|g| \leq 1$) for four different n_d/n (0.2, 0.4, 0.6, and 0.8). The horizontal dotted lines indicate the normal state

CP850, *Low Temperature Physics: 24th International Conference on Low Temperature Physics;*
edited by Y. Takano, S. P. Hershfield, S. O. Hill, P. J. Hirschfeld, and A. M. Goldman

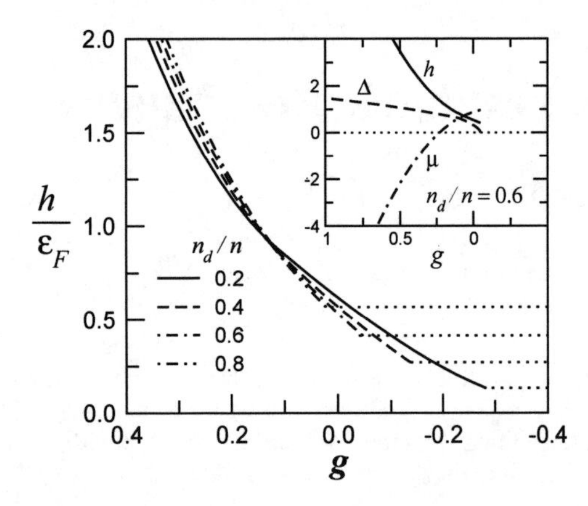

FIGURE 1. Chemical potential difference h versus the coupling constant g for four different values of n_d. Inset includes the results for the gap function Δ and the total chemical potential μ for $n_d/n = 0.6$.

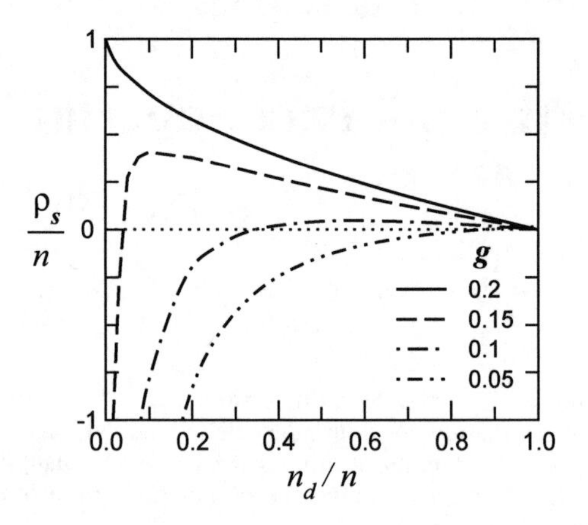

FIGURE 2. Superfluid density ρ_s versus n_d/n for fixed coupling constant $g's$.

in which the gap function is zero. The lines (for $\Delta \neq 0$) of h versus g for different n_d/n cross each other near $g \sim 0.15$. For $g \geq 0.17$, h increases with n_d for fixed g. On the other hand, h decreases as n_d increases for $g \leq 0.15$. Note that h plays the role of an effective external Zeeman field. A stable superfluid state corresponds to a positive slope of h (as a function of n_d for fixed g) [6, 7, 8].

The superfluid density ρ_s must be positive [6] to have a stable superfluid phase. The analytic result can be expressed as

$$\frac{\rho_s}{n} = 1 - \frac{\theta(\sqrt{\mu^2+\Delta^2} - h)\bar{k}_1^3 + \bar{k}_2^3}{2\sqrt{1-(\Delta/h)^2}}, \tag{5}$$

where $\bar{k}_{1,2} = [(\mu \mp \sqrt{h^2-\Delta^2})/\varepsilon_F]^{1/2}$ and $\theta(x)$ is the step function. In Fig. 2, we plot ρ_s as a function of n_d/n for several different g. Note that, for $n_d/n \to 0$, ρ_s changes sign exactly at $\mu = 0$. The second term in Eq. (5) is the paramagnetic response ρ_p, which is given by $\rho_p = -\frac{1}{6\pi^2 m}\int_0^\infty dk k^4 \delta(E_\uparrow(k))$ [6]. For $\mu < 0$, the solution to $E_\uparrow(k) = 0$ exists only when $h > \sqrt{\mu^2+\Delta^2}$ and takes place at $k = k_2$ with $k_2^2/2m = \mu + \sqrt{h^2-\Delta^2}$. For $n_d \to 0^+$, h is just slightly larger than $\sqrt{\mu^2+\Delta^2}$ and k_2 is small. Hence $\rho_p \to 0^+$ and $\rho_s \approx n > 0$. A typical case is shown in Fig. 2 for $g = 0.2$. For $\mu > 0$, $E_\uparrow(k) = 0$ happens when $h > \Delta$ and takes place at two k values: $k = k_{1,2}$ with $k_{1,2}^2/2m = \mu \pm \sqrt{h^2-\Delta^2}$. For $n_d \to 0^+$, h is just slightly larger than Δ and the $E_\uparrow(k) = 0$ points occur near $\xi(k) \approx 0$; hence $\partial E_\uparrow(k)/\partial k \to 0$. Since k_1 and k_2 are finite, $\rho_p \to -\infty$ and $\rho_s < 0$. The dashed lines in Fig. 2 (with $g = 0.15$) give a typical behavior of ρ_s in this case. For smaller g, ρ_s is less than zero for all nonzero n_d, indicating that the superfluid is unstable in this regions.

In summary, we have investigated the stability of a fermion mixture with unequal concentrations under a Feshbach resonance. We found that both the superfluid density and the slope of $h(n_d)$ are positive only for sufficiently strong coupling. This suggests that a uniform superfluid is stable in this regime. For weak interactions the normal state is the only stable uniform state.

ACKNOWLEDGMENTS

This research was supported by the National Science Council of Taiwan under grant numbers NSC93-2112-M-194-002/93-2120-M-194-002 (CHP), NSC93-2112-M-194-019 (STW) and NSC93-2112-M-001-016 (SKY).

REFERENCES

1. C. A. Regal *et al.*, *Phys. Rev. Lett.* **92**, 040403 (2004); M. W. Zwierlein *et al.*, *ibid* **92**, 120403 (2004); J. Kinast *et al.*, *ibid* **92**, 150402 (2004); C. Chin *et al.*, *Science* **305**, 1128 (2004).
2. E. Tiesinga *et al.*, *Phys. Rev.* A**47**, 4114 (1993); S. Inouye *et al.*, *Nature***392** 151 (1998); Ph. Courteille *et al. Phys. Rev. Lett.* **81**, 69 (2004); J.L. Roberts *et al.*, *ibid* **81**, 5109 (1998).
3. P. Fulde and R. A. Ferrell, *Phys. Rev.* **135**, A550 (1964); A. I. Larkin and Yu. N. Ovchinnikov, *Zh. Éksp. Teor. Fiz.* **47**, 1136 (1964) [*Sov. Phys. JETP* **20**, 762 (1965)].
4. T. Mizushima *et al.*, *Phys. Rev. Lett.* **94**, 060404 (2005); C. Mora and R. Combescot, cond-mat/0412449.
5. C. A. R. Sá de Melo, M. Randeria, and Jan R. Engelbrecht, *Phys. Rev. Lett.* **71**, 3202 (1993).
6. See, e.g., S.-T. Wu and S.-K. Yip, *Phys. Rev.* A **67**, 053603 (2003).
7. P.F. Bedaque *et al.*, *Phys. Rev. Lett.* **91**, 247002 (2003).
8. C.-H. Pao *et al.*, cond-mat/0506437.

Universal Thermodynamics of a Strongly Interacting Fermi Gas

J. E. Thomas[1], J. Kinast, and A. Turlapov

Physics Department, Duke University, Durham, NC 27708-0305, USA

Abstract. We study the properties of an optically-trapped, Fermi gas of ^{6}Li atoms, near the center of a broad Feshbach resonance, where strong interactions are observed. Strongly interacting Fermi gases exhibit universal behavior, and are of interest as models of exotic systems ranging from high temperature superconductors to neutron stars and quark-gluon plasmas. We measure the frequency and damping rate of the radial breathing mode of the trapped gas and consider quantum viscosity as a damping mechanism. We also demonstrate that the virial theorem holds and measure the heat capacity. In the experiments, energy is precisely added to the gas and an empirical temperature is determined from the spatial profiles of the cloud. Transitions are observed in both the damping rate and the heat capacity, measured as functions of the empirical temperature. Recent theory, using a pseudogap formalism, enables the first temperature calibration, and shows that the observed transition temperatures are close to that predicted for the onset of superfluidity in a strongly attractive Fermi gas.

Keywords: Fermi gas, strongly interacting, optical trap
PACS: 32.80.Pj, 03.75.Ss

1. INTRODUCTION

Since the first observation of a degenerate, strongly interacting Fermi gas [1], the field of interacting Fermi gases has made spectacular progress. Strongly interacting Fermi gases are produced in an optical trap [2], by using a magnetic field to tune a mixture of spin-up and spin-down atoms to a Feshbach resonance, where the zero energy scattering length is large compared to the interparticle spacing [1–4]. Strongly interacting Fermi gases can exhibit universal behavior and scale invariance, where the only natural length scale is the interparticle spacing. Under these conditions, the ratio of the interaction energy to the local kinetic energy is a universal parameter, denoted β [1,5]. This parameter was originally explored theoretically in the context of nuclear matter [5,6], and has now been measured by several groups [1,7–10], and found to be in reasonable agreement with recent predictions [11,12]. In the close proximity of a broad Feshbach resonance, the local thermodynamic properties of the trapped gas are believed to obey the universal hypothesis [13], i.e., they are independent of the details of the interparticle interactions, and are functions only of the density and temperature.

In the vicinity of a Feshbach resonance, pairing of spin-up and spin-down atoms occurs in the so-called crossover regime, part way between a Bose-Einstein condensate (BEC) and a Fermi superfluid comprising Cooper pairs. By tuning below resonance, Bose-Einstein condensates (BECs) of molecular dimers have been produced from a two-component strongly interacting Fermi gas [14–17].

In contrast to studies of stable molecular BEC's, which are produced below resonance, the study and proof of superfluidity just above resonance, in the strong Cooper pairing or strongly attractive regime, has been less straightforward. Over the past two years, however, substantial evidence for superfluidity has been obtained in a variety of experiments.

Macroscopic measurements provide evidence for superfluidity in a strongly attractive Fermi gas, and provide important information on the equation of state of this universal quantum system. Evidence for superfluid hydrodynamics has been obtained in observations of anisotropic expansion after release of the cloud [1,17] and in studies of the temperature and magnetic field dependence of the frequency and damping of collective modes [18–21]. Measurements of the heat capacity [10] and collective mode damping [21] as a function of empirical temperature reveal transitions in behavior, close to the predicted superfluid transition temperature [10,12,22,23]. Recently, the observation of vortices [24] in a strongly attractive Fermi gas has provided what appears to be a definitive proof of superfluidity.

Microscopic studies of strongly attractive Fermi gases have concentrated on the detection and probing of fermionic atom pairs. Pairs were first observed by projection onto a molecular BEC [25,26]. The pair binding energy has been probed in measurements of the pairing gap, by radiofrequency spectroscopy [27] and by modulating the interaction strength [28]. In the region of a

[1] E-mail: jet@phy.duke.edu

CP850, *Low Temperature Physics: 24th International Conference on Low Temperature Physics;*
edited by Y. Takano, S. P. Hershfield, S. O. Hill, P. J. Hirschfeld, and A. M. Goldman

Feshbach resonance, the pair wavefunction can contain both a dominant triplet contribution in the open collision channel, and a much smaller singlet contribution from the closed molecular channel [29,30]. Recently, molecular spectroscopy in the singlet manifold has been used to probe the molecular amplitude of the fermionic atom pairs and the superfluid order parameter throughout the Feshbach resonance region [30].

2. EXPERIMENTAL SYSTEM

We prepare a highly degenerate, strongly interacting Fermi gas of ^{6}Li. This is accomplished using evaporation of an optically-trapped, 50-50 mixture of spin-up/down states at 840 G, just above the center of a broad Feshbach resonance [1,10,18,20,21]. To reduce the temperature, we do not employ a magnetic sweep from a BEC of molecules, in contrast to several other groups [9,17,19,24,26,27]. Instead, we evaporate directly in the strongly attractive, unitary regime: We simply exploit the large collision cross section and the rapid vanishing of the heat capacity with decreasing temperature, which is especially effective in the superfluid regime. These properties make the gas easier to cool.

In the forced evaporation, the depth of the CO_2 laser optical trap is reduced to 1/580 of its maximum value, and then recompressed to 4.6% of the maximum trap depth for most of the experiments. From the trap frequencies measured under these conditions and corrected for anharmonicity, we obtain: $\omega_\perp = \sqrt{\omega_x\omega_y} = 2\pi \times 1696(10)$ Hz, $\omega_x/\omega_y = 1.107(0.004)$, and $\lambda = \omega_z/\omega_\perp = 0.045$. Then, $\bar{\omega} = (\omega_x\omega_y\omega_z)^{1/3} = 2\pi \times 589(5)$ Hz is the mean oscillation frequency. For most of the data reported, the total number of atoms is $N = 2.0(0.2) \times 10^5$. The corresponding Fermi temperature at the trap center for a noninteracting gas is $T_F = (3N)^{1/3}\hbar\bar{\omega}/k_B \simeq 2.4\,\mu$K, small compared to the final trap depth of $U_0/k_B = 35\,\mu$K (at 4.6% of the maximum trap depth). The coupling parameter of the strongly interacting gas at $B = 840$ G is $k_F a \simeq -30.0$, where $\hbar k_F = \sqrt{2mk_BT_F}$ is the Fermi momentum, and $a = a(B)$ is the zero-energy scattering length estimated from the measurements of Bartenstein et al. [31]

3. UNIVERSAL THERMODYNAMICS

As noted above, at a Feshbach resonance, a strongly interacting gas obeys the universal hypothesis, where the interparticle spacing, and hence the density n, determines the natural length scale. The local thermodynamic properties are then functions only of the density and temperature, the same variables that describe a noninteracting Fermi gas. The universal hypothesis has directly measurable consequences, some of which we will describe briefly.

3.1. Spatial Distribution and Fermi Temperature

At zero temperature, the local energy per particle for a resonantly interacting gas is just $(3/5)(1+\beta)\,\varepsilon_F(n)$ where $\varepsilon_F(n) = \hbar^2(3\pi^2 n)^{2/3}/(2m)$ is the local Fermi energy of a noninteracting Fermi gas and β is a universal constant [1,5,7].

For such a zero temperature gas, the net effect of the interactions is then equivalent to changing the bare mass m to an effective mass [6,10], $m^* = m/(1+\beta)$. The equation of state then yields precisely a zero temperature Thomas-Fermi profile, $n_0(\mathbf{x})$, for which the Fermi radii are altered from those of a noninteracting gas by a factor of $(1+\beta)^{1/4}$ [7]. Hence, the spatial distribution of the trapped gas is determined quite generally for a very low temperature cloud. We find that the measured spatial profiles of the cloud assume nearly the shape of a zero-temperature Thomas-Fermi profile [1,7,32]. Measurements of the cloud radii can then be used to determine β [7,9,10].

The Fermi temperature for a harmonically trapped, noninteracting gas, is given by $T_F = (3N)^{1/3}\hbar\bar{\omega}/k_B$. Since the effective mass for the strongly interacting gas is given by $m^* = m/(1+\beta)$, the effective oscillation frequency is altered by a factor of $\sqrt{1+\beta}$, and the Fermi temperature for the strongly interacting gas is given by

$$T_F' = T_F\sqrt{1+\beta}. \qquad (1)$$

3.2. Universal Hydrodynamics

At zero temperature, the local pressure of the trapped gas differs from the Fermi pressure of a noninteracting gas by a factor of $1+\beta$ and the pressure then scales as $n^{5/3}$. In this case, upon release from a harmonic trap, the expansion dynamics are governed precisely by a scale transformation, where the density evolves according to $n(\mathbf{x},t) = n_0(\tilde{\mathbf{x}})/\Gamma$, where $\tilde{x} = x/b_x(t)$, $\Gamma = b_xb_yb_z$, and $b_i(t)$ is a hydrodynamic expansion factor [1,33]. The predicted hydrodynamic expansion for release from a cigar-shaped trap is highly anisotropic, and independent of β, i.e., the gas expands rapidly in the originally narrow direction while remaining nearly stationary in the long direction as observed in experiments [1].

The breathing mode frequencies take on universal values when the local pressure scales as $n^{5/3}$, and are independent of β. For a cylindrically-symmetric trap, the radial breathing mode frequency for a zero temperature gas is given by

$$\omega_{hydro} = \sqrt{\frac{10}{3}}\,\omega_\perp = 1.83\,\omega_\perp. \qquad (2)$$

For the conditions of our trap, which deviate slightly from cylindrical symmetry, exact diagonalization of the linearized hydrodynamic equations yields $\omega_{hydro} = 1.84\,\omega_\perp$. This result is in very good agreement with our measurements [18,20,21], as shown in § 6.

One can generalize these arguments to show that they hold at all temperatures where the gas expands hydrodynamically *under isentropic conditions* [34].

3.3. Virial Theorem

According to the universal hypothesis, the local pressure P must be a function of the local density and temperature [13]. In this case, a strongly interacting Fermi gas must obey the virial theorem for a noninteracting gas at all temperatures, as we now show. One can readily show by elementary thermodynamic arguments that if $P = P(n,T)$, then

$$P = \frac{2}{3}\varepsilon(\mathbf{x}), \qquad (3)$$

where $\varepsilon(\mathbf{x})$ is the local energy density, i.e., the sum of the local kinetic and interaction energies [13,34].

Balance of the pressure and trapping forces in a harmonic potential requires that $N\langle U\rangle = (3/2)\int d^3\mathbf{x}\,P(\mathbf{x})$, where $\langle U\rangle$ is the average potential energy per particle. Using $\int d^3\mathbf{x}\,\varepsilon(\mathbf{x}) = E - N\langle U\rangle$, one obtains [34]

$$N\langle U\rangle = \frac{E}{2}. \qquad (4)$$

This result is remarkable: Analogous to an ideal noninteracting gas, a trapped, strongly interacting, unitary gas, comprising condensed superfluid pairs, noncondensed pairs, and unpaired atoms, should obey the virial theorem. Since $\langle U\rangle \propto \langle x^2\rangle$, the mean square transverse radius $\langle x^2\rangle$ of the trapped cloud should scale linearly with the total energy, as verified in our experiments, see § 4.2.

4. TOOLS FOR THERMODYNAMIC MEASUREMENTS

Equilibrium thermodynamic properties of the trapped gas, as well as dynamical properties, can be measured as functions of the temperature or of the total energy. The temperature is changed by adding energy to the gas at fixed total atom number and fixed magnetic field, starting from the lowest temperature samples. In the following, we describe first a method for precisely adding a known energy to the gas. Then we describe a method for associating an empirical temperature with the spatial profile of the gas, and a temperature calibration method using theoretically predicted spatial profiles [35].

4.1. Precision Energy Input

Energy is added to the gas by abruptly releasing the cloud and then recapturing it after a short expansion time t_{heat}. During the expansion time, the total kinetic and interaction energy is conserved. When the trapping potential $U(\mathbf{x})$ is reinstated, the potential energy of the expanded gas is larger than that of the initially trapped gas, increasing the total energy. After waiting for the cloud to reach equilibrium, the sample is ready for subsequent measurements.

After recapture, the increase in the total energy, ΔE, is given by

$$\Delta E = \int d^3\mathbf{x}\,[n(\mathbf{x},t_{heat}) - n_0(\mathbf{x})]\,U(\mathbf{x}), \qquad (5)$$

where n_0 is the initial spatial distribution, and n is the spatial distribution after expansion during the time t_{heat}, as described in § 3.2.

For a harmonically trapped cloud which is initially at nearly zero temperature, the total energy is close to that of the ground state, which is $3/4$ of the Fermi energy per particle, i.e., $E_0 = (3/4)k_BT_F\sqrt{1+\beta}$. The energy after expansion and recapture is given by

$$E = E_0\left[\frac{2}{3} + \frac{b_x^2(t_{heat}) + b_y^2(t_{heat})}{6}\right]. \qquad (6)$$

Equation 6 has a simple physical interpretation. After release from a harmonic trap, and subsequent recapture after a time t_{heat}, the potential energy in each transverse direction is increased as the square of the expansion factors, b_x and b_y, where $b_z(t_{heat}) \simeq 1$, for the conditions of our experiments. The total potential energy is half of the total energy, since the unitary gas obeys the virial theorem for an ideal gas at all temperatures, as shown in § 4.2. Hence, the initial potential energy in each direction is $1/6$ of the total energy. Note that, by using Eq. 5, the corrections to the energy change arising from trap anharmonicity are readily determined [10].

4.2. Test of the Virial Theorem

To test the virial theorem prediction, the gas is evaporatively cooled to the lowest temperature and then the energy is increased as described above. For each value of t_{heat}, E is calculated according to Eq. 6. For each final energy, the gas is released and the transverse radius of the cloud is measured after a *fixed* expansion time of 1 ms. The observed linear scaling of $\langle x^2 \rangle$ with the calculated E, Fig. 1 verifies the virial theorem prediction. The linear scaling of $\langle x^2 \rangle$ with E also confirms that the expansion dynamics is closely hydrodynamic at all temperatures, since $\langle x^2 \rangle$ is measured after a fixed expansion time for all energies, which implies that the expansion factor must be nearly the same over the range of temperatures studied.

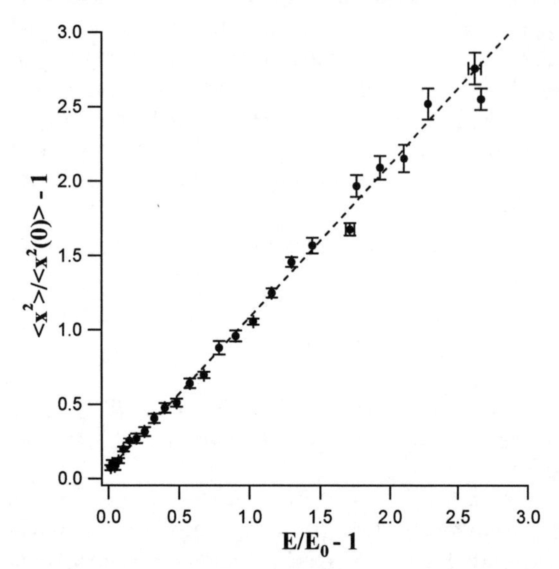

FIGURE 1. $\langle x^2 \rangle / \langle x^2(0) \rangle$ versus E/E_0 for a unitary gas of ^{6}Li, showing linear scaling, and verifying the virial theorem prediction. Here $\langle x^2 \rangle$ is the measured transverse mean square size. E is the total energy, calculated using Eq. 6. E_0 and $\langle x^2(0) \rangle$ denote the ground state values.

4.3. Empirical Temperature Measurement

Measurement of temperature in a noninteracting or weakly interacting Fermi gas is readily accomplished by fitting a Thomas-Fermi (T-F) distribution to the spatial profile of the cloud either in the trap, or after ballistic expansion, which alters the profile by a scale factor [1,33]. We normally integrate the measured column density of the expanded cloud over the axial dimension, and obtain the spatial profile in one transverse dimension, $n_{TF}(x; \sigma_x, T/T_F)$. The spatial profile is taken to be a function of two parameters, the Fermi radius σ_x, i.e., the cloud radius at zero temperature, and the reduced temperature T/T_F, i.e., the ratio of the Boltzmann temperature T to the trap Fermi temperature for a noninteracting gas, T_F.

One can consider $\sigma_x = \sqrt{2k_B T_F/(m\omega_x^2)}$ to set the length scale of the spatial profile and T/T_F as a shape parameter. At low T/T_F, the shape approaches a zero temperature T-F profile, $\propto (1 - x^2/\sigma_x^2)^{5/2}$, while at high T/T_F, the profile approaches a Maxwell-Boltzmann shape $\propto \exp[-m\omega_x^2 x^2/(2k_B T)] = \exp[-(x^2/\sigma_x^2)(T_F/T)]$.

In the latter case, only the product of T/T_F and σ_x^2 appears. Hence, for determination of the reduced temperature, it is convenient to determine the Fermi radius from the lowest temperature data, and then to hold this radius constant, i.e., to take $\sigma_x = c_x N^{1/6}$ in subsequent measurements at higher temperature, where c_x is held constant. In this way, the reduced temperature T/T_F is uniquely correlated with (and can be used to parametrize) the shape of the spatial profile.

For a unitary gas, the spatial profile is not precisely known, and there are no simple analytical formulae except at $T = 0$, where the equation of state assures that the shape of the cloud must take the zero temperature T-F form, with $\sigma_x \to \sigma_x'$, where $\sigma_x' = \sigma_x(1+\beta)^{1/4}$, as discussed above [7]. We obtain β by comparing the transverse radius of the trapped cloud for the interacting gas with that of the noninteracting gas [7]. For the noninteracting gas, we use either the calculated σ_x or the radius measured after ballistic expansion. For the interacting gas, we obtain σ_x' after hydrodynamic expansion for 1 ms. We find that $\beta = -0.49(0.04)$ (statistical error only). Similar results are obtained by measurements on the axial dimension of the trapped cloud without expansion [9] and by direct measurements of the interaction energy [8]. The discrepancy between the measurements may arise from the sensitivity of β to the precise location of the Feshbach resonance [9], which in ^{6}Li has been most recently measured by radiofrequency methods [31].

Although the spatial profile of a unitary gas is not precisely known, we observe experimentally that the binned, one-dimensional shape is closely approximated by a T-F profile for a noninteracting gas. Further, recent theoretical predictions of the spatial profile [35] show that the shape is nearly of the T-F form at all temperatures, as a consequence of the existence of preformed pairs. Hence, to provide a parametrization of the spatial profiles, we define an *empirical* reduced temperature $\tilde{T} \equiv (T/T_F)_{fit}$, and take the one dimensional spatial profile of the cloud to be of the form $n_{TF}(x; \sigma_x', \tilde{T})$.

In general, the empirical reduced temperature does not directly determine the reduced temperature T/T_F. However, at $T = 0$, the T-F shape is exact, so that $\tilde{T} = 0$ coincides with $T/T_F = 0$. Hence, the procedure for determining σ_x' from the data at very low temperature, where $\tilde{T} \simeq 0$, is consistent, i.e., we take $\sigma_x' = c_x' N^{1/6}$, where c_x' is a constant.

Further, at sufficiently high temperature, the

cloud profile must have a Maxwell-Boltzmann form $\propto \exp[-m\omega_x^2 x^2/(2k_B T)] = \exp[-(x^2/\sigma_x'^2)(1/\tilde{T})]$. This determines the natural reduced temperature scale, $\tilde{T}_{nat}$,

$$\tilde{T}_{nat} = \frac{T}{T_F\sqrt{1+\beta}}, \tag{7}$$

which follows from the interacting gas Fermi radius, $\sigma_x' = \sigma_x(1+\beta)^{1/4}$, and the definition of the noninteracting gas Fermi radius σ_x.

The empirical temperature scale is therefore exact at $T = 0$ and at high temperature, for a fixed interacting gas Fermi radius (which determines β). To calibrate $\tilde{T}$ more generally, we fit profiles of the form $n_{TF}(x;\sigma_x',\tilde{T})$ to the spatial profiles predicted as a function of T/T_F using a pseudogap formalism [10,35]. The value of σ_x' is determined from the lowest temperature theoretical profile, and $\tilde{T}$ is determined for all of the predicted profiles.

If the natural temperature were the correct scale at all T, then one would expect $T/T_F = \tilde{T}\sqrt{1+\beta} = 0.71\tilde{T}$. Remarkably, above the predicted superfluid transition temperature, where $T_c/T_F = 0.29$, i.e., for $\tilde{T} \geq 0.45$, the natural temperature scale is in close agreement with predictions [10], even though noncondensed pairs are believed to exist up to at least $T/T_F \simeq 0.6$, and are present in the predicted profiles. However, below the transition, for $0 < T \leq 0.29$, i.e., for $0 \leq \tilde{T} \leq 0.45$, we find that there is a systematic deviation: Here, $T/T_F = 0.54\tilde{T}^{2/3}$, and the natural temperature scale underestimates the reduced temperature [10]. This is reasonable, since the energy of the unitary gas, and hence the mean square cloud size, increases as a higher power of T/T_F than quadratic. The full empirical temperature calibration is shown in the inset in Fig. 3.

5. HEAT CAPACITY

The techniques of precision energy input and empirical temperature measurement provide a method for exploring the heat capacity [10,32] of a strongly interacting Fermi gas. In the experiments, the ^{6}Li gas is cooled to very low temperature, $\tilde{T} \simeq 0.04$, by forced evaporation at 840 G, just above the center of the Feshbach resonance, as described above. Then, the gas is heated by adding a known energy. Finally, the gas is released from the trap and allowed to expand for 1 ms. As observed above, the gas expands hydrodynamically by a scale factor, so that the shape of the expanded cloud closely approximates that of the trapped cloud, enabling a determination of $\tilde{T}$.

Figure 2 shows that the reduced energy of the gas, E/E_0, scales with empirical temperature $\tilde{T} = (T/T_F)_{fit}$ in much the same way as that of an ideal, noninteracting, Fermi gas. However, closer examination reveals that the

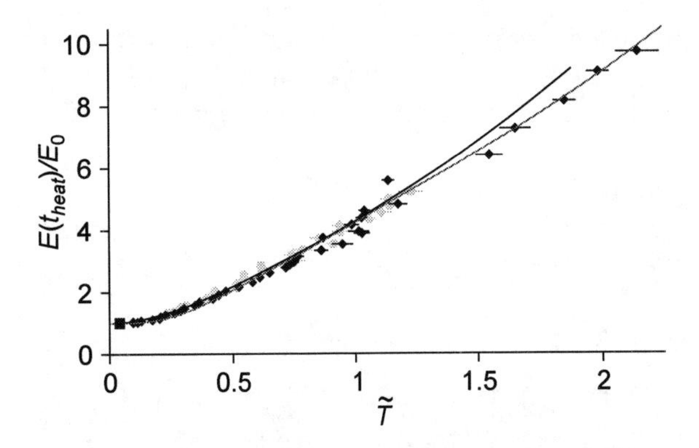

FIGURE 2. Total energy versus temperature. For each heating time t_{heat}, the temperature parameter $\tilde{T}$ is measured from the cloud profile, and the total energy $E(t_{heat})$ is calculated from Eq. 6 in units of the ground state energy E_0. Circles: noninteracting Fermi gas data; Diamonds: strongly interacting Fermi gas data. Upper solid curve: predicted energy versus reduced temperature for a noninteracting, trapped Fermi gas, $E_{ideal}(\tilde{T})/E_{ideal}(0)$; Lower solid curve: predicted energy versus $\tilde{T}$ for the unitary case. No temperature calibration is applied since $\tilde{T} \approx \tilde{T}_{nat}$ over the broad temperature range shown. Note that the lowest temperature point (solid square) is constrained to lie on the upper noninteracting gas curve.

low temperature data is better fit by a power law in $\tilde{T}$ than by the ideal gas scaling. The same data on a $log-log$ plot shows a transition in behavior [10,32].

By using the temperature calibration, and replotting the raw data as in Fig. 3, we find that the transition occurs at $T/T_F = 0.27$, in very good agreement with the prediction for the superfluid transition, $T_c/T_F = 0.29$ [10]. We also find that the behavior of the energy with temperature is in very good agreement with the predictions [10].

By fitting a power law in T/T_F to the data above and below the transition temperature, we obtain analytic approximations to the energy $E(T/T_F)$, from which the heat capacity is calculated using $C = (\partial E/\partial T)_{N,U}$, where the number N and trap depth U are constant in the experiments. For $T/T_F \leq 0.27$, we obtain $E/E_0 - 1 = 97.3\,(T/T_F)^{3.73}$, while for $T/T_F \geq 0.27$, $E/E_0 - 1 = 4.98\,(T/T_F)^{1.43}$. By differentiating the energy in each region with respect to T, we find that the heat capacity exhibits a jump at the transition temperature, comparable in size to that expected for a transition between a superfluid and a normal fluid [10].

The appearance of a transition in the behavior in the heat capacity, i.e., in the behavior of the energy versus temperature, is model-independent, as it appears in the empirical temperature data, $E(\tilde{T})/E_0$, without calibration [32]. However, the estimate of the transition temperature T_c/T_F and the magnitude of the jump in heat capacity are model-dependent, since the temperature estimates are obtained by calibration using the theoretical spatial profiles.

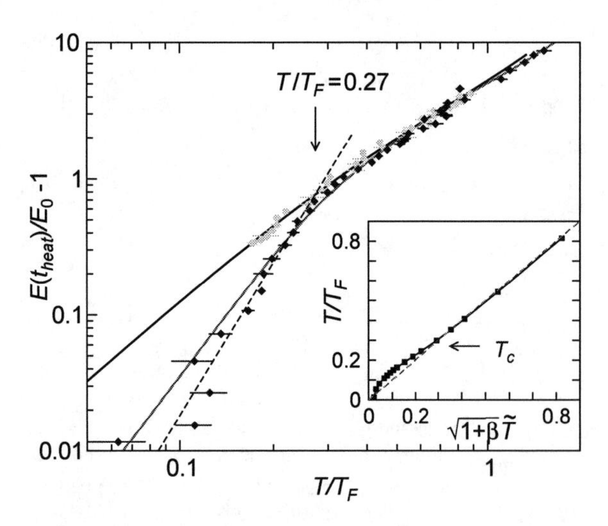

FIGURE 3. Energy input versus temperature from Fig. 2 after temperature calibration, on a $log - log$ scale. The strongly interacting Fermi gas shows a transition in behavior near $T/T_F = 0.27$. Circles: noninteracting Fermi gas data; Diamonds: strongly interacting Fermi gas data; Lower (upper) solid curves: prediction for a unitary (noninteracting), trapped Fermi gas, calculated at trap depth $U_0/E_F = 14.6$ as in the experiments; Dashed line: best fit power law $97.3\,(T/T_F)^{3.73}$ to the calibrated unitary data for $T/T_F \leq 0.27$. The inset shows the calibration curve, which has been applied to the unitary data (diamonds). The diagonal dashed line in the inset represents $T/T_F = \sqrt{1+\beta\tilde{T}}$. Here $E_0 \equiv E(T=0)$, and $E_F = k_B T_F$ is the noninteracting gas Fermi energy.

6. COLLECTIVE OSCILLATIONS

As the determination of the transition temperature in the crossover regime is of great interest, we look for corresponding transitions in the mechanical properties of the gas. In this section, we describe our comprehensive measurements of the temperature dependence of the frequency and damping of the radial breathing mode [21]. The temperature is increased by adding energy to the gas. Then, the empirical temperature is measured from the spatial profiles of the released cloud as described above.

The radial breathing mode is excited by releasing the gas from the trap for a short time and then recapturing the cloud. In contrast to the method used to add energy, the gas is not allowed to thermalize, and the expansion time, $25\,\mu$s, is so short that the energy increase is negligible. After recapture, the cloud is allowed to oscillate in the trap for a variable time, t_{hold} after which it is released and imaged as described above.

The width of the cloud, Fig. 4, oscillates at a frequency ω. The oscillation amplitude decays at a rate $1/\tau$. To determine ω and τ, we fit a damped sinusoid to the mean square width as a function of t_{hold}.

From the measured values of the trap oscillation frequencies, we predict the radial breathing frequency for a noninteracting gas $\omega_{nonint} = 2\,\omega_x = 2.10\,\omega_\perp$ and the hydrodynamic frequency for a strongly interacting (unitary) gas, $\omega_{hydro} = 1.84\,\omega_\perp$, as given in § 3.2.

Figure 5 shows the measured frequency ω in units of $\omega_\perp$, as a function of temperature. Remarkably, after correction for anharmonicity, the frequency is very close to the hydrodynamic value, and far from the collisionless (ballistic) value over the entire range of temperatures explored. This behavior suggests that the gas oscillates under conditions which are close to locally isentropic [34].

In contrast to the frequency, the damping rate, Fig. 6, shows a transition in behavior at $\tilde{T} \simeq 0.5$. For empirical temperatures in the range $0 \leq \tilde{T} \leq 0.5$, the data is well fit by a line (0.998 correlation coefficient), while above 0.5, the damping rate behaves quite differently, exhibiting non-monotonic behavior. The value of $\tilde{T} = 0.5$ lies just above the predicted superfluid transition temperature, where $\tilde{T} \simeq \tilde{T}_{nat}$ is a good approximation. Using Eq. 7, we find that $\tilde{T} = 0.5$ corresponds to $T/T_F = 0.35$. This is quite close to the value measured for the transition in the heat capacity, $T/T_F = 0.27$, and is consistent with recent predictions, $T_c/T_F = 0.29$ [22,32], $T_c/T_F = 0.31$ [36], and $T_c/T_F = 0.30$ [23].

The damping rate also appears to have a plateau and a further increase near $\tilde{T} = 1.0$, i.e., $T/T_F = 0.71$, close to the region where the pairing gap is comparable to the collective mode quantum, $\hbar\omega$. This behavior may arise from the breaking of weakly bound pairs in this temperature region.

6.1. Quantum Viscosity

In a unitary Fermi gas, there is a natural unit of viscosity, η which is determined by the interparticle spacing, L. Viscosity has dimensions of momentum/area. Hence, $\eta \simeq \hbar/L^3 \propto \hbar n$, where n is the local density. Since $\eta \propto \hbar$, we consider this scale as the natural unit of quantum viscosity. Following Gelman et al. [37], we take

$$\eta = \alpha\hbar n, \quad (8)$$

where α is a dimensionless constant.

It is instructive to determine α from the lowest damping rates observed in measurements of the breathing mode. For the axial mode measured by the Innsbruck group [19], the axial damping ratio is found to be very small, $1/(\omega_z\tau_z) = 1.5\times10^{-3}$. This corresponds to observed axial damping times of several seconds, since $\omega_z = 2\pi\times22.5$ Hz. For the radial breathing mode measured by the Duke group [20,21], the damping ratio is $1/(\omega_\perp\tau_\perp) = 1.3\times10^{-2}$, corresponding to damping times of up to seven milliseconds. Similar results are obtained by the Innsbruck group [19].

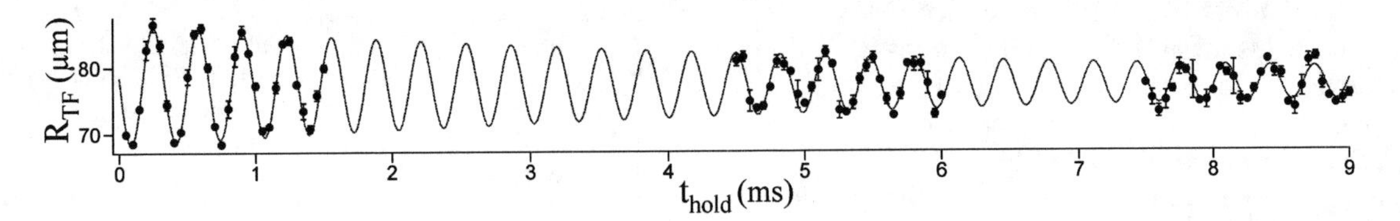

FIGURE 4. Radial breathing mode amplitude versus hold time.

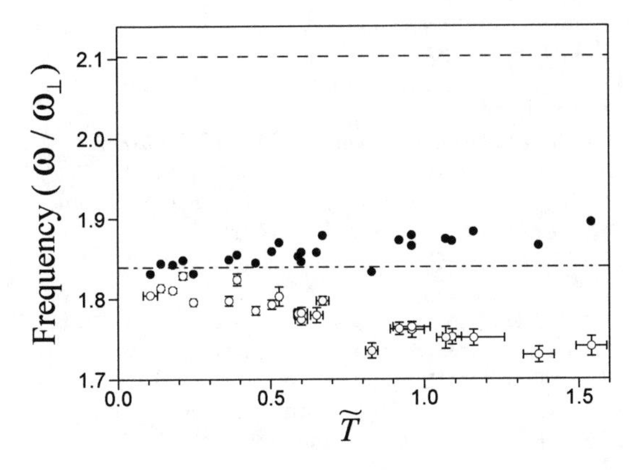

FIGURE 5. Frequency ω versus empirical reduced temperature $\tilde{T}$. Open circles–measured frequencies; Black dots–after correction for anharmonicity using a finite-temperature Thomas-Fermi profile. The dot-dashed line is the unitary hydrodynamic frequency $\omega_H = 1.84\,\omega_\perp$, for our trap parameters. The dashed line at the top of the scale is the frequency $2\,\omega_x = 2.10\,\omega_\perp$ observed for a noninteracting gas at the lowest temperatures.

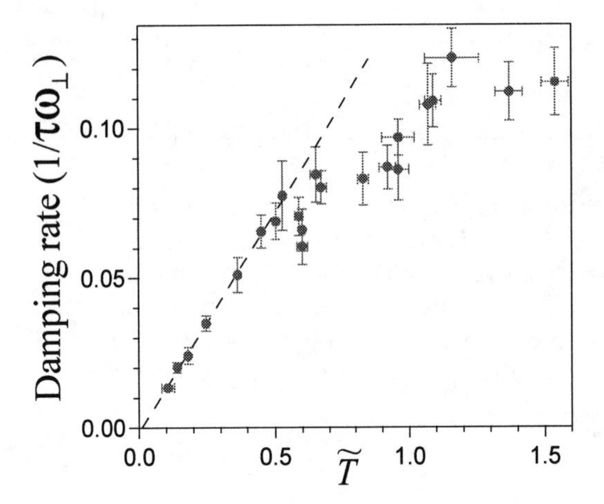

FIGURE 6. Temperature dependence of the damping rate for the radial breathing mode of a trapped ^{6}Li gas at 840 G, showing a transition in behavior. The dashed line shows fit to a line which extrapolates close to zero at zero temperature.

To determine α, we introduce a pressure term which arises from the shear viscosity [38], into the hydrodynamic equations for a compressible fluid. For low damping, the gas can be assumed to oscillate under nearly isentropic conditions at all temperatures [34]. In this case, the local stream velocity components are of the form $u_i = x_i \dot{b}_i / b_i$, where $b_i(t)$ is a scale factor [33,39]. The spatial derivatives of the stream velocity, which determine the shear pressure, are therefore spatially independent, and the gradient of the viscosity determines the spatial dependence of the shear pressure. The equations of motion for the b_i are readily solved and yield for the radial mode,

$$\frac{1}{\omega_\perp \tau_\perp} = \frac{4}{3} \frac{\alpha}{(3N\lambda)^{1/3}\sqrt{1+\beta}}, \tag{9}$$

and for the axial mode,

$$\frac{1}{\omega_z \tau_z} = \frac{16}{5} \frac{\alpha\lambda}{(3N\lambda)^{1/3}\sqrt{1+\beta}}. \tag{10}$$

Equation 10 predicts that the axial damping ratio is smaller than that of the radial mode by a factor of $\lambda \equiv \omega_z/\omega_\perp << 1$. Assuming $\beta \simeq -0.5$ [32], and using the measurements for the damping ratio of axial mode by the Innsbruck group, where $N = 4 \times 10^5$, and $\lambda = 0.03$, we obtain from Eq. 10, $\alpha = 0.4$. Using the parameters for our radial mode experiments, where $N = 2 \times 10^5$, we obtain $\alpha = 0.2$. From these results, we conclude that the measured damping ratios are comparable to those expected for the quantum viscosity scale.

Eqs. 9 and 10 also predict that the damping ratios should decrease as $N^{-1/3}$ with increasing total atom number N. However, we find experimentally that the data are nearly independent of N for fixed nonzero $\tilde{T}$: Decreasing the number of atoms N by a factor of 3 produces damping ratios which lie on the linear extrapolation for N atoms [21]. Hence, the observed damping ratios are consistent with the quantum viscosity scale, but viscosity may not be the cause of the damping.

7. CONCLUSIONS

We have studied a highly-degenerate, strongly interacting Fermi gas, which is prepared by direct evaporation at a Feshbach resonance in an optical trap. By precisely adding energy to the gas, we have verified that the virial theorem holds, despite strong interactions. Using an empirical temperature parameter linked to the cloud profiles, we observe transitions in both the heat capacity and in the damping rate of the radial breathing mode. The observation of these transitions is model independent, but the method used to calibrate the temperature scale is model dependent. Nevertheless, the two measured transition temperatures are consistent, and in good agreement with predictions of the superfluid transition temperature in the unitary regime.

ACKNOWLEDGMENTS

This research is supported by the Army Research Office and the National Science Foundation, the Physics for Exploration program of the National Aeronautics and Space Administration, and the Chemical Sciences, Geosciences and Biosciences Division of the Office of Basic Energy Sciences, Office of Science, U. S. Department of Energy.

REFERENCES

1. K. M. O'Hara, S. L. Hemmer, M. E. Gehm, S. R. Granade, and J. E. Thomas, *Science*, **298**, 2179 (2002).
2. J. E. Thomas, and M. E. Gehm, *Am. Scientist*, **92**, 238 (2004).
3. M. Houbiers, R. Ferwerda, H. T. C. Stoof, W. I. McAlexander, C. A. Sackett, and R. G. Hulet, *Phys. Rev. A*, **56**, 4864 (1997).
4. M. Houbiers, H. T. C. Stoof, W. I. McAlexander, and R. G. Hulet, *Phys. Rev. A*, **57**, R1497 (1998).
5. H. Heiselberg, *Phys. Rev. A*, **63**, 043606 (2001).
6. G. A. Baker, Jr., *Phys. Rev. C*, **60**, 054311 (1999).
7. M. E. Gehm, S. L. Hemmer, S. R. Granade, K. M. O'Hara, and J. E. Thomas, *Phys. Rev. A*, **68**, 011401(R) (2003).
8. T. Bourdel, J. Cubizolles, L. Khaykovich, K. M. F. Magãlhaes, S. Kokkelmans, G. V. Shlyapnikov, and C. Salomon, *Phys. Rev. Lett.*, **91**, 020402 (2003).
9. M. Bartenstein, A. Altmeyer, S. Riedl, S. Jochim, C. Chin, J. H. Denschlag, and R. Grimm, *Phys. Rev. Lett.*, **92**, 120401 (2004).
10. J. Kinast, A. Turlapov, J. E. Thomas, Q. Chen, J. Stajic, and K. Levin, *Science*, **307**, 1296 (2005), published online 27 January 2005 (10.1126/science.1109220).
11. J. Carlson, S.-Y. Chang, V. R. Pandharipande, and K. E. Schmidt, *Phys. Rev. Lett.*, **91**, 050401 (2003).
12. A. Perali, P. Pieri, and G. C. Strinati, *Phys. Rev. Lett.*, **93**, 100404 (2004).
13. T.-L. Ho, *Phys. Rev. Lett.*, **92**, 090402 (2004).
14. S. Jochim, M. Bartenstein, A. Altmeyer, G. Hendl, S. Riedl, C. Chin, J. H. Denschlag, and R. Grimm, *Science*, **302**, 2101 (2003).
15. M. Greiner, C. A. Regal, and D. S. Jin, *Nature*, **426**, 537 (2003).
16. M. W. Zwierlein, C. A. Stan, C. H. Schunck, S. M. F. Raupach, S. Gupta, Z. Hadzibabic, and W. Ketterle, *Phys. Rev. Lett.*, **91**, 250401 (2003).
17. T. Bourdel, L. Khaykovich, J. Cubizolles, J. Zhang, F. Chevy, M. Teichmann, L. Tarruell, S. Kokkelmans, and C. Salomon, *Phys. Rev. Lett.*, **93**, 050401 (2004).
18. J. Kinast, S. L. Hemmer, M. E. Gehm, A. Turlapov, and J. E. Thomas, *Phys. Rev. Lett.*, **92**, 150402 (2004).
19. M. Bartenstein, A. Altmeyer, S. Riedl, S. Jochim, C. Chin, J. H. Denschlag, and R. Grimm, *Phys. Rev. Lett.*, **92**, 203201 (2004).
20. J. Kinast, A. Turlapov, and J. E. Thomas, *Phys. Rev. A*, **70**, 051401(R) (2004).
21. J. Kinast, A. Turlapov, and J. E. Thomas, *Phys. Rev. Lett.*, **94**, 170404 (2005).
22. Q. Chen, J. Stajic, and K. Levin, *Phys. Rev. Lett.*, **95**, 260405 (2005).
23. P. Massignan, G. M. Bruun, and H. Smith, *Phys. Rev. A*, **71**, 033607 (2005).
24. M. Zwierlein, J. Abo-Shaeer, A. Schirotzek, C. Schunck, and W. Ketterle, *Nature*, **435**, 1047 (2005).
25. C. A. Regal, M. Greiner, and D. S. Jin, *Phys. Rev. Lett.*, **92**, 040403 (2004).
26. M. W. Zwierlein, C. A. Stan, C. H. Schunck, S. M. F. Raupach, A. J. Kerman, and W. Ketterle, *Phys. Rev. Lett.*, **92**, 120403 (2004).
27. C. Chin, M. Bartenstein, A. Altmeyer, S. Riedl, S. Jochim, J. H. Denschlag, and R. Grimm, *Science*, **305**, 1128 (2004).
28. M. Greiner, C. A. Regal, and D. S. Jin, *Phys. Rev. Lett.*, **94**, 070403 (2005).
29. G. M. Falco, and H. T. C. Stoof, *Phys. Rev. Lett.*, **92**, 130401 (2004).
30. G. B. Partridge, K. E. Strecker, R. I. Kamar, M. W. Jack, and R. G. Hulet, *Phys. Rev. Lett.*, **95**, 020404 (2005).
31. M. Bartenstein, A. Altmeyer, S. Riedl, R. Geursen, S. Jochim, C. Chin, J. H. Denschlag, R. Grimm, A. Simoni, E. Tiesinga, C. J. Williams, and P. S. Julienne, *Phys. Rev. Lett.*, **94**, 103201 (2005).
32. J. Kinast, A. Turlapov, and J. E. Thomas, Heat capacity of a strongly-interacting Fermi gas (2004), arXiv:cond-mat/0409283.
33. C. Menotti, P. Pedri, and S. Stringari, *Phys. Rev. Lett.*, **89**, 250402 (2002).
34. J. E. Thomas, A. Turlapov, and J. Kinast, *Phys. Rev. Lett.*, **95**, 120402 (2005).
35. J. Stajic, Q. Chen, and K. Levin, *Phys. Rev. Lett.*, **94**, 060401 (2005).
36. A. Perali, P. Pieri, L. Pisani, and G. C. Strinati, *Phys. Rev. Lett.*, **92**, 220404 (2004).
37. B. A. Gelman, E. V. Shuryak, and I. Zahed, Cold strongly coupled atoms make a near-perfect liquid (2005), arXiv:nucl-th/0410067.
38. L. D. Landau, and E. M. Lifshitz, *Fluid Mechanics*, Pergamon Press, New York, 1975.
39. D. Guéry-Odelin, F. Zambelli, J. Dalibard, and S. Stringari, *Phys. Rev. A*, **60**, 4851 (1999).

One-Loop Correction to the Mean Field Theory for the Neutral BCS Superconductors

Tomio Koyama

Institute for Materials Research, Tohoku University, Sendai 980-8577, Japan

Abstract. To investigate the strong coupling effect in neutral BCS superfluids we construct a theory beyond the Hartree –Fock-Bogoliubov approximation, in which the self-energy correction is incorporated into the single-particle fermionic excitations. We consider the self-energy correction arising from the interaction between the Goldstone boson and the fermions. The Ward-Takahashi relations resulting from the U(1) symmetry are systematically utilized to ensure the consistency between the single- and two-particle channels.

Keywords: degenerate Fermi gas, neutral BCS super-fluid, self-energy correction.
PACS: 03.75.S, 72.20.Fg, 03.75.Kk

INTRODUCTION

Recent observation of superfluidity in trapped fermion atomic gasses has explored a new challenging field of physics[1-3]. In these systems two neutral fermions of different hyperfine states form a BCS-like Cooper-pair or a molecule which can be condensed into the BEC state at low temperatures. The interaction between fermions in this system is tunable and is strong near the Feshbach resonance. Then, it has a meaning to construct a theory beyond the Hartree-Fock-Bogoliubov (HFB) approximation in the BCS superfluids. In this paper we investigate the self-energy correction to the HFB approximation.

FORMULATION

In this paper we restrict ourselves to the uniform single-component superfluid system at $T=0$ described by the conventional BCS Hamiltonian. In the following we use the Nambu's notation for the field operators. The single-particle Green function, which is a 2 by 2 matrix, is defined as

$$S(x-y) = -i < T\Psi(x)\Psi^{+}(y) > , \qquad (1)$$

where $\Psi(x)$ is the Nambu doublet field. The Fourier transform of the Green function is generally expressed in the superfluid state as

$$S^{-1}(p_0,\bar{p}) = p_0 - \varepsilon(\bar{p})\tau_3 - UM\tau_1 - \Sigma(p), \qquad (2)$$

where τ_1 and τ_3 are the Pauri matrices, $\varepsilon(\bar{p})$ is the free fermion energy measured from the chemical potential, U is the bare coupling constant, M is the superconducting 'polarization' defined as

$$M = < \psi_{\uparrow}(x)\psi_{\downarrow}(x) >, \qquad (3)$$

and $\Sigma(p)$ is the self-energy function. In eq.(2) the order parameter is assumed real. In neutral superfluid fermionic systems the Goldstone mode appears in the region below twice the energy gap and brings about a pole singularity in the two-particle Green function,

$$D(x-y) = i < T\tau_2(x)\tau_2(y) >, \qquad (4)$$

where

$$\tau_2(x) = \Psi^{+}(x)\tau_2\Psi(x). \qquad (5)$$

At low temperatures the self-energy correction in the low-energy region is expected to arise from the interaction between the Goldstone mode and the fermions, since the energy gap in the fermionic excitations is large and the Goldstone mode is well defined as the low energy excitation. From the detailed study of the Dyson equation we can show that the self-energy function is expressed in the one-loop order as

CP850, *Low Temperature Physics: 24th International Conference on Low Temperature Physics;*
edited by Y. Takano, S. P. Hershfield, S. O. Hill, P. J. Hirschfeld, and A. M. Goldman

$$\Sigma(p) = -\frac{ig_r^{\ 2}}{2}\int\frac{d^4q}{(2\pi)^4}\tau_2 S(p+q)\tau_2 D(q), \quad (6)$$

where we introduce the renormalized coupling constant g_r, which is defined later. To evaluate eq.(6) we need the two-particle Green function. To obtain $D(q)$ one has to introduce the vertex function. We use the 3-point vertex function Γ defined graphically in Fig.1. From the study of the Ward-Takahashi relation resulting from the U(1) symmetry we can prove the following rigorous relation for the vertex function,

$$S^{-1}(p)\tau_3 - \tau_3 S^{-1}(p) = 2iM\Gamma(p;p;0). \quad (7)$$

This relation can be utilized to ensure the gapless nature of the Goldstone mode in the calculation of $D(q)$, that is, we construct the approximate equation for $D(q)$ that satisfies eq.(7). The details of the calculation for $D(q)$ will be published elsewhere. In solving the above equations it is convenient to introduce the renormalized quantities. We choose a point on the Fermi surface of non-interacting Fermi gas as the renormalization point. i.e., $p_F = (0, \vec{p}_F)$. We define the renormalized coupling constant as

$$\Gamma(p_F; p_F; 0) = g_r \tau_2, \quad (8)$$

and the renormalized gap function and the renormalized chemical potential are defined as

$$\Delta_r = UM + \frac{1}{2}tr[\tau_1 \Sigma(p_F)], \quad (9)$$

$$\mu_r = \mu - \frac{1}{2}tr[\tau_3 \Sigma(p_F)], \quad (10)$$

In this formulation we can also prove that eq.(9) is rewritten as

$$\Delta_r = g_r M. \quad (11)$$

Let us present the numerical results briefly. In this paper we compare the results with those in the HFB approximation. The numerical procedure is as follows. For given values of μ_r, g_r and Δ_r we calculate the bare coupling constant U .Then, using the obtained bare coupling constant, we calculate the gap in the MFB approximation, Δ_{MF}. In Fig.2 we present two cases as a function of the renormalized coupling constant.

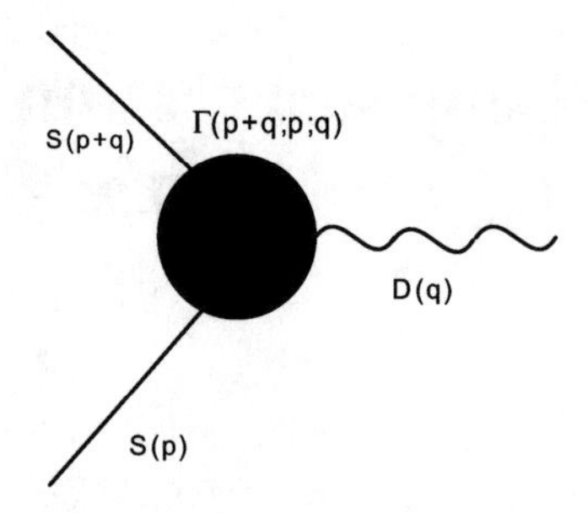

FIGURE 1. Definition of the 3-point vertex function. This vertex function describes the correction due to the interaction between the Goldstone mode and the fermions.

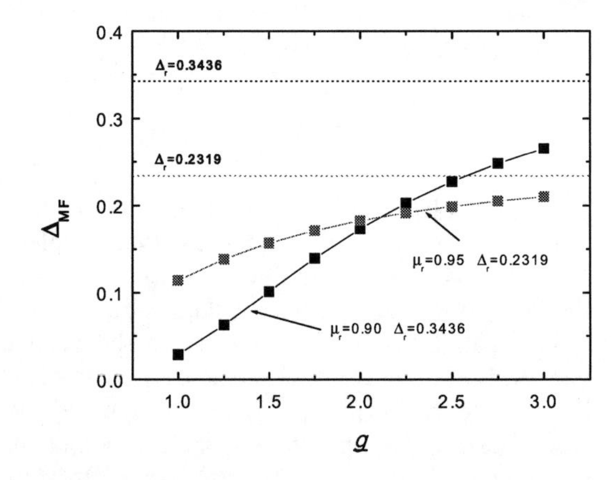

FIGURE 2. Comparison with the mean field results. The chemical potential and the gap are normalized by the chemical potential of the non-interacting system.

As seen in this figure, one understands that the self-energy correction due to the interaction between the Goldstone boson and the fermions enhances the gap function. However, as the renormalized coupling constant increases, the mean-field gap function approaches the renormalized one. This result indicates that in the strong coupling limit the mean-field approximation is justified. The details of the numerical results and discussions will be given in a separate paper.

REFERENCES

1. C. A.. Regal et al, *Phys. Rev. Lett.* **92**, 040403 (2004).
2. N. W. Zwierlein, et al, *Phys. Rev. Lett.* **91**, 250401 (2003).
3. S. Jochim, et al, *Science* **302**, 2101-2103 (2003).

Fulde-Ferrell-Larkin-Ovchinnikov States in Fermi Gases With Unequal Populations

T. Mizushima, K. Machida, and M. Ichioka

Department of Physics, Okayama University, Okayama 700-8530, Japan

Abstract. We theoretically investigate the microscopic structure of neutral Fermionic systems with unequal atomic populations. By using fully microscopic Bogoliubov-de Gennes framework, we study on the vortex structure in a single-vortex state with and without the Fulde-Ferrell-Larkin-Ovchinnikov (FFLO) modulation of the pairing field. It is found that the novel local density of states structure appears at intersection point of the FFLO nodal plane and the vortex line, which reflects the topological structure of the pairing field.

Keywords: Superfluidity, Quantized vortex, Quantum gas
PACS: 03.75.Hh, 03.75.Ss, 74.20Fg

INTRODUCTION

The possibility of the spatially modulated superfluid state, i.e., the Fulde-Ferrell-Larkin-Ovchinnikov (FFLO) state has been discussed for a long time in the various physics field, ranging from the condensed matter physics to the high energy community [1]. The FFLO state is favored in the system with mismatched Fermi surfaces of two species where the pairing has the intrinsic center-of-mass momentum. The recent achievement of Fermionic condensation in atomic gases [2] provides an opportunity to realize the FFLO state, because the system exhibits high flexibility of key parameters, such as the coupling constant and atomic population differences.

Here, we discussed the local density of states (LDOS) in a single-vortex state with and without the FFLO modulated pairing field from the fully microscopic point of view, by using the Bogoliubov-de Gennes (BdG) equation for the quasiparticle (QP) wave functions $u_{\mathbf{q}}(\mathbf{r})$ and $v_{\mathbf{q}}(\mathbf{r})$ [3]:

$$\begin{bmatrix} \frac{\nabla^2}{2}-\mu_\uparrow & \Delta(\mathbf{r}) \\ \Delta^*(\mathbf{r}) & -\frac{\nabla^2}{2}+\mu_\downarrow \end{bmatrix} \begin{bmatrix} u_{\mathbf{q}}(\mathbf{r}) \\ v_{\mathbf{q}}(\mathbf{r}) \end{bmatrix} = \varepsilon_{\mathbf{q}} \begin{bmatrix} u_{\mathbf{q}}(\mathbf{r}) \\ v_{\mathbf{q}}(\mathbf{r}) \end{bmatrix}. \quad (1)$$

The label $\mathbf{q}$ is the quantum number and the self-consistent equation is $\Delta(\mathbf{r}) = g\sum u_{\mathbf{q}}(\mathbf{r})v^*_{\mathbf{q}}(\mathbf{r})$ with the coupling constant g (< 0). To prepare the unequal spin population for up- and down-spins, the chemical potential is shifted as $\mu_{\uparrow,\downarrow} = \mu \pm \delta\mu$.

π-phase shift

Let us consider the microscopic structure in the rectangular pairing potential $\Delta(\mathbf{r}) = \Delta_0 \tanh(z/\xi)$. For simplicity, we remain only the z-axis along the modulation, and the other dimensions are neglected as redundant ones. Then, the one-dimensional BdG equation can be written in terms of $f_\pm = u \pm iv$ as $[v_z^2\nabla_z^2 + \varepsilon^2 - \Delta^2(z) \pm v_z\nabla_z\Delta(z)]f_\pm(z) = 0$ [4], where v_z the Fermi velocity along the z-axis. By using the hypergeometric functions, it is found that there always exists the bound state at the node of the pairing field, which has the zero energy, $\varepsilon = 0$. This results mean the appearance of the bound state, or soliton, due to the π-phase shifted pairing potential.

NUMERICAL RESULTS

Hereafter, we present the results of full numerical calculations of Eq. (1) in axially symmetric system: We write the eigenfunctions as $u_{\mathbf{q}}(\mathbf{r}) = u_{\mathbf{q}}(r,z)\exp[i(q_\theta - \frac{1}{2})\theta]$ and $v_{\mathbf{q}}(\mathbf{r}) = v_{\mathbf{q}}(r,z)\exp[i(q_\theta + \frac{1}{2})\theta]$ with $\Delta(\mathbf{r}) = \Delta(r,z)e^{-i\theta}$ for a single vortex where q_θ is an angular momentum of the eigenstates. We also use the same parameters with the previous work [5]. In addition, we consider the system with unequal populations induced by $\delta\mu$ which is crucial to stabilize the FFLO modulation.

Vortex structure

First, we consider the superfluid state having a single-vortex line along the z-axis with no FFLO modulation, where the order parameter has the continuous phase from 0 to 2π around the axis. Thus, the quasiparticle passing through the vortex line, schematically shown in Fig. 1 as the pass (i), experiences the π-phase shift of the pairing field which allows to form the bound state near the Fermi energy at the singularity line [6]. In Fig. 2, the upper panel shows the LDOS structure in a single-vortex state

CP850, *Low Temperature Physics: 24th International Conference on Low Temperature Physics;*
edited by Y. Takano, S. P. Hershfield, S. O. Hill, P. J. Hirschfeld, and A. M. Goldman

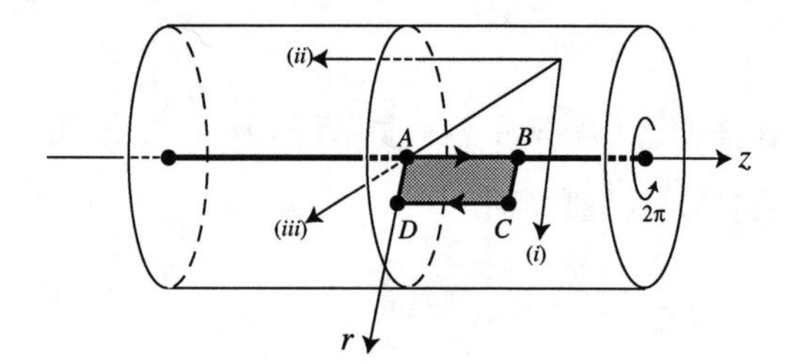

FIGURE 1. Schematic representation of the present system: $A \to \cdots \to D$ corresponds to the paths shown in Fig. 2.

with the finite population difference. Along the vortex line, corresponding to the pass $A \to B$, the state strongly bounded in the vortex core appears near the Fermi level. Away from the core site B toward the bulk area C, the bound state disappears and the bulk energy gap $\pm\Delta_0$ is recovered.

The presence of the finite population difference causes the shift of the Fermi level of each spin component, $\pm\delta\mu$. In this situation, while the core-bound state for QP's with the up-spin is formed below the Fermi energy, the bound state for down-spins is located outside it, i.e., the empty state. Then excess atoms generated by $\delta\mu$ can be one-dimensionally accumulated in the vortex core, which can be clearly observed via the Stern-Gerlach experiment.

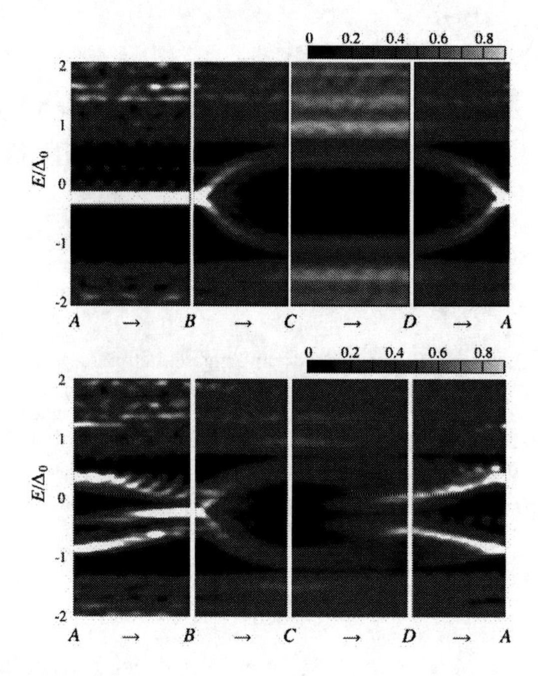

FIGURE 2. Local density of states of QP's with the up-spin in a single-vortex state without FFLO (upper panel) and with FFLO (lower panel) at $\delta\mu/\Delta_0 = 0.28$. In the energy scale, $E = 0$ corresponds to the Fermi energy.

Vortex structure in the FFLO state

The lower panel in Fig. 2 shows the LDOS in a single-vortex state with the FFLO modulation, where the FFLO nodal plane is located at $z = 0$. The novel topological structure of the pairing field causes the drastic change of the LDOS, compared to the conventional vortex state: In addition to the π-phase shift along the pass crossing the vortex, the QP's passing through the FFLO nodal plane also experience the π-phase shift. Then, the excess atoms are two-dimensionally localized at the FFLO nodal plane in addition to the vortex line. However, since the π shift due to the FFLO nodal plane is canceled by the phase shift from a vortex line at the intersection A, the QP's along the pass (iii) in Fig. 1 experience no phase shift.

This is also reflected in the splitting of bound states at the intersection point A in Fig. 2: Away from the intersection A toward the bulk area D, the splitting tends to merge. Similarly, along the pass $A \to B$, the single core-bound state is recover at the vortex core far away from the intersection. In the splitting bound states, the lower branch is located below the Fermi energy μ, while the another has the higher energy than μ. The same structure is also seen in the LDOS of down-spins. This results in the disappearance of the excess atoms at the intersection of the vortex line and the FFLO nodal plane.

Summary

Here, we have presented the novel structure of the local density of states in a single-vortex state in the presence of the FFLO modulation, which can be understandable from the simple topological point of view. The physical origin of the splitting of the core-bound state can be interpreted as the Doppler shift due to the non-vanishing center-of-mass momentum (see for the details, Ref. [5]).

REFERENCES

1. R. Casalbuoni and G. Nardulli, *Rev. Mod. Phys.* **76**, 263 (2004).
2. C.A. Regal, M. Greiner, and D.S. Jin, *Phys. Rev. Lett.* **92**, 040403 (2004).
3. T. Mizushima, K. Machida, and M. Ichioka, *Phys. Rev. Lett.* **94**, 060404 (2005).
4. K. Machida and H. Nakanishi, *Phys. Rev. B* **30**, 122 (1984).
5. T. Mizushima, K. Machida, and M. Ichioka, *Phys. Rev. Lett.* **95**, 117003 (2005).
6. N. Hayashi, T. Isoshima, M. Ichioka, and K. Machida, *Phys. Rev. Lett.* **80**, 2921 (1998).

Experimental Observation of Atomic Hydrogen Stabilized in Thin Films of Molecular H_2 at Temperatures ≤100 mK

S. Vasilyev*, J. Järvinen*, V. V. Khmelenko¶, and D. M. Lee¶

*Department of Physics, University of Turku, 20014 Turku, Finland
¶LASSP, Cornell University, Ithaca, NY 14853-2501, USA

Abstract. We report on experimental observation of hydrogen atoms embedded in 50 nm thick films of solid H_2. A 50 ppm mixture of H in H_2 is deposited on surfaces of the sample cell as a result of recombination of a polarized atomic hydrogen gas H↓. We did not see any noticeable recombination loss of the H atoms in the H_2 film during 4 days of observation at temperatures ≤100 mK.

Keywords: Atomic hydrogen, quantum diffusion.
PACS: 67.80.Mg, 67.65.+z

The supersolid behavior of ^{4}He recently discovered by Kim and Chan [1] has stimulated interest in studying the quantum behavior of other solids. Atomic hydrogen isolated in a solid H_2 matrix (H_M) is a system where quantum diffusion and quantum chemical reactions of atoms have been intensively studied experimentally and theoretically [2-5]. Typical methods to create the H_M include injection of the H and H_2 gas mixture into a cold finger [3], flash condensing it into superfluid ^{4}He [4], and irradiation of solid H_2 by γ-rays [5]. Maximum H concentrations of the order of 100 ppm can be reached with the first two methods. These studies were conducted in the temperature range 1.3–6 K. To our knowledge, there were no experimental attempts to cool the H_M to temperatures substantially below 1 K, where quantum effects will be more pronounced. In the present work we report on a novel preparation method of high density H_M samples discovered in our experiments with atomic hydrogen gas, which is suitable for cooling the H_M into the submillikelvin region by conventional dilution refrigerator and/or nuclear demagnetization techniques.

Solid layers of molecular H_2 are deposited on surfaces of the sample cell (Fig. 1*a*) designed for the experiments on the thermal compression of 2D atomic hydrogen gas (H↓) [6]. The gas of H↓ is generated in the cryogenic rf dissociator and accumulated in the sample cell located in the ≈5 T magnetic field at temperatures ≤300 mK with the walls covered by a superfluid helium film. High field (130 GHz) electron spin resonance (ESR) was used to monitor the H atom populations [6].

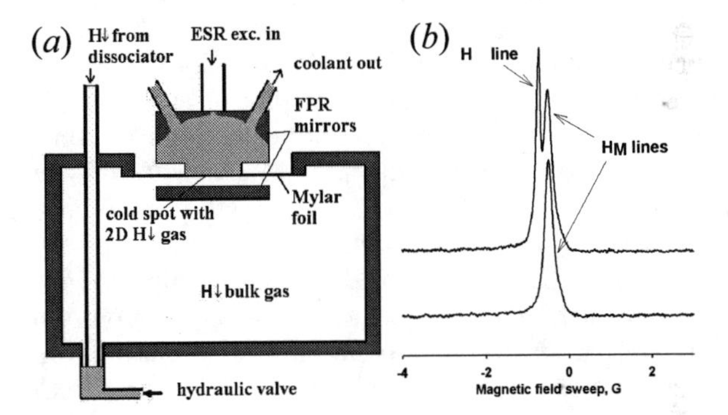

FIGURE 1. Schematic drawing of the sample cell with Fabry-Perot resonator (FPR) (*a*). ESR ***a→d*** transition line of the H atoms in the cell (*b*): H↓ and H_M samples (upper trace); H_M sample only, after the H↓ gas has been destroyed (lower trace).

The layers of H_2, grown beneath the helium film, increase the stability of the H↓ sample by reducing the relaxation rate from the doubly polarized hyperfine state ***b*** to the more reactive state ***a***. The *o*-H_2 molecules, resulting from the ***a***+***b***→H_2 recombination, appear in the bulk of the sample cell in highly excited rotational-vibrational states, and slowly relax down to the ground state, finally sticking to the walls. The rate of such deposition is very slow ~ 1–2 monolayers of H_2/hour, limited by the efficiency of the dissociator. We found (Fig. 1*b*) that hydrogen atoms are captured inside the H_2 layers. The atoms acquire a high enough kinetic energy to penetrate through the helium film, following inelastic collisions with highly excited molecules in the gas phase. We verified that the atoms do not diffuse into the sample cell from the dissociator along the H_2 film covering the filling line. Closing the

CP850, *Low Temperature Physics: 24th International Conference on Low Temperature Physics;* edited by Y. Takano, S. P. Hershfield, S. O. Hill, P. J. Hirschfeld, and A. M. Goldman

hydraulic valve at the cell bottom stopped the growth of the atomic ESR signals. The capture rate of atoms into the film increased with the bulk density of H↓ gas. To get a maximum yield of H_M atoms we kept the cell at a temperature of 300 mK, optimized for highest bulk density of H↓ ($\sim 10^{15}$ cm^{-3}). After one week of continuously filling the sample cell with H↓ gas we created ≈150 monolayers of H_2 film with ≈50 ppm relative concentration of atoms as detected by ESR. The thickness of the H_2 layer was measured by the quartz microbalance with an accuracy of 5 monolayers and agreed well with the estimate based on the known flux of H↓ from the dissociator. Both ESR lines of the H_M appeared to be shifted from that of the free atoms of the H↓ gas by ≈0.2 G towards the center of the spectrum. This can be explained by a decrease of the hyperfine constant due to the influence of the crystalline field. The H_M lines were well fitted by a Lorentzian lineshape, with the width gradually decreasing from 0.6 G to 0.2 G during three weeks of the measurement run. This effect can be explained by the natural ortho-para conversion process. Taking the known dependence of the H_M linewidth on the ortho-para content [5], we conclude that at the end of the measurement run we had very pure (≥99.9%) p-H_2 film. Little is known about the structure of the H_2 films obtained by the method described. We believe that the film is homogeneous, and has a smooth surface on a μm scale. In our studies of the H↓ gas adsorbed on the surface of the helium film above the H_2 films [6] we observe narrow (≤0.1 G) ESR lines, shifted from the bulk H↓ resonance due to the internal dipolar field in the 2D system (~4 G). The shift is anisotropic with respect to surface orientation in the main polarizing field, and a surface roughness (on a scale larger than helium film capillary radius) would lead to a 2D H↓ line broadening, not observed in the experiments [6].

We did not see any noticeable (≤10%) recombination loss of the atoms in the film during 4 days of observation at temperatures ≤100 mK. A slow decay of the sample was detected at higher temperatures, 300 and 480 mK (Fig. 2). Linearity of the data plots confirm that the decay process is of the second order, with the recombination rate constant $k_{rec} \approx 2 \cdot 10^{-25}$ cm^3/s, estimated at 480 mK. This is much smaller than the results of ref. [5] $k_{rec} \approx 10^{-22}$ cm^3/s (T=4.2 K, 99.9% p-H_2) and ref. [7] $k_{rec} \geq 2 \cdot 10^{-24}$ cm^3/s (T=1.3 K, *n*-H_2). To verify the influence of the main polarizing field on the recombination rate at T≤100 mK we decreased the field to zero for ~10 hours and then turned it back to 5 T. No decrease of H_M signal was detected, in agreement with the results of ref. [7]. Enhanced stability of the H_M samples at temperatures below 0.5 K can be explained by the energy level mismatch of neighboring atoms, which prevents them from approaching too closely and recombining [8].

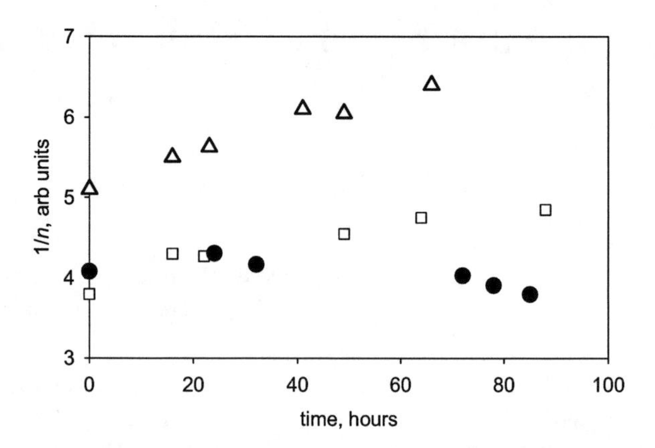

FIGURE 2. Decays of the H_M samples at temperatures of ≤100 mK (•), 300 mK (□) and 480 mK (Δ). Inverse concentration of the atoms is plotted as a function of time.

The rate of H_2 film growth can be increased by decreasing the sample cell area (~100 cm^2 at present). This will lead to a larger bulk density of H↓ and enhance the probability of capturing the H↓ atoms into the solid film and hence increase density of the H_M. The high stability of H in a molecular H_2 matrix at ultralow temperatures opens up the possibility of studying quantum diffusion as well as to search for possible collective quantum behavior of the embedded hydrogen atoms

We thank S. Jaakkola and K.-A. Suominen for illuminating discussions, the Finnish Academy of Sciences (proj. # 206109) and NSF (grant DMR-0504683) for financial support.

REFERENCES

1. E. Kim and M. H. W. Chan, *Science*, **305**, 1941 (2004).
2. Y. Kagan and A. Leggett, *Quantum Tunneling in Condensed Media*, Amsterdam: North-Holland, 1992, and references therein.
3. A. S. Iskovskih *et al.*, *Sov. Phys. JETP* **64**, 1085 (1986).
4. S. I. Kiselev, V. V. Khmelenko, and D. M. Lee, *Phys. Rev. Lett* **89**, 175301 (2002).
5. T. Kumada *et al*, *J. Chem. Phys.* **116**, 1109 (2002).
6. S. Vasilyev *et al.*, *Phys. Rev.* A **69**, 023610 (2004).
7. A. V. Ivliev *et al.*, *Sov. Phys. JETP Lett* **62**, 1268 (1985).
8. Yu. Kagan and L. A. Maksimov., *Sov. Phys. JETP* **52**, 688 (1981).

Electron Attachment Reaction Rates in 2D Atomic Hydrogen-Electron Mixed System on Liquid Helium Surface

Toshikazu Arai*, Tomoyuki Mitsui† and Hideki Yayama†

*LTM Center, Kyoto University, Kitashirakawa-Oiwakecho, Kyoto 606-8502, Japan
†Department of Physics, Kyushu University, 4-2-1 Ropponmatsu, Fukuoka 810-8560, Japan

Abstract. We have measured the temperature dependence of the electron attachment reaction rate of atomic hydrogen (H) on a liquid ^{4}He surface in applied magnetic fields of 0–5 T at 0.2–0.6 K. The measured surface state electron (SSE) losses are faster at lower temperatures for a given magnetic field. This behavior can be qualitatively understood, since the surface coverage of adsorbed H is large at low temperature and the collisions between H and SSE are frequent. However, the reaction is faster than expected based on the collision frequency argument. The measured reaction rate coefficient K_e is strongly temperature dependent. We observe that, as the temperature is lowered, K_e increases by several orders of magnitude. This indicates that some additional effect enhances electron attachment at low temperature. We discuss a possible reaction mechanism between H and SSE.

Keywords: atomic hydrogen, surface state electrons, electron attachment
PACS: 67.65.+z, 68

When surface state electrons (SSE) on liquid helium are exposed to a gas of hydrogen atoms (H), electron attachment (EA), H + H + e^- $\rightarrow$ H^- + H, occurs. This induces a loss of SSE [1]. In our previous work [2, 3], we measured the magnetic field dependence of the reaction rate, which provided a knowledge of spin-part contribution to the reaction. We found that the EA rate coefficient K_e is inversely proportional to square of applied magnetic field B. This result agreed with our model that a spin 'down' surface electron attaches to a so-called $|a\rangle$ state H atom which contains small admixture of electron spin 'up' part since H^- is electron spin singlet.

In this work, we measure the temperature dependence of EA reaction rates in applied magnetic fields. From this, we can elucidate the orbital contribution to the reaction mechanism.

We use a sample cell identical to the one described in Ref. [3]. A vibrating capacitor electrometer [4] is built in the sample cell to measure the SSE density. A magnetic field is applied perpendicular to the surface. We charge the surface with SSE and fill the sample cell with H at a constant flux ϕ until the SSE vanish. ϕ is turned on at $t=0$. Figure 1 shows SSE density reduction due to the EA reaction, measured at $B=3$ T. The reaction rate is larger at lower temperatures. This temperature dependence can be naively understood, because the surface coverage of H is higher at lower temperatures and the collision rate of e^- with H increases.

We obtained the reaction rate coefficient K_e by fitting the data to the solution of the rate equation for SSE,

$$\frac{dN_e}{dt} = -K_e N_e (\alpha N_H)^2, \qquad (1)$$

where N_e and N_H are the numbers of SSE and H, respectively. α is σ_H/N_H, where σ_H is the density of adsorbed H. H atoms are in equilibrium between the gas and adsorbed. The H density in the gas (n_H) and σ_H phases are related by the adsorption isotherm $\sigma_H = n_H \lambda_{th} \exp(E_a/k_B T)$, where $E_a/k_B = 1.0$ K is the surface binding energy of H on liquid ^{4}He, $\lambda_{th} = (2\pi\hbar^2/m_H k_B T)^{1/2}$ is the thermal de Broglie wavelength, and m_H is the atomic mass of hydrogen.

The rate equation for H is

$$\frac{dN_H}{dt} = \phi - K_e N_e (\alpha N_H)^2 - K_s^{\text{eff}} (\alpha N_H)^2. \qquad (2)$$

The effective surface recombination rate coefficient [5] K_s^{eff} for two-body recombination is written as $K_s^{\text{eff}} = AK_s + VK_v\lambda_{th}^{-2}\exp(-2E_a/k_B T)$. K_s and K_v are surface and volume recombination rate coefficients, respectively. These quantities and their magnetic field dependences are well established [5]. We used the values given by Arai *et al.* [6]

Neglecting the second term in Eq. 2 [2], the solution for N_e is

$$\frac{N_e(t)}{N_e(0)} = \exp\left[\frac{a}{b}\tanh(bt) - at\right], \qquad (3)$$

where $a = K_e\phi/K_s^{\text{eff}}$ and $b = \alpha\sqrt{\phi K_s^{\text{eff}}}$. All the measured SSE density decay curves are well fitted by Eq. 3 with two adjustable parameters, K_e and ϕ.

The temperature dependence of K_e is plotted in Fig. 2. For any given magnetic field, we observe an increase in K_e at lower temperatures.

CP850, *Low Temperature Physics: 24th International Conference on Low Temperature Physics;*
edited by Y. Takano, S. P. Hershfield, S. O. Hill, P. J. Hirschfeld, and A. M. Goldman

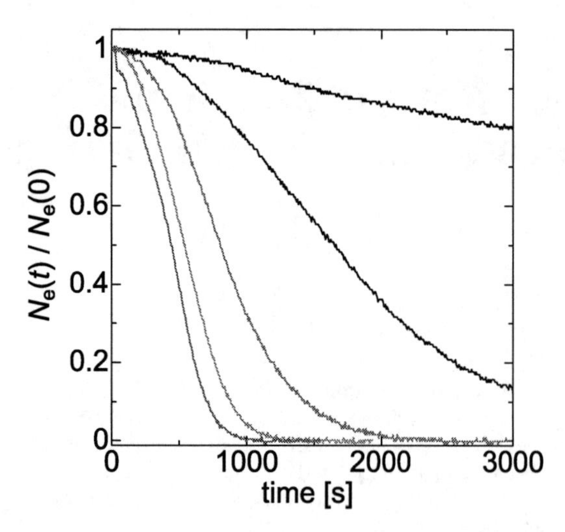

FIGURE 1. Measured SSE density as a function of time at $B = 3$ T at various temperatures, $T = 0.57$, 0.55, 0.51, 0.42, 0.39 K from upper to lower, respectively. The vertical axis is scaled by the initial density. The reaction rate is larger at lower temperatures.

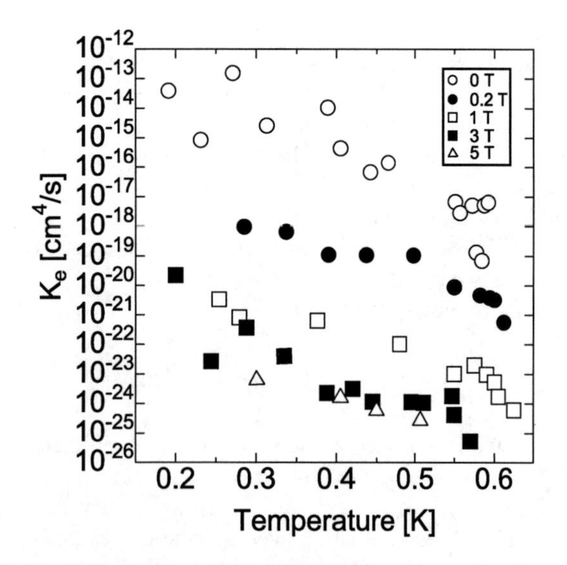

FIGURE 2. Measured EA rate coefficient K_e as a function of temperature. K_e increases as the temperature is lowered, at any magnetic field.

We can develop a simple model of SSE loss by assuming that the reaction occurs when three particles ($H + H + e^-$) meet together in a circle of diameter l_{eH}. We denote l_{eH} as the reaction cross section, which characterizes the microscopic physical mechanism of the reaction. Since our reaction is restricted on the two-dimensional surface, l_{eH} has a dimension of length. From simple scaling considerations, K_e may be roughly estimated as $K_e \sim l_{eH}^3 \overline{v_H}$, where $\overline{v_H} = (2\pi k_B T / m_H)^{1/2}$ is the two-dimensional thermal mean speed of a H atom. As shown in Fig. 3, l_{eH} is

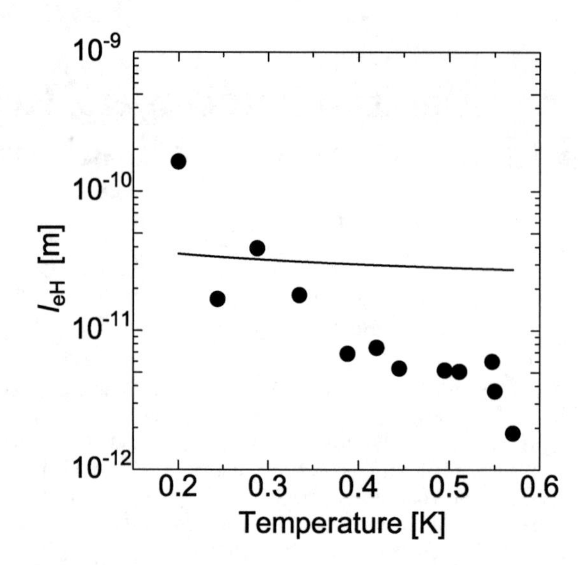

FIGURE 3. Temperature dependence of reaction cross section l_{eH}. Dots are derived from experimental data for $B = 3$ T. The measured l_{eH} increases at low temperatures. The line is proportional to $T^{-1/4}$ expected from the polarization potential.

strongly temperature dependent. l_{eH} becomes two orders of magnitude larger upon lowering the temperature from 0.6 K to 0.2 K. If there is an attractive interaction between e^- and H, this interaction can influence the motion of the slow H even at a long distance. We consider the interaction of induced electric dipole of H in the electric field near e^- (the polarization potential). Polarization potential $V(r)$ is proportional to r^{-4}. Again, simple scaling indicates $V(l_{eH}) \sim k_B T$, leading to $l_{eH} \sim T^{-1/4}$. The solid line in Fig. 3 is proportional to $T^{-1/4}$. It is evident that the measured temperature dependence is much stronger than the rough estimation. We conclude that a quantum mechanical treatment of polarization potential model is required.

This work is supported by a Grant-in-Aid for Scientific Research (B) from JSPS and a Grant-in-Aid for Exploratory Research from MEXT.

REFERENCES

1. T. Arai, and K. Kono, *Physica B*, **329-333**, 415–418 (2003).
2. T. Arai, T. Mitsui, and H. Yayama, *J. Low Temp. Phys.*, **138**, 445–450 (2005).
3. T. Arai, T. Mitsui, and H. Yayama, *J. Phy. Chem. Sol.* (2005), to be published.
4. T. Arai, A. Würl, P. Leiderer, T. Shiino, and K. Kono, *Physica B*, **284–288**, 164–165 (2000).
5. I. F. Silvera, and J. T. M. Walraven, "Spin-Polarized Atomic Hydrogen," in *Prog. in Low Temp. Phys. Vol. X*, edited by D. F. Brewer, Amsterdam: Elsevier, 1986, pp. 139–370.
6. T. Arai, M. Yamane, A. Fukuda, and T. Mizusaki, *J. Low Temp. Phys.*, **112**, 373–398 (1998).

Dynamics of Magnetically Induced Superflow in Spin-Polarized ^{3}He A_1

R. Masutomi[a], K. Kimura[b], S. Kobayashi[b], A. Yamaguchi[b], H. Ishimoto[b], and H. Kojima[a]

[a] *Serin Physics Laboratory, Rutgers University, Piscataway, New Jersey 08854, USA*
[b] *Institute for Solid State Physics, Tokyo University, Kashiwa, Chiba 277-8581, Japan*

Abstract. The spin relaxation time (τ) in superfluid ^{3}He A_1 phase is measured by observing the magnetic fountain pressure induced across a superleak. Preliminary data are reported on the dependence of τ on applied magnetic field (H), pressure and ^{4}He coverage. When $0.5 < H < 1$ tesla, τ increases linearly with H in accordance with Hammel-Richardson model. However, τ varies little with field when $2 < H < 8$ tesla. As pressure is decreased, τ increases. When the interior wall surfaces (including those of heat exchanger) are covered with five layers of ^{4}He, τ does *not* change significantly. Conflicting observations show that the driven spin flow phenomena are not well understood. A new oscillatory behavior in magnetic fountain effect at 29 bars is also described.

Keywords: superfluid ^{3}He, magnetic fountain effect, spin polarized system, spin current, spin relaxation
PACS: 67.57.-z, 67.60.-g, 72.25.-b, 85.75.-d

INTRODUCTION

Although the relaxation of magnetization of liquid ^{3}He in contact with other materials has been studied over many years, it is not yet generally well understood. We are making observations on the dynamics of magnetic fountain effect in superfluid ^{3}He A_1 phase to gain understanding of the magnetization relaxation phenomena. If ^{3}He magnetization relaxation time is assumed long, a simple theory implies that an applied magnetic field gradient is balanced by an induced fountain pressure gradient.[1] All of our previous experiments show that the induced pressure gradient relaxes over a relatively short time in the presence of steady field gradient.[2] The relaxation in fountain pressure is directly related to that in magnetization.

The experimental set up is the same as described previously.[2] Briefly, a small chamber equipped with a differential pressure sensor is connected via superleak channels to a larger reservoir immersed in liquid ^{3}He all in uniform magnetic field up to 8 tesla. When ^{3}He is cooled into A_1 phase and a ramped "square pulse" of magnetic field gradient is applied across the superleak, magnetic fountain pressure develops. The fountain pressure is measured by a capacitive differential pressure sensor. The fountain pressure relaxes exponentially (except at high pressures; see Fig. 3) with time constant τ when the field gradient is constant. This relaxation in pressure is interpreted as that arising from spin density relaxation.[3] Temperature and pressure dependence of τ was reported earlier.[2] Preliminary data of τ as a function of magnetic field and ^{4}He coverage of surfaces are presented in this report. A new observation on oscillations in the fountain pressure at 29 bars is also described.

RESULTS AND DISCUSSIONS

The dots in Fig. 1 show the dependence of measured τ on magnetic field at 21 bars in pure ^{3}He. Since τ does not vary much with temperature except near the

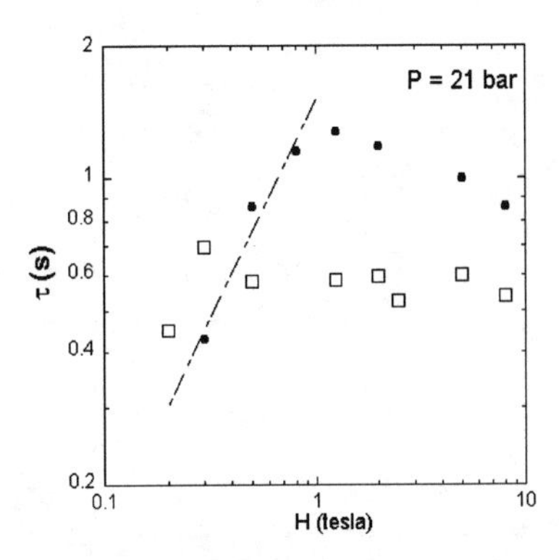

FIGURE 1. Magnetic field dependence of relaxation time.

CP850, *Low Temperature Physics: 24th International Conference on Low Temperature Physics;* edited by Y. Takano, S. P. Hershfield, S. O. Hill, P. J. Hirschfeld, and A. M. Goldman

"edges" of A_1 phase close to T_{c1} and T_{c2},[2] the average values near the middle of A_1 are plotted. Below 1 tesla, τ increases roughly linearly (shown by dashed line) with field. The linear field dependence is consistent with our earlier measurement below 1.5 tesla.[3] It is also consistent with the Hammel-Richardson(HR) model of spin relaxation.[4] Surprisingly, τ decreases at fields greater than 1.5 tesla. This is contrary to what would be expected from the linear field dependence of spin relaxation time observed by Schuhl et al.[5] in fields up to 8 tesla in *normal* liquid ^{3}He in contact with fluorocarbon particles. Apparently, the spin relaxation mechanism is different in our experiment in A_1 phase.

In the HR model, the spin relaxation is mediated by spin fluctuation within a dense solid-like ^{3}He layer adjacent to the cell/substrate wall surface. If the ^{3}He solid layer is replaced by ^{4}He, this mechanism should become ineffective. To test this, an amount sufficient to produce 5.0 layers of ^{4}He was introduced into the apparatus to preplate all surfaces including heat exchangers, and the experiment was repeated. The measured τ with ^{4}He preplating is shown by squares in Fig. 1. Surprisingly, there is no increase in τ except perhaps at low fields. The relaxation time no longer varies with magnetic field. Below 0.4 tesla, τ appears to increase with ^{4}He coating but it decreases above 0.7 tesla. The scatter in the data does not allow firm conclusion and more work is needed to clarify the effects of ^{4}He on spin relaxation in A_1 phase.

Fig. 2 shows the measured pressure dependence of τ in pure ^{3}He (dots) and in ^{4}He preplated cell surface(triangles) both in 8 tesla. The lines are guide to the eye. Both show the same trend with pressure, but the relative increase is variation in pure ^{3}He. The thickness of the solid-like layer adjacent to material walls decreases as pressure decreases. The observed pressure dependence in pure ^{3}He is consistent with the associated decrease in number of fluctuating spins, if HR model is applicable to our experiment. The lesser variation of τ with pressure with ^{4}He coating is in accord with this also. The effect of ^{4}He coverage on magnitude of τ discussed in the previous paragraph remains a puzzle.

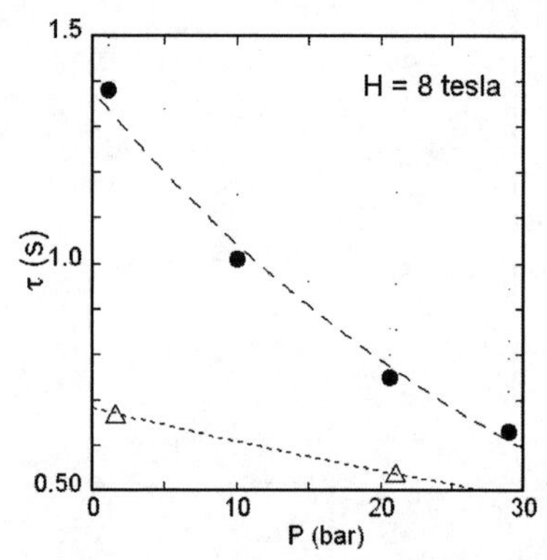

FIGURE 2. Pressure dependence of relaxation time.

When the pressure is increased to 29 bars, an oscillation behavior during the relaxation of fountain pressure is observed for the first time. An example is shown in Fig. 3. Starting at t = 0, the magnetic field gradient is increased linearly over 200 ms and is subsequently kept constant for 3.8 s. The plotted signal is proportional to the fountain pressure. The field gradient sequence is repeated for averaging. The oscillation frequency (6.6 Hz in Fig. 3) decreases as temperature increases. If the oscillation is a Helmoltz resonance with the superleak (length l and cross sectional area a) as the neck and the diaphragm (area b and tension σ) as the restoring force, the resonance frequency is given by $(1/2\pi)((\rho_s/\rho)^{1/2}(8\pi\sigma a/\rho l b^2)^{1/2}$, where ρ_s/ρ is the superfluid density fraction. This formula is in rough agreement with the observed frequency and its temperature dependence. Detailed analysis, however, requires more work to determine dependence on various parameters. The oscillation may give us another tool for studying the spin relaxation phenomenon.

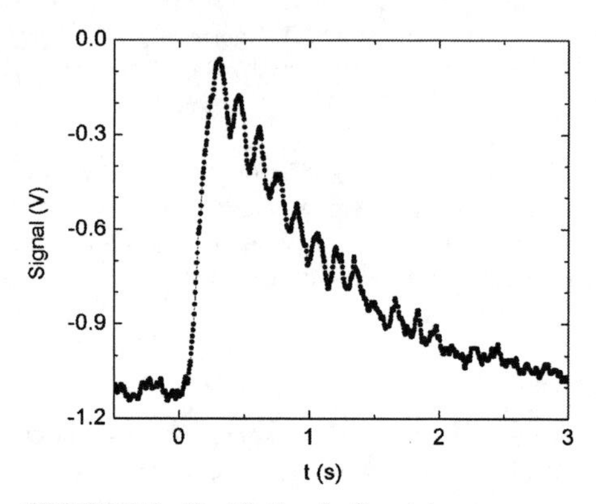

FIGURE 3. Oscillation in fountain pressure.

ACKNOWLEDGMENTS

Research is supported by NSF and Japan-US Collaborative Science under JSPS.

REFERENCES

1. M. Liu, *Phys. Rev. Lett.* **43**, 1740 (1979).
2. R. Masutomi, et al. *J. Low Temp. Phys.* **138**, 789 (2005).
3. S.T. Lu, Q. Jiang and H. Kojima, *Phys. Rev. Lett.* **89**, 1639(1989).
4. C. Hammel and R.C. Richardson, *Phys. Rev. Lett.* **33**, 1234 (1900).
5. A. Schuhl, et al., *Phys. Rev.B* **36**, 6811 (1987).

The Thermal Boundary Resistance of the Superfluid ^{3}He A-B Phase Interface in the Low Temperature Limit

D.I. Bradley, S.N. Fisher, A.M. Guénault, R.P. Haley, H. Martin, G.R. Pickett, J.E. Roberts and V. Tsepelin

Department of Physics, Lancaster University, Lancaster, LA1 4YB, United Kingdom

Abstract. We have constructed a vertical cylindrical cell in which we cool superfluid ^{3}He to the low temperature limit. At the top and bottom of this cylinder are pairs of vibrating wire resonators (VWRs), one to act as a heater and the other as a thermometer. Quasiparticle excitations are created by driving the heater VWRs. These excitations can only leave the cylinder via a small hole at the top. Using a shaped magnetic field, we can produce a layer of A phase across the tube, while maintaining low field B phase in the vicinity of the VWRs for reliable thermometry. Preliminary results show that the two A-B interfaces lead to a measurable extra resistance for quasiparticles between the top and bottom of the cylinder.

Keywords: superfluid, helium-3, interface
PACS: 67.57.Bc, 67.57.Np

INTRODUCTION

In recent years we have been investigating various properties of the ^{3}He A-B interface stabilized by a magnetic field in the zero temperature limit at low pressure [1]. Here we report a new method for measuring the thermal boundary resistance of the A-B interface at temperatures below $0.2T_c$, where the quasiparticles carrying the heat are in the ballistic regime. We use a profiled magnetic field to produce a layer of A phase separating two regions of B phase, thus creating two A-B interfaces that impede the flow of quasiparticles.

EXPERIMENT

The experiment is performed in a tailpiece attached to the inner cell of a Lancaster-style nested nuclear demagnetisation stage, illustrated in figure 1 and described in more detail in [2]. The cell consists of a vertical cylinder of epoxy-impregnated paper 8 mm in diameter and 45 mm long, closed at the bottom but open at the top through a 3 mm diameter orifice. At the top and bottom are pairs of vibrating wire resonators (VWRs). Each pair has a 13.5 μm wire that acts as a heater (source of quasiparticles); and a 4.5 μm wire that acts as a thermometer (a measure of quasiparticle density). The magnetic field required to produce and stabilise the A phase layer across the cylinder is generated by a set of superconducting solenoids, thermally anchored to the still of the dilution refrigerator. These provide the 340 mT field required to produce A phase from B phase [1], and compensation fields to maintain approximately 50 mT at the VWRs. This ensures the VWRs remain in low field B phase so

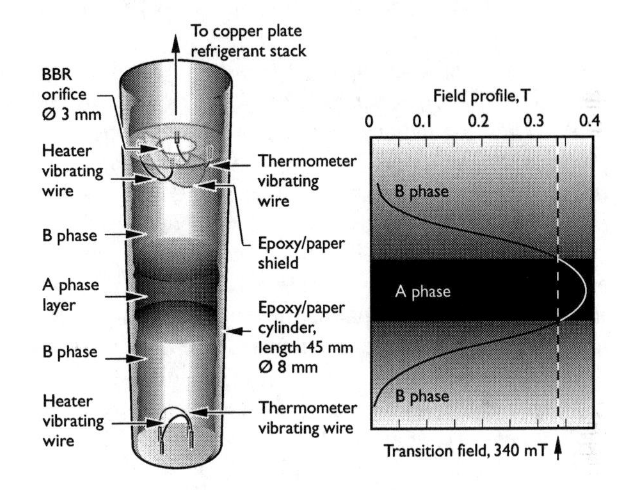

FIGURE 1. Cell schematic and magnetic field profile, see text. The vibrating wires are made from NbTi filaments.

that the frequency width or damping of their resonance, Δf_2, is proportional to the quasiparticle density, varying as $\exp(-\Delta_B/k_BT)$, where Δ_B is the isotropic B phase energy gap.

Each VWR is driven by an ac current. The voltage generated as the VWR moves through the field is measured by a lock-in amplifier. The bottom heater VWR is driven above the pair-breaking threshold to create quasiparticles which flow up the cylinder and exit through the small hole at the top. The two thermometer VWRs at the top and bottom are excited at low amplitude at their resonance frequencies and monitored continuously as the heater is turned on and off. Their resonance amplitude varies as $1/\Delta f_2$ and thus gives a record of the changes in quasiparticle density.

CP850, *Low Temperature Physics: 24th International Conference on Low Temperature Physics;*
edited by Y. Takano, S. P. Hershfield, S. O. Hill, P. J. Hirschfeld, and A. M. Goldman

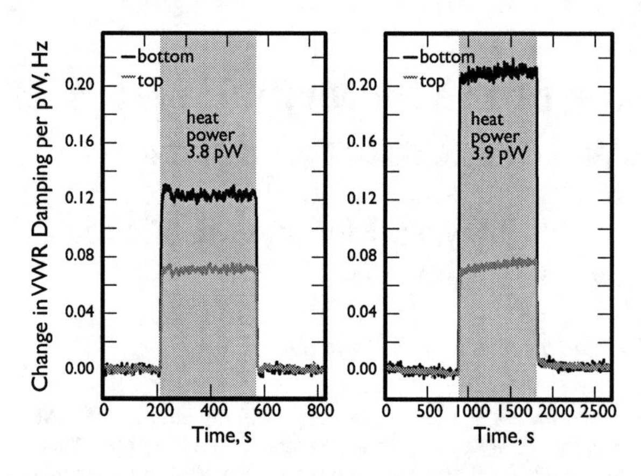

FIGURE 2. Change in vibrating wire damping (Δf_2) per picowatt at the top and bottom of the cylinder when heat is applied at the bottom. On the left the cell contains only B phase, whereas on the right it contains a layer of A phase.

RESULTS AND DISCUSSION

Figure 2 shows the change in Δf_2 of the top and bottom thermometer VWRs per pW for heater powers around 4 pW when the cylinder contains low field isotropic B phase everywhere (left) and a layer of A phase across the middle (right). A constant baseline arising from residual heat leaks and intrinsic damping of the VWR has been subtracted. When the cell contains only B phase, one can see that the applied heat leads to a substantial difference in the change in quasiparticle density between the top and bottom. This density gradient arises because the scattering at the cell walls is predominantly diffuse, and is discussed in [3]. At the lowest temperatures ($\sim 0.12\,T_c$) the ratio of the change in Δf_2 of the bottom VWR to that of the top is almost 1.8. Small mechanical differences make the bottom VWR slightly more sensitive to quasiparticle density than the top VWR by a factor of approximately 1.1. Thus the ratio of quasiparticle density change between bottom and top is ~ 1.7.

When the cell contains a layer of A phase, this (corrected) ratio rises to ~ 2.7, ie. the rise in quasiparticle density at the bottom of the cell is almost 3 times that of the top. This significant increase in density at the bottom indicates that the flow of quasiparticles is impeded by the layer of A phase. We have varied the thickness of the A phase layer from 1-2 cm down to 1-2 mm and have observed no change in this ratio. Therefore the extra impedance must come from the A-B boundary region. As quasiparticles approach the interface they are Andreev-reflected since the available states in the gap structures of A and B are very different at these low temperatures [4].

We now discuss complications that prevent us from using this extra impedance to make a straightforward calculation of the A-B boundary resistance. In large magnetic fields the two phases respond in different ways. The B phase gap is distorted by high field, reducing along the field direction and increasing perpendicular [5]. Additionally, there is a spin-dependent Zeeman splitting of the minimum energy of quasiparticles travelling along the field direction. Apart from aligning the gap nodes perpendicular to the field direction, the A phase is relatively unaffected by the field; there is a small difference in gap between the up-spin and down-spin pairs but no spin splitting of the excitation spectrum. Further, the A-B interface has an effect on both A and B textures. The A phase node direction is preferentially aligned along the interface. In our cell geometry the field and interface both align the nodes in the same direction so there is no competition. In the B phase the direction of gap suppression is aligned along the field but must bend away as it approaches the boundary region to lie parallel to the interface [6].

We made two measurements with a layer of highly distorted B phase at 320 mT, just below the transition field. First, we ramped up the field without ever having had the A phase layer in the cell. The ratio of quasiparticle density change rose from 1.7 to 1.9, showing that the distorted B phase itself adds a resistance to the flow of quasiparticles. Second, we ramped to high field to create the A phase layer (and thereby bend the B phase texture), then reduced the field to leave only B phase. This gave the surprising result of a density change ratio of 2.3. We are thus left to speculate that the B phase retains a memory of the bending of the texture by the A-B interface, or that the A phase leaves behind some other kind of textural defect. Work is in progress to extract a resistance that we can ascribe to the interface alone in order to compare with theoretical prediction [7].

In conclusion, we have implemented a new technique that uses ballistic quasiparticle flow to probe the A and B phase gap structures and textures in large magnetic fields, and discovered that the A-B interface presents a measurable impedance.

REFERENCES

1. see M. Bartkowiak, S. N. Fisher, A. M. Guénault, R. P. Haley G. R. Pickett and P. Skyba, *Phys. Rev. Lett*, **93**, 045301 (2004), and references therein.
2. D. I. Bradley et al., *A Levitated Droplet of Superfluid ^{3}He-B Entirely Surrounded by ^{3}He-A*, this conference.
3. D. I. Bradley et al., *Thermal Transport by Ballistic Quasiparticles in Superfluid ^{3}He-B in the Low Temperature Limit*, this conference.
4. D. J. Cousins, M. P. Enrico, S. N. Fisher, S. L. Phillipson, G. R. Pickett, N. S. Shaw and P. J. Y. Thibault, *Phys. Rev. Lett.*, **77**, 5245 (1996).
5. M. Ashida and K. Nagai, *Prog. Theor. Phys.* **74**, 949 (1985).
6. E. V. Thuneberg, *Phys. Rev. B*, **44**, 9685 (1991).
7. S. Yip, *Phys. Rev. B*, **32**, 2915 (1985).

Observation of A-B Phase Transition of Superfluid ^{3}He by Transverse Acoustic Response

M. Saitoh*, Y. Wada, Y. Aoki, R. Nishida, R. Nomura, and Y. Okuda

Department of Condensed Matter Physics, Tokyo Institute of Technology, 2-12-1 O-okayama, Meguro, Tokyo 152-8551, Japan

Abstract. By using an AC-cut quartz transducer at the frequencies of 9.56 and 28.7 MHz, we observed discontinuity in the temperature dependence of transverse acoustic impedance of superfluid ^{3}He at 24.9 bar. Though the transverse acoustic response of superfluid ^{3}He is not well understood, we can conclude that the discontinuity occurred at the *A-B* transition, from its temperature and frequency dependence. Since the density of states near the boundary within a coherence length dominates the momentum transfer between an oscillating wall and ^{3}He quasi-particles, the large difference in the transverse acoustic impedance between the *A* and *B* phases can be explained by a difference in the symmetry of the order parameters near the wall.

Keywords: superfluid ^{3}He, transverse acoustic impedance, Andreev bound states
PACS: 67.57.Np, 74.45.+c, 43.58.Bh

INTRODUCTION

Since the surface Andreev bound states (SABS) which form near the boundary of unconventional superconductors are strongly affected by the symmetry of the order parameter, they have provided a useful tool to investigate the structure of the order parameter [1]. SABS is detected as a significant change of transport properties due to the large modification of density of states at the Fermi level. SABS has been predicted to occur also in *p*-wave superfluid ^{3}He [2]. Recently, we have succeeded in observing SABS in superfluid ^{3}He at a pressure of 17 bar [3] by measuring the transverse acoustic impedance *Z*. Here, we report the behavior of *Z* at the *A-B* transition.

EXPERIMENTAL RESULTS AND DISCUSSION

By using an AC cut quartz transducer, *Z* was measured at the frequencies of 9.56 MHz (fundamental) and 28.7 MHz (third harmonics). The complex transverse acoustic impedance $Z = Z' + iZ''$ was deduced from the measured Q factor and resonance frequency $f_0 = \omega_0/2\pi$ by the relation [4] $Z' = \frac{1}{4}n\pi Z_q \Delta(1/Q)$, $Z'' = \frac{1}{2}n\pi Z_q \Delta f_0 / f_0$, where Z_q is the acoustic impedance of the quartz transducer, and n is the harmonics number. The measurements were made during a temperature sweep, by using each of the two frequencies for every other data point. Experimental details are presented in our previous papers [3, 5, 6].

The results for 9.56 MHz and 28.7 MHz are shown in Fig. 1 (a) and (b), respectively. Z' and Z'' are plotted by circles and triangles, respectively. The open and solid symbols represent the results for cooling at a rate of -0.45 μK/min and warming at a rate of 0.35 μK/min, respectively. The temperature was measured by a ^{3}He melting curve thermometer mounted on the nuclear stage, and was corrected for a temperature gradient between the sample and the thermometer by using pair breaking temperatures observed at 9.56 MHz at several pressures in B phase separately for cooling and warming.

Because the frequencies were in the limit of $\omega_0\tau >> 1$, where τ is the lifetime of the quasi-particles, neither Z' nor Z'' showed any temperature dependence in the region above the superfluid transition temperature T_c. Below T_c the energy gap began to open and the density of states at the Fermi level was

* present address: Low Temp. Phys. Lab., RIKEN, 2-1 Hirosawa Wako-shi, 351-0198, Japan

CP850, *Low Temperature Physics: 24th International Conference on Low Temperature Physics;* edited by Y. Takano, S. P. Hershfield, S. O. Hill, P. J. Hirschfeld, and A. M. Goldman

modified. The momentum transfer between the quasi-particles and the transducer caused Z to be temperature dependent. A peak in Z' appeared at 28.7 MHz, and there was a steep slope in Z'' at 9.56 MHz. This behavior agreed well with the microscopic quasi-classical calculation by Yamamoto *et al.* [7] The temperature dependences are well reproduced by the theory at both frequencies, above the temperature where a jump in Z occurred. The jump appeared at the temperature $0.92T_c$ in warming and $0.88T_c$ in cooling. It occurred simultaneously at the two frequencies, ruling out a possibility that it was an order-parameter collective mode. Moreover, the temperature at the jump in warming agreed well with the A-B phase transition temperature $T_{AB} = 0.93T_c$ at the pressure of 24.9 bar [8]. Therefore, it can be concluded that the jump occurred as the response of Z at the A-B phase transition. The hysteretic behavior can be understood as supercooling during cooling.

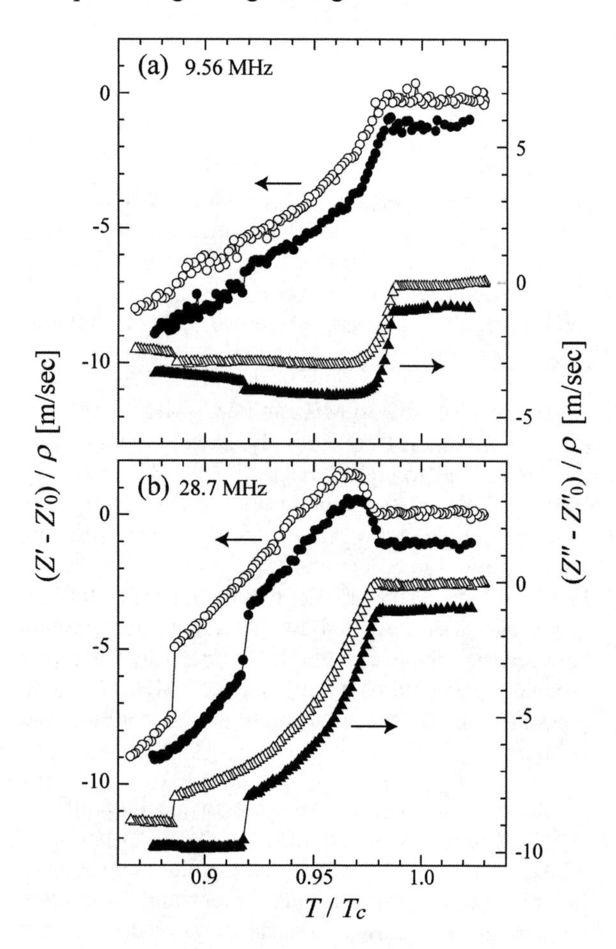

FIGURE 1. Results for Z' (open and solid circles) and Z'' (open and solid triangles) at 24.9 bar and at the frequencies of 9.56 (a) and 28.7 MHz (b). Open and solid symbols represent the results for cooling and warming process, respectively. The warming data have been shifted down by 1 m/sec for clarity.

CONCLUSION

In conclusion, the A-B phase transition in superfluid ^{3}He at the pressure of 24.9 bar was observed by the measurement of transverse acoustic impedance at the frequencies of 9.56 and 28.7 MHz. Although the transverse acoustic response at the transition is not yet understood, from the observation of simultaneous jumps in both Z' and Z'' at two different frequencies at the temperature known to be T_{AB}, it can be concluded that the A-B phase transition was observed as a jump in Z. A significant modification of SABS due to the change in the order parameter was sensitively detected by the transverse acoustic response.

ACKNOWLEDGMENTS

This work was partly supported by the Kurata Memorial Hitachi Science and Technology Foundation, a Grant-in-Aid for Scientific Research, and a 21st Century COE Program at Tokyo Tech "Nanometer-Scale Quantum Physics" by the Ministry of Education, Culture, Sports, Science and Technology of Japan.

REFERENCES

1. S. Kashiwaya and Y. Tanaka, *Rep. Prog. Phys.* **63**, 1641-1724 (2000).
2. Y. Nagato, M. Yamamoto, and K. Nagai, *J. Low. Temp. Phys.* **110**, 1135-1171 (1998).
3. Y. Aoki, Y. Wada, M. Saitoh, R. Nomura, Y. Okuda, Y. Nagato, M. Yamamoto, S. Higashitani, and K. Nagai, *Phys. Rev. Lett.* **95**, 075301 (2005).
4. W. P. Halperin, and E. Varoquaux, “Order-Parameter Collective Modes in Superfluid ^{3}He” in *Helium Three*, edited by P. Halperin and L. P. Pitaevskii, Amsterdam: North-Holland, 1990, pp. 353-522.
5. Y. Aoki, Y. Wada, A. Ogino, M. Saitoh, R. Nomura, and Y. Okuda, *J. Low Temp. Phys.* **138**, 783-788 (2005).
6. Y. Aoki, Y. Wada, M. Saitoh, R. Nomura, and Y. Okuda, *J. Phys. Chem. Solids* **66**, 1349-1351 (2005).
7. M. Yamamoto, Y. Nagato, S. Higashitani and K. Nagai, *J. Low Temp. Phys.* **138**, 349-354 (2005).
8. D. S. Greywall, *Phys. Rev. B* **33**, 7520-7538 (1986).

Pressure Dependence of the Transverse Acoustic Impedance of Superfluid ^{3}He-B

Y. Wada, Y. Aoki, M. Saitoh, R. Nishida, R. Nomura, and Y. Okuda

Department of condensed Matter Physics, Tokyo Institute of Technology, 2-12-1, O-okayama, Meguro, Tokyo, 152-8551, Japan

Abstract. We measured the pressure dependence of the complex transverse acoustic impedance of superfluid ^{3}He-*B* using an AC-cut quartz transducer. The measurements were performed by a CW bridge method at the third harmonics of the fundamental resonance frequency 9.56 MHz. We obtained the real (Z') and the imaginary (Z'') parts of the transverse acoustic impedance independently. Z' and Z'' did not change from the normal state values at T_c, but Z' started to increase at the pair breaking edge temperature T_{pb} upon cooling. The slope of the increase changed at a temperature defined as T^* which was lower than T_{pb}. With further cooling, it reached a maximum and then decreased slowly. Z'' had a small peak at T^* and decreased rapidly with decreasing temperature. These temperature dependences were possibly influenced by quasi-particle density of states within the energy gap originating from the surface Andreev bound states. At lower pressure the maximum of Z' and the small peak of Z'' moved to smaller T/T_c and became larger due to the pressure dependence of the energy gap.

Keywords: Transverse acoustic impedance, Superfluid ^{3}He-B, Andreev bound states
PACS: 67.57.Np, 74.45.+c, 43.58.Bh

INTRODUCTION

The surface Andreev bound states (SABS) can be observed at the surface of unconventional superconductors such as the high temperature superconductors by tunneling spectroscopy [1]. It has been a very important method to identify the symmetry of new superconductors. The existence of SABS was also predicted theoretically in *p*-wave superfluid ^{3}He [2]. SABS lead to quasi-particle density of states within the energy gap formed by the spatial variation of the order parameter in the vicinity of a rough surface. However, there has been no good method to investigate the surface property of superfluid ^{3}He. Recently, we obtained clear evidence of SABS by measuring the temperature dependence of the transverse acoustic impedance of superfluid ^{3}He using AC-cut quartz transducers [3]. In this paper, we report the pressure dependence of the transverse acoustic impedance Z.

EXPERIMENTAL RESULTS AND DISCUSSION

We used an AC-cut quartz transducer whose fundamental frequency was 9.56 MHz, and performed the measurements at the third harmonics of 28.7 MHz. The quality factor Q and the resonance frequency f_0 were determined from the acoustic response of the transducer. The complex transverse acoustic impedance $Z = Z' + iZ''$ is written as $Z' = \frac{1}{4}n\pi Z_q\Delta(1/Q)$, $Z'' = \frac{1}{2}n\pi Z_q\Delta f_0/f_0$, where Z_q is the acoustic impedance of the quartz, and n is the harmonic number of the transducer [4]. Details of our experiment were described in our previous papers [3, 5, 6].

Figure 1 shows the temperature dependence of Z' and Z'' at 17 bar. These plots show the changes from the normal liquid value $Z_0 = Z_0' + iZ_0''$ just above T_c. Neither component changed from Z_0 at T_c. At the pair breaking edge temperature T_{pb}, Z' started to increase but Z_0'' did not change. Z' had a kink at a temperature defined as T^* indicated by the vertical solid line, and Z'' had a small peak at the same temperature T^*. In the temperature range below T^*, Z' had a maximum and after that decreased, whereas Z'' decreased rapidly. Although we used a different transducer than in Ref. 3, these features agreed with the previous results. Therefore the measurements of Z did not depend on the transducer, confirming that Z was obtained correctly.

CP850, *Low Temperature Physics: 24th International Conference on Low Temperature Physics;*
edited by Y. Takano, S. P. Hershfield, S. O. Hill, P. J. Hirschfeld, and A. M. Goldman

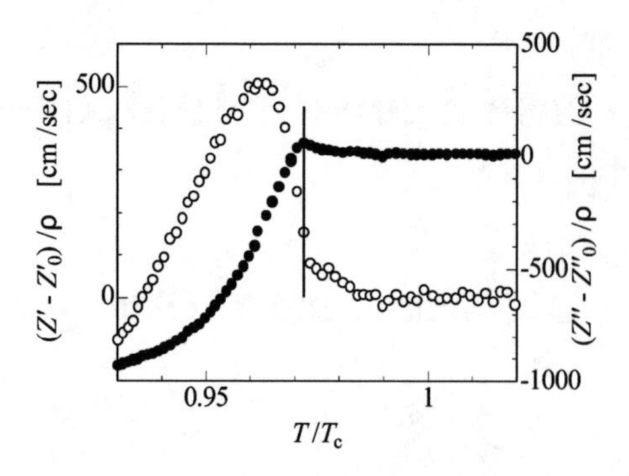

FIGURE 1. Temperature dependence of the transverse acoustic impedance at a frequency of 28.7 MHz at a pressure of 17 bar. Open circles are Z' and solid circles are Z''.

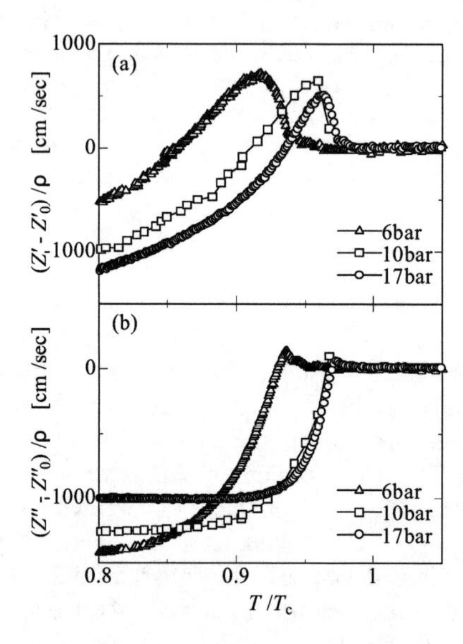

FIGURE 2. Temperature dependence of the real (a) and the imaginary (b) components of transverse acoustic impedance at a frequency of 28.7 MHz at 6, 10, and 17 bar; open triangles for 6 bar, open squares for 10 bar, open circles for 17 bar.

The pressure dependence of Z' and Z'' is shown in Fig. 2. Z' and Z'' at other pressures had a similar temperature dependence to that at 17 bar. At lower pressure, the features at T^* shifted to lower temperature. And the maximum of Z' and the small peak of Z'' also shifted to lower temperature and both of them became larger. In the previous experiment [3], these shifts were observed at higher frequencies. Lowering the pressure was found to be equivalent to increasing the frequency. We concluded in Ref. 3 that the kink of Z' and the peak of Z'' at T^* were anomalies due to the pair excitation of a surface bound state and a propagating Bogoliubov quasi-particle. We could confirm SABS formation by the pressure dependence of Z, considering the pressure dependence of the energy gap.

CONCLUSION

We measured the temperature dependence of the real and imaginary components of the transverse acoustic impedance of superfluid ^3He at a frequency of 28.7 MHz at 6, 10, and 17 bar. The pressure dependence was compared with previous results and attributed to quasi-particle density of states within the gap energy originating from the SABS.

ACKNOWLEDGMENTS

This work was partly supported by the Kurata Memorial Hitachi Science and Technology Foundation, a Grant-in-Aid for Scientific Research, and a 21st Century COE Program at Tokyo Tech "Nanometer-Scale Quantum Physics" by the Ministry of Education, Culture, Sports, Science and Technology, of Japan.

REFERENCES

1. L. H. Greene, M. Covington, M. Aprili, E. Badica, and D. E. Pugel, *Physica B* **280**, 159-164 (2000).
2. Y. Nagato, M. Yamamoto, and K. Nagai, *J. Low. Temp. Phys.* **110**, 1135-1171 (1998).
3. Y. Aoki, Y. Wada, M. Saitoh, R. Nomura, Y. Okuda, Y. Nagato, M. Yamamoto, S. Higashitani, and K. Nagai, *Phys. Rev. Lett.* **95**, 075301 (2005).
4. W. P. Halperin, and E. Varoquaux, "Order-Parameter Collective Modes in Superfluid ^3He," in *Helium Three,* edited by P. Halperin and L. P. Pitaevskii, Amsterdam: North-Holland, 1990, pp. 353-522.
5. Y. Aoki, Y. Wada, A. Ogino, M. Saitoh, R. Nomura, and Y. Okuda, *J. Low Temp. Phys.* **138**, 783-788 (2005).
6. Y. Aoki, Y. Wada, M. Saitoh, R. Nomura, and Y. Okuda, *J. Phys. Chem. Solids*, **66**, 1349-1351 (2005).

Thermal Transport by Ballistic Quasiparticles in Superfluid ^{3}He-B in the Low Temperature Limit

D.I. Bradley, S.N. Fisher, A.M. Guénault, R.P. Haley, H. Martin, G.R. Pickett, J.E. Roberts and V. Tsepelin

Department of Physics, Lancaster University, Lancaster, LA1 4YB, United Kingdom

Abstract. In the temperature range below $0.2T_c$, the gas of thermal excitations from the superfluid ^{3}He-B ground state is in the ultra-dilute ballistic regime. Here we discuss preliminary measurements of the transport properties of this quasiparticle gas in a cell of cylindrical geometry with dimensions much smaller than any mean free path. The vertical cylinder, constructed from epoxy-coated paper, has vibrating wire resonator (VWR) heaters and thermometers at the top and bottom, and a small aperture at the top which provides the only exit for quasiparticles. Using the thermometer VWRs, we measure the difference in quasiparticle density between the top and bottom of the tube when we excite the top or bottom VWR heater. This gives information about the transport of energy along the cylindrical ^{3}He sample and hence about the scattering behaviour involved when a quasiparticle impinges on the cylinder wall.

Keywords: superfluid, helium-3, ballistic
PACS: 67.57.Bc, 67.57.Hi

INTRODUCTION

In this paper we describe new experiments to investigate ballistic thermal transport in superfluid ^{3}He-B in low magnetic field (50 mT) where the order parameter is essentially isotropic. All the measurements are made at temperatures below $0.2T_c$, so that heat is carried by a very dilute gas of ^{3}He quasiparticles whose mean free paths in a bulk volume exceed 200 mm, much larger than the dimensions of the experimental cell.

EXPERIMENTAL

The measurements are made inside the cylindrical tailpiece of a Lancaster-style black body radiator (BBR) cell [1] illustrated in figure 1, and described in more detail in [2]. This is a cylindrical box 45 mm long with a uniform internal diameter of 8 mm. It is closed at the bottom but open at the top through a 3 mm diameter aperture that connects the ^{3}He contained in the BBR to the ^{3}He that is in contact with the nuclear refrigerant of silver-sintered copper plates. Quasiparticles created inside the BBR by heaters and residual heat leaks can only exit via this small orifice. The box is equipped with 4 vibrating wire resonators (VWRs) [1], one pair near the bottom of the cylinder and another close to the top but shielded from direct line of sight to the aperture. One VWR of each pair (wire diam. 13.5 μm) acts as heater, a source of quasiparticles. The other VWR (diam. 4.5 μm) is used as a thermometer, giving a measure of the quasiparticle density.

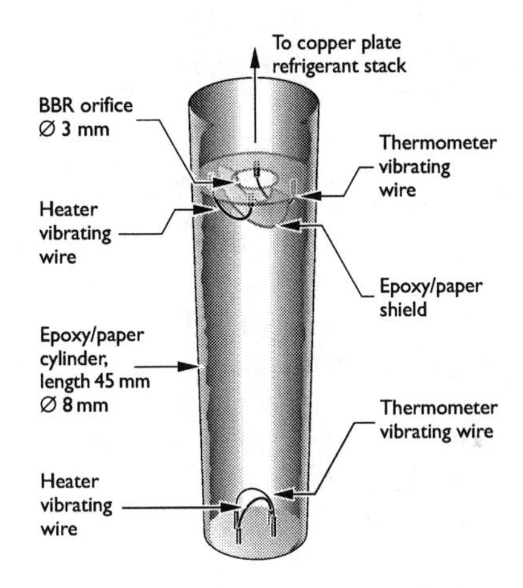

FIGURE 1. Cell schematic, see text. The vibrating wires are made from NbTi filaments.

Each VWR is driven by a current supplied by a waveform generator, and the voltage generated as the VWR moves in the 50 mT field is measured by a lock-in amplifier. The thermometer VWRs are excited at low amplitude at their resonance frequencies, and their resonant voltages are monitored continuously. From this we can infer the damping width, Δf_2, of the VWR resonance, which gives a measure of the quasiparticle damping. These two VWRs are tracked whilst power is applied using one of the heater VWRs, driven above the pair-breaking threshold to produce quasiparticles. Figure 2 shows the change in Δf_2 when a heat power of 4 pW is turned on and off. A constant baseline arising from residual heat leaks and the intrinsic damping of the VWRs has

CP850, *Low Temperature Physics: 24th International Conference on Low Temperature Physics;* edited by Y. Takano, S. P. Hershfield, S. O. Hill, P. J. Hirschfeld, and A. M. Goldman

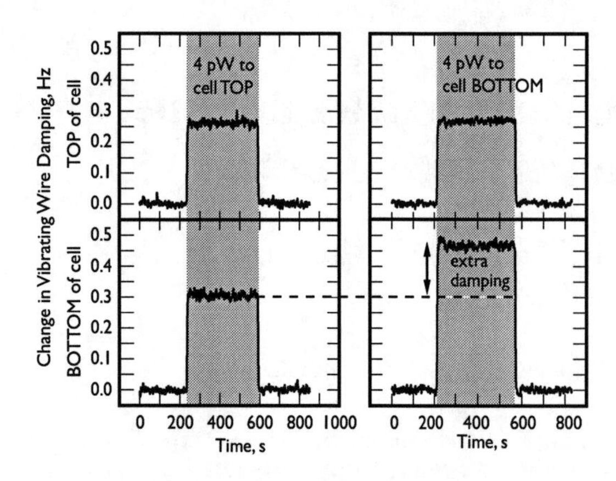

FIGURE 2. Change in vibrating wire damping (Δf_2) at the top and bottom of the cylinder when a heater power of 4 pW is applied to the top (left) and bottom (right).

been subtracted. The increase in damping is proportional to the increase in quasiparticle density in the vicinity of the thermometer VWRs.

We observe a marked difference between the response following excitation of the top heater and that following excitation of the bottom heater. The ratio of the change in Δf_2 of the bottom VWR to that of the top VWR is ~ 1.8 when quasiparticles are created by the heater at the bottom of the box. This ratio is only ~ 1.2 when the excitation is at the top. When heating outside the box is used (which heats the box uniformly) the ratio is ~ 1.1, allowing us to correct for the small difference in sensitivity of the two thermometer VWRs. So we see that the effect on the two thermometers is almost identical when the top heater is used. However, when the bottom heater is used the increase in quasiparticle density at the bottom is substantially larger, ~ 1.7 times that at the top. This ratio is insensitive to the heater power applied, which so far has ranged from 1 to 50 pW.

DISCUSSION

Consider the case of heating the bottom. We generate a continuous source of quasiparticles which then travel up the tube to escape through the hole at the top. It is instructive to compare this with the flow of gas through an open tube in the Knudsen (long mean free path) regime. The behaviour depends critically on the scattering boundary condition at the wall. For totally specular reflection, with no momentum transfer to the wall, there is zero particle density gradient along the tube. However, for completely diffuse reflection there is a gradient, equivalent to diffusive (viscous) transport with an effective mean free path of order the radius a of the cylinder. This analogy demonstrates that in our case we must have some degree of diffusive scattering of quasiparticles at the wall.

A simple kinetic theory expression [3] for pure diffuse scattering gives the number of particles per second passing along a long cylinder as $\frac{2}{3}\pi a^3 v \frac{dn}{dz}$ where $\frac{dn}{dz}$ is the constant density gradient along the tube and v is the mean speed. This assumes purely diffuse scattering at the wall. We know the quasiparticle current from the power applied to the heater VWR, and the quasiparticle average speed from a thermal average of the group velocity [1]. Hence we can find the density gradient in a long tube. We also know the density at the top from the effusion rate through the orifice close to the top thermometer. If we assume that our cell corresponds to a tube of length 45 mm, we can thus calculate the expected density at the bottom thermometer. This gives a ratio of 1.6, which compares quite well with the experimental 1.7. The discrepancy might arise from assuming we can use a long cylinder approximation, ignoring corrections for the tube ends. There may also be some degree of specular scattering. We are working on a more detailed calculation of quasiparticle transport in a tube with our geometry. This will allow us to say more about what fraction of the quasiparticles are diffusively scattered.

In conclusion, in conditions of ballistic transport we have observed a large quasiparticle gradient in our cell, which shows that the quasiparticle-wall scattering must be predominantly diffuse. Work is in progress to extend these measurements to higher temperatures.

ACKNOWLEDGMENTS

We thank M. Ward and I. Miller for their excellent technical support, the UK EPSRC for funding and the Royal Society for research support for RPH.

REFERENCES

1. S. N. Fisher, A. M. Guénault, C. J. Kennedy and G. R. Pickett, *Phys. Rev. Lett.*, **69**, 1073 (1992).
2. D. I. Bradley, S. N. Fisher, A. M. Guénault, R. P. Haley, H. Martin, G. R. Pickett, J. E. Roberts and V. Tsepelin, *A Levitated Droplet of Superfluid ^{3}He-B Entirely Surrounded by ^{3}He-A*, this conference.
3. E. H. Kennard, *Kinetic Theory of Gases*, McGraw Hill, London, 1938, pp. 302–305.

A Levitated Droplet of Superfluid ^{3}He-B Entirely Surrounded by ^{3}He-A

D.I. Bradley, S.N. Fisher, A.M. Guénault, R.P. Haley, H. Martin, G.R. Pickett, J.E. Roberts and V. Tsepelin

Department of Physics, Lancaster University, Lancaster, LA1 4YB, United Kingdom

Abstract. From our long experience of using profiled magnetic fields to stabilize and manipulate the A-B phase boundary in superfluid ^{3}He, we have constructed a cell in which we can create and move a droplet of B phase, levitated within A phase away from any walls at $T \sim 0.15T_c$. Uniquely, the A and B condensates are coherent across the A-B interface and at such low temperatures the superfluid is essentially pure, providing the most ordered phase boundary to which we have laboratory access. We configure the field so that within a bulk volume of superfluid, a region of high field (stabilizing the A phase) completely surrounds a region of lower field (stabilizing the B phase). Our preliminary measurements are at zero pressure and temperatures below $0.3T_c$ where the first-order transition from B to A phase is at 340 mT. We observe the formation of the droplet as we ramp the field, and we also study the transport of thermal excitations out of the droplet. Future plans include measurements at higher pressures where the A phase can be stabilized in low magnetic field at temperatures close to T_c. Upon cooling into the B phase we should then be able to make the first studies of nucleation uninfluenced by the presence of container walls.

Keywords: superfluid, helium-3, interface, droplet
PACS: 64.60.Qb, 67.57.Bc, 67.57Hi, 67.57.Np

INTRODUCTION

The first-order phase boundary between the A and B phases of superfluid ^{3}He is unique in providing the best-characterized and most ordered interface available for experimental study [1]. The system is inherently free of any defects or impurities. Here we describe a new method to create and control the interface in three dimensions, away from any influence of the container walls on the properties of the transition.

EXPERIMENTAL

The experiments are made inside a black-body radiator (BBR) [2] that forms the tailpiece of a Lancaster-style nuclear demagnetization stage [3] as shown in figure 1. The cell consists of a cylinder of epoxy-impregnated paper 8 mm in diameter and 44 mm long. At the top, a 3 mm diameter orifice provides a weak thermal link between the ^{3}He inside the BBR and the ^{3}He in contact with the silver-sintered copper plate refrigerant of the inner cell. The radiator volume contains pairs of heater and thermometer vibrating wire resonators (VWR) [2] at the top and bottom. Each VWR is driven at resonance by an ac current and the voltage generated is measured by a lock-in amplifier. The thermometer VWRs are excited at low amplitude so that the voltage is inversely proportional to the frequency width or damping of the resonance. The damping is proportional to the quasiparticle density in the vicinity of the VWR and varies as $\exp(-\Delta_B/k_BT)$ in the B phase, where Δ_B is the B phase energy gap. The heater VWRs are driven above the superfluid critical velocity, breaking pairs, and thus acting as a source of quasiparticles which can be used to calibrate the thermometers. Power also enters the radiator by residual heat leaks and from the latent heat released or absorbed by the motion of the A-B interface [3].

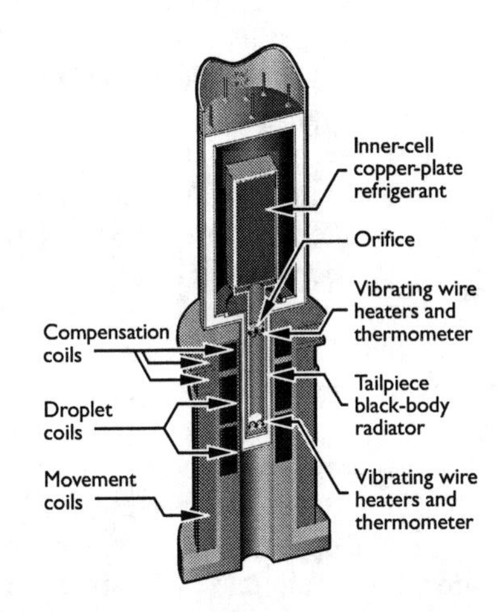

FIGURE 1. Black body radiator tailpiece surrounded by solenoid assembly, see text.

CP850, *Low Temperature Physics: 24th International Conference on Low Temperature Physics;* edited by Y. Takano, S. P. Hershfield, S. O. Hill, P. J. Hirschfeld, and A. M. Goldman

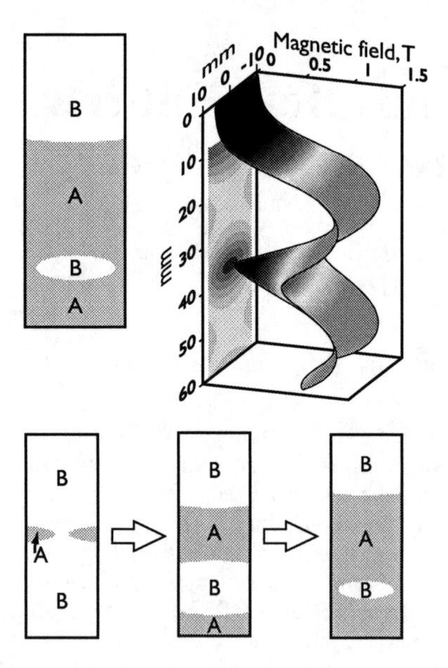

FIGURE 2. Field profile and process of B-phase droplet formation.

The magnetic field profiles we use to stabilise and manipulate the A-B interface inside the tailpiece are provided by a set of superconducting solenoids (see Fig. 1), thermally anchored to the still of the dilution refrigerator. At zero bar and temperatures below $0.3T_c$ the transition from B to A phase requires a field of 340 mT [4]. To create a droplet of B phase completely surrounded by A phase we use two identical co-axial solenoids carrying currents in opposite directions. As shown in figure 2 this produces large axial and radial field gradients, providing a field zero between the coils and a field magnitude increasing in all outward directions. We start with low field B phase everywhere in the cell, and ramp the current to the "droplet" solenoids. First, the A phase nucleates at the side walls and forms a layer across the cylinder that then grows thicker as the field is increased. Upon further ramping, the A phase enters at the bottom of the cell and grows upwards. Eventually the radial field grows large enough that the field at the side wall exceeds the transition field, and we form a droplet of B phase within A phase at the bottom of the cell. We are also able to move the B droplet up and down inside the A phase by ramping the current to the "movement" solenoid, which moves the zero field position along the vertical axis of the cell. Compensation coils ensure that the top of the cell always contains low field B phase for accurate thermometry.

We use thermodynamic methods to probe the behaviour of the A-B boundary. The latent heat released or absorbed by the moving interface leads to measurable changes in the quasiparticle density at the thermometer VWRs. The upward flow of quasiparticles produced by the bottom heater VWR also gives us much information about the order parameter structures that they must pass through [5]. In addition, the two VWRs at the bottom are crossed at right angles with the top of each loop centered on the cell axis and separated vertically by 0.43 mm. This allows for accurate positioning of the B-phase droplet inside the BBR, since we have previously measured an anomalous damping signature when the interface sits at the top of a VWR loop [6]. Our initial measurements show that we have created a B-phase droplet. We were also able to move the droplet on and off the thermometer and heater VWRs at the bottom of the cell and make rough measurements of quasiparticle transport. At the time of writing these measurements are still too preliminary for public display.

FUTURE

Previously, we discovered that the field-induced nucleation of B phase from A phase must be a surface effect since it could be triggered by mechanical shock but not by radiation, at odds with the accepted cosmic-ray "Baked Alaska" scenario [7]. We intend to use this new cell to investigate further by working at higher pressures where the A phase is stable at zero field. We will ramp the droplet solenoids, such that upon cooling, the superfluid at the container walls always remains in the A phase, thereby excluding any possibility of surface-induced A-to-B nucleation. It is then an open question as to whether the low field region isolated within bulk A phase will *ever* undergo a transition to B phase.

REFERENCES

1. A. J. Leggett and S. K. Yip, "Nucleation and Growth of ^{3}He-B in the Supercooled A-phase," in *Helium Three*, edited by W. P. Halperin and L. P. Pitaevski, Amsterdam: Elsevier, 1990, pp. 523–571.
2. S. N. Fisher, A. M. Guénault, C. J. Kennedy and G. R. Pickett, *Phys. Rev. Lett.*, **69**, 1073 (1992).
3. M. Bartkowiak, S. W. J. Daley, S. N. Fisher, A. M. Guénault, R. P. Haley, G. R. Pickett, G. N. Plenderleith and P. Skyba, *Phys. Rev. Lett.*, **83**, 3462 (1999).
4. I. Hahn, Y. H. Tang, H. M. Bozler and C. M. Gould, *Physica*, **194-196B**, 815 (1994).
5. D. I. Bradley, S. N. Fisher, A. M. Guénault, R. P. Haley, H. Martin, G. R. Pickett, J. E. Roberts and V. Tsepelin, *The Thermal Boundary Resistance of the Superfluid ^{3}He A-B Phase Interface in the Low Temperature Limit*, this conference.
6. M. Bartkowiak, D. I. Bradley, S. N. Fisher, A. M. Guénault, R. P. Haley and G. R. Pickett, *J. Low Temp. Phys.*, **134**, 345 (2004).
7. see M. Bartkowiak, S. N. Fisher, A. M. Guénault, R. P. Haley, G. R. Pickett, G. N. Plenderleith and P. Skyba, *Phys. Rev. Lett.*, **85**, 4321 (2000), and references therein.

Studies of Superfluid ^{3}He Confined to a Regular Submicron Slab Geometry, Using SQUID NMR

Andrew Casey, Antonio Córcoles, Chris Lusher, Brian Cowan and John Saunders

Department of Physics, Royal Holloway University of London, Egham, Surrey, TW20 0EX, U.K.

Abstract. The effect on the superfluid ground state of confining p-wave superfluid ^{3}He in regular geometries of characteristic size comparable to the diameter of the Cooper pair remains relatively unexplored, in part because of the demands placed by experiments on the sensitivity of the measuring technique. In this paper we report preliminary experiments aimed at the study of ^{3}He confined to a slab geometry. The NMR response of a series of superfluid samples has been investigated using a SQUID NMR amplifier. The sensitivity of this NMR spectrometer enables samples of order 10^{17} spins, with low filling factor, to be studied with good resolution.

Keywords: Superfluid ^{3}He films, SQUID NMR, confined ^{3}He
PACS: 67.57. Bc.

INTRODUCTION

The phase diagram of superfluid ^{3}He confined to a slab of thickness comparable to the superfluid coherence length remains to be established experimentally. At zero pressure and zero temperature the coherence length ξ_0 = 72 nm. In this experiment our objective is to study a single thick film resting on a planar surface. For a diffusively scattering boundary it is predicted that T_c is suppressed to zero for film thickness $t \sim \xi_0$, while T_{AB} is suppressed to zero at $t \sim 10\xi_0$ [1]. There is evidence that suppression of T_c is eliminated if the surface is specular. Exploration of the t - T plane should uncover new superfluid phases and a rich variety of new phenomena, particularly in the two dimensional superfluid regime, $t < \xi_0$.

The most convenient "fingerprint" of the superfluid phase is its NMR response. We have developed a SQUID NMR spectrometer of sufficient sensitivity to study a single slab as thin as 100 nm [2]. Previous NMR experiments [3-5] have involved confining ^{3}He into a stack of parallel plates of average thickness 0.3. 0.8 and 1.1 μm, using applied pressure to tune ξ_0.

EXPERIMENTAL DETAILS

The SQUID NMR spectrometer is similar to that reported previously, and has a bandwidth in flux-locked loop mode of 5 MHz. The sample coil, in series with a capacitor, Q-spoiler, and flip-chip SQUID input coil forms a tuned circuit resonating at 963 kHz with a Q of 100. Tipping pulses are applied using a separate orthogonal transmitter coil. The free induction decay following a 20° tipping pulse is Fourier transformed to give the signals shown here.

The sample cell consists of two highly polished silver plates, with surface roughness ± 10 nm, separated and sealed by a Kapton/Stycast 2850 gasket. This results in an inner plate separation of around 200 μm. The ^{3}He sample is condensed into this space through a heated fill line, and annealed at a few hundred mK, in order to grow a thick van der Waals film on the inner surfaces of the cavity. We note that the film on the upper surface has an upper limit to its thickness of less than 100 nm, due to gravity. Both silver plates are clamped to a silver mounting plate, which is thermalised to the nuclear demagnetisation stage. This minimises the temperature difference between the two plates, to eliminate the thermo-mechanical effects previously experienced [2]. The ^{3}He slab (thick film) is cooled through the film-silver plate interface. There seems to be no problem in cooling these small samples into the superfluid phases across this boundary with surface area of merely 1cm^2. In the following the temperature of the film is assumed to be the same as the temperature of the mounting plate, measured by pulsed NMR on platinum powder. The static magnetic field, **B**, of 27 mT is oriented parallel to the surface normal of the silver plates.

CP850, *Low Temperature Physics: 24th International Conference on Low Temperature Physics;*
edited by Y. Takano, S. P. Hershfield, S. O. Hill, P. J. Hirschfeld, and A. M. Goldman

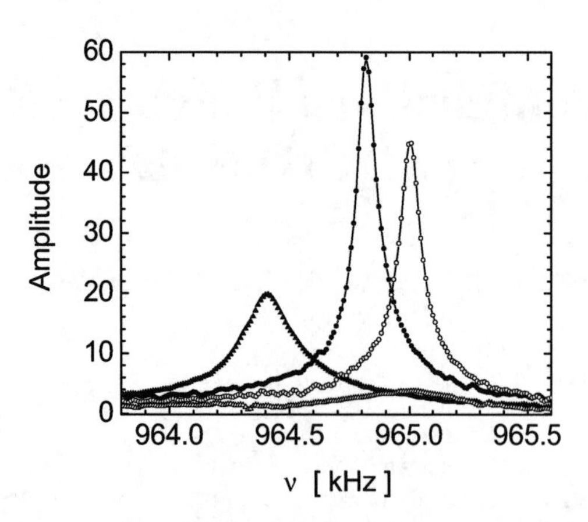

FIGURE 1. NMR signals from thin slab showing metastable A phase with negative frequency shift (filled symbols) and B phase at the Larmor frequency (open symbols). Triangles, $T = 0.38$ mk; circles, $T = 0.79$ mK.

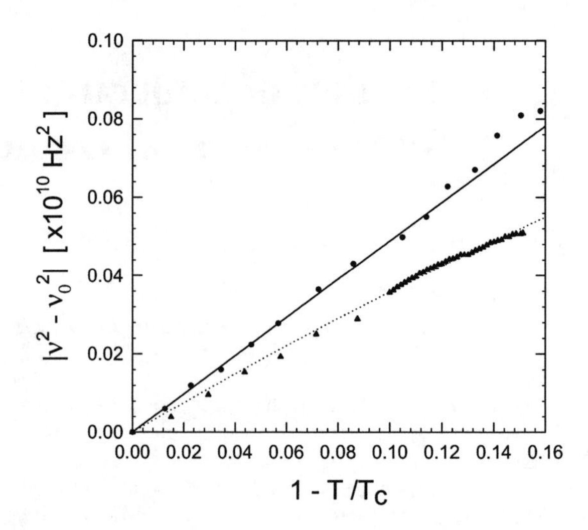

FIGURE 2. Frequency shift of bulk A phase droplets (circles), compared to results of ref [6] extrapolated to zero pressure (line). Negatively shifted signal from a 3.0 μm slab (triangles) with a guide-to-eye (dotted line).

RESULTS AND DISCUSSION

We have studied sample sizes in the range 3×10^{17} to 48×10^{17} spins, (where 3.2×10^{17} spins is equivalent to a uniform 100 nm film on each surface). In Fig. 1 we show signals from two coolings of a sample equivalent to a slab of thickness 3.0 μm on the lower surface. (We note that so far the cell has only been levelled to better than 0.1°, so that the sample is most likely wedge shaped). For such a thick film the equilibrium state is expected to be B phase, except for a very narrow temperature regime near T_c. The minimum temperatures were 0.38 mK for the first cooling and 0.75 mK for the second. In both cases we observe a strongly supercooled A-phase, with a negative frequency shift as expected for a slab with $\boldsymbol{l} \perp \boldsymbol{d}$. The magnetisation of this signal is equal to the normal state value, consistent with an equal spin pairing state. After around 20 hours in the superfluid state the line jumps to the Larmor frequency at 965 kHz, with reduced magnetisation, consistent with B-phase. On subsequent warming through T_c, no B → A transition has so far been resolved.

For samples which have been imperfectly annealed, the sample morphology probably includes bulk-like droplets. In this case the superfluid signatures depend on whether the sample has been annealed below T_c, and both A-like and B-like signals are observed. For these samples the A-like signature has a positive frequency shift. In Fig. 2 we show that the shift from an imperfectly annealed 1.8 μm sample is in reasonable agreement with previous data on the bulk A-phase shift [6], extrapolated to zero pressure. The temperature dependence of the frequency shift from the "3.0 μm slab" sample is shown for comparison.

CONCLUSIONS AND FUTURE WORK

We have successfully detected NMR signals from thin superfluid ^{3}He slabs of thickness around a few μm. The advantage of the present sample geometry is that in principle it allows continuous tuning of the film thickness. We have demonstrated sufficient sensitivity of our SQUID NMR spectrometer to allow measurements on films to thickness below 100 nm. Improvements in spectrometer sensitivity are possible and are the subject of current work. Alternative geometries in which the ^{3}He is confined within a single submicron slab of fixed, uniform spacing and where the reduced film thickness t / ξ_0 is tuned by pressure will be the subject of future work.

ACKNOWLEDGEMENTS

This work is funded by EPSRC EP/C522877/1, and is in collaboration with D. Drung and T. Schurig (PTB, Berlin) and Jeevak Parpia (Cornell University).

REFERENCES

1. See A. B. Vorontsov and J A Sauls, *Phys. Rev. B* **68**, 064508 (2003) and refs. to earlier work therein.
2. H. Dyball et al. *J. Low Temp. Phys.* **126**, 79 (2002).
3. M R Freeman and R C Richardson, *Phys. Rev. B* **41**, 11011 (1990).
4. K Kawasaki et al. *Phys. Rev. Lett.* **93**, 115301 (2004).
5. S. Miyawaki et al. *Phys. Rev. B* **62**, 5855 (2000).
6. P. Schiffer et al. *Phys. Rev. Lett.* **69**, 3096 (1992).

Studies Of Submicron ^{3}He Slabs Using A High Precision Torsional Oscillator

Antonio Corcoles*, Andrew Casey*, Jeevak Parpia¶, Roger Bowley#, Brian Cowan*, and John Saunders*

* *Department of Physics, Royal Holloway University of London, Egham, Surrey, TW20 0EX, U.K.*
¶ *LASSP, Department of Physics, Clark Hall, Cornell University, Ithaca, NY 14853, USA*
# *School of Physics and Astronomy, University of Nottingham, University Park, Nottingham, NG7 2RD, U.K.*

Abstract. A high precision torsional oscillator has been used to study ^{3}He films of thickness in the range 100 to 350 nm. In previous work we found that the films decoupled from the oscillator motion below 60 mK, in the Knudsen limit. This precluded observation of the superfluid transition. Here we report measurements using a torsional oscillator whose highly polished inner surfaces have been decorated with a low density of silver particles to act as random elastic scattering centres. This modification locks the normal film to the surface. A superfluid transition of the film is observed.

Keywords: Superfluid ^{3}He films, torsional oscillator, interfacial friction, Knudsen limit
PACS: 67.55.Fa, 47.75.Gx, 67.70.+n, 68.08.-p

INTRODUCTION

The study of flow in submicron ^{3}He films over a flat surface is expected to provide interesting insights into the problem of boundary scattering, since temperature provides a convenient method of tuning the inelastic mean free path relative to film thickness and surface roughness. Furthermore the superfluid transition of a film of thickness comparable to the superfluid coherence length is also of interest. We have developed a high precision torsional oscillator, with high frequency resolution and a very weak temperature dependent background, to study these phenomena [1].

DECOUPLING OF NORMAL ^{3}HE FILMS AND LOCKING THE FILM

In our first oscillator the ^{3}He film resided on two coin silver surfaces, mechanically polished to a measured surface roughness of approximately ± 10 nm. The films were observed to de-couple from the oscillator below around 60 mK, contrary to the naive expectation that a microscopically rough surface should lock the film [2]. Measurements of both the frequency shift and dissipation of the film could be described by a phenomenological model in which the film surface coupling is described by an interfacial friction. We found the surprising result that the momentum relaxation time between film and surface varied approximately as $1/T$.

Further analysis suggests that this result can be understood in terms of boundary scattering from a rough surface [3, 4]. For a model with surface of roughness amplitude l, with Gaussian correlations over length R, the film momentum relaxation time is given by

$$\frac{1}{\tau}=3\left(\frac{1}{\tau_\eta \tau_F}\right)^{1/2}\left(\frac{l}{R}\right)^2\frac{1}{k_F d} \tag{1}$$

where τ_η is the appropriate quasiparticle relaxation time, $\tau_F = \hbar/2E_F$, where k_F and E_F are the Fermi wavevector and Fermi energy, and d is the film thickness. This accounts for the temperature dependence of τ, since $\tau_\eta \propto 1/T^2$. We infer $R/l \sim 30$. Thus the slippage of the film is directly related to the surface profile, which can be independently characterised by modern surface probe techniques.

In order to lock the film we fabricated a new oscillator, in which the inner surfaces of the coin silver head on which the film resides were first polished, using the same procedures as previously, and

CP850, *Low Temperature Physics: 24th International Conference on Low Temperature Physics;*
edited by Y. Takano, S. P. Hershfield, S. O. Hill, P. J. Hirschfeld, and A. M. Goldman

subsequently decorated by silver particles, with a target spacing of order 10 μm. Assuming these particles act as elastic scattering centres, with temperture independent scattering time τ_{el}, we estimate that $\omega\tau_{el} \sim 3.10^{-3}$, sufficient to couple the film. The silver particles of nominal size 700 nm were deposited from an appropriately dilute suspension in ethanol. The surface was vacuum annealed at 750°C in order to bond the particles to the surface. This was followed by an ultrasonic cleaning step to dislodge weakly attached particles. Figure 1 shows a similarly treated surface.

Experiments on two ^{3}He films show that this surface treatment had the desired effect of locking the film to the surface (Fig. 2). No change in frequency shift is observed, as well as a much smaller film contribution to the dissipation.

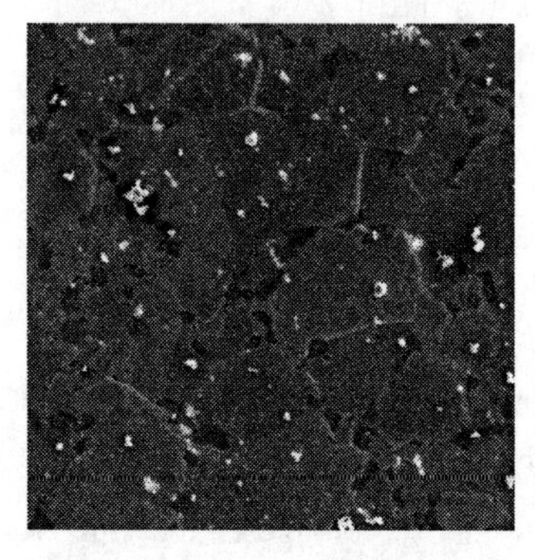

FIGURE 1. Polished silver surface, decorated by silver particle. Electron microscope picture (approx. 100 × 100 μm).

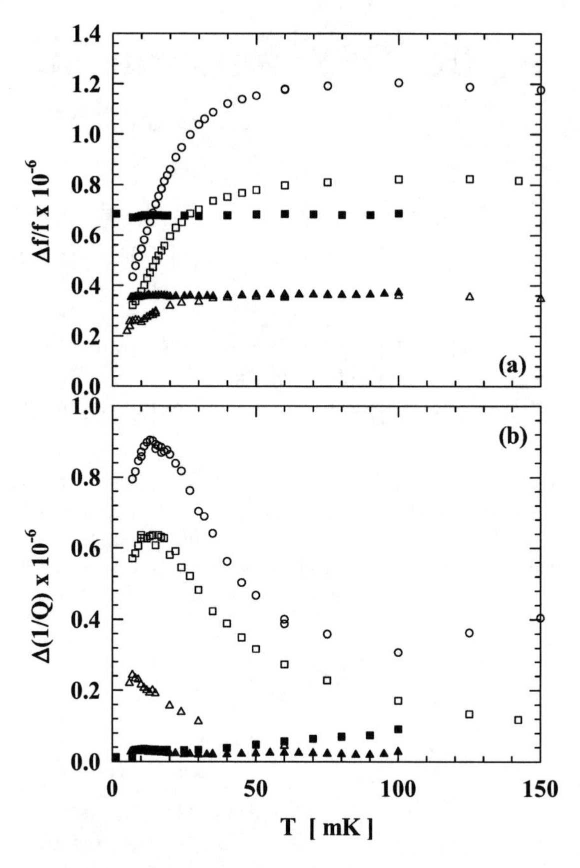

FIGURE 2. (a) Frequency shift due to film, and (b) film contribution to dissipation for two film thicknesses on the silver particle decorated surface (solid symbols), corresponding to frequency shifts of 0.9 and 1.8 mHz. In comparison, open symbols show previous data [2], obtained with a plain polished surface, for frequency shifts 0.9, 2.4, 3.5 mHz. Nominal film thickness is 100 nm/mHz for both oscillators.

SUPERFLUID TRANSITION IN FILM

With surfaces similar to that shown in Fig.1, we are able to observe a superfluid transition in the film, with superfluid transition temperature suppressed below bulk T_c, as expected for diffuse quasiparticle boundary scattering. Preliminary indications are that the superfluid density is similar to that measured in comparable films by third sound [5], but significantly greater than that found in earlier torsional oscillator measurements [6]. Further studies of the temperature dependence of the superfluid fraction, its suppression relative to bulk values and the suppression of T_c as a function of film thickness are underway.

ACKNOWLEDGMENTS

This work was supported by EPSRC grant GR/S20567/01.

REFERENCES

1. G. W. Morley, A. Casey, C.P. Lusher, B. Cowan, J. Saunders, and J. M. Parpia, *J. Low Temp. Phys.* **126**, 557-562 (2002).
2. A. Casey, J. Parpia, R. Schanen, B. Cowan and J. Saunders, *Phys. Rev. Lett.* **92**, 255301-4 (2004).
3. A. E. Meyerovich and A Stepaniants, *J. Phys.: Condens. Matter.* **12**, 5575-5597 (2000).
4. R. Bowley, unpublished.
5. A. M. R. Schecter, R. W. Simmonds, R. E. Packard and J. C. Davis, *Nature* **396**, 554-557 (1998).
6. J. Xu and B. C. Crooker, *Phys. Rev. Lett.* **65**, 3005-3008 (1990).

Heat Transfer Properties of Liquid ^{3}He below 1K

M. Katagiri*, M. Maeda†, K. Shinn*, T. Tsurutani*, Y. Fujii* and K. Hatanaka*

*Department of Applied Physics, Okayama University of Science, 1-1 Ridai-cho, Okayama 700-0005, Japan
†Tsukuba Magnet Laboratory, NIMS, 3-13 Sakura, Tsukuba 305-0003, Japan

Abstract. The characteristics of heat transfer from a flat copper surface to liquid ^{3}He have been studied under saturated vapor pressure using two types of sample cells in which the flows of liquid ^{3}He induced by convection were different from each other. The temperature difference between the heated copper surface and liquid ^{3}He was measured as a function of heat flux in steady states in the regions of the nonboiling state and nucleate boiling state. The characteristic curves are analyzed in terms of the convective flow along the heated surface and the bubble nucleation rates.

Keywords: Heat transfer, Liquid ^{3}He, Nucleate boiling state, Nonboiling state.
PACS: 67.55.Cx

The heat transfer characteristics of liquid ^{3}He have been studied previously from the nonboiling state to the film boiling one below 1 K by our group [1]. The temperature difference ΔT caused by the thin thermal boundary layer existing between a flat copper surface and bulk liquid ^{3}He was measured as a function of heat flux $\dot{q}$ in steady states. In the nucleate boiling state, the data could not be explained by Kutateladze's correlation model [2], and the measured ΔT was smaller by more than an order of magnitude than the value expected by the model. In order to further investigate the nucleate boiling state, we have observed the bubbles by a visualization method [3] using two types of sample cells, while the $\dot{q}$ -ΔT relations were measured. In this paper, we describe the effects of fluid flow induced by the convection on the heat transfer characteristics.

Figure 1a shows the sample cell I. Liquid ^{3}He stored in the cell was evaporated to maintain a steady state. The heater unit placed at the bottom of the cell comprised a copper block and a heater wire. The upper surface of the block, whose diameter was 7 mm, was the heat transfer surface polished with paper of a 0.3 μm roughness. To measure the surface temperature T_W, a germanium thermometer calibrated down to 0.3 K was embedded in the block. In the liquid ^{3}He, three thin carbon thermometers were placed as shown. The sides of the copper block were covered by epoxy resin Stycast 1266 to avoid bubble formation on them. Basically, this arrangement was the same as the cell used for the previous heat transfer experiment [1] except for the Pyrex glass tube used to pass the light. Figure 1b shows sample cell II, which was made to observe bubble generation on the entire copper surface. This cell had two flat fused-quartz windows. The surface of the copper block was made the same as in cell I, and a thin film strain gauge was glued onto the copper block as a heater. A calibrated Cernox thermometer was embedded in the block to measure T_W, and a RuO_2 thermometer was placed in liquid ^{3}He. In this cell, there was a free space filled with liquid ^{3}He between the block and the windows.

The measurements of the heat transfer characteristics have been performed with the following procedure. After a steady state was established for a given heat flux $\dot{q}$, all the temperatures were measured. The temperature of liquid ^{3}He, T_S, was measured with the carbon thermometer closest to the copper surface. The temperature gradient in the bulk liquid ^{3}He was measured with the three carbon thermometers, and the temperature just above the thermal boundary layer, T_L, was estimated. The temperature difference between the copper surface and liquid ^{3}He, ΔT, was determined from T_L and T_W by subtracting the temperature difference caused the Kapitza thermal resistance [1]. Since the temperature difference in the bulk liquid was negligible in comparison with ΔT, only one thermometer was placed in cell II and its reading was taken to be T_L. This allowed an unobstructred view of the copper surface.

Figure 2a and b show the $\dot{q}$ -ΔT relations at 0.7 K for cells I and II, respectively. The relation for cell I almost coincides with the one obtained previously [1]

CP850, *Low Temperature Physics: 24th International Conference on Low Temperature Physics;* edited by Y. Takano, S. P. Hershfield, S. O. Hill, P. J. Hirschfeld, and A. M. Goldman

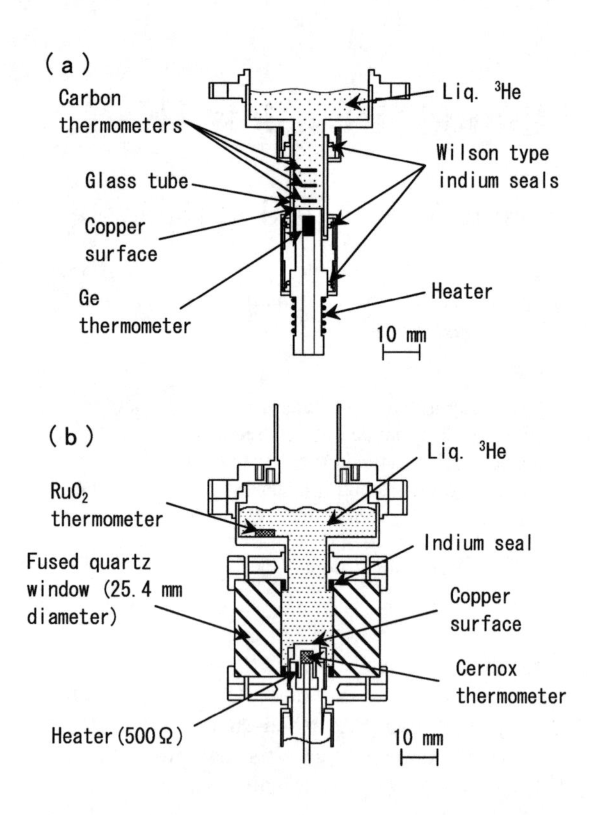

FIGURE 1. Experimental setup. (a) cell I, (b) cell II.

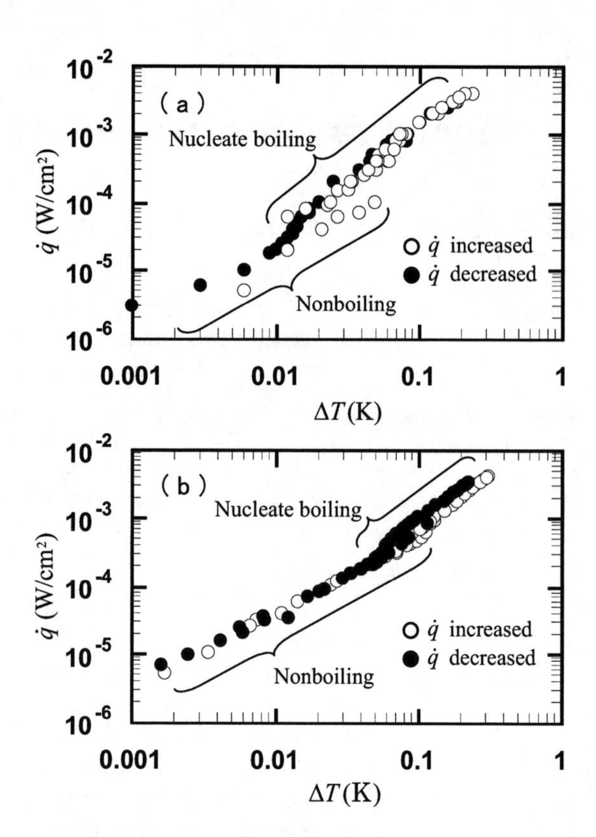

FIGURE 2. $\dot{q}$ -ΔT curves at 0.7 K. (a) cell I, (b) cell II.

for a same type of cell. The measurements were made with both increasing $\dot{q}$ and decreasing $\dot{q}$. While $\dot{q}$ was increased from zero, the nonboiling state was studied. And then, the state changed to the nucleate boiling state with a discontinuous decrease of ΔT. In the nucleate boiling state, the slope of the $\dot{q}$ -ΔT curve was initially steep, and then became gentle. In these experiments, the film boiling state was not reached because of the cooling power limit of the cryostat.

In the nonboiling state, ΔT in cell II was smaller than that in cell I for a given $\dot{q}$. On the other hand, the transition from the nonboiling state to the nucleate boiling state occurred a smaller value of $\dot{q}$ in cell I, and ΔT in the nucleate boiling state was smaller in this cell for a given $\dot{q}$. These features can be explained by the flow of bulk liquid ^{3}He. In cell II, there was a stronger flow of liquid induced by the convection along the heated copper surface because there was a free space between the copper block and the windows. This situation decreased ΔT in the nonboiling state and delayed the transition to the nucleate boiling state. The $\dot{q}$ -ΔT curves indicate that the agitation of the thermal boundary layer by bubbles nucleated at the heated surface was stronger in cell I, where permitted less flow along the heated surface. At large $\dot{q}$, the curves for the two cells approached each other, because the effect of fluid flow became small due to strong agitation by the bubbles, whose nucleation rate increased.

These results indicate that, to cool a hot plate effectively by the use of liquid ^{3}He, there should be no free space between the hot plate and the wall.

REFERENCES

1. M. Maeda, A. Beppu, Y. Fujii and T. Shigi, *Cryogenics* **40**, 713-719 (2000).
2. S. S. Kutateladze, *Heat Transfer in Condensation and Boiling*, Moscow: State Scientific and Technical Publishers of Literature on Machinery, 1952, pp. 114-133.
3. M. Katagiri, M. Maeda and Y. Fujii, *Physica B* **329-333**, 120-121 (2003).

Theory of Josephson Phenomena in Superfluid ^{3}He

Erkki Thuneberg

Department of Physical Sciences, University of Oulu, Finland

Abstract. Quite detailed theoretical description of superfluid ^{3}He is possible on length scales that are much larger than the atomic scale. We discuss weak links between two bulk states of ^{3}He-B. The current through the weak link is determined by the bound states at the link. The bound state energies are spin split, depending on the order parameters in the bulk. As a result, unusual current-phase relations with π states appear. Increasing Josephson coupling modifies the order parameter in the bulk. This leads to a stronger π state and to an additional current at constant pressure bias. The theoretical results are compared with experiments.

Keywords: superfluid ^{3}He, Josephson effect, current-phase relation, π state, Andreev bound state.
PACS: 67.57.De

INTRODUCTION

In spite of more than thirty years of studies of superfluid phases of ^{3}He, there are several active research directions. One of them is the study of Josephson phenomena in weak links between two volumes of bulk superfluid. Here we give a review of recent theoretical work on this subject [1-4]. We pay special attention to the role of bound quasiparticle states in the link, and explain how these determine the flow through the junction. In particular we study the equilibrium current-phase relation, $I(\phi)$, and the dc current at constant potential difference, $I_{\rm dc}(U)$. We compare the results with experiments. The experimental work has been reviewed in Ref. [5]. A theoretical review with a somewhat different emphasis has been given in Ref. [6].

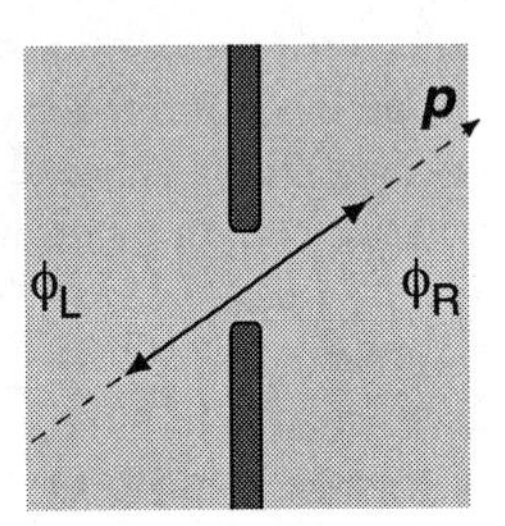

FIGURE 1. Sketch of a weak link between two bulk superfluids. The two bulk superfluids have phases ϕ_L and ϕ_R. One quasiparticle trajectory with momentum **p** is shown. When its energy is within the energy gap, the quasiparticle cannot escape to the bulk but is Andreev reflected as hole with essentially the same momentum. This in turn is Andreev reflected as particle on the other side of the link. Thus particle and hole type excitations bounce back and forth along the same trajectory.

WEAK LINKS

Let us consider two volumes of superfluid that are connected by a weak link. Here "superfluid" means generally a fermion superfluid, either superconducting metal or liquid ^{3}He. Our discussion of superconducting weak links is mainly aimed as an introduction for ^{3}He. Therefore we neglect all complications that arise from impurity or interfacial scattering in superconductors. Also we consider only superconductors of the conventional (s-wave) type. In the case of ^{3}He we limit to the superfluid B phase.

The geometry of the weak link is depicted in Fig. 1. The figure shows one quasiparticle trajectory with momentum **p**. In bulk superfluid such elementary excitations must have energy ε larger than the energy gap Δ of the superfluid. In the weak link, however, there exists energy eigenvalues within the energy gap, $|\varepsilon| < \Delta$. In such a state the quasiparticle cannot escape to the bulk but is Andreev reflected. In Andreev reflection a particle type excitation is converted to hole type and vice versa, with essentially unchanged momentum [7]. Thus the particles and holes travel the same trajectory back and forth, respectively. In one cycle, one Cooper pair, or equivalently two particles, are transmitted through the weak link.

The bulk superfluid states are characterized by phases ϕ_L and ϕ_R. The energies of the bound states depend essentially on the phase difference $\phi = \phi_R - \phi_L$ [8]. For superconductors this is depicted in Fig. 2. The states appear as pairs with positive and negative energies, $\pm|\varepsilon|$. Understanding that $d\varepsilon/d\phi$ plays a similar role as the group velocity, we can assign the states as propagating to the right ($\delta = +1$) and to the left ($\delta = -1$). These two states are not independent, however. Based on what was stated about Andreev reflection above, if a state with $\delta = +1$ is occupied, the corresponding $\delta = -1$ state must be empty. And vice versa. More generally, the occupations $f_\pm$ have to satisfy $f_+ + f_- = 1$. Really the states with

CP850, *Low Temperature Physics: 24th International Conference on Low Temperature Physics;*
edited by Y. Takano, S. P. Hershfield, S. O. Hill, P. J. Hirschfeld, and A. M. Goldman

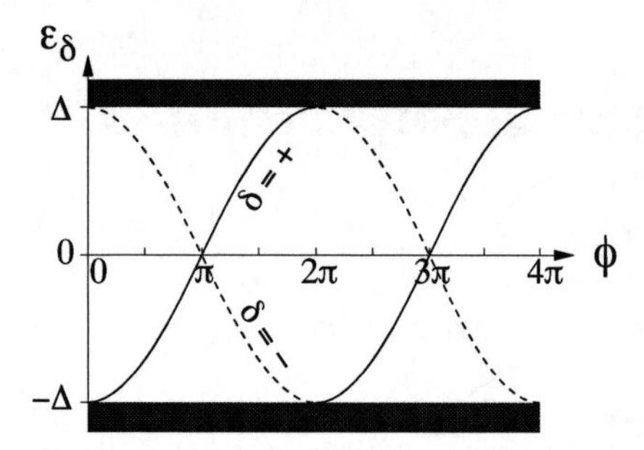

FIGURE 2. Bound state energies $\varepsilon_\delta(\phi)$ in a superconducting point contact. As ϕ increases, the states with transport to the right ($\delta = +1$) have increasing energy (solid lines), and the states with transport to the left ($\delta = -1$) have decreasing energy (dashed lines). The states at energies $|\varepsilon| > \Delta$ form a continuum.

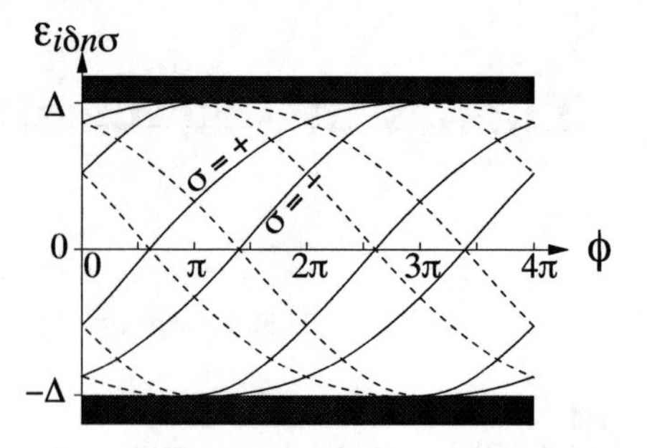

FIGURE 3. Example of bound state energies $\varepsilon_{i\delta n\sigma}(\phi)$ in a ^{3}He weak link. Compared to Figure 2, the bound state energies have smaller slopes and they are spin split ($\sigma = \pm 1$). Figure reprinted from Ref. [3].

positive and negative energies are the same bound state, whose energy relative to the ground state is the absolute value $|\varepsilon|$. Anyhow, we find it very useful to consider a bound state as a superposition of positive and negative energy states.

In the simplest case of a point contact, the bound state energies have the simple analytic expression $\varepsilon = \pm\Delta\cos\frac{\phi}{2}$ [8]. By a point contact, or a pinhole in case of ^{3}He, we mean a weak link whose dimensions are all small compared to the superfluid coherence length ξ_0.

It is an interesting observation that to a large extent the current through a weak link is determined by the bound states [9-12]. In the following we are interested in equilibrium or only small nonequilibrium, where the chemical potential difference $\mu_L - \mu_R \equiv U \ll \Delta \sim k_B T_c$. In this case the current has the form

$$I = \frac{2}{\hbar}\sum_i\sum_\delta \frac{d\varepsilon_{i\delta}}{d\phi} f_{i\delta}. \qquad (1)$$

Here $f_{i\delta}$ is the occupation of the state. In equilibrium it reduces to the Fermi distribution $f_{i\delta} = [\exp(\varepsilon_{i\delta}/k_B T) + 1]^{-1}$. The index i indicates different channels, which correspond to different directions and locations of the bound state (Fig. 1). In a point contact of area A, there are $M = k_F^2 A/4\pi$ channels, where k_F is the Fermi wave vector. In the normal state the current is $I = MU/\pi\hbar$ [13].

Kulik and Omel'yanchuk derived in 1975 the following formula for equilibrium current through a superconducting point contact [14]

$$I = \frac{M\Delta}{\hbar}\sin\frac{\phi}{2}\tanh\frac{\Delta\cos\frac{\phi}{2}}{2k_B T}. \qquad (2)$$

Their original derivation does not give much clue to understand the result. Now this result can be straightforwardly understood based on the formulas given above: all channels have the same bound state energies $\varepsilon = \pm\Delta\cos\frac{\phi}{2}$ and using the current formula (1) and the Fermi distribution gives the result (2).

As the temperature increases towards the superfluid transition temperature T_c, the Kulik-Omel'yanchuk result (2) reduces to sinusoidal form

$$I = I_c \sin\phi \qquad (3)$$

with critical current $I_c = M\Delta^2/4\hbar k_B T_c$.

^{3}HE WEAK LINKS

Most of the discussion of superconducting weak links applies also to weak links in ^{3}He. There are a few differences. In ^{3}He the Cooper pairs are in p-wave states, instead of the s-wave state considered above. Any scattering breaks these pairs. Thus the superfluid state is always suppressed near walls. This suppression also affects the bound states energies in a weak link, as depicted in Fig. 3. Compared to the superconducting case, the slopes $d\varepsilon/d\phi$ are smaller in magnitude. Another difference is that there is spin splitting of the energy states ($\sigma = \pm 1$). Whereas the states in the superconducting case were doubly degenerate, this degeneracy is generally broken in ^{3}He.

Because of the spin splitting the current formula (1) has to be generalized to the form

$$I = \frac{1}{\hbar}\sum_i\sum_\delta\sum_n\sum_\sigma \frac{d\varepsilon_{i\delta n\sigma}}{d\phi} f_{i\delta n\sigma}, \qquad (4)$$

where a factor of 2 is replaced by spin summation. There is also additional summation over index n, which takes into account that several bound states can occur at a given ϕ.

The spin splitting can lead to current-phase relationships that differ essentially from the standard sinusoidal form (3) [15]. Consider, for example, the case where the spin split states are shifted relative to each other by a

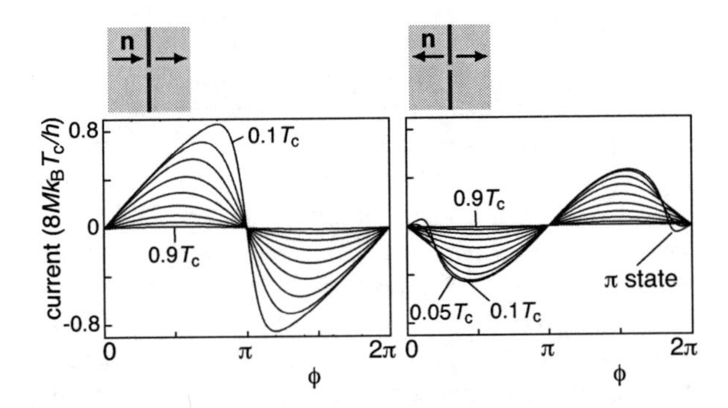

FIGURE 4. Current-phase relationships calculated for a pinhole. The left hand panel is for parallel $\hat{\mathbf{n}}$ vectors on the two sides of the junction. The right hand panel is for antiparallel $\hat{\mathbf{n}}$'s that are perpendicular to the wall. At high temperature the curves are sinusoidal (3). At lower temperature the curves become skew in the parallel case resembling the Kulik-Omel'yanchuk result (2). The antiparallel case has negative critical current I_c, or equivalently, is shifted by a phase difference π. At very low temperature it develops an additional kink that is known as π state. Figure adapted from Ref. [2].

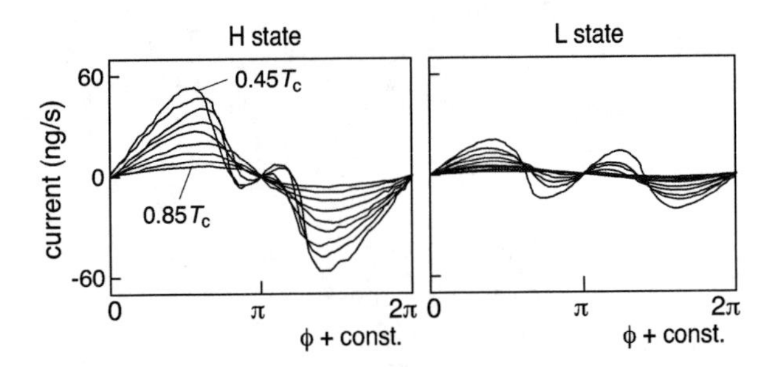

FIGURE 5. Measured current-phase relationships for a 65 × 65 array of 100 nm diameter apertures in 50 nm thick wall [17]. In cooling through T_c, the system randomly freezes to either H or L state. Only the relative value of the phase is determined experimentally. Figure adapted from Ref. [17].

phase difference $\approx \pi$. It follows that the leading sinusoidal components of the current from the two spin states [$\propto \sin(\phi+\psi_\sigma)$] cancel each other. What is then left are higher harmonics [$\propto \sin(n\phi+\psi_\sigma)$, $n=3,\ldots$]. Unusual current-phase relations are confirmed by calculations below.

The Cooper pairs of superfluid ^{3}He have p-wave form [16]. There are three orthogonal p-wave states: p_x, p_y, and p_z. The spin state is triplet. There are three triplet states, which conventionally are chosen as $-\uparrow\uparrow+\downarrow\downarrow$, $i\uparrow\uparrow+i\downarrow\downarrow$, and $-\uparrow\downarrow+\downarrow\uparrow$. Here we concentrate on the B phase, where p_x pairs have the first spin state, p_y pairs the second and p_z pairs the third. However, the spin coordinate axes (x',y',z') are rotated relative to the orbital axes (x,y,z). The rotation angle is fixed to 104°, but the axis $\hat{\mathbf{n}}$ of this rotation can vary.

In hydrodynamics of the B phase the mass (ϕ) and spin ($\hat{\mathbf{n}}$) degrees of freedom are independent. The weak link acts as a nonlinear element that couples mass and spin, as will be demonstrated below.

Isotextural Theory

Some current-phase relations calculated for a ^{3}He-B pinhole are shown in Fig. 4. Two cases are shown. First, the spin rotation axes $\hat{\mathbf{n}}$ on both sides of the junction are the same, and second, the rotation axes have opposite directions perpendicular to the wall of the pinhole. The former case resembles the Kulik-Omel'yanchuk result (2): at high temperatures $I(\phi)$ is sinusoidal but at lower temperatures it becomes skew towards $\phi=\pi$. The case with antiparallel $\hat{\mathbf{n}}$'s is also sinusoidal at high temperatures, but has negative I_c. (Such a case is called a π *junction* in superconductivity.) At temperatures around $0.1T_c$ it develops to a π *state*. This means that the slope of $I(\phi)$ is positive at both 0 and π.

Prior to theoretical calculations, the π state had been observed experimentally. The experimental results of Ref. [17] are shown in Fig. 5. The experimental $I(\phi)$ is sinusoidal at high temperature but develops a π state at low temperature. Moreover, two different metastable states with different sets of $I(\phi,T)$ curves was observed. These are called H and L, corresponding to high and low critical current. The two states appeared randomly in cooling through T_c, but the nucleated state remained stable in the superfluid state.

Let us consider a weak link under a constant chemical potential difference $U=\mu_L-\mu_R$. According to the Josephson relation

$$\frac{d\phi}{dt}=\frac{2U}{\hbar} \tag{5}$$

the phase ϕ increases linearly in time. This implies that the bound state energies (Fig. 3) are continuously shifting up or down. During shifting the occupations of these states can change in collisions with other quasiparticles. Since the collisions are rare, this thermalizes the occupations only at small $U<\hbar/\tau$, where τ is the scattering time. At larger bias, the occupations of the bound states remain essentially constant during the shift from $-\Delta$ to $+\Delta$, or from $+\Delta$ to $-\Delta$. The occupations of these states are then fixed to the thermal equilibrium occupations at the starting energies, $-\Delta$ or $+\Delta$.

It is now possible to calculate the current using Eq. (4). The resulting time-averaged current $I_{\rm dc}$ is plotted in Fig. 6. At large bias ($U\gg\hbar/\tau$) the occupations are determined by gap edges and the current therefore becomes independent of the bias U. At small bias ($U\ll\hbar/\tau$) the scattering has time to preserve a nearly thermal distribution, and the deviation from equilibrium distribution as well as the current is linear in U. Note that the equilib-

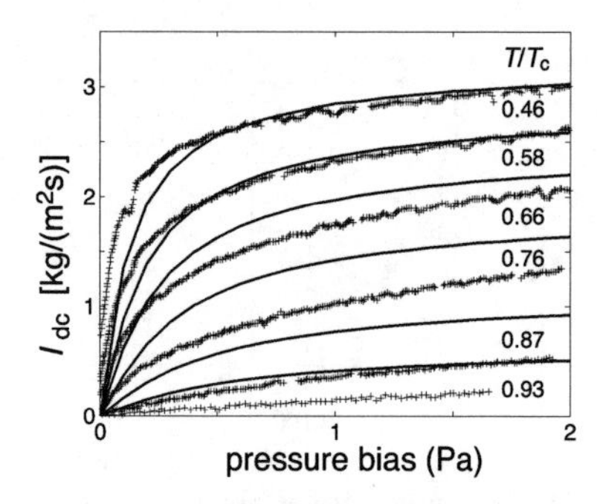

FIGURE 6. Average current vs. pressure bias. The data points are experimental results from Ref. [18], and the lines are theoretical results from Ref. [4]. The pressure bias of 1 Pa corresponds to $U = 4.7 \times 10^{-3} k_B T_c$. Figure reprinted from Ref. [4].

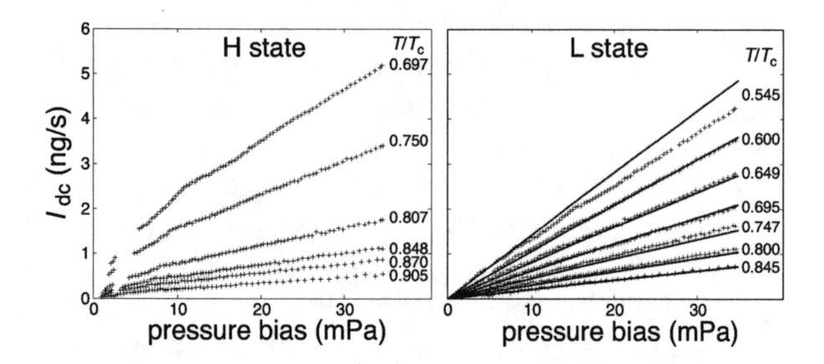

FIGURE 7. Average current vs. pressure at low pressure bias. The data points are experimental results from Ref. [19] for the same 65×65 array of apertures as the equilibrium currents in Fig. 5. The lines drawn on the right hand side panel are the results of isotextural theory, and fitted with a single parameter (the relaxation time) as in Fig. 6. According to isotextural theory, the same lines should fit also the H state, which clearly is not the case. Figure adapted from Ref. [3].

rium current (Fig. 4) is oscillating, and therefore does not contribute to the time-averaged current.

Fig. 6 shows also experimental data by Steinhauer et al. [18] The theory can reasonably be fitted to this data using the scattering time as the only fitting parameter. The agreement is surprisingly good taking into account that the single aperture in the experiment (dimensions 7.8 μm $\times$ 0.27 μm) is large compared to the superfluid coherence length $\xi_0 = 77$ nm, whereas the pinhole theory makes just the opposite assumption.

The current at finite bias was also measured for the same array of apertures as the equilibrium current in Fig. 5. The results are shown in Fig. 7. The current is different for the H and L states. The current in the L state is in reasonable agreement with theory using again the relaxation time as the only adjustable parameter.

As a summary thus far, we can say that theory and experiment are in reasonable agreement. Both show bistability (H and L state), π states, and similar dissipative currents $I_{\rm dc}(U)$. The temperature dependences agree as well as the order of magnitudes of all quantities.

In spite of this success, closer examination reveals some problems in the theory. 1) Only one of theoretical bistable states shows π state (Fig. 4). More detailed analysis reveals that this problem cannot be removed by considering other configurations of $\hat{\mathbf{n}}$'s [2]. 2) The theoretical π state occurs at a much lower temperature and is much weaker than in experiments — compare Figs. 4 and 5. 3) The theory does not predict any difference in $I_{\rm dc}(U)$ between the H and L states. This is because different $\hat{\mathbf{n}}$ configurations mainly shift the bound states energies $\varepsilon_{i\delta n\sigma}$ along the ϕ axis but do not much affect the shape of the $\varepsilon_{i\delta n\sigma}(\phi)$ curves. In the case that ϕ changes at constant rate, this only affects the instantaneous current $I(t)$, but the average $I_{\rm dc}$ is unchanged.

Anisotextural Theory

All the problems above can be explained by a single new concept. If the Josephson coupling is strong, it can change the spin-part of the order parameter on both sides. We call this *anisotextural* effect, since the $\hat{\mathbf{n}}$ orientations are commonly known as texture.

As an example consider the case of parallel $\hat{\mathbf{n}}$'s. Then there is no spin splitting and the bound state at phase difference $\phi = \pi$ lies at zero energy, $\varepsilon = 0$. The thermal occupation of this doubly degenerate state is $f = 1/2$. If we now allow spin splitting, the $\varepsilon = 0$ state can split into two with positive and negative energies (Fig. 3). Taking into account that the occupation of the negative energy state is larger than that of the positive energy state, this leads to a reduction of energy. Thus the spin splitting can take place spontaneously.

The anisotextural effect as a function of phase is shown in Fig. 8. Starting from the completely spin-symmetric situation at $\phi = 0$, the $\hat{\mathbf{n}}$ texture changes spontaneously as ϕ increases towards π. The reduction of the Josephson coupling energy $F_{\rm J}$ is associated with a change in $I(\phi)$ since

$$I = \frac{2}{\hbar}\frac{\partial F_{\rm J}}{\partial \phi}. \tag{6}$$

Thus a π state develops if the reduction of $F_{\rm J}$ is sufficient to produce a local minimum of $F_{\rm J}(\phi)$ at $\phi = \pi$.

An additional contribution to energy arises from the bending of $\hat{\mathbf{n}}$ (right hand panel of Fig. 8). In a simple model the total energy is written as

$$F[\eta] = F_J(\eta_0, \phi) + \tfrac{1}{2} K \int d^3 r |\nabla \eta|^2. \tag{7}$$

The first term is the Josephson coupling energy that depends on ϕ and on the tilting angle η_0 of $\hat{\mathbf{n}}$ at the weak link. The second term is a gradient energy for the tilting angle in the bulk. In comparison to experiments, the only free parameter is the tilting angle η_∞ on one side far away

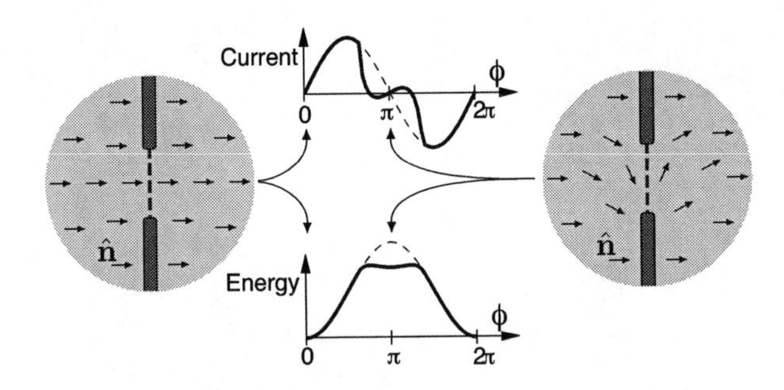

FIGURE 8. Principle of the anisotextural Josephson effect. The panels on the left and right hand sides depict an array of apertures and the configurations of the spin-rotation axis $\hat{\mathbf{n}}$. Assuming there is no spin structure ($\hat{\mathbf{n}} =$ constant) at the phase difference $\phi = 0$, the $\hat{\mathbf{n}}$ texture changes spontaneously when ϕ increases to π. This leads to a reduction of energy and to a positive slope of $I(\phi)$ at $\phi = \pi$, as shown by the change from dashed lines to solid lines in the current and energy plots.

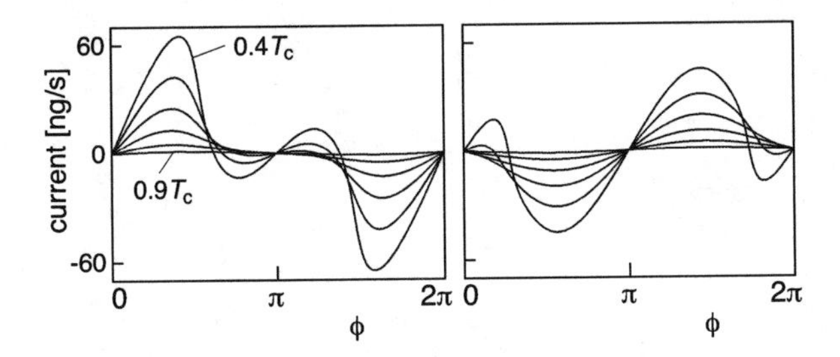

FIGURE 9. Current-phase relationships calculated using the anisotextural model (7). The theory contains one free parameter η_∞, and the result should be compared with the experimental one in Fig. 5. Figure adapted from Ref. [2].

from the junction, which is difficult to calculate because of the complicated shape of the experimental cell. Using the freedom to adjust η_∞, it is possible to generate the current-phase relations shown in Fig. 9. This should be compared with the experimental curves in Fig. 5.

Next we consider the anisotextural effect at constant bias U. Finite U means that the phase changes at a constant rate [Eq. (5)]. As a result of the anisotextural effect, the $\hat{\mathbf{n}}$ texture oscillates at the angular frequency $\omega = 2U/\hbar$. This oscillation of the spin structure generates spin waves, which radiate out of the junction.

In order to make an quantitative theory, we need to add one more contribution to the energy (7) so that the total energy is

$$F[\eta, S] = F_J(\eta_0, \phi) + \tfrac{1}{2}K \int d^3r |\nabla\eta|^2 + \tfrac{\gamma^2}{2\chi} \int d^3r S^2. \quad (8)$$

The last therm is the energy associated with net spin density S, where γ is the gyromagnetic ratio and χ the magnetic susceptibility [20]. Here S and η are conjugate variables. Writing the Hamiltonian equations of motion, this leads to a wave equation describing the spin waves and to a boundary condition for them. It turns out to be a quite standard radiation problem. The radiated power has the frequency dependence

$$P_{\rm rad} \propto \frac{\omega^2}{1 + (r_0\omega/c)^2}, \quad (9)$$

where r_0 is the radius of the Josephson array, and c the spin wave velocity. This expresses that there is little radiation for wave lengths longer than the source, but more at shorter wave lengths.

The dissipated power now has to come from the dc current, $P = UI_{\rm dc}$. Thus there is additional dc current due to the spin wave radiation. Including also the prefactors we get for it the expression

$$I_{\rm dc,rad} = \frac{P_{\rm rad}}{U} = \frac{2}{\hbar}\frac{[J_{\rm sp}(\eta_\infty)]^2}{4\pi cK}\frac{\omega}{1 + (\omega r_0/c)^2}. \quad (10)$$

Here the equilibrium spin current $J_{\rm sp}(\eta_\infty)$ plays a role of a coupling constant between mass and spin variables. Most interestingly, this coupling is different for the theoretical H and L states. Putting in numbers we see that the theory explains approximately one third of the observed H-L difference shown in Fig. 5. (The comparison is shown in Ref. [3].) This is not completely satisfactory, but we must remember that all parameters of the theory were fixed before the comparison.

Models to explain the observed $I_{\rm dc}(U)$ have also been suggested in Ref. [19]. These models differ essentially from the present ones. In particular, an A-phase-like distortion of the order parameter is associated with the linear part and a quasiparticle radiation mechanism with the nonlinear part of $I_{\rm dc}(U)$.

CONCLUSION

We have seen that coupling between mass and spin variables is essential for understanding the Josephson phenomena in superfluid ^{3}He. With spin splitting of bound states together with the anisotextural effect it is possible to explain most of the experimental results of both $I(\phi)$ and $I_{\rm dc}(U)$. The theory presented above uses the pinhole approximation. For more accurate results it is necessary to consider larger apertures. This would require selfconsistent determination of the order parameter within the aperture, which is calculationally demanding and has been done only in limiting cases [21, 22]. We expect that this might lead to qualitatively same type of results as the anisotextural effect for pinhole arrays, in particular, to enhanced π states and to a texture dependence of $I_{\rm dc}(U)$. This might explain the remaining differences between theory and experiment, and the experimental results in single apertures [23].

The anisotextural phenomena depend strongly on many parameters such as array size, geometry of the experimental cell and magnetic field. We hope that these could be tested in future experiments.

REFERENCES

1. J. K. Viljas and E. V. Thuneberg, *Phys. Rev. Lett.* **83**, 3868 (1999).
2. J. K. Viljas and E. V. Thuneberg, *Phys. Rev. B* **65**, 064530 (2002).
3. J. K. Viljas and E. V. Thuneberg, *Phys. Rev. Lett.* **93**, 205301 (2004).
4. J. K. Viljas, *Phys. Rev. B* **71**, 064509 (2005).
5. J. C. Davis and R. E. Packard, *Rev. Mod. Phys.* **74**, 741 (2002).
6. J. K. Viljas and E. V. Thuneberg, *J. Low Temp. Phys.* **136**, 329 (2004).
7. A.F. Andreev, *J. Eksp. Teor. Fiz.* **49**, 655 (1965) [*Sov. Phys. JETP* **22**, 455 (1966)].
8. I. O. Kulik, *Zh. Eksp. Teor. Fiz.* **57**, 1745 (1969) [*Sov. Phys. JETP* **30**, 944 (1970)].
9. C. Ishii, *Prog. Theor. Phys.* **44**, 1525 (1970).
10. A. Furusaki and M. Tsukada, *Physica B* **165&166**, 967 (1990).
11. C. W. J. Beenakker, *Phys. Rev. Lett.* **67**, 3836 (1991).
12. D. Averin and A. Bardas, *Phys. Rev. B* **53**, R1705 (1996).
13. S. Datta, *Electronic Transport in Mesoscopic Systems* (Cambridge, 1995).
14. I. O. Kulik and A. N. Omel'yanchuk, *Fiz. Nizk. Temp.* **3**, 945 (1977) [*Sov. J. Low Temp. Phys.* **3**, 459 (1977)].
15. S.-K. Yip, *Phys. Rev. Lett.* **83**, 3864 (1999).
16. A. J. Leggett, *Rev. Mod. Phys.* **47**, 331 (1975).
17. A. Marchenkov, R. W. Simmonds, S. Backhaus, K. Shokhirev, A. Loshak, J. C. Davis, and R. E. Packard, *Phys. Rev. Lett.* **83**, 3860 (1999).
18. J. Steinhauer, K. Schwab, Yu. M. Mukharsky, J. C. Davis, and R. E. Packard, *Physica B* **194-196**, 767 (1994).
19. R. W. Simmonds, A. Marchenkov, S. Vitale, J. C. Davis, and R. E. Packard, *Phys. Rev. Lett.* **84**, 6062 (2000).
20. A. J. Leggett, *Ann. Phys. (New York)* **85**, 11 (1974).
21. E. V. Thuneberg, J. Kurkijärvi, and J. A. Sauls, *Physica B* **165&166**, 755 (1990).
22. J. K. Viljas and E. V. Thuneberg, *J. Low Temp. Phys.* **129**, 423 (2002).
23. O. Avenel, Yu. Mukharsky, and E. Varoquaux, *Physica B* **280**, 130 (2000).

Superfluidity of Dense Neutron Matter with Spin-Triplet P-Wave Pairing in Strong Magnetic Field

Alexander N. Tarasov

Institute for Theoretical Physics, NSC Kharkov Institute of Physics and Technology, 61108, Kharkov, Ukraine

Abstract. A dense superfluid pure neutron matter with an effective Skyrme interaction, which depends on the density n of the neutrons with spin-triplet p-wave pairing similar to those of ^{3}He-A$_1$ and ^{3}He-A, is studied in a strong uniform static magnetic field H in the framework of a generalized non-relativistic Fermi-liquid theory. We present an analytic solution for a set of previously derived nonlinear integral equations for the components of the order parameter and an effective magnetic field H_{eff}. The obtained general formulas (valid for arbitrary parameterization of Skyrme forces) for the phase transition temperatures $T_{c1,2}$ of the neutron matter (from normal to a superfluid state of ^{3}He-A$_1$ type and then to a ^{3}He-A type state, respectively) and the expression for H_{eff} at T=0 are linear functions of strong magnetic field H and nonlinear functions of n. The gap equation is also solved at T=0.

Keywords: Superfluidity, Fermi liquid, Neutron matter, Skyrme forces
PACS: 05.30.-d; 67.57.-z; 97.60.Jd; 26.60.+c

INTRODUCTION

For a description of a superfluid pure neutron matter (SNM) with anisotropic spin-triplet p-wave pairing (existing inside the liquid core of a neutron star [1] at high densities $n>0.7n_0$, n_0=0.17 fm^{-3}) similar to those of ^{3}He-A$_1$ and ^{3}He-A [2] in a strong magnetic field, we have used a generalized non-relativistic Fermi-liquid theory [3]. We have obtained [4] a set of integral equations for the components of the order parameter (OP) and an effective (renormalized in the SNM) magnetic field (EMF) $\mathbf{H}_{eff}$. The OP has the form

$$\Delta_\alpha(\mathbf{p}) = \left(\Delta_+\hat{d}_\alpha + i\Delta_-\hat{e}_\alpha\right)\left(\hat{m}_j + i\hat{n}_j\right)\hat{p}_j, \quad \hat{\mathbf{p}} = \mathbf{p}/p. \quad (1)$$

Here $\Delta_\pm(T) \equiv \left(\Delta_\uparrow(T) \pm \Delta_\downarrow(T)\right)/2$; $\hat{\mathbf{d}}$, $\hat{\mathbf{e}}$ and $\hat{\mathbf{m}}$, $\hat{\mathbf{n}}$ are mutually orthogonal real unit vectors in the spin and orbital spaces, respectively (i.e., $\hat{\mathbf{d}}\cdot\hat{\mathbf{e}} = 0$, $\hat{\mathbf{d}}^2 = \hat{\mathbf{e}}^2 = 1$ and $\hat{\mathbf{m}}\cdot\hat{\mathbf{n}} = 0$, $\hat{\mathbf{m}}^2 = \hat{\mathbf{n}}^2 = 1$). The components of the OP can be written [4,5] as $\Delta_{\uparrow(\downarrow)}(T,\xi,p) = p\delta_{\uparrow(\downarrow)}(T,\xi)$ (with spin-up and spin-down projections on the quantization axis opposite to the magnetic field $\mathbf{H}$, see [2]) and $\xi(p)\mathbf{H}/H = |\mu_n|\mathbf{H}_{eff}(p,H)$ (here μ_n is the magnetic dipole moment of the neutron [6]).

We consider a dense SNM with an effective Skyrme interaction between neutrons, and the effective neutron mass $m_n^*(y)$ which depends on the density n is given by [5,7]

$$\frac{m}{m_n^*(y)} = 1 + \frac{myn_0}{4\hbar^2}\left[(1-x_1)t_1 + 3(1+x_2)t_2\right], \quad (2)$$

where $m=(m_p+m_n)/2$ is the mean value of the free nucleon masses, $y \equiv n/n_0$, and t_1, t_2, x_1 and x_2 are the Skyrme parameters.

Further we shall consider solutions (obtained by analytic methods) of nonlinear integral equations [4] (valid at an arbitrary temperature $0 \le T \le T_{c0}$) for the SNM in the magnetic field in two limiting cases: at temperatures near the phase transition (PT) temperature T_{c0} of the neutron matter (NM) from normal to superfluid state with triplet p-wave pairing in zero magnetic field and at T=0.

SNM IN MAGNETIC FIELD (NEAR T_{c0})

Near T_{c0} ($|T-T_{c0}| \ll T_{c0}$), by solving a linearized gap equation for the components $\delta_{\uparrow(\downarrow)}$, we have obtained [5] the following general formulas (valid for arbitrary parameterization of Skyrme forces) to the leading order in the small parameter $h_{ext} \equiv |\mu_n|H/\varepsilon_F \ll 1$ ($\varepsilon_F = p_F^2/2m_n^*$ is the Fermi energy):

$$t_{c1,2} \approx t_{c0}\left[1 \pm \frac{h_{ext}}{I_0}\left(AI_A + BI_B\right)\right], \quad (3)$$

where $t_{c0} \equiv T_{c0}/\varepsilon_F \ll a$. The functions $t_{c1,2}(a;h_{ext},y)$ depend on the cutoff parameter $a \equiv \varepsilon_{max}/\varepsilon_F - 1$ which is small, $a \ll 1$, in the present model of neutron Cooper pairing in a thin shell in the vicinity of the Fermi sphere. The reduced PT temperature t_{c0} has the following general form, valid for all Skyrme parameterizations:

$$t_{c0} = \frac{a}{2}\exp\left(\frac{\ell(a)}{2} + \frac{2}{c_3 n_0 y m_n^*}\right), \quad (4)$$

CP850, *Low Temperature Physics: 24th International Conference on Low Temperature Physics;*
edited by Y. Takano, S. P. Hershfield, S. O. Hill, P. J. Hirschfeld, and A. M. Goldman

$$\ell(a)=b_0+\frac{3}{8}a^2+\frac{3}{256}a^4+O\left(a^6\right), \qquad (5)$$

$$b_0=2\left(1-\frac{1}{9}+\frac{2}{75}\right)+4\sum_{k=1}^{\infty}(-1)^{k+1}Ei(-2k)\approx 1.64932, \qquad (6)$$

$$Ei(-x)=\int_{-\infty}^{-x}\frac{e^t}{t}dt, \qquad (7)$$

and $c_3\equiv t_2(1+x_2)/\hbar^2<0$ is the coupling constant of spin-triplet *p*-wave pairing of neutrons [4,5]. The integrals I_0, I_A and I_B in (3) are defined as

$$I_0(a;y)\equiv\int_{-a}^{a}dx\frac{\sqrt{(1+x)^3}}{x}\tanh\left(\frac{x}{2t_{c0}}\right)=\ell(a)+2\ln\left(\frac{a}{2t_{c0}}\right), \qquad (8)$$

$$I_A(a;t_{c0})\equiv\int_{-a}^{a}dx\sqrt{(1+x)^3}\frac{d}{dx}\left(\frac{\tanh\left(x/2t_{c0}\right)}{x}\right), \qquad (9)$$

$$I_B(a;t_{c0})\equiv\int_{-a}^{a}dx\sqrt{(1+x)^3}\,x\frac{d}{dx}\left(\frac{\tanh\left(x/2t_{c0}\right)}{x}\right). \qquad (10)$$

The functions $A(a;y,t_{c0})$ and $B(a;y,t_{c0})$ in (3) are of the same form as in Ref. [5] but with modified integrals $i_j(a;t_{c0})$ (j=1, 3, 5) defined as

$$i_j(a;t_{c0})\equiv\int_{-a}^{a}dx\sqrt{(1+x)^j}\frac{d}{dx}\tanh\left(\frac{x}{2t_{c0}}\right). \qquad (11)$$

SNM IN MAGNETIC FIELD (T=0)

Let us consider the SNM at T=0. In this case we have obtained from the integral equation [4,5] for the EMF on the Fermi surface the following solution to the first order in the small parameter $h_{ext}\ll a\ll 1$:

$$\gamma(y,H)\equiv\frac{|\mu_n|H_{eff}(p_F,H)}{\varepsilon_F(y)}=\frac{h_{ext}(H,y)}{1-\left(r+2sp_F^2\right)\nu_F}. \qquad (12)$$

Here $r\equiv t_0+(t_3/6)n^\alpha$ and $s\equiv(t_1-t_2)/(4\hbar^2)$ are combinations of the Skyrme parameters t_0, t_1, t_2 and t_3, $1/6\leq\alpha\leq 1/3$. The density of states $\nu_F=(m_n^* p_F)/(\pi^2\hbar^3)$ at the Fermi surface is

$$\nu_F(y)\approx 0.004190\frac{m_n^*(y)}{m_n}y^{1/3}\,(\text{MeV}^{-1}\text{ fm}^{-3}). \qquad (13)$$

It should be emphasized that the general approximate formula (12) for $H_{eff}(p_F,H)$ is valid for all parameterizations of the Skyrme forces admissible for an NM, and H_{eff} is independent of the cutoff parameter $a\ll 1$ and of the energy gap in the energy spectrum of neutrons in the SNM (to the first order).

We have also solved the equation [4,5] for the OP $\delta_0(a;y)=\delta_\uparrow(a;y,H=0)=\delta_\downarrow(a;y,H=0)$ of the SNM at H=0 and T=0 with a pairing of a ^{3}He-A type, and have derived the following value for the ratio of the maximal magnitude of the reduced anisotropic energy gap $g(a;y)\equiv p_F(y)\delta_0(a;y)/\varepsilon_F(y)$ to the PT temperature $t_{c0}(a;y)$ (see (4)), valid for arbitrary parameterization of the Skyrme forces (see (6) for b_0):

$$\frac{g(a;y)}{t_{c0}(a;y)}=2\exp\left(\frac{5}{6}-\frac{b_0}{2}\right)\approx 2.0174\cdot \qquad (14)$$

Note that, at T=0 in a sufficiently strong EMF such that

$$a^4\ll\gamma(y,H)\ll a\ll 1, \qquad (15)$$

the approximate analytic expressions for $g_\uparrow\neq g_\downarrow$ (which are by definition $g_{\uparrow(\downarrow)}(a;y,H)\equiv p_F(y)\delta_{\uparrow(\downarrow)}(a;y,H)/\varepsilon_F(y)$) can be found from integral equations [4] for the OP. But here we only remark that, according to our numerical estimates for the so-called RATP parameterization [7] (proposed for astrophysical purposes to describe NM properties in the core of a neutron star at high densities), the values $|g_{\uparrow(\downarrow)}-g|/g\lesssim 0.01$ are small even in sufficiently strong magnetic fields $H_{eff,RATP}<10^{17}$ G (see (15)) for such densities that $0.7\leq y\leq 1.8$.

CONCLUSION

By solving integral equations [4] for the OP and EMF, we have obtained analytic formulas (3), (4), (12) and (14) for the PT temperatures and the EMF valid for arbitrary parameterization of the Skyrme forces in a dense SNM (at subnuclear densities $n\geq 0.7n_0$ and supernuclear densities but $n<2.0n_0$), with anisotropic OP similar to those of ^{3}He-A$_1$ and ^{3}He-A in a sufficiently strong magnetic field in two limiting cases: at $|T-T_{c0}|\ll T_{c0}$ and at T=0. These results can be specified for concrete parameterizations of the Skyrme forces, as will be described in more detail elsewhere.

Note finally that the phenomenon of superfluidity in an NM at high densities $n>2n_0$ (inside the fluid core of a neutron star) should be investigated in the framework of a relativistic approach and with different interpretation of the hadron matter structure (including mesons, quarks, and other possible constituents).

REFERENCES

1. T. Takatsuka and R. Tamagaki, *Prog. Theor. Phys. Suppl*, **112**, 27-65 (1993).
2. D. Vollhardt and P. Wolfle, *The Superfluid Phases of Helium 3*, Taylor & Francis, London, 1990.
3. A.I. Akhiezer, V.V. Krasil'nikov, S.V. Peletminskii and A.A. Yatsenko, *Phys. Rep.*, **245**, 1-110 (1994).
4. A.N. Tarasov, *Physica B*, **329-333**, 100-101 (2003).
5. A.N. Tarasov, *Europhys. Letters,* **65**, 620-626 (2004).
6. Review of Particle Properties, *Phys. Rev. D,* **50**, 1233, 1673, 1680 (1994).
7. M. Rayet, M. Arnould, F. Tondeur and G. Paulus, *Astron. Astrophys.*, **116**, 183-187 (1982).

Pair Excitations in Fermi Fluids

Helga M. Böhm, Eckhard Krotscheck, Karl Schörkhuber, and Josef Springer

Institute for Theoretical Physics, Johannes Kepler University, A-4040 LINZ, Austria

Abstract. We present a theory of multi-pair excitations in strongly interacting Fermi systems. Based on an equations-of-motion approach for time-dependent *pair* correlations it leads to a *qualitatively* new structure of the density-density response function. Our theory reduces to both, i) the "correlated" random-phase approximation (RPA) for fermions if the two-pair excitations are ignored, and ii) the correlated Brillouin-Wigner perturbation theory for bosons in the appropriate limit. The theory preserves the two first energy-weighted sum rules. A familiar problem of the standard RPA is that its zero-sound mode is energetically much higher than found in experiments. The popular cure of introducing an average effective mass in the Lindhard function violates sum rules and describes the physics incorrectly. We demonstrate that the inclusion of correlated pair excitations gives the correct dispersion. As in ^{4}He, a modification of the effective mass is unnecessary also in ^{3}He.

Keywords: Fermi fluids, density response function, multi-pair excitations, He-3, correlated basis functions
PACS: 67.55.-s, 67.55.Jd, 67.55.Cx

MOTIVATION

Low-lying excitations in strongly interacting quantum fluids can be characterized by the quantum numbers of a single "quasi"-particle. This concept is insufficient to describe higher-lying excitations, in particular the roton in ^{4}He. The problem is solved by introducing "back-flow" correlations [1], which are a special case of pair excitations. Such a theory has been systematically developed [2] and applied [3] with remarkable agreement between theory and experiment and it preserves the two first energy-weighted sum rules.

We take the Fermi liquid density-density response function to be of the usual correlated RPA (CRPA) form

$$\chi^{\rm CRPA}(q,\omega) = \chi_0(q,\omega)/\left(1-\widetilde{V}_{\rm p\text{-}h}(q)\,\chi_0(q,\omega)\right) \quad (1)$$

where χ_0 is the Lindhard function, and $\widetilde{V}_{\rm p\text{-}h}$ a suitable effective interaction. Fixing $\widetilde{V}_{\rm p\text{-}h}$ through the sum rules

$$\frac{\hbar^2 q^2}{2m} = \frac{\hbar^2}{\pi}\int d\omega\,\omega\,\Im{\rm m}\,\chi(q,\omega)\,, \quad (2)$$

$$S(q) = \int d\omega\, S(q,\omega) \equiv \frac{\hbar}{\pi}\int d\omega\,\Im{\rm m}\,\chi(q,\omega)\,, \quad (3)$$

leads to a collective mode that is energetically much higher than found in experiments. The problem could be cured by introducing an average effective mass m^* in χ_0, but this procedure is unsatisfactory as it violates the above sum rules (3). Also, in recent measurements[4] in ^{3}He layers a collective mode was found *inside* the particle-hole continuum. An effective m^* in a RPA formula (1) lowers *both* the continuum and the collective mode, instead of moving the latter into the continuum.

Further strong evidence that introducing m^* does not describe the right physics comes from ^{4}He: There the roton position is corrected by introducing fluctuations at wavelengths comparable to the interparticle distance. It appears appropriate to improve the description of excitations in ^{3}He in the same manner.

STRONGLY INTERACTING FERMIONS

We describe strongly interacting fermions with a non-orthogonal basis of Jastrow-Feenberg states [5]

$$|\Psi_{\mathbf{m}}(1,\ldots,N)\rangle = \frac{F(1,\ldots,N)\,|\Phi_{\mathbf{m}}(1,\ldots,N)\rangle}{\langle\Phi_{\mathbf{m}}|F^\dagger F|\Phi_{\mathbf{m}}\rangle^{1/2}}\,, \quad (4)$$

where $\Phi_{\mathbf{m}}$ is a determinant of plane waves, and F is the correlation operator [5]. The excited state wave function has a generalized time-dependent Hartree-Fock form including single-pair ("c") and *two-pair* ("d") amplitudes:

$$|\Psi(t)\rangle = \frac{F\,|\Phi(t)\rangle}{\langle\Phi(t)|\,F^\dagger F\,|\Phi(t)\rangle^{1/2}}\,, \quad (5)$$

$$|\Phi(t)\rangle = e^{-iH_{\mathbf{oo}}t/\hbar}\, e^{\frac{1}{2}U(t)}\,|\Phi_0\rangle\,, \quad (6)$$

$$U(t) = c_{ph}(t)\,a_p^\dagger a_h + \tfrac{1}{2}\,d_{pp'hh'}(t)\,a_p^\dagger a_{p'}^\dagger a_{h'} a_h\,, \quad (7)$$

where $H_{\mathbf{oo}}$ is the energy of the ground state $|\Psi_{\mathbf{0}}\rangle$.

We first demonstrate the procedure for a weakly interacting Hamiltonian where the correlations F can be ignored. The time-dependent particle-hole amplitudes $c_{ph}(t)$ and $d_{pp'hh'}(t)$ are determined by a stationarity principle, treating the external field as a first order perturbation. Upon linearizing the equations of motion, the procedure produces a density fluctuation $\delta\rho(\mathbf{r};t)$.

To simplify the calculation, we (a) neglect exchange terms and (b) retain the same terms as in the Bose theory. We also assume a *local* pair excitation operator, *i.e.*

$$d_{pp'hh'}\,a_p^\dagger a_{p'}^\dagger a_{h'} a_h \longrightarrow \sum_{qq'} d(\mathbf{q},\mathbf{q}')\left[\hat\rho_{\mathbf{q}}\hat\rho_{\mathbf{q}'} - \hat\rho_{\mathbf{q}+\mathbf{q}'}\right]\,, \quad (8)$$

CP850, *Low Temperature Physics: 24th International Conference on Low Temperature Physics;*
edited by Y. Takano, S. P. Hershfield, S. O. Hill, P. J. Hirschfeld, and A. M. Goldman

where $\hat{\rho}_{\mathbf{q}}$ is the density operator. The c-d coupling is further simplified by replacing the amplitudes $c_{ph}(t)$ by their Fermi-sea averages

$$c_{h+q,h}(t) \longrightarrow c(\mathbf{q};t) = N \frac{\sum_{\mathbf{h}} n_h \bar{n}_{\mathbf{h}+\mathbf{q}} c_{h+q,h}(t)}{\sum_h n_h \bar{n}_{\mathbf{h}+\mathbf{q}}} \tag{9}$$

(n_k is is the Fermi function and $\bar{n}_k \equiv 1-n_k$). Introducing

$$\chi_0^{(\pm)}(q,\omega) = \pm\frac{1}{N}\sum_h \frac{n_h \bar{n}_{\mathbf{h}+\mathbf{q}}}{\hbar\omega \mp e_{h+q,h} + i0^+}, \tag{10}$$

where $e_{h+q,h} \equiv e_{\mathbf{h}+\mathbf{q}} - e_h$ are differences of particle and hole energies, we derive the following closed-form expression for the density-density response function:

$$\chi^{\text{Pair}}(q,\omega) = \frac{\kappa(q,\omega)}{1-\kappa(q,\omega)\widetilde{V}_{\text{p-h}}(q)} \tag{11}$$

$$\kappa(q,\omega) = \frac{\chi_0^{(+)}}{1-\chi_0^{(+)}\mathscr{W}(q,-\omega)} + \frac{\chi_0^{(-)}}{1-\chi_0^{(-)}\mathscr{W}(q,\omega)}, \tag{12}$$

$$\mathscr{W}(q,\omega) = \frac{1}{2}\sum_{\mathbf{q}'} \frac{\overline{\mathscr{W}}_3(\mathbf{q}-\mathbf{q}',\mathbf{q}')\overline{\mathscr{W}}_3(\mathbf{q}-\mathbf{q}',\mathbf{q}')}{\hbar\omega + \varepsilon(\mathbf{q}-\mathbf{q}') + \varepsilon(q')} \tag{13}$$

where $\overline{\mathscr{W}}_3$ is a three-"phonon" vertex, and $\varepsilon(q) = \hbar^2 q^2/(2mS(q))$ is the Feynman spectrum.

In order to apply the theory to strongly interacting systems, we next include the correlation operator F. The result in this ("correlated basis functions = CBF") approach is almost of the same structure as Eq. (11), containing a "multi-phonon" correction $V_{\text{mp}}(q;\omega)$

$$\chi^{\text{CBF}}(q,\omega) = \frac{\kappa(q,\omega)}{1-\kappa(q,\omega)\widetilde{V}_{\text{p-h}}(q) + V_{\text{mp}}(q,\omega)}. \tag{14}$$

The only term, in which the fully correlated χ^{CBF} deviates from the structure of the "weakly interacting" χ^{Pair}, is the multi-phonon term $V_{\text{mp}}(q,\omega)$. As its inclusion in the actual computations gives only marginal improvement [6], it can be omitted.

We conclude this section by noting that it is easily proved that $\chi(q,\omega)$ obeys the sum rules (3), independent of the specifics of $\mathscr{W}(q,\omega)$. This means that $\widetilde{V}_{\text{p-h}}(q)$ is *uniquely* defined by the static structure function.

RESULTS

In Fig. 1 we show the results for the dynamic structure factor $S(k,\omega)$ of ^{3}He. Input to all our calculations was the static $S(k)$ obtained in Ref. [7]. The "correlated RPA" (= omitting pair excitations) is shown on the left; as mentioned above the collective mode is higher than found experimentally. The right picture gives the CBF result (= including pair fluctuations). Obviously, the dispersion is significantly improved and a weak pair-excitation background can be seen outside the patricle-hole continuum. As the calculated zero-sound mode is practically identical with the experimental one, a modification of the effective mass is unnecessary to obtain this result.

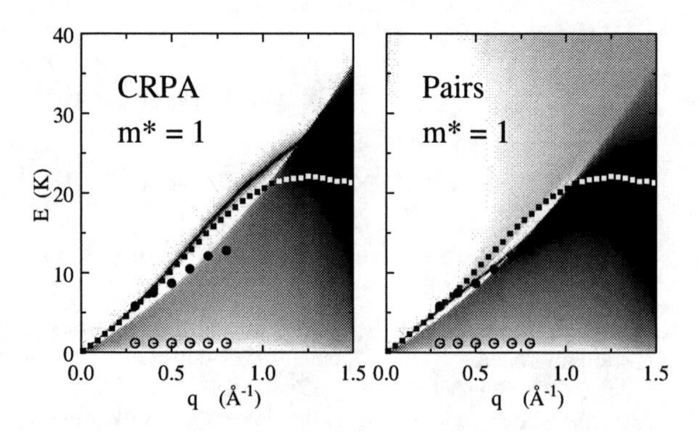

FIGURE 1. Saturation density ^{3}He dynamic structure factor. Dark line: theoretical zero-sound mode. Filled [open] dots: experimental phonon [magnon] modes (Ref. [8]). Squares: RPA.

In summary, the introduction of dynamic multi-pair correlations, so successful in ^{4}He, also proves necessary for obtaining the correct phonon dispersion in ^{3}He. A promising extension is the inclusion of spin fluctuations, which holds the potential of understanding the large effective mass in ^{3}He.

As a matter of course, the theory can also be applied to other strongly interacting Fermi fluids and to electrons.

ACKNOWLEDGMENTS

This work was supported by the FWF project P15083-N08. We thank H. Godfrin for providing the data [4].

REFERENCES

1. R. P. Feynman and M. Cohen, *Phys. Rev.*, **102**, 1189–1204 (1956).
2. H. W. Jackson, *Phys. Rev. A*, **9**, 964–975 (1974).
3. V. Apaja and M. Saarela, *Phys. Rev. B*, **57**, 5358 (1998).
4. H. Godfrin, M. Meschke, and H. J. Lauter, (private communication).
5. E. Feenberg, *Theory of Quantum Fluids*, New York: Academic Press, 1969.
6. K. E. Schörkhuber, "Multi–Particle–Hole Excitations in Many–Body Fermi Systems", Ph.D. thesis, Johannes Kepler University, Linz, 2002.
7. E. Krotscheck, *J. Low Temp. Phys.*, **119**, 103 (2000).
8. H. R. Glyde, B. Fåk, N. H. van Dijk, H. Godfrin, K. Guckelsberger, and R. Scherm, *Phys. Rev. B*, **61**, 1421–1432 (2000).

Force on a Slow Object in a Fermi Liquid in the Ballistic Limit

Timo H. Virtanen and Erkki V. Thuneberg

Department of Physical Sciences, University of Oulu, Finland

Abstract. We study normal Fermi liquids at low temperatures where the collisions of quasiparticles with each other can be neglected. We calculate the force exerted by the liquid on a slowly moving body. Our calculation is exact within the Fermi liquid theory, so that it takes into account all Fermi-liquid interaction effects as well as the effect caused by superfluid ^{4}He, when applied to mixtures of ^{3}He and ^{4}He. Our main result is for specular scattering, but we consider also diffuse scattering of quasiparticles off the surface.

Keywords: Ballistic limit, Fermi Liquid, ^{3}He-^{4}He mixture, Vibrating wire.
PACS: 67.60

INTRODUCTION

The mechanical properties of quantum liquids have been extensively studied using moving objects like vibrating wires. At high temperatures the force on the object is determined by hydrodynamic equations. At low temperatures the mean free path λ of elementary excitations grows, and hydrodynamic theory is no more applicable when λ approaches the size of the object. Especially, in mixtures of ^{3}He and ^{4}He, it is possible to study the properties of normal Fermi liquids over the whole range of mean free paths up to the ballistic limit, where the quasiparticle-quasiparticle collisions play no role [1, 2].

Consider a circular cylinder of radius a moving with a slow velocity $\mathbf{u}$ (perpendicular to the axis of the cylinder) through a normal Fermi liquid. We consider the force $\mathbf{F}$ that the liquid exerts on the cylinder in the ballistic limit. It has been expressed as [3]

$$\mathbf{F} = -Aan_3p_\mathrm{F}\mathbf{u}, \tag{1}$$

where n_3 is the number density of Fermi particles, p_F the Fermi momentum, and A a dimensionless factor. Ref. [4] cites an unpublished result by Bowley, according to whom $A = 2.36$ for reflection from a smooth wire, and $A = 2.58$ for a rough wire. More precise value $A = 3\pi/4$ in the specular case is given in Ref. [5].

The purpose of this work is to check the accuracy of Bowley's result. In particular, we are interested in finding if there are corrections to the result caused by Fermi-liquid interactions and the background ^{4}He superfluid. For specular scattering of quasiparticles, we find that Bowley's result is unaffected by Fermi-liquid interactions. For diffuse scattering our prefactor $A = 2.81$ differs slightly from Bowley's result.

FERMI LIQUID THEORY

Our starting point is Landau's Fermi liquid theory. The quasiparticle energy near the Fermi surface can be written

$$\varepsilon_\mathbf{p}(\mathbf{r},t) \;=\; \xi_\mathbf{p} + \delta\varepsilon(\hat{\mathbf{p}},\mathbf{r},t). \tag{2}$$

Here $\xi_\mathbf{p} = v_\mathrm{F}(p - p_\mathrm{F})$ and the Fermi velocity $v_\mathrm{F} = p_\mathrm{F}/m^*$ with effective mass m^*. Instead of the full distribution function $n(\mathbf{p},\mathbf{r},t)$, it is sufficient to consider the distribution that is integrated over energy,

$$\phi(\hat{\mathbf{p}},\mathbf{r},t) = \int d\xi_\mathbf{p}[n(\mathbf{p},\mathbf{r},t) - n_0(\mathbf{p})], \tag{3}$$

where n_0 is the unperturbed Fermi distribution. As a consequence ϕ depends only on the direction $\hat{\mathbf{p}}$ of momentum in addition to $\mathbf{r}$ and t. The correction $\delta\varepsilon$ of the quasiparticle energy is

$$\delta\varepsilon(\hat{\mathbf{p}},\mathbf{r},t) = Dp_\mathrm{F}\hat{\mathbf{p}}\cdot\mathbf{v}_s(\mathbf{r},t) + \langle F(\hat{\mathbf{p}},\hat{\mathbf{p}}')\phi(\hat{\mathbf{p}}',\mathbf{r},t)\rangle_{\hat{\mathbf{p}}'}, \tag{4}$$

where

$$D = 1 - \frac{m_3}{m^*}\left(1 + \frac{F_1^s}{3}\right), \quad F(\hat{\mathbf{p}},\hat{\mathbf{p}}') = \sum_{l=0}^{\infty} F_l^s P_l(\hat{\mathbf{p}}\cdot\hat{\mathbf{p}}'), \tag{5}$$

and m_3 is the bare fermion mass. The latter term in Eq. (4) is the standard Fermi-liquid interaction term with the average over the unit sphere of $\hat{\mathbf{p}}'$ denoted by $\langle\ldots\rangle_{\hat{\mathbf{p}}'}$. The former term comes from the flow velocity $\mathbf{v}_s$ of the superfluid ^{4}He in the case of ^{3}He-^{4}He mixture [6]. In the case of pure ^{3}He, $D = 0$. Note that the present theory makes no assumption about the diluteness of the ^{3}He component in mixtures.

The distribution function ϕ is determined from the kinetic equation and boundary conditions. The linearized

CP850, *Low Temperature Physics: 24th International Conference on Low Temperature Physics;*
edited by Y. Takano, S. P. Hershfield, S. O. Hill, P. J. Hirschfeld, and A. M. Goldman

Landau-Boltzmann kinetic equation for ϕ is

$$\frac{\partial \phi}{\partial t} + v_{\mathrm{F}}\hat{\mathbf{p}} \cdot \frac{\partial}{\partial \mathbf{r}}(\phi + \delta\varepsilon) = I, \tag{6}$$

where I is the collision term. The boundary condition appropriate for specular scattering is

$$\phi(\hat{\mathbf{p}}) = \phi(\hat{\mathbf{p}} - 2\hat{\mathbf{n}}(\hat{\mathbf{n}} \cdot \hat{\mathbf{p}})) + 2m_3 v_{\mathrm{F}}(\hat{\mathbf{n}} \cdot \hat{\mathbf{p}})(\hat{\mathbf{n}} \cdot \mathbf{u}), \tag{7}$$

where $\hat{\mathbf{n}}$ is the surface normal (pointing to the liquid). For diffusive scattering we get

$$\phi(\hat{\mathbf{p}}_{\mathrm{out}}) = 2\langle |\hat{\mathbf{n}} \cdot \hat{\mathbf{p}}_{\mathrm{in}}| \phi(\hat{\mathbf{p}}_{\mathrm{in}}) \rangle_{\hat{\mathbf{p}}_{\mathrm{in}}} + m_3 v_{\mathrm{F}}(\hat{\mathbf{p}}_{\mathrm{out}} + \frac{2}{3}\hat{\mathbf{n}}) \cdot \mathbf{u} \tag{8}$$

with an average over half of the unit sphere ($\hat{\mathbf{n}} \cdot \hat{\mathbf{p}}_{\mathrm{in}} < 0$). A consistency check for the boundary conditions is that the fermion particle current

$$\mathbf{J} = Dn_3\mathbf{v}_s + \frac{3n_3}{p_{\mathrm{F}}}(1 + \frac{F_1^s}{3})\langle \hat{\mathbf{p}}\phi(\hat{\mathbf{p}}) \rangle_{\hat{\mathbf{p}}} \tag{9}$$

behaves as expected: $\hat{\mathbf{n}} \cdot \mathbf{J} = n_3 \hat{\mathbf{n}} \cdot \mathbf{u}$. The force exerted by quasiparticles on a surface element da is obtained from the formula

$$\frac{d\mathbf{F}}{da} = -n_3\langle \{(1+F_0^s)\hat{\mathbf{n}} + (1+\frac{F_2^s}{5})[3(\hat{\mathbf{n}} \cdot \hat{\mathbf{p}})\hat{\mathbf{p}} - \hat{\mathbf{n}}]\}\phi(\hat{\mathbf{p}}) \rangle_{\hat{\mathbf{p}}}. \tag{10}$$

BALLISTIC LIMIT

In the ballistic limit $I = 0$. We assume the liquid is in equilibrium at infinity, and study an arbitrary surface element of a convex body. The kinetic equation (6) is easily solved for any incoming trajectory: $\phi(\hat{\mathbf{p}}_{\mathrm{in}}) = -\delta\varepsilon(\hat{\mathbf{p}}_{\mathrm{in}})$. In order to find ϕ for all directions, we define an auxiliary function $s(\hat{\mathbf{p}})$ by the conditions

$$\phi(\hat{\mathbf{p}}) = s(\hat{\mathbf{p}}) - \langle F(\hat{\mathbf{p}}, \hat{\mathbf{p}}')\phi(\hat{\mathbf{p}}') \rangle_{\hat{\mathbf{p}}'}, \tag{11}$$

$$s(\hat{\mathbf{p}}_{\mathrm{in}}) = -Dp_{\mathrm{F}}\mathbf{v}_s \cdot \hat{\mathbf{p}}_{\mathrm{in}}. \tag{12}$$

The functions ϕ and s on the surface are now fully determined by these conditions and the boundary condition (7) or (8). These can be solved by expanding ϕ and s using spherical harmonics. The force is obtained by substituting ϕ to the force expression (10).

For specular scattering we find the force

$$\frac{d\mathbf{F}}{da} = -\frac{3}{4}n_3 p_{\mathrm{F}}(\hat{\mathbf{n}} \cdot \mathbf{u})\hat{\mathbf{n}}. \tag{13}$$

In particular, there is no effect of Fermi-liquid interactions or the ^{4}He background on this *quasiparticle* force. For diffusive scattering we get

$$\frac{d\mathbf{F}}{da} = -\frac{1}{48}n_3 p_{\mathrm{F}}[9\mathbf{u} + 25(\hat{\mathbf{n}} \cdot \mathbf{u})\hat{\mathbf{n}}], \tag{14}$$

at least in the absence of Fermi-liquid interactions and ^{4}He (which implies $m^* = m_3$).

The total force on a body of arbitrary convex shape can now be obtained by integrating over its surface. For example, for a sphere of radius a we find

$$\mathbf{F}_{\mathrm{spec}} = -\pi a^2 n_3 p_{\mathrm{F}}\mathbf{u}, \tag{15}$$

$$\mathbf{F}_{\mathrm{diff}} = -\frac{13\pi}{9}a^2 n_3 p_{\mathrm{F}}\mathbf{u} \tag{16}$$

for specular and diffuse scattering, respectively. These are in agreement with the general result for a small spherical object

$$\mathbf{F} = -\sigma_{\mathrm{tr}} n_3 p_{\mathrm{F}}\mathbf{u}, \tag{17}$$

where σ_{tr} is the transport cross section. For a circular cylinder moving perpendicular to its axis we get

$$\mathbf{F}_{\mathrm{spec}} = -\frac{3\pi}{4}an_3 p_{\mathrm{F}}\mathbf{u}, \tag{18}$$

$$\mathbf{F}_{\mathrm{diff}} = -\frac{43\pi}{48}an_3 p_{\mathrm{F}}\mathbf{u}. \tag{19}$$

The specular scattering result agrees with Bowley's, but for diffuse scattering the present result is slightly larger.

The Fermi-liquid corrections in the diffusive case as well as the force exerted by the ^{4}He component will be discussed elsewhere.

REFERENCES

1. A. M. Guénault, V. Keith, C. J. Kennedy, and G. R. Pickett, *Phys. Rev. Lett.* **50**, 522 (1983).
2. J. Martikainen, J. Tuoriniemi, T. Knuuttila, and G. Pickett, *J. Low Temp. Phys.* **126**, 139 (2002).
3. D. C. Carless, H. E. Hall, and J. R. Hook, *J. Low Temp. Phys.* **50**, 605 (1983).
4. A. M. Guénault, V. Keith, C. J. Kennedy, S. G. Mussett, and G. R. Pickett, *J. Low Temp. Phys.* **62**, 511 (1986).
5. R. M. Bowley, and J. R. Owers-Bradley, *J. Low Temp. Phys.* **136**, 15 (2004).
6. I. M. Khalatnikov, *Sov. Phys. JETP* **28**, 1014 (1969).

Measurements of Superfluid ^{4}He Flow Through Sub-15 nm Aperture Arrays

J. A. Hoffmann[a], B. Hunt[a], M. Wang[a], C. T. Black[b], and J. C. Davis[a]

[a]*LASSP, Department of Physics, Cornell University, Ithaca, NY 1485, USA*
[b]*IBM T. J. Watson Research Center, Yorktown Heights, NY 10598, USA*

Abstract. We have constructed an experiment designed to study Josephson phenomena in ^{4}He. Motivated by reports [1,2] and our ideas for novel silicon nanofabrication techniques [3], we designed the experiment to study the possibility of a transition from stochastic to coherent phase slippage. Here we briefly describe the nanofabrication of sub-15 nm aperture arrays and show preliminary data. For temperatures below 2 K, these data show temperature dependence of the superfluid critical velocity through the weak link which is characteristic of phase-slip limited flow.

Keywords: superfluid ^{4}He, Josephson effects, nanofabrication.
PACS: 67.40.Rp

INTRODUCTION

The body of observations of Josephson behavior in ^{4}He has grown in the last several years with contributions from experiments exploiting advances in silicon technology for the fabrication of weak links. Sukhatme et al. observed the disappearance of a staircase pattern in the response of a hydrodynamic resonator as the temperature approached T_λ, a disappearance interpreted as a cross-over from a hysteretic to a non-hysteretic current-phase relation in the weak link [1]. Observation of a staircase response introduced the possibility of the coherent phase slips in their weak-link array. Hoskinson et al. measured superfluid oscillations through their aperture array with frequency predicted by the Josephson-Anderson phase evolution equation [2] but at temperatures seemingly too low for a non-hysteretic phase relation. This further hinted at a cross-over from stochastic to coherent phase slips. To further explore Josephson behavior in superfluid ^{4}He, we designed an experiment using arrays of much smaller apertures and, accordingly, with more agressive vibrational and acoustic noise suppression to compensate for the reduced conductance (and thus reduced signal) of the weak links.

DESIGN

Fabrication of Weak Links

In the past, the resolution of electron-beam lithography has defined the lower limit for the fabrication of silicon-based weak link devices. We use a novel technique (inspired by [4]) that exploits anisotropic etching to push back this lower limit. We use commercially-available silicon-on-insulator (SOI) wafers with single crystal, <100>-oriented silicon device layers. The finished devices are 5 mm chips with 130 μm suspended silicon membranes, onto which we pattern approximately 7600 apertures using e-beam lithography. Anisotropic KOH etching scales the ~150 nm e-beam features to sub-15 nm apertures [3].

Experimental Design

Figure 1 shows a schematic of our experimental cell, a hydrodynamic resonator with the inductive element formed by the aperture array and the capacitive element formed by a flexible Kapton membrane. Together, the weak link and membrane define two volumes of superfluid, connected only through the weak link.

The flexible membrane is coated with a superconducting Al-Pb-Al film and functions as the moveable portion of a SQUID-based displacement transducer [5]. We drive the membrane capacitively via a rigid electrode.

CP850, *Low Temperature Physics: 24th International Conference on Low Temperature Physics;*
edited by Y. Takano, S. P. Hershfield, S. O. Hill, P. J. Hirschfeld, and A. M. Goldman

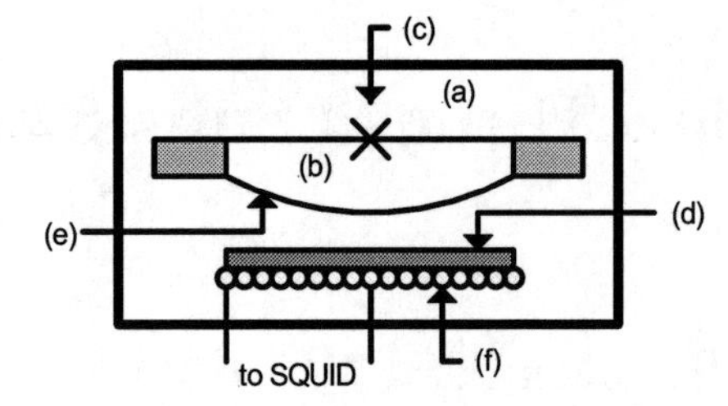

FIGURE 1. Schematic of experimental cell: (a) outer volume (b) inner volume (c) weak link (d) rigid electrode (e) flexible, superconducting membrane (f) SQUID pick-up coil.

Our experimental space is designed with two layers of passive vibration isolation. The inner lab space is mounted on a 25 ton concrete slab which is isolated from the building by six airsprings. The cryostat itself is then mounted on a lead-filled table supported by another set of three airsprings. Two layers of acoustic isolation, Sonex foam in the building room and a commercial sound room on the concrete slab, complete the low-noise infrastructure. Careful design of the cryostat is required to preserve the low-noise environment [6].

DATA AND DISCUSSION

Figure 2 shows the typical motion of the membrane after a pressure step is applied across the aperture array described previously. The membrane relaxes to a new equilibrium position, around which the hydrodynamic Helmholtz oscillations are visible. We extract the velocity of the membrane v_{mem} by a linear fit to the transient motion, which is limited by the critical velocity of superflow through the weak link array, v_{sf}. Assuming negligible normal flow, we obtain v_{sf} from $\rho_s a v_{sf} = \rho A v_{mem}$ where ρ_s is the superfluid density, a is the total area of the aperture array, ρ is the fluid density, and A is the membrane area. Determining the size of the apertures is difficult through imaging techniques due to the small size and the thinness of the silicon. Thus, we plot the product of the total aperture area (which is temperature independent) and the superfluid velocity in order to examine the temperature dependence of the superfluid velocity.

The scaled critical velocities show the linear temperature dependence observed/predicted at low temperatures [7] with increasing deviation from linear behavior at the temperature increases, particularly above 1.9 K. This differs from previous observations of linear temperature dependence above 2 K in experiments with apertures on order of 200 nm [8]. We postulate that the deviation from linear temperature dependence is indicative of movement toward non-hysteretic Josephson behavior.

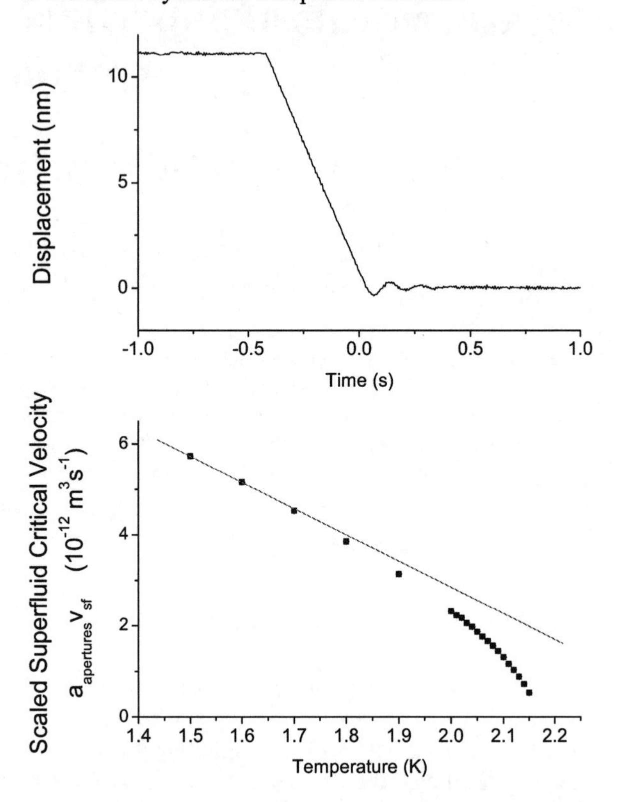

FIGURE 2. Top: Sample transient measured at 1.5 K after a 20 V pressure step. Bottom: Scaled superfluid critical velocity as a function of temperature. The absolute value of the velocity is a function of the total area of the apertures, a (see text). The line is a guide to the eye, based on the two lowest temperature points.

ACKNOWLEDGEMENTS

This work was supported by the NSF and Cornell University. JAH gratefully acknowledges support from an IBM Graduate Fellowship. BH gratefully acknowledges support from an NSERC Fellowship.

REFERENCES

1. K. Sukhatme, Y. Mukharsky, T. Chui, and D. Pearson, *Nature* **411**, 280 (2001).
2. E. Hoskinson, T. M. Haard, and R. E. Packard, *Nature* **433**, 376 (2005).
3. J. A. Hoffmann et al., in preparation.
4. N. N. Gribov, S. J. C. H. Theeuwen, J. Caro, and S. Radelaar, *Microelecton. Eng.* **35**, 317 (1997).
5. H. J. Paik, *J. Appl. Phys.* **47**, 1168 (1976).
6. J. A. Hoffmann, Ph.D. thesis, UC Berkeley (2005).
7. J. C. Davis et al., *Phys. Rev. Lett.* **69**, 323 (1992).
8. J. C. Davis, personal communication.

A Chemical Potential "Battery" for Superfluid ^{4}He Weak Links

E. Hoskinson*, Y. Sato*, K. Penanen† and R. E. Packard*

*Department of Physics, University of California, Berkeley, CA 94720, USA
†Jet Propulsion Laboratory, California Institute of Technology, Pasadena, CA 91109, USA

Abstract. Research and development of superfluid weak links has been hindered by the absence of a source of dc chemical potential, similar to a simple battery or voltage source for analogous superconducting devices. We describe here a method for generating a dc chemical potential difference, $\Delta\mu$, across a weak link array in superfluid ^{4}He. The presence of a $\Delta\mu$ forces quantum oscillations at a Josephson frequency, selectable by the adjustment of input power to a heater. We discuss a case in which the frequency locks onto a resonance feature where it exhibits remarkable stability, and amplitude magnification by a factor of 40.

Keywords: weak link, chemical potential, superfluid
PACS: 67.40.Rp, 67.40.Hf

Transient superfluid ^{4}He oscillations can be generated in a array of sub-micron apertures with the application of a pressure step, ΔP [1], or a temperature step ΔT [2]. The frequency of the oscillations f_J obeys the generalized Josephson frequency relation, $f_J = \Delta\mu/h$. Here $\Delta\mu = m_4(\Delta P/\rho - s\Delta T)$ is the chemical potential across the array, m_4 is the helium-4 atomic mass, ρ is the fluid density, and s its entropy per unit mass. Here we describe and demonstrate a method for producing continuous oscillations at a constant frequency, first proposed by Penanen and Chui [3]. A heater is used to maintain the constant $\Delta\mu$ which drives the oscillations. For reasons we do not yet fully understand, the system exhibits a high degree of stability at certain discrete frequencies. We propose a simple mathematical framework to describe the phenomenon.

Our experimental cell has been described in detail elsewhere [2]. A small volume of superfluid ^{4}He, itself immersed in a superfluid ^{4}He bath, is bounded on one side by the array (4225 apertures, nominally 70 nm in diameter, spaced on a 3 μm square lattice, in a 50 nm thick silicon nitride membrane), and on the other side by a flexible diaphragm. Inside the small volume there is a heater. The displacement of the diaphragm, detected using a SQUID-based transducer, indicates both the pressure ΔP across the array, and fluid flow through it.

If a small power W is applied to the heater, the inner volume warms, causing a net current to flow into it until steady state is reached. At this point $\Delta\mu = 0$ and $\Delta P = \rho s\Delta T$. Normal current I_n continues to flow, according to

$$I_n = -\frac{\rho_n\beta}{\eta}\left(\rho_n\frac{\Delta P}{\rho} + \rho_s s\Delta T\right). \tag{1}$$

This reduces to $I_n = \rho_n\beta\Delta P/\eta$ when $\Delta\mu = 0$. Here ρ_n and ρ_s are the normal and superfluid densities, η is the normal fluid viscosity, and β is a geometrical factor. To maintain zero net current at steady state, a super current flows in the opposite direction: $I_s = -I_n$. The steady state ΔT is reached when the heater power is balanced by thermal conduction through the walls of the inner volume, heat carried out by normal flow, and heat required to "warm" incoming superflow:

$$W = \frac{\Delta T}{R} + sT\left(I_s - \frac{\rho_s}{\rho_n}I_n\right). \tag{2}$$

Here R is the thermal resistance between the helium inside the inner volume at temperature $T + \Delta T$, and the helium outside, at constant temperature T.

If the heater power is slowly increased, ΔT, ΔP, I_n and I_s all increase according to the above relations, and $\Delta\mu = 0$ is maintained. I_s, however, cannot exceed the superfluid critical current I_c. Instead, when W exceeds a critical value, $\Delta\mu$ becomes non-zero and Josephson frequency oscillations, or "whistling," begins. A new steady state can be achieved in which the oscillations continue, $\Delta\mu = hf_J \neq 0$, the mean (dc) superflow $I_{s,\mathrm{dc}}$ replaces I_s in eq. 2, and $I_n = -I_{s,\mathrm{dc}}$.

Such a sequence of events is plotted in fig. 1. This data was taken at $T_\lambda - T = 1$ mK. For $t < 0$, a constant 63 nW is applied to the heater, and the measured pressure $\Delta P = 0.25$ Pa agrees with that predicted for when $\Delta\mu = 0$: $\Delta P = \rho sRW/\left(1 + \rho^2 s^2 TR\beta/\eta\right)$. Beginning at $t = 0$ the heater power is ramped linearly to a final 112 nW over 16 seconds. Initially ΔP rises to maintain $\Delta\mu = 0$ as the heater power rises, but at $t \simeq 1.5$ sec, the superfluid begins to oscillate. One might expect that transition to the whistling state to occur when I_s reaches I_c, at a fountain pressure $\Delta P_c = \eta I_c/\rho_n\beta_n$. An indepen-

CP850, *Low Temperature Physics: 24th International Conference on Low Temperature Physics;*
edited by Y. Takano, S. P. Hershfield, S. O. Hill, P. J. Hirschfeld, and A. M. Goldman

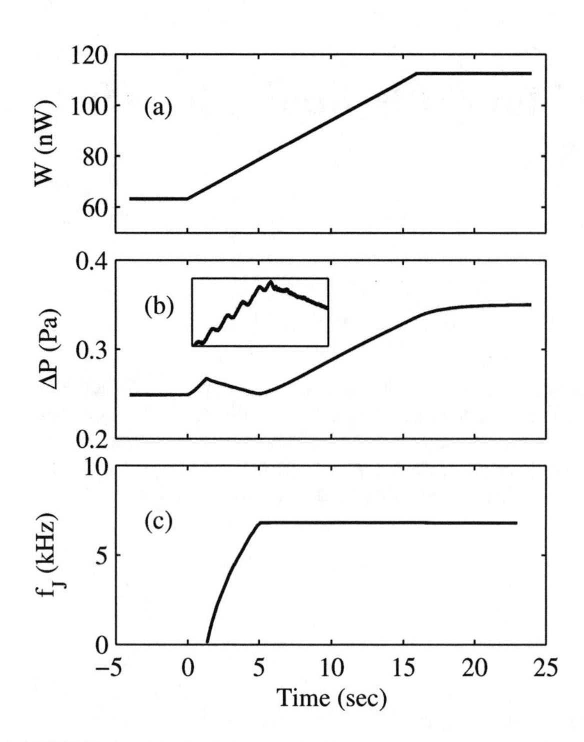

FIGURE 1. Method for producing a constant chemical potential difference and steady superfluid ^{4}He oscillations in a sub-micron aperture array. (a) Power delivered to the heater. (b) Evolution of the pressure across the array. The inset is a 0.4 sec wide section of data centered on the transition into the oscillating state that occurs at $t \simeq 1.5$ sec. (c) The oscillation frequency, shown locking into a stable frequency at $t \simeq 5$ sec.

dent measurement yielded $I_c = 5.6 \times 10^{-12}$ kg/sec. The corresponding ΔP_c is 2.0 Pa, almost a factor of 10 larger than the $\Delta P = 0.27$ Pa at which the system actually began to whistle. This suggests that the system is metastable at fountain pressures well below ΔP_c, and fluctuations can cause it to switch into a whistling state. Indeed, a close look at the transition at $t \simeq 1.5$ sec in the $\Delta P(t)$ plot of fig. 1b (see inset) reveals 23 Hz Helmholtz mode oscillations (not the same as Josephson frequency whistling) of increasing amplitude just before ΔP begins to drop and the superfluid begins to whistle.

As shown in fig. 1c, after the whistling begins, its frequency (and therefore $\Delta\mu$) rises steadily as the heater power continues to increase and ΔP drops. At $t \simeq 5$ sec, the frequency abruptly plateaus and ΔP begins rising again. For the next 11 seconds, ΔP continues to rise roughly linearly as the heater power rises, until $t = 16$ sec where the heater power reaches its final value and ΔP relaxes to a steady state value. From $t \simeq 5$ sec to $t = 16$ sec, the whistle, at 6.816 kHz with a width of only 0.25 Hz, drifts (downward) only 0.7 Hz, even while the heater power and pressure are increasing by 50%. We do not yet understand the reason for this remarkable stability. At $t > 24$ sec, the system remains in a steady state as long as the heater power is maintained. It seems likely that the 6.8 kHz value the whistle frequency locks onto is associated with a resonance feature of the system, but we have not yet determined the nature of this resonance. The amplitude of the 6.8 kHz signal is unexpectedly large – almost 40 times larger than would result from a current oscillation of amplitude I_c. From $t \simeq 5$ sec to $t = 16$ sec, the amplitude of the whistle varies (increases) by only 2%.

We have observed the system lock on to other frequencies as well. The frequency can be manipulated by varying the heater power – the steady state frequency achieved is dependent on the history of the heater power and on how fast it is changed.

The main unknown in the steady state formalism presented above is the mean superflow $I_{s,\text{dc}}$. We believe this quantity will depend primarily on the whistle frequency f_J, and possibly its phase ϕ, with the spectrum $I_{s,\text{dc}}(f_J, \phi)$ determined by the interaction of the superfluid oscillations with the resonant behavior of the overall system, perhaps similar in nature to the Fiske or Shapiro effects in superconductors [4], and in superfluid ^{3}He [5]. The intersection of this spectrum with the function $f_J(I_{s,\text{dc}}, W)$ derived from the steady state formalism will determine an allowed set of Josephson frequencies with unstable, metastable, and stable branches which can be traversed by manipulation of the heater power W. If the nature of the $I_{s,\text{dc}}(f_J, \phi)$ spectrum can be understood, it may be possible to design a cell to optimize stability and signal-to-noise.

We have demonstrated a method with which we have produced superfluid ^{4}He aperture array oscillations with a highly stable frequency and considerable amplitude magnification. The nature of the resonance behavior remains to be explained. This technique may prove to be an ideal method of operating a ^{4}He weak link device analogous to the dc-SQUID, which would be highly sensitive to rotation.

This work is supported in part by the NSF-DMR and NASA.

REFERENCES

1. E. Hoskinson, R. E. Packard, and T. M. Haard, *Nature* **433**, 376 (2005).
2. E. Hoskinson and R. E. Packard, *Phys. Rev. Lett.* **94**, 155303 (2005).
3. K. Penanen and T. Chui, *APS March meeting* L15.007 (2004).
4. T. Van Duzer and C. W. Turner, *Superconductive Devices and Circuits*, Upper Saddle River: Prentice Hall, 1999, pp. 185-187, 206-210.
5. J. C. Davis and R. E. Packard, *Rev. Mod. Phys.* **74**, 741 (2002).

Calibration Technique for Superfluid ^{4}He Weak-Link Cells Based on the Fountain Effect

E. Hoskinson and R. E. Packard

Department of Physics, University of California, Berkeley, CA 94720, USA

Abstract. Studies of superfluid ^{4}He weak-links require calibration constants which permit the determination of the pressure and temperature differences which drive Josephson oscillations. We describe a technique for calibrating ^{4}He weak-link cells in which a heater is used to induce fountain pressures detected by the deflection of a diaphragm. The technique determines the diaphragm spring constant, the inner cell volume, and the thermal conductance of the inner cells walls. This information is used to convert the measured deflection of the diaphragm into the total chemical potential difference across the weak link.

Keywords: weak-link, superfluid, calibration
PACS: 67.40.Rp, 67.40.Hf

Low frequency Helmholtz resonator-type cells for the study of superfluid helium flow through small apertures have been used in a variety of experiments. These cells utilize a flexible diaphragm as a piston to drive and detect fluid flow through the apertures. Flow can be driven electrostatically by application of a voltage between the metalized diaphragm and a nearby electrode. It can also be driven via the fountain effect with a heater. Displacement of the diaphragm, indicating both fluid flow (which directly displaces the diaphragm), and pressure (because the diaphragm has a spring constant), can be measured capacitively or with a SQUID-based superconducting displacement transducer. We have used such a cell to study superfluid ^{4}He quantum oscillations in an array of submicron sized apertures [1, 2]. Here we describe the method we use for measuring several dynamical quantities of the cell, and present data from such measurements.

A schematic of our cell is shown in figure 1. It consists of a cylindrical container of inner diameter 8 mm and height 0.6 mm, with walls and bottom constructed out of aluminum 7075. The top boundary is a 8 μm thick flexible Kapton diaphragm, with a 400 nm layer of superconducting lead evaporated onto the top surface as part of the diaphragm displacement transducer. The array, mounted on the bottom plate, consists of 4225 nominally 60 nm diameter apertures in a square array with 3 μm spacing, in a 50 nm thick 200 μm $\times$ 200 μm silicon-nitride membrane. This membrane is supported on a 3 mm $\times$ 3 mm, 0.5 mm thick silicon chip, glued into the bottom plate using Stycast 2850 FT with catalyst 24LV. The heater is a 54 mΩ length of CuNi wire, flattened and roughened to increase surface area. Unclad 50 μm superconducting NbTi wire are used for the heater leads to minimize thermal conduction along them and ensure all power delivered is dissipated inside the inner volume.

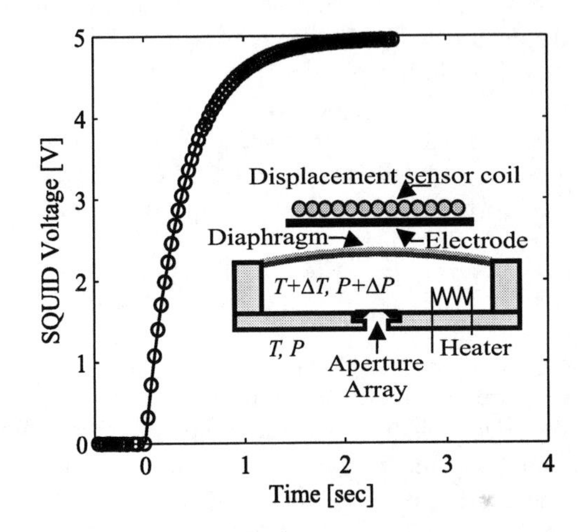

FIGURE 1. Displacement sensor voltage output versus time and a schematic of the experimental cell. The heater is turned on at time t = 0 sec. The temperature across the aperture array relaxes to a steady state value ΔT_f, determined mainly by thermal conduction through the inner cell walls. The diaphragm relaxes to a new position corresponding to the fountain pressure $\Delta P_f = \rho s \Delta T_f$.

Our method is based on measurement of heater driven relaxation transients. One such transient is shown in figure 1. When power W is suddenly applied to the heater inside the inner cell, a temperature difference $\Delta T(t)$ across the aperture array grows, driving superfluid current I_s into the inner cell (the fountain effect), causing the diaphragm to bulge and the pressure $\Delta P(t)$ to rise.

The diaphragm displacement transducer output signal is a voltage, ΔU, proportional to displacement, $x = \alpha \Delta U$. Displacement is proportional to ΔP : $kx = \Delta P A$, where k is the diaphragm spring constant and A is its area. Thus $\Delta P = k\alpha\Delta U/A \equiv \gamma\Delta U$. The constant γ is de-

CP850, *Low Temperature Physics: 24th International Conference on Low Temperature Physics;* edited by Y. Takano, S. P. Hershfield, S. O. Hill, P. J. Hirschfeld, and A. M. Goldman

termined from the Josephson frequency $f_j = \Delta\mu/h$ measured at the beginning of a pressure driven transient (excited electrostatically) where $\Delta T = 0$ [1]. Here $\Delta\mu = m_4(\Delta P/\rho - s\Delta T)$ is the chemical potential difference across the aperture array, h is Plank's constant, m_4 is the ^{4}He atomic mass, ρ is the fluid mass density and s is the fluid entropy per unit mass.

The total current $I_t = I_s + I_n$ into the inner cell, where I_s and I_n are the superfluid and normal components, is related to the diaphragm motion and ΔP through: $I_t = \rho A \partial x/\partial t = \rho(A^2/k)\partial\Delta P/\partial t$.

The evolution of $\Delta T(t)$ is determined by a balance of heat flows [2]:

$$C_p \frac{d\Delta T}{dt} = -sT\left(I_s - \frac{\rho_s}{\rho_n} I_n\right) - \frac{\Delta T}{R} + W. \quad (1)$$

Here ρ_s and ρ_n are the superfluid and normal fluid densities, R is the thermal resistance between the ^{4}He inside the inner cell and the ^{4}He outside it, and $C_p = c_p V$, where c_p is the heat capacity per unit volume of ^{4}He and V is the volume of the inner cell.

The size of our apertures is such that the viscous normal flow, while small in comparison with the superflow, is not entirely negligible. Viscous normal flow follows

$$I_n = -\frac{\rho_n \beta}{\eta}\left(\frac{\rho_n}{\rho}\Delta P + \rho_s s \Delta T\right), \quad (2)$$

where β is a geometrical factor and η is the viscosity. We measure β directly from the flow response to an electrostatically induced ΔP just above the superfluid transition temperature T_λ.

If the heater power applied is sufficiently small, no Josephson oscillations are excited and $\Delta\mu = 0$ is maintained throughout the transient. In this case ΔP will relax exponentially to a steady state fountain pressure

$$\Delta P_f = \frac{\rho s R W}{(1 + \rho^2 s^2 T R \beta/\eta)}, \quad (3)$$

with time constant

$$\tau_f = \frac{\Delta P_f}{\rho s W}\left(C_p + s^2 \rho^2 T \frac{A^2}{k}\right). \quad (4)$$

The constants A^2/k and V (recall $C_p = c_p V$) are determined from a fit to $\tau_f/\Delta P_f$ as a function of T (with τ_f and ΔP_f determined from transients such as the one shown in fig. 1). Alternatively, if V is already known, but the pressure calibration constant γ is not, the fit can determine A^2/k and γ. Published values are used for s, ρ_s, ρ_n, η, and c_p [3]. For the data shown in fig. 2, the parameters values were determined to be, with estimated uncertainty in the last digit specified in brackets, $\gamma = 0.0313(6)$ Pa/V, $k/A^2 = 1.88(4) \times 10^{12}$ N/m^5, $V = 2.45(5) \times 10^{-8}$ m^3, $\beta = 4.8(3) \times 10^{-20}$ m^3.

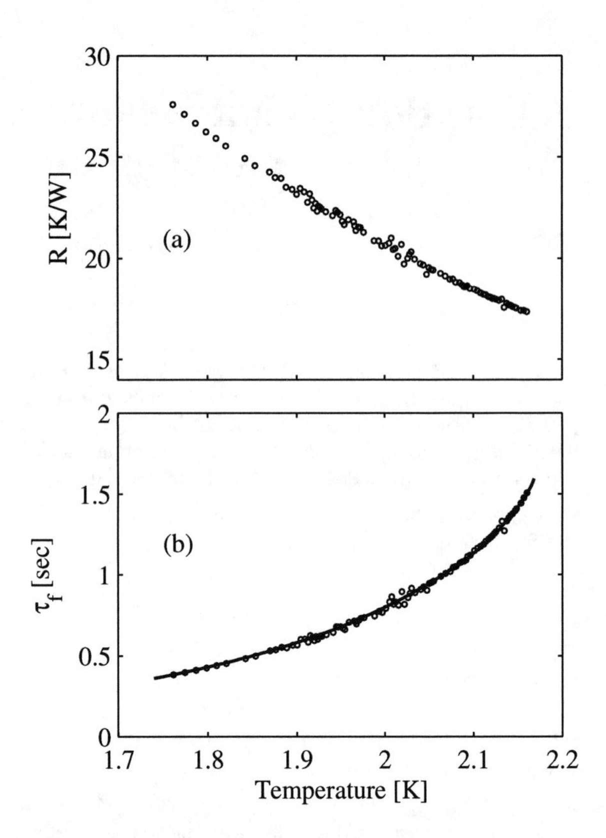

FIGURE 2. (a) Thermal resistance R from the measured fountain pressure ΔP_f versus temperature. R is likely determined mainly by conduction through the thin, large area diaphragm. Heat transport by normal flow through the apertures is taken into account separately, and is small compared to thermal conduction. (b) Fountain transient time constant τ_f versus temperature, data (circles), and fit (line), from which A^2/k and V are determined.

With the calibration constants determined using this method, eq. 1 can be numerically integrated, giving $\Delta T(t)$ and thus $\Delta\mu(t)$ for any measured $I_t(t)$, $\Delta P(t)$. The quantum mechanical phase $\Delta\phi$ across the aperture array can be determined by integration of the Josephson-Anderson phase evolution equation $d\Delta\phi/dt = -\Delta\mu/\hbar$.

This work is supported in part by the NSF-DMR and NASA.

REFERENCES

1. E. Hoskinson, R. E. Packard, and T. M. Haard, *Nature* **433**, 376 (2005).
2. E. Hoskinson and R. E. Packard, *Phys. Rev. Lett.* **94**, 155303 (2005).
3. R. J. Donnelly and C. F. Barenghi, *J. Phys. Chem. Ref. Data* **27**, 1217 (1998).

^{4}He Versus ^{3}He Josephson Effect: Vibration Decoherence

Sergey V. Pereverzev

Department of Physics and Astronomy, Rutgers University, New Jersey, USA

Abstract. Several on-going experiments searching for the Josephson effect in ^{4}He close to the λ-transition employ experimental cells with a weak link in the form of an array of submicron holes, with the size of the array and the cell dimensions very close to those used for ^{3}He-B. In the same environment, the ^{4}He experiment is more prone to decoherence by mechanical vibrations. The problem is due to the shift of the maximum of the vibration response of the experiment to low frequencies (0.1 Hz or less) and to the increase of the power spectrum density of the seismic velocities with decreasing frequency in this frequency range. To avoid decoherence, one needs to lower the cut-off frequency of the vibration isolation or to use an array with a larger open area. The latter option is briefly discussed.

Keywords: Superfluid, λ-point, Josephson effect, decoherence, mechanical vibrations.
PACS: 67.57.-z, 67.40.-w, 74.30.+r, 03.75.

INTRODUCTION

Reliable experimental observation of the Josephson effect in superfluid ^{4}He remains a challenging problem. If successful, such an experiment could shed light on the nature of the λ-transition and on the microscopic picture of superfluidity in ^{4}He, which is of fundamental importance. It would be easier to use ^{4}He at temperature near 2 K than ^{3}He cooled below 1 mK for applications to a superfluid gyroscope. In several on-going experiments searching for the Josephson effect in ^{4}He close to the λ-transition [1,2], experimental cells with a weak link in a form of an array of submicron holes are used, with the size of the array and cell dimensions very close to those used in ^{3}He-B experiments. In this paper, we compare the decoherence effect of mechanical vibrations on ^{4}He and ^{3}He experiments and show that seismic motion can cause decoherence in ^{4}He experiments.

MODEL AND ANALYSIS

In a typical experiment, a Helmholtz resonator of a small inner volume (V) with one flexible wall (membrane) of area A and effective spring constant k is placed inside a large external container filled with superfluid. The inner volume is connected to the large volume through an array of small holes (the flow channel) with total cross section a and effective channel length l. Potential energy in the Helmholtz oscillations is due to the membrane's elastic deformation X: $U=kX^2/2$. The holes are sufficiently small that the normal component is clamped inside the flow channel and the mass transfer and kinetic energy T in the Helmholtz oscillations are due to the superfluid velocity V_S in the channel. The mass conservation requires $d/dt(X)A\rho=-\rho_s aV_s$, where ρ is the liquid helium density and ρ_s is the density of the superfluid component (we suppose constant temperature everywhere in the cell). Then $T=\rho_s al(V_s)^2=\rho_s al((A\rho/a\rho_s)d/dt(X))^2$, leading to the Helmholtz resonance frequency

$$\omega_0 \cong (\rho_S ka / \rho^2 A^2 l)^{1/2} . \quad (1)$$

Acceleration of the apparatus causes a change in the hydrostatic pressure at the position where the resonator is located and generates superflow V_S in the flow channel (due to non-zero liquid compressibility κ). While Vs is below some critical velocity (in other words, for small quantum phase differences across flow channel), V_S is related to the amplitude of cell vibrations V_{Cell} and frequency ω by the formula [3]

$$V_S(\omega) = \frac{k\kappa VL}{l\rho A^2} \frac{\omega^2 V_{Cell}(\omega)}{\omega_0^2 - \omega^2 + i\omega\omega_0 / Q}. \quad (2)$$

Here Q is the quality factor of the Helmholtz resonance. Parameter L is an effective distance between the centers of gravity of the inner and outer containers (see [3] for details). We can use (2) to define the response function F(ω) of the experiment to small cell velocities: $V_s(\omega)=F(\omega)V_{Cell}(\omega)$. The frequency-dependent part of $F(\omega)$ has a maximum (with height Q) around ω_0, and is close to 0 at low frequencies and 1 at high frequencies. In a typical ^{3}He

CP850, *Low Temperature Physics: 24th International Conference on Low Temperature Physics;*
edited by Y. Takano, S. P. Hershfield, S. O. Hill, P. J. Hirschfeld, and A. M. Goldman

experiment, f_0 is in the range 3–100 Hz ($\omega_0=2\pi f_0$). For a λ-transition ^{4}He experiment (with the same cell), f_0 will be lower by a factor of $(\rho_s/\rho)^{1/2}$ (see (1)), lying in the 0.05–1 Hz range.

We need the power spectrum density of the cell velocities $S_V(\omega)$ which is strongly site-dependent. We assume that the apparatus is decoupled from building and pump vibrations, and what is left is given by the product of the vibration isolation transmissibility $T_I(\omega)=|V_{Cell}(\omega)/V_{Ground}(\omega)|$ and the power spectrum density of the ground velocities $S_{VG}(\omega)$. We assume that (as for a critically damped suspension) $T_I(\omega)$ is equal to 1 for frequencies below the cut-off frequency $\omega_C=2\pi f_C$ of the suspension and that, for $\omega>\omega_C$, $T_I(\omega)$ drops with a slope of 20 dB/octave (or faster for a multistage isolation). In a typical ^{3}He experiment, f_C is about 1 Hz and less than f_0.

Ultra-low frequency vibrations are important in special applications like the LIGO project or large particle accelerators. One can find $S_{VG}(f)$ at different locations in corresponding reports. We will refer here to the "low noise model" (described in [4]) — a minimum of geophysical observations worldwide. To avoid confusions, $S_{VG}(f)$ is measured in $(\mu m/s)^2 Hz^{-1}$, and is related to the power spectrum density of the displacements S_X by the formula $S_{VG}(f)=S_X(f)(2\pi f)^2$. $S_{VG}(f)$ in the low noise model has a wide maximum in the $5\cdot 10^{-2}$–10^0 Hz region due to large ocean waves. We can ignore this maximum; then $S_{VG}(f)\sim 1/f^\gamma$ with $\gamma\approx 2$ in the frequency range 10^{-3}–10^2 Hz.

For the RMS value of noise-induced superfluid velocity V_s^* in the flow channel we have

$$(V_s^*)^2 = \int_0^\infty |F(\omega)T_I(\omega)|^2 S_{VG}(\omega)d\omega. \qquad (3)$$

The RMS value of the quantum phase noise across the channel is proportional to V_s^*. We need a lower estimate of the change in V_s^* with the decrease of ω_0; so we can assume even slower increase of $S_{VG}(\omega)$ with decreasing frequency, namely $S_{VG}(\omega)= S_{V0}/\omega$, where $S_{V0}\equiv const$. With this substitution [5] we get

$$(V_s)^2 = \int_0^\infty |F(\omega)T_I(\omega)|^2 S_{V0} d(\ln(\omega)). \qquad (4)$$

This means that, for two experiments, one needs to compare the areas under the curves $|F(\omega)T_I(\omega)|^2$ plotted in a semi-logarithmic scale. To avoid an increase of the RMS value of V_s in ^{4}He experiments with respect to that in (same cell geometry) ^{3}He-B experiments, one needs to decrease the cut-off frequency of the vibration isolation ω_C by the same factor as the Helmholtz frequency ω_o is decreased [6]. A faster decrease of $T(\omega)$ for $\omega > \omega_C$ will not help without changing the ω_C itself (especially if f_0 falls below f_c).

DISCUSSION

It is possible to have the cut-off frequency of the insulation as low as 0.1 Hz or even 0.01 Hz, but this would be expensive. It will be more cost-effective to prepare the weak link for a λ-point ^{4}He experiment with a factor of 100 (or more) larger critical current (with a larger total open area of the holes). While increasing the total critical current of the array, one needs to keep constant the product of the array linear size (the lateral extent of the array) and the average current density (or below some critical value, see [7]). In the opposite case, the effective inductance of the spreading flow rises and, more dangerous, a macroscopic flow with a quantized circulation can be trapped in the array (see [7]). So, to get a 100 fold increase in the total current, one needs a 100 fold increase in the array linear size. This is a challenge for a 40 nm thick free-standing membrane! Consideration in the spirit of ref. [7] shows that it is possible to avoid trapping a quantized circulation without an excessive increase of the size of the array by making the superfluid density in the bulk higher with respect to that in the holes. One can suppress the superfluid transition inside the holes by a few mK with respect to the transition temperature everywhere else in the cell by strong electric field inside holes (see [8]).

ACKNOWLEDGMENTS

This paper originates from numerous discussions with Professor Georg Eska during our work on the ^{3}He Josephson effect project in Bayreuth University, Germany.

REFERENCES

1. E. Hoskinson, R. E. Packard and Thomas M. Haard, *Nature,* **433**, 376 (2005).
2. Joan Audrey Hoffmann, "Superfluid ^{4}He: On sinφ Josephson Weak Links and Dissipation of Third Sound", dissertation, University of California at Berkeley, 2005.
3. S.V. Pereverzev and J.C. Davis, *Czechoslovak Journal of Physics*, **46**, 109-110 (1996), suppl. S1 (Proceedings of the LT21, Prague, 1996).
4. J. Peterson, USGS Open-File Report 93-322, Albuquerque, NM (1993).
5. Eq. (4) will be exact for substitution $S_{V0}(\omega)=S_{VG}(\omega)/\omega$.
6. In our experience, coherent phenomena become unobservable in ^{3}He experiment for temperatures near T_c, when ω_0 drops below 5–10% of its zero temperature value (likely due to vibration decoherence).
7. S. V. Pereverzev and G. Eska, *Physica B,* **284-288**, 85-86 (2000).
8. H. A. Notarys, *Phys. Rev. Lett.*, **20**, 258-259 (1968).

Determination Of The Bulk Helium Critical Exponents Using Confined Helium

Mark O. Kimball, Manuel Diaz-Avila, and Francis M. Gasparini

Department of Physics, University at Buffalo, The State University of New York, Buffalo NY 14260, USA

Abstract. The specific heat of helium homogeneously confined in one or more dimensions is expected to collapse onto a scaling function which depends only on the ratio of the smallest dimension of confinement to the correlation length, written as L/ξ. This may be rewritten to explicitly show the temperature dependence of the correlation length as $L/\xi_0 t^{-\nu}$, where the constant ξ_0 is the prefactor of the correlation length, t is a dimensionless temperature difference from the superfluid transition, and ν is the critical exponent associated with the correlation length. Thus, in principle, one should be able to obtain the exponent ν from the scaling of thermodynamic measurements of confined helium for various L's. This would represent an independent determination of ν distinct from what is obtained using the behavior of the bulk superfluid density, or via the bulk specific heat and the hyperscaling relation. In practice, this analysis is hampered by the lack of a theoretical expression for the scaling function. We present preliminary results of analyses of specific heat data for planar confinement which spans a range of about 1200 in L and yields the exponent ν. The data are fit to an empirical equation which is obtained so that it has the proper asymptotic temperature dependence for large and small values of the scaling variable, which we take as $tL^{1/\nu}$. Results are compared with theoretical and other experimental determinations of ν.

Keywords: helium, critical exponents, scaling
PACS: 67.20.+k, 68.15.+e

Finite-size scaling theory states the specific heat of confined helium will scale as a function of $(L/\xi)^{1/\nu}$ where L is the spatial confinement, ξ is the correlation length, and ν is the bulk correlation-length critical exponent [1, 2]. Typically, this exponent is determined by a thermodynamic measurement such as the superfluid density [3] or derived from the specific heat [4] of bulk helium. However, the reverse may be done; one may ask which exponent provides the best collapse of specific heat data measured for a number of planar confinements.

In this analysis, we use specific heat data from measurements of helium confined to eight different planar confinements. The seven smallest confinements [5–7], which range from 48.3 nm to 986.9 nm in L, may be measured on Earth since effects due to a pressure gradient across the sample are not evident within the temperature resolution of the experiments. The largest confinement, which has a spatial separation between its plates of 57000 nm, was measured in a near-Earth orbit to reduce the effects of the pressure gradient across the thick slab of helium [8, 9]. The full range of confinements is nearly a factor of 1200 from smallest to largest. Figure 1 shows the $T > T_\lambda$ specific heat measured for ^{4}He confined in the eight different planar confinements. To attempt to scale the data, one uses

$$[C_p(\infty,t) - C_p(L,t)]\, t^{\alpha} = g_2(y), \qquad (1)$$

where $y = tL^{1/\nu}$ and $t = T/T_\lambda - 1$. α is the critical exponent of the specific heat. Once the specific heat data is

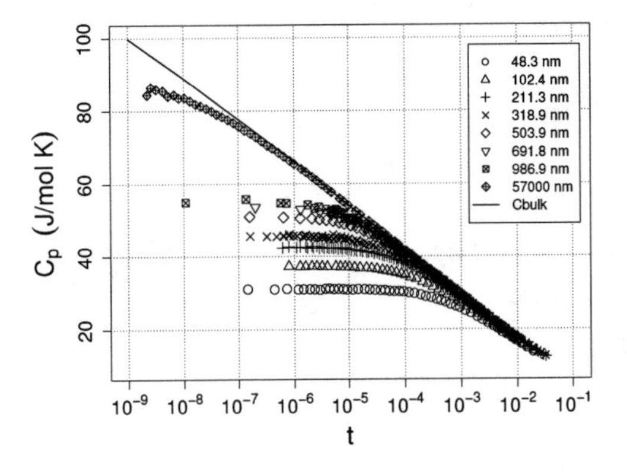

FIGURE 1. Specific heat data from helium confined to eight different planar geometries. This is for temperatures greater than T_λ.

cast in scaling form using Eq. 1, one needs a specific expression to use in fitting the data to determine the goodness of collapse for a given ν. Since there is not a known theoretical expression for the scaling function, we use an empirically-derived function chosen to have the proper limits for large and small values of the dimensionless scaling variable, $t\,(L/a_0))^{1/\nu}$ where a_0 is 3.56 Å. This is given by [10]

$$g_2(y) = \frac{A/\alpha}{1+ay^{\nu}} + \frac{by^{\alpha}}{1+cy^{\alpha+\nu}}. \qquad (2)$$

CP850, *Low Temperature Physics: 24th International Conference on Low Temperature Physics;*
edited by Y. Takano, S. P. Hershfield, S. O. Hill, P. J. Hirschfeld, and A. M. Goldman

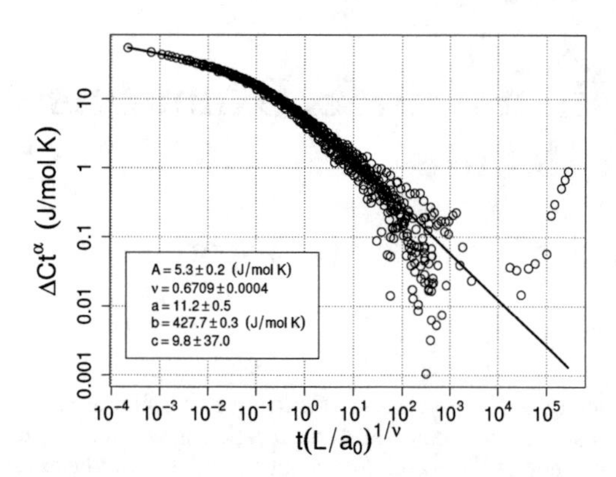

FIGURE 2. Best collapse of the scaled data. The solid line is described by Eq. 2 and the best collapse parameters shown on this plot.

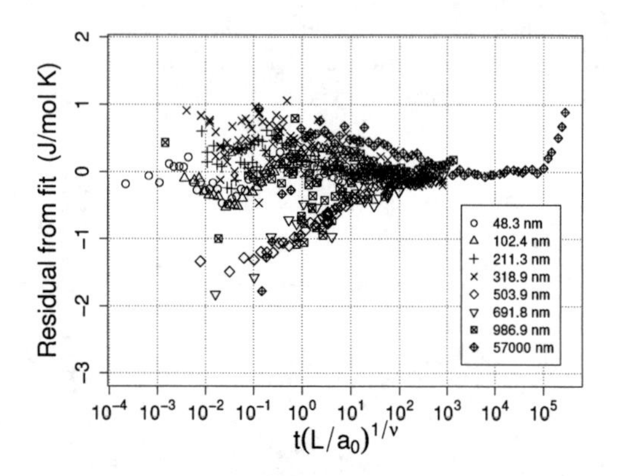

FIGURE 3. Residuals between the solid line and the scaled data in Fig. 2.

The two critical exponents are related by the hyperscaling equation

$$\alpha = 2 - 3\nu, \tag{3}$$

therefore Eq. 2 may be rewritten using only a single critical exponent as a fit parameter. A is the leading amplitude of the specific heat, and along with ν, a, b, and c is varied to best fit ΔC to $t^{(2-3\nu)}g_2(y)$. This procedure yields $\nu = 0.6709 \pm 0.0004$. This compares to 0.67155 ± 0.00027, 0.6709 ± 0.0001, and 0.6705 ± 0.0006, which correspond to the latest theoretical estimate [11], the latest value derived from Eq. 3 and the bulk specific heat [4], and the latest value from the superfluid density of bulk helium [3] respectively. *We find the experimental results remarkably close considering the variety of experiments used to determine* ν. The scaled data plotted using the exponent which provides the best collapse are shown in Fig. 2. Figure 3 shows the difference between this data and the line described by Eq. 2 and the best collapse parameters.

Ultimately, our analysis relies upon the function used to perform the least-squares minimization of the data in scaling form. It is certain that another function would provide different values for the fit parameters, including ν. Ideally, a method of determining the best collapse without relying on a specific function should be used. One method we have considered is to scale the data with a fixed ν and divide this data into a fixed number of equal-width bins based upon the scaling variable. We then would perform a linear regression fit of the data within each bin and record the residual error of the fit. The sum of the errors for all bins would be the measure of the goodness of collapse for a particular ν. We next repeat the procedure using different values for ν until we determine a neighborhood of values for ν which provide the smallest total residual error indicating the best collapse. While this will predict the value of ν which provides the best collapse, factors such as method used to weight the data and number of bins affect the result. We will continue to explore this method as a means to determine ν using confined helium.

ACKNOWLEDGMENTS

We wish to thank the National Science Foundation (DMR-9972285, and DMR-0242246) for their support of this work and the Cornell NanoScale Science & Technology Facility (526-94).

REFERENCES

1. M. E. Fisher, "The theory of critical point singularities," in *Critical Phenomenon, Proc. 51st "Enrico Fermi" Summer School, Varenna, Italy*, edited by M. Green, Academic Press, NY, 1971.
2. M. E. Fisher, and M. N. Barber, *Phys. Rev. Lett.*, **28**, 1516–1519 (1972).
3. L. S. Goldner, N. Mulders, and G. Ahlers, *J. Low Temp. Phys.*, **93**, 125–176 (1993).
4. J. A. Lipa, J. A. Nissen, D. A. Stricker, D. R. Swanson, and T. C. P. Chui, *Phys. Rev. B*, **68**, 174518 (2003).
5. S. Mehta, M. O. Kimball, and F. M. Gasparini, *J. Low Temp. Phys.*, **114**, 467–521 (1999).
6. M. O. Kimball, S. Mehta, and F. M. Gasparini, *J. Low Temp. Phys.*, **121**, 29–51 (2000).
7. M. Diaz-Avila, F. M. Gasparini, and M. O. Kimball, *J. Low Temp. Phys.*, **134**, 613–618 (2004).
8. J. A. Lipa, D. R. Swanson, J. A. Nissen, Z. K. Geng, P. R. Williamson, D. A. Stricker, T. C. P. Chui, U. E. Israelsson, and M. Larson, *Phys. Rev. Lett.*, **84**, 4894–4897 (2000).
9. J. A. Nissen, and J. A. Lipa, Numerical data provided via private communication (2005).
10. S. Mehta, and F. M. Gasparini, *Phys. Rev. Lett.*, **78**, 2596–2599 (1997).
11. M. Campostrini, M. Hasenbusch, A. Pelissetto, P. Rossi, and E. Vicari, *Phys. Rev. B*, **63**, 214503 (2001).

Testing The Universality Of The Lambda Transition Using Confined Helium Mixtures

Mark O. Kimball and Francis M. Gasparini

Department of Physics, University at Buffalo, The State University of New York, Buffalo NY 14260, USA

Abstract. The universality of phase transitions is an important prediction of theories of critical behavior. Simply stated, microscopically different systems near a critical point may be described by universal quantities if their dimensionality is the same and the order parameter has the same degrees of freedom. One way to test this idea is to measure the thermodynamic response of a set of systems to an input where the response changes with a variation in some quantity like spatial confinement, impurity concentration, or even pressure. While the response of each system is different, the behavior may still be described by a common critical exponent if the idea of universality is correct. Confined mixtures of ^{3}He-^{4}He and pure ^{4}He are believed to satisfy these requirements. Here, amplitudes such as the magnitude of the correlation length and the temperature of the transition both depend upon the concentration of the mixture and the extent of confinement. However, universality predicts the value of the critical exponent that describes the behavior near the transition should not be affected. Therefore, mixtures confined to a planar film should scale with the same critical exponent as pure ^{4}He, regardless of concentration or magnitude of film thickness. We used two different planar confinement cells: one at 48.3 nm and a second at 986.9 nm. We compare specific heat data taken from seven different concentrations split between the two cells and use finite-size scaling theory in an attempt to collapse all the data onto a universal curve using a single critical exponent.

Keywords: helium, mixtures, scaling, specific heat, universality
PACS: 64.70.-p, 67.60-g, 68.35.Rh

The superfluid transition of helium is an example of a continuous phase transition. Near the transition temperature, thermodynamic fluctuations dictate the behavior of the free energy. This transition has been studied quite extensively and the specific heat and superfluid density are well known in the thermodynamic limit. The behavior of helium confined to be homogeneously small in one dimension has also been studied [1–3]. Here, finite-size effects become evident near the transition since the temperature-dependent correlation length becomes comparable to the spatial confinement. Less well studied is the behavior of confined mixtures of ^{3}He-^{4}He. Compared to ^{4}He, mixtures have a larger correlation length and lower transition temperature. This reduction in transition temperature is a function of concentration and defines a locus of transitions known as the λ-line. Mixtures are expected to belong to the same universality class as pure helium and therefore should be described by the same critical exponents. Thus, mixtures of ^{3}He-^{4}He may be used to test the ideas of universality along a line of continuous transitions.

The cells used to confine helium to a planar geometry have been discussed in detail elsewhere [1, 2]. Briefly, these cells consist of two silicon wafers directly bonded together. One wafer has a particular thickness of oxide grown on it. The majority of this oxide is removed leaving a 4 mm wide ring on the outer circumference and an array of 0.2 mm square posts inside this ring. Upon bonding, the two wafers and border form a closed cell with the border and posts separating the two wafers by a distance defined by the oxide thickness. Helium is introduced into the cell through a hole in one of the wafers.

The small sample size requires the use of an ac technique to measure the heat capacity [4, 5]. A periodic voltage is applied to a heater on the cell bottom. This produces temperature oscillations which are measured using thermometers affixed on top of the cell. The heat capacity is related to the magnitude of the heat input and resulting temperature oscillations.

Since ^{3}He is a mobile impurity in the system, the heat capacity measured at constant pressure and concentration is renormalized and is not expected to show true critical behavior [6]. Therefore, one must perform a conversion to a specific heat at constant pressure and difference in chemical potential between ^{3}He and ^{4}He, $\phi = \mu_3 - \mu_4$. The conversion involves derivatives taken along the lambda line and is given by [7, 8]

$$C_{p\phi} = \frac{C_{px}\left(T\left.\frac{\partial x}{\partial T}\right|_{p,t}\left.\frac{\partial \phi}{\partial T}\right|_{p,t} - T\left.\frac{\partial s}{\partial T}\right|_{p,t}\right) + \left(T\left.\frac{\partial s}{\partial T}\right|_{p,t}\right)^2}{T\left.\frac{\partial s}{\partial T}\right|_{p,t} + T\left.\frac{\partial x}{\partial T}\right|_{p,t}\left.\frac{\partial \phi}{\partial T}\right|_{p,t} - C_{px}}. \quad (1)$$

where s is the molar entropy and the derivatives in this conversion are tabulated in [7].

Finite-size scaling theory [9, 10] predicts that the heat capacity of confined helium, $C_{p\phi}(L,\theta)$, will scale as a

CP850, *Low Temperature Physics: 24th International Conference on Low Temperature Physics;*
edited by Y. Takano, S. P. Hershfield, S. O. Hill, P. J. Hirschfeld, and A. M. Goldman

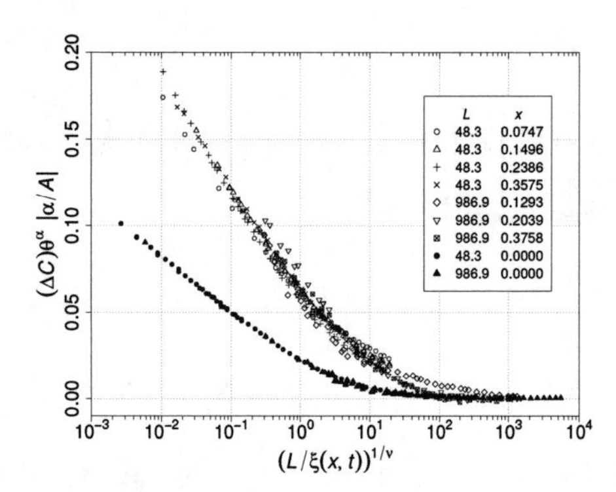

FIGURE 1. Here, all data are scaled using an α determined from each particular concentration.

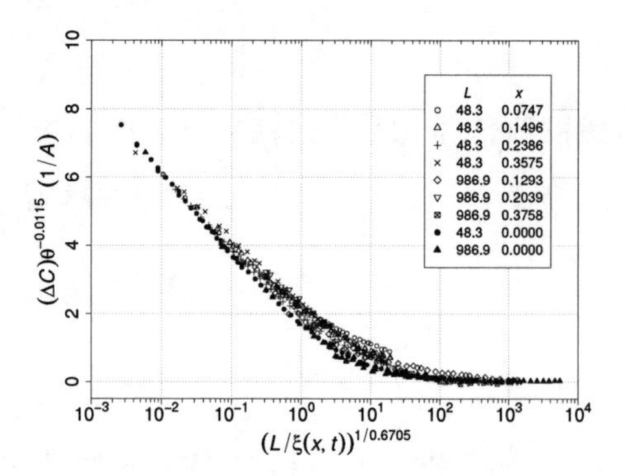

FIGURE 2. Scaled data for seven concentrations measured in two different confinement cells at $T > T_\lambda$. Here, the same α is used for the pure system and the mixtures.

function of L/ξ and is given by

$$\Delta C\theta^{\alpha}\left(\frac{\alpha}{A(x)}\right) = g_2\left(\left[\frac{L}{\xi(x,\theta)}\right]^{1/\nu}\right). \tag{2}$$

Here, $\Delta C = \left[C_{p\phi}(\infty,\theta) - C_{p\phi}(L,\theta)\right]$, $\theta = |T - T_\lambda|/T_\lambda$ is a dimensionless reduced temperature appropriate for $C_{p\phi}$[7], L is the spatial confinement length, ξ is the correlation length, $A(x)$ is the amplitude of the bulk specific heat, and α and ν are the critical exponents associated with the specific heat and correlation length respectively. The ratio $(\alpha/A(x))$ removes the leading concentration dependent amplitude of the heat capacity and allows the mixtures and pure system to be compared on the same scale.

Figure 1 shows 9 sets of data for $T > T_\lambda$ scaled according to equation 2. Here, the α used for each mixture in the scaling equation is unique and comes from fitting bulk data at each concentration. While the mixture data collapse, it is obvious that the pure data (solid symbols) stand apart. This is due to the fact that the bulk mixture data yield an exponent α which is close to -0.025, a factor of $\sim$2 more negative than for pure ^{4}He.

Another analysis is shown in figure 2. Here, data are scaled using an universal $\alpha = -0.0115$ derived from superfluid density measurements of pure ^{4}He [11]. For this analysis, one must assume the exponent which is determined from fitting the bulk $C_{p\phi}(x)$ is an effective exponent and the proper exponent used to scale the data is the one determined from the $x = 0$ data. However, the amplitudes A(x) come from fits of the data where α is allowed to take on its best fit value, ~ -0.025. While the collapse of all the data in Fig. 2 is perhaps satisfactory, one must remember this is not a self-consistent analysis since the value of α used to scale the data is not the same as that used to determine the magnitude $A(x)$ of the specific heat of the mixtures.

In summary, we have tested the ideas of universality using confined mixtures of helium. This includes seven different concentrations split between two different spatial confinements. Data for $x \neq 0$ scale amongst themselves but there is not a self-consistent means to allow the $x = 0$ to scale with the $x \neq 0$ data. One may question if these results indicate that the point at $x = 0$ on the phase diagram is special. This would be consistent with the determination of a different exponent obtained for bulk pure ^{4}He and mixtures of ^{3}He-^{4}He. However, the uniqueness of this point is not expected on theoretical grounds.

ACKNOWLEDGMENTS

We wish to thank the National Science Foundation (DMR-0242246) for their support and the Cornell NanoScale Science & Technology Facility (526-94).

REFERENCES

1. S. Mehta, M. O. Kimball, and F. M. Gasparini, J. Low Temp. Phys. **114**, 467 (1999).
2. M. O. Kimball, S. Mehta, and F. M. Gasparini, J. Low Temp. Phys. **121**, 29 (2000).
3. J. A. Lipa et al., Phys. Rev. Lett. **84**, 4894 (2000).
4. P. F. Sullivan and G. Seidel, Phys. Rev. **173**, 679 (1968).
5. S. Mehta and F. M. Gasparini, J. Low Temp. Phys. **110**, 287 (1998).
6. M. E. Fisher, Phys. Rev. **176**, 257 (1968).
7. F. M. Gasparini and M. Moldover, Phys. Rev. B **12**, 93 (1975).
8. G. Ahlers, Phys. Rev. A **10**, 1670 (1974).
9. M. E. Fisher, The theory of critical point singularities, in *Critical Phenomenon, Proc. 51st "Enrico Fermi" Summer School, Varenna, Italy*, edited by M. Green, Academic Press, NY, 1971.
10. M. E. Fisher and M. N. Barber, Phys. Rev. Lett. **28**, 1516 (1972).
11. L. S. Goldner, N. Mulders, and G. Ahlers, J. Low Temp. Phys. **93**, 125 (1993).

Measurements Of The Superfluid Fraction Of ^{4}He In 9.4 nm Channels, 19 μm Wide And 2000 μm Long

Manuel Diaz-Avila and Francis M. Gasparini

Department of Physics, University at Buffalo, The State University of New York, Buffalo NY 14260, USA

Abstract. We report measurements of the superfluid fraction of ^{4}He confined in small channels 9.4 nm high, 19 μm wide by 2000 μm long. This confinement corresponds to a film of finite lateral extent. The data show a shift in the transition to a lower temperature which is larger than the logarithmic dependence expected from finite-size scaling and Berezinskǐ -Kosterlitz-Thouless theory. This shift however is smaller than the one proposed by Sobnack and Kusmartsev for this kind of geometry. When examining the behavior of the shift for confinement at several widths, we found that the shift favors a power law with a larger exponent than predicted by Sobnack and Kusmartsev.

Keywords: superfluidity, finite-size effects, confined helium.
PACS: 67.40.-w

The superfluid properties of ^{4}He films have been studied extensively for many years. However, the properties of films which in addition are laterally constrained have not been explored in as much detail. We report measurements of the superfluid fraction of ^{4}He confined in small channels 9.4 nm high, 19 μm wide by 2000 μm long. We compare these data to earlier results.

To achieve the film geometry for this experiment, we confine helium between two silicon wafers, 5 cm in diameter, directly bonded at a separation determined by a lithographically-formed pattern of SiO_2. One wafer, with a center hole, is patterned with a 4000 μm wide outer ring, a 2000 μm wide inner ring, and a series of SiO_2 posts which provide a 310.6 nm separation between the two wafers. The outer ring is used to seal the cell and the inner ring defines two reservoirs, one that is immediately below the ^{4}He filling line located at the center of the cell and the other deeper into the cell. The second wafer has radially patterned onto it 72 channels 9.4 nm high, 19 μm wide and 2000 μm long which connect both reservoirs. When these wafers are bonded, the two reservoirs are connected by the narrow channels. Superfluid helium can be driven in resonance across the channels by using a film heater which is deposited on one of the wafers. The resulting temperature oscillations are detected using a biased germanium thermometer. The technique of adiabatic fountain resonance (AFR) has been described previously [1]. The superfluid density is obtained by fitting the frequency response to the excited AFR lineshape.

In Fig.1 we have plotted the superfluid fraction of ^{4}He and dissipation in two different channels. One result corresponds to the 9.4 nm high by 19 μm wide channels and the other corresponds to 10 nm high by 8 μm wide channels previously measured [2].

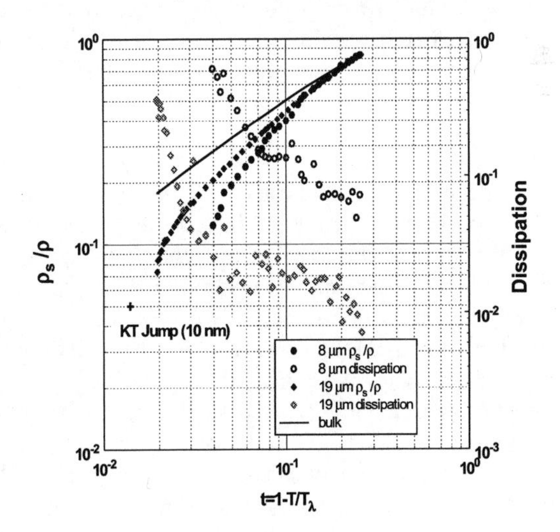

FIGURE 1. Superfluid fraction and dissipation of ^{4}He confined in two different films of finite lateral extent as function of the reduced temperature $t=1-T/T_\lambda$.

CP850, *Low Temperature Physics: 24th International Conference on Low Temperature Physics;* edited by Y. Takano, S. P. Hershfield, S. O. Hill, P. J. Hirschfeld, and A. M. Goldman

The cross indicates the expected Berezinskǐ-Kosterlitz-Thouless (BKT) transition [3,4] for the superfluid fraction. The temperature (T_c) at which the transition should take place is based on the behavior of the superfluid onset for planar films of effectively infinite lateral extent. This temperature is well described by the exponent ν expected from the 3D correlation length as $t_c=(1-T_c/T_\lambda)\sim L^{-1/\nu}$. Furthermore, given the behavior of the 2D correlation length, and finite-size scaling, one expects that there would be a further shift given by $\Delta t_c=[T_c(L,\infty)-T_c(L,W)]/T_c(L,\infty)=[2\pi/b\ln(W/\xi_o)]^2$ where ξ_o is the effective vortex core radius and b, a non-universal constant. We have estimated this shift to be of the order of 2×10^{-4} for these sets of data. However, we observed that, for both channels, the value of t_c at which the transition occurs is shifted to a much lower temperature than expected from finite size-scaling and BKT. This shift is larger for the 8 μm wide channels than for the 19 μm wide ones. This experiment, in addition to previous data obtained in our laboratory for 2D finite films [2,5], supports the idea that there is a new class of transition for thin films of finite lateral extent as predicted by Sobnack and Kusmartsev [6]. Their theory states that, for this kind of films, the shift in t_c should behave as a power law such that $\Delta t_c=(2\xi_o/W)^{1/2}$. This power-law behavior has not been confirmed yet for superfluid films. But, there is some suggestion from our most recent and preliminary data [7], in addition to the ones presented in this paper, that the power is larger than 1/2. These data corresponds to 9.4 nm channels with varying width (3 μm, 5 μm and 10 μm). By plotting the corresponding shift of the superfluid onset temperature versus the width of the channels and doing a least square fit, our data are *best represented by a power of 1.4±0.1.*

In general, one could argue that it is the dissipation which prevents ρ_s from reaching its expected value. But that is not the case. The general trend of the dissipation is that, far away from the transition, it slowly increases as one moves towards the transition, and then there is a sudden rise as the transition is approached. The rapid rise of the dissipation as ρ_s vanishes is qualitatively consistent with the intrinsic mechanism associated with vortex-pair unbinding characteristic of the two-dimensional superfluid. We studied the effect on the resonance frequency ω_0 as a function of excitation power in the heater. We observed, as shown in Fig. 2, that there is a mild dependence of ω_0 on power. Far from the transition, ω_0 decreases with increasing power. Closer to the transition this is reversed. Within this trend, one could reasonably extrapolate to ω_0 at zero power. The data of ρ_s plotted in Fig. 1 were obtained at 0.12 μW. Extrapolating to zero power would affect the values by ~2%. This is smaller than the size of the symbols used in the figure.

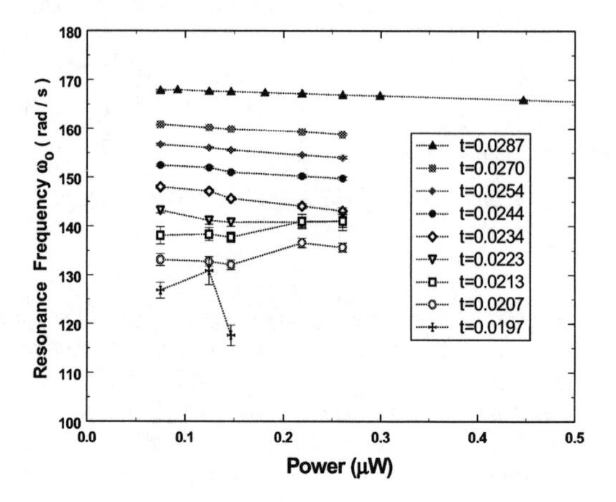

FIGURE 2. Resonance angular frequency plotted as function of power. The error bars represent the uncertainty with which the resonance frequency can be obtained from the signal's lineshape.

This new study supports the fact that in laterally confined 2D films the superfluid transition is shifted to lower temperatures than expected from BKT theory. However, it cannot verify the power-law behavior predicted by the Sobnack-Kusmartsev theory.

ACKNOWLEDGMENTS

We want to thank the National Science Foundation for its support, DMR-0242246, and the Cornell Nanofabrication Facility, project 526-94.

REFERENCES

1. Gasparini, F.M., Kimball, M.O.,and Mehta, S., *J. Low. Temp. Phys.* **125**, 215-238 (2001).
2. Diaz-Avila, M., Kimball, M.O., and Gasparini, F.M., *J. Low. Temp. Phys.* **134**, 613-618 (2004).
3. Berezinskǐ, V., *Sov. Phys. JETP* **32**, 493-500 (1971).
4. Kosterlitz, J.M., and Thouless, D.J., *J. Physics C* **6**, 1181-1203 (1973); and Kosterlitz, J.M., *J. Physics C* **7**, 1046-1060 (1974).
5. Kimball, M.O., Diaz-Avila, M. and Gasparini, F.M., *Physica B*: **329-333**, 248-249 (2003).
6. Sobnack, M.B., and Kusmartsev, F.V., *J. Low. Temp. Phys.* **126**, 517-526 (2002).
7. Diaz-Avila, M., and Gasparini, F.M., to be published.

Specific Heat of Helium in 2 μm^3 Boxes, Coupled or Uncoupled?

K.P. Mooney, M.O. Kimball, and F.M. Gasparini

Department of Physics, University at Buffalo, The State University of New York, Buffalo NY 14260, USA

Abstract. We report on recent measurements of the specific heat of helium confined in pill-boxes 2 μm across and 2 μm deep made lithographically on a silicon wafer. The experimental cells distribute liquid from a bulk reservoir to $\sim 10^8$ boxes by an array of very shallow fill-channels (0.019 μm and 0.010 μm) which represent a negligible volume compared to that of the boxes. Since the channels are so shallow, the helium in them becomes superfluid at a much lower temperature than the liquid in the boxes. Therefore, during the course of the heat capacity measurements, the liquid in the channels in always normal, and the cell would be expected to behave as a system of uncoupled boxes. We compare these measurements with one previously made of a cell where the confinement was to 1 μm boxes with an equivalent fill arrangement. While the shift in the position of the specific heat maximum relative to the 1 μm cell is what one would expect on the basis of finite-size scaling, there are discrepancies in the specific heat amplitude between the 2 μm cell utilizing different depth fill-channels, and with the 1 μm cell. It is possible that the channels, even though normal and of negligible volume, provide a weak coupling between the boxes leading to a collective rather than single-box behavior.

Keywords: finite-size effects, superfluidity
PACS: 67.40.Kh, 64.60.Fr

The specific heat of confined ^{4}He near the superfluid transition deviates substantially from bulk behavior as the correlation length ξ grows to be comparable to the the smallest confining spatial length scale L. This has been the subject of both experimental and theoretical work for some time [1].

Using a technique involving photolithography and direct silicon wafer bonding, we have been able to construct experimental cells to enclose helium in a specific geometry. Previously, we have used this method to build cells confining the helium to thin films or long narrow channels of square cross section. Details of cell construction, diagnostics, and mounting procedures can be found in prior publications [2, 3]. Our current cells make use of box-like structures to confine the helium in all three dimensions. Therefore, as the transition is approached, the behavior of the helium crosses over from three dimensional behavior (3D) to zero dimensional (0D).

Measurements where the liquid was confined to boxes of 1 μm have previously been reported [4]. We have constructed two new cells where the box size was 2 μm to investigate finite-size scaling for 0D crossover.

Each cell has on the order of 10^8 boxes. One needs a method of delivering helium from the fill line above the cell to each of the boxes. This is accomplished by patterning the top wafer with very shallow filling channels, 0.019μm in the case of both the 1 μm and first 2 μm cells, and 0.010μm in the case of the second 2 μm cell.

The shallowness of the channels serves two purposes. First, it minimizes the volume of helium contained in the channels compared to that in the boxes. In the case of the

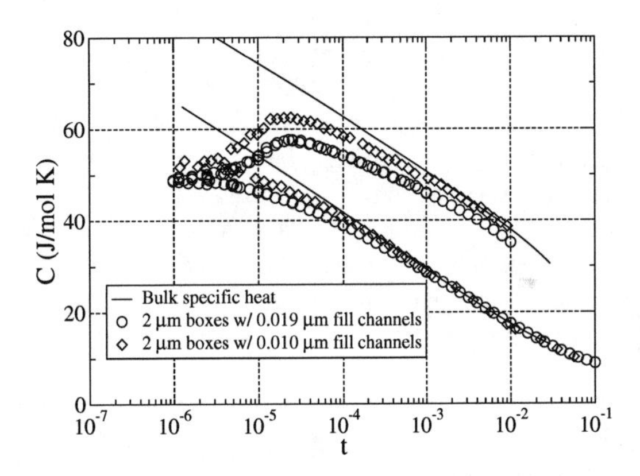

FIGURE 1. Specific heat of helium confined in boxes 2 μm in size but with different sized feed channels. The upper branch is for temperatures less than the bulk T_λ while lower branch is for data with $T > T_\lambda$. The solid lines is the specific heat of bulk helium.

1 μm cell, it represents about 1.8% of the total cell volume. The magnitude of the signal from this contribution is also drastically reduced due to finite-size effects. This minimizes unwanted heat capacity signal from the liquid in the channels. The superfluid transition temperature of the liquid in the channels is at a much colder temperature than the bulk T_λ. At all points during the measurements, the helium contained in the channels is *normal*. We can make a conservative estimate of the transition tempera-

CP850, *Low Temperature Physics: 24th International Conference on Low Temperature Physics;*
edited by Y. Takano, S. P. Hershfield, S. O. Hill, P. J. Hirschfeld, and A. M. Goldman

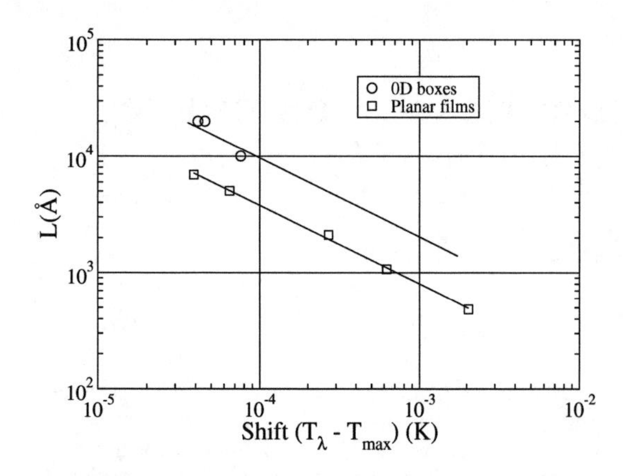

FIGURE 2. Confinement size in angstroms plotted as a function of the shift of C_m for both planar films and boxes.

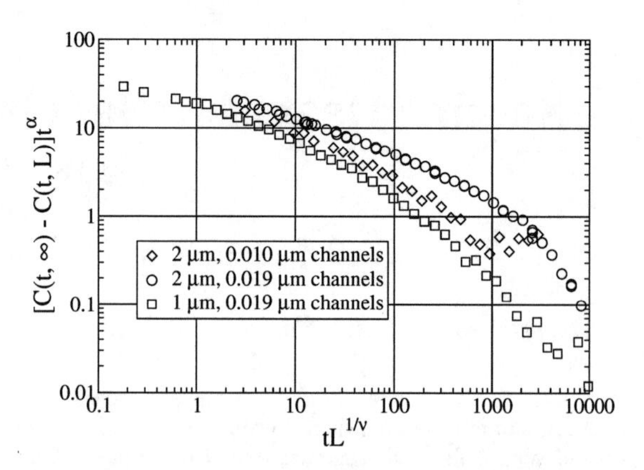

FIGURE 3. Data for both 2 μm cells and the 1 μm cell plotted according to Eq. 1. This is for $T > T_\lambda$ with L expressed in angstroms in the scaling variable.

ture in the channels by approximating the geometry as a 2-dimensional film of the same thickness, and taking the onset of superfluidity to be at the temperature of C_m, the maximum of the specific heat in the channels. We denote this temperature as T_m, and define a corresponding reduced temperature $t_m = (T_\lambda - T_m)/T_\lambda$. Using the data in Ref. [3] and the relation $t_m = a_0 L^{(-1/\nu)}$, gives roughly 10^{-2} for 100 Å thick films. Direct measurements of the superfluid fraction in the channels shows an onset in the vicinity of $t = 2 \times 10^{-2}$. It is desirable to have the superfluid transition colder than the temperature range at which one takes specific heat data so as to ensure the boxes are not coupled via a superfluid link.

Specific heat measurements are done using an AC calorimetry technique. An AC voltage is applied to a thin film heater evaporated onto the bottom of the cell, and the resulting temperature oscillations are detected by one of two germanium thermometers. The magnitude of the oscillations is inversely proportional to the heat capacity.

Figure 1 shows the specific heat for two different cells both patterned with 2 μm boxes. The most striking feature is that the data for the cell with 0.019 μm deep fill channels lies substantially below the data for the cell with the shallower channels. It also begins to deviate from the bulk at a warmer temperature, and never merges back into the bulk at colder temperatures.

The position of the specific heat maximum is the same for both cells. The difference in the position of the maximum between these cells, and the previous 1 μm cell is reasonably consistent with what one would expect on the basis of finite-size scaling. A plot of the confinement size versus the temperature shift in C_m should give a straight line on a log-log plot. This has been done for data taken where the helium was confined to a thin film. We would expect the data for the boxes to define a straight line parallel to the data for the films. While more points are clearly needed in the case of the boxes, the existing data is reasonably consistent with expectations.

On the basis of finite-size scaling, we should be able to cast the data in scaling form:

$$[C(t,\infty) - C(t,L)]t^\alpha = g_2(tL^{1/\nu}), \tag{1}$$

where ν and α are respectively the correlation length and specific heat exponents, and g_2 is a universal function. Clearly the data plotted in Fig. 1 will not scale since the confinement size is the same while the specific heat is different. When these two data sets are plotted along with data for the 1 μm cell, it is clear that none of these data scale at all. This contrasts markedly with data for the thin films of Ref. [3] which scale very well over many decades of reduced temperature.

It may be that our assumption of isolated boxes is incorrect, and there is some coupling, perhaps via fluctuations, between them. Future work needs to investigate this possibility, and will involve measurements where the size of the fill-channels distance are varied in a systematic way.

REFERENCES

1. M. N. Barber, *Phase Transitions and Critical Phenomena*, edited by C. Domb and J. L. Lebowitz, New York: (Academic Pres, 1983, pp 146-259).
2. I. Rhee, D. J. Bishop, A. Petrou, and F. M. Gasparini, *Rev. Sci. Instrum.*, **61**, 1528 (1990).
3. S. Mehta, M. O. Kimball, and F. M. Gasparini, *J. Low Temp. Phys.*, **114** (Nos.) 5/6, 467 (1999).
4. M. O. Kimball, M. Diaz Ávila, and F. M. Gasparini, *Physica B*, **329-333**, 286 (2003).

Specific Heat of Helium at Constant Volume along the Lambda Line

J. A. Lipa, J. A. Nissen, D. Avaloff and Suwen Wang

Hansen Experimental Physics Laboratory, Stanford University, Stanford, CA 94305, USA

Abstract. We report new measurements of the constant-volume specific heat of helium along the lambda line from 0.15 to 24.4 bars. The pressure in the cell was also recorded as a function of temperature using a gauge with a superconducting readout. This data can be used to convert the results to the constant-pressure specific heat along isobars. The constant-volume data compare well with earlier results and extend the temperature range of the measurements much closer to the lambda line. A preliminary conversion to $C_p(T,P)$ indicates good agreement with universality.

Keywords: Helium, Specific Heat, Lambda Line, Superfluid.
PACS: 65.40.Hq, 67.40.Kh

INTRODUCTION

Accurate measurements of the specific heat of helium near the lambda lines as a function of pressure or ^{3}He concentration are important for testing the predictions of universality that are a central feature of the theory of second order phase transitions. Universality predicts that the exponents characterizing the divergence of the specific heat and the singular behavior of the superfluid density are independent of pressure, P, or ^{3}He concentration along the respective lambda lines, and that some ratios of coefficients show similar behavior. These predictions have been tested with moderate precision in a number of experiments [1-4] and some small discrepancies have been noted [5]. Here we report new measurements of the constant-volume specific heat, C_V, in which the fluid pressure was also recorded as a function of temperature. This information enables us to compute the constant-pressure specific heat, C_P, to which universality should apply. It will also allow us to compute the superfluid density from earlier second sound velocity measurements [6] more accurately than has been possible to date. The present results extend significantly closer to the transition than previous work [1-3], allowing a better determination of the asymptotic behavior.

Here we describe some preliminary results of the measurements using earlier pressure measurements obtained by Ahlers [2] to convert from C_V to C_P. Our own pressure measurements have not yet been analyzed, but these are expected to result in significantly more accurate C_P data close to T_λ.

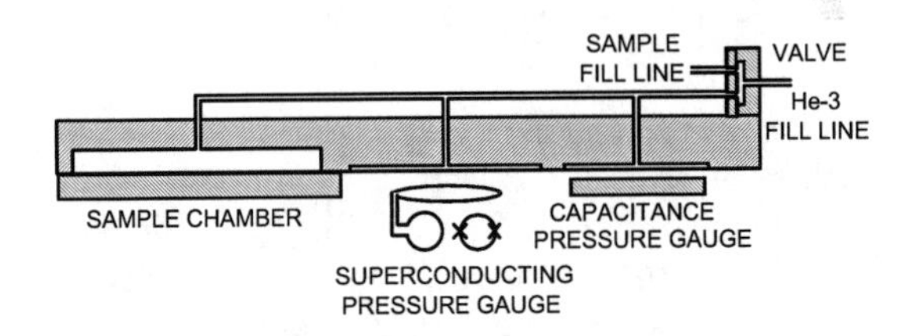

FIGURE 1. Sketch of the calorimeter assembly used in the measurements (not to scale).

APPARATUS

The basic configuration of the apparatus is shown in Fig. 1. It consists of a thin, disc-shaped sample chamber of height 1 mm and volume 1.5 cc made from annealed copper on the top and cylindrical wall. The base was a Be-Cu plate backed by a brass flange. The chamber was connected to a superconducting pressure gauge [7] via a small tube. Due to the stiffness of the flexure plate in the gauge, the thermodynamic trajectory of the helium during the measurements was very close to a constant volume path. A paramagnetic salt thermometer with a resolution of $< 10^{-9}$ K and a leak tight valve were attached to the copper top along with a heater and germanium thermometer. A second, smaller pressure gauge with a capacitance readout was also attached to the sample chamber near the valve. This was used to transfer an external pressure calibration to the superconducting gauge during the

CP850, *Low Temperature Physics: 24th International Conference on Low Temperature Physics;*
edited by Y. Takano, S. P. Hershfield, S. O. Hill, P. J. Hirschfeld, and A. M. Goldman

measurements. The assembly was installed in a three-stage thermal isolation system with a miniature cooler.

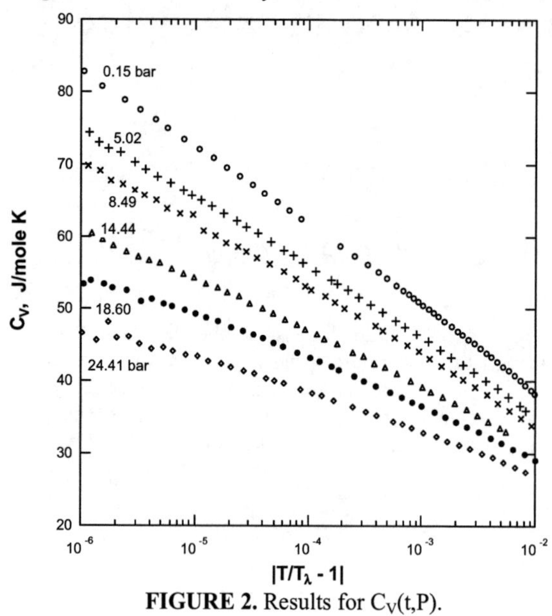

FIGURE 2. Results for $C_V(t,P)$.

RESULTS

The measurements were performed along six isochors at pressures extending from the 0.15 to 24.4 bars and the results for C_V below the transition with $t = |T/T_\lambda - 1| > 10^{-6}$ are shown in Fig. 2. The data have been corrected for the heat capacity of the empty calorimeter that was measured separately. The temperature of the calorimeter was found to be somewhat unstable during the measurements due to the use of high-pressure ^{3}He to close the valve. Because the line to the valve extended to the top of the cryostat, the internal pressure fluctuated with room temperature. This had the effect of adding noise to most of the measurements and limited the resolution of the experiment to about the level where gravity effects become important, at $t \sim 10^{-7}$. Additional data obtained for $t < 10^{-6}$ will be reported separately.

The results were converted to C_p along isobars intersecting the lambda line at the same pressure as the raw measurements using the method described by Ahlers [2]. Figure 3 shows the results obtained as a function of t and P. We find good agreement with previous measurements [1-3] except at 14.4 bars, where our results appear to be about 5% higher than expected in the region of overlap ($t > 10^{-5}$). We fitted some of the data with the asymptotic function:

$$C_p = A\, t^{-\alpha}\,(1 + a_0\, t^{0.529} + a_1\, t^{1.058}) + B, \quad T < T_\lambda \quad (1)$$

where A, a_0, a_1 and B are pressure-dependent adjustable constants. Since the present pressure corrections are only approximate, the exponent α was tentatively set equal to the value recently reported at the SVP for very wide range data [8]. The resulting fits are shown by the solid lines in Fig. 3 for two pressures. It can be seen that there is good agreement in both cases. The fit is slightly worse for 24.4 bars, but within the present uncertainties. Corrections from C_V to C_P are large at this pressure and need to be carefully evaluated. A more detailed analysis is in progress.

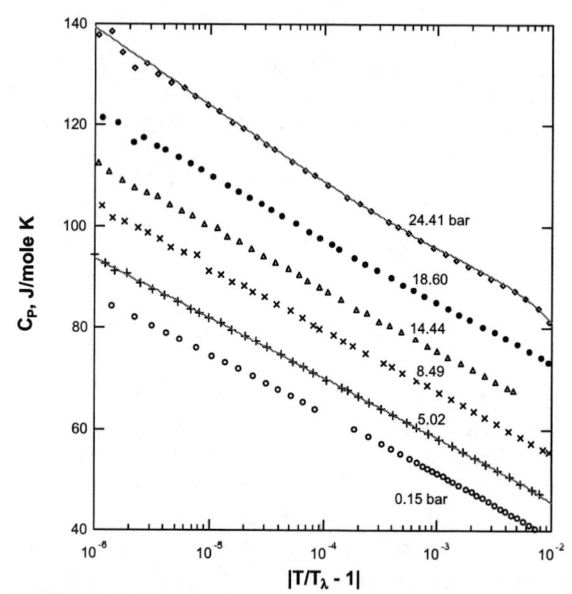

FIGURE 3. Results for $C_P(t,P)$. Data sets for $P > 0.15$ bar have been raised in increments of 10 J/mole K for clarity. Lines show fits to Eq. 1.

ACKNOWLEDGMENTS

This work was supported by NASA grant NAG3-2873.

REFERENCES

1. T.–H. McCoy and E. H. Graf, *Phys. Lett.*, **38A**, 287 (1972).
2. G. Ahlers, *Phys. Rev. A*, **8**, 530 (1973).
3. M. Okaji and T. Watanabe, *J. Low Temp. Phys.*, **32**, 555 (1978).
4. F. M. Gasparini and M. R. Moldover, *Phys. Rev. B*, **12**, 93 (1975).
5. F. M. Gasparini and A. A. Gaeta, *Phys. Rev. B*, **17**, 1466 (1978).
6. J. A. Nissen, D. R. Swanson, Z. K. Geng, K. Kim, P. Day and J. A. Lipa, *Physica B*, **284-288**, 51 (2000).
7. C. Edwards, L. Marhenke and J. A. Lipa, *Czech. J. Phys.* **46**, Suppl. S1, 2755 (1996).
8. J. A. Lipa, J. A. Nissen, D. A. Stricker, D. R. Swanson, and T. C. P. Chui, *Phys. Rev. B*, **68**, 174518 (2003).

Effect of Inhomogeneous Heat Flow on the Enhancement of Heat Capacity in Helium-II by Counterflow near T_λ

STP Boyd[a], AR Chatto[b], RAM Lee[b], RV Duncan[a,b], and DL Goodstein[b]

[a]*Department of Physics and Astronomy, University of New Mexico, Albuquerque NM 87131, USA*
[b]*Condensed Matter Physics, California Institute of Technology, Pasadena CA 91125, USA*

Abstract. In 2000 Harter *et al.* reported the first measurements of the enhancement of the heat capacity $\Delta C_Q \equiv C(Q)-C(Q=0)$ of helium-II transporting a heat flux density Q near T_λ. Surprisingly, their measured ΔC_Q was ~7-12 times larger than predicted, depending on which theory was assumed. In this report we present a candidate explanation for this discrepancy: unintended heat flux inhomogeneity. Because $C(Q)$ should diverge at a critical heat flux density Q_c, homogeneous heat flow is required for an accurate measurement. We present results from numerical analysis of the heat flow in the Harter *et al.* cell indicating that substantial inhomogeneity occurred. We determine the effect of the inhomogeneity on ΔC_Q and find rough agreement with the observed disparity between prediction and measurement.

Keywords: helium, superfluid, heat capacity, lambda transition.
PACS: 67.40.Kh, 67.40.Pm, 64.60.Ht

In order to evaluate the idea that unintended inhomogeneity of the heat flow in the Harter *et al.*[1] experiment might account for the discrepancy between measurement and predictions[2,3] of ΔC_Q, we must estimate the heat flow field $\boldsymbol{Q}(\boldsymbol{r})$ in the helium-II. It is not difficult to show[4] that thermal counterflow in helium-II can be solved simultaneously with the diffusive heat flow in the enclosing experimental cell using a standard finite-element solver[5], if the helium-II is nondissipative, nonvortical, nearly isothermal, and free of net mass flow ($\boldsymbol{J}=0$). These conditions should have been well-approximated in the Harter *et al.* experiment.

Such a numerical model has been constructed and solved for the Harter *et al.* cell. The model geometry is shown in Fig. 1. Not visible at this scale is the model for the Kapitza boundary resistance R_K: an artificial thin envelope of thickness δ=25 μm and thermal conductivity $\kappa_{RK}=\delta/R_K$ interposed between the helium and the cell walls.

For best accuracy, the helium-II diffusion coefficient should be modeled as $\kappa_{He}=\alpha(\rho_s/\rho_n)$, where α is a large constant required to reduce the variation of the scalar superfluid velocity potential function[4]. However, Harter *et al.* had a very short cell (0.64 mm) and did not approach closer to T_λ than ~0.5×10^{-6} K, limiting the maximum variation of ρ_s/ρ_n over the height of their cell to ~10%. To reduce the number of required computations we have approximated ρ_s/ρ_n as constant and set $\kappa_{He}=10^6$ W/cmK. Test reductions of κ_{He} to 10^5 W/cmK changed calculated enhancements by only ~0.01%, verifying that κ_{He} is sufficiently large.

To within their measurement noise, Harter *et al.* found that $t^{-\alpha}\Delta C_Q$ was linear in $(Q/Q_c)^2$, where $t=(T_\lambda-T)/T_\lambda$ is the reduced temperature. Keeping only the

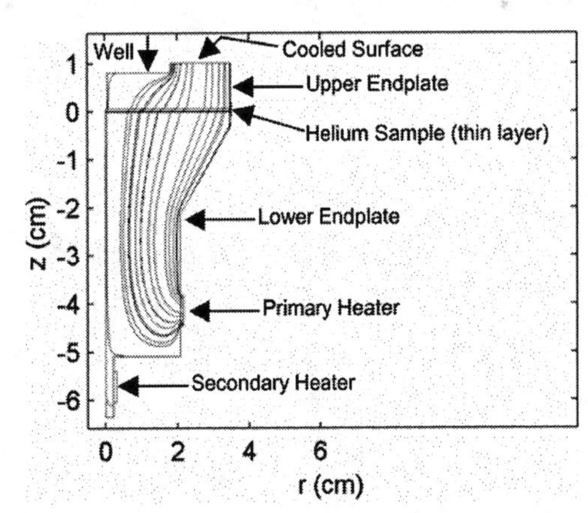

FIGURE 1. Cell model geometry with heat flow streamlines. The model is axisymmetric about r=0. Streamlines show heat flowing from the primary heater to the cooled surface. The principal cause of inhomogeneity of the heat flow in the helium is the "well" cut into the upper endplate to accommodate the diaphragm valve.

CP850, *Low Temperature Physics: 24th International Conference on Low Temperature Physics;* edited by Y. Takano, S. P. Hershfield, S. O. Hill, P. J. Hirschfeld, and A. M. Goldman

lowest-order $(Q/Q_c)^2$ term of the expansion for ΔC_Q[2], and neglecting the variation of reduced temperature t over the height of the cell, the fractional enhancement of ΔC_Q by inhomogeneity is

$$E = \frac{\int t^{-\alpha} \left(Q(\mathbf{r})/Q_c \right)^2 d\mathbf{r}}{\int t^{-\alpha} \left(Q_{\text{nom}}/Q_c \right)^2 d\mathbf{r}} = \frac{\int Q^2(\mathbf{r}) d\mathbf{r}}{Q_{\text{nom}}^2 V_{\text{helium}}} \qquad (1)$$

where the integrals are taken over the helium volume, and Q_{nom} is the "nominal" heat flux density (corresponding to that reported by Harter *et al.*) that would have been obtained for homogeneous heat flow.

In solutions of the linear heat flow equation for a given mixed boundary condition, the distribution of heat flux $\boldsymbol{Q}(\boldsymbol{r})$ is unaffected if all conductivities are scaled by the same factor. We have deliberately set κ_{He} so high that it is effectively infinite, thus $\boldsymbol{Q}(\boldsymbol{r})$ can depend only on the ratio of R_{K} to the endplate thermal conductivity κ_{Cu}. The calculations confirm this scaling: values of E agree to within ~0.1% or better for scenarios where the product $R_{\text{K}}\kappa_{Cu}$ is equal.

Although it was impossible to deduce an accurate R_{K} from the Harter *et al.* data, extensive measurements[6] exist of the value and reproducibility of R_{K} for Cu surfaces and helium-II near T_λ. Those measurements, together with others made by us at the University of New Mexico and Caltech, show that an estimate of R_{K}=1.0±0.2 cm^{2}K/W should be very reliable. We determined κ_{Cu} from published fits of κ_{Cu}(RRR)[7] and measurements of the RRR of several "core samples" cut from the bottom endplate of the Harter *et al.* cell by electrical discharge machining. These samples yielded RRR=240-260, thus κ_{Cu} =6.9-7.4 W/cmK.

The calculated E is shown in Fig. 2. Using R_{K}=1 cm^{2}K/W and κ_{Cu}=7.2 W/cm·K yields E=3.0, compared to the observed anomalous enhancement of ~7-12. Given the complexity and approximations involved in this post-experiment analysis, this level of agreement seems quite good.

Also shown in Fig. 2 are calculated maximum values of Q_{nom}/Q_c. Harter *et al.* found that above a maximum Q_{nom}/Q_c~0.3 (their "β" point) additional thermal resistance appeared between the bottom and top endplates. They proposed that this happened when the coherence length grew to exceed the surface roughness of the bottom endplate, effectively decreasing the bottom endplate area and increasing the apparent Kapitza resistance. Our present analysis provides another candidate explanation: the β point might be occurring when the maximum value of $|Q(\boldsymbol{r})|$

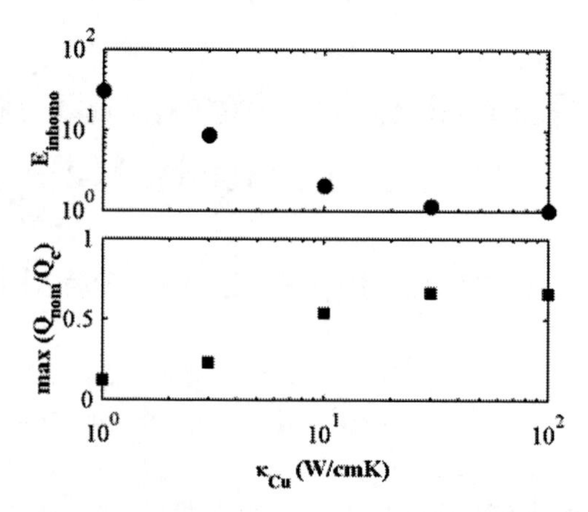

FIGURE 2. Calculated values of E and the β point over the expected range of κ_{Cu}, assuming R_{K}=1 cm^{2}K/W.

becomes comparable to Q_c, causing a local breakdown of superflow. In this case $\max(Q_{\text{nom}}/Q_c)$ should equal the calculated $Q_{\text{nom}}/\max(|Q(\boldsymbol{r})|)$, which for our estimated values of R_{K} and κ_{Cu} is 0.43, in reasonable agreement with observation. A recent reanalysis of some of the Harter *et al.* data by one of us (ARC) has shown that this new explanation may fit the data better than the correlation length argument.

In summary, numerical estimates of $\boldsymbol{Q}(\boldsymbol{r})$ in the Harter *et al.* cell indicate significant inhomogeneity which might explain both the anomalously large ΔC_Q and the β point. There is a clear need for new measurements of ΔC_Q in a new cell with homogenous heat flow. Such a cell has been prepared at the University of New Mexico, and the cooldown is presently underway at Caltech. Work supported by NASA Fundamental Physics Discipline NAG3-1763 (STPB), NAG3-2900 (ARC, RAML, DLG), and JPL #960494 (RVD). RVD acknowledges support at Caltech as a Moore Distinguished Scholar.

REFERENCES

1. Harter, A. W., Lee, R. A. M., Chatto, A., Wu, X., Chui, T. C. P., and Goodstein, D. L., *Phys. Rev. Lett.* **84**, 2195-2198 (2000).
2. Chui, T. C. P. Goodstein, D. L., Harter, A. W., and Mukhopadhyay, R., *Phys. Rev. Lett.* **77**, 1793-1796 (1996).
3. Haussmann, R., and Dohm, V., *Czech. J. Phys.* **46-S1**, 171 (1996).
4. Boyd, S. T. P., and Goodstein, D. L., in preparation.
5. *e.g.* FEMLAB v2.3, COMSOL Inc, www.comsol.com.
6. Dingus, M., Zhong, F., and Meyer, H., *J. Low Temp. Phys.* **65**, 185-212 (1986).
7. F. Pobell, *Matter and Methods at Low Temperatures*, Berlin: Springer-Verlag, 1992, p. 52. The Wiedemann-Franz inaccuracy (~10% here) has negligible effect here.

CW Measurement of the Upward-Going Temperature Wave in the Helium-4 Self-Organized Critical State

STP Boyd, DA Sergatskov, and RV Duncan

Department of Physics and Astronomy, University of New Mexico, Albuquerque NM 87131, USA

Abstract. We describe the first continuous-wave (CW) measurements of the upward-going temperature wave in the self-organized-critical (SOC) state which forms in ^{4}He under conditions of downward heat flow near T_λ under gravity. The CW technique permits measurements with extremely low (<1 nK) excitation amplitudes, allows continuous measurement of the wave velocity as the SOC state grows, and has yielded the first quantitative measurements of the attenuation. The CW measurements appear to support predictions for the velocity but disagree with predictions for the attenuation. This new technique may help us understand the underlying mechanism of the SOC state.

Keywords: helium, superfluid, lambda transition, self-organized criticality
PACS: 64.60.Ht, 67.40.Pm, 05.65.+b

When ^{4}He transports a heat flux Q downward in the Earth's gravitational field at temperature near T_λ, it can self-organize thermally into a state with vertical temperature gradient equal to the $\partial T_\lambda/\partial y$ imposed by the hydrostatic pressure head[1]. This "Self-Organized Critical" (SOC) state is at present the only means by which a bulk sample of ^{4}He can be brought uniformly extremely close to T_λ ($T-T_\lambda$<1 nK) in the Earth's gravitational field. Analysis indicates[2,3] that the SOC state should support an upward-going temperature wave (UGW), with velocity v_{UGW}~$|Q|^{3.3}$ and damping D_{UGW}~$|Q|$. The upward-going wave provides a unique new dynamic probe of the SOC state and of ^{4}He physics near T_λ.

We have previously reported[3] results of pulse time-of-flight (TOF) measurements which confirmed the existence of the upward-going wave and provided rough confirmation of the scaling v_{UGW}~$|Q|^{3.3}$. However, the TOF technique had several drawbacks: 1) it was not in the low-amplitude limit; 2) quantitative damping measurement could not be performed; and 3) the strong damping may have caused systematic error in the measured velocity. All of these concerns can be addressed by the continuous-wave (CW) technique.

The experimental cell and thermal network have been described previously[3]. For CW measurements, excitation of the upward-going wave at frequency $2f$ was accomplished by applying a sinusoidal voltage at frequency f to a heater on the bottom endplate. The temperature response of the helium was sensed at two heights, 5.59 and 7.87 mm above the bottom endplate, by two high-resolution (~1×10^{-10} K) PdMn magnetic alloy thermometers with SQUID readout. The magnitude and phase of the responses were then calculated by post-experiment numerical analysis.

"Direct" measurements of v_{UGW} were obtained from the phase delay difference of the two thermometers. Additional "phase slope" measurements of v_{UGW} were obtained by plotting the phase delay to each thermometer versus the helium-II temperature as the cell warmed and the SOC/helium-II interface moved upwards. Because $\partial T_\lambda/\partial y$ is effectively constant, the location of the SOC/helium-II interface, and thus the length of the SOC region, is linear in the helium-II temperature. The inverse slope of phase delay versus

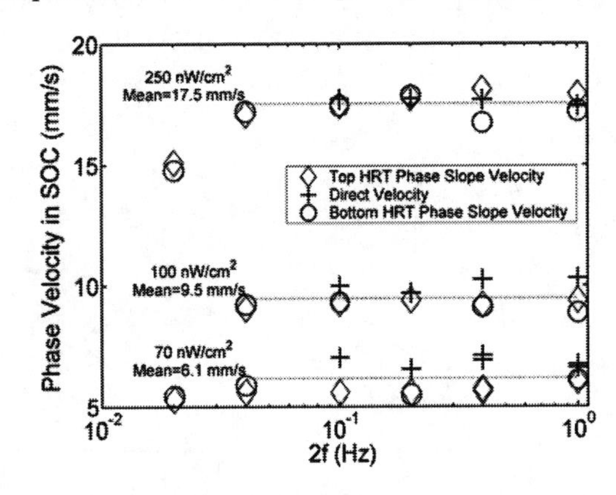

FIGURE 1. CW velocities of the SOC upward-going wave at three downward heat fluxes.

CP850, *Low Temperature Physics: 24th International Conference on Low Temperature Physics;* edited by Y. Takano, S. P. Hershfield, S. O. Hill, P. J. Hirschfeld, and A. M. Goldman

helium-II temperature thus gives v_{UGW}. This "phase slope" technique has provided the first confirmation that v_{UGW} remains constant as the SOC region grows, as expected.

The CW measurements of v_{UGW} are shown in Fig. 1. Mean velocities for each Q_{SOC} are indicated. The rolloff of v_{UGW} for $f < 0.04$ Hz is accompanied by distortions in the form of the data and appears to be an experimental artifact. The measurement precision can be seen from the two points where we have more than one data set: Q_{SOC}=70 nW·cm^{-2} at $2f = 0.4$ and 1 Hz. We see the direct velocity is somewhat larger than the phase slope velocity for lower Q_{SOC}. This discrepancy is unexplained at present. Lastly, we see no evidence of frequency dispersion for the data $0.04 \leqslant 2f \leqslant 1$ Hz, indicating that we are in the low-amplitude limit.

Figure 2 compares mean CW velocities to the TOF velocities previously reported[3]. The analysis of Ref. 2 indicates that the $v_{UGW} \sim Q^{3.3}$ scaling should only apply for Q_{SOC} less than approximately 100 nW·cm^{-2}. Figure 2 appears to support this conclusion. The line is a power-law fit to points with $Q_{SOC} \leqslant 100$ nW·cm^{-2}, yielding an exponent of 2.95. Additionally, we see that the point at 250 nW·cm^{-2} falls well below any Q^3 law. However, we can directly compare the CW and TOF values only at Q_{SOC}=70 nW·cm^{-2}, and there we find significant disagreement, by about a factor of two. Thus until more CW data can be obtained this agreement should be regarded as preliminary.

The CW attenuation measurements are shown in Fig. 3. The smooth curves are best fits for each Q_{SOC} using the simple linearized analysis of Ref. 3, with the fitted damping coefficient as indicated. The attenuation data disagree with predictions in two important ways. Firstly, for the two lower values of Q_{SOC} the low-frequency behavior is qualitatively

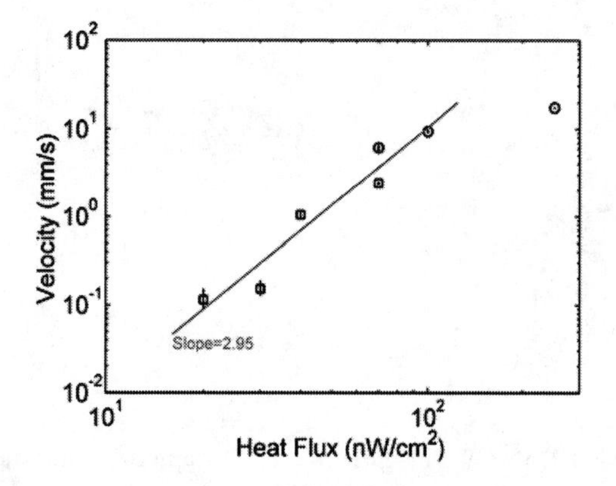

FIGURE 2. Comparison of mean CW velocities (circles) to the previously-reported TOF velocities (squares). Vertical bars indicate estimated error bars.

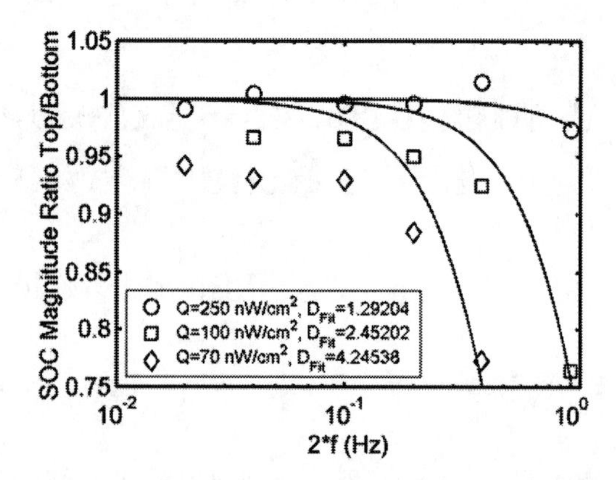

FIGURE 3. Amplitude ratios of the CW upward-going wave at the two thermometers.

different from the behavior of the smooth curves—the measured amplitude ratio shows no inclination to go to unity at these frequencies. Secondly, as shown, the fitted damping coefficients vary nearly as $1/Q_{SOC}$, rather than as the predicted Q_{SOC}. These results strongly suggest that the attenuation analysis of the simple linearized model is inadequate and needs to be revisited.

To summarize, the CW technique has been successfully implemented to study the upward-going wave of the SOC state. The new technique works well: it yields a new, very precise, "phase slope" velocity measurement; it achieves the low-amplitude limit; and it provides the first quantitative attenuation data. The initial group of CW measurements obtained in this first experiment appears to support predictions of v_{UGW} but disagrees strongly with predictions for attenuation. Additional CW data are urgently needed, and new measurements are presently underway at both UNM and Caltech.

We gratefully acknowledge discussions with D. L. Goodstein, R. A. M. Lee, and A. R. Chatto. This work supported by the Fundamental Physics Discipline of the NASA Office of Biological and Physical Research through JPL contract #960494. RVD acknowledges support at Caltech as a Moore Distinguished Scholar.

REFERENCES

1. Moeur, W. A., Day, P. K., Liu, F-C., Boyd, S. T. P., Adriaans, M. J., and Duncan, R. V., *Phys. Rev. Lett.* **78**, 2421 (1997), and references therein.
2. Weichman, P. B., and Miller, J., *J. Low Temp. Phys.* **119**, 155 (2000).
3. Sergatskov, D. A., Babkin, A. V., Boyd, S. T. P., Lee, R. A. M, and Duncan, R. V., *J. Low Temp. Phys.* **134**, 517 (2004), and references therein.

Measurement of the SOC State Specific Heat in ^{4}He

A. R. Chatto*, R. A. M. Lee*, R. V. Duncan†,*, P. K. Day** and D. L. Goodstein*

*Condensed Matter Physics, California Institute of Technology, Pasadena, CA 91125, USA
†Department of Physics and Astronomy, University of New Mexico, Albuquerque, NM 87131-1156, USA
**Jet Propulsion Laboratory, Pasadena, CA 91109, USA

Abstract. When a heat flux Q is applied downward through a sample of liquid ^{4}He near the lambda transition, the helium self organizes such that the gradient in temperature matches the gravity induced gradient in T_λ. All the helium in the sample is then at the same reduced temperature $t_{\mathrm{SOC}} = \frac{T_{\mathrm{SOC}}-T_\lambda}{T_\lambda}$ and the helium is said to be in the Self-Organized Critical (SOC) state. We have made preliminary measurements of the ^{4}He SOC state specific heat, $C_{\nabla T}(T(Q))$. Despite having a cell height of 2.54 cm, our results show no difference between $C_{\nabla T}$ and the zero-gravity ^{4}He specific heat results of the Lambda Point Experiment (LPE) [J.A. Lipa et al., *Phys. Rev. B*, **68**, 174518 (2003)] over the range 250 to 450 nK below the transition. There is no gravity rounding because the entire sample is at the same reduced temperature $t_{\mathrm{SOC}}(Q)$. Closer to T_λ, the SOC specific heat falls slightly below LPE, reaching a maximum at approximately 50 nK below T_λ, in agreement with theoretical predictions [R. Haussmann, *Phys. Rev. B*, **60**, 12349 (1999)].

Keywords: Self-Organized Criticality, Specific Heat, Helium, Superfluidity
PACS: 65.20+w,67.40.Kh,05.65+b

INTRODUCTION

In 1987, the Self-Organized Critical (SOC) state was predicted for the normal phase of ^{4}He in the presence of gravity [1]. Gravity creates a hydrostatic pressure gradient in the helium which creates a gradient in T_λ, the superfluid transition temperature, of $\nabla T_\lambda = 1.273\ \mu\mathrm{K/cm}$ [2]. The thermal conductivity $\kappa(t)$, where $t = \frac{T-T_\lambda}{T_\lambda}$, diverges as $t \to 0$ [3]. When a heat flux Q is applied downward through a sample of helium, the resulting temperature gradient parallels the gradient in T_λ, and the helium self organizes to satisfy the condition $Q/\kappa = \nabla T_\lambda$. Therefore, while there is a gradient in temperature, the temperature difference from T_λ is uniform throughout the entire sample, as shown in Fig. 1.

The SOC state in ^{4}He was first observed in 1997 by Moeur *et al.* [4]. They saw not only the expected normal phase SOC state, but also self-organization at temperatures below T_λ at higher heat flux. In addition, they found that κ diverged at $T_{\mathrm{DAS}}(Q)$, where $T_{\mathrm{DAS}}(Q)$ is the measured temperature at which perfect thermal conductivity of the superfluid state fails abruptly under a heat flux Q applied upwards through the helium [5]. This encouraged the interpretation that the heat flux was depressing the critical point T_λ to the lower $T_\lambda(Q) = T_{\mathrm{DAS}}(Q)$, and that the SOC state was therefore always on the 'normal' side of T_λ. In contrast, Weichman and Miller presented a theoretical model in one dimension that treated the high heat flux self organization as a superfluid with a series of phase slips in order to maintain the requisite temperature gradient [6].

In this paper, we report the first measurements of the specific heat of the SOC state [7].

EXPERIMENT

Our cell is constructed with two 2.3 cm diameter gold plated copper endplates epoxied to a cylindrical insulating Vespel® sidewall in order to give a 2.54 cm sample height. We have three high resolution thermometers (HRTs); one on each of the top and bottom endplates and a third on a 165 μm thick copper foil that penetrates the sidewall. This foil is positioned 0.64 cm above the bottom endplate and is in direct contact with the helium sample. We measured the size of our helium sample (0.389 moles) through a calibrated extraction and confirmed it with a traditional pulse heat capacity measurement in the superfluid phase.

Since the temperature of the SOC state is determined by the heat flux through the helium, one cannot put in a pulse of energy and measure the temperature change as in a conventional heat capacity measurement. Instead, the heat capacity is measured by establishing the SOC state throughout the sample at one heat flux, then switching to a new heat flux and measuring the amount of energy needed to re-establish the SOC state throughout the sample at its new temperature. This is shown schematically in Fig. 1. As heat is slowly added, the derivative of the top endplate thermometer is used to find the time when profiles 1 and 3 are reached, while the midplane thermometer measures the SOC temperature change.

CP850, *Low Temperature Physics: 24th International Conference on Low Temperature Physics;*
edited by Y. Takano, S. P. Hershfield, S. O. Hill, P. J. Hirschfeld, and A. M. Goldman

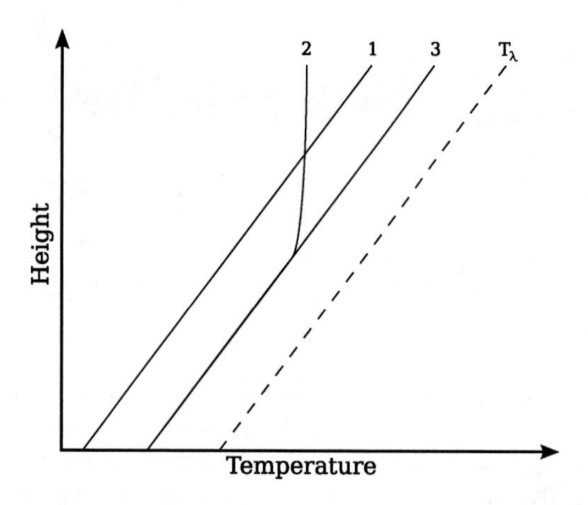

FIGURE 1. Profiles of helium temperature vs. sample height during the experimental procedure: (1) SOC state is fully established at the first heat flux; (2) Heat flux is decreased which raises the SOC temperature; (3) Energy is added to fully establish SOC state for the second heat flux. (Note: for $Q \lesssim 100\ \mathrm{nW/cm^2}$, the helium instead self organizes above T_λ.)

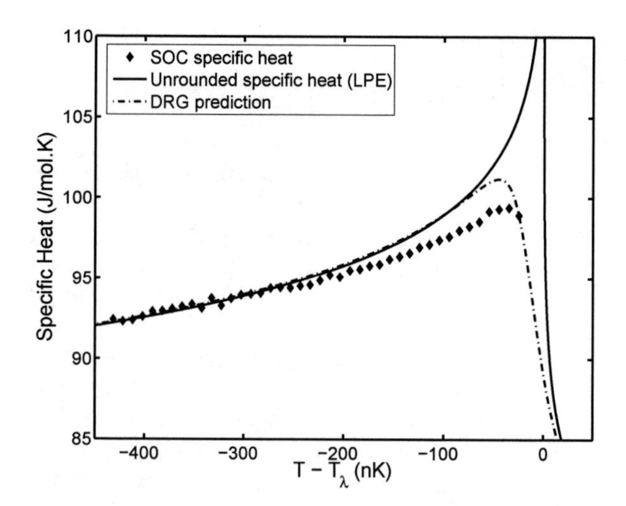

FIGURE 2. Specific heat of the SOC state

In traditional helium heat capacity measurements, where the helium is isothermal, the gradient in T_λ causes the heat capacity to be gravity rounded, (i.e. averaged over a range of reduced temperatures.) In contrast, in the SOC state, the temperature gradient is equal to ∇T_λ and the entire sample is equidistant from criticality. Therefore, there is no gravity rounding in our data, despite a sample height of 2.54 cm. Our SOC specific heat results shown in Fig. 2 are compared directly to a fit of the Lambda Point Experiment (LPE) results [8]. Haussmann's prediction for the SOC specific heat using Dynamic Renormalization Group (DRG) theory is also plotted [9].

We also measured the SOC temperature versus heat flux. Our results for 0.5 to 4.5 $\mu\mathrm{W/cm^2}$ are well fit by $t_{\mathrm{SOC}}(Q) = -(Q/Q_0)^{0.813}$ with $Q_0 = 760 \pm 10\ \mathrm{W/cm^2}$. Our Q_0 differs from Moeur *et al.* [4], but agrees well with a later experiment [10].

CONCLUSIONS

The data in Fig. 2 show no measurable difference from the LPE results in the range 250 to 450 nK below T_λ - i.e. the SOC state heat capacity is the same as that for the static superfluid. This may imply that the helium in this heat flux range of the SOC state is essentially in the superfluid phase. If so, this rebuts arguments that $T_{\mathrm{DAS}}(Q)$ is a depressed critical point, despite the fact that the thermal conductivity κ diverges at this temperature.

Closer to T_λ, we measure a slight depression in the specific heat, relative to LPE, which starts at approximately 250 nK below the transition. This differs from the prediction of the DRG theory, where the depression starts at approximately 100 nK below T_λ [9]. However, both theory and experiment reach a maximum at approximately 50 nK below T_λ.

ACKNOWLEDGMENTS

We would like to thank Dmitri Sergatskov, Steve Boyd, Alex Babkin, Alexander Churilov, and Talso Chui for helpful discussions and assistance with cell construction and cryovalve assembly and operation. This work was supported in part by the Fundamental Physics Discipline of the Microgravity Science Office of NASA.

REFERENCES

1. A. Onuki, *Jpns. J. Appl. Phys.*, **26-S3**, 365 (1987).
2. G. Ahlers, *Phys. Rev. Lett.*, **171**, 275 (1968).
3. Note: this is true only in the small Q limit.
4. W. A. Moeur, P. K. Day, F-C. Liu, S. T. P. Boyd, M. J. Adrianns, and R. V. Duncan, *Phys. Rev. Lett.*, **78**, 2421 (1997).
5. R. V. Duncan, G. Ahlers, and V. Steinberg, *Phys. Rev. Lett.*, **60**, 1522 (1988).
6. P. B. Weichman, and J. Miller, *J. Low Temp. Phys.*, **119**, 155 (2000).
7. Note: these measurements are very different from previous work [10] on the SOC/superfluid two-phase heat capacity.
8. J. A. Lipa, J. A. Nissen, D. A. Stricker, D. R. Swanson, and T. C. P. Chui, *Phys. Rev. B*, **68**, 174518 (2003).
9. R. Haussmann, *Phys. Rev. B*, **60**, 12349 (1999).
10. R. A. M. Lee, A. R. Chatto, D. A. Sergatskov, A. V. Babkin, S. T. P. Boyd, A. M. Churilov, T. D. McCarson, T. C. P. Chui, P. K. Day, R. V. Duncan, and D. L. Goodstein, *J. Low Temp. Phys.*, **134**, 495 (2004).

New Measurements of Wetting by Helium Mixtures

Ryosuke Ishiguro and Sébastien Balibar

Laboratoire de Physique Statistique de l'Ecole Normale Supérieure, associé aux Universités Paris 6 et 7 et au CNRS, 24 Rue Lhomond, 75231 Paris Cedex 05, France

Abstract. T. Ueno *et al.* found that the wetting of a wall by the interface between phase separated He mixtures was anomalous: the contact angle of the ^{3}He-^{4}He interface was non-zero near the tricritical point. This behavior was contradictory to "critical point wetting", and they proposed that it was due to critical Casimir forces, a consequence of fluctuations in a superfluid ^{4}He-rich film between the wall and the ^{3}He-rich bulk phase. In a new series of experiments, we repeated Ueno *et al.*'s measurements and extended them to lower temperature, looking also for a force associated with Goldstone modes. For this we changed the geometry of Ueno *et al.*'s cell and looked for possible artifacts in the interferometric images of the liquid meniscus near its contact line. We find that the contact angle is in fact zero at all temperatures : the ^{4}He-rich phase wets the wall completely. This does not mean that Casimir forces are absent, only that their amplitude is smaller than the van der Waals force, another long range force which favors wetting.

Keywords: wetting, Casimir forces, helium mixtures
PACS: 68.35.Rh, 67.60.-g, 64.60.Fr, 05.70.Jk, 64.60.Kw, 68.08.Bc

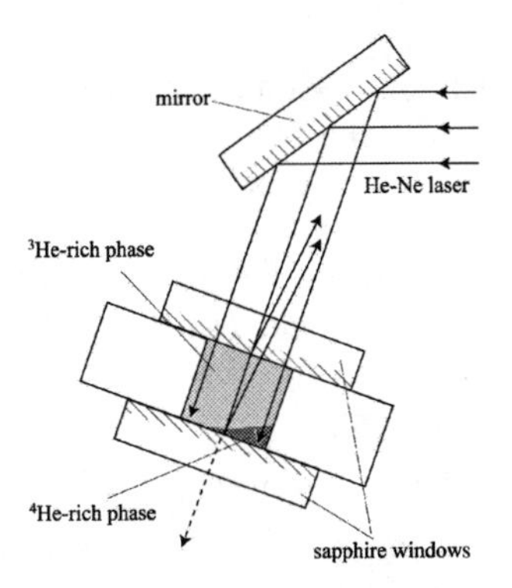

FIGURE 1. In this new experimental cell, the light beam is close to normal to the ^{3}He-^{4}He interface, so that refraction effects are negligible.

In a first experiment done in Kyoto, T. Ueno, M. Fujisawa, K. Fukuda, Y. Sasaki, and T. Mizusaki used MRI to measure the shape of the ^{3}He-^{4}He interface close to an epoxy wall [1]. They found that the contact angle increased from 20° at low temperature T to $40 \pm 40°$ as T approached the tri-critical temperature $T_t = 0.87$ K. This was surprising because the ^{4}He-rich phase should completely wet the wall close to T_t, according to Cahn's argument on critical wetting [2]. In order to check this, T. Ueno, S. Balibar, F. Caupin, T. Mizusaki and E. Rolley studied wetting with an optical interferometric technique [3]. Due to large refraction effect, their measurements were restricted to the temperature region from 0.81 to 0.86 K where the index difference between the concentrated phase and the diluted one is small. In this range, it was found that the contact angle increased with temperature from 15 to $55 \pm 15°$, a result which again indicated a possible exception to critical wetting.

In a third article [4], Ueno *et al.* explained that critical Casimir forces had the right sign and magnitude to explain the anomalous wetting behavior. These forces originate in the confinement of critical fluctuations between two surfaces [5-7]. In the case of He mixtures, Ueno *et al.* considered the fluctuations of superfluidity in a ^{4}He-rich superfluid film which might exist between the wall and the bulk ^{3}He-rich liquid.

It thus appeared worth checking Ueno *et al.*'s results by repeating their last experiment in a different geometry. For this, we rotated Ueno *et al.*'s cell by 90 degrees, so that the incidence of the light beam was close to normal on the ^{3}He-^{4}He interface. Refraction effects became negligible and the measurements could be extended to low T [8]. Our analysis of new interface profiles gave us values close to zero for the contact angle and values for the surface tension which were in reasonable agreement with previous measurements by Sato *et al.* [9] and by Leiderer *et al.* [10]. But it had rather large error bars. A better agreement for the surface tension was obtained when we assumed that the contact angle was zero. As shown in Fig. 2 we obtained precise agreement with previous measurements. In a further step, we then used Sato's value for the interfacial tension σ_i far from T_t [9] and Leiderer's value close to T_t [10] to obtain the contact angle (Fig. 3):

CP850, *Low Temperature Physics: 24th International Conference on Low Temperature Physics;* edited by Y. Takano, S. P. Hershfield, S. O. Hill, P. J. Hirschfeld, and A. M. Goldman

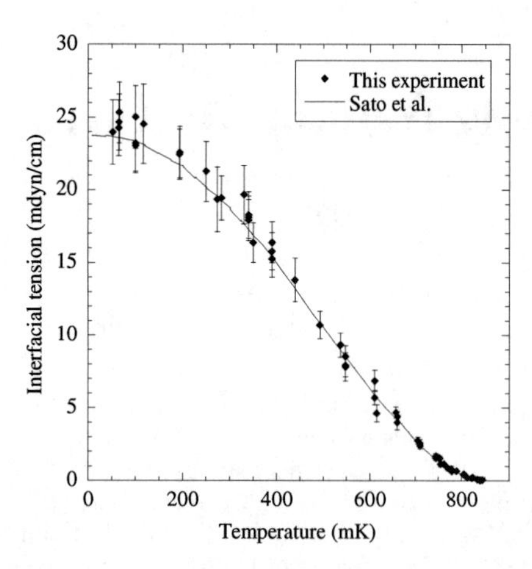

FIGURE 2. Assuming that the contact angle is zero, we find precise agreement between our measurements of the interfacial tension and previous measurements by Sato *et al*[9].

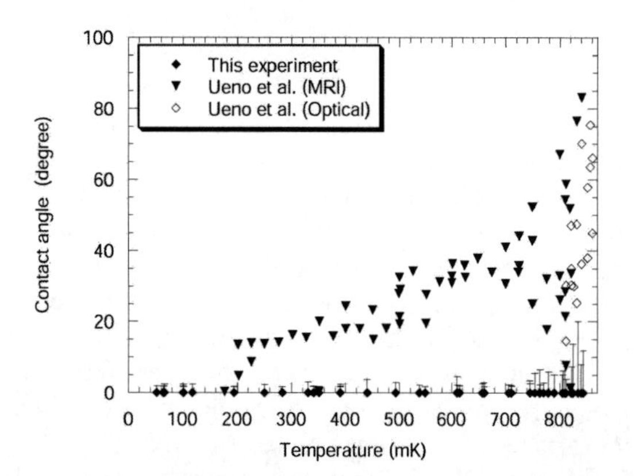

FIGURE 3. New measurements show that the contact angle of the ^{3}He-^{4}He interface is zero.

we found zero within 10 degrees in the whole temperature range.

There must have been some artifact in the experiment done by Ueno *et al.* in 2003. In these optical measurements, the interface profile is obtained from the difference between a reference pattern without interface and a pattern with an interface. In Ueno *et al.*'s experiment, the reference was extrapolated so that they could not account for the possible existence of defects. In the new experiment, the reference was measured.

Our results also call for some criticism of the theoretical interpretation by Ueno *et al.* [4]. We believe that these authors overestimated the magnitude of the critical Casimir forces by using the measurements by Garcia and Chan [8, 11]. This is because Garcia and Chan studied a pure ^{4}He superfluid film, while we are dealing with a mixture. It is also possible that Garcia and Chan's measurements themselves have overestimated the Casimir force, because the largest theoretical results are

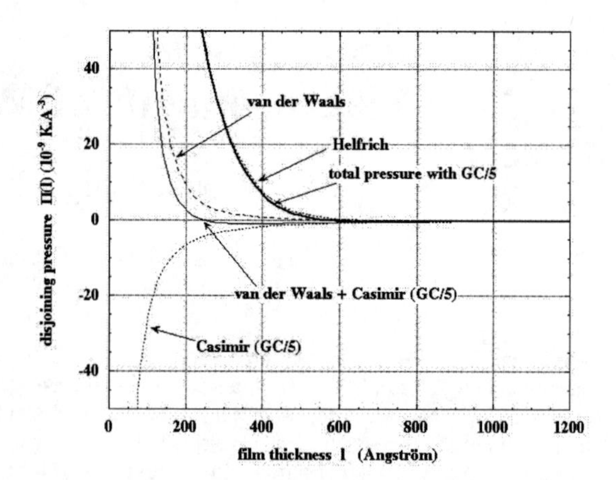

FIGURE 4. If the Casimir force is 5 times smaller than estimated by Ueno *et al.* from Garcia and Chan's measurements(GC/5), the disjoining pressure is nearly equal to the Helfrich repulsion [12] which is always positive, indicating complete wetting.

5 times smaller. If we assume that the Casimir force is 5 times smaller than assumed by Ueno *et al.*, we find that the van der Waals forces always dominate, so that complete wetting occurs, as shown by Fig. 4 where the disjoining pressure on the surface film is always positive. As for Goldstone modes, their effect on wetting must be again dominated by the van der Waals forces [8].

R. Ishiguro acknowledges support from JSPS for a Postdoctoral Fellowship for Research Abroad since April 2004.

REFERENCES

1. T. Ueno, M. Fujisawa, K. Fukuda, Y. Sasaki, and T. Mizusaki, *Physica B* **284-288**, 2057 (2000).
2. J. W. Cahn, *J. Chem. Phys.* **66**, 3667 (1977).
3. T. Ueno, S. Balibar, F. Caupin, T. Mizusaki and E. Rolley, *J. Low Temp. Phys.* **130**, 543 (2003).
4. T. Ueno, S. Balibar, T. Mizusaki, F. Caupin, and E. Rolley, *Phys. Rev. Lett.* **90**, 116102, (2003).
5. M. Fisher and P.G. de Gennes, *C.R. Acad. Sci. Paris*, **B 287**, 209 (1978).
6. M. Kardar and R. Golestanian, *Rev. Mod. Phys.* **71**, 1233 (1999).
7. M. Krech, *J. Phys.: Condens. Matter* **11**, R391 (1999).
8. R. Ishiruro and S. Balibar, *J. Low Temp. Phys.* **140**, 29 (2005).
9. A. Sato, K. Ohishi, and M. Suzuki, *J. Low Temp. Phys.* **107**, 165 (1997).
10. P. Leiderer, H. Poisel, and M. Wanner, *J. Low Temp. Phys.* **28**, 167 (1977); P. Leiderer, D. R. Watts, and W. W. Webb, *Phys. Rev. Lett.* **33**, 483 (1974).
11. R. Garcia and M.H. Chan, *Phys. Rev. Lett.* **83**, 1187 (1999).
12. W. Helfrich and R.M. Servuss, *Nuovo Cimento* **3D**, 137 (1984).

The Hydraulic Jump in Liquid Helium

Étienne Rolley*, Claude Guthmann*, Michael S. Pettersen† and Christophe Chevallier*

*Laboratoire de Physique Statistique, École Normale Supérieure, 75231 Paris, France
†Department of Physics, Washington and Jefferson College, Washington, PA 15301, USA

Abstract. We present the results of some experiments on the circular hydraulic jump in normal and superfluid liquid helium. The radius of the jump and the depth of the liquid outside the jump are measured through optical means. Although the scale of the apparatus is rather small, the location of the jump is found to be consistent with the assumption that the jump can be treated as a shock, if the surface tension is taken into account. The radius of the jump does not change when going down in temperature through the lambda point; we think that the flow is supercritical. A remarkable feature of the experiment is the observation of stationary ripples within the jump when the liquid is superfluid.

Keywords: hydraulic jump, liquid helium, capillary waves in shallow water
PACS: 47.15.-x, 47.20.Hw, 67.40.Hf

When a jet of liquid falls on a horizontal surface, as one often observes in the kitchen sink, a discontinuity may occur in the depth of the out-flowing fluid: at a certain distance R_j from the jet, there is an abrupt increase in the depth of the liquid, and a decrease in the average velocity of the liquid. This discontinuity is called the hydraulic jump, and was discussed in 1914 by Rayleigh.[1] Although the hydraulic jump is a popular undergraduate experiment,[2] the theory is challenging because of the free boundary, and it remains a problem of current theoretical interest.[2-7] Many experimental measurements of the jump radius R_j have been performed over the years (see the references cited in ref. [6]), mostly with fluids such as water or ethylene glycol. In this article, we report the results of observations of the hydraulic jump using liquid helium-4. Helium differs from the liquids that have previously been studied in having a much lower viscosity ($\nu \sim 2 \times 10^{-8}$ m^2/s), so jet Reynolds numbers up to $Re \sim 4 \times 10^4$ can be achieved. Another difference is the scale: typical values of R_j in our apparatus are a few mm, an order or two smaller than in previous experiments on the hydraulic jump. This has the effect of increasing the importance of the surface tension.

The cell used in the experiment was mounted in a pumped-helium optical cryostat, to permit direct imaging of the jump with a digital camera, as seen in Fig. 1. The jet was formed by admitting helium through a cupronickel capillary tube of 100 μm inner diameter. The capillary was aligned so that the jet would be perpendicular to the surface of impact. The circular symmetry of the jump attested to the correctness of the alignment. The substrate surface was a sapphire disc (chosen for its high thermal conductivity), optically flat and aluminized.

The flux of liquid, Q, was determined from the rate helium gas was admitted to the cell, measured with a flowmeter at room temperature. Four heat exchangers (copper capillaries 0.5 m long with a 1 mm inner diameter) cooled, liquefied and thermalized the gas. The resulting impedance limits the maximum flow rate in the experiment.

FIGURE 1. Image of the hydraulic jump viewed from an oblique angle.

The radius of the jump R_j was determined directly from the images taken by the digital camera. In order to permit measurement of the depth of the liquid, a horizontal wire was placed above the liquid. The image of the wire in the mirrored surface is displaced due to the index of refraction of the liquid on the surface. Measuring the displacement of the image relative to the wire thus permitted a measurement of the depth of the liquid outside the jump, d, to within a few μm.

Fig. 2 shows a comparison between the measurement of R_j above the lambda transition and various models. To analyze the data, we have considered three models. Two of the models, those of Watson[3] and of Bush and

CP850, *Low Temperature Physics: 24th International Conference on Low Temperature Physics;* edited by Y. Takano, S. P. Hershfield, S. O. Hill, P. J. Hirschfeld, and A. M. Goldman

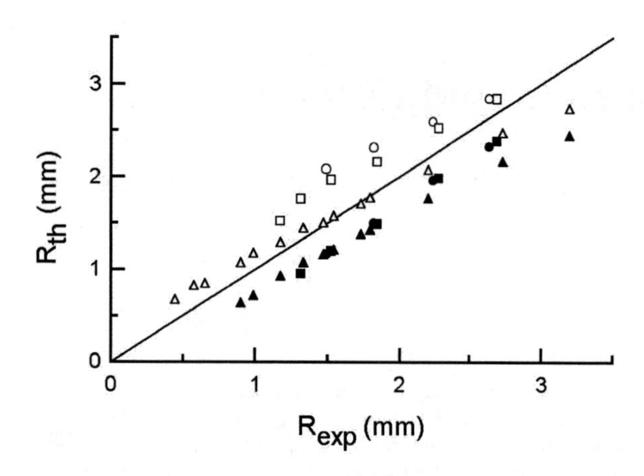

FIGURE 2. Experimental measurements of the hydraulic jump radius for flow rates ranging from Q=3 mm^3/s to 40 mm^3/s, compared with models of Watson (open symbols) and Bush (solid symbols). Circles: 2.45 K; squares: 3.0 K; triangles: 4.25 K.

Aristoff,[7] follow Rayleigh in treating the jump as a shock discontinuity, where mass and momentum fluxes are conserved. Watson's model improves on Rayleigh's in including the viscosity of the liquid. Near the jet impact, the fluid flow is modelled as a uniform flow near the free surface, and a growing boundary layer near the substrate; beyond the point where the boundary layer reaches the free surface, the flow is modelled with a similarity profile, which approximates the flow with the form $u(r,z) = U(r)f(z/H(r))$, with u the radial fluid velocity, and $H(r)$ is the depth of the fluid at distance r from the origin; the functions U and f are determined from the equations of motion and of continuity.[3] The matching of the velocity profiles where the two flow models join is not exact, but close. Bush and Aristoff's model is similar to Watson's, but includes the effect of the surface tension, which exerts a force on the curved surface of the jump and must be included in the momentum flux conservation condition. The model of Bohr *et al.*[4] differs in attempting to model the flow at the jump more accurately. It uses a one-parameter polynomial model of the flow profile; both the profile parameter and the height of the surface are variables, which vary continuously at the jump. This model is capable of representing the roll which is known to develop beyond the jump.[8]

In Fig. 2 it is seen that the predictions of Watson are slightly higher than the experimental results, and the predictions of Bush and Aristoff are slightly too low. It would appear that the effect of surface tension is important, but that the model of Bush and Aristoff is too crude, and overestimates the size of the effect. (Curiously, the experiments of Bush and Aristoff seem to indicate that their estimate of the effect of surface tension is not large enough.) The predictions of the model of Bohr *et al.* for our experiment are similar to those of Watson; this is not surprising in view of the fact that in practice, in all of our runs except at the highest temperatures and lowest flow rates, the jump is rather sharp.

Below the lambda transition, one might hope to observe a transition in the jump to the value predicted by Rayleigh in the inviscid case. However, we did not observe such a transition. We believe the fluid velocity in our experiment exceeds the critical velocity for superfluidity. For a fluid depth on the order of 10 μm, such as we expect within the jump,[3] the critical velocity is on the order of 50 mm/s,[9] corresponding to Q less than 1 mm^3/s. Unfortunately, at such low fluxes, the jet becomes unstable and drips instead. However, even if the helium in our experiments is normal below the lambda point, the viscosity of the normal fluid drops rather abruptly as the temperature is lowered from 2.17 K to 1.5 K. In this regime, we start to observe stationary ripples inside the jump, as seen in Fig. 1. These can be explained as waves generated at the jump and travelling upstream at the same velocity the fluid flows downstream. The ripple wavelength can be used to estimate the depth of the liquid inside the jump by equating the phase velocity of finite-depth capillary waves with the fluid velocity. The resulting estimate is in rough agreement with the prediction[3] and is much smaller than the depth outside the jump. The temperature dependence of the decay length is not yet completely understood. Presumably, the lower viscosity below the lambda point allows the ripples to propagate further upstream from the jump, so they become more prominent. Analysis of this phenomenon is ongoing.

ACKNOWLEDGMENTS

The Laboratoire de Physique Statistique de l'École Normale Supérieure is associated with Universités Paris 6 et Paris 7. M. S. Pettersen is grateful for the support of the École Normale Supérieure and the Université de Paris 7.

REFERENCES

1. Lord Rayleigh, *Proc. Roy. Soc.* A **90**, 324 (1914).
2. B. L. Blackford, *Am. J. Phys.* **64**, 164 (1996).
3. E. J. Watson, *J. Fl. Mech.* **20**, 481 (1964).
4. T. Bohr, P. Dimon and V. Putkaradze, *J. Fl. Mech.* **254**, 635 (1993).
5. P. Godwin, *Am. J. Phys.* **61**, 829 (1993).
6. S. Watanabe, V. Putkaradze and T. Bohr, *J. Fl. Mech.* **480**, 233 (2003).
7. J. W. M. Bush and J. M. Aristoff, *J. Fl. Mech.* **489**, 229 (2003).
8. I. Tani, *J. Phys. Soc. Japan* **4**, 212 (1949).
9. J. Wilks, *Properties of Liquid and Solid Helium*, Oxford: Oxford, 1987, pp 388-392.

Characterization of Scintillation Light produced in Superfluid Helium-4

G. Archibald*, J. Boissevain†, R. Golub**,‡, C.R. Gould‡, M.E. Hayden*, E. Korobkina**,‡, W.S. Wilburn† and J. Zou§

**Department of Physics, Simon Fraser University, Burnaby BC V5A 1S6, Canada*
†Los Alamos National Laboratory, Los Alamos NM 87545, USA
***Berlin Neutron Scattering Center, Hahn-Meitner Insitut, Berlin D-14109, Germany*
‡Department of Physics, North Carolina State University, Raleigh NC 27695, USA
§Department of Physics, Royal Holloway University of London, Egham, Surrey TW20 0EX, UK

Abstract. Experiments that place strict limits on the electric dipole moment of the neutron (n-EDM) represent one of the most promising opportunities to search for physics beyond the standard model. Over the last several years, a collaboration based at Los Alamos National Laboratory has been working toward the development of an apparatus that is expected to improve the current n-EDM limit by two orders of magnitude. A key feature of the proposed experiment involves the detection of ultraviolet scintillation light produced as a result of neutron capture on ^{3}He in superfluid liquid ^{4}He (ℓ-^{4}He). Typically such events give rise to a large prompt flash of light followed by a train of delayed photons ('after-pulses'). We recently performed an experiment to examine correlations between the amplitude of the initial light pulse and the number of after-pulses that are produced when a dilute mixture of ^{3}He in ℓ-^{4}He is irradiated with a neutron beam. Data from this experiment suggest that events associated with reaction products from neutron capture on ^{3}He can be distinguished from background events associated with other forms of ionizing radiation such as gamma rays. A sample of data from our experiment is presented.

Keywords: fluorescence, liquid helium, superfluid, neutron capture
PACS: 34.50.Gb, 33.50.-j, 78.70.-g, 67.60.Hr

Measurement of a non-zero value for the electric dipole moment (EDM) of an elementary particle would represent direct evidence for violation of time reversal symmetry and physics beyond the standard model, and would yield information crucial to understanding the baryon asymmetry that is observed in the universe. The anticipated sensitivity limit for a new experiment to search for the electric dipole moment of the neutron (n-EDM) that is being developed by a Los Alamos-based collaboration [1] is such that failure to observe a positive result will challenge our understanding of baryogenesis.

The new n-EDM search will make use of ultra-cold neutrons (UCN) produced in superfluid ^{4}He by down-scattering from a beam of 9Å neutrons. The liquid will be doped with a tiny admixture (fractional concentration $X \sim 10^{-10}$) of polarized ^{3}He intended to act as a magnetometer, as a neutron-polarization detector, and as a neutron spin-orientation analyzer. The latter functions are accomplished by (indirectly) monitoring scintillation light that is produced in the extreme ultraviolet (EUV) as products of neutron capture on ^{3}He are brought to rest within the liquid. This light is dominated by radiation from He_2^* molecules, of which those in the short-lived singlet-state give rise to a bright initial pulse (on ns timescales) and those in the long-lived triplet-state are responsible for a train of dimmer after-pulses. Since the cross section for neutron capture on ^{3}He is maximally spin-dependent, the gross features of the scintillation rate reflect the relative orientation of UCN and ^{3}He spins. More detailed discussions of the proposed n-EDM experiment and the production of scintillation light in liquid ^{4}He by ionizing radiation can be found in the references.

Our interest here is in addressing the efficiency with which the signature of direct neutron capture on ^{3}He in liquid ^{4}He can be distinguished from other forms of ionizing radiation, particularly at temperatures of order 300 mK where the proposed n-EDM search will take place. An attractive method for discriminating between these processes on the basis of after-pulse characteristics has been discussed in the literature for several years [2, 3]. It relies on the fact that the density of ionization events produced by neutron capture on ^{3}He is considerably larger than that produced by β particles from neutron decay or Compton scattering of γ-rays. This causes the dynamics of the excited dimers that are created along ionization tracks to be different, and ultimately influences the rates at which after-pulses are generated. One expects this behaviour to be reflected in the distribution of the number of single-photoelectron events that are detected during a short (few μs) time window following the initial scintillation pulse when plotted as a function of the total area of the main scintillation pulse. Encouraging re-

CP850, *Low Temperature Physics: 24th International Conference on Low Temperature Physics;*
edited by Y. Takano, S. P. Hershfield, S. O. Hill, P. J. Hirschfeld, and A. M. Goldman

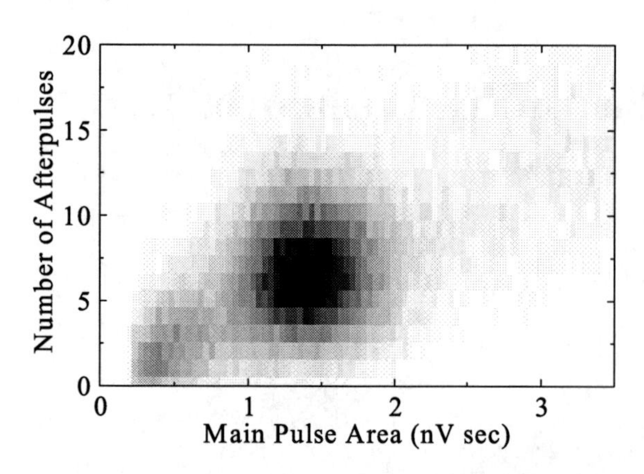

FIGURE 1. After-Pulse Scatter Plot: Scintillation events associated with neutron-capture on ^{3}He (dominant cluster near centre of figure) are readily distinguished from background γ- and β-induced events when the number of after-pulses in a 6.5 μs window is plotted as a function of the initial (main) scintillation pulse area. These data were acquired at 90 mK.

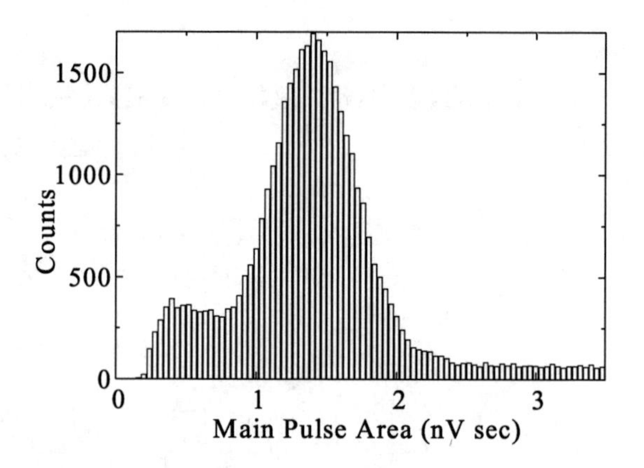

FIGURE 2. Scintillation Pulse-Area Histogram: The distinction between neutron-capture on ^{3}He and background γ- and β-induced events is less pronounced when one looks at a simple histogram of scintillation pulse areas. The geometry of our apparatus is such that the peak in this distribution occurs at a pulse-area equivalent to $\sim$ 20 single-photoelectron events.

sults have already been reported for temperatures of 1.8 Kelvin and above [2]. Reference [3] describes radiation-induced liquid ^{4}He scintillation light characteristics at lower temperatures, albeit without the use of neutrons.

We recently performed a series of experiments in which liquid ^{4}He containing small amounts of ^{3}He ($X \sim 10^{-4}$) was irradiated with a neutron beam. The liquid was confined to a 4.4 cm (inner) diameter by 22 cm long cylindrical acrylic cell coated with deuterated tetraphenyl butadiene (an organic fluor that down-converts EUV light into the visible) and aligned with its axis perpendicular to the beam. Independent light guides and photomultiplier tubes viewing either end of the cell allowed us to monitor scintillation events in coincidence. The cross sectional area of the beam ($\lesssim 1$ cm^2) was varied to provide detector count rates from 0.4 kHz (similar to the anticipated n-EDM search count rate) to 10 kHz, and the liquid temperature was varied from 0.09 - 1.9 K.

A full analysis of the data from these experiments is in progress. Nevertheless, we can already confirm that the implementation of a coincidence requirement is clearly advantageous for this type of experiment. Without it, we typically observed a 1200 Hz background event rate associated with γ-rays and β-particles (from ^{28}Al decay) relative to a 700 Hz signal from neutron capture on ^{3}He. With the coincidence requirement, which suppresses scintillation events that occur in the light guides, the background was reduced by a factor of four.

We can also confirm that the implementation of an event discrimination procedure based on an analysis of after-pulse characteristics leads to a further suppression of the background. Figure 1 illustrates the result of plotting the number of after-pulses as a function of the main pulse area. Two distinct peaks separated by a deep valley are visible. Neutron capture on ^{3}He is largely responsible for the peak at higher energies (pulse areas) while γ- and β-induced events account for the remainder of the distribution. Much of this contrast is lost when one looks at a simple histogram of scintillation pulse-areas, as shown in Fig. 2. Note that the event distributions presented in Figs. 1 and 2 have been extracted from the same data set.

An important feature of our experiment is that scintillation events were confined to a small volume of liquid at a fixed distance from the light guides and photomultiplier tubes. If this had not been the case, geometric variations in light collection efficiency throughout the cell would almost certainly have caused the neutron capture peak to broaden. This is an issue that should be given careful consideration in the design of similar experiments.

In conclusion, we find that significant levels of discrimination between scintillation events caused by neutron capture on ^{3}He in superfluid ^{4}He and other forms of ionizing radiation can be achieved by implementing relatively simple coincidence- and after-pulse characteristic requirements.

This work was supported by the Berlin Neutron Scattering Center (Hahn-Meitner Institut), Los Alamos National Laboratory (20040104DR), and NSERC (Canada).

REFERENCES

1. R. Golub and S. K. Lamoreaux *Phys. Rep.* **237**, 1 (1994) & Los Alamos Unclassified Report 02-2331 (2002).
2. K. Habicht, PhD Thesis, Technische Universtät Berlin (1998).
3. D. N. McKinsey *et al*, *Phys. Rev. A* **67**, 062716 (2003).

A Single Bubble Nucleated by Acoustic Waves in ^{3}He-^{4}He Mixtures

H. Abe[1], F. Ogasawara, Y. Saitoh, T. Ueda, R. Nomura and Y. Okuda

Department of Condensed Matter Physics, Tokyo Institute of Technology, 2-12-1, O-okayama, Meguro, Tokyo 152-8551, Japan

Abstract. We report the first study of a bubble nucleation in the dilute ^{3}He phase of phase-separated ^{3}He-^{4}He liquid mixtures. When an acoustic wave pulse of 1 msec duration was applied to the dilute phase at saturated vapor pressure, a single bubble was nucleated on an active area of the piezoelectric transducer. We succeeded in observing fast motion of the bubble by using a high-speed camera. We also observed bubbles nucleated in the ^{3}He rich phase and the pure superfluid ^{4}He by acoustic waves. Bubble shape in the dilute ^{3}He phase was similar to that in ^{3}He rich phase but quite different from that in pure superfluid ^{4}He.

Keywords: ^{3}He-^{4}He mixtures; acoustic waves; bubble; nucleation
PACS: 67.60. -g, 47.55. Dz, 43.35. -c

INTRODUCTION

Dynamics of a spherical bubble in a liquid has been studied extensively both theoretically[1] and experimentally[2, 3]. Such a bubble can be produced by focusing of pulsed laser beam in the normal liquids such as water and liquid nitrogen. One of the important features of the bubble dynamics is the collapsing process of a bubble near a solid boundary[4]. Many interesting phenomena have been observed in water such as a jet flow, vortex-ring formation and shock wave radiation. Benjamin and Ellis[5] predicted that a jet flow will create a hollow vortex ring, if it penetrates the surface of a spherical bubble axi-symmetrically. According to a recent theory, the collapse of a bubble which is generated by ultrasound in superfluids is accompanied by the nucleation of quantized vortex rings[6]. In this paper, we present the first visual observations of a sound-induced bubble in the dilute ^{3}He phase of phase-separated ^{3}He-^{4}He liquid mixtures.

EXPERIMENT AND RESULTS

The experiment was performed in a cell cooled by a ^{3}He-^{4}He dilution refrigerator. The refrigerator and the cell had optical windows on their sides. We could observe the inside of the cell by using a high-speed camera which operated at 1 ms/frame. The experimental setting in the cell can be seen in the first frame of Fig. 1. The distance between the two transducers was about 10 mm and the active area diameter was about 5 mm. The fundamental frequency was about 10 MHz. When the ^{3}He-^{4}He mixture was condensed into the cell, it separated into the ^{3}He rich phase and the dilute ^{3}He phase at low temperatures. The amount of ^{3}He was such that the interface between the two phases was always located above the upper transducer, and only the dilute phase was between the transducers. By increasing the voltage applied to the lower transducer, we found a single bubble to be nucleated on the active area of the transducer. The threshold voltage for the nucleation was 6.0 V at 300 mK. Figure 1 shows images of a bubble nucleated by a 10 MHz pulse, 6.7 V in amplitude and 1 ms in duration, at 300 mK. A single bubble was nucleated during the pulse and continued to expand after the pulse was turned off (1–10 ms). It reached the maximum radius and then gradually shrunk (10–13 ms). If the maximum radius was smaller than 0.5 mm, the bubble collapsed immediately and disappeared on the transducer. If it was larger than 0.5 mm, however, the bubble detached from the transducer while shrinking and continued to shrink (13–14 ms).

[1] E-mail: abe@ltp.ap.titech.ac.jp

CP850, *Low Temperature Physics: 24th International Conference on Low Temperature Physics;* edited by Y. Takano, S. P. Hershfield, S. O. Hill, P. J. Hirschfeld, and A. M. Goldman

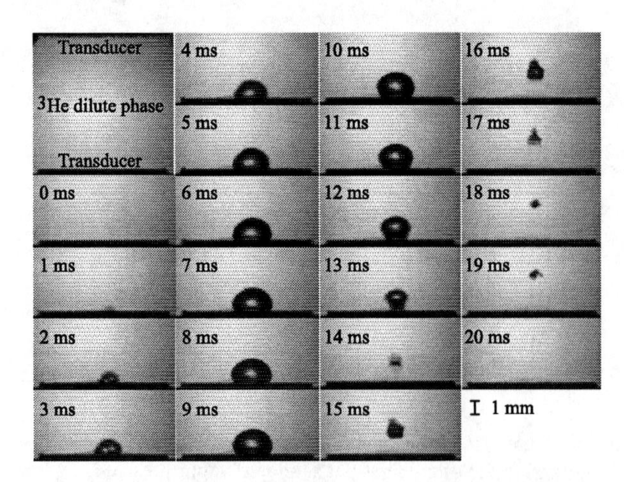

FIGURE 1. Images of a sound-induced bubble in the dilute ^{3}He phase at 300 mK. The pulse duration was 1 ms and the applied voltage was 6.7 V.

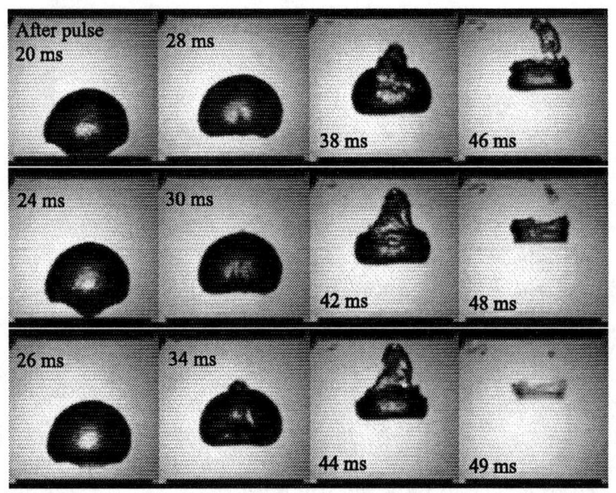

FIGURE 2. Images of the collapsing process of a bubble in the dilute ^{3}He phase at 200 mK. The pulse duration was 1 ms and the applied voltage was 10.8 V.

After the collapse the bubble had a small rebound and disappeared as it rose (14–20 ms). The collapsing process was clearly visible in the case of a larger bubble. Figure 2 shows a collapsing process for an applied voltage of 10.8 V at 200 mK. In this figure, we can see that an upward flow penetrated the bubble surface (34–44 ms). The behavior is characteristic of a liquid jet near a solid wall[5]. The transducers play a role of solid walls in the present experimental setup. The jet goes forward, due to the large liquid inertia, far from the collapsing bubble (46–49 ms). This is similar to a process which produces a hollow vortex ring in water [2].

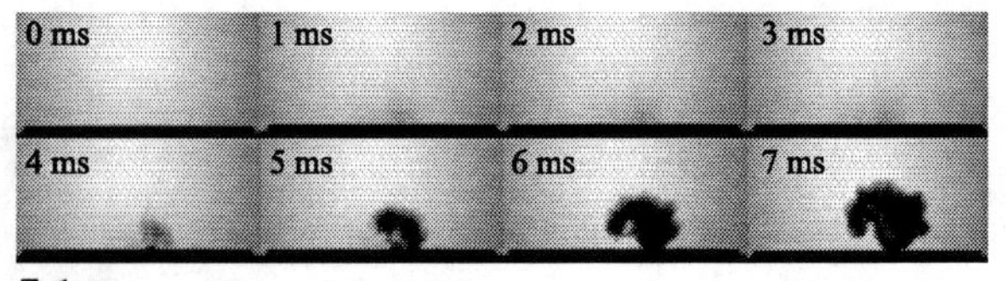

FIGURE 3. Images of a sound-induced bubble in superfluid ^{4}He at 200 mK. The pulse duration was 5 ms and the applied voltage was 10.2 V.

In order to investigate the shape of the bubbles, we performed the same experiments in the ^{3}He rich phase and in the pure superfluid ^{4}He. A single spherical bubble, similar to the one in the dilute ^{3}He phase, was nucleated in the ^{3}He rich phase. In contrast, bubbles were always non-spherical in pure ^{4}He. A typical sequence of image are shown in Fig. 3. The shape of the bubble was highly irregular, with an ill-defined surface. The threshold voltage in pure ^{4}He was 8.6 V at 300 mK. The acoustic wave power density applied to the transducer was twice as large as required in the dilute phase and the pulse duration was five times longer. Although ^{4}He is superfluid in both the pure liquid ^{4}He and the dilute ^{3}He phase, the bubble dynamics was quite different.

SUMMARY

We observed a single bubble nucleation by acoustic waves in the dilute ^{3}He phase. The process seemed to be controlled by the ^{3}He in superfluid ^{4}He. We found that the collapse of a bubble involved a liquid-jet formation. It is possible that the jet produces vortices or a vortex ring, as bubble collapses. Our observations may give a new prospect for the study of ^{3}He-^{4}He liquid mixtures.

REFERENCES

1. H. Lamb, *Hydrodynamics,* New York: Dover, 1932.
2. Y. Tomita and A. Shima, *Acustica Journal.* **71**, 161 (1990).
3. M. Tsubota, Y. Tomita, A. Shima, and I. Kano, *JSME Int. J., Ser. B* **39**, 257 (1996).
4. A. Vogel, W. Lauterborn, and R. Timm, *J. Fluid Mech.* **206**, 299-338 (1989).
5. T. B. Benjamin and A. T. Ellis, *Phil. Trans. R. Soc. Lond. A* **260**, 221-240 (1966).
6. N. G. Berloff and C. F. Barenghi, *Phys. Rev. Lett.* **93**, 090401 (2004); *J. Low. Temp. Phys.* **138**, 481 (2005).

Evaporative Isotopic Purification of Superfluid Helium-4

M.E. Hayden*, S.K. Lamoreaux† and R. Golub**

*Department of Physics, Simon Fraser University, Burnaby, BC V5A 1S6, Canada
†Physics Division, Los Alamos National Laboratory, Los Alamos, NM 87545, USA
**Department of Physics, North Carolina State University, Raleigh, NC 27695, USA

Abstract. We examine the feasibility of using evaporative techniques for isotopic purification of bulk liquid ^{4}He in the ultra-pure (sub ppb ^{3}He concentration) limit. At temperatures of order several-tenths of a degree Kelvin the ^{3}He-impurity diffusion coefficient and equilibrium vapour density are simultaneously large enough that one should be able to extract impurities simply by pumping on a free liquid surface. At somewhat lower temperatures the equilibrium distribution of impurities is dramatically modified by surface (Andreev) states, providing an alternate mechanism for purifying the bulk liquid.

Keywords: helium; isotopic purification; separation techniques
PACS: 67.60.-g, 64.75.+g, 81.20.Ym

Large quantities of exceptionally pure liquid ^{4}He (ℓ-^{4}He) are required for applications ranging from the production and containment of ultra-cold neutrons to studies of quantum evaporation. By far the most widely used and successful methods for producing this liquid are those based on the superfluid heat-flush effect. In particular, McClintock and collaborators [1] have developed a number of very robust heat-flush-based isotopic purification devices capable of delivering virtually unlimited quantities of liquid product with fractional ^{3}He concentrations $X < 5\times10^{-13}$ at temperatures $T \gtrsim 1$ K.

Our interest in this problem is associated with the design of a new experiment to search for the electric dipole moment of the neutron (n-EDM) with unprecedented sensitivity [2]. This experiment requires that ~ 10 litre volumes of ℓ-^{4}He with $X \sim 10^{-10}$ at $T \sim 0.3$ K be purified to $X < 10^{-12}$ on timescales of order 10^3 s [3]. One could certainly meet these specifications by heating the liquid, running it through a heat-flush-based purification system, and then cooling it back down to $T \sim 0.3$ K. On the other hand, significant reductions in both cost and complexity of the overall experiment may be possible if the liquid could be purified at lower temperatures. Here we present a systematic examination of two alternate approaches to this problem, both of which involve the introduction of a free liquid surface. In the first instance we consider the extraction of ^{3}He from ℓ-^{4}He by forced evaporation and in the second instance we look at the natural isotopic separation that occurs at temperatures $T \lesssim 0.1$ K when ^{3}He impurities are trapped in surface (Andreev) states [4]. Jewell *et al.* and Tulley [5] have used the former approach at $T > 1$ K and Safonov *et al.* [6] have used the latter to produce ℓ-^{4}He with $X \lesssim 10^{-9}$.

We begin by noting that the net flux of impurity atoms traversing a free liquid surface must be zero under equilibrium conditions. Likewise, the chemical potential of impurity atoms in the bulk liquid and the vapour phase must be equal. These constraints imply that [7]

$$\frac{n}{n^*} = \left(\frac{m}{m^*}\right)^{\frac{3}{2}} \exp\left(-E_B/kT\right) \tag{1}$$

and

$$\frac{\alpha_{lv}}{\alpha_{vl}} = \frac{m}{m^*} \exp\left(-E_B/kT\right) \tag{2}$$

where n (n^*) denotes the ^{3}He atom number density immediately above (below) the interface, α_{lv} (α_{vl}) is the thermally averaged probability that an impurity striking the surface from below (above) will pass into the vapour (liquid), $m^*/m \sim 2.4$ is the ^{3}He quasiparticle effective mass ratio, and $E_B \sim 2.8$ K is the binding energy for a single ^{3}He impurity in bulk ℓ-^{4}He. In deriving these expressions it is assumed that the ^{3}He behaves like an ideal gas that is independent of the ^{4}He background in both the liquid and the vapour.

Reducing n by pumping on the vapour will produce an imbalance in the equilibrium described above and will drive a net evaporation of ^{3}He at the liquid surface. In the limit where diffusion is rapid and n is successfully reduced to zero, n^* will decrease exponentially with a time constant $\tau_e = 4V/A\bar{v}^*\alpha_{lv}$ or

$$\tau_e = \frac{4V}{A\bar{v}\alpha_{vl}} \left(\frac{m^*}{m}\right)^{\frac{3}{2}} \exp\left(E_B/kT\right) \tag{3}$$

where V is the volume of liquid, A is the area of the liquid-vapour interface and $\bar{v}$ ($\bar{v}^*$) is the mean thermal speed of ^{3}He atoms (quasiparticles) in the vapour (liquid). The advantage of writing τ_e in the latter form is that the reflection coefficient for ^{3}He scattering from ℓ-^{4}He surfaces is small [8] from which we infer $\alpha_{vl} \sim 1$.

Diffusion will tend to further limit the rate at which ^{3}He can be extracted from the liquid by forced evaporation. We are thus interested in solutions to the diffusion

CP850, *Low Temperature Physics: 24th International Conference on Low Temperature Physics;*
edited by Y. Takano, S. P. Hershfield, S. O. Hill, P. J. Hirschfeld, and A. M. Goldman

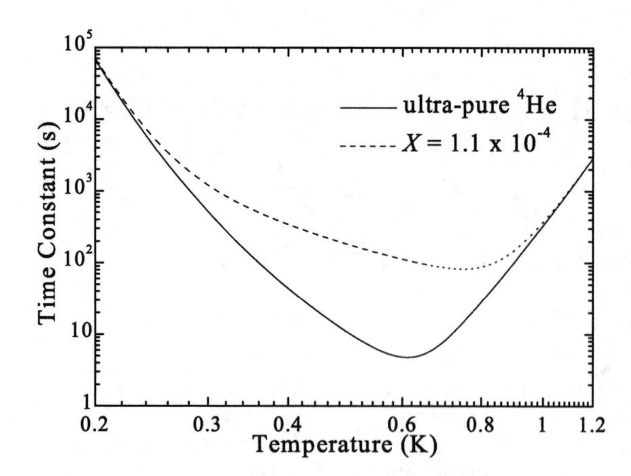

FIGURE 1. Time constant τ for removal of ^{3}He from a 10 litre volume of ℓ-^{4}He by forced evaporation through a 30 cm diameter free liquid surface (cf. Eq. (4)). Diffusion coefficients have been extracted from references [9], [10], and [11]. Note the influence of ^{3}He-^{3}He scattering when X is (relatively) large.

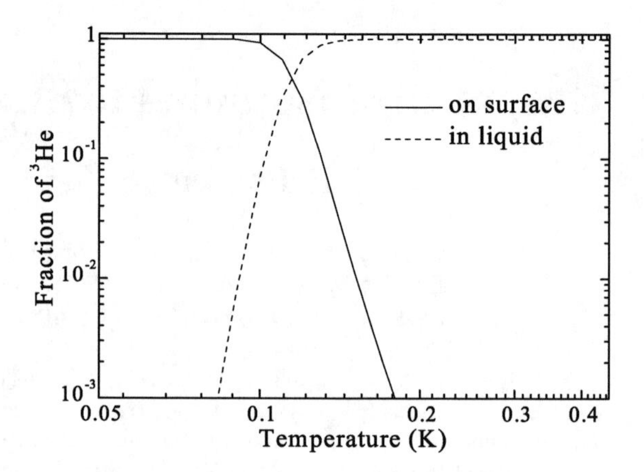

FIGURE 2. Distribution of ^{3}He impurities between the bulk liquid and the surface calculated for $X = 10^{-10}$ and using the same geometry described in Fig. 1. The maximum areal density of impurities σ here is of order 0.3 nm^{-2}, or roughly 8% of the density at which a second surface state starts to fill [12].

equation $D\,\nabla^2 n^* = \partial n^*/\partial t$, where D is the impurity diffusion coefficient. We take n^* to be uniform at $t = 0$ and set the impurity current density $j = D\,|\nabla n^*|$ to zero at the walls of the vessel and $j = \alpha_{lv} n^* \bar{v}^*/4 = Vn^*/A\tau_e$ at the free liquid surface. For a bath of uniform depth $L = V/A$, the time constant τ for removal of ^{3}He is dominated by the largest solution of

$$\tan\sqrt{\tau_D/\tau} = \sqrt{\tau_D \tau}/\tau_e \qquad (4)$$

where $\tau_D = L^2/D$. Figure 1 illustrates the manner in which τ varies as a function of temperature and ^{3}He concentration. In the low ^{3}He concentration limit a minimum value of τ is observed for $T \sim 0.6$ K. At higher temperatures the extraction process is limited by diffusion while at lower temperatures it is limited by thermal activation. Reasonable time constants are expected for temperatures as low as a few tenths of a degree Kelvin.

At yet lower temperatures, a significant fraction of the total number N of ^{3}He impurities in a closed system can end up forming a two-dimensional fermi-dirac gas bound to the liquid surface [4] by an energy $E_0 \sim 5$ K [12]. Equating the chemical potential of the ^{3}He in the surface state [13] to that in the bulk leads to the requirement

$$kT\ln\left[\frac{2\exp\left(\sigma\Lambda^2 m/2m_s\right)-2}{n^*\Lambda^3\left(m/m^*\right)^{3/2}}\right] = E_0 - E_B - \frac{\sigma V_0}{2} \qquad (5)$$

where σ is the areal density of ^{3}He atoms on the surface, $m_s/m \sim 1.5$ [12] is their effective mass ratio, $\Lambda = (2\pi\hbar^2/mkT)^{1/2}$ is the ^{3}He deBroglie wavelength, and $V_0/k \sim 0.23$ K$\cdot$nm^2 parameterizes the ^{3}He-^{3}He interaction [12]. With the further constraint $N = \sigma A + n^* V$ one can solve for the fraction $\sigma A/N$ of ^{3}He that resides on the surface at temperature T. This is illustrated in Fig. 2, which suggests that adequate levels of isotopic separation can be achieved simply by introducing a free liquid surface, cooling the liquid to temperatures $T \lesssim 0.1$ K, and removing the purified ℓ-^{4}He from below.

Clearly, the simple models and estimates presented here would benefit from further analysis and experimental validation. In particular, a microscopic theory that included coupling to ripplons and other excitations of the fluid may shed light on unforseen limitations of the proposed purification schemes.

REFERENCES

1. P. C. Hendry and P. V. E. McClintock, *Cryogenics*, **27**, 131 (1987) and references therein.
2. R. Golub and S. K. Lamoreaux, *Phys. Rep.* **237**, 1 (1994).
3. P. D. Barnes *et al.*, LANSCE Activity Report, p. 110 (2001) and LANL Tech. Report LA-UR 02-2331 (2002).
4. A. F. Andreev, *Sov. Phys. JETP* **23**, 939 (1966).
5. C. Jewell *et al.*, *Physica* **107B**, 587 (1981); C. Jewell, PhD Thesis, Lancaster University, p. 55 (1983); P. C. Tully, *U.S. Bureau of Mines Report 8054*, (Washington, 1975).
6. A. I. Safonov, *et al.*, *Phys. Rev. Lett.* **86**, 3356 (2001); ^{4}He purified in situ, S. A. Vasilyev (private communication).
7. M. W. Reynolds *et al.*, *J. Low Temp. Phys.* **84**, 87 (1991).
8. D. O. Edwards, *et al.*, *Phys. Lett.* **59A**, 131 (1976).
9. S. K. Lamoreaux *et al.*, *Europhys. Lett.* **58**, 718 (2002).
10. J. J. M. Beenakker *et al.*, *Physica* **18**, 433 (1952).
11. I. Adamenko *et al. J. Low Temp. Phys.* **111**, 145 (1998); R. Rosenbaum, *et al.*, *J. Low Temp. Phys.* **16**, 131 (1974).
12. J. P. Warren and C. D. H. Williams, *Physica B* **284-288**, 158 (2000).
13. D. O. Edwards *et al. Phys. Rev. B* **12**, 892 (1975).

A Density Functional for Liquid ^{3}He Based on the Aziz Potential

M. Barranco*, E.S. Hernández†, R. Mayol*, J. Navarro**, M. Pi* and L. Szybisz‡

*Departament ECM, Facultat de Física, Universitat de Barcelona, E-08028 Barcelona, Spain
†Departamento de Física, Facultad de Ciencias Exactas y Naturales, Universidad de Buenos Aires, 1428 Buenos Aires, and Consejo Nacional de Investigaciones Científicas y Técnicas, Argentina
**IFIC(CSIC-Universitat de València. Apartado 22085, E46071 València, Spain
‡Departamento de Física, CAC, Comisión Nacional de Energía Atómica, 1429 Buenos Aires, Argentina

Abstract. We propose a new class of density functionals for liquid ^{3}He based on the Aziz helium-helium interaction screened at short distances by the microscopically calculated two-body distribution function g(r). Our aim is to reduce to a minumum the unavoidable phenomenological ingredients inherent to any density functional approach. Results for the homogeneous liquid and droplets are presented and discussed.

Keywords: density functional, helium, pair correlation
PACS: 67.60.-g, 67.70.+n,61.46.+w

The total energy of a system consisting of N identical particles that interact pairwise can be universally expressed as

$$E = Tr_{1,\ldots,N}(H\rho_N) = Tr_1(h_1\rho) + \frac{1}{2}Tr_{1,2}(V\rho_2) \quad (1)$$

where Tr_i denotes tracing with respect to the quantum numbers of particle i, H is the N-body Hamiltonian with h_1 the one-body contribution and V the pair potential. Here ρ and ρ_2 are, respectively, the one- and two-particle density operators corresponding to the full density matrix ρ_N. The two-body density of particles 1 and 2 is usually expressed in terms of the pair correlation function $g(1,2)$ as $\rho_2(1,2) = \rho(1)\rho(2)g(1,2)$. Since Hohenberg-Kohn's theorem[1] states that the total energy of an N-particle system is a functional of ρ –and only ρ–, one of the main problems in many-body theory is to relate the pair correlation function to the one-body density. In a microscopic approach, this can be achieved by means of hierarchically coupled equations that contain summations of diagrams.

A phenomenological approach to circumvent the ignorance of the true pair distribution as a function of the single-particle (sp) density, preserving to some extent the many-body character of the system, is provided by Density Functional (DF) theories. These are mean field descriptions of the fully correlated system, where the interaction of two and more particles are included through parametrized functions of the sp density.

Recently, a DF has been proposed[2] for inhomogeneous ^{4}He that incorporates the available knowledge about the pair correlation function into a representation of the potential energy. In these works, one adopts the so-called "hybrid" DF

$$\begin{aligned} E &= \int d\mathbf{r}\,\tau(\mathbf{r}) + \int d\mathbf{r} \sum_{n=2}^{4} \frac{c_n}{n}\rho(\mathbf{r})\,[\tilde{\rho}(\mathbf{r})]^n \quad (2) \\ &+ \frac{1}{2}\int\int d\mathbf{r}d\mathbf{r}'\rho(\mathbf{r})\,g(|\mathbf{r}-\mathbf{r}'|)V_{LJ}(|\mathbf{r}-\mathbf{r}'|)\,\rho(\mathbf{r}') \\ &+ \frac{1}{2}\int\int d\mathbf{r}d\mathbf{r}'\rho(\mathbf{r})\,\mathcal{G}(|\mathbf{r}-\mathbf{r}'|)\,\rho(\mathbf{r}') \end{aligned}$$

where $\tau(\mathbf{r})$ is the kinetic energy density, $\tilde{\rho}(\mathbf{r})$ is a coarse-grained density obtained by averaging ρ with respect to the function $\rho_0 h(|\mathbf{r}-\mathbf{r}'|) = \rho_0\,[1-g(|\mathbf{r}-\mathbf{r}'|)]$, and the pair interaction term contains the full Lennard-Jones potential. The correlation kinetic energy is

$$\mathcal{G}(|\mathbf{r}-\mathbf{r}'|) = \frac{\hbar^2}{2m}\left[|\nabla\sqrt{g(|\mathbf{r}-\mathbf{r}'|)}|^2 + |\nabla'\sqrt{g(|\mathbf{r}-\mathbf{r}'|)}|^2\right] \quad (3)$$

After introducing an analytical form for the pair correlation function, that reproduces experimental data at saturation density ρ_0, we fit the coefficients c_n, n = 2 to 4 by using the energy per atom, density and incompressibility modulus at saturation. A good agreement ss obtained for density profiles of adsorbed films, as compared with Monte Carlo calculations, as well as a good figure for the surface tension and for various wetting data.

The purpose of the present work is to apply the above philosophy to construct a hybrid DF for inhomogeneous liquid ^{3}He. A complete DF for this system should be able to describe both the density (spin-symmetric or *S*) and the spin (spin-antisymmetric or *A*) channels of the homogeneous liquid; given the complexities of the *A*-channel, we postpone this issue for a later study and

CP850, *Low Temperature Physics: 24th International Conference on Low Temperature Physics;*
edited by Y. Takano, S. P. Hershfield, S. O. Hill, P. J. Hirschfeld, and A. M. Goldman

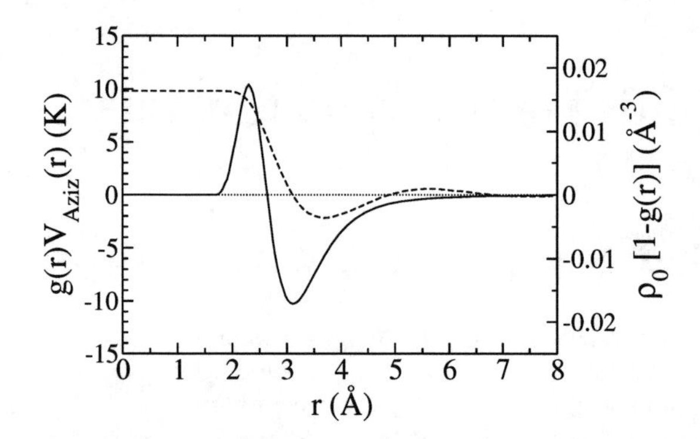

FIGURE 1. Screened Aziz potential $g(r)\,V_A(r)$ (full line) and direct pair correlation $h(r)$ (dashed line) as functions of radial distance between particles.

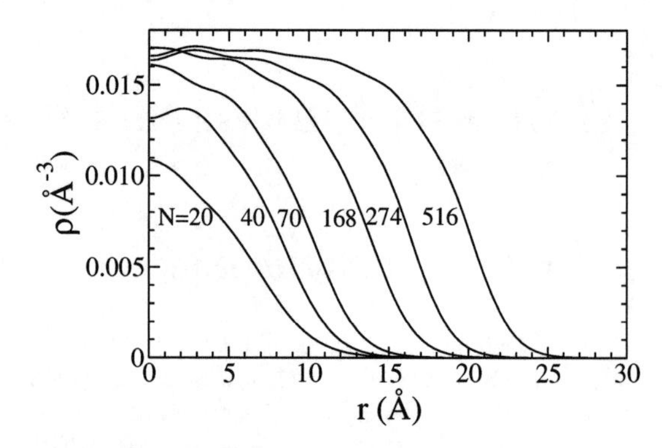

FIGURE 2. Radial density profiles of spherical ^3He_N clusters for various numbers of atoms N.

here concentrate in spin-saturated ^{3}He. Accordingly, we propose the following DF

$$
\begin{aligned}
E \quad &= \int d\mathbf{r}\, \frac{\hbar^2}{2m}\left(1-\frac{\tilde{\rho}(\mathbf{r})}{\rho_c}\right)^2 \sum_{i=1}^{N} |\nabla\phi_i(\mathbf{r})|^2 \qquad (4)\\
&+ \int d\mathbf{r}\, \frac{c}{2}\, \rho(\mathbf{r})\, [\tilde{\rho}(\mathbf{r})]^{1+\gamma}\\
&+ \frac{1}{2}\int\!\!\int d\mathbf{r}d\mathbf{r}'\, \rho(\mathbf{r})\, g(|\mathbf{r}-\mathbf{r}'|)\, V_A(|\mathbf{r}-\mathbf{r}'|)\, \rho(\mathbf{r}')\\
&+ \frac{f}{2}\int\!\!\int d\mathbf{r}d\mathbf{r}'\, \rho(\mathbf{r})\, \mathscr{G}(|\mathbf{r}-\mathbf{r}'|)\, \rho(\mathbf{r}')
\end{aligned}
$$

Here $\phi_i(\mathbf{r})$ are sp wave functions to be determined by solving the mean field Kohn-Sham (KS) equations [3] and the sp density is $\rho(\mathbf{r}) = \sum_{i=1}^{N} |\phi_i(\mathbf{r})|^2$. The two-body interaction V_A is the full Aziz HFD-B(HE) potential[5]. The density-dependent factor in the kinetic energy term and the repulsive correlation energy with strength c are chosen as in the Orsay-Paris (OP) DF,[4] and $f = 0.93$ is a constant parameter selected so as to preserve the OP value of the integral of the pairwise potential energy. In this way, the predicted equation of state coincides with the OP and experimental ones.

The main difficulty here is the lack of experimental information on the pair correlation; however, the key ingredients are the equal and opposite pair correlation functions $g_{\uparrow\uparrow}(|\mathbf{r}-\mathbf{r}'|)$ and $g_{\uparrow\downarrow}(|\mathbf{r}-\mathbf{r}'|)$ that can be computed within a diffusion Monte Carlo frame[6]. With these data, we numerically construct the density-density correlation function $g(|\mathbf{r}-\mathbf{r}'|) = [g_{\uparrow\uparrow}(|\mathbf{r}-\mathbf{r}'|) + g_{\uparrow\downarrow}(|\mathbf{r}-\mathbf{r}'|)]/2$, later employed as the input for the KS equations. In Fig. 1 we show the effective potential $g(r)\,V_A(r)$ and the direct pair correlation $h(r) = 1 - g(r)$, in full and dashed lines, respectively.

In order to assess the reasonability of this DF, we have computed density profiles and energetics of ^3He_N clusters for increasing number of atoms N. A byproduct of this calculation is the predicted value of the surface tension, which in this case reads 0.121 K Å^{-2}, to be compared with an experimental figure of 0.113 K Å^{-2}. Typical results are displayed in Fig. 2, where we plot density profiles for N = 20, 40, 70, 168, 274 and 516.

In summary, we are proposing a new DF for the spin-symmetric channel of nonhomogeneous liquid ^{3}He, which incorporates some desirable features of a true many-body system, namely the two-body correlation function, here determined by a rigorous simulation. This method permits to represent the two-body interaction by a realistic pair potential, avoiding the arbitrary core suppression of earlier DF's and eliminating both the core and the coarse-graining radius as parameters. Additionally, no gradients of the sp density need to be introduced through parametric functions. We believe that the present results encourage further research, especially, examining the dynamical response of the liquid and seeking to improve the description of the spin-antisymmetric channel

We are indebted to Jordi Boronat for kindly providing his numerical data on the pair correlation functions. This work was partially supported by grants 2001SGR00064 and FIS2004-00912, Spain, and PICT 03-08450 (ANPCYT), Argentina.

REFERENCES

1. P. Hohenberg and W. Kohn, *Phys. Rev.* **136**, B864 (1964).
2. L. Szybisz and I. Urrutia, *J. Low Temp. Phys.* **138**, 337 (2005) and *Phys. Lett.* **A 338**, 155 (2005).
3. W. Kohn and L. J. Sham, *Phys. Rev.* **140**, A1133 (1965).
4. J. Dupont-Roc, M. Himbert, N. Pavloff, and J. Treiner, *J. Low Temp. Phys.* **81**, 31 (1890).
5. R. A. Aziz, F. R. W. McCourt, and C. K. Wong, *Mol. Phys.* **61**, 1487 (1987).
6. J. Boronat, private communication.

Transport currents in Bose quantum liquids

V. Apaja, E. Krotscheck, A. Rimnac* and R. E. Zillich†

*Institut für Theoretische Physik, Johannes Kepler Universität, A 4040 Linz, Austria
†Fraunhofer ITWM, D-67663 Kaiserslautern, Germany

Abstract. Until now, most of what has been said about excitations in quantum liquids has concerned the dynamic structure function, which is observable by means of neutron scattering. The dynamic structure function can be calculated using standard linear response theory. However, at this level one needs only transition densities $\langle 0|\hat{\rho}|n\rangle$ or transition currents $\langle 0|\hat{\mathbf{j}}|n\rangle$, which are oscillatory in time and hence do not describe mass transport. In this work we go a step further and study transport currents in excited states, $\langle n|\hat{\mathbf{j}}|n\rangle$, which requires the calculation of $\langle n|\hat{\mathbf{j}}|n\rangle$ to second order. For that purpose, we take a well-tested microscopic theory of inhomogeneous quantum liquids and extend it to find the currents formed when helium atoms scatter off a helium slab or when excitations evaporate atoms (a setup experimented by A. F. G. Wyatt's group in Exeter). Current conservation was already a major theoretical problem encountered by R. Feynman and led him to introduce backflow corrections. We show that perfect current conservation is expected only for exact solutions of the time-dependent many-body Schrödinger equation. This is the first extensive theoretical study of transport phenomena in a quantum liquid based on an accurate microscopic theory.

Keywords: quantum sticking, quantum reflection, quantum evaporation
PACS: 67.20.+k,67.40.Db,67.40.Pm

THEORETICAL DESCRIPTION

Excitations in a condensed system can be described by studying the response to a weak, periodic external perturbation. In a quantum fluid like ^{4}He, it is useful to write the perturbed wave function as

$$\Psi(1,\ldots,N;t) = e^{-iH_{00}t/\hbar}\Phi(1,\ldots,N;t) \quad (1)$$

$$\Phi(1,\ldots,N;t) = \sqrt{\frac{1}{\mathcal{N}(t)}}e^{\frac{1}{2}\delta U_N(t)}\Psi_0(1,\ldots,N)$$

where $\Psi_0(1,\ldots,N)$ is the exact ground state, H_{00} is the ground state energy, and $\mathcal{N}(t)$ is the normalization integral. Without loss of generality, we can write the excitation operator as a series of multiparticle fluctuations:

$$\delta U_N(t) = \sum_{i=1}^{N}\delta u_1(\mathbf{r}_i;t) + \sum_{1=i<j}^{N}\delta u_2(\mathbf{r}_i,\mathbf{r}_j;t) + \ldots . \quad (2)$$

These fluctuations are determined by linearizing the stationarity principle [1]

$$\delta\int dt\,\langle\Psi(t)|H + U_{\rm ext}(t) - i\hbar\frac{\partial}{\partial t}|\Psi(t)\rangle = 0 . \quad (3)$$

The resulting equations of motion are continuity equations for the *first-order* currents and densities,

$$\delta\rho(\mathbf{r};t) = \langle\Psi_0|\hat{\rho}(\mathbf{r})\delta U_N(t)|\Psi_0\rangle + \text{c.c.} \quad (4)$$

$$\mathbf{j}(\mathbf{r};t) = \langle\Psi_0|\hat{\mathbf{j}}(\mathbf{r})\delta U_N(t)|\Psi_0\rangle + \text{c.c.} \quad (5)$$

where $\hat{\rho}(\mathbf{r})$ and $\hat{\mathbf{j}}(\mathbf{r})$ are the density and current operators. The theory leads to the plausible form

$$\chi(\mathbf{r},\mathbf{r}',\omega) = \sum_{st}\delta\rho_s(\mathbf{r})\,[G_{st}(\omega) + G_{st}(-\omega)]\,\delta\rho_t(\mathbf{r}') \quad (6)$$

for the density-density response function, where the $\delta\rho_s(\mathbf{r})$ are Feynman phonon densities, and

$$G_{st}(\omega) = [\hbar(\omega - \omega_s + i\varepsilon)\delta_{st} + \Sigma_{st}(\omega)]^{-1} \quad (7)$$

$$\Sigma_{st}(\omega) = \frac{1}{2}\sum_{mn}\frac{V^{(s)}_{mn}V^{(t)}_{mn}}{\hbar(\omega_m + \omega_n - \omega - i\varepsilon)} \quad (8)$$

is the Greens's function, with the self-energy correction $\Sigma_{st}(\omega)$. The $V^{(t)}_{mn}$ are three-phonon matrix elements. The theory provides the raw material for further studies.

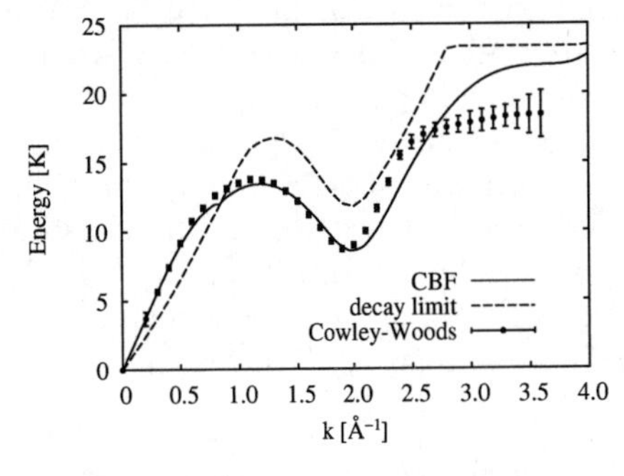

FIGURE 1. The figure shows a comparison between the calculated phonon-roton spectrum (solid line, labeled CBF) and the experimental spectrum [2]. Also shown is the *decay-limit* (dashed line) for phonons.

CP850, *Low Temperature Physics: 24th International Conference on Low Temperature Physics;* edited by Y. Takano, S. P. Hershfield, S. O. Hill, P. J. Hirschfeld, and A. M. Goldman

The first order current (5) vanishes in the time average. Hence, it does not describe particle transport. For that we must calculate second-order quantities,

$$\delta\rho^{(2)}(\mathbf{r};t) = \frac{1}{4}\langle\Psi_0|\,\delta U_N^*(t)\hat{\rho}(\mathbf{r})\delta U_N(t)\,|\Psi_0\rangle \quad (9)$$

$$\hat{\mathbf{j}}^{(2)}(\mathbf{r};t) = \frac{1}{4}\langle\Psi_0|\,\delta U_N^*(t)\hat{\mathbf{j}}(\mathbf{r})\delta U_N(t)\,|\Psi_0\rangle\,. \quad (10)$$

Already Feynman noted that these quantities satisfy a continuity equation only in an exact theory. The evaluation of the second-order current turns out to be a quite complicated task [3] which leads, however, to a reasonably simple result. The total transport current is a sum of an "elastic" and an "inelastic" contribution, $\mathbf{j}^{(2)}(\mathbf{r};t) = \mathbf{j}^{(2)}_{\rm el}(\mathbf{r};t) + \mathbf{j}^{(2)}_{\rm inel}(\mathbf{r};t)$; the inelastic current is a sum of "channel currents" $\mathbf{j}_m(\mathbf{r})$ into Feynman statem m. The "inelastic" current decays when the self-energy (8) has a non-trivial imaginary part, which describes, among others, the *decay* of quasiparticles. The decay rate is

$$\frac{d}{dt}\mathbf{j}^{(2)}_{\rm inel}(\mathbf{r};t) = \sum_{ms}\mathbf{j}_m(\mathbf{r})\left|\sum_t V^{(t)}_{ms}x_t\right|^2 2\pi\delta(\omega-\omega_m-\omega_s)\,. \quad (11)$$

Here $V^{(t)}_{ms}$ are the three-phonon matrix elements, and x_t are the expansion coefficients in Feynman states of the driving density fluctuation. Eq. (11) is recognized as "Fermi's golden rule". Microscopic many-body theory provides unambiguous expressions for the interaction strengths. We have shown in Ref. [4] that inelastic processes amount to about 90 percent of the physics in ^{4}He scattering off ^{4}He surfaces.

APPLICATIONS

Phonons have an anomalous dispersion at energies below about 10 K [5]; hence they have a finite penetration depth and decay in an angle of about 20° [6]. Fig. 2 shows the theoretical decay length for phonons in bulk ^{4}He and for phonons generated by a current of free particles impinging on a ^{4}He surface. Phonons with intermediate energy decay rapidly into low-energy phonons with $\hbar\omega < 1$ K, and only these and the very high-energy phonons traverse appreciable distances of order 1 mm. The angular distribution of decay products shows that most of them go to a narrow angle in the forward direction. A ^{4}He atom with energy 11.9 K impinging on a helium surface creates mostly ripplons, as Fig. 3 shows.

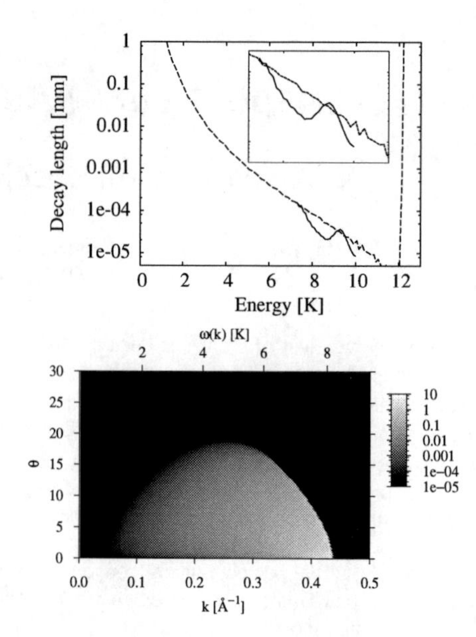

FIGURE 2. Upper panel: The decay length of phonons. The inset shows a magnified view of the decay length of ^{4}He atoms hitting a free surface from outside. Lower panel: Angular distribution of phonon decay products in bulk ^{4}He. Lighter shade of gray corresponds to a larger probability.

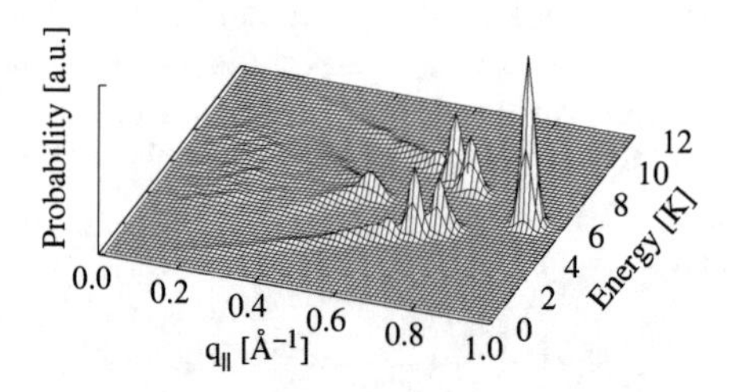

FIGURE 3. The figure shows the probability that a particle impinging on a ^{4}He slab with energy 11.9 K decays into two lower energy excitations that have wave vector $q_{\|}$ parallel to the slab and energy shown in the figure.

ACKNOWLEDGMENTS

This work was supported, in part, by the Austrian Science Fund (FWF) under project P15083-N08. We thank A. F. G. Wyatt for discussions.

REFERENCES

1. V. Apaja and E. Krotscheck, "A Microscopic View of Confined Quantum Liquids," in *Microscopic Approaches to Quantum Liquids in Confined Geometries*, edited by E. Krotscheck, and J. Navarro, World Scientific, Singapore, 2002, pp. 205–268.
2. R. A. Cowley and A. D. B. Woods, *Can. J. Phys.*, **49**, 177–200 (1971).
3. V. Apaja, E. Krotscheck, A. Rimnac, and R. Zillich, *Phys. Rev. B* (2005), to be published.
4. C. E. Campbell, E. Krotscheck, and M. Saarela, *Phys. Rev. Lett.*, **80**, 2169 (1998).
5. R. C. Dynes and V. Narayanamurti, *Phys. Rev. B*, **12**, 1720–1730 (1975).
6. R. V. Vovk, C. D. H. Williams, and A. F. G. Wyatt, *Phys. Rev. B*, **12**, 134508 (2003).

Critical Rayleigh Number Controversy For Liquid ^{4}He

Matthew J. Lees, Michael S. Thurlow, James R. T. Seddon and Peter G. J. Lucas

School of Physics and Astronomy, University of Manchester, Manchester, M13 9PL, United Kingdom

Abstract. We present new measurements of the critical Rayleigh number R_C for Rayleigh-Bénard convection in ^{4}He over the temperature range 2.2K to 3.1K in a cylindrical cell with aspect ratio Γ=16.6. A study of the convective flow pattern evolution as the Rayleigh number is increased above R_C has been reported by us earlier. The interesting feature of these new data is that R_C is close to 1800 at 2.2K, rising monotonically to about 2500 as the temperature is increased to 3.1K. This is at odds with previous work where R_C is generally below 1708, the expected value for a pure fluid with infinite Γ. Possible explanations for the difference between this earlier work and our data are discussed.

Keywords: Rayleigh-Bénard, convection, instability, helium.
PACS: 47.54.+r, 47.20.Bp, 47.27.Te

INTRODUCTION

Measurements[1] of the critical Rayleigh number R_C in convecting liquid ^{4}He lie between 1200 and 1750, generally lower than the ideal value[2] of 1708 for a cylindrical cell with fixed temperature upper and lower boundaries and infinite lateral extent. The value of R_C is affected by the cell aspect ratio[3] Γ (radius/height), the ratios[4] L of the thermal conductivities of the helium to the boundary material and D of the boundary thickness to the cell height d, and the fixed heat current lower boundary condition employed in the experiments[1]. However although experimentally[1] Γ>5, L~10^{-3} and D~1, calculations[3,4] show that the difference between these and the ideal conditions represented by Γ»1, L=0 and D»1 should change R_C from 1708 by no more than 1%. Here we present new data on R_C for ^{4}He where for the first time the flow pattern has been made visible[5,6], which challenge the speculation[4] that imperfect knowledge of the fluid parameters might be the source of the controversy.

MEASURING R_C

Our measurements of R_C in liquid helium were made using the cylindrical cell described by us earlier[5,6]. Key features of this cell are: (a) the radius and vertical height d are 9.13±0.01mm and 0.55±0.01mm respectively so that the aspect ratio Γ is 16.6±0.3, (b) for the plane copper lower boundary $L=1.4\times10^{-5}$ and D=18, while for the plane sapphire upper boundary $L=3\times10^{-4}$ and D=5.5, (c) the upper boundary is maintained at a constant temperature T_U which can be set between T_λ=2.172K and 4.2K, where the 1962 ^{3}He vapour pressure scale[7] is being used, (d) the lower boundary achieves a temperature T_L which is higher than T_U by the dissipation of a constant known heat current W, (e) the lateral sidewalls[5,6] are an annular Vespel[8] SP-22 spacing ring of rectangular cross-section with height d and horizontal thickness 1mm, surrounded by an annular region of horizontal thickness 5mm containing liquid helium at the mean temperature of the convecting fluid but with a smaller vertical spacing to prevent convection in this region. Critical Rayleigh numbers were measured at the five mid-plane temperatures given in Table 1, by noting the departure by two standard deviations from $\Delta T \propto W$ in the conduction regime. W was incremented in very small equally spaced steps in the vicinity of the convection onset, waiting for $\Delta T = T_L - T_U$ to stop changing before making the next increment. ΔT_C is not measurably different when the experimental protocol is changed to decrementing W. Critical Rayleigh numbers are then calculated using the formula $R_C = g d^3 \beta_T \Delta T / \nu D_T$ where g is the gravitational field 980.67cms^{-2}, β_T the thermal expansion coefficient, ν the kinematic viscosity and D_T the thermal diffusivity for normal liquid ^{4}He. We used the tables of Donnelly, Riegelmann and Barenghi[9] for data on the fluid parameters since these use the 1958 ^{4}He vapour pressure scale[10], which coincides with the 1962 scale[7] used by us in our experiments. Although the tables of Donnelly and Barenghi[11] are more recent, the data sources are the same[9]; the only difference being that the 1990 helium vapour pressure scale[12] is used.

CP850, *Low Temperature Physics: 24th International Conference on Low Temperature Physics;*
edited by Y. Takano, S. P. Hershfield, S. O. Hill, P. J. Hirschfeld, and A. M. Goldman

RESULTS

Our data on the temperature dependence of the critical Rayleigh number R_C are shown in Figure 1, Table 1 giving the mean values and fluid parameters used.

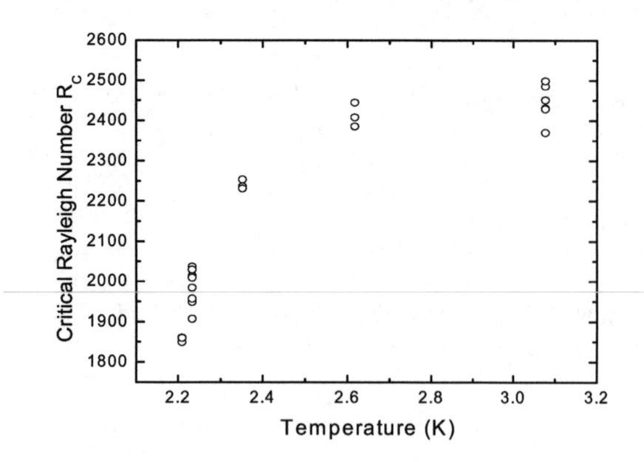

FIGURE 1. Temperature dependence of the critical Rayleigh number R_C, scatter indicating experimental errors.

TABLE 1. Temperature dependence of fluid parameters and critical Rayleigh number.

T (K)	$10^2\ \beta_T$ (K^{-1})	$10^4 D_T$ (cm^2s^{-1})	$10^4\nu$ (cm^2s^{-1})	ΔT_C (mK)	R_C
2.209	1.266	2.681	1.862	44.9	1857
2.233	1.716	2.986	1.919	40.8	1992
2.352	2.956	3.950	2.120	38.9	2240
2.617	4.447	4.720	2.342	36.8	2412
3.077	6.544	4.745	2.503	27.2	2443

All the R_C values lie above 1708, in contrast to those from the earlier workers[1]; and while R_C is close to the ideal value of 1708 near T_λ, its value increases monotonically to about 2500 as the temperature is increased to above 3K. Just below the thermal convection onset a small patch of straight parallel rolls is seen in the centre of the cell which grows[5,6] to fill the whole cell when $R = R_C$.

CONCLUSIONS

A flow pattern of straight parallel rolls is known[2] to be the preferred pattern for an ideal laterally infinite system. An explanation for this pattern being observed in our cell is that the annular layer of non-convecting helium acts as a "soft" sidewall boundary, minimizing the effect of sidewall forcing on the pattern so that the convecting layer simulates a laterally infinite system. The Vespel spacer is thermally "invisible" since its thermal response time of 10^{-3}s is short on the 10s time scale for the creation of a roll. However if our cell simulates an infinite system R_C should be close to 1708 at all temperatures, whereas we only find this to be the case just above T_λ. The data sources for the fluid parameters used in this and the earlier work[1] to calculate R_C all derive from the University of Oregon spline-fits[9,11,13] and the differences between these data sets are too small to account for our very different values of R_C. The d^3 dependence in the expression for R_C is a possible source of error, but the spacer height and the layer gap of the assembled cell at 300K both agree to within 0.01mm. Thermal contraction of the Vespel from 300K to 2K will reduce R_C by no more than 1.6%, and at 2K the value of d is consistent with the measured roll spacing. The absence of a Newton's ring pattern rules out the possibility of deformation of the sapphire upper boundary. We calculate that non-Boussinesq effects[2] are negligible, and a hexagonal roll pattern was not observed. We conclude therefore that while our attention to the sidewall boundary conditions has led to R_C values higher than those of the earlier workers[1] and closer to the ideal value of 1708 at T_λ, the trend for R_C to become larger as the temperature is increased above T_λ is not understood.

REFERENCES

1. R. W. Walden and G. Ahlers, *J. Fluid Mech.* **109**, 19-114 (1981); R. P. Behringer and G. Ahlers, *J. Fluid Mech.* **125**, 219-258 (1982); J. M. Pfotenhauer, P. G. J. Lucas and R. J. Donnelly, *J. Fluid Mech.* **145**, 239-252 (1984); R. W. Motsay, K. E. Anderson and R. P. Behringer, *J. Fluid Mech.* **189**, 263-286 (1988).
2. M. C. Cross and P. C. Hohenberg, *Rev. Mod. Phys.* **65**, 851-1112 (1993).
3. G. S. Charlson and R. L. Sani, *Int. J. Heat and Mass Transfer* **13**, 1479-1496 (1970).
4. G. P. Metcalfe and R. P. Behringer, *J. Low Temp. Phys.* **78,** 231-246 (1990).
5. M. J. Lees, M. S. Thurlow, J. R. T. Seddon and P. G. J. Lucas, *Phys. Rev. Lett.* **93**, 144502 (2004).
6. M. J. Lees, "Visualising Flow pattern Instabilities in Low Temperature Convection", Ph.D. Thesis, University of Manchester, 2003.
7. R. H. Sherman, S. G. Sydoriak and T. R. Roberts, *J. Res. NBS* **68A**, 547-588 (1964).
8. R. C. Richardson and E. N. Smith, *Experimental Techniques in Condensed Matter Physics at Low Temperatures*, New York: Addison-Wesley, 1988, pp. 120-121.
9. R. J. Donnelly, R. A. Riegelmann and C. F. Barenghi, "The Observed Properties of Liquid Helium at the Saturated Vapor Pressure", report, University of Oregon (1992).
10. F. G. Brickwedde, H. van Dijk, M. Durieux, J. R. Clement and J. K. Logan, *J. Res. NBS* **64A**, (Phys. & Chem.), 1-16 (1960).
11. R. J. Donnelly and C. F. Barenghi, *J. Phys. Chem. Ref. Data,* **27**, 1217-1274 (1998).
12. H. Preston-Thomas, *Metrologia*, **27** 3-10 (1990).
13. C. F. Barenghi, P. G. J. Lucas and R. J. Donnelly, *J. Low Temp. Phys.* **44**, 491-504 (1981).

Observation of a New Type of Negative Ion in Superfluid Helium

Ambarish Ghosh and Humphrey J. Maris

Department of Physics, Brown University, Providence, RI 02912,USA

Abstract. In recent work, we have developed a new technique for the study of the properties of electron bubbles (negative ions) in liquid helium. We use ultrasound to measure the critical negative pressure P_c at which an electron bubble becomes unstable and explodes. The value of P_c is affected, for example, by the quantum state of the electron and is reduced if the bubble is attached to a quantized vortex. In the present experiments, we have discovered a new type of object that appears to be larger than the usual electron bubble. We will consider possible explanations of these observations.

Keywords: Electron bubble, Critical negative pressure
PACS: 67.40.H, 43.35

INTRODUCTION

An electron injected into liquid helium forces open a spherical cavity or radius R approximately 19 Å in which there are no helium atoms.[1] The size of this so–called electron bubble is determined by a balance between the outward pressure exerted by the electron, the surface tension α, and the pressure P in the liquid. The equilibrium radius is found by minimizing the total energy E given by

$$E = \frac{h^2}{8mR^2} + 4\pi R^2 \alpha + \frac{4\pi}{3} R^3 P, \qquad (1)$$

where $h^2/8mR^2$ is the ground state energy of the electron confined in the bubble (m is the electron mass). A number of different techniques have been used to study electron bubbles including mobility measurements and optical studies.[1,2] These experiments are consistent with the radius found from Eq. 1. However, several groups have detected other negatively charged objects in superfluid helium. These objects include the "fast ion"[3] and the "exotic ions".[4,5] The size of these objects can be estimated from their mobility, and it appears that their radius lies in the range between 10 to 16 Å. This is much larger than the radius for negatively-charged impurity ions, and there are major difficulties with other possible explanations.[4] The purpose of this paper is to report a new experiment in which we detect for the first time another unidentified electron object (UEO). This appears to be an electron bubble that is larger than the normal electron bubble.

EXPERIMENTAL SETUP

We have constructed an experimental cell which can be used for optical and ultrasonic experiments down to 0.6 K (see Fig. 1). To inject electrons into the helium we use a field emission tungsten tip. In some experiments this tip was replaced by a ^{63}Ni β– source with an activity of approximately 5 mCi. The maximum energy of the emitted electrons is 67 keV giving a range in liquid helium of less than 1 mm. After entering the liquid, each electron forms a bubble which then moves through the liquid under the influence of the local electric field. An externally applied field is produced by applying a negative dc voltage to the tungsten tip (or radioactive source) and keeping the lower surface of the ultrasonic transducer at ground potential. In addition, each electron bubble is acted on by the space charge field produced by the other electron bubbles in the liquid. The number density of electrons in the liquid can be modified by changing the externally applied field. The density also changes with temperature because as the temperature is lowered the number of thermal excitations (phonons and rotons) decreases rapidly resulting in a reduction in the drag force on a moving bubble and a large increase in mobility.

CP850, *Low Temperature Physics: 24th International Conference on Low Temperature Physics;*
edited by Y. Takano, S. P. Hershfield, S. O. Hill, P. J. Hirschfeld, and A. M. Goldman

We have developed an ultrasonic method that can be used to detect single electron bubbles and to determine some information about their characteristics.[6] If the pressure in the liquid is decreased a bubble becomes larger, and at a critical negative pressure P_c, the bubble becomes unstable and begins to grow without limit. If the electron inside the bubble is in the ground state (1S), calculations predict that this pressure should be -1.89 bars. Measurements give results in good agreement with this estimate.[6] For an electron in an excited state, e.g., the 1P state, the outward pressure exerted by the electron is larger and so P_c is smaller.[7] For an electron bubble attached to a quantized vortex, the pressure in the liquid near to the vortex is reduced relative to the ambient pressure due to the Bernoulli effect, and so the magnitude of the critical pressure is reduced relative to the magnitude of P_c for electron bubbles in bulk liquid.[6]

In the experiments reported here, a hemispherical ultrasonic transducer was used to generate sound pulses at a frequency of frequency 1.35 MHz (see Fig. 1). To a good approximation one can assume that the pressure oscillation at the acoustic focus is proportional to the amplitude of the ac voltage V_{ac} driving the transducer. If there is an electron bubble close to the acoustic focus, it will explode as soon as the pressure becomes less than P_c, i.e., more negative than P_c. After the bubble becomes unstable it grows rapidly. A He-Ne laser beam is sent through the region around the acoustic focus and the light scattered from the bubble is detected through the use of a photo-multiplier tube (PMT in Fig.1). In the experiment, a sequence of a few hundred acoustic pulses is generated and the number of times that an exploding bubble is detected is counted. From the results, the probability S of cavitation per sound pulse is calculated.

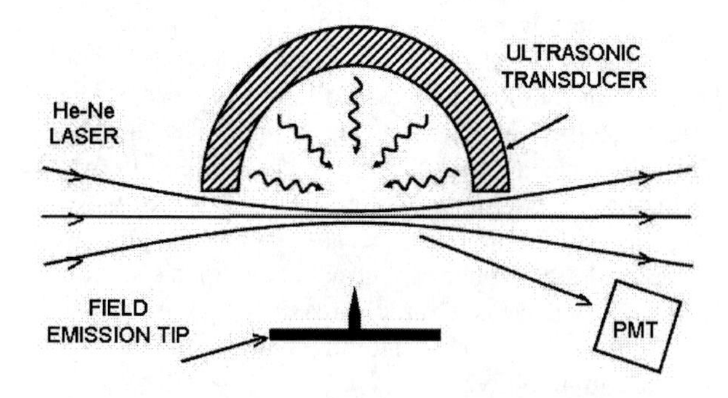

FIGURE 1. Schematic diagram of the experiment.

EXPERIMENTAL RESULTS

At high temperatures (e.g., 2 K) there are no electron bubbles attached to vortices. Consequently, the probability of cavitation is zero until the transducer voltage reaches a critical value V_{c1} such that the negative pressure swing at the focus reaches the critical pressure P_c. For this transducer voltage, there has to be an electron bubble precisely at the acoustic focus in order for an explosion to occur. When the voltage is increased above V_{c1} electron bubbles in a region near to the focus can explode and so the cavitation probability increases rapidly. From the variation of S with V_{ac}, the number density of the electron bubbles in the liquid can be determined.

At high temperatures, measurements show that there is a single threshold voltage at which cavitation begins to be seen. However, at low temperatures the experiment reveals multiple cavitation thresholds, thus indicating the presence of different types of electron bubbles with different critical pressures. Sample data showing this is in Fig. 2, taken at 0.72 K. Two distinct thresholds can be seen at 18.5 and 20.6 V. The solid curve in this figure is a fit to the data based on the assumption that two distinct objects are present, and suing the same type of analysis as in ref. 7.

To investigate these objects in more detail, we have made measurements of the cavitation probability as a function of voltage, temperature and applied electric field. From these data, it is clear that there are three distinct types of electron bubble present. The number density of each of these objects varies considerably with temperature and electric field and so for most

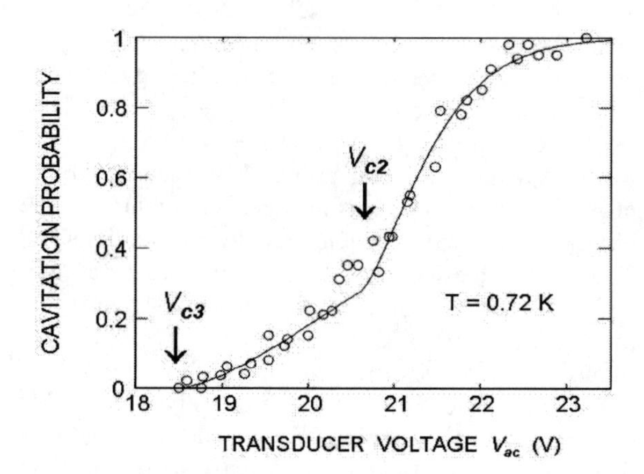

FIGURE 2. Probability of cavitation S as a function of the driving voltage V_{ac} applied to the transducer showing the two thresholds at V_{c2} and V_{c3}.

combinations of temperature and fields the thresholds for only one or two of the objects can be seen. The pressure at which each of these objects explodes is shown as a function of temperature in Fig. 3. We are confident that we have correctly identified two of these objects, but we do not know the nature of third object.

The first object #1 exploding at P_{c1} is the "normal" electron bubble in bulk liquid. This identification is based on the result that it was the only object seen when the radioactive source was used to inject electrons, the temperature was high ($T \geq 1.1$ K), and a small electric field (<100 V cm^{-1}) was applied to direct the electrons towards the transducer. Under these conditions, the electron bubbles should move through the liquid too slowly to produce vortices. When the temperature is lowered, the mobility of the electron bubbles increases and the number density of the type #1 objects decreases rapidly. For $T \leq 1.1$ K, the number density became too low for the cavitation threshold to be detected. This variation of the number density with temperature is consistent with object #1 being an electron bubble in bulk liquid.

For larger electric fields and at temperatures below 1.95 K, the second object #2 is seen with threshold at V_{c2}. These objects appear to be electron bubbles that are attached to quantized vortices. Evidence in favor of

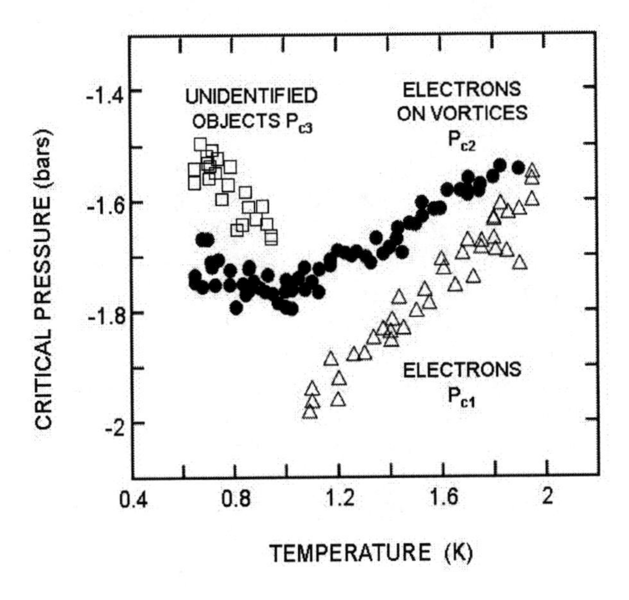

FIGURE 3. Measured negative pressure required to explode an electron in liquid helium as a function of temperature. Distinct thresholds are found for electron bubbles moving in bulk liquid P_{c1}, bubbles attached to a vortex P_{c2}, and for the new unidentified objects P_{c3}

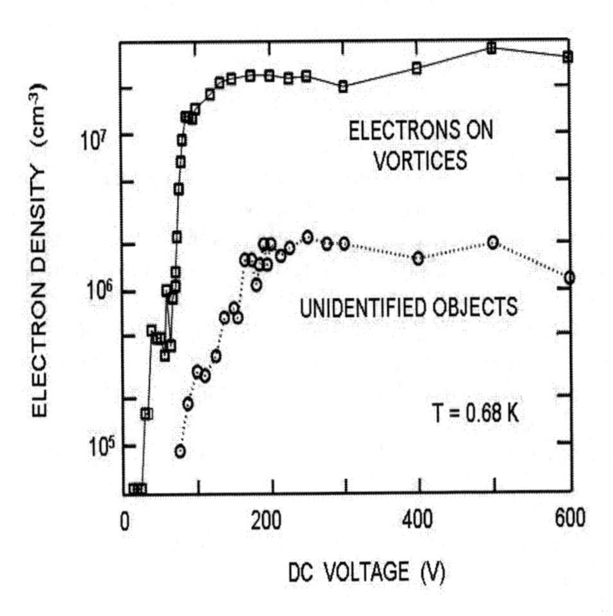

FIGURE 4. Measured density of electrons on vortices and unidentified electron objects as a function of the magnitude of the applied dc voltage (negative) that is applied to the radioactive source. These measurements were made at 0.62 K.

this assignment comes from a measurement of the number density of the objects as a function of electric field . When the radioactive source is used to inject electrons, the #2 objects are only seen when the electric field is sufficiently large to give the electron bubbles the critical velocity needed for the nucleation of quantized vortices. When the field emission tip is used instead of the radioactive source, the situation is a little more complicated. Near to the tip, the electric field is very large and vortices will be created even when the electric field in the bulk of the cell is not high enough to give electron bubbles the critical velocity. The vortices created near the tip, together with attached electrons, can travel across the cell losing energy as they go and may be able to reach the acoustic focus.

The new unidentified electron objects #3 (UEO's) appear only when the temperature is below 1 K. They can be detected when electrons are injected from either the tip or the radioactive source. With the radioactive source the UEO's appear only when a sufficiently large electric field is applied. This critical field varies with temperature. Figure 4 shows measurements of the number density of the UEO as a function of the magnitude of the negative voltage applied to the radioactive source. The UEO's first appear at a voltage that is roughly a factor of two larger than the voltage needed to produce electrons on vortices. The density of the UEO's is always at least one order of magnitude less than the density of electrons on vortices.

DISCUSSION

We have been unable to establish the physical nature of these objects. The observation that they are only seen when the electric field is above a critical value, suggests that they must have some connection with quantized vortices. The bubbles explode at a negative pressure that has a smaller magnitude than the explosion pressure for a normal electron bubble. This means that the objects are probably larger than standard bubbles. The pressure enters into the expression for the total energy as the term PV (see Eq. 1) and so for a larger bubble the effect of an applied negative pressure will be larger, thereby giving a critical pressure for explosion that has a smaller magnitude.

One possibility is that the UEO's are electron bubbles that are moving at high velocity just after having escaped from a vortex line. These bubbles would be accelerated by the electric field, and quickly reach the critical velocity at which a new vortex ring can be nucleated. The electron bubble would then again be trapped. While the bubble is moving at high speed, the pressure at the surface of the bubble will be reduced below the pressure in the bulk liquid due to the Bernoulli effect. As a result, the bubble should explode at a critical pressure of the bulk liquid which is of a smaller magnitude than the magnitude of the critical pressure for a stationary bubble. The critical velocity v_c for vortex nucleation is around 30 m s^{-1}, and so the order of magnitude of the Bernoulli pressure is $\rho v_c^2/2 = 0.65$ bars. This is of the right order of magnitude to explain the shift in the critical pressure that we detect. However, since electron bubbles are continually escaping from vortices at a certain rate, if it is these bubbles that are the UEO's we would expect that the probability that cavitation occurs should be proportional to the length of time for which the negative pressure is applied. To look for this effect we measured the cavitation probability S as a function of the number of cycles N in the sound pulse. However, we found no significant variation of S when N was changed, and thus we think that the UEO cannot be fast moving bubbles.

Since the UEO's appear to be related to the presence of vortices it is possible that they are electron bubbles that are attached to two vortex lines or to a single vortex line that has a circulation around it of $2h$. As far as we know, there has been no previous experimental support for the existence of doubly-quantized vortices, although Dalfovo has performed calculations of their energy and structure.[8]

ACKNOWLEDGMENTS

We thank D.O. Edwards and W. Guo for helpful discussions. This work was supported in part by the National Science Foundation through grant DMR-0305115

REFERENCES

1. The properties of ions in helium have been reviewed by A.L. Fetter, in *The Physics of Liquid and Solid Helium*, editor K.H. Benneman and J.B. Ketterson (Wiley, New York, 1976), chapter 3.
2. C.C. Grimes and G. Adams, *Phys. Rev.* **B41**, 6366 (1990) and **B45**, 2305 (1992); A. Ya. Parshin and S.V. Pereversev, *Pis'ma Zh. Eksp. Teor. Fiz.* **52**, 905 (1990) [*JETP Lett.* **52**, 282 (1990)].
3. C.S.M. Doake and P.W.F. Gribbon, *Phys. Lett.* **30A** 251 (1969).
4. G.G. Ihas and T.M. Sanders, *Phys. Rev. Lett.* 27, 383 (1971) and in *Proceedings of the 13th International Conference on Low Temperature Physics*, editors K.D. Timmerhaus, W.J. O'Sullivan and E.F. Hammel, (Plenum, New York, 1972), Vol. 1, p. 477; G.G. Ihas, Ph.D. thesis, University of Michigan, 1971.
5. V.L. Eden and P.V.E. McClintock, Phys. Lett. 102A, 197 (1984). V.L. Eden, M. Phil. thesis, University of Lancaster, 1986; C.D.H. Williams, P.V.E. McClintock, and P.C. Hendry, in Elementary Excitations in Quantum Fluids, Springer, Heidelberg, 1989), p. 192 and *Jap. J. Appl. Phys.* 26-3, DJ13 (1987).
6. J. Classen, C.-K. Su, M. Mohazzab and H.J. Maris, *Phys. Rev.* **57**, 3000 (1998).
7. D. Konstantinov and H.J. Maris, *Phys. Rev. Lett.*, **90**, 025302 (2003).
8. F. Dalfovo, *J. Low Temp. Phys.* **89**, 425 (1992).

Charge Transport in Liquid Helium at Low Temperatures

B. Sethumadhavan, W. Yao, H. Eguchi, A. Ghosh, Y H. Huang, Y. H. Kim, R. E. Lanou, H. J. Maris, A. N. Mocharnuk-Macchia and G. M. Seidel

Department of Physics, Brown University, Providence, RI 02912, USA

Abstract. In an experiment to investigate the possibility of using superfluid helium as a detection medium for low energy solar neutrinos, we have studied the currents produced by a radioactive source in a helium cell having a liquid/vacuum interface at 50 mK. A number of phenomena have been observed that appear not to have been described in the literature. These include the following: 1) The current at very low voltages in a cell having a free surface can be 100 times greater than in a filled cell. This additional current is associated with Penning ionization of metastable triplet dimers in surface states. 2) There is a large amplification of current in modest electric fields with a free surface present in the cell. This is the result of charges accelerated across the vacuum having sufficient energy to produce ionization and additional free charges upon hitting a liquid surface. The amplification becomes sufficiently large that breakdown occurs at potential differences across the vacuum of less than 1000 V. The dependence on ^{3}He concentration of these phenomena has been studied.

Keywords: superfluid helium, metastables, surface, solar neutrino
PACS: 51.50.+v, 26.65.+t, 67.40.Jg

INTRODUCTION

A single electron, extracted from liquid helium and accelerated by a field to an energy of several hundred eV, can be detected calorimetrically at low temperatures on striking a reasonably sized wafer [1]. In a preliminary experiment to study if such a process could be used to detect electrons that are produced by the scattering of low energy p–p neutrinos from the Sun, we have investigated the current resulting from the presence of a radioactive source in a cell partially filled with helium at 50 mK.

EXPERIMENT AND DISCUSSION

A 1 mCi radioactive beta source, ^{63}Ni, the mean energy of the emitted electrons being 17 keV, was placed at the bottom of a 2.5 cm diameter, 0.4 cm high cell. The helium level in the cell was adjustable and its position could be measured with a capacitance gauge. The current was measured upon applying a potential between a source electrode at the bottom of the cell and a collector electrode at the top. The range of a 17 keV electron in liquid helium is less than 100 microns.

It takes 43 eV, on average, for an energetic electron to produce an ionizing event in liquid helium. The ions and electrons, for the most part, rapidly undergo geminate recombination to atoms in either singlet or triplet excited states. The excited state atoms form He_2^* dimers with ground state helium atoms. The singlets radiatively decay to the dissociated ground state with the emission of EUV photons of 16 eV energy in less than 10^{-9}s. The triplet dimers have a long radiative lifetime of about 13 s.

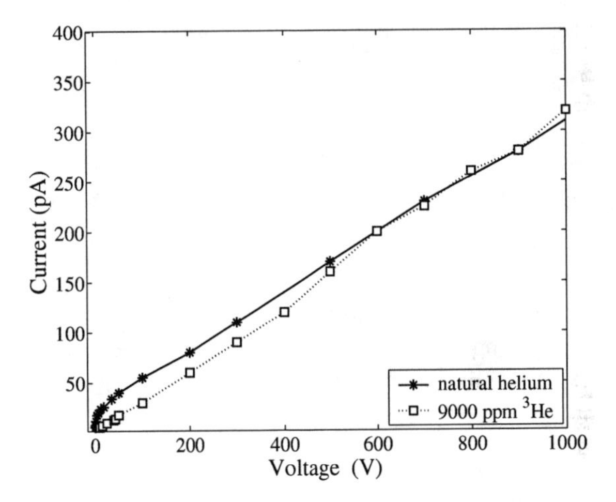

FIGURE 1. Current as a function of potential applied across the cell filled with natural helium and ^{4}He with 0.9% ^{3}He.

The current in the completely filled cell is small. The field dependence is shown in Fig. 1. At very low voltages there is a component, depending on the sign of the voltage, that is due to the primary electrons. As the voltage rises a small but increasing number of the electron/ion pairs are separated by the field and do not recombine. The manner in which the current varies with field depends on the initial spatial distribution of electron/ion pairs and upon the nature of the motion of the electron bubble and ion snowball in the combined Coulomb and applied field. The addition of ^{3}He atoms at 50 mK produces scattering of the moving charges, affects the formation of vortices and decreases the current, as shown in Fig. 1. To obtain a significant change in current, ^{3}He

CP850, *Low Temperature Physics: 24th International Conference on Low Temperature Physics;* edited by Y. Takano, S. P. Hershfield, S. O. Hill, P. J. Hirschfeld, and A. M. Goldman

concentrations of greater than 0.1% are required. A quantitative analysis and discussion of these results will be presented in a more comprehensive paper.

When the cell is partially filled with natural helium, the current is dramatically different as shown in Fig. 2. At very low fields the current rises to a plateau where it is 100 or more times larger than in the filled cell. Since the primary ionizing electrons are stopped with 100 microns or less of the bottom electrode, the production of electron/ion pairs cannot be influenced by the presence of a free surface, approximately 2 mm higher. The origin of this large excess current is due to the metastable triplet dimers in the $a^3\Sigma_u^+$ state. At low temperatures the triplet dimers travel ballistically and, in the filled cell, upon encountering a solid surface, nonradiatively decay. In the partially-filled cell having the geometry of a thin disc, roughly half of the dimers reach the free surface to which they become bound. Because of their large diameter outer electron orbits, these neutral dimers form bubbles in the liquid and have a lower energy on the surface. When two dimers interact, they are known to undergo a thresholdless Penning ionization process

$$He_2(a^3\Sigma_u^+) + He_2(a^3\Sigma_u^+) \rightarrow 3\,He(1^1S) + He^+ + e^- \quad (1)$$

or

$$He_2(a^3\Sigma_u^+) + He_2(a^3\Sigma_u^+) \rightarrow 2\,He(1^1S) + He_2^+ + e^-. \quad (2)$$

This process occurs in the bulk, but the probability of occurrence is small since the concentration of dimers is low. Also, upon ionization in the bulk the separation of the electron and ion is small, and the charges tend to recombine immediately producing no contribution to the current. However, on the free surface the density builds up to the point where one dimer is likely to encounter another. When Penning ionization does occur, one of the two charged particles is likely to enter the vacuum, and depending on the polarity of the applied potential, either is accelerated across the vacuum or is returned to the bulk liquid at a substantial distance from where it originated. In either case the Penning process at the free surface results in a current. The magnitude of this current is consistent with estimates of the number of triplet dimers formed by electrons stopped in liquid helium[2] and will be discussed in detail elsewhere.

This explanation of the large excess current is supported by the observed influence of small amounts of ^{3}He. The introduction of less than 10 ppm of ^{3}He reduces the current in the partly filled cell to approximately that observed in the filled cell. The ^{3}He scatters the triplet dimers sufficiently often that in undergoing a random walk from their origin very close to the bottom electrode, they are much more likely to encounter the bottom surface and be destroyed rather than to reach the free surface.

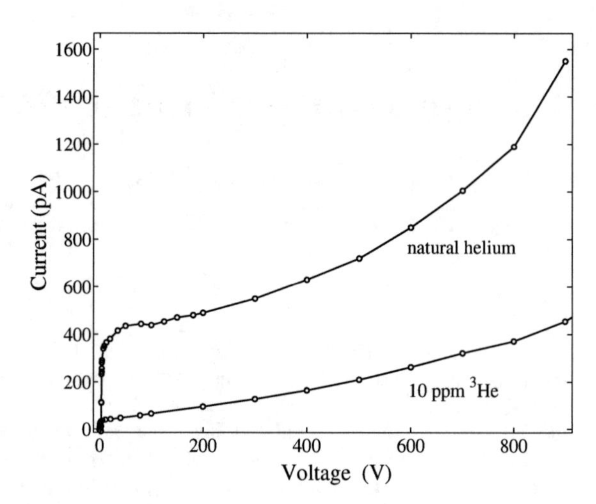

FIGURE 2. Current as a function of potential applied across the cell half full of helium. The sign of the potential is such that electrons travel upwards through the liquid and vacuum to the electrode on the top of the cell.

At higher voltages the total current is not simply the sum of the current from charge separation of secondaries in the bulk plus the current from surface Penning ionization. The current is a nonlinear function of field and, at potentials of less than 1000 V across the vacuum, electrical breakdown occurs. This amplification phenomenon must be the result of charged particles, accelerated across the vacuum gap, producing additional ionization events and dimers upon striking the film on the top electrode or the bulk liquid depending on sign of the charge. The helium atom density in the vacuum at 50 mK is far too small to be of significance. Amplification and breakdown occur for either sign of the potential applied to the cell, but there are differences that do depend on polarity. These details are presumed to be associated with the mechanisms by which electron bubbles and positive ion snowballs escape through the liquid surface. These pheonomena are still being explored with the aid of a ring electrode, which has been added to the cell.

ACKNOWLEDGMENTS

This work is supported in part by U.S. Department of Energy grant DE-FG02-88ER40452.

REFERENCES

1. B. Sethumadhavan, *et al*, *Nucl. Instr. and Meth. A*, **520**, 142 (2004).
2. J. Adams, Energy Deposition by Electrons in Superfluid Helium, Ph.D. Thesis, Brown University, 2000.

Properties of Moving Electron Bubbles in Superfluid Helium

Wei Guo and Humphrey J. Maris

Department of Physics, Brown University, Providence, Rhode Island 02912, USA

Abstract. It is well known that the Bernoulli effect modifies the shape of gas bubbles moving through a liquid. In this paper we investigate the influence of the Bernoulli pressure on the shape of electron bubbles moving through superfluid helium. We show that an electron bubble moving through liquid at zero pressure becomes unstable when its velocity reaches approximately 47 m s^{-1}. In addition, the change in shape contributes significantly to the variation of the bubble mobility with velocity.

Keywords: superfluid helium, electrons, Bernoulli.
PACS: 67.40, 78.20c

When a bubble moves through a liquid of low viscosity, the Bernoulli effect results in a variation of the liquid pressure over the surface of the bubble. If the liquid is treated as incompressible and the bubble is, for the moment, taken to be spherical it is straightforward to show that the liquid velocity is

$$\vec{v} = \frac{R^3}{2r^3}[3\hat{r}(\vec{V}\cdot\hat{r}) - \vec{V}] \qquad (1)$$

where $\vec{V}$ is the velocity of the bubble, R is the bubble radius, and $\vec{r}$ is the distance from the center of the bubble. As a result of this flow of the liquid there is a pressure distribution over the surface of the bubble given by

$$P = P_0 + \frac{1}{8}\rho V^2 (9\cos^2\theta - 5) \qquad (2)$$

where θ is the angle between the velocity $\vec{V}$ and the location of a point on the bubble surface, ρ is the liquid density and P_0 is the pressure in liquid that is at rest and far away from the bubble. As a result of this pressure variation, the bubble undergoes a change in shape such that the waist increases. The shape is determined by the Weber number We defined as

$$We = \frac{2\rho V^2 R}{\alpha} \qquad (3)$$

where α is the surface tension. At a critical Weber number We_c of 3.37, a gas bubble becomes unstable against growth of the waist.[1]

Consider now an electron bubble moving through superfluid helium at low temperatures where the drag on the bubble due to the normal fluid is small. It is important to recognize that the elastic properties of an electron bubble are different from those of a gas bubble. To a very good approximation a gas bubble of macroscopic size (e.g., ~ 1 cm) deforms in a way such that its volume remains constant. The extent to which the shape of the bubble changes due to the variation of the Bernoulli pressure over the bubble surface is determined by a balance between the pressure and the surface tension. For an electron bubble, on the other hand, the energy of the electron inside the bubble is a function of both the volume and the shape of the bubble.

For small velocity, the shape change can be calculated by perturbation theory. If the bubble is nearly spherical, the pressure is as given by Eq. 2, and the change in the shape can readily be calculated from the results for the stiffness of the bubble against deformation given in ref. 2. For larger velocity, we have developed a numerical method. We start with a guess at the bubble shape and then calculate the electron wave function and determine the pressure the electron exerts on the bubble surface. The flow of the liquid outside the bubble is next calculated and the variation of the Bernoulli pressure over the

CP850, *Low Temperature Physics: 24th International Conference on Low Temperature Physics;* edited by Y. Takano, S. P. Hershfield, S. O. Hill, P. J. Hirschfeld, and A. M. Goldman

surface is determined. We next add in the effect of surface tension and determine the net pressure imbalance ΔP at different locations on the surface. We then move the surface by a distance which is proportional to ΔP, and repeat this until we find the equilibrium shape. Note that in this calculation the liquid is treated as incompressible.

In Fig. 1 we show the pressure at which bubbles explode as a function of their velocity. At zero pressure bubbles become unstable and explode at a critical velocity V_{ex} of 47 m s^{-1}. Figure 2 shows the distance from the center of the bubble to the pole and to the equator for a bubble moving in liquid at zero pressure.

When electron bubbles move through the liquid at a sufficiently high velocity, a vortex can be nucleated and the bubble is then trapped on the ring. The critical velocity V_v for vortex nucleation [3] is approximately 44 m s^{-1}. Thus, by coincidence, V_{ex} and V_v are almost identical. From Fig. 2 it can be seen that when the velocity of the bubble is 44 m s^{-1}, there is a very significant change in the size (and shape) of the bubble compared to the size at zero velocity. This change is not taken into account in the Muirhead-Vinen-Donnelly [4] theory of vortex nucleation and it would be interesting to know how inclusion of this change in size would modify the predictions of this theory.

Finally, we mention that the increase in the radius of the waist of the bubble will increase the drag on the bubble due to scattering of thermal excitations. Thus, the change in shape should lead to

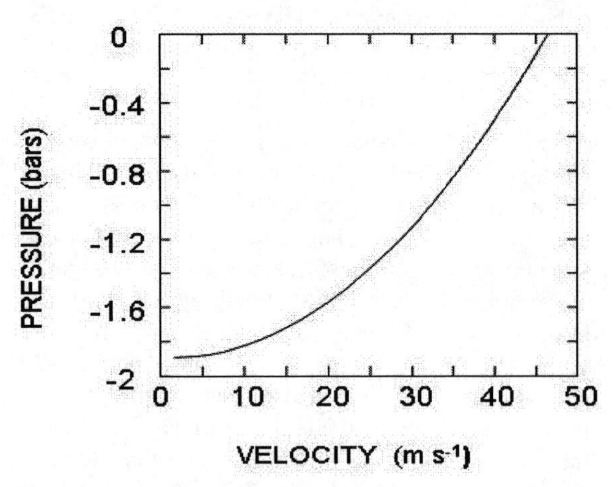

FIGURE 1. Pressure at which an electron bubble explodes as a function of the bubble velocity.

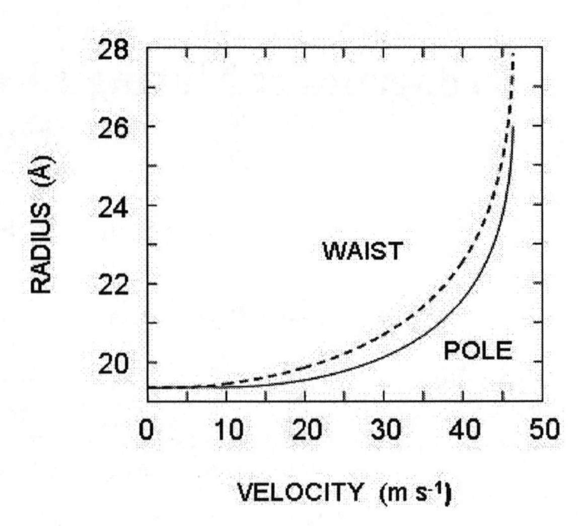

FIGURE 2. Radius to the poles and to the waist of an electron bubble as a function of velocity. Calculations are for $P = 0$.

a mobility that decreases with velocity. At 30 m s^{-1}, the change in the area of the equator is 14 %, corresponding to a decrease in mobility relative to the low field value of 12 %. The measured decrease in mobility[5] at this velocity is 18%, in surprisingly good agreement, considering that there may be other factors that contribute to the change in mobility with velocity.

This work was supported in part by the National Science Foundation through Grant No. DMR-0305115.

REFERENCES

1. V.I. Kusch, A.S. Sangani, P.D.M. Spelt and D.L. Koch, *J. Fluid. Mech.* **460**, 241 (2002).
2. H.J. Maris and W. Guo, *J. Low Temp. Phys.* **137**, 491 (2004).
3. The cited velocity is for isotopically pure ^{4}He. For a discussion, see R.J. Donnelly, *Quantized Vortices in Helium II* (Cambridge, Cambridge, 1991), pp 291-292. The estimate of 44 m s^{-1} is obtained by extrapolation of the data at higher pressure (R.M. Bowley, P.V.E. McClintock, F.E. Moss and P.C.E. Stamp, (*Phys. Rev. Lett.* **44**, 161 (1980)) to $P = 0$.
4. C.M. Muirhead, W.F. Vinen and R.J. Donnelly, *Phil. Trans. Roy. Soc.* **A311**, 433 (1984).
5. V.L. Eden and P.V.E. McClintock, *Phys. Lett.* **102A**, 197 (1984).

Experiments to Study Photoemission of Electron Bubbles from Quantized Vortices

Denis Konstantinov, Matthew Hirsch and Humphrey J. Maris

Department of Physics, Brown University, Providence, Rhode Island 02912, USA

Abstract. At sufficiently low temperatures, electron bubbles (negative ions) can become trapped on quantized vortices in superfluid helium. Previously, the escape of electron bubbles from vortices by thermal excitation and through quantum tunneling has been studied. In this paper, we report on an experiment in which light is used to release bubbles from quantized vortices (photoemission). A CO_2 laser is used to excite the electron from the 1S to the 1P state, and it is found that each time a photon is absorbed there is a small probability that the bubble containing the electron escapes from the vortex.

Keywords: electrons, bubbles, helium, photons.
PACS: 67.40, 78.20c

Electrons injected into liquid helium create a cavity in the liquid from which helium atoms are excluded. These so called "electron bubbles" have been studied in many low temperature experiments, mainly through measurements of their mobility.[1] If an electron bubble moves above a critical velocity, a vortex ring forms and the bubble helium can become trapped on the vortex with a binding energy U_0.[2] Electrons can escape from a vortex as a result of thermal excitation or by quantum tunneling. In this paper, we investigate another mechanism by which electrons can escape from a vortex, namely via optical excitation.

The apparatus is shown schematically in Fig. 1. A sharp tungsten tip is used to introduce electrons into a region of liquid between a metal perforated disk (PD) and a collector plate. The disk is located 5 mm below the point of the tip and has a number of holes in it to allow ions to pass through. The disk is separated from the collector by a cylindrical nylon spacer of height 5 mm and with holes at either side to allow light into the liquid that is between the PD and the collector. A constant voltage of -2 kV is applied to the tip to inject negative ions into the liquid, the voltage on the perforated disk is -100 V, and the collector plate is connected to ground through a 100 kΩ resistor.

Because of the high electric field in the region above the PD, electrons emitted from the tungsten tip quickly become trapped on vortices. Some of these vortex rings pass through the PD into the region of lower electric field between the PD and the collector. A light pulse from a CO_2 laser is applied to the region between the PD and the collector. Some fraction of the trapped electrons escape from the vortices, and give a pulse of current arriving at the collector a few hundred μs later. Through an analysis of the variation of the size of the current pulse with the energy in the light pulse the probability that the absorption of a photon leads to electron escape can be determined as described below.

The laser beam has a Gaussian profile with a waist diameter of about 3.6 mm. Typically a series of 1000 light pulses at intervals of 1 s were applied, and the average signal at the collector was measured using a digital oscilloscope. All of the measurements were

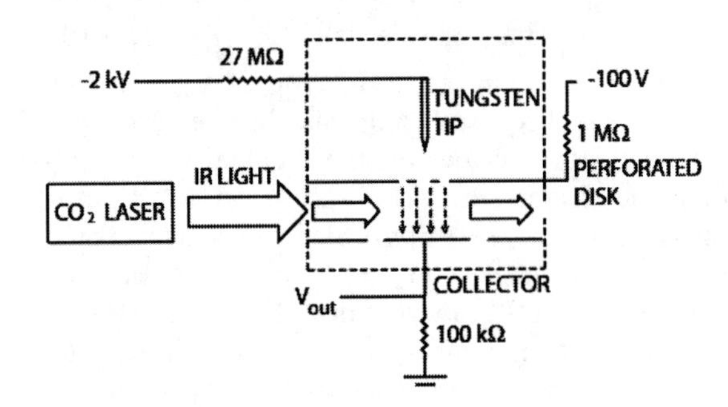

FIGURE 1. Schematic diagram of the apparatus.

CP850, *Low Temperature Physics: 24th International Conference on Low Temperature Physics;*
edited by Y. Takano, S. P. Hershfield, S. O. Hill, P. J. Hirschfeld, and A. M. Goldman

conducted at a static pressure of 1 bar where the cross-section for light absorption[3] has its maximum value of 4.6×10^{-15} cm^{-2}, and at 1 K where almost all ions should be trapped on vortices.

Let σ be the absorption cross-section for the 1S→1P transition at the CO_2 wavelength, and p be the escape probability. Then the rate at which ions escape from vortices due to a light pulse is given by

$$\frac{dN}{dt} = (N_0 - N) I \sigma p \qquad (1)$$

where N is the number of untrapped ions, N_0 is the total number of ions in the volume illuminated by the light, and I is the flux of photons per unit area per unit time. If a light pulse of length τ is introduced, the number of liberated ions is

$$N(I,\tau) = N_0[1 - \exp(-I\sigma\tau p)]. \qquad (2)$$

These released electrons give a pulse of current $J(t)$ at the collector, which is much larger than the current without illumination. The integral Q of the detector signal over the pulse is proportional to $N(I,\tau)$, and hence a measurement of Q for a series of values of I or τ can be used to find the probability p. In this analysis we use the calculated absorption cross-section for the 1S→1P transition.

In Fig. 2 we show examples of the collector signal as a function of time when the length of the light pulse and the intensity were varied. From these data, we calculate Q and then determine the value of p that gives the best fit to the data when Eq. 2 is used. An example of this fit is shown in Fig. 3.

This procedure gives a value for p of 1.3×10^{-4}. As far as we are aware, there is no theory with which to compare this result. Immediately after the light is absorbed, the shape of the bubble undergoes a dramatic change because of the change in the form of the electron wave function. Calculations[4] show that the bubble may break into two smaller bubbles which could be ejected away from the vortex line. In addition, the motion of the bubble wall results in energy dissipation in the surrounding liquid, and it is possible that this energy enables the bubble to escape from the confining potential. To attempt to distinguish between these possibilities we are currently making measurements of the escape probability as a function of temperature and electric field.

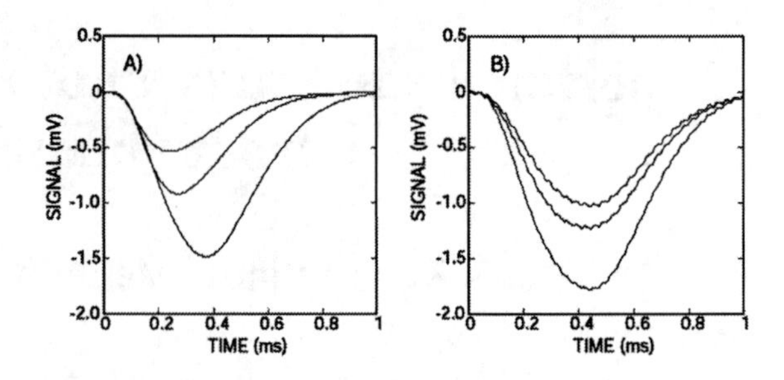

FIGURE 2. Signal at the collector as a function of time resulting from the application of a light pulse. A) Results for pulses of the same intensity length and length 50, 100 and 300 μs. B) Results for 400 μs pulses of intensity 35 %, 50 % and 100 % of the maximum laser output power.

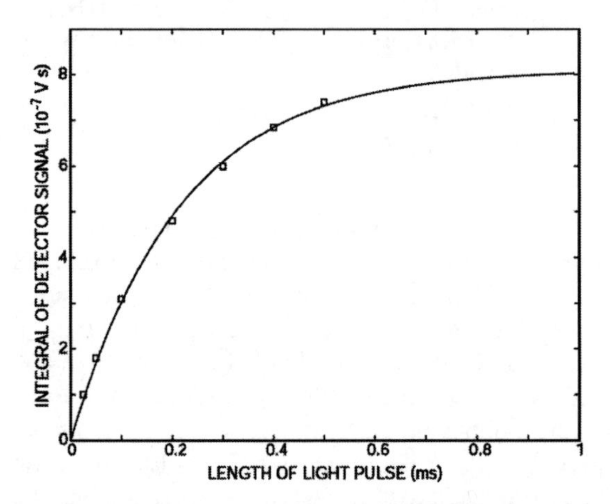

FIGURE 3. Plot of the integral of the detector signal as a function of the length of the light pulse to illustrate the fitting procedure. The signal saturates when all electrons have escaped from vortices.

We thank A. Ghosh, W. Guo, and G.M. Seidel for helpful discussions. This work was supported in part by the National Science Foundation through Grant No. DMR-0305115.

REFERENCES

1. For a review, see A.L. Fetter, in *The Physics of Liquid and Solid Helium*, edited by K.H. Benneman and J.B. Ketterson, New York: Wiley, 1960.
2. See R.J. Donnelly, *Quantized Vortices in Helium II*, London: Cambridge, 1991.
3. H.J. Maris and W. Guo, *J. Low Temp. Phys.* **137**, 491 (2004).
4. H.J. Maris, *J. Low Temp. Phys.* **120**, 173 (2000).

Calculation of the Cross-Section for the 1S→2P Transition of an Electron Bubble in Helium II

Wei Guo and Humphrey J. Maris

Department of Physics, Brown University, Providence, Rhode Island 02912, USA

Abstract. In a recent paper, we have calculated the effect of thermal and zero-point fluctuations on the 1S to 1P optical transition of the electron bubble in superfluid helium. We obtained a line shape that was in good agreement with the experimental results of Grimes and Adams (*Phys. Rev.* **45**, 2305 (1992)). Here we extend these calculations to consider the line shape for the 1S to 2P transition and compare the results with the data of Zipfel and Sanders.

Keywords: electrons, bubbles, helium, photons.
PACS: 67.40, 78.20c

The optical properties of electron bubbles in superfluid helium have been studied in a number of experiments. The photon energy required to cause the electron to make a transition from the 1S to the 1P state has been measured by Grimes and Adams,[1, 2] and Parshin and Pereversev.[3] The 1S→2P transition has been studied by Zipfel and Sanders.[4] At zero pressure the 1S→1P transition requires a photon of energy 0.1 eV, and the 1S→2P transition energy is 0.5 eV.

The line shape of these transitions is determined by the fluctuations in the shape and size of the electron bubble. Both zero-point and thermal fluctuations contribute to the line shape. As a consequence, measurements of the line shape as a function of temperature and pressure provide an interesting test of calculations of the fluctuations of a helium surface. In a recent paper[5] we have calculated the amplitude of these fluctuations and have determined the line shape for the 1S→1P transition. The results were in excellent agreement with the measurements of Grimes and Adams.[2] In this note, we apply the same method to calculate the shape of the 1S→2P transition.

Here, we describe the essential physics of the method; the calculation follows along the same general lines as already presented for the 1S→1P transition in ref. 5. The total energy E_b of the electron bubble is taken as

$$E_b = E_{el} + \alpha A + PV , \quad (1)$$

where E_{el} is the energy of the electron, α is the surface tension, P is the applied pressure, A is the surface area, and V is the volume of the bubble. For a mode of vibration in which the surface displacement varies as $Y_{lm}(\theta,\phi)$, the vibrational frequency ω_l is found. The statistical distribution of amplitudes for each mode l,m is then determined. For a set of mode amplitudes, we calculate the difference in energy between the 1S and 2P levels. Averaging over the statistical distribution of mode amplitudes then gives the line shape. Since the amplitude of the fluctuations of the bubble surface is small, it is sufficient to consider that the shift in the energy levels is proportional to the displacement. Within this approximation, the shift in the 1S level depends only on the normal mode with $l=0$, and the 2P level is shifted by the $l=0$ mode, and by the five degenerate modes with $l=2$.

In the previous calculation for the 1S→1P transition, the energy of the electron was sufficiently far below the potential barrier provided by the wall of the bubble that the penetration of the wave function into the helium could be neglected. However, for the 1S→2P transition it is necessary to allow for the finite height (~1 eV) of the barrier, and to do this we followed the method used by Grimes and Adams.[1] The results for the line shape at different pressures and temperatures are shown in Fig. 1. It can be seen that the line width increases with pressure and temperature,

CP850, *Low Temperature Physics: 24th International Conference on Low Temperature Physics;*
edited by Y. Takano, S. P. Hershfield, S. O. Hill, P. J. Hirschfeld, and A. M. Goldman

and hence the peak value of the cross-section decreases with an increase in these parameters.

The only experimental study of the 1S→2P transition is the work of Zipfel at a temperature of approximately 1.3 K. She made measurements of the line shape over a range of pressure up to 15 bars. In Fig. 2 we compare her results for the full width at half maximum with the results of our calculations. It can be seen that, allowing for the scatter in the experimental results and the approximations that have been made in the theory, the agreement between experiment and theory is good.

This work was supported in part by the National Science Foundation through Grant No. DMR-0305115.

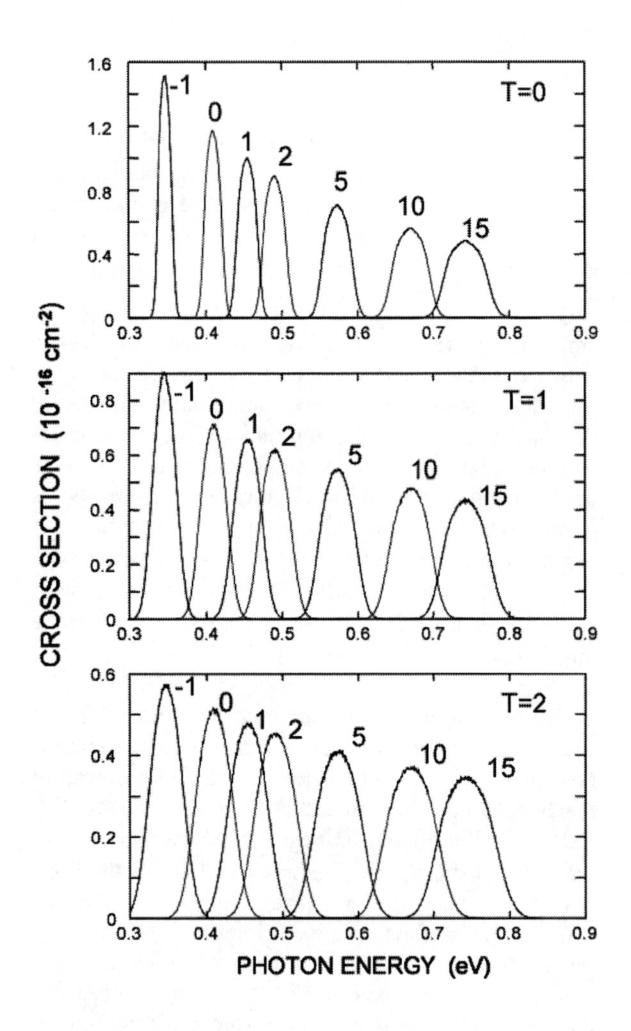

FIGURE 1. Calculated shape of the absorption cross-section for the 1S→2P optical transition as a function of pressure at temperatures of 0, 1 and 2 K. The different curves are labeled by the pressure in bars.

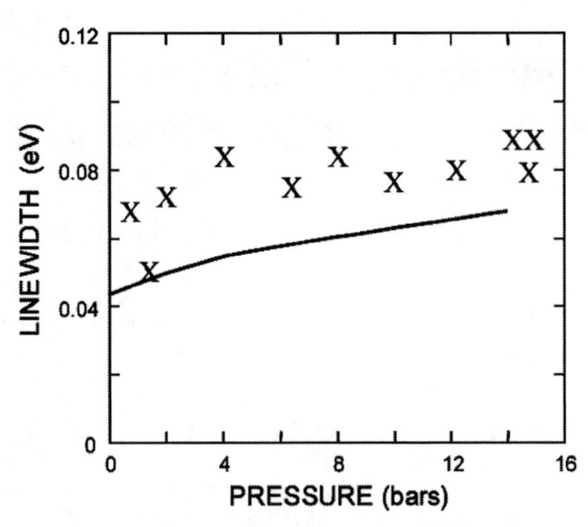

FIGURE 2. Full width at half height of the 1S→2P cross-section as a function of pressure at 1.3 K. The crosses are the measurements of Zipfel (ref. 4) and the solid curve is the calculated width.

REFERENCES

1. C.C. Grimes and G. Adams, *Phys. Rev. B* **41**, 6366 (1990).
2. C.C. Grimes and G. Adams, *Phys. Rev. B* **45**, 2305 (1992).
3. A. Ya. Parshin and S.V. Pereversev, Pis'ma Zh. ksp. *Teor. Fiz.* **52**, 905 (1990) [*JETP Lett.* **52**, 282 (1990)].
4. C.L. Zipfel, Ph.D. thesis, University of Michigan, 1969, unpublished; C.L. Zipfel and T.M. Sanders, in *Proceedings of the 11th International Conference on Low Temperature Physics*, edited by J.F. Allen, D.M. Finlayson, and D.M. McCall (St. Andrews University, St. Andrews, Scotland, 1969), p. 296.
5. W. Guo and H.J. Maris, *J.Low Temp. Phys* **137**, 491 (2004).

Small Multielectron Bubbles in Bulk Superfluid ^{4}He

Lauri Lehtovaara and Jussi Eloranta

Department of Chemistry, University of Jyväskylä, P.O.Box 35, 40014 Jyväskylä, Finland

Abstract. A computational method for describing multielectron bubbles in bulk superfluid helium (^{4}He) is described. The developed method can be used to study their stability and spectroscopic properties in both singlet and triplet manifolds. Our initial results indicate meta-stability of the spherical two-electron bubble in its singlet ground state.

Keywords: superfluid, helium, multielectron, bubble
PACS: 67.40.Yv

INTRODUCTION

Previous experimental and theoretical studies have established that excess electrons in superfluid ^{4}He reside in large solvation cavities ("bubbles") [1]. Geometry of the electron bubble is mainly dictated by the following factors: (i) mostly repulsive interaction between an electron and helium, (ii) a large zero-point motion of the electron, and (iii) surface tension of the liquid. In the ground state (0 K and saturated vapor pressure), the bubble has spherical geometry with an approximate radius of 18 Å.

If more than one electron occupies the bubble, the electron-electron interaction must be taken into account. The electron-electron repulsion pushes electrons away from each other and causes greater pressure towards the bubble edge. In the case of two electrons, it is not clear whether the repulsion is sufficiently strong to push the electrons far enough that the bubble would split into two one-electron bubbles. Provided that the two-electron bubble is energetically stable, it would be of interest to provide theoretical estimates of the bubble geometry and the electronic configuration.

Two-electron bubbles have been previously studied by Maris [2]. He treated the electronic degrees of freedom essentially exactly using a simple diffusion algorithm. This method can be applied to find only the symmetric lowest energy solution and therefore it was not possible to model the electronic triplet state. The liquid interface, providing the external potential for the electronic problem, was approximated by a hard-wall with overall shape given by spherical harmonics. This clearly prevents any electron density from penetrating the liquid. The superfluid contribution to the total energy was then approximated using a classical surface tension term, which ignores the quantum effects related to the diffuse bubble edge. Due to these approximations, this model, for example, overestimates the ground state one-electron bubble edge position by ~ 1 Å [1, 2].

COMPUTATIONAL METHOD

The interaction between an electron and a single helium atom is obtained through the pseudo-potential developed by Jortner *et al.* [3]. The total interaction energy is then calculated by convoluting the pseudo potential with the electron or liquid densities [1]. These convolutions can be evaluated efficiently by the Fast Fourier Transform (FFT) techniques [4].

The coupled electron-superfluid problem is split into two parts, which are solved iteratively. The superfluid helium part is evaluated using the bosonic density functional theory (DFT) of Dalfovo *et al.* [5]. To find the ground state of the superfluid, we use the imaginary time method, where bosonic Kohn-Sham equation is propagated in imaginary time. This approach works only when the ground-state wave function does not have any nodes. The electronic part is solved using the standard Hartree-Fock method employing a Gaussian basis set. The interaction potential between the electron and the liquid is obtained by mapping the basis functions onto a 3D grid, where the FFT convolution algorithm can be applied. Details of the method will be described in a separate publication. Solution of

CP850, *Low Temperature Physics: 24th International Conference on Low Temperature Physics;*
edited by Y. Takano, S. P. Hershfield, S. O. Hill, P. J. Hirschfeld, and A. M. Goldman

both the electronic and the liquid problems with updated particle densities is iterated until convergence is reached.

RESULTS AND DISCUSSION

To verify the validity of our numerical implementation, we have first recalculated the one-electron bubble problem. The resulting helium and electron density profiles for the one-electron bubble are shown in Fig. 1. The bubble edge distance is 17.6 Å, which is 0.3 Å shorter than in the earlier calculations of Eloranta *et al.* [1]. This difference probably originates from choice of the discretization scheme (1D spherical vs. 3D Cartesian grid). In both calculations the electron density penetrates into the liquid by *ca.* 5 Å.

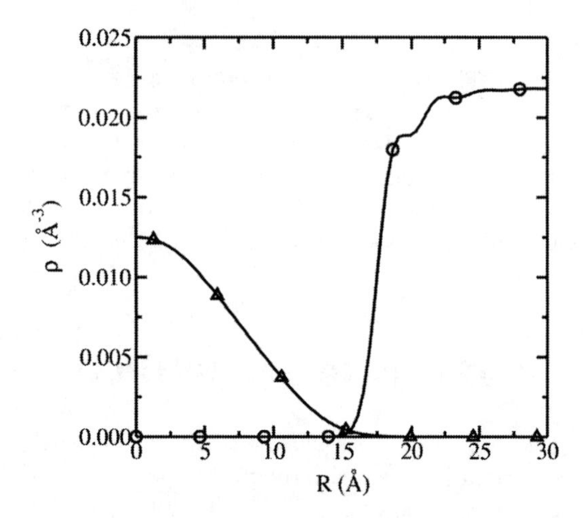

FIGURE 1. Density profiles of helium (circles) and the electron (triangles) for the one-electron bubble (electron density is scaled by a factor of 50).

For the two-electron bubble problem, we have calculated the singlet ground state. The helium and electron density profiles for this case are shown in Fig. 2. The spherical bubble edge is 33 Å and the electron density penetrates slightly over 5 Å into helium. Oscillations seen in the electron density profile are most probably artificial and caused by the Gaussian basis set (*s*-type Gaussians). When *d*-type Gaussians were included in the calculation and the initial bubble geometry (e.g. non-spherical) was chosen to favor splitting of their degeneracy, the calculations always converged to the spherical solution. Provided that this result will still hold when electron correlation is included, it would suggest stability of spherical two-electron bubbles. In addition to ground states, we are planning to extend our calculation to excited (both singlet and triplet) states using configuration interaction (CI) and electronic DFT methods. The use of CI and DFT methods would also allow us to take into account the electron correlation, which we have neglected here.

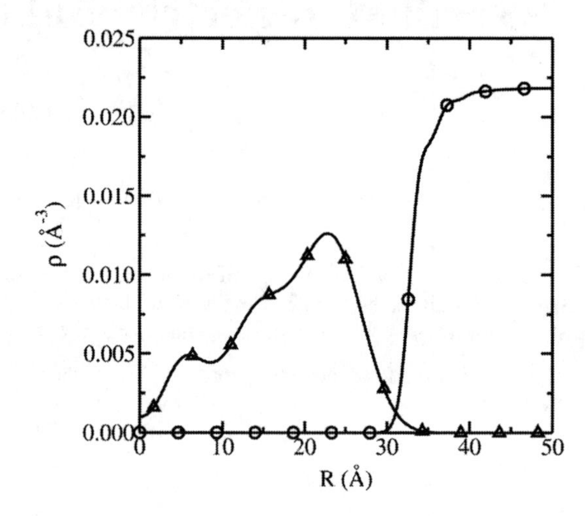

FIGURE 2. Density profiles of helium (circles) and the electrons (triangles) for the spherical two-electron bubble (electron density is scaled by a factor of 500).

ACKNOWLEDGMENTS

This research was funded by graduate school LASKEMO. The computational resources were granted by Finnish Center of Scientific Computing (CSC).

REFERENCES

1. J. Eloranta and A. Apkarian, *J. Chem. Phys.* **117**, 10139-10150 (2002) and references therein.
2. H. J. Maris, *J. Low Temp. Phys.* **132**, 77-94 (2003)
3. N. R. Kestner, J. Jortner, M. H. Cohen, and S. A. Rice, *Phys. Rev.* **140**, 56-66 (1965)
4. L. Lehtovaara, T. Kiljunen, and J. Eloranta, J. Comp. Phys. **194**, 78-91 (2004).
5. F. Dalfovo, A. Lastri, L. Pricaupenko, S. Stringari, and J. Treiner, Phys. Rev. B **52**, 1193-1209 (1995)

Quantum Turbulence: Where Do We Go From Here?

W. F. Vinen

School of Physics and Astronomy, University of Birmingham, Birmingham B15 2TT, UK

Abstract. It is now 50 years since the discovery of turbulence in the superfluid component of liquid ^{4}He, a form of turbulence that is strongly influenced by quantum effects and therefore often described as quantum turbulence. Early work focussed on turbulence associated with thermal counterflow in ^{4}He. The pioneering computer simulations of Schwarz led to considerable understanding of this type of turbulence, which has no classical analogue, but important aspects of it, both experimental and computational, remain unexplored. More recently, attention has turned to types of quantum turbulence, in both ^{4}He and ^{3}He-B, that have classical analogues, investigations (experimental, theoretical and computational) serving to emphasize similarities and differences between the classical and quantum cases and serving therefore to add to our understanding of both cases. It will be argued that progress is now hampered by the shortage of experimental techniques with which to probe quantum turbulence over a wide temperature range in the kind of detail possible in the classical case, and proposals are explored for addressing this shortage.

Keywords: Superfluids; quantized vortex lines; turbulence; Particle Image Velocimetry.
PACS: 67.57.-z, 67.40.-w, 47.37.+q, 67.40.Vs, 67.57.Fg, 67.40.Vs.

INTRODUCTION

The possible existence of turbulence in superfluid ^{4}He was first recognized in the 1950s. Feynman [1] suggested that flow of the superfluid through an aperture might lead to turbulence in the form of an irregular tangle of quantized vortex lines, and simultaneously and independently it was shown experimentally that the Gorter-Mellink force of mutual friction that limits heat transport in a thermal counterflow must almost certainly be due to turbulence in the superfluid component [2]. Soon it was recognized that the mutual friction was due to the scattering by vortex lines of the thermal excitations forming the normal fluid, and the subsequent observation of mutual friction in the uniformly rotating liquid provided the first convincing experimental demonstration of the existence and importance of vortex lines [3].

Detailed experimental study of counterflow mutual friction, both in the steady state and in transients, led to phenomenological equations describing the build-up and decay of turbulent vortex tangles in a heat current. Schwarz [4] realized that the behaviour of a tangle of vortex lines could be studied by computer simulations, based on what we now call the vortex filament model, in which the motion of the vortices is essentially classical, except for two factors: the force of mutual friction between a vortex core and the normal fluid; and the possibility of vortex reconnections when two vortices approach each other closely. Schwarz showed that the force of mutual friction, together with reconnections, could lead to the existence of a self-sustaining and essentially homogeneous tangle of vortices, with characteristics required to explain the experimental results.

CLASSICAL AND QUANTUM TURBULENCE

Thermal counterflow turbulence requires the forced relative motion of the normal and superfluid components of the helium, and therefore it has no obvious classical counterpart. Although observations were made on many different types of flow of critical velocities at which, presumably, turbulence was nucleated, there was practically no study until the mid 1990s of types of superfluid turbulence that might have classical analogues. This was strange in view of an obvious question: how are typical fully-developed classical turbulent regimes modified when the classical fluid is replaced by a superfluid? In other words, how do the severe quantum restrictions on rotational flow in a superfluid affect the turbulent flow? Or, how does quantum turbulence differ from analogous forms of classical turbulence?

CP850, *Low Temperature Physics: 24th International Conference on Low Temperature Physics;*
edited by Y. Takano, S. P. Hershfield, S. O. Hill, P. J. Hirschfeld, and A. M. Goldman

Two experiments relating to this question stand out, both carried out on ^{4}He at temperatures above 1K, where there is a significant fraction of normal fluid: that of Maurer & Tabeling [5] on pressure fluctuations in flow driven by counter-rotating discs; and that of Stalp *et al* [6] on the attenuation of second sound (due to mutual friction) in the wake of a moving grid. The first of these experiments showed rather directly that, at least on large length scales, there can exist in the helium a classical Richardson-Kolmogorov cascade, essentially indistinguishable from that in the corresponding classical regime. Interpretation of the second was less direct, but was nevertheless consistent with the same picture; in addition the moving grid experiment also showed that dissipation at the high wave number end of the cascade was of essentially the same form as that in a classical fluid with an effective kinematic viscosity of the same order of magnitude as that of helium I. A reasonable understanding of these experimental results now exists ([7] and references therein): both fluids are turbulent; turbulence in the superfluid component behaves in an essentially classical way on scales large compared with the average vortex line-spacing; on these large scales the two fluids are coupled by mutual friction, which leads to a common velocity field with a Kolmogorov energy spectrum; and dissipation on the scale of the line spacing mimics that due to viscosity in a classical fluid and is due to a combination of mutual friction and normal-fluid viscosity. In short, we see *quasi-classical behaviour* when we deal with turbulent superflows that have potential classical analogues.

Our comments so far relate to superfluid ^{4}He. Superfluid ^{3}He-B offers us another system in which quantum turbulence might be studied. Perhaps the most important difference lies in the viscosity of the normal fluid, the large value of which in ^{3}He-B means that turbulence in the normal component is practically impossible. We return later to recent results obtained with ^{3}He-B.

Our comments have also related largely to forms of fully-developed turbulence. There are still questions relating to the nucleation of turbulence and to the associated critical velocities, to which we shall also return later.

SHORTCOMINGS AND THE NEED FOR NEW EXPERIMENTAL TECHNIQUES

In spite of the success of the Schwarz simulations relating to thermal counterflow and of a good physical understanding of quasi-classical behaviour in ^{4}He at temperatures above 1K, our knowledge and understanding of quantum turbulence is far from satisfactory. There are two reasons.

First, very little is known from experiment about the potentially interesting and important case of turbulence at a very low temperature when there is a negligible fraction of normal fluid, and when therefore one might observe a very "pure" form of quantum turbulence. Much theoretical speculation exists in this case [7], but the only experiments relate to oscillating grids and wires, the measurements being either of the damping of the oscillating system or, in the case of ^{3}He-B, of Andreev scattering of thermal quasi-particles from the generated turbulence [8]. The results are often difficult to interpret, especially as the turbulence produced is complicated in form, ill-characterized, and not well-studied even in the classical case.

The second reason lies in the shortage of experimental techniques, which is the point of emphasis in this paper. Until very recently, the available techniques consisted of only the measurement of pressure and temperature gradients, forces on obstacles, the attenuation of second sound, and in one isolated case local pressure fluctuations [5]. The attenuation of second sound, applicable to ^{4}He above 1K, measures the length of vortex line per unit volume, averaged over the volume in which the second sound propagates, while pressure fluctuations, observed in a suitable way, can measure turbulent energy spectra over wave numbers less than the reciprocal of the spatial resolution of the transducer. These techniques are quite primitive, in comparison with those available in classical fluid mechanics, where direct visual observation and the use of various types of velocimeter allow the detailed velocity field to be observed and measured.

In view of the absence of detailed information about velocity fields in superfluid turbulence, it is remarkable how much progress has been made on the theory. However, much of the well-developed theory is confined to situations where the quantum turbulence is assumed to be homogeneous, so that the vortex-line density, for example, as measured with second sound, is also homogeneous. Even then, however, we remain ignorant of important aspects of the turbulence; for example, in the case of thermal counterflow, about the possibility of accompanying turbulence in the normal fluid (Schwarz assumed no normal-fluid turbulence). If we are to aim at an understanding of quantum turbulence comparable with that of classical turbulence, we need seriously to address this lack of detailed experimental data. It should also be emphasized that our existing well-established experimental techniques are not well-suited to the study of the initial development of turbulence from localized nucleation centres.

Further progress would be greatly facilitated by, and is probably dependent on, the development of new experimental techniques: especially to observe detailed velocity fields, both at high temperatures, where there is a significant fraction of normal fluid, and at low temperatures, where the normal-fluid fraction is negligible. In the low temperature regime, it would also be helpful to be able to generate and study homogeneous turbulence, produced, for example, by a steadily-moving grid. The development of the necessary mechanism for moving the grid, without significant frictional heating, is likely to be complete quite soon, and we need not dwell on it further.

The development of techniques that will allow us to map out velocity fields remains a major challenge. The importance of a direct visualization of these fields cannot be overemphasized. Much current research in classical turbulence is concerned with the existence coherent structures [9], which can hardly be seen except by direct visualization.

A potentially attractive technique is Particle Image Velocimetry (PIV), in which, in essence, the motion of seed particles within the liquid is followed photographically. There are two problems facing PIV: the practical choice of seed particle, which ought, ideally, to be neutrally-buoyant in a liquid with the low density of helium; and the theoretical problem of the velocity with which a neutrally buoyant particle moves in a superfluid, where there are velocity fields associated with the normal fluid, the superfluid component, and the vortex lines, and where in general these fields are not identical. Important progress on the choice of seed particle has been reported recently, and it seems that there need be no serious problem in choosing seed particles with sizes of order 1 μm or a little larger. And the beginnings of the necessary theory have also been reported [10], although they have served to emphasize that the calculation of the velocity with which a seed particle can be expected to move is not straightforward. Experimental confirmation that this is indeed the case has been provided by the recent PIV studies of thermal counterflow by Van Sciver and co-workers [11], who found, contrary to the simplest theory, that the seed particles seem to move at a velocity that is not equal to that of the normal fluid. The explanation may lie in the partial trapping of the seed particles by the vortex lines, but the details may be quite complicated.

Trapping of seed particles by vortex lines at a finite temperature in ^{4}He can be avoided by using sufficiently small particles. But very small particles are difficult to see. One possibility is to try to use ions, detected through their charge; however, imaging of individual ions is hardly possible in a turbulent superfluid, and even the tracking of groups of ions is difficult. Another possibility is to use the long-lived (~13 s) "neutral excitations" produced by high-energy electron bombardment and consisting of triplet He_2 molecules in small bubble states. McKinsey *et al* [12] have made the very interesting suggestion that these molecules can be detected by the visible fluorescence emitted after they are excited by two infra-red photons. With modest infra-red exciting lasers the fluorescence may be so strong that individual molecules could be imaged, although the tracking of relatively small numbers of molecules could be achieved more easily. In ^{4}He above 1K these molecules ought not to be trapped by vortex lines, so they are likely to track the normal fluid. At low temperatures, trapping is more likely, so that the molecules might then be used to track the motion of vortex lines. Successful development of the use of triplet molecules, or something similar, could make a major contribution to the study of quantum turbulence at all temperatures. A useful and important first step would be a measurement of cross-sections for trapping by vortices at low temperatures.

There seems to be no realistic way to use PIV to measure velocities of the superfluid component directly. A miniature Rayleigh disc might be used, but it is not clear how its operation would be affected by the presence of vortex lines.

Study of classical turbulence has often been facilitated by the use of local probes, such as hot-wire anemometers. This type of anemometer is inapplicable to a superfluid, but we have already noted the application to superfluid turbulence of another type of local probe, which measures pressure fluctuations. Existing measurements [5] used a pressure sensor with poor spatial resolution, but the development of modern nanofabrication techniques might facilitate the development of a pressure sensor with adequate sensitivity, adequate response time, and much better spatial resolution, perhaps approaching the smallest length scales found in quantum turbulence (typically 10-100 μm). Such a pressure sensor would be invaluable in the study of quantum turbulence at very low temperatures. It is significant that as yet we have no experimental evidence for intermittency in quantum turbulence, except from the isolated measurements of pressure fluctuations by Maurer and Tabeling; this is in spite of the importance of intermittency in current research in classical turbulence. Another local probe might be provided by a small second-sound resonator, with which it might be possible to measure local densities of vortex line along with the fluctuations in these densities; development of such a probe was described by Roche at a recent Workshop on New Experimental Techniques for the Study of Quantum Turbulence, held at the International Centre for Theoretical Physics, in Trieste.

STUDY OF ^{3}He-B

Special and novel techniques have been important in recent studies of turbulence in ^{3}He-B. NMR spectra allow the observation of changes of texture due to significant counterflow velocities at higher temperatures [13], and the Andreev scattering of thermal quasi-particles [8] has led to the possibility of "imaging" regions of turbulence in ^{3}He-B at very low temperatures. The former technique has led a recognition that a large mutual friction can inhibit the turbulent spin-up of the superfluid component in ^{3}He-B, while the latter technique has led to information about the form of turbulence generated by a vibrating wire or vibrating grid. However, these techniques, important though they are, along with others such as simple damping measurements on vibrating wires and grids in both ^{3}He and ^{4}He, do not provide unambiguous data relating directly to important characteristics of the quantum turbulence, such as turbulent velocity fields or energy spectra; as a result their interpretation is not straightforward and often depends on extensive computer modelling, with problems relating to inevitable oversimplification and to the uniqueness of the interpretation. Furthermore, these techniques tend to be applicable to only a limited range of turbulent flows, and often not to the simplest forms of turbulence. Nevertheless all these existing techniques will continue to be important, but we believe that they ought now to be supplemented by techniques that allow us the look at the details of a turbulent flow, such as those based on local probes and tracer particles.

SIMULATIONS

Simulations will continue to play an important role in the study of quantum turbulence, as they do in classical turbulence. With the use of increasingly powerful computers it will be possible to study coupled turbulence in both fluids, and it will be possible also to make greater use of a full (non-local) Biot-Savart treatment of vortex dynamics, instead of the local induction approximation (used by Schwarz [4]), which cannot describe long-range vortex interactions and which may often be inadequate even in thermal counterflow with high mutual friction [14]. Simulations based on the Non-Linear Schrodinger Equation (Gross-Pitaevskii equation), rather than the vortex filament model, used first by Schwarz, allow proper treatment of all-important reconnections [4], and are likely to become increasingly important (see, for example, [15]).

CONCLUSIONS

Existing experimental techniques, including those recently developed for special studies of turbulence in ^{3}He-B, will continue to play an valuable role. However, we believe that there is now an urgent need to develop techniques such as PIV and those based on local pressure sensors, which are applicable to a wide range of turbulent flows in both ^{4}He and ^{3}He, and which yield the type of detailed information that is available to those who study classical turbulence. Such a development will enhance greatly the prospects of a stimulating cross-disciplinary interaction with our colleagues in classical fluid mechanics.

ACKNOWLEDGEMENTS

I am grateful to friends and colleagues for many helpful discussions, and especially to the participants of a recent Workshop on New Experimental Techniques for the Study of Quantum Turbulence, held at the International Centre for Theoretical Physics, in Trieste.

REFERENCES

1. R. P. Feynman, in *Progress in Low Temperature Physics*, **1**, Amsterdam: North-Holland, 1955, ch.2.
2. W. F. Vinen, *Proc. Roy. Soc., Lond.* A**240**, 114 (1957), A**240**, 128 (1957), A**242**, 489 (1957).
3. H. E. Hall and W. F. Vinen, *Proc. Roy. Soc., Lond.* A**238**, 204, 215 (1956).
4. K. W Schwarz, *Phys. Rev.* B **38**, 2398 (1988).
5. J. Maurer and P. Tabeling, *Europhys. Letters* **43**, 29 (1998).
6. S. R. Stalp, L. Skrbek and R. J. Donnelly, *Phys. Rev. Letters* **82**, 4831 (1999).
7. W. F. Vinen and J. J. Niemela, *J. Low Temp. Phys.* **128**, 167 (2002).
8. D. I. Bradley *et al*, *Phys. Rev. Letters* **95**, 035302 (2005).
9. U. Frisch, *Turbulence*, Cambridge University Press, 1995.
10. D. R. Poole, C. F. Barenghi, Y. A. Sergeev and W. F. Vinen, *Phys. Rev.* B **71**, 064514 (2005).
11. T.Zhang and S. W. Van Sciver, *J. Low Temp. Phys.* **138**, 865 (2005)
12. D. N. McKinsey, W. H. Lippincott, J. A. Nikkel and W. G. Rellergert, *Phys. Rev. Letters*, in press
13. A. P. Finne *et al*, *Nature* **424**, 1022 (2003).
14. A. Mitani, M. Tsubota, and W. F. Vinen, to be published.
15. M. Kobayashi and M. Tsubota, *Phys. Rev. Letters* **94**, 065302 (2005).

The Generation Of Quantum Turbulence In ^{3}He-B By A Vibrating Grid At Low Temperatures.

D. I. Bradley, S. N. Fisher, A. M. Guénault, R. P. Haley, C. J. Matthews, G. R. Pickett, V. Tsepelin and K. Zaki

Department of Physics, Lancaster University, Lancaster, LA1 4YB, United Kingdom

Abstract. We have measured the onset of quantum turbulence generated by a vibrating grid resonator in ^{3}He-B. Our measurements were carried out in the low temperature regime where the normal fluid component is very dilute and can be described as a gas of ballistic quasiparticles. Consequently, the normal fluid component can not participate in turbulence generation. We have measured the onset of turbulence from the grid motion using two nearby vibrating wire resonators. The vibrating wires show a reduction in thermal quasiparticle damping due to Andreev reflection in the surrounding turbulent velocity field. Our measurements reveal a transition in the transient behavior of the onset of the vorticity signal at the vibrating wire resonators as a function of the grid velocity.

Keywords: superfluid; Helium-3; quantum turbulence; ultralow temperature.
PACS: 67.40.Vs, 67.57.Fg

INTRODUCTION

Quantum turbulence has been studied in superfluid ^{4}He at relatively high temperatures for many years and is found to behave surprisingly similar to classical turbulence[1, 2]. In those measurements turbulence is thought to occur in both the superfluid and the normal fluid components due to the coupling of the superfluid condensate to the normal fluid. The coupling between the components arises from interactions between the thermal excitations and vortex cores.

In contrast to ^{4}He, the normal fluid component in superfluid ^{3}He is so viscous that normal-component turbulence never develops under typical experimental conditions[3]. Furthermore, the measurements described below were made far below the transition temperature, where thermal quasiparticle excitations are so rare that their mean free paths are limited only by the container walls and vortices can move with little or no dissipation. Therefore, we have essentially pure superfluid with no, or very little, dissipation. These are the most ideal conditions for studying quantum turbulence.

The first measurements made with a vibrating grid in ^{3}He-B revealed that at low velocities, the grid response is linear, i.e. its velocity is proportional to the driving force, while at higher velocities, it becomes non-linear[4]. Furthermore, the force-velocity curve of a grid shows no saturation. This is in contrast to a vibrating wire resonator, where the response becomes very flat once the wire is driven above its pair-breaking critical velocity[5]. These results suggested that the grid generates turbulence at high velocities with negligible pair-breaking[4]. Therefore, a grid is an ideal generator for producing vortex lines alone. Below we describe measurements of transient behavior during the production of vortex lines by the grid.

MEASUREMENTS AND RESULTS

Our measurements were performed in the inner cell of a Lancaster-style nested double chamber cell attached to a demagnetization refrigerator[6]. The vibrating grid resonator is formed by a 5.1x2.8 mm mesh of 10 μm copper wires spaced by 50 μm, attached to a 5 mm square goal post made of 125 μm diameter Ta-wire. A layer of cigarette paper is used to insulate the mesh from the goal post. Two semicircular vibrating wire resonators are placed 1 and 2 mm in front of the grid. Both vibrating wires have 2.5 mm leg spacing and are

CP850, *Low Temperature Physics: 24th International Conference on Low Temperature Physics;*
edited by Y. Takano, S. P. Hershfield, S. O. Hill, P. J. Hirschfeld, and A. M. Goldman

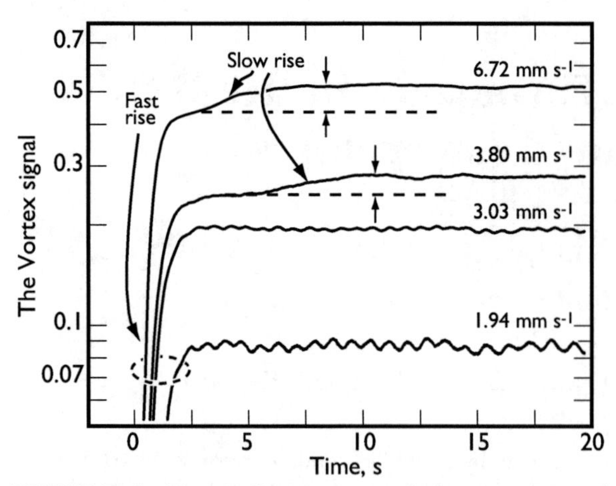

FIGURE 1. Onset of quantum turbulence generated by a vibrating grid at several grid velocities. Curves show behavior of the wire nearest to the grid and were measured in the temperature range of 165-190 μK and 0 bar. The definition of "vortex signal" is described in text.

made from 4.5 μm diameter NbTi wire. A similar wire, located 4 mm from the grid, acts as a thermometer and is enclosed in a cage made of the same mesh as used for the vibrating grid to prevent vorticity from reaching the thermometer wire.

In the low temperature ballistic regime the damping on a vibrating wire resonator is proportional to the thermal quasiparticle excitation density. To exchange momentum with the wire excitations must come from infinity, be normally reflected by the wire and return to infinity. Superfluid flow (from vorticity) can Andreev reflect incoming excitations thereby reducing the damping on the wire. This effect has been utilised to detect quantum turbulence with vibrating wire resonators and has also been measured directly[7].

The moving grid generates heating which causes increased damping on the thermometer wire. The near-by vibrating wires show a reduction in damping from local vorticity superimposed on the increased damping from the overall warming of the cell. This effect is removed from the responses by taking the ratio of the damping of the measurement wire to the damping of the thermometer wire. The fractional reduction of this damping ratio is a direct measure of the fraction of thermal quasiparticles Andreev-reflected by the surrounding vortex lines. This gives a measure of the local vortex line density around the wire, which we refer to as the "vortex signal".

Figure 1 illustrates the time development of vorticity around the vibrating wire closest to the grid, with the grid motion starting at zero. At low grid velocities, below 3 mm/s, the vortex signal rises rapidly to a constant value. The time constant of this rise is dominated by that of the vibrating grid resonator itself and the measurement equipment. This implies that at low grid velocities the generated vortex lines travel very quickly. A similar fast response is seen on decay, when the grid motion is halted[8]. We believe this represents a scenario where vorticity is produced as a gas of ballistic vortex rings[8].

However, once the grid velocity is increased above approximately 3 mm s^{-1} the initial rise of the vortex signal is followed by a further increase with a much slower time constant highlighted in the figure by double arrows. These two distinct regimes have also been observed in the decay of the turbulence[8]. We believe that the slower behavior corresponds to the development of quantum turbulence from the ballistic vortex rings[8]. These measurements clearly show that it takes several seconds for a vortex tangle to develop fully from vortex rings. The time scale of the slow rise is insensitive to temperature at these low temperatures.

ACKNOWLEDGMENTS

We acknowledge I.E.Miller and M.G.Ward for the excellent technical support. This research is supported by EPSRC grant GR/R54453/01

REFERENCES

1. W.F. Vinen, *Phys. Rev. B*, **61**, 1410–1420 (2000).
2. S.R. Stalp, L. Skrbek and R.J. Donnelly, *Phys. Rev. Lett.*, **82,** 4831–4834 (1999).
3. A.P. Finne, S. Boldarev, V.B. Eltsov, and M. Krusius, *J. Low Temp. Phys.*, **136**, 249-279 (2004).
4. D.I. Bradley, D.O. Clubb, S.N. Fisher, A.M. Guénault, C.J. Matthews, and G.R. Pickett, *J. Low Temp. Phys.*, **134**, 381–386 (2004).
5. D. I. Bradley, S. N. Fisher, A. M. Guénault, M. R. Lowe, G. R. Pickett, A. Rahm, and R. C. V. Whitehead, *Phys. Rev. Lett.* **93**, 235302 (2004).
6. G.R. Pickett and S.N. Fisher, *Physica B*, **329-333**, 75–79 (2003).
7. S.N. Fisher, A.J. Hale, A.M. Guénault and G.R. Pickett, *Phys. Rev. Lett.* **86**, 244-247 (2001).
8. D. I. Bradley, D. O. Clubb, S.N. Fisher, A.M. Guénault, R.P. Haley, C. J. Matthews, G.R. Pickett, V. Tsepelin, and K. Zaki, *Phys. Rev. Lett.* **95**, 035302 (2005).

The Decay of Quantum Turbulence Generated by a Vibrating Grid at Low Temperatures in Superfluid ^{3}He-B

D.I. Bradley, S.N. Fisher, A.M. Guénault, R.P. Haley, C.J. Matthews, G.R. Pickett, V. Tsepelin and K. Zaki

Department of Physics, Lancaster University, Lancaster, LA1 4YB, United Kingdom

Abstract. We describe measurements of the decay of quantum turbulence produced by a vibrating grid in superfluid ^{3}He-B at low temperatures. Two nearby NbTi vibrating wire resonators at different distances from the grid are used to detect the turbulence via a reduction in the quasiparticle damping owing to Andreev reflection by the flow around the vortex lines. The decay time constant of this turbulence, after stopping the grid motion, is longer at the wire furthest from the grid. This suggests that either the tangle evolves and moves away from the grid once the grid motion has been stopped, or that there is a faster decay mechanism in operation close to the grid.

Keywords: superfluidity; Helium-3B; quantum turbulence; vibrating grid
PACS: 67.57.Fg

INTRODUCTION

Recently there have been experiments investigating quantum turbulence in superfluid ^{3}He-B at low temperatures [1, 2]. The vibrating grid we use in the experiment detailed below is an ideal apparatus for generating such turbulence [3].

EXPERIMENT

The experimental set up is shown in figure 1 and is identical to that used in the measurements described in [2]. The grid assembly is constructed using 125 μm diameter tantalum wire bent into a 5 mm square goal post shape. Attached to this is a very fine copper mesh constructed of $\sim$ 10 μm rectangular wires 50 μm apart forming 40 μm square holes. The mesh is insulated from the tantalum wire with a thin layer of cigarette paper. In front of the grid are situated two vibrating wire resonators, each 2.5 mm in diameter, made from 4.5 μm diameter NbTi wire. The near wire is placed 1 mm from the grid and the far wire, 2 mm from the grid as shown. There is also a third wire resonator, not shown in figure 1, which is used for thermometry. This is situated $\sim$5 mm from the grid and enclosed in a cage made of the same mesh material used for the grid. The mesh cage is used to prevent vortex lines reaching the vibrating wire inside, to allow accurate thermometry of the cell.

The experiment forms part of the inner cell of a Lancaster style nuclear cooling stage [4] that is filled with liquid helium-3. The cell is cooled by adiabatic nuclear demagnetisation of copper to a base temperature of below 100 μK. All measurements discussed in this article were performed at zero pressure.

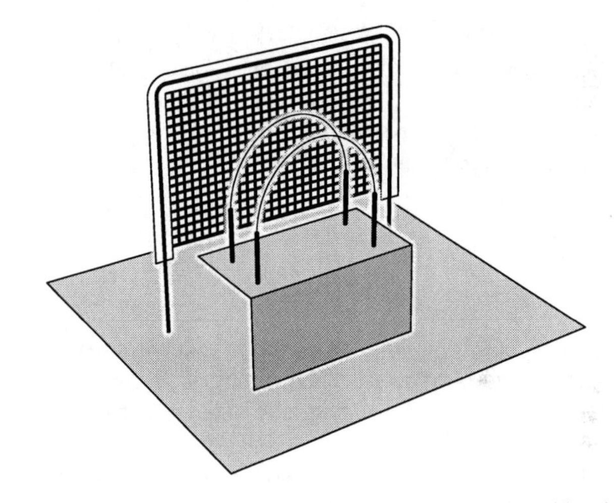

FIGURE 1. Experimental arrangement, showing the grid and the two nearby vibrating wires.

The grid is operated in the same way as a NbTi vibrating wire resonator; placed in a vertical magnetic field an AC current is passed through the tantalum loop setting up a horizontal Lorentz force, causing the grid to oscillate. The motion induces a Faraday voltage proportional to the grid velocity. This voltage is then measured with a lock-in amplifier. If the grid is driven above a velocity of 1 mm/s turbulence is produced [3].

CP850, *Low Temperature Physics: 24th International Conference on Low Temperature Physics;*
edited by Y. Takano, S. P. Hershfield, S. O. Hill, P. J. Hirschfeld, and A. M. Goldman

MEASUREMENTS

The turbulence produced by the grid is measured by the responses of the two nearby vibrating wires. The damping on a vibrating wire resonator gives a direct measure of the flux of thermal quasiparticle excitations reaching the wire. Vibrating wire resonators are very sensitive quasiparticle detectors since the damping is enhanced by several orders of magnitude by Andreev reflection processes. [5].

When vortices are present, the flow field around the vortex cores Andreev reflect a fraction of the incoming thermal quasiparticles, effectively shielding the vibrating wire. As a result a reduction in the damping on the vibrating wire is observed which gives a measure of the surrounding turbulence [1] [6].

The response of the three NbTi wire resonators is monitored as the grid is driven on resonance at a given velocity. When the drive to the grid is switched off, the transient responses of the wires give a measure of how the turbulence produced by the grid evolves and decays.

RESULTS

The thermal damping on each of the two wires near the grid is divided by the thermal damping on the thermometer wire to yield the damping ratio of the responses. This removes the effect of any temperature variation due to heating from the power input required to drive the grid.

Figure 2 shows the fractional reduction in the damping ratio, for the two vibrating wires close to the grid, plotted as a function of the time after the motion of the grid is stopped.

It is clear that the response of the near wire is qualitatively similar to that of the far wire. However, there are two important differences. The nearest wire indicates a higher initial vortex density closer to the grid, as one would expect. More interestingly, the decay of the vorticity shows a longer delay for the far wire and at late times there is a higher vortex density further away from the grid.

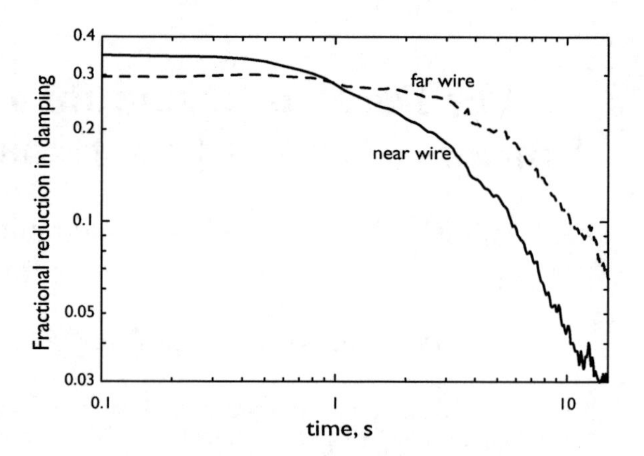

FIGURE 2. Fractional reduction in damping plotted as a function of the time after the motion of the grid is stopped.

CONCLUSION

There could be two explanations for this behavior. Either the tangle of turbulence moves away from the grid once its motion has been stopped, or (more likely) there is a faster decay mechanism in operation close to the grid. An enhanced vortex decay rate in proximity to a wall was also found in recent computer simulations [7] although the actual decay mechanism is unknown.

ACKNOWLEDGMENTS

We acknowledge the excellent technical support of I.E. Miller and M. Ward and useful discussions with W.F. Vinen, L.Skrbek and M. Tsubota. This work is supported by the UK EPSRC under grant GR/R54453

REFERENCES

1. D.I. Bradley, S.N. Fisher, A.M. Guénault, M.R. Lowe, G.R. Pickett, A.Rahm and R.C.V. Whitehead, *Phys. Rev. Lett.* **93**, 235302 (2004).
2. D.I. Bradley, D.O. Clubb, S.N. Fisher, A.M. Guénault, R.P. Haley, C.J. Matthews, G.R. Pickett, V. Tsepelin and K. Zaki, *Phys. Rev. Lett.* **95**, 035302 (2005).
3. D.I. Bradley, D.O. Clubb, S.N. Fisher, A.M. Guénault, R.P. Haley, C.J. Matthews and G.R. Pickett, *J. Low Temp. Phys.* **134**, 381 (2004).
4. S.N. Fisher and G.R. Pickett, *Physica B* **329-333**, 75-79 (2003).
5. S.N. Fisher, A.M. Guénault, C.J. Kennedy and G.R. Pickett, *Phys. Rev. Lett.* **63**, 2566 (1989).
6. S.N. Fisher, A.J. Hale, A.M. Guénault, and G.R. Pickett, *Phys. Rev. Lett.* **86**, 244 (2001).
7. M. Tsubota, T. Araki and W.F. Vinen, *LT23 Conf. Proc. Physica B* **93**, 235302 (2004).

Onset of Turbulence in Superfluid ^{3}He-B and its Dependence on Vortex Injection in Applied Flow

A.P. Finne*, R. Blaauwgeers†, S. Boldarev**, V.B. Eltsov*,**, J. Kopu* and M. Krusius*

*Low Temperature Laboratory, Helsinki University of Technology, P. O. Box 2200, FIN-02015 HUT, Finland
†Kamerlingh Onnes Laboratory, Leiden University, P. O. Box 9504, 2300 RA Leiden, The Netherlands
**Kapitza Institute for Physical Problems, Kosygina 2, 119334 Moscow, Russia

Abstract. Vortex dynamics in ^{3}He-B is divided by the temperature dependent damping into a high-temperature regime, where the number of vortices is conserved, and a low-temperature regime, where rapid vortex multiplication takes place in a turbulent burst. We investigate experimentally the hydrodynamic transition between these two regimes by injecting seed vortex loops into vortex-free rotating flow. The onset temperature of turbulence is dominated by the roughly exponential temperature dependence of vortex friction, but its exact value is found to depend on the injection method.

Keywords: ^{3}He-B, vortex, vortex dynamics, turbulence, onset of turbulence, NMR
PACS: 67.57.De, 47.37.+q

Introduction. Superfluid ^{3}He-B is the best medium to study the influence of friction in vortex dynamics. The friction arises when the superfluid vortex moves with respect to the flow of the normal component. It consists of the longitudinal dissipative and the transverse reactive contributions, characterized by the mutual friction parameters α and α' in the equation for the vortex line velocity $\mathbf{v}_L = \mathbf{v}_s + \alpha\hat{\mathbf{s}} \times (\mathbf{v}_n - \mathbf{v}_s) - \alpha'\hat{\mathbf{s}} \times [\hat{\mathbf{s}} \times (\mathbf{v}_n - \mathbf{v}_s)]$. Here $\hat{\mathbf{s}}$ is a unit vector parallel to the vortex line element. The velocities of the normal and superfluid fractions are $\mathbf{v}_n$ and $\mathbf{v}_s$, while the difference $\mathbf{v} = \mathbf{v}_n - \mathbf{v}_s$ is called the counterflow velocity, the hydrodynamic drive. An important feature of ^{3}He-B hydrodynamics is the large viscosity of the normal component. Transient processes in the normal flow decay quickly and can be neglected. Owing to this simplification, with some modification the results for rotating flow can be carried over to other types of flow. With careful design and preparation of the sample container the energy barriers preventing vortex formation can be maintained high, so that high-velocity vortex-free flow can be achieved.

The dynamics of quantized vortex lines can be explored if one injects vortex loops in this meta-stable high-energy state of vortex-free flow. The fate of the seed loops depends on temperature: At high temperatures the number of vortices (N) is conserved, the injected loops expand to rectilinear lines, and the flow relaxes only partially. At low temperatures below some onset temperature, rapid vortex proliferation from the seed loops takes place in a transient turbulent burst which leads to the formation of the equilibrium number of vortices ($N_{\rm eq} \sim 10^3$) and to a complete removal of the applied flow [1]. At sufficiently low temperatures ($T \lesssim 0.45\,T_c$) the turbulent burst always follows, even after the injection of a single vortex ring, independently of the injection method. However, in a narrow temperature regime around the onset of turbulence the situation is different: the injection may or may not lead to turbulence depending on the injection details and the velocity of the applied flow.

In this report we examine the dependence of the onset temperature on the injection properties. The motivation is the following: A number of different processes with their individual energy barriers have to be traversed sequentially before turbulence in the bulk volume becomes possible. The first is vortex nucleation. It is here avoided by the injection of the seed loops. Next follows a series of events which build up the vortex density locally somewhere in the bulk volume for turbulence to start. These processes act at the container wall. They have been explored in Ref. [2]. Compared to turbulence in viscous fluids, the path leading to turbulence in superfluids appears to be more straightforward to reconstruct.

Injection methods. In our experiment [3] the flow is created by rotating a cylindrical sample of 110 mm length and $R = 3$ mm radius around its symmetry axis. We employ four different techniques to inject vortex loops in the rotating flow, in order to study the transient evolution from the vortex-free state to a final stable state with a central vortex cluster consisting of rectilinear vortex lines. The final state is only meta-stable, unless the cluster contains the equilibrium number $N_{\rm eq}$ of vortex lines. With NMR techniques we measure the number of lines N close to both ends of the sample. At temperatures above the onset of turbulence one observes regular expansion of the injected loops to rectilinear lines, i.e. in the final state $N \ll N_{\rm eq}$. At temperatures below onset the

CP850, *Low Temperature Physics: 24th International Conference on Low Temperature Physics;*
edited by Y. Takano, S. P. Hershfield, S. O. Hill, P. J. Hirschfeld, and A. M. Goldman

injection evolves into a turbulent burst, which results in a large increase in N, so that in the final state $N \approx N_{eq}$.

The different injection techniques are compared in Fig. 1 and Table 1. The number of injected vortex loops, their size, proximity, and the initial vortex density at the injection site vary from one injection technique to the next. We would like to answer the question whether these properties, besides temperature and drive velocity $v = \Omega R$, influence the onset to turbulence.

The first two injection methods are the most reproducible. They are based on the properties of the first order interface between the A and B phases of superfluid ^{3}He. The A phase is stabilized with a magnetic barrier field over a short section in the middle of the long sample. The AB interface undergoes an instability of the Kelvin-Helmholtz type when flow is applied parallel to it. As a result of the instability vortex loops are tossed across the interface from the equilibrium vortex state in A phase to the vortex-free B phase. The KH instability can be triggered (1) by sweeping the rotation velocity Ω up to a well-defined critical threshold $\Omega_c(T,P)$ or (2) by sweeping at constant $\Omega > \Omega_c$ the magnetic field H from a low value up to where the A phase is suddenly nucleated with some magnetic hysteresis at $H > H_{AB}(T,P)$.

The third injection method makes use of rapid localized heating in a neutron absorption event. From the overheated volume, which is $\sim 50\,\mu$m in diameter, one or more vortex rings may evolve above a critical threshold $\Omega_{cn}(T,P)$. The number of rings depends on the applied flow $v = \Omega R$ [4]: Just above Ω_{cn} only one vortex loop is created per absorption event, but with increasing Ω the average number of loops per injection event increases, reaching $\langle N \rangle \approx 5$ at $\Omega/\Omega_{cn} = 4$.

In the fourth method we start from an existing remnant vortex and create the flow later. In a strict sense this is not injection as in the three earlier cases, but in practice it achieves the same result of placing a curved vortex loop in applied flow. With decreasing temperature the last one or two vortices require an ever longer time to annihilate at $\Omega = 0$ because of the rapidly reducing vortex damping. Thus it becomes possible to catch a remnant vortex before its annihilation in a random location at the cylindrical container wall. When Ω is suddenly increased to some final stable value, we may monitor the stability of the remnant vortex in the applied flow [2].

The variations in the injection properties prove to be larger between the different methods than the variability within one method from one run to the next. We therefore list the perturbations, which the different injection techniques generate, in an order from the strongest to the weakest: **(1)** First comes the nucleation of ^{3}He-A during a magnetic field sweep, where the injected loop number is typically in the few hundreds. Next comes **(2)** the KH instability of the AB interface, **(3)** followed by neutron absorption at higher flow velocities. **(4)** Neutron absorption at lower velocities $\Omega \gtrsim \Omega_{cn} \sim 1$ rad/s generates only one ring. The weakest perturbation is **(5)** a remnant vortex which can be studied down to low flow velocities. The absolute limit is the velocity at which a loop with a radius of curvature $\lesssim R$ is able to expand: $v \gtrsim \kappa/(4\pi R)\ln(R/\xi)$ which corresponds to $\Omega \sim 10^{-2}$ rad/s. (Here κ is the circulation quantum and ξ the coherence length.) In practice the limit is set by our NMR detection which at present requires $\Omega \gtrsim 0.4$ rad/s [3].

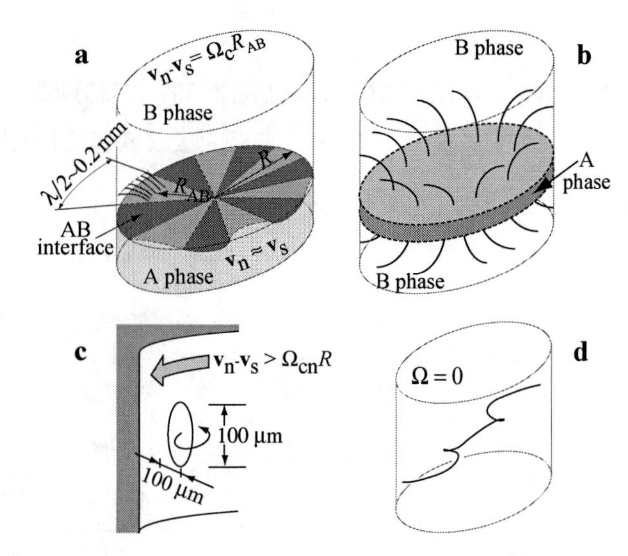

FIGURE 1. Vortex loop configurations at injection in a rotating cylinder: **(a)** Two-phase sample with roughly the equilibrium number of vortices in A phase and vortex-free B phase. The A phase vortices curve at the AB boundary on the interface, forming there a surface layer of vorticity. In the KH instability the A-phase vortices in the deepest trough of the interface wave are tossed on the B-phase side in the region between $R_{AB} \approx 2.6$ mm and the cylinder wall at $R = 3$ mm. **(b)** When the barrier field is slowly swept up (at constant Ω, T, P) the nucleating A-phase layer is unstable and a massive number of vortex loops escapes to the B phase. **(c)** In a neutron absorption event close to the cylinder wall vortex rings are extracted by the applied flow from the volume, heated by the reaction. **(d)** A remnant vortex, which has not yet had sufficient time to annihilate at $\Omega = 0$, forms a seed for new vortex formation when the applied flow velocity $v = \Omega R$ is increased from zero.

Onset of turbulence. Let us examine with our different injection methods the probability of turbulence within the transition regime. We first consider the situation at a higher temperature of $0.53\,T_c$: **(1)** Here the nucleation of ^{3}He-A with magnetic field always gives the equilibrium vortex state. **(2)** Also KH injection has a high probability $p_{AB} = 0.96$ to start turbulence ($P = 29$ bar, $\Omega = 0.8$–1.6 rad/s) [5]. In contrast, **(3)** for neutron absorption at $\Omega = 2.32\,\Omega_{cn}$ the probability is $p_n = 0.09$ (Fig. 3). **(4)** No turbulent bursts have been observed with neutron injection, if $\Omega < 2\Omega_{cn}$. **(5)** Similarly vortex multiplication is not triggered by a remnant vortex at this temperature

TABLE 1. Characterization of vortex injection methods in rotating superfluid ^{3}He-B.

Method	Trigger	Number of loops	Length scale	Location in sample container
Kelvin-Helmholtz instability of AB interface [3]	Slow Ω sweep across $\Omega_c(T,P,H)$ $\sim 0.8-1.6$ rad/s	smooth distribution $N \sim 3-30$ (peak $\sim$ 8, long tail up to 30)	$0.1-1$ mm, a wavelength of ripplon	At AB interface close to cylindrical wall at $R_{AB} \approx 2.6\,\mathrm{mm} < R$
Nucleation of ^{3}He-A in magnetic field from ^{3}He-B	Slow sweep of barrier field up to $H_{AB}(T,P)$	$1 \ll N < N_{eq}$	Circumference of AB interface along cylindrical wall	At newly forming AB interface
Neutron absorption [4, 7]	Neutron absorption event	$N \sim 1-5$ depends on Ω/Ω_{cn}	$\sim 100\,\mu$m (diameter of largest vortex ring)	Random location close to cylindrical wall
Remnant vortex [2]	Rapid increase of Ω from zero	$N \sim 1$	$\sim R$, size of remnant vortex loop at $\Omega = 0$	Random and distributed along sample

(at $P = 29$ or 10 bar). The different injection processes thus yield different onset temperatures for turbulence.

When temperature and mutual friction damping decrease, the magnitude of the flow perturbation, which is needed to start turbulence, decreases as well. At $0.45\,T_c$ ($P = 29$ bar), vortex injection via **(1)** the nucleation of the A phase or **(2)** the KH instability always result in a turbulent burst. **(3)** Similarly neutron absorption at high flow $\Omega > 2\Omega_{cn}$ also has unit probability to initiate turbulence. **(4)** Even at lower flow velocity $\Omega \gtrsim \Omega_{cn}$, where only a single vortex ring can be injected from the neutron bubble, the probability of obtaining turbulence is $p_n = 0.9 - 1$. **(5)** A remnant vortex gives a probability of about 0.9 ($P = 10$ bar) [2]. We concede that at low temperatures even the smallest perturbation, a single quantized vortex loop, will evolve to a turbulent tangle.

As seen from Table 1, the various injection methods differ in a variety of ways. The intensity of the perturbation, which they present to vortex-free flow, arises from a combination of different properties. A most important characteristic is the number of injected loops. This is illustrated as a probability distribution in Fig. 2 for two particular cases of injection, namely via KH instability and neutron absorption. Clearly the above examples from injection experiments at 0.53 and $0.45\,T_c$ support the simple notion that the more vortex loops are initially injected, the larger is the probability of turbulence. To explain these observations, we have to assume that at $0.53\,T_c$ one needs to inject 4 – 5 loops to achieve turbulence, while at $0.45\,T_c$ a single loop suffices. A characterization of the injection methods in Table 1 in terms of the number of injected loops would seem like a gross oversimplification which ignores other differences, like the size of the volume where the initial perturbation is localized. Still, in the light of our results it looks conceivable that the number of injected loops is a reasonable first measure of the intensity of the flow perturbation.

Discussion. With each injection method the transition from regular vortex dynamics at high temperatures to turbulent vortex dynamics at low temperatures occurs in a temperature interval of certain width. We attribute this width to the variability in the injection from one time to the next, since each injection method is characterized by a certain distribution of configurations of injected loops. Only some of these configurations result in turbulence at higher temperatures. With decreasing temperature more configurations become effective and eventually at low enough temperatures all configurations produced by a given injection method result in turbulence. The typical width of the transition for a given injection method is $\sim 0.06\,T_c$ [5, 2]. This width is generally smaller than the difference between the average onset temperatures measured with different injection methods.

Stability considerations of steady-state turbulence lead to the conclusion that on average the transition between regular and turbulent vortex motion is given by a condition on the ratio of the mutual friction coefficients: $q = \alpha/(1-\alpha') \sim 1$ [1, 6]. Our measurements with different injection methods and at different external conditions of T, P, Ω show that the onset temperature is approximately predicted by this criterion, but the exact value of temperature does not correspond to any universal critical value of q. Whether q may still have a universal critical value in the particular case of sustained homogeneous well-developed turbulence is unclear, since no such measurements exist in the relevant range $q \sim 1$.

The dependence of the onset on the intensity of the flow perturbation resembles that observed in recent measurements [8] on the flow of a classical viscous liquid in a circular pipe. Here a perturbation of finite magnitude ε needs to be injected in the flow, to turn it from laminar to turbulent. A scaling law connects the smallest possible perturbation ε and the Reynolds number: $\varepsilon \propto \mathrm{Re}^{-1}$, i.e. the minimum perturbation decreases with increasing flow velocity. In superfluids the analog of the Reynolds number Re is q^{-1} [1, 3]. The magnitude of q^{-1} increases monotonically with decreasing temperature, but does not depend on flow velocity.

The hydrodynamic transition from laminar (regular) to turbulent flow in viscous liquids and in superfluids differs from a usual first order phase transition, such as *eg.* the transition from supercooled meta-stable A phase to

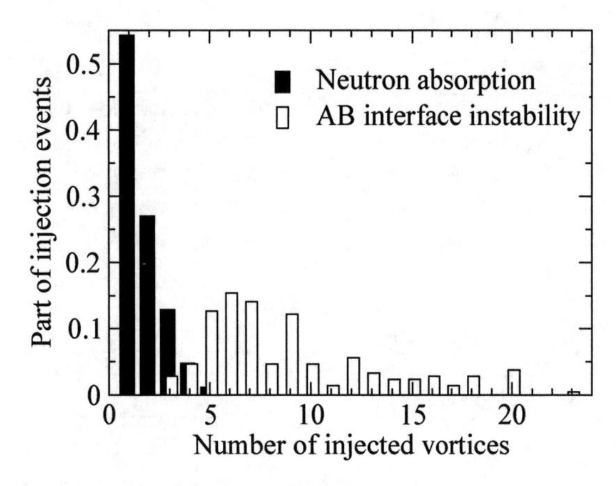

FIGURE 2. Two histograms of the number of events (vertical scale) in which a specified number of vortex loops (horizontal scale) is injected in rotating vortex-free B-phase flow. These histograms have been measured above the onset temperature of turbulence in the regime of regular vortex dynamics. The example of KH injection from the AB interface has been measured at $P = 29$ bar, $T = 0.77T_c$, and $\Omega = 1.34$ rad/s. Our measurements indicate that this distribution is not strongly temperature or velocity dependent. The example for injection via neutron absorption represents an interpolation for $\Omega = 2.32\Omega_{cn}$ from data measured at $T \approx 0.95T_c$ and $P =$ 2–18 bar (from Ref. [4]). The number of loops in neutron injection is strongly velocity-dependent, but only weakly T and P dependent.

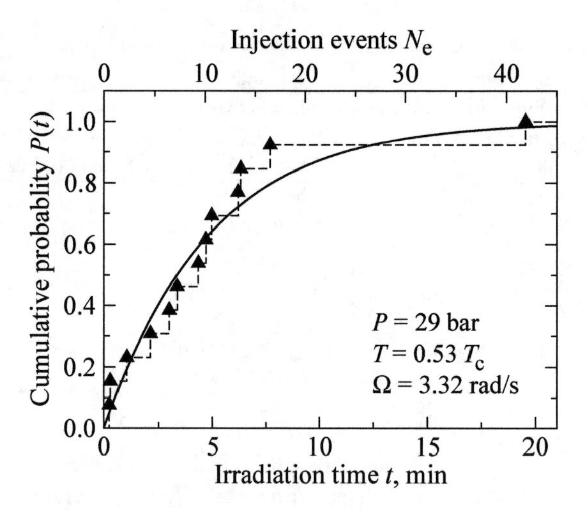

FIGURE 3. Probability of turbulence after vortex loop injection from neutron absorption events at $\Omega = 2.32\Omega_{cn}$. Each triangle represents a continuous neutron irradiation session of an originally vortex-free sample which, after a number of non-turbulent vortex injection events, terminates in a turbulent burst at the marked time. The triangles are distributed equidistantly along the vertical axis between 0 and 1 to form approximation (dashed line) for the probability $P(t)$ to observe turbulence within the time span t after starting irradiation. If each injection event has a probability p_n to generate turbulence independently, then $P(t) = 1 - (1 - p_n)^{\dot{N}_e t}$. Here the average rate of injection events $\dot{N}_e = 2.14\,\mathrm{min}^{-1}$ is determined from the measured vortex formation rate [7] and the average number of vortices formed in a neutron absorption event in Fig. 2. A fit of the measured data to the expected dependence $P(t)$ gives the fitted probability distribution (solid curve) with $p_n = 0.092$.

the stable equilibrium B state of superfluid ^{3}He. A single energy barrier, the creation of a sufficiently large seed bubble of B phase with a critical radius of order $\sim 1\,\mu$m, prevents the A→B phase transition. If such a seed bubble is injected, then the stable B phase is irreversibly created in all of the available volume. If the seed is too small, then it will shrink and disappear. In viscous liquid flow along a circular pipe [8], an injected perturbation creates turbulent "puffs" and "slugs" in the laminar flow, which are limited in space and do not extend over the whole length of flow. In our superfluid experiments an analogous case appears if vortex multiplication in a turbulent burst stops before the vortex number has reached N_{eq} and large applied flow remains. Such intermediate events are rare, but have been observed in the onset regime: In the middle of the transition region their proportion among the well-characterized events (*i.e.* those which conserve the number of injected vortices and those where the meta-stable flow is completely removed) is at most a few percent. The existence of such intermediate behavior is a further demonstration that in hydrodynamic transitions often there is no single well-defined energy barrier which one should overcome to cause a spontaneous change from a meta-stable to a stable flow pattern [2].

Conclusion. The onset temperature of turbulence after the injection of vortex loops in meta-stable vortex-free flow depends on several variables, but the decisive factor appears to be the number of injected loops: When the number of loops increases they are more likely to become unstable towards turbulence. As a result the onset temperature of turbulence increases and the value of the friction parameter q^{-1} at onset decreases. This connection between the amplitude of the flow perturbation and the "Reynolds number" q^{-1} at onset resembles the scaling law of classical turbulence [8].

REFERENCES

1. A.P. Finne *et al.*, *Nature* **424**, 1022 (2003).
2. A.P. Finne *et al.*, preprint *arXiv:cond-mat/0502119*.
3. A.P. Finne *et al.*, *J. Low Temp. Phys.* **136**, 249 (2004).
4. V.B. Eltsov, M. Krusius, and G.E. Volovik, in *Prog. Low Temp. Phys.* Vol. XV (Elsevier, Amsterdam, 2005).
5. A.P. Finne *et al.*, *J. Low Temp. Phys.* **138**, 567 (2004).
6. N.B. Kopnin, *Phys. Rev. Lett.* **92**, 135301 (2004); V.S. L'vov *et al.*, *JETP Lett.* **80**, 479 (2004); W.F. Vinen, *Phys. Rev. B* **71**, 024513 (2005).
7. A.P. Finne *et al.*, *J. Low Temp. Phys.* **135**, 479 (2004).
8. B. Hof *et al.*, *Phys. Rev. Lett.* **91**, 244502 (2003).

NMR Response of a Vortex Tangle in Rotating ^{3}He-B

J. Kopu*, V. B. Eltsov*,†, A. P. Finne*, M. Krusius* and G. E. Volovik*,**

*Low Temperature Laboratory, Helsinki University of Technology, P. O. Box 2200, FIN-02015 HUT, Finland
†Kapitza Institute for Physical Problems, Kosygina 2, 119334 Moscow, Russia
**Landau Institute for Theoretical Physics, Kosygina 2, 119334 Moscow, Russia

Abstract. Following the injection of vortex loops into an originally vortex-free sample of superfluid ^{3}He-B at $T < 0.6T_c$, transient states appear which exhibit a modified response in transverse NMR experiments when compared with the response from the final equilibrium state of rectilinear vortices. Here we investigate the NMR response of a model state where a random vortex tangle is assumed to be superimposed on the equilibrium vortex state. We show that this model can qualitatively account for the experimental findings, provided that the tangle is anisotropic.

Keywords: Superfluid helium 3, B phase, NMR, vortices, turbulence
PACS: 67.57.Fg, 67.40.Vs

Introduction: In the B phase of superfluid ^{3}He the presence of quantized vortices can be detected using transverse NMR. The method is based on the orienting effect of vorticity on the B-phase order parameter. This in turn results in a shift of the resonance frequency. The orientational influence is due to two different effects. The first one is the reduction in the global (averaged over vortex lines) counterflow. Another small local influence arises from the quantized superfluid flow field circulating around a vortex, and the suppression of the order parameter from the bulk form inside its core. Here we focus on these local contributions, which become important if the global counterflow is small.

Recent experiments [1] have monitored the NMR response as a function of time in a situation where some seed vortices were injected in an originally vortex-free sample of ^{3}He-B in a rotating cylindrical container. Before the attainment of the equilibrium vortex state, transient states were observed, with substantially modified responses as compared with the initial and final states. Motivated by this finding, we study the NMR response of model states involving a random vortex tangle.

Textures and NMR: The NMR lineshape in ^{3}He-B is determined by the texture of the unit vector $\hat{\mathbf{n}}$ that parametrizes the B-phase order parameter. Mathematically, the texture problem is formulated in terms of a free-energy functional, the minimum of which is obtained with the desired texture $\hat{\mathbf{n}}(\mathbf{r})$. Each orienting influence, such as that originating from vortex lines, contributes a term to this functional (see review [2]). The local resonance frequency ν depends on the component of $\hat{\mathbf{n}}$ along the magnetic-field direction $\hat{\mathbf{h}}$ (here taken parallel to the rotation axis, $\hat{\mathbf{h}} \parallel \hat{\mathbf{z}}$) as

$$\nu \approx \nu_L + \frac{\nu_B^2}{2\nu_L}(1 - \hat{n}_z^2), \qquad (1)$$

where ν_L and ν_B are the Larmor frequency and the longitudinal resonance frequency of ^{3}He-B, respectively.

The local contribution of the vorticity $\vec{\omega}$ to the free energy can be written as [3]

$$\frac{F_{\rm vort}}{aH^2} = \frac{2}{5}\left(\frac{\lambda}{\Omega}\right)\int d^3r\, \frac{(\vec{\omega}\cdot\hat{\mathbf{l}})^2}{|\vec{\omega}|}, \qquad (2)$$

where the (positive) textural parameters a and λ depend on temperature and pressure, H is the magnetic-field strength, Ω the angular velocity of rotation, and $\hat{\mathbf{l}} \equiv \cos\theta\hat{\mathbf{h}} + (1-\cos\theta)(\hat{\mathbf{n}}\cdot\hat{\mathbf{h}})\hat{\mathbf{n}} + \sin\theta\hat{\mathbf{h}}\times\hat{\mathbf{n}}$, with $\theta = \arccos(-1/4)$, defines the direction of orbital anisotropy. Here vortices are assumed to be locally polarized along the average vorticity $2\vec{\omega} = \langle\nabla\times\mathbf{v}_s\rangle$. The integration in Eq. (2) extends over the region with nonzero vorticity.

The equilibrium vortex cluster has $\vec{\omega} = \vec{\omega}_{\rm eq} = \Omega\hat{\mathbf{z}}$. In this case, the energy in Eq. (2) is minimized when $\hat{\mathbf{l}} \perp \hat{\mathbf{z}}$ or, equivalently, if $\hat{n}_z^2 = 1/5$. The effect of Eq. (2) is then to increase the NMR absorption at the *counterflow frequency* ν_c, obtained by substituting $\hat{n}_z^2 = 1/5$ in Eq. (1). If the vortex number is much less than the equilibrium one, this tendency is overcompensated by the reduction in global counterflow, which results in reduced absorption at ν_c. However, in the absence of appreciable global counterflow the local effect can be observed. Indeed, the parameter λ has been measured by comparing the NMR spectrum of a nonrotating vortex-free sample with that of a rotating sample in the equilibrium vortex state [4].

Experimental observations: In our experiments, a small number of vortices were injected into the (initially vortex-free) B-phase sample. At low enough temperatures, these seed vortices were seen to multiply until the equilibrium number was reached, whereafter the vortex configuration relaxed to the final equilibrium state of rectilinear vortex lines. During this evolution, the NMR

CP850, *Low Temperature Physics: 24th International Conference on Low Temperature Physics;*
edited by Y. Takano, S. P. Hershfield, S. O. Hill, P. J. Hirschfeld, and A. M. Goldman

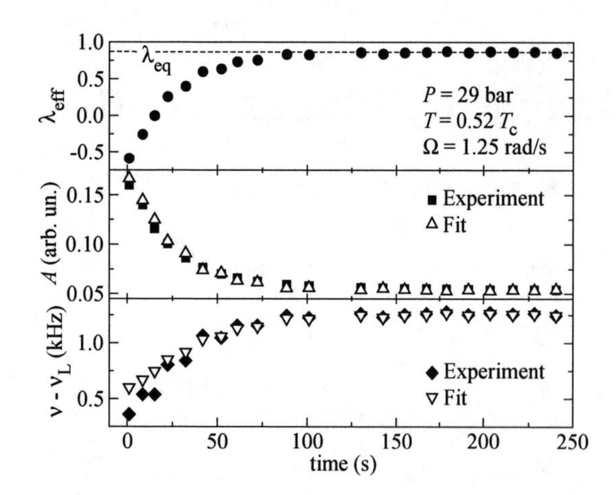

FIGURE 1. Slow relaxation of the overshoot in the NMR absorption in the Larmor region of the NMR lineshape. The maximum overshoot was reached at time $t = 0$. The uppermost graph shows the textural parameter λ_{eff}, treated as a fit parameter and obtained by comparing the measured NMR spectra with those calculated in a similar way as explained in Ref. [4]. The equilibrium-state value λ_{eq} measured in Ref. [4] (denoted with the dashed line) coincides remarkably well with the final-state value in the present measurements. The measured and calculated height (middle graph) of the NMR absorption peak and its frequency shift (lower graph) are also shown.

lineshape was monitored as a function of time.

Our observations can be summarized as follows. In the initial state of high counterflow, the NMR lineshape essentially consisted of a single peak at ν_c. After the injection, the absorption at ν_c abruptly decreased, and the absorption was shifted to the region close to the Larmor frequency ν_L. At the moment when the absorption at ν_c had dropped to the level corresponding to the NMR response of the equilibrium vortex cluster, the absorption in the Larmor region reached a maximum value well above that of the final state. This was followed by a slow relaxation process, during which the absorption near ν_L settled down to the final equilibrium value.

The most prominent signature of the transient vortex states is the substantially increased absorption near ν_L, while at the same time exhibiting negligible absorption at ν_c. Note that this 'Larmor overshoot' means that the effect of Eq. (2) for rectilinear vortex distributions is suppressed in these states. Interestingly, however, we found that a good fit between experimental and theoretical NMR lineshapes could be obtained for the transient states if the parameter λ in Eq. (2) is allowed to change in value, even including taking on *negative* values (see Fig. 1). In the following, we present one possibility how this seemingly unphysical situation can be realized.

Vortex-tangle model: To explain the measured NMR response, we relax the requirement of rectilinear vortices and consider a modified configuration with $\vec{\omega} = \vec{\omega}_{\text{eq}} + \vec{\omega}'$, where $\vec{\omega}'$ denotes a random vortex distribution with $\langle \vec{\omega}' \rangle = 0$ (the averaging is over different vortex configurations). This model state, which can be viewed as a random vortex tangle superimposed on an equilibrium cluster, corresponds to (on average) zero counterflow.

We estimate the orientational energy of such a vortex distribution by replacing the numerator and denominator of the integrand in Eq. (2) with their averaged counterparts. This procedure renormalizes the prefactor λ to

$$\lambda_{\text{eff}} = \lambda \frac{1 + (\omega_{\parallel}/\Omega)^2 - (\omega_{\perp}/\Omega)^2}{\sqrt{1 + (\omega_{\parallel}/\Omega)^2 + 2(\omega_{\perp}/\Omega)^2}}, \tag{3}$$

where the parameters $\langle \omega_x'^2 \rangle = \langle \omega_y'^2 \rangle = \omega_{\perp}^2$ and $\langle \omega_z'^2 \rangle = \omega_{\parallel}^2$ are assumed to be spatially constant. In averaging, we have used the fact that the order parameter varies slowly over length scales of the order of the intervortex distance.

For an isotropic tangle, $\omega_{\perp}^2 = \omega_{\parallel}^2 = \omega'^2/3$, we obtain a suppression of the parameter λ by a factor $\sqrt{1 + (\omega'/\Omega)^2}$. This expresses the fact that a superimposed random tangle reduces the orientational effect of the rectilinear vortex lines, and cannot fully explain the experimental observations. However, in the case of an anisotropic tangle, $\omega_{\perp}^2 \neq \omega_{\parallel}^2$, it is possible for λ_{eff} to change sign. The minimization of Eq. (2) would then give $\hat{\mathbf{l}} = \hat{\mathbf{z}}$ (or $\hat{n}_z^2 = 1$) which, via Eq. (1), implies absorption at the Larmor frequency. Furthermore, since the total vortex density $n \propto \Omega \sqrt{1 + (\omega_{\parallel}/\Omega)^2 + 2(\omega_{\perp}/\Omega)^2}$, Eq. (3) places a lower limit $n > \sqrt{3} n_{\text{eq}}$ for the vortex density in the situations corresponding to negative values of λ_{eff}.

Conclusions: We have demonstrated that, with a model consisting of a random vortex tangle superimposed on the equilibrium vortex state, the main features of the NMR lineshapes corresponding to transient rotating states in recent vortex-injection experiments [1] can be qualitatively explained. A necessary condition for this agreement is that the tangle exhibits uniaxial anisotropy. However, this is not the only possible model which explains the measured transient NMR response. The final judgement has to be based on further comparison with experiment and a full vortex-dynamics calculation. Still, our estimations based on Eq. (3) suggest that NMR measurement can be used to monitor stationary state turbulence, for instance around a vibrating grid [5].

REFERENCES

1. A. P. Finne *et al.*, *J. Low Temp. Phys.* **135**, 479 (2004).
2. E. V. Thuneberg, *J. Low Temp. Phys.* **122**, 657 (2001).
3. A. D. Gongadze *et al.*, *Fiz. Nizk. Temp.* **7**, 821 (1981) [*Sov. J. Low Temp. Phys.* **7**, 397 (1981)].
4. P. J. Hakonen *et al.*, *J. Low Temp. Phys.* **76**, 225 (1989).
5. D. I. Bradley *et al.*, *J. Low Temp. Phys.* **134**, 381 (2004).

Vortex State of ^{3}He-A Studied by NMR Linewidth

M. Kubota[a], Y. Kataoka[a], M. Yamashita[b], K. Izumina[a], and O. Ishikawa[c]

[a]*Institute for Solid State Physics, the University of Tokyo, Kashiwanoha, Kashiwa, 277-8581, Japan*
[b]*Graduate School of Science, Kyoto University, Kitashirakawa, Kyoto, 606-8502, Japan*
[c]*Graduate School of Science, Osaka City University, Sugimotocho, Osaka, 558-8585, Japan*

Abstract. The vortex state of superfluid ^{3}He-A has been studied under rotation by measuring the change of the width of the main NMR line at rotation speeds up to twice as fast as previously reported. We identify three temperature ranges according to the temperature dependence of the linewidth. These temperature ranges have been predicted by Fomin and Kamenskii. We also report the results of our preliminary NMR linewidth measurements for different rotational speed Ω's. These are consistent with previously reported results, where they overlap, and reveal a different Ω dependence for each temperature range.

Keywords: 3He-A, vortex state, NMR linewidth, spin diffusion, spin wave
PACS: 67.57.-z, 67.57.Fg, 67.57.Lm

INTRODUCTION

The vortex state of superfluid ^{3}He-A has been studied under rotation most commonly by a torsional oscillator and NMR. As has been known since the earliest NMR experiments, rotation leads to a broadening of the main NMR line in the A phase, as well as an appearance of a satellite peak due to vortices. The line broadening reveals the presence of a fluctuating effective field in addition to the usual fluctuations present in non-rotating superfluid. Fomin and Kamenskii[1] predicted that there are three temperature ranges in which the NMR linewidth due to rotation has different temperature dependences. In the temperature range near the superfluid transition temperature T_c, the temperature-dependent spin diffusion constant governs the variation of the linewidth, whereas the temperature dependence of the spin wave velocity plays a role in the lowest temperature range. The linewidth hardly changes in the intermediate temperature range.

Experimental work has been reported by the Helsinki group; see for example the review of Eltsov *et al.*[2] and references therein. As is discussed by these authors the most common vortex in ^{3}He-A is the continuously unlocked vortex, CUV. We report our preliminary NMR linewidth results at fast rotation speeds up to 1 rev/sec and refer to the three distinct temperature ranges where we observed new features under rotation.

EXPERIMENT

The experiment was performed using the ISSP ULT rotating cryostat[3]. The sample cell was made of Stycast 1266. It had a 3 mm inner diameter, was 12 mm long and was connected through a 1 mm diameter orifice to the ^{3}He column leading to the nuclear stage heat exchanger. The pressure of the ^{3}He was kept constant at 3.05 MPa. The experiment was performed in two modes. The constant Ω mode, as demonstrated in Fig. 1, consisted of taking data at various fixed temperatures at a steady rotation speed Ω. The other mode is a swept Ω mode, in which temperature was

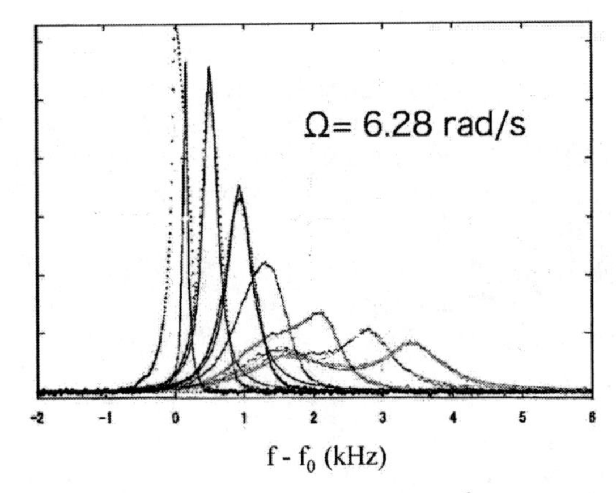

FIGURE 1. NMR spectra of superfluid ^{3}He-A (arbitrary unit) under rotation at P=3.05 MPa for temperatures, from right, 0.86, 0.89, 0.92, 0.95, 0.97, 0.99, 0.99, 1.02T_c.

CP850, *Low Temperature Physics: 24th International Conference on Low Temperature Physics;*
edited by Y. Takano, S. P. Hershfield, S. O. Hill, P. J. Hirschfeld, and A. M. Goldman

kept fixed while Ω was varied continuously. An example of some snapshots from such an Ω sweep are shown in Fig. 2 for $T=0.84T_c$. Quantitative evaluation of the linewidth was made by fitting the signal to Lorentzian curves where applicable.

By combining the data taken in the two modes we can determine the NMR linewidth change relative to the linewidth $\Delta\Gamma(0)$ at zero rotation, as a function of the reduced temperature $\tau = 1-T/T_c$ and angular speed Ω. We can analyze just the change under rotation. This is because the fluctuating field under rotation is random and its effect on the linewidth is just additive to those of Leggett and Takagi's mechanism and a field inhomogeneity, according to the idea given in Ref. 1. We have found that, without rotation, the linewidth becomes stable after some measurement runs. Preliminary data analysis indicates that we have a τ^2 dependence of $\Delta\Gamma(\Omega)$ very near T_c for $\Omega < \sim 3$ radian/sec in accordance with the expectation[1], when spin diffusion is the dominant effect that determines the linewidth. At higher rotational speed it becomes difficult to determine the width of the main peak near T_c because of its overlap with the vortex peak as seen in Fig. 1. In contrast, the data at lower temperatures indicate that the change in the linewidth depends linearly on τ. A linear dependence on τ seems to appear at the lowest temperature for all the data we took. In between, we observe a plateau where $\Delta\Gamma(\Omega)$ becomes independent of temperature for $\Omega < \sim 3$ radian/sec. At higher speeds this region disappears.

Quantitative analysis of the data and discussion of the physics will require a separate publication, which we plan for the near future. One of the complications of quantitative analysis is that there was a small temperature increase in the sample under fast rotation and we needed recalibration of the sample temperature against the NMR frequency shift of the sample itself. A continuous sweep of Ω at constant temperature reveals the Ω dependence of the linewidth. In the plateau region we find results that are consistent with those of Ref. 2. However, we observe that the Ω dependence is different in the three temperature regions.

This presence of different Ω dependences for the three temperature regions seems to differ from the prediction of Fomin and Kamenskii[1]. We are eager to complete the quantitative analysis and identify the origin of this effect.

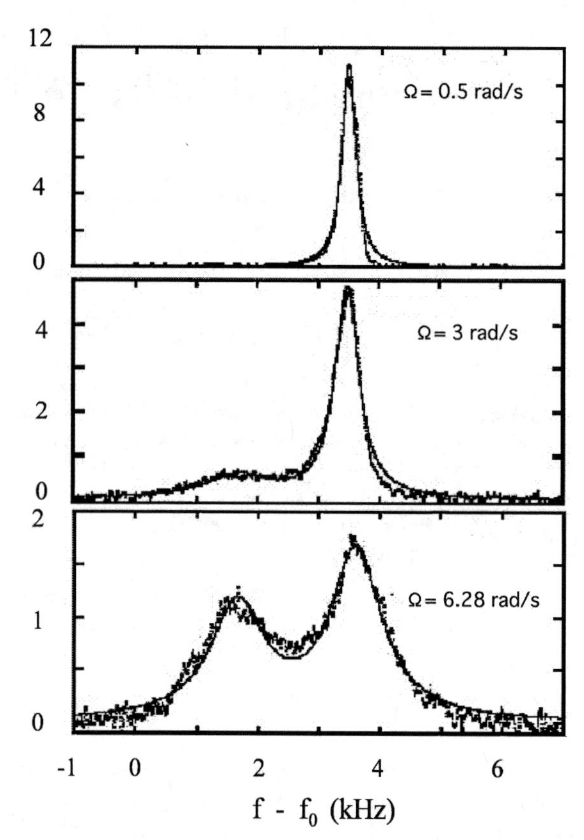

FIGURE 2. NMR spectra of ^{3}He-A at $T \sim 0.84T_c$ for rotation speeds $\Omega = 0.5$, 3, and 6.28 rad/sec. The temperature of the sample was determined from the main peak shift from the Larmor frequency. Vertical axes are scaled differently for comparison.

ACKNOWLEDGEMENTS

The authors thank I. A. Fomin for many communications and useful comments.

REFERENCES

1. I.A. Fomin and V.G. Kamenskii, *JETP Lett.* **35**, 302 (1982).
2. V.B. Eltsov, R. Blaauwgeers, M. Krusius, J.J. Ruohio, and R. Schanen, *J. Low Temp. Phys.* **124**, 123 (2001).
3. M. Kubota *et al.*, *Physica B* **329-333**, 1577-1581 (2003).

Rotating Superfluid ^{3}He-*A* in Parallel-Plate Geometry

Minoru Yamashita[a, b], Ken Izumina[a], Akira Matsubara[d], Yutaka Sasaki[d], Osamu Ishikawa[c], Takeo Takagi[e], Minoru Kubota[a] and Takao Mizusaki[b]

[a]*Institute for Solid State Physics, The University of Tokyo, Chiba 277-8581, Japan*
[b]*Department of Physics, Graduate School of Science, Kyoto University, Kyoto 606-8502, Japan*
[c]*Graduate School of Science, Osaka City University, Osaka 558-8585, Japan*
[d]*Research Center for Low Temperature and Materials Sciences, Kyoto University, Kyoto 606-8502, Japan*
[e]*Department of Applied Physics, Fukui University, Fukui 910-8507, Japan*

Abstract. We have measured NMR spectra of the ^{3}He-*A* phase restricted between parallel plates under rotation at a speed of $\Omega < 6.28$ rad/s, using a rotating cryostat at ISSP. The sample volume was divided by the plates into 110 disk-shaped spaces, each with a thickness of 12.5 μm and a radius of 1.5 mm. They were connected to a bulk superfluid through 0.3 mm-wide channels. Measurements were done by continuous wave (cw) NMR at 869 kHz. The axes of both H and Ω were perpendicular to the parallel plates. We observed a very narrow spectrum with a negative frequency shift, an indication that the gaps and the superfluid texture were well designed.

Keywords: superfluidity, helium3, vortex
PACS: 67.57.Fg

INTRODUCTION

The superfluid ^{3}He-*A* phase[1] in narrow channels[2, 3] has been widely studied. Since its energy gap is anisotropic, the orbital part of the order parameter, ***l***, must be perpendicular to the walls. The spin part, ***d***, can be directed by a magnetic field, H, to be $\boldsymbol{d} \perp H$. In bulk, ***l*** is parallel to ***d*** because of the dipole interaction. However, we can obtain a $\boldsymbol{l} \perp \boldsymbol{d}$ configuration if the superfluid is confined in a narrow gap on the order of the dipole coherence length, $\xi_D \sim 10$ μm, and a strong magnetic field ($H > 30$ G) is applied perpendicular to the gap. For this configuration, a vortex with a half quantum of circulation[4] is predicted to exist. The Helsinki group[3] tried to make parallel plates separated by a 19 μm gap. However, their gaps varied widely and the texture was not uniform.

PARALLEL PLATE CELL

Our sample cell consisted of 220 alternately stacked polyimide films of a 12.5 μm thickness and a 25 μm thickness. In each 12.5 μm film, we made a hole of a 3.0 mm diameter and a channel, 0.3 mm wide and 9.5 mm long, for introducing ^{3}He into the hole. By sandwiching the 12.5 μm film between the 25 μm films, we obtained a space 12.5 μm in thickness and 3.0 mm in diameter. These films were cut by a UV YAG laser,[5] which made accurate cuts with no burrs, so we were able to stack the films without additional spacing.

NMR SPECTRUM

We studied the superfluid in the parallel-plate geometry by cw NMR at 869 kHz at a pressure of 3.05 MPa. The parallel plates were orthogonal to both H and the rotation axis. For comparison, we also measured the NMR of bulk ^{3}He in a second cell, connected to the parallel-plate cell via the 0.3 mm wide channels. Both NMR spectra are shown in Fig. 1 for various temperatures. The liquid between the parallel plates shows a negative frequency shift, while the bulk liquid shows a positive frequency shift. A small peak near the Larmor frequency in the parallel plate spectra is attributed to solid ^{3}He because its magnetization increases with decreasing temperature.

In the *A* phase, the NMR absorption frequency, f, can be written as $f = f_L + f_A \cos 2\theta$ in the local

CP850, *Low Temperature Physics: 24th International Conference on Low Temperature Physics;*
edited by Y. Takano, S. P. Hershfield, S. O. Hill, P. J. Hirschfeld, and A. M. Goldman

approximation[1], where f_L is the Larmor frequency, f_A is the transverse frequency shift of the bulk *A* phase, and θ is the angle between ***l*** and ***d***. Since the dipole interaction forces θ to be 0 in the bulk liquid, we can determine f_A from the frequency shift of the bulk liquid. The negative frequency shift between the parallel plates shows that $\cos 2\theta = -0.93$. The broadening of the NMR spectra from that in the normal state, assuming that it is due to a misalignment of the parallel plates, indicates a nonuniformity of $\Delta\theta = 1.7°$. This result shows that gaps of our cell were well designed and that we could obtain a uniform texture. We don't know why we could not observe $\cos 2\theta = -1$. The misalignment due to our experimental setup is too small to explain the difference.

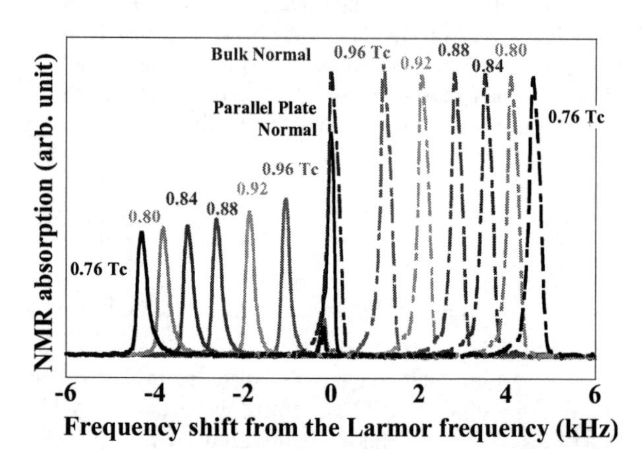

FIGURE 1. NMR spectrum in bulk (dashed lines) and between parallel plates (solid lines) at 3.05 MPa. The temperature for each NMR spectrum is indicated as a fraction of the transition temperature T_C (T_C = 2.46 mK).

The rotation dependence of the NMR spectrum was investigated after cooling the sample at rest through the superfluid transition. Below Ω_{Fr} = 1.0 rad/s, there was no change in the NMR spectrum. However, for $\Omega \geq \Omega_{Fr}$, a satellite signal was observed near f_L as shown in Fig. 2. This signal increased as the rotation speed increased for $\Omega_{Fr} \leq \Omega \leq \Omega_C$ = 1.8 rad/s. Above Ω_C, the satellite signal decreased as Ω was increased up to our maximum rotation speed 6.28 rad/s. It decreased rapidly as the rotation speed decreased and disappeared below 5.5 rad/s.

We attribute the satellite signal to the Fréedericksz transition[6], a textural transition of the ***l*** texture by rotation. The normal flow, v_N, induced by the rotation tends to orient ***l*** parallel to the plates and varies the dipole potential, which can trap a spin wave. The reduction of the satellite signal above Ω_C can be attributed to an existence of vortices. Since vortices form a cluster and give rise to a macroscopic superflow, v_S, the counterflow, $v_N - v_S$, is reduced by them. However, we have not observed any spin wave signals trapped by vortices as in bulk liquid[7]. A detailed explanation for the spin wave and vortices in the parallel-plate geometry will be published elsewhere.

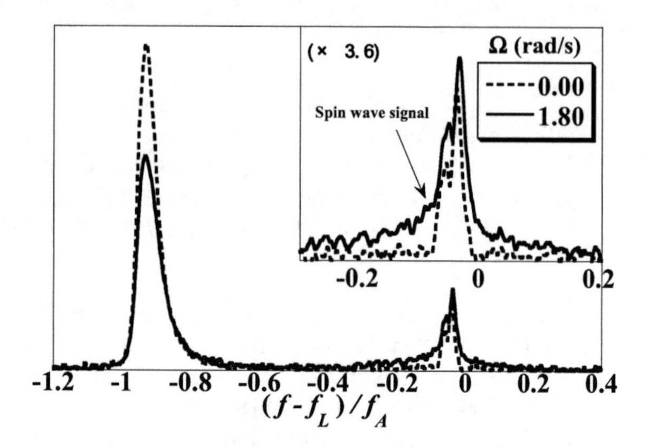

FIGURE 2. NMR spectrum between the parallel plates at 0 rad/s and 1.8 rad/s at $T = 0.81T_C$. The horizontal axis (frequency shift from f_L) is normalized by f_A. At 1.8 rad/s, a spin wave signal was observed near the Larmor frequency. The inset is an enlarged view near the spin-wave resonance.

ACKNOWLEDGMENTS

M. Yamashita acknowledges the support of a special research fellowship from the JSPS. The authors thank the Materials Design and Characterization Laboratory of the Institute for Solid State Physics, the University of Tokyo. This work was partially supported by a Grant-in-Aid for Scientific Research from MEXT and by a Grand-in-Aid for the 21st Century COE ``Center for Diversity and Universality in Physics".

REFERENCES

1. D. Vollhardt and P. Wölfle, *The Superfluid Phases of Helium 3*, London: Taylor & Francis, 1990.
2. R. Ishiguro, *et. al., Phys. Rev. Lett.*, **93**, 125301 (2004).
3. P. J. Hakonen, *et. al., Phys. Rev. Lett.*, **58**, 678 (1987).
4. M. M. Salomaa and G. E. Volovik, *Phys. Rev. Lett.*, **55**, 1184 (1985).
5. Will Co., Ldt. Hatori 3-17-17, Fujisawa, 251-0056, Japan, and Shinozaki Manufacturing Co., Ltd. Kitashinagawa 3-6-2, Tokyo, 140-0001, Japan.
6. P. M. Walmsley, D. J. Cousins, and A. L. Golov, *Phys. Rev. Lett.*, **91**, 225301 (2003).
7. P. J. Hakonen, O. T. Ikkala, and S. T. Islander, *Phys. Rev. Lett.*, **49**, 1258 (1982).

Quantum Turbulence at Very Low Temperatures: Status and Prospects

D. Charalambous*, P.C. Hendry*, P.V.E. McClintock* and L. Skrbek†

*Department of Physics, University of Lancaster, Lancaster LA1 4YB, UK
†Joint Low temperature Laboratory, Institute of Physics ASCR and Faculty of Mathematics and Physics, Charles University, V Holešovičkách 2, 180 00 Prague, Czech Republic

Abstract. The theory of how turbulent energy decays via a Richardson cascade is well-established for classical fluids, and it also seems to apply to the case of so-called co-flowing He II turbulence in the range from the superfluid transition temperature T_λ down to ~1 K, where its behaviour is similar to that of a classical fluid. For pure superfluids, e.g. He II in the mK range or ^{3}He-B in the μK range, where the normal fluid density ρ_n is near zero, the mode(s) through which quantum turbulence (QT) might decay have been much less clear because of the absence of viscosity to dissipate the turbulent energy on small length scales. Recent advances made in the theory of QT in this $T \to 0$ limit are consistent with such experimental evidence as is available, but new experiments supported by new techniques for the production and detection of QT are urgently required. The experimental situation is reviewed and prospects for further advances are considered.

Keywords: helium, superfluidity, turbulence, oscillators
PACS: 47.27.Cn, 47.37.1q, 67.40.Vs, 47.15.Cb

INTRODUCTION

Quantum turbulence (QT) [1, 2] in a superfluid (^{4}He or ^{3}He-B) can be thought of as a seemingly disordered collection of quantized vortex lines. The vortex lines themselves have been well-understood almost since their original discovery [3, 4] half a century ago. To reach an understanding of their collective motion in QT, however, represents a problem in complexity where major challenges remain to be tackled. It must be emphasized that the vortices are the *components* of QT. One can no more expect to predict the properties of QT from an understanding of these components than one could e.g. predict second sound or the other superfluid properties of liquid ^{4}He from a detailed knowledge of the ^{4}He atom; likewise, neither could one infer the periodic table, or chemistry, from a detailed knowledge of protons, neutrons and electrons. In each case, there are emergent phenomena that could hardly have been anticipated from an understanding of the microscopic components of the system. Note also that, in reality, and contrary to their appearance, the arrangement of vortex lines in QT is *not* random.

Below its superfluid transition temperature T_λ, liquid ^{4}He of density ρ can be regarded [5] as a mixture two components: a normal fluid component with viscosity, carrying the whole entropy of the liquid, and an inviscid superfluid carrying no entropy, with densities ρ_n and ρ_s respectively such that $\rho_n + \rho_s = \rho$. Quantum mechanics places severe restrictions on the motion of the superfluid, in that the circulation is quantized

$$\oint \mathbf{v}_s \cdot \mathbf{d}\ell = h/m_4 \tag{1}$$

where $\mathbf{v}_s$ is the fluid velocity, m_4 is the helium atomic mass and the integral is taken around a loop enclosing the vortex core; there is no such restriction on motions in the normal fluid component, which are classical. In the case of ^{3}He, the m_4 is replaced by $2m_3$, the mass of a Cooper pair.

The relative proportions of the two fluids are such that the normal fluid density $\rho_n/\rho = 1$ at $T = T_\lambda$ and $\rho_n/\rho = 0$ at $T = 0$; correspondingly, the superfluid density $\rho_s/\rho = 0$ at $T = T_\lambda$ and $\rho_s/\rho = 1$ at $T = 0$. In a typical experiment, QT is created by flow or by motion of an object in the liquid. In what follows, we will assume conventional isothermal flow; we will not consider superfluid/normal fluid counterflow (which, although often discussed in the literature, is actually a rather specialised kind of flow, peculiar to superfluids). Because the vortex lines are metastable states of the liquid, they may be expected to decay away with time. The main aim of the theory is to identify the underlying decay mechanisms, explore the corresponding physics, and characterise the form and timescale of the decay for comparison with experiment. As we will see, the theoretical situation is much clearer in the temperature range where the normal fluid density ρ_n is appreciable, say for $1 < T < T_\lambda = 2.17$ K. The main unsolved questions relate to the $T \to 0$ limit, by which we mean the temperature range in which $\rho_n \simeq 0$ so that the effect of viscous dissipation is negligible.

CP850, *Low Temperature Physics: 24th International Conference on Low Temperature Physics;*
edited by Y. Takano, S. P. Hershfield, S. O. Hill, P. J. Hirschfeld, and A. M. Goldman

The aim of this paper is to introduce QT in a manner accessible to non-specialists, to outline succinctly what is already understood, and to identify the major challenges still to be tackled. In particular, we describe the emergent theoretical picture describing QT in the $T \to 0$ limit, consider some of the experiments carried out to date in this temperature range, and discuss the requirements of the new generation of experiments now needed to test the theory. Our emphasis will be on experiments, but first we consider the theoretical background in order to set the context of the work.

Here we take no account of intermittency, despite its being a keyword of today's conventional turbulence research. We note that there is experimental evidence [9] strongly suggesting that intermittency occurs in QT as well, at least in the case of co-flowing He II turbulence generated between counterrotating discs [9]. Although its role in QT has not yet been widely recognized, there is little doubt that intermittency will become an important issue in future QT studies.

THEORETICAL BACKGROUND

Classical turbulence

In order to appreciate the problems presented by QT, it is first necessary to understand the accepted picture of classical turbulence (CT) [6, 7] and the manner in which it decays. The flow of the fluid is governed by the Navier-Stokes equation [8]

$$\frac{\partial \mathbf{v}}{\partial t} + (\mathbf{v}.\nabla)\mathbf{v} = -\frac{1}{\rho}\nabla p + \nu\nabla^2\mathbf{v} \qquad (2)$$

where $\mathbf{v}$ is the flow velocity, p is the pressure, ρ is the density, and ν is the kinematic viscosity. The ratio of the nonlinear inertial term $(\mathbf{v}.\nabla)\mathbf{v}$ to the dissipative term $\nu\nabla^2\mathbf{v}$ gives the Reynolds number $\mathrm{Re} = \ell u/\nu$. Here, u and ℓ are respectively a characteristic velocity and length describing the flow. In turbulent flows, Re is large and so the effect of dissipation can be ignored.

In typical flows, energy is fed into the turbulence at large length scales determined e.g. by the dimensions of a flow tube or the mesh size of a grid. Because the nonlinear term in (2) couples together motion on different length scales, turbulent energy flows without dissipation towards smaller and smaller length scales. Eventually the little eddies become so small that $\mathrm{Re} \sim 1$: the dissipative term in (2) then becomes important, and the energy is dissipated by viscosity. Within the inertial range of this Richardson cascade, the spectral energy density distribution is described by the Kolmogorov spectrum

$$E(k) = C\varepsilon^{2/3}k^{-5/3} \qquad (3)$$

where k is the wave number of the eddies, $E(k)$ is the energy per unit mass contained in eddies of wavenumber between k and $k + dk$, the Kolmogorov constant C is of order unity, and ε determines the rate at which energy flows down the cascade. Note that the fluid dynamics community commonly uses a notation where the turbulent energy is specified per unit mass, so it is of dimension $\mathrm{m}^2/\mathrm{s}^2$.

Quantum turbulence

We first consider QT at finite T. In mechanically generated co-flowing He II, the normal and superfluid components move together on large length scales, being strongly coupled by the action of the mutual friction force. The latter arises because, although vortex lines consist of an ordered flow in the *superfluid* component, the vortex cores around which the flow rotates are part of the *normal* fluid component[1]. Consequently, flows of this kind e.g. grid turbulence are characterised by a hydrodynamics that is essentially classical on large length scales, closely resembling conventional turbulence. On smaller scales, however, the quantum condition (1) will dominate and non-classical behaviour is then to be anticipated. In particular, the decay of the vortex tangle is expected to be driven by line *reconnections* [10], as sketched in Fig. 1. They create relatively sharp cusps in the reconnected lines, and the fast self-induced superflow velocity causes rapid motion of the vortex core through the viscous normal fluid component, leading to line shrinkage. Experiments above 1 K have studied QT created between two

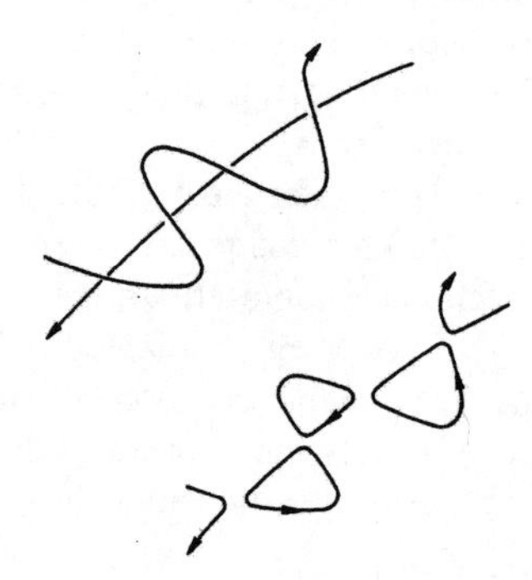

FIGURE 1. Sketch showing how quantized vortices are believed to reconnect when they cross, after Schwarz [10].

[1] In ^{3}He-B, considered later, the cores of the quantized vortices are macroscopic objects about hundred times larger than those of He II vortices, with complex structure that undergoes phase transitions. Nevertheless, the concept of the mutual friction force between normal and superfluid components holds generally.

counter-rotating disks [9], and the decay of QT created by a grid drawn once through a column of He II [11, 12]. In each case, the inferred energy distribution of the turbulence was consistent with the Kolmogorov energy spectrum (3) [13]. At least in this respect, therefore, QT is (perhaps unexpectedly) similar to classical turbulence.

In the $T \to 0$ limit, the absence of normal fluid component means that reconnections are no longer dissipative through this mechanism (though some small dissipation due to this process may still be expected). So, given that the cascade to smaller scales cannot be terminated by viscous damping, what happens? We can even wonder whether the QT will *ever* decay. Physical intuition suggests that it must in fact decay, given that vortices represent metastable states of the liquid – and experiments (see below) confirm this tentative conclusion. The theory of QT in the zero T limit and the decay mechanism in this regime [1, 15, 16, 17, 18, 19, 20] is due especially to Vinen, Barenghi, Brachet, Svistunov, Tsubota and others. We draw attention to the ever-growing role of numerical studies in such investigations. They can now be based on a full Biot-Savart simulation of the motion of vortex filaments [18, 21], as well as on the non-linear Schrödinger equation (Gross-Pitaevskii equation) [22, 23]. These studies follow the influential work of Schwarz [10] who recognized the importance of reconnections, but in an era when the available computer power forced him to limit his simulations to the so-called local induction approximation, thus neglecting any possibility of non-local interaction. Thus e.g. stretching of vortex lines, which is believed to play a central role in classical turbulence, could not be included.

The picture that is emerging involves a Richardson cascade from the largest length scales to successively smaller scales. It cannot end in viscous dissipation, as in the classical case, but it is probably followed by a *Kelvin wave cascade*. Reconnections generate a high density of Kelvin waves (vortex waves) on the lines, which mutually interact to generate Kelvin waves of successively higher frequency. Those of sufficiently high frequency can radiate phonons, thus terminating the cascade. What experimental evidence is there to support this general picture?

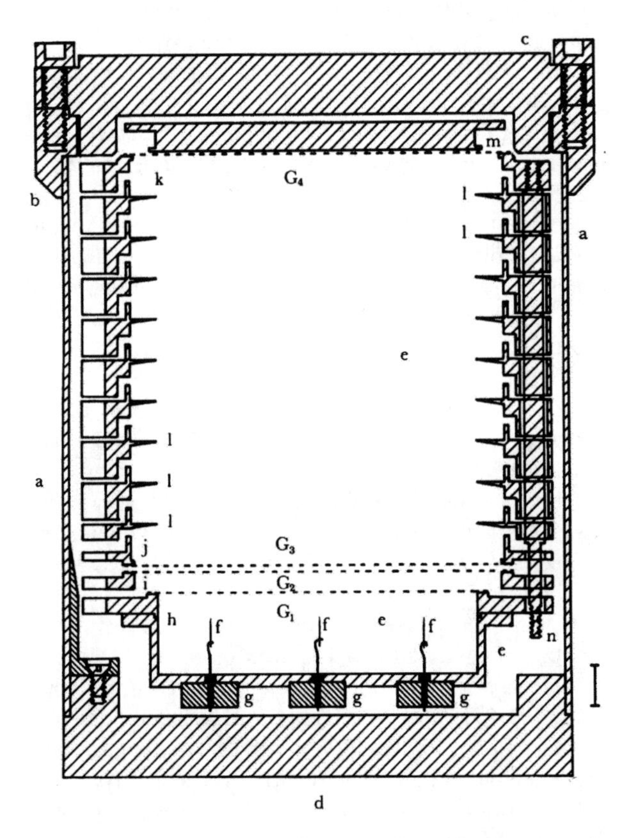

FIGURE 2. Schematic diagram of cell used for measurement of the Landau critical velocity [14]. The field emission tips f were used to inject negative ions, whose drift velocity across the space between grids G_3 and G_4 was then measured. The 10 mm vertical bar outside the cell on the lower right-hand side is to indicate the scale.

Early results

Interest in the decay of QT preceded the development of the theory. The experimental determination of the Landau critical velocity for roton creation [14] involved the use of the large experimental cell shown in Fig. 2. Negative ions were injected from the field emission tips f, and their transit time across the relatively low electric field region between grids G_3 and G_4 was measured. It is known that charged vortex rings and QT are created in the region of very high electric field near field emission tips. Negative ions are trapped on vortex lines so that, in the absence of a QT decay mechanism, a high density of spacecharge would have built up within the cell and would have prevented the experiment from working. That this did not happen [14] demonstrated that a QT decay mechanism must exist even in the absence of normal fluid component.

Fig. 3 illustrates a preliminary experiment [24] to investigate the decay of QT in superfluid ^{4}He in the $T \to 0$ limit. A fine-mesh grid, stretched to its yield point on a circular holder, is driven on-resonance by an electrostatic field. Because of the high Q of the oscillator, and the fact that it is vibrating in a dissipation-free superfluid, the amplitude rises until the critical velocity for QT generation is attained. After the grid has been driven for a few seconds, the drive is switched off. When a pulse of negative ions is passed through the QT that has been created, some of the ions get trapped on the vortices, leading to an attenuated signal at the collector. By propagating a sequence of such pulses, the recovery of the signal can be observed, corresponding to the decay of the QT cloud as shown in Fig. 4. Although direct quantitative comparison of the results of this experiment with the emerging theoretical or computational models is difficult, it clearly illustrates the decay of QT on a timescale of seconds.

EXPERIMENTS ON QT AT VERY LOW *T*

Techniques for creating QT

In addition to the vibrating grid mentioned above, several other techniques have also been used for creating QT in the low *T* limit, illustrated in Fig. 5.

The most widely used of these is the oscillating wire loop (top) [25, 26, 27]. The loop is formed from a single NbTi superconducting filament etched out of the copper matrix of standard magnet wire, typically x–y μm in diameter. A magnetic field is applied parallel to its legs. When an oscillatory current is passed through the loop, it responds to the resultant force as shown, moving in the direction perpendicular to the magnetic field and the wire. The induced voltage arising from the movement can be measured directly, there being no voltage due to the current in the static wire on account of its superconductivity (and we ignore the possibility of flux motion within the wire, associated with its movement).

At small driving amplitudes the response is usually linear, corresponding to laminar flow. Under these conditions the superfluid makes no contribution to the damping, but just provides an added hydrodynamic mass (see below). When the driving amplitude is such that a critical velocity is attained there is a sudden change in the response, attributable to the creation of QT [2]. In the crossover regime, a variety of interesting phenomena can appear, including hysteresis and intermittent switching between laminar and turbulent flow.

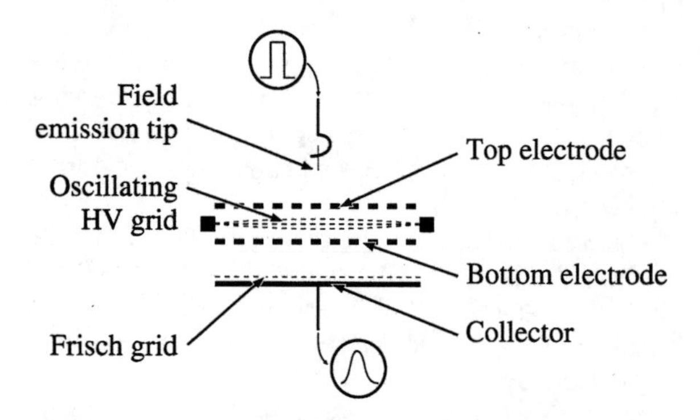

FIGURE 3. Schematic diagram showing the apparatus used for the preliminary experiment seeking evidence of QT decay at very low temperatures [24]. The electrode structure was immersed in isotopically pure He II with the cell illustrated in Fig. 2.

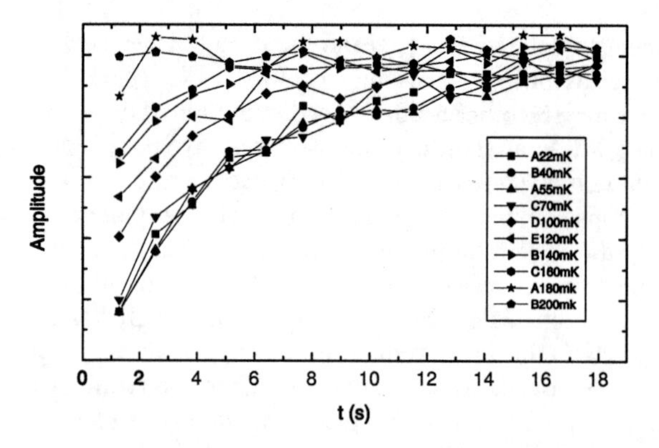

FIGURE 4. Amplitudes of a sequence of negative ion pulses after passage through the QT cloud [24]. The amplitude increases with time because the QT decays, reducing the extent of ion trapping on quantized vortices.

The oscillating microsphere (Fig. 5, second from top) behaves in many respects rather similarly [28, 29, 30, 31]. The magnetic sphere is levitated in He II between Nb electrodes. It carries an electrostatic charge, and it can therefore be driven by an applied oscillatory electric field. The resultant oscillations can be detected and measured. Again, investigation of the variation of the response amplitude with the driving force reveals a well-defined critical velocity at which there is a marked change in behaviour from linear response (laminar flow) to a much slower increase in response with drive (because of the creation of QT): see Fig. 6.

A recent innovation by Bradley *et al* [32] (Fig. 5, third from top) has been the creation of QT in ^{3}He-B by means of a grid mounted on a loop of superconducting wire. The grid material is identical to that used in the earlier work in He II [24, 33, 34], but it is moved magnetically rather than electrostatically. The authors report convincing evidence (see below) of QT production in ^{3}He-B when a critical velocity of the grid is exceeded.

A very recent innovation is the use of an oscillating quartz tuning fork [35]. The legs of the fork (Fig. 5, bottom) are about 3.4 mm in length, each with a cross-section of 400×400 μm^2. They oscillate in antiphase at a frequency of $\sim$30 kHz. Preliminary studies suggest that the behaviour of the fork in superfluid is generally similar to that of other vibrating objects in that it creates QT once

2 In ^{3}He-B there is also a critical threshold associated with reaching the pair-breaking velocity [26]. The Landau critical velocity for roton creation in He II [14] is many orders of magnitude larger than those at which QT is generated, and can be ignored in the present context.

a critical velocity has been exceeded. It is characterised by robustness, small size, and a high Q of up to 10^5, and there may be other advantages. These devices are under active investigation in Helsinki, Košice, Lancaster and Prague in terms of their potential for thermometry and particle detection, as well as for QT creation, both in He II and superfluid phases of ^{3}He.

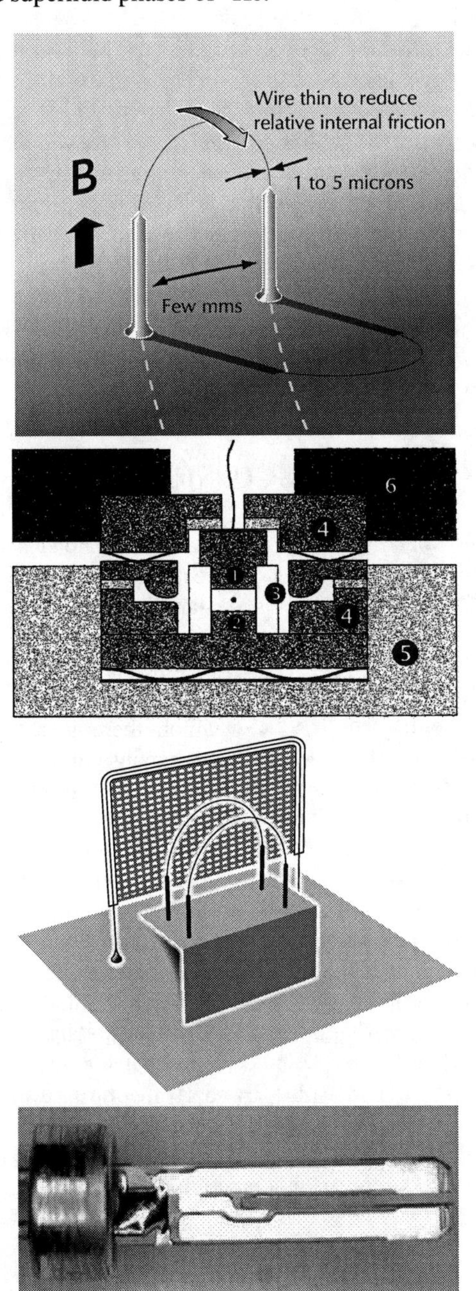

FIGURE 5. Other techniques used for creation of QT in superfluid ^{4}He and ^{3}He at very low T. From the top: vibrating wire loop [25, 26, 27]; oscillating levitated sphere [28, 29, 30, 31]; vibrating grid, on wire loop (with two additional loops as detectors) [32]; vibrating quartz fork [35].

All of the techniques shown in Figs. 3 and 5 involve the oscillation of a submerged object of some kind within the superfluid. Despite obvious cryogenic advantages this is not, in fact, ideal as a method of creating QT (see Discussion below) in that one aims to create a well-defined turbulent flow that can be approximated as homogeneous and isotropic in order to compare it with available theoretical models.

Hydrodynamic mass in a superfluid

The superfluid is inviscid and, provided that there are no free surfaces on which waves can be created, it causes no drag on an object moving at low velocity. Nonetheless, there are inertial effects associated with the need to push the superfluid out of the way. This is the origin of the classical hydrodynamic mass enhancement of

$$\Delta M = \beta V \rho \tag{4}$$

where β is a geometrical constant, V is the volume of the moving object, and ρ is the density of the fluid. The dynamics of a moving object in the superfluid in the $T \to 0$ limit are therefore identical to what they would be in a vacuum, except for the modified effective mass. Experiments with negative ion bubbles [36] confirm this picture.

For oscillating material objects submerged in He II, however, there are indications [37, 38] that the effective mass is enhanced by more than just the classical ΔM, even within the linear regime of low oscillatory amplitude. In the case of the oscillating grid, the constant β cannot reliably be evaluated because of the ill-defined though approximately rectangular cross-section of the grid wires, but Charalambous *et al* [37] arrive at this conclusion through investigation of the density (pressure) dependence of the resonant frequency. The wire used by Yano *et al* [38] is smooth and circular in cross-section, so that the value of $\beta = 1$ for an infinite cylinder can be assumed. The radius of the wire calculated from the measured hydrodynamic mass is significantly larger than the actual radius as determined by electron microscopy, i.e. the effective mass is larger than expected. It seems extremely likely that the physics underlying the anomalous mass enhancement is the same in both experiments [37, 38]. The origin of the effect has yet to be established, but one possibility is that surfaces exposed to superfluid become covered in a "fur" of small vortex loops that can then add [39] to the inertial effective mass. The general picture would be qualitatively similar to that considered by Kusmartsev [40] in trying to account for critical velocities in the flow of rotating superfluid ^{4}He. He envisaged the creation by fluctuations of a plasma of half-vortex rings at the wall. In terms of this picture, the concept of remanent vorticity [41], consisting of a few relatively long vortices pinned to protuberances, must be supplemented by the idea of the dense array of microscopic loops. But many problems need to be resolved (e.g. the stability of pinned vortex half-rings) before any firm conclusion can be drawn.

The seemingly anomalous effective mass of objects oscillating in the superfluid does not bear directly on QT as such, but the underlying physics may be highly relevant to the critical velocity that must be exceeded to initiate QT creation.

Techniques for detecting QT

Experiments on QT in He II above 1 K [11, 12] have used the attenuation of second sound to detect the vortices, almost universally. It is an extremely sensitive technique: using a resonant cavity, it can even enable the detection of individual vortex lines. Unfortunately, however, it is applicable neither in the low T range for He II because second sound (involving antiphase oscillation of the normal and superfluid components) cannot exist in the absence of normal fluid component, nor in superfluid phases of ^{3}He where second sound is heavily damped.

Most experiments on QT in He II in the $T \rightarrow 0$ limit have detected its creation indirectly, by observing the oscillatory dynamics of the moving object. In particular, the response as a function of the amplitude of the driving force usually undergoes a marked change at the transition between laminar flow and QT creation: see e.g. the oscillating microsphere data of Fig. 6. Although this approach provides evidence of QT production, it does not allow comparison with the theory described above. Exceptions include the preliminary ^{4}He experiment [24] where ion trapping was used as a detection method.

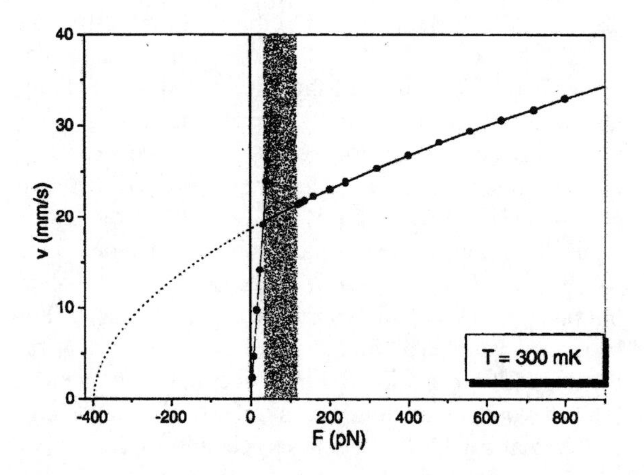

FIGURE 6. Plot of response as a function of forcing amplitude for the magnetically levitated microsphere [30].

QT in ^{3}He can be detected by application of the powerful technique of nuclear magnetic resonance (NMR). Indeed, the temperature induced transition from laminar to turbulent flow in ^{3}He-B at $\sim 0.6\ T_c$ was observed by NMR in a rotating sample when one or few vortex seed loops were injected into the originally vortex-free Landau state [42]. However, the NMR technique used is sensitive mainly to the large-scale counterflow rather than to the signal from vortex cores; and its use at lower temperatures becomes increasingly difficult. In experiments by Bradley *et al* [32] QT in ^{3}He-B generated by the electromagnetically oscillating grid (see above) was detected as a *decrease* in damping sensed by the nearby vibrating wire: the underlying physics is based on Andreev scattering (the incoming quasiparticle is back-scattered as a quasihole and vice versa) of ballistically propagating quasiparticles. The Andreev scattering channel appears through Galilean modification of the quasiparticle dispersion relation due to circulating superflow around the vortex core [43].

DISCUSSION

Several experiments show that QT can be created at low T, and there are two [24, 32] which clearly demonstrate that it decays on a timescale of several seconds. It has not yet, however, been possible to make a detailed comparison with theory. The main problem is that submerged oscillating objects create QT that is not formed at a well-defined initial length scale, because the object repeatedly traverses the QT that it has created, thus chopping it up into ill-defined smaller elements. The other problems are that the QT is not homogeneous: it may stay localised near the oscillating object, so that its density then decreases sharply with distance. Another source of inhomogeneity is that flow across the moving object often occurs at different speeds in different positions. A further complication is that the transition to turbulence due to submerged oscillating objects represents a problem which is as yet only partly understood, even for classical viscous flows. Experiments with flows due to e.g. a transversally oscillating cylinder [44], or an oscillating grid in water [45], displayed very rich and complex flow phenomena. This makes comparison of classical and quantum flows, and the problem of understanding the latter, extremely difficult.

The ideal QT experiment would somehow emulate the techniques used earlier above 1 K [11], in which a grid was drawn once through a column of liquid to create well-characterised homogeneous QT which was then detected with enormous sensitivity by use of second sound. For obvious reasons, this is extremely difficult to accomplish at very low T. QT creation by linear mechanical movement must be accomplished without frictional heating – a considerable technical challenge – and, as mentioned above, second sound cannot be used for detection because there is no normal fluid component.

Efforts to create the necessary mechanism for a drawn grid are currently in progress. To avoid friction, the grid is to be moved on magnetic bearings. The result is the levitated, superconducting, linear, stepper motor[3] sketched in Fig. 7. It is hoped that this device will avoid many of the disadvantages of the earlier oscillatory QT creation methods.

New techniques will also be required to detect the QT. There are two problems associated with the use of negative ions. First, although the capture cross-section in He II has not yet been measured in the $T \to 0$ limit, it is certainly very small [46] and probably highly dependent on electric field. So ions can only be suitable for the detection of QT at relatively high density and will probably not allow coverage of the wide range of densities needed for a detailed comparison with experiment. Secondly, they must be to some extent *invasive* probes in that any electric field, including their own spacecharge field, will apply a force to the QT, an effect that could be significant under certain circumstances.

The use of ions for QT detection in the superfluid phases of ^{3}He seems even more difficult, in view of a much larger size of the vortex cores leading to smaller substitution energy and presumably a smaller capture cross-section. The use of NMR and especially the novel detection technique based on Andreev scattering (see above) seem promising here. There is a clear call for complementary experiments probing comparable flows in He II and ^{3}He in the $T \to 0$ limit.

Further detection techniques currently under consideration include the use of: (a) calorimetry, which is sensitive but where it will be difficult to analyse the data for comparison with the theory; (b) tracers, e.g. excited He II molecules; and (c) miniature pressure transducers. The options are discussed by Vinen in these Proceedings [47].

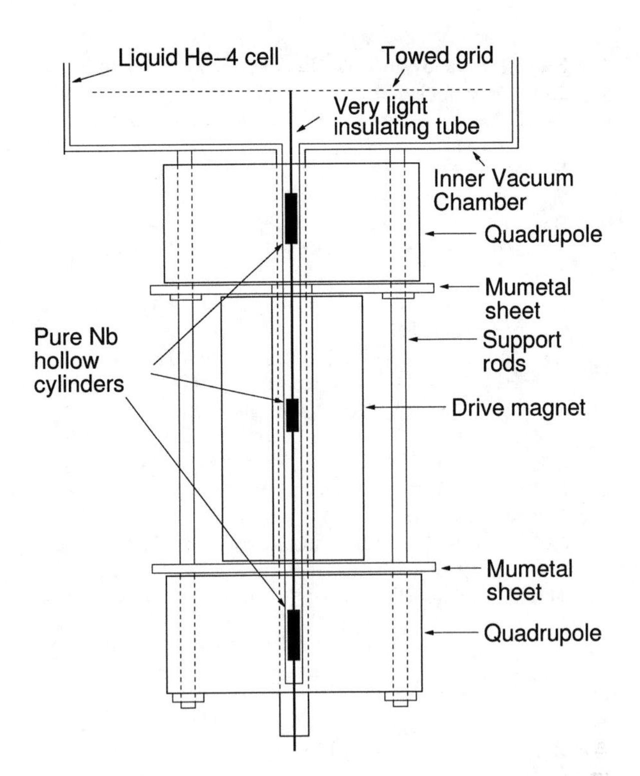

FIGURE 7. Schematic diagram of the linear mechanism being developed for used for creation of well-characterised homogeneous QT at very low temperatures. It uses magnetic bearings and propulsion, and the moving elements should not make any physical contact with static components of the cell.

CONCLUSIONS

It is evident that the theory of QT for $T \to 0$ has overtaken the experiments and moved far ahead. The study of QT in this range not only brings intrinsic interest of turbulence in a pure superfluid, but also represents a major intellectual challenge. As a consequence of the quantized circulation in superfluids it introduces *new physics*, e.g. the Kelvin wave cascades. Although adequate tests of the theory require the introduction of new experimental techniques for the creation and detection of QT, there are several experiments under way in different laboratories as well as many promising new ideas ready to be tried. It seems to us that the prospects for rapid progress towards an understanding of QT in the low temperature limit are excellent.

ACKNOWLEDGEMENTS

We acknowledge valuable discussions with W F Vinen and G G Ihas and we are grateful to G R Pickett for providing the oscillating loop diagrams in Fig. 5. The work was supported by the Engineering and Physical Sciences Research Council (UK) and by the Ministry of Education of the Czech Republic under research plan MS 002162083.

REFERENCES

1. Vinen, W. F., and Niemela, J. J., *J. Low Temp. Phys.*, **128**, 167–231 (2002).
2. Skrbek, L., *Physica C*, **404**, 354–362 (2004).

[3] Designed and built in collaboration with W.F. Vinen, but yet to be tested.

3. Hall, H. E., and Vinen, W. F., *Proc. Roy. Soc. (Lond.) A*, **238**, 204–214 (1956).
4. Vinen, W., *Progress in Low Temperature Physics*, **III**, Chapter 1 (1961).
5. Wilks, J., *The Properties of Liquid and Solid Helium*, Clarendon Press, Oxford, 1967.
6. Batchelor, G. K., *The Theory of Homogeneous Turbulence*, Cambridge University Press, Cambridge, 1953.
7. Frisch, U., *Turbulence*, Cambridge University Press, Cambridge, 1995.
8. Landau, L. D., and Lifshitz, E. M., *Fluid Mechanics*, Butterworth and Heinemann, Oxford, 1987.
9. Maurer, J., and Tabeling, P., *Europhys. Lett.*, **43**, 29–34 (1998).
10. Schwarz, K. W., *Phys. Rev. B*, **38**, 2398–2417 (1988).
11. Stalp, S. R., Skrbek, L., and Donnelly, R. J., *Phys. Rev. Lett.*, **82**, 4831–4834 (1999).
12. Skrbek, L., Niemela, J. J., and Donnelly, R. J., *Phys. Rev. Lett.*, **85**, 2973–2976 (2000).
13. Skrbek, L., Niemela, J. J., and Sreenivasan, K. R., *Phys. Rev. E*, **64**, 067301 (2001).
14. Ellis, T., and McClintock, P. V. E., *Phil. Trans. R. Soc. (Lond.) A*, **315**, 259–300 (1985).
15. Svistunov, B. V., *Phys. Rev. B*, **52**, 3647 (1995).
16. Tsubota, M., Araki, T., and Nemirovskii, S. K., *J. Low Temperature Phys.*, **119**, 337–342 (2000).
17. Barenghi, C. F., and Samuels, D., *Phys. Rev. Lett.*, **89**, 155302 (2002).
18. Vinen, W. F., Tsubota, M., and Mitani, A., *Phys. Rev. Lett.*, **91**, 135301 (2003).
19. Nore, C., Abid, M., and Brachet, M. E., *Phys. Rev. Lett.*, **78**, 3896– (1997).
20. Nore, C., Abid, M., and Brachet, M. E., *Phys. Fluids*, **9**, 2644 (1997).
21. Vinen, W. F., Tsubota, M., and Mitani, A., *J. Low Temperature Phys.*, **134**, 457–462 (2004).
22. Koplik, J., and Levine, H., *Phys. Rev. Lett.*, **71**, 1375 (1993).
23. Kobayashi, M., and Tsubota, M., *Phys. Rev. Lett.*, **94**, 065302 (2005).
24. Davis, S. I., Hendry, P. C., and McClintock, P. V. E., *Physica B*, **280**, 43–44 (2000).
25. Morishita, M., Kuroda, T., Sawada, A., and Satoh, T., *J. Low Temp. Phys.*, **76**, 387 (1989).
26. Bradley, D. I., *Phys. Rev. Lett.*, **84**, 1252 (2000).
27. Yano, H., Handa, A., Nakagawa, H., Obara, K., Ishikawa, O., Hata, T., and Nakagawa, M., *J. Low Temperature Phys.*, **138**, 561–566 (2005).
28. Jäger, J., Schuderer, B., and Schoepe, W., *Phys. Rev. Lett.*, **74**, 566–569 (1995).
29. Niemetz, M., Kerscher, H., and Schoepe, W., *J. Low Temp. Phys*, **126**, 287–296 (2002).
30. Niemetz, M., and Schoepe, W., *J. Low Temperature Phys.*, **135**, 447 (2004).
31. Schoepe, W., *Phys. Rev. Lett.*, **92**, 095301 (2004).
32. Bradley, D. I., Clubb, D. O., Fisher, S. N., Guénault, A. M., Haley, R. P., Matthews, C. J., Pickett, G. R., Tsepelin, V., and Zaki, K., *Phys. Rev. Lett.*, **95**, 035302 (2005).
33. Nichol, H. A., Skrbek, L., Hendry, P. C., and McClintock, P. V. E., *Phys. Rev. Lett.*, **92**, 244501/1–4 (2004).
34. Nichol, H. A., Skrbek, L., Hendry, P. C., and McClintock, P. V. E., *Phys. Rev. E*, **70**, 056307 (2004).
35. Eltsov, V. B., Hosio, J., and Skrbek, L., in preparation (2006).
36. Ellis, T., and McClintock, P. V. E., *Phys. Lett. A*, **89**, 414–416 (1982).
37. Charalambous, D., Hendry, P. C., McClintock, P. V. E., Skrbek, L., and Vinen, W. F., "Vibrating Grid as a Tool for Studying the Flow of Pure He II and its Transition to Turbulence," in these *Proceedings*.
38. Yano, H., Handa, A., Nakagawa, M., Obara, K., Ishikawa, O., and Hata, T., "Study on the turbulent flow of superfluid ^{4}He generated by a vibrating wire," in these *Proceedings*.
39. Vinen, W. F., Skrbek, L., and Nichol, H. A., *J. Low Temperature Phys.*, **135**, 423–445 (2004).
40. Kusmartsev, F. V., *Phys. Rev. Lett.*, **76**, 1880–1883 (1996).
41. Awschalom, D. D., and Schwarz, K. W., *Phys. Rev. Lett.*, **52**, 49–52 (1984).
42. Finne, A. P., Araki, T., Blaauwgeers, R., and et al, *Nature*, **424**, 1022 (2003).
43. Bradley, D. I., Fisher, S. N., Guénault, A. M., Lowe, M. R., Pickett, G. R., Rahm, A., and Whitehead, R. C. V., *Phys. Rev. Lett.*, **93**, 235302 (2004).
44. Honji, H., *J. Fluid Mech.*, **107**, 509–520 (1981).
45. de Silva, I. P. D., and Fernando, H. J. S., *Phys. of Fluids*, **6**, 2455–2464 (1994).
46. Ostermeier, R. M., and Glaberson, W. I., *J. Low Temp. Phys.*, **20**, 159 (1975).
47. Vinen, W. F., "Quantum turbulence: where do we go from here?," in these *Proceedings*.

Study on the Turbulent Flow of Superfluid ^{4}He Generated by a Vibrating Wire

H. Yano*, A. Handa*, M. Nakagawa†, K. Obara*, O. Ishikawa* and T. Hata*

*Graduate School of Science, Osaka City University, Osaka 558-8585, Japan
†Department of Physics, Hokkaido University of Education, Kushiro 085-8580, Japan

Abstract. We have studied the flow of superfluid ^{4}He generated by a vibrating wire. As the drive force increases, the velocity of the wire grows in the laminar-flow regime, until it suddenly drops at the onset of the turbulent-flow regime. As the drive force decreases, the turbulence disappears at a critical velocity. This result suggests that the vortices on the wire are confined within a finite size, even in turbulence. We have measured the critical velocity of seven vibrating wires, whose resonance frequencies range from 0.5 kHz to 9 kHz, at 1.4 K and found that the critical velocity is almost constant below an oscillation frequency of 2 kHz and increases above this frequency. We have also observed the response of a vibrating wire in superfluid ^{4}He at a low temperature of 30 mK. We find that the resonance frequency jumps upward at the same moment as the entry of the flow to a turbulent state. The frequency jump may be caused by vortex dynamics such as expansion, entanglement, and reconnection occurring in the turbulence.

Keywords: helium, superfluidity, turbulence, quantized vortex, vibrating wire
PACS: 67.40.Vs,47.27.Cn,47.15.Cb

INTRODUCTION

The vortex of superfluid ^{4}He has a simple structure: a normal-fluid core with a diameter of the order of the superfluid coherence length, and a superfluid current around the core decaying in proportion to the inverse of the distance from the core. This structure arises from the quantization of circulation, which prevents the core from becoming a superfluid. The core, therefore, can be imagined as a string which forms into a ring shape without ends or attaches itself to a boundary. A ring-shaped string easily dissipates by the mutual friction of the normal-fluid component. However, a string attached to a boundary is stable for hours or longer [1]. Vortex strings may expand and produce quantum turbulence when a superfluid current exceeding a certain velocity is applied [2]. This means that a moving object can generate turbulence in superfluid ^{4}He, even at temperatures at which the normal-fluid component almost vanishes, if some vortex strings are attached to the object.

In recent years, quantized vortices and quantum turbulence have been investigated using oscillating objects, which dissipate energy by the creation of vortices and turbulence [3–9]. A study using a micro sphere has shown that the superfluid flow around the sphere is laminar at low velocities and develops into turbulence above a critical velocity [3]. An oscillating grid exhibits a more complicated motion which varies with the drive force. Although at low drive force the response of the grid exhibits a Lorentzian curve as a function of frequency, hysteresis appears in the response at high drive forces and the effective mass of the grid is enhanced [4]. The increase in the effective mass can be understood in terms of the adiabatic response of vortex strings attached to the grid [5]. An oscillator with a simple shape is more suitable for detecting vortex nucleation. In the micro-sphere studies, intermittent shakes due to nucleation were observed in the oscillating response [6, 7]. The oscillation of a very thin wire has been used to detect the nucleation of an individual vortex in superfluid ^{3}He [8]. In preliminary reports, we have described the response of a vibrating wire in superfluid ^{4}He exhibiting a behavior similar to that of a micro sphere [9, 10]. The response reflects the interaction between vortices attached to the wire and a normal-fluid component.

In the present paper, we describe the motion of a vibrating wire in superfluid ^{4}He as a function of frequency between 0.5 kHz and 9 kHz. Turbulence is generated by vibration at a temperature of 1.4 K, where there are many quasi particles in the superfluid and the system behaves hydrodynamically. We also describe the responses at a low temperature of 30 mK, where quasi particles are almost absent. At this temperature, the motion of the vibrating wire is expected to directly reflect the vortex motion. Throughout this paper, the vortex dynamics in the ac flow of the superfluid ^{4}He are discussed by choosing coordinates that move with the wire.

EXPERIMENTAL

In a preliminary report [10], we have described the details of the vibrating wires. Each vibrating wire was made from 2.5-μm diameter NbTi wire formed into a semicir-

CP850, *Low Temperature Physics: 24th International Conference on Low Temperature Physics;*
edited by Y. Takano, S. P. Hershfield, S. O. Hill, P. J. Hirschfeld, and A. M. Goldman

cular shape. The surface of the wire was smooth within 0.2 μm. We set the distance between the legs of each vibrating wire so as to give a range of resonance frequencies. The vibration of the wire was generated by an ac electric current at the resonance frequency in a magnetic field of 25 mT. We measured the vibrating velocity as an induced voltage between the legs of the wire. The peak velocity given in this paper is that of the peak of the curve of the vibrating wire.

The experiment at 1.4 K was performed with a conventional helium-4 cryostat. We measured the response of the vibrating wire in a vacuum chamber at 4.2 K, then cooled it down to a temperature lower than the superfluid transition temperature of 2.2 K and liquefied superfluid ^{4}He directly into the chamber. This procedure reduces the number of vortex strings attached to the wire [9]. We made seven vibrating wires and set two wires in the chamber at a time, and the responses of all wires were measured at 1.4 K. The resonances of the wires were found to be in the frequency range from 0.5 to 9 kHz. The experiment at 30 mK, employing one vibrating wire, was cooled with a dilution refrigerator. In vacuum at 4.2 K, the resonance frequency and the Q value of the wire were found to be 3.8 kHz and 1700, respectively.

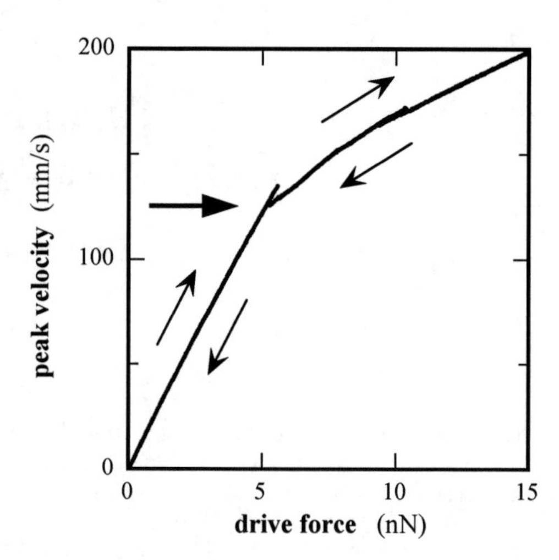

FIGURE 1. Peak velocity of a vibrating wire as a function of drive force at 1.4 K. The data were taken with up and down sweeps of the drive force. The velocity grows linearly at low drive force; however, the response exhibits a non-linear dependence above 125 mm/s. The critical velocity at which the non-linear dependence disappears in the down sweep process is indicated by the bold arrow.

RESULTS AND DISCUSSION

Critical Velocity

The responses of the vibrating wires indicate the existence of two types of flow. As the drive force is increased, the oscillation of a wire increases linearly and then at a certain drive force it drops suddenly and begins to increase non-linearly. The motion of the wire with an oscillation frequency of 8.8 kHz is plotted in Fig. 1, showing a typical response. The linear response indicates laminar flow around the wire, where the drag force is caused by the normal fluid [3]. In the non-linear regime, the oscillation expends extra energy, adding to the loss due to the normal fluid. The extra energy suggests that vortices are created in this regime. The vortices are generated by the expansion of vortex strings attached to the wire to form quantum turbulence [9]. A superfluid current causes a vortex string to expand unstably if it exceeds a velocity

$$v_s \approx (\kappa/4\pi r)[\log(8r/\xi) - 0.25], \qquad (1)$$

because of Glaberson–Donnelly instability [11]. Here r is the radius of the vortex loop, κ is the quantum of circulation, and ξ is the coherence length. The vortex strings on the wire are smaller than r derived from the equation at the maximum v_s of the linear region.

We also measured the response with decreasing drive force and found that the turbulence disappears at the intersection with the linear response, shown by the arrow in Fig. 1. If spontaneous expansion of the vortex strings occurs in the turbulent flow, the strings would grow infinitely and the critical velocity would be reduced. The disappearance of the turbulence at the intersection indicates that the strings do not grow in this way. In Fig. 1, the critical velocity at which the turbulence disappears is estimated to be 125 mm/s; consequently the strings must be confined within a radius of 0.7 μm, derived from Eq. (1). The dynamics of vortex strings is expected to confine the strings within a finite size. A large vortex string may easily entangle itself and, in this process, a ring vortex can become detached from the vortex string by reconnection. This entanglement and reconnection process is likely to limit the size of the strings on the wire in the turbulence.

The evolution of a vortex string may be influenced by the oscillation frequency of the superfluid current as well as its velocity. The frequency dependence of the critical velocity is useful for describing the vortex dynamics in turbulence. Figure 2 shows the measured critical velocity as a function of frequency for the vibrating wires. The critical velocity shows a remarkable frequency dependence: it is almost constant below 2 kHz and increases at high frequencies. This variation of the critical velocity should be caused by the motion of the vortex strings. The frequency dependence below 2 kHz indicates that vortex strings are confined within a finite size independent of flow oscillation. In other words, the strings may expand, entangle, and reconnect themselves completely within a period of oscillation. The period between reconnections

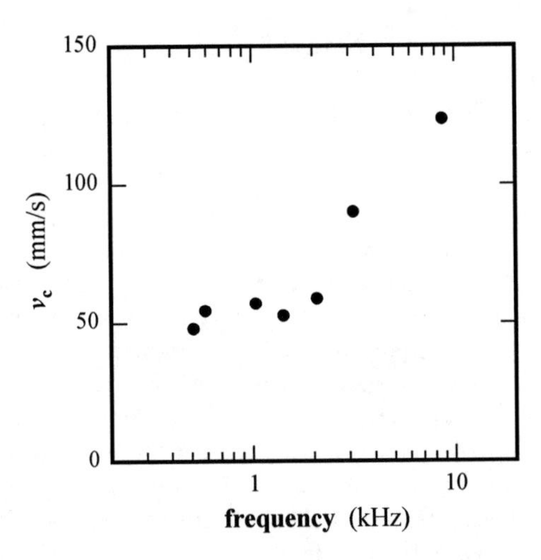

FIGURE 2. Critical velocity v_c at which the turbulence disappears for a decreasing drive force, as a function of frequency at 1.4 K. The velocity is almost constant below 2 kHz but increases with frequency at high frequencies.

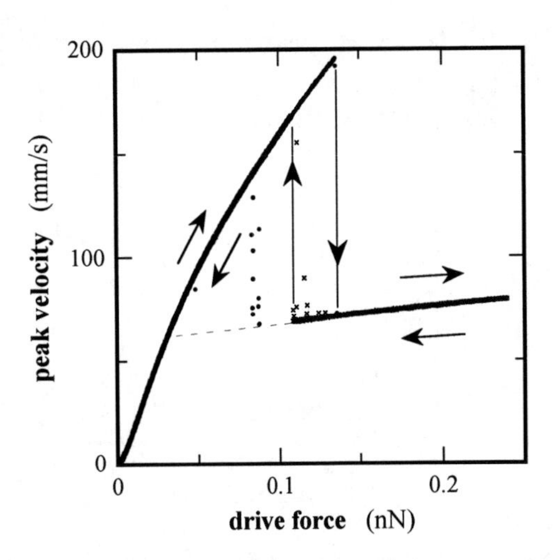

FIGURE 3. Velocity profile of vibrating wire at 30 mK. Laminar and turbulent flows appear at low and high drive forces, respectively. The data points between the two flow regimes are caused by intermittent switching (see text).

may determine the confinement size of the vortex strings. In a high-frequency flow, however, the strings may entangle and reconnect themselves before expanding sufficiently. Consequently the size of the strings is reduced and the critical velocity increases. From this frequency dependence a period of 0.5 ms can be derived, which is the inverse of 2 kHz, corresponding to the time between reconnections.

The present measurements were performed at a temperature of 1.4 K. Hence, the vortex strings must be influenced by the normal fluid, which makes up a fraction of the liquid at 1.4 K. The normal fluid reduces the size of the strings by the mutual friction between the normal fluid and the core of the vortices [2]. The critical velocity, therefore, is expected to increase with temperature in the hydrodynamic regime, in which normal fluid exists. Consequently, the critical velocity must depend on temperature as well as the flow frequency.

Wire Response At Low Temperatures

At temperatures where quasi particles are largely absent, the vortex dynamics can be observed more effectively. We prepared a vibrating wire with a resonance frequency of 3.8 kHz and observed the response in superfluid ^{4}He at 30 mK. The oscillation of the wire was similar to that obtained at high temperatures, but it varied more dramatically, as shown in Fig. 3. As the drive force is increased, the oscillation grows to almost 200 mm/s in the laminar flow and suddenly drops to 70 mm/s as the flow develops into turbulence. When the drive force is swept downwards, the peak velocity follows the same curve and jumps to a high value in the laminar flow. Since the upward jump in the velocity was not observed at high temperatures, as shown in Fig. 1, this behavior must be caused by the vortex dynamics directly. The size of the vortex strings attached to the wire can be derived from Eq. (1). In the laminar flow, the vortex strings are confined within 0.4 μm, which is estimated using the maximum velocity of 200 mm/s. In contrast, the size of the strings is larger than 1.3 μm in the turbulent flow, because continuous expansion of the strings occurs even at 70 mm/s. These estimates reveal that the strings reduce in size considerably when the flow around the wire changes from turbulent to laminar.

We occasionally observed unstable variations in the velocity between the laminar and turbulent flow regimes. We have plotted the averaged data in Fig. 3. This behavior is similar to unstable variations caused by intermittent switching between laminar and turbulent flow observed by Schoepe *et al.* [6, 7], though the switchings do not occur as often in our study. Schoepe *et al.* studied quantum turbulence using a micro sphere with a resonance frequency of 120 Hz. The intermittent switching might be influenced by the object shape and the oscillating frequency.

We also measured the resonance frequency of the vibrating wire by locking the phase to keep the quadrature at zero. Since the resonance frequency is not very sensitive to quasi particles in the ballistic regime [12], the motion of vortices attached to the wire is reflected effectively in the resonance frequency at low temperatures. Figure 4 shows resonance frequency data taken simultaneously with the peak velocity data shown in Fig. 3. The drive force dependence of the resonance frequency

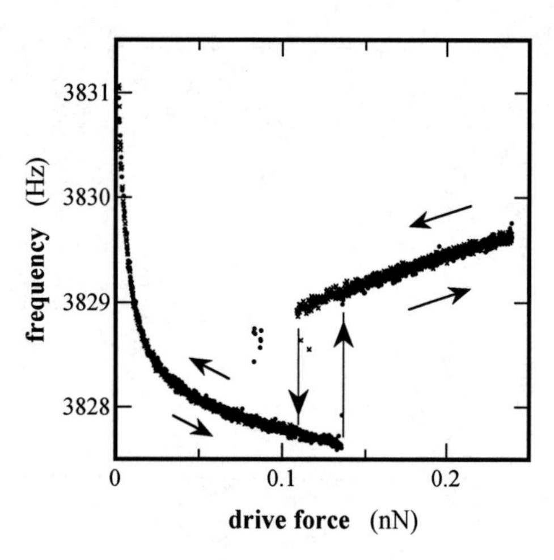

FIGURE 4. Resonance frequency of vibrating wire at 30 mK. The data were taken simultaneously with the peak velocity data shown in Fig. 3.

reveals a curious behavior: the frequency exhibits a sudden jump upward at the moment when the laminar flow develops into turbulence. The similar behavior between the resonance frequency and the peak velocity has been reported by Bradley *et al.* [13]. The upward jump in the frequency indicates a reduction in the moment of inertia, though the turbulence is likely to increase the mass of the wire.

By comparing these results with the resonance frequency in vacuum, we estimated the density of the wire in the superfluid to be 2.0 g/cm^3, which is much smaller than the expected value of 6.3 g/cm^3 for NbTi alloy [12]. This result suggests that the wire becomes thick in the superfluid due to the vortex strings attached to it. A vortex string on an oscillator behaves like a barrier against the superfluid current and increases the moment of inertia of the oscillator [5]. We have found that vortex strings on the wire are smaller in laminar flow than in turbulent flow. Hence, if the vortex string behaves as a barrier in the both types of flow, the moment of inertia would be greater in the turbulent flow. As mentioned above, the process of expansion, entanglement and reconnection may occur in turbulent flow. Entanglement may change the shape of the barrier presented by the string, allowing a superfluid current to pass through it. Reconnection can remove the mass of a ring vortex from the wire. Therefore, it is likely that the processes of entanglement and reconnection reduce the moment of inertia of the wire. Expansion of the string, however, may increase the mass of the wire. These effects combine and influence the response of the vibrating wire in turbulent flow in a complicated way. A theoretical picture is necessary for understanding the vortex dynamics in the ac flow of superfluid ^{4}He.

SUMMARY

We have described the response of a vibrating wire in superfluid ^{4}He. The turbulent flow generated by the vibrating wire exhibits various responses that vary with frequency and temperature. At 1.4 K, the turbulence vanishes at the intersection with laminar flow; on the other hand, at 30 mK the oscillation of the wire jumps upward when the turbulence ceases entirely. In both cases, entanglement and reconnection are expected to play an important role in the vortex dynamics, as well as expansion, though the interaction between the normal fluid and a vortex is expected to contribute at 1.4 K. The critical velocity at which the turbulence disappears at 1.4 K is almost constant below 2 kHz and increases with frequency above 2 kHz. The dependence might be due to the process of expansion, entanglement, and reconnection.

ACKNOWLEDGMENTS

We gratefully thank Prof. Tsubota and Prof. Vinen for various theoretical discussions.

REFERENCES

1. D. D. Awschalom and K. W. Schwarz, *Phys. Rev. Lett.* **52**, 49 (1984).
2. R. J. Donnelly, *Quantized Vortices in Helium II*, Cambridge University Press, Cambridge, 1991.
3. J. Jäger, B. Schuderer, and W. Schoepe, *Phys. Rev. Lett.* **74**, 566 (1995).
4. H. A. Nichol, L. Skrbek, P. C. Hendry, and P. V. E. McClintock, *Phys. Rev. Lett.* **92**, 244501 (2004).
5. W. F. Vinen, L. Skrbek, and H. A. Nichol, *J. Low Temp. Phys.* **135**, 423 (2004).
6. M. Niemetz, H. Kerscher, and W. Schoepe, *J. Low Temp. Phys.* **126**, 287 (2002).
7. W. Schoepe, *Phys. Rev. Lett.* **92**, 095301 (2004).
8. D. I. Bradley, *Phys. Rev. Lett.* **84**, 1252 (2000).
9. H. Yano, A. Handa, H. Nakagawa, K. Obara, O. Ishikawa, T. Hata, and M. Nakagawa, *J. Low Temp. Phys.* **138**, 561 (2005).
10. H. Yano, A. Handa, H. Nakagawa, M. Nakagawa, K. Obara, O. Ishikawa, and T. Hata, *J. Phys. Chem. Solids* **66**, 1501 (2005).
11. W. I. Glaberson and R. J. Donnelly, *Phys. Rev.* **141**, 208 (1966).
12. M. Morishita, T. Kuroda, A. Sawada, and T. Satoh, *J. Low Temp. Phys.* **76**, 387 (1989).
13. D. I. Bradley, D. O. Clubb, S. N. Fisher, A. M. Guénault, R. P. Haley, C. J. Matthews, G. R. Pickett, and K. L. Zaki, *J. Low Temp. Phys.* **138**, 493 (2005).

Cryogenic Buoyancy-Driven Turbulence

J.J. Niemela*, K.R. Sreenivasan* and R.J. Donnelly†

*ICTP, Strada Costiera 11, 34014 Trieste, Italy
†Department of Physics, University of Oregon, Eugene, Oregon 97403 USA

Abstract. Fluid turbulence is of considerable importance both fundamentally, as a paradigm for all nonlinear systems with many degrees of freedom, and in applications. In recent years there has been considerable effort to take advantage of some unique properties of low temperature liquid and gaseous helium. In particular, studies of turbulent thermal convection in conventional fluids have been aided by the use of low temperature helium which principally allows the limit of large Reynolds and Rayleigh numbers to be attained under controlled conditions. We discuss some directions and recent progress in these studies.

Keywords: convection, turbulence
PACS: 47.20.Bp;47.27.Jv;47.27.Te

INTRODUCTION

Fluid turbulence is an important consideration in the context of industrial applications such as the dispersal of pollutants, mass and heat transfer, and flows around ships and aircraft. The problem is also a paradigm for strongly nonlinear systems, distinguished by the interaction of a large number of degrees of freedom. In either context there remain many open questions, which are of consequence to a number of closely related problems such as interstellar energy transport [1], weather prediction and planetary magnetic fields [2], and, more indirectly, perhaps even market fluctuations [3]. Because of the inherent complexity of these problems, progress depends on a substantial input from experiment. This has led to a search for optimal test fluids for laboratory work in fluid turbulence, which has in turn pointed to the use of low temperature helium. The adaptation of low temperature technology to the study of classical turbulence is not, however, without difficulty, but the benefits we believe outweigh the limitations, which are mostly related to the adaptation of measurement technology.

FLUID EQUATIONS

We briefly introduce here the fluid equations written for a coordinate system fixed in space, and assuming the applicability of continuum mechanics. Let us consider a volume of fluid of density ρ and subject to a velocity $\mathbf{u}$. Conservation of mass takes the form

$$\frac{\partial \rho}{\partial t} + \nabla \cdot (\rho \mathbf{u}) = 0, \tag{1}$$

which, in the case of constant density, reduces to

$$\nabla \cdot \mathbf{u} = 0. \tag{2}$$

For a Newtonian fluid (having a linear relation between stress and strain tensors), the momentum equation reduces to

$$\frac{D\mathbf{u}}{Dt} = -\frac{1}{\rho}\nabla p + \nu \nabla^2 \mathbf{u} + \mathbf{F_{ext}}, \tag{3}$$

in terms of the pressure p, kinematic viscosity ν which is the ratio of the shear viscosity μ divided by the density ρ. For convenience of notation we have used the convective derivative $\frac{D}{Dt} = \frac{\partial}{\partial t} + \nabla \cdot \mathbf{u}$. An unspecified external body force term is represented by $\mathbf{F_{ext}}$.

For thermally driven flows, we have the buoyancy force of magnitude $F_{ext} = g\alpha\Delta T$ in the direction of the gravitational acceleration g, where α is the coefficient of thermal expansion and ΔT is the temperature difference across a layer of fluid in the direction of gravity. With the exception of variations in density that produce the buoyancy force, a simplifying Boussinesq approximation assumes that all other fluid property variations occurring as a result of the imposed thermal gradient can be neglected. The equation for energy conservation in the Boussinesq approximation is

$$\frac{\partial T}{\partial t} = \kappa \nabla^2 T - \mathbf{u} \cdot \nabla T. \tag{4}$$

Here κ is the thermal diffusivity of the fluid at temperature T. Additionally to these partial differential equations we must also supply the appropriate initial and boundary conditions; e.g., rigid or "stress-free" surfaces, conducting or insulating boundaries, rough or smooth surfaces, etc.

To motivate the principal dynamical control parameter for turbulent flows– the Reynolds number *Re*– we briefly consider isothermal flows driven solely by pressure gradients. Rescaling all velocities by some characteristic velocity $\mathbf{U}$, all lengths by some characteristic length L, times by L/U and normalizing the pressure by

CP850, *Low Temperature Physics: 24th International Conference on Low Temperature Physics;*
edited by Y. Takano, S. P. Hershfield, S. O. Hill, P. J. Hirschfeld, and A. M. Goldman

a factor ρU^2 the momentum equation (3) without the external force term can be re-written as

$$\frac{\partial \mathbf{u}}{\partial t} + \mathbf{u} \cdot \nabla \mathbf{u} = -\nabla p + \frac{1}{Re} \nabla^2 \mathbf{u}, \qquad (5)$$

where the Reynolds number is identified with $Re = UL/\nu$. It may be thought that when *Re* is very large (i.e. for turbulent flow) the viscous term in the momentum equation (the term $\nu\nabla^2\mathbf{u}$ in equation 3) becomes unimportant compared to the inertial term (the term $\mathbf{u} \cdot \nabla\mathbf{u}$). The Reynolds number can be defined of course on any scale, not just the largest. If we consider that at some scale the locally defined *Re* becomes of order unity, allowing for the domination of viscosity, we can understand conceptually the existence of a smallest scale in a turbulent flow.

It is evident from equation 5 that we do not have to take into account particular dimensioned values of the viscosities, lengths, velocities, etc. This is what allows us to connect in a general way diverse phenomena ranging from the flow of hydrogen and helium gas at astrophysical scales to, say, the the flow of blood in tiny capillaries. In the case where the geometrical factors are the same, *Re* becomes a dynamical similarity parameter and its equivalence between flows with otherwise different length and velocity scales, or different viscosities, is the principle behind wind-tunnel model testing.

In those flows where buoyancy derived from thermal gradients plays a role–which evidently occurs in nature more often than not, and in which we are mostly interested here–another control parameter, namely the Rayleigh number

$$Ra = g\alpha\Delta T L^3/\nu\kappa, \qquad (6)$$

emerges in the fluid equations. Physically, the Rayleigh number measures the ratio of the rate of potential energy release due to buoyancy with the rate of its dissipation due to thermal and viscous diffusion. Additionally, a parameter composed only of fluid properties, the Prandtl number,

$$Pr = \nu/\kappa, \qquad (7)$$

appears. Physically, *Pr* is a measure of the ratio of time scales due to thermal diffusion ($\tau_\theta = L^2/\kappa$) and momentum diffusion ($\tau_\nu = L^2/\nu$).

WHY HELIUM?

Turbulent flows occur in the limit of large *Re* and *Ra*, although there does not always exist any sharp boundary beyond which a flow becomes turbulent as one of these parameters is increased. Considering that *Re* or *Ra* in fluid turbulence are measured on a logarithmic scale, the attainment of the highest possible control parameter values is a useful target for experimental research. If suitably high *Re* or *Ra* were attainable using common fluids such as air or water, there would be little motivation for pursuing the difficult task of pushing low-temperature technology. However, this does not appear to be the case. The principal advantage of helium, as advertised numerous times (see for example ref. [4]) is its very small kinematic viscosity, even in the gas phase, leading to large *Re* without the need for large apparatus dimensions or of large velocities. Of course, it can be made to have a rather large viscosity as well in the gas phase (for very low densities). Similarly the ratio of fluid properties appearing in the Rayleigh number (see equation 6) can be orders of magnitude larger in cryogenic helium than for conventional fluids, and, conversely, can also be made rather small (also in the gas phase for very low densities). We have emphasized the capacity of low temperature helium gas in particular to produce a wide variation in the fluid properties (hence also *Re*, *Ra*) because we are often seeking to accurately measure scaling relations of some property of the turbulent flow, assumed invariant in some asymptotic limit of high *Re* or *Ra*. For this we need decades of the control parameter in the turbulent regime, which are better obtained in a single apparatus having consistent boundary conditions, protocol, etc.

RAYLEIGH-BÈNARD CONVECTION

Turning our attention now to buoyancy-driven flows, we note at the outset that it is not realistic to suppose that we can solve all real problems, which may have complicated boundary conditions, multiple phases and chemical species, magnetic fields, and so on. Consequently, most work focuses on a simpler problem for which reasonable progress can be made, but one that still contains the essential physics of the real problem in which we are ultimately interested. In the case of thermally-driven turbulence, this is the so-called Rayleigh-Bènard convection (RBC). In standard RBC, a thin, laterally infinite fluid layer is contained between two surfaces (either rigid or "stress-free") held at constant temperature. Usually the expansion coefficient is positive and so, when the fluid is heated from below (temperature decreasing from bottom to top), a mechanically unstable density gradient forms. The applied stress is measured in terms of the Rayleigh number, *Ra*, defined above. With increasing *Ra* the dynamical state goes from a uniform and parallel roll pattern at the onset of convection ($Ra \sim 10^3$) to turbulent flow at $Ra \sim 10^7 - 10^8$. For turbulent flows, the Prandtl number *Pr* determines the nature and relative sizes of the viscous and thermal boundary layers that are established on the solid surfaces.

Threlfall [6] is usually credited as being the first to recognize the advantages of using low temperature helium gas to investigate the thermal turbulence problem in the laboratory. Later low temperature work by Libchaber and co-workers [7] considerably broadened the awareness of the problem.

EXPERIMENTAL FEATURES

We briefly note some salient features of the low temperature experiments, which do not vary in substantial respects from one experiment to the next (the reader is invited to look in the various original publications [7, 8, 9, 10] for more detailed information). To approximate the constant temperature top and bottom boundaries, annealed OFHC copper is used (thermal conductivity near 1 kW/m-K at helium temperatures). Suitable heaters are attached to both plates which vary in design from experiment to experiment. On the bottom plate a constant heat current is applied while the heater at the top plate (in contact with a helium bath) is used to regulate its temperature. Typically, small (nominally 200-300 micrometer on a side) semiconductor crystals, either doped germanium or silicon, are placed in the flow to measure temperature fluctuations within the gas which is confined laterally by thin wall stainless steel walls. The sample cell is surrounded by thermal shields at various graded temperatures.

In laboratory experiments there is an additional length scale imposed by the need to laterally contain the fluid, the width-to-height aspect ratio $\Gamma = D/H$, where here D is the diameter for the more typical case of cylindrical containers. As the dynamics of thermal convection depend explicitly on H, they are significantly affected only in the limit in which the ratio of width to height is small. Having both Γ and Ra large, however, is difficult (note that H appears in the numerator of one and the denominator of the other). For absolute heights large enough to obtain high Ra, making D much larger would involve a technical and economic challenge. An exception is with the use of cryogenic helium gas, where a diameter of reasonable dimension can be coupled to a much smaller height without sacrificing too much of the upper limit of attainable Ra (recalling its H^3 dependence). Such an experiment has recently been performed [11] in aspect ratio 4. Note that while this does not represent any record for large Γ, it does represent the first attempt to obtain very high Ra in a moderately large Γ experiment. We will briefly discuss some of the preliminary observations in the context of turbulent heat transfer.

HEAT TRANSFER AND SCALING

The benefits of low temperature experiments to obtain scaling relations has been clearly evident in the context of dimensionless heat transfer, represented by the Nusselt number Nu defined as

$$Nu = \frac{q}{q_{cond}} = \frac{qH}{\lambda \Delta T}, \tag{8}$$

where q is the total heat flux, q_{cond} is the value the heat flux would have in the absence of convection, and λ is the thermal conductivity of the fluid. Nu represents the ratio of the effective turbulent thermal conductivity of the fluid to its molecular value and can reach values of over 10^4 in helium experiments [8] thus demonstrating the enormous enhancement of thermalization (one of the motivations for Threlfall's pioneering work in cryogenic turbulent convection–see ref. [6]).

THE CLASSICAL RESULTS

To partially motivate the push to larger Γ it is useful to briefly consider the expected scaling between Nu and Ra. The "classical" prediction (see, e.g., refs. [12, 13]) is a power law relation $Nu \sim Ra^{\beta}$ with $\beta = 1/3$. This value for the exponent can be easily motivated by considering that the thermal gradients occur only in very thin diffusive boundary layers near each horizontal heated surface, and that the intervening fluid, being fully turbulent, acts more or less like a thermal short circuit (the reader might be convinced of this scaling by taking the general power law relation above and assuming the physical heat flux Q appearing in Nu to have no implicit height dependence). In the limit of infinite Pr at least, where the problem becomes more tractable, the exponent $\beta = 1/3$ has recently been rigorously established [14]. The other so-called "classical" relation is a power law (albeit with logarithmic corrections) with $\beta = 1/2$, mostly due to the work of Kraichnan [15]. A modern theory [16] presents a $Ra - Pr$ phase portrait with different power law relations in neighboring regions combining to influence Nu, including separate relations with the two values of β given above.

SCALING RESULTS FROM EXPERIMENTS

Experiments have in fact generally measured exponents less than 1/3; for instance, Libchaber and co-workers championed the exponent 2/7 and subsequent theory which seemed to pin it down [7], although this is no longer considered the correct asymptotic limit [16]. Others have observed larger scaling exponents closer to 1/2 at high Ra [9], albeit for operating points quite close to the critical point of the fluid. The highest Ra obtained to date was in an experiment of Niemela *et al.* [8], in which, to lowest order, an exponent of 0.31 was observed over 10 decades of Ra up to $Ra \sim 10^{17}$. The main experimental results of each of these groups were obtained in $\Gamma = 1/2$ containers, which more or less represents an historically accepted minimum. However, Γ may play a significant role, at least for small values of it, through the action of a robust and organized "wind" which sweeps through the entire container [17, 18, 19]. Low temperature experiments have subsequently been performed [10] in a container of $\Gamma = 1$ for Rayleigh numbers up to $Ra \sim 10^{15}$. One of the interesting results at $\Gamma = 1$ was that at high enough Ra for the mean wind to have become significantly disordered, but low enough for the Boussinesq approximation to remain valid, roughly a decade of power-law scaling with $\beta = 1/3$ emerged.

The furthest step in the direction of fully turbulent RBC in a laterally extended system has only been done recently [11], in which a $\Gamma = 4$ container was used to obtain Ra up to 10^{13} (note that the highest Ra is sacrificed with increase in aspect ratio as the latter is practically effected by lowering the height while maintaining the diameter fixed). In this $\Gamma = 4$ experiment a mean wind also existed but was considerably less robust than in the smaller aspect ratio containers. A scaling exponent of approximately 0.31 was observed at low Ra (albeit in the turbulent regime) as in the case of Niemela *et al.* [8], but was followed by a transition to almost 2 decades of scaling consistent with $\beta = 1/3$ over the ultimate range of Ra for which the Boussinesq approximation could be assumed valid. Corresponding to this transition in scaling, the long-time correlation between temperature probes separated by a diameter vanished. Thus it appears as though the $\Gamma = 4$ experiments may approach the kind of conditions postulated for observing the $\beta = 1/3$ regime; namely, a more disordered bulk flow that separates the action of the two thermal boundary layers. Presumably, the heat transfer is influenced less by the mean wind, which otherwise sweeps more robustly through the entire cell in small-Γ containers, affecting Nu by its coupling to gradients in the imperfectly insulating sidewalls [20, 21, 10, 22]. It should also be mentioned that the exponent $\beta = 1/3$ is consistent with the corresponding region of the $Ra - Pr$ phase space according to ref. [16].

DISCUSSION

The $\Gamma = 4$ experiment in turbulent convection discussed above represents a different direction from that pursued in the past for helium experiments; namely the simultaneous attainment of both large Ra and large Γ, thus examining more accurately the problem of turbulent RBC. It should be possible to expand even further in this direction.

It is worth pointing out, however, that recent advances in numerical modelling have enabled the effects of finite Γ to be properly accounted for at high Ra. We note that there continues to be steady and substantial progress in this direction (R. Verzicco, personal communication).

REFERENCES

1. Spiegel, E.M. *Ann. Rev. Astron. Astrophys.* **9**, 323-352 (1971).
2. Glatzmaier, G.A., Coe, R.C., Hongre, L. & Roberts, P.H. *Nature* **401**, 885-890 (1999).
3. Mandelbrot, B.B. A Multifractal Walk Down Wall Street *Scientific American* February, 77-73 (1999).
4. Niemela, J.J. "Thermal turbulence in cryogenic helium gas" *Physica B* **329-333**, 429-430 (2003).
5. Frisch, U. *Turbulence, the Legacy of A. N. Kolmogorov*(Cambridge University Press, Cambridge, England) (1995).
6. Threlfall, D.C. *J. Fluid. Mech.* **67** 17-28 (1975).
7. Castaing, B., Gunaratne, G., Heslot, F., Kadanoff, L., Libchaber, A., Thomae, S., Wu, X.Z., Zaleski, A. & Zanetti, G. *J. Fluid Mech.* **204**, 1-30 (1989).
8. Niemela, J.J., Skrbek, L., Sreenivasan, K.R. & Donnelly, R.J. *Nature* **404**, 837-840 (2000).
9. Chavanne, X., Chilla, F., Chabaud, B., Castaing, B. & Hebral, B. *Phys. Fluids* **13**, 1300-1320 (2001).
10. Niemela J.J. & Sreenivasan K.R. *J.Fluid Mech.* **481** 355 (2003) .
11. Niemela, J.J. & Sreenivasan K.R. *to appear in J. Fluid Mech.* (2005).
12. Malkus, W.V.R. *Proc. R. Soc. London* **A 225**, 196-212 (1954).
13. Howard, L.N. *Proc. Eleventh International Congress on Applied Mechanics* (Ed. H. Görtler) Berlin: Springer, pp. 1109-1115 (1966).
14. Constantin, P. & Doering, C. R., *J. Stat. Phys.* **94**, 159Ű172 (1999).
15. Kraichnan, R.H. *Phys. Fluids* **5**, 1374-1389 (1962).
16. Grossmann, S. & Lohse, D. *Phys. Rev. E.* **66**, 016305 (2002).
17. Niemela, J.J., Skrbek, L., Sreenivasan, K.R. & Donnelly, R.J. *J. Fluid Mech.* **449**, 169-178 (2001).
18. K. R. Sreenivasan, A. Bershadskii and J. J. Niemela, Phys. Rev. E **65**, 056306 (2002).
19. Verzicco R. & Camussi, R. *J. Fluid Mech.* **477**, 19 (2003).
20. Ahlers, G. *Phys. Rev. E.* **63**, art. no. 015303 (2001).
21. Roche, P., Castaing, B., Chabaud, B., Hebral, B. & Sommeria, J. *Euro. Phys. J.* **24**, 405-408 (2001).
22. Verzicco, R. *J. Fluid Mech.* **473**, 201 (2002).

PIV Measurements of He II Counterflow Around a Cylinder

S. Fuzier[1], S. W. Van Sciver[1,2], and T. Zhang[3]

[1] National High Magnetic Field Laboratory, Florida State University, Tallahassee, FL 32310 USA
[2] Mechanical Engineering Department, FAMU-FSU College of Engineering, Tallahassee, FL 32310 USA
[3] GE Global Research Center, One Research Circle, EP 123, Niskayuna, NY 12309 USA

Abstract. The induced flow field of counterflow He II across a circular cylinder has been quantitatively studied using the particle image velocimetry (PIV) technique. Two different size cylinders (6.35 mm and 2 mm in diameter) were used and placed in a 20 mm wide rectangular channel. In these experiments, large-scale eddy motion generated by the He II counterflow was observed both in front of and behind the cylinder, an effect which has no analogue in classical fluids.

Keywords: superfluid helium, thermal counterflow, flow visualization
PACS: 67.40.Vs

INTRODUCTION

The two-fluid model describes He II as if it were consisting of two interpenetrating fluid components, the viscous normal fluid and the inviscid superfluid [1]. The normal fluid component contains the thermal excitations while the superfluid component carries no entropy. The relative concentration of the two fluid components varies continuously between T_λ = 2.176 K, where the fluid is all normal component, and absolute zero, where it is pure superfluid.

A heat current in He II is carried by the normal fluid flows at a velocity, v_n, given by the relation,

$$q = \rho s T v_n \qquad (1)$$

with q being the applied heat flux, ρ being the total fluid density and s the specific entropy. A special case is that of zero net mass flow where the bulk fluid is at rest and the normal fluid flow is balanced with flow of the superfluid component in the opposite direction. Under this condition, the superfluid velocity v_s is,

$$v_s = -\frac{\rho_n}{\rho_s} v_n \qquad (2)$$

where ρ_n and ρ_s are respectively the normal and superfluid densities. This mechanism is called thermal counterflow.

Of interest are experiments that can visualize thermal counterflow in He II in order to understand the velocity fields of the two fluid components.

EXPERIMENT

Recently, we have been applying the particle image velocimetry (PIV) technique to the study of thermal counterflow in He II. With PIV, one seeds the He II with micron sized particles and tracks their motion to obtain a full field map of the flow. For classical fluids, the particle motion is closely correlated with the bulk fluid velocities. However, in thermal counterflow, it appears that the particle mainly tracks the normal fluid flow, but may also be affected by the turbulent superfluid component. For counterflow in a one dimensional channel, this complex interaction is evident by the particles moving at roughly half the normal fluid velocity [2].

The present work is an extension of our recent study [3] of counterflow in more complex geometries. Here we place circular cylinders in a one dimensional rectangular channel, which is 200 mm long and 38.9×19.5 mm^2 in cross section. The channel is immersed in a constant temperature He II bath with the top end open to the bath while the lower end is closed by a Nichrome thin film heater. The side walls of the channel have optical windows for illumination of the particles with a laser sheet. The front wall has a 45×20 mm^2 optical window for imaging the particles with a CCD camera. A transparent cylinder is located inside the channel spanning the full width and orthogonal to the flow. In the widest part of the cylinder, the channel cross section is therefore reduced. Two different size circular cylinders have been used, 6.35 mm and 2 mm in diameter. Thus, for the larger cylinder the flow cross section of the channel is reduced by about 1/3

CP850, *Low Temperature Physics: 24th International Conference on Low Temperature Physics;*
edited by Y. Takano, S. P. Hershfield, S. O. Hill, P. J. Hirschfeld, and A. M. Goldman

while for the smaller cylinder by only 10%. The resulting flow fields are compared to determine how they are affected by the ratio of cylinder to channel dimension. In particular, we are interested in whether the large turbulent eddies seen both in front of and behind the larger cylinder are also present for the smaller cylinder [3].

RESULTS

Figure 1 contains examples of the observed counterflow field around the two cylinders. These are representative data which actually fluctuate in time and are therefore only typical. They were acquired at 2.03 K and for a heat flux of 7.2 kW/m^2 directed upward, which corresponds to v_n = 23 mm/s. Clearly, in both cases, turbulent structures occur both in front of and behind the cylinder. However, a noticeable effect for the small cylinder, which is less evident for the large cylinder, is the tendency for the large vortices in front of the cylinders to cling more closely to the channel walls than do those behind the cylinder. Although we have no theoretical explanation for these structures, it is speculated that the large vortices in front of the cylinder are the result of some form of flow separation occurring in the turbulent superfluid component. However, since the normal fluid component interacts more strongly with the suspended particles [2], it is possible that the transverse velocity component of v_n as it flows around the cylinder could push these vortices toward the channel walls. Further work is required to confirm this suggestion.

In classical fluid flow, the Reynolds number is normally used to scale the turbulence. For thermal counterflow, the Reynolds number is normally defined as,

$$\mathrm{Re}_D = \frac{\rho v_n D}{\mu_n} \tag{3}$$

where μ_n is the normal fluid viscosity and, D is the cylinder diameter in this case. For the results in Figure 1, the corresponding Reynolds numbers are 13288 and 4185 respectively, scaling with the cylinder diameter. It is worth noting that these values of Reynolds number in classical fluids would result in vortex shedding, where the vortices detach from the cylinder and produce a wake [4]. No such effect is seen in the present results suggesting a fundamental difference between classical and counterflow turbulence around a cylinder.

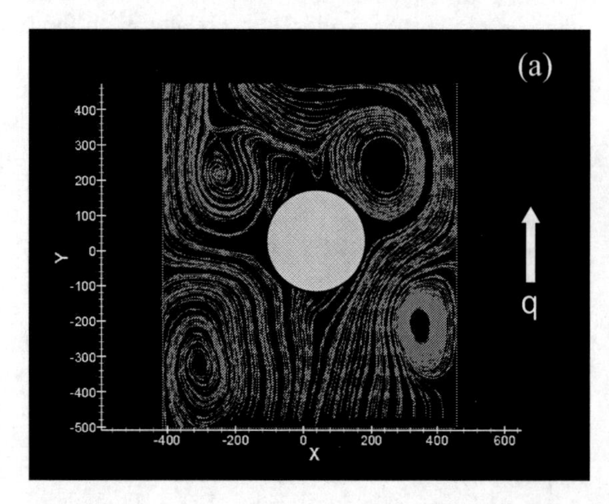

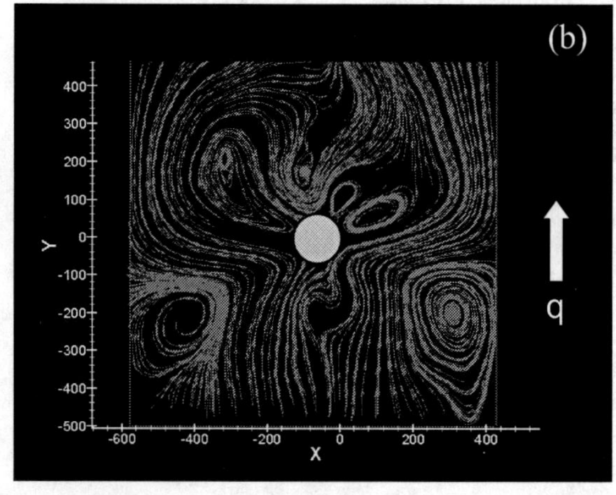

FIGURE 1. Comparison of turbulent structures for two different cylinder diameters (a) D = 6.35 mm and (b) D = 2 mm. Both cases are at T = 2.03 K and q = 7.2 kW/m^2 of channel cross section which corresponds to normal fluid velocity of v_n = 23 mm/s and Reynolds numbers of 13288 and 4185 respectively.

ACKNOWLEDGMENTS

The authors would like to acknowledge the help of Dr. L. Lourenco and Mr. S. Maier during these experiments. This work is supported by the Department of Energy and the Florida State University – Office of Research.

REFERENCES

1. D. Landau, *J. Phys.* (*U.S.S.R.*) **5**, 71 (1941).
2. T. Zhang and S. W. Van Sciver, *J. Low Temp. Phys.* **138**, 865-870 (2005).
3. T. Zhang and S. W. Van Sciver, *Nature Physics* **1**, 36-38 (2005).
4. H. Schlichting, *Boundary Layer Theory*, New York: McGraw-Hill, 1951, pp. 28-32.

Vibrating Grid as a Tool for Studying the Flow of Pure He II and its Transition to Turbulence

D. Charalambous*, P.C. Hendry*, L. Skrbek†, P.V.E. McClintock* and W.F. Vinen**

*Department of Physics, University of Lancaster, Lancaster LA1 4YB, UK
†Faculty of Mathematics and Physics, Charles University, Ke Karlovu 3, 121 16 Prague, Czech Republic
**School of Physics and Astronomy, University of Birmingham, Birmingham B15 2TT, UK

Abstract. We report a detailed experimental study of the flow of isotopically-pure He II, generated by a vibrating grid. Our measurements span a wide range of temperatures ($50\,\mathrm{mK} \leq T \leq 1.37\,\mathrm{K}$) and pressures ($2\,\mathrm{bar} \leq p \leq 15\,\mathrm{bar}$). The response of the grid was found to be of a Lorentzian form up to a sharply-defined threshold value. This threshold value does not change appreciably with pressure; the form of the resonant response of the grid is qualitatively the same for all temperatures while the threshold value is a monotonically increasing function of temperature. We discuss the measured variation of the resonant frequency of the grid as a function of applied pressure (density) of He II and relate this to a hydrodynamic effective mass of the grid. These measurements extend our previously reported studies [Nichol *et al*, *Phys. Rev. E* **70**, 056307 (2004)] and form an integral part of a series of experiments aimed at providing a better understanding of classical and quantum turbulence.

Keywords: helium, superfluidity, turbulence, oscillators
PACS: 47.27.Cn, 47.37.1q, 67.40.Vs, 47.15.Cb

The study of turbulence in quantum fluids [1] has attracted considerable attention, both from experimental and theoretical physicists. It has already been established that, although quantum turbulence (QT) involves a tangle of quantized vortex lines, the macroscopic flow of He II exhibits classical features such as the existence of an energy decay spectrum consistent with that proposed by Kolmogorov for turbulence in classical fluids [2].

We have recently demonstrated that a vibrating grid in He II can be used to study the transition to turbulence, in the zero temperature limit [3]. Here we present further measurements on the flow of He II induced by a vibrating grid (using a different grid but of the same specifications as that used in [3]), for a range of temperatures and pressures. Details of the experimental setup and procedure used can be found in [3].

The response of the grid was first studied in vacuum in order to establish that a constant temperature had been reached, as well as to compare the resonance characteristics with those obtained with the grid immersed in isotopically-pure He II. The resonance characteristic in vacuum was found to be of a Lorentzian shape, centered at a frequency of 1036.9(1) Hz and exhibiting a quality factor of $Q \sim 4000$ [4].

The resonance characteristics of the grid were also found to be Lorentzian in He II for low drive amplitude, but with the resonant frequency reduced by approximately 40 Hz—more about this below. When the response of the grid exceeds a pressure-independent threshold, however, the resonance curves broaden markedly, signaling the onset of nonlinearity and additional dissipation in the system. Shown in Fig. 1 is the variation of the response of the grid at resonance as a function of the applied driving voltage, for temperatures ranging from 350 mK to 1370 mK. All the data were obtained at a pressure of 15 bar. It can readily be seen that, for low driving voltages, the response of the grid is proportional to the driving force.

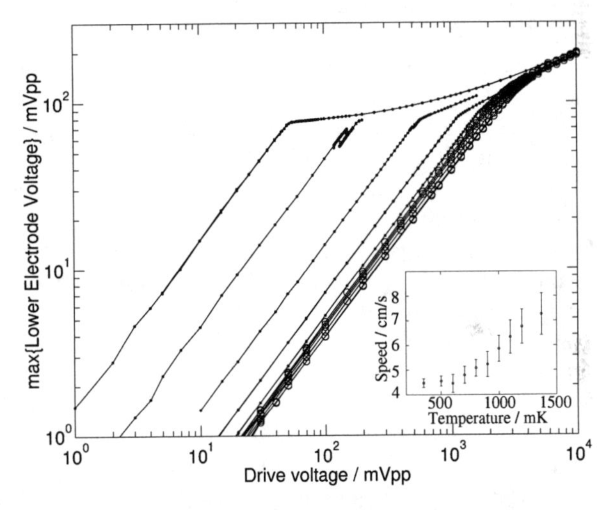

FIGURE 1. The variation of the response of the grid at resonance as a function of the driving voltage, for a range of temperatures at a pressure of 15 bar. Shown in the inset is the resonant response at the onset of dissipation, extracted from the data plotted in the main graph.

CP850, *Low Temperature Physics: 24th International Conference on Low Temperature Physics;*
edited by Y. Takano, S. P. Hershfield, S. O. Hill, P. J. Hirschfeld, and A. M. Goldman

In relation to the nonlinear regime, recall that the drag coefficient C_D is defined as [5]

$$C_D = \frac{F}{\frac{1}{2}\rho u^2 A}, \qquad (1)$$

where F is the magnitude of the drag force on the grid, A is the area of the grid, ρ the density of He II and u is the grid speed which can be calculated [3] from the amplitude of the signal picked up on the lower electrode. It is also known that for steady turbulent flow of a classical fluid at very large Reynolds number, the drag coefficient is roughly constant and hence the grid speed should be proportional to the square root of the drag force. The data plotted in Fig. 1 indicate that this might be the case in the present system, but we are unable to provide more conclusive evidence as this would require the application of considerably larger drive voltages to the grid, with the risk of causing electrical breakdown and hence damage to the system. An unexpected and unexplained observation (Fig. 1 inset) is that the maximum grid speed at onset of dissipation is a monotonically increasing function of temperature. Further experiments are in progress in order to further explore this phenomenon.

We have also measured the resonant frequency as a function of pressure p and hence density ρ of He II. Before discussing the results, we note that any submerged body moving in a fluid acquires an effective mass equal to its "bare" mass plus a quantity Δm proportional to the mass of fluid displaced [5]. We may thus write $\Delta m = \beta V \rho(p)$, where β is a mass enhancement factor, V is the volume occupied by the body and $\rho(p)$ is the density of the fluid at pressure p. The factor β can be calculated exactly for bodies of certain geometrical shape; for instance, it can be shown that $\beta = \frac{1}{2}$ for a sphere, $\beta = 1$ for an infinite circular cylinder and $\beta = 3$ for an elliptical cylinder of axes ratio 3, oscillating in the direction of its minor axis. For the present experimental system, it is straightforward to show [4] that

$$\frac{1}{f_1^2(p)} \approx \frac{1}{f_0^2}\left(\frac{\beta}{\rho_{\mathrm{Ni}}}\rho(p) + 1\right), \qquad (2)$$

where f_1 (f_0) refers to the resonant frequency of the nickel grid in He II (vacuum) and ρ_{Ni} is the density [6] of the grid. Shown in Fig. 2 is a plot of $1/f_1^2$ against ρ within the linear regime of weak forcing in the temperature-independent range near $T \sim 20$ mK. Also shown are two straight-line fits to the data: the continuous line fit was obtained by taking all data points into account whereas the dashed line fit was calculated after excluding the point for which $\rho = 0$ (vacuum). Shown in the inset is a magnified view of the region around the points for which $\rho \neq 0$, for clarity. Note that the continuous line fit gives $\beta \approx 4.4(2)$ and the corresponding value for the dashed line fit is $\beta \approx 3.1(1)$. It can be seen

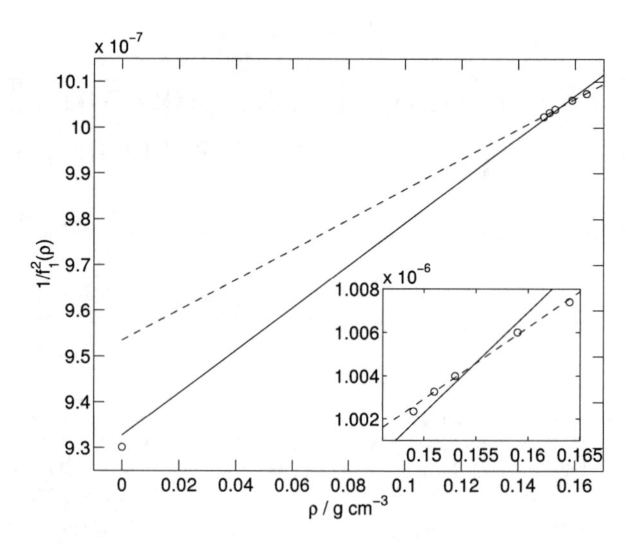

FIGURE 2. The measured resonance frequency of the grid as a function of the density of He II; the value of the density was extracted from the applied pressure using the data provided by Brooks & Donnelly [6]. The dashed line is a straight-line fit to the points excluding the one at zero density; the continuous line is a similar fit taking into account all data points. Note that the errorbars are too small to show on this scale.

that the data points of Fig. 2 do not all lie on a straight line, as one might expect from Eq. (2). Instead, the points for which $\rho \neq 0$ do appear to form an approximately straight line; if this is indeed the case, then one might conclude that the vibrating grid exhibits an additional effective mass renormalization which cannot be accounted for by the above analysis. We conjecture that this additional effective mass enhancement might be due to an effective "boundary layer" formed by vortex loops pinned to the vibrating grid.

We would like to thank A.M. Guénault & M. Giltrow for their invaluable assistance during the experiments; this work is part of the research plan MS 002162083 financed by the Ministry of Education of the Czech Republic and the EPSRC.

REFERENCES

1. Vinen, W. F., and Niemela, J. J., *J. Low Temp. Phys.*, **128**, 167–231 (2002).
2. Skrbek, L., and Stalp, S., *Phys. Fluids*, **12**, 1997–2019 (2000).
3. Nichol, H. A., Skrbek, L., Hendry, P. C., and McClintock, P. V. E., *Phys. Rev. E*, **70**, 056307 (2004).
4. Charalambous, D., Skrbek, L., Hendry, P. C., McClintock, P. V. E., and Vinen, W. F., submitted to *Phys. Rev. E* (2005).
5. Landau, L. D., and Lifshitz, E. M., *Fluid Mechanics*, Butterworth-Heinemann, Oxford, 1984, 2nd edn.
6. Brooks, J. S., and Donnelly, R. J., *J. Phys. Chem. Ref. Data*, **6**, 51 (1977).

Is Quantized Vorticity in Pure He II at Low Temperature Directly Related to Cavitation and Spinodal Pressure?

L. Skrbek

Joint Low Temperature Laboratory, Institute of Physics ASCR and Charles University, V Holesovickach 2, 180 00 Prague, Czech Republic

Abstract. We argue that the critical velocity for intrinsic nucleation of quantized vortices in isothermal flow of He II at low temperature can be viewed as approaching the spinodal limit in pressure and breakdown of superfluidity as a consequence of the Bernoulli equation. Breaking the liquid by cavitation that changes the topology from simply to multiply connected seems an essential requirement for intrinsic vortex nucleation and serves as an additional criterion of superfluidity, of the form $V_c = [2(p_{ext} - p_{sp})/\rho_s]^{1/2}$, where p_{ext} is the external pressure, p_{sp} denotes the spinodal limit, and ρ_s stands for the superfluid density. This criterion can be viewed as additional to the well-known Landau criterion for breakdown of superfluidity due to emission of quasiparticles.

Keywords: superfluidity, He II, quantized vortices
PACS: 67.40.Vs, 47.55.Bx

INTRODUCTION

The flow of He II can be described using the two-fluid model, assuming He II consists of two interpenetrating fluids: a viscous normal fluid of density ρ_n and velocity $\vec{v}_n$ carrying all the entropy and an inviscid superfluid of density ρ_s and corresponding velocity $\vec{v}_s$; the total density of He II is $\rho = \rho_s + \rho_n$. In case of isothermal flow and absence of quantized vortices the velocity fields $\vec{v}_n$ and $\vec{v}_s$ may be assumed to be independent. Under such conditions the flow of the superfluid obeys the Euler equation, which for steady flow (or relatively steady flow, such as motion of a sphere through a stationary fluid that can be regarded as steady by including constant velocity superimposed on the system [1]) becomes the Bernoulli equation

$$\frac{1}{2}\rho_s v_s^2 + p = \text{const}, \tag{1}$$

where p denotes the pressure. Therefore, for a superfluid at rest and in the absence of gravity we have that $p = \text{const}$. At very low temperature, where the normal fluid density is negligibly small, one can consider the flow of the single-component inviscid superfluid with quantized circulation. The circulation is zero in a simply connected region and is quantized in units of $\kappa \cong 0.997 \times 10^{-3}$ cm^2/s in multiply connected regions; the latter is a consequence of the condensate wave function being single-valued. The vorticity, defined as $\vec{\nabla} \times \vec{v}_s$, is zero everywhere in the bulk except inside the cores of quantized vortices that are believed to be of Ångstrom size and where superfluidity is broken [2]. We attempt to provide additional insight on the breakdown of superfluidity as well as on the extremely small size of the vortex core.

NEGATIVE PRESSURES AND QUANTIZED VORTICITY

Let us consider a singly-quantized rectilinear vortex in the zero temperature limit. The superfluid streamlines are concentric circles, the superfluid velocity being given by $v_s(r) = \kappa/(2\pi r)$, where r is the distance from the vortex axis. We ask how close, based on the Bernoulli Eq. 1, one can approach the axis before the liquid breaks.

At low temperature, say below 0.5 K, liquid He II exists at external pressures ranging from the saturated vapour pressure ($p_{sv} \approx 0$ bar) up to the solidification pressure ($p_{sol} \approx 25$ bar). Let us suppose that the external pressure is close to p_{sv}. Approaching the core of the vortex, the Bernoulli equation requires that the pressure should become negative, at a finite distance from the core. Do we expect something drastic to happen? In fact, no, because He II is a liquid, not a gas.

In liquid He II the local pressure can be both positive and negative—this occurs in many other liquids [3]. Negative pressure (which is nothing but a positive stress) corresponds to a metastable state: if a bubble of sufficiently large size forms, it will be able to grow without limit and the liquid breaks—this is known as *cavitation*. This metastable state exists thanks to the energy barrier that opposes nucleation of the gas phase and is a consequence of the liquid-gas transition being of first order. The maximum negative pressure at which the liquid ultimately breaks is called the *spinodal limit*. How close one

CP850, *Low Temperature Physics: 24th International Conference on Low Temperature Physics;* edited by Y. Takano, S. P. Hershfield, S. O. Hill, P. J. Hirschfeld, and A. M. Goldman

can approach the spinodal limit depends on various factors such as temperature, the purity of the liquid and/or the presence of boundaries; unless special care is taken liquids usually break much earlier [5].

For details on cavitation in the quantum liquids ^{4}He and ^{3}He, and for references to original works, see the excellent review by Balibar [3]. In the zero temperature limit the spinodal limit is believed to be around −9 bar. For now, let us accept this value and continue our discussion on the rectilinear singly quantized vortex.

Assuming naively that the superfluid density stays constant close to the vortex core, equal to $\rho \cong 0.145$ g/cm^3 at the saturated vapour pressure $p \simeq 0$; and $\rho \cong 0.172$ g/cm^3 at the highest available external pressure $p \simeq 25$ bar [4], the isotopically pure liquid may not break until the spinodal pressure $p_{sp} \cong -9$ bar is reached at the vortex core radius, which can be estimated as

$$r_c(p) \cong \frac{\kappa}{2\pi}\sqrt{\frac{\rho_s(p,r)}{2(p-p_{sp}-p_{gas})}} \approx \frac{\kappa}{2\pi}\sqrt{\frac{\rho_s(p)}{2(p-p_{sp})}} \quad (2)$$

At the zero temperature limit which is of our concern here we can safely neglect the gas pressure p_{gas}. Within the available pressure range, we get that the core size is bound between $r_c(0) \approx 1.42$ Å down to $r_c(25\text{ bar}) \approx 0.73$ Å, in surprising agreement with the commonly accepted values [2]. Note that a better approximation must take into account that the superfluid density decreases with decreasing external pressure and it is likely that it will behave so even in the range of negative pressures, although the exact behaviour is at present not known.

Breaking He II in the T → 0 limit

So far we have considered the existing quantized vortex and showed that if its core is of the commonly accepted size the liquid there must break. The next step is to ask ourselves how to create a vortex in an initially vortex-free sample of pure He II; in other words, what are the conditions for intrinsic vortex nucleation? The crucial requirement is to change the topology of the sample from simply connected to multiply connected. To do so, it is necessary to break the liquid along some path. This will *certainly* happen if the pressure along such a path reaches the spinodal pressure. For a flow that satisfies the Bernoulli equation this pressure corresponds to a velocity v_{max}^{crit} given by

$$v_{max}^{crit} \approx \sqrt{\frac{2(p_{ext}-p_{sp})}{\rho_s(p)}}. \quad (3)$$

For a pure sample of He II in the zero temperature limit this condition supplements the Landau criterion for superfluidity, that is the (phenomenological) condition for creating elementary excitations (rotons and phonons) which is based on the form of the energy spectrum. Note that for the entire range of external pressures this critical velocity satisfies 111 m/s $\leq v_{max}^{crit} \leq$ 199 m/s. It represents the upper limit for *intrinsic* vortex nucleation for a pure sample of He II moving with respect of a hydrodynamically smooth submerged surface. Any real surface in contact with He II, however, ought to be considered as rough because He II possesses an Ångstrom size healing length, characterizing the distance over which the macroscopic condensate wave function can appreciably change. The surface roughness causes additional (but unknown) enhancement of the flow velocity over its excrescences.

Let us stress here that the above approach is not applicable to superfluid ^{3}He; the latter is known to possess much larger vortex cores which can contain different superfluid phases. The critical velocities for intrinsic vortex nucleation are much lower, of the order of a few mm/s in superfluid ^{3}He-B [6]. It is also not clear at present up to what degree this approach might be used for He II at higher temperature, especially in the vicinity of the lambda-point, due to the two-fluid behaviour of He II.

To conclude, we have considered the striking similarity between physical phenomena of intrinsic vortex nucleation in pure He II at low temperature and homogeneous cavitation, needed to break the liquid in order to change the topology of the originally vortex-free sample from simply connected to multiply connected. Based on the Bernoulli equation, this simple idea leads to a criterion of superfluidity (Eq. 3) that supplements the famous Landau criterion for superfluidity.

ACKNOWLEDGEMENTS

The author acknowledges stimulating discussions with a number of colleagues, especially S. Balibar, D. Charalambous, P.C. Hendry, P.V.E. McClintock, W.F. Vinen and G.E. Volovik. This research is supported by the Institutional Research Plan AVOZ10100520 and by GAČR 202/05/0218.

REFERENCES

1. L.M. Milne-Thomson, *Theoretical hydrodynamics*, New York: Dover Publications, 1996.
2. R.J. Donnelly, *Quantized vortices in He II*, Cambridge: Cambridge University Press, 1991.
3. S. Balibar, *J. Low Temp. Phys.* **129**, 363–421 (2002).
4. J.S. Brooks and R.J. Donnelly, *J. Phys. Chem. Ref. Data* **6**, 51 (1977).
5. H. Maris and S. Balibar, *Physics Today* **53**, 29 (2000).
6. V.M.H. Ruutu, U. Parts, J.H. Koivuniemi, N.B. Kopnin and M. Krusius, *J. Low Temp. Phys.* **107**, 93 (1997).

Turbulence in He II Generated by Superflow

T.V. Chagovets, M. Rotter, J. Sindelar, F. Soukup, and L. Skrbek

Joint Low Temperature Laboratory, Institute of Physics ASCR and Charles University, Faculty of Mathematics and Physics, V Holesovickach 2, 180 00 Prague, Czech Republic

Abstract. We report experimental investigations of He II turbulence and its decay, generated by pure superflow through a channel of square crossection. Both ends of the channel are covered by sintered-silver superleaks, which prevent any net normal fluid flow through the channel. The steady-state turbulence is generated by applying a power to the heater immersed in He II in a volume adjacent to one of the superleaks having an outlet above the helium bath level in the glass cryostat. When the power is switched off, the superflow through the channel and the vortex line density in the middle of the channel decay. The steady-state vortex line density and the form of its decay is continuously measured by second sound attenuation and analyzed within available theoretical models.

Keywords: quantum turbulence, superfluidity
PACS: 67.40.Vs

INTRODUCTION

The He II turbulence generated either by a towed grid or in a counterflow channel by applying a heat pulse to its closed end as well as its subsequent decay has been a subject of investigation by many authors (see reviews [1, 2] and references therein). A purely classical phenomenological spectral model [3, 4] was shown to describe most of the decaying He II turbulence in a channel of finite size. Perhaps surprisingly, this feature was experimentally found not only for the decaying grid turbulence [4], but also for late decay of the counterflow turbulence [5, 6].

Although the character of the decay does not appreciably change with temperature while the normal fluid to superfluid density ratio varies by a factor of ten or so, the role of the normal fluid (e.g., its profile in counterflow) remains largely unknown. In order to shed more light on this issue, we designed and utilized an experimental apparatus where He II turbulence is generated by superflow, inside a channel 13 cm long and of square cross-section (6×6 mm^2). Both ends of the channel are conical and smoothly connected (via a thin In o-ring) to sintered silver superleaks, about 2 mm thick and 16 mm in diameter. They have been sintered *in situ* inside a short brass tube; their large diameter and filling factor of about 1/2 ensure that superfluid can freely enter and exit the flow channel. He II turbulence is thus generated by a pure superflow through the channel; there is no net normal fluid flow through it. This superflow is set by applying a heat to heater H1 (see Fig. 1). The upward moving superfluid is partly converted into the normal fluid above the channel and leaves the heater volume as a fountain streaming into the helium bath.

The schematic block diagram of the apparatus is shown in Fig. 1. To detect the vortex line density we use the second sound attenuation technique based on the gold plated membranes [5]. The set of preliminary results we are reporting here has been obtained in a glass helium cryostat that allowed simple visual control of the helium bath level with respect to the outlet of the flow channel.

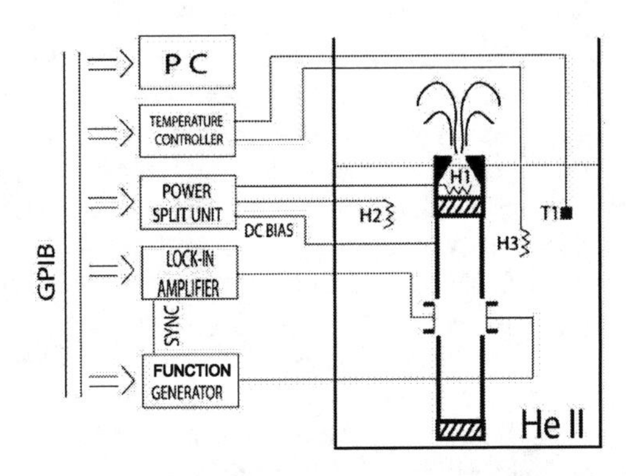

FIGURE 1. Schematic block diagram of the experiment, showing the main components used in measurements. Besides the second sound transducer/reciever across the middle of the channel, there is a thermometer and heater H3 used to control the temperature of the helium bath. The overall power distributed by PC via the home-made electronic power split unit between heaters H1 and H2 is kept constant, in order to maintain the temperature stability of the helium bath.

RESULTS

Figure 2 displays the steady-state and decaying vortex line densities measured at T=1.72 K and calculated assuming that the vortex tangle is random [6].

When a power of order 0.5 W is applied to heater

CP850, *Low Temperature Physics: 24th International Conference on Low Temperature Physics;* edited by Y. Takano, S. P. Hershfield, S. O. Hill, P. J. Hirschfeld, and A. M. Goldman

H1, the originally high second sound amplitude abruptly (within ≈ 0.2 s) decreases and stabilizes to a reproducible steady-state value. We keep the heater on for typically 10-15 s (the last two seconds of this state is included in Fig. 2) and at the instant marked as $t = 0$ the power is switched off.

After the initial fast decay, irrespectively of the steady state starting value of the vortex line density L, the subsequent decay can be described as exponential, of the form $L \propto \exp(-t/t_0)$ with the characteristic decay time $t_0 \approx 2$ s. We have not yet investigated how this decay depends on temperature; this particular set of decay data has been taken at experimental conditions when the bath temperature was held at T=1.72 K.

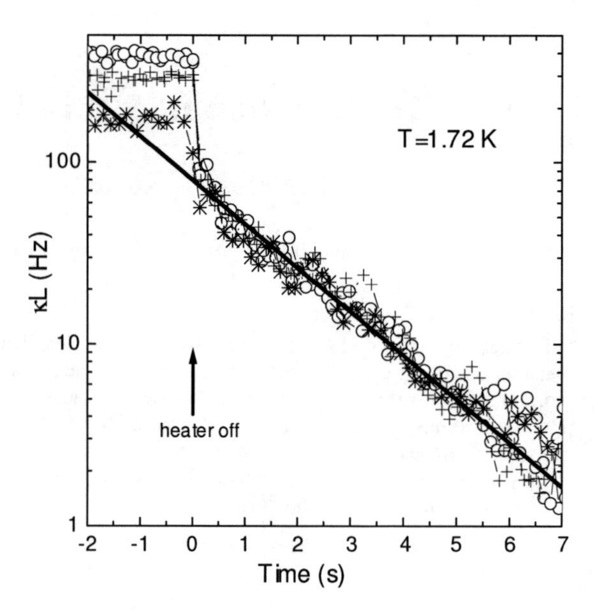

FIGURE 2. Steady state and decaying vortex line density (assuming that the vortex tangle is random) generated by pure superflow measured in the middle of the channel of a 6×6 mm^2 crossection. The steady state heat inputs 0.484 W (o), 0.388 W (+) and 0.295 W ($*$) applied to the heater placed above the upper superleak are switched off at $t = 0$. The solid line is a plot of an exponential function 80 e^{-t/t_0} with $t_0 = 1.8$ s.

DISCUSSION

The observed exponential decay is distinctly different from the decay of He II turbulence generated in counterflow, where the decay of L over most of the time can be characterized by a power law with an exponent of -3/2 [5, 6]. An exponential decay of L in He II turbulence was previously observed, but only as a very late stage – a low vortex line density part – of the complex decay of the grid generated turbulence, consisting of four distinctly different regimes [4]. In a later paper [7] it was shown that the exponential decay is consistent with a spectral energy density of the form $\Phi(k) = C\varepsilon\kappa^{-1}k^{-3}$, if the energy containing length scale is assumed to be limited by the size of the channel. Here C is the Kolmogorov constant, ε is the energy decay rate and κ denotes the quantum of circulation. Beyond the quantum scale $\ell_q = 2\pi(\varepsilon/\kappa^3)^{-1/4}$, the normal and superfluid eddies cannot be matched, due to quantized circulation in the superfluid. It seems plausible therefore that the exponential decay (Fig. 2) suggests that normal and superfluid eddies do not match even for larger length scales and much higher values of L.

In some cases it turned out to be favourable to use the home-made low-noise current to voltage converter as a preamplifier together with the SR 830. It eliminates the parasitic capacitance of the coaxial cable, while keeping the current noise level at 0.1 pA/$\sqrt{\text{Hz}}$.

Theoretical considerations of superfluid turbulence in a stationary normal fluid [8-10] lead to a steeper energy spectrum than the classical Kolmogorov $k^{-5/3}$, characterized in this case by a roll-off exponent up to -3. A steeper energy spectrum is generally caused by the dissipative mutual friction force, acting on all relevant length scales. According to the predicted flow phase diagram for helium superfluids [8, 11], depending on the way how the superfluid turbulence is generated, two different turbulent phases with different decay mechanisms may exist. This idea seems consistent with several types of experimentally observed turbulent states, historically marked as T1, T2, and T3 [2], in which the normal component may or may not be turbulent. Ongoing experiments with various channels at a range of temperatures ought to shed more light on this issue.

We thank J. Dupak for help with sintering the superleaks and W.F. Vinen for discussion. This research is supported by the Institutional Research Plan AVOZ10100520 and by GACR 202/05/0218.

REFERENCES

1. W.F. Vinen and J.J. Niemela, *J. Low Temp. Phys.* **128**, 167 (2002).
2. J.T. Tough, "Superfluid turbulence", in *Prog. in Low Temp. Phys.*, Vol. 8, North-Holland Publ. Co., (1982).
3. L. Skrbek and S.R. Stalp, *Phys. Fluids* **12**, 1997 (2000).
4. L. Skrbek, J.J. Niemela and R.J. Donnelly, *Phys. Rev. Lett.* **85**, 2973 (2000).
5. L. Skrbek, A.V. Gordeev and F. Soukup, *Phys. Rev.* **E67**, (2003) 047302.
6. A.V. Gordeev, T.V. Chagovets, F. Soukup and L. Skrbek, *J. Low Temp. Phys.* **138**, 549 (2005).
7. L. Skrbek, J.J. Niemela and K.R. Sreenivasan, *Phys. Rev.* **E64** (2001) 067301.
8. G.E. Volovik, *J. Low Temp. Phys.* **136**, 309 (2004).
9. V.S. L'vov, S.V. Nazarenko and G.E. Volovik, *JETP Letters* **80**, 479 (2004).
10. W.F. Vinen, *Phys. Rev.* **B71**, 024513 (2005).
11. L. Skrbek, *JETP Letters* **80**, 484 (2004).

Decay of Capillary Turbulence on the Surface of a Semiquantum Liquid

M. Yu. Brazhnikov*, G. V. Kolmakov*, A. A. Levchenko*, P. V. E. McClintock† and L. P. Mezhov-Deglin*

*Institute of Solid State Physics RAS, Chernogolovka, 142432, Russia
†Department of Physics, Lancaster University, Lancaster, LA1 4YB, UK

Abstract. We study the free decay of capillary turbulence on the charged surface of liquid hydrogen. Contrary to expectations based on the existing self-similar theory of nonstationary wave turbulent processes in ideal liquid, we find that decay begins from the *high frequency* end of the spectral range, while most of the energy remains localized at low frequencies. We show that finite damping of the waves changes qualitatively the character of the turbulent decay. Numerical calculations based on this idea agree well with the experimental data.

Keywords: Quantum liquids, surface waves, turbulence
PACS: 47.27.-i, 47.25.+i

It is known that the nonlinear interaction of capillary waves on the surface of liquid hydrogen is relatively strong, and that liquid hydrogen is a perfect fluid for studies of the turbulence on liquid surface [1, 2]. The possibility of driving the charged surface of liquid hydrogen directly through the application of electrical forces provides an additional advantage.

We report below the results of our studies of the decay of turbulence in a system of capillary waves on the surface of liquid hydrogen. We have observed that decay of the stationary turbulent spectrum starts in the high frequency domain, with the energy remaining localized in the low frequency range of the turbulent spectrum, i.e. near the driving frequency. At low frequencies the turbulent spectrum remains close to its unperturbed shape for a relatively long period of time after the driving force is switched off. This observation is in a sharp contrast to predictions of the self-similar theory of non-stationary wave turbulent processes (see the review [3]), where evolution of the spectrum is considered in the range of frequencies, at which viscous damping can be neglected. According to this theory, the decay of capillary turbulence should start with the damping of low-frequency surface waves, caused by the cascade transfer of energy to higher frequency scales.

The experimental arrangements were similar to those used in our earlier studies of steady-state turbulence on the charged surface of liquid hydrogen [4]. The measurements were made using an optical cell inside a helium cryostat. Hydrogen was condensed into a cup formed by a bottom capacitor plate and a guard ring 60 mm in diameter and 6 mm high. The layer of liquid was 6 mm thick. The top capacitor plate (a collector 60 mm in diameter) was located at a distance of 4 mm above the surface of the liquid. A two-dimensional positively charged layer was created just below the surface of the liquid. The temperature of the liquid was 15.5 K. The waves on the charged surface were excited by a periodic driving voltage applied between the guard ring and the upper electrode. The surface oscillations were detected from the variation of the total power $P(t)$ of a laser beam reflected from the surface, which was measured with a photodetector, sampled with an analogue-to-digital converter, and stored in a computer. Given the size of the light spot, the correlation function $I_\omega = \langle |\eta_\omega|^2 \rangle$ of the surface elevation $\eta(\mathbf{r},t)$ in frequency representation is directly proportional to the squared modulus of the Fourier transform of the detected signal, $I_\omega = \mathrm{const}\, P_\omega^2$ at frequencies above 50 Hz. The instantaneous power spectrum P_ω^2 of the nonstationary surface oscillations was calculated by application of a short-time Fourier transform [5] to the measured signal $P(t)$.

To establish the steady turbulent state at the surface of the liquid, an ac driving voltage at a frequency ω_p was applied for $\sim$10 s (on Fig. 1 the driving frequency is $\omega_p/2\pi = 97$ Hz). After the driving voltage was switched off, we could observe relaxation oscillations of the surface (Fig. 1 (b) and (c)).

It is clearly evident from Fig. 1 that, during decay of the turbulence, it is the *high frequency* components of the power spectrum that are damped first. The amplitude of the main peak at the driving frequency remains larger than the amplitudes of peaks at the harmonics all the time, both before and after the driving force is switching off; i.e. the surface of the liquid continues to oscillate mainly at the driving frequency. The wave amplitude in the turbulent distribution decreases homogeneously during the decay, and the power-law dependence of the spectrum persists, even for low frequencies down to 100 Hz.

CP850, *Low Temperature Physics: 24th International Conference on Low Temperature Physics;*
edited by Y. Takano, S. P. Hershfield, S. O. Hill, P. J. Hirschfeld, and A. M. Goldman

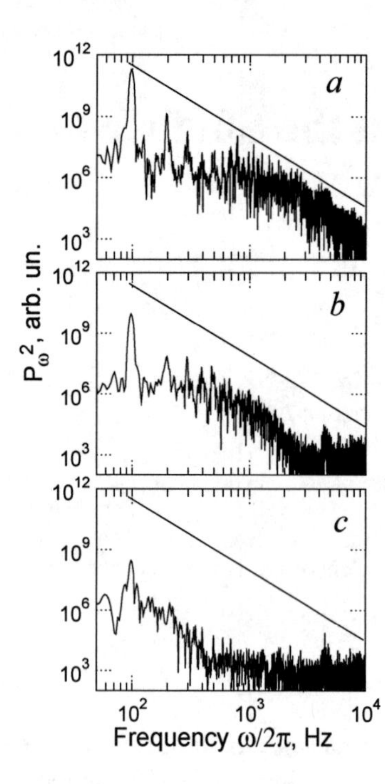

FIGURE 1. Instantaneous spectra of surface oscillations calculated for different moments of time: (a) $t = -0.18$ s, i.e. before the removal of the driving force, (b) $t = 0.44$ s and (c) $t = 1.07$ s. Time $t = 0$ corresponds to the moment when the driving force was switched off. The straight line corresponds to the power-law dependence $P_\omega^2 \sim \omega^{-7/2}$.

To understand better the reason of the apparent discrepancy between our observation and the theoretical predictions we have performed numerical calculations of the capillary turbulence decay, using the kinetic equations for waves (see [3]), in which a viscous damping of waves at all frequencies has been taken into account in addition to the kinetic intergal.

Figures 2 (a) and (b) show the results of our numerical calculations of decay of the turbulent spectra after removal of the narrow-band driving force, for two moments of time corresponding to experimental observations shown in plots (a) and (b) in Fig. 1.

From the results obtained we can claim that, for understanding of nonstationary turbulent processes at the surface of liquid hydrogen, viscous losses in a turbulent system are of central importance at all frequencies above the driving frequency. We have seen, both in experiments and in computations, that finite damping of the waves changes qualitatively the character of the turbulent decay: rather than a propagation of perturbation from low to high frequencies, caused by the cascade transfer of energy (the scenario considered in [3]), there is a relatively fast decay of the high frequency domain of the spectrum. Note that these considerations do *not* apply to stationary turbulent phenomena, where dissipation can be neglected completely over an inertial range of frequencies [6].

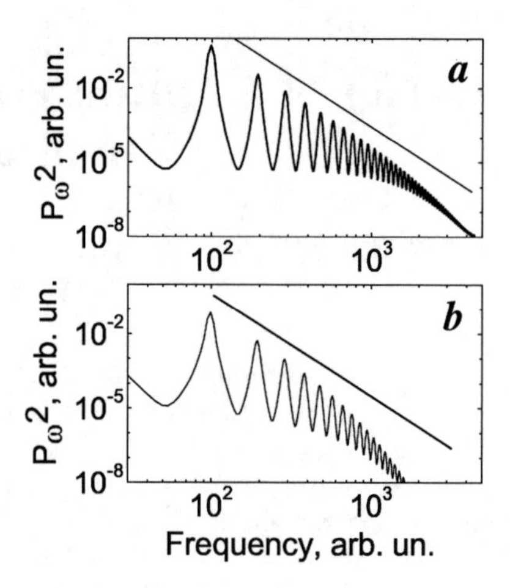

FIGURE 2. Results of our numerical calculations of decay of the turbulent spectra after removal of the narrow-band driving force. Plots (a) and (b) are calculated for moments of time corresponding to the experimental observations shown in plots (a) and (b) on Fig. 1.

ACKNOWLEDGMENTS

The authors are very grateful to V.E. Zakharov, E.A. Kuznetsov, M.T. Levinsen and A.N. Silchenko for many useful discussions, and to V.N. Khlopinskii for assistance in preparation of the experiments. The investigations are supported in part by RFBR (grant 05-02-17849), by the Ministry of Education and Science of RF (program "Quantum macrophysics"), and by the Engineering and Physical Sciences Research Council, UK. M.Yu.B. acknowledges a support from the Science Support Foundation, Russia.

REFERENCES

1. M.Y. Brazhnikov, G.V. Kolmakov, A.A. Levchenko, and L.P. Mezhov-Deglin, *JETP Lett.* **73**, 398 (2001).
2. G.V. Kolmakov, A.A. Levchenko, M.Yu. Brazhnikov, L.P. Mezhov-Deglin, A.N. Silchenko, and P.V.E. McClintock, *Phys. Rev. Lett.* **93**, 074501 (2004).
3. V. Zakharov, V. L'vov, and G. Falkovich, *Kolmogorov Spectra of Turbulence*, Vol. 1, Berlin: Springer, 1992.
4. M.Y. Brazhnikov, A.A. Levchenko, and L.P. Mezhov-Deglin, *Instrum. Exp. Tech.* **45**, 758 (2002).
5. S. Mallat, *A Wavelet Tour of Signal Processing*, New York: Academic Press, 1997.
6. A. N. Pushkarev and V. E. Zakharov, *Phys. Rev. Lett.* **76**, 3320 (1996).

Shielded Superconducting Linear Motor for Towed-Grid Studies of Quantum Turbulence

Shu-chen Liu, Yihui Zhou, and Gary G. Ihas

Department of Physics, University of Florida, Gainesville, FL 32611-8440, USA

Abstract. A motor is described which can pull a grid through a channel of pure superfluid ^{4}He to produce homogeneous isotropic turbulence. The motor is composed of a superconducting solenoid inside a superconducting shield to minimize Joule and eddy current heating of the liquid helium. Computer simulations show the design to be feasible.

Keywords: Quantum turbulence, towed-grid, superconducting linear motor, superconducting shield.
PACS: 67.40.Vs

Quantum turbulence produced by a towed grid in superfluid ^{4}He at very low temperature is proposed to decay not by viscosity (lacking normal fluid), or by the mutual friction between the superfluid and normal fluid, but by phonon radiation when energy flows into the smaller length scale of a Kelvin-wave cascade on the quantized vorticity [1, 2]. We propose a calorimetric technique to explore this energy decay mechanism.

To produce homogeneous isotropic turbulence (HIT) by towing a grid through superfluid helium, we have designed a superconducting linear motor in a superconducting shield. This design can avoid eddy current heating in the cell walls and other parts, and Joule heating from the solenoid coil which is immersed in the helium. To produce the HIT the grid must accelerate from v=0 to v=1 m/s over 1 mm, and move for the distance of 10 mm at a nearly constant speed of 1.0 m/s, then decelerate over 1 mm.

The motor is shown schematically in Fig. 1, which is made of three coaxial parts: one superconducting shield with the radius b and length h, one superconducting solenoid with the inner radius r and length l, and a light insulating rod with two niobium cylinders attached and separated by distance ΔS. The grid is attached to the end of the rod and hence is pushed by it.

By applying a properly shaped current pulse to the solenoid, the magnetic interaction with the niobium cylinders produce the required motive forces, with the upper cylinder accounting for the initial acceleration and the lower cylinder the deceleration of the grid.

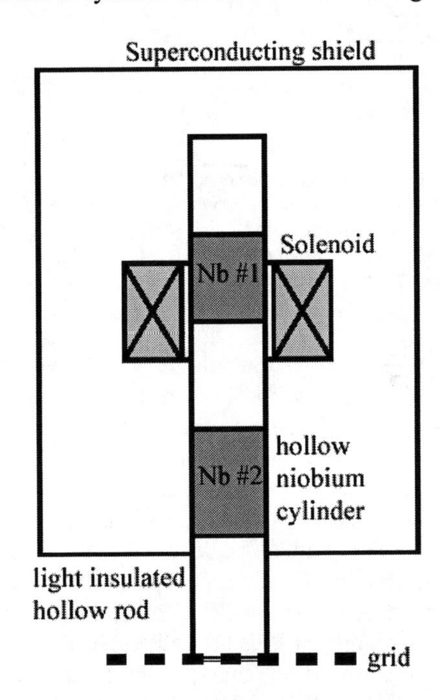

FIGURE 1. Shielded superconducting motor.

In order to find the best motor design, we have developed a LabView simulator program, in which the various design parameters may be varied while maintaining the required motion of the grid. Optimization in general yielded the lowest voltage and current applied to the solenoid, hence minimizing the magnetic field produced.

CP850, *Low Temperature Physics: 24th International Conference on Low Temperature Physics;*
edited by Y. Takano, S. P. Hershfield, S. O. Hill, P. J. Hirschfeld, and A. M. Goldman

The z component of the magnetic field at the position (ρ, θ, z) near the solenoid is derived as the following by generalizing the problem in Griffiths [3]:

$$B_z = \sum_{m=1}^{l/d} \sum_{n=1}^{Nd/l} \frac{\mu_0 IR}{4\pi} \int_0^{2\pi} d\phi \times$$

$$\frac{R - x\cos\phi - y\sin\phi}{\left[\rho^2 + R^2 + (z - z_0)^2 - 2xR\cos\phi - 2yR\sin\phi\right]^{3/2}} \quad (1)$$

$$\text{where } x = \rho\cos\theta, y = \rho\sin\theta, \quad (2)$$

$$R = r + \left(\frac{2n-1}{2}\right)d, z_0 = -\frac{\ell}{2} + \left(\frac{2m-1}{2}\right)d \quad (3)$$

For the magnetic field distribution inside the solenoid enclosed by the superconducting shield, we cite Eq. 12 in Sumner [4]. This process yielded the optimized parameters of Table 1 (units in mm).

TABLE 1. Optimal parameters* for the motor design (mm).

Parameter	Value	Parameter	Value	Parameter	Value
d	0.2	N	1500	r	5.0
l	10.0	z_0	3.5	r_{o1}	4.0
r_{o2}	4.0	l_1	9.0	l_2	11.0
ΔS	21.0	h	50.0	b	20.0

* d: the diameter of the superconducting wire; N: the number of turns of the solenoid; z_0: the initial position of Nb cylinder 1; r_{o1}: the outer radius of Nb cylinder 1; r_{o2}: the outer radius of Nb cylinder 2; l_1: the length of Nb cylinder 1; l_2: the length of Nb cylinder 2; ΔS: the distance between niobiums; the mass of the whole system 2.0 g.

Figure 2(a) shows the required current versus time curves for the unshielded and superconducting shielded solenoid with a sine function acceleration (velocity is the sine function of the niobium position). The curves show three peaks: the first and the third peaks are to accelerate and decelerate the niobium cylinders. The middle peak is due to the almost balanced magnetic forces on the two niobium cylinders at that position since each niobium cylinder is almost equidistant from the ends of the solenoid. We can run the simulator removing the middle peak (o-symbols) in Fig. 2(a), and see that the velocity versus position curve in Fig. 2(b) (again o-symbols) is virtually unaffected. The droop in velocity after z=10 mm is unavoidable since all forces, including gravity, are directed downward for the rest of the stroke.

Note that the superconducting shield requires a slightly higher current (0.14 A greater) to produce the same motion. The effect is small because the shield is significantly larger than the solenoid.

In the velocity versus position curves in Fig. 2(b), we see that the motion is as expected. The niobium cylinders travel at almost constant speed, 1 m/s, for 10 mm, but start to slow down when the middle current peak occurs, which is quite reasonable. After the first niobium passes z=10 mm, the second niobium is closer to the solenoid and experiences a stronger magnetic force in the opposite direction, resulting in the slight deceleration. Therefore, applying the third pulse produces the desired deceleration to rapidly stop the grid. The evaluation results prove that our superconducting linear motor is a very feasible design.

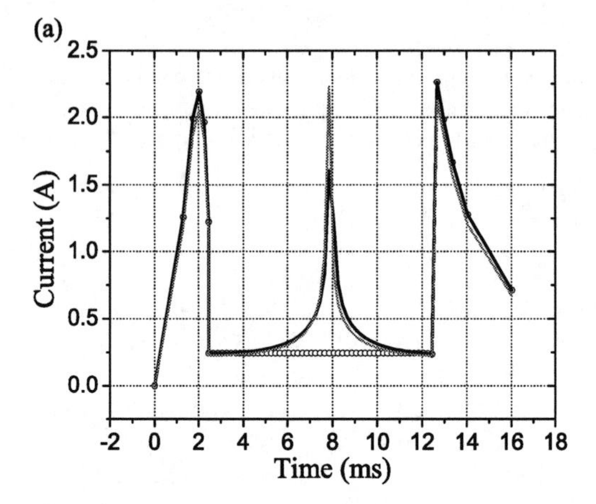

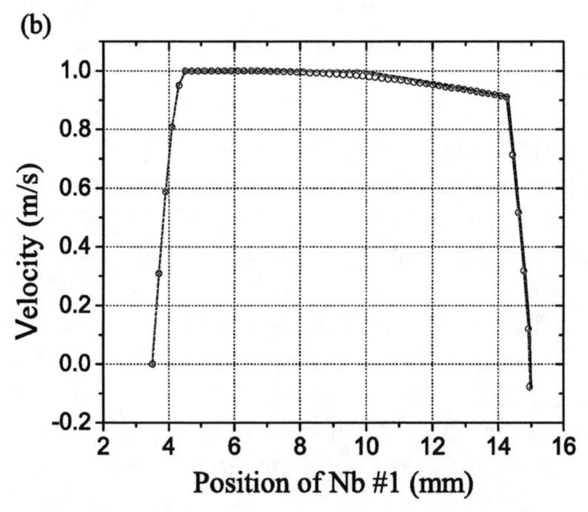

FIGURE 2. Unshielded (gray curves) and superconducting shielded (black curves) solenoid under the sine function acceleration: (a) I(t) curve; (b) v(z) curve. (○: modified data).

REFERENCES

1. W. F. Vinen and J. J. Niemela, *J. Low Temp. Phys.*, **128**, 167 (2002).
2. W. F. Vinen, M. Tsubota and A. Mitani, *J. Low Temp. Phys.*, **134**, 457 (2004).
3. D. J. Griffiths, *Introduction to Electrodynamics, 2nd ed.*, New Jersey: Prentice-Hall, 1989, pp. 213.
4. T. J. Sumner, *J. Phys. D: Appl. Phys.*, **20**, 692 (1987).

Experimental Studies of Decay and Formation of Capillary Turbulence on the Surface of Liquid Hydrogen

M. Yu. Brazhnikov, G. V. Kolmakov, A. A. Levchenko and L. P. Mezhov-Deglin

Institute of Solid State Physics RAS, Chernogolovka, 142432, Russia

Abstract. We present the results of investigations of transformation of the stationary turbulent cascade of capillary waves on the surface of liquid hydrogen after switching off the additional pumping force with the frequency lying below the frequency of the main pumping. It was observed that the additional pumping suppressed the high-frequency turbulent oscillations.

Keywords: Quantum liquids, capillary waves, turbulence
PACS: 47.27.-i, 47.25.+i

It is known that the turbulent behavior of capillary waves on the surface of a liquid can be described within the framework of weak turbulence theory [1]. In this approach the statistical properties of an ensemble of surface waves are characterized by "occupation numbers" n_ω for wave modes of frequency ω, obeying the kinetic equation

$$\frac{\partial n_\omega}{\partial t} = St[n_\omega] - 2\gamma_\omega n_\omega + F_{\text{drive}}(t) \qquad (1)$$

Here $St[n_\omega]$ is the collision integral, $\gamma_\omega = 2\nu(\rho/\sigma)^{1/3}\omega^{4/3}$ is the coefficient of viscous damping of the capillary waves, $F_{\text{drive}}(t)$ is the external low frequency driving force, ν is the coefficient of viscosity, σ is the surface tension, and ρ is the density of the liquid. For capillary waves, whose dispersion relation $\omega = (\sigma/\rho)^{1/2}k^{3/2}$ is of a decay type, it is the interaction of three surface waves that makes the main contribution to the collision integral (i.e. the decay of one wave into two waves, and the opposite process of confluence of two waves into one wave).

In the inertial frequency range (i.e. much higher than the driving frequency, but lower than the frequency scale where the dissipation plays the dominant role in the wave dynamics) the turbulent behavior of capillary waves is determined mainly by their nonlinear interaction. For wide-band pumping of the liquid surface the Kolmogorov turbulence spectrum $n_\omega \sim Q^{1/2}\omega^{-15/6}$ becomes established, where the energy flux Q is directed towards the high frequency domain. In the case of narrow-band pumping the frequency dependence should be changed to $n_\omega \sim \omega^{-19/6}$. It might be of definite interest to study the dynamics of transformation of the Kolmogorov spectrum by changing the frequency spectrum of the driving force. And this question is one of the main goals of our present studies.

It is known that the nonlinear interaction of capillary waves on the surface of liquid hydrogen is relatively strong, and that liquid hydrogen is a perfect fluid for studies of the turbulence on liquid surface. Results of our previous investigations of the stationary turbulent spectrum and decay of capillary turbulence on the surface of liquid hydrogen were presented in [2, 3].

We report below the results of our studies of evolution of the shape of spectrum and formation of a new stationary distribution of the capillary turbulence after switching off the additional pumping force with the frequency lying below the frequency of the main pumping. The additional pumping at low frequencies increases the nonlinearity and the energy flux from low frequencies to the high frequency domain. So, according to the predictions of the theory [1], we could expect the broadening of the inertial interval of capillary turbulence. But our further experimental investigations and numerical simulations have shown that the additional pumping leads to *suppression* of high-frequency turbulent oscillations and *narrowing* of the inertial interval.

The experimental arrangements were similar to those used in our earlier studies of steady-state turbulence on the charged surface of liquid hydrogen [4]. The measurements were made using an optical cell placed inside a helium cryostat. Hydrogen was condensed into a cup formed by a bottom capacitor plate and a guard ring 60 mm in diameter and 6 mm high. The layer of liquid was 6 mm thick. The top capacitor plate (a collector 60 mm in diameter) was located at a distance of 4 mm above the surface of the liquid. A radioactive target was attached to the lower plate of the capacitor. Positive electric voltage applied to the lower plate extracts positive ions from the ionized fluid layer adjacent to the target surface and draws them to the surface of liquid hydrogen, so that a quasi-twodimensional charged layer is formed under the surface. The temperature of the liquid was 15.5 K. The waves on the charged surface were excited by a periodic driving voltage applied between the guard ring and the upper electrode. The possibility of driving the charged surface of liquid hydrogen directly

CP850, *Low Temperature Physics: 24th International Conference on Low Temperature Physics;*
edited by Y. Takano, S. P. Hershfield, S. O. Hill, P. J. Hirschfeld, and A. M. Goldman

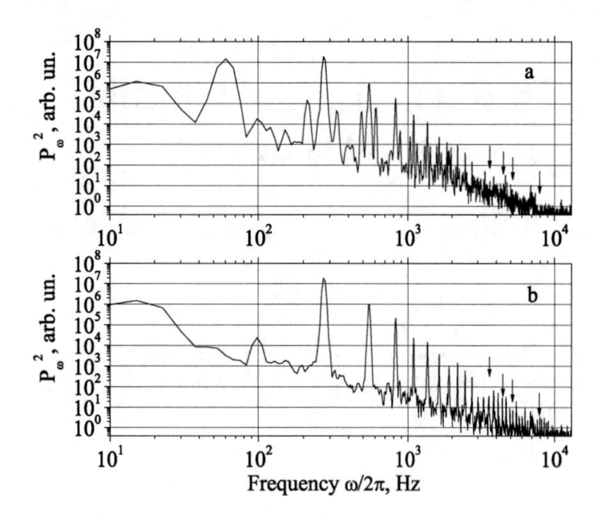

FIGURE 1. Stationary spectra of surface oscillations: (a) simultaneous drive at 61 and 274 Hz, (b) drive at a single frequency of 274 Hz.

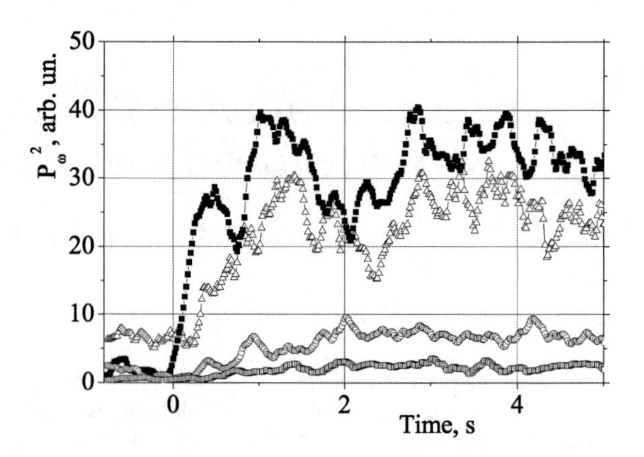

FIGURE 2. Time dependences of the squared amplitude of high-frequency surface oscillations. Triangles – 3.57 kHz; filled squares – 4.39 kHz; circles – 5.19 kHz; squares – 7.96 kHz. The additional drive was switched off at $t = 0$.

with an electrical force provides an advantage for our method capillary-wave generation.

The surface oscillations were detected from the variation of the total power $P(t)$ of a laser beam reflected from the surface, which was measured with a photodetector, sampled with an analogue-to-digital converter, and stored in a computer. Given the size of the light spot, the correlation function $I_\omega = \langle|\eta_\omega|^2\rangle$ of the surface elevation $\eta(\mathbf{r},t)$ in frequency representation is directly proportional to the squared modulus of the Fourier transform of the detected signal, $I_\omega = \mathrm{const}\, P_\omega^2$ at frequencies above 100 Hz. The instantaneous power spectrum P_ω^2 of the nonstationary surface oscillations was calculated by application of a short-time Fourier transform to the measured signal $P(t)$.

In this study, a steady turbulent state at the surface of the liquid was generated first by an ac driving voltage consisting of two frequencies $\omega_1/2\pi = 61$ Hz and $\omega_2/2\pi = 274$ Hz (Fig. 1a). Then the driving force at 61 Hz (additional pumping) was switched off, while the amplitude of pumping at 274 Hz was held constant, and the turbulent cascade was observed to relax to a new steady distribution (Fig. 1b). By comparing the two spectra, one can see an appearance of new high frequency oscillations in the range of ≈7–10 kHz after the additional drive was turned off.

Figure 2 shows evolution of the high-frequency turbulent oscillations (marked by arrows in Fig. 1) with time after the additional drive was turned off. The amplitude of the oscillations grows to several times its original value.

The increase in the amplitudes of the high-frequency turbulent oscillations can be qualitatively attributed to a decrease in the density of states (number of harmonics excited in unit frequency range) involved in the nonlinear energy transfer from low- to high-frequency oscillations in the inertial frequency range. When the additional pumping is turned off, the amplitudes of the oscillations at this frequency, as well as at multiple harmonics and combination frequencies, decrease rapidly. This damping gives rise to the redistribution of oscillation energy over frequencies, which is manifested in the evolution of the turbulent spectrum and in a noticeable increase in the amplitudes of waves with frequencies that are multiples of the main pumping frequency ω_2.

ACKNOWLEDGMENTS

The authors are very grateful to V.E. Zakharov and E.A. Kuznetsov for many useful discussions, and to V.N. Khlopinskii for assistance in preparation of the experiments. This work is supported in part by RFBR (grant 05-02-17849), by the Ministry of Education and Science of RF (program "Quantum macrophysics"), and by the Engineering and Physical Sciences Research Council, UK. M.Yu.B. acknowledges a support from the Science Support Foundation, Russia.

REFERENCES

1. V. Zakharov, V. L'vov, and G. Falkovich, *Kolmogorov Spectra of Turbulence*, Vol. 1, Berlin: Springer, 1992.
2. M.Y. Brazhnikov, G.V. Kolmakov, A.A. Levchenko, and L.P. Mezhov-Deglin, *JETP Lett.* **73**, 398 (2001).
3. G.V. Kolmakov, A.A. Levchenko, M.Yu. Brazhnikov, L.P. Mezhov-Deglin, A.N. Silchenko, and P.V.E. McClintock, *Phys. Rev. Lett.* **93**, 074501 (2004).
4. M.Y. Brazhnikov, A.A. Levchenko, and L.P. Mezhov-Deglin, *Instrum. Exp. Tech.* **45**, 758 (2002).

Simple Pinning Model for Vibrating Wire Turbulence in Superfluid Helium at Zero Temperature

R. Hänninen, A. Mitani and M. Tsubota

Department of Physics, Osaka City University, Sugimoto 3-3-138, Sumiyoshi-ku, Osaka 558-8585, Japan

Abstract. We present here a simple model to examine the effect of oscillating superfluid velocity on a quantized vortex. This situation can be realized in vibrating wire or vibrating grid experiments where the vortices are attached to the oscillating wires. Within this model we determine the critical velocity, above which the vortex motion becomes irregular and new vortices are created. We also determine the vortex creation rates.

Keywords: superfluid helium, vortex, turbulence, vibrating wire
PACS: 67.40.Vs

INTRODUCTION

Superfluid turbulence has been under active investigation during the last years. Traditionally, experimental work has concentrated on the counter-flow turbulence but recently there have been new measurements using vibrating wire, grids or spheres, where one generates a rapidly oscillating flow around the moving object [1–4]. Quantized vortices (few of them always present in the superfluid) move with the flow and, when the velocities are high enough, develop into a turbulent vortex tangle. Unfortunately there are no numerical simulations that would describe the behavior of vortices under oscillating flow and their effect on the measured quantities. Partly this is because the modeling of the experiment is not so simple and the resulting numerics take a lot of time. Here we would like to present a simple pinning model as a first step towards more realistic models.

THEORY AND MODEL

Quantized vortices in superfluid helium are very stable topological defects. One may consider them as one-dimensional line defects, since the radius of the vortex core is so small, $a \approx 0.1$ nm for He-II and $a \approx 10$ nm for ^{3}He-B [5]. In principle, the movement of these vortex lines gives all the required information to interpret the experiments.

The calculations presented here have been done using the vortex filament model that has been described by Schwarz [6, 7]. We consider He-II at temperatures below 1 K where the normal component is practically absent and mutual friction can be neglected. In this case any point on the vortex, denoted by $\mathbf{s}$, moves at the local superfluid velocity given by the familiar Biot-Savart equation

$$\dot{\mathbf{s}} = \frac{\kappa}{4\pi}\int \frac{(\mathbf{s}_1 - \mathbf{s}) \times d\mathbf{s}_1}{|\mathbf{s}_1 - \mathbf{s}|^3} + \mathbf{v}_a(t). \qquad (1)$$

Here $\kappa = 2\pi\hbar/m_4$ is the quantum of circulation and $\mathbf{v}_a$ is the applied, oscillating superfluid velocity.

Spatially homogeneous oscillating flow has no real effect on bulk vortices. However, in experimental situations there are two effects that make the vortex behave unexpectedly. First, a vortex near the wire/grid feels inhomogeneous flow. For example, frictionless flow around a cylinder has two stagnation points and maxima on the cylindrical surface. The second important fact is that a vortex must terminate at a boundary or form a closed loop. A vortex that ends on wire is not allowed to move totally freely, since the end point must stay on the wire.

Our model concentrates on this pinning effect. As a limiting case we fix the starting and ending points of the vortex and observe how the vortex responds to the oscillating flow, applied perpendicular to the line determined by these points. This corresponds to the limit where a vortex is put between two spheres, and we take the limit where the radius of the spheres is reduced to zero, or more generally, is much smaller than their separation.

RESULTS

The first task is to choose a correct separation D between the pinning points, since this greatly affects the obtained critical velocities. We have done calculation by using the values of 5 μm and 50 μm, both in principle appropriate for vibrating wire and grid experiments [8]. Simple scaling arguments can be used to obtain approximate results for other values [7].

CP850, *Low Temperature Physics: 24th International Conference on Low Temperature Physics;* edited by Y. Takano, S. P. Hershfield, S. O. Hill, P. J. Hirschfeld, and A. M. Goldman

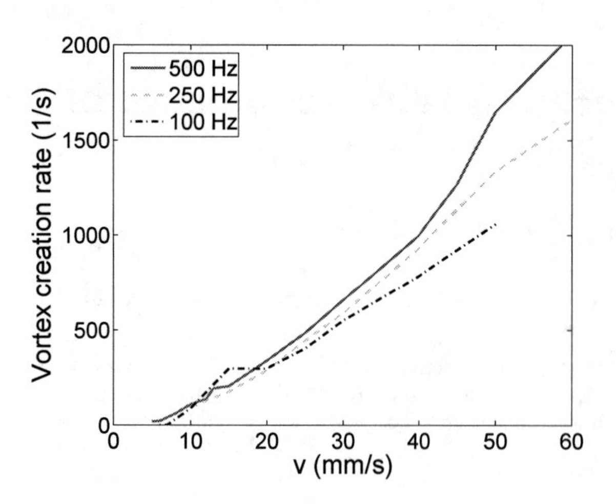

FIGURE 1. Vortex creation rate as a function of driving velocity for different driving frequencies when the separation between the pinning points is 50 μm (corresponding to a Kelvin resonance frequency of 60 Hz).

These calculations were performed using the full Biot-Savart equation and using a space resolution of $0.02D$. Reconnections were assumed to occur when the vortices approach each other closer than the space resolution. For the $D = 5\ \mu$m case, the frequencies used were below the lowest Kelvin resonance frequency, 4.9 kHz. For $D = 50\ \mu$m we used frequencies of 100, 250 and 500 Hz which are above the main resonance frequency, 60 Hz.

At low driving velocities the vortex just oscillates between the pinning points, trying to make a rotational motion around the axis defined by these points. The frequency of this rotational motion depends on the drive. At low drive frequencies it is the Kelvin-resonance frequency of the lowest Kelvin wave, determined by the vortex length. When the drive frequency is higher than this, it is typically the frequency of one of the upper modes. The important point is that, since the vortex length is constantly changing, the critical velocity remains finite.

For higher driving velocities vortex motion becomes more irregular and higher modes are excited. Above the critical velocity, reconnections start to occur, producing new vortices that quickly fly away. The vortex production rate is an approximately linear function of the drive, but at high frequencies one may observe some nonlinearity, as seen in Fig.1.

The critical velocity is mainly determined by the pinning distance D, which determines the characteristic length scale and hence the typical self-induced velocity for the vortex. However, the critical velocity will drop if the drive frequency is close to one of the resonance frequencies. For example, for $D = 5\ \mu$m the minimum critical velocity $v_{c,\min} \approx 20$ mm/s is obtained at 2 kHz whereas it is slightly below 30 mm/s at 4.9 kHz. The latter critical velocity is also obtained when using a non-oscillating constant drive. For $D = 50\ \mu$m we observed critical velocities of 6 mm/s and 5 mm/s at 100 Hz and 500 Hz, respectively. Especially for the smaller pinning distance, the critical velocities are not so clear. Sometimes one observes a creation of few new vortices also at somewhat lower drive but, after that, the vortex just oscillates for several seconds. This may be due to numerics, but is hard to check due to chaotic behavior of the vortex motion.

Above the critical velocity reconnections create new vortex loops. These new loops have quite a large size distribution but their average size is almost independent of the drive velocity. This average length goes down slightly with the frequency, being about 12 μm at 1 kHz and about 6 μm at 4.9 kHz for $D = 5\ \mu$m. For comparison, a time independent drive of 50 mm/s produces loops that are 60 μm in length, flying along the drive. With $D = 50\ \mu$m the average length of the new loops is approximately 46 μm for 100 Hz and 23 μm for 500 Hz. New vortex loops typically fly in random directions. Only at velocities just above the critical one, and at low frequencies, is the direction distribution no more isotropic, and one may observe some polarization along the drive.

The results presented here give some new information about the vortex motion in an oscillating flow. In real experiments the vortices are not totally pinned and reconnections with the wire may be important. Therefore it is important to check the effect of the inhomogeneous, oscillating flow on the vortex, without any pinning. We should also try to relate the vortex motion with the measured damping of the wire.

ACKNOWLEDGMENTS

We would like to thank L. Skrbek, W. F. Vinen and H. Yano for fruitful discussions and instructions.

REFERENCES

1. S. N. Fisher, A. J. Hale, A. M. Guénault, and G. R. Pickett, *Phys. Rev. Lett.*, **86**, 244–247 (2001).
2. D. I. Bradley, D. O. Clubb, S. N. Fisher, A. M. Guénault, C. J. Matthews, and G. R. Pickett, *J. Low Temp. Phys.*, **134**, 381–386 (2004).
3. H. Yano, A. Handa, H. Nakagawa, K. Obara, O. Ishikawa, T. Hata, and M. Nakagawa, *J. Low Temp. Phys.*, **138**, 561–566 (2005).
4. J. Jäger, B. Schuderer, and W. Schoepe, *Phys. Rev. Lett.*, **74**, 566–569 (1995).
5. R. J. Donnelly, *Quantized Vortices in Helium II*, Cambridge: Cambridge University Press, 1991.
6. K. W. Schwarz, *Phys. Rev. B*, **31**, 5782–5804 (1985).
7. K. W. Schwarz, *Phys. Rev. B*, **38**, 2398–2417 (1988).
8. W. F. Vinen, L. Skrbek, and H. A. Nichol, *J. Low Temp. Phys.*, **135**, 423–445 (2004).

Inertial Range and the Kolmogorov Spectrum of Quantum Turbulence

Makoto Tsubota* and Michikazu Kobayashi*

*Faculty of Science, Osaka City University, Sugimoto 3-3-138, Sumiyoshi-ku, Osaka 558-8585, JAPAN

Abstract. Quantum turbulence is numerically studied by solving the Gross-Pitaevskii equation. By introducing both energy dissipation at small scales and energy injection at large scales, we succeed in obtaining steady turbulence sustained by a balance between energy injection and energy dissipation. The energy spectrum in the inertial range takes the form of the Kolmogorov spectrum and the energy flux is consistent with the energy dissipation rate at small scales. These results reveal the properties of quantum turbulence in the inertial range, allowing us to propose a prototype of turbulence much simpler than conventional classical turbulence.

Keywords: superfluid turbulence, quantized vortices, Kolmogorov spectrum
PACS: 67.40.Vs, 47.37.+q, 67.40.Hf

INTRODUCTION

Quantum turbulence (QT), consisting of quantized vortices, has become an important subject not only for investigating the issues of superfluidity in low temperature physics, but also for studying a prototype turbulence simpler than classical turbulence (CT). The steady state of fully developed CT in an incompressible fluid follows a characteristic statistical law for the energy spectrum known as the Kolmogorov law[1]. The energy, injected into the fluid at large scales in the energy-containing range, is transferred in the inertial range to smaller scales without being dissipated. In the inertial range the energy spectrum follows the Kolmogorov law $E(k) = C\varepsilon^{2/3}k^{-5/3}$. Here the energy spectrum $E(k)$ is defined by $E = \int dk E(k)$, where E is the kinetic energy per unit mass and k is the wave number obtained from the Fourier transformation of the velocity field. The energy transferred to the energy-dissipative range is dissipated by viscosity with a dissipation rate ε equal to the energy flux Π in the inertial range. The Kolmogorov constant C is a dimensionless parameter of order unity.

The inertial range is believed to be sustained by the self-similar Richardson cascade process, in which large eddies are broken up into smaller eddies through vortex reconnections. However, the Richardson cascade has never been confirmed clearly in CT because it is impossible to definitely identify each eddy. In contrast, quantized vortices are definite and stable topological defects with quantized circulation. Consequently QT may give us a prototype of turbulence simpler than CT [2]. Some experimental [3, 4] and theoretical works [5, 6] have studied the Kolmogorov law for QT, but it was difficult to control the energy dissipation rate ε.

Recently we numerically simulated the Gross-Pitaevskii (GP) equation, which describes a quantum fluid at zero temperature [2]. In that work, we introduced a dissipative term that works only at smaller scales than the healing length, in order to remove compressible short-wavelength excitations created through vortex reconnections which may hinder the cascade process. We thus obtained the Kolmogorov spectrum for decaying turbulence starting from a random configuration of the velocity field. In this work, we also inject energy into the fluid at large scales by moving the random potential, thus obtaining steady turbulence sustained by the balance of energy injection and energy dissipation. The spectrum of the incompressible kinetic energy still obeys the Kolmogorov law, and the energy flux Π is almost independent of scale in the inertial range and consistent with the energy dissipation rate in the energy-dissipative range. These results reveal the properties of the inertial range of QT.

STEADY QUANTUM TURBULENCE

We study numerically the Fourier transformed GP equation,

$$[\mathrm{i} - \gamma(k)]\frac{\partial}{\partial t}\tilde{\Phi}(\mathbf{k},t) = [k^2 - \mu(t)]\tilde{\Phi}(\mathbf{k},t) + \frac{g}{L^6}\sum_{\mathbf{k}_1,\mathbf{k}_2}\tilde{\Phi}(\mathbf{k}_1,t)\tilde{\Phi}^*(\mathbf{k}_2,t) \times \tilde{\Phi}(\mathbf{k}-\mathbf{k}_1+\mathbf{k}_2,t) + \frac{1}{L^3}\sum_{\mathbf{k}_1}\tilde{V}(\mathbf{k}_1,t)\tilde{\Phi}(\mathbf{k}-\mathbf{k}_1,t). \quad (1)$$

CP850, *Low Temperature Physics: 24th International Conference on Low Temperature Physics;*
edited by Y. Takano, S. P. Hershfield, S. O. Hill, P. J. Hirschfeld, and A. M. Goldman

Here $\tilde{\Phi}(\mathbf{k},t)$ is the Fourier transformation of the macroscopic wave function $\Phi(\mathbf{x},t) = f(\mathbf{x},t)\exp[i\phi(\mathbf{x},t)]$, g is the coupling constant, and L the system size. The dissipation is given by

$$\gamma(k) = \gamma_0\theta(k - 2\pi/\xi), \tag{2}$$

which works only on scales smaller than the healing length $\xi = 1/f\sqrt{g}$ in order to remove compressible short-wavelength excitations. The chemical potential $\mu(t)$ depends on time to conserve the total number of particles $N = \int d\mathbf{x}\, f(\mathbf{x},t)^2$. The Fourier transformation of the moving random potential $V(\mathbf{x},t)$ is $\tilde{V}(\mathbf{k},t)$, which moves so that it satisfies the two-point correlation

$$\langle V(\mathbf{x},t)V(\mathbf{x}',t')\rangle = V_0^2 \exp\Big[-\frac{(\mathbf{x}-\mathbf{x}')^2}{2X_0^2} - \frac{(t-t')^2}{2T_0^2}\Big],$$

where V_0, X_0 and T_0 are the characteristic potential strength, space scale and time scale, respectively. Thus $\tilde{V}(\mathbf{k},t)$ injectes the energy into the fluid at the scale X_0 to set the energy-containing range.

We numerically solve Eq. (1) in a periodic box with spatial resolution containing 256^3 grid points. We consider the case of $g = 1$. As numerical parameters, we use a spatial resolution $\Delta x = 0.125$ and $L = 32$, where the length scale is normalized by the healing length ξ. The parameters of the moving random potential are $V_0 = 50$, $X_0 = 4$ and $T_0 = 6.4 \times 10^{-2}$. The numerical integral is calculated by using a pseudo-spectral method in space and the Runge-Kutta-Verner method in time with a time resolution $\Delta t = 1 \times 10^{-4}$. We start from an initial configuration in which both the density and the phase of the wave function are uniform. The system reaches a steady state after $t \simeq 25$ when both the total energy E and the incompressible kinetic energy $E^{\rm i}_{\rm kin}$, given by

$$E = \frac{1}{N}\int d\mathbf{x}\Phi^*[\nabla^2 + \frac{g}{2}f^2]\Phi,$$

$$E^{\rm i}_{\rm kin} = \frac{1}{N}\int d\mathbf{x}[\{f\nabla\phi\}^{\rm i}]^2,$$

are almost constant in time. Here $\{\cdots\}$ denotes the incompressible part, $\nabla \cdot \{\cdots\} = 0$, of the vector fields. In the steady state we can obtain the dissipation rate ε from the time derivative of the incompressible kinetic energy $\partial E^{\rm i}_{\rm kin}/\partial t$ after switching off the moving random potential.

In this steady turbulence, we can clearly define the energy-containing range $k < 2\pi/X_0$ in which the system takes the energy from the moving random potential, the energy-dissipative range $k > 2\pi/\xi$ in which the energy is dissipated by $\gamma(k)$, and the inertial range $2\pi/X_0 < k < 2\pi/\xi$ between these two ranges. Figure 1 (a) shows the energy spectrum of the incompressible kinetic energy $E^{\rm i}_{\rm kin}(k)$, defined by $E^{\rm i}_{\rm kin} = \int dk\, E^{\rm i}_{\rm kin}(k)$. In

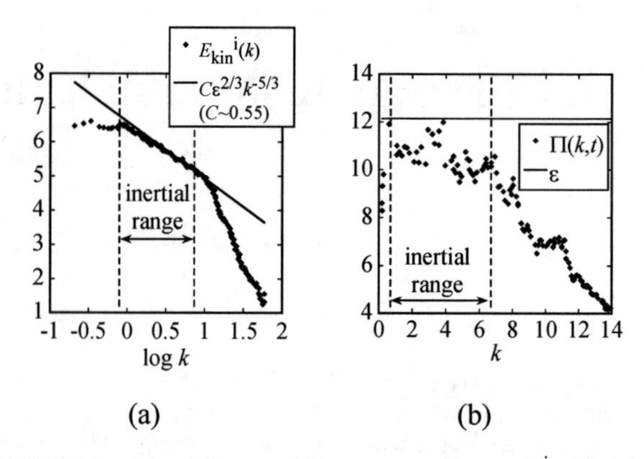

FIGURE 1. Dependence of the energy spectrum $E^{\rm i}_{\rm kin}(k)$ (a) and the energy flux $\Pi(k,t)$ (b) of the incompressible kinetic energy on the wave number k. Solid lines refer to the Kolmogorov law $C\varepsilon^{2/3}k^{-5/3}$ in (a) and the energy dissipation rate ε in (b), respectively.

the inertial range, the numerically obtained spectrum is quantitatively consistent with the Kolmogorov law using the ε obtained from the steady state. Figure 1 (b) shows the flux $\Pi(k)$ of the incompressible kinetic energy which is given by the energy budget equation derived from Eq. (1). The energy flux $\Pi(k)$ is almost independent of k in the inertial range and is close to the energy dissipation rate ε. This result supports strongly the scenario that QT is caused by quantized vortices; the energy-containing range, the inertial range and the energy-dissipating range are, respectively, caused by vortex nucleation, a vorttex Richardson cascade and the decay of vortices to compressible excitations, as believed for eddies in CT.

Detailed studies of this issue are in progress and will be reported shortly elsewhere.

REFERENCES

1. U. Frisch, *Turbulence* (Cambridge University Press, Cambridge, 1995).
2. M. Kobayashi and M. Tsubota, Phys. Rev. Lett. **94** (2005) 065302.
3. J. Maurer and P. Tabeling, Europhys. Lett. **43** (1), 29 (1998).
4. S. R. Stalp, L. Skrbek and R. J. Donnelly, Phys. Rev. Lett. **82**, 4831 (1999)
5. C. Nore, M. Abid and M. E. Brachet, Phys. Rev. Lett. **78**, 3896 (1997); Phys. Fluids. **9**, 2644 (1997).
6. T. Araki, M. Tsubota and S. K. Nemirovskii, Phys. Rev. Lett. **89**, 145301 (2002).

Critical Velocity for Superfluid Turbulence at High Temperatures

Akira Mitani, Risto Hänninen and Makoto Tsubota

Department of Physics, Osaka City University, Sumiyoshi-ku, Osaka 558-8585, Japan

Abstract. In rotating superfluid ^{3}He-B, a sharp transition to turbulence which is insensitive to the fluid velocity is observed in contrast to classical turbulence. Instead, this transition is controlled by an intrinsic parameter, $q \equiv \alpha/1-\alpha'$, of the superfluid, where α and α' are mutual friction parameters depending on temperature. However, rotating superfluid ^{3}He-B is the only system which shows this kind of transition, and many unsolved mysteries yet remain. Hence, we study numerically the quantum vortex dynamics in thermal (unidirectional) counter flow systems especially at high temperatures, with the goal of proposing a new system which also shows a turbulence transition controlled by q. It is shown that the critical velocity above which the system can sustain a vortex tangle increases sharply at $q \approx 0.4$. This means that no vortex tangle state can be stable for $q \gtrsim 0.4$. This may be closely relevant to the dramatic vortex dynamics transition caused by q in rotating superfluid ^{3}He-B.

Keywords: Superfluid, vortex dynamics, vortex tangle
PACS: 67.40.Vs, 03.75.Lm,47.37.+q

Since the pioneering experiment done by Finne *et al.* [1] on a velocity-independent criterion for superfluid turbulence, there has been a growing interest in the role of an intrinsic parameter, $q \equiv \alpha/(1-\alpha')$, where α and α' are mutual friction parameters. They observed a sharp transition to turbulence in rotating superfluid ^{3}He-B which is insensitive to the fluid velocity, in contrast to our knowledge about the transition for turbulence in classical fluids. Instead, this transition is controlled by the parameter q, which causes damping of the vortex motion [2]. Unfortunately, rotating superfluid ^{3}He-B is the only system which shows this kind of transition, and many unsolved mysteries yet remain. Since this kind of transition is purely governed by the parameter q, the rotation of the system must be unnecessary for this kind of transition. Hence, to observe similar transitions in a different system, we study the quantum vortex dynamics in thermal (unidirectional) counter flow systems without rotation. We also investigate the q dependence of the critical velocity v_c, above which the system can sustain a vortex tangle state, especially at high temperatures.

The vortex dynamics are calculated numerically using the localized induction approximation (LIA). We adopt LIA as a first step to find the role of large q in thermal counterflow, although we need the full Biot-Savart law to calculate the vortex dynamics accurately. The details of the vortex dynamics and the numerical scheme used are similar to the ones described in our previous paper [3], except that we adopted a new algorithm of vortex-vortex reconnection processes based on considerations of crossing lines [4]. This algorithm mimics what happens in the real system more accurately than our previous method.

We now describe briefly how to perform the numerical calculation to obtain the critical velocity v_c. The computation sample is taken to be a cube of size 1 cm. One set of faces is subjected to periodic boundary conditions. The other two sets are treated as smooth, rigid boundaries. Vortices approaching the boundaries reconnect to them and their ends can move smoothly, along the boundaries. We apply a uniform counter flow field, $\mathbf{v_{ns}} = \mathbf{v_n} - \mathbf{v_s}$ along the direction of the periodic boundary condition. The simulation starts with dilute vortex tangles; the results hardly depend on the initial configuration. For simplicity, we assume the system is superfluid ^{4}He so that we can neglect α', namely $q = \alpha$.

We find v_c at a finite temperature by observing the time development of the line length density of the vortex tangle with various counter flow velocities. If we compare the time development in the case of $v_{ns} \lesssim v_c$ and in the case of $v_{ns} \gtrsim v_c$, the two are similar in the early stages. However, they become different after a certain time. All vortices disappear with the counter flow velocity slightly below v_c. However, with the velocity slightly above v_c, the line length density gradually goes up and will presumably reach a steady value. We identify v_c at finite temperatures by observing this sudden transition. By repeating this calculation with various mutual friction parameters, we can obtain v_c as a function of q.

The result is given in Fig. 1. For comparison, the white circles, which indicate v_c at low q obtained by Schwarz [5], are also shown. We can see that the critical velocity increases sharply at $q \approx 0.4$. No vortex tangle states are established at $q > 0.41$, even with $v_{ns} = 1.0$ cm/s. This result suggests that v_c diverges and no vortex tangle state can be stable at $q \gtrsim 0.4$. We have to mention a discrep-

CP850, *Low Temperature Physics: 24th International Conference on Low Temperature Physics;*
edited by Y. Takano, S. P. Hershfield, S. O. Hill, P. J. Hirschfeld, and A. M. Goldman

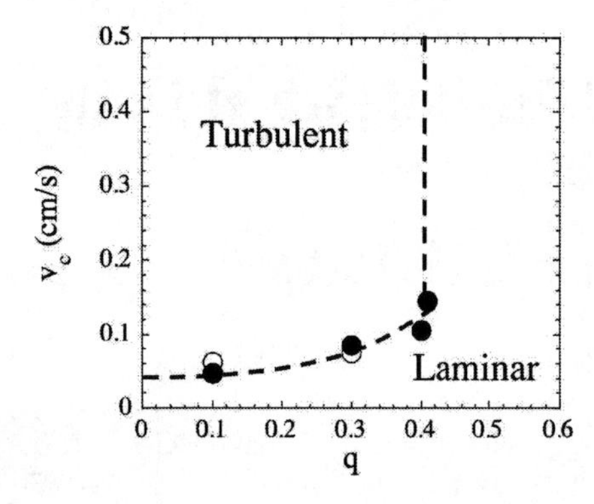

FIGURE 1. Computed values of v_c. Black and white circles represent our and Schwarz's [5] results, respectively. The dotted line is drawn to guide the eye.

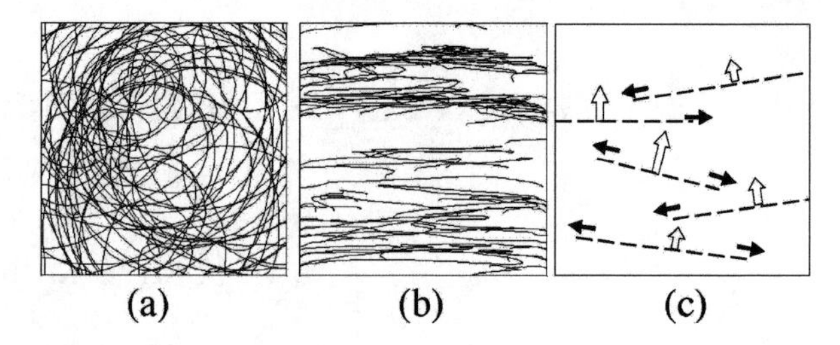

FIGURE 2. Illustration of the typical internal structure of the vortex tangle with large mutual friction coefficients (q>0.4). Frame (a) shows the vortex tangle viewed along the flow direction. Frame (b) shows the vortex tangle viewed across the flow direction. Frame (c) shows a schematic picture of the vortex tangle viewed across the flow direction. Black and white arrows show the direction of expansion and the direction of motion, respectively, of the vortex rings (dashed lines).

ancy between our result and Schwarz's result, which obtained stable vortex tangle even at $q \geq 0.4$ [6]. This discrepancy comes from the artificial mixing process used by Schwarz. This process reduces the anisotropy of the vortex tangle that plays an important role in the critical velocity as we will discuss in the following paragraph. This divergence of v_c may be closely relevant to the transition caused by q observed in rotating superfluid ^{3}He-B.

It is important to ask why v_c suddenly diverges in the high q region. Figs. 2(a) and (b) show the typical internal structure of the vortex tangle near v_c. Fig. 2 (c) shows a schematic illustration of the internal structure. As can be seen in these figures, the regions of high curvature and the associated three-dimensional random motion of the vortex line are damped out because of the high q value [2], whereas the two-dimensional frictional growth becomes more dominant [6]. We can assume, in this situation, the vortex tangle is an assembly of vortex rings which proceed in almost the same direction as $\mathbf{v}_{ns}$. Two origins of vortex-vortex reconnections, which act as a trigger for a transition to a turbulent state, can be listed. One is that a vortex ring collides with another during its expansion because the vortex rings are slightly out of alignment with each other (Origin A). The other is that a vortex ring overtakes another because each vortex ring has different self-induced velocity depending on its radius (Origin B). A simple analysis can show that α makes vortex rings perpendicular to $\mathbf{v}_{ns}$. As α, namely q, becomes large, the vortex tangle becomes more anisotropic. As a result of the high anisotropy, the probability of vortex-vortex reconnections due to Origin A decreases, because vortex rings can expand with few reconnections and then annihilate at the sidewalls. This is the reason why v_c increases rapidly in the high q region.

Furthermore, although we assume $\alpha' = 0$, a simple analysis can show that vortex rings of various radii have increasingly similar velocity as $1/(1-\alpha')$ becomes large. Consequently, the probability of vortex-vortex reconnections due to Origin B is reduced as $1/(1-\alpha')$ becomes large, namely as q becomes large.

A concrete analysis of the role of q in the critical velocity and a detailed numerical calculation with finite α' will be discussed in a later paper.

In summary, we have reported numerical results for the quantum vortex dynamics in thermal (unidirectional) counter flow systems, especially at high temperatures. Our results show that the critical velocity increases sharply at $q \approx 0.4$. This is due to the severe anisotropy of the vortex tangle caused by the large q value and means that no vortex tangle state can be stable at $q \gtrsim 0.4$. This phenomenon can be relevant to the vortex dynamics transition caused by the intrinsic superfluid parameter q showed by Finne *et al.* [1]

We are grateful for stimulating conversation with S.K. Nemirovskii and W.F. Vinen.

REFERENCES

1. A.P. Finne *et al.*, *Nature* **424**, 1022–1025 (2003).
2. C.F. Barenghi, R.J. Donnelly and W.F. Vinen, *Phys. Fluids* **28**, 498–504 (1985).
3. M. Tsubota, T. Araki and S.K. Nemirovskii, *Phys. Rev. B* **62**, 11751–11762 (2000).
4. L. Kondaurova and S.K. Nemirovskii, *J. Low Temp. Phys.* **138**, 555-560 (2005).
5. K.W. Schwarz, *Phys. Rev. Lett.* **50**, 364–367 (1983).
6. K.W. Schwarz, *Phys. Rev. B* **38**, 2398–2417 (1988).

Numerical Study of Stochastic Vortex Tangle Dynamics in Superfluid He

Luiza Kondaurova and Sergey K. Nemirovskii

Kutateladze Institute of Thermophysics SB RAS, 630090 Novosibirsk, Russia

Abstract. This paper is devoted to numerical simulation of stochastic vortex tangle dynamics in the Langevin formalism. A scale separation scheme is applied to evaluate a Biot-Savart integral entering into a full equation of motion. This procedure reduces the number of operations for integral computation. A new algorithm, which is based on consideration of crossing lines, is used for vortex reconnection processes. Calculations are performed for an open space with no walls. One of the most startling results is a formation and disappearance of regions of very high vorticity. These regions suddenly disappear and appear again in different places, with a resemblance to the intermittency phenomenon in classical turbulence. The distributions of loops with respect to their lengths and the spectrum of energy are calculated.

Keywords: superfluidity, vortices, turbulence.
PACS: 67.40.Vs, 47.32.Cc, 05.40.-a

A vortex tangle (VT), which consists of chaotic vortex loops, influences many properties of superfluid helium. Quantized vortices exist either in a nonequilibrium state (the superfluid turbulence, see, e.g., [1-4]) or in a thermodynamic equilibrium state (see, e.g., [5-6]). In order to describe various phenomena related to the presence of vortices, it is necessary to know the statistical description of VT. Evolution of vortex lines is calculated on the basis of the full Biot-Savart (BS) law. In addition, we use a new algorithm for vortex reconnection, introduce random forces stirring up the system, and apply a scale separation scheme for computation of the BS integral.

The calculation of the full BS integral is computationally expensive whenever a large number of mesh points are required to represent vortex lines. That is why the scale separation scheme is employed to solve the equation of motion for vortex lines containing the full BS integral [9]. The numerical scheme is as follows. Firstly, all space occupied by vortex lines is divided into a three-dimensional (3D) mesh. Secondly, the "center of mass" $\mathbf{R}_\nu$ and the total tangential vector $\mathbf{g}_\nu = \sum_{j=1}^{K} \mathbf{s}'_j \, d\xi_j$ are determined for each cell. Here $\mathbf{s}$ is the position of a point on a vortex line, ξ is the arc length, $\mathbf{s}'$ is the derivative with respect to the arc length, and K is the number of mesh points in cell ν. Thus the BS integral is split into three parts: the local term (LC), the nonlocal contribution of the neighboring environment (NLC), and the summation over the cells:

$$\dot{\mathbf{s}}_B = \mathrm{LC} + \mathrm{NLC} + \frac{k}{4\pi}\sum_\nu \frac{(\mathbf{R}_\nu - \mathbf{s})\times \mathbf{g}_\nu}{|\mathbf{R}_\nu - \mathbf{s}|^3}, \quad (1)$$

where $\dot{\mathbf{s}}_B$ is the propagation velocity of the vortex filament at a point $\mathbf{s}$, defined by the BS law. The width of the 3D mesh and the number of neighboring cells are selected experimentally according to the prescribed accuracy of the calculation. With this numerical scheme, the calculation time is reduced by a factor of more than 10. Taking into account the friction between the vortices and the normal component of helium, and a rapidly fluctuating random term, we obtain the equation for the dynamics of a point on a vortex line:

$$\dot{\mathbf{s}} = \dot{\mathbf{s}}_B + \alpha \mathbf{s}' \times (\mathbf{v}_n - \dot{\mathbf{s}}_B) - \alpha' \mathbf{s}' \times [\mathbf{s}' \times (\mathbf{v}_n - \dot{\mathbf{s}}_B)] + \mathbf{f}, \quad (2)$$

where α and α' are the friction coefficients, $\mathbf{v}_n$ is the velocity of the normal component, and $\mathbf{f}$ is the Langevin force. In the present study $\mathbf{f}$ is chosen in the form of white noise with both temporal and spatial variables. Its correlator satisfies the equation $\langle \mathbf{f}_i(\xi_1,t_1)\, \mathbf{f}_j(\xi_2,t_2)\rangle = D\, \delta_{ij}\, \delta(t_1 - t_2)\, \delta(\xi_1 - \xi_2)$, where i and j are the spatial components, t_1 and t_2 are arbitrary times, ξ_1 and ξ_2 define any points on the vortex line, and D is the strength of the Langevin force. Further we will take $\mathbf{v}_n = 0$ and neglect the term with α'.

CP850, *Low Temperature Physics: 24th International Conference on Low Temperature Physics;*
edited by Y. Takano, S. P. Hershfield, S. O. Hill, P. J. Hirschfeld, and A. M. Goldman

In the modeling of a reconnection process, two pairs of points are selected. It is assumed that the line segment between each pair moves at a constant velocity ($\mathbf{V}_i$ and $\mathbf{V}_j$ for the first and second pairs, respectively) during the time step Δt (see Fig. 1). From the solution of equation $\mathbf{s}_i + \mathbf{V}_i h + (\mathbf{s}_{i+1} - \mathbf{s}_i) l_1 = = \mathbf{s}_j + \mathbf{V}_j h + (\mathbf{s}_{j+1} - \mathbf{s}_j) l_2$, where $0 \le l_1 \le 1$, $0 \le l_2 \le 1$, and $0 \le h \le \Delta t$, the meeting of these line segments during Δt is determined. Here subscripts i, i+1, j, and j+1 relate to the first and second pairs of points. If the line segments has met, the reconnection occurs. Thus, if originally the points belong to the same loop, a pair of new loops is generated. Otherwise the confluence of the loops occurs.

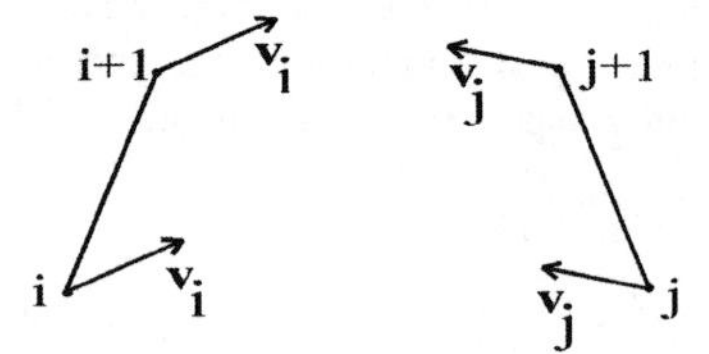

FIGURE 1. Elements of vortex lines may reconnect.

The simulation was performed with a temporal step $\Delta t = 5 \cdot 10^{-8}$s. The initial step along the vortex rings was $4\pi \cdot 10^{-5}$cm and the radii of the rings were $2 \cdot 10^{-3}$cm. The parameters were α=0.0098 and $D = 4 \cdot 10^{-4}$ cm/s.

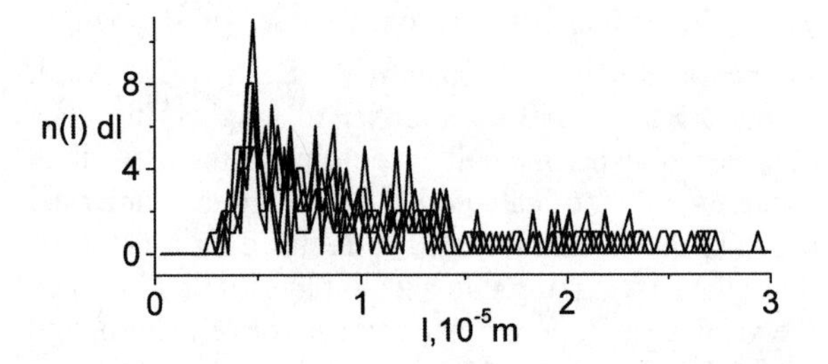

FIGURE 2. Distribution of the vortex loops on their length at t=1.1445 ms.

Animation of the numerical experiment shows that the system of rings evolves into a highly chaotic tangled inhomogeneous vortex structure with strong fluctuations. During the evolution, vortex tangles arise and disappear in different places. It resembles the intermittency phenomenon in classical turbulence. In our simulation, a dynamical equilibrium was reached, e.g., the total length of the VT fluctuated steadily around an average value. For this state we calculated the distribution of loops n(l) with respect to their lengths l (see Fig. 2). The distribution is described by the following relation: $n(l)dl \propto l^{-1}$. This distribution differs from $n(l)dl \propto l^{-2,5}$[5], which is believed to take place in an equilibrium state. We are not in a position to explain this discrepancy.

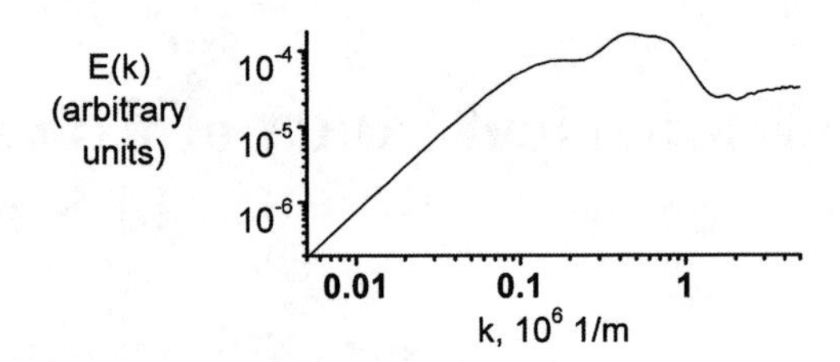

FIGURE 3. Energy spectrum at t=1.1445 ms.

For the isotropic case, the spectral density of the average kinetic energy [3] is expressed as follows:

$$E(k) = \frac{\rho_s k^2}{(2\pi)^2} \int_0^L \int_0^{L'} \frac{\sin(k|\mathbf{s}(\xi_1) - \mathbf{s}(\xi_2)|)}{k|\mathbf{s}(\xi_1) - \mathbf{s}(\xi_2)|} (d\mathbf{s}(\xi_1) \cdot d\mathbf{s}(\xi_2)), \quad (3)$$

where ρ_s is the superfluid density. In the region of small wave numbers, $E(k) \propto k^2$ (see Fig. 3). This law agrees with equipartition of energy, and points out indirectly that the system is close to a thermodynamic equilibrium.

As for quantitative results we cannot claim that an equilibrium state has been reached for sure. It is possible that we have observed some transient processes. Our preliminary simulation demonstrates that the Langevin approach is a very promising method for the study of chaotic vortex structures.

ACKNOWLEDGMENTS

It is a pleasure to acknowledge the fruitful discussions with B. Svistunov. The investigations were supported by RFBR (03-02-16179 and 05-08-01375).

REFERENCES

1. R. J. Donnely, *Quantized Vortices in Helium II*, Cambridge: University Press, 1991.
2. J. T. Tough, in *Progress in Low Temperature Physics*, edited by D. F. Brewer (North Holland, Amsterdam) **8**, 133 (1982).
3. S. K. Nemirovskii and W. Fiszdon, *Rev. Mod. Phys.* **67**, 37 (1995).
4. W.F. Vinen, *Phys. Rev. B.* **61**,1410 (2000).
5. N. D. Antunes and L. M. A. Bettencourt, *Phys. Rev. Lett.* **82**, 2824 (1999).
6. S. K. Nemirovskii, *Theoretical and Mathematical Physics* **141**, 141 (2004).
7. K. W. Schwarz, *Phys. Rev.* **38**, 2398 (1988).
8. T. Araki, M. Tsubota, and S. K. Nemirovskii, *Phys. Rev. B.* **62**, 11751 (2002).
9. Evgeny Kozik and Boris Svistunov, cond-mat/0408241.

NMR Studies of Texture in the B-like Phase of ^{3}He in Aerogel

V.V. Dmitriev*, N. Mulders†, V.V. Zavjalov*, and D.E. Zmeev*

*Kapitza Institute for Physical Problems, 2 Kosygina Str., Moscow 119334, Russia
†Department of Physics and Astronomy, University of Delaware, Newark, Delaware 19716, USA

Abstract. Continuous wave NMR experiments have been performed in the B-like phase of ^{3}He in a 98% open aerogel. The results allow us to get information about the texture of the ^{3}He order parameter inside the aerogel. We conclude that in the B-like phase the boundary between the bulk ^{3}He and the aerogel orients the vector of orbital anisotropy (**L**) parallel to it. Two stable textures and textural transition between them were observed in our sample.

Keywords: Superfluid ^{3}He, Aerogel.
PACS: 67.57.Pq, 67.57.Lm

INTRODUCTION

In the superfluid B-phase of "usual" bulk ^{3}He the spatial distribution of the order parameter (texture) is determined by minimization of the total free energy of the sample and depends on the cell geometry, magnetic field **H** and temperature [1-3]. The most important are the following four terms of the free-energy functional: the density of bulk-field orientation energy,

$$f_{BH} = -a(\mathbf{n}\cdot\mathbf{H})^2; \quad (1)$$

the density of the surface field energy,

$$f_{SH} = -d(\mathbf{H}\cdot\mathbf{R}\cdot\mathbf{s})^2 = -dH^2(\mathbf{L}\cdot\mathbf{s})^2; \quad (2)$$

the density of the gradient energy,

$$f_G = cf(\mathbf{n}, \nabla\mathbf{n}); \quad (3)$$

and the density of the surface dipole energy f_{SD}, which can be neglected in fields H≥100 Oe and will not be considered. Here $\mathbf{R}=\mathbf{R}(\mathbf{n}, \theta_0)$ is the order parameter matrix describing a rotation by an angle $\theta_0 \approx 104°$ around the unit vector **n**, $\mathbf{L}=\mathbf{R}^{-1}(\mathbf{n}, \theta_0)\mathbf{S}/S$ is the unit vector of orbital anisotropy, **S** is the spin density and **s** is the surface normal. It has been established that a and d in Eqs. 1 and 2 are positive and therefore **L** tends to be perpendicular to the walls of the sample cell. Far away from the walls, f_{BH} tends to align **n** and **L** parallel to **H**. Reorientation of **n** and **L** takes place at a distance of the order of the magnetic healing length $R_H \sim 1$ mm, which is determined by a competition between f_G and f_{BH}. Standard low excitation transverse continuous wave (CW) NMR does not disturb the texture and may be used for its study, because the frequency shift $\Delta\omega$ from the Larmor value ω_L depends on the angle φ between **H** and **n**, in other words on the angle ψ between **H** and **L**:

$$\Delta\omega = \frac{\Omega_B^2}{2\omega_L}\sin^2\varphi = \frac{2\Omega_B^2}{5\omega_L}(1-\cos\psi), \quad (4)$$

where Ω_B is the Leggett frequency. If sample dimensions are not much less than R_H then the NMR line has the following form: a peak at ω_L (signal from regions with $\varphi \approx 0°$) and a high frequency tail which drops to zero at some value of $\Delta\omega = \Delta\omega_0$, corresponding to regions with maximal φ. Usually φ is maximal near the cell walls oriented parallel to **H**. Here f_{SH} reaches its minimal value, if **L**⊥**H**, $\varphi \approx 63°$, and

$$\Delta\omega_0 = \frac{2}{5}\frac{\Omega_B^2}{\omega_L}. \quad (5)$$

Soliton-like textural defects with $\varphi = 90°$ are not energetically favorable, but can be formed and be stable [3]. These defects result in an additional peak in the NMR line at the maximal possible frequency shift $\Delta\omega_{max} = \Omega_B^2/2\omega_L$ [3]. It is known that the B-like phase of ^{3}He in aerogel is analogous to "usual" bulk ^{3}He-B [4-6]. However, as shown below, the texture of the B-like phase is different from that in a bulk ^{3}He-B sample of the same geometry.

CP850, *Low Temperature Physics: 24th International Conference on Low Temperature Physics;* edited by Y. Takano, S. P. Hershfield, S. O. Hill, P. J. Hirschfeld, and A. M. Goldman

EXPERIMENTAL SETUP

Experiments were done at pressures from 11.0 to 29.3 bar in magnetic fields from 81 to 1062 Oe (corresponding to NMR frequencies from 262 kHz to 3.4 MHz). Three cells with aerogel samples were used (Fig. 1). All samples had a cylindrical form with the axis oriented along **H**. Cells 1 and 2 were parts of the same experimental chamber and their aerogel samples were cut from the same piece. The cells are described in more detail in [6,7]. The temperature was measured with a vibrating wire calibrated by a Pt NMR spectrometer at pressures of 15.5, 17.5, 19.5 and 25.5 bar (at other pressures we used an extrapolation of these calibrations along with the computer program from [8]). In order to avoid the paramagnetic signal from solid ^{3}He on the surface of aerogel strands, all experimental chambers were preplated by the same amount of ^{4}He (we estimate that it corresponds to 2.5–4 monolayers of ^{4}He on the surface of the aerogel strands). Consequently no Curie-Weiss behavior of spin susceptibility was observed in our experiments.

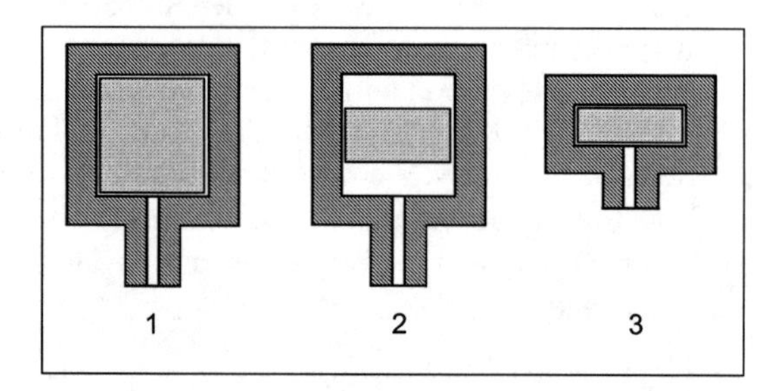

FIGURE 1. Experimental cells. Aerogel samples are shown in gray. The diameters of all samples are 5 mm. The heights of the aerogel samples are: 1 – 5 mm; 2 – 2.4 mm; 3 – 1.5 mm. The inner diameters of the cells are ~5.3 mm. The internal heights of the cells are: 1 – 5.3 mm; 2 – 5.3 mm; 3 – 1.7 mm. NMR coils are not shown.

EXPERIMENTS WITH CELLS 1 AND 2

The CW NMR signals from cells 1 and 2 obtained under the same conditions are shown in Fig. 2. Note that the signal from cell 2 (dashed line) is a sum of signals from bulk ^{3}He-B and the B-like phase of ^{3}He in aerogel. However, peaks “b” and “c” in Fig. 2 are entirely associated with the B-like phase of ^{3}He in aerogel because, on warming, they gradually move to the left and reach the Larmor position at $T=T_{ca}$. Peak “a” is situated at the Larmor position and, at least in part, is due to bulk ^{3}He-B. The high frequency edge of the signal from bulk ^{3}He-B is not seen; it becomes visible and appears in the given frequency range only at $T>T_{ca}$. The NMR line from cell 1 (solid line) looks qualitatively similar to that expected for bulk ^{3}He-B. However the maximal frequency shift observed in cell 2 is much larger than in cell 1, i.e. at least for cell 1 the edge of the high frequency tail does not correspond to Eq. 5. Moreover, the main peak in the signal from cell 1 is shifted from the Larmor value by about 100 Hz. These results show that the texture inside the aerogel does not correspond to the texture of bulk ^{3}He-B. If it did, the lines in Fig. 2 would have the high frequency edge at the same place.

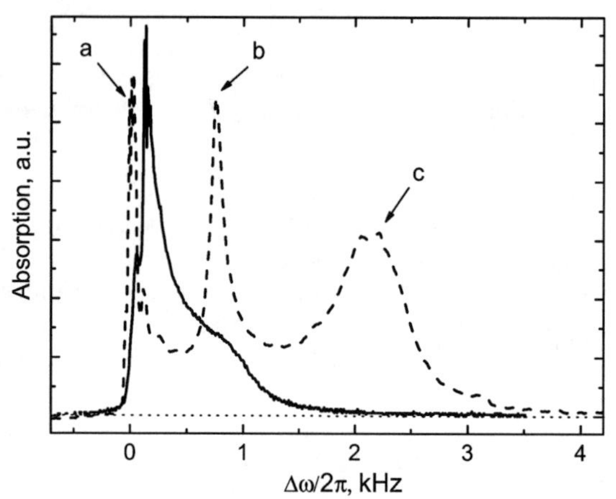

FIGURE 2. CW NMR absorption signals from cell 1 (solid line) and cell 2 (dashed line) obtained under the same conditions: P=25.5 bar, H=284 Oe, $T=0.71T_{ca}$. Here T_{ca} is the superfluid transition temperature of ^{3}He in aerogel, which we define as a temperature where the mean frequency shift of the line becomes zero (for these samples at a given pressure $T_{ca}\approx 0.76T_c$, where T_c is the superfluid transition temperature of the bulk ^{3}He-B).

EXPERIMENTS WITH CELL 3

NMR lines from the B-like phase in cell 3 are shown in Fig. 3. Two distinct peaks (1 and 2) are seen. Note that in contrast to the case of bulk ^{3}He-B no signal is seen at the Larmor frequency. Such a form of the NMR lines is typical for all pressures studied (see e.g. Fig. 1 from Ref. 10). We assert that peak 2 corresponds to a textural defect with φ=90°. This statement is based on our texture independent measurements of the Leggett frequency in the B-like phase of ^{3}He in aerogel in the same cell [7]. For this purpose we have used the homogeneously precessing domain (HPD) inside which the texture is known to be homogeneous. Measurements of the frequency of spatially homogeneous oscillations of the HPD made it possible to determine Ω_B [7,9] and then from Eq. 4 to find the value of φ for each feature in the NMR line. Consequently, as is illustrated by Fig. 4, the position of peak 2 can be used for measurements of the Leggett frequency [10]. The position of peak 1 is also well defined. At $T<0.95T_{ca}$ the corresponding angle is

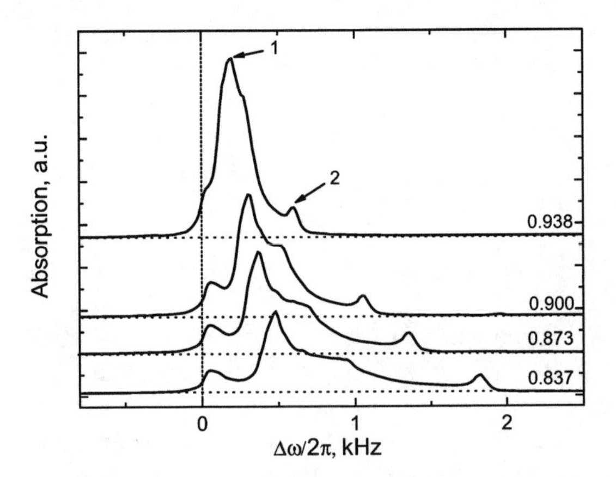

FIGURE 3. CW NMR absorption signals at different temperatures, with the zero levels shown by dotted lines. The temperature in terms of T/T_{ca} is shown near each curve. P=25.5 bar, H=532 Oe. For this sample at given conditions $T_{ca} \approx 0.805 T_c$.

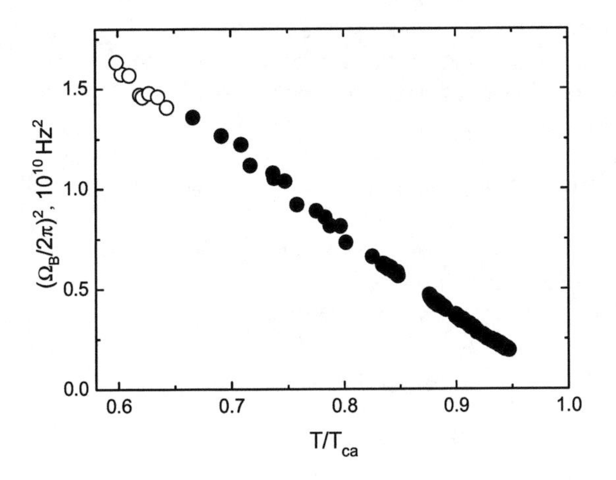

FIGURE 4. Leggett frequency in the B-like phase measured by the HPD techniques [7] (open circles) and by CW NMR on assumption that peak 2 corresponds to $\varphi = 90°$ (filled circles). P=25.5 bar. H=284 Oe for the HPD measurements and H=532 Oe for CW NMR.

$\varphi \approx 31°$, practically independent of temperature and very weakly dependent on pressure and magnetic field (φ grows by about 1° if the field is increased up to 1 kOe, or the pressure decreased from 29.3 to 11 bar).

Peak 2 has appeared in most of the experiments while cooling from the normal phase. However, in some of the runs it was not observed. We have not studied how the probability of the creation of such a textural defect depends on cooling rate, pressure and field. It seems that at low pressures the probability is lower (this may explain why we have not seen peak 2 at pressures of 13.0 and 14.5 bar in spite of several attempts). On warming from the state with the textural defect, we always observed a textural transition; very close to T_{ca}, peak 2 disappears (Fig. 5). After the

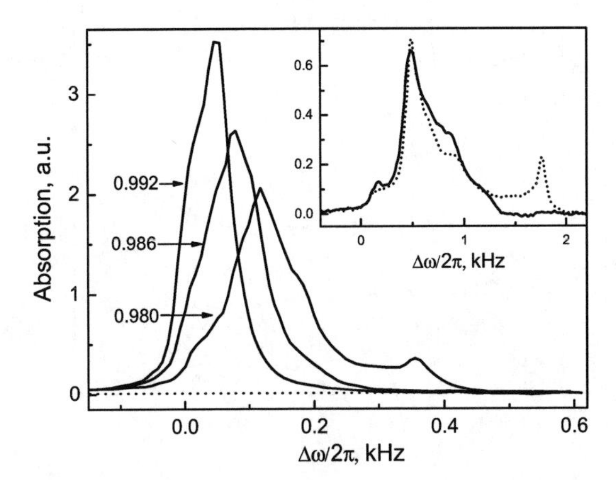

FIGURE 5. CW NMR absorption signals recorded close to T_{ca} during slow warm up. The temperature in terms of T/T_{ca} is shown near each curve. P=17.5 bar, H=142 Oe. The textural transition occurs at $0.980 < T/T_{ca} < 0.986$. Inset: CW NMR lines from states with and without the textural defect. Solid line – without the defect; dashed line – with the defect. $T = 0.946 T_{ca}$, P=17.5 bar, H=142 Oe.

transition the sample remains in the B-like phase, but subsequent cooling down to temperatures well below the transition temperature shows no signs of the defect. In this case, the NMR lineshape coincides with the one observed when no defect had formed upon cooling from the normal phase. The NMR lines recorded under the same conditions with and without the defect are shown in the insert of Fig. 5. Note that the lines are almost identical (except peak 2 associated with the defect); e.g. the position of peak 1 is practically the same. The state with the defect is not favorable, but the transition to the non-defect state has an associated energy barrier. This barrier is defined by a competition between the energies (Eqs. 1–3), which change with temperature in different ways [1,2]. Very close to T_{ca} the bulk field-orientation energy dominates and this can result in the transition.

We have also performed experiments with different orientations of H with respect to the axis of cell 3 (Fig. 6). Our main observations are the following: i) the NMR line does not depend on whether we cool down from the normal phase in a given orientation of H or rotate H to this orientation after the sample is cooled well below T_{ca}; ii) the lineshape does not depend on the sign of the angle η between H and the cell axis; iii) the frequency shift of peak 1 decreases with increasing η and disappears at $\eta = 90°$. Thus we can conclude that, in our sample, the pinning of the texture by local inhomogeneities of aerogel is not the main factor. The existence of the textural defect and the textural transition also support this conclusion.

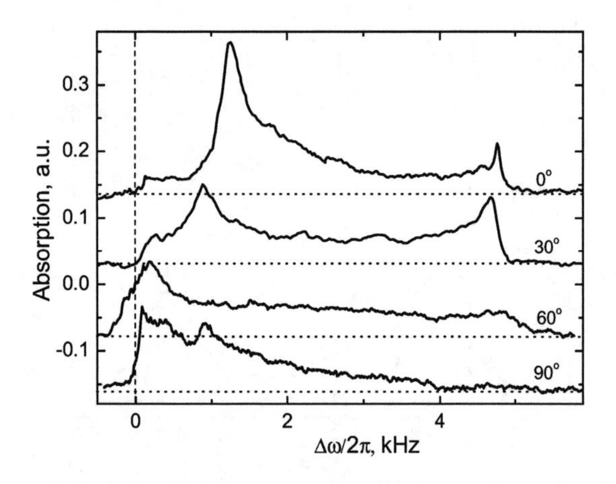

FIGURE 6. NMR absorption signals for different angles η between **H** and the axis of cell 3. Broad edges of the curve for η=60° are due to residual inhomogeneity of the field in this experiment. P=26 bar, H=142 Oe, $T \approx 0.89T_{ca}$.

DISCUSSION

In all our experiments, the NMR lines in the B-like phase are different from (or in some sense are opposite to) the lines expected for bulk ^{3}He-B in the same geometry. For example, for the disc geometry of cells 2 and 3, we would expect the main part of the line to be concentrated near the Larmor value. In contrast, for cell 1 which has the form of a long cylinder, the shifted part of the signal should be more prominent. However, the observed NMR lineshapes can be explained, if we assume that in the B-like phase the coefficients *a*, *c* and *d* in Eqs. 1–3 are changed. The bulk-term coefficients *a* and *c* should not change except for a renormalization according to the change of the gap, transition temperature etc.). But the surface-term coefficient *d*, even its sign may change. If *d* is negative, then L near the boundary tends to be parallel to the surface. This conjecture explains all the NMR lines observed in our experiments. This seems to be possible also for the following reason. Note that all our samples are surrounded by at least a 0.1 mm gap filled with bulk ^{3}He–B which has a similar order parameter. At the boundary of the aerogel, the amplitude of the order parameter changes from one value to another (and it is smaller inside the aerogel). This is not the case in a bulk ^{3}He–B sample, in which the amplitude of the order parameter drops to zero at the boundaries (i.e. on the cell walls). Preliminary results of computer simulations of the texture in cell 3 (with a negative *d*) are in a good agreement with the observed NMR spectra [11].

In order to check our conjecture against the experimental results of other groups, we should know whether the aerogel sample is surrounded by bulk ^{3}He–B or not. If the sample is in a glass tube (where it was grown), then near the wall we expect **L** to be normal to the surface. But in some places gaps can exist between aerogel and the glass, and here we expect **L** to be parallel to the aerogel surface. This can make the picture quite comlicated.

To summarize, we list the main results of our studies.
(1) We suggest that in the B-like phase the boundary between the aerogel and bulk ^{3}He orients **L** parallel to itself.
(2) Soliton-like textural defects with **n⊥H** can exist in the B-like phase. A textural transition (or "annealing" of such a defect) was observed.

ACKNOWLEDGMENTS

We thank I. Kosarev, who took part in an initial stage of the experiments, J. Kopu for providing us the results of computer simulations of the texture, I. Fomin for useful discussions and the authors of Ref. 8 for the computer program for vibrating wire calibration. This work was supported by CRDF (grant #RUP1-2632-MO04), RFBR (grant 03-02-17017), and the Ministry of Education and Science of Russia. V.V.Z. and D.E.Z. thank the Russian Science Support Foundation for financial support.

REFERENCES

1. W.F. Brinkman, H. Smith, D.D. Osheroff, and E.I. Blount, *Phys .Rev .Let.* **33**, 624 (1974).
2. H. Smith, W.F. Brinkman, and S. Engelsberg, *Phys .Rev B* **15**, 2199 (1977).
3. P.J. Hakonen, M. Krusius, M.M. Salomaa, R.H. Salmelin, J.T. Simola, A.D. Gongadze, and G.E.Vachnadze, G.A.Kharadze, *J. Low Temp. Phys.* **76**, 225 (1989).
4. H. Alles, J.J. Kaplinsky, P.S. Wootton, J.D. Reppy, J.H. Naish, and J.R.Hook, *Phys. Rev .Lett.* **83**, 1367 (1999).
5. B.I. Barker, Y. Lee, L. Polukhina, D.D. Osheroff, L.V. Hrubesh, and J.F. Poco, *Phys. Rev .Lett.* **85**, 2148 (2000).
6. V.V. Dmitriev, V.V. Zavjalov, D.E. Zmeev, I.V. Kosarev, and N. Mulders, *JETP Lett.* **76**, 321 (2002).
7. V.V. Dmitriev, V.V. Zavjalov, D.E. Zmeev, and N.Mulders, *JETP Lett.* **79**, 499 (2004).
8. C.B. Winkelmann, E. Collin, Yu.M. Bunkov, and H. Godfrin, *J.Low Temp. Phys.* **135**, 3 (2004).
9. V.V. Dmitriev, V.V. Zavjalov, and D.E. Zmeev, *J. Low Temp. Phys.* **138**, 765 (2005).
10. V.V. Dmitriev, N. Mulders, V.V. Zavjalov, and D.E. Zmeev, "CW NMR ...", in these Proceedings.
11. J. Kopu, private communication and to be published.

CW NMR Measurements of the Leggett Frequency in ^{3}He-B in Aerogel

V.V. Dmitriev*, N. Mulders†, V.V. Zavjalov* and D.E. Zmeev*

*Kapitza Institute for Physical Problems, 2 Kosygina Str., Moscow 119334, Russia
†Department of Physics and Astronomy, University of Delaware, Newark, Delaware 19716, USA

Abstract. The Leggett frequency in the B-like phase of ^{3}He in 98% silica aerogel has been measured in a wide range of temperatures and pressures by means of CW NMR.

Keywords: Superfluid ^{3}He, Aerogel.
PACS: 67.57.Pq, 67.57.Lm

INTRODUCTION

It is now established that the low-temperature superfluid phase of ^{3}He in aerogel is analogous to the B-phase of bulk ^{3}He [1-3]. An important question is how quantities that define the behavior of the superfluid are modified by aerogel strands. One such quantity is the Leggett frequency that characterizes the dipole interaction in superfluid ^{3}He. In bulk ^{3}He (i.e. in the absence of impurities) the Leggett frequency Ω_B in the B-phase can be measured using longitudinal or transverse CW NMR. In both cases it is necessary to have some knowledge about the spatial distribution (texture) of the order parameter, which determines the form of the NMR line. In the case of transverse NMR the frequency shift $\Delta\omega$ from the Larmor value ω_L is determined by the angle φ between the steady magnetic field **H** and the order parameter vector **n** [4]:

$$\Delta\omega = \frac{\Omega_B^2}{2\omega_L}\sin^2\varphi. \qquad (1)$$

A maximal possible frequency shift $\Delta\omega_{max}=\Omega_B^2/2\omega_L$ can be observed in the presence of soliton-like textural defects [5]. Such defects are not energetically favorable. So the maximal observed shift is usually determined by the boundary condition for **n** at the cell walls oriented along **H**, which corresponds to $\sin^2\varphi = 4/5$ at the walls. Far away from the walls (i.e. at distances much larger than the so-called magnetic healing length $R_H \sim 1$ mm) **n** is parallel to **H** and the shift is zero. Correspondingly the NMR line has the following characteristic form: a sharp peak at ω_L and a high frequency tail which drops to zero at some value of $\Delta\omega$. This value can be used to determine Ω_B [5]. In ^{3}He-B in aerogel the boundary conditions for **n** are different from those in the bulk ^{3}He-B [6], so the measurements of the Leggett frequency cannot be done analogously to the measurements in bulk ^{3}He. However, our previous experiments [7] made it possible to establish relationships between the CW NMR frequency shifts and Ω_B in ^{3}He-B in aerogel.

EXPERIMENT

Our sample of 98% aerogel has the form of a thin disc (diameter of 5 mm, height of 1.5 mm) with the axis oriented along **H** (the cell is described in more detail in [7]). Aerogel was coated with two atomic layers of ^{4}He. The experiments were done at pressures from 11.9 to 26 bar in magnetic fields from 81 Oe up to 1062 Oe (the obtained values of Ω_B did not depend on the magnitude of **H**). Standard linear CW NMR techniques were used. Temperature was measured with a vibrating wire calibrated by Pt NMR spectrometer at pressures of 15.5, 17.5, 19.5 and 25.5 bar (at other pressures we used extrapolation of these calibrations along with the computer program from [8]).

Typical CW NMR lines from the B-like phase in our cell are shown in Fig. 1. The peaks are attributed to the dipole frequency shift caused by specific texture of the order parameter. In previous experiments with the same sample [7] we have studied oscillations of the homogeneously precessing domain (HPD). Inside the HPD the texture is homogeneous and measurements of the frequency of the oscillations made it possible to determine Ω_B. We were able to make such

CP850, *Low Temperature Physics: 24th International Conference on Low Temperature Physics;*
edited by Y. Takano, S. P. Hershfield, S. O. Hill, P. J. Hirschfeld, and A. M. Goldman

measurements only in a narrow range of temperatures, where the HPD could be formed. However, values of Ω_B obtained with the HPD techniques allow us to find values of φ corresponding to the peaks in NMR lines obtained at the same conditions. It was found that the high frequency peak (peak 2) in Fig. 1 corresponds

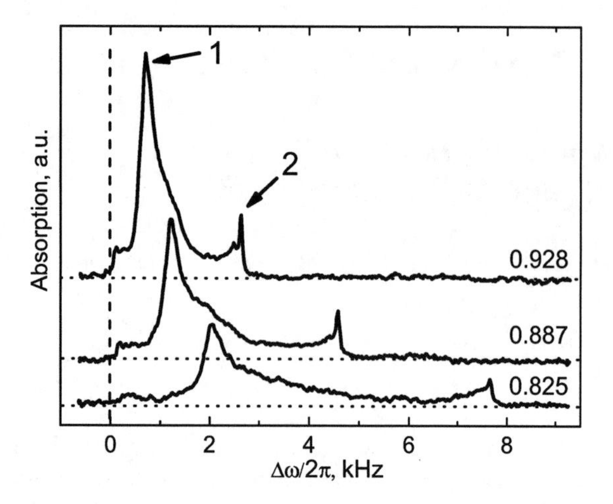

FIGURE 1. CW NMR lines at different temperatures (the corresponding zero levels are shown by dotted lines). Two peaks (1 and 2) characterize the texture. According to (1) peak 1 corresponds to $\varphi \approx 31°$ and peak 2 corresponds to $\varphi = 90°$. Note that no signal is seen at the Larmor frequency unlike the case of bulk ^{3}He-B. Temperatures (in terms of T/T_{ca}) are shown near the corresponding curves. Here T_{ca} is the superfluid transition temperature of ^{3}He in aerogel, which we define as a temperature where mean frequency shift of the line becomes zero. P=26.0 bar, H=142 Oe.

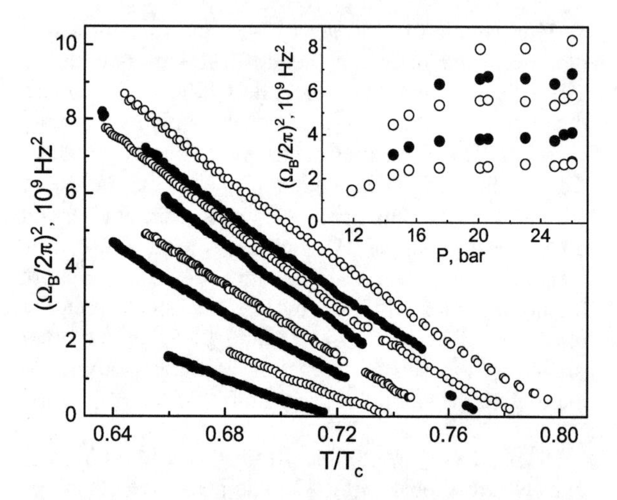

FIGURE 2. The Leggett frequency Ω_B as a function of T/T_c at different pressures (from left to right and from bottom to top: 11.9, 13.0, 14.5, 15.5, 17.5, 20.1, 23.0 and 26.0 bar). Here T_c is the superfluid transition temperature of bulk ^{3}He at corresponding pressure. At the insert the pressure dependence of Ω_B for different T/T_{ca} is shown (from bottom to top: T/T_{ca}= 0.92; 0.89; 0.86; 0.83 and 0.80). Note that in a wide range $\Omega_B=\Omega_B(T/T_{ca})$ only weakly depends on pressure.

to the textural defect with φ =90° (i.e. it corresponds to $\Delta\omega_{max}$). Such a defect appeared in most of the used pressures on cooling from the normal phase. The shift of peak 2 allowed us to determine Ω_B in a wide range of temperatures and pressures (Fig. 2). On warming we observed a textural transition: peak 2 abruptly disappeared very close to T_{ca} [6]. In a number of runs (i.e. at 13.0 and 14.5 bar), peak 2 was not observed at all. However we have found that at temperatures $T < 0.95T_{ca}$ the ratio of the frequency shifts for peaks 2 and 1 does not depend on temperature and weakly depends on **H** and pressure (e.g. for H=284 Oe this ratio grows from 3.6 at 11.9 bar to 3.74 at 26.0 bar). This allowed us to use also peak 1 for determination of Ω_B.

The ratio of energy gaps in the B-like phase and in the bulk ^{3}He-B can be estimated from our data if we assume that $\Delta^2=k\Omega_B^2/\chi$ (where Δ is the energy gap, χ is magnetic susceptibility and factor k is the same for the bulk ^{3}He-B and for ^{3}He-B in aerogel). Then we obtain that $\Delta_{aero} \approx 0.54\Delta$ at 26 bar and $\Delta_{aero} \approx 0.62\Delta$ at 11.9 bar. However such an assumption appears to be incorrect, because the obtained dependence of Δ_{aero}/Δ on pressure does not agree with the results of [9].

ACKNOWLEDGEMENTS

We thank the authors of [8] for the computer program for vibrating wire calibration. This work was supported by CRDF (grant #RUP1-2632-MO04), Russian Foundation for Basic Research (grant 03-02-17017) and by the Ministry of Education and Science of Russia. V.V.Z. and D.E.Z. thank the Russian Science Support Foundation for financial support.

REFERENCES

1. H. Alles, J.J. Kaplinsky, P.S. Wootton, J.D. Reppy, J.H. Naish and J.R. Hook, *Phys. Rev. Lett.*,**83**, 1367 (1999).
2. B.I. Barker, Y. Lee, L. Polukhina, D.D. Osheroff, L.V. Hrubesh and J.F. Poco, *Phys. Rev. Lett.*,**85**, 2148 (2000).
3. V.V. Dmitriev, V.V. Zavjalov, D.E. Zmeev, I.V. Kosarev and N. Mulders, *JETP Letters*, **76**, 321 (2002).
4. D.D. Osheroff and W.F. Brinkman, *Phys. Rev. Lett.*, **32**, 584 (1974).
5. P.J. Hakonen, M. Krusius, M.M. Salomaa, R.H. Salmelin, J.T. Simola, A.D. Gongadze, G.E. Vachnadze and G.A. Kharadze, *J. of Low Temp. Phys.*, **76**, 225 (1989).
6. V.V. Dmitriev, N. Mulders, V.V. Zavjalov and D.E. Zmeev, "NMR Studies…", in *these Proceedings*.
7. V.V. Dmitriev, V.V. Zavjalov, D.E. Zmeev and N. Mulders, *JETP Letters*, **79**, 499 (2004).
8. C. B. Winkelmann, E. Collin, Yu.M. Bunkov and H. Godfrin, *J. of Low Temp. Phys.*, **135**, 3 (2004).
9. G. Lawes, J.M. Parpia, *Phys. Rev. B*, **65**, 092511 (2002).

Hydrodynamic Property of Oscillating Superfluid ^{3}He in Aerogel

K. Obara*, Y. Nago*, H. Yano*, O. Ishikawa*, T. Hata*, H. Yokogawa† and M. Yokoyama†

*Graduate School of Science, Osaka City University, Osaka 558-8585, Japan
†Advanced Technology Research Laboratory, Matsushita Electric Works, Ltd., Japan

Abstract. The investigation of the superfluidity of liquid ^{3}He in aerogel of 97.5% and 98.5% porosities using the fourth sound resonance technique revealed two distinct observations. First, the superfluid transition temperature T_C and the superfluid density ρ_s/ρ of ^{3}He in aerogel are greatly suppressed. Second, the sound attenuation does not depend on temperature at higher temperatures, but monotonically diminishes with decreasing temperature at lower temperatures.

Keywords: superfluidity, aerogel, acoustic property, impurity effect
PACS: 67.57.-z,67.57.De,67.57.Pq,74.81.-g

Aerogel provides an interesting system to investigate superfluidity of liquid ^{3}He, because the average distance between the silica strands is comparable to the superfluid coherence length ξ_0, and the diameter of each silica strand is minute. Therefore, aerogel dose not act as a wall but an impurity, and does not destroy the superfluidity completely but does suppress the transition temperature T_C and the superfluid density ρ_s/ρ. Among many methods to investigate superfluidity in aerogel, we used two methods: cw and pulsed NMR to mesasure magnetization [1] and the fourth sound resonance technique. This report focuses on the fourth sound measurement. The superfluid density can be derived from the velocity of the fourth sound. Moreover, the hydrodynamic properties such as the viscous penetration depth δ_v and the viscous mean free path l_v can be calculated from the quality factor Q of the resonance line shape. We also performed a small angle neutron scattering experiment for our aerogel to get structural information [2]. The result shows the existence of weak spatial periodicity in our aerogel.

Previously, we reported the fourth sound velocity and $1/Q$ of 98.5% porosity aerogel at 13 bar [3]. This time, we prepared sound cells with 97.5% and 98.5% porosity aerogel, and a bulk ^{3}He sound cell which is used as a reference. These three resonators are cylindrical in shape. The diameter of each resonator is 8 mm, and the length is 15 mm. At both ends of each sound resonator, we place a piezo electrode as a pressure transducer to generate and detect a sound resonance.

The fourth sound resonance technique has one problem: aerogel is an elastic medium [4]. To prevent it from oscillating together with liquid, we have grown the aerogel directly between the particles of sintered silver powder (SSP) whose packing factor is 65%. The particle of SSP are almost spherical, with a diameter of about 70 μm. The aerogel is confined in so narrow a space that the characteristic frequency of elastic oscillation of an aerogel strand itself becomes higher than 1 MHz and decouples from the fluid oscillation whose frequencies are less than hundreds of herz [5]. The static pressure of liquid ^{3}He was adjusted at 22 bar. The temperature was measured by a Pt-NMR thermometer located in the bulk liquid.

Typical resonance lines are shown in the inset of Fig. 1. We can see four equally spaced resonance peaks corresponding to the $m = 1,2,3,4$ modes and a small peak at the lowest frequency, a Helmholtz resonance caused by the inlet hole located at the middle of the side face of the cylindrical resonator. The fourth sound resonance frequency f_m and the full width at half maximum Δf_m can be obtained by a Lorentzian fit. The sound velocity C_4 and the superfluid fraction ρ_s/ρ were calculated by using the following formulae:

$$C_4 = 2nLf_m/m, \quad (1)$$
$$\rho_s/\rho = (C_4/C_1)^2. \quad (2)$$

Here, L is the length of the resonator, n is the acoustic refraction index that represents the effective length of wave propagation in ^{3}He, and C_1 is the first sound velocity [6] whose temperature dependence is negligibly small. The temperature dependence of the superfluid density is shown in Fig. 1. Since the SSP is known to affect neither the superfluid transition temperature T_C nor ρ_s/ρ [7], the suppression of T_C and ρ_s/ρ is caused by the aerogel itself. We can see the suppression of the superfluid transition temperature, $T_C = 0.88T_C^{bulk}$ in 98.5% aerogel. The suppression of ρ_s/ρ is stronger for the 97.5% porosity aerogel. Figure 2 shows the loss of the second mode $1/Q = \Delta f/f_2$. At higher temperatures above 1.0

CP850, *Low Temperature Physics: 24th International Conference on Low Temperature Physics;*
edited by Y. Takano, S. P. Hershfield, S. O. Hill, P. J. Hirschfeld, and A. M. Goldman

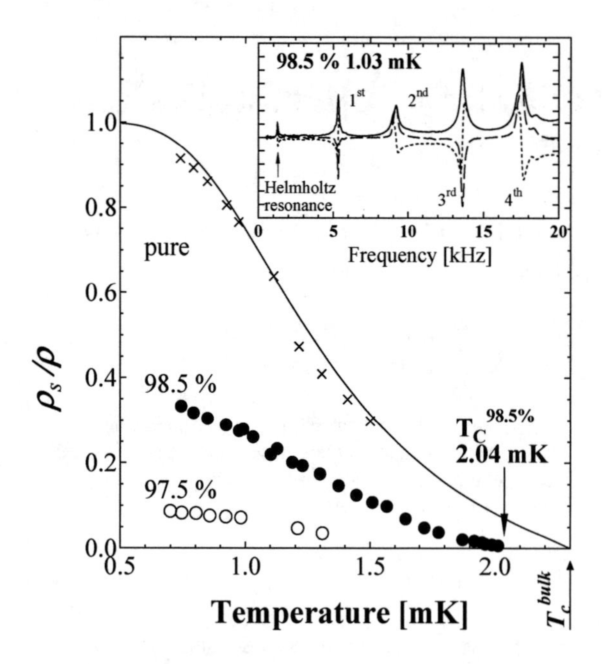

FIGURE 1. Superfluid density ρ_s/ρ as a function of temperature at 22 bar. ○, • and × represent the values for the 97.5% porosity aerogel, 98.5% porosity aerogel and pure ^{3}He cells, respectively. The inset shows the fourth sound resonance in the 98.5% porosity aerogel cell at 1.03 mK. Solid, dashed and dotted lines represent the amplitude, absorption and dispersion, respectively. Solid line represents the bulk value calculated from the torsional oscillator data [8].

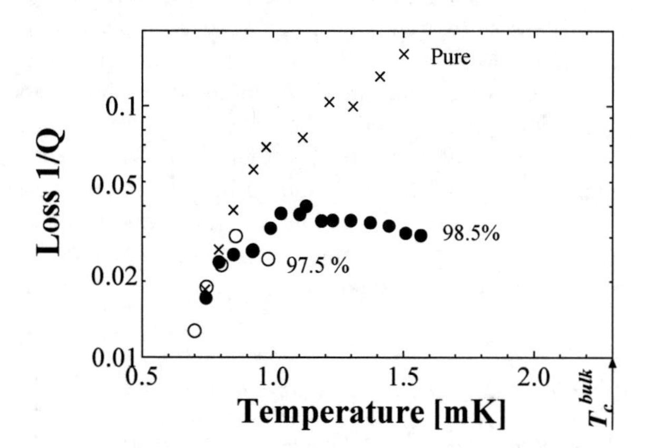

FIGURE 2. Loss of the second mode of the fourth sound resonance at 22 bar. The symbols are the same as in Fig. 1. The loss in ^{3}He in aerogel is always smaller than the loss in pure ^{3}He, but it merges with the pure ^{3}He value at the lowest temperatures. At higher temperatures, above 1.6 mK, it is hard to determine Δf_2 because the resonance intensity is too weak to be fitted to a Lorentzian.

mK, the loss in the pure ^{3}He cell shows an almost linear temperature dependence, but those in the aerogel cells do not depend on temperature, as reported in our previous paper [3]. The losses in the aerogel cells are always smaller than the loss in pure ^{3}He. This means that aerogel locks the normal fluid component more effectively. This can also explain the porosity dependence of the loss. At lower temperatures, however, the losses decrease monotonically with decreasing temperature and approach the pure ^{3}He value. This behaviour is not understood at this moment. We will do more detailed experiments to obtain the hydrodynamic properties such as the viscous penetration depth δ_v and the viscous mean free path l_v.

We thank Dr. Aizawa and Dr. Morii of Neutron Science Research Center of Japan Atomic Energy Research Institute who made the useful neutron scattering structure analysis of the aerogel.

REFERENCES

1. H. Nakagawa, K. Obara, H. Yano, O. Ishikawa, T. Hata, H. Yokogawa, and M. Yokoyama, *J. Low Temp. Phys.* **138**, 159 (2005).
2. This work was originally motivated by the result of Matsumoto et al. (*Phys. Rev. Lett.* **79**, 253 (1997)). They found that aerogels of same porosity showed a sample dependence: T_C and ρ_s/ρ in aerogel sample differed from one sample to another, implying that the internal structure plays a key role.
3. K. Kotera, T. Hatate, H. Nakagawa, H. Yano, O. Ishikawa, T. Hata, H. Yokogawa, and M. Yokoyama, *Physica B* **329**, 316 (2003). *J. Low Temp. Phys.* **138**, 159 (2005).
4. A. Golov, D.A. Geller, J.M. Parpia, and N. Mulders, *Phys. Rev. Lett.* **82**, 3942 (1999).
5. S. Higashitani, T. Ichikawa, M. Yamamoto, M. Miura, and K. Nagai, *Physica B* **329**, 299 (2003).
6. J. C. Wheatley, *Rev. Mod. Phys.* **47**, 415 (1975)
7. T. Takebayashi, Master's Thesis, Osaka City University, 2000.
8. J.M. Parpia, D.G. Wildes, J. Sanders, E.K. Zeise, J.D. Reppy, and R.C. Richardson, *J. Low Temp. Phys.* **61**, 337 (1985).
9. R. Hänninen and E.V. Thuneberg, *Phys. Rev. B* **67**, 214507 (2003)

A-B Phase Transition and Pinning of Phase Boundary of Superfluid ^{3}He in Aerogel

Osamu Ishikawa[a], Ryusuke Kado[a], Hisashi Nakagawa[b], Ken Obara[a], Hideo Yano[a], Tohru Hata[a], Hiroshi Yokogawa[c], and Masaru Yokoyama[c]

[a]*Graduate School of Science, Osaka City University, Osaka 558-8585, Japan*
[b]*Metrology Institute of Japan, Advanced Industrial Science and Technology Tsukuba, 305-8563, Japan*
[c]*Matsushita Electric Works Ltd. Kadoma, Osaka, 571-0500, Japan*

Abstract. Phase transition in superfluid ^{3}He in aerogel has been studied by NMR. Above 19 bar, we have clearly observed the *A*-like and *B*-like phases by following changes in the NMR lineshapes and resonance frequencies. There is a wide temperature region in which the *A*-like phase and the *B*-like phase coexist, extending from near the superfluid transition temperature T_c^{aero} to the lowest temperature of coexistence, T_{AB}^{aero}, below which only the *B*-like phase exists. There are two temperature regions, only in which the phase conversion occurs. Both regions are a few tens of μK wide, the upper region being just below T_c^{aero} and the lower one just above T_{AB}^{aero}. In cooling down and warming up with the two phases in coexistence, no phase conversion occurs between the two regions. The phase boundary between the *A*-like phase and *B*-like phase cannot move in aerogel due to strong pinning by inhomogenities of aerogel.

Keywords: superfluidity, helium3, aerogel, phase transition
PACS: 67.57.Bc, 67.57.Pq

INTRODUCTION

Superfluid ^{3}He in low density aerogel gives us an opportunity to study the effect of impurities on the anisotropic superfluid state of p-wave paring, in a very clean limit.[1,2] The suppression of the superfluid component and the superfluid transition temperature is a common feature in all experiments, and it is qualitatively understood in the same way as the suppression in an s-wave superconductor with random impurities. NMR and sound experiments revealed the coexistence of two phases.[3,4] One is thought to be an equal spin paring state similar to the bulk *A* phase and the other an isotropic sate similar to the bulk *B* phase. Here we call them the *A*-like phase and the *B*-like phase, respectively. We report on an NMR experiment, which examines the transition between the *A*-like and *B*-like phases and the coexistence of these states in 97.5% porosity aerogel.

AEROGEL

We used aerogel made by Matsushita Electric Works Ltd. Figure 1 shows the result of the neutron scattering experiment on an aerogel sample taken from the same batch. Here the differential cross section is plotted as a function of wave number. The data show one peak, implying that the distances between aerogel strands have a weak periodicity. The average distance corresponding to the peak wave number is 53.6 nm. There exists, however, a wide distribution of distances, indicating an inhomogeneous structure, but the inhomogeneity may be isotropic. To avoid a gap around it, we made aerogel directly in a glass cylinder, which is 4 mm in diameter and 10 mm in height. We coated the silica strands with 2.5 atomic layers of ^{4}He in advance.

NMR SPECTRUM IN COEXISTING STATE

The operating NMR frequency was 923 kHz. All the data presented here were taken at 23.5 bar. On cooling from the normal state, the *A*-like phase first appeared below the superfluid transition temperature T_c^{aero}=1.412 mK. The magnetization of the *A*-like phase was nearly the same as that of the normal state, and the NMR spectra showed an overall positive frequency shift and a change in the line shape.

CP850, *Low Temperature Physics: 24th International Conference on Low Temperature Physics;*
edited by Y. Takano, S. P. Hershfield, S. O. Hill, P. J. Hirschfeld, and A. M. Goldman

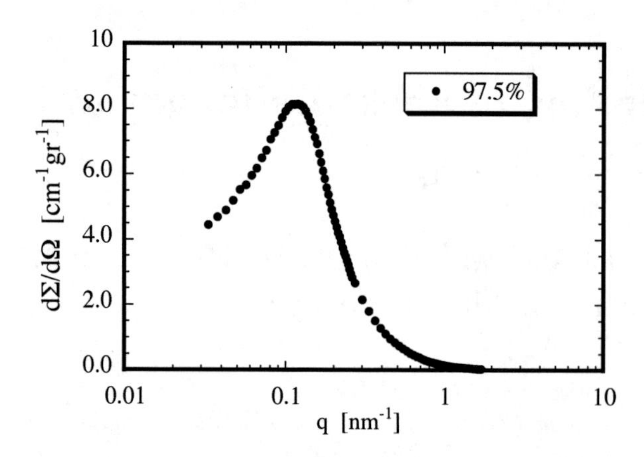

FIGURE 1. Differential cross section obtained by the neutron scattering experiment on 97.5% porosity aerogel.

This change is attributed to a collective spin motion.[5] On further cooling, the *B*-like phase appeared and coexisted with the *A*-like phase. The absorption signal of the *B*-like phase had one broad peak at a frequency higher than that of the *A*-like phase. This signal indicates that the n-texture is quite different from that in bulk liquid. Then the *A*-like signal disappeared below the supercooled *A*-*B* transition temperature, T_{AB}^{aero}. We call the coexistence temperature region just above T_{AB}^{aero} region 2. The conversion of the *A*-like phase to the *B*-like phase occurred in region 2, only with decreasing temperature. On warming from the *B*-like phase, the *A*-like phase appeared within a narrow temperature region, which we call region 1, just below T_c^{aero}. In this region, the phase conversion occurred only with increasing temperature.

Figure 2 shows the absorption signals when the liquid was warmed to region 1 after it had been cooled from the normal state to 1.273 mK, which is in region 2. At low temperatures, we can clearly recognize two signals showing positive frequency shifts. The signal with the smaller frequency shift is due to the *A*-like phase and that with the larger frequency shift is due to the *B*-like phase. Both signals approach the Larmor frequency with increasing temperature and merge near region 1. We can calculate the fraction of each phase at lower temperatures as well as near region 1 by assuming that the shape of the signal from the *A*-like phase does not change. This assumption was verified by a turn-around experiment from region 1, where we reversed the direction of temperature change from warming to cooling. We find that the fraction of each phase remained constant during the warming process depicted in Fig. 2. This means that no phase conversion occurs between regions 1 and 2.

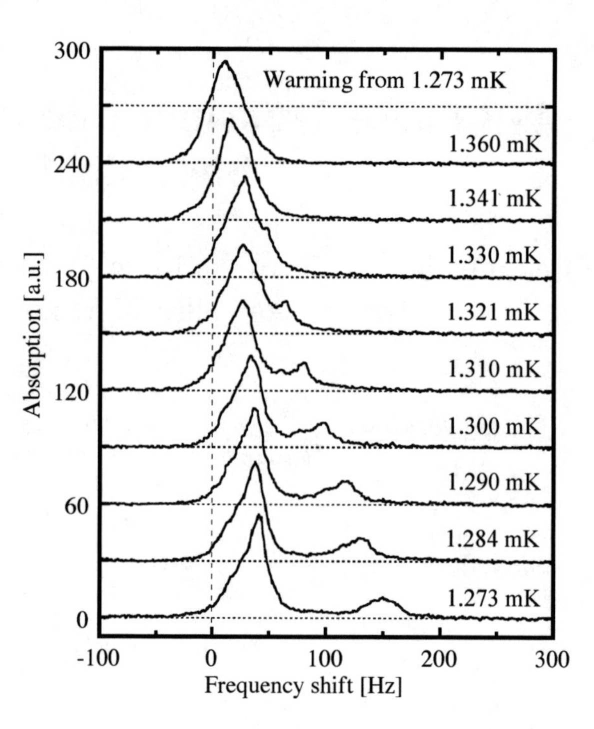

FIGURE 2. NMR absorption signals during warming from 1.273 mK, which is in region 2.

We have also observed no phase conversion during cooling from region 1, with the two phases in coexistence. These results suggest strong pinning of the *A*-*B* phase boundary by inhomogeneities in aerogel.

REFERENCES

1. J. V. Porto and J. M. Parpia, *Phys. Rev. Lett.* **74**, 4667 (1995).
2. D. T. Sprague *et al.*, *Phys. Rev. Lett.* **75**, 661 (1995).
3. E. Nazaretski, N. Mulders, and J. M. Parpia, *J. Low Temp. Phys.* **134**, 763 (2004).
4. J. E. Baumgardner, Y. Lee, and D.D. Osheroff, *Phys. Rev. Lett.* **93**, 055301 (2004).
5. H. Nakagawa *et al.*, to be published.

Pulsed NMR Measurements in Superfluid ^{3}He in Aerogel of 97.5 % Porosity

Osamu Ishikawa[a], Ryusuke Kado[a], Hisashi Nakagawa[b], Ken Obara[a], Hideo Yano[a], Tohru Hata[a], Hiroshi Yokogawa[c], and Masaru Yokoyama[c]

[a]Graduate School of Science, Osaka City University, Osaka 558-8585, Japan
[b]Metrology Institute of Japan, Advanced Industrial Science and Technology Tsukuba, 305-8563, Japan
[c]Matsushita Electric Works Ltd. Kadoma, Osaka, 571-0500, Japan

Abstract. Aerogel is made of thin SiO_2 strands of a few nanometer diameter. Since the coherence length of superfluid ^{3}He is much longer than the silica strand diameter and is nearly the same as the mean distance between silica strands, aerogel gives us a chance to study the effects of an impurity in superfluid ^{3}He. To investigate what superfluid states are formed in aerogel, we performed a pulsed NMR experiment. Both the A-like and B-like phases show a tipping angle dependent frequency shift in the FID signal after an rf pulse. The dependence in the A-like phase is well explained by an expectation based on the "robust phase" introduced by Fomin, while the FID frequencies in the B-like phase behave similarly to those observed in the bulk B phase in a slab geometry with the initial condition of a non-Leggett configuration.

Keywords: superfluidity, helium3, aerogel, pulsed NMR
PACS: 67.57.Lm, 67.57.Pq

INTRODUCTION

Low density aerogel is composed of silica strands of a few nanometer diameter with the average distance between them of the same order as the coherence length of superfluid ^{3}He. This situation allows aerogel to behave as an impurity in superfluid ^{3}He.[1] So far, two superfluid phases in aerogel have been recognized, with hysteretic behaviors in NMR and sound experiments.[2, 3] On cooling from the normal state, the A-like phase appears below the superfluid transition temperature T_c^{aero}. The magnetization is nearly the same as that of the normal state, which supports equal spin paring like the bulk A phase. With further cooling the B-like phase appears, showing a gradual decrease of magnetization. This supports non-equal spin pairing like the bulk B phase. Below T_{AB}^{aero}, the A-like phase disappears. But on warming from the B-like phase, the characteristic features of the A-like phase do not appear until near T_c^{aero} in a magnetic field. Without the field, no A-like phase is observed on warming from the B-like phase.[4] Recent pulsed NMR experiments on the B-like phase have shown that there exists a characteristic tipping angle beyond which the FID signal exhibits a frequency shift similar to that of the bulk B phase.[5] This suggests the structure of the order parameter in the B-like phase might be that of the bulk B phase. Here we report the results of continuous-wave (cw) and pulsed NMR in superfluid ^{3}He in 97.5 % porosity aerogel in both the A-like and B-like phases.

EXPERIMENTS AND RESULTS

We used aerogel made by Matsushita Electric Works Ltd, which was directly grown in a glass cylinder to avoid a gap. We coated the silica strands with 2.5 layers of ^{4}He film in advance to eliminate the surface magnetization of ^{3}He. The observed magnetizations below 24 bar show no increase due to the Curie-Weiss behavior of the surface magnetization at low temperatures. But at 32 bar, the magnetization increases at lower temperatures, which is well explained with a Curie-Weiss magnetism. We think that this difference comes from the film state of ^{4}He; at low pressures the ^{4}He film is in the superfluid state and inhibits ^{3}He from solidifying onto the surface of the silica strands. But at 32 bar the ^{4}He film itself probably solidifies and no more prevents a formation of a surface ^{3}He solid.

CP850, *Low Temperature Physics: 24th International Conference on Low Temperature Physics;*
edited by Y. Takano, S. P. Hershfield, S. O. Hill, P. J. Hirschfeld, and A. M. Goldman

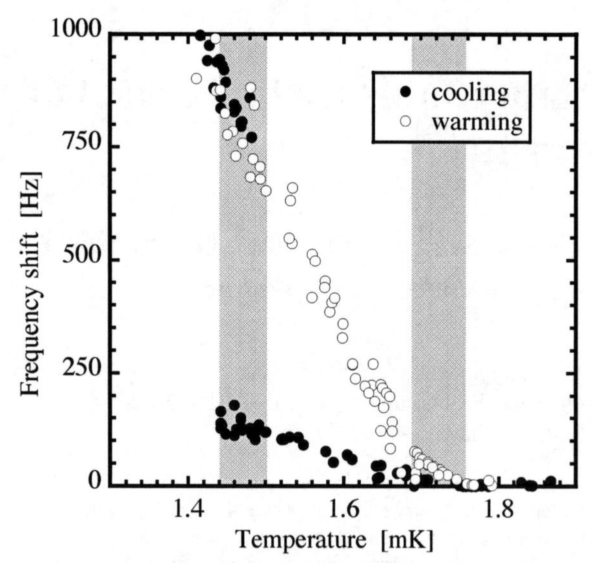

FIGURE 1. NMR frequency shift as a function of temperature at 32 bar. Solid circles are for cooling from the normal state and open circles are for warming from a complete B-like state.

Figure 1 shows the result of cw NMR at 32 bar. The measurements were made at 923 kHz by sweeping the magnetic field, and the field of resonance has been converted to a frequency shift relative to the Larmor frequency. The superfluid transition temperature T_c^{aero} was 1.755 mK. On cooling from the normal state, the A-like phase first appeared with a positive frequency shift with nearly linear temperature dependence. Below 1.500 mK, the B-like phase appeared in the large positive frequency region with one broad peak and the A-like and B-like phases coexisted within the 50 μK wide shaded region on the left in Fig. 1. The observed B-like phase signal with one peak is quite different from that in a previous experiment.[2] By using this feature, we could distinguish two signals from the two phases in coexistence. Below T_{AB}^{aero} of 1.440 mK, the A-like phase disappeared. On warming from the B-like phase, the A-like phase did not appear until the right shaded region, which was confirmed by cooling back from this region.[6] Such a hysteresis was observed at all pressures above 19 bar.

We kept the temperature at 1.550 mK on cooling and studied frequencies of FID signals in the A-like phase. The tipping angles were calibrated using the normal state of liquid ^{3}He. Figure 2 shows the frequency of the FID signal after an rf pulse at 32 bar. The frequencies are given relative to the Larmor frequency as in Fig. 1. In bulk A phase it is well known that the FID frequency depends on the tipping angle, as presented with a broken line in Fig. 2.[7] The obtained data in the aerogel experiment are similar to this curve. But, the prediction of Miura *et al.*,[8] using the order parameter proposed as the "robust phase"

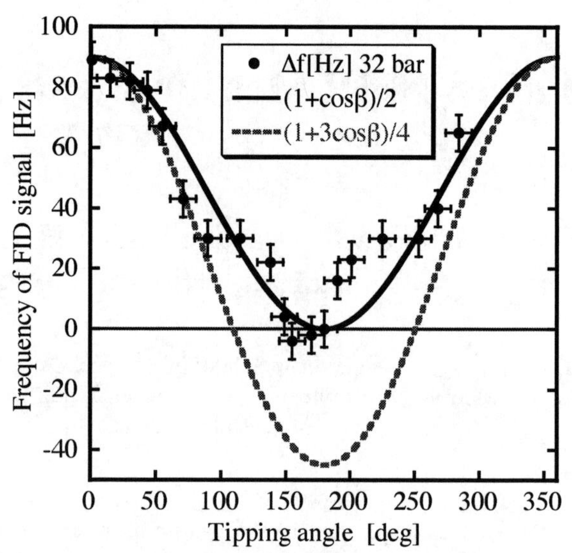

FIGURE 2. Frequency of the FID signal as a function of tipping angle at 32 bar. Broken and solid lines are expectations in bulk liquid and in aerogel, respectively.

by Fomin,[9] explains quite well the data as shown with the solid line.

We have observed the same tipping angle dependence also at 24 bar. This means that the surface magnetization is not the cause of this phenomenon.

We have also measured the frequency of the FID in the B-like phase. The results are very similar to those observed in the B phase confined in a parallel plate geometry.[10] This means that the order parameter of the B-like phase is very similar to that of the bulk B phase. From the result of the cw and pulsed NMR experiments, we conclude that the order parameter of the B-like phase in aerogel forms a non-Leggett configuration as in a slab geometry.[10]

REFERENCES

1. J. V. Porto and J. M. Parpia, *Phys. Rev. Lett.* **74**, 4667 (1995).
2. B. I. Baker *et al., Phys. Rev. Lett.* **85**, 2148 (2000).
3. G. Gervais *et al., Phys. Rev. Lett.* **87**, 035701 (2001).
4. E. Nazaretski *et al., J.Low Temp. Phys.* **134**, 763 (2004).
5. V.V. Dmitriev *et al., JETP Lett.* **76**, 312 (2002).
6. To be published elsewhere.
7. A. J. Leggett, *Rev. Mod. Phys.* **47**, 331 (1975)
8. M. Miura *et al., J. Low Temp. Phys.* **138**, 153 (2005).
9. I. A. Fomin, *JETP Lett.* **77**, 240 (2003).
10. O. Ishikawa *et al., J. Low Temp. Phys.* **75**, 35 (1989).

Compressed Silica Aerogels for the Study of Superfluid ^{3}He

J. Pollanen[a], H. Choi[a], J.P. Davis[a], S. Blinstein[a], T.M. Lippman[a], L.B. Lurio[c], N. Mulders[b], and W.P. Halperin[a]

[a] *Department of Physics and Astronomy, Northwestern University, Evanston, IL 60208, USA*
[b] *Department of Physics and Astronomy, University of Delaware, Newark, DE 19716, USA*
[c] *Department of Physics, Northern Illinois University, DeKalb, IL 60115, USA*

Abstract. We have performed Small Angle X-ray Scattering (SAXS) on uniaxially strained aerogels and measured the strain-induced structural anisotropy. We use a model to connect our SAXS results to anisotropy of the ^{3}He quasiparticle mean free path in aerogel.

Keywords: superfluidity, helium 3, aerogel
PACS: 67.57.Pq, 61.10.Eq

INTRODUCTION

Measurements of the low temperature phase diagram of superfluid ^{3}He in 98% aerogel indicate a stable *B*-phase and a metastable *A*-like phase [1-3]. Vicente *et al.* proposed that the relative stability of these phases can be attributed to local anisotropic scattering of the ^{3}He quasiparticles by the aerogel network [4]. This network consists of silica strands with a diameter of ~30 Å and average separation $\xi_a \approx 300$ Å. Vicente *et al.* also proposed using uniaxial strain of the aerogel to produce global anisotropy [4]. We have performed SAXS on two uniaxially strained aerogels and found that strain introduces anisotropy on the ~100 Å length scale. We relate this to anisotropy of the quasiparticle mean free path, λ.

EXPERIMENT

SAXS studies of two aerogel samples (97.1% and 98% porosity) were performed at Sector 8 of the Advanced Photon Source (APS) at Argonne National Laboratory, using a photon energy of 7.5 keV. The 97.1% sample was grown at Northwestern University and the 98% sample was grown at the University of Delaware. The samples were cylinders with diameter to height ratios of 1.53 (97.1%) and 0.65 (98%). For the sample grown at Northwestern radial shrinkage of ~10% was observed after supercritical drying. The porosity of the sample was measured after drying.

The samples were uniaxially strained along the cylinder axis and oriented such that the strain axis was perpendicular to the x-ray beam. For the 98% aerogel, SAXS was performed with nominally zero and 28% strain. For the 97.1% sample, the sample conditions were fully relaxed and a series of increasing strains in the range 3.5-52.8%. Beyond ~30% the sample showed significant damage in the form of cracks and the results were not analyzed. For each value of strain the scattered x-ray intensity, $I(q)$, was obtained, where q is the scattered x-ray wave vector. $I(q)$ was binned for various values of the azimuthal angle, ϕ, defined with respect to the incident x-ray beam in the plane of the CCD camera. $\phi = 90°$ is parallel to the strain axis.

RESULTS AND DISCUSSION

Our analysis of the scattering curves is based on a phenomenological scattering function used to fit $I(q)$,

$$I(q) = \frac{C\xi^d}{\left(1+q^4\xi^4\right)^{d/4}} \frac{\left(1+q^2\xi^2\right)^{1/2}}{q\xi} \times \sin\left[(d-1)\tan^{-1}(q\xi)\right], \quad (1)$$

where C, d, and ξ are fit parameters. Eq. (1) is a modified version of the structure factor described by Freltoft *et al.* for a fractal structure [5]. In Eq. (1), C is a constant, d is approximately the fractal dimension of the aerogel, and ξ is associated with the upper length

CP850, *Low Temperature Physics: 24th International Conference on Low Temperature Physics;* edited by Y. Takano, S. P. Hershfield, S. O. Hill, P. J. Hirschfeld, and A. M. Goldman

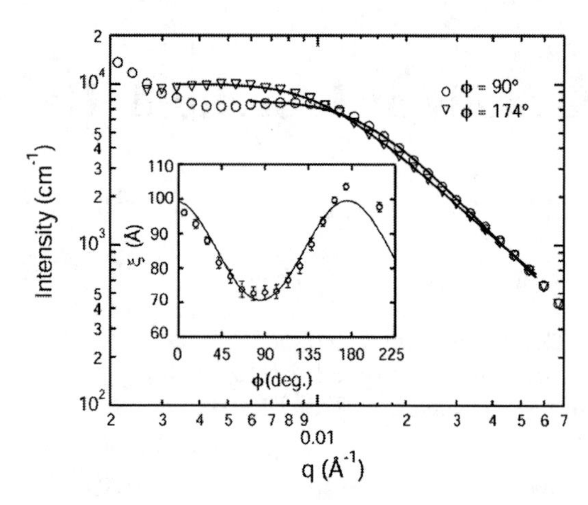

FIGURE 1. $I(q)$ fit with Eq. (1) (ϕ = 90°, 174°) for the 97.1% aerogel strained by 21.1%. The inset depicts $\xi(\phi)$ for this strain.

scale at which the aerogel ceases to be fractal in nature. This length scale is of the order of ξ_a. In the limit $q\xi$ >>1, Eq. (1) is proportional to q^{-d}. This procedure produced fits that match the data well. Fig. 1 presents $I(q)$ for a strain of 21.1% (ϕ = 90°, ϕ= 174°) along with fits produced by Eq. (1).

For each value of strain we plotted ξ vs. ϕ, and found it to vary as $\xi(\phi) = \xi_0 + \xi_1\sin(2\phi + \theta)$. The results for the 97.1% sample strained by 21.1% are presented in the inset of Fig. 1. For this sample we found θ= 99.4° ± 1.3° close to the expected symmetry axis of θ= 90°. ξ_0 ranged from 85-98 Å. Similar ϕ-dependence was found for C and d. A consistent sinusoidal behavior was observed for the 98% porosity aerogel compressed by 28%. We found that θ= 96.4° ± 1.5° for this sample. Note, the error estimates on θ are statistical. Previous SAXS studies on isostatically strained aerogels also indicate a decrease in ξ with strain [6]. The increase of ξ_1 with strain, shown in Fig. 2, demonstrates that anisotropy can be introduced into the aerogel.

For the 97.1% sample we discovered intrinsic anisotropy present with the sample unstrained. $\xi(\phi)$ was ~180° out of phase relative to the strained case. We suspect that this intrinsic anisotropy was produced during the synthesis of the aerogel. Slight anisotropy in the nominally unstrained 98% sample was observed but might be accounted for by the strain from the sample holder. For the unstrained 97% sample there was no stress from the sample holder.

We have used a simple model of the aerogel network to connect the anisotropy measured in ξ with the anisotropy of the quasiparticle mean free path [7].

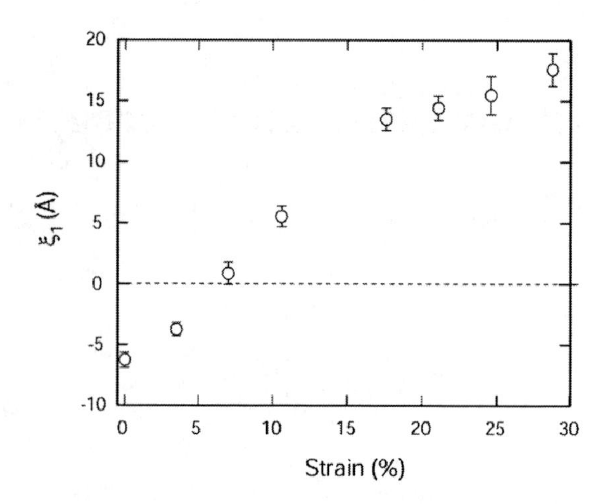

FIGURE 2. ξ_1 vs. strain for the 97.1% aerogel.

We find, $\lambda_\perp/\lambda_\parallel = 2(\xi_\perp/\xi_\parallel)/((\xi_\perp/\xi_\parallel)+1)$, where $\lambda_\perp(\lambda_\parallel)$ is the mean free path perpendicular (parallel) to the strain axis and $\xi_\perp \equiv \xi_0+\xi_1$ and $\xi_\parallel \equiv \xi_0-\xi_1$. At a strain of 28% $\lambda_\perp/\lambda_\parallel$ = 1.33 for the 98% sample and $\lambda_\perp/\lambda_\parallel$ = 1.21 for the 97.1% sample. In conclusion, uniaxial strain produces significant global anisotropy in the structure of the aerogel and may be used to test the relative stability of the A and B phases of superfluid ^{3}He. Preliminary measurements to this effect by J.P. Davis *et al.* have been conducted in 98% aerogel grown at Northwestern [8].

ACKNOWLEDGMENTS

The authors would like to thank J.A. Sauls for valuable theoretical insights. We are grateful to G.W. Scherer and J.F. Poco for their advice regarding aerogel shrinkage and fabrication. We also acknowledge the support of the National Science Foundation, DMR-0244099.

REFERENCES

1. G. Gervais, K. Yawata, N. Mulders, and W.P. Halperin, *Phys. Rev. B* **66**, 054528 (2002).
2. B.I. Barker, Y. Lee, L. Polukhina, and D.D. Osheroff, *Phys. Rev. Lett.* **85**, 2148 (2000).
3. E. Nazaretski, N. Mulders, and J.M. Parpia, *J. Low Temp. Phys.* **134**, 763 (2004).
4. C.L. Vicente, H.C. Choi, J.S. Xia, W.P. Halperin, N. Mulders, and Y. Lee, *Phys. Rev. B*, to appear.
5. T. Freltoft, J.K. Kjems, and S.K. Sinha, *Phys. Rev. B* **33**, 269 (1986).
6. T. Woignier, I. Beurroies, P. Delord, V. Gibiat, R. Sempere, and J. Phalippou, *Eur. Phys. J. AP* **6**, 267 (1999).
7. J.A. Sauls, (private communication).
8. J.P. Davis, H. Choi, J. Pollanen, and W.P. Halperin, in these Proceedings.

Acoustic Spectroscopy of Superfluid ^{3}He in Aerogel

J.P. Davis, H. Choi, J. Pollanen, and W.P. Halperin

Department of Physics and Astronomy, Northwestern University, Evanston, IL 60208, USA

Abstract: We have designed an experiment to study the role of global anisotropic quasiparticle scattering on the dirty aerogel superfluid ^{3}He system. We observe significant regions of two stable phases at temperatures below the superfluid transition at a pressure of 25 bar for a 98% aerogel.

Keywords: Superfluidity, Helium 3, Aerogel
PACS: 67.57.Pq, 67.57.Bc, 64.60.My

INTRODUCTION

Ultrasonic spectroscopy has proven to be a powerful tool in the study of ^{3}He. The acoustic impedance for transverse sound exhibits anomalies at phase transitions that mark the superfluid phase diagram of ^{3}He in 98% porosity silica aerogel [1]. The scattering of ^{3}He quasiparticles from the silica aerogel strands suppresses T_c and stabilizes the B-phase. An A-like phase is found to be metastable in zero field with large supercooling [1]. This is consistent with NMR [2] and low-frequency sound velocity measurements [3].

More recent acoustic tracking experiments by Vicente *et al.* [4] and NMR by Osheroff *et al.* [5] (99.3% aerogel) reveal that the A-like phase is in fact stable in a small temperature window near T_c at high pressure. Vicente *et al.* suggest that this stabilization is due to the *local* anisotropic scattering from the aerogel strands. Furthermore, they propose introducing *global* anisotropy by uniaxial compression of the aerogel to study the effect.

THEORY

Sauls [6] and Thuneberg *et al.* [7] have shown that local anisotropy can stabilize the axial state of superfluid ^{3}He within aerogel. The relative stability of the axial (A) and isotropic (B) phases can be expressed as the difference between the beta parameters. The beta parameters are the coefficients of the fourth order terms in the Ginzburg-Landau expansion of the free energy in powers of the order parameter and are proportional to the difference in the heat capacity jumps.

Sauls also noted [6] that large length scale correlations, or global anisotropy, in the aerogel might also favor phases with the orbital wavefunction perpendicular to the anisotropy axis, namely the planar or axial phases.

EXPERIMENT

In order to study the role of anisotropy, one needs a probe that is both directional and extremely sensitive to phase transitions in the ^{3}He. Transverse acoustic impedance has been shown to give a clear signature of all phase transitions in ^{3}He [1]. The magnetic field dependence of the phase diagram allows us to assign which phases are equal spin pairing (ESP), like the A-phase or non-ESP, like the B-phase.

We designed and built a cell to compress a pair of aerogel samples that sandwich an *ac*-cut quartz acoustic transducer, as shown in Fig. 1. The electrical impedance was measured with a continuous wave impedance bridge [8]. A melting curve thermometer (MCT) was used as the primary thermometer based on the Greywall temperature scale [9].

The 98.2% aerogel in this experiment was grown at Northwestern using a two-step synthesis with rapid supercritical extraction (RSCE); and ~10% shrinkage was observed. Similar aerogels were studied by small-angle x-ray scattering (SAXS) as a function of compression [10]. Anisotropy increases systematically with uniaxial compression.

CP850, *Low Temperature Physics: 24th International Conference on Low Temperature Physics;*
edited by Y. Takano, S. P. Hershfield, S. O. Hill, P. J. Hirschfeld, and A. M. Goldman

Additionally, there is evidence of some intrinsic anisotropy [10].

In our preliminary work, we performed temperature sweeps, Fig. 2, to determine how the ^{3}He might be affected by this aerogel before removing the spacers and compressing the samples; this is the data that we present here.

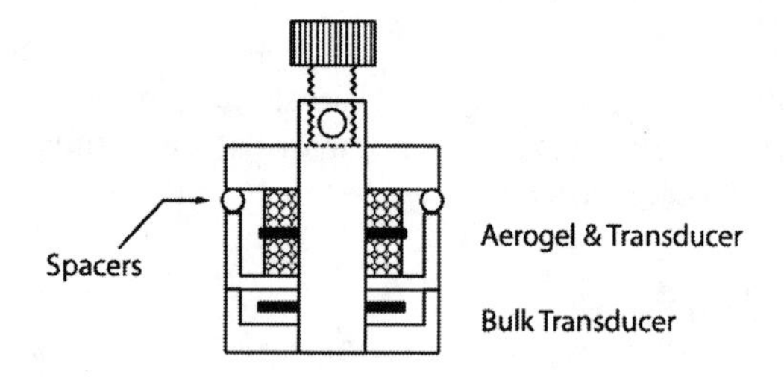

FIGURE 1. Compression cell. Spacers ensure transducer contact to the nominally uncompressed aerogel.

DATA AND RESULTS

At a pressure of 25 bar the bulk transition is at 2.36 mK, and is indicated by a separate bulk-transducer, Fig. 1. This trace is not shown in Fig. 2. We see no evidence for a bulk transition with the aerogel-sample transducer. In addition, the superfluid transition temperature in aerogel is less suppressed than previously observed by Gervais for a comparable porosity aerogel [1] (T_{ca} = 1.91 mK). The transition from normal to superfluid appears to be in two parts, the superposition of a broad transition and a narrow transition. At lower temperatures there are also two distinct features in the acoustic impedance. On warming one of these is exceedingly sharp ($\Delta T \approx 2$ μK) and it exhibits a small hysteresis that can be associated with a first order transition. All of these features have been reproduced on multiple temperature sweeps.

The double transitions can most naturally be associated with there being two, non-identical, aerogel samples with which the transducer is in contact. Tentatively we associate the two low temperature features as transitions from *B* to *A*-like phases on warming, based on: a) previous studies of transverse impedance experiments [1], and b) their supercooling. The stability of the *A*-like phase might be a consequence of intrinsic global anisotropy [10], or possibly anisotropy introduced by nominal strain from the sample holder. Further work at different pressures and as a function of compression and magnetic field should help to clarify this situation.

CONCLUSIONS

Preliminary studies of ^{3}He at 25 bar in 98% aerogel grown at Northwestern suggest that an *A*-like phase can be stabilized, likely due to global anisotropy induced in the aerogel sample.

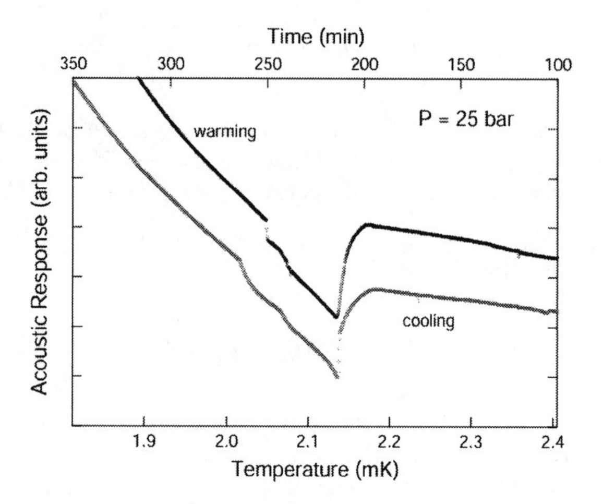

FIGURE 2. Acoustic impedance measurements of phase transitions for ^{3}He within a 98.2% aerogel. The major feature is the transition to superfluid in aerogel; smaller features are discussed in the text.

ACKNOWLEDGMENTS

We would like to thank J.A. Sauls, N. Mulders, and Y. Lee for helpful discussions and acknowledge support from the National Science Foundation, DMR-0244099.

REFERENCES

1. G. Gervias, K. Yawata, N. Mulders, and W.P. Halperin. *Phys. Rev. B* **66**, 054528 (2002).
2. K. Matsumoto, J. V. Porto, L. Pollack, E. N. Smith, T. L. Ho, and J. M. Parpia. *Phys. Rev. Lett.* **79**, 253 (1997).
3. E. Nazaretski, N. Mulders, and J.M Parpia. *J. Low Temp. Phys.* **134**, 763-768 (2004).
4. C.L. Vicente, H.C. Choi, J.S. Xia, W.P. Halperin, N. Mulders, and Y. Lee. *Phys. Rev. B*, **72** 094519 (2005).
5. D.D. Osheroff and J.E. Baumgardner. *J. Low Temp. Phys.* **138**, 117 (2005).
6. J.A. Sauls (unpublished internal communication).
7. E.V. Thuneberg, S.K. Yip, M. Fogelstrom, and J.A. Sauls. *Phys. Rev. Lett.* **80**, 2861 (1998).
8. P.J. Hamot, H.H. Hensley, and W.P. Halperin. *J. Low Temp. Phys.* **77**, 429 (1989).
9. D.S. Greywall. *Phys. Rev. B.* **33**, 11 (1986).
10. J. Pollanen, H. Choi, J.P. Davis, S. Blinstein, T.M. Lippman, L.B. Lurio, N. Mulders and W.P. Halperin. In these Proceedings.

Specific Heat of Disordered ^{3}He

H. Choi[a], J.P. Davis[a], J. Pollanen[a], N. Mulders[b], and W.P. Halperin[a]

[a]*Department of Physics and Astronomy, Northwestern University, Evanston, IL 60208, USA*
[b]*Department of Physics and Astronomy, University of Delaware, Newark, DE 19716, USA*

Abstract. Porous aerogel is a source of elastic scattering in superfluid ^{3}He and modifies the properties of the superfluid, suppressing the transition temperature and order parameter. The specific heat jumps for the *B*-phase of superfluid ^{3}He in aerogel have been measured as a function of pressure and interpreted using the homogeneous and inhomogeneous isotropic scattering models. The specific heat jumps for other *p*-wave states are estimated for comparison.

Keywords: specific heat, superfluid, helium 3, aerogel.
PACS: 67.57.Bc, 67.57.Pq

INTRODUCTION

Porous silica aerogel imbibed with liquid ^{3}He provides a source of scattering and modifies the superfluid phases. Torsional oscillator[1], NMR[2,3], and specific heat experiments[4] performed on ^{3}He in aerogel show substantial suppression of the transition temperature and order parameter. Thuneberg *et al.*[5] describe a model for such a system in the framework of the Ginzburg-Landau theory, where the free energy of various superfluid ^{3}He phases is expressed with different combinations of the β coefficients[6]. As a result, the specific heat jump of ^{3}He at the superfluid transition temperature can be calculated for the possible *p*-wave states of the superfluid.

According to the homogeneous isotropic scattering model (HISM), the impurities, i.e. the silica balls in the aerogel, are assumed to be distributed homogeneously in space. Then only one parameter, the quasiparticle mean free path, is needed to describe the effects of impurities in the strong scattering limit. The HISM, however, systematically underestimates the transition temperature. Sharma and Sauls introduced a second parameter in the inhomogeneous isotropic scattering model (IISM) the strand-strand correlation length ξ_a, to make the pressure-temperature phase diagram consistent with other superfluid properties[7]. The mean free path was estimated to be 150 nm from previous measurements of the phase diagram[1], consistent with transport measurements of spin diffusion[8] and thermal conductivity[9,10] of liquid ^{3}He in aerogel. Using this mean free path at zero magnetic field, the *B*-phase is predicted to be stable[5] and experiments agree[11]. We investigated the predictions of both the HISM and IISM on the specific heat jump for *B*-phase of superfluid ^{3}He in aerogel and compared the calculation with our experiment[4].

EXPERIMENT

The sample cell used in this experiment was made of high purity silver. It was cooled by adiabatic demagnetization of $PrNi_5$ after precooling with a dilution refrigerator. The cell was thermally isolated with a cadmium heat switch; then an adiabatic heat pulse technique was used to measure the specific heat. An LCMN thermometer was calibrated with a melting curve thermometer[4,12]. The measurement was performed in the pressure range of 1 to 29 bar and the temperature range of 0.75 to 5 mK.

RESULTS AND DISCUSSION

Fig. 1 shows the phase diagram obtained from the specific heat measurement. Both the transition temperature and the specific heat jump are suppressed in aerogel ^{3}He; this is a direct manifestation of the suppression of the order parameter. We present here a detailed analysis of the specific heat jump to show that this is, in fact, the superfluid *B*-phase.

In the Ginzburg-Landau theory, the specific heat jump for a given phase, ΔC_p, is calculated in terms of the α and β coefficients, $\Delta C_p = (\alpha'(T_c))^2/\beta_p$, where the subscript p denotes the superfluid phases of ^{3}He,

CP850, *Low Temperature Physics: 24th International Conference on Low Temperature Physics;*
edited by Y. Takano, S. P. Hershfield, S. O. Hill, P. J. Hirschfeld, and A. M. Goldman

axial (A), isotropic (B), polar (P), and planar (PL) phases. For each of these phases, β_p is defined as $\beta_A = \beta_{245}$, $\beta_B = \beta_{12} + \beta_{345}/3$, $\beta_P = \beta_{12} + \beta_{345}/2$, and $\beta_{PL} = \beta_{12} + \beta_{345}$, respectively.

For the HISM, the α and β coefficients are modified to include the pair breaking parameter $x = \hbar v_F/2\pi k_B T\lambda$ where λ is the mean free path[5]. The specific heat jump for the B-phase is calculated and compared with the experiment in the inset of Fig. 2. Within the HISM, the calculation is most consistent with the experiment for $\lambda = 160$ nm (solid curve). The same calculation is performed within the IISM by substituting x with $\tilde{x} = x/(1+(\xi_a/\lambda)^2/x)$ where ξ_a is the strand-strand correlation length of the aerogel[7]. For the IISM, $\lambda = 160$ nm and $\xi_a = 40$ nm best matched the experiment (dashed curve).

The relative stability of the B-phase is determined from the magnitude of β coefficients. If $\beta_B < \beta_p$, the B-phase is stabilized over a p-phase and vice versa. This stability condition enters into the specific heat jump through the relation:

$$\frac{\Delta C_B - \Delta C_p}{\Delta C_B} = \frac{\beta_B^{-1} - \beta_p^{-1}}{\beta_B^{-1}} = \frac{\beta_p - \beta_B}{\beta_p}. \qquad (1)$$

According to the HISM or IISM, the B-phase is stable up to 30 bar. (See Fig. 2.) Despite the stability condition for the B-phase, a metastable A-phase is observed in the superfluid ^{3}He in aerogel[11,13].

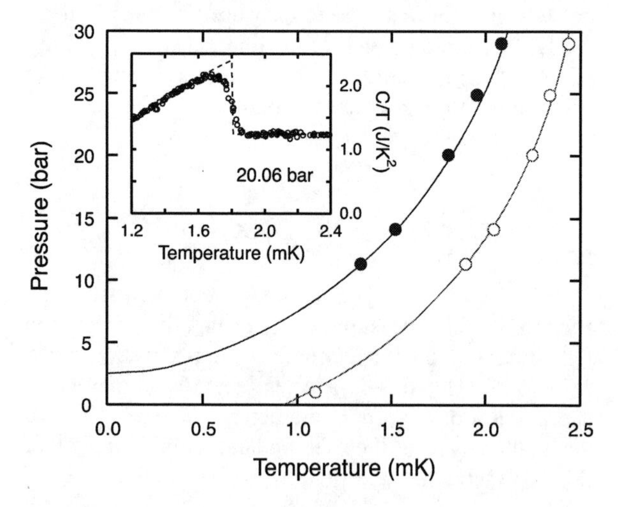

FIGURE 1. Phase diagram obtained from the specific heat measurement. Open circles and solid circles mark bulk superfluid transition and aerogel superfluid transition, respectively. For the aerogel superfluid transition, the IISM with $\lambda = 190$ nm and $\xi_a = 50$ nm is used to generate the curve. The inset is the C/T for ^{3}He in aerogel near T_{ca}. The specific heat jump, $\Delta C_B/C$, at T_{ca} is 0.92 as opposed to $\Delta C_B/C = 1.82$ for bulk ^{3}He at 20.06 bar.

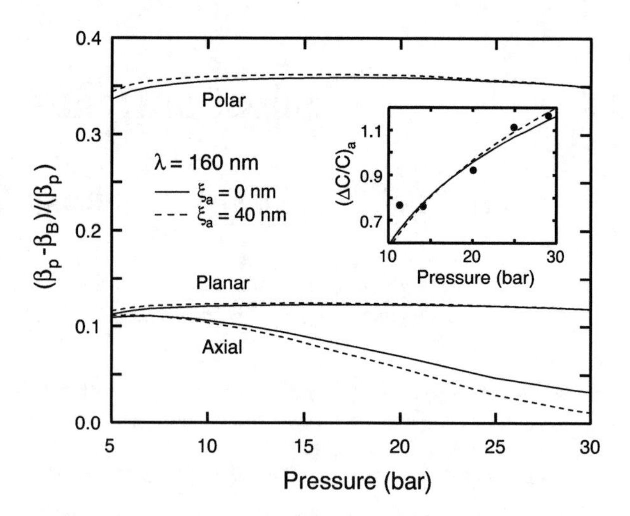

FIGURE 2. $(\beta_p-\beta_B)/\beta_p$ is calculated for the pressure range of 5 to 30 bar with $\lambda = 160$ nm, $\xi_a = 0$ nm and 40 nm for the HISM and IISM respectively. These parameters were obtained from the comparison between the measurement and calculation of the B-phase specific heat jump (inset).

The difference between the specific heat jump of the B-phase and the metastable phase should be observable and might be helpful in confirming its identity.

ACKNOWLEDGMENTS

We would like to thank T.M. Haard and K. Yawata for their valuable contribution to the project and J.A. Sauls and Y. Lee for helpful discussions. This work was supported by NSF, Grant No. DMR-0244099.

REFERENCES

1. K. Matsumoto *et al., Phys. Rev. Lett.* **79**, 253 (1997).
2. D.T. Sprague *et al., Phys. Rev. Lett.* **75**, 661 (1995).
3. D.T. Sprague *et al., Phys. Rev. Lett.* **77**, 4568 (1996).
4. H. Choi *et al., Phys. Rev. Lett.* **93**, 145301 (2004).
5. E.V. Thuneberg *et al., Phys. Rev. Lett.* **80**, 2861 (1998).
6. D. Vollhardt and P. Wölfle, *The Superfluid Phases of Helium 3*, Bristol, PA, USA: Taylor and Francis Inc., 1990, pp. 113-118.
7. J.A. Sauls and P. Sharma, *Phys. Rev. B* **68**, 224502 (2003).
8. E. Collin, Ph.D. thesis, Université Joseph Fournier Grenoble (2002).
9. P. Sharma and J.A. Sauls, *J. Low Temp. Phys.* **125**, 115 (2001).
10. S.N. Fisher *et al., Phys. Rev. Lett.* **91**, 105303 (2003).
11. G. Gervais *et al., Phys. Rev. B* **66**, 054528 (2002).
12. T.M. Haard, Ph.D. thesis, Northewstern University, Evanston, IL (2000).
13. Barker *et al., Phys. Rev. Lett.* **85**, 2148 (2000).

Coupling between Solid ^{3}He on Aerogel and Superfluid ^{3}He in the Low Temperature Limit

D.I. Bradley, S.N. Fisher, A.M.Guénault, R.P. Haley, G.R. Pickett, P. Skyba¶, V. Tsepelin, and R.C.V. Whitehead

Department of Physics, Lancaster University, Lancaster, LA1 4YB, UK
¶also Institute of Experimental Physics, Watsonova 47, 04353 Košice, Slovakia

Abstract: We have cooled liquid ^{3}He contained in a 98% open aerogel sample surrounded by bulk superfluid ^{3}He-B at zero pressure to below 120 μK. The aerogel sample is placed in a quasiparticle blackbody radiator cooled by a Lancaster-style nuclear cooling stage to ~200 μK. We monitor the temperature of the ^{3}He inside the blackbody radiator using a vibrating wire resonator. We find that reducing the magnetic field on the aerogel sample causes substantial cooling of all the superfluid inside the blackbody radiator. We believe this is due to the demagnetization of the solid ^{3}He layers on the aerogel strands. This system has potential for achieving extremely low temperatures in the confined fluid.

Keywords: superfluidity, aerogel, solid ordering, ^{3}He
PACS: 67.57, 67.90

INTRODUCTION

Low density silica aerogels have been widely used in recent years to study the effects of impurity scattering on superfluid ^{3}He [1]. The spacing between the aerogel strands is comparable to the superfluid coherence length and therefore suppresses superfluidity, and can induce gapless superfluidity under certain conditions [2]. The silica strands are coated in solid ^{3}He which, owing to the very large surface area, can have a very significant heat capacity at low temperatures and in modest magnetic fields. This opens up the possibility of using the solid ^{3}He in aerogel as a refrigerant to cool the fluid via nuclear demagnetization. Below, we discuss some of our initial findings.

EXPERIMENTAL SETUP

We use a typical Lancaster style nuclear cooling stage to cool a cell containing ^{3}He at zero pressure to ~110 μK. The experimental chamber consists of a blackbody radiator [3] formed from a sapphire tube and sealed at the top by a piece of stycast impregnated paper with a 0.23 mm hole to thermally connect the superfluid ^{3}He-B in the tube to the bulk ^{3}He-B (at ~110 μK) outside. The sapphire tube, of internal diameter 4.4 mm, contains 0.78 cm^3 of ^{3}He with 3.7 mg of 98% open aerogel at its base. At zero pressure, we note that the liquid ^{3}He inside the aerogel is always in the normal state.

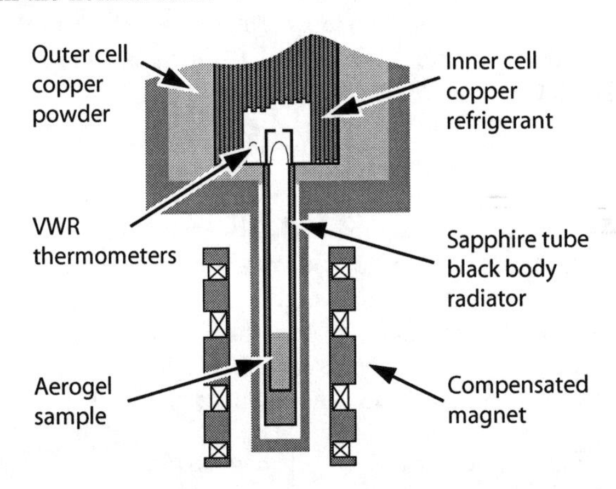

FIGURE 1. Schematic of the experimental cell showing the sapphire blackbody radiator containing the aerogel sample. The copper demagnetization final field magnet is not shown.

Inside the radiator tube are two NbTi vibrating wire resonators (VWRs), one as a thermometer and the other as a heater. Measurement of the damping of the thermometer wire allows us to infer the quasiparticle density and therefore the temperature. A third wire in the bulk ^{3}He allows us to check that the temperature outside the radiator remains constant at ~110 μK.

CP850, *Low Temperature Physics: 24th International Conference on Low Temperature Physics;* edited by Y. Takano, S. P. Hershfield, S. O. Hill, P. J. Hirschfeld, and A. M. Goldman

The magnetic field over the blackbody radiator is produced by a combination of two magnets — only one is shown in Figure 1. The final field magnet for the copper demagnetization (not shown) has a relatively flat profile over the aerogel sample and the vibrating wires. A second magnet provides a localized field on the aerogel sample but no significant field on the wires or the copper refrigerant.

The key point about a blackbody radiator is that it is a very sensitive power detector [3]. There is a simple relationship between the damping of a vibrating wire inside the radiator and the thermal power deposited in radiator which is balanced by the power leaving through the radiator hole. Therefore a measurement of damping permits the net instantaneous power into the radiator to be determined. The heat leak from the sapphire tube walls is easily determined from the steady state damping of the vibrating wire thermometer. This is typically a few pW once the experiment has been running a few weeks, and constant throughout the measurements below.

With the bulk ^{3}He-B at ~110 μK, we increase the field on the aerogel and wait overnight for the ^{3}He-B in the blackbody radiator to cool to ~ 200 μK. We then reduce the magnetic field over ~5 minutes to the final value and observe the temperature of the ^{3}He-B in the blackbody radiator. Throughout this process, the bulk ^{3}He-B temperature remains constant at ~110 μK.

RESULTS

We see a very dramatic change in the temperature of the ^{3}He-B in the blackbody radiator when we reduce the field on the aerogel sample. There is almost instantaneous cooling as the field reduces and the temperature of the ^{3}He-B in the blackbody radiator drops from ~200 μK to a minimum of around 118 μK.

Figure 2 shows several demagnetizations with the same initial conditions to different final fields. First we note that there is the dramatic cooling. Secondly, we measure the same minimum ^{3}He-B temperature independent of the final field. This is due to a balance between the heat leak into the radiator and the effective thermal boundary resistance between the ^{3}He-B and the normal ^{3}He liquid inside the aerogel (as the pressure is zero) [2] limiting the heat flow into the aerogel sample. The temperature of the ^{3}He inside the aerogel, which we cannot measure, must be very much lower and the aerogel is effectively acting as a 'black hole' absorbing quasiparticle excitations. Thirdly, we note that the higher the final field, the slower the warming rate (due to the heat leak) indicating a large field dependent heat capacity, as we observed earlier [4]. The heat leak also determines the final liquid temperature of ~190 μK. This temperature would also ultimately be achieved by a very long precool.

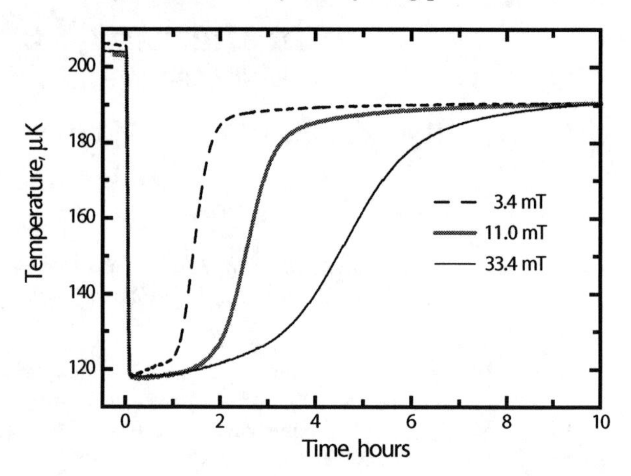

FIGURE 2. Temperature evolution of the ^{3}He-B in the blackbody radiator following three demagnetizations from an initial field of 124 mT to the final fields indicated.

Further measurements to lower fields and analysis to determine better the thermal properties of the system are currently in progress. But clearly these preliminary results confirm that solid ^{3}He in aerogel can be very effective as a nuclear refrigerant for the surrounding fluid. Also owing to the relatively poor thermal conductivity through the aerogel [2], the aerogel-confined fluid effectively decouples from the outside environment (and heat leaks) at low temperatures. This suggests that this technique could generate extremely low temperatures in the aerogel-confined fluid.

ACKNOWLEDGEMENTS

We thank N. Mulders for the aerogel sample and acknowledge the excellent technical support of I.E. Miller and M. Ward. This work is supported by the UK EPSRC under grant GR/S62109. P.S. acknowledges the support of the VEGA agency.

REFERENCES

1. Porto, J.V., and Parpia, J.M., *Phys. Rev. Lett.* **74**, 4667 (1995).
2. Fisher, S. N., Guénault, A. M., Mulders, N., and Pickett, G. R., *Phys .Rev. Letters* **91**, 105303–105306 (2003).
3. Fisher, S. N., Guénault, A.M., Kennedy, C. J., and Pickett, G. R., *Phys. Rev. Lett.* **69**, 1073 (1992).
4. Bradley, D. I., Fisher, S. N., Guénault, A. M., Haley, R. P., Mulders, N., Pickett, G. R., Skyba, P., and Whitehead, R. C. W., *J. Low Temp. Phys.* **138**, 129–134 (2005).

Effect of a Quenched Disorder on the Order Parameter of Superfluid ^{3}He

I. A. Fomin

P. L. Kapitza Institute for Physical Problems, ul. Kosygina 2, 119334 Moscow, Russia

Abstract. As a consequence of continuous degeneracy of the order parameter of the superfluid ^{3}He quenched disorder in a form of aerogel gives rise both to a disruption of the orientational long-range order of the condensate and to a significant change of the order parameter itself. There exist a class of quasi-isotropic order parameters which are not sensitive to the disorienting effect of aerogel. While the BW order parameter belongs to this class the ABM does not. A possible candidate for the order parameter of the observed A-like phase is discussed.

Keywords: superfluidity, quenched disorder
PACS: PACS numbers: 67.57.-z

The average distance ξ_a between the strands of high porosity aerogels introduced as impurities in the superfluid ^{3}He [1, 2] does not differ much from the superfluid coherence length ξ_0. This property is in conflict with the condition of applicability of the conventional theory of superconducting alloys [3] $n\xi_0^3 \gg 1$, where n is the average density of impurities. In aerogels $n\xi_0^3 \sim 1$ and the fluctuations of $n\xi_0^3$ are significant. In the vicinity of the transition temperature T_c, the effect of fluctuations can be described phenomenologically by an introduction of random terms in the Ginzburg and Landau (GL) functional. For the scalar order parameter ψ such an analysis has been made by Larkin and Ovchinnikov [4]. According to this analysis, the most singular corrections to the order parameter at $T \to T_c$ originate from the term $F_\eta = N(0)\int \eta(\mathbf{r})|\psi|^2 d^3r$ (here $N(0)$ is the density of states and $\eta(\mathbf{r})$ – a random scalar), which describes local variations of T_c. The generalization of this approach to the 3×3 complex matrix order parameter $A_{\mu j}$ is straightforward. The additional term in the GL functional has a form:

$$F_\eta = N(0)\int \eta_{jl}(\mathbf{r})A_{\mu j}A^*_{\mu l}d^3r. \qquad (1)$$

Now $\eta_{jl}(\mathbf{r})$ is a random tensor with zero average value. Since the strength of $t \to -t$ invariance tensor $\eta_{jl}(\mathbf{r})$ is real and symmetrical, its isotropic part $\eta^{(i)}_{jl}(\mathbf{r}) = \frac{1}{3}\eta_{mm}(\mathbf{r})\delta_{jl}$ describes local variations of $T_c = T_c(\mathbf{r})$ due to fluctuations of the density of scatterers; it couples only to the overall amplitude of the order parameter $A_{\mu l}A^*_{\mu l} \equiv \Delta^2$. The effect of $\eta^{(i)}_{jl}(\mathbf{r})$ is similar to that for the scalar case – broadening of the transition and a suppression of the absolute value of the order parameter. New properties originate from the anisotropic part $\eta^{(a)}_{jl}(\mathbf{r}) \equiv \eta_{jl}(\mathbf{r}) - \frac{1}{3}\eta_{ll}(\mathbf{r})\delta_{jl}$ which describes a local splitting of T_c for different projections of angular momentum because of breaking of spherical symmetry by the aerogel strands. The contribution of $\eta^{(a)}_{jl}(\mathbf{r})$ to F_η depends on the particular form of the order parameter and on its orientation with respect to the aerogel. Correspondingly $\eta^{(a)}_{jl}(\mathbf{r})$ can both influence the form of the order parameter and locally change its orientation. To make a qualitative guess of a possible effect of $\eta^{(a)}_{jl}(\mathbf{r})$ for different order parameters, one can start with a uniform order parameter and substitute it into F_η. No essential difference with respect to the scalar case is expected if $\eta^{(a)}_{jl}A_{\mu j}A^*_{\mu l} = 0$. This is the case for the BW order parameter of bulk B phase: $A^{BW}_{\mu j} = \Delta e^{i\varphi}R_{\mu j}$, where $R_{\mu j}$ is a real orthogonal matrix, but this is not the case for the bulk A phase (ABM). Its order parameter has a form $A^{ABM}_{\mu j} = (\Delta/\sqrt{2})\hat{d}_\mu(\hat{m}_j + i\hat{n}_j)$ with $(\hat{m}_j, \hat{n}_j) = 0$; then $\eta^{(a)}_{jl}(\mathbf{r})A_{\mu j}A^*_{\mu l} \sim -\eta^{(a)}_{jl}l_j l_l \neq 0$, where $l_j = e_{jik}m_i n_k$. This property of ABM order parameter is of importance in connection with the identification of the A-like phase which has been observed in liquid ^{3}He in aerogel just below T_c [5, 6].

The disorienting effect of the random tensor $\eta_{jl}(\mathbf{r})$ on the ABM order parameter has been emphasized by Volovik [7]. Following the arguments of Larkin [8] and Imry and Ma [9] he concluded that there may be no long-range order with the ABM order parameter and suggested instead a glass state. In that state, the order parameter preserves the ABM form locally but its orientation changes on a scale $L_{IM} \sim \xi_0/\eta^2$. The characteristic value of η^2 can be estimated e.g. from the measured width of the specific heat jump. This width is determined by the parameter $\alpha \equiv \eta^2/\sqrt{|\tau|}$ where $\tau = (T - T_c)/T_c$. A region where fluctuations dominate is determined by the condition $\alpha \sim 1$ [4]. For ^{3}He, using the data of Ref. [10] for 22.5 bar, one arrives at $\eta^2 \sim 1/5$ and $L_{IM} \sim 5\xi_0 \sim 10^{-5}cm$.

CP850, *Low Temperature Physics: 24th International Conference on Low Temperature Physics;*
edited by Y. Takano, S. P. Hershfield, S. O. Hill, P. J. Hirschfeld, and A. M. Goldman

Except for the disruption of the orientational order $\eta_{jl}^{(a)}$ influences a form of the order parameter. For the scalar case only the amplitude of the order parameter can change and its relative change is of the order of $\alpha \ll 1$. For the triplet case the decrements of different components of $A_{\mu j}$ can be different. Moreover their change can be not small in case of degeneracy. For a triplet Cooper pairing there are altogether 18 extrema of the free energy [11] and some of them can come close in energy. The contribution of fluctuations to the energy, being small in comparison with the full condensation energy, can be comparable to or larger than the small difference of the energies of the competing states. As a result fluctuations can shift the equilibrium and determine the choice of the order parameter. Let us consider this possibility in more details. With the random term, the GL functional takes the form:

$$F_{GL} = N(0) \int d^3r[(\tau + \frac{1}{3}\eta_{ll}(\mathbf{r}))A_{\mu j}A_{\mu j}^* +$$

$$(\eta_{jl}(\mathbf{r}) - \frac{1}{3}\eta_{mm}(\mathbf{r})\delta_{jl})A_{\mu j}A_{\mu l}^* + \frac{1}{2}\sum_{s=1}^{5}\beta_s I_s +$$

$$\frac{1}{2}\left(K_1\frac{\partial A_{\mu l}}{\partial x_j}\frac{\partial A_{\mu l}^*}{\partial x_j} + K_2\frac{\partial A_{\mu l}}{\partial x_j}\frac{\partial A_{\mu j}^*}{\partial x_l} + K_3\frac{\partial A_{\mu j}}{\partial x_j}\frac{\partial A_{\mu l}^*}{\partial x_l}\right)], \tag{2}$$

where I_s - 4th order invariants in the expansion of the free energy over $A_{\mu j}$. The coefficients $\beta_1,...\beta_5$ and K_1, K_2, K_3 are phenomenological constants.

Because of the presence of the **r**-dependent tensor $\eta_{jl}(\mathbf{r})$ in the functional (2), the solutions of the corresponding GL equations also depend on **r**. It is convenient to separate the order parameter into its average $\bar{A}_{\mu j}$ and a fluctuating part $a_{\mu j}(\mathbf{r})$ so that $A_{\mu j}(\mathbf{r}) = \bar{A}_{\mu j} + a_{\mu j}(\mathbf{r})$. Below T_c $\quad \bar{A}_{\mu j} \neq 0$. The terms proportional to $\eta_{jl}(\mathbf{r})$ in Eq. (2) dominate at $T \to T_c$. Within a region $|\tau| \leq \eta^4$ or $\alpha \geq 1$, $\quad a_{\mu j}(\mathbf{r})$ gives at least as important a contribution to the physical quantities as $\bar{A}_{\mu j}$. It is difficult to analyze solutions of GL equations within this region. An analysis is possible when $\alpha \ll 1$ and $|a_{\mu j}| \ll |\bar{A}_{\mu j}|$. In this region of parameters the iteration procedure of Ref. [4] renders equations for $\bar{A}_{\mu j}$:

$$\tau\bar{A}_{\mu j} + \frac{1}{2}\sum_{s=1}^{5}\beta_s[\frac{\partial I_s}{\partial \bar{A}_{\mu j}^*} + \frac{1}{2}(\frac{\partial^3 I_s}{\partial A_{\mu j}^*\partial A_{\nu n}\partial A_{\beta l}} < a_{\nu n}a_{\beta l} > +$$

$$2\frac{\partial^3 I_s}{\partial A_{\mu j}^*\partial A_{\nu n}^*\partial A_{\beta l}} < a_{\nu n}^* a_{\beta l} >)] = - < a_{\mu l}\eta_{lj} >; \tag{3}$$

and for $a_{\mu j}(\mathbf{r})$:

$$\tau a_{\mu j} + \frac{1}{2}\sum_{s=1}^{5}\beta_s[\frac{\partial^2 I_s}{\partial A_{\mu j}^*\partial A_{\nu n}}a_{\nu n} + \frac{\partial^2 I_s}{\partial A_{\mu j}^*\partial A_{\nu n}^*}a_{\nu n}^*] -$$

$$\frac{1}{2}K\left(\frac{\partial^2 a_{\mu j}}{\partial x_l^2} + 2\frac{\partial^2 a_{\mu l}}{\partial x_l \partial x_j}\right) = -\bar{A}_{\mu l}\eta_{lj}. \tag{4}$$

Equations (3) and (4) have the same structure as in the scalar case. The linear equation for $a_{\mu j}$ is solved by Fourier transformation and its solution has the following structure:

$$a_{\mu j}(\mathbf{k}) \sim \frac{\bar{A}_{\mu l}\eta_{lj}(\mathbf{k})}{(\kappa^2 + k^2)}, \qquad \kappa \sim 1/\xi. \tag{5}$$

Then for the average binary products entering Eq. (3) we have

$$< a_{\mu j}a_{\nu n} > \sim \bar{A}_{\mu l}\bar{A}_{\nu m}\int\frac{< \eta_{lj}(\mathbf{k})\eta_{mn}(-\mathbf{k}) >}{(\kappa^2 + k^2)^2}\frac{d^3k}{(2\pi)^3}. \tag{6}$$

These terms in Eq. (3) give corrections of a relative order of $\alpha \ll 1$ except for the directions in which $\kappa = 0$ (Goldstone directions) when the integral diverges. This is the case for the increments of $\bar{A}_{\mu j}$ and $\bar{A}_{\mu j}^*$ at the infinitesimal rotation θ_n:

$$\omega_{\mu j} = \theta_n e^{jnr}\bar{A}_{\mu r}, \quad \omega_{\mu j}^* = \theta_n e^{jnr}\bar{A}_{\mu r}^*. \tag{7}$$

The divergency means that in the Eq. (3) the contribution of the averaged products of the projections of $a_{\beta l}$ in the directions determined by $\omega_{\mu j}$ is of a relative order of α^λ, $\lambda < 1$. The value of λ depends on the mechanism of cutting off the divergency. This mechanism is provided by anharmonicity. The previously suggested evaluation of the diverging integral from the condition $|a_{\beta l}| \sim |\bar{A}_{\beta l}|$ [12] overestimates it and renders $\lambda = 0$. A better estimation is obtained by keeping the anharmonic (third order) terms in Eq. (4). The obtained nonlinear equation can be treated within a mean-field approach, i.e. the third-order terms are substituted as $aaa \to a < aa >$, where a stands for any of component of $a_{\mu j}$ and only diverging binary averages $< aa >_{sing}$ are kept. The resulting linear equation has solution of a symbolic form (all indices are suppressed):

$$a \sim \frac{Q\eta}{\Delta[(\xi_0 k)^2 + \beta < aa >]}, \tag{8}$$

where $\beta \sim 1/T_c^2$ is a typical value of β-coefficients in the GL expansion and Q is a tensor $Q_{rl} = \bar{A}_{\mu r}\bar{A}_{\mu l}^* + \bar{A}_{\mu l}\bar{A}_{\mu r}^*$. For binary averages in the Goldstone directions we have

$$< aa >_{sing} = const.\frac{Q^2}{\Delta^2}\int\frac{< \eta\eta > k^2 dk}{[(\xi_0 k)^2 + \beta < aa >_{sing}]^2}. \tag{9}$$

This is an equation for $< aa >_{sing}$. Its solution has a form

$$< aa >_{sing}^{3/2} = const.\frac{Q^2 < \eta\eta >_0}{\Delta^2\sqrt{\beta}\xi_0^3}. \tag{10}$$

Here $< \eta\eta >_0 = \int < \eta(0)\eta(\mathbf{r}) > d^3r$. With the aid of Eq. (8) $< aa >_{sing}$ can be estimated as $< aa >_{sing} \sim (\Delta Q^2 \alpha)^{2/3}$. At $\alpha \to 0$ the singular terms in the GL equation give a principal part of the contribution of fluctuations to this equation. GL-equation now has a structure:

$$\frac{\delta}{\delta A^*_{\mu j}}[F^{(0)} + F^{fl}_{reg} + F^{fl}_{sing}] = 0. \quad (11)$$

Consider a set of competing states which are nearly degenerate with ABM:

$$\left| \frac{F_c - F_{ABM}}{F_{ABM}} \right| \equiv \varepsilon \ll 1, \quad (12)$$

then, in Eq. (11) the first term is of the order of ε, the second of the order of α and the third of the order of $(\alpha)^{2/3}$. The character of the solution depends on the relative values of the small parameters ε and $(\alpha)^{2/3}$. If $\varepsilon \gg (\alpha)^{2/3}$ the order parameter is determined as a minimum of the unperturbed GL functional $F^{(0)}$ (supposedly ABM state) and fluctuations give small corrections to its form. In the opposite limit $\varepsilon \ll (\alpha)^{2/3}$ the form of the order parameter is determined by the fluctuations:

$$\frac{\delta}{\delta A^*_{\mu j}}[F^{fl}_{sing}] = 0. \quad (13)$$

This equation is satisfied if $Q \sim \delta_{jl}$:

$$A_{\mu j}A^*_{\mu l} + A_{\mu l}A^*_{\mu j} = const. \cdot \delta_{jl}, \quad (14)$$

or, equivalently:

$$\eta^{(a)}_{jl} A_{\mu j}A^*_{\mu l} = 0. \quad (15)$$

As has been discussed above, for the order parameters that meet condition (15), there is no disorienting effect of the random tensor $\eta^{(a)}_{jl}$ and long-range order is preserved. Generally this condition specifies a degenerate family of so-called "robust", or quasi-isotropic, states. For partial lifting of the degeneracy the remaining terms in the free energy have to be minimized

$$\frac{\delta}{\delta A^*_{\mu j}}[F^{(0)} + F^{fl}_{reg}] = 0 \quad (16)$$

with Eq. (14) as a constraint. Eq. (16) determines in particular the overall amplitude Δ. It has to be verified then that the found solutions satisfy condition (12). If it is the case there exists a temperature region where condition $\varepsilon \ll (\alpha)^{2/3}$ is met as well since ε does not depend on temperature but α grows as $1/\sqrt{|\tau|}$ at $T \to T_c$. Within this temperature region the above procedure of finding solution of GL equation which are determined by fluctuations is justified.

In case of A-like phase the competing states have to be of equal spin pairing (ESP) type. It follows from the experimental observation that magnetic susceptibility of the emerging phase coincides with that of normal and ABM phases [5]. A general form of the order parameter for ESP state is:

$$A^{ESP}_{\mu j} = \Delta \frac{1}{\sqrt{3}}[\hat{d}_\mu (m_j + i n_j) + \hat{e}_\mu (l_j + i p_j)], \quad (17)$$

where $\hat{d}_\mu$ and $\hat{e}_\mu$ are mutually orthogonal unit vectors, m_j, n_j, l_j, p_j – arbitrary real vectors. The order parameter $A^{ESP}_{\mu j}$ meets criterion (14) if vectors m_j, n_j, l_j, p_j satisfy the equation

$$m_j m_l + n_j n_l + l_j l_l + p_j p_l = \delta_{jl}. \quad (18)$$

A detailed description of this solution is presented elsewhere [13]. If additional constraint of the absence of spontaneous magnetization is imposed the obtained solution can be rewritten as:

$$A^R_{\mu j} = \Delta \frac{1}{\sqrt{3}} e^{i\psi}[\hat{d}_\mu (\hat{b}_j + i\cos\gamma \hat{c}_j) + \hat{e}_\mu (\hat{a}_j + i \sin\gamma \hat{c}_j)], \quad (19)$$

where $\hat{a}_j, \hat{b}_j, \hat{c}_j$ – mutually orthogonal unit vectors, and ψ and γ are parameters.

Now the condition (12) has to be verified. For the order parameter Eq. (19) ε is expressed in terms of coefficients β_i as:

$$\varepsilon = \frac{\beta_{13} - 4\beta_{45}}{9\beta_2 + \beta_{13} + 5\beta_{45}}, \quad (20)$$

where $\beta_{13} = \beta_1 + \beta_3$ etc. For the weak coupling values of β-coefficients $\varepsilon = 1/19$ i.e. small, but in that case BW phase is more favorable than ABM. The most important are strong coupling corrections to the combination β_{45} since the weak coupling value $\beta_{45} = 0$ determines a boundary between ABM and axiplanar phases in the space of β-parameters [14]. For the moment it is difficult to make definite conclusions about the value of β_{45} realized in liquid ^{3}He. Combinations of $\beta_1, ...\beta_5$ deduced from the analysis of thermodynamic data [15] do not form a complete set and cannot define ε without ambiguity. On the other hand the values of $\beta_1, ...\beta_5$ found theoretically from model assumptions [16] do not agree with the experimental findings, in particular about β_5.

Eq. (19) gives solution of GL equation of a zeroth order in the small parameter $\varepsilon/\alpha^{2/3}$. In the real ^{3}He this ratio is not necessarily very small so that the finite deviation from "robust" state occurs and disorienting effect of the random tensor $\eta_{jl}(\mathbf{r})$ comes into play. The outlined procedure offers in that case only a local form of the order parameter. Principal contribution to the integral Eq. (9) comes from the wave vectors estimated as $(\xi_0 k)^2 \sim \beta < aa >_{sing}$, or from the length scales

$L_{c-off} \sim \xi/(\alpha q^2)^{1/3}$, where $q^2 = Q^2/\Delta^2$ is the normalized deviation of the order parameter from the "robust" form. The local form of the order parameter is preserved if it changes its orientation in space on a length scale much greater than L_{c-off}. This is the case for Imry-Ma disorienting effect: $L_{IM}/L_{c-off} \sim 1/(\alpha q^2)^{2/3} \gg 1$. The resulting glass-type state is based on a "nearly robust" state and the characteristic size of the "domains" is much greater than in the case of the ABM-based glass state. The most important difference between the two states is that the order parameter $A^R_{\mu j}$, unlike the $A^{ABM}_{\mu j}$, does not have the combined gauge-orbital symmetry. As a consequence the property of superfluidity for $A^R_{\mu j}$ is preserved even in the glass-type state. Formally it means that the averages of a type $< A_{\mu j}(\mathbf{r}) A_{\nu l}(\mathbf{r}) >$ are finite and can be used as the order parameter of the "superfluid glass" state [17].

So, one can conclude that for a triplet Cooper pairing quenched disorder of a type of "fluctuations of the transition temperature" $\eta_{jl}(\mathbf{r})$ influences both the form of the order parameter and its local orientation. There are "robust" classes of order parameters, for which disorienting effect disappears. If the form of the order parameter required by the minimization of the initial free energy is not "robust" there is a mechanism of adjustment which tends to bring the order parameter closer to the "robust" form and to decrease the disorienting effect of the random tensor $\eta_{jl}(\mathbf{r})$. Both disorienting effect and the mechanism of adjustment originate from degeneracy of the unperturbed free energy with respect to orbital rotations of the order parameter.

I acknowledge the hospitality of the Laboratory of Atomic and Solid State Physics of Cornell University, where the final part of this work has been done. This research was supported in part by RFBR grant (no. 04-02-16417), by Ministry of Education and Science of Russian Federation and by CRDF grant RUP1-2632-MO04.

REFERENCES

1. J.V.Porto and J.M.Parpia, *Phys. Rev. Lett.*,**74**, 4667 (1995).
2. D. Sprague, T. Haard, J. Kysya, M. Rand, Y. Lee, P. Hamot, and W. Halperin *Phys. Rev. Lett.*,**75**, 661 (1995).
3. A.A. Abrikosov and L.P. Gorkov, *ZhETF*, **39**, 1781 (1961) [*Sov. Phys. JETP* **12**, 1243 (1961)].
4. A.I. Larkin and Yu.N. Ovchinnikov, *ZhETF* **61**, 1221 (1971).
5. B.I. Barker, Y. Lee, L. Polukhina, D.D. Osheroff, L.W. Hrubesh, and J.F. Poco, *Phys. Rev. Lett.*, **85**, 2148 (2000)
6. V.V. Dmitriev, V.V. Zavjalov, D.E. Zmeev, I.V. Kosarev, and N. Mulders, *JETP Lett.*, **76**, 371 (2002).
7. G.E. Volovik, *Pis'ma ZhETF* **63**, 281 (1996) [*JETP Lett.* **63**, 301 (1996)]
8. A.I. Larkin, *ZhETF* **58**, 1466 (1970) [*Sov. Phys. JETP* **31**, 784 (1970)].
9. Y. Imry and S. Ma, *Phys. Rev. Lett.*, **35**, 1399 (1975)
10. Jizhong He, A.D. Corwin, J.M. Parpia, and J.D. Reppy, *Phys. Rev. Lett.*,**81**, 115301 (2002).
11. V.I. Marchenko, *ZhETF*, **93**, 141 (1987) [*Sov. Phys. JETP* **66**, 79 (1987)].
12. I.A. Fomin, *Pis'ma ZhETF* **81**, 362 (2005) [*JETP Lett.*, **81**, 285 (2005)].
13. I.A. Fomin, *ZhETF*, **125**, 1115 (2004) [*JETP*, **98**, 974 (2004)]
14. N. D. Mermin and G. Stare, Report **2186**, Materials Science Center, Cornell University (1974)
15. C.M. Gould and H.M. Bozler, *J. Low Temp. Phys.*, **113**, 661 (1998).
16. J.A. Sauls and J.W. Serene, *Phys. Rev.*, **B 24**, 183 (1981)
17. G.E. Volovik and D.E. Khmel'nitskii, *Pis'ma ZhETF* **40**, 469 (1984) [*JETP Letters*, **40**, 1298 (1984)].

Impurity Effect on the Order Parameter Collective Mode in Superfluid ^{3}He in Aerogel

S. Higashitani*, M. Miura† and K. Nagai*

*Faculty of Integrated Arts and Sciences, Hiroshima University, Kagamiyama 1-7-1, Higashi-hiroshima 739-8521, Japan
†Graduate School of Advanced Sciences of Matter, Hiroshima University, Kagamiyama 1-3-1, Higashi-hiroshima 739-8530, Japan

Abstract. In liquid ^{3}He confined in aerogel, the ^{3}He quasiparticles are scattered by random distribution of short-range potentials provided by the aerogel. We discuss the impurity effect on the dynamics of a p-wave order parameter on the basis of the linear response theory in the Keldysh formulation. Among the various types of order parameter fluctuations, we focus on the squashing mode which is known to couple to longitudinal sound.

Keywords: superfluid, aerogel, impurity, sound, collective mode
PACS: 67.57.Jj, 67.57.Pq

INTRODUCTION

Acoustics has been a powerful tool for probing the order parameter collective modes in superfluid ^{3}He. In pure superfluid ^{3}He, some of the collective modes have been observed clearly as sound attenuation peak. Nomura *et al.* [1] performed high-frequency acoustic measurements in liquid ^{3}He confined in 98% porous aerogel over a wide temperature range across the transition temperature T_c. Unlike the pure superfluid, no attenuation peak was observed below T_c.

Recently, the acoustic experiment [1] has been analyzed [2] using a two-fluid hydrodynamic theory [3] in which phenomenological parameters are evaluated within the s-wave-unitarity-limit scattering approximation and by assuming superfluid ^{3}He to be in the BW state. It was shown that the two-fluid theory gives reasonable agreement with the experiment in both the normal and superfluid phases. The quasiparticle mean free path $l = v_F\tau$ in the aerogel, as a fitting parameter in that theory, was estimated to be 56 nm ($\tau T_c k_B/\hbar = 0.27$). Such a strong impurity scattering gives rise to considerable smearing of the gap structure in the density of states and even makes it completely gapless (see the inset in Fig. 1).

In the two-fluid theory, the collective-mode degrees of freedom are not taken into account, except for the fluctuation of the global phase of the order parameter. In the BW state, the so-called squashing mode is known to couple to sound. The acoustic experiment [1] suggests that the impurity scattering by aerogel causes large damping of the squashing mode. The damping of the collective mode is closely related to the structure of the quasiparticle density of states. The purpose of this paper is to study the impurity effect on the dynamics of the squashing mode in the case where superfluid ^{3}He is in the gapless regime.

FORMULATION

Our theory is based on the quasiclassical Green's function method for p-wave superfluid with impurities [2]. In this formulation, the non-equilibrium quantities of interest are calculated from the Keldysh quasiclassical Green's function δg^K defined as a 4×4 Nambu-space matrix. We are interested in the order parameter fluctuation induced by a longitudinal sound with wave vector $\mathbf{q}$ and frequency ω, given by

$$\delta\Delta_{\hat{p}}(Q) = -N(0)\int\frac{d\hat{p}'}{4\pi}v_{\hat{p},\hat{p}'}\int\frac{d\varepsilon}{4i}\delta g^{K12}(\hat{p}',\varepsilon,Q), \quad (1)$$

where $Q = (\mathbf{q},\omega)$, δg^{K12} denotes the upper-right 2×2 submatrix of δg^K, $N(0)$ is the density of states in the normal state at the Fermi level, and $v_{\hat{p},\hat{p}'}$ is the paring interaction. One can describe the dynamics of the order parameter by calculating δg^{K12} as the linear response to the effective perturbation (see Ref. [2])

$$\delta H_{\hat{p}}(Q) = \begin{pmatrix} 0 & \delta\Delta_{\hat{p}}(Q) \\ \delta\Delta^\dagger_{\hat{p}}(-Q) & 0 \end{pmatrix} \quad (2)$$

and then substituting the resulting δg^{K12} into Eq. (1).

We evaluate the impurity scattering effect on δg^K within the self-consistent t-matrix approximation with only the s-wave scattering channel retained [4]. In this approximation, the impurity effect is parameterized by the mean free time τ and the normalized cross section

CP850, *Low Temperature Physics: 24th International Conference on Low Temperature Physics;*
edited by Y. Takano, S. P. Hershfield, S. O. Hill, P. J. Hirschfeld, and A. M. Goldman

$\bar{\sigma}$ ($0 < \bar{\sigma} < 1$; $\bar{\sigma} = 1$: unitarity limit; $\bar{\sigma} \ll 1$: Born approximation).

In the BW state, the equilibrium order parameter has the form $\Delta_{\hat{p}} = d_i \sigma_i i\sigma_y$ with $d_i = \Delta \hat{p}_i$, where σ_i's are the Pauli matrices. One can express the fluctuation of the p-wave order parameter as $\delta\Delta_{\hat{p}}(Q) = \delta d_{ij}(Q)\hat{p}_j \sigma_i i\sigma_y$. The fluctuations coupled to the sound are $\delta d_0 \propto \delta d^-_{ii}$ and $\delta d_{\rm sq} \propto \delta d^-_{xx} + \delta d^-_{yy} - 2\delta d^-_{zz}$, where $\delta d^-_{ij}(Q) = [\delta d_{ij}(Q) - \delta d^*_{ij}(-Q)]/2$. The δd_0 describes the global phase fluctuation and $\delta d_{\rm sq}$ is the squashing mode.

IMPURITY EFFECT ON THE SQUASHING MODE

In pure superfluid ^{3}He, Eq. (1) in the long wavelength limit $q \to 0$ gives the well known result

$$L(\omega)\delta d_{\rm sq} = 0, \quad L(\omega) = \omega^2 - \frac{12}{5}\Delta^2. \qquad (3)$$

In the similar way as in the pure ^{3}He case, one can derive the corresponding equation in the impure system. A difference in the equation comes from the impurity self-energy yielding a renormalization of the energy variable ε. In addition, $\delta H_{\hat{p}}$ has a vertex correction; however, it does not contribute to Eq. (1) for the squashing mode in the limit $q \to 0$. We obtain, for impure superfluid,

$$L(\omega) = \omega^2 + \frac{i\omega}{\tau} a_1(\omega) - \frac{12}{5}\Delta^2 a_0(\omega) \qquad (4)$$

with

$$a_0(\omega) = \frac{\omega \int d\varepsilon \lambda^K}{\int d\varepsilon [\lambda \bar{\omega}]^K}, \quad a_1(\omega) = \frac{\int d\varepsilon [\lambda D \bar{\omega}]^K}{\int d\varepsilon [\lambda \bar{\omega}]^K}, \qquad (5)$$

where

$$\lambda = \frac{1}{\tilde{x}_+ \tilde{x}_- (\tilde{x}_+ + \tilde{x}_-)}, \quad \tilde{x}_\pm = \sqrt{\Delta^2 - \tilde{\varepsilon}^2_\pm} \qquad (6)$$

and $\bar{\omega}$ and D are defined by $\bar{\omega} = \tilde{\varepsilon}_+ - \tilde{\varepsilon}_- = \omega + iD/\tau$ with $\tilde{\varepsilon}_\pm$ the renormalized energy variable at $\varepsilon \pm \omega/2$. The Keldysh functions denoted by the superscript K are constructed from the corresponding retarded, advanced, and anomalous functions using the prescription given in Ref. [2]. The above equations are valid for both the Born and unitarity scattering regimes.

Let us discuss the collective mode dynamics described by the function $L(\omega)$. In Fig. 1, we show the numerical result of $L(\omega)$ at zero temperature. The parameters τ and $\bar{\sigma}$ are chosen to be the same as those estimated in Ref. [2] from comparison with the acoustic experiment [1]. The imaginary part of $L(\omega)$ shows a linear ω dependence up to $\omega \sim 2\Delta$. The squashing mode dynamics is, therefore, well described by a damped oscillator equation in the

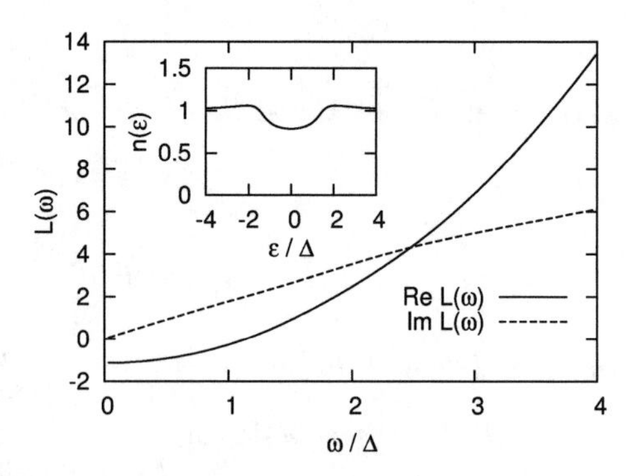

FIGURE 1. Real (solid line) and imaginary (dashed line) parts of $L(\omega)$ for impure BW state at zero temperature. The impurity scattering is assumed to be in the unitarity limit with $\tau\Delta/\hbar = 0.5$, corresponding to $\tau T_c k_B/\hbar = 0.27$ which was used in Ref. [2] to reproduce the acoustic experiment by Nomura *et al.*[1] Inset: the density of states for the same impurity parameter.

frequency range of interest. The physical origin of the ω-linear imaginary part of $L(\omega)$ is as follows: in the gapless regime, there are quasiparticle excitations that lead to the damping of the collective mode for arbitrarily small ω; the density of states available for the excitations increases linearly with ω. This is in contrast to the weak scattering regime, the Born limit with $\tau\Delta/\hbar \gg 1$. In that case, the density of states has a gap in the energy range $|\varepsilon| < \varepsilon_g$ (ε_g is equal to Δ in pure BW state but is less than Δ in the impure system). As a result, there is no excitation process for $\omega < 2\varepsilon_g$, so that the damping of the collective mode cannot occur at the low frequency $\omega < 2\varepsilon_g$.

The above observation implies that the sound attenuation from the squashing mode in aerogel spreads over a wide frequency range.

ACKNOWLEDGMENTS

This work is supported in part by a Grant-in-Aid for Scientific Research (No. 16540322) and a Grant-in-Aid for COE Research (No. 13CE2002) from the Ministry of Education, Culture, Sports, Science and Technology of Japan.

REFERENCES

1. R. Nomura, G. Gervais, T.M. Haard, Y. Lee, N. Mulders, and W.P. Halperin, *Phys. Rev. Lett.* **85**, 4325 (2000).
2. S. Higashitani, M. Miura, M. Yamamoto, and K. Nagai, *Phys. Rev. B* **71**, 134508 (2005).
3. M. Miura, S. Higashitani, M. Yamamoto, and K. Nagai, *J. Low Temp. Phys.* **134**, 843 (2004).
4. L.J. Buchholtz and G. Zwicknagl, *Phys. Rev. B* **23**, 5788 (1981).

Capillary Condensation of Liquid ^{4}He in Aerogel on Cooling Through λ Point

W. Miyashita, K. Yoneyama, H. Kato, R. Nomura[1] and Y. Okuda

Department of Condensed Matter Physics, Tokyo Institute of Technology, 2-12-1, O-okayama, Meguro, Tokyo 152-8551, Japan

Abstract. Capillary condensation of liquid ^{4}He in silica aerogel with a 90% porosity was investigated visually. The initial condition of the experiment was such that liquid ^{4}He was present in the sample cell but not in the aerogel. This situation was realized by introducing the liquid into the cell at a fast rate to avoid liquefaction in the aerogel. The free surface of the liquid rose up in the cell with filling and eventually reached the bottom of the aerogel. Then, the aerogel absorbed the liquid by capillary condensation. The height of the liquid in the aerogel rose with time t roughly as $t^{1/2}$ in the normal fluid phase. This behavior was consistent with the Washburn model. When the system was cooled through the λ point during the condensation, the liquid height started to rise faster in the superfluid phase with a constant velocity of about 0.3 mm/sec. The dynamics of capillary condensation was strongly dependent on whether the liquid ^{4}He was in the normal or the superfluid phase.

Keywords: liquid ^{4}He, superfluidity, capillary condensation, aerogel, Washburn model
PACS: 47.55.Mh, 67.40.-w, 68.03.-g

INTRODUCTION

The influence of disorder on phase transitions has attracted much attention. The gas-liquid phase transition of ^{4}He in silica aerogel was investigated by several groups.[1-3] Hysteretic behaviors in adsorption and desorption of the liquid were observed, and the hysteretic loop of a vapor-pressure isotherm strongly depended on both the porosity of the aerogel and temperature. These behaviors were recently reproduced by a local mean field theory of a fluid in contact with a gel network generated by a diffusion-limited cluster-cluster aggregation model.[4,5] So far most of the experiments were performed very slowly to study the phenomena in the so-called rate-independent limit. We filled ^{4}He in a sample cell very fast to see condensation dynamics in aerogel under a highly non-equilibrium condition. Aerogel is optically transparent and suitable for studying the dynamics in it. When the free surface of the liquid reached the bottom of the aerogel, it sucked up the liquid with a visible liquid-gas interface in it. The interface rose with time t roughly as $t^{1/2}$ in the normal fluid phase. This behavior was consistent with the Washburn model. When the system was cooled through the λ point during the condensation, the surface began to rise faster in the superfluid phase with a constant velocity of about 0.3 mm/sec. The way of capillary condensation depended on whether the liquid ^{4}He was in the normal or the superfluid phase.

EXPERIMENTAL AND RESULTS

The sample cell had two windows and we could observe the inside from room temperature. The cell was illuminated through the back window at room temperature and a video camera recorded the images at the front. Temperature was measured by a RuO_2 resistance thermometer inside the cell. Silica aerogel with a 90% porosity was set on a stage on the bottom of the cell. The volume of the aerogel was 0.62 cm^3 and the height 12 mm. We controlled a filling rate of ^{4}He gas into the cell by a needle valve of a gas handling system at room temperature.

At a low filling rate of the ^{4}He gas, the aerogel became almost uniformly opaque during the condensation before the bulk liquid appeared outside of the aerogel, as was reported by Lambert *et al.*[3] At a

[1] E-mail: nomura@ap.titech.ac.jp

CP850, *Low Temperature Physics: 24th International Conference on Low Temperature Physics;* edited by Y. Takano, S. P. Hershfield, S. O. Hill, P. J. Hirschfeld, and A. M. Goldman

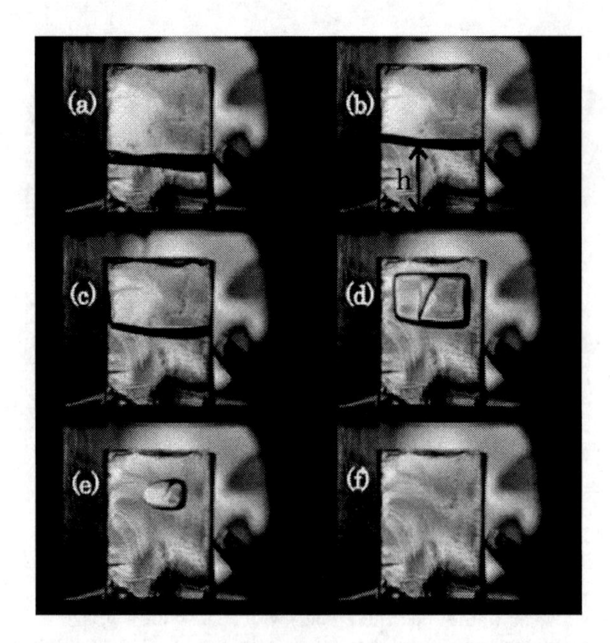

FIGURE 1. Successive images of the interface in aerogel. The images were taken at (a) 23 sec, (b) 52 sec, (c) 55 sec, (d) 57 sec, (e) 60 sec, and (f) 63 sec. $t = 0$ was the moment at which the interface appeared in the aerogel. Frame (c) was taken immediately after He became superfluid.

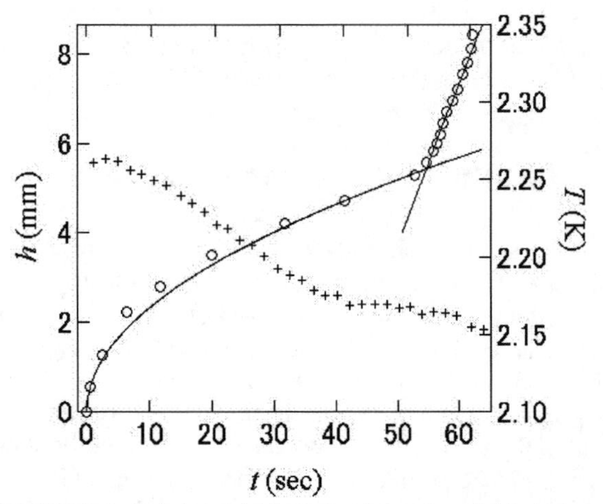

FIGURE 2. Height of the interface h in aerogel (circles) and temperature (crosses) as a function of time t. In the normal fluid phase $h \propto t^{1/2}$, and in the superfluid phase $h \propto t$.

high rate, however, the bulk liquid appeared outside of the aerogel before the aerogel became opaque. The liquid-gas interface rose in the cell as the condensation proceeded and finally touched the bottom of the aerogel. Then the aerogel absorbed the liquid with a visible interface in it. The height of the interface in the aerogel h was proportional to $t^{1/2}$ as is known as the Washburn model. We have already reported this behavior of normal fluid ^{4}He and determined the effective diameter of the pores to be 19 nm.[6]

To repeat the experiment in the superfluid phase, the initial temperature was adjusted to just above the λ point. The system was cooled down slowly so that the liquid became superfluid when the interface was in the middle of the aerogel. Successive images of the interface are shown in Fig. 1, and h and temperature are plotted against t in Fig. 2. In the normal fluid phase the interface rose from the bottom and the dynamics was almost consistent with the Washburn model. In the superfluid phase the interface appeared on all sides and moved with a constant velocity of about 0.3 mm/sec. A superfluid film must have climbed the outer surface of the aerogel rapidly to supply the liquid into the aerogel from all sides. We cannot clearly distinguish it in Fig. 1 but a dimly visible line rose very quickly, at about 2 mm/sec, just before the interface appeared on all sides. The line was probably the front of the film that had climbed the outer surface.

SUMMARY

The dynamics of capillary condensation of superfluid ^{4}He was very different from that of the normal fluid. The interface appeared from the bottom in the normal fluid phase and rose in aerogel with nearly $t^{1/2}$. This behavior was consistent with the Washburn model. The interface appeared on all sides of the aerogel in the superfluid phase and proceeded with a constant velocity.

ACKNOWLEDGMENTS

We are grateful to Y. Lee for helping us to prepare aerogels. This study was supported in part by a 21st Century COE Program at Tokyo Tech "Nanometer-Scale Quantum Physics" and by Grants-in-Aid for Scientific Research from the Ministry of Education, Culture, Sports, Science and Technology.

REFERENCES

1. A. P. Y. Wong and M. H. W. Chan, *Phys. Rev. Lett.* **65**, 2567-2570 (1990).
2. J. R. Beamish and T. Herman, *Physica B* **329**, 340-341 (2003).
3. T. Lambert *et al. J. Low Temp. Phys.* **134**, 293-302 (2004).
4. F. Detcheverry *et al. Phys. Rev. E* **68**, 061504-1-11 (2003).
5. F. Detcheverry *et al. Langmuir* **20**, 8006-8014 (2004).
6. W. Miyashita *et al. J. Phys. Chem. Solids*, **66**, 1509-1511 (2005).

Adsorption of Helium in Silica Aerogel Near the Critical Point

Tobias Herman, James Day, and John Beamish

Department of Physics, University of Alberta, Edmonton, Alberta, Canada, T6G 2J1

Abstract. When fluids are adsorbed in small pores, capillary forces cause the pores to fill before bulk liquid appears, and isotherms usually show hysteresis between filling and emptying. The tenuous structure of aerogels provides a unique medium in which to study the effect of dilute impurities on liquid-vapor properties, particularly near the critical point where it may be possible to observe equilibrium behavior. We have used a capacitive technique to study adsorption in silica aerogel of two different densities near the liquid-vapor critical point of helium. We compare the disappearance of hysteresis in helium adsorption isotherms in different density aerogels to the behavior in vycor glass.

Keywords: aerogel; helium; adsorption isotherms; liquid-vapor critical point
PACS: 64.60.Fr, 64.70.Fx, 68.03.Cd

INTRODUCTION

The liquid-vapor phase transition in porous media has considerable fundamental and practical interest. It is often described within the framework of capillary condensation since, in geometries like pores or channels, fluids tend to condense more readily than in bulk. Capillary condensation is characterized by very deep metastable states during adsorption and desorption, which lead to hysteresis along adsorption isotherms. However, near the liquid-vapor critical point (LVCP) this picture breaks down as thermal density fluctuations grow to macroscopic scale.

Aerogels provide an opportunity to study this transition in a medium without a well-defined pore shape and could be a model system in which to study liquid-vapor critical behavior in the dilute impurity limit. The disappearance of hysteresis as the temperature is raised has not been quantitatively studied for fluids in aerogel in the static limit. The existence of an equilibrium liquid-vapor transition for fluids in low density aerogels, necessary for there to be true liquid-vapor critical behavior, is still an open question.

In this paper, we show a series of adsorption and desorption isotherms of helium in two silica aerogels of different densities and in vycor. Our results show that in aerogels hysteresis in adsorption isotherms persists to temperatures very close to the LVCP, and that the temperature evolution of the hysteresis loops changes qualitatively with aerogel density.

RESULTS

Aerogel samples were synthesized from tetramethyl orthosilicate (TMOS) using a base catalyzed single step procedure and dried supercritically. Two aerogels were used, with densities of 110 kg/m^3 (**B110**) and 51 kg/m^3 (**B51**), corresponding to porosities of about 95% and just less than 98% respectively. The density of the adsorbed helium was measured using a capacitive technique. Thin ($\sim$ 0.5 mm) slices of aerogel and vycor had electrodes directly plated onto their surfaces. Since aerogel acts as a dilute impurity the dielectric constant of the fluid-aerogel system varies linearly with helium density, but for the vycor sample nonlinearity results in a 10% uncertainty in the absolute helium density values. Pressures were measured with an *in situ* capacitive pressure gauge and normalized to the bulk saturation pressure (P_0). Isotherms were collected over a range of temperatures from 4.4 K to 5.7 K[1, 2]. All points represent "equilibrated" values — the system was allowed to rest at each point until temperature, pressure, and density were stable.

Hysteresis in dense porous glasses such as vycor (porosity $\sim$ 29%) disappears at temperatures far below the bulk LVCP, as shown in Fig. 1. The loop becomes narrower and the low density closure moves upwards as the temperature is raised. The high density closure remains close to the bulk liquid density (121 kg/m^3 at 4.4 K). The loop appears to "zip" closed from the low density side, disappearing about 0.6 K below the bulk ^{4}He LVCP (T_c=5.195 K).

The disappearance of hysteresis in B110 is shown in Fig. 2. Hysteresis disappears at 5.155 K, only 40 mK below T_c. The onset of capillary condensation is sharper than in vycor, occurring over a narrow pressure range (note the different pressure scales in the three figures), but the roughly triangular shape of the isotherms in B110 is similar to the vycor isotherms. The temperature evolution of the hysteresis loop is also similar — the high density edge of the loop remains roughly equal to the bulk liquid density at each temperature while the low density side of the hysteresis loop shifts upwards until the loop disappears.

CP850, *Low Temperature Physics: 24th International Conference on Low Temperature Physics;*
edited by Y. Takano, S. P. Hershfield, S. O. Hill, P. J. Hirschfeld, and A. M. Goldman

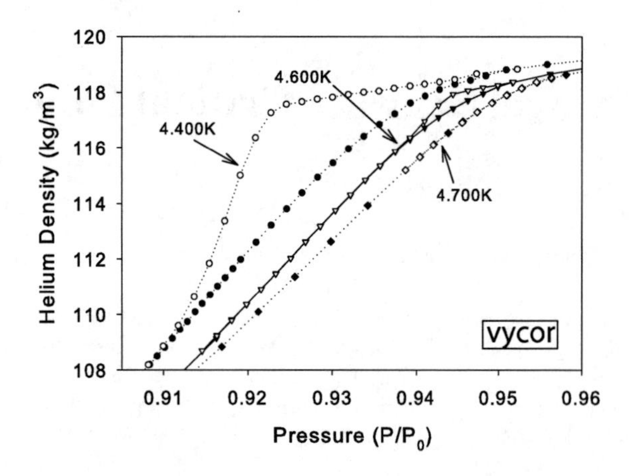

FIGURE 1. Temperature evolution of hysteresis loops in adsorption (solid symbols) and desorption (open symbols) isotherms of helium in vycor (at 4.400 K, 4.600 K, and 4.700 K).

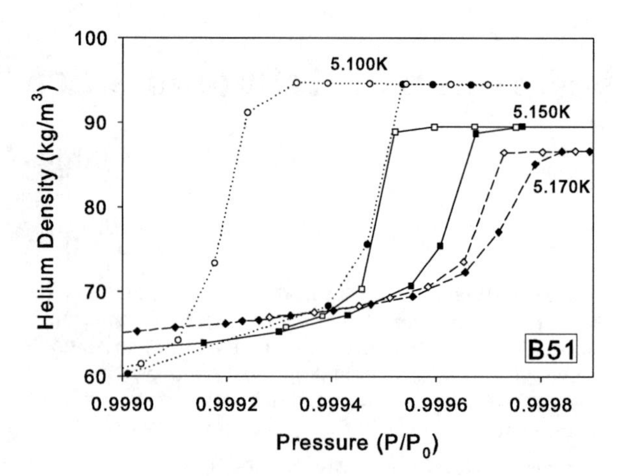

FIGURE 3. Temperature evolution of hysteresis loops in adsorption (solid symbols) and desorption (open symbols) isotherms of helium in aerogel B51 at T=5.100 K, 5.150 K and 5.170 K.

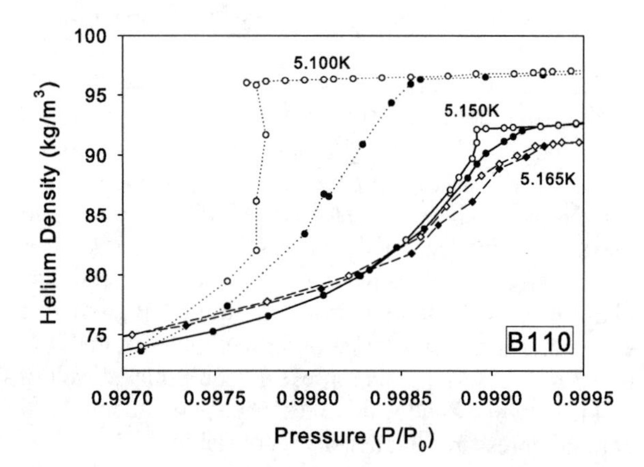

FIGURE 2. Temperature evolution of hysteresis loops in adsorption (solid symbols) and desorption (open symbols) isotherms of helium in aerogel B110 at T=5.100 K, 5.150 K and 5.165 K.

Figure 3 shows the disappearance of hysteresis in the lower density aerogel, B51. Capillary condensation is very sharp, giving rise to rectangular hysteresis loops, even close to the bulk LVCP. Hysteresis disappears between 5.170 K and 5.180 K, only ~ 20 mK below T_c. The evolution of the hysteresis loop from 5.100 K to 5.170 K is qualitatively different than for B110 — instead of the loop closing from the low density side, it narrows more symmetrically, until it covers a pressure range which is too small for us to resolve (at temperatures above 5.170 K).

Whereas capillary condensation (such as seen in vycor) can be viewed as a collection of first order transitions in a variety of microscopic environments, and B110 appears to exhibit this same behavior, aerogel B51 may be showing evidence of an underlying macroscopic transition. The roughly rectangular loops which persist until close to the LVCP may indicate that the equilibrium state of the system experiences a discontinuous change in fluid density, characteristic of a macroscopic liquid-vapor transition. It is difficult to experimentally test for the existence of an underlying equilibrium liquid-vapor transition, since the system is mired in a series of metastable states along the edge of the hysteresis loop, but simulations[3] of adsorption in aerogels indicate that there may be an equilibrium macroscopic transition within these loops.

ACKNOWLEDGMENTS

This work was supported by a grant from NSERC.

REFERENCES

1. T. Herman, *Liquid-vapor critical behavior in silica aerogel*, Ph.D. thesis, University of Alberta (2005).
2. T. Herman, J. Day, and J. Beamish, cond-mat/0505430 (2005).
3. F. Detcheverry, E. Kierlik, M. L. Rosinburg, and G. Tarjus *Langmuir*, **20**, 8006-8014 (2004).

Deformation of low density silica aerogel by helium

Tobias Herman, James Day, and John Beamish

Department of Physics, University of Alberta, Edmonton, Alberta, T6G 2J1, Canada

Abstract. Liquid-vapor interfaces are present throughout any porous medium during adsorption and desorption below the liquid-vapor critical point. Surface tension across these curved interfaces produces the pressure difference between the liquid and vapor phases which is responsible for capillary condensation; this pressure difference can also compress the medium. Aerogels have extremely small elastic moduli, so surface tension induced deformation can be significant and sometimes destructive. We present measurements of the dilation and compression of silica aerogel during adsorption and desorption of helium and discuss the implications for possible damage to aerogels.

Keywords: aerogel; compression; helium; surface tension;
PACS: 61.43.Gt, 62.20.Fe, 68.03.Cd

INTRODUCTION

Porous media have large surface areas, and consequently interfacial energy contributes significantly to their behavior. A porous medium exposed to vapor at low pressure collects a thin film on the surface of the pores which may reduce the interfacial energy and allow the matrix to expand [1]. As vapor pressure is increased this film thickens until the fluid undergoes capillary condensation. The curved liquid-vapor interface present during capillary condensation creates a pressure difference between the liquid and vapor phases which may deform elastically compliant materials like aerogel. Compression during liquid nitrogen adsorption and desorption has been measured [2] for some higher density aerogels – the samples exhibited large changes in volume and were permanently damaged.

We have measured[3] the macroscopic strain in low density aerogels during adsorption and desorption of low surface tension fluids (neon near its critical point, and helium). We found that the compression during capillary condensation is proportional to the surface tension of the adsorbate and inversely proportional to the aerogel's bulk modulus, which could be directly extracted from the adsorption and desorption isotherms. Since aerogels can suffer permanent damage when compressed more than about 5%, our results provide guidelines for the safe filling and emptying of aerogels.

RESULTS

Helium was studied in an aerogel with a density of 110 kg/m^3 (**B110**), corresponding to a porosity of 95%. The aerogel was synthesized from tetramethoxysilane (TMOS) using a standard one-step base catalyzed procedure followed by supercritical drying. The sample was cut into a 1 cm long cylinder and a linear variable differential transformer (LVDT) allowed high precision, non-contact measurement of changes in the aerogel length.

Isotherms for helium in B110 are shown in Fig. 1. The pressure of the cell was controlled by a low temperature ballast, so the isotherms are plotted with length change as a function of the cell pressure. Figure 1a shows the adsorption and desorption branches at 4.2 K. At low partial pressures the thin film of helium on the aerogel strands produced an expansion of about 0.08%. At higher pressures capillary forces contracted the aerogel, with the maximum change in length occurring during desorption. The linear portions of the isotherms closest to the bulk saturation pressure, P_0, have identical slope for the emptying and filling branches of the isotherm; in this region the gel is full of liquid and the aerogel matrix is only responding to the pressure difference across the liquid-vapor interface present at the surface of the aerogel sample. Figure 1b shows three desorption isotherms taken at 2.4 K, 4.2 K and 5.0 K. The temperature dependence of the maximum compression during capillary condensation is accounted for by the temperature dependence of surface tension.

Adsorption and desorption isotherms were also collected for neon at 43 K in an aerogel with a density of 51 kg/m^3 (**B51**), corresponding to a porosity of slightly less than 98%; neon at this temperature has the same surface tension as ^{4}He at 3.8 K. During filling, the aerogel exhibited a maximum linear compression of about 2.5%; once the gel was full it re-expanded to its original size. Upon removal of the neon the aerogel was compressed by almost 5% before emptying ($\sim$ 15% volume change). The gel showed evidence of a slight ($\sim$ 0.3%) permanent linear compression after desorption.

With knowledge of the bulk modulus of the gel, the breakthrough radius of the liquid-vapor interface during desorption, and the surface tension of the fluid, it is

CP850, *Low Temperature Physics: 24th International Conference on Low Temperature Physics;*
edited by Y. Takano, S. P. Hershfield, S. O. Hill, P. J. Hirschfeld, and A. M. Goldman

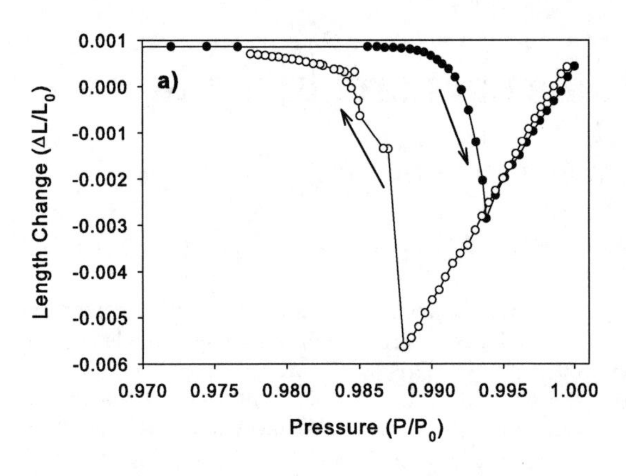

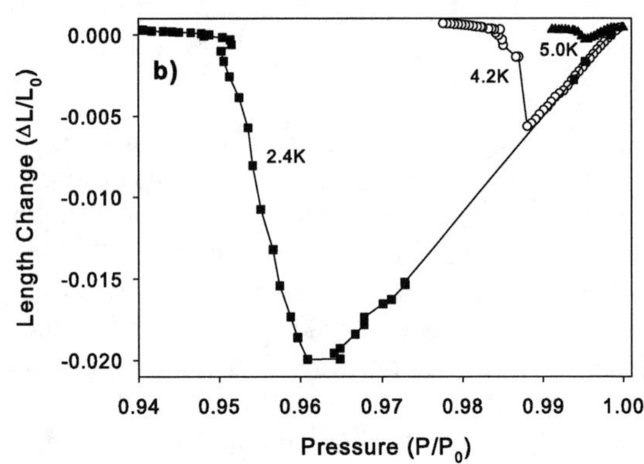

FIGURE 1. (a) Deformation of B110 during adsorption (solid symbols) and desorption (open symbols) of helium at 4.2 K. (b) Desorption isotherms at 2.4 K (squares), 4.2 K (circles), and 5.0 K (triangles).

possible to predict the deformation of the aerogel. The degree of deformation is very sensitive to the gel density because of the very strong dependence of elastic moduli on density ($K \propto \rho_{gel}^{3.7}$)[4], but the breakthrough radius is also important. Our isotherms allow us to determine the breakthrough radius for our aerogels (22 nm and 85 nm for B110 and B51 respectively) — to generalize our results we must extrapolate breakthrough radii for other densities of aerogel, so for simplicity we assume that radius also follows a power law in density.

Predicted maximum deformations of aerogels are plotted in Fig. 2 for a series of emptying temperatures. When the volumetric compression is kept at less than 5%, aerogels tend to behave elastically; however, for greater compression they can begin to show plastic deformation resulting in permanent damage. This cut-off is chosen somewhat arbitrarily since there are no definitive data on the location of the boundary between elastic and plastic behavior in aerogels. To avoid damage to a 98% gel ($\rho \approx$ 45 kg/m^3), for example, the temperature should not be

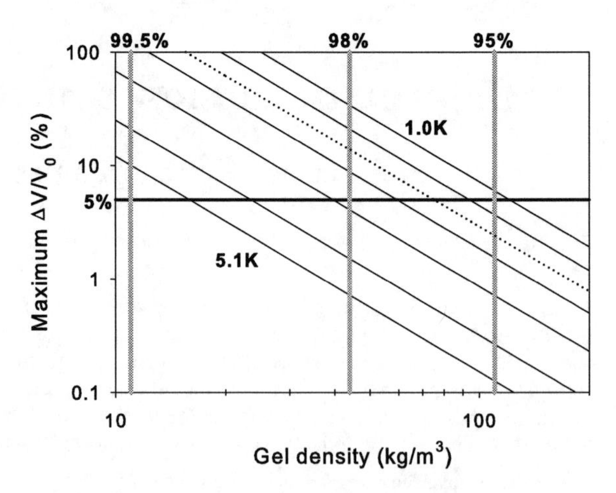

FIGURE 2. Predicted volumetric compression of aerogels as a function of density, for removing ^{4}He at various temperatures. The solid diagonal lines, from right to left, correspond to predicted maximum deformation at 1.0 K, 3.0 K, 4.2 K, 4.7 K, 5.0 K, and 5.1 K. The dotted line is the predicted deformation for removing ^{3}He at 1K. Grey vertical stripes have been added at densities corresponding to gels with porosities of 95%, 98%, and 99.5%. For compressions of less than 5%, indicated by the horizontal black bar, the gel should behave elastically.

allowed to dip below 4.6 K during filling and emptying. For comparison, the surface tension of liquid ^{3}He below 1 K is approximately the same as ^{4}He at 3.5 K, meaning that only gels with porosities below $\sim$ 96% should be filled with, or emptied of, ^{3}He at low temperatures.

The plastic deformation of aerogels affects their microstructure[5]. During compression the largest pores collapse and the long range correlations in the aerogel structure are affected — the gel correlation length decreases and the fractal dimension may change. If low density aerogels are filled or emptied at low temperature important details of their microstructure, as well as their macroscopic density, may be changed.

This work was supported by a grant from NSERC.

REFERENCES

1. P. Thibault, J. J. Préjean, and L. Puech, *Physical Review B*, **52**, 17491 – 17500 (1995).
2. G. Reichenauer, and G. W. Scherer, *Colloids and Surfaces a — Physicochemical and Engineering Aspects*, **187**, 41–50 (2001).
3. T. Herman, J. Day, and J. Beamish, to be published (2005).
4. T. Woignier, and J. Phalippou, *Journal de Physique (Paris)*, **50**, C4179–C4184 (1989).
5. I. Beurroies, L. Duffours, P. Delord, T. Woignier, and J. Phalippou, *Journal of Non-crystalline Solids*, **241**, 38–44 (1998).

Tortuosity of ^{4}He Films on Aerogel

C. E. Ashton*, N. Mulders† and A. I. Golov*

*School of Physics and Astronomy, University of Manchester, Manchester, M13 9PL, UK
†Department of Physics and Astronomy, University of Delaware, Newark, DE 19716, USA

Abstract. A torsional oscillator has been used to study the flow of liquid ^{4}He through silica aerogel. The tortuosity and dissipation of the flow in the fractal aerogel environment has been measured, for both capillary condensed films and a fully saturated cell. The scaling of tortuosity as a function of filling fraction has been investigated. We compare the results of our 88%-porous aerogel to a 92%-porous sample where a transverse sound technique was used. Our more sensitive technique has reduced the scatter in both the frequency and dissipation data. Like in the previous transverse sound experiment, we find tortuosity as a function of filling fraction to scale with an exponent of ~ -1.1. Further experiments are planned, to find the dependence of the exponent on the fractal dimension and porosity of aerogel.

Keywords: superfluidity, aerogel, fractal
PACS: 67.40.HF, 67.40.Pm, 61.43.Gt, 61.43.Hv

INTRODUCTION

This Paper reports our progress towards using tortuosity of superfluid ^{4}He films adsorbed onto aerogel, as a method of characterising the microscopic structure of aerogel. This experiment considers capillary condensed films [1], where initial depositions closely follow the morphology of the aerogel's silica matrix, while successive depositions of helium progressively smooth out the small scale irregularities of the film's profile. This effectively de-fractalises the aerogel until it is fully saturated. The scaling of the film's tortuosity with volume of adsorbed helium – or largest filled aerogel pore – can be used as an independent method to characterise aerogel, or indeed other porous media. It may be possible to use this technique to characterise either fractal or non-fractal materials [2].

An inviscid liquid accelerated along a tortuous path while passing round an obstacle exerts force on the obstacle. This force, reflecting the tortuosity of the streamlines, is proportional to their "tortuosity" α, which is sometimes expressed in terms of the "drag factor" χ where $\chi = 1 - \alpha^{-1}$ [3]. It is straightforward to determine the tortuosity of a film of superfluid helium adsorbed inside a porous material by measuring the moment of inertia of a piece of this material with helium inside it.

THE EXPERIMENT

We used a cylindrical (diameter 12.7 mm, height 7.7 mm) aerogel sample made by one of us (NM), of density $\rho_a = 0.269\,\mathrm{g/cm^3}$. Comparison with silica density 2.2 g/cm^3 leads to the aerogel porosity $\phi_a = 0.88$. SAXS revealed mass fractal correlations between the wave numbers $0.016\,\text{Å}^{-1}$ and $0.032\,\text{Å}^{-1}$, and a fractal dimension, $D = 2.0$.

The aerogel is mounted in a typical Manchester torsional oscillator (for more details see [4]). The aerogel was epoxied into a closely fitting cylindrical head made from Stycast 1266, which also had a BeCu torsion stem. A one piece BeCu vibration filter and torsion stem, was fastened to a brass base, from where all temperature regulation was performed. The head unit was made detachable from the filter, to aid manufacture, but primarily to allow easy sample change for future experiments. The filter was driven close to the resonant frequency of the cell head. The high Q-value and disparity between the inertial masses of the oscillator allow the following approximations to be made for the head resonant frequency, ν_R, and bandwidth, ν_B:

$$\nu_R = \frac{1}{2\pi}\left(\frac{K_1}{I_1}\right)^{1/2}, \quad \nu_B = \frac{1}{2\pi}\frac{\zeta_1 + \zeta_{12}}{I_1}, \tag{1}$$

where K_1 is the torsion constant of the cell head; I_1 the moment of inertia of the cell head; and ζ_1 and ζ_{12} are the damping coefficients of the head's torsion stem and helium in aerogel respectively. We found both ν_R and ν_B for an empty cell to be temperature independent, $\nu_{R0} = 248.626\,\mathrm{Hz}$ and $\nu_{B0} = 6.8\,\mathrm{mHz}$.

For a given helium filling fraction, ϕ_{He}, it is possible to deduce the density of coupled helium, ρ_C [4], and through

$$\alpha(\phi_{He}) = \left(1 - \frac{\rho_{Cs}(\phi_{He}) - \rho_0}{\rho_{Cn}(\phi_{He}) - \rho_0}\right), \tag{2}$$

determine a value for α, where ρ_{Cs} and ρ_{Cn} are the coupled helium density when helium is in the superfluid and normal phase respectively; ρ_0 is the density of he-

CP850, *Low Temperature Physics: 24th International Conference on Low Temperature Physics;*
edited by Y. Takano, S. P. Hershfield, S. O. Hill, P. J. Hirschfeld, and A. M. Goldman

lium solidified on the silica resulting from van der Waals attraction.

EXPERIMENTAL RESULTS AND DISCUSSION

The layer of solid helium on silica makes a temperature independent contribution to I_1; the number of moles in this layer is proportional to the surface area of the aerogel and was estimated, using a critical coverage of 35 μmoles/m^2 [5], to be $\rho_0 = 0.013\,\mathrm{g\,cm^{-3}}$. Further admissions of helium result in liquid films, the profile of which is determined by the van der Waals and capillary terms in the chemical potential [1]. The liquid makes a temperature dependent contribution to I_1 below T_c, reaching a minimum at $T < 0.5\,\mathrm{K}$ when all liquid is superfluid. Above T_c, this contribution is approximately temperature independent.

For a given filling fraction of helium, ϕ_{He}, the contribution to I_1 can be calculated by tracking $\nu_{\mathrm{R}}(T)$. Above T_c the ν_{R} is temperature independent, but as the temperature is decreased past T_c, ν_{R} increases in a manner which is consistent with the normal component decreasing. However, even taking into account the inertia of the solid helium layer on the silica strands, ν_R never recovers its empty cell value of $\nu_{\mathrm{R0}} = 248.626\,\mathrm{Hz}$. Below 0.5 K the normal fluid fraction is essentially zero, and the remaining inertia of the superfluid is attributed to tortuosity [6, 7]. A more detailed discussion of this analysis for our saturated 88%-porous sample can be found in [4].

In Fig. 1 we plot $\alpha(\phi)$ for the capillary condensed regime, alongside data for 92%-porous and 93.5%-porous aerogels. We fitted the function $\alpha = \phi_{\mathrm{He}}^{-\varepsilon}$ to our data. This is consistent with the physical definition of α, that for a bulk ($\phi = 1$) sample $\alpha = 1$ [3]. We found $\varepsilon = 1.060 \pm 0.013$, to be compared with $\varepsilon = 1.10 \pm 0.02$ for the 92%-porous aerogel. For comparison, the measured $\alpha(\phi)$ for liquid saturating non-fractal packed powders [8] and fused glass beads [3] of various porosities ϕ, give exponents between $\varepsilon = 1/2$ and $\varepsilon = 2/3$. The aerogels have a markedly different ε to the non-fractal porous media. We found the 88%-porous aerogel to have the same ε as the 92%-porous ($D = 2.0$) and 93.5%-porous aerogels within the experimental error, although the 92%-porous and 93.5%-porous aerogels were made by a different manufacturer (Airglass, Sweden). Work is currently under way to determine $\alpha(\phi_{\mathrm{He}})$ for a 98%-porous aerogel.

The values of dissipation for all ϕ_{He} were much smaller than the temperature-independent value of $\nu_{\mathrm{B0}} = 6.8\,\mathrm{mHz}$ for the empty cell.

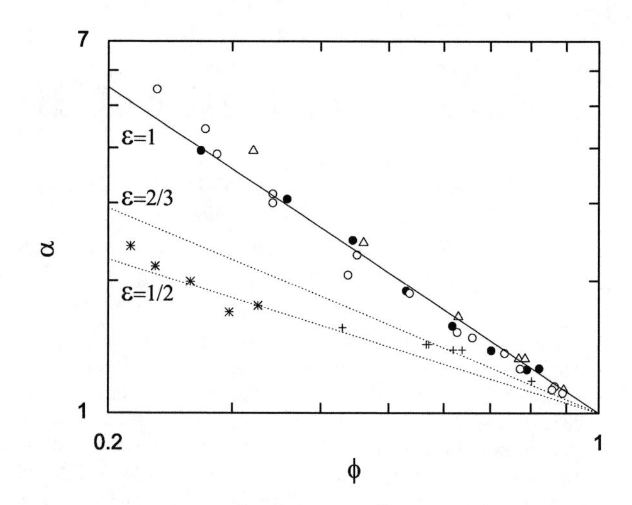

FIGURE 1. Tortuosity vs. helium filling fraction: (•) our data; (∘) data for a 92%-porous sample which employed a transverse sound technique [2]; (△) and for 93.5%-porous aerogel [9]. For comparison helium saturating (∗) fused glass beads [3] and (+) packed powders [8] are also shown. Solid lines are produced using an effective-medium theory for saturated porous media [10], for ($\varepsilon = 1$) self-similar needles orientated perpendicular to the flow, ($\varepsilon = 2/3$) self-similar needles randomly orientated, and ($\varepsilon = 1/2$) self-similar spheres. The solid line is a fit to our 88%-porous aerogel data.

ACKNOWLEDGMENTS

We are grateful to S. May for technical support. This work has been supported by EPSRC through grant GR/S27559. We thank D. Cousins, I. B. Berkutov and J. G. Grossmann for SAXS characterization of the aerogel, performed at CLRC Daresbury.

REFERENCES

1. A. I. Golov, I. B. Berkutov, S. Babuin, and D. J. Cousins, *Physica B*, **329-333**, 258 (2003).
2. S. Babuin, *Transverse Sound In Liquid Helium-4 In Aerogel*, Ph.D. thesis, University of Manchester (2004).
3. D. L. Johnson, T. J. Plona, and C. Scala, *Phys. Rev. Lett.*, **49**, 1840 (1982).
4. C. Ashton, S. Babuin, and A. I. Golov, *J. of Low Temp. Phys.*, **138**, 183 (2005).
5. P. A. Crowell, J. D. Reppy, S. Mukherjee, M. H. W. Chan, and D. W. Schaefer, *Phys. Rev. B*, **51**, 12721 (1995).
6. S. Babuin, and A. I. Golov, *J. Low Temp. Phys.*, **134**, 133 (2004).
7. G. K. S. Wong, P. A. Crowell, H. A. Cho, and J. D. Reppy, *Phys. Rev. B*, **48**, 3858 (1993).
8. M. Kriss, and I. Rudnick, *J. Low Temp.*, **3**, 339 (1970).
9. R. Dolesi, M. Bonaldi, and S. Vitale, *J. Low Temp. Phys.*, **118** (2000).
10. D. L. Johnson, and P. N. Sen, *Phys. Rev. B*, **24**, 2486 (1981).

Two-Dimensional Classical Wave Localization in a Third Sound System

D.R. Luhman, J.C. Herrmann and R.B. Hallock

Laboratory of Low Temperature Physics, Department of Physics
University of Massachusetts, MA 01003, USA

Abstract. Patterned calcium fluoride deposited on glass creates an effective two-dimensional scatterer of third sound propagating on a thin ^{4}He film. We have utilized a substrate with a periodic arrangement of scatterers and a substrate with a random arrangement of scatterers to investigate classical wave propagation and localization using third sound. We observe pass bands on the periodic substrate while only low frequency modes are observed in the disordered case. The presence of high frequency modes on the ordered substrate and the absence of high frequency modes on the disordered substrate indicates we are observing localization on the disordered substrate. We compare the disordered data to the two-dimensional localization theory of Cohen and Machta [1] and find reasonable agreement.

Keywords: localization, superfluid helium
PACS: 71.55.Jv, 43.20.+g, 67.40.Pm, 67.70.+n

INTRODUCTION

Localization is a phenomena whereby a wave is spatially confined and extended states are not allowed. The time averaged wave amplitude, $|\psi|$, exponentially decreases with distance, x, as $|\psi| \sim e^{-x/\xi}$. The localization length, ξ, sets the length scale of confinement and is a primary quantity of interest.

Localization has been studied in a wide variety of systems ranging from quantum electronic transport to classical acoustics. Localization has been observed using third sound waves in one-dimension [2, 3]. Here we report additional results [4] from ongoing experiments on continuous wave measurements of third sound in a random collection of two-dimensional scatterers and present strong evidence for the presence of localization.

Cohen and Machta [1] have calculated the localization length for a two-dimensional third sound system. The main assumption of the theory is that the characteristic times and lengths associated with the scatterers are much shorter than those of third sound. Their coarse-grain approach allows them to characterize the system by the index of refraction through the scatterers which is defined by $n_s = C_0/C_s$ where C_0 is the speed of third sound on a clean smooth substrate and C_s is the speed of third sound on the disordered substrate. They find that the localization length in such a system is strongly frequency dependent and given by $\xi(E) = l(E)e^{(E_0/E)^2}$ where E_0 is a constant, $l(E)$ the scattering mean free path and $E = 2\pi f$, where f is the observable third sound frequency. The theory is only valid for $E < E_0$. In terms of the experimentally accessible parameters the localization length can be written as

$$\xi(f) = \frac{C_0^3 n_s \bar{\rho}_0}{4\pi^3 f^3 (n_s^2 - 1)^2} \exp\left\{ \frac{\bar{\rho}_0}{2\pi} \left(\frac{C_0 n_s}{f(n_s^2 - 1)} \right)^2 \right\}, \quad (1)$$

where $\bar{\rho}_0$ is the average areal density of scatterers.

EXPERIMENTAL DETAILS

The substrates used to investigate localization were a plain glass substrate, a uniformly CaF_2 coated substrate, a glass substrate with a periodic array of scatterers and a glass substrate with a random, disordered collection of scatterers. Third sound was generated using a resistive silver strip and detected using thin-film aluminum superconducting transition-edge bolometers. The periodic substrate had three detectors spaced 3 mm apart and the disordered substrate had five detectors, each separated by 3 mm. The scatterers were formed by evaporating 300 nm of CaF_2 onto the substrate through a thin mask. The mask had 0.255 mm radius holes drilled in either a periodic or disordered fashion. The effectiveness of such scatterers has been previously shown [5].

The substrates were mounted in a ^{4}He pumped-bath dewar and brought to an operating temperature of $T = 1.650$ K. The third sound transmission data across the substrates was determined by applying a continuous sinusoidal voltage to a third sound generator and observing the voltage drop across a selected current-biased detector using an EG & G 5208 lock-in amplifier operating in the "$2f$" mode. Measurements were taken for several values of the ^{4}He film thickness, d.

CP850, *Low Temperature Physics: 24th International Conference on Low Temperature Physics;*
edited by Y. Takano, S. P. Hershfield, S. O. Hill, P. J. Hirschfeld, and A. M. Goldman

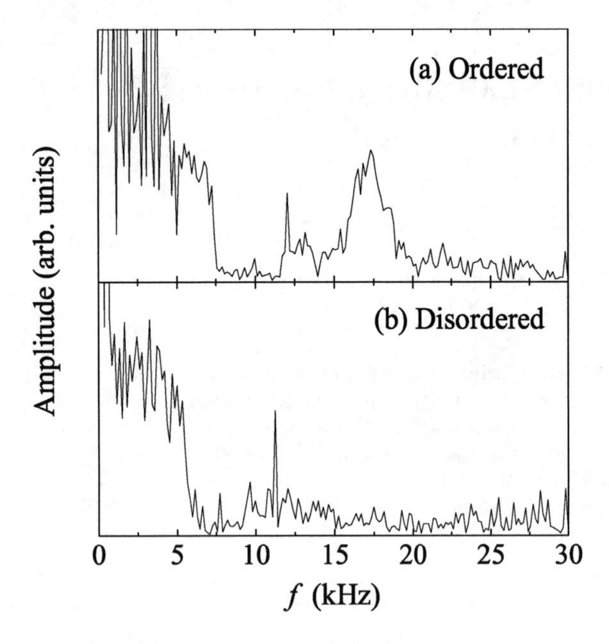

FIGURE 1. Third sound amplitude on the (a) ordered substrate and (b) disordered substrate versus frequency. The generator-detector distance was 3 mm and The helium film thickness is $d = 6.32$ layers for these data.

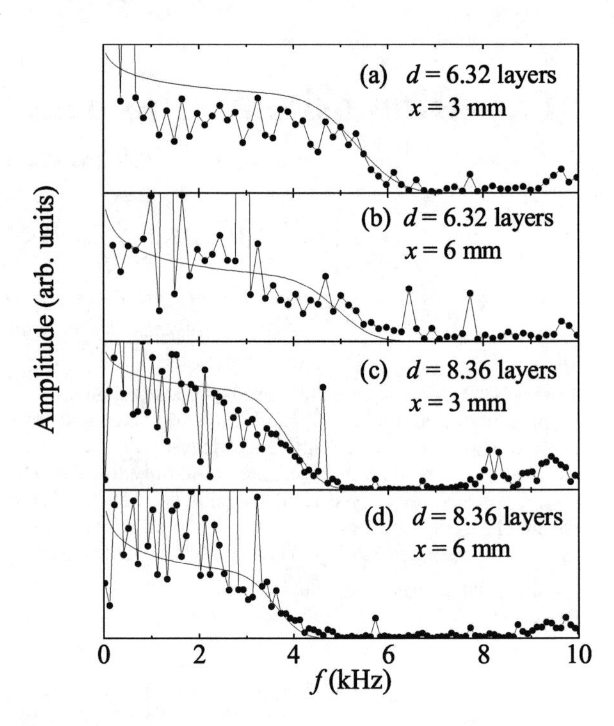

FIGURE 2. Third sound amplitude measured as a function of frequency for the disordered substrate for (a) $d = 6.32$ layers and $x = 3$ mm, (b) $d = 6.32$ layers and $x = 6$ mm, (c) $d = 8.36$ layers and $x = 3$ mm, (d) $d = 8.36$ layers and $x = 6$ mm. The lines connecting adjacent data points are a guide to the eye and the smooth solid lines are comparison to theory.

RESULTS

Qualitative evidence for localization can be seen by comparing data obtained from the periodic and disordered substrates as shown in Figure 1. The pass band near 17.5 kHz is clearly evident in the data from the ordered substrate and is greatly suppressed in the disordered data. Throughout the entire frequency range the amplitude on the disordered substrate does not significantly increase after the initial drop at low frequency. The absence of appreciable high frequency modes on the disordered substrate and the presence of them on the ordered substrate is evidence for classical wave localization.

In order to quantitatively compare our results to Eq. (1) we also need to include ordinary third sound attenuation. To model attenuation in our experiment we compare the data on the ordered substrate with Bergman's calculated result for the attenuation coefficient, α_B [6]. We find that our attenuation can be characterized by $\alpha \approx 0.01 d^3 \alpha_B$ where d is in layers. To compare our data to Eq. (1) we normalize the data such that the trend of the amplitude is toward unity as $f \to 0$ and compare it to $|\psi| = e^{-x(\xi^{-1}+\alpha)}$ where x is the distance from the third sound driver. Figure 2 shows a comparison of data from the disordered substrate to theory with n_s as a free parameter for two values of d and x. The theory is shown by a smooth solid line, which has two main features. The first is a gradual drop with increasing frequency followed by a second more abrupt drop. The abrupt drop is due to localization while the gradual drop is due to normal third sound attenuation. The theory is in reasonable agreement with the data, particularly with regard to the features specific to localization.

ACKNOWLEDGMENTS

We thank J. Machta for productive discussions. This work was supported by the National Science Foundation under grants DMR-0138009 and DMR-0213695 (MRSEC) and also by research trust funds administered by the University of Massachusetts Amherst.

REFERENCES

1. S. M. Cohen and J. Machta, *Phys. Rev. Lett.* **54**, 2242 (1985).
2. D.T. Smith, C.P. Lorenson, R.B. Hallock, K.R. McCall, and R.A. Guyer, *Phys. Rev. Lett.* **61**, 1286 (1988).
3. K. Kono and S. Nakada, *Phys. Rev. Lett.* **69**, 1185 (1992).
4. D.R. Luhman, J.C. Herrmann, and R.B. Hallock, *Phys. Rev. Lett.* **94**, 176401 (2005).
5. J.C. Herrmann and R.B. Hallock, *J. Low Temp. Phys.* **126**, 361 (2002).
6. D. Bergman, *Phys. Rev.* **188**, 170 (1969).

Model for Cyclically Astable Third Sound Resonance

F. M. Ellis, Ian Carbone, and Hoan Dang

Department of Physics, Wesleyan Unversity, Middletown, CT 06459, USA

Abstract. Third sound in a circular resonator is capable of rich behavior due to the coincidence of several characteristics: a high quality factor; the capability for wave-induced circulation changes; vortex pinning on the substrate; and the Doppler shifting of modes. One dimensional models for each of these components have previously been used to reproduce highly nonlinear third sound CW lineshapes using a steady-state approximation. We now include oscillator transients to account for cyclical amplitude modulations of third sound recently observed under conditions of a drive force where both the amplitude and frequency are fixed.

Keywords: superfluidity, circulation, vortices, pinning, third sound
PACS: 67.40 Hf, 67.40 Vs

INTRODUCTION

Third sound is a unique tool for the study of vortices in superfluid helium films. Large amplitude wave agitation can be used to create and destroy circulatory persistent current states in a resonator and the Doppler effect can be used to determine the corresponding net vorticity.[1] The persistence of flow concurrent with intermingled vortices demonstrates a remarkably strong pinning of the vortices to the substrate.

The high Q typical of third sound at low temperatures allows small changes in the resonant frequency to have dramatic consequences. There are two dominant sources of amplitude dependent frequency shifts that contribute to distorted resonance lineshapes. First, there is a quadratic mode coupling interaction associated with the AC Bernoulli pressure[2], and second, high amplitude wave agitation can induce changes in any persistent current present in the resonator. Both of these effects are illustrated in Figure 1.

Consider first a scan up through a resonance. The amplitude rises as resonance is approached, but the mode coupling effect shifts the frequency down. At a high enough drive amplitude, this results in a catastrophic jump of mode to an above-resonant situation. The scan then proceeds along the above-resonant tail to a lower amplitude. The resonance peak is completely bypassed. A scan down is qualitatively different. The mode coupling shift moves the resonance further down and elongates the eventual reaching of a peak. Just past the peak, the steady-state condition is unstable. The amplitude drops concurrent with the resonant frequency shifting back up, and the scan continues from a point on the below-resonant tail.

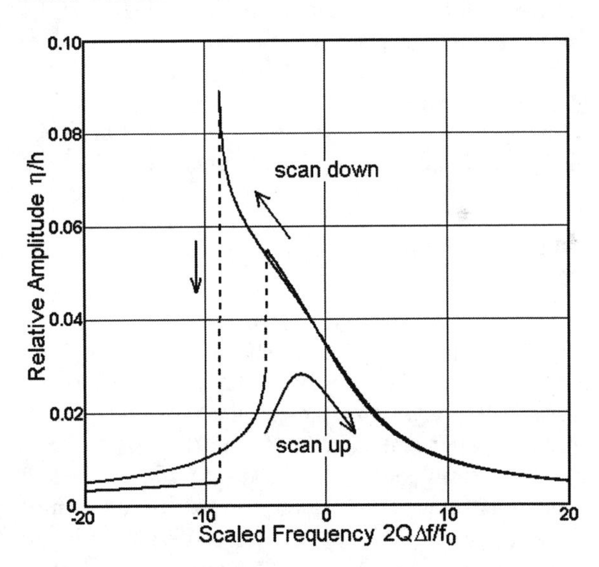

FIGURE 1. Distortion and catastrophic jumps are a consequence of amplitude dependent frequency shifts. The drive frequency is Δf relative to the linear resonance at f_0.

Induced circulation adds another source of distortion to the lineshape. As the amplitude increases beyond a threshold for de-pinning, the wave agitation first diminishes the circulation. At higher amplitudes, new vorticity is induced and the circulation increases.

CP850, *Low Temperature Physics: 24th International Conference on Low Temperature Physics;*
edited by Y. Takano, S. P. Hershfield, S. O. Hill, P. J. Hirschfeld, and A. M. Goldman

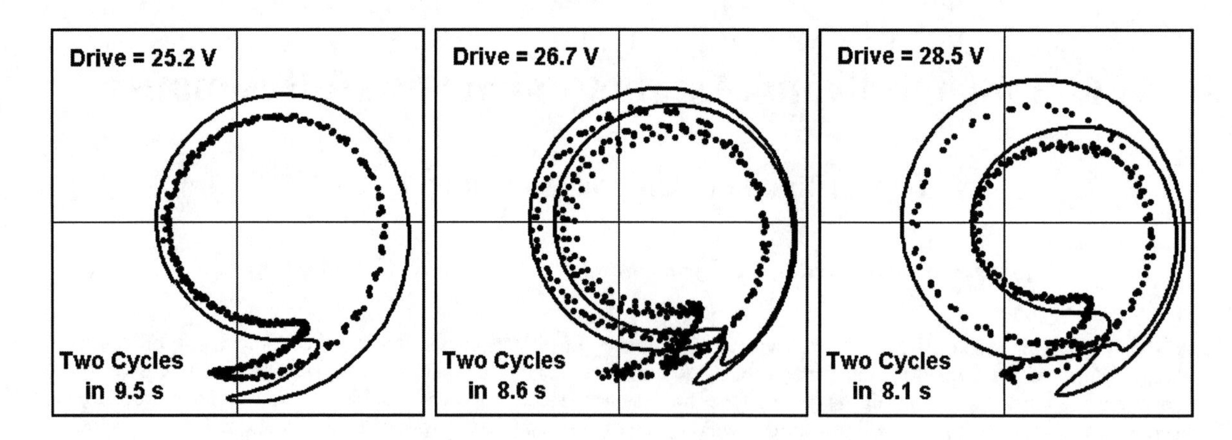

FIGURE 2. Real and imaginary parts of the cyclical data and model at different drives. Full scale is $\eta/h=0.15$ in all plots.

The sharp peak of the down scan results as the increasing circulation moves the mode faster into resonance. The jump to the below-resonant tail leaves the mode permanently shifted up to about $Q\Delta f/f_0=11$.

An interesting situation arises which has been experimentally observed[3]. With a fixed drive, it is possible to have a temporally unstable response. With the resonance above the drive, moderate wave agitation causes the circulation to decay. This shifts the mode through resonance, where the higher wave amplitude causes the circulation to increase. The mode is consequently shifted back up above the drive, initiating a new cycle. This repetition is the basis for the cyclically astable third sound resonance.

MODEL AND RESULTS

A steady state model for the interplay between resonance and circulation has been successfully applied to reproduce highly distorted lineshapes[4]. It captures the physics of the circular resonator and its spatially dependent circulation with two parameters, the wave-amplitude to film-thickness ratio, $z=\eta/h$, and a radially averaged circulation flow, scaled to the third sound speed, $\chi=\langle v_{circ}\rangle/c_3$. Flow changes are the result of wave-flow coupling dependent on both χ and z. The amplitude z is then self-consistently determined from the steady state response shifted by the mode coupling (amplitude z) and Doppler effects (flow χ).

This model is now extended to include the transient behavior that dominates the cyclically astable resonances. The amplitude z is taken to be complex and a discretized stepping of the simple harmonic oscillator with transients is performed, updating the oscillator parameters (resonance frequency and Q) appropriate to the flow χ and amplitude magnitude $|z|$. The transient aspect of the model is approximate only to the extent that the parameters change during the time steps.

Figure 2 shows experimental data together with the model behavior for a 12.3 mm dia. resonator with a gold substrate. Shown are the real and imaginary parts of the complex amplitude driven at 1670 Hz with c_3=21 m/s at 0.6 K. The paths cycle about the origin with the times for two revolutions shown. Three drive amplitudes are shown with all other conditions fixed. The flow was approximately χc_3=11 cm/s.

The model allows us to verify exactly what is happening during the cycling. Most of the cycle is spent with the third sound mode frequency above the drive. These are the major circular sections in the figures, dominated by a free oscillator transient and a slowly decaying circulation. The hook near the negative imaginary axis corresponds to where the mode moves down through the drive. The amplitude builds up and the circulation then increases moving the mode back above resonance. No longer driven near resonance, the amplitude decays and the circulation reverts to a slow decay. The model shows a flow variation of about 20% over the course of the cycling. The model also indicates that vortex de-pinning occurs with an exponential activation[4] of the form $\exp(bv_{flow})$ with $b=12\pm5$ s/m.

REFERENCES

1. F. M. Ellis and C. Wilson, *J. Low Temp. Phys.* **113**, 411-416 (1998).
2. R. Baierlein, F. M. Ellis, and H. Luo, *J. Low Temp. Phys.* **108**, 31-52 (1997).
3. F. M. Ellis, *J. Low Temp. Phys.* **134,** 97 (2004).
4. F. M. Ellis and Seungwook Ma, *Physica* **B284-288**, 129-130 (2000).

Third Sound Amplification and Detailed Balance

J. D. Eddinger and F. M. Ellis

Department of Physics, Wesleyan University, Middletown, CT 06459, USA

Abstract. Condensation of atoms from the vapor into a third sound resonance is expected to be capable of acoustic amplification. This results from normal to superfluid conversion that coherently accommodates atoms into the third sound velocity field. Consideration of third sound in light of the equilibrium detailed balance between vapor particles and the superfluid film provides further evidence that acoustic amplification is attainable.

Keywords: superfluidity, quantum condensation, third sound.
PACS: 67.40 Db, 67.40 Hf

INTRODUCTION

Condensation of vapor atoms into the superfluid He II is a phase transition that involves a change in order. The gas to liquid transition involves an increase in order, but the quantum degeneracy condition characterizing the superfluid requires that condensing atoms also take on the momentum order associated with the condensed state. This conversion of disordered vapor into ordered kinetic energy has been previously observed for the case of a film in a persistent current state. Henkel et al[1], referred to it as a "crystal-like growth" of the superfluid state. This direct conversion of disordered atoms into ordered kinetic energy is an interesting example of a heat engine illustrated schematically in Fig 1(A). The flow of heat through the system (driving the condensation process) results in mechanical energy (an increase in macroscopic kinetic energy). Alternatively, the process can also be interpreted as an example of a "matter laser" — particles coherently join the macroscopic deBroglie wave structure upon condensing.

In a more recent work, a similar coherent condensation process has been proposed.[2] Schematically shown in Fig. 1(B), the ordered kinetic energy state consists of the macroscopic film flow of a third sound oscillation. The mechanical "output" in this case is the acoustic energy of the third sound oscillations. Here, the analogy of coherent condensation to a matter laser is more obvious. Condensing atoms take up the role of transitions within an inverted population pumped by a temperature difference. The transitions involve adding particles into a macroscopically occupied cavity mode, in this case a third sound wave mode instead of a photon mode. The presence of multiple acoustic modes separates the acoustic energy gain process from the recycling of mass out of the film. This separation allows for a continuous gain similar to CW laser operation.

In this paper we discuss the consequences of the equilibrium film-vapor particle exchange to the coherent condensation process.

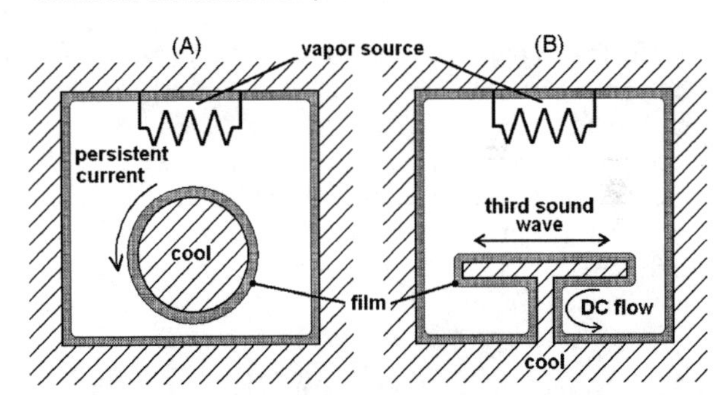

FIGURE 1. Kinetic energy of a persistent current (A) or a third sound mode (B) is generated by the condensation of atoms into the macroscopic quantum state of the film. In principle, moving slabs of film could be extracted from (A) and acoustic energy from (B).

DISCUSSION

With the assumption that condensing atoms coherently join the instantaneous macroscopic kinetic energy configuration of the third sound wave, an intuitive explanation of the self-oscillation condition is

CP850, *Low Temperature Physics: 24th International Conference on Low Temperature Physics;* edited by Y. Takano, S. P. Hershfield, S. O. Hill, P. J. Hirschfeld, and A. M. Goldman

obtained. The particles need to condense and be recycled through the cavity at a rate that exchanges all of the particles in the resonance region of the film during a time on the order of the third sound decay. The rate of energy input is determined by the average kinetic energy of the mode, and the rate of energy loss is the natural damping of the third sound.

$$\frac{dN}{dt} = \frac{N}{\tau} \frac{\langle E \rangle_{resonator}}{\langle KE \rangle_{exposed}}. \qquad (1)$$

This expression is exact with the kinetic energy average appropriately weighted by the spatial distribution of condensing atoms. The assumption that condensing atoms coherently join the local kinetic energy state of the film is crucial. Henkel's experiment demonstrated this for a persistent current state, but the validity of the assumption for a non-persistent macroscopic third sound flow is not entirely clear. Although the time and size scales of atomic condensing events are far removed from the corresponding time and size scale of the third sound, one would like a rigorous demonstration that the condensing atoms interact with the moving third sound film in the same way as with a persistent current. Here we make the case using the concept of detailed balance applied to the experimental fact that third sound takes a relatively long time to decay.

Consider the rate at which gas particles are accommodated into the film at equilibrium. At low temperatures ($0<T<1$ K), the vapor pressure (P_{film}) becomes quite small, particularly above a thin film. The degeneracy of the gas is totally negligible so classical kinetic theory can be applied. The approximate time for exchanging all atoms composing the film is given by

$$\tau = \frac{\rho h}{\alpha m} \frac{\sqrt{2\pi mkT}}{P_{film}} \qquad (2)$$

where α is the absorption probability for particles impinging on the film, ρ is the liquid density, and h is the film thickness. Theoretical[3] and experimental[4] evidence shows that α is close to unity until the perpendicular wavenumber of the impacting atom approaches zero. With wavenumbers characteristic of 0.2 K, α has dropped to 90%. In spite of the low vapor pressure, the film exchange time including the accomadation factor is quite short, as shown in Fig. 2.

What is remarkable about this result is that, in equilibrium at most temperatures, all of the atoms in the film are replaced on a time scale much shorter than the typical decay time of a third sound wave, shown as the dashed line in the Fig. 2. The exchange at elevated temperatures is even more impressive: at 0.6 K, refreshing the film's particles on the order of one cycle of the third sound wave oscillation. We take this as a demonstration that condensing atoms are accommodated into the kinetic motion of the third sound resonance. Any other fate of the atoms would contribute to an anomalously large sound attenuation.

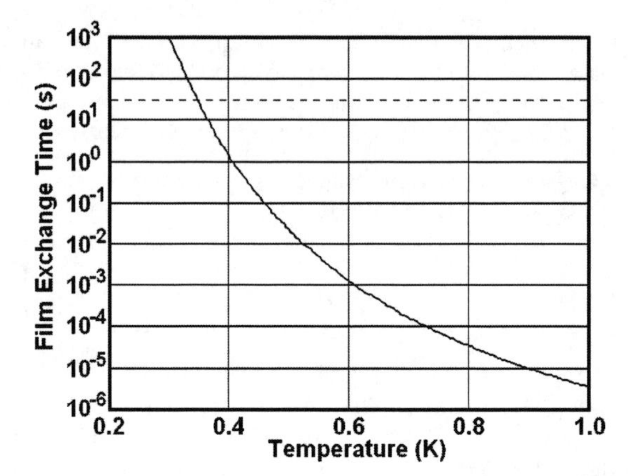

FIGURE 2. Characteristic time for the exchange of all atoms in an equilibrium film. The time scale for the decay of a high Q third sound resonance is shown as the dashed line.

The amplification by stimulated condensation is the product of a small imbalance of the equilibrium exchange, with the imbalance creating the gain as additional particles join the macroscopic quantum flow state. It is effectively a normal to superfluid conversion process that also increases the number of particles in the system. The liquid-vapor interface is, in some sense, acting as a semi-permeable membrane allowing the normal fluid excitations (as actual particles in the vapor) through while confining the superfluid state.

We are engaged in an experimental effort to achieve third sound amplification through the condensation of a vapor flux onto a resonating film. Our efforts are currently focused on mechanical issues.

REFERENCES

1. R.P. Henkel, E.N. Smith, J.D. Reppy, *Phys. Rev. Lett.* **23**, 1276-1281 (1969).
2. S. A. Jerebets and F. M. Ellis, *J. Low Temp. Phys.* **121**, 321 (2000).
3. D. R. Swanson and D. O. Edwards, *Appl. Phys. Rev. B* **37**, 1539-1549 (1988).
4. Mark Brown and Adrian F. G. Wyatt, *J. Phys.: Condens. Matter* **15**, 4717-4738 (2003).

Third sound and stability of ^{3}He-^{4}He mixture films

R. H. Anderson*, E. Krotscheck† and M. D. Miller†,*

*Department of Physics, Washington State University, Pullman, WA 99164-2814, USA
†Institut für Theoretische Physik, Johannes Kepler Universität, A 4040 Linz, Austria

Abstract.
We study third sound and the interaction between ^{3}He adatoms in two thin ^{3}He-^{4}He mixture films from a first-principles, microscopic theory. Utilizing the variational, hypernetted-chain Euler-Lagrange (HNC-EL) theory as applied to inhomogeneous boson systems, we calculate chemical potentials for both the ^{4}He superfluid film and the physisorbed ^{3}He. Numerical density derivatives of the chemical potentials lead to the sought-after third sound speeds that clearly reflect a layered structure of at least seven oscillations. In this paper, we report third sound on model substrates: Nuclepore, and sodium. We find that the effect of the ^{3}He depends sensitively on the particular ^{4}He film coverage. Our most important result is that, with the addition of ^{3}He, the third sound speed can *either* increase or decrease. In fact, in some regimes, the added ^{3}He destabilizes the film and can drive "layering transitions", leading to fairly complicated geometric structures of the film in which the outermost layer is predicted to consist of phase-separated regions of ^{3}He and ^{4}He.

Keywords: Third sound, ^{3}He-^{4}He mixtures, films
PACS: 67.20.+k,67.40.Db,67.40.Pm

INTRODUCTION

Low-temperature ^{4}He films can support propagating hydrodynamic modes despite being much thinner than the viscous penetration depth. These modes, termed third sound, are surface-type, density fluctuations which are neither purely longitudinal nor purely transverse in character. We are aware of three third sound experiments for ^{3}He-^{4}He mixture thin films. Two experiments were reported by Hallock and co-workers. The first, by Valles, Heinrichs, and Hallock, [1] was on a borosilicate glass substrate. The second, by Sheldon and Hallock, [2] was on a Nuclepore substrate. Noiray, Sornette, Romagnan, and Laheurte [3] reported results for third sound on both Nuclepore and glass. Third sound is a sensitive measure of film structure and impurity distributions.

The square of the third sound speed c_3 in the mixture film as a function of ^{3}He coverage is linear for small coverages. In the small ^{3}He coverage limit, we can write the total incompressibility as $m_4c_3^2 = m_4c_{30}^2(1-\Delta\theta)$, where $\theta = (\sigma_3/\sigma_{3\ell})$ is the ^{3}He coverage in units of monolayers, $\sigma_{3\ell} \equiv 1/\ell_3^2 = 0.065$ Å^{-2} is the areal density at "conventional" monolayer completion, and c_{30} the third sound speed for zero ^{3}He concentration

By inspection, the experimental slope in the Nuclepore case, [2] is $\Delta_{np} \approx 0.5$. The linear regime in Nuclepore extends out to 0.6 ℓ_3. For glass, the results qualitatively resemble older thick ^{3}He film results of Ref. [4]: *little or no linear regime at low coverages*. In this paper, we shall restrict ourselves to discussing two substrates: Nuclepore and sodium. We emphasize that our results for sodium are predictions since, unlike Nuclepore, no third sound measurements have yet been made on it.

We describe both the ^{4}He background and the ^{3}He impurity by the familiar Feenberg variational wave function, including triplet correlations, and allowing for the breaking of translational invariance and isotropy as demanded by the external field. The ^{4}He and ^{3}He densities $\rho_i(\mathbf{r})$ and the pair distribution functions $g^{(i,j)}(\mathbf{r},\mathbf{r})$ are determined by minimization of the total energy subject to the constraint of fixed particle number N. We note the coverage is defined as $\sigma_i = \int_{-\infty}^{\infty} dz \rho_i(\mathbf{r})$, where $i = 3,4$.

The central quantity of interest in this work is the ^{3}He and ^{4}He coverage dependence of the third sound speed, c_3. The third sound speed can be calculated in two different ways, namely as the hydrodynamic derivative $m_4c_3^2 = \sigma_4 (d\,\mu_4/d\,\sigma_4)$ of the ^{4}He chemical potential mu_4, and also from the long-wavelength limit of the low-lying excitations. The quantity Δ discussed above is proportional to a second density derivative of the ^{4}He chemical potential. For details of the calculation, we refer the reader to Refs. [5, 6].

RESULTS

We have carried out extensive studies of the coverage dependence of third sound for helium films on glass, Nuclepore, sodium, and lithium substrates.

We have fitted the numerical data for two and more layers with six parameters $\alpha_s, c, d, e, k, \gamma$ by the function

$$\mu_i(\sigma_4) = \mu_i(\infty) - \frac{\alpha_s\bar{\rho}^3}{\sigma_4^3} + \frac{c}{\sigma_4^4} + \frac{d}{\sigma_4^5} + \frac{e\cos(k\sigma_4-\gamma)}{\sigma_4^4}. \tag{1}$$

CP850, *Low Temperature Physics: 24th International Conference on Low Temperature Physics;*
edited by Y. Takano, S. P. Hershfield, S. O. Hill, P. J. Hirschfeld, and A. M. Goldman

Because all substrates exhibit layering transitions, the monolayer results are not well represented by the form (1). From the fits, we immediately obtain the bulk limit of the ^{4}He chemical potential, approximately -7.30 K for each of the substrates, which is in good agreement with the experimental value. We also obtain the energy of the Andreev state in the bulk limit of -5.2 K for all substrates, again in good agreement with experiment.

The coefficients representing the van der Waals term and the oscillatory part of the density are directly related to the properties of the ^{4}He film, and are reasonably consistent for the Nuclepore and sodium substrates. The simple van der Waals form (*i.e. the first two terms* of (1)), expected to be valid for structureless films, is not in good agreement with the calculated, microscopic chemical potentials. We conclude therefore that the films under consideration here are still far from the thickness where the chemical potential is dominated by the van der Waals term.

The most interesting aspect of our calculations is the effect of a small amount of ^{3}He on the third sound speed. In Figs. 1 and 2, we show incompressibilities for the Nuclepore and sodium substrates, respectively. To the extent that the linearization of the ^{3}He dependence of the incompressibility is legitimate, the results are quite interesting. Depending on the ^{4}He coverage, the incompressibility may *increase or decrease* with ^{3}He concentration. In the vicinity of layering transitions, the effect can be strong enough to make the film unstable. The effect is particularly pronounced in the triple-layer films with coverage between 0.165 Å^{-2} and 0.22 Å^{-2}. In the lower half of that coverage regime, the addition of ^{3}He would have the tendency to *stabilize* the films, whereas it *destabilizes* the translationally invariant configuration at coverages above 0.19 Å^{-2}. This is consistent with the drop in the speed of sound ($\Delta > 0$) seen in the Nuclepore experiment of Sheldon and Hallock [2]. Of course, inspection of the Nuclepore data, Fig. 1, shows clearly that the change in third sound speed depends quite sensitively on the coverage, and a quantitative comparison is not feasible.

Our calculations suggest a rather interesting scenario: as the coverage is further increased, the ^{4}He film can undergo a layering transition which is *driven by the presence of* ^{3}*He*. Our results indicate that this happens at the modest ^{3}He concentration of 0.25 monolayers, where the linearization should still be permitted, and a ^{4}He coverage of about 0.21 Å^{-2}. The configuration of the film in this regime would be a mixture of regions that would consist of three layers of ^{4}He without ^{3}He in Andreev states, and double layers of ^{4}He with ^{3}He in surface states. It is intriguing to note that the anomolous mixture film system investigated by Bhattacharyya and Gasparini [7] had a coverage ≈ 0.26 Å^{-2}; this value is located in a possibly unstable region as shown in Fig. 1.

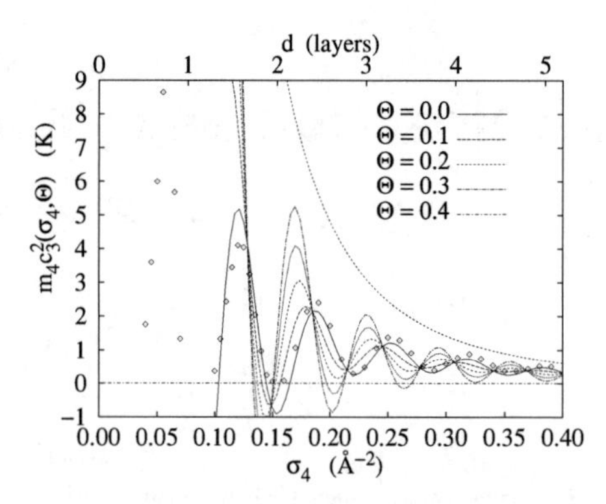

FIGURE 1. The figure shows the total incompressibility of the mixture film on Nuclepore as a function of ^{3}He coverage Θ in the low concentration regime. The short-dashed line shows the incompressibility from the Van der Waals term alone.

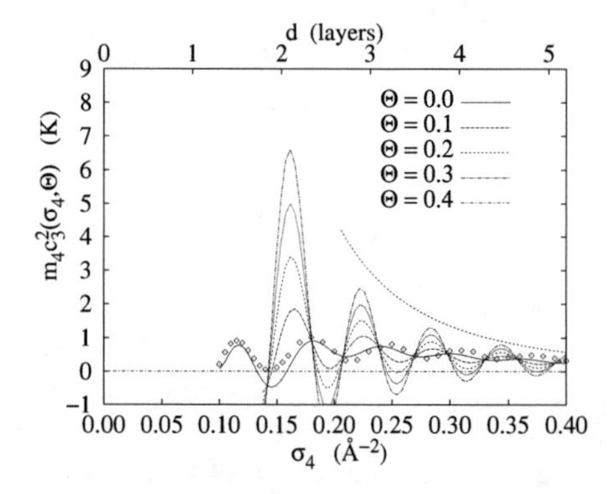

FIGURE 2. Same as Fig. 1 for the sodium substrate.

ACKNOWLEDGMENTS

This work was supported, in part, by the Austrian Science Fund (FWF) under project P15083-N08.

REFERENCES

1. J. M. Valles, R. M. Heinrichs, and R. B. Hallock, *Phys. Rev. Lett.*, **56**, 1704 (1986).
2. P. A. Sheldon and R. B. Hallock, *Phys. Rev. B*, **50**, 16082 (1994).
3. J. C. Noiray, D. Sornette, J. P. Romagnan, and J. P. Laheurte, *Phys. Rev. Lett.*, **53**, 2421 (1984).
4. F. M. Ellis, R. B. Hallock, M. D. Miller, and R. A. Guyer, *Phys. Rev. Lett.*, **46**, 1461 (1981).
5. V. Apaja and E. Krotscheck, "A Microscopic View of Confined Quantum Liquids," in *Microscopic Approaches to Quantum Liquids in Confined Geometries*, edited by E. Krotscheck, and J. Navarro, World Scientific, Singapore, 2002, pp. 205–268.
6. E. Krotscheck and M. D. Miller (2005), phys. Rev. B and J. Low Temp. Phys, in press.
7. B. Bhattacharyya and F. M. Gasparini, *Phys. Rev. Lett.*, **49**, 919 (1982).

Kosterlitz-Thouless Transition in ^{4}He Films Adsorbed to Rough Calcium Fluoride

D.R. Luhman and R.B. Hallock

Laboratory of Low Temperature Physics, Department of Physics
University of Massachusetts, Amherst, USA 01003

Abstract. Previous measurements in our lab have shown that the onset of superfluidity at the KT transition, typically seen as a sharp change in the frequency of a smooth-surface quartz crystal microbalance, becomes less identifiable in the presence of increasing surface roughness or disorder, while the peak in the dissipation is unchanged[1]. Using a series of microbalances coated with increasingly rough CaF_2, we have extended our measurements to lower ^{4}He film coverages and thus lower temperatures. We find at lower ^{4}He coverages that the presence of disorder on the substrate has a diminished effect on the frequency shift.

Keywords: Kosterlitz-Thouless transition, superfluid helium
PACS: 68.35.Ct, 64.70.Ja, 67.40.Pm, 68.15.+e

INTRODUCTION

The superfluid transition in thin helium films has been shown to be well described by the theory of Kosterlitz and Thouless (KT)[2]. The onset of superfluidity in static films is abrupt with a discontinuous jump in the areal superfluid density, σ_s. In practice the discontinuous step in σ_s at the onset of superfluidity is slightly rounded due to frequency and velocity effects in the adsorbed film. In addition to the sharp onset of superfluidity there is also a peak in the dissipation at the transition temperature, T_{KT}, that is also characteristic of the superfluid transition in two-dimensional films.

A number of experiments have investigated the effect of disorder on the KT transition through the use of three-dimensional multiply connected substrates, such as vycor[3] and packed alumina powder[4]. However there has been little work done on the effect of disorder in quasi-two-dimensional systems. In a previous experiment[1] we used a series of quartz crystal microbalances (QCMs), each with a different surface roughness to investigate the effect of surface roughness (i.e. disorder) on the superfluid transition. Both the shift in frequency and the dissipation were monitored as the helium film thickness was increased at a constant temperature of $T = 1.672$ K. It was shown that as the surface roughness increases the observed shift in frequency at the transition becomes less well defined and less identifiable with a sharp step in σ_s while the peak in the dissipation remains essentially unchanged. For the roughest substrate a shift in frequency at the transition was barely visible. An earlier experiment[5] on even rougher surfaces did not observe a shift in frequency at the onset of superfluidity. In another experiment, utilizing a QCM with porous gold electrodes near the capillary condensation transition, a dissipation peak was observed at the superfluid transition, however, a shift in frequency was not seen[6].

Here we report the results of further observations utilizing a series of QCMs with different surface roughness at lower temperatures and thinner helium films.

EXPERIMENTAL DETAILS

The experiment utilized five gold-plated AT-cut QCMs operated in their third harmonic (15 MHz). Four of the crystals were coated with different nominal thicknesses of CaF_2, t, to produce differing surface roughness. The surface roughness of CaF_2 increases as t increases; the structures increase in size while the porosity stays relatively constant ($\phi \approx 0.64$) with increasing t[7]. The remaining crystal was left uncoated. For a particular QCM, the same amount of CaF_2 was thermally deposited on each side. The values of t used in the experiment were 30, 60, 90, and 120 nm, covering the same range of CaF_2 thicknesses as in Ref. [1].

The crystals, along with a plain glass substrate with third sound generators and Zn bolometers, were mounted in a brass sample can attached to a dilution refrigerator. ^{4}He was slowly bled in the sample chamber and allowed to equilibrate. The temperature was then brought to $T = 0.820$ K. The helium film thickness was determined by measuring the speed of third sound in the ^{4}He film on glass and solving $C_3^2 = (\langle\rho_s\rangle/\rho)Fd(1+TS/L)^2$ for the film thickness d. $\langle\rho_s\rangle/\rho$ is the effective superfluid fraction in the film, F is the restoring force due to the van der Waals interaction, S is the specific entropy, and L

CP850, *Low Temperature Physics: 24th International Conference on Low Temperature Physics;*
edited by Y. Takano, S. P. Hershfield, S. O. Hill, P. J. Hirschfeld, and A. M. Goldman

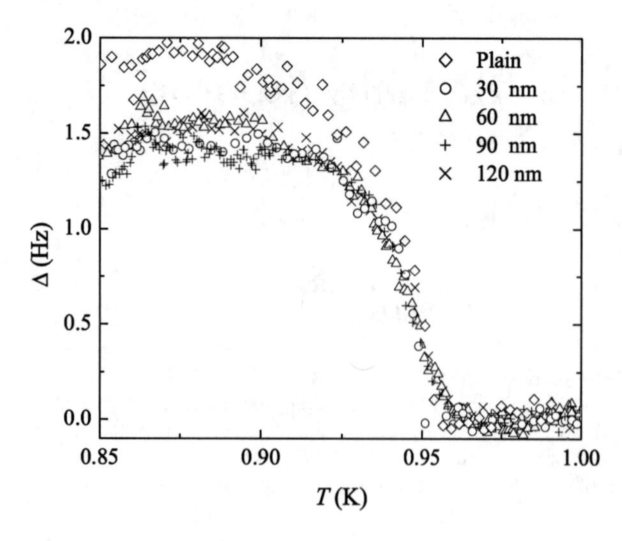

FIGURE 1. Frequency data with the background subtracted as a function of T. The large drop in frequency indicates the superfluid transition. The film thickness measured at $T = 0.820$ K was $d = 2.79$ layers for these data.

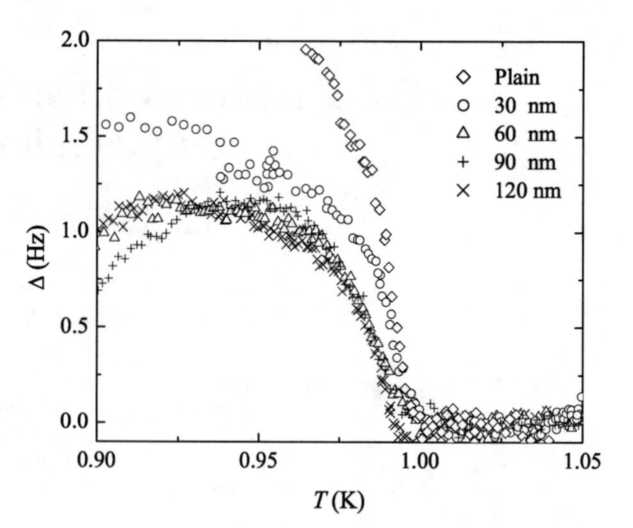

FIGURE 2. Frequency data with the background subtracted as a function of T. The large drop in frequency indicates the superfluid transition. The film thickness measured at $T = 0.820$ K was $d = 3.19$ layers for these data.

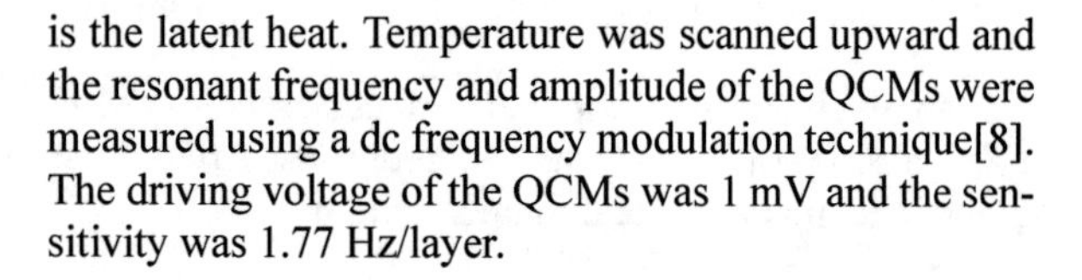
is the latent heat. Temperature was scanned upward and the resonant frequency and amplitude of the QCMs were measured using a dc frequency modulation technique[8]. The driving voltage of the QCMs was 1 mV and the sensitivity was 1.77 Hz/layer.

RESULTS

Figures 1 and 2 show frequency data with the background subtracted for $d = 2.79$ layers and $d = 3.19$ layers as measured at $T = 0.820$ K respectively. As the temperature was increased during data collection the film thickness decreased slightly as atoms moved from the film to the vapor. Due to this process, we estimate the thickness of the film at $T_{KT} = 0.950$ K (Fig. 1) is a few a tenths of a layer thinner than the value measured at $T = 0.820$ K and near 0.5 layers thinner at $T_{KT} = 0.992$ K (Fig. 2). In Fig. 1 the general shape and size of the mass coupling at the transition is virtually identical for all the substrates coated with CaF_2. The frequency shift on the plain QCM was slightly larger than on the CaF_2 substrates but essentially the same shape. For the thicker film data shown in Fig. 2 the substrates with larger values of t show a somewhat broader transition. This broadening is likely a crossover to the behavior seen in the earlier experiment at increased temperature and film thickness where the frequency shift was observed to be less distinct as t increased[1]. This general behavior is likely due to an increase in helium film thickness. At larger film thicknesses there is additional helium adsorbed to the disordered substrates (e.g. via capillary condensation) than at thinner helium film thicknesses. This will cause a disordered film thickness that varies greatly with position. We speculate that this disorder in the film thickness itself may lead to suppression in the distinctness of the frequency shift at the KT transition at larger film coverages.

ACKNOWLEDGMENTS

We thank N. Prokof'ev for productive discussions. This work was supported by the National Science Foundation under grants DMR-0138009 and DMR-0213695 (MRSEC) and also by research trust funds administered by the University of Massachusetts Amherst.

REFERENCES

1. D.R. Luhman and R.B. Hallock, *Phys. Rev. Lett.* **93**, 086106 (2004).
2. J. M Kosterlitz and D. J. Thouless, *J. Phys. C* **6**, 1181 (1973).
3. J.E. Berthold, D.J. Bishop, and J.D. Reppy, *Phys. Rev. Lett.* **39**, 348 (1977).
4. V. Kotsubo and G.A. Williams, *Phys. Rev. Lett.* **53**, 691 (1984).
5. J.C. Herrmann and R.B. Hallock, *Phys. Rev. B* **68**, 224510 (2003).
6. R.J. Lazarowick, P. Taborek, and J.E. Rutledge, *Bull. Am. Phys.* **49**, 378 (2004).
7. D. R. Luhman and R. B. Hallock, *Phys. Rev. E* **70**, 051606 (2004).
8. M.J. Lea, P. Fozooni, and P.W. Retz, *J. Low Temp. Phys.* **54**, 303 (1984).

Sensitive Measurement of the Kosterlitz-Thouless Transition

Shalva Ben-Ezra and William I. Glaberson

The racah Institute of Physics, The Hebrew University 91904, Jerusalem, Israel

Abstract. We have developed a new technique for probing the Kosterlitz - Thouless transition in thin helium films. The technique is based on operation of a single quartz crystal microbalance at two different harmonic frequencies simultaneously. The technique is insensitive to surface effects, and provides the means for monitoring the heating of the film. By suppressing heating and surface effects we were able to systematically compare the results with the existing theories of ANHS, Bowley, Armour and Benedict, and Armour and Bowley. Substantial agreement with the last model of Armour and Bowley was observed. To the best our knowledge this is the first time that the experimental data in the frequency range 10-30MHz is compared to theoretical models.

Keywords: Helium films, Quartz crystal microbalance, Kosterlitz - Thouless transition.
PACS: 67.40.-w, 67.40.Rp, 67.40.Vs

INTRODUCTION

It is generally believed that the superfluid transition in thin helium films is well characterized by the Kosterlitz-Thouless[1] theory. In this approach, thermally activated bound vortex-antivortex pairs are the dominant fluctuations and they mediate the transition from the superfluid to the normal state. The Kosterlitz-Thouless approach, suitable to a situation where the system is probed at zero frequency, leads to the characteristic discontinuous jump in superfluid density at the transition. The AHNS[2] theory described the vortex dynamics in a phenomenological way, using expressions for the superfluid density and vortex fugacity obtained from the static recursion relations of Kosterlitz. Significant contributions to the understanding of the KT transition were achieved by Bowley and co-authors[3,4]. They described a new theoretical approach based on the equivalence of the Coulomb-gas model and the sine-Gordon model for a fictitious field φ. The authors proposed an equation of motion to describe the relaxation of the fictitious field back to equilibrium. Their approach has the important advantage over previous theories in that the dynamic behavior is obtained directly from a renormalization treatment of the equation of motion for the system. It also has the advantage that it leads to agreement with the main features of the experimental results[5] at low frequencies on Mylar substrates.

EXPERIMENTAL TECHNIQUE

We have developed a new technique for probing the Kosterlitz-Thouless transition in thin helium films. A quartz crystal microbalance is simultaneously operated in two harmonic resonant modes - the fundamental, at a frequency of ~3.5 MHz, and ninth at a frequency of ~31.5 MHz[6,7]. Stability of the quartz crystal microbalance at the first overtone was 0.001Hz and at the ninth overtone 0.006Hz. We independently and simultaneously monitor the resonant frequencies and Q values of the two modes as the adsorbed film thickness is varied through the superfluid transition (see reference 7). Because the Kosterlitz-Thouless transition is frequency dependent, the transitions as observed in the two modes occur at different film thicknesses. In a perfectly linear system, the transition as observed in one mode should be independent of the amplitude of the other mode. In all of the measurements, the fundamental overtone was used to monitor any heating effects, if present. No changes in the first overtone Q^{-1}, $\Delta P/P$ and transition temperature T_c, were observed as the ninth harmonic mode amplitude was varied in the measurements reported here. At larger amplitudes, where the superfluid velocity is larger than ~*4cm/sec*, shifts in the first overtone T_c begin to be observable, indicating the onset of heating effects.

CP850, *Low Temperature Physics: 24th International Conference on Low Temperature Physics;*
edited by Y. Takano, S. P. Hershfield, S. O. Hill, P. J. Hirschfeld, and A. M. Goldman

Results and Discussion

In this section the results from quartz microbalance experiments are compared with the predictions of the dynamic theories. The results from torsional oscillator or quartz microbalance experiments are usually presented in the form of plots of $2\Delta P/P$ and ΔQ^{-1} against T. However, we choose instead a parametric plot so that the real and imaginary parts of $K_0\varepsilon^{-1}(\omega)$ are (see ref. 3) plotted against each other. The parametric plot has the advantage that it avoids the need to model the temperature dependence of $2\Delta P/P \sim K_0\mathrm{Re}[\varepsilon^{-1}(\omega)]$ and $\Delta Q^{-1} \sim K_0\mathrm{Im}[\varepsilon^{-1}(\omega)]$ and so it leads to a more direct comparison of theory and experiment. This type of plot is used in the theory described by Bowley and co-authors[3]. At first we compare our experiments with the dynamic renormalization group theory described by Armour and Bowley[3] (AB) and then with the refined phenomenological theory described by Bowley, Armour and Benedict[4] (BAB). In addition we graph parametric plots according the AHNS theory. In order to obtain a fit of the AB theory to the measured data, we repeat the calculation procedure described by AB. For this we tuned the values of two parameters from the theory: we varied $Y(0)$ until there was good agreement at the low temperature end (large $K_0\mathrm{Re}[\varepsilon^{-1}(\omega)]$), and then we varied $Y(l_c)$ to fit the average peak in the dissipation $K_0\mathrm{Im}[\varepsilon^{-1}(\omega)]$. In Fig. 1 we draw the theoretical curve $K_0\varepsilon^{-1}(\omega)$ with $Y(l_c)=1$ and for different values of the parameter $Y(0)$. In our case, best fit was obtained with $Y(0) \approx 2.0$ and $Y(0) \approx 0.75$ (line). There is some deviation between the new theory and experiment at the high temperature end (small $K_0\mathrm{Re}[\varepsilon^{-1}(\omega)]$). This is to be expected: this is the strong coupling regime, which the present theory describes with a very simple harmonic approximation to the pinning potential. There is some systematic deviation from the predicted curve near the peak in $2K_0\mathrm{Im}[\varepsilon^{-1}(\omega)]$. Looking at Fig.1 it is clear that the method presented by Armour and Bowley leads to a parametric curve of the dielectric function with a shape very close to that of the measurements. The second (dash line) theoretical curve describes the refined phenomenological theory by BAB. In order to obtain a fit with the measured data, we use the ratio of ΔQ^{-1} to $2\Delta P/P$ at the peak in the dissipation values as parameters. The corresponding quantity, for the different values of the thickness of the helium films in our experiments, is the peak of ratio $K_0\mathrm{Im}[\varepsilon^{-1}(\omega)]$ to $K_0\mathrm{Re}[\varepsilon^{-1}(\omega)]$ and has a measured value of ~0.64. This should be compared with the predictions of previous dynamical theories: Minnhagen's[8] theory gives a value of $2/\pi$. The BAB theory takes into account the more accurate recursion relations developed by Timm[9]. The ambiguity over paired and unpaired vortices, which arises in the AHNS work, was also resolved. An accurate numerical technique to calculate the associated linear response was used. The refined model gives predictions, which are consistent with the alternative theoretical picture of Minnhagen[8] and Wallin[10]. Nevertheless, there is still only a qualitative agreement between theory and experiment. The agreement between our experiment and the AB theory is good: it is substantially better than that

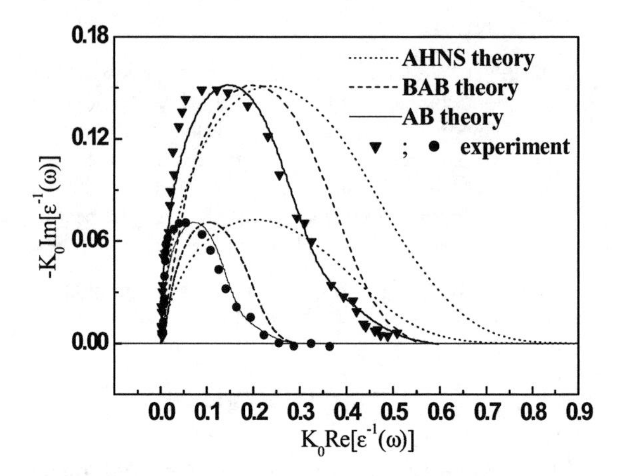

FIGURE 1. Parametric plot of $-K_0\mathrm{Im}\,[\varepsilon^{-1}(\omega)]$ against $K_0\mathrm{Re}[\varepsilon^{-1}(\omega)]$ according to AB theory (line), BAB theory (dash line) and AHNS theory (dot line). Experimental data for thin helium films on gold plated quartz crystal substrates are shown as circles and triangles.

obtained using the refined phenomenological theory by Bowley and co authors, which itself is more accurate than AHNS. The model described by AB does not yet account for the variation in dynamic response with film thickness and so we cannot use it to understand the systematic variation in the dissipation peak that is observed.

REFERENCES

1. J.M. Kosterlitz and D.J. Thouless, *J. Phys. C* **6**, 1181 (1973).
2. V. Ambegaokar, B.I. Halperin, D.R. Nelson and E.D. Siggia, *Phys. Rev. Lett.* **40**, 783 (1978).
3. A.D. Armour and R.M. Bowley, *Phys. Rev. B* **60**, 12389 (1999).
4. R.M. Bowley, A.D. Armour, and K.A. Benedict, *J. Low Temp. Phys.* **113**, 71 (1998).
5. D.F. McQueeney, Ph.D. thesis, Cornell University, 1988.
6. R. Brada, H. Chayet and W.I. Glaberson, *Phys. Rev. B* **8**, 12 874 (1993).
7. S. Ben-Ezra and W.I. Glaberson, *Physica B* **284-288**, 119-121 and 122-124 (2000).
8. P. Minnhagen, *Rev. Mod. Phys.* **59**, 1001 (1987).
9. C. Timm, *Physica C* **265**, 31 (1996).
10. M. Wallin, *Phys. Rev. B* **41**, 6575 (1990).

Interfacial Friction of ^{4}He Films Adsorbed on Grafoil Preplated with Kr

N. Hosomi*, S. Takizawa*, A. Tanabe*, Y. Aoki*, J. Taniguchi*, M. Hieda† and M. Suzuki*

*Department of Applied Physics and Chemistry, University of Electro-Communications, Chofu, Tokyo 182-8585, Japan

†Department of Physics, Nagoya University, Chikusa, Nagoya 464-8602, Japan

Abstract. We report results for interfacial friction of nonsuperfluid ^{4}He films adsorbed on Grafoil (exfoliated graphite) preplated with one atomic layer of Kr (^{4}He/Kr/Gr). In the present experiment, the quartz-crystal microbalance (QCM) technique was adopted, and the friction was measured down to 0.5 K for several oscillating amplitudes, ranging from 0.1 to 1 nm at 5 MHz. Above about the first layer completion, a decrease in the friction was clearly observed for a large amplitude below a certain temperature T_S. This decrease occurs abruptly at around the second layer completion. Although ^{4}He/Kr/Gr has a lower T_S than ^{4}He adsorbed on bare Grafoil when ^{4}He is two atoms thick, the areal density dependence of T_S is similar for the two systems.

Keywords: Interfacial friction, ^{4}He films, Grafoil
PACS: 67.20.+k, 81.40.Pq

The field of research in atomic-scale friction has been developing as the nanotribology[1]. Interfacial friction between a physisorbed film and substrate is one of the interesting topics in this field. The quartz-crystal microbalance (QCM) technique has enabled us to measure the interfacial friction. It has been applied to noble-gas films adsorbed on noble metals[2, 3].

The surface of graphite is atomically flat and the interaction of He atom is small. Therefore, ^{4}He films adsorbed on graphite (^{4}He/Gr) are ideal systems for a study of the interfacial friction. We have measured the interfacial friction of ^{4}He/Gr using the QCM technique [4]. So far, we have found that the friction for several amplitudes decreases below a certain temperature T_S, and T_S is intimately related to ^{4}He areal density. This friction has been predicted to depend strongly on the interaction between the physisorbed film and substrate[5]. In this paper, we report the interfacial friction of ^{4}He films adsorbed on graphite preplated with one atomic layer of Kr (^{4}He/Kr/Gr) and compare it with that of ^{4}He/Gr.

In the QCM technique, information on the friction is obtained from changes in the resonant frequency and Q-value of the quartz crystal. When the friction decreases, an increase in the resonant frequency and a change in the Q-value are observed. The resonator used for the QCM was an AT-cut quartz crystal with a fundamental resonant frequency of 5.0 MHz. A piece of Grafoil (exfoliated graphite), which was baked in vacuum at 900 °C for 3 hours, was pasted uniformly on the Ag electrodes of the QCM. Although the Q-value of the crystal was slightly lowered by the pasting, it remained better than 10^4. A large effective surface area and highly oriented c-axis means that Grafoil is a suitable substrate for a study of the interfacial friction. When ^{4}He films are locked to the oscillating substrate, the decrease of the resonant frequency is estimated to be 4.2 Hz/(atoms/nm^2). To preplate the Grafoil, 6.4 atoms/nm^2 of Kr were adsorbed at LN_2 temperature. The sample was then cooled down to

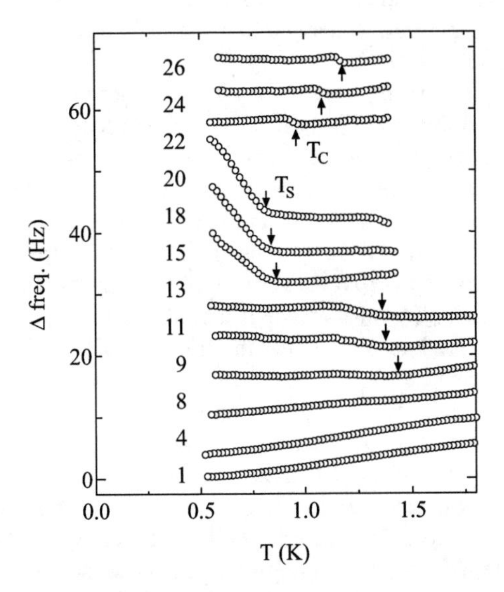

FIGURE 1. Changes in the resonant frequency as a function of temperature for an oscillating amplitude of 0.5 nm. Numbers represent the areal density in the unit of atoms/nm^2. The arrows indicate the friction decrease temperature T_S and the superfluid onset T_C. For clarity, the data sets have been shifted vertically.

CP850, *Low Temperature Physics: 24th International Conference on Low Temperature Physics;* edited by Y. Takano, S. P. Hershfield, S. O. Hill, P. J. Hirschfeld, and A. M. Goldman

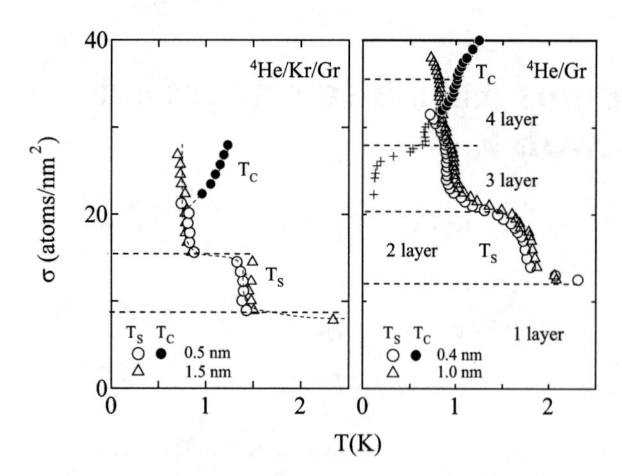

FIGURE 2. T_S and T_C as a function of ^{4}He areal density Left: ^{4}He/Kr/Gr; right: ^{4}He/Gr. Open symbols indicate T_S, and solid ones T_C. Crosses for ^{4}He/Gr represent T_C, which has been measured using a torsional oscillator[6]. Dashed lines correspond to the layer completion.

LHe temperature. In the present experiment, the Kr areal density is close to the commensurate ($\sqrt{3} \times \sqrt{3}$) phase on graphite.

We measured the resonant frequency and the Q-value down to 0.5 K for ^{4}He areal densities up to 26 atoms/nm^2. Figure 1 shows the temperature dependence of the resonant frequency for several ^{4}He areal densities. The oscillating amplitude of the quartz crystal was $\sim$0.5 nm. Above 9 atoms/nm^2, when the temperature is lowered below T_S, the resonant frequency increases accompanied by an increase of about 2% in the Q-value. This behaviour means that the friction on ^{4}He films decreases below T_S. For ^{4}He areal densities between 9 and 14 atoms/nm^2, T_S depends only weakly on ^{4}He areal density and is $\sim$1.4 K. As the density increases above 15 atoms/nm^2, T_S drops down to $\sim$0.9 K and the resonant frequency increases more strongly at low temperatures. Up to 22 atoms/nm^2, T_S is almost constant. Above 24 atoms/nm^2, the decrease in the friction below T_S disappears at an amplitude of $\sim$0.5 nm. In contrast, the resonant frequency increases rapidly at T_C, corresponding to the superfluid onset of the ^{4}He film.

It is interesting to compare the decrease in the friction at low temperatures between ^{4}He/Kr/Gr and ^{4}He/Gr. Figure 2 shows a comparison of the areal density dependence of T_S and T_C. The superfluidity for ^{4}He/Kr/Gr is observed at a lower areal density than for ^{4}He/Gr, which means that the effective adsorption potential is weaker. However, the areal density dependence of T_S and T_C shows a similar behaviour overall. In particular, T_S for ^{4}He/Kr/Gr was observed after the first layer completion and changes stepwise as seen in ^{4}He/Gr. On the other hand, there is a quantitative difference. For two-atom thick films, T_S for ^{4}He/Kr/Gr is $\sim$1.4 K while that for ^{4}He/Gr is $\sim$1.8 K. This difference almost vanishes for three-atom thick films, and T_S is $\sim$0.9 K. That is, T_S is affected strongly by the substrate mostly for two-atom thick films.

Finally, we mention the nonlinear behaviour of the interfacial friction of ^{4}He/Kr/Gr and ^{4}He/Gr. In the case of ^{4}He/Gr, the increase in the resonant frequency below T_S depends on the oscillating amplitude, and the resonant frequency increases greatly as the amplitude increases [7]. The amplitude dependence demonstrates that the friction depends nonlinearly on the sliding velocity. Furthermore, we found that the low-friction state below T_S at large amplitudes is metastable. The fiction below T_S remains small after dropping the amplitude. This low-friction metastable state may be related to a change in the structure of the ^{4}He film. We have found a similar behaviour in ^{4}He/Kr/Gr. This suggests that the behaviour may be a common phenomenon for ^{4}He films.

In summary, we measured the friction of ^{4}He films adsorbed on Grafoil preplated with one atomic layer of Kr (^{4}He/Kr/Gr). Above about the first layer completion, it was found that the friction decreases below T_S. T_S remains almost constant for two-atom thick films, and drops down at about the second layer completion. Although the ^{4}He areal density dependence of T_S shows a similar behaviour to that of bare Grafoil, T_S of ^{4}He/Kr/Gr is lower than that of ^{4}He/Gr for two-atom thick films. This difference of T_S almost disappears above three-atom thick films. We found that the effect of the substrate is manifest most sensitively for two-atom thick films.

REFERENCES

1. B. N. J. Persson, *Sliding Friction*, Springer: Berlin, 1998.
2. C. Mak and J. Krim, *Phys. Rev. B* **58**, 5157-5159 (1998).
3. L. Bruschi, A. Carlin, and G. Mistura, *Phys. Rev. Lett.* **88**, 046105-1-4 (2002); A. Carlin, L. Bruschi, M. Ferrari and G. Mistura, *Phys. Rev. B* **68**, 045420-1-6 (2003).
4. N. Hosomi, M. Suzuki, and M. Hieda, *J. Low Temp. Phys.*, **134**, 37-42 (2004); N. Hosomi, A. Tanabe, M. Hieda, and M. Suzuki, *J. Low Temp. Phys.* **138**, 361-366 (2005).
5. H. Matsukawa and H. Fukuyama, *Phys. Rev. B* **49**, 17286-17292 (1994).
6. P. A. Crowell and J. D. Reppy, *Phys. Rev. B* **53**, 2701-2718 (1996).
7. N. Hosomi, A. Tanabe, M. Hieda, and M. Suzuki, *J. Phys. Chem. Solids*, to be published.

Quantum Phase Transition of ^{4}He Confined in Nanoporous Media

Keiya Shirahama

Department of Physics, Keio University, Yokohama 223-8522, Japan

Abstract. ^{4}He confined in nanoporous media is an excellent model system for studying a strongly correlated Bose liquid and solid in a confinement potential. We studied superfluidity and liquid-solid phase transition of ^{4}He confined in a porous Gelsil glass that had nanopores 2.5 nm in diameter. The obtained pressure-temperature phase diagram is fairly unprecedented: the superfluid transition temperature approaches zero at 3.4 MPa, and the freezing pressure is enhanced by approximately 1 MPa from the bulk one. These features indicate that the confined ^{4}He undergoes a superfluid-nonsuperfluid-solid quantum phase transition at zero temperature. The nonsuperfluid phase may be a localized Bose-condensed state in which global phase coherence is destroyed by a strong correlation between the ^{4}He atoms or by a random potential.

Keywords: superfluidity, nanoporous media, quantum phase transition, strongly correlated system
PACS: 67.40.-w

INTRODUCTION

Recently, bosons in an attractive potential have been a subject of great interest. A number of exotic ground states, such as a Mott insulator, Bose glass, and supersolid, have been predicted [1]. Experimental studies have been performed employing cold alkali gas atoms in optical lattices [2], superconducting Josephson junction arrays [3], and granular superconducting films [4]. ^{4}He in porous media is a model system for a strongly correlated Bose liquid and solid in a confinement potential. Owing to recent progress in material synthesis, a great variety of nanoporous materials have become available. In porous materials, one can control (1) the density and correlation of ^{4}He atoms, from dilute Bose gas [5] to dense liquid or solid, (2) the dimensionality of the system, (3) spatial periodicity and disorder, and (4) the topology of the system. The degree of controllability is fairly unique and is unavailable in the other abovementioned experimental systems.

The present work is motivated by past studies of ^{4}He confined in porous Vycor glass, which have three-dimensionally interconnected nanopores with a diameter of 7 nm [6-8]. In this glass, the freezing pressure increases by about 1 MPa from the bulk freezing curve, while the superfluid transition temperature T_c decreases by about 0.2 K from the bulk λ line. From these results, it is inferred that the confinement of ^{4}He in smaller pores may lead to a further suppression of superfluidity, particularly at high pressures. We examine the phase diagram of ^{4}He confined in a nanoporous Gelsil glass with a nominal pore diameter of 2.5 nm [9].

PHASE DIAGRAM

The superfluidity of confined ^{4}He is mainly studied by using the torsional oscillator technique, which has been described elsewhere [9]. In our experiment, the Gelsil glass is placed inside a torsional oscillator and ^{4}He is introduced to it. We measure the shift in the resonance frequency ($f \sim 1956$ Hz) of the torsional oscillator, which is proportional to the superfluid density, as a function of temperature.

The frequency data are shown in Fig. 1. In Fig. 1(a), in addition to the small frequency shift that is due to the superfluid transition of bulk ^{4}He inside the oscillator cell, the superfluid transition of ^{4}He confined in the nanopores is clearly observed. T_c is approximately 1.43 K near 0 MPa. This value is already considerably lower than the value of the bulk λ point, 2.17 K. Both T_c and the frequency shift decrease as the pressure increases; the superfluidity is suppressed by pressurization.

Above the bulk freezing pressure P_f ($T < 0.8$ K) $\sim$ 2.53 MPa, the entire bulk ^{4}He in the cell solidifies. The bulk solid is formed by the capillary blocking method. Here, the cell pressure is estimated from the frequency [9]. Even under the blocked capillary condition, the frequency shifts are clearly observed as shown in Fig.1(b); in other words, liquid ^{4}He exhibits superfluidity even at pressures above P_f. T_c and the frequency shift are further suppressed with increasing pressure. In our observations, the lowest T_c is 38 mK at $P = 3.33$ MPa.

Since the torsional oscillator technique is not sensitive to the solidification of ^{4}He inside the pores, we studied the L-S transition by measuring the isochoric pressure. This measurement was conducted with a Straty-Adams type capacitance pressure gauge [11], which contained

CP850, *Low Temperature Physics: 24th International Conference on Low Temperature Physics;*
edited by Y. Takano, S. P. Hershfield, S. O. Hill, P. J. Hirschfeld, and A. M. Goldman

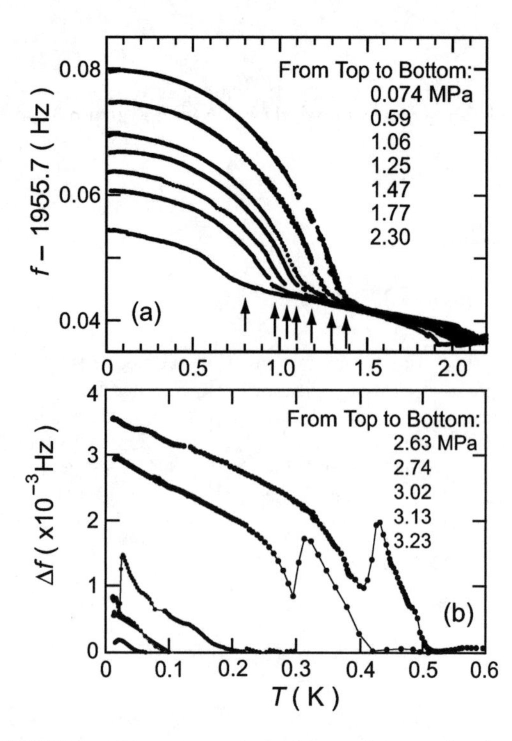

FIGURE 1. Temperature dependence of the torsional oscillator frequency. (a) Data at $P < P_f = 2.53$ MPa (bulk freezing pressure). To clarify the superfluid transitions, the ordinates of the data are shifted so that they overlap at $T = 1.5$ K. Small upturns seen around 2 K are due to the superfluid transition of the bulk liquid in the interspace between the porous glass sample and cell wall. Arrows indicate the superfluid transitions inside the nanopores. (b) Frequency shift data above P_f. The *n*-shaped anomalies are anti-crossing resonances resulting from the coupling to the superfluid fourth sound.

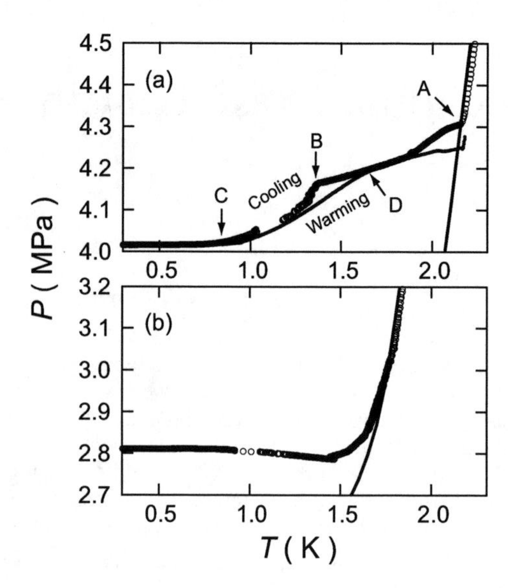

FIGURE 2. Pressure isochores for two typical starting pressures. Solid lines denote the bulk L-S boundary. (a) High pressure data. Circles and dots are the data taken during cooling and warming, respectively. In the cooling run, freezing starts at Point B (P_F, T_F) and ends at Point C. In the warming run, melting ends at Point D (P_M, T_M). (b) Cooling data starting at a low pressure. No freezing onset is observed, and a pressure minimum exists at about 1.4 K. See text.

three porous glass disks.

In Fig. 2, we show typical pressure data sets. The isochoric condition is realized by a blocked capillary method. Initially, we prepare liquid ^{4}He at high temperatures (2.5 K) and high pressures (5 MPa $< P <$ 7 MPa), and then slowly cool the system. After the entire bulk ^{4}He in the cell solidifies, the isochore departs from the bulk L-S boundary. This is indicated by Point A. The pressure abruptly drops at Point B. As in the case of ^{4}He in Vycor [6], this pressure drop indicates the freezing inside the pores. We denote Point B as (P_F, T_F). The freezing proceeds in the finite temperature range from Point B to C. Below C, the pressure becomes nearly independent of temperature down to 20 mK. Upon warming, the pressure traces a different curve, showing hysteresis. Melting appears to be completed at Point D (P_M, T_M), above which the warming data coincide with the cooling data, except near the bulk L-S boundary. Therefore, we conclude that ^{4}He clearly undergoes an L-S transition in the 2.5-nm pores and determine the freezing and melting points, as in the case of ^{4}He in the 7-nm Vycor pores.

As the starting pressure decreases, the pressure drop becomes less prominent and eventually disappears (Fig. 2(b)). Instead, a pressure minimum appears below about 3.8 MPa. We will comment on the pressure minimum later.

From the torsional-oscillator and pressure data, we obtain the *P-T* phase diagram that is shown in Fig. 3. T_c approaches 0 K at a critical pressure $P_c \sim 3.4$ MPa. As is obvious in Fig. 1, the frequency shift, which is proportional to the superfluid density ρ_s, also decreases continuously to zero as pressure approaches P_c [9].

The points of the onset of freezing onset and completion of melting are also summarized in Fig. 3. The freezing curve is increased to about 3.7 MPa at 1 K. It is difficult to record the freezing point below 3.65 MPa with our constant volume apparatus due to its weak temperature dependence. However, no freezing is observed at 3.42 MPa at least down to 20 mK. We therefore conclude that the freezing curve is located within a narrow pressure range of $3.42 < P < 3.65$ MPa. Although the freezing pressure at zero temperature is close to P_c, our experiments are currently not accurate enough to determine whether they coincide. Simultaneous measurements of superfluid density and pressure will resolve this issue.

It is remarkable that the overall shape of the L-S boundary is similar to that of bulk ^{4}He; in particular, the freezing pressure is nearly independent of temperatures below 1 K. Nevertheless, the liquid phase adjacent to the

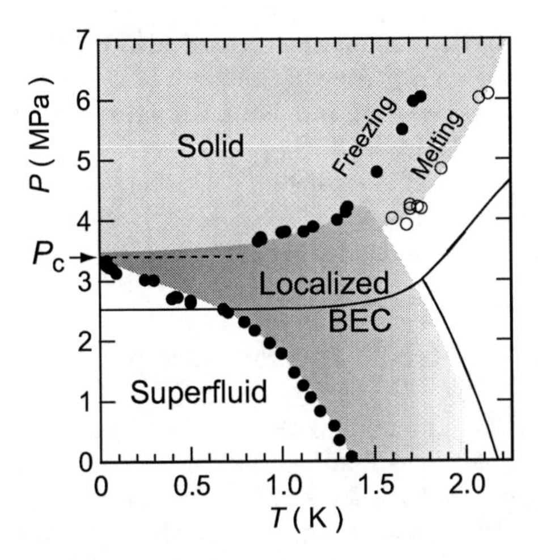

FIGURE 3. P-T phase diagram determined by T_c (solid circles), (P_F, T_F) (solid circles), and (P_M, T_M) (open circles). The solid line shows the bulk phase diagram. P_c denotes the critical pressure of 3.4 MPa, where the superfluidity is suppressed to zero temperature. The horizontal dashed line is the tentative lower bound of the freezing pressure, 3.42 MPa, below which freezing is not observed.

solid is not a superfluid. According to the Clapeyron-Clausius relation at the L-S coexistence, the entropy of this nonsuperfluid phase is very close to that of the solid phase. Consequently, the nonsuperfluid phase has a very small entropy, and it is possibly a novel ordered state.

QUANTUM PHASE TRANSITION AND LOCALIZED BEC

The continuous suppressions of T_c and ρ_s to zero are rather unprecedented; they strongly suggest that ^{4}He confined in the nanoporous glass undergoes a continuous quantum phase transition (QPT) near P_c and at zero temperature. In other words, the superfluid-nonsuperfluid-solid transition, which is not a first-order phase transition, is driven at zero temperature by pressure [12].

The nature of the nonsuperfluid but low-entropy phase adjacent to the superfluid and solid phases is interesting. We propose that this phase is a localized Bose-Einstein condensate (BEC) state. The idea of a localized BEC was first proposed by Glyde et al. [13] in their interpretation of inelastic neutron scattering results for liquid ^{4}He confined in Vycor and Gelsil glasses at zero pressure (the "fullpore" case). They observed clear roton peaks, which indicate a BEC or superfluidity, even at temperatures higher than T_c determined by a torsional oscillator or ultrasound. Since the pore size of the porous glasses is not uniform, the superfluid (or BEC) transition temperature has a spatial distribution. When the temperature is decreased below the bulk T_λ, the BECs grow from large pores. The size of the BECs is roughly limited to the pore size; hence, they exist locally. The system has no global phase coherence and does not exhibit superfluidity that can be detected by macroscopic measurements. As the temperature is decreased, the localized BECs grow continuously. Macroscopic superfluidity is realized when most of the BECs coalesce.

We attribute the pressure-induced suppression of superfluidity to formation of localized BECs. As the pressure increases, the spatial exchanges of ^{4}He atoms, which are indispensable for phase coherence, can be suppressed in the narrow pores because of the hard-core nature of He atoms. The suppression of atom exchanges prevents condensate growth, thereby causing the suppression of T_c.

In the above scenario, the small size of the pores is the only factor that is essential to the superfluid suppression. Disorder or randomness in the porous structure and pore walls may play an additional or alternative role in the localization of the condensates. That is, the nonsuperfluid phase may be a Bose glass state, which has been predicted for a Bose-Hubbard model [1].

The possibility of a quantum phase transition in ^{4}He was first proposed by Crowell et al. for ^{4}He films adsorbed on a Vycor glass [14]. They attributed the existence of the nonsuperfluid ^{4}He layer adjacent to the pore walls to a quantum localization of ^{4}He by a random substrate potential. This may also be the case for our ^{4}He-nanopore system. In Fig. 4, we show the global phase diagram by displaying the phase diagram of adsorbed film states with the P-T phase diagram. The superfluid phase exists within a ^{4}He density range defined by two quantum critical points.

The abovementioned idea of localized BEC-superfluid transition is analogous to a model for the superconducting-insulating transitions in Josephson junction arrays [3, 12] and granular metal films [4]. In both systems, the suppression of Cooper pair tunneling by Coulomb blockade results in a phase fluctuation. A phase diagram similar to our P-T phase diagram has been presented in [12].

The global phase diagram of Fig. 4 is very similar to the phase diagram of high-T_c cuprates. Emery and Kivelson proposed that, in the pseudogap state in the underdoped regime, superconductivity is destroyed by phase fluctuation, although the amplitude of the order parameter has a nonzero value [15]. Our picture of superfluid suppression is quite similar to their proposal, and the phase diagrams resemble each other. Although this similarity may be superficial, it lends a novel aspect to our ^{4}He-nanopore system.

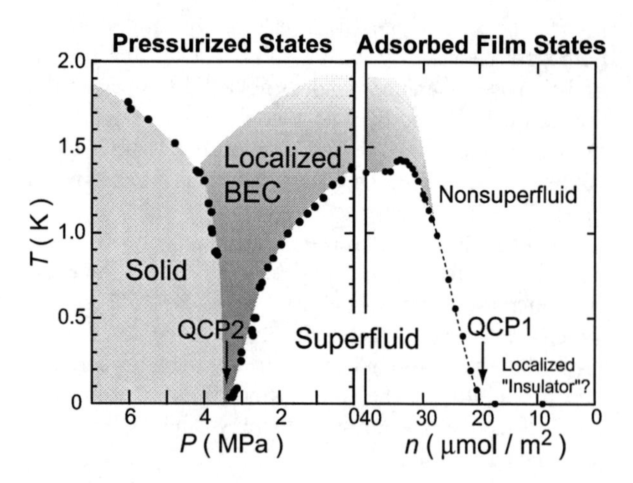

FIGURE 4. Global phase diagram from the thin film states to the compressed solid. The phase diagram contains two quantum critical points (QCP 1 and 2).

OPEN QUESTIONS

These experiments have led to some interesting questions. First, in the zero-pressure (fullpore) studies conducted by us and other groups, significant differences were observed in the superfluid transition temperature in the nominally 2.5-nm porous Gelsil samples. Miyamoto and Takano determined that T_c was approximately 0.9 K under a nearly fullpore condition [16]. More surprisingly, Mulders and Glyde observe no superfluid transition in their ultrasound studies, although they did observe clear phonon and roton signals even near bulk T_λ for the same sample [17]. These observations lead us to conclude that T_c is extremely sensitive to either a slight difference in the pore size or some unknown characteristics in the pore structure.

Second, the superfluid properties observed in our preliminary ultrasound experiment [18] show some differences from the torsional oscillator results. Although T_c is found to be identical to the torsional oscillator T_c at zero pressure ($T \sim 1.4$ K), superfluidity is not suppressed by pressure, i.e. T_c approaches 0.8 K near 3.5 MPa. Further, the superfluid density is not suppressed, instead showing a roughly T-linear behavior below T_c, down to zero temperature, at all pressures. This suggests that observation of superfluidity by dynamical means is strongly dependent on frequency. Further studies including simultaneous measurements of ultrasound and torsional oscillator for a single Gelsil sample are underway.

It is important to determine the localized Bose condensation temperature. The pressure minimum shown in Fig. 2(b) may be associated with the formation of the localized BECs, since such a pressure minimum is observed at the bulk λ transition [10]. However, it has been pointed out that such a minimum can be caused by the frost heave phenomenon [19]. We will conduct a heat capacity measurement, which will provide a definitive evidence for the localized BEC.

Our finding of the quantum phase transition of ^{4}He in nanoporous media provides a new perspective for studying the physics of a strongly correlated Bose system.

ACKNOWLEDGMENTS

The author is grateful to many collaborators, especially K. Yamamoto and Y. Shibayama for the torsional oscillator and pressure studies, and T. Kobayashi and M. Suzuki for the ultrasound experiment. Thanks are also due to M. Kobayashi, M. Tsubota, H. R. Glyde, and N. Mulders for stimulating discussions. This work is supported by a Grant-in-Aid for Scientific Research (A) and for Priority Area from the Ministry of Science and Education, Japan.

REFERENCES

1. M. P. A. Fisher *et al.*, *Phys. Rev.* **B40**, 546 (1989).
2. M. Greiner *et al.*, *Nature* **415**, 39 (2002).
3. L. J. Geerligs *et al.*, *Phys. Rev. Lett.* **63**, 326 (1989).
4. L. Merchant *et al.*, *Phys. Rev.* **B63**, 134508 (2001).
5. J. D. Reppy *et al.*, *Phys. Rev. Lett.* **84**, 2060 (2000).
6. E. D. Adams *et al.*, *Phys. Rev. Lett.* **52**, 2249 (1984).
7. J. R. Beamish *et al.*, *Phys. Rev. Lett.* **50**, 425 (1983).
8. L. -Z. Cao *et al.*, *Phys. Rev.* **B33**, 106 (1986).
9. K. Yamamoto *et al.*, *Phys. Rev. Lett.* **93**, 075302 (2004).
10. K. Yamamoto *et al.*, these Proceedings.
11. G. C. Straty and E. D. Adams, *Rev. Sci. Instrum.* **40**, 1393 (1969).
12. S. L. Sondhi *et al.*, *Rev. Mod. Phys.* **69**, 315 (1997).
13. H. R. Glyde *et al.*, *Phys. Rev. Lett.* **84**, 2646 (2000); O. Plantevin *et al.*, *Phys. Rev.* **B65**, 224505 (2002).
14. P. A. Crowell *et al.*, *Phys. Rev. Lett.* **75**, 1106 (1995).
15. V. J. Emery and S. A. Kivelson, *Nature* **374**, 434 (1995).
16. S. Miyamoto and Y. Takano, *Czech J. Phys.* **46**, 137 (1996); S. Miyamoto, Thesis, University of Florida (1995).
17. N. Mulders and H. R. Glyde, private communication.
18. T. Kobayashi *et al.*, these Proceedings.
19. M. Hiroi *et al.*, *Phys. Rev.* **B40**, 6581 (1989); T. Mizusaki, private communication.

Ultrasonic Study of Superfluidity of ^{4}He in a Nanoporous Glass

T. Kobayashi*, S. Fukazawa*, J. Taniguchi*, M. Suzuki* and K. Shirahama†

*Department of Applied Physics and Chemistry, University of Electro-Communications, Tokyo 182-8585, Japan
†Department of Physics, Keio University, Yokohama 223-8522, Japan

Abstract. Recently, Yamamoto et al. have carried out torsional oscillator measurements for ^{4}He filled in a nanoporous glass containing pores 2.5 nm in diameter (Gelsil), and reported a large suppression of superfluidity at high pressures. In order to study the mechanism of the superfluid suppression, we have carried out ultrasonic measurements for ^{4}He filled in the same nanoporous glass. We have clearly observed an increase in the sound velocity due to decoupling of the superfluid component. At zero pressure, the superfluid transition temperature T_C is suppressed to 1.4 K from the bulk lambda point, 2.17 K. The pressure dependence of T_C and the temperature dependence of the superfluid fraction are very different from those reported by Yamamoto et al.

Keywords: superfluid ^{4}He, porous glass, ultrasound
PACS: 67.40.-w, 43.35.Cg

Superfluidity of ^{4}He in porous materials has attracted the attention of many researchers for several decades. Recently, Yamamoto et al. have carried out torsional oscillator measurements for ^{4}He filled in a nanoporous glass containing 2.5 nm diameter pores (Gelsil).[1] They reported a large suppression of superfluidity. The superfluid transition temperature T_C drops down to 1.4 K at zero pressure. Furthermore, it decreases monotonically with increasing pressure, and approaches zero temperature at a critical pressure P_C of 3.4 MPa. This behavior differs from that of ^{4}He in porous Vycor glass, in which the T_C line terminates at the freezing curve.[2–4] Their results cannot be explained in terms of the conventional concept of the size effect for superfluidity. Thus motivated, we have performed ultrasonic measurements for ^{4}He filled in the same porous glass in order to study the mechanism of the suppression of superfluidity.

Ultrasound was used to study the superfluidity of ^{4}He in porous Vycor glass.[2] Decoupling of the superfluid component from the oscillating substrate caused an increase in the sound velocity below T_C. This increase was related to the superfluid density ρ_s as $\Delta v/v_0 = (1 - \chi_s)\, \rho_s/\,(2\rho)$, where χ_s is the fraction of the superfluid component which remains locked to the substrate, and ρ is the total density.

In the present experiment, we have prepared a Gelsil sample from the same batch used in the torsional oscillator measurements of Yamamoto et al.[5] The sample is cylindrical with dimensions of 5.2 mm in diameter, and 2.11 mm in thickness. The faces of the sample are polished and 10 MHz transverse $LiNbO_3$ transducers are then bonded to both ends with silicone adhesive. Before closing the sample cell made of Be-Cu, the sample is heated at 100°C to remove water.

For the ultrasonic measurement, we employ a continuous wave (cw) resonance technique. An alternating voltage is applied to one of the transducers, and the output of the other is fed into an rf lock-in amplifier to detect the standing wave in the sample. When the resonance condition is satisfied, the quadrature phase component becomes maximum whereas the in-phase one becomes zero. The frequency is controlled to keep the in-phase component zero, and the change in the sound velocity is calculated from the change in the resonance frequency as $\Delta v/v_0 = \Delta f/f_0$. The resonance amplitude is estimated from the output of the lock-in amplifier, and is controlled in the range between 0.06 – 6 μm/s. The sound velocity v_0 can be obtained form the resonance curve and the thickness of the sample. It is 1.98×10^3 m/s at 2 K. The bottom wall of the sample cell acts as a diaphragm pressure gauge.[6]

The inset of Fig. 1 shows the variation of the sound velocity for a ^{4}He film. The areal density of the film is 22 μmol/m^2. An increase in the sound velocity due to superfluidity is observed at T_C.

The main frame of Fig. 1 shows the variation in the sound velocity for full-pore ^{4}He at several pressures. The velocity amplitude of the transducer is 0.6 μm/s. At pressures above the freezing pressure of bulk ^{4}He, 2.5 MPa, we have used the blocked capillary method to set the sample pressure at a desired value. Upon cooling, the sound velocity starts to increase at temperature T_C. We believe that this increase indicates an onset of superfluidity of liquid ^{4}He in the Gelsil. The increase is independent of the amplitude from 0.06 to 6 μm/s; T_C is 1.4 K at 0.04 MPa. As the pressure increases, T_C decreases mono-

CP850, *Low Temperature Physics: 24th International Conference on Low Temperature Physics;*
edited by Y. Takano, S. P. Hershfield, S. O. Hill, P. J. Hirschfeld, and A. M. Goldman

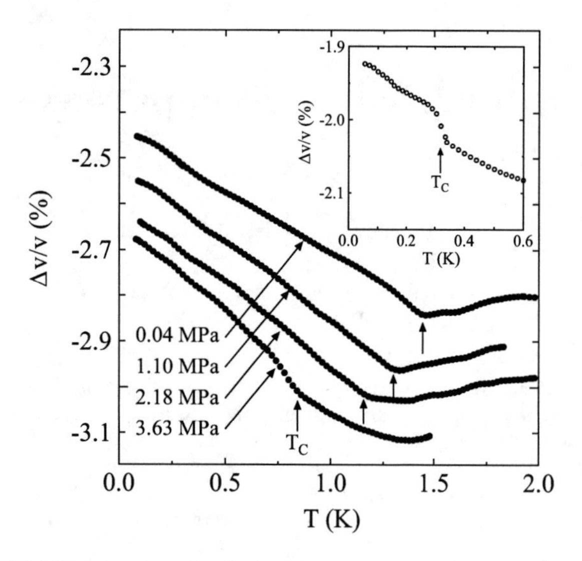

FIGURE 1. Sound velocity of the Gelsil sample with ^{4}He filled at several pressures. The vertical axis is the decrease in the sound velocity from that of empty Gelsil. The arrows indicate T_C at each pressure. Inset: sound velocity of the sample with a ^{4}He film.

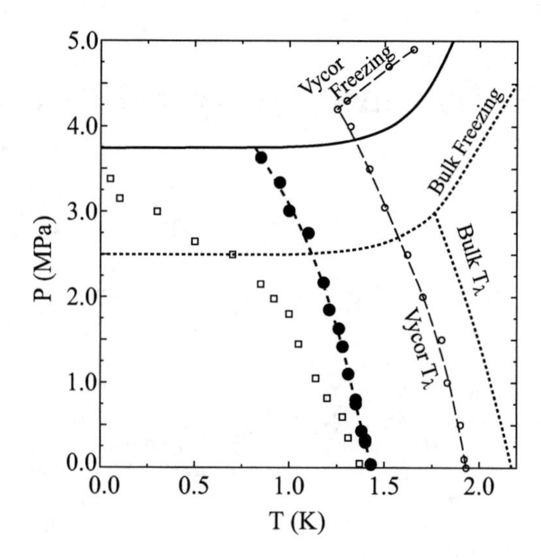

FIGURE 2. T_C data (closed circles), together with T_C (squares) of Yamamoto et al. by means of a torsional oscillator [1]. The solid line is the freezing curve, which was determined from the behavior of the pressure of the sample cell in the present experiment.

tonically. Above the freezing pressure of bulk ^{4}He, the increase in the sound velocity remains even above T_C. This suggests that part of ^{4}He in the Gelsil is frozen due to a pore-size distribution.[7]

We estimate the fraction of the decoupled mass relative to the total amount of *liquid* ^{4}He. Since the decrease in the sound velocity due to the inert layer of ^{4}He is about 1.8%, the fraction reaches about 0.35 at 0.04 MPa and at zero temperature. This value does not depend strongly on pressure.

It is interesting to compare the present result with the torsional oscillator result. In Fig. 2, T_C is plotted together with T_C of Yamamoto et al.[1] T_C of the present experiment at 0.04 MPa is suppressed from the bulk lambda point to 1.4 K, and coincides with T_C of Yamamoto et al. On the other hand, the pressure dependence is quite different: T_C in the present experiment shows a much weaker pressure dependence, and terminates at the freezing curve. On the other hand, T_C of Yamamoto et al. shows a strong pressure dependence, and is suppressed drastically at high pressure.

At present, the origin of this difference is not clear. We have used a different Gelsil sample from Yamamoto et al.'s, although it is form the same batch. The heat treatment was slightly different, and this might have altered the pore structure. However, the following features should be noted. The pressure dependence of T_C in the present experiment seems to be similar to that of ^{4}He in porous Vycor glass, although it is strongly suppressed.[2-4] On the other hand, the temperature dependence of the decoupling below T_C is different. In the present experiment, the decoupling increases linearly with temperature, whereas in Vycor it has a much weaker temperature dependence.[2]

In summary, we have carried out ultrasonic measurements for ^{4}He in Gelsil. We observed that T_C is suppressed to 1.4 K at 0.04 MPa, but its pressure dependence is quite different from the previous report of Yamamoto et al. Further study is needed to clarify the cause of the difference.

REFERENCES

1. K. Yamamoto, et al., *Phys. Rev. Lett.* **93**, 075302 (2004).
2. J. R. Beamish et al., *Phys. Rev. Lett.* **50**, 425 (1983).
3. E. D. Adams et al., *Phys. Rev. Lett.* **52**, 2249 (1984).
4. C. Lie-zhao et al., *Phys. Rev. B* **33**, 106 (1986).
5. The Gelsil sample was manufactured by Geltech Co., 3267 Progress Drive, Orlando, FL 32826, USA.
6. G. C. Straty and E. D. Adams, *Rev. Sci. Instrum.* **40**, 1393 (1969).
7. In the case of Vycor the sound velocity increases when ^{4}He freezes (see Ref. 2).

Interfacial Friction of ^{4}He Films Adsorbed on Porous Gold

J. Taniguchi*, K. Wataru*, K. Hasegawa*, M. Hieda† and M. Suzuki*

*Department of Applied Physics and Chemistry, The University of Electro-Communications, Chofu, Tokyo 182-8585, Japan
†Department of Physics, Nagoya University, Chikusa, Nagoya 464-8602, Japan

Abstract. We measured the interfacial friction of nonsuperfluid ^{4}He films adsorbed on porous gold using the quartz-crystal microbalance (QCM) technique. Above about the first layer completion, decoupling due to a decrease in the interfacial friction was observed for a large oscillation amplitude of 1.6 nm below a certain temperature T_S, which depends on the areal density of ^{4}He. The completion of the first layer and the inert layer was found to strongly affect the interfacial friction.

Keywords: Interfacial friction, ^{4}He film
PACS: 67.20.+k, 81.40.Pq

Friction is a common but poorly understood force from a microscopic point of view. In physisorbed films, the interfacial friction between the film and substrate has become an interesting topic of "nanotribology" [1]. The interfacial friction of Kr films on a Au substrate (Kr/Au) has been studied using the quartz-crystal microbalance (QCM) technique[2–4]. Mistura and coworkers reported that Kr films are locked on the substrate at small oscillation amplitudes, while slipping weakly above a certain amplitude[4]. They interpreted this phenomenon as a dynamic transition caused by the de-pinning of Kr films.

Recently, the interfacial friction of ^{4}He films on Hectorite[5] and Grafoil[6] was measured, and was found to be quite small. It is very interesting to study the friction of ^{4}He films adsorbed on a Au substrate. Thus motivated, we prepared a porous gold substrate, which has a highly uniform porous structure[8], and measured the interfacial friction of ^{4}He films[7]. It was found that ^{4}He films on porous gold (He/PG) also slip easily. In this paper, we report the interfacial friction of He/PG.

In the QCM technique, a change in interfacial friction is measured by recording the changes in the resonant frequency and the Q-value of the quartz crystal. The decrease in the frictional force causes an increase in the resonant frequency. At the same time, a change of Q^{-1} occurs, corresponding to the change in energy loss due to the frictional force.

In the present experiment, we used a porous gold substrate which had been cleaned by heating to 300°C. The Q-value of the QCM was $\sim 3.5\times10^4$ for a 4 MHz fundamental mode. The sensitivity for ^{4}He films was estimated to be 1.1 ± 0.1 Hz/(atoms/nm^2), which is about 20 times larger than with the conventional flat electrodes. To easily control the ^{4}He areal density on the porous gold electrodes, porous gold flakes were added to the sample cell. The total surface area of the cell was 10.0 m^2.

The measurements were performed between 0.7 K and 4.0 K with an oscillation amplitude of 1.6 nm. Figure 1 shows the variation of the resonant frequency on cooling as a function of temperature for various ^{4}He areal densities (σ). For clarity, we represent the decrease of the resonant frequency (ΔF) from that in vacuum at the lowest temperature (3958341.07 Hz).

As the temperature decreases, the resonant frequency decreases slightly and then a gentle increase appears below a certain temperature, T_S. (Here, we define T_S as the intersection of the extrapolation from high temperatures and the extension of the steepest increase on cooling.) When all of the adsorbed ^{4}He is locked to the oscillating substrate, the resonant frequency should decrease in proportion to the areal density at the estimated rate given above. However, the decrease in the resonant frequency is suppressed, even at the maximum decrease. This suppression indicates that the ^{4}He film partially slides. We define the decoupling fraction (D) due to the slippage as the ratio of the suppressed decrease in the resonant frequency to the expected decrease when all of ^{4}He is locked. D at the maximum frequency decrease is plotted as a function of areal density in the upper inset of Fig. 1. D is small at $\sim$2 atoms/nm^2, and increases by up to $\sim$30% with increasing areal density. In addition, it has a dip at 9 atoms/nm^2 and becomes small again above 16 atoms/nm^2. The former areal density is close to that of the first layer completion. This correspondence suggests that the interfacial friction of ^{4}He films becomes large when the first layer becomes dense. The latter areal density is near the inert layer completion. These results demonstrate that ^{4}He films on a Au substrate slide to some extent, and that the interfacial friction is strongly governed by the density of the non-superfluid layers.

Next, we focus on the increase of the resonant frequency below T_S. D at the lowest temperature is also

CP850, *Low Temperature Physics: 24th International Conference on Low Temperature Physics;*
edited by Y. Takano, S. P. Hershfield, S. O. Hill, P. J. Hirschfeld, and A. M. Goldman

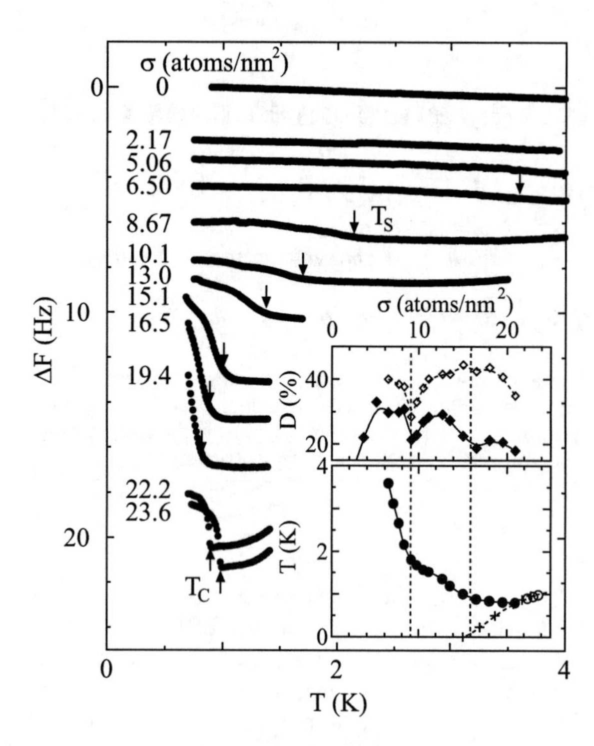

FIGURE 1. Variation of the resonant frequency at an oscillating amplitude of 1.6 nm. For clarity, we show the decrease of resonant frequency (ΔF) from the resonant frequency in vacuum at the lowest temperature (3958341.07 Hz). Numbers represent the ^{4}He areal densities. Upper inset: decoupling fraction due to slippage (*D*) at the maximum decrease of the resonant frequency (solid diamonds) and at the lowest temperature (open diamonds). Lower inset: T_S (solid circles) and T_C (open circles). Crosses indicate T_C measured by Csáthy et al. (Ref. 9)

plotted in the upper inset of Fig. 1. At 13.0 atoms/nm^2, it is larger than the value at the maximum decrease by $\sim$10%. At higher areal densities, it reaches more or less 40%, and seems to further increase below the lowest temperature of the experiment.

Above 22.2 atoms/nm^2, instead of the gentle increase, an abrupt increase in the resonant frequency was observed at a certain temperature, T_C. This is due to the onset of superfluidity, and the areal density dependence of T_C agrees well with that of superfluid onset obtained by Csáthy et al.[9] (See the lower inset of Fig. 1). Assuming that the increase at T_C is the universal jump due to the Kosterlitz-Thouless (KT) transition, we find that about 40% of the superfluid component is detected.

The lower inset of Fig. 1 shows T_S as a function of areal density together with T_C. T_S decreases steeply with increasing areal density until 9 atoms/nm^2, and then decreases slowly. On further increasing the areal density, the decrease becomes slow again above 16 atoms/nm^2. These two areal densities coincide with those at which the slippage above T_S is suppressed. It is concluded that T_S depends strongly on the film structure.

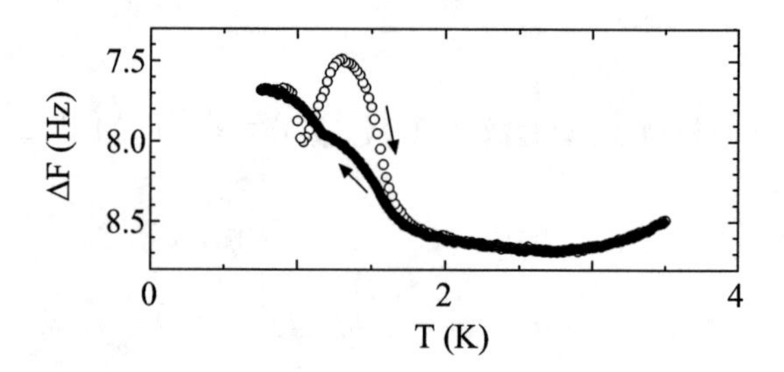

FIGURE 2. Temperature dependence of ΔF on cooling (solid circles) and warming (open circles) at 10.1 atoms/nm^2.

Finally, we report a hysteresis in the resonant frequency observed between cooling and warming at temperatures below T_S. As an example, the temperature dependence of ΔF at 10.1 atoms/nm^2 is shown in Fig. 2. On cooling, the resonant frequency increases monotonically. On warming, it stays constant at first and then dips at $\sim$1 K. After the dip, it becomes higher again than on cooling. The interfacial friction has a different temperature dependence between warming and cooling, and its temperature dependence on warming is rather complicated. The origin of this singular hysteresis has not yet been identified.

In summary, we measured the interfacial friction of nonsuperfluid ^{4}He films on porous gold using the QCM technique. It was found that the interfacial friction of He/PG is small. It is reduced at temperatures below T_S, accompanied by an anomalous hysteresis. Furthermore, it is strongly affected by the change in the film structure, as seen at the completion of the first layer and the inert layer.

REFERENCES

1. B. N. J. Persson, *Sliding Friction*, Berlin: Springer, 1998.
2. J. Krim, D. H. Solina, and R. Chiarello, *Phys. Rev. Lett.* **66**, 181–184 (1991).
3. C. Mak and J. Krim, *Phys. Rev. B.* **58**, 5157–5159 (1998).
4. L. Bruschi, A. Carlin and G. Mistura, *Phys. Rev. Lett.* **88**, 046105–1–4 (2002).
5. M. Hieda, T. Nishino, M. Suzuki, N. Wada, and K. Torii, *Phys. Rev. Lett.* **85**, 5142–5145 (2000).
6. N. Hosomi, M. Suzuki and M. Hieda, *J. Low Temp. Phys.* **134**, 37–42 (2004).
7. K. Wataru, J. Taniguchi, M. Hieda, and M. Suzuki, *J. Phys. Chem. Solid*, to be published.
8. M. Hieda, R. Garcia, M. Dixon, T. Daniel, D. Allara and M. H. W. Chan, *Appl. Phys. Lett.* **84**, 628–630 (2004).
9. G. A. Csáthy and M. H. W. Chan, *J. Low Temp. Phys.* **121**, 451–458 (2000).

Superfluid Transition of ^{4}He in Porous Gold Studied with Quartz Crystal Resonator

Tsuyoshi Kato*, Mitsunori Hieda*, Kazuhisa Wataru†, Masaru Suzuki†, Taku Matsushita* and Nobuo Wada*

*Department of Physics, Nagoya University, Furo-cho, Chikusa-ku, Nagoya 464-8602, Japan
†Department of Applied Physics and Chemistry, UEC, Chofugaoka, Chofu, Tokyo 182-8585, Japan

Abstract. We measured the superfluid density of full pore ^{4}He in porous gold with pores 390 Å in diameter and 70% in porosity by a quartz crystal resonator. The superfluid density critical exponent is found to be 0.72 ± 0.03 which is very close to that of bulk ^{4}He, 0.6705 ± 0.0006. The superfluid transition temperature T_C shows a reduction of 6 ± 2 mK from T_λ, the transition temperature of bulk ^{4}He. This reduction is in agreement with the pore-size dependence of T_C of ^{4}He confined in various porous media.

Keywords: liquid ^{4}He, porous gold, restricted geometry
PACS: 64.60.-i, 67.40.-w, 67.40.Yv

There is considerable interest in the superfluid transition of liquid ^{4}He confined in various porous media. The effect of quenched disorder due to porous media on the superfluid transition has been studied by torsional oscillator and heat capacity measurements [1, 2]. The superfluid density of ^{4}He in Vycor and aerogel shows a simple power law dependence on the reduced temperature $t = (T_C - T)/T_C$ (T_C is the transition temperature). The critical exponent of the superfluid density in porous gold (PG), as obtained from a power low fit including a high order term [2], is consistent with that of pure bulk ^{4}He.

A quartz crystal resonator (QCR), which is frequently used as a microbalance, is another possible tool to measure the superfluid density. QCR is found to be particularly effective after the surfaces of the crystal have been coated with a thin film of porous material. When the pore radius is less than the order of the viscous penetration depth, 210 Å at 12 MHz and at T_λ, mass loading of the normal component viscously trapped within the porous structure can be detected. In this paper we report on measurements of superfluid density in a thin PG sample using the QCR technique.

The PG sample is made by selectively leaching out silver from a 1 μm thick Ag-Au alloy film sputtered onto a quartz disk [3]. The pore diameter and porosity are found to be 390 Å and 70%, respectively. The QCR with PG was operated at $f = 12$ MHz (the third overtone mode), and the quality factor (Q) was 3.4×10^4 when immersed in liquid ^{4}He at 2 K. The temperature was measured by a RuO_2 resistance thermometer calibrated against the vapor pressure of liquid ^{4}He. The typical temperature stability was ± 0.1 mK. For comparison, an experiment using a QCR with a flat piece of gold (FG) was carried out separately.

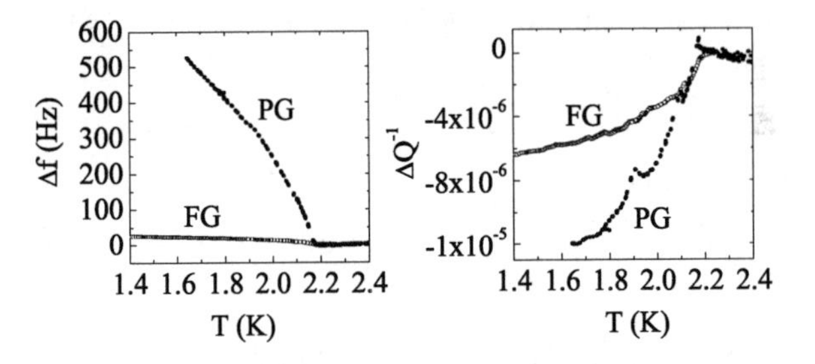

FIGURE 1. Frequency shift and inverse Q change of QCRs for PG and FG versus temperature.

Figure 1 shows the frequency shift Δf and the inverse Q change ΔQ^{-1} of the QCRs for PG and FG versus temperature. In the case of FG, Δf and ΔQ^{-1} are described [4] by the viscous coupling of the QCR to the surrounding fluid, and are given by

$$\Delta f = -\frac{f}{2}\Delta Q^{-1} = -\frac{2f}{n\pi Z_q}(\pi f \eta \rho)^{\frac{1}{2}}, \qquad (1)$$

where Z_q is the transverse acoustic impedance of quartz, 8.86×10^6 kg/m^{2}s, n is the harmonic acoustic number ($n = 1,3,5\cdots$), η is the viscosity, and ρ is the fluid density. From the difference of ΔQ^{-1} between PG and FG, the viscous coupling of PG is estimated to be twice as large. However, the anomalously large enhancement of Δf by PG cannot be explained by Eq. (1).

In the case of PG, the contribution of the mass locked in the pores is added to the frequency change due to the viscous coupling mentioned above. When the pore size is much less than the viscous penetration depth σ, the mass loading causes the frequency change without dissipation,

CP850, *Low Temperature Physics: 24th International Conference on Low Temperature Physics;* edited by Y. Takano, S. P. Hershfield, S. O. Hill, P. J. Hirschfeld, and A. M. Goldman

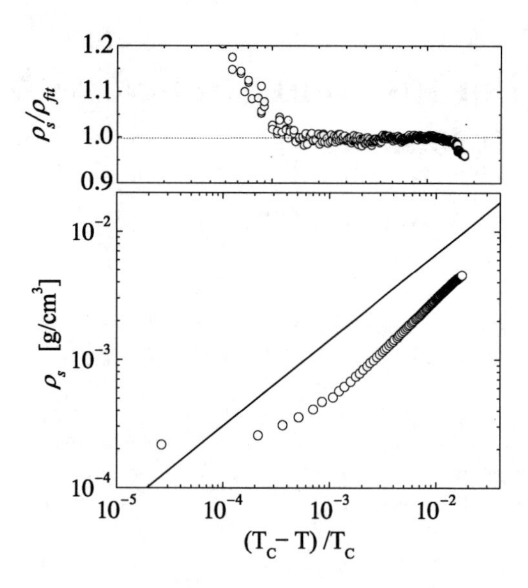

FIGURE 2. Superfluid density near T_C as a function of reduced temperature. Solid line represents the data of bulk helium. Upper panel shows the fractional deviation from the fit to Eq. (3).

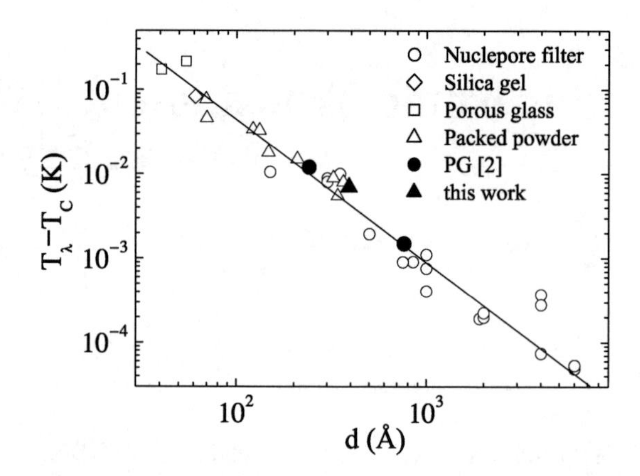

FIGURE 3. Pore-size dependence of the reduction of T_C from T_λ for various porous media [2, 7]. The straight line is a guide for the eye.

and is given by

$$\Delta f_m = -\frac{4f^2}{nZ_q}\frac{\Delta m}{A}, \tag{2}$$

where Δm is the mass of the fluid in the pores and A is the geometric area of the electrode. In this experiment, though the pore size is on the same order as the penetration depth σ, the observation that the enhancement of Δf is much larger than that of ΔQ^{-1} indicates that Eq. (2) is still valid.

Figure 2 shows the superfluid density near T_C calculated by Eq. (2). The superfluid density is expressed according to a course grain definition [1], *i.e.*, the superfluid density per unit volume of the entire experimental system including both PG and liquid helium. A slight deviation from the simple power law $\rho = \rho_0 t^\zeta$ of the bulk is seen and the deviation is larger than the data in Ref. 2.

To analyze critical properties near the critical temperature, one frequently uses a power law form with a high order term. The superfluid density is analyzed in the form [2, 5]

$$\rho_s(t) = \rho_{s0} t^\zeta (1 + bt^\Delta). \tag{3}$$

The second term represents the correction to a simple power law by a confluent singularity which is carried to allow for higher-order effects in the region far from the transition. The same procedure as in Ref. 2 is adopted to analyze our data. The parameter Δ in the nonlinear least squares fit is fixed at 0.5, and the best fit is obtained in the range $1.5 \times 10^{-3} < t < 10^{-2}$. The parameters are found to be $\rho_{s0} = 0.065 \pm 0.010$ g/cm^3, $\zeta = 0.72 \pm 0.03$, $b = 2.8 \pm 0.5$, and $T_C = 2.1695 \pm 0.0006$ K. The superfluid density exponent is found to be 0.72 ± 0.03 which is close to that of bulk ^{4}He, 0.6705 ± 0.0006 [6].

The transition temperature T_C shows a reduction of 6 ± 2 mK from T_λ, the transition temperature of bulk ^{4}He. Figure 3 compares the result of this experiment with the previously noted dependence of T_C on the pore size for various porous materials [2, 7]. The reduction of T_C is coincident with the pore-size dependence observed for ^{4}He confined in various porous media.

In summary, we measured the superfluid density of full pore ^{4}He in a PG using a QCR. The superfluid density exponent is found to be 0.72 ± 0.03 which is close to that of bulk ^{4}He, 0.6705 ± 0.0006. The transition temperature T_C shows a reduction of 6 ± 2 mK from T_λ of bulk ^{4}He.

We acknowledge the technical support provided by T. Kurokawa.

REFERENCES

1. P. A. Crowell *et al.*, *Phys. Rev. B*, **51**, 12721–12736 (1995).
2. J. Yoon and M. H. W. Chan, *Phys. Rev. Lett.*, **78**, 4801–4804 (1997).
3. M. Hieda *et al.*, *Appl. Phys. Lett.*, **84**, 684–687 (2004).
4. M. J. Lee, P. Fozooni, and P. W. Retz, *J. Low Temp. Phys.*, **54**, 303–331 (1984).
5. D. S. Greywall and G. Ahlers, *Phys. Rev. A*, **7**, 2145–2162 (1973).
6. L. Goldner, N. Mulders, and G. Ahlers, *J. Low Temp. Phys.*, **93**, 131–182 (1993).
7. N. Wada *et al.*, *Phys. Rev. B*, **52**, 1167–1175 (1995) and references therein.

Dynamics of Quantized Vortices in a Torsional Oscillator under Rotation: Proposed Experiments in Supersolid ^{4}He

Minoru Kubota[a], Muneyuki Fukuda[a], Toshiaki Obata[a], Yuji Ito[a], Andrey Penzev[a], Tomoki Minoguchi[b], and Edouard Sonin[c]

[a]*Institute for Solid State Physics, the University of Tokyo, Kashiwanoha 5-1-5, Kashiwa, Chiba, 277-8581 Japan*
[b]*Institute of Physics, University of Tokyo, Komaba, Meguro-ku, Tokyo 153-8902 Japan*
[c]*Racah Institute of Physics, Hebrew University of Jerusalem, Givat Ram, Jerusalem 91904, Israel*

Abstract. Recently there have been reports of superfluidity in solid ^{4}He and possibly in other solids. One of the common features of these systems is their small superfluid density. This would imply a rather long three dimensional (3D) coherence length. Sub-monolayer superfluid ^{4}He films condensed on 3D connected surfaces show 3D superfluid transitions, in some cases with a rather small 3D superfluid density and a long 3D coherence length, which implies a large 3D vortex core size. We have previously detected 3D vortex lines in a 3D superfluid made of a sub-monolayer ^{4}He film condensed on the pore surface of large pore diameter porous glass, by measuring the energy dissipation in a torsional oscillator under rotation. We propose application of this method to the new supersolids.

Keywords: . superfluid, supersolid, rotationally induced vortex lines, energy dissipation, vortex pairs, vortex rings
PACS: 67.40.Vs, 67.40.Rp, 67.80.-s

INTRODUCTION

Study of superfluidity in the newly reported solid He[1], as well as in other systems[2] with low superfluid density, demands special techniques to probe their unique properties. We propose a sensitive method of studying vortex line properties by a torsional oscillator (TO), by measuring a new energy dissipation peak under rotation as reported in [3] for He films formed on the 3D connected surfaces of porous glasses in addition to a static peak. The rotation induced dissipation peak in a He film is caused by the dynamics of two-dimensional (2D) vortices under the influence of the circulating superflow of vortex lines. We expect a long 3D coherence length[4] and thus a larger vortex core in some of the low superfluid density systems. We further expect that large-core vortices can be detected by a TO under rotation, where their flow fields couple with thermal excitations. Especially, those systems with a small 3D superfluid density would have low energy thermal excitations, which in some cases manifest themselves as a dissipation peak in a TO experiment under static condition, in the vicinity of the superfluid transition. A supersolid experiment reported by Kim and Chan is such an example[1], where a dissipation peak near T_c is reported.

3D SUPERFLUID OF MONOLAYER HE

We first review here sub-monolayer superfluid He films condensed on the 3D connected surfaces of porous media. An interesting feature of this 3D superfluid is that it provides a unique combination of two-dimensional (2D) and 3D physics. The system shows behaviors similar to those of a 2D film such as the superfluid density being proportional to the transition temperature T_c with an energy dissipation peak around T_c. Simultaneously, the system has 3D features such as a critical exponent of superfluid density ρ_s[4] close to 2/3, which is found in the λ transition of bulk ^{4}He. Thermally activated vortex-antivortex pairs (VAP's) play a crucial role in 2D Kosterlitz-Thouless (KT) transitions. In superfluid films that are 3D connected, 3D vortex movement is connected with 2D VAP's motion as has been quantitatively calculated using a model[5].

The vortex lines threading through such a system (see a pore vortex, Fig. 1) manifest themselves as a dissipation peak in addition to the static peak at T_c in a TO experiment under rotation, as recently reported by Fukuda *et al.*[3] It arises from 2D vortex pair dynamics, typical for KT superfluids[6] under the influence of the superfluid velocity field V_s circulating

CP850, *Low Temperature Physics: 24th International Conference on Low Temperature Physics;*
edited by Y. Takano, S. P. Hershfield, S. O. Hill, P. J. Hirschfeld, and A. M. Goldman

around the 3D vortex cores. The temperature at which this peak appears is determined by the vortex core size, in other words, by the minimum 3D coherence length. It is a unique feature of this system that the flow fields of 3D vortices interact with 2D thermal excitations. An increasing rotational speed increases the number of vortex lines, but V_s is constant for a given core size because of the quantization condition. We find some similarities with 3D superfluids in which vortex rings are thermally created near the superfluid transition temperature[7].

SUPERSOLIDS

Now we discuss the possibility of studying vortex lines threading through other superfluids with low energy excitations like supersolid He[1]. The reported superfluid density of bulk solid He, as well as solid He in porous media, is on the order of 1% of the total solid He density. From an estimation of the Josephson length[4] we can have a rough idea of the vortex core dimension, which is inversely proportional to the superfluid density. So we have on the order of a few tens of nm core size for supersolid He. This large size may explain the similar results of the TO experiments for a bulk toroidal cell and those containing porous media[1]. The core is larger than or comparable to the pore sizes used in the experiment.

In bulk liquid He immersed in Vycor, there has been an interesting experiment by Reppy and Tyler[8] using a Helmholz resonator, in which they observed a peak in the vicinity of T_λ. By changing the excitation amplitude by more than an order of magnitude, they found that the dissipation peak changed in size and position in temperature, quite similar to the behavior of He films formed in large pore porous glass[3], except for the very narrow temperature range for the bulk liquid ^{4}He.

Torsional oscillator dissipation data[1] for solid He are somewhat different from the result of the Helmholz resonance experiment and that for He films in porous media. Namely with the increase of the excitation amplitude the dissipation peak amplitude does not increase for the solid. Nevertheless, we hope to detect vortices in a supersolid under rotation by the dissipation signal change, since such an observation would shed light on this very poorly understood solid.

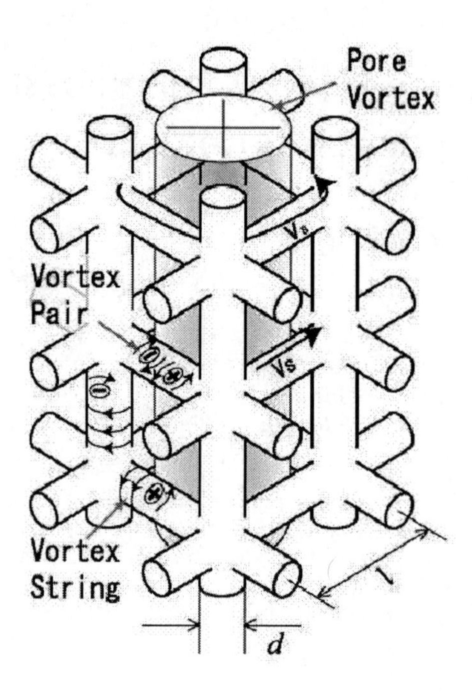

FIGURE 1. Model of a He film formed in a porous medium. A 3D vortex line called a pore vortex threads through the sample and produces quantized flow with V_s, which interacts with VAP's and causes energy dissipation in the system. In a supersolid, vortex rings with minimum sizes comparable to the 3D vortex core diameter would play the role of VAP's to produce extra dissipation under rotation.

REFERENCES

1. E. Kim and M. H. Chan, *Nature (London)* **427**, 225 (2004); *Science* **305**, 1941 (2004) and in these Proceedings.
2. See for example, H. Araki *et al.*, *J. Low Temp. Phys.* **134**, 1145-1151 (2004).
3. M. Fukuda, M.K. Zalalutdinov, V. Kovacik, T. Minoguchi, T. Obata, M. Kubota, and E.B. Sonin, *Phys. Rev. B* **71**, 212502-1-4 (2005).
4. N. P. Mikhin, V. E. Syvokon, T. Obata, and M. Kubota, *Physica B* **329-333**, 272-273 (2003).
5. T. Obata and M. Kubota, *Phys. Rev. B* **66**, 140506(R) -1-4 (2002); T. Obata, J.D. Reppy, and M. Kubota, *J. Low Temp. Phys.* **134**, 547-552 (2004). .
6. V. Ambegaokar, B. I. Halperin, D. R. Nelson, E. D. Siggia, *Phys. Rev. B* **21**, 1806-1826 (1980).
7. G. A. Williams, *Phys. Rev. Lett.* **82**, 1201 (1999); **68**, 2054 (1992).
8. J. D. Reppy and A. Tyler, in *Excitations in Two-Dimensional and Three Dimensional Quantum Fluids*, edited by A. G. F. Wyatt and H. J. Lauter, New York: Plenum Press, 1991, pp. 291-300.

Simultaneous Measurements of Heat Capacity and Superfluid Density of ^{4}He Adsorbed on Nanopores with Three-Dimensional Network

Ryo Toda, Mitsunori Hieda, Taku Matsushita and Nobuo Wada

Department of Physics, Nagoya University, Nagoya 464-8602, Japan.

Abstract. To investigate the superfluid onset of ^{4}He films formed in a three-dimensional network of nanopores 2.7 nm in diameter, we made simultaneous measurements of the superfluid density and the heat capacity. Errors in the ^{4}He coverage and temperature were reduced by mounting two experimental cells on the same cryostat. We observed a superfluid onset at 280 mK for a coverage of 1.53 atomic layers, and an anomaly in the heat capacity at the same temperature within an experimental error of 30 mK. The coincidence indicates that the superfluid transition of the ^{4}He film is macroscopic.

Keywords: superfluidity, ^{4}He film, ordered three-dimensional nanopores
PACS: 67.40.-w, 67.40.Kh, 67.70.+n

INTRODUCTION

The Bose-Einstein condensation and the superfluid onset of ^{4}He depend strongly on the dimensionality. In bulk ^{4}He, a peak in the heat capacity and the onset of superfluidity are observed at the same transition temperature. The critical exponents of the heat capacity and superfluid density take values characteristic of a three-dimensional (3D) transition. In contrast, ^{4}He films formed on flat solid surfaces show no heat capacity anomaly at the superfluid onset temperature observed by a torsional oscillator. The lack of heat capacity anomaly is in accordance with the Kosterlitz-Thouless theory for two-dimensional (2D) superfluids.

For ^{4}He films formed on Vycor glass in which pores 5–7 nm in diameter are randomly connected in three dimensions, a peak in the heat capacity was observed just at the superfluid onset temperature [1]. This suggests that the superfluid onset of ^{4}He films on Vycor glass is 3D rather than 2D. To further explore 3D superfluid transitions, study of ^{4}He films formed in smaller pores connected in three dimensions seems of value.

Many kinds of nanopores with regular pore structures have recently been synthesized. The substrate HMM-2 [2] has pores 2.7 nm in diameter with a regular 3D structure. According to our vapor-pressure measurements for adsorption [3], a uniform layer of ^{4}He is formed on the nanopore wall up to 2.16 atomic layers. Above about 1.4 atomic layers, we have observed heat capacity that is qualitatively different from that of ^{3}He of the same coverage, indication that a Bose fluid is formed [4]. Above the same coverage, superfluidity has been observed by a torsional oscillator [5].

In this article, we report measurements of the heat capacity and the superfluid density at the same coverage and temperature in order to examine whether there is a heat capacity anomaly just at the onset temperature of the superfluid. We developed an apparatus for simultaneous measurements of heat capacity and superfluid density, and optimized the cooling procedure for ^{4}He adsorption. A 3D character of the transition was indicated by the agreement between the heat capacity anomaly and the superfluid onset within the present experimental error.

APPARATUS AND RESULTS

HMM-2 used in the experiment is a powder whose particle size is about 0.1–10 μm. An organic-inorganic hybrid material forms the framework of an ordered 3D network of nanopores with an hcp structure (a = 5.54 nm, c = 8.86 nm) [2]. The mean pore diameter is estimated to be 2.7 nm.

The powder was placed in two experimental cells, a heat capacity cell and a torsional oscillator cell for superfluid density. The adsorption area of each cell was estimated from an N_2 adsorption isotherm, with an error of about 10%. The experimental apparatus is schematically shown in Fig. 1. Both the heat-capacity cell and torsional-pendulum cell were mounted on a copper block thermally linked to the mixing chamber of a dilution refrigerator. The temperature of the oscillator cell and the Cu block was monitored by the thermometer on the vibration isolator. Since a semi-adiabatic method was em-

CP850, *Low Temperature Physics: 24th International Conference on Low Temperature Physics;*
edited by Y. Takano, S. P. Hershfield, S. O. Hill, P. J. Hirschfeld, and A. M. Goldman

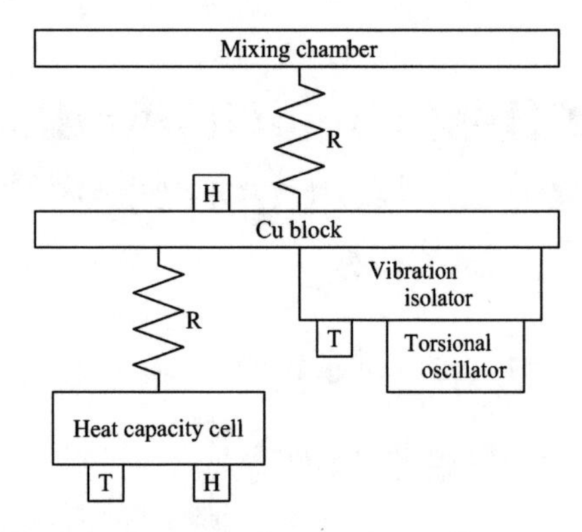

FIGURE 1. Diagram of experimental setup for simultaneous measurements of heat capacity and superfluid density. T, H, and R indicate thermometers, heaters, and weak thermal links, respectively.

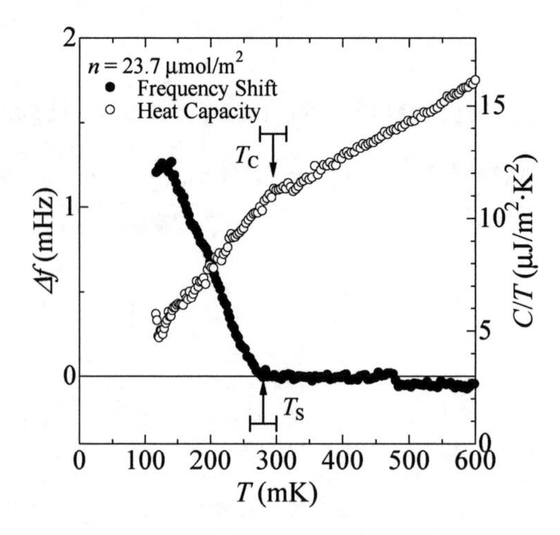

FIGURE 2. Heat capacity and superfluid density at the coverage 23.7 μmol/m^2. C/T (∘) shows an anomaly at $T_C = 295 \pm 20$ mK. The frequency shift of the torsional oscillator (•) shows a superfluid onset at $T_S = 280 \pm 20$ mK.

ployed for the calorimetry, the heat-capacity cell was connected to the Cu block through a weak thermal link. A typical thermal relaxation time between this cell and the Cu block was about 200 sec at 0.15–0.6 K.

The sample gas inlet tubes of the two cells were connected together at the Cu block. After introducing ^{4}He gas to the cells, the entire apparatus was slowly cooled down to about 1 K by controlling the heater on the block so as to keep a thermal equilibrium between the two thermometers within 0.5%. At each temperature of specific-heat and superfluid-density measurements, we waited until the temperature difference between the two experimental cells became less than 1%. After that, we measured the frequency of the torsional oscillator, and then applied a heat pulse to the heat capacity cell, raising its temperature by 2–3%.

The results of the simultaneous measurements at the coverage 23.7 μmol/m^2 (1.53 atomic layers) are shown in Fig. 2. The heat capacity C divided by T is shown by open circles. It shows a kink at $T_C = 295$ mK with an estimated uncertainty of $\pm$ 20 mK as shown by the error bar in the figure. The frequency shift Δf of the torsional oscillator, representing the superfluid density, is shown by solid circles. The scatter in Δf is $\pm$ 50 μHz. The result indicates an onset of superfluidity at $T_S = 280 \pm 20$ mK. Within the experimental error, T_C agrees with T_S. Measurements at other coverages are now in progress.

We observed a heat capacity anomaly at the superfluid onset temperature of ^{4}He films formed in 3D nanopores 2.7 nm in diameter, which is much smaller than the pore diameter of the usual Vycor glass studied so far. The relation between the 3D property and the pore size should be studied. It is also important to study the superfluid and heat capacity in one-dimensional nanopores of a similar pore diameter.

ACKNOWLEDGMENTS

We are grateful to S. Inoue, M. Katsuki and T. Kurokawa for technical support. We also thank S. Inagaki and Y. Fukushima for providing the substrate. This research was partly supported by a Grant-in-Aid for JSPS Fellows and a Grant-in-Aid for Scientific Research from the Ministry of Education, Culture, Sports, Science and Technology, Japan.

REFERENCES

1. S. Q. Murphy and J. D. Reppy, *Physica B* **165-166**, 547–548 (1990).
2. S. Inagaki, S. Guan, Y. Fukushima, T. Ohsuna, and O. Terasaki, *J. Am. Chem. Soc.* **121**, 9611–9614 (1999).
3. R. Toda, J. Taniguchi, R. Asano, T. Matsushita, and N. Wada, *J. Low Temp. Phys.* **138**, 177–182 (2005).
4. R. Toda, T. Yamada, J. Taniguchi, T. Matsushita, and N. Wada, *Physica B* **329-333**, 282–283 (2003).
5. T. Yamada, R. Toda, Y. Matsushita, T. Matsushita, and N. Wada, *J. Low Temp. Phys.* **134**, 601–606 (2004).

Localization of Bose-Einstein Condensate and Disappearance of Superfluidity of Strongly Correlated Bose Fluid in a Confined Potential

Michikazu Kobayashi and Makoto Tsubota

Faculty of Science, Osaka City University, Sugimoto 3-3-138, Sumiyoshi-ku, Osaka 558-8585, Japan

Abstract. We develop a model for a Bose fluid in a confined potential to study a new quantum phase caused by the localization of the Bose-Einstein condensate and the disappearance of superfluidity which was recently observed in liquid ^{4}He in porous glass at high pressures. The critical pressure for the transition to this phase can be defined by our new analytical criterion, which assumes that the size of the localized Bose-Einstein condensate is comparable to the scale of confinement. The critical pressure is quantitatively consistent with observations without free parameters.

Keywords: Bose-Einstein condensate, superfluidity, disorder, strong correlation
PACS: 01.30.Cc, 67.40.-w, 64.60.Cn

INTRODUCTION

The study of Bose condensed systems in random environments is a very important problem not only because of interest in the effects of impurities, but also for clarifying the effects of long-range correlations. Experimental works have studied liquid ^{4}He in porous glass and have observed many interesting phenomena [1–3]. In particular, Yamamoto *et al.* recently observed the disappearance of superfluidity of liquid ^{4}He confined in porous Geltech silica with very small pore size, 25Å at high pressures and very low temperatures [3]. More recently, a quasiparticle excitation spectrum was observed in the system at high pressures by using neutron scattering [4]. Yamamoto *et al.* suggested that there was a new quantum phase transition to localized Bose-Einstein condensate under the effect of strong correlation.

In previous work, motivated by Yamamoto *et al.*'s experiment, we qualitatively investigated the behavior of the critical temperature of the Bose-Einstein condensate (BEC) and the superfluidity of a Bose fluid under the effect of confinement and showed that the superfluid critical temperature goes to zero with a finite BEC critical temperature at high pressures. In this work, limiting our calculations to the case of zero temperature, we *quantitatively* investigate the behavior of the BEC, in particular, the localization of the BEC at high pressures. We use the three-dimensional Bose fluid model in a confined potential [6], and introduce our new criterion for finding the localization of the BEC. Assuming the BEC is localized, we calculate the energy of the system as a function of the size of the localized BEC and then minimize the energy. The resulting size gives the order of the scale of confinement above a critical pressure when the superfluid density disappears. The critical pressure is quantitatively consistent with the experimental one without free parameters.

THE MODEL OF BOSE FLUID IN A CONFINED POTENTIAL

The grand canonical Hamiltonian $\hat{H}-\mu\hat{N}$ of a Bose fluid in a confined potential is given by

$$\begin{aligned}\hat{H}-\mu\hat{N} =& \sum_{\mathbf{k}}[\varepsilon(\mathbf{k})-\mu]\hat{a}^\dagger(\mathbf{k})\hat{a}(\mathbf{k}) \\ &+\frac{1}{V}\sum_{\mathbf{k}_1,\mathbf{k}_2} U(\mathbf{k}_1-\mathbf{k}_2)\hat{a}^\dagger(\mathbf{k}_1)\hat{a}(\mathbf{k}_2) \\ &+\frac{1}{2V}\sum_{\mathbf{k}_1,\mathbf{k}_2,\mathbf{q}} g_0(\mathbf{q})\hat{a}^\dagger(\mathbf{k}_1+\mathbf{q})\hat{a}^\dagger(\mathbf{k}_2-\mathbf{q}) \\ &\times\hat{a}(\mathbf{k}_2)\hat{a}(\mathbf{k}_1),\end{aligned} \quad (1)$$

where $\hat{a}^\dagger(\mathbf{k})$ and $\hat{a}(\mathbf{k})$ are the free boson creation and annihilation operators with the wave number $\mathbf{k}$, $\varepsilon(\mathbf{k})=\hbar^2\mathbf{k}^2/2m$ is the kinetic energy of a particle of mass m, and μ is the chemical potential. $U(\mathbf{k})$ is the external confining potential, $V=L^3$ the volume of the system of size L, and $g_0(\mathbf{k})$ the interaction between two particles. The second term of the right-hand side represents the interaction between one particle and the confining potential; the second-order perturbation yields the Green's function

$$G^{\mathrm{R}}(k)=\frac{1}{\hbar^2V^2}\sum_{\mathbf{k}_1}|U(\mathbf{k}-\mathbf{k}_1)|^2G^0(\omega,\mathbf{k})^2G^0(\omega,\mathbf{k}_1). \quad (2)$$

CP850, *Low Temperature Physics: 24th International Conference on Low Temperature Physics;*
edited by Y. Takano, S. P. Hershfield, S. O. Hill, P. J. Hirschfeld, and A. M. Goldman

The third term refers to the interparticle interaction. Using the ring-approximation, we obtain

$$g(q) = \frac{g_0(\mathbf{q})}{1 + \frac{g_0(\mathbf{q})}{\hbar V} \sum\!\!\!\!\!\!\int_{\mathbf{k}} \frac{dk_0}{2\pi i} G^0(q) G^0(k+q)}. \quad (3)$$

Here $G^0(k)$ is the noninteracting Green's function

$$G^0(k) = G^0(\omega, \mathbf{k}) = \frac{\hbar}{\hbar\omega - [\varepsilon(\mathbf{k}) - \mu]}, \quad (4)$$

with frequency ω. Following the Bogoliubov method, we separate $\hat{a}(\mathbf{k})$ into the condensed part $a(\mathbf{k}_c) = \sqrt{N_c}$ with the smallest wave number of $\mathbf{k}_c = (2\pi/L, 2\pi/L, 2\pi/L)$ and the noncondensed part $\hat{a}(\mathbf{k} \neq \mathbf{k}_c)$. Then we calculate the Green's function $G^I(k)$ including the interparticle interaction $g(k)$ by using the Bogoliubov theory, finally obtaining the total Green's function $G(k) = G^R(k) + G^I(k)$. The superfluid component N_s, based on the two fluid model, can be given by linear response theory as [7]

$$N_s = N - \frac{\hbar}{6\pi m i} \sum\!\!\!\!\!\!\int_{\mathbf{k}} dk_0 \, k^2 \det[G(k)], \quad (5)$$

where N is the total number of particles. We assume a gaussian confinement $U(\mathbf{k}) = U_0 \exp(-k^2/2k_p^2)$ so that the wave number $k_p = 2\pi/r_p$ can be connected with the pore size r_p of the porous glass. Here the strength U_0 is estimated from the experimental critical coverage below which superfluidity vanishes [6]. For the interparticle interaction, we use $g_0(\mathbf{k}) = v_0(\sigma\sqrt{2\pi})^3 \exp(-k^2\sigma^2/2)$, where parameters v_0 and σ are determined by the comparison with the potential proposed by Aziz *et al.* [8]

To consider the phase of the localized BEC, we introduce a new criterion. First we assume a localized BEC of size L_g and calculate the energy E_g by replacing volume V in Eq. (1) with $V_g = L_g^3$. The total energy of the volume V should be proportional to $E_g V/V_g$, because the number of localized BECs is proportional to V/V_g. The ideal or weakly interacting Bose gas has E_g proportional to V_g. In the strongly correlated Bose fluid, however, E_g is no longer a simple linear function of V_g. Then, E_g/V_g depends on V_g. Calculating E_g/V_g as a function of V_g, we find $V_g = V_{g,\min}$ which minimizes E_g/V_g. Thus the localization of the BEC can be defined by the new criterion: if $V_{g,\min}$ exceeds V, the system is a non-localized BEC state. When $V_{g,\min}$ is reduced to being comparable to r_p^3, we judge that the BEC is localized. Since $V_{g,\min}$ is a function of N and pressure P, we can obtain the phase boundary of the transition to localization by using this criterion.

Figure 1 shows the dependence of N_s and $V_{g,\min}$ on the pressure, where all numerical parameters are fixed by References [3, 8] and there remain no free parameters. At the pressure $P \simeq 4.2$ MPa, the superfluid component

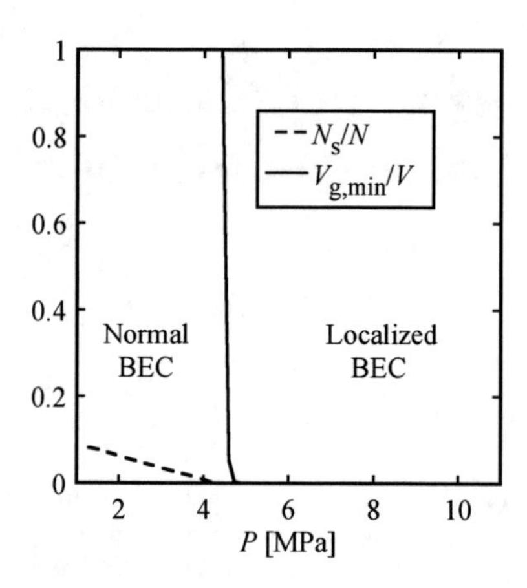

FIGURE 1. Dependence of the superfluid component N_s and the volume $V_{g,\min}$ of the localized BEC on the pressure. The pressure P can be obtained by the thermodynamic relation $P = -\partial E/\partial V$.

N_s disappears and $V_{g,\min}$ is reduced dramatically to the order of r_p^3. We can therefore define this pressure as the critical pressure P_c. Our critical pressure $P_c \simeq 4.2$ MPa is quantitatively consistent with the experimental one $P_c \simeq 3.5$ MPa [3], and we finally conclude that the experimentally observed disappearance of superfluidity at high pressures is caused by a transition to a localized BEC. For the case of porous glass with a larger pore size $r_p \simeq 70$ Å, we obtain a much larger critical pressure $P_c \simeq 9$ MPa which is too high for a BEC to be localized against solidification of liquid ^{4}He. This is also consistent with experimental results for ^{4}He in Vycor glass [3].

REFERENCES

1. J. D. Reppy, *J. Low. Temp. Phys.*, **87**, 205–245 (1992).
2. O. Plantevin, H. R. Glyde, B. Fåk, J. Bossy, F. Albergamo, N. Mulders, and H. Schober, *Phys. Rev. B*, **65**, 224505 (2002).
3. K. Yamamoto, H. Nakashima, Y. Shibayama, and K. Shirahama, *Phys. Rev. Lett.*, **93**, 075302 (2004).
4. K. Shirahama, private communication.
5. M. Kobayashi and M. Tsubota, *JLTP*, **138**, 189 (2005).
6. M. Kobayashi and M. Tsubota, *Phys. Rev. B*, **66**, 174516 (2002).
7. P. C. Hohenberg and P. C. Martin, *Ann. Phys.*, **34**, 291–359 (1965).
8. R. A. Aziz, A. R. Janzen, and M. R. Moldover, *Phys. Rev. Lett.*, **74**, 1586–1589 (1995).

One-Dimensional ^{4}He and ^{3}He Quantum Fluids Realized in Nanopores

Nobuo Wada*, Taku Matsushita*, Ryo Toda*, Yuki Matsushita*, Mitsunori Hieda*, Junko Taniguchi† and Hiroki Ikegami**

*Department of Physics, Nagoya University, Nagoya 464-8602, Japan
†Department of Applied Physics and Chemistry, University of Electro-Communiations, Chofu 182-8585, Japan
**Low Temperature Physics Laboratory, RIKEN, Wako 351-0198, Japan

Abstract. We have studied ^{4}He and ^{3}He adsorbed onto FSM-16, which has straight one-dimensional (1D) nanopores 18 or 28 Å in diameter. Fluid film tubes are formed on the nanopore walls covered with about one atomic layer of inert (solid) helium. The low-temperature heat capacity of ^{4}He fluid tubes is linear in temperature. This is due to 1D phonons, whose contribution becomes observable at temperatures below 1/10 of the energy-level separation between the ground state and the first excited state quantized in the azimuthal direction. The superfluidity of ^{4}He adsorbed in the nanopores was studied by a torsional oscillator. When the nonopores are preplated with ^{4}He layers, ^{3}He adatoms at a low density show a large heat capacity characteristic of a fluid. At high temperatures, the heat capacity is as large as that of a Boltzmann gas. With decreasing temperature, the heat capacity shows a Schottky-like peak followed by an asymptotic, linear dependence on temperature. A model calculation for a dilute ^{3}He gas in an adsorption potential of a nanopore suggests a dimensional crossover from two dimensional to 1D around the Schottky-like peak. The calculation also suggests a degenerate 1D state at sufficiently lower temperatures than the energy gap for the motion in the azimuthal direction.

Keywords: low-dimensional quantum fluids, helium 4, helium 3, nanopores
PACS: 67.40.Db, 67.55.Cx, 68.65.-k, 67.40.Kh

INTRODUCTION

Helium adsorbed on flat solid surfaces form two-dimensional (2D) quantum fluids. 2D ^{4}He fluid shows a Kosterlitz-Thouless superfluid transition in which the two dimensionality plays an essential role. New features of 2D Fermi fluids have been found in ^{3}He films adsorbed on graphite [1]. In one dimension, no quantum fluid has been realized with either ^{4}He or ^{3}He, to our knowledge, until our recent studies [2, 3]. In this paper, we review studies on the one-dimensional (1D) quantum fluids of ^{4}He and ^{3}He formed in FSM-16, which has 1D nanopores 18 Å or 28 Å in diameter.

For these new 1D substrates, the first layer completion and a uniform layer formation of the adatoms on the nanopore walls were determined from the vapor pressure of the adsorption. Obvious differences in the low-temperature heat capacities between ^{4}He and ^{3}He indicate that quantum fluid layers are formed on the inert (solid) helium layers above about one atomic layer. For the tubes formed by ^{4}He fluid layers, a Bose fluid region in the temperature vs. coverage diagram was experimentally determined by the heat capacity. A 1D phonon state of the Bose fluid was observed at sufficiently low temperatures. Superfluidity of the fluid was examined by a torsional oscillator. ^{3}He fluid films were made on the 1D nanopores preplated with ^{4}He layers. The heat capacity of the ^{3}He films at a low density was semiquantitatively explained by a dimensional crossover from a 2D Boltzmann gas at high temperatures to a 1D gas at low temperatures.

ADSORPTION ON 1D NANOPORES

The 1D nanopores of FSM-16 are formed by a honeycomb framework made of silica, as shown in the inset of Fig. 1(a) [4]. We used two varieties of the material, with pore diameters 18 Å and 28 Å. These diameters have an uncertainty of 2 Å, because a method of pore diameter measurement has not yet been well established. The mean length of the pores was estimated to be about 3000 Å from the grain size of the FSM-16 powder used in the experiment. After dehydration at 200°C for about two hours, the adsorption area S was determined by measuring a N_2 pressure isotherm at 77 K and by fitting it to the Brunauer-Emmett-Teller equation at $5 < P < 15$ kPa [2]. The coverage n is described by the adsorbed amount of

CP850, *Low Temperature Physics: 24th International Conference on Low Temperature Physics;* edited by Y. Takano, S. P. Hershfield, S. O. Hill, P. J. Hirschfeld, and A. M. Goldman

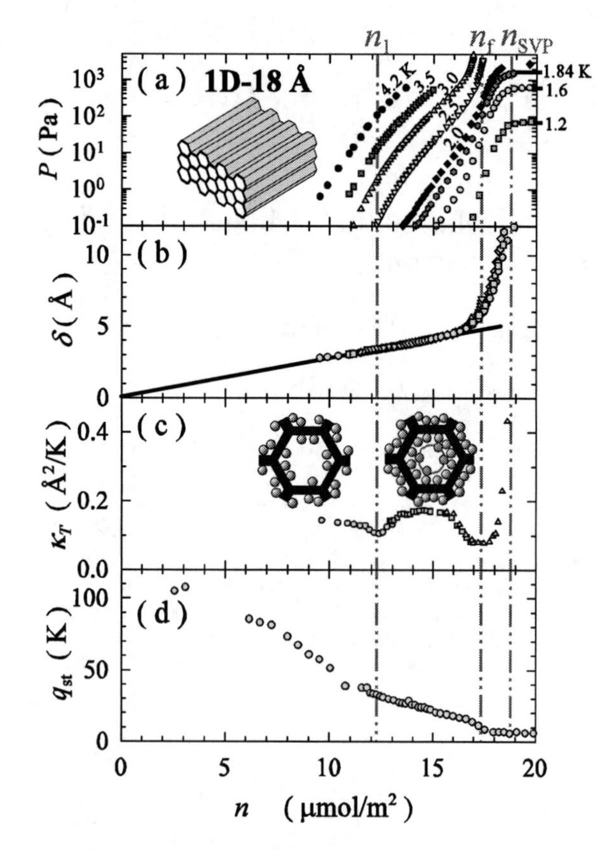

FIGURE 1. (a) Vapor pressure isotherms of ^{4}He adsorbed on 1D pores of FSM-16 18 Å in diameter. Complete filling with ^{4}He is indicated by the saturated vapor pressure of bulk liquid (solid line) above the coverage n_{SVP}. (b) Thickness δ of ^{4}He film on the pore wall, estimated by the FHH model. Linear increase up to n_f indicates a uniform layer formation. (c) Compressibility κ_T. Minimum at n_l indicates a first layer completion. (d) Isosteric heat of sorption q_{st}.

helium divided by S.

To investigate how helium adsorbs on the nanopores, we measured the vapor pressure $P(T,n)$ of adsorption, as shown in Fig. 1(a) for the case of ^{4}He adsorbed on the 18 Å pores [5]. Complete filling of pores with ^{4}He occurs at the coverage n_{SVP}, above which the pressure agrees with the saturation vapor pressure of bulk liquid (solid line). The chemical potential $\mu_S(T,n)$ of the adatoms is equal to that of the vapor, which is calculated from the pressure data. According to the Frenkel-Halsey-Hill (FHH) model [6], $\mu_S(T,n)$ of a film of thickness δ is a sum of the van der Waals potential $-A/\delta^3$ and the chemical potential of the bulk liquid. Then, δ of the adsorbed film is obtained as shown in Fig. 1(b), where we have chosen $A = 1100$ KÅ^3. The almost linear dependence up to n_f indicates a formation of a uniform layer on the nanopore walls up to this coverage. From the pressure isotherm, the compressibility κ_T of the film is obtained as shown in Fig. 1(c) by using $\kappa_T = 1/(n^2RT)\,[\partial \ln P/\partial n]_T^{-1}$, where R is the gas constant. The minimum of κ_T at coverage n_l corresponds to a first layer completion. Isosteric heat of sorption given by $q_{st}(T) = -\,[\partial \ln P/\partial(1/T)]_n$ is a work function for bringing an adatom to the gas state. Figure 1(d) shows $q_{st}(T)$ that is consistent with the assumption for $\mu_S(T,n)$ in the FHH model.

In the 18 Å and 28 Å pores, uniform layers were formed up to the coverages $n_f = 17.3$ $(1.4\,n_l)$ and 31.8 μmol/m^2 $(2.0\,n_l)$, respectively, for ^{4}He adsorption. The heat capacities of the adsorbed ^{4}He and ^{3}He were measured down to about 70 mK. In contrast to the classical regime [7], the heat capacities in the quantum regime are qualitatively different between ^{4}He and ^{3}He. This is due to the different quantum statistics of bosons and fermions.

^{4}HE FLUIDS IN 1D NANOPORES

Bose Fluid Film Tubes

Isotherms of the heat capacities of ^{4}He and ^{3}He adsorbed on the 18 Å pores are shown in Fig. 2(a) and Fig. 3, respectively [8]. At low temperatures, e.g., 0.1 K, the heat capacities are small roughly below n_l, indicating localization of the adatoms. At each coverage, an onset of localization is suggested by a drop in C/T below a specific temperature, which is marked by a solid circle in Fig. 2(b). Above n_l, the heat capacity isotherms are qualitatively different between ^{4}He and ^{3}He [3]. The ^{3}He isotherms increase monotonically with n, due to the large heat capacity of the ^{3}He fluid layer in the quantum regime [7] formed on the inert (solid) layer. In contrast, each ^{4}He isotherm has a maximum. A similar maximum is found in ^{4}He isotherm for the 28 Å pores, as shown in Fig. 4(a).

These isotherms for ^{4}He suggest that a Bose fluid appears above the coverage n_B of the isotherm maximum. Plotting n_B (solid square) as a function of T, we define this region in the T-n diagram, as shown in Fig. 2(b) and Fig. 4(b) for the 18 Å and 28 Å pores, respectively. This region appears above coverage $n_c = 1.15\,n_l$ (18 Å) and $1.4\,n_l$ (28 Å) in the present measurement down to 70 mK. In the case of the 28 Å pores, C/T vs. T at a fixed coverage shows a kink at temperature T_B (open triangle), which is on the boundary of the Bose fluid region. Taking the thickness δ at n_c to be that of an inert (solid) layer (Fig. 1(b) in the case of the 18 Å pores), we find that the Bose fluid forms a nanotube about 10 Å and 17 Å in diameter in the 18 Å and 28 Å pores, respectively.

It is known that 3D and 2D ^{4}He liquids show phonon heat capacities at low temperatures. The heat capacity of bulk ^{4}He superfluid shows a T^3 dependence with

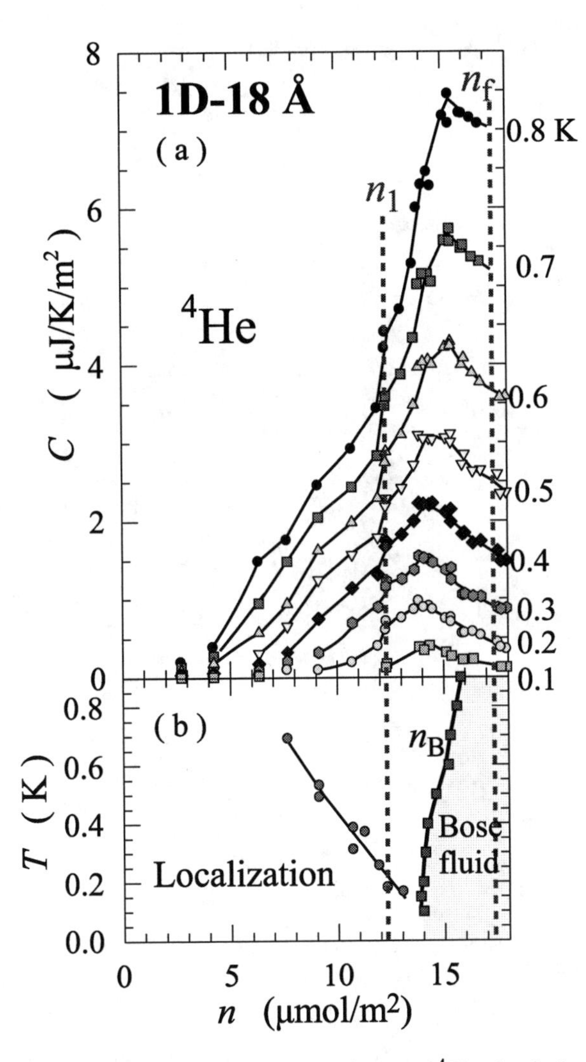

FIGURE 2. (a) Heat capacity isotherms of ^{4}He adsorbed on 1D 18 Å pores. It is suggested that a Bose fluid state appears above the coverage n_B of the isotherm maximum. (b) Phase diagram proposed from the heat capacity. Localized state is in the region below the solid circles. Bose fluid region is defined from n_B (solid squares).

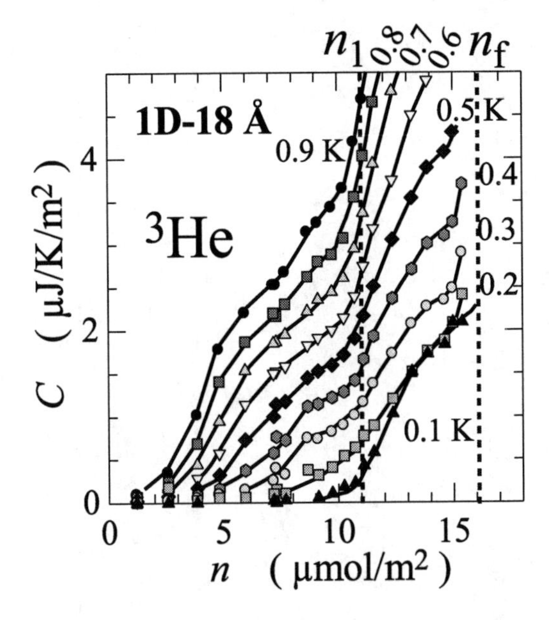

FIGURE 3. Heat capacity isotherms of ^{3}He adsorbed on 1D 18 Å pores.

the phonon velocity $v_I = 239$ m/sec below about 0.7 K at the saturated vapor pressure. ^{4}He fluid films formed on the 2D substrate of Hectorite show a T^2 dependence with $v_I = 100$–200 m/sec below about 0.2–0.3 K, which depends on the ^{4}He coverage [11]. The specific heat of ^{4}He adsorbed on the 1D 18 and 28 Å pores is plotted in Figs. 5(a) and (b), respectively, where C is divided by the amount $n - n_c$ of ^{4}He in the fluid layer and the gas constant R [2, 10]. For the 18 Å pores, C is linear in T below 0.10–0.15 K, which depends on n. The data for the 18 Å pores at higher temperatures and those for the 28 Å pores in the whole measured region show stronger temperature dependences, between T and T^2.

The heat capacities can be explained by the phonon heat capacities calculated for a nanotube of a fluid [9, 12]. Along the axial direction, we assume a continuous wave vector k with velocity v_I. In the azimuthal direction, discrete eigenstates are expected. If the same phonon velocity v_I is assumed in this direction, the energy of the first excited state (doublet) is given as $\Delta_{01} = 2\hbar v_I/d$ with respect to the ground state energy followed by higher energy levels at $2\Delta_{01}$, $3\Delta_{01}$ and so on. Here d is the diameter of the fluid tube. Thus, the phonon energy in the tube is given by $\varepsilon_{kl} = [(\hbar v_I k)^2 + (\Delta_{01}\ell)^2]^{1/2}$, where ℓ is an integer. The calculated phonon heat capacity shows a T-linear dependence below $0.1\Delta_{01}/k_B$, where the deviation is 0.6 %. The deviation is caused by thermal excitations from the ground state to the first excited states. By adjusting v_I and Δ_{01}, we fit the calculation to the data for $C/(n-n_c)R < 0.05$ as shown by solid lines in Figs. 5(a) and (b) [9].

In the case of the 18 Å pores, v_I was found to be 100–200 m/sec, depending on the coverage. These velocities are comparable to those of 3D and 2D ^{4}He liquids. Deviations from the T-linear behavior above 0.10–0.15 K (Fig. 5(a)) are caused by thermal excitations between the ground state and the first excited states in the azimuthal motion with $\Delta_{01}/k_B = 0.9$–1.4 K. In the 28 Å pores (Fig. 5(b)), no clear T-linear behavior is observed down to 70 mK, the lowest temperature of the experiment, indicating that Δ_{01}/k_B is 0.7 K or lower.

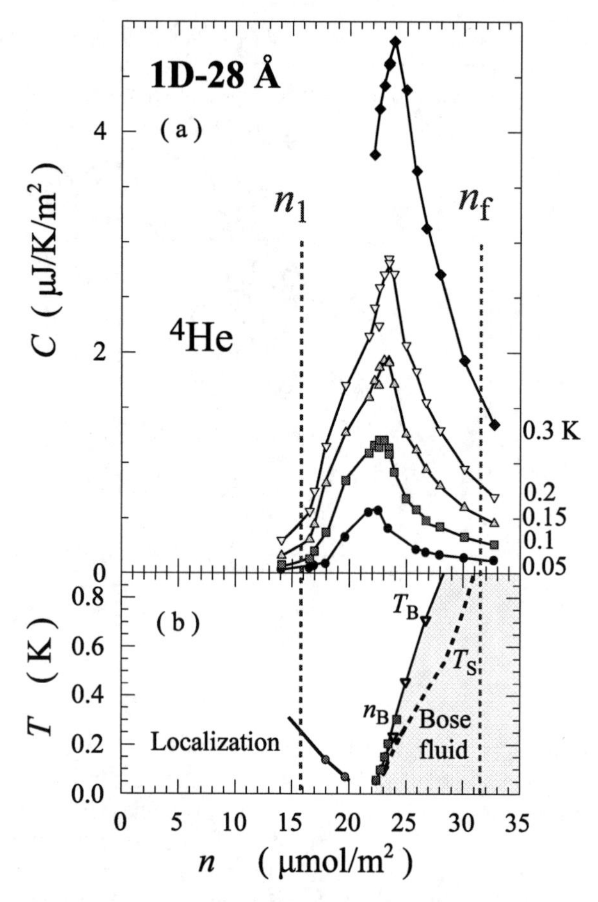

FIGURE 4. (a) Heat capacity isotherms of ^{4}He adsorbed on 1D 28 Å pores. (b) Phase diagram proposed from the heat capacity. Bose fluid region is defined from the coverage n_B (solid square) at the isotherm maximum and an anomaly of the heat capacity at T_B (open triangle). Dashed line shows the onset temperature T_S of the superfluidity.

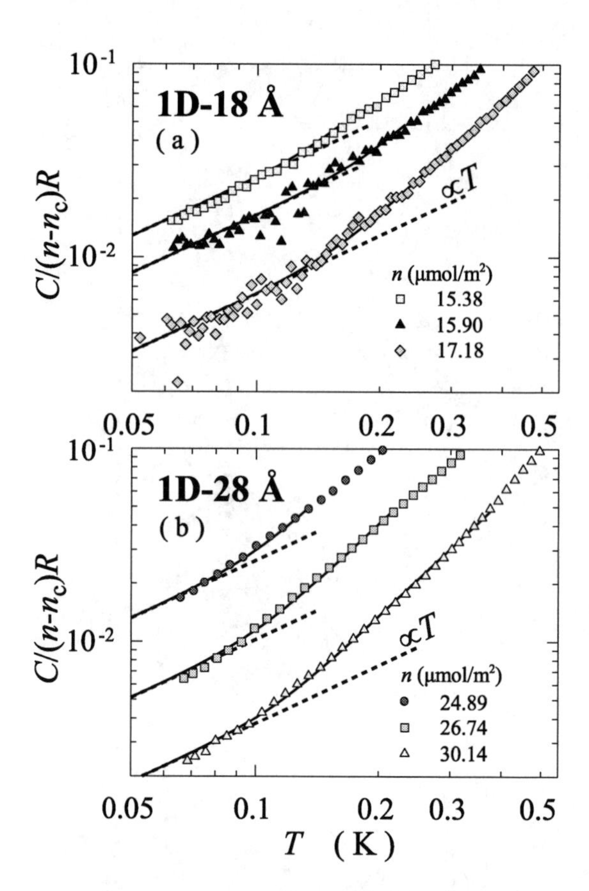

FIGURE 5. Specific heats $C/(n-n_c)R$ of ^{4}He adsorbed on 1D 18 Å and 28 Å pores, (a) and (b), respectively. T linear dependence at low temperatures suggests a 1D phonon state of the ^{4}He Bose fluids.

Superfluid in 1D Nanopores

The most interesting question is whether or not superfluidity occurs in ^{4}He inside the 1D nanopores. This was examined by a torsional oscillator, into which we pressed FSM-16 powder mixed with silver powder [13, 14]. At the measurement frequency of about 1 kHz and the superfluid sound velocity on the order of 100 m/sec, a frequency shift by superfluids in the nanopores could be observed only when they were connected by superfluid films on the grain surfaces of the substrates and the silver powder. For this reason, we first took data with ^{4}He in the cell. Next, we repeated the measurements after filling the nanopores with N_2 and then introducing ^{4}He into the cell. The difference in the frequency shifts due to superfluidity between the two cases gave the superfluid fraction of the ^{4}He adatoms inside the nanopores.

Under the ^{4}He-only and N_2-filled conditions, frequency shift ΔF due to superfluidity was observed in the 18 Å pores as shown in Fig. 6(a) [14]. The onset temperatures $T_S = 1.17$ and 1.18 K, respectively, are almost the same. Just below T_S, ΔF shows a steep increase, which suggests a Kosterlitz-Thouless transition of ^{4}He films on the silver powder and the grain surfaces of the substrates. The figure also shows a small but clear difference of ΔF below T_S, as indicated by the shaded region. The difference is due to superfluidity of the ^{4}He adatoms in the nanopores. The superfluidity was observed roughly above the coverage n_f in Fig. 2(b).

In the 28 Å pores, a large difference in ΔF was observed as shown in Fig. 6(b). Under the ^{4}He-only condition, superfluidity was observed above $n_c = 22\ \mu$mol/m^2. The superfluid onset temperature T_S increases with n as shown by the dashed line in Fig. 4(b). In comparison with the Bose fluid boundary (solid line) determined from the heat capacity anomaly, T_S is much lower at a given coverage.

Superfluidity in the 18 Å pores was observed at higher coverages (Fig. 6(a)) than those for which the 1D phonon

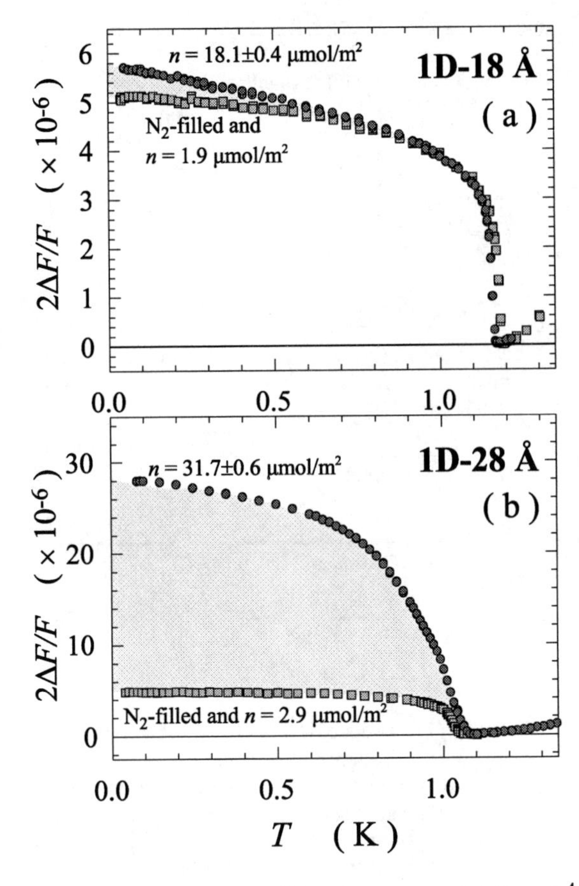

FIGURE 6. Frequency shifts due to superfluidity of ^{4}He adsorbed on 1D 18 and 28 Å pores, (a) and (b), respectively, measured in two conditions. One condition is that only ^{4}He was adsorbed. The other one is that the nanopores were filled with N_2. The differences of the frequency shift indicated by the shaded regions show superfluidity of the ^{4}He adatoms in the nanopores.

heat capacity was seen (Fig. 5(a)). In the 28 Å pores, superfluidity was observed above about 50 mK, which may not be yet in the 1D phonon region(Fig. 5(b)).

Superfluid in 3D Nanopores

The onset of superfluidity in adsorbed ^{4}He films has been studied in Vycor glass, whose pores are typically about 70 Å in diameter and randomly connected in three dimension. Superfluidity appears at coverages above 1.6 atomic layers. In a simultaneous experiment with a torsional oscillator experiment, heat capacity showed an anomaly of a peak just at the onset temperature of superfluidity [15]. This is clear evidence for a 3D phase transition. In a 2D Kosterliz-Thouless transition, superfluid onset occures without any anomaly in the heat capacity at temperatures lower than the Gintzburg-Landau temperature at which some heat capacity anomaly has been observed [16].

To investigate the effect of the dimensionality of pore connection on the superfluid onset, we studied ^{4}He adsorbed on HMM-2 whose pores, 27 Å in diameter, are connected three-dimensionally with a regular structure [17]. From the vapor pressure of adsorption, we found the first layer completion at $n_1 = 15.5\,\mu\text{mol/m}^2$ and a uniform layer formation up to $n_f = 33.5\,\mu\text{mol/m}^2 (2.16 n_1)$ [18]. Superfluidity was observed above $n_c \approx 21\,\mu\text{mol/m}^2$ [19, 20]. The onset temperature T_S increased with increasing n. At $n = 23.7\,\mu\text{mol/m}^2$ between n_c and n_f, we measured the heat capacity as well as the superfluid density as shown in Fig. 7. Here, we carefully made the coverage be the same for the two measurements [20]. The superfluid onset temperature was $T_S = 0.280 \pm 0.020$ K. C/T shows a change in the slope at $T_B = 0.295 \pm 0.020$ K. The two temperatures agree within the experimental uncertainties. In the T-n diagram (Fig. 4(b)) of the 1D 28 Å pores, which are very similar to the pores of HMM-2 in diameter, the superfluid onset temperature $T_S = 0.3$ K is significantly lower than $T_B = 0.6 \pm 0.1$ K of the Bose fluid boundary.

The heat capacity and superfluid density results for the 1D and 3D nanopores indicate a strong dependence of the superfluid onset on the dimensionality of pore connection. Superfluidity inside the 1D nanopores was not observed throughout the Bose fluid regions determined by the heat capacity. In contrast, in the 3D nanopores, the superfluid onset coincided with the heat capacity anomaly within the experimental error. This is similar to the superfluid transition observed for the Vycor glass [15] with a larger pore diameter.

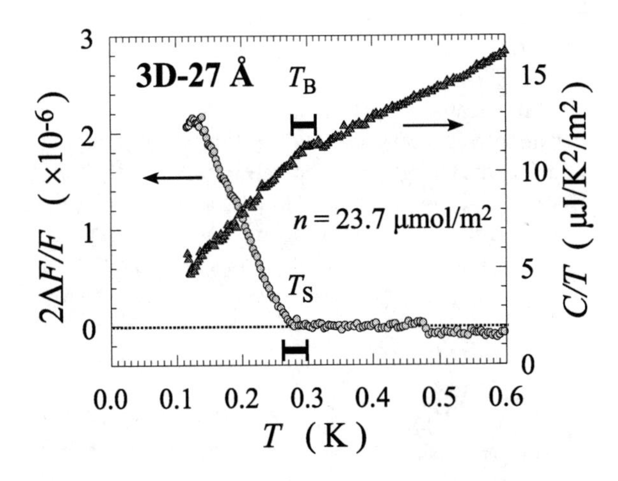

FIGURE 7. Frequency shift due to superfluidity and heat capacity of ^{4}He adsorbed on HMM-2 whose nanopores, 27 Å in diameter, are connected in three-dimension. Superfluid onset at T_S and anomaly of C/T at T_B occur at the same temperature within the experimental uncertainties.

^{3}HE FLUID FILM TUBES

Dimensional Crossover

Similar to ^{4}He, ^{3}He also forms a nanotube of a fluid film surrounded by a inert (solid) layer in each 1D pore of FSM-16, as shown schematically in Fig. 8. To realize a degenerate 1D state, the ^{3}He fluid density should be sufficiently low so that the Fermi energy is lower than the energy-level separation between the ground state and the first excited states for azimuthal motion. There is also an experimental demand that the heat capacity of the inert layer is as small as possible. Therefore, we preplated the nanopores with ^{4}He before introducing ^{3}He.

A molar heat capacity of ^{3}He adatoms as large as the gas constant R was in this manner as shown in Fig. 9(a), where the 18 and 28 Å pores had been preplated with 1.1 and 1.47 atomic layers of ^{4}He, respectively [21, 3]. The diameter of the ^{3}He fluid tube is estimated to be 10 and 17 Å, respectively, using the thickness of ^{4}He obtained from the vapore pressure of adsorption. At high temperatures, the molar heat capacity C/n is equal to or a little larger than $C/n = R$ of the 2D Boltzmann gas. With decreasing T, C/n at the lowest coverage shows a Schottky-like peak at $T \approx 0.3$ K for the 18 Å pores and 0.15 K for the 28 Å pores, followed by an approach to a linear T dependence. The peak height decreases with increasing ^{3}He coverage n_3.

To understand the heat capacity data, we solve the Schrödinger equation for a ^{3}He atom in an adsorption potential of a 1D nanopore, and calculate the heat capacity of ^{3}He gas in the pores [3, 21]. The energy $E_{\ell m}$ of a ^{3}He atom in the potential within a cross section orthogonal to the axial direction is labeled with quantum numbers ℓ and m for the radial and azimuthal degrees of freedom, respectively. The ground state energy is E_{00}. The wavefunction shows a maximum of probability density near the wall, because of a deep potential there. The first excited doublet is a 2D state with $\ell = 0$ and with energy E_{01}. We define the energy-level separation as $\Delta_{01} = E_{01} - E_{00}$. 3D states with $\ell \neq 0$ appear at much higher energies than E_{01}, due to the potential profile in the nanopore. Assuming a free translation along the pores, we calculate the heat capacity of a ^{3}He gas as shown in Fig. 9(b). In the 28 Å pores preplated with ^{4}He, the 2D energy-level separation is calculated to be Δ_{01}/k_B =0.328 K by assuming a reduced pore diameter of 18.2 Å [3]. At temperatures higher than Δ_{01}/k_B, C/nR is about 1, as expected for a 2D Boltzmann gas, and slowly increases with temperature due to thermal excitations to the 3D state located at Δ_{10}/k_B =2.46 K. At $T \approx 0.3\Delta_{01}/k_B$, C/nR shows a Schottky peak followed by the heat capacity of an ideal 1D ^{3}He gas below $0.1\Delta_{01}/k_B$. C/nR of a non-interacting 1D gas with a Fermi temperature $T_F = 0.042$ K is shown by the solid line. A Schottky peak appears only at a low density where T_F is much lower than Δ_{01}/k_B. The peak disappears when T_F exceeds the 2D level separation Δ_{01}/k_B. The Schottky peak indicates a dimensional crossover from a 2D Boltzmann gas to a 1D gas with decreasing temperature. In the case of the 18 Å pores, we assume a solid ^{4}He layer 3.78 Å in thickness, and obtain $\Delta_{01}/k_B = 1.20$ K and $\Delta_{10}/k_B = 7.11$ K. A 1D-2D crossover peak appears at $T \approx 0.35$ K.

The calculated heat capacities (Fig. 9(b)) semiquantitatively reproduce the experimental results (Fig. 9(a)). Thus, the Schottky-like peak at $T \approx 0.15$ K observed for the 28 Å pores indicates $\Delta_{01}/k_B \approx 0.5$ K between

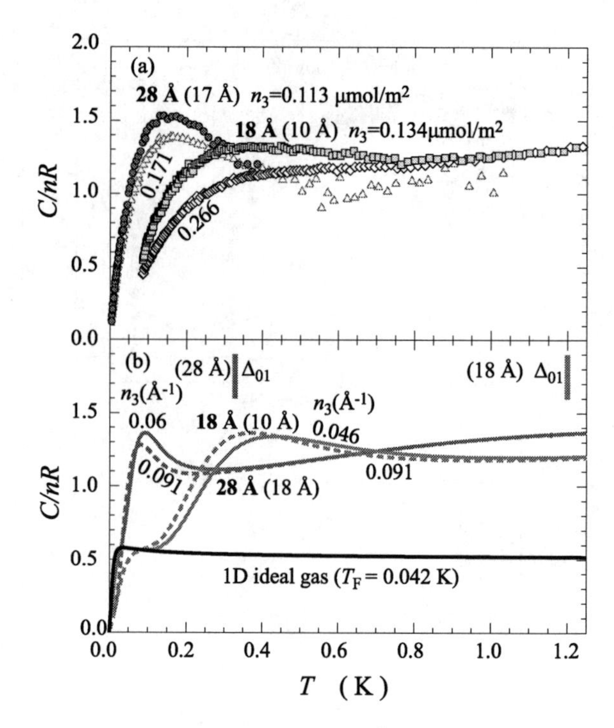

FIGURE 9. (a) Heat capacities of ^{3}He adsorbed on the 1D 18 and 28 Å pores preplated with 1.1 and 1.47 atomic layers of ^{4}He. (b) Calculated heat capacities of non-interacting dilute ^{3}He gas in the 1D nanopores. A Schottky peak appears at $T \approx 0.3\Delta_{01}/k_B$, which indicates a dimensional crossover. Heat capacity of a 1D gas appears at $T < 0.1\Delta_{01}/k_B$.

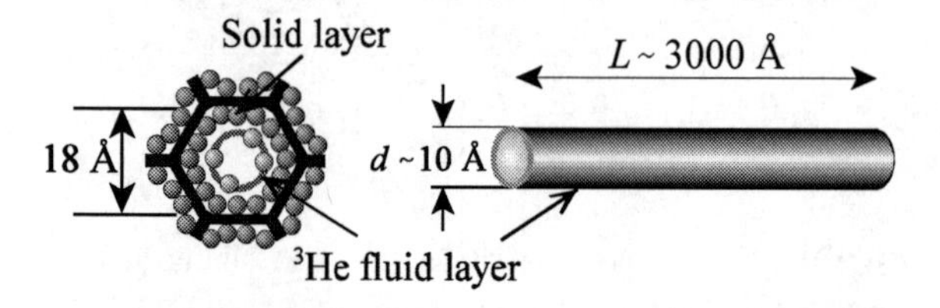

FIGURE 8. Adsorption model of helium atoms in 1D 18 Å pores. Fluid nanotube, about 10 Å in diameter, is formed in the 18 Å pore preplated with an inert solid layer.

the ground state and the first excited state. The maximum at $T = 0.3 - 0.4\,\mathrm{K}$ for the 18 Å pores indicates $\Delta_{01}/k_B \approx 1 - 1.3\,\mathrm{K}$. The level separation Δ_{01} increases with decreasing pore diameter, in agreement with the calculation.

Degenerate ^{3}He in Nanopores

At temperatures below about a few tens of mK, heat capacity of ^{3}He in the 28 Å pores appears to approach a T-linear behavior (Fig. 9(a)), suggesting a degenerate state of ^{3}He. The heat capacities below 50 mK at several densities are shown in Fig. 10(a). According to the calculation, the 1D condition is satisfied by the three lowest coverages at low temperatures, i.e., calculated Fermi temperatures ($T_F = 18$, 72, and 164 mK) are lower than the level separation $\Delta_{01}/k_B = 328\,\mathrm{mK}$.

The temperature dependences are fitted to an empirical formula, $C = \beta_0 + \gamma T + \beta_2 T^2$. The parameters β_0, γ, and β_2 are shown in Fig. 10(b)-(d). The γT term likely corresponds to the Sommerfeld term of a degenerate Fermi fluid. In an ideal degenerate 1D Fermi gas, γ decreases with increasing density as shown in Fig. 10(c), because the 1D density of states is proportional to $1/\sqrt{\varepsilon}$, where ε is the particle kinetic energy. In contrast to the ideal 1D Fermi gas, the observed γ value (Fig. 10(c)) increases with increasing n_3, although the three lowest densities satisfy the 1D condition in the calculation.

To explain the experimental result, there are two possibilities. One is a heterogeneity of the nanopores [22]. If there is a variation in the pore diameter, the lowest energy E_{00} for motion in the cross section also varies. Then, the averaged density of states deviates from $1/\sqrt{\varepsilon}$, and may even increase with ε. The other possibility is the effect of the interaction between ^{3}He atoms [23, 24]. In the 3D free space, the binding energy of ^{3}He atoms is estimated to be of the order of 1 mK. When the wave functions of ^{3}He atoms are confined in a nanopore, the binding energy increases to several tens of mK depending on the pore diameter. This energy, comparable to the Fermi energy of about a few tens of mK, may change the n_3 dependence of γ.

CONCLUSION

In 1D nanopores 18 Å and 28 Å in diameter, ^{4}He and ^{3}He fluid tubes are realized at coverages n above about one atomic layer. In the case of ^{4}He, we determined the Bose fluid region in a T-n diagram from the heat capacities. The T-linear heat capacities observed in the Bose fluid region suggest a completely 1D phonon state in the sense that motion in the cross section is in the

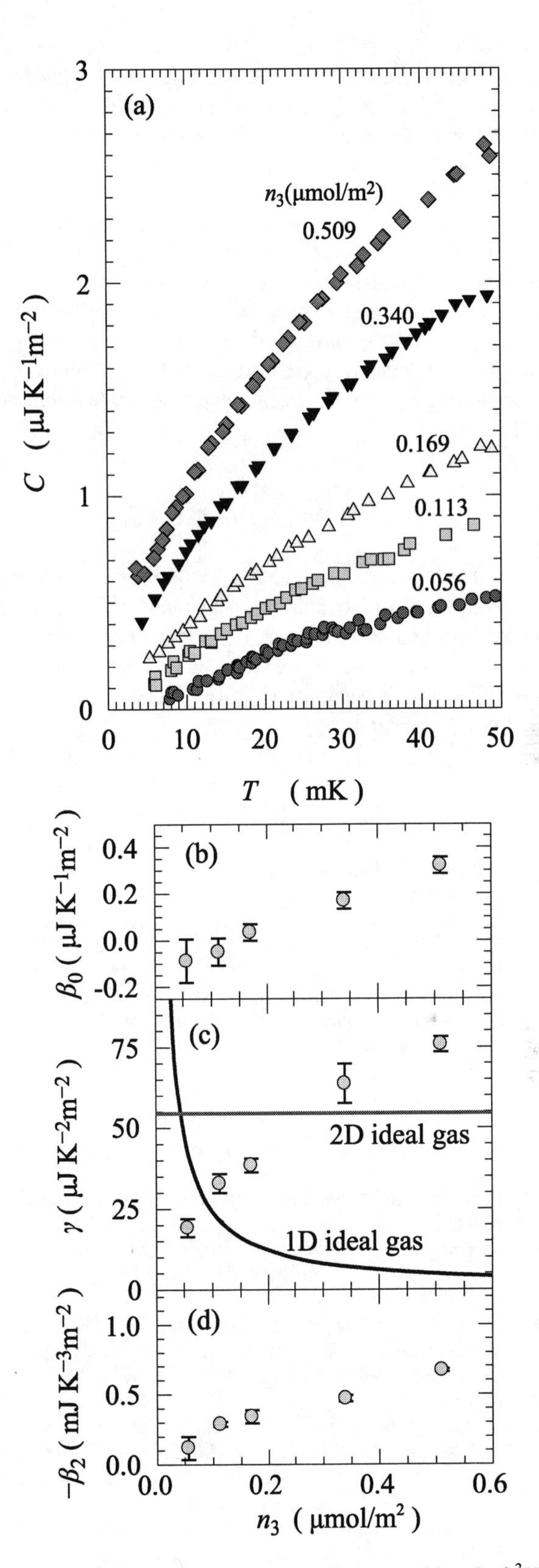

FIGURE 10. (a) Low temperature heat capacities C of ^{3}He adsorbed on 1D 28 Å pores preplated with ^{4}He layer. (b-d) Fitting parameters of $C = \beta_0 + \gamma T + \beta_2 T^2$. γT term likely corresponds to the Sommerfeld term of a degenerate ^{3}He fluid.

ground state and that there remain only thermal phonon excitations along the pores. Superfluidity of ^{4}He adatoms inside the pores was observed by a torsional oscillator. The superfluidity is not observed throughout the Bose fluid region.

^{3}He fluid tubes were made in the 1D nanopores preplated with ^{4}He layers. At a low density, a characteristic Schottky-like peak indicates a dimensional crossover from a 2D Boltzmann gas at high temperatures to a 1D fluid at low temperatures. A degenerate state of the ^{3}He fluid was suggested by the heat capacity asymptotically approaching a linear T dependence below a few tens of mK.

ACKNOWLEDGMENTS

We thank T. Kurokawa and S. Inoue for help with the experiments. We also thank S. Inagaki and Y. Fukushima for providing the nanopores. This research was partly supported by a Grant-in-Aid for Scientific Research from the Ministry of Education, Culture, Sports, Science and Technology of Japan.

REFERENCES

1. F. Ziouzia, H. Patel, J. Nyéki, B.P. Cowan, and J. Saunders, *J. Low Temp. Phys.*, **134**, 79–84 (2004).
2. N. Wada, J. Taniguchi, H. Ikegami, S. Inagaki, and Y. Fukushima, *Phys. Rev. Lett.*, **86**, 4322–4325 (2001).
3. J. Taniguchi, A. Yamaguchi, H. Ishimoto, H. Ikegami, T. Matsushita, N. Wada, S.M. Gatica, M.W. Cole, F. Ancilotto, S. Inagaki, and Y. Fukushima, *Phys. Rev. Lett.*, **94**, 065301 (2005).
4. S. Inagaki, Y. Fukushima, and K. Kuroda, *J. Chem. Soc. , Chem. Commun.*, 680–682 (1993).
5. H. Ikegami, T. Okuno, Y. Yamato, J. Taniguchi, N. Wada, S. Inagaki, and Y. Fukushima, *Phys. Rev. B*, **68**, 092501 (2003).
6. E. Cheng, and M.W. Cole, *Phys. Rev. B*, **38**, 987–995 (1988).
7. C. Kittel and H. Kroemer, *Thermal Physics*, W.H. Freeman and Company, San Fransisco, 1980, pp. 152–182.
8. N. Wada, J. Taniguchi, T. Matsushita, R. Toda, Y. Matsushita, H. Ikegami, M. Hieda, A. Yamaguchi, and H. Ishimoto, *J. Phys. Chem. Solids*, **66**, 1513–1516 (2005).
9. Y. Matsushita, R. Toda, M. Hieda, T. Matsushita, and N. Wada, *J. Phys. Chem. Solids*, **66**, 1521–1522 (2005).
10. Y. Matsushita, J. Taniguchi, R. Toda, T. Matsushita, M. Hieda, and N. Wada, These proceedings.
11. N. Wada, A. Inoue, H. Yano, and K. Torii, *Phys. Rev. B*, **52**, 1167–1175 (1995).
12. M.W. Cole and E.S. Hernández, *Phys. Rev. B*, **65**, 092501 (2002).
13. Y. Yamato, H. Ikegami, T. Okuno, J. Taniguchi, and N. Wada, *Physica B*, **329-333**, 284–285 (2003).
14. H. Ikegami, Y. Yamato, T. Okuno, J. Taniguchi, and N. Wada, *J. Low Temp. Phys.*, **138**, 171–176 (2005).
15. S.Q. Murphy and J.D. Reppy, *Physica B*, **165 & 166**, 547–548 (1990).
16. J. Yuyama, and T. Watanabe, *J. Low Temp. Phys.*, **48**, 331–348 (1982).
17. R. Toda, T. Yamada, J. Taniguchi, T. Matsushita, and N. Wada, *Physica B*, **329-333**, 282–283 (2003).
18. R. Toda, J. Taniguchi, R. Asano, T. Matsushita, and N. Wada, *J. Low Temp. Phys.*, **138**, 177–182 (2005).
19. T. Yamada, R. Toda, Y. Matsushita, T. Matsushita, and N. Wada, *J. Low Temp. Phys.*, **134**, 601–606 (2004).
20. R. Toda, M. Hieda, T. Matsushita, and N. Wada, These proceedings.
21. Y. Matsushita, T. Matsushita, R. Toda, M. Hieda, and N. Wada, These proceedings.
22. M.W Cole, F. Ancilotto, and S.M. Gatica, *J. Low Temp. Phys.*, **138**, 195–200 (2005).
23. Y Okaue and D.S. Hirashima, *J. Phys. Chem. Solids*, **66**, 1525–1526 (2005).
24. Y. Okaue, Y. Saiga, and D.S. Hirashima, These proceedings.

^{3}He Fluid Formed in One-Dimensional 1.8 nm Pores Preplated with ^{4}He Layer

Yuki Matsushita, Taku Matsushita, Ryo Toda, Mitsunori Hieda and Nobuo Wada

Department of Physics, Nagoya University, Nagoya 464-8602, Japan

Abstract. In one-dimensional (1D) nanopores 1.8 nm in diameter preplated with 1.2 atomic layers of ^{4}He, we have made ^{3}He film tubes whose diameter is estimated to be about 1.0 nm. At a low ^{3}He density, the heat capacity shows that of a 2D Boltzmann gas at high temperatures followed by a Schottky-like peak at about 300 mK. This result can be understood in terms of dimensional crossover from 2D to 1D with decreasing temperature. The peak appears at a temperature about 0.3 times the energy gap between the ground state and the first excited states of the azimuthal motion in the ^{3}He fluid tubes.

Keywords: low-dimensional quantum fluid, degenerate helium 3, nanopores
PACS: 67.55.Cx, 67.70.+n, 68.65.-k

INTRODUCTION

Two-dimensional (2D) ^{3}He films have been made on the free surface of bulk ^{4}He liquid and on a flat solid surface preplated with a ^{4}He layer. The heat capacity of these films shows a typical temperature dependence of either a 2D Boltzmann gas or a degenerate Fermi liquid. The two dimensionality is associated with the motion perpendicular to the plane being in the ground state. Recently, ^{3}He film tubes have been realized in one-dimensional (1D) nanopores, 2.8 and 2.2 nm in diameter, preplated with a ^{4}He layer [1–3]. The observed heat capacities at low densities are understood in terms of dimensional crossover from a 2D Boltzmann gas to a 1D gas with decreasing temperature. The crossover temperature depends on the pore diameter. To further examine the effect of the pore size on the thermodynamics of ^{3}He, we have employed 1D nanopores of a smaller diameter, 1.8 nm, in the present study.

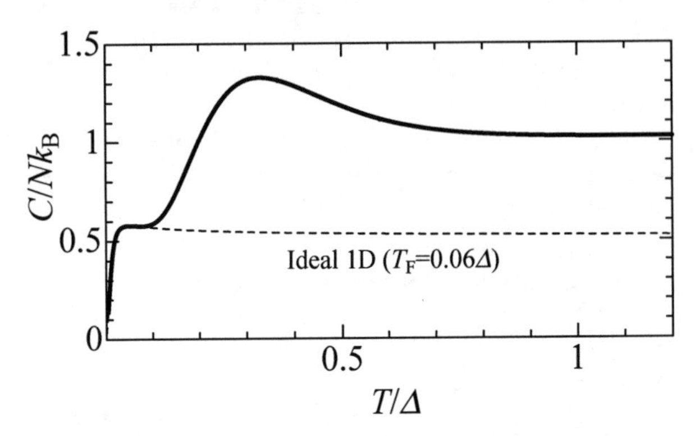

FIGURE 1. Specific heat C/Nk_B calculated for a non-interacting ^{3}He gas adsorbed on a 1D nanopore. $k_B\Delta$ is the energy gap between the ground state and the first excited states of the azimuthal motion. At a low density where the Fermi temperature becomes $T_F = 0.06\Delta$, C approaches that of a 2D Boltzmann gas at $T > \Delta$. The peak appearing around 0.3Δ indicates a dimensional crossover from 2D to 1D with decreasing T. At $T < 0.1\Delta$, C becomes that of an ideal 1D Fermi gas.

DIMENSIONAL CROSSOVER

FSM-16 is a silicate containing uniform 1D nanopores whose diameter can be selected between 1.5 and 4.8 nm during the synthesis [4]. The length of the pores is about 300 nm. For pores with diameters larger than 1.8 nm, we found a fomation of a uniform layer of ^{4}He by analyzing the pressure for adsorption [5]. In a previous work involving the 2.8 nm diameter nanopores [2], the pores were preplated with 23.4 μmol/m^2 (1.4 atomic layer) of ^{4}He. Since the thickness of the ^{4}He layer was estimated to be 0.55 nm, the ^{3}He films were tubes about 1.7 nm in diameter.

The heat capacity of a ^{3}He film tube is semiquantitatively explained by a calculation for a non-interacting ^{3}He gas adsorbed on a 1D nanopore [2, 3]. In the cross section of the nanopore, the motion of a ^{3}He atom has discrete energy levels. We define the gap energy between the ground state and the first excited states as $k_B\Delta$, where k_B is the Boltzmann constant. The heat capacity of the ^{3}He gas at a low density ($T_F = 0.06\Delta$ where T_F is the Fermi temperature) is shown in Fig. 1. At $T > \Delta$, the heat capacity becomes that of a 2D Boltzmann gas.

CP850, *Low Temperature Physics: 24th International Conference on Low Temperature Physics;*
edited by Y. Takano, S. P. Hershfield, S. O. Hill, P. J. Hirschfeld, and A. M. Goldman

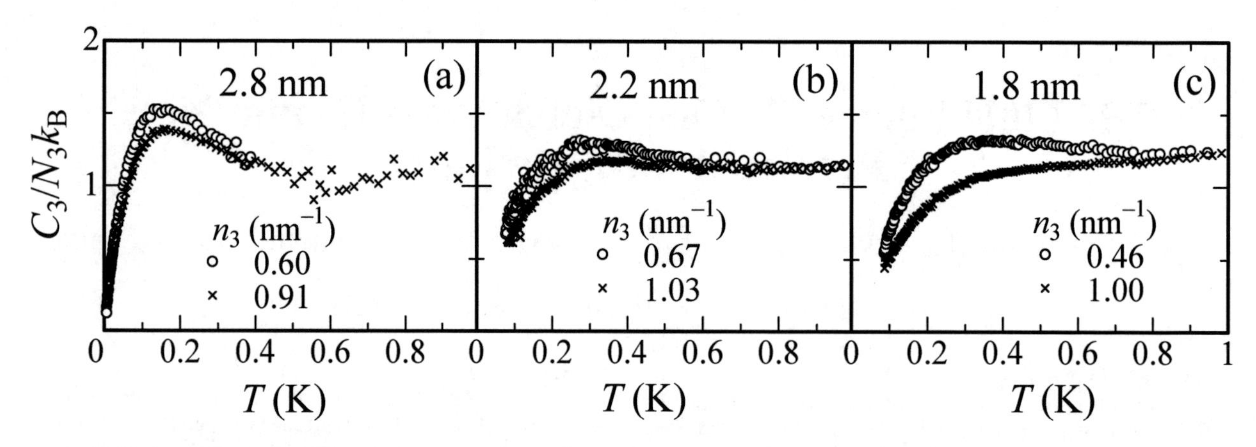

FIGURE 2. Specific heats of ^{3}He adsorbed on 1D nanopores preplated with ^{4}He layers. The pore diameters are (a) 2.8 nm, (b) 2.2 nm, and (c) 1.8 nm, respectively. n_3 is the linear density of ^{3}He atoms along the pore.

With decreasing T, the heat capacity shows a peak at $T = 0.3\Delta$ and, at $T < 0.1\Delta$, becomes that of an ideal 1D Fermi gas. The Schottky-like peak indicates a dimensional crossover. In the experimental results for the 2.8 nm pores shown in Fig. 2(a), the 2D Boltzmann gas heat capacity at $T > \Delta$ probably explains the data above about 0.5 K. The Schottky-like peak at $T = 0.3\Delta$ corresponds to the peak at 120 mK. From the peak position, the gap energy Δ is estimated to be about 400 mK. At high densities for which $T_F > \Delta$, the peak disappears because the discrete levels of the azimuthal motion are occupied even at absolute zero.

Figure 2(b) shows the results for the 2.2 nm pores preplated with 16.9 ± 0.3 μmol/m^2 (1.1 atomic layer) of ^{4}He. The diameter of the ^{3}He film tube is estimated to be about 1.3 nm, from the ^{4}He thickness 0.44 nm. In comparison with the results for the 2.8 nm pores, the Schottky-like peak has shifted to about 280 mK, reflecting a larger Δ in the smaller pores.

These results lead to a question: is there a dimensional crossover of ^{3}He gas for pores of an even smaller diameter than these.

RESULTS FOR 1.8 NM PORES

We have studied ^{3}He adsorbed on 1.8 nm pores of FSM-16 preplated with ^{4}He. In these pores, ^{4}He forms a uniform layer up to a coverage of 1.4 atomic layers [5]. When the amount of ^{4}He was less than one atomic layer, the observed ^{3}He heat capacity became much smaller than that of a gas state, indicating localization of the ^{3}He atoms.

When the pores are preplated with 14.5 μmol/m^2 (1.2 atomic layers) of ^{4}He, the heat capacity of ^{3}He is as shown in Fig. 2(c). In the pores, the ^{3}He adatoms form a tube 1.0 nm in diameter which is estimated from the ^{4}He thickness of 0.38 nm. The ^{3}He nanotube behaves like a gas and shows a Schottky-like peak in the heat capacity [Fig. 2(c)]. On the other hand, the peak temperature is not much higher than found in the 2.2 nm pores. To explain these results more quantitatively, we have to take into account the role of interaction.

In summary, we have observed a gas-like state of the ^{3}He adsorbed on the 1.8 nm pores preplated with ^{4}He. It would be of interest to study the degenerate state of the ^{3}He gas at low temperatures where the 1D condition is well satisfied.

ACKNOWLEDGMENTS

We thank S. Inoue and T. Kurokawa for technical support. This research was partly supported by a Grant-in-Aid for Scientific Research from the Ministry of Education, Culture, Sports, Science and Technology, Japan.

REFERENCES

1. J. Taniguchi, H. Ikegami, and N. Wada, *Physica B*, **329–333**, 274–275 (2003).
2. J. Taniguchi, A. Yamaguchi, H. Ishimoto, H. Ikegami, T. Matsushita, N. Wada, S. M. Gatica, M. W. Cole, F. Ancilotto, S. Inagaki, and Y. Fukushima, *Phys. Rev. Lett.*, **94**, 065301–1–4 (2005).
3. Y. Matsushita, J. Taniguchi, A. Yamaguchi, H. Ishimoto, H. Ikegami, T. Matsushita, N. Wada, S. M. Gatica, M. W. Cole, and F. Ancilotto, *J. Low Temp. Phys.*, **138**, 211–216 (2005).
4. S. Inagaki, Y. Fukushima, and K. Kuroda, *J. Chem. Soc., Chem. Commun.*, **22**, 680–681 (1993).
5. H. Ikegami, T. Okuno, Y. Yamato, J. Taniguchi, N. Wada, S. Inagaki, and Y. Fukushima, *Phys. Rev. B*, **68**, 092501–1–4 (2003).

Vapor Pressure Measurement for ^{4}He Films Adsorbed on 2D Mesoporous Hectorite

Ryota Asano, Ryo Toda, Yuki Matsushita, Mitsunori Hieda, Taku Matsushita and Nobuo Wada

Department of Physics, Nagoya University, Furo-cho, Chikusa-ku, Nagoya 464-8602, Japan

Abstract. The vapor-pressure measurement for adsorbed films is equivalent to the measurement of the chemical potential. By the measurement of ^{4}He films adsorbed on two-dimensional (2D) mesoporous Hectorite, we obtained the 2D isothermal compressibility, the isosteric heat, and the effective thickness deduced from the FHH model, mainly above a coverage of 15 μmol/m^2. The compressibility shows two dips at $n_1 = 17.5\pm0.5$ μmol/m^2 and $n_2 = 22.7\pm0.5$ μmol/m^2 which correspond to the first layer completion and appearance of the quantum-Bose-fluid layer, respectively. For the quantum-Bose-fluid layer, the phonon velocity deduced from the compressibility is on the order of 100 m/s, and it reasonably agrees with that obtained from the 2D phonon heat capacity.

Keywords: vapor pressure for adsorption, ^{4}He film, Hectorite
PACS: 67.70.+n, 68.03.-g, 67.40.Kh

Atomically thin ^{4}He and ^{3}He films on flat substrates, such as graphite and Hectorite, have been studied extensively for the study of two-dimensional (2D) quantum fluids system [1-4]. As the 2D nature relates intimately to the state of the films, it is important to obtain basic information about the film growth. One of excellent methods to study film growth is vapor-pressure measurement [5] for adsorbed film which is equivalent to the measurement of the chemical potential. From the analysis, we can obtain various useful quantities, *e.g.* 2D isothermal compressibility, isosteric heat of adsorption, and film thickness. In this paper, we report pressure measurements of ^{4}He adsorbed on Hectorite.

Hectorite (OH-02) used in this experiment is a 2D porous material made of layered silicate [1]. It has plane spaces of 17-20 Å between smectite layers of 9.6 Å thickness, which are supported by smectite fragments. The sample cell containing Hectorite is the same as that built for the heat-capacity measurements in the past experiment [1]. The BET surface area S by N_2 adsorption isotherm at 77 K is 255 ± 10 m^2. The vapor pressure of ^{4}He is measured by an in-situ capacitive pressure gauge with a metalized 125 μm-thick Kapton diaphragm [5]. The resolution was 0.5 Pa at low temperature. The pressure was calibrated against the saturated vapor pressure of ^{4}He.

Figure 1 shows various quantities obtained from the measurement of the vapor pressure P. The 2D isothermal

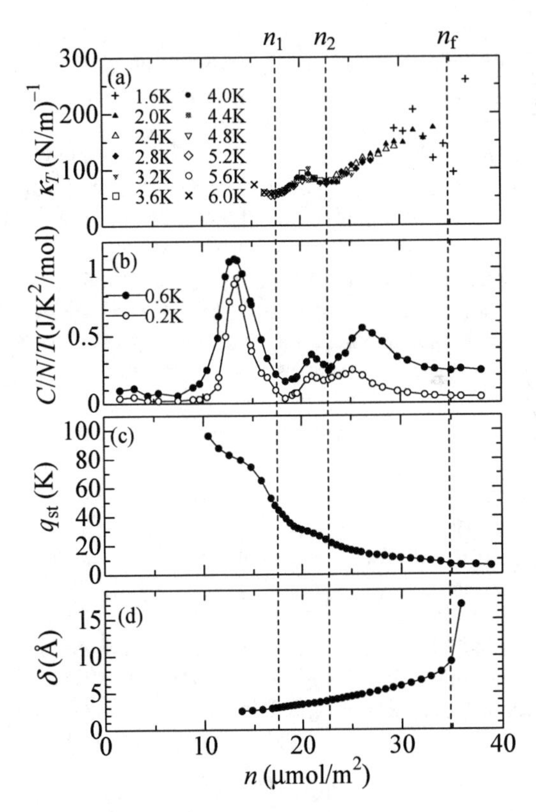

FIGURE 1. Various quantities of ^{4}He adsorbed on Hectorite. (a) Two-dimensional isothermal compressibility. (b) Isothermal heat capacities in the previous study [1]. (c) Isosteric heat of adsorption. (d) Film thickness estimated by the FHH model. n_1, n_2, and n_f indicate the coverage of the first layer completion, the appearance of the quantum-Bose-fluid layer, and the full pore.

CP850, *Low Temperature Physics: 24th International Conference on Low Temperature Physics;*
edited by Y. Takano, S. P. Hershfield, S. O. Hill, P. J. Hirschfeld, and A. M. Goldman

compressibility is defined by

$$\kappa_T = \frac{1}{n^2}\left(\frac{\partial n}{\partial \mu_a}\right)_T = \frac{1}{n^2 k_B T}\left(\frac{\partial n}{\partial \ln P}\right)_T, \quad (1)$$

where n is the coverage determined by dividing total adsorbed amount N by S, μ_a is the chemical potential of adsorbate, and k_B is the Boltzmann constant. As shown in Fig. 1(a), the compressibility shows two dips at $n_1 = 17.5 \pm 0.5$ μmol/m^2 and $n_2 = 22.7 \pm 0.5$ μmol/m^2 which correspond to the first layer completion and the appearance of a fluid layer, respectively. The heat capacity isotherm in the past experiment [1] also shows a minimum at n_2 as shown in Fig. 1(b), and above n_2 a qualitative difference between ^{4}He and ^{3}He due to the difference of the quantum statics was observed. Thus n_2 is the coverage of the appearance of a quantum-Bose-fluid layer. (Note that the quantum Bose fluid does not mean the superfluid. n_2 is slightly below the critical coverage of the superfluid $n_{on} = 24.3 \pm 0.3$ μmol/m^2 determined by the torsional oscillator study [1].)

Another thermodynamic quantity calculated from the pressure is the isosteric heat of adsorption q_{st} defined by $q_{st} = -\partial \ln P/\partial(1/T) - 5T/2$. q_{st} shows a small substeps at n_1 and n_2, as shown in Fig. 1(c). At n_1, a similar value of q_{st} is reported on graphite [6]. q_{st} approaches the constant value of 6 K at $n_f = 34.8 \pm 0.4$ μmol/m^2, which is almost equal to 7 K of the bulk ^{4}He liquid. This indicates that the bulk liquid appears above n_f.

Figure 1(d) shows the effective film thickness calculated by the Frenkel-Halsey-Hill (FHH) model for an ideal planar substrate, which is given by $\delta = \{a/[T\ln(P_0/P)]\}^{1/3}$ where P_0 is the saturated vapor pressure of ^{4}He at T. We have assumed that the adsorption coefficient a is the value for a glass substrate 1100 KÅ^3 [7]. At n_f, a sudden increase of the thickness is observed. This is understood that the pores are filled with ^{4}He at n_f above which the dosage adsorbs mostly on the outside surface of the grain having much smaller area than that inside pores.

When the quantum fluid layer is assumed to be formed independently on the underlying inert layer, the phonon velocity in the fluid layer is deduced from the 2D isothermal compressibility of the fluid κ_{Tf}, which is given by

$$v_p = \sqrt{\frac{\gamma}{\kappa_{Tf}\rho_f}}, \quad (2)$$

where $\gamma = \kappa_{Tf}/\kappa_{Sf}$ (κ_{Sf} is the adiabatic compressibility of the fluid.) and ρ_f is the areal mass density of the fluid. κ_{Tf} can be calculated by the replacement of n with $n - n_2$ in eq. (1). For the fluid layer, γ has a value in the range of $1 \leq \gamma \leq 2$. The phonon velocity in the case $\gamma = 1$ is shown in Fig. 2. When $\gamma = 2$ in the case of the 2D ideal gas, the velocity indicates slightly larger value by a factor of 1.4.

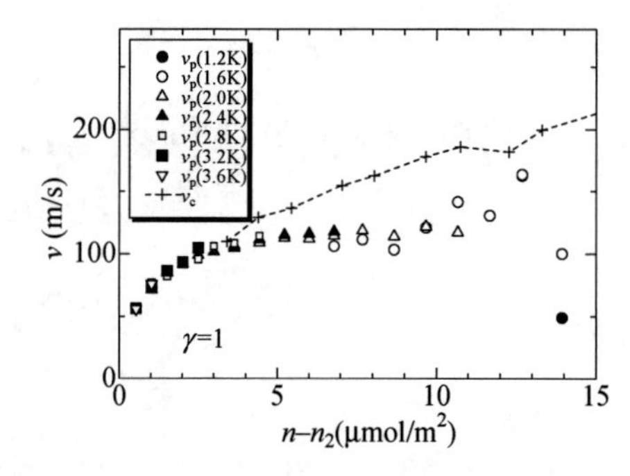

FIGURE 2. Phonon velocity v_p of the quantum Bose fluid layer deduced from the two-dimensional compressibility . For comparison, the phonon velocity v_c in the superfluid layer estimated from the 2D phonon heat capacity [1] is also shown.

Therefore we conclude that the phonon velocity deduced from κ_{Tf} is on the order of 100 m/s.

For comparison, the phonon velocity in the superfluid layer estimated from the data of the past heat-capacity measurements [1] by the 2D Debye model is also shown in Fig. 2. The velocity v_c is given by

$$v_c = \sqrt{\frac{3\zeta(3)Sk_B{}^3}{\pi\hbar^2\beta}}, \quad (3)$$

where ζ is the Riemann zeta function, and β is the coefficient of T^2 term of the heat capacity. v_c reasonably agrees with that obtained from κ_{Tf}. This indicates the validity of the analysis for the phonon velocity deduced from κ_{Tf}.

In this paper, we presented the vapor-pressure measurements of ^{4}He adsorbed on 2D-mesoporous Hectorite. From the analysis, the various useful quantities were obtained. The coverage of the first layer completion, the appearance of the quantum-Bose-fluid layer, and the full pore were identified. The validity of the analysis for the phonon velocity deduced from κ_{Tf} was indicated.

We acknowledge M. Katsuki and T. Kurokawa for technical support. We also acknowledge K. Torii for providing the substrate.

REFERENCES

1. N. Wada *et al.*, *Phys. Rev. B*, **52**, 1167–1175 (1995).
2. N. Wada, H. Yano, and Y. Karaki, *J. Low Temp. Phys.*, **113**, 317–322 (1998).
3. D. Tsuji *et al.*, *J. Low Temp. Phys.*, **134**, 31–36 (2004).
4. A. Casey *et al.*, *Phys. Rev. Lett.*, **90**, 115301 (2003).
5. H. Ikegami *et al.*, *Phys. Rev. B*, **68**, 092501 (2003).
6. D. S. Greywall and P. A. Busch, *Phys. Rev. Lett.*, **67**, 3535–3538 (1991).
7. E. Cheng and M. W. Cole, *Phys. Rev. B*, **38**, 987–995 (1988).

The First Layer of ^{4}He, H_2, and Ne Adsorbed on HiPco™ Carbon Nanotube Bundles

O. E. Vilches*, S. Ramachandran*, T. A. Wilson*†, and J. G. Dash*

*Department of Physics, University of Washington, Seattle, WA 98195-1560, USA
†Department of Physics and Astronomy, University of Massachusetts, Amherst, MA 01003, USA

Abstract. We summarize results from AC and DC heat capacity measurements of ^{4}He, H_2, and Ne adsorbed on HiPco™ purified, closed-end single-wall carbon nanotube bundles (SWNTB) for the first adsorbed layer. We find two regions in the coverage domain: below ≈1/3 monolayer the adsorbate occupies high binding energy sites, mostly the external grooves of the bundles, while above ≈1/3 monolayer the external graphene surface is covered. No phase transitions have been observed at any temperature for all the adsorbates, a range of T where two-dimensional phases and phase transitions are seen for the same adsorbates deposited on exfoliated graphite.

Keywords: Carbon Nanotubes, Adsorption, Heat Capacity, One and Two Dimensions
PACS: 05.70 Np, 67.70 +n, 68.43 Fg

INTRODUCTION

Single wall carbon nanotube bundles (SWNTB) have been proposed and are being used as substrates for physisorption of atoms and molecules [1-8]. Interest stems from the possibility of realizing on them one- and two-dimensional (1D, 2D) forms of matter and studying their evolution as a function of temperature (T, in K) and coverage (n_{ads}, in mmole per gram of SWNTB). ^{4}He, H_2 and Ne are some of the less massive and interacting atoms/molecules, and their single layer phases and phase transitions when adsorbed on graphite, a surface closely related to SWNTB, have been extensively studied. ^{4}He and H_2 have prominent commensurate phases with the substrate [9-12], while Ne is incommensurate with a solid-liquid-vapor triple point at 13.5 K [13,14]. Here we report on the similarities on the adsorption pattern and the lack of phase transitions as a function of T for the three substances on SWNTB at all coverages.

EXPERIMENTAL

We have measured the AC heat capacity (^{4}He and H_2) and the DC heat capacity (Ne) of many films on SWNTB. We used purified HiPco™ SWNTB as adsorber [15], pressed between soldered thin wall copper calorimeter shells. The AC cell had 106 mg and the DC cell had 413 mg of SWNTB from the same batch, Cernox thermometers and electric heaters; they were connected to an external gas dosing system. Initial AC measurements were done at 1 Hz [16]. The majority of the ^{4}He and all of the H_2 measurements were done at 0.1 Hz [17]. The earlier 1 Hz data can be scaled well with the 0.1 Hz data by multiplying the 1 Hz heat capacity for all n_{ads} and T by 3.8. Our AC measurements do not yield an absolute value of heat capacity (C) due to the internal time constant of the complex calorimeter cell. The DC measurements give the absolute value of C.

Helium. Earlier results from volumetric isotherms and AC calorimetry at f = 1 Hz have been published [16]. Additional measurements have been performed at both 1 Hz and 0.1 Hz in the 1.7 K<T<7 K range. The general behavior of the heat capacity as a function of T and n_{ads} are identical at both frequencies for equal or very similar coverages. At the lowest densities (≈0.02 monolayer) the heat capacity of the film increases almost linearly with T below 4 K, but with a negative intercept if extrapolated to 0K. For n_{ads}>8.4 mmole/g the heat capacity between 1.7 K and 4 K follows a T^2 behavior. Monolayer completion, from adsorption isotherms, occurs at n_{ads}≈10.9 mmole/g.

Hydrogen. Experiment spanned the first layer and the beginning of the second layer, for 1.7 K<T<8 K for most runs. For surface n_{ads} below 0.02 of a monolayer

CP850, *Low Temperature Physics: 24th International Conference on Low Temperature Physics;*
edited by Y. Takano, S. P. Hershfield, S. O. Hill, P. J. Hirschfeld, and A. M. Goldman

the heat capacity of the film below 3 K is essentially zero. By ≈1.1 mmole/g deposited C grows exponentially with T, then bends towards a constant value above 6 K. For n_{ads}>7.2 mmole/g the low T heat capacity is proportional to T^2. Monolayer completion is at ≈9.5 mmole/g.

Neon. The DC measurements span 1.7 K<T<18 K, and 0.22 mmole/g<n_{ads}<13.5 mmole/g, near monolayer completion. The low T heat capacity increases at a higher than linear rate from the lowest to the highest n_{ads}, almost saturating at 2.5<C/nR<3.2 for T>16 K. It is quite remarkable that for all doses C/nR tends to about 3 rather than 0.5 (1D classical gas), 1 (1D solid) or 2 (2D classical solid). There is no evidence of the 2D melting transition at 13.5 K [13,14] for what we would expect to be the 2D-like coating of the external graphene surface of the bundles. Monolayer completion has not been determined yet.

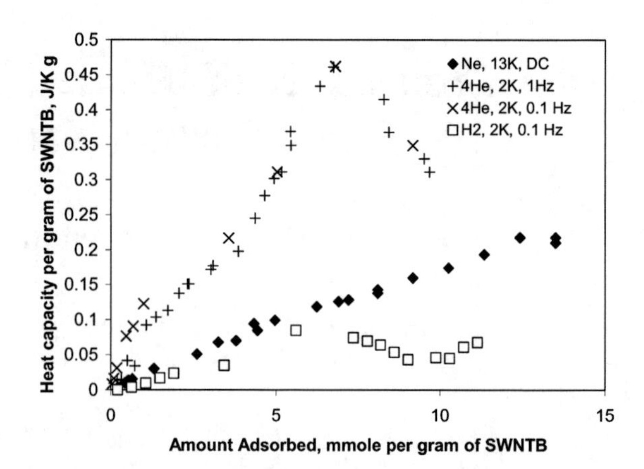

FIGURE 1. Heat capacity isotherms from C-T-n_{ads} grids. Vertical axis corresponds to Ne data. ^{4}He data at 1 Hz has been multiplied by 380, ^{4}He and H_2 data at 0.1 Hz by 100.

We acknowledge support from NSF DMR 0245423 and from the Bosack-Kruger Foundation.

DISCUSSION

The heat capacity data for ^{4}He, H_2, and Ne have many similarities. We show in Fig. 1 one total heat capacity isotherm *vs.* n_{ads} for each species, normalized to the mass of SWNTB in the respective cells. Each point in Fig. 1 comes from a C vs. T run. The y-axis corresponds to the Ne data. The isotherms for ^{4}He and H_2 (at lower T) have been scaled as stated in the caption. The "dip" in the H_2 data at 9.5 mmole/g corresponds to monolayer completion.

The heat capacity isotherms clearly show the same pattern of growth for the three species studied. First, higher binding sites are filled until all "linear" sites are occupied. This occurs a little below 1/3 monolayer coverage for ^{4}He and H_2. Graphite-like adsorption starts after the inflexion at 4 mmole/g for He [16], slightly less for H_2 [17] and after 7.5 mmole/g for Ne. The n_{ads} ratio for this feature for Ne to ^{4}He is rather large and should be studied by scattering techniques. The low temperature T^2 dependence for ^{4}He and H_2 starts after a maximum heat capacity in the isotherms of Fig. 1 has been passed; this is similar to adsorption on graphite, where the 2D quantum solids are highly compressible and the decrease in C with surface density is indicative of a rapid growth in their Debye temperatures (Θ_D) with n_{ads}. The low T heat capacity of Ne is ~T^2 at high n_{ads}, which yields Θ_D≈43 K, lower than the 52 K found for 2D Ne/graphite at monolayer completion. Still, there is no evidence of a melting transition up to 18 K in this first layer. This may be the result of dimensionality, or other properties of adsorption on this complex surface [18-20].

REFERENCES

1. M. M Calbi *et al.*, *Rev. Mod. Phys.* **73**, 857 (2001).
2. A. D. Migone and S. Talapatra, in *Encyclopedia of Nanoscience and Nanotechnology*, H. S Nalwa, Ed., American Scientific Publishers, **4**, 749-767 (2004).
3. G. Stan and M. W. Cole, *J. Low Temp. Phys.*,**110**, 539 (1998).
4. M. C. Gordillo, J. Boronat and J. Casulleras, *Phys. Rev. B*, **61**, R878 (2000), **68**, 125421 (2003).
5. M. Aichinger *et al., Phys. Rev. B*, **70**, 155412 (2004).
6. W. Teizer *et al., Phys. Rev. Lett.*, **82**, 5305 (1999), **84**, 1844(E) (2004).
7. L. C. Lasjausnias *et al., Phys. Rev. Lett.*, **91**, 025901 (2003).
8. M. Bienfait *et al., Phys. Rev. B*, **70**, 035410 (2004).
9.. R. E. Ecke *et al.,Phys. Rev. B*, **31**, 448 (1985).
10. D. S. Greywall, *Phys. Rev. B*, **47**, 309 (1993).
11. M. E. Pierce and E. Manousakis, *Phys. Rev. B*, **62**, 5228 (2000).
12. H. Wiechert, in Landolt-Börnstein *Numerical Data and Functional Relationships in Science and Technology, Group III: Condensed Matter*, **42A**, 283 (2003).
13. G. B. Huff and J. G. Dash, *J. Low Temp. Phys.*, **24**, 155 (1976).
14. R. E. Rapp, E. P. de Souza and E. Lerner, *Phys. Rev. B*, **24**, 2196 (1981).
15. Carbon Nanotechnologies Inc.
16. T. Wilson and O. E. Vilches, *Physica B*, **329-333**, 278 (2003).
17. S. Ramachandran *et al., J. Low Temp. Phys.* **134**, 115 (2004).
18. J. M. Phillips and J. G. Dash, *J. Stat. Phys, (in print).*
19. M. M. Calbi *et al.,, J. Chem. Phys.***115**, 9975 (2001).
20. F. Ancilotto *et al., Phys. Rev. B*, **70**, 165422 (2004).

Sound Velocity Measurements of Nuclear-Ordered Solid ^{3}He in the Low Field Phase

Satoshi Sasaki*, Daisuke Takagi*, Atsuyoshi Nakayama*, Yutaka Sasaki*,†
and Takao Mizusaki*

*Department of Physics, Graduate School of Science, Kyoto University, Kyoto 606-8502, Japan
†Research Center for Low Temperature and Materials Sciences, Kyoto University, Kyoto 606-8502, Japan

Abstract. We have measured, along the melting curve, the magnetic-field dependences of the velocities of the longitudinal and two transverse sound modes in the low field phase (U2D2) of nuclear-ordered solid ^{3}He crystals with a single magnetic domain. We determined the elastic-stiffness tensors $c_{ij}(T,B)$ along the melting curve at $T = 0.5$ mK. Since the U2D2 phase has a tetragonal symmetry, we extracted six components of the nuclear-spin elastic-stiffnesses tensor $\Delta c_{ij}^{\rm N}(0.5\ {\rm mK},B)$. We obtained the generalized Grüneisen constants $\Gamma_{ij}^{1/\chi}$ for the magnetic susceptibility χ from $\Delta c_{ij}^{\rm N}(0.5\ {\rm mK},B)$, which were proportional to B^2.

Keywords: U2D2 solid ^{3}He; Ultrasound; Nuclear magnetism; Multiple spin exchange
PACS: 62.20.Dc; 67.80.Cx; 67.80.Jd

The magnetism of nuclear-ordered solid ^{3}He in the low field phase (U2D2) and the high field phase (CNAF) is described by the multiple-spin exchange model (MSE model) [1]. Sound experiment is very useful in investigating the nuclear-spin system of solid ^{3}He at temperatures below the nuclear-ordering temperature, which is 0.93 mK in zero field at the melting pressure, because the ultrasound couples strongly to the nuclear spins through the large Grüneisen constants $\gamma^J (= d\ln J/d\ln V)$ for the exchange frequencies J.

We measured the change of sound velocity $\Delta v/v$ for both longitudinal and transverse sounds along the melting curve for a single domain crystal as a function of temperature T and magnetic field B. The experimental method and setup of the sample cell are the same as those described in [2]. For samples with different crystal orientations with respect to a sound propagation direction, we evaluated the elastic-stiffness tensors $c_{ij}(T,B)$, which can be divided into three parts as

$$c_{ij}(T,B) = c_{ij}^{\rm L} + c_{ij}^{\rm N} = c_{ij}^{\rm L} + c_{ij}^{\rm N}(0,0) + \Delta c_{ij}^{\rm N}(T,B) \qquad (1)$$
$$= c_{ij}^{0} + \Delta c_{ij}^{\rm N}(T,B),$$

where L and N represent the lattice and nuclear-spin contributions, respectively. $\Delta c_{ij}^{\rm N}(T,B)$ are the temperature- and field-dependent deviations of the nuclear-spin elastic-stiffness tensors from $c_{ij}^{\rm N}(0,0)$ ($\ll c_{ij}^{\rm L}$). We determined six kinds of $\Delta c_{ij}^{\rm N}(T,B)$ for the U2D2 phase of solid ^{3}He crystal from the temperature- and field-dependent changes of sound velocity $\Delta v/v$ along the melting curve.

In the U2D2 phase, $\Delta c_{ij}^{\rm N}(T,B)$ are related to the generalized Grüneisen constants $\Gamma_{ij}^{X} (\equiv (\partial^2 X/\partial\varepsilon_i\partial\varepsilon_j)/X)$ for X and the temperature- and field-dependent nuclear-spin exchange energy $\Delta U_{ex}^{\rm N}(T,B)$ as follows:

$$\Delta c_{ij}^{\rm N}(T) \equiv \Delta c_{ij}^{\rm N}(T, B\simeq 0) = \Gamma_{ij}^{c}\Delta U_{ex}^{\rm N}(T, B\simeq 0), \qquad (2)$$

$$\Delta c_{ij}^{\rm N}(B) \equiv \Delta c_{ij}^{\rm N}(T\simeq 0, B) = \Gamma_{ij}^{\chi^{-1}}\Delta U_{ex}^{\rm N}(T\simeq 0, B), \qquad (3)$$

where c is the spin wave velocity and χ is the magnetic susceptibility. Once we know $\Delta U_{ex}^{\rm N}$ together with the measured $\Delta c_{ij}^{\rm N}$, then Γ_{ij}^{c} and $\Gamma_{ij}^{1/\chi}$ can be evaluated. They are related to each other according to the MSE model and should be compared. Previously, we measured $\Delta c_{ij}^{\rm N}(T)$ at a constant field of $B = 0.03$ T $\simeq 0$ and obtained Γ_{ij}^{c} for c [3]. $\Delta c_{ij}^{\rm N}(T)$ were proportional to T^4. In the present work, we measured $\Delta c_{ij}^{\rm N}(B)$ at a fixed temperature, found that $\Delta c_{ij}^{\rm N}(B)$ were proportional to B^2, and then determined $\Gamma_{ij}^{1/\chi}$ for χ.

The magnetic-field dependences of $\Delta v/v$ in the U2D2 phase at a fixed temperature of $T \simeq 0.5$ mK for all sound modes (one longitudinal and two transverse modes) are proportional to B^2 as shown in Fig.1 and are analyzed as $\Delta v(B)/v = a_{\rm p}(\theta,\phi)B^2$, where $a_{\rm p}(\theta,\phi)$ depends on sound mode p and the crystal orientation (θ,ϕ). Here, θ is the angle between the anisotropy axis $\hat{\ell} \parallel \hat{z}$=[001] of U2D2 crystal and the sound propagation direction $\hat{k}$, and ϕ is the azimuthal angle of $\hat{k}$ measured from $\hat{x}$=[100].

The melting pressure changes as a function of T and B and thus the molar volume of solid, V_s, changes along the melting curve. The change of molar volume, ΔV_s, is given as follows:

$$\frac{\Delta V_s}{V_s} = \frac{\kappa\chi}{2\mu_0}\left(\frac{V_s}{\Delta V} - \gamma^{\chi}\right)B^2, \qquad (4)$$

CP850, *Low Temperature Physics: 24th International Conference on Low Temperature Physics;*
edited by Y. Takano, S. P. Hershfield, S. O. Hill, P. J. Hirschfeld, and A. M. Goldman

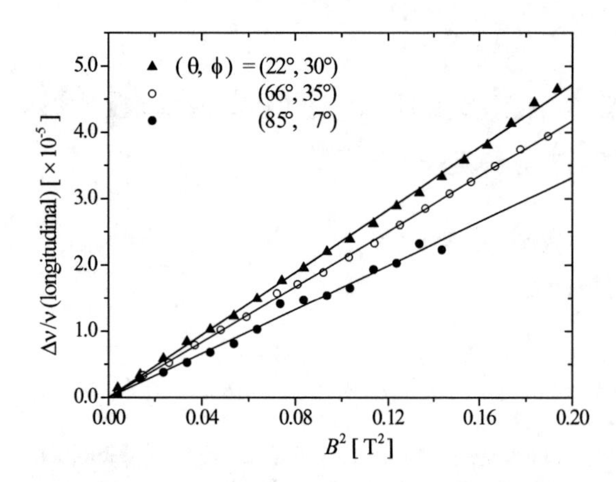

FIGURE 1. Changes of sound velocities $\Delta v(B)/v$ as a function of B^2 for the longitudinal mode for different crystal orientations. The crystal orientation (θ, ϕ) is defined in the text.

where κ is the compressibility, γ^χ is the Grüneisen constant for χ, and ΔV is the molar volume difference between solid and liquid. In the U2D2 phase in the zero temperature limit, $\Delta V_s/V_s = 6.0\times10^{-7}B^2$, where B is in T and $\gamma^\chi = 18$, $\mu_0 = 4\pi\times10^{-7}$ [H/m], $\kappa = 5.3\times10^{-8}$ Pa^{-1}, $V_s = 24.2\times10^{-6}$ [m^3/mol], $\Delta V = 1.314\times10^{-6}$ [m^3/mol] [4] and $\chi = 6.80\times10^{-5}$ (SI unit) [5].

In principle, the change in the sound velocity along the melting curve contains the uninteresting lattice contribution $\Delta v^{\rm L}(\Delta V_s)/v$ due to ΔV_s, in addition to the nuclear-spin contribution $\Delta v^{\rm N}/v$. This lattice contribution is isotropic and is given as follows:

$$\frac{\Delta v^{\rm L}(\Delta V_s)}{v} = (-\gamma^{\rm L} + \frac{1}{3})\frac{\Delta V_s}{V_s} = -1.1\times10^{-6}B^2, \qquad (5)$$

where the Grüneisen constant for Debye temperature $\gamma^{\rm L} = 2.2$ [6] and Eq. 4 has been used. This is only about 1% of the observed $\Delta v/v$. Therefore, $\Delta v/v$ can be regarded to be entirely the nuclear-spin contribution $\Delta v^{\rm N}/v$, with no need for a correction for $\Delta v^{\rm L}/v$. The six independent components of the field-dependent $\Delta c_{ij}^{\rm N}(B)$ are obtained from $\Delta v^{\rm N}/v$ for various samples and sound modes, and are formulated as $\Delta c_{ij}^{\rm N}(B) = {\rm c}_{ij}^{\rm B}B^2$. We obtained the coefficients ${\rm c}_{ij}^{\rm B}$, where i and j denote the direction of the strain (1: xx, 2: yy, 3: zz, 4: yz, 5: xy, 6: zx), as ${\rm c}_{11}^{\rm B} = 5.9$, ${\rm c}_{33}^{\rm B} = 8.6$, ${\rm c}_{12}^{\rm B} = 7.3$, ${\rm c}_{13}^{\rm B} = 7.9$, ${\rm c}_{44}^{\rm B} = 7.0$ and ${\rm c}_{66}^{\rm B} = 4.5$ [$\times10^3$ J/T^{2}m^3] within our error of 10%.

Next, in order to relate $\Delta c_{ij}^{\rm N}(B)$ with $\gamma_i^{\rm J}$, we divide the internal energy at $T \simeq 0$ into three parts as follows:

$$U = U_0^{\rm L} + U_0^{\rm N} + \Delta U_0^{\rm N}(B) = U_0 + \Delta U_0^{\rm N}(B), \qquad (6)$$

where $U_0^{\rm L}$ is the lattice part, and $U_0^{\rm N}$ and $\Delta U_0^{\rm N}(B)$ are the field-independent and field-dependent nuclear-spin parts, respectively. In the U2D2 phase, $\Delta U_0^{\rm N}(B)$ is given as

$$\Delta U_0^{\rm N}(B) = \Delta U_{\rm ex}^{\rm N} + \Delta U_{\rm z}^{\rm N} = \frac{\mu_0}{2\chi}M^2 - MB, \qquad (7)$$

where $\Delta U_{\rm ex}^{\rm N}$ is the field-dependent exchange energy due to the magnetization M at finite B, and $\Delta U_{\rm z}^{\rm N}$ is the Zeeman energy. The c_{ij} in Eq. 1 at $T \simeq 0$ are given as

$$c_{ij} = \left.\frac{\partial^2 U_0}{\partial\varepsilon_i\partial\varepsilon_j}\right|_S + \left.\frac{\partial^2\Delta U_0^{\rm N}(B)}{\partial\varepsilon_i\partial\varepsilon_j}\right|_S = c_{ij}^0 + \Delta c_{ij}^{\rm N}(B), \qquad (8)$$

where ε_i is the strain tensor in the direction of i (i = 1, 2, ..., 6) and the suffix S stands for a derivative under an adiabatic condition. M is constant with respect to ε_i under an adiabatic condition, that is, $\partial M/\partial\varepsilon_i|_S = 0$. Therefore, the generalized Grüneisen constants $\Gamma_{ij}^{1/\chi}$ are related to $\Delta c_{ij}^{\rm N}(B)$ as

$$\Delta c_{ij}^{\rm N}(B) = \left.\frac{\partial^2\Delta U_0^{\rm N}(B)}{\partial\varepsilon_i\partial\varepsilon_j}\right|_S = \left.\frac{1}{\chi^{-1}}\frac{\partial^2\chi^{-1}}{\partial\varepsilon_i\partial\varepsilon_j}\right|_S \cdot \Delta U_{\rm ex}^{\rm N}$$
$$= \Gamma_{ij}^{1/\chi}\Delta U_{\rm ex}^{\rm N} = \Gamma_{ij}^{1/\chi}\frac{\chi}{2\mu_0}B^2, \qquad (9)$$

where M in $\Delta U_{\rm ex}^{\rm N}$ is converted to B with a relation $M = \chi B/\mu_0$. From the values of ${\rm c}_{ij}^{\rm B}$ and Eq. 9, we obtain $\Gamma_{11}^{1/\chi} = 217$, $\Gamma_{33}^{1/\chi} = 319$, $\Gamma_{12}^{1/\chi} = 270$, $\Gamma_{13}^{1/\chi} = 291$, $\Gamma_{44}^{1/\chi} = 259$ and $\Gamma_{66}^{1/\chi} = 165$. These values and the relation $\Gamma_{ij}^{1/\chi} \equiv \gamma_i^\chi\gamma_j^\chi$ (assuming $\partial\gamma_j^\chi/\partial\varepsilon_i = 0$) give four independent Grüneisen constants $\gamma_i^\chi(\equiv(\partial\chi/\partial\varepsilon_i)/\chi)$ for χ as $\gamma_1^\chi = 16$, $\gamma_3^\chi = 18$, $\gamma_4^\chi = 16$, $\gamma_6^\chi = 13$. The analysis of each γ_i^χ and $\gamma_i^{\rm J}$ based on the MSE model in the same way as in Ref. [3], as well as the results and analysis in the high-field CNAF phase, will be published elsewhere.

We gratefully acknowledge useful discussions with Prof. Ohmi and Ms. Okamoto. This work is partly supported by the grants-in-aid for Scientific Research from the MEXT, Japan.

REFERENCES

1. M. Roger, J. H. Hetherington, and J. M. Delrieu, *Rev. Mod. Phys.* **55**, 1 (1983); M. Roger, *Phys. Rev. B* **30**, 6432 (1984).
2. S. Sasaki *et al.*, (to be published in *J. Phys. Chem. Solids* 2005).
3. M. Yamaguchi *et al.*, *Phys. Rev. Lett.* **91**, 115301 (2003).
4. W. Ni, J. S. Xia, and E. D. Adams, *Phys. Rev. B* **50**, 336 (1994).
5. D. D. Osheroff, H. Godfrin, and R. R. Ruel, *Phys. Rev. Lett.* **58**, 2458 (1987).
6. R. Wanner, *Phys. Rev. A* **3**, 448 (1971).

Nucleation and Growth of Stable Phase during the Magnetic Field Induced Phase Transition between U2D2 ^{3}He and CNAF ^{3}He

T. Tanaka*, H. Ito*, Y. Sasaki*,† and T. Mizusaki*,†

*Department of Physics, Graduate School of Science, Kyoto University, Kyoto, 606-8502, Japan
†Research Center for Low Temperature and Materials Sciences, Kyoto University, Kyoto, 606-8502, Japan

Abstract. Nucleation and growth of the stable phase were investigated in the magnetic field induced transition between the U2D2 phase and the CNAF phase of nuclear ordered solid ^{3}He. Time evolution of the stable phase, after sweeping the magnetic field B through the critical field B_{C1} to $B = B_{C1} + \Delta B$, was investigated by pulsed NMR. The volume fraction of the metastable phase showed two stage evolution and decreased exponentially in time during the first stage. In this stage, the rate of the evolution increased almost linearly with ΔB and decreased slightly at higher temperatures and was almost independent of the direction of the phase transition.

Keywords: Solid ^{3}He, U2D2, CNAF, first order phase transition, MRI
PACS: 64.60.Qb, 67.80.-s, 67.80.Jd, 68.08.-p, 76.60.Pc.

Nucleation and growth mechanisms during a first-order phase transition near $T = 0$ have attracted much interest and have been studied in various first-order transition in He such as the liquid-solid, gas-liquid, and superfluid ^{3}He A-B transitions, phase separation in a ^{3}He-^{4}He mixture, and structural transitions in solid ^{4}He. In bcc solid ^{3}He, a nuclear antiferromagnetic ordering at the melting pressure occurs at $T \simeq 0.9$ mK when the applied magnetic field B=0, where we express magnetic field H by $B = \mu_0 H$. The $B-T$ phase diagram at the melting pressure is separated into three regions. Below the critical field $B_{C1} \simeq 0.45$ T, the spin structure of nuclear ordered phase is U2D2[1], which has uniaxial symmetry. Above B_{C1} and below the upper critical field B_{C2}, there is another nuclear ordered phase. This high field phase has a CNAF spin structure[2], which has cubic symmetry. The both phases are antiferromagnetic. However, the spin symmetries of the two phases are quite different. The magnetization of the CNAF phase is as big as a half of the saturated magnetization $\frac{1}{2}\gamma\hbar$. Thus the phase transition between these two phases is first-order. When the magnetic field is swept through B_{C1}, phase transition occurs and the stable phase appears in the metastable phase. The evolution of the stable phase after this nucleation was studied recently[3, 5]. In this paper, we report an extension of these measurements to clarify the mechanism of the phase evolution.

The experimental method was similar to the one described in our previous report[3]. A U2D2 single crystal was produced at the melting pressure and was examined for its shape and domain distribution by MRI[4]. The temperature of the sample was obtained from B_{C1}, which was determined by pulsed NMR while sweeping B slowly. After achieving thermal equilibrium at B just below B_{C1}, B was quickly stepped up to $B = B_{C1} + \Delta B$ with time constant τ_B. The time evolution of the volume fraction of the U2D2 phase was obtained from intensities of free induction decay signals (FID) after small tipping angle ($\sim 1°$) pulses. The quantity $\delta m \equiv |M_{\rm CNAF} - M_{\rm obs}(t)|/|M_{\rm CNAF} - M_{\rm U2D2}|$ gives the volume fraction of the metastable U2D2 phase, where $M_{\rm CNAF}$ ($M_{\rm U2D2}$) is the signal intensity when the whole solid is in the CNAF (U2D2) phase, and $M_{\rm obs}(t)$ is the observed signal intensity at time t after the stepping of the field. After the transition was completed in the CNAF phase, the reversed process to the U2D2 phase was measured in the same manner with $\Delta B < 0$. In the previous experiment[3, 5], τ_B was as large as 20 sec, while in the present measurements it was only 1 sec. Also, the time interval between FID measurements was as short as 0.2 sec in the present experiment. Due to these improvements, we could observe much faster phenomena than in our previous observation.

As shown in Figure 1, the volume fraction of the metastable phase showed two stage evolution and decreased exponentially in time, as $\exp(-\gamma_1 t)$, in the first stage. Then at $\delta m = \delta m_{tr}$, the first stage terminated and the slower second stage followed. Figure 2 shows a typical ΔB dependence of the first stage rate γ_1 obtained at various temperatures. The γ_1 increased almost linearly with $|\Delta B|$ and decreased slightly with increasing temperature. Figure 3 shows a typical ΔB dependence of δm_{tr}. Neither quantities depended much on the size and the shape of the samples. The exponential time evolution

CP850, *Low Temperature Physics: 24th International Conference on Low Temperature Physics;*
edited by Y. Takano, S. P. Hershfield, S. O. Hill, P. J. Hirschfeld, and A. M. Goldman

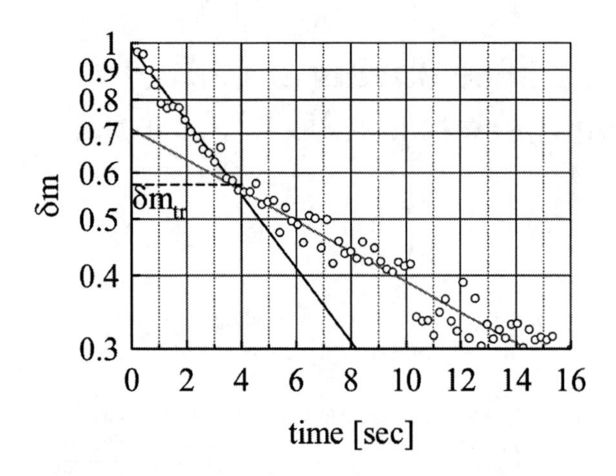

FIGURE 1. Time evolution of the volume fraction, δm, of the metastable U2D2 phase at 0.54 mK and for $\Delta B = +3$ mT.

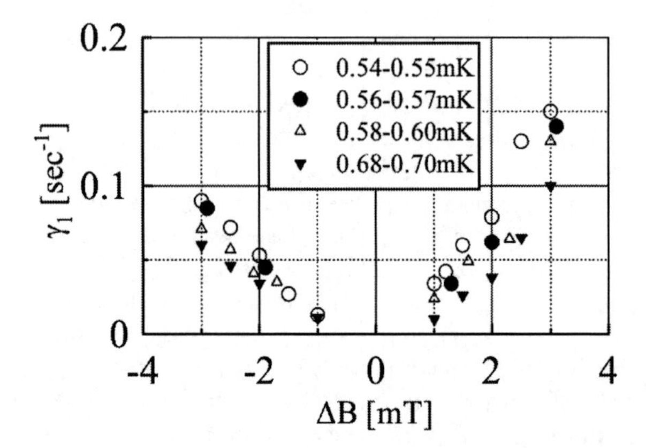

FIGURE 2. ΔB dependence of γ_1 obtained from a crystal of 5 mm^3.

in the first stage suggests that this process is governed by the nucleation of seeds of the stable phase. It is not governed by the growth process, which should typically give power-law time evolution. The weak ΔB dependence of the first stage rate γ_1 suggests that this process is not controlled by the homogeneous nucleation overcoming an energy barrier of the interfacial tension between the two phases, which should give an exponential dependence of γ_1 on ΔB. Moreover the estimated barrier height from naive argument[5] was so large that homogeneous nucleation should not take place in this system. We propose that this process is controlled by some heterogeneous nucleation, possibly occurring at nucleation sites such as magnetic defects embedded throughout the crystal[6] or magnetic domain walls in the U2D2 phase. Another important feature is that the ΔB dependences of γ_1 and δm_{tr} are not much influenced by the direction of

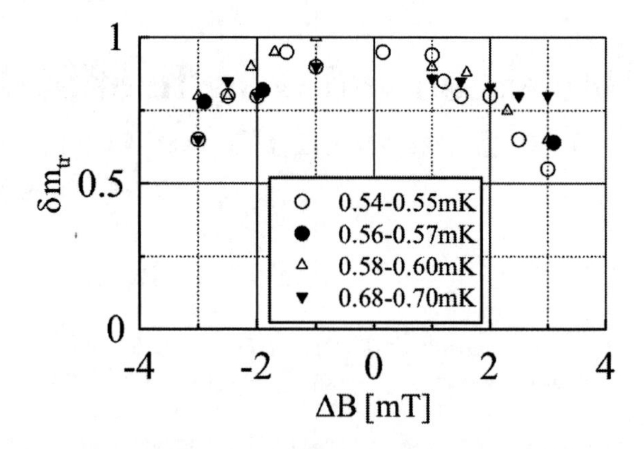

FIGURE 3. ΔB dependence of δm obtained from a crystal of 5 mm^3.

the phase transition. This may appear inconsistent with heterogeneous nucleation, since any sort of heterogeneity which favors one phase should be unfavorable to the other phase, and thus should lead to asymmetric ΔB dependences. Thus a heterogeneity like a liquid-solid interface can be ruled out. However, planar magnetic defects, which appear as a consequence of dislocations, might play a role, since their spin structure in each phases could alter to a configuration suitable for the nucleation of the other phase.

The second stage behavior was rather complicated and depended on the size, temperature and phase of the sample. We have not yet performed systematic studies on this stage. So far our knowledge of this phase transition is incomplete, but we suggest that this process might be controlled by an existence of many planar magnetic defects embedded in the crystal, whose density is somehow similar in the U2D2 and CNAF phases. Further detailed study is necessarily to test this hypothesis.

REFERENCES

1. D.D. Osheroff, M.C. Cross, and D.S. Fisher, *Phys. Rev. Lett.*, **44**, 792–795 (1980)
2. M. Roger, J.H. Hetherington, and J.M. Delrieu, *Rev. Mod. Phys.*, **55**, 1–64 (1983)
3. T. Tanaka, H. Ito, Y. Sasaki, and T. Mizusaki, *J. Low Temp. Phys.*, **138**, 847–852 (2005)
4. Y. Sasaki, E. Hayata, T. Tanaka, H. Ito, and T. Mizusaki, *J. Low Temp. Phys.*, **138**, 911–916 (2005)
5. T. Tanaka, H. Ito, Y. Sasaki, and T. Mizusaki, to be published in *J. Phys. Chem. Sol.*, (2005)
6. Y.P. Feng, P. Schiffer, and D.D. Osheroff, *Phys. Rev. B*, **49**, 8790–8796 (1994)

Nuclear Ordered Phases of Solid ^{3}He in Silver Sinters

Erwin A. Schuberth, Matthias Kath, and Simone Bago

Walther Meissner Institut, D-85748 Garching, Germany

Abstract. To determine the exact spin structure of the nuclear magnetic ordered phases of solid ^{3}He, the U2D2 low field and the high field phases above 0.4 T, a European Research and Training Network for neutron scattering from the ordered solid was established which consisted of a collaboration with the Hahn Meitner Institute, Berlin, and other European and US groups. For this experiment it is crucial to grow a single crystal within the sinter needed for cooling the solid to temperatures of the order of 500 μK and to keep it cold long enough to measure a magnetic neutron diffraction. The sinter is also necessary to absorb the major part (> 90%) of the heat generated by the neutron capture and decay reaction of the ^{3}He nucleus. In this work we studied the growth of crystals in Ag sinters of different pore sizes and with different growth speeds to find an optimal way to obtain single crystalline samples, or at least samples with only a few grains. We used SQUID magnetometry and NMR to measure the magnetization in the ordered phases. They were indicated by the known drop of the intensity, both in the NMR signal and in the dc magnetization, for the U2D2 phase, and by an increase of about 30% for the high field phase. The best results for cooling were obtained with sinters made from 700 Å "Japanese powder" with a packing fraction of 50% which were annealed at 130 °C after sintering and then had a calculated particle size of about 4200 Å. In the dc magnetization we found a paramagnetic surface contribution from a few monolayers of ^{3}He down to 500 μK in addition to the bulk magnetization.

Keywords: Solid ^{3}He, Nuclear Spin Ordered Phases, Magnetization
PACS: 67.80 Gb - 67.80 Jd - 75.30 Kz

We measured the dc magnetization and the NMR resonance of solid ^{3}He in Ag sinters well into the nuclear ordered phases. The first was done with a homemade SQUID detection system [1] in the cell which is shown in Fig. 1. The solid at the end of the extension is positioned in the SQUID pickup coil and is cooled through the sinter. The NMR setup was published previously [2] and only a short description is given here. The H_1 (and pickup) coil was wound around a cylindrical Ag sinter which was formed around a central Ag cooling rod. This arrangement was immersed in a pressure cell body, also filled with Ag sinter. Pulsed NMR measurements were performed at frequencies between 200 kHz and 3 MHz where we searched for a line splitting expected in the U2D2 structure [3].

We formed several sinters with varying packing fractions and heat treatments to find an optimal way to obtain single crystalline samples. A first attempt with 100 Å Pt black sinter failed (no solid was formed) and all of our following sinters were made of 700 Å "Japanese powder" which was "presintered" at 120 to 130 °C before pressing.

From the measured surface areas we estimated that in the presinter process we cluster the 700 Å particles to 2700 Å ones. Two sinters were formed by just pressing this powder; a third one was additionally annealed and the particle size had increased to 4200 Å. This third sinter with a 50% packing fraction proved most suitable for obtaining cold and well ordered solid.

The temperature of the solid inside the sinter can be determined only above T_{Neel} by applying a modified

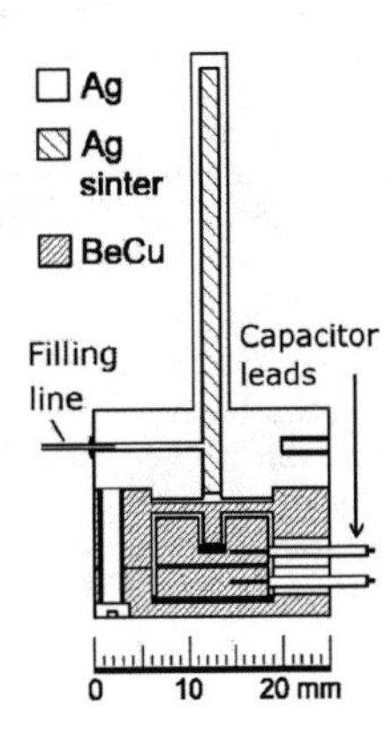

FIGURE 1. Drawing of the pressure cell (Ag) with an extension to fit into SQUID magnetometer and with a capacitive Straty-Adams transducer.

Curie-Weiss law [4] to the measured magnetization. Below T_{Neel} it can only be calculated. We employed a three-stage thermal model to obtain T_{solid} from the temperature of the nuclear stage and the sinter temperature using independent determinations of $R_{Kapitza}$ at various temperatures and the heat capacity data of Greywall and Busch [5]. For all sinters $R_{Kapitza}$ followed a $T^{-2.5}$ law with very small variations of the exponent. We believe that the given temperatures are correct within the error of less than 10%.

A line splitting of about 10 kHz on the high frequency side of the Larmor line (956 kHz) in the low field phase

CP850, *Low Temperature Physics: 24th International Conference on Low Temperature Physics;*
edited by Y. Takano, S. P. Hershfield, S. O. Hill, P. J. Hirschfeld, and A. M. Goldman

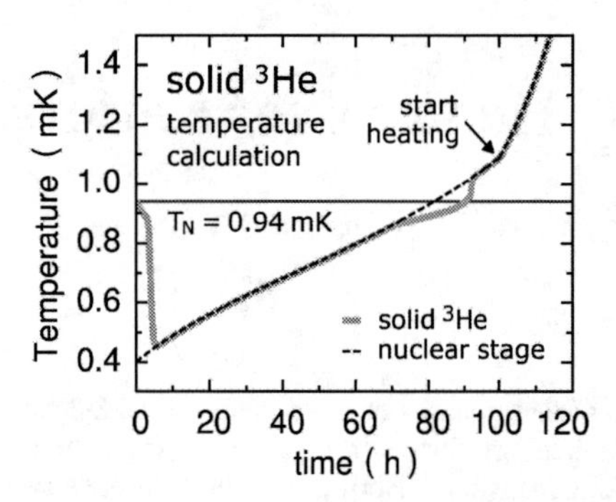

FIGURE 2. Calculated solid ^{3}He temperatures for the data of Figs. 3 and 4 vs time after the end of the demagnetization. The initial plateau and the one between 70 and 90 h are due to the latent heat at the first order phase transition from the paramagnetic to low field phase and out of it respectively.

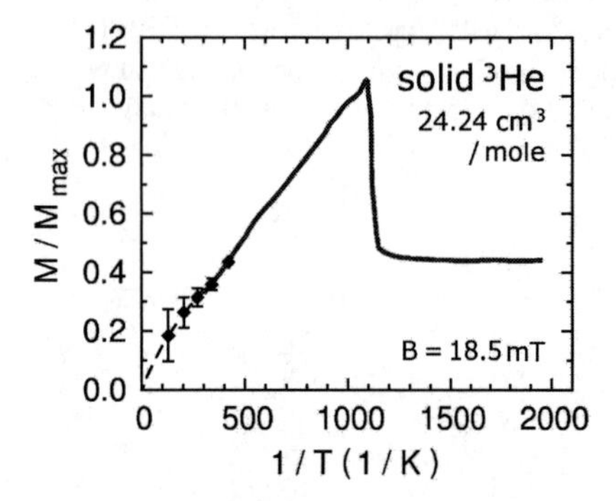

FIGURE 3. Normalized magnetization through the nuclear ordering transition of solid ^{3}He vs inverse temperature in sinter #3. The pressure was just above the melting pressure. Here and in Fig. 4 some special points on the continuous curves were selected as examples for the error bars in the respective T region.

was observed only after a slower growth than used in the first runs, holding the sample at the melting curve for 6 h, and continuing cooling over 12 h. But the amplitude of the whole signal was very low, only 2% of the maximum amplitude just above T_N. Other NMR lines could not be found in the range from 200 kHz to 3 MHz. That a small amplitude at the Larmor line is left could be due to surface layers since we found a paramagnetic contribution of similar size in the SQUID magnetization, see below. The small split-off lines can be attributed to two crystallites with a favorable orientation. The missing of further lines could be due to a large line broadening by a variety of crystallites with different orientations in the sinter pores or by a two magnon spin relaxation [6] favored by the numerous scattering centers provided by the sinter. In the high field phase an increase of the NMR intensity by about 30% is found, as expected in this "weak ferromagnetic" phase.

For the SQUID magnetization signals the background

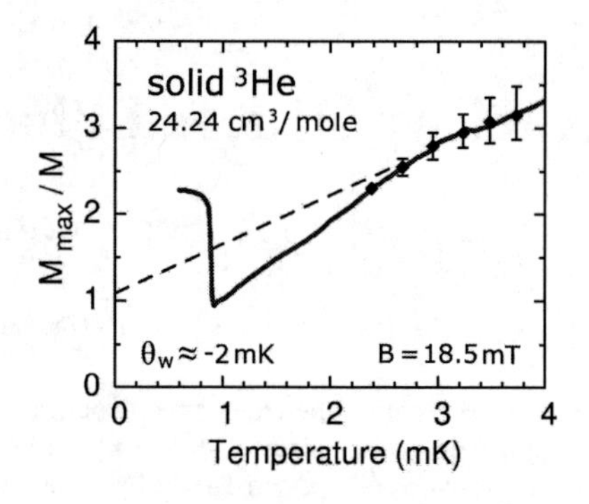

FIGURE 4. Data of Fig. 3 plotted as inverse magnetization of solid ^{3}He vs temperature.

of the 700 Å Ag powder proved to be a major problem, since its magnetization by far dominated the total signal. Therefore, the background had to be determined very accurately in each run for subtraction. Also, a paramagnetic surface layer contribution $\propto T^{-1}$ was subtracted. Above 4 mK our SQUID data follow a Curie-Weiss law with a negative Weiss constant of about -2.0(5) mK, consistent with the observed ordering temperature of 0.94 mK. In Fig. 3 the inverse magnetization between 1 mK and 3 mK deviates from the Curie-Weiss law towards a pure Curie behavior. Our data are in full agreement with earlier ones by Hata *et al.*[4] including the decrease of the solid magnetization in the low field phase down to 40% of its maximum value which is fully consistent with the U2D2 spin structure.

As for the growth and cooling of ^{3}He crystals for neutron scattering experiments, our present result with the annealed third sinter is encouraging. With even slower growth rates, a single crystal with a small number of magnetic domains may be obtained and the pore size above 4000 Å seems to be favorable for this. Indeed, the Berlin group found sharp structural reflections with similarly grown samples.

We gratefully acknowledge funding from EU, contract HPRN-CT-2000-00166.

REFERENCES

1. E. A. Schuberth, *Intern. J. Modern Phys. B* **10**, 357 (1996).
2. E. A. Schuberth, C. Millan-Chacartegui, and S. Schöttl, *J. Low Temp. Phys.* **134**, 637 (2004).
3. D. D. Osheroff, M. C. Cross, and D. S. Fisher, *Phys. Rev. Lett.* **44**, 792 (1980).
4. T. Hata, S. Yamasaki, T. Kodama, and T. Shigi, *J. Low Temp. Phys.* **71**, 193 (1988).
5. D. S. Greywall and P. A. Busch, *Phys. Rev. B* **36**, 6853 (1987).
6. Y. Sasaki, T. Matsushita, T. Mizusaki, and A. Hirai, *Phys. Rev. B* **44**, 7362 (1991).

Spin Wave and Sound in the High Field Phase of Solid ^{3}He

Y. Okamoto and T. Ohmi

Graduate School of Science, Kyoto University, Kyoto, 606-8502, Japan

Abstract. Correction to the spin wave velocity due to thermal magnons and magnon decay rate are calculated for the high field phase (HFP) of solid ^{3}He using Holstein-Primakoff 1/S expansion. Sound velocity is also obtained as a function of magnetic field and temperature, according to the methods similar to those developed by Khalatnikov et al. The anisotropy of the crystal is taken into consideration as was done in the U2D2 phase; the kinetic equation and the Boltzmann equation are expanded in terms of the strain tensor. Sound attenuation which stems from four-magnon processes is also evaluated with the use of a collision-time approximation.

Keywords: Solid ^{3}He, Spin wave, Sound
PACS: 67.80.Cx,67.80Jd

SPIN WAVE 1/S EXPANSION

Exchange Hamiltonian for $\frac{1}{2}$-spins which includes up to planar and folded four-atom exchange is given by

$$H_{ex} = -2\sum_{n=1}^{3} J_n \sum_{i<j} S_i \cdot S_j - 4 \sum_{\alpha=P,F} K_\alpha \sum_{i<j<k<l}^{\alpha} G_{ijkl} \quad (1)$$

$$G_{ijkl} = (S_i \cdot S_j)(S_k \cdot S_l) + (S_i \cdot S_l)(S_k \cdot S_j) - (S_i \cdot S_k)(S_j \cdot S_l). \quad (2)$$

The spin Hamiltonian is expanded with the spin wave operators using the Holstein-Primakoff (H-P) method for antiferromagnets. In the HFP, up and down spins are canted toward the direction of the magnetic field $H \parallel z$ by angle θ, and the quantization axis of the H-P transformation is taken to be canted accordingly. We ignore the contributions from the optical modes, which are about a few mK higher in energy than the acoustic mode. Three-magnon interactions are combined into four-magnon interactions to obtain the effective four magnon interaction V_4^{EFF} (Fig.1). V_4^{EFF} has the form

$$V_4^{EFF} = \frac{1}{\sqrt{\lambda_{k_1}\lambda_{k_2}\lambda_{k_3}\lambda_{k_4}}} W(J_n, K_\alpha, k_1, k_2, k_3, k_4), \quad (3)$$

where W is a function of J_1, J_2, J_3, K_P, K_F, k_1, k_2, k_3 and k_4, and λ_k is the energy eigenvalue for the acoustic mode. The terms in W that are of order $(ka)^0$ disappear, since those which comes from the three magnon interactions and the original four magnon interactions cancel out each other. The terms which are of order $(ka)^2$ do not have any contribution to the spin wave velocity v_s nor magnon decay rate, when energy and momentum conservation laws are applied. For our convenience, we divide W into two parts: one which is reducible with the propagator of the three magnon interactions $\lambda_{k_i} + \lambda_{k_j} - \lambda_{|\mathbf{k_i}+\mathbf{k_j}|}$ etc. (W_{RED}) and one which is not (W_{IRD}). Expansions are done up to the fourth order in ka of W.

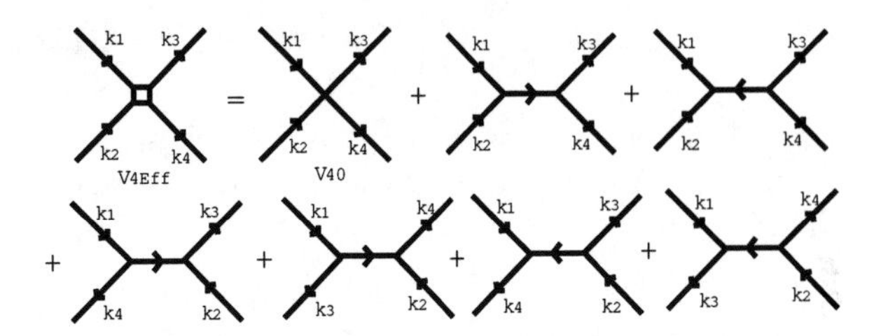

FIGURE 1. Effective four-magnon interactions.

Correction for the Spin Wave Velocity

The self-energy diagram of the lowest order in 1/S that gives spin wave velocity correction Δv_s is given in Fig. 2(a). The temperature dependence of Δv_s is given by

$$\Delta v_s = \int x^3 dx \left(\frac{k_B T}{\hbar v_s} a\right)^4 \frac{a}{\hbar}\Big\{X(J_1,J_2,J_3,K_P,K_F) - Y(J_1,J_2,J_3,K_P,K_F)\ln|\gamma\left(\frac{k_B T}{\hbar v_s}\right)^2 a^2x^2|\Big\}\tilde{n}(x), \quad (4)$$

where X and Y are polynomial functions of J_n and K_α corresponding to W_{RED} and W_{IRD} respectively, and have the energy dimension. $\tilde{n}(x) = 1/(\exp(x) - 1)$, and γ is the dispersion of the acoustic mode; $\lambda_k = v_s k(1 + \gamma(ka)^2)$. The expressions for X and Y will be given elsewhere [1]. The magnitude of Δv_s can be estimated from the exchange parameters obtained by the path integral Monte Carlo method [2]. Δv_s can be up to several cm/s.

Note that, although there appears to be a divergence on the order $1/k$ in V_4^{EFF} when $k_1 \to 0$, it is deceptive, and the divergence from the original four magnon interactions (V_{40}) and that from the three-magnon interactions always cancel out.

CP850, *Low Temperature Physics: 24th International Conference on Low Temperature Physics;*
edited by Y. Takano, S. P. Hershfield, S. O. Hill, P. J. Hirschfeld, and A. M. Goldman

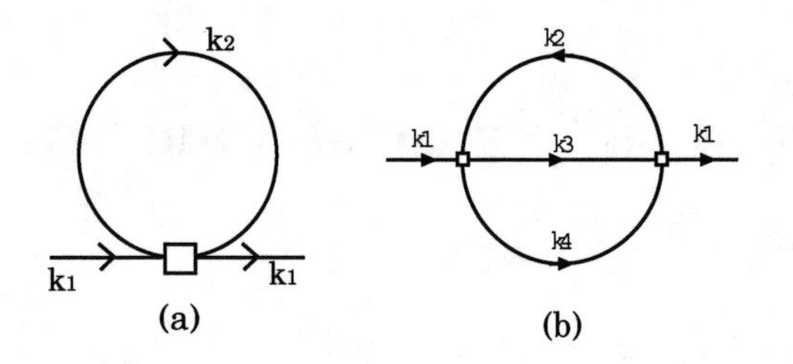

FIGURE 2. (a)The Diagram which contributes to Δv_s. (b)The Diagram which contributes to the decay rate

Magnon Decay Rate: Γ

Dominant terms in Γ come from the irreducible part (W_{IRD}) of V_4^{EFF}. They are of the order $1/\gamma$.

$$\Gamma = \frac{Z(J_n, K_\alpha)^2}{\hbar v_s}\left(\frac{k_B T}{\hbar v_s}\right)^7 a^8 \frac{1}{\gamma}, \tag{5}$$

where Z is a function of exchange parameters with the dimension of energy. The diagram considered is given in Fig. 2(b).

SOUND IN THE HFP OF SOLID ^{3}HE

We start with the the Boltzmann equation for the magnon number n and the momentum conservation law [3, 4]. As was done in the U2D2 phase, the anisotropy of the crystal is taken into consideration. Here, we are not able to ignore the contribution from the ground state energy E_0 in the stress tensor σ_{ik}, since it depends on the magnetic field through the canting angle θ. $\sigma_{ik} = \lambda_{iklm}u_{lm} + \frac{\partial \tilde{E}_0}{\partial u_{ik}}$; u_{ij} is the strain tensor for the lattice, λ_{iklm} is the elastic constant at $T = 0$, and $\tilde{E}_0 = E_0 +$ [zero point energy]. E_0 for the U2D2 and HFP are given by

$$\begin{aligned} E_0^{U2D2} &= \frac{N}{4}(-2J_2 + 4J_3 + 2K_F + 6K_P) - 2H\theta \\ &\quad -4(2J_1 + J_2 + 4J_3 + 3K_P)\theta^2 \qquad (6) \\ E_0^{HFP} &= -\frac{N}{2}(3J_2 + 6J_3 - 4J_1\cos 2\theta \\ &\quad +3(K_F + K_P)\cos 4\theta + H\sin\theta). \qquad (7) \end{aligned}$$

The equations are linearized in terms of u_{ik} and the magnon number n, and the sound velocity s and attenuation Δk are obtained by solving them. They are integral equations and not easy to solve. From the calculation for the U2D2 phase, we estimate that the collision time approximation gives fair results [4] and approximated the collision integral $I(n)$ as $I(n) = -\delta n/\tau$. Here, attenuation caused by thermal magnons (four-magnon scattering) is considered, and the life time τ in $I(n)$ is replaced by that of the thermal magnons: $\tau_{th} = \hbar/\Gamma$. Calculations with $I(n)$ which conserves energy and momentum will be discussed elsewhere [1].

Sound Velocity Changes

Velocity change of the longitudinal sound wave propagating in the direction z is given by

$$\frac{\Delta s}{s_0} = \frac{\hbar}{2ms_0^2}\left\{\frac{\partial^2 \Delta\tilde{E}_0}{\partial u_{zz}\partial u_{zz}} + \frac{\pi^2 v_s}{30}\left(\frac{k_B T}{\hbar v_s}\right)^4 a^3 \left(C_{zz}^2 + 2C_{zz} + \frac{4}{3}C_{zz}^2\left(\frac{v_s}{s_0}\right)^2 \frac{\omega^2\tau_{th}^2 - 1}{(\omega^2\tau_{th}^2+1)^2}\omega^2\tau_{th}^2\right)\right\}, \tag{8}$$

where $\Delta\tilde{E}_0$ is the change in $\tilde{E}_0$ due to H and C_{ij} are the Grüneisen constants for v_s defined by $C_{ij} \equiv \frac{1}{v_s}\frac{\partial v_s}{\partial u_{ij}}$. The condition $\frac{\partial \theta_0}{\partial u_{ij}} = 0$ is always used during the calculation.

1) Temperature Dependence
The temperature dependence of the sound velocity change is obtained from the second and third terms in (8), which are ascribed to the excitation of magnons. They are on the order of $10^{-4} - 10^{-5}$ and proportional to T^4.

2) Magnetic Field Dependence
The sound velocity change due to magnetic field is mostly attributed to $\Delta\tilde{E}_0$, the first term in Eq. (8). $\Delta s/s_0$ is on the order of $10^{-3} - 10^{-4}$, and is larger than the temperature dependence.

Attenuation of Sound

The attenuation α of the longitudinal wave propagating in the direction z is given by

$$\alpha = \frac{8}{3}\frac{\pi^2}{30}C_{zz}^2\frac{\hbar}{2s_0^2\rho}\left(\frac{v_s}{s_0}\right)^3 \frac{\omega^4\tau_{th}^3}{(\omega^2\tau_{th}^2+1)^2}\left(\frac{k_B T}{\hbar v_s}\right)^4. \tag{9}$$

In the hydrodynamic region ($\omega\tau_{thermal} \ll 1$), $\alpha \propto \omega^4 T^{-20}$, and in the collisionless region ($\omega\tau_{th} \gg 1$) $\alpha \propto \omega^0 T^{11}$.

We wish to thank Prof. T. Mizusaki and Dr. S. Sasaki for useful discussions and comments.

REFERENCES

1. Y. Okamoto, to be published.
2. D. M. Ceperley, and G. Jacucci, *Phys. Rev. Lett.*, **58**, 1648 (1987).
3. I. M. Khalatnikov, *An Introduction to the Theory of Superfluidity*, W. A. Benjamin, Inc., 1962, chap. 18-22.
4. Y. Okamoto, and T. Ohmi, *J. Low Temp. Phys.*, **134**, 151 (2004).

NMR Measurements on New Quantum Phases in 2D ^{3}He

S. Murakawa, H. Akisato, Y. Matsumoto†, D. Tsuji, K. Mukai, H. Kambara and Hiroshi Fukuyama

Department of Physics, University of Tokyo, 7-3-1 Hongo, Bunkyo-Ku, Tokyo 113-0033, Japan

Abstract. Monolayer ^{3}He is an ideal experimental system for studying strongly correlated two dimensional fermions. We measured magnetic properties of the second layer of ^{3}He adsorbed on ^{4}He-preplated graphite by NMR in a wide temperature (70 μK $< T <$ 300 mK) and density ($4.9 < \rho < 9.8$ nm^{-2}) ranges. The measured magnetization isotherms can be classified into four distinct regions depending on density. The present NMR results support the scenario given by the recent heat capacity measurements that the region just below the density for the 4/7 commensurate phase is a hole-doped Mott localized phase.

Keywords: Two dimensional system, Helium three, Magnetization.
PACS: 67.70.+n, 67.80.Jd, 71.10.Ay, 71.10.Fd

Helium three (^{3}He) adsorbed on a graphite substrate is known to be the ideal realization of a strongly correlated two dimensional (2D) fermion system. In the second layer of this system, there is a registered phase at a four seventh density ($\rho_{4/7}$) of the underlying first layer, called the 4/7 phase. This phase has very interesting properties [1] due to competition among the ferromagnetic and antiferromagnetic multiple exchange interactions and geometrical frustration of its triangular lattice structure. It has been proposed that the 4/7 phase has a gapless spin- liquid ground state [1]. On the other hand, it had been long believed that a fluid-solid two-phase coexistence region exists at densities just below $\rho_{4/7}$. However, it is becoming clear through recent heat capacity measurements [2] that there exists a novel phase in this region, i.e. a hole-doped Mott localized phase, possibly behaving as a uniform quantum fluid containing zero-point vacancies.

We have carried out continuous wave NMR measurements in a magnetic field of 170 mT and in a wide density range above and below $\rho_{4/7}$ including this peculiar region down to 70 μK. The sample cell containing a Grafoil substrate with a surface area of 53.6 m^2 is installed directly on a nuclear demagnetization stage. Further details of the experimental setup were described elsewhere [3]. For the first layer, the graphite surface is plated with ^{4}He (11.78 nm^{-2}) at 4.2 K. The next step consists of plating with ^{3}He at 2.7 K. We assume that the first layer density is fixed regardless of the amount of ^{3}He which forms the second layer and higher layers.

Figure 1 shows magnetization isotherms of the second layer ^{3}He at several different temperatures. The inset displays the temperature dependence of magnetization at several densities. In the density region we studied, the magnetic behavior separates into four distinct regions depending on density. At the lowest density (region I), the temperature dependence of magnetization shows a Fermi fluid behavior, i.e., a temperature-independent magnetization at low temperatures ($T \leq 100$ mK). In region II, whose density is just below $\rho_{4/7}$, the Fermi fluid behavior disappears gradually and the magnetization increases non-linearly with density. The density dependence of magnetization then switches to a moderate one in region III where the densities are slightly above $\rho_{4/7}$. In region IV, the highest density region, the magnetization reveals a ferromagnetic tendency (see the inset of Fig. 1).

Region II had been long believed to be the region where both the fluid and registered phases coexist. However, this conclusion was drawn from the measurements at only three different densities in this region [4]. In contrast, a much more rapid magnetization increase with density than a linear one is observed in this work, which was done in a finer density grid. Such a non-linear variation cannot be described by the conventional two-phase coexistence. Indeed, our results are consistent with the recent heat capacity measurements [2] which indicate that the

CP850, *Low Temperature Physics: 24th International Conference on Low Temperature Physics;*
edited by Y. Takano, S. P. Hershfield, S. O. Hill, P. J. Hirschfeld, and A. M. Goldman

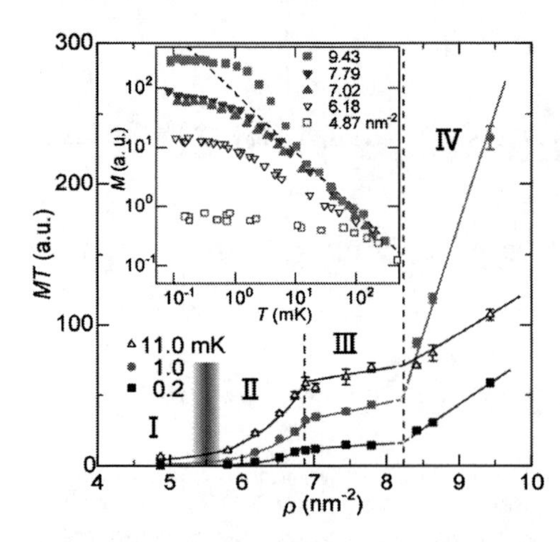

FIGURE 1. Isotherms of magnetization (M) multiplied by temperature (T). The solid lines are guides to the eye. The vertical dashed lines are boundaries between the four distinct regions. The inset shows the temperature dependences of M. The dashed line shows a Curie-law behavior.

system in region II is a uniform phase with unusual properties.

The boundary between regions II and III is 6.7–6.8 nm^{-2} where the density dependence of magnetization clearly changes. This density is close to 4/7 of the first layer density. We therefore conclude that the boundary is the 4/7 commensurate density ($\rho_{4/7}$). We thus claim that, if the 4/7 phase can be considered to be a Mott localized phase, region II should be a hole-doped Mott localized phase. The temperature dependences of magnetization are weaker than that of Curie's law ($\propto 1/T$). As the hole concentration increases, the magnetization decreases in the whole temperature range. This tendency is in agreement with the numerical calculations for the t-J model for a triangular lattice [5]. We speculate that this decrease of M is due to the release of the geometric frustration by the hole doping. The hole may form an antiferromagnetic polaron with a nontrivial spin configuration [6]. The magnetization deviates from Curie's law below about 100 mK and almost saturates below 300 μK. This temperature dependence seems to be related to the double peak structure observed in the heat capacity measurements [2].

Figure 2 shows temperature dependences of the line widths of NMR absorption spectra. The line widths increase below 10 mK and saturate below 300 μK. Although this peculiar behavior cannot be explained fully at present, it provides the first microscopic information on spin dynamics in these novel quantum phases. It should be noted that, in region II, the line width at low temperatures increases with increasing density. This change cannot be described by the conventional two-phase coexistence

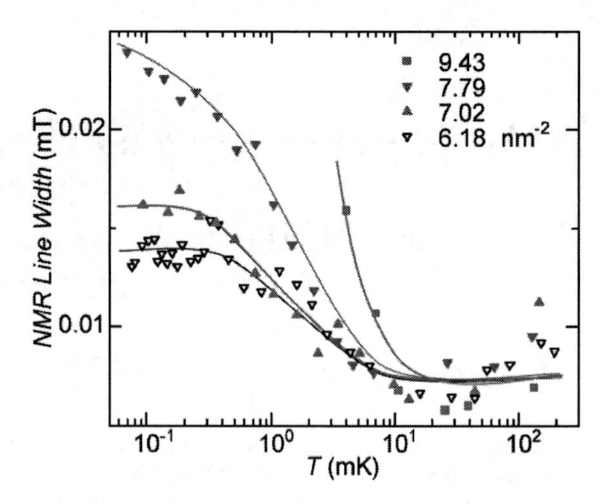

FIGURE 2. Temperature dependences of full widths at half maximum of NMR absorption spectra. The solid lines are guides to the eye. The contribution from a static magnet field inhomogeneity has been already subtracted.

either, supporting the assumption of a uniform phase for region II.

In region IV, the magnetization increases rapidly and linearly with density (see Fig.1). A ferromagnetic tendency is seen in the temperature dependence of M (see the inset of Fig. 1). This behavior indicates now a two-phase coexistence, between a ferromagnetic phase and the phase of region III. A possible origin of this first order phase transition is the demotion of ^{3}He to the second layer in region IV. This will increase the second layer density and change the nature of the exchange interaction to a ferromagnetic one.

This work was financially supported by Grants-in-Aid for Scientific Research from the MEXT, Japan and by the ERATO Project of JST. S. M. and Y. M. acknowledge financial supports from the JSPS.

REFERENCES

†Present address: Institute for Solid State Physics, University of Tokyo, 5-1-5 Kashiwanoha, Kashiwa, Chiba 277-8581, Japan

1. K. Ishida et al., *Phys. Rev. Lett.* **79**, 3451 (1997); H. Fukuyama and M. Morishita, *Physica B* **280**, 104 (2000); R. Masutomi et al., *Phys. Rev. Lett.* **92**, 025301 (2004).
2. Y. Matsumoto et al., *J. Low Temp. Phys.* **138**, 271 (2005).
3. S. Murakawa et al., *Physica B* **329**, 144 (2003); H. Akisato et al., *J. Low Temp. Phys.* **138**, 265 (2005).
4. D. S. Greywall, *Phys. Rev. B* **41**, 1842 (1990).
5. T. Koretsune and M. Ogata, *Phys. Rev. Lett.* **89**, 116401 (2002).
6. M. Héritier and P. Lederer, *Phys. Rev. Lett.* **42**, 1068 (1979); K. Machida and M. Fujita, *Phys. Rev. B* **42**, 2673 (1990).

^{3}He Films as Model Strongly Correlated Fermion Systems

Michael Neumann, Andrew Casey, Jan Nyéki, Brian Cowan and John Saunders

Dept. of Physics, Royal Holloway University of London, Egham, Surrey, TW20 0EX, U.K.

Abstract. Helium films on graphite are atomically layered. This allows a wide variety of studies of strong correlations in two dimensions with density as a continuously tunable parameter. Studies of a monolayer of ^{3}He adsorbed on graphite plated by a bi-layer of HD find a divergence of effective mass with increasing density, corresponding to a Mott-Hubbard transition between a 2D Fermi liquid and a quantum spin liquid phase. While the Fermi liquid survives in 2D, non-Fermi liquid features remain at finite T; recent theories find that this correction arises from the spin component of the backscattering amplitude. In another experiment a ^{3}He film is grown on graphite plated by a bi-layer of ^{4}He. The first ^{3}He layer only solidifies in the presence of an overlayer. However in the regime in which the system comprises a ^{3}He fluid bilayer, we observe a striking maximum in the temperature dependence of both heat capacity and magnetization. This feature is driven towards $T = 0$ with increasing film coverage, suggestive of a quantum critical point. Well below the maximum a linear temperature dependence of the heat capacity is recovered; the coverage dependence of the effective mass identifies a (bandwidth driven) Mott-Hubbard transition at 9.8 nm^{-2}.

Keywords: 2D ^{3}He, strongly correlated fermions, quantum criticality, Mott-Hubbard
PACS: 67.70.+n, 67.55.-s, 71.10.Ay. 71.30.+h

INTRODUCTION

Bulk liquid ^{3}He is the paradigm for strongly correlated fermion systems, and is well described at sufficiently low temperatures by Landau's theory of Fermi liquids. Does this theory survive in two dimensions [1]? What new phenomena might we find there?

It appears that ^{3}He films adsorbed on graphite provide model 2D systems to investigate these questions. This is made possible by a number of features. (1) The surface of graphite is atomically flat so that the helium films which grow on it are atomically layered. It is therefore possible to study a single fluid atomic layer in the absence of significant interlayer coupling. (2) In several systems studied there is no gas-liquid transition of a two dimensional ^{3}He monolayer (although ^{4}He does condense due to its lower zero-point motion). This allows the surface density of a 2D ^{3}He fluid to be varied over a rather wide range, tuning the strength of correlations in a convenient manner.

This system is relatively simple compared to strongly correlated electron systems, such as the wide range of intermetallic compounds and metallic oxides currently under active investigation. This 2D Fermi system is intrinsically isotropic and the effective mass ratio arises solely from correlation effects. There are no crystal field effects. The spin-orbit interaction is the ^{3}He nuclear dipole-dipole interaction, of order μK, and hence is negligible. The fluid ^{3}He atoms are subject to a periodic potential arising from the Van der Waals interaction with the underlying compressed solid layer, which turns out to play an important role.

This paper reviews the solidification of a ^{3}He monolayer as an example of a Mott-Hubbard transition from a strongly correlated 2D fluid to a quantum spin liquid. Our main aim, however, is to discuss a new ^{3}He bi-layer system, comprising two fluid layers, one of which is on the verge of solidification. On tuning the total coverage, it appears to exhibit a quantum critical point. This new system appears to provide an intriguing analogue of a heavy fermion metal, with an almost localised layer on the verge of solidification ("f-electrons"), coupled to a fluid overlayer ("s-electrons").

In this work we use exfoliated graphite with total surface area A = 182 m^2 as the substrate, providing sufficiently large samples, of order 10^{20} atoms, for thermodynamic measurements to below 1 mK. The heat capacity is measured by the adiabatic pulsed method, while the ^{3}He nuclear magnetization is measured selectively by continuous wave NMR at 27 mT.

CP850, *Low Temperature Physics: 24th International Conference on Low Temperature Physics;*
edited by Y. Takano, S. P. Hershfield, S. O. Hill, P. J. Hirschfeld, and A. M. Goldman

MOTT-HUBBARD TRANSITION OF A ^{3}HE MONOLAYER

In previous work [2] a ^{3}He monolayer was adsorbed on the surface of graphite, preplated with a bilayer of solid HD. Solidification of this monolayer occurs at a ^{3}He coverage, $n_3 = 5.2$ nm^{-2}. The 2D ^{3}He solid forms a triangular lattice in $\sqrt{7}\times\sqrt{7}$ commensuration with the HD underlayer. As the coverage is tuned towards this critical coverage, we observe an apparent divergence of the effective mass ratio, Fig. 1, which leads us to identify this as a Mott-Hubbard transition. In this work the heat capacity is fit to $c = \beta + \gamma T + \Gamma T^2$, where $\gamma = \pi k_B^2 A m^* / 3\hbar^2$. The magnetization enhancement relative to that of an ideal Fermi liquid also tracks this effective mass increase. It appears that as the effective mass diverges, the Landau parameter F_0^a tends to saturate at $\approx -3/4$, close to the value observed in bulk liquid ^{3}He and consistent with the picture of liquid ^{3}He as an almost localized system. Measurements of a ^{3}He monolayer adsorbed on graphite plated by a monolayer of ^{4}He [3] are also consistent with this scenario. In this case the $\sqrt{7}\times\sqrt{7}$ phase occurs at $n_3 = 6.8$ nm^{-2}.

The temperature region over which $(c-\beta)/T$ is constant collapses rapidly as the Mott transition is approached. Here the heat capacity appears to be "non-Fermi liquid-like" except at the very lowest temperatures. For $n_3 = 5.0$ nm^{-2} a plateau in $(c-\beta)/T$ develops when $T/T_F^* < 0.05$, where $T_F^* = T_F/(m^*/m)$ and $T_F = \pi\hbar^2 n_3 / k_B m$ is the ideal 2D Fermi gas temperature.

The wide range of effective mass ratios accessible in this system allows an interesting study of the higher order terms in the temperature dependence of the heat capacity. For bulk ^{3}He it is well established [4] that there is a $\Gamma T^3 \ln(T/T_0)$ term in the heat capacity beyond the leading order linear term. Recent theory [5] finds that in 2D the corresponding term is ΓT^2, with

$$c(T) = \gamma_0(1 + \tfrac{1}{2}F_1^s)T - a[A_s^2(\pi) + 3A_a^2(\pi)]T^2 \quad (1)$$

In this expression γ_0 is the ideal 2D Fermi gas value of γ, F_1^s (where $m^*/m = 1 + \frac{1}{2}F_1^s$) is the usual Landau parameter, and the factor a is proportional to the effective mass. The scattering amplitude has been decomposed into "charge" and spin components, A_s and A_a, see ref. 5 for details.

The unusual feature of this expression is that Γ depends on the symmetric and antisymmetric scattering amplitudes at a particular angle, not averaged over the Fermi surface. Just forward and backscattering (i.e. one dimensional) terms contribute, and comparison with our data [2, 5] leads to the conclusion that the spin component of the scattering dominates.

Elsewhere it has been proposed that, in the large effective mass limit, an isotropic Fermi system may be unstable with respect to a distortion in the quasiparticle dispersion relation near the Fermi energy [6]. This ultimately leads to the formation of a "fermion condensate". Anomalous temperature dependences are predicted for thermodynamic properties. Indeed anomalies are observed experimentally. In the present work the "β-term" in the heat capacity is observed to increase as the Mott transition is approached. And the magnetization increases above the Pauli magnetization at the lowest temperatures. The challenge is distinguish intrinsic phenomena from nuisance background effects arising from weak substrate heterogeneity, and more work remains to be done.

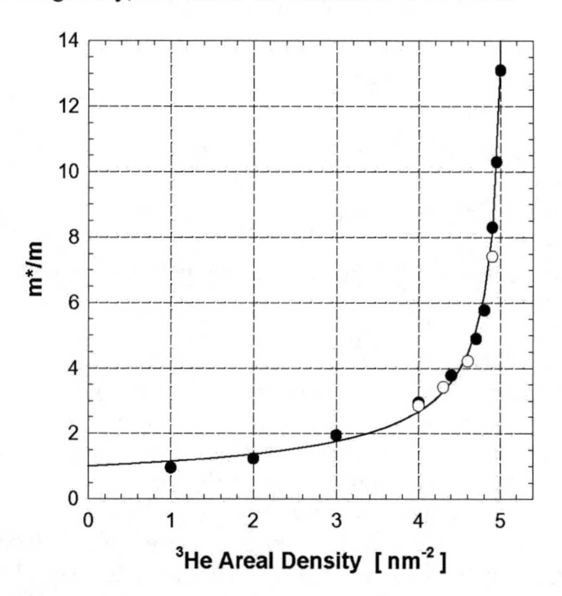

FIGURE 1. Effective mass of a ^{3}He fluid monolayer absorbed on graphite plated by a bi-layer of HD, showing apparent divergence on approaching density of $\sqrt{7}\times\sqrt{7}$ commensurate solid (quantum spin liquid phase). Heat capacity data (closed symbols); magnetic susceptibility data (open symbols), see [2].

NEW QUANTUM CRITICAL POINT IN A ^{3}HE FLUID BILAYER

In more recent experiments [7] we have investigated ^{3}He films adsorbed on graphite preplated by a solid bilayer of ^{4}He. This work was originally motivated by a desire to investigate the structure of helium multilayer films on graphite. Heat capacity and magnetization measurements show that for pure ^{3}He films on graphite only two layers solidify at saturated vapour pressure. The well known differences between the thermodynamic behaviour of localized S = 1/2 fermions coupled by exchange interactions, and a quantum degenerate fluid provide very clear experimental signatures of solid and fluid phases. In the case of pure ^{4}He films there is no conclusive evidence for solidification of the third layer, although Path Integral Monte Carlo calculations suggest this is possible [8]. Differences in behaviour of the two isotopes would arise because of the smaller zero point energy of ^{4}He, for motion both parallel and perpendicular to the plane of the substrate.

The properties of a ^{3}He bilayer on this composite substrate (graphite + solid ^{4}He bilayer) are extremely rich. As the ^{3}He coverage is increased we find the first monolayer is a 2D Fermi fluid with the effective mass ratio increasing to $m^*/m \sim 3.5$ at a coverage of 6.5 nm^2, at which point a second ^{3}He layer starts to form. [For reference we note that this density is equal to that of a bulk-liquid-layer.]

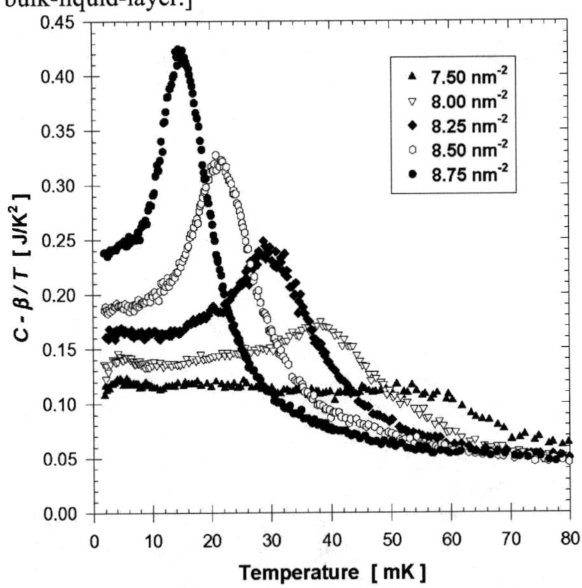

FIGURE 2. Temperature dependence of $\gamma(T)=(c-\beta)/T$ for a ^{3}He bilayer adsorbed on graphite plated by a bilayer of ^{4}He, for ^{3}He coverages shown.

In the present system we find that the first ^{3}He layer (L1) is solidified only in the presence of a fluid overlayer (L2). The onset of this process occurs at a total coverage n_3 = 9.2 nm^{-2}, with complete solidification at 9.85 nm^{-2}. A simple picture is that the formation of the fluid overlayer restricts the zero-point motion of the layer beneath, causing it to solidify. We suggest that this may be regarded as a bandwidth-driven Mott-Hubbard transition; L2 tunes the ratio of bandwidth to effective on-site repulsion (t/U in Hubbard model terminology) in L1, by increasing the effective U. In contrast in the second layer of ^{3}He on graphite, as well as ^{3}He on graphite preplated by a monolayer of ^{4}He, solidification into the $\sqrt{7}\times\sqrt{7}$ phase occurs prior to promotion. This is the density-driven Mott-Hubbard transition discussed previously

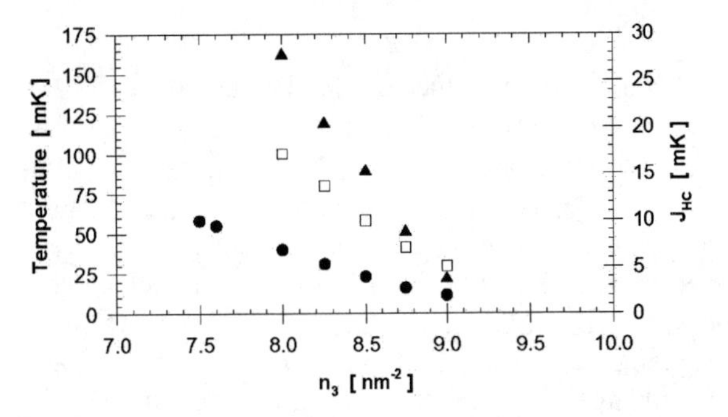

FIGURE 3. Coverage dependence of parameters characterising the heat capacity maximum. Left scale: T_0 (circles), Δ (triangles). Right scale: J (squares)

The properties of the 2D solid L1, which is a frustrated S = 1/2 magnet on a triangular lattice, are discussed elsewhere in these proceedings. In the rest of this paper we concentrate on the properties of the ^{3}He fluid bilayer found at coverages in the range 7.0 to 9.2 nm^{-2}.

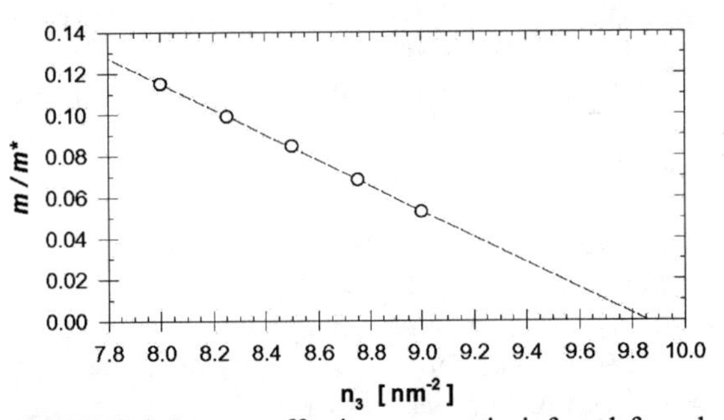

FIGURE 4. Inverse effective mass ratio inferred from heat capacity data below T_0

Heat capacity measurements are shown in Fig. 2, where we plot the temperature dependence of $\gamma(T)=(c-\beta)/T$. The most pronounced feature is a distinct maximum, at T_0, which is tuned to lower temperatures with increasing coverage. Above T_0, $\gamma(T)$ is strongly temperature dependent (non-Fermi liquid behaviour). Well below T_0, $\gamma(T)$ tends to a constant, indicative of a recovery of the Fermi liquid.

Below T_0 the heat capacity is well described by

$$c = \beta + \gamma_1 T + \gamma_2 \exp(-\Delta / T) \qquad (2)$$

We find that both T_0 and Δ decrease towards zero with total coverage n_3, Fig. 3. This identifies a new and unexpected quantum critical point. Above T_0 we fit the up-turn in the heat capacity by an exchange term δ/T^2, with $\delta=9N_s k_B J^2/4$, where N_s is the number of particles in L1. The parameter J is also plotted in Fig. 3. The behaviour of T_0, Δ and J indicate that the system is characterised by a single energy scale, tunable by total coverage.

Magnetization isotherms identify the coverage 9.2 nm^{-2} as that which L1 starts to solidify. The heat capacity between here and 9.85 nm^{-2}, at which point L1 is completely solid, is more complex. At 9.85 nm^{-2} the heat capacity of the 2D solid exhibits an exchange maximum at 8 mK, with ~$0.19k_B$ per particle. (The density of the solid L1 is established as 6.3 nm^{-2} from magnetization measurements). Thus the energy scale of exchange within the layer L1 masks the tuning of T_0 of the fluid bilayer towards zero. Furthermore from 9.5 to 9.75 nm^{-2} we can identify a secondary peak in the heat capacity, which tunes towards zero at an extrapolated coverage of 9.9 nm^{-2}. This may arise from zero-point vacancies.

Below T_0 we infer an effective mass from the parameter $\gamma(T << T_0) = \gamma_1$. The inverse effective mass ratio is plotted in Fig. 4 as a function of total coverage. These values extrapolate to indicate a mass divergence at total coverage 9.85 nm^{-2}. As previously discussed this is the lowest coverage at which L1 forms a uniform solid (with density 6.3 nm^{-2}). This supports the model of a Mott-Hubbard transition of L1, driven by changing the density of L2.

The magnetic susceptibility is determined by continuous wave NMR measurements on selected samples. These also show a maximum at T_0. For $T \sim 0.3T_0$, the susceptibility enhancement over the ideal Fermi gas value is consistent with a Wilson ratio of 4 i.e. $F_0^a \sim -3/4$, using the values of m*/m determined here. This saturation of F_0^a is as observed in 2D monolayers and bulk liquid. Above T_0 the magnetization approaches Curie law for the estimated density of L1.

How does the density of the fluid overlayer L2 act as a tuning parameter for L1? We have already discussed the idea that L2 tunes t/U of L1. Another potentially important process is competing exchange interactions between L1 and L2. To the extent that L1 is on the verge of localisation one might treat it in a lattice-gas model. In one such model [9] the competition between an indirect antiferromagnetic exchange, J_H, process and an RKKY interaction, J_{RKKY}, is evaluated as a function of the coverage of the fluid overlayer. With increasing coverage J_H/J_{RKKY} strongly decreases, reminiscent of the Doniach phase diagram [10, 11].

According to this picture the peak at T_0 would be a Kondo peak, accounting for the heavy fermion behaviour of the bilayer observed at $T<<T_0$ and the local moment behaviour of L1 well above T_0. This "Kondo temperature" is tuned to zero at the quantum critical point.

In conclusion the discovery of this new heavy fermion system may give the opportunity to gain fresh insights into the nature of quantum criticality and the elementary excitations in the quantum critical regime. The relative simplicity of these 2D ^3He systems provide new paradigms for strongly correlated Fermi systems in which the complexities of intermetallic compounds are stripped away. As such they should be an interesting challenge for theorists.

ACKNOWLEDGMENTS

This work was supported by EPSRC grant GR/S20567/01. We would like to thank Piers Coleman, Andrew Green, Malte Grosche, Chris Hooley, Andrew Schofield and Nic Shannon for interesting discussions.

REFERENCES

1. P. W. Anderson, *Phys. Rev. Lett.* **65**, 2306 (1990).
2. A. Casey et al. *Phys. Rev. Lett.* **90**, 115301 (2003).
3. Y. Matsumoto et al. *J. Low Temp. Phys.* **138**, 271(2005).
4. D. S. Greywall, *Phys. Rev.* B **27**, 2747 (1983).
5. A. V. Chubukov et al. *Phys. Rev.* B 71, 205112 (2005) and cond-mat/0502542.
6. J. W. Clark, V. A. Khodel and M. V. Zverev, *Phys. Rev.* B **71**, 012401 (2005).
7. M. Neumann et al. *J. Low Temp. Phys.* **138**, 391,(2005).
8. M. Pierce and E. Manousakis, *Phys. Rev.*B **63**, 144524 (2001).
9. S. Tasaki, *Prog. Theor. Phys.* **79**, 1311 (1988).
10. S. Doniach, *Physica* **91B**, 231 (1977).
11. P. Fazekas, *Lecture Notes on Electron Correlation and Magnetism,* World Scientific (1999).

^{3}He Bilayer Film Adsorbed on Graphite Plated with a Bilayer of ^{4}He: a New Frustrated 2D Magnetic System

Michael Neumann, Ján Nyéki, Brian Cowan and John Saunders

Department of Physics, Royal Holloway University of London, Egham, Surrey, TW20 0EX, U.K.

Abstract. The heat capacity and NMR response of a ^{3}He bilayer adsorbed on graphite plated with a bilayer of ^{4}He have been measured over the temperature range 1–80 mK. We find that the first ^{3}He layer requires the presence of a ^{3}He fluid overlayer before it solidifies. Solidification is completed at a total coverage close to 9.85 nm^{-2}. On further increasing the coverage the heat capacity maximum grows from 'antiferromagnetic-like' (AFM-like) to 'ferromagnetic-like' (FM-like). On the other hand, when the ^{3}He layer first solidifies, it has a low temperature saturation magnetisation corresponding to a significant fraction of full polarisation, and this increases with increasing coverage. Furthermore the effective exchange constant inferred from the high temperature magnetisation data is always ferromagnetic. The effective exchange constants inferred from the heat capacity and magnetisation are significantly larger than those observed in the second layer of pure ^{3}He films adsorbed on bare graphite. Otherwise there are strong similarities in the coverage dependence of the heat capacity and magnetisation, providing fresh insights into how the magnetic ground state of such 2D magnets evolves as the frustration is tuned with increasing coverage.

Keywords: 2D ^{3}He, frustrated magnetism.
PACS: 67,70.+n, 67.80.Jd, 75.40.Cx

INTRODUCTION

This study concerns ^{3}He films formed on graphite, which has been first preplated by two atomic layers of ^{4}He. We found previously that in this sytem solidification of the first ^{3}He layer is induced by the presence of a ^{3}He fluid overlayer [1]. A simple picture of the mechanism is that the formation of the fluid overlayer restricts the zero point motion normal to the surface of the layer beneath. This behaviour should be contrasted with the properties of ^{3}He films on bare graphite, for which the second layer solidifies prior to layer promotion, and subsequent layers are found to remain fluid. The main focus of this report is to compare the magnetic properties of the solid layer in the present system with those of the second layer of ^{3}He on bare graphite (hereafter referred to as the "second layer"), with a view to gaining further insight into the magnetic ground state.

RESULTS

We have performed measurements of heat capacity and magnetisation on the same sample, at a series of total ^{3}He coverages. The magnetisation is determined by continuous wave NMR. From these data we can identify the onset of the solidification of the first ^{3}He layer, L1 at 9.2 nm^{-2} and complete solidification at 9.85 nm^{-2}. This solid has a ferromagnetic Curie-Weiss constant, a low temperature magnetisation of order 0.3 M_0, where M_0 is the estimated saturation magnetisation of L1. This large magnetisation gives rise to a strong dipolar frequency shift of the NMR line at low temperatures.

We fit the high temperature magnetisation data to a high temperature series expansion for L1, plus a term to account for the magnetisation of the fluid overlayer L2. This enables us to infer the density of the solid layer L1, and the effective exchange constant J_χ, which is FM-like. We find that both these quantities depend rather weakly on total coverage, see Fig. 1. To compare with the "second layer", where the density of the 4/7 "quantum spin liquid phase" is 6.4 nm^{-2}, and the density at the FM anomaly is around 7.5 nm^{-2}, the solid density in the present system remains close to 6.3 nm^{-2}. It is interesting to note that this density is consistent with a 13/19 close-packed lattice commensurate with the underlayer (^{4}He solid of density 9.2 nm^{-2}), just as at the FM anomaly in the second layer [2]. The AFM "quantum spin liquid phase" is not observed in the present system.

CP850, *Low Temperature Physics: 24th International Conference on Low Temperature Physics;* edited by Y. Takano, S. P. Hershfield, S. O. Hill, P. J. Hirschfeld, and A. M. Goldman

The heat capacity data show an evolution with increasing coverage similar to that observed in the "second layer" [3]. There is a characteristic low temperature maximum due to exchange at $T \sim J_c$ arising from L1, as well as a contribution linear in T from the fluid overlayer L2. With increasing coverage the exchange maximum evolves in a similar way to that observed in the "second layer". The spin heat capacity at the maximum increases from ~ 0.19 k_B per particle (AFM-like) to ~ 0.41 k_B per particle (FM-like).

DISCUSSION

As is well established, the magnetic properties of 2D solid ^{3}He is governed by atomic ring exchanges, which together with the triangular lattice result in a strong magnetic frustration [3-6]. The most direct way to experimentally characterise this frustration is to determine the ratio of the different exchange constants which are inferred from the high temperature behaviour of the magnetisation and heat capacity, J_χ/J_c;

$$M = \frac{A}{T - 3J_\chi} \text{ and } c = \frac{9}{4} N_S k_B \left(\frac{J_c}{T}\right)^2 \quad (1)$$

The values of J_c from the heat capacity fits are shown in Fig.1. (The total heat capacity is fit, in a way self consistent with the inferred J_c, to the form $c(T) = \beta + a/T^2 + b/T^3 + \gamma T$ [3]). We also plot the temperature of the heat capacity maximum, J_p, which as expected is similar in magnitude and coverage dependence to J_c.

The range of J_χ/J_c is similar to that found in the "second layer". The simplest model capturing the essential physics of these systems is the J-J_4 model: $J = J_2 - 2J_3 < 0$ is an effective Heisenberg exchange constant (FM due to dominance of 3 particle ring exchange J_3 over 2 particle exchange J_2), frustrated by four particle exchange J_4. Thus as $J_4/J < 0$ tends to zero J_χ/J_c approaches unity. Our data clearly show that with increasing coverage the system becomes more weakly frustrated; a FM ground state is expected, and this is consistent with the observed magnetisation, which is close to full polarisation at the highest coverage.

An important open question is how the transition from FM to quantum spin liquid occurs with increasing frustration. Classically the FM ground state is unstable for $J_4/J < -1/4$, bounded by the point at which the FM spin wave velocity vanishes. In this regime the "second layer" remarkably shows both AFM-like properties in the heat capacity, and FM like properties in the magnetisation. Our main new finding is that very similar behaviour is also found in the system discussed here.

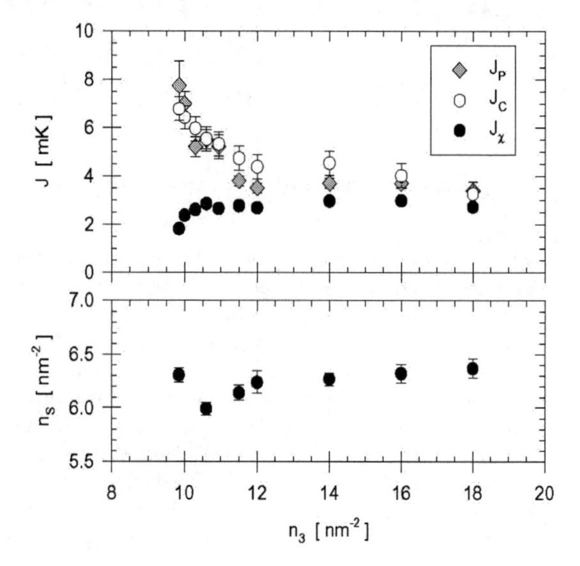

FIGURE 1. Effective exchange constants inferred from heat capacity and magnetisation, and inferred density of solid layer as a function of total ^{3}He coverage.

In particular if we plot the saturation magnetisation against the value of heat capacity maximum, a break identifies a phase boundary at $J_\chi/J_c = 0.52$. In the J-J_4 model this corresponds to $J_4/J = -0.27$. In an alternative model including six spin exchange J_6, where we assume $J_4 = J_6$, this corresponds to $J_4/J = -0.19$. This boundary is the same as identified in our results for the "second layer" [3,5].

Thus comparison of the present system with previous work on the "second layer" strengthens evidence for a new magnetic phase of a 2D spin-½ system on a triangular lattice with frustrated ring exchange. This may correspond to the gapless quantum phase, bordering on the FM phase, with mixed AFM and FM characteristics, recently proposed theoretically [7].

REFERENCES

1. M. Neumann et al. *J. Low Temp. Phys.* **138**, 391 (2005).
2. M. Roger et al., *J. Low Temp. Phys.* **112**, 451 (1998).
3. M. Siqueira et al., *Phys. Rev. Lett.* **78**, 2600 (1997).
4. M. Roger, *Phys. Rev. Lett.* **64**, 297 (1990).
5. M. Siqueira et al., *Czech. J. Phys. Suppl. S1* **46**, 407 (1996).
6. M. Roger et al., *Phys. Rev. Lett.* **80**, 1308 (1998).
7. T. Momoi and N. Shannon, *Prog. Theor, Phys.,* to be published.

Magnetism of Two-Dimensional Films of ^{3}He on Highly Oriented Graphite

H.M. Bozler, Jinshan Zhang, Lei Guo, Yuliang Du, C.M. Gould

Dept. of Physics and Astronomy, Univ. of Southern California, Los Angeles, CA 90089-0484, USA

Abstract. What is the effect of the structural length scale on the ordering of ^{3}He films? NMR experiments on the magnetism of second layer ^{3}He on Grafoil in the low field limit found ferromagnetic ordering for coverages over 20 atoms/nm^2. Finite temperature phase transitions are prohibited in 2D when only Heisenberg interactions are present. However ordering of a two-dimensional magnetic film can be a result of a phase transition caused by weak anisotropy and/or dipolar interactions, or could be a less interesting manifestation of finite size effects. By replacing Grafoil with ZYX grade highly oriented graphite, we can study the magnetism of two-dimensional films with a substantially increased structural coherence length and test the importance of finite size effects. Our new experiments find a region of coverages where the second layer ^{3}He films become ferromagnetic at temperatures above 1 mK, with no evidence for an increased suppression of the ordering due to increasing the coherence length. We show the results for the magnetism at a wide range of coverages as well as the effect of varying the magnetic field in the ferromagnetic cases. Our results support the interpretation in terms of a phase transition occurring at finite temperature.

Keywords: magnetism, NMR, He-3, 2-D
PACS: 67.70.+n, 76.60.-k

INTRODUCTION

Studies of two-dimensional magnetism in diverse systems provide evidence for ferromagnetic order at finite temperatures [1, 2, 3] even when the dominant exchange process has continuous symmetry. This observation appears to contradict assumptions that two-dimensional phase transitions in Heisenberg systems are suppressed due to long wavelength spin fluctuations [4]. However, two-dimensional magnetism is extremely sensitive to weak long range forces and finite temperature order can be stabilized by the influence of dipolar coupling [5, 6] and weak anisotropy [1, 2, 3]. Our previous NMR experiments on the magnetism of second layer ^{3}He on GTY Grafoil found ferromagnetic ordering for coverages over 20 atoms/nm^2 in the low field limit. However Grafoil is not a perfect substrate and the finite size effects compete with more interesting weak but long range interactions. We report here preliminary results from a study of ^{3}He films on high quality ZYX graphite that has the advantage of reducing the influence of finite size.

Compared to GTY Grafoil, Highly Oriented Pyrolytic Graphite (HOPG) grade ZYX has a dramatically greater structural coherence length increasing from 20 nm to 190 nm. The orientation misalignment is also reduced from $\pm 15^o$ to $\pm 5^o$ [7, 8, 9, 10]. The advantages of using ZYX graphite as a substrate are somewhat offset by the reduction of surface area which is typically 10 times smaller than Grafoil, resulting in 10 times less signal. However, with improvements in our SQUID NMR techniques, we managed to obtain good signals even at fields as low as 0.24 mT. We used pulsed NMR for signals with small magnetization (temperatures above 2 mK and/or coverages below 20 atoms/nm^2) and CW NMR when polarizations became significant.

If we assume that the observed onset of ferromagnetic order observed in Grafoil was entirely due to a finite size cutoff caused by the structural coherence, then ZYX graphite should show a suppression of this order. Based on estimates for the magnetic coherence length for a pure two-dimensional Heisenberg ferromagnet [11], we would estimate that the increased coherence length with ZYX would reduce the ferromagnetic onset by 25%.

2D ^{3}HE MAGNETIZATION ON ZYX GRAPHITE

Our estimates of ^{3}He coverage are obtained by a ^{3}He isotherm at 4.2 K. We found that the relaxation time for pressure equilibrium in the sample cell was quite long, and eventually needed to allow as much as 24 hours to obtain an accurate measurement of the equilibrium pressure with our *in situ* pressure gauge. By scaling to the known ^{3}He isotherm [12], we calculated the surface area of our ZYX graphite to be 0.24 m^2.

Magnetic isotherm measurements of ^{3}He on ZYX graphite over a wide range of coverages and temperatures of 3 mK and 6 mK are shown in Figure 1. All of these data are at 1.4 mT external field. These data illustrate the basic features of the magnetism that are quite similar to earlier studies performed by Godfrin and

CP850, *Low Temperature Physics: 24th International Conference on Low Temperature Physics;*
edited by Y. Takano, S. P. Hershfield, S. O. Hill, P. J. Hirschfeld, and A. M. Goldman

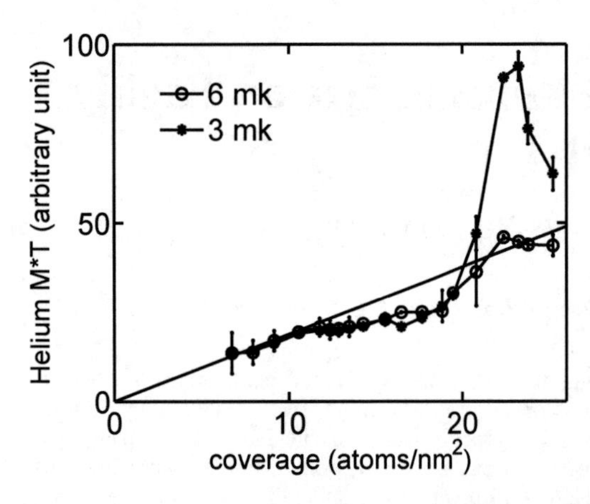

FIGURE 1. Magnetization times temperature for different ^{3}He film coverages on ZYX graphite at temperatures of 3 mK and 6 mK. The straight line indicates values for free spins assuming that the spins are free and localized.

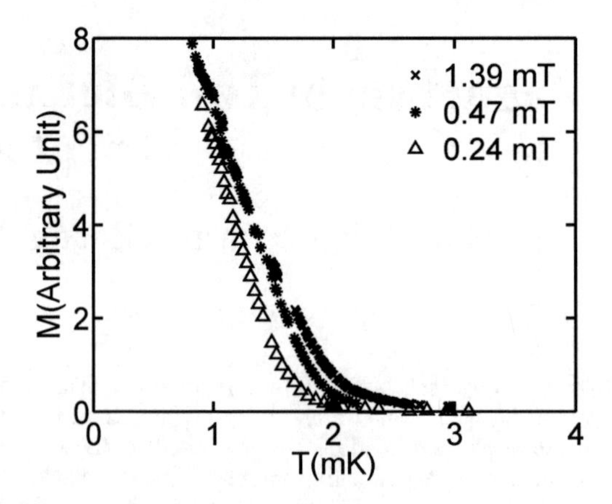

FIGURE 2. Magnetizations of ^{3}He film on ZYX graphite for different static fields at coverage 23.3 atoms/nm^2.

coworkers [7] on conventional Grafoil. A more detailed discussion of lower coverages will be discussed elsewhere, however the most obvious feature of Figure 1 is the rapid increase in ferromagnetic tendency at around 20 atoms/nm^2. The maximum ferromagnetic exchange occurs between 22.4 atoms/nm^2 and 23.3 atoms/nm^2.

DISCUSSION

Figure 2 shows the magnetization of second layer ^{3}He atoms at coverage of 23.3 atoms/nm^2 and applied fields of 1.39, 0.47 and 0.24 mT. We can see from figure 2 that the magnetization increases rapidly below an onset temperature at all of the fields and becomes nearly independent of the applied field at temperatures below 1 mK. This result is qualitatively similar to results with Grafoil at the same coverage with the apparent ordering temperature (obtained from extrapolating the steepest slope to zero as a straight line) remaining at least as large as in the Grafoil data. The curvature of the temperature data remains in the ZYX case. The reasons for this remain to be explored.

We found that the increase in structural coherence of ZYX graphite substrate does not greatly affect the onset temperature of ferromagnetic order. This result supports the interpretation that the observed ordering of these two-dimensional magnets is intrinsic, and not the result of finite size effects. Are there differences between the results using ZYX or conventional Grafoil? To date, the most striking effect of the increased coherence length is the resolution of multiple components in the cw NMR absorption lines. The analysis of NMR lineshapes on Grafoil and ZYX substrate will be discussed elsewhere [13].

ACKNOWLEDGMENTS

We would like to thank M. Bretz for providing us the ZYX graphite sample. This research is supported by the National Science Foundation grant DMR-0307382.

REFERENCES

1. H. M. Bozler, Y. Gu, J. Zhang, K. S. White, and C. M. Gould, *Phys. Rev. Lett.*, **88**, 065302 (2002).
2. D. L. Mills, *Ultrathin Magetic Structures 1*, p. 91 (1994), edited by J. A. C. Bland and B. Heinrich (Springer-Verlag, Berlin, Heidelberg).
3. M. A. Kastner, R. J. Birgeneau, G. Shirane, and Y. Endoh, *Reviews of Modern Physics*, **70**, 897 (1998).
4. N. D. Mermin, and H. Wagner, *Phys. Rev. Lett.*, **11**, 1133 (1966).
5. Y. Yafet, J. Kwo, and E. M. Gyorgy, *Phys. Rev. B*, **33**, 6519 (1986).
6. L. J. Friedman, A. L. Thomson, C. M. Gould, H. M. Bozler, and P. B. Weichman, *Phys. Rev. Lett.*, **62**, 1635 (1989).
7. H. Godfrin, and H. J. Lauter, *Progress in Low Temperature Physics*, **14** (1995), edited by W. P. Halperin (Elsevier, Amsterdam).
8. S. Murakawa, Y. Niimi, Y. Matsumoto, C. Bauerle, and H. Fukuyama, *Physica B*, **329-333**, 144 (2003).
9. Y. Niimi, S. Murakawa, Y. Matsumoto, H. Kambara, and H. Fukuyama, *Rev. Sci. Instru.*, **74**, 4448 (2003).
10. M. Bretz, *Phys. Rev. Lett.*, **38**, 501 (1977).
11. P. Kopietz, P. Scharf, M. S. Skaf, and S. Chakravarty, *Europhys. Lett.*, **9**, 465 (1989).
12. M. Bretz, J. G. Dash, D. C. Hickernell, E. O. McLean, and O. E. Vilches, *Phys. Rev. A*, **8**, 1589 (1973).
13. L. Guo, J. Zhang, Y. Du, C. M. Gould, and H. M. Bozler, *submitted to LT24 conference proceedings* (2005).

Evidence for Split NMR Lines in Ferromagnetic ^{3}He Films

Lei Guo, Jinshan Zhang, Yuliang Du, C.M. Gould and H.M. Bozler

Dept. of Physics and Astronomy, Univ. of Southern California, Los Angeles, CA 90089-0484, USA

Abstract. In earlier experiments on ferromagnetic ^{3}He films, we observed a complex lineshape due in part to the dipolar field generated by polarization of the ^{3}He nuclei. Much of the complex lineshape can be explained by the known distribution of the Grafoil platelets. However, there remained some evidence for a split NMR line at some temperatures. In our new experiments on ZYX grade exfoliated graphite where the size of individual platelets is much larger and the angular distribution is three times smaller, this splitting has become more evident over a wider range of temperatures. Now it is clear that the complex lineshape includes two peaks along with remaining orientation effects. We also find that roughly 20% of our signal comes from randomly oriented platelets. We present the details of our model for analyzing these lineshapes and the experimental results for the line splitting at several coverages in the ferromagnetic range. We discuss the possible sources of this line splitting.

Keywords: He-3, NMR, 2-D, graphite
PACS: 67.70.+n, 76.60.-k

Ferromagnetic order has been observed in two-dimensional ^{3}He films on grafoil [1] and more recently highly oriented ZYX grade exfoliated graphite [2]. A consequence of the ferromagnetism is high polarization of the ^{3}He spins. This high polarization results in a substantial frequency shift in the NMR signal due to the dipolar field of the spins. This shift has been observed in many experiments, but has been analyzed in more detail by Schiffer *et al.* [3] and more recently at USC in very low applied fields [1]. In both cases, the lineshape was qualitatively consistent with a rapidly broadening NMR line from graphite planes that were predominately oriented but also with large portions (25 - 50%) randomly oriented. The resulting NMR lineshape was a broad peak shifted towards lower frequency (here the applied field is normal to the oriented surfaces) with a background from the randomly oriented regions extending to positive frequencies. Figure 1 shows a set of NMR lines from Grafoil at a series of temperatures [1]. We see the line center shift as well as the broadening and the line components extending to positive shifts. A complete analysis taking into account the spin dynamics of disordered sheets of these lines found a reasonable model that included the spin dynamics of two-dimensional sheets with around 75% of the observed spins in a single NMR line [1]. However, the lineshapes observed do not fit well at all temperatures, and in fact there is an intermediate range of temperatures where there appears to be additional structure to the NMR line. Compared to Grafoil, ZYX has a much greater structural coherence and is more highly oriented [4, 5]. Recently we have studied the magnetism of ^{3}He films on ZYX graphite and have found that the features we observed in our original Grafoil experiment were much more pronounced and required further examination. [2]

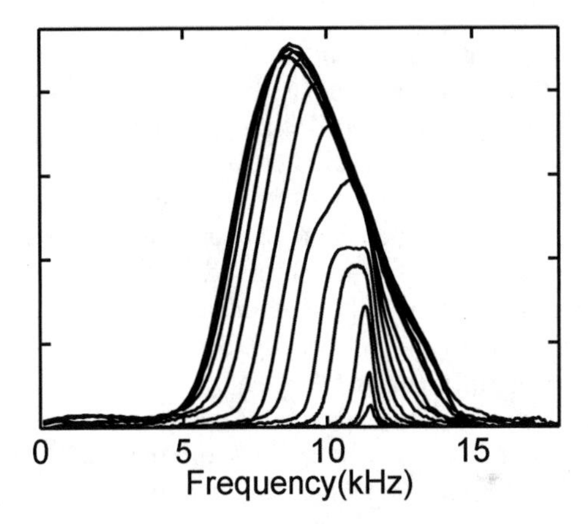

FIGURE 1. Grafoil cw lines from 2 mK (smallest amplitude) to 0.3 mK (largest amplitude) for a coverage of 22.5 atoms/nm^2 at 0.35 mT applied field

Figure 2 shows typical cw NMR absorption lines for ^{3}He films on a ZYX substrate at a coverage of 23.3 atoms/nm^2 and 0.47 mT applied field from 2.2 mK to 0.7 mK. We observed a double peak structure on the left of central frequency and a small hump on the right starting at temperature 1.07 mK. We have observed a similar set of NMR lines at both larger fields (1.39 mT) and smaller fields (0.24 mT). As with Grafoil, a portion of the NMR line appears with a positive frequency shift, however is more pronounced. Although similar features occur with conventional Grafoil, the double peaked structure is much more obvious with ZYX and the single central peak is no longer adequate to describe the data. Finally, the steepness of the lineshape on the low frequency side is inconsistent with a single resonance, but more

CP850, *Low Temperature Physics: 24th International Conference on Low Temperature Physics;*
edited by Y. Takano, S. P. Hershfield, S. O. Hill, P. J. Hirschfeld, and A. M. Goldman

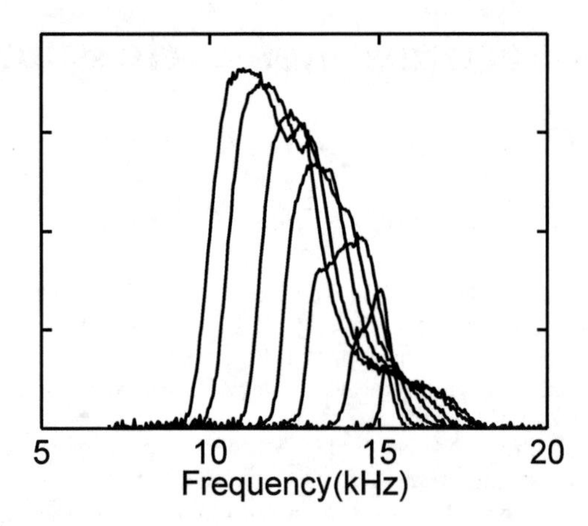

FIGURE 2. ZYX cw lines from 2.2 mK (smallest amplitude) to 0.7 mK (largest amplitude) for a coverage of 23.3 atoms/nm^2 at 0.47 mT applied field

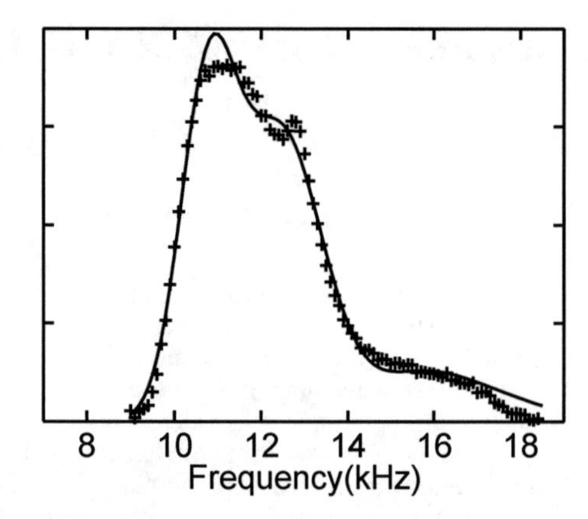

FIGURE 3. cw lineshape at 0.69 mK in 0.47 mT. Pluses are observations. The curve is the result of the spin wave model augmented with misoriented platelet contributions.

consistent with a multicomponent spectrum.

When we consider the possibilities for the existence of multiple peaks in the NMR line components, bound spin waves appear to be a good candidate. A simple dimensional analysis argument including the exchange energy within the two-dimensional sheets gives the correct order of magnitude for the observed frequency shifts. We assume that defects at the boundaries of each uniform crystallite pin the spins locally. Since the shape of each platelet is different, the frequency spectrum of the entire sample is a superposition of many spectra. The relative positions of resonances in two-dimensional ellipses are not strongly dependent upon ellipticity, and aside from scale factors and offsets, are qualitatively similar to simple one-dimensional spin-wave resonances [6], so we have indexed each resonance accordingly.

Figure 3 shows the raw cw absorption line at 0.69 mK (plus) the result of fitting (solid line). From the relative spacing of the peaks, the corresponding spin wave modes are labeled as 1, 3 and 5. For this fit we assumed that the distribution of orientation of the platelets is a narrow Gaussian distribution around normal and included an isotropic distribution to account for disoriented regions. The frequency shift of each spin wave spectrum is due to both the dipolar field and spin wave exchange. The angle of magnetization and then the center frequency were calculated using spin dynamics for two-dimensional sheets [1]. The resulting lineshapes are fit better when Gaussian rather than Lorenzian lines are used.

Figure 4 shows the average dipolar field vs temperature obtained from fitting our NMR lines. So far the fits using a spin wave model are qualitative, but certainly suggestive. By appropriately averaging over two-dimensional shapes, this model can be improved.

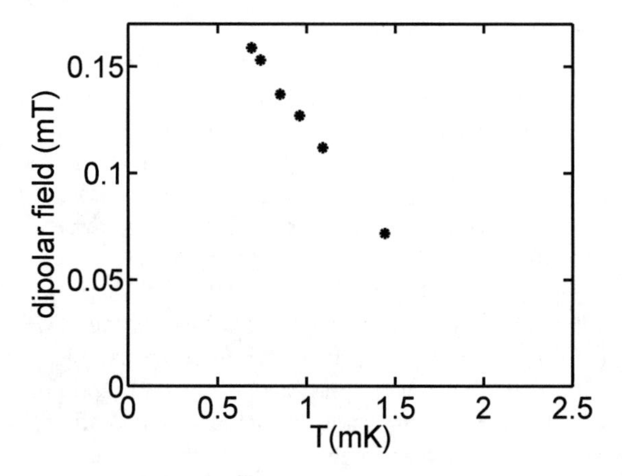

FIGURE 4. Average dipolar field as a function of temperature

ACKNOWLEDGMENTS

This research is supported by NSF through grant DMR-0307382.

REFERENCES

1. H. M. Bozler, Y. Gu, J. Zhang, K. S. White, and C. M. Gould, *Phys. Rev. Lett.*, **88**, 065302 (2002).
2. H. M. Bozler, J. Zhang, L. Guo, Y. Du, and C. M. Gould, *submitted to LT24 conference proceedings* (2005).
3. P. Schiffer, *J. Low Temp. Phys.*, **94**, 489 (1994).
4. M. Bretz, *Phys. Rev. Lett.*, **38**, 501 (1977).
5. H. Godfrin, and H. J. Lauter, *Progress in Low Temperature Physics*, **14** (1995), edited by W.P.Halperin (Elsevier, Amsterdam).
6. C. Kittel, *Introduction to Solid State Physics*, John Wiley and Sons Inc, 2005, 8 edn.

Magnetization Measurements of ^{3}He Monolayer Film on Graphite

Akira Yamaguchi, Hirofumi Nema, Yuichi Tanaka and Hidehiko Ishimoto

Institute for Solid State Physics, The University of Tokyo, 5-1-5 Kashiwanoha, Kashiwa, Chiba 277-8581, Japan

Abstract. A newly developed Faraday-type magnetometer is demonstrated through measurement of the nuclear magnetization of ^{3}He monolayer films adsorbed on Grafoil at temperatures down to 1.1 mK and in magnetic fields of up to 7 T. The magnetic force is measured capacitively by monitoring the displacement of a wire-suspended copper plate holding a stack of diffusion-bonded copper foils interleaved with Grafoil. The strong background signal due to the Grafoil and copper is suppressed by a double gradient coil system that produces opposite field gradients in two regions equidistant from the field center. The upper region contains Grafoil and copper, and the lower region contains non-exfoliated graphite sheets and copper. Preliminary results of magnetization measurements are reported for the first ^{3}He monolayer in the paramagnetic phase below 10 mK.

Keywords: Helium 3, Frustration, Magnetization, Two-dimensional magnets.
PACS: 67.70.+n, 67.80.Jd, 75.70.Ak

INTRODUCTION

A low-density solid ^{3}He film adsorbed on a graphite surface is one of the most ideal two-dimensional (2D) quantum spin systems on a triangular lattice, with a nuclear spin of $S = 1/2$.[1] Due to the hard-core potential between ^{3}He atoms, higher-order multiple-spin exchange processes play an important role in this system, in addition to two-particle exchange. The exchange of an even number of particles is an antiferromagnetic (AFM) process, while the exchange of an odd number is ferromagnetic (FM). The competition between these two processes, coupled with the geometric frustration inherent in a triangular lattice, causes strong frustration in the solid ^{3}He film. Recent experiments suggest that the AFM phase of the second atomic layer of the film has a quantum spin liquid ground state with very small or near-zero spin gap.[2,3] In the spin liquid state, some theoretical studies predict the appearance of a magnetic plateau in the magnetization curve.[4,5] Experimental verification of this phenomenon is therefore strongly desired. Recently, our group has undertaken the development of a Faraday magnetometer capable of measuring the nuclear magnetization of monolayer solid ^{3}He films in high magnetic fields. This type of magnetometer has been successfully used for magnetization measurements of intermetallic compounds and molecular magnets.[6]

EXPERIMENTS

Faraday magnetometric measurements rely on the fact that a material with a magnetic moment M_z in a field with gradient dH/dz experiences a force of $M_z dH/dz$. In the proposed Faraday cell, the force is measured as a capacitance change arising from the displacement of a wire-suspended movable part. The cell consists of adsorbed ^{3}He atoms, graphite substrates, and copper foils and a copper plate as thermal links. Specifically, the substrate is exfoliated graphite (Grafoil GTY grade, 76 μm thick).[7] Two Grafoil sheets (8 mm × 10 mm) are diffusion-bonded on both sides of well-annealed copper foil (10 μm thick). A total of 50 copper-Grafoil sandwiches are stacked and the copper foil tabs extended from the end are diffusion-bonded to a copper plate (1.0 mm thick). Four separate copper foils (25 μm thick) are also diffusion-bonded to the plate to serve as a thermal link to a rigid silver cage. The Grafoil and copper amount to 60.4 mmol and 35 mmol, respectively, much larger than that of the first-layer solid ^{3}He (0.24 mmol). This value is estimated from the Grafoil surface area (12.2 m^2). The strong background signal produced by the diamagnetism of the Grafoil and the nuclear paramagnetism of the copper makes it difficult to discriminate even saturated ^{3}He nuclear magnetization.

CP850, *Low Temperature Physics: 24th International Conference on Low Temperature Physics;*
edited by Y. Takano, S. P. Hershfield, S. O. Hill, P. J. Hirschfeld, and A. M. Goldman

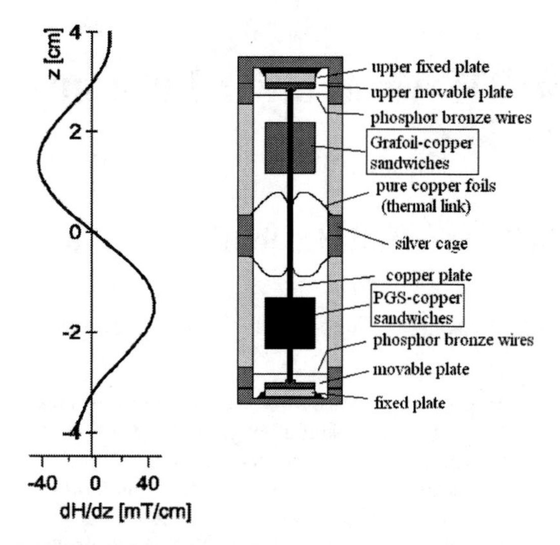

FIGURE 1. Schematic of the Faraday cell and corresponding field gradient profile.

To overcome this difficulty, a double-gradient coil system[8] is employed, as shown in Fig. 1. The coil consists of three NbTi superconducting solenoids, the two outer of which are wound in opposite directions to the central solenoid to produce opposite field gradients in two regions equidistant from the field center. The maximum field gradient is about 10 T/m, and is limited by the critical current of NbTi corresponding to the static field. The movable part of the Faraday cell is designed such that equal amounts of graphite and copper foils are present in each field gradient region. In the bottom field gradient region, non-exfoliated graphite sheet (PGS; 100 μm thick)[9] is used to eliminate the surface area. ^{3}He atoms are adsorbed on the exfoliated Grafoil in the upper field gradient region. By careful adjustment of the quantities of PGS, Grafoil and Cu foils, most of the background signal can be removed.

The movable part is suspended from a rigid silver cage by four phosphor bronze wires (100 μm in diameter) stretched crosswise at both ends. A silicon electrode with sputtered silver surface on one side is glued to the bottom end to form a parallel plate capacitor with a corresponding silver plate fixed to the rigid silver cage. The rigid cage is tightly fastened to a silver thermal link connected to the nuclear demagnetization stage. The entire Faraday cell is placed in a titanium can, which is then sealed with an indium O-ring. The Faraday cell and gradient coil are installed at the center of a 15 T superconducting magnet that produces a homogeneous field over the entire cell.

Figure 2 shows some preliminary results of magnetization measurements for the first ^{3}He monolayer in the paramagnetic phase (areal density of 10.5 atoms/nm^2) at 1.1 mK and 9.3 mK. Two magnetization curves can be reproduced within experimental error by a Brillouin function with a single adjustable parameter of saturation magnetization.

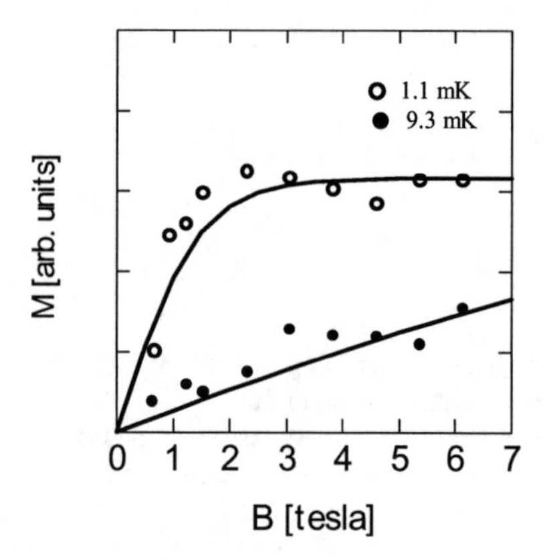

FIGURE 2. Magnetization of the first ^{3}He monolayer in the paramagnetic phase at 1.1 mK and 9.3 mK. The solid curves are the Brillouin function.

CONCLUSION

The nuclear magnetization of a solid ^{3}He monolayer film was successfully measured in magnetic fields of up to 7 T and at temperatures below 10 mK using a newly developed Faraday cell. Further improvements in accuracy are required, and modification of the device for measurements in magnetic field of up to 15 T is currently being pursued through development of a Nb_3Sn gradient coil system.

REFERENCES

1. Godfrin, H. and Rapp, R. E., *Adv. Phys.* **44**, 113-186 (1995).
2. Masutomi, R., Karaki, Y. and Ishimoto, H., *Phys. Rev. Lett.* **92**, 025301 (2004).
3. Collin, E., Triqueneaux, S., Harakaly, R., Roger, T., Bauerle, C., Bunkov. Y. M. and Gofrin, H., *Phys. Rev. Lett.* **86**, 2447-2450 (2001).
4. Misguich, G., Bernu, B., Lhuillier, C. and Waldtmann, C., *Phys. Rev. Lett.* **81**, 1098-1101 (1998).
5. Momoi, T., Sakamoto, H. and Kuno, K., *Phys. Rev. B* **59**, 9491-9499 (1999).
6. Sakakibara, T., Mitamura, H., Tayama, T. and Amitsuka, H., *Jpn. J. Appl. Phys.* **33**, 5067-5072 (1994).
7. Grafoil is a product of Graf Tech International Ltd.
8. Zhang, D., Probst, C. and Andres, K., *Rev. Sci. Instrum.* **68**, 3755-3760(1997).
9. PGS is a product of Matsushita Electric Industrial Co., Ltd.

Thermal Conductance between ^{3}He Solid Film and Graphite Substrate in Magnetic Field

Masashi Morishita

Graduate School of Pure and Applied Sciences, University of Tsukuba, Tsukuba, Ibaraki 305-8571, Japan

Abstract. Thermal conductance between a ^{3}He solid film and a graphite substrate was measured in magnetic fields up to 400 gauss. The thermal conductance is strongly affected by the magnetic field. Previous measurements in zero magnetic field show that heat is transferred only by some local spots between ^{3}He film and graphite substrates, and suggest the magnetic Kapitza conductance as the heat transfer mechanism. The results in the magnetic field support the magnetic Kapitza mechanism.

Keywords: ^{3}He film, thermal conductance, magnetic Kapitza conductance.
PACS: 67.70.+n, 67.80.Gb

INTRODUCTION

^{3}He films adsorbed on graphite substrates are subjects of vigorous research. However, the heat transfer mechanism between ^{3}He films and graphite substrates is not clearly understood. In a bulk helium system heat is thought to be transferred through an interface by phonons which come almost perpendicularly to the interface. However, such phonons cannot transfer heat in films. In bulk ^{3}He, magnetic interaction also contributes to heat transport through an interface (magnetic Kapitza conductance). The results of my previous thermal conductance measurements indicate that heat is transferred only by local spots between ^{3}He films and graphite substrates, and inside ^{3}He films heat is transferred by some magnetic mechanism.[1,2] Additionally, in the second layer the replacement of the first layer with a monolayer of ^{4}He reduces the thermal conductance.[2] These observations suggest magnetic Kapitza conductance as the heat transfer mechanism between ^{3}He films and graphite substrates. In this paper the results of thermal conductance measurements in magnetic fields are reported. The thermal conductance is strongly affected by the magnetic field. The results support the magnetic Kapitza mechanism.

EXPERIMENTAL

The thermal conductance is measured by the relaxation method analyzed by Shepherd.[3] The total surface area of the graphite (Grafoil grade GTA) is about 390 m^2. The magnetic field is applied by a superconducting magnet with a 6 inch bore diameter. Although a niobium tube with a 32 mm diameter near the sample cell distorts the magnetic field, the inhomogeneity is estimated to be smaller than several percent.

RESULTS AND DISCUSSION

Thermal conductance κ_b between a solid ^{3}He monolayer film and a graphite substrate was measured in magnetic fields up to 400 gauss in the temperature range between 0.2 and 1 mK. The areal density of the ^{3}He film was 7.3 nm^{-2}. Selected results are shown in Fig. 1. Isotherms of κ_b as a function of the magnetic field are also shown in Fig. 2 to show clearly the magnetic field dependence. In a higher temperature range κ_b decreases rapidly with the increasing magnetic field up to 100 gauss. My previous work suggests that heat is transferred between ^{3}He films and graphite substrates by the magnetic coupling between ^{3}He nuclear spins and magnetic impurities in the graphite substrates such as Fe and Ni, and the measured κ_b is dominated by the internal thermal conductance of ^{3}He films, not by the interfaces between the films and substrates.[1,2] The magnetic field reduces the influence of the active magnetic impurities. This means an increase of the effective path length of heat flow inside the ^{3}He films. So, the application of the magnetic field should decrease the

CP850, *Low Temperature Physics: 24th International Conference on Low Temperature Physics;*
edited by Y. Takano, S. P. Hershfield, S. O. Hill, P. J. Hirschfeld, and A. M. Goldman

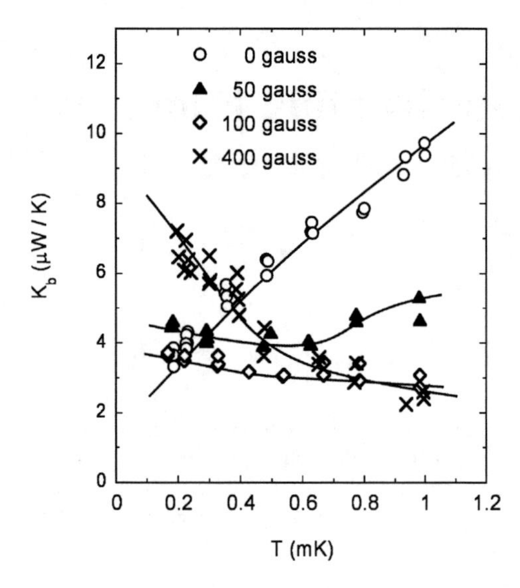

FIGURE 1. Selected measured thermal conductance between the solid ^{3}He film of 7.3 nm^{-2} and the graphite substrate in magnetic fields. The lines are guides for the eye.

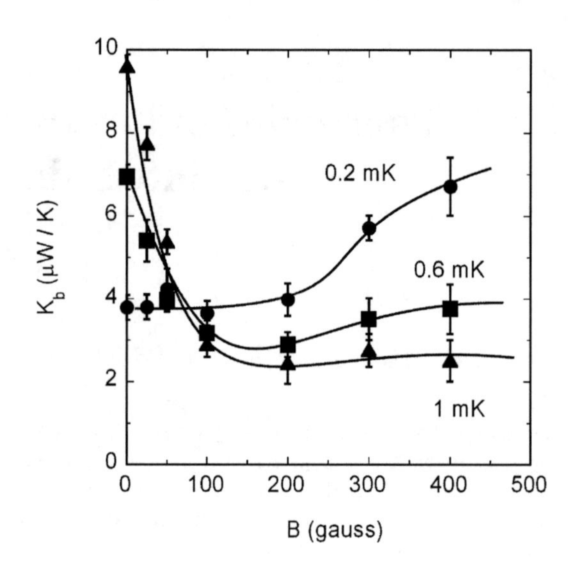

FIGURE 2. Isotherms of the measured thermal conductance between the solid ^{3}He film and the graphite substrate as a function of magnetic field. The lines are guides for the eye.

thermal conductance. This mechanism also explains the decrease of the thermal conductance in the second layer of ^{3}He by the replacement of the first layer with a monolayer of ^{4}He observed in previous measurements.[2] If the amount of ^{4}He is slightly too large to complete the first atomic layer, a part of the amorphous ^{3}He in the second layer is also replaced by ^{4}He. Magnetic impurities should be located at substrate heterogeneities on which amorphous ^{3}He exists. The replacement of the amorphous ^{3}He with ^{4}He also increases the effective path length of heat flow inside the ^{3}He films, and decreases the thermal conductance.

It might be expected that in a sufficiently high magnetic field the magnetic Kapitza mechanism does not work and κ_b should go to zero. The observed behavior seems to saturate at a finite value in a high magnetic field. Similar behavior has been observed at interfaces of bulk helium systems.[4-6] Some other heat flow paths might exist. In contrast, κ_b at the lowest temperature increases with the magnetic field. This increase should be attributed to the increase of the thermal conductivity of the ^{3}He film itself and explained as follows. The heat capacity (C) of the ^{3}He film simultaneously measured with κ_b obviously shifts to higher temperatures with increasing magnetic field. The observed temperature shifts of C are 20 times larger than the Zeeman energy. At 400 gauss the spin entropy change within the measured temperature range is estimated to be 40% of the entropy value in the high temperature limit, while in a zero magnetic field it is estimated to be 10%. The magnetic field considerably promotes a short-range order of the spins. The efficiency of heat transport inside ^{3}He films should increase with the growth of the short-range spin order.

CONCLUSION

Magnetic field dependence of the thermal conductance between a monolayer ^{3}He film and a graphite substrate has been measured. The observed behavior supports the magnetic Kapitza mechanism as the heat transfer mechanism between them.

ACKNOWLEDGMENTS

The author is grateful to A. Yamaguchi for a stimulating discussion. This research is supported by Grants-in-Aid for Scientific Research from MEXT, Japan, and the Cryogenics Center of University of Tsukuba.

REFERENCES

1. M. Morishita, *J. Low Temp. Phys.* **138**, 367-372 (2005).
2. M. Morishita, *J.Phys. Chem. Solids* (in press).
3. J. P. Shepherd, *Rev. Sci. Instrum.* **56**, 273-277 (1985).
4. T. Perry *et al.*, *Phys. Rev. Lett.* **48**, 1831-1834 (1982).
5. D. D. Osheroff and R. C. Richardson, *Phys. Rev. Lett.* **54**, 1178-1181 (1985).
6. S. N. Ytterboe *et al.*, *Phys. Rev. B* **42**, 4752-4755 (1990).

Heat Capacity of Dilute ^{3}He -^{4}He Fluid Mixture Monolayer Films on Graphite

Masashi Morishita

Graduate School of Pure and Applied Sciences, University of Tsukuba, Tsukuba, Ibaraki 305-8571, Japan

Abstract. Heat capacity of dilute ^{3}He-^{4}He mixture submonolayer films was measured. The excess heat capacity previously observed in pure ^{3}He fluid films disappears by the addition of a small amount of ^{4}He. The measured heat capacity changes slightly with the ^{4}He areal density at high temperatures. This behavior supports the coexistence of two-dimensional gas and liquid phases of ^{4}He. However, the variation of the heat capacity with the ^{4}He areal density in the present study is completely different from that obtained in previous measurements of the second atomic layer.

Keywords: ^{3}He-^{4}He film, amorphous, 2D gas-liquid coexistence.
PACS: 67.70.+n, 67.80.Gb

INTRODUCTION

In previous heat capacity (C) measurements of pure ^{3}He fluid films, an excess heat capacity (C_{ex}) has been observed.[1,2] This is believed to be a contribution from amorphous ^{3}He adsorbed on substrate heterogeneities, and to be almost independent of temperature in a certain range.[3] The upper cut-off temperatures of this range are reported to be several mK for the submonolayer and several tens of mK for the second layer.[4,5] However, no direct information has been obtained with regard to the lower cut-off temperatures. On the other hand, in the previous measurements of the second atomic layer dilute ^{3}He-^{4}He mixture films by Ziouzia *et al.*, linear decreases in the C isotherms have been observed and interpreted as evidence for two-dimensional (2D) gas-liquid phase separation of the ^{4}He film.[6] These measurements have not yet been performed for the submonolayer. In this paper, I report the results of C measurements of dilute ^{3}He-^{4}He fluid mixture submonolayer films and discuss the lower cut-off temperature of C_{ex} and the coexistence of 2D gas and liquid phases of ^{4}He.

EXPERIMENTAL

The apparatus and experimental techniques used in this work are the same as those in previous works.[4,5] The total surface area of the graphite substrate was 390 m^2. The sample films were annealed for 10 hours at about 30 K, where the vapor pressure is several mbar.

RESULTS AND DISCUSSION

The measured C values of dilute ^{3}He-^{4}He mixture submonolayer films are shown in Fig. 1. The areal density of ^{3}He was fixed at 0.2 nm^{-2} and that of ^{4}He was increased from 0.80 to 2.9 nm^{-2}. The apparent excess heat capacity is not distinguishable in any of the ^{3}He-^{4}He films. This disappearance of C_{ex} is due to the replacement of amorphous ^{3}He by ^{4}He. The variation in the measured C with the ^{4}He areal density was not significant. However, C increases slightly at high temperatures by about 5%–7% with an increase in the ^{4}He areal density from 0.8 to 1.1 nm^{-2}, and then it decreases. These variations in C are larger than the scatter of the data or the relative error. The first increase is due to further replacement of amorphous ^{3}He by ^{4}He or by an increase of ^{3}He in the fluid. These observations imply that the heterogeneous surface is entirely covered by an amount of ^{4}He that corresponds to an areal density of about 0.85 nm^{-2}. Under the assumption that the areal densities of amorphous ^{3}He and ^{4}He that replaces it are almost identical, the amount of amorphous ^{3}He in our sample cell is estimated to be 0.54 mmole. The spin entropy of this amount of ^{3}He at high temperatures is 3.1 mJ/K. Furthermore, assuming that C_{ex} is independent of temperature below 0.1 mK, the temperature variation of the spin entropy of the amorphous ^{3}He can be evaluated from the previously measured C_{ex}.[4,5] By comparing these two entropy values, the lower cut-off temperature of the amorphous ^{3}He contribution to the

CP850, *Low Temperature Physics: 24th International Conference on Low Temperature Physics;*
edited by Y. Takano, S. P. Hershfield, S. O. Hill, P. J. Hirschfeld, and A. M. Goldman

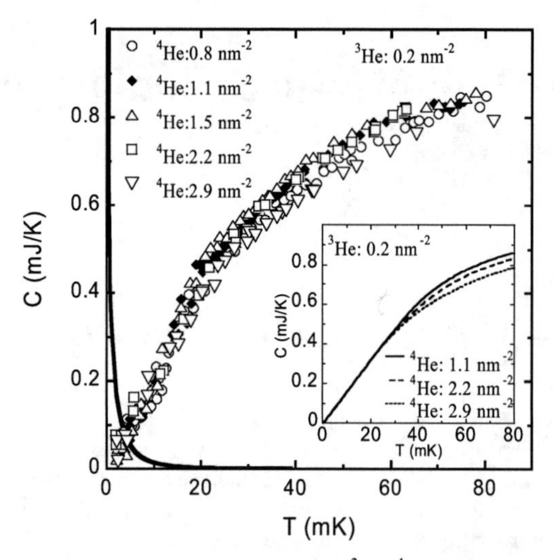

FIGURE 1. Heat capacity of dilute ^{3}He-^{4}He mixture films. The kinks around 20 mK are within the experimental errors due to an uncertainty in the addendum. The solid line shows C_{ex} observed in pure ^{3}He submonolayer films. The inset shows the estimated behaviors by assuming that the ^{3}He areal density in the 2D liquid phase of ^{4}He is 0.14 nm^{-2}.

heat capacity is estimated to be about 50–60 μK. Golov and Pobel predicted a value of 1.5 μK from the reported lower cut-off exchange frequency on porous glass.[3,7] The large estimated value of 50–60 μK means that the distance between ^{3}He atoms is not so small, even in the most dense regimes of amorphous ^{3}He.

The fluid film of ^{4}He on graphite is thought to separate into 2D gas and 2D liquid phases.[8,9] Ziouzia *et al.* have provided evidence for 2D gas-liquid separation in the second layer of ^{4}He by measuring the heat capacity of dilute ^{3}He-^{4}He fluid films.[6] The heat capacity of the 2D Fermi fluid is proportional to temperature ($C = \gamma T$) at sufficiently low temperatures, and the coefficient γ is proportional to the surface area where ^{3}He atoms can move around, not to the amount of ^{3}He. If the 2D gas-liquid separation occurs and the ^{3}He atoms dissolve only in the 2D gas phase of ^{4}He, then γ must show a linear decrease with increasing ^{4}He areal density. The results of Ziouzia *et al.* are in agreement with this argument. With regard to the submonolayer, a distinct areal density variation of the low temperature heat capacity is not seen in Fig. 1. However, 2D gas-liquid separation in the ^{4}He submonolayer cannot be ruled out solely on the basis of this observation. If the solubility of ^{3}He in the 2D liquid phase of ^{4}He is not low, then the low temperature heat capacity does not change with the ^{4}He areal density. Even in such a case, the high temperature behavior of the heat capacity must change with the ^{4}He areal density, provided the Fermi temperatures of ^{3}He in the 2D gas and liquid phases of ^{4}He are sufficiently different. For such cases, the total heat capacities of dilute ^{3}He-^{4}He fluid films are estimated as shown in the inset of Fig. 1. Here, the ^{3}He areal density in the 2D liquid phase of ^{4}He is assumed to be 0.14 nm^{-2}. The estimated results show a ^{4}He areal density dependence that is similar to the experimental results; this should indicate that the ^{4}He fluid film separates into 2D gas and liquid phases. The question thus arises as to why there is a significant difference in the solubility of ^{3}He in ^{4}He between the submonolayer and the second atomic layer. One possibility is the difference in the ^{3}He areal density. Our ^{3}He areal density is twice that of Ziouzia *et al.*[6] However, this possibility can be ruled out since a recent preliminary measurement with a ^{3}He areal density of 0.1 nm^{-2} confirmed that C behavior does not change significantly with the ^{4}He areal density. Stronger adsorption forces in the first atomic layer might play an important role.

CONCLUSION

It is confirmed that the excess heat capacity observed in pure ^{3}He fluid films completely disappears by the addition of a small amount of ^{4}He. Supporting evidence for 2D gas-liquid separation in ^{4}He fluid films is also obtained for a submonolayer film. However, the solubility of ^{3}He in ^{4}He is considerably different between the submonolayer and second layer.

ACKNOWLEDGMENTS

This research is supported by Grants-in-Aid for Scientific Research from MEXT, Japan, and the Cryogenics Center of the University of Tsukuba.

REFERENCES

1. D. S. Greywall and P. A. Busch, *Phys. Rev. Lett.* **65**, 64–67 (1990).
2. M. Morishita *et al.*, *J. Low Temp. Phys.* **110**, 351–356 (1998).
3. A. Golov and F. Pobell, *Phys. Rev. B* **56**, 12647–12650 (1996).
4. M. Morishita, H. Nagatani, and H. Fukuyama, *Physica B* **284–288**, 228–229 (2000).
5. M. Morishita, H. Nagatani, and H. Fukuyama, *Phys. Rev. B* **65**, 104524 - 1-7 (2002).
6. F. Ziouzia, J. Nyéki, B. Cowan, and J. Saunders, *Physica B* **329–333**, 252–253 (2003).
7. Y. Kondo *et al.*, *J. Low Temp. Phys.* **75**, 289–306 (1989).
8. D. S. Greywall and P. A. Busch, *Phys. Rev. Lett.* **67**, 3535–3538 (1991).
9. M. Pierce and E. Manousakis, *Phys. Rev. B* **59**, 3802–3814 (1999).

A Phase Transition in Solid ^{4}He

John M. Goodkind

University of California San Diego, La Jolla California 92093-0319, USA

Abstract. We have performed experiments using acoustic waves and heat waves that revealed a phase transition in solid ^{4}He with $T_C \sim .17$ K. Unusual properties of non-phonon excitations are measured by the experiments.

Keywords: Solid helium.
PACS: 67.80.-s, 67.90.+z

We have used the propagation of acoustic waves and heat waves[1,2,3,4] to search for evidence of a Bose condensation or superfluidity in solid ^{4}He. By adding a few ppm of ^{3}He we discovered a continuous phase transition with $T_C \sim 0.17$ K at molar volume 20.983 cm^3 (T_{melt} = 0.86 K). Within experimental uncertainties, this is the temperature interpreted as the onset of "non-classical moment of inertia" by the Penn State group[5] so that it is likely that the experiments are observing the same phenomena with different tools.

We infer the existence of the transition from a sharp attenuation (α) peak and velocity (v) shift of the acoustic waves as a function of T (Figure 1). The phenomenon is well described by a relaxation process in which the α peak and the midpoint of the shift in v occurs when $\omega\tau = 1$, where ω is the frequency of the acoustic waves (9, 27, and 45 MHz for our work). The data is fit by a relaxation time, $\tau = \tau_0/(T-T_C)$. τ_0 determines the intrinsic width of the anomaly. The frequency dependence predicted by the relaxation process is not observed (Figure 2) apparently due to the very narrow intrinsic width. The width of the anomaly is apparently influenced by ^{3}He concentration, dislocation density, molar volume and, perhaps, crystal orientation. Dislocations can be present simultaneously with the anomaly and produce a broad peak in α centered around 0.5 K. The dislocation density can be measured using the amplitude of the peak. Two crystals grown from the mixture with the highest ^{3}He concentration (220 ppm) showed two transitions that can be fit approximately by the relaxation model. The upper, small one, fits with values close to the others; $\tau_0 = 4\text{x}10^{-10}$ K sec and $T_C = 0.15$ K. The lower one has 4 times larger changes in α and v than the others with $\tau_0 = 0.9\text{x}10^{-10}$ K sec. and $T_C = 0.08$ K. The latter T_C is consistent with an extrapolation of the phase separation curve[6]. At higher density (20.972 cm^3, T_{melt} = 1.2 K) this mixture showed an upper anomaly with $\tau_0 = 0.3\text{x}10^{-10}$ K sec. and $T_C = 0.136$ K. A crystal with 14 ppm ^{3}He, and $V = 20.972$ cm^3 was fit by $\tau_0 = 2\text{x}10^{-10}$ K sec, T_C = 0.085 K so that in two cases we found T_C decreased with decreasing V. The influence of the relaxation process on the acoustic velocity and attenuation is comparable to or larger than other influences only when $\omega\tau \sim 1$ and for very low acoustic power. For pure crystals, the anomaly is too small to observe except in the heat pulse experiments discussed below. That experiment indicates that $\omega\tau >> 1$ for all T below melting in pure crystals.

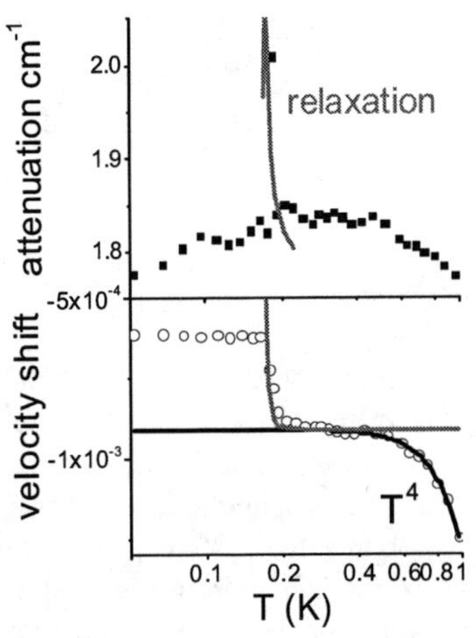

FIGURE 1. Crystal XI12 grown at T=0.86 K from mixture with 14 ppm ^{3}He. V = 20.983 cm^3. Data is at 9 MHz. Smooth curves are for $\tau = 5\text{x}10^{-11}/(T - .17)$.

The time constant for approach to the equilibrium value of α and v, after a change in T is of order 10^4 seconds. But immediately after a change in T, the

CP850, *Low Temperature Physics: 24th International Conference on Low Temperature Physics;*
edited by Y. Takano, S. P. Hershfield, S. O. Hill, P. J. Hirschfeld, and A. M. Goldman

acoustic properties change rapidly in the opposite direction[2]. That is, an increase of T above the α peak causes an immediate increase in α before decreasing toward equilibrium. Below the peak the fast response is a decrease in α. The model we adopted to describe this assumes that the excitations are created and destroyed at a very slow rate when the walls of the container change T, but that the T of the excitations follows the container T rapidly[2,3]. Thermodynamic analysis, assuming that the transition is continuous and that the heat capacity is finite at the critical temperature, requires that the heat capacity is higher just above T_C than just below. This implies that an order parameter is finite above T_C. Our attempts to describe that peculiar conclusion in terms of a Bose condensation[2] required invoking an equally peculiar condensation of thermal excitations. A number of other potentially viable explanations were suggested but have not been analyzed.

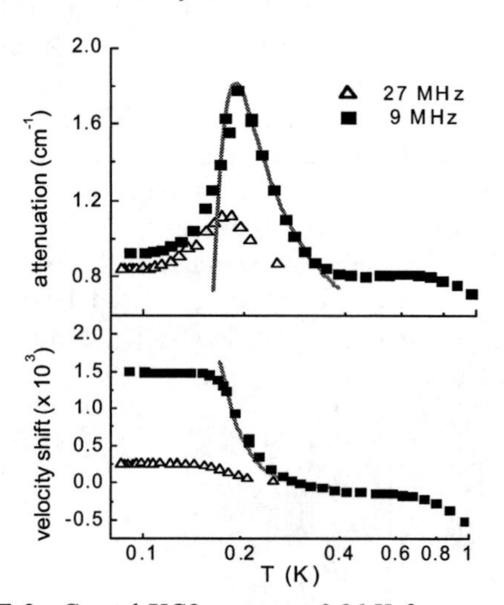

FIGURE 2. Crystal XG2 grown at 0.86 K from a mixture with 27 ppm ^{3}He. V = 20.983 cm^3. The smooth curves are computed for $\tau_0 = 5.2\times10^{-10}$ K sec, $T_C = 0.158$ K.

The experiments also measured a variety of phenomena arising from the non-phonon excitations identified by the T dependence of α. A resonant response of the acoustic propagation, as a function of temperature[1] was found to be consistent with an interaction between the acoustic wave and another traveling wave with a temperature dependent velocity. We assumed this was a wave traveling on the excitation gas and set constraints on its dispersion relation[1]. We detected these waves directly by measuring the interaction of acoustic waves with waves propagated by a heat pulse[3]. Heat waves were launched along a path intersecting the path of the acoustic pulses at right angles. The α and v were measured as a function of the delay time between the

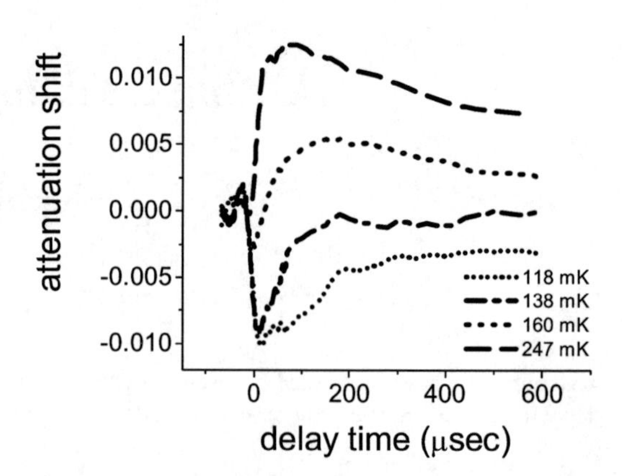

FIGURE 3. Shift of α as a function of delay time between the start of the acoustic pulse and start of the heat pulse.

start of the heat pulse and of the acoustic pulse. This provided a measure of the velocity of waves generated by the heat pulses as well as the magnitude and sign of their effect on the acoustic waves. Below the attenuation peaks shown in Figs. 1 and 2, the waves cause a decrease in α, consistent with the fast response to a change in T described above. However, the bolometer opposite the heater detected no wave at a velocity corresponding to the shortest delay time, indicating that a traveling wave generated by a heat pulse interacts very weakly with the bolometer and container walls, consistent with the above model. We also found that: 1) A wave propagated at 226 m/sec by the heat pulse and detected by the bolometer, is inconsistent with any phonon mode. 2) The shift in the acoustic properties due to the interaction with the heat pulses increases exponentially with the power in the heat pulses above T_C. This signifies an energy gap in the spectrum of these excitations that disappears or becomes small below T_C. 3) In pure solid the interaction with heat waves is measurable only in crystals with high dislocation density. In that case, the response is of the same sign as the response below the peak in α of Fig. 3, but much smaller. From this we conclude that the transition and associated relaxation process are present in pure crystals but $\omega\tau >> 1$ at 9 MHz[4] and the anomaly is masked by other effects.

REFERENCES

1. G. Lengua and J. M. Goodkind, *J. Low Temp. Phys.* **79**, 251 (1990).
2. P.-C. Ho, I. P. Bindloss and J. M. Goodkind, *J. Low Temp. Phys.* **109**, 409 (1997).
3. John M. Goodkind, *Phys. Rev. Lett.* **89**, 095301 (2002).
4. John M. Goodkind, in preparation.
5. E. Kim and M. H. W. Chan, *Science* **305**, 1941 (2004)
6. D. O. Edwards, A. S. McWilliams and J. G. Daunt, *Phys. Rev. Lett.* **9**, 195 (1962)

Pressure Induced Flow of Solid ^{4}He

James C. Day, Tobias K. Herman, and John R. Beamish

Department of Physics, University of Alberta, Edmonton, Alberta, Canada T6G 2J1

Abstract. Recent torsional oscillator measurements on solid ^{4}He in the pores of Vycor and in bulk have demonstrated non-classical rotational inertia, interpreted in terms of a transition to a supersolid phase. In the interest of determining whether or not solid ^{4}He exhibits any of the other unusual flow properties of a superfluid, we have used a piezoelectrically driven diaphragm to study the pressure-induced flow of solid helium into the Vycor pores. We also present preliminary results of the pressure-induced flow of bulk solid helium through an array of microchannels.

Keywords: helium; supersolid
PACS: 65.40.-b, 65.80.+n, 66.30.-h, 66.35.+a, 67.80.-s

INTRODUCTION

Non-classical rotational inertia (NCRI) has recently been seen in torsional oscillator measurements with solid helium confined in the pores of Vycor[1] and subsequently with bulk helium[2]. Mass transport can occur through motion of crystalline defects like dislocations or grain boundaries, and it has been suggested[3] that these may be essential to supersolidity. Since such extended defects would be pinned by a porous matrix, the observation of NCRI in Vycor constrains possible explanations of supersolidity. Our recent experiments[4] examined some alternative explanations that could mimic the decoupling seen in the Vycor study. We have eliminated those that involve a redistribution of mass due to some other phase transition, and have demonstrated the unlikelihood of those that involve a persistent liquid layer. It is thus interesting to see whether solid helium exhibits any of the other unusual flow properties associated with a superfluid.

We have used a capacitive technique to measure the flow of solid helium into the pores of Vycor in response to an external pressure change. Close to melting we saw a slow thermally activated flow, but it essentially disappeared below 1 K. We did not see any additional flow at temperatures as low as 30 mK, which allowed us to put an upper limit on possible pressure-induced superflow. We have also made preliminary bulk measurements, using a capacitive pressure gauge to detect flow of solid helium through an array of capillaries. The absence of flow at temperatures as low as 75 mK puts stringent constraints on possible superflow in bulk solid helium.

RESULTS

We grew our helium crystals in a beryllium copper cell using the blocked capillary technique. The cell's volume could be changed by using an external piezoelectric stack to drive a flexible diaphragm.

For our first set of measurements[4], the cell contained a thin disc of porous Vycor surrounded by bulk helium which we "squeezed" using the diaphragm. Flow of helium into the pores was detected by monitoring the capacitance between copper electrodes evaporated directly onto either side of the disc. Figure 1 shows the capacitance change when we squeezed a Vycor sample containing solid helium (which froze at 2.05 K and a pressure of 57 bar). The top curve shows the response at 1.3 K and is typical of the behavior above about 1 K. The initial jump (0.13 fF within a few seconds) simply reflects the geometric change in capacitance due to elastic compression of the Vycor sample. The subsequent, slower increase is due to changes in dielectric constant associated with flow of solid helium into the pores to equalize the pressures within and around the Vycor. This flow occurred more rapidly at higher temperatures, indicating that it involved a thermally activated process, presumably diffusion of vacancies along the pore wall or in the solid helium, with an activation energy around 9 K. At lower temperatures, flow became too slow to detect and only the elastic response remained, as the middle curve shows for a temperature of 500 mK.

If there is a supersolid and it responds to pressure gradients, then we would expect to see additional flow in the temperature range below 175 mK, where Kim and Chan saw decoupling from their Vycor sample in a torsional oscillator experiment[1]. The lowest curve in Fig. 1 shows the result of a squeeze at 75 mK, but there is no indication of such flow. This result was typical for longer compressions (up to 4 hours) and for all temperatures between 1 K and our lowest temperature (30 mK).

With there never being any detectable density change after the initial capacitance jump, this allows us to put upper limits on the speed at which solid helium can flow at low temperatures in response to a pressure gradient.

CP850, *Low Temperature Physics: 24th International Conference on Low Temperature Physics;*
edited by Y. Takano, S. P. Hershfield, S. O. Hill, P. J. Hirschfeld, and A. M. Goldman

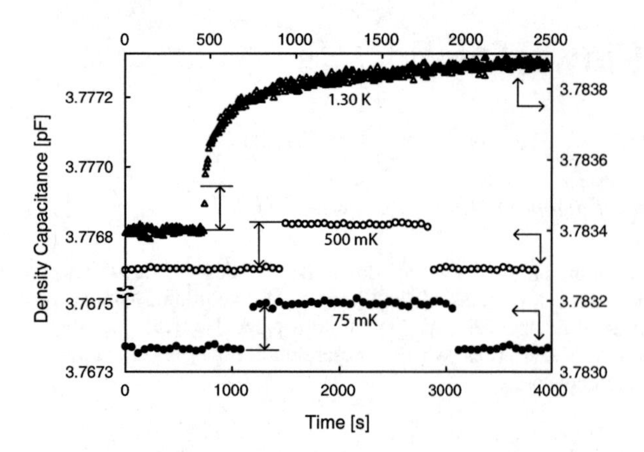

FIGURE 1. Pressure induced flow of solid helium into the pores of Vycor. The density capacitance is proportional to the amount of helium in the Vycor. At temperatures near the melting point of the confined helium (triangles), flow of solid helium into the Vycor was observed and the rate was temperature dependent. At temperatures below half the melting point of the confined helium (open circles) and in the range where decoupling was observed (closed circles), there was no observable pressure induced flow. Note the different scales on both axes.

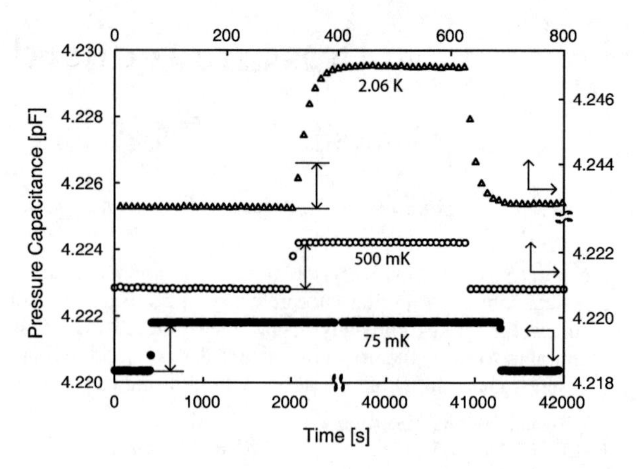

FIGURE 2. Pressure induced flow of bulk solid helium through an array of microchannels. The pressure capacitance is directly proportional to the pressure of the helium in the second chamber. At temperatures near the melting point of the helium (triangles), flow of solid helium through the channels was observed and the rate was temperature dependent. At temperatures below half the melting point of the helium (open circles) and in the range where decoupling was observed (closed cirlces), there was no observable pressure induced flow. Note the different scales on both axes.

The average speed must be less than about 0.015 nm/s. If we assume that 0.5% of the helium participates in superflow (the fraction that decouples in the torsional oscillator experiments[1]), then this would imply a critical velocity less than 3 nm/s.

For our second set of measurements, the cell contained two chambers separated by an array of roughly 40,000 glass capillaries (25 μm diameter, 3 mm long). Helium in the first, larger chamber was compressed and flow through the capillaries was detected by measuring the pressure rise in the second, smaller chamber with an *in situ* capacitive gauge. Figure 2 shows the pressure response in the second chamber due to compression and decompression (pressure changes of about 100 mbar) in the first chamber. There is an immediate pressure response which is, in this case, due to the capillary array that separates the two chambers flexing by about 30 nm (producing a pressure change of roughly 30 mbar in the second chamber). Close to the melting point (at a temperature of 2.13 K and a pressure of 43 bar), this initial jump is followed by a slower pressure rise due to flow of helium through the capillaries, as shown in the upper curve at 2.06 K. The middle curve shows the behavior at 500 mK, where no flow is seen after the initial 1.4 fF jump. The lowest curve shows essentially identical behavior at 75 mK, with no measurable flow over 12 hours. From this and the volume of the second chamber, we can put an upper limit of about 10^{-5} nm/s on the average velocity of pressure-induced flow at low temperatures. This is about 3 orders of magnitude lower than the limit from Greywall's experiments[5]. Even with a supersolid fraction of 1%, this would correspond to a critical velocity less than 10^{-3} nm/s.

Our measurements put limits on pressure induced flow velocities in confined and bulk solid helium that are 3 or more orders of magnitude lower than the critical velocities for NCRI in the torsional oscillator measurements[1, 2]. If helium forms a supersolid, it does not respond to pressure gradients by flowing at speeds comparable to the critical velocity associated with NCRI.

ACKNOWLEDGMENTS

This work was supported by the Natural Sciences and Engineering Research Council of Canada (NSERC).

REFERENCES

1. E. Kim, and M. H. W. Chan, *Science* **305**, 1941-1944 (2004).
2. E. Kim, and M. H. W. Chan, *Nature* **427**, 225-227 (2004).
3. N. Prokof'ev, and B. Svistunov, *Physical Review Letters* **94**, 155302 (2005).
4. J. Day, T. Herman, and J. Beamish, *Physical Review Letters* **95**, 035301 (2005).
5. D. S. Greywall, *Physical Review B* **16**, 1291-1292 (1977).

Search of Superfluidity of Solid ^{4}He in a Porous Vycor Glass by Means of the Ultrasound Technique

T. Kobayashi*, S. Fukazawa*, J. Taniguchi*, M. Suzuki* and K. Shirahama†

*Department of Applied Physics and Chemistry, University of Electro-Communications, Tokyo 182-8585, Japan
†Department of Physics, Keio University, Yokohama 223-8522, Japan

Abstract. Kim and Chan have reported that solid ^{4}He shows a reduction of the rotational moment of inertia below 0.2 K, which suggests an onset of superfluidity. Ultrasound should be sensitive to mass decoupling caused by superfluidity. If a superfluid component exists, the sound velocity of a porous material filled with solid ^{4}He could increase. We have carried out ultrasonic measurements for a porous Vycor glass filled with solid ^{4}He. Since the reported "critical velocity" is very low, we have adopted a continuous wave resonance technique which realizes the oscillating velocity less than 1×10^{-7} m/s. The resolution of the sound velocity is 10^{-5} for small oscillating velocities, and is enough to detect the expected mass decoupling. Although the present experimental conditions are rather limited, no signature of supersolid has been observed.

Keywords: supersolid, Vycor glass, ultrasound
PACS: 67.40.-w, 43.35.Cg

Recently, Kim and Chan performed torsional oscillator measurements of bulk solid ^{4}He and solid ^{4}He filled in a porous Vycor glass, and reported a reduction of the rotational moment of inertia below 0.2 K. This suggests a supersolid state of solid ^{4}He.[1] A supersolid of ^{4}He will open up a new field in superfluid physics. The features of their observation are as follows: (1) The superfluid onset T_C does not depend on pressure and is about 0.2 K. (2) The superfluid fraction at low temperatures is on the order of 10^{-2}. (3) The critical velocity is extremely small, and the superfluid fraction is suppressed above 10 μm/s.

Ultrasound is expected to be a useful probe for a supersolid. Two groups have already carried out precise ultrasound measurements of bulk solid ^{4}He.[2, 3] We report here ultrasound measurements for a porous Vycor glass filled with solid ^{4}He under a low amplitude conditions. Decoupling of a superfluid component from the oscillating substrate should cause an increase in the ultrasound velocity below T_C. The increase is related to the superfluid density as $\Delta v/v_0 = (1-\chi_s)\,\rho_s/\,(2\rho)$, where χ_s is the fraction of the superfluid component which remains locked to the substrate due to tortuosity of the pores, ρ_s the superfluid density and ρ the total density.

We use both longitudinal and transverse waves. The porous Vycor samples are ~15 mm in diameter and 13.12 mm in length for the longitudinal-wave measurements and ~15 mm in diameter and 8.42 mm in length for the transverse-wave measurements. The bottom and top faces of each sample are polished, and $LiNbO_3$ transducers, 10 MHz for the longitudinal sound and 3 MHz for the transverse sound, are then bonded to both ends with silicone adhesive. The area of the electrode is 13 mm^2 and 113 mm^2 for the longitudinal and transverse transducers. The porosity of the sample is about 28%. Before closing the sample cell, the sample is heated to 100°C to remove water.

In order to reduce the amplitude of ultrasound, we use a continuous wave (cw) resonance technique with an rf lock-in amplifier. An alternating voltage is applied to one of the transducers, and the output of the other is fed into the lock-in amplifier to detect the standing wave in the sample. The equivalent circuit for the transducer is shown in Fig. 1. The amplitude of the standing wave is estimated from the output voltage. The current I is related to the voltage V and the oscillating velocity of the transducer face $\dot{u}$ as

$$I = j\omega C_d V + 2\Phi\dot{u}, \tag{1}$$

where C_d is the electrostatic capacity, and Φ the coupling constant between the electric and mechanical systems. Compared with C_d, the impedance $Z_0 = 50\ \Omega$ is small, so that the current is proportional to the oscillating velocity.

For precise measurements of the sound velocity, a phase-locked loop is used. When the resonance condition is satisfied, the quadrature phase component becomes maximum whereas the in-phase one becomes zero. The frequency of the input voltage is controlled to keep the in-phase component zero, and the change in the sound velocity is calculated from the change in the resonance frequency as $\Delta v/v_0 = \Delta f/f_0$. The resolution of the sound velocity is 10^{-5} even for small oscillations. The absolute value of v_0 can be obtained from the resonance curve and the thickness of the sample. The resonance curve with the 3 MHz transverse transducers is shown in Fig. 1. The transverse sound velocity of the sample is 1.65×10^3 m/s at 4.2 K.

CP850, *Low Temperature Physics: 24th International Conference on Low Temperature Physics;*
edited by Y. Takano, S. P. Hershfield, S. O. Hill, P. J. Hirschfeld, and A. M. Goldman

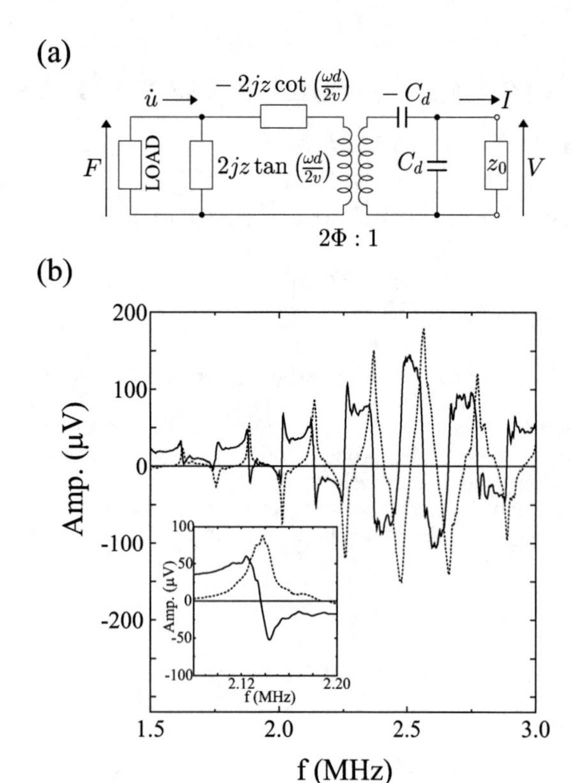

FIGURE 1. (a) Equivalent circuit for the transducer in the present experiment. Here $z = \rho v S$ is the product of the acoustic impedance and the area of the electrode, d the thickness of the transducer, and $\Phi = eS/d$ the coupling constant between the electric and mechanical systems (e is the piezoelectric constant). (b) A resonance curve of the porous Vycor sample using 3 MHz transverse transducers. Inset: enlarged view of the resonance curve.

The inset of Fig. 2 shows the resonance frequency and amplitude for the longitudinal wave with high pressure liquid ^{4}He. The sound velocity was obtained to be 3.59×10^3 m/s at 2 K. In this measurement ^{4}He inside the Vycor sample remains liquid, while the outer ^{4}He solidifies. The frequency increases below 1.72 K due to the superfluid decoupling, and the amplitude decreases at the same temperature. The increase in the frequency at zero temperature reaches about 40 kHz. Compared with the porosity, the observed increase corresponds to $\chi_s \sim$ 0.68. Presupposing the same χ_s for the possible supersolid, the frequency shift Δf due to the "supersolid transition" would be about 400 Hz.

The main panel of Fig. 2 shows the resonance frequency of the longitudinal wave with solid ^{4}He for various oscillating velocities. These velocities correspond to voltage amplitudes of 0.5 – 210 μV. To make solid ^{4}He in the Vycor sample, we pressurized liquid ^{4}He up to 12.5 MPa at 4.2 K, and cooled it down slowly. From the variation of the sound velocity, freezing of ^{4}He in the Vycor sample was confirmed.[4] The expected increase due to a 1% supersolid component is indicated by the vertical line in the figure. Because of the two-level tunneling systems in the Vycor glass, the sound velocity shows a maximum around 0.25 K and has a large temperature dependence in the low temperature region.[5] The velocity of the transducer is smaller than the reported "critical velocity". Therefore, it is expected that the resonance frequency will increase below about 0.2 K due to mass decoupling. As seen in the figure, we do not observe such an increase within the present accuracy. Similarly, in the transverse-wave data, we saw no anomaly within the experimental accuracy.

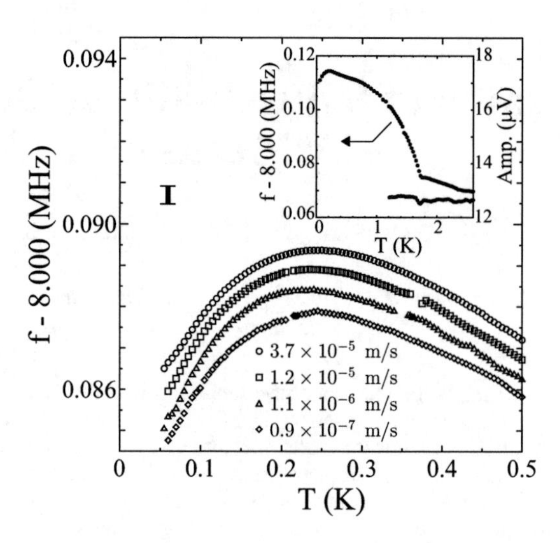

FIGURE 2. Resonance frequency of the porous Vycor sample with solid ^{4}He for the longitudinal wave. The starting pressure at 4.2 K is 12.5 MPa. The different symbols correspond to different oscillating velocities. For clarify, the data sets are shifted vertically by 0.0005 MHz with respect to each other. The vertical line in the figure indicates an expected change due to a 1% supersolid component. Inset: resonance frequency and amplitude for liquid ^{4}He in Vycor with a starting pressure of 9.1 MPa at 4.2 K.

In summary, we have made cw ultrasound studies for a porous Vycor glass filled with solid ^{4}He. The present experimental conditions were rather limited, but we could see no signature of the supersolid. Further studies are under way.

REFERENCES

1. E. Kim and M. H. W. Chan, *Nature*, **427**, 225 (2004); *Science*, **305**, 1941 (2004).
2. I. Iwasa et al., *J. Phys. Soc. Jpn.*, **46**, 1119 (1979).
3. G. A. Lengua et al., *J. Low Temp. Phys.*, **79**, 251 (1990).
4. J. R. Beamish et al., *Phys. Rev. Lett.*, **50**, 425 (1983).
5. N. Mulders et al., *Phys. Rev. B*, **48**, 6293 (1983).

Off-Diagonal Long-Range Order in Solid ^{4}He

D.E. Galli, M. Rossi and L. Reatto

Dipartimento di Fisica, Università degli Studi di Milano, Via Celoria 16, 20133 Milano, Italy

Abstract. We have found the first microscopic evidence for Bose-Einstein condensation (BEC) in crystalline ^{4}He based on a quantitative variational theory of solid ^{4}He: the shadow wave function technique. By computing the one-body density matrix ρ_1 in solid ^{4}He at $T = 0$ K, we find that off-diagonal long-range order is present for a perfect solid ^{4}He for a range of densities above the melting one, at least up to 54 bars. The key process giving rise to BEC is the formation of vacancy-interstitial pairs (VIPs). Such processes have a finite probability to be present in the ground state of the system, and we have confirmed this also by an exact calculation. Due to exchange or other processes these VIPs are unbound; however they are not permanent excitations but simply fluctuations of the perfect crystal induced by the large zero-point motion.

Keywords: Supersolid, Bose–Einstein condensation, solid helium
PACS: 67.80.-s

There is a renewed interest in the supersolid state of matter beginning from the recent observations of non-classical rotational inertia (NCRI) effects in bulk solid ^{4}He[1]. In the first theoretical discussions about a solid phase in which Bose-Einstein condensation (BEC) or superfluidity could be observed, the presence of delocalized ground state vacancies was recognized as the possible origin of these phenomena[2, 3]. Evidence for such zero point vacancies is still lacking and arguments against bulk superfluidity have appeared in literature[4, 5]. Leggett showed that BEC, at least at $T = 0$ K, is a sufficient condition for the observation of NCRI effects[6]. So a central quantity to compute is the one-body density matrix $\rho_1(\vec{r}, \vec{r}')$, whose Fourier transform represents the momentum distribution. A non-zero limit at large distance of ρ_1 (off-diagonal long-range order, ODLRO) implies BEC and then NCRI. In ref. [7] we discuss the results for ρ_1 computed with the shadow wave function (SWF) in a perfect hcp and in a bcc ^{4}He crystal. We found ODLRO for a range of densities above the melting one. For example at $\rho = 0.031$Å^{-3}, which corresponds to a pressure of about 54 bars, we find a condensate fraction of $n_c = (2.0 \pm 0.4) \times 10^{-6}$. These results constitute the first microscopic indication that BEC can be present in perfect solid ^{4}He. By perfect solid, we mean a solid where the number of maxima in the local density $\rho(\vec{r}) = \rho_1(\vec{r}, \vec{r})$ is equal to the number of ^{4}He atoms.

Here we present new results for the calculation of ρ_1 for a perfect solid ^{4}He at $T = 0$ K using the SWF variational technique. With the SWF, the full Bose symmetry is always preserved, and the crystalline order is an effect of the spontaneously broken translational symmetry [8-11]. In this way, all the microscopic processes that might be present due to the large zero-point motion of the ^{4}He atoms are in principle allowed. The SWF technique has obtained a number of important successes not only in the description of the ground state properties of both the liquid and the solid phase of ^{4}He, but, for example, it is able to give also an accurate description of longitudinal acoustic phonons[12] and the excitation spectrum of a vacancy[9].

Recently, we have introduced a new exact projector method, the shadow path integral Ground State (SPIGS), which uses a SWF as a trial wave function[13]. Presently, we have tested the accuracy of the SWF with this method only on diagonal properties. By comparing the static structure factor $S(\vec{k})$ and the local density $\rho(\vec{r})$ with the exact ones, we have found that within the SWF description the ^{4}He atoms are more localized. In fact, the Bragg peaks in $S(\vec{k})$ are 7% higher respect to the SPIGS values and the oscillations in $\rho(\vec{r})$ are more marked than in the exact case, as shown in Fig. 1. This stronger localization of the atoms given by SWF compared to the exact results suggests that the present SWF might underestimate the condensate. In our previous computation of the solid with one vacancy[12], we found an improved description if one extra-shadow variable was introduced. We follow this idea also for the perfect solid and introduce additional extra-shadows, then the SWF now reads

$$\psi_{\text{new}}^{\text{SWF}}(R) = \int dSdS_{\text{v}}\, F(R,S)L(S,S_{\text{v}}) \qquad (1)$$

with the same notation of ref. [12]. More precisely, if N is the number of ^{4}He atoms, we introduce $N + N_{\text{extra}}$ subsidiary variables. N of these are coupled to the real variables in the standard way, and N_{extra} are only coupled to the other N subsidiary variables via a pseudopotential which is separately optimized. We find that these extra-shadows indeed lower the energy and bring $S(\vec{k})$ and $\rho(\vec{r})$ in better agreement with SPIGS results (see Fig. 1). We find that the expectation value of the Hamiltonian has a minimum as a function of N_{extra}, and, at the density of the

CP850, *Low Temperature Physics: 24th International Conference on Low Temperature Physics;*
edited by Y. Takano, S. P. Hershfield, S. O. Hill, P. J. Hirschfeld, and A. M. Goldman

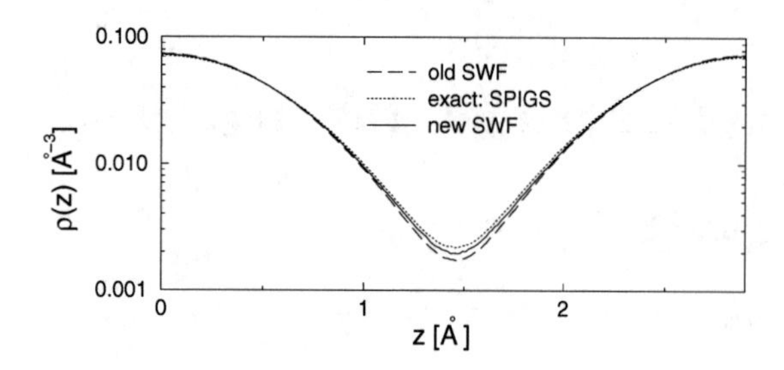

FIGURE 1. Average local density along the z direction for an *hcp* crystal at a global density of 0.031 $Å^{-3}$ computed with the SWF and the SPIGS methods.

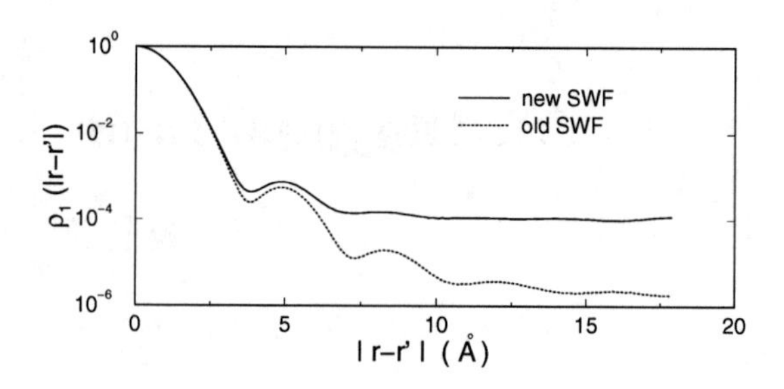

FIGURE 2. One-body density matrix computed in an perfect *hcp* solid ^{4}He at a density of 0.031 $Å^{-3}$ with the old SWF[7] and with the new SWF.

computation, $\rho = 0.031Å^{-3}$ the optimal N_{extra} is $N/20$ with a lowering of the energy of 1%.

With this improved SWF, we have computed also ρ_1 in a perfect *hcp* solid ^{4}He, which is plotted in Fig. 2. The lower degree of order in the structure of the crystal is observable in the oscillations of the plateau which reflect the crystalline order in the system; they are still present but very much damped. The value of the condensate fraction obtained with the new SWF at $\rho = 0.031Å^{-3}$ increases up to $n_c = (1.1 \pm 0.3) \times 10^{-4}$. So the improved description of the crystalline structure has a very relevant effect on ρ_1 and on the value of n_c. It is possible that, by using the new SWF at higher densities, the pressure range where ODLRO is present could be enlarged. As shown in ref. [7], from the behavior of the oscillations in the plateau of ρ_1, it is possible to gain informations on the microscopic mechanism at the basis of ODLRO in the solid. By analyzing the particle configurations which contribute to the plateau of ρ_1, we found that at least one of the coordinates $\vec{r}$ and $\vec{r}\,'$ refers to an interstitial position, i.e., a vacancy-interstitial pair (VIP) is present in the system[7]. The presence of VIP is the key process that gives rise to BEC in solid ^{4}He. Due to the fact that VIPs can appear with a small, but finite, probability at any lattice points in the perfect solid, one can destroy a particle in any lattice position and create it at any far distance with a non-zero probability due to the presence of holes in the crystal. Since the concentration of VIPs is finite the destruction-reconstruction process at arbitrary large distance is a zero energy cost process due to the indistinguishability of the atoms. Differently from the defected solid case, the microscopic processes that give rise to a finite probability to find an unoccupied (distorted) lattice position are part of the zero-point motion and are not excitation of the system. The condensate fraction is small but is finite; in a macroscopic sample, this would lead to a macroscopic occupation of the zero momentum state. The exchange of atoms and VIPs are evident not only in ρ_1 but also in $|\Psi|^2$ as seen in the configurations generated with SWF and with the exact SPIGS method[7]. It is worthy to stress that the name "pairs" is given to the vacancy-interstitial fluctuation of the crystal only to underline the origin of this microscopic process. Due to the exchange between atoms or other zero-point processes, in fact, these VIPs are unbound and interstitials can annihilate far away from the original partners, with vacancies coming from other VIPs.

The success of SWF in describing diagonal properties like the energy or the structure of solid ^{4}He does not strictly ensure that the off-diagonal properties of ρ_1 are accurately reproduced. However, since BEC implies NCRI, our results, if confirmed by an exact calculation, are very relevant, in that they provide the basic explanation of effects recently measured[1]. Moreover these NCRI effects can only be increased by the presence of disorder, if the experimental bulk sample contains some defects. Then, even if the bulk sample would be a perfect solid, the observed NCRI effects should be present in any case, perhaps at a different transition temperature and with a different superfluid fraction than the ones measured in ref. [1]. This indicates the relevance of computing ρ_1 also by an exact Quantum Monte Carlo method. The computation of ρ_1 by means of an exact projector method, such as SPIGS, is in progress.

REFERENCES

1. E. Kim and M.H.W. Chan, *Science* **305**, 1941 (2004).
2. A.F. Andreev and I.M. Lifshitz, *Soviet Phys. JETP* **29**, 1107 (1969).
3. G.V. Chester, *Phys. Rev. A* **2**, 256 (1970).
4. D.M. Ceperley and B. Bernu, *Phys. Rev. Lett.* **93**, 155303 (2004).
5. N. Prokof'ev and B. Svistunov, *Phys. Rev. Lett.* **94**, 155302 (2005).
6. A.J. Leggett, *Physica Fennica* **8**, 125 (1973).
7. D.E. Galli *et al.*, *Phys. Rev. B* **71**, 14056 (2005).
8. F. Pederiva *et al.*, *Phys. Rev. B* **56**, 5909 (1997).
9. D.E. Galli *et al.*, *J. Low Temp. Phys.* **134**, 121 (2004).
10. S.A. Vitiello *et al.*, *Phys. Rev. Lett.* **60**, 1970 (1988).
11. F. Pederiva *et al.*, *Phys. Rev. Lett.* **72**, 2589 (1994).
12. D.E. Galli *et al.*, *Phys. Rev. Lett.* **90**, 175301 (2003).
13. D.E. Galli and L. Reatto, *Mol. Phys.* **101**, 1697 (2003).

On the Growth Dynamics of ^{4}He Crystals near the First Roughening Transition.

I. A. Todoshchenko*, H. Alles*, H. J. Junes* and A. Ya. Parshin†

*Low Temperature Laboratory, Helsinki University of Technology P. O. Box 2200, FIN-02015 HUT, Finland
†P. L. Kapitza Institute, Kosygina 2, Moscow 119334, Russia

Abstract. The first roughening transition of a surface of hcp ^{4}He crystals was carefully studied in Paris in 1980s. By investigating the growth dynamics of the (0001) facet, the free energy of an elementary step was measured in the close vicinity of the transition and a good agreement was found with the theory of critical fluctuations developed by Nozières and Gallet. We believe, however, that the interpretation of the growth data near the roughening transition made by the Paris group is not self-consistent. We argue that with the step energies they obtained, assuming that the growth is due to the process of 2D-nucleation of terraces, another growth mechanism provided by screw dislocations should be much more effective.

Keywords: helium, crystals, roughening, growth
PACS: 68.35.Md 68.08.-p 67.80.-s

Facets exist on a crystal surface at low temperatures as smooth planes which are stabilized by the periodic lattice potential. As temperature increases, thermal fluctuations tend to destroy the facets, until the surface enters the rough state at the so-called roughening transition temperature T_R. The order parameter, which determines the size of the facet, is the free energy β of the elementary step on the facet.

However, measurements of the equilibrium facet size are very difficult. Another way to investigate the free energy of the elementary step is to study the growth kinetics of the facet. The growth rate of the facet strongly depends on the step free energy which can be thus measured provided that the growth mechanism and the corresponding kinetic coefficients are known.

Here we will discuss experiments and results reported in two papers, by Wolf *et al.* [1] and by Gallet *et al.* [2] In both papers similar techniques were used to investigate the growth dynamics of the (0001) facet of a hcp ^{4}He crystal near its roughening transition at $T_R = 1.3$ K. The chemical potential difference between liquid and solid phases $\Delta\mu$ was obtained by measuring the height difference H between the sample crystal and the reference crystal with much larger horizontal surface: $\Delta\mu = (\rho_s - \rho_l)gH/\rho_s$. Above T_R the velocity of the (0001) orientation is proportional to the driving force, $u \propto \Delta\mu$, which is the property of the rough surface. Below T_R the dependence of u on $\Delta\mu$ becomes nonlinear, indicating facet formation.

Wolf *et al.* [1] and Gallet *et al.* [2] suggested that the dominant growth mechanism of the facet is spontaneous nucleation of seeds of a new layer, and the growth velocity of the facet varies as $\Delta\mu \exp[-\pi\beta^2/(3d\rho_s\Delta\mu k_B T)]$, where d is the step height. The values of $\ln(u/\Delta\mu)$ plotted against $1/\Delta\mu$ fall indeed onto a straight line, and from the slope of the line the step free energy was obtained at different temperatures down to 1.130 K.

There is another mechanism of growth which is provided by the steps originating at screw dislocations. If driving force is applied, these steps form spirals which rotate around the dislocations, and the velocity of the facet varies as square of the driving force:

$$u_{sp} = \frac{N\rho_s d^2 k_{st}\Delta\mu^2}{19\beta}. \quad (1)$$

Here k_{st} is the mobility of the step and N is the number of layers produced by one turn around the dislocation. For such a growth there is a certain threshold because the step typically connects two dislocations of opposite sign, forming the so-called Frank-Read source, and cannot move away until the chemical potential difference compensates the largest possible curvature of the step, $2/l$ (l is the size of the Frank-Read source). The corresponding threshold is found from the average distance between dislocations $\langle l\rangle$: $\Delta\mu_c = 2\beta/(d\rho_s\langle l\rangle)$.

Wolf *et al.* and Gallet *et al.* evaluated the velocity of the spiral growth by using Eq. (1). They estimated that under the conditions of their experiments the spiral growth mechanism provides smaller facet velocities than they measured and is thus not relevant. In their evaluation they assumed that the mobility of the step is equal to the mobility of the rough surface. However, the assumption that the mobility of the step is the same as the mobility of the rough surface seems not correct. Only if the step is very sharp, so that the interface changes its height by d over the distance of a few interatomic spacings, then the step mobility can indeed be taken to be approximately equal to the mobility of the rough interface, disregarding

CP850, *Low Temperature Physics: 24th International Conference on Low Temperature Physics;*
edited by Y. Takano, S. P. Hershfield, S. O. Hill, P. J. Hirschfeld, and A. M. Goldman

the peculiarities of the scattering of quasiparticles by the microscopic step. Near the transition, where the step width ξ diverges, the step can be thought as the rough interface tilted by a very small angle $\sim d/\xi$ with respect to the facet orientation [3]. When the chemical potential difference is applied, the velocity of the step *along* the facet is larger by a factor of $\sim \xi/d$ than the *normal* velocity of the rough surface.

In the theory of critical fluctuations developed by Nozières and Gallet [3], the step width ξ and the step mobility k_{st} are given by

$$\xi = \frac{2}{\pi^2}\frac{\gamma d^2}{\beta}, \quad k_{st} = \frac{\pi^2}{2}\frac{\xi}{d}k_R = \frac{\gamma d}{\beta}k_R, \tag{2}$$

where $\gamma \approx 0.24\,\mathrm{erg/cm^2}$ [1] is the surface stiffness. We can estimate that even at 1.130 K, the lowest temperature of the measurements, the reported value of the step free energy $\beta(1.130\,\mathrm{K}) = 1.4\cdot 10^{-11}$ erg/cm suggests the step width $\xi = 4\cdot 10^{-6}$ cm and the factor of 600 in the mobility of the step compared to the mobility of the rough interface. This factor becomes even larger when approaching the roughening transition, but it was not taken into account by Wolf *et al.* [1] and Gallet *et al.* [2] in their estimation of the velocity of the spiral growth. It is thus clear that the possible role of the spiral growth in these experiments should be revised, as it was already pointed out in the recent review on helium crystal surfaces [4].

To obtain the velocity of the facet growing due to screw dislocations, we merge the mobility of the step given by Eq. (2) into the basic relation (1) and find $u_{sp} = N\rho_s d^3\gamma k_R \Delta\mu^2/(19\beta^2)$. We will take the value of $N = 2$ as the lowest possible for (0001) orientation and use the same value of the rough surface mobility, $k_R = 3.1\cdot 10^{-6}e^{7.8\mathrm{K}/T}$ s/cm, as was used by Wolf *et al.* and Gallet *et al.* The values of the step free energies are taken as they were reported by these authors.

We have found that, at all temperatures and at all driving forces, the velocity of the facet growing by screw dislocations is much higher than the measured velocities. Figure 1 presents three sets of original raw data obtained at 1.130 K, 1.173 K (Wolf *et al.* [1]) and 1.205 K (Gallet *et al.* [2]) in $\sqrt{u}$–$\Delta\mu$ coordinates. The velocities of the spiral growth with the values of the step energy reported in these works, namely $\beta(1.130\,\mathrm{K}) = 1.4\cdot 10^{-11}$ erg/cm, $\beta(1.173\,\mathrm{K}) = 6.3\cdot 10^{-12}$ erg/cm, and $\beta(1.205\,\mathrm{K}) = 8.4\cdot 10^{-13}$ erg/cm, are shown by straight lines which are connected by arrows with the corresponding experimental data sets. One can see that these lines lie always higher than the experimental points, and in some cases the spiral growth with the reported step energies gives an order of magnitude faster growth than it was actually measured.

We can also evaluate the values of the threshold for spiral growth using the average distance between dislocations measured in the same work by Wolf *et al.* [1],

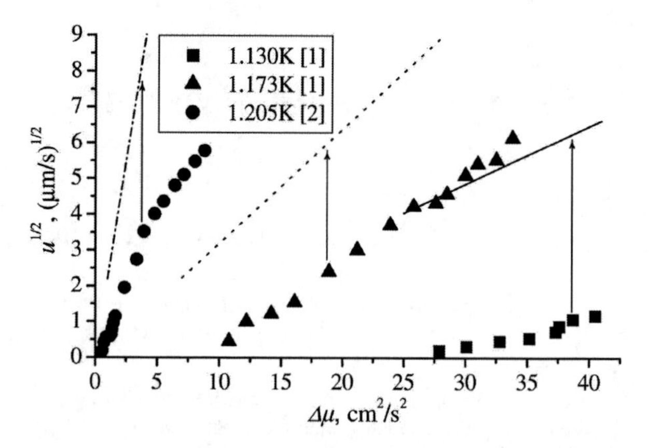

FIGURE 1. Velocities of the (0001) facet measured by Wolf *et al.* [1] and by Gallet *et al.* [2] (symbols) and evaluated velocities of the spiral growth (lines). See text for details.

$\langle l\rangle \approx 0.01$ cm. We find $\Delta\mu_c = 0.5\,\mathrm{cm^2/s^2}$ at 1.130 K, $0.2\,\mathrm{cm^2/s^2}$ at 1.173 K, and $0.03\,\mathrm{cm^2/s^2}$ at 1.205 K, so that $\Delta\mu_c$ is smaller by two orders of magnitude than the actual driving forces applied in the experiment (see Fig. 1).

We conclude that the values of the step free energies found in the experiments by Wolf *et al.* and by Gallet *et al.* suggest much larger facet velocities than what was measured, and hence the results reported in these two papers are not self-consistent. Thus, 20 years after these pioneering works, the adequate understanding of the nature of the roughening transition is still absent, and we would like to renew the attention to this interesting phenomenon.

ACKNOWLEDGMENTS

We are grateful to S. Balibar for sending us the raw data and for very fruitful discussions.

REFERENCES

1. P. E. Wolf, F. Gallet, S. Balibar, and E. Rolley, *J. Physique* **46**, 1987–2007 (1985).
2. F. Gallet, S. Balibar, and E. Rolley, *J. Physique* **48**, 369–377 (1987).
3. P. Nozières, and F. Gallet, *J. Physique* **48**, 353–367 (1987).
4. S. Balibar, H. Alles, and Parshin, A. Ya., *Rev. Mod. Phys.* **77**, 317–370 (2005).

Homogeneous Nucleation of Solid ^{4}He

Ryosuke Ishiguro, Frédéric Caupin and Sébastien Balibar

Laboratoire de Physique Statistique de l'Ecole Normale Supérieure, associé aux Universités Paris 6 et 7 et au CNRS, 24 Rue Lhomond, 75231 Paris Cedex 05, France

Abstract. We report the possible observation of the homogeneous nucleation of solid ^{4}He from highly pressurized liquid ^{4}He in the absence of walls. We have used a new spherical geometry for the transducer which focuses an ultrasound burst (1.3 MHz) in bulk superfluid ^{4}He. Cavitation and possible crystallization events are detected with a laser beam. From the signal amplitude and an analysis of nucleation times, we distinguish two regimes, depending on the static pressure in the cell. At low static pressure (0 to 5 bar), cavitation occurs during the negative pressure swings of the wave, as previously observed with a hemispherical transducer. At large static pressure (20 to 25 bar), the crystallization is shifted by half a period, indicating crystallization by the positive swings. We now need to understand the surprising dependence of the crystallization voltage on static pressure, which is possibly related to non-linear effects in the sound focusing. We also need to measure the instantaneous pressure at the acoustic focus, possibly by Brillouin scattering of the laser beam.

Keywords: nucleation, solid helium
PACS: 67.80.-s, 43.35.+d, 64.60.-i

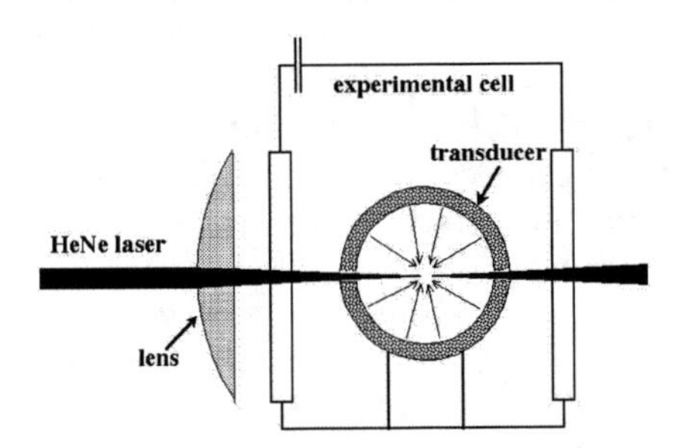

FIGURE 1. Experimental cell with the spherical transducer.

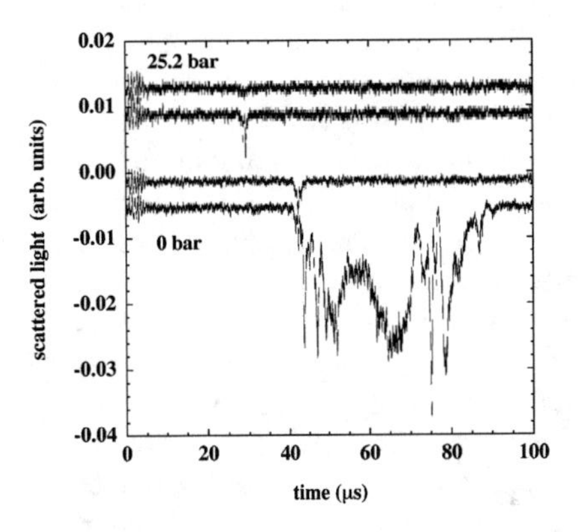

FIGURE 2. Recorded signals at the threshold for nucleation. Top two traces: the static pressure in the cell is 25.2 bar; bottom two traces: 0 bar.

As one increases the pressure of a liquid beyond the liquid-solid equilibrium pressure P_m, one expects the crystalline phase to appear. The liquid phase can be metastable with respect to the solid phase because the liquid-solid transition is first order, but the metastability is usually restricted to a small pressure region above P_m because of the influence of walls or impurities.[1] Even with a pure system like ^{4}He, this is what has been observed in the past. By focusing bursts of ultrasound waves, we can eliminate the influence of walls and look for the intrinsic metastability limits of liquids. In a first series of experiments, we investigated the liquid-gas transition in ^{4}He. We showed that, at low enough temperature, bubbles nucleate in the negative swings of the acoustic bursts and that the cavitation threshold pressure is close to -9.5 bar, the spinodal limit beyond which the liquid is unstable with respect to the gas.[2] After this, we looked for nucleation of crystals in the positive swings of the sound bursts. For a calibration of the sound amplitude, we focused the acoustic wave on a glass plate.[3] We showed that the nucleation of crystals was easy to detect by measuring the scattering of a laser beam at the acoustic focus. The crystallization took place 4.3 bar above P_m = 25.3 bar, the liquid-solid equilibrium pressure. This overpressure was 1000 times larger than in ordinary cells. However, we showed that the crystallization was still favoured by one particular defect on the glass plate. Eventually, we removed the glass plate and looked for homogeneous crystallization in the liquid.[4]

In the experiment by Werner et al. [4], we used a hemispherical transducer which focused 1 MHz sound bursts

CP850, *Low Temperature Physics: 24th International Conference on Low Temperature Physics;* edited by Y. Takano, S. P. Hershfield, S. O. Hill, P. J. Hirschfeld, and A. M. Goldman

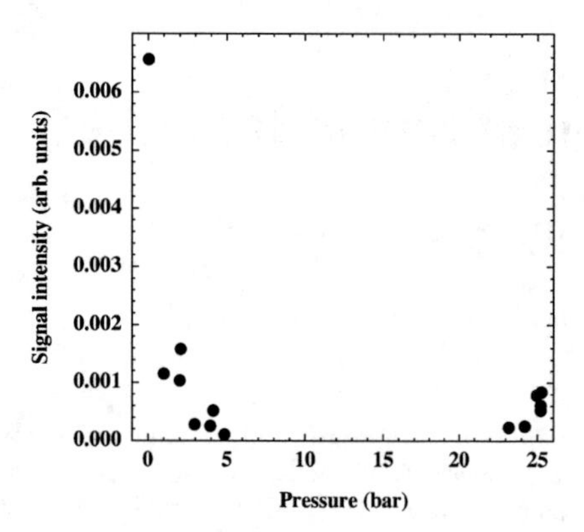

FIGURE 3. Signal intensity increases near the two limits of the stability region of liquid helium.

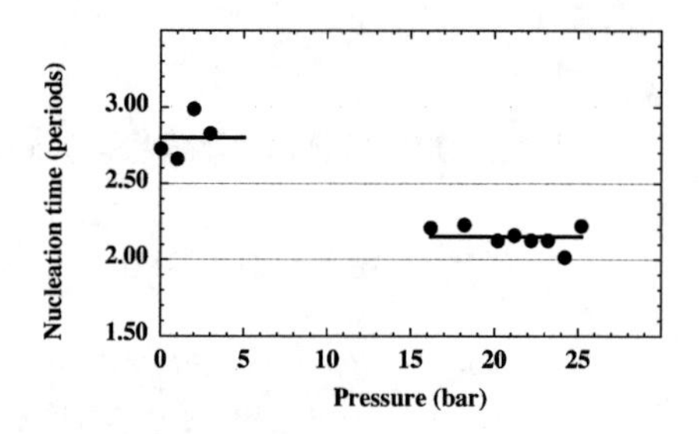

FIGURE 4. Nucleation occurs about half a period earlier at high static pressure than at low pressure. This shift indicates that we observe bubbles nucleating in the negative pressure swings at low pressures, but crystals nucleating in the positive pressure swings at high pressure.

at its center. We studied nucleation as a function of P_{stat}, the static pressure in the cell. In the range $18 < P_{stat} < 25$ bar, we found that the nucleation threshold voltage increased linearly with P_{stat}, and extrapolated to -9.5 bar at zero voltage, so that we assumed that the nucleated objects were bubbles. It allowed us to calibrate our transducers from the known cavitation pressure (-9.5 bar). From the maximum voltage without nucleation we estimated that we had pressurized liquid helium up to 163 ± 20 bar. In order to reach even higher pressures, we then added a second hemispherical transducer and obtained a spherical geometry (Fig. 1). Here we present preliminary results from this experiment. Figure 2 shows two pairs of recordings of the scattered light versus time at two different pressures: 0 and 25.2 bar. In both cases, the upper recording shows light scattering without nucleation but some scattering by acoustic wave, and the lower one shows nucleation. The probability of recording either one or the other type of signal was adjusted to about 0.5 by choosing appropriate acoustic amplitude. At zero bar, the nucleated objects are bubbles whose lifetime is large because the liquid is close to the liquid-vapor equilibrium line. At 25.2 bar, the signals are much smaller. This could be because the bubbles have a much shorter lifetime, consequently a much smaller size in a high pressure environment. However the acoustic wave may nucleate crystals which grow smaller and scatter light less efficiently than bubbles. Figure 3 shows the signal amplitude as a function of P_{stat}. There are two regimes. At low static pressure, the signal decreases as one goes away from the liquid-gas equilibrium. This is of course because the bubbles get smaller. If the signals were always due to bubbles, the amplitude should *decrease* continuously in the whole pressure range from 0 to 25 bar where the liquid is stable. Instead of this, we lose the signal between 5 and 23 bar, and the signal *increases* at high static pressure as the liquid-solid equilibrium line is approached.

Also in favor of crystallization at large P_{stat} is an analysis of nucleation times. Figure 4 shows the time, at which nucleation occurs, in units of the sound period (0.75 μs), after substraction of the flight time from the transducer surface to its center ($R = 9.43 \pm 0.01$ mm). Here, we excited the transducers with bursts of three periods, so that nucleation took place about three periods after the flight time. Despite some scatter due to noise, Fig. 4 shows a delay of about half a period between low and high pressure nucleation. This is consistent with bubbles nucleating in the negative swings and crystals in the positive swings.

We also found that the nucleation voltage increases with P_{stat}, as in ref. [4] but with a different slope. This might be due to non-linear effects which are known [5] to be important in a spherical geometry. As P_{stat} goes away from the spinodal limit, non-linear effects are less important, and the positive swings of the sound wave might be smaller. This could be checked in the near future by combining Brillouin light scattering with our acoustic technique.

R. Ishiguro acknowledges support from JSPS for a Postdoctoral Fellowship for Research Abroad since April 2004.

REFERENCES

1. For a review, see S. Balibar, T. Mizusaki, and Y. Sasaki, *J. Low Temp. Phys.* **120**, 293 (2000).
2. F. Caupin and S. Balibar, *Phys. Rev. B* **64**, 064507 (2001).
3. X. Chavanne, S. Balibar, and F. Caupin, *J. Low Temp. Phys.* **125**, 155 (2001).
4. F. Werner, G. Beaume, A. Hobeika, S. Nascimbene, C. Herrmann, F. Caupin, and S. Balibar *J. Low Temp. Phys.* **136**, 93 (2004).
5. C. Appert, C. Tenaud, X. Chavanne, S. Balibar, F. Caupin, and D. d'Humières, *Eur. Phys. J.* **B 35**, 531 (2003).

Observation and Simulation of Stress Driven Instability on Solid Helium Surface

H. Kojima[a] and P. Grinfeld[b]

[a] Serin Physics Laboratory, Rutgers University, Piscataway, NJ 08854 USA
[b] Massachusetts Institute of Technology, Cambridge, MA 02139 USA

Abstract. Preliminary data on observation of stress driven rearrangement instability on the surface of solid ^{4}He in contact with superfluid at its melting pressure are presented. When the applied strain exceeds a threshold value, the surface becomes unstable and forms irregular deformations. Threshold effects are also seen in numerical simulations.

Keywords: stress driven instability, solid helium, fracture, crack, crystal shape
PACS: 62.20.-x, 67.80.-s, 68.35.-p

INTRODUCTION

Imagine an undisturbed solid material initially in equilibrium with its liquid melt and consider what happens to the shape of the solid when an external stress is applied to it. It was shown theoretically that, if the stress exceeds threshold, the surface of the solid becomes unstable and forms corrugation.[1] The first systematic experimental study of this stress driven rearrangement instability (SDRI) on solid ^{4}He was reported by Torii and Balibar.[2] They indeed observed deformations and corrugations on solid ^{4}He surfaces. Our goal is to extend their study to include effects of orthogonally applied stresses, temperature dependence, dynamics of formation and decay of the corrugations and introduction of impurities into the solid.

The phenomenon of SDRI is related to effects of both practical and fundamental importance. The initial stages of cracking and fracturing are thought to be related to SDRI.[3] The self-assembly process of quantum dots *might* be understood in terms of SDRI where the stress arises between two films with different material composition and lattice constants .[4]

OBSERVATTIONS

A simple cryogenic optical interferometer apparatus has been constructed. A single mode optical fiber guides an incident He-Ne laser beam from room temperature to a beam expander in a vacuum can at low temperature. The expanded beam is incident onto a splitter plate. The horizontally reflected beam from the splitter is absorbed into a light dump. The vertically transmitted part from the splitter is incident onto a partially reflecting window which forms the lower wall of the solid growth chamber (its cross section is a 17.8×10.1 mm^2 rectangle and the depth is 20 mm). The light entering the chamber travels up and reflects at the mirror, which acts as the upper wall of the chamber, and returns back to the splitter. The transmitted and reflected lights from the chamber interfere and the resulting pattern is focused onto the end of an image conduit, which extends up into the room temperature region where it is recorded by a CCD video camera. There is one fringe shift per 93 μm change in solid height and our current resolution is about ±5 μm.

Strains parallel to the long side of the rectangular cell are applied onto solid helium by a movable piston attached to a piezoelectric transducer (PZT). When a voltage V_{pzt} is applied to the PZT, the piston imparts a calibrated strain S (= 5.8×10^{-8} V_{pzt} V^{-1}) onto the solid. Voltages can be applied as high as 5000 V.

To form solid in the cell, the liquid pressure is raised to about 24.5 bars and the temperature is stabilized at 1.20 K. The pressure is then slowly increased until a solid seed forms. The first time a solid is grown, a pressure greater than the equilibrium melting pressure (25 bars) is usually needed before a seed appears. Fig. 1 shows an example of interference pattern with a small seed solid (oval region at left). The presence of a solid meniscus around the seed is apparent.

CP850, *Low Temperature Physics: 24th International Conference on Low Temperature Physics;*
edited by Y. Takano, S. P. Hershfield, S. O. Hill, P. J. Hirschfeld, and A. M. Goldman

The seed can be easily manipulated to increase or decrease in size. The movable piston is located at the left side of the figure. The pattern outside of the seed is determined by the background interference.

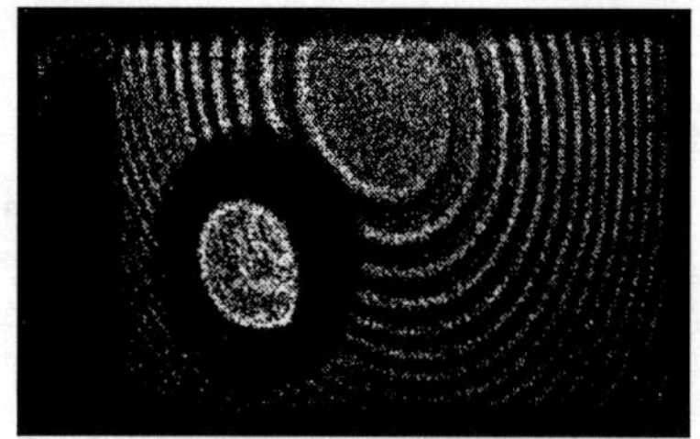

FIGURE 1. Solid seed.

The seed is shrunk to a small size and "annealed" for an hour and then grown back slowly over two to three hours to a vertical thickness of 3 ~ 4 mm over the entire bottom surface. Subsequently, strain is applied to the solid. While V_{pzt} is less than a threshold value (1800 ~ 2300 V), the interference pattern shows locally uneven but smooth variation of height. In general, the solid height decreases in the region near the piston. The height changes by lesser amounts or even increases in the region far away from the piston depending on the particular solid sample. According to the linear equilibrium response between liquid and solid under strain, the solid height is expected to fall globally. Our observations indicate that the applied strain on the solid is not uniformly transmitted across the entire length of the sample. The boundary condition between the solid helium and the chamber wall (machined copper at present) requires further studies. If V_{pzt} does not exceed the threshold value for the sample solid, the interference pattern reverts reversibly to the original pattern when V_{pzt} is decreased back to zero.

When V_{pzt} exceeds the threshold value for a given sample, spectacular localized patterns develop. This onset of instability is reminiscent of previous observations.[2] An example for V_{pzt} = 4000 V is shown in Fig. 2. The high values of V_{pzt} induce several regions

FIGURE 2. Deformed and irregular fringe pattern beyond instability threshold.

to form deep valleys whose nascent stages can be seen by the interference line pattern curling up into high line density regions. The interference pattern relaxes while V_{pzt} is held constant. This interesting relaxation effects will be investigated further in the future. The interference patterns are no longer reversible as V_{pzt} is decreased, and different pattern develops if same time sequence of V_{pzt} is repeated. The threshold and interference pattern are different for each solid samples. The orientation of crystal axis of our samples relative to relative to PZT is not controlled. This might explain the oblique angles of the deformations in Fig. 2.

We have also carried out numerical simulation studies based the formulations[5] for the equilibrium shape of deformable crystals. Our "sample" has dimensions **4x4y1z** discretized into a grid of 193×193×49.[6] Strains are applied in **x** and **y** directions and the vertical displacement is zero at **z** = 0 surface. An example of final surface profile is shown in Fig. 3 for a material (Poisson's ratio = 0.3) with unit Young's modulus and surface tension = 0.05 in reduced units. The intricate height profile does not show a regular pattern. Experiments are being planned to

FIGURE 3. Simulated final configuration.

apply strains onto solid ^{4}He from two orthogonal directions to compare with the simulation results.

ACKNOWLEDGMENTS

We are grateful to S. Balibar for very useful discussions. We thank R. Masutomi and T. Elkholy for development of our optical apparatus. The research is supported by NASA (NAG3-2868).

REFERENCES

1. M.A. Grinfeld, *Dokl. AN SSSR* **265**, 1358 (1982). R.J. Asaro and W.A. Tiller, *Metallurgical Transactions* **3,** 1789 (1972).
2. R.H. Torii and S. Balibar, *J. Low Temp. Phys.* **89**, 391 (1992).
3. K. Kassner and C. Misbah, *Europhys. Lett.* **28**, 245 (1999).
4. C. Teichert, *Applied Phys.* **A76**, 653 (2003)
5. M.A Grinfeld, *Thermodynamic Methods in the Theory of Heterogeneous Systems*, 32, Sussex: Longman, 1991.
6. M.A. Grinfeld, et al., MRS Fall 2004 Proceedings.

Kinetics of Phase Transition at the BCC–HCP–Liquid Triple Points of ^{4}He

E. Rudavskii, N. Mikhin, A. Polev, and Ye. Vekhov

Department for Quantum Fluids and Solids, B. Verkin Institute for Low Temperature Physics and Engineering 61103, Lenin Ave. 47, Kharkov, Ukraine

Abstract. The kinetics of BCC-HCP phase transition along the ^{4}He melting curve is studied by precise pressure measurements in the temperature range of 1.25 – 2.0 K in the vicinity of the triple points (BCC-HCP-He II and BCC-HCP-He I). It is found that the time dependence of the pressure change along the melting curve far from the triple point can be fitted to an exponential. During the crossing of the triple points, an extra contribution to the kinetics of pressure is found in addition to the exponential one. These anomalies can be explained by assumption a re-melting of the crystal during the structural phase transition.

Keywords: ^{4}He, BCC–HCP Phase Transition, Melting Curve, Triple Points, Kinetics.
PACS: 67.80.-s, 68.08.-p, 67.90.+z

INTRODUCTION

The special feature of the ^{4}He phase diagram is an existence of a very narrow region of the BCC phase near the melting curve, between the low triple point (LPT) and upper triple point (UTP) [1-3]. The phase transition kinetics has been recently studied by optical [4] and NMR [5,6] techniques. The optical experiments show that the phase transition occurs at LTP and UTP in different ways. At LTP, a new crystalline phase is grown from superfluid, and at UTP the phase transition is similar to a Martensitic transition. The NMR investigations of the BCC-HCP transition in dilute ^{3}He-^{4}He solid mixtures [5,6] have identified an additional diffusion process which can be caused by liquid droplets appearing during the phase transition. To date, we have no reliable explanation of the BCC-HCP phase transition kinetics.

EXPERIMENTS

The experiments were carried out using a copper cell [6] of a 10 mm diameter and 1.5 mm height, with a weak thermal contact (via a cold finger) with the 1 K pot. The temperature of the cold finger was stabilized and the sample was grown by the blocking capillary technique. During the experiments, stepwise cooling (heating) of the sample was employed. The state of the system was investigated by a precise pressure measurement *in situ*.

RESULTS AND DISCUSSION

The experiments made it clear that, far from triple points, the pressure change after heating (cooling) along the melting curve can be approximated by exponential dependence. An anomalous behavior of pressure and temperature was observed during crossing through the triple points. The pressure and temperature behavior of the sample during a cooling through UTP is plotted in Fig. 1. One can see an additional contribution to the pressure besides the exponential one. This contribution is caused by the HCP-BCC transition and is a non-monotonic function of time with a maximum (Fig. 1c). During a cooling and heating through LTP, the pressure change is approximately the same. The additional pressure change can be explained by a partial or full re-melting during the structural phase transition. This conclusion is in a good agreement with the results of the NMR experiments [5,6] and does not contradict the optical observations [4] in which the observed opacity of the crystal (at UTP) can result from the formation of small liquid droplets. Note that the re-melting effect during the BCC-HCP phase transition was earlier observed in acoustic experiments [7].

CP850, *Low Temperature Physics: 24th International Conference on Low Temperature Physics;* edited by Y. Takano, S. P. Hershfield, S. O. Hill, P. J. Hirschfeld, and A. M. Goldman

During a cooling through UTP and during a cooling and heating through LTP, the liquid is superfluid. In this case the heat exchange between the BCC and HCP solids is very fast due to the presence of superfluid. The anomalous pressure change appears without evident pressure lag, and the pressure changes smoothly. The observed dependence P(t) during a heating through UTP (Fig. 2) shows a different behavior, because the liquid phase is He I with a small heat conductivity. In this case the phase transition starts after overheating up to 0.1 K and manifests itself as a sudden pressure drop (Fig. 2b). At the same moment, the heat absorption is observed (Fig. 2a). Such unusual kinetics can be explained by a re-melting of the crystal during the phase transition. We have estimated that the heat absorption due to the BCC-He I transition is more then 1.5 times the heat release due to the He I-HCP transition.

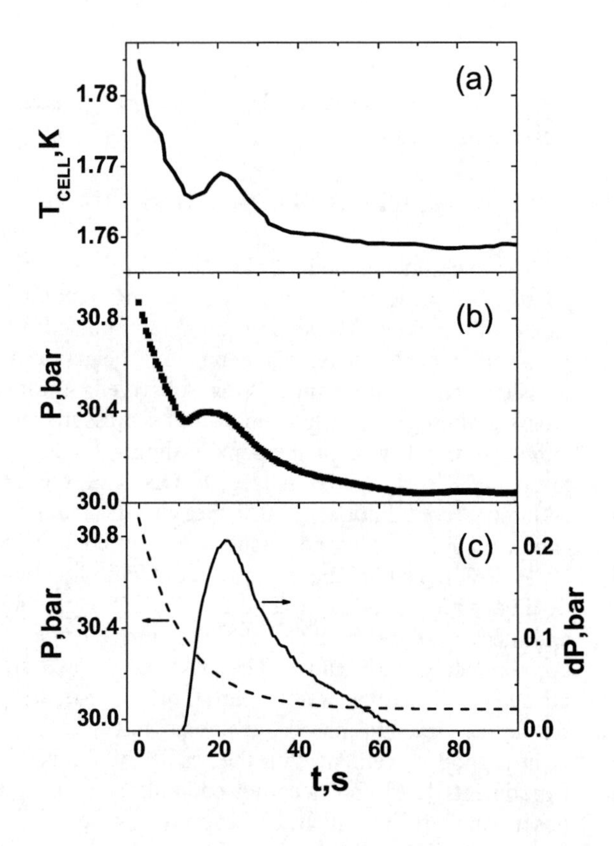

FIGURE 1. Typical time dependences of the temperature (a) and the pressure (b) during a cooling through UTP. (c) – A decomposition of the pressure data to an exponential behavior (the broken line) and an additional contribution (the solid line).

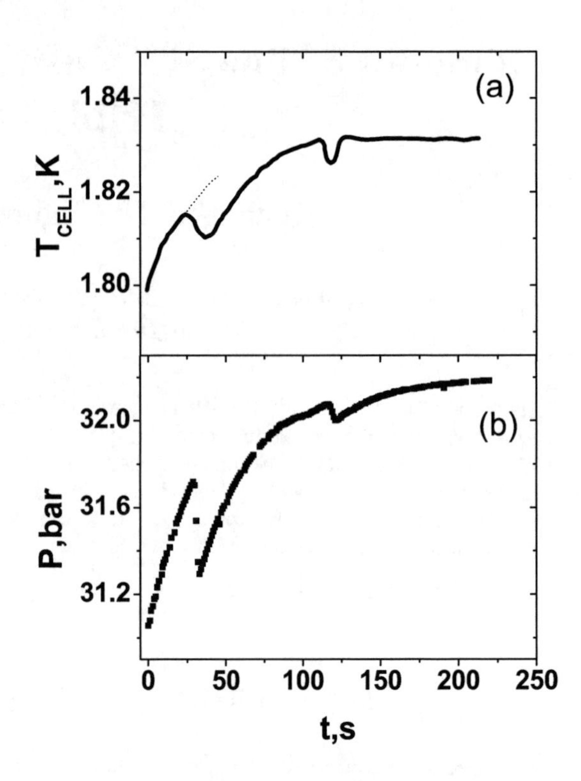

FIGURE 2. Typical time dependences of the temperature (a) and the pressure (b) during a heating through UTP.

ACKNOWLEDGMENTS

This study was supported in part by the Ukrainian Government Foundation for Basic Research (Project 02.07.00391).

REFERENCES

1. Edwards, D.O., and Pandorf, R.C., *Phys. Rev.* **144**, 143-151 (1966).
2. Grilly, E.R., *J. Low Temp. Phys.* **11**, 33-52 (1973).
3. Hoffer, J.K., Gardner, W.R., Waterfield, C.G., and Phillips, N.F., *J. Low Temp. Phys.* **23**, 63-102 (1976).
4. Okuda, Y., Fujii, H., Okumura, Y., and Mackana, H., *J. Low Temp. Phys.* **121**, 725-730 (2000).
5. Mikhin, N., Polev, A., and Rudavskii, E., *JETF Lett.* **73**, 470-473 (2001).
6. Polev, A., Mikhin, N., and Rudavskii, E., *J. Low Temp. Phys.* **127**, 279-287 (2002).
7. Dumin, N., Grigor'ev, V., and Svatko, S., *Sov. J. Low Temp. Phys.* **15**, 142-148 (1989).

Solid-Liquid Interface Motion of ^{4}He Induced by Heat Pulse

Y. Saitoh, T. Ueda, F. Ogasawara, H. Abe[1], R. Nomura and Y. Okuda

Department of Condensed Matter Physics, Tokyo Institute of Technology, 2-12-1, O-okayama, Meguro, Tokyo 152-8551, Japan

Abstract. Displacement of the solid-liquid interface of ^{4}He induced by a heat pulse of 2 ms duration was measured by a high-speed camera. Either crystallization or melting was induced at low temperatures depending on whether the heat pulse was applied to the interface from the solid side or the liquid side. The heat pulse had qualitatively the same effect on the interface as acoustic waves reported in R. Nomura *et al.*, Phys. Rev. B **70** 054516 (2004). However, the effect was smaller and a larger power was needed to induce an interface motion than acoustic waves. Another difference between them is that the heat pulse induced no interface motion at all above 0.8 K, where acoustic waves induced melting.

Keywords: solid ^{4}He, crystal growth, heat pulse
PACS: 67.80.-s, 81.10.Aj, 47.27.Te

INTRODUCTION

It is known from our recent research that acoustic waves can induce both crystallization and melting of ^{4}He at low temperatures depending on whether the waves are applied to the interface from the solid side or the liquid side.[1,2] We attribute this effect to acoustic radiation pressure which pushes the interface unidirectionally. Acoustic waves induce only melting regardless of their direction at high temperatures above about 0.8 K which depend on the orientation of solids. This high-temperature behavior cannot be explained quantitatively by the acoustic radiation pressure model, and heat flow accompanying the acoustic waves has been suspected to be the cause of the melting. Wolf *et al.* have already reported crystallization induced by heat flow.[3] We report here the effect of a heat pulse on the interface.

EXPERIMENTAL AND RESULTS

We performed the experiment in a sample cell cooled by a ^{3}He-^{4}He dilution refrigerator. The refrigerator and the cell had optical windows on their sides and the inside of the cell was observable through the windows at room temperature. Interface motion was visualized by a high-speed camera which took images at rates up to 1 frame/msec. A Mn-Ni wire 120 μm in diameter and 14 cm long was placed in the cell and used as a heater. It had a resistance of 5.2 Ω. It was wound around a plastic plate which had a rectangular shape, 5 mm × 4 mm × 0.5 mm. Superconducting wires were used as leads to avoid spurious heating in other places than the heater. Applied voltage was measured by a digital oscilloscope and power density I was calculated from the area of the moved interface. The high-speed camera was synchronized with the heat pulse by a trigger pulse.

The position of the interface was adjusted above or below the heater so as to apply heat pulses from the solid side or the liquid side. Crystallization induced by a heat pulse from the solid side is shown in Fig. 1a and melting by a heat pulse from the liquid side in Fig. 1b. The displacement induced by a heat pulse of a 2 ms duration is shown in Fig. 2. The circles are for a heat pulse applied from the liquid side with $I = 2100$ W/m^2 and the triangles are from the solid side with $I = 2700$ W/m^2. A positive displacement represents crystallization and a negative displacement melting. The interface moved toward the direction of the heat flow at low temperatures as in the acoustic wave case. However, the interface did not respond to the heat pulse at all above 0.8 K, where acoustic waves induced melting regardless of their direction.[1,2]

[1] E-mail: abe@ltp.ap.titech.ac.jp

CP850, *Low Temperature Physics: 24th International Conference on Low Temperature Physics;* edited by Y. Takano, S. P. Hershfield, S. O. Hill, P. J. Hirschfeld, and A. M. Goldman

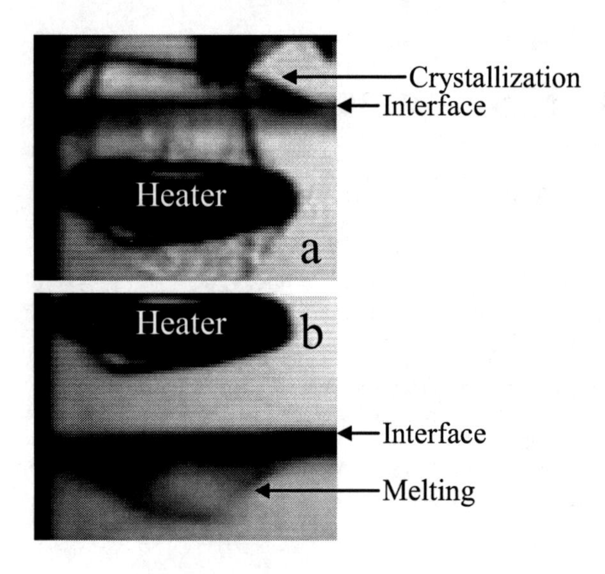

FIGURE 1. Interface motion induced by a heat pulse from the solid side (a) and the liquid side (b) at low temperature.

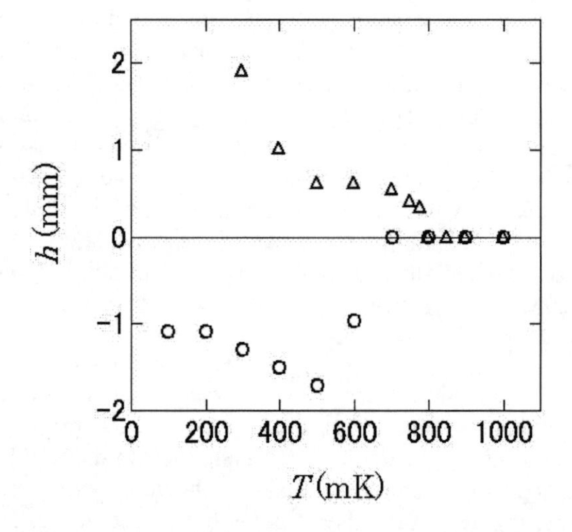

FIGURE 2. Temperature dependence of the interface displacement by a heat pulse. Circles are for a pulse applied from the liquid side and triangles are from the solid side.

Acoustic radiation pressure is a second order acoustic effect and can be regarded as momentum transfer of phonons to the interface.[4] The melting at temperatures above 0.8 K can be explained qualitatively by the acoustic radiation pressure model, which takes into account the large temperature dependence of the reflection coefficient of acoustic wave at the interface.[5] However, the observed amount of melting was too large to be explained by the model quantitatively[1,2] and thus a heating effect accompanying the acoustic waves has been suspected to be the cause of the melting. As is shown in Fig. 2, a heat pulse could induce no interface motion at all in the same temperature range. It is now certain that the observed melting in the acoustic experiment was caused not by mere heating but by a purely acoustic effect.

Because of the negative slope of the melting pressure of ^{4}He at low temperatures, warming of the system is expected to induce only crystallization. It is very intriguing that a heat pulse induces both crystallization and melting as an acoustic wave does. At low temperatures, a heat pulse can be regarded as a phonon beam in the ballistic regime. It is probable that the interface was moved by receiving the momentum of ballistic phonons as an acoustic radiation pressure. In the hydrodynamic regime at high temperatures heat is transferred by an Umklapp process and does not induce the interface motion.

SUMMARY

A heat pulse could induce both crystallization and melting of ^{4}He like an acoustic wave at low temperatures. A heat pulse induced neither crystallization nor melting above 0.8 K where an acoustic wave induced melting only. They acted on the solid-liquid interface similarly at low temperatures and differently at high temperatures.

ACKNOWLEDGMENT

This study was partly supported by Grants-in-Aid for Scientific Research from the Ministry of Education, Culture, Sports, Science and Technology, Japan.

REFERENCES

1. R. Nomura, Y. Suzuki, S. Kimura, and Y. Okuda, *Phys. Rev. Lett.* **90,** 075301 (2003).
2. R. Nomura, S. Kimura, F. Ogasawara, H. Abe, and Y. Okuda, *Phys. Rev. B* **70,** 054516 (2004).
3. P. E. Wolf, D. O. Edwards, and S. Balibar, *J. Low. Temp. Phys.* **51,** 489 (1983).
4. M. Sato and T. Fujii, *Phys. Rev. E* **64,** 026311 (2001).
5. S. Balibar, H. Alles, and A. Y. Parshin, *Rev. Mod. Phys.* **77,** 317 (2005).

Solidification and Melting of ^{4}He in Aerogel Observed by Ultrasound Propagation

Koichi Matsumoto*, Keiichi Yoshino*, Satoshi Abe*, Haruhiko Suzuki*, and Koji Tajiri¶

*Graduate School of Natural Science and Technology, Kanazawa University, Kanazawa 920-1192, Japan
¶Ceramics Research Institute, AIST, Nagoya 463-8560, Japan

Abstract. Helium in small confined geometries remains a liquid at considerably higher pressures than bulk melting. In the present work, we have studied solidification and melting of ^{4}He in aerogel whose structure is different from that of Vycor glass, for which the p-T phase diagram has been established. The velocity and attenuation of longitudinal ultrasound were measured to detect solidification and to determine the solidification pressure. The solidification pressure in 94% porous aerogel is elevated by about 0.3 MPa from that of bulk. This value is considerably smaller than in Vycor glass. We have observed smaller hysteresis on melting and freezing in aerogel than that in Vycor glass. Transmission of sound through the solid-liquid interface in aerogel has been observed.

Keywords: ^{4}He, crystallization, aerogel, ultrasound.
PACS: 67.40.-W, 87.80.Cx

INTRODUCTION

^{4}He confined in porous media has been intensively studied because interatomic interaction can be controlled by pore structure, size and ^{4}He density. The pressure-temperature phase diagram is significantly altered when ^{4}He is confined in narrow pores[1-4]. For ^{4}He in Vycor glass, the freezing pressure is elevated by about 1.5 MPa at about 1.5 K, and the superfluid transition is suppressed by about 0.2 K from that of bulk liquid. Vycor glass has three-dimensionally connected random pores 6 nm in diameter. The melting pressure elevation is ascribed to the inhibition of solid nucleation in narrow pores.

Silica aerogel has porosity up to 99.8%, and its structure varies with porosity and fabrication method. It is thought that aerogel consists of a disordered silica network which is different from Vycor. Superfluid transition in aerogel also differs from that of bulk ^{4}He [5]. We have studied longitudinal ultrasound of liquid ^{4}He in various aerogels at saturated vapor pressure and elevated pressure [6,7].

RESULTS AND DISCUSSION

We observed 10 MHz longitudinal ultrasound to study sound velocity and attenuation of ^{4}He in aerogel. Ultrasonic measurements were made using a standard pulse transmission and a phase sensitive detection technique. The sound cell is described elsewhere [6,7]. The pressure of ^{4}He was measured by a sensor at room temperature. Then, the pressure at the sample cell was determined using the sound velocity when the fill line was blocked. The pressure determination method is described in ref. 7.

Figure 1 shows representative velocity and attenuation measurements. The cell was cooled at a constant pressure from about 3K and then kept at the melting point of bulk ^{4}He in order to grow bulk solid in the void volume of the cell. After the fill line was blocked, the cell was cooled along the melting curve of bulk ^{4}He. When the bulk ^{4}He was completely frozen, the pressure of liquid ^{4}He in the aerogel became higher than the bulk melting pressure, and the sound velocity deviated from that on the melting curve. When the sample was cooled further, attenuation increased abruptly. We identified this change in attenuation as the initiation of freezing of the ^{4}He in the aerogel and attributed this large attenuation to the solid-liquid interface. For the low pressure run (3.2 MPa in Fig.1), the characteristic attenuation peak which corresponded to the superfluid transition of liquid ^{4}He in aerogel was observed and is shown in the figure. The transition temperature was

CP850, *Low Temperature Physics: 24th International Conference on Low Temperature Physics;*
edited by Y. Takano, S. P. Hershfield, S. O. Hill, P. J. Hirschfeld, and A. M. Goldman

lower than that at the bulk melting pressure by several millikelvins due to the higher liquid pressure.

The freezing pressure of ^{4}He in aerogel was determined using the sound velocity at the freezing point [7]. It agreed with the value calculated from the expected thermal contraction of the liquid and the pressure at which the bulk ^{4}He in the void volume was completely frozen. Since the pressure was higher than that of the bcc-hcp transition in bulk solid ^{4}He, we concluded that it was unrelated to that transition in the solid ^{4}He in the void volume. The freezing pressure elevation was about 0.3 MPa in the temperature range of 1.1 K to 1.8 K. The resulting melting curve for ^{4}He in aerogel is roughly parallel to the bulk meting curve.

The sound signal disappeared from approximately 1.6 K to 0.8 K in runs shown in Fig. 1. One reason for this may be large attenuation due to imperfect crystal structure in the cell. The cell was maintained at 1.4 K for 1 day to achieve an annealing effect in the solid. However, no sound signal was observed at this temperature. When the sample was cooled further, the sound signal was observed at low temperatures as shown in Fig. 1. The sound velocity depends on the blocking pressure which determines the ratio of solid to liquid. ^{4}He in the cell was partially solidified in this experiment so the sound velocity of the composite material of solid ^{4}He and aerogel could not be obtained. The fraction of the solid was estimated to be roughly 75%, 55%, and 7% for the runs at 3.7 MPa, 3.5 MPa, and 3.2 MPa, respectively, assuming the sound velocity of the composite material of solid ^{4}He and aerogel was the same as that of the bulk solid.

The attenuation at low temperatures has no systematic dependence on the ratio of solid to liquid. This indicates that the attenuation mainly took place at the solid-liquid interface. In the bulk case, the transmission of sound through the interface is small, because melting and freezing of bulk ^{4}He is a very rapid process[8]. However, it is not clear that the same situation occurs in this system. Acoustic properties of a solid-liquid interface in aerogel are still not well known. Although further experiments are necessary, the present work suggests that sound experiment is a powerful tool to study solid-liquid interface properties in aerogel.

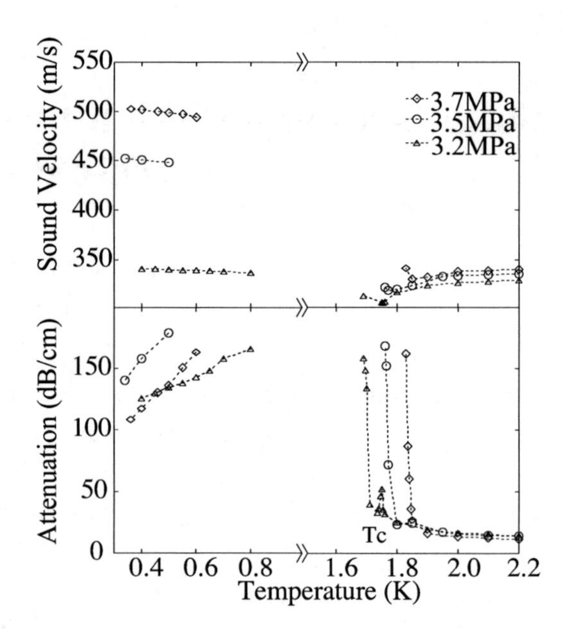

FIGURE 1. Sound velocity and attenuation in the solidification experiment. The data were taken upon cooling, with the initial pressures of 3.7 MPa, 3.5 MPa, and 3.2 MPa. In the 3.2 MPa run, the attenuation peak due to superfluid transition was observed (shown as Tc). Solidification of ^{4}He in aerogel was identified as an abrupt increase in attenuation.

ACKNOWLEDGMENTS

This work was supported by a Grant-in-Aid for Scientific Research from the Ministry of Education, Culture, Sports, Science and Technology of Japan.

REFERENCES

1. E. D. Adams, Y. H. Taung, K. Uhig, and G. E. Haas, *JLTP* **66**, 85 (1987).
2. J. R. Beamish, A. Hikata, L. Tell, and C. Elbaum, *Phys. Rev. Lett.* **50**, 425 (1983).
3. K. Yamamoto, H. Nakashima, Y. Shibahara, and K. Shirahama, *Phys. Rev. Lett.* **93**, 075302 (2004).
4. E. B. Molz, and J. R. Beamish, *JLTP* **101**, 1055 (1995).
5. J. Yoon, D. Sergatskov, J. Ma, N. Mulders, and M. H. W. Chan, *Phys. Rev. Lett.* **80**, 1461 (1998).
6. K. Matsumoto, Y. Matsuyama, D. A. Tayurskii, and K. Tajiri, *JETP Lett.*, **80**, 118 (2004).
7. K. Matsumoto, M. Nishikawa, K. Yoshino, S. Abe, H. Suzuki, D. A. Tayurskii and K. Tajiri, *J. Phys. Chem. Solids*, **66**, 1486 (2005). M. Nishikawa, K. Yoshino, S. Abe, H. Suzuki, K. Matsumoto, D. A. Tayurskii and K. Tajiri, *J. Phys. Chem. Solids*, **66**, 1506 (2005).
8. J. Roitrenaud, and P. Legros, *Europhys. Lett.* , **8**, 651-656 (1989).

Suppression of Freezing and Emergence of a Novel Ordered State in ^{4}He Confined in a Nano - porous Material

K. Yamamoto, Y. Shibayama, and K. Shirahama

Department of Physics, Keio University, Yokohama, 223-8522, Japan

Abstract. Confinement of ^{4}He in a porous material with nanometer - size pores suppresses both the freezing and superfluidity. In our previous investigation of superfluid density of ^{4}He confined in a porous Gelsil glass which has pores of 2.5 nm in diameter, it was demonstrated that the superfluidity is greatly suppressed by pressurization. In order to explore the overall $P-T$ phase diagram, we study the liquid - solid coexistence line. The freezing pressure is elevated up to about 3.4 MPa and independent of temperature below 1.3 K. Along with the previous measurement of superfluid density these features indicate that a nonsuperfluid phase exists next to the solid phase. The flat freezing curve indicates that this nonsuperfluid phase has small entropy as well as that of solid. Therefore the nonsuperfluid phase is possibly a novel ordered state, in which the global phase coherence is destroyed by strong correlation between ^{4}He atoms and/or by random potential.

Keywords: superfluidity, nano porous media, quantum phase transition, strongly correlated system
PACS: 67.40.-w

^{4}He confined in nano porous media offers an ideal example as a Bose system in random potential. Confining ^{4}He in a porous material, we can control the dimension and randomness of the system. In addition, the correlation between ^{4}He atoms can be easily varied over the wide range by changing the density, from the thin film to the pressurized solid.

Although there are considerable amounts of investigations for superfluid film states adsorbed on porous materials, the pressurized states have not been so interested. It was reported that the superfluid transition of ^{4}He in a porous Vycor glass having 6-nm pores was suppressed about 0.2 K from that of bulk [1, 2]. Our recent investigation employing a porous Gelsil glass with 2.5-nm pores [3] has revealed that the superfluidity is drastically suppressed; T_c is about 1.4 K at SVP and approaches 0 K at $P_c \sim 3.4$ MPa. The feature around 0 K suggests a continuous quantum phase transition (QPT) of ^{4}He restricted in a 2.5 nm space at 0 K and P_c. It is interesting and important to explore the overall $P-T$ phase diagram, in particular the liquid - solid coexistence line, in this system. In order to determine the phase diagram of ^{4}He restricted in the nano - space, we have carried out the measurements of the isochores of ^{4}He in the porous Gelsil glass.

The present Gelsil sample with 2.5 nm pores is taken from the same batch as that employed in our previous work [3]. The measurements of the isochores are carried out with a Straty-Adams type *in situ* capacitive pressure gauge [4] down to 20 mK.

Figure 1 shows typical isochores. The isochoric condition is realized by the capillary blocking method. Initially the sample cell is pressurized at $T > 2.5$ K and at

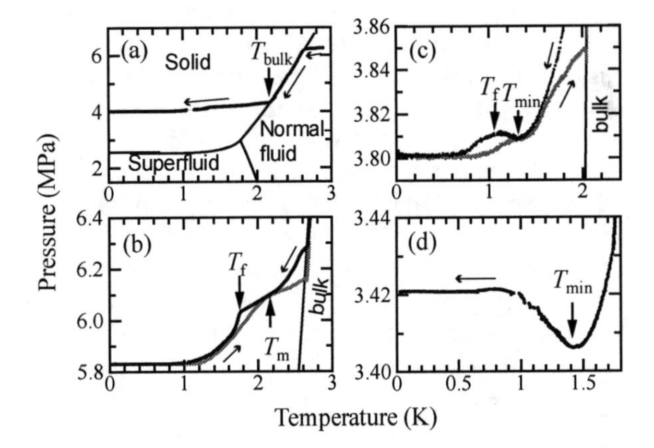

FIGURE 1. Typical isochores of ^{4}He in Gelsil at various density. (a) The temperature dependence of pressure of the sample cell. At T_{bulk} the freezing of bulk ^{4}He in the sample cell is completed. (b) The isochores which is cooled from $P_{ini} = 6.9$ MPa. In the cooling process (shown by black), freezing starts at T_f and freezing completion is not clear. T_m indicates the melting completion in warming run (shown by gray). (c) $P_{ini} = 5.41$ MPa. A pressure minimum at T_{min} is observed in addition to the freezing at T_f. (d) $P_{ini} = 5.33$ MPa.The indication of the freezing is not observed any more and there is only a pressure minimum at T_{min}

5 MPa $< P_{ini} <$ 7 MPa, and then the cell is cooled slowly by operating the dilution refrigerator (Fig. 1(a)). When the bulk ^{4}He that inevitably exists in the open space of the sample cell starts to freeze, the pressure drops following the bulk L - S phase boundary. After the bulk freezing completion at T_{bulk}, the pressure gradually decreases due to the density change of both bulk solid and supercooled liquid in the pores. The features of the

CP850, *Low Temperature Physics: 24th International Conference on Low Temperature Physics;*
edited by Y. Takano, S. P. Hershfield, S. O. Hill, P. J. Hirschfeld, and A. M. Goldman

isochore are classified into three types according to P_{ini}.

Fig. 1(b) shows the isochore at high P_{ini}, 6.9 MPa. In the cooling process, a sharp drop in pressure at T_f is observed. It is the onset of the freezing of ^{4}He in the nanopores as determined in the investigation of ^{4}He in Vycor [1]. Since the pressure decrease is so gradual below T_f, the completion of freezing cannot be identified. In the warming, the onset of melting in Gelsil is also not clear. At a temperature denoted as T_m the cooling and the warming curve intersect, meaning the melting is completed. The hysteresis between cooling and warming curve is also observed in ^{4}He in Vycor [1].

In Fig. 1(c), the isochore at P_{ini} = 5.41 MPa is shown. In the cooling run, a pressure minimum appears at T_{min}, then a sharp drop in pressure due to the freezing of ^{4}He in Gelsil is observed. As the pressure minimum is similar to the feature of bulk ^{4}He near T_λ [5], we suppose that the pressure minimum is associated with superfluidity of ^{4}He in Gelsil (See below). In Fig. 1(d), where P_{ini} = 5.33 MPa, only the pressure minimum is observed.

In Fig. 2, we summarize the freezing onset and the melting completion of ^{4}He in Gelsil with bulk $P-T$ phase diagram. The freezing pressure is elevated up to about 3.7 MPa at 1 K. It is difficult to measure the freezing curve below 1 K with our constant volume apparatus because the freezing curve seems to be independent of temperature. The indication of the freezing of ^{4}He in Gelsil is not observed at 3.42 MPa down to 20 mK (Fig. 1(d)). Therefore it is concluded that the freezing curve is nearly independent of temperature below 1.3 K. Along with our previous superfluid density measurements [3], this indicates that there is a nonsuperfluid phase adjacent to the solid phase. Since the freezing pressure is nearly independent of temperature, the entropy of the nonsuperfluid phase is very close to the solid phase from the Clapeyron - Clausius relation. Consequently, the nonsuperfluid phase has very small entropy, and is possibly a novel ordered state.

In Fig. 2, we also plot the pressure minimum on the $P-T$ phase diagram. The position of the pressure minimum are nearly parallel to the bulk λ-line, but much higher than the superfluid transition temperature T_c determined by the torsional oscillator measurement [3].

A possible picture for the above mentioned feature is that the nonsuperfluid phase is a localized Bose-Einstein Condensation (LBEC) state, which has phase coherence only at length scale of the pore size [6]. The LBEC state is possibly caused by the distribution of the pore size in Gelsil glass because the Bose condensates may grow from the large pores. The LBEC state cannot be detected by torsional oscillator, since LBEC's in larger pores are separated by normal fluid in thinner pores so that no phase coherence exists between the LBEC's. Phase coherence grows as the temperature decreases and macroscopic superfluidity is eventually realized at T_c.

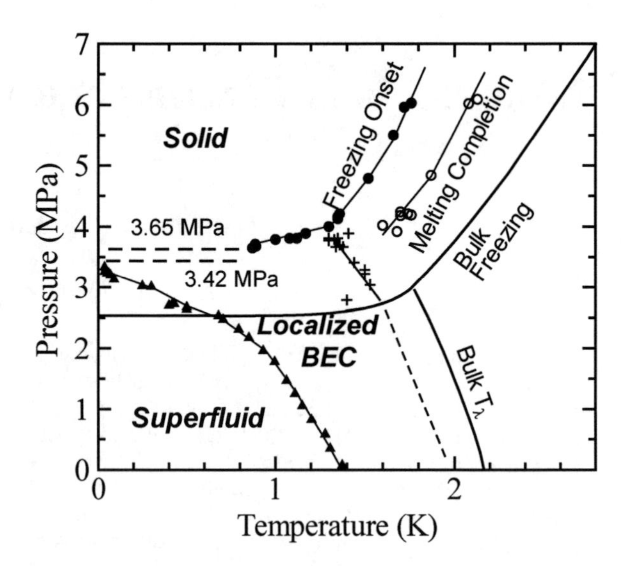

FIGURE 2. $P-T$ phase diagram. The closed and open circle are the freezing onset and the melting completion, respectively. The triangle is the superfluid transition temperature determined by our previous superfluid density measurement [3]. The cross indicates T_{min}. Solid lines through the data are guides to the eye.

It should be mentioned that the pressure minimum might be due to frost heaving [7]. Further detailed study is needed to elucidate the problem.

In summary, we have measured the isochores of ^{4}He in Gelsil and determine the liquid-solid coexistence line. The freezing pressre is nearly independent of temperature below 1.3 K. This fact indicates the existence of nonsuperfluid phase next to the solid phase. The nonsuperfluid phase is possibly a localized BEC state, which has phase coherence only with a short length scale of the pore size. These features could be due to the strong correlation between ^{4}He atoms, confinement, and random potential.

This work is supported by the Grant-in-Aid for Scientific Research (A) from JSPS. K. Y. is supported by Research Fellowships of the JSPS for Young Scientists.

REFERENCES

1. E. D. Adams *et al.*, *Phys. Rev. Lett.* **52**, 2249 (1984).
2. L-Z. Cao *et al.*, *Phys. Rev. B* **33**, 106 (1986).
3. K. Yamamoto *et al.*, *Phys. Rev. Lett.* **93**, 075302 (2004).
4. G. C. Straty *et al.*, *Rev. Sci. Instrum.* **40**, 1393 (1969).
5. J. Wilks, *The Properties of Liquid and Solid Helium*, Oxford: Clarendon press, 1967, pp. 12.
6. H. R. Glyde *et al.*, *Phys. Rev. Lett.* **84**, 2646 (2000); O. Plantevin *et al.*, *Phys. Rev. B* **65**, 224505 (2002).
7. M. Hiroi *et al.*, *Phys. Rev. B* **40**, 6581 (1989); T. Mizusaki (private communication).

Effective Free Energy for Solidification of Superfluid He-4 under Pressure: An Improvement of Previous Model

Tomoki Minoguchi

Institute of Physics, Graduate School of Arts and Sciences, University of Tokyo, Komaba 3-8-1, Meguro-ku, Tokyo 153-8902, Japan

Abstract. It is shown that the model effective free energy previously proposed for solidification of superfluid He-4 can be greatly improved with simple alteration. The model not only describes the pressure dependence of the roton parameters satisfactorily, but also numerically reproduces the superfluid equation of state (Maris' equation of state) surprisingly well. With this model, the overpressurized region can be more safely treated, and the freezing spinodal pressure is predicted as 139 bar, while it is 82 bar in the previously proposed model.

Keywords: superfluid, freezing, solidification, helium, He-4, 4He
PACS: 64.70.Dv, 67.40.-w, 67.80.-s, 68.08.-p, 64.30.+t, 64.60.My, 64.60.Qb

INTRODUCTION

Intensive efforts have been made for quantitative observation of the solid nucleation in overpressurized superfluid He-4 over the last 10 years [1,6]. The issue is deeply connected to some frontiers in physics: the first order phase transition and the macroscopic quantum tunneling. For comparison with experiment, a reliable theoretical model for the superfluid-solid transition is desired, by which the solid nucleation probability is quantitatively calculated.

THE EFFECTIVE FREE ENERGY

Previously Proposed Model

Following a modern concept of freezing, that is freezing is a process in which liquid deforms at a set of finite wave numbers in a restricted degree of freedom [2], I proposed in [3] a two order parameter description for the density deformation of superfluid He-4:

$$\frac{\delta\rho(\vec{r})}{\rho_l}=\xi(\vec{r})+\eta(\vec{r})\sum_{n=1}^{M}e^{i\vec{k}_n\cdot\vec{r}}, \tag{1}$$

where $\delta\rho=\rho-\rho_l$ is the deviation from the mean density ρ_l, and ξ and η the slowly varying order parameters respectively in the phonon and roton domains. k_n is the roton-minimum wave number. I assume six base vectors M=6 in a plane to realize a simple hexagonal reciprocal lattice, which may approximate hcp solid into which superfluid He-4 freezes. The superfluid and solid are characterized respectively by $(\xi,\eta)=(0,0)$ and $(\Delta\xi, \Delta\eta)$.

The zero temperature free energy for the density deformation up a fourth order in $\delta\rho$ was proposed to be

$$H=H_0-\frac{\tilde{b}}{3}\int d^3\vec{r}\,\delta\rho(\vec{r})^3+\frac{\tilde{u}}{4}\int d^3\vec{r}\,\delta\rho(\vec{r})^4, \tag{2}$$

where H_0 is of the quadratic deformation which describes the sound propagation with the phonon-roton spectrum $\omega_{\vec{k}}$:

$$H_0=\frac{m}{2\rho_l V}\sum_{\vec{k}}\frac{|\delta\dot{\rho}_{\vec{k}}|^2}{k^2}+\frac{m}{2\rho_l V}\sum_{\vec{k}}\frac{\omega_{\vec{k}}^2}{k^2}|\delta\rho_{\vec{k}}|^2. \tag{3}$$

The second and third terms in (2) are phenomenologically added to induce instability. Putting aside the kinetic term (the first term) in H_0, where the dot denotes the time derivative, I regard the combination of the second term in H_0 and the second and third terms in H as the grand potential (the T-μ function) W. Substituting (1) for W, I found the effective free energy functional

$$W=\int d^3\vec{r}\,w(\xi,\eta), \tag{4}$$

$$w(\xi,\eta)=\varphi(\xi,\eta)+\lambda(\nabla\xi)^2+\lambda'(\nabla\eta)^2, \tag{4-1}$$

where φ is a quartic polynomial in ξ and η, and λ and

CP850, *Low Temperature Physics: 24th International Conference on Low Temperature Physics;* edited by Y. Takano, S. P. Hershfield, S. O. Hill, P. J. Hirschfeld, and A. M. Goldman

λ' are determined by the energy dispersion $\omega_{\bar{k}}$ [4,8] along with the model assumption M=6.

At the melting pressure P_m, "$\varphi(\xi,\eta)$ should have two minima for the superfluid state and the solid state with the same value φ=0." By using an experimental value $\Delta\xi$ [5] along with values which characterize $\omega_{\bar{k}}$ and ρ_l at 0 K [9], the unknown parameters become just three: $\tilde{b}$, $\tilde{u}$ and $\Delta\eta$. By the condition given above in the double quotes, these three parameters are fixed and the free energy profile (4) is determined.

At pressures $P\neq P_m$, I employed

$$\varphi(\xi,\eta;\Delta P)=\varphi(\xi,\eta)-\Delta P\xi, \tag{5}$$

where $\Delta P=P-P_m$. The superfluid is now characterized by $(\xi,\eta)=(\kappa\Delta P,0)$ with the compressibility $\kappa=4.3\times 10^{-3}$ bar^{-1} at P_m. The roton parameters at P are extracted from the coefficient of the η^2 term in (5) and reproduce experimental values [4] for pressures between 0 bar and a value close to P_m=25.3 bar.

Here I claim, however, that the compressibility extracted from the coefficient of the $(\xi-\kappa\Delta P)^2$ term in (5) is correct only in the vicinity of P_m. In fact, the pressure dependence of the superfluid density determined by

$$\frac{\partial}{\partial\xi}\varphi(\xi,0;\Delta P)=0 \tag{6}$$

is wrong for $|\Delta P/P_m|>>1$ (see Fig. 2) because (5) is a linear approximation. Then the pressure dependence of the roton gap $\Delta(P)$ for $|\Delta P/P_m|>>1$ is not reliable and the freezing spinodal pressure P_c=82 bar estimated by (5), at which $\Delta(P_c)$=0 (Fig. 1), can be doubted.

Improvement

The full contribution of ΔP originates in

$$+\Delta P\int d^3\vec{r}\left(\frac{\rho_l}{\rho(\vec{r})}-1\right), \tag{7}$$

which should be added to W, and (5) should be replaced by

$$\varphi(\xi,\eta;\Delta P)=\varphi(\xi,\eta)+\Delta P\left(\frac{1}{1+\xi}-1\right). \tag{8}$$

With (8), the superfluid equation of state (SEOS) (6) gives surprisingly improved values as shown in Fig. 2. P_c is predicted to be 139 bar (see Fig. 1). This value seems to contradict the focused phonon experiment [6], in which no freezing was seen up to 160 bar at the focus. The contradiction can be a very interesting paradox.

This work was partially supported by the Ministry of Education, Science, Sports and Culture, Grant-in-Aid for Scientific Research (C), 17540313, 2005.

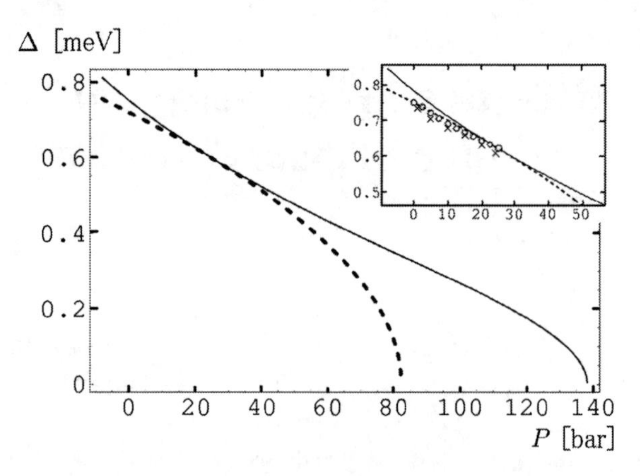

FIGURE 1. Calculated roton gap as a function of pressure with (5) (dotted line), and with (8) (solid line), at 0 K. The crosses and open circles are respectively the experimental data [4] at 1.3 K and their extrapolation to 0 K [9].

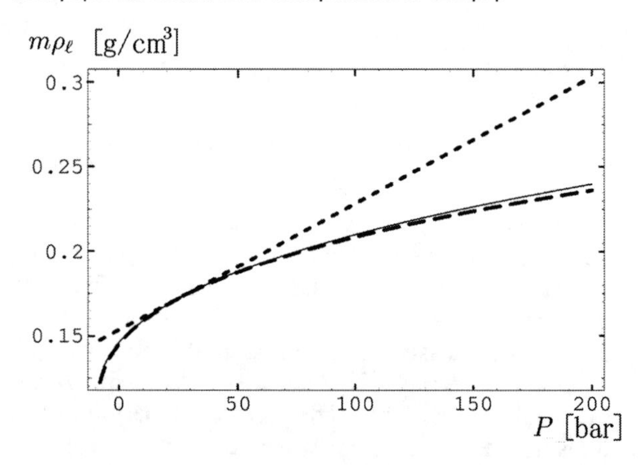

FIGURE 2. SEOS with (5) (dotted line) and with (8) (solid line). The latter is numerically very similar to Maris' equation of state (dashed curve) [7] which empirically fits experimental values (0 to 25 bar) quite well.

REFERENCES

1. J.P. Ruutu, P.J. Hakonen, J.S. Penttila, A.V. Babkin, J.P. Saramaki and E.B. Sonin, *Phys. Rev. Lett.*, **77**, 2514 (1996); Y. Sasaki and T. Mizusaki, *J. Low Temp. Phys.*, **110**, 491 (1998); X. Chavanne, S. Balibar and F. Caupin, *J. Low Temp. Phys.*, **126**, 615 (2002).
2. See for instance, P.M. Chaikin and T.C Lubensky, *Principles of Condensed Matter Physics,* Cambridge: Cambridge University Press, 1997, pp. 188-198.
3. T. Minoguchi, *J. Low Temp. Phys.*, **126**, 627 (2002).
4. O. W. Dietrich, et al., *Phys. Rev. A*, **5**, 1377 (1972).
5. E. R. Grilly, *J. Low Temp. Phys.*, **11**, 33 (1973).
6. F. Werner, G. Beaume, A. Hobeika, S. Nascimbene, C. Herrmann, F. Caupin and S. Balibar, *J. Low Temp. Phys.*, **138**, 93 (2004).

7. H. J. Maris, *Phys. Rev. Lett.*, **66**, 45 (1991); see also Eq. (3) in Ref. 6.
8. N. E. Phillips, et al., *Phys. Rev. Lett.,* **25**, 1260 (1970).
9. Data at T=0 (obtained by extrapolating data of Ref. 4) listed in Table VI in R. J. Donnelly and P. H. Roberts, *J. Low Temp. Phys.,* **27**, 687 (1977).

Transverse Phonon Frequencies in bcc Solid ^{4}He

G. Mazzi, D.E. Galli and L. Reatto

Dipartimento di Fisica, Università degli Studi di Milano, Via Celoria 16, 20133 Milano, Italy

Abstract. We have computed the excitation spectrum of transverse phonons in bcc solid ^{4}He near the melting density with the shadow wave function technique. The representation of a transverse phonon is obtained by taking the wave vector of the excitation operator (a density fluctuation written in terms of the shadow variables) as a combination of a reciprocal lattice vector $\vec{G}$ and a wave vector $\vec{k}$ orthogonal to $\vec{G}$; in this way one is able to compute almost pure transverse phonon excitation energies when $\vec{k}$ is not too large. The computed transverse phonon frequencies are in good agreement with recent experimental results.

Keywords: Solid helium, excitation, phonon
PACS: 67.80.-s

Due to the highly anharmonic dynamics revealed in inelastic neutron scattering (INS) experiments, the study of the excited states of solid helium is still one of the most interesting problems in solid state physics. Solid helium is a prototypical quantum solid, a solid in which the zero–point motion of the atoms covers a significant fraction of the lattice spacing. For this reason the theoretical description of the excited states in this system is quite difficult. Different methods have been applied in order to evaluate the lattice dynamics of solid helium: the self–consistent phonon theory (SCP)[1], the density functional approach[2], the shadow wave function (SWF) technique[3, 4] and, more recently, the path integral Monte Carlo method[5]. Experimental observations of unexpected excitations are also present in literature[6, 7]; this indicates that the physics of solid helium is still not completely understood.

The phonon frequencies reproduced by the earlier theoretical approaches[1] are in fair agreement with experimental data except for the bcc phase of solid ^{4}He where the transverse phonon branch T_1 along the $[\xi,\xi,0]$ direction turns out to be overestimated by almost a factor of 2. Recently we have applied the SWF technique to the calculations of longitudinal phonon frequencies in hcp[3] and bcc[4] solid ^{4}He, finding a very good agreement with the available experimental data. We have now extended the SWF technique to the calculation of transverse phonon frequencies and have applied this method to bcc solid ^{4}He. In the SWF variational theory[8] interparticle correlations are introduced to all orders through the coupling to the subsidiary (shadow) variables $S = (\vec{s}_1,\ldots,\vec{s}_N)$, one for each of the atoms $R = (\vec{r}_1,\ldots,\vec{r}_N)$. The standard representation of an excited state with SWF is obtained via an extension of the Feynman ansatz[9]:

$$\Psi_{\vec{q}}(R) = \int dSF(R,S) \times \sum_j e^{i\vec{q}\cdot\vec{s}_j} \quad , \tag{1}$$

where the explicit and implicit ground-state interparticle correlations contained in $F(R,S)$ have been multiplied by a density fluctuation written in terms of the shadow variables. With the SWF in eq.(1) one can usually describe longitudinal phonons[3] in the solid phase. In order to describe transverse phonons, we have taken the wave vector of the excitation $\vec{q}$ as a sum of two wave vectors: a reciprocal lattice vector $\vec{G}$ plus a wave vector $\vec{k}$ orthogonal to $\vec{G}$. In this way if one assumes that, as in harmonic solids, it is possible to reduce all vectors to the vectors in the first Brillouin zone, the excited state (1) represents a state in which the reduced wave vector of the excitation $\vec{k} = \vec{q} - \vec{G}$ is perpendicular to $\vec{q}$ when $\vec{k} \to 0$; so, in this limit, state (1) represents a pure transverse phonon. This can be seen by noting that the density fluctuation becomes $\sigma_k = \sum_j e^{i(\vec{k}+\vec{G})\cdot\vec{s}_j}$; if one writes $\vec{s}_j$ as a lattice site $\vec{R}_j$ plus a displacement $\vec{u}_j$, by making a Taylor expansion, one obtains

$$\sigma_k = \sum_j (\vec{G}+\vec{k}) \cdot \vec{u}_j e^{i\vec{k}\cdot\vec{R}_j} \quad . \tag{2}$$

Equation 2 shows that the density fluctuation used as excitation operator, for small displacements $\vec{u}_j$, has the same form as in the one–phonon approximation[1] for a transverse phonon when $\vec{k} \to 0$. In principle one could obtain transverse phonons for any $\vec{k}$, if $\vec{q} = \vec{G} + \vec{k}$ is chosen perpendicular to $\vec{k}$. However, the state (1) has to respect the periodic boundary conditions in the simulation box so that we can access a discrete set of q vectors and, apart from special cases, it is not possible to choose $\vec{q}$ exactly perpendicular to $\vec{k}$. In contrast, it is always possible to take $\vec{k}$ perpendicular to a chosen reciprocal lattice vector $\vec{G}$; however, in this case one obtains a good representation of a transverse phonon only when $|\vec{k}| << |\vec{G}|$. As $|\vec{k}|$ increases, the lack of orthogonality between $\vec{q}$ and

CP850, *Low Temperature Physics: 24th International Conference on Low Temperature Physics;*
edited by Y. Takano, S. P. Hershfield, S. O. Hill, P. J. Hirschfeld, and A. M. Goldman

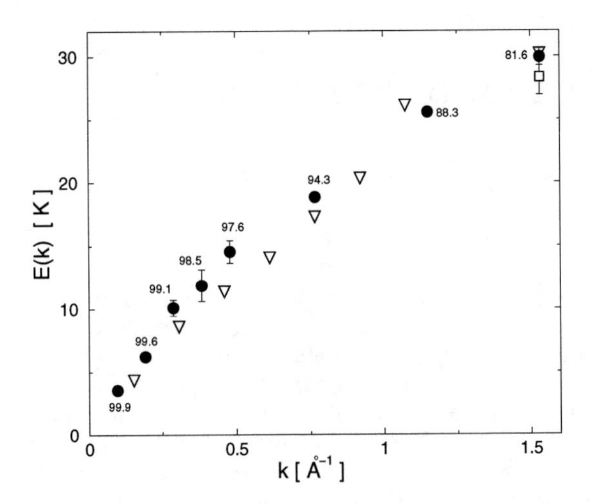

FIGURE 1. Circles: transverse phonon energies along the $[\xi,0,0]$ direction. Open squares: pure transverse phonon energy computed as described in the text. Triangles: experimental data from INS[7]. The numbers on the graph are the percentage weigths of the transverse component.

$\vec{k}$ implies that what we compute is actually a mixture of transverse and longitudinal phonons.

The transverse phonon frequencies computed at density of $\rho = 0.029$ Å^{-3} for bcc solid ^{4}He are shown in Fig.1 through 3. At $T = 0$ K the stable state is hcp as in experiment but with a suitable choice of the simulation box we can stabilize the bcc phase which experimentally is stable only at finite T. At small wave vector $\vec{k}$ we find always good agreement with experiments along the principal directions of the lattice. In Fig.1, which is for the $[\xi,0,0]$ direction, we have shown also the unique pure transverse phonon energy (the open square at half zone) computed with $\vec{G}+\vec{k}$ orthogonal to $\vec{k}$; in this case the agreement with the experimantal data is very good and this result gives a strong indication of the quality of the SWF in the description of the transverse phonon modes in solid helium. For the other wave vectors, when $\vec{k}$ is perpendicular to $\vec{G}$, it is possible to obtain a quantitative information on the degree of trasversality of the phonon mode by computing the ratio between the component of $\vec{k}$ perpendicular to $\vec{G}+\vec{k}$ and $|\vec{k}|$. This is simply the cosine of the angle between $\vec{G}+\vec{k}$ and $\vec{G}$; the numbers shown in Fig.1 through 3 near the computed data represent this ratio as a percentage.

We notice a good overall agreement of the computed phonon transverse energies with experiment; this is specially noteworthy for the low branch in the $[\xi,\xi,0]$ direction which is not well represented by SCP and by density functional. On the other hand a significant discrepancy is present in the higher energy branch in the same direction. This suggests that a more complex representation of the wave function is needed in this case.

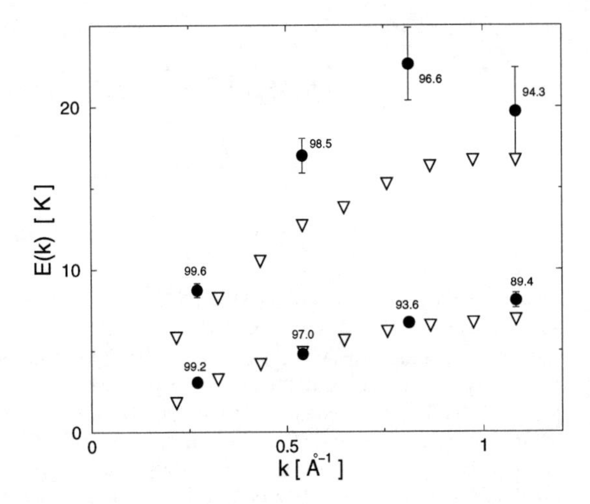

FIGURE 2. Circles: transverse phonon energies along the $[\xi,\xi,0]$ direction. Triangles: experimental data from INS[7]. The numbers on the graph are the percentage weigths of the transverse component.

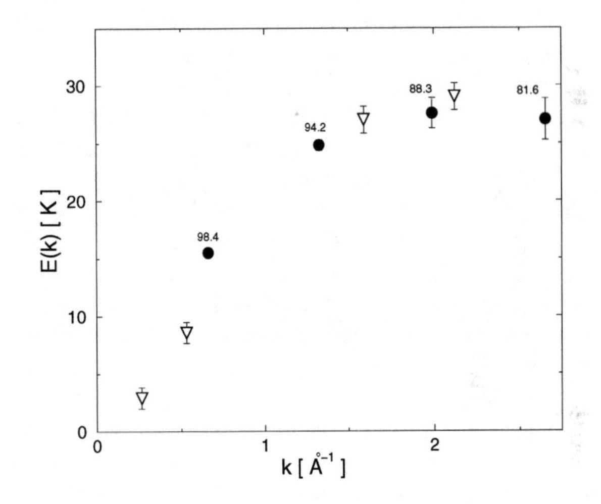

FIGURE 3. Circles: transverse phonon energies along the $[\xi,\xi,\xi]$ direction. Triangles: experimental data from INS[10]. The numbers on the graph are the percentage weigths of the transverse component.

REFERENCES

1. H.R. Glyde, *Excitation in Liquid and Solid Helium*, Claredon Press, Oxford, 1994, pp. 74-98.
2. V. Tozzini and M.P. Tosi, *Physica B* **262**, 369 (1999).
3. D.E. Galli and L. Reatto, *Phys. Rev. Lett.* **90**, 175301 (2003).
4. D.E. Galli *et al.*, *J. Low Temp. Phys.* **134**, 121 (2004).
5. V. Sorkin, E. Polturak, and J. Adler, cond-mat/0502108.
6. J.M. Goodkind, *Phys. Rev. Lett.* **89**, 095301 (2002).
7. T. Markovich *et al.*, *J. Low. Temp. Phys.* **129**, 65 (2002).
8. S.A. Vitiello *et al.*, *Phys. Rev. B* **42**, 228 (1990).
9. W. Wu *et al.*, *Phys. Rev. Lett.* **67**, 1446 (1991).
10. E.B. Osgood *et al.*, *Phys. Rev. A* **5**, 1537 (1972).

Solid ^{4}He in Narrow Porous Media

M. Rossi, D.E. Galli and L. Reatto

Dipartimento di Fisica, Università degli Studi di Milano, via Celoria 16, 20133 Milano, Italy

Abstract. We give a variational Monte Carlo description at $T = 0$ K of ^{4}He filling a porous glass within the shadow wave function technique. The confining media is modeled by a smooth cylindrical pore, and we have considered two different pore radii: $R = 13$ Å and $R = 22.5$ Å. The radial density profiles show a strong layering of the ^{4}He atoms. In the $R = 13$ Å pore, as the density is increased, solidification takes place layer by layer, starting from the pore wall, and the radius is too small to allow a bulk-like solid to nucleate in the small liquid region at the center of the pore. For the $R = 22.5$ Å pore we find an hcp lattice accommodated in the central region surrounded by four layers which turn out to be solid at the density of the computation. Computing the one-body density matrix (in the $R = 13$ Å case) we are able to estimate the Bose-Einstein condensate fraction, which is still non-zero even when the whole system is in the solid phase.

Keywords: solid helium, Bose Einstein condensate, porous media
PACS: 67.80.Mg, 67.90.+z, 67.40.Db

In this paper we present a variational Monte Carlo (VMC) study of the ground state properties of ^{4}He filling a porous media. We have modeled the confining media with a smooth cylindrical pore[1] and the He-pore potential parameters are chosen to reproduce the He-Si interaction[2]. We have considered two different radii, $R = 13$ Å, a size comparable to the nominal pore size of Gelsil glass employed in ref.[3], and $R = 22.5$ Å, which is comparable to the Gelsil glass employed in ref.[4]. Our trial wave function is a shadow wave function[5] modified in order to allow different degrees of correlations between ^{4}He atoms depending on the distance from the pore wall but still preserving the full Bose symmetry[6].

We find that ^{4}He forms a distinct layered structure near the pore wall. The atoms in the layer adjacent to the pore wall are very localized under all conditions of a filled pore. The properties of this layer (0th layer) turn out to be almost independent of the total density of the system. The 0th layer is always solid in the considered density range and its areal density is 0.12 Å^{-2} for the both considered radii. In the study of the inner layers we have then chosen to fix the positions of the atoms in this layer treating them as a part of the substrate. This choice is dictated by the possibility of simplified and less time consuming simulations. We give our results in terms of effective density ρ_{eff} in which we have excluded the high density 0th layer.

We have performed VMC simulations with a pore of radius $R = 13$ Å and length $L = 21.62$ Å, with periodic boundary conditions along the pore axis lying in the z direction, and with a number of dynamical ^{4}He atoms between $N = 132$ and $N = 231$, which correspond to effective densities $\rho_{\text{eff}} = 0.02296$ Å^{-3} and $\rho_{\text{eff}} = 0.04070$ Å^{-3}, respectively. The radial density profiles show that at all densities the dynamical atoms arrange themselves in layers around the axis of the cylinder, and the layered structure of the ^{4}He atoms in the pore is made evident by Fig. 1. As we increase the effective density of the system the peaks in the radial density profile become sharper and the minima between adjacent layers become lower, starting from the adsorbed layer. This feature suggests a layer by layer localization and solidification starting from the outermost layer. This is confirmed by the energy per particle vs effective density ρ_{eff} plot, reported in Fig. 2. Since the radial density profile is not enough to characterize if and which kind of solid order is present, we have also computed the static structure factor and the density contour separately for each layer in order to characterize its microscopic state. We find four different phases: in phase I all the dynamical system is in the liquid phase, in phase II the first layer (the layer adjacent the adsorbed, locked layer) becomes solid while the other two layers remain liquid, and in phase III only the third (central) layer is still liquid. It is interesting to note that

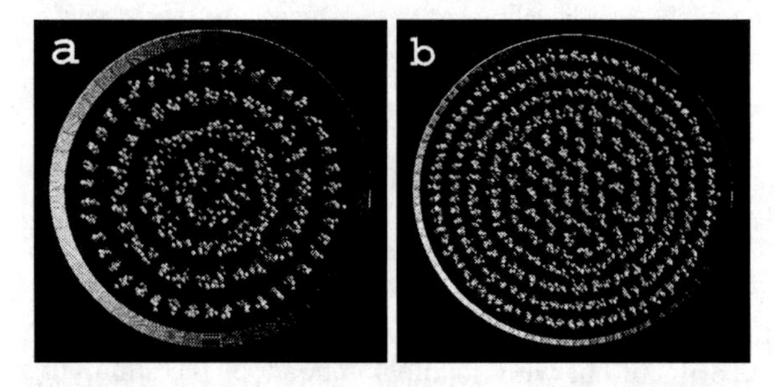

FIGURE 1. Snapshot along the pore axis of the ^{4}He atom and shadow particle coordinates (they cannot be distinguished from each other in the present scale) during a simulation run of (a) the $R = 13$ Å pore ($\rho_{\text{eff}} = 0.04070$ Å^{-3}) and (b) the $R = 22.5$ Å pore ($\rho_{\text{eff}} = 0.04182$ Å^{-3}).

CP850, *Low Temperature Physics: 24th International Conference on Low Temperature Physics;*
edited by Y. Takano, S. P. Hershfield, S. O. Hill, P. J. Hirschfeld, and A. M. Goldman

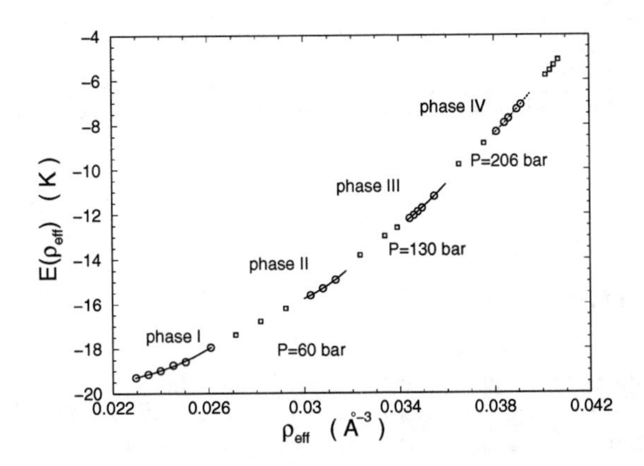

FIGURE 2. Energy per particle as a function of ρ_{eff} (circles: stable points; squares: unstable points) and phase boundaries from Maxwell construction for the $R = 13$ Å pore. The solid lines show the third-degree polynomial fit[5] of the VMC results interrupted at the phase boundaries. The dotted line shows the polynomial fit for phase IV which cannot be unambiguously characterized. Pressures at the phase coexistence are also displayed.

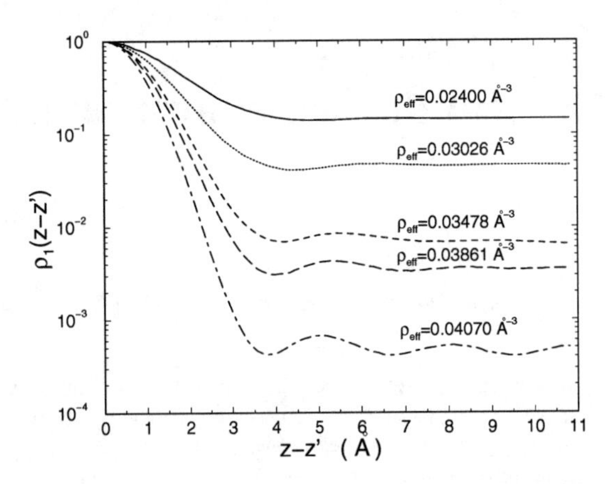

FIGURE 3. One-body density matrix $\rho_1(z-z')$ along the axial direction for the central layer in the $R = 13$ Å pore at different effective densities.

the solidification of both the first and the second layers occurs at the same areal density of about 0.09 Å^{-2} which is substantially larger than the freezing density found in a pure 2D system[7]. The solid layers have a triangular lattice with some defects. For ρ_{eff} higher than 0.03548 Å^{-3} it is not possible to unambiguously characterize the microscopic phase of the third layer. At $\rho_{\text{eff}} = 0.03548$ Å^{-3} the areal density of the third layer is equal to the freezing value 0.09 Å^{-2} of the other two layers. Then we might expect it to become solid (phase IV), but the static structure factor shows a solid third layer only at $\rho_{\text{eff}} = 0.04070$ Å^{-3}. Furthermore, at the highest ρ_{eff} the system undergoes a structural transition from a configuration with three layers to one with three layers plus a filled center.

The major features that emerge from our simulations are that the solidification takes place starting from the pore wall and that the system does not enter a crystalline ordered phase (all the layer solidified) up to very high pressures. The widely accepted mechanism for solidification in pores is that the solid nucleates within the liquid region at the center of the pore[8]. However there must be a transition region between the amorphous layers adsorbed at the pore wall and the crystalline solid in the center[9], and we conclude that our narrow pore, because of its reduced size, accommodates only this "transition region". In fact, with a larger pore radius, we have found a bulk-like solid only in the central region. We have performed a VMC simulation with a $R = 22.5$ Å and $L = 15.57$ Å pore with $N = 699$ dynamical ^{4}He atoms, which corresponds to $\rho_{\text{eff}} = 0.04182$ Å^{-3}. We find that there is a defected hcp lattice at the center of the pore surrounded by three solid layers plus the 0th one which constitute the transition region. In Fig. 1(b) we show a snapshot of the ^{4}He atom coordinates taken during the simulation run. By changing the pore length L (for the both R values) we have verified that there are no size effects due to the small value of L on the microscopic structure of the system.

For the $R = 13$ Å pore we have studied also the presence of Bose-Einstein condensation by computing the single layer contribution to the one-body density matrix $\rho_1(z-z')$ along the pore axis. The most important feature is that ρ_1 has a non-zero plateau (called off-diagonal long range order, ODLRO) for a very wide pressure range. Even at a pressure of order 250 bar the inner two layers have a small but finite condensate fraction even when the considered layer is in the solid phase, as shown in Fig. 3. By studying the long distance tail oscillations of ρ_1, whose maxima are registered with the layer crystalline lattice, it is possible to argue that the dominant contribution to ODLRO is given by the presence of defects such as vacancies rather than the vacancy-interstitial pairs creation due to the zero point motion of the ^{4}He atoms, as is the case for bulk ^{4}He[10].

REFERENCES

1. G. Stan *et al.*, *Phys. Rev. B* **62**, 2173 (2000).
2. G. Vidali *et al.*, *Surf. Sci. Rep.* **12**, 133 (1991).
3. K. Yamamoto *et al.*, *Phys. Rev. Lett.* **93**, 075302 (2004).
4. J. V. Pearce *et al.*, *Phys. Rev. Lett.* **93**, 145303 (2004).
5. T. MacFarland *et al.*, *Phys. Rev. B* **51**, 13577 (1994).
6. M. Rossi *et al.*, *Phys. Rev. B* **69**, 2125210 (2004).
7. P. A. Whitlock *et al.*, *Phys. Rev. B* **38**, 2418 (1988).
8. C. Lie-Zhao *et al.*, *Phys. Rev. B* **33**, 106 (1986).
9. M. Schindler *et al.*, *Phys. Rev. B* **53**, 11451 (1996).
10. D. E. Galli *et al.*, *Phys. Rev. B* **71**, 140506 (2005).

Critical Phase Separation in Solid Helium Isotopic Mixtures

M. Poole, J. Saunders and B. Cowan

Millikelvin Laboratory, Royal Holloway University of London, Egham, Surrey, TW20 0EX, UK.

Abstract. We report measurements of critical phase separation and spinodal decomposition in solid helium isotopic mixtures. Pressure measurements are found to be in good agreement with Mullin's model calculation of the excess volume. We also see clearly the signature of the bcc to hcp transition in the dilute ^{3}He component. NMR measurements give an indication of the spatial modulation of the mixture concentration, characteristic of spinodal decomposition. The non-exponential relaxation profiles – indicative of a concentration distribution – are interpreted in terms of a stretched exponential relaxation function.

Keywords: solid helium, phase transitions.
PACS: 67.80.Jd, 68.35.Rh

INTRODUCTION

The phase diagram of a two-component system has certain generic features that are manifest, particularly, in the phase separation process. The essential details of such a system are shown in Fig. 1.

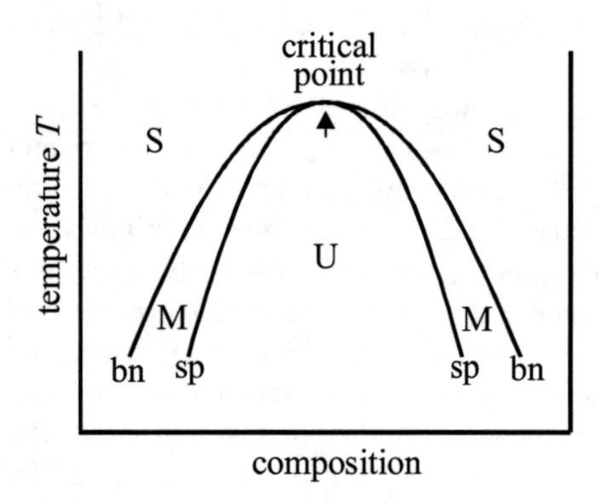

FIGURE 1. Generic phase diagram for a two-component system.

The system exists as a homogeneous mixture in the stable region labeled S. In the unstable region U and the metastable region M, it is energetically favourable for the system to form as coexisting regions of different composition, the two equilibrium concentrations being given by the binodal line, labeled bn. In the metastable region the system remains stable with respect to small composition fluctuations, so that the growth of the new phase is reliant on the occurrence of fluctuations of adequate magnitude. Such fluctuations provide 'nuclei' for the evolution of the new phase; thus in the metastable region the new phase grows through a nucleation-growth mechanism [1].

In the unstable region of the phase diagram, separated from the metastable region by the spinodal line sn, the system is unstable with respect to infinitesimal variations in composition. Thus *any* fluctuation will become magnified and the system will evolve by a sort of 'uphill diffusion'. This was clarified by Cahn and Hilliard [2]; essentially the particle flux is driven by a gradient in the chemical potential, rather than the concentration, in the general case. This mechanism is known as spinodal decomposition. Phase separation evolves through the spatial modulation of the mixture composition. And by solving the Cahn-Hilliard equation one finds a dominant wavelength for these modulations; short wavelength variations are too costly in interfacial energy, while long wavelength fluctuations are unfavourable as they require diffusion over large distances. Thus the occurrence of an optimum, intermediate, length scale.

We have studied these phase separation phenomena in solid mixtures of ^{3}He and ^{4}He. Relevant experimental details are given in [3]; in particular we grow good quality mixture crystals whose phase separation is investigated by a combination of NMR and pressure measurements. The pressure increase upon phase separation is a consequence of the excess volume of the homogeneous mixture [4]. The initial pressure of the crystal considered here was 36.7 bar.

CP850, *Low Temperature Physics: 24th International Conference on Low Temperature Physics;*
edited by Y. Takano, S. P. Hershfield, S. O. Hill, P. J. Hirschfeld, and A. M. Goldman

NMR is a particularly useful probe of the system. NMR sees only the ^{3}He but, importantly, the relaxation times T_1 and T_2 are sensitive functions of the ^{3}He concentration. Moreover, NMR can measure the self-diffusion coefficient of the ^{3}He atoms and the characteristic dimensions of ^{3}He-rich regions (droplet size) [5].

Away from the critical point, upon cooling through the transition line (into the metastable region), growth of the new phase is clearly seen as the appearance of a second component to the NMR relaxation signals with a different relaxation time. This has enabled us to follow the growth of the new phase droplets and the depletion of the homogeneous background [3, 5].

The situation is different when one cools through the critical point, reported here. For the helium mixture system this corresponds to an initial ^{3}He concentration $x_3 = 0.5$. Only then can one move directly into the unstable region, avoiding the metastable region. If the system were to evolve simply by moving along the coexistence (binodal) curve, this would be reflected in the NMR by the appearance of two distinct signals, the one corresponding to the depleted, $x_3 < 0.5$, regions and the other corresponding to the enriched, $x_3 > 0.5$, regions. This is not what is observed.

When cooling through the critical point the spin echo relaxation profile is observed to become non-exponential. It is found that the relaxation profile is fitted well by a 'stretched exponential' or Kohlsrausch-Williams-Watts function [6]

$$f(t) = f(0)e^{-(t/\tau)^n}. \qquad (1)$$

This may be understood as arising from a distribution of relaxation times τ – this a consequence of a distribution of ^{3}He concentrations. Here the index n, varying between 0 and 1 gives an indication of the breadth of the distribution: $n = 1$ indicates a single component, while $n \to 0$ indicates a very broad spread of concentrations.

We emphasise that the adoption of the stretched exponential profile, Eq. (1), is but a convenient parameterisation of the experimental data. There may well be good reasons why this is an appropriate representation in some 'central limit' way; that is discussed elsewhere [7].

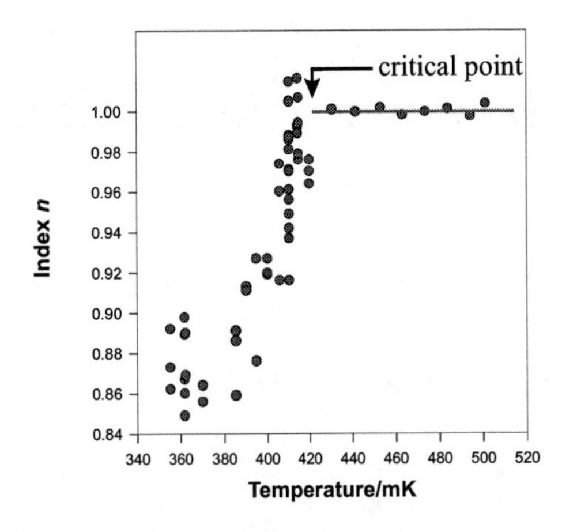

FIGURE 2. Variation of index n upon cooling through the critical point.

The variation of the index n is shown in Fig. 2. Above the critical point, $n = 1$, indicating a homogeneous mixture. Upon stepwise cooling through the critical point the drop in n reflects the emergence of a spread of relaxation rates – a broadening of the distribution of ^{3}He concentrations around $x_3 = 0.5$. This indicates a (smooth) spatial variation of concentration, showing that in this region the phase separation takes place through spinodal decomposition.

Although we can measure the diffusion coefficient of the ^{3}He below the critical point, the NMR relaxation indicates that the motion of the atoms is *unbounded.* In other words the ^{3}He can travel throughout the entire specimen. This corroborates the conclusion of a spatial modulation of the concentration throughout the crystal, rather than distinct regions of relatively uniform composition.

REFERENCES

1. V. V. Slezov and J. Schmelzer, *Phys. Solid State* **39**, 2210 (1997).
2. J. W. Cahn and J. E. Hilliard, *J. Chem. Phys.* **31**, 688 (1959).
3. A. Smith, V. A. Maidanov, E. Ya. Rudavskii, V. N. Grigor'ev, V. V. Slezov, M. Poole, J. Saunders and B. Cowan, *Phys. Rev. B* **67**, 245314 (2003).
4. W.J. Mullin, *Phys. Rev. Lett.* **20**, 254 (1968).
5. S. Kingsley, V. Maidanov, J. Saunders and B. Cowan, *J. Low Temp. Phys.* **113**, 1017 (1998).
6. G. Williams and D. C. Watts, *Trans. Faraday Soc.* **66**, 80 (1970).
7. B. Cowan (to be published).

Observation of Stages of Homogeneous Nucleation of Liquid Droplets in Solid Isotopic Helium Mixtures

M. Poole, J. Saunders and B. Cowan

Millikelvin Laboratory, Royal Holloway University of London, Egham, Surrey, TW20 0EX, UK.

Abstract. We have made simultaneous pressure and NMR measurements during the evolution of phase separation in solid helium isotopic mixtures. The experiments reported were performed at initial pressures and concentrations where the phase-separated ^{3}He inclusions formed as liquid droplets. Our observations indicate clearly all three stages of the homogeneous nucleation process: creation of nucleation sites; growth of the new-phase component at these nucleation sites; and finally dissolution of sub-critical droplets with the consequent further growth of the super-critical droplets, otherwise known as 'Ostwald Ripening'.

Keywords: solid helium, phase transitions.
PACS: 67.80.Jd, 68.35.Rh

Upon cooling, a solid solution of ^{3}He and ^{4}He will phase separate into regions rich in ^{3}He and regions rich in ^{4}He. The phase diagram for such a system is well established [1]. In this paper we report experiments performed on *dilute* solutions of ^{3}He in ^{4}He. Such mixtures separate into droplets of almost pure ^{3}He embedded in a solid matrix of almost pure ^{4}He. And the ^{3}He droplets may form either as a liquid or a solid depending upon the ambient pressure.

Our experiments have involved simultaneous NMR and pressure measurements during step-wise cooling through the phase separation transition. Simple NMR spin-spin and spin-lattice measurements yield characteristic double exponential relaxations from which the amounts of ^{3}He in the two evolving phases may be determined [2]. But in the presence of a magnetic field gradient, NMR spin echo measurements also allow determination of the size of the ^{3}He droplets during the phase separation process [2, 3].

Phase separation experiments at constant volume (as are ours) result in a pressure increase during the separation process. This is a consequence of the excess volume [4] associated with the process; there is a nonlinear term in the mixture molar volume as a function of composition. Thus pressure measurements provide an alternative way of following the phase separation and in particular it allows a determination of the characteristic time constant of the separation process.

The theory of homogeneous nucleation in solid mixtures has been formulated by Slezov and Schmelzer [5]. Central to the theory of nucleation is the identification of three stages of the process. Upon initial cooling through the transition the system is in a metastable state where it is energetically favourable for regions of higher ^{3}He concentration to exist. Thermodynamic fluctuations then result in such regions forming; the creation of these nucleation sites is the first stage of the process.

Once the nucleation sites have been formed, further cooling will result in the growth of the ^{3}He–rich droplets on these sites at the expense of the ^{3}He remaining in the dilute background; this is the second stage of the process.

Finally, when the background is almost completely depleted and the system is moving close to its true equilibrium state, the critical size of the droplets becomes very large. The sub-critical droplets are energetically disfavoured and so they will dissolve. The super-critical droplets are stable and so they continue to grow at the expense of the evaporated smaller droplets. The true equilibrium state is then one with a single ^{3}He–rich domain. This last stage of the process takes place on a long time scale as it relies upon diffusion of the very dilute ^{3}He atoms to the sites of the growing droplets. This is a highly non-linear process characterised by a $t^{1/3}$ evolution of the droplet diameter [6]. This last stage of the phase separation process is known as Ostwald ripening. Another

CP850, *Low Temperature Physics: 24th International Conference on Low Temperature Physics;*
edited by Y. Takano, S. P. Hershfield, S. O. Hill, P. J. Hirschfeld, and A. M. Goldman

possible mechanism for the ripening process, leading to a $t^{1/3}$ evolution, is the bodily diffusion of entire droplets towards each other. But this happens over a much longer time scale.

We report measurements on a 2% ^{3}He mixture crystal grown at a pressure of 28 bar. At this pressure the temperature at which phase separation starts was found to be 295 mK and new-phase regions form as ^{3}He-rich liquid droplets. We studied this crystal in two separate cooling cycles. In the first we lowered the temperature from above the transition in small steps of ~5 mK. And in the second we cooled in one step to 100 mK. But note that this final temperature is still significantly above the spinodal temperature, which would be about 40 mK for this crystal.

NMR allows the determination of the size and the ^{3}He concentration of the new-phase clusters. And from this the density of the clusters may be found. In the small-step coolings we found this concentration to be approximately constant after the first two steps below the transition. Thus we conclude that the *nucleation* stage of the phase separation process occurs in these two steps. Following this, upon further cooling, the droplets are observed to grow but their number remains constant; this is the *growth* stage of the process. This growth stage may also be followed through the observed pressure increase in the crystal. The time constant for this diffusive growth is found to be ~ 500s by both methods of determination.

The theory of homogeneous nucleation gives the concentration of new-phase clusters in terms of the degree of supersaturation. Within the framework of the Slezov-Schmelzer theory [5], the droplet surface tension σ may be expressed in terms of the cluster concentration or in terms of the growth time constant. Thus it may be found independently from NMR measurements (cluster concentration) or from pressure measurements (growth time constant) [3]. In both cases the determinations of σ are found to agree; this supports the identification of homogeneous nucleation. The low temperature variation gives $\sigma \sim A - BT^2$, consistent with Fermi liquid behaviour, and the zero-temperature value of σ is approximately 1.5×10^{-5} J m^{-2}.

The treatment of the early stages of the separation process is based upon the assumption that the total volume of the new-phase droplets is sufficiently small that it has negligible effect on the degree of supersaturation. This allows a linearised approach to the kinetics where the droplet growth is independent of the state of the other droplets. This breaks down in the latter stages of the separation process where the degree of supersaturation becomes small.

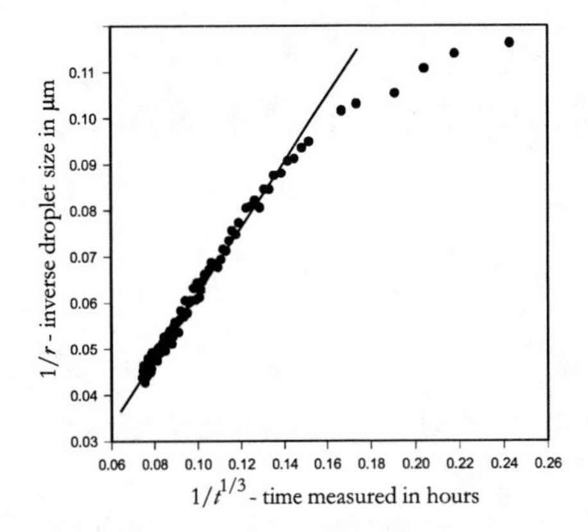

FIGURE 1. The Ostwald ripening process. The sample, of 2% ^{3}He concentration and molar volume 20.9 cm^3 was quenched to a temperature of 100 mK.

Then the sub-critical droplets will evaporate, providing more ^{3}He for the growth of the super-critical droplets. But at the same time the critical droplet size increases. So eventually the ultimate equilibrium state will consist of *one* macroscopic droplet of the new phase. Clearly this growth process is highly nonlinear. The kinetics of the process is described by Lifshitz and Slezov [6], although the asymptotic $t^{1/3}$ behaviour may be justified by dimensional arguments. The droplet curvature ~ $r(t)^{-1}$ will lead to concentration gradients of magnitude ~ $\sigma/r(t)^2$ which will drive a diffusive flux ~ $D\sigma/r(t)^2$. It is this flux that leads to the growth of the supercritical droplets so that

$$\frac{\mathrm{d}r(t)}{\mathrm{d}t} \sim \frac{D\sigma}{r(t)^2} \qquad (1)$$

which has solution $r(t) \sim (\sigma Dt)^{1/3}$. The full solution [6] gives the correct prefactor. This behaviour is shown in Fig. 1. We have plotted r^{-1} against $t^{-1/3}$ so that the Lifshitz-Slezov asymptotic behaviour is indicated by the approach to linearity on the left hand side. The slope of the asymptote is 7.23×10^{10} s$^{1/3}$ m^{-1}.

REFERENCES

1. D. O. Edwards and S. Balibar, *Phys. Rev.* B **39**, 4083 (1989).
2. S. C. J. Kingsley, I. Kosarev, L. Roobol, V. Maidanov, J. Saunders and B. Cowan, *J. Low Temp. Phys.* **110**, 400 (1998).
3. A. Smith, V. A. Maidanov, E. Ya. Rudavskii, V. N. Grigor'ev, V. V. Slezov, M. Poole, J. Saunders and B. Cowan, *Phys. Rev.* B, **67**, 245314 (2003).
4. W. J. Mullin, *Phys. Rev. Lett.* **20**, 254 (1968).
5. V. V. Slezov and J. Schmelzer, *Phys. Solid State*, **39**, 1971 (1997).
6. I. M. Lifshitz and V. V. Slezov, *Soviet Physics JETP*, **35**, 331 (1959).

Experimental Evidence For Phase Separation In Quench-Condensed Films Of Normal Hydrogen

Nina V. Krainyukova

Institute for Low Temperature Physics and Engineering NASU, 47 Lenin ave., Kharkov 61103, Ukraine

Abstract. A transmission high energy electron diffraction study of quench-condensed films of normal hydrogen in the temperature interval 3–5 K is reported. The difference between the measured structure of nH_2 films and the known bulk structure is ascribed to a periodic ortho-para layering that entails a new characteristic sequence of basal planes. Our analysis compares experimental diffractograms and calculations based on structure modelings.

Keywords: Quantum Solids, Hydrogen.
PACS: 67.70.+n; 67.80.-Cx; 61.14.-x

INTRODUCTION

Numerous structural studies[1-5] (x-ray, electron and neutron diffraction) of hydrogen samples prepared by condensation from the gas phase on a substrate held at liquid helium temperatures have revealed the essential difference of the observed structures from the hcp structures expected in the relevant regions of the currently accepted phase diagram.[1] The transition to the low temperature cubic phase was found to be incomplete.[3,4] Intermediate close packed structures were predicted[4-6] in hydrogen and deuterium.

EXPERIMENTAL

A high energy (50 keV) electron diffraction study of quench-condensed films (200–300 Å in thickness) of normal hydrogen on a polycrystalline aluminum substrate in the temperature interval 3–5 K is presented. The films were prepared by depositing hydrogen gas, which is precooled to a liquid nitrogen temperature, at low (0.5–2 Å/s) rates to avoid heating. Discontinuous films composed of nanocrystals are usually produced by such a preparation. The precise digital data recording was performed at deposition temperatures (with no preceding heat treatment) with the step in S units ~0.015 Å^{-1} (S=4πsinθ/λ, θ is the diffraction angle and λ is the wavelength of electrons) and showed no difference in diffraction patterns within a sample. The substrate intensities were recorded before the deposition and then subtracted from the total intensities. The diffracted intensity $I_{calc,k}$ for each film fragment k was calculated using the Debye formulae. The total intensity $I_{calc}=\Sigma w_k I_{calc,k}$ (Fig. 1) was considered as a superposition of diffracted intensities from different film fragments k with their relative weights w_k (the total $\Sigma w_k=1$), which were used as independent variables in the minimization of the R factor $R=\Sigma|I_{exp}-I_{calc}|/\Sigma(I_{exp}+I_{calc})$.[7]

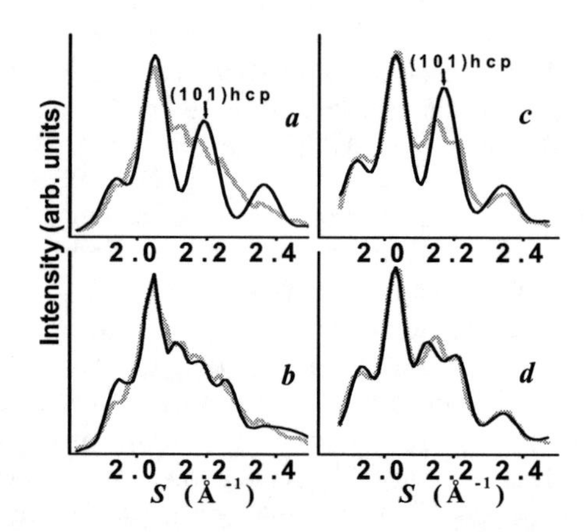

FIGURE 1. Experimental I_{exp} (light gray) and calculated (thin solid line) diffracted intensities: T=4.6 K (*a* and *b*), T=3.4 K (*c* and *d*); *a* and *c*– the best-fit simple mixture of fcc and hcp structures; *b* and *d* – the same but applying all models including intermediate layered structures.

RESULTS AND DISCUSSION

At lower temperatures we observe a time dependent transformation of diffraction patterns, which is attributed to Ostwald's rule. High disorder in the condensate, typical of a frozen liquid, changes to

CP850, *Low Temperature Physics: 24th International Conference on Low Temperature Physics;*
edited by Y. Takano, S. P. Hershfield, S. O. Hill, P. J. Hirschfeld, and A. M. Goldman

the hcp structure characteristic of the high temperature region. Later in time, the new structure forms, and Ostwald's rule dictates that structure is more favorable than hcp at the temperature of observation. No time-dependent transformations were revealed at higher temperatures, apparently because faster processes occur in this case.

A simple mixture of fcc and hcp (even for their best fit combination, see Fig. 1) obviously does not fit the experiment. The suppression of hcp reflections and also a highly reproducible split of the hcp peak (101) (that may not be caused by random stacking faults) are obvious. This split indicates that some new crystal order arises and is presumably connected with intermediate phases (see below).

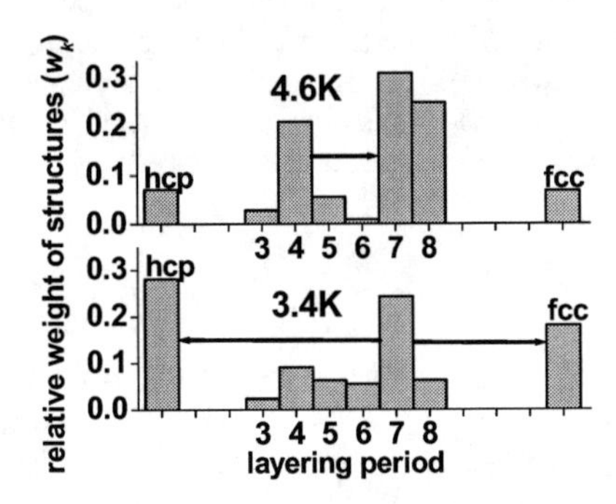

FIGURE 2. Distribution functions over modeling structures at 4.6K and 3.4 K.

Including the effect of orientational ordering of orthomolecules in the fcc crystallites does not improve the agreement with the data. Furthermore, in the temperature region of the experiment, we expect only dynamically correlated rotational motion of these molecules, instead of orientational ordering. Finally, we succeed with models that are based on a new periodic sequence of basal planes. This periodicity implies that there are two types of molecules, which tend to separate.[8,9] If the period involves n basal planes, the first n_1 planes (one after other) comprise only one type of molecule, e.g., orthomolecules with fcc-like stacking (ABCABC...); the next one plane consists of a random mixture of two types; while the rest (n-n_1-1) are composed only of the other molecule type, e.g., paramolecules with hcp-like stacking (ABABAB...). Such structural models (the space group $P3m1$) with a period ranging from 3 up to 8 basal planes were involved in the analysis together with hcp and fcc structures. We included fcc-like sequences of basal planes, because the fcc phase was shown in numerous experiments to be dominant in hydrogens prepared at liquid He temperatures. Probably the martensitic transition to the fcc phase is facilitated in quench condensates by a well developed free surface at higher temperatures.

Two periodic structures with 4 and 7(8) layers are dominant in the higher temperature distribution function (Fig. 2). At lower temperatures the 4-layered structure disappears first; the relative weight of the 7(8) layered structure decreases, replaced by the hcp structure (presumably enriched by paramolecules) and the fcc structure (with dominant orthomolecules). This change can imply a tendency toward complete separation (although it is not realized as yet). It is remarkable to note that the period of the second observed layered structure is nearly twice the period of the first structure. It could be understood if we assume that the 4-layered structure forms at higher temperatures; at lower temperatures the simplest way to increase the degree of separation is to double the period. Our observations are in some qualitative agreement with phase diagram[9] that predicts the phase separation into two hcp-like phases in the temperature interval and concentration under study; but two phases are replaced in our case by hexagonal structures with a step-wise increase in the layering period (or degree of separation) at lowering temperatures. The low-temperature phase transition in the separated system obviously has to be incomplete,[3,4] since the pH_2-subsystem may not transform to the cubic phase.

The author is deeply grateful to V. G. Manzhelii, M. A. Strzhemechny, A. I. Prokhvatilov and M. I. Bagatskii for fruitful discussions of the problem, she is also greatly indebted to B. W. van de Waal for software used in film fragment modeling and thank Yasu Takano for numerous comments on the paper.

REFERENCES

1. Manzhelii, V. G., and Strzhemechny, M. A., "Quantum molecular crystals", in *Physics of Cryocrystals*, edited by V. G. Manzhelii et al., NY: AIP Press, Woodbury, 1996, pp. 3-211 (and references therein).
2. Collins, G. W., Unites, W. G., Mapoles, E. R., and Bernat, T. P., *Phys. Rev. B* **53**, 102-106 (1996).
3. Barret, C. S., Meyer, L., and Wasserman, J., *J. Chem. Phys.* **45**, 834-837 (1966).
4. Yarnell, J. L., Mills, R. L., and Schuch, A. F., *Sov. J. Low Temp. Phys.* **1**, 366-371 (1975).
5. Silvera, I. F., *Rev. Mod. Phys.* **52**, 393-452 (1980).
6. Conradt, R. N. J., Albrecht, U., Herminghaus, S., and Leiderer, P., *Physica B* **194-196**, 679-680 (1994).
7. Krainyukova, N. V., and van de Waal, B. W., *Thin Solid Films* **459**, 169-173 (2004).
8. Stepanova, G. I., *Fiz. Tverd. Tela* [*Sov. Phys.-Solid State Physics*] **4**, 1263-1269 (1962).
9. Prokhvatilov, A. I., *Sov. J. Low Temp. Phys.* **16**, 722-725 (1990).

Simulated NMR Line Shapes for Random Ordering in Solid Hydrogen Films

Kiho Kim*, J. R. Bodart†, and N. S. Sullivan¶

*Department of Physics, University of South Alabama, Mobile, AL 36688, USA
†Department of Natural Science, Chipola College, Marianna, FL 32446, USA
¶Department of Physics, University of Florida, Gainesville, FL 32611-8440, USA

Abstract. Random disorder was seen for ortho-H_2 concentrations $0.48 < x[\text{o-}H_2] < 0.69$ at temperatures [~200 mK ≤ T ≤ ~600 mK] in submonolayer H_2 films adsorbed on hexagonal boron nitride. The NMR line shapes for the adsorbed H_2 samples were distinct from those of bulk samples because a crystal field is introduced by the substrate in addition to the electrostatic quadrupole-quadrupole interactions between two adjacent H_2 molecules. Computer simulations have been carried out to determine the NMR line shapes of solid H_2 films for different local order parameters and Guassian broadening functions. The simulated NMR line shapes are presented and discussed.

Keywords: NMR, H_2
PACS: 05.70 Ln, 05.70 Jk, 64.

INTRODUCTION

Diatomic H_2 molecules form a "frustrated" system when adsorbed as a commensurate layer on boron nitride (BN) because of the lattice symmetry of the hexagonal substrate. Previous studies[1] have shown that the frustration of H_2 molecules plays an important role in the orientational ordering in 2D. The combined effect of frustration and disorder gives rise to a glassy behavior for $0.48 < x[\text{o-}H_2] < 0.69$ at temperatures [~200 mK ≤ T ≤ ~600 mK] in submonolayer H_2 films adsorbed on BN. In 2D very distinctive NMR line shapes (LS), quite different from the typical Pake-doublet NMR LS in 3D, are observed in thick and submonolayer H_2 films on BN. Since extensive theoretical work concerning the NMR LS of H_2 samples in 2D has not been reported elsewhere, we carried out computer simulations (CS) in order to understand the peculiar NMR LS observed in H_2 films on BN.

COMPUTER SIMULATIONS

The CS allow us to consider the effect of different order parameters (σ) corresponding to different environments on the BN substrate, because the effective field determining the molecular orientations varies with $x[\text{o-}H_2]$ and the competition between crystal field effects and intermolecular interactions. The CS for the final NMR LS are based on two assumptions: (i) The fluctuations of the molecular orientation are very fast compared to the relevant frequency for the NMR LS measurements, i.e. the intramolecular dipolar frequency d. In this case, the NMR LS depends only on a single-particle thermodynamic average, which can be determined from the local reduced density matrices. (ii) In the absence of interactions that break time-reversal symmetry, the dipole moments are all zero. The assumed vanishing of the dipole moments allows us to neglect the spin-orbit interaction in the calculation of the NMR LS for thin films of H_2. The Hamiltonian giving the NMR spectrum can be found in Ref. [1]. With this assumption, for a powdered sample, the NMR spectrum is given by a sum of distinct Pake–powder spectra $\Pi(\sigma, \omega)$ with a distribution $P(\sigma)$ where $-0.5 < \sigma < 1$. The NMR LS $S(\omega)$ is given as

$$S(\omega) = \int_{-0.5}^{1} d\sigma P(\sigma)\left[\int \Pi(\sigma,\omega) g(\omega-\omega',\Delta) d\omega'\right] \quad (1)$$

where $\Pi(\sigma, \omega)$ denotes the well-known Pake-doublet of frequency span $6\sigma d$. The factor $g(\omega, \Delta)$ is an appropriate convolution function which takes into account the intermolecular dipolar broadening

CP850, *Low Temperature Physics: 24th International Conference on Low Temperature Physics;*
edited by Y. Takano, S. P. Hershfield, S. O. Hill, P. J. Hirschfeld, and A. M. Goldman

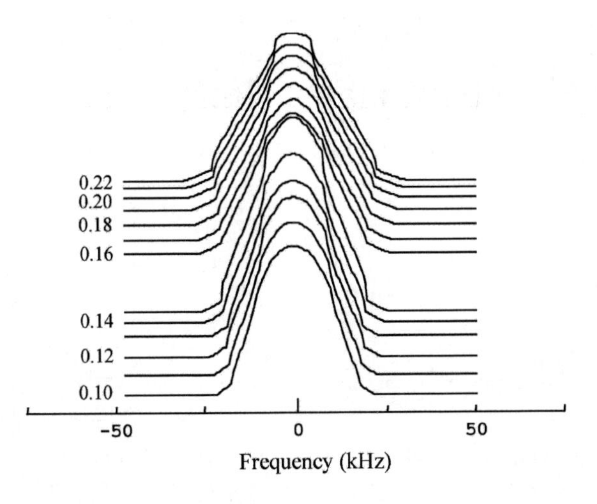

FIGURE 1. Simulated NMR LS for $0.10 \leq \sigma \leq 0.22$.

specified by Δ. For the sake of simplicity, we have approximated this factor as the product of a Lorentzian and a Gaussian of appropriate widths:

$$g(\omega) \propto \frac{\exp(-\frac{\omega^2}{3.1\Delta^2})}{1+(\frac{\omega}{\Delta})^2}. \quad (2)$$

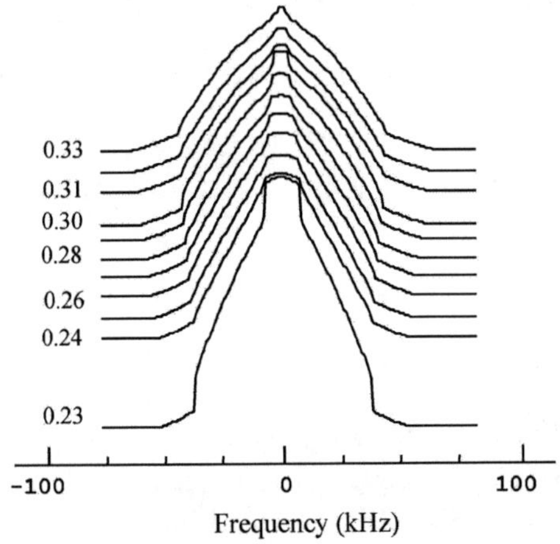

FIGURE 2. Simulated NMR LS for $0.23 \leq \sigma \leq 0.33$.

Δ is given in unit of the Lorentzian width with $(\Delta_0/2\pi) = 7.25$ kHz for pure ortho molecule in the totally disordered para-orientational state. (Para-H_2 molecules are not orientable.) $d = 57.67$ kHz for H_2. CS programs were written to examine the effect of differing σ on the quadrupolar line shapes. The line shapes were generated by allowing the order parameter to vary over the allowed range for fixed Gaussian broadening as a function of Δ which controls the broadening effect. Simulated NMR line shapes are presented in Figs. 1–3 for $0.10 \leq \sigma \leq 0.40$ under fixed broadening $\Delta = 33$ kHz.

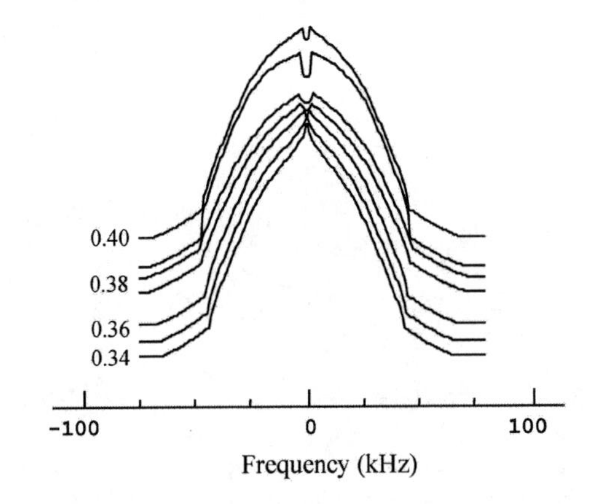

FIGURE 3. Simulated NMR LS for $0.34 \leq \sigma \leq 0.40$.

One can directly compare individual NMR LSs with different σ under the same broadening parameter. Structural changes in the NMR spectra have served as a guide for mapping out the phase diagram in our previous experimental NMR studies.[1] These LS changes may not result only from thermodynamic phase changes but from competition between the effects of variations in σ and in Δ. The NMR LS with the lowest σ in Fig. 1 has the narrowest central line. We can still see the slanted shoulder for $\sigma = 0.10$. As σ increases (that is, H_2 molecules enter more ordered states), dramatic changes are observed in the NMR LS, especially near $\sigma = 0.15$. The central Gaussian-like peak decreases and the separation of the slanted shoulders becomes larger.

ACKNOWLEDGMENTS

This work is supported by the special faculty-development fund and the presidential travel award at the University of South Alabama.

REFERENCES

1. K. Kim and N. S. Sullivan, *J. Low Temp. Phys.* **114**, 173 (1999); K. Kim and N. S. Sullivan, *Phys. Rev. B* **55**, R664 (1997); Kiho Kim and N. S. Sullivan, *Phys. Rev. B* **57**, 12595 (1998).

Study Of Structural Effects Of Vacancies In Solid p-H_2 With Shadow Wave Functions

Francesco Operetto and Francesco Pederiva

Dipartimento di Fisica and INFN, Università di Trento, Via Sommarive, 14 I-38050 Povo, Trento, Italy

Abstract. We employ the shadow wave functions in connection with variational Monte Carlo calculations in order to study the local environment of crystalline para-H_2 in which a vacancy is present. In particular we look at the distribution of molecules surrounding the empty site, in order to study possible deformations in the local structure of the solid. We present some of the results obtained on the densities and angular distributions of the molecules.

Keywords: defective solid para-hydrogen, vacancies.
PACS: 67.40.Yv, 67.80.-s

INTRODUCTION

Shadow wave functions (SWF) are one of the most powerful tools[1] to understand the behavior of defective quantum crystals. In fact, they provide a way to describe an ordered system without the need of imposing *a-priori* equilibrium positions of the atoms or molecules of the system itself. Calculation of the vacancy formation energy at T=0 was already presented in a recent paper[2]. We show here some of the results of a detailed analysis performed in solid para-H_2, in which we explore the structural properties of the defective quantum solid, and in particular the effects of the a vacancy on the structure and the local order of the system.

SHADOW WAVE FUNCTIONS

Shadow wave functions are a particular form of trial solution for the Hamiltonian

$$H = -\frac{\hbar^2}{2m}\sum_{i=1}^{N}\nabla_i^2 + \sum_{i<j} v(r_{ij}) \tag{1}$$

where the $\mathbf{r}_i$ are the coordinates of N molecules, and v is the Silvera-Goldman[3] intermolecular potential. The SWF assumes the following form:

$$\Psi_T(R) = \int K(R,S)\varphi(S)dS \tag{2}$$

where $R=\{\mathbf{r}_1\ldots\mathbf{r}_N\}$ and $S=\{\mathbf{s}_1\ldots\mathbf{s}_N\}$ are the coordinates of the molecules and of a set of auxiliary variables, respectively. The kernel K describes the quantum mechanical correlations due to the interaction between molecules, and the connection between the real degrees of freedom and the auxiliary degrees of freedom. This term is typically of the form of a Jastrow product, multiplied by a product of Gaussians:

$$K(R,S) = \prod_{i<j} f(r_{ij}) \prod_{i=1}^{N} e^{-C(r_i - s_i)^2} \tag{3}$$

The function φ describes the correlations among the auxiliary degrees of freedom (the “shadows”). The detailed analytical form of this functions can be found elsewhere[2]. The SWF contains a set of variational parameters which are minimized by following the variational principle, which requires that the expectation value of the Hamiltonian

$$E_T = \frac{\langle \Psi_T | H\Psi_T \rangle}{\langle \Psi_T | \Psi_T \rangle} \tag{4}$$

be minimized with respect to the choice of the parameters in the wavefunctions. Simulations were performed in the fcc crystal, using a box including 108 lattice sites filled with 107 molecules. We fixed the density of the system ρ, and determined the size of the simulation box accordingly. After optimizing the wavefunction, runs consisted in generating 2x10^7 configurations of the system. The quantities of interest were computed by post-processing a number of such configurations, saved each 2000 steps, in order to

CP850, *Low Temperature Physics: 24th International Conference on Low Temperature Physics;*
edited by Y. Takano, S. P. Hershfield, S. O. Hill, P. J. Hirschfeld, and A. M. Goldman

avoid as much as possible the effects of autocorrelations in the generation of the random walk.

RESULTS

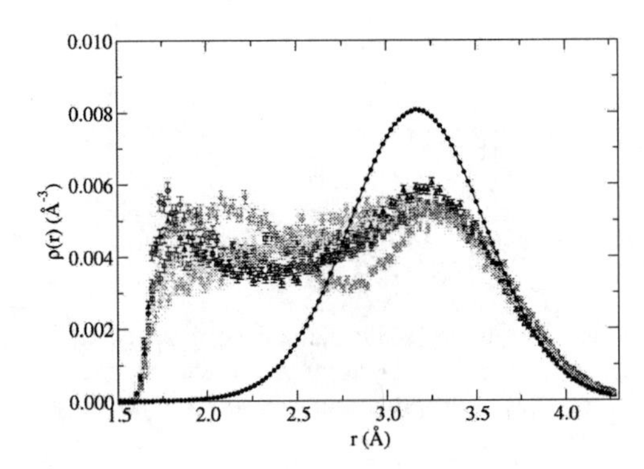

FIGURE 1. Density of molecules around a vacant site at density ρ=0.043 $Å^{-3}$. The Gaussian curve joining the points represents the density of nearest neighbors of a filled site. The other curves are estimates of the density of nearest neighbors of a vacant site obtained in different runs.

The simulations were performed at three values of density: ρ=0.0215 $Å^{-3}$ (at negative pressure), ρ=0.026 $Å^{-3}$ (saturation density), and ρ=0.043 $Å^{-3}$ corresponding to a pressure of about 0.4 GPa. The quantities computed in order to study the local order of the system were the one-body density of the nearest neighbors of filled and empty sites $\overline{\mathbf{r}}$, defined as

$$\rho(r) = \left\langle \sum_{<nn>} \delta(r - |\mathbf{r}_i - \overline{\mathbf{r}}|) \right\rangle, \qquad (5)$$

and the angular distribution:

$$\Theta(\cos(\theta)) = \left\langle \sum_{<nn>} \delta(\cos(\theta) - \frac{\mathbf{r}_i \cdot \mathbf{r}_j}{r_i r_j}) \right\rangle. \quad (6)$$

By combining the information extracted from these two curves, it is possible to reconstruct the position of the neighbors of a vacancy and look for possible changes in the local structure of the crystal. The position of the vacancy, which is not determined *a priori* in a SWF simulation, is determined by dividing the crystal into Wigner-Seitz cells. A vacancy is defined in this case as an empty site, which has no doubly occupied nearest neighbor WS cells.

In Figs. 1 and 2 we show the results obtained for the density and the angular distribution near a vacant site at density ρ=0.043 $Å^{-3}$. Distortion effects would be most likely expected at low densities, where the space available to each single molecule is larger. At saturation density we actually found that the density of the nearest neighbors of a vacancy tends to move into the empty site, as already observed in calculations for ^{4}He crystals[4]. However, at this higher density, it looks clear not only that the molecules tend to occupy the vacant site, but that the icosahedral structure of the molecules surrounding the empty site tends to be heavily deformed. This is perfectly visible if one examines the angular correlations between shadows, hich in the crystal are more rigid degrees of freedom, and which represent a sort of mean position for the molecules.

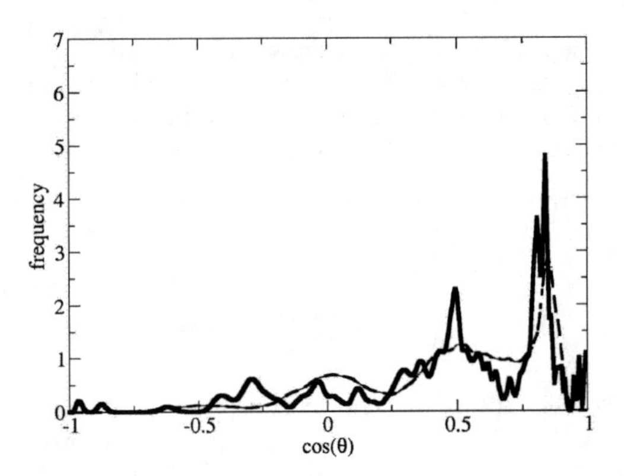

FIGURE 2. Angular distribution of shadow degrees of freedom around a filled (dashed curve) and a vacant site (solid line) at density ρ=0.043 $Å^{-3}$.

We also performed calculations on the diffusion of vacancies throughout the crystal. Although our calculations were performed at T=0, we can give an estimate of the mobility of vacancies due to the exchange of molecules induced by quantum fluctuations. We observe a non-zero diffusion at all the densities considered, although it tends to decrease when the density is increased.

We thank G.V. Chester and G. Galli for useful conversations about the subject of this paper. Calculations were in part performed at CINECA with an INFM/ICP grant.

REFERENCES

1. S.A. Vitiello. K.A. Runge, and M.H. Kalos, *Phys. Rev. Lett.* **60**, 1970 (1988); L. Reatto and G.L. Masserini, *Phys. Rev. B* **38**, 4516 (1988).
2. F. Operetto and F. Pederiva, *Phys. Rev. B* **69**, 024203 (2004).
3. I. .F. Silver and V.V. Goldman, *J. Chem. Phys.* **69**, 4209 (1978).
4. B. Chaudhuri, F. Pederiva, and G.V. Chester, *Phys. Rev. B* **60**, 3271 (1999).

Variational Monte Carlo Study Of Isotopic Impurities In Solid Molecular Para-Hydrogen

Francesco Operetto and Francesco Pederiva

Dipartimento di Fisica and INFN, Università di Trento, Via Sommarive 14, I-38050 Povo, Trento, Italy

Abstract. We employed shadow wave functions in combination with variational Monte Carlo calculations to study the properties of ortho-D_2 molecules in a para-H_2 crystal. In particular we computed the excess energy due to the presence of impurities. We also computed the energies for clustered and isolated impurities at different concentrations, trying to determine a limit for the solubility of o-D_2 in p-H_2.

Keywords: ortho-deuterium, impurities, para-hydrogen.
PACS: 67.40.Yv, 67.80.-s

INTRODUCTION

The equations of state of solid para-Hydrogen (p-H_2) and ortho-Deuterium (o-D_2) have been extensively studied in the past by experiments[1]. It has been shown that the same intermolecular potential can be safely used for reproducing their main features[2]. In particular, the both species carry a total angular momentum J=0, and can be treated from the theoretical point of view as point objects interacting via a spherically symmetric potential. In this paper we show some preliminary results we have obtained by means of shadow wave function (SWF) calculations[3,4] and variational Monte Carlo (VMC) methods on the properties of impurities of o-D_2 embedded in a solid p-H_2 matrix. In particular, we computed the excess energy as a function of the concentration of the o-D_2 molecules. Moreover, we investigated the variation of the excess energy when the impurities are all kept close to each other, and when they are separated by at least a shell of p-H_2 molecules. From the computed results, we try to infer some limits for the solubility of o-D_2 in p-H_2.

METHOD

In our SWF-VMC calculations we seek for a trial solution for the following Hamiltonian:

$$H = -\frac{\hbar^2}{2m_H}\sum_{i=1}^{N_H}\nabla_i^2 - \frac{\hbar^2}{2m_D}\sum_{i=1}^{N_D}\nabla_i^2 + \sum_{i<j} v(r_{ij}) \quad (1)$$

which describes a system of N_H p-H_2 molecules and N_D o-D_2 molecules, interacting by means of the Silvera-Goldman potential[2]. The SWF used in this work is an extension of the standard form[3]:

$$\Psi(R) = \phi(R)\int K(R,S)\phi_S(S)dS \quad (2)$$

where the correlation functions are adapted to deal with the coexistence of two molecular species in the system. In particular, the function

$$\phi(R) = \prod_{i<j}^{N_H} f_{HH}(r_{ij}) \prod_{i<j}^{N_D} f_{DD}(r_{ij}) \prod_{i,j}^{N_H N_D} f_{HD}(r_{ij}) \quad (3)$$

is a Jastrow product describing the quantum correlations among p-H_2 molecules, o-D_2 molecules and p-H_2 and o-D_2 molecules. The functions f are expansions of the correlations in terms of a suitable base, and contain up to 20 parameters each. The function $\phi_S(S)$ is a similar product, but is evaluated on the auxiliary degrees of freedom S. In this case the correlation functions f are of the form

$$f(s_{ij}) = e^{-\delta v(\alpha s_{ij})}, \quad (4)$$

where v is the intermolecular potential and α and δ are variational parameters. The Kernel K, connecting particles and shadows, has the form:

$$K = \prod_{i=1}^{N_H} e^{-C_H(\mathbf{r}_i - \mathbf{s}_i)} \prod_{i=1}^{N_D} e^{-C_D(\mathbf{r}_i - \mathbf{s}_i)}, \quad (5)$$

CP850, *Low Temperature Physics: 24th International Conference on Low Temperature Physics;*
edited by Y. Takano, S. P. Hershfield, S. O. Hill, P. J. Hirschfeld, and A. M. Goldman

TABLE 1. Energy per molecule ε_{imp} (in K) of a p-H_2 crystal containing o-D_2 impurities.

N_D	ε (close)	ε (far)
0	-87.045(5)	-
1	-87.388(4)	-
2	-87.763(5)	-87.762(5)
4	-88.527(11)	-88.547(11)
6	-89.279(12	-89.274(12)
8	-90.002(12)	-90.048(13)
10	-90.746(12)	-90.786(12)
18	-93.855(11)	-93.772(13)
27	-97.284(11)	-97.001(12)

where C_H and C_D are variational parameters.

We performed simulations at a fixed density of ρ=0.026 Å^{-3}, in a periodic box accommodating a fcc lattice with 108 total molecules. The parameters of the wavefunction have all been determined by minimizing a linear combination of the expectation value of the Hamiltonian and its variance. We used different numbers of p-H_2 and o-D_2 molecules. The initial positions of the o-D_2 impurities can be chosen initially in such a way to have them clustered or diluted into the crystal, i.e. arranged in such a way that each impurity has a full shell of p-H_2 surrounding it. Expectation values of operators of interest have been computed by means of the Monte Carlo method.

RESULTS

Due to the finite number of molecules used in the simulations, we could study only a discrete set of concentrations of o-D_2 impurities in p-H_2. In Table 1 we report the results we obtained for the energy per molecule ε_{imp} as function of the number of impurities N_{imp}. In Fig. 1 we plot the excess energy per impurity, defined as

$$E_{imp} = \frac{N}{N_{imp}}\left(\varepsilon_{imp} - \varepsilon\right), \qquad (6)$$

where ε is the energy per molecule of the system without impurities. As it can be seen, the excess energy per impurity remains almost constant in the range of concentrations considered. A rough estimate of the excess energy of a single o-D_2 impurity in the p-H_2 crystal can be obtained by considering, as a perturbation term in the Hamiltonian, the difference in kinetic energy due to the difference in the mass of the two isotopes. Such an evaluation gives an excess binding energy of about 35 K, not far from the value of about 40 K obtained from the VMC simulations. This fact suggests that the contribution of the mass effect to the excess energy of the impurity is dominant.

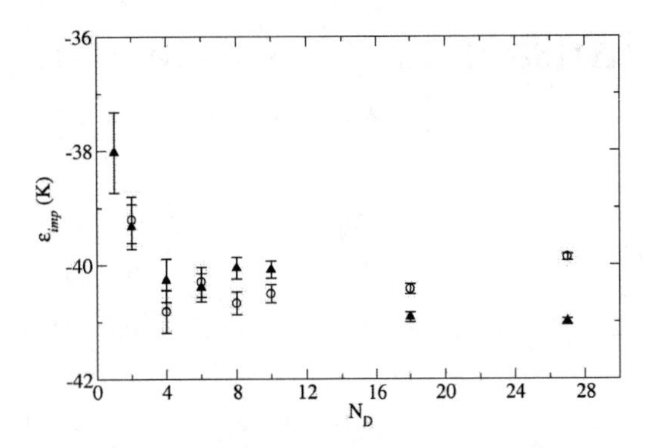

FIGURE 1. Excess energy per impurity (in K) in the p-H_2 crystal as a function of the number of o-D_2 impurities. Filled triangles: close impurities, empty circles: far impurities.

It is interesting to observe that the excess energy for clustered impurities and diluted impurities agree within error bars for concentrations up to 10%. Below this threshold we are led to conclude that the o-D_2 is soluble in the p-H_2 crystal. For higher concentrations, however, the excess energy per impurity for clustered impurities becomes lower than the energy for diluted impurities well outside error bars. This phenomenon is probably due to the fact that o-D_2 impurities tend to be more localized than p-H_2 molecules, due to the higher mass. Clustered impurities tend to occupy less volume than the same number of p-H_2 molecules and therefore it is possible for the p-H_2 molecules surrounding the cluster to reduce their kinetic energy considerably. The use of SWF to study this kind of effects is crucial, because no *a-priori* lattice structure is given to the system (the SWF is manifestly translationally invariant), and therefore relaxation effects can be described without the need of re-optimizing the structure.

Calculations for this project were partly performed at the ECT* supercomputing facilities in Trento.

REFERENCES

1. A. Driessen, J.A. de Waal, and I.F. Silvera, *J. Low Temp. Phys.* **34,** 255 (1979).
2. F. Silvera and V.V. Goldman, *J. Chem. Phys.* **69**, 4209 (1978).
3. S.A. Vitiello, K.A. Runge, and M.H. Kalos, *Phys. Rev. Lett.* **60**, 1970 (1988); L. Reatto and G.L. Masserini, *Phys. Rev. B* **38**, 4516 (1988).
4. F. Operetto and F. Pederiva, *Phys. Rev. B* **69,** 024203 (2004).

Diffusion Monte Carlo Study Of The Equation Of State Of Solid p-H_2: Role Of Many-Body Interactions

Francesco Operetto and Francesco Pederiva

Physics Department and INFN, University of Trento, Via Sommarive 14, I-38050 Povo, Trento, Italy

Abstract. We present a Diffusion Monte Carlo calculation of the equation of state of solid para-hydrogen in a range of pressures from around 0 up to 0.4GPa. Two interactions among the ones available have been used, namely the Silvera-Goldman and Buck potentials, and results have been compared with experimental data. A preliminary study of the effect of three-body potentials is also included.

Keywords: Solid para-Hydrogen, Equation of State, Diffusion Monte Carlo,
PACS: 67.80-s

INTRODUCTION

The physics of solid para-hydrogen has recently gained interest among experimental and theoretical physicists. In particular, there are two main issues that have been raised. The first is the occurrence of a transition to a kind of superfluid state in small droplets of p-H_2 embedded in ^{4}He and mixed ^{3}He-^{4}He droplets[1]. The second is the possibility of the occurrence of a phase transition of solid p-H_2 to a fluid dissociated state at very low temperatures and at pressures slightly below the transition to the metallic phase[2]. In this context it is interesting to have an idea of the accuracy of the models used for describing molecular para-hydrogen. Diffusion Monte Carlo methods provide a clean way of comparing model potentials in many-boson systems. We propose here some results of an accurate analysis we performed on two of the most used intermolecular interaction available, the Buck[3] and the Silvera-Goldman[4] potentials.

COMPUTATIONAL METHODS

We consider a periodic system of N molecules of p-H_2 described by the following Hamiltonian:

$$H = \frac{\hbar^2}{2m}\sum_{i=1}^{N}\nabla_i^2 + \sum_{i<j} v(r_{ij}) \qquad (1)$$

where $\{\mathbf{r}_i\}$ are the coordinates of the molecules. The size of the simulation box is determined by the density of the system ρ. The potential v has been taken equal either to the Buck or the Silvera-Goldman forms. The Diffusion Monte Carlo algorithm uses an imaginary time propagation of a density of walkers in the configuration space to project the exact ground state of the many-boson problem (i.e. the lowest symmetric eigenstate), starting from an arbitrary trial wavefunction[5]. The computed eigenenergy for the N particles therefore depends only on the choice of the interaction in the Hamiltonian. For technical reasons it is better to start projecting from a wavefunction which is as close as possible to the correct eigenfunction. Such wavefunction is also used for guiding the population of walkers in the imaginary time propagation. As trial function we chose the popular Jastrow-Nosanow form:

$$\Psi = \prod_{i<j} f(r_{ij}) \prod_{i=1}^{N} e^{-C(\mathbf{r}_i - \mathbf{R}_i)} \qquad (2)$$

where the $\{\mathbf{R}_i\}$ are the coordinates of the lattice sites around which each molecule has confined. Despite the fact that this wavefunction is not symmetric under exchange of molecules, it can be safely used to describe a quantum solid where actual exchange processes in the perfefct crystal are rare. The function f in the Jastrow is an expansion in term of an appropriate basis, whose coefficients have been determined by a variational Monte Carlo calculation in which the expectation value of the Hamiltonian is

CP850, *Low Temperature Physics: 24th International Conference on Low Temperature Physics;*
edited by Y. Takano, S. P. Hershfield, S. O. Hill, P. J. Hirschfeld, and A. M. Goldman

minimized with respect to the expansion coefficients. The maximum number of terms used in such expansion was 40. Calculations were performed using a face-centered-cubic (fcc) lattice.

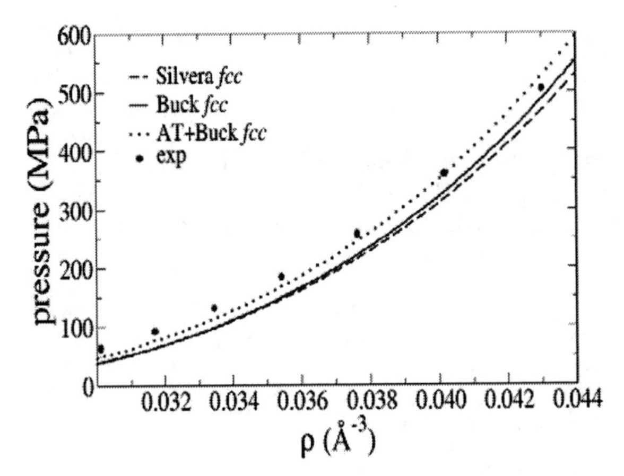

FIGURE 1. Computed value of the pressure with the Buck (solid line) and Silvera-Goldman (dashed line) potentials in the fcc crystal, compared with the experimental data of Silvera (points). The dotted curve represents the pressure curve obtained from the DMC values corrected with the inclusion of an Axilrod-Teller three-body force.

The number of molecules used in the periodic box was N=108. The coefficients for the expansion of the function f and the parameter C of Eq. (2) have been variationally recomputed for all the densities considered. The following DMC projection was performed using different values of the imaginary time-step and for different populations.

RESULTS

In Fig. 1 we show the results obtained for the pressure of the p-H_2 hcp crystal as a function of the density of the system for the Buck and Silvera-Goldman potentials. Pressures were obtained by fitting the DMC results with a Murnaghan curve, and computing the derivative with respect to the density:

$$P = \frac{\partial \varepsilon}{\partial \rho} \rho^2 \tag{3}$$

where ε is the energy per molecule. The theoretical predictions are compared with the experimental results of Ref. [6]. As it can be seen neither of the two interactions is actually giving a very satisfactory description of the pressure, neither near the saturation density, nor at higher densities. In particular, pressures appear to be underestimated. The situation tends to slightly improve at high pressures. The Silvera-Goldman potential includes a term $1/r^9$ which should effectively describe a triple-dipole interaction between the molecules. In order to check whether the effect of three body forces, when explicitly included in the intermolecular potential, can improve the description of the system, we tried to add to the Hamiltonian a three-body force of the Axilrod-Teller form:

$$V_3 = \alpha \left(\frac{1 + 3\cos(\varphi_1)\cos(\varphi_2)\cos(\varphi_3)}{r_{12}^3 r_{13}^3 r_{23}^3} \right) \tag{4}$$

where the φ are the angles described by the three molecules. The contribution of V_3 to the total energy was estimated perturbatively, and added to the energies evaluated with the use of the Buck potential, which does not include the effective triple-dipole term. The parameter $\alpha = 4.22\times10^4$ KÅ^{-9} was fitted against the experimental values of the pressure. The value of the coefficient is much larger than the corresponding C_9 coefficient in the SG potential[4]. As it can be seen from Fig. 1, the inclusion of the three body force considerably improves the description of the pressure over all the range of densities considered. In particular, the discrepancy observed at densities slightly higher than the saturation density is about halved. Also the agreement at high pressures is much more satisfactory. In conclusion, we showed how the currently used two-body intermolecular interactions do not give a fully realistic description of the equation of state of p-H_2. Inclusion of explicit three-body (and possibly many-body) potentials seems to be a necessary step to improve the results.

This work was partially supported by the INFM-DEMOCRITOS laboratory. Calculations were partly performed on the ECT* computing facilities. We thank M.H. Kalos, G.V. Chester and G. Galli for useful discussions.

REFERENCES

1. S. Grebenev, B. Sartakov, J.P. Toennis, and A.F. Vilesov, *Science* **289,** 1532 (2000).
2. S.A. Bonev, E. Schwegler, T. Ogitsu, and G. Galli, *Nature* **431**, 669 (2004).
3. U. Buck, F. Huisken, A. Kohlhase, D. Otten, and J. Shaefer, *J. Chem. Phys.* **78**, 4439 (1982).
4. I.F. Silvera and V.V. Goldman, *J. Chem. Phys* **69**, 4209 (1978).
5. P.J. Reynolds, D.M. Ceperley, B.J. Alder, and W.A. Lester, Jr, *J. Chem. Phys.* **77**, 5593 (1982).
6. J. van Straten and I.F. Silvera, *Phys. Rev. B.* **37**, 1989 (1988).

Two-Pulse Electron Spin Echo Study of Deuterium-Helium Solids

E. P. Bernard[a], V. V. Khmelenko[a], E. Vehmanen[b], P. P. Borbat[c], J. H. Freed[c], and D.M. Lee[a]

[a]Laboratory of Atomic and Solid State Physics, Cornell University, Ithaca, NY 14853, USA
[b]Department of Chemistry, University of Jyväskylä, P.O. Box 35, FIN-40351, Jyväskylä, Finland
[c]ACERT, Baker Laboratory, Department of Chemistry, Cornell University, Ithaca, NY 14853, USA

Abstract. We measured electron spin echoes from deuterium atoms within deuterium-helium solids with an X-band pulse ESR spectrometer. Our two-pulse electron spin echo envelope modulation (ESEEM) measurements are well described by a model that places 50% – 60% of the deuterium atoms at the interface between the molecular deuterium nanoclusters and the superfluid liquid helium. The remainder of the atoms lie in substitutional sites within the nanoclusters. We also report the spin-lattice relaxation times T_1 for these atoms.

Keywords: electron spin resonance, atomic deuterium, impurity-helium solids.
PACS: 76.30.Rn, 76.30.-v, 82.70.Gg, 81.07.-b

INTRODUCTION

Deuterium-helium (D_2-He) solids form when a mixture of D_2 and He gases is injected into superfluid helium at 1.5 K. If the mixture is subject to an RF discharge immediately prior to injection, deuterium atoms are produced that become embedded in the resulting solid. Studies of D_2-He solids by x-ray scattering show that these solids are composed of nanoclusters of molecular deuterium that are loosely aggregated into a gel within the superfluid helium.[1] A typical cluster is 90 Å in diameter and contains a blend of FCC and random close packed structures, with several stacking faults. In samples made with an RF discharge, deuterium atoms are present at average concentrations of ~ 10^{18} cm^{-3}. Measurements by CW ESR show that these atoms reside within the molecular clusters, so that local radical concentrations exceed average concentrations by at least a factor of ten[1]. Our recently constructed X-band pulse ESR spectrometer allows for electron spin echo envelope modulation (ESEEM) to provide a sensitive probe of the nuclear environment immediately surrounding the D atoms[2].

EXPERIMENT

We formed two D_2-He samples with stabilized atoms from gas mixtures with the ratio D_2:He = 1:20 according to the method described in our previous work.[1] We used a 90°-τ-180°-τ-echo primary echo sequence, with a 90° pulse time of 40 ns. For T_1 measurements, τ was held fixed at 800 or 1312 ns while the recovery time t between subsequent pulse pairs was varied from 500 μs to 50 s. For ESEEM measurements, τ was varied from 460 ns to 12 μs, with t held constant at 400 ms. All three deuterium hyperfine lines were separately excited at a frequency of 8930 MHz and magnetic fields of 3105.5, 3182.5 and 3261.1 Gauss.

RESULTS

Our recovery time data are modeled by the Bloch-type recovery function $V(t) \propto 1-\exp(-t/T_1)$, where V is the echo amplitude and T_1 is the spin-lattice relaxation time.[3] Fits to the data from the middle field hyperfine line give T_1 values of 64 ms and 87 ms for samples 1 and 2, respectively. The other hyperfine lines give similar times. The fit for sample 2 is inset in figure 1. Our sampled echo signals are processed by a blend of boxcar and phase corrected methods as described by our group and Astashkin *et al.*[2,4] The resulting ESEEM data is fit by the product of a sum of two decaying exponentials and a sum of two oscillatory modulation terms (see figure 1):

CP850, *Low Temperature Physics: 24th International Conference on Low Temperature Physics;*
edited by Y. Takano, S. P. Hershfield, S. O. Hill, P. J. Hirschfeld, and A. M. Goldman

$$E(\tau) = A\left(Qe^{-\tau/T_{2Q}} + (1-Q)e^{-\tau/T_{2L}}\right)\left(BE_B(\tau) + (1-B)E_S(\tau)\right)$$

where A, Q, T_{2Q} and T_{2L} parameterize the echo decay and B weights the modulation functions E_B and E_S. The function E_B simulates the ESEEM signal of D atoms *within* the solid clusters by assuming they are substituted in an FCC lattice of D_2 molecules. The function E_S is produced by the same model, but with the population of each coordination shell reduced by one-half. This simulates D atoms at the *surfaces* of the clusters in contact with the helium. We incorporate work by Kumada *et al.*[5] showing that D atoms reside in substitutional sites within solid D_2, that they have a 70 kHz isotropic coupling to their nearest molecular neighbors, and that the D atoms do not distort their immediate environment from the x-ray determined nearest neighbor spacing of 3.60 Å.[6] We calculate E_B and E_S using the "spherical approximation" of Mims *et al.*[7], which simplifies powder averaging over all crystalline orientations by ignoring the angular structure of the molecular lattices surrounding the radicals. At the 3.60 Å nearest neighbor distance this causes minimal distortion of the ESEEM signals.[8] Least-squares fitting finds the parameters A, Q, T_{2Q}, T_{2L} and B. Table 1 shows the four relevant parameters.

TABLE 1. Optimum fitting parameters of the two-population model for the three deuterium ESR lines.

Sample, hyperfine line	Q	$T_{2Q}/\mu s$	$T_{2L}/\mu s$	B
1, low field	0.648	1.12	5.26	0.394
1, middle field	0.672	1.13	5.51	0.403
1, high field	0.647	1.11	5.24	0.377
2, low field	0.793	0.689	2.96	0.477
2, middle field	0.792	0.732	3.04	0.457
2, high field	0.774	0.649	2.85	0.504

DISCUSSION

The failure of single population models—those that place all of the D atoms either in the interior or at the surface of the clusters—to fit the data necessitates the complexity of the two-population model described above. The presence of faults and randomly stacked structures in the clusters implies a great variety of environments around the deuterium atoms within the clusters. However, the angular insensitivity of this ESEEM method suggests that all these environments produce modulation signals very similar to E_B. Similarly, we assume that the great variety of environments surrounding the D atoms on the cluster surfaces produce ESEEM signals that, when averaged, resemble E_S. The parameter B then estimates the fraction of atoms located *within* the clusters to be 40% - 50%. The form of the echo decay term was found empirically and the fitting parameters T_{2Q} and T_{2L} should not be associated with either the bulk or surface atom populations. The much less concentrated samples of Kumada *et al.*[5] showed much longer decay times, which suggests that our signal decay is dominated by interactions among the D atoms.

Pulse ESR methods offer the first indication that a large fraction of the atomic radicals within impurity-helium solids reside on the cluster-helium interface, as suggested by Gordon.[9]

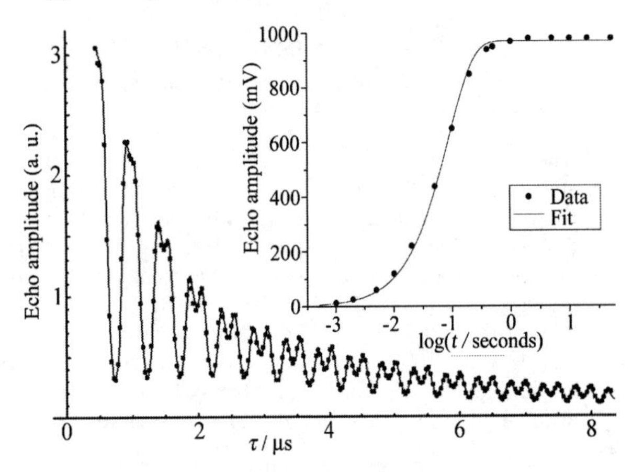

FIGURE 1. Two-pulse ESEEM data from the low field hyperfine line of sample 1. Dots show every fourth data point. The line shows the best fit of the two population model $E(\tau)$; see also table 1. Inset: Fit of $V(t)$ to recovery time data from the middle hyperfine line of sample 2.

This work is supported by NASA grant NAG 3-2871 and NIH/NCRR grant RR016292.

REFERENCES

1. E. P. Bernard, R. E. Boltnev, V. V. Khmelenko, V. Kiryukhin, S. I. Kiselev and D. M. Lee, *Phys. Rev. B* **69**, 104201 (2004).
2. E. P. Bernard, V. V. Khmelenko, E. Vehmanen, P. P. Borbat, J. H. Freed and D. M. Lee, these proceedings.
3. A. Schweiger and G. Jeschke, *Principles of Pulse Electron Paramagnetic Resonance*, Oxford: New York, 2002, p. 225.
4. A. V. Astashkin, V. V. Kozlyuk and A. M. Raitsimring, *J. Mag. Res.* **145**, 357-363 (2000).
5. T. Kumada, T. Noda, J. Kumagai, Y. Aratono and T. Miyazaki, *J. Chem. Phys.* **111**, 10974-10978 (1999).
6. A. F. Schuch and R. L. Mills, *Phys. Rev. Lett.* **16**, 616-618 (1966).
7. W. B. Mims, J. Peisach and J.L. Davis, *J. Chem. Phys.* **66**, 5536-5550 (1977).
8. M. Iwasaki, K. Toriyama and K. Nunome, *J. Chem. Phys.* **86**, 5971-5982 (1987).
9. E. B. Gordon, *Low Temp. Phys.* **30**, 756-762 (2004).

ESR Investigations of Spin-Pair Radicals in Nitrogen-Helium Solids

E. Vehmanen[*,†], V.V. Khmelenko[*], E.P. Bernard[*], H. Kunttu[†] and D.M. Lee[*]

** Laboratory of Atomic and Solid State Physics, Cornell University, Ithaca, NY 14853, USA*
† Department of Chemistry, University of Jyväskylä, P.O. Box 35, FIN-40351, Jyväskylä, Finland

Abstract. We have investigated nitrogen-helium solids by X-band CW-ESR. Our samples, containing molecules and atoms of either the ^{14}N or the ^{15}N isotope, were produced from N_2:He gas mixtures with different nitrogen impurity concentrations ranging from 0.25% to 1%. Samples were formed in a helium cryostat at 1.5 K by applying a radio-frequency discharge to a gas mixture jet that was then introduced into superfluid ^4He. We achieved average concentrations of stabilized nitrogen atoms as high as $2 \cdot 10^{19}$ atoms/cm^3. We observed a strong ESR signal of stabilized nitrogen atoms at ~3235 Oe. For the first time we also detected weak ESR signals corresponding to the $\Delta M_S = 2$ transitions of $^{14}N \cdots {}^{14}N$ or $^{15}N \cdots {}^{15}N$ spin-pair radicals in nitrogen-helium solids at ~1617 Oe. The intensities of these signals were ~1000 or more times smaller than those of the main atomic signals. We investigated the thermal stability of the nitrogen-helium solids and observed the spin-pair radical signals in dry samples at $T = 3.1$ K. After explosive disintegration of the samples at ~3.3 K, the spin-pair radical signal disappeared.

Keywords: atomic nitrogen, impurity-helium solids, spin-pair radicals
PACS: 67.40.Yv, 76.30.-v, 67.80.Mg

INTRODUCTION

Injection of atoms and molecules into superfluid helium leads to the creation of impurity-helium solids (gels) with high concentrations of stabilized atoms. When two atoms are separated in a solid matrix by only a small distance (5-10 Å), they form a spin-pair radical with coupled electronic spins. The first observations of N···N spin-pair radicals in a neon matrix were made by Knight *et al.*[1] In nitrogen-helium solids local concentrations of stabilized atoms in N_2 nanoclusters as high as 10^{21} cm^{-3} are achieved,[2] with atoms typically separated by 10 Å on the average. In this work we make the first observations of N···N spin-pair radicals in impurity-helium solids with high concentrations of nitrogen atoms.

EXPERIMENTAL TECHNIQUES

We formed our nitrogen-helium solid samples in a Janis cryostat at 1.5 K by sending a N_2:He gas mixture jet through a radiofrequency discharge and then into superfluid ^4He. We used a Varian X-band (f = 9.07 GHz) continuous wave spectrometer to record ESR spectra of allowed transitions of atomic nitrogen and of forbidden $\Delta M_S = 2$ transitions of spin-pair radicals. Average atomic nitrogen concentrations were calculated by comparing the nitrogen main line intensities to a reference ruby crystal with a known number of spins. This crystal was attached to the bottom of the TE_{011} resonant cavity into which a quartz beaker containing each sample was inserted. Some spin-pair radical signals were near the detection limit of the spectrometer, but the sensitivity was improved by applying higher microwave power. After measurements at temperatures 1.25 K and 1.35 K were performed, the samples were warmed with a computer-controlled heater to 3 K in order to evaporate the liquid helium and investigate the main atomic line and spin-pair radicals in dry samples. Then the temperature was slowly increased in small steps until the samples disintegrated explosively. Measurements were made using N_2:He gas mixtures 1:100, 1:200 and 1:400 on both isotopes ^{14}N and ^{15}N to see how the gas mixture ratio affects spin-pair radical formation.

RESULTS AND DISCUSSION

For the first time the $\Delta M_S = 2$ ESR signals of N···N spin-pair radicals for both the ^{14}N and ^{15}N isotopes in impurity-helium solids were observed. Using a

CP850, *Low Temperature Physics: 24th International Conference on Low Temperature Physics;*
edited by Y. Takano, S. P. Hershfield, S. O. Hill, P. J. Hirschfeld, and A. M. Goldman

$[N_2]$:[He]=1:200 gas mixture we formed our nitrogen-helium sample and achieved average atomic nitrogen concentrations of 2×10^{19} cm^{-3}. We observed in the ESR spectrum of the sample not only a strong atomic nitrogen peak at ~3235 Oe but also a three orders of magnitude weaker $\Delta M_S = 2$ spin-pair radical line at ~1617 Oe. The strong dipolar interaction broadened signals of individual N atoms, making the $\Delta M_S = 1$ transitions of spin-pair radicals unobservable. CW-ESR spectra of atomic main lines and magnified spin-pair radical lines are shown in Fig. 1 for samples containing either ^{14}N or ^{15}N.

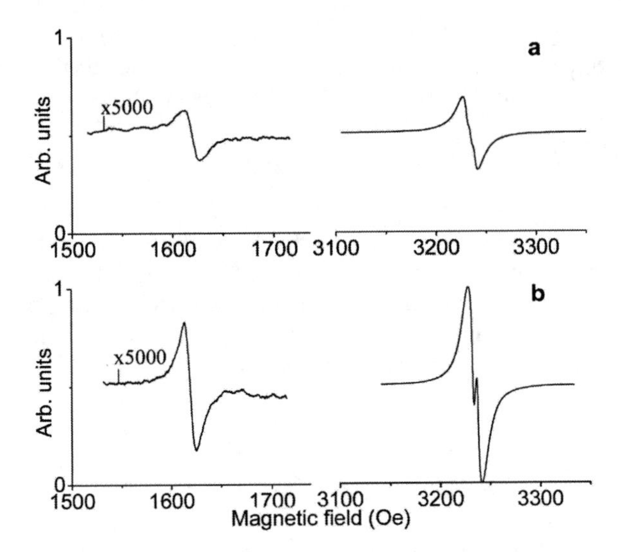

FIGURE 1. ESR spectra of nitrogen atoms and spin-pair radicals in nitrogen-helium solids at T=1.35K. a) $^{14}N\cdots{}^{14}N$ and ^{14}N, b) $^{15}N\cdots{}^{15}N$ and ^{15}N. Spectra obtained in samples which were formed using $[N_2]$:[He]=1:100 gas mixture.

The intensity of the spin-pair radical $\Delta M_S = 2$ transition was strongly affected by the choice of gas mixture, even though the main ESR line of individual N atoms remained essentially unchanged. After normalizing each signal with our ruby reference we conclude that use of the 0.5%-mixture favors spin-pair radical formation more than 0.25% or 1%-mixtures. This is shown in Fig. 2, where signal intensities are compared with the signal of the $[^{15}N_2]$:[He]=1:200 mixture. The relative intensity between main and spin-pair lines varied by large factors from a bit more than 1000 in the ^{15}N sample formed from a 0.5% gas mixture to almost 10 000 (see Fig. 1b) in most of the other samples where the spin-pair radicals were observed.

Spin-pair radical lines were observed at almost exactly half the field as that for the main atomic line in every sample. Large differences in line widths exist however. While the atomic line was 31-34 Oe in width

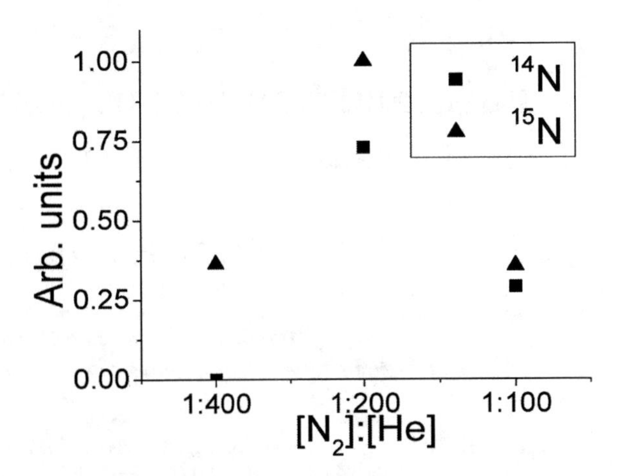

FIGURE 2. Spin-pair radical line intensities in samples formed by using different gas mixtures. Intensities are relative to that of the $[^{15}N_2]$:[He]=1:200 signal.

for both $^{14}N_2$ and $^{15}N_2$ samples in superfluid regardless of nitrogen content in the gas mixture, spin-pair radical line widths varied from 20-34 Oe under the same conditions. By changing applied microwave power we observed the ESR peak amplitude increase but no saturation effect on line width was seen. At 3.0 K the dried samples became more compact and both atomic lines and spin-pair radical lines were ~20% more intense. Further increase of the temperature by 0.1-0.2 K did not make any significant difference. At 3.3 K the samples exploded and no spin-pair radical signals were detected.

CONCLUSIONS

For the first time we have observed the $\Delta M_S = 2$ N···N spin-pair radical ESR spectra in nitrogen-helium solid samples and also demonstrated the effects of different nitrogen content in the gas mixtures on the spin-pair radical intensities. These studies suggest that it will be possible to perform saturation experiments on the $\Delta M_S = 2$ line to determine the average distance between atoms in spin-pair radicals. This work was supported by NASA grant NAG3-2871 and Academy of Finland.

REFERENCES

1. L.B. Knight, Jr., B.A. Bell, D.P. Cobranchi and E.R. Davidson, *J. Chem. Phys.* **111**, 3145-3154 (1999).
2. E.P. Bernard, R.E. Boltnev, V.V. Khmelenko and D.M. Lee, *J. Low Temp. Phys.* **134**, 199-204 (2004).

Pulse and Continuous Wave Electron Spin Resonance Investigations of H and D Atoms in Impurity-Helium Solids

V.V. Khmelenko[*], E.P. Bernard[*], E. Vehmanen[*,†] and D.M. Lee[*]

** Laboratory of Atomic and Solid State Physics, Cornell University, Ithaca, NY 14853, USA*
† Department of Chemistry, University of Jyväskylä, P.O. Box 35, FIN-40351, Jyväskylä, Finland

Abstract. By using pulse and continuous wave (CW) X-band ESR methods the kinetics of the tunneling exchange chemical reaction $D+HD \rightarrow D_2+H$ in HD-D_2-He solids has been studied. It was found that in the course of this reaction the number of HD molecules in the vicinity of D atoms decreased while the surroundings of H atoms did not change.

Keywords: atomic hydrogen, impurity-helium solids, tunneling reactions
PACS: 67.40.Yv, 67.80.Mg, and 76.30.Rn

INTRODUCTION

Impurity-helium (Im-He) solids are porous gel-like materials created by injecting a mixed beam of helium gas and gaseous impurity atoms and molecules into superfluid ^{4}He.[1] Im-He solids are utilized for the matrix isolation of high concentrations of stabilized atoms and for studying exotic tunneling chemical reactions at low temperatures.[2] In this work we use CW and pulse ESR spectroscopy to study the kinetics of a tunneling exchange chemical reaction of hydrogen isotopes and for monitoring the changes in the immediate molecular environment of the H and D atoms during the course of this reaction.

EXPERIMENTAL TECHNIQUES

Im-He solids with H and D atoms stabilized in molecular (HD-D_2) nanoclusters are produced *in situ* in a Janis cryostat by sending gaseous helium with small (5%) admixtures of impurities (H_2 and D_2) through a high frequency discharge into superfluid helium at T~1.5 K contained in a small sample beaker. After sample collection, the beaker is lowered into a TE_{011} microwave resonant cavity. We use two different low temperature inserts in the cryostat. The first is for detection of CW ESR signals by an X-band homodyne spectrometer (Varian E-4) and employs a high (~3000) quality factor cavity. Atomic concentrations are found by comparing the intensity of atomic CW ESR signals with the intensity of signals from a ruby crystal with a known number of spins. The second insert is for the detection of electron spin echo signals by a pulse X-band homodyne spectrometer and has a low (~700) quality factor cavity.

RESULTS AND DISCUSSION

We investigated the tunneling reaction $D+HD \rightarrow D_2+H$ in Im-He solids created from the gas mixture $[H_2]:[D_2]:[He]=1:4:100$. The concentrations of the H and D atoms obtained by CW ESR measurements show the progress of the reaction in Fig. 1. Within the accuracy of our measurements, the concentration of H atoms remains almost constant while the concentration of D atoms decreases over the course of the tunneling reaction. Each of the allowed CW ESR lines of the H and D atoms is accompanied by two satellite lines, which are separated from the main line by ~4.5 gauss.[2] These satellite lines are associated with ESR transitions involving a spin flip of an electron in either an H or a D atom and a simultaneous spin flip of a proton in a neighboring HD molecule.[2,3] Trammel et al.[3] show the ratio of the intensities of the satellite lines to their respective main lines to be proportional to the number of HD nearest neighbors around the H or D atoms. Assuming a constant distance from these atoms to the HD molecules allows changes in the number of HD neighbors to be followed. We found that the satellite

CP850, *Low Temperature Physics: 24th International Conference on Low Temperature Physics;*
edited by Y. Takano, S. P. Hershfield, S. O. Hill, P. J. Hirschfeld, and A. M. Goldman

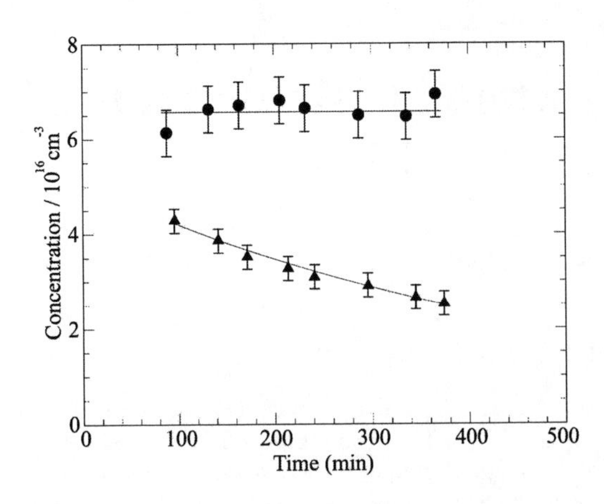

FIGURE 1. Time dependence of the concentrations of H (circles) and D (triangles) atoms in HD-D_2-He solid.[2]

line intensity ratio for H atoms is 0.11 and that this value does not change as the reaction proceeds.[2,4] For D atoms the initial satellite line intensity ratio of 0.20 decreases to 0.14 in 14 hours.[4] These results show that the number of HD molecules surrounding H atoms does not change while the number of HD neighboring D atoms decreases by 30%. An analogous study of D_2 neighbors is not possible because the resulting satellite lines cannot be resolved from the main line.

Pulse ESR measurements of electron spin echo envelope modulation (ESEEM) signals provide complementary information about the HD and D_2 neighbors of the H and D atoms. In general, the depth of the echo modulation increases with the number of deuterons neighboring the excited atom. The ESEEM signal of H atoms in an HD-D_2-He solid is shown in Fig. 2a. The ESEEM signals of H atoms taken at the beginning of the experiment and six hours later were identical, showing that the number of deuterons surrounding H atoms to be constant. We model the modulation of D atom echo signals as due to two populations of D atoms, one residing in the interior of the cluster with twelve nearest neighbors, the other at the cluster-helium surface with six nearest neighbors. In clusters formed from only deuterium and helium gases, this analysis showed that 55% of D radicals reside in the interior of the clusters, with the remaining 45% at the cluster surface.[5] If we assume that H and D atoms in our hydrogen containing sample are similarly distributed, we can fit the number of HD and D_2 nearest neighbors in our model to the ESEEM signals. The best fit of our model to the ESEEM signal of H atoms, presented as the solid line in Fig. 2a, was obtained for the configuration in which ten deuterium and two HD molecules surround a typical H atom in the interior of the clusters. As the number of D atoms decreases with time, their echo amplitudes are correspondingly reduced (Fig. 2b and 2c). Fitting our ESEEM modulation model to signals of D atoms

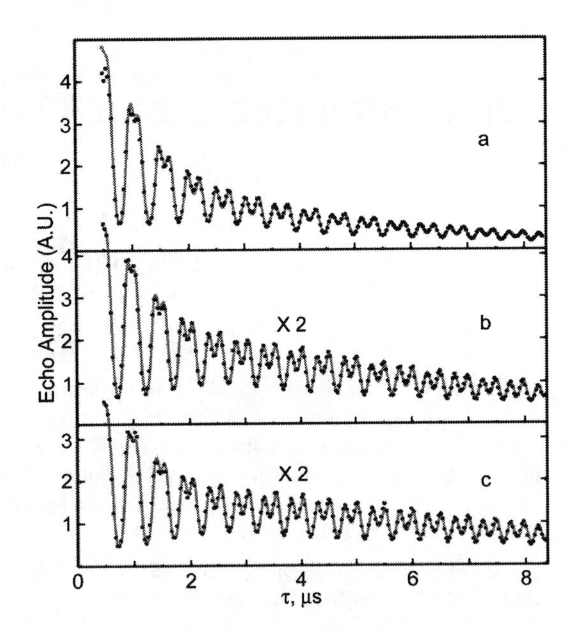

FIGURE 2. Two pulse ESEEM signals (dots) for H and D atoms in HD-D_2-He solid at 1.35 K taken at 8.93 GHz: (a) - low field hyperfine (hf) component of H atoms at 2910.1 gauss; (b, c) - low field hf components for D atoms at 3105.5 gauss at the beginning of the experiment and six hours later, respectively. Solid lines are the best fits of model calculations to experimental data (see text).

obtained at the beginning of the experiment (solid line in Fig. 2b) shows that a typical D atom in the interior of the clusters is surrounded by nine D_2 and three HD molecules. Fitting the same model to data obtained six hours later (solid line, Fig. 2c) shows that ten D_2 and two HD molecules surround a typical D atom within the cluster.

In conclusion, CW and pulse ESR methods provide complementary information about trapping sites and changing environment of atoms participating in low temperature tunneling reactions. This information is important for elucidation of the mechanisms of tunneling reactions. This work was supported by NASA grant NAG 3-2871.

REFERENCES

1. E.B. Gordon, V.V. Khmelenko, A.A. Pelmenev, E.A. Popov and O.F. Pugachev, *Chem. Phys. Lett.* **155**, 301-304 (1989).
2. S.I. Kiselev, V.V. Khmelenko, and D.M. Lee, *Phys Rev. Let.* **89**, 175301 (2002).
3. G.T. Trammel, H. Zeldes, and R. Livingston, *Phys. Rev.* **110**, 630-634 (1958).
4. E.P. Bernard, R.E. Boltnev, V.V. Khmelenko and D.M. Lee, *J. Low Temp. Phys.* **138**, 829-834 (2005).
5. E.P. Bernard, V.V. Khmelenko, E. Vehmanen, P.P. Borbat, J.H. Freed, and D.M. Lee, these proceedings.

Nuclear Spin Relaxation Times for Methane-Helium "Slush" at 4 MHz using Pulsed NMR

J. A. Hamida and N. S. Sullivan

University of Florida, Gainesville, FL 32611, USA

Abstract. We report measurements of the nuclear spin-lattice relaxation times (T_1) and spin-spin relaxation times (T_2) for small grains of methane suspended in liquid helium (methane-helium "slush") for temperatures 2 K<T<5 K. The observed spin-spin relaxation rate $1/T_2$ is consistent with internal diffusion as opposed to surface scattering, which has been shown to be dominant for hydrogen-helium "slush". The most interesting feature observed for methane-helium mixtures is the existence of three different time scales for samples aged at 4.2 K. The possible origins of this distribution of relaxation times are discussed.

Keywords: Helium, Methane, Slush, NMR
PACS: 67.80.Jd, 76.60.-k, 76.60.Es

INTRODUCTION

One of the most promising large-scale advances in rocket propulsion is the use of atomic propellants (such as atomic H) stabilized in cryogenic matrices to inhibit or delay their re-combination into molecules. We reported earlier[1] the studies of a suitable cryogenic matrix consisting of a slush of small grains of solid hydrogen suspended in liquid helium. For hydrogen slush measurements showed that the thermal relaxation was dominated by scattering at grain boundaries, which makes the matrix unsuitable. As an alternative matrix we have studied methane-helium slush. The interest in methane is the absence of the quadrupolar surface interactions that dominate the relaxation for hydrogen matrices. The results for the measurements of the relaxation for methane-helium slush are compared to the relaxation rates observed for hydrogen-helium mixtures.

RESULTS AND DISCUSSIONS

The nuclear spin-spin relaxation times (T_2) and spin-lattice relaxation times (T_1) were measured using a coherent pulsed NMR spectrometer operating at 4 MHz. T_2 values were determined from the decay of solid echoes, and the T_1 values were obtained from repeated $\pi/2$-τ- $\pi/2$ sequences. T_2 studies for freshly prepared methane-helium slush at 4.2 K revealed the existence of two different relaxation times. Methane has a very complex ground state at low temperature, consisting of a mixture of different molecular species (A, T and E, corresponding to different orientational symmetries[2]).

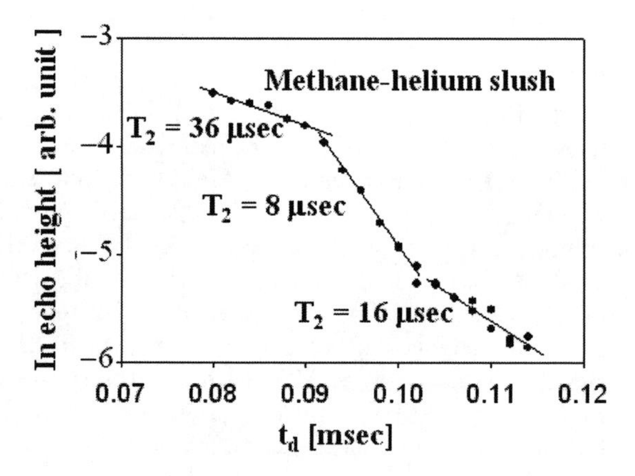

FIGURE 1. Experimental results for the spin-spin relaxation times T_2 for a methane-helium slush sample at 4 K aged for three weeks. t_d is the delay time between two pulses used to obtain a solid echo at time $2t_d$ and thus determine the T_2 processes.

We have carried out measurements on methane-helium slush as the sample was aged to understand the roles played by the different species. Measurements of T_2 for samples 3 weeks old at 4 K revealed three different time scales (see Fig. 1) instead of the two typically associated with (i) internal diffusion and (ii) surface scattering. For the nuclear spin-lattice relaxation time, T_1, we found a single exponent irrespective of the sample's age (see Fig. 2). This latter observation would indicate that at very short times (~μsec) there is a single fluctuating field that dominates the process of thermal relaxation. The complex evolution for T_2 would be consistent with a

CP850, *Low Temperature Physics: 24th International Conference on Low Temperature Physics;*
edited by Y. Takano, S. P. Hershfield, S. O. Hill, P. J. Hirschfeld, and A. M. Goldman

sequential relaxation at longer times (~100 μsec) with a possible bottleneck in the relaxation in some cases.

RELAXATION PROCESSES

The relaxation process involves different mechanisms associated with the different molecular species. The E species have total nuclear spin I=0 and do not participate except with their slow conversion into other species: T (I=1) and A (I=2). The energy difference[3,4] between the species is a few K and spin-spin interactions between different species (A and T) are significantly reduced because they do not conserve energy at low temperatures. Because of their spherical spatial symmetry and thus reduced dipole-dipole interactions, the A species form a narrow central component of the NMR line shape. Their relaxation is expected to be driven by relatively slow spin-spin and cross-relaxation with the T species. Because of their strong dipole-dipole interactions, the T species contribute a broad NMR line shape with only the central component overlapping with that of the A species. Thermal fluctuations of the T species are much more effective in relaxation because they create fluctuating local magnetic fields that can induce rapid spin flips between neighboring spins. There can, however, be a bottleneck in the relaxation if the intrinsic A species relaxation is slower than the cross-relaxation with the T species.

The appearance of three different relaxation rates (see Fig. 1) after aging for three weeks is not understood. This time period for aging is comparable to that for the conversion of the species, principally the decay of the E and T species to A species. The reduced fractional concentration of T species relative to the A species would lead to a faster relaxation rate but this would be offset by local ordering of the T molecular species due to their intermolecular electrostatic octupolar interactions. The relaxation of ordered molecules would be expected to be very slow at 4 K.

In this model, the fast relaxation component observed for T_2 would be related to the characteristic reorientational fluctuation rate τ_R for the T species. The value of τ_R can be estimated by noting that the ratio T_1/T_2, if dominated by the T species fluctuation, would be of the order of $(\omega\tau_R)^2/6$ yielding τ_R ~ 8 μsec. We can also estimate τ_R from T_2 using the known second moment of methane and this yields τ_R ~ 6 μsec, which is consistent.

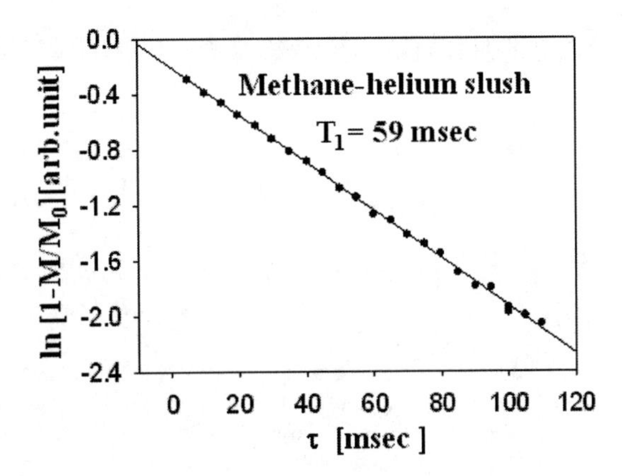

FIGURE 2. Spin-lattice relaxation time T_1 for methane-helium slush at 4.2 K measured form the magnetization decays using repeated $\pi/2$- τ–π /2 pulse trains.

CONCLUSIONS

Measurements of the nuclear spin-spin and spin-lattice relaxation times for finely divided methane particles suspended in liquid helium reveal a complex relaxation process that depends on the spin-spin interactions of the different molecular species and the age of the sample. The results suggest that the fast reorientational rates of the T species dominate the relaxation leading to an initial bottleneck for spin-spin relaxation, which has not previously been observed. Cross-relaxation and diffusion dominate the relaxation for methane "slush" in contrast to the surface relaxation reported for hydrogen "slush" formed under similar conditions.

ACKNOWLEDGMENTS

We gratefully acknowledge many helpful discussions with Gary Ihas and K. A. Muttalib. Pradeep Bhupathi is also thanked for his help. Work supported by the NASA grant no. NAG3-2750.

REFERENCES

1. M. Matusiak, J. A. Hamida, G. G. Ihas and N. S. Sullivan, *J. Low Temp. Phys.* **134**, 775 (2004).
2. B. Wilson, *J. Chem. Phys.* **3**, 276 (1935).
3. W. Press and A. Kollmar, *Solid State Commun.* **17**, 405 (1975).
4. H. Glattli, A. Sentz and M. Eisenkramer, *Phys. Rev. Lett.* **28**, 871 (1972).

Search for New Tool for Production of Ultracold Neutrons

L.P. Mezhov-Deglin[a], V.B. Efimov[a], A.A. Levchenko[a], G.V. Kolmakov[a], A.V. Lokhov[a] and V.V. Nesvizhevsky[b]

a *Institute of Solid State Physics RAS, Chernogolovka, Moscow region, 142432 Russia;*
b *Institute Laue-Langevin, Grenoble, France*

Abstract. The structure and properties of highly porous nanocluster condensates in superfluid helium — impurity gels, whose dispersive system or frame is formed by impurity clusters surrounded by a layer of solidified helium — are under wide investigations now. We are going to verify the hypothesis that gels formed in He-II by deuterium or heavy water clusters (a matter with low rate of neutron absorption) and cooled below a few mK can be used for lowering the energy of cold neutrons down to the energy range of ultracold neutrons. The first step in this direction includes the neutron studies of the structure of deuterium or heavy water gel samples prepared in a wide quartz glass cell filled with He-II at temperatures above 1.2 K. The measurements are planned to be performed at ILL in the new optical cryostat designed at ISSP specially for this studies.

Keywords: Nanocluster condensate, impurity gel, superfluid helium
PACS: 67.40.Yv, 61.82.Rx

INTRODUCTION

The study of impurity-helium condensates or impurity gels, formed by impurity nanoclusters in superfluid He-II, is an exciting area in current condensed matter and soft matter physics [1,2]. The nature and composition of the impurity gels is not presently understood in its entirety. The information derived from X-rays, ESR and ultrasound studies [3] should allow a much needed critical analysis of their structure, its evolution with time and temperature, their physical properties, and chemical reactions involving the gel molecules and their radicals.

We have proposed [4,5] that ultracold neutrons could be produced at a few mK temperaures by inelastic collision of cold neutrons with a highly porous condensate of nanoparticles formed in He-II by nanoparticles of low neutron absorption. The main goal of our studies is to verify the feasibility of this proposal. Ultracold neutrons (UCN) are useful in the fundamental physics of elementary particles and in many other fields. The great interest in UCN arises from the fact that they are reflected totally from a clean solid surface, which makes it possible to confine them in a closed container for a period of time comparable to their beta-decay lifetime (~15 min). This long confinement time allows high precision measurements of the properties of neutrons and their interaction with matters and fields. However, available UCN densities are low. This limits significantly any applications of UCN.

The new approach we have proposed for a large increase in UCN density is based on equilibrium thermalization of neutrons during their diffusive motion in a large sample of impurity-helium gel, which consists of nanoparticles of heavy water, deuterium, oxygen, or other low-absorption molecules in He-II cooled below a few mK. This process is analogous to the thermalization of neutrons by a moderator in a nuclear reactor. Nanoparticles should provide sufficiently large cross-section of coherent scattering. However, the distances between the scattering centers in an impurity gel are greater by a factor of 100, which should enable this process at lower energies (larger wavelengths) close to that of UCN (about a few mK).

CURRENT STATE OF THE STUDIES

Our first step is a design and construction of a special optical cryostat suitable for neutron studies of the structure of macroscopic gel samples in He-II at temperatures above 1.2 K. The sample cell and the metal cryostat have been prepared and tested at the Institute of Solid State Physics in Chernogolovka. The neutron studies will be performed at ILL, Grenoble, by a joint team including researches from

CP850, *Low Temperature Physics: 24th International Conference on Low Temperature Physics;*
edited by Y. Takano, S. P. Hershfield, S. O. Hill, P. J. Hirschfeld, and A. M. Goldman

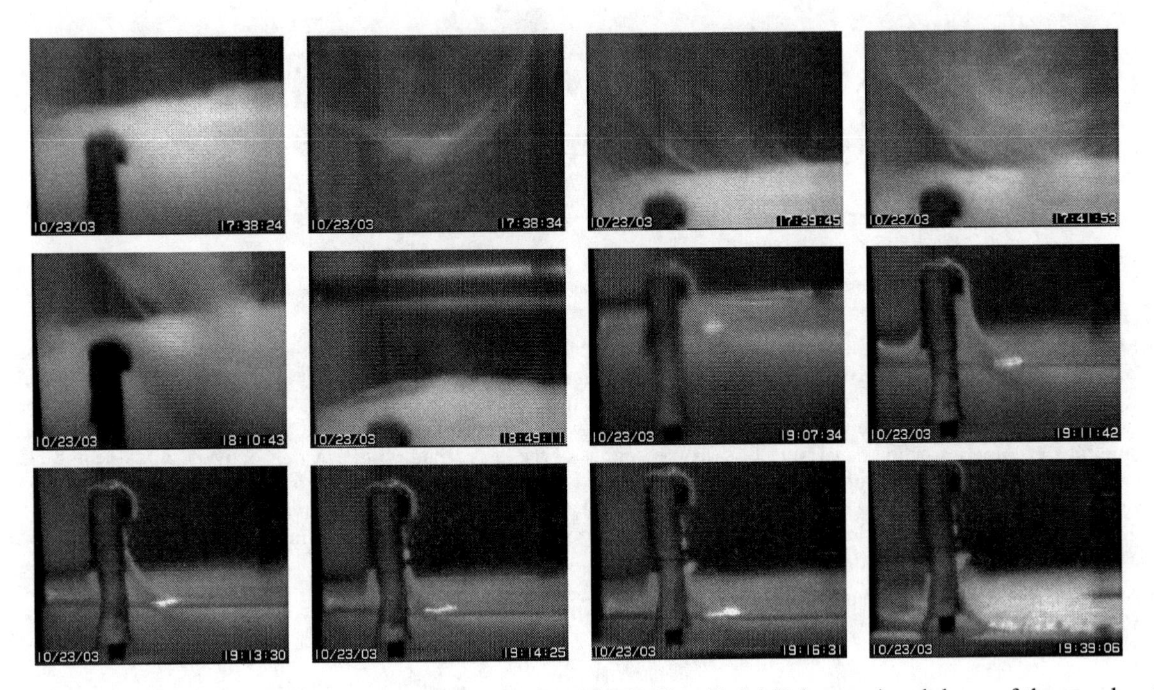

FIGURE 1. Formation of the deuterium gel sample in superfluid He-II at T=1.4 K (top row) and decay of the sample with lowering of the level of He-II (middle and bottom rows). The diameter of the Teflon plate at the bottom of the cell is ~3 cm, and the height of the column supporting the resistive thermometer placed above the plate is ~1 cm.

ISSP, ILL and JINR at Dubna. This work is in progress.

The frames shown in Fig.1 illustrate the process of formation and decay of a deuterium gel sample in the glass cell with a Teflon plate placed at the bottom.

ACKNOWLEDGMENTS

The study was supported in part by the Ministry of Education and Science of the Russian Federation as part of the Government Science and Technology program “Topical Problems in Condensed Matter Physics” and by the Presidium of RAS under the program “Macrophysics”.

REFERENCES

1. L.P. Mezhov-Deglin and A.M. Kokotin, *JETP Lett.* **70**, 756 (1999).
2. L.P. Mezhov-Deglin and A.M. Kokotin, *Instruments and Experimental Techniques* **2**, 279 (2001).
3. E.P. Bernard, R.E. Boltnev, V.V. Khmelenko, V. Kiryukhin, S.I. Kiselev, and D.M. Lee, *Phys. Rev. B* **69**, 104201 (2004).
4. V.V. Nesvizhevsky, *Physics of Atomic Nuclei* **65**, 409 (2002).
5. L.P. Mezhov-Deglin, V.V. Nesvizhevsky, and A.V. Stepanov, *Physics-Uspekhi* **46**, 89 (2003).

Laser Induced Fluorescence Detection Of Metastable He_2 Molecules In Superfluid Helium

W.G. Rellergert, W.H. Lippincott, J.A. Nikkel, and D.N. McKinsey

Department of Physics, Yale University, New Haven, CT 06520, USA

Abstract. Ionization events in superfluid helium result in the copius production of metastable He_2 molecules with lifetimes of roughly 10 seconds. Because of these extremely long lifetimes, individual molecules can be detected using laser induced fluorescence (LIF). By using a cycling transition, each molecule can be probed many times over its lifetime resulting in a high signal to noise ratio. We describe an approach that uses two pulsed infrared lasers to drive the cycling transition, which results in a position resolution equal to or less than the radius of the laser beam profile. A radioactive source, focused laser beam, or electric discharge in superfluid helium could be used to create He_2 triplet molecules which would then be tracked by intersecting beams. Rastering the beams throughout the volume of interest would allow imaging in three dimensions. This approach also has particle physics applications, as it could allow for the detection of neutrinos, weakly interacting massive particles (WIMPs), and ultracold neutrons.

Keywords: superfluidity, helium, turbulence, fluorescence
PACS: 95.55.Vj, 29.40.Gx, 33.50.-j

In the 1950s and 1960s it was discovered that ionizing radiation, in addition to making helium scintillate brightly in the ultraviolet [1, 2], also produces long lived excitations in liquid helium [3]. These were later attributed to the creation of He_2 excimer molecules, and studies showed that about 50% of the energy of an electron passing through the liquid went into the production of the lowest lying singlet $He_2(A^1\Sigma_u^+)$ and triplet $He_2(a^3\Sigma_u^+)$ states [4, 5]. It was found that about 13000 $He_2(a^3\Sigma_u^+)$ states and 19000 $He_2(A^1\Sigma_u^+)$ states are created per MeV [6].

Both the $He_2(A^1\Sigma_u^+)$ and $He_2(a^3\Sigma_u^+)$ emit 80 nm photons when they decay [7, 8]. However, the lifetime of the $He_2(a^3\Sigma_u^+)$ is about 13 s [9] whereas the lifetime of the $He_2(A^1\Sigma_u^+)$ is only around 1 ns [10]. This is because the transition of the $He_2(a^3\Sigma_u^+)$ to the dissociative ground state $He_2(X^1\Sigma_g^+)$ is forbidden, requiring a spin flip. Of the noble gases, this is unique to helium, as the lifetimes of the lowest triplet state of neon, argon, krypton and xenon are 6.6 μs, 3.2 μs, 350 ns, and 50 ns respectively [11]. This can be explained qualitatively by noting that the spin orbit coupling scales as Z^4, and that the excited helium atom is in an S state whereas other excited noble gas atoms are left in P states. They therefore contain an intrinsic orbital angular momentum that contributes to the spin orbit coupling, while the helium atom does not.

In addition to radiative decays, the molecules can also decay through reactions with other He_2 molecules [12] and collisions with the walls of their container. The movement of the $He_2(a^3\Sigma_u^+)$ molecules is limited by scattering with rotons, phonons and ^3He impurities (see Figure 1). By adjusting the temperature or ^3He concentration, the distance a molecule can travel in its radiative lifetime can be varied over several orders of magnitude.

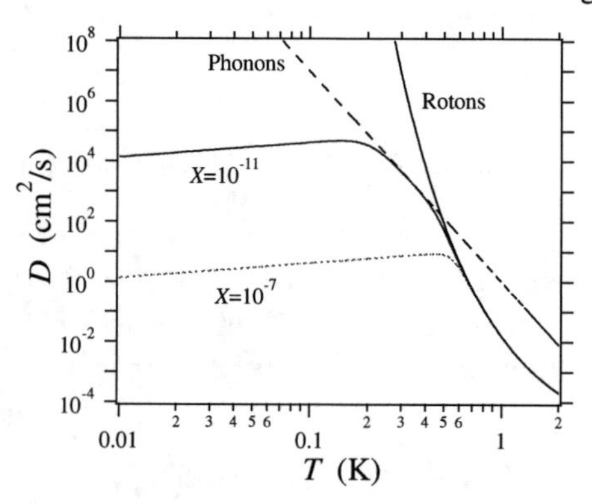

FIGURE 1. The diffusion constant,D, vs. T for He^3 concentrations in commercially available liquid helium (10^{-7}) and purified liquid helium (10^{-11}). Also shown are the roton and phonon dominated curves.

Because the radiative lifetime of the $He_2(a^3\Sigma_u^+)$ is 13 s, it should be possible to detect an individual molecule using laser induced fluorescence, since one molecule could be cycled many times over its lifetime. Many detailed spectroscopic studies of the triplet molecules have been performed, and optical absorption frequencies are well established [4, 5, 12, 13]. Figure 2 shows the energy levels of the lowest lying triplet states of the He_2 molecule and one possible cycling transition using two infrared lasers. The first laser would be used to send in a pulse of 910 nm light. This would excite molecules from $a^3\Sigma_u^+ \rightarrow c^3\Sigma_g^+$. A second pulse of 1040 nm light would immediately follow, exciting

CP850, *Low Temperature Physics: 24th International Conference on Low Temperature Physics;*
edited by Y. Takano, S. P. Hershfield, S. O. Hill, P. J. Hirschfeld, and A. M. Goldman

the molecules from $c^3\Sigma_g^+ \rightarrow d^3\Sigma_u^+$. From there approximately 90% of the time the molecule will decay to the $b^3\Pi_g$ giving off a photon at 640 nm, and 10% of the time it will decay to the $c^3\Sigma_g^+$. Both the $c^3\Sigma_g^+$ and the $b^3\Pi_g$

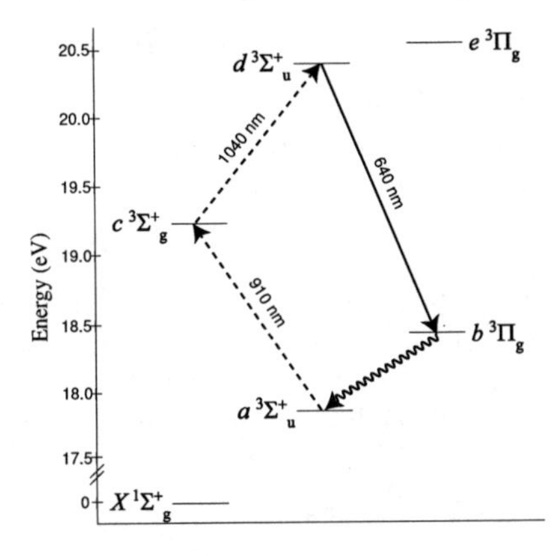

FIGURE 2. Energy levels of the He_2 molecule below 21 eV. Also shown is the cycling transition described in the text.

then decay back to the ground state $a^3\Sigma_u^+$. One can therefore detect the presence of helium molecules by detecting the 640 nm radiation given off during the transtition. The cross section of the $a^3\Sigma_u^+ \rightarrow c^3\Sigma_g^+$ transition has been measured to be $2.4 \times 10^{-15}\,\text{cm}^2$ [13], and if one assumes that the $c^3\Sigma_g^+ \rightarrow d^3\Sigma_u^+$ transition will have a comparable cross section, lasers emitting 250 $\mu\text{J}/\text{cm}^2$ would be enough to drive the transition with a 90% efficiency. This scheme is advantageous because the infrared laser light is out of the detectable range of the light detector, and because it allows for position resolution as molecules will only be excited where the beams overlap.

When driving cycling transitions in molecules one must give careful consideration to the rotational and vibrational structure of the energy levels. Any time a molecule decays from one level to another, there is a small chance that it will fall to a different rotational or vibrational level. This can be problematic because the molecules might fall into a state with a long relaxation time where they are blind to the pump lasers. Eltsov *et al.* determined the vibrational relaxation time for the $a^3\Sigma_u^+$ to be 140±40 ms, and demonstrated that the rotational relaxation times were much faster [13]. In the scheme mentioned above, the only transition with a significant probability of falling to the wrong vibrational state is the $c^3\Sigma_g^+ \rightarrow a^3\Sigma_u^+$ transition. Through calculation of Franck-Condon factors and branching ratios, we expect about 1.2% of the molecules undergoing the cycling transition to fall to the first vibrational state, $a(1)$, instead of the zeroth, $a(0)$. Those molecules are then effectively lost for subsequent cycles. However, they can be regained by adding another laser operating at 1070 nm to pump the molecules in the $a(1)$ to the $c(0)$ where they will decay back to $a(0)$ 94% of the time.

There are advantages to the vibrational levels, however. In some experiments it may be desirable to cause molecules created by one event to be undetectable for a following event. In that case, another laser can force the molecules into a vibrational state so that they are blind to the pump lasers. Also, this gives rise to another method to detect the molecules using inexpensive CW laser diodes. One laser at 800 nm would excite the molecule from $a(0) \rightarrow c(1)$ which would decay to the $a(1)$. That laser could then be turned off and a second laser at 1070 nm would drive the molecule from $a(1) \rightarrow c(0)$, where it would decay back to $a(0)$ emitting a detectable photon at 920 nm.

The laser linewidth requirements for these excitation schemes are not very stringent because the absorption lines are broadened by the liquid to a width of about 120 cm^{-1}. While this also reduces the absorption cross sections for the transitions, we calculate that the measured value of $2.4 \times 10^{-15}\,\text{cm}^2$ in superfluid helium is sufficient to ensure good signal to noise with modest laser power.

One application of this technique is imaging flows and turbulence in superfluid helium. A radioactive source, for instance, could be used to create He_2 molecules which could then be imaged to determine their paths. In addition, the molecules may become trapped on vortex lines allowing for the possibility of imaging the shape and dynamics of vortices. Currently, neutron absorption tomography [14] is useful in imaging superfluid flows, but can image only in two dimensions. With two infrared lasers one would be able to image in three dimensions.

In conclusion, laser induced fluorescence promises to be a useful method to detect He_2 molecules, which can be used as probes to study superfluid helium.

REFERENCES

1. H. Fleishman, H. Einvinder, and C. S. Wu, *Rev. Sci. Inst.* **30**, 1130 (1959).
2. E. H. Thorndike and W. J. Shlaer, *Rev. Sci. Inst.* **30**, 838 (1959).
3. C. Surko and F. Reif, *Phys. Rev* **175**, 229 (1968).
4. W. S. Dennis *et al.*, *Phys. Rev. Lett.* **23**, 1083 (1969).
5. J. C. Hill *et al.*, *Phys. Rev. Lett.* **26**, 1213 (1971).
6. J. S. Adams, "Energy Depositions by Electrons in Superfluid Helium", Ph.D. Thesis, Brown University (2001).
7. C. M. Surko, R. E. Packard, G. J. Dick, and F. Reif, *Phys. Rev. Lett.* **24**, 657 (1970).
8. M. Stockton, J. W. Keto, and W. A. Fitzsimmons, *Phys. Rev. Lett.* **24**, 654 (1970).
9. D. N. McKinsey *et al.*, *Phys. Rev. A* **59**, 200 (1999).
10. P. Hill, *Phys. Rev. A* **40**, 5006 (1989).
11. T. Oka *et al.*, *J. Chem. Phys.* **61**, 4740 (1974).
12. J. W. Keto, F. J. Soley, M. Stockton, and W. A. Fitzsimmons, *Phys. Rev. A* **10**, 887 (1974).
13. V. B. El'tsov *et al.*, *Sov. Phys. JETP* **81**, 909 (1995).
14. M. E. Hayden *et al.*, *Phys. Rev. Lett.* **93**, 105302 (2004).

On the Formation Mechanism of Impurity Helium Solids

Evgeny Popov[a,b], Jussi Ahokas[a], Jussi Eloranta[a] and Henrik Kunttu[a]

[a] *Department of Chemistry, University of Jyväskylä, Survontie 9, 40500 Jyväskylä, Finland*
[b] *Institute of Energy Problems of Chemical Physics (Branch), Chernogolovka, Moscow Region, Russia*

Abstract. In this study we present new data on cooling of seeded (N_2 and NO) helium jets prior to penetration into the bulk helium. In this case, the fine structure originating from the molecular excited states is comparable to kT and therefore the corresponding ultraviolet and visible (UV/VIS) emissions can be used to estimate the temperature of the surroundings of the molecules. Cooling down of the clusters inside the bulk helium occurred slowly, presumably due to extra heat generated by atom-atom recombination.

Keywords: impurity-helium solids, molecular clusters, low temperature luminescence spectra, N_2, NO.
PACS: 67.80.-s, 61.46.+w

INTRODUCTION

Formation of porous materials in superfluid ^{4}He ("impurity helium solids") has been a subject of extensive experimental studies. In most experiments, gas mixtures of N_2 and He were passed through a radio-frequency discharge tube and then directed onto the surface of the superfluid. Upon prolonged gas deposition, the impurity helium solid forms in the superfluid [1, 2]. It has been shown that extensive clustering in the gas phase occurred and further aggregation of the clusters took place in the bulk helium [3].

In this study we present new data on cooling of seeded (N_2 and NO) helium jets prior to penetration into the bulk helium. In this case, the fine structure originating from the molecular excited states is comparable to kT and therefore the corresponding UV/VIS emissions can be used to estimate the spin-orbit temperature of the clusters. The NO molecular probe produced reasonable estimate of the jet temperature whereas the N_2 system could be analyzed only qualitatively. However, this demonstrated that cooling down of the clusters inside the bulk occurred slowly. This is presumably due to extra heat generated by atom-atom recombination processes.

EXPERIMENTAL

Our experimental setup consists of a liquid ^{4}He bath cryostat equipped with a set of quartz windows for optical observations. By pumping the reservoir with a one-stage mechanical pump the lowest accessible temperature was *ca.* 1.5 K. A cryogenic discharge source, a gas nozzle, and an optical cell were placed inside the cryostat. An induction coil was installed coaxially around the liquid-nitrogen cooled discharge tube. The distance between the discharge region and the nozzle was 2 cm. A radio-frequency (RF) generator operating at 40 MHz and at 10–70 W output power was used to drive the induction coil. The optical cell was located directly under the nozzle with separation between the nozzle opening and the surface of He II varied from 2 cm to 10 cm. A fountain pump was used to continuously supply superfluid helium from the main reservoir to the optical cell. The N_2/He, N_2/O_2/He, or NO/He gas mixture was passed through the discharge tube and the resulting excited gas jet was directed towards the liquid surface in the cell. Additional details of the experimental setup can be found in our previous publications [3].

RESULTS AND DISCUSSION

The relevant electronic states and optical transitions of NO are summarized in Fig. 1. Due to spin-orbit splitting in the upper (B $^2\Pi_{3/2}$ and $^2\Pi_{1/2}$) and lower (X $^2\Pi_{3/2}$ and $^2\Pi_{1/2}$) states, the corresponding emission spectrum consists of a doublet structure. Experimental emission spectrum, originating from the gas jet, *ca.* 5 cm above superfluid liquid, is shown in Fig. 2. Notice the intensity difference between the lines in the doublet.

CP850, *Low Temperature Physics: 24th International Conference on Low Temperature Physics;*
edited by Y. Takano, S. P. Hershfield, S. O. Hill, P. J. Hirschfeld, and A. M. Goldman

Typical molecular spin-orbit relaxation rates due to collisions with rare gas atoms have been measured to be in the range of 10^{-11} to 10^{-12} s^{-1} cm^3 $molecule^{-1}$ [4]. Under the present conditions (gas density ~ 10^{19} atoms cm^{-3}) this would lead to thermalization times on the order of 100 ns. The radiative lifetime of the B $\rightarrow$ X transition is 3 μs in the gas phase [5] and 1 μs in an argon matrix [6]. Hence the excited spin-orbit split states have sufficient time to thermalize in the jet before emitting photons. In practice this means that the intensities of the lines in the doublet should reflect the temperature of the jet and that the molecular probe could be used as a molecular thermometer. By analyzing the intensity difference according to the Boltzmann distribution law, we obtained the jet temperature of 20 K at 5 cm above the superfluid level.

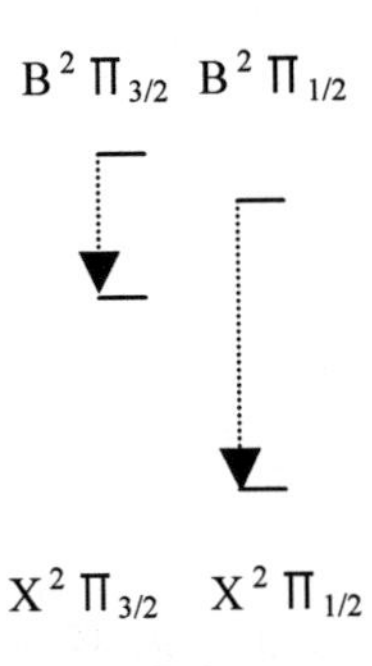

FIGURE 1. Simplified energy level diagram of the NO ($B\ ^2\Pi$ - $X\ ^2\Pi$) transitions showing the doublet splitting in the B and X states (28.7 cm^{-1} and 121.1 cm^{-1}) [7].

In order to find a suitable molecule for probing the temperature of the system, the following issues must be considered: (i) suitable splitting of the upper state comparable to kT, (ii) sufficient splitting in the ground state to eliminate a need for high spectral resolution, (iii) radiative lifetime that is much longer that the spin-orbit relaxation time, and (iv) suitable ratio of the spin-orbit constant to rotational constant (e.g. Hund's case a).

In another set of experiments we attempted to use N_2 ($B\ ^3\Pi_{g\ 0,1,2} \rightarrow A\ ^3\Sigma_u$) triplet set of lines of N_2 as an indicator of temperature. However, due to very rich rotational structure and missing spin-orbit splitting in the A state, it was not possible to obtain temperature quantitatively. It was still possible to obtain relative temperature by judging the intensities of the rotational and spin-orbit split bands. The essential outcome was that the clusters forming the condensate must have had higher temperature than 77 K. Alternatively, it is possible that the lifetime of the excited state became much shorter than the spin-orbit and rotational relaxation times.

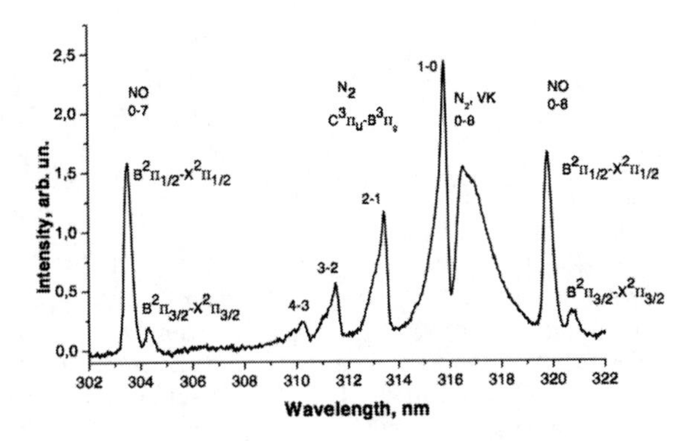

FIGURE 2. Luminescence spectrum of a mixture 1% N_2 and 0.1% O_2 in gaseous He jet, which was passed through RF discharge and directed toward the He II surface. Emission was collected 5 cm above He II. The temperature of the jet was estimated to be 20 K. Note that NO is a reaction product of the N and O atoms formed in the discharge.

ACKNOWLEDGMENTS

Support from the Academy of Finland is greatly acknowledged.

REFERENCES

1. E. B. Gordon, V. V. Khmelenko, A. A. Pelmenev, E. A. Popov, O. F. Pugachev, and A. F. Shestakov, *Chem. Phys.* **170**, 411 (1993).
2. E. P. Bernhard, R. E. Boltnev, V. V. Khmelenko, V. Kiryukhin, S. I. Kiselev, and D. M. Lee, *J. Low Temp. Phys.* **134**, 133 (2004).
3. E. A. Popov, J. Eloranta, J. Ahokas, and H. Kunttu, *Low Temp. Phys.* **29**, 510 (2003).
4. S. D. Le Picard, B. Bussery-Honvault, C. Rebrion-Rowe, P. Honvault, A. Canosa, J. M. Launay, and B. R. Rowe, *J. Chem. Phys.* **109**, 10319 (1998).
5. J. Brzozowski, N. Elander, and P. Erman, *Phys. Scr.* **9**, 99 (1974).
6. J. Eloranta, K. Vaskonen, H. Häkkänen, T. Kiljunen, and H. Kunttu, *J. Chem. Phys.* **109**, 7784 (1998).
7. G. Herzberg, *Molecular Spectra and Molecular Structure. I. Spectra of Diatomic Molecules* (2nd ed.), Princeton: Van Nostrand, 1963, pp. 269.

Applicability of Density Functional Theory to Model Molecular Solvation in Superfluid ^{4}He

Teemu Isojärvi, Lauri Lehtovaara, and Jussi Eloranta

Department of Chemistry, University of Jyväskylä, Survontie 9, 40500 Jyväskylä, Finland

Abstract. Density functional theory (DFT) has been extensively applied to model solvation of atomic and molecular species ("impurities") in superfluid helium. The interaction between the impurity and the surrounding liquid may range from purely repulsive (e.g. alkali metal atoms and most diatomic excimers) to deeply bound potentials (e.g. aromatic compounds). In order to apply DFT to model processes relevant to low temperature chemistry in superfluid ^{4}He, it is essential to obtain the limits of applicability of the theory. For purely repulsive potentials, the spatial gradient of the liquid density remains small and DFT is expected to produce accurate results. This has been verified previously by comparing DFT results to quantum Monte Carlo calculations. For strong binding potentials, however, DFT fails since the individual He atoms tend to localize about the potential minimum. The present work tests the accuracy of DFT between the weakly and strongly bound regimes. This is done by comparing DFT and quantum Monte Carlo (diffusion Monte Carlo) results for various realistic model potentials with varying degree of binding.

Keywords: Density functional theory, solvation, Monte Carlo, helium.
PACS: 67.40.Yv, 67.40.Db

INTRODUCTION

The Orsay-Trento (OT) density functional theory (DFT) of liquid helium-II is a phenomenological model, specifically calibrated to reproduce several bulk parameters of superfluid helium (density, compressibility etc.) as well as its static response function $\chi(q)$ [1]. Therefore it can be expected that the model describes most correctly systems that resemble bulk conditions (i.e. no rapid spatial oscillations in the density). A situation of particular interest, in which large oscillations may emerge, is the packing of helium atoms in layers around a solvated molecule [2, 3]. A system with a sufficiently attractive molecular center will most probably be poorly described by the OT DFT. Examples of such attractive molecular centers are benzene (binding energy ~ 95 K) and various excimers (binding energies some hundreds of K) [3, 4].

At present, there seem to be no good quantitative estimates about the limits of the validity of DFT in such a situation. To find such an estimate we calculate the radial helium density distributions of doped helium droplets with both OT DFT and importance-sampled diffusion Monte Carlo (DMC).

COMPUTATIONAL METHODS

We choose to simulate a cluster of 40 atoms of ^{4}He in all calculations. To model the presence of an impurity, we use a spherically symmetric external potential of the form

$$V(r) = Ae^{-br} - (c/r)^6 \tag{1}$$

which has a hard repulsive core, a minimum at an intermediate distance and an 'r^{-6}-behavior' at large distances from the origin (*cf.* Fig. 1). In the DMC calculation, we use the Aziz potential for the He-He pair interaction [5].

To find an approximate point of breakdown of the OT DFT model, we use (at present) two sets of parameters in (1): $A = 1.457 \times 10^6$, $b = 3.15$, $c = 1.519377$, and $A = 1.457 \times 10^7$, $b = 3.15$, $c = 2.230140$. All parameter values are given in atomic units. These model potentials produce wells of similar widths of about 1 Å and locations of minima at about 4 – 4.5 Å from the origin, but different well depths of 10 K and 100 K, respectively.

CP850, *Low Temperature Physics: 24th International Conference on Low Temperature Physics;*
edited by Y. Takano, S. P. Hershfield, S. O. Hill, P. J. Hirschfeld, and A. M. Goldman

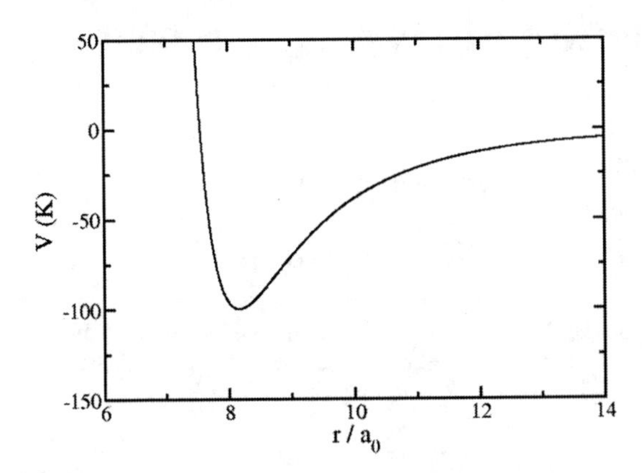

FIGURE 1. A plot of the potential function $V(r)$ for a well depth of 100 K. Bohr's constant is denoted by a_0.

For the DMC, we choose to use a trial function of the form (r_i is the position vector of the i-th particle)

$$\Psi_T(r_1, r_2, \cdots, r_n) = \prod_{i=1}^{n} \theta(r_i) \prod_{i<j} \phi(|r_i - r_j|) \quad (2)$$

with the single-particle- and pair-correlation parts

$$\theta(r_i) = \exp\left(-a|r_i| - (b/|r_i|)^5\right) \quad (3)$$

$$\phi(r_i, r_j) = \exp\left(-\alpha|r_i - r_j| - \beta \ln|r_i - r_j|\right) \times \exp\left(-(\gamma/|r_i - r_j|)^5\right) \quad (4)$$

The trial function parameters are chosen as follows (in atomic units): $a = 0.18$, $b = 8.5$, $\alpha = 0.005$, $\beta = 0.03$, $\gamma = 4.8$ (10 K well depth), and $a = 0.26$, $b = 9.0$, $\alpha = 0.02$, $\beta = 0.03$, $\gamma = 4.8$ (100 K well depth).

For the DFT results, imaginary time propagation of the relevant nonlinear Schrödinger dynamics is applied [6].

RESULTS AND DISCUSSION

Figures 2 and 3 show the calculated radial density distributions (DFT and DMC with second order approximation) for the 10 K and 100 K potentials.

Both figures clearly exhibit the formation of two distinct solvent layers. A surprising point is that, although the DFT method is known to tend to overlocalize the helium atoms to the bottom of deep potential wells, the discrepancy between the DMC and DFT results at 100 K is mostly seen in the second solvent layer. This possibly implies that the trial function parameters in (3) and (4) still have to be further optimized using variational Monte Carlo (VMC). However, no gross differences between the results is observed for either value of well depth. Noteworthy at this point is that OT DFT seems to reproduce fairly well the details of the first solvent layer, which is especially important considering the modeling of atomic and molecular spectra in liquid ^{4}He. Further calculations involving deeper potentials have to be made to find the point of failure of the OT DFT and to establish clear conditions for its applicability.

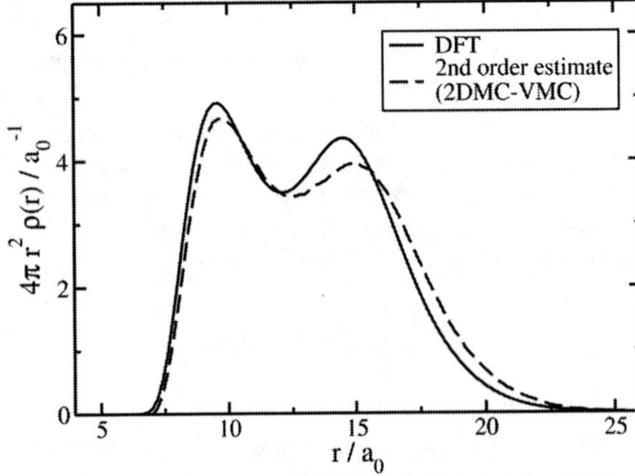

FIGURE 2. The DMC and OT DFT radial density distributions for the potential well of depth 10 K.

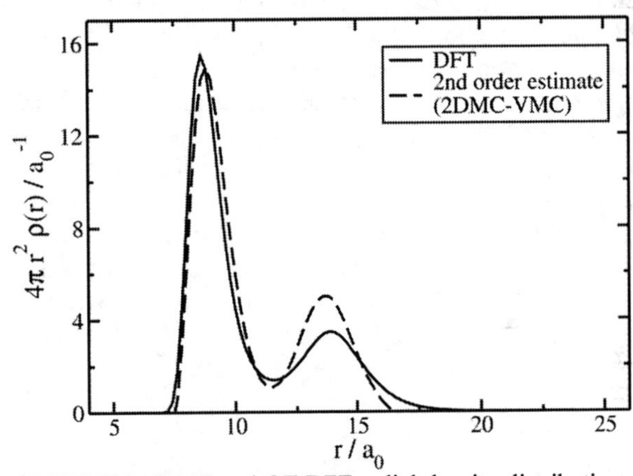

FIGURE 3. DMC and OT DFT radial density distributions in the 100 K case.

REFERENCES

1. F. Dalfovo, A. Lastri, L. Pricaupenko, S. Stringari, and J. Treiner, *Phys. Rev. B* **52**, 1193-1209 (1995)
2. T. Kiljunen, L. Lehtovaara, H. Kunttu, and J. Eloranta, *Phys. Rev. A* **69**, 012506 (2004).
3. P. Huang and K. B. Whaley, *Phys. Rev. B* **67**, 155419 (2003).
4. J. Eloranta and V. A. Apkarian, *J. Chem. Phys.* **115**, 752-760 (2001).
5. A. R. Janzen and R. A. Aziz, *J. Chem. Phys.* **107**, 914-919 (1997)
6. L. Lehtovaara, T. Kiljunen, and J. Eloranta, *J. Comp. Phys.* **194**, 78(2004)

Adsorbed Xenon and the Production of Hyperpolarized ^{129}Xe

E. V. Krjukov, J. D. O'Neill and J. R. Owers-Bradley

School of Physics and Astronomy, University of Nottingham, Nottingham, NG7 2RD, England

Abstract. We show that it is possible to induce a large nuclear spin polarization in ^{129}Xe adsorbed on a silica gel substrate using ^{3}He as a relaxant, in a reasonable time. We also describe the complete production cycle for the generation of large quantities of hyperpolarized xenon from xenon gas by a cryogenic method for use in MRI and material science.

Keywords: Hyperpolarized, Xenon, Adsorbed, NMR
PACS: PACS numbers; 76.60.-k, 76.60.Es, 67.55.-s

INTRODUCTION

The possibility of using prepolarized noble gases, ^{3}He and ^{129}Xe, as contrast agents in magnetic resonance studies has excited scientists for some years. Laser optical pumping is currently the only method of production. However, in this paper we shall describe an alternative cryogenic method which has the potential to produce large quantities of hyperpolarized ^{129}Xe within a reasonable time frame. The brute force concept is very simple. Take the ^{129}Xe nuclear spin system to the extreme conditions of very low temperatures and high magnetic field where the equilibrium state corresponds to a high degree of spin polarization, allow the system to reach this equilibrium and then quickly remove the polarized nuclei from the cryostat before relaxation processes destroy the polarization. This technique demands that we satisfy conflicting requirements. On one hand, nuclear relaxation time should be long enough to preserve polarization at high temperatures outside the cryostat (this is possible for S=1/2 spin systems). On the other hand, the nuclear spin system should reach equilibrium at very low temperatures within a practical timescale. Relaxation times in bulk solid xenon increase dramatically with cooling and become impractically long at temperatures around 10 mK, even when using oxygen as a relaxant [1, 2]. Biškup et al [3] proposed a way to solve this problem and following their work we show here that a high degree of ^{129}Xe polarization is achievable and that the polarization can be retained at higher temperatures.

METHOD OF PRODUCTION

We first adsorb the xenon onto a suitable substrate and use ^{3}He as a relaxant over the xenon layer, allowing the polarization to grow in a finite time. There is sufficient coupling between the mobile ^{3}He atoms and the uppermost xenon layer to induce relaxation to near equilibrium in a few hours. We use a high surface area silica gel substrate onto which a single monolayer of xenon is adsorbed at temperatures of around 160 K. Following cooling to 1 K, ^{3}He is added and the system is further cooled to the base temperature of 10 mK. The ^{3}He also improves thermal contact between the xenon and the mixing chamber. The degree of polarization is detected using pulsed nuclear magnetic resonance (NMR). The magnetization is destroyed by a comb of rf pulses and the subsequent growth of the NMR signal is monitored by recording free induction decay signals induced by tipping pulses of a few degrees. We have previously shown [1] that, for the first xenon monolayer which is in direct contact with ^{3}He, the spin-lattice relaxation is much faster than for the lower layers. An Oxford Instruments top loading dilution refrigerator with a 15.5 tesla superconducting magnet provides the high B/T >1.5 mK T^{-1}. Under these conditions, the equilibrium polarization for ^{129}Xe is 41%.

RESULTS

Fig. 1 shows a graph of polarization degree against time in a magnetic field of 14.0 T. It can be seen that the majority of the relaxation occurs during the first three hours. Notice that we have not achieved the equilibrium polarization at the lowest temperature; this is because a quantity equating to only 2 monolayers of ^{3}He were added above the xenon surface. This is sufficient to induce ^{129}Xe relaxation but, from our Aluminium NMR thermometer in the cell, we know it is not enough to ensure good thermal contact between the xenon and the walls of the cell.

CP850, *Low Temperature Physics: 24th International Conference on Low Temperature Physics;*
edited by Y. Takano, S. P. Hershfield, S. O. Hill, P. J. Hirschfeld, and A. M. Goldman

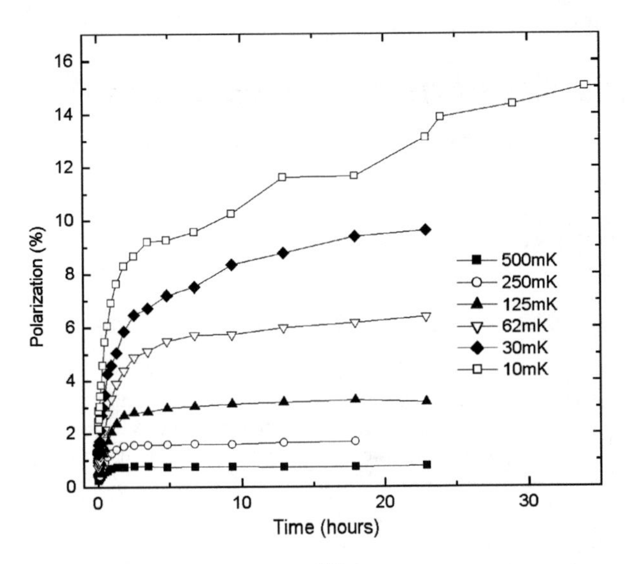

FIGURE 1. The growth of ^{129}Xe polarization for various cryostat temperatures.

REMOVING THE HYPERPOLARIZED XENON

Before the xenon sample is removed from the cryostat and warmed to generate hyperpolarized gas, it is essential to break the coupling between the ^{3}He and the xenon [1]. We add the equivalent of a monolayer of ^{4}He to the cell at the lowest temperatures. The ^{4}He is preferentially adsorbed onto the xenon, displacing the ^{3}He thereby decreasing the relaxation rate by about a factor of 20 [1]. With the coupling reduced between the ^{3}He and ^{129}Xe, the sample can then be warmed to above 3 K in order to evaporate the ^{3}He. One potential drawback with this technology is that ^{3}He becomes contaminated with ^{4}He. We have solved this problem by application of a purification method which was proposed by Dmitriev [4]. We are able to take some ^{3}He directly from the condensing line of our dilution fridge where it has about 12% of ^{4}He impurity and then pump it through charcoal at 4.2 K. We collect ^{3}He with only 0.05% of ^{4}He in good agreement with [4]. This recovered ^{3}He is still an effective relaxant despite the small amount of ^{4}He. After xenon polarization we return the ^{3}He and ^{4}He gases back to the dilution refrigerator.

To prevent coupling between the ^{129}Xe and ^{131}Xe and subsequent rapid relaxation to equilibrium, the sample must remain within a magnetic field. To this end there is a 4.5 T superconducting magnet, located above the main magnet, with a minimum field of 1 T between them, and above this there is a 0.03 T permanent magnet to hold the polarized ^{129}Xe prior to a transfer to a 77 K storage vessel.

DISCUSSION

We now estimate how much hyperpolarized Xe gas we can generate. A full cycle of our xenon polarization lasts about one day. Firstly, condensation of the xenon and helium takes about half a day. Secondly, cooling down the mixing chamber with the sample in it takes about 6 hours. During the third stage the xenon itself cools and the polarization grows. The first and third stages could be improved dramatically by technical development. The cell size can be as large as 100 cm^3. The density of silica gel = 0.48 g cm^{-3} and the surface area = 550 m^2 g^{-1}, so the surface area is over 25,000 m^2. This corresponds to around 0.25 moles or just over 5 litres of hyperpolarized STP xenon gas for a single monolayer coverage of xenon. If natural abundance xenon is used, the concentration of ^{129}Xe is 26% compared to 86% for enriched xenon; the latter however is rather more expensive. The rate of production is poorer than the best low pressure indirect optical pumping systems but better than typical high pressure xenon optical pumping systems. The advantages of our technique are that the xenon produced is already in its frozen state and free from contamination by alkali metals. We are presently attempting to improve the thermal contact between the cell and the mixing chamber by using more ^{3}He. To increase the potential production rate we also plan to examine plating with multiple monolayers of xenon, the use of higher surface area substrates and cryostat designs with larger capacity cells.

CONCLUSIONS

It is hoped that, with further development of this technique, large scale production of hyperpolarized xenon by the brute force method described here will be achievable. Our priority is to use the hyperpolarized gas to supplement the lung studies already under way using ^{3}He. Xenon is more plentiful than helium and is likely to be the best choice for future work in this area despite the lower NMR sensitivity of the ^{129}Xe. Another potential advantage of the technique is that other noble gas isotopes may be polarized as well as ^{31}P and ^{13}C.

REFERENCES

1. E. V. Krjukov, J. D. O'Neill, and J. R. Owers-Bradley, *J. Low. Temp. Phys.* **140**, 397-408 (2005).
2. A. Honig, X. Wei, A. Lewis, E. ter Haar, and K. Seraji-Bozorgzad, *Physica B* **284-288**, 2049-2050 (2000).
3. N. Biškup, N. Kalechofsky, and D. Candela, *Physica B* **329**, 437-438 (2003).
4. V. V. Dmitriev, L. V. Levitin, V. V. Zavjalov, and D. Ye. Zmeev, *J. Low. Temp. Phys.* **138**, 877-880 (2005).

Structure, Growth Mechanisms And Characterization Of Noble Gas Clusters Formed In Multiporous Confinement

Nina V. Krainyukova

Institute for Low Temperature Physics and Engineering NASU, 47 Lenin ave., Kharkov 61103, Ukraine

Abstract. Size-dependent structures and growth mechanisms of noble gas nanoclusters inside multiporous amorphous carbon films were studied by low temperature high energy electron diffraction. Multiply twinned particles such as icosahedra and decahedra (with a five-fold symmetry) were shown to prevail. Atomic dynamics in smallest clusters grown in the porous confinement was found to be essentially suppressed.

Keywords: Confined Geometry, Porous Medium, Nanoclusters.
PACS: 61.46.+w; 81.05.Rm; 81.10.Bk; 61.14.-x

INTRODUCTION

The structure and growth mechanisms of clusters in a confined geometry are essentially different from those in the bulk. Free clusters,[1-5] clusters grown in pores[6-8] and impurity-helium solids[9] exhibit behaviors characteristic of restricted geometries. The common feature of all these clusters is the crystallization *inward* from their surfaces due to the heat removal through pore walls or superfluid ^{4}He, or due to the evaporation of single atoms in the case of free clusters. The requirement of surface energy minimization determines the formed structures and leads to formation of multiply twinned particles (MTPs)[10] such as icosahedra and decahedra with a five-fold symmetry.

EXPERIMENTAL AND ANALYSIS METHODS

Neon and argon clusters grown from the gas in multiporous confinement of amorphous carbon[11] films were studied by high energy (50 keV) electron diffraction with the help of a liquid helium cryostat. The effective film thickness δ characterizes the pore filling and was determined from the analysis of electron beam absorption. Intensities I_{exp} were recorded with the step in s ~0.015 Å^{-1}. Here $s=4\pi\sin\theta/\lambda$, where θ is the diffraction angle, and λ is the electron wavelength (Fig. 1). The diffraction intensity $I_{calc,N}$ for each cluster N was calculated using the Debye formulae that involve the Debye-Waller factor $\exp(-\langle u^2\rangle_N s^2)$ allowing for a mean-square atomic displacement $\langle u^2\rangle_N$. The total intensity $I_{calc}=\Sigma w_N I_{calc,N}$ (Fig. 1) was taken to be a superposition of diffraction intensities from different clusters N with their relative weights w_N (the total $\Sigma w_N=1$) chosen as fitting parameters in the minimization of the R-factor $R=\Sigma|I_{exp}-I_{calc}|/\Sigma(I_{exp}+I_{calc})$.[7]

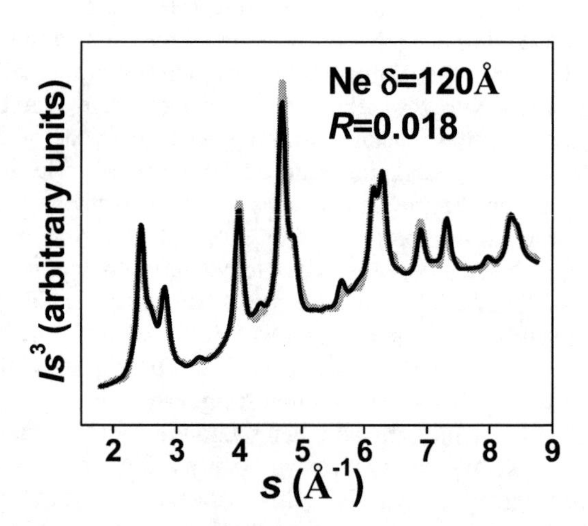

FIGURE 1. Experimental (gray) and calculated (black) diffractograms from Ne clusters formed in an amorphous carbon substrate. The substrate contribution was subtracted.

The distribution functions of cluster sizes and structures are obtained by this fitting of each experiment diffractogram to a set of calculated intensities $I_{calc,N}$ for individual model clusters.

CP850, *Low Temperature Physics: 24th International Conference on Low Temperature Physics;*
edited by Y. Takano, S. P. Hershfield, S. O. Hill, P. J. Hirschfeld, and A. M. Goldman

RESULTS AND DISCUSSION

The obtained distribution functions of sizes and structures reveal an absolute prevalence of MTPs (~60–90%) among clusters grown in the pores of the amorphous carbon.

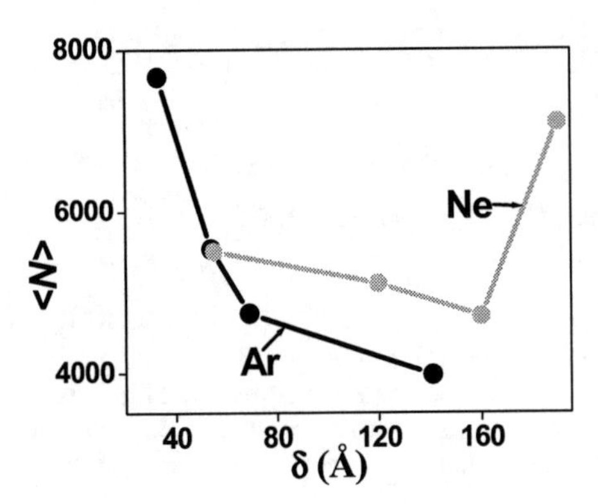

FIGURE 2. Dependence of the average cluster size <*N*> on the effective sample thickness for Ar and Ne.

Unexpected at first sight, the decrease of the average cluster size <*N*> (the number of atoms) with δ (Fig. 2) can be explained on the basis of diffusion-like filling mechanisms in smaller pores. Indeed this δ dependence can imply two stages in the filling process. At first, large pores are easily filled with the gas flowing inside the porous carbon, and then the gas atoms diffuse into smaller pores, resulting in the decrease of the average cluster size. We ascribe the strong increase of <*N*> when the effective film thickness reaches approximately 160 Å to an unhindered growth of clusters outside the pores, after they are completely filled.

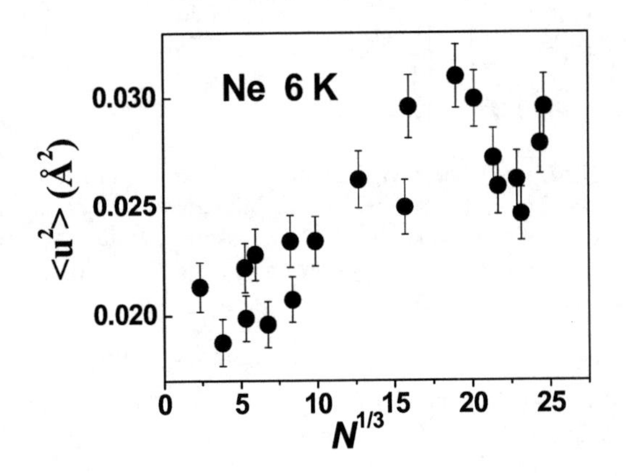

FIGURE 3. Mean-square atomic displacement $<u^2>$ in Ne samples at 6 K as a function of cluster size.

Atomic dynamics of clusters in pores differs considerably from that of the bulk as well as free clusters. The open geometry of the free cluster surface leads to an increasing[4,5] mean-square atomic displacement $<u^2>$ with the decreasing cluster size. In clusters grown in the pores, we can see (Fig. 3) the opposite. The mean-square atomic displacement is close to the bulk value for large clusters but falls off noticeably for small clusters. This behavior indicates that the interaction of cluster atoms with pore walls is much stronger than the interaction between themselves.

The method of analysis presented in this paper is applicable to a wide variety of objects that undergo a growth process typical of confined geometries.

ACKNOWLEDGMENTS

The author is deeply grateful to V. G. Manzhelii, M. A. Strzhemechny, A. I. Prokhvatilov and M. I. Bagatskii for fruitful discussion and is greatly indebted to B. W. van de Waal for valuable comments and also for his computer programs applied to cluster modeling.

REFERENCES

1. Van de Waal, B. W., *The FCC/HCP Dilemma*, Enschede: B. V. Febodruk, 1997, pp. 1-162.
2. Reinhard, D., Hall, B. D., Ugarte, D., and Monot, R., *Phys. Rev. B* **55,** 7868-7881 (1997).
3. Torchet, G., de Feraudy, M.-F., and Loreaux, Y., *J. Mol. Struct.* **485-486**, 261-267 (1999).
4. Eremenko, V. V., Kovalenko, S. I., and Solnyshkin, D. D., *Low Temp. Phys.* **29**, 353-355 (2003).
5. Verkhovtseva, E. T., Gospodarev, I. A., Gryshayev, O. V., Kovalenko, S. I., Solnyshkin, D. D., Syrkin, E. S., and Feodosyev, S. B., *Low Temp. Phys.* **29**, 386-393 (2003).
6. Silva, D. E., Sokol, P. E., and Ehrlich, S. N., *Phys. Rev. Lett.* **88**, 155701-1-4 (2002).
7. Krainyukova, N. V., and van de Waal, B. W., *Thin Solid Films* **459**, 169-173 (2004).
8. Huber, P., and Knorr, K., *Phys. Rev. B* **60,** 12657-12665 (1999).
9. Kiselev, S. I., Khmelenko, V. V., Lee, D. M., Kiryukhin, V., Boltnev, R. E., Gordon, E. B., and Keimer, B., *Phys. Rev. B* **65,** 024517-1-12 (2001).
10. Hofmeister, H., "Fivefold Twinned Nanoparticles", in: *Encyclopedia of Nanoscience and Nanotechnology*, edited by H. S. Nalwa, V. 3, Los Angeles: American Scientific Publishers, 2004, pp. 431-452.
11. Townsend, S. J., Lenosky, T. J., Muller, D. A., Nickols, C. S., and Elser, V., *Phys. Rev. Lett.* **69,** 921-924 (1992).

Temperature and Pressure Control of Cold Helium Gas above Liquid Helium for Laser Spectroscopy of Atoms and Molecules at 3–30 K

Yukari Matsuo*, Yoshimitsu Fukuyama*, and Yoshiki Moriwaki†

*RIKEN Discovery Research Institute, Wako Saitama 351-0198, Japan

† Department of Physics, Toyama University, Toyama City, Toyama 930-8555, Japan

Abstract. We have developed a novel and simple technique to control the temperature and pressure of cold helium gas independently in a designated space above liquid helium. The technique functions through adjustment of both the pumping speed of helium gas and the evaporation of liquid helium using a heater immersed in the liquid helium. This technique enables us to investigate the peculiar temperature dependence of collision cross sections of Ba^{+*} and He, or Ba^{+*}-He and He at temperatures of 3–30 K through observation of laser-induced fluorescence spectra of Ba^{+*}-He exciplexes.

Keywords: cold helium gas, barium, laser spectroscopy
PACS: 33.50.Dq, 34.30.+h, 39.90.+d, 67.40.Yv:

INTRODUCTION

Laser spectroscopy of impurity atoms and molecules in the cold environments of gaseous, liquid and solid helium reveals not only the characteristics of the impurities in these environments; it also reveals the properties of the helium matrices [1]. We have observed laser-induced fluorescence spectra of Ba^{+*}-He exciplexes that are formed by Ba^{+} produced by laser ablation in cold helium gas and surrounding helium atoms. Those observations yielded valuable information on the interatomic potential of Ba^{+}-He and collision cross sections at low temperatures [2]. Studying impurities in helium gas is advantageous because the temperature and pressure can be changed over wide ranges. Nevertheless, in a conventional cryostat, it is difficult to control both the temperature and pressure of cold helium gas arbitrarily because the temperature determines the pressure of the cold gas above liquid helium. Merely slowing the pumping speed will increase the temperature and pressure. This paper presents a new and simple technique using a heater immersed in liquid helium. We report control of the temperature and pressure by adjusting both the helium-gas pumping speed and liquid helium evaporation.

EXPERIMENTAL SETUP

The experimental setup for laser spectroscopy of impurities in cold helium environment using a cryostat is shown in Fig. 1. The apparatus comprises a cryostat chamber, ablation laser, excitation laser, and optical detection system. Details of the system for laser spectroscopy are explained elsewhere [2].

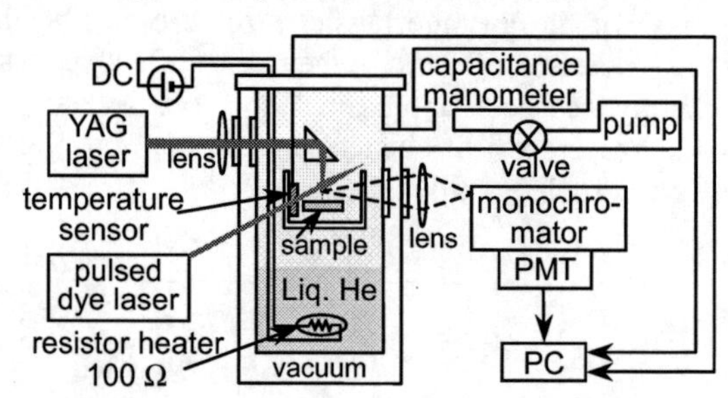

FIGURE 1. Experimental setup. A 100 Ω resistor heater is immersed in the bottom part of the liquid helium. Typically, a voltage of 0–1 V is applied to the resistor. Conductance of the vacuum valve is adjusted manually to control the pressure.

The cryostat chamber contains liquid helium in the bottom part, where a 100 Ω resistor heater is immersed. The space above the liquid helium is filled with cold helium gas evaporated from liquid helium. Typically, a

CP850, *Low Temperature Physics: 24th International Conference on Low Temperature Physics;*
edited by Y. Takano, S. P. Hershfield, S. O. Hill, P. J. Hirschfeld, and A. M. Goldman

voltage of 0–1 V is applied to the resistor. The temperature is monitored using a commercial temperature sensor held in an open-top quartz container, which is fixed above the liquid level to prevent the disturbance of direct He gas flow. The pressure is measured using a capacitance manometer. The essential point of the system is that the temperature of the cold helium gas decreases when the heater is turned on, because the supply of colder helium gas is accelerated by vaporization. Therefore, the temperature and pressure of cold helium gas in a designated space can be controlled.

RESULTS

Pressure, temperature, and heat applied to the liquid helium are depicted in Fig. 2 for a series of typical experimental conditions used for laser spectroscopy of impurities in cold helium gas. When the heater is turned on, the temperature decreases because of the colder helium vapor rising from the liquid. Adjustment of the helium-gas pumping speed can control the pressure. In this manner, temperatures and pressures of 3–30 K and 2000–27000 Pa are realized and held constant for at least half an hour.

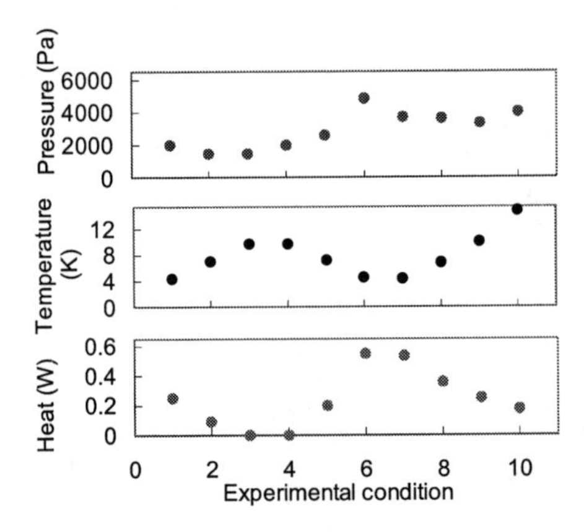

FIGURE 2. Pressure, temperature, and heat applied to the liquid helium. The horizontal axis represents different conditions in which a desired temperature and pressure are obtained. Generally, heating lowers the temperature. Pressure can be changed for a constant temperature (points 6 and 7); temperature can be changed for a constant pressure (points 7–10).

Using this technique, we observed laser-induced fluorescence spectra of Ba^{+*}-He and Ba^{+*} at 3–30 K and 2000–27000 Pa. Typical spectra observed in several conditions are shown in Fig. 3. Fluorescence spectra observed over a frequency region of 20000–22000 cm^{-1} show a multi-peak broad band of Ba^{+*}-He (vibrational bands v=0, 1, 2 and 3) and a sharp D1 emission line of Ba^{+}. The intensity of broad fluorescence bands increases at higher temperatures and lower pressures. Analysis of the intensity engenders formation and dissociation cross sections of Ba^{+*}-He. Details of the analysis are described elsewhere [3].

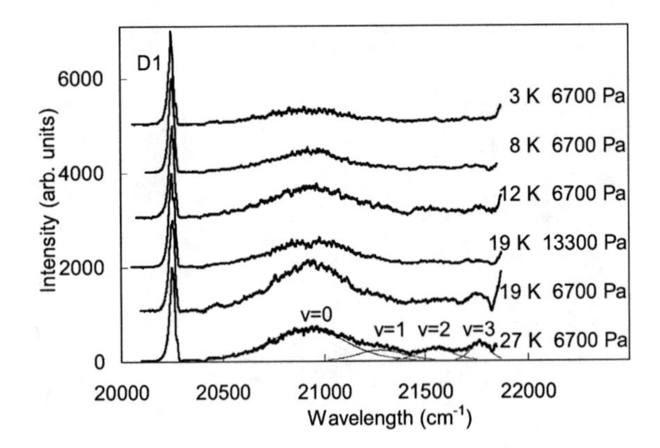

FIGURE 3. Observed laser-induced fluorescence spectra of Ba^{+*}-He and Ba^{+} (solid lines). Fitted Gaussian curves (dashed lines) correspond to emissions from four vibrational levels.

CONCLUSION

We developed a simple and novel technique to control the temperature and pressure of the cold helium gas independently in a designated space above the liquid helium surface. The technique uses a heater immersed in liquid helium to adjust both the helium-gas pumping speed and liquid helium evaporation. Using this technique, we observed a peculiar temperature dependence of cold collision cross sections between Ba^{+*} and He, or Ba^{+*}-He and He at temperatures of 3–30 K [3]. This simple but effective technique enables observation of the characteristics of impurities in wide ranges of temperatures and pressures of cold helium gas.

REFERENCES

1. B. Tabbert, H. Günther, and G. zu Putlitz, *J. Low Temp. Phys.* **109**, 653-707 (1997).
2. Y. Fukuyama, Y. Moriwaki, and Y. Matsuo, *Phys. Rev.* A **69**, 042505 (2004).
3. Y. Fukuyama, Y. Moriwaki, and Y. Matsuo, submitted to *Phys. Rev.* A.

SUPERCONDUCTIVITY

Two-Gap Features from Tunneling Studies on Trilayered Cuprates, $HgBa_2Ca_2Cu_3O_{8+\delta}$ with T_C~132K

N. Miyakawa[*], K. Tokiwa[†], T. Watanabe[†], A. Iyo[‡] and Y. Tanaka[‡]

[*]*The Center of General Education and Humanity, Tokyo University of Science, Suwa, 5000-1 Toyohira, Chino, Nagano 391-0292, Japan*
[†]*Department of Applied Electronics, Tokyo University of Science, Chiba 278-8510, Japan*
[‡]*National Institute of Advanced Industrial and Science and Technology, Tsukuba 305-8531, Japan*

Abstract. We present point contact tunneling spectroscopy measured on nearly optimally-doped $HgBa_2Ca_2Cu_3O_y$ (Hg1223) with T_C~132K. The tunneling conductances on Hg1223 exhibit two kinds of gaps that originate from crystallographically inequivalent inner four-fold CuO_2 plane (IP) and outer five-fold CuO_2 planes (OP). The averaged-gap magnitude at OP, Δ_{av}(OP), is 37.8 meV but that at IP, Δ_{av}(IP), is 55.9 meV. This result shows that in multilayered cuprates such as Hg1223 the superconducting density of states at IP is different from that at OP, suggesting the existence of two Fermi surfaces.

Keywords: multilayered cuprates, tunneling spectroscopy, two-gaps
PACS: 74.72.Jt, 74.81.-g, 74.50.+r

INTRODUCTION

Tunneling spectroscopy is a powerful tool to investigate the quasiparticle density of states (DOS) in both the pairing and normal states with high energy resolution (≤meV). Tunneling studies on high-T_C cuprates (HTCs) have been focused on the doping dependence of superconducting gap [1], pseudogap [2-3], and inhomogeneity in the CuO_2 plane [4] for mono- or bi-layered cuprates. However the reports of physical properties in the superconducting states for multilayered cuprates (MLCs), that have more than three sheets of CuO_2 planes in a unit cell, have not been sufficient. MLCs have two crystallographically inequivalent kinds of CuO_2 planes, i.e. inner plane(s) (IP) with four-fold oxygen coordination and outer planes (OP) with five-fold oxygen coordination. We have recently reported the tunneling results of $TlBa_2Ca_2Cu_3O_{10-\delta}$ (Tl1223) [5] as well as $(Cu,C)Ba_2Ca_3Cu_4O_{12+\delta}$ (Cu1234) [6], in which we have observed two distinct kinds of gaps that are not observed for mono- or bi-layered cuprates. From comparison with results of the previous extensive studies on MLCs by NMR [7] as well as those of our tunneling studies on Bi2212 as a function of doping [1, 3], we determined that the spectra displaying the large (small) gap magnitude correspond to the DOS of IP(OP).

In this paper, we report that two-gap features can also be observed on other trilayered cuprate $HgBa_2Ca_2Cu_3O_{8+\delta}$ (Hg1223) and they are most probably one of the characteristic common features for MLCs. In addition, we found that IP and OP in the MLCs have quite different electronic structures, suggesting the existence of two Fermi surfaces.

EXPERIMENTAL

Hg1223 polycrystalline samples were prepared by the high-temperature and high-pressure synthesis technique. The superconducting temperature T_C was determined as 132K from zero-resistance temperature as well as onset temperature of ac-susceptibility. SIN junctions were prepared by a point contact technique using a Au-tip, and dI/dVs were measured by standard ac lock-in technique.

RESULTS AND DISCUSSION

From our tunneling studies on trilayered Hg1223, we observed wide spread of gap magnitude Δ, from

CP850, *Low Temperature Physics: 24th International Conference on Low Temperature Physics;*
edited by Y. Takano, S. P. Hershfield, S. O. Hill, P. J. Hirschfeld, and A. M. Goldman

28.9 to 70.0 meV as determined from peak voltage. Then, we plot the histogram showing the statistical distribution of the observed superconducting gap, Δ of Hg1223 in Fig. 1(a). From Fig. 1(a) we found that the averaged gap magnitude with smaller gap was 37.8meV and that with larger gap was 55.9 meV. Similar two-gaps for Hg1223 with T_C~132K was observed by Jeong *et al.* [8]. Figure 1(b) shows the representative tunneling conductances, where #Hg-S1 is a representative spectrum for smaller gap region in Fig.1(a) and #Hg-L1 is that for larger gap region, but some spectra showed two-gaps in a tunneling spectrum and their results will be shown in separate paper.

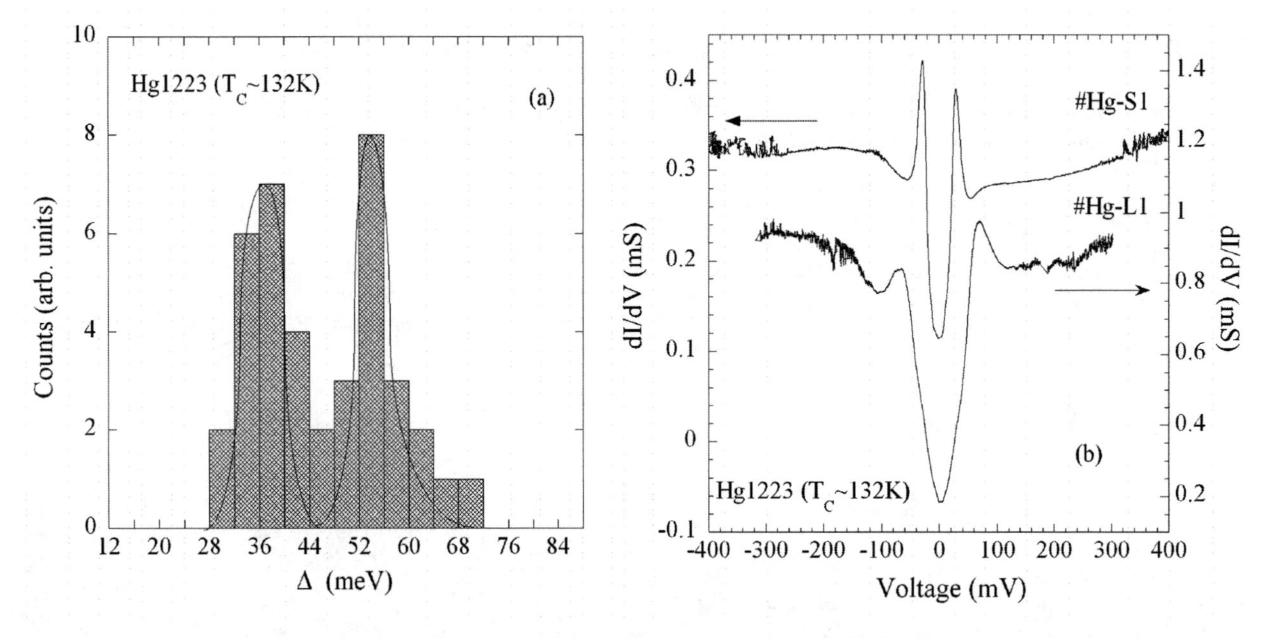

FIGURE 1. (a) Histogram showing the statistical distribution of superconducting gap Δ of Hg1223 measured at 4.2K through our tunneling studies. (b) Representative tunneling conductance on Hg1223 with T_C~132K measured at 4.2K.

Let us discuss about the origin of two kinds of gaps that were observed on Hg1223. A similar distribution of gaps was observed on other MLCs, Tl1223 with T_C~91K [5] and Cu1234 with T_C~117K [6], but it has never observed on mono- and bi-layered cuprates. Therefore the two kinds of gaps observed by tunneling studies may be a characteristic common feature for multilayered cuprates with n≥3. The main difference of crystallographical structure between mono- or bi-layered cuprates and MLCs with n≥3, is that MLCs have two kinds of inequivalent CuO_2 planes (i.e. IP and OP). Therefore it is plausible to consider that these two kinds of gaps originate from these two kinds of CuO_2 planes. In addition, NMR studies reported that the carrier concentration at IP is always lower than that at OP [7], and we reported that $\Delta(p)$ in hole-doped high-T_C cuprates increases with decreasing hole concentration, p [1, 3]. In addition, temperature dependence of dI/dV showing large Δ on Tl1223, Cu1234 and Hg1223 showed pseudogap at T>T_C and it disappeared at T^* of IP measured by site selecting NMR studies [7]. Therefore by linking with these results we assign that the spectra showing large (small) gap reflect the DOS at IP (OP) of Hg1223. This result suggests the existence of two Fermi surfaces.

REFERENCES

1. N. Miyakawa *et al., Phys. Rev. Lett.* **83**, 1018 (1999), N. Miyakawa *et al., Phys. Rev. Lett.* **80**, 157 (1998).
2. T. Timusk and B. Statt, *Rep. Prog. Phys.* **62**, 61 (1999).
3. N. Miyakawa *et al., Physica C* **364-365**, 475 (2001), N. Miyakawa *et al., Trends in Applied Spectroscopy* **4**, 47 (2002), N. Miyakawa *et al., Physica C* **357-360**, 126 (2001).
4. S.H. Pan *et al., Nature (London)* **413**, 282 (2001).
5. N. Miyakawa *et al., Int. J. Mod. Phys. B* **19,** 225 (2005).
6. N. Miyakawa *et al., Int. J. Mod. Phys.* B **17**, 3612 (2003).
7. H. Kotegawa *et al., J. Phys. Chem. Solids* **62**, 171 (2001), H. Kotegawa *et al., Phys. Rev B* **64**, 064515 (2001), H. Kotegawa *et al., Phys. Rev B* **65**, 184504 (2002).
8. G.T. Joeng *et al., Phys. Rev. B* 49, 15416 (1994).

130-K Superconductivity in the Hg-Ba-Ca-Cu-O System

A. Salem[1], G.A. Gamal[2] , G. Jakob[1], and H. Adrian[1]

1Institute of Physics, Johannes Gutenberg University, 55099 Mainz, Germany
2 South Valley University, Qena, Egypt

Abstract. $(Hg_{0.9}Re_{0.1})Ba_2Ca_2Cu_3O_{8+\delta}$ (HgRe-1223) HTSC thin films have been prepared by pulsed laser deposition (PLD) of a precursor on (100)-oriented $SrTiO_3$ substrates followed by annealing in a controlled mercury atmosphere. The critical temperature is 129 K determined by the temperature dependence of AC susceptibility. Resistance measurements show a superconducting transition at T_c= 130 K with transition width $\Delta T_c \approx 3$ K.

Keywords: Hg(Re)-1223; AC susceptibility; Resistance measurement.
PACS: R74.72.-h; 74.72.Jt; 74.78.Bz.

INTRODUCTION

Mercury superconductors possess excellent superconducting properties, even above 77 K. Critical current densities of the order 10^5 A/cm^2 at 77 K and irreversibility fields of $\mu_0 H_{irr}$ (100 K) = 0.5 T demonstrate the great application potential of these superconductor materials. Several researchers [1, 2, 3, 4, 5] have reported T_c above 130 K for the Hg series of compounds $HgBa_2Ca_{n-1}Cu_nO_{2n+2+\delta}$, where n is the number of CuO_2 layers per unit cell, sometimes with Pb doping for Hg. The highest known T_c at atmospheric pressure is about 135 K in $HgBa_2Ca_2Cu_3O_{8+\delta}$ [3] and 138 K in $Hg_{0.8}Tl_{0.2}Ba_2Ca_2Cu_3O_{8+\delta}$ [6]. The Hg-based family has several important features, such as relatively simple structures, the possibility to vary T_c in a wide range by reducing/oxidizing gaseous treatment etc. Therefore, they can be considered as excellent objects to study superconductivity phenomena in Cu-based oxides. In this work we report the growth and the transport properties by measuring the resistivity and magnetization of HgRe-1223 films.

EXPERIMENTAL RESULTS AND DISCUSSION

The preparation of $Hg_{0.9}Re_{0.1}Ba_2Ca_2Cu_3O_y$ thin films involves three main steps, the preparation of the precursor target, the ablation of the precursor by pulsed laser deposition on (100) $SrTiO_3$, and the formation of $Hg_{0.9}Re_{0.1}Ba_2Ca_2Cu_3O_y$ by gas/solid diffusion. The precursor material with the nominal composition $Hg_{0.9}Re_{0.1}Ba_2Ca_2Cu_3O_y$ was prepared from stoichiometric mixture of ReO_2 and commercially available multiphase $Ba_2Ca_2Cu_3O_y$. The resulting material after sintering in an oxygen atmosphere served as a target for pulsed laser deposition (PLD). In an oxygen atmosphere ($\approx$. 0.3 mbar) precursor thin films with an optically smooth surface were obtained after deposition onto (100) aligned $SrTiO_3$ substrates at a pulse frequency of 8 Hz for 1 h at room temperature with a laser flux of 650 mJ/pulse. After the deposition the substrates with the precursor films were sealed into quartz tubes sandwiched by a HgRe-223 pellet on the film side and a Re-223 pellet on the substrate side. The HgRe-223 pellet provides the Hg atmosphere to transform the precursor film into the HgRe-1223 film. The ratio between Re-223 and HgRe-223 was approximately 1:3. The total quantity of the reaction powder of 0.641 g (consisting of 0.5 g $Re_{0.1}Ba_2Ca_2Cu_3O_y$ and 0.141 g HgO) was arbitrarily selected and was not of crucial importance for the production of superconducting films. The relationship of the partial pressure is also independent of the total quantity of the powder. The samples annealed in close proximity of the stoichiometric bulk pellet showed excellent superconducting characteristics while those placed as little as a centimeter away were not superconducting at all. A fused-quartz rod (7 mm OD, 50 mm length) was inserted in the quartz tube in order to reduce the free volume. In preparation for heat treatment, the quartz tubes (8 mm ID, 12 mm OD) were evacuated to 1 mbar and sealed off at a length of about 12 cm by a hydrogen torch. The evacuated tubes were then fast

CP850, *Low Temperature Physics: 24th International Conference on Low Temperature Physics;*
edited by Y. Takano, S. P. Hershfield, S. O. Hill, P. J. Hirschfeld, and A. M. Goldman

heated to 750 °C in 1 h (to overcome the problem of $CaHgO_2$ formation), and then to 850 °C in 30 min and held at this temperature for 1 h. Finally the samples were cooled to room temperature in 6 hours. In samples that were heat treated below 850 °C , HgRe-1201 and Ca_2CuO_3 are the dominate phases. In the annealing temperature range from 850 °C to 890 °C, the heat treatment led to the information of the HgRe-1223 phase.

The AC susceptibility measurement is very useful in studying the magnetic properties of materials, especially of the magnetic phase transitions where the response of magnetic moments in low field is of great importance. Resistivity, on the other hand, is easier to measure, and can be a better guide for applications. Generally, the T_c value determined from the resistivity drop to zero occurs at a somewhat higher temperature than its susceptibility counterpart. Figure 1 shows a drop in resistance of HgRe-1223 thin film that occurs at $T_c \approx 130$ K. The T_c values that are determined in the present work of the HgRe-1223 thin films are ordinarily midpoint values at which R (T) has decreased by 50 % below the normal state resistivity. The point at which the first derivative of the resistance curve, shown in the lower inset of Fig. 1, reaches its maximum value could be selected as defining T_c, since it is the inflection point of the original curve. No annealing in argon or oxygen was necessary to enhance T_c, which suggests that as prepared samples had close to optimal hole concentrations.

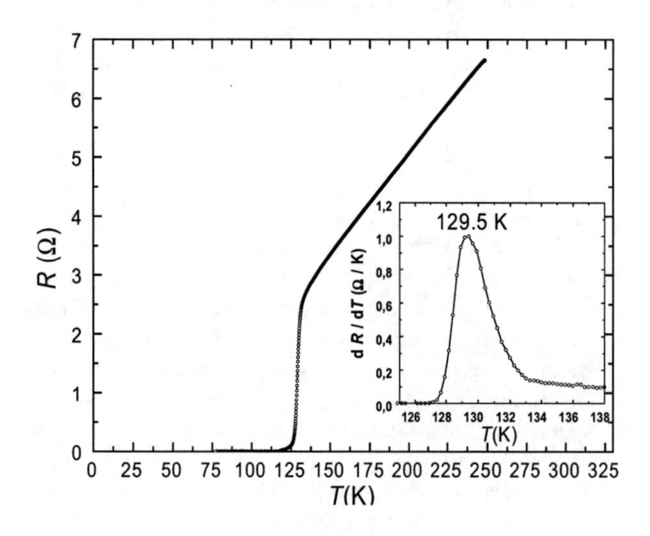

FIGURE 1. Temperature dependence of the electrical resistance for a HgRe-1223 thin film measured in zero field. The lower inset is a relation between the temperature derivative of the resistive transition and temperature.

We can see in Fig. 2 that there is an analogous drop in susceptibility at $T_c \approx 129$ K. This happens because any tiny part of the material going superconductive loses its resistance, and $R = 0$ when one or more continuing superconducting paths are in place between the measuring electrodes. In contrast, AC susceptibility measurements depend on macroscopic current loops to shield the B field from an appreciable fraction of the sample material, and this happens when full superconducting current paths become available.

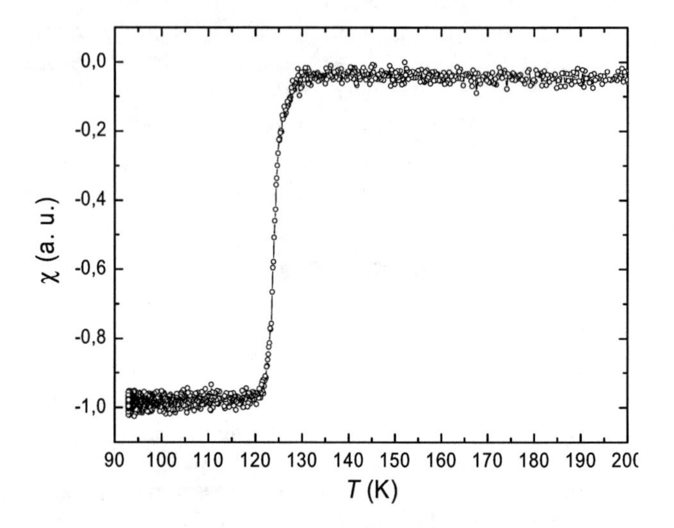

Figure 2. AC susceptibility for as-prepared HgRe-1223 films.

ACKNOWLEDGMENTS

We acknowledge support by the Government of Egypt, the Deutsche Forschungsgemeinschaft (grant AD87-2), the Materials Science Research Center of the University of Mainz, and LT24.

REFERENCES

1. C. W. Chu, L. Gao, F. Chen, Z. J. Huang, R. L. Meng, and Y. Y. Xue, *Nature* **365**, 323 (1993).
2. Z. Iqbal, T. Datta, D. Kirven, A. Lungu, J. C. Barry, F. J. Owens, A. G. Rinzler, D. Yang, and F. Reidinger, *Phys. Rev. B* **49**, 12322 (1994).
3. A. Schilling, M. Cantoni, J. D. Guo, and H. R. Ott, *Nature* **363**, 56 (1993).
4. A. Schilling, O. Jeandupeux, J. D. Guo, and H. R. Ott, *Physica C* **216**, 182 (1993).
5. A. Schilling, O. Jeandupeux, S. Buchi, H. R. Ott, and C. Rossel, *Physica C* **235-240**, 229 (1994).
6. P. Dai, B. C. Chakoumakos, G. F. Sun, K. W. Wong, Y. Xin, and D. F. Lu, *Physica C* **243,** 201 (1995).

Quasi-two-dimensional Transport Properties of Layered Superconductors $Nd_{2-x}Ce_xCuO_{4+\delta}$ and $Ca_{2-x}Sr_xRuO_4$.

T.B. Charikova[a], A.I. Ponomarev[a], N.G. Shelushinina[a], A.O. Tashlykov[a], A.V. Khrustov[a], and A.A. Ivanov[b]

[a] *Institute of Metal Physics Ural Division RAS, 620219 Ekaterinburg, Russia*
[b] *Moscow Engineering Physics Institute, 115410 Moscow, Russia*

Abstract. The results of an investigation of the in-plane $\rho_{ab}(T)$ and out-of-plane $\rho_c(T)$ resistivities ($1.5K \le T \le 300K$) of $Nd_{2-x}Ce_xCuO_{4+\delta}$ single crystal films (x=0.12; 0.15; 0.17) with (001) and (110)orientations and $Ca_{2-x}Sr_xRuO_4$ single crystals ($0 \le x \le 2.0$) are presented. For stoichiometric reduced $Nd_{2-x}Ce_xCuO_4$ and Ca-doped $Ca_{2-x}Sr_xRuO_4$ compound the conduction mechanism appears to be different between ab- and c - directions: a combination of in-plane metallic ($d\rho_{ab}/dT>0$) and out-of-plane nonmetallic ($d\rho_c/dT<0$) behavior. These layered quasi-two-dimensional systems are Anderson conductors with a very anisotropic localization length ($R_{loc}^{ab} \gg R_{loc}^{c}$). The increase of the non-stoichiometric oxygen in Nd-system and the increase of the Ca-content in $Ca_{2-x}Sr_xRuO_4$ leads to a decrease of the anisotropy coefficient. So in both quasi-two-dimensional systems the destruction of the stoichiometric order with oxygen in Nd-system and with Ca-content in Ru-system demonstrates Anderson-type disorder-induced transition.

Keywords: Layered superconductors, transport properties, anisotropic behavior.
PACS: 74.25.Fy, 74.70.Pq, 74.72.Jt, 74,78.-w

INTRODUCTION

The layered perovskite Ce-doped cuprate $Nd_{2-x}Ce_xCuO_{4+\delta}$ and Ca-doped rutenate $Ca_{2-x}Sr_xRuO_4$ have attracted interest because of existence of the transition from metallic to non-metallic behavior in both of the systems. In the compound $Ca_{2-x}Sr_xRuO_4$ the metal-non-metal transition appears after Ca-doping [1]. In Nd-compounds the change of the cerium content and nonstoichiometric disorder leads to the metal-non-metal transition [2]. There are discussions about anomalous transport properties in the normal phase of layered quasi-two-dimensional perovskite-like crystal structure: a coexistence of in-plane metallic ($d\rho_{ab}/dT>0$) and out-of-plane nonmetallic ($d\rho_c/dT<0$) behavior [3]. The effect of nonstoichiometric disorder and impurity doping on the anisotropic transport properties and the difference of the conduction mechanism between *ab*- and *c*-directions in $Nd_{2-x}Ce_xCuO_{4+\delta}$ films and $Ca_{2-x}Sr_xRuO_4$ single crystals are presented in this paper.

EXPERIMENT AND DISCUSSION

Flux separation techniques [2] were used for growing two types of the single crystal films $Nd_{2-x}Ce_xCuO_{4+\delta}$ with x=0.12; 0.15; 0.17: films with the standard orientation (001): *c*-axis is perpendicular to the $SrTiO_3$ (100)-substrate; and films with the orientation (110): *c*-axis is parallel to the $SrTiO_3$ (110)-substrate. The films were annealed under various conditions: annealed in O_2 ("oxidized"") and annealed in vacuum ("reduced"). The films thickness was 1200-2000 Å. The single crystals of $Ca_{2-x}Sr_xRuO_4$ ($0 \le x \le 2.0$) with dimensions ($3 \times 2 \times 0.2$) mm^3 were grown by floating-zone method [1].

Fig. 1 shows $\rho_{ab}(T)$ and $\rho_c(T)$ dependencies for optimally reduced $Nd_{1.85}Ce_{0.15}CuO_4$ film. In the temperature range (300÷25)K ρ_{ab} of this reduced film has a metallic behavior with $\rho_{ab}(T) \sim T^2$. On the contrary, $\rho_c(T)$ is non-metallic: it increases with decreasing temperature ($\rho_c(T) \sim 1/T$ at (300÷100)K and then $\sim -\ln T$ in the range (70÷25)K. Fig.2 shows

CP850, *Low Temperature Physics: 24th International Conference on Low Temperature Physics;*
edited by Y. Takano, S. P. Hershfield, S. O. Hill, P. J. Hirschfeld, and A. M. Goldman

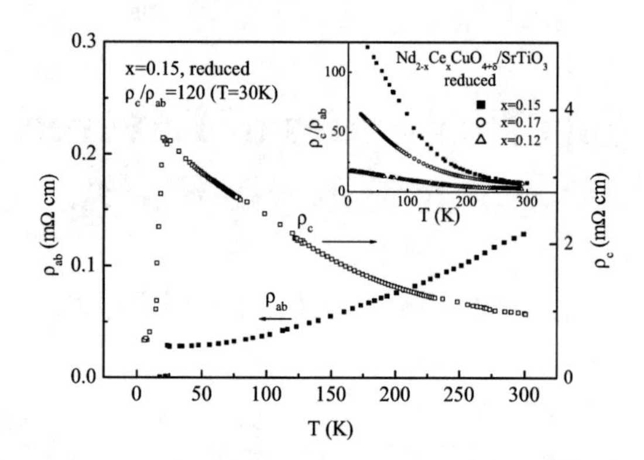

FIGURE 1. Temperature dependencies of the ρ_{ab} and ρ_c for reduced $Nd_{1.85}Ce_{0.15}CuO_4$ films. The insert shows the anisotropy coefficient $\rho_c/\rho_{ab}(T)$.

$\rho_{ab}(T)$ and $\rho_c(T)$ dependencies for single crystal $Ca_{2-x}Sr_xRuO_4$ with x=0.5. As for Nd-system in-plane and out-of-plane temperature dependencies of the resistivity are different: $\rho_{ab}(T) \sim T^{1.8}$ – metallic behavior in $\Delta T=(300\div30)$K and $\rho_c(T) \sim -\ln T$ in the range $(300\div190)$K – non-metallic behavior.

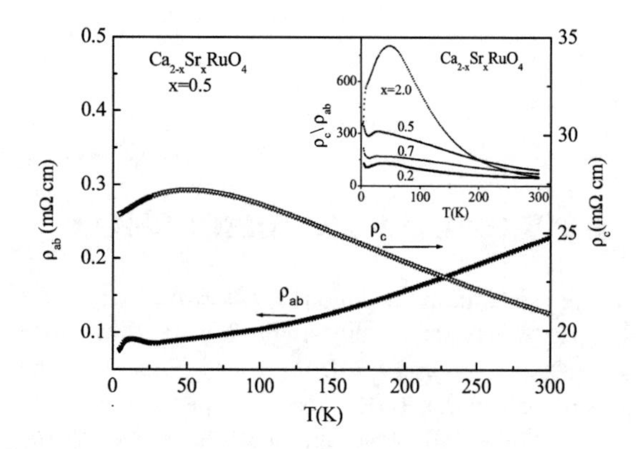

FIGURE 2. Temperature dependencies of the ρ_{ab} and ρ_c for single crystal $Ca_{2-x}Sr_xRuO_4$ with x=0.5. The insert shows the anisotropy coefficient $\rho_c/\rho_{ab}(T)$.

According to [3] these disordered layered quasi-two- dimensional systems are Anderson conductors with a very anisotropic localization length ($R_{loc}^{ab} >> R_{loc}^{c}$). Let us compare the *ab*-plane localization length $R_{loc}^{ab} = \ell\exp(\pi/2 \times k_F\ell)$ with the inelastic scattering length $L_\varphi^{ab} = (D_{ab} \times \tau_\varphi)^{1/2}$ (ℓ-mean free path, k_F -Fermi wave vector, D_{ab} -diffusion coefficient, τ_φ -phase coherence time) and in *c*-direction - $R_{loc}^{c} \cong c/\ln(\tau_{esc}/\tau)$ with c=6 Å (τ_{esc} -the time of the transfer between the conducting planes, τ -transport relaxation time, c-the distance between the CuO_2 - and RuO_2 -planes). These parameters were estimated from the experimental dates [2].

For $Nd_{2-x}Ce_xCuO_{4+\delta}$ films with different cerium and oxygen content we have: for "oxidized" film (x=0.12) $L_\varphi^{ab} \cong 35.6$ Å > $R_{loc}^{ab} \cong 17$ Å – non-metallic behavior in *ab*-plane; $R_{loc}^{c} \cong 1.36$ Å < c = 6 Å – non-metallic behavior in *c*-direction. For "oxidized" film (x=0.15) $L_\varphi^{ab} \to \infty$ > $R_{loc}^{ab} \cong 10$ Å – non-metallic behavior in *ab*-plane; $R_{loc}^{c} \cong 1.46$ Å < c = 6 Å – non-metallic behavior in *c*-direction. In the "reduced" film (x=0.15 – maximum of the anisotropy coefficient (fig.1,insert)) $L_\varphi^{ab} \cong 1200$ Å << $R_{loc}^{ab} \to \infty$ – metallic behavior in *ab*-plane, but in *c*-direction $\tau_{esc}/\tau \sim 1$ – non-metallic coherent transport at low temperature (fig.1). Thus for stoichiometric reduced $Nd_{1.85}Ce_{0.15}CuO_4$ a combination of in-plane metallic $\rho_{ab}(T)$ and out-of-plane nonmetallic behavior $\rho_c(T)$ takes place, as an intrinsic property of the quasi-two-dimensional systems. However, with increasing of disorder due to non-stoichiometric oxygen and decreasing of Ce-content both in-plane and out-of-plane resistivity become non-metallic.

For the single crystals $Ca_{2-x}Sr_xRuO_4$ with x=0.5 and 0.7 we have: (x=0.5) $L_\varphi^{ab} \cong 4$ Å << $R_{loc}^{ab} \to \infty$ – metallic behavior in *ab*-plane and in *c*-direction $R_{loc}^{c} \cong 2.2$ Å < c = 6 Å –nonmetallic behavior; (x=0.7) $L_\varphi^{ab} \cong 4.7$ Å << $R_{loc}^{ab} \to \infty$ – metallic behavior and in *c*-direction $R_{loc}^{c} \cong 3.8$ Å < c = 6 Å – nonmetallic behavior. We have found that the increase of the Ca-content in $Ca_{2-x}Sr_xRuO_4$ (fig.2,insert) leads to a decrease of the anisotropy coefficient.

So in both quasi-two-dimensional systems the destruction of the stoichiometric order with oxygen in Nd-system and with calcium in Ru-system demonstrate Anderson-type disorder-induced metal-non-metal transition.

The authors thank Y.Maeno for supplied samples of $Ca_{2-x}Sr_xRuO_4$. This work was supported by the RFBR-Ural grant 04-02-96084 and State Contract No.40.012.1.1.11.46(12/04).

REFERENCES

1. Nakatsuji, S., and Maeno, Y., *Phys. Rev. Letters* **84**, 2666-2669 (2000).
2. Ponomarev, A.I., Harus, G.I., Charikova, T.B., et al, *Modern Phys. Letters B* **17**, 701-707 (2003).
3. Sadovskii, M.V., "High-T_c Superconductors," in "*Superconductivity and Localization*", Singapore: World Scientific Publishing Co. Pte. Ltd., 2000, pp.192-223.

Doping and Momentum Dependence of Charge Dynamics in $Nd_{2-x}Ce_xCuO_4$ (x = 0, 0.075, and 0.15) Studied by Resonant Inelastic X-ray Scattering

K. Ishii*, K. Tsutsui†, Y. Endoh*,**, T. Tohyama†, S. Maekawa†, M. Hoesch*, K. Kuzushita*, T. Inami*, M. Tsubota*, K. Yamada†, Y. Murakami‡,* and J. Mizuki*

*Synchrotron Radiation Research Center, Japan Atomic Energy Research Institute, Hyogo 679-5148, Japan
†Institute for Materials Research, Tohoku University, Sendai 980-8577, Japan
**International Institute for Advanced Studies, Kyoto 619-0025, Japan
‡Department of Physics, Tohoku University, Sendai 980-8578 Japan

Abstract. We report a resonant x-ray scattering (RIXS) study at Cu K-edge in the n-type cuprate superconductor $Nd_{2-x}Ce_xCuO_4$ from the half-filled Mott insulator to the T_c = 25 K superconductor. The Mott gap is clearly seen in the spectra of x = 0, while both interband excitation across the Mott gap and intraband excitation in the upper Hubbard band are observed at electron-doped x = 0.075 and 0.15. The momentum dependence of x = 0.075 is presented and the dispersion relation of x = 0.075 is qualitatively similar to that of x = 0.15. We find in the doping dependence that the intraband excitation at low energy increase in intensity with carrier concentration (x).

Keywords: resonant inelastic x-ray scattering, high-T_c superconductor, charge excitation
PACS: 74.25.Jb, 74.72.Jt, 78.70.Ck

Exploration of the electron dynamics in high-T_c superconducting Cu oxides is very important not only to understand the mechanism of superconductivity but also to clarify the motion of electrons under strong Coulomb interaction. While spin dynamics is extensively studied by neutron scattering, charge dynamics can be elucidated by photon due to the coupling to the electron charge. Angle-resolved photoemission spectroscopy gives plenty of information on the momentum dependence of the occupied state through one-particle excitation spectra [1]. On the other hand, resonant inelastic x-ray scattering (RIXS) in the hard x-ray regime can measure two-particle excitaions spectra with momentum resolution, unlike conventional optical methods. RIXS at Cu K-edge has been applied to some high-T_c materials and dispersion relation of the excitation across the Mott gap, more precisely the gap between the Zhang-Rice band and the upper Hubbard band (UHB), was reported [2, 3, 4, 5]. Recently we have succeeded to distinguish the intraband excitation in the upper Hubbard band (UHB) from the interband excitation across the Mott gap in the n-type superconductor $Nd_{1.85}Ce_{0.15}CuO_4$, taking advantage of n-type over p-type to observe the intraband excitation [6]. As a function of momentum transfer, the intraband excitation shifts to higher energy up to 2-2.5 eV at the Brillouin zone boundary, accompanied by an increase of the spectral peak width. Theoretical calculation demonstrated that these features of the intraband excitation are qualitatively similar to the dynamical charge correlation function ($N(\mathbf{q},\omega)$) in the two-dimensional Hubbard model. Here we extend the RIXS study for $Nd_{2-x}Ce_xCuO_4$ and report the momentum dependence of x = 0.075 and the doping dependence of $Nd_{2-x}Ce_xCuO_4$ from the half-filled Mott insulator (x = 0) to the optimum doping (x = 0.15).

The RIXS experiments at Cu K-edge were carried out at BL11XU at SPring-8 [7]. A Si (400) channel-cut monochromator and a spherically-bent Ge (733) analyzer were used and the overall energy resolution is about 400 meV estimated from the full width at half maximum of the elastic scattering. The polarization of incident x-ray is π, and the surface of the single crystal is normal to the **c** axis, which was kept in the scattering plane. The energy of incident x-ray (E_i) is fixed at 8991 eV where low energy excitations below 3 eV are resonantly enhanced. Because of the strong two-dimensionality of the CuO_2 plane, the momentum dependence along the $\mathbf{c}^*$ axis is small. Then we select absolute momentum transfer ($\mathbf{Q}$) to be $\mathbf{Q} = (h,k,12.5)$ with $0 \leq h,k \leq 0.5$ where the scattering angle (2θ) is close to 90 °. Because the elastic intensity is roughly proportional to $\cos^2 2\theta$, the elastic intensity is significantly reduced at the momenta. It is quite important to measure excitations at low energy.

We show the contour plot of the RIXS spectra of $Nd_{1.925}Ce_{0.075}CuO_4$ (x = 0.075) in Fig. 1. The elastic tail and the higher-energy component is subtracted from the raw data in the same manner in the previous paper for x = 0.15 [6]. There are two characteristic excitations in the

CP850, *Low Temperature Physics: 24th International Conference on Low Temperature Physics;*
edited by Y. Takano, S. P. Hershfield, S. O. Hill, P. J. Hirschfeld, and A. M. Goldman

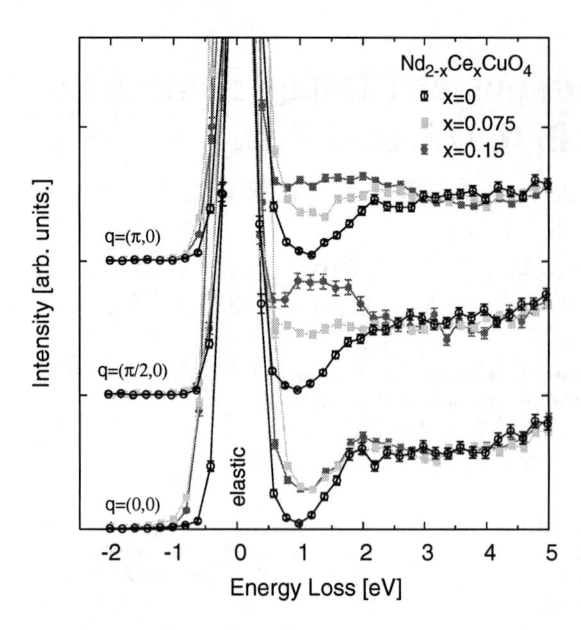

FIGURE 1. Doping dependence of RIXS spectra along the $(0,0)-(\pi,0)$ direction.

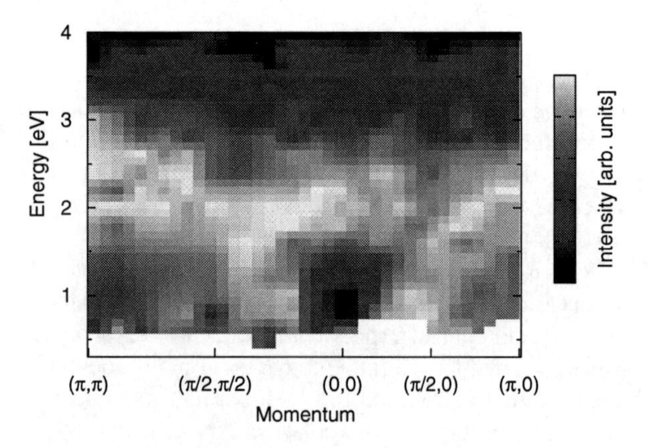

FIGURE 2. RIXS spectra of x = 0.075. The elastic tail and the higher energy components are subtracted from the raw data.

spectra. One is the excitation at 2 eV and the intensity is concentrated near the zone center. The other is the dispersive excitation whose peak energy shifts to higher energy as a function of $|\mathbf{q}|$, where $\mathbf{q}$ is the reduced momentum in the CuO_2 plane. These characteristics are qualitatively similar to that of x = 0.15, even though the separation of them is rather indistinct at x = 0.075. The former excitation is assigned to be an interband excitation across the Mott gap and the latter is the intraband excitation in UHB.

The doping dependence of RIXS spectra of $Nd_{2-x}Ce_xCuO_4$ along the $(0,0)-(\pi,0)$ direction are shown in Fig. 2. At x = 0, Mott gap is clearly seen at all the momenta. The magnitude of the gap at $\mathbf{q} = (0,0)$ is consistent with that observed in the optical conductivity [8]. When electrons are doped in the UHB, the excitation below the Mott gap increases at finite $|\mathbf{q}|$ and the intensity roughly proportional to the electron concentration x. As mentioned in the first paragraph, the intraband excitation in the RIXS spectrum is qualitatively similar to $N(\mathbf{q},\omega)$. When carrier concentration increases, theoretically calculated $N(\mathbf{q},\omega)$ increases in intensity. Therefore we confirm again that the intensity developed below the Mott gap is the intraband excitation.

Since the RIXS spectra exhibit characteristics of dynamical charge correlation function $N(\mathbf{q},\omega)$ including the dispersion relation, our results demonstrates that RIXS is a useful tool to study the charge dynamics with momentum resolution. It is noted that $N(\mathbf{q},\omega)$ is directly observed in non-resonant inelastic x-ray scattering. However its application is limited to materials with light elements due to the small scattering cross section and it is still weak to observe electronic excitations in high-T_c superconductors.

In summary, we have carried out a RIXS study for $Nd_{2-x}Ce_xCuO_4$ (x = 0, 0.075, and 0.15) and observed both the interband excitation across the Mott gap and the intraband excitation in the UHB. The momentum dependence of x = 0.075 is qualitatively similar to that of x = 0.15, that is, the interband excitation across the Mott gap is concentrated at the Brillouin zone center and the intraband excitation shows almost linear dispersion relation. The intensity of the intraband excitation increases with the electron concentration x.

ACKNOWLEDGMENTS

The authors thank T. Uefuji for supplying a crystal of $Nd_{1.925}Ce_{0.075}CuO_4$. K. T., T. T., S. M., and K. Y. were supported by the Japanese Ministry of Education, Culture, Sports, Science and Technology, Grant-in-Aid for Scientific Research. K. T., T. T., and S. M. were also supported by CREST, NAREGI Nanoscience Project. M. H. acknowledges support from the Japanese Society for the Promotion of Science.

REFERENCES

1. A. Damascelli, et al., *Rev. Mod. Phys.*, **75**, 473 (2003).
2. M. Z. Hasan, et al., *Science*, **288**, 1811 (2000).
3. Y. J. Kim, et al., *Phys. Rev. Lett.*, **89**, 177003 (2002).
4. Y.-J. Kim, et al., *Phys. Rev. B*, **70**, 094524 (2004).
5. K. Ishii, et al., *Phys. Rev. Lett.*, **94**, 187002 (2005).
6. K. Ishii, et al., *Phys. Rev. Lett.*, **94**, 207003 (2005).
7. T. Inami, et al., *Nucl. Instrum. Method Phys. Res., Sect. A*, **467-468**, 1081 (2001).
8. Y. Onose, et al., *Phys. Rev. B*, **69**, 024504 (2004).

Interlayer Transport Properties Observed Using Small Mesa Structures for Electron-doped $Sm_{1.86}Ce_{0.14}CuO_{4-\delta}$

Tsuyoshi Kawakami*, Takasada Shibauchi*, Yuhki Terao* and Minoru Suzuki*

*Department of Electronic Science and Engineering, Kyoto University, Kyoto 615-8510, Japan

Abstract. Interlayer transport properties are measured for electron-doped $Sm_{1.86}Ce_{0.14}CuO_{4-\delta}$ by using 30 nm-high small mesa structures to elucidate intrinsic characteristics of this material without influence of the partial epitaxial decomposition phase parallel to the layers. We find that the mesas show some properties quite diffrent from those of the bulk single crystals, providing intriguing insight into the electronic states in electron-doped cuprates analogous to overdoped $Bi_2Sr_2CaCu_2O_{8+\delta}$.

Keywords: Electron-doped cuprates, Interlayer transport, Critical current density
PACS: 74.25.Fy, 74.25.Sv, 74.50.+r, 74.72.Jt

INTRODUCTION

High-T_c cuprates have layered structures, which consist of alternation of superconducting layers and charge reservoir layers. Among them, hole-doped $Bi_2Sr_2CaCu_2O_{8+\delta}$ (BSCCO), where charge reservoir layers are insulator-like, has very large anisotropy and is understood to possess a stack of high-quality tunnel-type Josephson junctions [1]. Electron-doped cuprates also show highly anisotropic conduction, whose anisotropy parameter is as large as 10^4 [2], comparable to BSCCO. Owing to the large anisotropy, the interlayer transport is expected to be dominated by tunneling conduction. Thus it is of interest to compare interlayer transport properties of electron-doped cuprates with those of hole-doped BSCCO.

In reality, the bulk single crystals of electron-doped cuprates were recently found to be plagued with the phase decomposition and chemical inhomogeneity, which are produced as quasi-two-dimensinal epitaxial structure of the secondary phase parallel to the layers [3]. The decomposition proceeds during the oxygen reduction annealing, which is essential for superconductivity in electron-doped cuprates. The secondary phase exists at an average of 300 nm intervals perpendicular to layers, so that the ordinary bulk measurement of the interlayer transport is affected. Here we report intrinsic interlayer (c-axis) transport measurements of 30 nm-high small mesa structures in an electron-doped cuprate $Sm_{2-x}Ce_xCuO_{4-\delta}$ (SCCO), which are much thinner than the typical separation of the secondary phase.

EXPERIMENTAL

Single crystals of SCCO are grown by a self-flux method [2, 4], having plate-like shapes with flux-free and shiny surfaces perpendicular to c-axis, which are requisite for the mesa fabrication. The Ce concentration $x \sim 0.14$ was determined by EDS. The as-grown crystals were annealed under reducing atmosphere (flowing Ar gas) at 950°C for 20 hours, and then the mesa structures with a four-terminal configuration were fabricated on the surfaces of the annealed crystals with conventinal photolithography and Ar ion milling technique [1]. Typical size of the small mesas was 30 nm in height and 8 μm × 12 μm in lateral area; the height is sufficiently smaller than the typical separation of the secondary phase, and thus the effects of the secondary phase can be eliminated. We examined the interlayer transport of the SCCO mesas through the c-axis resistivity (ρ_c) and the current-voltage (I–V) characteristics.

RESULTS AND DISCUSSIONS

Inset of Fig. 1 shows the temperature (T) dependence of ρ_c in the mesa and bulk single crystal. The SCCO bulk single crystal was grown and oxygen-reduced under identical conditions used for the mesa fabrication, and additionally annealed under flowing O_2 gas at 500°C for 10 hours after oxygen reduction for further homogenization of the oxygen deficiency in the bulk crystal [5]. This treatment under oxygenating atmosphere can sharpen the superconducting transition with little affecting the shape of ρ_c–T curve above the T_c. The mesa samples reproducibly exhibit ρ_c values considerably smaller than those of SCCO bulk crystals and metallic temperature dependence with a residual resistivity ratio much greater than that of bulk crystals. They also show a sharper resistive transition than the bulk crystals, which reflects exceedingly high homogeneity within the mesa structure. The reproducibility and high homogeneity for the mesa samples demonstrate that the secondary phase, which can

CP850, *Low Temperature Physics: 24th International Conference on Low Temperature Physics;*
edited by Y. Takano, S. P. Hershfield, S. O. Hill, P. J. Hirschfeld, and A. M. Goldman

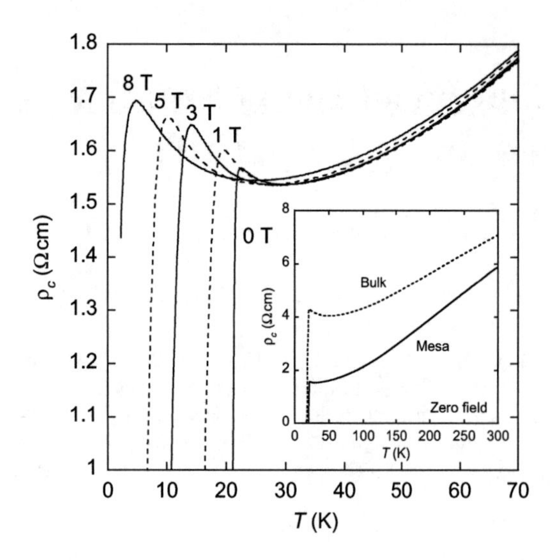

FIGURE 1. Temperature dependence of ρ_c in magnetic fields along to the c-axis for an SCCO mesa. Inset: ρ_c–T curves for the mesa and an SCCO bulk single crystal in zero field.

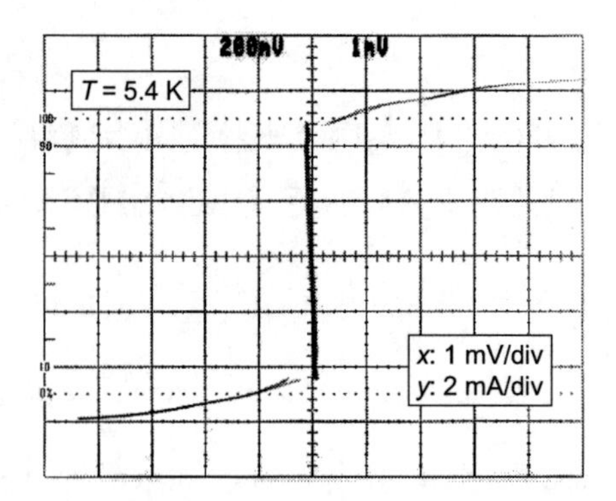

FIGURE 2. Oscilloscope image of current (y-axis) versus voltage (x-axis) characteristics for the mesa. Negative resistance observed on the supercurrent can be attributed to geometry of the electrodes.

cause the excess resistivity in the bulk crystals, is able to be successfully avoided by the mesa fabrication.

Let us now turn to the ρ_c behavior of the mesa in the presence of magnetic fields (see Fig. 1). Without field ρ_c exhibits a clear upturn just above T_c, and in fields we find the suppression of this ρ_c upturn and significant negative magnetoresistance at low temperatures, which we note is quite analogous to the case especially for overdoped BSCCO in the pseudogap region [6, 7]. Moreover, the ρ_c–T curves in fields show a parallel shift of the transition, which is in contrast to the resistivity broadening observed in underdoped hole-doped cuprates and is rather similar to the behavior in the overdoped regime.

Figure 2 exhibits the I–V characteristics of the mesa at 5.4 K. Distinct critical current and slight hysteresis are observed. The critical current density J_c for the mesa is found to be 4.7×10^3 A/cm^2, which is as large as that of overdoped BSCCO; however, large hysterisis and resistive branches observed for the BSCCO mesas cannot be seen. Meanwhile, the J_c for the mesa is much larger than the reported result for a $Pr_{1.85}Ce_{0.15}CuO_{4+\vartheta}$ bulk single crystal, where $J_c = 50$–200 A/cm^2 at 4.2 K [8]. The discrepancy of J_c is attributed to the avoidance of the secondary phase as well as decrease of the Joule heating and suppression of the effects of self magnetic field. Furthermore, we also found the obtained J_c for our mesas to be consistent with Josephson plasma frequencies reported for a series of high-T_c cuprates [9]. These results suggest that the genuine J_c value in SCCO is expected to be closer to our results in the mesas, demonstrating significance of thinning samples below several dozen nanometers in studies on electron-doped cuprates.

CONCLUSIONS

We extracted the intrinsic interlayer transport by fabricating the small mesa structures in an electron-doped cuprate. The SCCO mesas differ significantly from the bulk single crystals in ρ_c and J_c; the differences can result mainly from the secondary phase. The T-dependence of ρ_c in fields for the mesas quanlitatively resembles that in overdoped BSCCO with the pseudogap. The J_c for the mesa is found to be 4.7×10^3 A/cm^2 at 5.4 K, which should be close to a intrinsic value in SCCO.

ACKNOWLEDGMENTS

This work is partly supported by the 21st Century COE Program (Grant No. 14213201). T. Kawakami is supported by a JSPS Research Fellowship.

REFERENCES

1. M. Suzuki, T. Watanabe, and A. Matsuda, *Phys. Rev. Lett.*, **82**, 5361–5364 (1999).
2. Y. Hidaka and M. Suzuki, *Nature*, **338**, 635–637 (1989).
3. P. K. Mang *et al.*, *Phys. Rev. B*, **70**, 094507 (2004).
4. J. L. Peng, Z. Y. Li, and R. L. Greene, *Physica C*, **177**, 79–85 (1991).
5. M. Matsuda, Y. Endoh, K. Yamada, H. Kojima, I. Tanaka, R. J. Birgeneau, M. A. Kastner, and G. Shirane, *Phys. Rev. B*, **45**, 12548–12555 (1992).
6. T. Shibauchi, L. Krusin-Elbaum, G. Blatter, and C. H. Mielke, *Phys. Rev. B*, **67**, 064514 (2003).
7. T. Kawakami, T. Shibauchi, Y. Terao, M. Suzuki, and L. Krusin-Elbaum, *Phys. Rev. Lett.*, **95**, 017001 (2005).
8. K. Schlenga *et al.*, *Physica C*, **235-240**, 3273–3274 (1994).
9. H. Shibata and T. Yamada, *Physica C*, **293**, 191–195 (1997).

Anomalous Change of Hall Coefficient in Overdoped $La_{2-x}Sr_xCu_{1-y}Zn_yO_4$ around $x = 0.2$

Jun Tonishi*, Takao Suzuki ,†,* and Takayuki Goto*

*Department of Physics, Sophia University, 7-1 Kioi-cho, Chiyoda-ku, Tokyo 102-8554, Japan
†Advanced Meson Science Laboratory, RIKEN, 2-1 Hirosawa, Wako, Saitama 351-0198, Japan

Abstract. The Hall coefficient (R_H) has been measured in 0.5% Zn-doped $La_{2-x}Sr_xCu_{0.995}Zn_{0.005}O_4$ under high magnetic fields up to 12 T. With decreasing temperature, R_H increases and begins to decrease below a temperature T_{R_H}. This characteristic temperature T_{R_H} has the local maximum around $x = 0.195$, and this Sr-concentration coincides with that the superconducting transition temperature is slightly suppressed. This behavior is quite similar to the phenomena observed in the stripe phase in $x \sim 0.12$. These results suggest that the anomalous decrease of R_H around $x = 0.195$ observed in this study is responsible for the "1/4"-anomaly [as reported by Kakinuma *et al.*, Phys. Rev. B **59**, 1491 (1999).].

Keywords: $La_{2-x}Sr_xCuO_4$; Hall effect
PACS: 74.72.Dn,74.25.Fy

In $La_{2-x}Sr_xCuO_4$ around $x \sim 0.115$, of which is so-called "1/8"-concentration, the suppression of the superconductivity and several anomalies in physical properties have been reported [1, 2], and these phenomena have been discussed in terms of the "stripe" modulation of the spin and/or charge densities [3]. Recently, Kakinuma *et al.* investigated transport properties of the Zn-doped $La_{2-x}Sr_xCu_{1-y}Zn_yO_4$ (LSCZO) in the overdoped region. They reported that a slight suppression of the superconducting transition temperature T_c and anomalously less-metallic behaviors of the thermoelectric power around $x = 0.2$, and these anomalies are enhanced by the Zn-doping [4]. In addition, the slowing down of the Cu-spin fluctuations is suggested by results of muon-spin-relaxation measurements in this concentration [5]. However, the origin of these anomalies reported in the concentration of $x \sim 0.2$ is unknown yet.

The purpose of this study is to investigate the charge transport in this concentration, so that we examined the temperature dependence of the Hall coefficient in the Zn-doped $La_{2-x}Sr_xCu_{0.995}Zn_{0.005}O_4$.

Sintered samples used in this study were prepared by the usual solid-state reaction method. All the samples were confirmed to be single phase by X-ray diffraction. The temperature dependence of the magnetization was measured at the applied field of 10 Gauss under the field cooling (FC) condition using a Quantum Design SQUID magnetometer. The Hall coefficient was measured by a dc method with decreasing temperature from 100 K in 12 T using a six-probe geometry. The typical size of the sample is $2 \times 6 \times 0.2$ mm^3 and the electric current is 30 mA. The intrinsic Hall voltage is obtained by calculating the mean value for all four permutations of the direction of applied magnetic field and the electric current.

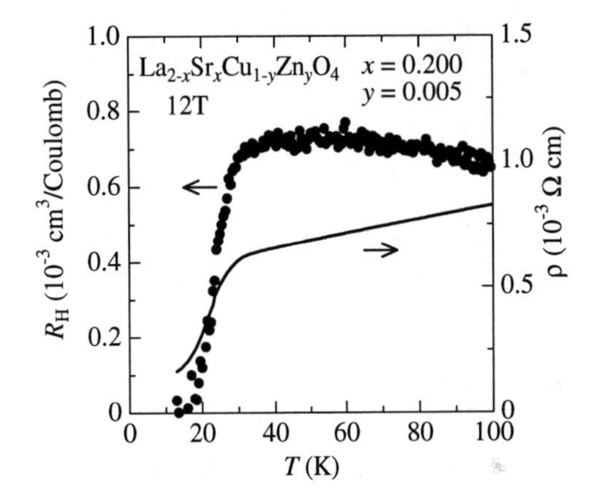

FIGURE 1. Temperature dependence of the Hall coefficient (R_H) and the electric resistivity (ρ) for $La_{1.8}Sr_{0.2}Cu_{0.995}Zn_{0.005}O_4$ in 12 T. Closed circles are R_H and solid line is ρ.

Figure 1 shows the temperature dependence of the Hall coefficient (R_H) and the electric resistivity (ρ) in case of $x = 0.200$ and $y = 0.005$ in 12 T as a typical case. With decreasing temperature, R_H increases monotonically and begins to decrease below a temperature around $\sim$ 60 K in the normal conducting state far above T_c. The anomalous decrease of R_H indicates the modification of the electronic state. With the view of investigating the quantitative Sr concentration dependence of this characteristic temperature where the change of the electronic state occurs, the Hall angle ($\cot\theta_H$) is calculated by the following expression:

$$cot\,\theta_H = \rho / H R_H \tag{1}$$

Figure 2(a) shows the temperature dependence of $\cot\theta_H$

CP850, *Low Temperature Physics: 24th International Conference on Low Temperature Physics;*
edited by Y. Takano, S. P. Hershfield, S. O. Hill, P. J. Hirschfeld, and A. M. Goldman

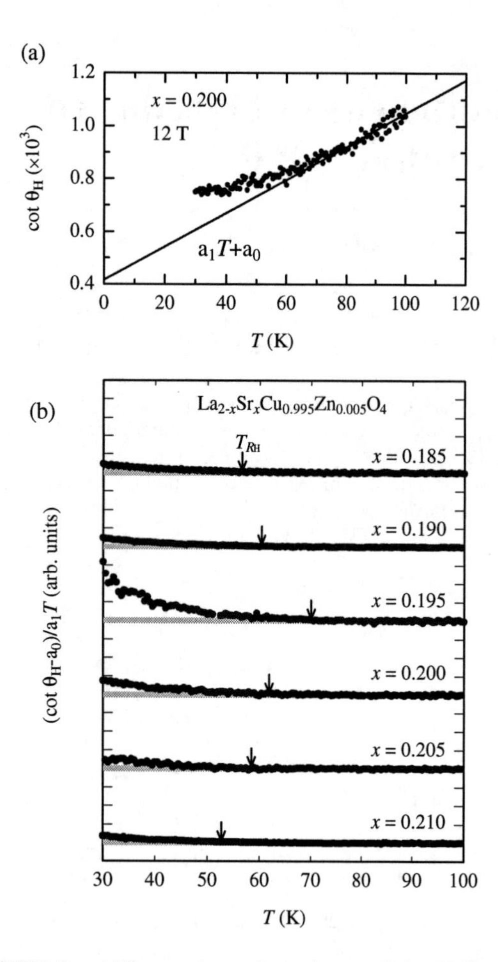

FIGURE 2. (a)Temperature dependence of the Hall angle ($\cot\theta_{\rm H}$) . Solid line is fitted linear function of $a_1T + a_0$ in the temperature range between 70 K and 100 K. (b)Plot of $(\cot\theta_{\rm H} - a_0)/a_1T$ versus T in each sample. The arrows indicate the temperature $T_{R_{\rm H}}$.

in case of $x = 0.200$. With decreasing temperature, $\cot\theta_{\rm H}$ decreases linearly and begins to deviate from the linear temperature dependence in the lower temperature region. The linear function of $a_1T + a_0$ is fitted for the data between 70 K and 100 K. In order to make clear and to define the temperature where this deviation occurs, Fig 2(b) shows the plot of $(\cot\theta_{\rm H} - a_0)/a_1T$ versus T for all samples measured in this study. The characteristic temperature $T_{R_{\rm H}}$ is defined as the temperature where $(\cot\theta_{\rm H} - a_0)/a_1T$ deviates from the horizontal line. The arrows indicate the temperature $T_{R_{\rm H}}$. It is apparent that the deviation is the most significant in $x = 0.195$. Figure 3 shows the phase diagram for $T_{R_{\rm H}}$ and $T_{\rm c}$ in $La_{2-x}Sr_xCu_{0.995}Zn_{0.005}O_4$. The superconducting transition temperature $T_{\rm c}$ is defined as the onset temperature in the magnetization curve. Obviously, the characteristic temperature $T_{R_{\rm H}}$ has the local maximum and the superconducting transition temperature $T_{\rm c}$ has the local minimum in $x = 0.195$. That is to say, the Sr concentration in which the electronic state is changed significantly coincides with that $T_{\rm c}$ is lowered. This result suggests that the anomalous decrease of $R_{\rm H}$ around $x = 0.195$ observed in this study is responsible for the slight suppression of the superconductivity and for the "1/4"-anomaly reported by Kakinuma *et al* [4]. Finally, we mention that the relation between $T_{R_{\rm H}}$ and $T_{\rm c}$ found out in this study has a striking resemblance to the well known phase diagram in the concentration of x around 0.12.

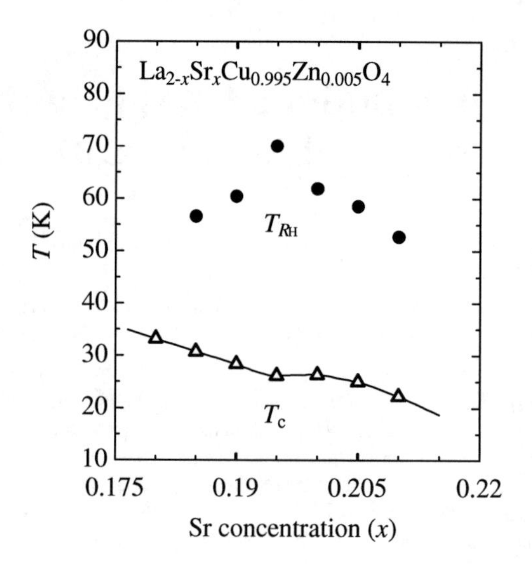

FIGURE 3. Phase diagram for $T_{R_{\rm H}}$ and $T_{\rm c}$ in LSCZO for $y = 0.005$. Closed circles are $T_{R_{\rm H}}$ and open triangles are $T_{\rm c}$. Solid line is guide for the eyes.

In summary, the temperature dependence of Hall coefficient $R_{\rm H}$ was measured in 0.5% Zn-doped $La_{2-x}Sr_xCu_{0.995}Zn_{0.005}O_4$, and the anomalous decrease of $R_{\rm H}$ was observed below the temperature $T_{R_{\rm H}}$ far above $T_{\rm c}$. This characteristic temperature $T_{R_{\rm H}}$ has the local maximum in $x = 0.195$ where $T_{\rm c}$ shows the local minimum. These results suggest that the electronic state appeared below $T_{R_{\rm H}}$ is responsible for the reported slight superconducting suppression around $x = 0.195$.

ACKNOWLEDGMENTS

We are grateful to T. Naito for useful suggestions in the construction of the sample rotating system in magnetic fields.

REFERENCES

1. H. Takagi *et al.*, *Phys. Rev. B* **40**, 2254 (1989).
2. M. Sera *et al.*, *Solid State Commun.* **69**, 851 (1989).
3. J.M. Tranquada *et al.*, *Nature* (*London*) **375**, 561 (1995).
4. N. Kakinuma *et al.*, *Phys. Rev. B* **59**, 1491 (1999).
5. I. Watanabe *et al.*, *Phys. Rev. B* **62**, 11985 (2000).

Increase of the Sound Velocity by Magnetic Fields in $La_{2-x}Sr_xCuO_4$ around $x = 0.220$

Takao Suzuki[*,†], Takaya Ota[†], Jun Tonishi[†] and Takayuki Goto[†]

[*]*Advanced Meson Science Laboratory, RIKEN, 2-1 Hirosawa, Wako, Saitama 351-0198, Japan*
[†]*Department of Physics, Sophia University, 7-1 Kioi-cho, Chiyoda-ku Tokyo, 102-8554, Japan*

Abstract. The sound velocity V_s of $La_{2-x}Sr_xCuO_4$ around $x = 0.220$ has been measured in magnetic fields up to 12 T. The increase of V_s by applying magnetic fields was observed in case of $x = 0.220$, and resembles the enhancement of the increase of V_s observed in the concentration of x around 0.115. The magnetic field dependence of V_s in $x = 0.220$ scales with that observed in $x = 0.120$. This result strongly suggests that the observed magnetic field effect on the sound velocity has the same origin with that in $x \sim 0.115$.

Keywords: High-Tc, sound velocity
PACS: 74.72.Dn, 74.25.Ld

It is well known in $La_{2-x}Sr_xCuO_4$ that the superconducting transition temperature T_c is slightly lowered in the concentration of x around 0.115 compared with that of the neighbouring region [1]. In this concentration, transport properties show an glimpse of anomalous changes above T_c [2], and gradual increase of the sound velocity at low temperatures, which is interpreted as the precursor of the structural phase transition to the TLT phase [3]. The relation between these anomalies (1/8-anomalies) has been investigated intensively over the last decade.

Recently, Kakinuma *et al.* investigated transport properties of the Zn-doped $La_{2-x}Sr_xCu_{1-y}Zn_yO_4$ in the overdoped region, and reported that anomalously less-metallic behaviors of the electrical resistivity and the thermoelectric power, and that the enhancement of the anomaly by the Zn-doping [4]. In addition, the slowing down of the Cu-spin fluctuations with Zn doping is observed by the μSR measurement in those concentrations [5]. These rsesults in the overdoped region remind us the anomalous behavior observed in $x \sim 0.115$, and the formation of an ordering of holes and/or spins is discussed on the analogy of the stripe order around $x \sim 0.115$. However, the enhancement of the anomalies by other factor, which is the structural phase transition or the applying magnetic fields[6], has not been investigated yet. In this study, we report the experiment results of the sound velocity of the overdoped $La_{2-x}Sr_xCuO_4$ in magnetic fields up to 12 T.

Sintered samples used in this study were prepared by the usual solid-state reaction method. All the samples were confirmed to be of the single phase by the powder x-ray diffraction. The sound velocity V_s was measured using the phase comparison method with $\sim$10MHz longitudinal waves generated by PZT transducers. Ultrasonic measurements in magnetic fields were carried out under the configuration of $\mathbf{k} \parallel \mathbf{u} \parallel \mathbf{H}$ in order to exclude the contribution of the vortex lattice. Temperature dependence of the sound velocity was measured with decreasing temperature in each magnetic field.

Figure 1 shows the temperature dependence of the sound velocity in 0 T and 12 T. The relative change $\Delta V_s(T,H)/V_s(T = 3.5\text{K}, H = 0\text{T})$ is plotted in each sample. In the inset, a typical temperature dependence of $\Delta V_s/V_s$ in zero field in the overdoped region is shown. The sound velocity V_s is quite sensitive to the change of the crystal structure and to the distortion of the crystal lattice because V_s directly corresponds to the elastic constant of the crystal lattice c through the formula of $c = \rho V_s^2$ (ρ: the density). With increasing x, as shown in Fig.1, the softening and the hardening associated with the structural phase transition are weakened in connection with the disappearance of this transition. The sound velocity in 12 T is the same temperature dependence as V_s in 0 T within the resolution except for the case of $x = 0.220$ below 16 K. This result means that the structural phase transition from the high-temperature tetragonal phase to the orthorhombic phase is not affected by the magnetic field. In case of $x = 0.220$, the sound velocity in 12 T increases at low temperatures compared with that in 0 T although the change by the magnetic field is not observed in the neighboring concentration. The magnetic field dependence of the sound velocity at 3.5 K is shown in Figure 2. Each data is normalized by V_s in 0 T. All measurements were carried out with decreasing the magnetic field after cooling from 45 K in 12 T. Evidently, the monotonous increase of the sound velocity by applying magnetic fields is observed only in case of $x = 0.220$ within the resolution. It should be emphasized that this value $x = 0.220$ coincides with x where the dip-

CP850, *Low Temperature Physics: 24th International Conference on Low Temperature Physics;*
edited by Y. Takano, S. P. Hershfield, S. O. Hill, P. J. Hirschfeld, and A. M. Goldman

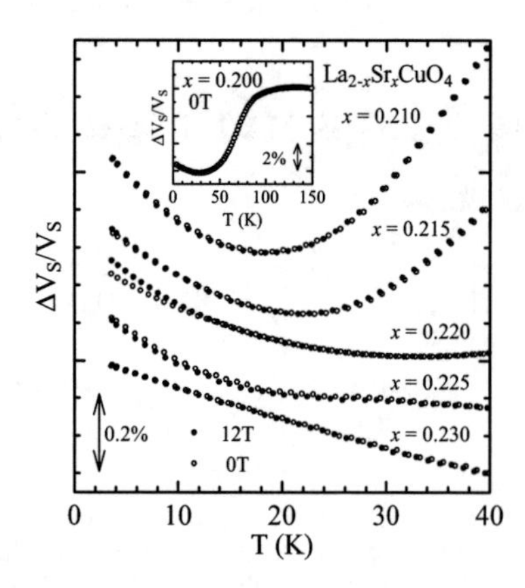

FIGURE 1. Temperature dependence of the sound velocity of $La_{2-x}Sr_xCuO_4$ in 0 T and 12 T. The relative change of the sound velocity $\Delta V_s/V_s$ is plotted in each sample. The inst shows the temperature dependence of $\Delta V_s/V_s$ in case of $x = 0.200$ in 0 T as a typical behavior in the overdoped region.

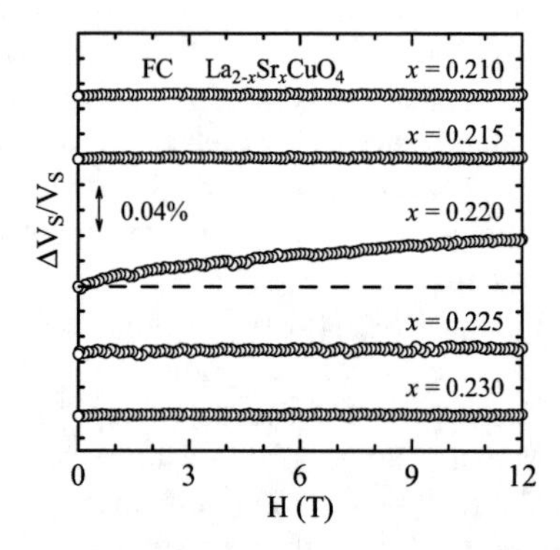

FIGURE 2. Magnetic field dependence of $\Delta V_s/V_s$ at 3.5 K. All measurements were carried out with decreasing magnetic field after cooling in 12 T from 45 K to 3.5 K.

like anomaly in the $T_c - x$ curve is reported [4]. Thus, we find a magnetic field effect on the sound velocity in this characteristic concentration of $x = 0.220$. The range of x where the magnetic field effect is observed is quite narrow compared to that in case of $x \sim 0.12$ [6]. Figure 3 shows the Magnetic field dependence of $\Delta V_s/V_s$ in $x = 0.220$ at 3.5 K and in $x = 0.120$ at 5 K. The data of $x = 0.12$ was measured in the previous work [6]. Both data are normalized by V_s in 0 T. The field dependence of V_s in $x = 0.220$ scales with that in $x = 0.12$ in the region of magnetic fields from 0 T to 12 T although the value of the velocity change in $x = 0.220$ is one order of magnitude smaller than the change in $x = 0.12$. This result strongly suggests that the observed magnetic field effect on the sound velocity has the same origin with that in $x \sim 0.115$.

Finally, we discuss a possible scenario of the increase

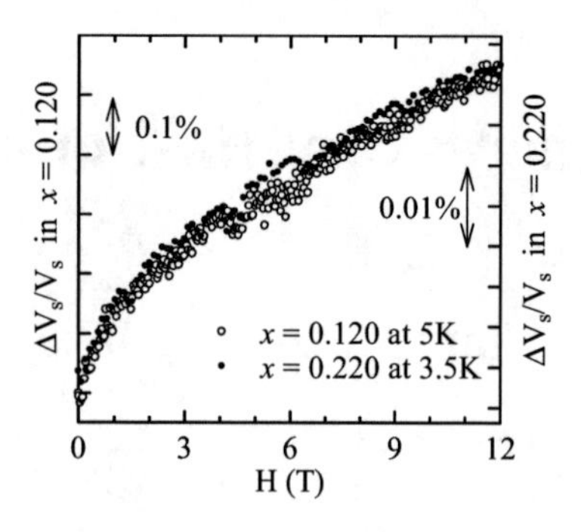

FIGURE 3. Comparison of the Magnetic field dependence of $\Delta V_s/V_s$ in $x = 0.220$ at 3.5 K(the right side vertical axis, closed circles) with that in $x = 0.120$ at 5 K (the left side vertical axis, open circles).

of the sound velocity by magnetic field in $x = 0.220$. The enhancement of the increase of V_s in $x \sim 0.115$ is interpreted as the stabilization of the low-temperature-tetragonal-like local distortion associated with the development of the stripe order by magnetic fields, because the static stripe order favors the low-temperature tetragonal phase. In $x = 0.220$, the crystal structure determined by the diffraction measurement is the high-temperature tetragonal (I4/mmm) at 3.5 K [4]. The local distortion of the crystal lattice, as a precursor of the structural change to the low-temperature phase, will appear associated with the development of the quasi-static spin order if the frequency of the Cu-spin fluctuation is lowered by magnetic fields. The increase of the sound velocity observed in this study will be accompanied with that local distortion.

In summary, we measured the sound velocity V_s of the overdoped $La_{2-x}Sr_xCuO_4$ in magnetic fields. The increase of V_s by magnetic fields is confirmed in case of $x = 0.220$, and the magnetic field dependence scales with that observed in the sample of $x = 0.120$ which shows the 1/8-anomalies. From these results, it is suggested that the origin of the increase of V_s by magnetic fields is the same with that in $x \sim 0.115$.

ACKNOWLEDGMENTS

The authors would like to thank I. Watanabe, T. Adachi, Y. Koike and K. Yamada for the useful discussion.

REFERENCES

1. H. Takagi *et al*, *Phys. Rev . B* **40**, 2254 (1989).
2. M. Sera *et al*, *Solid State Commun.* **69**, 851 (1989).
3. T. Fukase *et al*, *Physica B* **165-166**, 1289 (1990).
4. N. Kakinuma *et al*, *Phys. Rev. B* **59**, 1491 (1999).
5. I. Watanabe *et al.*, (private communication).
6. T. Goto *et al*, *Physica B* **246-247**, 572 (1998).

STM/STS Study on Local Electronic States of $La_{2-x}Sr_xCuO_4$

N. Momono, T. Goto, A. Hashimoto, K. Takeyama, Y. Ichikawa, M. Oda, M. Ido

Division of Physics, Graduate School of Science, Hokkaido University, Sapporo 060-0810, Japan

Abstract. We performed scanning tunneling microscopy/spectroscopy (STM/STS) on overdoped $La_{1.81}Sr_{0.19}CuO_4$ (T_c=33K). We can clearly see oxygen atoms on the cleaved surface of Cu-O plane. The tunneling spectrum measured on the cleaved surface is similar to the so-called zero temperature pseudogap (ZTPG), which was recently reported in lightly doped cuprates by McElroy et al.[Phys. Rev. Lett. **94**, 197005 (2005)] and by Hanaguri et al. [Nature **430,** 1001 (2004)]. The ZTPG-like spectrum indicates that the electronic states of the topmost Cu-O plane, where STS spectrum was measured, will be seriously modified on account of the marked reduction of the doping level.

Keywords: STM/STS, La-based cuprate, pseudogap
PACS: 74.25.Jb, 74.50.+r, 74.72.Dn

The local electronic states in high-T_c cuprates have intensively been studied by scanning tunneling microscopy and spectroscopy (STM/STS) so far [1-7]. In particular, in $Bi_2Sr_2CaCu_2O_{8+y}$ (Bi2212), the tunneling spectrum, which is proportional to the local density of states (LDOS), has been investigated in detail as a function of spatial position and temperature. It has been reported that the tunneling spectrum for underdoped Bi2212 is extremely inhomogeneous; various types of the energy gap at E_F ranging from a d-wave SC type with different gap sizes to the so-called zero temperature pseudogap (ZTPG) appear when the STM/STS tip is swept over the surface [3]. It was recently reported in lightly doped Bi2212 and $Na_xCa_{2-x}CuO_2Cl_2$ (NCCOC) that the electronic 2D superstructure, which looks like a 'checkerboard', appears in the region where the ZTPG is observed [3-6]. The checkerboard-like charge order was also found in the pseudogap state above T_c [7]. Thus the checkerboard-like charge order is a potential candidate of an order hidden in the pseudogap state although the origin of the checkerboard-like charge order is still unclear. To clarify the relationship between the checkerboard-like charge order and the pseudogap, it will be important to clarify whether or not the checkerboard-like charge order is inherent in high-T_c cuprates, though the pseudogap is well established in many high-T_c cuprates [8-10]. However STM/STS study has so far been limited only to Bi-based cuprates, YBCO and NCCOC because it is rather difficult in other cuprates to obtain a flat, clean and stable cleaved surface, which is crucial for STM/STS measurements. In the present study, we report atomic-resolution STM image observed successfully in overdoped $La_{2-x}Sr_xCuO_4$ (La214) and STS spectrum measured just after imaging.

The La214 crystal with x=0.19 was grown by the traveling solvent floating zone method. The superconducting transition temperature (T_c) is about 33K. To obtain a flat and clean surface of La214, the sample was cleaved in-situ at T=9K in ultrahigh vacuum before approach of STM tip to the surface.

Figure 1(a) shows a typical STM image of the cleaved surface over an area of ~7nm × ~3nm, acquired with a sample bias voltage of -0.9V and a tunneling current of 0.3nA. In Fig. 1(a), one can clearly see bright atomic rows and some defects of atoms. However no checkerboard-like charge order was observed in the STM image of the present overdoped La214 sample. The spacing between neighboring atoms along the atomic row is ~3 Å. This spacing is very close to the distance between nearest-neighbor oxygen atoms on Cu-O plane (~2.8 Å). However, the spacing between the neighboring atomic rows is ~6 Å, which is almost double the oxygen-spacing of Cu-O plane. It suggests that atomic rows of oxygen may be removed between bright atomic rows. On the other hand, we observed dark atoms between bright atomic rows in a different region of the identical cleaved surface (Fig. 1 (b)). The atomic spacing between the dark atoms is also ~3 Å, indicating that the dark atom will also correspond to the oxygen atom on Cu-O plane. These STM images of oxygen atoms on the Cu-O plane indicate that the present crystal is cleaved between Cu-O and La (Sr)-O planes and the cleavage causes many atomic defects.

CP850, *Low Temperature Physics: 24th International Conference on Low Temperature Physics;*
edited by Y. Takano, S. P. Hershfield, S. O. Hill, P. J. Hirschfeld, and A. M. Goldman

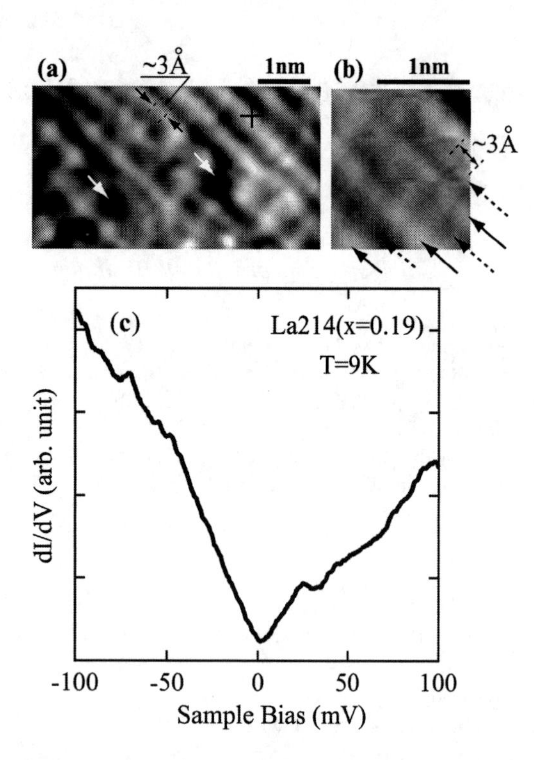

FIGURE 1. (a) Atomic-resolution STM image of overdoped La214 (x=0.19), which was acquired under V_s=-0.9V and I_t=0.3nA. The bright lines indicate atomic rows of oxygen on the Cu-O plane. The white arrows indicate the defects of atoms. (b) STM image acquired on a different region of the same cleaved surface as that of (a). The image is much zoomed for viewability. Solid-line arrows indicate the bright atomic rows, and broken-line arrows the dark atomic rows between the bright ones. (c) STS spectrum measured at the position indicated by the cross symbol in the STM image (a).

Here we focus on a possible reason why the dark and bright atomic rows of oxygen appear alternatively, as shown in Fig 1(b). The La214 crystal with x=0.19 takes orthorhombic structure at low temperatures (T<~130K), with the buckled Cu-O plane where atomic rows of oxygen along $(110)_{HTT}$ axis are displaced upward and downward alternatively (HTT is the acronym for "high temperature tetragonal"). Since the buckling of the Cu-O plane mainly originates from mismatch in lattice constants between La(Sr)-O and Cu-O planes, the buckling will take place on the topmost Cu-O plane owing to its lower La(Sr)-O layer even though the upper La(Sr)-O layer is removed in the process of cleaving the crystal. Therefore, in STM measurements on the cleaved surface of Cu-O plane, the two kinds of atomic rows of oxygen are expected to emerge as bright and dark lines in the STM image, because the tunneling current strongly depends on the height of atom in the vertical position

Figure 1(c) shows a typical STS spectrum at 9K (<<T_c), which was acquired on the cleaved surface where the STM images were taken (Fig. 1(a)). An asymmetric large V-shaped gap was observed instead of d-wave superconducting gap, although the sample shows bulk superconductivity. The present shape of the STS spectra reminds us of the ZTPG reported for lightly doped Bi2212 and NCCOC samples, suggesting that the doping level p will be largely reduced, at least, in the topmost Cu-O plane compared with the bulk p-value.

We suggest two origins for the reduction of the doping level in the topmost surface. One is the defect of oxygen atoms on the topmost surface, which will be caused by cleaving. The other is that the top La (Sr)-O layer, which is carrier-reservoir layer for Cu-O plane, is cleaved off and the Cu-O plane is exposed to vacuum. In La214, La (Sr)-O layer exists both above and below Cu-O plane, and the carriers (holes) are introduced into Cu-O plane from both La (Sr)-O layers. Then if upper La (Sr)-O layer is removed by cleaving, the doping level of the topmost Cu-O plane will be markedly reduced. These two origins may be inevitable in the process of cleaving in La214.

In summary, we observed the STM images of La214 with atomic resolution and measured the STS spectra. In the present study, STS data suggest that the doping level of the topmost Cu-O plane will be markedly reduced. To clarify whether such a reduction of doping level is inevitable on the cleaved surface of La214 is important for discussing the data of surface-sensitive probes such as STM/STS and photoemission spectroscopy.

This work was supported in part by Grants-in-Aid for Scientific Research and the 21st century COE program "Topological Science and Technology" from the Ministry of Education, Culture, Sports, Science and Technology of Japan.

REFERENCES

1. Ch. Renner and O. Fischer, *Phys. Rev. B* **51**, 9208 (1995).
2. M. Oda et al., *Phys. Rev. B* **53**, 2253 (1996).
3. K. McElroy et al., *Phys. Rev. Lett.* **94**, 197005 (2005).
4. C. Howald et al., *Phys. Rev. B* **67**, 14533 (2003).
5. T. Hanaguri et al., *Nature* **430**, 1001 (2004).
6. N. Momono, et al., *J. Phys. Soc. Jpn* **74**, 2400 (2005).
7. M. Vershinin et al., *Science* **33**, 1995 (2004).
8. N. Momono et al., *J. Low. Temp. Phys.* **117**, 353 (1999).
9. T. Matsuzaki et al., *J. Phys. Soc. Jpn.* **73**, 2232 (2004).
10. T. Nakano et al., *J. Phys. Soc. Jpn.* **67**, 2622 (1998).
11. N. Yamada and M. Ido, *Physica C* **203**, 240 (1992).

Raman and Optical Reflection Studies of Electronic States in $La_{2-x}Sr_xCuO_4$

S. Sugai*, J. Nohara*, Y. Takayanagi*, N. Hayamizu*, T. Muroi*, K. Obara* and K. Takenaka†

*Department of Physics, Faculty of Science, Nagoya university, Furo-cho, Chikusa-ku, Nagoya 464-8602, Japan
†RIKEN (The Institute of Physical and ChemicalResearch), Wako 351-0198, Japan

Abstract. The electronic states in $La_{2-x}Sr_xCuO_4$ are systematically investigated by Raman scattering, infrared-ultraviolet reflection spectroscopy, and electric resistivity. A narrow quasi-particle band is created at E_F and sharply grows up by collecting the density of states from the high energy region at 1-1.5 eV as temperature decreases near the insulator-metal transition. The relaxation time of conducting carriers is limited by the quasi-particle band width near the insulator-metal transition, but it drastically increases in the overdoped phase. The quasi-particle band gives the anomalous electronic properties.

Keywords: High T_c superconductor, $La_{2-x}Sr_xCuO_4$, quasi-particle band, Raman scattering, infrared-ultraviolet spectroscopy
PACS: 74.25.Gz,74.25.Jb,74.72.Dn,72.15.Lh

The high T_c superconductivity has been extensively investigated. The charge states, however, include still many unresolved issues about the insulator-metal transition, the formation of the k-dependent quasi-particle band, the k-dependent pseudo-gap and superconducting gap, the non-Fermi liquid behavior, the anomalous temperature dependent resistivity, and so on. The present experiments using the complementary selection rule of Raman scattering and infrared-ultraviolet reflection spectroscopy in cooperation with the electric resistivity disclosed that the development of the sharp quasi-particle band at the Fermi energy (E_F) is important for the physical properties in the underdoped phase.

Single crystals were synthesized by a traveling-solvent floating-zone method in an infrared radiation furnace. The superconducting transition temperatures are 13 K ($x = 0.06$), 27 K (0.08), 33 K (0.1), 33 K (0.115), 42 K (0.15), 32 K (0.2), and 24 K (0.22), and 13 K (0.25).

The upper panels in Fig. 1 show the temperature dependence of the low energy parts of B_{2g} Raman spectra. The B_{2g} spectra present the electronic excitations around $(\pi/2,\pi/2)$. The averaged intensity increases from $x = 0.035$ to 0.6 and then gradually decreases above $x = 0.06$, while the B_{1g} intensity gradually increases as carrier density increases. It indicates that the electronic density of states increases first around $(\pi/2,\pi/2)$ as carriers are doped and then shifts into the $(\pi,0)$ as carrier density increases [1]. In the B_{2g} spectra at $x = 0.06$ the central peak at energy zero sharpens and the intensity reaches the maximum as temperature decreases from 300 K to 100 K. And then the peak intensity decreases from 100 K to 40 K. The temperature dependence is caused by the growth of the quasi-particle band at E_F from 300 K to 100 K, the formation of the pseudo-gap from 100 K to 40 K and the formation of the superconducting gap below T_c as shown in Fig. 2. The temperature dependence decreases as carrier density increases. The clear superconducting coherent peak (pair breaking peak) is observed at 14.1 and 14.5 meV at $x = 0.1$ and 0.15 in the B_{2g} spectra at 5 K, respectively. The superconducting peak is created within the quasi-particle band below 0.1 eV. The present carrier density dependence is consistent with the angle resolved photo-emission spectroscopy (ARPES) [2].

The lower panels in Fig. 1 show the temperature dependence of the in-plane optical conductivity. The optical conductivity increases with the increase of the doped carrier density x. The temperature dependence is very different in the underdoped phase and the overdoped phase. In the underdoped phase the Drude peak increases with nearly keeping the width as temperature decreases. It is quite different from the normal Drude response that the increase of the peak height inevitably causes the decrease of the width, because the integrated intensity gives the carrier density from the f-sum rule,

$$\int_0^{\omega} \sigma(\omega')d\omega' = \frac{\pi n(0-\omega)e^2}{2m}. \tag{1}$$

The width of the Drude peak that is the inverse relaxation time is almost the same as that of the central Raman peak near the insulator-metal transition at $x = 0.06$. So that the inverse relaxation time of conducting carriers is determined by the narrow quasi-particle band width. The Drude width is almost unchanged in the underdoped phase in spite of the gradual increase of the quasi-particle band width as x increases. In the optimum and overdoped phases the Drude width becomes very narrow and de-

CP850, *Low Temperature Physics: 24th International Conference on Low Temperature Physics;*
edited by Y. Takano, S. P. Hershfield, S. O. Hill, P. J. Hirschfeld, and A. M. Goldman

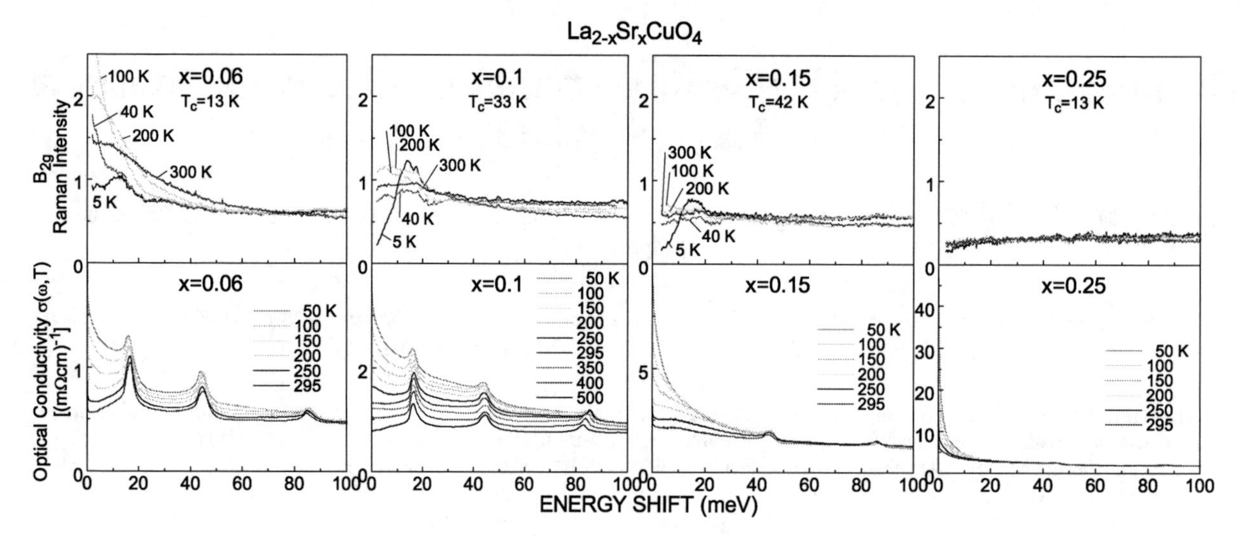

FIGURE 1. Temperature dependence of the B_{2g} Raman spectra (upper panels) and optical conductivity obtained from the reflection spectra (lower panels).

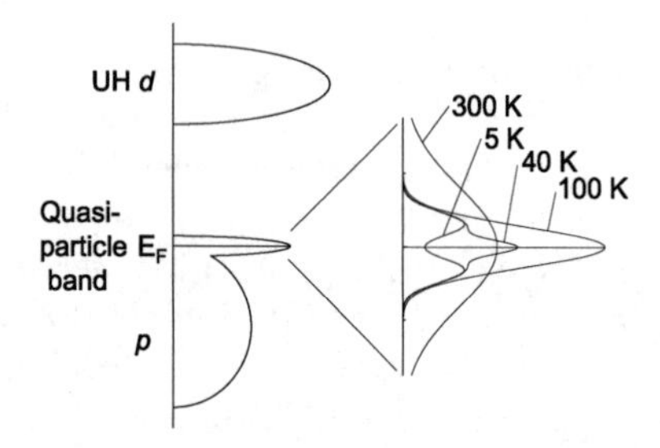

FIGURE 2. Temperature dependence of the quasi-particle band near the insulator-metal transition. This temperature dependence is derived from the Raman spectra at $x = 0.06$.

creases as temperature decreases. The electronic state approaches the normal metal in which the inverse lifetime of carriers is much smaller than the band width.

In order to clarify the temperature and doping dependence of the carrier density in the narrow quasi-particle band, the carrier density from $E = 0$ to 0.05 eV is obtained using the Eq. 1. Fig. 3 shows the result. The open squares are the experimental points. In the underdoped phase the temperature dependent component which gives the main conducting component is about 0.2 times the doped carriers. The growth fo the quasi-particle band on decreasing temperature is causes by collecting the electronic density of states from the high energy region at 1-1.5 eV. This is the common process in the reconstruction of electronic states at the insulator-metal transition in high T_c superconductors.

At present many models are based on the constant carrier density on temperature. However, the sharp growth of the quasi-particle band causes the anomalous electronic properties in high T_c superconductors. The resis-

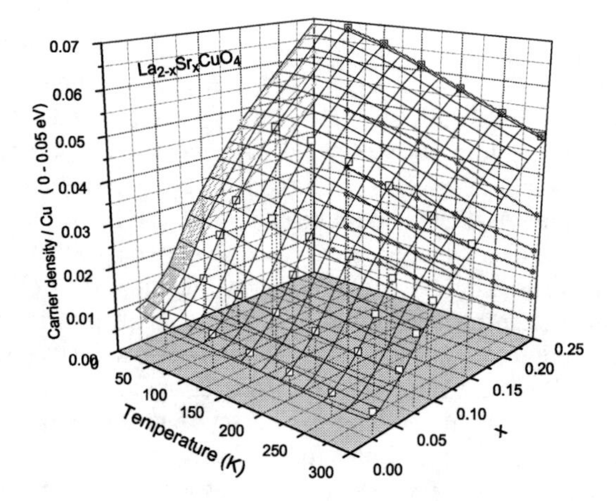

FIGURE 3. Temperature and Sr concentration dependence of the carrier density in the quasi-particle band from E_F to 0.05 eV, which is calculated by Eq. 1.

tivity can be expressed by the power law temperature dependence $\rho = a + bT^c$ above 100 K. The exponent c changes linearly from 2.1 at $x = 0.01$ to 1 at $x = 0.16$ and then increases to 1.4 at $x = 0.25$. The x dependence is related to the change of the relaxation time wich is limited by the quasi-particle band width to the much longer relaxation time as x increases and the change of large density of states from $(\pi/2, \pi/2)$ to $(\pi, 0)$ at $x = 0.16$.

REFERENCES

1. S. Sugai, H. Suzuki, Y. Takayanagi, T. Hosokawa, and N. Hayamizu, *Phys. Rev. B* **68**, 184504 (2003).
2. T. Yoshida *et al.*, *Phys. Rev. Lett.* **91**, 027001 (2003).

Dynamic Scaling Analyses of AC Fluctuation Microwave Conductivity in Superconducting $La_{2-x}Sr_xCuO_4$ Thin Films

H. Kitano[1], T. Ohashi[1], A. Maeda[1], and I. Tsukada[2]

[1)] *Department of Basic Science, University of Tokyo, 3-8-1 Komaba, Meguro-ku, Tokyo 153-8902, Japan*
[2)] *Central Research Institute of Electric Power Industry, 2-11-1 Iwadokita, Komae, Tokyo 201-8511, Japan*

Abstract. We studied the ac fluctuating microwave conductivity, $\sigma_{FL}(\omega)$, as a function of swept-frequency for superconducting $La_{2-x}Sr_xCuO_4$ (LSCO) thin films. In the vicinity of T_c, we found that the magnitude of $\sigma_{FL}(\omega)$ showed a power law dependence and that the phase of $\sigma_{FL}(\omega)$ was nearly independent of ω, showing a good agreement with the behavior expected in the dynamic scaling theory. Moreover, the dynamic scaling analysis of both the magnitude and the phase of $\sigma_{FL}(\omega)$ suggested a possibility that the critical fluctuation of underdoped LSCO belongs to the 2D-XY universality class.

Keywords: fluctuation, dynamic scaling, microwave conductivity, cuprate superconductor, LSCO, 2D-XY.
PACS: 74.25.Nf, 74.40.+k, 74.72.Dn, 74.78.Bz

INTRODUCTION

The large thermal fluctuation effects of the superconducting order parameter in high-T_c cuprates, which are enhanced by the short coherence lengths and the quasi two-dimensionality, enable the exploration of the fluctuation-dominated critical regime very close to T_c. In such a critical regime, the ac fluctuating conductivity, $\sigma_{FL}(\omega)$, is predicted to scale as [1],

$$\sigma_{FL}(\omega) \approx \xi^{z+2-d} S(\omega\xi^{z}), \tag{1}$$

where ξ is the coherence length diverging at T_c, z is a dynamic critical exponent, d is an effective spatial dimension, and S is a complex universal scaling function, respectively. In order to check this scaling relation directly, the measurement of $\sigma_{FL}(\omega)$ as a function of swept-frequency is crucially required.

In this paper, we studied the critical fluctuations of superconducting $La_{2-x}Sr_xCuO_4$ (LSCO) thin films by measuring the dynamic microwave conductivity, $\sigma(\omega) = \sigma_1(\omega) - i\sigma_2(\omega)$, as a function of the swept-frequency.

EXPERIMENTAL

The highly c-axis oriented LSCO thin films with x=0.07 were deposited on $LaSrAlO_4$ (LSAO) substrates by a pulsed laser deposition technique using pure ozone [2]. The use of LSAO substrate has the following advantages in this study: (1) Compressive strain gives rise to an increase of T_c compared with LSCO bulk crystals. (2) The tetragonal symmetry of LSAO substrate supports the CuO_2 planes with very flat square-lattices, free from disorders due to corrugations and twin boundaries.

To obtain the frequency dependence of $\sigma(\omega)$ directly, a non-resonant broadband technique was applied [3]. By measuring the sample and three known standards very carefully, we succeeded in obtaining $\sigma(\omega)$ for LSCO thin films from 0.1 GHz to 10 GHz at several temperatures down to 10 K.

RESULTS AND DISCUSSION

Figures 1(a) and 1(b) show the frequency dependence of the magnitude, $|\sigma_{FL}(\omega)|$, and the phase, $\phi_\sigma(=\tan^{-1}[\sigma_2/\sigma_1])$, of $\sigma_{FL}(\omega)$ for LSCO with x=0.07 at several temperatures, respectively. Here, $\sigma_{FL}(\omega)$ was obtained by subtracting the normal-state microwave conductivity $\sigma(\omega)$ at 32 K, which was found to be almost independent of frequency. In the vicinity of the resistive T_c (~19 K), we found that $|\sigma_{FL}(\omega)|$ showed a power law dependence similar to $1/\omega$ and that ϕ_σ was nearly independent of ω with a constant value, $\phi_\sigma \sim (\pi/2)\times 0.9$, as shown in Fig. 1. These results were found to agree with the behavior of Eq. (1) in the limit of $T \rightarrow T_c$., assuming that d is 2 [4].

We examined the scaling behavior of both $|\sigma_{FL}(\omega)|$ and ϕ_σ for temperatures above T_c, and found that both were scaled successfully over a wide range of frequencies, as shown in Fig. 2. In addition, we found that the obtained scaling functions of $|\sigma_{FL}(\omega)|$ and ϕ_σ

CP850, *Low Temperature Physics: 24th International Conference on Low Temperature Physics;*
edited by Y. Takano, S. P. Hershfield, S. O. Hill, P. J. Hirschfeld, and A. M. Goldman

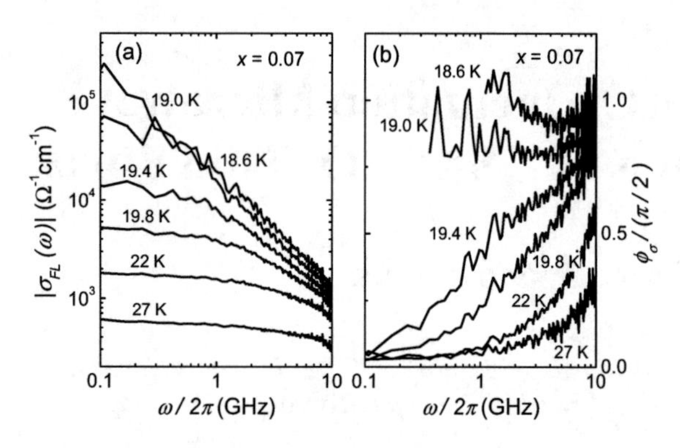

FIGURE 1. The frequency dependence of (a) the magnitude and (b) the phase of the fluctuating conductivity for LSCO thin film with x=0.07, respectively.

were similar to the 2D Gaussian form, except for the very vicinity of T_c. This shows a sharp contrast to the previous results for the optimally doped $YBa_2Cu_3O_y$, in which the 3D-like behavior was suggested [5,6].

The scaling parameters, ω_0 ($\sim\xi^{-z}$) and σ_0 ($\sim\xi^{z+2-d}$), which include information on the critical exponents, were obtained independently. As shown in Figs. 2(c) and 2(d), we found that both ω_0 and σ_0 greatly deviated from the behavior of the Gaussian form (dashed-dotted curve) below ε~0.04 (Here, $\varepsilon \equiv T/T_c$-1). Rather, they were similar to a characteristic behavior (solid curve) predicted by the Berezinskii-Kosterlitz-Thouless (BKT) theory [7], where ξ diverges as ~ exp[$b/\varepsilon^{1/2}$] in close to T_c (Here, b is a constant) and the dynamic critical exponent z is 2. Moreover, we found that the whole behavior of ω_0 and σ_0 could be successfully fit to an interpolation formula (dashed

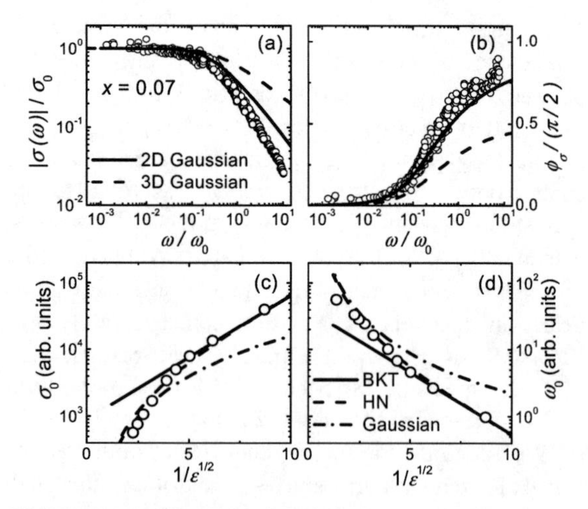

FIGURE 2. Scaling behavior of (a) the magnitude and (b) the phase of $\sigma_{FL}(\omega)$ for the same material. Solid and dashed lines are 2D and 3D Gaussian scaling functions, respectively. (c) σ_0 and (d) ω_0 as a function of $1/\varepsilon^{1/2}$. See the literature for details of three curves.

curve) between BKT and Gaussian forms, proposed by Halperin and Nelson [8]. These results strongly suggest a possibility that the BKT transition occurs in underdoped LSCO. Indeed, we also observed the strong frequency dependence of the phase stiffness energy, which was very similar to the previous report for underdoped $Bi_2Sr_2CaCu_2O_y$ [9]. Thus, all of the obtained results implied that the critical fluctuation of underdoped LSCO belongs to the 2D-XY universality class.

In the conventional critical phenomena, the critical fluctuation is considered to be universal. However, some recent theoretical models [10], which are based on the quantum criticality for competing ground states underlying the phase diagram of the high-T_c cuprates, suggest a possibility that the critical behavior can vary with hole doping. Thus, it is interesting to investigate whether the universality class in LSCO is changed by hole doping or not, which is now in progress.

CONCLUSION

We have measured the ac fluctuating microwave conductivity over a wide range of frequency in close to T_c for underdoped LSCO thin film with x=0.07. All of the obtained results suggested a possibility that the critical fluctuation of underdoped LSCO belongs to the 2D-XY universality class.

ACKNOWLEDGMENTS

This work was partly supported by the Grant-in-Aid for Scientific Research (13750005, 14340101 and 15760003) from the Ministry of Education, Science, Sports and Culture of Japan.

REFERENCES

1. D. S. Fisher, M. P. A. Fisher, and D. A. Huse, *Phys. Rev. B* **43**, 130-159 (1991.
2. I. Tsukada, *Phys. Rev. B* **70**, 174520 (2004).
3. H. Kitano *et al.*, *Physica C* **412-414**, 130 (2004).
4. A. T. Dorsey, Phys. Rev. B **43**, 7575 (1991).
5. J. C. Booth *et al.*, Phys. Rev. Lett. 77, 4438 (1996).
6. R. A. Wickham and A. T. Dorsey, Phys. Rev. B 61, 6945 (2000).
7. V. L. Berezinskii, *Sov. Phys. JETP,* **32**, 493 (1970); J. M. Kosterlitz and D. J. Thouless, *J. Phys. C* **6**, 1181 (1973).
8. B. I. Halperin and D. R. Nelson, J. Low Temp. Phys. **36**, 599 (1979).
9. J. Corson *et al.*, Nature **398**, 221 (1999).
10. S. Sachdev, *Science* **288**, 475 (2000).

Tunneling Spectroscopy on Underdoped $La_{2-x}Sr_xCuO_4$

T. Kato, T. Maruyama, T. Noguchi, T. Pyon, and H. Sakata

Department of Physics, Tokyo University of Science, 1-3 Kagurazaka, Shinjuku-ku, Tokyo 162-8601, Japan

Abstract. We report the results of low-temperature tunneling spectroscopy on underdoped $La_{1.94}Sr_{0.06}CuO_4$ single crystal with T_c = 14 K. The tunneling spectra have been obtained on the *ab* plane at 4.2 K indicating gap structure with the asymmetric V-shaped background. While we have observed the spectra with gap value Δ consistent with previous ARPES data of the underdoped samples, a few spectra exhibit almost the same magnitude of Δ observed on the optimally doped sample in our former tunneling study. This result suggests the existence of the strong gap inhomogeneity in under doped $La_{2-x}Sr_xCuO_4$.

Keywords: tunneling spectroscopy, La-based cuprates
PACS: 74.50+r, 74.72.Dn

INTRODUCTION

The electronic properties of the underdoped high-temperature superconductors (HTSCs) have been attracting much attention because of their unusual behavior such as pseudogap state [1], strong gap inhomogeneity [2,3], stripe order [4], and increasing gap magnitude against the suppression of T_c as doping decreases [5,6]. Scanning tunneling spectroscopy (STS) using low-temperature scanning tunneling microscopy (LT-STM) has been used for investigating low-energy electronic structure in the HTSCs. Since STM is a surface-sensitive technique, STM study have been performed mostly on $Bi_2Sr_2CaCu_2O_{8+\delta}$ (BSCCO), which has good cleavage plane. However, it is difficult to obtain highly over- or under-doped BSCCO single crystal because hole doping in BSCCO is achieved by non-stoichiometric oxygen atoms. In contrast to this, substituting cations in $La_{2-x}Sr_xCuO_4$ (LSCO) realizes wide-range doping control. However, since LSCO has no cleavage plane, it is unexplored material for LT-STM/STS experiments to date.

We have performed tunneling spectroscopy using LT-STM on underdoped LSCO single crystal. The target Sr composition *x*, i.e. hole concentration, is set to 0.06, which is just above the superconductor-insulator transition.

EXPERIMENT

The underdoped LSCO (x=0.06) single crystal used in this study was grown by traveling-solvent floating zone technique [7], and annealed in air at 900°C for 50 h then 500°C for 50 h. The samples for the measurements were cut out from the crystal after the determination of the crystallographic axes by Laue method. The superconducting transition temperature T_c determined by magnetization measurements was 14 K. The tunneling spectroscopy measurements were performed by a laboratory-build LT-STM/STS system. The sample was mounted in the LT-STM, cooled down to 4.2 K in pure helium gas, and then mechanically fractured perpendicular to the *c* axis in order to expose the surface. Electrochemically etched Pt-Ir wire was used as a STM tip, which was parallel to the *c* axis of the sample. Typical tunneling parameters for the tunneling spectroscopy measurements were 60 mV and 90 pA giving tunneling conductance of 1.5 nS. The experimental details have been described elsewhere [8,9].

RESULTS AND DISCUSSION

The surface topography observed by the LT-STM shows highly flat but rather narrow terraces on the exposed *ab* plane. The step height corresponds to a half of the unit cell height (~ 6–7 Å) reflecting the layered structure of LSCO along the *c* axis [9]. The

CP850, *Low Temperature Physics: 24th International Conference on Low Temperature Physics;*
edited by Y. Takano, S. P. Hershfield, S. O. Hill, P. J. Hirschfeld, and A. M. Goldman

tunneling spectroscopy measurements were performed on such terraces and we have collected over 20,000 tunneling spectra at different points far from step edges. Since coherence peaks were not apparent in most of the measured spectra as described below, we identified the energy giving the negative peak in the second derivative of the tunneling spectra as the gap value Δ.

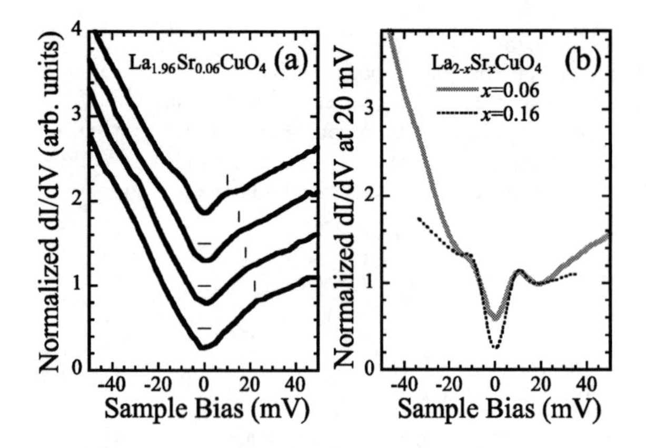

FIGURE 1. (a) Typical tunneling spectra of $La_{1.94}Sr_{0.06}CuO_4$ at 4.2 K sorted by the magnitude of Δ denoted by the vertical bars in the positive bias side. Each curve is shifted by 0.5 for clarity. (b) Representative spectrum having apparent coherence peaks. The dashed line is taken from x=0.16 sample for comparison [9].

The value of Δ in the obtained spectra was varied from location to location on the length scale of ~4 nm. Fig. 1a shows representatives of the spectra. The measured Δ distributed from 10 to 20 meV or more. The value of $\Delta \approx 20$ meV is in agreement with the previously reported ARPES (angle-resolved photoemission spectroscopy) data for the underdoped region [6], whereas the lower limit of $\Delta \approx 10$ meV is almost the same magnitude as that observed in the optimally doped sample in our tunneling study [9,10]. This result suggests the existence of the strong inhomogeneity of the electronic states in the underdoped LSCO.

While the asymmetric V-shaped background, which has been also observed in the optimally doped sample, seems to be common characteristics of the tunneling spectra of LSCO, there are obvious changes in the spectra in comparison with the optimally doped sample. First, the bias asymmetry becomes strong in the underdoped sample. Similar situation is seen in $Ca_{2-x}Na_xCuO_2Cl_2$ [11]. Second, the gap feature in the underdoped LSCO becomes very weak: Most of the spectra exhibit no apparent coherence peaks. Although there are a few spectra with coherence peaks, their zero-bias conductance are remarkably high as compared with that of the optimally doped sample as shown in Fig. 1b. This broad spectral characteristic is in agreement with ARPES results on the underdoped cuprates [6] indicating short quasiparticle lifetime.

CONCLUSION

In summary, low-temperature tunneling spectroscopy has been performed on an underdoped LSCO (x=0.06) single crystal at 4.2 K. The obtained spectra exhibit weak gap feature and strong bias asymmetry. The spectra having both small Δ comparable to that of optimum doping and clear coherence peaks were also observed suggesting presence of the strong inhomogeneity of the electronic states.

REFERENCES

1. Ch. Renner, B. Revaz, J. -Y. Genoud, K. Kadowaki, and Ø. Fischer, *Phys. Rev. Lett.* **80**, 149 (1998).
2. S. H. Pan, J. P. O'Neal, R. L. Badzey, C. Chamon, H. Ding, J. R. Engelbrecht, Z. Wang, H. Eisaki, S. Uchida, A. K. Guputa, K. –W. Ng, E. W. Hudson, K. M. Lang, and J. C. Davis, *Nature (London)* **413**, 282 (2001).
3. K. M. Lang, V. Madhavan, J. E. Hoffman, E. W. Hudson, H. Eisaki, S. Uchida, and J. C. Davis, *Nature (London)* **415**, 412 (2002).
4. J. M. Tranquada, B. J. Sternlieb, J. D. Axe, Y. Nakamura, and S. Uchida, *Nature (London)* **375**, 561 (1995).
5. N. Miyakawa, P. Guptasarma, J. F. Zasadzinski, D. G. Hinks, and K. E. Gray, *Phys. Rev. Lett.* **80**, 157 (1998).
6. A. Ino, C. Kim, M. Nakamura, T. Yoshida, T. Mizokawa, A. Fujimori, Z.-X. Shen, T. Kakeshita, H. Eisaki, and S. Uchida, *Phys. Rev. B* **65**, 094504 (2002).
7. I. Tanaka, K. Yamane, and H. Kojima, *J. Cryst. Growth* **96**, 711 (1989).
8. H. Sakata, M. Oosawa, K. Matsuba, N. Nishida, H. Takeya, and K. Hirata, *Phys. Rev. Lett.* **84**, 1583 (2000)
9. T. Kato, S. Okitsu, and H. Sakata, *Phys. Rev. B* **72**, 144518 (2005)
10. T. Kato, H. Morimoto, A. Katagiri, S. Okitsu, and H. Sakata, *Physica C* **392-396**, 221 (2003).
11. T. Hanaguri, C. Lupien, Y. Kohsaka, D.-H. Lee, M. Azuma, M. Takano, H. Takagi, and J. C. Davis, *Nature (London)* **430**, 1001 (2004).

Correlated Decrease between the Superconducting Volume Fraction and T_c and Possible Phase Separation in the Overdoped Regime of $La_{2-x}Sr_xCuO_4$

Y. Tanabe, T. Adachi, T. Noji, H. Sato and Y. Koike

Department of Applied Physics, Graduate School of Engineering, Tohoku University, 6-6-05, Aoba, Aramaki, Aoba-ku, Sendai 980-8579, Japan

Abstract. In order to prove the possible phase separation in the overdoped high-T_c superconductors, suggested from the muon-spin-relaxation measurements, the superconducting (SC) volume fraction has been investigated in detail in the overdoped regime of $La_{2-x}Sr_xCuO_4$. From measurements of the magnetic susceptibility, χ, using a single crystal in which the Sr concentration, x, continuously changes in the overdoped regime, both the absolute value of χ at 2 K, $|\chi_{2K}|$, on field cooling reflecting the SC volume fraction and T_c have been found to decrease with increasing x. Moreover, it appears that $|\chi_{2K}|$ has a roughly linear relation to T_c. It has been concluded that phase separation into SC and normal-state regions takes place in the overdoped regime of $La_{2-x}Sr_xCuO_4$.

Keywords: magnetic susceptibility, phase separation, $La_{2-x}Sr_xCuO_4$, overdoped regime,
PACS: 74.25.Ha, 74.62.Dh, 74.72.Dn

The inhomogeneity of the superconductivity in the CuO_2 planes is one of the recent central issues in the study of high-T_c superconductivity. Scanning-tunneling-microscopy (STM) experiments in the optimally doped $Bi_2Sr_2CaCu_2O_{8+\delta}$ have revealed that both the local density of states and the superconducting (SC) gap are spatially inhomogeneous [1]. However, STM is very sensitive to the surface electronic state. Therefore, the inhomogeneity in the CuO_2 plane is still controversial and it is necessary to confirm it from measurements reflecting the bulk nature of a sample.

In the overdoped regime of $Tl_2Ba_2CuO_{6+\delta}$, it has been found from the muon-spin-relaxation (μSR) maeasurements that the SC carrier density over the effective mass decreases with an increase in the hole concentration, suggesting possible microscopic phase separation into SC and normal-state regions [2]. From measurements of the magnetic susceptibility, χ, in $La_{2-x}Sr_xCuO_4$ (LSCO), on the other hand, it has been pointed out that the shielding volume fraction estimated from χ at 4.2 K on zero-field cooling using bulk samples is almost 100 % in the range $0.07 \leqq x \leqq 0.27$ [3]. In general, however, the shielding volume fraction is larger than the real SC volume fraction in the presence of inhomogeneity such as the microscopic phase separation.

In this paper, we have investigated the hole-concentration dependence of the SC volume fraction from the χ measurements on field cooling in the overdoped regime of LSCO [4].

A single-crystal rod with the Sr concentration, x, changing continuously in the overdoped regime was grown by the traveling-solvent floating-zone method. Several pieces of single crystal obtained by slicing the single-crystal rod were used for the χ measurements. The x value of each single-crystal piece was estimated by ICP. The details have been described elsewhere [4]. The χ measurements were carried out at low temperatures down to 2 K, using a superconducting quantum interference device magnetometer.

Figure 1 shows the temperature dependence of χ in a magnetic field of 10 Oe on field cooling for powdered or non-powdered samples of single-crystal LSCO pieces with $x = 0.198 - 0.273$, together with the data of polycrystalline powder samples of LSCO with $x = 0.18$, 0.20 and 0.22 [5]. The SC transition looks broad for the powdered samples, which is ascribed to the temperature dependence of the penetration depth. With increasing x, the absolute value of χ at 2 K, $|\chi_{2K}|$, as well as T_c decreases and becomes almost zero for $x > 0.273$. As for the data of non-powdered samples of

CP850, *Low Temperature Physics: 24th International Conference on Low Temperature Physics;* edited by Y. Takano, S. P. Hershfield, S. O. Hill, P. J. Hirschfeld, and A. M. Goldman

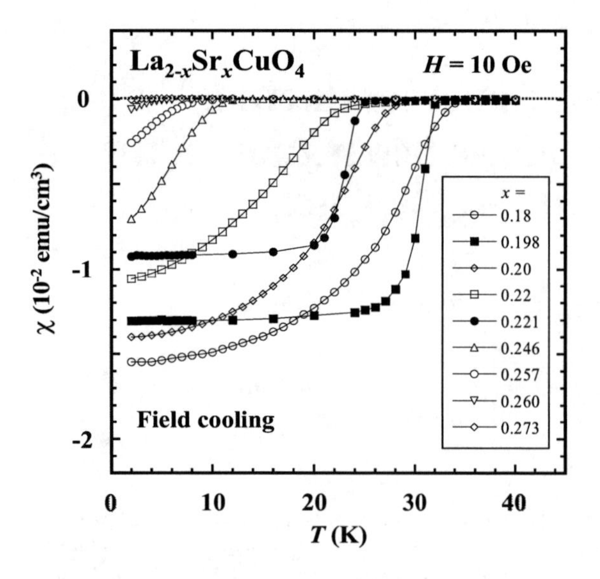

FIGURE 1. Temperature dependence of the magnetic susceptibility, χ, in a magnetic field of 10 Oe on field cooling for single-crystal $La_{2-x}Sr_xCuO_4$ pieces with $x = 0.198 - 0.273$. Open symbols indicate the data of powdered samples obtained from each single-crystal piece, while closed ones are the data of non-powdered samples of single-crystal pieces in a field parallel to the c axis. The data of polycrystalline powder samples with $x = 0.18$, 0.20, 0.22 are also plotted for reference [5].

$x = 0.198$ and 0.221, the SC transition is sharp owing to the good quality of the single-crystal pieces, and both T_c and the value of $|\chi_{2K}|$ decrease with increasing x [6]. These results mean the decrease of the SC volume fraction with increasing x in the overdoped regime of LSCO, strongly suggesting that phase separation into SC and normal-state regions takes place.

Figure 2 displays the plot of $|\chi_{2K}|$ vs T_c for powdered samples of single-crystal LSCO pieces. The data of the polycrystalline powder samples of LSCO are also plotted [5]. Both T_c and $|\chi_{2K}|$ seem to decrease cooperatively with increasing x and disappear at $x \sim 0.273$ in the overdoped regime of LSCO. Supposing that the microscopic phase separation suggested from the STM experiment occurred [1], the proximity effect due to the increase of the normal-state regions with increasing x would result in the decrease in T_c.

There are two possible origins of the phase separation in the overdoped regime. One is related to the decrease of the SC condensation energy with increasing x [7]. In this case, phase separation into the hole-poor SC region with $x \sim 0.19$ and the hole-rich normal Fermi-liquid state region with $x \sim 0.30$ will probably take place. The other is due to doping of holes into the Cu$3d$ orbital, producing free Cu spins

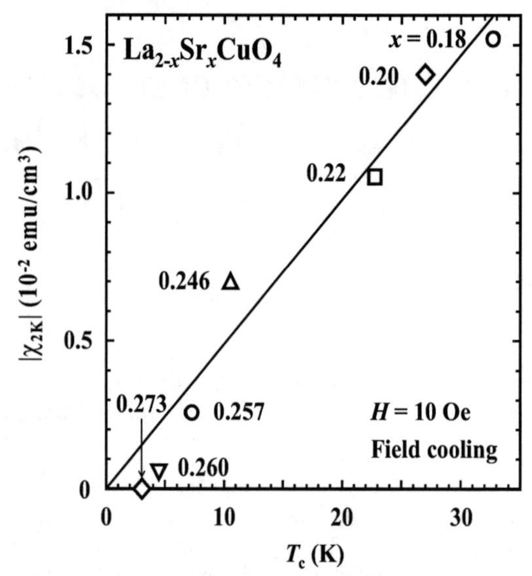

FIGURE 2. $|\chi_{2K}|$ vs T_c for powdered samples of single-crystal LSCO pieces. The data of polycrystalline powder samples with $x = 0.18$, 0.20, 0.22 are also plotted [5].

around the holes [8] and/or disturbing the antiferromagnetic correlation between Cu spins [9]. Both of them will bring about the local destruction of superconductivity around holes, generating normal-state regions in the SC sea.

In summary, it has been found from the χ measurements in the overdoped regime of LSCO that both T_c and $|\chi_{2K}|$ decrease with increasing x, strongly suggesting the phase separation into SC and normal-state regions.

This work was supported by the Iketani Science and Technology Foundation and also by a Grant-in-Aid for Scientific Research from the Ministry of Education, Culture, Sports, Science and Technology, Japan.

REFERENCES

1. S. H. Pan *et al.*, *Nature (London)* **413,** 282 (2001).
2. Y. J. Uemura, *Solid State Commun.* **120,** 347 (2001).
3. T. Nagano *et al.*, *Phys. Rev. B* **48,** 9689 (1993).
4. Y. Tanabe *et al.*, *J. Phys. Soc. Jpn.* **74,** 2893 (2005).
5. N. Oki, T. Adachi, S. Yairi and Y. Koike, unpublished.
6. In order to avoid the difference in the effect of the demagnetizing field between samples, we formed each single-crystal piece into the same rectangular shape within the error of ± 4 %.
7. T. Matsuzaki *et al.*, *J. Phys. Soc. Jpn.* **73,** 2232 (2004).
8. M. Oda *et al.*, *Physica C* **183,** 234 (1991).
9. S. Wakimoto *et al.*, *Phys. Rev. Lett.* **92,** 217004 (2004).

Buckling of the CuO_2 Plane in Single Crystals of La-Based High-T_C Cuprates Observed by NMR

Takayuki Goto[a], Masanori Ueda[a], Hidemitsu Sumikawa[a], Takao Suzuki[a], Masaki Fujita[b], Kazuyoshi Yamada[b], Tadashi Adachi[c], and Yoji Koike[c]

[a]*Faculty of Science and Technology, Sophia University, Tokyo 102-8554 Japan*
[b]*Institute for Materials Research, Tohoku University, Sendai 980-8577, Japan*
[c]*Graduate School of Engineering, Tohoku University, Sendai 980-8579, Japan*

Abstract. The buckling of CuO_2 plane in single crystals of La-based high-T_C cuprates LSCO (x=0.15) and LBCO (x=0.08) was directly observed by Cu-NMR. In both the cases, buckling patterns obtained by NMR disagree with those expected in the averaged structure at the vicinity of the structural phase transitions.

Keywords: NMR, local structure, La-based cuprate.
PACS: 74.72.Dn, 76.60.-k, 61.72.Hh

INTRODUCTION

La-based high-T_C cuprates show the structural phase transition from *I4/mmm* phase to *Cmca* at the temperature T_{d1}. Among them, $La_{2-x}Ba_xCuO_4$ (LBCO) and Nd-doped $La_{2-x}Sr_xCuO_4$ (LSCO) show the successive transition to $P4_2/ncm$ at still lower temperature T_{d2}[1]. These two structural phase transitions are associated with the change in the flatness of CuO_2 plane. In *I4/mmm* phase, the plane is flat, and CuO_6 octahedra stand vertically. In low temperature phases, there appear a characteristic buckling pattern in the plane. Especially, the pattern in the $P4_2/ncm$ phase is believed to pin the dynamically fluctuating stripe and stabilize it. This static stripe order suppresses the superconductivity as is well known in Nd-doped LSCO and LBCO[2,3].

However, there is a disagreement in reported experimental results on the bucking in the CuO_2 plane. Billinge *et al.* reported by the study on the pair distribution function of the neutron scattering that the CuO_6 octahedra in LBCO tilt randomly even in the *I4/mmm* phase[4]. They claim that the local structure is quite different from the averaged one. Some other reports by techniques of XAFS, XANES *etc.*, also support the existence of the local structure, while still others do not [5~9]. The purpose of this work is to investigate the existence of the local structure in single crystals by Cu-NMR, which is a local probe, and is expected to detect local structures. We also expect that this study will give a further insight to the role of the structure of CuO_2 plane in the stripe order.

EXPERIMENTAL

The single crystals of LSCO (x=0.15, T_C=37K) and LBCO (x=0.08, T_C=24K) were prepared by the conventional floating zone method [3,10]. Cu-NMR "angle-swept spectra" were obtained by recording the spin-echo amplitude while rotating a single crystal in a constant magnetic field.

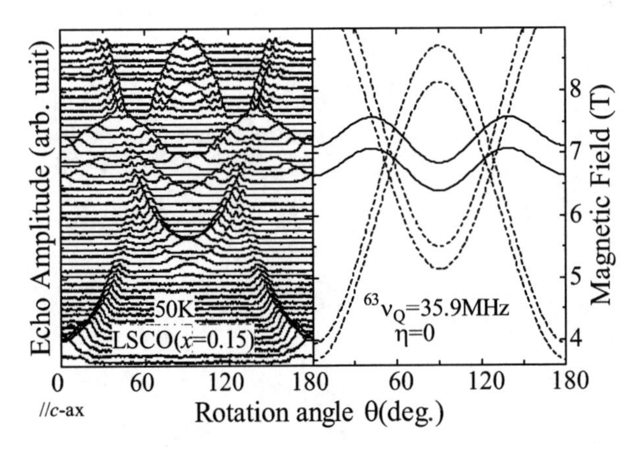

FIGURE 1. Angle-swept spectra of $^{63/65}$Cu-NMR in LSCO (left), and calculated curves with parameters of $^{63}\nu_Q$ = 35.9MHz, $K \approx 0.5\%$ and $\eta = 0$ (right). The field is tilted from c-axis to (cos29°, sin29°,0)$_t$ in the tetragonal notation.

CP850, *Low Temperature Physics: 24th International Conference on Low Temperature Physics;*
edited by Y. Takano, S. P. Hershfield, S. O. Hill, P. J. Hirschfeld, and A. M. Goldman

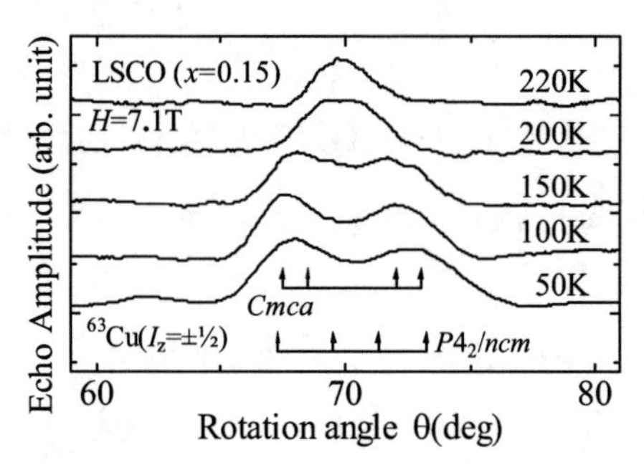

FIGURE 2. The temperature dependence of profile of angle-swept spectra in LSCO. Arrows show peak positions calculated assuming that the principal axis of the EFG tensor is tilted from c-axis by 3.5°.

RESULTS AND DISCUSSION

Figure 1 shows the rotational pattern of Cu-NMR spectra in LSCO. Each peak shows a slight split when the direction of the applied field is tilted from c-axis. We made a calculation of the quadrupolar shift with a second order perturbation assuming that the principal axis of the electric field gradient is parallel with the direction of the apical oxygen. In *Cmca* phase, there are the four tilting directions of the octahedron $(\pm1,\pm1,0)_t$, for the sample has a twin structure. In $P4_2/ncm$ phase, the tilting directions are $(\pm1,0,0)_t$ or $(0,\pm1,0)_t$. The calculated peak positions are shown by arrows in Fig. 2. One can see that *Cmca* with the tilting angle 3.5° reproduces the observed splitting at 50K. This indicates that the local structure observed by NMR is identical with the averaged structure which is reported to be *Cmca.*

When the temperature is raised, the split shrinks in the high temperature phase *I4/mmm.* Figure 3 shows the temperature dependence of the splitting width and the ultrasonic velocity. The latter is a good probe for the structural phase transition. In general, the lattice shows the softening around the transition temperature. In La-based cuprates, T_{d1} is determined from the

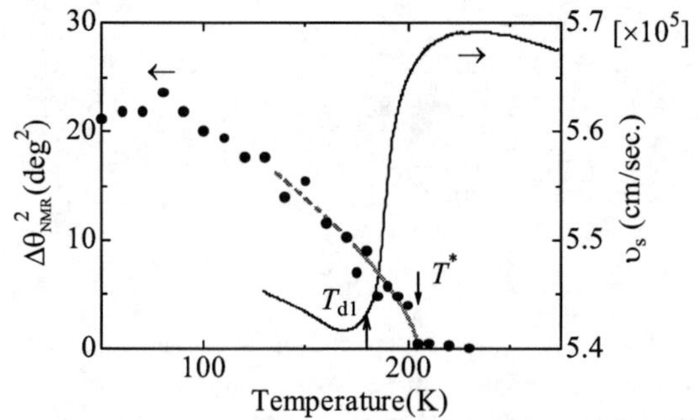

FIGURE 3. Temperature dependence of NMR-splitting width (solid-symbols), and the ultrasound velocity measured by longitudinal wave of 9MHz, C_{11} mode. The definition of T_{d1} is stated in the text. T^* is the onset of the peak splitting. The gray curve is a guide to eyes.

velocity data as the crossing point of linear extrapolations from the both sides of the transition[11]. Thus determined T_{d1} agrees with the onset of the orthorhombicity observed by X-ray diffraction. Note that onset of the NMR peak split is T^*=205K much higher than T_{d1}=180K. In the temperature region between T_{d1} and T^*, where the averaged structure is *I4/mmm*, what NMR sees is the local structure of *Cmca*. This local structure is static in time, because the characteristic time scale of NMR is very slow. As the temperature is decreased to T_{d1}, the static but non-periodic buckling pattern is first formed at T^*, and then it is aligned periodically at T_{d1} .

Finally, we show in Fig. 4 the rotational profile of ^{63}Cu-NMR in LBCO (x=0.08), T_{d2} of which is reported to be around 10K[12]. Calculated peak positions based on the local structure of $P4_2$/ncm and *Cmca* are shown by arrows. The observed profile of a single spired peak suggests that the local structure is likely to be the former up to 100K, while the averaged one is the latter. The existence of the buckling pattern of $P4_2/ncm$ can be related with the recently reported field-induced stripe order in LBCO with x slightly apart from 1/8[10].

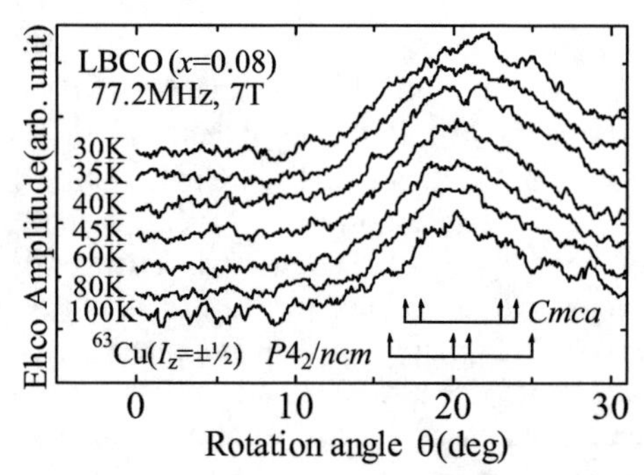

FIGURE 4. The temperature dependence of profile of angle-swept spectra in LBCO. The field is tilted from c-axis to $(\cos35°,\sin35°,0)_t$. Arrows show the calculated position of peaks expected for local structures of *Cmca* and $P4_2/ncm$ with a tilting angle of 4.5°

REFERENCES

1. J. D. Axe *et al.*, *Phys. Rev. Lett.* **62**, 2751–2754 (1989).
2. J. M. Tranquada *et al.*, *Nature* **375**, 561, (1995).
3. M. Fujita *et al.*, *Phys. Rev. B* **70**, 104517(2004), *ibid.* *B***66**, 184503 (2002).
4. S. J. L. Billinge *et al.*, *Phys. Rev. Lett.* **72**, 2282 (1994).
5. D. Haskel *et al.*, *Phys. Rev. Lett.* **76**, 439 (1996).
6. A. Bianconi *et al.*, *Phys. Rev. Lett.* **76**, 3412 (1996).
7. S.-W. Han *et al.*, *Phys. Rev. B* **66**, 094101 (2002).
8. C. Friedrich *et al.*, *Phys. Rev. B* **54**, R800–R803 (1996).
9. M. Braden *et al.*, *Phys. Rev. B* **63**, 140510 (2001).
10. T. Adachi *et al.*, *Phys. Rev. B* **71**, 104516 (2005).
11. T. Hanaguri, Dr. Thesis, Tohoku Univ. (1993).
12. T. Fukase *et al.*, *JJAP Series* 7, **213** (1992); *Physica B* **165&166**, 1289 (1990).

Muon Spin Relaxation Study of 1/8 Anomaly in High-T_c Superconductor $La_{2-x-y}Eu_ySr_xCuO_4$

J. Arai[a], S. Kaneko[a], T. Goko[a], S. Takeshita[a], K. Nishiyama[b] and K. Nagamine[b]

[a] *Department of Physics, Tokyo University of Science, Noda, Chiba 278-8510, Japan*
[b] *Institute of Materials Structure Science, KEK, Tsukuba, Ibaraki 305-0801, Japan*

Abstract. In order to investigate the origin of the so-called 1/8-anomaly in high-T_c La(Sr,Ba)-214, we have carried out the measurements of electric resistivity, magnetic susceptibility and muon spin relaxation (μSR) for the samples $La_{2-x-y}Eu_ySr_xCuO_4$. Eu(y=0.1)-substitution stabilizes a low temperature tetragonal (LTT) phase and consequently, around x=0.125 (1/8) the magnetic ordering temperature T_m shows the maximum value while the superconducting critical temperature T_c is most suppressed, indicating a strong competition between them, and over the whole x range T_c are entirely lowered. On the other hand, with increasing y up to y~0.05 at x=0.125 (1/8), T_m steeply increases and T_c rapidly decreases, indicating transition to LTT, but both the temperatures gradually decrease in LTT phase above y~0.05. In all Eu-substituted samples, the resisitivity upturn is observed at low temperature, implying carrier localization. The anomalous suppression of T_c in Eu-substituted LSCO can be attributed to the strong carrier localization in LTT phase and the subsequent peculiar magnetic order such as spin and charge stripe-order.

Keywords: high-T_c superconductivity, 1/8 anomaly, muon spin relaxation
PACS: 74-72-Dn; 76.75.+i:

The superconducting critical temperature T_c in $La_{2-x}M_xCuO_4$ (M=Ba, Sr) is known to be strongly suppressed for M=Ba (LBCO) but slightly suppressed for M=Sr (LSCO), which is called 1/8-anomaly and its origin has not been elucidated [1]. At low temperature, a structural transition from a low temperature orthorhombic (LTO) phase to a low temperature tetragonal (LTT) phase takes place in LBCO samples while in LSCO samples no structural transition occurs. Neutron diffraction experiments have revealed the existence of peculiar orders such as spin density wave (SDW) and charge density wave (CDW) in Nd-doped $La_{1.48}Nd_{0.4}Sr_{0.12}CuO_4$ which has LTT phase and 1/8 anomaly as well as in LBCO [2]. Thus, superconductivity is supposed to be closely related to the structures such as LTO and LTT phase, and its peculiar magnetic and charge orders. In order to clarify the relation among them, we carried out the measurements of electric resistivity, magnetic susceptibility and muon spin relaxation (μSR) in $La_{2-x-y}Eu_ySr_xCuO_4$.

The polycrystalline samples of $La_{2-x-y}Eu_ySr_xCuO_4$ were prepared by a usual solid state reaction method and were confirmed to be single phase by X-ray diffraction. The magnetic susceptibility was measured by a SQUID magnetometer. A μSR experiment was carried out at the Muon Science Laboratory of High Energy Accelerator Research Organization in Japan.

Figs.1(a) and (b) show the representative μ-spin relaxation curves for x=0.125 (1/8) and x=0.150 in the samples with y=0.1. The oscillation can be observed

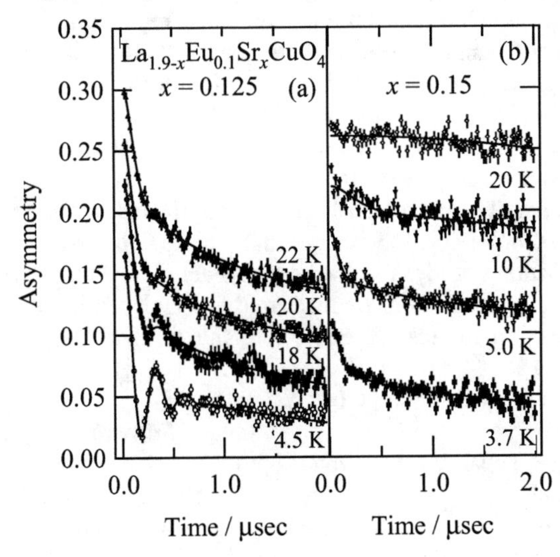

FIGURE 1. μ-spin relaxation curves for (a) x=0.125 (1/8) and (b) x=0.150 in the samples with y=0.1.

CP850, *Low Temperature Physics: 24th International Conference on Low Temperature Physics;*
edited by Y. Takano, S. P. Hershfield, S. O. Hill, P. J. Hirschfeld, and A. M. Goldman

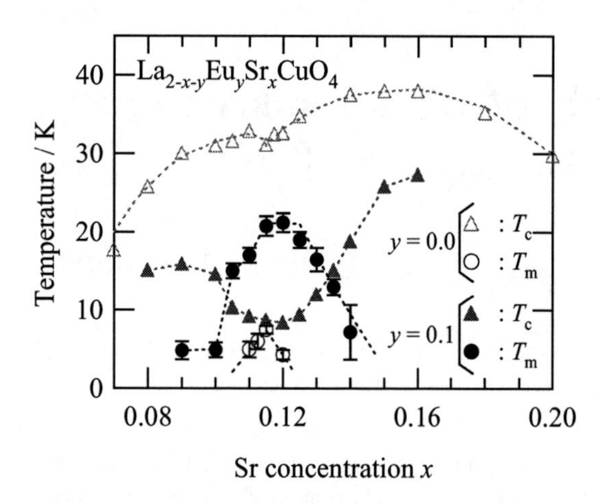

FIGURE 2. Sr(x) concentration dependence of T_m and T_c for Eu-substituted (y=0.1) samples and LSCO (y=0.0).

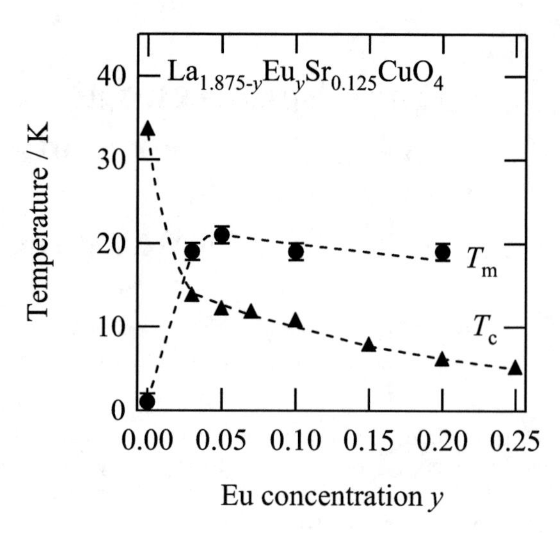

FIGURE 3. Eu(y) dependence of T_m and T_c for x=0.125 (1/8).

below about 20 K in (a) but no oscillation even at 3.7 K in (b). We defined the magnetic ordering temperature T_m as the onset temperature where an oscillation starts. In Fig.2 the temperatures, T_m and T_c for the sample with y=0.1 are plotted against Sr concentration (x), where T_m and T_c for y=0 (LSCO) are also plotted for a comparison. A phase diagram of Eu-substituted LSCO has been reported; LTT phase for y=0.1 and LTO for y=0 [3].

In Fig.2, around x=0.125 (1/8) T_m shows maximum value and T_c is most suppressed, indicating a strong competition between them while in LSCO (y=0) with LTO phase a slight T_c-suppression and lower T_m can be seen. Here, it should be noted that in the optimally doped sample (x=0.15) with y=0.1 there exists no magnetic order and T_c is rather high despite being in LTT-phase but is suppressed from that of LSCO (y=0). And furthermore, a resisitivity upturn can be observed before the onset of superconductivity in the temperature dependence of the electric resistivity for all samples with y=0.1, which is not shown here, indicating carrier localization. To clarify the effect of the Eu-substitution, we measured Eu (y) dependence of the temperatures, T_c and T_m in the samples keeping x=0.125 (1/8), which is shown in Fig.3. With increasing y up to y~0.05, T_m and T_c change drastically but decrease slowly above y=0.05 and furthermore, the more enhanced resisitivity-upturn is also observed, which is not shown here. This indicates that the development of LTT phase around y=0.05 destroys the superconductivity and dramatically give rise to the magnetic order, but in the LTT phase above y~0.05 the increase of y hardly affect T_c and T_m. Fig.2 reveals that in LTT phase caused by Eu-substitution, T_c is generally suppressed over whole x range compared with that of LSCO (y=0) even above x=0.15 where no magnetic order is present. As well known, CuO_2 plane in LTT structure is not electronically uniform because of strong buckling due to the tilt of CuO_6 octahedra and furthermore, potential randomness due to the difference of ion radius between La and Eu, which tends to localize hole carriers. As can be seen in Fig.2, a slight T_c-suppression and magnetic order can be observed around x=1/8 even in LSCO without LTT phase and furthermore, we have reported that the increase of randomness on CuO_2 plane due to a slight Zn-substitution enhances T_m and suppresses T_c severely [4, 5]. Therefore, the hole concentration 1/8 is supposed to be essential for the peculiar order such as spin and charge stripe-order, and the carrier localization caused by the randomness on CuO_2 plane works to enhance further the magnetic order. In conclusion, the anomalous suppression of T_c in Eu-substituted LSCO can be attributed to the strong carrier localization in LTT phase and the subsequent peculiar magnetic order such as spin and charge stripe-order.

REFERENCES

1. A. R Moodennbaugh et al., *Phys. Rev. B* **38**, 4596 (1988).
2. J. M. Tranquada et al., *Nature* **375**, 561 (1995).
3. R. Werner et al., *Phys. Rev. B* **62**, 3705 (2000).
4. J. Arai et al., *Physica B* **289~290**, 347 (2000).
5. T. Ishiguro et al., *J. Mag. Mag. Mat.* **226~230**, 275 (2001).

Scanning Tunneling Microscopy and Spectroscopy on $La_{1.68}Nd_{0.2}Sr_{0.12}CuO_4$

Ryo Saito, Tetsuro Noguchi, Tesu Pyon, Takuya Kato, and Hideaki Sakata

Department of Physics, Tokyo University of Science, 1-3 Kagurazaka, Shinjuku-ku, Tokyo 162-8601, Japan

Abstract. We report the results of scanning tunneling microscopy and spectroscopy measurements on $La_{1.68}Nd_{0.2}Sr_{0.12}CuO_4$ single crystal with T_c = 13 K. A topographic image has revealed a narrow terrace structure on a surface prepared at 4.2 K. An observed superconducting gap value 2Δ has been distributed from 14 meV to 24 meV. A mean value of 2Δ is 18 meV, which corresponds to $2\Delta/k_BT_c \approx 16$.

Keywords: Scanning tunneling microscopy, Tunneling spectroscopy, La-based cuprates
PACS: 74.50.+r, 68.37.Ef, 74.72.Dn

INTRODUCTION

Since static charge/spin order was discovered in $La_{2-x-y}Nd_ySr_xCuO_4$ (LNSCO) by elastic neutron scattering [1], there has been increasing interest in its electronic states. One of the most powerful techniques to investigate the electronic states of a sample is scanning tunneling microscopy and spectroscopy (STM/STS), which provides both a surface structure and local density of states of the sample. A large number of STM/STS measurements on Bi- and Y-based high-T_c superconductors have been performed and obtained an atomic image of the surface structure and a superconducting energy gap [2-4]. On the other hand, there have been few reports of the STM/STS study on $La_{2-x}Sr_xCuO_4$ (LSCO) family [5] because of a difficulty in preparing a clean surface required for the STM/STS measurements. In this paper, we report the results of STM/STS measurements on $La_{1.68}Nd_{0.2}Sr_{0.12}CuO_4$ single crystal whose surface was prepared by cracking of the sample at 4.2 K just before the measurements. We have succeeded in obtaining a surface topography. An observed superconducting gap value was distributed from 14 meV to 24 meV. A mean gap value was 18 meV, which corresponds to $2\Delta/k_BT_c \approx 16$.

EXPERIMENT

$La_{1.68}Nd_{0.2}Sr_{0.12}CuO_4$ single crystal was grown by a traveling-solvent-floating-zone method [6]. A superconducting transition temperature T_c was determined to be 13 K from magnetization measurements by SQUID. STM/STS measurements were performed by using a laboratory built scanning tunneling microscope. To obtain a clean and fresh surface required for the STM/STS measurements, a sample of 1.4×1.4×4 mm^3 was cracked to expose the *ab*-plane at 4.2 K in helium gas just before the measurements. Chemically etched Au wire was used as a STM tip, which is perpendicular to the exposed *ab*-plane of the sample. STM measurements were performed in constant current mode. Typical tunneling parameters are tunneling current I of 0.07 nA and sample bias voltage V of 0.5 V, thus tunneling conductance G is 0.14 nS. Tunneling spectra were obtained by numerical differentiation of measured current-voltage (I-V) curves.

RESULTS AND DISSCUSSION

Fig. 1 shows a typical STM image of the *ab*-plane, which is exposed by cracking of the sample at 4.2 K. Many steps and narrow terraces can be seen. Although we have not identified the surface layer because of lack of atomic resolution, each terrace is thought to be La (or Nd) O or CuO_2 layer. This feature is sometimes destroyed by scan with high-tunneling conductance $G \approx 1$ nS. Similar terrace structure has been observed in our previous study on $La_{1.84}Sr_{0.16}CuO_4$ single crystal [5]. This structure is common characteristic of the cracked surface of LSCO family.

CP850, *Low Temperature Physics: 24th International Conference on Low Temperature Physics;*
edited by Y. Takano, S. P. Hershfield, S. O. Hill, P. J. Hirschfeld, and A. M. Goldman

FIGURE 1. A typical STM image of a cracked surface of LNSCO at 4.2 K (8.8×8.8 nm^2). Sample bias voltage is 444 mV and tunneling current is 0.070 nA. Many steps and narrow terraces can be seen.

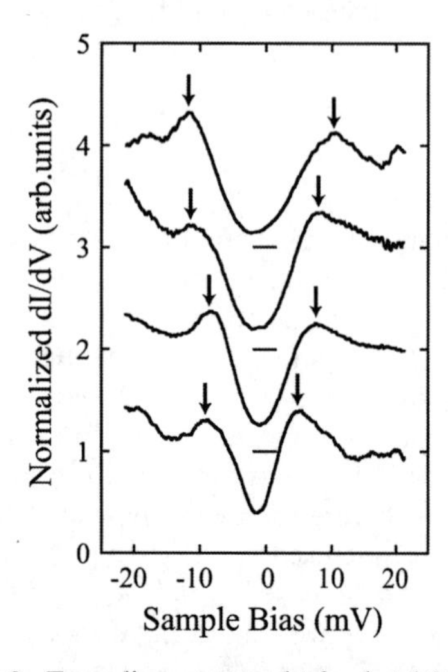

FIGURE 2. Tunneling spectra obtained at 4.2 K at different locations on the surface. Each spectrum is normalized by the value of dI/dV at sample bias voltage V of 20 mV and shifted vertically for clarity. The arrows indicate positions of coherence peaks. Superconducting gap value 2Δ is distributed from 14 meV (bottom) to 24 meV (top).

Fig. 2 shows representative tunneling spectra, which were taken at a number of points chosen randomly on the surface. Each spectrum is normalized by the value of dI/dV at sample bias voltage of 20 mV. These spectra exhibit clear coherence peaks and an asymmetric background conductance, which is slightly larger value at negative bias than the value at positive bias. This asymmetric background conductance similarly observed in those of $La_{1.84}Sr_{0.16}CuO_4$ [5]. Superconducting gap value 2Δ defined as a voltage between the coherence peaks is distributed from 14 meV (bottom of Fig. 2) to 24 meV (top of Fig. 2). A mean value of 2Δ is 18 meV, which corresponds to $2\Delta/k_BT_c \approx 16$. This is much larger than that of LSCO (x = 0.16, $2\Delta/k_BT_c \approx 4.8$) [5].

The origin of this distribution of the gap value is not clear at the present time. There are two possible explanations for the distribution of this superconducting gap value: an inhomogeneity of the electronic states observed in Bi2212 and LSCO [7-9] or a spatial change of the electronic states due to a charge order. To investigate the origin of this distribution, further measurements of spatial variation of tunneling spectra with an atomic resolution are required.

REFERENCES

1. J. M. Tranquada, B. J. Sternlieb, J. D. Axe, Y. Nakamura, and S. Uchida, *Nature* **375**, 561 (1995).
2. H. L. Edwards, J. T. Markert, and A. L. de Lozanne, *Phys. Rev. Lett.* **69**, 2967 (1992).
3. Ch. Renner and Ø. Fischer, *Phys. Rev. B.* **51**, 9208 (1995).
4. S. Kaneko, N. Nishida, K. Mochiku, and K. Kadowaki, *Physica C* **298**, 105 (1998).
5. T. Kato, H. Morimoto, A. Katagiri, S. Okitsu, and H. Sakata, *Physica C* **392-396**, 221 (2003).
6. Y. Nakamura and S. Uchida, *Phys. Rev. B* **46**, 5841 (1992).
7. S. H. Pan, J. P. O'Neal, R. L. Badzey, C. Chamon, H. Ding, J. R. Engelbrecht, Z. Wang, H. Eisaki, S. Uchida, A. K. Gupta, K. –W. Ng, E. W. Hudson, K. M. Lang, and J. C. Devis, *Nature* **413**, 282 (2001).
8. K. M. Lang, V. Madhavan, J. E. Hoffman, E. W. Hudson, H. Eisaki, S. Uchida, and J. C. Davis, *Nature* **415**, 412 (2002).
9. T. Kato, S. Okitsu, and H. Sakata, to be published.

Pressure Effect on Superconductivity and Magnetic Order in $La_{2-x}Ba_xCuO_4$ with x = 0.135

T. Goko*, K. H. Satoh*, S. Takeshita*, J. Arai*, W. Higemoto†, K. Nishiyama** and K. Nagamine**

*Department of Physics, Tokyo University of Science, Noda, Chiba 278-8510, Japan
†Advanced Science Research Center, JAERI, Tokai, Ibaraki 319-1195, Japan
**Institute of Materials Structure Science, KEK, Tsukuba, Ibaraki 305-0801, Japan

Abstract. We have performed muon spin relaxation (μSR), resistivity and magnetic susceptibility measurements under high pressure for $La_{2-x}Ba_xCuO_4$ with $x = 0.135$. An oscillation component is observed in the μSR spectra at low temperatures under ambient pressure and pressure of 1.1 GPa. This result suggests that the low-temperature tetragonal (LTT) structure is dispensable for the appearance of the magnetic order, since several experimental results indicate that the LTT structure vanishes at 1.1 GPa. The superconducting temperature T_c increases drastically with increasing pressure, while the magnetic ordering temperature T_m hardly depends on pressure. The suppression of T_c therefore has no correlation with T_m. However, by applying pressure, the magnetic ordering region is replaced by the superconducting region. We speculate that the suppression of T_c correlates with the magnetic volume fraction.

Keywords: high-T_c superconductor; 1/8 problem; pressure effect; muon spin relaxation
PACS: 74.72.Dn; 76.75.+i; 74.62.Fj

An anomalous suppression of the superconducting transition temperature T_c is observed in a narrow range of $x \sim 1/8$ for $La_{2-x}M_xCuO_4$ ($M =$ Ba, Sr), which is called "1/8 problem". This is one of the most important issues to understand the relation between superconductivity and magnetism in high-T_c cuprates. Structural instability is inherent in La-214 compounds. With decreasing temperature, La-214 compounds undergo a well-known phase transition from a high-temperature tetragonal structure to a low-temperature orthorhombic (LTO) structure except for the overdoped region. In addition, $La_{2-x}Ba_xCuO_4$ (LBCO) displays a second transition to a low-temperature tetragonal (LTT) structure around $x = 1/8$ at low temperatures. Because superconductivity is strongly suppressed in LBCO around $x = 0.125$ and slightly suppressed in $La_{2-x}Sr_xCuO_4$ (LSCO) around $x = 0.115$, it is widely believed that the LTT structure plays an important role in suppressing superconductivity around $x \sim 1/8$. On the other hand, in La-214 compounds around $x \sim 1/8$, a SDW-like magnetic order has been observed by neutron scattering experiments and muon spin relaxation (μSR) experiments. These results suggest the close relation between the suppression of superconductivity and the appearance of magnetic order.

In oder to clarify the relation among superconductivity, the magnetic order and the LTT structure, we have performed zero-field μSR, resistivity and magnetic susceptibility measurements under high pressure in LBCO with $x = 0.135$, which is a little away from the hole concentration of 1/8. The LTT phase of LBCO with $x = 0.125$ is suppressed by applying pressure up to $\sim$ 0.6 GPa [1]. In La-214 compounds, the carrier concentration hardly depends on pressure [2] in contrast to other high-T_c cuprates. Then we can suppress the LTT structure by applying pressure without changing the carrier concentration.

The polycrystalline sample of LBCO with $x = 0.135$ was prepared by a solid-state reaction method. To increase the homogeneity of Ba concentration, we sintered the sample at 1100 °C for 140 hours. We carried out resistivity measurements under pressure up to 1.4 GPa and magnetization measurements under pressure up to 0.8 GPa. In the μSR measurement, hydrostatic pressure up to 1.1 GPa was generated by using a piston-cylinder type pressure cell made of a non-magnetic alloy Ni-Co-Cr-Mo (MP35N). The dimension of the cylindrical sample is about $\phi 7 \times 10$ mm^3, which is much smaller than that in conventional μSR measurement. To increase the ratio of the sample signal to the background, we used dc muon beam and the defining counter which is as small as the sample. We performed the zero-field μSR measurement under high pressure by using decay muons with high momentum of $\sim$ 100 MeV/c at the M9B beam line at TRIUMF.

In Fig. 1, the values of T_c determined by zero resistivity are plotted as a function of pressure for LBCO with $x = 0.135$. With increasing pressure, T_c increases linearly with an initial rate of $dT_c/dP =$ +23.7 K/GPa. At $P_d \sim 1.0$ GPa, the pressure dependence of T_c changes to a small rate of +2.1 K/GPa. The value of dT_c/dP above P_d

CP850, *Low Temperature Physics: 24th International Conference on Low Temperature Physics;*
edited by Y. Takano, S. P. Hershfield, S. O. Hill, P. J. Hirschfeld, and A. M. Goldman

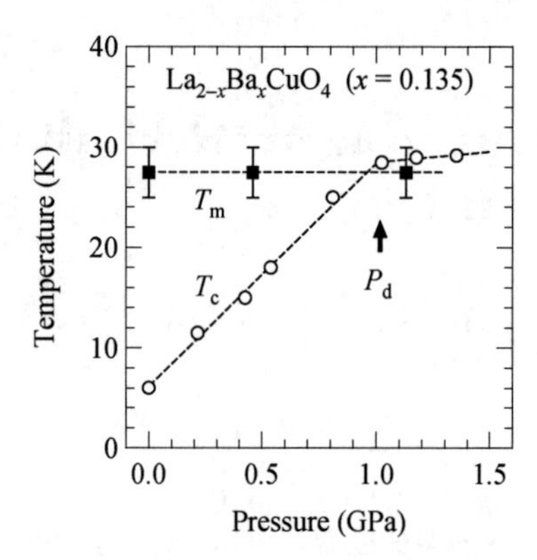

FIGURE 1. Pressure dependence of magnetic ordering temperature T_m and superconducting temperature T_c.

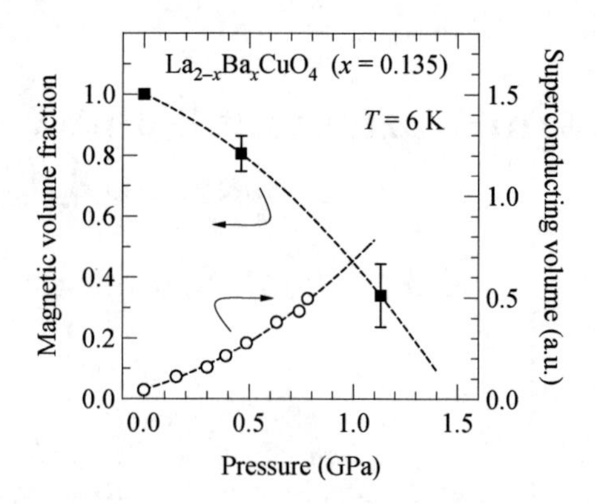

FIGURE 3. Pressure dependence of magnetic and superconducting volume at 6 K.

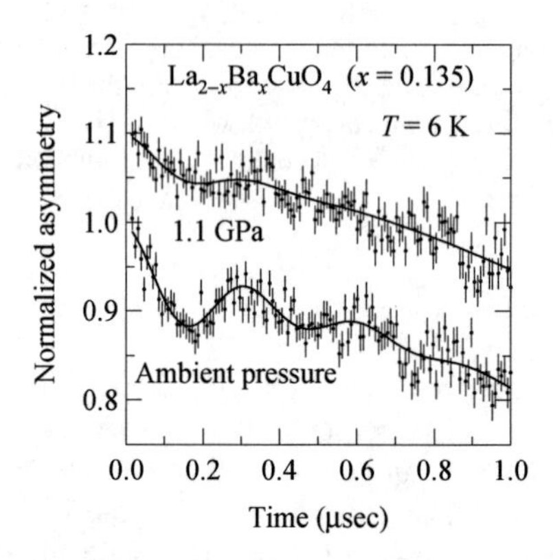

FIGURE 2. ZF-μSR time spectra under ambient pressure and 1.1 GPa at 6 K. The spectrum for 1.1 GPa is vertically offset for clarify.

is nearly equal to that in the LTO phase of LSCO [3], suggesting that P_d corresponds to the LTT-LTO phase transition pressure. Moreover, P_d agrees with the pressure where the semiconductor-like upturn observed in the resistivity curve of the LTT phase disappears. The abrupt increase in T_c below P_d therefore is attributed to the suppression of the LTT structure. The value of dT_c/dP below P_d is much larger than that for LBCO with $x = 1/8$, although P_d's have similar values for both samples. Furthermore, by applying pressure above P_d, T_c for $x = 0.135$ exhibits almost complete recovery, while T_c for $x = 1/8$ remains suppressed [3]. These results indicate that the 1/8 hole concentration is essential factor in the suppression of T_c and the LTT structure plays a role to enhance the suppression, which is similar to the Zn substitution effect on LSCO.

Figure 2 shows the μSR spectra for LBCO with $x = 0.135$ under ambient pressure ($< P_d$) and 1.1 GPa ($> P_d$) at $T = 6$ K. These signals include contributions from the sample and the pressure cell. The ratio of the sample signal to the background is roughly 1 : 10. The slow relaxation observed over a wide temperature range is due to the pressure cell and is reproduced by the Gaussian function with $\sigma \sim 0.5$ μsec^{-1}. An oscillation component is observed at low temperatures under both pressures. This observation means that the LTT structure is dispensable for the appearance of the magnetic order in LBCO with $x = 0.135$.

The pressure dependence of the magnetic ordering temperature T_m determined by the appearance of the oscillation component is shown in Fig. 1. With increasing pressure, T_m hardly changes, although T_c drastically increases. Similar behavior is observed in LBCO with $x = 1/8$ [4]. We therefore conclude that the suppression of T_c has no correlation with T_m. This is quite different from the relation between T_c and T_m seen in the hole concentration dependence in LBCO and LSCO.

In order to estimate the magnetic volume fraction, we analyzed the μSR time spectra. The details are described in Ref. [4]. The magnetic volume fraction is reduced from $\sim$ 100% to $\sim$ 30% at 6 K by applying pressure up to 1.1 GPa, as shown in Fig. 3. On the other hand, the superconducting volume estimated from the magnetic susceptibility increases steadily by applying pressure. This result indicates that the magnetic ordering region is replaced by the superconducting region by applying pressure. We speculate that the suppression of T_c correlates with the magnetic volume fraction rather than T_m.

REFERENCES

1. S. Katano, S. Funahashi, N. Môri, Y. Ueda, and J. A. Fernandez-Baca, *Phys. Rev. B*, **48**, 6569 (1993).
2. N. Tanahashi, Y. Iye, T. Tamegai, C. Murayama, N. Môri, S. Yomo, N. Okazaki, and K. Kitazawa, *Jpn. J. Appl. Phys.*, **28**, L762 (1989).
3. N. Yamada, and M. Ido, *Physica C*, **203**, 240 (1992).
4. T. Goko, K. H. Satoh, S. Takeshita, J. Arai, W. Higemoto, K. Nishiyama, and K. Nagamime, *to be published* (2005).

In-Plane Electrical Resistivity under Strong Magnetic Fields up to 27 T in $La_{2-x}Ba_xCuO_4$ and $La_{2-x}Sr_xCuO_4$ around $x = 1/8$

T. Adachi[1], T. Kawamata[1], Y. Koike[1], K. Kudo[2], T. Sasaki[2] and N. Kobayashi[2]

[1] *Department of Applied Physics, Tohoku University, 6-6-05, Aoba, Aramaki, Sendai 980-8579, Japan*
[2] *Institute for Materials Research (IMR), Tohoku University, Katahira 2-1-1, Sendai 980-8577, Japan*

Abstract. Magnetic-field effects on the so-called charge-spin stripe order, namely, the charge stripe order and spin stripe order in the La-214 system have been investigated from the in-plane electrical-resistivity, ρ_{ab}, measurements. In $La_{2-x}Ba_xCuO_4$ (LBCO) with $x = 0.10$ and $La_{2-x}Sr_xCuO_4$ with $x = 0.115$ where the incommensurate charge peaks are weak and unobservable in zero field from the elastic neutron-scattering measurements, respectively, the normal-state value of ρ_{ab} at low temperatures markedly increases with increasing field up to 27 T. As for $x = 0.11$ in LBCO where the charge stripe order is fairly stabilized in zero field, the increase in ρ_{ab} with increasing field is negligibly small. In conclusion, the magnitude of the increase in ρ_{ab} by the application of magnetic field is well correlated with the stability of the charge stripe order in zero field for the La-214 system around $x = 1/8$. Our understanding is as follows. When the charge-spin stripe order is not fully stable in zero field, magnetic field operates to stabilize the charge-spin stripe order. The value of ρ_{ab} increases with increasing field depending on the stability of the charge stripe order.

Keywords: stripe order, magnetic-field effect, La-based high-T_c superconductor, electrical resistivity
PACS: 74.25.Fy, 74.62.Dh, 74.72.Dn

In recent years, the magnetic-field effects on the so-called charge-spin stripe order, namely, the charge stripe order and spin stripe order in the La-214 system have been one of the most interesting issues. Elastic neutron-scattering measurements in magnetic fields for $La_{2-x}Sr_xCuO_4$ (LSCO) with $x = 0.10$[1] have revealed the enhancement of the incommensurate (IC) magnetic peaks corresponding to the spin stripe order by the application of magnetic field. On the other hand, the enhancement is quite small for $x = 0.12$,[2] which is believed to be due to the fairly good stability of the spin stripe order even in zero field at $x \sim 1/8$. As for the magnetic-field effect on the charge stripe order, no enhancement of the IC charge peaks is observed for $La_{1.6-x}Nd_{0.4}Sr_xCuO_4$ with $x = 0.15$[3] where the charge-spin stripe order is stabilized on the tetragonal low-temperature (TLT) structure (space group: $P4_2/ncm$).

In order to address this issue, we have investigated the in-plane electrical resistivity, ρ_{ab}, for $La_{2-x}Ba_xCuO_4$ (LBCO) with $x = 0.08$, 0.10 and 0.11 in magnetic fields parallel to the c axis up to 15 T.[4] It has been found that the superconducting transition curve shows a broadening in magnetic fields for $x = 0.08$, while it shifts toward the low-temperature side in parallel with increasing field for $x = 0.11$ where the charge-spin stripe order is formed at low temperatures even in zero field.[5] As for $x = 0.10$, where the intensity of the IC charge peaks is weak but that of the IC magnetic peaks is almost the same as that in $x \sim 1/8$,[5] the broadening is observed in low fields and it changes to the parallel shift in high fields above 9 T. Moreover, the normal-state value of ρ_{ab} at low temperatures markedly increases with increasing field up to 15 T. It has been inferred that these pronounced features of $x = 0.10$ are due to the magnetic-field-induced stabilization of the charge stripe order.

In this paper, we have expanded the applied magnetic field up to 27 T in the ρ_{ab} measurements of LBCO with $x = 0.08$, 0.10, 0.11 and also measured ρ_{ab} for LSCO with $x = 0.115$ where the superconductivity is a little suppressed, in order to clarify the relation between the magnetic field and the stability of the charge stripe order.

Single crystals of LBCO and LSCO were grown by the traveling-solvent floating-zone method.[6] The ρ_{ab} measurements were carried out up to 27 T using a hybrid magnet at the High Field Laboratory for Superconducting Materials (HFLSM), IMR, Tohoku University.

CP850, *Low Temperature Physics: 24th International Conference on Low Temperature Physics;*
edited by Y. Takano, S. P. Hershfield, S. O. Hill, P. J. Hirschfeld, and A. M. Goldman

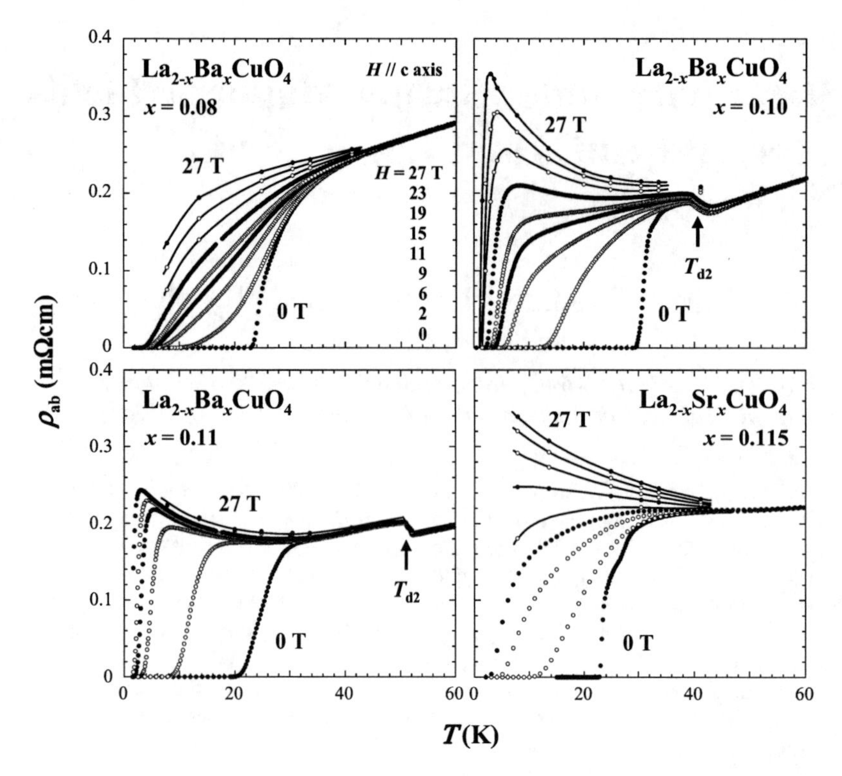

FIGURE 1. Temperature dependence of the in-plane electrical resistivity, ρ_{ab}, in various magnetic fields parallel to the c axis for $La_{2-x}Ba_xCuO_4$ with x = 0.08, 0.10, 0.11 and $La_{2-x}Sr_xCuO_4$ with x = 0.115. The temperature T_{d2} corresponds to the structural phase-transition temperature between the orthorhombic mid-temperature (space group: *Bmab*) and tetragonal low-temperature ($P4_2/ncm$) phases. Solid lines are to guide the reader's eye.

Figure 1 displays the temperature dependence of ρ_{ab} in various magnetic fields for LBCO with x = 0.08, 0.10, 0.11 and LSCO with x = 0.115. For x = 0.08, the broadening continues up to 27 T, which is an usual behavior in the underdoped high-T_c cuprates with a huge upper critical field, H_{c2}. For x = 0.10, the normal-state value of ρ_{ab} at low temperatures continues increasing up to 27 T and ρ_{ab} exhibits an insulating behavior in high fields. These suggest that the charge stripe order is further stabilized by the application of magnetic field up to 27 T. For x = 0.11, on the contrary, the increase of ρ_{ab} with increasing field is negligibly small up to 27 T. That is, ρ_{ab} seems stiff to the application of magnetic field, indicating the nearly perfect stabilization of the charge stripe order even in zero field at $x \sim 1/8$.

As for LSCO with x = 0.115, the broadening is observed at 2 T and changes to be like a parallel shift at 9 T. Moreover, ρ_{ab} at low temperatures increases with increasing field and shows an insulating behavior in high fields. These suggest that the magnetic-field-induced localization of carriers takes place also for LSCO around x = 1/8.

The behavior of ρ_{ab} for LSCO with x = 0.115 in each field appears to be quite similar to that for LBCO with x = 0.10. Elastic neutron-scattering measurements for LSCO with $x \sim 1/8$ in the orthorhombic mid-temperature phase (space group: *Bmab*) have revealed that no IC charge peaks are observed in zero field, though the spin stripe order is fairly stabilized in zero field.[7] Accordingly, it is possible that the charge stripe order is stabilized by the application of magnetic field also for LSCO with x = 0.115. The similarity of the response of ρ_{ab} to magnetic field between LBCO with x = 0.10 and LSCO with x = 0.115 implies the similar stability of the charge stripe order in zero field, as suggested from the thermal-conductivity measurements also.[8,9]

In conclusion, the present results suggest that the magnitude of the increase in ρ_{ab} by the application of magnetic field is well correlated with the stability of the charge stripe order in zero field for the La-214 system around x = 1/8. Our understanding is as follows. When the charge-spin stripe order is not fully stable in zero field, magnetic field operates to stabilize the charge-spin stripe order. The value of ρ_{ab} increases with increasing field depending on the stability of the charge stripe order.

REFERENCES

1. B. Lake *et al.*, *Nature (London)* **415**, 299 (2002).
2. S. Katano *et al.*, *Phys. Rev. B* **62**, R14677 (2000).
3. S. Wakimoto *et al.*, *Phys. Rev. B* **67**, 184419 (2003).
4. T. Adachi *et al.*, *Phys. Rev. B* **71**, 104516 (2005).
5. M. Fujita *et al.*, *Physica C* **426-431**, 257 (2005).
6. T. Adachi *et al.*, *Phys. Rev. B* **64**, 144524 (2001).
7. H. Kimura *et al.*, *Phys. Rev. B* **59**, 6517 (1999).
8. K. Kudo *et al.*, *Phys. Rev. B* **70**, 014503 (2004).
9. T. Kawamata *et al.*, *Proc. of LT24* (in this issue).

Field-Induced Magnetic Order and Thermal Conductivity in $La_{2-x}Ba_xCuO_4$

T. Kawamata*, N. Takahashi*, M. Yamazaki*, T. Adachi*, T. Manabe*, T. Noji*, Y. Koike*, K. Kudo† and N. Kobayashi†

*Department of Applied Physics, Tohoku University, 6-6-05, Aoba, Aramaki, Aoba-ku, Sendai 980-8579, Japan
†Institute for Materials Research, Tohoku University, Katahira 2-1-1, Aoba-ku, Sendai 980-8577, Japan

Abstract. The thermal conductivity in the ab plane, κ_{ab}, of $La_{2-x}Ba_xCuO_4$ ($x = 0.06, 0.08, 0.10, 0.11$) single crystals has been measured in magnetic fields parallel to the c axis up to 14 T. By the application of magnetic field, κ_{ab} has been found not to be suppressed for $x = 0.06$ and 0.08, while it has been found to be suppressed at low temperatures below about the superconducting transition temperature for $x = 0.10$ and 0.11. The suppression of κ_{ab} is smaller for $x = 0.11$ than for $x = 0.10$. These results strongly support the stripe-pinning model for the field-induced magnetic order where the dynamically fluctuating stripes of spins and holes are regarded as being pinned by vortex cores, leading to the development of the static stripe order.

Keywords: Thermal conductivity, $La_{2-x}Ba_xCuO_4$, Stripe order, Field-induced magnetic order
PACS: 74.25.Fy, 74.62.Dh, 74.72.Dn

In the La-based high-T_c cuprates, it is well known that the superconductivity is anomalously suppressed at p (the hole concentration per Cu) $\sim 1/8$, which is called the $1/8$ anomaly. The $1/8$ anomaly is marked in $La_{2-x}Ba_xCuO_4$ (LBCO) with the tetragonal low-temperature (TLT) structure (space group: $P4_2/ncm$) [1, 2] and also in the Zn- or Ni- doped $La_{2-x}Sr_xCuO_4$ (LSCO) [3]. Accordingly, the $1/8$ anomaly is understood in term of the stripe-pinning model, where dynamically fluctuating stripes of spins and holes are regarded as existing and being pinned by the TLT structure or impurities to be statically stabilized, leading to the suppression of superconductivity [4, 5].

Our previous thermal conductivity measurements in magnetic fields for LSCO have suggested that the field-induced magnetic order reported from the neutron scattering experiments [6, 7] is due to pinning of the dynamically fluctuating stripes by vortex cores [8]. Furthermore, considering the results that the suppression rate of the thermal conductivity in the ab plane, κ_{ab}, by the application of magnetic field for $x = 0.115$ is smaller than those for $x = 0.10$ and 0.13 and that the suppression rate becomes small for $x = 0.13$ when the static stripe order is developed by Zn- or Ni-doping, it has been inferred that the suppression rate is related to the magnitude of the development of the static stripe order [9].

In this paper, in order to confirm the above-mentioned inference, we have measured κ_{ab} of LBCO single crystals in magnetic fields.

Single crystals of LBCO with $x = 0.06, 0.08, 0.10$ and 0.11 were grown by the traveling-solvent floating-zone method. Thermal-conductivity measurements were carried out by the conventional steady-state method. Magnetic fields up to 14 T were applied parallel to the c axis.

Figure 1 shows the temperature dependence of κ_{ab} of LBCO in magnetic fields parallel to the c axis. For $x = 0.06$ and 0.08, κ_{ab} monotonically decreases with decreasing temperature in zero field. For $x = 0.10$ and 0.11, on the other hand, κ_{ab} monotonically decreases with decreasing temperature similarly to $x = 0.06$ and 0.08 down to T_{d2}, which is the structural phase transition temperature between the orthorhombic mid-temperature (OMT) phase (space group: $Bmab$) and the TLT phase, jumps at T_{d2} and then exhibits a peak below T_{d2} in zero field. The monotonic decrease of κ_{ab} with decreasing temperature observed for $x = 0.06$ and 0.08 implies that the mean free path of heat carries is very short. Since Ba^{2+} ions of which the ionic radius is much larger than that of La^{3+} ions are expected to induce local TLT structure regions even in the OMT phase [10], it is likely that the local TLT structure regions scatter heat carries so strongly as to make the mean free path very short. Accordingly, the peak of κ_{ab} for $x = 0.10$ and $x = 0.11$ is interpreted as being due to the increase of the mean free path of heat carries on account of the extension of the TLT structure regions all over the sample below T_{d2}.

For clarity of the suppression of κ_{ab} by the application of magnetic field, the suppression rate of κ_{ab}, $\left[\kappa_{ab}(0) - \kappa_{ab}(H)\right]/\kappa_{ab}(0)$, is plotted as shown in Fig. 2. The suppression rate for $x = 0.06$ and 0.08 is zero within the experimental accuracy, while the suppression rate for $x = 0.10$ and 0.11 starts to increase with decreasing temperature at about the superconducting transition temperature, T_c. The starting temperature is almost independent

CP850, *Low Temperature Physics: 24th International Conference on Low Temperature Physics;*
edited by Y. Takano, S. P. Hershfield, S. O. Hill, P. J. Hirschfeld, and A. M. Goldman

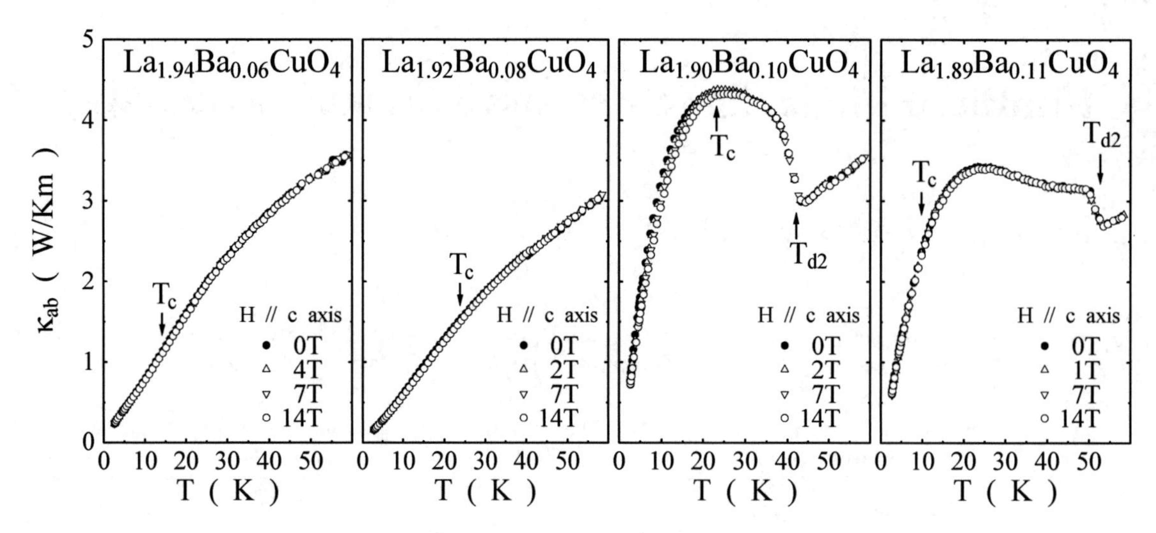

FIGURE 1. Temperature dependence of κ_{ab} in magnetic fields parallel to the c axis for $La_{2-x}Ba_xCuO_4$ ($x = 0.06$, 0.08, 0.10, 0.11).

of the magnitude of the applied magnetic field. These behaviors are very similar to our previous ones observed in LSCO, though the suppression rate is smaller than that in LSCO [8, 9]. Moreover, it is found that the suppression rate is smaller for $x = 0.11$ than that for $x = 0.10$. According to the stripe-pinning model by vortex cores, the small suppression rate indicates that the static stripe order is already developed considerably in zero field. Therefore, it follows that the static stripe order is developed in zero field in LBCO than in LSCO. This is reasonable, because the pinning by the TLT structure takes place in zero field in LBCO [11]. Moreover, it follows that the static stripe order is developed in zero field for $x = 0.11$ more than for $x = 0.10$. This is also reasonable, because p in $x = 0.11$ is closer to $1/8$ than that in $x = 0.10$. Accordingly, the present results strongly support the stripe-pinning model by vortex cores.

As for no change of κ_{ab} by the application of magnetic field in $x = 0.06$ and 0.08, it appears that κ_{ab} does not change even if the static stripe order is developed by vortex cores, because the average distance between Ba^{2+} ions (~ 11 Å and ~ 9.5 Å in $x = 0.06$ and 0.08, respectively), namely, the mean free path of heat carriers is much shorter than the average distance between vortex cores (~ 130 Å at 14 T).

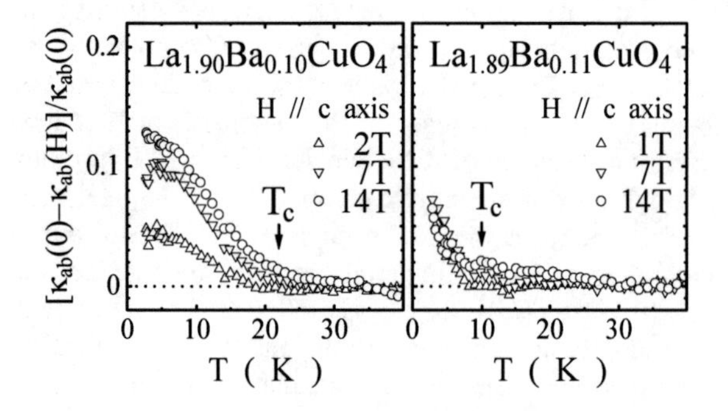

FIGURE 2. Temperature dependence of the suppression rate of κ_{ab} by the application of magnetic field parallel to the c axis for $La_{2-x}Ba_xCuO_4$ ($x = 0.10$, 0.11).

In conclusion, the field-induced magnetic order observed around $p \sim 1/8$ is due to pinning of the dynamically fluctuating stripes by vortex cores. Furthermore, the field effect is marked when the static stripe order is not so developed in zero field.

The thermal conductivity measurements were performed at the High Field Laboratory for Superconducting Materials, Institute for Materials Research, Tohoku University. This work was supported by a Grant-in-Aid for Scientific Research from the Ministry of Education, Culture, Sports, Science and Technology, Japan. One of the authors (T. K.) was supported by the Japan Society for Promotion of Science.

REFERENCES

1. K. Kumagai *et al.*, *J. Magn. Magn. Mater.* **76-77**, 601 (1988).
2. A. R. Moodenbaugh *et al.*, *Phys. Rev. B* **38**, 4596 (1988).
3. Y. Koike *et al.*, *Solid State Commun.* **82**, 889 (1992).
4. J. M. Tranquada *et al.*, *Nature (London)* **375**, 561 (1995).
5. T. Adachi *et al.*, *Phys. Rev. B* **70**, 060504(R) (2004).
6. S. Katano *et al.*, *Phys. Rev. B* **62**, R14677 (2000).
7. B. Lake *et al.*, *Nature (London)* **415**, 299 (2002).
8. K. Kudo *et al.*, *Phys. Rev. B* **70**, 014503 (2004).
9. T. Kawamata *et al.*, *Proc. of ISS2004* (Physica C, in press).
10. M. Sera *et al.*, *J. Phys. Soc. Jpn.* **66**, 765 (1997).
11. M. Fujita *et al.*, *Proc. of ISS2004* (Physica C, in press).

Magnetic and Electric Properties of $RuSr_2GdCu_2O_8$

R. Cobas [a], S. Muñoz Pérez [a] and J. Albino Aguiar [b]

[a] *Pós-Graduação em Ciência de Materiais, CCEN, UFPE, 50670-901, Recife, PE, Brasil*
[b] *Laboratório de Supercondutividade e Materiais Avançados, Departamento de Física, Universidade Federal de Pernambuco, 50670-901, Recife, PE, Brasil*

Abstract. We have performed an experimental study of magnetic and magneto-transport properties on a polycrystalline sample of $RuSr_2GdCu_2O_8$. (Ru-1212). Two peaks in the derivative of the resistive curves were identified in the temperature T_1= 43 K and T_2 =34 K at zero magnetic field. The magnetization presents an increase, at the same temperature T_2 of the resistive curves, followed by a decrease at lower temperatures, which leads to the appearance of a peak in the magnetization versus temperature plots upon entering of superconducting state. We discuss how the anomalies observed in the magnetic and electric measurements could be originated by fault in the stoichiometric of Ru-1212.

Keywords: Ruthenates, Superconducting ferromagnet.
PACS: 74.70.Pq, 74.70.–b, 74.72.–h.

The coexistence of superconductivity (SC) and ferromagnetism (FM) in chemically homogeneous Ru-1212 opened a controversial line of study because it has been demonstrated[1] that a long-range FM order can not coexist with a uniform SC order. Recently, a question of phase purity of Ru-1212 samples has arisen again.[2-4] In this work we note that it is possible to explain some anomalies observed in resistive and magnetic measurements in terms of unknown phase finely dispersed in Ru-1212.

A polycrystalline sample of Ru-1212 was prepared following the Reference [7]. X-ray diffraction measurements were performed using a Rigaku powder difractometer with Cu $K\alpha$ radiation. The magnetoresistance was carried out using a standard four-probe method operating at 80 Hz in the temperature range 2 < T < 300 K at applied magnetic field ranging from 0 to 9 T. Ac susceptibility measurements were done using a Quantum Design SQUID magnetometer. Zero-field cooling (ZFC) curves were obtained at 200 Hz for the same range of temperature and applied magnetic field, looking for a correlation between transport and magnetic properties.

X-ray diffraction pattern and Rietveld simulation of Ru-1212 are shown in Fig. 1. The pattern was assigned to the tetragonal P4/mmm type structure and lattice constant, a = 3.836(4) Å c = 11.54(1) Å, in good agreement with values reported previously.[5] Some extra peaks marked with circles indicated the presence of a few percent of an impurity phase. According to Bauernfeind[6] and Lorentz[7] this impurity phase can be attributed to $SrRuO_3$. However, we can not rule out the existence of other phases such as Gd_2CuO_4[8], or a solid solution $(Sr_{1-x}Gd_x)(Ru_{1-y}Cu_y)O_{3-\delta}$[4] as recently reported.

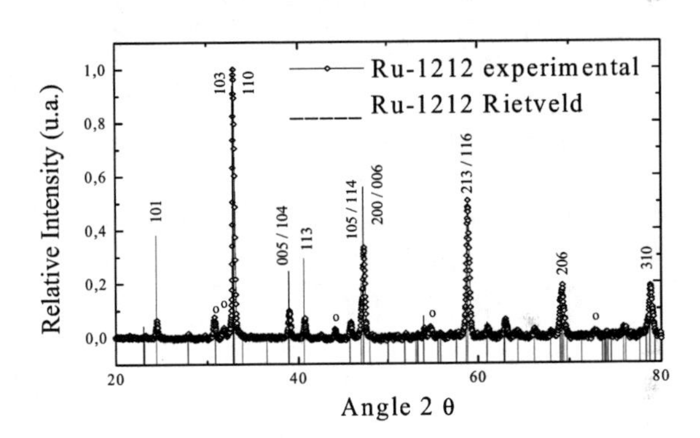

FIGURE 1. X-ray (Cu$K\alpha$) diffraction pattern and Rietveld simulation of the Ru-1212 sample.

The majority Ru-1212 phase forms already at 1030°C.[2] However SC could only be obtained in samples annealed for a long time at 1060°C.[4,7,9] On the other hand, it is known that at 1060°C in oxygen occurs the decomposition of Ru-1212, by the partial evaporation of RuO_2 accompanied by redistributions between Ru and Cu. These redistributions lead to local deviations from stoichiometric composition.[2]

This scenario is harmonious with a phase separation in AFM and FM domains[10] where the SC might be present producing a crypto-superconducting fine structure[11] (probably AFM) within the insulating FM Ru-1212.[2,4,11]

CP850, *Low Temperature Physics: 24th International Conference on Low Temperature Physics;*
edited by Y. Takano, S. P. Hershfield, S. O. Hill, P. J. Hirschfeld, and A. M. Goldman

The resistivity measurements (not shown) at lower temperatures reveal a slight upturn of the resistance above the onset of SC at ~ 50 K, while the zero resistivity state is observed in the range 30 K to 5 K for magnetic field applied up to 9 T. The temperature derivative of the resistivity (Fig. 2) clearly shows two broad overlapped peaks indicating that the resistive transition proceeds in two steps.[7] Both peaks were identified as the inter-grain T_2 and intra-grain T_1 transition temperatures. However the anomalous dependence with magnetic field, observed for inter-grain peak T_2, is not typical of single-phase conventional cuprate (see Fig. 2).

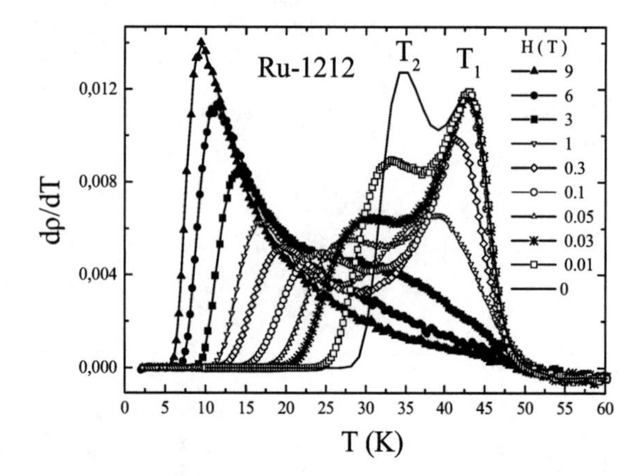

FIGURE 2. Derivative dρ/dT of the resistive curves for Ru-1212.

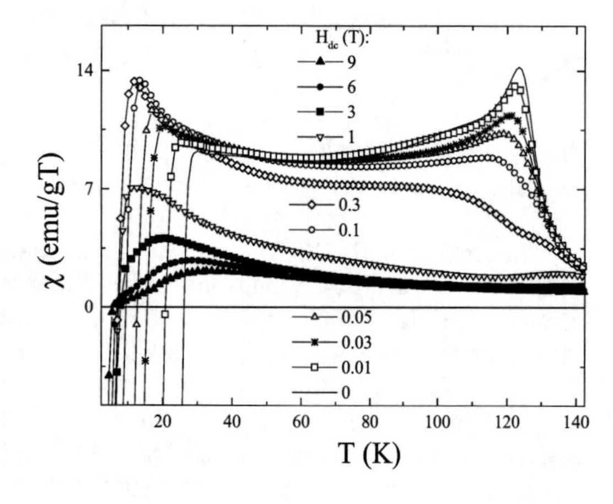

FIGURE 3. Temperature dependence of ac magnetic susceptibility for Ru-1212.

The results of ZFC ac magnetization measurements (for different superimposed dc magnetic fields) are shown in Fig. 3. For $H_{dc} = 0$ a peak occurs at 130 K.[9] Traditionally, the existence of one peak has been interpreted as AFM order of Ru lattice.[12] The widths and temperatures of these transitions depend on the applied field. For $H \geq 0.3$ T disappears the AFM peak and FM ordering develops. The presence of a FM component at high field was indicated by hysteresis loops in *M-H* curves (not shown). The disappearance of this peak with increasing applied magnetic field may be related to the spin-reorientation of the Ru lattice.

At low temperature the magnetization curves show a clear increase (in the same intervals of temperatures T_2 of the resistive curves) followed by a decrease at lower temperatures. The close relation in temperature between derivative and magnetic peaks indicated that both have the same origin. On the other hand these anomalies can not be attributable to single phase Ru-1212. Recently has been shown that these magnetic transitions can be attributable to impurities present in Ru-1212 sintering at 1060°C in flowing oxygen.[13]

In conclusion, the susceptibility as well as the derivative dρ/dT curves of Ru-1212 exhibits not expected peaks close to the onset of SC. The possible presence of an enigmatic structure in the chemically uniform Ru-1212 could be responsible for the anomalies observed. Further experiments to determine the influence of local deviations from stoichiometric composition are needed.

REFERENCES

1. L. N. Bulaevskii, A. I. Buzdin, M. L. Kulie and S. V. Panjukov, *Adv. Physica* **34**, 175 (1985).
2. A.T. Matveev, A.N. Maljuk, A. Kulakov, C.T. Lin and H.U. Habermeier, *Physica C* **407**, 139-146 (2004).
3. A.L. Vasiliev, M. Aindow, Z.H. Han, J.I. Budnick, W.A. Hines, P.W. Klamut, M. Maxwell and B. Dabrowski, *Appl. Phys. Letters* **85**, 3217 (2004).
4. A.T. Matveev, E. Sader, A. Kulakov, A. Maljuk C.T. Lin and H.U. Habermeier, *Physica C* **403**, 231-239 (2004).
5. I. Felner, U. Asaf, S. Reich and Y. Tsabba, *Physica C* **311**, 163 (1999).
6. L. Bauernfeind, W. Widder and H. Braun, *J. of Low Temperature Physic* **105**, 5-6 (1996).
7. B. Lorenz, R.L. Meng, Y.S. Wang, J. Lenzi, Y.Y. Xue and C.W. Chu, *Physica C* **363**, 251-259 (2001).
8. M. Hrovat, A. Benčan, J. Holc, Z. Samardžija and D. Mihailovič, *J. Mater. Sci. Lett.* **19**,1423 (2000).
9. P.W. Klamut, B. Dabrowski, M. Maxwell, J. Mais, O. Chmaissem, R. Kruk, R. Kmiec and C.W. Kimball, *Physica C* **341-348**, 455-456 (2000).
10. Y.Y. Xue, B. Lorenz, A. Baikalov, F. Chen, R.L. Meng and C.W. Chu, *Physica C* **408-410**, 638-640 (2004).
11. C.W. Chu, Y.Y. Xue, S. Tsui, A.K. Heilman, B. Lorentz and R.L. Meng, *Physica C* **335**, 231-238 (2000).
12. G.V.M. Williams and S. Krämer, *Phys. Rev. B* **62**, 4132 (2000).
13. T.P. Papageorgiou, T. Herrmannsdörfer, R. Dinnebier, T. Mai, T. Ernst, M. Wunschel and H.F. Braun, *Physica C* **377**, 383-392 (2002).

SQUID Based DC Magnetometry and AC Susceptibility of $RuSr_2GdCu_2O_8$ Under 10 kbar of Pressure

J. R. O'Brien [a, b], D. Bird [c], S. Gomez [a], H. Oesterreicher [b] and M. S. Torikachvili [c]

[a] *Quantum Design, 6325 Lusk Boulevard, San Diego, CA 92121, USA*
[b] *University of California, San Diego, Dept. of Chemistry, La Jolla, CA, 92093-0317, USA*
[c] *San Diego State University, Dept. of Physics, San Diego, CA 92182-1233, USA*

Abstract. The compound of $RuSr_2GdCu_2O_8$ shows distinct magnetic transitions, which occur on independent sub-lattice. Comparison is made between measurements in the cell at no pressure and maximum 10 kbar of quasi-hydrostatic pressure. The chain site ruthenium oxide layer has a weakly ferromagnetic transition of 133 K and increases to 139 K under pressure. The plane site copper oxide layer has an intra-granular superconducting critical temperature of 49 K and increases to 51 K under pressure. The response to pressure is related to anisotropic changes in lattice parameters. The bulk or inter-granular superconducting transition also shows a significant increase with applied pressure. With low frequency (0.2Hz) AC susceptibility, the peak in the imaginary component increases from 21 K to 24 K under pressure.

Keywords: Superconductors, Cuprate, Pressure
PACS: 74.25.Ha; 74.62.Fj; 74.72.-h

INTRODUCTION

The discovery of ruthenium based superconducting copper oxide compounds [1] opens up new areas of research into the effect of strong magnetic interactions between chain site metal ions. [2] In particular, the title compound is a direct analog to the $YBa_2Cu_3O_7$ material, with the 6-fold oxygen coordination of Ru substituting for the chain site Cu with a 4-fold square planar oxygen coordination. [3] The difficulty in synthesis prevents a systematic study of changes to the lattice parameters through chemical substitution. Thus, the influence of external pressure can provide a valuable tool into determining the effect of decreasing the unit cell parameters. The magnetic coupling of the Ru sub-lattice should be strongly influenced by changes in the lattice parameters, particularly along the a-axis. For superconductivity, it is the ratio of the c-axis to a-axis, in particular the exact positioning of the apical oxygen, which determines the hole content on the copper oxide planes. [4] The lattice parameters of the grain boundaries and edges are also expected to quickly decrease to closer to bulk value, resulting in sharper transitions.

The single-phase sintered bulk compound is prepared [5] using the Sr_2GdRuO_6 precursor and CuO phases with a final annealing at 1060 C in oxygen. Lattice parameters are a=3.838 A and c/3 = 3.844 A. The pressure cell is prepared by inserting tin wire as the internal pressure standard. The transmitting fluid fills the cylindrical Teflon container, then the 21.4 milligram bar shaped sample is inserted and the cap is put into place. The zero pressure magnetometer analysis is performed before and after pressurization. Once assembled, pressure is applied up to the maximum rating of about 10 kbar. The SQUID based magnetometer is a Quantum Design MPMS XL-5AC. The remanent field is below 0.01 Oe for the cooling process. Once at 1.8 K, a field of 10 Oersted is applied and the magnetization versus temperature is acquired, as shown in figure 1. The pressure is stable over 10 thermal cycles and weeks of measurements.

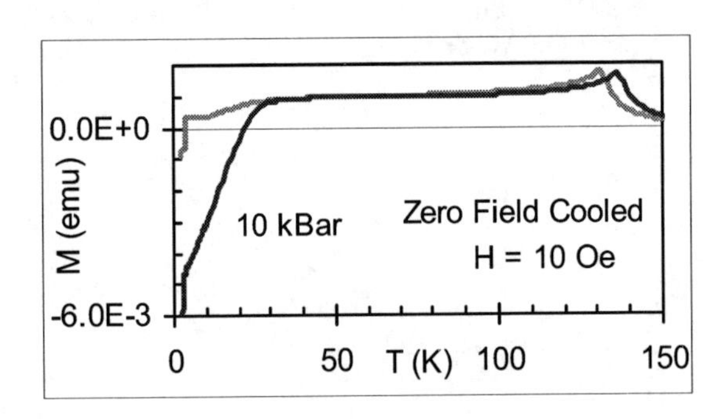

FIGURE 1. Magnetization at zero and 10 kbar of pressure. There is an increase in the bulk superconducting properties and the Ru ordering temperature as the pressure is increased.

CP850, *Low Temperature Physics: 24th International Conference on Low Temperature Physics;*
edited by Y. Takano, S. P. Hershfield, S. O. Hill, P. J. Hirschfeld, and A. M. Goldman

RESULTS

The lowest temperature transition observed in figure 1 is the tin internal pressure reference standard. In the DC magnetization experiment with a bias field of 10 Oe, the critical transition temperature decreases from 3.66 K to 3.14 K under pressure. This change in T_c of tin corresponds to a pressure of 10.6 kBar. The bulk superconducting critical temperature near 30 K increases with pressure, but more dramatic is the larger diamagnetic moment. High T_c ceramics are known for the sensitivity of grain boundary junctions. Increasing pressure clearly enhances the inter-granular contact, thus improving the shielding currents for the ZFC state. If figure 1 is magnified from 30 K to 50 K, the improved superconducting volume fraction while under pressure is observed up to the intra-granular T_c. The inverse magnetization versus temperature is analyzed from 45 K to 65 K to determine the onset temperature of bulk superconductivity. Above the superconducting critical temperature, Curie-Weiss behavior for the paramagnetic Gd^{3+} ion is expected to dominate until the region of the Ru ordering is approached. The effect of the pressure is to shift the Ru ordering to higher temperatures, making the region from 51 K to 70 K more like a paramagnetic response. The magnetic ordering temperature of the Ru lattice is determined from the plot of moment squared versus temperature. The slope is extrapolated to zero and shifts from 133 K to 139 K with increase in pressure. From other data not shown, the magnetic transition defined as the peak in the real component of the AC susceptibility increased from 130 K to 136 K under increasing pressure.

The SQUID based AC susceptibility is operated in the frequency range below 10 Hz to minimize inherent phase shift caused by the large metallic pressure cell. Figure 2 shows the results for zero DC field and using a 0.2 Hz, 1 Oe AC drive signal.

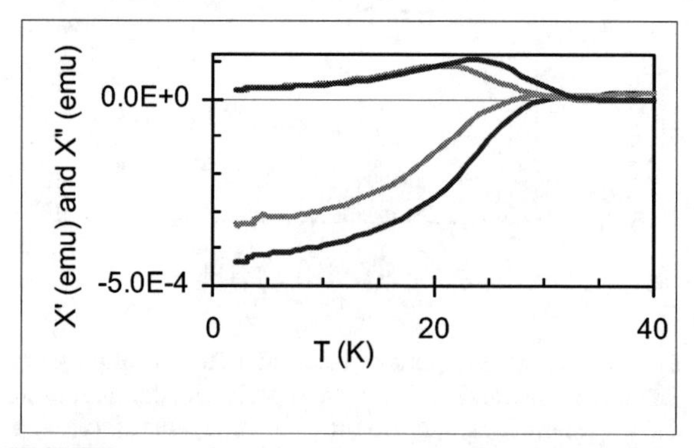

FIGURE 2. Real and imaginary parts of AC susceptibility at ambient and 10 kbar of pressure. There is a clear increase in the X" peak and midpoint of X' transition temperatures.

The AC susceptibility data presented in figure 2 shows the tin transition in zero DC biased field. For both ambient and maximum cell pressure, higher values of the tin transition temperatures are obtained. The difference still confirms the pressure of 10.6 kBar. Figure 2 shows the increase in peak temperature for the imaginary (X") component from 21 K to 24 K. The higher bulk transition temperature under pressure is seen in the real (X') component. Interestingly, the amplitude of the AC susceptibility response is much closer than in the DC data, suggesting AC analysis techniques are not as limited by grain boundary issues. An experiment which removes the interference from the tin transition may yield information on the anti-ferromagnetic transition of the Gd ions around 3 K.

CONCLUSIONS

The $RuSr_2GdCu_2O_8$ system is useful to explore the influence of pressure on magnetic transitions in various independent sub-lattice. An increase in the magnetic ordering temperature of the ruthenium lattice with higher pressure likely corresponds to a substantial decrease in the a-axis lattice parameter. The slight increase for the onset of superconductivity suggests decreases in both the c-axis and a-axis result in few net holes to be created as pressure is applied. The increase in bulk superconducting properties with increasing pressure reflects a more mechanical solution to the issue of short coherence length, poor granular contact in ceramic materials. Future research suggests finding the synthesis conditions for the smaller rare earth element substitution for Gd, like Dy or Er to increase the magnetic ordering temperature and possibly the superconducting temperature as well.

ACKNOWLEDGMENTS

The pressure cell is obtained from easyLab, Ltd. Support from NSF Grant No. DMR-0306165 (MST and DB) is gratefully acknowledged.

REFERENCES

1. A. Ono, *Jpn. J. Appl. Phys.* **34**, L1121 (1995).
2. I. Felner, U. Asaf, Y. Levi and O. Millo, *Phys. Rev. B* **55**, R3374 (1997).
3. B. Lorenz, R. L. Meng, Y. Y. Xue and C. W. Chu, *Physica C* **383**, 337-342 (2003).
4. J. R. O'Brien and H. Oesterreicher, *J. Solid State Chem.* **135**, 307-311 (1997).
5. M. S. Torikachvili, I. Bossi, J. R. O'Brien, F. C. Fonseca, R. Muccillo and R.F. Jardim, *Physica C* **408**, 195 (2004).

Infrared Study of $YBa_2Cu_3O_{7-\delta}$ in High Magnetic Fields

Minghan Chen,[a] D. B. Tanner,[a] and Y. J. Wang[b]

a Department of physics, University of Florida, Gainesville, FL 32611-8440, USA
b National High Magnetic Field Laboratory, Tallahassee, FL 32310-3076, USA

Abstract. The magnetic-field dependence of the far-Infrared transmittance was measured for $YBa_2Cu_3O_{7-\delta}$ thin films. The measurement was done at 4.2K for $YBa_2Cu_3O_{7-\delta}$ films on both sapphire and MgO substrates. The results show no significant change in the transmittance spectra as the field was applied. This implies very low loss for the vortex state at low temperatures.

Keywords: Infrared, magnetic field, transmittance
PACS: 32.30.Bv

INTRODUCTION

Far-infrared spectroscopy has been widely applied to investigate the vortex dynamics in high T_c superconductors. Karrai *et al.*[1] measured the transmittance of $YBa_2Cu_3O_{7-\delta}$ thin films, finding an increase in transmittance below ~125 cm^{-1} with increasing field. This effect was attributed to dipole transitions associate with bound states in the vortex cores. Measurements of the far-infrared reflectivity R of a superconducting $YBa_2Cu_3O_{7-\delta}$ thin film by Eldridge *et al.*[2] found a strong dependence on magnetic field, suggesting that the increase in transmission at low wavenumber in the experiment of Karri *et al.*[1] was mainly due to a decrease in reflectivity. In their experiment they also observed the phonon mode at far-infrared region in the high magnetic field, which is usually not expected in the *ab*-plane films at zero field.

In contrast, Liu *et al.*[3] measured $YBa_2Cu_3O_{7-\delta}$ films at various temperatures in magnetic fields up to 30 T. Their data clearly indicates that at low temperatures (below 50 K) there is no significant field dependence to either reflectance or transmittance. But at higher temperatures, such as 60 K and 72 K, there was a significant increase of the far-infrared transmittance (below ~120 cm-1) with increasing magnetic fields.

EXPERIMENT

We studied superconducting and non-superconducting films on several substrates. The substrates used for the superconducting $YBa_2Cu_3O_{7-\delta}$ films were MgO and sapphire, with a CeO buffer layer. The films are about 400 Å in thickness, with T_c around 85-90 K. A non-superconducting $YBa_2Cu_3O_{6+x}$ film on silicon was also studied.

Our far-infrared transmittance measurements were done at the National High Magnetic Field Laboratory, Tallahassee FL. The experiment used a Bruker 113v spectrometer and light pipe in conjunction with an 18-Tesla superconducting magnet.[4] The sample insert allowed exchange of sample and reference, so that absolute transmittance measurements could be made.

RESULTS

The transmittance spectra of the films taken at 4.2 K at different magnetic fields are shown in Fig. 1. We observed practically no influence of the magnetic field on the far infrared transmittance spectra of any of the films. The noise around 140 and 230 cm^{-1} is due to poor beam-splitter efficiency at these frequencies. Outside of these two regions, the variation in transmission with field is $\Delta T/T < \pm 8\%$, set by the signal to noise ratio in the data. Moreover, note that for the two superconducting samples (Fig. 1, panels a and b) the low frequency transmittance tends to zero, as expected for a sample where the far infrared properties are dominated by the inductive response (σ_2). Were the loss (σ_1) significant, there would be finite transmittance at low frequencies. Thus, we can conclude that with the external field perpendicular to the superconducting $YBa_2Cu_3O_{7-\delta}$ film, no far-infrared

CP850, *Low Temperature Physics: 24th International Conference on Low Temperature Physics;*
edited by Y. Takano, S. P. Hershfield, S. O. Hill, P. J. Hirschfeld, and A. M. Goldman

magnetoresistance was detected at 4.2 K and in the high field regime. The non-superconducting sample (Fig. 1, panel c) also showed no field-dependent absorption. (The oscillations in this sample are due to multiple internal reflections in the Si substrate.)

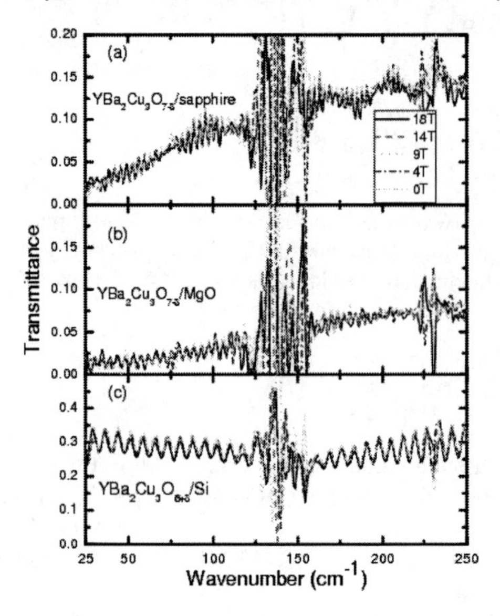

FIGURE 1. At 4.2 K, transmittance of samples at different magnetic fields.

DISCUSSION

Because our infrared electric field is parallel to the *ab* plane, the vortices oscillate within their pinning potential. We model the vortex state as a parallel mixture of vortex cores (normal state) surrounded by superconductor. The fractional area of the cores is $H/H_{c2}(T)$ with H_{c2} the upper critical field. The dielectric function of this model is:

$$\varepsilon(\omega)=\frac{\omega_{ps}^2}{\omega(\omega+i0^+)}\left[1-\frac{H}{H_{c2}}\right]-\frac{\omega_{ps}^2}{\omega(\omega+i/\tau_s)}\left[\frac{H}{H_{c2}}\right]+\varepsilon_{ir} \qquad (1)$$

Where ω_{ps} is the superfluid plasma frequency, $1/\tau_s$ is the damping constant inside the vortex, and ε_{ir} is the non-superconductor part of the dielectric function. (A more complete model would include a normal fluid, midinfrared absorption, and high-frequency processes. Here, these are all lumped into ε_{ir}.)

In order to explain the absence of changes in our optical spectrum, consider the change of optical conductivity σ as the vortex density is increased: $\sigma_s(\omega)=\omega_{ps}^2[1-H/H_{c2}]/4\pi i\omega$; $\sigma_n(\omega)=\omega_{ps}^2[H/H_{c2}]S(\omega)/4\pi$. Here, $S(\omega)$ is the frequency dependence of the vortex conductivity and is initially dominated by $1/(i\omega-1/\tau_s)$. The change in $\sigma(\omega)$ due to a conversion of super to normal fluid is given by $\Delta\sigma(\omega)=\omega_{ps}^2(H/H_{c2})[S(\omega)-i/\omega]/4\pi$. Specifically, $\Delta\sigma(\omega)$ is maximum at zero frequency and decreases rapidly with frequency for $\omega > 1/\tau_s$. In our experiment, the measurements are limited to $\omega > 25\ cm^{-1}$ and the quasi-particle scattering rate $1/\tau_s$ of our films is very small. Moreover, the change in $\Delta\sigma(\omega)$ is expected to be not big for fields up to our maximum field of 18 T when H_{c2} is about 240 T. Thus, it is possible that any change in the spectra should be relatively small in our far-infrared frequency and magnetic field range.

CONCLUSION

The transmittances of $YBa_2Cu_3O_{7-\delta}$ films with different substrate were measured at 4.2 K in the high magnetic field up to 18 T. The transmittance spectra did not show significant changes with the changing of the magnetic field. This result may indicate a very low loss for the vortex state at low temperature.

ACKNOWLEDGEMENTS

Supported by the National Science Foundation, DMR-0305043, and the Department of Energy, DE-AI02-03ER46070.

REFERENCES

1. K. Karri, E. J. Choi, F. Dunmore, S. Liu and H. D. Drew, *Physical Review Letters* **69**, 152-155 (1992).
2. J. E. Eldridge, M. Dressel, D. J. Matz, B. Gross, Q. Y. Ma, and W. N. Hardy, *Physical Review B* **52**, 4462-4466 (1995).
3. H. L. Liu, A. Zibold, D. B. Tanner, Y. J. Wang, M. J. Burns, K. A. Delin, M. Y. Li, and M. K. Wu, *Solid State Communications* **109**, 7-12 (1999).
4. H. K. Ng, Y. J. Wang, in: Z. Fisk et al. (Eds.) *Proceedings of Physical Phenomena at High Magnetic Fields-II*, Singapore: World Scientific Press, 1996, pp. 729.

Disorder Driven Localization in Charge Neutrally Doped 123 Superconductors

S. R. Ghorbani*, M. Andersson† and Ö. Rapp†

*Department of Physics, Tarbiat Muallem University of Sabzevar, P. O. Box 397, Sabzevar, Iran
†Solid State Physics, IMIT, Kungliga Tekniska Högskolan 229, 10044 Stockholm, Sweden

Abstract. Charge neutral dopings with equal amounts of CaTh or CaPr on rare earth site in Nd-123 have been investigated by studies of the normal state electrical resistivity, ρ, the Seebeck coefficient S, and the Hall coefficient R_H. In spite of comparable rates of depression of the superconducting T_c significant differences are observed in the doping concentration dependence of ρ, S, and R_H between the two dopings. These results are discussed in terms of disorder induced localization.

Keywords: 123 compounds, charge neutral dopings, electronic reansport, localization, disorder
PACS: 74.72.Jt, 74.62.Dh, /4.25.Fy

INTRODUCTION

A well established view of high-T_c superconductors is that they are stable in a small range of charge concentrations, with rapid, and usually parabolic depression of the superconducting T_c on both sides of optimal doping. However, similarly strong deterioration of the superconducting state can be produced by charge neutral dopings. Examples are Ca_xPr_x and Ca_xTh_x doping on rare earth site in 123-compounds for x in a range up to about 0.1. In that concentration range Pr and Th are both +4 valent, and equal amounts of the doping elements can be dissolved in the orthorhombic phase up to, or for CaTh slightly below this limit [1]. This observation raises the question what is the cause for a strong depression of T_c in these alloy systems. We have previously investigated this problem from resistivity, ρ, [2], thermoelectric power, S, [3] and Hall coefficient R_H, [4]. In the present paper the room temperature transport properties are compared in doped Nd-123, to further elucidate the different roles of Ca_xPr_x and Ca_xTh_x doping in the depression of T_c.

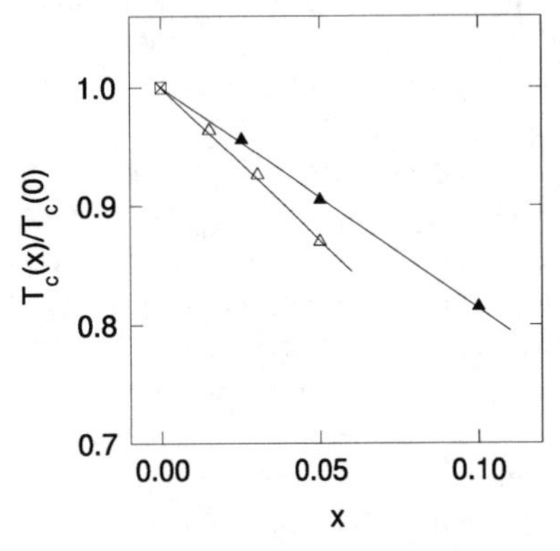

FIGURE 1. Superconducting T_c *vs* doping concentration x in $Nd_{1-2x}Ca_xTh_xBa_2Cu_3O_{7-\delta}$ (open symbols) and in $Nd_{1-2x}Ca_xPr_xBa_2Cu_3O_{7-\delta}$ (closed symbols). T_c is linear for both dopings with - d $\ln T_c/dx \approx 2.5$ and 2.3 respectively.

SAMPLES AND EXPERIMENTS

Polycrystalline samples were prepared by standard methods as described before [1]. Phase purity, oxygen stoichiometry, and the expected distribution of elements on rare earth site have been verified with X-ray and neutron diffraction, and bond valence sum calculations.

Transport measurements were made on bar shaped samples. Standard dc arrangements with electrical contacts by silver paint was used for ρ and Hall studies. R_H was evaluated from measurements in magnetic fields up to 8 T, with switching of the electric and magnetic fields. The thermopower was studied on similar bars, using small reversible temperature gradients up to 1.5 K.

Fig. 1 shows that the depression of T_c is similarly strong in both alloy systems, and linear in doping concentration. This indicates that other mechanisms than charge variation cause destruction of superconductivity.

RESULTS

The Hall coefficient at 295 K is shown in Fig 2 *vs* the reduction of T_c. There is a strong increase of $R_H(x)$ with depression of T_c for CaTh doping, while $R_H(x)$ is almost constant for CaPr doping. Analyses of $R_H(x,T)$ in terms of a two band model with a narrow peak in the density of states [5], and in the Anderson model [6], have shown that both dopings introduce electronic disorder [4]. However, the results in Fig. 2 are not straightforwardly related

CP850, *Low Temperature Physics: 24th International Conference on Low Temperature Physics;*
edited by Y. Takano, S. P. Hershfield, S. O. Hill, P. J. Hirschfeld, and A. M. Goldman

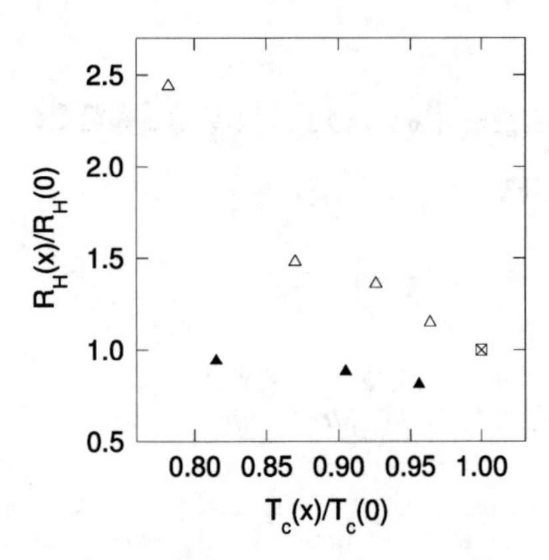

FIGURE 2. Normalized Hall coefficient at 295 K for CaTh doped Nd-123 (open symbols) and CaPr doped Nd-based 123 (filled symbols) *vs* reduction of T_c.

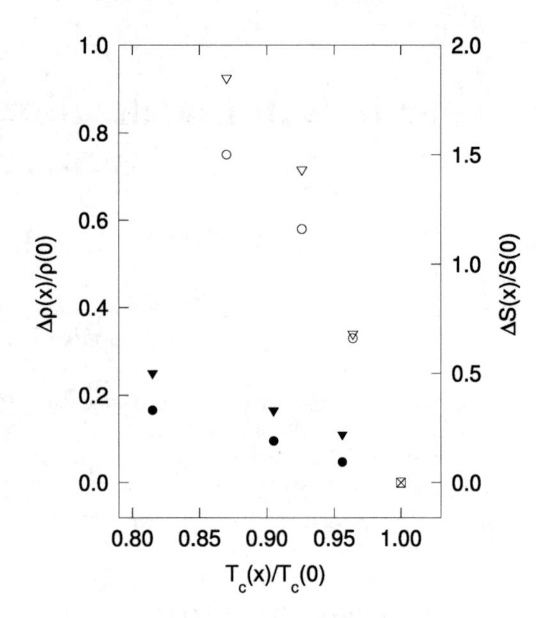

FIGURE 3. $\Delta\rho$(295 K) and ΔS(295 K) vs reduction of T_c. Circles: $\Delta\rho$, left hand scale. Triangles: ΔS, right hand scale. Open symbols: CaTh doping, filled symbols: CaPr doping.

to disorder. It is therefore interesting to observe that there is qualitative agreement with the results in Fig 3, with a strong effect on a transport property for CaTh doping, and a weak effect for CaPr doping.

Fig. 3 shows ΔS and $\Delta\rho$, both at 295 K, *vs* the depression of T_c. ΔS=$S(x)$-$S(0)$ and $\Delta\rho$=$\rho(x)$-$\rho(0)$. A decrease of T_c is accompanied by a strong increase in S and ρ for CaTh doping and a much weaker effect for CaPr doping. Fig. 3 shows that the scaling factor between $\Delta\rho$ and ΔS is almost the same for CaPr and CaTh doping.

For S it is well known that the reduction of T_c from its maximum value in an alloy system is a function of S(290 K) [7]. This result concerns charge doped alloys over several orders of magnitude of S. It can be noted that the relation in Fig 3 for charge neutrally doped alloys is qualitatively similar. The hole concentration was estimated in [7] by other methods, e.g. bond valence sum calculations. With this analogy, our results can be interpreted as a beginning hole localization in the CuO_2 planes.

$\Delta\rho$ in Fig. 3 reflects directly a strongly increasing disorder with CaTh doping, and a weaker one with CaPr doping. In fact, taking $\Delta\rho$ as a measure of the change of $k_F l$ (Fermi wave vector and mean free path), ΔT_c in 123 compounds with CaTh doping, but not with CaPr, could be qualitatively described by weak localization and electron-electron interaction effects on T_c [2].

There has been little theoretical and experimental work on disorder corrections to S. However, Ting et al [8] found that weak localization effects should be absent, as for the Hall effect. Interaction effects are small at room temperature. In their picture observable disorder effects also at room temperature can be understood from a free electron-like $\sigma_o = ne^2\tau/m$, and $S = \pi^2 k_B^2 T/3e\varepsilon_F$. The increase of $\Delta\rho$ with reduced T_c then results from an increased scattering rate, presumably combined with a decreased charge density, n, while the increase of S arises from a decrease of the Fermi energy ε_F. These three parameter changes all reflect increased electronic disorder.

Effects on transport properties are seen to be weaker for CaPr- than for CaTh-doping. Yet the rates of depression of T_c are similar. Detailed analyses of $S(x,T)$ have suggested that in addition to the weaker disorder effect in CaPr doped samples, there is a contribution from localization of charge carriers in the CuO_2 planes [3].

In the present work we have compared transport properties at room temperature in charge neutrally doped Nd-123. Disorder effects can be clearly seen in S and ρ. Differences between CaPr and CaTh doping are of the same nature in all transport properties, with strong changes for CaTh doping and weak changes for CaPr doping.

REFERENCES

1. P. Lundqvist, Ö. Rapp, R. Tellgren, and I. Bryntse, *Phys. Rev. B* **56**, 2824 (1997).
2. B. Lundqvist, P. Lundqvist, and Ö. Rapp, *Phys. Rev. B* **57**, 2824 (1998).
3. S. R. Ghorbani, M. Andersson, and Ö. Rapp, *Phys. Rev. B* **66**, 104519 (2002).
4. S. R. Ghorbani, M. Andersson, and Ö. Rapp, *Physica C*, in press (2005).
5. V. E. Gasumuyants, V. I. Kaidanov, and E. V. Vladimirskaya, *Physica C* **248**, 255 (1995).
6. P. W. Anderson, *Phys. Rev. Lett.* **67**, 2092 (1991).
7. S. D. Obertelli, J. R. Cooper, and J. L. Tallon, *Phys. Rev. B* **46**, 14928 (1992).
8. C. S. Ting, A. Houghton, and J. R. Senna, *Phys. Rev. B* **25**, 1439 (1982).

In-Plane Superfluid Density of Highly Underdoped $YBa_2Cu_3O_{6+x}$

D. M. Broun*, P. J. Turner*, W. A. Huttema*, S. Özcan†, B. Morgan†, Ruixing Liang**, W. N. Hardy** and D. A. Bonn**

*Department of Physics, Simon Fraser University, Burnaby, BC, V5A 1S6, Canada
†Cavendish Laboratory, Madingley Road, Cambridge, CB3 0HE, United Kingdom
**Department of Physics and Astronomy, University of British Columbia, Vancouver, BC, V6T 1Z1, Canada

Abstract. The two most prominent states of matter in the phase diagram of the high temperature superconductors are the Mott insulator and d-wave superconductor. Important clues for connecting the two states can be obtained by studying how the fundamental excitations evolve with doping. Using a breakthrough in the preparation of high purity $YBa_2Cu_3O_{6+x}$ crystals, we can tune into a regime of the underdoped cuprate phase diagram in which the transition temperature can be varied continuously between $T_c = 25$ K and zero, without changing cation disorder. A combination of microwave spectroscopy and measurements of the lower critical field H_{c1} has been used to map out the temperature-dependent superfluid density as a function of doping, with surprising results. The superfluid density becomes anomalously small in this doping range but shows no sign of a Kosterlitz-Thouless transition. In addition, the temperature slope of the superfluid density is roughly proportional to T_c, suggesting a strong doping dependence of the charge-current renormalization factor. In contrast to this, electron currents between the CuO_2 layers are not renormalized, implying that the in-plane quasiparticles are fundamentally different from the physical electrons that tunnel between layers.

Keywords: cuprate, high temperature superconductivity, superfluid density, lower critical field, surface impedance, penetration depth
PACS: 74.72.Bk, 74.25.Nf, 74.25.Bt, 74.25.Ha

The superfluid density ρ_s is one of the most fundamental physical properties of a superconductor. It is both the energy scale for fluctuations in the phase of the superconducting order parameter [1] and a thermodynamic probe of the spectrum of excitations from the superconducting condensate. In a clean d-wave superconductor nodal quasiparticles deplete ρ_s linearly with temperature [2]. This effect was first observed in measurements of the magnetic penetration depth λ on optimally doped $YBa_2Cu_3O_{6.95}$ and provided some of the earliest evidence for unconventional superconductivity in the cuprates [3]. Similar measurements have since been used to probe the nature of the nodal quasiparticles at other hole dopings and in different materials [4]. Here we report measurements of the superfluid density in the most mysterious part of the cuprate phase diagram — where superconductivity gives way to antiferromagnetism. The measurements find a sublinear correlation between T_c and $\rho_s(T \rightarrow 0)$, and show that charge currents *within* the CuO_2 layers are renormalized downwards as the d-wave superconductor is underdoped. They also reveal an unconventional role for phase fluctuations, with no vortex unbinding transition despite near two-dimensionality.

Recent advances in the preparation of $YBa_2Cu_3O_{6+x}$ single crystals now give us an unprecedented level of control over hole doping within an individual sample, *without* changing its oxygen content [5]. Near the composition $YBa_2Cu_3O_{6.35}$ it is possible to tune T_c over a 10 K range using room temperature ordering under ambient pressure [6], and over approximately 15 K using hydrostatic pressures up to 20 kbar. The process is illustrated in Fig. 1. Ordering of oxygen atoms in the CuO chain layers at room temperature into chainlets of increasing length gradually removes electrons from the CuO_2 planes, slowly increasing the hole doping [7, 8]. Below 0°C the oxygen atoms are frozen in place, allowing low temperature experiments to be carried out at fixed dopings.

In the $YBa_2Cu_3O_{6+x}$ system we have previously applied the continuous doping method to measuring λ_c, the screening length for currents flowing along the crystal c axis [6]. The data, plotted as $1/\lambda_c^2$, are shown in Fig. 2 and reveal the power of the technique to carry out doping studies in a *single* sample . These experiments confirm that d-wave pairing persists throughout the superconducting phase [9], but are less than ideal as a means of studying the superconductivity. λ_c is an indirect probe of the superfluid density — its interpretation requires a model of how electrons tunnel between CuO_2 layers — and measurements of λ_c may not be sensitive to subtle electronic correlations occurring *within* the CuO_2 planes, such as charge-current renormalization.

In the conventional geometry for penetration depth measurements a uniform magnetic field is applied in the

CP850, *Low Temperature Physics: 24th International Conference on Low Temperature Physics;*
edited by Y. Takano, S. P. Hershfield, S. O. Hill, P. J. Hirschfeld, and A. M. Goldman

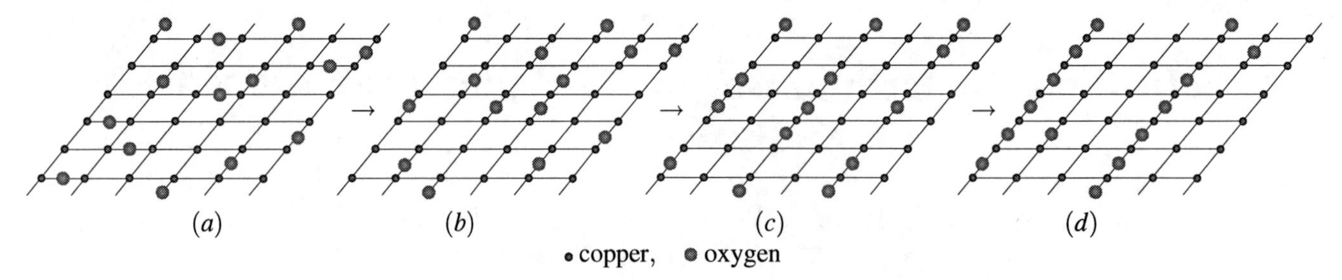

FIGURE 1. Illustration of oxygen ordering in the CuO chain layers of $YBa_2Cu_3O_{6.333}$. (a) For $T \gg 150°C$ the material is tetragonal and oxygen occupies a- and b-axis sites with equal likelihood. (b) Below $T \approx 150°C$ the material adopts the Ortho-I structure, with oxygen randomly distributed amongst b-axis sites. Doping is very low and $T_c = 0$. (c) On annealing at room temperature for 3 to 6 weeks, the oxygen orders into the Ortho-II structure, with every second chain empty. Chainlets lengthen, doping holes and making T_c small but finite. (d) High hydrostatic pressure ($p \approx 20$ kbar) promotes further ordering at room temperature. This figure shows the approach to the sought-after Ortho-III′ phase, as yet unrealized in the laboratory.

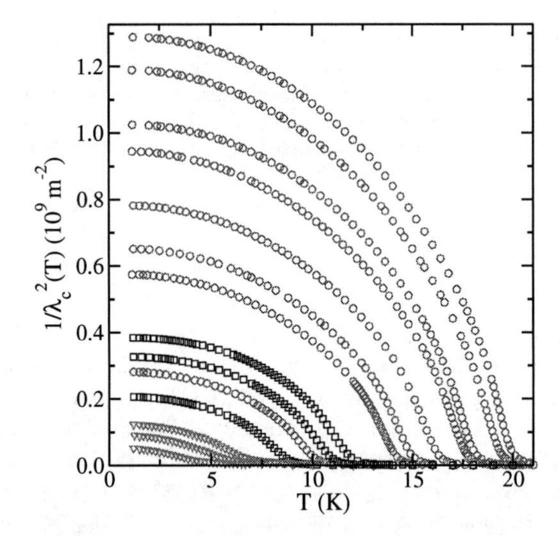

FIGURE 2. Evolution of the c-axis superfluid density with doping during gradual oxygen ordering at room temperature. Sets of curves denoted by different symbols correspond to slightly different oxygen contents near $YBa_2Cu_3O_{6.35}$. (Reproduced from [6] with permission.)

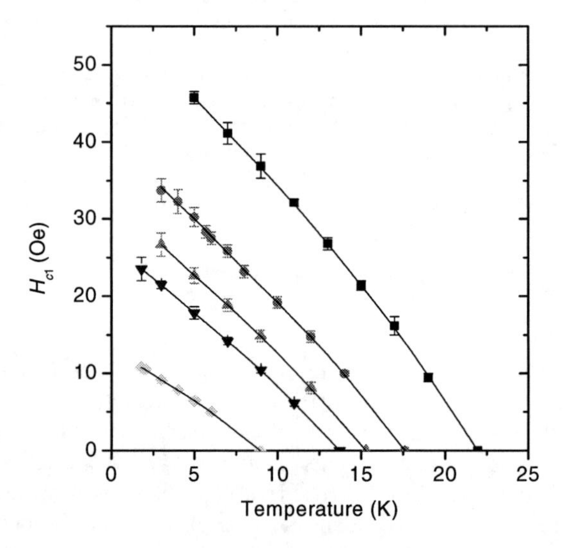

FIGURE 3. H_{c1} for an $YBa_2Cu_3O_{6.333}$ ellipsoid as doping is tuned by room temperature annealing. (Reproduced from [10] with permission.)

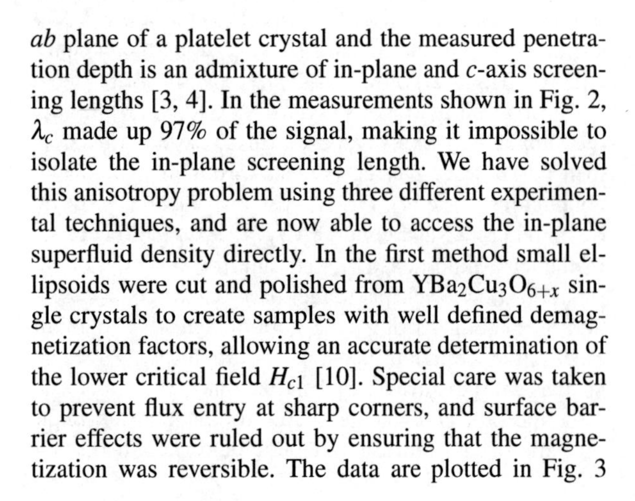

ab plane of a platelet crystal and the measured penetration depth is an admixture of in-plane and c-axis screening lengths [3, 4]. In the measurements shown in Fig. 2, λ_c made up 97% of the signal, making it impossible to isolate the in-plane screening length. We have solved this anisotropy problem using three different experimental techniques, and are now able to access the in-plane superfluid density directly. In the first method small ellipsoids were cut and polished from $YBa_2Cu_3O_{6+x}$ single crystals to create samples with well defined demagnetization factors, allowing an accurate determination of the lower critical field H_{c1} [10]. Special care was taken to prevent flux entry at sharp corners, and surface barrier effects were ruled out by ensuring that the magnetization was reversible. The data are plotted in Fig. 3 at five dopings, with T_c tuned by room-temperature annealing. In Ginzburg–Landau theory $H_{c1} = h[\ln(\kappa) + 0.5]/(8\pi e \lambda_{ab}^2)$, where κ is the ratio of penetration depth to coherence length. κ is weakly doping and temperature dependent in the cuprates [11], making H_{c1} a direct measure of ρ_s. Below 0.6 T_c, H_{c1} is linear in temperature, indicating the presence of line nodes in the energy gap. The temperature slope and zero-temperature extrapolation of $H_{c1}(T)$ are plotted in Fig. 4. (In both cases, as we report below, the H_{c1} results are supported by very similar data from microwave measurements of the penetration depth.) Figure 4 shows a sublinear correlation between T_c and $\rho_s(T \to 0)$, in disagreement with the long-established Uemura phenomenology [12]. It also reveals a decline in slope, a sign that paramagnetic quasiparticle currents are becoming ineffective at depleting the superfluid density as $T_c \to 0$. In a d-wave supercon-

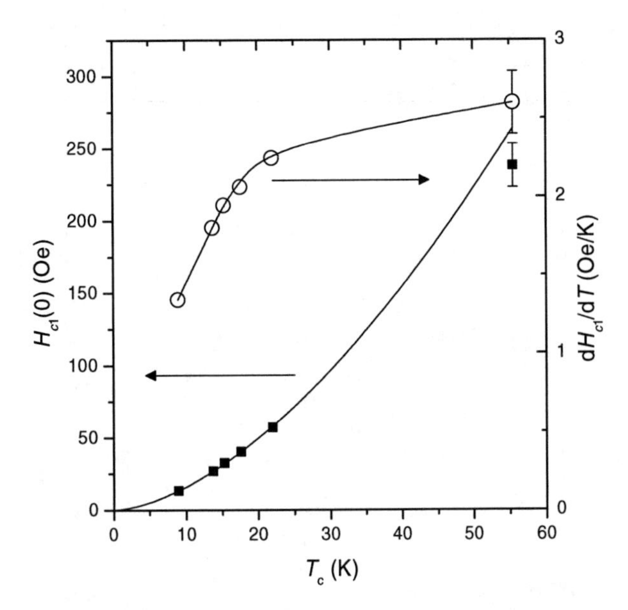

FIGURE 4. $H_{c1}(T \to 0)$ and $-dH_{c1}/dT$ as functions of T_c for an $YBa_2Cu_3O_{6.333}$ ellipsoid. (Reproduced from [10] with permission.)

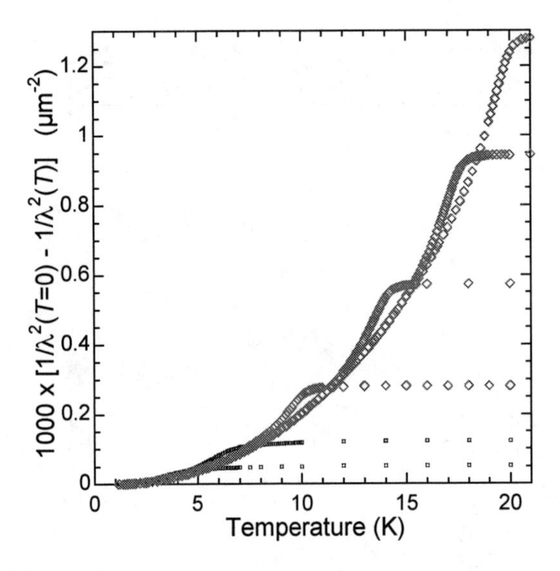

FIGURE 5. Paramagnetic part of the superfluid response along the c direction. Data are taken from reference [6] and for clarity not all dopings are shown. $1/\lambda_c^2(T=0) - 1/\lambda_c^2(T)$ probes the low energy quasiparticle excitations in the vicinity of the d-wave gap nodes. Away from the immediate vicinity of T_c, the c-axis data all collapse onto a single curve, indicating a nodal spectrum and charge that do not vary with hole doping.

ductor $-dH_{c1}/dT \propto \alpha^2 v_F/v_\Delta$, where α is the charge-current renormalization factor, v_F is the Fermi velocity, and v_Δ is the group velocity due to the slope of the energy gap near the nodes. In the cuprates, v_F and v_Δ are known to depend only weakly on doping [13, 14, 15], so $-dH_{c1}/dT$ reflects the evolution of α^2 with doping. In Fermi liquid theory charge-current renormalization is due to interaction-induced backflow reducing the electrical current carried by quasiparticles [16, 17, 18, 19]. In more exotic theories, such as the gauge theories of the t–J model, the d-wave quasiparticles evolve continuously with underdoping into the spinon excitations of an RVB spin liquid, with a corresponding decline in effective charge [20, 16]. Those models, however, predict a much stronger doping dependence of α than we find. An interesting contrast is observed when we examine quasiparticle currents along the c-direction. This can be seen most clearly in a plot of $1/\lambda_c^2(T=0) - 1/\lambda_c^2(T)$, the paramagnetic part of the c-axis superfluid density. Away from the immediate vicinity of T_c, the c-axis data all collapse onto a single curve, indicating no renormalization of the effective charge and showing that the physical electrons that tunnel between CuO_2 layers are very different from the electron quasiparticles within them.

In addition to H_{c1} measurements, we have used two microwave cavity perturbation techniques to probe the in-plane penetration depth directly. The first of these is surface cavity perturbation, and a schematic of this apparatus is shown in Fig. 6. Here the sample, a $3.6 \times 1.6 \times 0.16$ mm^3 single crystal, is mounted on an independently heated sapphire thermal stage above a movable microwave resonator, with the temperature difference supported across optical fibres. Microwave fields applied to the sample through a small hole in the end wall of the resonator drive screening currents in the crystal that are confined to flow in the CuO_2 planes. The sensitivity is calibrated using the principle that an increase in the sample's penetration depth by an amount $\Delta\lambda$ will produce the same frequency shift as moving the resonator a distance $\Delta\lambda$ away from the sample. We have found this method to give good results for the temperature dependence of the penetration depth, but it has proven difficult to calibrate, despite the use of a precision capacitance read-out of the resonator–sample distance. A second approach, using cavity perturbation at 5.48 GHz on small ellipsoid samples, has circumvented the calibration problem. Details of the apparatus will be published elsewhere, but the method entails mounting an $YBa_2Cu_3O_{6+x}$ ellipsoid on an independently heated sapphire rod and positioning it at the centre of a cylindrical rutile dielectric resonator, with the microwave field accurately oriented along the crystal c direction. In both techniques the absolute penetration depth is obtained by requiring that the microwave conductivity in the normal state be purely real. Data from a sample with $T_c = 16.8$ K are plotted in Fig. 7, with data at seven other dopings to be published elsewhere. The microwave data are very similar to the H_{c1} data, with a sublinear correlation between T_c and $\rho_s(T \to 0)$, and a temperature slope that declines on underdoping. A striking feature of the data

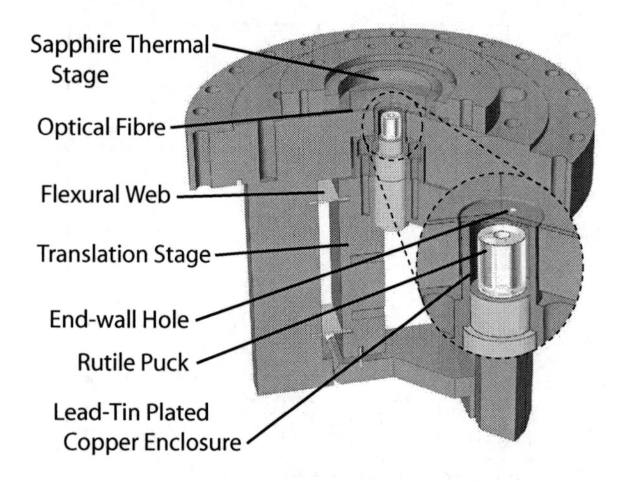

FIGURE 6. Surface cavity perturbation apparatus. The sample is positioned in the evanescent microwave fields above a hole in the end-wall of a rutile-dielectric-loaded 6.6 GHz resonator, inducing *ab*-plane screening currents. The sample is attached to a sapphire thermal platform that is thermally isolated from the resonator housing by optical fibres. The spacing between the sample and the movable resonator is varied in order to calibrate the sensitivity of the microwave experiment, with resonator position monitored by a precision capacitance transducer.

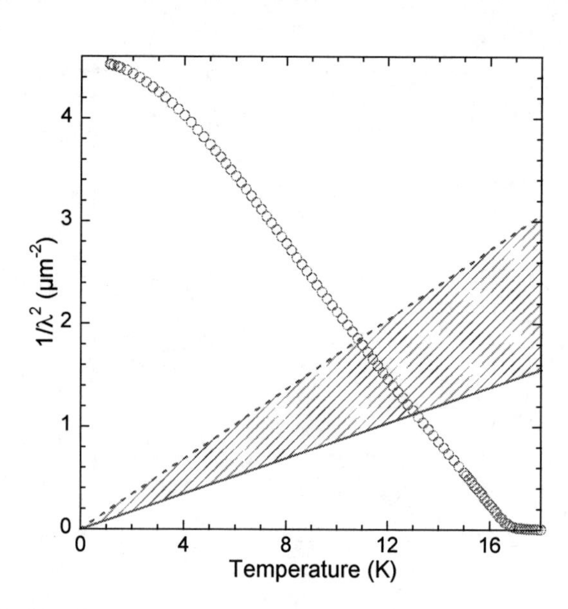

FIGURE 7. In-plane superfluid density of an $YBa_2Cu_3O_{6.333}$ ellipsoid with $T_c = 16.8$ K. The vortex unbinding transition, which is absent, should occur in the hatched region for a 2D superconductor. The upper bound corresponds to $\rho_{2D} = (2/\pi)T$ in each CuO_2 plane, the lower bound to $\rho_{2D} = (2/\pi)T$ in each CuO_2 bilayer.

in Fig. 7 is the wide range of linear temperature dependence, extending from T_c down to $T = 5$ K. Underdoped $YBa_2Cu_3O_{6+x}$ is electronically highly two dimensional, with $\lambda_c/\lambda_{ab} \approx 100$. As a result, a vortex unbinding transition should occur when the 2D superfluid density in a layer of thickness d, $\rho_{2D}(T) = \hbar^2 d/4k_B e^2 \mu_0 \lambda^2(T)$, falls to $(2/\pi)T$ [21]. This minimum 2D superfluid density is shown in Fig. 7 for both isolated CuO_2 planes and for bilayers, as the dashed and solid lines respectively. There is no sign of vortex unbinding, or of 3D-XY critical fluctuations. Related work on thin films *has* reported the observation of vortex unbinding, but the effective thickness of the fluctuating 2D layers is the *entire* film thickness [22]. Our experiments on bulk samples support this, with a mean-field-like transition in the superfluid density.

In conclusion, we have used room-temperature annealing of CuO-chain order to continuously tune hole doping in highly underdoped $YBa_2Cu_3O_{6+x}$. Using measurements of H_{c1} and microwave cavity perturbation we have probed the temperature and doping dependence of the in-plane superfluid density. We find a sublinear correlation between T_c and $\rho_s(T \to 0)$; a decline in the effectiveness of quasiparticle currents to deplete the superfluid density; and a mean-field like superfluid transition that shows neither vortex unbinding nor 3D-XY critical fluctuations.

We acknowledge useful discussions with M. J. Case, M. Franz and I. F. Herbut, and support from the National Science and Engineering Research Council of Canada and the Canadian Institute for Advanced Research.

REFERENCES

1. V. J. Emery, and S. A. Kivelson, *Nature* **374**, 434 (1995).
2. M. Prohammer, and J. P. Carbotte, *Phys. Rev. B* **43**, 5370 (1991).
3. W. N. Hardy, et al., *Phys. Rev. Lett.* **70**, 3999 (1993).
4. D. A. Bonn, and W. N. Hardy, "Microwave Surface Impedance of High Temperature Superconductors," in *Physical Properties of High Temperature Superconductors*, edited by D. M. Ginsberg, World Scientific, Singapore, 1996, vol. V, pp. 7–98.
5. R. Liang, et al., *Physica C* **383**, 1 (2002).
6. A. Hosseini, et al., *Phys. Rev. Lett.* **93**, 107003 (2004).
7. J. Zaanen, et al., *Phys. Rev. Lett.* **60**, 2685 (1988).
8. B. W. Veal, et al., *Phys. Rev. B* **42**, 6305 (1990).
9. D. E. Sheehy, T. P. Davis, and M. Franz, *Phys. Rev. B* **70**, 054510 (2004).
10. R. Liang, et al., *Phys. Rev. Lett.* **94**, 117001 (2005).
11. K. E. Gray, et al., *Phys. Rev. B* **45**, 10071 (1992).
12. Y. J. Uemura, et al., *Phys. Rev. Lett.* **62**, 2317 (1989).
13. X. J. Zhou, et al., *Nature* **423**, 398 (2003).
14. J. M. Harris, et al., *Phys. Rev. B* **54**, 15665 (1996).
15. Ch. Renner, et al., *Phys. Rev. Lett.* **80**, 149 (1998).
16. X. G. Wen, and P. A. Lee, *Phys. Rev. Lett.* **80**, 2193 (1998).
17. A. J. Millis, et al., *J. Phys. Chem. Solids* **59**, 1742 (1998).
18. A. C. Durst, and P. A. Lee, *Phys. Rev. B* **62**, 1270 (2000).
19. L. B. Ioffe, and A. J. Millis, *J. Phys. Chem. Solids* **63**, 2259 (2002).
20. P. A. Lee, and X. G. Wen, *Phys. Rev. Lett.* **78**, 4111 (1997).
21. I. F. Herbut, and M. J. Case, *Phys. Rev. B* **70**, 094516 (2004).
22. Y. Zuev, et al., cond-mat/0407113 (2004).

Momentum Dependence of Mott Gap Excitations in Optimally Doped $YBa_2Cu_3O_{7-\delta}$ Studied by Resonant Inelastic X-ray Scattering

K. Ishii*, K. Tsutsui†, Y. Endoh*,**, T. Tohyama†, K. Kuzushita*, T. Inami*, K. Ohwada*, S. Maekawa†, T. Masui‡, S. Tajima‡, Y. Murakami§,* and J. Mizuki*

*Synchrotron Radiation Research Center, Japan Atomic Energy Research Institute, Hyogo 679-5148, Japan
†Institute for Materials Research, Tohoku University, Sendai 980-8577, Japan
**International Institute for Advanced Studies, Kyoto 619-0025, Japan
‡Superconducting Research Laboratory, ISTEC, Tokyo 135-0062, Japan
§Department of Physics, Tohoku University, Sendai 980-8578 Japan

Abstract. Momentum dependence of Mott gap excitations in the twin-free $YBa_2Cu_33O_{7-\delta}$ at optimal doping has been studied by the resonant inelastic x-ray scattering at the Cu *K*-edge. We have succeeded to distinguish between the excitations from CuO chain and that from the CuO_2 plane. The excitation from the chain is enhanced at 2 eV near the zone boundary of the $\mathbf{b}^*$ direction, while the excitation from the plane is broad at 1.5-4 eV and almost independent of the momentum transfer. Theoretical calculation based on the one-dimensional and two-dimensional Hubbard models reproduces these features when we assume a smaller value of on-site Coulomb repulsion of the chain site than that of the plane site.

Keywords: resonant inelastic x-ray scattering, high-T_c superconductor, charge excitation
PACS: 74.25.Jb, 74.72.Bk, 78.70.Ck

$YBa_2Cu_3O_{7-\delta}$(YBCO) is a unique material among high-T_c superconductors because it has a characteristic crystal structure, that is, one-dimensional CuO chain runs along the crystalline **b** axis in addition to the double two-dimensional CuO_2 planes. While the CuO_2 plane is the most important unit in the superconductivity, the role of the CuO chain has not been clarified yet. Anisotropic characters in both normal [1, 2] and superconducting states [3, 4] are observed experimentally, which indicate the contribution of the CuO chain to the electronic properties. In order to separate the intrinsic electronic structure of the CuO chain from that of the CuO_2 plane, we applied the resonant inelastic x-ray scattering (RIXS) method using hard x-rays. It has a great advantage that electronic excitations can be observed with momentum resolution. In the RIXS studies of Mott insulators at the *K*-edge of transition metal elements, peak features of the Mott gap are observed at 2-3 eV [5, 6, 7, 8]. When holes are doped, the excitation across the Mott gap robustly persists and the excitation below the Mott gap develops simultaneously [9, 10]. Here we report a RIXS study of optimally doped YBCO (T_c = 93 K). From the momentum dependence of a twin-free crystal, we can successfully distinguish the Mott gap excitation of the CuO chain from that of the CuO_2 plane. In this paper, we give a brief summary of our previous letter [11] in additon to some supplementary information.

The RIXS experiments at Cu *K*-edge were carried out at BL11XU at SPring-8 [12]. A Si (400) channel-cut monochromator and a spherically-bent Ge (733) analyzer were used and the overall energy resolution is about 400 meV estimated from the full width at half maximum of the elastic scattering. The polarization of incident x-ray is π and is kept in the *ab* plane of the sample. We determined the incident energy (E_i) at 8990 eV where resonant peak of 2 eV was found at the scattering vector of $\mathbf{Q} = (4.5,0,0)$. All the data were collected at room temperature.

In Fig. 1, we show RIXS spectra of a twin-free YBCO crystal near the Brillouin zone center and at the zone boundaries. We found two characteristics in the low energy region which are considered to be excitations across the Mott gap. One is a momentum-independent broad feature at 1.5-4 eV and the other is an excitation at 2 eV which is prominent at the zone boundary of $\mathbf{b}^*$ axis, $(0,\pi)$ and (π,π). It is noted that the spectra of the zone boundary were measured at two different Brillouin zone and these characteristics are independent of the Brillouin zone. The momentum dependence of the Mott gap excitations can be seen more clearly in the contour plot in Fig. 2 where the elastic scattering and the higher-energy excitation at 5.5 eV are subtracted from the raw data. The excitation along the $\mathbf{a}^*$ and $\mathbf{b}^*$ axes should be equivalent in the CuO_2 plane. On the other hand, the momentum dependence along the $\mathbf{b}^*$ axis is larger than that along the $\mathbf{a}^*$ axis in the CuO chain be-

CP850, *Low Temperature Physics: 24th International Conference on Low Temperature Physics;*
edited by Y. Takano, S. P. Hershfield, S. O. Hill, P. J. Hirschfeld, and A. M. Goldman

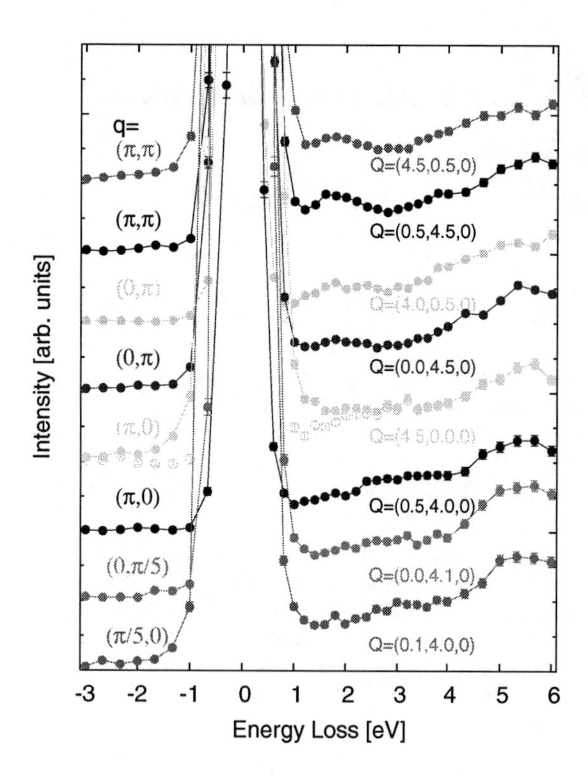

FIGURE 1. RIXS spectra of twin-free $YBa_2Cu_3O_{7-\delta}$ crystal at the optimal doping. **Q** and **q** denote the absolute momentum transfer and reduced wave vector in the *ab* plane, respectively. Filled circles are raw data and open circles of $\mathbf{Q} = (4.5, 0, 0)$ shows the data in which the elastic scattering is subtracted.

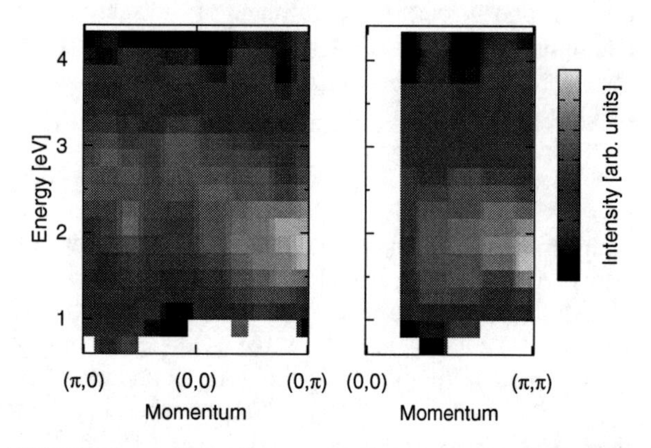

FIGURE 2. Momentum dependence of Mott gap excitations of $YBa_2Cu_3O_{7-\delta}$.

cause the chain runs along the **b** axis. Therefore we can assign the intensity enhanced at the zone boundary of $\mathbf{b}^*$ axis originates from the Mott gap excitation of the chain and the momentum-independent broad excitation comes from the plane. Smaller dispersion in the hole-doped CuO_2 plane than that of the undoped one [5, 6] was predicted in a theoretical calculation [13].

In order to confirm the assignment of the excitations more accurately, we calculated the RIXS spectra by the numerically exact diagonalization technique on small clusters [11]. One-dimensional and two-dimensional Hubbard models are used for the chain and the plane, respectively, where the Zhang-Rice band is regarded as the lower Hubbard band. The two characteristics mentioned above are well reproduced. A smaller value of the Coulomb repulsion (U) for the chain than that of the plane is needed to explain the peak energy of the Mott gap excitations, which suggests that the charge transfer gap of the chain is smaller than that of the plane.

In summary, we have carried out a RIXS study for optimally doped $YBa_2Cu_3O_{7-\delta}$. From the momentum dependence of a twin-free crystal, we could separate the Mott gap excitation in the CuO chain from that in the CuO_2 plane. Combined with the theoretical calculation performed simultaneously, the Mott gap of the chain is found to be smaller than that of the chain.

ACKNOWLEDGMENTS

K. T., T. T., and S. M., were supported by the NAREGI Nanoscience Project and a Grant-in-Aid for Scientific Research from the Ministry of Education, Culture, Sports, Science and Technology of Japan. The numerical calculations were performed in the supercomputing facilities in ISSP, Univ. of Tokyo and IMR, Tohoku University. The crystal growth was supported by the New Energy and Industrial Technology Development Organization (NEDO) as the Collaborative Research and Development of Fundamental Technologies for Superconductivity Applications.

REFERENCES

1. B. Koch, et al., *Solid State Comm.*, **71**, 495 (1989).
2. K. Takenaka, et al., *Phys. Rev. B*, **50**, 6534 (1994).
3. D. N. Basov, et al., *Phys. Rev. Lett.*, **74**, 598 (1995).
4. R. Gagnon, et al., *Phys. Rev. Lett.*, **78**, 1976 (1997).
5. M. Z. Hasan, et al., *Science*, **288**, 1811 (2000).
6. Y. J. Kim, et al., *Phys. Rev. Lett.*, **89**, 177003 (2002).
7. T. Inami, et al., *Phys. Rev. B*, **67**, 045108 (2003).
8. Y.-J. Kim, et al., *Phys. Rev. Lett.*, **92**, 137402 (2004).
9. Y.-J. Kim, et al., *Phys. Rev. B*, **70**, 094524 (2004).
10. K. Ishii, et al., *Phys. Rev. B*, **70**, 224437 (2004).
11. K. Ishii, et al., *Phys. Rev. Lett.*, **94**, 187002 (2005).
12. T. Inami, et al., *Nucl. Instrum. Method Phys. Res., Sect. A*, **467-468**, 1081 (2001).
13. K. Tsutsui, et al., *Phys. Rev. Lett.*, **91**, 117001 (2003).

Possible Formation of Bosonic Stripes in Temperature Range up to 1200 K: Verification of the Predicted Five-Level T-δ Diagram for $YBa_2Cu_3O_{6+\delta}$

Alexander V. Mitin

P. L. Kapitza Institute for Physical Problems, RAS, Kosygin Str. 2, Moscow 119334, Russian Federation

Abstract. Data of recent experiments evidence that the superconducting order does not disappear inside the pseudogap region of underdoped cuprates and can persist, e.g., within segregated hole-rich stripes. As follows from the author's scenario, the quantized width of dominant bosonic stripes decreases from $6a$ to $2a$ with lowering δ in $YBa_2Cu_3O_{6+\delta}$, where a is the mean Cu-Cu distance in CuO_2 layers. Such a shrinkage is accompanied by rise of local critical temperatures up to 1200 K. Experimental results are found to be in agreement with the predicted five-level T-δ diagram for $YBa_2Cu_3O_{6+\delta}$.

Keywords: Superconductivity, cuprates, pseudogap, bosonic stripes, transport properties.
PACS: 74.20.Mn, 74.25.Dw, 74.25.Fy, 74.72.Bk, 74.81.-g

Comparative studies of underdoped oxides allowed to conclude that a pseudogap genesis can be explained in a self-consistent way by presence of segregated hole-rich superconducting filaments with local critical temperatures $T^*_{ci} \sim \hbar^2/2k_B m_e d_B^2$ up to 800 K [1], where k_B is the Boltzmann constant, m_e is the electron rest mass, and $1/d_B^2$ is the boson density ($d_B \geq 2a$). As a development of this idea, it was proposed the string model [2] for cuprate electron spectra (ES) which is based on the direct overlap of O $2p$-like states. Besides, the model takes into account coherent displacements of Cu^{2+} with regard to the oxygen sublattice. This results in a fragmentation of CuO_2 layers into domains with violation of rotational symmetry and dimerization of O-O bonds.

Both factors promote a partial bosonization of ES with generation of extended coherent states along strings composed of overlapping antisymmetric O $2p(\sigma^*)$ orbitals. For fully filled O $2p$ shells, the lowest values of self-frequency ω_o (zero mode) and binding energy $E^*_B \approx \hbar\omega_o = \hbar^2/2m_e r_o^2 \approx 2.05$ eV of these states near the top of $2p(\sigma)$ subband are mainly depending on the spacing $2r_o \approx 2.72$ Å of O^{2-} positions [2]. Accordingly, the bottom of $2p(\sigma)$ subband is determined by the nearest mode $E_B \approx \hbar\omega_o(1+2n) \approx 6.15$ eV. These modes were recently observed in $Ca_2CuO_2Cl_2$ with the rather perfect CuO_2 layers [3]. The middle of the interval $w_\sigma = E_B - E^*_B$ is occupied by states from the narrower $2p(\tau)$ subband of width $w_\tau \approx 2$ eV. In addition to Cu^{2+} displacements, the composite $2p(\sigma^*)$ bosons are coupled by exchange interaction $E^*_{ex} \approx -E^*_B t_\sigma^2/w_\sigma^2 \approx -128$ meV, where $|t_\sigma| = w_\sigma/4$ is the hopping integral. The zero mode ω_o of the $2p(\sigma^*)$ boson strings assigns the vacuum of the Hilbert space for hole excitations in CuO_2 layers.

Insertion of oxygen into $YBa_2Cu_3O_{6+\delta}$ basal planes produces excitations in the form of quantized hole orbitals, whose hierarchy is defined by rank η (an analog of the principal quantum number in atoms). At $\delta < 1/8$, the CuO_2 layers are occupied mainly by second-rank hole orbitals or, in other words, by fermion-like curls of convolved strings involving $4\eta-1$ correlated $2p$-like states with the binding energy $E'_{B\eta} = \hbar\omega_o/2\eta \approx 0.5$ eV [2]. The bound state of the dopant ion and hole curl can be viewed as a pseudo-atom [2]. The further rise of δ leads to a crystallization of pseudo-atom "nuclei" with supercells $2\sqrt{2}\times2\sqrt{2}$ and 1×8 [4] (in units of the Cu^{2+} lattice). The latter implies emergence of $\{CuO_3\}_n$ chains creating potential wells where the curls are collectivized into extended strings (crooked as two-dimensional springs of width $w_\eta = \eta a$) carrying ordered hole pairs. These bosonic stripes (BS) have to be stable up to $T^*_{c\eta}(\delta) = C^*_\eta D^*_\eta \hbar\omega_o/2k_B(2\eta^2+\eta) \approx 1200$ K at $\eta = 2$ [2], where $C^*_\eta \leq 1$ is the factor of BS compatibility with potential wells, $D^*_\eta = 1-(1-\delta/\delta_\eta)^2$, and δ_η is the optimal doping for stripes of given η. To achieve $T^*_{c\eta} \approx 1200$ K, the space between BS of $\eta = 2$ must be filled by fermion-like curls of second rank that yields $\delta_\eta = 1/8 + 6/8^2 \approx 0.22$. The BS and surrounding curls are identified with the peak and hump in the famous "peak-dip-hump" structure of ES. At that, the boson states are located at $E''_{B\eta} \approx \hbar\omega_o/8\eta$, while the curl states are distributed around $E'_{B\eta} = 4E''_{B\eta}$.

CP850, *Low Temperature Physics: 24th International Conference on Low Temperature Physics;*
edited by Y. Takano, S. P. Hershfield, S. O. Hill, P. J. Hirschfeld, and A. M. Goldman

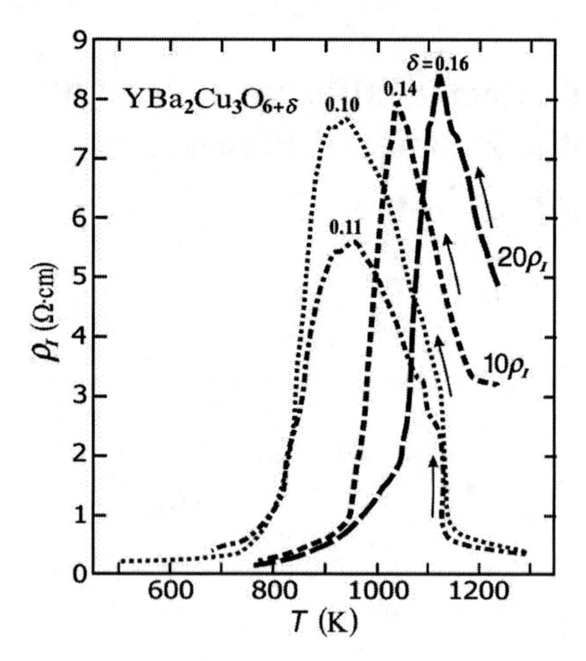

FIGURE 1. Change of the integrated resistivity of underdoped samples $YBa_2Cu_3O_{6+\delta}$ measured under cooling in argon flow.

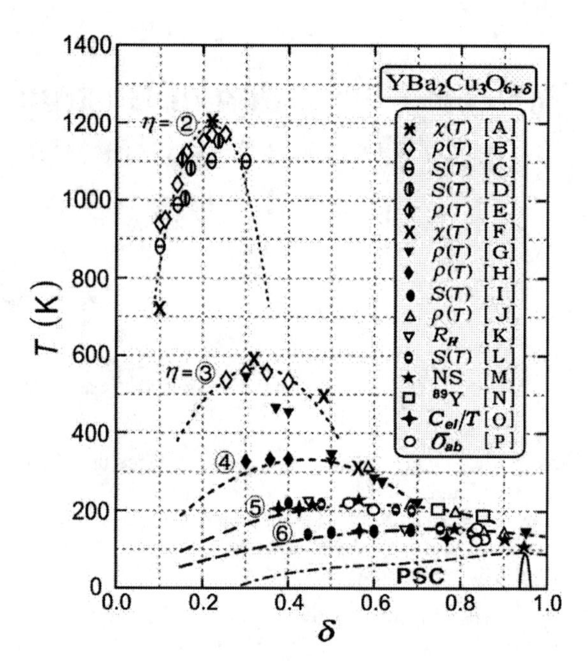

FIGURE 2. Comparison of the predicted five-level T-δ diagram for $YBa_2Cu_3O_{6+\delta}$ (dash lines) with experimental data.

Formation of frustrated superconducting channels composed of BS with a gap $|\Delta_{s\eta}| = E''_{B\eta} \approx 128$ meV should be detected as a decline in resistivity $\rho(T)$ and thermopower $S(T)$ below $T^*_{c\eta}(\delta)$ under cooling samples with $\delta < 1/4$. Moreover, the samples with $\delta < 1/8$ have to demonstrate a maximum in $\rho(T)$ at $T^*_{c\eta}(\delta)$, since the formation of BS network should occur after condensation of curls into pancakes initiating a dielectrization of spacings between ones with drastic growth in $\rho(T)$.

To verify these predictions, the simultaneous measurements of $S(T)$ and "integrated resistivity" $\rho_I(T)$ were recently performed [5]. The use of the term $\rho_I(T)$ is caused by nonlinearity of I-V characteristics [1]. A set of $\rho_I(T)$ curves obtained at the same current density J=0.1 A/cm^2 for samples with $\delta \leq 0.16$ is shown in Fig. 1. They resemble the sharp maxima in $\rho(T)$ of $La_2CuO_{4+\delta}$ which had been interpreted in terms of filamentary superconductivity [6]. It is worth to note that the variation of $\rho_I(T)$ at $T > 1150$ K for $\delta \leq 0.12$ obeys the thermally activated hopping law $\rho_I(T) = \rho_o \exp(\varepsilon_A/k_B T)$ with expected $\varepsilon_A \approx E'_{B\eta} \approx 0.5$ eV. The locations of high-temperature maxima in $\rho_I(T)$ and in $S(T)$ curves are depicted in Fig. 2.

With growing δ, the huge peaks in $\rho_I(T)$ near 1200 K transform into kinks. Moreover, kinks and local maxima appear in the range 500 K $< T <$ 580 K indicating, within the proposed scenario, a formation of BS with η=3 (symbols B in Fig. 2). Symbols C relate to maxima in $S(T)$. Fig. 2 also includes the T^* values from papers cited in Ref. [5]. One can see that almost all data group near five dashed lines defined by the unified formula $T^*_{c\eta}(\delta) = D^*_\eta \hbar\omega_o / 2k_B(2\eta^2+\eta)$ with $\eta = 2 \div 6$.

The region of percolation superconductivity (PSC) is bounded by dash-dotted line in Fig. 2. The most uniform superconducting state takes place at $\delta = 0.96$, when dynamically ordered segments of sixth-rank BS with a size $6a \times 48b$ (in the optimal case $b = 1.015a$) overlay CuO_2 layers. Quantum broadening of momentum across stripes $\varepsilon_q = \hbar^2/m_e(\eta a)^2 \approx 14$ meV leads to anisotropic gaps $|\Delta_{s\eta}|_b = E''_{B\eta} = \hbar\omega_o/8\eta \approx 43$ meV and $|\Delta_{s\eta}|_a = E''_{B\eta} - \varepsilon_q \approx 29$ meV along and perpendicularly to $\{CuO_3\}_n$ chains in accordance with the ARPES data [7]. Besides, extended diagonal strings with loosely-coupled pairs imitate a suppression of $|\Delta_{s\eta}|_d$ along the O-O bonds.

In summary, the obtained results may help to explain a controversial nature of pseudogap manifestations and their relation to the superconducting ordering in cuprates.

This work was supported by the Branch of Physical Sciences of RAS under Program for Basic Research "Strongly Correlated Electrons in Semiconductors, Metals, Superconductors, and Magnetic Materials" (Project 3.5) and by the RFBR (Grant No. 05-08-50074).

REFERENCES

1. A. V. Mitin et al., *JETP* **80**, 1075-1089 (1995).
2. A. V. Mitin, *Proc. 14-th Ural Int. Winter School on the Physics of Semiconductors.* Ekaterinburg, 2002, L10.
3. F. Ronning et al., *Phys. Rev. B* **67**, 165101-12 (2003).
4. A. A. Aligia et al., *Phys. Rev. B* **57**, 1241-1247 (1998).
5. A. V. Mitin, *Bulletin of the RAS. Physics* **69**, 576-579 (2005).
6. P. M. Grant at al., *Phys. Rev. Lett.* **58**, 2482-2485 (1987).
7. D. H. Lu at al., *Phys. Rev. Lett.* **86**, 4370-4373 (2001).

Superfluid Density and Residual Conductivity in Optical Spectra of YBCO

T.Kakeshita*, T.Masui† and S.Tajima†

*Superconductivity Research Laboratory, ISTEC, 1-10-13 Shinonome, Koto-ku, Tokyo 135-0062, Japan
†Dept. of Physics, Osaka University, Machikaneyama 1-1, Toyonaka, Osaka 560-0043, Japan

Abstract. We report on superfluid density estimated from optical spectra of $YBa_2Cu_3O_{7-\delta}$(YBCO). Some recent reports have suggested that optical measurements generally underestimate the superfluid density compared to μSR. We discuss the origin of the suppression of the superfluid density with nano-scale inhomogeneous picture on CuO_2 plane.

Keywords: superfluid density, residual conductivity, inhomogeneous, nano-scale, $YBa_2Cu_3O_{7-\delta}$
PACS: 71.27.-a,74.25Gz,74.72Bk

INTRODUCTION

Now it is focused whether inhomogeneity in CuO_2 plane is essential for high-Tc superconductors. Some cuprates exhibit such an inhomogeneous behavior[1, 2, 3]. The scanning tunneling spectroscopy (STM) measurement[1] demonstrated an existence of superconducting and nonsuperconducting domains in plane whose size is nano-scale (~nm) and therefore the inhomogeneity is, so-called, "nano-scale" phase separation. The number of superconducting domains is increasing with doping in the underdoped region and the behavior is qualitatively similar to a universal relation between superfluid density and Tc as seen by μSR measurements[4].

It is known that optical spectra can also estimate superfluid density from a sum rule of optical conductivity. Also in case of optical spectra, the behavior is roughly consistent with μSR and STM, however, some recent optical measurements on cuprates indicate that the superfluid density estimated from optical spectra is considerably smaller than that of μSR measurements[5, 6]. As the reason, we consider the difference of the probes. An optical measurement is a macro-scale probe applying electric field , while the measurement to estimate superfluid density by μSR is a micro-scale probe in the presence magnetic field. We discuss a relation between superfluid density and residual conductivity, considering the difference.

EXPERIMENTAL RESULTS AND DISCUSSION

Single crystals were grown by a conventional flux method. We use twin-free crystals of YBCO with Tc of 77K(underdoped). Fig.1 shows the reflectivity for the underdoped composition. For the a-axis polarization, the spectra exhibit typical behavior seen in 90~130K-class cuprates[7, 8, 9, 10] . A parabolic shape emerges below ~1000cm^{-1} in the superconducting state and a dip-like behavior is clearly seen around the frequency. The parabolic behavior already appears in the normal state and the appearing temperature seems consistent with the pseudogap temperature[11]. The reflectivity of the optimum doping is higher than that of the underdoped one, however, the position of the dip-like behavior is not so changed. On the other hand, the b-axis spectrum is different from the a-axis. The b-axis spectra of the underdoped composition have three features which are not seen for the a-axis polarization. First, some optical phonons due to oxygen deficiencies in the chain structure appear with decreasing temperature. The most remarkable mode exists at ~230cm^{-1}. The behavior is consistent with optical reflectivity using untwinned crystal[12] and Raman spectra of underdoped composition[13]. Second, the superconducting response is fairly suppressed compared to a-axis. Despite that the reflectivity of b-axis spectra is higher than that of a-axis in far-infrared region in the normal state, it is lower in the superconducting state. Third, the dip-like behavior is weak compared to a-axis. It is considered that the behavior is screened by the normal state carriers in the chain structure.

Fig.2 shows the optical conductivity obtained from Kramers-Kronig transformation. The missing area in the conductivity spectra between just above Tc (80K) and the lowest temperature (5K) corresponds to a superfluid density at ω=0. It seems that the optical conductivity has two components at low temperatures; Drude term of quasiparticle in the nodal direction $(\pi/2, \pi/2)$ and d-wave symmetric gap in antinodal direction $(\pi, 0)$[14]. The total b-axis conductivity is totally larger than that of the a-axis due to the carriers in the chain structure. The missing area of b-axis is narrower than that of a-axis. Fig.3 shows a comparison to the universal line by

CP850, *Low Temperature Physics: 24th International Conference on Low Temperature Physics;*
edited by Y. Takano, S. P. Hershfield, S. O. Hill, P. J. Hirschfeld, and A. M. Goldman

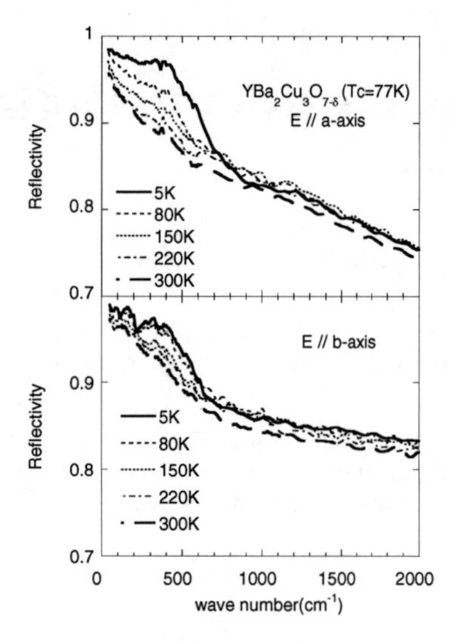

FIGURE 1. Reflectivity of underdoped YBCO.

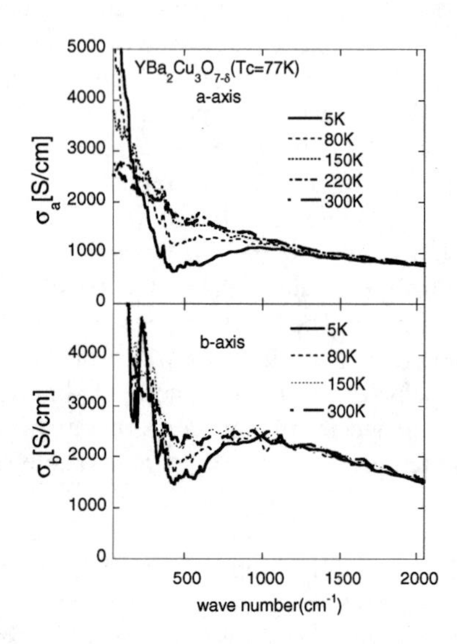

FIGURE 2. Optical conductivity of underdoped YBCO.

μSR[4]. The values obtained by optical spectra are a factor of 2$\sim$3 smaller than those of μSR.

Here, we discuss the origin of the strong suppression of the superfluid density in optical spectra. It is suggested that a disorder state in CuO_2 plane results in the residual conductivity[15]. We suggest that the discrepancy between μSR and optical measurement is essential to the nano-scale disorder correlated with one of the most remarkable feature of cuprates, extremely short coherent length, $\sim$nm. μSR is a micro-scale probe and obtain an local information of inhomogeneous CuO_2 plane. On the other hand, optical measurement is a macro-scale probe and observe the average of the plane. We think such nanoscale inhomogeneity is the origin of the difference of superfluid density between μSR and optical measurements. We also plot the value from Zn-YBCO[16]. It suggests that an increase of residual conductivity is closely connected with suppression of superfluid density.

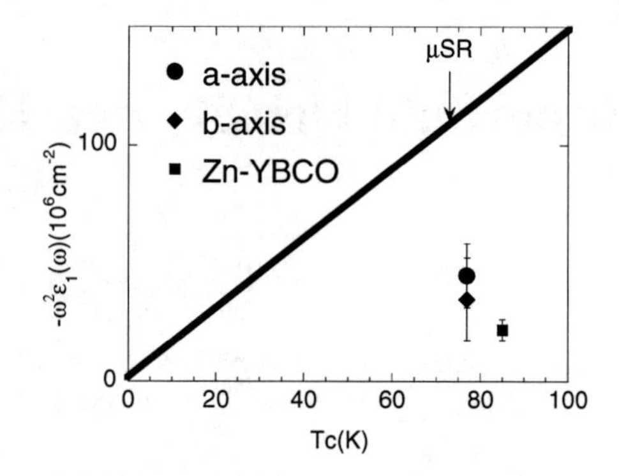

FIGURE 3. Plot of superfluid density against Tc.

SUMMARY

We have discussed the relation between superfluid density and residual conductivity. We suggest that the difference between μSR and optical data results from scale difference in probing nano-scale inhomogeneity in the CuO_2 plane.

ACKNOWLEDGMENTS

This work was supported by the New Energy and Industrial Technology Development Organization(NEDO) through ISTEC as the Collaborative Research and Development of Fundamental Technologies for Superconductivity Applications..

REFERENCES

1. K. M. Lang *et al., Nature* **415**, 412 (2002).
2. T. Hanaguri *et al., Nature* **430**, 1001 (2004).
3. J. M. Tranquada *et al., Nature* **375**, 561 (1995).
4. Y. J. Uemura *et al., Phys. Rev. Lett.* **62**, 2317 (1989).
5. C. C. Homes *et al., Nature* **430**, 539 (2004).
6. S. Tajima *et al., Phys. Rev. B* **71**, 094508 (2005).
7. D. N. Basov *et al., Phys. Rev. Lett.* **77**, 4090 (1996).
8. N. L. Wang *et al., Phys. Rev. B* **68**, 054516 (2003).
9. J. J. McGuire *et al., Phys. Rev. B* **62**, 8711 (2000).
10. J. J. Tu *et al., Phys. Rev. B* **66**., 144514 (2002).
11. K. Takenaka *et al., Phys. Rev. B* **50**, 6534 (1994).
12. N. L. Wang *et al., Phys. Rev. Lett.* **89**, 87003 (2002).
13. M. F. Limonov, *et al., Phys. Rev. B* **61**, 12412 (2000).
14. T. Yoshida *et al., Phys. Rev. Lett.* **91**, 027001 (2003).
15. S.Barabash *et al., Phys. Rev. B* **61**, 14924 (2000).
16. N. L. Wang *et al., Phys. Rev. B* **57**, 11081 (1998).

Charge Dynamics of the High-Temperature Superconductor $YBa_2Cu_3O_y$

Kozo Okazaki, Koichi Yonetani, Nobuyuki Kaji and Shunji Sugai

Department of Physics, Nagoya University, Nagoya 464-8602, Japan

Abstract. We have studied the charge dynamics of the high-temperature superconductor $YBa_2Cu_3O_y$ (YBCO) by measuring optical reflectivity at various temperatures from the insulating sample ($y = 6.3$) to the optimally-doped sample ($y = 6.91$). We have found that the spectral weight of the optical conductivity increases in the narrow low-energy region ($\hbar\omega < 0.06$ eV) with decreasing temperature. This behavior is attributed to an an evolution of the coherent quasiparticle band with temperature.

Keywords: $YBa_2Cu_3O_y$, $La_{2-x}Sr_xCuO_4$, optical conductivity, quasiparticle band
PACS: 74.25.Gz, 74.72.Bk, 78.20.-e, 72.80.Ga

The high-T_c superconductors have been extensively studied since their discovery and some anomolous metallic behaviors have been found in the normal state of the underdoped or optimally doped region, such as a T-linear resistivity [1] or a temperature-dependencet Hall coefficient [2]. However, the origins of these anomolous behaviors have not been fully understood yet. These problems should be important to understand how the superconducting state realizes with carrier doping. Systematic measurments of optical properties with a wide composition and temperature rage are expected to provide fruitful informations to study these problems.

In this study, we have performed systematic measurements of the optical reflectivity of $YBa_2Cu_3O_y$ (YBCO) at various temperatures from the insulating sample (y = 6.3) to the optimally-doped sample (y = 6.91) and deduced the optical conductivity. We have observed an evolution of a Drude-like componnet of the optical conductivity with hole doping and found that the spectral weight of the Drude-like component increases in the narrow low-energy region ($\hbar\omega < 0.06$ eV) with decreasing temperature rather than its width decreases expected from a simple Drude theory. This behavior is attributed to an evolution of the coherent quasiparticle band with temperature and should be important to understand the charge dynamics of YBCO.

Single crystals of YBCO were grown by a conventional self-flux method with Y_2O_3 crucibles. The obtained single crystals were sealed in quartz tubes with ceramics quenched from various temperatures and annealed at the same temperatures. The oxygen concentration was determined from the quenched temperature [3]. The amounts of oxygen and quenched temperatures are 6.30 (900 °C), 6.37 (800 °C), 6.47 (750 °C), 6.72 (600 °C), and 6.91 (400 °C). The superconducting transition temperatures (T_c) determined by measurements of dc resistivity were 0 K (y = 6.30), 20 K (y = 6.37), 47 K (y = 6.47), 67 K (y = 6.72), and 91 K (y = 6.91). The hole concentration per Cu atom in the CuO_2 plane, p is estimated according to Tallon *et al.* [4].

Near-normal incidence in-plane ($E \perp c$) reflectivity spectra $R(\omega)$ were measured using a Fourier-type interferometer (0.006 - 1.2 eV) and a grating spectrometer (0.8 - 6.25 eV). We have deduced the optical conductivity $\sigma(\omega)$ via Kramers-Kronig transformation. For the extrapolation in the low-energy region, we have assumed a Hagen-Rubens formula. In the high-energy region, we have used room-temperature data in Ref.[5] below 31 eV and extrapolated assuming $R(\omega) \propto \omega^{-4}$ above 31 eV.

Figure 1 shows the optical conductivity of $YBa_2Cu_3O_y$ at various temperatures above T_c. We can see that Drude-like components peaked around $\omega = 0$ evolve with hole doping and decreasing temperature except for $y = 6.30$. In the spectrum of y = 6.30, a broad structure is observed in the midinfrared (mid-IR) region ($\sim$ 0.5 eV). This is a so-called "mid-IR absorption", which is also observed in $La_{2-x}Sr_xCuO_4$ (LSCO) [6]. This mid-IR absorption becomes weaker with hole doping and almost vanishes at y = 6.47. This doping dependence is in contrast to LSCO. For the case of LSCO, the mid-IR absorption is commonly observed for unperdoped samples [6]. This may reflect a different evolution of the elctronic structure with hole doping between YBCO and LSCO. For $6.37 \leq y \leq 6.72$, the Drude-like component has a broad peak centered around 30 meV above 200 K. Charge transport may become incoherent above this temperature in these compositions [7].

The inset of each panel in Fig. 1 shows the effective carrier number per unit cell, $N_{\mathrm{eff}}(\omega)$ defined as,

$$N_{\mathrm{eff}}(\omega) \equiv \frac{2m_0 V}{\pi e^2} \int_0^{\omega} \sigma(\omega') d\omega', \qquad (1)$$

where m_0 is a bare-electron mass and V is the unit-cell

CP850, *Low Temperature Physics: 24th International Conference on Low Temperature Physics;*
edited by Y. Takano, S. P. Hershfield, S. O. Hill, P. J. Hirschfeld, and A. M. Goldman

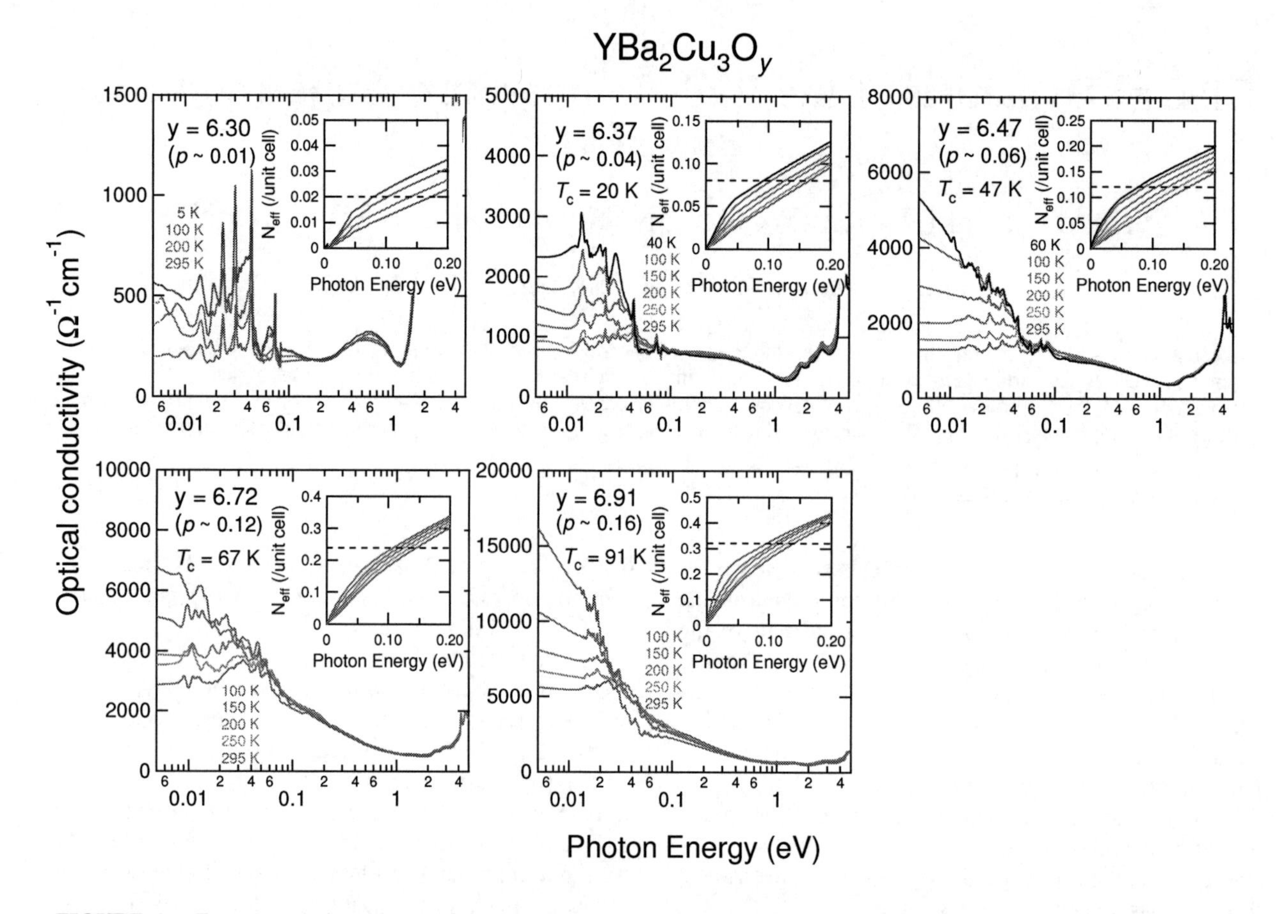

FIGURE 1. Temperature dependencc of the optical conductivity of $YBa_2Cu_3O_y$. The inset of each panel shows temperature dependence of $N_{eff}(\omega)$. Hole density per Cu atom in the CuO_2 plane is indicated by p and the dashed lines in the inset indicate $2p$.

volume. We note that the unit cell of $YBa_2Cu_3O_y$ includes two CuO_2 planes and one CuO chain. The dashed lines indicate $2p$ (number of the CuO_2 planes per unit cell × hole density per Cu atom). This corresponds to the f-sum rule when the holes are doped only in the CuO_2 planes. It is interesting that the sum rule is almost fulfilled within 0.1 eV. This means that most of the Drude-like component is contributed from the conducting carriers in the CuO_2 planes. Furthermore, temperature dependence of $N_{eff}(\omega)$ is almost restricted in the narrower energy region ($\hbar\omega < 0.06$ eV). This can be understood that the spectral weight of the coherent quasiparticle band in the CuO_2 planes evolve with decreasing temperature but its life time or scattering rate does not depend on temperature significantly. This behavior is much different from the behavior expected from a simple Drude theory. In this theory, the temperature dependence of conductivity should originate from the temperature variation of the scattering rate. This behavior should be important to understand charge dynamics of YBCO.

This work was supported by the 21st Century COE program of Nagoya University and Grant-in-Aid for Scientific Research from the Ministry of Education, Culture, Sports, Science and Technology, Japan.

REFERENCES

1. M. Gurvitch and A. T. Fiory, *Phys Rev. Lett.* **59**, 1337 (1987).
2. T. Nishikawa, J. Takeda and M. Sato, *J. Phys. Soc. Jpn.* **63**, 1441 (1994).
3. K. Kishio, J. Shimoyama, T. Hasegawa, and K. Kitazawa, *Jpn. J. Apl. Phys.*, **26**, L1228 (1987).
4. J. L. Tallon, C. Bernhard, H. Shaked, R. L. Hitterman, and J. D. Jorgensen, *Phys. Rev. B* **51**, 12911 (1995).
5. S. Tajima, H. Ishii, T. Nakahashi, T. Takagi, S. Uchida, M. Seki, S. Suga, Y. Hidaka, M. Suzuki, T. Murakami, K. Oka, and H. Unoki, *J. Opt. Soc. Am. B* **6**, 475 (1989).
6. S. Uchida, T. Ido, H. Takagi, T. Arima, Y. Tokura, and S. Tajima, *Phys. Rev. B* **43**, 7942 (1991).
7. K. Takenaka, J. Nohara, R. Shiozaki, and S. Sugai, *Phys. Rev. B* **68**, 134501 (2003).

Transport Properties of Zn-doped $Y_2Ba_4Cu_7O_{15-\delta}$

A. Matsushita,[a] Y. Yamada,[b,d] S. Sekiya,[c] T. Aoyagi,[a] K. Fukuda,[c,d] and F. Ishikawa[c,d]

[a] *National Institute for Materials Science, 1-2-1 Sengen, Ibaraki 305-0047, Japan*
[b] *Department of Physics, Niigata University, Niigata 950-2181, Japan*
[c] *Graduate School of Science and Technology, Niigata University, Niigata 950-2181, Japan*
[d] *Center for Transdisciplinary Research, Niigata University, Niigata 950-2181, Japan*

Abstract. We have measured the electrical resistivity and Hall coefficient of $Y_2Ba_4(Cu_{1-x}Zn_x)_7O_{15-\delta}$ (x = 0 - 0.05) polycrystalline samples synthesized using a high-oxygen-pressure technique. The sign of the Hall coefficient was positive above the superconducting transition temperature T_c for non-doped $Y_2Ba_4Cu_7O_{15-\delta}$ (Y247). The Hall coefficient dramatically changed after 4% Zn doping; the sign of the Hall coefficient changed from positive to negative below about 120 K and still showed superconductivity in 4%Zn-doped Y247. These behaviors could be explained by considering another conduction path, the so-called CuO double chain, instead of the CuO_2 plane.

Keywords: transport properties, $Y_2Ba_4Cu_7O_{15}$, Zn doping, CuO double chains
PACS: 74.25.Fy, 74.62.Dh, 74.72.Bk

INTRODUCTION

After a report of high-T_c superconductivity in the CuO double chains in reduced $Pr_2Ba_4Cu_7O_{15-\delta}$ (Pr247)[1], several studies have confirmed it [2,3]. Pr247 is a compound of the Pr123 family. This compound has both CuO double chains and CuO single chains and shows metallic conductivity at low temperatures owing to the metallic conduction in the CuO double chains. The structure of the CuO double chains is so rigid that it has no oxygen deficiency. Because of the one-dimensional structural features of the CuO double chains, their physical properties are very interesting. Some researchers have suggested that the CuO double chain might be in the Tomonaga-Luttinger liquid state [4-6]. On the other hand, CuO single chains easily absorb or release oxygen atoms. Therefore, the carrier density can be varied by controlling the amount of oxygen deficiency in CuO single chains.

In Pr247, the charge carriers in the CuO_2 planes are considered to be removed owing to Pr 4f - O 2p hybridization [7]. Therefore, the conduction path through CuO_2 planes is eliminated, and the transport property of the CuO double chains appears. It is well known that the Zn substitution in $YBa_2Cu_3O_7$(Y123) significantly suppresses the superconductivity. The Zn atoms reside in the CuO_2 planes, and the electrical resistivity becomes semiconductor-like. Thus, Zn-doping is also expected to eliminate the conduction in the CuO_2 planes and may demonstrate the transport properties or superconductivity of CuO double chains. In this paper, we report the transport properties of Zn-doped Y247 to elucidate the transport properties of CuO double chains.

EXPERIMENTAL

The precursor of the sample was prepared using a polymerized complex method with appropriate amounts of Y_2O_3, CuO, $Ba(NO_3)$, and ZnO. These were decomposed at 800-880ºC in air for 24 hours. The precursor was pressed into a pellet and then reacted in pure oxygen gas of 10 atm at 986ºC for 30 hours (the as-sintered sample). The reduced sample was obtained by post-annealing in a vacuum at 400°C for 3-24 hrs. The final polycrystalline samples were characterized by X-ray diffraction with Cu Kα radiation. Samples were confirmed to be the Y247 structure containing a small amount of the $Ba(Cu,Zn)O_2$ phase and the ZnO phase. The Zn concentrations were determined with an electron probe microanalyzer (EPMA). The Hall coefficient was measured with the conventional 5-probe method. The electrical resistivity and magnetoresistance up to 7 T were measured with the conventional 4-probe method.

CP850, *Low Temperature Physics: 24th International Conference on Low Temperature Physics;*
edited by Y. Takano, S. P. Hershfield, S. O. Hill, P. J. Hirschfeld, and A. M. Goldman

RESULTS AND DISCUSSION

Figure 1 shows the electrical resistivity for the non-doped Y247, $Y_2Ba_4(Cu_{0.96}Zn_{0.04})_7O_{15-\delta}$ (4%Zn-Y247), and the reduced 5%Zn-Y247. The values of T_c were 85 K and 25 K for Y247 and 4%Zn-Y247, respectively, while the reduced 5%Zn-Y247 showed semiconducting behavior. Figure 2 shows the Hall coefficients of Y247 and 4%Zn-Y247. The Hall coefficient of Y247 was positive through the normal state. On the other hand, the behavior of the Hall coefficient of 4%Zn-Y247 was quite different. It was positive at high temperatures but rapidly changed to negative around 110 K as decreasing temperature. The reduced 5%Zn-Y247 showed similar behavior.

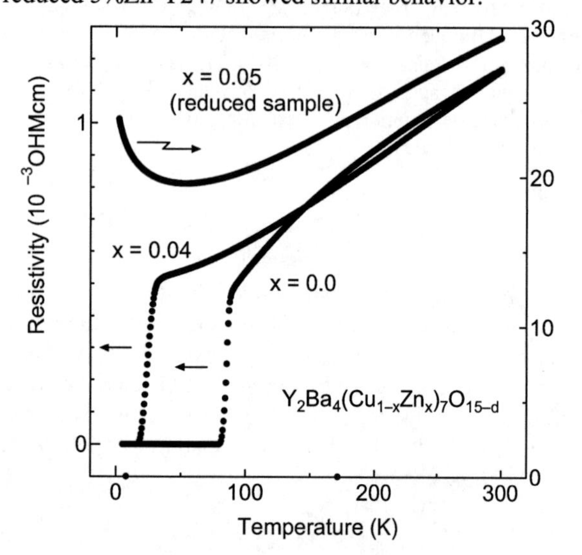

FIGURE 1. Electrical resistivity as a function of the temperature.

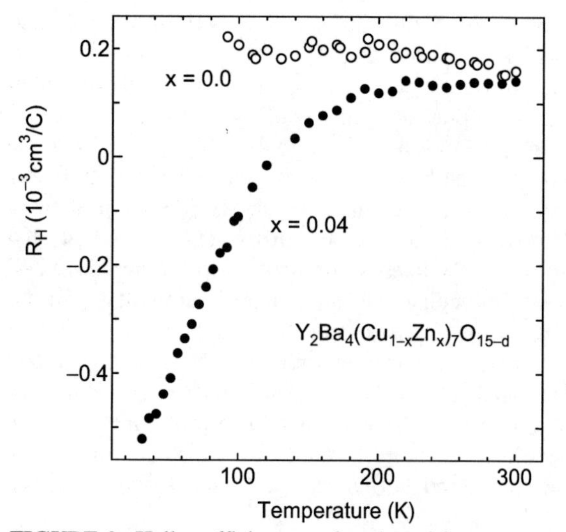

FIGURE 2. Hall coefficient as a function of the temperature.

In all of the high-T_c cuprate superconductors reported so far, the central structural unit is the CuO_2 plane, where superconductivity occurs. The sign of the charge carriers in the CuO_2 planes depends on the existence of the so-called apical oxygen. Most of the high-T_c cuprate superconductors have apical oxygen, and its carrier is a hole. The exception is the $(Nd,Ce)_2CuO_4$ compound with a so-called T' structure, which does not have apical oxygen atoms and whose carrier is an electron. The positive Hall coefficient of Y247 is consistent with this prediction, since the structure of Y247 contains apical oxygen. However, the negative Hall coefficient observed in 4%Zn-Y247 constituted unexpected behavior and is considered to be the transport property of the CuO double chains. In fact, similar behavior was observed in the reduced $Pr_2Ba_4Cu_7O_{15-\delta}$ (Pr247) [2]. In Pr247, the negative Hall coefficient was observed in reduced samples, while the as-sintered Pr247 showed a positive Hall coefficient in the entire temperature region. In contrast, a reduction treatment was not necessary to obtain the negative Hall coefficient in Zn-doped Y247. In as-sintered Pr247, a broad peak was observed in the temperature dependence of resistivity, which has been considered to be a characteristic behavior of CuO double chains [8]. However, the broad peak was not observed in any Zn-doped Y247. Thus, there are slight differences in the transport properties between Pr247 and Zn-doped Y247. Whether the superconductivity in the Zn-doped Y247 occurs in the CuO double chains is not clear at present.

REFERENCES

1. M. Matsukawa et al., *Physica C* **411**, 101-106 (2004).
2. Y. Yamada et al., *Physica C*, in press.
3. S. Watanabe et al., *Physica C*, in press.
4. T. Mizokawa et al., *Phys. Rev. B* **65**, 193101 (2002).
5. S. Fujiyama et al., *Phys. Rev. B* **67**, 060404(R) (2003).
6. K. Takenaka et al., *Phys. Rev. Lett.* **85**, 5428 (2000).
7. R. Fehrenbacher and T. M. Rice, *Phys. Rev. Lett.* **70**, 3471- 3474 (1993).
8. S. Horii et al., *Phys. Rev. B* **61**, 6327-6333 (2000).
9. I. Terasaki et al., *Phys. Rev. B* **54**, 11993-11996 (1996).

Carrier Localization And Ion-cluster Effect Affected By Oxygen Content For The Fe-Doped YBCO Superconductors

Shixun Cao, Lingwei Li, Jincang Zhang, and Chao Jing

Department of Physics, Shanghai University, Shanghai, 200444, PR China

Abstract. The oxygen content, Hall coefficient and the superconductivity are systematically studied for the Fe-doped cuprate high-T_c superconducting $YBa_2Cu_{3-x}Fe_xO_y$ (x=0, 0.1, 0,2) system. The results show that the oxygen content has great influence on the transport and transfer of carriers, the suppression of superconducting transition temperature T_c caused by Cu(1) site substitution could be weakened by the increase of oxygen content. The carrier density on CuO_2 plane plays a key role to the superconductivity. It proves that with the increase of Fe-doping concentration or the decrease of oxygen content (for a same doping concentration), the effective oxygen vacancy increases, which enhances the ion-cluster effect and the carrier localization, resulting in the decrease of carrier density that participates in superconducting transport on the CuO_2 plane, and accordingly, the decreases of T_c.

Keywords: oxygen content, carrier localization, ion-cluster
PACS: 74.25.Jb; 74.72.Bk; 74.62.Bf

INTRODUCTION

Extensive studies have been made on $YBa_2Cu_{3-x}M_xO_y$ (M=Fe,Co) systems to understand the mechanism of high-T_c superconductivity.[1,2] The main conclusions concerning its structure and superconductiing properties are as follows: 1) the Fe/Co atoms locate primary at Cu(1) site; 2) an orthorhombic to tetragonal structure phase transition occurs at x=0.12~0.15; 3) T_c decreases and oxygen content y increases with the increase of Fe/Co content. It is well known that the oxygen content affects the crystal structure, electron transport properties and superconductivity of the YBCO system. Therefore, study on the oxygen content in these systems is one of the most important subjects.

EXPERIMENTAL

Samples of $YBa_2Cu_{3-x}Fe_xO_y$ (x=0, 0.1, 0.2) were prepared by solid-state reaction from appropriate amount of dried high-purity Y_2O_3, Ba_2CO_3, CuO and Fe_2O_3. To obtain samples with different oxygen content, the precursor were divided in to several groups, and annealed under high oxygen pressure at different temperatures for 10hrs, then quenched into liquid nitrogen. The samples were characterized by powder X–ray diffraction on the Riguka D/max-2550 diffractometer (18KW, Cu-Kα). Oxygen content was determined by thermo-gravimetric technique. Temperature dependences of resistivity (standard four-probe technique) and hall coefficient (using the five-probe method) were conducted on PPMS-9 (quantum design).

RESULTS AND DISCUSSION

Table 1 listed the Fe-doping concentration x, oxygen content y and superconducting transition temperature T_c of the samples. For the sample prepared in air (x=0.1, y=6.99; x=0.2, y=7.02), the change of resistivity vs. temperature, and lattice parameters are consistent with the previous reports.[1,2] After high oxygen pressure treatment, the oxygen content increases and T_c was obviously improved. Previous studies show that for the undoped $YBa_2Cu_3O_y$, the oxygen content y can approach 7 prepared in air with slow cooling rate, so that each Cu(1) has four oxygen coordination. However, Co and Fe doped YBCO systems cannot get enough oxygen content in air because of their equilibrium oxygen pressure is higher than that of the pure YBCO. Sydow *et al.*[3] obtained higher oxygen content and T_c by treating them in high pressure O_2 or in O_3. Recently, Ren *et al.*[4] and Liu *et al.*[5] use the high pressure synthesis method and makes the highly doped semi-conductivity $R(Ba_{1-x}Sr_x)_2Cu_{2.5}Fe_{0.5}O_y$ (R=Y or rare earth; x=0, 0.5, 1)

CP850, *Low Temperature Physics: 24th International Conference on Low Temperature Physics;*
edited by Y. Takano, S. P. Hershfield, S. O. Hill, P. J. Hirschfeld, and A. M. Goldman

system have a T_c as high as 50 K. Hence, the suppression of T_c caused by Cu(1) site substitution could be weakened by the increase of oxygen content.

TABLE 1. The doping concentration x, oxygen content y and superconducting transition temperature T_c of $YBa_2Cu_{3-x}M_xO_y$ systems

x	T_c	y
0	92	6.96
0.1	74	6.99
0.2	58	7.07
0.2	45	7.02
0.2	34	6.87
0.2		6.53

The hall coefficients R_H as a function of temperature are shown in Fig. 1 and Fig. 2. Previous hall studies on high-T_c superconductors show that, R_H in polycrystalline samples is close to the R_{Habc} (field H parallel to c, current I in the a-b planes) obtained in single crystals.[6] Therefore, measures the R_H of polycrystalline samples also give us information of the carriers participating the transport on the CuO_2 planes. It can be seen that, above T_c, R_H was positive and decreases with the increases of temperature. The magnitude and temperature dependence are consistent with previous reports in the related systems. With the increase of Fe-doping concentration or the decrease of oxygen content (for a same doping concentration), the R_H at room temperature and its maximum was increased, i.e., the carrier density $n_H(\sim 1/eR_H)$ on CuO_2 planes decreases. The change of n_H are consistent with T_c, which proves that the carrier density on CuO_2 planes plays a key role on superconductivity. It is known that the doping ions on Cu site form into clusters especially at higher doping concentration or low oxygen pressure.[7,8] The ion-cluster effect will cause the carrier localization. Hence, the decrease of carrier density on the CuO_2 plane with the increase of Fe-doping concentration or the decrease of oxygen content (for a same doping concentration), may caused by the enhancement of the ion-cluster effect and carrier localization, and accordingly, resulting in the suppression of superconductivity.

In summary, the oxygen content, Hall coefficient and superconductivity are studied for the Fe-doped YBCO system. The oxygen content has great influence on the transport properties and superconductivity. It proves that with the increase of Fe-doping concentration or the decrease of oxygen content (for a same doping concentration), the effective oxygen vacancy increases, which enhance the ion-cluster effect and carrier localization, resulting in the decrease of carrier density that participates in superconducting transport on the CuO_2 plane, and accordingly, the decreases of T_c.

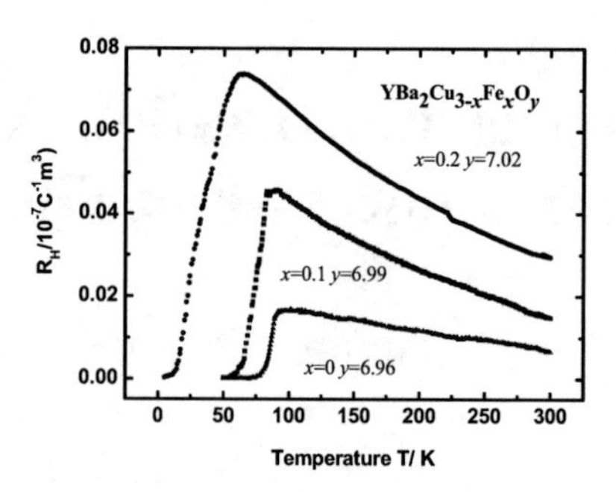

FIGURE 1. Hall coefficient R_H as a function of temperature $YBa_2Cu_{3-x}Fe_xO_y$ (x= 0, 0.1, 0.2) system.

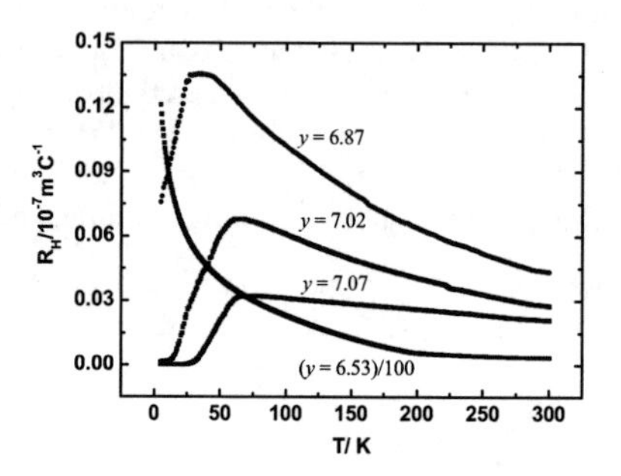

FIGURE 2. Hall coefficient R_H as a function of temperature $YBa_2Cu_{2.8}Fe_{0.2}O_y$ ($6.53 \leqslant y \leqslant 7.07$) system.

ACKNOWLEDGMENTS

This work is supported by the National Foundation of Natural Science of China (No.10274049), the "Shuguang" Project (No.03SG35) and the Leading Academic Discipline of Shanghai (No.T0104).

REFERENCES

1. J. M. Tarascon, P. Barboux, P. F. Miceli, L. H. Greene, G. W. Hull, M. Eibcshule, and S. A. Sunshine, *Phys. Rev. B* **37**, 7458-7469 (1988).
2. J. Zhang, L. Liu, C. Dong, J. Li, H. Chen, X. Li, and G. Cheng, *Phys. Rev. B* **65**, 054513-7 (2002).
3. J. P. Sydow, R. A. Buhrman, and B. H. Moeckly, *Appl. Phys. Letters* **72**, 3512-3514 (1998).
4. Z. Ren, G. Che, Y. Ni, C. Dong, H. Chen, S. Jia, and Z. Zhao, *Phys. Rev. B* **69**, 014507-6 (2004).
5. Y. Liu, G. Che, K. Li, W. Huang, H. Chen, and Z. Zhao, *Physica C* **411**, 47-63 (2004).
6. G. Kallias, I. Panagiotopoulos, and D. Niarchos, *Phys. Rev. B* **48**, 15992-15998 (1993).
7. M. S. Islam, and C. Ananthamohan, *Phys. Rev. B* **44**, 9492-9499 (1991).
8. P. Li, J. Zhang, G. Cao, C. Jing, and S. Cao, *Phys. Rev. B* **69**, 224517-7 (2004).

Superconductivity of Y-123/$Ba_{0.1}CuO_z$/Y-123 Thin Film System

Shinichiro Koba[*1] and Yoshinori Hakuraku[†]

[*]*Yatsushiro National College of Tech., 2761 Hirayama-shinmachi, Yatsushiro, Kumamoto 866-8501, Japan*
[†]*Faculty of Engineering, Kagoshima University, 1-21-40 Koorimoto, Kagoshima, Kagoshima 890-0065, Japan*

Abstract. By measuring the ac susceptibility of $YBa_2Cu_3O_{7-\delta}$/Ba_xCu-O_z/$YBa_2Cu_3O_{7-\delta}$ (Y-123/Ba_xCuO_z/Y-123) thin films, a variation of Tc with time has been observed. The optimal offset transition temperature (Tce) of a simple Y-123 film in this study was 88.7K. However, the superconducting transition of Y-123/$Ba_{0.1}CuO_z$/Y-123 trilayers reached an onset temperature of 96.8K, followed by Tce at 93.5K.

Keywords: YBCO thin film, doped copper oxide, Tc variatons.
PACS: 74.78.Bz, 74.62.Dh

INTRODUCTION

We report here the fabrication of bilayer and trilayer thin film systems, using $YBa_2Cu_3O_{7-\delta}$ (Y-123) as an electrode, in order to investigate the physical properties and superconducting characteristics of superconductor /normal-type (S/N) junctions such as Y-123/Pr-Ba-Cu-O (a S/N superlattice) [1]. Trilayer systems were initially fabricated using a mid-layer film (the N layer) comprised of copper oxide [2, 3] or A-Cu-O [A=Sr, (Sr,Ca)] [4]. In the case of CuO, the off-set temperature (Tce), as determined by ac magnetic susceptibility measurements, decreased from 92.5 K to 86.0 K as the thickness of the mid-layer increased from 30 to 50 nm. However, a higher value of Tce than that of the Y-123 electrode was observed in these samples [2, 3]. XPS data showed that Ba diffuses across the interface from the Y-123 layer, as far as 20 nm into the mid-layer. In this report, we show that the value of Tc for Y-123/$Ba_xCuO_{2-\delta}$/Y-123 trilayer systems, where $x = 0.1$ in the sputtering target composition, is higher than that of the other trilayer systems mentioned above. This shift of Tc towards a higher temperature suggests the presence of Cu-Ba-Ca-Cu-O and Ba-Ca-Cu-O phases, which are reported to have values of Tc of at least 100 K [5]. However, the reported examples were fabricated using a high-pressure method. If a phase with Tc=100 K or greater in the Y-123/$(Ba,Ca)_xCuO_z$/ Y-123 tri-layer thin film system can be stabilized, then a synthesis route avoiding high pressures would be realized.

EXPERIMENTAL

Y-123 targets were prepared by calcining mixtures of starting materials at 900 °C for 10 h. The mixtures were then pressed into plates and sintered at 900 °C for 5 h. Targets for the mid-layers were prepared by calcining mixtures containing the desired ratio of $BaCO_3$ and CuO at 600 °C for 10 h, pressing them into plates, and sintering at 600 °C for 10h. Samples were deposited by dc magnetron sputtering onto $SrTiO_3$ (100) substrates (STO) using two in-situ targets. The two Y-123 thin-film electrodes were deposited at 800 °C under a flow of 99.9999%-pure oxygen and Ar gas at partial pressures of 0.15 and 0.62 Torr, respectively. The thickness of each Y-123 film was 150nm [2]. The $Ba_xCuO_{2-\delta}$ film was sputter-deposited at 600 °C under a flow of oxygen with a partial pressure close to that used in the Y-123 deposition process. The thickness of the $Ba_{0.1}CuO_z$ film was optimized for highest Tc by varying between 20 nm and 50 nm. The surface area of the Y-123/$Ba_xCuO_{2-\delta}$/Y-123 samples was 8 mm^2. The ac magnetic susceptibility was measured at 800Hz under a field of 10 mOe, applied along the tri-layer direction. The temperature was varied at a rate of 1.5 K/min. The accuracy of these measurements was checked by using a carefully prepared, standard Y-123 bulk specimen.

[1]corresponding author E-mail address: koba@as.yatsushiro-nct.ac.jp

CP850, *Low Temperature Physics: 24th International Conference on Low Temperature Physics;*
edited by Y. Takano, S. P. Hershfield, S. O. Hill, P. J. Hirschfeld, and A. M. Goldman

RESULTS AND DISCUSSION

The composition of the targets for deposition of the mid-layer can be envisaged as CuO containing a small quantity of Ba-cuprate impurity. Therefore, the XRD peaks shown in Fig. 1 (a) exhibit a mixture of CuO and Ba-cuprate (Ba-Cu-O). However, the high intensity and sharpness of the CuO peaks indicate that this is the main phase. All Y-123 films are c-axis oriented, as shown in Fig. 1 (b)-(d). The XRD pattern of a $Ba_{0.1}CuO_z$ thin film deposited to a thickness of 200 nm on Y-123 is shown in Fig. 1(b). The (002) and (111) peaks of CuO are observed in addition to the peaks of Ba-Cu-O. The intensity of the Ba-Cu-O peaks is very low compared with that of the (002) CuO peak. These results indicate that $Ba_{0.1}CuO_z$ has been deposited in the form of Ba-doped CuO films. Peaks corresponding to Ba-Cu-O are not visible in the XRD pattern of Y-123/$Ba_{0.1}CuO_z$/Y-123 (Fig. 1(d)) because the $Ba_{0.1}CuO_z$ film is only 30 nm thick. However, the existence of the $Ba_{0.1}CuO_z$ film was verified by the presence of the (110), (002), and (111) CuO peaks. The ac magnetic susceptibility curves of the sample measured in Fig. 1 (d) are shown in Fig. 2 (b). The gap between Tce and the onset temperature (Tco) is denoted by ΔTc. The as-prepared trilayer sample (Fig. 2(b)) shows a superconducting transition of Tce = 93.5K (ΔTc = 3.3K), an enhancement of 4.8 K over the highest measured Tce (88.7 K) of 30 simple Y-123 film samples. However, a significant decrease of Tc with time was observed on exposure to the atmosphere for six weeks, as shown in Figs. 2(b)-(e). The value of Tce for the trilayer system decreased by 12.5 K

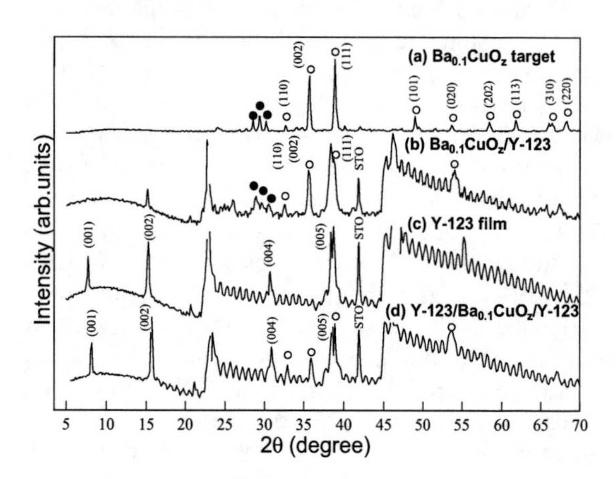

FIGURE 1. XRD patterns of (a) $Ba_{0.1}CuO_z$ bulk used for targets, (b) $Ba_{0.1}CuO_z$/Y-123 film system (thickness of $Ba_{0.1}CuO_z$ = 200nm), (c) Y-123 film, (d) Y-123/$Ba_{0.1}CuO_z$/ Y-123 film system (thickness of $Ba_{0.1}CuO_z$ = 30nm). The intensity axes of (b)-(d) are scaled in logarithmic fashion. Open circles denote CuO peaks. Solid circles indicate Ba-Cu-O peaks. Indices beside peaks are for CuO (with open circles) and Y-123.

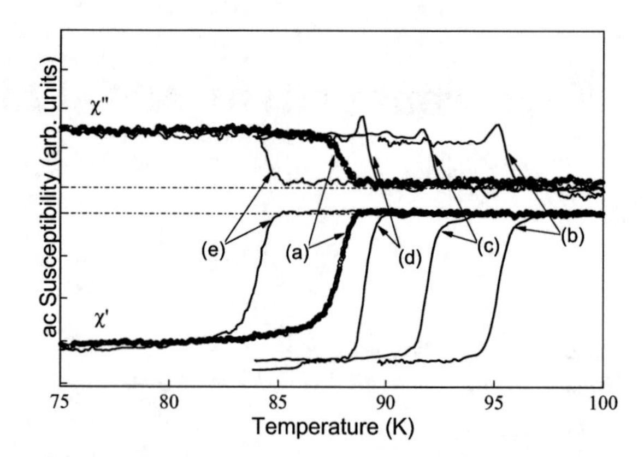

FIGURE 2. Time evolution of ac magnetic susceptibility for (a) simple Y-123 film (Tce=86.9K, ΔTc=1.7K) exposed to the atmosphere for 6 weeks (open circles) and (c)-(e) Y-123/$Ba_{0.1}CuO_z$/Y-123 film system (thickness of $Ba_{0.1}CuO_z$ = 30 nm) exposed to the atmosphere for: (c) 2 weeks (Tce=90.7K, ΔTc =2.1K), (d) 4 weeks (Tce=88.2K, ΔTc =1.7K), (e) 6 weeks (Tce=81.0K, ΔTc=3.5K). (b) The as-prepared Y-123/$Ba_{0.1}CuO_z$/Y-123 film system (Tce=93.5K, ΔTc=3.3K). ΔTc is the difference between Tco and Tce.

after 6 weeks, in contrast to a decrease of only 1.8 K for the simple Y-123 film. A comparison of Figs. 2(a) and (e) shows that the value of Tce for the trilayer system was 5.9 K lower than that of the simple Y-123 film after 6 weeks. We speculate that the high level of degradation of Tc of the trilayer system compared with the simple film is due to the presence of $Ba_{0.1}CuO_z$ and/or the interface between it and Y-123. However, further investigation will be required to confirm this hypothesis. It is notable that, in the course of our experiments, we observed a transition temperature Tce > 100 K for the trilayer system. However, this was an unstable phenomenon with a probability of only 0.02. In summary, we have observed an enhancement of Tce from 88.7 K for a simple Y-123 film to 93.5 K (ΔTc = 3K) for Y-123/$Ba_{0.1}CuO_z$/Y-123, a difference of 4.8 K. The value of Tce decreased with time by as much as 12.5 K for the trilayer system.

REFERENCES

1. D. P. Norton, D. H. Lowndes, S. J. Pennycook and J. D. Budai, *Phy. Rev. Lett.* **67**, 1358-1361 (1991).
2. T. Ogushi, M. Hirose, S. Koba, S. Higo and I. Kawano, *Mod. Phys. Lett. B* **9**, 1059-1067 (1995).
3. S. Koba, S. Higo, R. Nagata, Y. Ozono, M. Onishi, T. Sakamoto, K. Shibata, I. Kawano and T. Ogushi, *Czech. J. of Phys.* **46**, 1511-1512 (1996).
4. S. Koba, M. Saito, Z. Mori, T. Doi and Y. Hakuraku, *Physica C* **388-389**, 443-444 (2003).
5. C. Q. Jin, S. Adachi, X. J. Wu, H. Yamauchi and S. Tanaka, *Physica C* **223**, 238-242 (1994).

Magnetization and Resistance of Melt-Textured Growth YBCO Near T_c and at Low Magnetic Fields

W. M. Tiernan, N. S. Bingham, and J. C. Combs

Department of Physics, Mesa State College, Grand Junction, CO, USA

Abstract. We report measurements of resistance and magnetization vs. temperature for melt-textured-growth YBCO performed at selected magnetic fields, $0.05 < B < 15$ mT, and at temperatures near the superconducting transition. The lack of a traditional Meissner effect during field cooling is attributed to a stable non-equilibrium vortex state. Unusual flux relaxation observed near the resistive transition on warming may be due to sample inhomogeneities and suggests that the resistive transition may be affected by percolation effects.

Keywords: superconductivity, cuprate, magnetization, resistance
PACS: 74.72.Bk

INTRODUCTION

Measurements of the field-cooled magnetization of $YBa_2Cu_3O_7$ (YBCO) in magnetic fields below 15 mT and with magnetic field aligned with the *c* crystalline axis have yielded two quite different results. Some investigators have reported a partial Meissner effect accompanying the resistive transition [1]. Others have observed no magnetic response at the resistive transition but have observed a Meissner effect depressed 6 K or more below the transition, with larger temperature depression for larger fields [2]. Here we describe resistance and magnetization vs. temperature measurements done on a melt-textured growth sample of YBCO down to temperatures of 77 K and at selected magnetic fields from 0.05 mT to 15 mT. The magnetic response for field-cooled and zero field-cooled histories and for warming and cooling are investigated.

SAMPLE DETAILS AND EXPERIMENTAL RESULTS

The sample is a melt-textured growth YBCO hexagonal disk obtained from SCI in Columbus, OH and is composed of a matrix of *c*-axis aligned YBCO grains with insulating green phase inclusions. The disk is 6 mm thick and 21 mm across. Four lead resistance contacts were made using silver paint. Magnetic field was applied perpendicular to the disk and parallel to the aligned YBCO *c* axes. Magnetic field perpendicular to the disk was measured with a Hall sensor placed next to the sample. Warming and cooling runs were done at a rate of 40 mK/min. The sample had a room temperature resistivity of 9×10^{-6} Ω m. The superconducting resistive transition had a 90%-10% width of 0.6 K, similar to that of good quality single crystals. We were unable to detect any field-induced broadening of the resistive transition from 0.05 mT to 15 mT.

We performed data runs at selected applied magnetic fields between 0.05 mT and 15 mT. Figure 1 shows a typical set of data runs at an applied field of 4.0 mT, including *R* vs. *T* data and measured field vs. *T* data for three histories. The measured field runs include a field-cooled run, a warming run made in (nominal) zero field after field cooling, and a warming run with the field turned on after zero field cooling. There is no observable magnetic response for the field-cooled run and none was seen to our cryostat low temperature limit of 77 K for any of our magnetic fields. The warming magnetic runs show the escape or in-migration of magnetic field as the sample warms through the transition. The escape of trapped flux and in-migration of excluded flux both appear to have similar temperature dependences.

Figure 2 shows log-log plots of fluctuation conductivity vs. $(T-T_c)/T_c$ for this sample. The normal state resistance was estimated by performing a linear fit to the sample resistance between 180 and 240 K. For figure 2a, T_c was chosen as 92.0 K, near the bottom of the transition. The observed slopes and

CP850, *Low Temperature Physics: 24th International Conference on Low Temperature Physics;*
edited by Y. Takano, S. P. Hershfield, S. O. Hill, P. J. Hirschfeld, and A. M. Goldman

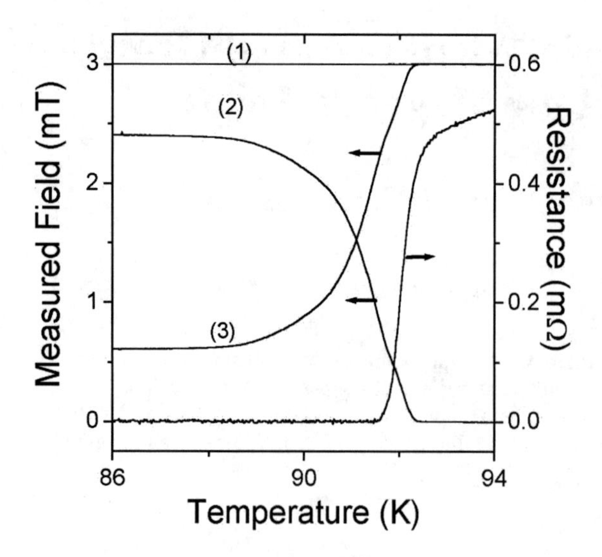

FIGURE 1. Resistance and measured field vs. temperature at 4.0 mT. Measured field curves: (1) field-cooled, (2) $B = 0$ warming after field cooling, (3) warming at 4.0 mT after zero field cooling.

"break" in the curve are similar to data for YBCO single crystals [3]. In figure 2b, T_c is chosen as 92.5 K. This temperature coincides with the "break" in the fluctuation conductivity data in (a) and also coincides closely with the temperature that magnetic flux completely escapes from the sample when it is warmed. In 2(b) the fluctuation conductivity data forms a straight line with slope 0.6, close to the 3D Aslamasov-Larkin prediction of 0.5.

CONCLUSIONS

The absence of a field-cooled Meissner effect, combined with the variety of field-cooled responses seen by others in YBCO, suggests that our observations are due to a non-equilibrium but stable vortex state. As the sample is field cooled it passes through the vortex state region of its phase diagram. Vortices formed here may become trapped, preventing the passage to a Meissner state. Weaker pinning could explain the different results seen by others.

The flux relaxation seen when warming through the resistive transition in figure 1 is not complete until the resistive transition is almost over, $R \approx 90\% R_N$. In a homogeneous superconductor, magnetic relaxation on warming should be almost complete when the resistive transition *begins*. Our observations may be due to inhomogeneity in the sample, which could produce a range of transition temperatures in different parts of the sample. A mix of superconducting and non-superconducting regions could result in a percolation transition, with zero resistance occurring when the superconducting regions reach the percolation threshold, $p_c \approx 0.25$ in 3D. This is in accord with our observation that flux relaxation on warming is about 30% complete at $R = 0$. The lack of homogeneity in this sample may be due to melt-textured growth. Alternatively, the similarity of our *R(T)* data to that seen in single crystals suggests that inhomogeneity may be an inherent property of YBCO single crystals.

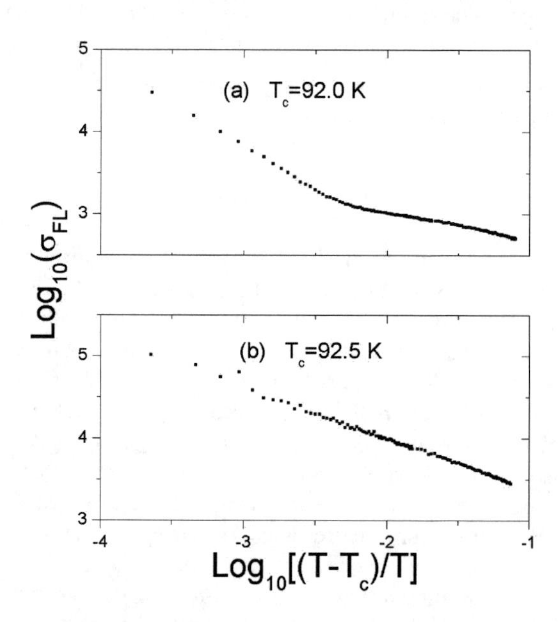

FIGURE 2. Log-log plots of fluctuation conductivity: (a) $T_c = 92.0$ K. (b) $T_c = 92.5$ K.

REFERENCES

1. A. P. Malozemoff, L. Krusin-Elbaum, D. C. Kronemeyer, Y. Yeshurin, and F. Holtzberg, *Phys. Rev. B* **38**, 6490-6499 (1988).
2. H. Safar, H. Pastoriza, F. de la Cruz, D. J. Bishop, L. F. Schneemeyer, and J. V. Waszcak, *Phys. Rev. B* **43**, 13610-13613 (1991).
3. T. A. Friedmann, J. P. Rice, J. Giapintzakis, and D. M. Ginsberg, *Phys. Rev. B* **39**, 4258-4266 (1989).

Transport Properties for YBCO Films Prepared by In-field CVD Process

Satoshi Awaji*, Kazuhiro Sugimoto*, Yanwei Ma† and Kazuo Watanabe*

*High Field Laboratory for Superconducting Materials, IMR, Tohoku University, Sendai 980-8577, Japan
†Applied Superconductivity Lab.,Institute of Electrical Engineering, Chinese Academy of Sciences, Beijing 100080, China.

Abstract. We measured the high field transport properties as functions of temperature, magnetic field and field angles for the YBCO films deposited in the 0 T and the 8 T by the in-field CVD process . The FWHM values in ϕ-scan of 103 reflection are about 3.6 ° for 0T sample and 2.8 ° for 8T sample. Both films show the dip structure at B//c in the angular dependence of resistivity in a magnetic field. This means that the c-axis correlated disorder works effectively in these samples. The dip in the angular dependence of resistivity appears below a certain temperature, which depends on the magnetic field. With decreasing temperature, it disappears once and then appears again for the 0T sample. However, this reentrant behavior is not observed in the 8T sample. Therefore we found that the possibility of two separated regions of the pinned vortex liquid phase and those may be related with the in-plane texture through the grain boundary pinning.

Keywords: CVD, pinned vortex liquid, c-axis correlated disorder, angular dependence, resistivity
PACS: 74.25.Qt, 74.72.Bk, 74.25.Fy

INTRODUCTION

It is well known that the magnetic field orientation effects on the basis of the magnetic anisotropy is effective for the texture control. In addition, the other magnetic field effect such as the morphology change was also reported [1, 2]. A texture control is one of the key technologies for the development of coated conductors. In that case, the in-plane texture is obtained by the aligned buffer templates. From both a practical and a basic research points of view, low angle grain boundaries are very interested as a pinning center in the aligned polycrystalline materials [3]. In particular, it is expected that the grain boundaries are effective as c-axis correlated pinning centers, which improve an irreversibility field [4, 5]. In this study, we prepared YBCO films by in-field CVD process, controlling the in-plane texture by the magnetic field, and measured the transport properties in high magnetic fields up to 17 T. We discuss dissipation in vortex liquid states.

EXPERIMENTAL

YBCO films were grown on single crystalline MgO substrates in the absence of a magnetic field and at 8 T by a chemical vapor deposition (in-field CVD). The detail of the film synthesis has been published previously [2]. These films have the in-plane texture. The full width of half maximum of ϕ-scan for the 103 peak is about 3.6 ° for the 0 T sample and 2.8 ° for the 8 T sample [6]. The sample was mounted on the rotated sample holder with a cernox and a capacitance thermometers. The sample temperature was controlled by both He gas flow in a temperature variable cryostat and the heater placed on the sample holder. Magnetic fields were applied using a 20 T superconducting magnet at the High Field Laboratory for Superconducting Materials (HFLSM), Institute for Materials Research (IMR), Tohoku University. The magnetic field angle of B//c-axis was defined as $\theta = 0°$ and transport currents were always perpendicular to the magnetic field and c-axis.

RESULTS AND DISCUSSIONS

The dip structures on the angular dependence of resistance at B//c-axis are observed for both samples (not shown here). This means that the c-axis correlated disorder works as a dominant pinning center for B//c. In order to study the detailed behavior of the dip, we compare the temperature dependence of the resistivity between $\theta = 0°$ (B//c) and 12° off the dip. Figures 1 and 2 show the measured resistivity and its difference at the various magnetic fields up to 17 T for the 8 T and the 0 T samples, respectively. Note that the resistivity difference corresponds to the magnitude of the dip. The resistivity difference for the 8 T sample indicates a single peak in the low temperature tail of the superconducting transition as shown in Fig. 1. The peaks grow and shift to lower temperature with increasing magnetic fields up to 9 T and saturate above 9 T. This behavior is similar with the typical nature, in the case of the c-axis correlated dis-

CP850, *Low Temperature Physics: 24th International Conference on Low Temperature Physics;*
edited by Y. Takano, S. P. Hershfield, S. O. Hill, P. J. Hirschfeld, and A. M. Goldman

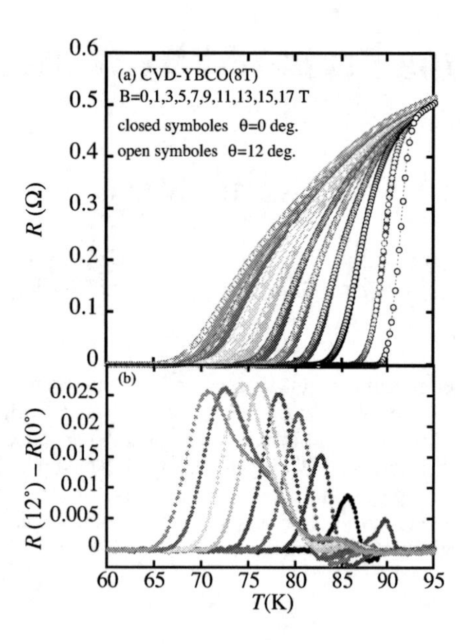

FIGURE 1. Temperature dependence of resistivity (a) and resistivity difference between $\theta = 0$ and 12° (b) for the 8 T sample.

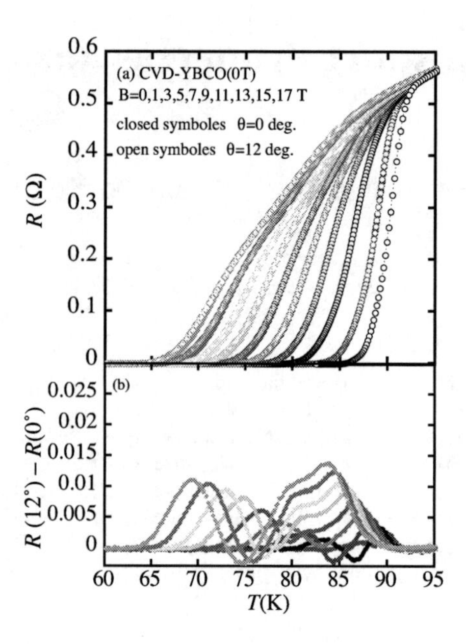

FIGURE 2. Temperature dependence of resistivity (a) and resistivity difference between $\theta = 0$ and 12° (b) for the 0 T sample.

order [5].

On the other hand, two peaks of the resistivity difference are observed for the 0 T sample as shown in Fig. 2. In other words, the dip structure of the angular dependent resistivity appears and then disappears once at a certain temperature with decreasing temperature. Further decreasing temperature, the dip appears again and increases. Since the dip of the angular dependent resistivity for B//c means the reduction of the dissipation due to the c-axis correlated disorders, the regions, where the dip exists, are recognized as a pinned vortex liquid or a partially entangled vortex liquid phase [7]. Therefore, it is suggested that there are two different regions of the pinned vortex liquid region in the 0 T sample.

If we compare the transport properties between the 0 T and 8 T samples each other, the peak of the resistivity difference of the 8 T sample is larger than that of the 0 T sample. Hence, the c-axis correlated disorders of the 8 T sample are more effective in comparison with the 0 T sample.

The 8 T sample has better in-plane texture than the 0 T sample as mentioned above, because the magnetic field at the CVD process improves not only c-axis orientation but also in-plane texture in case of the single crystalline substrates [6]. It is proposed that the low angle grain boundaries act as the pinning center [3]. These behaviors in terms of the c-axis correlated disorder are considerably related with the in-plane alignment through a grain boundary pinning.

In conclusion, we found that the possibility of two different regions of the pinned vortex liquid phase in the CVD-YBCO films. It is considered that the reentrant properties of the dissipation in the vortex liquid region are related with the in-plane texture through the grain boundary pinning.

ACKNOWLEDGMENTS

This work was partially supported by Grant-in-Aid for Young Scientists (B) from the Ministry of Education, Science and Technology, Japan.

REFERENCES

1. Y. Ma et al., *Appl. Phys. Lett.*, **77**, 3633–3635 (2000).
2. Y. Ma et al., *Jpn. J. Appl. Phys.*, **39**, L726–L729 (2000).
3. V. M. Pan et al., *IEEE Trans. Appl. Supercond.*, **13**, 3714–3717 (2003).
4. Crabtree et al., *Handbook of Superconducting Materials*, Eds. D. A Cardwell and D. S. Ginley, London: IOP, 2003, pp. 173–180.
5. S. Awaji et al., *Phys. Rev. B*, **69**, 214522-1–4 (2004) .
6. Y. Ma et al., *Physica C*, **353**, 283–288 (2001).
7. T. Puig et al., *Phys. Rev. B*, **60**, 13099–13106 (1999).

Precursors to Long-Range Superconductivity of $YBa_2Cu_3O_7$ Films :Thermal Fluctuations *vs* Onset Broadening

Detlef Görlitz, Dirk Dölling, Sven Skwirblies, and Jürgen Kötzler

Institut für Angewandte Physik und ZFM, Universität Hamburg, D-20355 Hamburg, Germany

Abstract. A previous study of the zero-field conductivity parallel to the CuO_2 planes of $YBa_2Cu_3O_{7-\delta}$ films is extended to the microwave regime and to temperatures between 50 K and 200 K. At 21.3 GHz, $\sigma(\omega,T)$ is well described by the two-fluid model comprising a Drude term and a 3-dimensional XY-type superfluid, except for temperatures very close to T_c. Upon decreasing frequency, a peak of σ' emerges at temperatures $T_\omega > T_c$, in qualitative agreement with the previous report at low frequencies. A detailed analysis of the peak positions T_ω indicates that this peak arises from a narrow gaussian distribution of transitions to an incoherent Ginzburg-Landau superfluid, rather than from thermally excited vortex-antivortex fluctuations as suggested earlier.

Keywords: Superconductivity, vortex fluctuations, cuprate.
PACS: 74.25.Nf, 74.40.+k, 74.72.Bk, 74.78.Bz

Due to the high transition temperature of the cuprate superconductors thermal fluctuations have been considered to influence strongly on the onset of long-range order, i.e. both the nucleation of a superfluid density at T_{GL} and of its phase coherence at $T_c < T_{GL}$[1]. Especially in thin films, vortex-antivortex (v/a) fluctuations were expected to give rise to a Kosterlitz-Thouless(KT) type of ordering, and characteristic peaks of the low-frequency conductivity $\sigma'(\omega,T)$ near T_c have been interpreted by this model (see Refs. in [2]). More recently, these σ'-peaks were found to occur above a thickness-dependent crossover temperature $T^* < T_c$ (see Fig. 1(b)) and tentatively related rather to the nucleation energy of v/a-pairs than to a KT-type unbinding [2]. Conductivity peaks were also observed at microwave frequencies [3,4], but in this context also an interpretation in terms of a broadening of T_{GL} has been proposed [5]. In order to shine more light into this problem here, we present and discuss $\sigma(\omega,T)$-data in the extremely wide range from 30 mHz to 21 GHz.

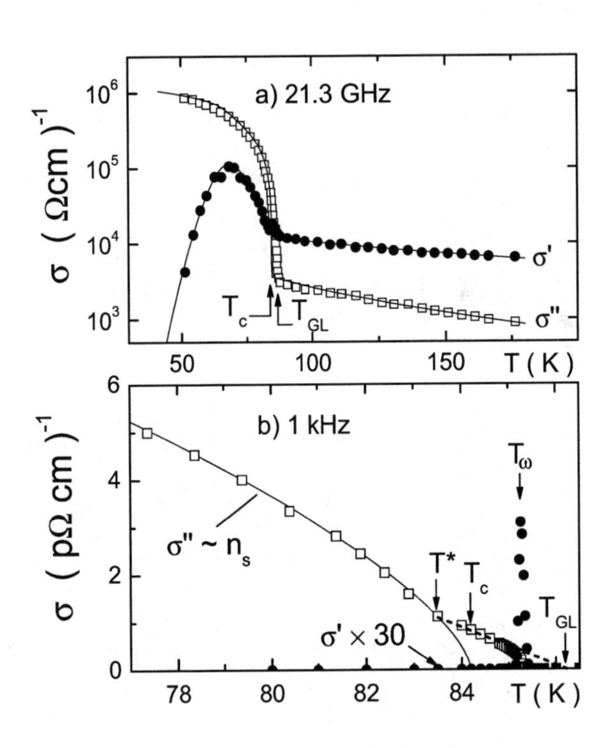

FIGURE 1. Dynamic conductivity $\sigma' - i\,\sigma''$ of a 50 nm thin YBCO-film measured at (a) 21.3 GHz and (b) 1 kHz[2]. The solid curves are fits explained in the text along with the different characteristic temperatures.

The loss and screening components of σ, determined from cavity perturbation measurements[6] at 21.3 GHz, are depicted in Fig. 1(a). The solid curves are fits to a two-fluid model, comprising a Drude-term $\sigma_D(\omega,T) = \sigma_0\,(1-n_s)\,/\,(1+i\omega\tau)$ with an exponentially increasing relaxation rate $\tau \sim \exp(n_s b T_c/T)$, and a 3D-

CP850, *Low Temperature Physics: 24th International Conference on Low Temperature Physics;* edited by Y. Takano, S. P. Hershfield, S. O. Hill, P. J. Hirschfeld, and A. M. Goldman

XY superfluid term $\sigma_S(\omega,T) = 1/(i\mu_0\omega\lambda^2(T))$ with $\lambda^{-2} \sim n_s(T) \sim (1-T/T_c)^{0.69}$ [1]. The evidence for the latter is also demonstrated by the low frequency results below T_c, where $\sigma=\sigma_s \sim n_s$, shown in Fig.1 (b), where in addition to T_c also the thickness-dependent vortex-loop blowout temperature T*, the σ'-peak temperature T_ω, and the mean onset temperature T_{GL} for the incoherent GL superfluid are indicated.

A closer view on the precursor regime above T_c is provided by Fig. 2(a). Upon increasing frequency, the locations of the conductivity peaks, T_ω, shift towards higher temperatures. The analysis of this variation in Fig. 2(b) shows an Arrhenius-type behavior at lower frequencies, $\omega / 2\pi < 1$ MHz, but strong deviations in the microwave regime: In Ref.[2] the energy barriers, defined at low frequencies, have been related to the nucleation energies of v/a cores and, moreover, using a phenomenological model for $\sigma(\omega,T)$, the shape of $\omega\sigma'$ (and also for $\omega\sigma''$, not shown here) was well reproduced, see broken curves in Fig. 2(a). However, both Fig. 2(a) and 2 (b), demonstrate that this v/a-fluctuation model becomes invalid above 1 MHz.

Referring to earlier arguments against v/a-fluctuations in the high-T_c cuprates triggered by microwave conductivities [5], we follow here the suggested alternative route and assume a Gaussian distribution of the GL (onset) temperatures T_0 about the mean T_{GL}:

$$P(t_0) = \frac{1}{\sqrt{2\pi}\,\delta t_0} \exp\left[-\frac{(t_0-1)^2}{2\,\delta t_0{}^2}\right], \qquad (1)$$

where $t_0 = T_0/T_{GL}$. Then, within an effective medium approach[3,5] and with $t \equiv T/T_0$, the dynamic conductivity reads:

$$\frac{1}{\sigma(\omega,T)} = \int_0^{T/T_{GL}} \frac{dt\,P(t_0)}{\sigma_D^+(\omega,t)} + \int_{T/T_{GL}}^{\infty} \frac{dt\,P(t_0)}{\sigma_D^-(\omega,t)+\sigma_{GL}(\omega,t)}, \qquad (2)$$

where $\sigma_{GL}(\omega,t) = \sigma_{GL}(\omega,0)(1-t)$ is the contribution by the GL-superfluid and σ^+_D and σ^-_D are Drude contributions on the normal fluid above and below T_0, respectively. The results of theses calculations [7] for

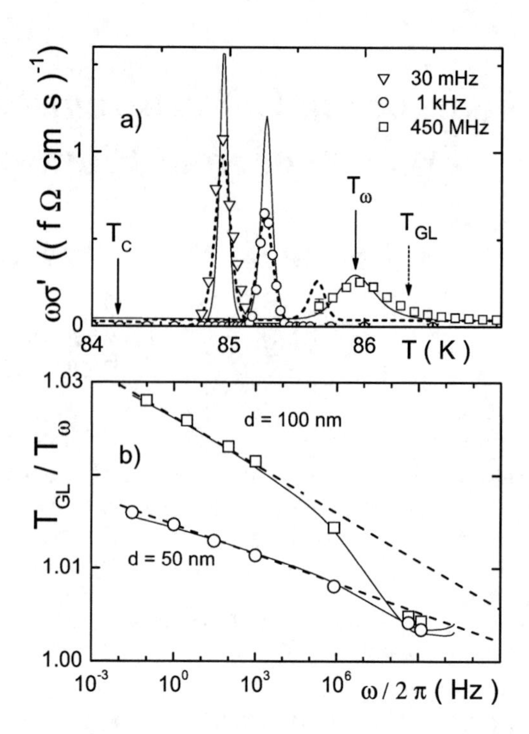

FIGURE 2. (a) Conductivity peaks of a 50 nm film between T_{GL} and T_c. (b) Arrhenius-plots of the peak temperatures for 50 nm and 100 nm film. Broken curves present nucleation model [2], solid curves T_{GL}-broadening model.

distribution widths $\delta t_0 = 0.0022$ for 50 nm and 0.004 for 100 nm are indicated by solid lines in Fig. 2(a), showing an excellent agreement for the peak positions. Some slight discrepancies in the amplitudes in Fig. 2(a) may be associated with the shape of the distribution function.

The present analysis, which for the first time comprises data from low frequencies up to the microwave regime, reveals that a stronger frequency variation of the peak-temperatures results from a broader distribution of GL-onset temperatures. More details will be given elsewhere [7].

REFERENCES

1. D.-S. Fisher *et al., Phys. Rev.* **B 43**, 130 (1991).
2. J. Kötzler *et al., Phys. Rev.* Lett. **87**, 127005 (2001); ibid. **89**,149704 (2002).
3. A. A. Golubov *et al., Physica* **C 213**, 139 (1993); J. R. Waldram *et al., Phys. Rev.* **B 59**, 1528 (1999).
4. A. Hosseini *et al., Phys. Rev.* **B 60**, 1349 (1999).
5. H. K. Olsson and R. H. Koch, *Phys. Rev. Lett.* **68**, 2406 (1992).
6. D. Görlitz, D. Dölling, and J. Kötzler, *Rev.Sci.Instr.* **75**, 1243 (2004).
7. D. Görlitz and J. Kötzler, unpublished.

Evidence for Current-Driven Phase Slips in $YBa_2Cu_3O_{7-\delta}$ Microstrips

P. Morales, M. DiCiano and J.Y.T Wei

Department of Physics, University of Toronto, 60 St.George Street, Toronto, ON M5S1A7, Canada

Abstract. We report distinctly nonlinear current-voltage characteristics in thin-film microstrips of high-T_c superconducting $YBa_2Cu_3O_{7-\delta}$, fabricated by a novel chemical-free technique. Both current-biased and voltage-biased measurements were made between 4.2K and $T_c \approx 91$K. Discontinuities were seen for the former and *S*-shaped non-linearities for the latter, in striking agreement with the 1D phase-slip phenomenology established for low-T_c nanowires. For our quasi-2D high-T_c microstrips, our observations indicate the formation of current-driven phase slip lines well below T_c.

Keywords: Phase Slips, High-T_c, Superconductivity, Cuprates
PACS: 74.40.+k, 74.78.Na, 74.25.Op, 74.78.-w

The ability of a superconductor to transport current with no resistance is one of its more important physical properties. Many proposed superconductive devices are designed around this fundamental property. However, the transition into the resistive state can vary vastly, depending on a superconductor's physical dimensions and chemical composition, as well as temperature and magnetic field. In reduced dimensions, a superconductor can become resistive through the occurence of phase slips [1–4]. For the current-driven phase slip process, the superconducting order parameter (OP) fluctuates between a finite value and zero, within a region spanning two coherence lengths 2ξ. As the OP fluctuates, its phase ϕ slips by 2π, while the superfluid momentum $p=\nabla\phi$ decreases by $2\pi/L$, where L is the length of the sample. While producing local dissipation, this decrease in momentum allows the superconducting condensate to remain below its critical velocity, thus preserving its long-range phase coherence.

It is well known that current-driven phase slips are manifested as distinct non-linearities in the current-voltage (*I-V*) characteristics [1]. Such nonlinear *I-V* characteristics have been well studied in conventional low-T_c superconductors, most recently in quasi-1D nanowires of Pb and Sn [5, 6]. For the high-T_c superconductors, several transport studies have been reported on chemically- or physically-etched cuprate thin films, ranging from 100nm to 20μm in width [7–13]. In these studies, highly non-linear *I-V* characteristics were seen below T_c, and attributed to either collective flux flow, vortex instability, phase slip regions or mesoscopic domains. All of these measurements were made under *current*-biasing. Here we present both *current*- and *voltage*-biased *I-V* measurements made on epitaxial $YBa_2Cu_3O_{7-\delta}$ (YBCO) microstrips, which were fabricated by a chemical-free method to ensure optimal doping.

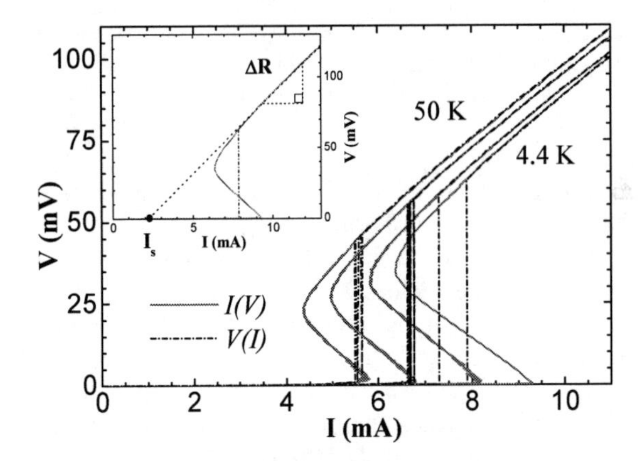

FIGURE 1. Current-voltage characteristics of an optimally-doped YBCO microstrip 1μm wide and 80μm long. Main panel shows data taken at 4.4, 30.0, 42.5 and 50.0 K, under both *current*-biasing (dashed) and *voltage*-biasing (solid). Inset shows the intersection (I_s) of the dV/dI (=ΔR) incline with the current axis.

Our chemical-free technique for fabricating YBCO microstrips is based on selective epitaxial growth [14]. Different regions of a YBCO film were physically and electrically isolated by amorphous $SrTiO_3$(STO) barriers patterned by electron-beam lithography. Atomic force microscopy (AFM) images of the YBCO deposited on top of the STO barriers indicated their amorphousness, while resistance versus temperature measurements confirmed their insulating behavior. *I-V* measurements were made in both voltage-biased $I(V)$ and current-biased $V(I)$ modes. The current and voltage signals were pulsed, with widths of 200μs and duty cycles of 5%, in order to minimize Joule heating.

Figure 1 plots representative *I-V* data taken on a 1μm

CP850, *Low Temperature Physics: 24th International Conference on Low Temperature Physics;*
edited by Y. Takano, S. P. Hershfield, S. O. Hill, P. J. Hirschfeld, and A. M. Goldman

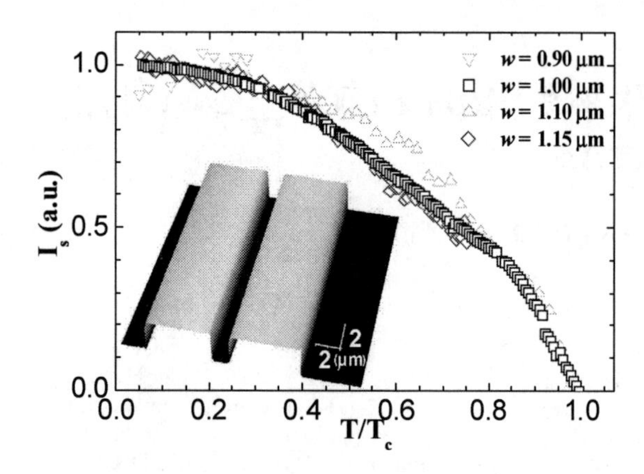

FIGURE 2. Intersection point (I_S) of the linear dV/dI incline with the current axis, plotted in arbitrary units versus reduced temperature T/T_C. I_S can be interpreted as the time-averaged value of the supercurrent through the phase slip region [1]. Main panel shows data for YBCO microstrips of various widths. Inset shows a perspective AFM image of a typical YBCO microstrip, with the light regions being the amorphous barriers and middle dark region being the epitaxial channel.

wide 80μm long microstrip at various temperatures. Each $V(I)$ shows an abrupt voltage jump at a critical level of current, while each $I(V)$ shows an s-shaped non-linearity which coincides with the voltage jump. This observation bears striking resemblance to the *I-V* measured on superconducting Pb and Sn nanowires, whose widths w are smaller than their coherence lengths ξ [5]. In these 1D geometries, the s-shaped non-linearities have been clearly identified as manifestations of phase slip centers using a time-dependant Ginzburg-Landau formalism [15]. In our YBCO microstrips, where $w>\xi$, the appearance of similar *I-V* non-linearities can be attributed to the formation of phase slip lines, which are 2D analogues of phase slip centers [16, 17].

According to the phase slip model of Skocpol, Beasley and Tinkham (SBT) [1], non-equilibrium quasiparticles due to the OP fluctuations should diffuse over a distance Λ_Q on either side of a phase slip region before recombining into the condensate, giving rise to an increment of resistance (ΔR) along the sample [1]. The linear dV/dI incline from above the voltage jump can be related to Λ_Q through the relation $2\Lambda_Q/L=\Delta R/R$, where L is the sample length and R is the normal-state sample resistance. From our data on YBCO, the quasiparticle relaxation length Λ_Q was determined to be 20.1μm at 4.2K. This experimental value varies little between samples, and shows a weak temperature dependence except near T_c where it expectedly diverges.

Furthermore, the OP fluctuations should cause the supercurrent to oscillate within the phase slip region [1]. Following the SBT treatment, the time-averaged value of the supercurrent (I_s) can be determined from the intersection of the $V(I)$ incline with the current axis (see inset of Fig.1). Figure 2 shows temperature dependance of the I_s extracted from our *I-V* data for YBCO microstrips of various widths. Plotted in arbitrary units, the different I_s curves appear to have a similar temperature dependence. This independence of I_s on sample width attests to the universality of the phase slip process governing the transport in our YBCO microstrips.

In summary, our measurement of the *I-V* characteristics in optimally-doped YBCO microstrips have provided clear evidence for current-driven phase slips well below T_C. Our results imply that the formation of phase slip lines (PSL), which are 2D analogues of phase slip centres (PSC), determines the transition of our YBCO microstrips into the resistive state. Further studies are needed to elucidate the analogy between PSC and PSL, particularly whether their formations are based on similar physical mechanisms.

ACKNOWLEDGMENTS

The authors acknowledge assistance by Liz Chia-Wei Chang, Stephanie Chiu and Eugenia Tam. This work was supported by funding from NSERC, CFI, OIT, MMO/EMK and the Canadian Institute for Advanced Research.

REFERENCES

1. W.J. Skocpol, M.R. Beasley and M. Tinkham, *J. Low. Temp. Phys* **16**, 145 (1974).
2. J.S. Langer and V. Ambegaokar, *Phys. Rev.*, **164**, 498 (1967).
3. D.E. McCumber and B.I. Halperin, *Phys. Rev. B*, **1**, 1054 (1970).
4. A. Bezryadin et al., *Nature* **404**, 971 (2000); C. N. Lau et al., Phys. Rev. Lett., **87**, 217003 (2001).
5. D.Y. Vodolazov et al., *Phys. Rev. Lett.*, **91**, 15 (2003).
6. M.L. Tian, et al., *Appl. Phys. Lett.* **83**, 1620 (2003).
7. M.J.M.E. de Nivelle, G.J. Gerritsma and H. Rogalla, *Phys. Rev. Lett.*, **70**, 1525-1528 (1993).
8. V.M. Dmitriev, I.V. Zolochevskii and E.V. Kristenko, *Physica C*, **235-240**, 1973-1974 (1994).
9. F.S. Jelila et al., *Phys. Rev. Lett.*, **81**, 9, (1998).
10. M.M. Abdelhadi and J.A. Jung, *Phys. Rev. B*, **67**, 054502 (2003).
11. J.A. Bonetti et al., *Phys. Rev. Lett.*, **93**, 087002 (2004).
12. S.G. Doettinger et al., *Phys. Rev. Lett.*, **73**, 1691 (1994).
13. M.N. Kunchur, *Phys. Rev. Lett.*, **89**, 137005 (2002).
14. P. Morales, M. DiCiano and J.Y.T. Wei, *Appl. Phys. Lett.*, **86**, 192509 (2005).
15. S. Michotte et al., *Phys. Rev. B*, **69**, 094512 (2004).
16. I.M. Dmitrenko, *Low Temp. Phys.* **22**, 648 (1996).
17. A.G. Sivakov et al., *Phys. Rev. Lett.*, **91**, 26 (2003).

Studies of the Zero-Bias Conductance Peak (ZBCP) in Thin-Film Superconducting $YBa_2Cu_3O_7$ Planar Tunnel Junctions: Detection and Modeling of ZBCP Splittings

L.H. Greene[1], P. J. Hentges[1*a], W. K. Park[1], G. Westwood[2*b], M. M. Pafford[2*c], and W.G. Klemperer[2]

Department of [1]Physics; [2]Chemisty; and The Frederick Seitz Materials Research Laboratory, University of Illinois at Urbana-Champaign, Urbana, IL 61801, USA

Abstract. Planar tunneling spectroscopy is performed on superconducting $YBa_2Cu_3O_7$ (YBCO) thin films which we grow in four different crystallographic orientations by off-axis sputter deposition. The tunneling insulator is formed via a patented solution-based zirconia deposition process, and measurements on these junctions can show remarkably sharp spectral features. A systematic study of the dependence of the tunneling conductance on applied magnetic fields, temperature and different counter-electrode deposition techniques, and comparison of these results with other published data, reveal some universal behaviors for when spontaneous and magnetic-field induced splittings of the ZBCP occur, or not. Our modeling, in which the values of the tunneling cone, surface faceting and quasiparticle lifetime are varied, accounts for the observed behaviors.

Keywords: High-temperature superconductivity, Unconventional superconductivity; Andreev bound states; Planar tunneling spectroscopy; Zero-bias conductance peak, Doppler effect in quasiparticles, Broken time-reversal symmetry.
PACS: 74.50.+r, 74.72.-h, 74.25.Fy

INTRODUCTION

Quasiparticle (QP) planar tunneling spectroscopy applied to superconductors allows momentum-resolved spectroscopic measurements of the near-surface QP density of states near the Fermi surface (FS) [1,2]. Planar tunneling into unconventional superconductors in which the order parameter (OP) changes sign at the FS can exhibit surface-induced Andreev bound states (ABS), which specifically arise from the sign-change of the OP at the FS [3-6]. These ABS are observed as a zero-bias conductance peak (ZBCP) in the tunneling conductance. Specific behaviors and evolutions of this ZBCP as a function of crystallographic orientation, temperature, magnitude and direction of applied magnetic field are reported ([7] and references therein). In particular, a literature search has revealed that the field-induced and spontaneous splittings of the ZBCP, attributed to the Doppler effect and broken time-reversal symmetry, respectively [5,6], which may be measured in junctions formed on optimally doped materials, occur concomitantly in a given junction [8]. We show how surface disorder [8,9] and the size of the tunneling cone [8,10] can affect the observed behaviors, thereby resolving much of the controversy of when splittings of the ZBCP are observed, or not.

EXPERIMENTAL DETAILS

Thin films of YBCO are grown in four crystallographic orientations, namely, (001), (100), (103) and (110), by magnetron sputter deposition [7]. Tunnel junctions are fabricated by several methods as reported earlier [5,7], including a novel method recently developed in which zirconia is deposited by a solution-based hydrolization and condensation technique, shown to be gentler to the thin-film surface than previous junction fabrication techniques [11-14].

ATOMIC-SCALE DISORDER EFFECTS

Spontaneous and field-induced splittings of the ZBCP have been observed in planar tunnel junctions fabricated on thin films ([4-5,7-22], for example). Splittings are not observed for junctions fabricated on disordered thin films [22-23], planar junctions fabricated on single crystals [24], ramp junctions [25] and grain-boundary junctions [26]. Field-induced and

CP850, *Low Temperature Physics: 24th International Conference on Low Temperature Physics;* edited by Y. Takano, S. P. Hershfield, S. O. Hill, P. J. Hirschfeld, and A. M. Goldman

spontaneous splittings occur together, or not, in a given junction, which we attribute to the effect of atomic scale disorder [8,9]. Impurity bound states (IBS) form in unconventional superconductors in the presence of such disorder, and it has been shown that these IBS may couple with the ABS [27]. Based on reported evolutions of the ZBCP with applied magnetic field [4-5,7-26] and discussions [28-29], we propose that the IBS decreases the QP lifetime, τ, thereby inducing inhomogeneous broadening to the ABS, which would then mask the field-induced splitting and quench the spontaneous splitting. This proposal is supported by our calculations based on the FRS model [6] in which we fit data with a varying τ and find that splittings disappear upon decreasing τ [12]. We are further investigating our proposal in the context of Kalenkov *et al.* [30] and note reported calculations of diffusive contacts to d-wave superconductors [31].

TUNNELING CONE EFFECTS

Junctions formed on our thin-films of YBCO using the solution-based zirconia technique [10-14] exhibit conductance features sharper than those ever previously reported in the literature and junctions exhibiting these sharp features do not exhibit splittings of the ZBCP. Materials microanalyses indicate that there exist regions on these coated film surfaces which exhibit atomic scale smoothness that extend over hundreds of nm [12]. Therefore, the notion of a tunneling cone can be addressed [32]. Furthermore, the characteristics observed in our tunneling conductance taken on different orientations indicate a momentum resolution. Fitting our data using the model of FRS [6] and varying the tunneling cone value indicates that we can achieve momentum resolution on the FS on the order of a few degrees. [8,10-12] As expected, neither spontaneous nor field-induced splittings are observed in junctions with narrow tunneling cones due to the filtering out of electrons with a large transverse momentum component.[10-12]

ACKNOWLEDGMENTS

We acknowledge valuable discussion and collaborations with H. Aubin, E. Badica, A. V. Balatsky, Yu. S. Barash, M. Covington, D. G. Hinks, D. M. Ginsberg, P. M. Goldbart, M. Fogelström, A. J. Leggett, T. Löfwander, M. Randeria, J. A. Sauls, D. Sheehey and N. Trevidi, and technical assistance from W. L. Feldmann. This work is supported by the U.S. DoE-DMR: DEFG02-ER9645439, through the FS-MRL, with additional support for LHG and PJH from NSF-DMR 99-72087. Materials characterization was carried out in the UIUC-CMM, partially supported by the U.S. DoE: DEFG02-91-ER45439.

REFERENCES

* Present addresses: [a]Hillsboro, Portland, OR; [b]Dept of Chem., Purdue, IN; [c]Rohm and Hass, Philadelphia, PA.

1. W. L. McMillan and J. M. Rowell, *Superconductivity*, R. D. Parks, ed., NY: Marshall-Dekkar, 1969. p. 561.
2. E. L. Wolf, *Principles of Electron Tunneling Spectroscopy,* London: Oxford University Press, 1985.
3. C-R Hu, *Phys. Rev. Lett.* **72**, 1526 (1994).
4. Y. Tanaka and S. Kashiwaya, *Phys. Rev. Lett.* **74**, 3451 (1995).
5. M. Covington *et al., Phys. Rev. Lett.* **79**, 278 (1997).
6. M. Fogelström *et al., Phys. Rev. Lett.* **79**, 281 (1997).
7. M. Covington and L. H. Greene, *Phys. Rev. B* **62**, 12440 (2000); L. H. Greene *et al., Physica B* **280**, 159 (2000), and references therein.
8. L. H. Greene *et al., Physica C* **387**, 162 (2003).
9. L. H. Greene *et al., Physica C* **408-410**, 804 (2004).
10. P. J. Hentges *et al., Physica C* **408-410**, 801 (2004).
11. P. J. Hentges *et al., IEEE Trans. Appl. Supercon.* **13**, 801 (2003).
12. P. J. Hentges, PhD Thesis, UIUC (2004); P. J. Hentges *et al., Phys. Rev. B* (in preparation).
13. US Patent **6,838,404 B2** (2005).
14. L. H. Greene *et al., J. Mater. Chem.* **14**, 3158 (2004).
15. J. Geerk *et al., Z. Phys. B*, **73**, 329 (1989).
16. S. Kashiwaya *et al., J. Phys Chem. Sol.* **59**, 2034 (1998).
17. J. Lesueur *et al., Proc. SPIE* **3481**, 419 (1998).
18. R. Krupke *et al., Phys. Rev. Lett.* **83**, 4634 (1999).
19. X. Grison *et al., Physica B* **284-288**, 559 (2000).
20. Y. Dagan and G. Deutscher, *Phys. Rev. Lett.* **87**, 177004 (2001).
21. R. Beck *et al., Phys. Rev. B* **69**, 144506 (2004).
22. E. Badica *et al.*, in *Superconducting and Related Oxides: Physics and Nanoengineering IV*, D. Pavuna, I. Bozovic, eds. (SPIE, Bellingham, 2000) **4058**, p 52.
23. M. Aprili *et al., Phys. Rev. B* **57**, R8139 (1998).
24. H. Aubin *et al., Phys. Rev. Lett.* **89**, 177001 (2002).
25. W. Wang *et al., Phys Rev. B* **60** 4272 (1999); I. Iguchi *et al., Phys Rev. B* **62**, R6131 (2000); I. Iguchi *et al., IEICE Trans. Electron* **E85-C**, 789 (2002).
26. L. Alff *et al., Phys. Rev. B* **58**, 11197 (1998); L. Alff *et al., Eur. Phys. J. B.* **5**, 423 (1998).
27. L. J. Buchholtz *et al., J. Low Temp. Phys.* **101**, 1099 (1995); A. V. Balatsky *et al., Phys. Rev. B* **51,** 15547 (1995); A. V. Balatsky and M. I. Salkola, Phys. *Rev. Lett.* **76,** 2386 (1996); I. Adagideli *et al., Phys. Rev. Lett.* **83,** 5571 (1999); I. Adagideli *et al., Phys. Rev. B* **66,** 140512R (2002); A. V. Balatsky and M. I. Salkola, *Phys. Rev. Lett.* **80,** 1117 (1998); M. Graf *et al., Phys Rev. B.* **61,** 3255 (2000); K. V. Samokhin and M. B. Walker, *Phys. Rev. B* **64**, 172506 (2001); W. K. Neils and D. J. Van Harlingen, *Phys. Rev. Lett.* **88**, 47001 (2002).
28 J. A. Sauls, M. Fogelström, (*private communications*).
29. Yu. S. Barash, (*private communications*).
30. M. S. Kalenkov *et al., Phys. Rev. B* **70**, 184505 (2004).
31. Y. Tanaka *et al., Phys Rev. B* **69**, 144519 (2004); Y. Tanaka *et al., Phys Rev. B* **71**, 094513 (2005).
32. M. B. Walker, *Phys. Rev. B* **60**, 9283 (1999).

Tunneling into YBCO Superconductor at High Magnetic Field

R. Beck[a], Y. Dagan[a,b], G. Leibovitch[a], G. Elhalel[a] and G. Deutscher[a]

[a] *School of Physics and Astronomy, Faculty of Exact Science, Tel-Aviv University, 69978, Tel-Aviv, Israel*
[b] *Center for Superconductivity Research, Department of Physics, University of Maryland, College Park, Maryland 20742, USA*

Abstract. We studied the tunneling density of states of *YBCO* films at high magnetic field up to 15 Tesla parallel to the films' surface and perpendicular to the CuO_2 planes. We observed a transition in the tunneling conductance at high fields. At 6 Tesla in increasing magnetic fields, the gap-like feature shifts discontinuously from 15meV to a lower bias of 11meV, and becoming more pronounced as the field increases. We found the effect to be anisotropic. We discuss its origin in term of current and/or magnetic field.

Keywords: Tunneling, High magnetic field, Superconductivity, YBCO thin films.
PACS: 74.50.+r, 74.25.Ha

INTRODUCTION

Tunneling into a *d-wave* superconductor is far richer than into a conventional superconductor. Hu[1] predicted that tunneling along the node direction will result in zero energy bound states, resulting from the π phase difference at adjacent lobes. Modification of the phases, for example by a supercurrent parallel to surface, will result in a splitting of the zero bias peak, proportional to the superfluid velocity[2]. Moreover, even in nodal directed tunneling, when the junction size exceeds the roughness of the films' surface[2], one can also measure the superconducting gap value as it appears as an additional peak in tunneling differential conductance near the gap value.

We investigate the tunneling conductance of *In/insulator/YBCO* junctions when a magnetic field is applied parallel to the film and perpendicular to the CuO_2 planes. We observe that for slightly underdoped samples the gap-like feature shifts discontinuously at about 6 Tesla in increasing magnetic field. We discuss the origin of the shift in terms of current and/or field.

EXPERIMENT

We measure the tunneling conductance of slightly underdoped thin films of $YBa_2Cu_3O_{7-\delta}$ oriented along the node direction. The crystallographic orientation of the films is shown in Fig 1a. The advantage of this configuration is that it enables one to apply magnetic fields perpendicular or parallel to the CuO_2 plains while keeping the field parallel to the surface. Forming a planar tunneling junction perpendicular to the film's surface allows studying the influence of the magnetic field on the superconductor with little influence of vortices. A detailed description of the films preparation and junction configuration can be found in reference[3].

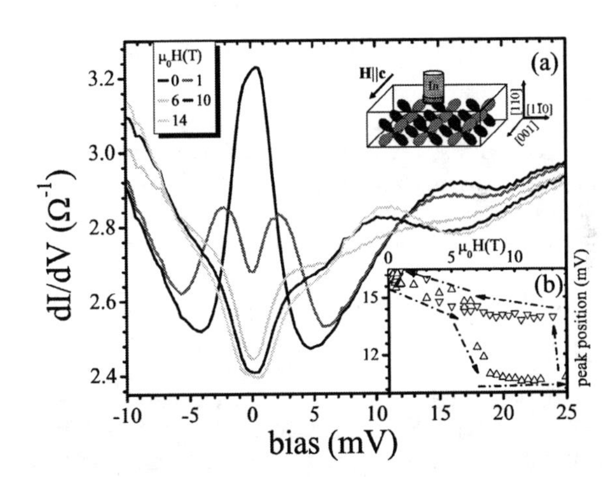

FIGURE 1. Differential conductance of the tunneling junction at 2.5K in increasing magnetic fields. a) Illustration of the film crystallographic orientation and experimental configuration. b) Gap-like feature bias value in increasing (Δ) and decreasing (∇) fields.

A typical tunneling spectrum is shown in Fig 1. At zero magnetic field the zero-bias peak and the gap-like feature, located at 16.3 *meV*, are clearly shown. Following the evolution of the tunneling spectrum with

CP850, *Low Temperature Physics: 24th International Conference on Low Temperature Physics;*
edited by Y. Takano, S. P. Hershfield, S. O. Hill, P. J. Hirschfeld, and A. M. Goldman

increasing magnetic field, reveals two opposite behaviors. While the zero bias peak is continually splitting with magnetic field, the gap feature reverses its trend at some point. At low fields the gap-like feature is slightly reduced down to 15 *meV*, until at 6 Tesla it can not be detected any more. Further increasing the magnetic field results in the appearance of a new gap-like feature, at 11 *meV*, whose amplitude grows with increasing magnetic fields. Next we decrease the magnetic field, we note that by reducing it from 14 to 13 Tesla the position of the gap-like feature shifts **continuously** from 11 to 14 *meV* [see also Ref. 4]. The bias values of the gap-like feature in both, increasing and decreasing fields, present the hysteretic behavior shown in Fig. 1b. This effect is highly reproducible, and can be measured up to 25K.

We repeated the experiment with films having different crystallographic orientations. In particular, films oriented with the a-axis perpendicular to the surface and the c-axis parallel to the film surface, and the magnetic field parallel to the c-axis as before. In such case, the gap-like feature is hardly modified with fields up to 15 Tesla. The same holds for the magnetic field applied parallel to the CuO_2 planes for any crystallographic configuration.

DISCUSSION AND CONCLUSIONS

From the hysteretic behavior of the gap-like feature we deduce that the bias value is strongly affected by induced currents rather than by the field itself. We notice that the gap-like feature is continuously shifting when reducing the field from its maximum value by 1T. Hence, we infer that the gap-like feature at 11 meV is indeed the successor of the zero field gap-like feature.

A possible explanation for the discontinuous shift at 6 Tesla could be a sudden change in the induced current, for example due to a phase transition in the vortex matter[5]. However, the low bias region is highly sensitive to the surface currents, via the Doppler shift effect on the Andreev-Saint-James bound states[2]. Therefore, one would expect to see the same evidence for the discontinuous shift in the low bias region as well. But, as shown in the contour plot in Fig 2, the low bias region is continuous over the whole range of magnetic field, in contrast to the high bias region, were states are shifted abruptly to the region of the new gap-like feature.

An alternative origin for the discontinuous shift is the magnetic field alone. However, in such a case the modification of the gap-like feature with field should be present in a-axis films as well, which is inconsistent with our measurements.

Therefore, with no other alternative, we conclude that the anisotropic discontinuous shift taking place at 6 Tesla is the result of a combination of strong currents and high magnetic field. To the best of our knowledge, no theoretical model has predicted such effect.

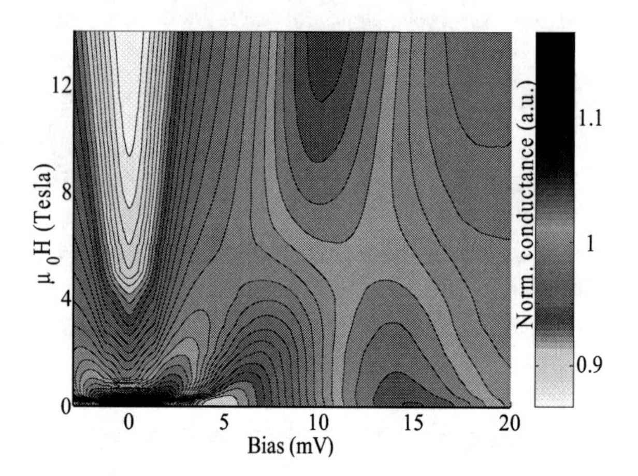

FIGURE 2. Contour plot for the conductance for all fields studied. light (dark) color represents low (high) bias. Note that the low bias region changes continuously with increasing fields, while the high bias region is highly discontinuous at 6 Tesla.

In summary, we have shown that the tunneling density of states is highly modified by the presence of high magnetic fields – high currents. The gap-like feature, at high bias, shifts **discontinuously**, while the low bias region evolves **continuously**. We conclude that the gap-like feature bias depends strongly on currents, but the shift is anisotropic and due to a combination of both strong currents and high magnetic fields.

ACKNOWLEDGMENT

This work was supported by the Heinrich Hertz Minerva center for High Temperature superconductivity, the Israel Science Foundation, the Oren Family Chair of Experimental Solid state physics.

REFERENCES

1. C.R. Hu , *Phys.Rev. Lett.* **72**, 1526 (1994).
2. M. Fogelström *et al.*, *Phys.Rev. Lett.* **79**, 281 (1997).
3. R. Beck *et.al.*, *Phys.Rev.B* **69**, 144506 (2004).
4. R. Beck *et.al.*, *Phys.Rev.B* **72**, 104505 (2005).
5. Y. Radzyner, A. Sheulov and Y. Yeshurun, *Phys.Rev. B* **65**, 100513(R) (2002).

Homogenous Crack-Free Large Size YBCO/YSZ/Sapphire Films for Application

B. Almog, M. Azoulay and G. Deutscher*

*School of Physics and Astronomy, Faculty of Exact Science, Tel-Aviv University, 69978, Tel-Aviv, Israel

Abstract. $YBa_2Cu_3O_{7-\delta}$(YBCO) films grown on Sapphire are highly suitable for applications. The production of large size (2-3") homogeneous, thick($d \geq 600nm$) films of high quality is of major importance. We report the growth of such films using a buffer layer of Yttrium-stabilized ZrO_2(YSZ). The films are highly homogeneous and show excellent mechanical properties. They exhibit no sign of cracking even after many thermal cycles. Their critical thickness exceeds 1000nm. However, because of the large lattice mismatch there is a decrease in the electric properties(increases R_s, decreases j_c).

Keywords: YBCO, YSZ, Sapphire
PACS: 85.25.Am, 74.25.Fy, 74.78.Bz, 74.78.Fk, 62.20.Mk

INTRODUCTION

Using superconductors for commercial applications requires high quality large size films. $YBa_2Cu_3O_{7-\delta}$(YBCO) on $(1\bar{1}02)$-oriented Al_2O_3(r-cut Sapphire) is one of the best choices because it is available in various sizes and shapes at low cost and has good electric properties($\varepsilon \approx 10$, $\tan\delta \approx 10^{-8}$). However, such films are usually of poor quality due to the large mismatch between the lattice constant of YBCO and Sapphire and the chemical reaction of Al impurities with the superconductor. These can be improved by introducing a buffer layer. Common choices are MgO, CeO_2 and Yttrium-stabilized ZrO_2(YSZ).

Another problem caused by the incompatible substrate is the appearance of micro-cracks above a critical thickness. The thickness of the films is of major importance for current carrying devices and for microwave applications (the thickness should be at least twice the penetration distance to avoid losses from the substrate[1]). Largest critical thickness values reported up today were between 300nm[2, 3] and 700nm[4].

In this article we report the growth of high quality, large size, crack free films of YBCO on top of r-cut Sapphire with a buffer layer of YSZ.

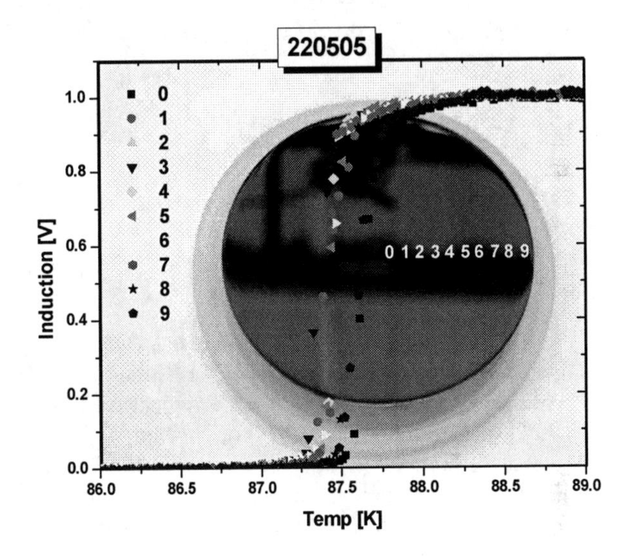

FIGURE 1. Typical 2" YBCO Film. The homogeneity is verified by the small changes of T_c measured inductively at different locations along the radius.

SAMPLE PREPARATION AND CHARACTERIZATION

Sample preparation. Both YSZ and YBCO layers were grown by off axis dc magnetron sputtering from stoichiometric single targets. The YSZ layer was sputtered at 5mTorr of Ar with the substrate at $955°C$. The thickness of the YSZ was 1000Å at a deposition rate of 0.01nm/sec. During the sputtering of YBCO the heater temperature was $930°C$. The substrate was heated by IR radiation and its temperature was estimated as $750°C$. The gas mixture was 50/50 percent Oxygen and Argon at 200mTorr. The dc power on the target was 140W, yielding a sputtering rate of $\sim 0.03nm/sec$. An annealing process in pure Oxygen was performed for 16 Hours at approximately $500°C$.

Film characterization. The crystallographic structure of the films was tested by X-Ray diffraction(XRD). All of the films showed the existence of only (001) orientation of YBCO. The c-axis value was calculated by

CP850, *Low Temperature Physics: 24th International Conference on Low Temperature Physics;*
edited by Y. Takano, S. P. Hershfield, S. O. Hill, P. J. Hirschfeld, and A. M. Goldman

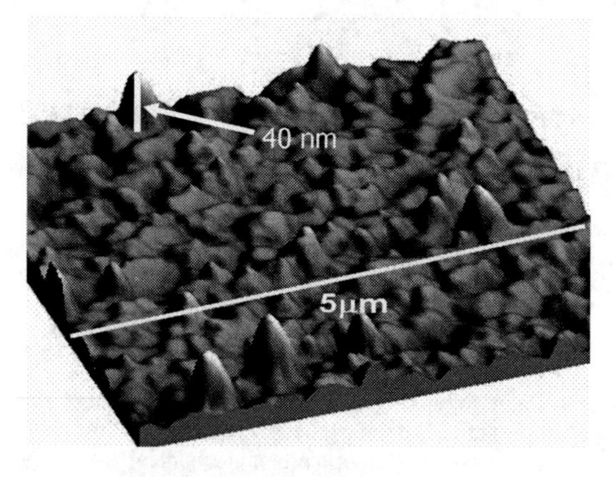

FIGURE 2. Typical AFM Picture. The roughness of the film is 5-10nm(RMS) with a small density of large outgrowth particles(20x20nm and 40nm height).

TABLE 1. Electric Properties of Selected Films.

Sample ID	Thickness(nm)	$j_c(MA/cm^2)$	$R_s(m\Omega)$*
160304	700	0.3	1.83
240304	700	0.2	1.56
300304	700	0.2	1.59
190505	400	0.37	
100205	150	0.6	
130205	150	0.9	

* measured only for films thicker than 600nm

the (0010) peak to be 11.68 ± 0.01Angstrom. The surface morphology was examined by an Atomic Force Microscope (AFM). Figure 2 shows a typical AFM image with a surface roughness(RMS) of 5-10nm. A few large outgrowth particles with typical sizes of 20x20nm and 40nm height are also visible. An important result which is clearly seen in all the AFM pictures is the absence of any micro-cracks.

Growing homogenous large size films is usually problematic. In order to check the homogeneity we made inductive measurements of the critical temperature (T_c) in different areas of the film. The inductive coils were 1mm in diameter. Figure 1 shows the high homogeneity of the films - less than 0.2K change in T_c along the radius of the sample. The critical temperature is 87.5K which is a typical value for high quality YBCO films on Sapphire.

The critical current density(j_c) was measured inductively by measuring the magnetic field penetration profile. A description of this technique is given elsewhere[5]. Typical j_c values are given in table 1. The j_c dependence on the film's thickness shows the already reported[6] feature of $\frac{1}{\sqrt{d}}$ as expected from the collective pinning model. This is shown in figure 3. The surface resistance(R_s) was measured with a dielectric resonator similar to that described by Zuccaro *et al.*[7].

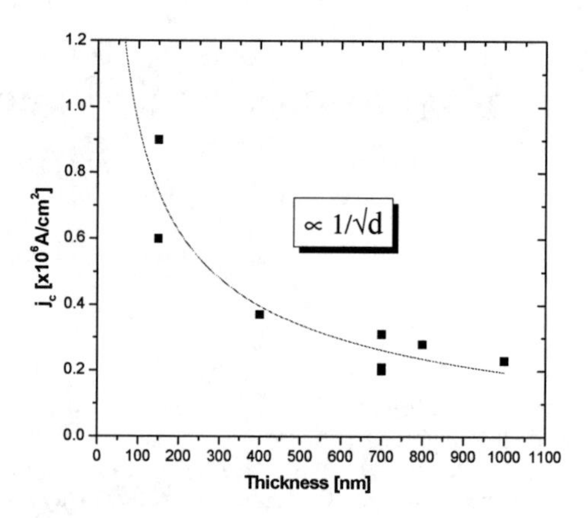

FIGURE 3. Critical Current Density Versus Thickness. The solid line represents the $j_c \propto \frac{1}{\sqrt{d}}$ behavior expected from the collective pinning model.

CONCLUSIONS AND DISCUSSION

We described the growth procedure of YBCO films on Sapphire with a buffer layer of YSZ. The electric properties of the films were slightly lower than the highest quality films reported on Sapphire. However, the use of YSZ as a buffer turned out to be advantageous for the mechanical properties. The films showed no micro-cracking even after many thermal cycles and their critical thickness exceeded 1000nm. Another feature of our films which is imperative for application was their homogeneity. The use of YSZ presented a tradeoff between electric properties and mechanical strength. This reduction in electric properties may be improved in the future by introducing a second buffer layer with a better lattice matching to YBCO(e.g. CeO_2).

Acknowledgments. This work was supported by the Heinrich Hertz Minerva Center for High Temperature Superconductivity, the Israel Science Foundation and the Oren Family Chair of Experimental Solid state physics.

REFERENCES

1. N. Klein *et al.*, *J. Appl. Phys.* **67**, 6940 (1990).
2. G. Kästner *et al.*, *Phys. Stat. Sol. (a)* **150**, 381 (1995).
3. A. G. Zaitsev *et al.*, *Applied Superconductivity: Inst. Phys. Conf. Ser.* **158**, 25 (1997).
4. A. G. Zaitsev *et al. J. Supercond.* **11**, 361 (1998).
5. B. Almog, M. Azoulay and G. Deutscher, to be published.
6. R. Wördenweber, *Supercond. Sci. Technol.* **12**, 86 (1999).
7. C. Zuccaro *et al.*, *IEEE Trans. Appl. Supercond.* **1**, 295 (1997).

The Pseudogap Value of Y-123 and Bi-2223 Thin Films Obtained from Temperature Dependence of Resistivity

D.D. Prokofiev, M.P. Volkov, and Yu.A. Boikov

A.F.Ioffe Physical-Technical Institute RAS, Polytekhnicheskaya 26, Saint-Petersburg 194021, Russia

Abstract. The temperature dependence of resistivity of underdoped and optimally doped high -Tc superconductors can be used not only for the determination of the pseudogap opening temperature T*, but as well for the determination of temperature dependence of pseudogap. In this communication we performed the analysis for $YBa_2Cu_3O_{7-x}$ and $Bi_2Sr_2Ca_2Cu_3O_{10}$ thin films and compared the results with the prediction of the BSC – BEC crossover theory. Some published data on resistivity of $YBa_2Cu_3O_{7-x}$ single crystals and films were also included in the analysis. The temperature dependencies of excess conductivity $\Delta\sigma = 1/\rho - 1/\rho_n$, where ρ - measured and ρ_n – extrapolated normal resistivities can be described by the relation $\Delta\sigma = A(1-T/T^*)\exp(\Delta^*/T)$. We argued that the Δ^* in this relation is the pseugogap, observed in different experiments in underdoped high –Tc superconductors. The obtained temperature dependence of pseudogap is discussed in terms of BSC - BEC crossover theory

Keywords: resistivity; pseudogap; $YBa_2Cu_3O_{7-x}$; $Bi_2Sr_2Ca_2Cu_3O_{10}$; thin films.
PACS: 74.25 Dw, 74.72.Bk, 74.72.Hs, 74.78 Bz

INTRODUCTION

The pseudogap in underdoped and optimally doped high-Tc superconductors remains a subject of intense study [1, 2]. The pseudogap and superconducting gap can be of the same origin according to ARPES data [1]. The temperature dependence of resistivity is used usually for the determination of the pseudogap opening temperature T^* as the temperature of the sudden change in the resistivity decreasing rate at temperature decreasing [3]. We proposed earlier a method for the determination of the pseudogap value using temperature dependence of resistivity [4]. We used this method for $YBCO_{6.85}$ thin films and compared the obtained temperature dependence of pseudogap with the predictions of the theory of the BCS-BEC crossover [5]. This method was discussed in more details in Ref. [6] and it was shown that it based on the measurements of ρ(T) with high precision. In this communication we present the results of pseudogap determination for $Bi_2Sr_2Ca_2Cu_3O_{10}$ and $YBa_2Cu_3O_{7-\delta}$ thin films and single crystals using this approach.

EXPERIMENTAL RESULTS

The pseudogap evaluation is based on the analysis of the temperature dependence of the additional or excess conductivity $\Delta\sigma=1/\rho-1/\rho_n$, where ρ is measured resistivity (at $T<T^*$) and ρ_n is the extrapolation of the normal resistivity from the region $T>T^*$ to lower temperatures. This extrapolation is shown on the insert to Fig.1 (curve 2) for the YBCO thin film with $7-\delta=6.85$ on $SrTiO_3$ substrate. The difference $\rho_n-\rho$ (insert to Fig.1, curve 3) is used to determine the pseudogap opening temperature T^*. It can be seen from Fig. 1 that the dependence of $\ln\Delta\sigma$ on inverse temperature 1/T is about linear in the temperature region 100 – 140 K (curve 1) and deviates from the linearity at higher temperatures. The relation

$$\Delta\sigma = A(1 - T/T^*)\exp(\Delta^*/T)\text{.........}(1)$$

describes the experimental results in a wider temperatute region 100 K – T^*. The relation (1) is shown on Fig. 1 as curve 2, where parameters Δ^* and A were determined by the least-square fitting. The obtained value of parameter $\Delta^* = 520$ K is close to the pseudogap value ~ 500 K obtained in Ref. [7] for YBCO films with the same doping level δ from the analysis of photoinduced decrease in the light transmission. So we suppose that the parameter Δ^* in the relation (1) is the pseudogap value for underdoped high–T_c superconductor.

The deviation of experimental results (curve 1) from the relation (1) (curve 2) at high temperature can be treated as the manifestation of the temperature dependence of Δ^* at $T \to T^*$ and can be used for the

CP850, *Low Temperature Physics: 24th International Conference on Low Temperature Physics;*
edited by Y. Takano, S. P. Hershfield, S. O. Hill, P. J. Hirschfeld, and A. M. Goldman

determination of this dependence. The temperature dependences of Δ^* obtained by fitting of experimental results with the relation (1) (where $\Delta^* = \Delta^*(T)$) are presented on Fig. 2. Our experimental results are shown for $Bi_2Sr_2Ca_2Cu_3O_{10}$ thin film (curve 1) and for $YBa_2Cu_3O_{7-\delta}$ thin films with $7-\delta=6.85$ (curve 3) and 6.96 (curve 6). The experimental results from Ref. [3] for single crystals with $7-\delta=6.88$ and $7-\delta=6.78$ are shown by curve 2 and curve 5, respectively, the results from [9] for thin film with $7-\delta=6.85$ are shown by curve 4.

As is seen from Fig.2 for $Bi_2Sr_2Ca_2Cu_3O_{10}$ and YBCO samples with nearly optimal doping ($7-\delta=6.88$ and 6.85) the pseudogap Δ^* doesn't depend on temperature far from T^* and falls drastically at $T\rightarrow T^*$. This falling part of the $\Delta^*(T)$ can be approximated by $(T-T^*)^{1/2}$ up to $T=T^*$ (solid lines on Fig.2). The Δ^* value for $Bi_2Sr_2Ca_2Cu_3O_{10}$ thin film is higher than the values for YBCO samples and the ratio (Δ^*Bi-2223)/($\Delta^*YBCO_{6.88}$) = 1.11, this ratio is close to the ratio of critical temperatures (~1.16).

For YBCO samples with lower doping Δ^* changes with temperature more smoothly (curve 5 for $7-\delta = 6.78$), that resembles the temperature dependence of pseudogap, obtained by the theory of BCS-BEC crossover [5]. Curve 7 and 8 present the $\Delta^*(T)$ dependences obtained by this theory with the crossover parameter $\mu/\Delta^*(0)=-2$ and -10 respectively, where μ - chemical potential and $\Delta^*(0)$ is taken from the low temperature part of the $\Delta^*(T)$ curve.

The relation (1) can be analysed taking into account that T^* can be considered as the mean-field temperature of the superconducting transition [8] and superconducting pairs exist below T^* leading to excess conductivity $\Delta\sigma$. Then it is natural to suggest that $\Delta\sigma$ is proportional to the number of pairs (n_s~$1-T/T^*$) and inversely proportional to the number of pairs broken by thermal motion (n_b~$\exp(-\Delta^*/T)$), that leads to relation (1). At temperatures near T^* the pseudogap $\Delta^*(T)$ can be described by square root relation $(T-T^*)^{1/2}$ that also follows from mean-field approximation.

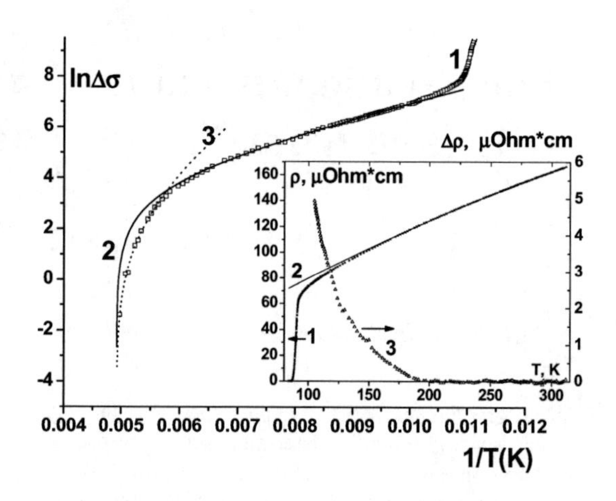

FIGURE 1. Curve 1 – the dependence of lnΔσ on inverse temperature 1/T for YBCO thin film. Curve 2 – the function $\ln[A(1-T/T^*)\exp(\Delta^*/T)\}$, where $\Delta^*=520$ K. Curve 3 – the same function where $\Delta^* \rightarrow \Delta^*(T) = k\,\Delta^*(1-T/T^*)^{1/2}$, k=2.6. On the insert temperature dependence of resistivity ρ(T) (curve 1) is shown along with the extrapolation of normal resistivity $\rho_n(T)$ from high temperature region (curve 2), which has Sd~0.02% in the 200-300 K range. Curve 3 – the temperature dependence of $(\rho_n - \rho)$.

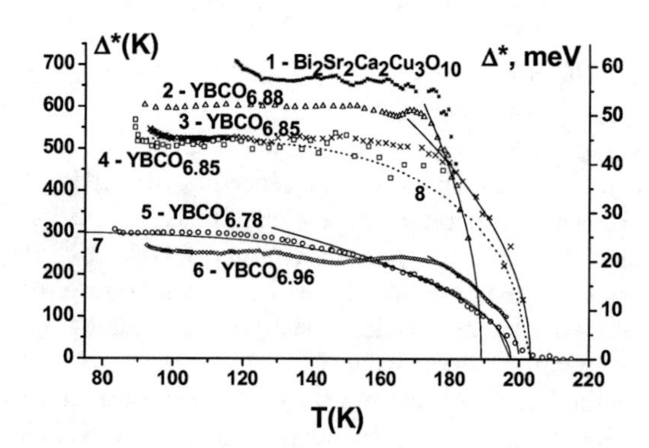

FIGURE 2. Temperature dependence of Δ^* for Bi- and Y-based HTSC thin films and single crystals.

ACKNOWLEDGMENTS

This work was supported by the President's Grant Program "Leading scientific school" 2200.2003.2.

REFERENCES

1. T. Timusk and B. Staat, *Rep. Progr. Physics* **62**, 61-112 (1999).
2. V. I. Belyavski and Yu. V. Kopaev, *Physics – Uspekhi* **174**, 457-466 (2004).
3. K. Takenaka, K. Mizuhashi, H. Takagi, and S. Ushida, *Phys. Rev.* B **50**, 6534-6537 (1994).
4. D. D. Prokof'ev, M. P. Volkov, and Yu. A. Boikov, LT23, Kyoto, Japan, LT2680, (2002).
5. E. Babaev and H. Kleinert, *cond-mat/9804206*, (1998).
6. D. D. Prokof'ev, M. P. Volkov, and Yu. A. Boikov, *Physics of the Solid State*, **45**, 1168-1176 (2003).
7. V. V. Kabanov, J. Demsar, B. Podobnic, et al. *Phys. Rev.* B **59**, 1497-1510 (1999).
8. V. Emery and S. A. Kivelson, *Nature* **374**, 434-438 (1995).
9. A. P. Solov'ev, H.-U. Habermeier, and T. Haage, *Low Temperature Physics* **28**, 24-35 (2002).

Pulsed Laser Deposition of T, T* and T' $La_{1.85}Y_{0.15}CuO_4$

S. C. Wimbush and E. Takayama-Muromachi

National Institute for Materials Science, International Center for Young Scientists,
1-1 Namiki, Tsukuba, Ibaraki 305-0044, Japan

Abstract. Following a report of superconductivity in nominally undoped T' lanthanum cuprates which, if verified, will have wide-ranging implications for the understanding of the mechanism of high-temperature superconductivity, an attempt has been made to independently reproduce or otherwise elucidate the reported results. Utilising a different preparation technique from the original report, thin film samples of nominally undoped $La^{3+}_{1.85}Y^{3+}_{0.15}CuO_4$ have been prepared by pulsed laser deposition in all three structural variations reported for the system, labeled T, T* and T'. The unique nature of pulsed laser deposition as a highly non-equilibrium process has been exploited in the fabrication of the metastable T* and T' phases, allowing a significant improvement in sample quality over that obtainable by the use of chemical pressure and epitaxial strain alone.

Keywords: T'-214 system, thin films, pulsed laser deposition
PACS: 74.72.-h, 74.10.+v, 68.55.-a, 81.15.Fg

INTRODUCTION

The prototypical high-T_c cuprate superconductor, as discovered by Bednorz and Müller [1], was a 214 phase of the form A_2CuO_4, with A formed from a solid solution of La and Ba, the latter constituent serving to provide the charge carrier doping believed essential for superconductivity to occur. This material and many others like it, subsequently discovered, form in the so-called T structure. By the use of techniques such as chemical pressure or epitaxial strain, a metastable T' structure of the same system can instead be formed. In a recent paper [2], Tsukada *et al.* report that such a T'-214 cuprate can be superconducting without charge carrier doping. This work aims to corroborate or otherwise explain their finding.

EXPERIMENTAL TECHNIQUE

Thin film samples (thickness ~200 nm) were deposited on single crystal $SrTiO_3$ (100) substrates by pulsed laser deposition in a 30 Pa oxygen atmosphere. $La_{1.85}Y_{0.15}CuO_4$ targets were prepared by standard ceramic processing techniques from powders of the constituent oxides. The substrate was heated during deposition to a range of temperatures, and the influence of both the deposition temperature and the laser fluence on the resultant sample structure was investigated by x-ray diffraction.

RESULTS AND DISCUSSION

X-ray diffraction of the prepared samples reveals exclusively (00*l*) peaks of the phases of interest, indicating a high degree of *c*-axis orientation of the grown films. In some samples, impurity peaks that cannot be attributed to $La_{1.85}Y_{0.15}CuO_4$ are also seen. Examining the full set of samples, the T peaks of the pure La_2CuO_4 phase [3] are clearly seen to split into three more or less evident peaks, attributable to the T, T* and T' phases, in order of increasing 2θ value (decreasing *c*-axis lattice parameter). The attribution of peaks to the T' phase follows the work of Tsukada *et al.* [4], which states the lattice parameters $c(\mathrm{T}) = 1.315$ nm and $c(\mathrm{T'}) = 1.255$ nm, in broad agreement with observation here. The speculative attribution of peaks to a T* phase is, at this stage, a best guess, based on the appearance of all the expected (00*l*) peaks in the right places and the fact that its lattice parameter should lie between those of the other two structures, it being an ordered combination of the two. This phase has yet to be isolated and studied in detail, and work is ongoing to confirm this supposition.

To elucidate the effect of the varying deposition conditions on the resultant structure of the films, figures 1 and 2 show an analysis of the x-ray data for two series' of deposited films, varying deposition temperature and laser fluence respectively. The integral peak intensities of the strong (004) peak group

CP850, *Low Temperature Physics: 24th International Conference on Low Temperature Physics;*
edited by Y. Takano, S. P. Hershfield, S. O. Hill, P. J. Hirschfeld, and A. M. Goldman

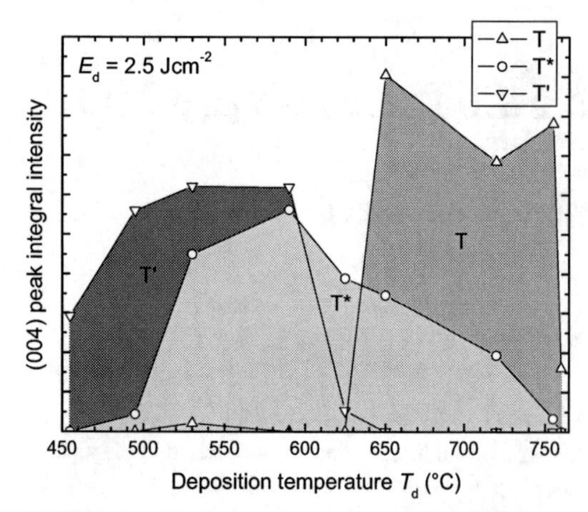

FIGURE 1. Variation in resultant sample structure on changing the deposition temperature. Curve areas are normalised since absolute x-ray intensities cannot be compared between different phases.

(T, T* and T' peaks), which also lies well clear of any substrate peaks, are plotted against the deposition parameter. From these results, a number of observations regarding the phase formation of the three phases can be made.

From figure 1, we see that there exist well-separated temperature regimes in which the T' and T phases are preferentially formed, with the equilibrium T phase dominating at higher temperatures, as might be expected. However, we also observe the unexpected formation of the third T* phase throughout the temperature range, in coexistence with the other phase.

Since the T' phase is of primary interest here, a second series of films was deposited at low temperature, this time varying the laser fluence with the aim of reducing or eliminating the T* fraction. The results are shown on figure 2, where we see that the T phase is suppressed in all cases at this low temperature, while the relative proportions of T* and T' can be altered by altering the laser fluence, high fluences favouring T'. Unfortunately, the deposition setup is limited to the fluence just less than 5 Jcm^{-2} used here, although experiments are in progress to increase this. Similarly at the low end of the scale shown, where T* is almost exclusively formed.

From these results, we may summarise the optimum conditions for obtaining samples forming each of the three structures as follows:

TABLE 1. Optimum Deposition Conditions.

Phase	Temperature	Energy
T	high (~750°C)	any (not shown)
T*	intermediate (~625°C)	low (< 4 Jcm^{-2})
T'	low (~500°C)	high (> 3.5 Jcm^{-2})

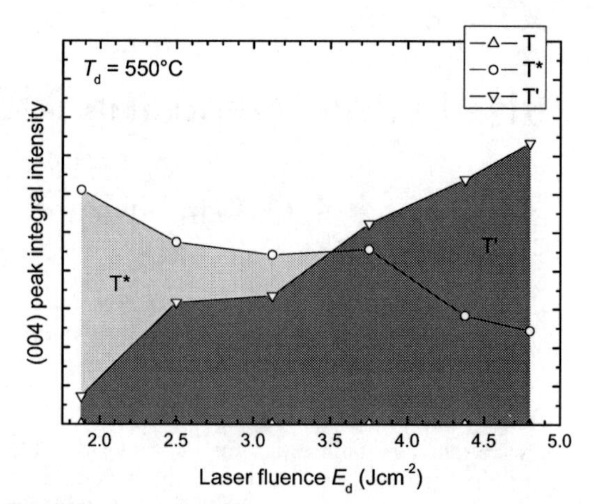

FIGURE 2. Variation in resultant sample structure on changing the laser fluence. Curve areas are normalised since absolute x-ray intensities cannot be compared between different phases.

Under these conditions, samples approaching phase purity (comparable integral intensities of the dominant phase in each case, with less than 10% absolute integral intensity of combined other phases) have been obtained.

CONCLUSION

The pulsed laser deposition of metastable T' lanthanum cuprates has been investigated and phase purity of the resultant samples approached. The unexpected formation of a second metastable phase, tentatively identified as T* has been observed. No superconductivity has been observed in any nominally undoped samples prepared to date.

ACKNOWLEDGMENTS

This work was performed using Special Coordination Funds for Promoting Science and Technology from the Ministry of Education, Culture, Sports, Science and Technology of the Japanese Government.

REFERENCES

1. J. G. Bednorz and K. A. Müller, *Z. Phys. B* **64**, 189-193 (1986).
2. A. Tsukada, Y. Krockenberger, H. Yamamoto, and M. Naito, *Physica C* in press. doi:10.1016/j.physc.2003.12.023.
3. JCPDS PDF number: 70-0449.
4. A. Tsukada, T. Greibe, and M. Naito, *Phys. Rev. B* **66**, 184515 (2002).

High Quality $YB_2C_3O_{7-\delta}$ Films with Controllable In-Plane Orientations Grown on YSZ Substrate

P. A . Lin[a,b], H. H. Hung[d], R. L. Lo[a] , M. K. Wu[a,c], and C. C. Chi[a,b]

a. *Department of Physics National Tsing Hua University, Hsinchu, Taiwan 300, Republic of China*
b. *Materials Science Center National Tsing Hua University, Hsinchu, Taiwan 300, Republic of China*
c. *Academia Sinica Institute of Physics, Taipei Taiwan 11529, Republic of China*
d. *National Synchrotron Radiation Research Center, Hsinchu, Taiwan, Republic of China*

Abstract. We use pulsed-laser deposition (PLD) technique to grow High T_C superconducting $YB_2C_3O_{7-\delta}$ films onto virgin and ion-bombarded YSZ substrates. We have found that 0-degree in-plane orientated high quality thin films can be grown on either substrate when the growth temperature is about 810℃. The T_C of the thin film grown in this condition is about 90.3K, and the J_C is about $4x10^6$ Acm^{-2} at 77K. At a lower temperature around 690℃, on the virgin substrates, we observe pure 45-degree in-plane orientation; while on the ion-bombarded substrates, small amount of 0-degree domain is present with the majority of 45-degree domain. Despite of the difference of the in-plane orientations, the T_C and J_C of the films grown at low temperature are as good as films grown at high temperature. This investigation is essential for making biepitaxial grain boundary junctions.

Keywords: High T_c & J_c, biepitaxial grain boundary junctions
PACS: 74.78.Bz

High T_c superconducting grain-boundary Josephson junctions are important for superconducting devices. There are several methods to fabricate Josephson junctions. The earliest grain boundary junction was natural grain boundary found in polycrystalline films [1]. In 1988, IBM developed the first bi-crystalline grain boundary junctions. According to Mannhart et al, the original goal of bi-crystalline grain boundary junctions was to study the influence of grain boundaries on superconducting transport properties [2,3,4]. However, the goal shifted to the application for making devices later. Unfortunately, bicrystal technique has flaws such as the difficulty to extend to IC processing and expensiveness of its substrates. To eliminate these flaws, Char et al, at Conductus, developed a new type of junction, called biepitaxial grain boundary junction [5]. This kind of junction composes of two layer-structures, YBCO/ CeO2/ MgO/ SrTiO3 and YBCO/ SrTiO3, on a single crystal substrate. The location of these junctions can be artificially defined by photolithography. With this advantage, bi-epitaxial grain boundary junctions become essential [6,7]. All the above biepitaxial junctions require a number of different materials, which makes deposition processes more complicated. To remedy this, a simpler fabrication process was recently developed in our lab. The junction fabricated from this process is called homo-bi-epitaxial grain boundary junction, whose structure is YBCO (high-T)/ YBCO (low-T)/ YSZ and YBCO (high-T)/ YSZ [8]. Preliminary studies of this process in our lab showed minor instability at high temperatures and some contradictory results [9]. This might be due to the ion-milling processing for patterning. The purpose of this paper is to resolve these problems and to improve the fabrication process.

The virgin substrate is referred as a single crystal with the 10% yttria content and is etched by an ion source to create a rough surface. From AFM images (not shown), we found smooth surface on the virgin substrate and many islands and holes on the rough substrate.

To grow YBCO thin films, an excimer laser with the wavelength of 248nm was used. First, we vary laser energy density, oxygen pressure, and the distance between target and holder to optimize the YBCO superconducting properties on virgin substrate. Then, we compare the films deposited on ion-bombarded substrates with those on the smooth substrates in the deposition temperature range of 660℃ to 810℃. In this temperature range, all films deposited on either kind of substrates are c-axis films. All other conditions are fixed with the oxygen pressure at 320 mTorr and the distance between target and holder at 39mm with laser energy density at $1.5J/cm^2$. Figure 1 shows x-ray φ-scan for films deposited at several different temperatures. At the high end of our experimental temperature range, we found no difference between virgin substrates and ion-bombarded substrates; they all have pure 0-degree orientations. At 790℃, there is a very small 45-degree orientation appeared on the

CP850, *Low Temperature Physics: 24th International Conference on Low Temperature Physics;*
edited by Y. Takano, S. P. Hershfield, S. O. Hill, P. J. Hirschfeld, and A. M. Goldman

virgin substrate. But on the ion-bombarded substrate, the peaks broaden greatly, with ±9-degree sub-peaks locating around the 0-degree peak. At 730℃, the 0-degree domain is almost the same with 45-degree domain on virgin substrate, but on ion-bombarded substrate, the 45-degree orientation domain is prominent. When temperature lowers to about 690℃, on the virgin substrate, pure 45-degree orientation is found; on the other hand, on the ion-bombarded substrate, small 0-degree peak was found.

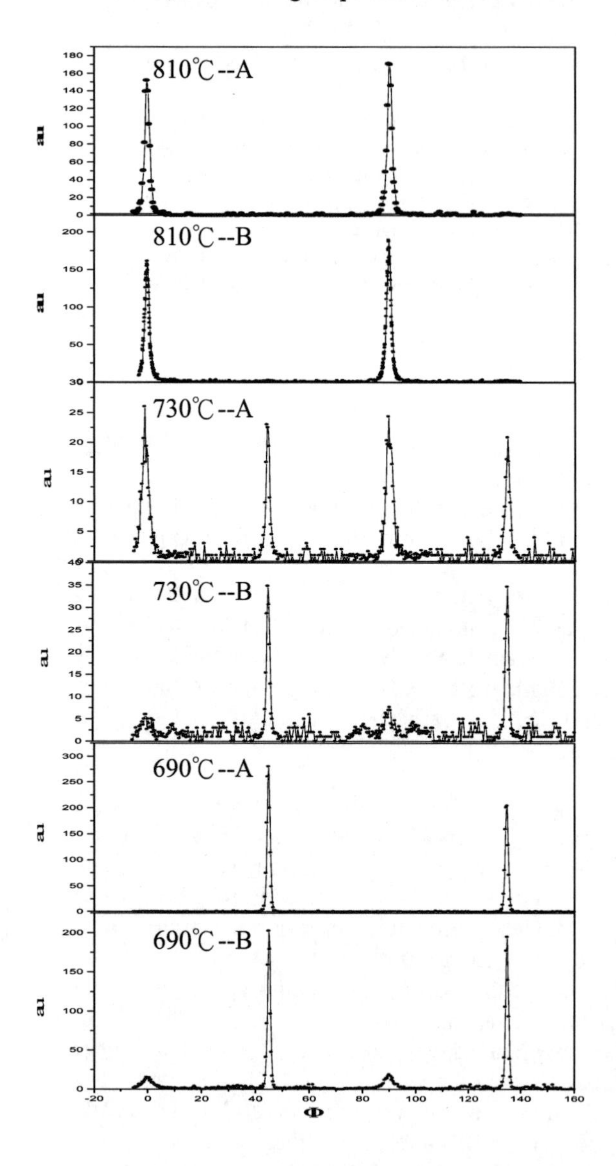

FIGURE 1. Thin films deposited on virgin substrate--A, and on ion-bombarded substrate--B at different temperatures.

It is interesting that, at lower temperature 660℃ a small 0-degree peak reappear on the virgin substrate (not shown). The reason for the reappearance of 0-degree peak is unclear.

We use SQUID magnetometer to measure M-H curves, and Bean model to estimate films' J_c values [10,11]. The geometry of our samples is almost the same as 0.4cmx0.4cmx3x10^{-5} cm in length, in width, and in thickness. The resulting J_c are tabulated in Table 1.

TABLE 1. List Jc at different temperature.

	Virgin surface		Etching surface	
	J_c (5k) Acm^{-2}	J_c (77k) Acm^{-2}	J_c (5k) Acm^{-2}	J_c (77k) Acm^{-2}
810℃	$4.1X10^7$	$4.1\ X10^6$	$4.8X10^7$	$4.9X10^6$
790℃	$4X10^7$	$3.5X10^6$	$6.4\ X10^5$	$3X10^5$
730℃	$1.2x10^5$	$1.3x10^3$	$4x10^5$	$3.3x10^3$
690℃	$3.1X10^7$	$2\ .3X10^6$	$3.6X10^6$	$4.3X10^5$
660℃	$1x10^7$	$2.3x10^5$	$4.7x10^5$	$3.1x10^3$

At high temperature about 810℃, the films deposited on ion-bombarded substrates are as good as those on virgin substrates. This indicates that damaged surface can heal itself. Therefore, we can have a stable pure 0-degree orientation at high temperature. Our transport measurements obtain high T_c and J_c for pure 0-degree and 45-degree in-plane orientated films. Thus we have found the most favorable conditions for fabricating be-epitaxial grain boundary junctions, which we intend to carry out in the future.

We thank financial support provided by NSC-93-2112-M-007-032 and NSRRC, Hsinchu, for providing in-house four-circle x-ray diffractometer.

REFERENCES

1. Koch.R. H., C. P. Umbach, G. J. Clark, P. Chaudhari, and R. B. Laibowitz, *Appl. Phys. Lett.* **51**,200–202 (1987)
2. H. Hilgenkamp, and J. Mannhart, *Rev of Mod Phys.* **74**, 485 (2002)
3. Chaudhari, P., J. Mannhart, D. Dimos, C. C. Tsuei, J. Chi, M.M. Oprysko, and M. Scheuermann, *Phys. Rev. Lett.* **60**, 1653 (1988)
4. Dimos. D, P. Chaudhari, J. Mannhart, and F. K. LeGoues, *Phys. Rev. Lett.* **61**.219 (1988)
5. Char, K., M. S. Colclough, S. M. Garrison, N. Newman, and G.Zaharchuk, *Appl. Phys. Lett.* **59**, 733–735 (1991)
6. M.Y. Li, et. al, *Physica C* **235**, 589 (1994)
7. S.H. Tsai, M.K.Wu, and C.C. Chi *Chinese J. Phys.* **36**, 355 (1998)
8. S.H. Tsai, M.K.Wu, and C.C. Chi *Physica C* **339**, 155 (2000)
9. Thesis of Ko-Hsin Lee (2002)
10. E.M. Gyorgy, R.B. van Dover, K.A. Jackson, L.F. Schneemeyer, and J.V. Waszczak, *Appl. Phys. Lett.* **55**, 283 (1989)
11. E. Zeldov, J. R. Clem, M. McElfresh, and M. Darwin, *Phys. Rev. B* **49** ,9802 (1994)

Reentrant Irreversibility and Magnetic Transition in Strongly Underdoped $Y_{0.47}Pr_{0.53}Ba_2Cu_3O_{7-\delta}$ Single Crystals

V. Sandu,[1,*] P. Gyawali,[1] T. Katuwal,[1]
B. J. Taylor,[2] M. B. Maple,[2] and C. C. Almasan[1]

[1]*Department of Physics, Kent State University, Kent, OH 44242, USA*
[2]*Department of Physics, University of California at San Diego, La Jolla, CA, USA*

Abstract. Magnetization measurements on $Y_{0.47}Pr_{0.53}Ba_2Cu_3O_7$ single crystal in the vicinity of the superconducting transition temperature $T_c = 7$ K revealed a reentrant irreversibility for applied magnetic fields higher than the irreversibility field H_{irr}. The transition to the second hysteresis loop occurs through an irreversible jump in magnetization. The temperature range of this irreversibility extends up to 40 K. Generally, this double hysteretic regime has been associated in low dimensional systems with the pinning of spin and/or charge density waves.

Keywords: magnetization, $Y_{1-x}Pr_xBa_2Cu_3O_7$, density waves, fluctuations.
PACS: 74.25.Ha, 74.25.Op, 74.72.Bk, 75.30.Fv

Underdoped cuprates seem to be an inexhaustible source of new phenomena whose origin is in the low charge carrier density and low dimensional charge conducting substructures. These features represent a very strong hint that an extended fluctuation regime has to accompany the superconducting (SC) to normal phase transition. Recently, Panagopoulos *et al.* [1] reported thermal hysteresis in $La_{2-x}Sr_xCuO_4$ single crystals, which extends up to 290 K, suggesting an association with superconducting fluctuations (SF). However, the question that still remains is: If the low charge carrier density enhances the superconducting fluctuations, why there is no continuous extension of the fluctuation regime with increasing underdoping? It was found that close to the superconducting to antiferromagnetic limit, the range of the superconducting phase fluctuations shrinks (but does not vanish) [2-3]. Therefore, it is of interest to check more carefully the range of carrier concentrations where the superconducting phase is very close to the superconductor to insulator transition. In our investigations, we used the antidoping effect of praseodymium, which partially substitutes for yttrium as $Y_{1-x}Pr_xBa_2Cu_3O_{7-\delta}$ and drives the system to the Mott antiferromagnetic state (AF) for $x \geq 0.55$. μSR investigations [4] have revealed a smooth crossover and small overlap for $x < 0.55$ between SC and AF orders of both Pr and planar Cu spins. In these highly underdoped materials we have found a double hysteretic behavior as a function of field. Specifically, a second loop opens close to T_c and remains distinct up to temperatures as high as $5 \times T_c$.

Magnetization M vs H ($H \| c$), M vs T, and magnetic relaxation measurements were performed on $Y_{0.47}Pr_{0.53}Ba_2Cu_3O_7$ ($T_c = 7$ K) single crystals using a SQUID magnetometer. These measurements give a superconducting volume fraction close to 100% at 4 K. Up to 6 K, the *M(T,H)* curves are typical for cuprates. With further increasing T, the opening of a new irreversible loop appears after a jump in magnetization. With increasing T, the new hysteresis loop, including the jump, shifts to lower H values. Both loops eventually merge even though they still remain distinct (see Fig. 1 for $T = 6.7$ K). The second irreversible loop is still visible up to T as high as 40 K (Inset to Fig. 1). Relaxation data, taken above T_c, show an extremely small average relaxation rate, $\Delta M/\Delta t = 1.2 \times 10^{-11}$ emu/sec, at 12 K and 0.03 T (data not shown).

M(T) plots show thermal hysteresis (a difference between zero-field-cooled and field-cooled magnetization curves, see Fig. 2) above T_c, which overlaps the expected paramagnetism.

To understand this behavior, we recall that the proximity to antiferromagnetism in strongly underdoped cuprates generates strong antiferromagnetic fluctuations, which, at microscopic level, coexist with the superconducting ordering. Neutron scattering experiments on $La_{2-x}Sr_xCuO_4$, have shown that these remnant antiferromagnetic excitations are consistent with spin density waves [5]. If spin (or charge) modu-

CP850, *Low Temperature Physics: 24th International Conference on Low Temperature Physics;*
edited by Y. Takano, S. P. Hershfield, S. O. Hill, P. J. Hirschfeld, and A. M. Goldman

lation is present in the superconducting state, then, above the critical temperature, the competing density wave (DW) type order is dominant since superconductivity is present only through fluctuations of the phase of the order parameter [6].

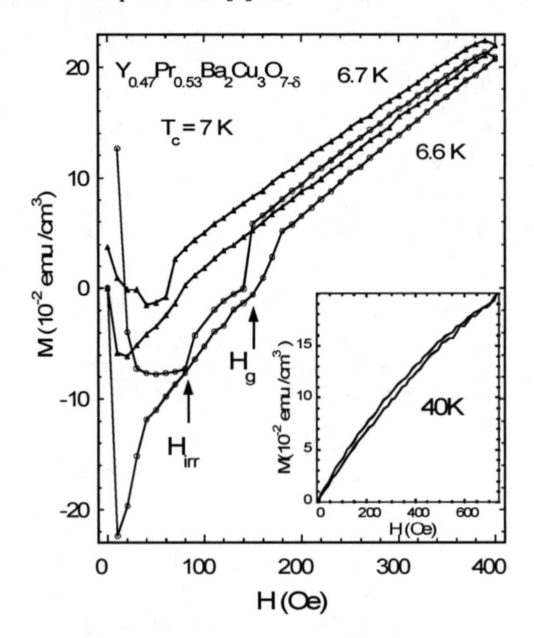

FIGURE 1. Plots of magnetization M vs magnetic field H in the superconducting regime of the $Y_{0.47}Pr_{0.53}Ba_2Cu_3O_{7-\delta}$ single crystal for T = 6.6 (open symbols) and 6.7 K (solid triangles). The two arrows mark the irreversibility field H_{irr} and the opening of the second hysteresis H_g. Inset: the hysteresis curve at 40 K.

The low dimensionality of the cuprates makes them sensitive to modulated ordering [7]. In the case of the present experiments, it is less likely that SF are responsible for the second irreversibility. The pinning becomes negligible once the reversibility is reached and a recovery of irreversibility at higher fields as a result of SF is not physically acceptable at temperatures as high as $5 \times T_c$, where thermal fluctuations become much stronger than the pinning energy. On the other hand, it was shown that both d-wave SC and d-wave DW promote local AF state in the presence of impurities [8]. At low hole concentrations, the carriers are expelled from the AF regions creating nano-scale charge and spin inhomogeneities. Therefore, it is expected the coexistence of AF both with SC, at low temperature, and with DW at high temperature. In the latter T-range, the irreversibility appears due to the infinite number of metastable states of the DW in the presence of impurities, which makes the local magnetic order history dependent, hence, the total moment irreversible.

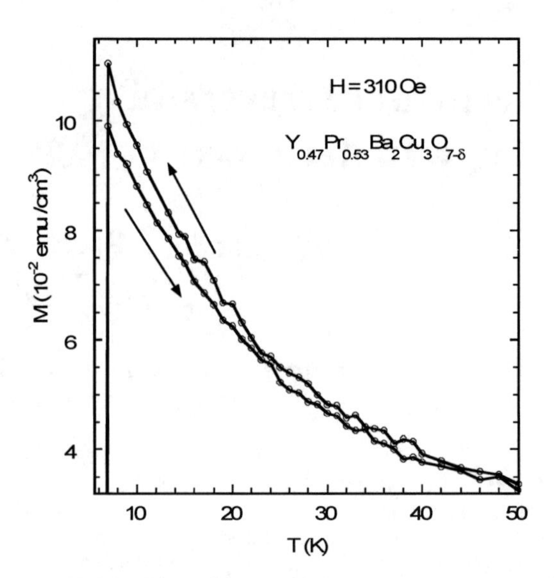

FIGURE 2. Plots of magnetization M vs temperature T for the $Y_{0.47}Pr_{0.53}Ba_2Cu_3O_{7-\delta}$ single crystal measured in a field of 310 Oe.

In summary, we found that in strongly underdoped but still superconducting Pr-doped cuprates there is a second magnetic hysteresis and, associated, a thermal hysteresis in the normal state. We argue that this new irreversibility is consistent with the existence of local antiferromagnetic regions induced by d-wave type density waves in the presence of impurities.

ACKNOWLEDGMENTS

This research was supported by the National Science Foundation under Grant No. DMR-0406471 at KSU and the US Department of Energy under Grant No. DE-FG02-04ER-46105 at UCSD.

* Permanent address: *National Institute of Materials Physics, Bucharest-Magurele, 01725 Romania.*

REFERENCES

1. C. Panagopoulos, *et al., Phys. Rev. B* **69**, 144508 (2004).
2. Y. Wang, *et al., Phys. Rev. Lett.* **88**, 257003 (2002).
3. T. Katuwal, *et al., Phys Rev B* **72**, 174501 (2005).
4. D. W. Cooke *et al., Phys. Rev. B* 41 4802 (1990).
5. S. Wakimoto, *et al., Phys. Rev. B* **60**, R769 (1999).
6. S. Sachdev and E. Demler, *Phys. Rev. B* **69**, 144504 (2004).
7. B. Grévin, *et al., Phys. Rev. Lett.* **80**, 2405 (1998).
8. Y. Osashi, *Phys Rev B* **60** 15388, (1999).

Temperature and Frequency Dependence of Complex Conductance of Ultrathin $YBa_2Cu_3O_{7-x}$ Films: Observation of Vortex-Antivortex Pair Unbinding

V.A. Gasparov*, G.E. Tsydynzhapov*, I.E.Batov† and Qi Li¶

*Institute of Solid State Physics RAS, 142432, Chernogolovka, Russian Federation

†University of Erlangen-Nьrnberg, D-91058 Erlangen, Germany

¶Department of Physics, Pennsylvania State University, University Park, PA 16802, USA

Abstract. We have studied the temperature dependencies of the complex sheet conductance, $\sigma(\omega,T)$, of 1-3 unit cell (UC) thick $YBa_2Cu_3O_{7-x}$ films sandwiched between semiconducting $Pr_{0.6}Y_{0.4}Ba_2Cu_3O_{7-x}$ layers at high frequencies. We have found: (i) quadratic temperature dependence of the kinetic inductance, $L_k^{-1}(T)$, with a break in slope at T^{dc}_{BKT}, (ii) the maximum of real part of conductance and large shift of the onset temperature and the peak of $\omega\sigma_1(T)$ to higher temperatures with increasing ω. We obtain the universal ratio $T^{dc}_{BKT}/L_k^{-1}(T^{dc}_{BKT})$ = 25, 25, and 17 nH K for 1, 2 and 3-UC films, respectively in relation with theoretical prediction of 12 nH K for vortex-antivortex unbinding transition.

Keywords: superconductivity, vortex-antivortex pairs, Berezinskii-Kosterlitz-Thouless transition.
PACS: 74.80.Dm, 74.25.Nf, 74.72.Bk, 74.76.Bz

INTRODUCTION

Although, many observations of the Berezinski – Kosterlitz - Thouless (BKT) transition in cuprates compounds have been reported, detailed comparison of the experimental data with the theory showed disagreements possibly due to inhomogeneity and vortex pinning [1]. It was suggested [2] that a precondition for the BKT transition to occur in a superconductor, i.e. the sample size $L<\lambda_{eff}$, is not satisfied even in ultra thin YBCO films. At high frequency however, the electromagnetic response of a 2D superconductor is dominated by those bound pairs that have the correlation length $\xi_+(T)$ equal to the vortex diffusion length $l_\omega(T) \propto \omega^{-1/2}$. This implies that it is possible to detect the response of vortex-antivortex pairs with short separation lengths at high frequencies even though the dc BKT transition is not present. Here we report the frequency and temperature dependences of the complex sheet conductance, $\sigma(\omega,T)$, of 1- UC to 3-UC thick YBCO films sandwiched between semiconducting $Pr_{0.6}Y_{0.4}Ba_2Cu_3O_{7-x}$ layers [1] at high frequencies, as a indication of observation dynamic BKT transition.

RESULTS AND DISCUSSION

Ultrathin YBCO layers sandwiched between 10 nm buffer and 15 nm cover layers of $Pr_{0.6}Y_{0.4}Ba_2Cu_3O_{7-x}$ were grown epitaxially on atomically flat and well - lattice - matched (100) $LaAlO_3$ substrates using a multitarget pulsed-laser deposition system [1]. The $\sigma(\omega,T)$ at RF in thin films was investigated employing a single coil mutual inductance technique. The high frequency (100 MHz – 1 GHz) measurements were performed using the spiral coil cavity. The MW losses were measured using a resonant cavity technique with the gold-plated copper cylindrical cavity operated at 30 GHz [1].

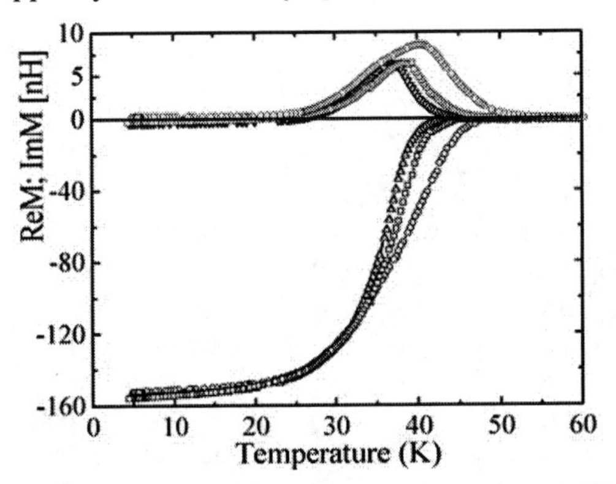

FIGURE 1. ReM(T) and ImM(T) curves for a 2-UC S2 film at different frequencies: 3 MHz (triangles), 26 MHz (squares) and 500 MHz (circles).

Figure 1 displays the ReM(T) and ImM(T) of complex mutual inductance M between the coil and the film curves for a 2UC film at three different

CP850, *Low Temperature Physics: 24th International Conference on Low Temperature Physics;* edited by Y. Takano, S. P. Hershfield, S. O. Hill, P. J. Hirschfeld, and A. M. Goldman

frequencies from 3 MHz to 500 MHz as measured by: LC circuit and spiral coil resonator. The most noticeable feature of these data is rather high shift of the onset point of ReM(T) transition and maximum ImM(T) position with frequency, not observed in such measurements on 200 nm thick YBCO films. The ReM(T) and ImM(T) data are converted to $L_k^{-1}(T) = \mu_0 \lambda^2 / d$ and $\mathrm{Re}\sigma(T) \equiv \sigma_1(T)$ using the mathematical inversion procedure [1]. Fig. 2 shows the $L_k^{-1}(T)$ curves at very low perpendicular magnetic fields, and zero field $\omega\sigma_1(T)$ for the 1-UC and 2-UC films (S1). We found that $L_k^{-1}(T)$ fit well below characteristic temperature, which we define as T^{dc}_{BKT}, by a parabolic dependence: $L_k^{-1}(T) = L_k^{-1}(0)[1-(T/T_{c0})^2]$, shown as thin solid lines. We determined the T^{ω}_{BKT} as the peak position of $\sigma_1(T)$ point, which coincides with the temperature where $l_\omega \approx \xi_+$.

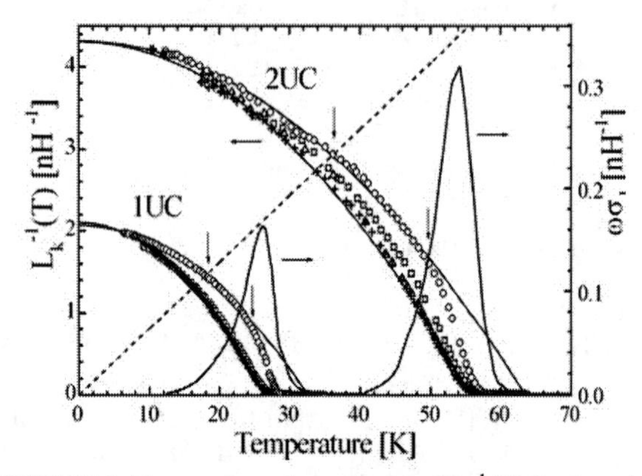

FIGURE 2. Temperature dependence of $L_k^{-1}(T)$ for 1- and 2- UC S1 films at 8 MHz and different magnetic fields: 0 mT (circles), 2 mT (squares), 3 mT (triangles) and 4 mT (crosses). The solid lines curves shows $\omega\sigma_1(T)$ at zero field.

There are three major features observed in our RF and MW measurements: (i) the large frequency dependence of onset temperature of $L_k^{-1}(T)$ and maximum position of $\omega\sigma_1(T)$; (ii) a foot jump in temperature dependence of the $L_k^{-1}(T)$, which is destroyed in weak magnetic fields; and (iii) a maximum in $\omega\sigma_1(T)$ with the onset of transition at higher temperatures than that of the $L_k^{-1}(T)$. The qualitative explanation of these features is as following. By probing the system at finite frequencies, the observed bound-pair response is dominated by those pairs with the correlation length $\xi_+(T) \approx l_\omega \propto (14D/\omega)^{1/2}$, which determines the BKT transition temperature at a given frequency T^{ω}_{BKT}. We plot theoretical BKT function $L_k^{-1}(T)$ as dashed straight line on Fig. 2 derived from the universal relationship: $\hbar^2c^2d/16\pi e^2\lambda^2k_BT_{BKT} = 2/\pi$, predicted by theory for *dc* case. It is obvious from the Fig. 2 that T^{dc}_{BKT} determined from the intercept of theoretical line with $L_k^{-1}(T)$ is lower then the peak position of $\sigma_1(T)$. Nevertheless, we found almost constant ratio of $T^{dc}_{BKT}/L_k^{-1}(T^{dc}_{BKT})$ equal to 25, 25, and 17 nH K for 1-, 2- and 3-UC S1 films, respectively. It is obvious from Fig.3, that all $\lambda^{-2}(T)$ data at 8 MHz fall on the same scaling curve, which proof of our definition of T^{dc}_{BKT} as the break point.

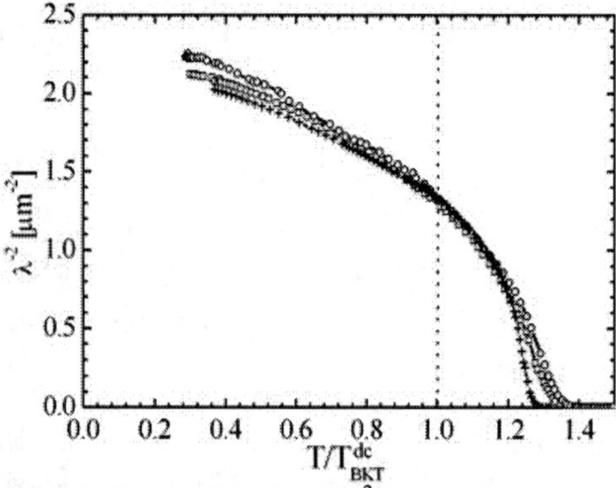

FIGURE 3. Penetration depth $\lambda^{-2}(T)$ derived from $L_k(T)$ vs normalized temperature T/T^{dc}_{BKT} for: 1UC (circles), 2UC (squares) and 3 UC (crosses) films.

It is easy to show from [3] that the ratio $L_k^{-1}/\omega\sigma_1(T) = \pi(Y-1)/2Y\ln Y$ in high frequency limit (here $Y = (l_\omega/\xi_+)^2$). From the Y(T) data, we thus found that the vortex diffusion constant D(T) follow with activation T dependence due to pinning of vortex core [1]. We found also that the $Y(T^{\omega}_{BKT}/T)$ deviates from exponential temperature dependence close to T^{ω}_{BKT} and collapse into one curve at $T > T^{\omega}_{BKT}$, indicating that the temperature dependence of $\xi_+(T^{\omega}_{BKT}/T)$ is the same for all samples and frequencies.

Notice that a peak of $\sigma_1(T)$ and its frequency dependence is not due to the skin effect [1]. We obtained $\delta = 0.13$ µm at RF and 5.4 µm at MW for the peak temperature from resistivity of the 3UC sample, far too small compared to a film size and too high compared with a thickness.

In summary, we have compared complex conductivity of ultrathin YBCO films with the dynamic theory for BKT transition and found that the unbinding of the vortex-antivortex pairs with short correlation length were observable at high frequencies.

ACKNOWLEDGMENTS

We are grateful S.E. Korshunov for discussions. This work was supported by Russian Government (Grant MSh-2169.2003.2), INTAS (Grant No. 01-0617) and NSF DMR-0405502.

REFERENCES

1. V.A. Gasparov *et al., J. Low Temp. Phys.* **139**, 49 (2005).
2. J.M. Repaci *et al., Phys. Rev. B* **54**, 9674 (1996).
3. P. Minnhagen *et al., Rev. Mod. Phys.* **59**, 1001 (1987).

Angular Dependence Of Critical Current Density In Ca-Doped YBCO Epitaxial Thin Films

A. Augieri[1], G. Celentano[1], L. Ciontea[2], J. Halbritter[3], V. Galluzzi[1], U. Gambardella[4], A. Mancini[1], T. Petrisor[2], A. Rufoloni[1], and A. Vannozzi[1]

[1] *Superconductivity section, ENEA Research Center, Frascati, 00044, Italy*
[2] *Technical University of Cluj, Str. C. Daicoviciu 15, 3400 Cluj-Napoca, Romania*
[3] *Institut fur Material Forschung, FZK, Postfach 3640, 76021 Karlsruhe, Germany*
[4] *INFN Frascati National Laboratory, Frascati, 00044, Italy*

Abstract. The angular dependence of the critical current density J_c was analyzed in $Y_{0.9}Ca_{0.1}Ba_2Cu_3O_{7-\delta}$ epitaxial thin films grown by PLD technique on $SrTiO_3$ and CeO_2 buffered Al_2O_3 substrates. The latter shows a broad peak in the J_c vs θ curves at θ=0° ($H // c$), revealing a directional character of the pinning mechanisms. This feature is affected both by temperature and magnetic field intensity emerging at intermediate T and H values. The origin of the c-axis correlated pinning is discussed and compared with the previously reported studies on the directional pinning.

Keywords: correlated pinning, critical current, angular dependence
PACS: 74.25.–q, 74.25.Fy, 74.25.Ha, 74.25.Qt, 74.25.Sv, 74.70.Dd, 74.78.Bz

INTRODUCTION

The applications of the coated conductors depend on the grain boundaries (GB) properties, as well as on the GB and intragrain (IG) pinning mechanisms. It has been shown that the introduction of Ca as partial substitute of Y in YBCO thin films increases the critical current density J_c in large angle (>10°) GB [1,2], while the same evidence is not so clear in low angle GB (<10°) [3,4], typical for coated conductors. The improvement of the coated conductors critical current can also be pursued by the introduction of controlled defects correlated on one direction, creating an anisotropic pinning force, as in the case of heavy ion irradiated samples [5,6]. More recently evidences of correlated pinning have been also observed in YBCO thin films grown on CeO_2 buffered substrates and have been ascribed to dislocations grown normal to the film surface [7,8]. In this work we report on the *dc* current transport characterization performed on $Y_{0.9}Ca_{0.1}Ba_2Cu_3O_{7-\delta}$ epitaxial thin films grown by PLD technique, either on CeO_2 buffered r-cut Al_2O_3 or (100) $SrTiO_3$ substrates. While the Ca doped YBCO film grown on CeO_2 buffered Al_2O_3 substrate showed directional features of the pinning mechanism, the film grown on (100) $SrTiO_3$ showed only the usual angular dependence peaked at $H \perp c$, associated to the isotropic pinning sites in the layered crystal structure of the cuprates.

CRITICAL CURRENTS

The critical current is determined from the voltage-current (V-I) characteristics, at an electrical field of 1 μV/cm, on patterned striplines 30÷50 μm wide, 2 mm long. The current is always perpendicular to the applied magnetic filed H, in order to achieve the maximum Lorentz force configuration, while the angle between the c-axis and the applied magnetic field, θ, can be varied continuously in the range ±110°. At θ=0°, i.e. H parallel to the c-axis, the comparison between the film grown on (100) $SrTiO_3$ substrate and the one grown on CeO_2 buffered r-cut Al_2O_3 substrate does not show significant differences. The $J_c(H)$ curves shown in Fig. 1 measured at reduced temperature $t=T/T_c$=0.85 are very similar, and the sample on $SrTiO_3$ exhibits a slightly higher critical current up to a 5 T field. Both films show the usual power law decay at low field followed by an exponential decay at high field.

CP850, *Low Temperature Physics: 24th International Conference on Low Temperature Physics;*
edited by Y. Takano, S. P. Hershfield, S. O. Hill, P. J. Hirschfeld, and A. M. Goldman

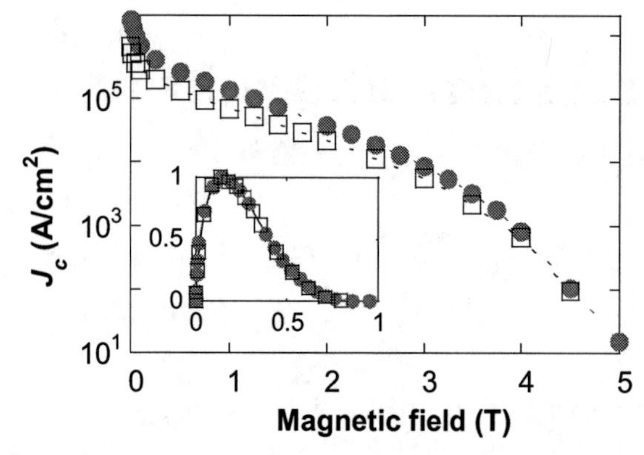

FIGURE 1. $J_c(H)$ comparison between samples at the same reduced temperature t=0.85. Open symbols: film on CeO_2/Al_2O_3. Solid symbols: film on $SrTiO_3$. The inset shows the reduced pinning force vs the reduced critical field.

The behavior of the reduced pinning force $f_p=F_p/F_{pmax}$, summarized in the inset of Fig. 1, indicates that the same pinning mechanisms are active in this angle configuration, either at low field or at higher fields.

On the other hand, there are substantial differences when, at fixed magnetic field, the critical current is studied as a function of θ. In Fig. 2 the normalized $J_c(\theta)$ behavior at different H values is shown in order to emphasize the angle dependence occurring in both samples. The film on $SrTiO_3$ (solid symbols) shows the standard behavior with two marked peaks at θ=±90°, corresponding to the so called intrinsic pinning, occurring when the field is parallel to the $a \bullet b$ planes. This fact implies that, apart from the intrinsic pinning, the pinning mechanisms active in this film have an isotropic character [6].

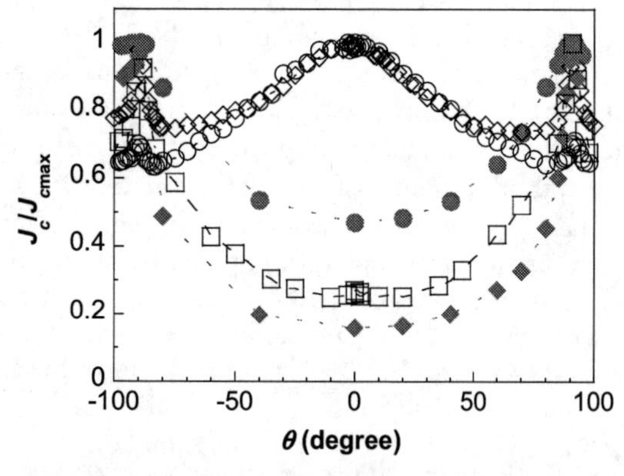

FIGURE 2. $J_c(\theta)/J_{cmax}$ curves measured at the same reduced temperature t=0.85 and different H values. Open symbols: film on CeO_2/Al_2O_3 (◊ 200 mT, ○ 870 mT, □ 2.5T). Solid symbols: film on $SrTiO_3$ (● 100 mT, ♦ 1T).

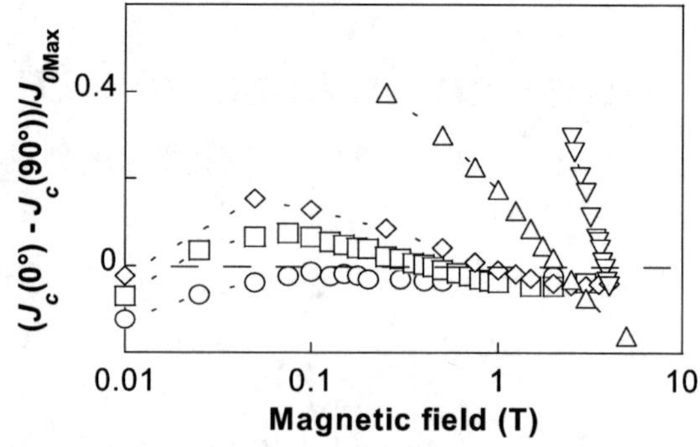

FIGURE 3. Difference between $J_c(0°)$ and $J_c(90°)$ normalized to the zero-field critical current density J_{0Max} for several temperatures: T=72 K (○), T=69 K (□), T=66 K (◊), T=55K (Δ), T=48 K (∇).

On the contrary, films grown on CeO_2/Al_2O_3 substrate (open symbols) exihibit a c-axis correlated pinning revealed by an additional broad peak centered at θ=0° [7]. The presence of this θ=0° peak, as well as its amplitude is affected by the temperature and the magnetic field strength. In Fig. 3 the $J_c(0°)$-$J_c(90°)$ normalized to the zero-field J_c as a function of H for different temperatures are reported. This function is proportional to the difference of the pinning forces between the θ=0° and 90° configurations. Below T= 72 K (t=0.92), the c-axis correlated pinning becomes dominant over the intrinsic pinning for increasing H as the temperature is reduced. The $J_c(0°)$-$J_c(90°)$ maximum is reached for a magnetic field strength value, which seems to decrease sligthly with temperature. This behaviour cannot be ascribed to the presence of calcium in the YBCO system but is more likely due to disorder induced in the film structure also by the diffusion of the Ce from the CeO_2 buffer layer in the YBCO film, with formation of Ce based compounds, during the growth process. Though the most valued evidence for structural disorder is considered to be the presence of dislocation, it is worth noting that also GB size and faceting could be affected [9] giving rise to an anisotropic behavior of pinning force.

REFERENCES

1. G. Hammerl *et al.*, *Nature* **407**, 162 (2000).
2. R. F. Klie *et al.*, *Nature* **435**, 475 (2005).
3. A. Augieri *et al.*, *Physica C* **401**, 320 (2004).
4. A. Weber *et al.*, *Appl. Phys. Lett.* **82**, 772 (2003).
5. L. Civale, *Supercond. Sci. Technol.* **10**, A11 (1997).
6. L. Civale *et al.*, *Appl. Phys. Lett.* **84**, 2121 (2004).
7. H. Yamada *et al.*, *Supercond. Sci. Technol.* **17**, 58 (2004).
8. J.C. Nie *et al.*, *Supercond. Sci. Technol.* **17**, 845 (2004).
9. J. Halbritter, to be published.

Effect of Ca Substitution at the Tm Site in the $TmBa_2Cu_3O_y$ Superconductor

Y. H. Lee,[1] Y. J. Chen,[1] J. J. Pan,[2] D. C. Ling,[2] and H.-C. I. Kao[1]

[1]*Department of Chemistry, Tamkang University, Tamsui 251, Taiwan*
[2]*Department of Physics, Tamkang University, Tamsui 251, Taiwan*

Abstract. A series of samples with composition $(Tm_{1-x}Ca_x)Ba_2Cu_3O_y$ were prepared by the conventional solid-state reaction method with $0 \leq x \leq 0.150$. Unit cell *a*- and *c*-axis increase and *b*-axis decreases with increase of the amount of Ca substitution. As a result, orthorhombicity decreases as x increases. Both T_c and y decrease with x, whereas hole concentration (p/Cu) increases with increasing Ca content up to 0.34/Cu for x = 0.075 and then decreases dramatically with x. The highest T_c (89 K) is found for x = 0 sample.

Keywords: $(Tm_{1-x}Ca_x)Ba_2Cu_3O_y$, superconductivity, structure, hole concentration, oxygen stoichiometry.
PACS: 74.72.Bk, 74.62.Bf

INTRODUCTION

$YBa_2Cu_3O_y$ and $RBa_2Cu_3O_y$ (R = rare earth, except Ce, Pr, Tb) are 90 K superconductors [1,2]. They are abbreviated as R123. Partial substitution of the R, Ba or Cu site always causes a decrease of T_c. It indicates that the un-substituted R123 compound has an optimal hole concentration, leading to either a hole doping [3] or a hole filling effect [4] caused by substitution. In general, orthorhombicity (defined as $2(b - a)/(a + b)$) of the unit cell decreases with increasing the amount of dopant [5], that is also related to the T_c of the R123 compounds [5]. Substitution of the trivalent R site with a divalent cation usually gives rise to the hole doping effect in the lighter rare earth R123 or Y123 [6,7]. However, not many studies on the heavier rare earth R123 compounds are reported. In order to elucidate the effect caused by the Ca substitution in Tm123, a series of $(Tm_{1-x}Ca_x)Ba_2Cu_3O_y$ is prepared, unit cell parameters, hole concentration, oxygen stoichiometry and T_c are investigated to understand the role of the Ca substitution in the system.

EXPERIMENTAL

Bulk samples of $(Tm_{1-x}Ca_x)Ba_2Cu_3O_y$ were prepared by the conventional solid-state reaction method. The purity of the samples was characterized by an x-ray diffraction (XRD) with Cu K_α radiation and a TG-DTA. GSAS program was employed for the Rietveld analysis [8]. T_c was found from the resistivity versus temperature curve measured by a standard 4-probe method. Oxygen stoichiometry and the hole concentration of the compound were determined by an iodometric titration method [9].

RESULTS AND DISCUSSION

Polycrystalline samples of $(Tm_{1-x}Ca_x)Ba_2Cu_3O_y$ were prepared with x = 0, 0.025, 0.050, 0.075, 0.100, 0.125 and 0.150. All of them are single-phase compounds with orthorhombic crystal system. Purity of the samples were carefully examined by XRD patterns in the range of $28° \leq 2\theta \leq 32°$. Impurity phases, $BaCuO_{2+x}$ related compounds, were not found. TG-DTA was also employed for the same analysis, no eutectic transition was observed at temperature lower than the peritectic transition of the sample so that no second phase was observed.

Unit cell axes as a function of x are plotted in Fig.1. Both the *a*- and *c*-axis increases with increasing x, while the *b*-axis decreases with increasing x. As a result, the orthorhombicity is linearly dependent on x following a mathematic relationship with y = 0.0182(2) – 0.023(3) x and R = –0.97. It strongly suggests that substitution of the Ca ions onto Tm site is successful.

It has been found by iodometric titration that the oxygen stoichiometry (y) of these compounds is higher than that of Y123 [10]. y decreases with increasing x from 7.00 for x = 0 to 6.85 for x = 0.150 as shown in

CP850, *Low Temperature Physics: 24th International Conference on Low Temperature Physics;*
edited by Y. Takano, S. P. Hershfield, S. O. Hill, P. J. Hirschfeld, and A. M. Goldman

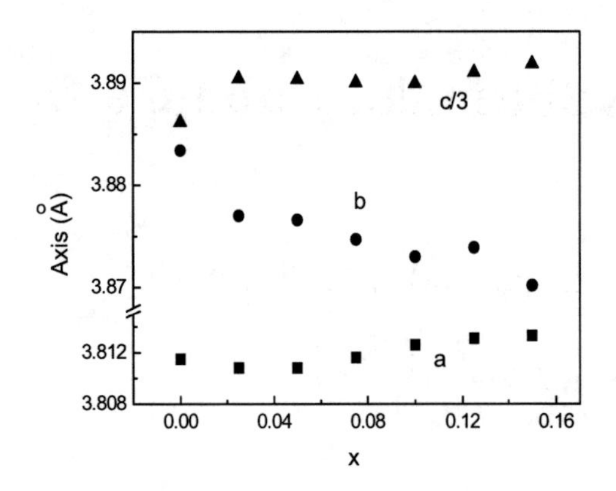

FIGURE 1. Unit-cell axes dependence with x.

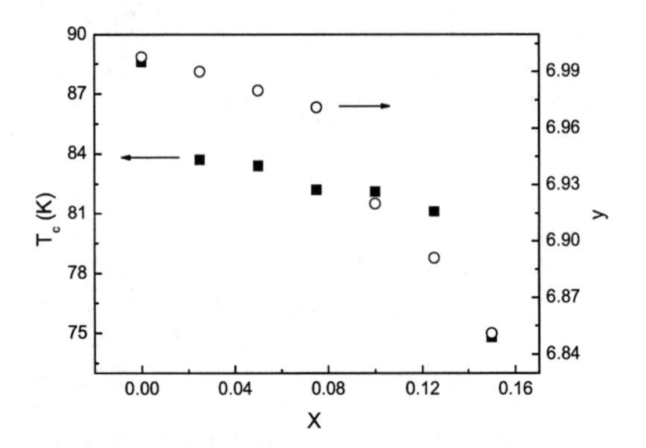

FIGURE 2. T_c and y dependence with x.

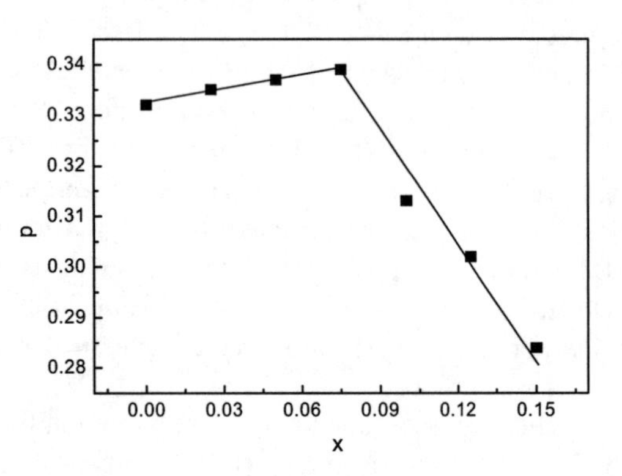

FIGURE 3. p/Cu dependence with x.

Fig. 2. T_c of the samples is also plotted in Fig. 2. Variation of T_c with x is similar to that of y with x. The average hole concentration (p/Cu) found in the compounds is in the range of $0.28 \leq p/Cu \leq 0.34$. For the samples with $x \leq 0.075$, p/Cu increases with increasing the amount of Ca that is common for the Y123 compound [6,7]. It is surprising that p dramatically decreases with increasing x for $x > 0.075$

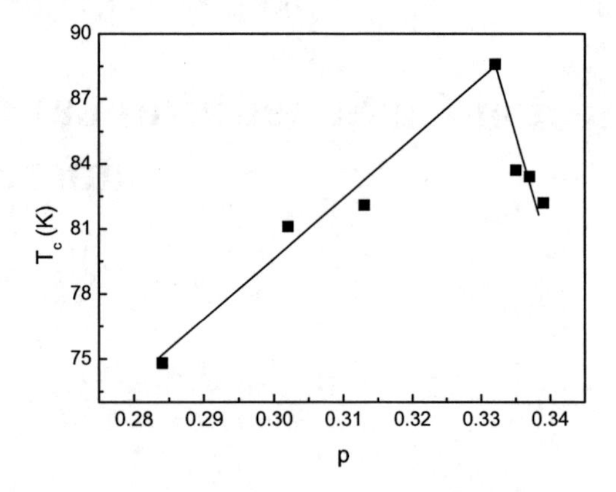

FIGURE 4. T_c dependence with p.

as illustrated in Fig. 3, indicative of a hole filling effect. This result is supported by the O-2p XANES spectra. A dome-like T_c versus p plot is shown in Fig. 4, which is commonly observed in the high-T_c cuprates. The un-substituted Tm123 has the highest T_c at 89 K with p/Cu = 0.340. More work about the distribution of the holes in the CuO_2 layers or in the CuO chains is under investigation.

ACKNOWLEDGMENTS

This work is financially supported by the National Science Council of the R. O. C. under the contract no. NSC93-2113-M-032-011.

REFERENCES

1. M. K. Wu, J. R. Ashburn, C. J. Torng, P. H. Hor, R. L. Meng, L. Gao, Z. J. Huang, Y. Q. Wang and C. W. Chu, *Phys. Rev. Lett.* **58**, 908 (1987).
2. P. Hor, R. L. Meng, Y. Qu Wang, L. Gao, Z. J. Huang, J. Bechtold, K. Forster and C. W. Chu, *Phys. Rev. Lett.* **58**, 1981 (1987).
3. H.-C. I. Kao, C. H. Chin, R. C. Huang and C. M. Wang, *Physica C* **341–348**, 623 (2000).
4. H. M. Luo, S. Y. Ding, G. X. Lu, B. N. Lin, H. C. Ku, C. H. Lin, H.-C. I. Kao and J. C. Ho, *Supercond. Sci. Technol.* **14**, 320 (2001).
5. C. H. Chin and H.-C. I. Kao, *Physica C* **388–389**, 381 (2003).
6. H.-C. I. Kao and C. H. Lin, *Physica C* **364–365**, 575 (2001).
7. V. P. S. Awana, et al., *Physica C* **262**, 272 (1996).
8. A. C. Larson and R. B. von Dreele, Report La-UR-86-748, Los Alamos National Lab. Los Alsmos, NM, USA (1990).
9. H. E. Appelman, et al., *Inorg. Chem.* **26**, 3237 (1987).
10. M. A. Beno, L. Soderholm, D. W. Capone, D. G. Hinks, J. D. Jorgensen, J. D. Grace, and I. K. Schuller, *Appl. Phys. Lett.* **51**, 57 (1987).

Distinctive Features of the Crystal Structure and the Superconductivity of RE-123 (RE=Gd and Sm) Prepared under a Magnetic Field

Yasukuni Matsumoto[1], Masatoshi Tomomatsu[1], Yusuke Terasaki[1], Shougo Kato[1], and Akihiko Nishida[2]

[1]*Department of Electrical Engineering, Fukuoka University, 8-19-1 Nanakuma, Fukuoka 814-0180, Japan*
[2] *Department of Applied Physics, Fukuoka University, 8-19-1 Nanakuma, Fukuoka 814-0180, Japan*

Abstract. The melting point (T_M) of $REBa_2Cu_3O_{7-y}$ (RE=Gd and Sm, RE-123) under a magnetic field up to 10T has been investigated by using X-ray diffraction and SEM-EDX analyses. Fired samples were directly immersed into liquid nitrogen in order to quench in the structural features. In the case of Gd-123, the T_M increased by about 5°C applying a magnetic field of 10T as compared with the case of zero field. In the case of Sm-123, there was no remarkable change of the T_M with magnetic field. The magnetic field effect on the unit cell parameters *a*, *b* and *c* has been investigated by the Rietveld refinement of the X-ray diffraction pattern. The T_c of Gd-123 prepared under a magnetic field of 10T was about 10K lower than that in the case of zero field.

Key words: melting point, Rietveld analysis, lattice constant
PACS: 74.72Jt, 74.62.Bf

INTRODUCTION

There has been a continuing effort to increase the critical current density J_c of high-T_c materials. One distinctive problem is the poor connection of the CuO_2 plane between the grains. One method to overcome the problem is material preparation by the melt-growth technique under a magnetic field. Such a method has been attempted in Bi-2212[1], Bi-2223[2] and YBCO[3].

It is important to know the melting point T_M of relevant material under magnetic field because the heat treatment profile is dependent on T_M . We are interested in the preparation of Gd-123 materials by the melt-growth technique under a magnetic field because the Gd atom has a large magnetic moment. Here, we report the experimental results of a T_M estimation, and the change of the crystal unit cell and superconducting transition temperature T_c of Gd-123 under a magnetic field. The results on Sm-123 are also reported.

EXPERIMENTAL

At first, the Gd-123 and Sm-123 powders were prepared by a conventional sintering method. Raw materials of $Gd_2O_3(Sm_2O_3)$, $BaCO_3$, and CuO were weighed and mixed to the stoichiometric ratio and calcined at 960°C for 24h in air. Finally, the powders were sintered at 940°C for 20h in air. The sintered powders were pressed into rectangular shape pellets with a pressure of 200MPa. A prepared sample was then contained in a Pt box and was heated at a relevant temperature for 1 hour followed by immersion within 10 seconds into liquid nitrogen. The temperature fluctuation within the furnace was less than 0.5°C. X-ray diffraction patterns and SEM observation including EDX analysis were used to determine T_M.

RESULTS and DISCUSSION

It is well known that the $GdBa_2Cu_3O_{7-y}$ phase will change to Gd_2BaCuO_5 (Gd-211) and a liquid at T_M, i.e. if the Gd-211 phase is detected then melting of Gd-123 has occurred. Based upon this fact, we examined the X-ray diffraction patterns and performed SEM-EDX analyses of the samples in order to estimate T_M.

One distinctive feature of Gd-123 sample quenched at 1020°C appears in the crystal structure: The intensity of 013 peak is remarkably greater than that of 103 peak, while the usual Gd-123 phase showing superconductivity has an inverse relation to

CP850, *Low Temperature Physics: 24th International Conference on Low Temperature Physics;*
edited by Y. Takano, S. P. Hershfield, S. O. Hill, P. J. Hirschfeld, and A. M. Goldman

the one mentioned above. Therefore, we estimated the intensity ratio of *I*(131) from Gd-211 phase to *I*(013) from Gd-123 phase as an indicator of melting. This circumstance is the same in case of Sm-123.

Experimental results are shown in Fig.1. In this figure, the abscissa is the heat treatment temperature, and the ordinate is the intensity ratio. In case of Gd-123 under zero magnetic field, there is no peak due to Gd-211 at 1020°C and 1023°C, but the Gd-211 peaks obviously appear at 1025°C. When a magnetic field of 10T is applied, the Gd-211 peaks were not observed even at 1030°C. If the magnetic field is 5T, there is no Gd-211 peak at 1028°C, but those are observed at 1030°C with the intensity ratio of 0.04. Such a circumstance is the same in the SEM-EDX observations. The experimental results indicate that the melting point, T_M, of Gd-123 phase is about 5°C higher when a magnetic field of 10T is applied.

On the other hand, there is no remarkable change of the T_M of Sm-123 phase whether or not a magnetic field is applied, as shown in Fig.1. The difference in the magnetic field effect on T_M between Gd-123 and Sm-123 may be due to the difference of the atomic magnetic moment of Gd and Sm.

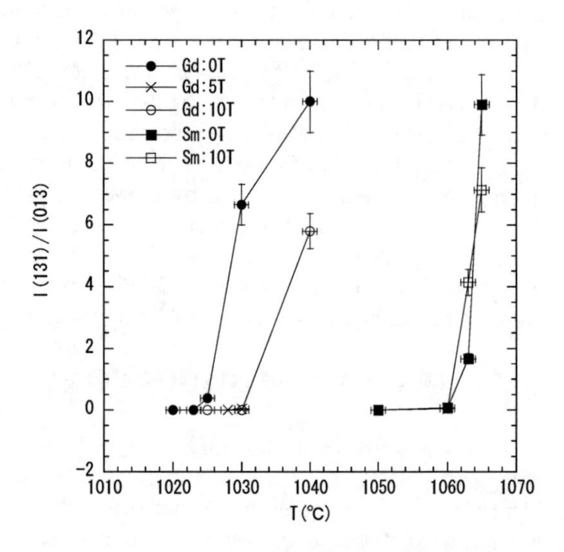

FIGURE 1. Determination of the T_M for the samples heat-treated under a magnetic field or not.

We examined the feature of the X-ray diffraction pattern by a Rietveld analysis of samples heat-treated at 1020°C either under the magnetic field of 10T (10T sample) or of zero field (ZF sample) followed by quenching. The resultant lattice constants *a*, *b* and *c* are 3.8730 Å, 3.8808 Å and 11.815 Å for the ZF sample and are 3.8744 Å , 3.8821 Å and 11.836 Å for the 10T sample. It should be noticed that the increment in the *c*-axis length of 10T sample is appreciable. There is no remarkable change of the lattice constant in Sm-123 fired whether under a magnetic field or not.

Oxygen occupation of O(5) site is not zero in the quenched sample, which is, as well known, but completely zero in the standard superconducting Gd-123, where site naming is owing to the work by Jorgensen *et al.*[4]. If we do not consider the occupation of O(5) site, refinement shows a fatal error. Therefore, the total oxygen number for the unit cell becomes larger than 7.

The magnetic field effect on T_c was examined in differently processed samples which were prepared by gradual cooling from 960°C under a magnetic field of 10T or zero field. The magnetically processed sample with a *c*-axis length of 11.745 Å had a T_c of 65K, while the zero field sample with *c*-axis of 11.730 Å had a T_c of 75K. This result is in accord with the common knowledge that the larger the *c*-axis length, the lower is T_c. On the other hand, the 10T sample and the ZF sample showed nearly the same T_c of 94K after annealing at 450°C for 100h in an O_2 gas flow of 100ml/min. Long time annealing may have caused the loss of the magnetic field effect associated with material fabrication. However, we are not aware of any mechanism for these results at present stage.

In conclusion, firstly, in case of Gd-123 the melting point, T_M, and the unit cell volume both increased with the application of a magnetic field. But in case of Sm-123 there was no remarkable change in T_M or the unit cell volume whether a magnetic field is applied or not. Secondly, oxygen occupation of O(5) site has to be considered in the cases of both of the 10T sample and the ZF sample, which may be affected by quenching the high temperature phase. The application of a magnetic field in the fabrication of Gd-123 or Sm-123 materials results in an additional oxygen content in the unit cell.

The authors acknowledge Dr. M. Takumi for his help in carrying out the Rietveld analysis.

REFERENCES

1. W.P.Chen, H.Maeda, K.Kakimoto, P.X.Zhang, K.Watanabe, and M.Motokawa, *Physica C* **320,** 96-100 (1999).
2. J.G.Noudem, J.Beille, D.Bourgault, D.Chateigner, and R.Tournier, *Physica C* **264,** 325-330 (1996).
3. S.Awaji, K.Watanabe, M.Motokawa, A.Kuramochi, T.Fukase, and K.Kimura, *IEEE Trans. Appl. Supercond.* **9,** 2014-2017 (1999).
4. J.D.Jorgensen, B.W.Veal, A.P.Paulikas, L.J.Nowicki, G.W.Crabtree, H.Claus, and W.K.Kwok, *Phys. Rev. B* **41,** 1863-1877 (1990).

Negative Expansion Of $Eu(Ba_{1-x}La_x)_2Cu_3O_{7-d}$ Compounds

Victor Eremenko*, Sergiy Feodosyev*, Igor Gospodarev*, Vladimir Ibulaev*, R. William McCallum+, Mikhailo Shvedun*, Valentyna Sirenko*

*Institute for Low Temperature Physics & Engineering NAS of Ukraine, Kharkov, 61103, Ukraine
+Ames Laboratory-USDOE, Iowa State University, Ames, Iowa 50011

Abstract. Variation of crystal-lattice parameters with temperature was studied by means of X-ray diffraction technique on the perovskite-like structures with Ba- substitution for light rare earth. Anisotropic negative thermal expansion was observed and explained within the "membrane effect" model proving its lattice origin.

Keywords: thermal expansion, cuprate superconductor, membrane effect
PACS: 65.40.De , 43.35.Gk ,74.25.-q t

INTRODUCTION

It was shown recently that distortion of $Eu(Ba_{1-x}R_x)Cu_3O_{7+d}$ structures produced by rare earth (R = La, Nd, Pr) on Ba site is relevant to crystallographic transition from orthorhombic to tetragonal symmetry, alteration of "buckling angle" at CuO_2 plane layer, and changes in ionic band lengths [1,2]. It gives rise to suppression of high temperature superconductivity in these compounds and a number of specific effects in the vicinity of superconducting transition seeking for a comprehensive study. In the presented work effect of negative thermal expansion (NTE) is considered for such material family members.

EXPERIMENTAL PART

The ceramic samples of $Eu(Ba_{1-x}La_x)Cu_3O_{7+d}$ compounds were studied by means of X-ray diffraction measurements in the temperature range 4.2 – 300 K. Characterization of the samples has shown that procedure of their preparation provided perfect location of the substituting La atoms in Ba position. The observed tendency of the studied materials to expand negatively in the c axis direction is illustrated by figure 1.

DISCUSSION

In the case of strongly anisotropic atomic interactions the amplitude of atomic vibrations along direction of weak coupling, which is normal to either planes or chains, is substantially higher than in the strong-coupling direction. In result of atomic displacement along the weak-coupling direction, the spacing between the in-layer atoms is thereby increased for the magnitude Δl, quadratic in this displacement. Eventually, the in-layer compressing force proportional to Δl arises. This is the reason for the anisotropic negative expansion similar to the "membrane effect" [3]. The relevant microscopic description of such a compressing force was conceived here with the results illustrated by figure 2. In [4, 5], the temperature variation of the linear thermal expansion coefficient (LTEC) in the strong-coupling direction is readily described by the following expression:

$$\alpha_l(T) \approx A\partial_T \left\langle u_l^2 \right\rangle_T [\delta - \Delta(T)], \qquad (1)$$

with $\Delta(T)$, the temperature derivatives ratio of the atomic root mean square displacements (RMSD) normal $\partial_T \left\langle u_\perp^2 \right\rangle_T$ and parallel $\partial_T \left\langle u_l^2 \right\rangle_T$ to l-direction.

CP850, *Low Temperature Physics: 24th International Conference on Low Temperature Physics;* edited by Y. Takano, S. P. Hershfield, S. O. Hill, P. J. Hirschfeld, and A. M. Goldman

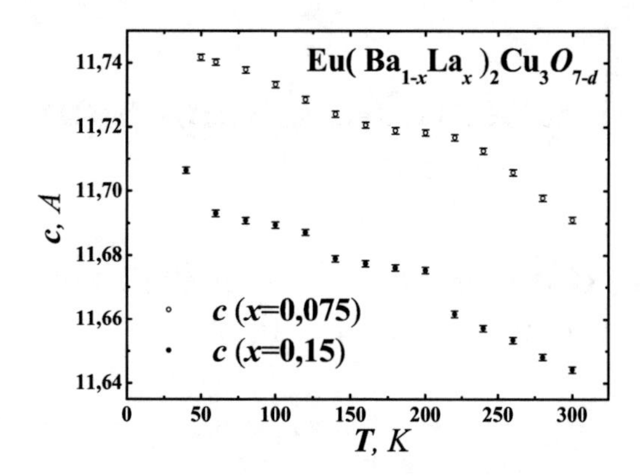

FIGURE 1. Negative expansion of the compounds $Eu(Ba_{0.85}La_{0.15})_2Cu_3O_{7-d}$ and $Eu(Ba_{0.925}La_{0.075})_2Cu_3O_{7-d}$ along the crystallographic c-axis direction.

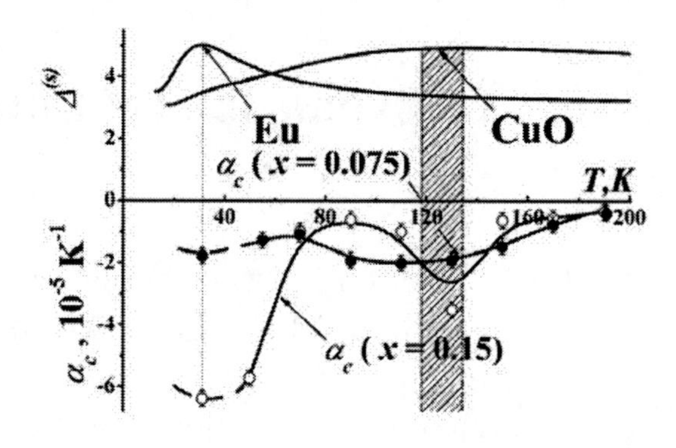

FIGURE 2. Temperature dependences of $\Delta^{(s)}(T)$ values for the layers of Eu and Cu-O of the parent $EuBa_2Cu_3O_7$, compound and the measured temperature dependences of LTEC.

The position of maximum in the dependence $\Delta^{(s)}(T)$ of strongly anisotropic layered crystal is compatible with the position of minimum in the temperature dependence of LTEC measured parallel to the layers.

In a number of multilayer high-T_c superconductors, like $Bi_2Sr_2CaCu_2O_x$, a pronounced anisotropy of elastic properties common to the layer crystals was revealed so far. The negative thermal expansion in the layers observed in this case is in good agreement with the above presented expression. Otherwise, in multilayer structures similar to high-T_c superconductors of the 1-2-3 type, the interlayer coupling is of the same order for all the layers while in-layer interactions vary substantially for different layers. Local anisotropy of the «chain type» is pertinent to the layers characterized by the weak in-layer interaction, e.g., the layers of rare earth and Cu-O chains. The in-plane RMSD of the atom pertained to such a layer noticeably exceeds RMSD of the same atom in the perpendicular direction approaching the classical limit at sufficiently low temperature. There can, therefore, exist the temperature range where the raise of temperature is accompanied by contraction of the interlayer spacing between the CuO_2 – Eu – CuO_2 and BaO - CuO - BaO layers, which apparently explains observations of the negative LTEC values along the ***c***-axis (figure 2). Remarkably, the change in $\Delta^{(s)}(T)$ of the Cu-O layers with temperature is rather smooth, and position of the minimum on the relevant measured dependences is evidently sensitive to variation of *d* and *x*. The dashed area in figure 2 outlines the specified temperature range. The CuO_2 and BaO layers do not contribute to the negative thermal expansion. As is clearly demonstrated by figure 1, the lattice parameter *c* is rapidly decreased with increasing *x*. Local anisotropy of Eu and Cu-O layers is then enhanced giving rise to deepening of minimum in LTEC dependence (figure 2). In conclusion, the negative thermal expansion along the c-axis of the compounds under study is explained in terms of the "membrane effect" indicating to lattice origin of observed peculiarities.

ACKNOWLEDGMENTS

This work was supported in part by grant of the MES of Ukraine #M/257 and of the Ukrainian Fundamental Research Foundation #10/50.

REFERENCES

1. Xu Youwen, M. J. Kramer, K. W. Dennis, H. Wu, A. O'Connor, R. W. McCallum, S. K. Malik, W. B. Yelon, *Physica C*, **341-348**, 613-14.2 (2000).
2. V. V. Eremenko, S. B. Feodosyev, I. A. Gospodarev, V. A. Sirenko, M. Yu. Shvedun, R. W. McCallum, M. Tovar, *Low Temperature Physics* **31** 350-355 (2004).
3. I. M. Lifshits, *Zh. Exp. Teor. Fiz.* **22**, 475 (1952).
4. S. B. Feodosyev, I. A. Gospodarev, and E. S. Syrkin, *Phys. Status Solidi* **B150**, K19 (1988).
5. S. B. Feodosyev, E. S. Syrkin, I. A. Gospodarev, V.P. Popov, A. A. Gurskas, and N.M. Nesterenko, *Fiz. Tverd. Tela* **31**, 186 (1989).

Pressure Effects on the Superconductivity in $Pr_2Ba_4Cu_7O_{15-\delta}$ Oxide Superconductor

K. Fukuda [a,c], A. Kaeriyama [a,c], F. Ishikawa [a,c], Yuh Yamada [b,c], A. Matsushita [d] and H. Takahashi [e]

[a] *Graduate School of Science and Technology, Niigata University, Igarashi, Niigata 950-2181, Japan*
[b] *Department of Physics, Niigata University, Igarashi, Niigata 950-2181, Japan*
[c] *Center for Transdisciplinary Research, Niigata University, Igarashi, Niigata 950-2181, Japan*
[d] *National Institute for Materials Science, Sengen, Tsukuba, Ibaraki 304-0047, Japan*
[e] *College of Humanities and Sciences, Nihon University, Sakurajousui, Setagaya, Tokyo 156-8550, Japan*

Abstract. We have investigated the pressure dependence of the superconducting transition temperature of $Pr_2Ba_4Cu_7O_{15-\delta}$ (Pr247) oxide superconductor. We synthesized Pr247 using a powder sintering method under high pressure oxygen. Occurrence of a superconducting phenomenon at approximately $T_c = 10$ K was confirmed after reduction treatment at 923 K in argon gas for 4 days. Under high pressure the superconducting transition disappeared and the temperature dependence of electrical resistivity behaved like the one of as-sintered sample. Experimental results are well explained by the theory on the basis of Tomonaga-Luttinger liquid model.

Keywords: $Pr_2Ba_4Cu_7O_{15-\delta}$, Double-chains, Pressure effects, one-dimensional superconductivity
PACS: 74.62.Fj, 74.72.Jt, 74.25.Fy

One of the most famous high temperature superconductors discovered so far is $YBa_2Cu_3O_{7-\delta}$ (Y123). It is well known that the Pr-substitution for Y-site in Y123 and $YBa_2Cu_4O_8$ (Y124) compounds dramatically suppresses superconducting transition temperature T_c and superconductivity in CuO_2 planes disappears [1]. The suppression of electronic conduction in CuO_2 plane is due to the strong hybridization between Pr 4*f* and O 2*p* orbitals [2]. We recently succeeded in synthesizing a $Pr_2Ba_4Cu_7O_{15-\delta}$ (Pr247) oxide which is a compound with an alternative repetition of the single and double chains along the *c*-axis, isostructural with superconducting $Y_2Ba_4Cu_7O_{15-\delta}$ [3,4]. As-sintered Pr247 shows metallic conductivity at low temperatures in the CuO double chains. The resistivity in Pr247 was investigated by changing the oxygen content, and superconductivity with $T_c = 10$ K was found after a reduction treatment [3]. By the reduction treatment, the amount of oxygen deficiency in the CuO double chains are controlled, and the carrier density is varied. The NQR experiment revealed that the superconductivity is realized at the CuO double chains [5]. These experiments suggest that the material shows the possibility of novel one-dimensional 1D superconductivity. We previously reported the effects of pressure on the electrical resistivities of as-sintered Pr124 and Pr247 [6]. In this paper, we report pressure effects on superconductivity in Pr247.

Polycrystalline samples of Pr247 which is the same sample as used for the previous report [3], were synthesized using a powder sintering method under high pressure oxygen, followed by annealing at 923 K for 4 days in argon gas [1,3]. The electrical resistivity under pressure was measured by a conventional four-probe method using Ni-Cr-Al piston-cylinder clamp cell.

Figure 1 shows temperature T dependence of the electrical resistivity ρ of Pr247 sample at various pressures up to 3.0 GPa. The superconducting transition evidently appears around 10 K and the zero-resistivity temperature is determined to be 3.8 K at ambient pressure. The value of ρ at ambient pressure is proportional to T^2 below 150 K, which is the same temperature dependence of ρ along the *b*-axis of Pr124 single crystal [7]. With increasing pressure, the electrical resistivity increased in all temperature region. The ρ - T curve at 1.0 GPa shows a drop of the electrical resistivity owing to a superconducting fluctuation was observed, but the temperature region of zero resistivity vanished. The T dependence of the resistivity shows broad maximum above 2.0 GPa. The

CP850, *Low Temperature Physics: 24th International Conference on Low Temperature Physics;* edited by Y. Takano, S. P. Hershfield, S. O. Hill, P. J. Hirschfeld, and A. M. Goldman

superconducting transition disappeared and the temperature dependence of electrical resistivity behaved like the one of as-sintered sample which also shows a broad maximum around $T = 150$ K [3].

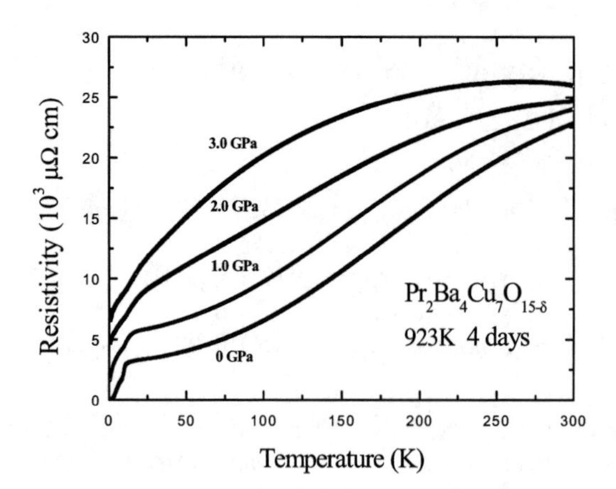

FIGURE 1. Temperature dependence of the electrical resisitivity of Pr247 at various pressures up to 3.0 GPa.

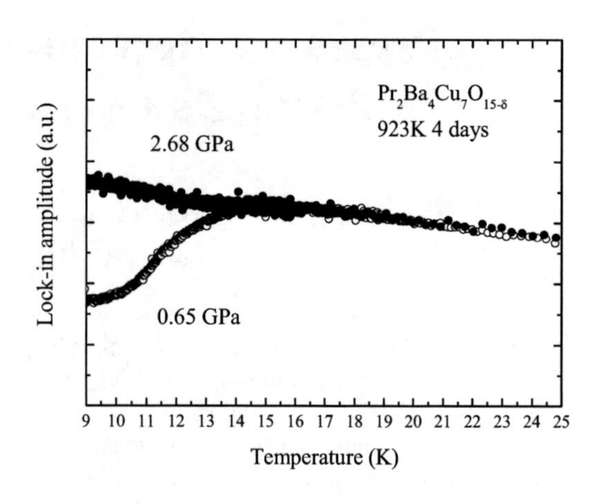

FIGURE 2. Temperature dependence of the ac susceptibility of superconducting Pr247 at $P = 0.65$ GPa and 2.68 GPa.

Figure 2 shows the temperature dependence of the ac susceptibility of superconducting Pr247 sample. A drop owing to Meissner effect is observed below 13 K at pressures of 0.65 GPa. The superconducting transition disappeared at pressures of 2.68 GPa.

The experimental results show that applying pressure suppresses the appearance of the superconducting phase. Recently, Sano *et al.* discussed the one-dimensional conducting mechanism in CuO double chain of Pr247 on the basis of Tomonaga - Luttinger liquid theory [8]. According to their phase diagram which is obtaind in the weak coupling mechanism, the CuO double chain of Pr247 is expected to be in the boundary region between the normal phases and the superconducting phase, and the applying pressure may lead to the phase transition from the superconducting phase to the normal phases. This agrees well with present experimental results.

In conclusion, we have measured the ρ - T curves at various pressures for Pr247. The superconducting transition was observed at ambient pressure but applying pressure leads to the suppression of the superconducting transition and no evidence for superconductivity was observed above 2.0 GPa. This experimental result was well explained by the theory on the basis of Tomonaga - Luttinger liquid model.

ACKNOWLEDGMENTS

The work was partially supported by a Grant in Aid for Scientific Research from the Ministry of Education, Science and Culture, Japan.

REFERENCES

1. S. Horii, Y. Yamada, H. Ikuta, N. Yamada, Y. Kodama, S. Katano, Y. Funahashi, S. Morii, A. Matsushita, T. Matsumoto, I. Hirabayashi and U. Mizutani, *Physica C* **302**, 10-22 (1998).
2. R. Fehrenbacher and T. M. Rice, *Phys. Rev. Lett.* **70**, 3471-3474 (1993).
3. M. Matsukawa, Y. Yamada, M. Chiba, H. Ogasawara, T. Shibata, A. Matsushita and Y. Takano, *Physica C* **411**, 101-106 (2004).
4. Y. Yamada and A. Matsushita *Phisica C* **426**, 213-219 (2005).
5. S. Watanabe, Y. Yamada and S. Sasaki, *Physica C* **426**, 473-477 (2005).
6. A. Matsushita, Y. Yamada, N. Yamada, S. Horii and T. Matsumoto, *Physica C* **242**, 381-384 (1995).
7. S. Horii, U. Mizutani, H. Ikuta, Y. Yamada, J. H. Ye, A. Matsushita, N. E. Hussey, H. Takagi and I. Hirabayashi, *Phys. Rev. B* **61**, 6327-6333 (2000).
8. K. Sano, Y. Ōno and Y. Yamada, *J. Phys. Soc. Jpn.* **74**, 2885-2888 (2005).

Evidence for Antiferromagnetic Order Within the CuO_2 Planes in Superconducting $Pr_2Ba_4Cu_7O_{14.5}$

Shinji Watanabe,[a] Yuh Yamada[b] and Susumu Sasaki[c]

[a]*Graduate School of Science and Technology, Niigata University, Ikarashi2-8050, Niigata950-2181, Japan*
[b]*Department of Physics, Niigata University, Ikarashi2-8050, Niigata 950-2181, Japan*
[c]*Department of Materials Science and Technology, Faculty of Engineering, Niigata University, Ikarashi2-8050, Niigata 950-2181, Japan*

Abstract. We present NQR experiments for the planar Cu sites in superconducting $Pr_2Ba_4Cu_7O_{14.5}$ (T_c = 15 K). The planar signals observed around 30 MHz at 300K disappear below 285 K, which indicates that a large internal magnetic field is present at the planar sites due to antiferromagnetic ordering below T_N = 285 K. Indeed, we observe NQR signals at 50 - 130 MHz, well below T_N. From the lineshape analysis, we found that there are two chemically inequivalent planar Cu sites with internal magnetic fields of 7.1 T and 9.6 T. This confirms that the planar Cu sites are insulating, while metallic conductivity is due to the CuO double chains. Thus, it is very likely to say that the superconductivity in $Pr_2Ba_4Cu_7O_{14.5}$ is driven by the CuO double chains, not by the CuO_2 planes.

Keywords: superconductivity, double chain, NQR, AF order, Pr247
PACS: 74.72.Bk; 76.60.Gv

INTRODUCTION

Since the discovery of high-T_c cuprate superconductors, extensive studies on strongly correlated electron system have been in progress on the basis of the physical properties of the two-dimensional CuO_2 planes. Moreover, from the viewpoint of low-dimensional physics, particular attention has been paid to the physical role of one-dimensional (1D) CuO chains included in the high- T_c cuprates such as the Y-based superconductors with the transition temperature T_c of approximately 90 K. It is well known that Pr-substitution for the Y-sites in the $YBa_2Cu_3O_{7-y}$ (Y123/7-y) and $YBa_2Cu_4O_8$ (Y124) compounds drastically suppresses T_c, and further substitution destroys the superconductivity driven by the CuO_2 planes. In the nonsuperconducting phase, antiferromagnetic order in the CuO_2 planes has been clarified by NMR/NQR experiments on $PrBa_2Cu_3O_7$ (Pr123/7) [1] and $PrBa_2Cu_4O_8$ (Pr124) [2]. The Pr-substitution effect on T_c has been ascribed to the localization of carriers in the CuO_2 planes, caused by the hybridization of Pr-*4f* and O-*2p* in the CuO_2 planes [3]. In contrast, the double chains remain conducting along the chain axis down to low temperatures, which has been studied from the viewpoint of strongly correlated quasi-1D electrons [4].

The mixed structure of Pr123 (only single chains) with Pr124 (only double chains) is realized in the $Pr_2Ba_4Cu_7O_{15-y}$ (Pr247/15-y) compound which consists of a single and a double chain (parallel to the *c*-axis) in the unit cell [5]. Recently, we found that the y = 0.5 compound (Pr247/14.5) shows superconductivity below 15 K. Quite surprisingly, it is suggested from measurements such as electrical resistivity [5], that the superconductivity is realized in the double chains, not in the CuO_2 planes. To experimentally clarify this suggestion, we performed a site-selective NQR experiment in the absence of an external magnetic field for the superconducting sample Pr247/14.5. In this paper, we report the results of NQR experiments for the planar Cu sites in superconducting Pr247/14.5.

EXPERIMENTAL

Polycrystalline powder samples of Pr247/15-y were synthesized using a powder sintering method under high-oxygen pressure. The details are described in Ref. 5. The NQR experiments were carried out using home-built pulsed spectrometers. The resonance signals were obtained by a phase alternating add-subtract spin-echo technique. The NQR spectra were obtained by integrating the signal intensity and by changing the frequency point by point.

CP850, *Low Temperature Physics: 24th International Conference on Low Temperature Physics;*
edited by Y. Takano, S. P. Hershfield, S. O. Hill, P. J. Hirschfeld, and A. M. Goldman

RESULTS AND DISSCUSSION

Fig.1 shows the NQR spectrum of Pr247/14.5 at 1.7 K. As shown in Fig.1, this spectrum is observed at much higher frequencies than the typical spectra of high-T_c cuprates (~ 30 MHz). This indicates that a large internal magnetic field is present at the planar sites due to antiferromagnetic order.

To understand the spectrum, we consider the Hamiltonian for spin 3/2 nuclei in a magnetically ordered state in the presence of an electric field gradient (EFG), which is written as

$$\hat{H} = -\gamma\hbar I \cdot H + \frac{1}{6} h\nu_Q \left(3I_z^2 - I(I+1)\right), \qquad (1)$$

where the coordinates refer to the principal axes of the EFG. Here, γ and H are the nuclear gyromagnetic ratio and the hyperfine field, respectively. The value of ν_Q is the NQR frequency in the paramagnetic state and we assume the asymmetry parameter η is equal to 0 as in the case of parent materials of Pr247. As expected from the lineshape of this spectrum, we considered the case in which the Zeeman interaction is greater than the electric quadrupole interaction. From equation (1), three resonance lines are expected for one isotope, I = 3/2, and the frequency interval between the first-order satellite lines is given as $\nu_Q(3\cos\theta\text{-}1)$, where θ is the angle between the maximal principal axis of the EFG and the direction of the hyperfine field. The three resonance lines correspond to the transitions $I = 3/2 \leftrightarrow 1/2$, $1/2 \leftrightarrow -1/2$, $-1/2 \leftrightarrow -3/2$. In Pr247, there are two planar sites, Cu(2) and Cu(3), and Cu has two isotopes, ^{63}Cu and ^{65}Cu, so that it is expected that at least twelve resonance lines are observed. The solid line in Fig.1 is the fit to the experimental spectrum with the parameters for ^{63}Cu, H = 7.1 T, θ = 75°, with a width of 5 MHz, and H = 9.6 T, θ = 61°, with 2 MHz, in for the two inequivalent planar sites. The ν_Q for both Cu(2) and Cu(3) are about 30 MHz. This spectrum provides direct evidence for an antiferromagnetic ordering of both planar Cu spins. This confirms that the planar Cu sites are insulating and metallic conductivity is due to the CuO double chains. Thus, it is very likely to say that the superconductivity in Pr247/14.5 is driven by the CuO double chains, not by the CuO_2 planes.

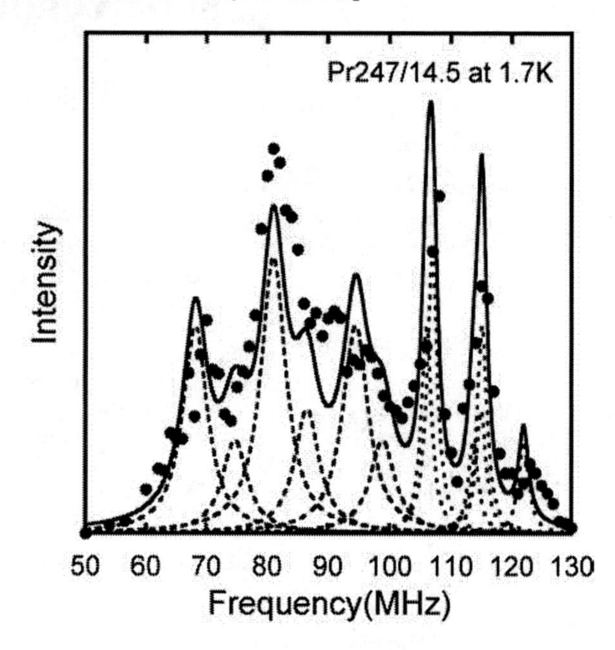

FIGURE 1. Planar site spectra of Pr247/14.5 at 1.7 K in zero-magnetic field. Full circles indicate the experimental data. The dotted [broken] lines indicate simulations of the Cu(2) [Cu(3)] signals which include the two isotopes ^{63}Cu and ^{65}Cu. The fitting parameters are given in the text.

REFERENCES

1. M. Matsumura M. Horiike, T. Yokoyama and H. Yamagata, *J. Phys. Soc. Jpn.* **68**, 748 (1999).
2. S. Fujiyama, M. Takigawa, T. Suzuki, N. Yamada, S. Horii and Y. Yamada, *Phys. Rev. B.* **67**, 060404 (2003).
3. R. Fehrenbacher and T. M. Rice, *Phys. Rev. Lett.* **70**, 3471 (1993).
4. S. Fujiyama, M. Takigawa and S. Horii, *Phys. Rev. Lett.* **90**, 147004 (2003).
5. M. Matsukawa, Y. Yamada, M. Chiba, H. Ogasawara, T. Shibata, A. Matsushita and Y. Takano, *Physica C* **411**, 101 (2004).

Paraelectric Permittivity And Temperature Dependence Of Resistivity And Hall Coeficient In High-T_c Metal Oxides

A. I. Golovashkin, A. L. Karuzskii, A. N. Lykov, V. N. Murzin and A. V. Perestoronin

P. N. Lebedev Physical Institute of RAS, Leninsky pr. 53, Moscow 119991, Russia

Abstract. Local electric fields in the metal oxide superconductors, which are generally close to the metal-insulator phase transition and possess the very short length of the mean free path for current carriers, order of a lattice constant, can result in the paraelectric permittivity for bound charges (soft dipoles), which obeys the Curie law. The Hall effect and dc resistivity temperature behaviours are explained here by the model of the paraelectric crystal close to the point of the Mott-Hubbard instability, in the ground state of which the current is carried by a liquid of boson-like pairs of carriers in upper and lower Hubbard bands. Fermion-like carriers, temperature excited over the energy of boson-like pair dissociation (pseudo gap), explain the temperature behaviour of Hall effect. Available data are compared with the model.

Keywords: HTSC, resistivity, Hall effect.
PACS: 74.72.Bk, 77.84.Bw, 71.30.+h, 71.10.Fd, 77.22.Ch, 77.80.Bh, 72.15.Lh, 72.20.Dp, 72.20.My.

Recently it has been argued [1] that in high-T_c metal oxide superconductors (HTSC) with relevant to these compounds perovskit-like layered structure and the metal-insulator type instability the effects of the local (acting) electric fields [2,3] should result in the paraelectric permittivity for bound charges (soft dipoles), which would obey the Curie law in the normal state, and in the very short length of the mean free path for current carriers, which could be order of a lattice constant, as revealed by microwave measurements [4-7]. Static properties such as the dc resistivity and Hall effect temperature behaviours can be explained as well by the model of the paraelectric crystal close to the point of the Mott-Hubbard instability, in the ground state of which the current is carried by a liquid of boson-like pairs of carriers in upper and lower Hubbard bands. The pair has been assumed [1,5,7] to be formed by the charge transfer mechanism between the oxygen atoms in neighboring atomic cells that gives oxygen state O^{2-} as a negatively charged boson-like empty (for holes) state in the lower Hubbard band and the state O^0 as a positively charged boson-like occupied (for holes) state in the upper Hubbard band. The Mott-Hubbard instability corresponds to the order-of-lattice-constant length of the mean free path and results in the temperature insensitivity of Drude conductivity. Nearly linear on temperature T resistivity results from the Curie law via the local (acting) electric field as shown previously [1,5]. When the length of the mean free path for current carriers is order of a lattice constant the local electric fields are coming in the semi classical equation of current carrier dynamics for HTSC via a replacement of the electric field **E** by $\mathbf{E}+\hat{b}\mathbf{P}/\varepsilon_0$, where **P** is the polarization, ε_0 is the vacuum permittivity, $\hat{b}$ is the operator, which characterizes the Lorentz contribution $\hat{b}\mathbf{P}/\varepsilon_0$. Here and in the following we concern with the normal state, in which the paraelectric susceptibility for bound charges obeys [1,5,7] the Curie law $\chi=A(T-T_0)^{-1}$, where A and T_0 are some constants. Taking into account the relation between **P** and χ, the expression $\sigma_D=ne^2\tau/m$ for Drude conductivity, in which the scattering time should be temperature independent because of the very short length of the mean free path, one obtains [1,5] for the dc resistivity

$$\rho(T)=\frac{E(0)}{j(0)}=\frac{\sigma_D^{-1}}{(1+\dfrac{b_0A(0)}{T-T_0})}\sim(T-T_0)\,,\quad(1)$$

which occurs to be proportional to T at not very low values of static components b_0 of $\hat{b}$ and $A(0)$ of χ due to the paraelectric properties. The Ohm law in the Eq. (1) includes the usually omitted for metals influence of the displacement currents via effects of the local (acting) electric fields. Fermion-like carriers, temperature excited over the energy of boson-like pair dissociation (pseudogap Δ_1), explain the temperature behaviour of Hall effect. We refer these fermion-like carriers to the

CP850, *Low Temperature Physics: 24th International Conference on Low Temperature Physics;*
edited by Y. Takano, S. P. Hershfield, S. O. Hill, P. J. Hirschfeld, and A. M. Goldman

oxygen state O^- composed of the O^{2-} and valence band hole states. Here the Hall effect is considered in more details and available data are compared with the model.

The Hall voltage under above assumptions will be generated in a magnetic field **B** by the transverse motion of the fermion-like carriers O^- exclusively. Boson-like carriers O^0 and O^{2-} are strictly correlated in adjacent cells of a crystal lattice that corresponds to the nearly half-filled Hubbard bands and could be considered as holes and electrons with equal density and mobility. So there should be no contribution to the Hall voltage from these states. From the equality of transverse current flows produced by the Lorenz force $j^x_L \equiv j^y_{O^-}\varphi_{O^-} = j^y_{O^-}\mu_{O^-}B_z = en_{O^-}\mu^2_{O^-}B_zE_y(1-b_0+b_0\varepsilon_{p0})$, here φ_{O^-} is the Hall angle for the fermion-like carriers O^-, and by Hall voltage $j^x_H = \sigma_D E_x(1-b_0+b_0\varepsilon_{p0}) = e(2n_O{}^0\mu_O{}^0 + n_{O^-}\mu_{O^-})E_x(1-b_0+b_0\varepsilon_{p0})$ the Hall coefficient $R_H = E_x/j^yB_z$, where the total current equals $j^y = \sigma_D E_y(1-b_0+b_0\varepsilon_{p0})$ and ε_{p0} is the static relative permittivity of bound charges, can be derived at not very high magnetic fields ($\omega_c\tau \ll 1$, $\omega_c = eB/m$) as

$$R_H = \frac{n_{O^-}\mu^2_{O^-}}{e(2n_{O^0}\mu_{O^0} + n_{O^-}\mu_{O^-})^2(1-b_0+b_0\varepsilon_{p0})}. \quad (2)$$

The combined conditions of independence on temperature for the total carrier density $n = n_{O^-}(T) + 2n_O{}^0(T) = n \times\exp(-\Delta_1/kT) + n(1-\exp(-\Delta_1/kT)) = const$ and for Drude conductivity $\sigma_D = e(2n_O{}^0\mu_O{}^0 + n_{O^-}\mu_{O^-})$ in a first order on density approximation, as it will be shown elsewhere, lead to the solution for mobilities as functions of respective densities $\mu_{O^-} = e\tau^0_{O^-}n/(m_{O^-}n_{O^-}) + \mu_i$ and $\mu_O{}^0 = e\tau^0{}_O{}^0 n/(m_O{}^0 n_O{}^0) + \mu_i$, where tau symbols and μ_i are some constants. These give the temperature dependence of $R_H(T)$ and Hall mobility, assuming $\mu_i = 0$ and $m_O{}^0 = 2m_{O^-}$,

$$R^{ef}_H(T) = R_H(T)(1-b_0+b_0\varepsilon_{p0}(T)) = \frac{(\tau^0_{O^-})^2 e^{\frac{\Delta_1}{kT}}}{en\tau^2}, \quad (3)$$

$$\mu_H(T) = R_H(T)\sigma(T) = e(\tau^0_{O^-})^2\exp(\Delta_1/kT)/m_{O^-}\tau, \quad (4)$$

which occur to be inversely proportional to the fermion-like carrier density. Fitting these temperature dependencies to the experimental data presented in Ref. [1] (see also Fig. 1) demonstrates a good applicability of the used first order approximation. The model developed here allows to estimate a mean value of the pseudogap averaged over the fermion-like carrier population at a given temperature from experimental data. Comparisons of the pseudogap variations from Hall data are shown in Fig. 1 for optimally oxygen doped $HoBa_2Cu_3O_{7-\delta}$ from [8] and for oxygen underdoped $YBa_2(Cu_{1-z}Zn_z)_3O_{6.78}$ with different zinc doping from [9]. Data of Fig.1 show that the higher (room) temperature mean value of Δ_1 increases from 367 K for the optimally oxygen doped sample with the superconducting critical temperature order of 90 K up to 420 K for the oxygen underdoped samples with zinc impurity.

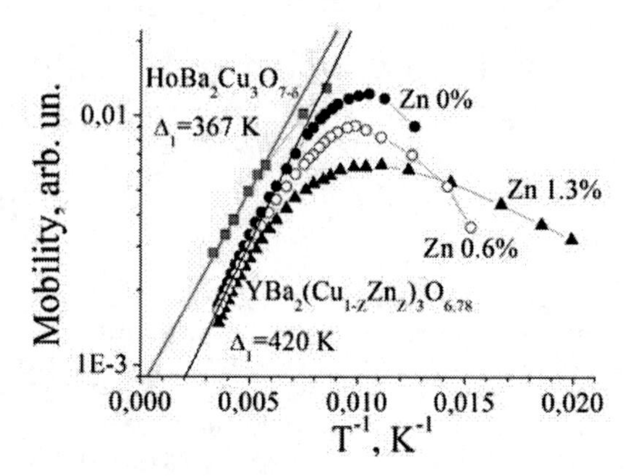

FIGURE 1. Hall mobility vs. reciprocal T for optimally oxygen doped $HoBa_2Cu_3O_{7-\delta}$ derived from data of Ref. [8] and for oxygen underdoped $YBa_2(Cu_{1-z}Zn_z)_3O_{6.78}$ derived from data of Ref. [9] (symbols) compared with dependencies (lines) predicted by Eq.(4)

The found trend to increase a pseudogap with decreasing T_c also correlates with ARPES data. The lower (200–100 K) temperature mean value of Δ_1 decreases with decreasing the superconducting critical temperature from 75 K down to 50 K in samples with higher zinc doping that can be explained by the decrease of the averaged over wave vector superconducting gap and pseudogap in these samples. The seen in the figure at temperature lower than 100 K decrease of experimental data for underdoped samples may results from the superconducting fluctuations near T_c and from the inhomogeneity of the oxygen underdoped samples with zinc impurity. The superconducting fluctuations are not accounted by presented model.

Work is supported by GNTP "Superconductivity".

REFERENCES

1. A. I. Golovashkin et al., *Physica B* **359-361**, 551-553 (2005).
2. V. N. Murzin, *Trudy FIAN* **48**, 145 (1969).
3. V. L. Ginzburg. *Contemp. Phys.* **33**, 15 (1992).
4. A. S. Shcherbakov et al., *Fiz.Met.Met.* **64**, 735 (1987).
5. A. L. Karuzskii et al., *Preprint FIAN* **No.215**, (P. N. Lebedev Physical institute, Moscow, 1988).
6. V. Müller et al., *Solid State Commun.* **67,** 997 (1989).
7. A. L. Karuzskii, *Physica C* **282-287**, 1581 (1997)
8. A. I. Golovashkin et al., *Pisma ZhETP [JETP Lett.]* **48,** 27 (1988).
9. Y. Abe, K. Segawa, Y. Ando, *Phys. Rev.* B **60,** R15055 (1999).

Isovalent Substitution and Heat Treatment Control of Chain Oxygen Disorder, Structure and T_c in $Y_{1-x}Sm_xSrBaCu_3O_{6+z}$

A. Nafidi, B. Bouallal, A. El Kaaouachi, M. Bellioua and H. Sahsah

Condensed Matter Physics Laboratory, Faculty of Sciences, B.P 8106, Hay Dakhla, University Ibn Zohr, 80000 Agadir, Morocco

Abstract. We report here on the preparation, X-ray diffraction (XRD) with Rietveld refinement, AC susceptibility measurements and effect of heat treatments in $Y_{1-x}Sm_xSrBaCu_3O_{6+z}$. Each sample was subject to two types of heat treatment: oxygen annealing [O] and argon annealing followed by oxygen annealing [OAO]. For the samples [O], as x increases from 0 to 1, the orthorhombicity ε decreases, together with T_c. However, for the samples [OAO], T_c generally increases with x as ε decreases. This last evolution of T_c and ε in opposite senses is unusual and it's observed for the first time. For a given x, the argon treatment increases ε (for $0 \leq x \leq 1$), T_c (for $x \geq 0.4$) and the distance d[Cu(1)-(Sr/Ba)] for x<0.5 and decrease it for x>0.5. Remarkable correlations were observed. A combination of several factors such as decrease in d[Cu (1)-O(1)]; increase in cationic and chain oxygen ordering; the mobile hole content in the $Cu(2)O_2$ plans (p_{Sh}) and in-phase purity for the [OAO] samples may account for the observed data.

Keywords: Isovalent substitution, heat treatments, T_c, Rietveld refinement structure, chain oxygen order-disorder, phase transition, $Y_{1-x}Sm_xSrBaCu_3O_{6+z.}$ ceramic cuprites.
PACS: 74.72.Bk, 74.62.Bf, 74.62.Dh, 61.10.Nz

INTRODUCTION

It is well-know that $YBa_2Cu_3O_{6+z}$ with $z \approx 1$ is superconducting below 92 K and characterised by double $Cu(2)O_2$ layers (oriented along the a-b plane) responsible for carrying the supercurrent and Cu(1)O chains (along the b direction) which provide a charge reservoir for these planes [1]. We have investigated the structural and superconducting properties of $Y_{1-x}Sm_xSrBaCu_3O_{6+z}$ (x = 0, 0.2, 0.4, 0.5, 0.6, 0.8 and 1). We found that the influence of argon heat treatment on these properties depends on Sm content, x.

EXPERIMENTAL

Our samples were polycrystalline prepared by solid state sintering. Each sample was subject to two types of heat treatment: (i) the sintered sample was annealed in oxygen at 450°C for 3 days, this sample was denoted as [O] and (ii) the same sample was heated in argon at 850°C for 1 day, cooled and later annealed in oxygen at 450°C for 3 days [OAO]. For each sample, XRD data were collected with Philips diffractometer and refined with Rietveld refinement. T_c was checked by measuring the real part χ′(T) of the AC susceptibility and confirmed by the measure of the resistivity ρ(T).

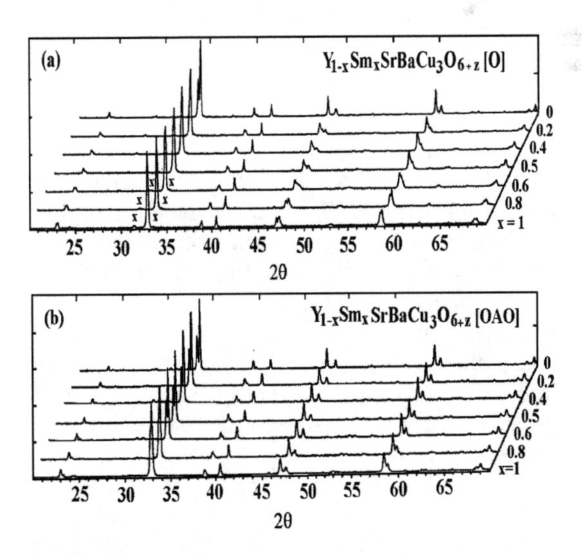

FIGURE 1. XRD pattern of $Y_{1-x}Sm_xSrBaCu_3O_{6+z}$ as a function of x. (a) samples [O] (b) samples [OAO].

CP850, *Low Temperature Physics: 24th International Conference on Low Temperature Physics;* edited by Y. Takano, S. P. Hershfield, S. O. Hill, P. J. Hirschfeld, and A. M. Goldman

RESULTS AND DISCUSSION

Fig. 1 allows the clear identification of the orthorhombic splitting, as well as the observation that some weak unidentified impurity peaks are eliminated in the samples [OAO]. For each sample, our iodometry measurements indicate $6+z = 6.94 \pm 0.04$. As seen in Fig. 2, for the samples [O], as x increases from 0 to 1, $\varepsilon = (b-a)/(b+a)$ decreases, together with T_c. However, for the samples [OAO], T_c generally increases with x as ε decreases. Note that T_c increases by 6 K to 85 K for x=1 [2]. The lattice parameter b is constant but a (and c) increase indicating an increase of the number of oxygen atoms per chain leading to a decrease of ε (T_c) [O] and a tetragonal structure. For a given x, the [OAO] treatment increases ε (for $0 \leq x \leq 1$), T_c (for $x \geq 0.4$) and the distance d[Cu(1)-(Sr/Ba)] (decrease T_c) for x<0.5 and decrease it (increase T_c) for x>0.5.

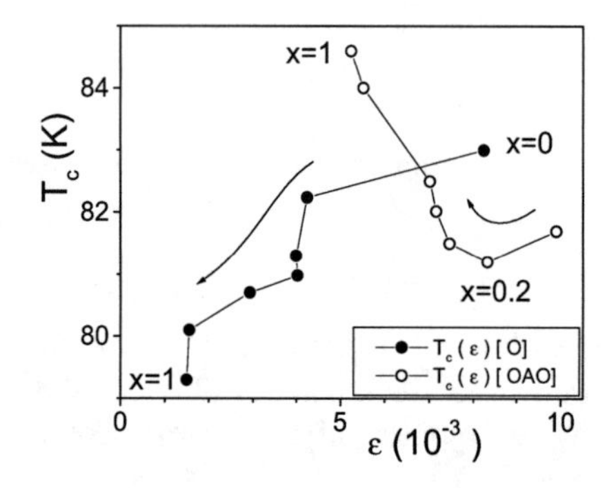

FIGURE 2. Variation of the orthorhombicity ε as a function of T_c and heat treatments of $Y_{1-x}Sm_xSrBaCu_3O_{6+z}$.

In the samples [OAO], the thermal parameter of the apical oxygen O(1) decreased from 2.02 Å^2 to 0.27 Å^2. It is possible that cationic disorder along the c axis decreased for $x \geq 0.2$. The heat treatment in argon at 850 °C removes the oxygen from the structure. That increases the atomic diffusion and the Sm-Sr/Ba-Sm order. When Sm ion occupies in Ba (or Sr) site, the same amount of Ba (or Sr) cation is pushed into Y site. Sm is a three-valence ion. It increases the positive charge density around Ba (or Sr) site and the attractive force with oxygen anion. As a result, oxygen vacancies O(5) along the a-axis in the basal plane have higher chance to be filled. On the other hand, Ba^{2+}(or Sr^{2+}) in Y^{3+} site make decrease the attractive force with oxygen anion in Cu(2) plane. This increases the buckling angle Cu(2)-O(3)-Cu(2) along the a axis. When x increased from 0 to 1, the two changes of cation sites increase the parameter a. The two arguments (cationic and anionic disorders) are justified here by the remarkable correlations observed between $T_c(x)$, d[Cu(2)- Y/Sm] (x) in "Fig. 3" and p_{sh} (x) (deduced from the under saturation zone of the universal relation T_c/T_{cmax} (p_{Sh}) [3]) in "Fig. 4"; and between $\delta T_c(x) = T_c[OAO]-T_c[O]$ and $\delta\varepsilon(x)$. So the structural and superconducting properties are correlated with the effect of argon heat treatment.

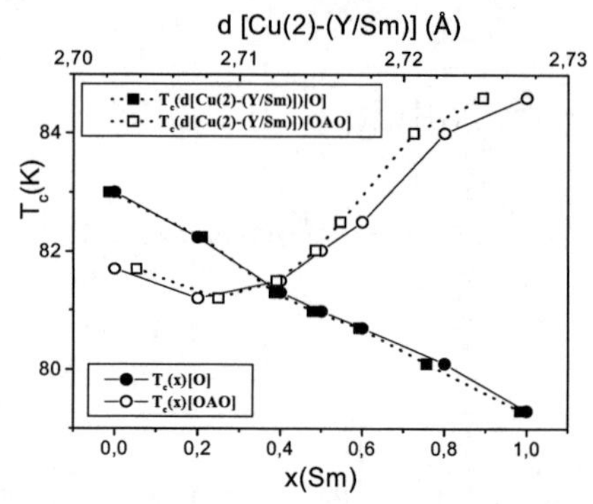

FIGURE 3. Correlation between d[Cu(2)-Y/Sm] distance and T_c as a function of x and heat treatment.

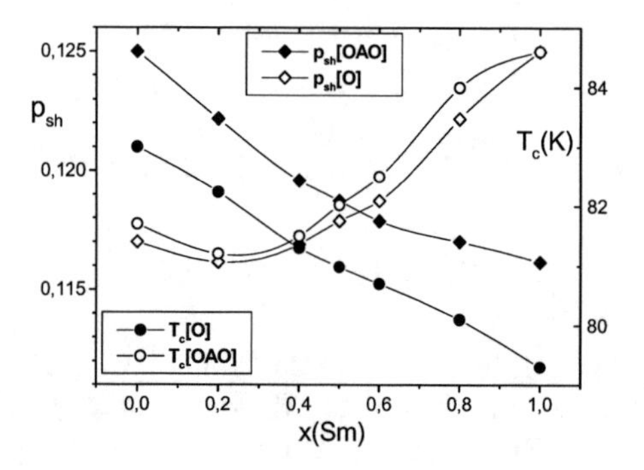

FIGURE 4. Correlation between p_{Sh} and T_c as a function of x and heat treatment of $Y_{1-x}Sm_xSrBaCu_3O_{6+z}$.

CONCLUSIONS

These results are the outcome of interplay between cationic disorder along the c axis (in agreement with the predictions of [4]) and oxygen disorder in basal plane. A combination of several factors such as decrease in d[Cu(1)-O(1)]; increase in cationic and chain oxygen ordering; p_{Sh} and in-phase purity for the samples [OAO] may account for the observed data.

REFERENCES

1. Tokura, Y., et al, *Nature* **337**, 345 (1989).
2. Bellioua, M., Nafidi, A., et al, *Physica C* **383**, 183-190 (2002). *Comptes Rendus Physique.* **5**, 285-292 (2004).
3. Zhang, H. and Sato, H., *Phys. Rev. Lett.* **70**, 11 (1993).
4. Wada, T., et al, *Phys. Rev. B* **39**, 9126 (1989).

Enhancement of T_C (~ 130K) in $TlBa_2Ca_2Cu_3O_y$ Synthesized under Ambient Pressure

S. Mikusu [a], S. Takami [a], K. Tokiwa [a], K. Takeuchi [b], A. Iyo [c], Y. Tanaka [c] and T. Watanabe [a]

[a] *Department of Applied Electronics, Tokyo University of Science, Noda, Chiba, Japan*
[b] *Faculty of Industrial Science &Technology, Tokyo University of Science, Oshamambe, Hokkaido, Japan*
[c] *National Institute of Advanced Industrial Science and Technology(AIST), Tsukuba, Ibaraki, Japan*

Abstract. We have succeeded to obtain the $TlBa_2Ca_2Cu_3O_y$ (Tl-1223) with the superconducting transition temperature T_C ~130K under ambient pressure. By controlling oxygen contents adequately in the starting mixture, we found that maximum T_C values of each samples prepared by a conventional method are quiet different. It is suggested that adjusting oxygen contents in the starting composition is one of the important factors to obtain Tl-1223 samples with high T_C. The details of synthesis condition and the results of neutron diffraction studies are presented.

Keywords: $TlBa_2Ca_2Cu_3O_y$; ambient pressure; T_C~130K
PACS: 74.62.Bf; 74.62.-c; 74.72.Jt; 74.72.-h

INTRODUCTION

Recently, $TlBa_2Ca_2Cu_3O_y$ (Tl-1223) samples synthesized under high pressure have been found to have T_C > 130K, which is comparable with that of $HgBa_2Ca_2Cu_3O_y$ (Hg-1223) that is known as the highest T_C superconductor [1,2].

With a conventional synthesis method, we have successful prepared the Tl-1223 samples with T_C~130K by controlling oxygen contents in the starting composition. In this paper, we have reported the synthesis condition to obtain the sample with T_C~130K.

EXPERIMENTAL

The samples were prepared from mixtures of Tl_2O_3, BaO, BaO_2, Ca_2CuO_3 and CuO by solid state reaction. These materials were weighed at nominal compositions of $TlBa_2Ca_2Cu_3O_y$ or $Tl_{0.9}Ba_2Ca_2Cu_3O_y$. Then, controlling oxygen contents (y=9.1-9.7) in the starting composition was carried out by varying the ratio between BaO and BaO_2. The samples sealed in quartz tube were sintered at 750°C-850°C for 2h and then quenched in cold water. The powder x-ray diffraction (XRD) patterns of the as-prepared Tl-1223 sample indicated nearly single phase. In order to obtain optimum-doped samples, post-annealing was done in Ar atmosphere because all as-prepared samples are in over-doped state. No changes were observed in the XRD pattern of annealed samples.

The superconducting properties were measured by the electrical resistivity and the dc magnetic susceptibility. The neutron diffraction measurement for the annealed samples was carried out using the general purpose of powder diffract-meter (GPPD) at the intense pulsed neutron source in the Argonne National Laboratory. The diffraction date was analyzed using Rietveld analysis computer program, GSAS [3].

RESULTS AND DISCUSSION

All as-prepared samples we obtained showed almost same T_C at around 110K in spite of the difference of oxygen contents in the starting composition. Moreover, we found that the samples optimized a doping state by post-annealing show a different maximum T_C (The T_C of the samples of y=9.1, 9.35, 9.5, and 9.7 were 123, 126, 127, and 117K, respectively). The synthesis temperature, which can obtain single-phase samples, is also related to oxygen

CP850, *Low Temperature Physics: 24th International Conference on Low Temperature Physics;* edited by Y. Takano, S. P. Hershfield, S. O. Hill, P. J. Hirschfeld, and A. M. Goldman

contents of starting compo-sition, and lower synthesis temperature tends to enhance T_C after post-annealing. Therefore, we point out that adjusting oxygen contents in the starting com-position is one of the most important factors to obtain Tl-1223 samples with high T_C. Finally, by reduced Tl composition in the starting mixture, i.e. in the nominal composition of $Tl_{0.9}Ba_2Ca_2Cu_3O_{9.5}$, we successfully prepared Tl-1223 samples with T_C∽130K (FIGURE 1). Although the sample surface was deteriorated by annealing above 700°C, the XRD pattern of this sample was essentially identified with that of as-prepared sample.

We have not obtained the Tl-1223 samples with T_C ∽ 130K without reducing Tl contents, as present. However, the T_C of Tl-1223 samples with stoichio-metric Tl composition is seemed to improve by adjustment of starting oxygen contents and synthesis temperature.

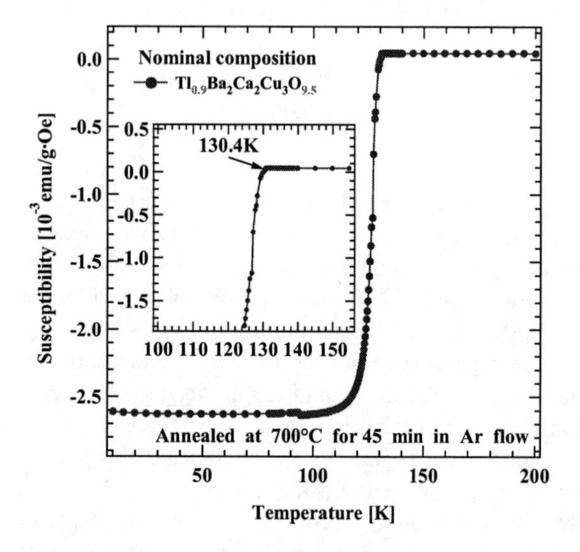

FIGURE 1. Temperature dependence of the susceptibility for the Tl-1223 samples with T_C∽130K. The dc magnetic susceptibility was measured by a SQUID magnetometer (Quantum Design MPMS) under the field-cooling mode with a magnetic field of 5 Oe.

In neutron powder diffraction experiments for the sample with T_C∽ 127K (the sample annealed for $TlBa_2Ca_2Cu_3O_{9.5}$), we carried out refinement of the crystal structure. As a result of refinement, we found that the amount of Tl substitution for Ba and Ca-sites is very little. It has been suggested that the remark-able enhancement of *Tc* value for the high-pressure synthesized Tl-1223 samples arise by suppression of Tl substitution for Ba- and Ca-site [1], which inheren-tly occurs in samples prepared by conventional methods. However, the T_C in the sample prepared by reducing Tl content in starting composition was higher than that of the sample, which contains no Tl substitu-tion for Ba or Ca sites. Moreover, in the crystal structure which was determined by our neutron powder diffraction study for the high-pressure synthe-sized (HPS) samples [4], we found that Tl substitutes for about 3% of Ca-site in spite of having the high T_C value (T_C∽132K). Then we have attempted to do the precise comparison of crystal structure between our HPS samples and conventional ones reported by B.Morosin *et al* [5]. The data of crystal structure for their conventional Tl-1223 samples (T_C∽120K) was obtained from single crystal X-ray measurements. We found that the amount of Tl substitution for Ba and Ca-sites in HPS samples is quite similar to that of conventional samples. Therefore, it seems that slight Tl substitution for Ba- and Ca-sites gives no essential effect on the difference of T_C in these two samples. Further studies on another key factors for the enhancement of T_C -value is progressing.

SUMMARY

Tl-1223 samples synthesized under ambient pressure successfully showed high T_C (∽130K) by optimizing synthesis conditions. We found that, in particular, it is necessary to control oxygen contents in the starting composition in order to obtain Tl-1223 samples with high T_C. In addition, from a result of neutron powder diffraction studies, it is suggested that a small amount of Tl substitution for Ba- and Ca-sites gives no serious influence on T_C enhancement.

ACKNOWLEDGMENTS

This work was supported by Tanaka Kikinzoku Kogyo Ltd, Japan.

REFERENCES

1. A. Iyo, Y. Tanaka, Y. Ishiura, M. Tokumoto, K. Tokiwa, T. Watanabe and H. Ihara, *Supercond. Sci. Technol.* **14**, 504 (2001).
2. A. Iyo, Y. Aizawa, Y. Tanaka, M. Tokumoto, K. Tokiwa, T. Watanabe and H. Ihara, *Physica C* **357-360**, 324 (2001).
3. A. C. Larson and R. B. von Dreele, General Structure Analysis System (GSAS), *Los Alamos National Laboratory Report LAUR* 86-748 (2004).
4. S. Mikusu et al., unpublished.
5. B. Morosin and E. L. Venturini and D. S. Ginley, *Physica C* **183** 90-98 (1991).

Phase Formation and Superconductivity in Co-doped $GaSr_2(Tm,Ca)Cu_2O_z$ Cuprate

H.K. Lee and Y.H. Kim

Department of Physics, Kangwon National University, Chunchon 200-701, Republic of Korea

Abstract. The effect of Co substitution for Ga on the phase formation and superconducting properties of $(Ga_{1-x}Co_x)Sr_2(Tm_{0.7}Ca_{0.3})Cu_2O_z$ ($0 \leq x \leq 0.5$) prepared under ambient pressure was investigated. X-ray diffraction revealed that the phase purity of the samples increases as the Co-doping concentration increases and single-phase materials could be obtained for samples with $x \geq 0.35$. Contrary to the Co-free sample, superconductivity could be induced by partially substituting Co for Ga in the $GaSr_2(Tm_{0.7}Ca_{0.3})Cu_2O_z$ system. The superconducting behavior of the Co-doped samples is discussed in connection with thermoelectric power measurements.

Keywords: $GaSr_2(Tm_{0.7}Ca_{0.3})Cu_2O_z$, Co substitution, superconductivity, ambient pressure synthesis
PACS: 74.72.Jt, 74.10.+v, 74.62.Bf

INTRODUCTION

The Ga-1212 type cuprate $GaSr_2YCu_2O_z$ was independently developed in several laboratories [1,2,3]. The crystal structure is closely related to that of $CuSr_2YCu_2O_7$ (Cu-1212) and can be described as replacement of the square-planar copper (chain copper) with tetrahedral gallium. The $GaSr_2YCu_2O_z$ cuprate becomes superconducting below about 40 K when the Y-site is doped by Ca in the composition of $GaSr_2(Y_{1-x}Ca_x)Cu_2O_z$ at a doping level $x > 0.15$-0.2, but only after synthesis under elevated oxygen pressure. Cava et al. [2] reported that it is difficult to find conditions which make single-phase $GaSr_2(R_{1-x}Ca_x)Cu_2O_z$ samples superconducting at oxygen pressures of 25 atmospheres or below. In connection with the preparation of single-phase Ga-1212 samples, it is recalled that, although $CuSr_2YCu_2O_7$ can be prepared only at a high pressure above 2 Gpa [4], the Cu-1212 structure can be stabilized at ambient pressure with partial substitution of the chain copper with elements of valence greater than +2 [5,6]. Therefore, it would be interesting to extend the study on the synthesis of single-phase superconducting Ga-1212 samples at ambient pressure by the introduction of an appropriate ion in the Ga site. In this paper, we report the effect of Co substitution for Ga on the phase formation and superconducting properties of $GaSr_2(Tm_{0.7}Ca_{0.3})Cu_2O_z$ prepared under ambient pressure.

SAMPLE PREPARATION

Polycrystalline samples with a nominal composition of $(Ga_{1-x}Co_x)Sr_2(Tm_{0.7}Ca_{0.3})Cu_2O_z$ ($0 \leq x \leq 0.5$) were prepared by using a solid-state reaction technique. Starting materials with appropriate ratios of pure (≥ 99.9%) Ga_2O_3, Co_3O_4, $SrCO_3$, Tm_2O_3, $CaCO_3$, and CuO powders were thoroughly mixed and calcined at 900 °C for 12 h in air. The resultant powders were palletized, sintered at 1010 °C for 24 h in flowing oxygen, and slowly cooled to below 100 °C. These pellets were finally annealed in flowing oxygen at 400 °C for 2 h and 350 °C for 5 h, and slowly cooled to room temperature

RESULTS AND DISCUSSION

Figure 1 presents X-ray diffraction (XRD) patterns of $(Ga_{1-x}Co_x)Sr_2(Tm_{0.7}Ca_{0.3})Cu_2O_z$ at room temperature. The diffraction lines are indexed on the basis of an orthorhombic unit cell of the space group Ima2 [7]. The XRD data indicate that the phase purity of the $(Ga_{1-x}Co_x)Sr_2(Tm_{0.7}Ca_{0.3})Cu_2O_z$ samples

CP850, *Low Temperature Physics: 24th International Conference on Low Temperature Physics;*
edited by Y. Takano, S. P. Hershfield, S. O. Hill, P. J. Hirschfeld, and A. M. Goldman

increases with increasing Co-doping content x and the samples with x ≥ 0.35 exhibit a single-phase nature.

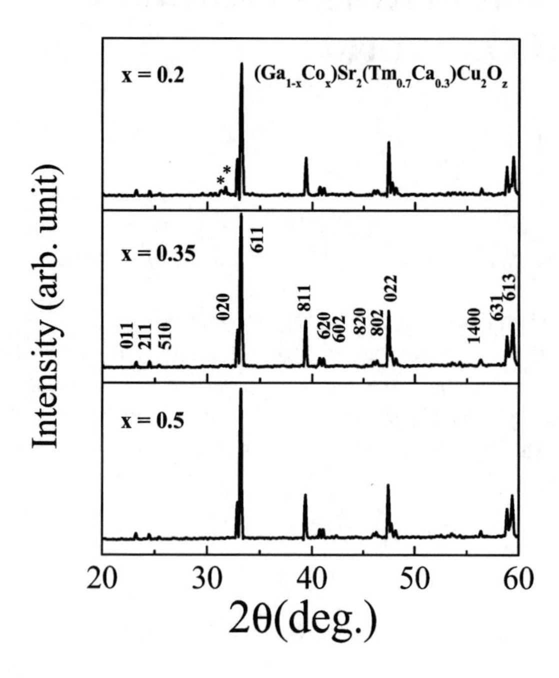

FIGURE 1. Powder X-ray diffraction patterns for $(Ga_{1-x}Co_x)Sr_2(Tm_{0.7}Ca_{0.3})Cu_2O_z$ samples. Impurity lines are shown with asterisks.

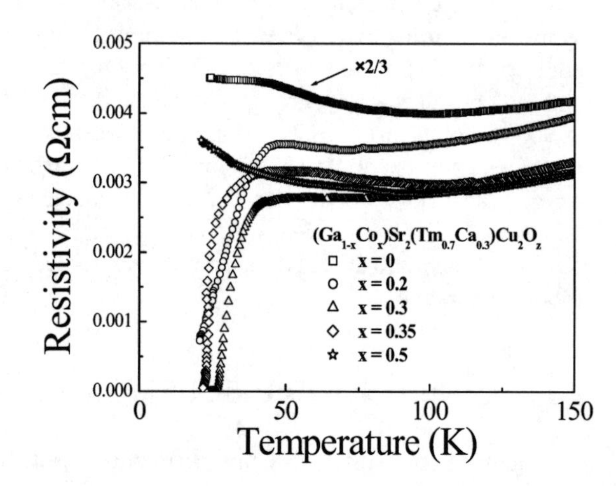

FIGURE 2. Temperature dependence of electrical resistivity of $(Ga_{1-x}Co_x)Sr_2(Tm_{0.7}Ca_{0.3})Cu_2O_z$ (x = 0-0.5) samples.

Figure 2 shows the temperature dependence of electrical resistivity for $(Ga_{1-x}Co_x)Sr_2(Tm_{0.7}Ca_{0.3})Cu_2O_z$ samples with x = 0, 0.2, 0.3, 0.35 and 0.5. On progressive replacement of Ga by Co, the superconducting transition temperature of the samples increases up to x = 0.3 and then decreases thereafter. The samples with x = 0.2 and 0.3 show a superconducting drop at about 45 K.

For p-type cuprates, it is possible to use room-temperature thermoelectric power (TEP) as a probe to determine the hole-doping state of the cuprates based on a systematic correlation between the room-temperature TEP and the hole concentrations in the superconducting planes [8]. The room-temperature TEP for the $(Ga_{1-x}Co_x)Sr_2(Tm_{0.7}Ca_{0.3})Cu_2O_z$ samples with x = 0, 0.1, 0.2, 0.3. 0.4 and 0.5 are 41.7 μV, 34.2 μV, 24.5 μV, 18.4 μV, 19.9 μV and 21.9 μV, respectively. The larger positive value of 18.4 μV for the x = 0.3 sample suggests that all samples are in an underdoped state. The decrease of TEP at a Co-doping content up to x = 0.3 implies that the hole density on the superconducting planes increases. Similarly, the increase of TEP at a higher Co-doping content above x = 0.4 can be attributed to a decrease of hole density of the sample. Therefore, the change of the superconducting properties with increasing Co-doping content for $(Ga_{1-x}Co_x)Sr_2(Tm_{0.7}Ca_{0.3})Cu_2O_z$ samples can be explained in broad terms by the change of hole density of the samples.

ACKNOWLEDGMENTS

This work was supported by Grant No. (R05-2003-000-10434-0) from the Korea Science and Engineering Foundation.

REFERENCES

1. J.T. Vaughey, J.P. Thiel, E.F. Hasty, D.A. Groenke, C.L. Stern, K.R. Poepplemeier, B. Dabroski, D.G. Hinks and A.W. Mitchell, *Chem. Mater.* **3,** 935-940 (1991).
2. R.J. Cava, R.B. Van Dover, B. Batlogg, J.J. Krajewski, L.F. Scheemeyer, T. Siegrist, B. Hessen, S.H. Shen, W.F. Peck and L.F. Rupp, *Physica C* **185-189,** 180-183 (1991).
3. B. Dabrowski, P. Radaelli, D.G. Hinks, A.W. Mitchell, J.T. Vaughey, D.A. Groenke and K.R. Poeppelmeier, *Physica C* **193,** 63-67 (1992).
4. B. Okai, *Jpn. J. Appl. Phys.* **29,** L2180-L2182 (1990).
5. P.R. Slater and C. Greaves, *Physica C* **180,** 299-306 (1990).
6. T. Den and T.Kobayashi, *Physica C* **196,** 141-152 (1992).
7. E. Kandyel, *Physica* C **415,** 1-8 (2004).
8. S.D. Obertelli, J.R. Cooper and J.L. Tallon, *Phys. Rev. B* **46,** 14928-14931 (1992).

Crystal Growth of Superconducting La2126 without HIP Treatment

Kazuhiro Koike*, Takashi Noji*,†, Tadashi Adachi*,† and Yoji Koike*,†

*Department of Applied Physics, Tohoku University, 6-6-05, Aoba, Aramaki, Sendai 980-8579, Japan
†CREST, Japan Science and Technology Corporation (JST)

Abstract. We have succeeded in the crystal growth of the superconducting $La_{2-x-y}Ba_xSr_yCaCu_2O_6$ of the La2126 phase by only the traveling-solvent floating-zone (TSFZ) method without the hot-isostatic-pressure (HIP) treatment. In the TSFZ growth, we added a small amount of B_2O_3 to the solvent for the stabilization of the growth and used high-pressure oxygen of $\sim$ 10 atm as a growth atmosphere for the suppression of oxygen defects. We have obtained single crystals with the size of the order of $2\times2\times3$ mm^3 and determined the composition to be $La_{1.80}Ba_{0.07}Sr_{0.19}Ca_{0.97}Cu_{1.97}O_6$ by the inductively-coupled-plasma atomic-emission-spectrometry. The powder x-ray diffraction has revealed that the crystals are of the single phase of La2126 and include no impurity phases such as $La_{2-x}M_xCuO_4$ (M: alkaline earth). The as-grown single crystals have shown superconductivity of the bulk with the transition temperature T_c = 20 K.

Keywords: superconductivity, cuprate, La2126, crystal growth, traveling-solvent floating-zone method
PACS: 81.10.Fq, 74.72.Dn, 74.25.-q, 74.62.Dh

The compound $La_{2-x}M_xCaCu_2O_6$ (M: alkaline earth) of the so-called La2126 phase has a pair of pyramidal Cu-O planes in the unit cell, which is the only electrically active part [1]. Because of the simple crystal structure, La2126 is expected to be a suitable system to extract the important information about the electronic state in the Cu-O plane.

As for single crystals of the superconducting La2126, the hot-isostatic-pressure (HIP) treatment, namely, post-annealing of obtained single-crystals above 1000 °C under an oxygen pressure of $\sim$ 300 atm is inevitable [2]. Therefore, we have aimed to obtain large-sized single crystals of the superconducting La2126 by only the traveling-solvent floating-zone (TSFZ) method without the HIP treatment.

Feed rods for the TSFZ growth were prepared by the solid-state reaction. High-purity powders of La_2O_3, $BaCO_3$, $SrCO_3$, $CaCO_3$ and CuO were mixed and sintered. After cycles of grinding and calcininig in air at 900-1050 °C for 24 h, the powder mixtures were isostatically cold-pressed into a rod of 6 mm in diameter and $\sim$100 mm in length and sintered in air at 1130-1150 °C for 36 h. A solvent rod was also prepared by solid-state reaction using high-purity powders of La_2O_3, MCO_3, $CaCO_3$ and CuO in the molar ratio of La : M : Ca : Cu = 2-x : x : 1 : 4. The powders were mixed and sintered in air at 900 °C for 12 h. After pulvarization, the powder mixtures were isostatically cold-pressed into a rod of 6 mm in diameter. The rod was sintered in air at 900 °C for 12 h. The sintered ros was sliced into pieces and a piece of $\sim$ 0.4 g was used as a solvent for the TSFZ growth. The TSFZ growth was performed under an oxygen pressure of $\sim$ 10 atm in an infrared heating furnace. The growth speed was 0.5 mm/h. Crystals thus obtained were evaluated by the powder x-ray diffraction and the x-ray back-Laue photography. The composition was analyzed by the inductively-coupled-plasma atomic-emission-spectrometry (ICP-AES). The magnetic susceptibility, χ, was measured using a SQUID magnetometer. The in-plane electrical resistivity, ρ_{ab}, was measured by the dc four-point probe method.

Table 1 displays the compositions of feed rods and obtained single-crystals analyzed by ICP-AES, lattice parameters, the superconducting transition temperature, T_c, and the superconducting volume fraction (SVF) estimated from χ at 2 K on zero-field cooling. First, the crystal growth using the Feed Rod I free of Ba and Sr was attempted. The growth was sufficiently stable that a crystal rod consisting of a lot of single crystals has been obtained. The size of one crystal (Crystal I) is the order of $1\times1\times2$ mm^3. The x-ray measurements have revealed that the Crystal I is a single crystal of the La2126 phase. The Crystal I has shown superconductivity with T_c = 18 K, but SVF is as small as $\sim$ 0.5 %. The small SVF may be due to oxygen defects.

Since ionic radii of Ba^{2+} or Sr^{2+} are larger than that of Ca^{2+}, it is expected that the substitution of Ba or Sr for Ca increases the lattice parameters so that oxygen is introduced into the crystal enough for the appearance of superconductivity with a large SVF. Then, the crystal growth using the Feed Rod II with Ba and Sr was tried. Although the growth was not so stable, a crystal rod made up of plenty of single crystals has been obtained. The size of one crystal (Crystal II) is also the order

CP850, *Low Temperature Physics: 24th International Conference on Low Temperature Physics;*
edited by Y. Takano, S. P. Hershfield, S. O. Hill, P. J. Hirschfeld, and A. M. Goldman

TABLE 1. Compositions of feed rods and obtained single-crystals analyzed by ICP-AES, lattice parameters a and c, T_c, and the superconducting volume fraction (SVF) for the La2126 crystals growth by the TSFZ method. T_c is defined as the onset temperature of the Meissner effect. SVF is estimated from χ at 2 K on zero-field cooling.

	La	Ba	Sr	Ca	Cu	a (Å)	c (Å)	T_c (K)	SVF (%)
Feed Rod I (Nominal)	1.7	-	-	1.3	2				
Crystal I (ICP)	1.89	-	-	1.20	1.91	3.823	19.41	18	~0.5
Feed Rod II (Nominal)	1.7	0.1	0.2	1	2				
Crystal II (ICP)	1.78	0.07	0.19	1.13	1.83	3.832	19.58	30	~1
Feed Rod III (Nominal)	1.7	0.1	0.2	1	2.2				
Crystal III (ICP)	1.80	0.07	0.19	0.97	1.97	3.827	19.54	20	~40

of $1\times1\times2$ mm^3. It has been revealed that the Crystal II is a single crystal of the La2126 phase and exhibits superconductivity with T_c = 30 K, but SFV is as small as that in the Crystal I. When SVF is very small, a suspicion comes up that the superconductivity is not due to the La2126 phase but due to a small amount of impurity phases such as $La_{2-x}M_xCuO_4$ of the so-called La214 phase. However, no impurity phases have been detected in the Crystal I nor in the Crystal II by the x-ray diffraction. From the composition analysis by ICP-AES, it has turned out that the Crystal II is short of Cu, which is guessed to be caused by the evaporation of Cu out of the molten zone in the TSFZ growth . Therefore, the small SVF may be due to defects of Cu.

Then, the crystal growth using the Feed Rod III with an excess Cu was tried. Here, 2 wt.% of B_2O_3 was added to the solvent so as to increase the viscosity of the solvent and stabilize the growth. The growth was sufficiently stable that a crystal rod consisting of many single-crystals has been obtained. The size of one crystal (Crystal III) is as large as the order of $2\times2\times3$ mm^3. The x-ray back-Laue photograph of the (001) plane has confirmed the single crystallinity as shown in Fig. 1, and the powder x-ray diffraction has revealed that the Crystal III is of the single phase of La2126 and includes no impurity phases such as $La_{2-x}M_xCuO_4$. The composition of the Crystal III has been determined to be $La_{1.80}Ba_{0.07}Sr_{0.19}Ca_{0.97}Cu_{1.97}O_6$ by ICP-AES. The Crystal III has shown superconductivity with T_c = 20 K and a large SVF, as shown in Fig. 2. These results can claim that the superconductivity is of the bulk of the La2126 phase. However, the superconducting transition is quite broad . The inhomogeneous distribution of cations and oxygen in the crystal may induce the inhomogenity of the hole concentration, leading to the broad transition.

In conclusion, it is effective for obtaining the superconducting La2126 crystals with a large SVF by only the TSFZ method without the HIP treatment to increase the lattice parameters by doping Ba and Sr and to lessen the defects of Cu. However, it will be necessary for the observation of a sharp superconducting transition to make the distribution of cations and oxygen in the crystal homogeneous.

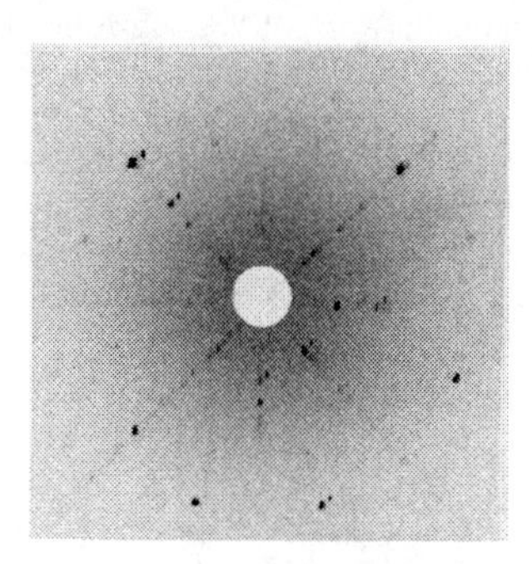

FIGURE 1. X-ray back-Laue photograph of the (001) plane of the single-crystal $La_{1.80}Ba_{0.07}Sr_{0.19}Ca_{0.97}Cu_{1.97}O_6$.

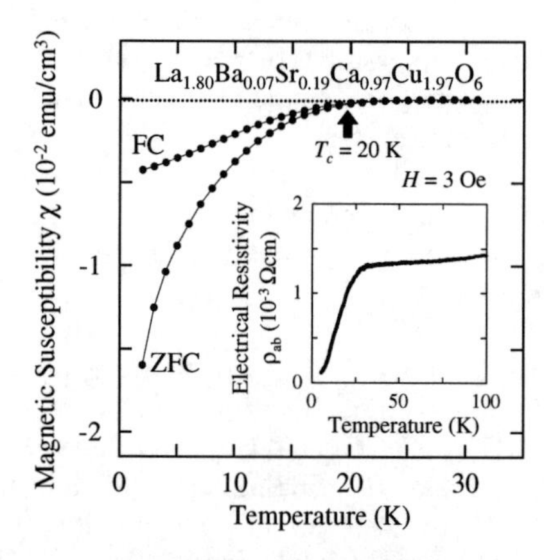

FIGURE 2. Tempreratur dependence of the magnetic susceptibility, χ, in a field of 3 Oe on zero-filed cooling (ZFC) and on field cooling (FC) for the single-crystal $La_{1.80}Ba_{0.07}Sr_{0.19}Ca_{0.97}Cu_{1.97}O_6$. The inset shows the temperature dependence of the in-plane electrical resistivity, ρ_{ab}.

This work was supported by a Grant-in-Aid for Scientific Research from the Ministry of Education, Culture, Sports, Science and Technology, Japan.

REFERENCES

1. N. Nguyen *et al.*, *Mater. Res. Bull.* **15**, 891 (1990).
2. K. Kinoshita *et al.*, *Phys. Rev. B* **46**, 9116 (1992).

Pseudogap in Pb-doped Bi2201 Studied by the Out-of-Plane Resistivity in Magnetic Fields up to 40 T

K. Kudo*, T. Sasaki*, E. Ohmichi†, T. Osada†, Y. Miyoshi* and N. Kobayashi*

*Institute for Materials Research, Tohoku University, 2-1-1 Katahira, Aoba-ku, Sendai 980-8577, Japan
†Institute for Solid State Physics, University of Tokyo, 5-1-5 Kashiwanoha, Kashiwa 277-8581, Japan

Abstract. We determine the onset temperature of the pseudogap corresponding to the precursor of the superconductivity by measuring the out-of-plane resistivity ρ_c of $Bi_{1.74}Pb_{0.38}Sr_{1.88}CuO_{6+\delta}$ in magnetic fields up to 40 T. It has been found that the semiconducting upturn of $\rho_c(T)$ below T^* is suppressed by the application of magnetic fields up to 40 T at low temperatures below T^{**} which is lower than T^*. That is, with decreasing temperature, the magnetic-field insensitive pseudogap is formed below T^*, and then the magnetic-field sensitive one participates below T^{**}. These results have confirmed that the pseudogap corresponding to the precursor of the superconductivity is formed below T^{**}.

Keywords: Pseudogap, Out-of-plane resistivity, Bi-based cuprates, Bi2201
PACS: 74.25.Dw, 74.25.Fy, 74.72.Hs

The origin of the pseudogap in high-T_c cuprates has not been settled yet, though it has been recognized as an important phenomenon to approach the superconducting mechanism. Both theoretical and experimental aspects are classified into two groups, such as the precursor of the superconductivity (precursor SC) and the competing energy gap or hidden order parameter to the superconductivity[1].

The pseudogap should be suppressed by the application of a magnetic field if it relates to the superconductivity, so that the magnetic field dependence of the pseudogap provides a significant information for the clarification of the origin. The c-axis magnetoresistance MR is useful in this issue, because the c-axis transport follows a tunneling nature which is sensitive to the depletion of ($\pm\pi$, 0) and (0, $\pm\pi$) states at the Fermi surface where the pseudogap first opens[2, 3]. In $Bi_2Sr_2CaCu_2O_{8+\delta}$[2], $Bi_2Sr_{2-x}La_xCuO_{6+\delta}$[3] and $Bi_{1.74}Pb_{0.38}Sr_{1.88}CuO_{6+\delta}$[4, 5], the negative MR being indicative of the suppression of the pseudogap has been reported. These results have supported the precursor scenario for the pseudogap.

In order to understand the phase diagram of the high-T_c cuprates, it is important to determine the temperature below which the precursor SC starts to develop. In this issue, there exist two aspects on the pseudogap temperature correlating to the precursor SC proposed from MR measurements[2, 3, 4, 5]. One is that the pseudogap corresponding to the precursor SC is formed below T^* below which the semiconducting upturn of $\rho_c(T)$ appears[2]. Another is that the pseudogap independent of the superconductivity is formed below T^* and then the pseudogap corresponding to the precursor SC develops below T^{**} which is lower than T^*[3, 4, 5]. Since the onset temperature of the pseudogap corresponding to the precursor SC has been determined as the temperature below which the negative MR appears, such the controversial arguments may arise from insufficient magnitude of applied magnetic-fields for detecting the field-dependent pseudogap clearly. In this study, therefore, we try to determine the onset temperature of the pseudogap corresponding to the precursor SC by measuring the out-of-plane resistivity ρ_c of Pb-doped Bi2201 in strong magnetic fields up to 40 T.

Single crystals of $Bi_{1.74}Pb_{0.38}Sr_{1.88}CuO_{6+\delta}$ were grown by the floating-zone method and then annealed under vacuum, flowing Ar, flowing O_2 or 7 atm O_2 atmosphere to control the hole concentration p[4, 5]. The achieved T_c's are found in the $T-p$ phase diagram in Fig. 2. p values of 0.21, 0.27 and 0.35 are determined form the area of the Fermi surface from ARPES[6]. In other samples, p values are estimated from the annealing conditions and the resulted T_c. Samples measured are the underdoped sample UD ($T_c = 16$K), the overdoped sample OD1 ($T_c = 11$K) and the heavily overdoped sample OD2 ($T_c = 0.5$K), as indicated by closed symbols in Fig. 2. The ρ_c was measured by a standard DC four-terminal method in magnetic fields up to 15, 27 and 40 T parallel to the c-axis, using a superconducting magnet, a hybrid magnet (HFLSM, IMR, Tohoku Univ.) and a pulsed magnet (ISSP, Tokyo Univ.), respectively.

Figures 1(a) and 1(b) show the temperature dependence of ρ_c in magnetic fields up to 40 T. At low temperatures, $\rho_c(T)$ exhibits the semiconducting upturn as a result of the depletion of ($\pm\pi$, 0) and (0, $\pm\pi$) states at the Fermi surface caused by the pseudogap formation. The onset temperature of the upturn T^* is defined as the local minimum of $\rho_c(T)$, as indicated by black arrows. By

CP850, *Low Temperature Physics: 24th International Conference on Low Temperature Physics;*
edited by Y. Takano, S. P. Hershfield, S. O. Hill, P. J. Hirschfeld, and A. M. Goldman

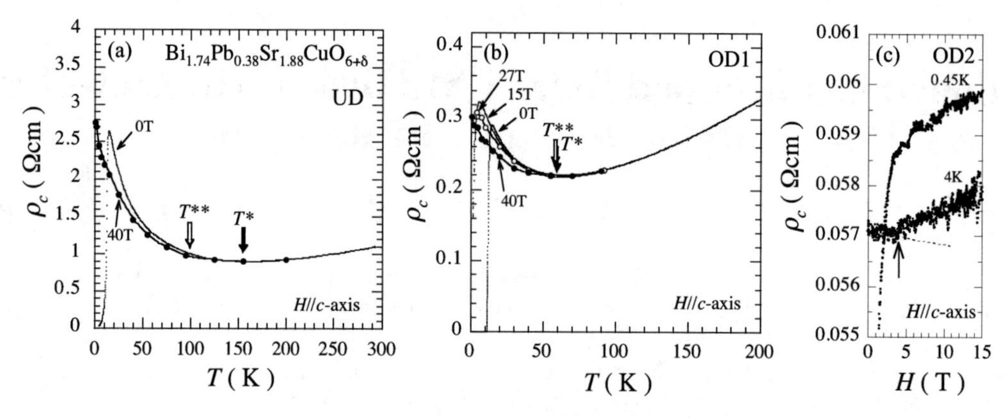

FIGURE 1. (a)(b)Temperature dependence of the out-of-plane resistivity ρ_c of $Bi_{1.74}Pb_{0.38}Sr_{1.88}CuO_{6+\delta}$ in magnetic fields up to 40 T parallel to the c-axis. (c)Magnetic field dependence of ρ_c of $Bi_{1.74}Pb_{0.38}Sr_{1.88}CuO_{6+\delta}$ in magnetic fields up to 15 T parallel to the c-axis. Broken line is a guide for eyes.

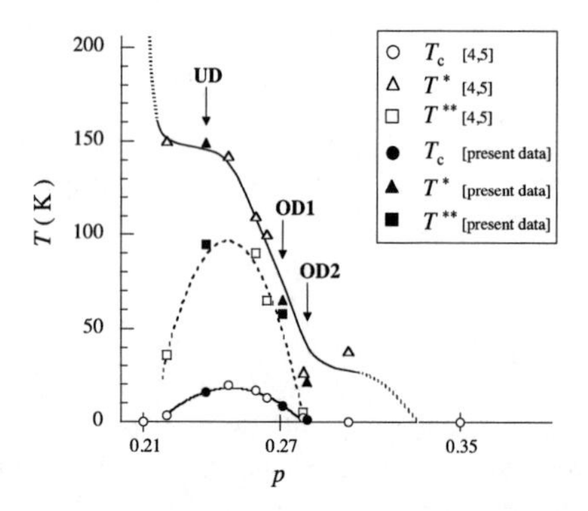

FIGURE 2. Hole-concentration p dependence of T_c, T^* and T^{**} in $Bi_{1.74}Pb_{0.38}Sr_{1.88}CuO_{6+\delta}$. Solid and open symbols indicate the present data and those from ρ_c measurements in 15 T[4, 5], respectively.

the application of magnetic fields, the upturn of $\rho_c(T)$ is suppressed. The suppression is attributed to the enhancement of the tunneling conductance parallel to the c-axis on account of the suppression of the pseudogap.

T^{**} is defined as the temperature where the negative MR ($\Delta\rho_c/\rho_c(0,T) = [\rho_c(H,T) - \rho_c(0,T)]/\rho_c(0,T)$) reaches 1 %, as indicated by white arrows in Figs. 1(a) and 1(b). It is found that T^* and T^{**} are close in the overdoped sample OD1 while they are far from each other in the underdoped sample UD. As shown in Fig. 2, this tendency resembles the $T-p$ phase diagram proposed from ρ_c measurements in 15 T[4, 5]. In Fig. 2, T^* decreases with increasing p and then comes close to the bell-like $T^{**}-p$ dome at the overdoped region. The present data determined in 40 T agree well with these curves, as shown by solid symbols. The result indicates that T^{**} is distinguished from T^* and recognized as another pseudogap temperature. That is, with decreasing temperature, the magnetic-field insensitive pseudogap is formed below T^*, and then the magnetic-field sensitive one participates below T^{**}. Accordingly, it is confirmed that the onset temperature of the pseudogap corresponding to the precursor SC is not T^* but T^{**}.

If a sufficient magnetic field was applied to close the pseudogap corresponding to the precursor SC, the negative MR is expected to change into the positive one above a critical field. This change may be observed in the magnetic field dependence of ρ_c in OD2. As shown by the arrow in Fig. 1(c), the slope of $\rho_c(H)$ is changed from negative to positive around 4 T at 4.2 K. At low temperatures below T_c, on the other hand, negative slope of $\rho_c(H)$ is not observed in OD2. This may be because the pseudogap is already closed when the resistive state appears. In addition, it is noted that the upturn of $\rho_c(T)$ of OD2 remains even in 27 T, which suggests that the pseudogap below T^* survives in high magnetic fields.

In conclusion, it has been found that the semiconducting upturn of $\rho_c(T)$ below T^* is suppressed by the application of magnetic fields up to 40 T at low temperatures below T^{**} which is lower than T^*. These results have confirmed that the onset temperature of the pseudogap corresponding to the precursor of the superconductivity is not T^* but T^{**}.

This work was partly supported by a Grant-in-Aid for Scientific Research (15340107) from the ministry of Education, Science, Sports, and Culture of Japan.

REFERENCES

1. T. Timusk and B. Statt, *Rep. Prog. Phys.* **62**, 61 (1999).
2. N. Morozov *et al.*, *Phys. Rev. Lett.* **84**, 1784 (2000).
3. A. N. Lavrov, Y. Ando and S. Ono, *Europhys. Lett.* **57**, 267 (2002).
4. K. Kudo *et al.*, *Physica C* **426-431**, 251 (2005).
5. K. Kudo *et al.*, *cond-mat/0405106*, unpublished.
6. T. Sato *et al.*, (private communications).

Electronic inhomogeneity as a function of out-of-plane disorder in Ln-doped Bi2201 superconductors

A. Sugimoto[a], S. Kashiwaya[a], H. Eisaki[a], H. Kashiwaya[a], H. Tsuchiura[b], Y. Tanaka[c], K. Fujita[d], S. Uchida[d,e]

[a] *Nanoelectronics Research Institute, National Institute of Advanced Industrial Science and Technology (AIST), Tsukuba, Ibaraki, 305-8568, JAPAN*
[b] *Department of Applied Physics, Tohoku University, Sendai 980-8579, Japan*
[c] *Department of Applied Physics, Nagoya University, Nagoya 464-8603, Japan*
[d] *Department of Advanced Material Science, The University of Tokyo, Bunkyo, Tokyo, 113-8656, JAPAN*
[e] *Department of Physics The University of Tokyo, Bunkyo, Tokyo, 113-8656, JAPAN*

Abstract. The local electronic properties of $Bi_2Sr_{1.6}Ln_{0.4}CuO_{6+y}$, (Ln= La, Gd) is investigated by scanning tunneling microscopy/spectroscopy (STM/STS). As increasing the magnitude of out-of-plane disorder by increasing the mismatch of Ln-ion doping, enlargement of deviation of Δ ($\sigma\Delta$) is detected. The value of $\sigma\Delta$ is linearly enhanced as decreasing T_c including the results of optimum $Bi_2Sr_2CaCu_2O_{8+y}$. These results suggest the out-of-plane disorder suppresses the superconductivity and enhances pseudogap regions in superconducting state efficiently.
Keywords: STM/STS, High-T_c superconductor, Inhomogeneity, Bi2201.
PACS: 68.37.Ef; 74.25.Jb.; 74.72.Hs;

INTRODUCTION

Recently, nanoscale-electronic inhomogeneous structure (coexistence of superconductivity (SC) and pseudo-gap (PG) state) has been widely reported in $Bi_2Sr_2CaCu_2O_{8+y}$ (Bi2212) [1-5]. However, the exact origin of the inhomogeneity is unclear at present. Not only the oxygen doping level, but also the lattice disorder such as impurity substitution by another type of cation, is considered to play an important role for superconductivity. In high T_c cuprate superconductors, CuO_2 planes mainly contribute superconductivity. However, the disorder outside of CuO_2 planes (out-of-plane disorder) also affects superconducting properties such as T_c [6-8]. The T_c of $Bi_2Sr_{2-x}Ln_xCuO_{6+y}$ (Ln-Bi2201) superconductor is systematically controllable by doping the different type of Lanthanide (Ln = La, Nd, Eu, Gd). The doped Ln ions replace the Sr ion at the SrO plane, next to the CuO_2 plane. T_c is monotonically decreasing with increasing mismatch of Ln-ion radius [6]. In this report, we show the STM/STS observation on Lanthanide doped $Bi_2Sr_2CuO_{6+y}$ (Ln-Bi2201) with different types of Ln (Ln = La, Gd) The two types of Ln samples would sort out the effect of the out-of-plane disorder while keeping the same carrier doping level. We investigate how nanoscale-electronic structures including inhomogeneous features is changing as the function of out-of-plane disorder and T_c.

EXPERIMENTAL

The Ln-Bi2201 (nominally $Bi_2Sr_{1.6}Ln_{0.4}CuO_{6+y}$) single crystals were fabricated by traveling solvent floating zone method [6,7]. The T_c is T_c = 34K for La-doped, T_c = 14K for Gd-doped, respectively. The samples were prepared to have optimum (OP) doping level by oxygen annealing with same condition. For comparison, the optimum doped Bi2212 superconductor (nominally $Bi_{2.1}Sr_{1.9}CaCu_2O_{8+y}$, T_c = 93K) was also investigated. A home-built STM unit with Pt/Ir tip was stored in a low temperature chamber cooled by liquid He. The observation temperature is 9K. The detail of the observing condition is described in Ref. [5].

CP850, *Low Temperature Physics: 24th International Conference on Low Temperature Physics;*
edited by Y. Takano, S. P. Hershfield, S. O. Hill, P. J. Hirschfeld, and A. M. Goldman

RESULTS AND DISCUSSION

Figures 1 (a) and (c) show the topographic images of Ln-Bi2201 ((a) is the La-Bi2201 and (b) is the Gd-Bi2201). The supermodulation structure on the BiO plane is clearly observed in each sample. Figs. 1 (b) and (d) are the Δ–map on the same field of (a) and (c), respectively. The Δ is defined as the half of peak-to-peak voltage in each spectrum data as shown in Fig. 2(a). When the gap peaks were unclear at the occupied-state side, especially in Ln-Bi2201, Δ was defined the peak voltage at the unoccupied (positive bias) side. As shown in Fig. 1 (b) and (d), the spatial inhomogeneous gap distributions are clearly observed in each Ln-Bi2201, as same as Bi2212.

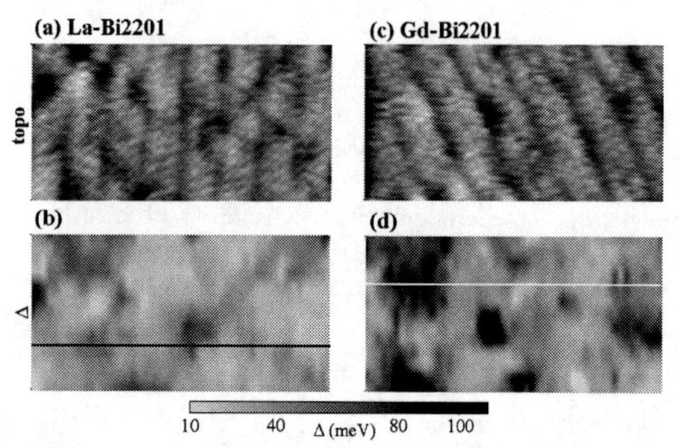

FIGURE 1. (a) Topographic and (b) Δ image of La-Bi2201. (c) Topographic and (d) Δ image of Gd-Bi2201. (20 nm x 10 nm, with the interval of x = 0.4 nm and y = 0.8 nm in Δ images.)

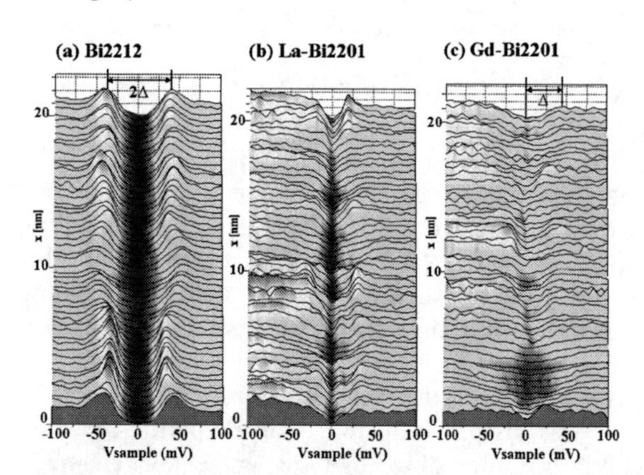

FIGURE 2. The scanning tunneling spectrum along the length of 20nm. (a) Bi2212, (b) La-Bi2201, (along the black line in Fig. 1(b)) (c) Gd-Bi2201. (the white line in Fig. 1(d))

Figs. 2 show the typical conductance spectrum along the length of 20 nm with the interval of 0.4 nm. In the case of Bi2212 (fig. 2 (a)), the spectra show clear peaks and smooth variation of Δ. On the other hand, in the case of the Ln-Bi2201 (b) and (c), spectra show more complex structure and ambiguous peaks. Compared to the spectra on Bi2212 (fig. 2(a)), the remarkable features on Ln-Bi2201 as follows: (1) Δ are widely distributed (10 to 110 meV); (2) The gap coherence peaks are significantly suppressed; (3) The zero bias conductance are enhanced and not uniformly distributed.

TABLE 1. Physical properties of observed samples

Sample	T_c	Δ_{ave}	σΔ
Bi2212 (OP)	93K	42meV	7 meV
La-Bi2201 (OP)	34K	36meV	17meV
Gd-Bi2201 (OP)	14K	48meV	22meV

The distribution of Δ is evaluated by the standard deviation of Δ (σΔ) and average value of Δ (Δ_{ave}). The parameters are listed in Table. 1. As shown in Fig. 1 (d) (Gd-Bi2201), the area with large Δ is broadened toward high energies, compared to La-Bi2201. As the consequence, σΔ of Gd-Bi2201 (22meV) is larger than that of La-Bi2201 (17meV). The PG area with large Δ is enhanced and SC area is destroyed as increasing of out-of-plane disorder. Compared σΔ as a function of T_c, including Bi2212, σΔ is linearly enhanced as decreasing T_c, while Δ_{ave} is not directly related to T_c. The σΔ, as the nanoscale-microscopic parameter of inhomogeneity induced by out-of-plane disorder, is strongly related T_c, as the macroscopic property of superconductors, under the comparison within the OP condition.

SUMMARY

The inhomogeneous Δ structures in Ln-Bi2201 samples are observed by using STM/STS. As increasing the magnitude of out-of-plane disorder, enlargement of deviation of Δ (σΔ) are detected. The out-of-plane disorder enhances pseudogap regions in superconducting state efficiently.

This work was supported by Grant-in-Aid for Japan Society for Promotion of Science (JSPS) Fellows.

REFERENCES

1. K. M. Lang *et al., Nature* **415**, 412 (2002).
2. K. McElroy *et al., Phys. Rev. Lett.* **94**, 197005 (2005).
3. C. Howald *et al., Phys. Rev. B.* **64**, 100504 (2001).
4. A. Matsuda *et al., Physica C* **388**, 207 (2003).
5. A. Sugimoto *et.al., Physica C* **412**, 270 (2004).
6. H. Eisaki *et al., Phys. Rev. B* **69**, 064512 (2004).
7. K. Fujita *et al., Phys. Rev. Lett.* **95,** 097006 (2005).
8. T. Noda *et al., Science* **286**, 265 (1999).

Scanning Tunneling Microscopy/Spectroscopy of Heavily Overdoped $Bi_2Sr_2CuO_y$ Single Crystals

Hideki Mashima [*], Noritaka Fukuo [*], Go Kinoda [†], Taro Hitosugi [**], Toshihiro Shimada [**], Takeshi Kondo [‡], Yoshinori Okada [‡], Hiroshi Ikuta [‡], Yuji Matsumoto [*] and Tetsuya Hasegawa [**]

[*] *FCRC, Tokyo Institute of Technology, Midori-ku, Yokohama 226-8503, Japan*
[†] *KAST, Sakato, Takatsu-ku, Kawasaki, 213-0012, Japan*
[**] *Department of Chemistry, University of Tokyo, Bunkyo-ku, Tokyo 113-0033, Japan*
[‡] *Department of Crystalline Materials Science, Nagoya University, Furo-cho, Chikusa-ku, Nagoya 464-8603, Japan*

Abstract. We have carried out scanning tunneling microscopy/spectroscopy (STM/STS) of heavily overdoped $Bi_2Sr_2CuO_y$ single crystals with T_c<2 K. STS spectra showed a pseudogap structure with dull coherence peaks. Furthermore, we found that the magnitude of pseudogap is spatially non-uniform in a nm scale and varies from 10 to 30 meV. This supports a scenario that the electronic inhomogeneity is a consequence of randomly distributed Coulomb potential in the CuO_2 layers induced by disorder of excess oxygen in the BiO layers.

Keywords: electronic inhomogeneity, Bi2201, STM/STS, pseudogap
PACS: 74.25.Jb

INTRODUCTION

Scanning tunneling spectroscopy (STS) has been extensively used for the elucidation of the pseudogap state in a high temperature superconductor (HTSC), $Bi_2Sr_2CaCu_2O_y$ (Bi2212) in various doping levels [1,2]. From variable temperature STS measurements, it has been clarified that the superconducting gap changes into the pseudogap almost continuously at T_c, and that the pseudogap state survives up to room temperature in underdoped or optimally doped compounds. Furthermore, recent STS studies have shown granular behavior of superconductivity that superconducting and pseudogap-like regions are mixed up in a nanometer scale [3,4,5]. As origin of the electronic inhomogeneity, unscreened Coulomb potential associated with excess oxygen atoms in the BiO layers has been frequently argued [3,5].

It is generally believed that HTSCs in the overdoped regime can be regarded as conventional metals, and less suffer from the electronic inhomogeneity. However, if the inhomogeneity is governed by oxygen disorder, it should be insensitive to carrier concentration itself. In this paper, we present the results of STM/STS measurements on $Bi_{2-x}Pb_xSr_2CuO_y$ (Pb-Bi2201) with x = 0.38 in the heavily overdoped regime. In this compound, carriers are introduced by Pb substitution for Bi, and the amount of excess oxygen is comparable to those of optimally-doped Bi2201. Therefore, it is expected that correlation between electronic inhomogeneity and oxygen disorder could be discussed more definitely.

EXPERIMENTAL

Pb-Bi2201 single crystals with x = 0.38 were grown by the floating-zone method. The as-grown crystals were further subjected to vacuum annealing. The annealed crystals did not show diamagnetic transition down to 2 K.

STM/STS measurements were carried out using an UHV-low temperature STM instrument equipped with a low temperature cleavage stage. The base pressure of the STM chamber was maintained at less than 2×10^{-10} Torr during the measurements. Single crystals were cleaved at 77 K to avoid oxygen loss from the surfaces. STM tips used in this study were mechanically sharpened Pt/Ir wires.

CP850, *Low Temperature Physics: 24th International Conference on Low Temperature Physics;*
edited by Y. Takano, S. P. Hershfield, S. O. Hill, P. J. Hirschfeld, and A. M. Goldman

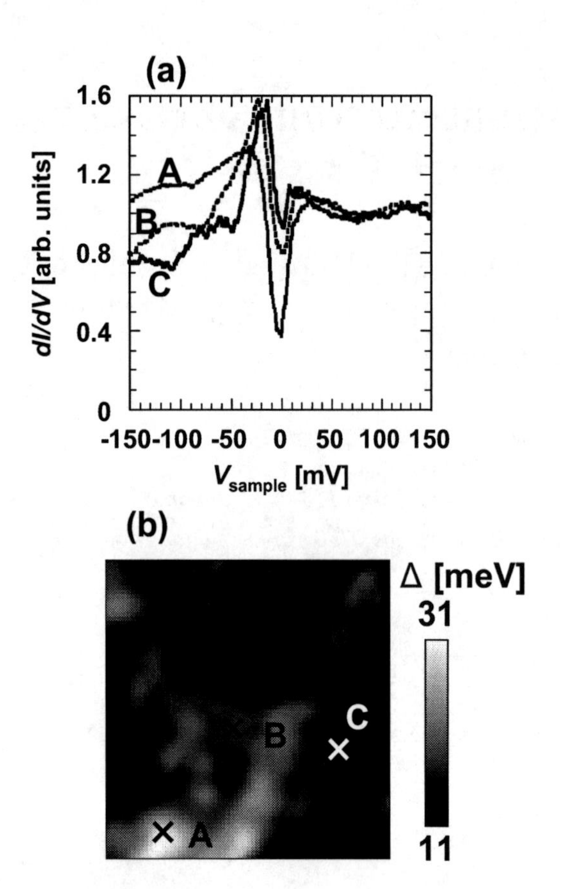

FIGURE 1. STS results of Pb-Bi2201 at 4.3 K (a) Typical tunneling spectra obtained at different locations, (b) Δ map observed on a 10 nm × 10 nm region. The spectra are normalized at 100 mV.

RESULTS AND DISCUSSION

STS results of Pb-Bi2201 are summarized in Fig. 1. Fig. 1(a) shows typical tunneling spectra observed at different spatial locations on a cleaved surface. Remarkably, each spectrum demonstrates a pseudogap structure near zero bias. This implies that carrier concentration is not an important factor for pseudogap formation. From Fig. 1(a), furthermore, it is noted that the magnitude of pseudogap is spatially non-uniform.

Fig. 1(b) is a spatial map of the gap value Δ. Values of Δ were deduced from conductance spectra as half the peak-to-peak energy separation. Cross symbols in the figure indicate the locations where spectra A, B and C in Fig. 1(a) were observed. Notably, smaller and larger gap regions, imaged with brighter and darker contrasts in Fig. 1(b), are smoothly connected by intermediate ones. This behavior quite

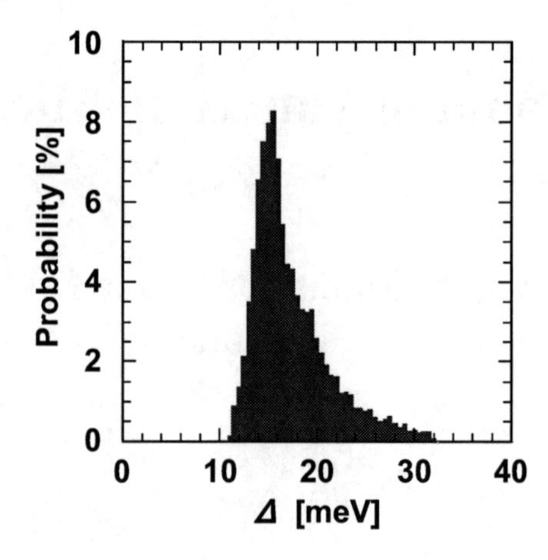

FIGURE 2. Δ histogram calculated from the Δ map in Fig. 1(b).

resembles that of electronic phase separation into superconducting and pseudogap states observed in Bi2212 [3,4].

Fig. 2 is a Δ histogram obtained from the Δ map in Fig. 1(b). As can be seen, pseudogap is distributed in a range of 10-30 meV. We found that the relative standard deviation of Δ, $\delta\Delta/\bar{\Delta}$, increases with increasing excess oxygen. This also is consistent with the local picture assuming disorder-driven inhomogeneity [5].

SUMMARY

We have performed low temperature STM/STS measurements on Pb-doped Bi2201 with T_c<2 K. The pseudogap value Δ evaluated from the STS spectra was found to be spatially non-uniform. This finding leads to a conclusion that electronic inhomogeneity arises from disorder of excess oxygen atoms in the BiO layers, which may induce the distribution of Coulomb potential in the CuO_2 planes.

REFERENCES

1. Ch. Renner *et al.*, *Phys. Rev. Lett.* **80**, 3606 (1998).
2. A. Matsuda *et al.*, *Phys. Rev. B* **60**, 1377 (1999).
3. S. H. Pan *et al.*, *Nature* **413**, 282 (2001).
4. K. M. Lang *et al.*, *Nature* **415**, 412 (2002).
5. G. Kinoda *et al.*, *Phys. Rev. B* **71**, 020502 (2005).

Optimization Of Pb And La Co-Doped $Bi_2Sr_2CuO_{6+\delta}$ Superconductors

Y. Arao*, M. Tange*, H. Ikeda*†, T. Koyano*† and R. Yoshizaki*†

*Faculty of Frontier Science, Graduate School of Pure and Applied Sciences, University of Tsukuba, Tsukuba, Ibaraki 305-8571, Japan

† Cryogenics Division, Research Facility Center for Science and Technology, University of Tsukuba, Tsukuba, Ibaraki 305-8577, Japan

Abstract. We obtained zero-resistance T_c of 43.8 K for the single crystals of Pb and La co-doped $Bi_2Sr_2CuO_{6+\delta}$. The superconductivity with the onset of about 75 K was observed, and the annealing effect was investigated.

Keywords: Bi-2201, high-T_c, doping effect, $Bi_2Sr_2CuO_6$, Pb and La co-doping.
PACS: 74.25.Ha, 74.62.-c, 74.72.Hs, 74.62.Bf

INTRODUCTION

The single layered Bi-2201 material provides one of the best candidates for understanding the mechanism of the superconductivity because of its relatively low T_c and simple crystal structure. In the study of Bi-2201, doping experiments have established general phenomenological trends for understanding the physical properties [1]. The optimized T_c is raised to 38 K with La doping [2] and to 34 K with Pb doping [3] defined by zero-resistance temperature. Recently T_c exceeds 40 K by co-doping of Pb and La in the Bi-2201 phase, $(Bi,Pb)_2(Sr,La)_2CuO_{6+\delta}$ (BPSLCO) [4]. Here we report the anisotropy of the optimal superconductivity of the BPSLCO single crystals and the superconductivity with the onset of about 75 K observed in a several crystals.

EXPERIMENTAL

We prepared two types of $Bi_{2-x}Pb_xSr_{1.9}La_{0.1}CuO_{6+\delta}$ compounds with x=0.3 and 0.4. Single crystals of BPSLCO are grown by a floating zone method in air. Annealing of the sample was carried out at 450 °C for 80 h in flowing Ar or O_2 gas or in air. An attention was paid to avoid the evaporation of Pb in the annealing process. The magnetic moment was measured with using a commercial magnetometer with a reciprocating sample option. The magnetic field was applied parallel to the crystal c axis.

RESULTS AND DISCUSSION

The temperature dependence of the in-plane (ρ_{ab}) and the out-of-plane (ρ_c) resistivity for the as-grown sample (x=0.4) is shown in Fig. 1. The zero-resistance T_c of 43.3 K and 43.8 K with the 10-90% transition width of 4.4 K and 3.5 K were obtained from the ρ_{ab} and ρ_c curves, respectively. The anisotropy parameter

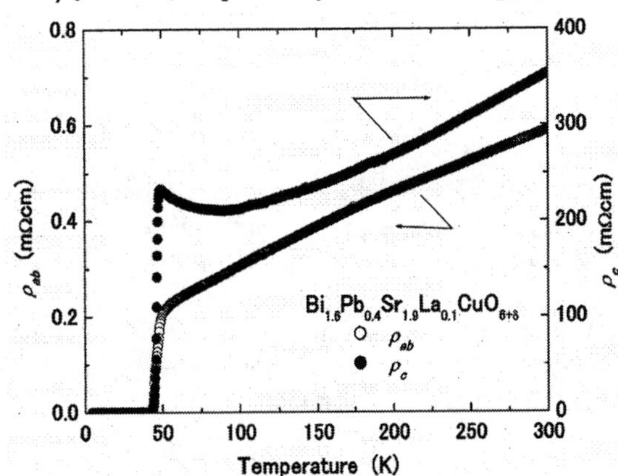

FIGURE 1. Temperature dependence of the in-plane resistivity ρ_{ab} (○) and the out-of-plane one ρ_c (●) for the as-grown sample with x=0.4.

CP850, *Low Temperature Physics: 24th International Conference on Low Temperature Physics;* edited by Y. Takano, S. P. Hershfield, S. O. Hill, P. J. Hirschfeld, and A. M. Goldman

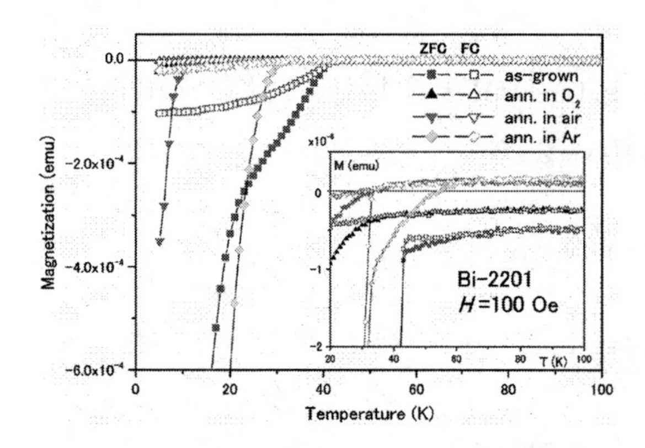

FIGURE 2. Temperature dependence of the magnetization measured for the sample with x=0.3. Inset: the high temperature region is displayed in a magnified scale.

γ was estimated from the ρ_c/ρ_{ab} ratio at 300 K and to be about 25. The reduction of the anisotropy with Pb doping is the normal aspect in the Pb-doped BSCCO materials in Bi-2212 phase [5].

The diamagnetic moment of the superconductivity for an as-grown sample with x=0.3 (thin slab-shape with m=0.51 mg) indicated the onset temperature of about 43 K observed in H=100 Oe as shown in the main panel of Fig. 2, where the solid symbols are the data in ZFC (zero-field-cooled) mode and the open symbols are the ones in FC (field-cooled) mode. T_c of this bulk superconductivity varied with the annealing condition. Bulk superconductivity disappeared by annealing in O_2 due to doping of carriers. T_c restored by reducing the hole carriers with succeeding annealing in air and in Ar after the first annealing in O_2.

Here we observed the higher onset temperature of superconductivity in magnified scales as shown in the inset of Fig. 2. Although the volume of this superconductivity was very small, the irreversibility temperatures were varied with the annealing condition like the bulk superconductivity as mentioned above. The onset temperature of the small superconductivity is estimated to be 75 K for the as-grown sample from the irreversibility line in various magnetic fields. It is noted that the features are reproducible in other samples. We observed the high-T_c superconductivity for the samples from two rods in the five prepared rods with x=0.3.

There are three possible explanations as the origin of this 75 K superconductor observed in the Bi-2201 phase. One is a scenario that the spatial inhomogeneities in the gap magnitude observed in Bi-2212 samples [6] exist also in the Bi-2201 phase, since the inhomogeneity is enhanced in the underdoped region. The superconductivity in the present case is also increased by the annealing in Ar after the succeeding annealing in O_2 and in air. The second scenario is that the modulation–free portions exist as intergrowth layers and induce the high-T_c superconductivity referring to the 85 K superconductivity observed in Tl-2201 phase [7]. The relaxation of the modulation is, however, observed by Pb doping in Bi-2201 [4], but not for the content of excess oxygen. The third one is the possible intergrowth of the 2212 phase in the sample, since La atom is able to occupy at the oxygen deficient A-site in-between the double CuO_2 layers observed in YBCO structure [8]. This possibility was confirmed by preparing the BPSLCO samples from the Bi-2212 composition, but the results did not show the 75 K superconductivity.

In summary, we observed the superconducting properties for the $Bi_{2-x}Pb_xSr_{1.9}La_{0.1}CuO_{6+\delta}$ compounds with x=0.3 and 0.4. The anisotropy parameter was estimated to be 25 from the in-plane and the out-of-plane resistivity at 300 K. We found that the existence of 75 K superconductivity in the Pb an La co-doped Bi-2201 phase, and the annealing effect on the superconductivity was measured.

REFERENCES

1. K. Remschnig et al., *Phys. Rev. B* **43**, 5481-5488 (1991).
2. S. Ono and Y. Ando, *Phys. Rev. B* **67**, 104512- (2003).
3. Z. Jianwu et al., *Supercond. Sci. Technolo.* **14**, 599-602 (2001).
4. T. Amano et al., *Physica C* **412-414**, 230-234 (2004).
5. T. Motohashi et al., *Phys. Rev. B* **59**, 14080-1486 (1999).
6. S. H. Pan et al., *Nature* **413**, 282-285 (2001).
7. Y. Shimakawa et al., *Phys. Rev. B* **40**, 11400-11402 (1989).
8. R. Yoshizaki et al., *Jpn. J. Appl. Phys.* **26**, L1703- L1706 (1987).

Doping Evolution of the Electronic Structure in the Single-layer Cuprate $Bi_2Sr_{2-x}La_xCuO_{6+\delta}$

M. Hashimoto[a], K. Tanaka[b], T. Yoshida[a], A. Fujimori[a], D. H. Lu[b], Z.-X Shen[b], S. Ono[c] and Y. Ando[c]

[a]*Department of Physics and Department of Complexity Science and Engineering, University of Tokyo, Kashiwa, Chiba 277-8561, Japan*
[b]*Department of Applied Physics and Stanford Synchrotron Radiation Laboratory, Stanford University, Stanford, California 94305, U.S.A.*
[c]*Central Research Institute of Electric Power Industry, Komae, Tokyo 201-8511, Japan*

Abstract. The evolution of the electronic structure with hole doping in the single-layer cuprate $Bi_2Sr_{2-x}La_xCuO_{6+\delta}$ (Bi2201) has been studied by angle-resolved photoemission spectroscopy measurements. Results for lightly-doped Bi2201 are compared with the results on other families of cuprate superconductors.

Keywords: Angle-resolved photoemission; High-T_c cuprates.
PACS: 71.30.+h; 74.72.Hs; 79.60.-i

How the electronic structure evolves from the antiferromagnetic insulator to the superconductor with hole doping is a fundamentally important issue in understanding the high-T_c cuprates. Angle-resolved photoemission spectroscopy (ARPES) studies have so far revealed various aspects of the doping evolution of the electronic structure in $La_{2-x}Sr_xCuO_4$ (LSCO) [1-3], $Ca_{2-x}NaSr_xCuO_2Cl_2$ (Na-CCOC) [4-6], and $Bi_2Sr_{2-x}La_xCaCuO_{8+y}$ (Bi2212) [7]. In LSCO, with hole doping, a quasi-particle (QP) appears as "in-gap" states in the nodal direction to form a Fermi arc, spectral weight is transferred from the lower Hubbard band (LHB) to the "in-gap" states for further doping, and the "in-gap" states evolve into the complete Fermi surface for higher doping level [2]. In Bi2212 and Na-CCOC, on the other hand, the LHB is shifted upward upon hole doping and the Fermi arc, where a QP peak crosses E_F, appears where the top of the LHB touches E_F around the node. The question of why there are two types of evolution, however, is still an open issue. It has been shown theoretically that the next-nearest-neighbor hopping integral t' plays an important role in the different behaviors of different cuprate families [8, 9]. t' is primarily influenced by the position of the apical oxygen [9]. A comparative ARPES study of Bi2212 and LSCO has supported this idea [7]. Empirical correlation between t' and the T_c at optimum doping $T_{c,max}$ has also been pointed out [8, 9].

$Bi_2Sr_{2-x}La_xCuO_{6+\delta}$ (La-doped Bi2201) is a Bi-based single-layer high-T_c cuprate system. Thus comparison of Bi2201 with the single layer cuprates LSCO and Na-CCOC would give useful information to understand the systematic changes in the electronic structure among the different families of cuprates. In order to see to which type the evolution of the electronic structure with hole doping Bi2201 belongs, to discuss the relationship between t' and ARPES spectra, and to identify differences from the other cuprates, we have carried out ARPES study of Bi2201 focusing on the lightly-doped region.

High quality single crystals of Bi2201 were grown by the floating zone method [10, 11]. Samples in the lightly-doped to underdoped regions, x = 0.60, 0.80 and 0.96 were prepared and had the hole concentrations of p = 0.12, 0.10 and 0.05, respectively. Only the p = 0.12 sample showed superconductivity with T_c = 17 K. ARPES measurements were performed at beamline 5-4 of Stanford Synchrotron Radiation Laboratory (SSRL) using a SCIENTA SES-200 analyzer with the total energy resolution of 15 meV and the angular resolution of 0.3 degree. Measurements were performed at 10 K. Samples were cleaved *in situ* under an ultrahigh vacuum of 10^{-11} Torr to obtain clean surfaces.

CP850, *Low Temperature Physics: 24th International Conference on Low Temperature Physics;*
edited by Y. Takano, S. P. Hershfield, S. O. Hill, P. J. Hirschfeld, and A. M. Goldman

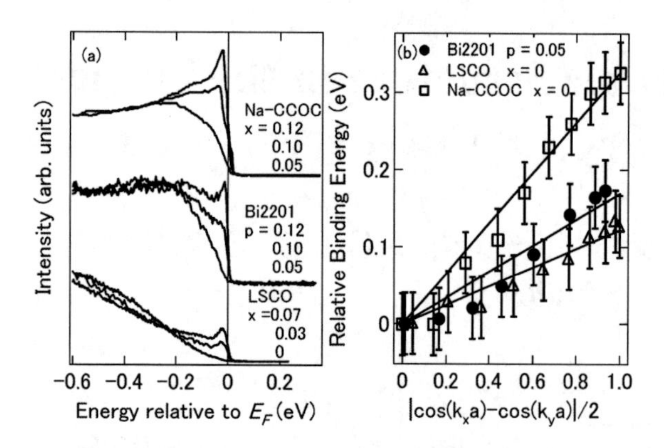

FIGURE 1. Comparison of the ARPES results for Bi2201, Na-CCOC [5], and LSCO [2]. (a) Spectra near $\mathbf{k} = (\pi/2, \pi/2)$, (b) Peak dispersion along the "underlying" Fermi surface as a function of the d-wave order parameter $|\cos k_x a - \cos k_y a|/2$ for undoped or lightly-doped samples.

Comparison of the spectra near $\mathbf{k} = (\pi/2, \pi/2)$ with those of Na-CCOC [5] and LSCO [2] are shown in Fig. 1(a). In the spectra for $p = 0.05$ and 0.10, one can hardly see spectral weight at E_F, corresponding to the insulating behavior in the lightly-doped region. This is similar to Na-CCOC and contrasted to LSCO. In the spectrum for $p = 0.05$, there is a single dispersive peak around ~ -0.4 eV corresponding to the LHB, which becomes closest to E_F near $(\pi/2, \pi/2)$. One can clearly see that the LHB moves towards E_F with hole doping. In the spectrum for $p = 0.12$, which shows superconductivity, one can see a QP peak crossing E_F. In LSCO, the LHB stays at ~ -0.4 eV and a QP peak appears at E_F upon hole doping in contrast to Bi2201 and Na-CCOC.

Figure 1(b) shows the band dispersion along the "underlying" Fermi surface as a function of the d wave order parameter $|\cos k_x a - \cos k_y a|/2$ for undoped ($x = 0$) LSCO [2] and Na-CCOC [4], and Bi2201 for the most lightly-doped ($p = 0.05$) samples. From this comparison, one can see that the slope for Bi2201 is intermediate between those for LSCO and Na-CCOC. Because the slope is considered to represent the magnitude of the next-nearest-neighbor hopping t' [12], one can say that the magnitude of t' of Bi2201 is larger than that of LSCO and is smaller than that of Na-CCOC, indicating a systematic increase of t' from LSCO to Bi2201 to Na-CCOC. This agrees with the theoretical prediction [9] that the magnitude of t' increases with decreasing influence of the apical oxygen, because the Cu-apical oxygen distance is larger in Bi2201 than in LSCO and Na-CCOC does not have an apical oxygen.

In summary, the ARPES study of Bi2201 shows the doping evolution similar to Na-CCOC. The band dispersion along the "underlying" Fermi surface shows intermediate behavior between LSCO and Na-CCOC, which indicates that the next-nearest-neighbor hopping t' is smaller than that for Na-CCOC but is larger than that for LSCO. This agrees with the behavior predicted by the theoretical calculation that the magnitude of t' increases with decreasing influence of the apical oxygen.

This work was supported by the US-Japan joint Research Program from JSPS and a Grant-in-Aid for Scientific Research in Priority Area "Invention of Anomalous Quantum Materials" from MEXT, Japan. SSRL is operated by the Department of Energy's Office of Basic Energy Science, Division of Chemical Sciences and Material Sciences.

REFERENCES

1. A. Ino, C. Kim, M. Nakamura, T. Yoshida, T. Mizokawa, Z.-X. Shen, A. Fujimori, T. Kakeshita, H. Eisaki, and S. Uchida, *Phys. Rev. B* **62**, 4137 (2000).
2. T. Yoshida, X.J. Zhou, T. Sasagawa, W.L. Yang, A. Fujimori, H. Eisaki, Z.-X. Shen, T. Kakeshita, and S. Uchida, *Phys. Rev. Lett.* **91**, 027001 (2003).
3. A. Ino, T. Mizokawa, A. Fujimori, K. Tamasaku, H. Eisaki, S. Uchida, T. Sasagawa, and K. Kishio, *Phys. Rev. Lett.* **79**, 2101 (1997).
4. F. Ronning, T. Sasagawa, Y. Kohsaka, K.M. Shen, A. Damascelli, C. Kim, T. Yoshida, N. P. Armitage, D. H. Lu, D.L. Feng, L.L. Miller, H. Takagi, and Z.-X. Shen, *Science* **282**, 2067 (1998).
5. K. M. Shen, F. Ronning, D.H. Lu, W.S. Lee, N.J.C. Ingle, W. Meevasana, F. Baumberger, A. Damascelli, N.P. Armitage, L.L. Miller, T. Kohsaka, M. Azuma, M. Takano, H. Takagi, and Z.-X. Shen, *Phys. Rev. Lett.* **93,** 267002 (2004) .
6. Y. Kohosaka, T. Sasagawa, F. Ronning, T. Yoshida, C. Kim, T. Hanaguri, M. Azuma, M. Takano, Z.-X. Shen, and H. Takagi, *J. Phys. Soc. Jpn.* **72**, 1018 (2003).
7. K. Tanaka, T. Yoshida, A. Fujimori, D.H. lu, Z.-X. Shen, X.-J. Zhou, H. Eisaki, Z. Hussain, S. Uchida, Y. Aiura, K. Ono, T. Sugaya, T. Mizuno, and I. Terasaki, *Phys. Rev. B* **70**, 092503 (2004).
8. R. Raimondi, J. H. Jeferson, and L. F. Feiner, *Phys. Rev. B* **53**, 8774 (1996).
9. E. Pavarini, I. Dasgupta, T. Saha-Dasgupta, O. Jepsen, and O. K. Andersen., *Phys. Rev. Lett.* **87**, 047003 (2001).
10. S. Ono and Y. Ando, *Phys. Rev. B* **67**, 104512 (2003).
11. Y. Ando, Y. Hanaki, S. Ono, T. Murayama, K. Segawa, N. Miyamoto, and S. Komiya, *Phys. Rev. B* **61**, R14956 (2000).
12. C. Kim, P.J. White, Z.-X. Shen, T. Tohyama, Y. Shibata, S. Maekawa, B.O. Wells, Y.J. Kim, R.J. Birgeneau, and M. A. Kanster, *Phys. Rev. Lett.* **80**, 4245 (1998).

Local Enhancement of Zero Bias Conductance in $Bi_2Sr_{1.6}Nd_{0.4}CuO_y$

T. Machida, N. Tsuji, Y. Kamijo, K. Harada, T. Kato, R. Saito, and H. Sakata

Department of physics, Tokyo University of Science, 1-3Kagrazaka, Shinjuku-ku, Tokyo, Japan

Abstract. Low temperature scanning tunneling microscopy and spectroscopy on $Bi_2Sr_{1.6}Nd_{0.4}CuO_y$ (T_c=26.5 K) at 4.2 K, reveals the existence of a region with locally enhanced zero bias conductance whose radius is about 10 Å. This enhancement is due to an additional state appears at the energy around -5meV. It is thought that this local enhancement of ZBC arises from an impurity or a defect in the CuO plane.

Keywords: Scanning tunneling microscopy, Scanning tunneling spectroscopy, cuprate.
PACS: 74.72.Hs, 74.50.+r.

INTRODUCTION

A number of scanning tunneling microscopy and spectroscopy (STM/STS) measurements has been performed on $Bi_2Sr_2CaCu_2O_y$ (Bi2212). Since the STM/STS experiments have the ability to investigate the local electronic structure at atomic scale, the local enhancement of zero bias conductance (ZBC) which arises from the vortex core state [1-3], inhomogeneous electronic structure [4,5] and electronic structure perturbed by impurity in CuO plane [6-9] has been observed in Bi2212. There are few reports with respect to the electronic structure in $Bi_2Sr_2CuO_y$ (Bi2201) having the most fundamental structure in Bi based cuprates. In this study, we have performed STM measurement on the cleaved surface of $Bi_2Sr_{1.6}Nd_{0.4}CuO_y$ (Bi2201-Nd) (T_c=26.5 K) at 4.2 K and observed the local enhancement of ZBC.

EXPERIMENT

The sample used in this report is $Bi_2Sr_{1.6}Nd_{0.4}CuO_y$ (Bi2201-Nd) single crystal grown by floating zone method and annealed at 750°C in air. This crystal has superconducting transition temperature (T_c) of 26.5 K characterized by SQUID magnetmeter.

We have performed STM/STS experiments immediately after the sample cleavage in He gas at 4.2 K. The Pt-Ir (80%-20%) wire was mechanically shaped as STM tip which was perpendicular to the exposed *ab*-plane. The bias voltage was applied to the sample, namely the negative and positive bias corresponds to the occupied and unoccupied states, respectively. The STM measurements were performed in the constant current mode. The tunneling spectra (dI/dV) were obtained by the numerical differentiation of the *I-V* characteristics.

RESULTS AND DISCUSSION

Fig. 1a shows a topograph of the sample surface. Since the cleavage is achieved between the two adjacent BiO layers, this topography is demonstrating the BiO plane. We have performed the STS measurement on this surface. Fig. 1b shows the observed tunneling spectra at the point A and B in Fig. 1a. The spectrum at point B reveals the superconducting gap about 15 meV with rather low ZBC. Fig. 1c shows the ZBC map in gray scale. In overall region, ZBC shows the almost same value as that of point B, except near point A. The spectrum at point A shown in Fig. 1b indicates the ZBC whose magnitude is about 2.5 times greater than that of the spectrum at point B.

CP850, *Low Temperature Physics: 24th International Conference on Low Temperature Physics;*
edited by Y. Takano, S. P. Hershfield, S. O. Hill, P. J. Hirschfeld, and A. M. Goldman

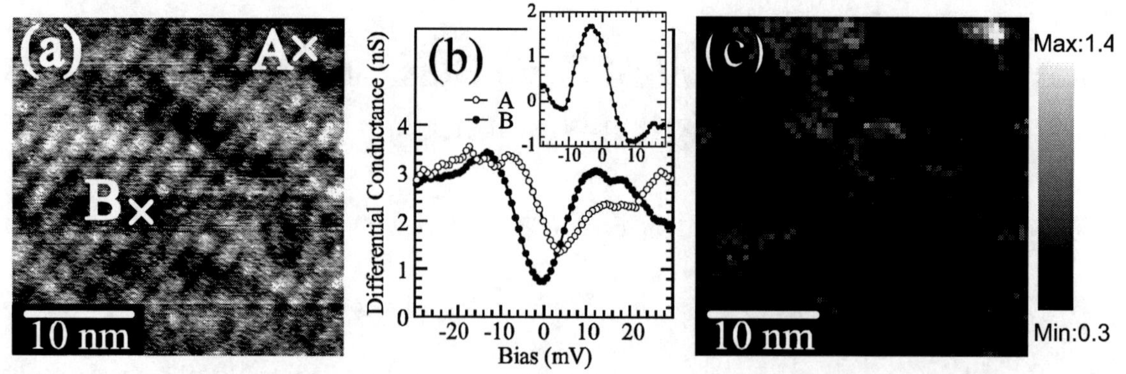

FIGURE 1. (a) Topographic image (32×32 nm) of the surface BiO plane which is exposed after cleavage of a Bi2201-Nd single crystal (T = 4.2 K, I = 172 pA, V_{sample}= 300 mV). (b) Tunneling spectra taken at the point A and B shown in (a) (open and filled circle, respectively). Inset is difference between two spectra. (c) A zero bias conductance map (32×32 nm) taken on the same surface shown in (a).

The length scale of the region corresponding to the extremely bright region is about 10 Å in diameter in Fig. 1c. The difference between the local density of states (LDOS) at point A and B, shown in inset of Fig. 1b, indicates that this ZBC enhancement is due to the extra density of states in negative bias around –5 meV.

In the STM experiment on Bi2212 substituting certain impurity atoms for copper atoms, an additional state with the length scale about 10 Å have been observed in the tunneling spectrum at an impurity site [6-9]. Since the additional state observed in this study is similar to that reported in Bi2212 qualitatively, thus it is thought that this phenomena is due to the defect or impurity in CuO_2 plane. In this study, impurities are not substituted for copper atoms in the sample intentionally, however, strontium atom is substituted by neodymium atom. It is convincing that neodymium ion substituted for copper atom in CuO_2 plane. Thus this additional state may due to the impurity resonance at neodymium atom in CuO_2 plane.

CONCLUSION

We have performed scanning tunneling microscopy measurement on the surface in $Bi_2Sr_{1.6}Nd_{0.4}CuO_y$ (T_c =26.5 K) at 4.2 K and observed the region with the radius of about 10 Å showing a local enhancement of zero bias conductance (ZBC). An additional state appears around –5 meV. This behavior is consistent with that in $Bi_2Sr_2CaCu_2O_y$ substituting a nonmagnetic impurity or defect for Cu atom qualitatively. It is thought that this local enhancement of ZBC arises from an impurity or defect in the CuO_2 plane.

REFERENCES

1. Ch. Renner, B. Revaz, K. Kadowaki, I. Maggio-Aprile, and Ø. Fischer, *Phys. Rev. Lett.* **80**, 3606 (1998).
2. S. H. Pan, E. W. Hudson, A. K. Gupta, K.-W. Ng, H. Eisaki, S. Uchida, and J. C. Davis, *Phys. Rev. Lett.* **85**, 1536 (2000).
3. K. Matsuba, H. Sakata, N. Kosugi, H. Nishimori, and N. Nishida, *J. Phys. Soc. Jpn.* **72**, 2153 (2003).
4. K. M. Lang, V. Madhavan, J. E. Hoffman, E. W. Hudson, H. Eisaki, S. Uchida, and J. C. Davis, *Nature* **415**, 412 (2002).
5. S. H. Pan, J. P. O'Neal, R. L Badzey, C. Chamon, H. Ding, J. R. Engelbrecht, Z. Wang, H. Eisaki, S. Uchida, A. K. Gupta, K.-W. Ng, E. W. Hudson, K. M. Lang, and J. C. Davis, *Nature* **413**, 282 (2001).
6. H. W. Hudson, K. M. Lang, V. Madhavan, S. H. Pan, H. Eisaki, S. Uchida, and J. C. Davis, *Nature* **411**, 920 (2001).
7. S. H. Pan, E. W. Hudson, K. M. Lang, H. Eisaki, S. Uchida, and J. C. Davis, *Nature* **403**, 746 (2000).
8. Ali. Yazdani, C. M. Howald, C. P. Lutz, A. Kapitulnik, and D. M. Eigler, *Phys. Rev. Lett.* **83**, 176 (1999).
9. H. W. Hudson, S. H. Pan, A. K. Gupta, K. -W. Ng, and J. C. Davis, *Science* **285**, 88 (1999).

Crystal Structure Features of HTSC Cuprates and Relative AF Phases

Svetlana G. Titova[a], Stepan V. Pryanichnikov[a], Olga M. Fedorova[a], Vladimir F. Balakirev[a], and Ivan Bobrikov[b]

[a]*State Institute of Metallurgy UrD RAS, Amundsen St. 101, 620016 Ekaterinburg, Russia*
[b]*Frank Laboratory of Neutron Physics, JINR, 141980 Dubna, Russia*

Abstract. We report X-ray and neutron diffraction study results for HTSC cuprates $Hg_{0.8}Tl_{0.2}Ba_2CaCu_2O_x$, $Bi_2Sr_2CaCu_2O_y$ and compare them with earlier studied AF phases CuO and Y_2BaCuO_5 in temperature range 100 - 300 K. For both HTSC systems at concentration of charge carriers near optimal doping level there is a local minimum of unit cell volume at $T_2 \approx 260$ K, which we explain as a state with strongest charge carrier localization and strongest lattice distortion owing to polaron band collapse. For both HTSC systems the inhomogeneous state has been observed between $T_1 \approx 160$ K and T_2. Obtained results are explained as instability of electronic homogeneous state during polaron band collapse, when polaron band is less than half-filled.

Keywords: Cuprates, Polaron, Polaron Band Collapse.
PACS: 74.72, 74.62.B, 64.75, 71.38

INTRODUCTION

A number of physical properties (ultrasonic attenuation, heat capacity and crystal structure) show similar features at $T_1 \sim 160$ K and $T_2 \sim 260$ K for HTSC cuprates and relative antiferromagnetic phases with Cu^{2+} ions as CuO and Y_2BaCuO_5 [1]. As the temperature of these features is a function of neither of charge carrier concentration nor the type of crystal structure, we have concluded that they are determined by a short range state of CuO-clusters and connected with polaron charge carriers. To investigate the origin of T_1 and T_2 anomalies we have performed X-ray and neutron diffraction study of HTSC cuprates $Hg_{0.8}Tl_{0.2}Ba_2CaCu_2O_x$ (I) and $Bi_2Sr_xCaCu_2O_y$ (II) at temperature range 100-300 K.

EXPERIMENTAL

Ceramic sample synthesis for (I) and (II) systems is described in refs. [2, 3]. X-ray diffraction study was performed using Ni-filtered Cu Kα radiation, internal standard SiO_2. Time of flight neutron diffraction data were collected as a function of temperature by the fixed angle backscattering detector at the HRPD instrument (ISIS, RAL) for (I) and in HRFD machine (IBR-2, FLNP JINR) for (II). The GSAS package [4] was used for crystal structure determination. The structural parameters were refined with $R_{wp} \sim$ 2-8 %, χ^2 ~1-3.

For studied materials (I) and (II), as well as for earlier investigated CuO and Y_2BaCuO_5, a slop of temperature dependence of unit cell parameters was observed at T_2 ~260 K. For HTSC near optimally doped phases a local minimum of the cell volume has been observed at T_2 [3,5]. In this case lattice volume increases on cooling for temperature range T_1-T_2. Moreover, in this temperature range both neutron and X-ray diffraction peaks demonstrate a broadening, see Figures 1 and 2. We explain this broadening as an appearance of an inhomogeneous state with small variation of lattice parameters (and charge carrier concentration).

For interpretation of physical properties of layered intercalation compounds based on titanium dichalcogenides the concept of inhomogeneous state appearance in case of polaron band collapse, when the Fermi level is situated near the bottom of the band, has been worked out [6,7]. Narrowing of the polaron band leads to formation of two fractions; for one of them Fermi level is situated near the middle of the band,

CP850, *Low Temperature Physics: 24th International Conference on Low Temperature Physics;*
edited by Y. Takano, S. P. Hershfield, S. O. Hill, P. J. Hirschfeld, and A. M. Goldman

while second fraction is insulating [6,7]. We interpret observed features for HTSC cuprates using the same concept; when T_2 corresponds to polaron band collapse – strongest electron-lattice interaction, strongest lattice distortion and lowest charge carrier mobility.

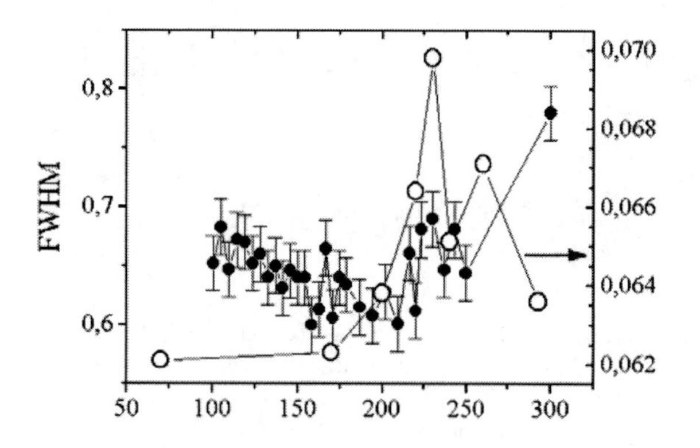

FIGURE 1. The diffraction line width (FWHM) as a function of temperature for $Bi_2Sr_2CaCu_2O_y$; the data for X-rays (full circles, Δ2ϑ, deg.) and neutron scattering (open circles, Δd, nm).

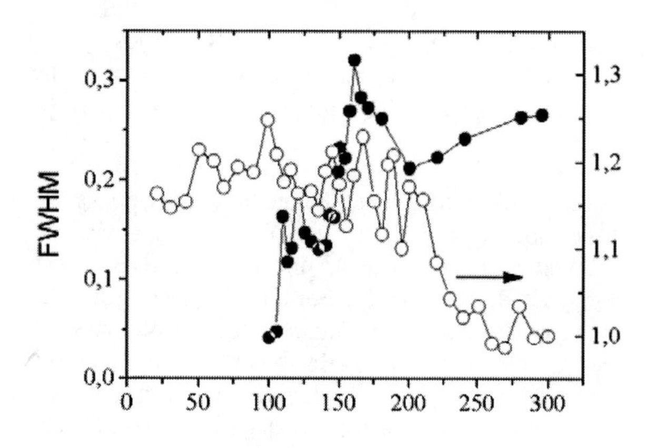

FIGURE 2. The diffraction line width (FWHM) as a function of temperature for $Hg_{0.8}Tl_{0.2}Ba_2CaCu_2O_x$; the data for X-rays (full circles, Δ2ϑ, deg.) and neutron scattering (open circles, Δd, rel. units).

In temperature range T_1-T_2 the system becomes inhomogeneous, as we observe from reflex broadening. Inhomogeneous state is a common feature and should be visible from other experiments. For example, in the same temperature range, the NMR lines for relative Pb,Cu-1212 HTSC cuprates demonstrate significant broadening due to the same reason [8]. As this inhomogeneity is similar for (I) and (II) cuprates, it could not be connected with structural modulations, typical for Bi-based HTSC compounds.

Both negative thermal expansion and maximum of FWHM in temperature range T_1-T_2 may be connected with redistribution of fraction volume, when cooling leads to a growth of the fraction with higher lattice volume. For both (I) and (II) systems, increase of lattice volume is almost linearly connected with reduction of charge carrier concentration [2,3]. If the density of state at the Fermi level for the fraction with Fermi level situated at the middle of the polaron band increases on cooling, it will lead to a growth of the insulating fraction. At T_1 point we will have the final situation with narrow polaron band and high density of states at Fermi level for "optimal" fraction and relatively large quantity of "insulating" fraction.

We would like to express our gratitude to Prof. Ingrid Bryntse, Stockholm University, for high quality ceramic samples of (I) system. We thank Prof. J.T.S. Irvine (St. Andrews University, Scotland) and Dr. K. Knight (ISIS, RAL) for the help with neutron diffraction experiment and ISIS for the possibility to perform high-resolution neutron diffraction. The work is supported by RFBR, grant NSc-468.2003.3.

REFERENCES

1. S. G. Titova, A. N. Titov, D. O. Shorikov, D. I. Kochubey et al., *Crystallogr. (rus.)* **47**, 934-938 (2002).
2. S. Titova, I. Bryntse, J. Irvine, B. Mitchell and V. Balakirev, *Journal of Superconductivity*, **11**, 471-480 (1998).
3. O. M. Fedorova, S. G. Titova, A. M. Yankin, and V.F. Balakirev, *Izvestia RAN. Ser. Phys. (rus.)*, **69**, 1050-1054 (2005).
4. A .C. Larson and R. B. Von Dreele LANSCE, MS-H805. Los Alamos National Laboratory, Los Alamos, 1986. NM 875.
5. S. G. Titova, V. F. Balakirev, Y. Ohishi, I. Bryntse, and D. I. Kochubey, *Physica C,* **388-389**, 215-216 (2003).
6. A. N. Titov and A. V. Dolgoshein, *Phys. Solid State*, **40**, 1081-1083 (1998).
7. A. N. Titov, Yu. M. Yarmoshenko, S. G. Titova, L. S. Krasavin, and M. Neumann, *Physica B,* **328**, 108-110 (2003).
8. D. P. Tunstall, S. G. Titova, J. T. S. Irvine and B. J. Mitchell, *Phys.:Condens. Matter.*, **10**, 2539-2550 (1998).

The Phase-sensitive c-Axis Twist Experiments on $Bi_2Sr_2CaCu_2O_{8+\delta}$ and Their Implications

Richard A. Klemm

Department of Physics, Kansas State University, Manhattan, KS 66506, USA

Abstract. There are presently three sets of c-axis twist experiments on $Bi_2Sr_2CaCu_2O_{8+\delta}$ (Bi2212): The bicrystal experiments of Li *et al.*, the artificial cross-whisker experiments of Takano *et al.*, and the natural cross-whisker experiments of Latyshev *et al.*. We summarize these experiments and the extensive theoretical analyses of their possible implications. All three experiments can only be understood in terms of a substantial s-wave superconducting order parameter component for $T \leq T_c$. The Li and Latyshev experiments are also consistent most other data that the c-axis tunneling is strongly incoherent.

Keywords: superconductivity, cuprate, order parameter symmetry, incoherent tunneling
PACS: 74.50.+r, 74.60.Jg, 74.72.Hs, 74.80.Dm

INTRODUCTION

There has long been a raging debate over the orbital symmetry of the superconducting order parameter (OP) in the high temperature superconductors [1]. Here we summarize the latest phase-sensitive tests of OP symmetry from three c-axis twist experiments using $Bi_2Sr_2CaCu_2O_{8+\delta}$ (Bi2212), the subject of a recent review [2].

BICRYSTAL TWIST JOSEPHSON JUNCTIONS

Li *et al.* made extraordinarily perfect Bi2212 bicrystal junctions twisted an angle ϕ_0 about the c axis [3]. These junctions were extensively characterized using high resolution transmission electron microscopy (HRTEM), electron energy loss spectroscopy, and low energy electron diffraction, etc. [4], and more recently, by off-axis electron holography [5], providing compelling evidence for their remarkable atomic perfection. Li *et al.* measured near to the transition temperature T_c the critical current densities J_c^J and J_c^S across the twist junction and constituent single crystals for 12 samples with different ϕ_0 values. They found $J_c^J(\phi_0)/J_c^S = 1$ at $T = 0.9T_c$ [3]. In Fig. 1, the solid diamonds represent $\log_{10}[KJ_c^J(\text{A/cm}^2)]$ for these junctions, where $K = 4.00$ to simulate the low temperature T values, $J_c(T=0)/J_c(0.9T_c) = 4.00$ [6]. From extensive theoretical analysis, the c-axis quasiparticle tunneling is strongly incoherent, and an s-wave OP component completely dominates J_c^J for all $T \leq T_c$ [3, 7, 8, 9].

ARTIFICIAL CROSS-WHISKER JUNCTIONS

Takano *et al.* sintered two Bi2212 whiskers with insulating surfaces together, obtaining artificial cross-whisker junctions (ACWJ's) [10]. The measured $J_c^J(\phi_0)$ at 4.2K varies substantially with ϕ_0, but is flat for $30° \leq \phi_0 \leq 60°$, as shown by the open circles in Fig. 1 [10]. Fits to these data indicated that the OP at 4.2 K is most likely predominantly s-wave, but the c-axis quasiparticle tunneling would have to be coherent [11]. However, the current-voltage ($I-V$) characteristics showed irregular branching, suggesting a complicated recrystallization of the interface region.[10, 12] The larger junction resistance $R(T, 60°)$ for $T_c \leq T \leq 105$ K than above 105 K also suggests an extra insulating interface barrier [11, 12]. More recently, the $J_c^J(45°)$ data on several junctions are non-vanishing, and consistent with each other [13]. In addition, HRTEM for two ACWJ's indicated flat junctions, and Fraunhofer and Shapiro step analyses for one $\phi_0 = 45°$ ACWJ demonstrated the tunneling to be weak and first order [11, 13]. $J_c^J(\phi_0)$ was also independent of T for $4.2\text{K} \leq T \leq 60\text{K}$ [13], ruling out any possible d-wave interpretation based upon the assumption of strong tunneling and Andreev scattering at the twist junction [14], as noted recently [2, 15]. These combined experiments provide strong supporting evidence that the OP is predominantly s-wave for $T \leq 60$K [11, 13], and that the strong $J_c^J(\phi_0)$ dependence outside $30° \leq \phi_0 \leq 60°$ is most likely extrinsic [2, 11].

CP850, *Low Temperature Physics: 24th International Conference on Low Temperature Physics;*
edited by Y. Takano, S. P. Hershfield, S. O. Hill, P. J. Hirschfeld, and A. M. Goldman

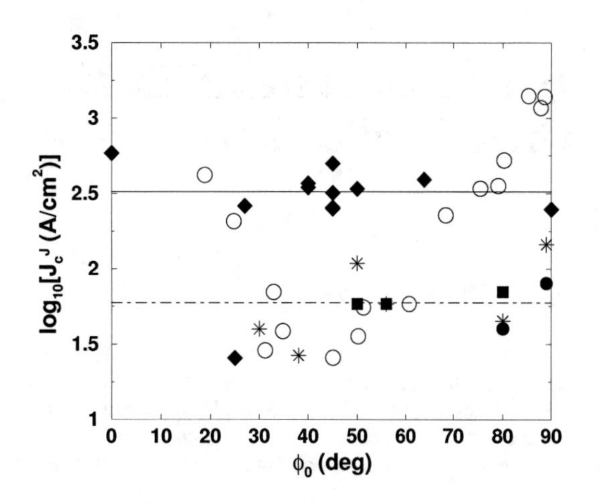

FIGURE 1. A comparison of $\log_{10}[J_c^J(\text{A/cm}^2)]$ versus ϕ_0 data at 4.2 K of the bicrystal data multiplied by 4.00 as described in the text,[2] (solid diamonds), with the ACWJ data,[9] (open circles), and the NCWJ data [13], measured with methods (a) (stars), (b) (solid circles), and (c) (solid squares). Reprinted with permission from Philosophical Magazine, http://www.tandf.co.uk.

NATURAL CROSS-WHISKER JUNCTIONS

Latyshev *et al.* annealed naturally-formed Bi2212 cross whiskers, obtaining natural cross-whisker junctions (NCWJ's) [16]. The measured room temperature resistance per square of the NCWJ's $R_{\text{sq}}^J(\phi_0)$ and intrinsic junctions R_{sq}^S in bulk Bi2212 satisfy $R_{\text{sq}}^J(\phi_0)/R_{\text{sq}}^S \approx 60$, independent of ϕ_0, as are the $J_c^S/J_c^J(\phi_0) \approx 40$ at 4.2K. $J_c^J(\phi_0) = I_c^J(\phi_0)/S$ was measured in three ways: (a) from its zero magnetic induction B value, (b) from the oscillatory part of the Fraunhofer-like $I_c(B)$, and (c) from the sharp rise in the $I-V$ characteristics at $eV = 2\Delta$. These data are presented as the stars, solid circles, and solid squares in Fig. 1, respectively. The ϕ_0-independent J_c^J data of Latyshev *et al.* are consistent with those of Takano *et al.* for $30° \leq \phi_0 \leq 60°$. $J_c^J(T,\phi_0)/J_c^J(4.2\text{K},\phi_0)$ for 4.2K$\leq T \leq$ 68K were about the same for $\phi_0 = 38°, 86°$. As a minimum, the Latyshev *et al.* data provide strong evidence that the tunneling at the weak junctions is incoherent, allowing at least a small (≥ 0.03) s-wave relative fraction of the OP to pass. However, since $J_c^J(\phi_0)R_{\text{sq}}^J(\phi_0) \approx J_c^S R_{\text{sq}}^S$, independent of ϕ_0, the Latyshev *et al.* data are also consistent with the Li *et al.* data [3, 16], and with a predominant s-wave OP for $T \leq$ 68K$\approx T_c$ [2].

CONCLUSIONS

The combined Bi2212 twist experiments provide strong evidence for a substantial s-wave OP component for $T \leq T_c$, and that the c-axis quaisparticle transport is incoherent.

ACKNOWLEDGMENTS

The author would like to thank Yu. Latyshev, Q. Li, K. Scharnberg, M. Tachiki, and Y. Takano for helpful discussions.

REFERENCES

1. K. A. Müller, *Phil. Mag. Lett.* **82**, 279 (2002).
2. R. A. Klemm, *Phil. Mag.* **85**, 801 (2005).
3. Qiang Li, Y. N. Tsay, M. Suenaga, R. A. Klemm, G. D. Gu, and N. Koshizuka, *Phys. Rev. Lett.* **83**, 4160 (1999).
4. Y. Zhu, Q. Li, Y. N. Tsay, M. Suenaga, G. D. Gu, and N. Koshizuka, *Phys. Rev. B* **57**, 8601 (1998).
5. M. A. Schofield, L. Wu, and Y. Zhu, *Phys. Rev. B* **67**, 224512 (2003).
6. V. Ambegaokar and A. Baratoff, *Phys. Rev. Lett.* **10**, 486 (1963); *ibid.* **11**, 104 (1963).
7. R. A. Klemm, C. T. Rieck, and K. Scharnberg, *Phys. Rev. B* **58**, 1051 (1998); *ibid.* **61**, 5913 (2000).
8. A. Bille, R. A. Klemm, and K. Scharnberg, *Phys. Rev. B* **64**, 174507 (2001).
9. G. B. Arnold and R. A. Klemm, *Phys. Rev. B* **62**, 661 (2000).
10. Y. Takano, T. Hatano, M. Ohmori, S. Kawakami, A. Ishii, S. Arisawa, S.-J. Kim, T. Yamashita, K. Togano, and M. Tachiki, *J. Low Temp. Phys.* **131**, 533 (2003).
11. R. A. Klemm, *Phys. Rev. B* **67**, 174509 (2003).
12. Y. Takano, T. Hatano, A. Fukuyo, M. Ohmori, P. Ahmet, T. Naruke, K. Nakajima, T. Chikyow, K. Ishii, S. Arisawa, K. Togano, and M. Tachiki, *Sing. J. Phys.* **18**, 67 (2002).
13. Y. Takano, T. Hatano, S. Kawakami, M. Ohmori, S. Ikeda, M. Nagao, K. Inomata, K. S. Yun, A. Ishii, A. Tanaka, T. Yamashita, and M. Tachiki, *Physica C* **408-410**, 296 (2004).
14. K. Maki and S. Haas, *Phys. Rev. B* **67**, 020510 (2003).
15. G. B. Arnold, R. A. Klemm, W. Körner, and K. Scharnberg, *Phys. Rev. B* **68**, 226501 (2003).
16. Yu. I. Latyshev, A. P. Orlov, A. M. Nikitina, P. Monceau, and R. A. Klemm, *Phys. Rev. B* **70**, 094517 (2004).

STM/STS Study on 4a×4a Electronic Charge Modulation of $Bi_2Sr_2CaCu_2O_{8+\delta}$

A. Hashimoto, Y. Kobatake, S. Nakamura, N. Momono, M. Oda and M. Ido

Department of Physics, Hokkaido University, Sapporo 060-0810, Japan

Abstract. We performed low-bias STM measurements on lightly doped Bi2212 crystals, and confirmed the development of two-dimensional (2D) superstructure with a periodicity of four lattice constants (4a) within the Cu-O layer at $T < T_c$. This 4a×4a superstructure orients along the Cu-O bonding direction, and is nondispersive. The nondispersive 4a×4a superstructure was clearly observed within the ZTPG or d-wave gap, while it tended to fade out outside the gaps.

Keywords: STM/STS, pseudogap, electronic charge ordering, 4a×4a superstructure, Bi2212
PACS: 74.25.Jb, 74.50.+r, 74.72.Hs

Recently STM/STS studies on the pseudogap state of $Bi_2Sr_2CaCu_2O_{8+\delta}$ (Bi2212) at $T > T_c$ have revealed a nondispersive ~ 4a×4a superstructure[1]. The nondispersive 4a×4a superstructure, electronic in origin, was also reported in the LDOS maps taken in the zero temperature pseudogap (ZTPG) regimes of lightly doped $Ca_{2-x}Na_xCuO_2Cl_2$ (Na-CCOC) and Bi2212 samples[2, 3]. A similar 4a×4a superstructure was first observed around the vortex cores of Bi2212 exhibiting a pseudogap-like V-shaped spectrum with no coherence peak[4]. Such 4a×4a superstructure has attracted much attention because this superstructure can be a possible electronic hidden-order in the pseudogap state. Furthermore, Fang *et al.* and Howald *et al.* found a nondispersive ~ 4a×4a superstructure with anisotropy in the superconducting (SC) state of Bi2212 in addition to a dispersive one[5, 6], which was first found by Hoffman *et al.* and McElroy *et al.* and has been successfully explained in terms of a quasiparticle scattering interference[7, 8]. However, the nondispersive ~ 4a×4a superstructure was not confirmed in later LDOS measurements on Bi2212 at $T < T_c$[1, 7].

In the present study, we performed low-bias STM measurements on lightly doped Bi2212 samples, and succeeded in observing nondispersive 4a×4a electronic superstructure with a substructure having a periodicity of 4a/3. The present 4a×4a structure clearly appears within the ZTPG or d-wave paring gap.

The single crystal of Bi2212 in the present study was grown by the traveling solvent floating zone method. We performed STM/STS experiments at $T \sim 9$ K on two samples A and B cut from the same single crystal with $T_c \sim 72$ K ($p \sim 0.11$). In the present STM/STS experiments, the sample was cleaved in an ultra-high vacuum at ~ 9 K just before the approach of the STM tip to the cleaved Bi-O layer in situ.

Shown in Figs. 1(a) and 1(b), are typical STM images measured at ~ 9 K on sample A under both low-bias voltage ($V_s = 30$ mV) and high-bias ($V_s = 600$ meV). Since there exists a semiconducting gap E_g of the order of 100 meV at E_F in the electronic structure of the Bi-O layer, electron tunneling occurs predominantly between the STM tip and the Cu-O layer in STM imaging under bias voltages lower than E_g ($V_s < E_g/e$), while it occurs predominantly between the STM tip and the Bi-O layer in STM imaging under bias voltages higher than E_g[9]. Thus low-bias STM images will provide information about the Cu-O layer, while high-bias STM images will select the Bi-O layer. In fact, the 1-D superlattice,

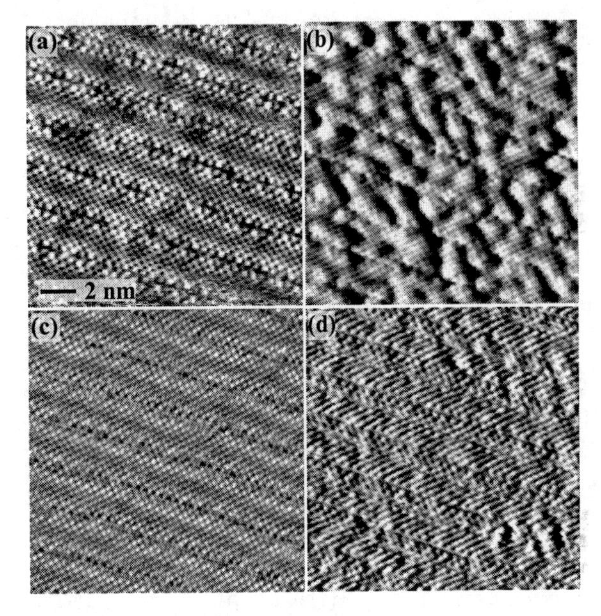

FIGURE 1. STM images of sample A under bias voltage of (a) 600 mV, and (b) 30 mV and those of sample B under bias voltage of (c) 600 mV, and (d) 10 mV. The resolutions of these images are 220 × 220 pixels (a, b) and 350 × 350 pixels (c, d).

CP850, *Low Temperature Physics: 24th International Conference on Low Temperature Physics;*
edited by Y. Takano, S. P. Hershfield, S. O. Hill, P. J. Hirschfeld, and A. M. Goldman

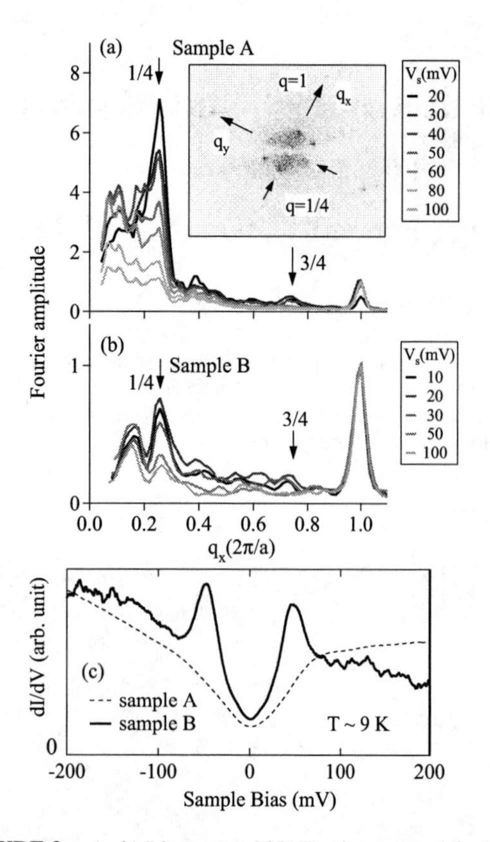

FIGURE 2. (a, b) Line cuts of 2D Fourier maps of the STM images for sample A and B along q_x axis at various bias voltages. The inset is a 2D Fourier map of Fig. 1(b). (c) A line average of the STS spectra measured along a straight line of 40 nm in sample A (broken line), and a typical d-wave like spectrum of sample B (solid line).

inherent in the Bi-O layer, appears more clearly in high-bias STM image at 600 mV than in low-bias STM one at 30 mV, as seen in Figs. 1(a) and 1(b).

In Fig. 1(b), we can identify a Cu-O bond-oriented, 2D superstructure over the cleaved surface. A 2D Fourier map $F(q_x, q_y)$ of the STM image at $V_s = 30$ mV for sample A shows that the 2D superstructure has periodicity of 4a×4a (Fig. 2). Line cuts of the 2D Fourier map along q_x axis show that the Fourier peaks corresponding to the 4a×4a superstructure are most intense at the lowest bias, 20 mV, and weakened rapidly as V_s increases. However, their position and width are independent of V_s, indicating the nondispersing nature of the superstructure, which enables us to observe the superstructure in STM masurements clearly. The superstructure tends to fade out above $V_s \sim 80$ mV. The weak Fourier peaks corresponding to the periodicity of 4a/3×4a/3 also appear, indicating that the nondispersive 4a×4a superstructure has an internal structure.

The STM image, taken at $V_s = 10$ mV, for sample B is shown in Fig. 1(d). we can identify the 2D superstructure, not throughout the cleaved surface, but over local regions on a nanometer scale. The line cuts of 2D Fourier map along the (π/a, 0) direction indicates that the periodicity is 4a×4a and nondispersive (Fig. 2(b)). The 4a×4a peaks appear below $V_s = 50$ mV, and are much weaker than those for sample A, which is consistent with the difference between features of real-space 4a×4a superstructures observed for sample A and B by STM.

The nondispersive 4a×4a uperstructure observed in the present study corresponds with findings of Hanaguri *et al.* for Na-CCOC, and Fang *et al.* and Howald *et al.* for the SC state of Bi2212[2, 5, 6].

The STS measurements reveal that the energy gap of sample A is of ZTPG type while the gap of sample B is of d-wave type, and the gap width, Δ_0, is ~ 80 meV on the positive bias side in sample A and ~ 60 meV in sample B (Fig. 2(c)). It should be noticed here that the bias voltage of ~ 80 mV for sample A and ~ 50 mV for sample B, where the 4a×4a peaks tend to fade out, roughly correspond to the gap widths Δ_0's of both samples. This suggests the possibility that electronic states within the paring gap, namely the hole pairs, may be associated with the formation of the 4a×4a superstructure.

As shown in Fig. 2(c), sample A and B exhibit different gap shape and gap width each other, which indicates that doping level p's of both samples are rather different from , though both samples were cut from the same single crystal: p of sample A is lower than that of sample B. It is worth while pointing out here that the cleaved surface of sample A has more missing atom rows, within the cleaved Bi-O layer than that of sample B, as seen in Figs. 1(a) and 1(c). If the missing atom rows are introduced in the Bi-O layer on cleaving, excess oxygen atoms within Bi-O layer will also be removed around the missing atom rows at the same time, which leads to the reduction of the average doping level within the Cu-O layer. Thus, the difference of the density of missing atom rows between A and B samples may give the reason why the doping level is lower in the sample A than in the sample B.

This work was supported in part by Grants-in-Aid for Scientific Research and the COE program "Topological Science and Technology" from the Ministry of Education, Culture, Science and Technology of Japan.

REFERENCES

1. M. Vershinin *et al.*, *Science*, **303**, 1995 (2004).
2. T. Hanaguri *et al.*, *Nature*, **430**, 1001 (2004).
3. K. McElroy *et al.*, *Phys. Rev. Lett.*, **94**, 197005 (2005).
4. J. E. Hoffman *et al.*, *Science*, **295**, 466 (2002).
5. A. Fang *et al.*, *Phys. Rev. B*, **70**, 214514 (2004).
6. C. Howald *et al.*, *Phys. Rev. B*, **67**, 014533 (2003).
7. J. E. Hoffman *et al.*, *Science*, **297**, 1148 (2002).
8. K. McElroy *et al.*, *Nature*, **422**, 592 (2003).
9. C. Manabe *et al.*, *J. Phys. Soc. Jpn.*, **66**, 1776 (1997).

Influences of Energy Dependent Quasiparticle Lifetime Effect on Tunneling Spectra of High-T_C Superconductors

Satoshi Kashiwaya[a], Akira Sugimoto[a], Hiromi Kashiwaya[a], Tetsuro Matsumoto[a], Hiroshi Eisaki[a], Masao Koyanagi[a], Hiroki Tsuchiura[b] and Yukio Tanaka[c]

[a]*Nanoelectronics Research Institute of AIST, Tsukuba, Ibaraki, 305-8568, Japan*
[b] *Department of Applied Physics, Tohoku University, Aobaku, Sendai, 980-8579, Japan*
c Department of Applied Physics, Nagoya University, Nagoya, 464-8603, Japan

Abstract. The local electronic states of optimally doped $Bi_2Sr_2CaCu_2O_{8+\delta}$ (Bi2212) and Ln-doped $Bi_2Sr_{1.6}Ln_{0.4}CuO_{6+\delta}$ (Ln-Bi2201, Ln=La, Gd) were measured by low-temperature scanning tunneling spectroscopy. Spatially inhomogeneous electronic states were detected for all samples, and the gap amplitudes were distributed in the range between 40-140meV in the case of Bi2212 and 30-200meV in the case of Ln-Bi2201.The spectra showed enhanced coherence peaks with large residual conductance when the gap amplitude was relatively small, while the peaks in the spectra were gradually broadened as the gap amplitude became larger. We demonstrate that a series of spectra with various gap amplitudes obtained on a single sample is systematically reproduced by the conductance spectra for *d*-wave BCS density of states which take account of the energy dependent quasiparticle lifetime broadening factor. This fact indicates the presence of serious influences of an intrinsic broadening effect due to the strong electron correlation in tunneling spectroscopy of high-T_c superconductors.

Keywords: superconductivity, tunneling spectroscopy, pseudogap
PACS: 74.50.+r, 74.72.Hs

INTRODUCTION

The presence of the intrinsic inhomogeneity of electronic states in Bi2212 cuprates is one of crucial issues of high-T_c superconductor physics[1]. Spatially resolved conductance spectra showed the coexistence of a wide variety of the density of states with various gap amplitudes in nano-meter patch pattern in Bi2212. However, the origin and the universality of this phenomenon are unclear at present. In the case of Ln-doped Bi2201, varying the ion radius of doped Ln ions to A-sites controls the magnitude of out-of plane disorder, thus the influence of the impurity doping on the intrinsic inhomogeneity can be systematically studied in this material [2]. Although the effects of the out-of plane disorder on the electronic states are known to be quite different from those of impurities onto the Cu-O plane based on resistivity measurements, the real space information is not clear yet. In this paper, we report the observation of the intrinsic inhomogeneity under the presence of out-of plane disorder in Ln-doped Bi2201. The gap structure evolution as the function of gap amplitude is analyzed in terms of the energy dependent quasiparticle lifetime effect.

EXPERIMENTAL

The local electronic states of Bi22121 and Bi2201 are observed by scanning tunneling spectroscopy (STS) at about 9K for all data presented here. The samples were grown by the traveling solvent floating zone method and the surfaces of the samples were cleaved in-situ at low temperature. The samples were optimally doped Bi2212, La-2201, Gd-2201 of which Tcs were 93K, 32K and 14K, respectively. Tunneling spectroscopy has been performed under the conditions of the work function larger than 2eV and the resistance of the junction larger than 100MΩ, and with clear atomic image in topography mode.

RESULTS AND DISCUSSION

The STS results on Bi2201 and Bi2212 samples showed spatial inhomogeneity with a few nano-meter size patch structure for all cases. There seemed no prominent patch size dependence due to the impurity

CP850, *Low Temperature Physics: 24th International Conference on Low Temperature Physics;*
edited by Y. Takano, S. P. Hershfield, S. O. Hill, P. J. Hirschfeld, and A. M. Goldman

doping in gap maps. On the other hand, the gap amplitude and the gap structure showed clear response to the doping. In the case of Bi2212, the gap amplitudes 2Δ were distributed in the range between 40-80meV, while they were in the range between 40-200meV in the case of Ln-doped Bi2201. This fact means the gap amplitude is tend to be enlarged as the result of impurity doping. For more details about experimental results including spatial distribution information are presented in the other report[3].

We found a systematic behavior of the gap structure evolution as the function of 2Δ common to all the samples used here. One of the examples for the gap-averaged spectra obtained on La-Bi2201 is shown in Fig. 1. The gap structure shows enhanced coherence peaks when 2Δ is relatively small, while the coherence peaks are gradually suppressed as 2Δ becomes larger. The peaks almost disappear when the 2Δ exceeds 100meV. The gap structure without coherence peaks observed below the T_c is usually referred to as zero-temperature pseudo gap whose origin is unclear. The residual conductance inside the gap is increased as 2Δ becomes smaller.

In order to explain the origin of this gap structure evolution, we focus on the quasiparticle lifetime effect that induces a broadening effect on tunneling spectra. Although the single broadening factor originally introduced by Dynes, et al[4], is widely used, here we extend it to include the energy dependence $\Gamma(E)$[5]. This idea is based on the fact that the lifetime of the quasiparticles of high-T_c superconductors is dominated by electron-electron scattering, and the quasiparticle lifetime has strong energy dependence as detected in optical measurements[6]. Referring to the energy dependence of scattering rates in infrared spectroscopy, the energy dependence of $\Gamma(E)$ is assumed to have energy dependence as described in the inset of Fig.2. The dependence means that the broadening effects become rapidly enhanced as the energies of the quasiparticles become larger. A series of spectra based on BCS type *d*-wave density of states calculated for various gap amplitudes with a fixed $\Gamma(E)$ is shown in Fig.2. It is quite clear that above mentioned features of spectra detected in the experiments are well reproduced in the calculation. Moreover, we can demonstrate that the other sets of spectra obtained on Bi2212 and Gd-Bi2201 are also reproduced in the calculation only by using corresponding energy dependence of $\Gamma(E)$ based on the same formulation. The influence of out of plane disorder appears only in the energy dependence of $\Gamma(E)$. That is, the scattering rate of the quasiparticles *i.e.* the influence of the broadening effect on tunneling spectra is enhanced in accordance with the impurity doping. Since the energy dependences of $\Gamma(E)$ used in the fitting can be extracted from the scattering rates of infrared spectroscopy for all three cases, no parameters except 2Δ are required for the present fitting. The consistency between experiment and calculation clearly suggests that the gap structure is seriously modified by the quasiparticle lifetime effect, and that the origin of the peudogap is regarded as the largely broadened gap structure due to the strong electron-electron correlation in the framework of the present scenario.

We would like to thank Prof. S. Uchida and K. Fujita for fruitful discussion.

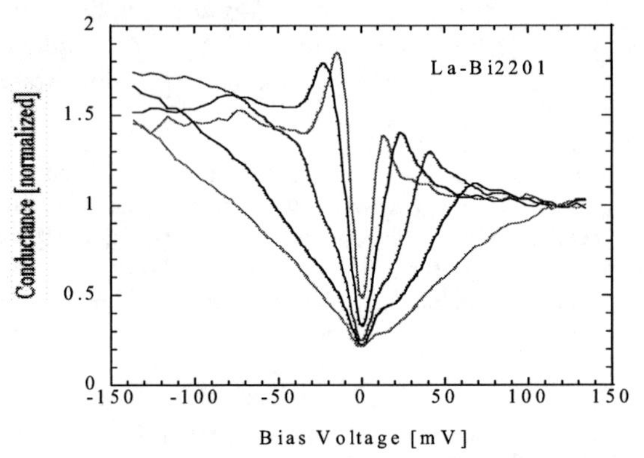

FIGURE 1. Gap-averaged spectra obtained by STS on La-Bi2201 at 9K are plotted.

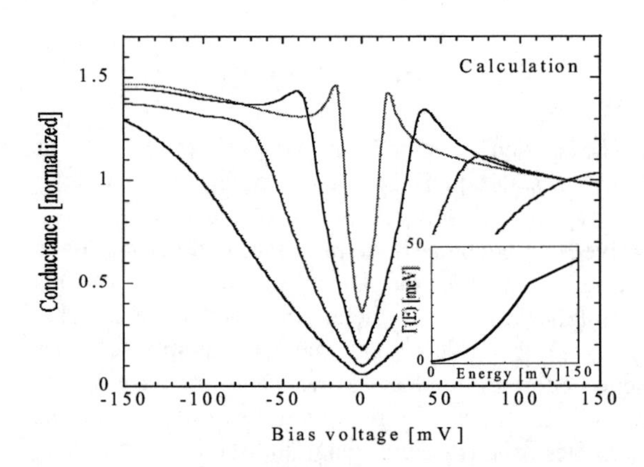

FIGURE 2. Theoretical spectra for various gap amplitudes are plotted. The energy dependence of $\Gamma(E)$ is shown in the inset.

REFERENCES

1. S. H. Pan, *et al.*, *Nature* **413**, 282 (2001); K. McElroy, *et al.*, *Cond-mat*/0404005.
2. H. Eisaki, *et al*, *Phys. Rev.* ***B*** **69**, 064512 (2004).
3. A. Sugimoto, *et al.*, Proceeding of this conference.
4. R. C. Dynes, *et al.*, *Phys. Rev. Lett.* **41**, 1509 (1978).
5. Similar calculation has also been presented by D. Coeffy, *et al*, *Phys. Rev. Lett.* **70**, 1529 (1993).
6. T. Timusk and B. Statt, *Rep. on Prog. in Phys.* **62**, 61, (1999).

Evolution of Coherence and Superconductivity in Electron-Doped Cuprates

G. Blumberg*, M.M. Qazilbash†, B.S. Dennis* and R.L. Greene†

*Bell Laboratories, Lucent Technologies, Murray Hill, NJ 07974, USA
†Department of Physics, University of Maryland, College Park, MD 20742, USA

Abstract. The superconducting (SC) phase diagram of the electron-doped cuprates has been explored by Raman spectroscopy as a function of doping x, temperature T, and magnetic field H. The data is consistent with *nonmonotonic* SC order parameter (OP) of the d-wave form. The persistence of SC coherence peaks in the B_{2g} channel for all dopings implies that superconductivity is mainly governed by interactions in the vicinity of $(\pm\pi/2a, \pm\pi/2a)$ regions of the Brillouin zone. Effective upper critical field lines $H^{*}_{c2}(T,x)$ at which the superfluid stiffness vanishes and $H^{2\Delta}_{c2}(T,x)$ at which the SC amplitude is suppressed by field have been determined. The difference between the two quantities suggests the presence of phase fluctuations that increase for $x < 0.15$. It is found that the field suppresses the magnitude of the SC gap linearly at an anomalously large rate. $H^{2\Delta}_{c2}$ value that is about 10 T for optimally doped samples decreases below a Tesla for overdoped cuprates.

Keywords: Superconductivity, Raman Scattering, Magnetic Field
PACS: 74.25.Ha, 74.25.Gz, 74.72.-h, 78.30.Er

We use electronic Raman spectroscopy to study quasi-particle (QP) spectra renormalization in magnetic fields on single crystals and films of $Pr_{2-x}Ce_xCuO_{4-\delta}$ (PCCO) and $Nd_{2-x}Ce_xCuO_{4-\delta}$ (NCCO) with different Ce doping covering most of the SC phase diagram. We find that the SC gap magnitude is strongly suppressed in magnetic fields. From the temperature and doping dependence of the SC coherence peak, we extract an effective upper critical field line $H^{*}_{c2}(T,x)$ at which the superfluid stiffness vanishes. Field dependence of the measured SC gap value reveals an estimate of $H^{2\Delta}_{c2}(T,x)$, an upper critical field at which the SC amplitude is completely suppressed by field. We find that ξ_{GL} increases from 60 Å for optimal doping (OPT) to 220 Å for the most overdoped (OVD) sample with $T_c = 13$ K. For the latter case $k_F\xi_{GL} \approx 150$ is much larger than for p-doped cuprates but is still not reaching the regime of conventional BCS-like tightly overlapping Cooper pairs.

In the SC state, the strength of the low-frequency Raman scattering intensity in the normal state is reduced and the spectral weight moves to the 2Δ coherence peak resulting from excitations out of the SC condensate (Fig. 1). In the B_{2g} and A_{1g} channels, the "pair-breaking" SC coherence peaks appear for all dopings while in the B_{1g} channel these SC coherence peaks are negligibly weak in the underdoped (UND) and the most OVD films. For the OPT crystal, the SC coherence peak energy is larger in the B_{2g} channel compared with that in B_{1g}. The intensity below the SC coherence peaks vanishes smoothly without a threshold to the lowest frequency measured. The smooth decrease in the Raman response

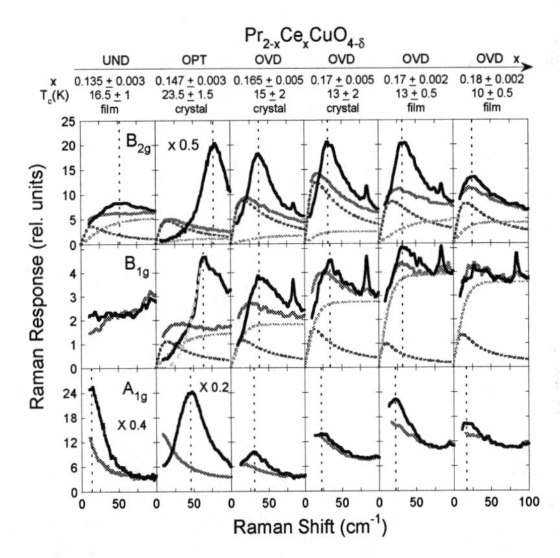

FIGURE 1. Doping dependence of the low energy electronic Raman response of PCCO single crystals and films for B_{2g}, B_{1g} and A_{1g} channels obtained with 647 nm excitation. The columns are arranged from left to right in order of increasing cerium doping. The light curves are the data taken just above the respective T_c of the samples. The normal state response in the B_{2g} and B_{1g} channels is decomposed into a coherent Drude-like contribution and an incoherent continuum (dotted lines). The dark curves show the data taken in the superconducting state at $T \approx 4$ K. The dashed vertical lines indicate positions of the SC coherence peaks.

below the SC coherence peak was interpreted in terms of a *nonmonotonic* d-wave OP with nodes along the (0,

CP850, *Low Temperature Physics: 24th International Conference on Low Temperature Physics;* edited by Y. Takano, S. P. Hershfield, S. O. Hill, P. J. Hirschfeld, and A. M. Goldman

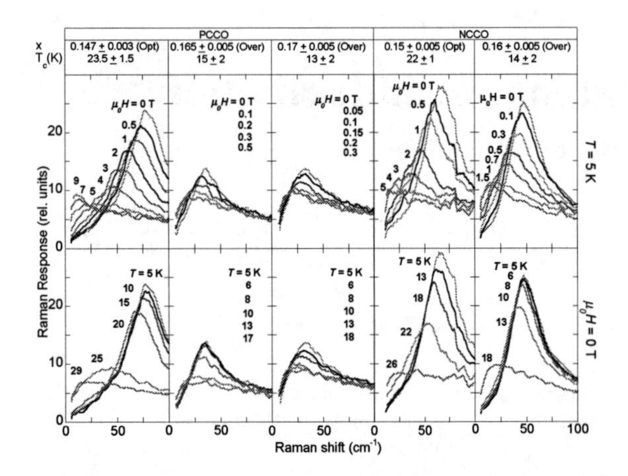

FIGURE 2. Raman response function for 647 nm excitation and right-left (RL) polarization for five single crystals of PCCO and NCCO with different Ce dopings x. The first row shows the disappearance of the 2Δ coherence peak in increasing magnetic field applied normal to the *ab*-plane of the crystals at 5 K. The second row shows the temperature dependence of the 2Δ peak in zero magnetic field.

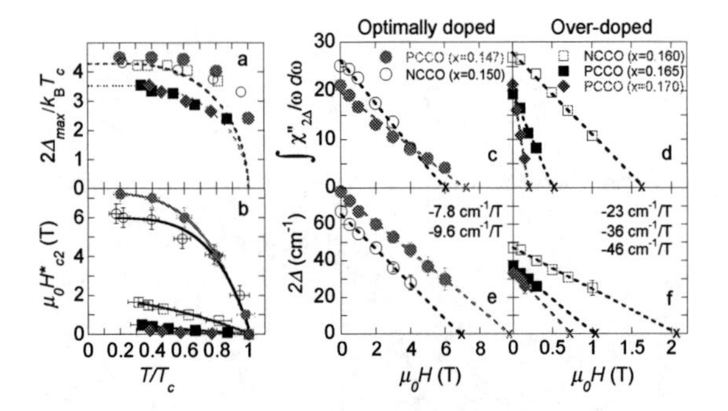

FIGURE 3. Temperature and field dependence of the SC OP amplitude and stiffness for two NCCO (open symbols) and three PCCO (filled symbols) single crystals. The temperature dependence (a) of the reduced SC gap magnitude at zero field, and (b) of the effective upper critical field H^*_{c2} that completely suppresses the SC coherence peak intensity at the given temperature. The field dependence at 5 K of the integrated coherence intensity $\int(\chi''_{2\Delta}(\omega)/\omega)\,d\omega$ is shown in panels (c) and (d) and the SC coherence peak energy (2Δ) in panels (e) and (f) for OPT and OVD crystals correspondingly. Approximated from the data effective upper critical fields H^*_{c2} and the values of the SC gap collapse $H^{2\Delta}_{c2}$ are indicated with "x". Panels (e) and (f) include also the rate of the gap suppression $\frac{d2\Delta(H)}{dH}$.

0)→(π/a, π/a) diagonal and the maximum gap being closer to this diagonal than to the BZ boundaries.

Fig. 2 exhibits the field and temperature dependence of the SC coherence peak at the maximum gap value ($2\Delta_{max}$) for the OPT (x $\approx$ 0.15) and OVD (x > 0.15) PCCO and NCCO crystals. The coherence peak loses intensity and moves to lower energies by either increasing the temperature or magnetic field. We define an effective

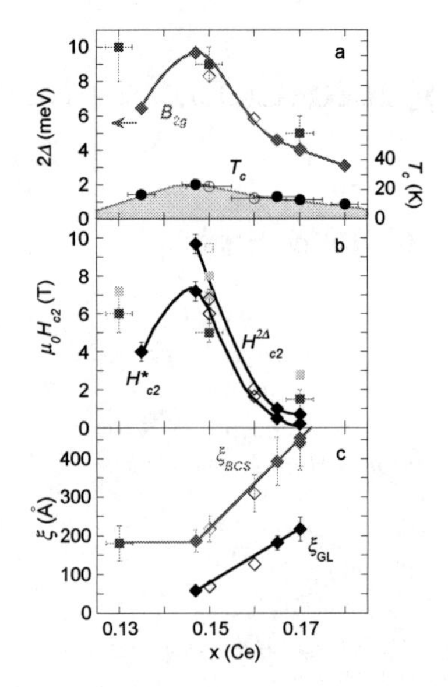

FIGURE 4. Phase diagram of PCCO (filled diamonds) and NCCO (open diamonds) superconductors explored by electronic Raman scattering in magnetic field. Panels show: (a) T_c (circles), the maximum energy of the SC 2Δ coherence peak and the distance between coherence peaks from point contact tunneling spectroscopy (squares); (b) The doping dependence at 5 K of the effective upper critical fields $H^*_{c2}(x)$ and the fields suppressing the gap amplitude $H^{2\Delta}_{c2}(x)$ is compared to upper critical fields obtained from other measurements (squares) including point contact tunneling, Nernst effect, and thermal conductivity; (c) the Ginzburg-Landau SC coherence length $\xi_{GL}(x) = \sqrt{\Phi_0/2\pi H^{2\Delta}_{c2}(x)}$ is compared to the BCS coherence length $\xi_{BCS}(x) = \hbar v_F/\pi\Delta_{max}(x)$ obtained from the Raman gap data. All solid lines are guides to the eye.

upper critical field, $H^*_{c2}(T,x)$, as the field that completely suppresses the coherence peak intensity (See Fig. 3). Reduced gap values and effective upper critical fields as a function of the reduced temperature for five single crystals of various doping levels are plotted in Figs. 3a-3b. For the lowest measured temperature, $2\Delta_{max}/k_BT_c$ values fall between 4.5 for the OPT crystals and 3.5 for the most OVD crystals. In Fig. 4 we show that the fields suppressing the gap amplitude, $H^{2\Delta}_{c2}(x)$, are related to the SC gap magnitude.

ξ_{GL} for the *n*-doped cuprates is significantly larger than for their *p*-doped counterparts and that leads to important differences. First, the size of the Cooper pair is much larger than the average inter-particle spacing: $k_F\xi_{GL}$ ranges between 40 and 150. Second, a larger Cooper pair size requires further pair interactions to be taken into account and leads to a more complicated non-monotonic momentum dependence of the SC gap than the simplest *d*-wave form $\Delta(\mathbf{k}) \propto \cos(k_{Fx}a) - \cos(k_{Fy}a)$ that well describes the gap function for *p*-doped cuprates with tight Cooper pairs.

Magneto-Optical Response of Electron Doped Cuprates $Pr_{2-x}Ce_xCuO_4$

Naveen Margankunte*, Alexandre Zimmers†, D.B. Tanner*, R.L. Greene† and Y. J. Wang**

*Department of Physics, University of Florida, Gainesville, FL 32611, USA
†Center for Superconductivity Research, Department of Physics, University of Maryland, College Park, MD 20742, USA
**National High Magnetic Field Laboratory, Florida State University, Tallahassee, FL 32306, USA

Abstract. We report mid-infrared transmission measurements of electron doped $Pr_{2-x}Ce_xCuO_4$ (PCCO) thin films for a wide range of dopings, in the large energy pseudogap regime both as a function of temperature and magnetic field. While the temperature dependent measurements show clear signatures of pseudogap, there is no magnetic field induced effect.

Keywords: PCCO, Magnetic field, Optics, Pseudogap
PACS: 74.25.Nf, 74.72.Jt, 78.20.Ls

INTRODUCTION

The normal state of cuprates shows many anomalous properties, attributed at present to the formation of a pseudogap. In the *ab*-plane, the pseudogap shows up as a drop in the optically defined scattering rate for hole-doped cuprates [1], whereas for electron-doped materials it appears directly in the conductivity [2, 3] spectra at around 0.15 eV. This high energy pseudogap has been attributed to changes in antiferromagnetic spin correlations [2] and spin density waves (SDW) [3] respectively. In this paper, we present magnetic-field and temperature dependent transmission data in the large energy pseudogap region.

EXPERIMENT

Thin Films of $Pr_{2-x}Ce_xCuO_4$ (PCCO) of several compositions were grown on $LaSrGaO_4$ (LSGO) substrates using pulsed laser deposition [4]. The films used in this experiment included a highly underdoped (x = 0.11) non-superconducting sample, an optimally doped (x = 0.15) sample and an overdoped (x = 0.18) sample. The mid-infrared studies in magnetic field were performed at the National High Magnetic Field Laboratory, using a Bruker 113v spectrometer with custom-built light-pipe optics to carry the mid-infrared radiation through the sample and on to a 4.2 K helium-cooled bolometer detector [5]. We employed a 30 T resistive magnet and measured the 4.2 K transmittance ratio, $\mathscr{T}(\mathscr{H})/\mathscr{T}(0)$. Two different sample holders were used, one where the *ab*-plane of the sample was $\perp$ to the magnetic field and one where the *ab*-plane was at an angle of $\theta = 25^0$ to the magnetic field, ensuring an in-plane $\mathscr{H}$ component.

ZERO FIELD DATA

Figure 1 shows the temperature dependent transmission of the three PCCO samples (with x = 0.11, x = 0.15, x = 0.18) at 0 T. The most prominent changes can be seen for the highly underdoped sample at around 1500 cm^{-1}. This behavior is in accord with the results of reflectance studies presented earlier [3]. In comparison, transmission changes in optimally doped and overdoped samples are small. This large increase in transmission (not due to the substrate) is observed as the temperature is lowered and the pseudogap is established.

MAGNETIC FIELD DATA

In contrast to its strong temperature dependence, the magneto-transmission of the underdoped PCCO, with magnetic field $\perp$ to the *ab*-plane did not show any change within the experimental signal to noise of $\pm 1\%$, as shown in Fig. 2. For the optimally doped sample, we measured the effect of having components of the magnetic field both along the *ab*-plane and along the *c*-axis. This oblique configuration has revealed doping-dependent properties in the DC resistivity and has been attributed to a quantum critical point in these materials [6]. No magneto-optical effect exceeding $\pm 2\%$ was observed. Finally, the overdoped sample with magnetic field $\perp$ to *ab*-plane also does not show any change in

CP850, *Low Temperature Physics: 24th International Conference on Low Temperature Physics;*
edited by Y. Takano, S. P. Hershfield, S. O. Hill, P. J. Hirschfeld, and A. M. Goldman

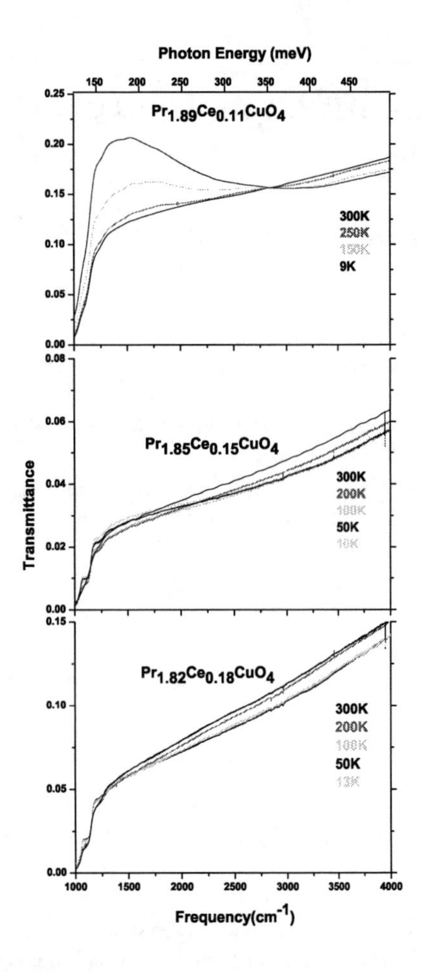

FIGURE 1. Transmission at various temperatures of PCCO at 0 T for $x = 0.11$, $x = 0.15$ and $x = 0.18$. The low frequency cutoff comes from the transmission cutoff of the LSGO substrate.

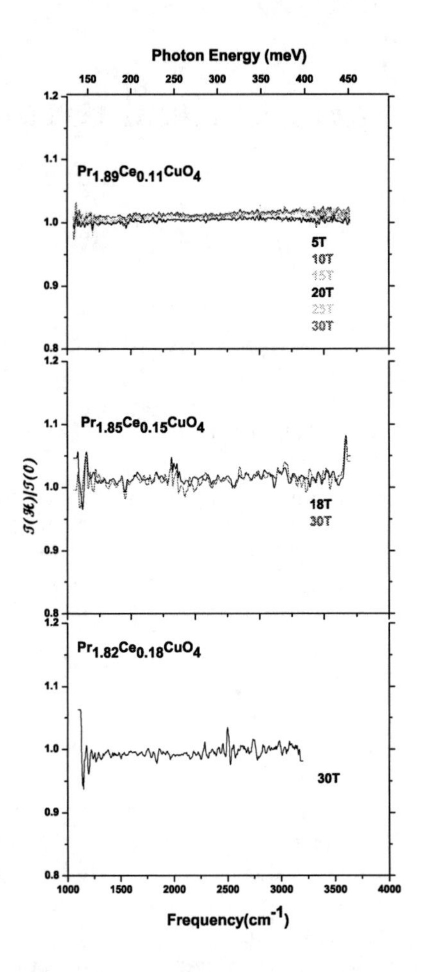

FIGURE 2. Normalized (to 0 T) magneto-transmission spectra at 4.2 K of PCCO for $x = 0.11$, $x = 0.15$ and $x = 0.18$.

transmission within experimental signal to noise resolution of $\pm 3\%$.

CONCLUSIONS

These experiments indicate that the high energy pseudogap is unaffected by magnetic field. This result is probably due to the fact that the Zeeman energy of the field (about 3 meV) is small compared to the energy scale of the antiferromagnetic correlation (0.12 eV), so the *ab*-plane spins are perturbed very little by the applied magnetic field.

ACKNOWLEDGMENTS

NM and DBT acknowledge support from the DOE. DE-AI02-03ER46070, and the NSF, DMR-0305043. The work at Maryland was supported by NSF Grant No. DMR-0352735. NHMFL is supported by NSF Cooperative Agreement No. DMR-0084173 and by the State of Florida.

REFERENCES

1. A.V. Puchkov, D.N. Basov, and T. Timusk, *J. Phys. Condens. Matter* **8**, 10049 (1996).
2. Y. Onose *et al., Phys. Rev. B.* **69**, 024504 (2004).
3. A. Zimmers *et al., Europhys. Lett.,* **70**, 225-231 (2005).
4. E. Maiser *et al., Physica C* **297**, 15 (1998).
5. H.K. Ng and Y.J. Wang, *Physical Phenomena at High Magnetic Fields-II*, Singapore: World Scientific, 1996, pp. 729.
6. Y. Dagan *et al., Phy. Rev. Lett.* **94**, 057005 (2005).

Spin Dynamics in an Electron-Doped High-Tc Superconductor

M. Fujita*, M. Matsuda† and K. Yamada*

*Institute for Materials Research, Tohoku University, Sendai, Miyagi 980-0821, Japan
†Quantum Beam Science Directorate, Japan Atomic Energy Agency, Tokai, Ibaraki 319-1195, Japan

Abstract. We have performed inelastic neutron-scattering measurements on an electron-doped superconductor, $Pr_{1-x}LaCe_xCuO_4$ to clarify the doping dependence of spin correlations. In measured samples with x=0.09, 0.11, 0.13, 0.15 and 0.18, the low-energy spin correlations are commensurate, unlike in hole-doped systems, which have incommensurate spin correlations. Upon electron doping, the commensurate peaks centered at the (π, π) reciprocal position are more drastically broadened at a higher energy, suggesting a reduction of slope of spin excitations ω/q, namely an effective nearest exchange constant J_{eff}.

Keywords: spin correlations, neutron scattering, electron-doped cuprate
PACS: 74.72.Jt, 74.25.Ha, 75.50.Ee

INTRODUCTION

Antiferromagnetic (AF) spin correlations are believed to play a crucial role in the superconducting (SC) mechanism. An extensive neutron-scattering study on hole-doped (p-type) $La_{2-x}Sr_xCuO_4$ (LSCO) has provided evidence for an intimate relationship between spin correlations and superconductivity. That is, incommensurate spin correlations with the modulation vector parallel to the Cu-O bond direction are observed throughout the entire SC region and the modulation period is inversely proportional to the SC transition temperature (T_c) in the underdoped region. [1] On the other hand, the electron-doped (n-type) superconductor, $Nd_{2-x}Ce_xCuO_4$, shows commensurate spin correlations.[2] These observed correlations in the p- and n-type systems suggest that electron-hole symmetry exists in the spin correlations upon doping. Thus, to clarify the universal spin correlations that are directly related to the high-T_c mechanism, a comprehensive study for the two systems is important. Hence, this paper uses neutron scattering measurements to investigate the doping dependence of the low-energy spin correlation in n-type $Pr_{1-x}LaCe_xCuO_4$ (PLCCO) as the effects of rare-earth moment are much weaker than those in the NCCO system.[3]

EXPERIMENTAL

Single crystals of PLCCO with x=0.09, 0.11, 0.13, 0.15, and 0.18 were grown by the traveling-solvent floating-zone method. All the as-grown single crystalline rods were cut into ~30 mm long pieces and annealed under argon gas flow between 950-960°C for 10-12 hours. Magnetic susceptibility was measured using a SQUID (superconducting quantum interference device) magnetometer. A superconducting shielding signal was observed in all samples and T_c gradually decreased upon doping from 26 K in x=0.09 to 10 K in x=0.18. Two samples with x=0.09 and 0.11, which locate in the vicinity of the AF and SC phase boundary, exhibit a fair AF order at low temperature, suggesting that the AF and SC phases coexist in the sample.

Inelastic neutron scattering experiments were performed on thermal triple-axis spectrometers, TAS-1 and TOPAN, installed in the JRR-3 at the JAEA in Tokai, Japan. Each crystal was mounted in the (h k 0) zone. All crystallographic indices in this paper are described in orthorhombic notation.

RESULTS AND DISSCUSSION

Figure 1 shows the magnetic excitations spectra for the x=0.09 sample measured at 8 K and (a) 10 meV, (b) 6 meV, (c) 3 meV. At each energy, the peak is centered at the commensurate position of h=1, which centered at the commensurate position of h=1, which corresponds to the AF zone center in the undoped system. Figure 2 shows a similar series of scans for the x=0.15 sample. The commensurate peaks are also observed at all measured ω. Thus, commensurate spin correlations appear to be common in low-energy spin fluctuation of the electron-doped systems. However, the peak-width in the overdoped (x=0.15) sample broadens faster than in the optimum doped (x=0.11) sample as the energy increases, suggesting that the spin correlations degrade upon electron doping. Figure 3 plots the resolution- cor-

CP850, *Low Temperature Physics: 24th International Conference on Low Temperature Physics;*
edited by Y. Takano, S. P. Hershfield, S. O. Hill, P. J. Hirschfeld, and A. M. Goldman

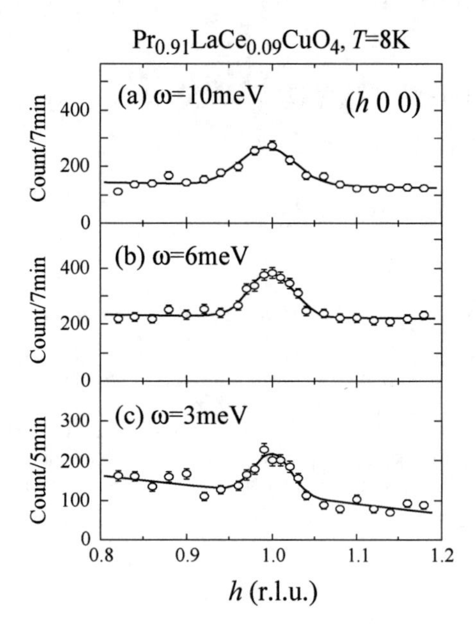

FIGURE 1. Inelastic neutron scattering spectra of $Pr_{0.91}LaCe_{0.09}CuO_4$ at (a) 10 meV, (b) 6 meV, and (c) 3 meV at 8 K. Solid lines are fits assuming a single Gaussian peak at h=1.

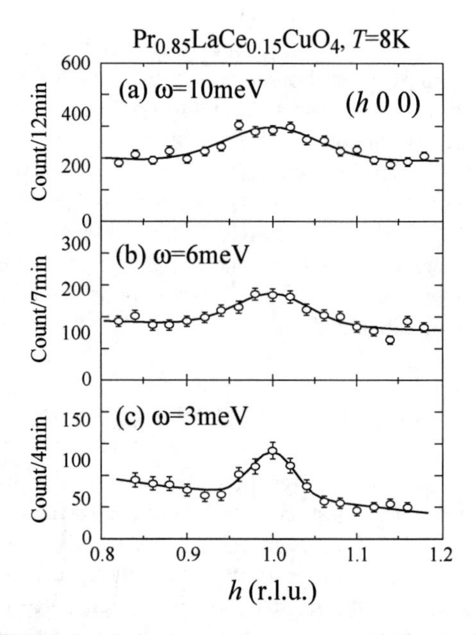

FIGURE 2. Inelastic neutron scattering spectra of $Pr_{0.85}LaCe_{0.15}CuO_4$ at (a) 10 meV, (b) 6 meV, and (c) 3 meV at 8 K. Solid lines are fits assuming a single Gaussian peak at h=1.

rected peak-width (κ) for five samples. At a low energy of 3-4 meV (closed circles), κ weakly increases as the Ce concentration increases. However, κ measured at 10-11meV (open circles) drastically increases with the Ce concentration. Broadening of the peak width at a higher

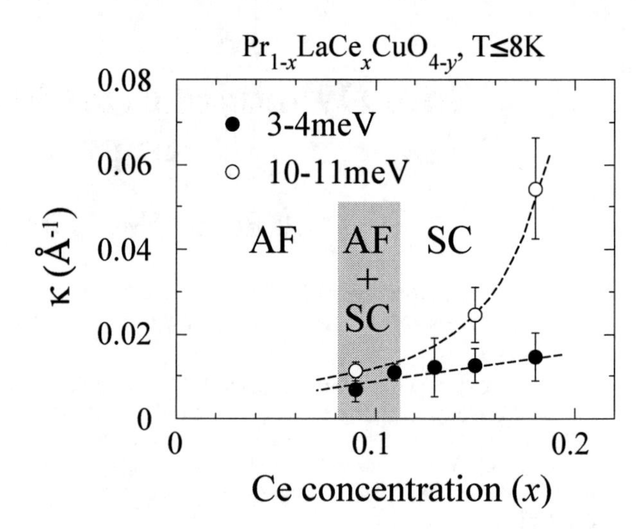

FIGURE 3. Ce concentration dependence of the resolution corrected peak-width κ for $Pr_{1-x}LaCe_xCuO_4$ at 3-4 meV (closed circles) and 10-11 meV (open circles). Solid lines are guides to the eye.

energy upon electron doping suggests that ω/q decreases in the vicinity of the AF zone center, namely the slope of spin excitation spectra in the low energy region, even in the SC phase. This result is consistent with a reduction in the average of the nearest neighbor exchange constant (J_{eff}) shown for the AF phase.[4] Therefore, the disappearance of the SC phase in a p-type system would be characterized by a reduced J_{eff}, which is different from the vanishment of the superconductivity in the overdoped region of LSCO for which evidence of separate SC and Fermi liquid phases has been reported.[5] To understand the differences in the spin correlations between these two systems and the universal relation between magnetism and high-T_c superconductivity, further experiments such as observations of the high-energy spin excitation are required.

ACKNOWLEDGMENTS

This work was supported in part by the Japanese Ministry of Education, Culture, Sports, Science and Technology, Grant-in-Aid for Encouragement of Young Scientists (A), 17684016, 2005.

REFERENCES

1. K. Yamada *et al.*, *Phys. Rev. B.* **57**, 6165 (1998).
2. K. Yamada *et al.*, *Phys. Rev. Lett.* **90**, 137004 (2003).
3. M. Fujita *et al.*, *Phys. Rev. B.* **67**, 0145145 (2003).
4. T.R. Thurston *et al.*, *Phys. Rev. Lett.* **65**, 263 (1990).
5. S. Wakimoto *et al.*, *Phys. Rev. Lett.* **92**, 217004 (2004).

Pressure Effects on The Superconductivity in $MoSr_2YCu_2O_{8-\delta}$ Oxide Superconductor

Yuh Yamada[a),c)], M. Ueno[b)], K. Fukuda[b),c)], F. Ishikawa[b),c)] and A. Matsushita[d)]

[a)]*Depatment of Physics, Niigata University, 8050 Igarashi-ninocyo, Niigata 950-2181, Japan*
[b)]*Graduate School of Science and Technology, 8050 Igarashi-ninocyo, Niigata 950-2181, Japan*
[c)] *Center for Transdisciplinary Research, 8050 Igarashi-ninocyo, Niigata 950-2181, Japan*
[d)]*Mateleals Engineering Laboratory, NIMS, Sengen 1-2-1, Tsukuba, Ibaraki 304-0047, Japan*

Abstract. We have investigated the pressure dependence of superconducting transition temperature T_c in $MoSr_2YCu_2O_{8-\delta}$ oxide superconductor (Mo1212). The superconducting Mo1212 samples were synthesized by solid-state reaction with multiple annealing process. The T_c -onset of this sample exhibited about 38 K at an ambient pressure. The pressure dependence of T_c was obtained about 7.1 K/GPa up to 2 GPa.

Keywords: Pressure effect, Oxide superconductor, Coexisting antiferromagnetism and superconductivity.
PACS: 74.62.Fj, 74.72.Jt, 74.25.Fy, 74.25.Ha

INTRODUCTION

Coexisting ferromagnetism and superconductivity have recently been discovered in the 1212-type layered cuprate superconductivity $RuSr_2GdCu_2O_8$ (Ru1212) [1]. This material is isostructural with $YBa_2Cu_3O_{7-\delta}$ with Y, Ba, and Cu1 (the chain copper atom) being completely replaced by Gd, Sr, and Ru, respectively and displays bulk superconductivity below T_c = 0 - 46K which is associated with the CuO_2 planes, and a Curie transition at T_f = 132 K.

We have reported that the pressure dependence of the superconducting and ferromagnetic transition temperature of the Ru1212 [2]. In these results, the behavior of pressure dependence in T_C is at a glance similar to optimum-doped oxide superconductor, like $YBa_2Cu_3O_7$. However, it is makes no sense at all that these samples are the optimum-doped oxide superconductor from the structural data. We would rather take account of the effect of ferromagnetic moments.

Recently, the $FeSr_2YCu_2O_{8-\delta}$ (Fe1212) compounds were found to show superconductivity by applying annealing process under moderately reducing conditions [3]. The Fe1212 could not be observed the magnetic order in magnetic susceptibility measurement at the temperature range 2 to 300 K. We have investigated the pressure dependence of T_c in Fe1212 superconductor which exhibited the T_c-onset of about 60 K. The pressure dependence of T_c was obtained to be about 2.1 K/GPa up to 2 GPa [4].

More recently, Felner and Galstyan [5] reported the superconducting and magnetic properties of $MoSr_2YCu_2O_{8-\delta}$ (Mo1212). This material displays the antiferromagnetic ordering transition below a superconducting transition temperature. It is interesting to study the superconducting and magnetic properties of this system with the pressure effects.

In the present work, we have investigated the pressure dependence of the superconducting transition temperature T_c for the Mo1212 compound.

EXPERIMENTAL PROCEDURE

A polycrystalline sample of Mo1212 was synthesized by solid-state reaction of a stoichiometric mixture of the Y_2O_3, CuO, $SrCO_3$ and Mo. These were decomposed at 750℃ and die-pressed into pellets before reaction in air at 850℃ for 24 hours. This was followed by reaction in flowing oxygen at 960℃ for 72 hours and annealed in 100 atm oxygen at 600℃ for 24 hours. The final product was characterized by X-ray diffraction with Cu Kα radiation.

Magnetization measurements in the temperature range 5 to 300 K in magnetic fields up to 1 T were made using a superconducting quantum interference device magnetometer (SQUID). The temperature dependence of the resistivity under high pressures was

CP850, *Low Temperature Physics: 24th International Conference on Low Temperature Physics;*
edited by Y. Takano, S. P. Hershfield, S. O. Hill, P. J. Hirschfeld, and A. M. Goldman

measured using the DC four-probe method up to P = 2.0 GPa. The sample was filled into a Teflon Cell with Daphne oil 7373 (Idemitsu Kosan Co., Japan) as pressure transmitting media and pressurized in a piston-cylinder-type clamp made by NiCrAl ally.

RESULTS AND DISCUSSION

Powder X-ray diffraction (XRD) measurements indicate that the final product is nearly Mo1212 single-phase (~ 90 %) and confirmed the tetragonal structure (P4/*mmm*). The XRD patterns left a few minor reflections, most of them belonging to the $SrMoO_4$ phase which shows Pauli-paramagnetism.

In figure 1, the temperature dependence of magnetization for the Mo1212 was shown which was measured under the field of 20 Oe and 1T. The Meissner effect was observed below 38 K confirming the occurrence of a superconducting transition.

An anomaly of magnetic susceptibility originated in antiferromagnetic ordering transition (T_N) is observed at about 14 K (see inset figure). The same results were reported and discussed by Felner *et al* [5].

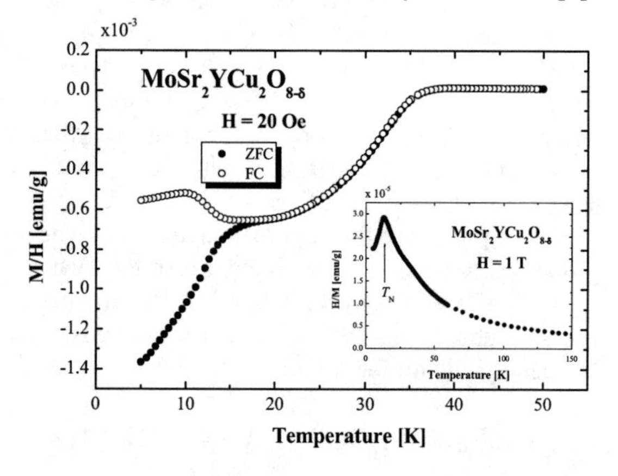

FIGURE 1. Zero field cooling (ZFC) and field cooling (FC) magnetization curves measured at 20 Oe of the Mo1212. The inset shows the temperature dependence of magnetization under the field of 1T.

Figure 2 shows the temperature dependence of resistivity measured at various pressures for Mo1212. The large changes with increasing pressure in the resistivity versus temperature curve are observed. The magnitude of resistivity in the normal state decreased up to 1.5 GPa and the value of T_c increased, with increasing pressure. It would appear that the enhancement of the magnitude of resistivity in 2.0 GPa arise from the decline of the hydrostatic pressure caused by the solidification of Daphne oil. The pressure coefficient of the T_c was 7.1 K/GPa up to 2 GPa which was one of the largest value in all oxide superconductors [6]. The similar positive large pressure effects were observed in the under-doped superconducting oxide frequently. One of the typical under-doped superconductor is $YBa_2Cu_4O_8$ compound (Y124) which indicates the large pressure effects on T_c [7]. Miyatake *et al.* reported that partially substituting Ca for Y in Y124 increases T_c [8]. If the same origin works, Ca-doped Mo1212 is supposed to enhance the T_c. We synthesized the $MoSr_2(Y_{0.9}Ca_{0.1})Cu_2O_{8-\delta}$ and investigated the value of T_c. However, the T_c on Ca-doped Mo1212 did not change greatly in comparison with that on non-doped Mo1212. In consideration of this result, there is a possibility that the change of T_c with pressure for Mo1212 was concerned with not only carrier concentration but also the magnetization. More detailed research is necessary for the examination.

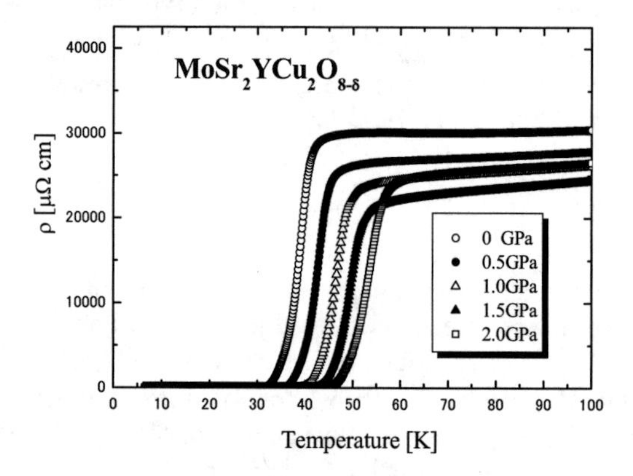

FIGURE 2. Temperature dependence of electrical resistivity of Mo1212 measured at various pressures.

ACKNOWLEDGMENTS

The work was partially supported by a Grant in Aid for Scientific Research from the Ministry of Education, Science and Culture, Japan.

REFERENCES

1. L. Bauernfeind *et al.*, *Physica C* **254**, 151-158 (1995).
2. Yuh Yamada *et al.*, *J. Phys. Cond. Mat.* **14**, 10763-10766 (2002).
3. J. Shimoyama *et al.*, *Physica C* **341-348**, 563-564 (2000).
4. H. Hamada *et al.*, *Physica C* **388-389**, 367-368 (2003).
5. I. Felner and E. Galstyan, *Phys Rev B* **69**, 024512 (2004).
6. H. Takahashi and N. Mori, *Studies of High Temperature Superconductors Vol. 16*, New York: Nova Science Publishers, 1996, pp. 1-64.
7. Y. Yamada *et al.*, *Physica C* **173**, 185-194 (1991).
8. T. Miyatake *et al.*, *Nature* **341**, 41-42 (1989).

Infrared Absorption Study of $Ca_{2-x}Na_xCuO_2X_2$ (X=Cl, Br)

Takumi Hasegawa[a], Norio Ogita[a], Toshihisa Kondo[b], Yuji Zenitani[c], Hirokazu Kawashima[c], Teruhiko Suzuki[c], Jun Akimitsu[c], and Masayuki Udagawa[a]

[a] *Faculty of Integrated Arts & Sciences, Hiroshima University, Higashi-Hiroshima, 739-8521, Japan*
[b] *Venture Business Laboratory of Hiroshima University, Higashi-Hiroshima, 739-8527, Japan*
[c] *Department of Physics, Aoyama-Gakuin University, Sagamihara, Kanagawa 229-8558, Japan*

Abstract. IR-active phonon spectra of $Ca_{2-x}Na_xCuO_2X_2$ (X=Cl, Br) have been measured by a CsI powder method in the energy region between 250 and 4000 cm^{-1} at room temperature. Two absorption peaks with the E_u symmetry have been clearly observed for the undoped crystals of $Ca_2CuO_2Cl_2$ and $Ca_2CuO_2Br_2$. However, the observed two peaks disappear for the Na-doped superconducting samples. From the comparison of the highest-energy E_u phonon, which is the Cu-O stretching vibration, the interaction of the Cu-O bond along the CuO_2 plane for the T-structure is stronger by 20 % than that of the T'-structure in the 2-1-4 family. To understand the effect of the apical ions, first-principles calculations of the E_u phonon energy for T- and T'-structure La_2CuO_4 is performed and the preliminary results agree with the experimental tendency.

Keywords: Infrared absorption, Phonon, Carrier doping, Superconductivity.
PACS: 74.72.-h, 74.25.Gz, 74.25.Kc, 78.30.-j, 71.15.Mb.

It has been believed that apical oxygen plays an important role for a microscopic mechanism of high-*Tc* superconductors, but the role is still unclear. In order to clarify this problem, the apical oxygen replacement by halogen ions in the K_2NiF_4 type structure (T-structure) is regarded with the suitable crystals. Superconductivity with the apex halogen system has been observed for $Sr_2CuO_2F_2$ (Tc = 46 K)[1], $(Ca,Na)_2CuO_2Cl_2$ (Tc = 28 K)[2], and $(Ca,Na)_2CuO_2Br_2$ (Tc = 18 K)[3]. However, the dynamical study has not been well proceeded because of difficulties of sample preparation and treatment. In addition, infrared absorption spectra of the apex halogen system have been measured only for the undoped crystals[4-6]. To clarify the apical ions effect for the lattice dynamical properties and doping effect for the infrared active phonon, the infrared absorption spectra of $Ca_{2-x}Na_xCuO_2X_2$ (X=Cl, Br) have been measured. Furthermore , to clarify the effect of the apical ions, first-principles calculations have been performed.

Polycrystalline samples were synthesized by the general solid-state reaction method. The detailed description of the sample preparation is reported in ref.3. Superconductivity appears at $x > 0.10$ for $Ca_{2-x}Na_xCuO_2Cl_2$ and at $x > 0.15$ for $Ca_{2-x}Na_xCuO_2Br_2$. Infrared absorption spectra were measured by a dual beam-type infrared spectrometer (Hitachi model 270-50) in an energy region between 250 and 4000 cm^{-1} at room temperature. The samples were pellets of CsI and $Ca_{2-x}Na_xCuO_2X_2$ powder. Since the samples are heavily damaged by a small amount of water, the pellet preparation was done in a dry box filled with Ar gas, and the pellet was set into a specially designed sample cell with KBr windows. The cell was also filled with dry Ar gas. The use of KBr crystal windows increases the available lowest energy from 250 to 330 cm^{-1}. Furthermore, a vacuum cell with the KBr windows was set in the reference beam line to compensate the KBr window effect.

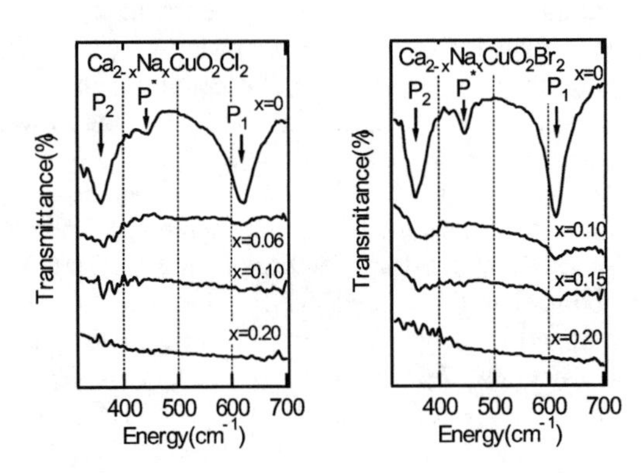

FIGURE 1. Na concentration dependence of infrared absorption spectra of $Ca_{2-x}Na_xCuO_2Cl_2$ and $Ca_{2-x}Na_xCuO_2Br_2$ measured at room temperature.

CP850, *Low Temperature Physics: 24th International Conference on Low Temperature Physics;* edited by Y. Takano, S. P. Hershfield, S. O. Hill, P. J. Hirschfeld, and A. M. Goldman

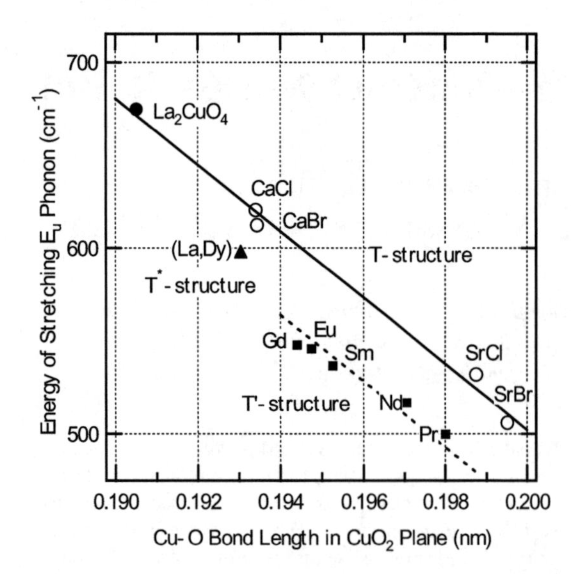

FIGURE 2. Bond length dependence of the energy of the stretching Cu-O vibration along the CuO_2 plane. The closed symbols are taken from ref. 4 and 8, and the open symbols denote the present result. For T'-structure (square symbols) and T^*-structure (triangle symbols), only rare earth ions are presented, and $R_2CuO_2X_2$ compounds are abbreviated by RX.

Figure 1 shows the representative IR-active phonon spectra of $Ca_{2-x}Na_xCuO_2X_2$ (X=Cl, Br) in an energy region between 330 and 700 cm^{-1}. For the low Na concentration samples, two peaks marked by P_1 and P_2 are clearly found. We assign the observed peaks as the E_u vibration, since the peak energy is similar with that of the in-plane phonons reported by Tajima *et al.*[4] P_1 and P_2 are the Cu-O stretching and bending vibrations in the CuO_2 plane, respectively. With increasing the Na-concentration, the absorption of P_1 and P_2 decreases and disappears at x = 0.10 for $Ca_{2-x}Na_xCuO_2Cl_2$ and at x = 0.20 for $Ca_{2-x}Na_xCuO_2Br_2$. According to the photoemission experiments[7], the carrier density of Br is lower than that of Cl even at the same concentration. Thus, the above result of the disappearance can be understood by the lower concentration of Br.

Next we discuss the Cu-O bond length dependence of the Cu-O stretching vibration in the CuO_2 plane. The bond length dependence of the P_1 energy is summarized in Fig. 2, where the open and closed marks denote the present and reported data[4,8], respectively. We referred to ref. 8 for La_2CuO_4 and to ref. 4 for the T'- and T^*-structure compounds. As shown in Fig. 2, the energy of the stretching vibration of T'-structure is systematically smaller by about 50 cm^{-1} than that of T-structure. Since the ionic displacement of P_1 is the same for T- and T'-structures, this difference is caused by existence of the apical ions. Thus, it is concluded that the atomic interaction of the Cu-O bond in the CuO_2 plane increases by about 20 % for the T-structure. This is the new experimental result of the effect of the apex ion and suggests that the apex ion changes the charge distribution in the CuO_2 plane.

To investigate this apical ions effect on the charge distribution, we have tentatively performed first-principles (local spin density approximation) calculations of phonon energies for T- and T'-structure La_2CuO_4 by ABINIT package. A detailed description of the calculations will be found elsewhere. In the calculations, the relaxed structures with the space group Bmab were used for both structures to produce an antiferromagnetic order in the CuO_2 plane. A difference between the relaxed T-structure and the experimental result[9] was lower than 3 % for lattice constants and lower than 0.012 for atomic positions. The relaxed T'-structure did not differ from the I4/mmm one. It is found that the phonon energy difference as shown in Fig. 2 is caused by a different charge distribution in the CuO_2 plane. The calculated results show that orbital mixing between $d_{x^2-y^2}$ and $d_{3z^2-r^2}$ is larger for the T-structure than that for the T'-structure. This suggests that the position of oxygen out of the CuO_2 plane affects on the charge distribution at the CuO_2 plane as discussed by Y. Ohta et al.[10] through the Madelung potential.

Hiroshima group was supported by a Grant-in-Aid for Scientific Research Priority Area "Skutterudite" (No. 15072205) and for COE Research (No. 13CE2002) of MEXT (Ministry of Education, Sports, Culture, Science and Technology). Aoyama-Gakuin group was also supported by 21st COE program from MEXT, 2002-2004 and Grant-in-Aid for Scientific Research on Priority Area of MEXT. The ABINIT package is a common project of the Universite Catholiqué de Louvain, Corning Incorporated, and other contributors (http://www.abinit.org/).

REFERENCES

1. M. Al-Mamouri et al., *Nature (London)* **369**, 382 (1994).
2. Z. Hiroi et al., *Nature (London)* **371**, 139 (1994).
3. Y. Zenitani et al., *Physica* C **419**, 32 (2005).
4. S. Tajima et al., *Phys. Rev. B* **43**, 10496 (1991).
5. A. Zibold et al., *Phys. Rev. B* **53**, 11734 (1996).
6. A. E. Lavat and E. J. Baran, *Mat. Letters* **63**, 296 (2002).
7. T. Takahashi (Private communication).
8. K. Ohbayashi et al., *J. Phys. Soc. Jpn.* **59**, 1372 (1990).
9. H. Takahashi et al., *Phys. Rev. B* **50**, 3221 (1994).
10. Y. Ohta et al., *Phys. Rev. Lett.* **66**, 1228 (1991).

Synthesis of New Electron-Doped Cuprates $Li_xSr_2CuO_2X_2$ (X= Cl, Br, I)

Tetsuya Kajita[a, b], Masatsune Kato[a, b], Takashi Noji[a, b] and Yoji Koike[a, b]

[a] *Department of Applied Physics, Tohoku University, 6-6-05 Aoba, Aramaki, Aoba-ku, Sendai 980-8579, Japan*
[b]*CREST, Japan Science and Technology Corporation (JST)*

Abstract. We have successfully synthesized a new electron-doped superconducting cuprate $Li_xSr_2CuO_2I_2$ by the electrochemical Li-intercalation technique. The electrochemical intercalation was carried out at room temperature under a constant potential of 500 mV vs. Li/Li^+ using a three-elecrode cell. It has been found that Li^+ ions tend to be intercalated readily with the increase of the ionic radius of X^- in $Sr_2CuO_2X_2$ (X = Cl, Br, I). The magnetic susceptibility measurements have revealed the appearance of superconductivity with the superconducting transition temperature T_c = 4.5 K in $Li_{0.25}Sr_2CuO_2I_2$, which is lower than T_c = 8.0 K in $Li_{0.15}Sr_2CuO_2Br_2$.

Keywords: Superconductor, Li-intercalation, Electron-doping, $Li_xSr_2CuO_2X_2$ (X = Cl, Br, I)
PACS: 74.72.-h, 82.45.Xy

In our previous works, we have succeeded in synthesizing new electron-doped cuprates $Li_xSr_2CuO_2X_2$ (X = Cl, Br) by the electrochemical Li-intercalation technique [1,2]. As shown in Fig. 1, the compound $Sr_2CuO_2X_2$ is a layered perovskite with the K_2NiF_4 structure and essentially isostructural to the well-known hole-doped high-T_c superconductor $(La,Sr)_2CuO_4$. In the rock-salt layer of Sr^{2+} and X^-, X^- ions shift away from the CuO_2 plane, which is due to the larger radius of X^- than that of O^{2-} and the smaller Coulomb attraction between Cu^{2+} and X^- than between Cu^{2+} and O^{2-}. This leads to the formation of X^--X^- double layers, so that Li^+ ions can be intercalated between the electronegative X^- layers, which are weakly bound through the van der Waals force. It has been found that $Li_xSr_2CuO_2Br_2$ shows superconductivity with the superconducting transition temperature T_c = 8.0 K, while no superconductivity appears in $Li_xSr_2CuO_2Cl_2$. The type of carriers doped into the CuO_2 plane changes from hole-like to electron-like with the increase of the *a*-axis length from 3.78 Å in $(La,Sr)_2CuO_4$ to 3.97-3.99 Å in $Sr_2CuO_2X_2$ (X = Cl, Br), indicating the large *a*-axis length would be important for the appearance of electron-doped superconductivity. Therefore, it is interesting to study whether the value of T_c increases for X = I with the larger *a*-axis length of 4.03 Å than those for X = Cl and Br.

In this paper, we report the synthesis of a new electron-doped cuprate $Li_xSr_2CuO_2I_2$ and discuss the superconductivity of $Li_xSr_2CuO_2X_2$ (X = Cl, Br, I).

Polycrystalline host samples of $Sr_2CuO_2I_2$ were synthesized as follows [3]. First, polycrystals of $SrCuO_2$ were prepared from stoichiometric amounts of $SrCO_3$ and CuO powders. The powders were mixed, ground and heated in air at 925 °C for 10 h. The products were then pulverized, pressed into pellets and sintered at 950°C for 20 h. Next, the obtained single-phase samples of $SrCuO_2$ were mixed with a stoichiometric amount of SrI_2 and pressed into pellets. Then the pellets were sintered at 580°C for 24 h in an evacuated quartz ampoule in order to suppress the volatilization of iodine, though no ampoule was necessary in the case of X = Cl and Br. The obtained samples of $Sr_2CuO_2I_2$ were highly unstable in air due to the hygroscopic nature. In the case of X = Cl and Br, the obtained samples of $Sr_2CuO_2X_2$ were mixed with naphthalene, pelletized and then sintered again in air to obtain porous samples which are suitable for the homogeneous intercalation of Li. In the case of X = I, however, porous samples were not prepared, because the volatilization of iodine was inevitable in the process of the volatilization of naphthalene. The electrochemical Li-intercalation was carried out at room temperature in an argon-filled glove box using a three-electrode cell. The working electrode was a pellet of $Sr_2CuO_2I_2$ which was covered with Ni meshes. Li sheets were used as the counter electrode and the reference electrode. As an electrolyte, 1.0 M $LiClO_4$

CP850, *Low Temperature Physics: 24th International Conference on Low Temperature Physics;*
edited by Y. Takano, S. P. Hershfield, S. O. Hill, P. J. Hirschfeld, and A. M. Goldman

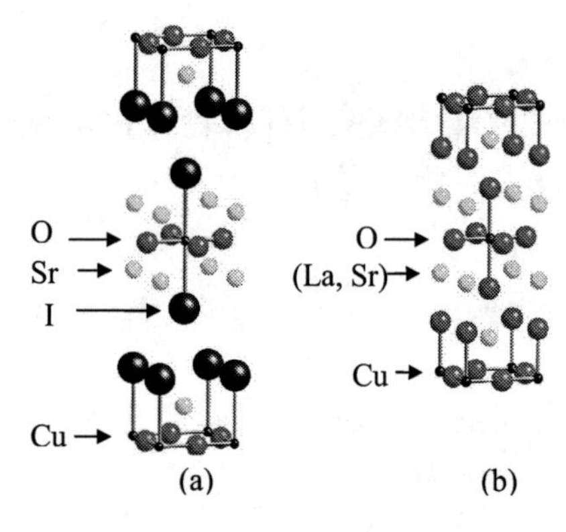

FIGURE 1. Crystal structures of (a) $Sr_2CuO_2I_2$ and (b) $(La, Sr)_2CuO_4$.

dissolved in propylene carbonate (PC) was used. The Li-intercalation was performed under a constant potential of 500 mV (vs. Li/Li^+) using a potentiostat. The total amount of Li intercalated into $Sr_2CuO_2I_2$ was estimated according to the simple Faraday law and also from the ICP analysis. The magnetic susceptibility was measured using a SQUID magnetometer in a magnetic field of 3 Oe.

$Li_xSr_2CuO_2I_2$ with $0.10 \leqq x \leqq 0.28$ were successfully obtained through the Li-intercalation within a half day. On the other hand, the maximum value of x in $Li_xSr_2CuO_2Cl_2$ was less than 0.10 despite the Li-intercalation for 1 week. Probably, Li^+ ions were intercalated readily owing to the large ionic radius of I^-.

The temperature dependence of the magnetic susceptibility, χ , of $Li_{0.25}Sr_2CuO_2I_2$ is shown in Fig. 2. A single-step diamagnetic response due to the Meissner effect is observed below 4.5 K. The superconducting volume fraction at 2 K, which is estimated from the χ value on zero-field cooling, is ~2 %. The value of T_c = 4.5 K is independent of x, while the superconducting volume fraction at 2 K is below 2 %.

The above results suggest that a phase separation takes place in the sample as in the case of $Li_xSr_2CuO_2Br_2$. The sample may include a pristine phase with $x = 0$, a superconducting phase with $x = x_{sc}$ and an amorphous phase of Li-Sr-Cu-O-I with $x > x_{sc}$. In particular, $Li_xSr_2CuO_2I_2$ may include a large amount of the pristine phase, because the surface area into where Li^+ ions are intercalated is small for the non-porous sample. In fact, non-porous samples of $Li_xSr_2CuO_2Br_2$ showed no superconductivity. The phase separation is also observed in several alkali-matal-intercalated superconductors such as $Li_xCsSr_2Nb_3O_{10}$ [4] and Na_xHfNCl [5]. One possible reason of the phase separation is as follows. When Li^+ ions are intercalated, the lattice is deformed. In order to minimize the energy loss due to the deformation, it is likely that Li^+ ions tend to be intercalated into the same layer, leading to the phase separation.

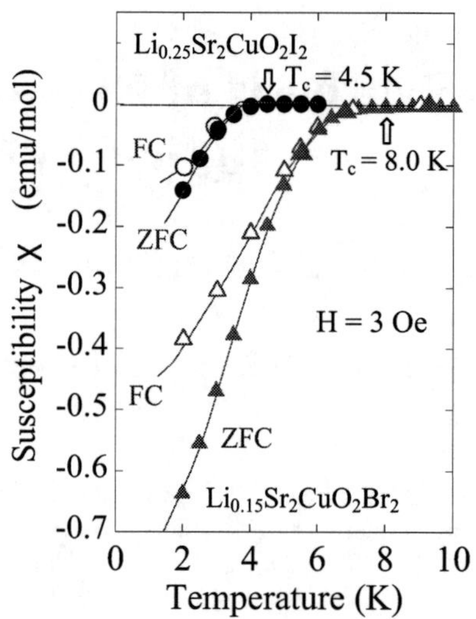

FIGURE 2. Temperature dependence of the magnetic susceptibility, χ , of $Li_{0.25}Sr_2CuO_2I_2$ (● , ○) and $Li_{0.15}Sr_2CuO_2Br_2$ (▲ , △)dmeasured in a magnetic field of 3 Oe on warming after zero-field cooling (ZFC) and on field cooling (FC).

The increase of the a-axis length, namely, the increase of the Cu-O bond length in the CuO_2 plane decrease of the Madelung potential at the Cu site, so that electronegative electron carriers tend to be readily introduced into the CuO_2 plane, as discussed in our previous reports [1, 2]. Accordingly, the Li-content x_{sc} for X = I is expected to be larger than that for X = Br. Therefore, it may follow that the superconducting phase in $Li_xSr_2CuO_2I_2$ is situated in the heavily overdoped regime, so that T_c = 4.5 K for X = I is lower than T_c = 8.0 K for X = Br. A higher T_c value may be obtained in the present system by controlling the carrier concentration.

In conclusion, a new electron-doped superconducting cuprate $Li_xSr_2CuO_2I_2$ with T_c = 4.5 K has successfully been synthesized by the electrochemical Li-intercalation technique. The suppression of the phase separation in the Li-intercalation compounds is an important issue to obtain a higher T_c value. Anyway, Li-intercalated oxyhalides with the K_2NiF_4 structure are good candidates for new electron-doped superconductors.

This work was supported by a Grant-in-Aid for Scientific Research from the Ministry of Education, Culture, Sports, Science and Technology.

REFERENCES

1. T. Kajita *et al., Jpn. J. Appl. Phys.* **43**, L1480 (2004).
2. T. Kajita *et al.*, Proc. of ISS2004, *Physica C*, in press.
3. C. S. Knee *et al., J.Mater.Chem.* **13**, 1507 (2003).
4. M. Kato *et al., Physica C* **388-389**, 445 (2003).
5. S. Shamoto *et al., Physica C* **402**, 283 (2004).

Specific Heat of Layered Ruthenates $Sr_2Ru_{1-x}Zr_xO_4$

N. Umeyama, S. I. Ikeda, I. Nagai, Y. Tanaka, Y. Yoshida, and N. Shirakawa

Nanoelectronics Research Institute (NeRI), National Institute of Advanced Industrial Science and Technology (AIST), Umezono 1-1-1, Tsukuba, Ibaraki, 305-8568, Japan

Abstract. We have measured specific heat (C_p) for polycrystalline single-phased $Sr_2Ru_{1-x}Zr_xO_4$ ($0.0 \leq x \leq 0.8$). The value of C_p at low temperatures is reduced with increasing x. Both of the Debye temperature (Θ) and the electronic specific heat coefficient (γ) estimated from the data decreased with increasing x for $x < 0.4$. The smaller γ towards x = 0.4 seems to blend into the insulating transport property for $x \geq 0.4$. We argue that the decrease of Θ with increasing x stems from the increased unit-cell volume.

Keywords: Ruddlesden-Popper phase, layered perovskite, 4d-electron, ruthenates, zirconates, specific heat
PACS: 65.40.Ba Heat capacity, 74.70.Pq Ruthenates

INTRODUCTION

The layered perovskite Sr_2RuO_4 is metallic and Pauli paramagnetic, and shows a spin-triplet superconductivity (T_c = 1.5 K) [1]. Since a spin-triplet superconductor has the internal degree of freedom, it may be controlled by a substitutional element or a pressure. We focused on Zr-substituted Sr_2RuO_4, which is predicted to have a strong structural strain by the large difference in the ionic radius between Ru^{4+} (0.062 nm) and Zr^{4+} (0.072 nm) [2]. The number of 4d electrons decreases to zero by substituting Zr for Ru. Therefore, Zr-substituted Sr_2RuO_4 is expected to show a metal-insulator transition and the decrease of the electronic specific heat coefficient (γ).

Recently, we have succeeded in the synthesis of polycrystalline single-phase $Sr_2Ru_{1-x}Zr_xO_4$ ($0.0 \leq x \leq 0.8$) (SRZ214) [3] by a conventional solid-state reaction method. The unit-cell volume increased by about 10 % by the Zr-substitution of x = 0.8. Here, we report the x-dependence of the specific heat in SRZ214.

EXPERIMENT

The measurements of the specific heat at constant pressure (C_p) were performed on a small piece of each sample of SRZ214 by a relaxation method (Quantum Design PPMS) in the temperature (T) range between 2 K and 30 K. The samples were checked for phase purity by a powder x-ray diffraction method. Since these polycrystalline-sample measurements might be affected by the grain boundaries, we prudently confirmed whether the time-relaxation curve of the sample measurements has a shape similar to that of addenda measurements, where the addenda is constructed with the wire, the Apiezon-N grease, and the sample platform. In addition, we have roughly checked the electrical resistivity of these samples by the four-probe technique at room temperature.

RESULTS and DISCUSSION

In the results for SRZ214, the value of C_p above about 15 K increased with increasing x, while that of C_p at ~2 K decreased with increasing x. For a detailed discussion, we present the data in Fig. 1 as a C_p / T vs. T^2 plot. For $x \leq 0.2$, T^2-dependences of C_p / T follow straight lines below 12 K. For $x \geq 0.4$, they follow straight lines only in the range, 4 K $\leq T \leq$ 12 K, and they seem to decrease towards zero below 4 K except for x = 0.6, which seems to have a finite γ.

The samples of $x \geq 0.4$ have an insulating conductance at room temperature, and the T-dependences of the magnetic susceptibility with varying x in the range, 2 K $\leq T \leq$ 350 K, have not shown a phase transition [3]. The samples of $x \geq 0.4$ should be insulators at low temperatures.

CP850, *Low Temperature Physics: 24th International Conference on Low Temperature Physics;*
edited by Y. Takano, S. P. Hershfield, S. O. Hill, P. J. Hirschfeld, and A. M. Goldman

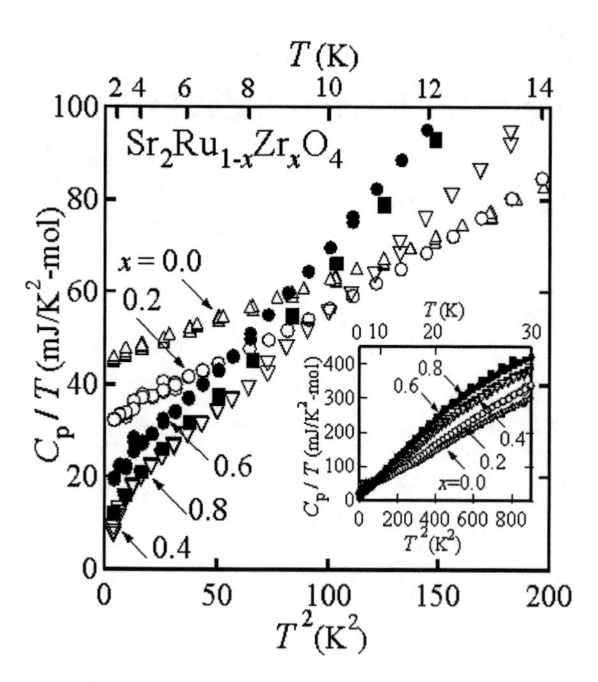

FIGURE 1. Low-temperature specific heat per SRZ214-mole with varying x plotted in C_p / T vs. T^2. The inset shows the data below 30K.

The Debye model at low-temperature limit ($T \ll \Theta$) is given by

$$C_p \approx C_v = \gamma T + \frac{12\pi^4}{5} N k_B \frac{1}{\Theta^3} T^3, \quad (1)$$

where C_v is specific heat at constant volume, Θ the Debye temperature, N the number of oscillators per unit cell, and k_B the Boltzmann's constant. The x-dependence of C_p / T and Θ obtained by fitting this formula to the data for $4\text{ K} \le T \le 12\text{ K}$ are displayed in Fig. 2, where C_p / T of $x \le 0.2$ can be regarded as γ.

For the results of $x = 0.0$ (Sr_2RuO_4), $\Theta = 417$ K and $\gamma = 45$ mJ/K^2-mol are in agreement with the results of the previous studies, which were $\Theta = 414$ K and $\gamma = 45.6$ mJ/K^2-mol [4] from the measurement for the polycrystal. The value of γ decreases with increasing x for $x < 0.4$, and tends to zero towards 0 K for $x = 0.4$ and 0.8. The results are consistent with the insulating transport property for $x \ge 0.4$. The sample for $x = 0.6$ is anomalous in that it seems to exhibit a finite density of states. Although an enhancement is seen in the susceptibility at 2 K for the same x, more detailed studies are necessary to sort out the nature of this anomaly.

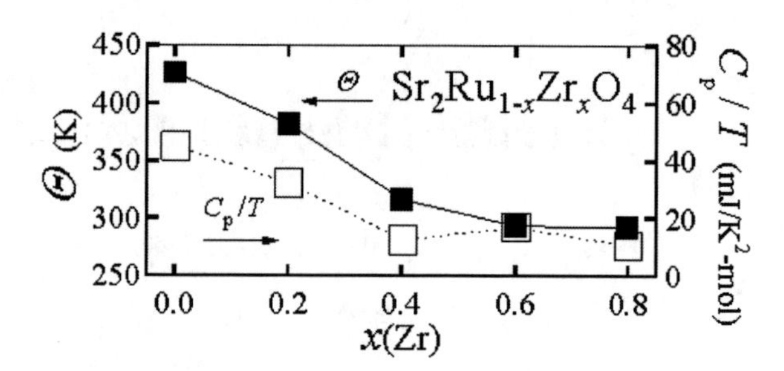

FIGURE 2. Debye temperature Θ (■) and C_p / T (□) with varying x, estimated from the data of SRZ214. Here, for $x \le 0.2$, C_p / T can be regarded as the electronic specific heat coefficient γ.

The value of Θ also decreases with increasing x, and is about 300 K for $x \ge 0.4$. In the Debye model, the Θ is given by,

$$\Theta = \frac{\hbar v}{k_B}\left(\frac{6\pi^2 N}{V}\right)^{1/3}, \quad (2)$$

where v is the sound velocity and V the volume per mole. We have found that V increases with increasing x in SRZ214 [3]. This lattice expansion may be the main cause of the decrease in Θ.

ACKNOWLEDGMENTS

We are grateful to I. Hase and T. Yanagisawa for discussing the physical properties for SRZ214.

REFERENCES

1. A. P. Mackenzie, and Y. Maeno, *Rev. Mod. Phys.* **75**, 657 (2003).
2. R. D. Shannon, *Acta Crystallogr. Sect.* **A 32,** 751 (1976).
3. N. Umeyama, S. I. Ikeda, I. Nagai, and Y. Yoshida, (unpublished).
4. J. J. Neumeier, M. F. Hundley, M. G. Smith, J. D. Thompson, C. Allgeier, H. Xie, W. Yelon, and J. S. Kim, *Phys. Rev. B* **50**, 17910-17916 (1994).

Scanning Tunneling Microscopy and Spectroscopy of Sr_2RuO_4

H. Kambara, Y. Niimi, K. Takizawa*, H. Yaguchi*, Y. Maeno*†, and Hiroshi Fukuyama

Department of Physics, Graduate School of Science, University of Tokyo, 7-3-1 Hongo, Bunkyo-ku, Tokyo 113-0033, Japan
**Department of Physics, Graduate School of Science, Kyoto University, Kitashirakawa, Sakyo-ku, Kyoto 606-8502, Japan*
†Kyoto University International Innovation Center, Kyoto 606-8501, Japan

Abstract. We report scanning tunneling microscopy/spectroscopy (STM/STS) studies of the spin-triplet superconductor Sr_2RuO_4 at temperatures down to 30 mK and in magnetic fields up to 6 T. Clean sample surfaces were prepared by cleaving the single crystals along the *ab* plane at 100 K in an ultrahigh vacuum environment. In most cases of the 1.5-K phase samples ($T_c \sim 1.5$ K), we observed a gap structure ($\Delta \sim 5$ meV) in the local density of states for cleaved surfaces which are likely the SrO planes. This gap is *not* related to the superconducting state and consistent with the data reported recently by C. Lupien *et al.* On the other hand, for the 3-K phase samples (Sr_2RuO_4–Ru eutectic system, $T_c \sim 3$ K), thin line defects of about 1 nm width were observed. The topography of the defects depends strongly on the bias voltage applied between sample and tip, which suggests that the electronic state varies largely near the defects.

Keywords: Sr_2RuO_4, Spin-triplet superconductor, STM/STS
PACS: 68.37.Ef, 74.20.Rp, 74.70.Pq

INTRODUCTION

In recent years, there has been a considerable progress in the understanding of the unconventional superconductor Sr_2RuO_4, which is considered to have the spin-triplet *p*-wave Cooper pairing symmetry [1]. A recent π-SQUID experiment demonstrated the odd-parity symmetry, in contrast with the *s* or *d*-wave pairing [2]. It is strongly demanded to investigate the peculiar electronic state of Sr_2RuO_4 much extensively, for example by using the scanning tunneling microscopy/spectroscopy (STM/STS) technique. This technique enables us to measure the local electronic density of states (LDOS), which had played an important role to determine the *d*-wave symmetry in the high-T_c cuprate, $Bi_2Sr_2CaCu_2O_{8+\delta}$ [3]. Very recently, an STM/STS study of the superconducting state of Sr_2RuO_4 was reported by Lupien *et al.* [4]. These authors observed two kinds of tunnel conductance spectra. One displays a large gap-like structure which is not originated from the superconducting state, while the other one does a narrow superconducting gap. The superconducting gap was observed only when the topmost layer happened to be a RuO plane.

In this report, we present results of our STM/STS studies for not only the 1.5-K phase samples but also the 3-K phase (Sr_2RuO_4–Ru eutectic [5]) ones. Single crystals of Sr_2RuO_4 were grown in an infrared image furnace by the floating zone method [6]. The T_c values for the 1.5-K phase samples were 1.30 ~ 1.35 K determined from AC magnetic susceptibility measurements. The experiments were carried out using an ultra-low temperature STM which works at temperatures down to 30 mK, in magnetic fields up to 6 T and in ultra-high vacuum (UHV) [7]. Samples were cleaved at 100 K and at pressures below 1×10^{-7} Pa before being transferred to the STM head while maintaining UHV conditions.

CP850, *Low Temperature Physics: 24th International Conference on Low Temperature Physics;*
edited by Y. Takano, S. P. Hershfield, S. O. Hill, P. J. Hirschfeld, and A. M. Goldman

RESULTS

Figures 1(a)(b) show an STM image and a differential conductance (d*I*/d*V*) curve at 30 mK for the 1.5-K phase sample. The surface atomic structure is clearly resolved in Fig. 1(a). In this experiment only this type of surface structure was observed. It is believed that the uppermost layer is the SrO plane because the interlayer bonding between Sr-O is weaker than that of Ru-O. We did not observe any surface reconstructions in contrast to the report by Matzdorf *et al.* [8]. The observed STM images are consistent with those reported by Barker *et al.* [9]. The dark spot in Fig. 1(a) is not an atomic vacancy but more likely an impurity atom embedded slightly below the surface. This is because the dark spot has a slightly higher LDOS at low bias voltages below -50 meV and the observed corrugation (0.03~0.06 nm) is too small for the vacancy.

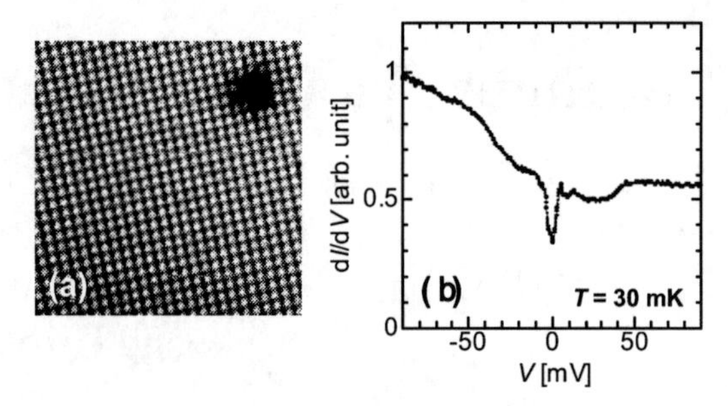

FIGURE 1. (a) STM image (9×9 nm^2, V = -0.1 V, I = 0.2 nA, T = 30 mK) and (b) Typical differential conductance (d*I*/d*V*) curve for the 1.5-K phase of Sr_2RuO_4.

The typical d*I*/d*V* curve shows an asymmetric spectrum within ±100 meV as well as the gap-like structure ($\Delta \approx$ 5 meV) (see Fig. 1(b)). The superconducting gap ($\Delta \sim$ 0.4 meV) feature was never observed inside the large gap. This result is consistent with the previous works [4, 9]. The large gap structure was observed even at temperatures higher than T_c or in magnetic fields much higher than H_{c2}, which clearly indicates that this feature is not related to superconductivity. We also found that this gap-like structure is not strongly influenced by the surface defect (dark spot) density which varies at each cleavage even for samples of the same batch. The physical origin of the gap is not known yet.

On the other hand, for the 3-K phase samples, it was more difficult to obtain smooth cleaned surfaces because of the existence of Ru-lamella inclusions. Nevertheless, we obtained STM images with line defects of atomic size width as shown in Fig. 2(a). Figures 2(b)-(d) show the sample-bias voltage (V) dependence of the STM images near the line defect (5×10 nm^2 each). The appearance of the defect changes greatly with V from –1.0 V to –0.15 V. There are no crystalline steps at these line defects which always go along the [110] direction (see Fig. 2(d)). This means that the variations of contrast near the line defects reflect those of the electronic states and not of the topography. We found that the electronic state on the SrO surface of Sr_2RuO_4 is widely homogeneous except around the few defects present (points or lines). This feature is rather different from what was observed in the high-T_c cuprates, for example $Bi_2Sr_2CaCu_2O_{8+\delta}$ [10].

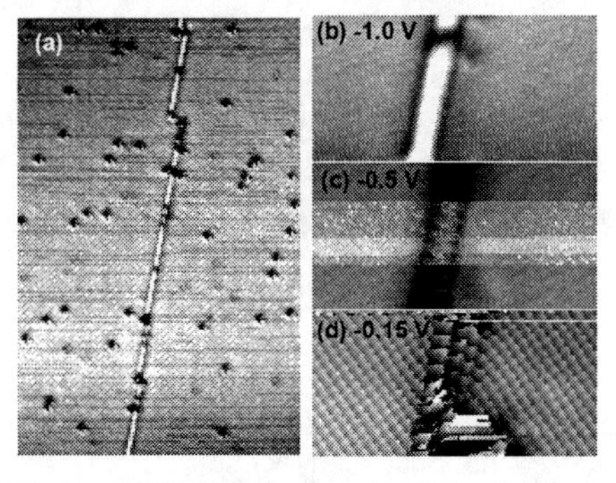

FIGURE 2. STM images for the 3-K phase sample of Sr_2RuO_4. (a) Line defects are running from top to bottom (75×50 nm^2, V = -1.0 V, I = 0.12 nA, T = 30 mK). (b)-(d) STM images (5×10 nm^2) at V = -1.0 V (b), -0.5 V (c), -0.15 V (d). The positions are not exactly the same.

ACKNOWLEDGEMENTS

We thank C. Winkelmann for useful discussions. This work was supported by Grants-in-Aid for Scientific Research from MEXT, Japan and ERATO Project of JST.

REFERENCES

1. Mackenzie, A. P. and Maeno, Y., *Rev. Mod. Phys.* **75**, 657 (2003).
2. Nelson, K. D. *et al.*, *Science* **306**, 1151 (2004).
3. Pan, S. H. *et al.*, *Nature* **403**, 746 (2000).
4. Lupien, C. *et al.*, cond-mat/0503317.
5. Maeno, Y. *et al.*, *Phys. Rev. Lett.* **81**, 3765 (1998).
6. Mao, Z. Q. *et al.*, *Mat. Res. Bull.* **35**, 1813 (2000).
7. Matsui, T. *et al.*, *Physica B.* **329-333**, 1653 (2003), Kambara, H. *et al.*, *J. Phys. Chem. Sol.* **66**, 1552 (2005).
8. Matzdorf, R. *et al.*, *Science* **289**, 746 (2000).
9. Barker, B. I. *et al.*, *Physica B.* **329-333**, 1334 (2003).
10. Pan, S. H. *et al.*, *Nature* **413**, 282 (2001).

Search for Spontaneous Field at the Ru Site in the Superconducting State of Sr_2RuO_4

Hiroshi Murakawa,[1] Kenji Ishida,[1] Kentaro Kitagawa,[1] Z.Q.Mao,[1†] and Yoshiteru Maeno[1,2]

[1]*Department of Physics.Graduate School of Scienec, Kyoto University, Kyoto 606-8502, Japan*
[2]*International Innovation Center, Kyoto University, Kyoto 606-8501, Japan*

Abstract. The linewidth of the Ru nuclear quadrupole resonance (NQR) spectrum was measured in the superconducting state of Sr_2RuO_4 in order to search for the spontaneous field at the Ru site. We considered that the large hyperfine field would be expected at the Ru site in the superconducting state if the spontaneous field revealed by the μSR measurement originated from the nonzero spin moment. The linewidth of the spectrum is unchanged within an experimental error, which gives that the maximum value of the spontaneous field at the Ru nucleus site is less than 5 G if any. Our result suggests that the spontaneous field originates from the orbital component of the spin-triplet pairs.

Keywords: Sr_2RuO_4, spin-triplet superconductivity, Ru NQR, time-reversal-symmetry
PACS: 74.70 Pq, 76.60.-k, 70.60.G

Since the discovery of superconductivity in Sr_2RuO_4 [1], numerous experimental and theoretical studies have been performed [2]. Until now, the following superconducting (SC) properties have been confirmed. (1) ^{17}O [3,4] and ^{99}Ru [5] Knight shifts in the field parallel to the RuO_2 plane reveal that the SC pairs are in a spin-triplet state with the spin component in the RuO_2 plane. (2) Muon spin relaxation (μSR) experiments suggest that time reversal symmetry is broken in the SC state [6]. (3) From measurements of the nuclear spin-lattice relaxation rate $1/T_1$ [7], the penetration depth [8], and angle-dependent specific heat in the SC state [9], the strongly anisotropic gap structure containing nodes or very deep gap minima is revealed. Taken these experimental results together, the spin triplet SC state expressed as $d(\boldsymbol{k}) = z\Delta_0(\sin k_x + i\sin k_y)$ is promising [9], for which the SC *d*-vector, perpendicular to the spin direction of the Cooper pair, points to the *c* axis.

Quite recently, it has been reported that ^{101}Ru Knight shift measurement in the magnetic field parallel to the *c* axis suggests that the SC *d*-vector can be rotated by a small magnetic field, and that the *d*-vector is in the RuO_2 plane in the field range $H > 200$ Oe [10]. This implies that the spin-orbit part of the pairing interaction, which keeps the spins in the RuO_2 plane, is very weak, about 25 mK. To identify the *d*-vector in the ground state (H=0), further measurements are still needed.

Another interesting feature of the SC properties in Sr_2RuO_4 is the appearance of a spontaneous field below T_c by the μSR measurement [6]. The magnitude of the spontaneous field is ~ 0.5 G, independent of the crystal direction. Although the observation of the spontaneous field is reproducible among the μSR measurements, it has not been detected by other experimental techniques. Recently, a similar spontaneous field was observed in μSR measurements in the Pr-filled skutterudite superconductor $PrOs_4Sb_{12}$; its magnitude is ~ 1.2 G [11]. Although the appearance of the spontaneous field is strong evidence of unconventional superconductivity in these compounds, the origin of the spontaneous field is not as yet well understood.

Probably, the spontaneous field originates from the spin or orbital part of the spin triplet Cooper pairs with $S = 1$, $L = 1$. We point out that the large hyperfine field would be induced at the Ru-nucleus site if the spontaneous field originated from the spin part of the pairs. This is because the hyperfine coupling between the Ru nucleus and electron spins (-250 kG/μ_B) [5] is two orders of magnitude larger than that between the muon and electron spins (~ 1 kG/μ_B), which is determined by the dipolar interaction. In this case, it is expected that the hyperfine field of about 120 G arises at the Ru site, which would broaden the Ru nuclear magnetic resonance (NQR) spectrum in the SC state. On the other hand, when the spontaneous field

CP850, *Low Temperature Physics: 24th International Conference on Low Temperature Physics;*
edited by Y. Takano, S. P. Hershfield, S. O. Hill, P. J. Hirschfeld, and A. M. Goldman

originates from the orbital part of the pairs, the spontaneous supercurrents induced in the vicinity of impurity surfaces, and/or domain walls give rise to the macroscopic spontaneous field, which would be the same at the Ru site and the implanted muon site (~ 0.5 G). Thus, the measurements of internal field at the Ru site should give us crucial evidence of the origin of the spontaneous field. Here we report the measurements of the linewidth of the Ru NQR spectrum under zero external field down to 84 mK.

We used three pieces of high-quality single crystals from the same batch, showing a high SC transition temperature of T_c = 1.5 K. ^{101}Ru-NQR-signal is observed at 3.28 MHz (corresponding to the ±1/2↔±3/2 transition), and at 6.56 MHz (the ±3/2↔±5/2 transition). We measured the full width at half maximum (FWHM) of the lower-frequency NQR signal. This is because the spectrum is sharper than the higher-frequency spectrum by a factor of two, implying that the small change of the linewidth can be precisely detected in the spectrum. When the internal field appears, it lifts the degeneracy of the (1/2↔3/2) and (-1/2↔-3/2) transitions, which makes the Ru spectrum broader by the ratio of $\gamma/2\pi$ = 0.22 kHz / G.

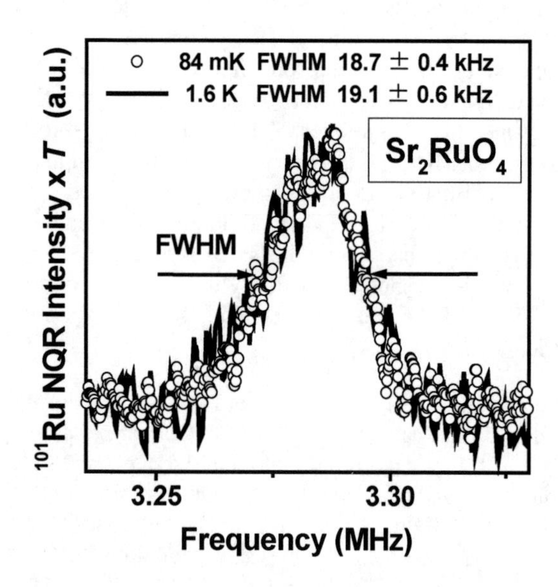

FIGURE 1. Ru NQR spectrum at 1.6 K and 84 mK, arising from the ±1/2↔±3/2 transition

Figure 1 shows the Ru-NQR spectra arising from the ±1/2↔±3/2 transition at 1.6 K and 84 mK. The FWHM at 1.6 K and 84 mK are 19.1 ± 0.6 kHz and 18.7 ± 0.4 kHz, respectively. The linewidth of the spectrum remains unchanged across T_c, showing that the appearance of the internal field cannot be detected within experimental errors (± 1 kHz ~ ± 5 G). If the spontaneous field originates from the spin part of the pairs, the appearance of the internal field is estimated to be ~ 120 G (~ 26 kHz) using the hyperfine coupling at the Ru site (-250 kG/μ_B) and that at the muon site (~ 1 kG/μ_B). Obviously, this is not the case. From the present measurement of the Ru NQR spectrum, we can say that the maximum value of the spontaneous field at the Ru nucleus site is less than 5 G if any. The tiny spontaneous field (~ 0.5 G) revealed by the μSR measurement is out of the resolution, and cannot be detected from the Ru NQR measurement. Therefore the absence of the broadening of the ^{101}Ru spectrum suggests that the spontaneous field originates from the orbital part of the pairs. However, the nature of the supercurrent which induces the spontaneous field is still unclear. What kind of impurities are effective or how large the domain size for a single spontaneous field is, should be clarified.

In conclusion, we measured the linewidth of the ^{101}Ru-NQR spectrum in the SC state to search for the spontaneous field at the Ru site. The linewidth of the spectrum is unchanged within an experimental error (~ 5 G), suggesting that the spontaneous field originates from the orbital component of the spin-triplet pairs.

ACKNOWLEDGMENTS

This work was partially supported by the 21 COE program on ``Center for Diversity and Universality in Physics" from MEXT of Japan, and by Grants-in-Aid for Scientific Research from the Japan Society for the Promotion of Science (JSPS) and MEXT.

REFERENCES

† present address :
Department of Physics, Tulane University

1. Y. Maeno *et al., Nature* (London) **372,** 532 (1994).
2. A. P. Mackenzie and Y. Maeno, *Rev. Mod. Phys.* **75,** 657 (2003).
3. K. Ishida *et al., Nature* (London) **396,** 658 (1998).
4. H. Mukuda *et al., J. Low. Temp. Phys.* **117,** 1587 (1999).
5. K. Ishida *et al., Phys. Rev. B* **63,** 060507 (2001).
6. G. M. Luke *et al., Nature* (London) **394,** 558 (1998).
7. K. Ishida *et al., Phys. Rev. Lett.* **84,** 5387 (2000).
8. I. Bonalde *et al., Phys. Rev. Lett.* **85,** 4775 (2000).
9. K. Deguchi *et al., Phys. Rev. Lett.* **92,** 047002 (2004).
10. H. Murakawa *et al., Phys. Rev. Lett* **93,** 167004 (2004).
11. Y. Aoki *et al., Phys. Rev. Lett.* **91,** 067003 (2003).

Spectroscopy of Sr_2RuO_4/Ru Junctions in Eutectic

H. Yaguchi [a], K. Takizawa [a], M. Kawamura [b,c], N. Kikugawa [a], Y. Maeno [a,d], T. Meno [e], T. Akazaki [b], K. Semba [b] and H. Takayanagi [b]

[a] *Department of Physics, Graduate School of Science, Kyoto University, Kyoto 606-8502, Japan*
[b] *NTT Basic Research Laboratories, NTT Corporation, Atsugi, Kanagawa 243-0198*
[c] *RIKEN, Wako, Saitama 351-0198, Japan*
[d] *International Innovation Center, Kyoto University, Kyoto 606-8501, Japan*
[e] *NTT-ATN Corporation, Atsugi, Kanagawa 243-0018, Japan*

Abstract. We have investigated the tunnelling properties of the interface between superconducting Sr_2RuO_4 and a single Ru inclusion in eutectic. By using a micro-fabrication technique, we have made Sr_2RuO_4/Ru junctions on the eutectic system that consists of Sr_2RuO_4 and Ru micro-inclusions. Such a eutectic system exhibits surface superconductivity, called the 3-K phase. A zero bias conductance peak (ZBCP) was observed in the 3-K phase. We propose to use the onset of the ZBCP to delineate the phase boundary of a time-reversal symmetry breaking state.

Keywords: Sr_2RuO_4, tunnelling spectroscopy, zero bias conductance peak, spin-triplet superconductivity, ruthenate
PACS: 74.45.+c, 74.70.Pq, 74.81.-g

It is now well established that the superconductor Sr_2RuO_4[1] is indeed a spin-triplet superconductor [2]. Importantly, an NMR experiment first revealed that the Knight shift of ^{17}O is not affected by the superconducting (SC) transition, indicative of the spin state of Cooper pairs is being triplet [3]. Subsequently, a muon spin relaxation measurement demonstrated that spontaneous magnetic moments accompany the SC transition, indicative of time reversal symmetry breaking in the SC phase [4]. Taken these results together, the basic form of the vector order parameter is constrained to be $\boldsymbol{d} = z(k_x + ik_y.)$, corresponding to the chiral state. However, this basic form is too simplified to explain existing experimental results. The SC gap structure recently proposed based on detailed experiments on the angle dependence of the specific heat has successfully reconciled the discrepancy between the basic form of the order parameter and the existing experimental results [5].

Amongst the many interesting superconducting properties of Sr_2RuO_4, the enhancement of the SC transition temperature in Sr_2RuO_4-Ru eutectic is rather surprising [6]. This eutectic system, consisting of Sr_2RuO_4 and micron-size Ru inclusions, shows a broad SC transition with an onset of approximately 3 K, called the 3-K phase. Several experiments suggest that 3-K phase superconductivity is filamentary and occurs in the Sr_2RuO_4 side of the interface with Ru [7].

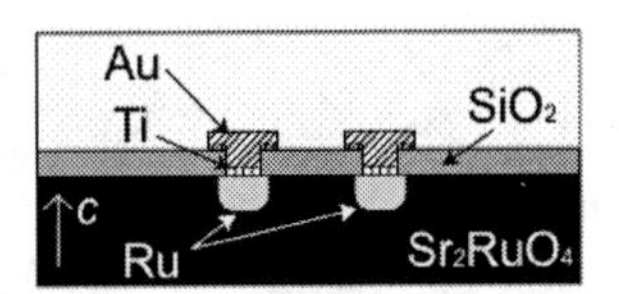

FIGURE 1. Schematic of the device fabricated on Sr_2RuO_4-Ru eutectic.

In order to measure the conductance at the interface between a single Ru micro inclusion and Sr_2RuO_4, we fabricated the device illustrated in Fig. 1 by using a micro-fabrication technique [8]. The device has a Ti/Au electrode directly attached to individual Ru micro-inclusions through 2 μm x 3 μm contact windows. The SiO_2 film is an insulating layer deposited on the polished ab-plane surface of Sr_2RuO_4-Ru eutectic. We were able to achieve good contacts only on Ru inclusions because of a non-superconducting layer forming on Sr_2RuO_4 surface. For actual measurements, therefore two Sr_2RuO_4/Ru junctions in series are inevitably involved.

Figure 2 shows the normalised differential conductance as a function of the bias voltage at temperatures between 0.3 K and 2.1 K. A clear zero bias conductance peak is seen, which is a hallmark of unconventional superconductivity and suggests a sign change of the order parameter on the Fermi surface [9]. As the ZBCP persists to a temperature well above

CP850, *Low Temperature Physics: 24th International Conference on Low Temperature Physics;*
edited by Y. Takano, S. P. Hershfield, S. O. Hill, P. J. Hirschfeld, and A. M. Goldman

1.5 K, the ideal T_c of bulk Sr_2RuO_4, the 3-K phase is responsible for the ZBCP. While the measurements involve *apparently* two junctions in series, we have confirmed only one of the two junctions *i.e.* a single interface of Sr_2RuO_4 and Ru is responsible for the ZBCP shown in Fig. 1 by the following way: We made three sets of measurements involving a third junction. Among them, one set of the measurements using the third junction gave rather featureless spectra, while the other two combinations gave the ZBCP. Thus we conclude that only one of the three junctions is responsible for the observed ZBCP. Whilst the spectra obtained in the present study are very similar to those previously reported on experiments using a break junction technique in Ref. 10, the width of the whole conductance peak in the present study is about the half that reported in Ref. 10. We suggest that this quantitative difference between the two works may be attributed effectively two junctions in series are involved in Ref. 10.

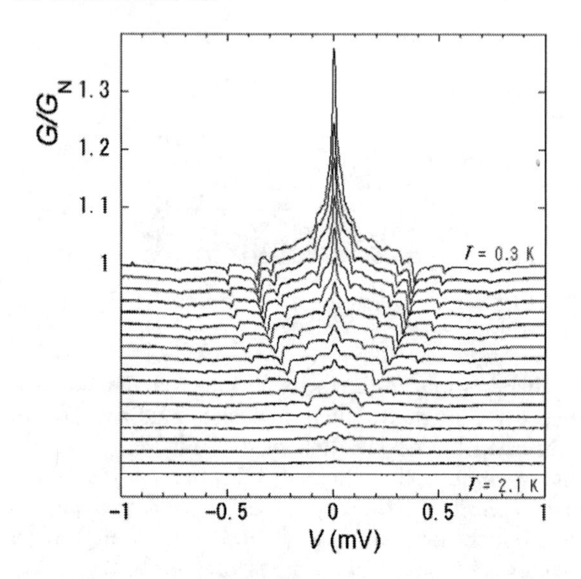

FIGURE 2. Normalised differential conductance at $H = 0$ as a function of the bias voltage at temperatures between 0.3 K and 2.1 K with a step of 0.1 K. A zero bias conductance peak is seen. Traces have been offset for clarity

We have also investigated the spectra in magnetic fields and observed the ZBCP. We have defined the onset magnetic field H^* at which the ZBCP commences. Figure 3 is a plot of such H^* as a function of temperature T, in addition to $H_{c2}(T)$ from Ref. 11. Noticeably, There is a clear difference between H^* and H_{c2} for H // ab or $H \sim 0$ whilst those fields match with each other rather well for H // c. Based on Sigrist and Monien's phenomenological theory, we suggest that the onset of the ZBCP H^* corresponds to the onset of the chiral state. For H // ab or $H \sim 0$, the nucleation of the state $\boldsymbol{d} = zk_x$ occurs at the onset of the 3-K phase and the additional k_y component with a relative phase of $\pi / 2$, necessary for a time-reversal symmetry breaking state, is induced at a lower temperature. For H // c, the onset of the 3-K phase already corresponds to a time-reversal symmetry breaking state due to the c-axis component of the applied field. A more detailed discussion with the help of Sigrist and Monien's theory [12] is described in Ref. 8.

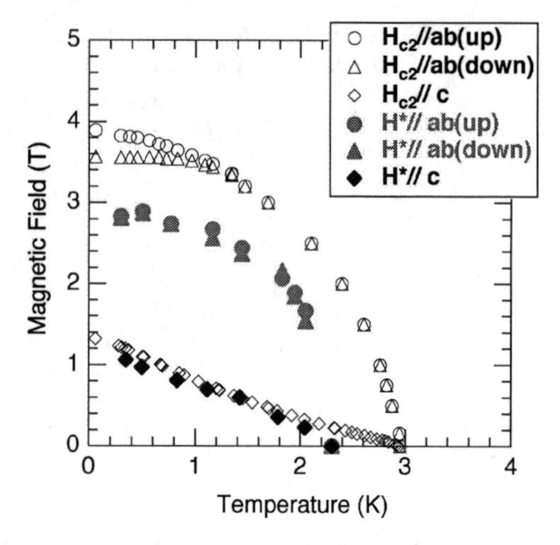

FIGURE 3. Onset of the ZBCP, H^* and the upper critical field H_{c2} as a function of temperature. There is a clear difference between H^* and H_{c2} for H // ab or $H \sim 0$.

ACKNOWLEDGEMENTS

This work was supported in part by a Grant-in-Aid for Scientific Research from the JSPS and 21COE program on "Center for Diversity and Universality in Physics" from the MEXT of Japan.

REFERENCES

1. Y. Maeno *et al.*, *Nature* **372**, 532 (1994).
2. A. P. Mackenzie and Y. Maeno, *Rev. Mod. Phys.* **75**, 657 (2003).
3. K. Ishida *et al.*, *Nature* **396**, 658 (1998).
4. G. M. Luke *et al.*, *Nature* **394**, 558 (1998).
5. K. Deguchi *et al.*, *J. Phys. Soc. Jpn.* **73**, 1313 (2004).
6. Y. Maeno *et al.*, *Phys. Rev. Lett.* **81**, 3765 (1998).
7. T. Ando *et al.*, *J. Phys. Soc. Jpn.* **68**, 1651 (1999). H. Yaguchi *et al.*, *Phys. Rev. B* **67**, 214519 (2003).
8. M. Kawamura *et al.*, *J. Phys. Soc. Jpn.* **74**, 531 (2005).
9. Y. Tanaka and S. Kashiwaya, *Phys. Rev. Lett.* **74**, 3451 (1995).
10. Z. Q. Mao *et al.*, *Phys. Rev. Lett.* **87**, 037003 (2001).
11. H. Yaguchi *et al.*, *Physica C* **388-389**, 507 (2003).
12. M. Sigrist and H. Monien, *J. Phys. Soc. Jpn.* **70**, 2409 (2001).

Structural Properties of $GdSr_2RuCu_2O_8$ under Strong Compression

Ravhi S. Kumar [a], Andrew L. Cornelius [a], J. Janaki [b], T. Geetha Kumary [b], and Malcolm. F. Nicol [a]

[a]High Pressure Science and Engineering Center (HiPSEC) and Department of Physics, University of Nevada Las Vegas, Las Vegas, NV 89154, USA
[b]Materials Science Division, IGCAR, Kalpakkam 603102, India

Abstract. The crystal structure of hybrid ruthenocuprate superconductor $GdSr_2RuCu_2O_8$ has been investigated up to 32 GPa using a diamond anvil cell (DAC) and synchrotron x-rays in the angle dispersive geometry. The tetragonal structure is found to be stable and no structural anomalies were found up to the experimental limit. The bulk modulus obtained by fitting the PV data with Birch-Murnaghan equation yielded $B_0 = 147$ (4) GPa with $B_0' = 8.5$. This value is higher than the one reported for the isostructural Y-123 compound and may be the result of partial replacement of rigid $Ru\text{-}O_2$ layers for $Cu\text{-}O_2$ layers.

Keywords: X-ray diffraction, High pressure, Equation of state, Diamond anvil cell.
PACS: 61.10.Nz, 62.50. + p, 64.30.+t

INTRODUCTION

The discovery of high temperature superconductivity ($T_c \approx 46K$) with ferromagnetic state ($T_N \approx 133K$) in the ruthenocuprate $GdSr_2RuCu_2O_8$ (Gd-1212) has drawn much attention in recent years [1,2]. The crystal structure of Gd-1212 is similar to $YBa_2Cu_3O_7$ (Y-123) and derived on replacing Y by Gd, Ba by Sr and $Cu\text{-}O_2$ by $Ru\text{-}O_2$ layers. Neutron diffraction experiments show a modified Y-123 structure with an enlarged unit cell resulting due to the rotation of $Ru\text{-}O_6$ octahedra and an abrupt change in the buckling of $Cu\text{-}O_2$ layers at ferromagnetic boundary manifesting itinerant ferromagnetism[3] residing in the $Ru\text{-}O_2$ layers. Further, recent Raman studies show doubling of unit cell due to $Ru\text{-}O_6$ octahedral rotations and a possible structural transition induced by magnetic ordering [4]. Application of pressure is found to enhance the ferromagnetic ordering and a competition between magnetic and superconducting phases has been observed in the transport measurements. [5, 6]. To elucidate the behavior of $Ru\text{-}O_2$ layers under pressure and to understand the structural response of this compound, we have investigated the crystal structure up to 32 GPa using synchrotron x-rays. We present the equation of state, variation of cell parameters and discuss the pressure effect on the crystal structure in the following sections.

EXPERIMENT

Samples were synthesized by solid state reaction described elsewhere and found to be in single phase with a tetragonal (P4/mmm) structure (a = 3.8400 (1) Å, c = 11.5700 (2) Å) [7]. Powdered sample is loaded with a ruby chip in a Merrill-Bassett type DAC. Silicone fluid was used as pressure transmitting medium. The experiments were performed at HPCAT, APS, using synchrotron x-rays ($\lambda = 0.3888$ Å) of typical beam size of 30x30 μm. The diffraction images were collected using an imaging plate for 10-20s and then integrated using Fit2D program. The Rietveld refinements were carried out using RIETICA (LHPM) software.

RESULTS AND DISCUSSION

The representative x-ray diffraction patterns at different pressures are shown in Fig.1 with a typical Rietveld refinement plot at 30 GPa. On increasing pressure the peaks shift gradually and we observed no pressure induced transitions up to 32 GPa. The PV data is fitted with Birch-Murnaghan equation shown below

CP850, *Low Temperature Physics: 24th International Conference on Low Temperature Physics;* edited by Y. Takano, S. P. Hershfield, S. O. Hill, P. J. Hirschfeld, and A. M. Goldman

$$P = 3/2\, B_0\, [\, (V/V_0)^{-7/3} - (V/V_0)^{-5/3}]\ \mathrm{x}\ \{\, 1 + \tfrac{3}{4}\, (B_0{'}-4)\ [\, (V/V_0)^{-2/3} - 1]\} \quad (1)$$

and yielded a bulk modulus value of B_0 = 147(4) GPa with B_0' = 8.5 (Fig.2.).

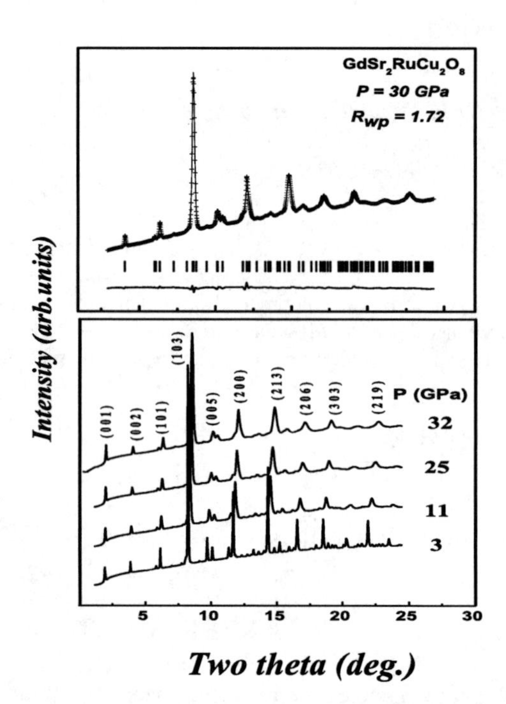

FIGURE 1. High pressure x-ray diffraction patterns of $GdSr_2RuCu_2O_8$ at different pressures. The upper panel shows a typical Rietveld refinement at 30 GPa. The difference and the phase markers are shown in the plot.

While comparing the pressure studies on the electrical resistivity of this compound with R-123 (R= Y and Pr) it is seen that the pressure co-efficient is much smaller and increasing ferromagnetic interactions under pressure [5,8]. As the ferromagnetic interactions are attributed by $Ru\text{-}O_2$ planes, compression of these planes strongly enhances the growth of magnetic phase. The gradual decrease of c/a ratio (inset of Fig.2) and the marginal changes observed in the position parameters of the atoms indicate uniform contraction and support the above observations. Our structural studies further show that the partial replacement of $Cu\text{-}O_2$ layers by $Ru\text{-}O_2$ leads to higher bulk modulus value in comparison with Y-123. Recent studies show intermediate valence (5+ and 4+) for Ru in this compound [9] and the origination of superconducting phase depends on the charge distribution between the $Ru\text{-}O_2$ and $Cu\text{-}O_2$ planes; it would be worth investigating the change in the valence states of Ru under pressure. Efforts are under way to study the scenario in detail.

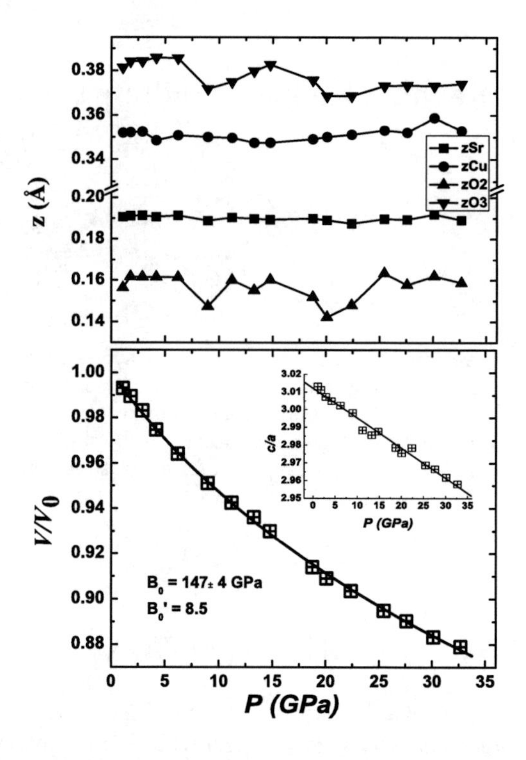

FIGURE 2. PV plot for $GdSr_2RuCu_2O_8$. The solid line is the fit with the Birch-Murnaghan equation of state. The inset shows the change of c/a with pressure. The top panel shows the variation of z parameter for all atoms.

ACKNOWLEDGEMENTS

Use of the HPCAT facility was supported by DOE-BES, DOE-NNSA (CDAC), NSF, DOD – TACOM, and the W.M. Keck Foundation.

REFERENCES

1. I. Felner, *et al., Phys. Rev. B* **55**, R3374 (1997).
2. C. Bernhard, *et al., Phys. Rev. B* **59**, 14 099 (1999).
3. O. Chmaissen *et al., Phys. Rev. B* **61**, 6401 (2000).
4. A. Fainstein *et al., Phys. Rev. B* **65**, 184517 (2002).
5. B. Lorenz *et al., Physics C* **383**, 337 (2003).
6. Jun Shibata *et al., Physica C* **388-389**, 235 (2003).
7. J. Janaki *et al., Mat. Chem. and Physics* **75**, 110 (2002).
8. Zhigang Zou *et al., Phys. Rev. Lett.* **80**, 1074 (1998).
9. R. S. Liu *et al., Phys. Rev. B* **63**, 212507 (2001).

Role Of LO Phonon In Perovskite-Type Superconductors

Eiko Matsushita and Shin-ichi Uno

Faculty of Engineering, Gifu University, Gifu 501-1193 Japan

Abstract. From the viewpoint of strong coupling of electrons with LO phonon in high transition-temperature (T_c) superconductors, the role of optical phonons is studied in the reduced Perovskite-type oxides. The ^{16}O-^{18}O isotope effect on T_c is explained by analyzing roles of optical phonons in good agreement with many experimental data; T_c- α relation ($T_c \propto M^{-\alpha}$) including both $\alpha > 0.5$ (for lower T_c) and $\alpha < 0.5$ (for higher T_c). In comparison with pressure-dependence of T_c, LO phonon mode to couple to electrons changes from Z.B. type (under-doped region) to Z.C. type (over-doped region) at a doped point x=1/8 in $La_{2-x}Ba_xCuO_4$. The superconductivity due to Bose-condensation of quasi-particles is suggested through the roles of LO phonon.

Keywords:. superconductivity, isotope effect, LO-phonon, theory, perovskite
PACS: 74.20.Fg, 74.25.Kc, 74.62.Fj , 74.72.-h

INTRODUCTION

Since the evidence for electron-phonon coupling was reported by the angle-resolved photoemission spectroscopy (ARPES) in copper-oxide superconductors, oxygen-isotope effect has been attracted the attention in Perovskite-like superconductors, $La_{2-x}Sr_xCuO_4$ (LSCO) and $Bi_2Sr_2CaCu_2O_8$ (Bi2212) with high transition-temperature (T_c) [1-3]. Although in an early stage for $La_{2-x}Ba_xCuO_4$ (LBCO), $YBa_2Cu_3O_7$ (YBCO), etc., an overall trend for anomalous $^{16}O \rightarrow {}^{18}O$ isotope effect on T_c was tried to explain by using small bipolarons [4], however, recent experimental results suggest the contribution of the longitudinal optical (LO) phonon associated with the movement of O-atoms.

In this paper, therefore, we study the role of optical phonons in perovskite-type $Ba(Pb_{1-x}Bi_x)O_3$ (T_c=13K) and $(Ba_{1-x}K_x)BiO_3$ (T_c=30K), and the reduced type oxides LSCO (T_c=40K), YBCO (T_c=90K) and Bi2212 (T_c=92K) with various T_c, and we explain O-isotope effect on T_c for many experimental data in these substances. Then we propose a model that LO phonon mode to couple to electrons changes from Z.B. type (under-doped region) to Z.C. type (over-doped region) at a certain doped point (x=1/8 in LBCO) in comparison with pressure-dependence of T_c, and furthermore, we discuss high-T_c superconductivity due to Bose-condensation of quasi-particles.

OXYGEN ISOTOPE EFFECT ON T_C

First, by extending the site representation formalism developed by Appel and Kohn [5] for the phonon-mediated superconductivity, we start from the generalized BCS-like equation for T_c derived successfully in Ref. [6]:

$$T_c \propto \theta \exp[-1/N(0)V_{eff}] \ , \tag{1}$$

$$V_{eff} = V + \frac{V_o - U^*}{1-(V_o - U^*)N(0)\ln(\omega_o/\omega_a)}, \tag{2}$$

$$U^* = \frac{U}{1+N(0)U\ln(\omega_F/\omega_o)}. \tag{3}$$

Within the weak-coupling theory, we introduced the effective electron-electron interaction vertex V_{eff} which is enhanced by optical phonon interaction V_o (cutoff frequency ω_o) and by a factor $\ln(\omega_o/\omega_a)$ in the denominator, and is reduced by Coulomb repulsion (cutoff frequency ω_F). It is easily guessed that optical phonon due to O vibration contributes to T_c, and the O-mass (M) dependence is introduced through only ω_o.

In Fig.1, T_c- α relation ($T_c \propto M^{-\alpha}$) is described for some parameters of U^*/V_o in good agreement with more than 120 experimental data for some kinds of high-T_c superconductors [7], where $N(0)V_o$=0.3 estimated for YBCO, ω_o=50 meV and ω_a=10 meV are assumed for the calculation of $\alpha = -\partial \ln T_c / \partial \ln M$. The solid line is the curve depicted for a parameter of

CP850, *Low Temperature Physics: 24th International Conference on Low Temperature Physics;*
edited by Y. Takano, S. P. Hershfield, S. O. Hill, P. J. Hirschfeld, and A. M. Goldman

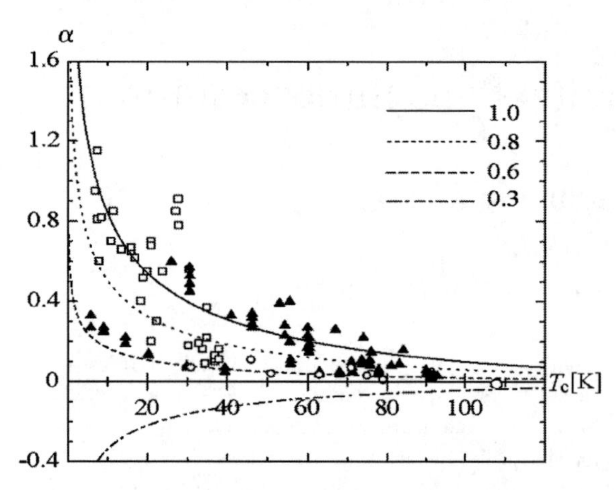

FIGURE 1. Theoretically predicted T_c- α relation for many perovskite-like oxide superconductors; solid line: U^*/V_o=1.0, dotted line: 0.8, broken line: 0.6, dash and dotted line: 0.3. Experimental data are plotted as the symbols □, ▲, ○ corresponding to LSCO, YBCO, Bi2212 / $EuBa_2Cu_{1-x}Zn_xO_7$, respectively.

U^*/V_o=1.0 which shows the best fit to most of the experimental data. Here U* is taken as a number. The parameters with $U^*/V_o>0.5$ are realized when strong electron-electron correlation exceeds the half of the contribution of optical phonon playing on T_c, and induce anomalous T_c- α relation with both $\alpha>0.5$ and $\alpha<0.5$. If U^*/V_o becomes smaller than 0.5 (dash and dotted line with $\alpha<0$ in Fig.1), the contribution of optical phonon disappears such that $\alpha(T_c)$ behavior becomes similar to Morel-Anderson curve in metal superconductors [8] where α reduces from 0.5 (BCS-value) with decreasing T_c. For U^*/V_o=0.5, no isotope effect would be expected for the appearance of T_c.

DISCUSSION

Now let us consider the role of optical phonon given for the relative movement of light O-atoms, and also Bose-condensation of quasi-particles due to LO phonon. We had the result that when U^*/V_o exceeds 0.5, α becomes positive. It means that the optical phonon makes the overall behavior of $\alpha>0.5$ (for lower T_c) and $0<\alpha<0.5$ (for higher T_c) in oxide superconductors with large Coulomb repulsion. Oxides with $U^*/V_o>0.5$ exhibits the importance of strong electron-electron correlation to enhance T_c.

In the present theoretical formula, high-T_c oxides have two aspects: (i) Optical phonon contributes to induce anomalous O-isotope effect. (ii) Two kinds of LO-phonon modes (Z.B. and Z.C. types) are made mainly at O-sites. Here, basing on the pressure-dependence of T_c, we present the prospect that the LO phonon mode to couple to electrons changes from Z.B. type in under-doped region to Z.C. type in over-doped region, at a doped point x=1/8 in LBCO. That is, in under-doped region, Z.B. type (Q_3 mode in Ref. [6]) LO phonon with $dT_c/dp>0$ couples to electrons, whereas in over-doped region, Z.C. type (Q_1 mode in Ref. [6]) LO phonon with $dT_c/dp<0$. This is the conclusion derived from the comparison with ferroelectric transition point T_a coming from various O-vibration modes in perovskite-oxides [9]. The x=1/8 point is a passing point with missing vibration, because the vibration mode of O-atoms exchanges from Z.B. into Z.C. type with increasing x, and vice versa. Therefore, an optimally doped x=1/8 takes the form of dip point of T_c. Thence, it will be expected to analogize the origin of superconductivity for both under-doped and over-doped regions. That is, in under-doped region, we would be able to explain that optical mode Q_3 plays a role to make charged boson at $T_c<T<T^*$ (pseudo-gap region), and condenses for $T<T_c$. In over-doped region, Q_1 mode plays and condenses simultaneously at T_c.

In this approach, the simplified extended BCS-like theory is in success in analyzing explicitly the role of optical phonon on T_c, from which it would be explained why Coulomb interaction U^* becomes large in perovskite-like oxide superconductors. However, as the appearance of T_c might be decided as Bose condensation of small bipolaron made by optical phonon, we have to exchange Debye temperature θ in Eq. (1) to a characteristic temperature forming small bipolaron, θ_{bp}, and also V in Eq. (2) to polaron-polaron attraction V_{bp} to describe charged boson. Thus we will introduce the strong electronic correlation in addition to the conventional mechanism of superconductivity, and furthermore, the derivation of T_c-equation for d-wave superconductivity will be published in the forthcoming paper.

ACKNOWLEDGMENTS

This work was supported in part by Grand-in-Aid for Scientific Research from The Ministry of education, Culture, Sports, Science and Technology in Japan.

REFERENCES

1. A.Lanzara et al., Nature **412**, 510-514 (2001).
2. J. Hofer et al., Phys. Rev. Lett. **84**, 4192-4195 (2000).
3. G.H. Gweon et al., Nature **430**, 187-190 (2004).
4. A.S. Alexandrov, Phys. Rev. B **46**, 14932-14935 (1992).
5. J.Appel and W.Kohn, Phys. Rev. B **4**, 2162-2174 (1971).
6. E. Matsushita, Ferroelectrics **105**, 27-32 (1990).
7. e.g. G. Zhao et al., Phys. Rev. B **54**, 14982-14985 (1996).
8. P.Morel and P.W.Anderson, Phys. Rev. **125**, 1263-1271 (1962).
9. G.A. Samara, T. Sakudo, and K. Yoshimitsu, Phys. Rev. Lett. **35,** 1767 -1769 (1975).

Spin-Phonon Coupling in High-T_c Copper Oxides.

T. Jarlborg

DPMC, University of Geneva, CH1211 Geneva 4, Switzerland

Abstract. Band calculations on $HgBa_2CuO_4$ and $La_{(2-x)}Sr_xCuO_4$ with phonon and spin-waves within the CuO planes show that partial gaps are created at various energies depending on wavelengths. Spin and phonon gaps appear at different energies when the modulations are along [1,1,0], while they are at the same energy for modulations along [1,0,0]. It is shown that the ability to form gaps and antiferromagnetic waves is correlated with the strength of the interaction parameter λ_{sf} for spin fluctuations. Many unusual properties of the high-T_C oxides can be understood from spin-phonon coupling.

Keywords: Superconductivity, cuprates, band theory
PACS: 74.20.Mn,74.25.Jb,74.72.-h

INTRODUCTION

Properties of high-T_C superconductors have been seen to depend both on phonons [1, 2] and magnetic fluctuations [3, 4]. Therefore it is expected that the mechanism for high-T_C superconductivity should involve phonons as well as magnetic fluctuations. Here, band results show that spin-phonon coupling (SPC) is important and many results are compatible with observations.

Band calculations on the high-T_C cuprates $HgBa_2CuO_4$ and $La_{(2-x)}Sr_xCuO_4$ show that gaps or pseudo gaps appear near the Fermi energy, E_F, when the potential has a modulation within the CuO-plane [5]. The calculations are made for long unit cells oriented either along [1,1,0] or [1,0,0] containing so-called half-breathing (bond-stretching) phonons. The planar O-atoms on both sides of a Cu-atom are alternatively displaced towards ("compressed") or away from the Cu ("diluted"). Calculations with anti-ferromagnetic (AFM) field on the Cu atoms are made to model spin-waves. When the calculations are made for co-existing phonon and spin waves, it turns out that the magnetic moments are largest on the "diluted" Cu-sites. The nodes of modulated spin waves coincide with "compressed" Cu-sites. The phonon mode creates a potential modulation of the potential (a Fourier component V_Q). Spin waves induce a modulation in the spin-polarized part of the potential and magnetic moments on the Cu. Gaps appear at different energies depending on the wave length of the modulation. The gain in kinetic energy is maximized when the gap is at E_F. The position of E_F is controlled by doping (calculations use the virtual crystal approximation). The wavelength of the spin wave is twice that of the phonon, because the spin can be up or down at "diluted" Cu-positions. Hence, for modulations along [1,1,0], the band structure has two gaps, one due to the phonon and one at a higher energy because of the spin wave, and there is no direct SPC for electrons at E_F. However, for modulations along [1,0,0] one realizes that two rows of CuO are required along the unit cell, so that the phase of the spin wave along each row will differ by π. The band results show in this case a constructive SPC, where the gap from separate phonon and AFM-modulations will open a gap at the same energy, i.e. at E_F for the correct doping. In addition, a phonon with large atomic displacements will increase the moments of the spin wave, and both waves contribute to a common gap.

The abovementionned results can be found in refs, [5]-[7], where it also is shown that a small correction to the band structure (which affects the potential and the localization) will stabilize a gapped AFM state instead of the metallic non-magnetic state for the undoped system. The same correction will increase the mass enhancement from spin fluctuations in doped cases [7].

RESULTS AND DISCUSSION

The calculated density of states (DOS) at E_F, $N \approx 1.5(Cu \cdot eV)^{-1}$. The atomic displacement $u^2 = 1.5\hbar\omega/K$ at low T and $u^2 = 3k_BT/K$ at high T (relative to the Debye temperature). The force constant $K = 25$ eV/Å^2 for a typical O phonon with $\hbar\omega \sim 80$ meV makes u about 0.06 Å at room temperature. Spin waves can, in analogy with phonons, be assigned a magnetic moment $m^2 = k_BT/K_m$, where $K_m = d^2E_{tot}/dm^2$. This gives $m \approx 0.3$ μ_B/Cu at room temperature for the co-existing case. A partial gap removes roughly half of the DOS within an energy $\Delta = 100$ meV around E_F. The gain in kinetic energy is approximately $\frac{1}{2}N\Delta^2 = 8$ meV/Cu, i.e. about 15-20 percent of the elastic energy $U = \frac{1}{2}K \cdot u^2$ at the average u. Thus, the phonon energy is expected to decrease by the this percentage because of the coupling to the spin wave, but only in the doped case when E_F is at the gap.

CP850, *Low Temperature Physics: 24th International Conference on Low Temperature Physics;*
edited by Y. Takano, S. P. Hershfield, S. O. Hill, P. J. Hirschfeld, and A. M. Goldman

These results compare resonably well with experiment [2]. Less doping implies coupling at longer wavelengths so that the softening moves away from the zone boundary and finally disappears. The softening should also disappear when the pseudo gap disappears. This is expected when the Fermi-Dirac occupation, f, becomes too wide to separate the occupied majority spins at $E_F - \Delta$ from the unoccupied minority spins at $E_F + \Delta$. As function of T, $m(T) = N \cdot \Delta \cdot (f(\frac{-\Delta}{k_B T}) - f(\frac{\Delta}{k_B T})$, while from the band calculations $\Delta \approx const. \cdot m$. By solving iteratively the two last equations as function of T, one finds a stable gap (and m) up to a $0.9 \cdot T_{max}$ after which Δ and m drop to zero at T_{max}, as is the case with the pseudo gap near T^*. The reason is the strong exponential T-dependence of f, and the feedback of m on Δ.

Isotope shifts are expected. The zero-point motion makes a mass (M) dependence $u \sim (K \cdot M)^{-1/4}$. A lower M will also decrease K because of the phonon softening due to the promotion of spin waves at larger u (see above), so the isotope effect will be stronger than what is suggested from a constant K. The effect on the pseudogap will be moderated at large T, when there is no explicit mass dependence of u, $u^2 \sim K^{-1}$. The effect on the superconducting $T_c \sim \omega exp(-1/\lambda)$ is complex. Pure phonon $\omega \sim \sqrt{K/M}$, but with SPC also K will drop when phonon amplitudes increase because of lower M. A pure λ_{sf} from spin fluctuations tend to increase at increased phonon amplitudes. The mass dependence of ω_{sf} and λ_{sf} may act oppositely on T_c. Anharmonic effects are expected, which will mix phonon and spin contribution. Experiments show small isotope effects on T_c [1].

The mass dependence of u is largest at low T, which suggests comparison with isotope shifts of the penetration depth Λ measured at low T [1]. Some assumptions lead to $\Lambda^{-2} \sim N \cdot v^2$, where v is the Fermi velocity. The band crossing E_F along Γ-M has larger v than at the crossing along X-M. Phonon and spin waves along [1,0,0] will produce the gaps in the latter region, so that v will increase. If O^{16} are substituted by O^{18} it will decrease u for half breathing phonons by 3 percent. The band results tell that the DOS is reduced roughly by a factor $\frac{1}{2}$ by spin fluctuations if a certain phonon distortion is present and by 20-25 percent if no distortions are present. By simple interpolation from these results one can estimate that reducing u by 3 percent will increase N by 1-2 percent. As $N \sim v^{-1}$ this suggests a decrease of Λ^{-2} by 1-2 percent. This is smaller but of the same order as in ref [1]. It can be noted that the gap starts in the X-M region from waves along [1,0,0] (and [0,1,0]). This fits to the observation of 'destruction' of the Fermi surface in this region of k-space [8], while it remains along Γ-M if no waves are oriented along [1,1,0].

Photoemission and tunneling spectra show the pseudogap and a 'dip' in the DOS about a tenth of an eV below E_F. The position of this dip relative to the main gap at E_F depends on doping according to tunneling data [9]. A projection of these data on to the calculated DOS suggests that the dip is located at a fixed energy (a fixed wavelength in our mechanism), while the main gap moves to higher energy for decreased doping. This is consistent with a type of perpendicular spin-phonon coupling as follows: Consider the shortest possible coexisting SPC along [1,0,0]. The half-breathing phonon creates maximum 'diluted' distortions for two out of four pairs of Cu, where condition for magnetism is optimal (while magnetic nodes fall on the two other pairs of Cu). The perpendicular rows of (optimally magnetized) Cu can develop spin waves along [0,1,0], possibly coupled with phonons along the same direction. The distortion along [1,0,0] has increased the ability for having strong spin fluctuations along [0,1,0]. While the extent of the former wave is fixed to 4 cells, it is possible to form longer waves for the latter so that the main gap is tied to E_F for the given doping. This perpendicular SPC is expected to have the same properties as discussed above for parallel coupling. Complete band calculations are more difficult because very large unit cells are needed to cover even the smallest configuration. However, band results for a simplified case, with only a half-breathing mode with large and small moments (no full spin wave) along [1,0,0] together with longer complete spin waves along [0,1,0], show that two gaps form simultanously at two different energies. At low evergy the gap is due to the short wave along [1,0,0], while at larger energy a second partial gap develops because of a longer spin wave along [0,1,0]. The doping dependence of the two gaps fits with data in ref. [9], but the energy scale is too large, almost by a factor of two. This could be explained if the DOS was larger, if all wavelengths were larger or if the real doping, x, is not what it is supposed to be.

In conclusion, it is shown that many properties of high-T_c materials are consistent with SPC, although more precise works are needed for quantitative results. More detailed results will be published elsewhere [10].

REFERENCES

1. J. Hofer, et al., *Phys. Rev. Lett.* **84**, 4192 (2000).
2. T. Fukuda, et al., *Phys. Rev. B* **71**, 060501 (2005).
3. M. A. Kastner, et al., *Rev. Mod. Phys.* **70**, 897 (1998).
4. H. Mook, P. Dai, and F. Dogan, *Phys. Rev. Lett.* **88**, 097004 (2002).
5. T. Jarlborg, *Phys. Rev. B* **64**, 060507(R) (2001).
6. T. Jarlborg, *Phys. Rev. B* **68**, 172501 (2003).
7. T. Jarlborg, *J. Phys.: Cond. Matter* **16**, L173 (2004).
8. M.R. Norman, et al., *Nature* **392**, 157 (1998).
9. J.F. Zasadzinski, et al., *Phys. Rev. Lett.*, **87**, 067005 (2001).
10. T. Jarlborg, *unpublished*, (2005).

Density-Matrix Renormalization Group Study of Phase Diagram in Systems with Strong Electron-Electron and Electron-Phonon Interactions

Masaki Tezuka*, Ryotaro Arita†,* and Hideo Aoki*

*Department of Physics, University of Tokyo, Hongo, Tokyo 113-0033, Japan
†Max-Planck-Institut für Festkörperforschung, Stuttgart, D-70569 Germany

Abstract. The problem of many-body systems that have both a strong electron-electron interaction and a strong electron-phonon interaction is studied. We focus on the regime where the phonon energy scale is comparable with the electron energy scale, for which adiabatic (and anti-adiabatic) approximations fail so that the electrons and phonons have to be treated on an equal footing. Here we employ the density-matrix renormalization group (DMRG) to calculate charge, spin and pairing correlation functions in the Holstein-Hubbard model in one dimension. For the half-filled electronic band, the degeneracy (as in the Hubbard model) of charge and on-site pair correlation exponents is shown to approximately persist. However, if we make the electronic structure electron-hole *a*symmetric (in a trestle = chain with second-neighbor hopping), a regime is shown to exist where a pairing is the single, most dominant correlation.

Keywords: Superconductivity, Electron-phonon interaction, Strongly Correlated Electrons, Density-Matrix Renormalization Group
PACS: 71.10.Hf; 71.38.-k; 74.20.Mn

Introduction. There has been an increasing fascination, both experimental and theoretical, with the many-body physics when the electron-electron (el-el) interaction and the electron-phonon (el-ph) interaction coexist. One crucial problem is whether and how superconductivity (SC) can arise in such a situation. This has been studied typically for the Holstein-Hubbard (HH) model at half-filling, for which competition between charge-density wave (CDW), spin-density wave (SDW) and SC phases is a key issue.

Here we have calculated, for the first time, various correlation functions, including the correlation of pairs with various symmetries, against real-space distance r for the one-dimensional Holstein-Hubbard model. We employ the density-matrix renormalization group (DMRG), with an improved algorithm for the warmup step when el-el and el-ph interactions are comparable. We have found the following. The SC correlation becomes, at best, nearly as dominant as the CDW correlation. Inspired by the fact that such a degeneracy between SC and CDW correlations occurs exactly for the purely electronic (Hubbard) model at half filling, and that the degeneracy can be lifted by destroying the electron-hole symmetry there, we then consider an electron-hole *a*symmetric system in a trestle (= chain with next-neighbor hopping), for which a regime is shown to exist where a pairing is the single, most dominant correlation.

Method. The Holstein-Hubbard model is

$$H = -t\sum_{i,\sigma}(c^\dagger_{i+1,\sigma}c_{i,\sigma}+\mathrm{H.c.})-t'\sum_{i,\sigma}(c^\dagger_{i+2,\sigma}c_{i,\sigma}+\mathrm{H.c.})$$

$$+ U\sum_i n_{i\uparrow}n_{i\downarrow}+g\sum_{i,\sigma}n_{i\sigma}(b_i+b_i^\dagger)+\hbar\omega\sum_i b_i^\dagger b_i, \quad (1)$$

where $c^\dagger_{i\sigma}$ creates an electron of spin σ on the ith lattice site, t is the nearest-neighbor hopping (which we take as the unit of energy hereafter), t' is the second-neighbor hopping, U is the on-site electron-electron repulsion, $a_i^\dagger$ creates an on-site (Einstein) phonon at i, g is the on-site el-ph interaction, and $\hbar\omega$ is the bare phonon energy.

Jeckelmann and White have introduced the *pseudo-site method*[3], which makes the application of DMRG to models with on-site phonons feasible. On top of an electron pseudo-site (open circles in Fig.1), $M = 2^N$ lowest phonon levels (0-boson, 1-boson, ..., 2^N-1-boson states) are considered for each site, which are expressed, in terms of fictitious N phonon "pseudo-sites" (circles in Fig. 1 with two horizontal bars representing

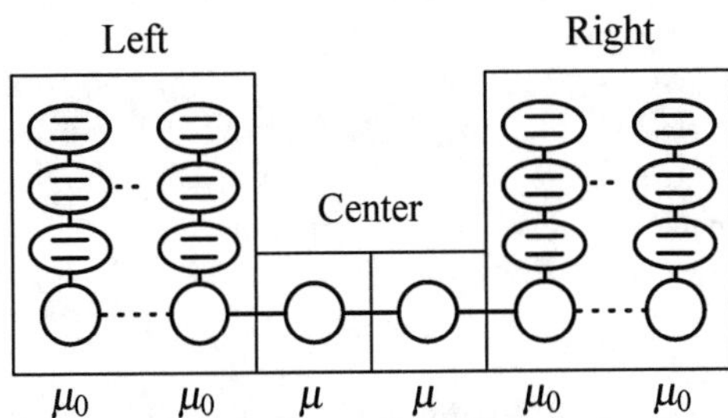

FIGURE 1. Schematic illustration of the pseudo-site method for an electron-phonon system. Open circles on the bottom and ellipses with two horizontal bars respectively represent the electron pseudo-sites and the phonon pseudo-sites.

CP850, *Low Temperature Physics: 24th International Conference on Low Temperature Physics;*
edited by Y. Takano, S. P. Hershfield, S. O. Hill, P. J. Hirschfeld, and A. M. Goldman

a binary expression). Thus the Hamiltonian and other operators in the original system are expressed in terms of operators defined for the pseudo-sites, which are, as in real sites in conventional DMRG, incorprated into the system one by one.

For the Holstein-Hubbard model, we find that, as the electron pseudo-site is added, the retained states tend to deviate significantly from those required to describe the system at later stages. This is because the electrons in the added electron pseudo-sites feel the bare repulsion there, while the repulsion at other sites is reduced due to the electron-phonon coupling, and this degrades the convergence. To remedy this we can modify the original chemical potential (μ_0 in Fig.1) to μ for the electrons for the added electron pseudo-sites in the calculation for the ground states (but not for the transformation of the Hamiltonian of the enlarged block), in such a way that the expectation value of the number of electrons at each of the new pseudo-sites equals the average number of electrons per site in the target state. This method, which we call the *compensation method*[4], has indeed given lower ground state energies and hence better ground states, throughout the warm-up stages and into the finite-method iterations.

In the conventional finite algorithm for the ground state with N electrons, when we consider states having $n_R \in [n_1, n_2] \equiv E_R$ electrons for a right block B_R, states with $n_L \in [N-n_2-2, N-n_1]$ electrons can be constructed for the enlarged left block B_L. When B_R is optimized, the possible range of the electron number of B_R is $[N-(N-n_1)-2, N-(N-n_2-2)] = [n_1-2, n_2+2]$ which is broader than E_R. This is not the case for the pseudo-site method and the range of the electron numbers cannot be enlarged. This implies that, if we want an accurate result, we need a large variety in electron numbers at the stage of the infinite algorithm. So we have kept $m = 200$ states per block and increase the number of retained states as the system grows before we go into the finite algorithm to have $m = 800$.

Results. We have calculated the correlation functions against real-space distance for on-site spin-singlet pair, nearest-neighbor singlet pair and nearest-neighbor triplet pair as well as for the charge and spin for a half-filled 40-site Holstein-Hubbard chain. Intuitively, U enhances the SDW correlation, while the electron-phonon interaction enhances the CDW, with the phonon-mediated attraction $\lambda \equiv 2g^2/\hbar\omega$ in the $\omega \to \infty$ limit. So the region of interest is the boundary between the SDW and CDW where $U \simeq \lambda$.

For the simple chain with $t' = 0$, we have found that, for $U \simeq \lambda$ the on-site pair correlation as well as charge and spin correlations decay with power laws, while for U sufficiently smaller than λ only the CDW and on-site pair correlations decay with power laws. If we examine the exponents (η in $r^{-\eta}$), we observe that $\eta_{\text{on-siteSC}}$ is slightly larger than η_{CDW}. This is reminiscent of the degeneracy due to the electron-hole symmetry in the Hubbard model [2]. On the other hand, the result is inconsistent with the TL result[1], in which $\eta_{\text{on-siteSC}} = 1/\eta_{CDW}$ in the spin-gapped metallic region so that $\eta_{CDW} > 1$ would imply $\eta_{\text{on-siteSC}} < 1$. So, while some authors assume that $\eta_{CDW} > 1$ would imply a dominant SC[5], this does not necessarily hold in the Holstein-Hubbard system.

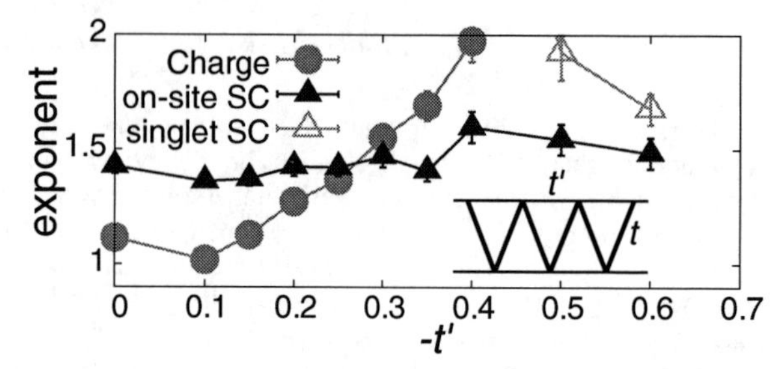

FIGURE 2. Exponents for the correlation functions for the Holstein-Hubbard model with $(U, g, \omega) = (2.0, 3.0, 4.5)$ against the second-neighbor hopping t'.

The above observation has led us to introduce t' to degrade the electron-hole symmetry. As $|t'|$ is increased, we can see in Fig.2 that η_{CDW} rapidly increases while $\eta_{\text{on-siteSC}}$ remains roughly constant, which makes the on-site pairing correlation the single, most dominant correlation for $0.3 \lesssim -t'$. We also notice that the exponent for the nearest neighbor singlet pair correlation approaches that for the on-site pair around $-t' \simeq 0.5$.

Discussion. The situation with $t' \neq 0$ may have a relevance to real systems such as the alkali-fullrides A_3C_{60} which have an fcc structure. As for the deviation from the Tomonaga-Luttinger behavior, we speculate that this may come from a strong energy-dependence in the electron-electron interaction as introduced by the coupling to phonons.

This work is in part supported by a Grant-in-Aid for Science Research on Priority Area from the Japanese Ministry of Education.

REFERENCES

1. H.J. Schulz, "Fermi Liquids and Non-Fermi Liquids", in *Mesoscopic Quantum Physics*, edited by E. Akkermans *et al.*, Amsterdam: Elsevier, 1995, pp. 533–603.
2. Y. Nagaoka, *Prog. Theor. Phys.* **52**, 1716 (1974).
3. E. Jeckelmann and S. R. White, *Phys. Rev. B* **57**, 5376 (1998).
4. M. Tezuka, R. Arita and H. Aoki, *Physica B* **359-361**, 708 (2005).
5. R.T. Clay and R.P. Hardikar, preprint(cond-mat/0505184).

Polaronic Effect in Lightly Doped High-Tc Cuprates

T. Sakai[a] and D. Poilblanc[b]

[a]Japan Atomic Energy Agency (JAEA), SPring-8, Mikazuki, Hyogo 679-5148, Japan and CREST JST
[b]Laboratoire de Physique Théorique, CNRS-UMR 5152, Université Paul Sabatier, F-31062 Toulouse, France

Abstract. Phonon effect of the in-plane oxygen atoms is investigated by numerical exact diagonalizations of the t-J Holstein model. The present study indicate that the breathing vibration mode gives rise to a polaronic effect which yields a broadening of the quasiparticle spectrum, as the one observed in angle-resolved photoemission spectroscopy.

Keywords: electron-phonon interaction, high-Tc cuprates, polaronic effect
PACS: 74.20.Mn, 71.27.+a, 71.38.+i

INTRODUCTION

The electron-phonon interaction in the high-Tc cuprates gives rise to various interesting phenomena. Recent angle-resolved photoemission spectroscopy (ARPES) measurements of $Ca_{2-x}Na_xCuO_2Cl_2$[1,2] indicate that the quasiparticle (QP) peak is quite broad and shifted from the chemical potential. This puzzling phenomena could be explained by polaronic effect of the in-plane breathing vibrations of oxygen atoms. Indeed, here we support this scenario and provide evidences for such a picture using a numerical investigation of the role of phonons in a *strongly correlated* electron model, namely an extended t-J-Holstein model. [3]

MODEL AND CALCULATION

In order to investigate the oxygen phonon effect in the CuO_2 plane, we consider an extended t-J-Holstein model defined by the Hamiltonian

$$\begin{aligned} H = & -t\sum_{\langle i,j\rangle,\sigma}(\tilde{c}^+_{j,\sigma}\tilde{c}_{i,\sigma}+\tilde{c}^+_{i,\sigma}\tilde{c}_{j,\sigma}) \\ & +J\sum_{\langle i,j\rangle}(\vec{S}_i\cdot\vec{S}_j-\frac{1}{4}n_in_j) \\ & +\sum_{i,\delta}(\frac{p^2_{i,\delta}}{2m}+\frac{1}{2}m\Omega^2u^2_{i,\delta}) \\ & +g\sum_{i,\delta}u_{i,\delta}(n^h_i-n^h_{i+\delta}) \end{aligned} \tag{1}$$

where $\tilde{c}_{j,\sigma}$ and $\tilde{c}^+_{j,\sigma}$ are the hole operators at each Cu site, and the last two terms describe the in-plane breathing vibration of each O atom between the nearest-neighbor Cu sites. The phonon terms can be rewritten in the boson representation

$$\begin{aligned} H_{\text{e-ph}} = & \Omega\sum_{i,\delta}(b^+_{i,\delta}b_{i,\delta}+\frac{1}{2}) \\ & +\lambda_0\sum_{i,\delta}(b_{i,\delta}+b^+_{i,\delta})(n^h_i-n^h_{i+\delta}) \end{aligned} \tag{2}$$

where $\lambda_0=\sqrt{1/2m\Omega}$. We approximately truncate the phononic Hilbert space to a finite number of bosonic states up to n_{ph} at each oxygen site. Within the one-phonon approximation (n_{ph}=1), we perform numerical diagonalization based on the Lanczos algorithm to study $\sqrt{8}\times\sqrt{8}$ unit-cell cluster with periodic boundary conditions with a total of 16 phonon modes, namely a Cu_8O_{16} cluster. We fix J/t=0.4 as a realistic parameter and take t as a unit of energy. The calculated expectation value of the phonon number per site is ~ 0.1 ($\ll n_{ph}$) for λ_0=0.6, which justifies the validity of the present approximation even such a strong coupling.

The dynamical spectral function of the single-hole QP is calculated by the form

$$\begin{aligned} P_1(\omega) = & \sum_l\left|\left\langle\Psi^{N-1}_l\middle|\tilde{c}_{k,\sigma}\middle|\Psi^N_0\right\rangle\right|^2 \\ & \times\delta(\omega-E^{(1)}_l+E^{(0)}_0) \end{aligned} \tag{3}$$

CP850, *Low Temperature Physics: 24th International Conference on Low Temperature Physics;*
edited by Y. Takano, S. P. Hershfield, S. O. Hill, P. J. Hirschfeld, and A. M. Goldman

where Ψ_0^N and Ψ_l^{N-1} are the antiferromagnetic ground state and one-hole excited states, respectively on the N-site cluster. The calculated spectral functions of the QP with $k=(\pi/2,\pi/2)$ for various electron-phonon coupling constants with fixed phonon frequency $\Omega=0.1$ are shown in FIGURE 1. It reveals that with increasing electron-phonon coupling the sharp QP peak is split into two parts; a lower energy structure and a higher broad peak. A gap-like structure also appears between them in the strong coupling regime. We suggest that the broad structure we find above this pseudo-gap is connected to the broad peak shifted from the chemical potential observed in the ARPES measurement of $Ca_{2-x}Na_xCuO_2Cl_2$[1,2]. We also calculate the QP weight (QPW)

$$Z_{1h}=\frac{\left|\left\langle\Psi_0^{N-1}\middle|\tilde{c}_{k,\sigma}\middle|\Psi_0^N\right\rangle\right|^2}{\left\langle\Psi_0^N\middle|\tilde{c}_{k,\sigma}^{+}\tilde{c}_{k,\sigma}\middle|\Psi_0^N\right\rangle} \tag{4}$$

shown in FIGURE 2. It indicates a crossover behavior characterized by a rapid decrease of the spectral weight with increasing electron-phonon coupling. This drastic reduction of the Z-factor is consistent with the previously discussed polaronic scenario characterized by a transfer of the QPW to higher energies. For large-enough phonon coupling, we therefore expect a very small residual QPW at the bottom of the hole-creation spectrum (i.e. the chemical potential) which might not be even detectable experimentally. Note that similar conclusions was obtained by a different approach based on a diagrammatic Monte Carlo calculation[4].

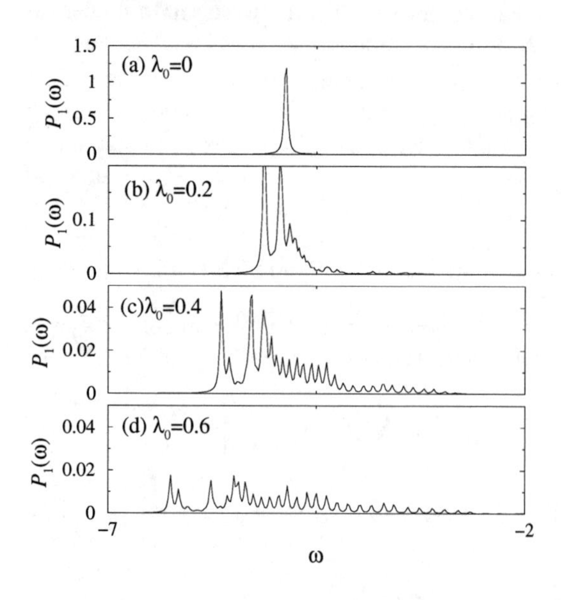

FIGURE 1. Dynamical spectral function of a single-hole QP on a Cu_8O_{16} cluster with momentum k=(π/2, π/2) vs frequency in the presence of in-plane oxygen modes for J=0.4, Ω=0.1, and λ_0=(a) 0, (b) 0.2, (c) 0.4, and (d) 0.6.

SUMMARY AND DISCUSSIONS

The present numerical diagonalization study on the extended t-J-Holstein model indicates that a polaronic effect of the in-plane breathing vibration of the oxygen atoms leads to a broadening of the QP spectrum as observed in ARPES of lightly doped cuprates. The detailed structure of the polaronic QP is to be investigated in near future.

ACKNOWLEDGMENTS

We wish to thank Prof. S. Ishihara for fruitful discussion. This work was partly supported by the Japanese Grants-in-Aid for Scientific Research (B)(No. 14340099). The numerical computation was done in part using the facility of the Supercomputer Center, Institute for Solid State Physics, University of Tokyo.

REFERENCES

1. K. M. Shen, F. Ronning, D. H. Lu, W. S. Lee, N. J. C. Ingle, W. Meevasana, F. Baumberger, A. Damascelli, N. P. Armitage, L. L. Miller, Y. Kohsaka, M. Azuma, M. Takano, H. Takagi and Z. –X. Shen, *Phys. Rev. Lett.* **93**, 267002 (2004).
2. F. Ronning, K. M. Shen, N. P. Armitage, A. Damascelli, D. H. Lu, Z. –X. Shen, L. L. Miller and C. Kim, *Phys. Rev. B* **71**, 094518 (2005).
3. T. Sakai , D. Poilblanc and D. J. Scalapino, *Phys .Rev. B* **55**, 8445 (1997).
4. A. S. Mishchenko and N. Nagaosa, *Phys. Rev. Lett.* **93**, 036402 (2004).

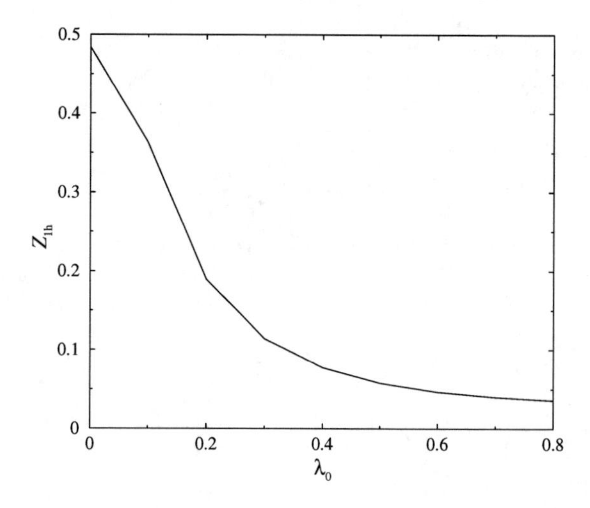

FIGURE 2. Quasiparticle weight of a single hole on a Cu_8O_{16} cluster with momentum k=(π/2, π/2) vs λ_0.

The True Nature of High-Temperature Superconductivity[†]

John D. Dow[a], Dale R. Harshman[a,b,c], and Anthony T. Fiory[d]

[a]Department of Physics, Arizona State University, Tempe, Arizona 85287-1504 U.S.A.,
[b]Physikon Research Corporation, P. O. Box 1014, Lynden, Washington 98264 U.S.A*.,
[c]Department of Physics, University of Notre Dame, Notre Dame, Indiana 46556 U.S.A.,
[d]Department of Physics, New Jersey Institute of Technology, Newark, NJ 07102 U.S.A.

Abstract. The question of whether high-temperature superconductivity is *s*-wave or *d*-wave has been resolved in favor of a nodeless gap function consistent with *s*-wave pairing [*Phys. Rev.* **B 69**, 174505 (2004)]. In the case of $YBa_2Cu_3O_7$, where the CuO_2 and the CuO layers both contain Cu *d*-bands, this result implies that the superconducting hole condensate resides in the BaO layers. Thus the theory for $YBa_2Cu_3O_7$ also describes systems without cuprate-planes, such as doped Ba_2YRuO_6 ($T_c \sim 93$ K) and materials such as $GdSr_2Cu_2RuO_8$ or $Gd_{2-z}Ce_zSr_2Cu_2RuO_{10}$ with onset T_c's near 45 K, whose cuprate-planes are either weak-ferromagnetic or antiferromagnetic, but whose BaO or SrO layers superconduct. Clearly high-temperature superconductivity (1) is in BaO or SrO layers, or in interstitial regions, rather than in cuprate-planes, (2) is *s*-wave, not *d*-wave, and (3) usually involves unit cells with both normal and superconducting planes.

Keywords: Electronic structure; Y-based cuprates; Type-II superconductivity
PACS: **74.25.Jb; 74.72.Bk; 74.60.-w**

INTRODUCTION

There are two basic schools of thought about high-temperature superconductivity: (i) that the superconducting condensate [1-6] resides in the BaO, SrO, or interstitial layers of the materials [5], and (ii) that the condensate resides in the cuprate-planes [7]. Most authors agree that in most materials, the superconductivity is *p*-type.

CUPRATE-PLANES

Advocates of cuprate-plane superconductivity have not adequately explained how the superconductivity can be *p*-type in a cuprate-plane that has a net charge of approximately -2: with about +2 for Cu, and -2 for each of two O. (The number of holes per cuprate plane is approximately +0.2 [8].) Most models are two-dimensional and treat only a single cuprate-plane layer of the unit cell as having a net negative charge near -1.8, which purportedly superconducts *p*-type because of holes in the CuO_2 bands that are otherwise nearly full of electrons.

[†] Research supported by the U.S. Army Research Office contract W911NF-05-1-0346.

Indeed, the experimental proof of cuprate-plane superconductivity is largely based on a single datum obtained by Cava *et al.* [9], through which a continuous curve with a step was drawn. Although Jorgensen *et al. claimed* to have confirmed the results of Cava *et al.*, they did not [10]. The data of Cava *et al.* and the data of Jorgensen *et al.* are given in Fig. 1, and are straight lines except for a single point from Cava *et al.* alone. (There is no such similar point from the data of Jorgensen *et al.* who claimed to have confirmed the results of Cava *et al.*) Hence, except for that single, unconfirmed point (which we believe is erroneous), the original evidence that the cuprate-planes superconduct is missing.

STM AND PHOTOEMISSION DATA

Authors who have studied the cuprate-planes, most notably by scanning tunneling microscopy (STM) [11] and by photoemission (PE) [12, 13], have concluded that those planes are *d*-like, with strong Cu-related signals. But STM and PE methods of studying the cuprate-planes do not directly determine that those planes are superconducting. In fact, muon data show that the cuprate-planes do not contribute to the observed *p*-type superconductivity [5]. Hence the *d*-

CP850, *Low Temperature Physics: 24th International Conference on Low Temperature Physics;*
edited by Y. Takano, S. P. Hershfield, S. O. Hill, P. J. Hirschfeld, and A. M. Goldman

like behavior of the cuprate-planes, while the superconductvity is *s*-like, is consistent with those planes being non-superconducting.

THE HOLES OF SUPERCONDUCTIVITY ARE IN THE BAO OR SIMILAR LAYERS

In non-magnetic materials, such as $YBa_2Cu_3O_7$, positive muon spin-rotation spectroscopy (μ^+SR) detects the vortex lattice, which is produced only by the superconducting bands. The determination that μ^+SR, after first being corrected for fluxon de-pinning, detects a signal that is *s*-like [5], means that the superconducting bands detected by muons are not Cu-like (as the cuprate-planes or the CuO layers are), but heavily oxygen-like, and consistent with a strong coupling two-fluid model of the superconductivity. According to recent band structure calculations [14], the superconductivity is associated with BaO bands.

THE IMPORTANCE OF FLUXON-DE-PINNING

For over a decade of muon studies, Sonier *et al.* [15] have been neglecting the established fluxon de-pinning and have been claiming that they find evidence of *d*-wave superconductivity, (without actually presenting a successful fit of the *d*-wave theory to the data). This never-proven *d*-wave pairing idea, together with the neglect of fluxon de-pinning, has left theorists claiming that the pairing is *d*-wave, and that the cuprate-planes superconduct [7]. Harshman *et al.* [5] did account for fluxon-de-pinning when fitting their muon data and found both that the superconductivity is *s*-wave, and is not *d*-wave. This discredits the claims of theorists who neglect fluxon de-pinning and argue that the pairing is *d*-wave in the cuprate planes [7].

SUMMARY

The μ^+SR data imply that the cuprate-planes *do not superconduct*; and that the BaO layers do superconduct (in $YBa_2Cu_3O_7$), and are over-whelmingly *s*-like. Our model can reasonably describe all known μ^+SR data on $YBa_2Cu_3O_7$.

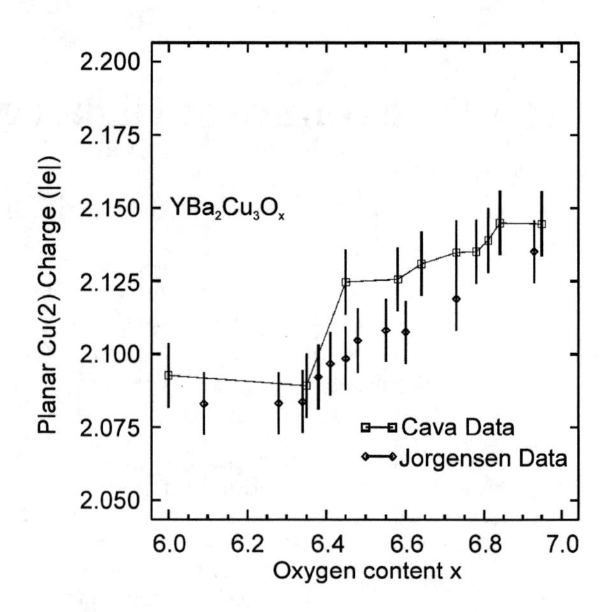

FIGURE 1. Cuprate-planar Cu(2) charges of $YBa_2Cu_3O_x$ as extracted [16] from data by Cava *et al.* [9] and by Jorgensen *et al.* [10]. The Cava data are systematically higher (due to different calibrations of the two monochromators). The Jorgensen data describe a straight line, while the Cava data have a *jump of a single point*, which is the basis of the dubious claim of cuprate-plane superconductivity.

References

1. D. R. Harshman, *et al.*, *Phys. Rev. B* **36**, 2386 (1987).
2. B. Pümpin, *et al.*, *Phys. Rev. B* **42**, 8019 (1990).
3. B. Pümpin *et al.*, *Physica C* **162-164,** 151 (1989).
4. D. R. Harshman, *et al.*, *Phys. Rev. B* **39**, 851 (1989).
5. D. R. Harshman, *et al.*, *Phys. Rev. B* **69**, 174505 (2004).
6. A. T. Fiory, *et al.*, *J. Electronic Mater.* **34**, 474 (2005).
7. D. J. Scalapino, *Phys. Rpts.* **250**, 1 (1995).
8. H. Zhang and H. Sato, *Phys. Rev. Lett.* **70**, 1697 (1993).
9. R. J. Cava, *et al.*, *Physica C* **156**, 523 (1988).
10. J. D. Jorgensen, *et al.*, *Phys. Rev. B* **41**, 1863 (1990).
11. K. McElroy, *et al.*, *Nature* **422**, 592 (2003).
12. J. C. Campuzano, *et al.*, "Physics of Conventional and Unconventional Superconductors," in *Physics of Superconductors*, *Vol. II*, edited by K. H. Benneman and J. B. Ketterson, Berlin: Springer, 2004, pp. 167-273.
13. A. Damascelli, *et al.*, *Rev. Mod. Phys.* **75**, 473 (2003).
14. P. de la Mora, *et al.*, "Superconductivity in the BaO Layers of $YBa_2Cu_3O_7$," to be published.
15. J. E. Sonier, *et al.*, *Phys. Rev. Lett.* **83**, 4156 (1999); *Rev. Mod. Phys.* **72**, 769 (2000).
16. J. D. Dow and H. A. Blackstead, unpublished.

* Present address

Study of d- and p-wave Pairing in the Hubbard Model Using the Dynamical Cluster Approximation

Ryotaro Arita and Karsten Held

Max Planck Institute for Solid State Research, Heisenbergstr. 1, Stuttgart 70569, Germany

Abstract. The Hubbard model is commonly considered to be relevant for unconventional superconductors like cuprates and ruthenates. For a more thorough insight into the physics of this model, we use a new projective quantum Monte Carlo (PQMC) algorithm within the framework of the dynamical cluster approximation (DCA). Employing this DCA(PQMC) approach, we calculate static pairing susceptibilities for various symmetries and study the crossover from $d_{x^2-y^2}$-wave to p_{x+y}-wave pairing as a function of doping.

Keywords: Superconductivity, Low Temperatures, Hubbard model, Dynamical Cluster Approximation, Quantum Monte Carlo Simulations
PACS: 71.10.Fd,71.27.+a,74.20.Rp

INTRODUCTION

The dynamical mean field theory (DMFT)[1] is a very successful method and has been applied to a variety of problems in the field of strongly correlated electrons. For example, by virtue of this method, our understanding of the Mott-Hubbard transition has greatly improved. However, DMFT neglects non-local correlations and wave-vector-dependent fluctuations. Hence, DMFT cannot describe anisotropic superconductivity, which is one of the most interesting phenomena in strongly correlated electron systems.

On the other hand, the dynamical cluster approximation (DCA) proposed by Hettler *et al.*[2] is a promising (cluster) extension of DMFT which systematically incorporates non-local correlations. By means of DCA, various properties of the Hubbard model on the square lattice, such as the possibility of $d_{x^2-y^2}$-pairing have been studied quite intensively[2, 3]. DCA and other methods such as the functional renormalization group indicate that $d_{x^2-y^2}$-wave instability becomes dominant in the t-t' Hubbard model when the electron density is near half-filling. But, it is still controversial which pairing symmetry is dominant for intermediate and low densities[4]. Especially, the possibility of the crossover from $d_{x^2-y^2}$-wave to p_{x+y}-wave is of great interest.

Our aim is to study this competition of different pairing channels. To this end, we employ a newly developed projective quantum Monte Carlo (PQMC) method [8] for solving the DCA equations at $T=0$. In the following two Sections, we first introduce the new DCA(PQMC) algorithm and then present some first results obtained by DCA(PQMC).

METHOD

The Hamiltonian of the t-t' Hubbard model is

$$H = -t \sum_{\langle i,j\rangle,\sigma}^{\text{nn}} (c^\dagger_{i\sigma}c_{j\sigma}+\text{h.c.}) + t' \sum_{\langle i,j\rangle,\sigma}^{\text{nnn}} (c^\dagger_{i\sigma}c_{j\sigma}+\text{h.c.}) + U\sum_i n_{i\uparrow}n_{i\downarrow} \quad (1)$$

in the standard notation. Here, t and t' denote the hopping amplitude between nearest neighbor (nn) and next nn (nnn) sites, respectively; in the following $t'=0.4t$ and a band width $D=4$ sets our unit of energy. It should be noted that the γ-band of Sr_2RuO_4 is modeled by $t'\sim 0.4t$. Hence, the possibility of triplet superconductivity for this case is of particular interest.

In DCA, the original reciprocal space with N **k**-points is divided into N_c cells, and the coarse-graining transformation is applied[2]. The original lattice model H is then mapped onto a self-consistently embedded cluster of Anderson impurities (denoted by H_{AIM} in the following)[3]. To solve H_{AIM}, the quantum Monte Carlo (QMC) algorithm by Hirsch and Fye[5] is the conventional tool[6]. However, this method is restricted to rather high temperatures T, since the numerical effort grows like $1/T^3$. Therefore, the study of superconducting phases at low-T is usually problematic[7, 3].

As a new path to low T, Feldbacher *et al.* recently developed a projective QMC (PQMC) algorithm[8], which allows for solving the Anderson impurity model in the $T\to 0$ limit. We extended[9] this PQMC approach to coupled impurities, i.e., to H_{AIM}. Thereby, ground state expectation values $\langle\Psi_0|\mathscr{O}|\Psi_0\rangle/\langle\Psi_0|\Psi_0\rangle$ of an observable $\mathscr{O}$ (one- and two-particle Green functions in our case)

CP850, *Low Temperature Physics: 24th International Conference on Low Temperature Physics;*
edited by Y. Takano, S. P. Hershfield, S. O. Hill, P. J. Hirschfeld, and A. M. Goldman

are calculated as:

$$\langle\mathscr{O}\rangle_0 = \lim_{\theta\to\infty}\lim_{\tilde\beta\to\infty} \frac{\mathrm{Tr}\left[e^{-\tilde\beta H_T} e^{-\theta H_{\mathrm{AIM}}/2}\mathscr{O} e^{-\theta H_{\mathrm{AIM}}/2}\right]}{\mathrm{Tr}\left[e^{-\tilde\beta H_T} e^{-\theta H_{\mathrm{AIM}}}\right]}. \quad (2)$$

Here, H_T is a "trial" Hamiltonian, given by the non-interacting part of H_{AIM} with a shifted impurity level[10]. For such a non-interacting H_T, the limit $\tilde\beta \to \infty$ can be taken analytically, see [8]. Then, the starting point of the PQMC algorithm is a zero-temperature non-interacting Green function $G_0(R,R',\tau,\tau')$. We choose a cluster size $N_c = 4\times 4$ for R,R' and τ,τ' are truncated to $0\le\tau,\tau'\le\theta = 48$ discretized into $L = 64$ slices. For the measurement of physical quantities, $\mathscr{L}$ central time slices are taken, and $\mathscr{P}$ time slices on the right and left side of the measuring interval are used for projection[8, 10].

FIRST RESULTS

Let us now turn to the dominant pairing symmetry in the t-t' Hubbard model. For discussing the pairing instability, we first calculate

$$\chi(Q,\tau) = \sum_{K_1,K_2} g(K_1)g(K_2) \langle c^\dagger_{K_1\uparrow}(\tau)c_{K_2\downarrow}(0)c^\dagger_{-K_2\downarrow}(0)c_{-K_1\uparrow}(\tau)\rangle \quad (3)$$

for the self-consistent 4×4 cluster[11]. Here, $c_{K\sigma} = \sum_X c_{X\sigma}\exp(iKX)$ and $g(K)$ is the form factor, e.g., $\sqrt{2}\sin(K_x)$ for p_x wave, $\sqrt{2}\sin(K_x+K_y)$ for p_{x+y} wave and $\cos(K_x)-\cos(K_y)$ for $d_{x^2-y^2}$ wave. Then, we calculate $\mathrm{Im}\chi(\omega)$ from $\chi(\tau)$ by the maximum entropy method and obtain the static susceptibility from the Kramers-Kronig relation, $\chi(\omega=0) = \int \mathrm{Im}\chi(\omega)/\omega d\omega$.

Fig.1 is a typical example, where the inverse static pairing susceptibilities for $U = 2t, n = 0.3, T = 0$ is plotted. Besides the DCA(PQMC) results at $T = 0$, conventional DCA(QMC) susceptibilities are shown for $T > 0$. We can see that p_{x+y}-wave becomes dominant for $n = 0.3$. This is in contrast to the nearly half-filled case.

In a more detailed study presented elsewhere [9], we compare $\chi(\omega=0)$ for various pairing symmetries in the parameter range $0.3 < n < 0.8$ and $2t < U < 4t$. We find that p_{x+y}- ($d_{x^2-y^2}$-) wave is dominant among triplet (singlet) pairing in this region. The crossover between $d_{x^2-y^2}$-wave and p_{x+y}-wave occurs at $n\sim 0.4$; a p_{x+y}-wave phase does not exist for intermediate densities.

ACKNOWLEDGMENTS

We acknowledge support by the Alexander von Humboldt foundation (RA) and the Emmy Noether program of the Deutsche Forschungsgemeinschaft (KH).

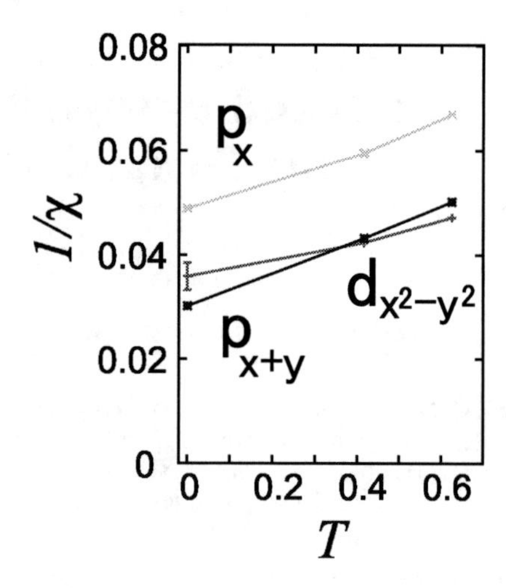

FIGURE 1. Inverse pair susceptibilities as a function of T for $U = 2t, n = 0.3$.

REFERENCES

1. W. Metzner and D. Vollhardt, *Phys. Rev. Lett.* **62**, 324 (1989); G. Kotliar and D. Vollhardt Physics Today **57**, 53 (2004); A. Georges *et al.*, *Rev. Mod. Phys.* **68**, 13 (1996).
2. M.H. Hettler *et al.*, *Phys. Rev. B* **58**, (1998) 7475; T. A. Maier *et al.*, cond-mat/0404055.
3. T. A. Maier *et al.*, cond-mat/0504529.
4. C. Honerkamp and M. Salmhofer, *Phys. Rev. Lett.* **87**, 187004 (2001); *Phys. Rev. B* **64**, 184516 (2001); A. A. Katanin and A. P. Kampf, *Phys. Rev. B* **68**, 19501 (2003); C. Honerkamp, D. Rohe, S. Andergassen, and T. Enss, *Phys. Rev. B* **70**, 235115 (2004), T. Nomura and K. Yamada, *J. Phys. Soc. Jpn.* **69**, 3678 (2000), K. Kuroki, T. Tanaka, K. Kimura and R. Arita, *Phys. Rev. B* **69** 214511 (2004).
5. J. E. Hirsch and R. M. Fye, *Phys. Rev. Lett.***56** 2521 (1986).
6. M. Jarrell, T. A. Maier, C. Huscroft, and S. Moukouri, *Phys. Rev. B* **64**, 195130 (2001).
7. R. Arita, S. Onari, K. Kuroki, and H. Aoki, *Phys. Rev. Lett.* **92** 247006 (2004).
8. M. Feldbacher, K. Held, and F. F. Assaad, *Phys. Rev. Lett.* **93**, 136405 (2004); for multi-orbital DMFT(PQMC) see R. Arita and K. Held, cond-mat/0504040.
9. R. Arita and K. Held, in preparation.
10. We shift the impurity potential in such a way that the Green functions of H_T and H have the same average impurity occupation and, hence, the same asymptotics at large imaginary time τ. This provides for better results for a *finite* number of projection time slice $\mathscr{P}$ (for more details, see [9]).
11. For simplicity, we look at the susceptibility on the self-consistent cluster. The lattice susceptibilities will be larger and divergent at phase transitions [2, 3]. However, we expect that the cluster susceptibility already captures the essential features of the competition between different pairing symmetries.

Superconductivity from a Long-Range Repulsive Interaction

S. Onari*, R. Arita†, K. Kuroki** and H. Aoki†

*Graduate School of Engineering, Nagoya University, Chikusa, Nagoya 464-8603, Japan
†Department of Physics, University of Tokyo, Hongo, Tokyo 113-0033, Japan
**Dept of Applied Physics & Chemistry, University of Electro-Communications, Chofu, Tokyo 182-8585, Japan

Abstract. The lattice model with short-range interactions (exemplified by the Hubbard model) is known to exhibit quite different features from those in the electron gas with the long-range Coulomb interaction. In order to explore how they cross over to each other, we have studied an extended Hubbard model which includes repulsions up to the 12th neighbors with the simplified fluctuation exchange (FLEX) approximation for the square lattice. We have found that (i) in the most dilute density region, spin and charge fluctuations become comparable, and *s*- and *p*-waves superconductivity become dominant, in agreement with the result for the electron gas by Takada, while (ii) the dominant spin fluctuation and its reflection on $d_{x^2-y^2}$ and d_{xy} pairing, both the effect of lattice structure, persists well away ($n \gtrsim 0.2$) from the half filling.

Keywords: superconductivity, Hubbard model, electron gas, charge fluctuation, spin fluctuation
PACS: 74.20.Mn

The current interest, as kicked off by the high-Tc cuprate, in the superconductivity arising from short-range electron-electron repulsive interactions leads naturally to the question of whether and how this would be related to another prototype for the electron system — the electron gas with the Coulomb ($1/r$) interaction. Study of superconductivity in the electron gas has in fact a very long history, in which Takada[1, 2] has elaborated the mechanism of pairing arising from the long-range Coulomb interaction.

There are two essential differences between the system with a short-range repulsion (typically the Hubbard model with the shortest possible (on-site) interaction) and the Coulomb gas: (i) For short range interactions, the spin fluctuation dominates in magnitude over the charge fluctuation, while the latter becomes significant for long-range interactions. (ii) While the electron gas is modeled in a continuous (jellium) space, the Hubbard model is defined on a lattice, where the band filling emerges as a crucial parameter. Specifically, the half filling is a special point, around which the Mott's metal-insulator transition occurs. Near half filling the (typically antiferromagnetic) spin fluctuation becomes dominant,which favors the pairing symmetry $d_{x^2-y^2}$, as studied with, e.g., the fluctuation exchange (FLEX) approximation[3, 4, 5, 6] for the model high-T$_c$ cuprates and also for more general shapes of the Fermi surface with varied filling and the lattice structure[7, 8]. So the study of the crossover between the two extreme (on-site and $1/r$) interactions is intriguing.

So, while there are intensive studies for the extended Hubbard model that extends the repulsion to the nearest-neighbor interaction V, the purpose of the present paper is to explore what happens to the pairing symmetry as we change the interaction from the on-site to long-ranged one ($\sim 1/r$) for a square lattice to compare the result with those for the electron gas. We adopt the simplified FLEX which includes bubbles and ladder diagrams like the RPA[9, 10]. For the FLEX study we have cut off the interaction at a finite distance (the 12-th neighbor), and we shall see whether the pairing in the lattice model with a finite range still resembles those in the electron gas.

We show that the pairing in the present model reproduces the pairing symmetries (*s*-wave, *p*-wave) of the electron gas for the dilute region, while the pairing symmetries ($d_{x^2-y^2}$, d_{xy}) that reflect the lattice structure persist well away ($n \gtrsim 0.2$) from the half filling. The key factors that govern the symmetry of the gap function are identified to be the structure of charge and spin susceptibilities especially (near half filling), and the size of the Fermi surface (in the dilute regime).

Let us first introduce a long-ranged Hubbard model,

$$\mathscr{H} = -t\sum_{i,j}^{\mathrm{nn}}\sum_{\sigma} c^{\dagger}_{i\sigma}c_{j\sigma} + U\sum_{i} n_{i\uparrow}n_{i\downarrow} + \frac{1}{2}\sum_{i,j}\sum_{\sigma\sigma'} V_{ij}n_{i\sigma}n_{j\sigma'}, \quad (1)$$

where off-site Coulomb repulsion $V_{ij} \propto 1/r_{ij}$ is extended as far as the 12th neighbor for the square lattice here (for which electrons can interact for the band filling as dilute as $n = 0.03$). Hence two parameters (the on-site U and the nearest $V \equiv V_1$) characterize the model.

It is difficult to include beyond the second-neighbor repulsion in the full FLEX calculation (since the inclusion of the third neighbor makes every quantity a 13×13 matrix). So we use here a simplified FLEX, where we consider full bubble diagrams and restricted ladder (the U term) diagrams, which are also adopted in the RPA[9, 10] with nearest-neighbor Coulomb interaction, for the effective interaction in a self-consistent

CP850, *Low Temperature Physics: 24th International Conference on Low Temperature Physics;*
edited by Y. Takano, S. P. Hershfield, S. O. Hill, P. J. Hirschfeld, and A. M. Goldman

manner. Both the FLEX and the simplified FLEX satisfy "conserving approximations" formulated by Baym and Kadanoff[11, 12]. In terms of the irreducible susceptibility, $\overline{\chi}(q) = -\frac{T}{N}\sum_k G(k+q)G(k)$, where $q \equiv (\mathbf{q}, \varepsilon_n)$ with $\varepsilon_n \equiv 2n\pi T$ being the Matsubara frequencies for bosons and $k = (\mathbf{k}, \omega_n)$ with $\omega_n = (2n-1)\pi T$ for fermions, the spin susceptibility is $\chi_{\rm sp}(q) = \overline{\chi}(q)/[1 - U\overline{\chi}(q)]$, while the charge susceptibility is $\chi_{\rm ch}(q) = \overline{\chi}(q)/\{1 + [U + 2V(q)]\overline{\chi}(q)\}$, where $V(q) = 2V_1(\cos q_x + \cos q_y) + 4V_2(\cos q_x \cos q_y) + 2V_3(\cos 2q_x + \cos 2q_y) + \cdots + (V_{12}$-term).

The gap function and T_c are obtained from Éliashberg's equation,

$$\lambda\phi(k) = -\frac{T}{N}\sum_{k'}\Gamma(k-k')G(k')G(-k')\phi(k'), \qquad (2)$$

where$\lambda = 1$ corresponds to $T = T_c$, and Γ is the pairing interaction (Γ_s in the singlet and Γ_t in the triplet channel). Here we take $N = 32 \times 32$ **k**-point meshes and the Matsubara frequencies ω_n from $-(2N_c - 1)\pi T$ to $(2N_c - 1)\pi T$ with $N_c = 16384$, $T = 0.01$, $U = 4.0$ and the unit of energy is the nearest-neighbor hopping integral $t = 1$.

We show in Fig.1 the eigenvalue of Éliashberg's equation λ, along with the charge ($\chi_{\rm ch}$) and spin ($\chi_{\rm sp}$) susceptibilities against the band filling n for $V = 1.0$. We see that the dominant pairing changes as $d_{x^2-y^2} \to d_{xy} \to p \to s$ as the band filling becomes dilute.

Let us compare this with the phase diagram obtained by Takada[1] for the electron gas, where the sole parameter is r_s. We see that p- and s-waves appear in the same order in both models in the dilute regime, where $\chi_{\rm ch}$ and $\chi_{\rm sp}$ are comparable. However, the onset values of r_s for s and p waves (which is defined as $r_s = \sqrt{2}/(a_B^* k_F)$, where a_B^* is the effective Bohr radius and indicated on the top axis in Fig.1) exhibit significant differences between the two. One factor should be that a 2D system is more favorable for the superconductivity than in 3D as expected from the plasmon mechanism[2].

If we turn to the region closer to the half filling, the effect of the lattice (spin fluctuations with well-defined structures in the spin susceptibility) is dominant to give d_{xy} (which is favored as $\chi_{\rm ch}/\chi_{\rm sp}$ increases) and $d_{x^2-y^2}$, and we see that this occurs well away from the half filling. We consider this as a phenomenon that interpolates between the electron gas and a lattice model.

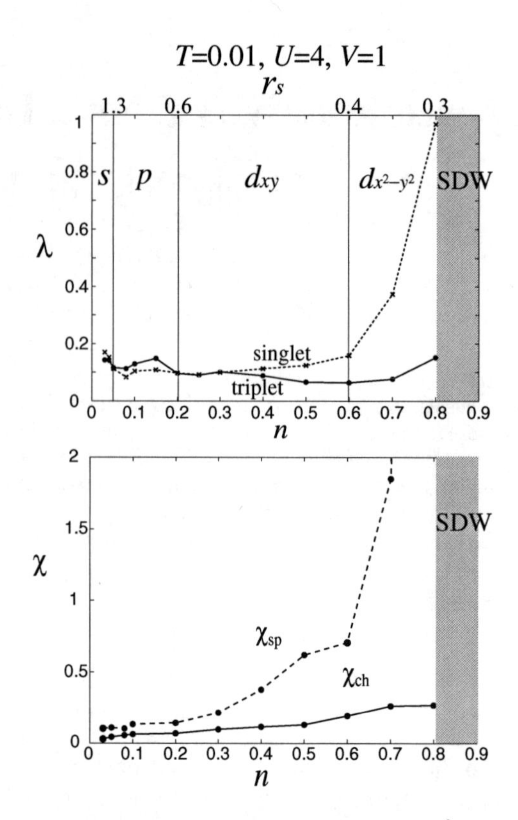

FIGURE 1. Top: The maximum eigenvalue of Éliashberg's equation λ for the triplet (solid line) and singlet (dotted) superconductivity with the dominant pairing symmetry and the r_s values indicated. The SDW phase is identified from the divergence of the spin susceptibility. Bottom: The charge (solid line) and spin (dotted) susceptibilities against n.

ACKNOWLEDGMENTS

This work is in part supported by a Grant-in-Aid for Science Research on Priority Area from the Japanese Ministry of Education. Numerical calculations were performed at the supercomputer center, ISSP.

REFERENCES

1. Y. Takada, *Phys. Rev. B* **47**, 5202 (1993).
2. Y. Takada, *J. Phys. Soc. Jpn.* **45**, 786 (1978).
3. N. E. Bickers, D. J. Scalapino, and S. R. White, *Phys. Rev. Lett.* **62**, 961 (1989).
4. N. E. Bickers and D. J. Scalapino, *Ann. Phys. (N.Y.)* **193**, 206 (1989).
5. T. Dahm and L. Tewordt, *Phys. Rev. B* **52**, 1297 (1995).
6. M. Langer, J. Schmalian, S. Grabowski, and K. H. Bennemann, *Phys. Rev. Lett.* **75**, 4508 (1995).
7. K. Kuroki and R. Arita, *Phys. Rev. B* **64**, 024501 (2001).
8. S. Onari, R. Arita, K. Kuroki, and H. Aoki, *Phys. Rev. B* **68**, 024525 (2003).
9. A. Kobayashi, Y. Tanaka, M. Ogata, and Y. Suzumura, *J. Phys. Soc. Jpn.* **73**, 1115 (2004).
10. Y. Tanaka, Y. Yanase, and M. Ogata, *J. Phys. Soc. Jpn.* **73**, 319 (2004).
11. G. Baym and L. P. Kadanoff, *Phys. Rev.* **124**, 287 (1961).
12. G. Baym, *Phys. Rev.* **127**, 1391 (1962).

Spinons, Holons and Flux Tubes in a SU(3) Theory of High T_c Superconductors

Stewart E. Barnes

Department of Physics, University of Miami, Coral Gables, FL 33124, USA

Abstract. Described is a flavour of the RVB theory of high T_c materials which uses the "slave boson" method. Holes are added by SU(3) rotation in the auxiliary particle space in a manner which does not violate the constraint $Q_i = 1$. The effective spinon hopping matrix element $J - 2xt$ is reduced by doping. The local SU(3) maps to S. C. Zhang's SO(5) on two sites but has different π-operators. Kondo like infrared divergences occur. The ARPES humb-dip-peak structure is reproduced.

Keywords: High T_c superconductors, $t-J$ and Hubbard models, highly correlated electrons
PACS: 71.30.+h74.20.-z74.20.Mn 74.25.Ha 74.72.-h

INTRODUCTION

It remains the case that there is no universally accepted theory for high temperature superconductivity. Outlined here[1, 2, 5] is a "slave boson" realisation of the RVB scenario[3]. The theory contains SO(5) rotations[4] in the space which accommodates both the anti-ferromagnetic (AF) magnetic super-conductive (SC) order parameters. Within the usual three dimensional auxiliary particle space spanned by the single hole boson $b_i^\dagger$ and two spin $f_{\sigma i}^\dagger$; $\sigma = \uparrow, \downarrow$ fermions are made SU(3) rotations which mix the hole and spin particles *but* which maintain the constraint $Q_i = 1$. The result is a coherent state in which the holes form a condensate. When two sites are considered, the SU(3) rotations contain those of SO(5), *however* the superconductivity stems from RVB and not BCS pairing and implies a new definition of the "π-operators". In the mean-field slave boson gauge theory[6], Bose condensation simply implies the thermal wavelength is of the order of the inter-particle distance. Here condensation requires the hole Bose level be re-normalised in energy to the physical chemical potential. This channel produces a "gap equation" which determines T_c. The new π-mode corresponds to a transition between the hole and AF Bose condensates of the theory. The neutron mode in SC state is an analogue of the ^{4}He roton mode and reflects the energy of an AF pseudo-magnon, with a given $\vec{Q}$ relative to the hole condensate.

THEORY

It is not possible to mix Bose and Fermi operators. For this reason, it is necessary to introduce a primitive supersymmetry where both Bose and Fermi versions of the same operator exist[2, 5]. This is the role of the unit flux tubes and the associated operators u_i where i is a site index. Thus the Bose version of the spin operators are $f_{\sigma i}^\dagger u_i$ while the Fermi hole operator is $b_i^\dagger u_i$ [1]. In, e.g., the AF state there is coexistence of *both* the fermionic $f_{\uparrow i}^\dagger$ and bosonic $f_{\uparrow i}^\dagger u_i$ spin operators[2]. In Fig. 1 is shown a comparison between the ARPES data and the *fermion* "spinon" sector of an AF-magnet predicted by the theory. The Bose sector comprises AF spin waves which lie in the shaded region and are not seen in this experiment.

The constraint, $Q_i = \sum_\sigma f_{\sigma i}^\dagger f_{\sigma i} + b_i^\dagger b_i = 1$, effectively defines a unit sphere and the allowed SU(3) rotation operators R are such that $Q_i = 1$, and, e.g.,

$$R f_{\sigma i}^\dagger R^{-1} \equiv \tilde{f}_{\sigma i}^\dagger = (1 - \frac{\psi_i^{*\sigma} \psi_i^\sigma}{2}) f_{\sigma i}^\dagger + \psi_i^\sigma b_i^\dagger u_i, \quad (1)$$

which coherently adds, to a given spin wave state, holes with the spin dependent wave function ψ_i^σ. Clearly the result has the nature of a coherent state, however since $Q_i = 1$ it is not possible to have a finite $\langle b_i^\dagger \rangle$ which is usually taken to indicate a condensate. However a coherent state, with a definite energy, in itself implies that holes do not contribute to the energy, i.e., that there is a condensate at the chemical potential.

That there be such a hole condensate requires that

$$i\hbar \frac{\partial b_i^\dagger}{\partial t} = [b_i^\dagger, \mathscr{H}] \quad (2)$$

have zero energy solutions and this reflects the Kondo physics. The $t-J$-model contains the near neighbour

[1] It is only possible to define an operator u_i which attaches a flux tube because this has unit magnitude.

[2] These are essentially equivalent to $S = 1/2$ Holstein-Primakoff bosons. The $f_{\sigma i}^\dagger$ over specify the problem. The $f_{\downarrow i}^\dagger$ generates the vacuum state while $f_{\uparrow i}^\dagger$ is the unique "spin particle". The method is simply an adaptation of the Jordan-Wigner transformation to higher dimensions.

CP850, *Low Temperature Physics: 24th International Conference on Low Temperature Physics;*
edited by Y. Takano, S. P. Hershfield, S. O. Hill, P. J. Hirschfeld, and A. M. Goldman

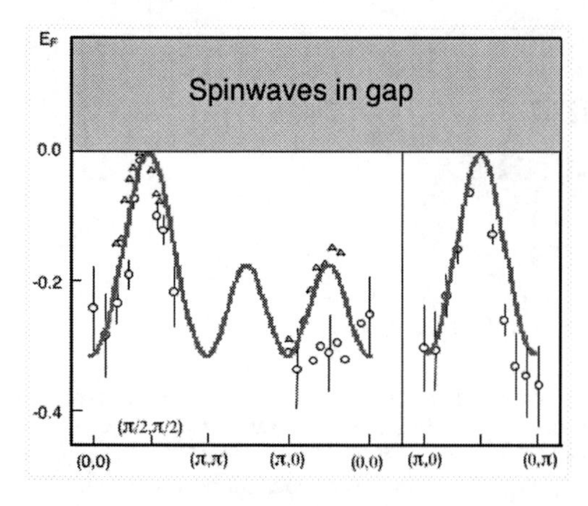

FIGURE 1. This are the spinon bands in "undoped" antiferromagnetic $Sr_2CuO_2Cl_2$. The open circles (Wells et al.) and open triangles, (LaRosa et al.) are experiment[7]. The lowest energy spin excitations are AF magnons with smaller energy than the spinons. The vertical scale is in eV.

kinetic energy $\mathscr{H}_T = -t\sum_{<ij>\sigma} b_i^\dagger b_j f_{\sigma i}^\dagger f_{\sigma j}$ which in $O(t^2)$, for Eqn. (2), contains infrared dangerous diagrams. There is an infrared cut-off since (i) in the insulating parent compound the Bose level must lie above the chemical potential and (ii) because the spinon dispersion $\sim J$ plays the role of the usual Kondo cut-off $k_B T_K$. Materials such as the original La-compounds which have the Bose level far from the chemical potential require $k_B T_K$ to be smaller and hence narrower spinon bands and lower T_c.

The same current-current coupling which leads to a magnon instability in ferromagnets[8] operates here. When high charge current states near $\pi/2, \pi/2$ become occupied the AF magnons become destabilised and spin currents are induced. Near to optimal doping these spin currents develop into the staggered flux phase realisation of the RVB scheme. Concurrently a Fermi surface crossing develops near $\pi/2, \pi/2$ in the spinon band of Fig. 1 and at T_c a uniform hole condensate appears. The spinon band now corresponds to both the super-conductive coherence peak and a "Kondo" resonance. The Kondo spinon band corresponds to good quasi-particles with energies $E_{\vec{k}} = \sqrt{\varepsilon_{\vec{k}}^2 + \Delta_{\vec{k}}^2}$ with $\varepsilon_{\vec{k}} = ((J-2xt)/2\sqrt{2})(\cos k_x a + \cos k_y a)$ and $\Delta_{\vec{k}} = (J/2\sqrt{2})(\cos k_x a - \cos k_y a)$. The formation of such a band requires condensed Bose holes and superconductivity. The presence of a concentration x of such holes reduces the effective exchange to $(J-2xt)$ in $\varepsilon_{\vec{k}}$ but *not* in the gap $\Delta_{\vec{k}}$. Thus, *if* d-wave superconductivity was absent ($\Delta_{\vec{k}} = 0$), there would be a quantum critical point for $(J-2xt) = 0$ for which both the spinons and AF magnons have zero dispersion.

Fermionic holes are manifest as the ARPES "hump".

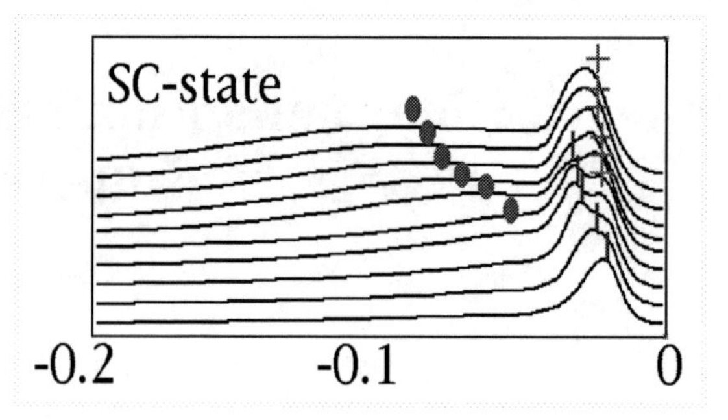

FIGURE 2. Theory to be compared with ARPES on the super-conductive state Bi2212[9]. The strongly dispersive "holon" bonding band is indicated by a filled circle. The | and + indicate the positions of the "spinon" coherence peaks associated with the bonding and anti-bonding bands respectively. The intensity is in arbitrary units and the energy is in eV.

This reflects poorly defined fermions which are a convolution of such a hole and a spinon propagator. The theory, Fig.2, has been compared with data on a bi-layer material. This hump band has a dispersion determined by t and hence a large bi-layer splitting while that of the spinon coherence peak is small consistent with its Kondo origin.

The phase diagram is consistent with the SO(5) classification. There is no RVB transition temperature and T_c always corresponds to hole Bose condensation for both under and over-doped materials.

ACKNOWLEDGMENTS

Many of these results stem from a collaboration with S. Maekawa and part of this work was performed at and supported through the Institute for Materials Research, Tohoku University, Japan.

REFERENCES

1. S. E. Barnes in *Physics of Transition Metal Oxides* S. Maekawa, et al., Berlin: Springer-Verlag, (2004).
2. S. E. Barnes and S. Maekawa, *Phys. Rev. B* **67**, 224513 (2003).
3. P. W. Anderson, *Science* **235**, 1196 (1987).
4. S.-C. Zhang, *Science* **275**, 1089 (1997).
5. S. E. Barnes and S. Maekawa *J. Phys. Cond. Matter* **14**, L19-L28 (2002).
6. See e.g., review: P. A. Lee et al., cond-mat/0410445.
7. B. O. Wells, Z. -X. Shen, A. Matsuura, D. M. King, M. A. Kastner, M. Greven, R. J. Birgeneau, *Phys. Rev. Lett.* **74**, 964 (1995); S. LaRosa, I. Vobornik, F. Zwick, H. Berger, M. Grioni, G. Margaritondo, *Phys. Rev. B* **56**, R525 (1997).
8. Y. B. Bazaliy, B. A. Jones and S.-C. Zhang, *Phys. Rev. B* **57**, R3213 (1998).
9. D. L. Feng et al., *Phys. Rev. Lett.* **86**, 5550 (2001).

Electronic Specific Heat of *s*-, *p*-, and *d*-wave Superconducting States

J. Samuel Millán[a], Luis A. Pérez[b] and Chumin Wang[c]

[a] *Facultad de Ingeniería, UNACAR, C.P. 24180, Cd. del Carmen, Campeche, México*
[b] *Instituto de Física, Universidad Nacional Autónoma de México (UNAM), A.P. 20-364, C.P. 01000, D.F., México*
[c] *Instituto de Investigaciones en Materiales, UNAM, A.P. 70-360, C.P. 04510, México D.F., México*

Abstract. Based on the BCS formalism, a comparative study of the electronic specific heat for *s*-, *p*, and *d*-symmetry superconducting states is performed in a square lattice described by a generalized Hubbard model, in which correlated-hopping interactions are included in addition to the repulsive Coulomb ones. The *p*- and *d*-wave superconducting states are respectively obtained at middle and high electronic density regimes, *i.e.* the later is formed by holes and in fact, its critical temperature is two orders of magnitude larger than the *p*-channel one. The electronic specific heat analysis shows a power law behavior in *p*- and *d*-wave superconducting states, instead of the exponential temperature dependence found in the *s* channel. Finally, a good agreement with experimental data from Sr_2RuO_4 is obtained.

Keywords: Superconducting gap symmetry, Hubbard model, two-dimensional systems.
PACS: 71.10.Fd, 74.20.Fg, 74.20.Rp, 74.70.Pq

One of the physical quantities that yields information about the symmetry of superconducting states is the electronic specific heat, which is highly sensitive to the low-energy excitations of the system. A fully gapped superconductor has an exponentially activated electronic specific heat, while an anisotropically gapped superconductor will have an electronic specific heat following some power law of the temperature, as occur in the cuprate superconductors [1] and in Sr_2RuO_4 [2]. Three-band Hubbard models have been proposed to describe the dynamics of the carriers on the CuO_2 and RuO_2 planes, and the electronic states close to the Fermi energy can be reasonably well described by a single-band tight-binding model with next-nearest-neighbor hoppings [3,4]. Recently, we have found that the second-neighbor correlated-hopping interaction (Δt_3) is essential in the $d_{x^2-y^2}$ wave superconductivity [5] and that a further small distortion of the right angles in the square lattice leads to *p*-wave superconductivity [6]. It is worth mentioning that a similar distortion has been observed at the surface of Sr_2RuO_4 [7].

In this work, we start from a single-band Hubbard model, in which first (Δt) and second (Δt_3) neighbor correlated-hopping interactions are considered in addition to the on-site (U) and nearest-neighbor (V) Coulomb interactions. The corresponding Hamiltonian ($\hat{H}$), using the same notations as in Ref. [6], is

$$\hat{H} = -t \sum_{<i,j>\sigma} c_{i\sigma}^{+} c_{j\sigma} - t' \sum_{<<i,j>>\sigma} c_{i\sigma}^{+} c_{j\sigma} + U \sum_{i} n_{i\uparrow} n_{i\downarrow} + \frac{V}{2} \sum_{<i,j>} n_i n_j + \Delta t \sum_{<i,j>,\sigma} c_{i\sigma}^{+} c_{j\sigma} (n_{i-\sigma} + n_{j-\sigma}) + \Delta t_3 \sum_{<<i,j>>,\sigma,<i,l>,<j,l>} c_{i\sigma}^{+} c_{j\sigma} n_l \, . \tag{1}$$

If we further consider a small distortion of the right angles in the square lattice, the second-neighbor interactions change and their new values are $-t'_{\pm} \equiv -t' \pm \delta$ and $\Delta t_3^{\pm} \equiv \Delta t_3 \pm \delta_3$, where $\pm$ refers to the $\hat{x} \pm \hat{y}$ direction. We apply the BCS formalism to Eq. (1) obtaining two coupled integral equations, which determine the superconducting gap (Δ_α) and the chemical potential (μ_α) at a given temperature (T) and electronic density (n) [6], where $\alpha = s, p$ or d indicates the superconducting gap symmetry. Hence, the electronic specific heat of superconducting states (C_α) is given by [8]

$$C_\alpha = \frac{2k_B\beta^2}{4\pi^2} \iint_{1BZ} f(E_{\mathbf{k}}^{\alpha})\left[1 - f(E_{\mathbf{k}}^{\alpha})\right]\left[\left(E_{\mathbf{k}}^{\alpha}\right)^2 + \beta E_{\mathbf{k}}^{\alpha} \frac{dE_{\mathbf{k}}^{\alpha}}{d\beta}\right], \tag{2}$$

where $\beta = 1/(k_B T)$, $f(E)$ is the Fermi-Dirac distribution, $E_{\mathbf{k}}^{\alpha} = \sqrt{\left[\varepsilon(\mathbf{k}) - \mu_\alpha\right]^2 + \Delta_\alpha^2(\mathbf{k})}$ is the single-particle

CP850, *Low Temperature Physics: 24th International Conference on Low Temperature Physics;*
edited by Y. Takano, S. P. Hershfield, S. O. Hill, P. J. Hirschfeld, and A. M. Goldman

excitation energy and $\varepsilon(\mathbf{k})$ is the mean-field dispersion relation [6]. To obtain the specific heat of the normal state (C_n) we take Δ_α equal to zero [8].

In figure 1, the electronic specific heat of *s*- (open circles), *p*- (open triangles), and *d*-wave (open squares) superconducting states is plotted as a function of the temperature, for any U, $V=\delta=0$, $t'=0.45t$, $\Delta t=0.5t$, $\Delta t_3=0.15t$, $\delta_3=0.1t$, with $n=1.0$ and 1.4 in the *p*- and *d*-channels, respectively. In the case of *s*-channel, the results are obtained by using $U=-1.3t$, $n=1.0$, and the rest of the parameters equal to zero. Inset shows the corresponding normalized superconducting gaps versus the temperature. Notice the non-BCS behavior of *p*- and *d*-wave superconducting-gap temperature dependence. In consequence, their specific heats present a power-law temperature dependence instead of the exponential behavior observed for the *s*-channel. In addition, the discontinuities in the specific heat at the critical temperature (T_c) are respectively 1.77, 0.75 and 0.8 for *s*-, *p*- and *d*-wave superconducting states, where the value for the *s*-channel is larger than 1.43, the expected BCS value, since in the attractive-Hubbard model all the electrons participate in the condensate.

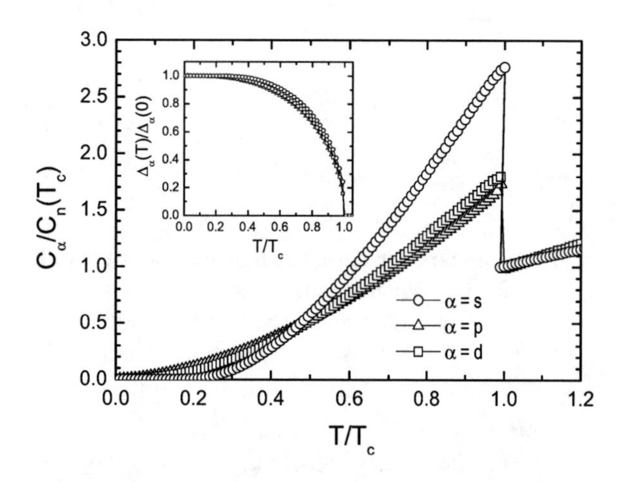

FIGURE 1. Superconducting electronic specific heat (C_α) normalized by that of the normal state at T_c [$C_n(T_c)$] versus temperature (T) for *s*-(open circles), *p*-(open triangles) and *d*-wave (open squares) states. Inset: Corresponding superconducting gaps versus temperature.

Figure 2 shows a comparison of the normalized *p*-wave specific heat (C_p) of Fig. 1 with the experimental data obtained from the spin-triplet superconductor Sr_2RuO_4 [9]. Notice the notable agreement in both, the discontinuity at T_c and the temperature dependence below T_c, as a consequence of the nature of the *p*-wave superconducting state obtained in the generalized Hubbard model.

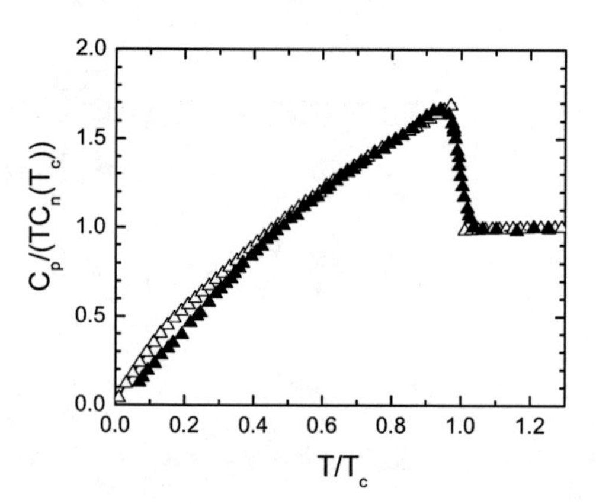

FIGURE 2. Normalized *p*-wave specific heat, as in Fig. 1, versus temperature in comparison with the experimental data (solid triangles) obtained from Sr_2RuO_4 [9].

In summary, the presence of nodes in the superconducting gap, induced by the correlated hoppings, modifies the electronic absorption of the thermal energy through the single-particle dispersion relation [6] and consequently has deep effects on the electronic specific heat. The results show a power law behavior in *p*- and *d*-wave superconducting states, instead of the exponential temperature dependence found in the *s* channel. Finally, an excellent agreement with experimental data from Sr_2RuO_4 is obtained.

ACKNOWLEDGMENTS

This work was partially supported by CONACyT-41492F and by UNAM-IN116605 and -IN110305. J.S.M. acknowledges PROMEP-UACAR-2001-01-01. Computations have been done at the Bakliz of UNAM.

REFERENCES

1. C.C. Tsuei and J.R. Kirtley, *Rev. Mod. Phys.* **72**, 969-1016 (2000).
2. A.P. Mackenzie and Y. Maeno, *Rev. Mod. Phys.* **75**, 657-712 (2003).
3. E. Dagotto, *Rev. Mod. Phys.* **66,** 763-840 (1994).
4. I.I. Mazin and D.J. Singh, *Phys. Rev. Lett.* **79**, 733-736 (1997).
5. L.A. Pérez and C. Wang, *Solid State Commun.* **121,** 669-674 (2002).
6. J.S. Millán, L.A. Pérez and C. Wang, *Phys. Lett. A* **335**, 505-511 (2005).
7. R. Matzdorf, Z. Fang, Ismail, J. Zhang, T. Kimura, Y. Tokura, K. Terakura, and E.W. Plummer *Science* **289** (2000) 746-748.
8. M. Tinkham, *Introduction to Superconductivity, 2nd Ed.*, New York: McGraw-Hill, 1996, pp. 64-66.
9. S.Nishizaki, Y. Maeno, and Z. Mao, *J. Phys. Soc. Jpn.* **69**, 572-578 (2000).

Physical Meaning of the Current Vertex Corrections: DC and AC Transport Phenomena in High-T_c Superconductors

Hiroshi Kontani

Department of Physics, Nagoya University, Nagoya, 464-8602, Japan

Abstract. Famous non-Fermi liquid-like behaviors of the transport phenomena in high-T_c cuprates (R_H, $\Delta\rho/\rho$, S, ν, etc) are caused by the current vertex corrections in neary antiferromagnetic (AF) Fermi liquid, which was called the backflow by Landau. We present a simple explanation why the backflow is prominent in strongly correlated systems. In nearly AF Fermi liquid, R_H is enhanced by the backflow because it changes the effective curvature of the Fermi surfaces. Therefore, the relaxation time approximation is not applicalbe to a system nearby a magnetic quantum critical point (QCP).

Keywords: transport phenomena, high-T_c superconductors, current vertex correction, backflow
PACS: 72.10.-d, 78.20.Bh, 71.10.Ay

In the present article, we explain that the backflow always plays important roles in Fermi liquids. In a Fermi liquid, excited quasiparticles (QP) interact with each other. Landau showed that the QP energy at $\mathbf{p}$ in the presence of QP excitations is expresed as [1]

$$\tilde{\varepsilon}_{\mathbf{p}} = \varepsilon_{\mathbf{p}} + \sum_{\mathbf{k}} f_{\mathbf{p},\mathbf{k}} \delta n_{\mathbf{k}}, \tag{1}$$

where $\varepsilon_{\mathbf{p}} = \mathbf{p}^2/2m^*$ and $\delta n_{\mathbf{k}}$ represents the deviation of the QP distribution function from the ground state. $f_{\mathbf{p},\mathbf{k}}$ is the QP interaction (i.e., Landau interaction function) which arises from the electron-electron correlations. Hereafter, we consider the paramagnetic state.

First, we discuss an isotropic Fermi liquid. Here, we consider to add a QP just above the Fermi surface (FS) at $\mathbf{k}$ as shown in Fig.1, whose lifetime τ is infinitesimally long. Then, the QP energy at $\mathbf{p}$ shifts by $f_{\mathbf{p},\mathbf{k}}$, which induces the change of QP velocity [1]. Its summation over the FS is given by $\sum_{|\mathbf{p}|<k_F} \nabla_{\mathbf{p}} f_{\mathbf{p},\mathbf{k}} = N \oint_{FS} dp_{\parallel} f_{\mathbf{k},\mathbf{p}} \mathbf{v}_{\mathbf{p}}$, which is called the backflow by Landau. In a spherical system, the QP velocity at $\mathbf{k}$ and the backflow are given by $\mathbf{v}_{\mathbf{k}} = \mathbf{k}/m^*$ and $\frac{1}{3}F_1\mathbf{k}/m^*$ (F_1 being a Landau parameter), respectively. Because $1+\frac{1}{3}F_1 = m^*/m$ due to Galilei invariance, the total current $\mathbf{J}_{\mathbf{k}}$ is $\mathbf{k}/m$ [1]. Thus, the backflow dominates the QP velocity in a strongly correlated Fermi liquid where $m^* \gg m$.

The backflow plays an important role in DC transport phenomena. To show this fact, we study the current vertex correction in the hydrodynamic limit ($\omega\tau \ll 1$) based on the microscopic Fermi liquid theory. The resistivity of interacting electrons without umklapp scatterings should be completely zero, as consequence of the momentum conservation law. However, the relaxation time approximation (RTA) gives a finite resistivity because $\rho^{\mathrm{RTA}} \propto \tau > 0$. Yamada and Yosida solved this discrepancy by taking the backflow (i.e., the current vertex correction) into account; they succeeded in reproducing $\rho = 0$ in the absence of the umklapp scatterings, even when τ is finite [2].

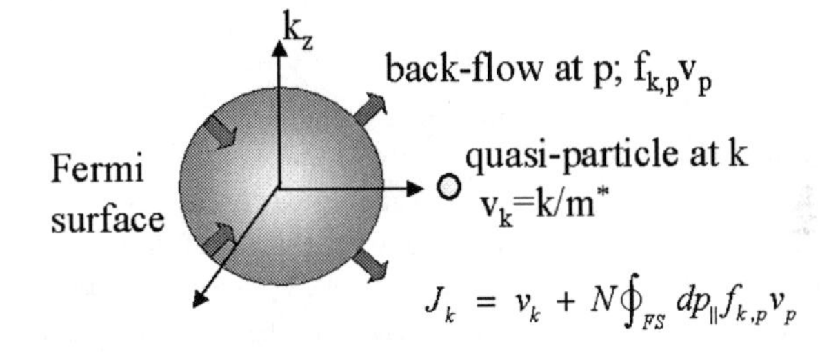

FIGURE 1. Backflow is the induced QP velocity over the FS due to other QP excitations, which is one of the most essential properties of Fermi liquids.

In 1999, we found that the backflow in nearly antiferromagnetic (AF) metals plays highly nontrivial roles; the total current $\mathbf{J}_{\mathbf{k}}$ is no more parallel to the QP velocity $\mathbf{v}_{\mathbf{k}}$ due to the backflow [2, 3]. The Bethe-Salpeter equation in the microscopic Fermi liquid theory is given by

$$\vec{J}_{\mathbf{k}} = \vec{v}_{\mathbf{k}} + \oint_{FS} dk_{\parallel} \Gamma(\mathbf{k},\mathbf{p}) \vec{J}_{\mathbf{p}}, \tag{2}$$

where $\Gamma(\mathbf{k},\mathbf{p})$ corresponds to the Landau interaction function in the hydrodynamic limit, and $k_{\parallel}$ is the momentum along the FS. In the FLEX approximation, $\Gamma(\mathbf{k},\mathbf{p}) = \int d\varepsilon[\mathrm{cth}(\varepsilon/2T) - \mathrm{th}(\varepsilon/2T)]\mathrm{Im}V^{\mathrm{FLEX}}_{\mathbf{k}-\mathbf{p}}(\varepsilon)\tau_{\mathbf{p}}$ and $V^{\mathrm{FLEX}}_{\mathbf{k}}(\varepsilon) \propto \chi^s_{\mathbf{k}}(\varepsilon) \propto \xi^2(1+\xi^2(\mathbf{k}-\mathbf{Q})^2 - i\varepsilon/\omega_{\mathrm{sf}})^{-1}$, where $\chi^s_{\mathbf{k}}$ is the spin susceptibility; ξ is the AF correlation length and $\mathbf{Q} = (\pm\pi,\pm\pi)$. In the SCR or FLEX approximation, $\xi^2 \propto T^{-1}$ and $\xi \gg 1$ at lower temperatures. Then, $\Gamma(\mathbf{k},\mathbf{p})$ takes a large value only when $|\mathbf{k}-\mathbf{p}-\mathbf{Q}| \lesssim \xi^{-2}$. As shown in Fig.2, a QP excitation at $\mathbf{k}$ causes the backflow through the QP interaction

CP850, *Low Temperature Physics: 24th International Conference on Low Temperature Physics;*
edited by Y. Takano, S. P. Hershfield, S. O. Hill, P. J. Hirschfeld, and A. M. Goldman

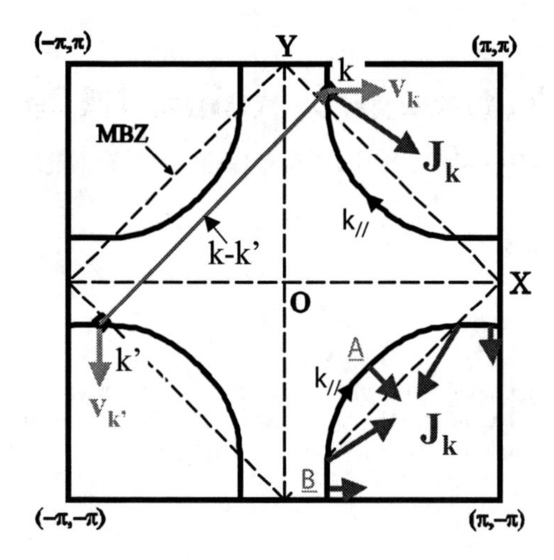

FIGURE 2. Schematic picture of $\mathbf{J_k}$ in high-T_c cuprates. Because the excited QP at $\mathbf{k}$ reduces the QP energy $\varepsilon_{\mathbf{k}'}$ only for $\mathbf{k}' \approx \mathbf{k}+\mathbf{Q}$, the induced beckflow is proportional to $\mathbf{v}_{\mathbf{k}+\mathbf{Q}}$.

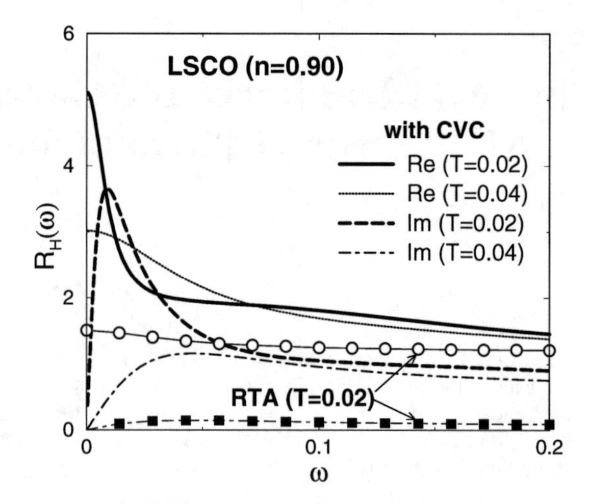

FIGURE 3. AC Hall coefficient obtained by the FLEX approximation. $R_H(\omega) \approx 1/ne$ within the RTA, whereas $R_H(\omega)$ by the conservation approximation shows a prominent ω-dependence, which is consistent with experiments.

Γ in eq. (2), only on the small portion of the FS around $\mathbf{k}' \approx \mathbf{k}+\mathbf{Q}$. Then, the total current becomes $\mathbf{J_k} \propto \mathbf{v_k} + \mathbf{v_{k'}}$, so $\mathbf{J_k}$ is not perpendicular to the FS except the points A and B, which are on the symmetric axes. In summary, the highly anisotropic Landau interaction function $\Gamma(\mathbf{k},\mathbf{k}')$ due to the AF fluctuations causes the nontrivial $\mathbf{k}$-dependence of $\mathbf{J_k}$.

We show that the nontrivial behavior of $\mathbf{J_k}$ shown in Fig. 2 leads to the enhancement of R_H. The expression for the Hall conductivity is given by [2]

$$\sigma_{xy}^{\rm RTA}/B_z = |e|^3 \oint_{\rm FS} dk_\parallel |\vec{v}_{\mathbf{k}} \tau_{\mathbf{k}}|^2 C_{\mathbf{k}}^v, \quad (3)$$

$$\sigma_{xy}/B_z = |e|^3 \oint_{\rm FS} dk_\parallel |\vec{J}_{\mathbf{k}} \tau_{\mathbf{k}}|^2 C_{\mathbf{k}}^J, \quad (4)$$

where $k_\parallel$ is the momentum parallel to the FS, $C_{\mathbf{k}}^v = -\partial\theta_{\mathbf{k}}^v/\partial k_\parallel$, $C_{\mathbf{k}}^J = -\partial\theta_{\mathbf{k}}^J/\partial k_\parallel$, $\theta_{\mathbf{k}}^v = \tan^{-1}(v_{\mathbf{k}x}/v_{\mathbf{k}y})$, $\theta_{\mathbf{k}}^J = \tan^{-1}(J_{\mathbf{k}x}/J_{\mathbf{k}y})$, and $\tau_{\mathbf{k}} = 1/2{\rm Im}\Sigma_{\mathbf{k}}(-i\delta)$ is the lifetime of QP. Because $C_{\mathbf{k}}^v$ represents the FS curvature (FSC) at $\mathbf{k}$, we recognize the well-known result of the RTA; *σ_{xy} is determined by the FSC.* However, this statement does not hold in nearly AF systems any more, because the "effective FSC" in eq.(4), $C_{\mathbf{k}}^J$, strongly deviates from the true FSC. In Fig. 2, $C_{\mathbf{k}}^J > 0$ around A whereas it is negative around B. In contrast, $C_{\mathbf{k}}^v > 0$ everywhere. Moreover, the "cold spot" where $\tau_{\mathbf{k}}$ takes the maximum value on the FS is around A (B) in the hole-doped (electron-doped) systems. As a result, σ_{xy} is positive (negative) in hole-doped (electron-doped) systems, as recognized in eq.(4). In fact, R_H in electron-doped systems is negative although its FSC observed by ARPES is positive. This fact cannot be explained by the idea, "deformation of the FS due to SDW transition", because $R_H < 0$ even above T_N. In Ref. [3], we solved this discrepancy by taking the fact that the effective FSC around the cold-spot becomes negative due to the backflow. We have shown that $C_{\mathbf{k}}^J \propto \xi^2$ in both hole-doped and electron-doped systems in terms of the conserving approximation, so R_H in under-doped system is strongly enhanced at lower temperatures in proportion to T^{-1}. Moreover, characteristic frequency and temperature dependences of the AC Hall coefficient are also reproduced very well by taking the backflow into account, as shown in Fig.3 [5].

In summary, the backflow is a natural consequence of the basic Fermi liquid equation, eq.(1), and it cannot be ignored in strongly correlated systems. The total current in nearly AF metals is no more perpendicular to the FS due to the backflow, because the Landau interaction function caused by the AF fluctuations has a prominent momentum dependence. This fact means that the RTA is totally broken down. The Hall coefficient shows a strong temperature dependence because the backflow changes the "effective FSC" around the cold spot. By taking the backflow into account, we succeeded in reproducing characteristic non-Fermi liquid-like behaviors in various DC and AC transport phenomena in high-T_c cuprates, even in the pseudo-gap region [2, 3, 4].

REFERENCES

1. D. Pines and P. Nozieres, *The Theory of Quantum Liquids*, Benjaminn, New York, 1966.
2. H. Kontani and K. Yamada, J. Phys. Soc. Jpn. **74**, 155 (2005), and references are therein.
3. H. Kontani et al., Phys. Rev. B **59**, 14723 (1999).
4. H. Kontani, Phys. Rev. B **67**, 014408 (2003).
5. H. Kontani, cond-mat/0507664; cond-mat/0511015.

Large Thermoelectric Effects and Inelastic Scattering in Unconventional Superconductors

Mikael Fogelström* and Tomas Löfwander†

*Applied Quantum Physics, MC2, Chalmers University of Technology, S-412 96 Göteborg, Sweden
†Institut für Theoretische Festkörperphysik, Universität Karlsruhe, 76128 Karlsruhe, Germany

Abstract. The thermoelectric coefficient $\eta(T)$ in unconventional superconductors is enhanced below T_c by intermediate strength impurity scattering that is intrinsically particle-hole asymmetric. We compute $\eta(T)$ for a strong-coupling d-wave superconductor and investigate the effects of inelastic scattering originating from electron-boson interactions. We show that $\eta(T)$ is severely suppressed at temperatures just below T_c by a particle-hole symmetric inelastic scattering rate. At lower temperatures inelastic scattering is frozen out and $\eta(T)$ recovers and regains its large amplitude. In the limit $T \to 0$, we have $\eta(T) \sim \eta_0 T + \mathcal{O}[T^3]$, where the slope η_0 contains information about the Drude plasma frequency, the details of impurity scattering, and the change in effective mass by electron-boson interactions. In this limit $\eta(T)$ can be used as a probe, complementary to the universal heat and charge conductivities, in investigations of the nature of nodal quasiparticles.

Keywords: Unconventional superconductivity, transport properties
PACS: 74.24.Fy,74.72.-h

Low-temperature transport measurements have provided a wealth of information about nodal quasiparticles in unconventional superconductors [1, 2]. Thermal conductivity is of particular importance, because theory predicts universality in the sense that the low-temperature asymptotic does not depend on the properties of the impurity potential [3]. This prediction has been confirmed experimentally [1]. However, there are some difficulties in analysing the low-temperature thermal conductivity. First, the leading T^2-dependence of the electronic contribution to $\kappa(T)/T$ is masked by a phonon contribution with the same T^2-dependence [2]. Second, experiments are often done on ultra-clean samples, which means that the T^2 power law of $\kappa(T)/T$ holds only in a small temperature bracket [4].

In a recent paper we discussed how isotropic elastic scattering by impurities of intermediate strength, *i.e* described by a phase shift $0 < \delta_0 < \frac{\pi}{2}$, gives rise to an electron-hole asymmetric scattering time and consequently a large non-universal thermoelectric response [5]. A careful study of the thermoelectric coefficient in the low-temperature regime would reveal information about the bare elastic scattering rate and potential strength. This information is hard to extract from thermal conductivity data. In this report we examine the interplay between elastic and inelastic scattering and show how it affects the thermoelectric coefficient.

The thermoelectric coefficient, $\eta(T)$, is defined as

$$\delta \vec{j}_e = -\eta(T)\nabla T = 2\mathcal{N}_f \int d\vec{p}_f \int \frac{d\varepsilon}{4\pi i} e\vec{v}_f \delta g^K. \quad (1)$$

The quasiclassical propagator $\delta \hat{g}^K$ has a closed form in which the self-consistently computed equilibrium Green function $\hat{g}_0^R$ and the self-energy $\hat{\Sigma}^R$ serve as input (see Graf *et al.* [3] and Ref. [5] for details). In the present study we consider a composite self-energy in particle-hole space $\hat{\Sigma}^R = \hat{\Sigma}^R_{\text{imp}} + \hat{\Sigma}^R_{\text{in}}$, where the impurity self-energy $\hat{\Sigma}^R_{\text{imp}}$ is diagonal with $\Sigma^R_{3,\text{imp}}$ and $\Sigma^R_{0,\text{imp}}$ being its particle-hole symmetric and anti-symmetric parts, respectively. The self-energy $\hat{\Sigma}^R_{\text{in}}$ includes the effects of inelastic electron-boson scattering, but we also assume that this interaction mediates the pairing. Below, $\hat{W}^R$ is the off-diagonal component, *i.e* the usual strong-coupling function related to the energy dependent gap as $\hat{\Delta}^R(\varepsilon) = \hat{W}^R(\varepsilon)/Z(\varepsilon)$. Contrary to the impurity self-energy the diagonal part of the inelastic self-energy is particle-hole symmetric as $\Sigma^R_{0,\text{in}} = 0$. Finally, $Z(\varepsilon)$ is the energy-renormalization function defined by the scattering renormalized energy $\tilde{\varepsilon}^R = Z^R(\varepsilon)\,\varepsilon = \varepsilon - \Sigma^R_{3,\text{imp}} - \Sigma^R_{3,\text{in}}$. This model is the same as we used in our study of the thermal conductivity [4]. With this input we use Ref. [5] and write down the response function

$$\eta_{ij}(T) = -\frac{e}{4T^2}\int d\varepsilon\, \varepsilon\, \text{sech}^2\frac{\varepsilon}{2T}\int d\vec{p}_f[v_{f,i}v_{f,j}] \times \frac{\mathcal{N}(\vec{p}_f,\varepsilon)\,\Im\,\Sigma^R_{0,imp}(\varepsilon)}{\left[\Re\,\Omega^R(\vec{p}_f;\varepsilon)\right]^2 - \left[\Im\,\Sigma^R_{0,imp}(\varepsilon)\right]^2}, \quad (2)$$

where $\mathcal{N}(\vec{p}_f,\varepsilon) = -\mathcal{N}_f \Im\left[\tilde{\varepsilon}^R/\Omega^R(\vec{p}_f;\varepsilon)\right]$ is the density of states, and $\Omega^R = \sqrt{|W^R(\vec{p}_f,\varepsilon)|^2 - (\tilde{\varepsilon}^R)^2}$. Note that $\eta(T)$ is directly proportional to the imaginary part of the particle-hole asymmetric part $\Im\,\Sigma^R_{0,imp}$, which is an odd function of energy.

CP850, *Low Temperature Physics: 24th International Conference on Low Temperature Physics;*
edited by Y. Takano, S. P. Hershfield, S. O. Hill, P. J. Hirschfeld, and A. M. Goldman

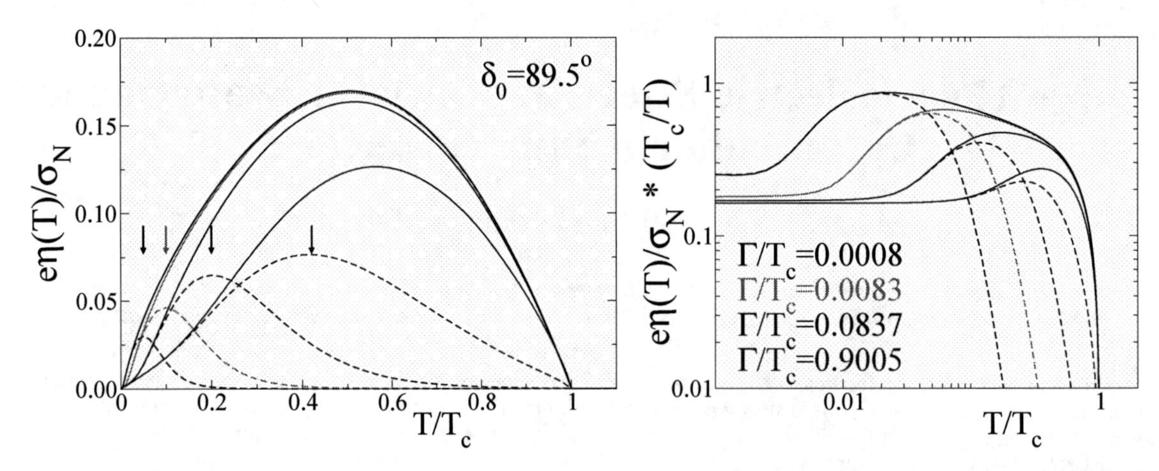

FIGURE 1. Thermoelectric coefficient, $\eta(T)$, calculated for four orders of magnitude in scattering rate, as indicated in the right panel. The dashed lines are results of a self-consistent calculation with the inelastic scattering model introduced in Ref. [4]. The full lines are the corresponding results obtained with an *effective* model where the inelastic part is included only as an effective mass via $m^*/m = 1+\lambda_{\text{in}}$ [4]. The temperature dependence obtained with the effective model is the same as within weak-coupling theory. In the left panel $\eta(T)$ is scaled by the normal state Drude conductance, $\sigma_N = e^2 v_f^2 \mathcal{N}_f / 2\Gamma$, where $v_f, \mathcal{N}_f$, and Γ are effective values of respective bare constants. The arrows indicate the cross-over, T^*, between the two scattering modes. In the right panel we show the temperature dependence of $\eta(T)/T$ with emphasis on the low-temperature regime where the leading $T-$dependence is $\eta(T)/T \approx \eta_0$. The constant, η_0, is non-universal since it explicitly depends on the ratio of $\partial \Im\Sigma^R_{0,imp}(\varepsilon)/\partial\varepsilon|_{\varepsilon=0}$ and $\Im\Sigma^R_{3,imp}(\varepsilon=0)$.

In Figure 1 we show $\eta(T)$ for a strong-coupling d-wave superconductor. The results are compared with an effective model-calculation where only the mass renormalization by inelastic scattering is accounted for, as described in Ref. [4]. As seen in Figure 1, inelastic scattering affects $\eta(T)$ much in the same way as it does the thermal conductivity reported in Ref. [4], *i.e* it dominates at temperatures $T \leq T_c$ but freezes out at low $T \ll T_c$ where instead elastic scattering limits the transport. We separate the two regimes where the two types of scattering dominate by introducing a cross-over temperature T^*, defined as the temperature where $\eta(T)$ has its maximium. The value of T^* depends on the elastic scattering rate Γ for a given electron-boson coupling spectra. The suppression of $\eta(T)$ in the interval $T^* \lesssim T \leq T_c$ comes from that in our model inelastic scattering does not break particle-symmetry and hence reduces the weight of the kernel in the integral in Eq. (2).

In the low-T regime $T \ll |\Im\Sigma^R_{3,imp}(0)|$, we obtain

$$\frac{\eta(T)}{T} = e\frac{\pi^2}{3}\frac{2\mathcal{N}_f v_f^2}{\pi\mu\Delta_0}\left|\frac{\partial_\varepsilon \Im\Sigma^R_{0,imp}(\varepsilon)}{\Im\Sigma^R_{3,imp}(\varepsilon)}\right|_{\varepsilon=0} + \mathcal{O}\left[T^2\right], \quad (3)$$

where Δ_0 is the spectroscopic gap, and $\mu = (1/\Delta_0)|d\Delta(\phi)/d\phi|_{\phi_{\text{node}}}$ is the opening rate of the gap function at the node. In analogy with the thermal conductivity [4], the remaining effect at $T \ll T^*$ of inelastic scattering is a modification of the $T \to 0$ asymptotic. When we write explicitly $\mathcal{N}_f \to \mathcal{N}_f^* = \mathcal{N}_f^0(1+\lambda_{\text{in}})$ and $v_f \to v_f^* = v_f^0/(1+\lambda_{\text{in}})$ in Eq. (3), one factor $1+\lambda_{\text{in}}$ remains in the denominator. Within the bare theory, this result can be traced back to that the spectroscopic gap Δ_0 in the weak-coupling limit is replaced by the strong coupling off-diagonal function $W(0)$.

In summary, we have discussed the temperature-dependence of the thermoelectric effect that results from the interplay of inelastic electron-boson scattering and elastic impurity scattering. At high temperatures, $T > T^*$, particle-hole symmetric inelastic scattering dominates and the thermoelectric coefficient is suppressed. On the other hand, at low temperatures, $T < T^*$, particle-hole asymmetric elastic impurity scattering dominates and the thermoelectric coefficient is enhanced.

We gratefully acknowledge financial support from the Swedish Research Council (M.F), and the EC under the Spintronics Network RTN2-2001-00440 (T.L).

REFERENCES

1. L. Taillefer *et al*, *Phys. Rev. Lett.* **79**, 483 (1997); S. Nakamae *et al.* , *Phys. Rev. B* **63**, 184509 (2001); J. Takeya, Y. Ando, S. Komiya, and X. F. Sun, *Phys. Rev. Lett.* **88**, 077001 (2002).
2. R. W. Hill, *et al*, *Phys. Rev. Lett.* **92**, 027001 (2004).
3. M. J. Graf, S.-K. Yip, J. A. Sauls, and D. Rainer, *Phys. Rev. B* **53**, 15147 (1996); A. Durst and P. A. Lee, *Phys. Rev. B* **62**, 1270 (2000).
4. T. Löfwander and M. Fogelström, *cond-mat/0503391*, accepted for publication in *Phys. Rev. Lett.*
5. T. Löfwander and M. Fogelström, *Phys. Rev. B* **70**, 024515 (2004).

Low-Temperature Thermal Conductivity of Superconductors With Gap Nodes

Tomas Löfwander* and Mikael Fogelström†

*Institut für Theoretische Festkörperphysik, Universität Karlsruhe, 76128 Karlsruhe, Germany
†Applied Quantum Physics, MC2, Chalmers University of Technology, S-412 96 Göteborg, Sweden

Abstract. We report results for the electronic thermal conductivity in d-wave superconductors at low temperatures, including effects of anisotropic impurity scattering.

Keywords: Unconventional superconductivity, transport properties, anisotropic impurity scattering
PACS: 74.24.Fy,74.72.-h

INTRODUCTION

At low temperatures nodal quasiparticles give the dominating contribution to the electronic response functions in the d-wave superconducting cuprates. We report results for the low-temperature thermal conductivity, with a focus on effects of anisotropic impurity scattering.

Impurity scattering is pair breaking and the density of nodal quasiparticles is enhanced with increasing impurity concentration. At low temperatures, the thermal conductance is due to these quasiparticles and has a linear temperature dependence. However, the slope of the conductance is independent of the properties of impurity scattering [1, 2, 3]. This universality has been exploited in a large body of experiments on the cuprates [4, 5, 6].

The next term in an expansion in temperature is proportional to T^3. This term is not universal and the details of impurity scattering is important.

Inelastic scattering is suppressed at low temperatures and the residual effect of coupling of the nodal quasiparticles to boson modes is only through the effective mass. This holds over a quite wide temperture range below a crossover temperature T^*, below which impurity scattering limits the quasiparticle lifetime (see [7] and original references therein).

RESULTS

We use the same model as in our recent report [7]. The thermal conductance has the form $\kappa(T) = \kappa_0 T + \kappa_1 T^3 + \mathscr{O}[T^5]$. The $T \to 0$ slope is $\kappa_0 = (\pi^2/3)(2\mathscr{N}_f v_f^2)/(\pi\mu\Delta_0)$, where v_f is the Fermi velocity, $\mathscr{N}_f$ is the density of states at the Fermi level, μ is the slope of the d-wave gap function at the node, and Δ_0 is the maximum spectroscopic gap. In the simplest model we have $\mathscr{N}_f = p_f/(2\pi v_f)$. The slope κ_0 has been measured experimentally for a variety of materials, recently with a focus on the doping dependence [6, 8, 9, 10, 11, 12]. The general trend is that κ_0 decreases monotonically with doping, from the optimally doped to the underdoped region. Given the nominally doping independence of the Fermi momentum p_f [13] and the Fermi velocity v_f [14] seen experimentally, the only remaining variable is the gap function through μ and Δ_0. Interestingly, the doping dependence then leads to that κ_0 follows the pseudogap if μ is assumed constant [8].

Much less is known about the T^3-term and we will now study the influence of anisotropic impurity scattering on κ_1. For isotropic scattering, κ_1 was first studied in [3] and later also in [15, 7]. We treat the impurities in the t-matrix approximation with the impurity potential written as

$$u(\vec{p}_f, \vec{p}_f') = \sum_{l=-\infty}^{\infty} u_l e^{il(\phi-\phi\prime)}, \tag{1}$$

where the angle $\phi \in [0, 2\pi]$ parameterizes the Fermi surface. We keep for simplicity the first three partial waves (i.e. we keep s-, $p-$, and $d-$wave scattering for $l =$ 0, 1, 2). We assume that $u_{-l} = u_l$ are real and introduce phase shifts $\delta_l = \arctan(\pi\mathscr{N}_f u_l)$.

There are two effects of anisotropic impurity scattering. First, the spectrum changes when the impurity bands are shifted in energy with respect to the corresponding situation for isotropic scattering. We illustrate this effect in Fig. 1. New bands can also form at higher energies, but that is of no importance for the low-T thermal conductivity. Second, there are vertex corrections to the conductivity. Only the odd partial waves (here the p-wave) couple to the driving force (the temperature gradient) and we obtain a vertex correction when the p-wave phase shift is non-zero.

In Fig. 2 we show the phase shift dependence of κ_1. The p-wave phase shift does not dramatically influence

CP850, *Low Temperature Physics: 24th International Conference on Low Temperature Physics;*
edited by Y. Takano, S. P. Hershfield, S. O. Hill, P. J. Hirschfeld, and A. M. Goldman

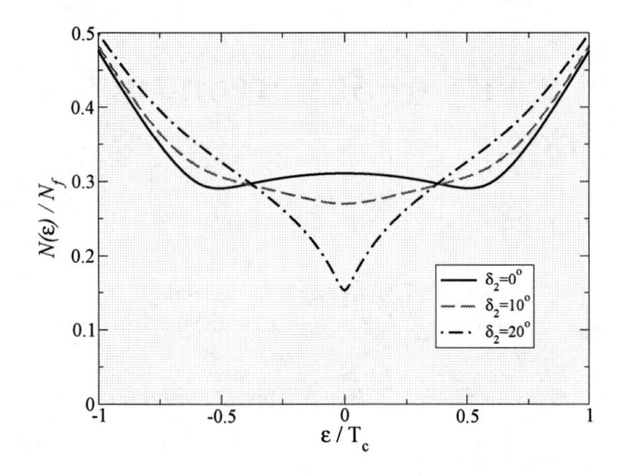

FIGURE 1. D-wave phase shift dependence of the low energy density of states for $\Gamma_0 = 0.1T_c$, $\delta_0 = 88^o$, and $\delta_1 = 0$.

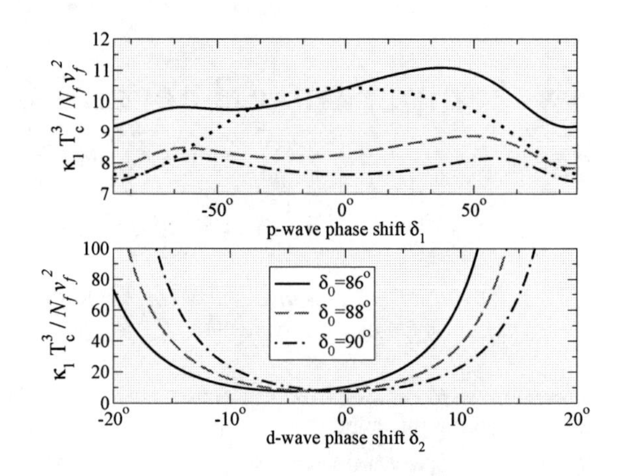

FIGURE 2. Phase shift dependence of the T^3 term of $\kappa(T)$. In (a) the p-wave phase shift varies while the d-wave is $\delta_2 = 0$. In (b) the d-wave phase shift varies while the p-wave is $\delta_1 = 0$. The dotted line in (a) is the result when vertex corrections are neglected for $\delta_0 = 86^o$. There are no vertex corrections in (b). In all cases $\Gamma_0 = 0.1T_c$.

the low-T thermal conductance, see Fig. 2(a). The effect on κ_1 is here about 10%, which is much larger than the $\sim 1\%$ influence on κ_0 (not shown in the figure). We also see that the vertex correction is important to get the correct numbers out. The most dramatic effect is obtained when there is scattering in the d-wave channel, see Fig. 2(b). This is solely due to the spectral changes: the impurity band that for isotropic scattering was at the Fermi level is shifted by the d-wave phase shift to a finite energy, which has a large impact on the conductivity. In fact, for large d-wave phase shifts the impurity band is so far from the Fermi level that the conductivity is not well described by an expansion in powers of T/γ, where γ is the quasiparticle lifetime at the node (impurity scattering contribution to the imaginary part of the energy). This is analogous to the situation for isotropic impurity scattering when the s-wave phase shift is not close to unitary limit [7]. When the d-wave phase shift approaches the unitary limit (90°), the impurity band approaches the Fermi level again (not shown in the figure).

SUMMARY

In summary, we report results for the influence of anisotropic impurity scattering on the low-temperature thermal conductivity in d-wave superconductors. We have shown that the largest impact on the conductivity is through the spectral changes that occur when there is a non-zero d-wave phase shift. A p-wave phase shift leads to vertex corrections, which have an impact on the T^3-term although it leaves the T-term unaffected.

ACKNOWLEDGMENTS

We acknowledge financial support from the Wenner-Gren foundations (T.L.), the EC under the Spintronics Network RTN2-2001-00440 (T.L.), and the Swedish Research Council (M.F.).

REFERENCES

1. P. A. Lee, *Phys. Rev. Lett.*, **71**, 1887 (1993).
2. Y. Sun, and K. Maki, *Europhys. Lett.*, **32**, 355 (1995).
3. M. J. Graf, S.-K. Yip, J. A. Sauls, and D. Rainer, *Phys. Rev. B*, **53**, 15147 (1996).
4. L. Taillefer *et al.*, *Phys. Rev. Lett.*, **79**, 483 (1997).
5. S. Nakamae *et al.* , *Phys. Rev. B*, **63**, 184509 (2001).
6. J. Takeya, Y. Ando, S. Komiya, and X. F. Sun, *Phys. Rev. Lett.*, **88**, 077001 (2002).
7. T. Löfwander, and M. Fogelström, condmat/0503391.
8. M. Sutherland *et al.*, *Phys. Rev. B*, **67**, 174520 (2003).
9. X. F. Sun, S. Komiya, J. Takeya, and Y. Ando, *Phys. Rev. Lett.*, **90**, 117004 (2003).
10. D. G. Hawthorn *et al.*, *Phys. Rev. Lett.*, **90**, 197004 (2003).
11. M. Sutherland *et al.*, *Phys. Rev. Lett.*, **94**, 147004 (2005).
12. X. F. Sun, K. Segawa, and Y. Ando, cond-mat/0502223.
13. A. Damascelli, Z. Hussain, and Z.-X. Shen, *Rev. Mod. Phys.*, **75**, 473 (2003).
14. X. J. Zhou *et al.*, *Nature*, **423**, 398 (2003).
15. T. Löfwander, and M. Fogelström, *Phys. Rev. B*, **70**, 024515 (2004).

Gap Distributions in Cuprate Superconductors

Ashot Melikyan, Brian M. Andersen, Tamara S. Nunner, and P. J. Hirschfeld

Department of Physics, University of Florida, Gainesville, FL 32611, USA

Abstract. Recent Scanning Tunneling Microscopy (STM) data on $Bi_2Sr_2CaCu_2O_{8+x}$ (BSCCO) [1] impose stringent restrictions on the spatial distribution of the superconducting order parameter and scalar potential due to impurities. In Ref. [2] it was shown that these distributions are inconsistent with a conventional mean-field approach where the inhomogeneities in the LDOS are driven by a scalar impurity potential. It was further demonstrated that the salient experimental features of the LDOS spatial modulations can be obtained if the pairing coefficient itself is assumed to be enhanced by the dopant atoms. Here, we report additional features of the LDOS that support the conclusions of Ref. [2].

Keywords: Superconductivity, BSCCO, BCS, impurities, STM
PACS: 74.72.-h, 74.25.Jb, 74.20.Fg

It is commonly believed that the cuprate superconductors near optimal doping are well described within the BCS framework. Recent STM experiments by McElroy *et al.* [1], who succeeded in mapping out individual dopant oxygen atoms in BSCCO, opened a possibility to test the applicability of the traditional d-wave BCS paradigm to the cuprates by comparing the measured tunneling conductance in the presence of impurities to the predictions of the BCS theory. This possibility is especially valuable in the study of the cuprate superconductors, where the microscopic theory is presently unknown, and the analysis of controlled perturbations of the superconducting state by impurities can impose significant restrictions on the class of possible underlying microscopic theories. In studying the relation of nano-scale inhomogeneities in BSCCO to the position of dopant atoms, it was found[1] that (i) in the vicinity of dopant oxygen atoms the spectral gap is enhanced and the height of the "coherence peaks" is suppressed, (ii) the peaks at negative and positive bias are symmetrically positioned around zero bias, and (iii) the charge modulations on the surface are of the order of a few percent or less, while the gap typically varies in the range of $35-80$ meV.

Within the mean-field approximation, the BCS state is described by the Hamiltonian $\mathscr{H}$:

$$\sum_{\mathbf{k}s}\xi_{\mathbf{k}}\hat{c}^{\dagger}_{\mathbf{k}s}\hat{c}_{\mathbf{k}s}+\sum_{\mathbf{r}s}V_{\mathbf{r}}\hat{c}^{\dagger}_{\mathbf{r}s}\hat{c}_{\mathbf{r}s}+\sum_{\langle\mathbf{r}\mathbf{r}'\rangle s}\left(\Delta_{\mathbf{r},\mathbf{r}'}\hat{c}^{\dagger}_{\mathbf{r}s}\hat{c}^{\dagger}_{\mathbf{r}'s}+\text{h.c.}\right), \quad (1)$$

where $\mathbf{r}$ denotes the lattice sites, $\langle\ldots\rangle$ stands for the nearest-neighbors, $V(\mathbf{r})$ is the impurity potential, and $\xi_{\mathbf{k}}=-2t(\cos k_xa+\cos k_ya)-4t'\cos k_xa\cos k_ya-\mu$, with $t'/t=-0.3$, $\mu/t=-1.0$. Diagonalization of $\mathscr{H}$ can be cast as an eigenvalue problem

$$\left[(\hat{\xi}+V(\mathbf{r}))\tau_3+\hat{\Delta}\tau_1\right]\psi(\mathbf{r})=E\psi(\mathbf{r}) \quad (2)$$

where τ_i are Pauli matrices, $\psi(\mathbf{r})=(u(\mathbf{r}),v(\mathbf{r}))^T$ is the Nambu spinor and $V(\mathbf{r})$ is the impurity potential. The action of the kinetic and the off-diagonal pairing operators on an arbitrary function $f(\mathbf{r})$ defined on lattice sites $\mathbf{r}$ is $\hat{\xi}f(\mathbf{r})=-t\sum_\delta f(\mathbf{r}+\delta)-t'\sum_{\delta'}f(\mathbf{r}+\delta')-\mu f(\mathbf{r})$ and $\hat{\Delta}f(\mathbf{r})=\sum_\delta\Delta_{\mathbf{r},\mathbf{r}+\delta}f(\mathbf{r}+\delta)$, where δ (δ') denotes the (next-) nearest neighbor sites. The bond variables $\Delta_{\mathbf{r},\mathbf{r}'}$ are determined from the Bogoliubov-de Gennes (BdG) self-consistency conditions. In the simplest model of attractive nearest neighbor density-density interactions, these conditions assume the following form: $\Delta_{\mathbf{r},\mathbf{r}'}=g_{\mathbf{r},\mathbf{r}'}\sum_{E_n>0}(v_n^*(\mathbf{r})u_n(\mathbf{r}')+v_n^*(\mathbf{r}')u_n(\mathbf{r}))$, where $(u_n(\mathbf{r}),v_n(\mathbf{r}))^T$ are the eigenfunctions from (2) with eigenenergy E_n. Traditionally, it is assumed that the coupling constant g is uniform in space, and that the impurities affect the modulations of quantities such as LDOS by acting primarily as sources of a scalar diagonal impurity potential $V(\mathbf{r})$. Note that by self-consistency, the bond variables $\Delta_{\mathbf{r},\mathbf{r}'}$ will also be modulated. The results of Ref. [2] suggest that this τ_3-channel scattering is incapable of describing the tunneling conductance experiments in BSCCO and reproducing properties (i)-(iii). It was found [2] that these properties arise naturally, if one assumes a dopant-enhanced pairing interaction $g_{\mathbf{r},\mathbf{r}'}$.

In order to model the effect of the oxygen dopants, we randomly generate $N_{imp}=pL^2$ impurities, where p is doping and L is the lattice size. The impurities located a distance r_z away from the CuO_2 plane, are assumed to modulate $V(\mathbf{r})=V_0\sum_{i=1}^{N_{imp}}e^{-|\mathbf{r}-\mathbf{r}_i|\lambda}$ (τ_3 channel) or $g_{\mathbf{r},\mathbf{r}'}=g_0+\delta g\sum_{i=1}^{N_{imp}}\left(e^{-|\mathbf{r}-\mathbf{r}_i|/\lambda}+e^{-|\mathbf{r}'-\mathbf{r}_i|/\lambda}\right)$ (τ_1 channel).

Our numerical computations[2] are summarized in Fig. 1. As shown in the upper panel, the standard scenario with $V(\mathbf{r})$ driving the modulations is in contradiction with experiment: (i) the height of the largest peak at negative bias increases with the gap, which is defined in

CP850, *Low Temperature Physics: 24th International Conference on Low Temperature Physics;*
edited by Y. Takano, S. P. Hershfield, S. O. Hill, P. J. Hirschfeld, and A. M. Goldman

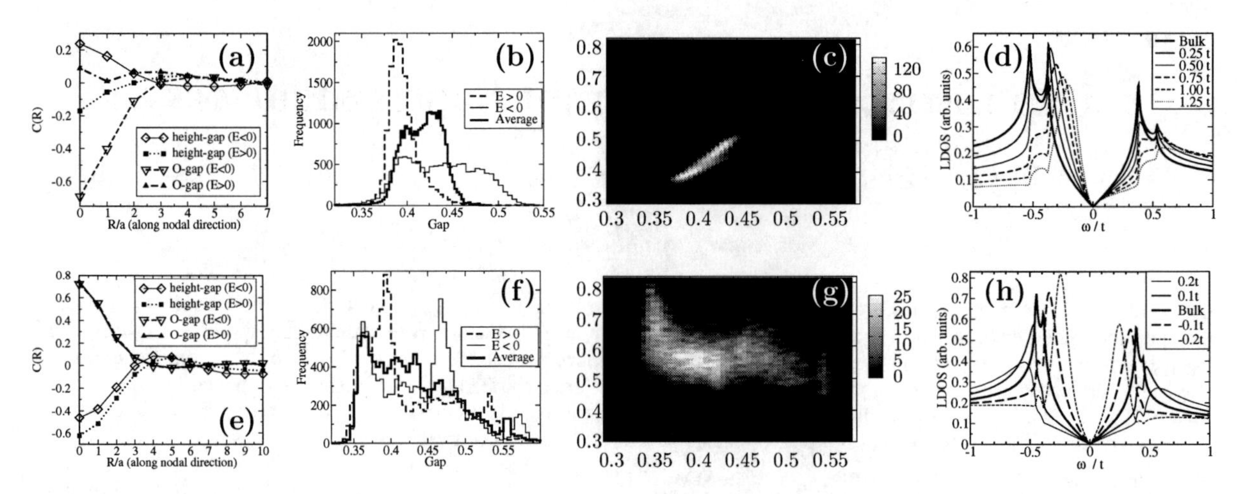

FIGURE 1. Upper and lower panels refer to τ_3- and τ_1- impurities respectively. The parameters used are $r_z = 0.8a$, $\lambda = 0.7a$, $V_0 = t$, $g_0 = 1.16$, and $\delta g = 0.5t$. *(a, e)*: The correlation function between the gap values and dopant atom locations (squares), and between the gap and the oxygen impurity locations. *(b, f)*: Histogram with the distribution of the gaps. *(c, g)*: Average gap – negative bias peak height distribution, where the gap (horiz.) is in units of *t* and the peak height (vert.) is in arb. units. *(d, h)*: the LDOS for a single point-like impurity problem; for τ_3 impurity, the perturbation is $H'(\mathbf{r}) = V\delta(\mathbf{r})$, while τ_1 potential is defined as a perturbation of the uniform order parameter $\Delta_{0,\delta} = \eta_\delta(\Delta_0 + \delta\Delta)$ on four bonds around $\mathbf{r} = 0$, where η_δ is $+1$ (-1) for $\delta = \pm\hat{a}x(\pm\hat{a}y)$.

experiments as half a distance between the largest peaks at negative and positive bias (see Fig. 1(c)); (ii) the position of the peaks in the LDOS at positive and negative bias are highly asymmetric: the peak height– gap correlations have opposite signs at $E > 0$ and $E < 0$ (see Fig. 1(a)); (iii) in order to reproduce the experimentally observed modulations of the gap by a factor of two, the impurity potential must be chosen so large that the charge density develops pronounced modulations. Even for V_0 used in the top row of Fig. 1, where the gap is modulated by only 20%, the charge density varies by a factor of 2.5, in contradiction with the STM experiments[1].

The results obtained by assuming the modulated pairing strength $g_{\mathbf{r},\mathbf{r}'}$ [2] are shown in the lower panel of Fig. 1. Note that (i) the peak height is anti-correlated with the gap for both positive and negative bias (Fig. 1(e)); (ii) the correlations shown in Fig. 1(e) are symmetric with respect to sign of bias: the location of impurity atoms is strongly correlated with the regions of large gap; (iii) in addition, the charge modulations in this case are extremely small, reaching at most 2%. Thus, the assumption of modulated paring $g_{\mathbf{r},\mathbf{r}'}$ resolves[2] the difficulties associated with the conventional potential scattering theory and satisfies the experimental criteria (i)-(iii).

Regarding the distribution of the LDOS gaps (Figs. 1(b) and 1(f)), we note that the distribution of the average gap (solid bold lines), has a positive skew for τ_1-impurities and follows closely the distribution of the potential $g_{\mathbf{r},\mathbf{r}'}$ (not shown), while for the τ_3-model the distribution of the gaps has a negative skew.

The distribution of only positive (negative) bias gap is shown in Figs. 1(b) and 1(f) as a thin (dashed) line. The bimodal character of the τ_1 distributions can be qualitatively understood from Fig 1(h). Consider e.g. $E > 0$: in a uniform system, the thermally broadened LDOS has two maxima as a function of energy: the "coherence" and the "van-Hove" peaks. In the presence of a single point-like impurity $\delta\Delta$, both the position and the height of the peaks evolve continuously as $\delta\Delta$ is increased adiabatically from zero. At a certain critical value of $\delta\Delta_c$ the height of the coherence peak becomes smaller than that of the van-Hove peak. The gap defined as the position of the highest maximum of the LDOS curve, therefore is a discontinuous function of $\delta\Delta$ at $\delta\Delta_c$. As a result, the distribution of the gaps acquires a "dip" between the gap value at $(\delta\Delta_c - 0)$ and at $(\delta\Delta_c + 0)$. For a τ_3-impurity the situation is similar, except the critical value V_c of the impurity potential is rather large at $E > 0$. Even for the relatively strong impurity potential used in Figs. 1(a-c), $V(\mathbf{r})$ is smaller than V_c, and the bimodality at the positive bias is absent (see Fig. 1(b)).

Although here we have only considered the extreme cases of pure τ_1 and τ_3 scatterers, real dopants certainly give rise to perturbations in both channels. We have argued that the gap histograms and the peak height-gap scatter plots show that the dominant scattering mechanism in BSCCO is of τ_1 origin.

REFERENCES

1. K. McElroy, H. Eisaki, S. Uchida, and J. C. Davis, *Bull. Am. Phys. Soc. March*, 2005 and unpublished.
2. T. S. Nunner, B. M. Andersen, A. Melikyan, and P. J. Hirschfeld, *cond-mat/0504693*.

Electronic Crystal State Reflects Bond Orderings and an Empirical Algorithm for Doping Curve Analysis

H. Oesterreicher

University of California, San Diego, Dept. of Chemistry and Biochemistry, La Jolla, CA, 92093-0317, USA

Abstract. STM and ARPES display a partly disordered electronic crystal state that is here shown to corroborate a bond order model. This model predicts characteristic plaid structures for bond centered pairs at $3a_0/2$. We show that indications for some of these predicted patterns are now seen in experiments. The elastic energy as given by the period of the electronic crystal is connected with a quantitative algorithm for T_c vs. holes. Accordingly $T_c=2f_a600/Pab$. In the bond order model parameters such as optimal holes, T_c, or pseudo-gaps are enhanced by high isolation of the planes (f_a). Accordingly high T_c superconductivity and the electronic crystal state are predictable purely on structural data.

Keywords: Electronic crystals, superconducting T_c predictions.
PACS: 74.62.-c; 74.72.-h

INTRODUCTION

It has been suggested [1] that 'all' high T_C superconductivity corresponds to charge transport in a universal bond system of exchange coupled 'local' pairs. For cuprates the 'local' pairs have been assumed to reflect bond centered doped charge at $3a_0/2$ distance, referred to as trijugate [3J or homo-conjugate] position. The pairs are arranged in bond orders [BO] in plaid formation. This 3J system has a combination of property of a mobile conjugate and an isolated system. Here we outline that charge modulation in STM and ARPES patterns [2,3] show aspects of partly disordered bond systems as predicted.

The period of the electronic crystal is here related to hole-concentration and to an empirical algorithm for doping curves (T_c vs. holes). This period Pab, where a and b are multiples of lattice parameter [e.g.P4x4 for $4a_0x4b_0$], gives the number of pairs n_p=1/Pab, representing the elastic energy [for the linear 'source' region, where all holes are transformed into pairs $2n_p$=h]. In the effective pair picture $T_c=2f_a600n_p=2f_a600/Pab$. Here, f_a represents the diminished correlation of the plane configurations, when the apical system interacts with the plane. It depends on the bond valence [1] in the c-direction. However, to first approximation f_a =1, 2/3, ½ for apical layer coordination of 0,1 and 2 [CuO_2, CuO_3, CuO_4] respectively. By depicting Pab, STM therefore ties in with quantitative phenomenological rules [1] that support a charge order view for high T_c superconductivity. These rules concern characteristic feature of doping curves with charge-lattice lock-ins, T_c dependence on period of charge modulations etc. They organize cuprates into 'musical' compound families. Empirical rules allow for compound specific theoretical T_c and doping curve predictions purely on structural parameters.

RESULTS

Superconductivity can occur as a result of a

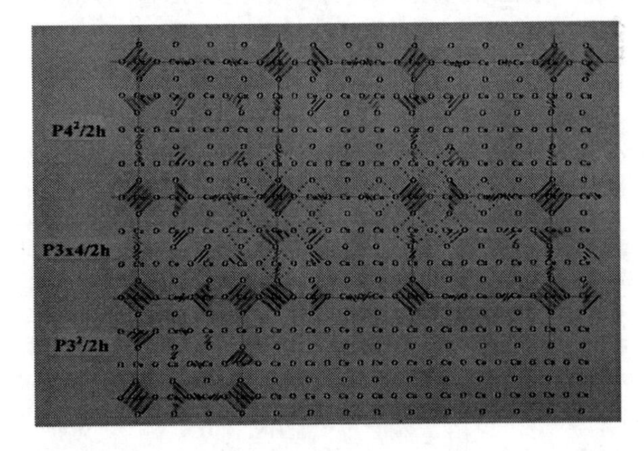

FIGURE 1. Schematic of various proposed superconducting Pab plaids. Alternating pair direction creates primary and secondary channels of charge. Half-tone shading indicates primary pair and bold shading the moment direction. The system, including the immediate neighbors to the pair is referred to as the pair-function [two selected pair functions are given as pointed].

CP850, *Low Temperature Physics: 24th International Conference on Low Temperature Physics;*
edited by Y. Takano, S. P. Hershfield, S. O. Hill, P. J. Hirschfeld, and A. M. Goldman

TABLE 1. Charge order periodicity and resulting characteristic hole numbers of selected bond patterns are the basis of a 'musical' T_c level scheme. T_{cmax} would be obtainable if f_a=1. h_{opK} are Knight shift data [bracketed are calculated]. OP and IP stand for outer and inner plane respectively. ^{o} stands for observed, ^{c} for calculated on $T_c = f_a T_{cmax}$. For $HgBa_2CaCu_3O_{8+x}$ OP, f_a =0.88>2/3 with T_c=132^{c}K is due to its unusually large distance to the apical O. Pab with a=b are preferred by tetragonal symmetry depicted Tet

h_{op}	Pab	T_{cmax}[K]=600h	Examples	h_{opK}	f_a	T_c[K]
0.250	$P4^2/4h$	150	$Bi_2Sr_2CaCu_2O_{8.25}$ (Tet)	0.25	2/3	100^c, 96°
			$HgBa_2Ca_2Cu_3O_{8+x}$ [OP] Tet	0.25	0.88	132^c, 133°
0.222	P3x6/4h	133	$YBa_2Cu_3O_{6.95}$	0.22	2/3	89^c, 94°
0.222	$P3^2/2h$	133	$HgBa_2Ca_2Cu_3O_{8+x}$ [IP] Tet	0.21	1	133^c, 133°
0.200	P4x5/4h	120	$YBa_2Cu_4O_8$ Tet	0.19	2/3	80^c, 80°
0.167	P3x4/2h	100	$YBa_2Cu_3O_{6.7}$	0.16	2/3	66^c, 63°
0.160	$P5^2/4h$	96	$La_{1.84}Sr_{0.16}CuO_4$	[0.16]	½	48^c, 38°
0.125	$P4^2/2h$	75	$RuSr_2GdCu_2O_8$ Tet	[0.12]	2/3	50^c, 48°

transformation [1] of 'Tranquada stripes' of singles into a BO of paired holes by a move of alternate holes into a position corresponding to a 3J bond around a super-exchange O [^{s}O], termed $P4^2/2h$ [FIG 1].

In order to compare with experiment we simplify these assumptions. The local pair will reside in 2 adjacent CuO_4 units within a Cu_2O_7 motif. These units are oriented in alternating directions by 90° to relax lattice pressure. This forms primary and secondary channels of charge, with the former collecting the parallel parts of the pairs together with half of the perpendicular pair. Flipping the orientation of the pairs transforms a primary into a secondary channel providing a mechanism of charge transport. $P4^2/2h$ can be seen as the basis of various related varieties at different hole-concentration. Examples for these patterns are given in Table 1 and FIG 1. The characteristic h_{op}=0.25 is straightforwardly explained by a doubling of holes corresponding to $P4^2/4h$.

Data [2,3] for $Ca_{2-x}Na_xCuCl_2O_2$ in the pseudo-gap region and near h=0.10 show features that remind of CuO_4, or Cu_2O_7 and their satellites as basic units in both renditions. They are arranged in plaid formations of various fragments of characteristic Pab as predicted.

A high degree of isolation of the CuO_2 planes is the major factor in increasing the magnitude of T_c, the optimal number of doped holes [h_{op}] and the magnitude of pseudo-gaps or the Neel temperature of the boundary phase. High plane isolation [Hg analogs] is related with record values of T_c and low isolation with absence of T_c [$YBa_2Cu_3O_{6.7}$ or $PrBa_2Cu_3O_7$ in special preparations]. h_{op} is also dictated by structure through f_a.

Knight shift has shown [1] optimal hole numbers to center on characteristic values such as h_{op}=0.16, 0.20, 0.22 and 0.25 per CuO_2. These values have been identified as related to characteristic patterns of local pairs according to $P5^2/4h$, P4x5/2h, $P3^2/2h$ and $P4^2/4h$ respectively. A fruitful language can be developed which deals with a buildup of complex quantum chemical features by bringing two holes into vicinity of a super-exchange O.

This approach is now also supported by STM data, where motifs of the BO model appear pictorially. It is plausible that the step from super-gaped into superconductivity will involve the full correlation of the undifferentiated type of charge blots into 3J bonds, possibly between O.

DISCUSSION

The occurrence and strength of high T_C superconductivity can now be related to detail of structure. The relative isolation of the CuO_2 plane dictates local super-exchange and other important properties. It therefore also extends to the filling degree and type of Pab providing a comprehensive phenomenological and microscopic understanding.

REFERENCES

1. H. Oesterreicher, *J. Superconductivity* **17**, 439 (2004).
2. T. Hanagurii, *Nature* **430**, 1001 (2004).
3. K. M. Shen, et al, *Science* **307**, 901-904 (2005).

Quantum Monte Carlo Optimization in Many Fermion Systems

Takashi Yanagisawa[1,2]

[1] *Condensed-Matter Physics Group, Nanoelectronics Research Institute AIST Central 2, 1-1-1 Umezono, Tsukuba 305-8568, Japan*
[2] *Graduate School of Science, Osaka University, Toyonaka, Osaka 560-8531, Japan*

Abstract. We have developed a Quantum Monte Carlo optimization (QMCO) method for many fermion systems. The ground state wave function is written as a linear combination of the basis states generated using the auxiliary field method The bases are chosen with weights that are determined by diagonalizing the Hamiltonian to lower the ground state energy. The QMCO method is free from a difficulty originating from the negative sign problem. We apply the QMCO method to the two-leg ladder and two-dimensional Hubbard model. We can reproduce the exact results for the Hubbard model on small clusters. The pair correlation function is indeed enhanced compared to that for U=0 as the strength of the repulsive Coulomb interaction increases.

Keywords: Quantum Monte Carlo method, auxiliary fields, diagonalization, Hubbard model, two-leg ladder
PACS: 74.20.-z, 71.10.Fd, 74.72.-h.

INTRODUCTION

The effect of strong correlation between electrons is important for many significant quantum critical phenomena such as unconventional superconductivity (SC) and metal-insulator transition. Recently the mechanisms of superconductivity in high-temperature superconductors and organic superconductors have been extensively studied using various two-dimensional (2D) models of electronic interactions. The two-leg ladder[1)-4)] and 2D Hubbard model[5-10)] are the simplest and most fundamental ones among such models.

The Quantum Monte Carlo (QMC) method is a method to treat the correlated electron systems numerically. In this paper we employ a Quantum Monte Carlo optimization (QMCO) method developed recently. The ground state wave function is written as a linear combination of the basis states that are generated using the auxiliary method based on the Hubbard-Stratonovich transformation. The weights of basis states are determined by diagonalizing the Hamiltonian to obtain the lowest energy state in the selected subspace.

MODEL AND THE METHOD

The Hamiltonian is the 2D Hubbard model given by

$$H = -t\sum_{\langle ij\rangle\sigma}\left(c_{i\sigma}^{+}c_{j\sigma} + h.c.\right) + U\sum_{i} n_{i\uparrow}n_{i\downarrow}, \qquad (1)$$

where $c_{j\sigma}^{+}$ ($c_{j\sigma}$) is the creation (annihilation) operator of an electron with spin σ at the j-th site. t is the transfer energy between the nearest-neighbor (n.n.) sites; the sites form a rectangular lattice. $\langle ij\rangle$ denotes summation over all the n.n. bonds. For the two-leg ladder we denote the transfer among the rung as t_d. U is the on-site Coulomb energy.

In a Quantum Monte Carlo simulation, the ground state wave function is given as

$$\psi = e^{-\lambda H}\psi_0, \qquad (2)$$

where ψ_0 is a initial one-particle state represented by a Slater determinant. The exponential operator projects out the ground state from the initial wave function. Using the standard method of QMC, the wave function in eq. (2) is written as

CP850, *Low Temperature Physics: 24th International Conference on Low Temperature Physics;* edited by Y. Takano, S. P. Hershfield, S. O. Hill, P. J. Hirschfeld, and A. M. Goldman

$$\psi = (e^{-\Delta\lambda H})^m \psi_0 \approx (e^{-\Delta\lambda K} e^{-\Delta\lambda V})^m \psi_0. \quad (3)$$

K and V are the first and second term of the Hamiltonian in eq. (1), respectively. Then we use the Hubbard-Stratonovich transformation:

$$\exp\left(-\alpha \sum_i n_{i\uparrow} n_{i\downarrow}\right) = \left(\frac{1}{2}\right)^N \sum_{s_i=\pm 1} \exp\left[2a \sum_i s_i (n_{i\uparrow} - n_{i\downarrow}) - \frac{\alpha}{2} \sum_i (n_{i\uparrow} + n_{i\downarrow})\right] \quad (4)$$

The wave function is expressed by the one-particle Slater determinants, leaving the summation over all the configurations of auxiliary fields $s_i=\pm 1$:

$$\psi = \sum_j c_j \varphi_j, \quad (5)$$

where φ_j is a Slater determinant corresponding to a configuration j of the auxiliary fields. We can generate the auxiliary fields by a random sampling. In Quantum Monte Carlo simulations, we must do an extrapolation to obtain the expectation values for the ground state wave function. Since an extrapolation in terms of the number of bases appears unstable, we use the variance method.[11,12] We have performed extrapolation with up to 2000 basis states.

RESULTS

In Table I we compare our results with the exact values and available ones obtained by the constrained path Monte Carlo method[13] on the 4×4 square lattice with periodic boundary conditions. In order to investigate a possibility of SC originating from the electronic interaction, we have computed pair correlation function for the two-leg ladder model. The correlation function of pairs within the rung is shown as a function of the distance in Fig.1; the pair correlation is obviously enhanced compared to that for U=0.

TABLE 1. Ground state energy per site for the 4×4 Hubbard model. The boundary conditions are periodic in both directions. The constrained path Monte Carlo (CPMC) results are from Ref. [13]. The column QMCO indicates the present results. Exact results were obtained by the diagonalization.

Ne	U	CPMC	QMCO	Exact
10	4	-1.2238	-1.2237	-1.2238
14	4	-0.9831	-0.9836	-0.9840
14	8	-0.728	-0.730(2)	-0.7418
14	10		-0.650(2)	-0.6754
14	12	-0.606	-0.617(5)	-0.6282

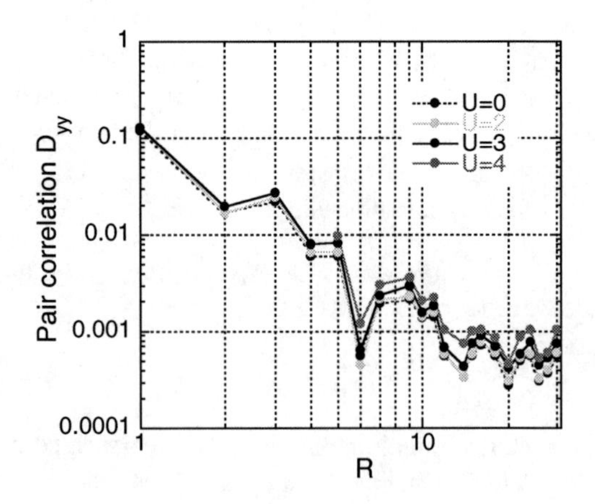

FIGURE 1. Pair correlation function as a function of the distance R for the two-leg ladder. The system is 60×2, t_d=1 in units of t, the number of electrons is 80 and the periodic boundary condition is imposed.

REFERENCES

1. K. Yamaji and Y. Shimoi, *Physica C* **222**, 349 (1994).
2. K. Yamaji et al., *Physica C* **235-240**, 2221 (1994).
3. S. Koike, K. Yamaji, and T. Yanagisawa, *J. Phys. Soc. Jpn.* **68**, 1657 (1999).
4. S. Koike, K. Yamaji, and T. Yanagisawa, *J. Phys. Soc. Jpn.* **69**, 2199 (1999).
5. T. Nakanishi, K. Yamaji and T. Yanagisawa, *J. Phys. Soc. Jpn.* **66**, 294 (1997).
6. K. Yamaji et al., *Physica C* **304**, 225 (1998).
7. T. Yanagisawa, S. Koike and K. Yamaji, *Phys. Rev. B* **64**, 184509 (2001).
8. T. Yanagisawa et al., *J. Phys.: Condens. Matter* **14**, 21 (2002).
9. T. Yanagisawa et al., *Phys. Rev. B* **67**, 132408 (2003).
10. M. Miyazaki et al., *J. Phys. Soc. Jpn.* **73**, 1643 (2004).
11. S. Sorella, *Phys. Rev. B* **64**, 024512 (2001).
12. T. Kashima and M. Imada, *J. Phys. Soc. Jpn.* **70**, 2287 (2001).
13. S. Zhang et al., *Phys. Rev. Lett.* **78**, 4486 (1997).

Disorder Operator In 2D Insulating States

Ryuichi Shindou[1,3], Ken-Ichiro Imura[2] and Masao Ogata[3]

1: Physics Department, University of California, Santa Barbara, CA, 93106-4030, USA
2: RIKEN, 2-1 Hirosawa, Wako, Saitama, 351-0198, Japan
3: Department of Physics, University of Tokyo, 7-3-1 Hongo, Bunkyo-ku, Tokyo, 113-0033, Japan

Abstract. In this brief report, we introduce a disorder operator or (dual) vortex field compatible with the periodic boundary condition (p.b.c.) and estimate its expectation value with respect to non-interacting wavefunctions (w.f.) and gapped mean-field ordered w.f. on a quite general footing. For metallic cases, its expectation value reduces to zero in the thermodynamic limit for almost arbitrary Fermi surfaces (F.S.). Meanwhile, it remains finite for the band insulating case/gapped mean-field order case, whose value is characterized by the localization length.

Keywords: doped Mott insulator, electronic disproportions, localization length, dual vortex field, lattice model
PACS: 71.27.+a, 71.10.Fd, 71.30.+h

Since the discovery of high-Tc cuprates, doped correlated insulators have been intensively studied both experimentally and theoretically. Among them, recent STM experiments have revealed that nano-scale electronic disproportions are spontaneously formed in doped Mott insulators such as lightly doped Bi-2212 and $CaNaCuO_2Cl_2$ [1, 2]. In the meanwhile, extensive Lorentz optical microscopy measurements on doped manganite have also revealed that charged ordered region and ferromagnetic metallic region are phase-separated in relatively larger length scale such as micrometer length scale [3]. Contemplating seriously on these different kinds of experimental observations, we may naturally ask whether there exists a universal classification of insulators against doping.

These questions have been previously studied in terms of the duality transformation [4,5]. The idea therein is based on the dual relation between insulator and superconductor (S.C.), where carrier doping in insulator is regarded as applied magnetic field in the latter. (See TABLE1.) However these preceding works did not necessarily give a complete answer on the above questions. For example, one might still wonder what are the dual counterparts of the *coherence length/magnetic penetration depth* in 2D correlated insulators (See TABLE1.).

In this brief report, we will introduce a more straightforward approach to these quantitative but important questions, where we construct their insulating order parameter (vortex field) *directly in terms of original electron operator.* Namely we consider the following non-local operator whose

TABLE1. Duality relation

2D insulator	2D S.C.
Doping	Applied magnetic field
Doped electron density	Magnetic induction
?	*Penetration depth*
?	*Coherence length*
Vortex field	S.C. order parameter

vortex represents a deviation of an electron's density from its mean value $\overline{n} = N/L^2$ (N: total electron number, L: linear dimension of our system size) [6]:

$$\hat{\eta}(r) = \exp\left\{\sum_{r'} \theta(r'-r)\cdot(\hat{n}_{r'} - \overline{n})\right\}, \quad (1)$$

$$\theta(r'-r) = \ln[(x'-x) + i(y'-y)]. \quad (2)$$

Here the summation w.r.t. $r' = (x', y')$ in eq. (1) is taken over entire systems but $r' = r$. Then, one can readily understand that the line integral of the gradient of $\arg \hat{\eta}(r)$ along a closed loop is always related to a total number of electrons inside this loop, measured from its mean value $\overline{n}$:

$$\oint_{\Gamma} d\vec{l} \cdot \vec{\nabla} \arg \hat{\eta}(r) = 2\pi i \cdot \sum_{r' \subset \Gamma} (\hat{n}_{r'} - \overline{n}). \quad (3)$$

Then 2π vortex of the vortex field at $r = r_0$ represents *one extra* electron/hole located at this point (see TABLE1.) and thus the above non-local operator is an appropriate 2D insulating order parameter in the context of their duality language.

CP850, *Low Temperature Physics: 24th International Conference on Low Temperature Physics;*
edited by Y. Takano, S. P. Hershfield, S. O. Hill, P. J. Hirschfeld, and A. M. Goldman

In order to understand the dual counterparts of the coherence length/magnetic penetration depth at *small* doping regions, we have only to calculate the 2-point/4-point correlation function of the above vortex field w.r.t. *non-doped* correlated insulating w.f., most of which are usually well-described by gapped *mean field* w.f. such as Slater's AF insulating w.f..

As a warm-up for the calculations of correlation functions, we will calculate, in this brief report, its expectation values w.r.t. band metal w.f. and band insulating w.f./mean-field gapped w.f. and confirm that our vortex operator indeed vanishes in metallic case, while remains finite in insulating case.

When trying to perform this lesson, one might immediately notice that the vortex field defined by eqs. (1) and (2) is ill-defined on a periodic lattice system. Then we employ, instead of eq. (2), the following alternative on a $L\times L$ periodic lattice:

$$\theta(r'-r)=\ln\left[-i\left(e^{i\frac{2\pi(x'-R_x)}{L}}-1\right)+\left(e^{i\frac{2\pi(y'-R_y)}{L}}-1\right)\right], \quad (4)$$

where (R_x,R_y) is the dual lattice site corresponding to $r=(x,y)$ on our original lattice. In the case of the square lattice with its lattice constant unit, it reads;

$$(R_x,R_y)\equiv(x,y)+(1/2,1/2).$$

Note that eq. (4) reproduces eq (2) in thermodynamic limit at $r'\approx r$ and also that eq. (4) is compatible with p.b.c..

Bearing in mind both non-interacting and gapped mean-field ordered w.f., we consider those ground state w.f. $|g\rangle$ which are described by a simple Slater determinant. Then some analytic manipulations reveal that the expectation value of our vortex field defined by eqs. (1) and (4) w.r.t. this type of w.f. can be expressed as the following determinant :

$$\langle g|\eta(r)|g\rangle = z\cdot\det\left[\hat{f}\cdot\hat{D}_z(r)+\hat{1}-\hat{f}\right], \quad (5)$$

where $\ln|z|$ can be evaluated by the complex integral [6]:

$$|z|=\exp[-N\times 0.4648...+\mathrm{O}(1)]. \quad (6)$$

$L^2\times L^2$ matrices $\hat{f}$ and $\hat{D}_z(r)$ are given as follows:

$$\left[\hat{f}\right]_{(n,k|n',k')}=\delta_{n,n'}\delta_{k,k'}f(\varepsilon_{n,k}), \quad (7)$$

$$\left[\hat{D}_z(r)\right]_{(n,k|n',k')}=-i\left[\hat{D}_x(r)\right]_{(n,k|n',k')}+\left[\hat{D}_y(r)\right]_{(n,k|n',k')}, \quad (8)$$

$$\left[\hat{D}_\mu(r)\right]_{(n,k|n',k')}\equiv\left[\hat{\partial}_\mu\right]_{(n,k|n',k')}+\left[A_{k,\mu}\right]_{n,n'}\delta_{k+\Delta_x,k'}, \quad (9)$$

$$\left[\hat{\partial}_\mu\right]_{(n,k|n',k')}\equiv\delta_{n,n'}\left(\gamma_\mu\,\delta_{k+\Delta_\mu,k'}-\delta_{k,k'}\right), \quad (10)$$

$$\left[A_{k,\mu}\right]_{n,n'}\equiv\gamma_\mu\left(\langle u_{n,k}|u_{n',k+\Delta_\mu}\rangle-\delta_{n,n'}\right), \quad (11)$$

$$\gamma_\mu\equiv\exp[-2\pi i\cdot R_\mu/M],\quad \Delta_\mu\equiv 2\pi e_\mu/M. \quad (12)$$

Here $f(\varepsilon)$ is the Fermi distribution function, in which informations about the F.S. or energy gap are encoded. Note that $\left[\hat{D}_z\right]$ is a discretized version of covariant derivative in momentum space, where $|u_{n,k}\rangle$ is the periodic part of the (magnetic) Bloch w.f. and thus $\left[A_{k,\mu}\right]$ is the gauge field in k-space.

In order to see in a simple way that eq. (5) indeed vanishes for metallic cases, one should consider those single-band metals within whose F.S. any non-contractible loop cannot be embedded. Then one readily notices that, after a proper arrangement of column and row, $\hat{f}\cdot\hat{D}_z+\hat{1}-\hat{f}$ in eq. (5) can be regarded just as an upper-triangle matrix. Thus, the product of diagonal matrix elements gives its determinant [6] :

$$\begin{aligned}|\langle g|\eta(r)|g\rangle| &= |z|\cdot\left|\det\left[\hat{f}\cdot\sqrt{2}+\hat{1}-\hat{f}\right]\right| \\ &= |z|\cdot\exp\left[N\cdot\frac{\ln 2}{2}\right],\end{aligned} \quad (13)$$

In comparison with eq. (6), the r.h.s. of eq. (13) indeed vanishes in the thermodynamic limit for an arbitrary shape of F.S. on condition that any non-contractible loop cannot be embedded within its F.S..

As for the band insulators, we evaluated eq. (5) by treating the gauge field, i.e. eq. (11), as a perturbation of the order of $1/M$ and obtained the followings [6]:

$$|\langle g|\eta(r)|g\rangle| = \exp\left[-\frac{1}{8}\lambda^2+\mathrm{O}\left(\frac{1}{M}\right)\right]. \quad (14)$$

Here λ denotes the localization length which is defined in terms of the so-called "quantum metric" $\hat{g}(k)$ [7]:

$$\lambda=\int_{\mathrm{B.z.}}\mathrm{Tr}[\hat{g}(k)]\cdot d^2k, \quad (15)$$

$$\begin{aligned}\mathrm{Tr}[\hat{g}]\equiv&\sum_{\mu=x,y}\sum_{n:\text{filled band}}\sum_{m:\text{empty band}}\\ &\times\langle\partial_{k_\mu}u_{n,k}|u_{m,k}\rangle\langle u_{m,k}|\partial_{k_\mu}u_{n,k}\rangle.\end{aligned} \quad (16)$$

This localization length is infinite (finite) when the system has a (no) Drude weight or peak [8]. Thus our vortex field remains finite when the system is band insulator/gapped mean-field ordered state.

REFERENCES

1. T. Hanaguri, et.al., *Nature* **430**, 1001-1005 (2004).
2. K. M. Lang, et.al., *Nature* **415**, 412-416 (2002).
3. M.Uehara, et.al., *Nature* **430**, 1001 (2004).
4. L. Balents, et.al., *Int. J. Mod. Phys. B* **12**, 1033 (1998).
5. Z. Tesanovic, *Phys.Rev. Lett.* **93**, 217004 (2004)
6. R.Shindou, et.al., in preparation.
7. N. Marzari et.al., *Phys. Rev. B* **56**, 12847-12865 (1997).
8. I. Souza, et.al., *Phys. Rev. B* **62**, 1668, (2000)

Self-Consistent Solution of the Bogoliubov-de Gennes Equation for a Single Vortex in *f*-wave Superconductors: Application to Sr_2RuO_4

Masaru Kato[*][†], Hisataka Suematsu[*][†], and Kazumi Maki[¶]

[*]*Department of Mathematical Sciences, Osaka Prefecture University, Sakai, Osaka 599-8531, Japan*
[¶]*Department of Physics and Astronomy, University of Southern California, Los Angeles, CA 90089-0484, USA*
[†]*JST-CREST, 4-1-8, Honcho, Kawaguchi, Saitama 332-0012, Japan*

Abstract. The paring sysmmetry of the triplet superconductor Sr_2RuO_4 is still controversial. Recently, the scanning tunneling microscopy measurement has been done on the quasi-particle structure in Sr $_2RuO_4$. They obtained the local density of states around a single vortex. Using the Bogoliubov-de Gennes equation in the differential form of f-wave ($\Delta\pm$ (k) = Δ**d** ($k_x \pm k_y$) cos (ckz)) superconductivity that has the horizontal nodes, we investigate the quasi-particle structure. As compared with the p-wave pairing, the local density of states of f-wave paring at the vortex core is much similar to the experimental data.

Keywords: Sr2RuO4, Bogoliubov-de Gennes equation, vortex, quasi-particle structure
PACS: 74.70.Pq, 74.20.Rp, 74.20.Fg, 74.25.Qt

INTRODUCTION

Since the discovery of the superconductivity in the layered ruthenate Sr_2RuO_4 in 1994 [1,2], its superconducting symmetry is still controversial. Initially, Rice and Sigrist proposed 2D $p_x \pm ip_y$-wave superconductivity, from the analogy of superfluid ^{3}He [3]. After that, spin-triplet pairing and related chiral symmetry-breaking was confirmed [4-5]. But the specific heat mesurement [6] and the superfluid density measurement [7] showed nodal structure in the superconducting order parameter.

A possible candidate for superconductivity that has the triplet paring with chiral symmetry-breaking and nodal structure is f-wave model [8,9],

$$\Delta_\pm(\boldsymbol{k}) = \Delta\hat{d}\left(k_x \pm ik_y\right)\cos\left(ck_z\right), \qquad (1)$$

where c is the lattice constant in the c-axis.

Recently, Lupien et al. measured the local density of states (LDOS) around a single vortex in Sr_2RuO_4 by the scanning tunneling microscope (STM) [10]. They showed the LDOS for the uniform superconducting state as well as that for the vortex states.

In this study, using the Bogoliubov de Gennes equation, we investigate the quasi-particle spectrum around a single vortex in an f-wave superconductor and compare it with the LDOS of Sr_2RuO_4. In the previous our study we use the unphysical parameter $k_F\xi$=3 [11,12], because of limited numerical capacity, but in present study we use somewhat reliable parameter $k_F\xi$=10.

MODEL AND RESULTS

We solve the Bogoliubov-de Gennes equation for f_+-wave superconductors in a differential form as follows [11];

$$\begin{aligned} E_n u_n(\mathbf{r}) &= \left[\frac{1}{2m}\left(p+\frac{e}{c}A\right)^2\right]u_n(\mathbf{r}) \\ &-\frac{1}{k_F}\int dz'dk_z\cos\left(ck_z\right)e^{ik_z z'} \\ &\times\left\{\frac{1}{2}\left[\left(i\partial_x+\partial_y\right)\Delta_+\left(\mathbf{r}_\parallel, z-\frac{z'}{2}\right)\right]\right. \\ &\left.+\Delta_+\left(\mathbf{r}_\parallel, z-\frac{z'}{2}\right)\left(i\partial_x+\partial_y\right)\right\}v_n\left(\mathbf{r}_\parallel, z-z'\right), \end{aligned} \qquad (2)$$

CP850, *Low Temperature Physics: 24th International Conference on Low Temperature Physics;*
edited by Y. Takano, S. P. Hershfield, S. O. Hill, P. J. Hirschfeld, and A. M. Goldman

$$
\begin{aligned}
E_n v_n(\mathbf{r}) = &-\left[\frac{1}{2m}\left(p-\frac{e}{c}A\right)^2\right]v_n(\mathbf{r}) \\
&-\frac{1}{k_F}\int dz'dk_z \cos(ck_z)e^{ik_z z'} \\
&\times\left\{-\frac{1}{2}\left[(i\partial_x-\partial_y)\Delta_+^*\left(\mathbf{r}_\parallel, z-\frac{z'}{2}\right)\right]\right. \\
&\left.-\Delta_+^*\left(\mathbf{r}_\parallel, z-\frac{z'}{2}\right)(i\partial_x-\partial_y)\right\}u_n(\mathbf{r}_\parallel, z-z'),
\end{aligned} \tag{3}
$$

where u_n and v_n are quasi-particle wave functions of n-the eigenstate and E_n is its the energy eigenvalue and $r_\parallel$ is a coordinate vector in the xy-plane. In order to solve the quasi-particle spectrum around a single vortex, we consider the cylindrical space with height L and radius R, and set the vortex at its centerline. Then we use the Fourier-Bessel expansion in the xy-plane and the Fourier expansion along the z-axis [11]. We solve the resulting discrete Bogoliubov-de Gennes equation and the self-consistent equation under the electron number conservation condition.

First we show the order parameter structures around a single vortex with winding number +1. For the vortex with winding number -1 show similar order parameter structure but there are small differences inside of the vortex core because the bound state structures are different between them.

In Fig.2 we show the LDOS for the f-wave and p-wave superconductors around the vortex with the winding number -1. Comparing Fig. 4(b) in Ref. [10], the f-wave data explain the experiment much better than the p-wave data. Especially at the center of the vortex, the LDOS around the bound state peak is similar to the experimental data. But, outside of the vortex core, the shoulder structures around the energy gap is somewhat different from Fig. 3(a) in Ref.[10]. In order to explain the results other mechanism, such as the multi-band model, may be needed [10].

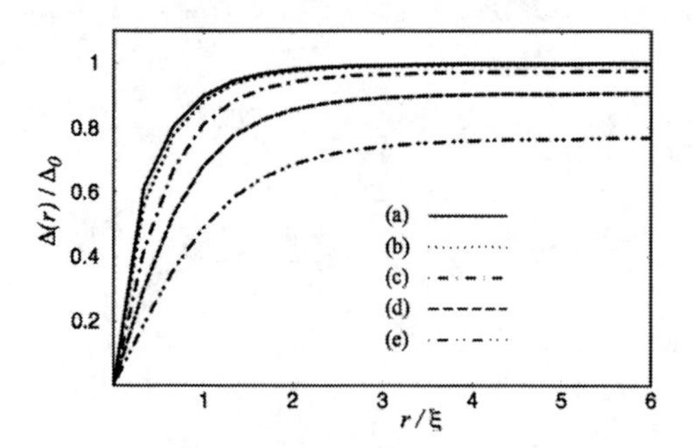

FIGURE 1. Order parameter structure around a single vortex in the f-wave superconductor. (a) T/T_c=0.1, (b) T/T_c=0.2, (c) T/T_c=0.4, (d) T/T_c=0.6 and (e) T/T_c=0.8.

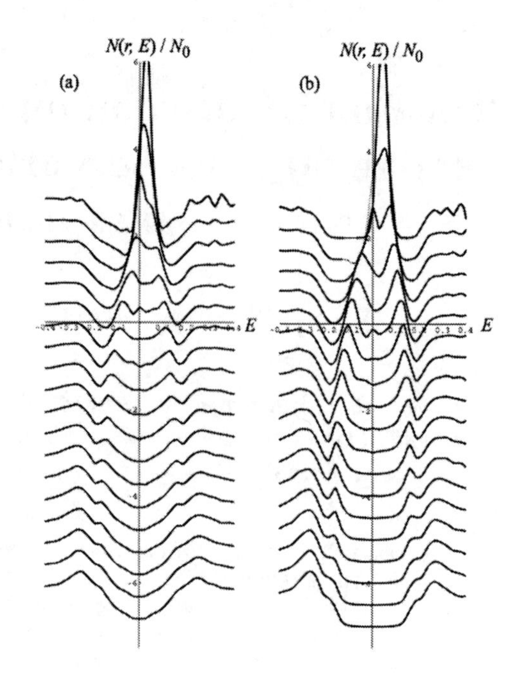

FIGURE 2. Local density of states for (a) f-wave and (b) p-wave superconductivity at various positions from the vortex center.

SUMMARY

We investigate the quasi-particle structure around a single vortex in an f-wave superconductor. The LDOS for f-wave superconductors at the vortex center is similar to the experimental data.

ACKNOWLEDGMENTS

Part of the numerical computation has been done on the computer facility of Japan Atomic Energy Research Institute.

REFERENCES

1. Y. Maeno et al., *Nature* **372**, 532 (1994).
2. A.P. Mackenzie and Y. Maeno, *Rev. Mod. Phys.* **75**, 657 (2003).
3. T.M. Rice and M. Sigrist, *J. Phys. Cond. Matt.* **7**, L643 (1995).
4. G.M. Luke et al., *Nature* **394**, 558 (1998).
5. K. Ishida et al., *Nature* **396**, 653 (1998).
6. S. Nishizaki et al., *J. Phys. Soc. Jpn.* **69**, 572 (2001).
7. I. Bonalde et al., *Phys. Rev. Lett.* **85**, 4775 (2000).
8. H. Won and K. Maki, *Europhys. Lett.* **52**, 427 (2000).
9. K. Izawa et al., *Phys. Rev. Lett.* **86**, 2653 (2001).
10. C. Lupien et al., cond-mat/0503317 (2005).
11. M. Kato, H. Suematsu and K. Maki, *Physica C* **408-410** 535-536 (2004).
12. M. Kato, H. Suematsu and K. Maki, *Phys. Chem. Solids* in press.

Raman Spectra of p-wave Superconductors with Non-Magnetic Impurities

M. Miura*, S. Higashitani†, M. Yamamoto† and K. Nagai†

*Graduate School of Advanced Sciences of Matter, Hiroshima University, Kagamiyama 1-3-1, Higashi-Hiroshima 739-8530, Japan
†Faculty of Integrated Arts and Sciences, Hiroshima University, Kagamiyama 1-7-1, Higashi-Hiroshima 739-8521, Japan

Abstract. We present a theory of impurity effect on the Raman spectra in spin-triplet p-wave superconductors. The contribution from the order parameter collective modes is taken into account. As an example of the p-wave paring, we consider a $k_x + ik_y$ state, which has been proposed as a candidate for Sr_2RuO_4. We show that impurity scattering gives rise to significant modification to the Raman spectra in the p-wave superconductor.

Keywords: p-wave superconductor, order parameter collective mode, Raman scattering
PACS: 74.20.Rp, 78.30.Er, 78.30.Ly

INTRODUCTION

In unconventional paring states, there are various collective modes reflecting the internal degrees of freedom of the order parameter. Some of the order parameter collective modes in p-wave states have been observed in superfluid ^{3}He by sound attenuation measurements. Raman scattering experiment provides a possibility for the observation of the order parameter collective modes[1, 2, 3] in superconductors. Kee *et al.* [4] have discussed the Raman scattering in a quasi-two-dimensional superconductor Sr_2RuO_4 [5]. They showed that a sizable peak due to the collective mode appears in the Raman spectrum assuming that the paring state has a p-wave symmetry $k_x + ik_y$ and the sample is free from impurities. It is expected, however, from the recent study [6, 7] of impurity effect on acoustics of superfluid ^{3}He confined in aerogel that the observability of the collective-mode peak is sensitive to disorder.

Devereaux and Einzel [8] have proposed a theory of the electronic Raman scattering in clean unconventional superconductors, which is based on the Landau transport equation generalized to superconducting state. Equivalent theory can be formulated in terms of the Keldysh quasiclassical Green's function. An advantage of the latter formulation is that the impurity effect can be treated by the well-established impurity-averaged Green's function technique. The quasiclassical Green's function theory for the impure p-wave state has been developed by the present authors [7] for studying the impurity effect in the superfluid ^{3}He-aerogel system. In this paper, we apply this theory to the present problem and calculate the Raman spectra in the $k_x + ik_y$ state in quasi-two-dimensional electron system.

RAMAN RESPONSE FUNCTION

There are massive collective modes in the $k_x + ik_y$ state, called real and imaginary clapping modes[2, 3]. To see the coupling of the clapping modes to the Raman response, it is convenient to write the Raman vertex $\gamma_{\mathbf{k}}$ as the sum of A_{1g}, B_{1g} and B_{2g} components (see Refs. [4] and [8]). From the analysis of the non-equilibrium gap equation, one can show that the real and imaginary clapping modes are coupled to the B_{2g} ($\gamma_{\mathbf{k}} \propto k_x k_y$) and B_{1g} ($\gamma_{\mathbf{k}} \propto k_x^2 - k_y^2$) channels, respectively.

The Raman spectra for the B_{1g} and B_{2g} channels are the same because of the isotropic gap Δ of the two-dimensional $k_x + ik_y$ state. We have calculated the corresponding Raman response function $\chi(\omega)$. Within the self-consistent Born approximation for δ-function scattering centers, the impurity effect on the Raman response is described by a complex energy $\tilde{\varepsilon}_\pm = \varepsilon \pm \omega/2 + (1/2\tau)\tilde{\varepsilon}_\pm/\tilde{x}_\pm$, where $\tilde{x}_\pm = \sqrt{\Delta^2 - \tilde{\varepsilon}_\pm^2}$ and $1/\tau$ is the impurity scattering rate. The response function $\chi(\omega)$ can be written as

$$\chi(\omega) = -\mathrm{Im}\left[\int \frac{d\varepsilon}{4i}[\lambda(\tilde{x}_+\tilde{x}_- + \tilde{\varepsilon}_+\tilde{\varepsilon}_- - \Delta^2)]^K + \frac{\Delta^2\left(\int \frac{d\varepsilon}{4i}[\lambda\bar{\omega}]^K\right)^2}{\int \frac{d\varepsilon}{4i}[\lambda(\bar{\omega}^2 - 2\Delta^2)]^K}\right], \quad (1)$$

where $\lambda = 1/\tilde{x}_+\tilde{x}_-(\tilde{x}_+ + \tilde{x}_-)$, $\bar{\omega} = \tilde{\varepsilon}_+ - \tilde{\varepsilon}_-$ and the Keldysh functions denoted by the superscript K are constructed from the corresponding retarded, advanced and anomalous functions using the prescription given in Ref. [7]. The second line in Eq. (1) is the contribution from the clapping mode. In the normal state, Eq. (1) is reduced to

CP850, *Low Temperature Physics: 24th International Conference on Low Temperature Physics;*
edited by Y. Takano, S. P. Hershfield, S. O. Hill, P. J. Hirschfeld, and A. M. Goldman

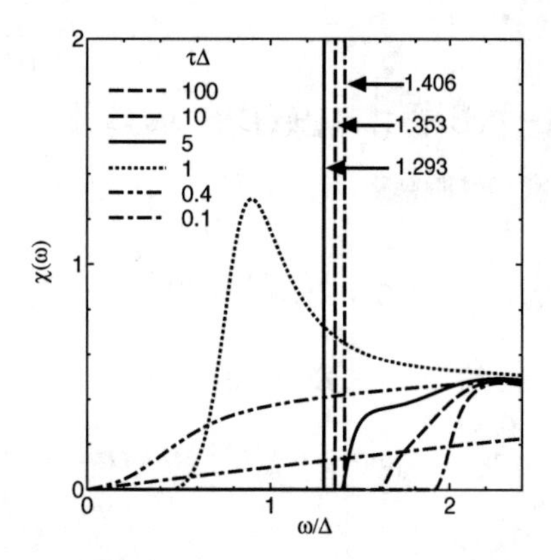

FIGURE 1. Raman response function in the impure $k_x + ik_y$ state at $T = 0$ [K] for various values of the impurity scattering parameter $\tau\Delta$. The impurity effect is evaluated within the self-consistent Born approximation. The vertical lines for $\tau\Delta = 5, 10, 100$ represent δ-function peaks due to the clapping mode. In the clean limit $\tau\Delta \to \infty$, the δ-function peak is located at $\omega/\Delta = \sqrt{2}$.

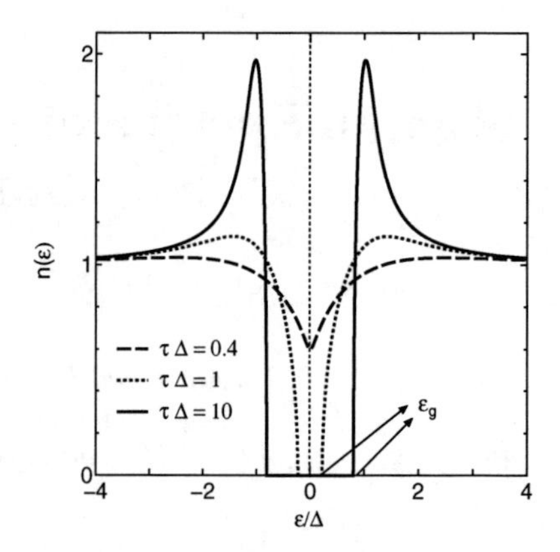

FIGURE 2. Quasiparticle density of states in the impure $k_x + ik_y$ state in the Born limit. There is a gap in the energy range $|\varepsilon| < \varepsilon_g = \Delta[1 - (1/2\tau\Delta)^{2/3}]^{3/2}$ for $\tau\Delta > 0.5$. The density of states becomes gapless for $\tau\Delta < 0.5$.

$\chi_n(\omega) = \omega\tau/[1 + (\omega\tau)^2]$. In the clean limit, Eq. (1) reproduces the result obtained by Kee *et al.*[4]

In Fig. 1, we show the numerical results of $\chi(\omega)$ at zero temperature in the Born limit with various values of $\tau\Delta$. In the weak scattering regime ($\tau\Delta = 5, 10, 100$), there is a δ-function peak denoted by a vertical line, which is due to the clapping mode. The peak position is reduced by the impurity effect. Above the peak position, $\chi(\omega)$ increases from a threshold $\omega = 2\varepsilon_g$, where ε_g denotes the gap edge in the density of states (see Fig. 2). The contribution to $\chi(\omega)$ for $\omega > 2\varepsilon_g$ comes from the quasiparticle excitations and the off-resonant coupling to the clapping mode. For large $\tau\Delta$, the threshold $2\varepsilon_g$ becomes less than the collective mode peak position, then quasiparticle excitations lead to damping of the clapping mode and consequently to the broadening of the peak. When $\tau\Delta$ is small enough to make the density of states gapless, the peak structure is smeared out. For the smallest value of $\tau\Delta$ in Fig. 1, $\chi(\omega)$ is well approximated by the normal-state result $\chi_n(\omega)$.

CONCLUSION

We have discussed the impurity effect on the Raman spectra in the $k_x + ik_y$ state in a quasi-two dimensional superconductor. The Raman spectra depend strongly on the quasiparticle mean free time τ. A well-defined collective mode peak appears in the Raman spectra in the weak scattering regime $\tau\Delta \gg 1$. However, small value of $\tau\Delta$ gives rise to substantial broadening of the peak. In particular, when $\tau\Delta$ is so small that the density of states becomes gapless, the collective mode peak cannot be observed.

In the numerical results presented here, the impurity effect is evaluated within the Born approximation. In the case of the unitarity limit scattering, a remarkable difference arises from the existence of the density of states around zero energy. The low energy states also contribute to the Raman spectra and yield another excitation processes leading to the collective mode damping. Detailed study of such a strong impurity scattering effect shall be reported elsewhere.

ACKNOWLEDGMENTS

This work is supported in part by a Grant-in-Aid for Scientific Research (No. 16540322) and a Grant-in-Aid for COE Research (No. 13CE2002) from the Ministry of Education, Culture, Sports, Science and Technology of Japan.

REFERENCES

1. L. Tewordt, *Phys. Rev. Lett.* **83**, 1007 (1999).
2. S. Higashitani and K. Nagai, *Physica B* **284–288**, 539 (2000).
3. S. Higashitani and K. Nagai, *Phys. Rev. B* **62**, 3042 (2000).
4. Hae-Young Kee, K. Maki, and, C.H. Chung, *Phys. Rev. B* **67**, 180504 (2003).
5. A. P. Mackenzie and Y. Maeno, *Rev. Mod. Phys.* **75**, 657 (2003).
6. R. Nomura, G. Gervais, T.M. Haard, Y. Lee, N. Mulders, and W.P. Halperin, *Phys. Rev. Lett.* **85**, 4325 (2000).
7. S. Higashitani, M. Miura, M. Yamamoto, and K. Nagai, *Phys. Rev. B* **71**, 134508 (2005).
8. T.P. Devereaux and D. Einzel, *Phys. Rev. B* **51**, 16336 (1995).

Toward the Thorough Microscopic Clarification of Superconducting Mechanism and Gap Structure in Sr_2RuO_4

Takuji Nomura

Synchrotron Radiation Research Center, Japan Atomic Energy Research Institute, Mikazuki, Sayo, Hyogo 679-5148, Japan

Abstract. A microscopic mechanism of triplet superconductivity and a possible superconducting gap structure in Sr_2RuO_4 are discussed on the basis of the standard Fermi liquid picture. We describe the Ru4d-electron system on the RuO_2 networks by using a three-band repulsive Hubbard model. Many-electron correlation is treated within the third order perturbation expansion in the on-site Coulomb interaction. The Eliashberg equation is solved numerically using the effective pairing interaction evaluated perturbatively. As a consequence, a highly anisotropic p-wave pairing state is obtained as the most probable pairing symmetry. The temperature dependences of specific heat, thermal conductivity and ultrasound attenuation rate are also analyzed, using the band- and momentum-dependent superconducting gap function calculated microscopically. It is shown that the theoretical results are consistent with the temperature dependences of those quantities, assuming the chiral pairing symmetry (i.e., $\Delta(\mathbf{k}) \sim k_x \pm ik_y$).

Keywords: Sr_2RuO_4, spin-triplet superconductivity, pairing mechanism, superconducting gap structure
PACS: 74.70.Pq, 74.20.Mn, 74.20.Rp

INTRODUCTION

The discovery of superconductivity in the quasi-two-dimensional ruthenium oxide Sr_2RuO_4 has stimulated condensed matter physicists to study intensively its intriguing electronic properties. Many excellent experimental and theoretical works in the last decade have elucidated that Sr_2RuO_4 is so ideal a compound for deepening our understanding of many-electron correlation as high-T_c copper oxides [1].

One of the most striking properties of Sr_2RuO_4 is unconventional superconductivity, most likely spin-triplet p-wave superconductivity [2, 3]. Much effort has been devoted to clarify the unconventional pairing mechanism. In addition to the superconducting mechanism, the gap structure has been also controversial. Rice and Sigrist singled out the pairing symmetry $\mathbf{d}(\mathbf{k}) \sim (k_x \pm ik_y)\hat{z}$ phenomenologically [4]. However the power-law temperature dependences observed in many physical quantities below T_c apparently contradict this pairing symmetry [1].

In the present short article, a possible solution for the superconducting mechanism and gap structure of Sr_2RuO_4 is proposed from a microscopic theoretical point of view. It is shown that the most probable pairing symmetry is triplet p-wave for the realistic electronic structure of Sr_2RuO_4, as a natural result from many-electron correlation [5, 6]. Specific heat, thermal conductivity and ultrasound attenuation rate are analyzed using the band- and momentum-dependent gap function calculated by the microscopic theory [7, 8]. Both the band- and momentum-dependences of the gap function are essential for explaining the temperature dependences of these quantities in Sr_2RuO_4.

MODEL AND THEORY

Since the electronic density of states near the Fermi level is dominated by the partial density of states of Ru4dε orbitals [9, 10], we take hopping parameters describing the transfers between the Ru4dε-like Wannier orbitals. We adopt the following non-interacting Hamiltonian (these Wannier orbitals are characterized by $\ell = \{xy, yz, xz\}$):

$$\begin{aligned} H_0 &= \sum_{\mathbf{k},\ell,\sigma} \xi_\ell(\mathbf{k}) c^\dagger_{\mathbf{k}\ell\sigma} c_{\mathbf{k}\ell\sigma} \\ &\quad + \sum_{\mathbf{k},\sigma} \lambda(\mathbf{k}) (c^\dagger_{\mathbf{k}yz\sigma} c_{\mathbf{k}xz\sigma} + c^\dagger_{\mathbf{k}xz\sigma} c_{\mathbf{k}yz\sigma}), \end{aligned} \quad (1)$$

where $c_{\mathbf{k}\ell\sigma}[c^\dagger_{\mathbf{k}\ell\sigma}]$ is the electron annihilation[creation] operator (the pseudo-momentum, orbital and spin states are denoted by $\mathbf{k}$, ℓ and σ, respectively), and the energy dispersions are

$$\begin{aligned} \xi_{xy}(\mathbf{k}) &= 2t_1(\cos k_x + \cos k_y) \\ &\quad + 4t_2 \cos k_x \cos k_y - \mu_{xy}, && (2) \\ \xi_{yz}(\mathbf{k}) &= 2t_3 \cos k_y + 2t_4 \cos k_x - \mu_{yz}, && (3) \\ \xi_{xz}(\mathbf{k}) &= 2t_3 \cos k_x + 2t_4 \cos k_y - \mu_{xz}, && (4) \\ \lambda(\mathbf{k}) &= 4t_5 \sin k_x \sin k_y. && (5) \end{aligned}$$

The hopping parameters $t_1, ..., t_5$ and the chemical potentials μ_ℓ's are determined to reproduce the Fermi surface

CP850, *Low Temperature Physics: 24th International Conference on Low Temperature Physics;*
edited by Y. Takano, S. P. Hershfield, S. O. Hill, P. J. Hirschfeld, and A. M. Goldman

topology. As easily seen by diagonalizing H_0, the Fermi surface consists of three circles (three cylinders in three-dimension), which are usually named α, β and γ (See Fig. 2(a)). We introduce the Coulomb interaction part:

$$\begin{aligned} H' &= \frac{U}{2}\sum_i\sum_\ell\sum_{\sigma\neq\sigma'} c^\dagger_{i\ell\sigma}c^\dagger_{i\ell\sigma'}c_{i\ell\sigma'}c_{i\ell\sigma} \\ &+\frac{U'}{2}\sum_i\sum_{\ell\neq\ell'}\sum_{\sigma,\sigma'} c^\dagger_{i\ell\sigma}c^\dagger_{i\ell'\sigma'}c_{i\ell'\sigma'}c_{i\ell\sigma} \\ &+\frac{J}{2}\sum_i\sum_{\ell\neq\ell'}\sum_{\sigma,\sigma'} c^\dagger_{i\ell\sigma}c^\dagger_{i\ell'\sigma'}c_{i\ell\sigma'}c_{i\ell'\sigma} \\ &+\frac{J'}{2}\sum_i\sum_{\ell\neq\ell'}\sum_{\sigma\neq\sigma'} c^\dagger_{i\ell\sigma}c^\dagger_{i\ell\sigma'}c_{i\ell'\sigma'}c_{i\ell'\sigma}, \end{aligned} \quad (6)$$

where the operator $c_{i\ell\sigma}[c^\dagger_{i\ell\sigma}]$ is the electron annihilation[creation] operator at i-th Ru site ($c_{\mathbf{k}\ell\sigma}$ is the Fourier transform of $c_{i\ell\sigma}$). The microscopic origin of H' is the Coulomb interaction between the Ru4d electrons. In the present calculation, $U = 3.385$, $U' = J = J' = 0.5U$ for the discussion of pairing mechanism, and $U = 4$, $U' = J = J' = 0.33U$ for the analysis of physical quantities. At least the qualitative features stated below about the calculated results are robustly unchanged between these parameter sets. The total Hamiltonian is $H = H_0 + H'$.

The superconducting order parameter (in other words, anomalous self-energy) is determined by solving the Eliashberg equation:

$$\begin{aligned} \lambda\Delta_{a,\sigma_1\sigma_2}(k) = &-\frac{T}{N}\sum_{a',k',\sigma_3\sigma_4} V_{a\sigma_1\sigma_2,a'\sigma_3\sigma_4}(k,k') \\ &\times|G_{a'}(k')|^2\Delta_{a',\sigma_4\sigma_3}(k'), \end{aligned} \quad (7)$$

where $k^{(\prime)} = (\mathbf{k}^{(\prime)}, i\omega_n{}^{(\prime)})$ $[\omega_n = (2n+1)\pi T]$, and band index $a^{(\prime)}$ takes α, β and γ. $G_a(k)$ is the Green's function for band a. The effective interaction between the Fermi liquid quasi-particles, $V_{a\sigma_1\sigma_2,a'\sigma_3\sigma_4}(k,k')$, is evaluated by the third order perturbation expansion in H'. The lengthy procedure of expansions and the numerical solutions of the Eliashberg equation (7) are given in the work [6].

P-WAVE PAIRING MECHANISM

The expected pairing symmetry is determined by the eigenvector $\Delta_a(k)$ giving the positive maximum eigenvalue of the eigenvalue equation (7). We can find indeed that a spin-triplet anisotropic p-wave pairing state gives the maximum eigenvalue of Eq. (7) [6].

The essence of the p-wave pairing is investigated by extracting the momentum dependence of the pairing interaction $V_{\gamma\sigma\sigma,\gamma\sigma\sigma}(k,k')$ on the γ band [5, 6]. In Fig. 1, $V_{\gamma\sigma\sigma,\gamma\sigma\sigma}(k,k')$ is depicted. We find that $V_{\gamma\sigma\sigma,\gamma\sigma\sigma}(k,k')$

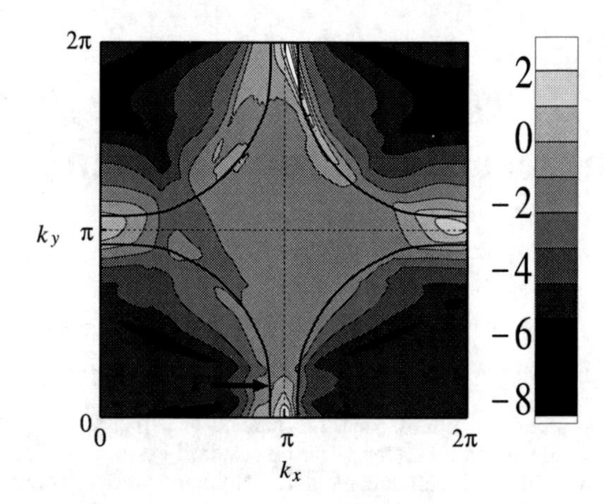

FIGURE 1. Calculated effective pairing interaction on the γ band, $V_{\gamma\sigma\sigma,\gamma\sigma\sigma}(\mathbf{k}i\pi T, \mathbf{k}'i\pi T)$, is shown as a function of $\mathbf{k}$ by two-dimensional contourplot, where $\mathbf{k}'$ is fixed as pointed by the arrow. The thick circle is the γ Fermi surface.

takes high values around $\mathbf{k} \approx -\mathbf{k}'$ and relatively low values around $\mathbf{k} \approx \mathbf{k}'$. According to Eq. (7), this momentum dependence favors opposite signs of $\Delta_\gamma(k)$ between these two points, in other words, is attractive in odd parity pairing channel. Among the odd parity pairing channels, the positive maximum eigenvalue is usually expected to be obtained for the pairing state yielding the smallest number of nodes.

It is interesting that the above essential momentum dependence originates from the third order vertex-corrected terms, and therefore could not simply be regarded as originating from the exchange processes of spin fluctuations [5]. Actually, enhancement of ferromagnetic spin fluctuations is not observed experimentally in Sr_2RuO_4 [11], and spin fluctuation contributions do not seem essential for the p-wave pairing for the case of Sr_2RuO_4. Some authors demonstrated that the same vertex-corrected terms essentially induce the p-wave pairing in two-dimensional repulsive Fermi gas in the weak coupling limit [12, 13]. Therefore we might consider that Sr_2RuO_4 is an analogous system to the isotropic repulsive Fermi gas in two-dimension.

GAP STRUCTURE

We assume naturally that the chiral symmetry $\Delta_a(k) \sim k_x \pm ik_y$ is realized, following Rice and Sigrist [4]. The gap structure and the density of states obtained in the present microscopic theory are presented in Fig. 2. The gap function takes the maximum value on the γ band. Thus we have succeeded in deriving the 'Orbital Dependent Superconductivity' [14] microscopically. There is a nodal structure on the γ Fermi surface along the directions [100] and [010], which is attributed to the 2π-

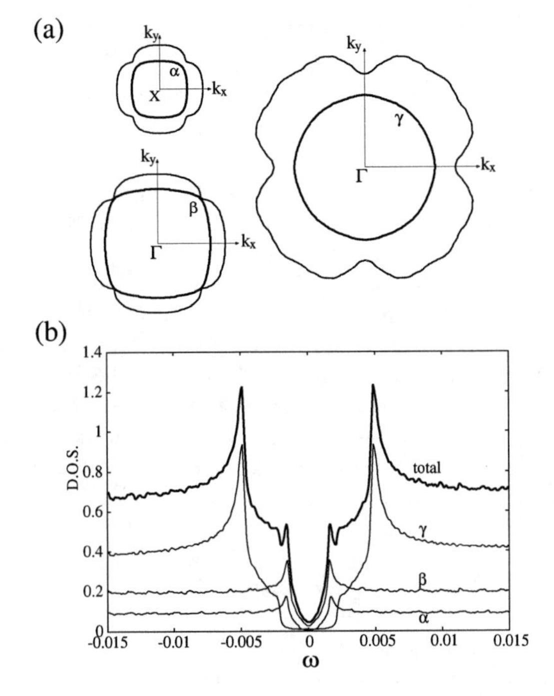

FIGURE 2. (a) The three Fermi surface sheets and the gap function. The Fermi surfaces are depicted by the thick solid lines. The dependence of the gap magnitude on in-plane direction is expressed by the distance from the Fermi circle along the direction. k_x and k_y axes correspond to [100] and [010] axes, respectively. (b)Calculated density of states in the superconducting state. The vertical axis represents the density of states (arb. units) and the horizontal axis represents the energy ω (the Fermi level corresponds to $\omega = 0$). The thick solid line denotes the total density of states, and the thin solid lines denote the contributions from each band. The unit of energy on the horizontal axis is about 660(K).

periodicity and odd parity of the order parameter [5, 15].

Our theory predicts additional nodal structures on the α and β Fermi surfaces on (*or* near) the diagonal lines [7, 8]. These nodal structures would be a result from the strong momentum dependence due to the incommensurate antiferromagnetic fluctuations.

ANALYSIS OF PHYSICAL QUANTITIES

We can analyze physical quantities, using the obtained band- and momentum-dependence of $\Delta_a(k)$ and determining the temperature dependence of the gap magnitude below T_c by the simple BCS equation [7, 8].

In Fig. 3, the result of specific heat is presented. The overall temperature dependence is consistent with the experimental data. The contribution from the γ band is dominant near T_c, since the partial density of states near the Fermi level is the largest on the γ band.

Thermal conductivity $\kappa_{\mu\nu}$ and ultrasound attenuation

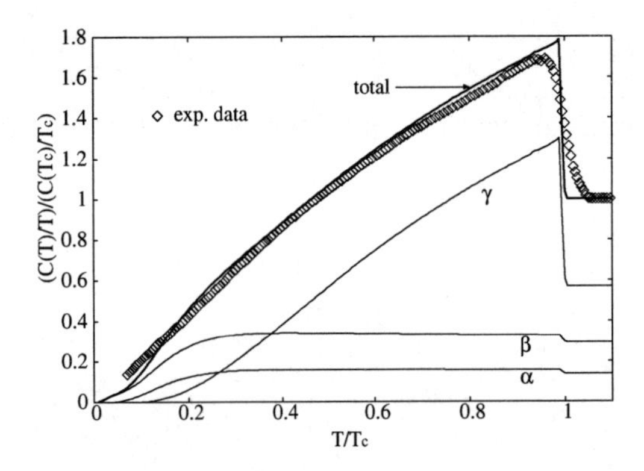

FIGURE 3. Calculated specific heat (thick solid curve) is compared with experimental results (◇, [16]). Contributions from each band are separately presented by the thin curves.

rate α are calculated, respectively, by the following formulas (8) and (9) [8]:

$$\kappa_{\mu\nu} = \frac{1}{8T^2} \sum_{a,\mathbf{k}_F} v_{\mathbf{k}_F a\mu} v_{\mathbf{k}_F a\nu} \int_{-\infty}^{\infty} dz \frac{z^2}{\cosh^2(\frac{z}{2T})} I_a(\mathbf{k}_F, z), \tag{8}$$

$$\alpha = \frac{\omega_0(\mathbf{q})}{8T} \sum_{a,\mathbf{k}_F} |\Lambda_{\mathbf{k}_F,\mathbf{q},a}|^2 \int_{-\infty}^{\infty} dz \frac{1}{\cosh^2(\frac{z}{2T})} I_a(\mathbf{k}_F, z), \tag{9}$$

with

$$I_a(\mathbf{k}_F, z) = \left| \frac{\partial \xi_a(\mathbf{k})}{\partial \mathbf{k}} \right|^{-1}_{\mathbf{k}=\mathbf{k}_F} \frac{1 + \frac{|\tilde{z}_a^R(\mathbf{k}_F,z)|^2 - |\Delta_a(\mathbf{k}_F)|^2}{|\tilde{z}_a^R(\mathbf{k}_F,z)^2 - |\Delta_a(\mathbf{k}_F)|^2|}}{\mathrm{Im}\sqrt{\tilde{z}_a^R(\mathbf{k}_F,z)^2 - |\Delta_a(\mathbf{k}_F)|^2}}, \tag{10}$$

where $\mathbf{k}_F$ is the Fermi wavenumber, $\xi_a(\mathbf{k})$ is the diagonalized band dispersion, $v_{\mathbf{k}_F a\mu}$ is μ-component of the Fermi velocity, $\Lambda_{\mathbf{k}_F,\mathbf{q},a}$ is the electron-phonon coupling matrix. $\tilde{z}_a(\mathbf{k},z) = z - \Sigma_a(\mathbf{k},z)$, using impurity self-energy $\Sigma_a(\mathbf{k},z)$. We have assumed that the impurity scattering is in the unitarity limit, and $\Sigma_a(\mathbf{k},z)$ is determined by the self-consistent T-matrix approximation.

The numerical results of thermal conductivity are shown in Fig. 4. There we can show also the contributions from each band separately (note that the quantity is separable into contributions from each band, since the formula (8) contains the summation with respect to band index a). The reason why the γ band does not dominantly contribute to the thermal transport is that the Fermi velocity is not large around the nodal structure on the γ band.

In order to calculate ultrasound attenuation rate, we approximate the electron-phonon coupling function $\Lambda_{\mathbf{k}_F,\mathbf{q},a}$ by five harmonic functions, as in [8]. The numerical results fit well the experimental results in the overall temperature region, where the Fourier coefficients for the five harmonics are used as the fitting parameters (Fig. 5).

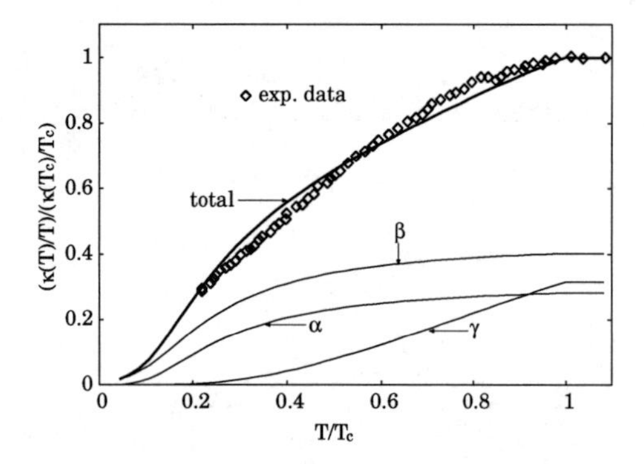

FIGURE 4. Calculated thermal conductivity $\kappa_{[100]}(T)$ (thick solid line) is compared with the experimental data (◇) read from [17]. The directions of the thermal current and the temperature gradient are both along [100] direction. The vertical and horizontal axes represent $\kappa_{[100]}(T)/T$ and temperature T normalized by the value at $T = T_c$, respectively. The contributions from each band are separately shown by the thin solid lines.

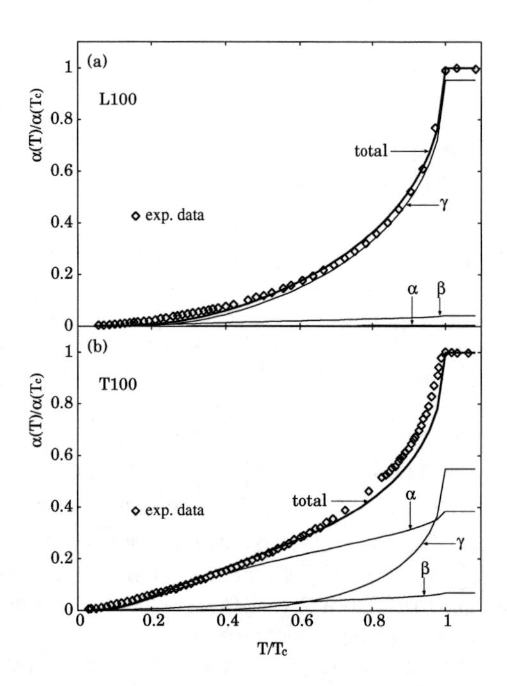

FIGURE 5. Calculated ultrasound attenuation rates (thick solid lines) are compared with the experimental data (◇) read from [18]. (a) For the longitudinal wave propagating along [100] direction. (b) For the transverse wave propagating along [100] direction. The vertical and horizontal axes represent $\alpha(T)$ and temperature T normalized by the value at $T = T_c$, respectively. The contributions from each band are separately shown by the thin solid lines.

We can reproduce the experimentally observed huge difference between the longitudinal and transverse modes along [100] direction [8]. This huge difference originates mainly from the momentum dependence of $\Lambda_{\mathbf{k}_F,\mathbf{q},\gamma}$ [8, 19].

CONCLUSION

In this short article, we have proposed a possible mechanism of the triplet superconductivity and a gap structure in Sr_2RuO_4 from the microscopic theoretical viewpoint on the basis of the Fermi liquid picture. Superconducting mechanism for other actual systems (high-T_c cuprates, organic superconductors, heavy Fermion superconductors ...) has been analyzed by the same formulation [20]. We consider that the mechanism of anisotropic pairing in various strongly correlated electron systems could be naturally understood under the unified view [20].

The author is grateful to Prof. Kosaku Yamada, Prof. Yoshiteru Maeno and Dr. Kazuhiko Deguchi for invaluable discussions and communications. The experimental data in Fig. 3 are provided by K. Deguchi and Y. Maeno.

REFERENCES

1. A.P. Mackenzie and Y. Maeno, *Rev. Mod. Phys.* **75**, 657–712 (2003), and references therein.
2. K. Ishida, H. Mukuda, Y. Kitaoka, K. Asayama, Z.Q. Mao, Y. Mori and Y. Maeno, *Nature* **396**, 658–660 (1998).
3. J. A. Duffy, S. M. Hayden, Y. Maeno, Z. Mao, J. Kulda and G. J. McIntyre, *Phys. Rev. Lett.* **85**, 5412–5415 (2000).
4. T.M. Rice and M. Sigrist, *J. Phys. Condens. Matter* **7**, L643–L648 (1995).
5. T. Nomura and K. Yamada, *J. Phys. Soc. Jpn.* **69**, 3678–3688 (2000).
6. T. Nomura and K. Yamada, *J. Phys. Soc. Jpn.* **71**, 1993–2004 (2002).
7. T. Nomura and K. Yamada, *J. Phys. Soc. Jpn.* **71**, 404–407 (2002).
8. T. Nomura, *J. Phys. Soc. Jpn.* **74**, 1818–1829 (2005).
9. T. Oguchi, *Phys. Rev. B* **51**, 1385–1388 (1995).
10. D.J. Singh, *Phys. Rev. B* **52**, 1358–1361 (1995).
11. Y. Sidis, M. Braden, P. Bourges, B. Hennion, S. NishiZaki, Y. Maeno and Y. Mori, *Phys. Rev. Lett.* **83**, 3320–3323 (1999).
12. A.V. Chubukov, *Phys. Rev. B* **48**, 1097–1104 (1993).
13. J. Feldman, H. Knörrer and E. Trubowitz, *Helv. Phys. Acta* **70**, 154–191 (1997).
14. D.F. Agterberg, T.M. Rice and M. Sigrist, *Phys. Rev. Lett.* **78**, 3374–3377 (1997).
15. K. Miyake and O. Narikiyo, *Phys. Rev. Lett.* **83**, 1423–1426 (1999).
16. K. Deguchi, Z.Q. Mao and Y. Maeno, *J. Phys. Soc. Jpn.* **73**, 1313–1321 (2004).
17. M.A. Tanatar, S. Nagai, Z.Q. Mao, Y. Maeno and T. Ishiguro, *Phys. Rev. B* **63**, 064505 (2001).
18. C. Lupien, W.A. MacFarlane, C. Proust, L. Taillefer, Z.Q. Mao and Y. Maeno, *Phys. Rev. Lett.* **86**, 5986–5989 (2001).
19. M.B. Walker, M.F. Smith and K.V. Samokhin, *Phys. Rev. B* **65**, 014517 (2001).
20. Y. Yanase, T. Jujo, T. Nomura, H. Ikeda, T. Hotta and K. Yamada, *Phys. Reports* **387**, 1–149 (2003).

Superconductivity in MgB_2: Magneto-Raman Measurements

G. Blumberg*, A. Mialitsin*, B.S. Dennis* and J. Karpinski†

*Bell Laboratories, Lucent Technologies, Murray Hill, NJ 07974, USA
†Solid State Physics Laboratory, ETH, CH-8093 Zürich, Switzerland

Abstract. Polarization-resolved Raman scattering measurements of MgB_2 superconductor as function of excitation, temperature and magnetic field have revealed four distinct superconducting features: a clean gap below 35 cm^{-1} and three superconducting coherence peaks at 50 and 110 cm^{-1} for the E_{2g} symmetry and at 75 cm^{-1} for the A_{1g} symmetry. Their temperature and field dependences have been established. Superconductivity induced renormalization of the E_{2g} phonon consistent with electron-phonon coupling $\lambda \approx 0.3$ has been observed.

Keywords: Superconductivity, Raman Scattering, Magnetic Field
PACS: 74.70.Ad, 74.25.Ha, 74.25.Gz, 78.30.Er

The two superconducting (SC) gaps nature of MgB_2 is experimentally established by a number of spectroscopies. The SC gaps have been assigned to distinctive (σ- and π-band) Fermi surface (FS) sheets by means of ARPES [1]. The larger gap, $\Delta_\sigma = 5.5-6.5$ meV, is attributed to σ-bonding states of the boron $p_{x,y}$ orbitals and the smaller gap, $\Delta_\pi = 1.5-2.2$ meV, to π-states of the boron p_z orbitals. Scanning tunneling microscopy (STM) has provided a reliable fit for the smaller gap, $\Delta_\pi = 2.2$ meV [2]. This value is close to the absorption threshold energy $2\Delta_0 = 31$ cm^{-1} obtained from magneto-optical far-IR studies [3]. The nominal upper critical field H_{c2}^{π} deduced from the coherence length $\xi_\pi = 49.6$ nm by vortex imaging is $H_{c2}^{\pi} \approx 0.13$ T [2] that is much smaller than the critical field $H_{c2}^{optical} \approx 5$ T found by magneto-optical measurements [3]. To explain near 40 K SC T_c in MgB_2 a large, in order of unity, electron-phonon coupling constant λ is required [4]. This coupling is expected to manifest itself in strong self-energy effects across the SC phase transition.

Raman scattering from the *ab* surface of MgB_2 single crystals with $T_c \approx 38$ K was performed in back scattering geometry using circularly polarized light. The data in magnetic fields was acquired with a continuous flow cryostat inserted in the horizontal bore of a superconducting magnet. We used the $\lambda_L = 482.5$ and 752.5 nm excitation lines of a Kr^+ laser and a triple-grating spectrometer for the analysis of the scattered light.

In Fig. 1 we show evolution of the SC coherence peaks for both the E_{2g} and the A_{1g} scattering channels across the SC transition for two cases: varying temperature at zero magnetic field (a, c) and varying the magnetic field at 8 K (b, d). The coherence peaks lose their intensity and move to lower energies by either increasing the temperature or field. In the SC state we observe a clean threshold of Raman intensity, $2\Delta_0 = 35$ cm^{-1} for all

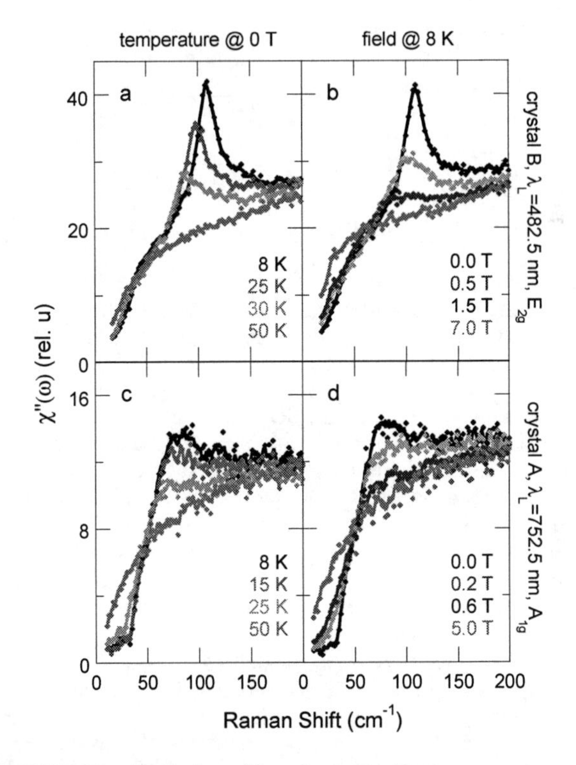

FIGURE 1. Evolution of low-frequency Raman response as function of temperature at zero field (a,c) and field at 8 K (b,d). The E_{2g} channel, right-left (*RL*) polarization, with $\lambda_L = 482.5$ nm excitation is shown in (a,b) and the A_{1g} channel, right-right (*RR*) polarization, with $\lambda_L = 752.5$ nm in (c,d).

scattering geometries. This threshold appears along with two SC coherence peaks in the E_{2g} scattering channel, $2\Delta_L^E = 110$ cm^{-1} and $2\Delta_S^E = 50$ cm^{-1}, and another peak in A_{1g} channel, $2\Delta^A = 75$ cm^{-1}. The energy scales Δ_0 and Δ_L^E are consistent respectively with Δ_π and Δ_σ from one-electron spectroscopies [1, 2].

CP850, *Low Temperature Physics: 24th International Conference on Low Temperature Physics;*
edited by Y. Takano, S. P. Hershfield, S. O. Hill, P. J. Hirschfeld, and A. M. Goldman

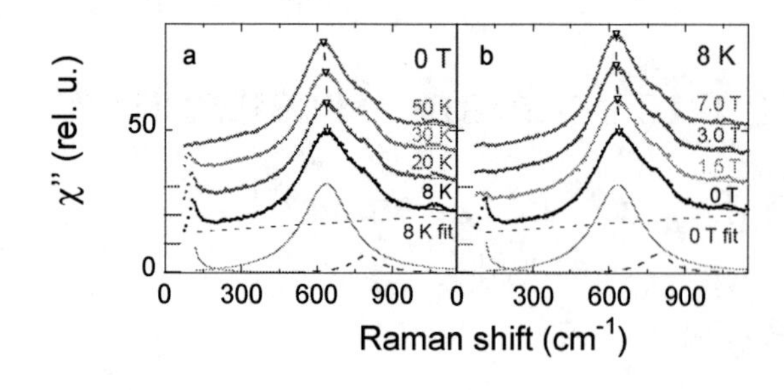

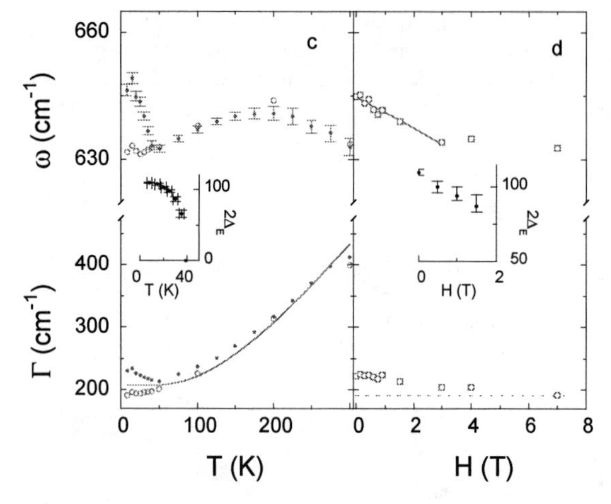

FIGURE 2. Phononic self-energy effect due to SC transition. Evolution of the Raman response in the E_{2g} channel (RL) across the SC transition as function of temperature in zero field (a) and as function of field applied parallel to the c-axis at 8 K (b), $\lambda_L = 482.5$ nm. The data (dots) are fitted (solid line) with two phononic oscillators and SC coherence peak on electronic continuum (decompositions for the lowest spectra are shown). The E_{2g} phonon frequency $\omega(T,H)$ and the damping constant $\Gamma(T,H)$ are drawn as functions of temperature (c) and field (d). In (c) zero field cooling (solid symbols) and 7 T cooling (empty symbols) are shown. The low temperature limit of ω_0^{SC} and ω_0^N are approximated for corresponding cooling. Solid line is fit in normal state to a model of anharmonic decay, $\Gamma(T) = \Gamma_0 + \Gamma_3[1+2n(\Omega/2)] + \Gamma_4[1+3n(\Omega/3)+3n^2(\Omega/3)]$. Here $\Omega(T) = hc\omega_h/k_BT$, harmonic frequency $\omega_h = 515$ cm^{-1} [5, 6], n is the Bose-Einstein distribution function, the internal temperature independent line width Γ_0 is small, $\Gamma_3 = 190$ and $\Gamma_4 = 20$ cm^{-1} are broadening coefficients due to the cubic and quartic anharmonicity. Since $\Gamma_3 + \Gamma_4 \gg \Gamma_0$ we conclude that the anharmonic decay is primarily responsible for the large damping constant of the E_{2g} phonon. Insets show temperature (b) and field (c) dependences of the $2\Delta_L^E$ SC coherence peak.

We study the effect of magnetic field on the superconductivity induced spectral features. The intensity threshold $2\Delta_0$ and the $2\Delta_S^E$ features smear out already at magnetic field as weak as 0.2 T, consistent with H_{c2}^{π} deducted from vortex imaging [2]. The $2\Delta^A$ SC coherence peak persists up to 0.6 T while the $2\Delta_L^E$ peak is suppressed only beyond 2 T. $2\Delta_L^E(T,H)$ are shown in the insets of Fig. 2, it exhibits BCS-like temperature dependence and a linear reduction in field with an anomalously rapid slope about -15 cm^{-1}/T. A linear continuation for the $2\Delta_L^E$ gap collapse approximates to 7 T, field higher than $H_{c2}^{optical}$ [3].

In Fig. 2 we summarize temperature dependence of the E_{2g} Raman response for MgB_2 crystal measured on cooling in zero- and 7 T field and as function of field at 8 K. We fit the data with a model containing two phononic oscillators on electronic Raman continuum including the coherence peak in the SC state. The temperature and field dependences of the E_{2g} phonon frequency $\omega(T,H)$ and the damping constant $\Gamma(T,H)$ are shown in panels (c-d).

Following Refs. [7, 8] we relate the electron phonon coupling constant λ to $\kappa = (\omega_0^{SC}/\omega_0^N) - 1$ as $\lambda = -\kappa(T)\Re(\frac{\sin u}{u})$, where $u \equiv \pi + 2i\cosh^{-1}(\omega^N/2\Delta_L^E)$. Using $2\Delta_L^E = 110$ cm^{-1} we estimate λ at about 0.3.

In summary, we have measured the polarization, excitation, temperature and field dependence of the Raman response for MgB_2 single crystal superconductor and observed four distinct superconductivity induced features: (1) a clean gap below 35 cm^{-1} consistent with the SC gap in π-states of the boron p_z orbitals; (2) a shoulder at 50 cm^{-1} in E_{2g} spectra; (3) a SC coherence peak of A_{1g} symmetry at 75 cm^{-1}; and (4) the largest gap in E_{2g} spectra at 110 cm^{-1} consistent with the SC gap in σ-bonding states of the boron $p_{x,y}$ orbitals. The features (1-3) are shown to be fragile to magnetic field applied along the c-axis. The largest gap (4) shows BCS-like temperature dependence and $2\Delta_L^E/k_BT_c$ ratio about 4 indicating a moderately strong coupling limit. While the $2\Delta_L^E$ coherence peak intensity sustains up to 2 T field the gap magnitude is suppressed at very rapid rate of -15 cm^{-1}/T. We study temperature dependence of the E_{2g} boron stretching phonon and conclude that the anharmonic decay is primarily responsible for the anomalously large damping constant of this mode. For this phonon we observe superconductivity induced self-energy effects from which we estimate the electron-phonon coupling $\lambda \approx 0.3$.

REFERENCES

1. S. Tsuda *et. al.*, *Phys. Rev. Lett.* **87**, 177006 (2001); S. Souma *et. al.*, *Nature* **423**, 65 (2003).
2. M. R. Esklidsen *et. al.*, *Phys. Rev. Lett.* **89**, 187003 (2002).
3. A. Perucchi *et. al.*, *Phys. Rev. Lett.* **89**, 097001 (2002).
4. A. Y. Liu, I. I. Mazin, and J. Kortus, *Phys. Rev. Lett.* **89**, 087005 (2001).
5. A. Shukla *et. al.*, *Phys. Rev. Lett.* **90**, 095506 (2003).
6. J. Kortus, *et. al.*, *Phys. Rev. Lett.* **86**, 4656 (2001).
7. R. Zeyher and G. Zwcknagl, *Z. Phys. B* **78**, 479 (2002).
8. C.O. Rodriguez *et. al.*, *Phys. Rev. B* **42**, 2692 (1990).

Superconductivity and Normal State In-Plane Resistivity of MgB_2: Screened Optical Phonon Approach

Dinesh Varshney[a], M. S. Azad[a], K. K. Choudhary[b], and R. K. Singh[c]

[a]*School of Physics, Vigyan Bhawan, Devi Ahilya University, Khandwa Road Campus, Indore-452017, India*
[b]*Department of Physics, S. V. I. T. S., Sanwer Road, Indore-453331, India*
[c]*Lakshmi Narain College of Technology, Raisen Road, Bhopal-462021, India*

Abstract. It is intended to trace the evolution of an effective two-dimensional dynamic interaction with s wave pairing embodying the screening of holes as carriers (σ band) by charge density fluctuations and by optical phonons leading to superconductivity in high-T_c MgB_2. Within the developed effective interaction potential the T_c of MgB_2 for σ (π) band electrons is estimated as 42 (19) K. The pressure derivative of T_c yields a negative value. Temperature dependence of resistivity of MgB_2 is then theoretically analyzed within the framework of electron–phonon i.e., Bloch–Gruneisen model. The estimated contribution to resistivity, when subtracted from single crystal data implies a quadratic temperature dependence over most of the temperature range.

Keywords: Borides, phonons, charge density fluctuations, resistivity, pressure effects.
PACS: 74.20.Mn, 74.25. -q, 74.25. Kc, 74.62. Fj

INTRODUCTION

The discovery of MgB_2 as a superconductor has attracted the scientific community due to the fact that its transition temperature T_c is higher and normal state properties are unusual compared to those previously known for the *A*15 inter-metallic compounds [1]. It is widely accepted that the B planes are conducting planes like CuO_2-planes in cuprates. Experimental probes like inelastic neutron scattering (INS) measurement [2] is a powerful probe related to vibrational modes in the crystal. INS results in an acoustic phonon of energy of about 36 meV and highly dispersive optic branches peaked at 54, 78, 89, and 97 meV.

The effect of boron substitution on the T_c of MgB_2 from magnetization measurement [3] was about 0.26 relative to the one-half BCS value. In subsequent work, the isotope effect for both B and Mg has been reported and found a B isotope effect of 0.30, and a small α_{Mg} of about 0.02 and points in a natural way towards a phonon based pairing.

Among non-conventional mechanisms, which have been invoked so far for MgB_2, collective charge fluctuations [4], electron-hole asymmetry [5] are important. Electron-energy-loss-spectroscopy (EELS) [6] has been successfully used to investigate the electronic structure. The charge fluctuation mechanism is emphasized in MgB_2 by its strong anisotropy and by the presence of a layer stacking sequence. Hence, it has become a key issue whether charge fluctuations can contribute to the pairing mechanism.

EFFECTIVE INTERACTION POTENTIAL

The effective interaction between each pair of holes in B planes is determined by screening of Coulomb repulsion via collective excitations in layered fluid of holes (plasmons) and in underlying ionic lattice (optical phonons). The effective interaction between a pair of carriers is [7] $V(q,\omega) = V_c(q)\varepsilon_{eff}^{-1}(q,\omega)$ to get:

$$\varepsilon_{eff}(q,\omega) = \varepsilon_\infty \frac{(\omega^2 - \Omega_1^2)(\omega^2 - \Omega_2^2)(\omega^2 - \Omega_3^2)}{(\omega^2 - A_1^2)(\omega^2 - A_2^2)(\omega^2 - A_3^2)} \quad (1)$$

Here, $\Omega_2^2 \cong \Omega_{ps,\pi}^2 + A_2^2$ is π electron 3D optical plasmon mode, $\Omega_3^2 \cong \Omega_{ps,\sigma}^2 + A_3^2$ is σ electron 2D acoustic plasmon mode in the long wavelength limit and screened optical phonon mode $\Omega_1^2 \cong \omega_{TO}^2$ in the adiabatic approximation. The evaluation of T_c requires information about hole-hole and hole-phonon coupling strengths. We define the renormalized Coulomb repulsive parameter $\mu^* = \mu[1+ \ln(\Omega_{2,3}/\omega_{TO})]^{-1}$. The

CP850, *Low Temperature Physics: 24th International Conference on Low Temperature Physics;*
edited by Y. Takano, S. P. Hershfield, S. O. Hill, P. J. Hirschfeld, and A. M. Goldman

Coulomb strength parameter is $\mu = N_{3D/2D}(\varepsilon_F)U$, where $N_{3D/2D}(\varepsilon_F)$ is the 3(2) dimensional density of states at Fermi energy and U is the static screened interaction averaged over Fermi sphere or circle

RESULTS AND DISCUSSION

In the calculation of the superconducting state parameters of MgB_2, the effective mass m^* of carriers is obtained as 3 (1) m_e from the jump ΔC in the specific heat at T_c and 3 (2) dimensional carrier density is $n \cong 22 \times 10^{21}$ cm^{-3} ($4 \times 10^{14} cm^{-2}$). The electron-phonon coupling strength $\lambda_{ph\sigma(\pi)} = 0.88$ (0.55). Other parameters of carriers are Fermi velocity $v_{F\sigma(\pi)} \cong 2.73$ ((9.0) $\times 10^7$cm/s, and $\varepsilon_{F\sigma(\pi)} \cong 0.63$ (2.2) eV. Turning to parameters related to ions, a reasonable value of the $\varepsilon_\infty \cong 3.5$ [6]. The optic-mode frequencies for an inverse-power r^{-s} overlap repulsion yields $\omega_{LO} \cong 89.4$ meV and $\omega_{TO} \cong 62.9$ meV. The estimated value of the static dielectric constant of lattice then is $\varepsilon_0 \cong 7$. The calculated value of the LO frequency is to be compared with measured [2] E_{2g} phonon mode of B.

Finally, we evaluate the $2k_F$ scattering of carriers on the Fermi sphere/circle. We find $\mu_{\sigma(\pi)} = 0.35$ (0.27) and $\mu^*_{\sigma(\pi)} = 0.19$ (0.14). In a true sense, the acoustic plasmon energy which is about 3.25 eV is considerably larger than $\varepsilon_{F\,\sigma} = 0.63$ eV and is logically inconsistent to ignore the plasmon energy to account for retardation effect. This is a constraint for all theories using pairing through excitations with energies exceeding or comparable to ε_F and is an unsolved problem.

For strong coupling we evaluate T_c following [8]

$$T_c^{\sigma(\pi)} = \frac{0.42\,\Omega_-^{\eta}\Omega_+^{\delta}}{(1.368)^{\eta}} \frac{\exp\left[\lambda_{pl}\{\lambda_{eff} + \lambda_{pl}\}^{-1}\right]}{[\exp(2/\lambda_{eff}) - 1]^{1/2}} \quad (2)$$

where λ_{eff} is effective coupling strength, $\eta = \lambda_{ph}/[\lambda_{ph} + \lambda_{pl}]$, and $\delta = \lambda_{pl}/[\lambda_{ph} + \lambda_{pl}]$. We find a $T_c^{\sigma(\pi)} \cong 42$ (19) K using λ_{pl} of about 0.1, in the range of measured values [1] T_c = 39 K. We define the isotope exponent $\alpha_\pm = 0.5 - \xi_\pm$, with In the adiabatic limit $[\Omega_{2,3} >> \omega_{TO}]$, We find ξ_+ (ξ_-) = 0.5 (0.0) and hence $\alpha_+ = 0.0$ and $\alpha_- = 0.5$. Foremost, the severely reduced value of α_{Mg} as earlier obtained [3] $\{\alpha_{Mg} \cong 0.02\}$ is predicted in a simple manner. The objective we have in mind is to see how we can retrace the sizeable α_B from the developed model. For resonance coupling ($\Omega_{2,3} \cong \omega_{LO}$), we find $\alpha_+ = \alpha_- = 0.25$, yielding an isotope exponent of 0.25, independent from the individual mode-coupling strengths. We thus comment that the estimates of α_B of about 0.25 agree well with the reported value [3] of 0.26.

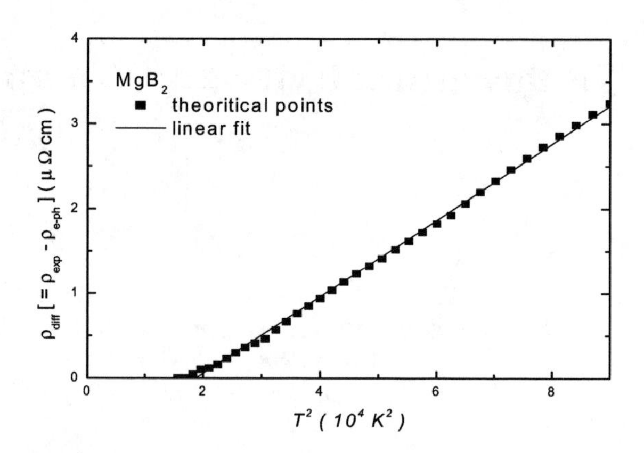

FIGURE 1. Variation of ρ_{diff} [= ρ_{exp} - (ρ_0 + $\rho_{e\text{-}ph}$ {ρ_{ac} + ρ_{op}})] (m Ω cm) with T^2 (10^4 K^2)

The analysis of the pressure dependence of T_c needs the understanding of the change in the electronic density of states, the phonon frequencies and the screening parameter with the pressure. We refer to Equation (2), of which the pressure derivative is obtained. The resulting values of the slope dT_c/dP of about −1.5 K/GPa agree well with that determined earlier [8] of −1.1 K/GPa. We follow Bloch-Gruneisen formula in estimating the electron-phonon contributions and notice that contribution from acoustic as well optical phonon together with the zero temperature-limited resistivity is smaller than the published data on the single crystal [10]. The difference between measured ρ and calculated ρ (= ρ_0 + $\rho_{e\text{-}ph}$) leads to a T^2 dependence and is interpreted in terms of electron-electron scattering (see Figure 1) in MgB_2.

ACKNOWLEDGMENTS

Financial assistance from DRDO, New Delhi and AICTE, New Delhi is gratefully acknowledged.

REFERENCES

1. Nagamatsu, J., *et al., Nature* **410**, 63 (2001).
2. Osborn, R., *et al., Phys., Rev. Lett.* **87**, 017005 (2001)
3. Budko, S. L., *et al., Phys. Rev. Lett.* **86**, 1877 (2001); Hinks, D. J., *et al., Nature* **411**, 457 (2001).
4. Sharapov, S. G., *et al., Eur. Phys. J. B.* **30**, 45 (2002).
5. Hirsch, J. E., *Phys., Lett.* **A 282**, 392 (2001).
6. Yu, R. C., *et al., Physica* **C 363**, 184 (2001).
7. Zeleny, V., *et al., Physica* **C 388/389**, 129 (2003).
8. Varshney, D., *et al., Sup. Sci. Tech.* **17**, 1 (2004).
9. Goncharov, A. F., and Struzhkin, V. V., *Physica* **C 385**, 117 (2003).
10. Sologubenko, A. V., *et al., Phys. Rev. B* **66**, 014504 (2002).

Ultraviolet Photoemission Spectroscopy Study of As-grown MgB_2 Film

Y. Harada*, Y. Nakanishi†, S. Tsuda**, T. Takahashi†, H. Iriuda†, M. Kuroha†, S. Shin** and M. Yoshizawa†

*Iwate Industy Promotion Center, Iiokashinden 3-35-2, Morioka, 020-0852, Japan
†Graduate School of Frontier Materials Function Engineering, Iwate University, Ueda 4-3-5, Morioka, 020-8551, Japan
**Institute for Solid State Physics, University of Tokyo, Kashiwa Chiba 277-8581, Japan

Abstract. The intermetallic superconductor MgB_2 is one of the promising candidates for use in superconducting electronic devices because of its high critical temperature of 39K. The application requires the developments of a high-quality film fabrication process. Recently we succeeded to deposit the as-grown MgB_2 films on MgO (100) substrates by using a molecular beam epitaxy (MBE) method. To evaluate the film property microscopically, the electric structure of the MgB_2 films was studied by means of ultraviolet photoemission spectroscopy (UPS). A clear peak was observed at 8-12 eV and 5-8eV binding energy corresponding to the B $2p_\sigma$ and $2p_\pi$ derived state. These features are in good agreement with those of bulk MgB_2 reported previously.

Keywords: MgB_2, MBE, UPS
PACS: 74.70.Ad, 74.78.-w, 78.40.-q

The intermetallic superconductor MgB_2 is one of the promising candidates for use in superconducting electronics device because of its high critical temperature (T_c) of 39K [1]. The application requires the developments of a high-quality film fabrication technique.

Recently we succeeded to deposit the as-grown MgB_2 films on MgO (100) substrate in the co-evaporation conditions of low growth temperature and low growth rate by using a molecular beam epitaxy (MBE) apparatus [2]. In this process as-grown technique was utilized, which can deposit the high-quality and very smooth MgB_2 films at relatively low substrate temperature (T_s) without after-annealing process. Thus this deposition technique is essentially of importance and available for the fabrication.

The photoemission study is one of the powerful tool to investigate the filled electric states directly. However, a few study have succeeded to investigate that of film [3] and those of bulk's [4, 5, 6, 7] apart from the high-resolution measurements of the superconducting gap [8, 9]. The main reason is the difficulty of having good samples for electron spectroscopies. This technique is surface sensitive and the surface of MgB_2 samples can be very different from the bulk, as well as unstable, due to the tendency of Mg to migrate toward the surface and to form oxides. In this study, we report the electric structure of as-grown MgB_2 film by means of ultraviolet photoemission spectroscopy (UPS).

These films for this study were deposited at 200°C on MgO (100) substrate without post-annealing process. The fabrication procedures for the as-grown MgB_2 films by MBE have been described elsewhere in detail [2]. The films thickness was fixed at 5000Å. Furthermore, thin Mg cap layer was deposited on the MgB_2 films for protection of the surface of the film to avoid oxidation. The structure and crystallinity were characterized by XRD. Fig.1 shows a 2θ-θ XRD scan of these MgB_2 film. Peaks from the (001) planes of MgB_2 were observed in addition to the peak from the substrate. This result implies these films shows c-axis oriented.

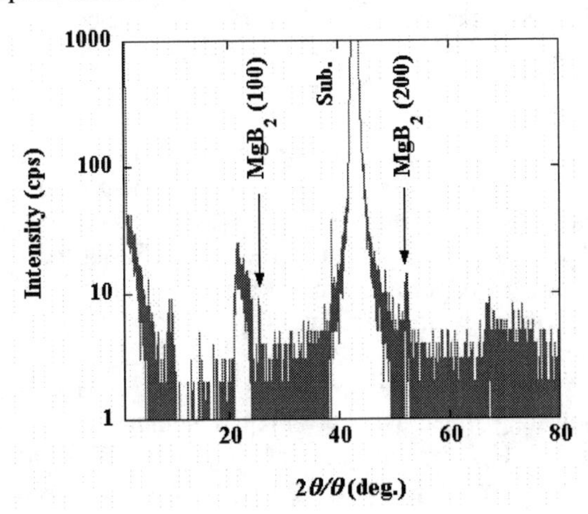

FIGURE 1. XRD pattern of MgB_2 film grown on MgO (100) substrate.

The superconducting properties of the films were studied by SQUID magnetometer. Fig.2 shows the zero-field-cooled (ZFC) and the field-cooled (FC) dc magnetization

CP850, *Low Temperature Physics: 24th International Conference on Low Temperature Physics;* edited by Y. Takano, S. P. Hershfield, S. O. Hill, P. J. Hirschfeld, and A. M. Goldman

(M-T) curves of this film in a 10 Oe field applied normal to the film surface. A clear diamagnetic transition was observed at 34K.

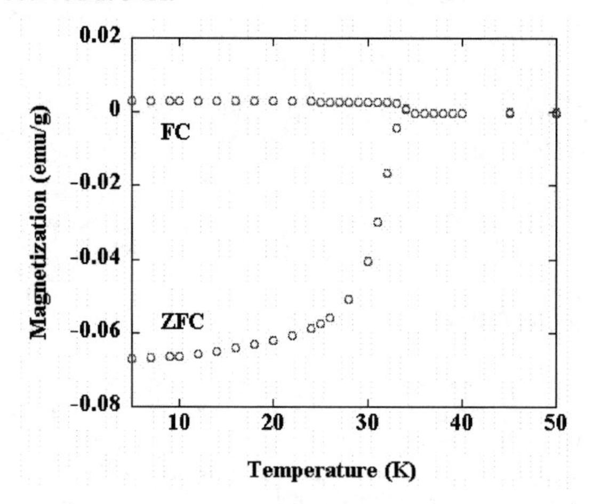

FIGURE 2. Dc magnetization curves (M-T) of MgB_2 film on MgO (100) substrate in the field of 10 Oe applied perpendicular to the films surface.

To evaluate the film property microscopically, the electric structure of the MgB_2 films was studied by means of UPS. The base pressure was 3.0×10^{-11} Torr and all measurements were performed at room temperature. To remove the Mg and MgO cap layers of the samples were cleaned *in situ* by Ar ion sputtering with the gun operating at 4keV and a vacuum of about 1.0×10^{-5} Torr for about 20 minutes. Fig.3 shows the valence-band spectra of MgB_2 film obtained by a He II (40.8 eV) resonance line at room temperature.

A clear peak was observed at 8 - 12 eV and 5 - 8 eV binding energy. These features are in good agreement with those of MgB_2 samples reported previously [3, 7]. We think the former corresponds to the B 2s derived state and the latter corresponds to the B $2p_\sigma$ and $2p_\pi$ derived state, comparing with band-structure calculations [7]. However, 8 - 12 eV binding energy was well known to be influenced by Oxygen, for example, leaving MgO that Mg cap was oxidized and diffusing from interface between MgO substrate and film. Then we need more detailed structural analysis, such as XRD, XPS and electron diffraction.

Moreover, the spectra do not show a clear Fermi edge, indicating the roughness or/and inhomogeneity of its surface. Probably, Ar ion sputtering damages the surface morphology and electric structure. From this result, a new technique how to keep the surface clean in *ex-situ* condition or how to remove the Mg and MgO cap layer completely to obtain clean surface.

We examined the electric state of as-grown MgB_2 film by means of UPS. A characteristics feature of UPS spectra as same as those of MgB_2 was observed. A clear peak was observed at 8 - 12 eV and 5 - 8 eV

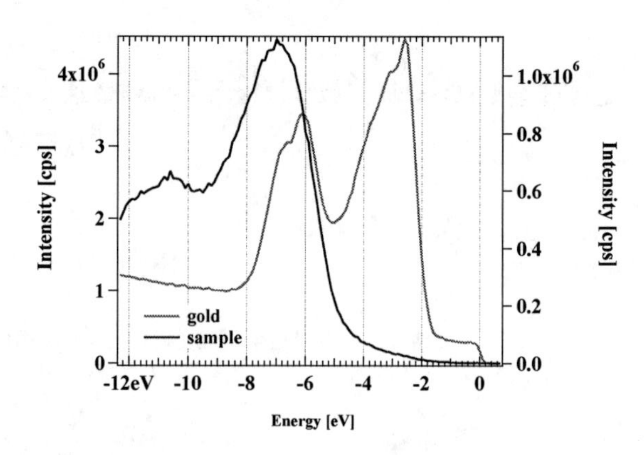

FIGURE 3. The valence-band spectra of MgB_2 film obtained by He II resonance line at room temperature.

binding energy. These features are in good agreement with those of MgB_2 samples reported previously. A lack of Fermi edge indicates that the surface of the present film is not sufficient for the present requirement of the measurement. A new technique how to keep the surface clean in *ex-situ* condition is needed.

REFERENCES

1. J. Nagamatsu, N. Nakagawa, T. Muranaka, Y. Zenitani, and J. Akimitsu, *Nature*, **410**, 63 (2001).
2. Y. Harada, M. Uzuka, Y. Nakanishi, and M. Yoshizawa, *Physica C*, **412**, 1383 (2004).
3. K. Ueda, H. Yamamoto, and M. Naito, *Physica C*, **378**, 225 (2002).
4. R. P. Vasquez, C. U. Jung, M.-S. Park, H.-J. Kim, J. Y. Kim, and S.-I. Lee, *Phys. Rev. B*, **64**, 052510 (2001).
5. H. Uchiyama, K. M. Shen, S. Lee, A. Damascelli, D. H. Lu, D. L. Feng, Z. X. Shen, and S. Tajima, *Phys. Rev. Lett.*, **88**, 157002 (2002).
6. A. Goldoni, R. Larciprete, S. Lizzit, S. La Rosa, A. Bianco, and M. Bertolo, *Phys. Rev. B*, **63**, 132503 (2002).
7. C. Jariwala, A. Chainani, S. Tsuda, T. Yokoya, S. Shin, Y. Takano, K. Togano, S. Otani, and H. Kito, *Phys. Rev. B*, **68**, 174506 (2003).
8. T. Takahashi, T. Sato, S. Souma, T. Muranaka, and J. Akimitsu, *Phys. Rev. Lett.*, **86**, 4915 (2001).
9. S. Tsuda, T. Yokoya, T. Kiss, Y. Takano, K. Togano, H. Kitou, H. Ihara, and S. Shin, *Phys. Rev. Lett.*, **87**, 177006 (2001).

Nonequilibrium Response of Superconducting MgB_2 Meander Line against Pulse Laser Irradiation

Takekazu Ishida [a,f], Daisuke Fujiwara [a,f], Shigehito Miki [a,f], Kazuo Satoh [b,f], Tsutomu Yotsuya [b,f], Hisashi Shimakage [c,f], Zhen Wang [c,f], Masahiko Machida [d,f], and Masaru Kato [e,f]

[a] *Department of Physics and Electronics, Osaka Prefecture University, 1-1 Gakuen-cho, Sakai, Osaka 599-8531, Japan*
[b] *Technology Research Institute of Osaka Prefecture, 2-7-1 Ayumino, Izumi, Osaka 594-1157, Japan*
[c] *Kansai Advanced Research Center, National Institute of Information and Communications Technology, 588-2 Iwaoka-cho, Nishi-ku, Kobe, Hyogo 651-2429, Japan*
[d] *CCSE, Japan Atomic Energy Research Institute, 6-9-3 Higashi-Ueno Taito-ku, Tokyo 110-0015, Japan*
[e] *Department of Mathematical Sciences, Osaka Prefecture University, 1-1 Gakuen-cho, Sakai, Osaka 599-8531, Japan*
[f] *JST-CREST, 4-1-8, Honcho, Kawaguchi, Saitama 332-0012, Japan*

Abstract. We performed 20-ps pulse laser irradiation experiments to a superconducting MgB_2 meander line to study a thermal-relaxation process for designing an MgB_2 neutron detector. We observed a thermal-relaxation signal resulting from pulse laser irradiation. The response time was faster than 1 μs, meaning that the detector would be capable of counting events at a rate of more than 10^6 events per second. The nonequilibrium process was realistically simulated by means of the time-dependent Ginzburg-Landau equation.

Keywords: nonequilibrium superconductivity; MgB_2 thin films; meander line; pulse laser; computer simulation
PACS: 74.40.+k; 74.70.Ad

INTRODUCTION

We proposed a superconducting detector using a high-quality ^{10}B-enriched MgB_2 thin film at higher operating temperatures [1]. This can be used as a neutron detector. The principle for detecting a nonequilibrium phenomenon is to observe a transient resistance change induced by the nuclear reaction of neutron and ^{10}B in MgB_2. The thermal energy causes a partial destruction of superconductivity, i.e., an electric resistance variation of the MgB_2 thin film. It is helpful to study the basic properties of an MgB_2 detector by means of pulse laser irradiation. It is also powerful to simulate a nonequilibrium process with the aid of the first principle calculations of dynamical superconducting properties.

In this paper, we report the fabrication method of a membrane-type structured MgB_2 device and a thermal-relaxation process by irradiating it with a 20-ps pulse laser as a thermal energy source. The combination of the time-dependent Ginzburg-Landau equation, the Maxwell equation, and the energy conservation law successfully reproduce this process.

EXPERIMENTAL

We first fabricated the front side of MgB_2 detectors on the 380-μm-thick Si substrates that had been coated with 500-nm-thick chemical vapor deposited (CVD) SiN films. The MgB_2 detector consisted of a 200-nm-thick MgB_2 thin-film meander line, a 300-nm-thick SiO protective layer, and 150-nm-thick Al electrodes. The fabrication process for the MgB_2 detector was already reported in our preceding publication [2]. After fabricating the MgB_2 device on the front side, anisotropic Si etching by ethylene diamine pyrocatechol (EDP) was applied on the backside to create a SiN membrane [3]. The critical temperature of the film was 27.5 ± 0.5 K.

A pulse laser with a pulse width of 20 ps, a wavelength of 1547 nm was used to simulate a nonequilibrium process caused by nuclear reaction. A single laser pulse has 4.1 MeV, of which the output was controlled by an erbium-doped fiber amplifier (EDFA) and a computer-controlled attenuator. We observed a clear transient signal by using a digital

CP850, *Low Temperature Physics: 24th International Conference on Low Temperature Physics;*
edited by Y. Takano, S. P. Hershfield, S. O. Hill, P. J. Hirschfeld, and A. M. Goldman

oscilloscope (frequency range: 1 GHz, max sample rate: 4 GS/s).

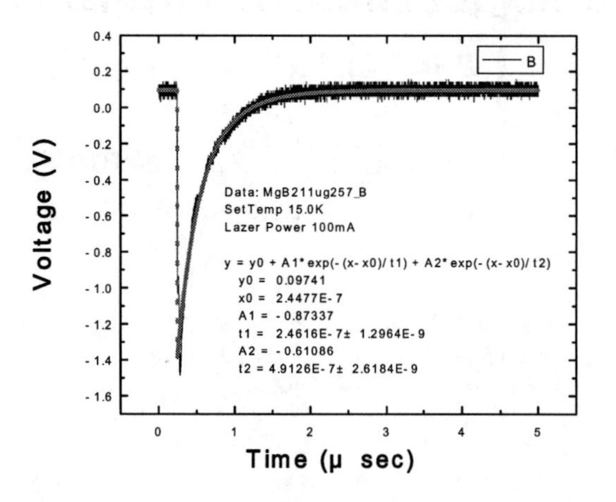

FIGURE 1. Nonequilibrium responses of the MgB_2 detector at 14 K. Data are fitted by the double-exponential model. .

In Fig. 1, we show typical thermal-relaxation signals from the MgB_2 devices. These results clearly indicate that these signals were related to the superconductivity of the MgB_2 detector and the pulse laser irradiation. We fitted the data with the double exponential model, and obtained relaxation times of 0.246 μs and 0.491 μs. This is very important because a response time of our detector is much faster than 1 μs, confirming an excellent performance of the detector for potential pulsed neutrons.

NUMERICAL SIMULATIONS

In order to simulate the nonequilibrium dynamics of the superconducting state caused by the pulse laser irradiation to MgB_2 strip line, we solve the time-dependent Ginzburg-Landau equation coupled with the Maxwell equation and the energy conservation law [3-5] to describe the dynamics of the superconducting order parameter and the vector potential, respectively. The energy conservation law takes care of quasiparticle behavior (energy dissipation) as the Joule heat and the heat diffusion both in superconducting and normal states. We track the time evolution of local temperature. The heat released by the pulse laser caused a hot spot region. In the present simulation, the superconducting order parameter Δ, the vector potential A, and the temperature T can simultaneously be computed so as to be able to describe the destruction of superconductivity and thermal diffusion of the hotspot. In Fig. 2, we show the snapshot of the sequential changes of the order parameter when the pulse laser is irradiated from the top side.

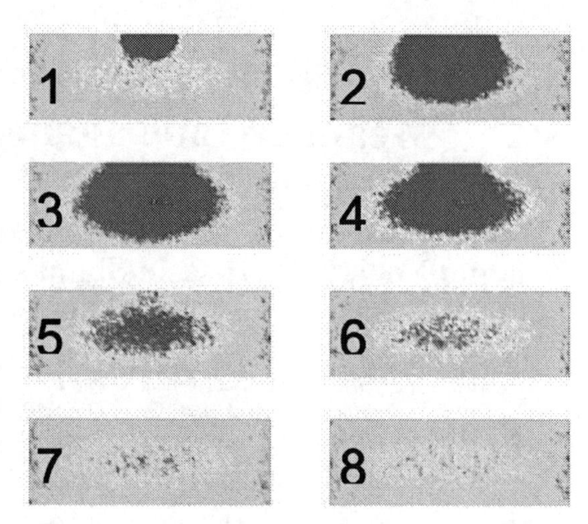

FIGURE 2. Sequential evolution (in $4\tau_0$ steps) of superconducting order parameter $|\Delta|$ in MgB_2 detector obtained by large-scale computer simulation. Here, τ_0 is an intrinsic relaxation time of elementary microscopic process, and is supposed to be 10^{-11} to 10^{-10} s. These display the hot-spot dynamics nucleated by the pulse laser irradiation to the MgB_2 meander line from the top.

CONCLUSION

We succeeded in observing clear output signals induced by pulse laser irradiation to MgB_2 detector. These nonequilibrium phenomenons are well reproduced by a computer simulation.

ACKNOWLEDGMENTS

This work was partly supported by a Gant-in-Aid for Scientific Research from the Ministry of Education, Culture, Sports, Science and Technology of Japan (Grant No. 16360477).

REFERENCES

1. K. Takahashi, K. Satoh, T. Yotsuya, S. Okayasu, K. Hojou, M. Katagiri, A. Saito, A. Kawakami, H. Shimakage, Z. Wang, and T. Ishida, *Physica C* **392-396**, 1501 (2003).
2. S. Miki, K. Takahashi, D. Fujiwara, H. Shimakage, Z. Wang, K. Satoh, T. Yotsuya, K. Moriwaki, H. Fukuda, S. Okayasu, M. Katagiri, Y. Morii, K. Hojou, N. Niimura, and T. Ishida, *Nucl. Instr. Meth. A* **529**, 405 (2004).
3. M. Machida and H. Kaburaki, *Phys. Rev. Lett.* **71**, 3206 (1993).
4. M. Machida and H. Kaburaki, *Phys. Rev. Lett.* **75**, 3178 (1995).
5. M. Machida, T. Koyama, M. Kato, and T. Ishida, *Nucl. Instr. Meth. A*, **529**, 409 (2004).

Influence of Al Doping on the Critical Fields in Magnesium Didoride Single Crystals

T. Klein*, C.Marcenat†, L. Lyard*, J. Marcus*, B. Kang**, H-J. Kim**, H-S. Lee** and S-I. Lee**

*Laboratoire d'Etudes des Propriétés Electroniques des Solides, Centre National de la Recherche Scientifique, B.P. 166, F-38042 Grenoble Cedex 9, France
†CEA - Grenoble, DRFMC-SPSMS, 17 rue des Martyrs, 38054 Grenoble Cedex 9,France
**NVCRICS and department of Physics, Pohang University of Science and Technology, Pohang 790-784 Republic of Korea

Abstract. The lower (H_{c1}) and upper (H_{c2}) critical fields of $Al_xMg_{1-x}B_2$ single crystals have been deduced from specific heat and magnetization measurements (with $x = 0$, 0.1 and 0.2). We show that H_{c1} and H_{c2} are both decreasing with increasing doping content. The corresponding anisotropy parameter $\Gamma_{H_{c2}}(0)$ value then also decreases from ~ 5 in pure MgB_2 samples down to ~ 1.6 for $x = 0.2$ whereas $\Gamma_{H_{c1}}(0)$ remains on the order of 1 in all measured samples.

Keywords: superconductivity, mixed state
PACS: 74.25.Dw, 74.25.OP, 74.70.Ad

INTRODUCTION

It is now well established that MgB_2 belongs to an original class of superconductors in which two weakly coupled bands with very different anisotropies coexist. This unique behaviour has been rapidly confirmed by spectroscopy [1] and specific heat [2] measurements which both revealed the existence of two distinct superconducting gaps. One of the main consequence of this two band superconductivity is a strong temperature dependence of the anisotropy of the upper critical field [3] $\Gamma_{H_{c2}} = H_{c2}^{ab}/H_{c2}^{c}$ (H_{c2}^{ab} and $/H_{c2}^{c}$ being the upper critical fields parallel to the ab-planes and c-axis, respectively). On the other hand, the lower critical field (H_{c1}) is almost isotropic at low temperature and the corresponding anisotropy $\Gamma_{H_{c1}} = H_{c1}^{c}/H_{c1}^{ab}$ slightly increases with temperature merging with $\Gamma_{H_{c2}}$ for $T \to T_c$ [4].

The influence of chemical doping and/or impurity scattering on the physical properties has then been widely addressed in both carbon ($Mg(B_{1-x}C_x)_2$) and aluminum ($Mg_{1-x}Al_xB_2$) doped samples. Whereas all studies agree on a significant increase of the upper critical field in both directions in C doped samples [5], no concensus was met in the latter system. Indeed, magnetization measurements in single crystals suggested an increase of $H_{c2}^{c}(0)$ [6] associated with a progressive change in the curvature of the $H_{c2}^{c}(T)$ line which has been attributed to a decrease of the diffusivity in the $p-$band (substitution on the Mg site is expected to have a greater impact on the diffusivity of the 3D $p-$ band then on the one of the $s-$orbitals which are localized in the boron planes). This effect has not been confirmed in polycrystalline samples [7] in which various measurements rather suggested a decrease of both H_{c2}^{c} and H_{c2}^{ab} with increasing doping. The influence of Al doping has then been related to the shift of the Fermi level and to the related weakening of the electron-phonon coupling [7]. As some uncertainty might be related to the determination of H_{c2} from either magnetic or transport measurements, we performed specific heat measurements in Al doped single crystals ($x = 0$, 0.1 and 0.2 in both directions). We confirm that $H_{c2}^{c}(0)$ and $H_{c2}^{ab}(0)$ both decrease with increasing x and that the correponding anisotropy $\Gamma_{H_{c2}}(0)$ also decreases (being equal to 5.6, 3.4 and 1.6 for $x = 0$, 0.1 and 0.2 respectively). We also deduced H_{c1} from local Hall probe magnetization measurements. As expected from the decrease of the carrier density with doping, H_{c1} decreases with x in both directions and $\Gamma_{H_{c1}}(0)$ remains on the order of 1 for all measured samples.

H_{C2} MEASUREMENTS

Specific heat measurements have been performed on small single crystals using an ac-technique for magnetic fields up to 8T both parallel and perpendicular for the $ab-$planes. For $x = 0$ and $x = 0.1$ well defined sharp specific heat jumps were observed for all fields allowing an unambiguous determination of H_{c2} (see for instance [3] for $x = 0$). For $x = 0$, H_{c2}^{ab} exeeds 8 T for $T < 20$ K and the low temperature values were deduced from high field magnetotransport data [3]. For $x = 0.2$ the zero field anomaly is much broader and led to a T_c value (~ 20 K) significantly lower than the one deduced from the onset

CP850, *Low Temperature Physics: 24th International Conference on Low Temperature Physics;*
edited by Y. Takano, S. P. Hershfield, S. O. Hill, P. J. Hirschfeld, and A. M. Goldman

of the diamagnetic signal ($\sim$ 24 K). This suggests that, for those high doping levels, the Al content of the surface is lower than that of the bulk [8]. Note however that complete screening is only obtained around 18 K which coincides with the mid-point of the specific heat jump. The H_{c2} values deduced from the onset of the specific heat anomaly anomaly, AC transmittivity measurements ($\sim$ 1% screening) and/or magnetotransport measurements have been reported on Fig.1 for $x = 0$, 0.1 and 0.2. As shown in Fig.1, the $H^c_{ab}(T)$ is nearly linear for $x = 0.2$ and did not reproduce the upward curvature previouly reported by Kang *et al.* [6]. The low temperature anisotropy decreases from ~ 5 in the pure sample to ~ 3 for $x = 0.1$ and finally ~ 1.6 for $x = 0.2$.

As discussed in ref [7], in the clean limit, the $H_{c2}(0)$ values are expected to be mainly determined by the parameters of the $s-$ band . The decrease of both $H_{c2}(0)$ values and $\Gamma_{H_{c2}}$ can then be, at least qualitatively, attributed to the decrease of the superconducting gap Δ_s as the reduction of the density of state due to electronic doping weakens the electron-phonon coupling. This decrease is only partially compensated by a decrease of the in-plane Fermi velocity (the Fermi velocity along the c-axis remaining approximatively constant) thus leading to a decrease of both $H^c_{c2} \propto (\Delta_s/v_F^{ab})^2$, $H^{ab}_{c2} \propto (\Delta_s)^2/v_F^{ab}v_F^c$ and $\Gamma_{H_{c2}} \propto v_F^{ab}/v_F^c$. The influence of both inter and intra band scattering however still has to be studied in details

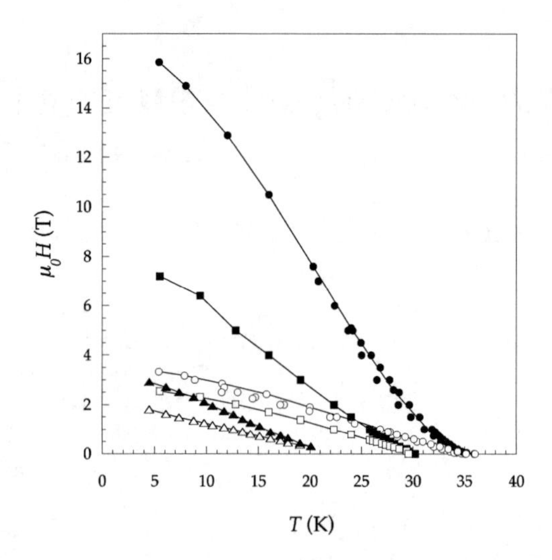

FIGURE 1. Temperature dependence of the upper critical field for $H//c$ (open symbols) and $H//ab$ (closed symbols) in $(Mg_{1-x}Al_x)B_2$ single crystals with $x = 0$ (circles), $x = 0.1$ (squares) and $x = 0.2$ triangles

H_{C1} MEASUREMENTS

The first penetration field (H_p) has been defined as the field for which a non zero magnetic field is detected by a miniature Hall probe located close to the center of the sample (as shown in [4] this value is not affected by the position of the probe due the absence of significant bulk pinning). The H_{c1} values have then been deduced from H_p taking into account the influence of geometrical barriers as discussed in [4]. The corresponding values have been reported in Fig.2 together with the H_{c2} values. As shown H_{c1} decreases with increasing x reflecting the decrease of the superfluid density as the bands are progressively filled up by electronic doping. The anisotropy parameter here remains on the order of 1 in all three samples suggesting that the average of the squared Fermi velocity over the entire Fermi surface (i.e. taking into account both bands) is independent on doping.

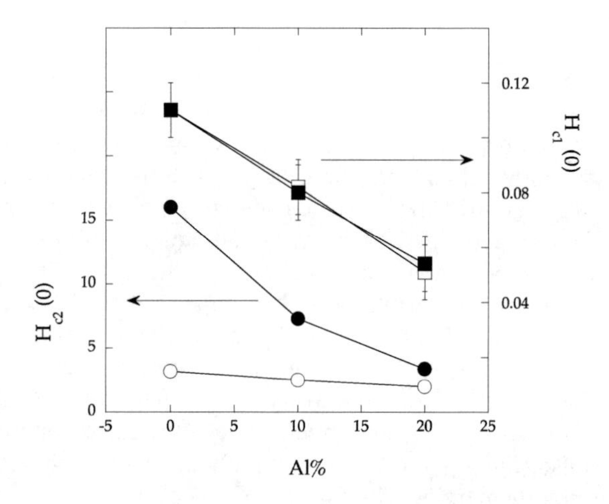

FIGURE 2. H_{c2} (circles) and H_{c1} (squares) value as a function for x for $H//c$ (open symbols) and $H//ab$ (closed symbols)

REFERENCES

1. P. Szabo *et al. Phys. Rev. Lett.*, **87**, 137005 (2001).
2. F. Bouquet *et al. Phys. Rev. Lett.*, **87**, 047001 (2001).
3. see for instance L. Lyard *et al. Phys. Rev. B*, **66**, 180502(R) (2001).
4. L. Lyard *et al. Phys. Rev. Lett.*, **92**, 057001 (2004); L. Lyard *et al. Phys. Rev. B*, **70**, 180504(R)(2004).
5. R.R. Ribeiro *et al. Physica C*, **384**, 227 (2003); Z. Holanova *et al. Physica C*, **404**, 195 (2004); R.H. Wilke *et al. Phys. Rev. Lett.*, **92**, 217003 (2004).
6. B. Kang *et al.* cond-mat/0409496 (unpublished).
7. M. Putti *et al. Phys. Rev. B*, **71**, 144505 (2005); M. Angst *et al. Phys. Rev. B*, **71**, 144512 (2001).
8. Note that T_c values on the order of 24 K (from magnetic measurements) have also been reported in polycrystals [7] for $x = 0.2$.

Development of Two Superconducting Energy Gaps in the Aluminum Doped MgB_2

P. Samuely*, P. Szabó*, Z. Hoľanová*, S. Bud'ko#, M. Angst#, and P. C. Canfield#

*Centre of Low Temperature Physics, IEP Slovak Academy of Sciences & P.J.Šafárik University, Watsonova 47, SK-04353 Košice, Slovakia

#Ames Laboratory and Iowa State University, Ames, IA 50011, USA

Abstract. Point-contact spectroscopy measurements on the aluminum substituted $Mg_{1-x}Al_xB_2$ samples reveal a retention of two superconducting energy gaps in the doping range for x = 0, 0.1 and 0.2. T_c's of those samples are decreased proportionally to the electron doping from 39 K down to 24 K. The large gap on the σ-band is decreased proportionally with decreasing T_c. For 10 % Al doped sample MgB_2 the small gap is almost unchanged compared to MgB_2, but for 20 % Al with T_c = 24 K it is reduced by approximately 15 %. The results are compared with those obtained on carbon doped MgB_2. A larger tendency to approach two gaps is found for Al doped MgB_2 but the merging of the gaps and the single gap superconductivity is still not realized in the studied doping range.

Keywords: MgB_2, two band/two gap superconductivity
PACS: 74.50.+r, 74.70.Ad, 74.62.Dh

Two different energy gaps are the key element of the unusual superconductivity in magnesium diboride, the system with the highest transition temperature of 40 K among intermetalics [1]. Strongly coupled quasiparticles in the σ band with a large energy gap only weakly interact with the weakly coupled π-band quasiparticles with a small gap. Substitutional studies started early after discovery of superconductivity in MgB_2, but the only proved *on site* substitutions are aluminum for magnesium and carbon for boron. An important issue which can be affected by partial substitution is the scattering. Significant *interband* scattering will lead to averaging of two gaps (decreasing Δ_σ and increasing Δ_π), canceling the interference effect responsible for high T_c which will drop to ~ 25 K [2]. In our previous studies on carbon doped MgB_2 [3] we have shown that in spite of the strong decrease of T_c = 22 K in the heavily doped systems (10 % C), two gaps are preserved, being decreased proportionally to T_c. Erwin and Mazin [4] have shown theoretically that due to different spatial symmetry of σ and π orbitals it is difficult to increase the interband scattering and that only out-of-plane distortions could make it. Since both orbitals reside in the boron plane, carbon replacing boron is ineffective in interband scattering. A different situation could happen when magnesium is replaced by aluminum. We show here by the point-contact measurements that two gaps approach more in Al-doped MgB_2 but merging is not accomplished up to 20 % Al doping.

Al-doped polycrystalline MgB_2 samples have been prepared by a two step synthesis at high temperatures with a special care to avoid as much as possible the sample inhomogeneities [5]. Powder x-ray measurements show no phase separation. Shifts in the lattice constants are close to literature values and a moderate peak broadening indicates a little variation of Al content throughout the samples. Magnetization curves reveal narrow transitions with T_c = 39, 31 and 25 K for the undoped, 10 % Al and 20 % Al MgB_2 samples [5]. Point-contact spectroscopy has been performed in a way described in Ref. 3. The point-contact spectrum - differential conductance versus voltage between a metallic tip and a superconductor can be compared with the Blonder, Tinkham and Klapwijk (BTK) theory using as input parameters the energy gap Δ, the barrier parameter z and a parameter Γ for the quasi-particle lifetime broadening. In the case of two-gap superconductor the overall conductance can be expressed as a weighted sum of two partial BTK conductances from the quasi two-dimensional σ band (with a large gap Δ_σ) and the 3D π band (with Δ_π). The weight factor α for the π-band contribution can vary from 0.65 for the tunneling/point-contact current strictly in the MgB_2 *ab*-plane to 0.99 of the *c*-axis current injection.

CP850, *Low Temperature Physics: 24th International Conference on Low Temperature Physics;* edited by Y. Takano, S. P. Hershfield, S. O. Hill, P. J. Hirschfeld, and A. M. Goldman

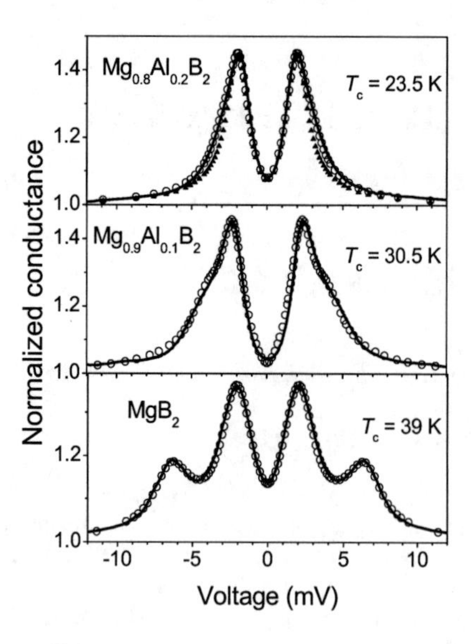

FIGURE 1. Point-contact spectra on Al-doped MgB_2 taken at 4.2 K, except for middle spectrum measured at 1.5 K. Solid lines are the data, open circles – the two-gap fits and solid triangles in the top part – the single-gap fit.

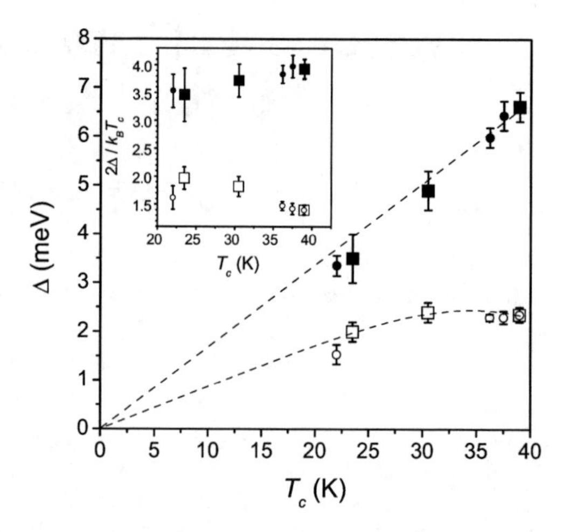

FIGURE 2. Energy of the gaps as a function of T_c in Al-doped MgB_2 – squares and C-doped MgB_2 – circles. Lines are guides to the eye.

In Fig. 1 representative point-contact spectra normalized to the conductance in the normal state, taken on samples with all three Al concentrations are presented. The zero- and 10% Al-doped samples clearly show two-gap spectra. These are not singular cases but several spectra with a contribution from both bands and small broadening parameter (Γ_i<0.2Δ_i) enabling to distinguish two gaps, have been collected. This is the first time to our knowledge that for the MgB_2 doped samples with T_c as low as 30 K two gaps are well revealed directly in the raw point contact/tunneling data. In the case of the 20 % Al-doped MgB_2 sample due to the close position of two gaps only peaks from the small gap and humps from the large gap at plus/minus 3.5 mV are visible. Fits to the two-gap BTK formula are shown by open symbols. Closed triangles in the upper part is a fit to a single gap BTK formula which is apparently failing. A behavior of the spectra in magnetic field which gives more evidence for two gap superconductivity in Al doped MgB_2 will be given elsewhere.

Fig. 2 summarizes the development of the superconducting energy gaps obtained from fitting the data shown in Fig. 1 and other data as a function of T_c. For a comparison also the gaps of the carbon-doped samples are indicated [3]. For both dopings the large gap in the σ band shows a linear decrease proportional to a decreased T_c. The small gap Δ_π is essentially the same for pure MgB_2 and 10 % Al-doped sample, then for 20 % Al-doped system it is suppressed by about 15 per cent. Again it is a qualitatively similar to the behavior of the carbon doped MgB_2, with a stronger tendency to approach two gaps in Al-doped MgB_2 (see also the reduced gaps in the inset of Fig. 2). Thus, even if the interband scattering is increased by 20 % Al doping, it is still not sufficient to merge two gaps.

Other measurements [6,7] of the gaps in Al-doped MgB_2 report similar decrease of the large gap with decreased T_c, but there is less agreement for the small gap development and an approaching of the gaps. Beside the scattering effect, both Al and C substitutions introduce the electron doping, resulting to a decrease of the σ-band hole DOS, and consequently to a decrease of T_c and both gaps. The recent theoretical model of Kortus *et al.* [8] considers both – the band filling and the interband scattering. The presented data could be used for a more quantitative analysis of the interplay of these two effects in MgB_2.

In conclusion, the development of two energy gaps upon Al doping in MgB_2 has been determined by point-contact spectroscopy showing approaching of two gaps but no merging up to 20 % Al doping.

The work was supported by the grant APVT 51-016604 and by the US Steel Košice, CLTP is operated as the Centre of Excellence of SAS.

REFERENCES

1. Review *Superconductivity in MgB_2*, eds. W. Kwok *et al.* [*Physica C* **385**, 1 (2003)].
2. A.Y. Liu, I.I. Mazin, and J. Kortus, *Phys. Rev. Lett.* **87**, 087005 (2001).
3. Z. Hoľanová *et al.*, *Phys. Rev. B* **70**, 064520 (2004).
4. S.C. Erwin and I.I. Mazin, *Phys. Rev. B* **68**, 132505 (2003).
5. M. Angst *et al.*, *Phys. Rev. B.* **71**, 144512 (2005).
6. R.S. Gonnelli *et al.*, cond-mat/0407267.
7. M. Putti *et al.*, *Phys. Rev. B.* **71**, 144505 (2005).
8. J. Kortus *et al.*, *Phys. Rev. Lett.* **94**, 027002 (2005).

π-band Goes Dirty by Carbon Doping in MgB_2?

P. Szabó*, P. Samuely,* Z. Hoľanová*, S. Bud'ko#, P. C. Canfield#, and J. Marcus$

*Centre of Low Temperature Physics, IEP Slovak Academy of Sciences & P.J.Šafárik University, Watsonova 47, SK-04353 Košice, Slovakia
#Ames Laboratory and Iowa State University, Ames, IA 50011, USA
$LEPES CNRS, Grenoble, France

Abstract. Point-contact measurements in a magnetic field on the pure and heavily carbon-doped MgB_2 samples are presented. Analysis of the magnetic-field behavior of the spectra indicates that the π band becomes dirty upon carbon doping.

Keywords: : MgB_2, two band/two gap superconductivity
PACS: 74.50.+r, 74.70.Ad, 74.62.Dh

Superconductivity in MgB_2 at $T_c \sim 40$ K represents a promising application potential. Prospects of applications depend on a success in increasing the values of the upper critical field H_{c2} and the critical current J_c. For that the material must be driven from the clean superconducting limit to the dirty one. This can be realized for example via substitution of the constituting elements [1,2]. An undesirable side effect in the case of a two-band/two-gap superconductor would be a significant increase of the interband scattering, leading to a merging of two gaps [3]. In our recent papers we have shown [4] that even in the heavily carbon-doped sample $Mg(B_{0.9}C_{0.1})_2$ both gaps are present and the interband scattering is not significantly changed.

In the present paper an effect of the applied magnetic field on the point-contact (PC) spectra is used for study of the intraband scatterings in the pure and heavily carbon-doped MgB_2. Experiments have been performed on the polycrystalline samples of MgB_2 with $T_c = 39$ K and $Mg(B_{0.9}C_{0.1})_2$ with $T_c = 22$ K. Experimental details can be found elsewhere [1,2,4].

The PC spectra measured between a normal metal M and a superconductor consists of the Andreev reflection (AR) contribution and the tunneling contribution. At $T = 0$ due to the excess current of AR the conductance inside the gap is twice higher than the value outside. The tunnelling contribution yields a peak at the gap edge and a reduced conductance inside. The Blonder-Tinkham-Klpawijk (BTK) theory interpolates between these two limits with an arbitrary barrier strength z as a parameter. When in an applied magnetic field the superconductor is in the mixed state, the excess current $I_{exc} \sim \Delta$ will be reduced by a factor of $(1-n)$, where n represents the normal state region created by the vortices at the contact area. Thus I_{exc} for the two-gap superconductor in the mixed state can be described by the sum $I_{exc}(H)=\alpha(1-n_\pi)\Delta_\pi + (1-\alpha)(1-n_\sigma)\Delta_\sigma$, where α is the weight of the π-band contribution and Δ_π and Δ_σ are the gap values of the π and σ bands, respectively. Considering that n_π and n_σ represent the separate filling of each bands with increasing applied field, their values can be identified with the values of the zero-bias DOS $N_\pi(0,H)$ and $N_\sigma(0,H)$. This model proposed in Refs. 5 and 6, enables a direct experimental determination of $N_\pi(0,H)$ in MgB_2. Using the theoretical calculations of Koshelev and Golubov [7] the field dependence of $N_\pi(0,B)$ can give a direct information about the ratio of the in-plane diffusivities in both bands D_σ/D_π.

Fig. 1 shows magnetic field dependencies of the normalized PC spectra obtained on the point-contacts of M-MgB_2 (left panel) and M-$Mg(B_{0.9}C_{0.1})_2$ (right panel). The zero-field spectrum of MgB_2 reveals a clear two-gap structure. At finite fields the peak corresponding to Δ_π is very rapidly suppressed. On the other hand, the zero field $Mg(B_{0.9}C_{0.1})_2$ spectrum shows only one pair of smeared maxima at Δ_π, but the different field sensitivity of each gap allows us to identify the large gap Δ_σ, too (see the humps at plus/minus 3 mV at 0.5 T).

CP850, *Low Temperature Physics: 24th International Conference on Low Temperature Physics;*
edited by Y. Takano, S. P. Hershfield, S. O. Hill, P. J. Hirschfeld, and A. M. Goldman

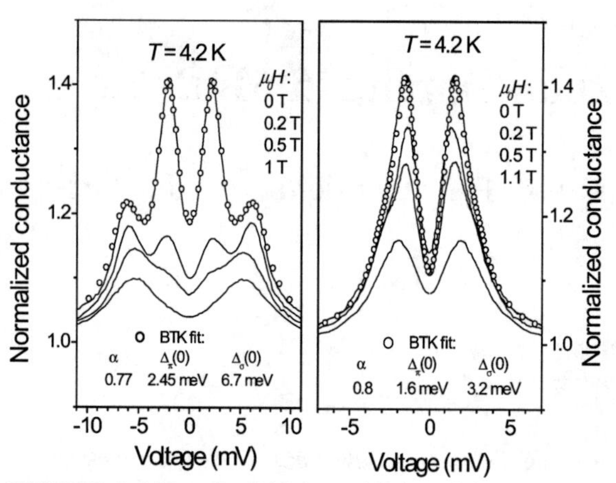

FIGURE 1. Magnetic field dependencies of the normalized PC spectra obtained on M-MgB_2 (left panel) and M-$Mg(B_{0.9}C_{0.1})_2$ junctions (right panel).

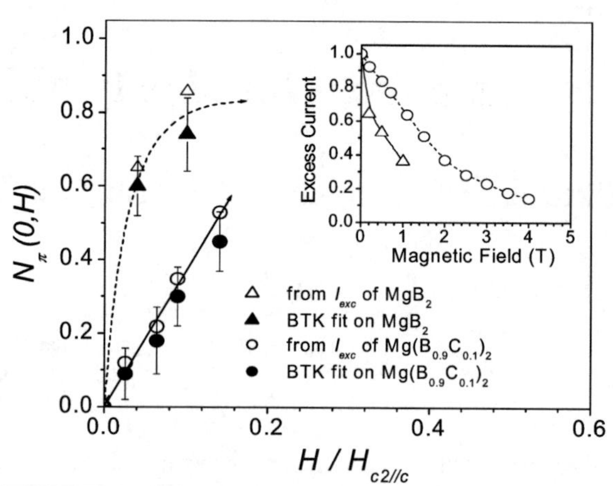

FIGURE 2. Field dependencies of the the zero-bias DOS $N_\pi(0,H)$ for samples MgB_2 (triangles) and $Mg(B_{0.9}C_{0.1})$ (circles). Inset – the field dependencies of the excess current determined from PC spectra.

The effect of an applied field on the excess currents $I_{exc}(H)$ of the junctions made on MgB_2 and $Mg(B_{0.9}C_{0.1})_2$ is shown in the inset of Fig. 2. I_{exc}'s have been calculated by integrating the normalized PC spectra (shown in Fig. 1) after subtracting an area below unity. Both $I_{exc}(H)$ dependencies reveal a two-gap behavior: a rapid quasi-linear fall at low magnetic fields is followed by a slower decrease at higher fields. This crossover defines the field H^*, below which the superconductivity is mostly determined by the 3D π-band. At higher fields the 2D σ–band dominates. If H_{c2} is large enough as compared to H^*, in the low field region $H<H^*$, the suppression of the σ-band contribution by the field can be neglected ($N_\sigma(0) = 0$). If we also neglect a small reduction of the energy gaps at these fields [8], then $I_{exc}(H)$ can be expressed as $I_{exc} = \alpha(1-n_\pi)\Delta_\pi(0) + (1-\alpha)\Delta_\sigma(0)$. This enables a simple estimate of $N_\pi(0)$'s in both samples at $H<H^*$ since the parameters α, $\Delta_\pi(0)$ and $\Delta_\sigma(0)$ have already been determined from the BTK fits at $H = 0$ (shown by circles in Fig.1). The resulting $N_\pi(0,H)$ for both samples are shown in the main panel of Fig. 2 by open symbols. The field scale for MgB_2 has been normalized to the value $H_{c2//c} = 5$ T (minimal upper critical field of our MgB_2 sample) and for $Mg(B_{0.9}C_{0.1})_2$ to the value $H_{c2//c} = 7.8$ T [2].

The correctness of this $N_\pi(0,H)$ estimate can be verified by the fitting of the normalized conductance spectrum to the two-band BTK formula in the mixed state [6], in which also the presence of the vortices is incorporated as $G=\alpha[N_\pi+(1-N_\pi)g_\pi]+ (1-\alpha)[N_\sigma+(1-N_\sigma) g_\sigma]$, with partial conductances $g_{\pi,\sigma}$. Here, all parameters are considered but they are so many that the previously determined N_π must be used as a guide. More details about the fitting procedure will be published elsewhere. The values of N_π determined from the BTK fits are shown in Fig. 2 by filled symbols. One can see a good agreement between these two determinations.

The steep increase of $N_\pi(0,H)$ in the pure MgB_2 can be ascribed within the framework of the Koshelev's model [7] to a small value of the ratio of the in-plane diffusivities $D_\sigma/D_\pi < 1$. It means a dirtier σ band than the π one in MgB_2. On the other hand the slower increase of $N_\pi(0)$ in $Mg(B_{0.9}C_{0.1})_2$ is indicating an increase of D_σ/D_π suggesting that the diffusivity in the π band decreases faster, than D_σ upon carbon doping. A similar conclusion about the dirtier π band in the carbon-doped MgB_2 has been recently drawn by Angst *et al.* [1], when $H_{c2}(T)$ dependencies have been analysed.

In conclusion, the analysis of the in-magnetic field behavior of the point-contact spectra indicates that the π-band becomes dirty upon carbon doping in MgB_2.

The work was supported by the Science and Technology Agency, the contract No. APVT-51-0166 and by the US Steel Košice. CLTP is operated as the Centre of Excellence of SAS.

REFERENCES

1. M. Angst, et al., *Phys. Rev. B* **71**, 144512 (2005).
2. A. Ribeiro et al., *Physica C* **384**, 227 (2003).
3. A. Y. Liu, et al., *Phys. Rev. Lett.* **87**, 087005 (2001).
4. P. Samuely et al., *Phys. Rev. B* **68**, 020505(R) (2003); Z. Hoľanová et al., *Phys. Rev. B* **70**, 064520.
5. Yu.G. Naidyuk et al., *cond-mat/0403324* (2004).
6. Y. Bugoslavsky et al., *cond-mat/*0502153 (2005).
7. A.E. Koshelev and A.A. Golubov, *Phys. Rev. Lett.* **90**, 177002 (2004).
8. Y. Bugoslavsky et al., *Phys. Rev. B* **69**, 132508 (2004).

Specific Heats of $Mg(B_{1-x}C_x)_2$: Two-Gap Superconductors

R. A. Fisher, N. Oeschler, N. E. Phillips, W. E. Mickelson, and A. Zettl

LBNL and Departments of Chemistry and Physics, University of California, Berkeley, CA 94720, USA

Abstract. Specific heats (C) of two polycrystalline samples of $Mg(B_{1-x}C_x)_2$ (nominal x = 0.1 and 0.2) were measured in magnetic fields (B) to 9 T and temperatures (T) from 1 to 40 K. Neither sample shows evidence of magnetic impurities nor do they have large non-superconducting fractions. The transition temperature (T_c) decrease monotonically with x, and the superconducting specific heats can be well fitted with two energy gaps (α) using the phenomenological α model.

Keywords: $Mg(B_{1-x}C_x)_2$, specific heat, two-gap superconductivity
PACS: 74.25.Bt; 74.60.Ec; 75.40.Cx

MgB_2 is an anisotropic superconductor with a transition temperature of ~39 K. Carbon is the only element that has been substituted for boron. The amount is not easily determined and the relation of x and T_c for $Mg(B_{1-x}C_x)_2$ samples is in dispute [1].

Polycrystalline samples were synthesized [2,3] with nominal x's of 0.1 and 0.2 for measurements of C from 1 to 40 K in magnetic fields to 9 T. The T_c's are 31.4 and 19.7 K, respectively. They have small residual Sommerfeld constants (γ_r) and a 90% or greater superconducting fraction (f_s)—see Table 1.

The electronic specific heat (C_e) is determined as follows: First, the low-temperature data from 1 to 4 K are globally fitted to $C(B) = D(B)/T^2 + \gamma_v(B)T + B_3T^3 + a(B)e^{-[b(B)/T]}$, where $D(B)/T^2$ represent the hyperfine components, $\gamma_v(B)$ the Sommerfeld coefficients for the mixed/vortex state EDOS, B_3T^3 the field-independent lattice term (the global variable), and the exponential terms are the B-dependent expressions for the low-T superconducting contributions. [For B = 0, $\gamma_v(0) \equiv \gamma_r$ and is assumed to be associated with normal material.]

Table 1. Parameters that characterize $Mg(B_{1-x}C_x)_2$.

x	0	0.1	0.2
T_c (K)	38.9	31.4	19.7
$f_s = (1 - \gamma_r/\gamma)$	0.99	0.94	0.90
γ (mJ K^{-2} mol^{-1})	2.53	1.26	1.43
Θ_D (K)	965	631	586
B_n (T)	~17	~13	~11
γ_1/γ	0.45	0.55	0.72
$\alpha_1 = \Delta_1(0)/k_BT_c$	0.6	0.8	0.9
γ_2/γ	0.55	0.45	0.28
$\alpha_2 = \Delta_2(0)/k_BT_c$	2.2	2.5	2.0

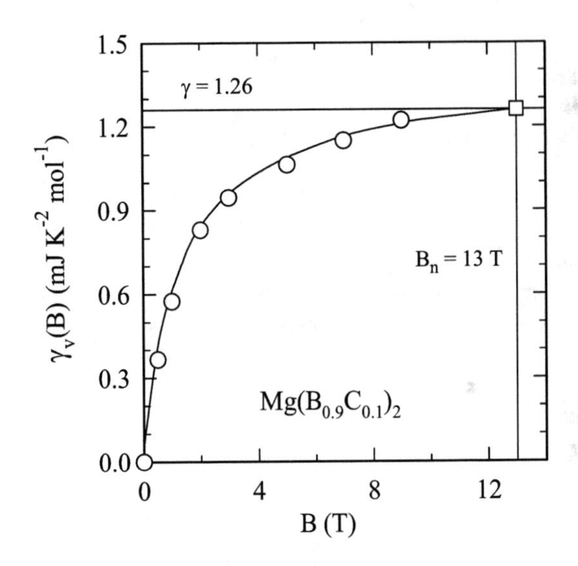

Figure 1. Evolution of γ_v with B for $Mg(B_{0.9}C_{0.1})_2$.

The fitted parameters are used to extrapolate C(B) to T = 0. Second, C-data for $T > T_c(B)$ for B = 0 and 9T are globally fitted to $C(B) = \gamma T + \sum B_nT^n$, where the sum B_nT_n represents the harmonic-lattice specific heat (C_{lat}) with n = 3, 5, 7 ··· for as many terms as necessary to achieve a good fit. B_3 is constrained to the value obtained from the low-T fit and γ is evaluated by requiring that $\int C(0)/TdT$ from T = 0 to T_c be equal to γT_c—the third law requirement. The electronic specific heat, uncorrected for the normal component, is given by $C_e'(B) = C(B) - [D(B)/T^2 + C_{lat}]$. It is

CP850, *Low Temperature Physics: 24th International Conference on Low Temperature Physics;*
edited by Y. Takano, S. P. Hershfield, S. O. Hill, P. J. Hirschfeld, and A. M. Goldman

normalized to $C_e(B)$, for one mole of superconducting material, using γ_r: $C_e(B) = [C_e'(B) - \gamma_r T]/(1 - \gamma_r/\gamma)$. Normalized $\gamma_v(B)$'s are plotted *vs* B in Figures 1 and 2

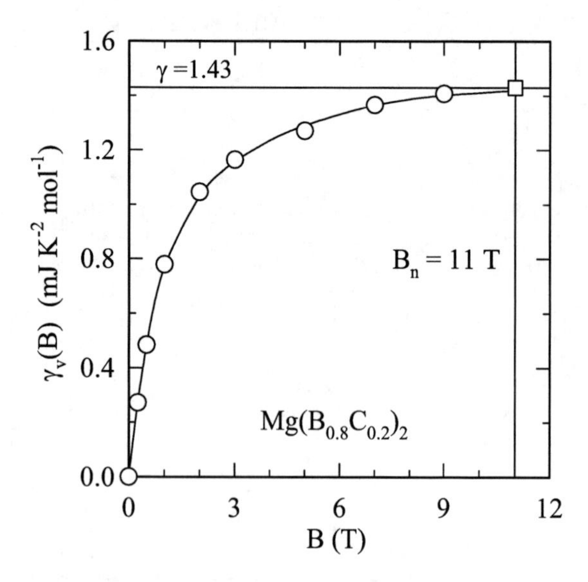

Figure 2. Evolution of γ_v with B for $Mg(B_{0.8}C_{0.2})_2$.

and extrapolated to γ at B_n where the samples are normal. The B_n decrease with increasing x and, except for MgB_2, are substantially less than $B_{c2}(0)$ determined by magnetization and resistivity [3], which may indicate the presence of filamentary superconductivity.

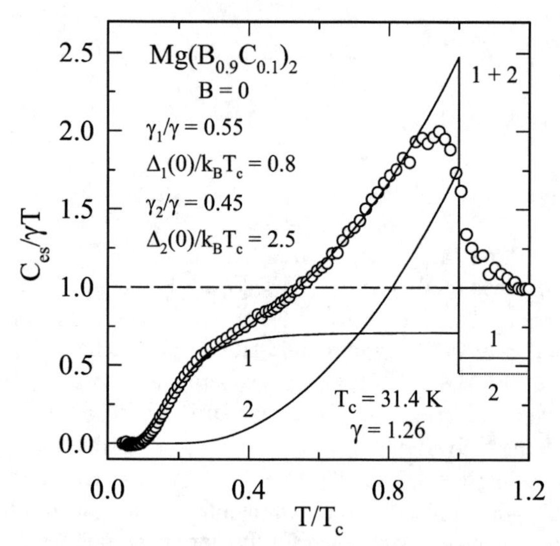

Figure 3. C_{es}/T *vs* T/T_c for $Mg(B_{0.9}C_{0.1})_2$.

Table 1 lists various parameters for x = 0, 0.1, and 0.2. The lattice softens (Θ_D decreases) as x increases and γ (the EDOS) decreases relative to MgB_2.

Figures 3 and 4 are scaled plots of $C_{es}/\gamma T$ *vs* T/T_c for B = 0, where C_{es} is the superconducting-state specific heat. C_{es} for x = 0.2 are similar to those previously published [4]. Two-gaps are required to fit

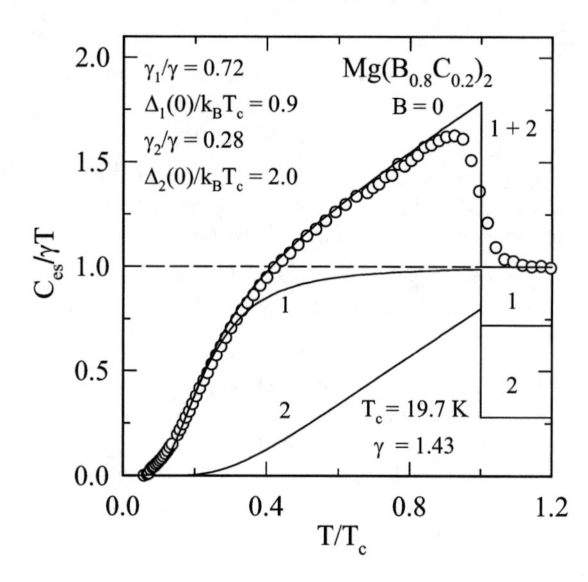

Figure 4. C_{es}/T *vs* T/T_c for $Mg(B_{0.8}C_{0.2})_2$.

C_{es} using the modified alpha model [5,6]. The gap components and parameters are given in the figures and listed in Table 1. Sizes of α_1 and α_2 vary with x for the three samples, but changes for the smaller gap are greater than those for the larger one. The fractional contribution of the larger gap decreases as x increases, and the ratio, α_2/α_1, decreases as x increases signaling their convergence. In all cases one gap (α_2) is larger than the BCS weak-coupling value and the other (α_1) smaller as theoretically predicted [7]. At T = 0 the fitted α values for all x are in reasonable agreement with those determined from point-contact spectroscopy [8] to within their combined experimental accuracy.

REFERENCES

1. S. M. Kazakov *et al.*, *Phys. Rev. B* **71**, 24533 (2005).
2. W. Mickelson *et al.*, *Phys. Rev. B* **65**, 52505 (2002).
3. M. Angst *et al.*, *Phys. Rev. B* **71**, 144512 (2005).
4. R. A. Riberio *et al.*, *Physica C* **384**, 227 (2003).
5. F. Bouquet *et al.*, *Phys. Rev. Lett.* **87**, 47001 (2001).
6. R. A. Fisher *et al.*, *Physica C* **385**, 180 (2003).
7. V. Z. Kresin and S. A. Wolf, *Physica C* **169**, 476 (1990).
8. Z. Hol'anová et al, *Phys. Rev. B* **70**, 64520 (2004).

Superconductivity in Porous MgB_2

V. Grinenko [*], E. Krasnoperov [*], V. Stoliarov [*], A. Bush [#], and B. Mikhajlov [+]

[*] *RRC Kurchatov Institute, 123182 Moscow, Russia*
[#] *MIREA, 119454, Moscow, Russia*
[+] *IMET, 119991, Moscow, Russia*

Abstract. Porous MgB_2 samples have been prepared by the heat treatment of a pressed mixture of Mg and MgB_2 powders. It was found that linked superconducting structure is formed down to the minimum density $\nu_c= d/d_o\cong 0.16$ (percolation threshold), where d is the density of MgB_2 averaged over the sample, $d_o= 2.62$ g/cm^3 is the X-ray density. The critical temperature of the porous phase decreases with decreasing ν and $T_{c2}\cong 32$K is minimal at $\nu_c\cong 0.16$.

Keywords: porous MgB_2, critical density, inter-grain
PACS: 74.25.Ha, 74.62.Bf

INTRODUCTION

The weak grain-boundary links limit substantially the critical current in ceramic high-temperature superconductors [1]. In the magnesium diboride, the inter-grain resistance is also observed after machining as consequence of the high hardness of MgB_2. This limits the critical current of wires and the tapes made using PIT method [2] in comparison with the bulk samples synthesized by hot isostatic pressing [3,4]. On the other hand scanning tunnel microscopy [5] of samples with a high Jc has revealed the presence of a normal inter-grain layer in the form of an amorphous region extending over 5 to 20 nm.

In this work, porous MgB_2 samples down to minimal average density were prepared. Porous phase and inter-grain superconductivity was investigated depending on the average density (porosity) of samples.

SAMPLES AND METHODS

Porous samples were prepared using a pressed mixture of MgB_2 and Mg powders. The initial weight proportion MgB_2:Mg was varied from pure magnesium diboride up to a ratio 1:7. The content of magnesium diboride in the volume of sample is characterized by the normalized average density $\nu= d/d_o$, where d is the density of MgB_2 averaged over sample, $d_o= 2.62$ g/cm^3 is the X-ray MgB_2 density. The tablets were heated up to a temperature of 900°C in a He atmosphere at a pressure of 1.5- 1.7 bar. The time of annealing was 1- 5 hours. After treatment the ν value was determined by direct measurements of weight and volume of a tablet with the assumption, that the quantity of boron remains constant. In this way, porous MgB_2 samples with a different content of pure magnesium from zero to 20% excess have been prepared. Using a Hall sensor the magnetic moment M(H) and diamagnetic susceptibility $<\chi(T)>= M(T)/H$ were measured after zero field cooling (ZFC).

RESULTS AND DISCUSSION

The temperature dependences of $-4\pi<\chi(T)>$ measured in the field H= 3 Oe (ZFC) before and after heat treatment are shown in Fig.1. Initial tablet($\nu\cong 0.24$ extruded) has a small magnetic moment which goes to

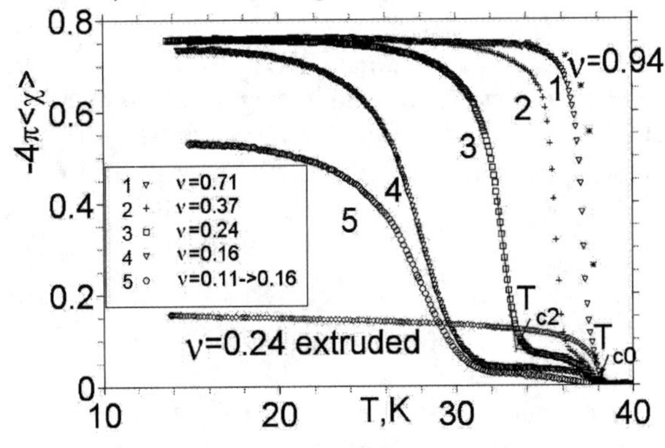

Figure 1. Temperature dependences of the susceptibility (after ZFC, H= 3 Oe) for porous magnesium diboride with the different ν.

CP850, *Low Temperature Physics: 24th International Conference on Low Temperature Physics;* edited by Y. Takano, S. P. Hershfield, S. O. Hill, P. J. Hirschfeld, and A. M. Goldman

zero at a temperature equal to T_{co} corresponding to a bulk sample [3, 5]. The weak diamagnetism in extruded sample is explained by the fact that superconducting MgB_2 particles have no superconducting interface. After annealing porous samples acquire a large diamagnetic moment at low temperatures, and $<\chi(T)>\sim -1/4\pi$ (demagnetization factor of tablets ~ 0.5). The value and behavior of M(T) depend on the annealing time. However, after 2-h annealing at T= 900C the behavior of M(T) was found virtually to be unchanged and looks like it is shown in Fig.1 for the samples with initial density $\nu > 0.16$. Moreover, composition of small ν samples might be changed during annealing to 20 % Mg excess or up to 15 % content of non-superconducting MgB_4 phase. But M(T) dependence is identical within the accuracy of measurements. As seen in fig.1 the dense samples (with $\nu>0.7$) exhibit a single superconducting transition. The moment of the dense sample [9] is labeled by asterisks. With increasing porosity (decreasing ν), two superconducting transitions are observed. The first (at T_{co}) is attributed to initial granules, and the second (with a higher magnetic moment) characterizes the disappearance of inter-grain currents at the temperature of T_{c2}.

Left curve in fig.1 shows the $-4\pi<\chi(T)>$ dependence for the sample with initial density $\nu\cong 0.11$ that, after the treatment, has increased up to $\nu\cong 0.16$. Critical temperature T_{c2} of this sample coincides with critical temperature of a sample with initial density $\nu\cong 0.16$ (curve 4). It is seen, that the magnitude of the susceptibility of this tablet is noticeably less (curve 5). This is due to the reduction of tablet size after treatment, and also because the density of the sample is near to percolation threshold. Below this threshold connected superconducting structure does not exist.

In order to determine $T_{c2}(0)$ in zero field, susceptibility was measured in magnetic field range 3 - 200Oe and results were extrapolated using square approximation $H\propto (1-T_{c2}/T_{c2}(0))^2$. The instance of this dependence is shown in inset in fig. 2. It is worth to note, the extrapolated temperature $T_{c2}(0)$ coincides with the temperature at that resistance become zero. The dependence of the relative critical temperature $T_{c2}(0)/T_{co}$ as function of ν is shown in fig. 2. With decreasing ν, additional (porous) phase with the lower critical temperature appears. At the further reduction of density T_{c2} decreases and one is minimal at $\nu_c\cong 0.16$.

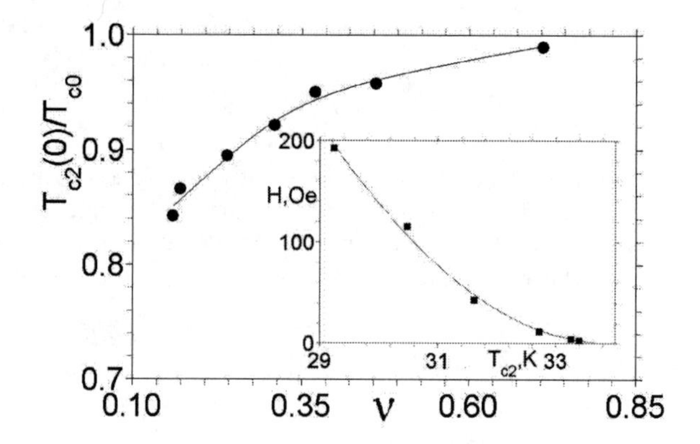

Figure 2. Normalized critical temperatures of porous phase $T_{c2}(0)/T_{co}$ as functions of the MgB_2 density ν. The inset shows experimental values $T_{c2}(H)$ and approximating curve for a sample with $\nu\cong 0.24$.

It is known that T_c decreases with lattice parameters reduction. It takes place with pressure [6] and atomic substitution [7]. In porous phase T_c can decrease as a result of defects that exist in this phase. X-ray analysis has shown that lattice parameters **a** and **c** are reduce with increasing porosity. At $\nu> 0.7$ they are the same as in the initial powder. In the sample with $\nu_c\cong 0.16$ the **a** is low by 0.33 % and **c** by 0.14 % in comparison with initial MgB_2. Most probably, the boron vacancies induce strain in crystal structure and reduce the lattice parameters.

The local critical current density J_c in porous medium was estimated from the magnetization curves for the sample with $\nu\cong 0.24$. Using a Bean model [8] and percolation theory for the random resistor networks proposed in [9] we estimate J_c at T= 20K as $J_c\approx 3\cdot 10^5$ A/cm^2.

REFERENCES

1. P. Chaudhary et al., *Phys. Rev. Lett.* **60,** 1653 (1988).
2. R. Flukiger, H.L. Suo, N. Musolino, C. Beneduce, P.Toulemonde and P.Lezza, *Physica C* **385,** 286–305 (2003).
3. J. Nagamatsu, N. Nakagawa, T. Muranaka, Y. Zenitani and J. Akimitsu, *Nature* **410,** 63 (2001).
4. T. C. Shields, K. Kawano, D. Holdom and J. S. Abell, *Supercond. Sci. Technol.* **15,** 202–205 (2002).
5. S. P. Anan'ev, V. A. Grinenko, V. E. Keilin, E. P. Krasnoperov, I. A. Kovalev, S. L. Kruglov, P. A. Levit and V. A. Stoliarov, *Supercond. Sci. Technol.* **17,** S274-S275 (2004).
6. Yang Shao et al., *J. Phys.: Condens. Matter* **16,** 1103-1113 (2004).
7. A. Bharathi et al., *Physica C* **370,** 211–218 (2002).
8. Bean C. P., *Phys. Rev. Lett.* **8,** 250 (1962).
9. M. Eisterer, M. Zehetmayer and H.W. Weber, *Phys. Rev. Lett.* **90,** 247002 (2003).

Nano Sized Powder Additives of SiC and Diamond to MgB_2 as Artificially Inductor of Pinning Force for the Dense Samples Obtained by High Pressure Technologies

A. Morawski[1], W. Pachla[1], D. Kuzmenko[1], T. Łada[1], A. Zaleski[2], O. Eibl[3], W. Haessler[4] and P. Kovac[5]

[1] *Institute of High Pressure Physics of the Polish Academy of Sciences (PASc), Warsaw, Poland*
[2] *Institute of Low Temperature Physics and Structural Researches, PASc, Wroclaw, Poland*
[3] *Institut für Angewandte Physik, Universität Tübingen, Germany*
[4] *IFW, Dresden, Germany*
[5] *IEE, SASc, Bratislava, Slovak Republic*

Abstract. The nano-powder additives of SiC (11nm) and diamond (6 nm) were mixed with Mg and amorphous B powders in the production route of MgB_2 bulk and wires samples. Special attention was paid to the precursor powders with the respect to chemical purity, grain morphology and secondary phases. Commercially available MgB_2 powders (ex-situ technology) and also mixtures of MgH_2 and amorphous B powders (in-situ technology) were applied as the starting precursors. Electron microscopy was used extensively for the characterization of these powders applying both SEM and TEM analysis. Superconducting bulks and wires samples have been produced, by hot pressing, hot isostatic pressing and hydrostatic extrusion processes. High grain refinement and density in the final wire samples have been achieved by the severe plastic deformation (reduction over 99.98%) with the use of the cumulative (multi-step) hydrostatic extrusion. Superconducting properties of these samples were measured, with particular attention paid to the $J_c(B)$ characteristics. Critical currents at lower and higher magnetic fields depended sensitively on the amount of the additives and grain refinement. TEM imaging and electron spectroscopic imaging were used to evaluate the size and distribution of the both of additives within the microstructure. The strain field of the additives resulted from the misfit in thermal expansion coefficient between the additives and the MgB_2 matrix was imaged in bright and dark field TEM and is particularly important for the flux-pinning properties of these materials. Influence of such synthesis parameters as pressure, temperature and time on the superconducting properties and the morphology of the products were examined and obtained results were compared. This allowed to choose the optimal conditions for the improvement of the bulk MgB_2 characteristics of the wire samples.

Keywords: MgB_2,SiC and diamond additives, critical current density, hot pressing, wire extrusion.
PACS: 74.70.-b , 74.72.-h

INTRODUCTION

The effect of strong decrease of critical current density J_c, as a function of magnetic field rising, at practically applicable temperature, not less than 20 K, is the main point to resolve in close application of MgB_2 superconductor material. Several ways were implicated to improve the J_c i.e.: the enhancement of the grain connectivity, by grains size nano scale engineering, by introducing effective pinning centers, inter-grain doping by elements, etc [1-6]. All the ways strongly depend on the synthesis and deformation process applied. For manufacturing metal claded MgB_2 superconductors, the great variety of the processes as: hot pressing, hot isostatic pressing, cold and hot rolling [1,3,5,6], rotary swaging [5], wire drawing [1], two-axial rolling [2,5], and hydrostatic extrusion [2], have been applied. The effective pinning centers are produced by introducing the defects into the grains, segregations of the nano sized grains by various nano sized additives elements placed in grain boundary, as nano-SiC [3,4,7,9], Si [4], nano-carbon [4], nano-borides, various silicides [3], or nano-MgO [6]. Doped samples usually, show higher J_c values than undoped ones [3,4,6,8]. In some cases they posed higher Tc-even up to 42,6 K, obtained by tangential stress between the core MgB_2 matrix surrounded by the SiC nano layer [7] or between the substrates (Al_2O_3

CP850, *Low Temperature Physics: 24th International Conference on Low Temperature Physics;*
edited by Y. Takano, S. P. Hershfield, S. O. Hill, P. J. Hirschfeld, and A. M. Goldman

or SiC) and MgB_2 layer [8]. Recently, Fujii et al.[9] have fabricated *in-situ* MgB_2 tapes starting from MgH_2 as the Mg source. MgH_2 is believed to prevent the oxidation of Mg, one of the main reason of J_c lowering in MgB_2 samples. The negative influence is the lower starting density, caused by the H_2 diffusion throughout the sample during the decomposition process of MgH_2 [1], so HP and HIP or HE methods are excellent for effective densification of the samples. In this paper we have reported an *in-situ* method applied for easy doping by nano-SiC and nano-diamond with combination of the High Pressure Technology (HPT) used for samples densification.

EXPERIMENTAL

The in-situ nano-SiC (11nm) and diamond (6 nm) additives were used for MgB_2 samples fabrication from MgH_2 and B amorphous powders. The hot pressing HP, hot isostatic pressing HIP and hydrostatic extrusion HE, were applied as the forming processes. The pressure range was up to 0.8 GPa for HIP and HE and up to 0.4 GPa for HP processes, typically lasted from 1 to 14 hours. The temperature was in the range from 620 to 670°C. For nano-powder processing a new, very effective, ultrasonic high-gas pressure mixing, crushing and cleaning method was applied [7]. All procedures were carried out in the glove box.

RESULTS AND DISCUSSION

The SEM analysis of doped samples indicated homogeneously distributed very small grains of SiC. The TEM showed that the grains were few nanometers in size. The densities of the HP-ed samples were high and especially high for HIP-ed samples, up to almost 97% of the theoretical density.

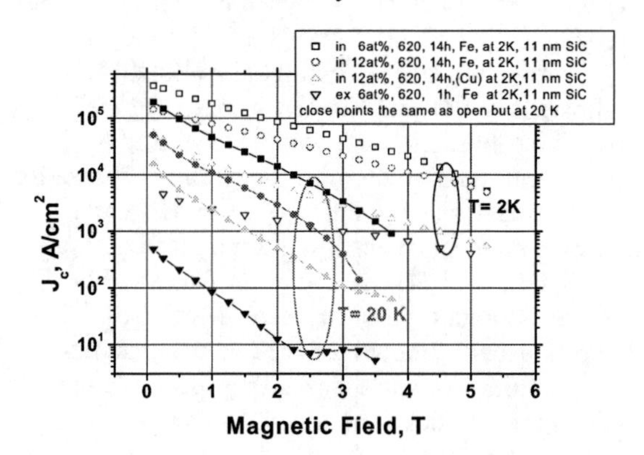

FIGURE 1. Critical current density by Bean model for various quantities of 11 nm SiC additives.

The *Jc* 10^5 at 20 K was obtained for SiC in-situ doped samples and the diamond doped were again much better at higher fields (see Fig. 1 and Fig 2).

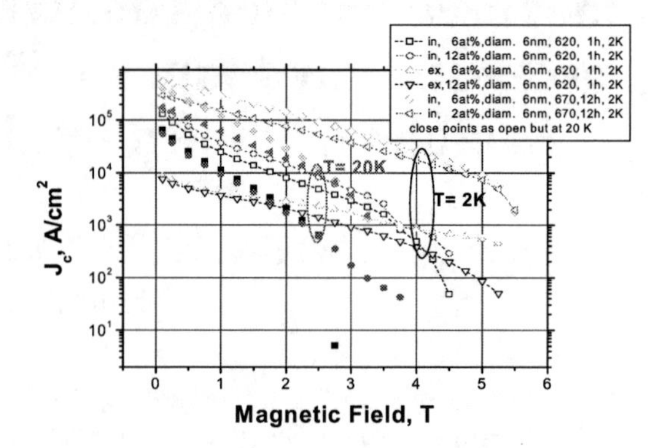

FIGURE 2. Critical current density estimated from Bean model for various quantities of 6 nm diamond additives.

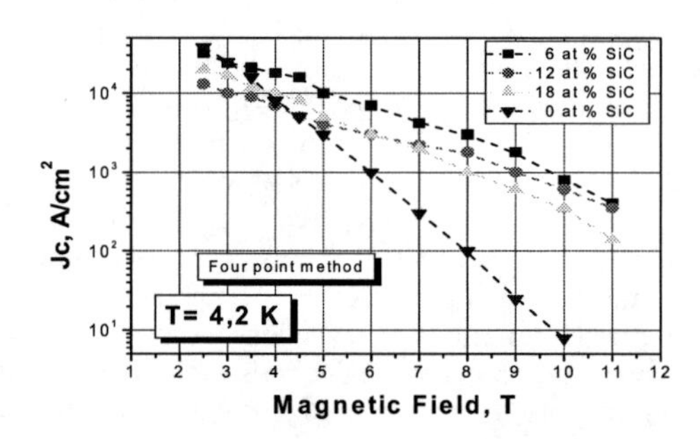

FIGURE 3. Critical current density Jc(B) vs. magnetic field dependence by four point contact method measurements, for pure and SiC doped, 1.6 mm^2 core, wire samples, after HE and 60 minutes annealing at 620°C in argon gas at ambient pressure.

The diamond additives and HP technology are very promising ways in applications and production of high current lids for magnets and high power systems.

ACKNOWLEDGMENTS

This work was supported by funding from the EU-FP6 STRP funded Specific Targeted Research Project HIPERMAG no. NMP3-CT-2004-505724.

REFERENCES

1. C.H. Jiang, et al.,*Supercond. Sci. T.* **18**, L17-L22 (2005).
2. W. Pachla, et al., *Supercond. Sci. T.* **18**, 552-556 (2005).
3. A. Matsumoto, et al., *Supercond. Sci. T.* **16**, 926-930 (2005).
4. X. L. Wang, et al., *Physica C* **408-410**,63-67 (2004).
5. P. Kovac, et al.,*Supercond. Sci. T* **17**, 1225-1230 (2005).
6. C. H. Jiang, et al., *Physica C* **423**, 45-50 (2005).
7. A. Morawski, et al., Prz. Elektr. **11**, 1125-1128 (2004).
8. A.V. Pogrebnyakov, et al.,*Nature Mater* **1**, 35-38 (2002).
9. H. Fuji, et al., *Supercond. Sci. T.* **15**, 1571-1576 (2005).

Tunneling Spectroscopy On Partially Deuterated κ-(BEDT-TTF)$_2$Cu[N(CN)$_2$]Br By STM

Koichi Ichimura, Kazushige Nomura and Atsushi Kawamoto

Division of Physics, Hokkaido University, Sapporo 060-0810, Japan

Abstract. The superconducting phase of organic superconductor κ-(BEDT-TTF)$_2$Cu[N(CN)$_2$]Br was studied by the electron tunneling using STM. Tunneling spectra were obtained at partially deuterated κ-(BEDT-TTF-d[3,3])$_2$Cu[N(CN)$_2$]Br. The tunneling conductance curve exhibits clear superconducting gap structure. Tunneling spectra with linear dependence on the energy are explained by *d*-wave pairing. We obtained the gap parameter as a function of the effective correlation strength, which can be controlled by partial deuteration. The large gap value suggests strong coupling superconductivity. Moreover, a zero bias conductance peak (ZBCP) was found in d[3,3] salt. *d*-wave pairing in κ-(BEDT-TTF)$_2$X is strongly supported.

Keywords: organic superconductor, tunneling spectroscopy, STM, *d*-wave pairing, gap parameter
PACS: 74.70.Kn, 74.50.+r, 74.25.Jb

The superconductivity in organic charge transfer salts κ-(BEDT-TTF)$_2$X is characterized as its superconducting phase adjoins the antiferromagnetic phase [1]. A lot of attention has been given to which mechanism brings about the superconductivity in the neighborhood of the Mott insulating phase. κ-(BEDT-TTF)$_2$Cu[N(CN)$_2$]Br is located near this phase boundary. Kawamoto *et al.* [2] revealed that the effective correlation can be controlled finely by partial deuteration of BEDT-TTF molecules.

In investigating the superconducting state, electron tunneling is useful since the electronic density of states can be obtained directly with high energy resolution. Tunneling spectroscopy using STM, *i.e.* STS, especially has an advantage because of its non-contacting tip configuration. Previously, we revealed *d*-wave pairing symmetry in κ-(BEDT-TTF)$_2$Cu(NCS)$_2$ by STS on lateral surfaces [3]. The nodal direction of the gap was determined as $\pi/4$ from the k_b and k_c axes. Our next interest is how the pairing symmetry and the gap parameter depend on the effective correlation. In this article, we present STS results on κ-(BEDT-TTF-d[3,3])$_2$Cu[N(CN)$_2$]Br, which is located closer to the Mott boundary.

Single crystals were grown by the standard electro-chemical method. Samples for d[3,3] salt were cooled slowly around 80 K with rate of about –0.05 K/min. The superconducting transition temperature was determined as T_c=12.0 K for d[3,3] salts from the magnetic transition with applying a magnetic field of 10 Gauss by SQUID magnetometer. As-grown surfaces of the *a-c* plane were investigated by low temperature STM. The tunneling current flows normal to the conducting layer in this configuration. The tunneling differential conductance was directly obtained by lock-in detection, in which 1 kHz ac modulation with amplitude of 0.1 mV was superposed in the ramped bias voltage.

Figure 1 shows a typical tunneling differential conductance curve for d[3,3] salt. The differential

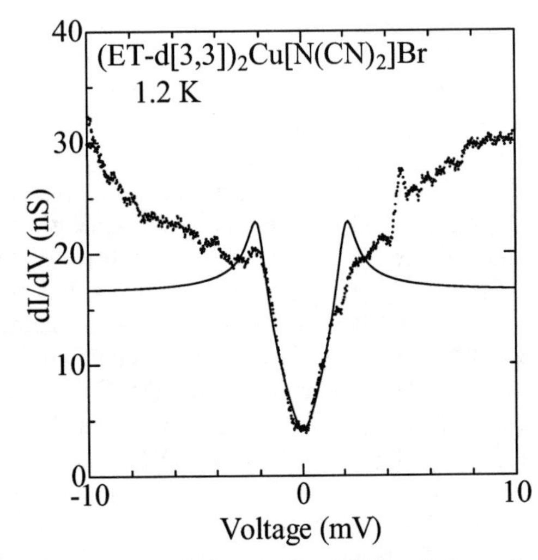

FIGURE 1. Typical tunneling differential conductance curve in the superconducting phase. The solid line represents the *d*-wave calculation.

CP850, *Low Temperature Physics: 24th International Conference on Low Temperature Physics;* edited by Y. Takano, S. P. Hershfield, S. O. Hill, P. J. Hirschfeld, and A. M. Goldman

TABLE 1. Gap parameter Δ and corresponding $2\Delta/kT_c$ for κ-(BEDT-TTF-d[n,n])$_2$Cu[N(CN)$_2$]Br.

	Δ (meV)	$2\Delta/kT_c$
d[0,0]	2.1-3.9	4.3-7.9
d[2,2]	3.0-4.8	5.8-9.3
d[3,3]	1.4-3.3	2.7-6.4

conductance around the zero bias voltage is well reduced compared with the normal conductance. Although the magnitude of the gap varies depending on sample, the functional form at low energy region, which reflects the symmetry of the pair wave function, is identical. The conductance inside the gap edges has energy dependence in contrast to the s-wave case. Gap anisotropy is thus suggested. We examine the *d*-wave symmetry with line nodes. The solid line in Fig. 1 represents the calculated conductance curve based on the simple *d*-wave symmetry described as, $\Delta(\boldsymbol{k})=\Delta\cos(2\phi)$ with Δ=2.1 meV and Γ=0.21 meV, where Γ is the lifetime broadening [4]. The linear dependence on the energy inside the gap is well explained by *d*-wave symmetry. For d[3,3] salt, the gap parameter is obtained by as Δ=1.4-3.3 meV by the above fitting. Correspondingly, the reduced gap is obtained as $2\Delta/kT_c$=2.7-6.4. The scattering in value is due to sample dependence.

Gap values for d[0,0], [5] d[2,2] and d[3,3] salts are listed in Table 1. These values are almost consistent with that for κ-(BEDT-TTF)$_2$Cu(NCS)$_2$ [6]. The effective correlation strength U/W, where U is the on-site Coulomb interaction and W is the band width, becomes large by the deuteration. The dependence of $2\Delta/kT_c$ on U/W is not monotonous. It is noteworthy that value of $2\Delta/kT_c$ in three salts is slightly larger than that of mean field theory. Strong coupling superconductivity is suggested.

It should be noted that anomalous enhancement of conductance around zero bias is observed occasionally as shown in Fig. 2. The tunneling conductance at zero bias voltage increases divergently in contrast to the case of the superconducting gap. We think that this sharp peak corresponds to the zero bias conductance peak (ZBCP) unique to anisotropic superconductors. The ZBCP has already been found in high-T_c cuprates. The present result is the first observation of the ZBCP in κ-(BEDT-TTF)$_2$X. The ZBCP is originated from the zero energy state, which is the bound state formed around gap node where the sign of the order parameter changes [6]. Therefore, the ZBCP is thought to be related to the presence of nodes of the superconducting gap. The ZBCP in our tunneling spectra supports *d*-wave pairing symmetry in κ-(BEDT-TTF)$_2$X.

Normally, one can hardly observe the ZBCP in the present tunneling configuration, in which electrons

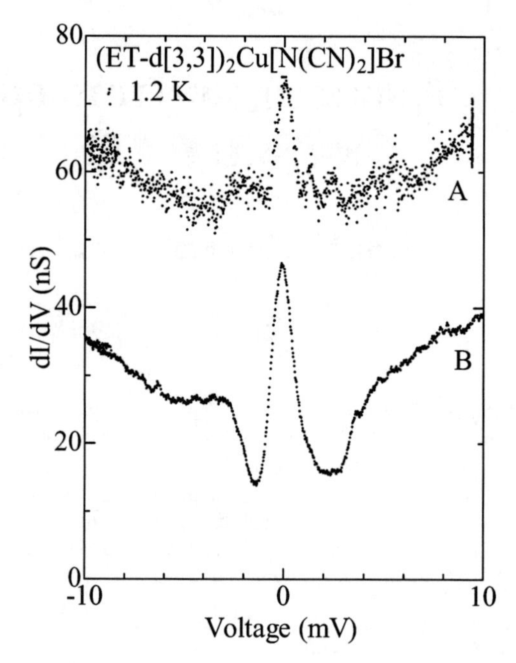

FIGURE 2. Tunneling conductance curve with the zero bias anomaly. Curves for two different samples are shown. The zero conductance line of each curve is shifted by 20 nS for clarity.

tunnel perpendicular to the conducting plane. The in-plane anisotropy would be averaged out. It is essential that electrons with wave number around the gap nodes contribute selectively to the tunneling in observing the ZBCP. We think that micro steps enabled us to observe the ZBCP. The in-plane component of the wave number can be resolved in the tunneling from step edges. However, we cannot know the nodal direction by the present tunneling configuration. STS measurement at lateral surfaces [3] in κ-(BEDT-TTF)$_2$Cu[N(CN)$_2$]Br is our future study to determine the direction of nodes.

REFERENCES

1. K. Kanoda, *Hyperfine Interact.* **104**, 253-249 (1997).
2. A. Kawamoto, H. Taniguchi and K. Kanoda, *J. Am. Chem. Soc.* **120**, 10984-10985 (1998).
3. T. Arai, K. Ichimura, K. Nomura, S. Takasaki, J. Yamada, S. Nakatsuji and H. Anzai, *Phys. Rev. B* **63**, 104518 (2001).
4. R. C. Dynes, V. Narayanamurti and J. P. Garno, *Phys. Rev. Lett.* **41**, 1509-1512 (1978).
5. K. Ichimura, K. Suzuki, K. Nomura and A. Kawamoto, *Synth. Metals* **133-134**, 213-214 (2003).
6. T. Arai, K. Ichimura, K. Nomura, S. Takasaki, J. Yamada, S. Nakatsuji and H. Anzai, *Solid State Commun.* **116**, 679-682 (2000).
7. S. Kashiwaya and Y. Tanaka, *Rep. Prog. Phys.* **63**, 1641-1724 (2000).

Sound Velocity Measurements near the Superconducting Transition Temperature of κ-(BEDT-TTF)$_2$Cu[N(CN)$_2$]Br under the Magnetic Field

Takayuki Simizu, Noriyuki Yoshimoto, Yoshiki Nakanishi and Masahito Yoshizawa

Graduate School of Engineering, Iwate University, 4-3-5 Ueda Morioka Iwate 020-8551, Japan

Abstract. Layered organic superconductor κ-(BEDT-TTF)$_2$Cu[N(CN)$_2$]Br shows a step-like softening in the sound velocity below the superconducting transition temperature. In this study, we have measured the sound velocity in the superconducting state of κ-(BEDT-TTF)$_2$Cu[N(CN)$_2$]Br in the applied magnetic field for various directions. The sound velocity as a function of the applied magnetic field shows remarkable elastic anomalies around $H = 2$ T. It was found that, the magnetic field parallel to the conducting planes of κ-(BEDT-TTF)$_2$Cu[N(CN)$_2$]Br, the sound velocity shows a sharp peak, which is considered to be related to the pinning by Josephson vortices.

Keywords: organic superconductor, sound velocity, vortex
PACS: 74.25.Ld; 74.70.Kn

INTRODUCTION

κ-(BEDT-TTF)$_2$Cu[N(CN)$_2$]Br is an organic superconductor which the superconducting transition temperature $Tc = 11.8$ K at ambient pressure [1]. It has layered structure with two dimensional conducting planes and insulating planes. It causes some anisotropic phenomena. The vortex behavior is one of the interest things. The magnetic field applied nearly parallel to the conducting planes of this salt, vortex shows the "lock-in state" [2]. This vortex state had observed in small angle range ~ ± 0.5 ° from conducting planes.

This salt shows that resistivity and susceptibility increase, depending on the cooling rate in the normal state. It was caused by ordering of the terminal ethylene groups in BEDT-TTF molecule [3, 4]. The control of the cooling rate plays an important role in the measurement on this salt.

EXPERIMENTAL

The sample was crystallized by electrochemical oxidation method [1]. The volume of this sample is $0.8 \times 0.7 \times 1.6$ mm^3 (for $a \times b \times c$). We had used the X-cut quartz piezoelectric transducers of 80 MHz (Valpey Fisher Co.) for high frequency ultrasound measurements. These transducers were glued to the parallel surfaces of this sample by liquid polymer Thiokol LP-32. The sound velocity measured by an ultrasonic apparatus with a phase comparison method. The longitudinal sound propagated to the interlayer direction of κ-(BEDT-TTF)$_2$Cu[N(CN)$_2$]Br for [010].

Cryogenic system had been constructed by using GM type cryo-cooler (Sumitomo heavy industries ltd.) combined with helium-free 5 T sprit-pair superconducting magnet (Toshiba Co.). This system has a cylindrical sample space for $\phi = 38$ [mm] and $l = 60$ [mm] in diameter and length, respectively, inside the triple radiation shields. Temperature was monitored by Cernox thermometer (Lake Shore Cryotronics Inc.) mounted near the sample. Sweep rate of the temperature had been controlled within 10 K / hour from room temperature down to 4 K by PID controller Model 34 (Cryogenic Control Systems Inc.). The sample mounted on a piezo-rotator (attocube systems AG) that is settled on the oxygen free copper cold stage. The rotation of this rotator is controlled by input voltage and repetition frequency for its piezo devices. The minimum driving angle step is about 0.0001 °.

CP850, *Low Temperature Physics: 24th International Conference on Low Temperature Physics;*
edited by Y. Takano, S. P. Hershfield, S. O. Hill, P. J. Hirschfeld, and A. M. Goldman

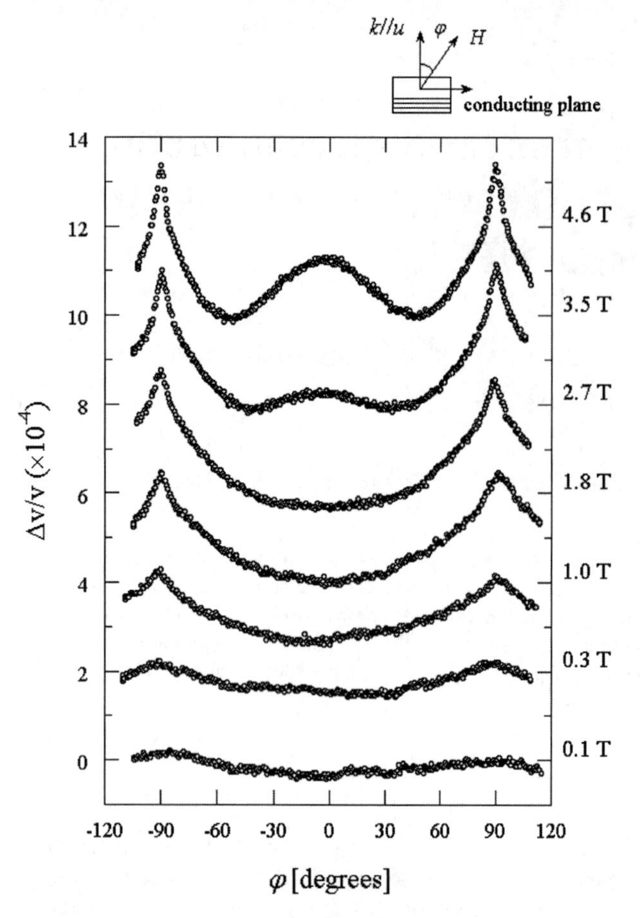

FIGURE 1. Angular dependence of the sound velocity in κ-(BEDT-TTF)$_2$Cu[N(CN)]$_2$Br at $T = 4.2$ K.

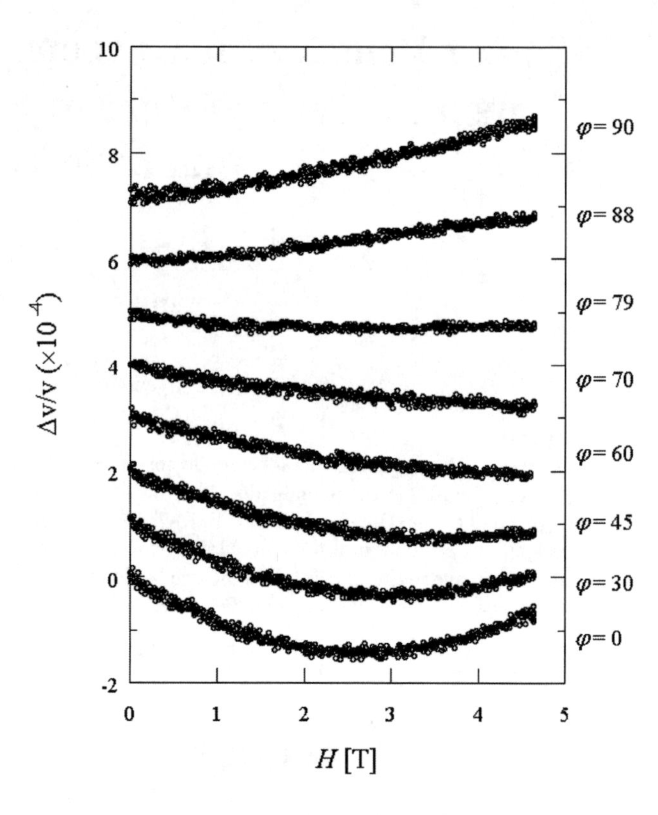

FIGURE 2. Magnetic field dependence of the sound velocity in κ-(BEDT-TTF)$_2$Cu[N(CN)]$_2$Br at $T = 4.2$ K.

RESULTS

Figure 1 shows the angular dependence of the sound velocity change under the magnetic field at 4.2 K. The sound velocity shows a remarkable φ-dependence, where φ is the angle of the magnetic field **H** to the interlayer direction or to the wave vector **k** and the polarization vector **u**. It shows a shallow minimum around $\varphi = 0$, in the magnetic field below 2.7 T. And an alternative maximum grows with increasing field. On the other hand, a sharp peak, around $\varphi = 90$ °, shows a gradual increase with the magnetic field. Figure 2 shows the magnetic field dependence of the sound velocity at 4.2 K. There shows the both of the increasing field and decreasing field process. Elastic softening appeared around $H = 2 \sim 3$ T for $\varphi < 45$ °. At 60 ° $< \varphi <$ 79 °, the sound velocity shows quite a weak dependence on the magnetic field. The magnetic field applied nearly parallel to the conducting planes, $\varphi = 88$ ° and 90 °, the sound velocity shows a completely change to gradually increasing.

The sound velocity measurements on κ-(BEDT-TTF)$_2$Cu(NCS)$_2$ had been reported in ref. 5. It shows the similar elastic softening in **u** // **H** configuration. It suggests that the common origin caused the magnetic field dependence of the sound velocity change in these quasi two dimensional layered organic superconductors. It has been known that, for $\varphi = 90$ ° configuration, Josephson vortices participate a pinning state in layered superconductors. Elastic strains couple with the flux pinning in superconducting state [6]. We are considering that the anomalous velocity enhancement near $\varphi \sim 90$ ° is attributed to the pinning of Josephson vortices.

REFERENCES

1. A. M. Kini *et al.*, *Inorganic Chemistry* **29**, 2555 (1990).
2. P. A. Mansky *et al.*, *Physical Review B* **50**, 15929 (1994).
3. M. A. Tanatar *et al.*, *Physical Review B* **59**, 3841 (1999).
4. H. Taniguchi *et al.*, *Physical Review B* **59**, 8424 (1999).
5. M. Yoshizawa *et al.*, *Solid State Communications* **89**, 701 (1994).
6. M. Yoshizawa *et al.*, *Journal of the Physical Society of Japan* **66**, 2355 (1997).

Superconductivity Emerging from Spin-Liquid Mott Insulator in Triangular Lattice System

Y. Shimizu[1,2,*], H. Kasahara[1], Y. Kurosaki[1], K. Miyagawa[1,3], K. Kanoda[1,3], M. Maesato[2], and G. Saito[2]

[1]*Department of Applied Physics, University of Tokyo, Hongo, Bunkyo-ku, Tokyo, 113-8656, Japan,*
[2]*Division of Chemistry, Graduate School of Science, Kyoto University, Oiwaketyo, Kitashirakawa, Sakyoku, Kyoto, 606-8502, Japan.*
[3]*CREST, Japan Science and Technology Corporation, Japan.*

Abstract. Mott transition and superconductivity in the triangular-lattice system κ-$(ET)_2Cu_2(CN)_3$ are investigated by resistivity and NMR measurements under pressure. The Mott insulator at ambient pressure is confirmed to undergo a transition at 0.4 GPa to the Fermi liquid, which shows superconductivity at low temperatures. The ^{13}C nuclear spin-lattice relaxation rate in the superconducting state shows the cubic temperature dependence without the coherence peak below T_C, which points to a highly-anisotropic superconducting order parameter.

Keywords: Superconductivity, triangular lattice, Mott transition, spin liquid
PACS: 71.30.+h, 74.25.Nf, 74.70.Kn

INTRODUCTION

Superconductivity on triangular lattice is of great interest since the spin frustration has been argued to make the electron pairing unconventional. The organic molecular conductor κ-$(ET)_2Cu_2(CN)_3$[1] is a model material of triangular-lattice superconductors, where ET denotes an organic molecule bis(ethylenedithio)tetrathiafulvalene. This is a Mott insulator without long-range magnetic order at ambient pressure, as evidenced by the NMR and magnetic susceptibility measurements.[2] Since the Mott transition occurs by applying a soft pressure, the system is a bare Mott insulator near the metal-insulator transition.[3] The experimental results are consistent with a theoretical prediction from the Hubbard model on triangular lattice, which suggests that the spin liquid state is realized near the Mott transition due to the spin frustration and charge fluctuations.[4] The superconducting state neighboring to the spin liquid state also attracts much interest, especially from the view of the resonating valence bond.[5]

Here we report transport and ^{13}C NMR studies of κ-$(ET)_2Cu_2(CN)_3$ under hydrostatic pressure. The Mott transition of this system is monitored by resistive measurements under hydrostatic pressure. The metallic state appearing above a critical pressure is demonstrated to have the Fermi liquid nature. In the superconducting state under pressure, the ^{13}C nuclear spin-lattice relaxation rate $1/^{13}T_1$ shows a power-law temperature dependence below T_C without the Hebel-Slichter coherence peak.

EXPERIMENTAL

The crystals were grown with the conventional electrochemical oxidation method. The in-plane resistivity was measured by the four-probe method. ^{13}C NMR measurements were performed on a single crystal prepared with ^{13}C-enriched ET molecules. The hydrostatic pressure was applied and cramped at room temperature using a BeCu cell. The superconducting diamagnetic shielding was monitored by the rf impedance measurements using the in-situ NMR coil. A magnetic field of 2 T was applied exactly parallel to the superconducting layer within 0.1°, which was attained by identifying the vortex lock-in state. The spin-echo signal was obtained after the $\pi/2$- τ -π-τ' pulse. The $^{13}T_1$ was evaluated from a single-

* Present address: RIKEN, Wako, Saitama 351-0198, Japan

CP850, *Low Temperature Physics: 24th International Conference on Low Temperature Physics;* edited by Y. Takano, S. P. Hershfield, S. O. Hill, P. J. Hirschfeld, and A. M. Goldman

exponential fit of the recovery of the spin-echo intensity.

RESULTS AND DISCUSSION

The temperature dependence of the resistance of κ-$(ET)_2Cu_2(CN)_3$ is shown in Fig. 1. The insulating behavior at ambient pressure is rapidly suppressed at 0.4 GPa with a resistance drop around 15 K. The discontinuous feature of the Mott transition indicates a first-order one as seen in Ref. 3. No indication of magnetic ordering was observed by ^{1}H NMR measurements up to this critical pressure of Mott transition, which evidences that the spin liquid phase neighbors to the metal or superconducting phase.[6] The onset of the superconducting transition is observed around 3.6 K under a zero magnetic field.

The temperature dependence of the ^{13}C nuclear spin-lattice relaxation rate $1/^{13}T_1$ at 0.4 GPa is shown in Fig.2. A linear temperature dependence is observed below 10 K down to 4 K, suggesting that the metallic phase under pressure has the Fermi liquid nature. This is consistent with the T^2 dependence in the resistivity at low temperatures below 7 K, shown in the inset of Fig. 1. The superconducting transition is seen as a drop of the susceptibility around 3.4 K under 2 T by the rf impedance measurements, as displayed in the inset of Fig. 2. The $1/^{13}T_1$ shows a kink around T_C and decreases with a power law, ~ T^3. We see no coherence peak below T_C. These features indicate that the superconducting order parameter is nodal or highly anisotropic. Our preliminary determination of the Knight shift and the spin-echo decay rate shows mysterious temperature dependence, which needs to be explored in the future work.[7]

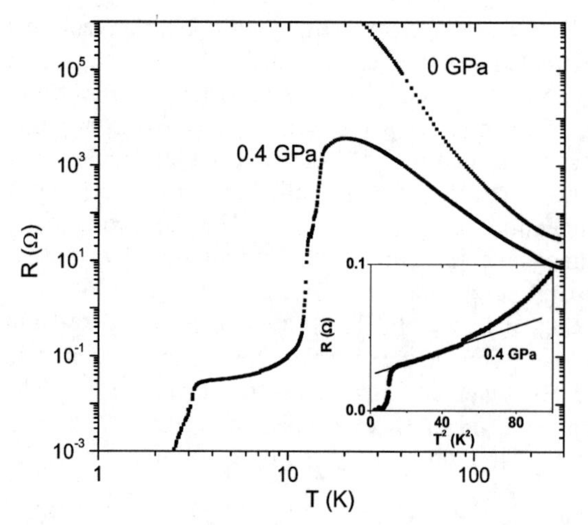

FIGURE 1. Temperature dependence of resistance of κ-$(ET)_2Cu_2(CN)_3$ at ambient pressure and 0.4 GPa. The inset shows the resistance at 0.4 GPa, plotted as a function of square of temperature.

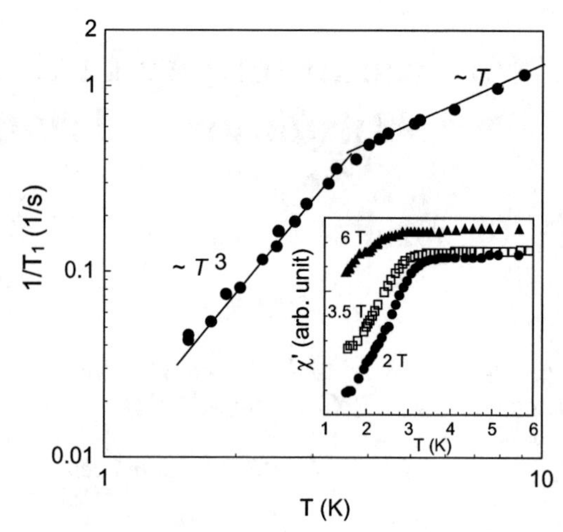

FIGURE 2. ^{13}C NMR $1/T_1$ of κ-$(ET)_2Cu_2(CN)_3$ at 0.4 GPa and 2 T. The inset shows the thermal variation of rf impedance under magnetic fields.

CONCLUSION

We have found that the superconductivity of κ-$(ET)_2Cu_2(CN)_3$ is highly anisotropic one possibly with nodes in the gap parameter. The Mott transition from the spin-liquid Mott insulator into the Fermi liquid was identified by the resistivity and the NMR spin-lattice relaxation rate.

ACKOWLEDGEMENT

This work was supported by MEXT KAKENHI on Priority Area of Molecular Conductors (No. 15073204) and by JSPS KAKENHI (No. 15104006).

REFERENCES

1. U. Geiser, *et al.*, *Inorg. Chem.* **30**, 2586-2587 (1991).
2. Y. Shimizu, K. Miyagawa, K. Kanoda, M. Maesato, and G. Saito, *Phys. Rev. Lett.* **91**, 107001 (2003).
3. T. Komatsu, N. Matsukawa, T. Inoue, and G. Saito, *J. Phys. Soc. Jpn.* **65**, 1340-1356 (1996).
4. H. Morita, S. Watanabe, and M. Imada, *J. Phys. Soc. Jpn.* **71**, 2109-2113 (2002).
5. J. Liu, J. Schmalian, and N. Trivedi, *Phys. Rev. Lett.* **94**, 127003 (2005); B. J. Powell and R. H. McKenzie, *ibid.* **94**, 47004 (2005).
6. K. Kurosaki, *et al.*, cond-mat/0504273(unpublished).
7. Y. Shimizu, *et al.*, to be published.

Two Band Fluctuation Exchange Study on the Pressure Dependence of the Superconducting Transition Temperature of β'-(BEDT-TTF)$_2$ICl$_2$

Tsuguhito Nakano*, Kazuhiko Kuroki* and Ryotaro Arita†

*Dept. of Applied Physics and Chemistry, Univ. of Electro-Communications, Chofu, Tokyo, 182-8585, Japan
†Dept. of Physics, Univ. of Tokyo. Hongo, Tokyo, 113-0033, Japan

Abstract. We performed a fluctuation exchange (FLEX) study on the original 3/4-filled multiband Hubbard model for a pressure-induced organic superconductor β'-(BEDT-TTF)$_2$ICl$_2$. Our study is motivated by the fact that a first principles calculation suggests that dimerization is not so strong under high pressure, so that the validity of adopting an effective single band model called the 'dimer model' is not obvious. Solving the linearized Eliashberg equation, a d_{xy}-wave like superconducting state with realistic superconducting transition temperature is obtained. Our results suggest that the band structure peculiar to this material may be playing an important role in the occurrence of superconductivity.

Keywords: superconductivity, pressure effects, β'-(BEDT-TTF)$_2$ICl$_2$
PACS: 74.62.Fj, 74.70.Kn

Organic charge-transfer (C-T) compounds exhibit a variety of electronic properties due to low-dimensionality and strong electron correlation. Especially, unconventional superconductivity is one of the most exciting issues in condensed matter physics.

In the title compound, β'-(BEDT-TTF)$_2$ICl$_2$, which has been known for its Mott insulating state and antiferromagnetic (AF) transition at $T_N = 22$K, superconductivity has recently been found at high pressure ($p >$ 8.2GPa).[1] The maximum value of the superconducting (SC) transition temperature $T_c = 14.2$K (at 8.2GPa) is the highest among the C-T salts. Since the SC phase seems to sit next to the AF insulating phase in the pressure-temperature phase diagram, there is a possibility of the pairing due to AF spin fluctuations. In the BEDT-TTF (ET) molecule layer, which is responsible for the electronic properties, the system is essentially a 3/4-filled two-band system since two ET molecules are packed in a unit cell with 0.5 holes per ET molecule (FIG.1(a)). At ambient pressure, however, because of the dimerization of two ET molecules connected by $p1$, a half-filled single band system is effectively realized (FIG.1(b)). It is considered that the structural change caused by applying high pressure results in the itineracy of the electrons.

To investigate the origin of SC, a fluctuation exchange (FLEX)[2] study has been performed on an effective single band Hubbard model, namely the 'dimer model' obtained in the strong dimerization limit.[3, 4] The hopping integrals of the dimer model are determined by a certain transformation from those of the original lattice, which are determined from a first principles calculation by Miyazaki *et al.*[5] The obtained phase diagram is

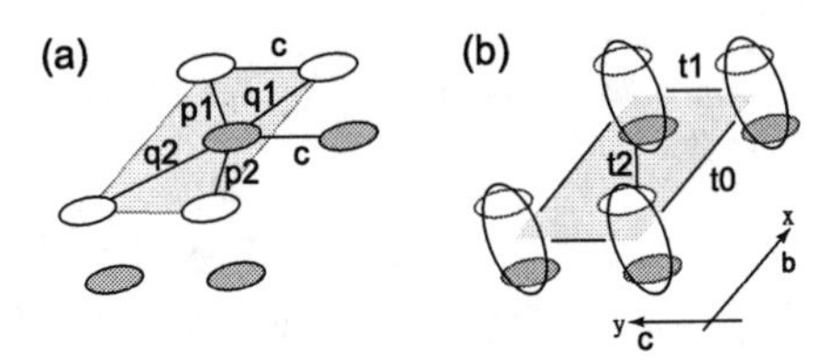

FIGURE 1. Schematic illustration of ET layer of β'-(ET)$_2$ICl$_2$. (a) Original two band lattice. (b) Effective two band lattice. Large ellipses represent dimers.

qualitatively similar to the experimental one although SC phase appears in a higher pressure regime.

However, the hopping integrals of the original lattice suggest that the dimerization is not so strong, so that there is a possibility that multiband effects may play an important role. In the present study, we obtain T_c and the gap function $\phi(k)$ by applying the two-band version of FLEX for the Hubbard model on the original two-band lattice using the hopping integrals determined by Miyazaki *et al.*[5] Realistic values of T_c with d_{xy}-wave like gap are obtained in a pressure regime similar to those in the single band approach.

In the present study, we adopt a standard Hubbard Hamiltonian having two sites in a unit cell, where each site corresponds to an ET molecule. $\mathscr{H} = \sum_{\nu,\nu'}\sum_{<i,j>,\sigma} t_{ij}\left(c_j^{\nu} c_i^{\nu'} + \mathrm{h.c}\right) + U\sum_{i,\nu} n_{i,\uparrow}^{\nu} n_{i,\downarrow}^{\nu}$, where ν and ν' denote the two sites in unit cell, and i,j denote the unit cells. The hopping t_{ij} corresponds to $p1, p2, \cdots, c$ shown in FIG.1(a). By using the two-band version of FLEX, which is a kind of self-consistent random phase approximation, the Green's function

CP850, *Low Temperature Physics: 24th International Conference on Low Temperature Physics;*
edited by Y. Takano, S. P. Hershfield, S. O. Hill, P. J. Hirschfeld, and A. M. Goldman

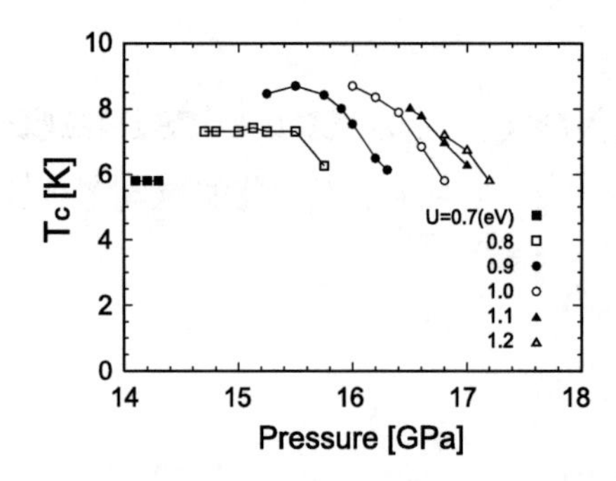

FIGURE 2. Transition temperature T_c as functions of pressure for several values of U.

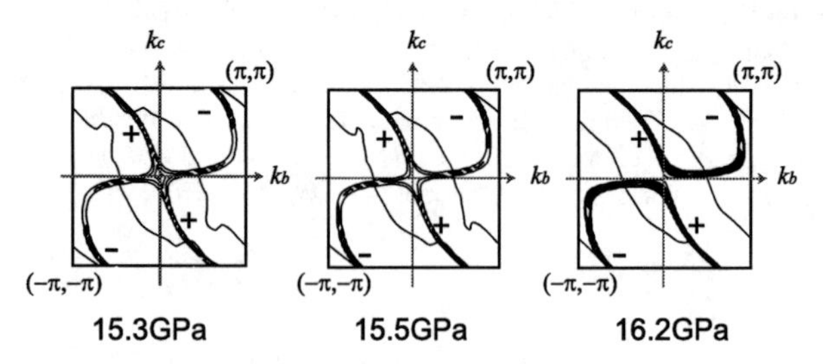

FIGURE 3. Contour plots of $|G(k)|^2$ and the nodes of SC order parameter $\phi(k)$.

$G_{\nu\nu'}(k, \varepsilon_n)$ and the pairing interaction $V^{(2)}_{\nu\nu'}$ are obtained, and we can calculate T_c and the gap function $\phi(k)$ by solving the linearized Eliashberg's equation. We use 64×64 k-point meshes and 16384 Matsubara frequencies in order to assure convergence at low temperature.

Now we move on to the results. The pressure dependence of T_c obtained for several values of U is shown in FIG.2. Qualitatively, FIG.2 is similar to the experimental phase diagram in that T_c appears at a certain pressure and decreases with increase of pressure.[1] The pressure where SC occurs, however, turns out to range from 14.2GPa to 17.2GPa, in rough agreement with the result of the dimer model approach.

In FIG.3, the nodal lines of $\phi(k)$ and the contour plots of $|G(k)|^2$ for several values of pressure are shown for $U = 0.9$eV. The ridges of $|G(k)|^2$ correspond to the Fermi surfaces (FS). As seen from the position of the FS and the gap nodes, the pairing symmetry is d_{xy}-wave like as in the single band approach. The peak position of the spin susceptibility χ_s, which has a large value at the nesting vector of FS, stays around $Q = (\pi, \pi/4)$ regardless of the pressure, but the peak value decreases with pressure corresponding to the dimensional crossover of the FS. Consequently, the pairing interaction $V^{(2)} \propto \chi_s$ becomes smaller, so that T_c becomes lower, with increasing pressure.

An important point in the present study is the value of T_c. In the present model, $T_c/W \sim 0.0006$ is obtained,

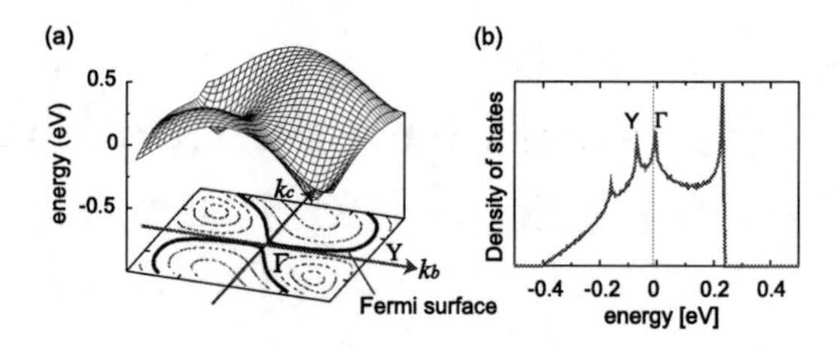

FIGURE 4. (a) $U = 0$ dispersion of the antibonding band. (b) DOS of the same band.

where W is the total band width. This value is not so small compared even to that for the Hubbard model on a two dimensional square lattice near half filling, namely a model for the high T_c cuprates. It is surprising that such a high T_c is obtained even in a 3/4-filled model, considering the fact that a similar 3/4-filled model for other organic compounds such as $\kappa-(ET)_2X$ [6] and $(TMTSF)_2X$[7] does not exhibit such high T_c within the same FLEX approach. The origin of this high T_c may be related to the band structure of this material. FIG.4(a) shows the $U = 0$ dispersion of antibonding at $p = 15.5$GPa. Two saddle points (SP) of the band dispersion are located at the Γ point and the Y point near the Fermi level E_F. The existence of SP results in the peaks of the density of states (DOS) (FIG.4(b)). For finite U, the energy of these SP become close to the E_F. Moreover, these saddle points are located at both the ends of the nesting vector $Q = (\pi, \pi/4)$. Thus, there are many states that contribute to the pair scattering. These characteristic features of the band structure may be related to the high T_c of this material.

To summarize, we have studied the superconducting phase diagram of β'-(BEDT-TTF)$_2$ICl$_2$ using two-band version of FLEX. Our results show that the high T_c in this material may be related to the band structure peculiar to this system, which results in a large DOS near the E_F.

Numerical calculations has been done at the Computer Center, ISSP, University of Tokyo. This study has been supported by Grants-in-Aid for Scientific Research from the Ministry of Education, Culture, Sports, Science and Technology of Japan, and from the Japan Society for the Promotion of Science.

REFERENCES

1. H. Taniguchi *et al.*, *J. Phys. Soc. Jpn.* **72**, 468 (2003).
2. N. E. Bickers *et al.*, *Phys. Rev. Lett.* **62**, 961 (1989).
3. H. Kontani, *Phys. Rev. B* **67**, 180503(R) (2003).
4. H. Kino *et al.*, *J. Phys. Soc. Jpn.* **73**, 25 (2004).
5. T. Miyazaki, and H. Kino, *Phys. Rev. B* **68**, 220511(R) (2003).
6. K. Kuroki *et al.*, *Phys. Rev. B* **65**, 100516(R) (2002).
7. K. Kuroki *et al.*, *Phys. Rev. B* **63**, 094509 (2001).

The Influence of Magnetic Fields on Resistivity of the Pressure Induced Superconductor, β'-$(ET)_2ICl_2$

Mika Kano*, Nobuyki Kurita*, Masato Hedo*, Yoshiya Uwatoko*, Masashi Miyashita†, Hiromi Taniguchi† and Stanley W. Tozer¶

*ISSP, University of Tokyo, 5-1-5 Kashiwanoha Kashiwa-shi, Chiba 277-8581, Japan
† Department of Physics, Saitama University, 225 Shimo-Ohkubo Sakura-ku Saitama-shi, Saitama 338-8570, Japan
¶NHMFL, Florida State University, 1800 E. Paul Dirac Dr. Tallahassee, FL 32310-3706, USA

Abstract. High pressure magnetotransport studies to 8.3 GPa using a diamond anvil cell are reported for the organic Mott-insulator, β'-$(ET)_2ICl_2$. Previous research shows that, at a pressure of 8.2 GPa, this material possesses the highest T_C among the organic conductors. A sharp drop observed in ρ(T) at 8.3GPa which we associate with the superconducting phase is completely suppressed in a magnetic field (H=18T).

Keywords: DAC, high pressure, organic superconductor, BEDT-TTF
PACS: 74.70.Kn

INTRODUCTION

It has been reported that the strongly insulating charge-transfer complex, β'-(BEDT-TTF)$_2$ICl$_2$ is metallized by application of extremely high pressure [1]. When the metallic state is stabilized, superconductivity with the highest T_C among organics appears, as shown in the Fig 1. The onset temperature for the M-I transition, which takes place around 6.5GPa, decreases with increasing pressure. On the other hand, the peak in the resistivity at low temperatures, which appears around 7.0 GPa, gradually moves up in temperature by applying pressure, and eventually becomes a robust superconducting transition within the metallic phase. Higher pressures drive the transition temperature down, with a maximum value in T_C of 14.2K (onset) being achieved at 8.2GPa which closely tracks the M-I phase boundary.

To acquire more information about this superconducting phase, we have measured the magnetotransport properties to 8.3 GPa in external magnetic fields up to 18 T.

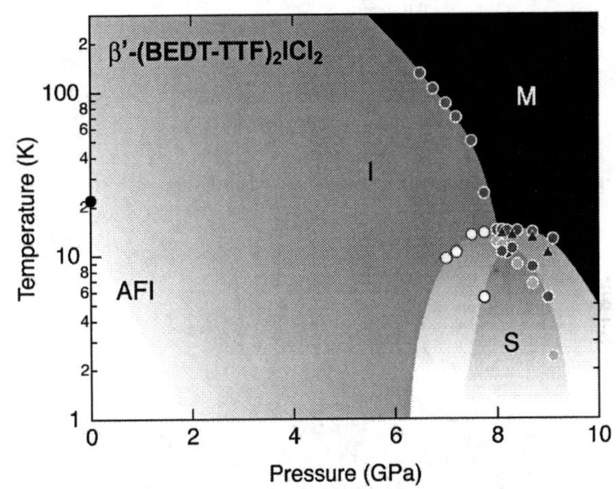

FIGURE 1. P-T phase diagram of β'-(BEDT-TTF)$_2$ICl$_2$.

EXPERIMENTAL

Pressure was generated using a turnbuckle type Diamond Anvil Cell (DAC). This DAC consists of two modified brilliant cut diamonds enclosed in a cell body of nonmagnetic NiCrAl alloy. The 0.9 mm diameter culets are separated by a BeCu gasket indented to 65 μm that is coated with a mixture of diamond powder and Stycast 1266 epoxy. Four electrical lead wires of pressed gold are attached to the

CP850, *Low Temperature Physics: 24th International Conference on Low Temperature Physics;*
edited by Y. Takano, S. P. Hershfield, S. O. Hill, P. J. Hirschfeld, and A. M. Goldman

sample using graphite paint. The sample is placed inside a cavity in the gasket with a liquid pressure medium, glycerin. The DAC is then assembled, and a load is applied at room temperature using a hydraulic press. Prior to removing the cell from the press, the pressure is locked in by turning the body of the turnbuckle. The pressure is calibrated by the fluorescence of a ruby chip located in the cavity near the sample. The electric resistivity was measured by the standard four-probe method with DC current parallel to the c-axis(I=1μA), and external magnetic fields were applied perpendicular to the conducting plane.

RESULTS AND DISCUSSION

The temperature dependence of the resistivity under several pressures and the effect of magnetic fields are shown in Fig. 2. At 6.5 GPa, the resistivity in insulating region shows a slight kink around 6 K, which is suppressed in an external magnetic field of 18T. At 8.3GPa, the insulating state is suppressed and a steep decrease (peak structure) in the resistivity is observed at 12.4K, but it is also completely destroyed in a magnetic field of 18 T.

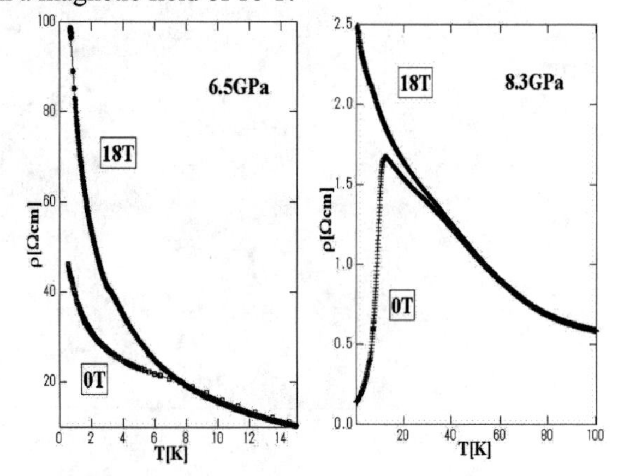

FIGURE 2. Magnetic field effect of the resistivity under 6.5GPa and 8.3GPa.

We also measured the magnetic field dependence of the resistivity at a pressure of 8.3 GPa and a temperature of 0.6 K, which is shown in figure 3. $\Delta\rho$ is defined as $\Delta\rho=\rho-\rho_0$, where ρ_0 represents the resistivity with no external magnetic field. $\Delta\rho/\rho_0$ displays a positive magnetoresistance with an S-shaped curve and a tendency toward saturatation above 18T. This result and the resistance recovery under magnetic field proves that the peak structure observed at 12.4K was due to superconductivity.More interestingly, the strong magnetic-field effect was still observed above the temperature of peak structure. Because of this kind of behavior, the presence of a superconducting phase even at higher temperature or some magnetic contribution is predicted. Details are not known yet, but we are continuing these studies.

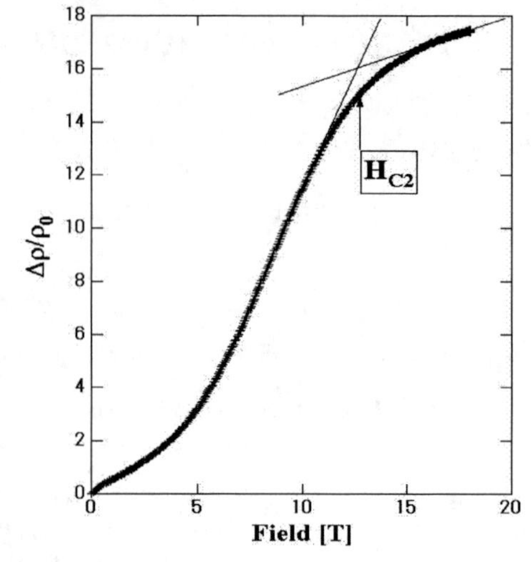

FIGURE 3. Magnetic field dependence of the resistivity at 8.3GPa.

In summary, we have measured the electrical resistivity of β'-$(BEDT\text{-}TTF)_2ICl_2$ in magnetic fields up to 18T at several pressures, and the results support the idea that the peak structures observed in lower temperature is associated with superconductivity. We were motivated to use the turnbuckle DAC in order to facilitate the study of the superconducting transition in β'-$(BEDT\text{-}TTF)_2ICl_2$, which has the highest T_C among the organic charge-transfer complexes. We were able to successfully measure the resistivity in magnetic fields under high pressures. Using this combination of techniques will allow us to study pressure induced phase-transitions and the associated mechanisms in other materials in the hope of discovering new pressure induced superconductors.

ACKNOWLEDGMENTS

One of the authors, Kano, wishes to thank Ms. Kazuko Nakazawa (Koyama) for supporting us on pressure calibration of the turnbuckle DAC. This work was partially supported by Grants-Aid for Scientific Research (No. 16038209) from the Ministry of Education, Culture, Sports, Science and Technology.

REFERENCES

1. M. Miyashita, K. Uchiyama K. Satoh, N. Mori, H. Okamoto, K. Miyagawa, K. Kanoda, M. Hedo, and Y. Uwatoko, *J. Phys. Soc. Jpn.* **72**, 468-471 (2003).

Competition between Singlet and Triplet Pairing Superconductivity when Spin and Charge Fluctuations Coexist

Kazuhiko Kuroki* and Yukio Tanaka†

*Dept. of Applied Physics and Chemistry, Univ. of Electro-Communications, Chofu, Tokyo, 182-8585, Japan
†Dept. of Applied Physics, Nagoya University, Nagoya, 464-8603, Japan

Abstract. We study the competition between spin singlet and triplet pairings in a quasi-one-dimensional organic superconductor $(TMTSF)_2PF_6$, where coexistence of $2k_F$ spin and $2k_F$ charge fluctuations along with the disconnectivity of the Fermi surface favors spin-triplet f-wave-like pairing. Comparison with the situation for a certain model of a cobaltate superconductor $Na_xCoO_2 \cdot yH_2O$ shows that whether the charge fluctuations favor singlet or triplet pairing depends on the character of the underlying spin fluctuations.

Keywords: superconductivity, $(TMTSF)_2X$, spin and charge fluctuations, singlet and triplet pairings
PACS: 74.70.Kn,74.20.-z,74.20.Rp

Correlation between spin fluctuations and the pairing symmetry in unconventional superconductors has been of great interest especially after the discovery of d-wave pairing near antiferromagnetic ordering in the cuprates. In some of the possible unconventional superconductors, the character of the spin fluctuations and pairing symmetry do not seem to go hand in hand at first glance. For example, in a quasi-one-dimensional organic superconductor $(TMTSF)_2PF_6$, superconductivity lies next to the $2k_F$ spin density wave (SDW) phase in the pressure-temperature phase diagram, which naively suggests that spin-singlet d-wave-like pairing mediated by antiferromagnetic (in the wide sense of the term) spin fluctuations. However, several experiments suggest that the pairing occurs in the spin-triplet channel.[1, 2]

As a possible solution for this puzzle, we have proposed that if $2k_F$ $(= \pi/2$ because the band is quarter-filled) charge fluctuations coexist with $2k_F$ spin fluctuations, as suggested from an experimental fact that CDW actually coexists with SDW[3, 4], triplet 'f-wave' like pairing, which is essentially a fourth neighbor pairing having a gap of the form $\sin(4k_x)$ (Fig.1(b)) may become competitive against singlet 'd-wave' like pairing, which is a second neighbor pairing with a gap form $\cos(2k_x)$ (Fig.1(a)).[5] This close competition arises because (i) the number of nodes of the f-wave gap on the Fermi surface is the same as that of the d-wave because of the disconnectivity of the Fermi surface, and (ii) the pairing interactions due to spin and charge fluctuations have the form $V^s = \frac{3}{2}V_{\rm sp} - \frac{1}{2}V_{\rm ch}$ for singlet pairing and $V^t = -\frac{1}{2}V_{\rm sp} - \frac{1}{2}V_{\rm ch}$ for triplet pairing, where $V_{\rm sp}$ and $V_{\rm ch}$ are contributions from spin and charge fluctuations, respectively, so that the absolute values of V_s and V_t become comparable when $V_{\rm sp} \simeq V_{\rm ch}$.

Here we adopt a microscopic model, $H = -\sum_{<i,j>,\sigma} t_{ij} c^\dagger_{i\sigma} c_{j\sigma} + U\sum_i n_{i\uparrow} n_{i\downarrow} + \sum_{<i,j>} V_{ij} n_i n_j$, on a quasi-one-dimensional lattice to verify the above phenomelogical argument, where we take into account intra (t) and interchain $(t_\perp)$ hoppings and also repulsive interactions up to third nearest neighbors (U, V, V', V'') within the chains and nearest neighbor repulsions $V_\perp$ between the chains (Fig.2). We adopt random phase approximation (RPA) to obtain the pairing interaction, and then we solve the gap equation to obtain T_c, which is determined as the temperature where the eigenvalue of the gap equation λ reaches unity.[6]

In Fig.3, we show λ as functions of T for parameter values where $2k_F$ spin and charge fluctuations coexist with similar magnitude. It can be seen that f-wave and d-wave closely compete, but the former dominates over the latter. Similar results have recently been obtained by

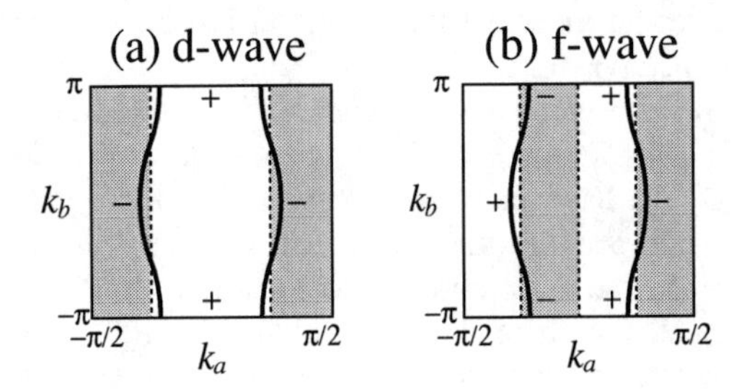

FIGURE 1. d-wave (a) and f-wave (b) Gap Functions Shown Schematically along with the Fermi Surface (Solid Curves). The dashed lines represent the nodes of the gap, whose k_b dependence is omitted for simplicity.

CP850, *Low Temperature Physics: 24th International Conference on Low Temperature Physics;*
edited by Y. Takano, S. P. Hershfield, S. O. Hill, P. J. Hirschfeld, and A. M. Goldman

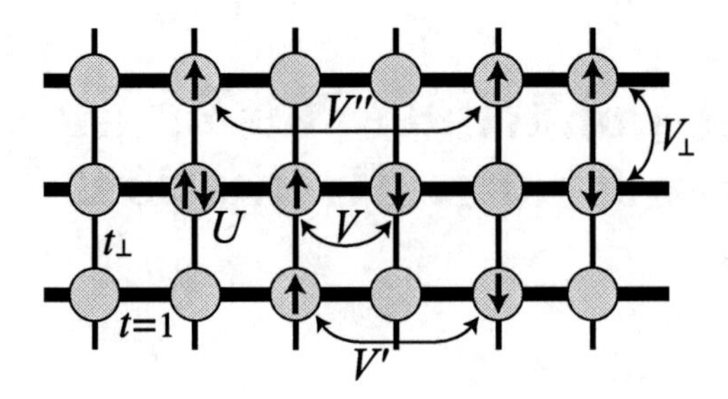

FIGURE 2. Model of the Present Study.

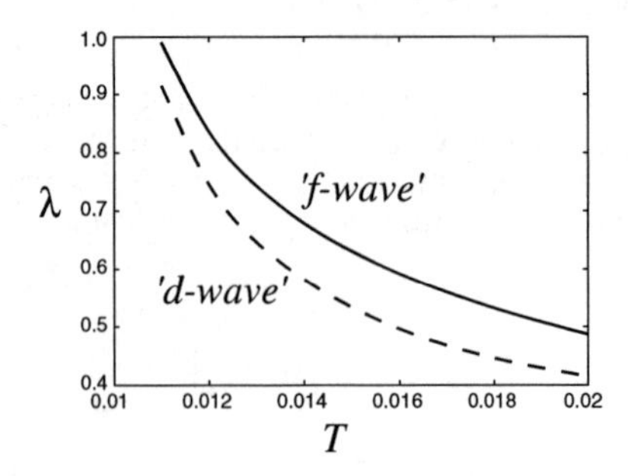

FIGURE 3. Eigenvalue of the Gap Equation Plotted as Functions of Temperature for d-wave and f-wave Pairings. $t_\perp = 0.2$, $U = 1.7$, $V = 0.8$, $V' = 0.45$, $V'' = 0.2$, and $V_\perp = 0.4$, all in units of t.

several groups using different approaches. [7, 8]

The present result can intuitively be understood in real space as follows. When $2k_F$ SDW and $2k_F$ CDW coexist, spin and charge are aligned in a pattern shown in Fig.4(a).[9] When this pattern 'melts' to become metallic, a pairing between parallel spins which are separated by four lattice spacings (f-wave) is likely to take place, while a pairing between antiparallel spins separated by two lattice spacings (d-wave) tends to be degraded by the next nearest neighbor repulsion V'. An important point here is that distant pairings are usually unfavorable because they result in additional nodes in the gap function, but in the present case the additional node in the f-wave gap does not intersect the Fermi surface because of the disconnectivity of the Fermi surface.

Now, it is interesting to compare the above situation to that in a model for a cobaltate superconductor $Na_xCoO_2 \cdot yH_2O$. We have previously proposed a model[10] that takes into account the disconnected pocket Fermi surfaces and the characteristic band structure found in the band calculation of this material.[11] In this model, a nearly ferromagnetic spin fluctuation arises due to a large density of states near the Fermi level located close to the top of the band. The ferromagnetic nature of the spin fluctuation induces spin-triplet f-wave pairing, but if we turn on sufficiently large off-site repulsions, triplet pairings give way to singlet pairings. What is happen-

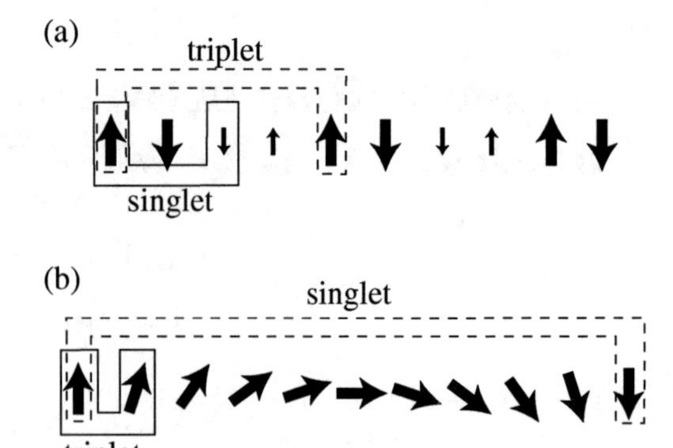

FIGURE 4. Spin and Charge Configuration when $2k_F$ SDW and CDW Coexist in a Quarter Filled System (a). Nearly ferromagnetic spin alignment (b). In both cases, dashed and solid lines represent possible pairings when the density waves 'melt' to become metallic.

ing here is just the opposite of what we have seen above. Namely, when spin fluctuations are nearly ferromagnetic (spin alignment with a long wave length), triplet pairs are formed at close distances (Fig.4(b)), which are degraded by off-site repulsions, giving way to singlet-pairings formed at further distances. Here again, the disconnectivity of the Fermi surfaces assists distant pairings.[10]

From the two cases considered here, we can conclude that the effect of the off-site repulsions, or the charge fluctuations, on the pairing symmetry competition depends on the charactor of the underlying spin fluctuations, and also on the shape of the Fermi surfaces.

Numerical calculations have been done at the Computer Center, ISSP, University of Tokyo. This study has been supported by Grants-in-Aid for Scientific Research from the Ministry of Education, Culture, Sports, Science and Technology of Japan, and from the Japan Society for the Promotion of Science.

REFERENCES

1. I.J. Lee *et al.*, *Phys. Rev. Lett.* **88**, 017004 (2002).
2. I.J. Lee *et al.*, *Phys. Rev. Lett.* **78**, 3555 (1997).
3. J. P. Pouget and S. Ravy, *J. Phys. I* **6**, 1501 (1996).
4. S. Kagoshima *et al.*, *Solid State Comm.* **110**, 479 (1999).
5. K. Kuroki *et al. Phys. Rev. B* **63**, 094509 (2001).
6. Y. Tanaka and K. Kuroki, *Phys. Rev. B* **70**, 060502 (2004); K. Kuroki and Y. Tanaka, *J. Phys. Soc. Jpn.* **74**, 1694 (2005).
7. Y. Fuseya and Y. Suzumura, J. Phys. Soc. Jpn. **74**, 1263 (2005).
8. J.C. Nickel *et al.*, cond-mat/0502614.
9. N. Kobayashi *et al. J. Phys. Soc. Jpn.* **67**, 1098 (1998).
10. K. Kuroki *et al. Phys. Rev. Lett.* **93**, 077001 (2004); *Phys. Rev. B* **71**, 024506 (2005).
11. D.J.Singh, *Phys. Rev. B* **61**, 13397 (2000).

Observation of a Non-Magnetic Impurity Effect in the Organic Superconductor $(TMTSF)_2ClO_4$

S. Takahashi,[a] S. Hill,[a,*] S. Takasaki,[b] J. Yamada[b] and H. Anzai[b]

[a]Department of Physics, University of Florida, Gainesville, FL 32611-8440, USA
[b]Department of Material Science, University of Hyogo, Kamigori-cho, Hyogo 678-1297, Japan

Abstract. We report a non-magnetic impurity effect on the superconductivity observed in $(TMTSF)_2ClO_4$. By varying the sample cooling rate in the vicinity of an anion ordering transition at T_{AO} = 24 K, we can systematically control the amount of disorder in the sample. We then use dc transport and microwave periodic orbit resonance measurements (closely related to cyclotron resonance) to determine the critical temperature T_c and the transport scattering time τ. We find a simple relationship between the two quantities, which strongly suggests that the superconductivity is suppressed by non-magnetic impurities, thus supporting the scenario for unconventional *p*-wave superconductivity in this material.

Keywords: triplet superconductivity, non-magnetic impurity effect, anion ordering, TMTSF and quasi-one-dimensional.
PACS: 74.70.Kn, 74.62.Dh, 74.20.Mn, 72.15.Gd

INTRODUCTION

The nature of the superconductivity in the quasi-one-dimensional (Q1D) organic superconductor $(TMTSF)_2X$ ($X=ClO_4$, PF_6 etc) remains an open-question even though it was first discovered 26 years ago [1]. It was believed for a long time that $(TMTSF)_2X$ was a conventional BCS superconductor. However, recent experiments have shown evidence for possible triplet superconductivity, *e.g.* a divergence of H_{c2} for X = ClO_4 and PF_6 [2], no NMR Knight shift through T_c for X = PF_6 [3], and a non-magnetic impurity effect for the X = $(ClO_4)_{1-x}(ReO_4)_x$ alloy [4]. Here we report on a non-magnetic impurity effect associated with disordered ClO_4 anions, which undergo an order-disorder transition at a temperature T_{AO} = 24 K in the X = ClO_4 salt. The degree of ordering is sensitive to the cooling rate at T_{AO}. Our investigation is focused on the effect of this disorder on the superconducting critical temperature T_c. We characterize the disorder using a periodic-orbit resonance (POR) technique, which enables us to determine both the Fermi velocity v_F and the quasiparticle scattering time τ in the normal state [5]. We find a direct correlation between T_c and τ, suggesting that pair-breaking due to non-magnetic impurities is responsible for the suppression of T_c.

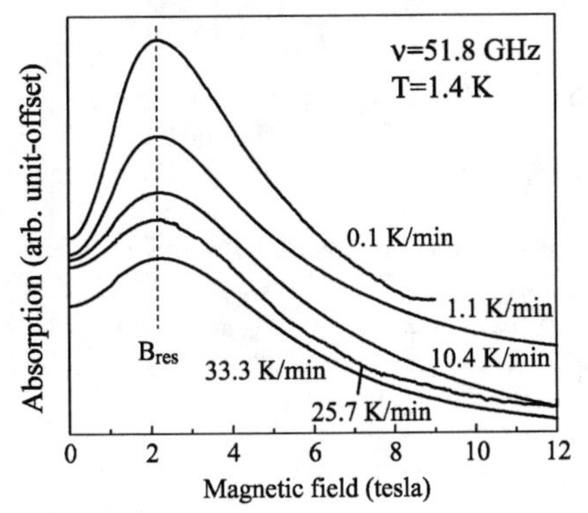

FIGURE 1. Microwave absorption as a function of magnetic field.

EXPERIMENTS

We separately performed POR and dc transport measurements in order to determine the cooling rate dependence of τ and T_c, respectively.

Study of the Periodic-Orbit Resonance

The POR measurements were performed using a millimeter vector network analyzer and a cavity perturbation technique [6]. A 17 T superconducting solenoid with a He flow cryostat was used for magnetic field and temperature control. The field orientation was varied using a rotating cavity [7]. We first used the angle-dependence of the POR to align the sample with the applied field parallel to c^* [8]. This was carried out in the relaxed state achieved via the slowest cooling rate (0.01 K/min). We then studied

CP850, *Low Temperature Physics: 24th International Conference on Low Temperature Physics;*
edited by Y. Takano, S. P. Hershfield, S. O. Hill, P. J. Hirschfeld, and A. M. Goldman

the cooling rate dependence of the POR, particularly the resonance linewidth from which we determine τ. Prior to each cooling, the sample was heated to 50 K, and then cooled to 15 K at the appropriate rate. The range of cooling rates achievable in the flow cryostat was 0.01 to 37 K/min. In addition to determining τ, we studied the angle-dependence of the POR to ensure that the cooling rate did not affect v_F, i.e. the band structure. Fig. 1 shows microwave absorption as a function of magnetic field at different cooling rates; the data were obtained at 1.4 K and 51.8 GHz. The Lorentzian-like POR is centered at ~2 T. As seen in the figure, the position of the POR is independent of the cooling rate. However, the resonance becomes broader with increasing cooling rate. This confirms that faster cooling leads to increased anion disorder and to an enhancement of the quasiparticle scattering rate. Thus, we can control the concentration of non-magnetic impurities in this way.

dc Transport Measurements

The interlayer resistance R_{zz} was measured for various cooling rates using a four-probe low-frequency (17 Hz) lock-in technique. In order to measure the superconducting critical temperature, T_c (onset $T_c \sim$ 1.2 K in the relaxed state), we employed a home-built ^{3}He cryostat [7]. We employed an identical cooling procedure as used for the POR measurements, achieving cooling rates between 0.03 and 28 K/min in the ^{3}He cryostat. The resistance was measured from 50 K down to 0.5 K. Each trace (not shown) displayed a kink at around 24 K indicative of the anion ordering. The resistance data were identical between 50 K and T_{AO} = 24 K for different cooling rates. However, they became distinguishable below T_{AO}, with higher resistances measured for faster cooling rates. The temperature dependence of R_{zz} was nearly linear below 8 K. Therefore, we estimated the residual resistance $R_{res}(T = 0\text{ K})$ using a polynomial fit. More importantly, T_c shifted to lower temperatures as the cooling rate was increased. We determined T_c at the completion temperature, where the maximum dR_{zz}/dT and zero resistance lines intersect; the transition width varied from 0.1 to 0.3 K. These tendencies were completely systematic and reproducible for all cooling rates.

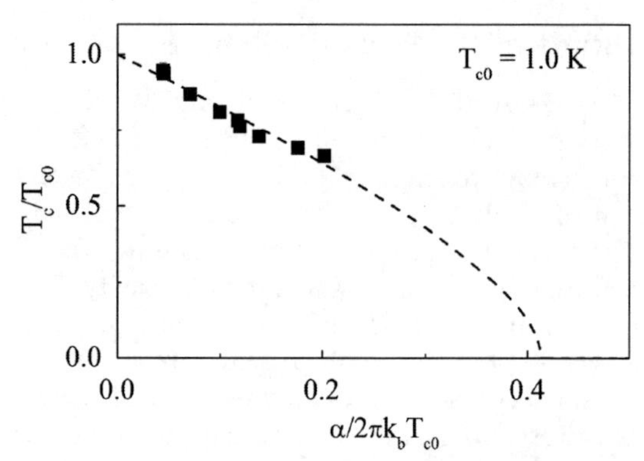

FIGURE 2. T_c/T_{c0} vs $\alpha/k_B T_{c0}$. The solid line is a fit to the universal function [9].

DISCUSSION AND SUMMARY

By comparing the dc transport and POR data, we find that changes in R_{zz} (and R_{res}) can be attributed to changes in the transport scattering time for all but the fastest cooling rates (we speculate that regions of the sample may become insulating for the fastest cooling rates, causing a departure from this behavior). We plot T_c as a function of the pair breaking strength α in Fig. 2. Here we make the assumption that the pair-breaking strength α (= $\hbar/2\tau_K$), which is inversely related to the pair breaking time τ_K, is proportional to the transport scattering rate, *i.e.* $\alpha \propto 1/\tau$. The data in Fig. 2 have been fit to the universal function in the Abrikosov-Gor'kov theory for the impurity effect on T_c [9],

$$\ln\frac{T_c}{T_{c0}} = \psi(\frac{1}{2}) - \psi(\frac{1}{2} + \frac{\alpha}{2\pi k T_c}), \qquad (1)$$

where ψ is the digamma function and T_{c0} is the critical temperature in the absence of impurities. The good agreement between the data and the fit suggests that the anion disorder suppresses T_c, supporting the scenario for unconventional spin-triplet (*p*-wave) superconductivity in $(TMTSF)_2ClO_4$.

ACKNOWLEDGMENTS

This work was supported by the National Science Foundation (DMR0196461 and DMR0239481).

REFERENCES

*Corresponding author: hill@phys.ufl.edu

1. T. Ishiguro, K. Yamaji and G.Saito, *Organic Superconductors*, Heidelberg: Springer-Verlag, 1998, pp. 4-11.
2. I. J. Lee, A. P. Hope, M. J. Leone and M. J. Naughton, *Synth. Met.* **70**, 747 (1995); I. J. Lee, P. M. Chaikin and M. J. Naughton, *Phys. Rev. B* **62**, R14669 (2000).
3. I. J. Lee et al., *Phys. Rev. Lett.* **88**, 017004 (2002).
4. N. Joo et al., *Eur. Phys. J. B* **40**, 43 (2004).
5. S. Takahashi et al., *Phys. Rev. B* **72**, 024540 (2005).
6. M. Mola and S. Hill, *Rev. Sci. Instrum.* **71**, 186 (2000).
7. S. Takahashi and S. Hill, *Rev. Sci. Instrum.* **76**, 023114 (2005).
8. A. E. Kovalev et al., *Phys. Rev. B* **66**, 134513 (2002).
9. R. D. Parks, *Superconductivity*, New York: M. Dekker, 1969, chap. 18 by K. Maki.

Fermi Surface and Electronic Properties of κ-$(BETS)_2FeCl_4$

T. Konoike[a], S. Uji[a], T. Terashima[a], M. Nishimura[a], T. Yamaguchi[a], K. Enomoto[a], H. Fujiwara[b], B. Zhang[c] and H. Kobayashi[c,d]

[a]*National Institute for Materials Science, Tsukuba, Ibaraki 305-0003, Japan*
[b]*Osaka Prefecture University, Sakai, Osaka 599-8570, Japan*
[c]*Institute for Molecular Science, Okazaki, Aichi 444-8585, Japan*
[d]*JST-CREST, Kawaguchi 332-0012, Japan*

Abstract. Shubnikov-de Haas (SdH) and angular dependent magnetoresistance oscillations are measured in the organic conductor κ-$(BETS)_2FeCl_4$. This salt has a quasi-two dimensional Fermi surface, which is almost consistent with the band structure calculation. The SdH oscillations suggest the presence of a large internal field of 10.6 T. The ratio of the interlayer to intralayer transfer integral is estimated as 1 / 146 from the width of the coherence peak.

Keywords: magnetic-field-induced superconductivity, organic conductor
PACS: 71.18.+y, 74.70.Kn

INTRODUCTION

The quasi-two-dimensional organic conductor κ-$(BETS)_2FeBr_4$ (Br salt) is known to show magnetic-field-induced superconductivity (FISC) [1]. For the stabilization of the FISC, both Zeeman and orbital effects should be suppressed. In the Br salt, the Zeeman effect is suppressed due to the large negative exchange interaction between the conduction π electrons on the BETS molecules and the localized Fe 3d spins (π-d interaction). The orbital effect is also strongly suppressed due to the low dimensionality of the system as long as the field is applied parallel to the conducting layers. The isostructural compound, κ-$(BETS)_2FeCl_4$ (Cl salt) shows successive antiferromagnetic and superconducting transitions (T_N = 0.45 K, T_c = 0.1 K) [2]. In order to obtain better understanding of the FISC, we have investigated the electronic state of the Cl salt by means of Shubnikov-de Haas (SdH) and angular dependent magnetoresistance oscillations (AMROs).

EXPERIMENT

The single crystals of the Cl salt were obtained by an electrochemical oxidation [3]. The resistance was measured by a conventional four-terminal ac technique with electric current perpendicular to the conducting *ac*-layers. The measurements were carried out by using a ^{4}He cryostat or a dilution refrigerator with superconducting magnets at NIMS.

RESULTS AND DISCUSSION

The inter-layer resistance as a function of magnetic field perpendicular to the conducting layers is shown in Fig. 1. There is no indication of superconductivity at zero field, though the Meissner signal is observed in the ac susceptibility measurement below 0.1 K [2]. The reason is unknown at present. A kink observed at 1.6 T is probably related to the canted antiferro-magnetic to paramagnetic transition of the Fe 3d spins. This structure is sample dependent, and is

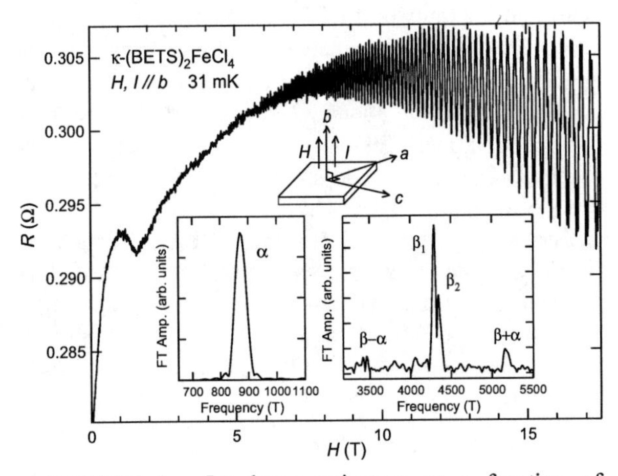

FIGURE 1. Interlayer resistance as a function of magnetic field. Insets show the FT spectra.

CP850, *Low Temperature Physics: 24th International Conference on Low Temperature Physics;* edited by Y. Takano, S. P. Hershfield, S. O. Hill, P. J. Hirschfeld, and A. M. Goldman

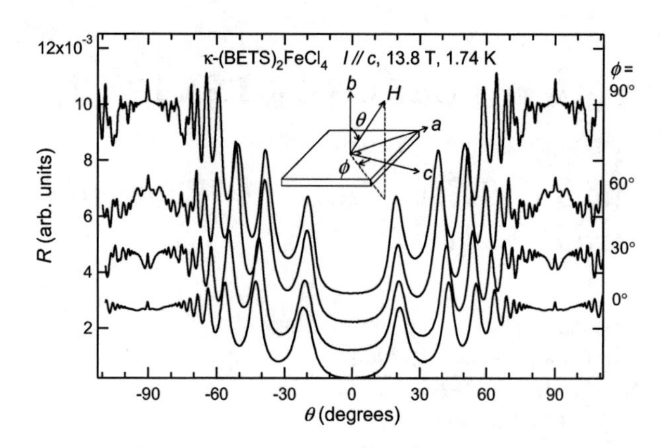

FIGURE 2. AMROs for various azimuthal angles ϕ.

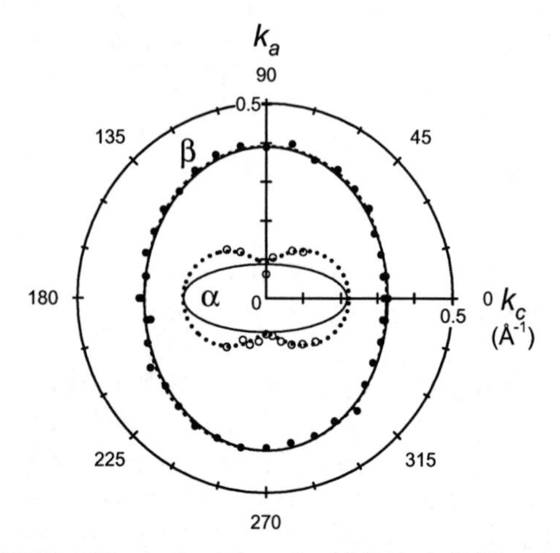

FIGURE 3. FS's obtained from the AMROs (solid lines).

inconsistent with the previous report, where the background resistance decreases with field monotonically [4].

In the figure, several SdH oscillations are evident. The Fourier transform (FT) spectra (insets of Fig. 1) show the two fundamental oscillations, α and β (β_1 and β_2) corresponding to the Fermi surface (FS) cross-sectional areas of about 20 and 100 % of the first Brillouin zone, respectively. These are almost consistent with the band structure calculation and the previous report [4]. As shown in the right inset, we observed a splitting of β (β_1 and β_2) for the first time. In the presence of a large exchange field H_J, two SdH frequencies corresponding to the up and down spin electrons are observed, as pointed out by Cepas *et al.* [5]. In this case, we can estimate the exchange field from the difference of the frequencies; $\delta F = (1/4)\cdot g\cdot(m_{eff}/m_0\cdot H_J)$, where g is the g-factor, m_{eff} is the effective mass and m_0 is the free electron mass. Assuming g=2, we obtain the large exchange field of 10.6 T, which suggests the large π-d coupling in this salt.

Figure 2 shows the resistance as a function of the field angle θ, defined in the inset. Since the observed oscillatory behavior is periodic in $\tan\theta$, and the peak positions do not change with field (not shown), we can ascribe the oscillation to AMRO, whose origin is interpreted in terms of the periodic motion of the electrons on the corrugated FS. Using the period in $\tan\theta$ for each azimuthal angle ϕ and assuming simply corrugated FS, we can construct the FS as shown by the solid lines in Fig. 3. The obtained FS is almost consistent with the band structure calculation, and quite similar to that of the Br salt.

At $\theta = \pm 90°$, peaks in the resistance are observed in Fig. 2. This structure arises from the interlayer coherent motion of electrons on the small closed orbits formed on the sides of the slightly corrugated FS [6]. The ratio of the interlayer to intralayer transfer integral is estimated from the peak width [6]; $t_\perp/t_{//} = 1/146$. This value is larger than 1/170 for the Br salt, which suggests a larger orbital effect in the Cl salt.

In summary, we observed SdH oscillations and AMROs in the Cl salt and obtained a similar FS to that of the Br salt showing FISC. The presence of a large internal field of 10.6 T suggests that the π-d interaction is also fairly large in the Cl salt. The analysis of the coherence peak shows that the orbital effect is larger in the Cl salt than in the Br salt.

ACKNOWLEDGMENTS

This work is supported by a Grant-in-Aid for Scientific Research from the Ministry of Education, Culture, Sports, Science and Technology (No. 15073225). One (T. K.) of the authors is supported by Research Fellowships of the Japan Society for the Promotion of Science for Young Scientist.

REFERENCES

1. T. Konoike *et al.*, *Phys. Rev. B* **70**, 094514 (2004).
2. T. Otsuka *et al.*, *J. Solid State Chem.* **159**, 407 (2001).
3. H. Kobayashi *et al.*, *J. Am. Chem. Soc.* **118**, 368 (1996).
4. N. Harrison *et al.*, *Phys. Rev. B* **57**, 8751 (1998).
5. O. Cepas *et al.*, *Phys. Rev. B* **65**, 100502 (2002).
6. N. Hanasaki, *et al.*, *Phys. Rev. B* **57**, 1336 (1998).

High Field FISDW State in Organic Superconductor $(DMET\text{-}TSeF)_2I_3$

Kokichi Oshima*,†, Michael J. Naughton†, Eiji Ohmichi**, Toshihito Osada** and Reizo Kato‡

*Graduate School of Natural Science and Technology, Okayama University, Okayama 700-8530, Japan
†Department of Physics, Boston College, Chestnut Hill, MA 02467-3809, USA
**ISSP, The University of Tokyo, Kashiwa 277-8581, Japan
‡Condensed Molecular Materials Laboratory, RIKEN, Wako 351-0198, Japan

Abstract. The Field Induced Spin Density Wave (FISDW) state has been studied for the title compound. The FISDW has been widely studied as one of the unique properties of the quasi one-dimensional system, but the FISDW state of the title compound has not been known until recently. We have reported the existence of an FISDW state above 25 T [1]. To decide the possible higher FISDW phase boundaries, we performed pulsed field transport measurements and magnetic torque measurements up to 40 T. And we have newly discovered another phase boundary around 35 T. It was found that the nested Fermi surface has a larger area than those of the other DMET-TSeF family salts. We discuss the origin of the high FISDW threshold field considering the crystal anisotropy, and propose a role of the inter-layer coupling as one of the possible origins.

Keywords: Quasi one-dimensional system, Organic metal and superconductor
PACS: 73.43.Qt,75.30.Fv

INTRODUCTION

The FISDW state is one of the unique properties of the quasi one-dimensional system [2]. We have reported the title compound as a superconductor with T_c around 0.5 K [3], but we could not identify any FISDW state up to the magnetic field of 12 T. Since other DMET-TSeF salts known to date with quasi one-dimensional Fermi surfaces at the lowest temperatures show FISDW states, it has been a puzzle that we cannot find any FISDW state in the title compound. We have recently reported the existence of a field induced state above 25 T [1]. To investigate the reason why the threshold is so high, we started pulsed field measurements. Another purpose is to confirm that the state is actually magnetic. We have found a new phase boundary using transport measurements. We ascribe one of the possible reasons of the high threshold field to a larger coupling along the b-axis.

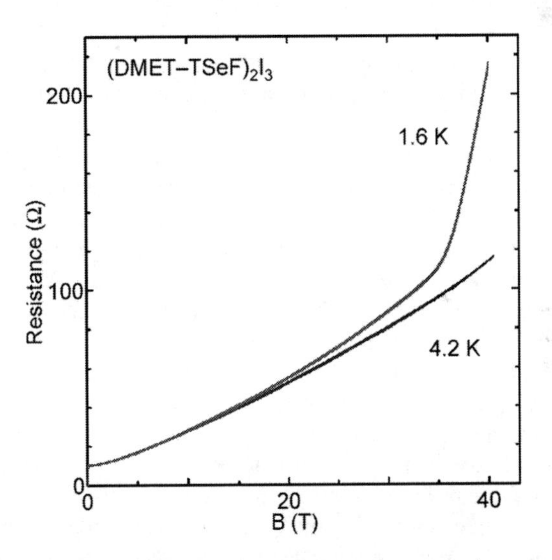

FIGURE 1. In-plane magnetoresistance at liquid ^{4}He temperature. The sample experienced several resistance 'jumps' before reaching the lowest temperature.

EXPERIMENTAL

The samples are made using the standard electrochemical oxidation of DMET-TSeF molecules [4]. The magnetoresistance has been measured using the 4-probe method in a cryostat which can be cooled to liquid ^{3}He temperatures. As reported in a recent paper, this compound shows clear Lebed-resonances [1]; therefore we can easily see that it has quasi one-dimensional Fermi surfaces even at the lowest temperatures. The crystal structure of the present salt is similar to those of the salts with anions AuI_2 and $AuBr_2$, both of which show FISDW states below 10 T at ambient pressure and under pressure, respectively. The magnetic field has been obtained using a pulsed field coil immersed in liquid nitrogen facilitated at the ISSP - Tokyo.

CP850, *Low Temperature Physics: 24th International Conference on Low Temperature Physics;*
edited by Y. Takano, S. P. Hershfield, S. O. Hill, P. J. Hirschfeld, and A. M. Goldman

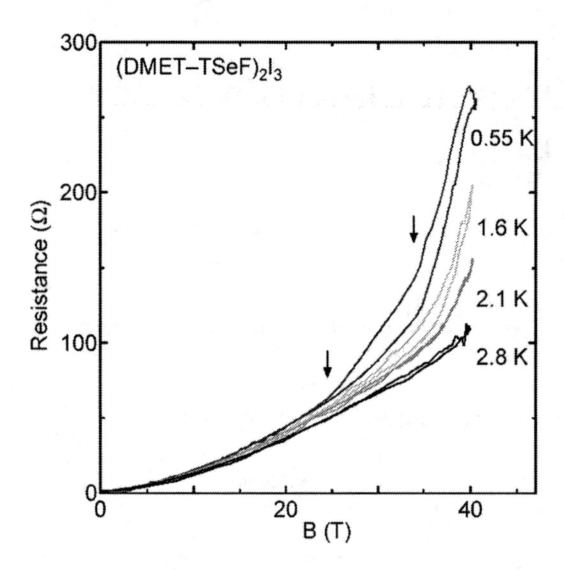

FIGURE 2. In-plane magnetoresistance down to liquid ^{3}He temperatures for a different sample from that in Fig. 1. Arrows indicate the possible phase boundaries.

RESULTS AND DISCUSSIONS

The results of magnetoresistance measurements are shown above in Fig. 1 and Fig. 2. The figures show upward and downward field sweeps. The samples have experienced several resistance 'jumps' before reaching the lowest temperatures. The new boundary around 35 T is clearly observed in Fig. 1 for 1.6 K data. Fig. 2 indicates the magnetoresistance for different temperatures down to liquid ^{3}He temperature. The arrows indicate the possible boundaries of the FISDW state. The origin of the behavior with hysteresis for the different sweep directions is not known at present. Unfortunately we could not confirm these states to be magnetic by the magnetic torque measurements, therefore we cannot say that this is really an FISDW state from our data. But, from the evidences on the family compounds and the result of the Lebed-resonances indicating the quasi one-dimensional Fermi surface, we can safely state that the boundaries are due to the FISDW state.

Therefore, the assumption which we made in a recent paper that 25 T may be the last boundary should be corrected. In these data we measured along the most conductive direction (a-axis) and applied field is along the crystal b-axis. We should note the b-axis in DMET-TSeF corresponds to that of c-axis in the TMTSF system. We should also note that we had no clear anomalous behavior indicating FISDW below 25 T in the measurement using a static magnetic field up to 27 T at 0.5 K. Therefore we can consider the threshold field to be 25 T. The FISDW frequency is about 88 T (period 0.0114 T^{-1}) if the boundaries can be considered to be periodic. Compared with the frequency for the (DMET-TSeF)$_2$AuCl$_2$ salt (about 45 T [5]), the frequency is nearly double the AuCl$_2$ value. On the contrary, in the AuI$_2$ salt case, the boundaries are not periodic. The value of the period for the I$_3$ salt is comparable for the separation between the N=0 and N=1 states in the AuI$_2$ salt (0.0133 T^{-1} [6]).

From the standard theory, an FISDW state is made by field dependent nesting of Fermi surfaces [2]. Then the boundaries are periodic in 1/B, and the produced pocket area perpendicular to the field is larger when the period is smaller. Therefore, the pocket for the I$_3$ should be larger than those of other family salts. We do not know if 35 T is the last boundary or not. To decide, we need higher field, around 60 T. Another possibility is to measure the Hall resistance for these states. As is known from the quantized behavior for the family salts, zero Hall resistance can be expected for the last FISDW state. The room temperature Fermi surfaces for the salts known to date in DMET-TSeF family are quasi one-dimensional, and which are comparable with the famous TMTSF salts [4]. And the transfer integrals of the family salts are not so different from each other as discussed in Ref. [3].

A low temperature crystal structure change is one possibility for the origin of the high threshold field. As noted before [1], we have examined the low temperature crystal structure, but found no structural change down to 90 K. Though not shown in this paper, recent measurements on ρ_{zz} revealed that this component decreases more rapidly than the other components as temperature decreases. This indicates that the Fermi surface nesting vector cannot simply be treated by the components in the two-dimensional plane, but the third component should be considered at the same time. This kind of effect is already shown in the TMTSF compounds [7][8]. In this context, we can speculate that the threshold field is dependent on the third directional coupling (in another words, warping) in the DMET-TSeF Fermi surfaces. Therefore, the difference in the size of the pockets cannot be simply understood by the two-dimensional standard theory using extended Hückel-type transfer integrals. This may be related to the aperiodic FISDW behavior in the (DMET-TSeF)$_2$AuI$_2$ salt.

REFERENCES

1. K. Oshima *et al*, *Synth. Met. in press* (2005).
2. T. Ishiguro *et al*, *Organic Superconductors*, Berlin:Springer, 1998, pp. 313–364.
3. K. Oshima *et al*, *Synth. Met.*, **70**, 861 (1995).
4. R. Kato *et al*, *Synth. Met.*, **61**, 199 (1993).
5. N. Biskŭp *et al*, *Phys. Rev. B*, **60**, R15005 (1999).
6. N. Biskŭp *et al*, *Phys. Rev. B*, **62**, 21 (2000).
7. T. Takahashi *et al*, *Physica B*, **143**, 417 (1986).
8. J. Delrieu *et al*, *Physica B*, **143**, 412 (1986).

Triplet Superconductivity in a Two-Chain Hubbard Model by the Ring-Exchange Mechanism

T. Shirakawa*, S. Nishimoto† and Y. Ohta**

*Graduate School of Science and Technology, Chiba University, Chiba 263-8522, Japan
†Institut für Theoretische Physik, Universität Göttingen, D-37077 Göttingen, Germany
**Department of Physics, Chiba University, Chiba 263-8522, Japan

Abstract. We calculate the binding energy of holes and pair correlation functions in the two-chain zigzag-bond Hubbard model by the density-matrix renormalization group (DMRG) method and show that the model can be superconducting in spin-triplet channel when we make an appropriate choice of the signs of hopping integrals for the ferromagnetic ring-exchange mechanism to work. We argue that the mechanism proposed may have possible relevance to the triplet superconductivity in $(TMTSF)_2X$.

Keywords: triplet superconductivity, ring exchange, DMRG, two-chain Hubbard model, zigzag ladder
PACS: 74.20.Mn, 71.10.Fd, 74.70.Kn, 75.10.-b

Spin-triplet superconductivity has attracted much attention in the field of strongly correlated electron systems. Among them are Bechgaard salts $(TMTSF)_2X$, for which recent experimental studies [1] have shown that the Cooper pairs are in the spin-triplet state. The low-energy electronic states of the system may be described by the coupled one-dimensional (1D) Hubbard model with weak interchain coupling. The hopping integrals have the unique structure as shown in Fig. 1 [2]; they show an alternating sign change along the zigzag bonds connecting two chains, while the sign along the 1D chain is always positive in the electron notation (i.e., for the bands of 3/4 filling of electrons). We hereafter use the hole notation for convenience; i.e., the system is in the quarter filling of holes. The signs of hopping integrals are then always negative along the 1D chain and are changing signs along the zigzag bonds.

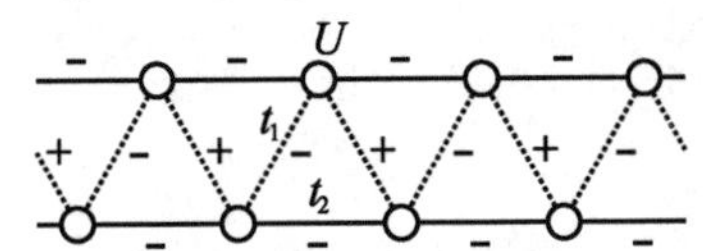

FIGURE 1. Schematic representation of the two-chain zigzag-bond Hubbard model in the hole notation for the signs.

Let us then notice that, in Hubbard models defined on such triangular-lattice related structures with proper signs of hopping integrals, the ring exchange mechanism of two spins on a triangle provides the system with ferromagnetic spin correlations [3, 4]. The structure of hopping integrals of $(TMTSF)_2X$ [2], in fact, satisfies the ferromagnetic sign rule $t_1t_2t_3 > 0$ (in the hole notation) for three hopping integrals of all the triangles.

We calculate the ground state of the relevant two-chain Hubbard model by the density-matrix renormalization group (DMRG) method [5] and show that indeed the system is metallic with short-range ferromagnetic correlations. Furthermore, we show that the attractive interaction acts between holes, being caused by the gain in kinetic energy due to ring exchange of holes. We also show that the superconducting pair correlations in the spin-triplet channel extend long-ranged in power-law length dependence, while the pair correlations in the singlet channel as well as the spin-density-wave (SDW) correlations decay exponentially, indicating that the system is in the state of spin-triplet superconductivity. We thus propose that the triplet superconductivity is realized in the two-chain Hubbard model by the ring-exchange mechanism [6].

The two-chain Hubbard model (see Fig. 1) is defined by the Hamiltonian

$$H = \sum_{<ij>\sigma} t_{ij}(c^\dagger_{i\sigma}c_{j\sigma} + \mathrm{H.c.}) + U\sum_i n_{i\uparrow}n_{i\downarrow} \qquad (1)$$

in the standard notations. We assume the quarter filling of holes. t_{ij} is the rearest-neighbor hopping integral; we take $t_2 = -1$ as the unit of energy and t_2 to change signs alternately as in Fig. 1. The signs of t_1 and t_2 can instead be taken all positive because the models where the product of the three hopping integrals of the triangles is positive are equivalent under canonical transformation. We apply the DMRG method for clusters of length L (containing $2L$ sites) with open boundary condition; we use L up to 128 with keeping up to $m \simeq 4500$ density-matrix eigenstates, whereby the discarded weights are typically of the order $10^{-7} - 10^{-6}$ and the ground-state energy is obtained in the accuracy of $\sim 0.001|t_2|$.

We first of all find that the system is metallic with vanishing charge gap $\Delta_c(L)$ in the thermodynamic limit

CP850, *Low Temperature Physics: 24th International Conference on Low Temperature Physics;*
edited by Y. Takano, S. P. Hershfield, S. O. Hill, P. J. Hirschfeld, and A. M. Goldman

$L \to \infty$ [7]. We then find that the spin gap $\Delta_s(L)$ vanishes (or becomes exponentially small if it exists) at $L \to \infty$ and that the spin correlation decays with nearly exponential length dependence; the apparent contradiction between these two (as well as the rapid decrease in the spin correlations at short distances) may be reconciled by the spin-triplet pairing present in our model.

We now calculate the binding energy of holes defined as $\Delta_b = \lim_{L\to\infty} \Delta_b^{\pm}(L)$ with $\Delta_b^{\pm}(L) = E_L(N \pm 2) + E_L(N) - 2E_L(N \pm 1)$, where $E_L(N)$ is the groundstate energy of the system with N holes. We first confirm that the compressibility evaluated from the calculations of $\Delta_c(L)$ and $\Delta_b^{\pm}(L)$ agrees well with each other as the gradient of the two curves (if the factor 4 is taken into account) for small 1 L regions agrees well. The positive gradient indicates also that the system is thermodynamically stable against phase separation. We then find that the extrapolated value Δ_b is negative; i.e., the attractive interaction acts between holes in the thermodynamic limit. The energy gain responsible for this may come from the motion of two holes running around the triangle, avoiding the on-site repulsion U and exchanging spins for triplet coupling, i.e., from the ring-exchange mechanism. The pairing mechanism is thus kinetic in origin.

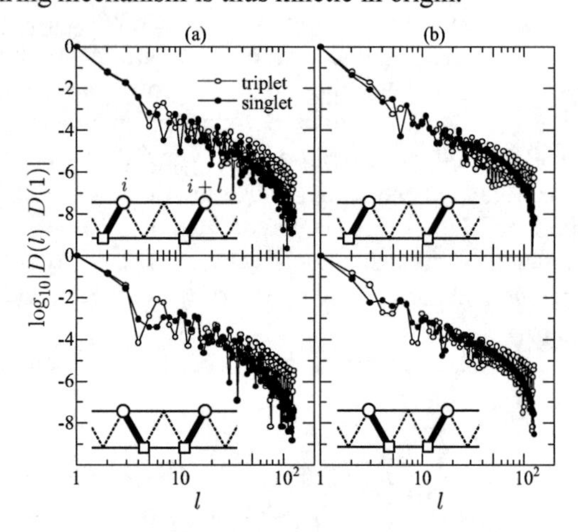

FIGURE 2. Pair correlation functions $D(l)$ calculated at $L = 128$ and $U = 10$ with (a) $|t_1| = 0.25$ and (b) $|t_1| = 0.5$.

We also calculate the pair correlation function defined as $D(l) = \langle \Delta_{i+l}^{\dagger} \Delta_i \rangle$ with $\Delta_i = c_{i\uparrow} c_{i+r\downarrow} - c_{i\downarrow} c_{i+r\uparrow}$ for singlet pairs and $\Delta_i = c_{i\uparrow} c_{i+r\downarrow} + c_{i\downarrow} c_{i+r\uparrow}$ for triplet pairs where $i + r$ denotes the neighboring sites of i. The results at $L = 128$ are shown in Fig. 2. We find that $D(l)$ shows the power-law length dependence for the interchain triplet pairing but decays exponentially for the singlet pairing as well as for the triplet pairing on the single 1D chain. These are the case also at $L = 64$, indicating that the size of the clusters used is sufficiently large. Then, combined with the attractive interactions of two holes shown above, our results indicate that the system should be in the state of spin-triplet superconductivity where the pairing of holes occurs between the interchain nearest-neighbor sites.

Finally, let us consider possible relevance of our results to Bechgaard salts. The appropriate model is a coupled sequence of the 1D chains. We have cut out two neighboring chains as a minimum model to seek for consequences of the interchain coupling. Since the ring-exchange mechanism works also for models of more than two chains [3] where the hopping integrals of all the triangles satisfy the ferromagnetic sign rule, we point out that the triplet pairing state obtained in the two-chain model may persist also in quasi-1D (or 2D) systems; future studies will be interesting. Also pointed out is that there is intriguing competition between the SDW and spin-triplet superconducting states in the experimental pressure-temperature phase diagram [8]. Because small intersite Coulombic repulsions exist in the materials [9], we examine their effects; we find that the inclusion of a realistic value $V_2 \simeq 1$ [9] does not change our results. However, if we include V_1 between the 1D chains as well, the triplet pair correlation becomes less long-ranged and thus the SDW correlation can be comparable with it. A true long-range order may be selected among these competing correlations when the two (or three) dimensionality is taken into account, although to predict which order is realized is beyond the scope of the present work. We hope that future quantitative analyses will help clarifying the issue.

ACKNOWLEDGMENTS

We thank G. I. Japaridze, R. M. Noack, K. Sano, Y. Suzumura, and M. Tsuchiizu for useful discussions. This work was supported in part by Grants-in-Aid for Scientific Research of Japan. A part of computations was carried out at the Research Center for Computational Science, Okazaki Research Facilities, and the Institute for Solid State Physics, University of Tokyo.

REFERENCES

1. I. J. Lee *et al.*, *Phys. Rev. B* **68**, 092510 (2003); *Phys. Rev. Lett.* **88**, 207002 (2002), and J. I. Oh and M. J. Naughton, *Phys. Rev. Lett.* **92**, 067001 (2004).
2. L. Ducasse *et al.*, *J. Phys. C: Solid State Phys.* **19**, 3805 (1986).
3. K. Penc *et al.*, *Phys. Rev. B* **54**, 4056 (1996).
4. S. Daul and R. M. Noack, *Phys. Rev. B* **58**, 2635 (1998).
5. S. R. White, *Phys. Rev. Lett.* **69**, 2863 (1992); *Phys. Rev. B* **48**, 10345 (1993).
6. Y. Ohta *et al.*, *Phys. Rev. B* (2005), in press.
7. S. Nishimoto and Y. Ohta, *Phys. Rev. B* **68**, 235114 (2003).
8. D. Jérome, *Science* **252**, 1509 (1991).
9. S. Nishimoto, M. Takahashi, and Y. Ohta, *J. Phys. Soc. Jpn.* **69**, 1594 (2000).

Magnetization of Spinel Compound $CuRh_2S_4$ Under Pressure

Masakazu Ito, Kiyotaka Ishii, Fumihiko Nakamura and Takashi Suzuki

Department of Quantum Matter, ADSM, Hiroshima University, Higashi-Hiroshima 739-8530 Japan

Abstract. Magnetization measurements have been carried out under quasihydrostatic pressure P up to 0.74 GPa on the chalcogenide spinel compound $CuRh_2S_4$ with the superconducting transition temperature T_c between 4 and 5 K. The shielding volume fraction does not depend on pressure below 0.74 GPa. The value of T_c increases linearly with dT_c/dP = 0.59 K/GPa. For whole pressure range we measured, the upper critical field H_{c2} shows linear temperature dependence below 15 kOe. We also found the zero-temperature upper critical field $H_{c2}(0)$ increases proportional to pressure with $dH_{c2}(0)/dP$ = 5.51 kOe/GPa.

Keywords: $CuRh_2S_4$; Chalcogenide spinel compound; High pressure; Pressure-induced superconductor-insulator transition; Magnetization.
PACS: 74.70.Dd, 74.25.Ha, 74.62.Fj

INTRODUCTION

The thiospinel compound $CuRh_2S_4$ is known for the superconductor with the superconducting transition temperature T_c between 4 and 5 K at 0.1 MPa [1]. Recently, we investigated electrical resistivity of $CuRh_2S_4$ under quasihydrostatic pressure up to 8.0 GPa and found a pressure-induced superconductor-insulator (SI) transition [2, 3]. With increasing pressure, T_c increases initially and then change to slightly decrease after having a maximum value 6.4 K at 4.0 GPa. With further compression, T_c of $CuRh_2S_4$ disappears abruptly at critical pressure P_{SI} between 5.0 and 5.6 GPa. In order to further investigate the effects of pressure on the superconducting properties of $CuRh_2S_4$, we have performed the magnetization measurement under quasi-hydrostatic pressure.

EXPERIMENTAL

A polycrystalline $CuRh_2S_4$ was prepared by a direct solid-state reaction as reported previously [2]. Temperature dependence of magnetization M in magnetic field H was measured with a Quantum Design MPMS SQUID magnetometer. The Magnetization was measured as a function of temperature in $2 \leq T \leq 6$ K during warm up in H after cooling to 2 K in the zero field (ZFC). Pressure up to 0.74 GPa was generated using a piston-cylinder Be-Cu clamp cell which can be attached with the sample rod of MPMS [4]. The equal volume mixture of Fluorinert FC70 and FC77 was used as the pressure transmitter.

RESULTS AND DISCUSSION

Figure 1 shows T dependence of magnetization divided by applying field (H = 50 Oe), M/H of $CuRh_2S_4$ at various pressures. For 0.1 MPa, the superconducting transition temperature is 4.46 K which is determined by the onset of M/H drop. The value of M/H at 2 K does not depend systematically on pressure. In other words, the shielding volume fraction of $CuRh_2S_4$ is not influenced by compression. The pressure dependence of T_c is shown in Fig. 2. T_c depends linearly on the pressure with the rate of dT_c/dP = 0.59 K/GPa. This value is consistent with the value of previous work by Shelton et al., ~0.5 K/GPa [5]. The pressure enhancement of T_c might arise from increasing of the Debye temperature θ_D by compression.

Figure 3 shows temperature dependence of the upper critical field H_{c2} determined from the M/H measurements under H up to 15 kOe at various pressures. H_{c2} show linear temperature dependence for all pressures and $(dH_{c2}/dT)_{Tc}$ can be obtained from the slope of the straight lines in Fig. 3. According to the theory of the type-II superconductor, the extrapolated

CP850, *Low Temperature Physics: 24th International Conference on Low Temperature Physics;* edited by Y. Takano, S. P. Hershfield, S. O. Hill, P. J. Hirschfeld, and A. M. Goldman

value of the zero-temperature upper critical field can be obtained from the relation [6, 7],

$$H_{c2}(0) = 0.693\ T_c\ (-dH_{c2}/dT)_{Tc}. \qquad (1)$$

Pressure dependence of $H_{c2}(0)$ is plotted in Fig. 4. With increasing pressure, $H_{c2}(0)$ increases linearly at the rate of $dH_{c2}(0)/dP$ = 5.51 kOe/GPa.

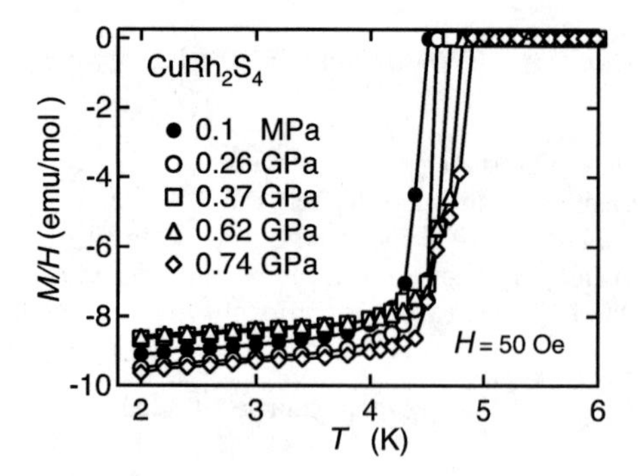

FIGURE 1. Temperature dependence of magnetization divided by applying field, M/H of $CuRh_2S_4$ for various pressures.

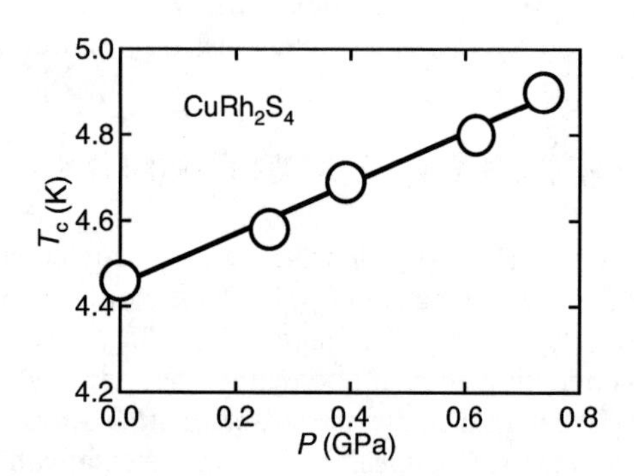

FIGURE 2. Pressure dependence of T_c. The solid line is guide to the eyes.

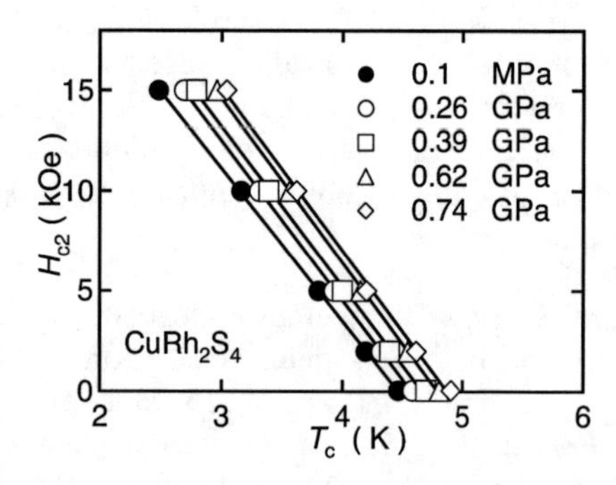

FIGURE 3. Temperature dependence of H_{c2} at various pressures. The straight lines are drawn through the data for each pressure to indicate dT_c/dP.

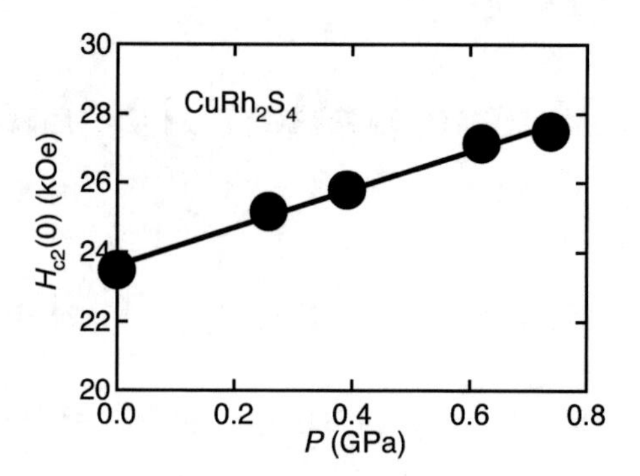

FIGURE 4. Pressure dependence of $H_{c2}(0)$. The solid line is guide to the eyes.

CONCLUSION

We have studied the magnetization of the chalcogenide spinel superconductor $CuRh_2S_4$ under pressure up to 0.74 GPa. The shielding volume fraction is independent of pressure below 0.74 GPa. With increasing pressure, T_c increases with the rate of dT_c/dP = 0.59 K/GPa. Furthermore, we found that the zero temperature upper critical field $H_{c2}(0)$ increases linearly with the rate of $dH_{c2}(0)/dP$ = 5.51 kOe/GPa by compression.

ACKNOWLEDGMENTS

This work was partially supported by the Grant-in-Aid for COE Research (No. 13CE2002), the Scientific Research (No. 16740205, No. 17340113) from MEXT of Japan and aid funds from Energia, Inc..

REFERENCES

1. N. H. Van Maaren, G. M. Schaeffer and F. K. Lotgering, *Physics Lett. A*, **25**, 238 (1967).
2. M. Ito, J. Hori, H. Kurisaki, H. Okada, A. J. Perez Kuroki, N. Ogita, M. Udagawa, H. Fujii, F. Nakamura, T. Fujita and T. Suzuki, *Phys. Rev. Lett.*, **91**, 077001 (2003).
3. M. Ito, H. Kurisaki, F. Nakamura, T. Fujita, T. Suzuki, J. Hori, H. Okada and H. Fujii, *J. App. Phys.*, **97**, 10B112 (2005).
4. Y. Uwatoko, T. Hotta, E. Matsumoto, H. Mori, T. Ohki, J. L. Sarrao, J. D. Thompson, N. Mori and G. Oomi, *Rev. High Pressure Sci. Technol.*, **7**, 1508 (1998).
5. R. N. Shelton, D. C. Johnston and H. Adrian, *Solid State Commun.*, **20**, 1077 (1976).
6. N. R. Werthamer, E. Helfand and P. C. Hohenberg, *Phys. Rev.*, **147**, 295 (1966).
7. R. R. Hake, *Appl. Phys. Lett.*, **10**, 189 (1967).

Specific Heat of Chalcogenide Superconductor TlV_6S_8

J. Hori*, A. Katai*, Y. Tange*, A. Furukawa*, Y. Fujii*,
T. Ohtani†, and M. Harada†

*Department of Applied Physics, Okayama University of Science, Ridai-cho, Okayama 700-0005, Japan
†Laboratory for Solid State Chemistry, Okayama University of Science, Ridai-cho, Okayama 700-0005, Japan

Abstract. The specific heat (C) of the chalcogenide compound superconductor TlV_6S_8 has been measured by the thermal relaxation method. The superconducting transition temperature (T_c) obtained from the C was ~ 2.5 K. The Sommerfeld coefficient γ estimated from the C in the normal state was 163 mJ / K^2 mol. γ of TlV_6S_8 is fairly large compared to that of the conventional superconductors. The large value of γ indicates that TlV_6S_8 is one of the strongly correlated electron systems. The value of C/T below T_c increased gradually with decreasing temperature and had a peak at ~ 2 K. We suspect that this behavior is derived from the quasi-one-dimensional structure.

Keywords: quasi-one-dimensional structure, chalcogenide superconductor, specific heat measurements
PACS: 65.40.-b

INTRODUCTION

The chalcogenide compound $A_xV_6S_8$ (A = In, Tl, K, Rb, Cs) belongs to the Nb_3Te_4-type hexagonal structure (space group $P6_3$). The structure of $A_xV_6S_8$ is constructed by face-sharing and edge-sharing VS_6 octahedra, where V-V zigzag chains are running parallel to the c-axis, composing a quasi-one-dimensional structure [1]. It has large hexagonal channels along the c-axis which are occupied by A atoms. TlV_6S_8 is a starting material to make $In_xV_6S_8$, $K_xV_6S_8$ $Rb_xV_6S_8$, and $Cs_xV_6S_8$. The superconducting transition of TlV_6S_8 was confirmed previously by the electric resistivity and ac-susceptibility measurements [2-5]. The ac-susceptibility measurements revealed that the powdery TlV_6S_8 shows the diamagnetism with two steps below superconducting transition temperature (T_c) [3-5]. The stepwise diamagnetism is explained by the inter-grain and intra-grain superconducting transition. The stepwise diamagnetism is not observed in the sintered samples. The electric resistivity and ac-susceptibility measurements were performed for other samples of $A_xV_6S_8$. However, the specific heat (C) has not been measured in $A_xV_6S_8$. In this work, we have measured the specific heat of TlV_6S_8 to investigate the thermodynamic properties in the superconducting state.

EXPERIMENTAL

The polycrystalline sample of TlV_6S_8 was prepared by a solid-state reaction method. The stoichiometric amount of Tl, V, and S elements were mixed in an evacuated quarts tube. The mixture of the elements was heated to 800°C in 17 days and sintered subsequently at this temperature for 5 days. Then the sample was cooled down to room temperature slowly. The powdery sample of TlV_6S_8 was pressed into pellets and sintered at 750 °C. The specific heat measurements were carried out by a thermal relaxation method from 0.4 to 3.0 K using a ^{3}He-^{4}He dilution refrigerator. The sample was fixed on silver addenda with a silver paste. The purity of silver for addenda was 99.98 %. The heat capacity of the addenda was below 10 % of the sample's heat capacity.

SPECIFIC HEAT MEASUREMENTS

Figure 1(a) shows the temperature (T) dependence of C/T for TlV_6S_8. A change in C due to the superconducting transition was observed at T_c ~ 2.5 K. However, the jump of C was not sharp. The plot of C/T as a function of T^2 is shown in Fig.1(b). C/T in the normal state (T > 2.5 K) is described by $C/T = \gamma + \beta T^2$ as shown in Fig.1(b) by the broken line, where γ is the Sommerfeld coefficient and β is a prefactor. The value of γ and β estimated by C/T in the normal state was ~ 160 mJ/K^2 mol and ~ 40 mJ/K^4 mol

CP850, *Low Temperature Physics: 24th International Conference on Low Temperature Physics;*
edited by Y. Takano, S. P. Hershfield, S. O. Hill, P. J. Hirschfeld, and A. M. Goldman

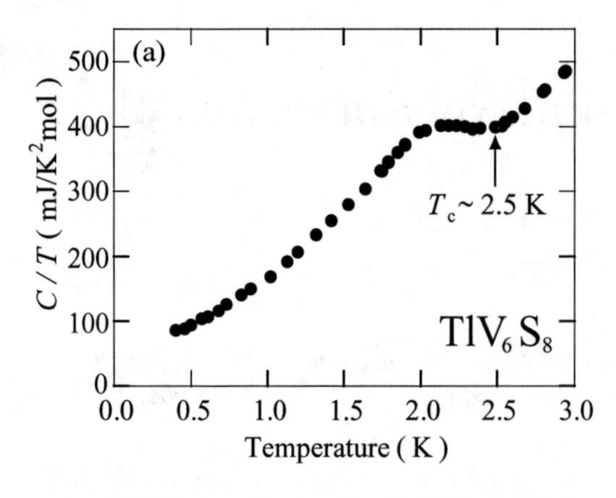

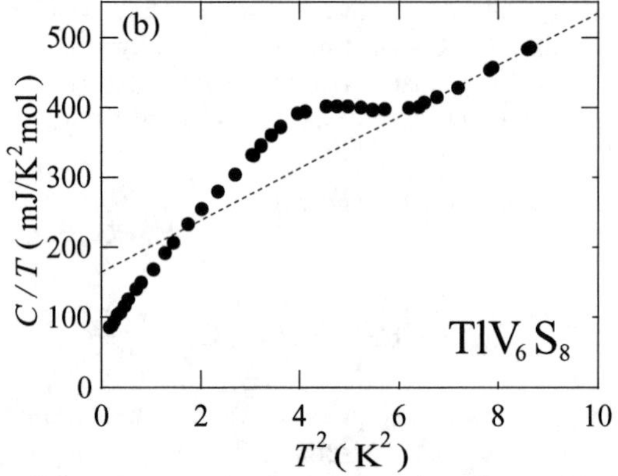

FIGURE 1. (a) Temperature dependence of C/T for TlV_6S_8. The change of C/T due to the superconducting transition was observed at ~ 2.5 K. (b) Plot of C/T as a function of T^2. The broken line is given by $C/T = \gamma + \beta T^2$ with γ ~ 160 mJ/K^2 mol and β ~ 40 mJ/ K^4 mol.

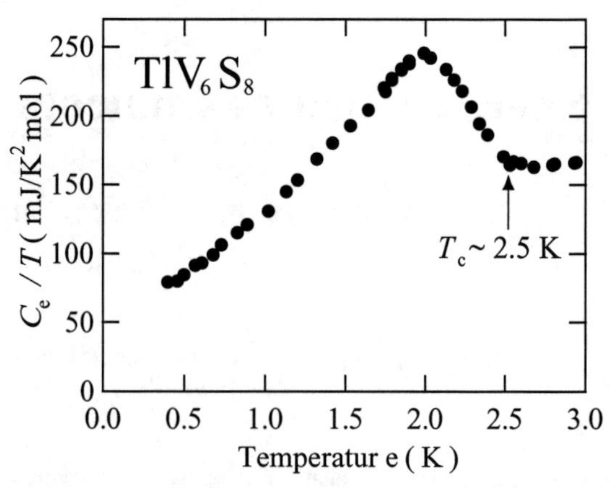

FIGURE 2. Temperature dependence of the electronic specific heat C_e devided by T for TlV_6S_8. The C_e had a peak at ~ 2 K.

respectively. The Debye temperature (Θ_D) calculated from β was ~ 90 K. Though the temperature range of normal state we have measured is very narrow and the estimated γ has a large margin of error, γ is at least lager than 100 mJ/K^2 mol. γ of TlV_6S_8 is fairly large compared to that of conventional superconductors. Even in the niobium chalcogenide superconductors which have the quasi-one-dimensional structures like TlV_6S_8, γ is 13.6, 21.4, and 18.7 mJ/K^2 mol for Nb_3S_4, Nb_3Se_4, and Nb_3Te_4 respectively[6]. The large value of γ indicates that TlV_6S_8 is one of the strongly correlated electron systems. Figure 2 shows the temperature dependence of electronic specific heat C_e divided by T. The value of C_e was estimated by the subtraction βT^3 from C. C_e/T below T_c increased gradually with decreasing temperature and had a peak at ~ 2 K. We suspect that the broad peak of C at T_c is due to the quasi-one-dimensionality of TlV_6S_8. Okamoto *et. al.* reported that Nb_3Te_4 did not show the jump of C at T_c because of the filamentary superconductivity in the sample [6]. The one-dimensional anisotropy in Nb_3Te_4 originated from the metallic chains of Nb atoms is in strong compared to Nb_3S_4 and Nb_3Se_4 [6,7]. The chains Nb_3Te_4 may work as weak-coupled Josephson-type junctions. The small volume fraction of filamentary superconducting regions may not be detected by the specific heat measurements. TlV_6S_8 also have one-dimensional chains of V atoms. In TlV_6S_8, if the small superconducting region is connected by the Josephson-type junctions as Nb_3Te_4 and distributed in the sample, the specific heat may show the broad peak.

SUMMARY

In summary, we have measured the specific heat of TlV_6S_8 and obtained the thermodynamic parameters of γ~160 mJ/K^2 mol, β~ 40 mJ/K^4, and Θ_D ~ 90 K. The large value of γ indicates that TlV_6S_8 is the material of the strongly correlated electron system. The broad peak of C due to the superconducting transition may arise from the quasi-one-dimensional structures.

REFERENCES

1. W. Bensch and J. Koy, *J. Alloys and Compounds* **178,** 193-204 (1992).
2. W. Bensch and J. Koy, *Solid State Commun.* **93**, 261-264 (1995).
3. T. Ohtani, Y. Miyoshi, Y. Fujii, T. Koyakumaru, T. Kusano, and K. Minami, *Solid State Commun.* **120**, 95-99 (2001).
4. Y. Fujii, T. Koyakumaru, T. Ohtani, Y. Miyoshi, S. Sunagawa, and M. Sugino, *Solid State Commun.* **121**, 165-169 (2002).
5. M. Sugino, T. Ueoka, Y. Fujii, T. Ohtani, and M. Yamaguchi, *Physica C* **388-389**, 575-576 (2003).
6. H. Okamoto, H. Taguti, and Y. Ishihara, *Phys. Rev. B* **53**, 384-388 (1996).
7. A. Oshiyama, *J. Phys. Soc. Jpn.* **52**, 587-596 (1983).

Heat Capacity Measurement on Li_2Pd_3B and Li_2Pt_3B

H. Takeya,[*] M.El Massalami,[#] R.E. Rapp,[#] K. Hirata,[*] K.Yamaura,[*] K. Yamada,[*] and K. Togano[*]

[*] *National Institute for Materials Science, Tsukuba, Ibaraki 305-0047, Japan*
[#] *Instituto de Fisica – UFRJ, CxP68-528, 21945-970, Rio de Janeiro, Brazil*

Abstract. Superconductivity has been recently found in two Li containing compounds, Li_2Pd_3B and Li_2Pt_3B. They show superconducting transition at the temperatures, 7.5 K and 2.17 K respectively. The structure analysis has been reported by Eibenstein et. al. in 1997 and they take the same cubic structure with the symmetry of $P4_332$. Heat capacity of these compounds was measured for confirming their bulk superconductivity and investigating superconducting properties. $\Delta C/T_c$ are evaluated to be around 18 mJ/mol K^2 for Li_2Pd_3B and 9.7 mJ/mol K^2 for Li_2Pt_3B. Electronic heat capacity (γ) and Debye temperature (θ_D) are derived from the normal state data and lead to 9.0 mJ/mol K^2 and 221 K for Li_2Pd_3B, 7.0 mJ/mol K^2 and 228 K for Li_2Pt_3B. Those physical parameters of the two compounds are discussed.

Keywords: Superconductor, Li_2Pd_3B, Li_2Pt_3B, Heat capacity, Ternary borides.
PACS: 74.70.Dd, 74.25.Bt

INTRODUCTION

Research of boride compounds with light elements has been encouraged since the discovery of MgB_2. Possibility of higher superconducting transition was suggested by some electronic band structure calculations for materials composed of light elements such as LiB_2, BeB_2 and LiBC. So far, there have not been any positive results on superconductivity of those materials.

We have been interested in multinary borides such as RRh_4B_4, RM_3B and RNi_2B_2C (R=Y, Lanthanoid Elements, M=Rh, Pd, Pt). They show various physical properties of superconductivity and magnetism. Since the discovery of MgB_2, we have been researching new superconducting materials composed of light elements and transition metals. Lithium and boron are recognized as interesting elements which react with many other metals. Those compounds are expected to reveal interesting physical properties. Superconductivity was recently confirmed in Li_2Pd_3B and Li_2Pt_3B around at 7.5 K and 2.17 K respectively [1]. Two of those compounds consist of distorted octahedral coordinations of Pd-B in the cubic lattice structure. In this report, heat capacity was measured and the fundamental parameters of those new superconductors are referred.

EXPERIMENTAL

A two-step arc-melting synthesis was performed for the preparation of materials. At the start, stoichiometric compounds of Pd_3B and Pt_3B were prepared in high purity argon atmosphere. Then, Li lumps were absorbed in Pd_3B and Pt_3B while they were heated by arc-flame. Further annealing process was performed at 973-1173 K for 24-48 h in a sealed Ta tube. Magnetic properties were measured using a SQUID magnetometer (Quantum Design Co., Ltd.). Chemical composition of Li:Pd(Pt):B was obtained by the induced coupled-plasma (ICP) analysis. The heat capacity measurements were carried out on two different types of calorimeters. A relaxation type calorimeter was operated down to 2 K. The other is a zero-field semi-adiabatic calorimeter operated down to 0.4 K.

RESULTS AND DISCUSSION

Li_2Pd_3B and Li_2Pt_3B gave the same X-ray powder diffraction patterns perfectly indexed by the cubic space group $P4_332$ as was reported by Eibenstein et. al [2]. The compositions of Li, Pd, Pt and B coincide in the stoichiometric ratio within 5 %.

CP850, *Low Temperature Physics: 24th International Conference on Low Temperature Physics;* edited by Y. Takano, S. P. Hershfield, S. O. Hill, P. J. Hirschfeld, and A. M. Goldman

The bulk superconductivity of Li_2Pd_3B and Li_2Pt_3B was confirmed by the heat capacity measurements. Figure 1 shows a heat capacity (C) vs. temperature (T) plot for the Li_2Pd_3B under the zero magnetic field and 7 T measured by the heat-pulse relaxation method. The normal state properties of Li_2Pd_3B are revealed after quenching the superconductivity by applying a magnetic field of 7 T over H_{c2}. The normal-state heat capacity is described by Sommerfeld (γT) and Debye (βT^3) terms as a conventional formula, $C_n=\gamma T+\beta T^3$. The fitting formula gave γ=9.0 mJ/mol K^2, β=1.08 mJ/mol K. The Debye temperature is θ_D=221 K. A superconducting jump appears at 7.5 K in 0 T. The value of $\Delta C/\gamma T$ at T_c is calculated to be 2.0. Figure 2 shows the heat capacity of Li_2Pt_3B measured by the semi-adiabatic method. This method was done only under zero magnetic field. The γ and β values are 7.0 mJ/mol K^2 and 0.98 mJ/mol K estimated by the fitting formula of $C_n=\gamma T+\beta T^3$ ($T>T_c$). θ_D=228 K, T_c =2.17 K, and $\Delta C/\gamma T_c$=1.39 are obtained in the same manner for Li_2Pd_3B. The above-mentioned physical parameters of Li_2Pd_3B and Li_2Pt_3B are listed in Table. 1.

TABLE 1. Normal and superconducting parameters of Li_2Pd_3B and Li_2Pt_3B

Compound	T_c [K]	γ [mJ/mol K^2]	θ_D [K]	$\Delta C/\gamma T_c$ [none]
Li_2Pd_3B	7.5	9.0	221	2.0
Li_2Pt_3B	2.17	7.0	228	1.39

attention to the similarity of the electronic configurations between Pd and Pt, the evaluation of their pseud-isotope effect o The γ of Li_2Pd_3B is almost 30 % higher than that of Li_2Pt_3B, while θ_D is 3% lower. γ is related to the electron density state of $N(E_F)$ expressed by γ=(2/3)$\pi^2k_B^2(1+\lambda)N(E_F)$, whereas k_B and λ denote the Boltzmann's constant and the electronic mass-enhancement. T_c is described by T_c=0.85$\theta_D$exp(-1/$N(E_F)V$), V:BCS interaction parameter in the BCS model [3]. Therefore, in case of using the same values of λ and V for the two compounds, the T_c of Li_2Pd_3B is calculated to be 5.7 K based on the measured T_c=2.17 K for Li_2Pt_3B. The measured T_c for Li_2Pd_3B, 7.5 K, is 32 % higher. Since $\Delta C/\gamma Tc$ by the BCS theory is 1.43, the value of 1.39 for Li_2Pt_3B is quite reasonable as a weakly-coupled BCS-type superconductor. The value of 2.0 for Li_2Pd_3B suggests the possibility of strong-coupling corrections, although it is not far from thevalue of the BCS theory.

It is difficult to control the accurate Li stoichiometry because of the loss during arc-melting. Since there is a strong dependence of T_c on the stoichiometry of the Li content, studying boron isotope effect on these materials is discouraged. Paying n superconductivity was performed. The relation ($T_{c,\,Pt}$ $\sqrt{M_{Pt}}$) / ($T_{c,\,Pd}$ $\sqrt{M_{Pd}}$) = 0.4 is 60 % away from the BCS standard model. The T_c of Li_2Pd_3B from this pseud-isotope effect is estimated to be only 2.93 K based on the measured T_c=2.17 K for Li_2Pt_3B. The measured T_c for Li_2Pd_3B, 7.5 K, is 255 % higher. Therefore, the difference of T_c between Li_2Pd_3B and Li_2Pt_3B is not just attributed to a modification of the lattice vibrations. Further investigations are still needed to understand the features of those two superconductors.

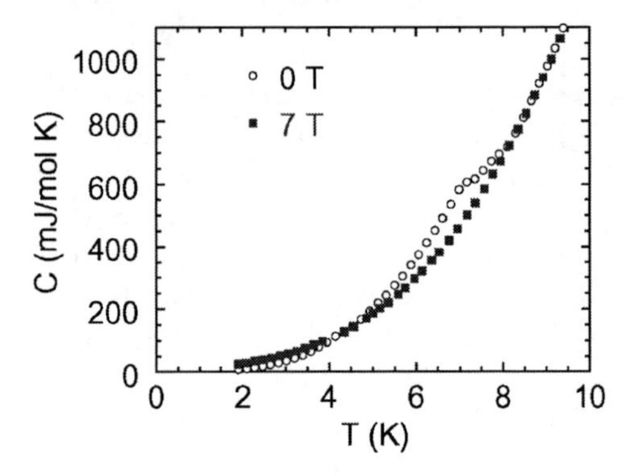

FIGURE 1. Heat capacity vs. temperature plot of Li_2Pd_3B.

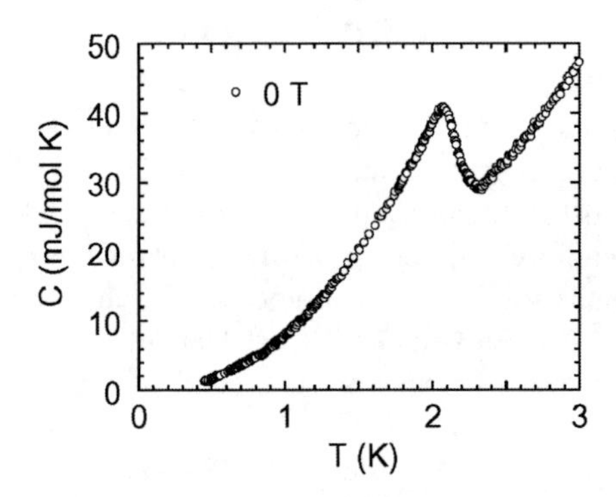

FIGURE 2. Heat capacity vs. temperature plot of Li_2Pt_3B.

ACKNOWLEDGMENTS

The partial financial supports from the Japan Society for the Promotion of Science and the Brazilizn agencies CNPq and Faperj are acknowledged.

REFERENCES

1. K.Togano, P. Badica, Y. Nakamori, S. Orimo, H. Takeya and K. Hirata, *Phys.Rev. Lett.* **93**, 247004 (2004).
2. U. Eibenstein and W. Jung, *J.Solid State Chem.* **133**, 21 (1997).
3. R. Meservey and B.B. Schwartz, "Equilibrium Properties," in *Superconductivity*, edited by R.D. Parks, New York: Marcel Dekker, Inc., 1969, pp.117-191.

A Penetration Depth Study on Li_2Pd_3B and Li_2Pt_3B

H. Q. Yuan[1], D. Vandervelde[1], M. B. Salamon[1], P. Badica[2,3], and K. Togano[2]

[1]*Department of Physics, University of Illinois at Urbana & Champaign, 1110 W. Green Street, Urbana, IL 61801, USA*
[2]*Institute for Materials Research, Tohoku University, 2-1-1 Katahira, Aoba-ku, Sendai, 980-8577, Japan*
[3] *National Institute of Materials Physics, Bucharest, POBox MG-7, RO-76900, Romania*

Abstract. In this paper we present a penetration depth study on the newly discovered superconductors Li_2Pd_3B and Li_2Pt_3B. Surprisingly, the low-temperature penetration depth $\lambda(T)$ demonstrates distinct behavior in these two isostructural compounds. In Li_2Pd_3B, $\lambda(T)$ follows an exponential decay and can be nicely fitted by a two-gap BCS superconducting model with a small gap Δ_1=3.2K and a large gap Δ_2=11.5K. However, linear temperature dependence of $\lambda(T)$ is observed in Li_2Pt_3B below $0.3T_c$, giving evidence of line nodes in the energy gap.

Keywords: superconductivity, penetration depth, Li_2Pd_3B, Li_2Pt_3B.
PACS: 74.25.Ha, 74.25.Nf, 74.70.Dd

1. INTRODUCTION

The recent discovery [1, 2] of superconductivity in the ternary lithium borides Li_2Pd_3B and Li_2Pt_3B has been attracting wide attention due to their rich physical properties. Most likely, this system provides a model example bridging unconventional superconductivity with the classic BCS superconductivity.

Li_2Pd_3B and Li_2Pt_3B crystallize in a perovskite-like cubic structure composed of distorted octahedral units of BPd_6 and BPt_6,[3] which resembles other metallic oxides, e.g., the high T_c cuprates and the sodium cobaltates. Measurements of thermodynamic and transport properties revealed a superconducting transition temperature T_c of 7-8K in Li_2Pd_3B, but around 2.5 K in Li_2Pt_3B.[1, 2] The large difference of T_c in these two compounds is difficult to understand within the conventional BCS theory. Up to now, only a very few reports can be found in the literature on the superconducting properties of Li_2Pd_3B and Li_2Pt_3B. Similar to $MgCNi_3$, the estimated superconducting parameters of Li_2Pd_3B are close to the BCS values.[1, 2] Electronic structure calculations showed that the Fermi surface of these two borides are very complicate and suggested that at least a two-band model is required to study their electronic properties.[4] Recent results of X-ray photoemission spectroscopy demonstrated that the electronic correlation in the Pd-compound does not play a role on the physical properties.[5] Nothing has been reported on the pairing mechanism of these two superconductors and further efforts are required to shed light on the superconducting nature. In this paper, we investigate the penetration depth in these two compounds, which is directly related to their superconducting gap structures.

2. EXPERIMENTAL METHODS

Precise measurements of penetration depth $\lambda(T)$ were performed utilizing a tunnel-diode based, self-inductive technique at 21 MHZ down to 90 mK in a dilution fridge. With this method, the change of $\lambda(T)$ is proportional to the resonant frequency shift $\Delta f(T)$, i.e., $\Delta\lambda(T)=G\Delta f(T)$, where the factor G is determined by sample and coil geometries. The magnetization M(T, H) was measured using a commercial SQUID magnetometer (MPMS, Quantum Design).

Polycrystalline Li_2Pd_3B and Li_2Pt_3B have been synthesized by using a two-step arc-melting method. First, alloys of Pd_3B and Pt_3B have been prepared from the mixtures of Pd (99.9%), Pt (99.99%) and B (99.5%). Then Li has been introduced into the melt. Powder X-ray diffraction identifies a single phase. To measure penetration depth, single-crystal-like pieces were selected from the smashed polycrystalline batches.

3. RESULTS AND DISCUSSSION

As an example, in Fig.1 the temperature dependence of the frequency shift $\Delta f(T)=f(T)-f(0)$ is shown for Li_2Pd_3B. For comparison, the corresponding

CP850, *Low Temperature Physics: 24th International Conference on Low Temperature Physics;*
edited by Y. Takano, S. P. Hershfield, S. O. Hill, P. J. Hirschfeld, and A. M. Goldman

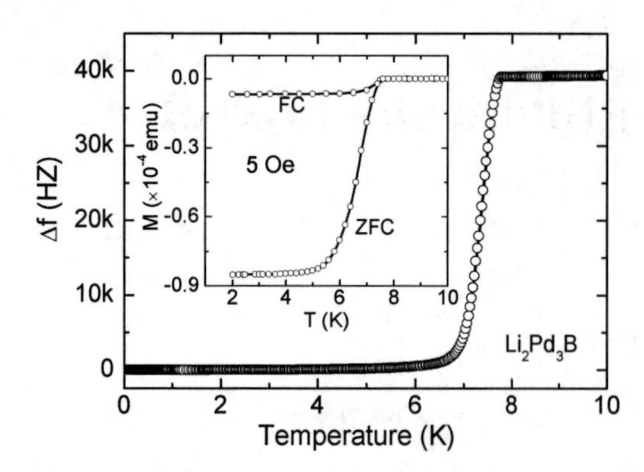

FIGURE 1. Temperature dependence of the frequency shift $\Delta f(T)$ for Li_2Pd_3B. Inset: magnetization vs. temperature measured in zero field cooling (ZFC) and field cooling (FC).

magnetization M(T) is plotted in the inset. Clearly, the results from different measurements are consistent and demonstrate a sharp superconducting transition around 7K, indicating good sample quality. Similar features are also observed in Li_2Pt_3B with $T_c \approx 2.4K$.

The low-temperature penetration depth $\Delta\lambda(T)$ of Li_2Pd_3B and Li_2Pt_3B, derived from $\Delta\lambda(T)=G\Delta f(T)$ where G=6.28 and 10.13 Å/Hz for Pd- and Pt-compounds respectively, is presented in Fig.2. At first glance, $\Delta\lambda(T)$ shows a much weaker temperature dependence in Li_2Pd_3B than in Li_2Pt_3B. A detailed analysis revealed that $\Delta\lambda(T)$ of Li_2Pd_3B can be well fitted by a two-gap BCS model below $0.3T_c$, as described in Ref.[6] for MgB_2, with a small energy gap (Δ_1=3.23K) and a large energy gap (Δ_2=11.55K). The fraction c_1 contributed from the small gap is about 5%, indicating only a small fraction of carriers are responsible for the opening of the small gap. To explore other possible gap functions, fits to the weak coupling BCS model (dotted line) and to a power law (dash-dotted line) are also included in Fig. 2(a). Obviously, the two-gap model (solid line) is the best fit, suggesting that Li_2Pd_3B is a wholly gapped superconductor. On the other hand, $\Delta\lambda(T)$ of Li_2Pt_3B behaves quite differently at low temperature. Instead of an exponential decrease of $\Delta\lambda(T)$ observed in the Pd-compound, $\Delta\lambda(T)$ follows a linear temperature dependence in the Pt-compound, indicating the existence of line nodes in the superconducting energy gap. It is noted that the superfluid density can be fitted by a two-gap BCS model and a d-wave superconducting gap over the whole temperature range ($T<T_c$) for Li_2Pd_3B and Li_2Pt_3B, respectively. Unconventional superconductivity is usually discussed beyond the phonon pairing mechanism. Recently, Agterberge *et al* argued that exotic superconductivity may arise from the conventional phonon mechanism in the systems with a multi-pocket Fermi surface located at some symmetry points, due to the competition of phonon and Coulomb interactions.[7] Considering the

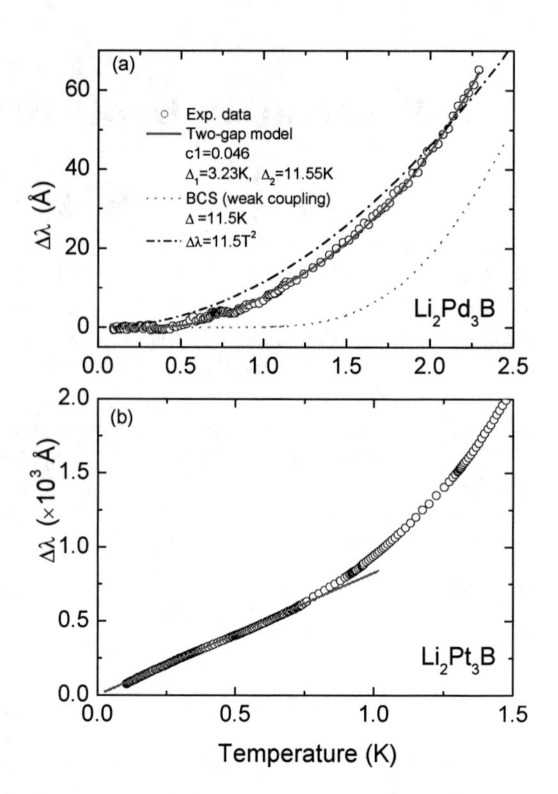

FIGURE 2. Temperature dependence of the penetration depth Δλ (T) for (a) Li2Pd3B and (b) Li2Pt3B.

complicate Fermi surface consisting of multiple pockets and the van Hove singularity around the Fermi energy in Li_2Pd_3B and Li_2Pt_3B,[4] the evolution of superconductivity and the unconventional superconducting features could be understood within this model; details will be published elsewhere.

ACKNOWLEDGMENTS

We are grateful for the useful discussion with D. F. Agterberg. This work is supported by the Department of Energy. HQY acknowledges the ICAM postdoctoral fellowship.

REFERENCES

1. Togano, K., Badica, P., Nakamori, Y., Orimo, S., Takeya, H., and Hirata, K., *Phys. Rev. Lett.* **93,** 247004 (2004).
2. Badica, P., Kondo, T., and Togano, K., *J. Phys. Soc. Jpn.*, **74**, 1014-1019 (2005).
3. Eibenstein, U., and Jung, W., *J. Solid. State. Chem.*, **133** 21-24 (1997).
4. Chandra, S., Mathi Jaya, S., and Valsakumar., M. C., cond-mat/0502525.
5. Yokoya, T., Muro, T., Hase, I., Takeya, H., Hirata, K., and Togano, K., *Phys. Rev. B* **71**, 092507 (2005).
6. Kim, M. S., Skinta, J. A., and Lemberger, T. R., *Phys. Rev., B* **66**, 064511 (2002).
7. Agterberg, D. F., Barzykin, V., and Gor'kov, L. P., *Phys. Rev. B* **60**, 14868-14871 (1999).

Low Temperature Properties and Superconductivity of YB_6 and YB_4

K. Flachbart[*], S. Gabáni[*], J. Kačmarčík[*], T. Mori[#], S. Otani[#], and V. Pavlík[*]

*Centre of Low Temperature Physics, Slovak Academy of Sciences, Watsonova 47, SK-04353 Košice, Slovak Republic

#Advanced Materials Laboratory, National Institute for Materials Science, Tsukuba, Ibaraki 305-0044, Japan

Abstract. Two borides, cubic YB_6 and tetragonal YB_4, have been investigated at low temperatures. The electrical resistivities of YB_6 and YB_4 as a function of temperature exhibit typical metallic behavior. Moreover, in YB_6 an abrupt decrease of resistivity at 7.5 K, incident for a transition to superconducting state, was observed. Further resistivity and specific heat measurements in magnetic field have shown that this compound is a BCS-type superconductor with a critical magnetic field of $B_C \approx 0.6$ T. On YB_4, on the other hand, no superconductivity down to 0.05 K was observed.

Keywords: YB_6, YB_4, superconductivity, electrical resistivity, specific heat
PACS: 74.25.Fy, 72.15.Eb, 65.40.Ba

INTRODUCTION

Compounds between yttrium (Y) and boron (B) form a rather wide series of boron-rich binary borides. Among them the metallic YB_2, YB_4, YB_6 and YB_{12} compounds have became attractive both from scientific and technical point of view. For example, YB_6 and YB_{12} are superconductors [1], and YB_6 has been considered as a possible candidate for high temperature thermoelectric materials [2]. Comparison between different classes of boron rich binary borides [3] has shown that superconductivity is found above all among low boron containing compounds (with B/M $\leq 2 - 2.5$, where M denotes a metallic atom). On the other hand, superconductivity is much less typical for high-boron borides (MB_4, MB_6, MB_{12}). From this point of view the investigation of low temperature or/and superconducting properties of high boron yttrium borides presents an interesting topic.

In recent years some of the properties of YB_6 and YB_4 have been intensely investigated, above all the electronic band structure [4, 5], and the optical [6] and NMR spectra [5]. There is, however, still little known above all about their low temperature and superconducting properties. In this paper, we present specific heat and electrical resistivity investigations of these compounds.

EXPERIMENTAL

YB_6 and YB_4 single crystals were grown by the rf-heated floating zone method in an atmosphere of ambient argon at a pressure of 0.5 MPa. Further details of sample preparation are described in [7]. Specific heat was measured by a typical transient heat pulse method with a small temperature increase relative to the system temperature. Resistance measurements between 1.6 and 300 K were performed in a conventional ^{4}He cryostat using standard dc four-terminal method. A standard ac lock-in technique at 17 Hz was used to measure the temperature and magnetic field dependence of the resistance below 1.6 K in a ^{3}He-^{4}He - dilution refrigerator. In all measurements the magnetic field was applied perpendicular to the electrical current. Magnetization measurements were realized in a Quantum Design SQUID magnetometer.

RESULTS AND DISCUSSION

In Fig. 1 the specific heat $C(T)$ of YB_6 below 10 K is displayed. In zero magnetic field the $C(T)$ dependence has a discontinuity at $T_C = 7.5$ K, clearly indicating a second order phase transition into the

CP850, *Low Temperature Physics: 24th International Conference on Low Temperature Physics;* edited by Y. Takano, S. P. Hershfield, S. O. Hill, P. J. Hirschfeld, and A. M. Goldman

superconducting state. With increase of magnetic field T_C gets reduced. In the normal state the specific heat can be described using $C(T) = \gamma T + \beta T^3$, where the first term with $\gamma \approx 4$ mJK^{-2}mole^{-1} represents the electronic and the second one with $\beta \approx 0.2$ mJK^{-4}mole^{-1} the phonon contribution to the heat capacity. From these data we obtained a ratio $\Delta C(T_C) / \gamma T_C \approx 1.4$ for the specific heat discontinuity relative to the normal state electronic specific heat at transition temperature which points to BCS-type superconductivity in this material.

On YB_4, on the other hand, no $C(T)$ anomaly down to 2 K was observed. In this case the specific heat at lowest temperatures can be described with $C(T) = \gamma T + \beta T^3$, where $\gamma \approx 1.45$ mJK^{-2}mole^{-1} and $\beta \approx 0.014$ mJK^{-4}mole^{-1}. The ratio between coefficients $\gamma(YB_6)/\gamma(YB_4)$ of YB_6 and YB_4 corresponds well with the ratio $D(YB_6,E_F)/D(YB_4,E_F)$ between calculated electronic densities of states at the Fermi level of these two compounds [5].

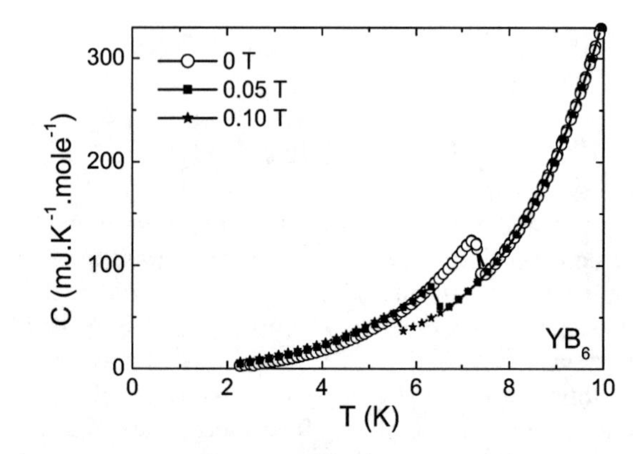

FIGURE 1. Specific heat of YB_6 as a function of temperature in various magnetic fields.

The electrical resistivities of YB_6 and YB_4 as a function of temperature show below 300 K a typical metallic behavior. Their residual resistivity ratios were found to be 4.3 for YB_6 and 37 for YB_4. In YB_6 at 7.5 K an abrupt decrease of resistivity, incident for a transition to the superconducting state and confirming the specific heat results, was observed. Consecutive resistivity measurements as a function of magnetic field at various temperatures below T_C have shown that this compound has a critical magnetic field of $B_{c2} \approx 0.6$ T. The corresponding B_{c2} dependence as a function of temperature obtained from magneto-resistive measurements is displayed in Fig. 2. From *dc* magnetization measurements the critical field B_{c1} was determined to be about 8 mT. On YB_4 sample, on the other hand, no superconductivity was observed down to 0.05 K.

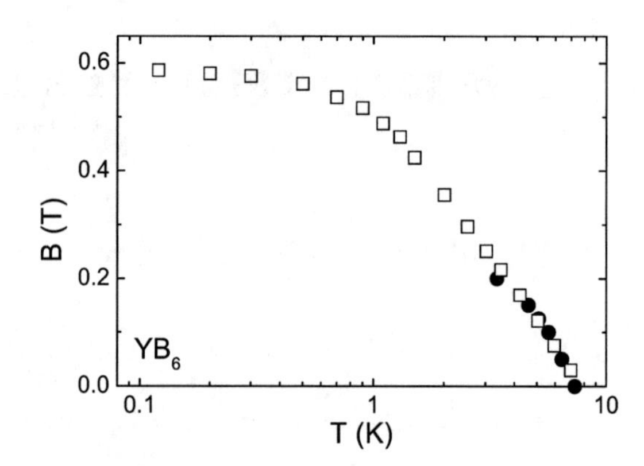

FIGURE 2. Critical magnetic field B_{c2} as a function of temperature. (♦) - data received from specific heat measurements, (□) - data received from magneto-resistive measurements.

The reason for the different behavior of YB_6 and YB_4 is most probably the disparity of the electron - phonon interaction function in these materials, which can be associated with their distinct Fermi surfaces and / or phonon spectra. To resolve this question, further (above all spectroscopy) investigations will be needed.

ACKNOWLEDGEMENTS

The work was supported by the Slovak Grant Agency VEGA, under contract 2-4061, by the Slovak Science and Technology Agency APVT, contract 51-0166, and by PRESTO, Japan Science and Technology Agency. The authors also acknowledge the sponsorship of a part of liquid nitrogen by US Steel Košice.

REFERENCES

1. B.T. Matthias, T.H. Geballe, K. Andres, E. Corenzwit, G.W. Hull, and J.P. Maita, *Science* **159**, 530 (1968).
2. Y. Imai, M. Mukaida, M. Ueda, and A. Watanabe, *Intermetallics* **9**, 721 (2001).
3. C. Buzea, and T. Yamashita, *Superconductors, Science and Technology* **14**, R115 (2001).
4. I.R. Shein, S.V. Okatov, N.I. Medvedeva, and A.L. Ivanovskii, *cond-mat / 0202015* (2002).
5. B. Jäger, S. Paluch, W. Wolf, P. Herzig, O.J. Zogal, N. Shitsevalova, and Y. Paderno, *J. Alloys Comp.* **383**, 232 (2004).
6. S. Kimura, H. Okamura, T. Nanba, M. Izekawa, S. Kunii, F. Iga, N. Shimizu, and T. Takabatake, *J. Electr. Spectr. Relat. Phenom.* **101-103**, 761 (1999).
7. S. Otani, M.M. Korsukova, T. Mitsuhashi, and N. Kieda, *J. Cryst. Growth.* **217**, 378 (2000).

Electron Transport and Superconducting Properties of ZrB_{12}, ZrB_2 and MgB_2

V.A. Gasparov* and A. Suslov†

*Institute of Solid State Physics RAS, 142432, Chernogolovka, Russian Federation

†National High Magnetic Field Laboratory, Tallahassee, Florida 32310, USA

Abstract. We report on the measurements of the temperature dependences of the resistivity, ρ(T), the penetration depth, λ(T), the upper critical magnetic field $H_{c2}(T)$, for single crystal samples of ZrB_{12}, ZrB_2 and thin films of MgB_2. Although in the normal state ZrB_{12} behaves like a simple metal, the resistive Debye temperature, 300 K, is three times less relative to the specific heat data (800-1200 K). Quadratic ρ(T) term was observed in ZrB_{12} and ZrB_2 below 25 K and 100 K, respectively, indicating the possible importance of the electron-electron interaction. Unconventional behavior of the ZrB_{12} superfluid density with a shoulder at T/T_c =0.65 was observed. We found a linear dependence of $H_{c2}(T)$ for ZrB_{12} from T_c down to 0.35 K. Both the λ(T) and $H_{c2}(T)$ dependences can be explained by a two band BCS model with *two superconducting gaps and T_c's.*

Keywords: superconductivity, two gap model, ZrB_{12}, ZrB_2, MgB_2.
PACS: 74.70.Ad, 74.25.Nf, 72.15.Gd, 74.25.Ha

INTRODUCTION

The discovery of superconductivity in MgB_2 initiated a search for superconductivity in other borides (see references in [1]). Yet, only nonstoichiometric boride compounds ($MoB_{2.5}$, $NbB_{2.5}$, Mo_2B, W_2B, $BeB_{2.75}$) demonstrate such transition (see references in [2]). It was suggested (see references in [3]) that the superconductivity in ZrB_{12} is due to the effect of a cluster of boron atoms. While the superconductivity in ZrB_{12} was discovered a while ago (see references in [2]), there has been little and controversial efforts devoted to study the electron transport in this compound. Here, we report unusual superconducting properties of ZrB_{12}. Comparative data from ZrB_2 single crystals and MgB_2 thin films are also present.

RESULTS AND DISCUSSION

Under ambient conditions, ZrB_{12} crystallizes in the UB_{12} type *fcc* structure of the UB_{12} type. In this structure, Zr atoms are located among the close-packed B_{12} clusters [2]. In contrast, ZrB_2 consists of two-dimensional graphite-like monolayers of boron atoms intercalated with Zr monolayers. The ZrB_{12} and ZrB_2 single crystals have been grown by the floating-zone method by the Kiev group (see references in [3]). Metallographic and X-ray analysis indicated that some parts of the ZrB_{12} ingot contained needle like phase of non-superconducting ZrB_2 [5]. Thus special care was taken to cut the ZrB_2-free <100>-oriented rectangular ZrB_{12} bars. Two highly crystalline, superconducting films of MgB_2 with T_{c0}'s of 38 K and 39 K, were grown on *r*-plane sapphire substrates by the Oak Ridge group (see references in [2]).

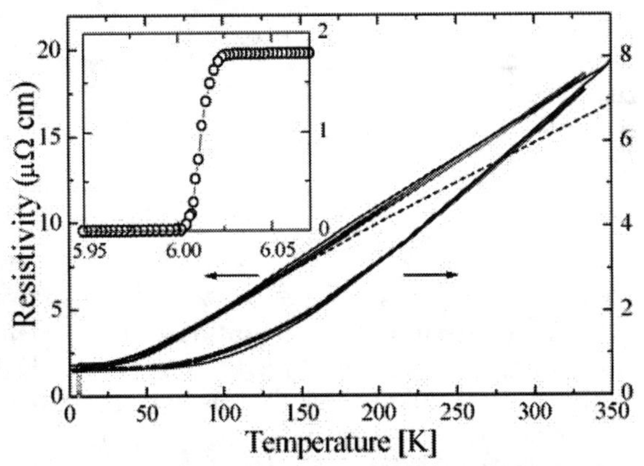

FIGURE 1. Temperature dependence of ρ(T) for ZrB_{12} (upper curve) and ZrB_2 (lower curve). The solid lines - the BG t^5 model, the dashed line is t^3 BG fit from [6].

Fig. 1 shows the temperature dependence of the resistivity of ZrB_{12} and ZrB_2 single crystal samples. We did not observe a superconducting transition in ZrB_2 down to 1.3 K, even though such transition occurred at 5.5 K in polycrystalline samples [1]. It was suggested (see references in [2]) that this contradiction could be associated with a non-

CP850, *Low Temperature Physics: 24th International Conference on Low Temperature Physics;*
edited by Y. Takano, S. P. Hershfield, S. O. Hill, P. J. Hirschfeld, and A. M. Goldman

stoichiometry in the zirconium sub-lattice. The transition temperature in ZrB_{12} (T_{c0}=6.0 K) is consistent with the previously reported values and is larger than that of ZrB_2 polycrystalline samples (5.5 K) [1].

Above 25 K the Bloch - Grüneisen (BG) model describes the ρ(T) dependence of ZrB_{12} and ZrB_2 fairly well [3] and shows that the electron-phonon interaction dominates at this temperature range. The resistive Debye temperature T_R, is equal to 300 K and 700 K respectively. The former one is very close to T_R=280 K observed on polycrystalline ZrB_{12} samples [2]. At the same time, T_D for ZrB_{12} calculated from the specific heat C(T) data on a rather large sample prepared by one of us (VAG) [4], is three times higher, possibly due to presence of ZrB_2 phase. The deviations of ρ(T) from the BG model at low temperatures relates to possible effect of electron-electron interaction on conductivity [3].

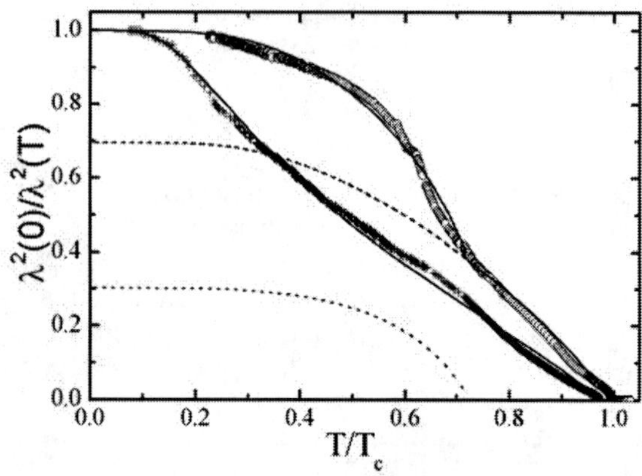

FIGURE 2. Superfluid density of the ZrB_{12} (circles), and the MgB_2 thin film (crosses). The two band model calculations are shown by the dashed (p-band term), dotted (d-band term), and solid (total) lines.

The radio frequency LC techniques (see references in [2]) have been used for the penetration depth λ(T) measurements. Fig. 2 shows $\lambda^2(0)/\lambda^2(T)$ versus reduced temperature T/T_c for the ZrB_{12} single crystal and MgB_2 film as determined from these techniques [2,5]. The unconventional behavior of ZrB_{12} superfluid density with pronounced shoulder at T/T_c=0.65 is easily noticed. This feature can be explained by the two band model [5], which assumes the existence of two independent *p*- and *d*-bands in electron structure of ZrB_{12} with the different gaps, Δ's, and transition temperatures, T_c's [5]. The total conductivity $1/\lambda^2$ is sum of the partial bands conductivities: $1/\lambda_p^2+1/\lambda_d^2$. Fitting of the experimental data gives the following *p*- and *d*- band parameters: T^p_c= 6.0 K, $\Delta_p(0)$= 0.73 meV, $\lambda_p(0)$= 170 nm and T^d_c=4.35 K, $\Delta_d(0)$= 1.2 meV, $\lambda_d(0)$= 260 nm [5].

Fig. 3 presents the $H_{c2}(T)$ dependence obtained from the onsets of the ρ(H) and λ(H) in the normal state. In contrast to the BCS theory, the $H_{c2}(T)$ dependence is linear over an extended temperature range with no evidence of a saturation down to 0.35 K. Linear extrapolation of $H_{c2}(T)$ to T=0 gives $H_{c2}(0)$ = 162 mT. We used this value to obtain the coherence length ξ(0), by employing the relations $H_{c2}(0) = \phi_0/2\pi\xi^2(0)$. The latter yields ξ(0) = 45 nm. We explain linear $H_{c2}(T)$ in the frames of the two band model described above. From the λ(T) data, we found that the diffusivities in the p- and d- bands are equal to D_p=56 cm²/sec and D_d=17 cm²/sec, respectively. According to [7], the large ratio D_p/D_d ≈ 3 leads to enhancement of the zero-temperature $H_{c2}(0)$ value. Thus the limiting value of $H_{c2}(0)$ is dominated by the *d*-band with the lower diffusivity, while the derivative dH_{c2}/dT in the vicinity of T_c is governed by the *p*-band with the larger diffusivity. Although the observed two-gap behavior of λ(T) in ZrB_{12} is similar to that in MgB_2, *discovery of two different T_c in these bands is unconventional.*

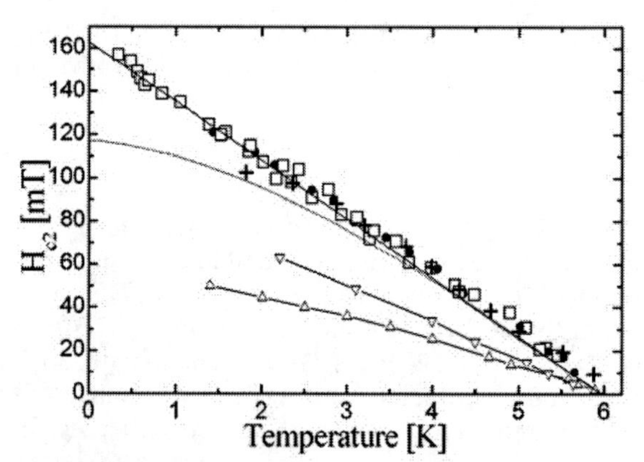

FIGURE 3. Temperature dependence of H_{c2} in ZrB_{12} extracted from: *λ(H)* - bold points; ρ(H) -squares; contact spectroscopy [6] - crosses; C(T) - up triangles and *ρ(H)* - down ones [4]. The dotted line represents the BCS dependence.

ACKNOWLEDGMENTS

We are grateful V.B. Filipov, A.B. Lyashenko and Yu.B. Paderno for the ZrB_2 and ZrB_{12} single crystals, and to H.-Y. Zhai, H.M. Christen, M.P. Paranthaman and D.H. Lowndes, for the MgB_2 films. This work was supported by the Russian Grant MSh-2169.2003.2 and INTAS Grant. 01-0617. NHMFL is supported by the NSF and State of Florida.

REFERENCES

1. V.A. Gasparov et al., *JETP Letters* **73**, 532 (2001).
2. V.A. Gasparov et al., *JETP* **101**, 98 (2005).
3. V.A. Gasparov et al., *JETP Letters* **80**, 330 (2004).
4. R. Lortz et al., PRB 72, 024547 (2005).
5. V.A. Gasparov et al., cond-mat/0508151.
6. D. Daghero et al., *Supercond. Sci. Technol.* **17**, S250 (2004).
7. A. Gurevich, *Phys. Rev. B* **67**, 184515 (2003).

Superconducting Transition Temperature of $(Nb_{1-x}Zr_x)_{0.8}B_2$

K. Nishimura, K. Mori, K. Ohya, and E. Ikeda

Faculty of Engineering, Toyama University, Toyama 930-8555, Japan

Abstract. The superconducting transition temperatures T_C of the nonstoichiometric $(Nb_{1-x}Zr_x)_{0.8}B_2$ pseudobinary system have been investigated via resistivity, magnetization and specific heat. Powder X-ray diffraction measurements revealed that the lattice parameters of the *a*- and *c*-axes monotonically increased as the substitution rate of Zr for Nb was increased. From resistivity and magnetization measurements, T_C's were observed at about 4 K with x from 0 to 0.3; the superconductivity disappeared at $x = 0.5$ and 0.7; T_C were around 6 K with $x = 0.8$, 0.9 and 1.0. The electronic specific heat coefficients estimated from the specific heat decreased as x was increased, which is consistent with the result of a theoretical work.

Keywords: superconductivity, $Nb_{0.8}B_2$, $Zr_{0.8}B_2$.
PACS: 74.25.Dw, 74.70.Ad

INTRODUCTION

It is widely accepted that a number of physical properties of metallic materials can be scaled against the electron-to-atom ratio (*e/a* ratio). The superconducting transition temperature of metals and alloys is one of those properties, as proposed by Matthias et al.[1] According to Gladstone et al.[2], the highest T_C is expected to appear around the *e/a* ratio of 4.5 among 4d transition metals and alloys, which takes place between Zr^{4+} and Nb^{5+}. This work aims to investigate a variation of T_C of the pseudobinary compound, $(Nb_{1-x}Zr_x)_{0.8}B_2$. Due to the recent discovery of T_C in MgB_2 at 39 K[3], intensive studies of superconductivity (SC) of transition metal diborides with the AlB_2-type crystal structure have been carried out experimentally and theoretically. From the experimental studies of the $Nb_{1-x}B_2$ system, it was found that T_C largely depends on a sample preparation process, indicating instability of the crystal structure[4-6]. Although improving a fabrication technique of materials to raise T_C is an important subject, we would focus on a variation of T_C by substitution of Zr for Nb in $(Nb_{1-x}Zr_x)_{0.8}B_2$ in this work. The latest experimental studies have reported that the highest T_C of $Nb_{1-x}B_2$ is found with a nonstoichiometric sample of about $1-x = 0.8$[4-6].

RESULTS AND DISCUSSION

All the samples were prepared by argon arc melting. No heat treatment was made for the samples. Power X-ray diffraction was employed to determine phase purity, and little impurity phase was detected at each sample. The lattice parameters of the samples were estimated from the diffraction peaks as shown in Fig. 1. The *a*- and *c*-axes increase as the Zr concentration x increases, reflecting the larger ion radius of Zr^{4+} than that of Nb^{5+}. Rather smooth expansion of the cell volume with x indicates that the substitution of Zr for Nb was successfully achieved in these samples.

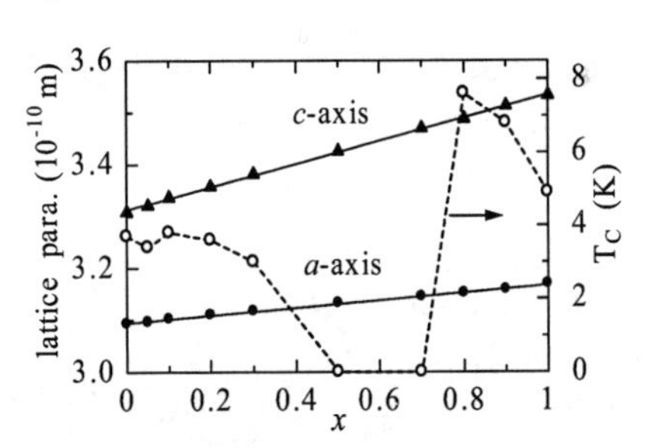

FIGURE 1. Lattice parameters (closed symbols) and T_C (open circles) of $(Nb_{1-x}Zr_x)_{0.8}B_2$.

The SC of the present samples was examined with the resistivity and magnetization. The results of DC resistivity measurements are plotted in Fig. 2, where the data for $x = 0.05$ and 0.1 were omitted to avoid confusion since they almost coincided with those for $x = 0.2$. There are two intriguing features; the samples of $x = 0.5$ and 0.7 did not show the superconducting transition above 2 K. This implies that T_C of $(Nb_{1-x}Zr_x)_{0.8}B_2$ are not in a simple relation with the *e/a* ratios. Another characteristic aspect is that the resistivity at the normal state above T_C decreased roughly with increasing x; that is , the residual

CP850, *Low Temperature Physics: 24th International Conference on Low Temperature Physics;*
edited by Y. Takano, S. P. Hershfield, S. O. Hill, P. J. Hirschfeld, and A. M. Goldman

resistivity of $(Nb_{1-x}Zr_x)_{0.8}B_2$ does not follow the Nordhiem's rule, $\rho \sim x(1-x)$, for alloys[7]. The DC magnetization measurements using a SQUID (Quantum Design MPMS) were made with an external field of 50 Oe above 2 K. Definite diamagnetism due to the SC was observed with the samples except for those of $x = 0.5$ and 0.7, with which no indication of SC was found above 2 K. The resultant relationship between the observed T_C and the *e/a* ratios is drawn in Fig. 1, using the open circles with the broken line. The present result of $T_C = 3.7$ K for $Nb_{0.8}B_2$ agrees well with that of $T_C = 4$ K reported for the arc-melting sample [5], but rather lower than about 9.5 K reported for the samples prepared by the different methods [4,6]. T_C of ZrB_2 has been reported to be 5.5 K [8], which is rather close to the present value of 5.0 K for $Zr_{0.8}B_2$.

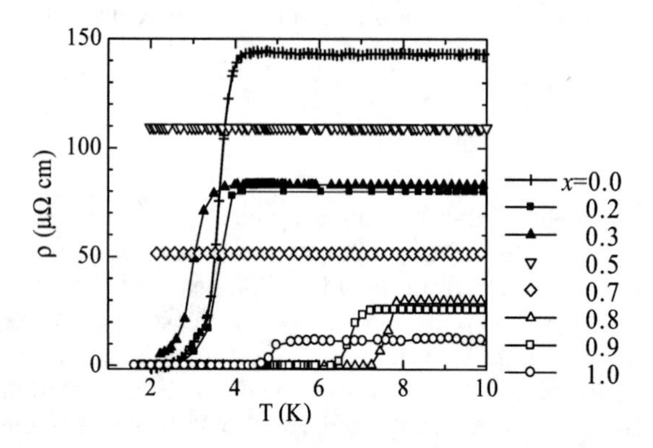

FIGURE 2. Resistivity of $(Nb_{1-x}Zr_x)_{0.8}B_2$ with the samples of *x* indicated.

The disappearance of SC at $x = 0.5$ and 0.7 may be due to band broadening by alloying. The *e/a* measure for T_C is only accessible when a rigid-band picture for the electronic structure is valid; i.e. the substitution of Zr for Nb just moves the Fermi level without changing density of states (DOS). Since the electronic specific heat coefficient γ is directly associated with DOS at the Fermi level, we carried out the specific heat (SH) measurements with the chosen samples. The measured SH data using PPMS system (Quantum Design) are plotted in Fig. 3. The SC contributions to SH are apparent with the sample of $x = 0.0$, but rather faint with those of $x = 0.8$ and 1.0 due to their small volume fractions of the superconducting states. The *C/T* values of the samples of $x = 0.5$ and 0.7 slightly rose at lower temperatures; its origin is not clear at this stage. The deduced values of γ and Debye temperature Θ_D for $x = 0.0$, 0.5, 0.7, 0.8 and 1.0 are 2.7, 2.0, 1.6, 1.2 and 0.7 mJ/mol K^2, and 810, 850, 920, 920 and 950 K, respectively. These γ and Θ_D values are comparable with the reported ones[9,10], and smoothly decrease as *x* increases. The recent band calculation predicts that DOS for NbB_2 is about three times larger than that of ZrB_2[11]. The observed tendency of γ in Fig. 3, therefore, supports the theoretical prediction. We could not find any hint to explain why the SC disappears at $x = 0.5$ and 0.7 from the present results of the specific heat measurements. A different sample preparation process could give us deeper understanding of the SC of $(Nb_{1-x}Zr_x)_{0.8}B_2$ system, which is in a planning stage.

This work is partially supported by the Takahashi Industrial and Economic Research Foundation.

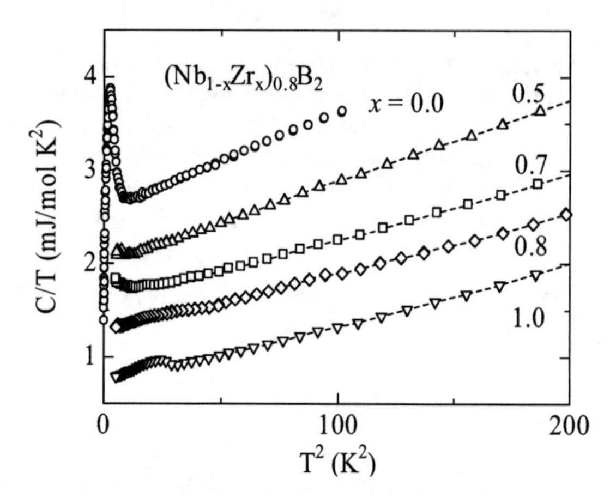

FIGURE 3. Specific heat of the samples of *x* indicated. The data marked by open symbols were taken without field; those marked by closed symbols were taken with 4T.

REFERENCES

1. B. T. Matthias, T. H. Geballe, V. B. Compton, E. Corenzwit, and G. W. Hull Jr., *Rev. Mod. Phys.* **36**, 155-156 (1964).
2. G. Gladstone, M. A. Jensen, and J. R. Schriefer, "Superconductivity in the Transition Metals" in *Superconductivity*, edited by R. D. Parks, Marcell Dekker, New York, 1969, pp. 665-816.
3. J. Nagamatsu, N. Nakagawa, T. Muranaka, Y. Zenitani, and J. Akimitsu, *Nature* **410**, 63(2001).
4. A. Yamamoto, C. Takao, T. Masui, M. Izumi, and S. Tajima, *Physica C***383**, 197-206 (2002).
5. H. Takeya, K. Togano, Y. Sung, T. Mochiku, and K. Hirata, *Physica C***408**, 144-145 (2004).
6. R. Escamila, O. Lovera, T. Akachi, A. Duran, R. Falconi, F. Morales, and R. Escudero, *J. Phys. Condens. Matter* **16**, 5979-5990 (2004).
7. H. A. Fairbank, *Phys. Rev.* **66**, 274-281 (1944).
8. V. A. Gasparov, N. S. Sidorov, I. I. Zver'kova, and M. P. Kulakov, *JETP Lett.* **73**, 532 (2001).
9. B. Fisher, K. B. Chashka, L. Patlagon, and G. M. Reisner, *Physica C***384**, 1-10(2003).
10. G. L. E. Muzzy, et al., *Physica C***382**, 153-165 (2002).
11. P. Mora, M. de la, Castro, and G. Tavizon, *J. Solid State Chem.* **169**, 168-175 (2002).

Structural and Electrical Properties of $CaBe_{0.75}Si_{1.25}$

Fumihiko Sano, Yumiko Takahashi, Kouichi Takase, Yoshiki Takano and Kazuko Sekizawa

Department of Physics, College of Science and Technology, Nihon University, 1-8 Kanda-Surugadai, Chiyoda-ku, Tokyo 101-8308, Japan

Abstract. Ternary silicide compounds, $Ca(Be_xSi_{1-x})_2$ ($0.25 \leq x \leq 0.5$) have been synthesized by a sealed tantalum ampoule method. When $x = 0.375$, a single phase of an AlB_2-type structure is observed, whereas some sub-phases are observed when $x \neq 0.375$ by X-ray diffraction measurements. $CaBe_{0.75}Si_{1.25}$ ($x = 0.375$) is isostructural to the high-temperature superconductor MgB_2 (AlB_2-type structure) and similar to $Ca(Al,Si)_2$. However, a detailed analysis reveals superlattice lines, indicating the ordering of Be and Si atoms. Whereas $Ca(Al,Si)_2$ is a superconductor with a transition temperature T_C of 7.7 K, the electrical resistivity of $CaBe_{0.75}Si_{1.25}$ shows metallic behavior above 4.2 K and is not a superconductor. In $CaBe_{0.75}Si_{1.25}$, the ordering of Be and Si atoms may play an important role in the suppression of superconductivity.

Keywords: Calcium disilicide, AlB_2-type structure
PACS: 74.62.Dh

INTRODUCTION

The discovery of superconductivity at 39 K in MgB_2 led to a growing interest in closely related systems with a AlB_2-type structure. $CaSi_2$ has an AlB_2-like structure under high pressure (16.5 GPa) and shows superconductivity below 14 K though $CaSi_2$ with a trigonal structure at ambient pressure is a superconductor with T_c = 1.58 K [1]. This result indicates that the flat Si plane of the AlB_2-like structure causes the rise of T_c. On the other hand, substitution of another element for a part of Si resulted in a realization of the AlB_2-type structure of $CaSi_2$ compounds at ambient pressure. For instance, $Ca(Al, Si)_2$ of the AlB_2-type structure (T_c = 7.7 K) has been reported[2].

In this study, Be was chosen for the substitution element [3], because CaBeSi is suggested to have electronic states similar to those of MgB_2 near E_F [4]. Then, $Ca(Be_xSi_{1-x})_2$ ($0.25 \leq x \leq 0.5$) was synthesized for the investigation of the superconductivity of AlB_2-type $CaSi_2$ compounds.

EXPERIMENTAL

Polycrystalline samples were prepared by a sealed tantalum ampoule method. The Ca granules (99.5%), Be chips (99.9%) and Si granules (99.999%) were used as the starting materials for the synthesis of $Ca(Be_xSi_{1-x})_2$ samples. The elements were weighted in a stoichiometric ratio, mixed and placed in a Ta ampoule 6.5 mm in diameter and 120 mm in length. Then ampoule was closed by a hand vice and sealed by arc melting. The Ta ampoule was heated up to 1350 K at a rate of 400 K/h, held for 1 h, and then cooled to room temperature. In order to protect each sample from oxidization, all processes were carried out in an Ar atmosphere.

The structure and lattice parameters of each sample were characterized by the synchrotron X-ray powder diffraction on BL02B2 at the Spring-8. The wavelength was 0.04968 nm. The Experimental data

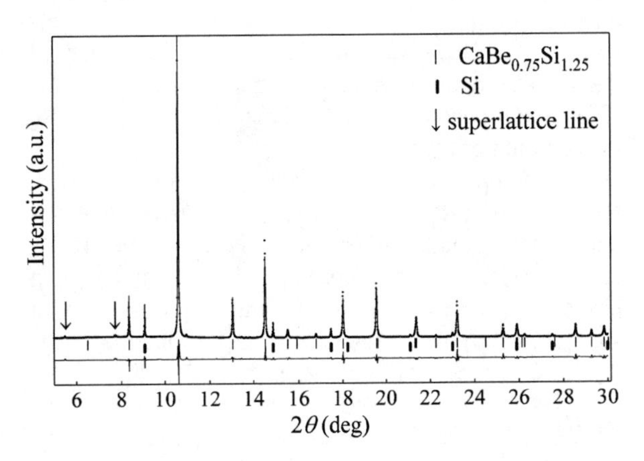

FIGURE 1. Rietveld refinement pattern of the synchrotron X-ray powder diffraction data for $CaBe_{0.75}Si_{1.25}$.

CP850, *Low Temperature Physics: 24th International Conference on Low Temperature Physics;* edited by Y. Takano, S. P. Hershfield, S. O. Hill, P. J. Hirschfeld, and A. M. Goldman

were analyzed by the Rietveld method with RIETAN2000 program [5]. The temperature dependences of the electrical resistivity $\rho(T)$ of the samples were measured by employing a standard direct-current (DC) four-probe method. The temperature range was 300 K to 4.2 K.

RESULTS AND DISCUSSION

Figure 1 shows the Rietveld refinement pattern of the synchrotron X-ray powder diffraction data for $CaBe_{0.75}Si_{1.25}$. From this, it was found that the sample was an almost single phase of AlB_2-type structure (*P6/mmm*, a = 3.9342Å , c = 4.3757Å) with a small quantity of Si impurities. For $x \neq 0.375$ other impurity phases were observed. The amount of impurity phases increased as x became apart from 0.375. Table 1. shows the results of the Rietveld analysis for $CaBe_{0.75}Si_{1.25}$ (x = 0.375). It shows the Wyckoff position WN, an occupation factor g and fractional coordinates x, y, z of each atom. We analyzed the crystal structure assuming that Be and Si have different fractional coordinates at the same site, although they should generally occupy the same site. As a result, the reliable factor of the analysis was improved. The crystal structure obtained from this analysis is shown in Fig. 2 The fractional coordinates x and y of Be and Si shifted a little bit from their ideal positions of AlB_2-type structure. This result indicated that the honeycomb structure which is constructed by Be and Si in the xy plane has a little distortion or Be and Si have the thermal vibration parameters with an anisotropic properties. Furthermore, small peaks denoted by arrows at small 2θ in Fig. 1 suggest an existence of a superlattice. These lines indicated that the lattice constants a and b became three times larger than the original axis. The ordering of Be and Si atoms and stacking behavior are now being examined in detail. The better refinement obtained by using the above mentioned assumption is considered to be due to the ordering of them.

The temperature dependence of the electrical resistivity of $CaBe_{0.75}Si_{1.25}$ shows a metallic behavior and superconductivity is not observed above 4.2 K (Fig. 3). The modulation of the AlB_2-type structure in $CaBe_{0.75}Si_{1.25}$ may prevent the appearance of superconductivity. However, the value of the resistivity at room temperature and the residual resistivity ratio ($R_{300}/R_{4.2}$ = 1.17) are close to those of high T_c superconductors such as YBCO [6]. If the single phase compound of the AlB_2-type structure is obtained with different x, there is a possibility that $Ca(Be_xSi_{1-x})_2$ becomes a superconductor.

TABLE 1. Structure parameters for $CaBe_{0.75}Si_{1.25}$ obtained by Rietveld analysis of the synchrotron X-ray powder diffraction data.

	WN	g	x	y	z
Ca	1*a*	1.0	0.0	0.0	0.0
Be	12*q*	0.0625	0.3375	0.6719	0.5
Si	12*q*	0.1041	0.3330	0.6662	0.5

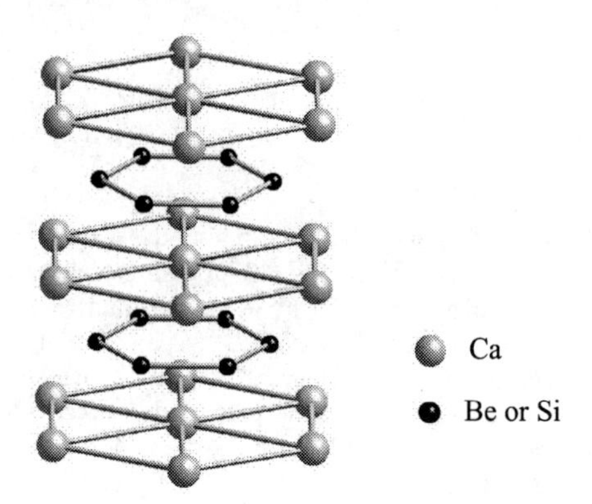

FIGURE 2. Crystal structure of $CaBe_{0.75}Si_{1.25}$.

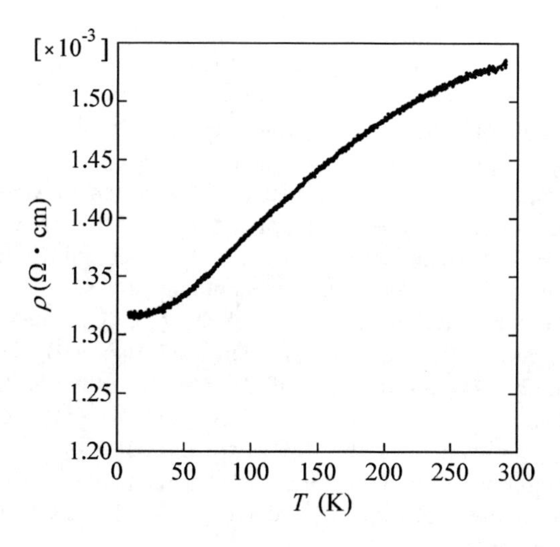

FIGURE 3. Temperature dependence of the electrical resistivity of $CaBe_{0.75}Si_{1.25}$.

REFERENCES

1. P. Bordet et al., *Phys. Rev. B* **62**, 11 392 - 11 397 (2000).
2. M. Imai, et al., *Appl. Phys. Lett.* **80**, 1019 - 1021 (2002).
3. N. May et al., *Z.Naturforsh* **29b**, 325 - 327 (1974).
4. G. Satta et al., *Phys. Rev. B* **64**, 104507-1 - 104507-9 (2001).
5. F. Izumi et al., *Mater. Sci. Forum* **321-324**, 198 - 203 (2000).
6. T. Wuernisha et al., *J. of Alloys and Compounds* **377**, 48 -54 (2004).

Correlation Of The Superconductivity With The Multi-Stack Structure In MgB_2-Type Superconductor CaAlSi

Hajime Sagayama[1], Yusuke Wakabayashi[1], Hiroshi Sawa[1], Takashi Kamiyama[1], Akinori Hoshikawa[2], Stefanus Harjo[2], Kentaro Uosato[3], Ajay K. Ghosh[3], Masashi Tokunaga[3] and Tsuyoshi Tamegai[3]

[1]*Institute of Materials Structure Science, High Energy Accelerator Research Organization, Tsukuba 305-0801, Japan*
[2]*Neutron Science Research Center, Japan Atomic Energy Research Institute, Tokai, Ibaraki, 319-1195, Japan*
[3]*Department of Applied Physics, The University of Tokyo, Hongo, Bunkyo-ku, Tokyo 113-8656, Japan*

Abstract. CaAlSi is a superconductor with the AlB_2-like structure. In this system, there are two types of multi-stack structures along *c*-axis, 5-fold structure and 6-fold structure. The 5-fold structure sample and the 6-fold structure sample have T_C of 6 K and 8 K, respectively. To investigate the correlation of stack structure with the superconductivity, we determined the crystal structures accurately using synchrotron radiation X-ray diffraction and neutron diffraction measured at KEK. Al and Si atoms form slightly corrugated honeycomb layers and those are ordered in both stack structures. The difference of stacking structure is caused by phase relation among Al-Si layers.

Keywords: CaAlSi, superconductivity, multi-stack structure
PACS: 61.10.Nz, 61.12.Ld , 74.62.Bf

The discovery of superconductivity at 39 K in MgB_2 with AlB_2-type structure[1] stimulated a great number of studies on superconductivity in compounds which have the intercalated graphite structure. $CaSi_2$ has a rhombohedral structure (R-3m, Z=6) in which the silicon sp^3 layers are separated by Ca plane. A pressure-induced phase transition from the rhombohedral structure to the trigonal structure (P-3m1, Z=1) occurs around 8 GPa at room temperature[2]. In pressure range larger than 8 GPa, the bonding state of Si atoms in $CaSi_2$ varies from sp^3- to sp^2-type by applying pressure. Superconductivity appears around 12 GPa and transition temperature T_C increases rapidly up to 14 K with applying pressure[3]. Therefore, the superconductivity in $CaSi_2$ might be related to the form of Si network. Superconductivity is also found in the $CaAl_{2-x}Si_x$ for $0.6 < x < 1.2$ that crystallizes in AlB_2-type structure (a_P, c_P)[4,5]. Maximum T_C = 7.8K appears for x = 1, namely, in CaAlSi. The T_C of CaAlSi also increases slightly by applying pressure[6].

In CaAlSi system, there are two types of multi-stack structures along *c*-axis, $1a_P \times 1a_P \times 5c_P$ (5-fold) structure and $1a_P \times 1a_P \times 6c_P$ (6-fold) structure, observed by the transmission electron microscopy and the X-ray diffraction measurement at room temperature[7]. It is remarkable that the 5-fold and the 6-fold structure samples have T_C of 6K and 8K, respectively[8]. In these multi-stack samples, large anisotropy and cusp-like feature of H_{C2} were observed[7,9]. This unusual feature is characteristic of thin film superconductor, even though bulk superconductivity was confirmed by specific heat measurement. This contradiction is not solved yet.

Our primary interest in CaAlSi is correlation between superconductivity and the periodicity of multi-stack structure. To investigate this problem, the origin of the difference between stacking structures should be revealed. Information concerning the arrangement of Al and Si atoms is also necessary because the character of hexagonal planes might be a large factor dominating the feature of superconductivity in CaAlSi like in $CaSi_2$.

In case that Al and Si arrangement is disordered, free energy gaps between various periodic structures in CaAlSi are small; therefore, the periodicity might be sensitive to external pressure. To investigate the pressure dependence of the periodicity of multi-stack structure,we have performed synchrotron radiation X-ray diffraction experiment using two dimensional

CP850, *Low Temperature Physics: 24th International Conference on Low Temperature Physics;*
edited by Y. Takano, S. P. Hershfield, S. O. Hill, P. J. Hirschfeld, and A. M. Goldman

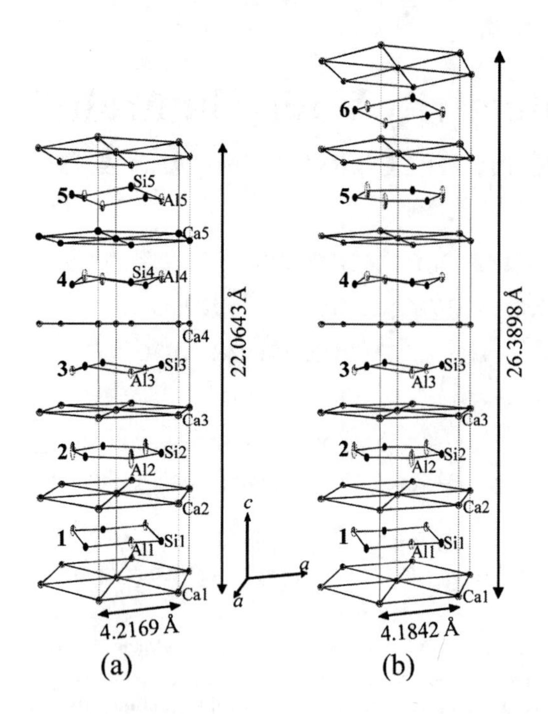

FIGURE 1. Multi-stack structures of CaAlSi, 5-fold (a) and 6-fold (b) structure with anisotropic displacement ellipsoids at room temperature. Gray, white and black ellipsoids correspond to Ca, Al and Si atoms, respectively.

(2D) cylindrical imaging-plate detector installed at BL-1 Photon Factory, KEK, Tsukuba, Japan. X-ray diffraction experiments in this work have been carried out on the high-quality single crystals of 5-fold and 6-fold structure samples chosen among fragments of crushed large single crystals. Details on the single crystal sample preparation can be found in references[7,9]. No change of the periodicity of 5-fold structure was found under high pressure up to 8 GPa. This result shows that the periodicity of multi-stack structure cannot be controlled applying pressure; therefore, a relation between multi-stack structure and T_C cannot be understood by pressure uniformly.

Consequently, to obtain information of the Al-Si network in the 5-fold and the 6-fold structures accurately, we have performed synchrotron radiation X-ray diffraction experiments. The DENZO program was used for 2D image processing. An incident X-ray beam with the wavelength of 0.6867 (0.6896) Å produced by Si(111) monochromator was used. The wavelengths were calibrated with CeO_2 powder diffraction pattern. The intensities of Bragg reflections were measured in a half-sphere of reciprocal space in the range $2\theta < 120^{\circ}$ at room temperature. The number of unique reflections with $I > 3\sigma(I)$ was 2422 (3532). The Crystal Structure program developed by RIGAKU was used for refinements.

Figure 1 shows the results of single crystal structure analyses on the 5-fold and the 6-fold structures of CaAlSi at room temperature. Space group for the 5-fold (6-fold) structure is P3 ($P6_3$). Lattice parameters were refined as a = 4.2169 (4.1842) ~ a_P, c = 22.0643 ~ $5c_P$ (26.3898 ~ $6c_P$) Å. Residuals factors are R = 0.045 (0.035) and R_W = 0.063 (0.044). Goodness-of-fit indicator is S = 1.012 (1.115). An Al-Si arrangement ordering in the 6-fold structure was confirmed from the powder neutron diffraction pattern measured at room temperature using a time-of-flight diffractometer VEGA, KENS, KEK.

Al and Si atoms form honeycomb layers and are ordered in *ab* plane in both multi-stack structures. The 5-fold structure is close to the 6-fold structure from which 5th Al-Si and Ca layers are removed. The 2nd Al-Si layers in both structures and the 5th layer in the 6-fold structure are almost flat. In these layers, anisotropic displacement ellipsoids of Al atoms are extended along *c*-axis. This tendency was shown in low temperature measurement. Therefore this anisotropy indicates existence of z position disorder of Al atoms, not of thermal oscillation. The planar defect due to the randomness of Al displacement might relate to the unusual angular dependence of H_{C2}. To discuss this problem exactly, information about local lattice defect is necessity. The corrugations of Al-Si layers are smaller than that of Si layers in $CaSi_2$ at ambient pressure. This fact suggests that the bonding between Al and Si atoms is a mixture of sp^3- and sp^2-type hybridization.

In conclusion, the crystal structures of the 5-fold and the 6-fold structures in CaAlSi were determined by synchrotron X-ray radiation diffraction and powder neutron diffraction. Al and Si form slightly corrugated honeycomb layers and are ordered. Ca atoms are intercalated at regular intervals between them. The difference of stacking structures is caused by phase relation between Al-Si layers.

REFERENCES

1. J. Nagamatsu *et al., Nature (London)* **410,** 63 (2001).
2. P. Bordet *et al., Phys. Rev. B Condensed Matter and Materials Physics* **62,** 11392 (2000).
3. S. Sanfilippo *et al., Phys. Rev. B Condensed Matter and Materials Physics* **61,** R3800 (2000).
4. M. Imai *et al., Appl. Phys. Lett.* **80,** 1019 (2002).
5. B. Lorenz *et al., Physica C* **383,** 191 (2002).
6. B. Lorenz *et al., Phys. Rev. B Condensed Matter and Materials Physics* **68,** 014512 (2003).
7. A. K. Ghosh *et al., Phys. Rev. B Condensed Matter and Materials Physics* **68,** 054507 (2003).
8. T. Tamegai (private communication).
9. A. K. Ghosh *et al., Physica C* **392-396,** 29-33 (2003).

Origin of Two-Dimensional Superconductivity in CaAlSi

T. Tamegai, K. Uozato, T. Nakagawa, and M. Tokunaga

Department of Applied Physics, The University of Tokyo, 7-3-1 Hongo, Bunkyo-ku, Tokyo 113-8656, Japan

Abstract. CaAlSi crystallized in AlB_2 structure and shows superconductivity at about 8 K. Although the anisotropy of the upper critical field (H_{c2}) is weak, its angular dependence shows an anomalous cusp-like feature when the field is applied parallel to the plane. Such a cusp-like angular dependence is predicted in the case of thin a film superconductor, when the coherence length is larger than the film thickness. It suggests that the superconductivity in CaAlSi is decoupled into two-dimensional superconducting units. The size of the two-dimensional superconducting unit is estimated from the temperature dependence of $H_{c2}(\theta)$. We also examined a possible defect structure using magneto-optical imaging.

Keywords: CaAlSi, Tinkham model, anisotropic GL model
PACS: 74.25.-q, 74.25.Op, 74.70.Dd

INTRODUCTION

Since the discovery of superconductivity at T = 39 K in MgB_2 with AlB_2 structure, interest has been renewed in layered intermetallic superconductors [1]. Several intermetallic superconductors with AlB_2 structure have been reported so far [2-5]. Among them, CaAlSi shows the highest superconducting transition temperature T_c = 7.8 K [4]. Reflecting the layered crystal structure, normal and superconducting properties in CaAlSi show anisotropy with anisotropy parameter $\gamma=(m_c/m_{ab})^{1/2}\sim 2$. Here, m_c and m_{ab} are effective masses along c-axis and ab-plane. Despite the small value of γ, the angular dependence of the upper critical field, $H_{c2}(\theta)$, does not follow the anisotropic GL model. Instead, it is well described by the Tinkham model for thin film superconductors [7], although the sample is a bulk crystal. On the other hand, $H_{c2}(\theta)$ in a related compound CaGaSi is well fitted by the anisotropic GL model [8]. By comparing CaAlSi and CaGaSi, we have shown that the anomaly in $H_{c2}(\theta)$ is closely related to the superstructure along c-axis, which is present only in CaAlSi [6,8]. The cusp-like anomaly of $H_{c2}(\theta)$ in a bulk crystal can be possible when the superconducting layers, thinner than the coherence length (ξ), are effectively decoupled by non-superconducting layers thicker than ξ. However, even with the superstructure, the real c-axis lattice parameter is only 24 A, which is much shorter than ξ. In this paper, we report the temperature dependence of $H_{c2}(\theta)$ in well-characterized CaAlSi single crystals, and estimate the size of possible defects. We also report magneto-optical observations of flux penetrations in ac-plane in order to check the presence of semi-macroscopic defects along ab-plane, which may also act to decouple superconducting layers.

EXPERIMENTAL

Single crystals of CaAlSi are grown by the floating-zone method as described in ref. [6]. After the growth, the crystals are annealed in order to remove residual strain and make the superconducting transition sharper. The crystal annealed at 750°C for three days in a evacuated quartz tube with a Ca chunk shows a sharp transition at T_c=7.3 K with ΔT_c less than 0.1 K. Electrical resistivity of the sample as a function of magnetic field at various temperatures and field-angles is measured by the four-probe method. The upper critical field is determined by the resistive criterion. In order to check the possible defect structures along ab-plane, the vortex penetration in ac-plane is visualized by the magneto-optical method using an in-plane magnetized garnet film as an indicator [9]. The ac plane of the crystal is polished before the observation and the garnet film is placed directly on top of it.

RESULTS AND DISCUSSION

Figure 1 shows angular dependence of the upper critical field $H_{c2}(\theta)$ for a sample with T_c=7.3 K. θ =0° corresponds to field parallel to a-axis. The ratio of H_{c2} for fields parallel to c-axis and a-axis is about 2 for all temperatures. At lower temperatures, $H_{c2}(\theta)$ shows a cusp-like feature when the field is applied parallel to

CP850, *Low Temperature Physics: 24th International Conference on Low Temperature Physics;*
edited by Y. Takano, S. P. Hershfield, S. O. Hill, P. J. Hirschfeld, and A. M. Goldman

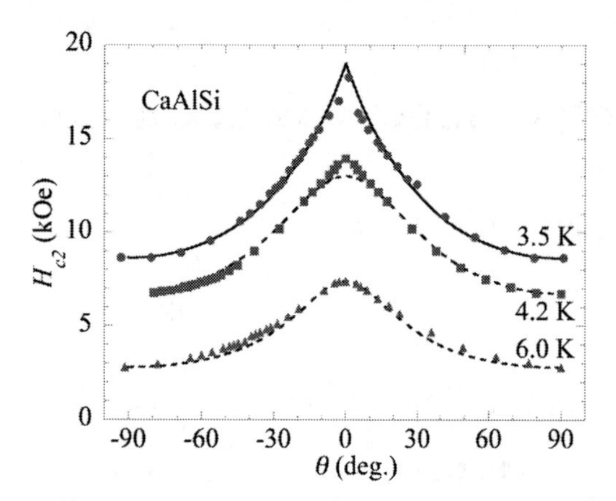

FIGURE 1. Angular dependence of the upper critical field $H_{c2}(\theta)$ for a CaAlSi single crystal at $T = 3.5$, 4.2 and 6.0 K. The solid curve is a fitting using the Tinkham model, while broken curves are fittings using the anisotropic GL model.

a-axis. However, the cusp-like feature becomes weaker at higher temperatures. Usually, $H_{c2}(\theta)$ is described by the anisotropic GL model unless γ is extremely large as in the case of $Bi_2Sr_2CaCu_2O_{8+y}$. $H_{c2}(\theta)$ with cusp-like feature has been observed in two-dimensional superconductors where the thickness of the sample is smaller than ξ [10], and is explained in ref. [7]. A similar cusp-like feature is also observed in a multilayer system consisting of successive stacking of superconducting and normal layers when the normal layer is thicker than ξ [11]. The presence of the cusp-like feature in a bulk crystal of CaAlSi requires built-in normal layers which decouple the two-dimensional superconducting layers. However, since ξ becomes longer near T_c, even such a system should show the anisotropic GL behavior near T_c. Actually, $H_{c2}(\theta)$ at 6.0 K in Fig. 1 is well fitted by the anisotropic GL model. If we assume that the thickness of the normal layer, d_n, is the same as ξ at 6.0 K, d_n is estimated as 180 A by using $\xi_c(0)$= 100 A [6].

In order to check whether there are macroscopic defects along ab-plane, vortex penetrations in ac-plane is visualized by the magneto-optical method. Figure 2 shows differential images constructed by subtracting a image at (H_a-5) Oe from that at H_a Oe. At very low field, the sample shows Meissner response except for two cracks running almost horizontally in Fig. 2(a). As the field is increased, vortices first penetrate the sample along the cracks as in Fig. 2(b). When the field is further increased, vortices start to penetrate the sample along c-axis as in Figs. 2(c)-(e). Finally, vortices penetrate the whole sample as in Fig. 2(f). Except for obvious cracks and minor inhomogeneities, the vortex penetration is normal and any additional channeling or pinning along ab-plane is not observed.

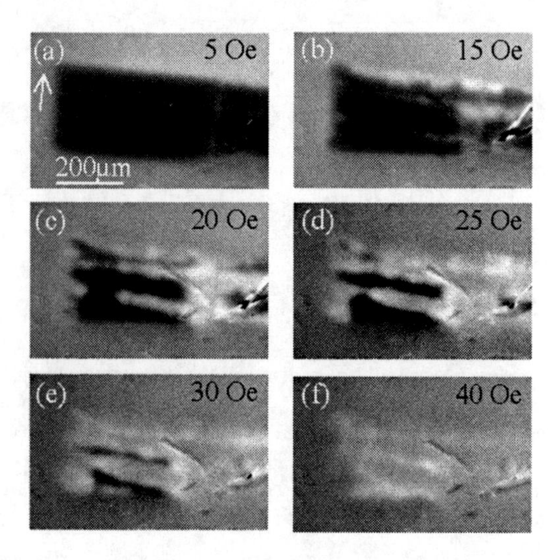

FIGURE 2. Differential magneto-optical images of vortex penetrations into ac-plane of a CaAlSi single crystal. Each image is constructed by subtracting an image taken at (H_a-5) Oe from another taken at H_a Oe, which is indicated in the figure. An arrow in (a) shows the direction of c-axis.

Presence of defects along ab-plane would affect the out-of-plane transport and may lead to SNS or SIS voltage-current characteristics. In order to check this possibility, voltage-current measurements in a small mesa fabricated by Ar-ion milling is now in progress.

In summary, we have found that the cusp-like angular dependence of H_{c2} in CaAlSi at low temperatures gradually becomes smooth close to T_c. From this result, we speculate that defect structures parallel to ab-plane with thickness of about 180 A are present in CaAlSi single crystals.

ACKNOWLEDGMENTS

This work is supported by Grant-in-aid for Scientific Research from the Ministry of Education, Culture, Sports, Science, and Technology.

REFERENCES

1. J. Nagamatsu *et al.*, *Nature* **410**, 63 (2001).
2. M. Imai *et al.*, *Phys. Rev. Lett.* **87**, 077003 (2001).
2. S. Majumdar *et al.*, *Phys. Rev. B* **63**, 172407 (2001).
4. M. Imai *et al.*, *Appl. Phys. Lett.* **80**, 1019 (2002).
5. H. Kito *et al.*, *Physica C* **377**, 185 (2002).
6. Ajay Kumar Ghosh *et al.*, *Phys. Rev. B* **68**, 054507 (2003).
7. M. Tinkham. *Phys. Rev.* **129**, 2413 (1963).
8. T. Tamegai *et al.*, *Int. J. Mod. Phys. B* **19**, 369 (2005).
9. A. Soibel *et al.*, *Nature* **406**, 282 (2000).
10. D. E. Morris *et al.*, *Phys. Rev. Lett.* **6**, 600 (1962).
11. C. S. L. Chun *et al.*, *Phys. Rev. B* **29**, 4915 (1984).

Search for Superconductivity in Layered Silicides

T. Tamegai, T. Nakagawa, K. Uozato, and M. Tokunaga

Department of Applied Physics, The University of Tokyo, 7-3-1 Hongo, Bunkyo-ku, Tokyo 113-8656, Japan

Abstract. Several layered silicide systems are studied in order to search for superconductivity. $CaAl_2Si_2$ and YAl_2Si_2, both having puckered double Si-Al layers, are found to be a semimetal and a typical metal, respectively. Despite the fact that these two compounds have the same crystal structure, it is difficult to obtain a homogeneous solid solution $Y_{1-x}Ca_xAl_2Si_2$. None of them show superconductivity down to 0.3 K. $LaAlSi_2$ is composed of alternate stacking of $LaSi_2$ and $LaAl_2Si_2$ layers. $LaAlSi_2$ polycrystalline sample with noticeable amount of impurities shows a trace of superconductivity starting from 0.7 K. However, it is found that $LaAlSi_2$ is not a superconductor, but the superconductivity originates from an unidentified impurity phase containing a part or all of La, Al, and Si.

Keywords: layered silicide, $CaAl_2Si_2$, YAl_2Si_2, $LaAlSi_2$
PACS: 72.15.-v, 74.10.+v

INTRODUCTION

The discovery of superconductivity in MgB_2 renewed our interest to layered intermetallic compounds [1]. Isotope effect study in MgB_2 indicates that the pairing mechanism is mediated by phonons [2]. The coexistence of two kinds of weakly interacting superconductivity in a two-dimensional σ band and a three-dimensional π band opens a new field of muti-band superconductivity [3]. Superconductivity in structurally similar layered intermetallic compounds has been found in $AE(M_{1-x}Si_x)_2$ (AE: alkaline earth elements, M: Al, Ga) [4,5]. Among these compounds, superconductivity in CaAlSi shows not only relatively high T_c (~8 K) [5], but also an anomalous angular dependence of the upper critical field [6]. These observations triggered our search for superconductivity in layered intermetallic silicides.

$CaAl_2Si_2$ has a hexagonal layered structure, where puckered double (Si,Al) layers are sandwiched by Ca layers forming triangular lattices. Ca in $CaAl_2Si_2$ can be replaced by Y and rare earth elements. On the other hand, by stacking $CaAl_2Si_2$ structure and AlB_2 structure, new structures like $LaAlSi_2$ and $La_3Al_4Si_6$ are reported [7]. In this paper, we have prepared polycrystalline samples of $CaAl_2Si_2$, YAl_2Si_2, and $LaAlSi_2$ and characterized their physical properties. None of them show superconductivity down to 0.3 K. However, in the case of La, we detected a trace of superconductivity originated from an unidentified impurity phase.

EXPERIMENTAL

Polycrystalline samples of $CaAl_2Si_2$, YAl_2Si_2, and $LaAlSi_2$ are prepared by arc-melting of the constituent elements in Ar atmosphere. Samples are annealed in evacuated quartz tube at the optimum temperature for as long as three weeks. Electrical resistivity, ρ, is measured by the four-probe method down to 0.3 K. Hall coefficient, R_H, is estimated by applying field up to 50 kOe. Magnetization is measured by SQUID magnetometer. Specific heat is measured by the relaxation method. In the case of $CaAl_2Si_2$, we grow a single crystal by the floating-zone method using an image furnace and characterized anisotropic properties.

RESULTS AND DISCUSSION

Figure 1(a) shows temperature dependence of ρ, for both $CaAl_2Si_2$ crystal and YAl_2Si_2. ρ in $CaAl_2Si_2$, for current parallel to *ab*-plane, is relatively large and its temperature dependence is weak, while that in YAl_2Si_2 shows typical metallic behavior with residual ρ of 5 $\mu\Omega$cm. ρ_c/ρ_{ab} in $CaAl_2Si_2$ is about 2. Temperature dependences of R_H in $CaAl_2Si_2$ and YAl_2Si_2 are shown in Fig. 1(b). R_H at high temperatures in $CaAl_2Si_2$ is negative but changes the polarity below about 70 K. This fact, together with the temperature dependence of ρ, indicates that $CaAl_2Si_2$ is a semimetal with both electrons and holes as carriers. On the other hand, R_H in YAl_2Si_2 is temperature independent and negative. Carrier density deduced from R_H for YAl_2Si_2 is 2.5×10^{22} cm^{-3}.

CP850, *Low Temperature Physics: 24th International Conference on Low Temperature Physics;*
edited by Y. Takano, S. P. Hershfield, S. O. Hill, P. J. Hirschfeld, and A. M. Goldman

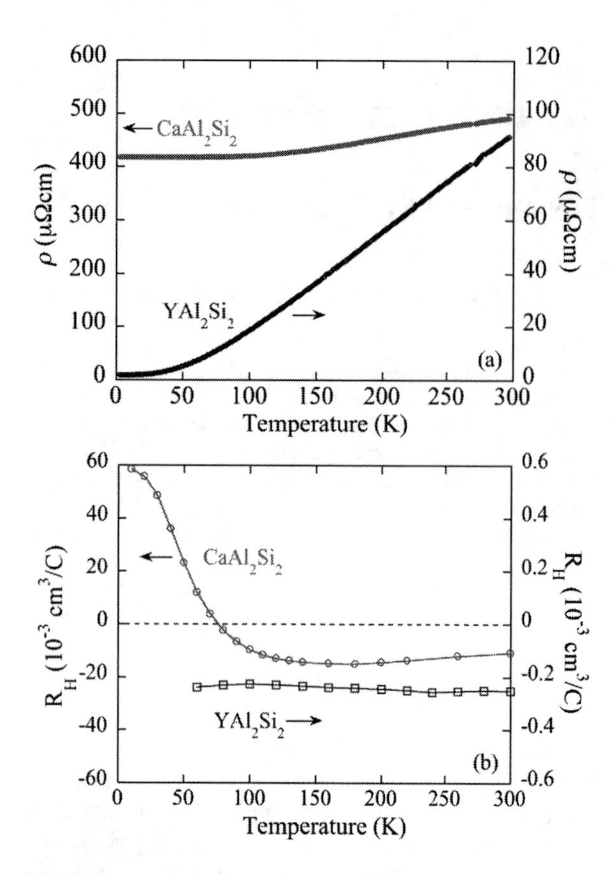

FIGURE 1. Temperature dependence of (a) resistivity and (b) Hall coefficient for $CaAl_2Si_2$ crystal and YAl_2Si_2.

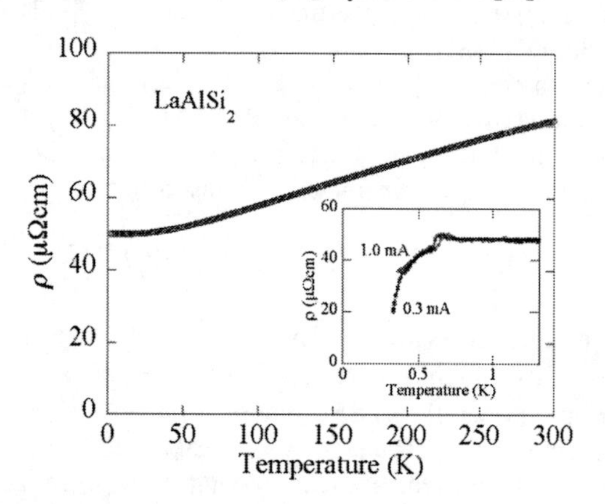

FIGURE 2. Temperature dependence of resistivity for $LaAlSi_2$. Inset shows the current dependence of resistivity at low temperatures.

Specific heat is measured at low temperatures. From the standard analysis we evaluated the electronic specific heat coefficients, γ, and the Debye temperature, Θ_D. Thus obtained values are γ = 0.52 mJ/mol K^2 and Θ_D = 390 K for $CaAl_2Si_2$ and γ = 5.9 mJ/mol K^2 and Θ_D = 390 K for YAl_2Si_2, respectively. One order of magnitude larger γ in YAl_2Si_2 is consistent with the transport data and recent band structure calculation [8]. We also prepare alloys of $Y_{1-x}Ca_xAl_2Si_2$. Electron probe microanalyses show that samples are inhomogeneous even after long-time annealing. One possible reason for the inhomogeneity is a large discrepancy of *c*-axis lattice constants in the two end members. At lower temperatures we observe a drop in resistance at around 1.2 K in YAl_2Si_2. However, the transition depends on the current density, and it suggests that the superconductivity is percolative. X-ray diffraction data shows a trace of aluminum, as much as 5 %, indicating that the observed superconductivity is caused by the aluminum, and YAl_2Si_2 is not a superconductor down to 0.3 K.

$LaAlSi_2$ can only be obtained in multi-phased form by arc melting followed by three weeks annealing. X-ray diffraction shows that the volume fraction of $LaAlSi_2$ is about 70 %. Figure 2 shows temperature dependence of ρ of the multi-phased $LaAlSi_2$. It shows metallic conduction with residual ρ of 50 μΩcm. The inset of Fig. 2 shows the low temperature part of $\rho(T)$. A weak onset of superconductivity can be identified at 0.7 K. However, the transition depends on the current density. T_c's of all possible known impurities (La: 4.9 K, 6.1 K, $LaAl_2$: 3.2 K, $LaSi_2$: 2.5 K) do not match the observed T_c. These facts indicate that the superconductivity originates from an unidentified impurity phase, and YAl_2Si_2 is not a superconductor.

In summary, we have studied several layered silicides searching for superconductivity. The isostructural $CaAl_2Si_2$ and YAl_2Si_2 are a semimetal and a metal, respectively. They do not show superconductivity down to 0.3 K. Multi-phased $LaAlSi_2$ sample show a trace of superconductivity at 0.7 K coming from an unidentified impurity phase, containing a part or all of La, Al, and Si.

ACKNOWLEDGMENTS

This work is supported by Grant-in-aid for Scientific Research from the Ministry of Education, Culture, Sports, Science, and Technology.

REFERENCES

1. J. Nagamatsu *et al.*, *Nature* **410**, 63 (2001).
2. S. L. Bud'ko *et al.*, *Phys. Rev. Lett.* **86**, 1877 (2001).
3. F. Bouquet *et al.*, *Phys. Rev. Lett.* **87**, 047001 (2001).
4. M. Imai *et al.*, *Phys. Rev. Lett.* **87**, 077003 (2001).
5. M. Imai *et al.*, *Appl. Phys. Lett.* **80**, 1019 (2002).
6. Ajay Kumar Ghosh *et al.*, *Phys. Rev. B* **68**, 054507 (2003).
7. N. Lyaskovska *et al.*, *J. Alloys Comp.* **367**, 180 (2004).
8. Q. Huang (private communications).

Superconductivity In $Y_2Pd(Ge_{1-x}Si_x)_3$ (x = 0-1)

Katsunori Mori, Yasunobu Koshi, and Katsuhiko Nishimura

Faculty of Engineering, Toyama University, 3190 Gofuku, Toyama 930-8555, Japan

Abstract. The electrical resistivity (2 - 300K), magnetization (2 - 10 K), and specific heat (0.5 - 10 K) were measured in $Y_2Pd(Ge_{1-x}Si_x)_3$, which is of AlB_2-type structure as the same as MgB_2. The superconducting critical temperature showed a maximum of T_c = 3.48 K near x = 0.05 and monotonically decreased with x increasing up to x = 0.7 in Si substitution. The variation of T_c with Si concentration is related to the electron-phonon coupling constant λ, which was calculated by use of the McMillan's equation. The variation of λ also exhibited a maximum near x = 0.05, and it monotonically decreased with x increasing up to x = 0.7 in Si substitution.

Keywords: superconductivity, $Y_2Pd(Ge_{1-x}Si_x)_3$, electron-phonon coupling constant.
PACS: 74.62.Bf, 74.70.Dd

INTRODUCTION

Since the recent discovery of the superconductivity in MgB_2 [1], many experimental and theoretical studies have been performed to understand the mechanism of superconductivity at comparatively high T_c, 39 K. The crystal structure of this material has been known as hexagonal (AlB_2 type). The characteristic boron honeycomb sheets are sandwiched between the Mg triangular sheets like an intercalated graphite. Recently, Y_2PdGe_3 with the same AlB_2- type structure as MgB_2 and superconductivity at 3 ~ 3.8 K has been reported [2, 3, 4]. By replacing Mg and B atoms in MgB_2 with Y and (Pd,Ge) respectively, the structure of Y_2PdGe_3 can be realized. In this study, Si atoms are substituted for Ge to investigate the impact of the lighter element in the honeycomb layers. $Y_2Pd(Ge_{1-x}Si_x)_3$ with different x has been studied by the electrical resistivity, specific heat, and magnetization measurements.

RESULTS AND DISCUSSION

The samples were prepared from stoichiometric amounts of elemental Y, Pd, Ge, and Si ingots. The ingots were melted in Ar atmosphere by arc furnace. The X-ray diffraction patterns were obtained employing Cu-K_α radiation. The electrical resistivity of the samples was measured in the temperature range from 2 to 273 K by use of DC four-probe method. The magnetization was measured in the temperature range from 2 to 20 K using SQUID magnetometer at a magnetic field of 50 Oe. The specific heat was measured by PPMS using heat relaxation method in the temperature range from 0.5 to 25 K.

Most peaks in the powder X-ray diffraction patterns of $Y_2Pd(Ge_{1-x}Si_x)_3$ between x = 0 and x = 0.4 are indexed with hexagonal AlB_2-type structure (a = 4.239 Å, c = 3.955 Å for the sample of x = 0). Increasing x further above 0.5, new peaks appeared and are also indexed with another hexagonal structure (a = 8.036Å, c = 8.646Å for the sample of x = 1). It is well known [5, 6] that a and c parameters of Y_2PdSi_3 are approximately twice those of the former corresponding hexagonal structure.

The specific heat, C/T, for the samples of x = 0,

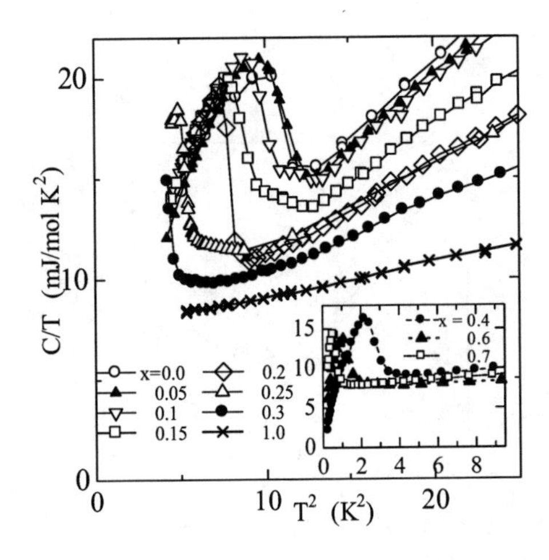

FIGURE 1. C/T vs. T^2. curves in $Y_2Pd(Ge_{1-x}Si_x)_3$.

CP850, *Low Temperature Physics: 24th International Conference on Low Temperature Physics;* edited by Y. Takano, S. P. Hershfield, S. O. Hill, P. J. Hirschfeld, and A. M. Goldman

0.05, 0.10, 0.15, 0.20, 0.25, 0.30, and 1.0 as a function of temperature, T^2, above $T = 2$ K is shown in Fig. 1 and that for the samples of $x = 0.40$, 0.60, and 0.70 is shown in the inset down to $T = 0.5$ K. Analyzing the data of specific heat shown in Fig. 1, the superconducting critical temperature, T_c, the electronic specific heat coefficient, γ, and the Debye temperature, θ_D are obtained. The Si concentration, x, dependence of γ and θ_D, are shown in Fig. 2, and that for T_c in Fig. 3. As seen in Fig. 2, the values of γ and θ_D monotonically increase with increasing x up to 0.4 and above 0.4 both exhibit almost constant.

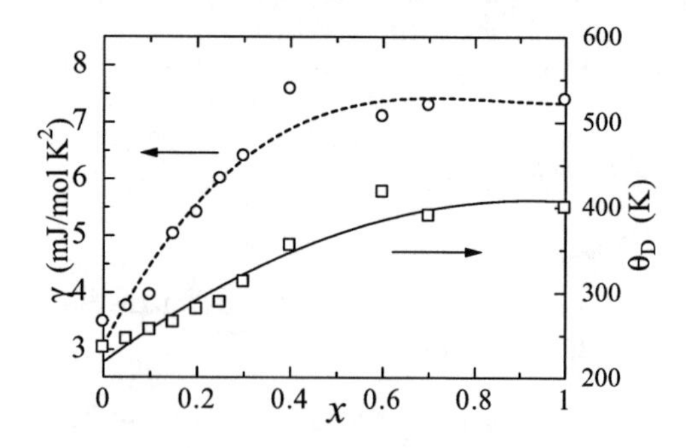

FIGURE 2. γ, θ_D vs. x curves in $Y_2Pd(Ge_{1-x}Si_x)_3$.

On the other hand as seen in Fig. 3, the variation of T_c with x determined by the specific heat measurement increases with x at lower concentration and then decreases with further increase in x. The variation of T_c with Si concentration is related to electron-phonon coupling constant λ, the equation of which is proposed by McMillan [7]. According to the McMillan's equation, λ is expressed as follows; $\lambda=\{1.04+\mu^*\ln(\theta_D/1.45T_c)\}/\{(1-0.62\mu^*)\ln(\theta_D/1.45T_c)-1.04\}$, where μ^* is the Coulomb coupling constant. Using the above expression and the Coulomb coupling constant, $\mu^*=0.10$, the value of λ is calculated and is

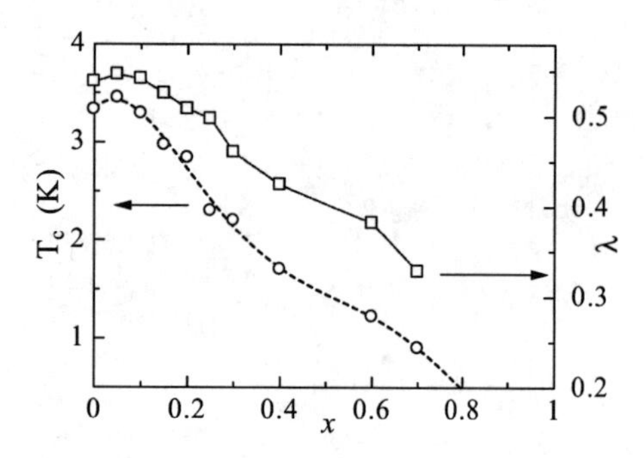

FIGURE 3. T_c, λ vs. x curves in $Y_2Pd(Ge_{1-x}Si_x)_3$.

shown in Fig. 3 as a function of Si concentration, x. As can be seen in Fig. 3, the value of λ shows the maximum around $x = 0.05$ and then decreases monotonically with increasing x up to 0.7. Therefore, the variation of T_c for $Y_2Pd(Ge_{1-x}Si_x)_3$ is controlled by the electron-phonon coupling strength.

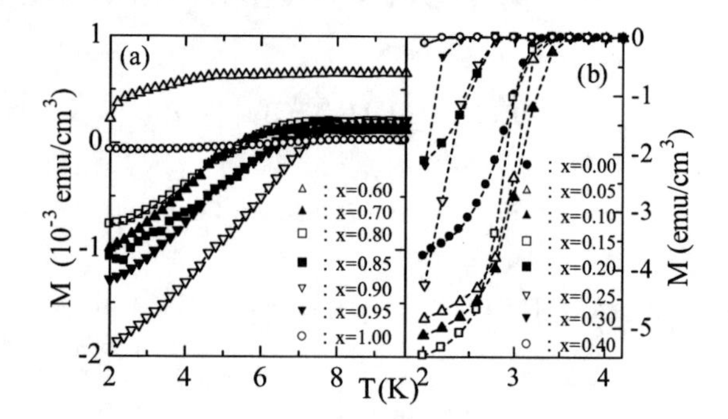

FIGURE 4. Plots of M vs. T under H=50 Oe.

Figure 4 (a) shows the temperature dependence of the magnetization, M, under $H = 50$ Oe for the samples of above $x = 0.6$. It is noted that the magnetization for the samples of $x > 0.6$ shows the negative value even though about three order magnitude smaller in comparison with that for the samples of $x < 0.5$ shown in Fig. 4(b). The temperature dependence of the resistivity, ρ for these samples does not go to zero but shows kink. Mallik and Sampathkumaran[6] have suggested a weak superconducting signal around 7 K for Y_2PdSi_3 compound. According to them, M and ρ for Y_2PdSi_3 exhibit a drop below 8 and 6 K respectively, though M does not become negative. Because of measuring the negative magnetization, we believe that a Y-Pd-Si phase undergoing superconducting transition at 8 K might exist and it is worthwhile to search for this phase.

REFERENCES

1. Nagamatsu, J., Nakagawa, N., Muranaka, T., Zenitani, Y., and Akimitsu, J., *Nature* **410**, 63-64 (2001).
2. Sampathkumaran, E. V., Majumdar, S., Schneider, W., Molodtsov, S. L., and Laubschat, C., *Physica B* **312-313**, 152-154 (2002).
3. Takano, S., Iriyama, Y., Kimishima, Y., and Uehara,M., *Physica C* **383**, 295-298 (2003).
4. Ghosh, A. K., Nakamura, H., Tokunaga, M., and Tamegai, T., *Physica C* **388 – 389**, 567-568 (2003).
5. Chevalier, B., Lejay, Etourneau, P., J., and Hagenmuller, P., *Solid State Comm.* **49**, 753-760 (1984).
6. Mallik, R., and Sampathkumaran, E. V., *J. Magn. Magn. Mater.* **164**, L13-L17 (1996).
7. McMillan, W. L., *Phys. Rev.* **167**, 331-344 (1968).

Local Field Measurements in $PrOs_4Sb_{12}$ with Broken Time-Reversal Symmetry

Shigeru Kasahara*,†, Tsuyoshi Tamegai*, Hitoshi Sugawara** and Hideyuki Sato‡

*Department of Applied Physics, The University of Tokyo, Bunkyo-ku, Tokyo 113-8656, Japan
†National Institute for Materials Science, Tsukuba, Ibaraki 305-0047, Japan
**Faculty of Integrated Arts and Sciences, Tokushima University, Minamijosanjima-cho, Tokushima 770-8502, Japan
‡Department of Physics, Tokyo Metropolitan University, Hachioji, Tokyo 192-0397, Japan

Abstract. We performed local magnetization measurements using micro-Hall probe in filled skutterudite $PrOs_4Sb_{12}$, which shows a superconducting state with broken time-reversal symmetry. It turns out that there appears a remarkable dip in the local magnetization hysteresis near zero field, which is characteristic of chiral superconductors, while temperature dependence of local magnetization shows no distinctive anomaly related to the spontaneous field. Possible situations of the spontaneous field is discussed from the results of local magnetization.

Keywords: $PrOs_4Sb_{12}$, Chiral superconductivity, Local magnetization, Hall probe
PACS: 74.70.Tx 74.25.Ha

INTRODUCTION

Coexistence of magnetism and superconductivity is one of the most fascinating issues in low temperature physics. Recently, a paring state with broken time-reversal symmetry has been reported in a filled skutterudite superconductor $PrOs_4Sb_{12}$ by μSR experiments [1]. To get further insight into the unusual superconductivity, we performed local magnetization measurements using micro-Hall probe. The local magnetization hysteresis shows a remarkable dip near zero field, which is characteristic of chiral superconductors. On the other hand, temperature dependence of local magnetization shows no distinctive anomaly related to the spontaneous magnetic field. We discuss possible situations of the source of spontaneous moment in the superconducting state from the view point of local magnetization measurements.

EXPERIMENTAL

The local magnetization measurement is a powerful technique which can give a semi-microscopic information on magnetic properties of a material. In this study, local magnetization measurements are performed using a micro-Hall probe with the active area of 30×30 μm^2. The details of the experiments are described elsewhere [2, 3]. Single crystals of $PrOs_4Sb_{12}$ were grown by the Sb-flux method [4]. Clear de Hass-van Alphen oscillations represent excellent quality of our samples [4]. A single crystal with typical dimensions of $500 \times 500 \times 500$ μm^3 was directly glued on the Hall-probe with the active area almost located at the center of the sample.

RESULTS AND DISCUSSION

Figure 1 demonstrates local magnetization hysteresis curves in $PrOs_4Sb_{12}$ measured by using micro-Hall probe. A magnetic field was applied along one of the crystal axes, which we call [001] direction [3]. It is worth noting that a clear dip appears in each hysteresis loop near the central magnetization peak field, namely first penetrating field of vortices, H_{cp}. The inset shows the local magnetization re-plotted as a function of local magnetic induction B_z. We see that the present anomalies appear when B_z is close to zero. The dip at $B_z \approx 0$ G traces the same curves for every measurement in the same setting. A similar anomaly near $B = 0$ G is also found in Sr_2RuO_4 [5, 6, 7] which is another superconductor with broken time-reversal symmetry [8, 9].

Figure 2 shows local magnetization curves in $PrOs_4Sb_{12}$ as a function of temperature. Here, the data are shifted vertically for clarity. We measured the local magnetization for both of zero field cooling and field cooling process. Clear double transitions at $T_{c1} \sim 1.80$ K and $T_{c2} \sim 1.75$ K are observed. We observe the diamagnetism of the high temperature phase is definitely weak, and easily suppressed in the field-cooling process. Another step appearing at 1.72 K may be related to the double minima structure in the flux flow resistively [10] or the difference in the double transition between the thermal expansion and the specific heat measurements [11, 12, 13]. On the other hand, despite

CP850, *Low Temperature Physics: 24th International Conference on Low Temperature Physics;*
edited by Y. Takano, S. P. Hershfield, S. O. Hill, P. J. Hirschfeld, and A. M. Goldman

a large anomaly observed in the hysteresis loops in Fig. 1, no additional anomaly suggesting the existence of spontaneous field was observed.

Currently, two models are discussed as a possible origin for the spontaneous field in the superconducting state. It depends on the ordering of either spin or orbital moments (or both) of superconducting pairs [14, 15]. (i) One arises from the undumped spontaneous current caused by the spatial inhomogeneities of the order parameter [14]. Such spontaneous current is considered to appear near impurities, surfaces, and domain boundaries between degenerate superconducting phases. In this case, a spontaneous moment is generated inhomogeneously in a range smaller than λ, the magnetic penetration depth. In this case, local field near the source is expected to be larger than that observed in μSR experiment, which is averaged over the sample. Spontaneous moment in Sr_2RuO_4 is considered to arise from this source. (ii) Another possibility is non-unitary odd-parity state originated from the ordered spin effect with triplet superconductivity [15] (similar to the A_1 phase in the superfluid ^{3}He). In the latter case, a three dimensional vector $\mathbf{d}(\mathbf{k})$, representing wave vector dependence of the order parameter, is complex, and $\mathbf{d}(\mathbf{k})\times\mathbf{d}^*(\mathbf{k})$ has a finite value, which gives rise to a uniform spin moment in a superconducting domain.

As shown in Fig. 2, the temperature dependence of the local magnetization has no anomaly except for the multiple transition. The result suggests either the spontaneous moment is absent at least in the active area of the Hall probe [model (i)], or the spontaneous moment is averaged out due to the domain size much small than the size of the Hall probe [model (i) or (ii)]. Further studies of spatial distribution on the local magnetic induction will give a clue to the source of spontaneous field.

Let us return to the anomaly in the hysteresis loop near $B = 0$ G. Recently, magnetization process in a chiral p-wave superconductor is theoretically discussed for both the multi-domains and the single-domain state [16]. When a magnetic field is applied to the degenerate multi-domains, the domain wall moves so that the unstable domains shrink to vanish. Since the local magnetic induction has a ridge along the domain wall, there can be a reduction of the negative magnetization for the crossing or appearance of domain walls on the active area of the Hall probe. Alternatively, difference of lower critical field H_{c1} between the degenerate superconducting domains is also discussed [16]. If H_{c1} for one of the chiral domains is small, vortices first penetrate this domain when the local B_z is reversed. At slightly higher field, vortices penetrate another domain with the opposite chirality. Such a difference of H_{c1} in the degenerate domains is another candidate to explain the anomaly in the magnetization hystereses.

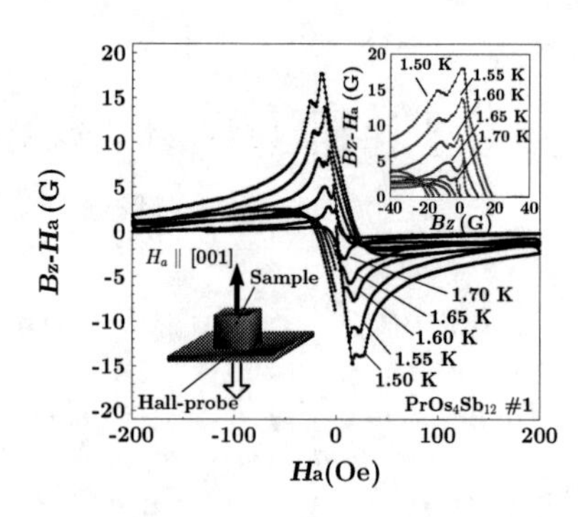

FIGURE 1. Local magnetization hysterises in $PrOs_4Sb_{12}$. The inset shows a re-plot as a function of local magnetic induction B_z.

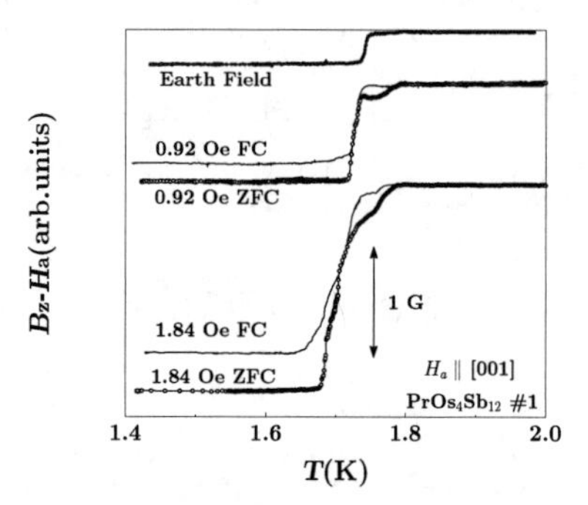

FIGURE 2. Local magnetization in $PrOs_4Sb_{12}$ as a function of temperature under several magnetic fields.

ACKNOWLEDGMENTS

This work is supported by a Grant-in-Aid for Science Research from Ministry of Education, Culture, Sports, Science and Technology.

REFERENCES

1. Y. Aoki *et al.*, *Phys. Rev. Letters*, **91**, 067003 (2003).
2. S. Kasahara *et al.*, *Phys. Rev. B*, **71**, 224501 (2005).
3. S. Kasahara *et al.*, *to be published in Physica C.*
4. H. Sugawara *et al.*, *Phys. Rev. B*, **66**, 220504(R) (2002).
5. T. Tamegai *et al.*, *Physica B*, **284-288**, 543 (2000).
6. K. Yamazaki *et al.*, *Physica C*, **378-381**, 537 (2002).
7. T. Tamegai *et al.*, *Physica C*, **388-389**, 499 (2003).
8. Y. Maeno *et al.*, *Nature (London)*, **372**, 532 (1994).
9. G. M. Luke *et al.*, *Nature (London)*, **394**, 558 (1998).
10. M. Kobayashi *et al.*, *J. Phys. Soc. Jpn.*, **74**, 1690 (2005).
11. R. Vollmer *et al.*, *Phys. Rev. Letters*, **90**, 057001 (2003).
12. K. Izawa *et al.*, *Phys. Rev. Letters*, **90**, 117001 (2003).
13. M.-A. Measson *et al.*, *Phys. Rev. B*, **70**, 064516 (2004).
14. M. Sigrist and K. Ueda, *Rev. Mod. Phys.*, **63**, 239 (1991).
15. M. Sigrist and T. M. Rice, *Phys. Rev. B*, **39**, 2200 (1989).
16. M. Ichioka *et al.*, *Phys. Rev. B*, **71**, 172510 (2005).

Crystalline Electric Field Effects in $Pr_{1-x}La_xOs_4Sb_{12}$

C. R. Rotundu and B. Andraka

Department of Physics, University of Florida, PO Box 118440, Gainesville, Florida 32611-8440, USA

Abstract. The specific heat of $Pr_{1-x}La_xOs_4Sb_{12}$ alloys was measured in fields to 14 T. CEF energies remain approximately constant across this system. The field-induced AFQ order is suppressed by the alloying and is not observed for x=0.2 above 0.4 K.

Keywords: CEF, heavy fermions, qudrupolar order, magnetic phase diagram
PACS: 65.40.Ba, 71.27.+a, 74.62.Dh

The discovery[1] of superconductivity and heavy fermion behavior in $PrOs_4Sb_{12}$ opens new frontiers for heavy fermion physics. A non-magnetic crystalline electric field (CEF) ground state of Pr seems to preclude a conventional mechanism leading to the formation of heavy electrons, based on a Kondo impurity model. Thus, the outstanding question is the nature of the heavy fermion state in this compound. Proximity of the low temperature state to the antiferro-quadrupolar (AFQ) order[2] suggests either fluctuations of this order parameter or the quadrupolar Kondo-effect as the primary candidates. Our recent investigation[3] of $Pr_{1-x}La_xOs_4Sb_{12}$ demonstrated that heavy fermion character is strongly reduced by relatively small amounts of La substituted for Pr, despite the rather weak, if any, effect of this substitution on accessible single-ion parameters. For instance, there is an unusually small variation of the lattice constant. Similarly, the temperature of the maximum in the susceptibility (approximately 4 K) does not show any systematic variation on x. This maximum is associated with excitations between the lowest CEF levels. The low temperature susceptibility hence is consistent with a CEF scheme of Pr, independent of the La content. Thus, the zero field alloying study implies a non-single impurity origin of the heavy fermion state.

The purpose of the current investigation is to determine how the field-induced AFQ order is affected by alloying and to provide further insight on how the single-ion parameters of $Pr_{1-x}La_xOs_4Sb_{12}$ depend on x. Specific heat for alloys with $x = 0.02$, 0.1, 0.2, and 0.6 was measured in fields to 14 T.

There are compelling evidences[4, 5] that the CEF ground state of $PrOs_4Sb_{12}$ is a singlet and the first excited state is a triplet at about 8 K. Magnetic field splits the triplet and lowers its lowest energy eigestate with respect to the singlet such that the two should cross in a field of about 9 - 10 T. Instead, when the two levels are sufficiently close in energy, the AFQ order is observed between 4 and 12 T. Thus, the long range order removes the degeneracy of a quasi-doublet in magnetic fields.

Figure 1 shows the magnetic phase diagram for $x = 0$, 0.02, and 0.1 for fields up to 14 T. Closed and open circles represent the Schottky and AFQ anomalies, respectively, for the undoped compound. When 2 % of La is substituted for Pr, both the size of the AFQ anomaly and its temperature are reduced with respect to $x = 0$ for identical fields. For instance, the reduction of the transition temperature in fields 8 and 9.5 T is 15 - 20 % (Fig. 1). This is a modest change in comparison to $PrPb_3$, another well studied Pr-material exhibiting AFQ order in the zero field[6]. In the latter case, 2 % of La is sufficient to suppress the long range order completely.

The f-electron specific heat for $x = 0.1$ is shown in Fig. 2. Broad maxima in $H = 6$, 12, and 14 T are clearly Schottky anomalies due to excitations between the lowest CEF levels. On the other hand, peaks at 8 and 10 T are somewhat sharper and of higher magnitude, although this magnitude is considerably reduced with respect to $x = 0$ and 0.02. These peaks most probably represent some short range order instead. However,since there is no clear distinction between anomalies due the field induced order and those for the Schottky effect, in Fig. 1 we used one symbol for all data points for corresponding to $x = 0.1$. We would like to stress that the temperatures of the Schottky maximum in high and small fields are not changed between $x = 0$ and 0.1.

These Schottky anomalies are also seen in $x = 0.2$ for $H = 13$ and 14 T (Fig. 3). Similarly, the specific heat for $x = 0.6$ (not shown) displays broad structures 13 and 14 T, at the same roughly coinciding with those for $x = 0$ and 0.1, and 0.2, for the same fields. Thus, the La-doping does not change the CEF scheme and energies of Pr as postulated from the zero field investigation[3]. Therefore, the observed sensitivity of electronic properties of $Pr_{1-x}La_xOs_4Sb_{12}$ on x have to be due to intersite effects, most probably AFQ interactions. There is no AFQ transition for $x = 0.2$ (and 0.6; not shown) down to 0.4 K in H=10 T. Instead, the f-electron specific heat in-

CP850, *Low Temperature Physics: 24th International Conference on Low Temperature Physics;*
edited by Y. Takano, S. P. Hershfield, S. O. Hill, P. J. Hirschfeld, and A. M. Goldman

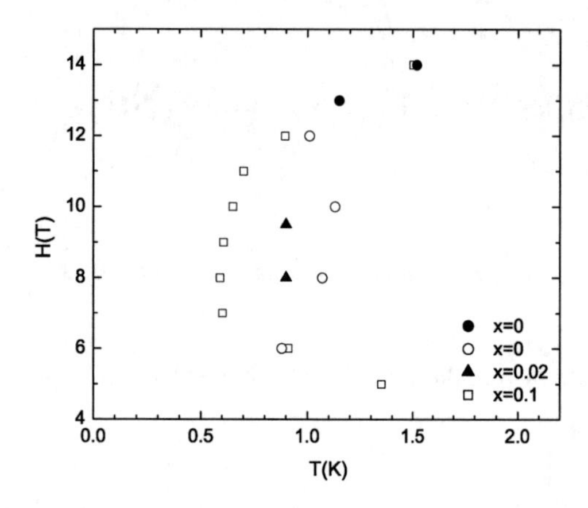

FIGURE 1. Magnetic field phase diagram for x=0, 0.02, and 0.1. Closed and open circles for x=0 represent Schottky and AFQ anomalies, respectively.

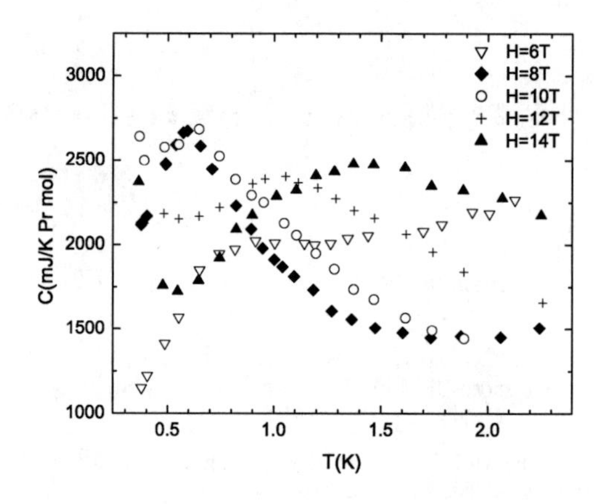

FIGURE 2. f-electron specific heat of $Pr_{0.9}La_{0.1}Os_4Sb_{12}$ in magnetic fields.

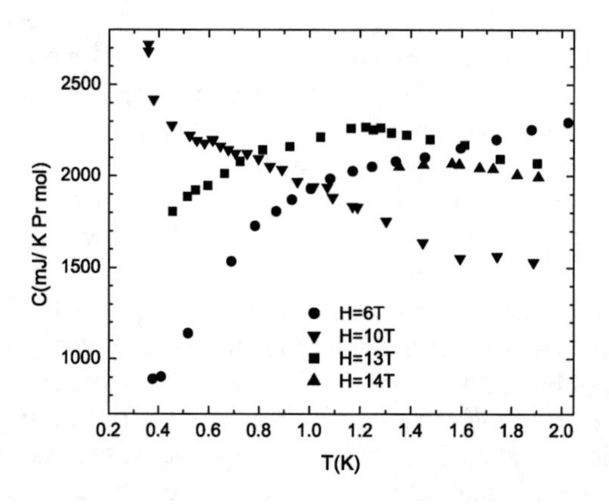

FIGURE 3. f-electron specific heat of $Pr_{0.8}La_{0.2}Os_4Sb_{12}$ in magnetic fields.

creases continuously with a decrease of temperature. Recall that 10 T corresponds roughly to the crossing field for $PrOs_4Sb_{12}$ and the maximal AFQ transition temperature is somewhere between 9 and 10 T. The low temperature tail in Fig. 3 is due to nuclear effects.

Thus, the magnetic field study of $Pr_{1-x}La_xOs_4Sb_{12}$ provides additional evidences that single-ion parameters of Pr do not depend on the La-content. The field-induced AFQ is relatively robust but no signatures of this long range order can be detected in $x = 0.2$ and 0.6. An interesting question is the nature of the low temperature state in a crossing field for alloys with $x > 0.2$. What is the mechanism responsible for removal of entropy associated with a quasi-doublet? Very strong nuclear effects prohibited us to measure the specific heat at lower temperatures, with sufficient accuracy, for moderately diluted alloys.

ACKNOWLEDGMENTS

This work was supported by the Department of Energy under Grant No. DE-FG02-99ER45748.

REFERENCES

1. E. D. Bauer, N. A. Fredrick, P.-C. Ho, V. S. Zapf, and M. B. Maple, *Phys. Rev. B* **65**, 100506(R) (2002).
2. Y. Aoki, T. Namiki, S. Ohsaki, S. R. Saha, H. Sugawara, and H. Sato, *J. Phys. Soc. Jpn.* **71**, 2098 (2002).
3. C. R. Rotundu, P. Kumar, and B. Andraka, cond-mat/0402599 (2004).
4. C. R. Rotundu, Y. Takano, B. Andraka, H. Sugwara, Y. Aoki, and H. Sato, *Phys. Rev. Lett.* **92**, 037203 (2004).
5. E. Goremychkin, R. Osborn, E. D. Bauer, M. B. Maple, N. A. Frederick, W. M. Yuhasz, F. M. Wooodard, and J. W. Lynn, *Phys. Rev. Lett.* **93**, 157003 (2004).
6. T. Kawae, M. Shimogai, M. Mito, K. Takeda, H. Ishii, and T. Kitai, *Phys. Rev. B* **65**, 012409 (2002).

Tunneling Spectroscopy On Epitaxial UNi_2Al_3 Thin Films

Andrey Zakharov, Martin Jourdan and Hermann Adrian

Johannes Gutenberg-University, Institute of Physics, Staudinger Weg 7, 55128Mainz, Germany

Abstract. We are presenting the first results of tunneling spectroscopy experiments performed on the heavy fermion superconductor UNi_2Al_3. Planar junctions consisting of an a*-axis oriented UNi_2Al_3 thin film as a base electrode, AlO_x insulating layer and a metal counter electrode were prepared employing an in vacuo process. The observation of the well-known superconducting density of states of the counter electrode (Pb) allows the evaluation of the junction quality. Although the junctions are in the tunneling regime, no features which could be associated with the superconducting state of the heavy fermion compound were observed.

Keywords: Unconventional superconductors; Tunneling spectroscopy; Heavy-fermion, UNi_2Al_3.
PACS: 74.70.Tx; 71.27.+a; 73.40.Gk

For the heavy fermion compound UPd_2Al_3 there is evidence for an unconventional magnetic mechanism of superconductivity [1, 2]. The compound UNi_2Al_3 [3] has the same crystal structure as UPd_2Al_3, but the magnetic and superconducting properties are quite different. Whereas UPd_2Al_3 is a simple antiferromagnet below $T_N \approx 14$ K with ordered moments of about $0.85\mu_B$, UNi_2Al_3 orders antiferromagnetically at $T_N \approx 5$ K with an incommensurable propagation vector and smaller moments of $\mu \approx 0.2\mu_B$ [4]. Furthermore, in contrast with UPd_2Al_3 there is evidence for a spin-triplet superconducting state in UNi_2Al_3 [5]. Thus a different pairing mechanism may be responsible for superconductivity of UNi_2Al_3. Tunneling spectroscopy could provide direct information about the order parameter and pairing mechanism of this compound.

(100)-oriented UNi_2Al_3 thin films were grown by molecular beam epitaxy (MBE) on a single crystalline $YAlO_3$ (112) substrate [6]. In order to obtain a 4-terminal cross-junction geometry a narrow stripe of the UNi_2Al_3 film was deposited employing a shadow-mask technique. After a thin layer of Al was deposited on top of the base electrode, an artificial insulating barrier was formed by natural oxidation in 0.8 mbar of dry oxygen. Finally, through a shadow mask the Pb counter electrode was evaporated. All junction preparation steps were performed in situ without breaking the vacuum.

In this paper we present experimental results obtained on the three junctions, prepared with different thicknesses of the Al layers (sample #1 – 6.5 nm of Al, sample #2 – 4 nm, sample #3 – 2 nm). The oxidation process was the same for all samples.

The quality of the UNi_2Al_3 thin films was investigated by $R(T)$-measurements using a standard four probe technique. All three samples have high residual resistance ratios ($RRR \approx 10$) and become superconducting at $T_c \approx 1$K with resistive transition width $\Delta T_c \approx 0.05$K (Fig. 1), proving the high quality of the UNi_2Al_3 base electrodes. Measurements of the contact resistances of the junctions above T_c^{Pb} showed only weak temperature dependence. The contact resistance continuously increased by a factor of ~2 with decreasing a temperature from 300 K to 7 K. This temperature dependence indicates a pinhole-free barrier consisting of an insulating material.

The measurements of the tunneling conductivity of the junctions were performed employing a standard AC-modulation technique. At temperatures below the superconducting critical temperature of the counter electrode T_c^{Pb}, but above T_c of UNi_2Al_3 (superconductor–insulator–normal-conductor [SIN] regime) the typical tunneling density of states of superconducting Pb including strong coupling-features was observed, proving the contacts to be in the tunneling regime. Below the superconducting critical temperature of the base electrode, UNi_2Al_3–AlO_x–Pb contacts are expected to be in the superconductor–insulator–superconductor [SIS] regime. Due to an opening of the superconducting energy gap of UNi_2Al_3 additional conductivity features as well as a reduction of the low bias conductivity are expected. Applying a magnetic field B = 0.3 T which is overcritical for Pb,

CP850, *Low Temperature Physics: 24th International Conference on Low Temperature Physics;*
edited by Y. Takano, S. P. Hershfield, S. O. Hill, P. J. Hirschfeld, and A. M. Goldman

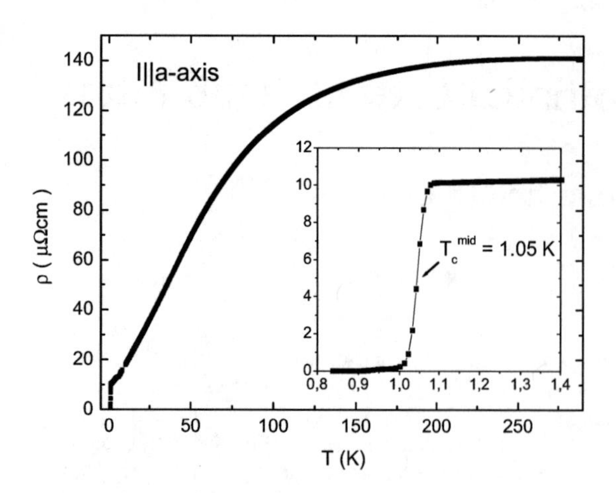

FIGURE 1. Electrical resistivity of a UNi_2Al_3 base electrode (sample #1) as a function of temperature. The inset shows the superconducting transition.

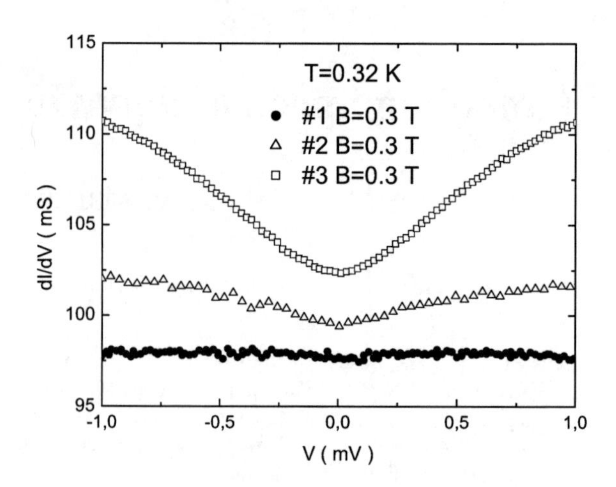

FIGURE 3. Differential conductivity at a magnetic field of B = 0.3T (curves for samples #2 and #3 are vertically displaced for clarity). No features related to superconductivity of UNi_2Al_3 are observed.

but not for UNi_2Al_3, the superconductivity of the counter electrode can be suppressed, and the influence of the density of states of the base electrode should be directly observable. However, none of these features was observed (Fig. 2, Fig. 3). The junction with the thickest AlO_x barrier can be described perfectly by a Dynes-fit for an SIN-junction with a small offset conductivity (S_{offset} = 2mS) and broadening parameter (Γ = 40μV). A reduced barrier thickness results in an increased offset current, but the Pb gap is still clearly visible. Additionally a V-shaped background is a characteristic feature of junction with small barrier thickness (Fig. 3).

There are several possibilities to explain the absence of any trace of superconductivity of UNi_2Al_3 in the tunneling data. First, the UNi_2Al_3 base electrode could be normal conducting in the bulk and superconducting only in some percolation path. However, the high critical current density measured for UNi_2Al_3 films ($I_{c,\,0.3K} \approx 10^4 A/cm^2$ [7]) corresponds to bulk superconductivity. Alternatively, superconductivity could be suppressed just in a degraded surface layer. In an unconventional superconductor, a third possibility is intrinsic pair breaking at the interface to the barrier. Finally, the insulating layer could be oxidized not completely, resulting the tunneling into non superconducting Al.

For a final conclusion concerning the prospect of UNi_2Al_3 tunneling junctions a careful characterization and optimization of the barrier properties will be necessary.

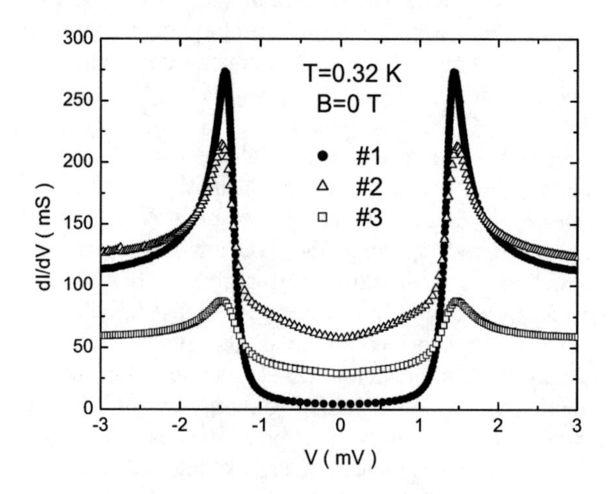

FIGURE 2. Differential conductivity of junctions with different barrier thickness at $T = 0.32$ K. The observed gap feature is entirely associated with the Pb electrode (see text).

Financial support by the German Research Foundation (DFG) is acknowledged.

1. M. Jourdan, M. Huth, and H. Adrian, *Nature* **398**, 47 (1999).
2. N. K. Sato, N. Aso, K. Miyake, R. Shiina, P. Thalmeier, G. Varelogiannis, C. Geibel, F. Steglich, P. Fulde, T. Komatsubara, *Nature* **410**, 340 (2001).
3. C. Geibel, S. Thies, D. Kaczorowski, A. Mehner, A. Grauel, B. Seidel, U. Ahlheim, R. Helfrich, K. Petersen, C. D. Bredel, and F. Steglich, *Z. Phys. B* **83**, 305 (1991).
4. A. Schröder, J. G. Lussier, B. D. Gaulin, J. D. Garret, W. J. L. Buyers, L. Rebelsky, and S. M. Shapiro, *Phys. Rev. Lett.* **72**, 136 (1994).
5. K. Ishida, D. Ozaki, T. Kamatsuka, H. Tou, M. Kyogaku, Y. Kitaoka, N. Tateiwa, N. K. Sato, N. Aso, C. Geibel, and F. Steglich, *Phys. Rev. Lett.* **89**, 037002 (2002).
6. A. Zakharov, M. Jourdan, M. Foerster, and H. Adrian *Physica B* **359-361**, 1108-1110 (2005).
7. M. Jourdan, A. Zakharov, M. Foerster, and H. Adrian, Phys. Rev. Lett. **93**, 097001 (2004).

Low Temperature Behavior of LaRhSn Superconductor

Matúš Mihalik[a], Vladimír Sechovský[a], Slavomír Gabáni[b] and Marián Mihalik[b]

[a]*Charles University, Faculty of Mathematics and Physics, Department of Electronic Structures, Ke Karlovu 5, 121 16 Prague 2, Czech Republic*
[b]*Slovak Academy of Sciences, Institute of Experimental Physics, Watsonova 47, 04353 Košice, Slovakia*

Abstract. We report on superconductive properties of LaRhSn single crystal studied by resistivity and specific heat measurements. Resistivity measurements revealed anisotropy between measurements performed along the *a*-axis the *c*-axis, but the critical temperatures are almost the same (T_c = 1.9 K for *a*-axis and T_c = 1.83 K for *c*-axis). The transition to the superconductive state is accompanied by a peak in the specific heat data. The critical field $\mu_0 H_c$ rises from 0.168 T at 1.5 K to 0.735 T at 0.11 K.

Keywords: Superconductivity, Rare earths
PACS: 74.70.Dd, 74.25.Fy, 74.25.Bt

In recent years, there has been a great deal of interest in the study of rare-earth containing intermetallic compounds of the type *RTX*, where *R* is rare earth, *T* is a *d*-metal and *X* is a *p*-metal. These compounds crystallize in a variety of structure types, depending upon the nature of *T* and *X* elements. LaRhSn crystallize in the hexagonal ZrNiAl-type structure (space group $P\bar{6}2m$). Electrical resistivity measurements on polycrystalline LaRhSn sample revealed superconducting phase with the critical temperature T_c = 2 K and with the critical field rising from 0.1 T at 1.6 K to 0.7 T at 0.2 K [1]. It is interesting that the magnetic susceptibility of LaRhSn is remarkably higher than that one of CeRhSn [2]. Up to now the single crystals of LaRhSn were used only as the reference of nonmagnetic analogue to the CeRhSn compound [3] but the superconducting transition in this compound was studied only on the polycrystalline samples. We decided to grow a single crystal and study specific heat and electric transport properties near the critical temperature T_c and along both main crystallographic axes.

Two single crystals (i) and (ii) of LaRhSn were grown by modified Czochralski method in a tri-arc furnace at the Departement of Electronic Structures in Prague. A melt consisting of the stoichiometric composition of the constituent elements with the high purity was used. Analysis of the X-ray powder diffraction did not reveal any impurity phase within the sensitivity of the method. The lattice parameters, refined by Rietveld method (*a* = 0.7479(3) nm; *c* = 0.4217(4) nm; x_{La} = 0.591(4) and x_{Sn} = 0.243(8)) are in good agreement with parameters published by Slebarski et al. [2]. Finally the Laue diffraction pattern confirmed good quality of the single crystals in both cases. The specific heat and electrical resistivity measurements were performed on the PPMS (Quantum Design) apparatus using ^{3}He insert for cooling below 2 K. The critical field H_c was determined from resistivity measurements performed in ^{3}He^{4}He dilution refrigerator (IEP SAS Košice).

The specific heat of LaRhSn (Fig. 1.) exhibits a lambda peak at 1.8 K. Based on the measurements of the temperature dependence of electrical resistivity we attributed this peak to the superconducting transition.

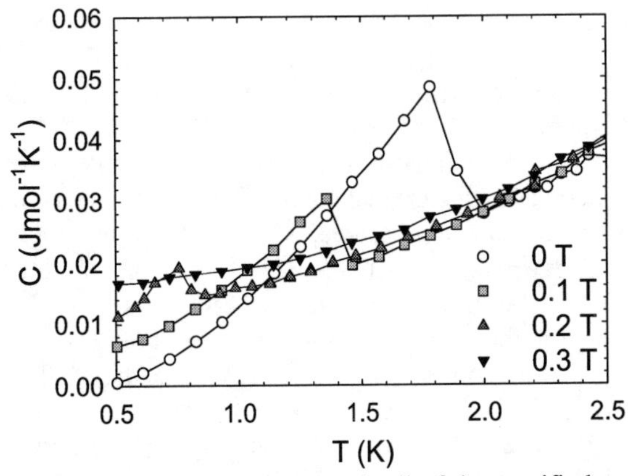

FIGURE 1. Low temperature detail of the specific heat shows an evolution of the superconducting transition due to applied magnetic field along the *c*-axis (crystal ii). The lines are only guides for the eyes.

CP850, *Low Temperature Physics: 24th International Conference on Low Temperature Physics;*
edited by Y. Takano, S. P. Hershfield, S. O. Hill, P. J. Hirschfeld, and A. M. Goldman

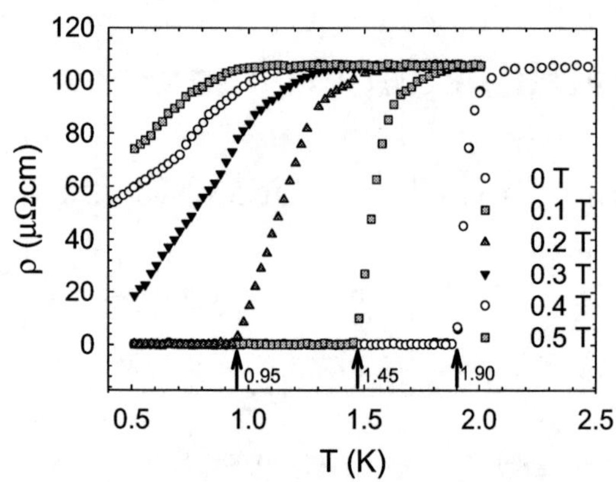

FIGURE 2. Resistivity measured along the *a*-axis. The field was applied along the *a*-axis (crystal ii).

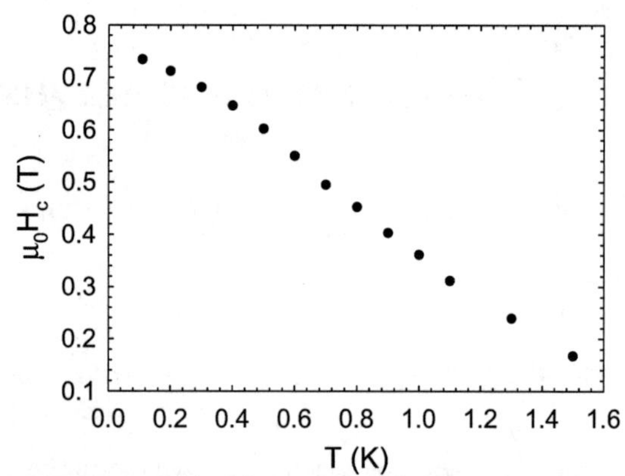

FIGURE 3. Critical field measured along the *a*-axis. The field was applied along the *c*-axis (crystal i).

An applied magnetic field along the *c*-axis shifts the peak to the lower temperatures and decreases its height. Finally the 0.3 T field wiped out the peak from measured temperature range. In the temperature range 2 – 15 K specific heat data fit equation $C = \gamma T + \beta T^3$ with parameters $\gamma = 8.78$ mJ mol^{-1}K^{-2} and $\beta = 7.9 \times 10^{-4}$ Jmol^{-1} K^{-4}. The γ coefficient is close to γ (=7.7 mJ mol^{-1}K^{-2}) obtained for another low temperature superconductor LaRhSb [4].

The resistivity drops to zero value at T_c =1.9 K for *a*-axis (Figs. 2) and at T_c = 1.83 K for *c*-axis. This drop clearly indicates that the LaRhSn is superconducting below T_c. With the field applied along the *a*-axis the superconducting transition shifts to the lower temperatures (Fig. 2) and finally in fields higher than 0.2 T the transition vanishes from the measured temperature region. The drop of resistivity in field 0.3 T is almost linear below 0.9 K. By extrapolating the resistivity to the zero we can estimate that at 0.3 T the transition to the superconducting state occurs near to 0.4 K. In the case of the current and field applied along the *c*-axis the value of T_c is first lower than in the case of the *a*-axis in fields 0 T (1.83 K), 0.1 T (1.38 K), 0.2 T (0.88 K) and then is higher for field 0.3 T (0.5 K). This behavior indicates differences of the temperature dependence of the critical field for the *a*- and *c*- axis. Results of our measurements performed on single crystal ii indicate lower critical fields and slightly lower critical temperatures than published in [1]. The reason for the different critical temperatures may be the different amount of impurities in the samples. The reason for the different critical fields may be the demagnetization factor, which changes the internal field in the sample.

A more detailed study of $H_c(T)$ was performed in ^{3}He^{4}He dilution refrigerator in geometry current applied along the *c*- axis and magnetic field applied along the *a*- axis. The temperature was first stabilized and then the field was applied. The results are presented in Fig.3 and are in good agreement with [1].

The present study of the superconducting phase transition of LaRhSn indicates different critical temperatures which evolutes in field completely different. This can suggest larger coupling of electrons to the Cooper's pairs parallel to the *a*-axis at zero field. An applied magnetic field changes coupling and in fields higher than 0.3 T the coupling is larger along the *c*-axis. To verify these assumptions a detailed study of $H_c(T)$ at temperatures down to 0.11 K has to be extended on the *a*- axis geometry.

ACKNOWLEDGMENTS

This work is a part of the research program MSM 0021620834 financed by the Ministry of Education of the Czech Republic. The work has also been supported by the Grant Agency of the Charles University (Grant #282/2004/B-FYZ/MFF). Measurements performed in ^{3}He^{4}He dilution refrigerator were supported by Center of Excellence of the SAS, Contract No. I/2/2003.

REFERENCES

1. P.-C. Ho, V. S. Zapf, A. Ślebarski and M. B. Maple, *Phyl. Mag.* **84**, 2119 (2004).
2. A. Slebarski, M. Radlowska, T. Zawada, M. B. Maple, A. Jezierski and A. Zygmunt, *Phys. Rew. B* **66**, 104434 (2002).
3. M. S. Kim, Y. Echizen, K. Umeo, S. Kobayashi, M. Sera, P. S. Salamakha, O. L. Sologub, T. Takabatake, X. Chen, T. Tayana, T. Sakakibara, M. H. Jung and M. B. Maple, *Pys. Rew B* **68**, 0554416 (2003).
4. S. K. Malik, H. Takeya and K. A. Gschneidner, *J. All. and Comp.* **207-208**, 237-240 (1994).

Atomic Substitution And Carbon Isotope Effect In Superconducting $MgCNi_3$

T. Klimczuk[1,2], V. Gupta[1], M. Avdeev[3], J.D. Jorgensen[3], and R.J. Cava[1]

[1] *Department of Chemistry, Princeton University, Princeton NJ 08544, USA*
[2] *Faculty of Applied Physics and Mathematics, Gdańsk University of Technology, Narutowicza 11/12, 80-952 Gdańsk, Poland*
[3] *Materials Science Division, Argonne National Laboratory, Argonne IL 60439, USA*

Abstract. The superconductor $MgCNi_3$ has been chemically doped by substitution of Ru and Fe for Ni and B for C. Neither ferromagnetism nor increasing T_C is observed, suggesting that $MgCNi_3$ has optimal superconducting properties at its intrinsic composition and electron count The effect of Carbon isotope substitution on Tc is reported: the carbon isotope effect coefficient is $\alpha_C = 0.54 \pm 0.03$. This indicates that carbon-based phonons play a critical role in the superconductivity.

Keywords: Alloys, superconductivity, chemical doping, superconducting isotope effect
PACS: 74.70.Ad, 74.62.Dh, 74.62Bf

INTRODUCTION

$MgCNi_3$ has provided a link between the perovskite-based oxides (HTSC) and intermetallic superconductors[1]. The calculated electronic structure suggested that $MgCNi_3$ is close to a ferromagnetic instability, and that a removing, as little as 0.1 electrons per formula unit should put the Fermi level near the peak in the expected electronic DOS, leading to a ferromagnetism[2]. However, this effect was not observed with the partial substitution of Co, Fe, Ru, Mn for Ni ($MgCNi_{3-x}M_x$) (Refs. 3-7) and B for C ($MgC_{1-x}B_xNi_3$) (Ref. 8) or with inducing carbon deficiency ($MgC_{1-x}Ni_3$) (Ref. 9).

In this paper we compare the effect of partial substitution on Ni and C site in $MgCNi_3$. The carbon isotope effect is also discussed.

EXPERIMENT

The series of samples with compositions $Mg_{1.2}C_{1.5-x}{}^{11}B_xNi_3$ (x=0, 0.05, 0.1, 0.15, 0.2 and 0.25), $Mg_{1.2}C_{1.5}Ni_{3-x}Fe_x$ (x=0, 0.005, 0.01, 0.015, 0.02, 0.025, 0.033, 0.05, 0.066 and 0.1) and $Mg_{1.2}C_{1.5}Ni_{3-x}Ru_x$ (x=0, 0.005, 0.01, 0.02, 0.033, 0.066, 0.1, 0.2, 0.3, 0.4, 0.5, and 0.6) were synthesized. The starting materials were bright Mg flakes (99% Aldrich Chemical), fine Ni powder (99.9% Johnson Matthey and Alfa Aesar), Ru powder (99.95% Alpha Aesar) and Fe powder (99.5% Alpha Aesar), glassy carbon spherical powder (Alfa Aesar) and enriched boron metal powder ^{11}B (99.5 At.% ^{11}B – Eagle-Picher Ind., Inc.). Previous studies on $MgCNi_3$ indicated the need to employ excess magnesium and carbon in the synthesis in order to obtain optimal carbon content[1,9]. More experimental details are described in Refs. 3,5,6,8.

The Fe and Ru concentration in $MgCNi_{3-x}Fe_x$ and $MgCNi_{3-x}Ru_x$ were assumed to be equal to the nominal concentration. In the case of $MgC_{1-x}B_xNi_3$ the time of flight neutron-powder-diffraction data were collected on the special environmental powder diffractometer (SEPD) at the intense pulsed neutron source (IPNS) at Argonne National Laboratory. The true boron concentration was calculated using the Rietveld refinement method with the GSAS (EXPGUI) suite. All details are described in Ref. 8.

RESULTS AND DISCUSSION

The cell parameter shows a linear relationship between concentration of dopant[5,6,8]. The solubility limits were determined as between 0.066 and 0.1 for $MgCNi_{3-x}Fe_x$, 0.5 and 0.6 for $MgCNi_{3-x}Ru_x$, and between 0.2 and 0.25 (nominal value) for $MgC_{1-x}B_xNi_3$. Considering the difference in covalent radius, the substantially higher solubility of Ru is surprising.

CP850, *Low Temperature Physics: 24th International Conference on Low Temperature Physics;* edited by Y. Takano, S. P. Hershfield, S. O. Hill, P. J. Hirschfeld, and A. M. Goldman

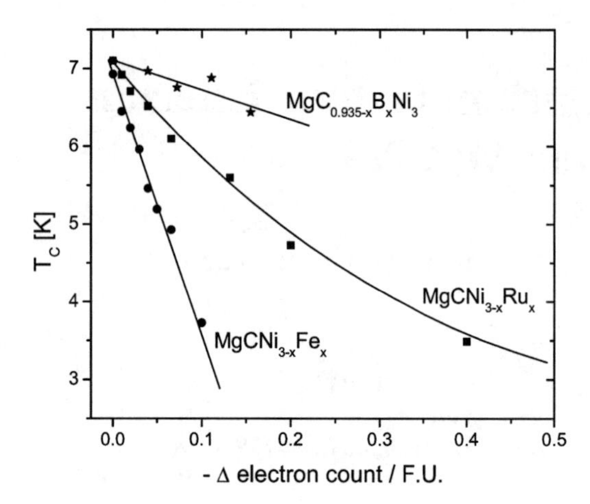

FIGURE 1. Superconducting critical temperature (T_C) in $MgCNi_{3-x}Ru_x$, $MgCNi_{3-x}Fe_x$ and $MgC_{0.935-x}B_xNi_3$ as a function of - Δ electron count. Data taken from Refs. 5,6,8.

As seen in Fig. 1, although T_C decreases systematically with the hole doping level, the rate of decrease is much smaller for B doping. This is because the substitution of boron for carbon is the least structurally and electronically disruptive, due to chemical similarity of B and C. The fastest decrease of T_C in the case of Fe (3*d*) substitution suggests substantial differences from the case of Ru (4*d*) substitution. The normal state susceptibility measurements, which show that χ is decreasing with Ru and increasing with Fe doping[6], suggests that 3*d* element substitutions have suppressed T_C by magnetic pair breaking and not through a band structure effect.

Doping neither the Ni nor the C site drives the system to a ferromagnetic state and superconductivity is suppressed in all cases. This suggests that $MgCNi_3$ has optimal superconducting properties at its intrinsic composition and electron count.

The carbon atom plays a special role for the presence of superconductivity in $MgCNi_3$ (Refs. 1,9). It is located in the center of Ni_6 octahedron and is the lightest element in the compound. Therefore carbon is the best candidate for a possible superconducting isotope effect.

Four independent groups of samples of $Mg^{12}CNi_3$ and $Mg^{13}CNi_3$ were synthesized. In each group, one ^{12}C and one ^{13}C (99% Cambridge Isotope Lab., Inc.) sample was made. The normalized AC susceptibility data (ZFC) are presented in the Fig. 2. As is expected from the BCS theory T_C is lower for $Mg^{13}CNi_3$. The average shift, $\Delta T_C = T_C(^{12}C) - T_C(^{13}C)$, was calculated to be 0.3K with the standard deviation value 0.015K, which gives the value of the carbon isotope effect coefficient $\alpha_C = 0.54(3)$, very close to what is expected for a phonon mediated superconductor, indicating that carbon-based phonons play a critical role in the superconductivity.

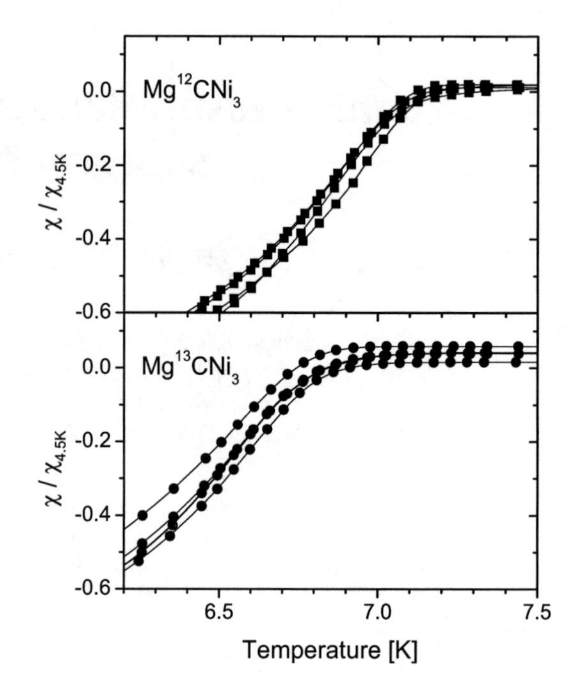

FIGURE 2. Normalized AC susceptibility (χ / $\chi_{4.5K}$) data for all $Mg^{12}CNi_3$ (upper panel) and $Mg^{13}CNi_3$ (lower panel) samples in the same temperature range 6.2-7.5K (Ref. 10)

REFERENCES

1. T. He, Q. Huang, A.P. Ramirez, Y. Wang, K.A. Regan, N. Rogado, M.A. Hayward, M. K. Haas, J.J. Slusky, K. Inumara, H.W. Zandbergen, N.P. Ong & R.J. Cava, *Nature* **411**, 54 (2001).
2. H. Rosner, R. Weht, M.D. Johannes, W.E. Pickett, and E. Tosatti, *Phys. Rev. Lett.* **88**, 027001 (2002).
3. M.A. Hayward, M.K. Haas, A.P. Ramirez, T. He, K.A. Regan, N. Rogado, K. Inumaru, and R.J. Cava, *Solid State Commun.* **119**, 491-495 (2001).
4. T.G. Kumary, J. Janaki, A. Mani, S.M. Jaya, V.S. Sastry, Y. Hariharan, T.S. Radhakrishnan, and M.C. Valsakumar, *Phys. Rev. B* **66**, 064510 (2002).
5. T. Klimczuk, V. Gupta, G. Lawes, A.P. Ramirez and R.J. Cava, *Phys. Rev. B* **70**, 094511 (2004).
6. T. Klimczuk and R.J. Cava, *Solid State Commun.* **132**, 379-382 (2004).
7. A. Das and R.K. Kremer, *Phys. Rev. B* **68**, 064503 (2003).
8. T. Klimczuk, M. Avdeev, J.D. Jorgensen and R.J. Cava, *Phys. Rev. B* **71**, 184512 (2005).
9. T.G. Amos and Q. Huang, J.W. Lynn and T. He and R.J. Cava, *Solid State Commun.* **121**, 73 (2002).
10. T. Klimczuk and R.J. Cava, *Phys. Rev. B* **70**, 212514 (2004).

The Specific Heat of $Na_{0.3}CoO_4 \cdot 1.3H_2O$: Possible Non-Magnetic Pair Breaking, Two Energy Gaps, and Strong Fluctuations in the Superconducting State

N. E. Phillips*, N. Oeschler*, R. A. Fisher*, J. E. Gordon†, M.-L. Foo‡, and R. J. Cava‡

*LBNL and Department of Chemistry, University of California, Berkeley, CA 94720, USA
†Physics Department, Amherst College, Amherst, MA 01002, USA
‡Department of Chemistry, Princeton University, Princeton, NJ 08544, USA

Abstract. Two samples of $Na_{0.3}CoO_2 \cdot 1.3H_2O$ have very different specific heats reflecting extreme sensitivity to minor differences in sample treatment. They have nearly the same superconducting transition temperatures with strong deviations from BCS behavior and evidence of two energy gaps that are markedly different for the two samples. Non-magnetic scattering centers may have a strong, but unusual, pair-breaking effect.

Keywords: specific heat, superconductivity, two energy gaps, $Na_{0.3}CoO_2 \cdot 1.3H_2O$
PACS: 74.20.De, 74.20.Rp, 74.25.Bt, 74.25.Ha

The superconductivity of $Na_{0.35}CoO_2 \cdot 1.3H_2O$ [1] is of interest for comparison with that of the cuprates. Theoretical work suggests that its superconductivity is different, and is very possibly unique. The conduction-electron, superconducting-state specific heat (C_{es}) gives information about the symmetry of the order parameter and clues to the nature of the pairing mechanism. A number of specific-heat measurements have been reported, but in many cases the interpretation is limited by contributions from magnetic impurities or lack of data at sufficiently low temperatures. We have measured the specific heat (C) of three samples. The results for sample 1, which showed a weak antiferromagnetic ordering near 6.5 K, but no superconductivity, will be reported elsewhere [2]. Measurements on samples 2 and 3, which showed clean transitions to the superconducting state, and no significant contribution from paramagnetic centers, are reported here. The data were analyzed as $C = C_e + C_l$, where C_e is the electron contribution and C_l is the lattice contribution. In the normal-state $C_e = C_{en}$

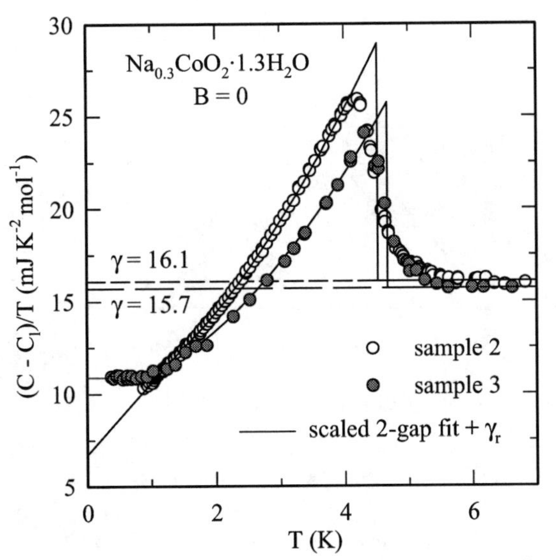

FIGURE 1. Total specific heat minus the lattice contribution for samples 2 and 3.

$C - C_l$ is shown in Fig. 1 for samples 2 and 3. Extrapolations to 0 K give non-zero intercepts, "residual" γ's (γ_r), of 6.67 and 10.9 mJ mol^{-1} K^{-2}, respectively, that measure a residual density of states (DOS). All specific-heat measurements on this material that permit reasonably unambiguous extrapolations to 0 K show contributions of this type. They have been interpreted as showing that a fraction of the sample (γ_r/γ) remains normal. On the basis of this interpretation C_e for one mole of superconducting material is shown in Figs. 2 and 3. For sample 2 the deviations from BCS behavior are remarkably similar to those for MgB_2 for which they are known to be

CP850, *Low Temperature Physics: 24th International Conference on Low Temperature Physics;* edited by Y. Takano, S. P. Hershfield, S. O. Hill, P. J. Hirschfeld, and A. M. Goldman

associated with the presence of two energy gaps. Two-gap fits characterized by the 0-K energy-gap parameters [$\Delta_i(0)$] and the fractional contributions to the normal-state density of states (γ_i/γ), i = 1 and 2, are included in the figure. The dotted curves represent a fit based on a T^2 behavior of C_{es} that is a signature of line nodes in the gap, which has also been reported by Yang et al. [3]. The gap parameters are $\Delta_1(0)/k_BT_c$ = 2.15, $\Delta_2(0)/k_BT_c$ = 1.00, γ_1/γ = 0.55, and γ_2/γ = 0.45. However, the T^2 behavior can also be regarded as an artifact associated with the superposition of the two contributions, and the solid curves represent a two-gap fit with no nodes and $\Delta_1(0)/k_BT_c$ = 2.20, $\Delta_2(0)/k_BT_c$= 0.70, γ_1/γ = 0.55, and γ_2/γ = 0.45. For sample 3, for which there is no evidence of a T^2 behavior, C_{es} can also be represented by a two-gap fit, with $\Delta_1(0)/k_BT_c$ = 2.30, $\Delta_2(0)/k_BT_c$= 1.10, γ_1/γ = 0.80, and γ_2/γ = 0.20.

The interpretation of the residual DOS as associated with non-superconducting regions of the sample is relatively plausible, but NQR measurements [4] show that a sample-dependent residual DOS of the same magnitude as that seen in C_e governs the low-T relaxation of the *same* Co nuclei that respond to the transition at T_c. It can be understood as well on the basis of a pair-breaking mechanism [5], which would also account for γ_r. However, it would be an unprecedented pair-breaking mechanism that does not affect T_c. With this interpretation, C_e obtained as described above would be only an approximation, but there are reasons to expect it to show the main features of the pure superconducting state. As shown in Fig. 4, the temperature of the onset of the transition to the mixed state is independent of field (B), evidence of unusually strong fluctuations.

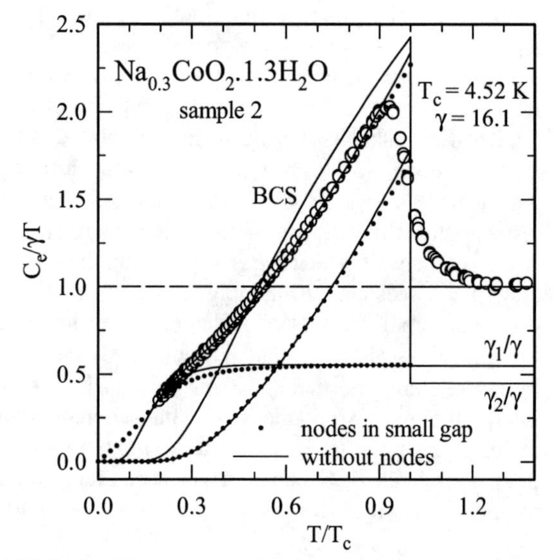

FIGURE 2. Electron contribution to the specific heat, sample 2, (see text).

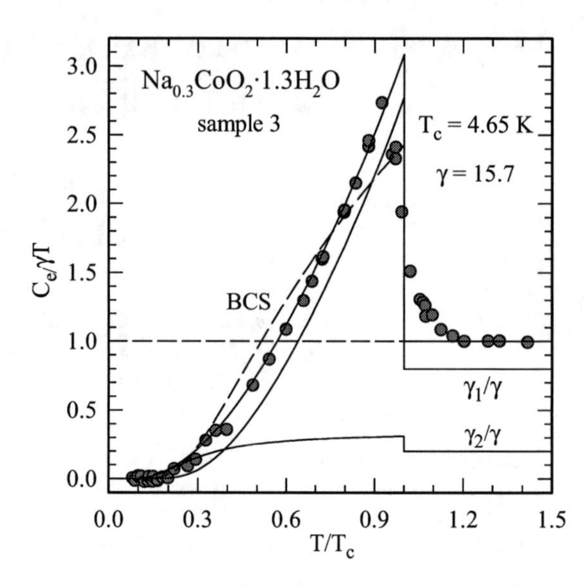

FIGURE 3. Electron contribution to the specific heat, sample 3 (see text).

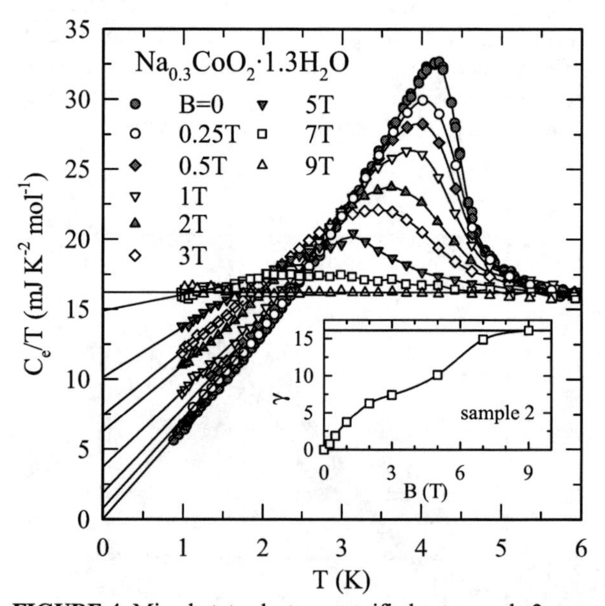

FIGURE 4. Mixed-state electron specific heat, sample 2.

REFERENCES

1. K. Takada *et al., Nature (London)* **422**, 53 (2003).
2. N. Oeschler *et al.*, to be published.
3. H. D. Yang *et al., Phys. Rev. B* **71**, 20504 (2005).
4. T. Fujimoto *et al., Phys. Rev. Lett.* **92**, 47004 (2004).
5. Y. Bang *et al., cond-mat/0307510.*

Co Nuclear-Quadrupole-Resonance Measurements on $Na_xCoO_2 \cdot yH_2O$

– Phase Diagram for Bilayered-Hydrate System –

Yoshihiko Ihara[a], Hideo Takeya[a], Kenji Ishida[a], Chishiro Michioka[b], Kazuyoshi Yoshimura[b], Kazunori Takada[c], Takayoshi Sasaki[c], Hiroya Sakurai[d], and Eiji Takayama-Muromachi[d]

[a]*Department of Physics. Kyoto University, Kyoto 606-8502, Japan*
[b]*Department of Chemistry, Kyoto University, Kyoto 606-8502. Japan*
[c]*Advanced Material Laboratory, National Institute for Materials Science, Ibaraki 305-0044, Japan*
[d]*Superconducting Materials Center, National Institute for Materials Science, Ibaraki 305-0044, Japan*

Abstract. Co nuclear quadrupole resonance (NQR) measurements were performed on several superconducting (SC) $Na_xCoO_2 \cdot yH_2O$ samples with different values of a SC transition temperature (T_c). The measurements were done for the investigation of the relationship between T_c and microscopic physical quantities derived from the Co NQR. We measured Co-NQR frequency ν_Q and the nuclear-lattice relaxation rate $1/T_1$ at the Co site, and found the relationship between ν_Q and the value of $1/T_1T$ at T_c: the higher-T_c sample possesses the higher ν_Q and larger value of $1/T_1T$ at T_c. When ν_Q exceeds a critical value (~12.45 MHz), magnetic ordering was observed. A possible phase diagram in $Na_xCoO_2 \cdot yH_2O$ is proposed, in which the NQR frequency ν_Q is the tuning parameter and determines the ground states. In this phase diagram, the SC phase is adjacent to the magnetic phase, suggesting that the superconductivity is intimately related to magnetic fluctuations. We found the coexistence of the superconductivity and the magnetic ordering in the same sample with the highest ν_Q.

Keywords: superconductivity, Hydrate Sodium Cobaltate, NQR.
PACS: 74.70.-b, 71.27.+a, 74.25.Dw, 76.60.Gv

After the discovery of superconductivity in $Na_xCoO_2 \cdot yH_2O$ with the bilayer-hydrate structure[1], much attention has been paid because of rich physics in this compound. The superconductivity that occurs on the CoO_2 block layer with a triangular structure is a contrasting example of the cuprate and ruthenate superconductors, in which superconductivity occurs on the square-lattice CuO_2 and RuO_2 layers. It becomes known that H_2O molecules situated between CoO_2 layers are easily removed under ambient conditions, resulting in a degradation of superconductivity[2]. The fragile hydrate condition causes large sample dependence, and makes certain types of experiments *e.g.* photoemission and resistivity measurements difficult, because the sample is to be kept in vacuum, or wiring is needed. A Co-NQR measurement is suitable for studying this system because it is free from such difficulties. The important results by NQR measurements have been reported since the discovery of the superconductivity[3,4].

We measured the Co-NQR frequency ν_Q corresponding to the $\pm 5/2 \leftrightarrow \pm 7/2$ transition, and the nuclear-lattice relaxation rate $1/T_1$ in various samples with different values of T_c[5,6]. It was found that ν_Q and the behavior of $1/T_1T$ in the normal state depend on sample. It was suggested that ν_Q and $1/T_1T$ at T_c are related to T_c[6]. From the neutron-diffraction measurement, it was reported that the insertion of H_2O largely changes the *c*-axis lattice parameters and distorts CoO_2 block layers[7]. We consider that ν_Q is a promising parameter to detect the distortion of CoO_6 octahedron microscopically since ν_Q is proportional to the electric field gradient at the Co site. Therefore we proposed a phase diagram of $Na_xCoO_2 \cdot yH_2O$ using ν_Q as a tuning parameter[6]. In this paper, we report the boundary region between the superconducting (SC)

CP850, *Low Temperature Physics: 24th International Conference on Low Temperature Physics;*
edited by Y. Takano, S. P. Hershfield, S. O. Hill, P. J. Hirschfeld, and A. M. Goldman

and magnetic phases in the phase diagram and the coexistence of superconductivity and magnetic ordering in a sample with a higher NQR frequency.

The samples we used in the present study were well characterized by inductively coupled plasma atomic emission spectroscopy, X-ray diffraction and redox titration measurements[8]. The Co valence of these samples is nearly the same and about +3.4. This result implies that the difference of ν_Q does not originate from the difference of the Co valence, but from the local symmetry around the Co nucleus, *i.e.*, distortion of CoO_2 block layers along the *c*-axis direction.

$1/T_1$ was measured from 150 K to 1.5 K. Figure 1 shows the temperature dependence of $1/T_1T$. The values of $1/T_1T$ at T_c increase with increasing T_c, when ν_Q is smaller than 12.45 MHz. It was suggested that the superconductivity is related to magnetic fluctuations. We found a magnetic ordering in the samples in which the magnetic fluctuations in the normal state are larger than that in the highest-T_c sample[6]. In these samples, $1/T_1T$ shows clear divergence at $T_M \sim 6$ K. It seems that the superconductivity in $Na_xCoO_2 \cdot yH_2O$ emerges in the vicinity of the magnetic phase as shown in Fig. 2. Since such a phase diagram is commonly observed in unconventional superconductivity reported so far[9], we claim that $Na_xCoO_2 \cdot yH_2O$ is classified to be an unconventional superconductor.

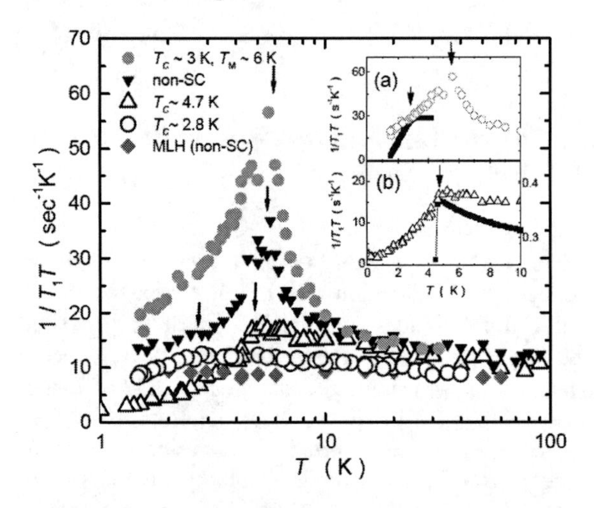

FIGURE 1. Temperature dependence of $1/T_1T$ measured on several samples. The arrows show T_c and T_M. Inset (a): $1/T_1T$ and AC susceptibility in the sample where superconductivity and magnetic ordering coexist. Inset (b): $1/T_1T$ and DC susceptibility in the highest T_c sample.

We show that ν_Q changes systematically and is related to the *c*-axis lattice parameter. According to the neutron diffraction measurements[7], a strong inverse correlation between the CoO_2 layer thickness and T_c is reported. A theoretical study suggests that such a distortion of the CoO_6 octahedron stabilizes six small hole-pockets near K point and gives rise to ferromagnetic (FM) fluctuations[10]. According to the theoretical scenario, the spin-triplet superconductivity induced by the FM fluctuations is expected. A spin susceptibility measurement in the SC state is needed to identify the symmetry of the SC pairs.

We found the coexistence of the magnetism and superconductivity in the sample with the highest $\nu_Q \sim$ 12.68 MHz, whose spectral width is almost the same as others. The magnetic ordering at $T_M \sim 6$ K was confirmed by the diverging behavior of $1/T_1T$ measured at 12.68 MHz and the superconductivity was identified from the observation of the Meissner signal below $T_c \sim 3$ K by an AC-susceptibility measurement. Although the decrease of $1/T_1T$ related to the opening of the SC gap was not observed at T_c (Fig.1(a)), considerably large Meissner signal suggests that the superconductivity is realized in the entire region of the sample. Further experiments are needed to reveal how the superconductivity and magnetism can coexist.

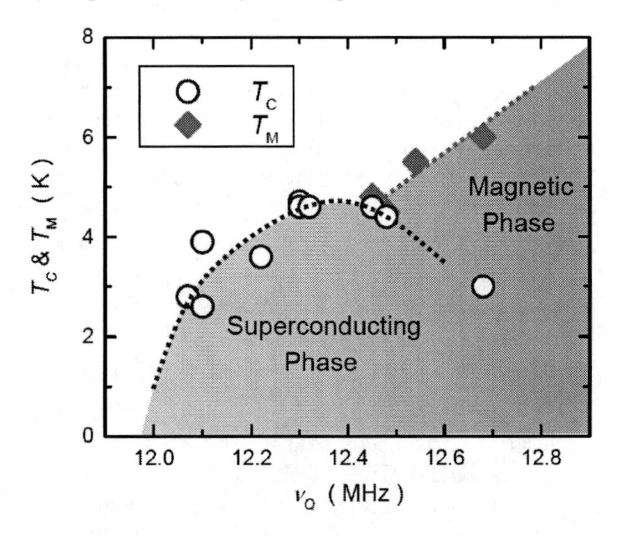

FIGURE 2. Phase diagram of $Na_xCoO_2 \cdot yH_2O$.

ACKNOWLEDGMENT

This work was partially supported by 21 COE program on "Center for Diversity and Universality in Physics" from the MEXT of Japan.

REFERENCES

1. K. Takada *et al.*, *Nature* **422**, 53 (2003).
2. J. D. Jorgensen *et al.*, *Phys. Rev. B* **68**, 214517 (2003).
3. K. Ishida *et al.*, *J. Phys. Soc. Jpn.* **72**, 3041 (2003).
4. T. Fujimoto *et al.*, *Phys. Rev. Lett.* **92**, 125 (2004).
5. Y. Ihara *et al.*, *J. Phys. Soc. Jpn.* **73**, 2069 (2004).
6. Y. Ihara *et al.*, *J. Phys. Soc. Jpn.* **74**, 867 (2005)
7. J. W. Lynn *et al.*, *Phys. Rev. B* **68**, 214516 (2003)
8. H. Sakurai *et al.*, *J. Phys. Soc. Jpn.* **73**, 9 (2004).
9. T. Moriya *et al.*, *Advance in Physics* **49**, 555 (2000).
10. M. Mochizuki *et al.*, *cond-mat*/0503233.

Superconductivity in $KMnO_4$-treated $Na_xK_z(H_2O)_yCoO_2$

S. Neeleshwar[1], Y. Y. Chen[1], J. C. Ho[1*], C. -J. Liu[2], C. -Y. Liao[2], W. -C. Hung[2], J. -S. Wang[2], and C.-J. C. Liu[2]

[1]Institute of Physics, Academia Sinica, Taipei, Taiwan
[2]Department of Physics, National Changhua University of Education, Changhua, Taiwan

Abstract. A series of sodium potassium cobalt oxyhydrates $Na_xK_z(H_2O)_yCoO_2$ (x=0.38-0.28 and z=0-0.08) were synthesized using aqueous $KMnO_4$ solution with various molar ratio of $KMnO_4$/Na with respect to the sodium content in the parent compound γ-$Na_{0.7}CoO_2$. Superconductivity was observed at the transition temperature $T_c \sim 4.5$ - 4.6 K in all fully-hydrated samples.

Keywords: Superconductivity, $KMnO_4$-treated $Na_xK_z(H_2O)_yCoO_2$
PACS: 74.70.–b

The general interest on superconductivity in sodium cobalt oxyhydrate, $Na_x(H_2O)_yCoO_2$[1], arises because of its close resemblance with that in high T_c cuprates, for which the mechanism leading to the intriguing phenomenon is yet to be fully elucidated. Both types of compounds exhibit two-dimensional layers (triangular CoO_2 and square CuO_2) in structure, and have spin-1/2 ions (Co^{4+} and Cu^{2+}) in electronic configuration. In contrast to the earlier report by Schaak et al.[2], a recent study by Milne et al.[3] on the superconducting phase diagram indicates a broader range of Na content (x ~ 0.29-0.37) corresponding to a narrow range of the superconducting transition temperature of $T_c \sim 4.3$-4.8 K. The sodium potassium cobalt oxyhydrates in this work were synthesized through a non-toxic route.

The superconductive $Na_x(H_2O)_yCoO_2$ was generally obtained by immersing the parent compound of γ-$Na_{0.7}CoO_2$ in Br_2/CH_3CN solution, followed by filtering and rinsing with acetonitrile. In this work, an aqueous $KMnO_4$ solution was used as a de-intercalating and oxidizing agent. With increasing aqueous $KMnO_4$ concentration, more Na^+ can be removed but a small amount of K^+ would enter the lattice, thus forming $Na_xK_z(H_2O)_yCoO_2$. For the $c \approx 19.6$ Å phase, the substitution of Na^+ by K^+ likely just has no significant effects on the electronic structure of the materials. Such an alternative approach was well developed by Liu and has been reported in Refs. 4-7. The contents of sodium and potassium of resulting products were determined by an inductively coupled plasma-atomic emission spectrometer (ICP-AES).

Knowing that the degree of hydration in this system can drastically affect T_c, we kept our samples in a covered chamber with saturated water vapor for a long period of time in order to achieve and maintain full hydration. Fig. 1 shows the temperature dependence of zero-field-cooled (ZFC) dc magnetization data, taken from a high Na content (x = 0.38, z = 0) sample with a SQUID magnetometer. Indeed, T_c drops from the fully-hydrated value of 4.5 K to 3.2 K for the partially-hydrated condition.

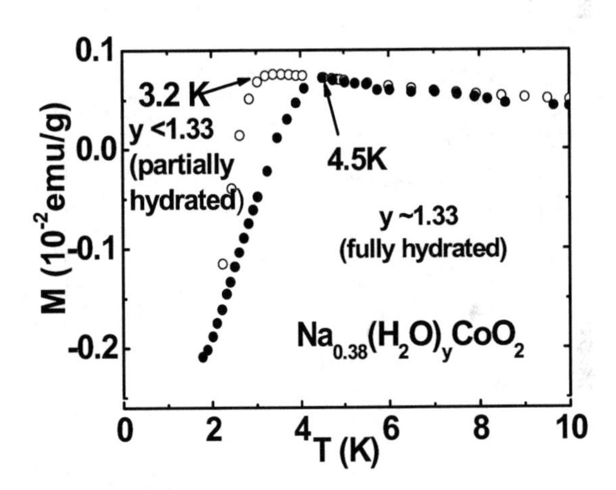

FIGURE 1. Temperature dependence of magnetization at 20 Oe for partially- and fully- hydrated samples with (x = 0.38, z = 0).

Similar results on three lower-Na content but K-containing samples with (x = 0.33, z = 0 02), (x = 0.31, z = 0.05), and (x = 0.28, z = 0.08), respectively, are shown in Fig. 2. The onset of superconducting transition occurs at virtually the same T_c of 4.5-4.6 K. Table 1 lists the T_c values for all samples. On the other hand, as in other reports,[1-3] the magnetically determined superconducting volume fraction is small in all cases.

CP850, *Low Temperature Physics: 24th International Conference on Low Temperature Physics;*
edited by Y. Takano, S. P. Hershfield, S. O. Hill, P. J. Hirschfeld, and A. M. Goldman

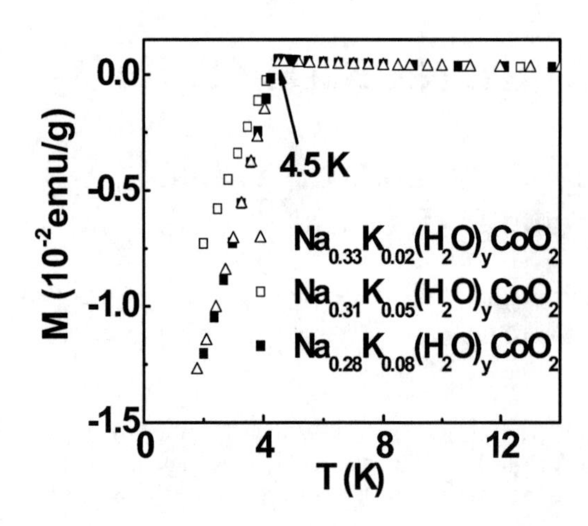

FIGURE 2. Temperature dependence of magnetization at 10 Oe for three fully-hydrated ($y\approx1.33$) samples with different Na and K contents.

The field dependence of magnetization at 1.8 K for two samples with (x = 0.38, z = 0) and (x = 0.31, z = 0.05) in Fig. 3 is similar to $Na_{0.31}CoO_2{\cdot}1.3H_2O$[8].

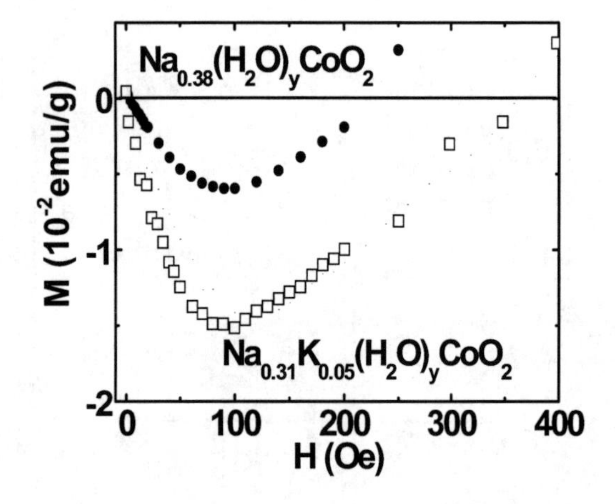

FIGURE 3. Field dependence of magnetization at 1.8 K for two fully-hydrated samples ($y\approx1.33$).

In conclusion, the $KMnO_4$-treated $Na_xK_z(H_2O)_yCoO_2$, when fully hydrated ($y\approx1.33$) has the superconducting transition temperature of T_c near 4.5-4.6 K for the (x, z) range of [(0.38, 0), (0.28, 0.08)].

TABLE 1. Superconducting transition temperature of fully-hydrated $Na_xK_z(H_2O)_yCoO_2$ ($y\approx1.33$).

x (Na)	z (K)	T_c (K)
0.38	0	4.5
0.34	0.01	4.5
0.33	0.02	4.6
0.31	0.05	4.5
0.28	0.08	4.5

ACKNOWLEDGMENTS

This work was supported by the National Research Council of the Republic of China under Grant Nos. NSC 93-2112-M-001-022 and 93-2112-M-018-003.

REFERENCES

* on leave from Wichita State University, Wichita, Kansas, USA.

1. K. Takada, H. Sakurai, E. Takayama-Muromachi, F. Izumi, R.A. Dilanian, and T. Sasaki, *Nature* **422**, 53-55 (2003).
2. R.E. Schaak, T. Klimczuk, M.L. Foo, and R.J. Cava, *Nature* **424**, 527-529 (2003).
3. C.J. Milne, D.N. Argyriou, A. Chemseddine, N. Aliouane, J. Veira, S. Landsgesell, and D. Alber, *Phys. Rev. Lett.* **93**, 247007- (2004).
4. C.-J. Liu, C.-Y. Liao, L.-C. Huang, C.-H. Su, S. Neeleshwar, Y.Y. Chen, and C.-J.C. Liu, *Physica C* **416**, 43-46 (2004).
5. C.-J. Liu, W.-C. Hung, J.-S. Wang, and C.-J.C. Liu, *J. Am. Chem. Soc.* **127**, 830-831 (2005).
6. C.-J. Liu, C.-Y. Liao, L.-C. Huang, C.-H. Su, W.-C. Hung, S. Neeleshwar, Y.Y. Chen, and C.-J.C. Liu, Ch*inese J. Phys.* **43**, 547-555(2005).
7. C.-J. Liu, C.-Y. Liao, *L.-C. Huang, C.-H. Su,* S. Neeleshwar, Y.-Y. Chen, *cond-mat/*0407420.
8. G. Cao, C. Feng, Yi Xu, W. Lu, J. Shen , M. Fang and Z.-an Xu, *J. Phys. Condens, Matter.* **15**, L519-L525, (2003).

Growth of the Single Crystal $BaBiO_3$ by the Floating-Zone Method

Yoshinori Imai*,†, Takashi Noji*,†, Masatsune Kato*,† and Yoji Koike*,†

*Department of Applied Physics, Tohoku University, 6-6-05, Aoba, Aramaki, Aoba-ku, Sendai 980-8579, Japan
†CREST, Japan Science and Technology Corporation (JST)

Abstract. A large-size single-crystal of the perovskite $BaBiO_3$ has been successfully grown by the floating-zone (FZ) method. The FZ growth was carried out under flowing O_2 gas at 2 atm in an infrared heating furnace equipped with the quartet ellipsoidal mirrors. The growth rate was 1 mm/h. The rotation speed of the upper and lower shafts was 10 rpm in the opposite direction. Diffraction spots in the x-ray back-Laue photograph of the grown crystal were very sharp, indicating the good quality of the grown crystal. The powder x-ray diffraction revealed that the grown crystal was of the single phase of $BaBiO_3$.

Keywords: $BaBiO_3$, perovskite, floating-zone method, single-crystal growth
PACS: 74.75.Dd, 81.10.Fq

INTRODUCTION

The perovskite $BaBiO_3$ is well known as a parent compound of the superconductor $Ba_{1-x}K_xBiO_3$ with the superconducting transition temperature $T_c \sim 30K$ [1], which is the highest among oxides free of copper. This family of compounds has neither quasi-two-dimensionality nor magnetic order, both of which are believed to be essential to the appearance of the high-T_c superconductivity in the Cu-based superconductors. The high T_c value in this system is believed to result from a strong coupling of electrons with the breathing phonon mode.

The perovskite structure of $BaBiO_3$ is a distorted one with the monoclinic symmetry. This distortion is due to the size mismatch between the shorter Ba-O bond length and the longer Bi-O bond length. In this compound, the average valence of Bi is +4 with the electronic configuration of $5d^{10}6s^1$. Accordingly, $BaBiO_3$ is expected to be a metal with one delocalized electron per Bi. In fact, however, the compound is not a metal but a CDW insulator because of the charge disproportionation of Bi^{4+} ions into Bi^{3+} and Bi^{5+}, which has been pointed out through some experiments, for example, the neutron diffraction [2], EXAFS [3] and the optical studies [4]. The CDW gap has been estimated to be 1.9 eV by the optical reflectivity mesurements [5].

As for $Ba_{1-x}K_xBiO_3$, several groups have succeeded in growing the single-crystal with $5.0 \times 5.0 \times 5.0$ mm^3 in size using the electrochemical method [6, 7]. In the electrochemical method, KOH is used as a flux, so that the growth of the single-crystal $BaBiO_3$ free of K using this method is difficult. Formerly, the single-crystal $BaBiO_3$ was grown by the hydrothermal method [8]. However, the size of the single crystal was on the order of 1 mm^3.

In this paper, the floating-zone (FZ) method is proposed for the growth of the large-size single-crystal $BaBiO_3$ of good quality.

EXPERIMENTAL

The procedure of the growth of single-crystal $BaBiO_3$ by the FZ method is as follows. In order to prepare the feed rod for the FZ growth, stoichiometric mixed powders of BaO_2(2N) and Bi_2O_3(3N) were heated in a flowing gas of O_2 at 800 °C for 48 hours with the intermediate grinding. After grinding for an hour, the powder was isostatically cold-pressed at 400 bar into a rod of 6 mm in diameter and ~120 mm in length. The rod was sintered at 830 °C for 24 hours in a flowing gas of O_2. The FZ growth was carried out using the obtained feed rod under the flowing gas of O_2 at 2 atm in an infrared heating furnace equipped with the quartet ellipsoidal mirrors (Crystal System Inc., Model FZ-T-4000-H). The growth rate was 1 mm/h. The rotation speed of the upper and lower shafts was 10 rpm in the opposite direction to secure the homogeneity of the liquid in the molten zone.

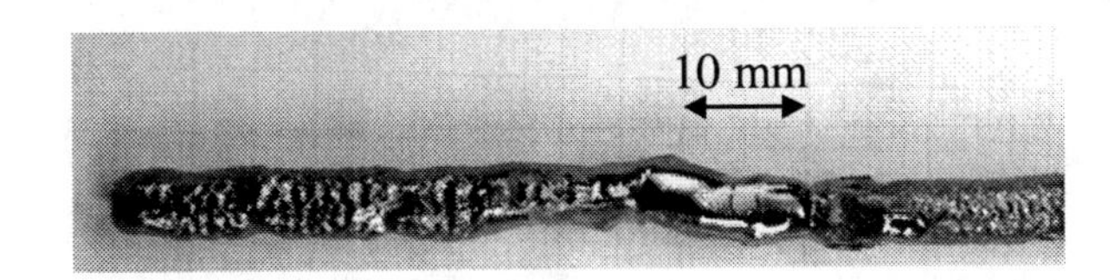

FIGURE 1. As-grown crystal of $BaBiO_3$ by the floating-zone method.

CP850, *Low Temperature Physics: 24th International Conference on Low Temperature Physics;*
edited by Y. Takano, S. P. Hershfield, S. O. Hill, P. J. Hirschfeld, and A. M. Goldman

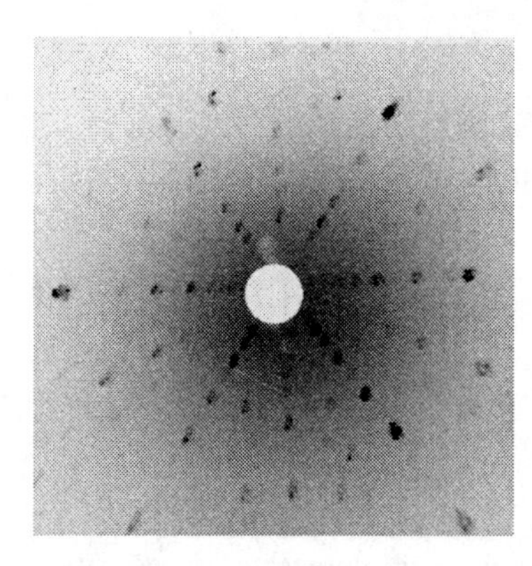

FIGURE 2. X-ray back-Laue photograph of the grown crystal of $BaBiO_3$ in the x-ray parallel to the [100] direction on the basis of the pseudo-cubic symmetry.

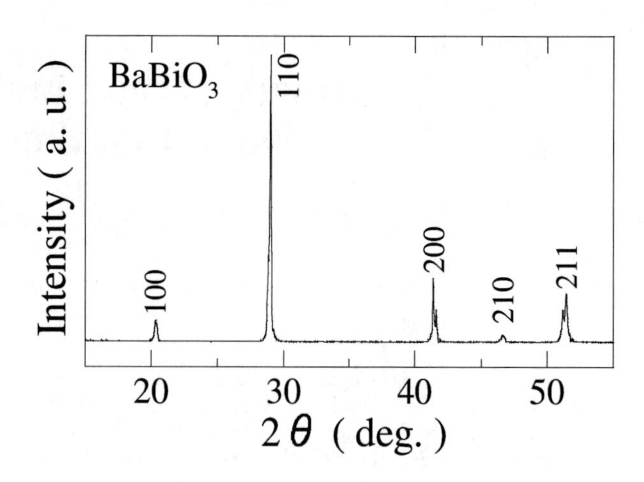

FIGURE 3. Powder x-ray diffraction pattern of the single-crystal $BaBiO_3$. Diffraction peaks are indexed on the basis of the pseudo-cubic symmetry.

The grown crystal was checked by the x-ray back-Laue photography, the powder x-ray diffraction using Cu Kα radiation and the inductively coupled plasma atomic emission spectrometry(ICP-AES).

RESULTS AND DISCUSSION

An as-grown crystal, which is golden bronze in color, is shown in Fig. 1. The dimensions are 6 mm in diameter and $\sim$ 100 mm in length. A typical x-ray back-Laue photograph of the grown crystal is shown in Fig. 2. Although the grown crystal is composed of several domains, the diffraction spots are very sharp. The typical dimensions of a single domain are $2 \times 2 \times 10$ mm^3, which are much larger than those grown by the hydrothermal method [8]. Fig. 3 shows the powder x-ray diffraction pattern of the grown crystal. There are seen Bragg peaks of the $BaBiO_3$ phase with the monoclinic symmetry without any impurity phases. The pseudo-cubic lattice parameter a_p is determined to be 4.354 Å, which is consistent with the value reported by [2]. The ICP-AES analysis has revealed that the molar ratio of Ba:Bi is 0.99:1.01. Accordingly,it is concluded that the quality of the grown single-crystal is very good.

SUMMARY

A large-size single-crystal of the perovskite $BaBiO_3$ has been successfully grown by the FZ method. The crystal is golden bronze in color. The typical dimensions of a single domain are $2 \times 2 \times 10$ mm^3. High quality of the grown crystal has been confirmed by the x-ray back-Laue photography, the powder x-ray diffraction and ICP-AES.

ACKNOWLEDGEMENTS

This work was supported by a Grant-in-Aid for Scientific Research from the Ministry of Education, Culture, Sports, Science and Technology, Japan.

REFERENCES

1. R. J. Cava, B. Batlogg, J. J. Krajewski, F. Farrow, L. W. Rupp Jr, A. E. White, K. Short, W. F. Peck and T. Kometani, *Nature* **332**, 814–816 (1988).
2. D. E. Cox and A. W. Sleight, *Solid State Commun.* **19**, 969–973 (1976).
3. A. Balzarotti, A. P. Menushenkov, N. Motta and J. Purans, *Solid State Commun.* **49**, 887–890 (1984).
4. S. Tajima, S. Uchida, A. Masaki, H. Takagi, K. Kitazawa, S. Tanaka and S. Sugai, *Phys. Rev. B* **35**, 696–703 (1987).
5. J. Ahmad, T. Nishio and H. Uwe, *Physica C* **388-389**, 455–456 (2003).
6. M. L. Norton, *Mat. Res. Bull.* **24**, 1391–1397 (1989).
7. T. Nishio, H. Minami and H. Uwe, *Physica C* **357-360**, 376–379 (2001).
8. R. D. Shannon and P. E. Bierstedt, *J. Am. Ceram. Soc.* **53**, 635–636 (1970).

Electrochemical Synthesis of the Perovskite $Ba_{1-x}Cs_xBiO_3$ from Molten Salts

Yoshinori Imai[*,†], Masatsune Kato[*,†], Takashi Noji[*,†] and Yoji Koike[*,†]

[*]Department of Applied Physics, Tohoku University, 6-6-05, Aoba, Aramaki, Aoba-ku, Sendai 980-8579, Japan
[†]CREST, Japan Science and Technology Corporation (JST)

Abstract. Single crystals of the perovskite $Ba_{1-x}Cs_xBiO_3$ with $x = 0.22$ have been successfully synthesized by the electrochemical method from molten salts. The electrodeposition of $Ba_{1-x}Cs_xBiO_3$ from $CsOH \cdot H_2O$ flux containing $Ba(OH)_2 \cdot 8H_2O$ and Bi_2O_3 was performed in a Teflon crucible under a water-saturated argon atmosphere at 270 °C. Single crystals of high quality were grown up to $1 \times 1 \times 1$ mm^3 in size. The crystal symmetry was cubic at room temperature. However, no superconductivity was observed above 2 K in the magnetic susceptibility measurements.

Keywords: $Ba_{1-x}Cs_xBiO_3$, electrochmemical method, molten salts, single crystal
PACS: 74.62.Bf, 74.75.Dd, 82.45.Aa

In the perovskite $BaBiO_3$ with the monoclinic symmetry, the BiO_6 octahedron is on the tilt because of the size mismatch between the shorter Ba-O bond length and the longer Bi-O bond length. According to the simple electron-band theory, $BaBiO_3$ has a half-filled band of 6s electrons of Bi^{4+}, leading to a metallic behavior. However, the valence state of Bi^{4+} with $6s^1$ is unstable, so that Bi^{4+} disproportionates into Bi^{3+} with $6s^2$ and Bi^{5+} with $6s^0$, leading to a CDW insulator [1].

The tilt of the BiO_6 octahedron is alleviated by the hole doping through the partial substitution of K for Ba in $BaBiO_3$. The crystal structure of $Ba_{1-x}K_xBiO_3$ changes from monoclinic to cubic via orthorhombic with incresing x. There is no tilt of the BiO_6 octahedron in the cubic phase. The metal-insulator transition coincides with the structural phase transition from orthorhombic to cubic. The cubic $Ba_{0.6}K_{0.4}BiO_3$ shows superconductivity with the superconducting transition temperature $T_c = 30$ K [2], which is the highest T_c among oxides free of copper.

So far, some carrier-doped compounds based on $BaBiO_3$ have been reported: for example, hole-doped superconductors, $Ba_{1-x}A_xBiO_3$ (A = K,Rb) [3], and electron-doped insulators, $Ba_{1-x}Bi_xBiO_3$ [4]. The purpose of this research is to synthesize a new hole-doped compound of $Ba_{1-x}Cs_xBiO_3$ through the partial substitution of Cs^+ for Ba^{2+} in $BaBiO_3$. The substitution of Cs is more effective in alleviating the tilt of the BiO_6 octahedron than that of K, because the ionic radius of Cs^+ is much larger than that of K^+. Therefore, a higher T_c than 30 K is expected in the cubic $Ba_{1-x}Cs_xBiO_3$ with a small x value because of the large density of states at the Fermi energy.

Single crystals of $Ba_{1-x}Cs_xBiO_3$ were synthesized by the electrochemical method from molten salts. All electrodepsitions were performed under water-saturated flowing argon gas using a three-electrodes cell. Pt wires with the diameter of 0.5 mm were used as the working electrode and the counter electrode. A rod of Bi (5N) was used as the referrence electrode. $CsOH{\cdot}H_2O$ (3N) was placed in a Teflon crucible and melted at 270 °C. The CsOH solution was stirred for about 2 h. $Ba(OH)_2{\cdot}8H_2O$ (2N) and Bi_2O_3 (3N) were then slowly added to the solution. The molar ratio of the melt was Cs : Ba : Bi = 17.5 : 0.33 : 1. After 24 h, cyclic voltammograms of the quiescent solution were run to determine the electrocrystallization potential. Upon application of the chosen potential, the electrocrystallization process was instantly initiated. Crystals were grown for 48 h. Grown crystals were taken out of the solution, washed with the distilled water and then dried.

The crystal structure and composition of the products were determined by the powder x-ray diffraction using Cu Kα radiation and the inductively coupled plasma atomic emission spectrometry (ICP-AES), respectively. Magnetic susceptibility measurements were performed using a SQUID magnetometer (Quantum Design, Inc).

The cyclic voltammogram obtained in the $Ba(OH)_2$-Bi_2O_3-CsOH solution is shown in Fig. 1. The initial potential of the voltammetric scan (400 mV vs. Bi) was set at a value where no apparent electrochemical processes occur; the potential range was swept first anodically up to 900 mV vs. Bi, cathodically to −400 mV vs. Bi and then anodically up to the initial potential. Around the peak at 560 mV vs. Bi, the oxidation from Bi^{3+} to Bi^{4+} is expected to occur. Taking the low diffusion velocity of the ions in the solution into consideration, the applied potential for the electrocrystallization of $Ba_{1-x}Cs_xBiO_3$

CP850, *Low Temperature Physics: 24th International Conference on Low Temperature Physics;*
edited by Y. Takano, S. P. Hershfield, S. O. Hill, P. J. Hirschfeld, and A. M. Goldman

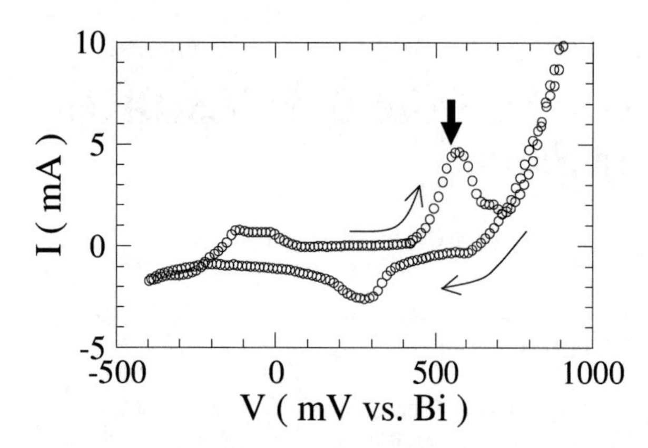

FIGURE 1. Cyclic Voltammogram from −400 mV vs. Bi to 900 mV vs. Bi in the $Ba(OH)_2$-Bi_2O_3-CsOH solution. The sweep rate is 5 mV/sec. The big downward arrow exhibits the potential in which the electrodeposition was performed.

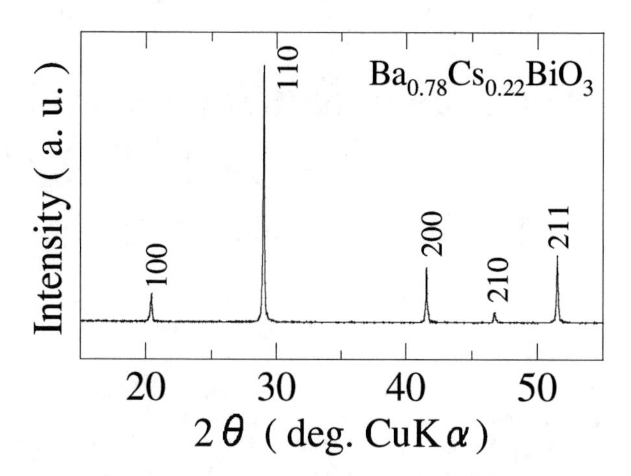

FIGURE 3. Powder x-ray diffraction pattern of the grown crystal $Ba_{0.78}Cs_{0.22}BiO_3$ at room temperature. The diffraction peaks are indexed on the basis of the cubic symmetry.

was determined to be 550 mV vs. Bi which was somewhat lower than the peak potential. Single crystals of $Ba_{1-x}Cs_xBiO_3$ were successfully electrodeposited on the Pt wire. As-grown crystals are shown in Fig. 2. The typical dimensions of a single crystal were $1 \times 1 \times 1$ mm^3.

The powder x-ray diffraction pattern of the grown crystal at room temperature is shown in Fig. 3. All the peaks can be indexed on the basis of the simple cubic $Pm\bar{3}m$ lattice, which is the same structure as the superconducting $Ba_{1-x}K_xBiO_3$ ($x \geq 0.37$). The Cs content x of the grown crystal is determined to be $x = 0.22$ by ICP-AES. This x value in the grown crystal of $Ba_{1-x}Cs_xBiO_3$ with the cubic symmetry is much smaller than the minimum K content ($x = 0.37$) required for the dissolution of the tilt of the BiO_6 octahedron in $Ba_{1-x}K_xBiO_3$ [2]. The lattice parameter is determined to be $a = 4.341$ Å, which is much larger than $a = 4.283$ Å in the superconducting $Ba_{0.6}K_{0.4}BiO_3$ [2]. However, the cubic $Ba_{0.78}Cs_{0.22}BiO_3$ showed no superconductivity above 2 K in the magnetic susceptibility measurements as opposed to the superconducting $Ba_{1-x}K_xBiO_3$ with the cubic symmetry.

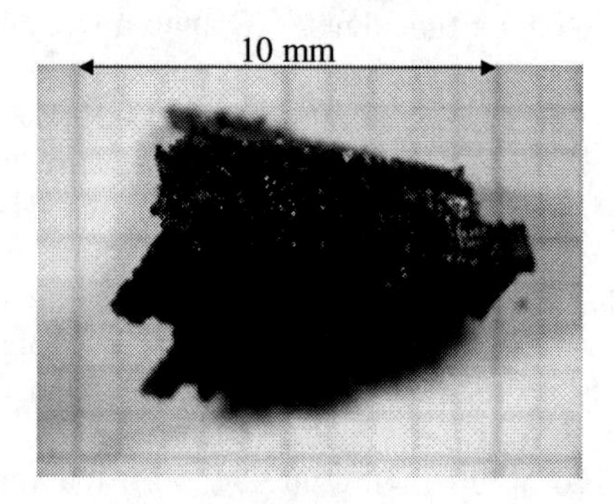

FIGURE 2. As-grown crystal of $Ba_{1-x}Cs_xBiO_3$ by the electrohemical method.

The possibility that the tilt of the BiO_6 octahedron exists at low temperatures is considered as a reason of non-superconductivity in the grown crystal, as follows. That is, in general, the thermal expansivity of the ionic bond tend to be larger than that of the covalent bond. Accordingly, the size mismatch in the present system would increase at low temperatures because the ionic Ba-O bond is expected to shrink more than the covalent Bi-O bond, leading to the tilt of the BiO_6 octahedron at low temperatures.

In summary, single crystals of the perovskite $Ba_{0.78}Cs_{0.22}BiO_3$ with $1 \times 1 \times 1$ mm^3 in size have been successfully synthesized by the electrochemical method from molten salts. The electrodeposition of $Ba_{1-x}Cs_xBiO_3$ from $CsOH \cdot H_2O$ flux containing $Ba(OH)_2 \cdot 8H_2O$ and Bi_2O_3 was performed in a Teflon crucible under a water-saturated argon atmosphere at 270 °C. The lattice parameter was 4.341 Å with the cubic symmetry at room temperature. However, no superconductivity was observed above 2 K.

This work was supported by a Grant-in-Aid for Scientific Research from the Ministry of Education, Culture, Sports, Science and Technology, Japan.

REFERENCES

1. D. E. Cox and A. W. Sleight, *Solid State Commun.* **19**, 969 (1976).
2. S. Pei *et al.*, *Phys. Rev. B* **41**, 4126 (1987).
3. R. J. Cava *et al.*, *Nature* **332**, 814 (1988).
4. Y. Imai *et al.*, *Physica C* **388-389**, 449 (2003).

Single-Crystal Growth of the Superconducting $Ba_{1-x}K_xBiO_3$ by the Floating-Zone Method

Takashi Noji,[a, b] Tatsuya Kato,[a] Yoshinori Imai[a, b] and Yoji Koike [a, b]

[a]*Department of Applied Physics, Graduate School of Engineering, Tohoku University, 6-6-5, Aoba, Aramaki, Aoba-ku, Sendai 980-8579, Japan*
[b]*CREST, Japan Sience and Technology Corporation (JST)*

Abstract. We have successfully grown single crystals of $Ba_{1-x}K_xBiO_3$ with the volume of the order of 1 mm^3 by the floating-zone method. It was important to use a high-pressure reducing atmosphere and the "twice-scanning" method for the successful growth. The crystal growth was performed under argon pressure of 1.0 MPa. The potassium content of each single crystal is much smaller than that of the feed rod. Single crystals of x = 0.32, 0.42, 0.50 (estimated from the ICP-AES analysis) annealed in flowing oxygen gas show superconductivity with the transition temperature T_c=29 K, 25 K, 15 K, respectively. From the powder x-ray diffraction analysis and electrical resistivity measurements, each obtained single-crystal has been found to be composed of at least two phases with different x values.

Keywords: $Ba_{1-x}K_xBiO_3$, single-crystal growth, floating-zone method
PACS: 74. 25. Ha, 74. 62. Bf, 74. 62. Dh

Among oxide superconductors not containing copper, $Ba_{1-x}K_xBiO_3$ (BKBO) exhibits the highest superconducting transition temperature, T_c = 30 K for x ~ 0.4, in spite of the low carrier concentration [1]. Single crystals of BKBO have been grown by the flux method and the electrochemical technique so far [2 - 4]. However, it has not yet been reported that the floating-zone (FZ) method is able to be applied to BKBO. In the FZ method, the control of the composition of a single crystal is comparatively easy. In general, the FZ method allows the growth of high-quality single crystals.

In this paper, we have tried the single-crystal growth of BKBO with various potassium contents by the FZ method. The superconducting properties have been investigated.

In order to prepare the feed rod for the FZ growth, first, we prepared the polycrystalline powder of $BaBiO_3$ by the solid-state reaction method. That is, the prescribed amount of BaO_2 and BiO_3 of 99.9 % purity was mixed, ground and prefired at 800 °C in oxygen gas for 12 h. Next, K_2CO_3 of 99.8 % purity and Bi_2O_3 were added to the prefired powder of $BaBiO_3$ in order to prepare a feed rod with various ratios of Ba : K : Bi. After 1 h grinding, the powder was isostatically cold-pressed into a rod of 5 - 6 mm in diameter and ~ 100 mm in length. Then, the rod was sintered at 650 °C in air for 12 h. Single crystals were grown in an infrared radiation-convergence type furnace with halogen lamp using a premelted feed rod. That is, the so-called "twice-scanning" method was used [5]. The high-density premelted feed rod was prepared by the first scan. In the first scan, the lamp power was elevated gradually under argon pressure of 1 MPa until the feed rod melted. The molten zone passed through at a speed of ~ 20 mm/h. The premelted feed rod and the as-sintered rod were rotated at ~ 20 rpm in the opposite direction. Next, the usual growing procedure was carried out under argon pressure of 1 MPa at the growth rate of 1.0 mm/h, using the premelted feed rod. Since the obtained rod was covered with potassium evaporating from the molten zone, single crystals were obtained by crushing the rod mechanically. In order to fill up oxygen vacancies and remove the strain, the as-grown crystals were post-annealed in flowing oxygen gas at 300 °C for 150 h.

The products were identified by the powder x-ray diffraction analysis and the back-Laue photography. The composition was determined by the inductively coupled plasma atomic emission spectroscopy (ICP-AES). The magnetic susceptibility was measured using a SQUID magnetometer. Electrical resistivity measurements were carried out by the dc four-point probe method.

CP850, *Low Temperature Physics: 24th International Conference on Low Temperature Physics;*
edited by Y. Takano, S. P. Hershfield, S. O. Hill, P. J. Hirschfeld, and A. M. Goldman

By using the "twice-scanning" method in a high-pressure reducing atmosphere, namely, under argon pressure of 1 MPa, a rod containing single crystals of BKBO was successfully obtained. This atmosphere was necessary to suppress the evaporation of potassium out of the molten zone and to keep the molten zone stable. It was hard to keep the molten zone stable in air and also in oxygen gas. The best growth rate was 0.8 – 1 mm/h. It was confirmed that single crystals were not obtained at the growth rate of 2 mm/h and that the molten zone was not keep stable at the rate slower than 0.5 mm/h on account of the penetration of the molten liquid into the feed rod.

The volume of the obtained single-crystals was the order of 1 mm^3. Many spots were clearly observed in the Laue photograph, but we could take no photograph with only a single set of symmetric spots. The powder x-ray diffraction peaks of the obtained single-crystals have revealed that these are of the single phase with the cubic perovskite structure. Looking at each peak in detail, however, it was slightly split. These mean that each obtained single-crystal is composed of at least two phases with different x values. Table 1 shows the nominal composition of the feed rods and the composition of the obtained single-crystals determined by ICP-AES. It is found that the average of the potassium content of each single crystal is much smaller than that of the feed rod. This is due to evaporation of potassium during the two scans.

TABLE 1. Nominal composition of the feed rods and composition of the obtained single-crystals determined by ICP-AES.

	Ba	K	Bi
Feed rod A (Nominal)	0.60	0.50	1.00
Single crystal (ICP)	0.64	0.32	1.04
Feed rod B (Nominal)	0.60	0.80	1.00
Single crystal (ICP)	0.57	0.42	1.01
Feed rod C (Nominal)	0.50	1.00	1.00
Single crystal (ICP)	0.49	0.50	1.01

Figure 1 displays the typical temperature dependences of the magnetic susceptibility on field cooling and the electrical resistivity for single crystals obtained from the feed rod A, B and C. It is found from the susceptibility that single crystals of x = 0.32, 0.42, 0.50 show superconductivity with the transition temperature T_c = 29, 25, 15 K, which are comparable to T_c values of polycrystalline samples of BKBO, respectively [1]. The superconducting volume fraction of the single crystal (x = 0.42) grown from the feed rod B is estimated from the Meissner effect to be 30 – 45 % of the perfect diamagnetism. The superconducting volume fraction of single crystals (x = 0.32, 0.50) grown from the feed rod A, C is smaller than that from B. Moreover, no resistance drop due to superconducting is observed for single crystals from the rod A and C, owing to the high resistivity value at low temperature. These results suggest that single crystals from the rod A and C are composed of a superconductor and an insulator with different x values so that superconducting paths are not connected with one another. It may be inevitable that single crystals of BKBO grown by the FZ method have a domain structure due to a kind of phase separation.

We would like to thank K. Takada and M. Ishiguro for their help in the ICP analysis.

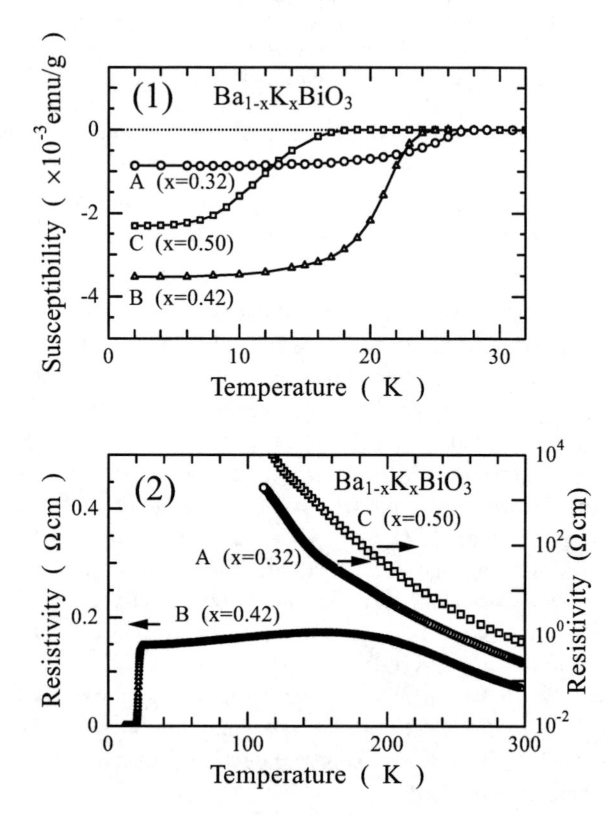

FIGURE 1. Temperature dependences of (1) the magnetic susceptibility on field cooling in a field of 3 Oe and (2) the electrical resistivity for single crystals of $Ba_{1-x}K_xBiO_3$ obtained from the feed rod A, B and C.

REFERENCES

1. S. Pei *et al., Phys. Rev. B* **41**, 4126 (1990).
2. J.P. Wignacourt *et al., Appl. Phys. Lett.* **53**, 1753 (1988).
3. L.F. Schneemeyer *et al., Nature* **335**, 421 (1988).
4. M.L. Norton, *Mat. Res. Bull.* **24**, 1391 (1989).
5. K. Oka and T. Ito, *Physica C* **227**, 77 (1994).

Magnetization Study of the Superconducting Pyrochlores AOs_2O_6 (A=Rb, K) under Pressure

K. Miyoshi, Y. Takamatsu, M. Miura, K. Fujiwara and J. Takeuchi

Department of Material Science, Shimane University, Matsue 690-8504, Japan.

Abstract. Dc magnetization has been observed as a function of temperature for superconducting β-pyrochlore oxides KOs_2O_6 (T_C=9.6 K) and $RbOs_2O_6$ (T_C=6.3 K) under pressure $p < 4$ GPa. It is found that T_C for KOs_2O_6 increases and shows a maximum of 10 K at p=0.5 GPa, whereas T_C for $RbOs_2O_6$ shows a monotonic increase and reaches 8.7 K at p=3.9 GPa. The T_C - p curves suggest that the relationship between T_C and lattice constant in these compounds is different from each other.

Keywords: β-pyrochlore, high-pressure, magnetization, osmates.
PACS: 74.62.Fj, 74.25.Ha, 74.70.Dd

The recent discovery of β-pyrochlore superconductor AOs_2O_6 (A=Cs, Rb, K) [1-3], in which only the Os atoms form a three-dimensional network of corner-sharing tetrahedra different from the conventional $A_2B_2O_7$ pyrochlores, has stimulated renewed interest in superconductivity on geometrically frustrated structures, such as in $LiTi_2O_4$ [4] and $Cd_2Re_2O_7$ [5-6]. Indeed, the superconductivity in AOs_2O_6 is currently studied extensively and a few experiments have revealed the characteristic features of unconventional superconductivity. For KOs_2O_6, an anisotropic superconducting gap is inferred from the anomalous magnetic field dependence of penetration depth measured by μSR [7], and also from the absence of the coherence peak in $1/(T_1T)$ below T_C in the NMR experiments [8]. For $RbOs_2O_6$, however, the recent magnetic field penetration depth [9] and specific heat measurements [10] in addition to the small coherence peak in $1/(T_1T)$ observed below T_C [8] suggest a conventional BCS-type superconductivity. The superconducting pairing mechanism, which should be universal among AOs_2O_6, is currently under debate.

In AOs_2O_6, T_C systematically increases from 3.3 K (A=Cs) to 9.6 K (A=K) with decreasing lattice constant a from 10.148 Å (A=Cs) to 10.099 Å (A=K), implying that T_C is increased by applying physical pressure (p) in contrast to the general trend in BCS-type superconductor. Actually, a positive pressure effect for T_C has been observed in AOs_2O_6 through the magnetic measurements under pressure up to 1.2 GPa [9, 11]. Also, electrical resistivity (ρ) for AOs_2O_6 has been measured under pressure up to 10 GPa using a cubic anvil press, and a characteristic p – T phase diagram for AOs_2O_6 has been proposed, determining T_C from the midpoint of the resistive drop [12]. However, it is difficult to obtain a precise phase diagram from the ρ(T) data because the resisitivity drop occurs over wide T range (ΔT=1-5 K) under pressure [12].

To gain more insight into the mechanism of the superconductivity in AOs_2O_6, it is important to establish the $T_C - p$ relations, which are worthy to be compared with the results of structural analysis under high pressure in the future study. In the present work, we have performed dc magnetization measurements for AOs_2O_6 (A=Rb, K) under high pressure up to 4 GPa and report the precise $T_C - p$ relations. In the measurements, a miniature diamond anvil cell (DAC) for $p < 4$ GPa and a piston-cylinder-type (PC-type) cell for $p < 1$ GPa with an outer diameter of 8 mm were used to generate high pressure and they are designed for used with a commercial SQUID magnetometer. The details of the DAC are given in Ref. [13]. Powder samples of AOs_2O_6 were synthesized by solid-state reaction as described in the literatures [1-3].

In Figs. 1(a) and 1(b), the temperature variations of zero-field-cooled dc magnetization M at various pressures for KOs_2O_6 and $RbOs_2O_6$ are shown. The data are shifted intentionally along vertical axis for clarity. In both figures, $M(T)$ exhibits a sharp drop below ~ 7 K at ambient pressure, corresponding to the

CP850, *Low Temperature Physics: 24th International Conference on Low Temperature Physics;*
edited by Y. Takano, S. P. Hershfield, S. O. Hill, P. J. Hirschfeld, and A. M. Goldman

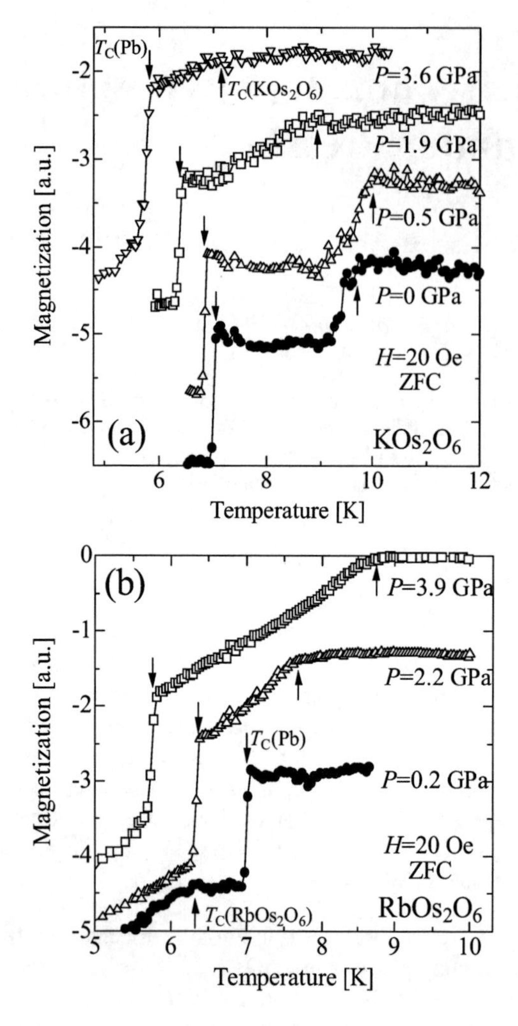

FIGURE 1. Temperature dependence of zero-field-cooled dc magnetization under pressure for KOs_2O_6 (a) and $RbOs_2O_6$ (b).

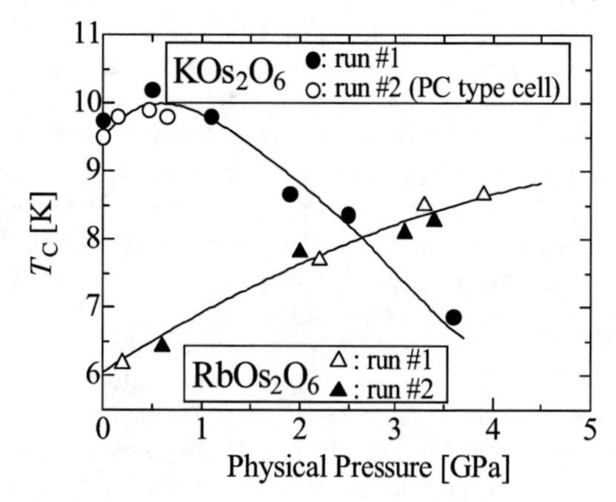

FIGURE 2. Plots of T_C versus pressure data for KOs_2O_6 and $RbOs_2O_6$.

superconducting transition of high-purity lead (Pb), which is loaded into the gasket hall in DAC together with AOs_2O_6 sample. *In situ* pressure measurements were done by determining p from T_C shift of Pb. As seen in Fig. 1(a), $M(T)$ at ambient pressure shows a sudden decrease below ~9.6 K, indicating the onset of diamagnetic susceptibility due to the superconducting transition in KOs_2O_6. In our experiments, T_C was estimated from the diamagnetic onset. In Fig. 1(a), T_C for KOs_2O_6 is increased to ~10 K at p=0.5 GPa, and then decreased by the further application of pressure up to 3.6 GPa. We also confirmed that T_C for KOs_2O_6 makes a maximum at p~0.5 GPa through the measurements using a PC-type cell. For $RbOs_2O_6$, a diamagnetic onset is observed at ~6.3 K at p=0.2 GPa in Fig. 1(b) and shifts toward higher temperatures as pressure is increased, reaching 8.7 K at p=3.9 GPa.

In Fig. 2, T_C versus pressure data for AOs_2O_6 are summarized. The behavior of the $T_C - p$ curve for KOs_2O_6 which shows a maximum at p~0.5 GPa agrees with that observed in earlier work [11]. On the other hand, the monotonic increase of T_C for $RbOs_2O_6$ is seen in Fig. 2. The $T_C - p$ curve for $RbOs_2O_6$ is in marked contrast to that decided from the midpoint of the resistive drop in the previous work, where it is suggested that the $T_C - p$ curve for $RbOs_2O_6$ shows a maximum at p~2 GPa [12]. It should be also noted that T_C for $RbOs_2O_6$ at p=3.9 GPa is still smaller than that for KOs_2O_6 at ambient pressure, even though the lattice constant for $RbOs_2O_6$ at p=3.9 GPa is expected to be much smaller than that for KOs_2O_6 at ambient pressure, considering the lattice compressibility for AOs_2O_6 under pressure calculated in Ref. [14]. This suggests that the relationship between T_C and lattice constant in these compounds is different from each other, although T_C systematically changes with A ion radius or lattice constant.

REFERENCES

1. S. Yonezawa *et al.*, *J. Phys.: Condens. Matter* **16**, L9 (2004).
2. S. Yonezawa *et al.*, *J. Phys. Soc. Japan* **73**, 819 (2003).
3. S. Yonezawa *et al.*, *J. Phys. Soc. Japan* **73**, 1655 (2004).
4. D. C. Johnston, *J. Low Temp. Phys.* **16**, 145 (1976).
5. H. Sakai *et al.*, *J. Phys.: Condens. Matter* **13**, L785 (2001).
6. M. Hanawa *et al.*, *Phys. Rev. Lett.* **87**, 187001 (2001).
7. A. Koda *et al.*, arXiv:cond-mat/0402400.
8. K. Arai *et al.*, arXiv:cond-mat/0411460.
9. R. Khasanov *et al.*, *Phys. Rev. Lett.* **93**, 157004 (2004).
10. M. Bruhwiler *et al.*, *Phys. Rev. B* **70**, 020530(R) (2004).
11. T. Muramatsu *et al.*, *J. Phys. Soc. Japan* **73**, 2912 (2004).
12. T. Muramatsu *et al.*, arXiV:cond-mat/0502490.
13. M. Mito *et al.*, *Jpn. J. Appl. Phys.* **40**, 6641 (2001).
14. R. Saniz *et al.*, arXiV:cond-mat/0506028.

Possible Superconductivity in $Ag_5Pb_2O_6$ Probed by AC Susceptibility

Shingo Yonezawa* and Yoshiteru Maeno*,†

*Department of Physics, Graduate School of Science, Kyoto University, Kyoto 606-8502, Japan
†International Innovation Center, Kyoto University, Kyoto 606-8501, Japan

Abstract. We will present the physical properties of a layered silver oxide $Ag_5Pb_2O_6$. The resistivity shows T^2 dependence in an unusually wide range of temperature, up to room temperature. Considering the specific heat, this T^2 dependence is not attributable to either strong electron correlation or strong electron-optic phonon interaction. In the AC susceptibility measurements we found a diamagnetic phase transition around 48 mK, which suggest the superconductivity of $Ag_5Pb_2O_6$. Positive peaks of the real part of the AC susceptibility, observed in magnetic fields, indicate reversibility of magnetic process. These observations implies that $Ag_5Pb_2O_6$ is a type-I superconductor, which are quite rare among oxides.

Keywords: $Ag_5Pb_2O_6$, silver oxide, type-I superconductivity
PACS: 74.10.+v, 74.70.Dd

INTRODUCTION

$Ag_5Pb_2O_6$ has a layered crystal structure containing a silver Kagome lattice parallel to its *ab* plane and silver chains along the *c* axis [1]. This silver oxide exhibits metallic conductivity. Band calculation by Brennan and Burdett [2] shows that its conductivity mainly comes from the Ag5s orbital, and that its Fermi surface has a quasi-three-dimensional character because both the silver chain and Kagome lattice contribute to the density of states at the Fermi level.

We here present the physical properties of single crystal $Ag_5Pb_2O_6$. Interestingly, its resistivity not only along the *c* axis but also in the *ab* plane shows T^2 dependence up to room temperature. What is more, we observed a new phase below 48 mK in the AC susceptibility measurement, which is possibly the type-I superconducting phase.

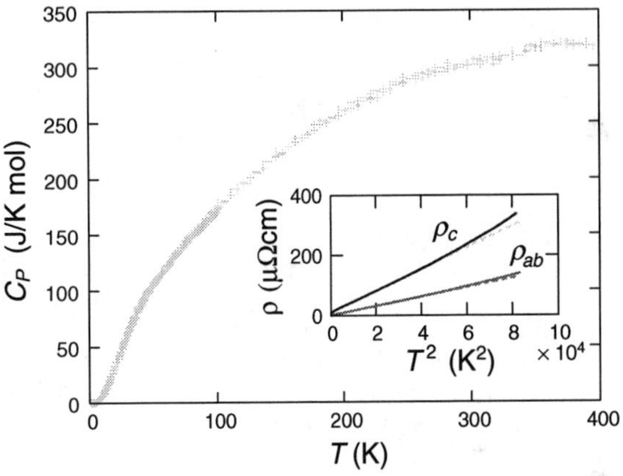

EXPERIMENTAL

In this work we used single crystals of $Ag_5Pb_2O_6$ grown by a self-flux method [3]. The resistivity was measured by a conventional four-probe method down to 4.2 K. The specific heat measurements between 0.3 K to 400 K were performed with a commercial calorimeter (Quantum Design, Model PPMS). The AC susceptibility is measured by a mutual inductance method. For the specific heat and the AC susceptibility measurements, we used a cluster of single crystals. The susceptibility measurements were performed with a ^{4}He-^{3}He dilution refrigerator (Cryoconcept, Model DR-JT-S-100-10), covering the temperatures as low as 38 mK.

FIGURE 1. (color online) Temperature dependence of the specific heat of $Ag_5Pb_2O_6$. The solid lines in the inset show the resistivity plotted against T^2. The broken lines are the results of fitting with $\rho(T) = \rho_0 + AT^2$ up to 220 K.

RESULTS AND DISCUSSION

The inset of Fig. 1 shows the resistivity along the *c* axis (ρ_c) and in the *ab* plane (ρ_{ab}) plotted against T^2. Apparent T^2 dependence was observed for both ρ_c and ρ_{ab}. If we approximate as $\rho(T) = \rho_0 + AT^2$, we obtained the coefficient of T^2 term $A_c = 3.59 \times 10^{-3}$ $\mu\Omega$cm/K^2 for $\rho_c(T)$ and $A_{ab} = 1.5 \times 10^{-3}$ $\mu\Omega$cm/K^2 for $\rho_{ab}(T)$ by fitting up to 220 K.

We now discuss some possible origins of the T^2 dependence of resistivity of $Ag_5Pb_2O_6$: One is the electron-electron scattering owing to strong correlation and another is the interaction between electrons and high-frequency optic phonons, which is suggested in

CP850, *Low Temperature Physics: 24th International Conference on Low Temperature Physics;*
edited by Y. Takano, S. P. Hershfield, S. O. Hill, P. J. Hirschfeld, and A. M. Goldman

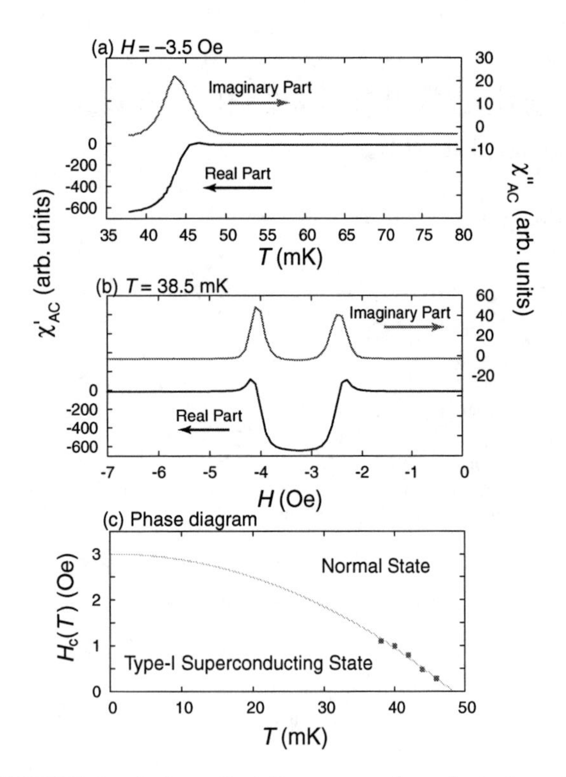

FIGURE 2. (color online) Temperature dependence (a) and field dependence (b) of the AC susceptibility of $Ag_5Pb_2O_6$. A new phase with a large diamagnetic signal is observed. The horizontal axis of (b) is the external applied field, however the exact zero is shifted due to the residual field in the instruments. (c) Estimated phase diagram of $Ag_5Pb_2O_6$. Here we defined $H_c(T) = |H_{p+} - H_{p-}|/2$, where H_{p+} and H_{p-} are the centers of the peak of χ'' in the field sweep data. The solid line shows the fitted parabolic law $H_c(T) = H_{c0}[1-(T/T_{c0})^2]$ with $H_{c0} = 3.0$ Oe and $T_{c0} = 48$ mK.

MgB_2 [4]. However, the specific heat shown in Fig. 1 is not compatible with these possibilities: The electronic specific heat coefficient $\gamma_e = 3.42$ mJ/K^{2}mol is too small to assume strong electron correlations. In addition, the specific heat at room temperature converges to the classical Dulong-Petit value, $C_{cl} = 3k_B r N_A = 324$ J/K mol, where $r = 13$ is the number of atoms in one formula unit. This fact indicates that any phonon mode with higher energy than room temperature does not exist.

We show in Fig. 2 the AC susceptibility χ_{AC} measured with a cluster of single crystals. A phase transition with a large diamagnetic signal is observed below 48 mK. This is probably the superconducting phase because of the following reason: The observed diamagnetic signal was as large as that of the superconducting transition of pure aluminum with similar volume, measured under the same condition. Since $Ag_5Pb_2O_6$ contains no magnetic ions, any magnetic phase transition with such a large signal should not occur.

Assuming the superconductivity of $Ag_5Pb_2O_6$, we deduce that $Ag_5Pb_2O_6$ is a type-I superconductor. One reason is that the estimated H_{c0} shown in Fig 2(c) is 3.0 Oe, and comparable to that of tungsten: $H_{c0} = 1.141$ Oe with $T_{c0} = 15.4$ mK [5]. Another is the positive peaks of χ' in Fig 2(b), indicating that $\partial M/\partial H$ is positive and that the magnetic process is reversible. In type-I superconductors, such peaks, which is called "the differential paramagnetic effect" (DPE) [6], are often observed in their intermediate states. On the other hand, DPE in type-II superconductors are relatively small ($|\chi'_{DPE}| \lesssim 0.01|\chi'_{dia}|$, where χ'_{dia} is the diamagnetic susceptibility in the full Meissner state)[7] because $\partial M/\partial H$ goes to zero near the transition.

CONCLUSION

In conclusion, we have measured the resistivity, the specific heat and the AC susceptibility of the silver layered oxide $Ag_5Pb_2O_6$. We found the T^2 dependence of resistivity up to room temperature. The origin of this behavior is still in a mystery. The new phase is observed and it is probably the type-I superconducting phase. However, much deeper studies are needed to confirm the superconductivity of $Ag_5Pb_2O_6$.

ACKNOWLEDGMENTS

One of the authors (SY) is supported by the Grant-in-Aid "Mochiduki Fund" from The Yukawa Memorial Foundation. This work has been supported by a Grant-in-Aid for the 21st Century COE "Center for Diversity and Universality in Physics" from Ministry of Education, Culture, Sports, Science and Technology (MEXT) of Japan.

REFERENCES

1. M. Jansen, M. Bortz, and K. Heidebrecht, *J. Less-Common Met.*, **161**, 17–24 (1990).
2. T. D. Brennan, and J. K. Burdett, *Inorg. Chem.*, **33**, 4794–4799 (1994).
3. S. Yonezawa, and Y. Maeno, *Phys. Rev. B*, **70**, 184523 (2004).
4. T. Masui, K. Yoshida, S. Lee, A. Yamamoto, and S. Tajima, *Phys. Rev. B*, **65**, 214513 (2002).
5. J. C. Wheatley, R. T. Johnson, and W. C. Black, *J. Low Temp. Phys.*, **1**, 641–667 (1969).
6. R. A. Hein, and R. L. Falge, Jr, *Phys. Rev.*, **123**, 407–415 (1961).
7. S. S. Banerjee, S. Saha, N. G. Patil, et al., *Physica C*, **308**, 25–32 (1998).

Anomalous Superconducting Properties in the Weak-Ferromagnetic Superconductors $RuSr_2GdCu_2O_8$ and $RuCa_2PrCu_2O_8$

B. C. Chang[*], C. Y. Yang[*], Y. Y. Hsu[†], B. N. Lin[¶], and H. C. Ku[*]

[*] *Department of Physics, National Tsing Hua University, Hsinchu, Taiwan 300, R.O.C.*
[†] *Institute of Physics, Academia Sinica, Taipei, Taiwan 115, R.O.C.*
[¶] *Electronics Research and Service Organization, ITRI, Hsinchu, Taiwan 310, R.O.C.*

Abstract. Anomalous superconducting properties are reported through transport, magnetic, and x-ray absorption near-edge spectroscopy (XANES) studies on the weak-ferromagnetic superconductors $RuSr_2GdCu_2O_8$ and $RuCa_2PrCu_2O_8$. For $RuSr_2GdCu_2O_8$ [T_m = 131 K, resistivity T_c(onset) = 56 K, T_c(zero) = T_c(dia) = 39 K], a spontaneous vortex state above T_0 = 30 K and a weak Meissner state below T_0 with a lower critical field $B_{c1}(0)$ = 12 G is observed due to the weak dipole field (~ 4 G) of the weak-ferromagnetic order. In $RuCa_2PrCu_2O_8$ (T_m = 47 K, possible T_c = 37 K), an anomalously large field-cooled diamagnetic shielding signal is observed below T_c. Both samples show mixed-valence $Ru^{4/5+}$ character from XANES data.

Keywords: High-T_c cuprates, magnetic superconductor, spontaneous vortex state.
PACS: 74.72.-h, 74.25.Ha

High-T_c superconductivity with anomalous magnetic properties was reported in the weak-ferromagnetic Ru-1212 system $RuSr_2RCu_2O_8$ (R = rare earths) [1–4]. Recently, possible superconductivity was observed in the Ca-substituted weak-ferromagnetic system $RuCa_2RCu_2O_8$ [5]. For the prototype compound $RuSr_2GdCu_2O_8$, the occurrence of superconductivity with T_c onset as high as 60 K is related to the quasi-two-dimensional CuO_5 bi-layers separated by a rare earth in the Ru-1212 structure. The weak-ferromagnetic order with an ordering temperature T_m above T_c is due to ordered Ru moments in the RuO_6 octahedron due to a strong Ru-$4d_{xy;yz;zx}$-O-$2p_{x;y;z}$ hybridization. Here we report the anomalous transport, magnetic, and x-ray absorption near-edge spectroscopy (XANES) results for two oxygen-annealed samples $RuSr_2GdCu_2O_8$ and $RuCa_2PrCu_2O_8$.

The ac electrical resistivity ρ(T) and volume magnetic susceptibility χ_V (T) are shown in Fig. 1 for $RuSr_2GdCu_2O_8$ under 1 G field-cooled (FC) and zero-field-cooled (ZFC) conditions. The resistivity shows a non-Fermi-liquid linear T-dependence behavior down to the weak-ferromagnetic ordering temperature T_m of 131 K, and changes to a T^2-dependence below T_m due to the magnetic order.

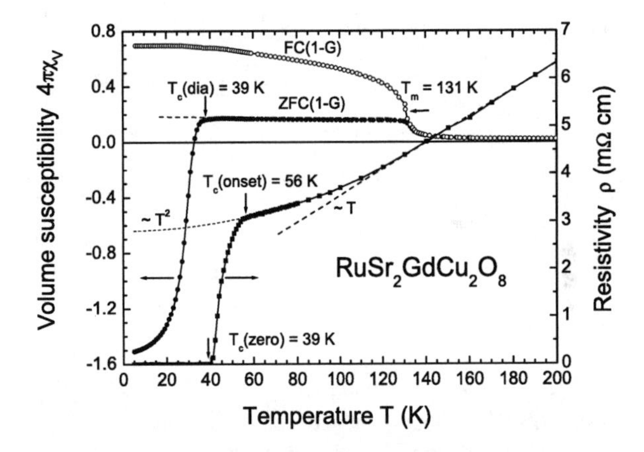

FIGURE 1. Electric resistivity ρ(T) and volume magnetic susceptibility χ_v(T) under 1 G field-cooled (FC) and zero-field-cooled (ZFC) conditions for $RuSr_2GdCu_2O_8$.

A superconducting onset temperature of 56 K with zero resistivity at 39 K is observed. The very broad transition is a common feature for all reported Ru-1212 resistivity data, and indicates that the superconducting Josephson coupling along the tetragonal c-axis, and Cu-O layers, may be partially blocked by the dipole field B_{dip} of the weak ferromagnetically ordered Ru moments. The diamagnetic T_c of 39 K was also observed in 1 G ZFC

CP850, *Low Temperature Physics: 24th International Conference on Low Temperature Physics;*
edited by Y. Takano, S. P. Hershfield, S. O. Hill, P. J. Hirschfeld, and A. M. Goldman

susceptibility measurement. A full Meissner shielding signal $4\pi\chi_V \sim 1.5$ was recorded at 5 K. From low field superconducting hysteresis measurements, the lower critical field B_{c1} observed from the maximum of initial magnetization curve can be fit to a simple parabolic equation $B_{c1}(T) = B_{c1}(0)[1 - (T/T_0)^2]$ with $B_{c1}(0) = 12$ G and $T_0 = 30$ K [6]. This indicates the existence of a spontaneous vortex state in zero applied magnetic field between 30 K and 56 K due to the weak dipole field $B_{dip} \sim 4$ G of the weak-ferromagnetic order, where 39 K is the melting temperature from the spontaneous vortex lattice/glass to the spontaneous vortex liquid state. No diamagnetic field expulsion signal can be detected below T_c in the 1 G field-cooled (FC) condition, due to the strong flux pinning where the superconductivity coexists with the weak-ferromagnetic order.

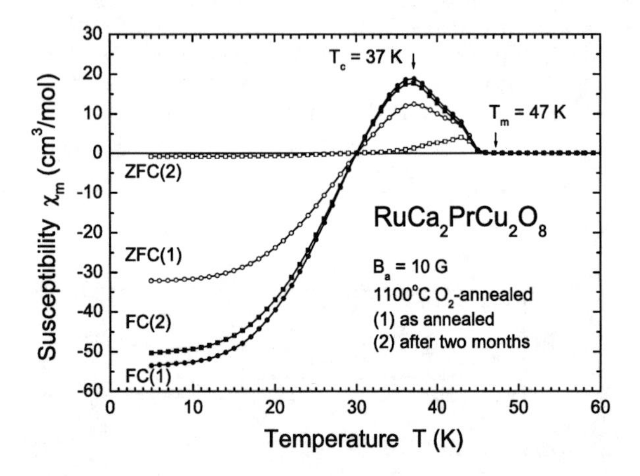

FIGURE 2. Temperature dependence of the molar susceptibility $\chi_m(T)$ of oxygen annealed $RuCa_2PrCu_2O_8$ under 10 G FC and ZFC conditions for (1) an as-annealed sample and (2) a sample after 2 months.

The temperature dependence of the molar susceptibility $\chi_m(T)$ of $RuCa_2PrCu_2O_8$ under 10 G FC and ZFC conditions for (1) an as-annealed sample as well as for (2) a sample after 2 months, are shown in Fig. 2. Below 47 K, a sharply increasing positive susceptibility indicates the onset of the weak-ferromagnetic order with an ordering temperatures $T_m \sim 47$ K. At lower temperature of around 37–39 K, the susceptibility starts to decrease and eventually becomes diamagnetic below 30 K, which is a sign of possible superconductivity. However, the resistivity is not zero in this sample. The Meissner shielding signal is smaller than the diamagnetic field expulsion signal. This anomalous behavior is probably due to the complex coexistence and interplay between the superconductivity and the weak-ferromagnetic order. For a sample measured after two months, a much smaller shielding signal is observed due to sample degradation or oxygen concentration change.

The normalized Ru L_3-edge (2p-4d dipole transition) XANES at room temperature for $RuSr_2GdCu_2O_8$ and $RuCa_2PrCu_2O_8$ as well as for two standards, RuO_2 (Ru^{4+}) and $RuSr_2GdO_6$ ($\sim Ru^{5+}$), are shown in Fig. 3. The almost identical threshold energy E_o for the two Ru-1212 samples, as compared with $RuSr_2GdO_6$ standard, indicate that Ru valence is close to 5+ with a similar RuO_6 environment. Peak A is the transition form $2p_{3/2}$ to 4d-t_{2g} and peak B is the transition form $2p_{3/2}$ to 4d-e_g. The energy separation $\Delta E = 2.6$ eV for $RuSr_2GdCu_2O_8$ and $RuCa_2PrCu_2O_8$ is larger than 2.0 eV for RuO_2 and smaller than 3.2 eV for $RuSr_2GdO_6$, which indicates a mixed-valence $Ru^{4/5+}$ character. The Ru self-doping with anisotropic hybridization may drive the resulting Ru^{4+}/Ru^{5+} mixed-valent system metallic and ferromagnetic via a double exchange interaction.

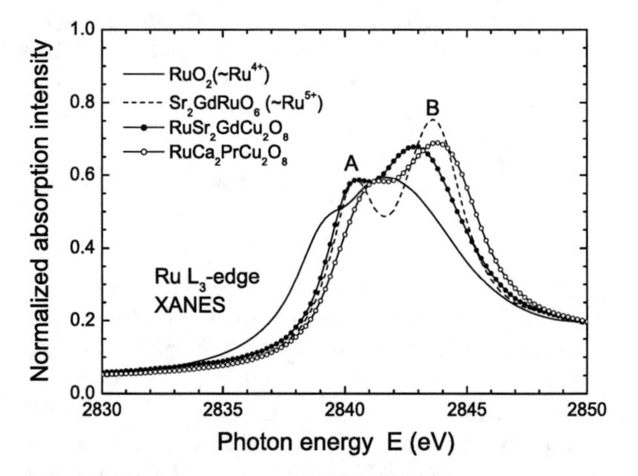

FIGURE 3. Ru L_3-edge XANES for $RuSr_2GdCu_2O_8$ and $RuCa_2PrCu_2O_8$.

ACKNOWLEDGMENTS

This work was supported by NSC of R.O.C. under contract No. NSC93-2112-M007-011.

REFERENCES

1. L. Bauernfeind, W. Widder, and H. F. Braun, *Physica C* **254**, 151 (1995).
2. C. Bernhard, J. L. Tallon, Ch. Niedermayer, Th. Blasius, et al., *Phys. Rev. B* **59**, 14099 (1999).
3. R. L. Meng, B. Lorenz, Y. S. Wang, J. Cmaidalka, Y. Y. Xue, and C. W. Chu, *Physica C* **353**, 195 (2001).
4. T. Nachtrab, D. Koelle, R. Kleiner, C. Bernhard, and C. T. Lin, *Phys. Rev. Lett.* **92**, 117001 (2004).
5. H. C. Ku, C. Y. Yang, B. C. Chang, B. N. Lin, Y. Y. Hsu, and M. F. Tai, *J. Appl. Phys.* **97**, 10B110 (2005).
6. C. Y. Yang, B. C. Chang, H. C. Ku, and Y. Y. Hsu, *Phys. Rev. B,* accepted (2005).

Pressure Study of Pure and Rh-doped $RuSr_2GdCu_2O_8$ Magnetic Superconductors

M. S. Torikachvili,[a] M. Steiger,[a] L. Harding,[a] D. Bird,[a] N. R. Dilley,[b] S. Gomez,[b] J. R. O'Brien,[b] and R. F. Jardim[c]

[a] *Department of Physics, San Diego State University, San Diego, CA 92182-1233, USA*
[b] *Quantum Design, 6325 Lusk Boulevard., San Diego, CA 92121, USA*
[c] *Instituto de Física, Universidade de São Paulo, SP, 05315-970, Brazil*

Abstract. We carried out an investigation of the effect of quasi-hydrostatic pressures up to 1.2 GPa in pure and Rh-doped $RuSr_2GdCu_2O_8$ compounds, by means of measurements of electrical resistivity in magnetic fields up to 9 T. The onset temperatures for superconductivity, and for magnetic order in the undoped $RuSr_2GdCu_2O_8$ compound are $T_c \approx 50$ K, and $T_m \approx 133$ K, respectively. The partial substitution of Rh for Ru lowers both these transitions temperatures. However, the effect of pressure for all compositions studied, up to 10% substitution, was to increase both T_c, and T_m. The effect of pressure on the upper critical magnetic field of the pure, and Rh-doped compounds is discussed.

Keywords: pressure; magnetic superconductors; critical field.
PACS: 74.25.Ha; 74.25.Op; 74.62.-c; 74.62.Fj.

INTRODUCTION

The coexistence of superconductivity (SC) with weak ferromagnetism (FM) in the rutheno-cuprates with general composition $RuSr_2LnCu_2O_8$ (Ln = Eu, Sm, and Gd) is quite remarkable.[1] For example, SC with onset at $T_c \approx 50$ K coexists with magnetic order ($T_m \approx 133$ K) in $RuSr_2GdCu_2O_8$, and the onset of SC doesn't affect the ordered state noticeably.[2]

The resistivity (ρ) transition to the SC state spans a quite broad T-range of 15 K or higher in these materials. These broad transitions have been attributed to cation disorder, self-induced vortices,[3] and granularity.[4] Lorentz et al. extracted the onset of the intra- and inter-granular SC, as well as of FM from the $d\rho/dT$ data, and they determined that $T_{c,inter}$, $T_{c,intra}$, and T_m all increased with pressure (P) up to about 2 GPa.[4]

In order to probe the SC, magnetic, and granular behavior of these materials, we studied the effect of pressure and magnetic field (H) in pure and Rh-doped $RuSr_2GdCu_2O_8$. The polycrystalline specimens for this study were synthesized by reacting CuO with $Sr_2(Ru,Rh)GdO_6$ precursors.[5] These measurements were performed in a Quantum Design 9-T measurement station (PPMS-9), using a 1.5 GPa self-clamping quasi-hydrostatic cell from EasyLab.

RESULTS AND DISCUSSION

The partial substitution of Rh for Ru up to about 25% can be accomplished while retaining phase purity. The substitution of Rh reduces T_c and T_m, while driving the SC behavior towards granularity. As shown in Fig. 1a, the SC transition evolves from a linear drop in ρ vs T for the pure compound to a 2-step drop in the 10% Rh-substituted material. The first and second drops represent the onset of intra- and inter-granular SC, respectively. The effect of the high pressure in the Rh-doped material is to raise the values of $T_{c,inter}$, and $T_{c,intra}$, as shown in Fig. 1b. The value of T_m = 123.7K is also raised with pressure at the rate of about 6.1 K/GPa (data not shown). The ρ vs T curves for the best undoped materials did not show noticeable anomalies at T_m, or $T_{c,inter}$, and a reliable determination of their values could not be made from these curves.

In order to determine the effect of the magnetic field on the SC state of the pressurized materials, we carried out measurements of magnetoresistivity in fields up to 9 T. The values of ρ/ρ_{300K} (T) for the unpressured, and pressured materials (with P_{max} = 1.1

CP850, *Low Temperature Physics: 24th International Conference on Low Temperature Physics;*
edited by Y. Takano, S. P. Hershfield, S. O. Hill, P. J. Hirschfeld, and A. M. Goldman

GPa, and 1.2 GPa, respectively) are shown in Fig. 2. Since the behavior of $\rho(T,H)$ does not depend strongly on P, only the isofield data for the samples pressurized with P_{max} are shown in Fig. 2. The magnetic field induces noticeable changes in the shape of $\rho(T)$. As the H increases, the SC transition becomes much broader near the onset of SC, and it sharpens up again near the zero-resistance state.

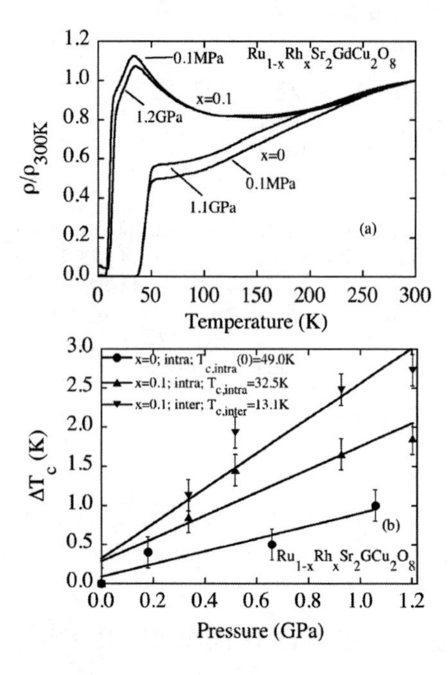

FIGURE 1. (a) Normalized electrical resistance ρ/ρ_{300K} versus T, in pure and Rh-doped $RuSr_2GdCu_2O_8$, in ambient P, and P_{max}; (b) pressure dependence of $T_{c,inter}$, $T_{c,intra}$, and T_m. The solid lines in (b) are guides to the eye.

In light of the broad and step-like transitions to the SC state, it is not trivial to determine the upper critical field H_{c2} vs T phase diagram. However, assuming the same T onset for SC in all fields, and taking the midpoint of the SC transitions as T_c, the upper limit for $H_{c2}(T)$ can be determined, as shown in Fig.3.[6] The magnitude of dH_{c2}/dT increases with H, reflecting the narrowing of the SC transition in higher fields. The positive curvature of $H_{c2}(T)$ is reminiscent of other high-T_c cuprates. The extrapolated value of $H_{c2}(T=0)$ can be estimated by using the WHH expression $H_{c2}(0) = -0.7(dH_{c2}/dT)T_c$. Assuming that for the pure material at ambient P $T_c = 44.3$ K, and using the $dH_{c2}/dT = -0.75$ T/K value extracted from the high H portion of Fig. 3, the yielded value for $H_{c2}(0)$ is ≈ 23.3 T. This value clearly increases with pressure.

In summary, our magnetoresistance measurements in $Ru_{1-x}Rh_xSr_2GdCu_2O_8$ under pressure show that 1) $T_{c,inter}$, $T_{c,intra}$, and T_m all increase with P; and 2) the $H_{c2}(T)$ curves are shifted to higher T with P.

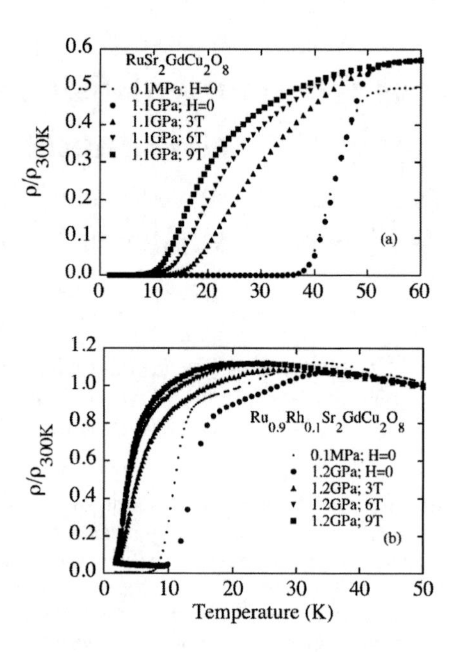

FIGURE 2. Normalized electrical resistance ρ/ρ_{300K} versus T, in (a) $RuSr_2GdCu_2O_8$; and (b) $Ru_{0.9}Rh_{0.1}Sr_2GdCu_2O_8$; for H = 0, 3, 6, and 9 T, at P_{max} = 1.1, and 1.2 GPa, respectively. ρ/ρ_{300K} (T, H=0, P=1atm) curves are shown for reference.

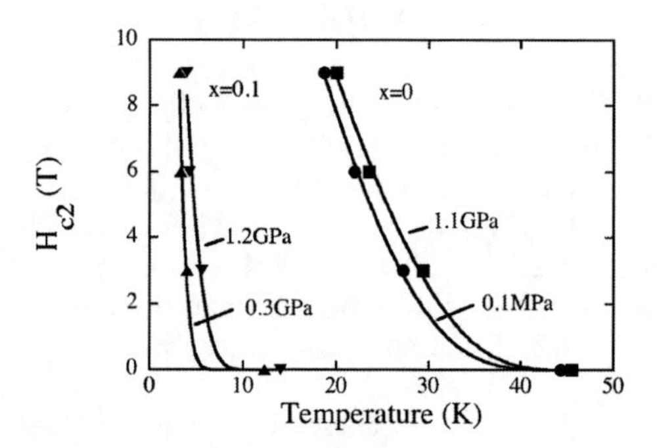

FIGURE 3. Curves of H_{c2} vs T for $Ru_{1-x}Rh_xSr_2GdCu_2O_8$. The solid lines correspond to fits to the phenomenological expression $H_{c2}(T) = H_{c2}(0)[1-(T/T_c)^2]^\alpha$.

ACKNOWLEDGMENTS

The support from NSF Grant No. DMR-0306165 (MST, MS, LH, and DB), Fapesp-Brazil Grant No. 99/10798-0 (RFJ), and CNPq-Brazil Grant No. 303272/04-0 (RFJ) are gratefully acknowledged.

REFERENCES

1. L. Bauernfeind et al., *J. Low Temp. Phys.* **105**, 1605 (1996).
2. C. Bernhard et al., *Phys. Rev. B* **59**, 14099-107 (1999).
3. Y. Tokunaga et al., *Phys. Rev. Lett.* **86**, 5767 (2001).
4. B. Lorenz et al., *Phys. C* **383**, 337-42 (2003).
5. T. P. Papageorgiou et al., *Phys. C* **377**, 383 (2002).
6. M. T. Escote et al., *Phys. Rev. B* **66**, 14503 (2002).

Transport, Heat Capacity and Magnetization of the Magnetic Superconductor $RuSr_2(Gd_{1.5}Ce_{0.5})Cu_2O_{10-\delta}$

D. G. Naugle[1], K. D. D. Rathnayaka[1], V. B. Krasovitsky[2], B. I. Belevtsev[2], M. P. Anatska[1], G. Agnolet[1], and I. Felner[3]

[1] *Department of Physics, Texas A&M University, College Station, TX 77843, USA*
[2] *B.Verkin Institute for Low Temperature Physics and Engineering, Kharkov, 61103, Ukraine*
[3] *Racah Institute of Physics, The Hebrew University, Jerusalem, 91904, Israel*

Abstract. Samples of $RuSr_2(Gd_{1.5}Ce_{0.5})Cu_2O_{10-\delta}$ have been studied. Resistivity, thermopower, heat capacity and magnetization were investigated in the temperature range 1.8-300 K under a magnetic field up to 8 T. The resistive transitions to the superconducting state are found to be determined by a granular structure, so that the intragranular, T_{c0}, and intergranular, T_{cg}, transition temperatures were found. Heat capacity, *C(T)*, shows a jump at the superconducting transition, corresponding to intragranular superconductivity (with $T_{c0} \approx 37.5$ K). A Schottky-like anomaly is found in *C(T)* below 20 K. The possible reasons for this anomaly (associated with rare-earth Gd ions) are considered.

Keywords: ruthenocuprates, magnetic superconductors
PACS: 74.72.Jt; 74.81.Bd; 74.25.Bt; 75.40.Cx

The ruthenocuprate $RuSr_2(Gd_{1.5}Ce_{0.5})Cu_2O_{10-\delta}$ belongs to the known family of magnetic superconductors. Superconductivity is associated with CuO_2 planes, while magnetic order is thought to be connected with RuO_2 planes. The exact nature of the magnetic order in this compound is still unknown, but it is conjectured that below 80-100 K weak-ferromagnetic order dominates. Superconductivity in this family of compounds is apparent below 50 K, where both superconducting and magnetic order coexist.

In this report, the transport, magnetic and thermal properties of samples of $RuSr_2(Gd_{1.5}Ce_{0.5})Cu_2O_{10-\delta}$ as prepared (by a solid-state reaction method) and annealed (12 hours at 845°C) in pure oxygen at 30, 62, 78 atm, will be presented. The measurements were made with a Quantum Design PPMS and a home-made thermoelectric power measuring system.

The samples behave as inhomogeneous (granular) superconductors. This manifests itself to the greatest extent in resistive properties, as can be seen, for example, in Fig. 1 for an annealed (62 atm) sample. The rather broad and shouldered *R(T)* curves in the region of the superconducting transition are indicative of the inhomogeneity effects. The most obvious inhomogeneity source is the granular structure (with a grain size of a few μm), determined by the polycrystalline structure. Non-homogeneous oxygen distribution can cause oxygen depletion of the grain-boundary regions and, hence, weak electrical connectivity between the grains, as is often the case in cuprates. Above the superconducting transition, the rather high resistivity (10^{-2} Ω cm) and the weak increase in resistance with decreasing temperature support this suggestion. The onset temperature of superconductivity is about 49 K with a zero-resistance temperature about 25 K. Derivatives *dR(T)/dT* reveal two peaks, the positions of which can be attributed to intragranular and intergranular superconducting transitions at temperatures T_{c0} and T_{cg}, respectively. The intergranular superconductivity is determined by Josephson coupling between the grains. T_{c0} and T_{cg} are equal to 37.5 K and 32.8 K, respectively, in zero field. At the maximum field 8 T used in this study, they reduce to 34.7 K and 12.4 K, respectively. Thus, the magnetic field has a weak influence on the intragranular transition temperature T_{c0}. The intergranular T_{cg} is far more sensitive to magnetic field, with the main variations occurring in the low field region $H < 0.5$ T.

The *R(T)* behavior of the as-prepared sample, which is expected to be the most depleted in oxygen, substantiates our assumptions. The *R(T)* curve of this sample, taken at $H = 0$, indicates that T_{c0} and T_{cg} are equal to 37 K and 18.5 K, respectively. Consequently,

CP850, *Low Temperature Physics: 24th International Conference on Low Temperature Physics;*
edited by Y. Takano, S. P. Hershfield, S. O. Hill, P. J. Hirschfeld, and A. M. Goldman

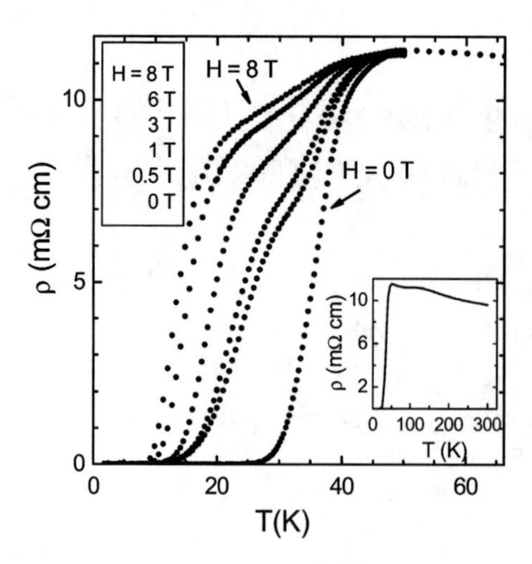

FIGURE 1. Resistivity $\rho(T)$ at different magnetic fields. The inset shows $\rho(T)$ in the range 2-300 K at zero field

high-pressure oxygen annealing affects primarily the intergranular properties. The intragranular properties are less affected.

The thermoelectric power (S) is sensitive to oxygen annealing as well. The magnitude of S is far (1.5 times) larger in the as prepared sample than that in the sample annealed at 62 atm of O_2. The temperatures of maximum slope in the $S(T)$ curves at superconducting transitions (which can be taken as T_c values) are 39 K and 21 K in 62-atm and as prepared samples, respectively.

In contrast to transport properties, the heat capacity was found to be nearly insensitive to the granular structure and oxygen annealing. The specific heat data are found to be practically the same for all 4 samples studied. The low temperature behavior of the non-phonon part of the heat capacity, $\Delta C(T) \approx C_{tot} - C_{ph}$, is shown in Fig. 2 for the as prepared sample (C_{tot} is as-measured data, $C_{ph}(T)$ is the lattice heat capacity). The ΔC is determined by electron and magnetic contributions. The extraction procedure for $\Delta C(T)$ will be detailed elsewhere. The main features in $\Delta C(T)$ are (i) the jump at the superconducting transition, and (ii) the upturn below 20 K (Schottky-like anomaly). No upturn in $C(T)$ for a similar sample with Gd replaced by Eu, which has no magnetic moment, is observed. A jump at T_c in the heat capacity of a ruthenocuprate with a similar chemical composition [$RuSr_2(Gd_{1.4}Ce_{0.6})Cu_2O_{10-\delta}$] was reported earlier [1], but no Schottky-like anomaly was previously reported.

In heat capacity studies, the temperature of the superconducting transition is usually associated with onset of the $C(T)$ jump (T_c^{onset}) or with the point of maximum slope of $C(T)$ in this region. In zero field

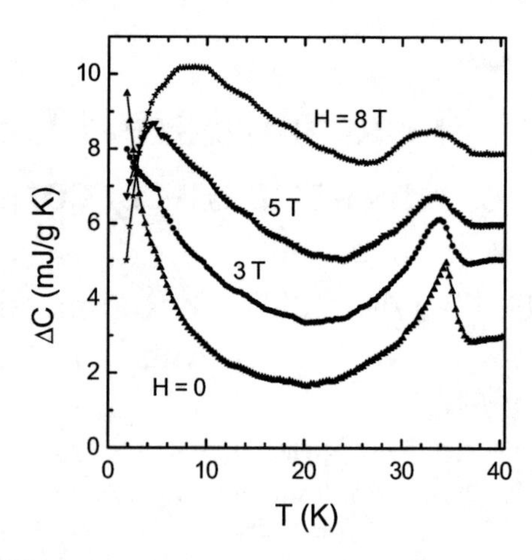

FIGURE 2. Temperature curves of the non-phonon part of the heat capacity, measured in different magnetic fields.

$T_c^{onset} \approx 37$ K which is very close to the intragranular T_{co} determined from the $R(T)$ curves. An external field up to $H = 8$ T hardly produces shifts in T_c^{onset} (Fig. 2). This correlates well with the very weak shift in T_{c0} in a magnetic field, and is evidence of enormous upper critical fields in ruthenocuprates.

The low temperature Schottky-like anomaly can be attributed to splitting of the ground term $^8S_{7/2}$ of paramagnetic Gd^{3+} ions. According to Kramers' theorem, the degenerate ground term should be split into 4 doublets in tetragonal symmetry. The sources of splitting can be crystal-electric-field effects, and the internal and external magnetic fields. The crystal-field effect can be ignored, in the first approximation, since Gd^{3+} has zero orbital angular momentum. Other sources of splitting cannot be, however, excluded. In particular, internal molecular fields can arise in the ruthenocuprate from both the Gd and Ru sublattices [2] and can coexist with superconductivity [3]. Even if a direct Gd-Gd exchange interaction is unlikely, these ions can be magnetically polarized by the *4d-4f* interaction. Magnetic data obtained suggest that Gd^{3+} ions do not behave as a simple system of non-interacting paramagnetic particles. More analytical work is necessary, however, to clarify sources of the heat capacity anomaly found.

This research was supported by the Robert A. Welch Foundation (A-0514, A-1386), NSF (DMR-0103455, DMR-0315478, DMR-0422949) and CRDF (UPI-2566-KH-03).

REFERENCES

1. Chen X. H. et. al, *Phys. Rev. B* **63**, 064506 (2001).
2. Tallon J. L. et al., *Phys. Rev. B* **61**, R6471 (2000).
3. Leviev G. I. et al., Preprint cond-mat/0311245.

Fermi Surface Topology of Borocarbide Superconductor YNi_2B_2C:Angle-resolved Photoemission Spectroscopy

T. Baba[a], T. Yokoya[b], S. Tsuda[a], S. Shin[a,c], T. Watanabe[a], M. Nohara[d], H. Takagi[d] and T. Oguchi[e]

[a]*Institute for Solid State Physics (ISSP), University of Tokyo, Kashiwa, Chiba 277-8581, Japan*
[b] *The Graduate School of Natural Science and Technology, Okayama University, Okayama, Okayama 700-8530, Japan*
[c]*The Institute of Physical and Chemical Research (RIKEN),Sayo-gun, Hyogo 679-5148, Japan*
[d]*Department of Advanced Materials Science, University of Tokyo, Kashiwa 277-8581, Japan*
[e]*Department of Quantum Matter, Greduate School of Advanced Sciences of Matter (ADSM), Hiroshima University, Kagamiyama, Higashi-Hiroshima 739-8526, Japan*

Abstract. We have performed high-resolution angle-resolved photoemission spectroscopy (ARPES) of the borocarbide superconductor YNi_2B_2C to study the Fermi surface (FS) topology of the anisotropic *s*-wave superconductor. We successfully observed experimental valence band dispersions and Fermi surfaces (FSs), and indicated two nested parts connected by the known nesting vector Q~(0.55,0,0).

Keywords: anistropic *s*-wave superconductor, angle-resolved photoemission spectroscopy, Fermi surface, nesting
PACS: 74.25.Jb, 71.18.+y, 74.70.Dd, 79.60.-i

INTRODUCTION

In borocarbide superconductors (Y,Lu)Ni_2B_2C, an extremely large superconducting (SC) gap anisotropy is observed in various experiments [1-6], despite the fact that isotope effect [7] and band calculations [8] indicate phonon-mediated superconductivity. But the type of the SC gap nodal structure seems to be controversial. This may be partly because these studies discuss SC gap anisotropy assuming an isotropic FS , which may be oversimplified compared to complicated anisotropic three dimensional electronic structure of (Y,Lu)Ni_2B_2C [9]. What is more, Kohn anomalies in phonon dispersion of (Y,Lu)Ni_2B_2C have been observed at a wave vector Q~(0.55,0,0), indicating the presence of FS nesting [10]. So, it is very useful to investigate the FS topology for not only clarifying the origin of large SC gap anisotropy but also understanding the general physics of superconductivity which coexists with FS nesting instability.

EXPERIMENT

Single crystals of YNi_2B_2C (T_c=15.1K) were grown by a floating zone method [11]. All the ARPES data presented here have been measured with a high-resolution hemispherical analyzer using a monochromatic HeIIα resonance line (40.814eV). Energy resolution was set to ~60meV and the measurements were performed in normal state (20K). Single crystal clean surfaces of YNi_2B_2C (001) were prepared *in situ* by repeating Ar-ion bombardment and flash heating.

RESULTS AND DISCUSSION

Band dispersion

In Fig 1, we show an ARPES spectral intensity map of YNi_2B_2C (001) measured along Γ (Z) – X(111) high symmetry directions as functions of binding energy and momentum *k*. We observed several dispersive bands within 6eV of the Fermi level (E_F). Since the Ni3*d* states make a large contribution to the HeIIα spectrum relative to the B and C 2*sp* due to the larger coross-section, the energy bands from E_F to ~4eV should have a mainly Ni3*d* character. On the other hand, weak dispersive bands with lower

CP850, *Low Temperature Physics: 24th International Conference on Low Temperature Physics;*
edited by Y. Takano, S. P. Hershfield, S. O. Hill, P. J. Hirschfeld, and A. M. Goldman

intensities at ~6eV have a B and/or C 2*sp* character. Near E_F, we found seven high intensity regions which move toward E_F as indicated by the arrows, indicating existence of FSs.

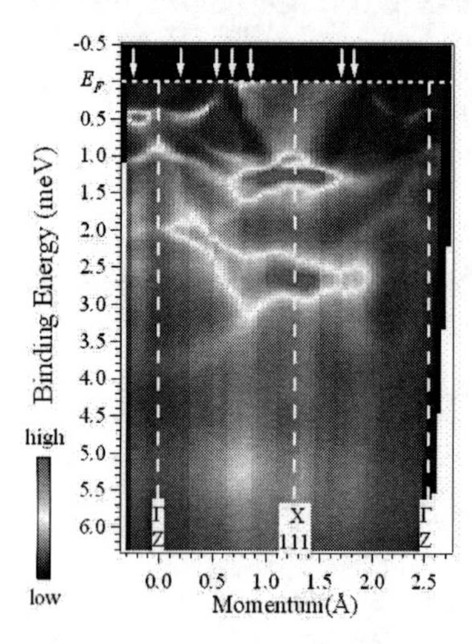

FIGURE 1. ARPES spectral intensity map of YNi_2B_2C (001) as functions of binding energy and *k*.

Fermi surface

To check these FSs, we show the experimental FSs of YNi_2B_2C (001) in Fig.2, which was obtained by integrating the ARPES spectral intensity within $E_F \pm$ 20meV, over the two dimensional Brillouin zone (BZ). Band structure calculations indicate that there are three electron-like FS sheets from 19th (or 27th), 18th (or 26th), 17th (or 25th) bands [9]. In our experimental results, at first, around the Γ (Z) point (0.0Å^{-1}), we found two FSs which have circular and star like FS sheets. These correspond to the 18th and 17th FS sheets near the k_z=0 plane, respectively. Second, two FSs around the X (111), which have square-like (outer) and ellipsoidal-like (inner) FS sheets, correspond to 17th and 18th FS sheets near the k_z=1/4 plane, respectively. In the next BZ around the Γ (Z) point (2.54Å^{-1}), a banana-like FS sheet which correspond to the 17th FS sheet, near the k_z=1/2 plane, is observed.

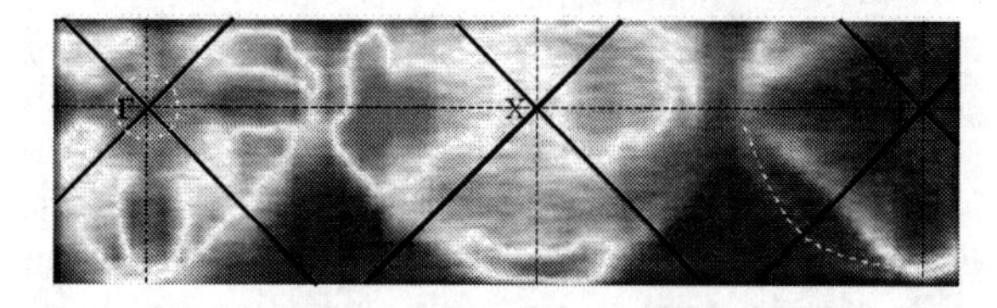

FIGURE 2. Experimental FS of YNi_2B_2C (001) measured with a HeIIα resonance line at 20K.

Nesting properties

To investigate the nesting properties, we show the symmetrized FS about the first BZ and the second BZ in Fig. 3 (a) and (b), respectively. In the first BZ, we found a nested part with a nesting vector Q~(0.55,0,0) on 17th FS indicated by black arrow in Fig. 3(a). The size of this nesting vector is in good agreement with the Kohn anomalies in phonon dispersion of (Y,Lu)Ni_2B_2C [10]. The position of nested part on FS is also the same as the band structure calculations [12] and as experimental results of two-dimensional angular correlation of electron-positron annihilation radiation in $LuNi_2B_2C$ [13]. Additionally, we also found the same value of nesting vector on 17th FS in the second BZ, as shown in Fig. 3(b) by a white arrow. To examine how the nested parts affect the SC gap, it is desirable to measure the SC gap at k_F points which are connected by the nesting vector. We are performing experiments to measure the *k*-dependent SC gap, and will discuss them in future publications.

(a) 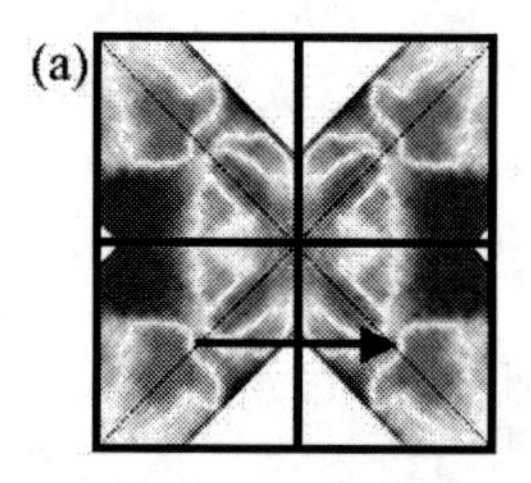 (b) 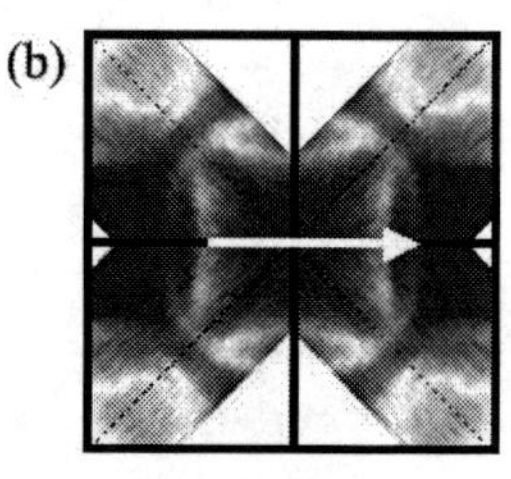

FIGURE 3. Symmetrized FSs about (a) 1st BZ and (b) 2nd BZ. Arrows show the nesting vector Q~(0.55,0,0).

ACKNOWLEDGMENTS

This work was supported by Grant-in-aid from the Ministry of Education, Science, and Culture of Japan. T. B. thanks center of excellence for applied physics on strong correlation for financial support.

REFERENCES

1. T. Kohara *et al., Phys. Rev. B* **51**, 3985 (1995).
2. M. Nohara *et al., J. Phys. Soc. Jpn.* **66**, 1888 (1997).
3. T. Yokoya *et al., Phys. Rev. Lett.* **85**, 4952 (2000).
4. K. Izawa *et al., Phys. Rev. Lett.* **89**,137006(2002).
5. T. Park *et al.,. Phys. Rev. Lett.* **90**, 177001(2003).
6. T. Watanabe *et al., Phys. Rev. Lett.* **92**,147002 (2004).
7. D. D. Lawrie *et al., Physica C* **245**, 159 (1995).
8. L. F. Mattheiss, *Phys. Rev. B* **49**, 13279 (1994).
9. K. Yamauchi *et al., Physca C* **412-414**, 225 (2004).
10. J. Zarestky *et al., Phys. Rev. B* **60**, 11932 (1999).
11. H. Takeya *et al.,Physica C* **256,** 220 (1996).
12. J. Y. Rhee *et al., Phys. Rev. B* **51**, 15585 (1995).
13. S.B. Dugdale *et al., Phys. Rev. Lett.* **83**, 4824 (1999).

Specific Heat of YNi_2B_2C Analyzed Using Two-gap Model

C. L. Huang[1], C. P. Sun[1], J. –Y. Lin[2], H. J. Kim[3], E. M. Choi[3], S. I. Lee[3], and H. D. Yang[1*]

[1] Department of physics, National Sun Yat-Sen University, Kaohsiung 804, Taiwan
[2] Institutre of Physics, National Chiao-Tung University, Hsinchu 300, Taiwan
[3] National Creative Research Initiative Center for Superconductivity and Department of physics, Pohang University of Science and Technology, Pohang 790-784, Republic of Korea

Abstract. New specific-heat data of YNi_2B_2C single crystal were measured and analyzed using two-gap model. The obtained T_c=13.77K, γ_n=19.74 mJ/mol K^2, Θ_D=533K, $\Delta C/\gamma_n T_c$=1.57 are consistent with those of previously reported indicating a good quality of crystal. In the superconducting state, the C_e can not be fitted to a simple isotropic-gap BCS relation. Instead, it can be well described by a two-gap model. The large gap Δ_L=2.67meV and small gap Δ_S=1.19meV can be obtained with 71% and 29% weighting for the density of states at Fermi surface, respectively. These results are compared with those taken by point-contact spectroscopy recently.

Keywords: Superconductivity, YNi_2B_2C, two-gap model, specific-heat.
PACS: 74.20.Rp, 74.25.Bt, 74.25.Jb, 74.70.Dd

INTRODUCTION

The RNi_2B_2C (R=Dy, Ho, Er, Tm, Lu and Y) have been among the most interesting superconductors during the last decade. There are a lot of varieties of physical properties, such as relatively high T_c ~ 15K, and coexistence of superconductivity and long-range magnetic order. At the earlier stage, YNi_2B_2C was considered to be a conventional isotropic *s*-wave superconductor [1]. However, recent angular dependent thermal conductivity, specific heat, and point contact spectroscopy seemly provide the point nodes in the superconducting gap function [2]. In this article, we present new specific-heat data and analyze them with BCS-like isotropic *s*-wave and two-gap models. We find that the two-gap model, which has been successfully applied to MgB_2 [3], to fit the data is better than the isotropic s-wave.

EXPERIMENTS

The YNi_2B_2C single crystal was grown by the high temperature flux method [4], and the low temperature specific heat *C(T,H)* has been measured in the temperature range 0.6K~20K under different magnetic fields (*H* = 0-8T) using heat-pulse thermal relaxation calorimeter [5]. The data were taken with the field being perpendicular to the c-axis of the crystal.

RESULTS AND DISCUSSION

Figure 1 shows the specific heat *C* (*T*, *H*) without and with the applied magnetic field (*H*=8T). T_c = 13.77K determined from the present measurement is consistent with previous works. The normal-state specific heat at the absence of magnetic field can be simply described by

$$C_n(T) = \gamma_n T + C_{lattice}(T) \tag{1}$$

where $\gamma_n T$ is the electronic term due to free charge carriers and $C_{lattice}(T) = \beta T^3 + \alpha T^5$ represents the phonon contribution which is assumed to be independent of the magnetic field. However, in order to achieve optimal fitting, the normal-state data with *H*=8 T at temperatures ranging from 3.5 K to 20 K are used but with adding a T^{-2} hyperfine contribution term to Eq. (1) which is thought to be due to a very low concentration of paramagnetic centers. It is found that γ_n = 19.74 ± 0.27 (mJ/mol K^2), β = 0.077 ± 0.003 (mJ/mol K^4) giving rise to Θ_D= 533 ± 7 K and α = 0.00018 ± 0.00001 (mJ/mol K^6) yielding the best fitting to experimental data. They are consistent with

CP850, *Low Temperature Physics: 24th International Conference on Low Temperature Physics;*
edited by Y. Takano, S. P. Hershfield, S. O. Hill, P. J. Hirschfeld, and A. M. Goldman

those of which have been measured by other groups. In fact, these parameters are justified to make the entropy balance for the second order superconducting normal phase transition.

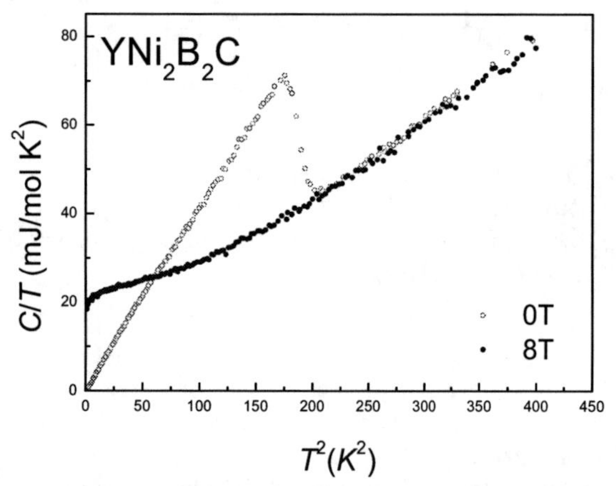

FIGURE 1. Temperature dependence of specific heat (C) data with (H=8T) and without the applied field for YNi_2B_2C.

The electronic specific heat in the superconducting state is given by $C_e(T)= C(T)-C_{lattice}(T)$, and we use the conventional BCS isotropic *s*-wave (single gap) and two-gap models to fit the data. Fitting results are shown in Fig. 2.

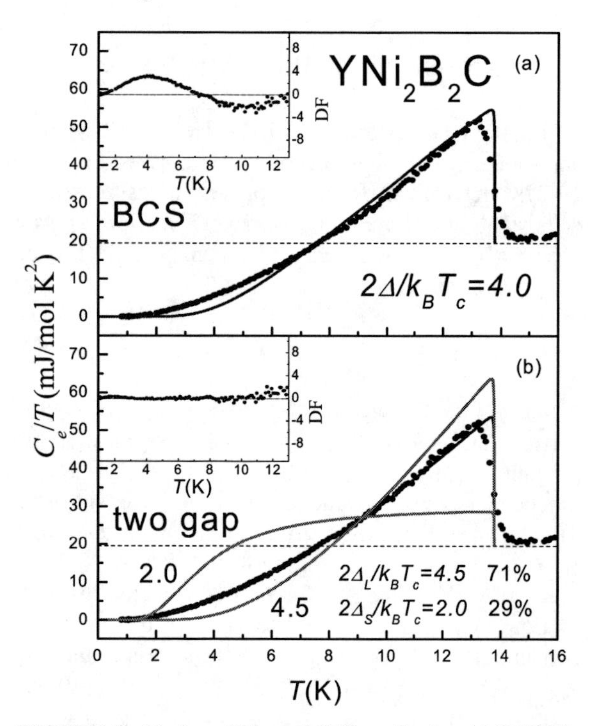

FIGURE 2. Various fitting of C_e/T vs. T using (a) BCS-like isotropic *s*-wave and (b) two-gap models. The goodness of the fit is shown in the inset, where the DF represents the deviation between the data and fitting line.

Apparently, one single gap using $2\Delta/k_BT_c = 4.0$ (Fig.2 (a)) can not describe the data. Instead, we can get an excellent fitting (Fig. 2 (b)) using two-gap model which has been successfully applied to MgB_2 [3]. Consequently, larger gap Δ_L=2.67meV and small gap Δ_S=1.19meV can be obtained with 71% and 29% weighting for density of states at the Fermi surface, respectively. The derived values of two gaps are fairly consistent with recent results from point-contact spectroscopy [6], where anisotropic gap (or a multiband) values with Δ_{max}=2.4 meV and Δ_{min} =1.5 meV were revealed.

Beyond these works, we also find other evidences supporting two gap model, such as, the similar positive curvature seen in $H_{c2}(T)$ vs. T at low fields [7] and $\gamma(H) \sim H^{\alpha}$ with $\alpha=\Delta_S/\Delta_L \sim 0.44$ based on the Bogoliubov-de Gennes framework [8]. The detailed analysis of $C(T,H)$ data for YNi_2B_2C is in progress and will be published elsewhere [9].

CONCLUSIONS

We have measured H and T dependence of specific heat of YNi_2B_2C and successfully used the two-gap model to describe the superconducting gap function. Based on this analysis and the related reports, YNi_2B_2C is inferred to be a possible two-gap superconductor in company of MgB_2.

ACKNOWLEDGMENTS

This research was supported by the National Science Council of Republic of China under contract Nos. NSC93-2112-M110-001 and NSC93-2112-M009-015.

REFERENCES

*Corresponding author, email: yang@mail.phys.nsysu.edu.tw

1. E. Johnston-Halperin *et al.*, *Phys. Rev. B* **51**, 12852 (1995).
2. P. Raychauhuri, *et al.*, *Phys. Rev. Lett.* **93**, 156802 (2004).
3. F. Bouquet *et al.*, *Europhys. Lett.* **56**, 856 (2001).
4. B. K. Cho *et al.*, *Phys. Rev. B* **52**, 3684 (1995).
5. H. D. Yang *et al.*, *Phys. Rev. Lett.* **87**, 167003 (2001).
6. S. Muhhopadhyay *et al.*, *Cond-mat* /0412486.
7. S. V. Shulga *et al.*, *Phys. Rev. Lett.* **80**, 1730 (1998).
8. N. Nakai *et al.*, *J. Phys. Soc. Jpn.* **71**, 23 (2002).
9. C. L. Huang *et al.*, unpublished.

A 1st Order Transition in $ErNi_2B_2C$

Y. Ishida*, T. Nagata†, H. Kawano-Furukawa†, H. Yoshizawa** and H. Takeya‡

*G.S.H.S., Ochanomizu Univ., Bunkyo-ku, Tokyo 112-8610, Japan
†Department of Physics, Ochanomizu University, Bunkyo-ku, Tokyo 112-8610, Japan
**Neutron Science Laboratory, I.S.S.P., The University of Tokyo, Ibaraki 319-1106, Japan
‡National Institute for Materials Science, Tsukuba, Ibaraki 305-0047, Japan

Abstract. The weak ferromagnetic (WFM) transition observed in $ErNi_2B_2C$ is believed to be of 2nd order because a corresponding anomaly is observed in the temperature dependence of the specific heat. Our detailed neutron diffraction measurements reveal, however, that the temperature dependence of the WFM order parameter shows a clear hysteresis near the WFM transition temperature, indicating that a 1st order process may also be involved in this transition.

Keywords: borocarbides, weak ferromagnetism, superconductivity
PACS: 74.70.Dd, 74.25.Ha, 75.25.+z

INTRODUCTION

It is known that superconductivity (SC) and ferromagnetism (FM) can coexist if the internal field (H_{int}) mediating the FM is less than the superconducting critical field (H_c) [1]. Furthermore, for type II superconductors, a spontaneous vortex phase is expected when $H_{\rm int}$ satisfies the inequality, $H_{\rm c1} < H_{\rm int} < H_{\rm c2}$, where $H_{\rm c1}$ and $H_{\rm c2}$ are the lower and upper critical field, respectively [2].

The superconductor family $RENi_2B_2C$ (where RE = Y or a rare earth) was discovered in 1994 [3]. In the case of $ErNi_2B_2C$ there is a coexistence of weak ferromagnetism (WFM) and SC [4, 5]. The WFM transition of this system was first reported by Canfield *et al.* from a small anomaly in the temperature (T) dependence of the specific heat and the appearance of a remanent magnetization at lower temperature [4]. Since then the WFM transition has been believed to be 2nd order. In the present study, we have carried out neutron diffraction measurements in this system in order to investigate the nature of the ordering process in the WFM phase. Our results show that the T - dependence of the WFM order parameter clearly exhibits hysteresis, indicating that the WFM transition in $ErNi_2B_2C$ includes a 1st order process.

EXPERIMENTAL PROCEDURE

A single crystal specimen of $ErNi_2{}^{11}B_2C$ was grown by the floating zone method. To reduce the high neutron absorption cross section of ^{10}B (20% natural abundance), boron enriched with ^{11}B was used to prepare this crystal. The superconducting, Néel and WFM transition temperatures of the crystal were $T_c \sim 8.6$ K, $T_N \sim 6.0$ K and $T_{WFM} \sim 2.6$ K (on heating), respectively.

Neutron diffraction measurements were performed using the triple axis spectrometer GPTAS (in two axis mode) installed at the JRR-3 reactor in JAERI, Tokai, Japan. Neutrons with a fixed initial neutron momentum of $k_i = 2.67$ Å^{-1}, and a combination of horizontal collimators of 40'-40'-40' and a PG filter before the sample position were utilized. The sample was set with a scattering plane (h 0 l). The lattice parameters of $ErNi_2{}^{11}B_2C$ in tetragonal notation are $a = b \sim 3.5$ Å and $c \sim 10.5$Å.

EXPERIMENTAL RESULTS AND DISCUSSION

Fig. 1(a) shows the T - dependence of the integrated intensity of the fundamental SDW peak at **Q** = (0.55 0 0). The intensity shows two step increases at $T_N = 6.0$ K and $T_{WFM} \sim 2.6$ K, corresponding to the SDW and WFM transitions, respectively. The inset shows an enlargement of the figure at around T_{WFM} and a small hysteresis is clearly observed.

To clarify the hysteresis behavior, we measured the detailed T - dependence of the peak intensities at **Q** = (1 0 0) and (0 0 1) where only the WFM order parameter contributes to the intensity. The results are depicted in Figs. 1(b) and (c). A hysteresis is clearly observed, which indicates that the WFM transition is 1st order. Furthermore, it is clear that both intensities at (1 0 0) and (0 0 1) show two step decreases at ~ 2.1 K and ~ 2.6 K on heating, suggesting that the WFM ordering may consist of 2 successive transitions. As mentioned above, the WFM transition has been believed to be a 2nd order transition because it was first reported from its corresponding anomaly in the specific heat [4]. Our results suggest that the transition to the WFM phase at around 2.3 K

CP850, *Low Temperature Physics: 24th International Conference on Low Temperature Physics;*
edited by Y. Takano, S. P. Hershfield, S. O. Hill, P. J. Hirschfeld, and A. M. Goldman

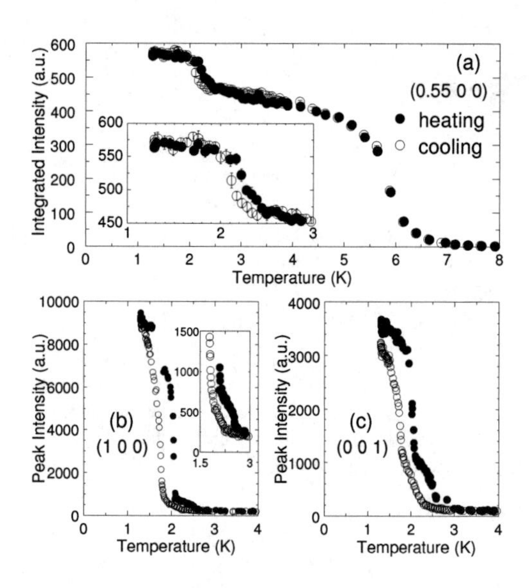

FIGURE 1. T - dependence of the intensities at $\mathbf{Q}$ = (a) (0.55 0 0), (b) (1 0 0) and (c) (0 0 1). The insets are enlargements at around $T_{WFM} \sim 2.6$ K

includes successive 1st and 2nd order transitions. To elucidate the detailed nature of the ordering process to the WFM phase, more detailed experiments are needed and such experiments are in progress.

Finally, we would like to mention that 3 different models for the WFM spin structure were reported by 3 independent groups [6-8]. In the WFM state, the strongest magnetic peak appears at the modulation vector $\mathbf{q} = 11/20\mathbf{a}^*$, indicating that the magnetic unit cell in the WFM phase is $20a \times b \times c$. In the $RENi_2{}^{11}B_2C$ system, the RE atoms occupy the body centered positions. Thus, the magnetic unit cell contains 40 Er atoms. Using a mean-field calculation, Jensen proposed a model expressed as $d(3p)d(5p)d(5p)d(5p)$. Here, the d and the p denote down spins and a pair of spins (up-up or down-down; for detailed explanations, please see the original paper ref. [6]). A similar model was proposed by Choi *et al.* [7] On the other hand, by assuming that a strong RKKY interaction with $\mathbf{q} = 11/20\mathbf{a}^*$ governs its spin structure, we proposed $d(5p)d(5p)d(4p)u(4p)$ as the WFM spin structure [8]. Here the u denotes up spins. The difference between the two models is only one spin among 40 Er spins in the magnetic unit cell, and it makes a difference in the expected net moment of a factor of two. Due to a screening effect of the superconductivity, however, it is not easy to measure the exact value of the net moment, which can make it difficult to test the two models by magnetization measurements.

We would like to point out here, however, that by taking into account the extinction rules in the diffraction measurements, the two models can easily be distinguished. As described above, the Er ions occupy the body centered tetragonal positions at the corners (x, y, z) = (0, 0, 0) and the body center positions (x, y, z) = (1/2, 1/2, 1/2) in the chemical unit cell. Here (x, y, z) indicates coordinates of atomic positions in the tetragonal notation. This leads to a structure factor for $\mathbf{q}$ = (1 0 0) of "0" with Jensen's and Choi's models, but "$2S_{Er}$" in our model. Here, S_{Er} is the magnitude of the Er spin, and $2S_{Er}$ corresponds to the difference between the total spin in the magnetic unit cell between the layers at $z = 0$ and $z = 1/2$ in our model. As clearly shown in Figs. 1(b) and 1(c), there appear magnetic intensities at $\mathbf{Q}$ = (1 0 0) and (0 0 1), and the intensity at $\mathbf{Q}$ = (1 0 0) is of the same order as that expected for a structure factor of "$2S_{Er}$". This strongly supports our conclusion that the magnetic structure of the WFM phase is $d(5p)d(5p)d(4p)u(4p)$.

SUMMARY

In summary, we performed detailed neutron diffraction measurements on $ErNi_2B_2C$ at low temperatures. Our results revealed that the scattering intensities at (1 0 0) and (0 0 1) show two-step changes and hysteresis. By combining the results of specific heat measurements, we concluded that the WFM transition at around 2.3 K consists of two successive transitions, one of 2nd order and the other of 1st order. We also discussed the models for the WFM spin structure and the experimental results clearly support that the spin arrangement in the WFM phase is $d(5p)d(5p)d(4p)u(4p)$ as we had proposed [8].

REFERENCES

1. V.L. Ginzburg, *Sov. Phys. JETP.*, **4**, 153 (1957).
2. E. L. Blout and C. M. Varma, *Phys. Rev. Lett*, **42**, 1079 (1979), M. Tachiki *et al.*, *Solid State Commun.*, **31**, 927 (1979); *ibid.* **34**, 19 (1980), H. S. Greenside *et al.*, *Phys. Rev. Lett*, **46**, 49 (1981).
3. *Rare Earth Transition Metal Borocabides (Nitrides): Superconductivity, Magnetic and Normal State Properties*, edited by K. H. Muller and V. Narozhnyi, Dordrecht: Kluwer Academic, 2001.
4. P.C. Canfield *et al.*, *Physica C* **262**, 249 (1996).
5. H. Kawano *et al.*, *Physica C* **60**, 1053 (1999).
6. J. Jensen, *Phys. Rev. B* **65**, 249 (2002).
7. S. -M. Choi *et al.*, *Phy. Rev. Lett.* **87**, 107001 (2001).
8. H. Kawano-Furukawa *et al.*, *Phy. Rev. B* **65**, 180508 (2002).

Superconductivity and Magnetism in $Dy_{1-x}Lu_xNi_2B_2C$ Compounds

E. Ikeda, K. Mori, and K. Nishimura

Faculty of Engineering, Toyama University, Toyama 930-8555, Japan

Abstract. We have measured the temperature dependencies of the electrical resistivity, magnetization, and specific heat of $Dy_{1-x}Lu_xNi_2B_2C$ compounds and determined the superconducting and antiferromagnetic transition temperatures for the samples of $x = 0.0$, 0.05, 0.1, 0.15, 0.2, 0.25, 0.3, 0.4, 0.5, 0.7 and 1.0. The magnetic entropies were found to vary almost monotonically against x. The effective magnetic moments decreased with increasing x up to $x = 0.7$. For the sample of $x = 0.4$, both T_c and T_N were observed at about 6 K. The superconductivity was not found for the sample of $x = 0.2$ above 2 K.

Keywords: superconductivity, antiferromagnetism, $DyNi_2B_2C$, $LuNi_2B_2C$
PACS: 74.25.Dw, 74.70.Dd

INTRODUCTION

Several quaternary intermetallic borocarbides RNi_2B_2C (R = rare earth elements) are magnetic superconductors in which superconductivity (SC) and antiferromagnetism coexist in some temperature ranges [1-6]. Comprehensive studies of the $Dy_{1-x}Lu_xNi_2B_2C$ system by Cho et al. [1] have revealed that de Gennes (dG) scaling did not hold for the T_c variation; the T_c values increased as the dG factor increased below $x = 0.10$. In the region above $x = 0.3$, the T_c variation appeared to follow an extrapolation of the Abrikosov-Gorkov theory for the dilute magnetic impurity limit. There was a non-SC region between the two; they claimed that the samples of $x = 0.25$ and 0.15 were not superconductors. An additional study, via specific heat measurements, by Michor et al. [2] indicated that $DyNi_2B_2C$ was a bulk superconductor. These authors also found that the critical concentration for the suppression of SC was $x = 0.15$, implying that a sample of $x = 0.2$ was a superconductor. Neither paper showed a variation of the antiferromagnetic (AF) transition temperatures T_N in the region above $x = 0.2$. This paper aims to elucidate the magnetic properties of $Dy_{1-x}Lu_xNi_2B_2C$, and resolve this discrepancy regarding the critical concentration for SC. We are also interested in the relationship between T_c and T_N as a function of x near the cross over point where T_c and T_N appear at the same time.

RESULTS AND DISCUSSION

Polycrystalline samples of $Dy_{1-x}Lu_xNi_2B_2C$ were prepared by arc melting in an argon gas atmosphere. The as-cast samples were annealed in vacuum at 1373K for 72h. Samples were then characterized by powder X-ray diffraction experiments and found to be single phase. The electrical resistivity was obtained using DC four-point method. The magnetization and susceptibility were observed with a SQUID (Quantum Design MPMS). The specific heat was measured by the thermal relaxation method using a PPMS system (Quantum Design).

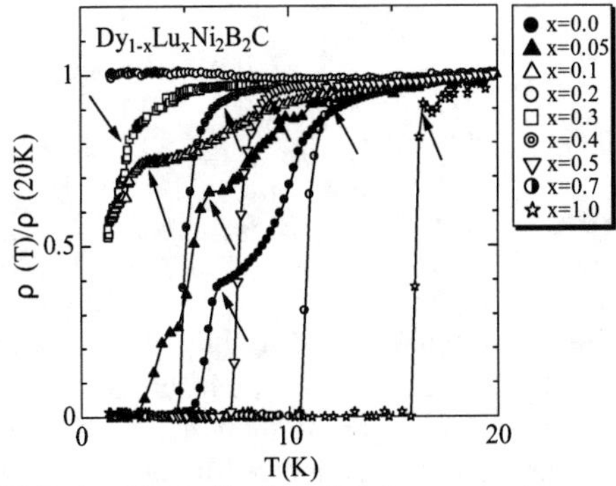

FIGURE 1. Normalized resistivity versus temperature of $Dy_{1-x}Lu_xNi_2B_2C$. Arrows indicate the T_c values.

Figure 1 shows the temperature dependence of normalized electrical resistivity of $Dy_{1-x}Lu_xNi_2B_2C$. The resistivity becomes zero for $x = 0.0$, 0.05 and 0.4-1.0 in the temperature range studied. For the samples of $x = 0.1$ and 0.3, the resistivity was not observed to

CP850, *Low Temperature Physics: 24th International Conference on Low Temperature Physics;*
edited by Y. Takano, S. P. Hershfield, S. O. Hill, P. J. Hirschfeld, and A. M. Goldman

go to zero down to 2K. Where a significant drop in resistivity was observed, this is attributed to the onset of SC. The resistivity for $x = 0.2$, however, appears to slightly increase as the temperature decreases. No sign of SC or T_N was observed at $x = 0.2$ via resistivity.

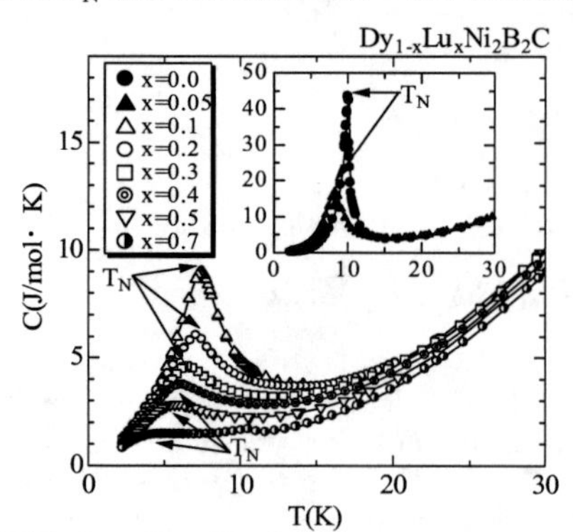

FIGURE 2. Specific heat versus temperature of $Dy_{1-x}Lu_xNi_2B_2C$.

Figure 2 shows the temperature dependence of specific heat (*C*) of $Dy_{1-x}Lu_xNi_2B_2C$ ($x = 0.0$-0.7). The large peaks of C are ascribed to the contributions from the AF transitions. The peak heights decrease and their positions shift toward lower T when x increases. The contributions to C from the superconducting transitions are too small to be seen. A small anomaly for $x = 0.7$ at about 10 K, however, is attributable to the superconducting transition since the temperature is consistent with that observed from resistivity.

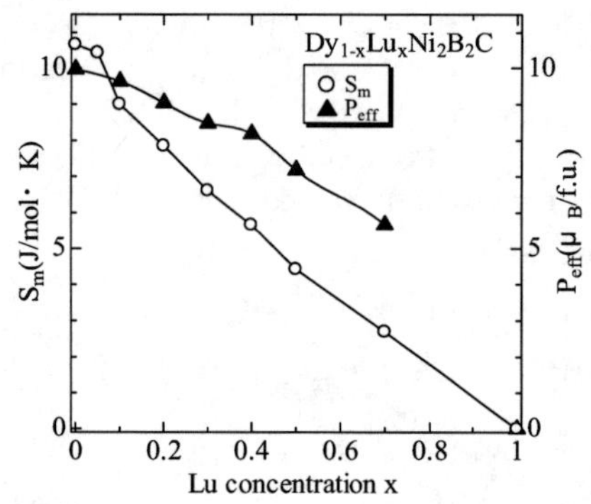

FIGURE 3. S_m and P_{eff} versus Lu concentration x of $Dy_{1-x}Lu_xNi_2B_2C$.

We also measured C of $LuNi_2B_2C$ applying an external magnetic field of 6T which suppressed the SC. The C values of $LuNi_2B_2C$ in the normal state enabled us to estimate the magnetic contributions of specific heats C_m by subtracting C ($LuNi_2B_2C$) values at 6 T from C ($Dy_{1-x}Lu_xNi_2B_2C$) ones at 0 T. The magnetic entropy S_m due to the magnetic Dy ions was evaluated by integrating the C_m/T vs T data up to 30K. The results of S_m evaluations are shown in Fig. 3. The S_m value of $DyNi_2B_2C$ is almost $R\ln4$, which means the magnetic ground state of Dy ions is a quartet state owing to the crystal electric field effect. This result supports previous work [7]. Figure 3 also shows the effective magnetic moments P_{eff} of $Dy_{1-x}Lu_xNi_2B_2C$ in the paramagnetic states deduced from the susceptibility data. The P_{eff} decreased with increasing x up to $x = 0.7$. The S_m varies monotonically against x. This implies that no pronounced change in the magnetic structure of Dy ions occurs across the $Dy_{1-x}Lu_xNi_2B_2C$ system.

The T_c and T_N values, taken from Figs. 1 and 2, are plotted against dG factors together with the reported T_c values [1] in Fig. 4. The present results for T_c are well correlated with the reported ones. The sample of $x = 0.2$ did not show SC. T_N decreases with increasing x, indicating that magnetic interactions between Dy ions are reduced gradually by the substitution of Lu for Dy. We found that both the superconducting and AF transitions occur at about 6 K with the sample of $x = 0.4$.

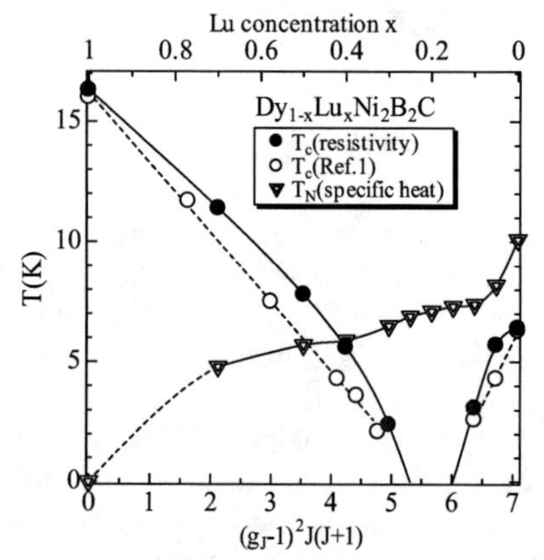

FIGURE 4. T_c and T_N versus dG factor of $Dy_{1-x}Lu_xNi_2B_2C$. The lines are drawn as a guide for the eye.

REFERENCES

1. B. K. Cho, P. C. Canfield, and D. C. Johnston, *Phys. Rev. Lett.* **77**, 163-166 (1996).
2. H. Michor, M. El-Hagary, R. Hauser, E. Bauer, and G. Hilscher , *Physica B* **259-261**, 604-605 (1999).
3. R. Nagarajan, C. Mazumber, Z. Hossain, S. K. Dhar, K. V. Gopalakrishnan, L. C. Gupta, C. Godart, B. D. Padalia, and R. Vijayaraghavan, *Phys. Rev. Lett.* **72**, 274-277 (1994).
4. R. J. Cava et al., *Nature (London)* **367**, 252 (1994).
5. R. J. Cava et al., *Nature (London)* **367**, 146 (1994).
6. H. C. Ku, C. C. Lai, Y. B. You, J. H. Shieh, and W. Y. Guan, *Phys. Rev. B* **50**, 351-353 (1994).
7. M. S. Lin et al., *Physica C* **249**, 403-408 (1995).

Possibility of Triplet Pairing in $Dy_{1-x}Y_xRh_4B_4$ Series

A.J. Zaleski*, A.V. Tswyashchenko†, E.P. Khlybov†,**, L.N. Fomicheva†, I.E. Kostyleva†,**, S.A. Lachenkov‡ and O.G. Zamolodchikov‡

*Institute of Low Temperature and Structure Research, PAS, Wrocław, Poland
†Institute for High Pressure Physics, RAS, Troitsk, Russia
**International Laboratory of High Magnetic Fields and Low Temperatures, Wrocław, Poland
‡Institute of Metallurgy, RAS, Moscow, Russia

Abstract. We have investigated magnetic and superconducting properties of the series of $Dy_{1-x}Y_xRh_4B_4$ samples by means of AC and DC magnetization measurements. It was established that this compound exhibits ferromagnetic order in temperatures below $T_N = 39$K for $x = 0$ and $T_N = 12$K for $x = 0.4$, and superconducting behavior for temperatures ranging from $T_c = 4.5$K for $x = 0$ to $T_c = 10$K for $x = 1$. At the same time the effective magnetic moment changes from about $\mu_{eff} = 9.5\mu_B$ for $x = 0$ to $\mu_{eff} = 0$ for $x = 1$. Magnetic hysteresis loops show Meissner behavior below the T_c, in spite of the substantial ferromagnetic component of hysteresis. $H_{c2}(T)$dependencies are similar to those observed earlier for HoN_2B_2C samples. Our experiments suggest that we are dealing with coexistence of ferromagnetism and superconductivity, thus they show possibility of triplet pairing in $Dy_{1-x}Y_xRh_4B_4$.

Keywords: superconductivity, ferromagnetism, coexistence
PACS: 74.25.Ha, 74.70.Dd, 75.50.Cc

INTRODUCTION

Among the different ternary rare earth rhodium borides the one basing on dysprosium was considered as ferromagnetic and not superconducting [1]. But this compound can exist in three different polymorphic variety. Most popular one, with the unit cell of dimenssions $5.3 \times 5.3 \times 7.4$ Å is stable and really not superconducting.

Interestingly, by using high pressure synthesis it is possible to prepare the other polymorphic phases which appear to be stable in normal conditions. It turned out that in synthesised this way $DyRh_4B_4$ both ferromagnetism and superconductivity coexist. The subject of our study was the investigation of the magnetic properties of $DyRh_4B_4$ and the whole series where, carrying magnetic moment, dysprosium was partially substituted by non-magnetic yttrium.

METHODS

Samples under study were prepared under high pressure of 8 GPa by the method described elsewhere [2].

X-ray analysis proved samples were single phased (space group I4/mmm) with lattice constant changing from $a = 7.453$Å, $c = 14.950$Å for $DyRh_4B_4$ to $a = 7.434$Å, $c = 14.934$Å for YRh_4B_4.

Magnetic measurements were carried out with the use of Quantum Design SQUID magnetometer system.

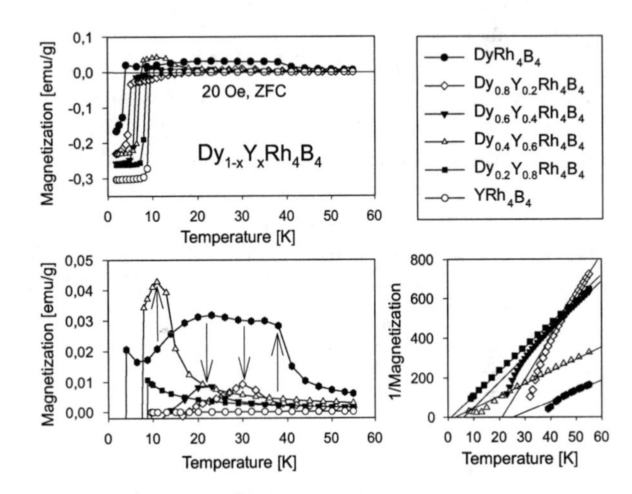

FIGURE 1. Magnetization vs. temperature for $Dy_{1-x}Y_xRh_4B_4$ series.

RESULTS AND DISCUSSION

Magnetization temperature dependence for the series of $Dy_{1-x}Y_xRh_4B_4$ for $x = 0, 0.2, 0.4, 0.6, 0.8, 1$ is presented in Fig.1. It is seen that for non-substituted $DyRh_4B_4$ sample evident transition to ferromagnetic state can be observed already at temperature $T_N =$ 39K while with decreasing temperature, superconducting transition can be seen at as high as $T_c = 4.5$K. Diluting the magnetic moments of dysprosium, by substituting them with yttrium, resulted in decreasing ferromagnetic transition temperature to $T_N = 14$K for $Dy_{0.4}Y_{0.6}Rh_4B_4$.

CP850, *Low Temperature Physics: 24th International Conference on Low Temperature Physics;* edited by Y. Takano, S. P. Hershfield, S. O. Hill, P. J. Hirschfeld, and A. M. Goldman

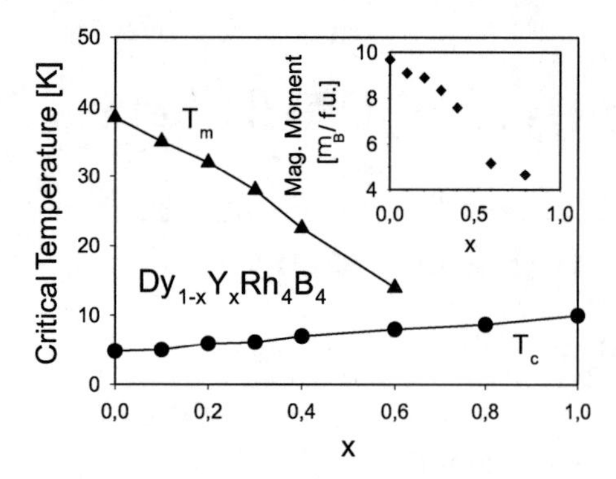

FIGURE 2. Superconducting and ferromagnetic transition temperature dependence vs. Dysprosium content. Inset: Effective magnetic moment of Dy ion for $Dy_{1-x}Y_xRh_4B_4$ series.

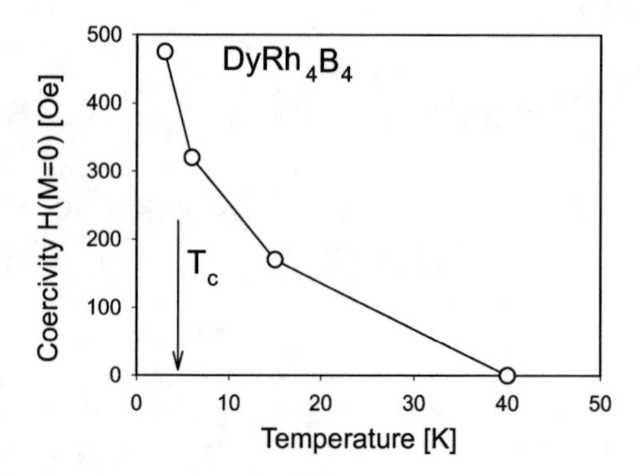

FIGURE 3. Coercivity vs. temperature for $DyRh_4B_4$. Superconducting critical temperature is marked on the plot.

Samples with lower dysprosium contents do not show any signs of magnetic ordering. Superconducting critical temperature changes from about $T_c = 4.5$K for $DyRh_4B_4$ to about $T_c = 10$K for YRh_4B_4.

Inverse magnetization temperature dependence is presented in Fig. 1 shows the change of Curie temperature with yttrium substitution.

Magnetic and superconducting transition temperatures for different yttrium substitution levels were summarized in Fig. 2. Superconducting transition temeperature varies almost linearily with yttrium content. It supports our previous finding that our samples are homogeneous. At inset of Fig.2 effective magnetic moment on Dy ion, derived from magnetization measurements, on composition dependence is presented. The change is non-monotonic. There is evident jump for the composition for which superconducting and ferromagnetic transition temperatures should be equal.

Magnetic hysteresis, measured for the whole series, shows, for low temperatures, superposition of superconducting-type and ferromagnetic-type loops. Plot of temperature dependence of coercivity (magnetic field, for which magnetization equals zero) is presented in Fig. 3 for $DyRh_4B_4$. Coercivity value for T_c (marked with arrow on Fig. 3) is high enough, that it may be expected, that spontaneously created mixed state is achieved during superconducting transition. Probably that is why Meissner part of hysteresis loop in virgin sample is visibly less than expected.

Ternary rhodium borides were discovered by Matthias et al. [3]. They found out, that with some rare earths compounds are superconducting and for the other - ferromagnetic. Of special interest was the one with erbium, for which superconductivity and ferromagnetism coexisted [4]. Although ferromagnetic order was observed for temperatures higher than superconductivity appearance, oscillatory character of its order parameter, with the characteristic length much smaller than the coherence length, enables both orders to coexist. The change of oscillatory character to the uniform one results in reentrant behavior. In the case of $ErRh_4B_4$ superconducting transition is observed for $T_{c1} = 8.7$K while it orders ferromagnetically (with uniform order parameter) at $T_{c2} = 0.9$K.

In the case of our sample, magnetic and resistivity measurements down to $T = 0.32$K did not show reentrant behavior.

SUMMARY

Series of ternary rhodium borides $Dy_{1-x}Y_xRh_4B_4$ was prepared under high pressure. We have found out that although all samples are superconducting, the ones with yttrium content less than $x = 0.3$ exhibit the signs of ferromagnetic ordering at temperatures higher than superconducting ordering, which suggests the coexistence of both orders. Non-substituted $DyRh_4B_4$ did not show reentrant behavior at temperatures above $T = 0.32$K .

The work was partially supported by grants of the Russian Foundation for Fundamental Research (04-02-16061, 03-02-16107a), Presidium RAS and SCENET-2.

REFERENCES

1. M.B. Maple and Ø. Fisher, *Superconductivity in Ternary Compounds*, Berlin: Springer Verlag, 1982.
2. A.V. Tsvyashchenko, L.N. Fomicheva, A.A. Sorokin, G.K. Ryasny, B.A. Komissarova, L.G. Shpinkova, K.V. Klementiev, A.V. Kuznetsov, A.P. Menushenkov, V.N. Trofimov, A.E. Primenko, and R. Cortes, *Phys. Rev. B* **65**, 174513 (2002).
3. B.T. Matthias, E. Corentzwit, J.M. Vanderberg, and H. Barz, *Proc. Natl. Acad. Sci. USA* **74**, 1334 (1977).
4. W.A. Fertig, D.C. Johnston, L.E DeLong, R.W McCallum, M.B. Maple, and B.T. Matthias, *Phys. Rev. Lett.* **38**, 987 (1977).

Magnetic Field-Induced Quantum Critical Point in $CeAuSb_2$

L. Balicas,[1] S. Nakatsuji,[2] H. Lee,[3] P. Schlottmann,[1] T. P. Murphy,[1] and Z. Fisk[3]

[1]*National High Magnetic Field Laboratory, Florida State University, Tallahassee-FL 32306, USA*
[2]*Department of Physics, University of Kyoto, Kyoto 606-8502, Japan*
[3]*Department of Physics, University of California Davis, Davis, California 95616, USA*

Abstract. Transport, magnetic and thermal properties at high magnetic fields (H) and low temperatures (T) of the heavy fermion compound $CeAuSb_2$ are reported. At H=0 this layered system exhibits antiferromagnetic order below T_N = 6 K. Applying B along the inter-plane direction, leads to a continuous suppression of T_N and a quantum critical point at $H_c \cong$ 5.4 T. Although it exhibits Fermi liquid behavior within the Neel phase, in the paramagnetic state the fluctuations associated with H_c give rise to unconventional behavior in the resistivity (sub-linear in T) and to a $TlnT$ dependence in the magnetic contribution to the specific heat. For $H > H_c$ and low T the electrical resistivity exhibits an unusual T^3-dependence.

Keywords: Quantum-criticality, Ce based heavy-Fermion compound.
PACS: 75.30.Mb, 75.20.Hr, 75.30.Kz, 75.40.-s

INTRODUCTION

Quantum criticality [1] is common to a large variety of very different phenomena ranging from low-dimensional quantum systems to high-temperature superconductivity, disorder-induced criticality (e.g. Griffiths phase) and heavy fermion compounds at the verge of antiferromagnetic (AF) order. For strongly correlated electrons a quantum critical point (QCP) is obtained when either (i) the long-range order is suppressed to T=0 (second order phase transition) or (ii) the critical end-point terminating a line of first-order transitions is depressed to T=0.[2] A QCP can be tuned by an external variable, such as pressure, chemical composition or the magnetic field H.[3] H is an ideal control parameter, since it can be reversibly and continuously tuned towards the QCP.[4] In alloys the disorder driven effects cannot be separated from the quantum criticality of the translational invariant system.[4] Hence, it is essential to consider stoichiometric systems. Two compounds with field-tuned QCP, $YbRh_2Si_2$ and $Sr_3Ru_2O_7$, reached prominence due to the non-Fermi liquid (NFL) behavior triggered by the quantum fluctuations associated with the QCP. In this manuscript we present a Ce-compound, $CeAuSb_2$, exhibiting a field-tuned QCP with unusual transport and thermodynamic properties. $YbRh_2Si_2$, $Sr_3Ru_2O_7$ and $CeAuSb_2$ have a field-tuned QCP as a common thread, yet their properties are considerably different. This points towards a lack of universality among the different systems as a fundamental component of quantum criticality.

RESULTS AND DISCUSSION

Here, we report on anomalous properties of the tetragonal metallic compound $CeAuSb_2$, which at H = 0 orders AF [5] with T_N = 6.0 K. For $T < T_N$, $\rho(T)$ has the typical AT^2 dependence of a FL and extrapolating C_e/T to T=0 yields $\gamma \sim 0.1$ J/mol.K^2. Hence, $CeAuSb_2$ can be considered a system of relatively light heavy-fermions. Above T_N, on the other hand, ρ(T) displays a T^α dependence with $\alpha \leq 1$ and, C_e/T has a $-lnT$ dependence, both characteristic of NFL behavior due to a nearby QCP. A magnetic field along the inter-plane direction leads to two subsequent MM transitions and the concomitant continuous suppression of T_N to T=0 at H_c = 5.3 ± 0.2 T. As the AF phase boundary is approached from the paramagnetic (PM) phase, γ is enhanced and the A coefficient of the resistivity diverges as $(H\text{-}H_c)^{-1}$. When T is lowered for $H \sim H_c$, the T-dependence of ρ is sub-linear and the one of C_e/T is approximately $-lnT$. At higher fields, $H >> H_c$, an unconventional T^3-

CP850, *Low Temperature Physics: 24th International Conference on Low Temperature Physics;*
edited by Y. Takano, S. P. Hershfield, S. O. Hill, P. J. Hirschfeld, and A. M. Goldman

dependence emerges in ρ and becomes more prominent as H increases.

The upper panel of Fig. 1 shows $C(T)/T$ as a function of T for $CeAuSb_2$ at H=5 T, and for its isostructural non-magnetic analog $LaAuSb_2$ at H=0 T. The large peak for $CeAuSb_2$ signals the AF transition. The subtraction of both curves yields the magnetic contribution to the heat capacity $C_e(T)/T$. For $3 < T <$ 20 K, $C_e(T)/T$ displays a $-lnT$ NFL-like dependence. The lower panel of Fig. 1 shows $C_e(T)/T$ as a function of H at $T = 1$ K which clearly indicate that the effective mass of the quasi-particles, as given by $C_e(T)/T$ for $T \to 0$ increases considerably as $H \to H_c$, although it remains finite. Because of the AFM order the effective mass cannot be defined precisely. In the PM phase the $-lnT$-dependence does not continue to very low T. This is similar to the behavior of the specific heat of $Sr_3Ru_2O_7$, where the $-ln(T)$ dependence does not continue to very low T and there is a cross-over to a constant C/T at the lowest T.

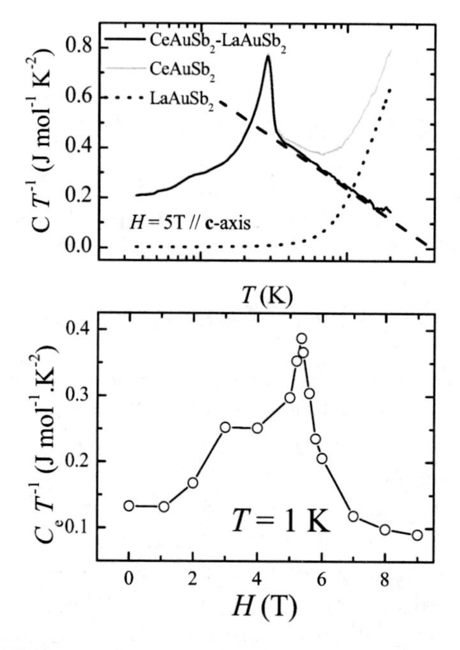

FIGURE 1. Upper panel: Heat capacity divided by temperature C/T vs. T down to 0.38 K for $CeAuSb_2$ with H=5 T applied along the c-axis (blue line), as well as for $LaAuSb_2$ at H=0 T (black line). The difference is the magnetic contribution to the heat capacity C_e/T (in magenta), which shows a lnT-dependence. Lower panel: C_e/T vs H at T=1 K. Notice the pronounced enhancement of C_e/T as H approaches H_c.

Fig. 2 depicts a qualitative sketch of the H-T phase diagram. It shows the dependence of the exponent $n \cong \partial ln(\rho(T) - \rho(0)/\partial lnT$ on H and T. Here different values of ρ_0 were used in the PM and AF phases. The PM phase is indicated by the blue region which is influenced by the QCP leading to the anomalous NFL value $n \leq 1$. This is analogous to the behavior of $YbRh_2Si_2$ and $Sr_3Ru_2O_7$. The FL state (in green) is recovered below the Néel temperature but is gradually suppressed as $H \to H_c$. A FL-like n=2 exponent is obtained above H_c but only over a limited range of T. Instead a value $n = 3$ is observed in the spin polarized PM phase at higher fields and lowest Ts.

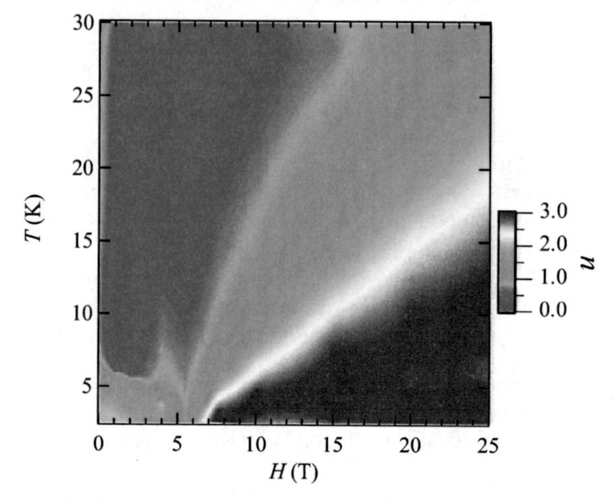

FIGURE 2. Exponent n of $\rho(T)$ in the T - H plane.

In conclusion, the field-tuned QCP systems represent a remarkable challenge from both the experimental and the theoretical perspectives, since the different compounds revealing some common aspects do *not* seem to belong to the same universality class.

ACKNOWLEDGEMENTS

This work is sponsored by the National Nuclear Security Administration under the Stewardship Science Academic Alliances program through DOE grant DE-FG03-03NA00066 and was performed at the NHMFL which is supported by NSF through DMR-0084173 and the State of Florida. LB acknowledges support from the NHMFL in-house program, SN from Grants-in-Aids for Scientific Research from JSPS and for the 21st Century COE Center for "Diversity Universality in Physics" from MEXT of Japan, and PS by grants (DOE) DE-FG02-98ER45707 and (NSF) DMR01-05431..

REFERENCES

1. S. Sachdev, *Quantum Phase Transitions*, Cambridge: Univ. Press, 1999.
2. A. J. Millis, A. J. Schofield, G. G. Lonzarich, and S. A. Grigera, *Phys. Rev. Lett.*, **88**, 217204 (2002).
3. G. R Stewart, *Rev. Mod. Phys.*, **73**, 797 (2001).
4. K. Heuser, E. W. Scheidt, T. Schreiner, and G. R. Stewart, *Phys. Rev. B*, **57**, R4198 (1998).
5. A. Thamizhavel, *et al.*, *Phys. Rev. B*, **68**, 054427 (2003).

$CePt_3Si$: Heavy Fermion Superconductivity and Magnetic Order without Inversion Symmetry

E. Bauer*, I. Bonalde†, A. Eichler**, G. Hilscher*, Y. Kitaoka‡, R. Lackner*, St. Laumann*, H. Michor*, M. Nicklas§, P. Rogl¶, E.W. Scheidt||, M. Sigrist†† and M. Yogi‡

*Institute of Solid State Physics, Vienna University of Technology, A-1040 Wien, Austria
†Centro de Física, Instituto Venezolano de Investigaciones Científicas, Caracas 1020-A Venezuela
**Institute of Applied Physics, Technical University of Braunschweig, D-38106 Braunschweig, Germany
‡Department of Materials Science and Technology, Osaka University, Osaka 560-8531, Japan
§Max Planck Institute for Chemical Physics of Solids, D-011187 Dresden, Germany
¶Institut für Physikalische Chemie, Universität Wien, A-1090 Wien, Austria
||Chemische Physik und Materialwissenschaften, Universität Augsburg, D - 86159 Augsburg, Germany
††Institute of Theoretical Physics, ETH-Hönggerberg, 8093 Zürich, Switzerland

Abstract. Ternary $CePt_3Si$ crystallizes in the tetragonal *P4mm* structure which lacks a center of inversion. Antiferromagnetic order sets in at $T_N \approx 2.2$ K followed by superconductivity (SC) below $T_c \approx 0.75$ K. Large values of $H'_{c2} \approx -8.5$ T/K and $H_{c2}(0) \approx 4$ to 5 T were derived, referring to Cooper pairs formed out of heavy quasiparticles. The mass enhancement originates from Kondo interactions with a characteristic temperature T_K of roughly 10 K. NMR and μSR results show that both magnetic order and SC coexist on a microscopic scale without having spatial segregation of both phenomena. The absence of an inversion symmetry gives rise to a lifting of the degeneracy of electronic bands by spin-orbit coupling. As a consequence, the SC order parameter may be composed of spin-singlet and spin-triplet components as indicated from a very unique NMR relaxation rate $1/T_1$ and a linear temperature dependence of the penetration depth λ. A helical modification of the order parameter would explain the absence of significant anisotropy of the upper critical field when comparing the field parallel and perpendicular to the $\vec{c}$ axis.

Keywords: heavy fermion superconductivity, inversion symmetry, magnetic order
PACS: 74.70.Tx, 71.27.+a, 75.30.Mb

INTRODUCTION

Among materials with strong electron correlations, those exhibiting superconductivity have attracted the most intense interest because a rich variety of unconventional SC states can be investigated in detail. Also because in such systems Cooper pairs are formed not by the interaction of electrons with phonons, resulting in general in the highly symmetric s-wave state, but rather by magnetic fluctuations. Theoretically, this can be ascribed to narrow electronic bands, inherent to highly correlated electron systems. As a consequence, the effective mass m^* of the charge carriers becomes very large ($m^* \approx 10$ to $1000 \times m_0$, m_0 ... free electron mass), thus the particle velocity is small. The retardation effect ($\propto W/(\hbar\omega_D)$, W ... electronic band width, ω_D ... Debye frequency) weakens because of the small bandwidth W. Hence, the electron phonon interaction may be insufficient in forming Cooper pairs.

Heavy fermion compounds naturally provide alternative possibilities for attractive forces between electrons, i.e, spin fluctuations, since many of such systems exhibit weak magnetic order or are in the proximity to a magnetic instability. Applying control parameters like pressure or substitution allows to tune these materials across a quantum critical point, where frequently superconductivity occurs and Cooper pairs are, most likely, formed by magnetic fluctuations [1]. Cooper pairing in such a scenario may happen also in a different angular momentum channel. This means that Cooper pairs may have either spin singlet or spin triplet configurations and the orbital angular momentum may lead to a highly anisotropic gap with lines or points as zero nodes. Furthermore, these Cooper pairs consist of heavy particles carrying the supercurrent. As a result, the upper critical field H_{c2} becomes exceptionally large.

It is generally believed that for spin singlet Cooper pairing time reversal invariance provides the necessary conditions, while spin triplet pairing needs additionally an inversion center [2]. The absence of inversion symmetry removes, however, the distinction of even and odd parity and leads immediately to a mixing of the spin channels [3]. Moreover, it induces peculiar band splittings, which are detrimental to certain kinds of Cooper

CP850, *Low Temperature Physics: 24th International Conference on Low Temperature Physics;*
edited by Y. Takano, S. P. Hershfield, S. O. Hill, P. J. Hirschfeld, and A. M. Goldman

pairing channels. In particular, spin triplet pairing was claimed to become unlikely under these circumstances [2].

A couple of SC without centrosymmetric crystal structures were found in recent years. Among them are binary R_2C_{3-y} with R = La or Y, which can be substituted by various other elements. Superconducting transition temperatures up to 18 K were observed for this family of compounds [4, 5, 6]. Another example is $Cd_2Re_2O_7$ with $T_c = 1$ K [7, 8]. Although the pyrochlore structure of $Cd_2Re_2O_7$ exhibits a center of symmetry, a series of low temperature structural phase transitions causes that superconductivity occurs in a non-inversion symmetric environment. Different to the previous case, strong spin-orbit coupling dramatically affects the electronic band structure. Nevertheless, normal state properties of these systems are quite simple and the charge carriers do no exhibit large effective masses.

All previously studied SC with strong electron correlations in the normal state region exhibit a center of inversion in its crystal structure. The first example of a heavy fermion superconductor (HFSC) without inversion symmetry became previously discovered $CePt_3Si$ (tetragonal space group P4mm) [9]. UIr [10] and $CeRhSi_3$ [11] have very recently expanded this family.

Some key experimental features already derived for $CePt_3Si$ are: i) Superconductivity ($T_c = 0.75$ K) is mediated by Cooper pairs, formed by heavy quasiparticles due to the presence of Kondo interactions; ii) Long range magnetic order ($T_N = 2.2$ K) coexists with superconductivity on a microscopic scale. The coupling between both phenomena, however, appears to be quite weak. iii) The upper critical field $H_{c2}(0) \approx 4$ to 5 T exceeds the Pauli-Clogston limit ($H_P \approx 1.1$ T) by far. iv) The $1/T_1$ relaxation rate of NMR measurements does not follow the simple T^3 power law of HFSC and additionally exhibits a sort of Hebel-Slichter peak below T_c. The latter has never been observed in HFSC. v) Thermal conductivity and penetration depth measurements [12, 13] performed at $CePt_3Si$ at very low temperatures reveal power laws which indicate line nodes in the electronic gap and thus evidence unconventional superconductivity.

These observations have triggered substantial theoretical work on the role of pairing symmetry in such a superconductor. The essential element to model non-centrosymmetric systems is the presence of antisymmetric spin-orbit coupling. An inherent feature is then the mixing of spin-singlet and spin-triplet Cooper-pairing channels which are otherwise distinguished by parity [3]. This mixing of pairing states, yielding a two-component order parameter, seems to be the appropriate scenario to account for the above indicated, mutually contradicting, experimental features observed in ternary $CePt_3Si$.

The aim of the present paper is to report on previously performed experimental studies on this material, as well as considering recent theoretical progress made for superconductivity without inversion symmetry.

RESULTS AND DISCUSSION

Normal State Properties of $CePt_3Si$

The structure type of $CePt_3Si$ derives from hypothetical $CePt_3$ with the cubic centrosymmetric $AuCu_3$ structure by filling the void with stabilizing Si. Whilst binary compounds $\{La, Ce, Eu\}Pt_3$ do not exist, $\{Pr, Nd, Sm, Gd\}Pt_3$ are reported to form a $AuCu_3$-type phase; in any case incorporation of Si acts as a stabilizer forming a series of ternary compounds with a significant tetragonal distortion of the unit cell to $c/a \approx 1.34$. Fig. 1 presents the structure of $CePt_3Si$ in three-dimensional view highlighting the Pt-Si units as tetragonal bi-pyramids within the tetragonal frame of the rare earth (RE) atoms. The lattice constants of $CePt_3Si$ (space group *P4mm* No. 99) are $a = 0.4072(1)$ nm and $c = 0.5442(1)$ nm. The lack of a mirror plane perpendicular to the four-fold axis and consequently the loss of a center of inversion is clearly seen from Fig. 1. The generating point group C_{4v} of the unit cell causes a broken inversion symmetry with an absence of the mirror plane $z \rightarrow -z$.

Physical properties of ternary $CePt_3Si$ are dominated by the onset of long range, presumably antiferromagnetic order below $T_N \approx 2.2$ K followed by SC below $T_c = 0.75$ K. This intriguing coincidence of two ordering

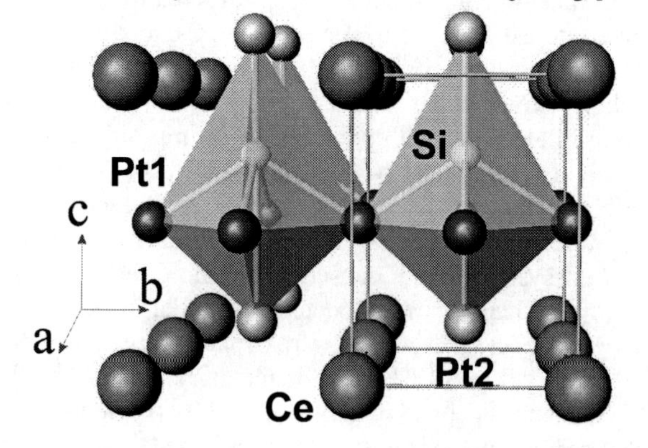

FIGURE 1. Crystal structure of $CePt_3Si$. Pt1 and Pt2 denote the independent crystallographic sites in 2(c) at 0.5; 0; 0.6504, and in 1(a) at 0; 0; 0 (fixed). For convenient comparison with the parent $AuCu_3$ structure, the origin is shifted by (0.5; 0.5; 0.8532).

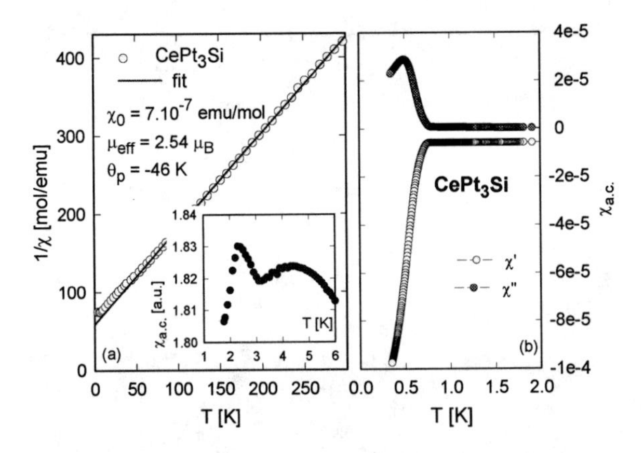

FIGURE 2. (a): Temperature dependent susceptibility χ of $CePt_3Si$ plotted as $1/\chi$ vs. T. The solid line is a least squares fit according to the modified Curie Weiss law. The inset shows the low temperature behavior deduced from an a.c. susceptibility study. (b): Temperature dependence of χ' and χ'' of $CePt_3Si$ below 2 K.

phenomena in $CePt_3Si$ have to be considered in the context of crystal electric field (CEF) splitting and Kondo interaction which substantially modify Hund's $j = 5/2$ ground state of the Ce ion. The response of the system associated with the mutual interplay of these mechanisms will be highlighted below.

Temperature dependent magnetic susceptibility provides information concerning the effective magnetic moments (μ_{eff}) involved in a particular system, the interaction strength between these moments via the paramagnetic Curie temperature (θ_p), and about phase transitions present in a certain sample. Results taken from SQUID measurements performed at $\mu_0 H = 1$ T are shown in Fig. 2(a) together with an a.c. susceptibility measurement in the inset of this figure. At elevated temperatures, $\chi(T)$ of $CePt_3Si$ exhibits a Curie Weiss behavior - and the anomaly around 2.2 K (inset, Fig. 2(a)) indicates the magnetic phase transition. To qualitatively account for the region above about 50 K, the modified Curie Weiss law,

$$\chi = \chi_0 + \frac{C}{T + \theta_p}, \tag{1}$$

was applied. χ_0 represents a temperature independent Pauli-like susceptibility and C is the Curie constant. Results of this procedure are shown in Fig. 2(a) as solid line. χ_0 is of the order of 10^{-8} emu/mol, and the effective magnetic moment deduced from the Curie constant C matches the theoretical value associated with the 3+ state of cerium, thereby inferring a rather stable magnetic moment. The paramagnetic Curie temperature $\theta_p \approx -46$ K is large and negative, indicative of strong antiferromagnetic interactions. In the Kondo picture, also necessary to explain the large Sommerfeld value γ (see below), the

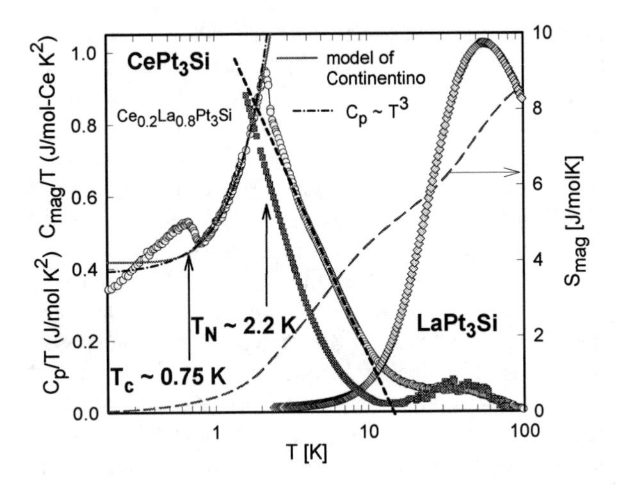

FIGURE 3. Temperature dependent magnetic contribution to the specific heat, C_{mag} of $CePt_3Si$ and of $Ce_{0.2}La_{0.8}Pt_3Si$ plotted as C_{mag}/T on a logarithmic temperature scale. The phonon contribution to define ΔC_p is taken from $C_p(T)$ of $LaPt_3Si$, which is added for purpose of comparison. The long-dashed line represents the magnetic entropy (right axis). The short-dashed line is a guide to the eyes and roughly indicates the non-Fermi liquid behavior. The solid and the dashed-dotted lines are fits described in ref. [16].

value of θ_p suggests a Kondo temperature of the order of 10 K ($T_K \approx \theta_p/4$) [14].

Large values of θ_p ($H \| [100]$ and $H \perp [100]$) were also deduced from a previous investigation of single crystalline $CePt_3Si$ [15], confirming the substantial antiferromagnetic interaction strength. Although c/a of the crystal structure is larger than one, the magnetic anisotropy of $\chi_{\|}$ with respect to $\chi_{\perp}$ is unexpectedly small [15].

The transition into a superconducting ground state around 0.75 K is clearly evidenced by $\chi_{a.c.}$ measurements. Results are presented in Fig. 2(b).

Substantial information concerning the magnetic and the paramagnetic properties of $CePt_3Si$ can be deduced from the temperature dependent magnetic contribution to the specific heat $C_{mag}(T)$. The latter may be defined by the difference ΔC_p between $CePt_3Si$ and isostructural non-magnetic reference $LaPt_3Si$, with $\Delta C_p \sim C_{mag}$. Plotted in Fig. 3 is C_{mag}/T vs. T of $CePt_3Si$ together with the raw data of $LaPt_3Si$. The low temperature behavior of the latter can be accounted for in terms of the Debye model with $\theta_D = 255$ K together with a Sommerfeld value $\gamma \approx 9$ mJ/molK^2. $C_{mag}(T)$ of $CePt_3Si$ defines three regimes: i) the SC state of $CePt_3Si$ below $T_c = 0.75$ K, ii) the magnetically ordered range below $T_N \approx 2.2$ K and iii) the paramagnetic region above T_N. This region is characterized by a Schottky like anomaly with a weak local maximum around 40 to 50 K. Such numbers imply a CEF level approximately 100 K above the ground state doublet. However, since $C_p(T)$ of $LaPt_3Si$ slightly exceeds the specific heat data of $CePt_3Si$, $C_{mag}(T) \sim C_{mag}(T)$ becomes negative, hence a reliable evaluation of CEF level scheme is not possible, at least, considering this quantity only. The integrated entropy S_{mag} (right axis, Fig. 3) reaches $R\ln 2$ around 25 K and the entropy of 8.7 J/molK integrated up to 100 K is slightly less than $R\ln 4 = 11.5$ J/molK. These results indicate that the ground state of Ce^{3+} ions is a doublet with the first excited level above about 100 K. The twofold degeneracy of the ground state doublet, however, is lifted by magnetic order as well as by Kondo type interaction spreading entropy to higher temperatures.

The second interesting aspect in the paramagnetic temperature range of $CePt_3Si$ is an almost logarithmic tail of $\Delta C_p/T$ above T_N, extrapolating to about 12 to 13 K. The logarithmic temperature dependence observed just above the magnetic transition may be considered as hint of non-Fermi liquid (nFl) behavior. Therefore, it is a unique observation at ambient pressure that non-Fermi liquid behavior, magnetic ordering and eventually a SC transition consecutively arises on the same sample upon lowering the temperature. To corroborate the nFl property deduced for $CePt_3Si$ from the specific heat data and to exclude short range order effects and inhomogeneities above the magnetic phase transition, $C_p(T)$ was studied for diluted $Ce_{0.2}La_{0.8}Pt_3Si$ as well. Results are shown in Fig. 3 as $\Delta C_p/T$ vs. $\ln T$ (filled squares). This diluted sample - without magnetic ordering - exhibits a similar logarithmic contribution to the specific heat, like parent $CePt_3Si$, and thus states this feature as intrinsic property. An extrapolation of the logarithmic tail reveals a temperature of about 7 K. The latter reflects the fact that, owing to an increase of the unit cell volume due to the Ce/La substitution, the hybridisation of the Ce $4f$ electrons with conduction electrons decreases, hence the Kondo interaction strength decreases as well.

Two models described in detail in Ref. [16] have been applied to the magnetic contribution to the specific heat data in the magnetically ordered temperature range, confirming antiferromagnetic ordering and a Sommerfeld value $\gamma \approx 0.4$ J/molK^2.

For better understanding of the ground state of the Ce^{3+} ion as well as of superconductivity, inelastic neutron scattering was performed at the time-of-flight instrument HET at ISIS [16] and at the JRR-3 reactor of the Japan Atomic Energy Research Institute, JAERI, Japan [15]. Results of both studies are rather contradictory. While we found a simple crystal electric field (CEF) scheme with doublets at 0, 152 and 232 K, the authors of Ref. [15] reported about CEF doublets at 0, 11 and 250 K. Of course, the ground state doublet becomes split due to magnetic order and/or the Kondo effect. The latter scheme would be responsible for a substantial contribution of the first excited CEF level to the ground state properties. Specific heat data within the paramagnetic temperature range provides the possibility to proof CEF effects as inferred from Schottky-like anomalies, as well as the level splitting considering magnetic entropy data. This would allow to roughly decide between the two sets of data derived from the inelastic neutron scattering [16, 15].

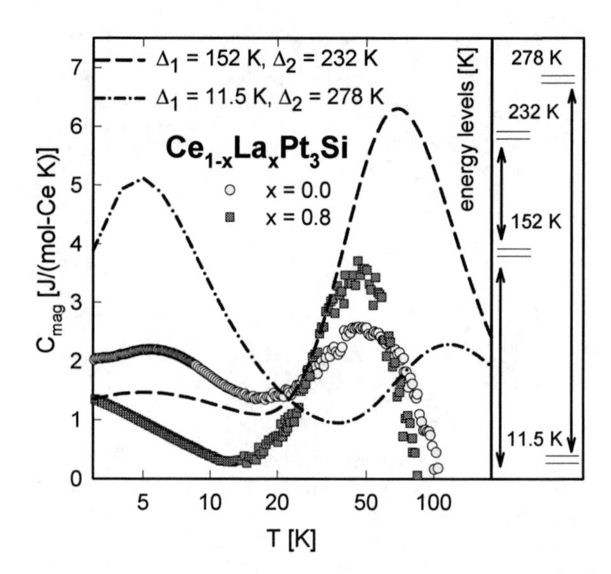

FIGURE 4. Temperature dependent magnetic contribution to the specific heat, C_{mag}, of $CePt_3Si$ and $Ce_{0.2}La_{0.8}Pt_3Si$. The dashed and the dashed-dotted lines are model calculations with CEF splitting sequences 0-152-232 and 0-11.5-278 K, respectively. The ground state degeneracy is lifted by Kondo interaction with $T_K = 10$ K. The right panel schematically shows the two CEF schemes

In Fig. 4 we show the magnetic contribution, $C_{mag}(T)$, to the specific heat in the paramagnetic temperature regime of $CePt_3Si$ and of $Ce_{0.2}La_{0.8}Pt_3Si$, together with a calculation of the Schottky contribution according to the above indicated sequences. In order to account for the lifting of the ground state doublet, a Kondo contribution to the specific heat with $T_K = 10$ K is added. Since the heat capacity of $LaPt_3Si$ exceeds those of the Ce based compounds at high temperature, $C_{mag}(T)$ crosses over to negative values. This denotes that the lattice dynamics of the Ce and the La based compounds behave increasingly different as the temperature rises.

A comparison between experimental and calculated data is thus reliable only at low - and becomes rather uncertain at high temperatures. Nevertheless, this comparison indicates that the system does not possess a low lying doublet (about 10 K above the ground state), which would require much higher C_{mag} at low -, but much smaller values at high temperatures. Hence, physics at low temperatures originates from the ground state doublet only. A similar conclusion can be drawn from the temperature dependent entropy S_{mag} derived from the

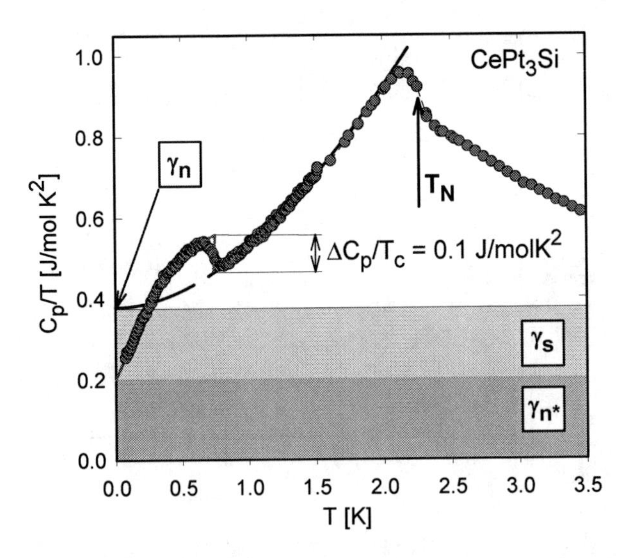

FIGURE 5. Temperature dependent specific heat C_p/T of $CePt_3Si$; the dashed line is a T^3 extrapolation of $C_p(T)$ at 0 T. The light and the dark shaded areas represent the Sommerfeld values γ_s and γ_n^* associated with electrons condensing into Cooper pairs, and normal state electrons, respectively.

data of Fig. 4 by integrating up C_{mag}/T. Qualitatively, the Kondo contribution at low temperatures seems to accurately account for the experimentally derived magnetic entropy when assuming the first excited level to be well separated from the ground state. A low lying level at $\Delta_1 = 11.5$ K, however, releases to much entropy and thus does not convincingly describe the heat capacity data.

Superconducting Properties of $CePt_3Si$

Signs of bulk SC of $CePt_3Si$ below $T_c = 0.75$ K are numerous: zero resistivity, diamagnetic signal in the susceptibility, a jump in the specific heat and NMR relaxation rate at T_c.

Substantial information concerning the superconducting state is provided by heat capacity data taken at low temperatures for $CePt_3Si$. Results are shown in Fig. 5 as C_p/T vs. T. The phonon contribution is negligible in the temperature range shown. Besides the already mentioned magnetic phase transition at $T_N = 2.2$ K and the logarithmic contribution above that temperature, the superconducting transition at $T_c \approx 0.75$ K is the most prominent feature.

The Sommerfeld coefficient $\gamma_n \approx 0.39$ J/molK^2 of $CePt_3Si$ at zero field, obtained from an extrapolation of the antiferromagnetically ordered region ($C_p \propto T^3$), evidences the large effective masses of the charge carriers involved. The extrapolation shown in Fig. 5 (dashed line) satisfies the basic requirement of superconductivity, the entropy balance between the SC and normal state regions. Another careful extrapolation of the heat capacity data in the superconducting temperature range towards zero yields about $\gamma_n^* \approx 210$ mJ/mol-K^2. This non-vanishing contribution within the SC state may hint at two mechanisms: i) not all electrons are involved in the SC condensate; rather, roughly half of them ($\gamma_s \approx \gamma_n^*$) are forming normal state long range magnetic order, which then co-exists with superconductivity on a microscopic scale. ii) A somewhat gapless superconducting state. This would cause finite values of the Sommerfeld constant and power laws of the thermodynamic and transport properties instead of an exponential behavior as in typical BCS systems. The coexistence of both states is evidenced from μSR spectroscopy and a superconducting state with nodes in the gap (gapless at certain sites) is supported by magnetic penetration depth $\lambda(T)$ measurements (see below).

The jump of the specific heat anomaly associated with superconductivity, $\Delta C_p/T|_{T_c} \approx 0.1$J/molK^2, leads to $\Delta C_p/(\gamma_n T_c) \approx 0.25$, which is significantly smaller than the figure expected from the BCS theory ($\Delta C_p/(\gamma T_c) \approx 1.43$). Even using the electronic specific heat coefficient in the SC state, $\gamma_s \approx 0.18(1)$ J/molK^2, we obtained $\Delta C_p/(\gamma_s T_c) \approx 0.55$ that is still below the BCS value.

Again, two scenarios may explain the substantial reduction of $\Delta C_p/\gamma T_c$ with respect to the BCS value; i) strongly anisotropic gaps yield a reduced magnitude of $\Delta C_p/(\gamma T_c)$ [17] and ii) not all electrons condense into Cooper pairs. This may imply that the electrons responsible for normal state features, such as antiferromagnetic order, coexist with those forming the Cooper pairs. In fact, the finite value of $\gamma_s \approx 0.18$ J/molK^2 provides evidence that even at $T = 0$ K a significant portion of the Fermi surface is still not involved in the SC condensate. It should be noted that $\Delta C_p/(\gamma T_c)$ of unconventional spin triplet superconductor Sr_2RuO_4 is similarily downsized as that of $CePt_3Si$ [17].

Microscopic evidence for the latter conclusion can be found from zero-field μSR spectroscopy data obtained in the magnetic phase below and above T_c in the magnetic phase [18]. At temperatures much above T_N, the μSR signal is characteristic of a paramagnetic state with a depolarization solely arising from nuclear moments. Below T_N the μSR signal indicates that the full sample volume orders magnetically. High statistic runs performed above and below T_c did not show any change of the magnetic signal, supporting the view of a microscopic coexistence between magnetism and SC.

As shown in Fig. 6, the application of magnetic fields reduces T_c, resulting in a rather large change of $dH_{c2}/dT \approx -8.5$ T/K. An extrapolation of $T_c(H)$ towards zero yields $H_{c2}(0) \approx 5$ T, well above the paramagnetic limiting field $H_p \approx 1.1$ T [9].

This order of magnitude for the upper critical field

is found from measurements on a single crystal as well [19]. It is interesting to note that the anisotropy deduced from measurements parallel and perpendicular to the $\vec{c}$-axis is rather small, i.e., $H_{c2}^{c}/H_{c2}^{ab} \approx 1.18$ for $T \rightarrow 0$. The large H_{c2} values imply that the Zeeman splitting must be non-negligible below T_c since the estimated paramagnetic limit $H_p \approx 1.1$ T. The lack of inversion symmetry in $CePt_3Si$ can explain the absence of paramagnetic limiting even in the case the pairing state would be *s*-wave like [20]. A theoretical model, however, predicts a distinct anisotropy for in-plane fields and fields along the $\vec{c}$-axis [21]. For fields along the $\vec{c}$-axis, H_{c2} is independent of the Zeeman field and thus there is no paramagnetic limiting. This follows from $\vec{g}_{\vec{k}}\vec{H} = 0$; $\alpha\vec{g}_{\vec{k}}\sigma$ introduces the antisymmetric spin-orbit coupling, where α is the coupling constant and $\vec{\sigma}$ are Pauli matrices [20]. In fact, band structure calculations revealed a significant coupling strength α with $\alpha > k_BT_c$ [22]. In the case of broken symmetry, $\vec{g}_{\vec{k}} = -\vec{g}_{-\vec{k}}$. The Rashba-type like spin-orbit coupling $\vec{g}_{\vec{k}}\vec{\sigma}$ would not account for the relatively large in-plane upper critical field. In order to get rid of the distinct reduction of the in-plane paramagnetic effect in $CePt_3Si$, a helical structure of the order parameter was assumed [21]. In the case of $\vec{H} \perp \vec{c}$, the field can induce a phase which gives rise to an additional phase factor in the many body wave function, thereby creating a helical vortex phase, able to explain the diminishing of paramagnetic limiting. For $\vec{H} \parallel \vec{c}$, the vortex phase appears to be quite conventional.

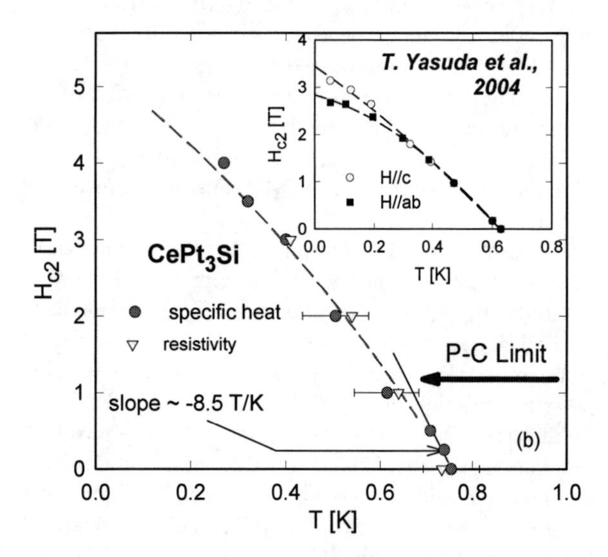

FIGURE 6. Temperature dependence of the upper critical field H_{c2}. The solid straight line yields $H_{c2}' \approx -8.5$T/K; the dashed line is a guide to the eyes (Ref. [9]). PC indicates the Pauli - Clogston limiting field. The inset shows measurements of H_{c2} on a single crystalline $CePt_3Si$ sample with fields along and perpendicular to the $\vec{c}$ axis. Data are taken from Ref. [19].

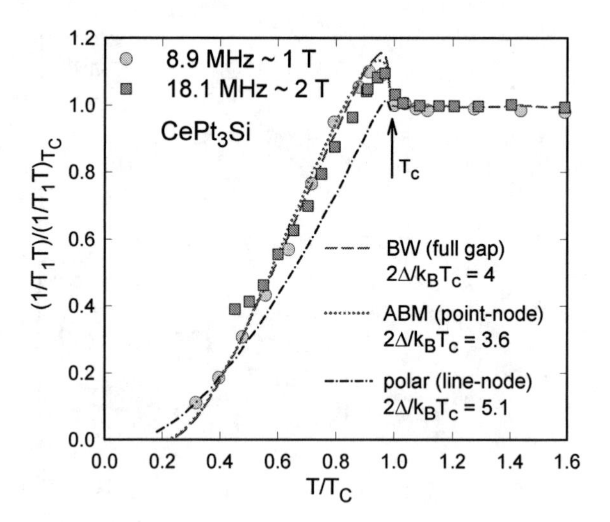

FIGURE 7. A plot of $(1/T_1T)/(1/T_1T)_{T_c}$ vs. T/T_c at 8.9 MHz ($H \sim 1$ T) and 18.1 MHz ($H \sim 2$ T). The dashed line is for the Balian-Werthamer model (BW isotropic triplet SC state) with a value of $2\Delta/k_BT_c = 4$. The dotted line assumes a point-node model with $2\Delta/k_BT_c = 3.6$ and the dashed-dotted line represents a fit by a line-node gap model with $2\Delta/k_BT_c = 5.1$.

Another microscopic information about the SC state can be obtained from the temperature dependent ^{195}Pt nuclear spin-relaxation rate $1/T_1$ [23]. Results are shown as a $(1/T_1T)/(1/T_1T)_{T_c}$ vs. T/T_c plot in Fig. 7 for 8.9 and 18.1 MHz. The relaxation behavior $1/T_1T$ of $CePt_3Si$ reminds to a kind of Hebel-Slichter anomaly [24] indicating coherence effects as in conventional BCS SC. The peak height, however, is significantly smaller than that observed for conventional BCS SC and, additionally, shows no field dependence at the 8.9 MHz ($H \sim 1$ T) and 18.1 MHz ($H \sim 2$ T) run. Notably, $CePt_3Si$ is the first HF SC that exhibits a peak in $1/T_1T$ just below T_c. Neither an exponential law nor the usual T^3 law, reported for most of the unconventional HF SC (see e.g., Ref. [25] and Refs. therein), is observed for the data down to $T = 0.2$ K ($\approx 0.3T_c$).

To account for the relaxation behavior below T_c, three unconventional models were adopted for a description of the temperature dependence of $1/T_1$ at $H \sim 1$ T. The dashed line in Fig. 7 represents a fit according to the Balian-Werthamer model (BW isotropic spin-triplet SC state) with $2\Delta/k_BT_c = 3.9$ [26]. Note that the peak of the BW model in $1/T_1T$ originates from the presence of an isotropic energy gap. The dashed-dotted line is a fit using a line-node model with $2\Delta/k_BT_c = 5.1$. The dotted line refers to a point-node model with $2\Delta/k_BT_c = 3.6$. The models used, however, failed to give satisfactory description of the observed temperature dependence of $1/T_1$ over the entire temperature range.

The data seem to start following the line-node model at the lowest measured temperatures. However, the data

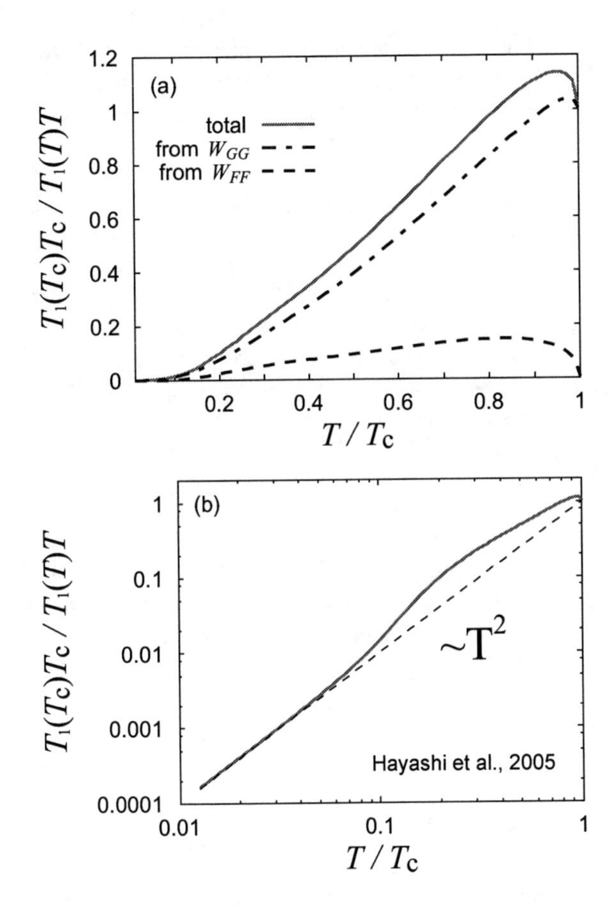

FIGURE 8. Temperature dependent nuclear spin- lattice relaxation rate $1/T_1T$ (solid lines). (a) Dashed line is the contribution of the anomalous Green functions W_{FF} related to the coherence effect. Dashed-dotted line is the contribution of the regular Green functions W_{GG} related to the density of states. (b) Plot of the same data on a double- logarithmic scale. Dotted line is a plot of T^2. From the plot, it is noticed that $1/T_1$ follows the T^3 law at low temperatures (after Ref. [27]).

are described reasonably well by the BW (nodeless) model just below T_c. The experimentally observed peak in $1/T_1T$ would, possibly, indicate the presence of an isotropic energy gap.

In fact, a recent study of the nuclear magnetic relaxation rate for a non-centrosymmetric superconductor [27] proposed a two-component order parameter composed of spin-singlet and spin-triplet pairing components. The Hamiltonian considered incorporates the lack of inversion symmetry through an antisymmetric spin-orbit coupling [20]. The pairing interaction is characterized by three coupling constants, taking into account interactions within each spin channel and scattering of Cooper pairs between the two channels.

Results concerning the nuclear spin-lattice relaxation rate $1/T_1$ are plotted in Figs. 8(a,b); data are taken from Ref. [27]. Using reasonable temperature dependencies

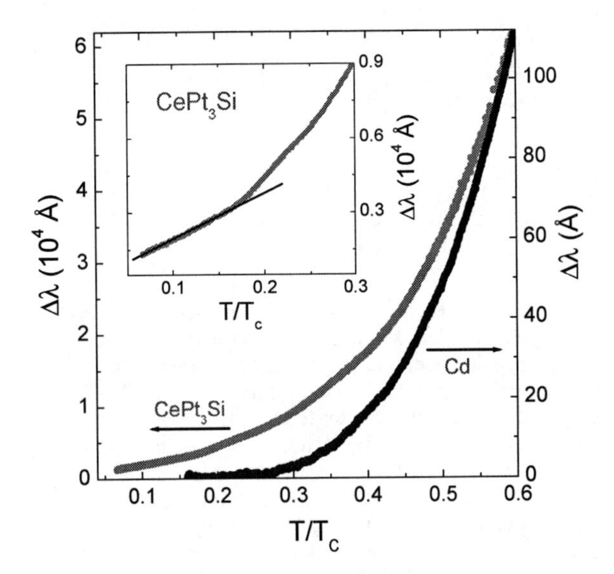

FIGURE 9. Low temperature behavior of $\Delta\lambda(T)$ in polycrystalline CePt$_3$Si and Cd samples. $\lambda(T)$ is independent of temperature below $0.28T_c$ in Cadmium, as expected for a s-wave superconductor. The inset displays a linear low temperature behavior of $\lambda(T)$ in CePt$_3$Si.

of the spin-singlet and the spin-triplet component of the order parameter [27], reveals two distinct features: i) slightly below T_c a coherence effect in terms of a Hebel-Slichter peak becomes obvious inferred from the contribution W_{FF}, where W_{FF} is related to coherence effects. ii) Well below T_c the $1/T_1T$ relaxation rate behaves proportional to T^2, a power law which is characteristic of line nodes in the gap. The respective function W_{GG} describes the contribution from the density of states. It should be noted that the nodes formed are a result from the superposition of spin-singlet and spin-triplet contributions. Each separately would not produce line nodes [27].

A comparison of the theoretically obtained results with the experimental data (see Fig. 7) reveals excellent qualitative agreement, and the two dominating features, i.e., the peak below T_c and the T^3 are well reproduced. This indicates that the superconducting order parameter consists of two components, the spin-singlet (s-wave) and a spin triplet (p-wave) component.

Fig. 9 shows the temperature dependent magnetic penetration depth $\lambda(T)$ deduced for polycrystalline CePt$_3$Si [13]. $\lambda(T)$ of Cadmium is added, being a classic s-wave superconductor, for which the low temperature dependence of $\lambda(T)$ is exponential. The inset to Fig. 9 is a close-up of $\Delta\lambda(T)$ vs T/T_c for CePt$_3$Si for temperatures $T < 0.3T_c$, where it can be clearly seen that the penetration depth follows a linear temperature behavior below $0.17T_c$. For line nodes in the energy gap the penetration depth is expected to be linear in the low temperature limit, where the temperature dependence of the

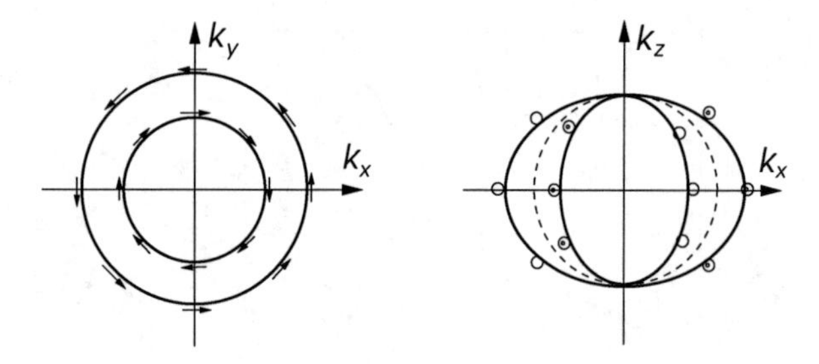

FIGURE 10. Fermi surface splitting. The lack of inversion symmetry gives rise to antisymmetric spin-orbit coupling which leads in general to a splitting of the spin degeneracy of the electronic states. The figure depicts schematically the splitting of the Fermi surfaces for the generating point group C_{4v} valid in $CePt_3Si$. The spinor states are k-dependent in the way that the spin quantization axes lie perpendicular to the Fermi surface and the z-axis with the spin pointing in opposite direction for the two Fermi surfaces. This feature is essential for the possible spin configuration of the pairing states which can be constructed from two electrons on the same Fermi surface.

energy gap can be neglected. Since $CePt_3Si$ is both a clean ($l > \xi_0$) and a local ($\lambda(0) > \xi_0$) superconductor [9], the penetration depth experimental result points out to the existence of lines of nodes in the structure of the superconducting pairing state and, hence, to unconventional superconductivity in $CePt_3Si$. The latter is in perfect agreement with the low temperature NMR $1/T_1$ data.

Summary

We summarize that non-centrosymmetric $CePt_3Si$ is a heavy fermion SC with $T_c = 0.75$ K that orders magnetically at $T_N = 2.2$ K. Specific heat, NMR and μSR studies indicate that superconductivity and long range magnetic order coexist on a microscopic scale and may be originated by two different sets of electrons. The NMR $1/T_1$ shows unexpected features which were neither found before in conventional nor in heavy fermion SC, indicative of very unusual shapes of the SC order parameter. Unconventional SC is backed also by penetration depth studies. In fact, the various theoretical scenarios developed for this compound support these conclusions.

The absence of inversion symmetry gives rise to antisymmetric spin-orbit coupling. In the case of $CePt_3Si$ the generating point group is C_{4v} which implies that there is no reflection symmetry $z \to -z$ where z is the fourfold tetragonal rotation axis. This is incorporated into band structure by a Rashba-like term

$$\varepsilon_{\vec{k}}\sigma_0 \to \varepsilon_{\vec{k}}\sigma_0 + \alpha(\hat{z} \times \vec{k}) \cdot \vec{\sigma} \tag{2}$$

with σ_0 as unit matrix and $\vec{\sigma}$ as the Pauli matrices in the electron spinor space. Due to this spin-orbit coupling, the electron bands split into two yielding two different Fermi surfaces with opposite spinor orientation as depicted in Fig. 10. Obviously there is now a clear restriction on the spin configurations for zero momentum Cooper pairs.

The mixing of the spin singlet and spin triplet channel for this electron band structure leads to a non-vanishing spin susceptibility at $T = 0$ in any case [3, 28, 29]. Thus the absence of paramagnetic limiting would be naturally explained. In fact, previously performed Knight shift measurements carried out parallel and perpendicular to the $\vec{c}$-axis [30] exhibit almost temperature independent data above and below T_c. This microscopically corroborates both the presence of spin-triplet components in the order parameter and a large value of the spin-orbit coupling constant α [28].

ACKNOWLEDGMENTS

Work supported by the Asutrian FWF P16370, P18054 and by FONACIT, Venezuela, under Grant No. S1-2001000639.

REFERENCES

1. N. D. Mathur et al., *Nature* **394**, 39 (1998).
2. P.W. Anderson, *Phys. Rev.* B **30**, 4000 (1984).
3. L.P. Gor'kov and E.I. Rashba, *Phys. Rev. Lett.* **87**, 037004 (2001).
4. M.C. Krupka et al., *J. Less Common Met.* **19**, 113 (1969).
5. A.L. Giorgi et al., *J. Less Common Met.* **2**, 131 (1970).
6. G. Amano et al., *J. Phys. Soc. Jpn.* **73**, 530 (2004).
7. M. Hanawa et al., *Phys. Rev. Lett.* **87**, 187001 (2001).
8. H. Sakai et al., *J. Phys.: Cond. Mat.* **13**, L785 (2001).
9. E. Bauer et al., *Phys. Rev. Lett.* **92**, 027003 (2004).
10. T. Akazawa et al., *J. Phys.: Cond. Mat.* **16**, L29 (2004).
11. N. Kimura, et al., *Phys. Rev. Lett.* **95**, 247004 (2005).
12. K. Izawa, et al., *Phys. Rev. Lett.* **94**, 197002 (2005).
13. I. Bonalde et al., *Phys. Rev. Lett.* **94**, 207002 (2005).
14. A. Hewson, *The Kondo Problem to Heavy Fermions*, Cambridge University Press, (1993).
15. N. Metoki et al., *J. Phys.: Cond. Mat.* **16**, L207 (2004).
16. E. Bauer et al., *Physica* B **359-361**, 360 (2005) .
17. A.P. Mackenzie and Y. Maeno, *Rev. Mod. Phys.* **75**, 657 (2003).
18. A. Amato, et al., *Phys. Rev.* B **71**, 092501 (2005).
19. T. Yasuda et al., *J. Phys. Soc. Jpn.* **73**, 1657 (2004).
20. P.A. Frigeri et al., *Phys. Rev. Lett.* **92**, 097001 (2004).
21. R.P. Kaur et al., Phys. Rev. Lett. **94**, 13702 (2005).
22. K.V.Samokhin et al., *Phys. Rev.* B **69**, 094514 (2004).
23. M. Yogi, et al., *Phys. Rev. Lett.* **93** , 027003 (2004).
24. L.C.Hebel and C.P. Slichter, *Phys. Rev.* **107**, 901 (1957).
25. H. Tou et al., *J. Phys. Soc. Jpn.* **64**, 725 (2003).
26. R. Balian and N. R. Werthamer, *Phys. Rev.* **131**, 1553 (1963).
27. Hayashi et al., *cond-mat*/0504176.
28. P.A. Frigeri et al., *New J. Phys.* **6**, 115 (2004).
29. L.N. Bulaevski et al., *Sov. Phys. JETP* **44**, 1243 (1976).
30. M. Yogi et al., *Physica* B (2005) in press.

Lines of Nodes in the Superconducting Gap of Noncentrosymmetric $CePt_3Si$

Ismardo Bonalde*, Werner Brämer-Escamilla* and Ernst Bauer†

*Centro de Física, Instituto Venezolano de Investigaciones Científicas, Apartado 21874, Caracas 1020-A, Venezuela
†Institut für Festkörperphysik, Technische Universität Wien, A-1040 Wien, Austria

Abstract. We report on magnetic penetration depth $\lambda(T)$ measurements in $CePt_3Si$ down to 50 mK. It is found that $\lambda(T)$ is proportional to T below $T < 0.16T_c$, which is interpreted as evidence for lines of nodes in the gap. A kink is observed in the penetration depth data around 0.5 K. We analyze the implications of these results on the symmetry of the superconducting pairing state.

Keywords: penetration depth, $CePt_3Si$, order parameter, symmetry
PACS: 74.20.Rp, 74.25.Nf, 74.70.Tx

INTRODUCTION

The recently discovered heavy-fermion superconductor $CePt_3Si$ ($T_c = 0.75$K) [1] has a number of interesting properties. It has no spatial inversion symmetry, which implies that the spatial component of the wave function has no definite parity. To preserve the antisymmetry of the pairing, the same applies to the spin component. The direct implication of the parity violation is that the pairing symmetry should, in principle, be a mixture of spin-singlet and spin-triplet states. Recent theoretical arguments [2, 3] indicate that such a mixture can be neglected under certain conditions; thus, the singlet and triplet pairing states can be considered separately as if parity is conserved.

In $CePt_3Si$ the upper critical field H_{c2} exceeds the paramagnetic limiting field [1], which readily points out to the occurrence of pure spin-triplet pairing in this material. This leads to a conflict because pure spin-triplet channels are known to be forbidden in the absence of inversion symmetry [4]. Frigeri *et al.* [2] showed, however, that spin-triplet states are not completely suppressed in the absence of an inversion center. Further theoretical support for an odd-parity state was given by Samokhin *et al.* [3], who showed that SOC in $CePt_3Si$ is strong and that this fact, along with the absence of inversion symmetry, always yields an odd-parity order parameter with line nodes for one-dimensional irreducible representations (IRs). In recent work, Sergienko and Curnoe [5] proposed a model for noncentrosymmetric superconductors with strong SOC in which the temperature dependence of the thermodynamic properties would be governed by an even-parity function. Such a function would transform according to the IRs of the point group. In the scenarios of Frigari *et al.* and Sergienko and Curnoe the superconducting gap is an isotropic and fully gapped even function.

To further complicate the physical picture, recent specific heat measurements suggest that the intrinsic properties of $CePt_3Si$ could be masked by sample-dependent results. Early specific heat data exhibit an anomaly -around 0.5 K- which suggested the existence of a second superconducting transition [6, 7]. Later, such an anomaly was attributed to an antiferromagnetic (AF) transition of a second phase existing in some samples [8]. More recently, measurements of the specific heat in single-crystals [9] show no second anomaly and the appearance of the superconductive phase at $T_c = 0.5$ K. All these controversial discussions make unclear the physics of the superconducting state of $CePt_3Si$.

Here we report measurements of the magnetic penetration depth $\lambda(T)$ in polycrystalline $CePt_3Si$. $\lambda(T)$ has proven to be a fundamental probe of the energy gap structure of unconventional superconductors. The data exhibit a linear temperature dependence below 130 mK, which is interpreted as a strong evidence for line nodes in the gap structure of the superconducting pairing state of $CePt_3Si$.

RESULTS AND DISCUSSION

The details of the sample and the experimental setup can be found elsewhere [10]. Figure 1(a) shows $\Delta\lambda(T)$ against T. Here $\Delta\lambda(T) = \lambda(T) - \lambda(T_{min})$, and T_{min} is the lowest temperature of the experiment. A strikingly broad transition is observed. A broad transition in polycrystalline samples is usually associated to inter-grain or proximity effects, but such effects are not expected to be relevant in the present case because of the very small measuring magnetic fields (about 5 mOe) [11]. The in-

CP850, *Low Temperature Physics: 24th International Conference on Low Temperature Physics;*
edited by Y. Takano, S. P. Hershfield, S. O. Hill, P. J. Hirschfeld, and A. M. Goldman

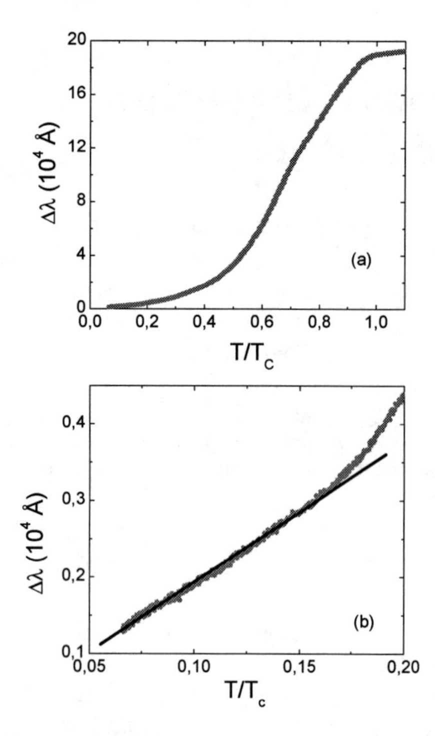

FIGURE 1. (a) $\Delta\lambda(T)$ vs T/T_c for a polycrystalline sample of $CePt_3Si$, and (b) the same data in the low temperature region.

flection point or kink around 0.53 K can then be interpreted as signaling a transition to either another superconducting state or an AF phase. $\lambda(T)$ falls rapidly below a superconducting transition, and peaks at the Neel temperature before falling exponentially at lower temperatures. If the kink in $\lambda(T)$ around 0.5 K is due to an AF transition, it would imply that such a transition would appear in the susceptibility data as a small peak. On the other hand, one can discern between AF and superconducting transitions by measuring the upper critical field $H_{c2}(T)$. We cannot carry out this type of measurement in our lab.

Figure 1(b) shows $\Delta\lambda(T)$ for $T < 0.2T_c$. We notice here that in this low temperature region $\lambda(T)$ is a response of the superconducting phase present in the sample. Any signal from an AF phase which emerges at a much higher temperature would be negligible. It can be clearly seen in Figure 1(b) that $\Delta\lambda(T) \propto T$ below $T \simeq 0.16T_c$. A T-linear dependence of the penetration depth in the low temperature region is expected for clean, local superconductors with line nodes in the gap function. Assuming a spherical Fermi surface (as it has been done in other studies [2, 3]), it is found [1] that both the mean free path $l = 800$ Å and $\lambda(0) = 11000$ Å are larger than the coherence length $\xi_0 = 81$ Å, implying that $CePt_3Si$ is a clean, local superconductor. Hence, the experimental penetration depth results point out to the existence of line nodes in the structure of the superconducting pairing state and, therefore, to unconventional superconductivity in $CePt_3Si$.

Some of the theories proposed [3, 5] for $CePt_3Si$ allow, without specifying a particular state, for the existence of line nodes in the pairing state. This is consistent with the penetration depth data presented here. The $\mathbf{d}(\mathbf{k}) = \hat{\mathbf{x}}k_y - \hat{\mathbf{y}}k_x$ pairing model [2] with no line nodes is not supported by the $\lambda(T)$ data. However, based on the theoretical scenarios suggested so far for the noncentrosymmetric $CePt_3Si$ [2, 3, 5] it is not even possible to discern the parity -whether it is singlet or triplet- of the pairing state.

CONCLUSIONS

Regardless of the pairing state multiplicity and the present of AF phases, the penetration depth data require the existence of line nodes in the superconducting gap structure of the studied $CePt_3Si$ sample.

ACKNOWLEDGMENTS

The work was supported by FONACIT, Venezuela, grant S1-2001000693, and the Austrian FWF P16370.

REFERENCES

1. E. Bauer, G. Hilscher, H. Michor, C. Paul, E. W. Scheidt, A. Gribanov, Y. Seropegin, H. Noël, M. Sigrist, and P. Rogl, *Phys. Rev. Lett.* **92**, 027003 (2004).
2. P. A. Frigeri, D. F. Agterberg, A. Koga, and M. Sigrist, *Phys. Rev. Lett.* **92**, 097001 (2004).
3. K. V. Samokhin, E. S. Zijlstra, and S. K. Bose, *Phys. Rev. B* **69**, 094514 (2004).
4. P. W. Anderson, *Phys. Rev. B* **30**, 4000 (1984).
5. I. A. Sergienko, and S. H. Curnoe, *Phys. Rev. B* **70**, 214510 (2004).
6. E. W. Scheidt, F. Mayr, G. Eickerling, P. Rogl, and E. Bauer, *J. Phys.: Condens. Matter* **17**, L121 (2005).
7. G. R. Stewart, Abstract:B39.00003:Two Superconducting Transitions in Heavy Fermion $CePt_3Si$, 2005 APS March Meeting.
8. J. S. Kim, D. J. Mixson, D. J. Burnette, T. Jones, P. Kumar, B. Andrake, G. R. Stewart, V. Cracium, W. Acree, H. Q. Yuan, D. Vandervelde, and M. B. Salamon, *Phys. Rev. B* **71**, 212505 (2005).
9. N. Tateiwa, Y. Haga, T. Matsuda, S. Ikeda, T. Yasuda, T. Takeuchi, R. Settai, and Y. Ōnuki, *J. Phys. Soc. Japan* **74**, 1903 (2005).
10. I. Bonalde, W. Brämer-Escamilla, and E. Bauer, *Phys. Rev. Lett.* **94**, 207002 (2005).
11. R. B. Goldfarb, M. Lelental, and C. A. Thompson, "Alternating-field susceptometry and magnetic susceptibility of superconductors", in *Magnetic Susceptibility of Superconductors and Other Spin Systems*, edited by R. A. Hein, New York: Plenum Press, 1991, pp. 49-80.

Neutron Scattering Studies On Stoner Gap In The Superconducting Ferromagnet UGe_2 By Using A Small Piston-Cylinder-Type Clamp Cell

N. Aso[*], Y. Uwatoko[†], T. Fujiwara[†], G. Motoyama[¶], S. Ban[‡], Y. Homma[§], Y. Shiokawa[§], K. Hirota[*], and N.K. Sato[‡]

[*] *Neutron Sci. Lab., ISSP, University of Tokyo, Tokai, Ibaraki 319-1106, Japan*
[†] *ISSP, University of Tokyo, Kashiwa 277-8581, Japan*
[¶] *Dept. of Mat. Sci., Graduate School of Material Science, University of Hyogo, Hyogo 678-1297 Japan*
[‡] *Department of Physics, Graduate School of Science, Nagoya University, Nagoya 464-8602, Japan*
[§] *Oarai Branch, Inst. for Mater. Research, Tohoku University, Oarai, Ibaraki 311-1313, Japan*

Abstract. The design and fundamental properties of a small copper-beryllium (CuBe) based piston-cylinder-type clamp cell for low-temperature (LT) neutron diffraction (ND) measurements are reported. The results obtained for the superconducting ferromagnet UGe_2 showed that the perfectly polarized ferromagnetic state is realized below P_x.

Keywords: pressure cell, UGe_2, neutron diffraction, superconductivity, ferromagnetism, Stoner model
PACS: 65.40.-b, 71.28.+d, 71.30.+h, 71.27.+a

PRESSURE CELL

During the last two decades, strongly correlated electrons systems (SCES) have been investigated extensively. Recently, high pressure (HP) investigations have been actively carried out since such SCESs have ground states which are quite sensitive to external pressure. We have used the McWhan-type HP cell [1,2] for neutron scattering experiments under HP in the reactor JRR-3M in JAERI, Japan but this cell is so big and heavy that it is not good for the LT measurements using a dilution refrigerator due to the limitation of the space. Thus, there is an actual need to develop a smaller and more convenient pressure cell for LT ND measurements. In this paper, we report the design and fundamental properties of a small CuBe-based clamp cell and illustrate its application to the superconducting ferromagnet UGe_2, as an example.

The HP cell was designed for use with an existing ISSP dilution refrigerator with the sample space of height ~ 65 mm and diameter ~ 30 mm. Figure 1(a) shows a schematic drawing of the cell. Hardened CuBe alloy was used for most of the cell to obtain the maximum pressure of 2 GPa. As an example of the pressurizing test for the cell, we illustrate in Fig. 1(b) the pressure dependence of linewidths of $\boldsymbol{Q}$=(2,0,0)

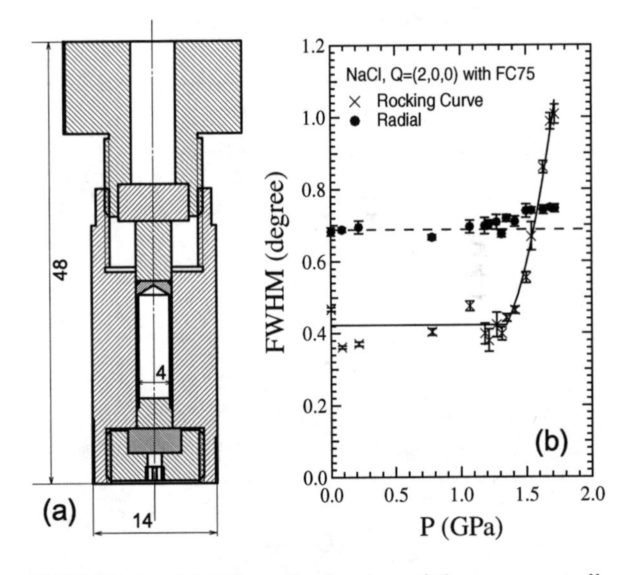

FIGURE 1. (a) Schematic drawing of the pressure cell. Hardened CuBe alloy is used for a cylinder. Tungsten carbide is used for a piston and a backup. The various kinds of inner cells made of teflon, copper and aluminum were tested. (b) Full widths at half maximum (FWHM) of Bragg profiles at $\boldsymbol{Q}$=(2,0,0) for single-crystalline NaCl as a function of pressure. The lines are guides to the eye.

CP850, *Low Temperature Physics: 24th International Conference on Low Temperature Physics;*
edited by Y. Takano, S. P. Hershfield, S. O. Hill, P. J. Hirschfeld, and A. M. Goldman

Bragg reflections of single-crystalline NaCl, where a fluorinert FC75 was used as a pressure transmitting media. The pressure was estimated by determining the change in lattice parameter of the NaCl itself. [3] The rocking curve linewidth keeps constant up to around 1.3 GPa and increases above the pressure. The radial scan linewidths, which roughly measure the pressure distribution in the cell, are unchanged within the accuracy of several % up to 1.3 GPa and then slowly increases. These findings indicate that there is a hydrostatic limit at around a critical pressure of P = 1.3 GPa for FC75.

APPLICATION TO UGE2

By using the above HP cell with a combination of FC75 and Al inner cell, a single crystal of UGe_2 can be successfully pressurized up to 1.6 GPa at LT. The rocking curve linewidth for UGe_2 does not change through this pressure range. Figure 2 shows the temperature-pressure phase diagram determined by the temperature dependence of ferromagnetic Bragg peak intensities at $\boldsymbol{Q}$ = (0,0,1). The detailed experimental set up is reported elsewhere. [4]

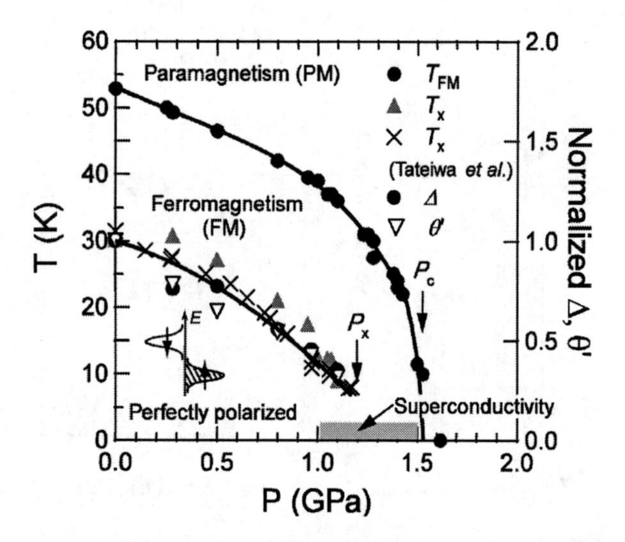

FIGURE 2. Phase diagram of UGe_2 determined by our ND measurements. The solid lines are guides to the eye. Pressure dependence of the obtained parameters Δ, and Θ' from the Stoner model are also plotted, which are normalized with respect to a respective value; Δ = 39.5 K and Θ' = 83.4 K. We also plot T_x taken from Ref. [5].

A Curie temperature (T_{FM}) is about 52 K at ambient pressure, and monotonically decreases with increasing pressure. The characteristic temperature T_x (~ 30 K at ambient pressure), which is characterized by a steep increase in the ferromagnetic Bragg peak intensities, also decreases with increasing pressure and becomes suppressed to zero another critical pressure P_x (~ 1.2 GPa). In Fig. 2, we also plot the pressure dependence of the obtained parameters Δ, and Θ' from the Stoner model [6], which is applied to the temperature dependence of ferromagnetic Bragg peak intensities at $\boldsymbol{Q}$ = (0,0,1) below T_x. In the Stoner model, the magnetization (square root of the neutron intensities) is expressed as follows;

$$M = M_0\left\{1 - \alpha T^{\frac{3}{2}} \exp(-\Delta / T)\right\}. \quad (1)$$

$$\alpha = \frac{3}{4}\sqrt{\pi}\left(\frac{1}{E_F}\right)^{\frac{3}{2}}, \Delta = 2E_F\left(\frac{\Theta'}{E_F} - 2^{-\frac{1}{3}}\right). \quad (2)$$

where M_0 indicates the magnetization at zero temperature, Δ a so-called Stoner gap, E_F a Fermi energy, and θ' is a molecular field coefficient. It should be noted that these quantities of Δ, Θ' and T_x lie on a single line, suggesting that the characteristic temperature T_x is related to the Stoner gap Δ (equivalently Θ') in the heavy quasiparticle band. These observations imply that the perfectly polarized ferromagnetic state is realized below P_x in UGe_2.

ACKNOWLEDGMENTS

The LT ND experiments were supported by ISSP, University of Tokyo (PACS No. 3432 and 3474). This work was supported by a Grant-in-Aid from the Ministry of Education, Culture, Sports, Science and Technology, Japan.

REFERENCES

1. McWhan, D. B. et al., *Phys. Rev.* **B 120,** 4612-4623 (1979).
2. Onodera, A. et al., *Jpn. J. Appl. Phys.* **26,** 152-156 (1987).
3. Skelton, E. F. et al., *Rev. Sci. Instrum.* **55,** 849-855 (1984).
4. Aso, N., et al., cond-matt/0505266 (2005).
5. Tateiwa, N. et al., *J. Phys. Cond. Matt.* **13,** L17-L23 (2001); *ibid* **13,** 6443 (2001).
6. Stoner, E. C. *Proc. Roy. Soc.* **A 165,** 372 (1938).

Spin Fluctuations and Weak Pseudogap Behaviors in $Na_{0.35}CoO_2$: Renormalization of Band Structure

Keiji Yada and Hiroshi Kontani

Department of Physics, Nagoya University, Nagoya, 464-8602, Japan

Abstract. We analyze the normal electronic states of $Na_{0.35}CoO_2$ based on the multi-orbital Hubbard model using the FLEX approximation. The fundamental electronic property of this system is drastically changed by the presence or absence of the small hole pockets associated with the e'_g orbital. This change of the Fermi surface topology may be caused by the crystalline electric splitting due to the trigonal distortion. When small hole pockets are absent, the weak pseudogap behaviors appear in the density of states and the uniform spin susceptibility, which are observed by recent experiments. We estimate the mass enhancement factor of quasiparticle $m^*/m \simeq 1.5 \sim 1.8$. This result supports the recent ARPES measurements.

Keywords: FLEX, Na_xCoO_2, spin fluctuations, pseudogap
PACS: 74.25.-q, 74.25.Ha, 74.25.Jb

$Na_{0.35}CoO_2 \cdot 1.3H_2O$ is the first Co-oxide superconductor with $T_c \sim 4.5K$[1]. While many theoretical and experimental studies are widely performed, the topology of Fermi surface (FS) and the low-energy electronic structure are still unresolved. To resolve these problems is very important to find out the mechanism of superconductivity.

In Na_xCoO_2, local density approximation (LDA) calculations[2] have predicted that Na_xCoO_2 has a large FS associated with the a_{1g} band and six small hole pockets corresponding to the e'_g band. However, such small pockets are not observed in recent ARPES measurements[3, 4]. ARPES measurements also observed that renormalized quasiparticle bandwidth is approximately half of that calculated by band calculation. In the present work, we study the normal electronic states in $Na_{0.35}CoO_2$ using the fluctuation exchange (FLEX) approximation to elucidate Fermi surface topology and renormalized band structure. In Na_xCoO_2, the topology of the FS is sensitively changed by the a_{1g}-e'_g splitting $3V_t$, whose value can be modified by the trigonal distortion of crystal. In this work, we study the many body effect for various values of $3V_t$.

We have reported[5] that density of states (DOS) on the Fermi energy and uniform spin susceptibility increase as temperature decreases when small pockets exist. On the other hand, when small pockets are absent, both of them decrease at lower temperatures. It is a weak pseudogap behavior due to magnetic fluctuations. In this case, the degree of reduction of DOS is greater when the top of the e'_g band is just below the Fermi level. In experimental measurements, both uniform spin susceptibility[6] and DOS in the photoemission spectroscopy[7] decrease at lower temperatures. These pseudogap behaviors are consistent with the latter result. As the pseudogap in the DOS is more prominent in bilayer hydrate samples than that in monolayer ones, the effect of intercalation of water is expected to raise the e'_g band slightly as expected by the analysis based on the point charge model. As a result, we have concluded that the small Fermi pockets do not exist or very small if any.

In this paper, we report the band structure of $Na_{0.35}CoO_2$ at $3V_t = 0.12$ (eV). In this case, small hole pockets are absent; they sink just below the Fermi energy. We find that the quasiparticle band is renormalized by the electronic correlation, and mass enhancement factor $m^*/m \simeq 1.5 \sim 1.8$. It agrees with the result of ARPES measurements.

The model Hamiltonian used in the present study is as follows.

$$H = H_0 + H', \tag{1}$$

$$\begin{aligned} H_0 &= \sum_{i,j,\sigma}\sum_{\ell\ell'} t_{ij}^{\ell\ell'} c^\dagger_{i\ell\sigma} c_{j\ell'\sigma} \\ &= \sum_{\mathbf{k},\sigma}\sum_{\ell\ell'} \varepsilon_{\mathbf{k}}^{\ell\ell'} c^\dagger_{\mathbf{k}\ell\sigma} c_{\mathbf{k}\ell'\sigma}, \end{aligned} \tag{2}$$

where ℓ represents $3d$ orbitals of Cobalt atoms (5 orbitals) and $2p$ orbitals of Oxygen atoms (2×3 orbitals), and $\hat{t}$ has the Slater-Koster's matrix form. H' is on-site Coulomb integrals in $t_{2g}(d_{xy}, d_{yz}, d_{zx})$ orbitals.

$$H' = H_U + H_{U'} + H_J + H_{J'}, \tag{3}$$

where U (U') is the intra-orbital (inter-orbital) Coulomb interaction, J is the Hund's coupling and J' represents the pair-hopping interaction. We put $U' = 1.3, J = J' = 0.13$, $U = U' + 2J = 1.56$ (eV) hereafter. In these parameters, deformation of the FS due to the interaction is very small.

CP850, *Low Temperature Physics: 24th International Conference on Low Temperature Physics;*
edited by Y. Takano, S. P. Hershfield, S. O. Hill, P. J. Hirschfeld, and A. M. Goldman

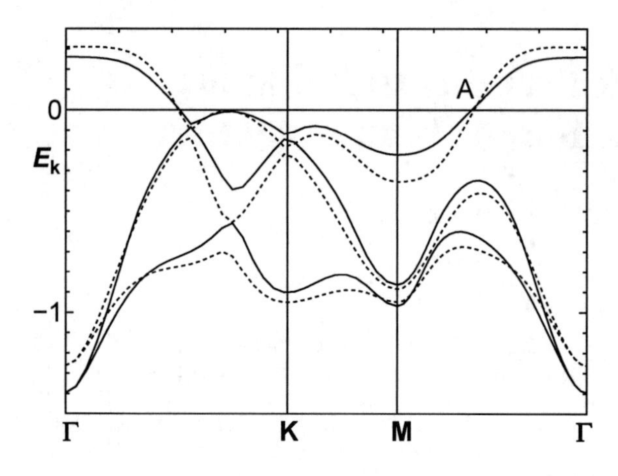

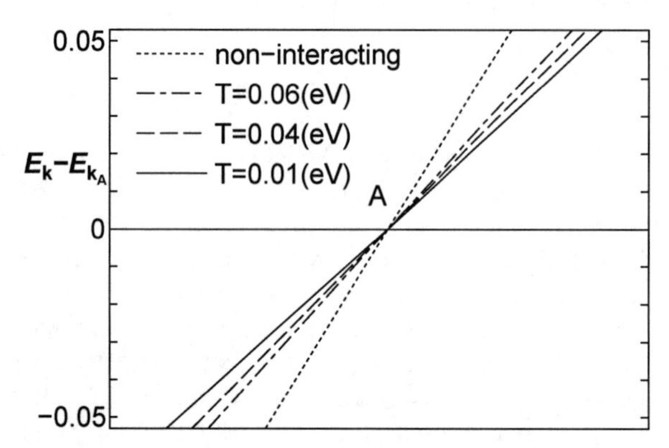

FIGURE 1. Top: The renormalized band dispersion (solid line) and the band dispersion without Coulomb interactions (dot line) at $3V_t = 0.12$. Bottom: Closeup of vicinity of point A at various temperatures.

We calculate the normal selfenergy $\Sigma(\mathbf{k}, \omega)$ in the FLEX approximation scheme. Then, the renormalized quasiparticle band dispersion $E_\mathbf{k}$ is decided by the following equation.

$$\det|(E_\mathbf{k} + \mu)\mathbf{1} - \hat{\varepsilon}_\mathbf{k} - \Sigma(\mathbf{k}, E_\mathbf{k})| = 0. \quad (4)$$

In this case, we take care that both $\hat{\varepsilon}_\mathbf{k}$ and $\Sigma(\mathbf{k}, E_\mathbf{k}) = \left(\hat{\Sigma}^R(\mathbf{k}, E_\mathbf{k}) + \hat{\Sigma}^A(\mathbf{k}, E_\mathbf{k})\right)/2$ are complex Hermitian matrices. Here we employed the mean value of retarded and advanced selfenergy on the selfenergy to cancel out the effect of damping of quasiparticle. By diagonalizing the Green function $\hat{G}$, we can switch to band-representation. In the band-representation, mass enhancement factor m^*/m is determined by following equation.

$$\begin{aligned} \frac{m^*_\alpha}{m_\alpha} &= \frac{d\varepsilon^\alpha_\mathbf{k}}{dk} \Big/ \frac{dE^\alpha_\mathbf{k}}{dk} \\ &= \left(1 + \frac{\partial \mathrm{Re}\Sigma^\alpha(\mathbf{k}, E^\alpha_\mathbf{k})}{\partial k} \Big/ \frac{d\varepsilon^\alpha_\mathbf{k}}{dk}\right)^{-1} \\ &\quad \times \left(1 - \frac{\partial \mathrm{Re}\Sigma^\alpha(\mathbf{k}, E^\alpha_\mathbf{k})}{\partial E^\alpha_\mathbf{k}}\right), \end{aligned} \quad (5)$$

where $k = \mathbf{k} \cdot \mathbf{n}_\perp$ ($\mathbf{n}_\perp$ is a unit vector perpendicular to the FS), and α indicates the band index. The first bracket in eq.5 is so-called k-mass and the second bracket is so-called ω-mass. In strongly correlated electron systems, ω-mass constitutes the main part of mass enhancement and it makes the effective mass heavy.

In Fig. 1, we show the calculated band dispersion of $Na_{0.35}CoO_2$. Near the Fermi level, we notice that the electronic band is renormalized and the slope of band dispersion decreases at lower temperatures. The mass enhancement factor at point A in Fig. 1 are 1.78, 1.73, 1.60 and 1.47 at $T = 0.01, 0, 02, 0.04$ and 0.06 (eV) respectively. The mass enhancement factor would increase further if one takes large U, U'. In the three bands we show in Fig. 1, the top band which across the Fermi energy is strongly renormalized in whole. The other two bands below the Fermi energy are also renormalized for $|E^\alpha_\mathbf{k} - \mu| \lesssim 1$ (eV). However, $|E^\alpha_\mathbf{k} - \mu|$ are greater than $|\varepsilon^\alpha_\mathbf{k} - \mu|$ for $|E^\alpha_\mathbf{k} - \mu| \gtrsim 1$ (eV), which is inconsistent with ARPES measurements. We conclude that the electronic structure for $|E^\alpha_\mathbf{k} - \mu| \lesssim 1$ (eV) is well reproduced. Higher energy spectrum by the FLEX approximation will be improved if one takes the vertex corrections. The FLEX approximation gives reliable results for lower energy scale.

In summary, we analyzed the multi-orbital Hubbard model for $Na_{0.35}CoO_2$ using the FLEX approximation, and we found the weak pseudogap behavior in the DOS and spin susceptibility when small hole pockets are absent. The electronic band is renormalized by the electronic correlation, and mass enhancement factor $m^*/m \simeq 1.5 \sim 1.8$. This result supports the recent ARPES measurements.

REFERENCES

1. K. Takada, H. Sakurai, E. Takayama-Muromachi, F. Izumi, R.A. Dilanian, and T. Sasaki, *Nature* **422**, 53 (2003).
2. D.J. Singh, *Phys. Rev. B* **61**, 13397 (2000).
3. H.-B. Yang, Z.-H. Pan, A.K.P. Sekharan, T. Sato, S. Souma, T. Takahashi, R. Jin, B.C. Sales, D. Mandrus, A.V. Fedorov, Z. Wang, and H. Ding, *Phys. Rev. Lett.* **95**, 146401 (2005).
4. M.Z. Hasan, D. Qian, Y. Li, A.V. Fedorov, Y.-D. Chuang, A.P. Kuprin, M.L. Foo, and R.J.Cava, cond-mat/0501530.
5. Keiji Yada, and Hiroshi Kontani, *J. Phys. Soc. Jpn.* **74**, 2161 (2005).
6. Mai Yokoi, Taketo Moyoshi, Yoshiaki Kobayashi, Minoru Soda, Yukio Yasui, Masatoshi Sato, and Kazuhisa Kakurai, *J. Phys. Soc. Jpn.* **74**, 3046 (2005).
7. T. Shimojima, T. Yokoya, T. Kiss, A. Chainani, S. Shin, T. Togashi, S. Watanabe, C. Zhang, C. T. Chen, K. Takada, T. Sasaki, H. Sakurai, and E. Takayama-Muromachi, *Phys. Rev. B* **71**, 020505(R) (2005).

Point Contact Spectroscopy Study of $ZrZn_2$

C.S. Turel*, M.A.Tanatar*, R.W.Hill†, E.A.Yelland**, S.M.Hayden** and J.Y.T. Wei*

*Department of Physics, University of Toronto, 60 St. George Street, Toronto, ON M5S1A7, Canada
†Department of Physics, University of Waterloo, ON N2L3G1, Canada
**H.H. Wills Physics Laboratory, University of Bristol, Bristol BS81TL, UK

Abstract. We have performed point contact spectroscopy on spark-cut single crystals of ferromagnetic $ZrZn_2$, using normal-metal tips in a dilution refrigerator down to 100mK. The differential conductance spectra show low-energy peak structures which evolve systematically with temperature below 1.1 K. We associate these state-conserving peak spectra with the surface superconductivity recently observed in $ZrZn_2$. Implications of our data on the electron pairing in $ZrZn_2$ are discussed.

Keywords: superconductivity, ferromagnetism, inter-metallics, tunneling spectroscopy, Andreev interference
PACS: 74.20.Rp, 74.50.+r, 74.70.Ad

Superconductivity and ferromagnetism tend to be competing types of order. The intermetallic ferromagnetic compound $ZrZn_2$ has recently been shown to display surface superconductivity under some circumstances [1, 2] in the ferromagnetic state, suggesting their possible coexistence and complex interplay [3, 4, 5]. A proposed scenario for this coexistence involves spin-mediated electron pairing, which could produce a spin-triplet, odd-parity superconducting order parameter (OP) with nodes in *k*-space [6].

Experimentally, the presence of OP nodes can be detected by tunneling spectroscopy, in the form of zero-bias conductance peaks, which are a manifestation of surface states due to Andreev interference resulting from the sign change about nodes in the OP [7, 8]. This Andreev phenomenology has been used to reveal OP nodes in several unconventional superconductors, ranging from $YBa_2Cu_3O_{7-\delta}$ [9, 10] to Sr_2RuO_4 [11, 12] and $CeCoIn_5$ [13].

To look for evidence of a nodal OP in $ZrZn_2$, we have performed point-contact spectroscopy (PCS) on single-crystal samples. The $ZrZn_2$ crystals were grown using a directional-cooling technique and cut to expose (111)-faces by spark-erosion [2]. These crystals show bulk ferromagnetic susceptibility below ~30K, and a resistive downturn below ~0.3K [1]. The PCS measurements were made in a dilution refrigerator, using either Pt-Ir or Au tips with a pulsed-signal technique to minimize Joule heating [13].

Typical dI/dV conductance spectra measured are plotted in Figures 1 and 2. Figure 1 plots the data taken with a Au tip, showing the spectral evolution with temperature up to 1.1K, above which the conductance becomes independent of bias voltage. Figure 2 plots the data for a Pt-Ir tip, showing a similar evolution up to 0.98K. The spectra shown in each plot are normalized relative to the highest

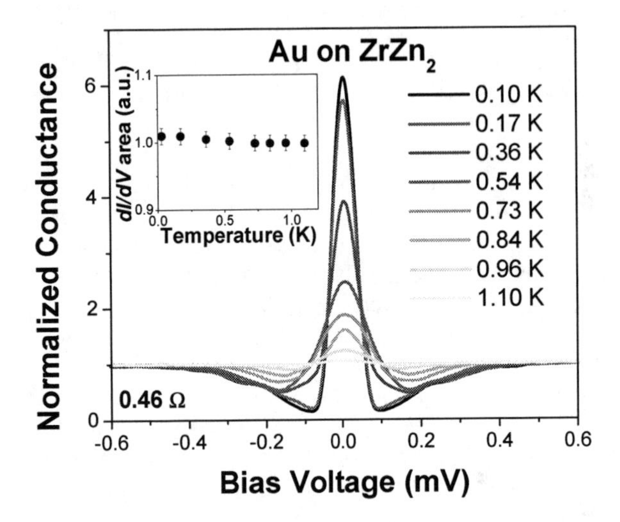

FIGURE 1. Normalized conductance *vs.* voltage as a function of temperature for a 0.46Ω Au/$ZrZn_2$ point-contact junction. Shown in the inset is the dI/dV spectral area, integrated between ±0.5 mV, as a function of temperature.

temperature data. The generic similarities between the two plots attest to the reproducibility of our measured spectra, as well as to their independence on tip material. At base temperature in each plot, a pronounced low-energy peak structure is observed, flanked by symmetric dip structures. With increasing temperature these spectral features evolve such that the total integrated dI/dV spectral area is conserved, as shown in the inset of each figure. The spectral details and their evolutions varied slightly between junctions and samples. In some cases, discernible peak-and-dip structures persisted to as high as ~1.8K.

The conductance spectra we observed provide further support for the existence of a superconducting sur-

CP850, *Low Temperature Physics: 24th International Conference on Low Temperature Physics;*
edited by Y. Takano, S. P. Hershfield, S. O. Hill, P. J. Hirschfeld, and A. M. Goldman

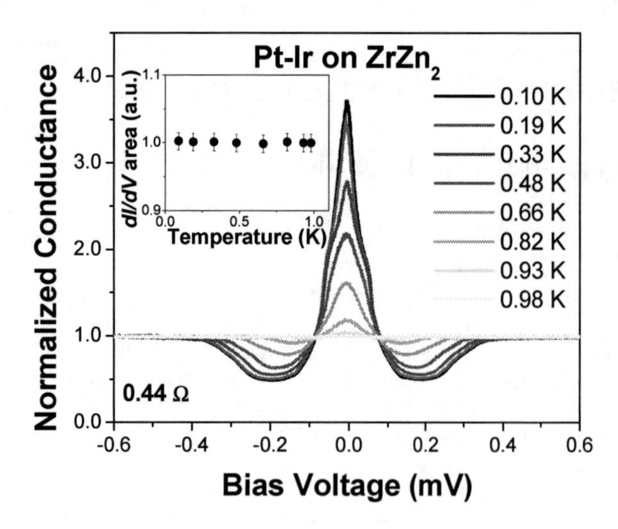

FIGURE 2. Normalized conductance *vs.* voltage as a function of temperature for a 0.44Ω Pt-Ir/$ZrZn_2$ point-contact junction. Shown in the inset is the dI/dV spectral area, integrated between ± 0.5 mV, as a function of temperature.

face layer in spark-cut $ZrZn_2$, although the composition of this layer is still unknown [2]. The observed spectral peaks can be interpreted as a distinct signature of Andreev surface states arising from a nodal OP. First, the spectral peaks were generally taller than twice the normalized spectral background, as is expected from the generalized Blonder-Tinkham-Klapwijk formalism [7, 8, 10, 14]. Second, the conservation in the dI/dV spectral area is entirely consistent with the conservation of quasiparticle states which mediate the Andreev interference process [7]. Third, the microscopic scale of our probe [15] largely rules out spurious scenarios based on material inhomogeneity, especially those involving conventional superconductivity with a non-nodal OP symmetry [5].

In conclusion, the apparent robust formation of Andreev surface states could be regarded as strong evidence for local pairing correlations on the surface in $ZrZn_2$. It is remarkable that these Andreev surface states seem to persist above the apparent resistive transition [1]. This observation suggests that in $ZrZn_2$ it is easier for the electrons to pair microscopically than for them to phase-condense macroscopically. This picture would be consistent with the fragile nature of the resistive transitions reported thus far [2]. More detailed studies correlating the spectral and transport data with surface preparation, in particular the effect chemical etching has on the surface [2], are currently under way to elucidate this picture.

ACKNOWLEDGMENTS

Work supported by NSERC, CFI/OIT, MMO/EMK and the Canadian Institute for Advanced Research in the Quantum Materials Program.

REFERENCES

1. C. Pfleiderer, M. Uhlarz, S.M. Hayden, R. Vollmer, H.v. Löhneysen, N.R. Bernhoeft and G.G. Lonzarich, *Nature* **412**, 58 (2001).
2. E. A. Yelland *et al.*, *cond-mat*/0502341.
3. S. S. Saxena, P. Agarwal, K. Ahilan, F. M. Grosche, R. K. W. Haselwimmer, M. J. Steiner, E. Pugh, I. R. Walker, S. R. Julian, P. Monthoux, G. G. Lonzarich, A. Huxley, I. Sheikin, D. Braithwaite and J. Flouquet, *Nature* **406**, 587 (2000).
4. Dai Aoki, Andrew Huxley, Eric Ressouche, Daniel Braithwaite, Jacques Flouquet, Jean-Pascal Brison, Elsa Lhotel and Carley Paulsen, *Nature* **413**, 613 (2001).
5. D. J. Singh and I.I. Mazin, *Phys. Rev. Lett.* **88**, 187004 (2002).
6. M. B. Walker and K.V. Samokhin, *Phys. Rev. Lett.* **88**, 207001 (2002).
7. C.-R. Hu, *Phys. Rev. Lett.* **72**, 1526 (1994).
8. Y. Tanaka and S. Kashiwaya, *Phys. Rev. Lett.* **74**, 3451 (1995); Y. Tanaka, Y. Tanuma, K. Kuroki and S. Kashiwaya, *J. Phys. Soc. Jpn.* **71**, 2102 (2002).
9. L. Alff, H. Takashima, S. Kashiwaya, N. Terada, H. Ihara, Y. Tanaka, M. Koyanagi and K. Kajimura, *Phys. Rev. B* **55**, 14757 (1997).
10. J. Y. T. Wei, N.-C. Yeh, D.F. Garrigus, and M. Strasik, *Phys. Rev. Lett.* **81**, 2542 (1998).
11. F. Laube, G. Goll, H.v. Löhneysen, M. Fogelström and F. Lichtenberg, *Phys. Rev. Lett.* **84**, 1595 (2000).
12. Z. Q. Mao, K.D. Nelson, R. Jin, Y. Liu and Y. Maeno, *Phys. Rev. Lett.* **87**, 037003 (2001).
13. P. M. C. Rourke, M.A. Tanatar, C.S. Turel, J. Berdeklis, C. Petrovic and J.Y.T. Wei, *Phys. Rev. Lett.* **94**, 107005 (2005).
14. For a recent review, see G. Deustcher, *Rev. Mod. Phys.* **77**, 109 (2005).
15. Using the Sharvin formula, $R = 4\rho\ell/3\pi a^2$, with $\rho \sim 0.6\ \mu\Omega\cdot$cm and $\ell \sim 55$ nm for $ZrZn_2$ [2], the point-contact radius for a 0.4 Ω junction is estimated to be $a \sim 20$ nm.

Non-Fermi-Liquid Behavior in $CeCoIn_5$ Near the Superconducting Critical Field

R. Movshovich, C. Capan*, F. Ronning, A. Bianchi**, P. G. Pagliuso♦, E. D. Bauer, and J. L. Sarrao

Los Alamos National Laboratory, Los Alamos, NM 87545, USA
**Department of Physics and Astronomy, Louisiana State University, Baton Rouge, Louisiana 70802, USA*
***Hochfeld-Magnetlabor Dresden, Forschungszentrum Rossendorf, 01328 Dresden, Germany*
♦Instituto de Fi ́sica "Gleb Wathagin", UNICAMP, 13083-970, Campinas, Brazil

Abstract. We measured specific heat and resistivity of heavy fermion $CeCoIn_5$ at and above the superconducting critical field with field in [001] (H_{c2} = 5 T) and [100] directions (H_{c2} = 12 T), and at temperatures down to 50 mK. At the critical fields the data show Non Fermi Liquid (NFL) behavior down to the lowest temperatures. With increasing field the data exhibit crossover from NFL to Fermi liquid behavior at the lowest temperatures. Analysis of the scaling properties of the specific heat, and comparison of both resistivity and the specific heat with the predictions of a spin-fluctuation theory suggest that the NFL behavior is due to incipient antiferromagnetism (AF) in $CeCoIn_5$, with the quantum critical point in the vicinity of H_{c2}. Below H_{c2} the AF phase which competes with the paramagnetic ground state is superseded by the superconducting transition. To separate the quantum critical point H_{QCP} from the superconducting critical field H_{c2} we performed a series of Sn-doping and hydrostatic pressure studies. Sn-doping appears not to be able to separate H_{QCP} and H_{c2}. Preliminary resistivity in magnetic field measurements under hydrostatic pressure indicate that the QCP point moves inside the superconducting phase in the H-P plane with increasing field.

Keywords: Superconductivity, correlated electrons, Quantum Critical Point, heavy fermions, thermodynamic properties.
PACS: 74.70.Tx, 71.27.+a, 74.25.Fy, 75.40.Cx

INTRODUCTION

Quantum Critical Points (QCP) in condensed matter systems has been active area of research for more than a decade[1]. At a QCP material appears to be at the boundary between two ground states at zero temperature, with some tuning parameter (be it pressure, doping, magnetic field, or other) capable of moving the compound between these two ground states.

Particular interest is generated by a number of the stoichiometric Ce and Yb-based compounds. $CeCoIn_5$ is one such compound. It is a heavy fermion superconductor (T_c = 2.3 K [2]), with a number of other fascinating properties, such as a first order nature of the superconducting phase transition close to the superconducting critical field H_{c2} [3-4], and a phase transition within a superconducting state close to H_{c2} [5,6], identified with the possible formation of the inhomogeneous Fulde-Ferrell-Larkin-Ovchinnikov (FFLO) superconducting state. In particular interest to present paper, superconductivity in $CeCoIn_5$ develops out of the NFL state, characterized by increasing Sommerfeld coefficient $\gamma = C/T$, where C is specific heat, with decreasing temperature, and close to linear-in-temperature resistivity. Upon suppressing T_c to zero at H_{c2} = 5 T for the field parallel to tetragonal c-axis, specific heat continues to increase logarithmically with temperature at $T \to 0$, and resistivity recovers a Fermi liquid $\rho \propto T^2$ behavior only for fields above H_{c2} [7,8]. These results indicate that a QCP was located very close to H_{c2}. The proximity of H_{QCP} and H_{c2} raises a question about the origin of the NFL behavior and a possibility of a superconducting quantum critical point. To try to address this question we undertook a series of anisotropy, doping, and pressure studies. Field anisotropy studies

The superconducting critical field for the field-in-plane orientation (H⊥c) is H_{c2} = 12 T, a factor of 2.4 greater than that for the out of plane field H ‖ c. Our measurements indicate that in this orientation $H_{QCP} \approx H_{c2}$ as well, with specific heat diverging logarithmically at Hc_2 down to the lowest temperature (see Fig. 1), and a non-T^2 resistivity [9].

CP850, *Low Temperature Physics: 24th International Conference on Low Temperature Physics;*
edited by Y. Takano, S. P. Hershfield, S. O. Hill, P. J. Hirschfeld, and A. M. Goldman

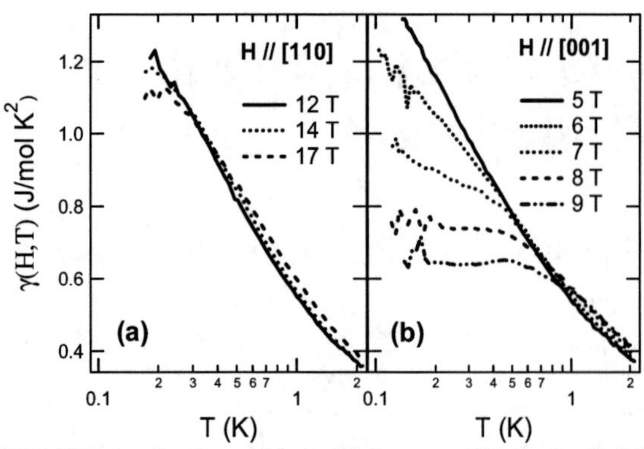

FIGURE 1. Sommerfeld coefficient $\gamma = C/T$ for both field orientations at and above the H_{c2}

DOPING STUDIES

Another possible line of attack is to use doping to separate H_{QCP} from H_{c2}. We used Sn-doping on the In site, and measured specific heat and resistivity of a number of $CeCoIn_{5-x}Sn_x$ samples[10]. Superconductivity is efficiently suppressed at a rate of $dT_c/dx = -0.6$ K/at. % Sn. However, for a significant range of doping $x \leq 0.12$ specific heat and resistivity behavior do not differ from the pure compound, with the strongest divergence in specific heat occurring at Hc_2 (see Fig. 2).

PRESSURE STUDIES

Finally, hydrostatic pressure provides a clean way of tuning a QCP, where possible effects of disorder, which necessarily complicate analysis and interpretation of the doping studies, are absent. Our preliminary resistivity measurements under pressure [11] indicate that pressure indeed leads to appearance of a significant Fermi Liquid regime for fields immediately above H_{c2}. This region is already observed at a pressure as low as 6 kbar, and extends to higher temperatures at higher pressures. We suggest that this behavior is due to a separation of the H_{QCP} from H_{c2} with pressure, with H_{QCP} moving inside the superconducting dome. This lends support to our previous suggestion that the NFL behavior exhibited by $CeCoIn_5$ at H_{c2} was not of the superconducting origin. Superconductivity is playing a parasitic role and superceding another ground state whose fluctuations are responsible for NFL behavior.

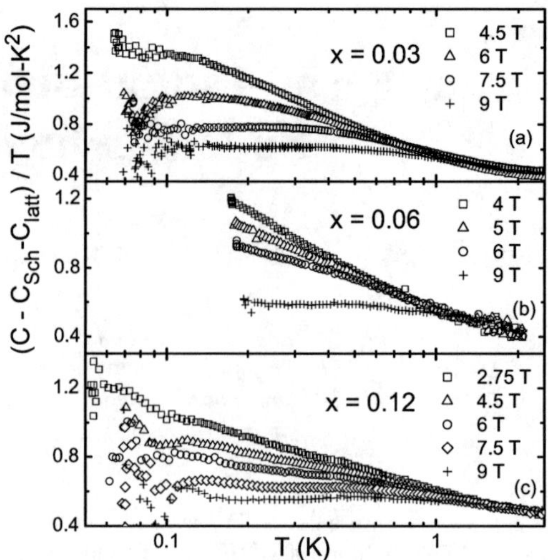

FIGURE 2. Electronic contribution to the specific heat $C_{el} = C - C_{Sch} - C_{latt}$ divided by temperature T of $CeCoIn_{5-x}Sn_x$ in magnetic fields $H \geq H_{c2}$, H||[001] for a) $x = 0.03$, b) $x = 0.06$, and c) $x = 0.12$, with the strongest divergences in specific heat occurring at H_{c2}.

ACKNOWLEDGMENTS

We want to thank I. Vekhter, P. Littlewood, P. Coleman, J. D. Thompson, and Z. Fisk for numerous discussions and contributions to this work.

REFERENCES

1. G. R. Stewart, *Rev. Mod. Phys.* **73**, 797 (2001).
2. C. Petrovic *et al.*, *J. Phys. Condens. Matter* **13**, L337 (2001).
3. K. Izawa et al., *Phys. Rev. Lett.* **87**, 057002 (2001).
4. A. Bianchi *et al.*, *Phys. Rev. Lett.* **89**, 137002 (2002).
5. H. Radovan *et al.*, *Nature (London)* **425**, 51 (2003).
6. A. Bianchi *et al.*, *Phys. Rev. Lett.* **91**, 18704 (2003).
7. J. Paglione *et al.*, *Phys. Rev. Lett.* **91**, 246405 (2003).
8. A. Bianchi *et al.*, *Phys. Rev. Lett.* **91**, 257001 (2003).
9. F. Ronning, *et al.*, *Phys. Rev. B* **71**, 104528 (2005).
10. E. D. Bauer *et al.*, *Phys. Rev. Lett.* **94**, 047001 (2005).
11. F. Ronning, unpublished.

Fulde-Ferrell-Larkin-Ovchinnikov Superconducting State in $CeCoIn_5$: New Evidence from Pressure Studies

C. F. Miclea*, M. Nicklas[1],*, J. L. Sarrao†, G. Sparn*, F. Steglich* and J. D. Thompson†

*Max Planck Institute for Chemical Physics of Solids, Nöthnitzer Str. 40, 01187 Dresden, Germany
†Los Alamos National Laboratory, Los Alamos, New Mexico 87545, USA

Abstract. $CeCoIn_5$ exhibits the highest superconducting transition temperature ($T_c = 2.3$ K) among the Ce-based heavy-fermion compounds. Power-law dependencies in specific heat (C), thermal conductivity (κ), and nuclear spin-lattice relaxation rate ($1/T_1$) for $T \ll T_c$ point to an unconventional superconducting state. Within the superconducting state specific-heat shows an anomaly in the vicinity of the upper critical field $H_{c2} = 11.6$ T for magnetic field parallel to the c-axis. This phase-transition is discussed as possible realization of a Fulde-Ferrell-Larkin-Ovchinnikov (FFLO) phase. On the other hand, $CeCoIn_5$ is located close to an antiferromagnetic (AFM) quantum critical point (QCP) at ambient pressure and underlying magnetic fluctuations may be responsible for the specific-heat feature. We present first results of a specific-heat study under pressure and in high magnetic fields to clarify the origin of this phase.

Keywords: heavy fermions, superconductivity, hydrostatic pressure
PACS: 71.27.+a; 74.25.Dw; 74.70.Tx

For a type-II SC in the clean limit and with a high Maki parameter [1] $\alpha > 1.8$ an inhomogenous FFLO [2, 3] SC phase at high fields is predicted to appear, between the normal and the vortex state. The heavy-fermion compound $CeCoIn_5$ is a clean limit superconductor [4, 5] with a very high Maki parameter $\alpha \geq 3.5$ [6] being a very good candidate for the realization of the FFLO state. Moreover, upon applying a magnetic field H perpendicular to the c-axis, the SC phase-transition changes from second to first order around 10 T [6] and a new low temperature phase, possibly the FFLO state, appears [7, 8]. $CeCoIn_5$ is situated close to a quantum critical point at ambient pressure [4] and this proximity may be responsible for the low-temperature phase. By applying pressure the system is driven away from the QCP [9, 10] and therefore the underlying AFM fluctuations are expected to become weaker and less important for the system. To probe the effect of underlying magnetic fluctuations we measured the specific-heat of $CeCoIn_5$ under pressure and in magnetic field.

The specific-heat measurements were carried out on high quality single crystals of $CeCoIn_5$ grown from excess In-flux. A miniature Cu-Be piston-cylinder type pressure-cell was utilized to generate pressures up to 1.4 GPa (Fig. 1). To guarantee the precise alignment of the samples insight the pressure cell, small discs fitting exactly in the pressure-cell were spark eroded from the single crystalline $CeCoIn_5$ platelets. A stack of three discs each separated by grease serving as pressure medium was placed in the cell, along with lead serving as pressure gauge. This procedure makes sure that the crystallographic [001]-plane is perpendicular to the axis of the pressure cell, allowing a precise orientation of the sample with respect to the direction of the magnetic field H. Figure 2 shows the specific-heat obtained at $P = 0.45$ GPa for different magnetic fields applied perpendicular to the c-axis. The total heat capacity (HC) of the loaded cell at zero field including addenda (pressure-cell, lead, grease) is depicted in the inset. The contribution of the sample to the total heat capacity reaches 54% on top of the superconducting phase-transition.

Here we present first results of specific-heat measurements under pressure on $CeCoIn_5$ with magnetic field applied in the [001]-plane. On applying a small pressure superconductivity is becoming more robust. T_c is increasing from $T_c = 2.24$ K at atmospheric pressure to $T_c = 2.43$ K at 0.45 GPa, in good agreement with resistivity data [9]. However, qualitatively the results at $P = 0.45$ GPa are similar to the findings at ambient pressure. In small magnetic fields, the specific-heat exhibits a mean-field like shape at the SC transition, typical for a second order phase-transition. With increasing magnetic field the anomaly moves to lower temperatures and sharpens up indicating the change from a second to a first order phase-transition at a characteristic temperature T_0. This is even more convincing, if one bears in mind that in high fields the $H_{c2}(T)$ phase-line is crossed

[1] Corresponding author: Max Planck Institute for Chemical Physics of Solids, Nöthnitzer Str. 40, 01187 Dresden, Germany. Phone: +49 (0)351 4646 2400, Fax: +49 (0)351 4646 2402, Email: nicklas@cpfs.mpg.de

CP850, *Low Temperature Physics: 24th International Conference on Low Temperature Physics;*
edited by Y. Takano, S. P. Hershfield, S. O. Hill, P. J. Hirschfeld, and A. M. Goldman

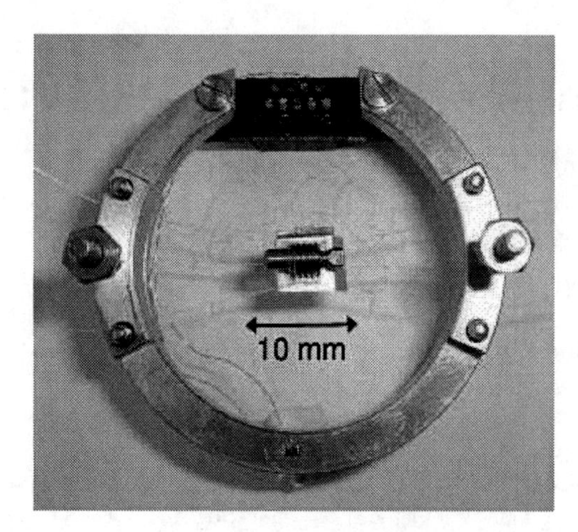

FIGURE 1. Miniature piston-cylinder type pressure cell on a silver platform for specific-heat experiments in a dilution refrigerator. The heater and the thermometer are attached underneath the platform. The length of the pressure cell is about 8 mm.

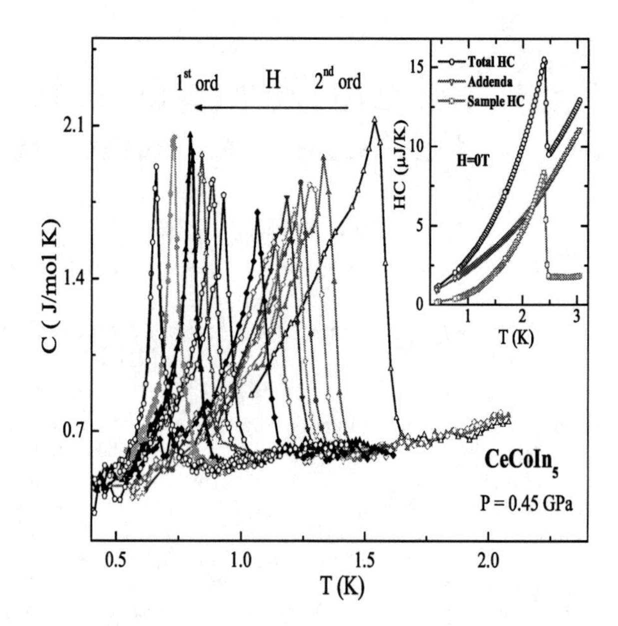

FIGURE 2. Specific-heat (C) of $CeCoIn_5$ in different magnetic fields at $P = 0.45$ GPa. The inset shows the total heat capacity (HC) of the loaded pressure cell (circles) and the separate contributions of sample (squares) and addenda (triangles).

in a glancing angle in a measurement at constant field, which would actually cause a broadening of the second order phase-transition anomaly. In addition to the change in shape, the height of the peak as function of magnetic field shows a minimum at T_0, typical for a crossover from a second to a first order phase-transition. T_0 is enhanced from $T_0(0\text{ GPa}) = (0.75 \pm 0.05)$ K to $T_0(0.45\text{ GPa}) = (1.1 \pm 0.1)$ K. Similar results for $T_0(P)$ were obtained from magnetization measurements under pressure [11]. The initial slope $\delta H_{c2}(0.45\text{ GPa})/\delta T\,|_{H=0} = 29.4$ T/K is nearly unchanged compared to the value at atmospheric pressure ($\delta H_{c2}(0\text{ GPa})/\delta T\,|_{H=0} = 30.5$ T/K) and the orbital limiting field $H_{orb} = 0.7 \cdot T_c \cdot |\delta H/\delta T|_{H=0}$ is only changing little from $H_{orb}(0\text{ GPa}) \approx 47.9$ T to $H_{orb}(0.45\text{ GPa}) \approx 49.9$ T. The Maki parameter $\alpha = \sqrt{2}H_{orb}/H_{Pauli}$ is almost constant and therefore also at $P = 0.45$ GPa much higher than the required value for the formation of the FFLO state ($\alpha = 1.8$). Here, we took the upper critical field $H_{c2}(T = 0)$ as Pauli limiting field H_{Pauli}.

At ambient pressure the specific-heat results show a second anomaly inside the superconducting state which in conjunction with the fulfilled prerequisites makes a strong case for the realization of the FFLO state. At $P = 0.45$, $CeCoIn_5$ is still Pauli limited $P = 0.45$ GPa with a high Maki parameter $\alpha \approx 5.7$ and the SC phase-transition is changing from second to first order at high magnetic fields. It is also unlikely that the mean-free path and the correlation length are changing significantly by applying hydrostatic pressure. Therefore, $CeCoIn_5$ can be still considered a clean superconductor satisfying all requirements for the existence of the FFLO state at $P = 0.45$ GPa. Preliminary results show an anomaly inside the SC state at $P = 0.45$ GPa. These results provide additional evidence for the realization of the FFLO SC phase in $CeCoIn_5$ and support the conclusion that the anomaly is not related to the vicinity of a QCP.

ACKNOWLEDGMENTS

We thank R. Borth and R. Koban for machining the pressure cell and for their help in the preparation process. We also thank A. D. Bianchi for stimulating discussions.

REFERENCES

1. K. Maki and T. Tsuneto, *Prog. Theor. Phys.* **31**, 945 (1964).
2. P. Fulde and R. A. Ferrell, *Phys. Rev.* **135**, A550 (1964).
3. A. I. Larkin and Y. N. Ovchinnikov, *Zh. Eksp. Teor. Fiz.* **47**, 1136 (1964) [*Sov. Phys. JETP* **20**, 762 (1965)].
4. C. Petrovic *et al.*, *J. Phys.: Condens. Matter* **13**, L337 (2001).
5. R. Movshovich *et al.*, *Phys. Rev. Lett.* **86**, 5152 (2001).
6. A. Bianchi *et al.*, *Phys. Rev. Lett.* **89**, 137002 (2002).
7. A. Bianchi *et al.*, *Phys. Rev. Lett.* **91**, 18 (2003).
8. H. A. Radovan *et al.*, *Nature* **425**, 51 (2003).
9. M. Nicklas *et al.*, *J. Phys.: Condens. Matter* **13**, L905 (2001).
10. E. Lengyel *et al.*, *High Pres. Res.* **22**, 185 (2002).
11. T. Tayama *et al.*, *J. Phys. Soc. Jpn.* **74**, 1115 (2005).

Andreev Reflection at the Normal-Metal / Heavy-Fermion Superconductor $CeCoIn_5$ Interface by Point-Contact Spectroscopy

Wan Kyu Park*, Laura H. Greene*, John L. Sarrao† and Joe D. Thompson†

*Department of Physics and the Frederick Seitz Materials Research Laboratory, University of Illinois at Urbana-Champaign, Urbana, IL 61801, USA
†Los Alamos National Laboratory, Los Alamos, NM 87545, USA

Abstract. Dynamic conductance between a normal-metal (N) Au tip and the heavy-fermion superconductor (HFS) $CeCoIn_5$ single crystal has been measured. The contact is shown to be in the Sharvin limit. Our clean conductance spectra are reproducibly obtained over a wide range of temperature (60 K to 400 mK). The background conductance shows a gradual development of asymmetry, starting at T^* ($\sim$ 45 K) down to T_c (2.3 K), which we attribute to the emergence of the heavy-fermion liquid. The enhanced sub-gap conductance observed below T_c, arising from Andreev reflection (AR), is an order of magnitude smaller ($\sim$ 13.3 % at 400 mK) than that observed for N/conventional superconductors but consistent with other N/HFS data reported. From analyses using an extended Blonder-Tinkham-Klapwijk model, we find the conductance data at 400 mK is consistent with strong coupling ($2\Delta/k_BT_c = 4.64$), and that the temperature dependence of the zero bias conductance is consistent with a *d*-wave order parameter, in agreement with the literature. However, attempts to fit full conductance curves as a function of temperature clearly show that this model cannot account for our data. We provide restrictions and suggestions for theoretical studies to account for AR at the N/HFS interface.

Keywords: heavy-fermion superconductor, Andreev reflection, Blonder-Tinkham-Klapwijk model, point-contact spectroscopy
PACS: 74.50.+r, 74.45.+c, 74.70.Tx, 74.20.Rp

Andreev reflection (AR) is an electronic transport phenomenon occurring at an interface between a normal metal (N) and a superconductor (S) [1]. In essence, it is a quantum mechanical scattering of a quasi-particle from a pairing potential, giving rise to a retro-reflected hole and a transferred Cooper pair at the same time. As a consequence, the conductance below the energy gap, Δ, is enhanced by 100 % in a pure metallic N/S contact. Blonder, Tinkham, and Klapwijk (BTK) [2] formulated a theory to explain a transitional behavior from AR to tunneling using an effective barrier strength, $Z_{eff} = [(1-r)^2/4r+Z_0^2]^{1/2}$, as a parameter. Here, Z_0 is due to an insulating tunnel barrier layer and r is the ratio of Fermi velocities in N and S.

The heavy-fermion superconductor (HFS) $CeCoIn_5$ has been drawing much attention due to the rich physical phenomena it displays [3], including reports of Fulde-Ferrell-Larkin-Ovchinikov [4] and quantum phase transitions [5]. Various measurements have been showing that the superconducting order parameter in $CeCoIn_5$ has line nodes, implying a *d*-wave symmetry [6, 7]. There have been reports claiming an unconventional pairing symmetry from point-contact spectroscopy (PCS) [8, 9], but without providing clear spectroscopic evidence [10].

We have measured dynamic conductance spectra of $CeCoIn_5$ [11] using the PCS technique we developed [11]. An electrochemically polished Au tip is brought into contact with a high quality single crystal of $CeCoIn_5$ using a combined (mechanical and piezoelectric) approach. The contact is formed on the *c*-axis face (*ab*-plane) of the crystal. Conductance spectra are taken using the standard four probe lock-in technique [11].

Figure 1 shows normalized conductance spectra taken over a wide temperature range from 60 K to 400 mK [11]. For $T > T_c$ (2.3 K), an asymmetry in the background conductance is observed to develop, gradually increasing with decreasing temperature. We attribute this to a signature of emerging coherent heavy-fermion liquid [12]. Below T_c, the asymmetry remains almost the same, whereas an enhancement of conductance is seen in the sub-gap region. Using the resistance at the highest bias ($\sim$ 1.1 Ω) and Wexler's formula [13], the diameter of the contact is estimated to be $\sim$ 460 Å [11], which is much smaller than the electronic mean free path for $T \leq T_c$ [6]. This indicates that the conductance data in Fig. 1 are taken from a contact in the Sharvin or ballistic limit, thereby ensuring the *spectroscopic* nature of our data. The maximum value of the applied current is $\sim$ 2 mA, generating negligible Joule heating compared to the cooling power of the ^{3}He cryostat. Therefore, the enhanced sub-gap conductance below T_c is due to AR at the Au/$CeCoIn_5$ interface. According to the BTK theory, no AR can occur at the N/HFS interface due to the disparity of the Fermi velocities. Instead, only an extreme tunneling behavior should

CP850, *Low Temperature Physics: 24th International Conference on Low Temperature Physics;*
edited by Y. Takano, S. P. Hershfield, S. O. Hill, P. J. Hirschfeld, and A. M. Goldman

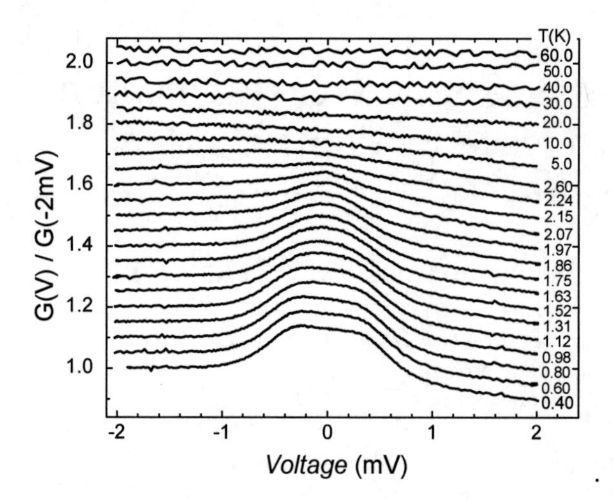

FIGURE 1. Conductance vs. voltage spectra of a ballistic Au/$CeCoIn_5$ point contact between 60 K and 400 mK. Curves are shifted vertically for clarity. After Ref. [11]

be observed ($Z_{eff} \geq 6$ in Au/$CeCoIn_5$ using the Fermi velocities reported [6]). However, AR is commonly observed in HFS point contacts [14, 15]. Deutscher and Noziéres [16] addressed this issue and proposed that the Fermi velocity entering in the ratio r is without a mass enhancement factor. The conductance curve at 400 mK, once normalized, shows an enhancement due to AR of 13.3 % at zero-bias, an order of magnitude smaller than that observed in conventional S/N contacts but consistent with other N/HFS contacts [15]. We note that AR could be observed along the c-axis of a d-wave superconductor because of large tunneling cones in metallic contacts.

Extended BTK models [17] were used to fit the conductance data with three parameters, Δ, Γ, and Z_{eff}. The zero-bias conductance vs. temperature data can be best fit using a d-wave pairing symmetry. A d-wave fit to the conductance vs. voltage data at 400 mK gives rise to $Z_{eff} = 0.365$, $\Gamma = 218\ \mu eV$, and $\Delta = 460\ \mu eV$, from which we obtain $2\Delta/k_BT_c = 4.64$, in agreement with strong coupling results [18]. However, extended BTK models cannot account for the temperature dependence of the full conductance data below T_c [11]. We attribute this to an intrinsic property of the N/HFS interface which is not taken into account in the extended BTK models. We have attempted to obtain better fits to our data by considering additional effects such as breakdown of Andreev approximation [19], mismatch of Fermi velocities and momenta [20], large and energy-dependent Γ [21], etc. These existing models do not reproduce our experimental data [11]. In addition, calculations taking into account spatial variations of the effective mass and the order parameter just give rise to usual BTK conductances with proper scalings of parameters [22]. We are presently investigating the applicability of the two-fluid model [12] to these results.

ACKNOWLEDGMENTS

We are grateful to A.J. Leggett, D. Pines, V. Lukic, and J. Elenewski for fruitful discussions and to B.F. Wilken, A.N. Thaler, P.J. Hentges, K. Parkinson, and W.L. Feldmann for experimental help. This work was supported by the U.S. DoE Award No. DEFG02-91ER45439, through the FS MRL and the CMM at UIUC.

REFERENCES

1. A. F. Andreev, *Sov. Phys. JETP* **19**, 1228 (1964); G. Deutscher, *Rev. Mod. Phys.* **77**, 109 (2005).
2. G. E. Blonder, M. Tinkham, and T. M. Klapwijk, *Phys. Rev. B* **25**, 4515 (1982); G. E. Blonder and M. Tinkham, *ibid.* **27**, 112 (1983).
3. J. D. Thompson *et al.*, *Physica B* **329-333**, 446 (2003).
4. R. A. Radovan *et al.*, *Nature* **425**, 51 (2003); A. Bianchi *et al.*, *Phys. Rev. Lett.* **91**, 187004 (2003).
5. V. A. Sidorov *et al.*, *Phys. Rev. Lett.* **89**, 157004 (2002); J. Paglione *et al.*, *ibid.* **91**, 246405 (2003).
6. R. Movshovich *et al.*, *Phys. Rev. Lett.* **86**, 5152 (2001).
7. K. Izawa *et al.*, *Phys. Rev. Lett.* **87**, 057002 (2001); M. R. Eskildsen *et al.*, *ibid.* **90**, 187001 (2003); H. Aoki *et al.*, *J. Phys.: Condens. Matter* **16**, L13 (2004).
8. G. Goll *et al.*, *Acta Physica Polonica B* **34**, 575 (2003).
9. P. M. C. Rourke *et al.*, *Phys. Rev. Lett.* **94**, 107005 (2005).
10. G. Sheet and P. Raychaudhuri, cond-mat/0502632; W. K. Park and L. H. Greene, cond-mat/0507489.
11. W. K. Park *et al.*, *Phys. Rev. B* **72**, 052509 (2005); W. K. Park and L. H. Greene, cond-mat/0510489.
12. S. Nakatsuji, D. Pines, and Z. Fisk, *Phys. Rev. Lett.* **92**, 016401 (2004).
13. G. Wexler, *Proc. Phys. Soc. London* **89**, 927 (1966).
14. Yu. G. Naidyuk and I. K. Yanson, *J. Phys.: Condens. Matter* **10**, 8905 (1998); H. v. Löhneysen, *Physica B* **218**, 148 (1996).
15. G. Goll, H. v. Löhneysen, I. K. Yanson, and L. Taillefer, *Phys. Rev. Lett.* **70**, 2008 (1993); G. Goll, C. Bruder, and H. v. Löhneysen, *Phys. Rev. B* **52**, 6801 (1995); C. Obermair *et al.*, *ibid.* **57**, 7506 (1998); Y. DeWilde *et al.*, *Phys. Rev. Lett.* **72**, 2278 (1994); *Physica B* **218**, 165 (1996); Yu. G. Naidyuk *et al.*, *Physica B* **218**, 161 (1996).
16. G. Deutscher and P. Nozières, *Phys. Rev. B* **50**, 13557 (1994).
17. Y. Tanaka and S. Kashiwaya, *Phys. Rev. Lett.* **74**, 3451 (1995); S. Kashiwaya, Y. Tanaka, M. Koyanagi, and K. Kajimura, *Phys. Rev. B* **53**, 2667 (1996); S. Kashiwaya and Y. Tanaka, *Rep. Prog. Phys.* **63**, 1641 (2000).
18. C. Petrovic *et al.*, *J. Phys.: Condens. Matter* **13**, L337 (2001).
19. A. Golubov and F. Tafuri, *Phys. Rev. B* **62**, 15200 (2000).
20. N. A. Mortensen, K. Flensberg, and A.-P. Jauho, *Phys. Rev. B* **59**, 10176 (1999).
21. F. B. Anders and K. Gloos, *Physica B* **230-232**, 437 (1997).
22. V. Lukic, J. Elenewski, (private communications).

Anomalous Resistivity of $CeCoIn_5$ Single Crystals

T. Hu,[1] H. Xiao,[1] T. A. Sayles,[2] M. B. Maple,[2] and C. C. Almasan[1]

[1]*Department of Physics, Kent State University, Kent OH 44242, USA*
[2]*Department of Physics, University of California at San Diego, La Jolla, CA 92903, USA*

Abstract. In-plane angular θ dependent resistivity ρ_{ab} (ADR) was measured on single crystals of $CeCoIn_5$ heavy fermion superconductor at temperatures $3\ K \leq T \leq 20\ K$ and in magnetic fields H up to 14 T. We find two different symmetries in ADR in the *H-T* phase diagram, one in the low field region, while the other in high field region. Both symmetries are in the non-Fermi liquid region. The boundary between the two symmetries of the ADR is the same as the crossover from $d\rho_{ab}(\theta=0)/dH > 0$ to $d\rho_{ab}(\theta=0)/dH < 0$, where $\rho_{ab}(\theta=0)$ is the resistivity when H is perpendicular to the *ab*-planes.

Keywords: CeCoIn5, angular dependent resistivity, torque.
PACS: 74.70.Tx, 74.25.Fy, 74.25.Op

The heavy fermion material $CeCoIn_5$ [1] is an unconventional superconductor with the highest superconducting transition temperature T_c of 2.3 K among heavy fermion compounds. Some measurements revealed *d*-wave parity pairing in this material [2, 3]. The superconductivity is near an antiferromagnetic quantum critical point, while the normal state exhibits pronounced non-Fermi liquid behavior [4]. Understanding the non-Fermi liquid behavior in the normal state of $CeCoIn_5$ can shed light on the nature of unconventional superconductivity of this exotic compound.

In this paper, we present angular θ dependent in-plane resistivity ρ_{ab} and torque τ measurements of $CeCoIn_5$, performed at different temperatures T and applied magnetic fields H. Two distinct symmetries were found in $\rho_{ab}(\theta,H,T)$ in the non-Fermi liquid region. Also, $\tau(\theta,H)$ shows paramagnetism with no evidence of a metamagnetic transition.

Single crystals of $CeCoIn_5$ ($T_c = 2.3$ K) were grown using the flux method. Typical size of crystals is $0.5\times0.5\times0.1 mm^3$. Both the out-of-plane ρ_c and in-plane ρ_{ab} resistivities were determined by using the electrical contact configuration of the flux transformer geometry [5]. The angular dependent resistivity $\rho_{ab}(\theta)$ and torque $\tau(\theta)$ were measured by rotating the single crystal from $H||c$-axis to $H||a$-axis, with θ the angle between H and the *c*-axis [see a sketch of the experimental configuration in the Inset to Fig. 1(b)].

Typical angular dependent resistivity curves of $CeCoIn_5$ are shown in Figs. 1(a) and 1(b), measured in low and high applied magnetic fields, respectively (the definition for high and low field regions is given in Fig. 2). Figure 1(a) shows that $\rho_{ab}(\theta)$ measured at 6 K and 3 T displays a four-fold-like symmetry while Fig. 1(b) shows that $\rho_{ab}(\theta)$ measured at 6 K and 8 T displays a six-fold-like symmetry. Note that the two distinct ADR behaviors shown in Figs. 1(a) and 1(b) have nodes at the same angle $\theta \approx 60^0$.

Figure 2 is the *H-T* phase diagram of $CeCoIn_5$, on which the $\rho_{ab}(H,T)$ data with four-fold-like and six-fold-like symmetry are represented as solid diamonds and solid stars, respectively. The solid squares represent the position of the maximum value of the resistivity, while the hatched line separates the $d\rho_{ab}(\theta=0)/dH < 0$ and $d\rho_{ab}(\theta=0)/dH > 0$ regions. Note that all $\rho_{ab}(H,T)$ with four-fold-like symmetry lies in the $d\rho_{ab}(\theta=0)/dH > 0$ region and the $\rho_{ab}(H,T)$ data with six-fold-like symmetry lies in the $d\rho_{ab}(\theta=0)/dH < 0$ region. Therefore, the change in the symmetry in $\rho_{ab}(\theta)$ correlates with the change in the field dependence of the resistivity measured with $H\perp a$-axis.

Figure 3 displays the angular dependence of the torque measured at 6 K and 3 T. These data are fitted well with $A\sin 2\theta$, with A a fitting parameter which depends on H and T.

Measurements of $\tau(\theta)$ at the same temperature and different values of the applied magnetic field ($1\ T \leq H \leq 14$ T) show that the parameter A displays an H^2 dependence (see Inset to Fig. 3). Hence the torque data follow:

$$\tau = -\frac{H^2}{2}(\chi_c - \chi_a)\sin 2\theta \qquad (1)$$

CP850, *Low Temperature Physics: 24th International Conference on Low Temperature Physics;*
edited by Y. Takano, S. P. Hershfield, S. O. Hill, P. J. Hirschfeld, and A. M. Goldman

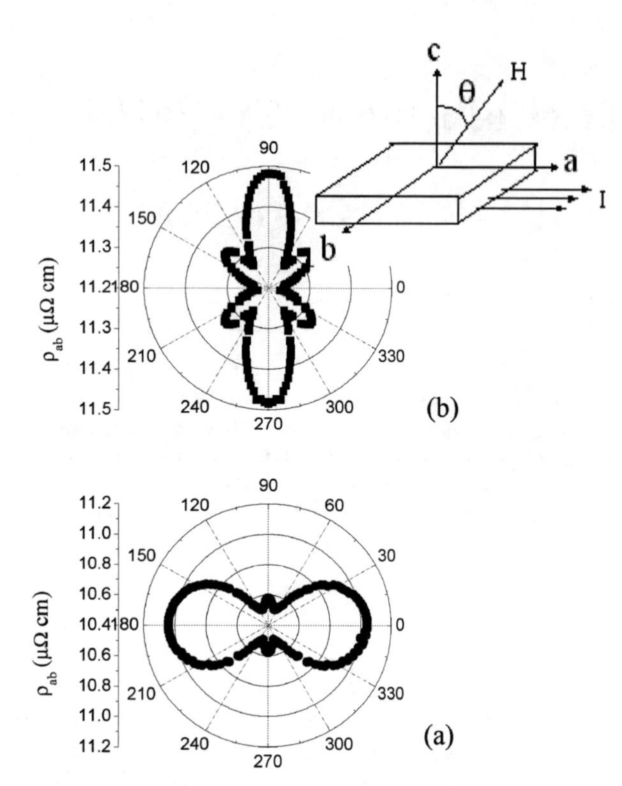

FIGURE 1. Polar plot of angular dependent resistvity of $CeCoIn_5$ measured at (a) 6 K and 3 T and (b) 6 K and 8 T. Plot (a) shows four–fold-like symmetry while plot (b) shows six-fold-like symmetry.

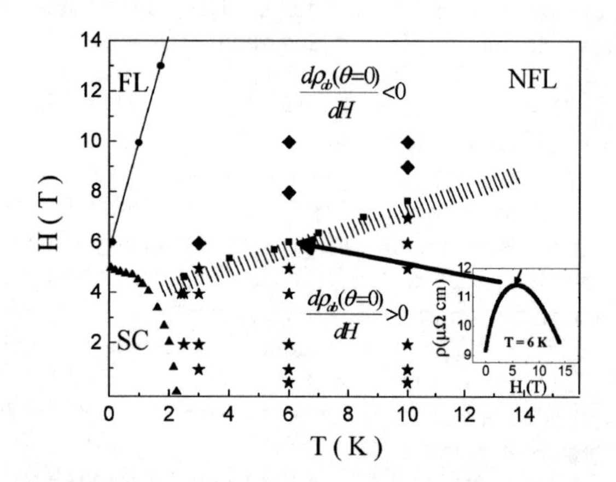

FIGURE 2. Field-temperature *H-T* phase diagram of CeCoIn5. The angular dependent resistivity data with four-fold-like (diamond) and six-fold-like (solid star) symmetries are shown. The solid squares represent the position at which $\rho_{ab}(\theta=0)$ is maximum.

This indicates anisotropic paramagnetism in the non-Fermi liquid region, which reflects the anisotropy between the *a*- axis and *c*- axis. Hence, there is no metamagnetic transition between 1 T and 14 T. This means that the change in the ADR behavior is not due to the spin ordering.

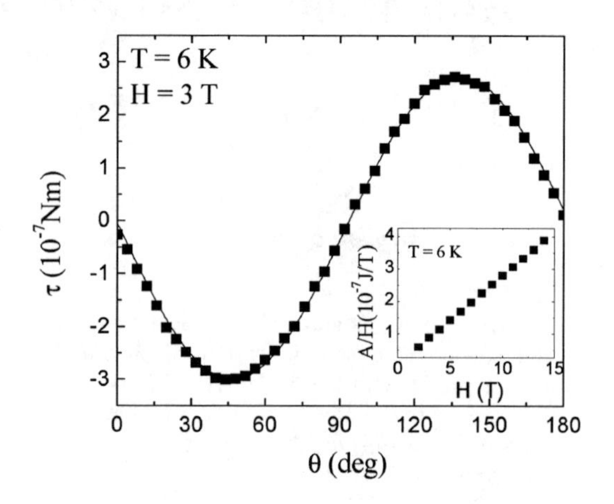

FIGURE 3. Plot of the angular θ dependent torque τ measured at 6 K in a magnetic field of 3 T. The solid line is a fit of the data with $sin2\theta$. Inset: Plot of A/H vs H.

In summary, we measured the angular dependent resistivity (ADR) and torque of $CeCoIn_5$. Two distinct ADR behaviors appear in the non-Fermi region, i.e., a four-fold-like symmetry and a six-fold-like symmetry, separated by the boundary between $d\rho_{ab}(\theta=0)/dH < 0$ and $d\rho_{ab}(\theta=0)/dH > 0$, where $\rho_{ab}(\theta=0)$ is the resistivity when H is perpendicular to the *ab*-planes. The normal state in the non-Fermi liquid region is paramagnetic, with no evidence for any metamagnetic transition. Therefore, the change in the behavior of the ADR is not due to spin ordering.

ACKNOWLEDGMENTS

This research was supported by the National Science Foundation under Grant No. DMR-0406471 at KSU and Grant No. DMR-0335173 at UCSD, and by the U. S. Department of Energy under Grant No. DE-FG02-04ER46105 at UCSD.

REFERENCES

1. C. Petrovic et al., *J. Phys. Condense. Matter* **13**, 337 (2001).
2. H. Aoki et al., *J. Phys. Condens. Matter* **16,** 13 (2004).
3. K. Izawa et al., *Phys. Rev. Lett.* **87**, 057002 (2001).
4. V. A. Sidorov, et al., *Phys. Rev. Lett.* **87**, 157004 (2002).
5. C. N. Jiang, et al., *Phys. Rev. B.* **55**, 3390 (1997).

Angular Resistivity Study in $CeCoIn_5$ Single Crystals

H. Xiao,[1] T. Hu,[1] T. A. Sayles,[2] M. B. Maple,[2] and C. C. Almasan[1]

[1] *Department of Physics, Kent State University, Kent, Ohio 44242, USA*
[2] *Department of Physics, University of California at San Diego, La Jalla, California 92903, USA*

Abstract. Angular dependent resistivity measurements were performed on $CeCoIn_5$ single crystals at a temperature T of 2.3 K in the low field region ($H \leq 1$ T). The resistivity curves scale below a certain critical angle θ_c ($\theta = 0^{\circ}$ when H is parallel to the c-axis). The critical angle θ_c is related with the anisotropy γ of the material. The explicit functional dependence of resistivity on field and angle is obtained based on the time-dependent Ginzburg-Landau theory. The scaling is consistent with this functional dependence.

Keywords: $CeCoIn_5$, scaling, flux-flow.
PACS: 74.70.Tx, 74.25.Fy, 74.25.Op,

Thermal conductivity measurements indicate that $CeCoIn_5$ is in the superclean regime, in which vortex viscosity may be greatly enhanced and leads to anomalous vortex dynamics [1]. It is of great interest to know how vortices behave in magnetic field and to compare this behavior with the one obtained in high transition temperature T_c superconductors.

Angular dependent resistivity measurements were performed at 2.3 K for $H \leq 1$ T. It is found that the resistivity curves scale with the perpendicular field component $H\cos\theta$, which is a result of flux-flow dissipation. We also obtained the explicit functional dependence of the resistivity on field and angle, which is consistent with the scaling observed experimentally.

The single crystal has a $T_{c0} = 2.3$ K in zero field. The in-plane resistivity ρ_{ab} was determined using an algorithm described elsewhere [2]. The samples were rotated in an applied magnetic field from $H||c$-axis ($\theta = 0^{\circ}$) to $H||I||a$-axis ($\theta = 90^{\circ}$).

Typical resistivity curves are shown in Fig. 1. The resistivity is largest at $\theta = 0^{\circ}$ and decreases monotonically as θ increases. At $\theta = 90^{\circ}$, the resistivity has a nonzero minimum, therefore there is still a small amount of dissipation in the system.

In another protocol, we scanned the field and measured ρ_{ab} at fixed angles. A plot of ρ_{ab} vs the perpendicular field component $H\cos\theta$, measured at different angles, is shown in Fig. 2. For angles smaller than a critical angle $\theta_c \approx 54^{\circ}$, the different resistivity curves overlap, i.e. the resistivity depends only on the perpendicular field component: $\rho(H,\theta) = \rho(H\cos\theta)$.

Similar scaling behavior in the mixed state was previously reported in $Bi_2Sr_2CaCu_2O_8$ [3] and MgB_2

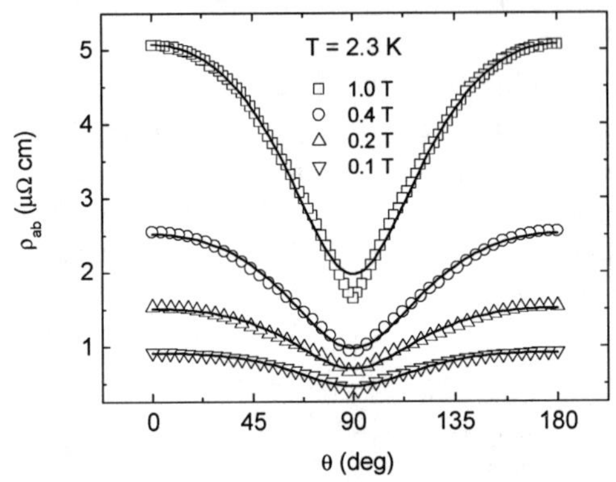

FIGURE 1. Resistivity ρ_{ab} vs angle θ measured at 2.3 K. The solid lines are fits of the data with Eq. (3).

[4]. In highly anisotropic superconductors like $Bi_2Sr_2CaCu_2O_8$, the magnetic field penetrates in the forms of two dimensional pancake vortices, therefore vortex dissipation depends only on, hence scales with, the field component perpendicular to the ab-planes [5]. On the other hand, based on the anisotropic Ginzburg-Landau theory, Blatter et al. [6] obtained the same scaling expression; i.e., $\rho(H,\theta) = \rho(H,\varepsilon_\theta) = \rho(H\cos\theta)$ for $\theta < \theta_c$, where $\varepsilon_\theta = (\cos^2\theta+\gamma^{-2}\sin^2\theta)^{1/2}$ for superconductors with anisotropy γ. Therefore, this general relationship explains the scaling of the resistivity observed in the mixed state of both highly anisotropic as well as less anisotropic superconductors.

The above scaling implies that the resistivity should depend on field and angle as $H\cos\theta$ for $\theta < \theta_c$. Also note that the data of Fig. 2 show that $\rho(H\cos\theta)$ is nonlinear. Next we obtain an explicit functional

CP850, *Low Temperature Physics: 24th International Conference on Low Temperature Physics;*
edited by Y. Takano, S. P. Hershfield, S. O. Hill, P. J. Hirschfeld, and A. M. Goldman

dependence for resistivity in the mixed state. Consider one coordinate frame *(x,y,z)* associated with the crystallographic axes in which the current I is along the y-axis. Another coordinate frame *(x',y',z')* is obtained by rotating the first one around the x-axis such that z' is always along the magnetic field H and θ is the angle between z and z', hence, the c-axis and H (see Inset to Fig. 3). Based on the dissipation energy conservation and the continuity equation, we write:

$$\frac{j_{yy}^2}{\sigma_{ab}^{(yy)}} = \frac{j_{y'y'}^2}{\sigma_{ab}^{(y'y')}} + \frac{j_{z'z'}^2}{\sigma_{ab}^{(z'z')}} \qquad (1)$$

and

$$\frac{1}{\sigma_{ab}^{(yy)}} = \frac{\cos^2\theta}{\sigma_{ab}^{(y'y')}} + \frac{\sin^2\theta}{\sigma_{ab}^{(z'z')}}, \qquad (2)$$

respectively. The first term on the right hand side of Eq. (2) is the flux-flow dissipation, while the second term is a result of quasiparticle dissipation and thermal activation, which explains the non-zero resistivity at $\theta = 90^{\circ}$. Based on the time-dependent Ginzburg-Landau theory, Kopnin calculated the flux-flow conductivity for anisotropic superconductors [7]: $\sigma^{(y'y')} = uaH_{c2}(\theta)\sigma_n(\theta)/2H$, with $u = \xi^2/l_E^2$ (ξ is coherence length and l_E is characteristic length which determines the scale of spatial variations of the gauge-invariant potential Φ), a is a coefficient given by numerical calculations using the vortex order parameter obtained by solving the GL equation, and $H_{c2}(\theta) = H_{c2}/(\cos^2\theta+\gamma^{-2}\sin^2\theta)^{1/2}$. With this expression for $\sigma^{(y'y')}$, Eq. (2) becomes:

$$\rho_{ab}^{(yy)} = \frac{m_1\cos^2\theta}{\sqrt{(\cos^2\theta+\gamma^{-2}\sin^2\theta)^{1/2}}} + \beta\sin^2\theta \quad , \qquad (3)$$

where $m_1 = 2H\rho_n/uaH_{c2}$ and β are fitting parameters; β represents the quasiparticle dissipation. The solid lines in Fig. 1 are fits of the data with Eq. (3). Note that Eq. (3) describes the experimental data very well except in the vicinity of 90°. This small discrepancy could be the result of the lock-in transition in which the dissipation is greatly reduced around 90°. The field dependence of m_1 is plotted in Fig. 3. Note that m_1 is linear in H, which is consistent with its definition.

Equation (3) is also consistent with the scaling $\rho(H\cos\theta)$. Indeed, it gives the scaling when the second term is substantially smaller than the first term and when $\gamma^{-2}\sin^2\theta << \cos^2\theta$, i.e. $\theta << \tan^{-1}\gamma \approx 63^{\circ}$ ($\gamma \approx 2$). The first condition is satisfied since in the mixed state the flux-flow dissipation is much larger than the quasiparticle dissipation. The second condition is experimentally satisfied, i.e., $\rho(H\cos\theta)$ scales for $\theta \leq 54^{\circ}$ (Fig. 2). This indicates the consistency between the the experimental data and the explicit functional dependence of resistivity given by Eq. (3).

In summary, angular dependent resistivity was measured in the mixed state, which shows

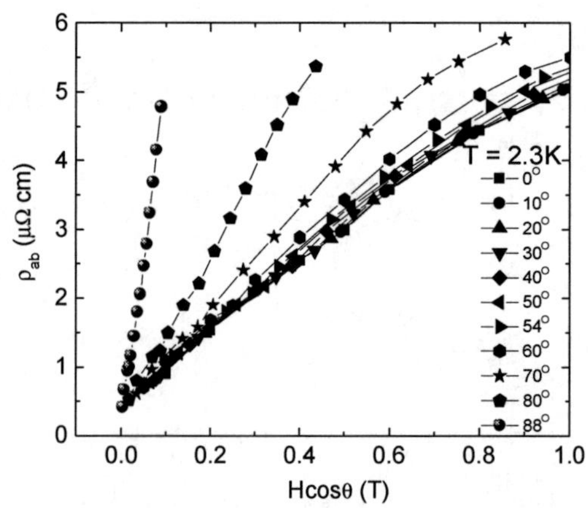

FIGURE 2. Resistivity ρ_{ab} vs the perpendicular field component $H\cos\theta$, measured at 2.3 K and fixed angles θ.

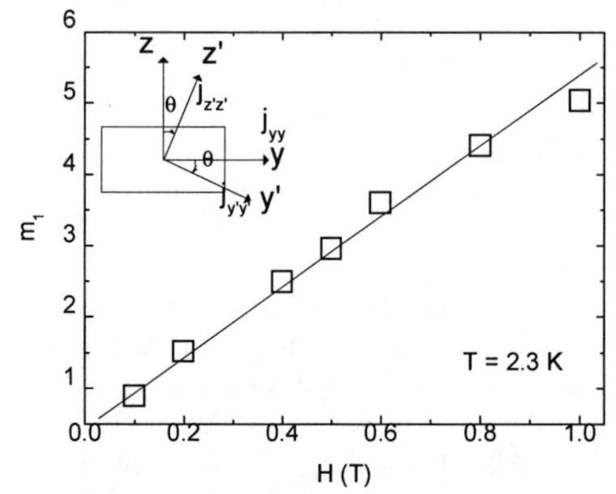

FIGURE 3. The field H dependence of the fitting parameter m_1. The solid line is a guide to the eye.

$H\cos\theta$ scaling, a result of flux-flow dissipation. The explicit functional dependence of ρ_{ab} on H and θ obtained is consistent with the $H\cos\theta$ scaling.

ACKNOWLEDGMENTS

This research was supported by the National Science Foundation under Grant No. DMR-0406471 at KSU and the U. S. Department of Energy under Grant No. DE-FG02-04ER46105 at UCSD.

REFERENCES

1. Y. Kasahara *et al.*, *cond-mat/0506071.*
2. C. N. Jiang *et al.*, *Phys. Rev. B* **55**, 3390 (1997).
3. H. Raffy *et al.*, *Phys. Rev. Lett.* **66**, 2515 (1991).
4. Ferdeghini *et al.*, *Eur. Phys. J. B* **30**, 147 (2002).
5. P. H. Kes *et al.*, *Phys. Rev. Lett.* **64**, 1063 (1990).
6. G. Blatter *et al.*, *Phys. Rev. Lett.* **68**, 875 (1992).
7. N. Kopnin, *Theory of Nonequilibrium Superconductivity,* Oxford: Clarendon press, 2001, pp. 238-245.

Microscopic Evidence for the FFLO State in $CeCoIn_5$ Probed by NMR

K. Kumagai*, K. Kakuyanagi*, M. Saitoh*, S. Takashima†, M. Nohara†, H. Takagi† and Y. Matsuda**

*Division of Physics, Graduate School of Science, Hokkaido University, Sapporo 060-0810, Japan
†Department of Advanced Materials Science, University of Tokyo, Kashiwa, Chiba 277-8581, Japan
**Department of Physics, Kyoto University, Kyoto 606-8502, Japan

Abstract. An anomalous temperature dependence of the NMR spectrum has seen found in the newly-discovered high field and low temperature superconducting phase of $CeCoIn_5$. The present NMR study provides direct microscopic evidence that the Fulde-Ferrel-Larkin-Ovchinikov (FFLO) state is realized in the heavy fermion superconductor, $CeCoIn_5$

Keywords: NMR, FFLO, heavy fermion superconductor, $CeCoIn_5$
PACS: 76.60Cq, 71.27.+a, 74.25Db, 74.70.Tx

Among possible exotic superconducting (SC) phases in the presence of a strong magnetic field, a spatially-nonuniform SC state has become the subject of intense investigation after the pioneering work by Fulde and Ferrel, as well as Larkin and Ovchinnikov (FFLO) in the mid-1960's [1]. In the FFLO state, pair-breaking due to the Pauli effect is reduced by the formation of a new pairing state $(k\uparrow, -k+q\downarrow)$. As a result, a new SC state with a spatially-oscillating order parameter and spin polarization should appear in the vicinity of the upper critical field H_{c2}. In spite of many efforts to find the FFLO state, it has never been observed in conventional superconductors. In the last decade, the heavy fermion superconductors such as $CeRu_2$ and UPd_2Al_3 have been proposed as candidates for the observation of the FFLO state, but subsequent research has called the interpretation of the data in terms of the FFLO state into question [2].

A new superconductor, $CeCoIn_5$ has aroused great interest, because several measurements have led to a renewed discussion of a possible FFLO state [3–6]. Pauli-limited superconductivity is supported by the fact that the phase transition from SC to normal metal at the upper critical fields is of the first order below $\sim$1.3 K [7, 8]. Heat capacity measurements revealed that a new phase appears within the SC state in the vicinity of H_{c2} with H parallel to the *ab*-plane at low temperatures [3, 4].

While these experimental and theoretical results make the FFLO scenario a very appealing one for $CeCoIn_5$, there is no direct experimental evidence so far which verifies the spatially-nonuniform SC state expected in the FFLO state. Therefore, a powerful probe of the quasiparticle excitations in the high field SC phase is required to shed light for this subject. NMR is particularly suitable for this purpose because NMR can monitor the low energy quasiparticle excitations sensitively. Here we present ^{115}In-NMR data to extract microscopic information on the quasiparticle structure [9].

Single crystals of $CeCoIn_5$ were grown by a flux method as described in Ref. [10]. NMR measurements were performed on a single crystal (T_c=2.3 K with a transition width less than 0.1 K) by using a phase-coherent pulsed NMR spectrometer. Experiments were always carried out with the magnetic field H parallel to the *a*-direction under the field-cooled condition. We report NMR results at the In(1) site which is located in the center of the square lattice of Ce atoms. The Knight shift ^{115}K was obtained from the central ^{115}In-line using a gyromagnetic ratio of $^{115}\gamma_N$=9.3295 MHz/T and by taking into account the electric quadrupole frequency, ν_Q=8.173 MHz and the asymmetric parameter, η=0.

Figure 1 shows ^{115}In-NMR spectra of $CeCoIn_5$ obtained for two magnetic fields: (a) H=11.8 T, (b) H=11.3 T. The line width in the normal state at H=11.8 T is 15 kHz and is temperature-independent. The peak position shifts to the higher frequency side, which indicates the Knight shift of ^{115}In(1) increases smoothly with decreasing temperature in the normal state. Compared to the NMR result at H=11.8 T, the NMR spectrum at H=11.3 T is anomalous, as shown in Fig. 1(b). The spectrum with a single peak is observed down to T=320 mK, and the peak frequency slightly decreases below T_c. Then the spectrum shows additional complex features just below the phase boundary at T=320 mK. The higher resonance line grows rapidly with T. A double peak structure shows up at T=240 mK, followed by a shoulder structure at T=260–300 mK.

We simulate this anomalous spectrum assuming a simple sinusoidal modulation of the gap function along the

CP850, *Low Temperature Physics: 24th International Conference on Low Temperature Physics;*
edited by Y. Takano, S. P. Hershfield, S. O. Hill, P. J. Hirschfeld, and A. M. Goldman

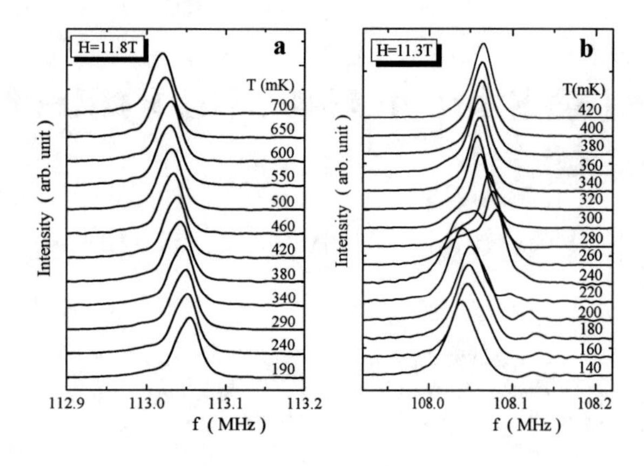

FIGURE 1. ^{115}In(1)-NMR spectra [central line of the ($\pm1/2 \leftrightarrow \mp1/2$) transition] as a function of frequency for various temperatures at (a) H=11.9 T (in the normal state) and (b) H=11.3 T (below the normal to SC phase boundary at $T\sim$700 mK, and into the new SC phase at $T\sim$300 mK).

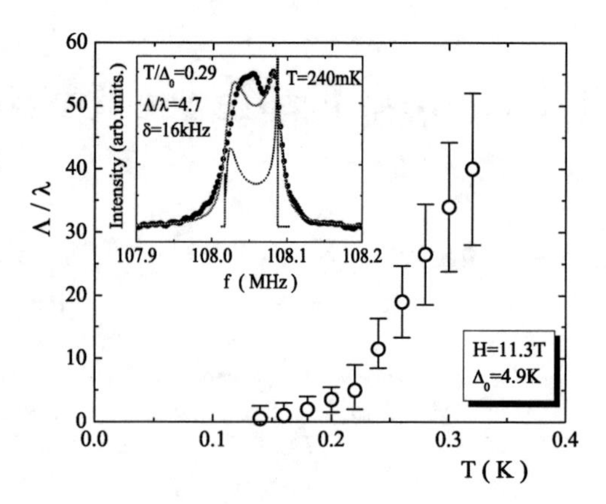

FIGURE 2. Temperature dependence of $\frac{\Lambda}{\lambda}$ for H=11.3 T and Δ_0=4.9 K. The inset shows the ^{115}In-NMR spectrum at T=240 mK. The solid red line and blue dotted line represent the simulated spectra with and without the convolution with a Lorentzian function (δ=16 kHz), respectively.

applied field, $\Delta(x) = \Delta_0 \sin x/\Lambda$, where Λ is the wave length [11]. The present measurement was done under the condition of very small H_1 and short pulse width τ_w (10 μs). This means a very small tipping angle of the nuclear spin by the NMR pulse, namely, $\theta = \gamma_N H_1 \tau_w \ll \pi/2$. In this case, the spin echo intensity is proportional to H_1^3 [12]. As the NMR spectrum depends on the Knight shift, and also on H_1, which are spatially-dependent in the FFLO state, the NMR spectrum is given by [9]

$$I(k) \propto \int_0^{\Lambda/2} \delta\left(K\left(\frac{\Delta(x)}{T}\right) - k\right) [H_1(x)]^3 \mathrm{d}x, \quad (1)$$

where $K(\Delta/T)$ is the Yoshida function for the Knight shift in the SC state [13]. We assumed a local DOS with a d-wave SC gap with line nodes. We obtain the functional form of the NMR intensity and then calculate the spectrum numerically to fit the experimental data. There are only two adjustable parameters, $\frac{\Lambda}{\lambda}$ and $\frac{T}{\Delta_0}$, where λ is the magnetic penetration depth.

After convolution with a Lorentzian shape due to inhomogeneous broadening, we show that the simulation reproduces well the observed spectra as shown in Fig. 2. This suggests that the evolution of the NMR spectrum with temperature is compatible with what is expected in the FFLO phase. In Fig. 2 we show the temperature dependence of $\frac{\Lambda}{\lambda}$ obtained from the fitting. As λ is expected to be nearly temperature-independent ($\sim$2000 Å) in the temperature range concerned ($T < 0.3$ K), the wave length of the spatial oscillation of the SC order parameter Λ decreases with lowering temperature. We would like to stress that, in spite of the very crude model, the evolution of the NMR spectrum with temperature is compatible with what is expected in the FFLO phase, and also that the presence of well-separated NMR lines implies that the quasiparticle excitation around the planar nodes is spatially localized. Finally we point out that the spatial dependence of the order parameter along the magnetic field may be Bloch wall-like or rectangular rather than sinusoidal far below $T_c(H)$ [11]. The present result calls for further investigations of the real space structure of the SC order parameter and the vortex core structure.

In summary, the ^{115}In-NMR spectrum in $CeCoIn_5$ exhibits a dramatic change in the vicinity of $H_{c2}^{\parallel}$. On the basis of the simulation of the NMR spectrum, we are able to establish clear evidence for a spatially-inhomogeneous SC state at high field and low temperatures, as expected in the FFLO state.

REFERENCES

1. P. Fulde and R.A. Ferrel, Phys. Rev. **135**, A550 (1964), A.I. Larkin and Y.N. Ovchinnikov, Sov. Phys. JETP **20**, 762 (1965).
2. For example, K. Gloos *et al.*, Phys. Rev. Lett. **70**, 501 (1993); M.R. Norman, *ibid.* **71**, 3391 (1993).
3. H.A. Radovan *et al.*, Nature **425**, 51 (2003).
4. A. Bianchi *et al.*, Phys. Rev. Lett. **91**, 187004 (2003).
5. T. Watanabe *et al.*, Phys. Rev. **B70**, 020506 (2004).
6. C. Capan *et al.*, Phys. Rev. **B70**, 134513 (2004).
7. A. Bianchi *et al.*, Phys. Rev. Lett. **89**, 137002 (2002), T. Tayama *et al.*, Phys. Rev. B **65**, 180504 (2002).
8. K. Izawa *et al.*, Phys. Rev. Lett. **87**, 057002 (2001).
9. K. Kakuyanagi *et al.*, Phys. Rev. Lett. **94**, 047602 (2005).
10. C. Petrovic, Europhys. Lett, **53**, 354 (2001) and J. Phys. Cond. Mat. **13**, L337 (2001).
11. M. Tachiki *et al.*, Z. Phys. B **100**, 369 (1996).
12. E. L. Hahn, Phys. Rev. **80**, 580 (1950)
13. K. Yoshida, Phys. Rev. **110**, 769 (1958)

Possibility of FFLO State in Organic Superconductor λ-$(BETS)_2FeCl_4$

S. Uji[1], T. Terashima[1], T. Yamaguchi[1], K. Enomoto[1], T. Konoike[1], M. Nishimura[1], S. Yasuzuka[2], H. Tanaka[3], M. Tokumoto[3], A. Kobayashi[4], C. Hengbo[5], H. Kobayashi[5], E. S. Choi[6], D. Graf[6], T. Tokumoto[6], and J. S. Brooks[6]

[1]*National Institute for Materials Science, Tsukuba, Ibaraki 305-0003, Japan*
[2]*Faculty of Science, Osaka City University, Osaka 558-8585, Japan*
[3]*Institute of Advanced Industrial Science and Technology and CREST JST, Tsukuba, Ibaraki 305-8568, Japan*
[4]*Research Centre for Spectrochemistry, Univ. of Tokyo, Bunkyo-ku, Tokyo 113-0033, Japan*
[5]*Institute for Molecular Science and CREST JST, Okazaki, Aichi 444-8585, Japan*
National High Magnetic Field Laboratory, Florida State University, Tallahassee FL 32306, USA

Abstract. A magnetic two dimensional organic conductor λ-$(BETS)_2FeCl_4$ has a superconducting phase only under high magnetic fields between 18T and 42 T. This field induced superconductivity is qualitatively understood by Jaccarino-Peter effect, and an inhomogeneous superconducting phase, the so-called FFLO phase, whose order parameter oscillates in space, is expected to be stabilized. Under high magnetic fields parallel to the conducting layers, characteristic features in the interlayer resistance are observed, suggesting that the Josephson vortices are strongly pinned at some fields. The results are explained in terms of a commensurability effect between the vortex lattice and the oscillating order parameter.

Keywords: Organic conductor, FFLO state, field induced superconductivity
PACS: 71.18.+y 71.20.Rv 74.70.Kn

INTRODUCTION

The two dimensional organic material λ-$(BETS)_2FeCl_4$, where BETS isbis(ethylenedithio)-tetraselenafulvalene, is a good example of low dimensional magnetic conductors where rich ground state properties appear [1,2]. In the absence of the magnetic field, λ-$(BETS)_2FeCl_4$ shows a transition from a paramagnetic metal (PM) to an antiferromagnetic insulator (AFI) around 8 K. The AFI phase is destabilized by the magnetic field of 10 T, and the PM phase is recovered at higher fields. When the magnetic field is applied parallel to the conducting layers (*ac* plane), a superconducting (S) phase is induced above 17 T below 1 K. The critical temperature T_c has a maximum (5 K) at 31 T. This superconductivity is easily destroyed when the magnetic field is tilted from the conduction layers (*ac* planes). This rich phase is ascribed to the exchange interaction between the Fe *3d* moments and the π electron spins on the BETS molecules.

The overall feature of the field induced S phase is well understood by Fischer theory [3], based on the Jaccarino-Peter (J-P) effect [4]. In the PM phase, the localized Fe moments are aligned along the external field (H). Because of a strong negative d-π exchange interaction J, the π electron spins experience a strong internal field ($-H_{int}$) created by the Fe *3d* moments, whose direction is antiparallel to H. Therefore, when the internal field is compensated with the external field ($H=H_{int}$), the Zeeman effect, one of the mechanisms that destroy superconductivity, is completely suppressed. Moreover, when H is parallel to the conducing layers, the orbital effect, which is the other destructive mechanism of superconductivity, is also suppressed. Therefore, superconductivity can be stabilized under high inplane fields. For this salt, an inhomogeneous superconducting state with a spatially modulated order parameter, the so-called a FFLO state [5], is expected to be stabilized [2,6,7]. The necessary conditions for the FFLO state are: quenched orbital effect, large paramagnetic susceptibility and clean

CP850, *Low Temperature Physics: 24th International Conference on Low Temperature Physics;*
edited by Y. Takano, S. P. Hershfield, S. O. Hill, P. J. Hirschfeld, and A. M. Goldman

limit superconductors. All the conditions are well fulfilled for λ-$(BETS)_2FeCl_4$. In order to investigate the possibility of the FFLO state in the field induced S phase, we have performed the resistance measurements under high magnetic fields up to 45 T.

EXPERIMENTS

Single crystals of λ-$(BETS)_2FeCl_4$ were synthesized electrochemically. The resistance measurements were performed by a conventional *ac* technique with the electric current along the *c* axis in the hybrid magnet at NHMFL.

RESULTS AND DISCUSSION

Figure 1 shows the interlayer resistance (I//*c*) in fields parallel to the *c* axis. At 0.5 K, the resistance linearly increases with field, and then shows an anomalous feature (successive dip structure) in the superconducting transition. Similar anomalies are also observable at high fields around 40 T. The dips are smeared out with increasing temperature, but the dip positions, denoted by arrows, do not change. The dips become more evident at higher currents. At 1.6 K, we observe no anomalous features. When the field is tilted from the *c* axis by a few degrees, the dips disappear. The resistance as a function of the field angle θ, which is the angle between the *c* axis and field is plotted in the inset of Fig.1. We note that a quite sharp peak is evident for H//*c*.

For λ-$(BETS)_2FeCl_4$, the Ginzburg-Landau coherence length perpendicular to the layers is less than the interlayer spacing, showing highly anisotropic superconductivity. For such anisotropic superconductors, it is known that a transition of flux line structures as a function of the magnetic field direction, a lock-in transition takes place. In magnetic fields sufficiently tilted from the layers, there exist conventional tilted flux lines or combined vortex structures made by Abrikosov (pancake) and Josephson vortices (JV), which penetrate into the superconducting layers perpendicularly, and lie in the insulating layers, respectively. When the field becomes almost parallel to the layers, a lock-in transition takes place: the Abrikosov vortices are excluded and only the JV are stabilized. Since the JV are weakly pinned in the insulating layers, the sufficiently high interlayer current depins the JV, and they move in the insulating layers. The motion, which is probably collective motion of the JV lattice, causes a large energy dissipation, *i.e.* finite resistance. Therefore, it is very likely that the sharp peak in the inset of Fig. 1 arises from the energy dissipation due to the depinned JV lattice. As the field is tilted, the lock-in transition occurs and the flux lines are strongly pinned in the superconducting layers (small energy dissipation). As the field is further tilted, the S phase is removed and the PM phase is recovered. Our systematic measurements show that the successive dip structure is ascribed to the strong pinning of the JV. As first predicted by Bulaevskii [8], the JV lattice in the FFLO state is strongly pinned by the periodic structure of the superconducting order parameter when the JV lattice constant is commensurate with the period of the order parameter oscillation. Since both periods change with field, such strong pinning may happen at some different fields as observed in Fig. 1. This is a novel commensurability effect between the JV lattice and the oscillating order parameter.

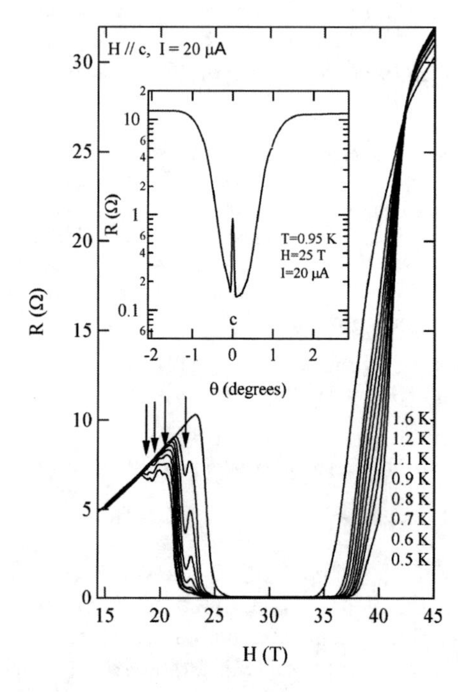

FIGURE 1. Interlayer resistance (I//*c*) in fields parallel to the *c* axis. The arrows show dips in the resistive transitions.

REFERENCES

1. S. Uji *et al.*, *Nature* **410**, 908 (2001); S. Uji *et al.*, *J. Phys. Soc. Jpn.* **72**, 369 (2003).
2. L. Balicas *et al.*, *Phys. Rev. Lett.* **87**, 067002 (2001).
3. O. Fischer, Helv. *Phys. Acta* **45**, 331 (1972).
4. V. Jaccarino and M. Peter, *Phys. Rev. Lett.* **9**, 290 (1962).
5. A. I. Larkin, and Yu. N. Ovchinnikov, *Sov. Phys. JETP* **20**, 762 (1962); P. Fulde, and R. A. Ferrell, *Phys. Rev.* **135**A, 550 (1964).
6. M. Houzet *et al.*, *Phys. Rev. Lett.* **88**, 227001 (2002).
7. H. Shimahara, *J. Phys. Soc. Jpn.* **71**, 1644 (2002).
8. L. Bulaevskii *et al.*, *Phys. Rev. Lett.* **90**, 067003 (2003).

Electronic Structure of Vortex in the FFLO Superconducting State

Masanori Ichioka, Hiroto Adachi, Takeshi Mizushima, and Kazushige Machida

Department of Physics, Okayama University, Okayama 700-8530, Japan

Abstract. Based on the quasiclassical theory, we investigate the vortex structure in the Fulde-Ferrell-Larkin-Ovchinnikov (FFLO) state suggested to exist for the high field phase of $CeCoIn_5$. The pair potential has 2π-phase winding around the vortex line, and π-phase shift at the nodal plane of the FFLO modulation. These topologies of the phase structure affect the distribution of paramagnetic moment and low energy electronic states. We also discuss the resonance line shape of the Knight shift in the FFLO state.

Keywords: Fulde-Ferrell-Larkin-Ovchinnikov state, vortex state, electronic structure, microscopic theory
PACS: 74.25.Op, 74.25.Jb, 74.70.Tx

INTRODUCTION

Much attention has been focused on a new superconducting phase, considered as a Fulde-Ferrell-Larkin-Ovchinnikov (FFLO) state, at high field in a quasi-two dimensional (Q2D) heavy fermion superconductor $CeCoIn_5$ [1, 2, 3, 4]. The FFLO state [5, 6] was proposed in the superconductors when the paramagnetic depairing effect is eminent in the mismatched Fermi surface, where the Fermi surfaces for up-spins and down-spin split due to the Zeeman effect under magnetic fields. In the mismatched Fermi surface, Cooper pairs of up- and down- spins are likely to have non-zero momentum for the center of mass coordinate of the Cooper pair. Then we expect the spatial modulation of the pair potential [7, 8, 9, 10, 11, 12]. The possible FFLO state is widely discussed in various research fields, ranging from superconductors in condensed matter, neutral Fermion superfluids in an atomic cloud [13], to color superconductivity in high energy physics [14].

In theoretical studies, many calculations for the FFLO state have been done by neglecting vortex structure. However, we have to include the vortex structure, because the FFLO state appears at high fields in the mixed states. When we consider vortex structure in the FFLO state, there are two possible cases, i.e., the modulation vector of the FFLO state is parallel or perpendicular to the applied magnetic field. In our study, the former case is investigated by the quasiclassical theory [8, 10, 15, 16]. In a previous quasi-classical study by Tachiki *et al.* [8] for this case, they reduce the FFLO states including Abrikosov vortex to a problem of the one dimensional system along the magnetic field direction, and the FFLO modulation is calculated. In our study, three dimensional structure of the vortex and the FFLO modulation are determined by the selfconsistent calculation based on the quasi-classical theory.

On the other hand, the vortex and nodal plane structure in the FFLO state were calculated by the Bogoliubov de-Gennes (BdG) theory for a single vortex in a cylindrical superconductor [17]. This study clarifies the importance of the topological structure in the pair potential. The pair potential has 2π-phase winding around the vortex line, and π-phase shift at the nodal plane of the FFLO modulation. These topologies of the phase structure affect the distribution of paramagnetic moment and low energy electronic states within the superconducting gap. The paramagnetic moment is not enhanced at the intersection point of a vortex and a nodal plane, while it is enhanced at the vortex core and the FFLO nodal plane. These paramagnetic structures are tightly related to the spectral evolution around the vortex and the nodal plane in the LDOS for up and down spin electrons.

The purpose of this paper is to investigate these characters of the vortex structure in the FFLO state by the quasiclassical theory. We calculate the spatial structure of the pair potential, paramagnetic magnetization (i.e., Knight shift in a magnetic field) and electronic states under given period of the FFLO modulation. And we compare these structures with those obtained by the BdG theory.

QUASI-CLASSICAL THEORY WITH PARAMAGNETIC EFFECT

Under magnetic fields, the Hamiltonian of superconductors is given by

$$\mathscr{H} - \mu_0 \mathscr{N} = \sum_{\sigma=\uparrow,\downarrow} \int d^3\mathbf{r}\, \psi^\dagger_\sigma(\mathbf{r}) K_\sigma(\mathbf{r}) \psi_\sigma(\mathbf{r})$$

CP850, *Low Temperature Physics: 24th International Conference on Low Temperature Physics;*
edited by Y. Takano, S. P. Hershfield, S. O. Hill, P. J. Hirschfeld, and A. M. Goldman

$$- \int d^3\mathbf{r}_1 \int d^3\mathbf{r}_2 \left\{ \Delta(\mathbf{r}_1,\mathbf{r}_2)\psi_\uparrow^\dagger(\mathbf{r}_1)\psi_\downarrow^\dagger(\mathbf{r}_2) + \Delta^*(\mathbf{r}_1,\mathbf{r}_2)\psi_\downarrow(\mathbf{r}_2)\psi_\uparrow(\mathbf{r}_1) \right\} \quad (1)$$

with

$$K_\sigma(\mathbf{r}) = \frac{\hbar^2}{2m}\left(\frac{\nabla}{\mathrm{i}} + \frac{2\pi}{\phi_0}\mathbf{A}\right)^2 + \sigma\mu_\mathrm{B}B(\mathbf{r}) - \mu_0 \quad (2)$$

and the flux quantum ϕ_0. $\sigma = \pm$ for up/down spin electrons. The superconductivity is suppressed by the paramagnetic depairing effect due to the Zeeman splitting term $\mu_\mathrm{B}B(\mathbf{r})$ in addition to the orbital depairing effect by the vector potential $\mathbf{A}$. Using the quasi-classical Green's functions $g(\omega_l + \mathrm{i}\tilde{\mu}B,\mathbf{k},\mathbf{r})$, $f(\omega_l + \mathrm{i}\tilde{\mu}B,\mathbf{k},\mathbf{r})$, $f^\dagger(\omega_l + \mathrm{i}\tilde{\mu}B,\mathbf{k},\mathbf{r})$, Eilenberger equations are given by [8, 10, 15, 16]

$$\left\{\omega_l + \mathrm{i}\mu_\mathrm{B}B + \hbar\mathbf{v}_\mathrm{F}\cdot\left(\frac{\nabla}{2} + \mathrm{i}\frac{\pi}{\phi_0}\mathbf{A}\right)\right\} f = \Delta(\mathbf{r},\mathbf{k})g,$$
$$\left\{\omega_l + \mathrm{i}\mu_\mathrm{B}B - \hbar\mathbf{v}_\mathrm{F}\cdot\left(\frac{\nabla}{2} - \mathrm{i}\frac{\pi}{\phi_0}\mathbf{A}\right)\right\} f^\dagger = \Delta^*(\mathbf{r},\mathbf{k})g,$$
$$\hbar\mathbf{v}_\mathrm{F}\cdot\nabla g = \Delta^*(\mathbf{r},\mathbf{k})f - \Delta(\mathbf{r},\mathbf{k})f^\dagger, \quad (3)$$

where $g = (1 - ff^\dagger)^{1/2}$, $\mathrm{Re}g > 0$, and $\Delta(\mathbf{r},\mathbf{k}) = \Delta(\mathbf{r})\phi(\mathbf{k})$. In Eq. (3), $\mathbf{r} = (\mathbf{r}_1 + \mathbf{r}_2)/2$ is the center of mass coordinate of the Cooper pair, and $\mathbf{k}$ is a relative momentum of the Cooper pair. Since we report the s-wave pairing case in this paper, the pairing function $\phi(\mathbf{k}) = 1$.

In the following, length, temperature, Fermi velocity, magnetic field and vector potentials are scaled by R_0, T_c, $\bar{v}_\mathrm{F}$, B_0 and B_0R_0, respectively. Here, $R_0 = \hbar\,\bar{v}_\mathrm{F}/2\pi k_\mathrm{B}T_\mathrm{c}$, $B_0 = \hbar c/2|e|R_0^2$ and $\bar{v}_\mathrm{F} = \langle v_\mathrm{F}^2\rangle_\mathbf{k}^{1/2}$ is an averaged Fermi velocity on the Fermi surface. $\langle\cdots\rangle_\mathbf{k}$ indicates the Fermi surface average. Energy, Δ and ω_l are scaled by $\pi k_\mathrm{B}T_\mathrm{c}$. The pair potential is selfconsistently calculated by

$$\Delta(\mathbf{r}) = g_0N_0T\sum_{0\le\omega_l\le\omega_\mathrm{cut}}\left\langle\phi^*(\mathbf{k})\left(f + f^{\dagger*}\right)\right\rangle_\mathbf{k} \quad (4)$$

with $(g_0N_0)^{-1} = \ln T + 2T\sum_{0\le\omega_l\le\omega_\mathrm{cut}}\omega_l^{-1}$. We use $\omega_\mathrm{cut} = 20k_\mathrm{B}T_\mathrm{c}$. The vector potential is selfconsistently determined by the paramagnetic moment and the supercurrent contributions;

$$\nabla\times(\nabla\times\mathbf{A}) = \nabla\times\mathbf{M}_\mathrm{para}(\mathbf{r}) - \frac{2T}{\tilde{\kappa}^2}\sum_{0\le\omega_l}\langle\mathbf{v}_\mathrm{F}\mathrm{Im}g\rangle_\mathbf{k}, \quad (5)$$

where

$$M_\mathrm{para}(\mathbf{r}) = M_0\left(\frac{B(\mathbf{r})}{\bar{B}} - \frac{2T}{\tilde{\mu}\bar{B}}\sum_{0\le\omega_l}\langle\mathrm{Im}\{g\}\rangle_\mathbf{k}\right) \quad (6)$$

is the paramagnetic moment in the vortex lattice state, with an average flux density $\bar{B}$, $\tilde{\kappa} = B_0/\pi k_\mathrm{B}T_\mathrm{c}\sqrt{8\pi N_0}$, $\tilde{\mu} = \mu_\mathrm{B}B_0/\pi k_\mathrm{B}T_\mathrm{c}$, and the normal state paramagnetic moment $M_0 = (\tilde{\mu}/\tilde{\kappa})^2\bar{B}$. N_0 is the density of states at the Fermi energy in the normal state. In the following, we see the case when $\tilde{\kappa} = 20$ and $\tilde{\mu} = 1.7$, giving $H_{c2}(T\sim 0)\sim 0.2$ in our unit.

As a model of Fermi surface in $\mathrm{CeCoIn_5}$, we use a Q2D Fermi surface with rippled cylinder-shape, and the Fermi velocity is given by $\mathbf{v}_\mathrm{F} = (v_a, v_b, v_c) \propto (\cos\theta, \sin\theta, \tilde{v}_z\sin k_z)$ [16]. In our calculation we set $\tilde{v}_z = 0.5$, so that the anisotropy ratio $\gamma = \xi_c/\xi_{ab} \sim \langle v_c^2\rangle_\mathbf{k}^{1/2}/\langle v_a^2\rangle_\mathbf{k}^{1/2} \sim 0.5$. A magnetic field is applied along a-axis direction in our calculation. Thus, (x,y,z) of the coordinate for vortices corresponds to (b,c,a) of the crystal coordinate.

We solve Eq. (3) and Eqs. (4)-(6) alternately, and obtain selfconsistent solution, by fixing a unit cell of the vortex lattice and a period L of the FFLO modulation [15, 16]. The unit cell of the vortex lattice is given by $\mathbf{r} = w_1(\mathbf{u}_1 - \mathbf{u}_2) + w_2\mathbf{u}_2$ with $-0.5 \le w_i \le 0.5$ (i=1, 2), $\mathbf{u}_1 = (a,0)$ and $\mathbf{u}_2 = (a/2, a_y)$. Reflecting anisotropy ratio γ, we set $a_y/a = \sqrt{3}\gamma/2$ with $\gamma = 0.5$. For the FFLO modulation, we assume $\Delta(x,y,z) = \Delta(x,y,z+L)$ and $\Delta(x,y,z) = -\Delta(x,y,-z)$. Then, $\Delta(\mathbf{r}) = 0$ at the FFLO nodal plane $z = 0$, and $\pm 0.5L$.

When we calculate the electronic state, we solve Eq. (3) with $i\omega_l \to E + i\eta$. The local density of states (LDOS) is given by

$$N_\sigma(\mathbf{r},E) = \langle\mathrm{Re}\{g(\omega_l + \mathrm{i}\sigma\tilde{\mu}B,\mathbf{k},\mathbf{r})|_{i\omega_l\to E+i\eta}\}\rangle \quad (7)$$

for each spin component. We typically use $\eta = 0.01$.

FFLO VORTEX STRUCTURE

The amplitude of the order parameter is shown by Fig. 1 at $\bar{B} = 0.15$ ($a = 9.8R_0$) and $T = 0.2$, when we set $L = 50R_0$. There we see that $|\Delta(\mathbf{r})|$ is suppressed near the vortex center at $x = y = 0$ and the FFLO nodal plane at $z = 0$, $\pm 0.5L$. Far from the FFLO nodal plane, $\Delta(\mathbf{r})$ shows a typical profile of the conventional vortex, as shown in Fig. 1(b).

Correspondingly, paramagnetic moment $M_\mathrm{para}(\mathbf{r})/M_0$ is presented in Fig. 2. From Fig. 2(a), we see that $M_\mathrm{para}(\mathbf{r})$ is enhanced at the vortex core and the FFLO nodal plane, but it is not enhanced at the intersection point of a vortex and a nodal plane. While $M_\mathrm{para}(\mathbf{r})$ has peak at the vortex center in the plane $z = 0.25L$ far from the nodal plane [Fig. 2(b)], it has minimum at the vortex center in the FFLO nodal plane $z = 0$ [Fig. 2(c)].

The reason of this $M_\mathrm{para}(\mathbf{r})$ structure can be understood if we see the LDOS spectrum for the up- and down-spin electrons presented in Fig. 3. In the quasi-

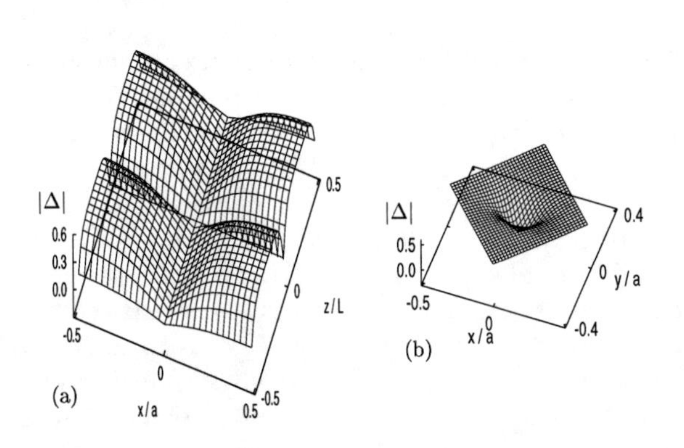

FIGURE 1. Amplitude of the pair potential, $|\Delta(\mathbf{r})|$ within a unit cell at $\bar{B}=0.15$ and $T=0.2T_{\mathrm{c}}$. (a) Profile in the xz plane at $y=0$. (b) Profile in the xy plane at $z=0.25L$. The vortex center runs along the z axis ($x=y=0$), and the FFLO nodal plane is at the plane $z=0$ and $\pm 0.5L$.

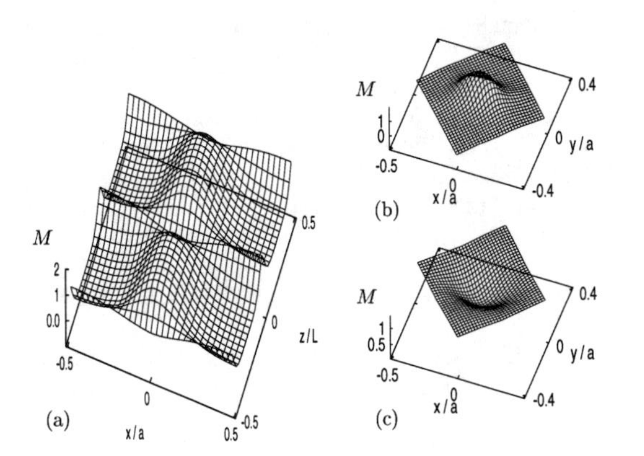

FIGURE 2. Paramagnetic moment $M_{\mathrm{para}}(\mathbf{r})/M_0$ within a unit cell at $\bar{B}=0.15$ and $T=0.2T_{\mathrm{c}}$. (a) Profile in the xz plane at $y=0$. Profile in the xy plane at $z=0.25L$ (b) and $z=0$ (c)

classical theory, $N_\uparrow(E,\mathbf{r})$ and $N_\downarrow(E,\mathbf{r})$ are symmetric by $E \leftrightarrow -E$. The spectra of the DOS for up- (down-) spin electrons is shifted to positive (negative) energy by $\tilde{\mu}\bar{B}$ due to the Zeeman term. At the vortex center far from the nodal plane [Fig. 3(a)], $N_\uparrow(E,\mathbf{r})$ and $N_\downarrow(E,\mathbf{r})$, respectively, have a "zero energy peak" at $E=\mu_+$ and $E=\mu_-$ with $\mu_\pm \equiv \mu_0 \pm \tilde{\mu}\bar{B}$. Since a vortex has phase winding 2π, along the trajectory through the vortex center $\Delta(\mathbf{r})$ changes sign by the phase change π at the vortex center. At the nodal point where $\Delta(\mathbf{r})$ changes sign, the zero-energy peak appears in the LDOS due to the low energy bound states. Since peak of up- (down-)spin electrons at $E>0$ ($E<0$) is empty (filled) states, there

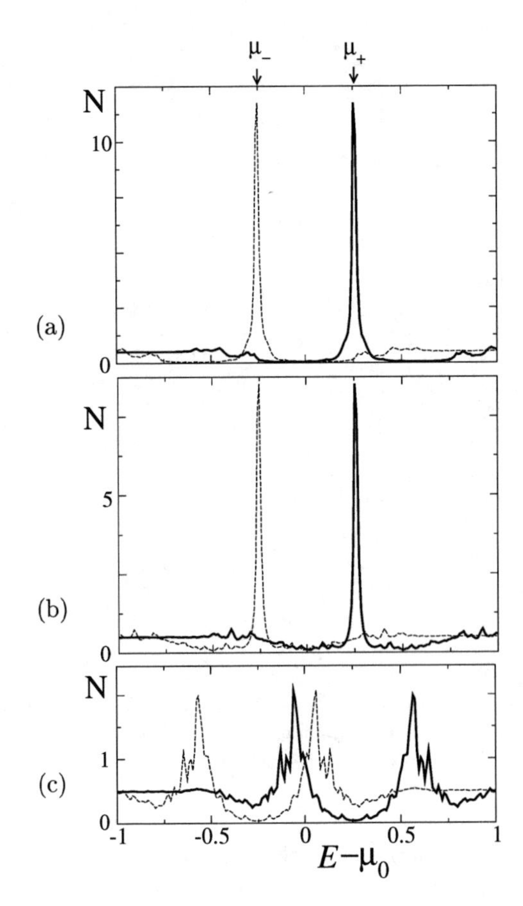

FIGURE 3. Local density of states for up-spin electrons $N_\uparrow(E,\mathbf{r})$ (solid lines) and down-spin electrons $N_\downarrow(E,\mathbf{r})$ (dashed lines) at three points $\mathbf{r}=(x,y,z)$. (a) At the vortex center far from the FFLO nodal plane, $x=y=0$ and $z=0.25L$. (b) At midpoint between vortices in the FFLO nodal plane, $x=0.5a$ and $y=z=0$. (c) At vortex center in the FFLO nodal plane, $x=y=z=0$.

appears enhanced paramagnetic moment $M_{\mathrm{para}}(\mathbf{r}) = -\mu_{\mathrm{B}}\int_{-\infty}^{0}(N_\uparrow(E,\mathbf{r})-N_\downarrow(E,\mathbf{r}))\mathrm{d}E$ at the vortex core. This situation is the same also at the FFLO nodal plane outside of vortex core. Since $\Delta(\mathbf{r})$ changes sign along the trajectory intersecting nodal plane, $N_\uparrow(E,\mathbf{r})$ and $N_\downarrow(E,\mathbf{r})$ have a peak at μ_+ and μ_- as shown in Fig. 3(b), and there appears large $M_{\mathrm{para}}(\mathbf{r})$ at the FFLO nodal plane.

However, along the trajectory through the intersection point of a vortex and a nodal plane, $\Delta(\mathbf{r})$ does not change the sign, because the phase shift is 2π by summing π due to vortex and π due to the nodal plane. Thus, the zero-energy peak does not appear in Fig. 3(c). Instead, $N_\uparrow(E,\mathbf{r})$ has two broad peaks at finite energy shifted upper or lower from μ_+. In this spectrum, one of the peaks is occupied ($E<0$) and the other is empty ($E>0$) both in $N_\uparrow(E,\mathbf{r})$ and $N_\downarrow(E,\mathbf{r})$. Then, since the unbalance of up- and down spin electron density is small, $M_{\mathrm{para}}(\mathbf{r})$ is suppressed at the vortex center in the nodal plane.

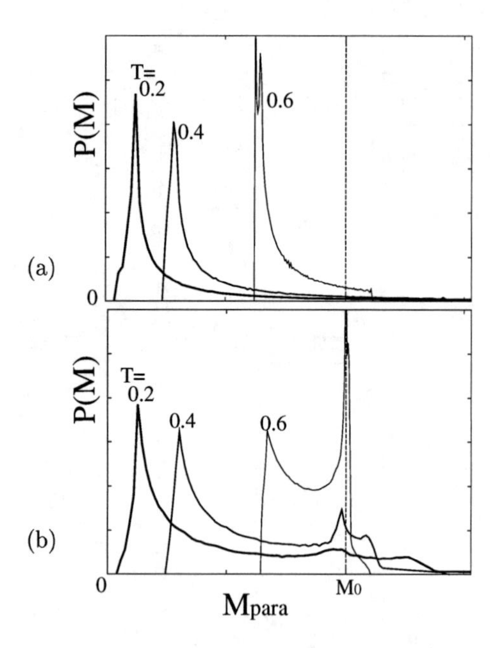

FIGURE 4. Calculated line shape $P(M)$ for the Knight shift as a function of $M_{\rm para}$ at $T/T_{\rm c}$=0.2, 0.4 and 0.6 without (a) and with (b) the FFLO modulation. $\bar{B}=0.15$ and $L=50R_0$. Vertical dashed lines show $M_{\rm para}=M_0$ at the normal states.

These features of the spatial structure of the paramagnetic moment and the LDOS are qualitatively consistent with those obtained by the BdG theory [17].

The temperature dependence of $M_{\rm para}(\mathbf{r})$ below $T_{\rm c}$ corresponds to the Knight shift in the NMR experiment. Outside the vortex core, $M_{\rm para}(\mathbf{r})$ is decreasing from M_0 on lowering T, as expected in the singlet pairing case. However, $M_{\rm para}(\mathbf{r})$ is not decreased at the vortex core or the FFLO nodal plane. From the spatial structure of $M_{\rm para}(\mathbf{r})$, we calculate the distribution function $P(M)=\int\delta(M-M_{\rm para}(\mathbf{r}))\mathrm{d}\mathbf{r}$, which corresponds to the resonance line shape observed by NMR experiments. In the conventional vortex lattice state without the FFLO modulation, shown in Fig. 4(a), the peak of the distribution function $P(M)$ comes from the signal from the outside of the vortex core. Thus, T-dependence of the peak position M corresponds to the Knight shift outside of the vortex core. In the vortex state with the FFLO modulation, as presented in Fig. 4(b), the line shape $P(M)$ becomes double peak structure. In addition to the peak seen in Fig. 4(a), a new peak coming from the FFLO nodal plane appears near $M_{\rm para}\sim M_0$. The volume contributions of the FFLO nodal plane is large because of the two dimensional sheet structure, compared with those by the one dimensional structure of the vortex line. The height of the new peak decreases on lowering T. These features of $P(M)$ are basically consistent with the observed NMR resonance line shape in the "FFLO" phase in $CeCoIn_5$ [4].

SUMMARY AND DISCUSSION

We investigated the structure of the vortex and the FFLO nodal plane in the FFLO state, based on the quasiclassical theory, calculating the spatial structure of the pair potential, paramagnetic moment and the electronic states under given period of the FFLO modulation. The topologies of the phase structure in the pair potential affect the distribution of paramagnetic moment, and low energy electronic states. These structures are consistent with those obtained by the BdG theory [17]. We also discuss the resonance line shape of the Knight shift in the FFLO state. We hope that these features will be observed in experiments to confirm the FFLO states

REFERENCES

1. A. Bianchi, R. Movshovich, C. Capan, P.G. Pagliuso, and J.L. Sarrao, *Phys. Rev. Lett.* **91**, 187004 (2003).
2. H. A. Radovan, N.A. Fortune, T.P. Murphy, S.T. Hannahs, E.C. Palm, S.W. Tozer, and D. Hall, *Nature (London)* **425**, 51 (2003).
3. T. Watanabe, Y. Kasahara, K. Izawa, T. Sakakibara, Y. Matsuda, C. J. van der Beek, T. Hanaguri, H. Shishido, R. Settai, and Y. Onuki, *Phys. Rev. B* **70**, 020506(R) (2004).
4. K. Kakuyanagi, M. Saitoh, K. Kumagai, S. Takashima, M. Nohara, H. Takagi, and Y. Matsuda, *Phys. Rev. Lett.* **94**, 047602 (2005).
5. P. Fulde and R.A. Ferrell, *Phys. Rev.* **135**, A550 (1964).
6. A.I. Larkin and Y.N. Ovchinnikov, *Sov. Phys. JETP* **20**, 762 (1965).
7. K. Machida and H. Nakanishi, *Phys. Rev. B* **30**, 122 (1984).
8. M. Tachiki, S. Takahashi, P. Gegenwart, M. Weiden, M. Lang, C. Geibel, F. Steglich, R. Modler, C. Paulsen, and Y. Onuki, *Z. Physik B* **100**, 369 (1996).
9. H. Shimahara, *Phys. Rev. B* **50**, 12760 (1994).
10. U. Klein, D. Rainer, and H. Shimahara, *J. Low Temp. Phys.* **118**, 91 (2000).
11. M. Houzet and A. Buzdin, *Phys. Rev. B* **63**, 184521 (2001).
12. H. Adachi and R. Ikeda, *Phys. Rev. B* **68**, 184510 (2003).
13. T. Mizushima, K. Machida, and M. Ichioka, *Phys. Rev. Lett.* **94**, 060404 (2005).
14. R. Casalbuoni and G. Nardulli, *Rev. Mod. Phys.* **76**, 263 (2004).
15. M. Ichioka, A. Hasegawa, and K. Machida, *Phys. Rev. B* **59**, 184 (1999); **59**, 8902 (1999).
16. M. Ichioka, K. Machida, N. Nakai, and P. Miranović, *Phys. Rev. B* **70**, 104510 (2004).
17. T. Mizushima, K. Machida, and M. Ichioka, to appear in *Phys. Rev. Lett.*; *cond-mat/0504665*.

Knight Shift in the FFLO State of a Two-Dimensional D-Wave Superconductor

Anton B. Vorontsov* and Matthias J. Graf†

*Department of Physics and Astronomy, Louisiana State University, Baton Rouge, LA 70803, USA
†Theoretical Division, Los Alamos National Laboratory, Los Alamos, NM 87545, USA

Abstract. We report on the Fulde-Ferrell-Larkin-Ovchinnikov (FFLO) state in two-dimensional d-wave superconductors with magnetic field parallel to the superconducting planes. This state occurs at high magnetic field near the Pauli-Clogston limit and is a consequence of the competition between the pair condensation and Zeeman energy. We use the quasiclassical theory to self-consistently compute the spatially nonuniform order parameter. Our self-consistent calculations show that the FFLO state of a d-wave order parameter breaks translational symmetry along preferred directions. The orientation of the nodes in real space is pinned by the nodes of the basis function in momentum space. Here, we present results for the Knight shift and discuss the implications for recent nuclear magnetic resonance measurements on $CeCoIn_5$.

Keywords: low-dimensional superconductivity, Fulde-Ferrell-Larkin-Ovchinnikov phase, d-wave superconductivity.
PACS: 75.60.-d, 76.60.Cq, 74.81.-g

The Fulde-Ferrell-Larkin-Ovchinnikov (FFLO) state of spin-singlet superconductors is the compromise between the pairing condensate, favoring anti-parallel spin alignment, and the Zeeman effect, favoring parallel spin alignment along the field [1, 2]. This compromise leads to a spatially inhomogeneous state of "normal" and "superconducting" regions, where the "normal" regions are defined by a spectrum of spin-polarized quasiparticles.

The FFLO phase of d-wave superconductors is modified by the anisotropy of the order parameter in momentum space compared to s-wave superconductors. The upper critical transition line, $B_{c2}(T)$, has a kink at low temperatures, $T^* \sim 0.06\,T_c$, corresponding to the discontinuous change in the modulation of the order parameter [3, 4, 5, 6]. Recent calculations of the spatial modulation of the order parameter in 2D near B_{c2} predicted that the energetically favored state at low T and high B forms a "square lattice" instead of the 1D stripe order [7, 8].

Here, we restrict our study to temperatures above this structural phase transition, $T > T^*$, and address the quasiparticle response in the FFLO phase between the lower critical field B_{c1} and the upper critical field B_{c2}. In addition, we assume that $\mathbf{B}$ is parallel to the superconducting planes. In this geometry the magnetic field affects the superconducting condensate only through the Zeeman coupling of the quasiparticle spin to the field. Furthermore, we assume a cylindrical Fermi surface.

Within the quasiclassical theory of superconductivity we calculate self-consistently the order parameter $\Delta(\mathbf{R},\hat{\mathbf{p}})$ and the quasiclassical Green's functions by solving Eilenberger's equation in a constant magnetic field $\mathbf{B}$. The Zeeman coupling of the quasiparticle spin with magnetic field enters through $\mu\mathbf{B}\cdot\sigma$, where σ_i are Pauli spin matrices and $\mu = (g/2)\mu_B$ is the absolute value of the magnetic moment of a quasiparticle with negative charge e; $\mu_B = |e|/2mc$ is the Bohr magneton. Note that the g-factor is a free material parameter in this calculation.

From the solutions of Eilenberger's equation we can calculate measurable quantities like the free energy, quasiparticle local density of states [6] and local magnetization $\mathbf{M}(\mathbf{R})$. Here, we consider spin-singlet order parameters that factorize into $\Delta(\mathbf{R},\hat{\mathbf{p}}) = \Delta(\mathbf{R})\cos 2\phi$, with a spatially dependent amplitude, $\Delta(\mathbf{R})$, and an angular dependent $d_{x^2-y^2}$-wave basis function.

The local magnetization is given by the paramagnetic response of the medium and the spin-vector component of the quasiclassical Matsubara Green's function [9]:

$$\mathbf{M}(\mathbf{R}) = 2\mu N_f\left[\mu\mathbf{B} + T\sum_{\varepsilon_n}\int\frac{d\hat{\mathbf{p}}}{2\pi}\,\mathbf{g}(\hat{\mathbf{p}},\mathbf{R};\varepsilon_n)\right], \quad (1)$$

with the normal-state density of states N_f at the Fermi level. The normal-state susceptibility, $\chi_N = 2\mu^2 N_f$, is defined by $M_N = \chi_N B$.

For comparison, we show in Fig. 1 the calculated temperature dependence of the magnetization in the uniform superconducting (USC) phase for three different values of B. Increasing B changes the T-dependence of the magnetization from linear to quadratic with a residual zero-temperature value due to the field induced shift in the spin-split density of states of the gapless d-wave superconductor. This result is in agreement with scaling arguments by Yang and Sondhi [10].

In Fig. 2, we show temperature scans of the minimum, average and maximum local magnetization for the stable FFLO phase, with spatial order-parameter modulation along the $\langle 110\rangle$ direction, i.e., along the nodal direction

CP850, *Low Temperature Physics: 24th International Conference on Low Temperature Physics;*
edited by Y. Takano, S. P. Hershfield, S. O. Hill, P. J. Hirschfeld, and A. M. Goldman

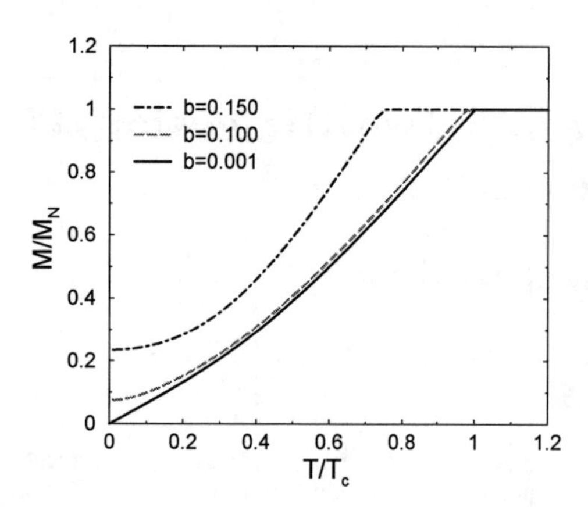

FIGURE 1. Magnetization M of the spatially uniform superconducting (USC) phase normalized by the normal-state magnetization M_N for different magnetic fields $b = \mu B/2\pi T_c$.

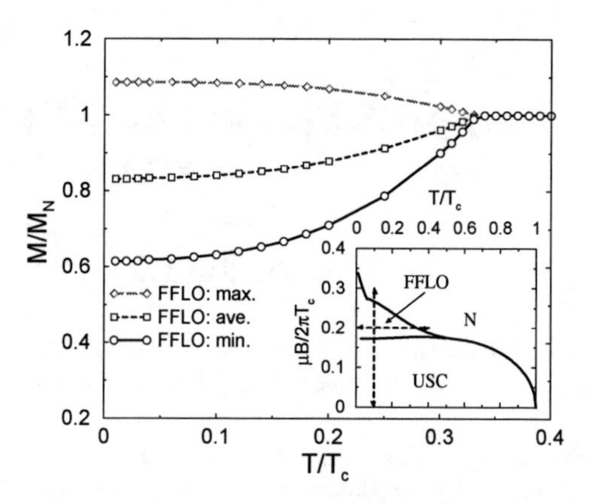

FIGURE 2. Temperature scan of the normalized local magnetization at fixed magnetic field $b = 0.2$ in FFLO phase for $\langle 110 \rangle$ orientation of spatial order-parameter modulation. Inset: Phase diagram with temperature and field scans across FFLO phase.

of the gap function. The Knight shift K is proportional to the change in the local magnetic field at the nucleus, thus it is directly proportional to the local magnetization. Since K is weighted by the field distribution, the largest contribution comes from areas where the derivative of $\mathbf{M}$ vanishes, which are at the minimum and maximum locations of $\mathbf{M}$. The calculated bifurcation between minimum and maximum local magnetization seen in Fig. 2 is in qualitative agreement with measurements of the Knight shift on $CeCoIn_5$ reported by Kakuyanagi et al. [11].

In Fig. 3, we show field scans of the local magnetization at $T/T_c = 0.1$ starting in the USC phase and into the FFLO $\langle 110 \rangle$ phase. It illustrates the nonlinear magnetic response of the quasiparticles due to an external field and the continuous second order transition at the lower critical field B_{c1}, which also is signaled by the appearance of a single domain wall. This finding clearly contradicts the claim by Yang and Sondhi [10] about a first order transition at B_{c1} between the USC and FFLO phase.

In addition, we calculated the spin-resolved local density of states in the FFLO state [6] (not shown). We found that the characteristic Andreev bound states, due to the periodic sign change of the order parameter, are responsible for the excess spin polarization of quasiparticles at the domain walls seen in Figs. 2 and 3. Therefore, the Andreev bound states should be clearly visible features in scanning tunneling spectroscopy measurements.

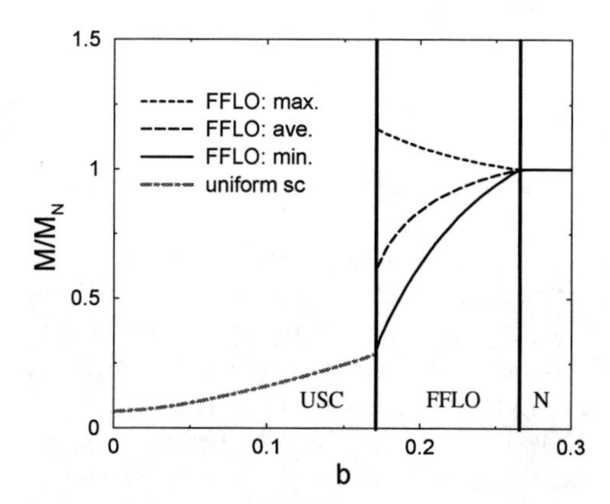

FIGURE 3. Field scan of the normalized local magnetization at fixed temperature $T/T_c = 0.1$ across USC and FFLO phases for $\langle 110 \rangle$ orientation of order-parameter modulation.

ACKNOWLEDGMENTS

We thank N. J. Curro, R. Movshovich and V. F. Mitrović for helpful discussions and A. V. Balatsky and J. A. Sauls for suggesting this problem. This research is supported by the Department of Energy, under contract W-7405-ENG-36, (MJG) and the LSU Board of Regents (ABV).

REFERENCES

1. P. Fulde and R. Ferrell, *Phys. Rev.* **135**, A550–A563 (1964).
2. A. I. Larkin and Y. N. Ovchinnikov, *Sov. Phys. JETP* **20**, 762–769 (1965).
3. K. Maki and H. Won, *Czech. J. Phys.* **46**, 1035–1036 (1996).
4. H. Shimahara and D. Rainer, *J. Phys. Soc. Japan* **66**, 3591–3599 (1997).
5. U. Klein, D. Rainer, and H. Shimahara, *J. Low Temp. Phys.* **118**, 91–104 (2000).
6. A. B. Vorontsov, J. A. Sauls, and M. J. Graf, *cond-mat/0506257* (2005).
7. H. Shimahara, *J. Phys. Soc. Japan* **67**, 736–739 (1998).
8. K. Maki and H. Won, *Physica B* **322**, 315–317 (2002).
9. J. A. X. Alexander, T. P. Orlando, D. Rainer, and P. M. Tedrow, *Phys. Rev. B* **31**, 5811–5825 (1985).
10. K. Yang and S. L. Sondhi, *Phys. Rev. B* **57**, 8566 (1998).
11. K. Kakuyanagi et al., *Phys. Rev. Lett.* **94**, 047602 (2005).

Conductance Characteristics between a Normal Metal and a 2D Fulde-Ferrell-Larkin-Ovchinnikov Superconductor

Qinghong Cui[1], C.-R. Hu[2], J. Y. T. Wei[3], and Kun Yang[1]

[1] *Department of Physics, Florida State University, Tallahassee, Florida 32306, USA*
[2] *Department of Physics, Texas A&M University, College Station, Texas 77843, USA*
[3] *Department of Physics, University of Toronto, 60 St. George St. Toronto, ON M5S1A7, Canada*

Abstract. We use the Blonder-Tinkham-Klapwijk formalism to calculate the conductance characteristics between a normal metal and a 2-dimensional Zeeman-field-driven, *s*- or *d*-wave, Fulde-Ferrell-Larkin-Ovchinnikov superconductor.

Keywords: tunneling, superconductivity, conductance characteristics, microconstriction.
PACS: 74.25.Fy, 74.50.+r

The Fulde-Ferrell-Larkin-Ovchinnikov (FFLO) state [1,2] has received renewed interest recently due to the experimental indication of its presence in $CeCoIn_5$ [3,4], a quasi 2-dimensional (2D) *d*-wave superconductor. However direct evidence of the spatial oscillation of the order parameter, as would be revealed by, for example, the phase-sensitive Josephson effect [5], does not yet exist. Another possible way to detect the phase structure directly is conduction through micro-constrictions, which has been used to probe the phase sensitive surface states of *d*-wave superconductors. Here we calculate the conductance characteristics between a normal metal (N) and a 2D *s*- or *d*-wave FFLO superconductor.

We begin with the Hamiltonian:

$$\mathcal{H} = \sum_{\mathbf{k}\sigma} \xi_{\mathbf{k}} c^{+}_{\mathbf{k}\sigma} c_{\mathbf{k}\sigma} + \sum_{\mathbf{k},\mathbf{k}',\mathbf{q}} V_{\mathbf{k}\mathbf{k}'} c^{+}_{\mathbf{k}+\mathbf{q}\uparrow} c^{+}_{-\mathbf{k}+\mathbf{q}\downarrow} c_{-\mathbf{k}'+\mathbf{q}\downarrow} c_{\mathbf{k}'+\mathbf{q}\uparrow} + \mu_0 B \sum_{\mathbf{k}} (c^{+}_{\mathbf{k}\uparrow} c_{\mathbf{k}\uparrow} - c^{+}_{\mathbf{k}\downarrow} c_{\mathbf{k}\downarrow}) \quad (1)$$

where $\xi_{\mathbf{k}} = \varepsilon_{\mathbf{k}} - \mu$, $\varepsilon_{\mathbf{k}}$ is the kinetic energy, μ is the chemical potential, B is the Zeeman magnetic field, μ_0 is the magnetic moment of the electron. For an *s*-wave superconductor, $V_{\mathbf{k}\mathbf{k}'} = -V_0$, with $V_0 > 0$, while for a *d*-wave superconductor, $V_{\mathbf{k}\mathbf{k}'} = -V_0 \cos(2\theta_{\mathbf{k}}) \cos(2\theta_{\mathbf{k}'})$. In this work we will focus on the Fulde-Ferrell (FF) version, namely states with pairing of electrons at a single total momentum $\mathbf{q}$, and the more general situation will be discussed elsewhere. In this case the mean field Hamiltonian is:

$$\mathcal{H}_{\mathrm{MF}} = \sum_{\mathbf{k}\sigma} (\xi_{\mathbf{k}} + \sigma \mu_0 B) c^{+}_{\mathbf{k}\sigma} c_{\mathbf{k}\sigma} + \sum_{\mathbf{k}} (\Delta_{\mathbf{kq}} c^{+}_{\mathbf{k}+\mathbf{q}\uparrow} c^{+}_{-\mathbf{k}+\mathbf{q}\downarrow} + \Delta^{*}_{\mathbf{kq}} c_{-\mathbf{k}+\mathbf{q}\downarrow} c_{\mathbf{k}+\mathbf{q}\uparrow}) \quad (2)$$

where $\Delta_{\mathbf{kq}}$ is the pairing potential and determined by the self-consistent condition: $\Delta_{\mathbf{kq}} = -\sum_{\mathbf{k}'} V_{\mathbf{k}\mathbf{k}'} \langle c_{-\mathbf{k}'+\mathbf{q}\downarrow} c_{\mathbf{k}'+\mathbf{q}\uparrow} \rangle$. Diagonalize the Hamiltonian (2), we get the quasi-particle spectra for up- and down-spin ($\sigma = \pm 1$): $E_{\mathbf{k}\sigma} = E_{\mathbf{k}} + \sigma(\xi^{(a)}_{\mathbf{k}} + \mu_0 B)$ where $E_{\mathbf{k}} = (\Delta^2_{\mathbf{kq}} + \xi^{(s)2}_{\mathbf{k}})^{1/2}$, $\xi^{(s)}_{\mathbf{k}} = (\xi_{\mathbf{k}+\mathbf{q}} + \xi_{\mathbf{k}-\mathbf{q}})/2$, $\xi^{(a)}_{\mathbf{k}} = (\xi_{\mathbf{k}+\mathbf{q}} - \xi_{\mathbf{k}-\mathbf{q}})/2$.

In the presence of a normal metal-superconductor interface, the quasi-particles obey the Bogoliubov-de Gennes equations:

$$Eu(\mathbf{x}) = \hat{h}_0 u(\mathbf{x}) + \int d\mathbf{x}' \Delta(\mathbf{s}, \mathbf{r}) v(\mathbf{x}')$$
$$Ev(\mathbf{x}) = -\hat{h}_0 v(\mathbf{x}) + \int d\mathbf{x}' \Delta^{*}(\mathbf{s}, \mathbf{r}) u(\mathbf{x}') \quad (3)$$

where $\mathbf{s} = \mathbf{x} - \mathbf{x}'$, $\mathbf{r} = (\mathbf{x} + \mathbf{x}')/2$, $\hat{h}_0 = -\nabla^2/2m - \mu$, and $\Delta(\mathbf{s}, \mathbf{r}) = \int d\mathbf{k}\, e^{i\mathbf{k}\cdot\mathbf{s}} \tilde{\Delta}(\mathbf{k}, \mathbf{r}) e^{i2\mathbf{q}\cdot\mathbf{r}}$; E will be shifted by $-\sigma \mu_0 B$ depending on the spin of the quasiparticle. In our model, we assume $\mathbf{q}$ is parallel to the N/S interface ($x = 0$) and its value is determined by minimize the Hamiltonian (1). Neglect the proximity effect at the N/S interface, we have $\tilde{\Delta}(\mathbf{k}, \mathbf{r}) = \Delta(\mathbf{k})\, \Theta(x)$, where $\Theta(x)$ is the step function, and $\Delta(\mathbf{k})$ is the pairing potential in the bulk superconductor. Following the method by Blonder-Tinkham-Klapwijk [6,7], the normalized tunneling conductance at zero temperature is given by $G = G_{ns} / G_{nn}$ (with average over spin implied), and

CP850, *Low Temperature Physics: 24th International Conference on Low Temperature Physics;*
edited by Y. Takano, S. P. Hershfield, S. O. Hill, P. J. Hirschfeld, and A. M. Goldman

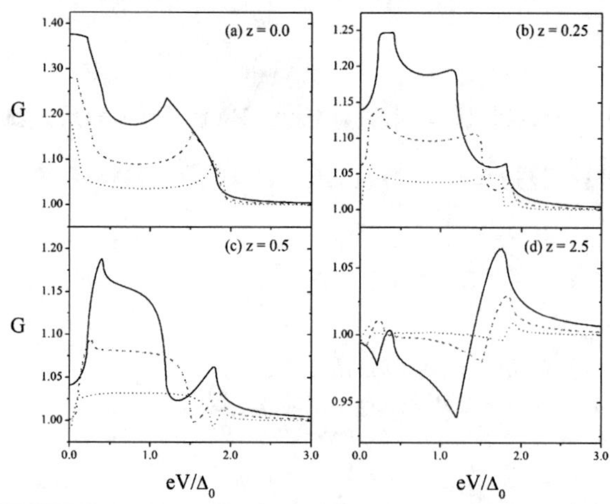

FIGURE 1. Normalized tunneling conductance vs voltage for (normal metal)-(s-wave FF superconductor) junction: (a) $z = 0$, (b) $z = 0.25$, (c) $z = 0.5$, (d) $z = 2.5$. Solid, $H = 0.704$, $Q = 0.804$; dash, $H = 0.8$, $Q = 0.892$; dot, $H = 0.9$, $Q = 0.952$, where $H = \mu_0 B / \Delta_0$, $Q = v_F q / \Delta_0$, Δ_0 is the no-Zeeman-field pairing potential of normal BCS state.

$$
\begin{aligned}
G_{ns} &= -\frac{e^2}{\pi}\int_{-\pi/2}^{\pi/2} d\theta\ (1+|a(-E,\theta)|^2 - |b(E,\theta)|^2) \\
G_{nn} &= -\frac{e^2}{\pi}\int_{-\pi/2}^{\pi/2} d\theta\ (1-|b(+\infty,\theta)|^2)
\end{aligned}
\tag{4}
$$

where

$$
a(E,\theta) = \frac{\cos^2\theta}{\eta_+(z^2+\cos^2\theta)-\eta_- z^2}, \tag{5}
$$

$$
b(E,\theta) = -\frac{z(z+i\cos\theta)(\eta_+ - \eta_-)}{\eta_+(z^2+\cos^2\theta)-\eta_- z^2}, \tag{6}
$$

$$
\eta_\pm = \frac{(E-\frac{qk_{Fy}}{m}) \pm \sqrt{(E-\frac{qk_{Fy}}{m})^2 - \Delta(\mathbf{k}_F)^2}}{\Delta^*(\mathbf{k}_F)}, \tag{7}
$$

θ is the angle between $\mathbf{k}_F$ and x-axis, $z = mU/k_F$, U is the barrier intensity. With strong spin-exchange field, the ordinary BCS state is destroyed, while the FF state becomes the energetically favorable state against the normal state. In Fig. 1-3, we calculate the normalized tunneling conductance under various Zeeman fields and barrier intensities for both s- and d-wave superconductors.

ACKNOWLEDGMENTS

Work supported by NSF DMR-0225698.

REFERENCES

1. P. Fulde and R.A. Ferrell, *Phys. Rev.* **135**, A550 (1964).

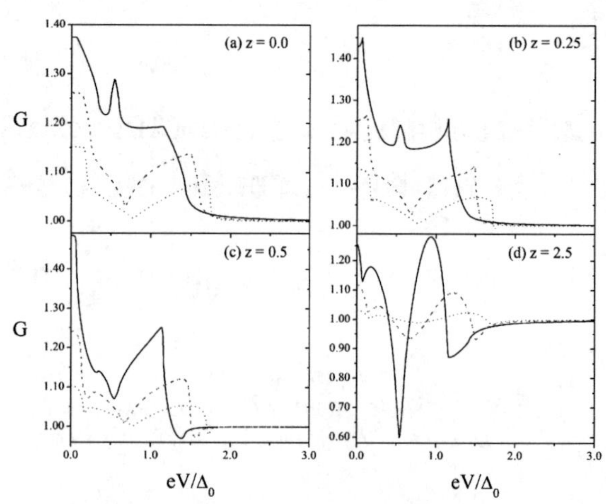

FIGURE 2. Normalized tunneling conductance vs voltage for (normal metal)-(d-wave FF superconductor) junctions with $\mathbf{q}$ along the nodal direction. Solid, $H = 0.544$, $Q = 0.608$; dash, $H = 0.68$, $Q = 0.816$; dot, $H = 0.776$, $Q = 0.94$.

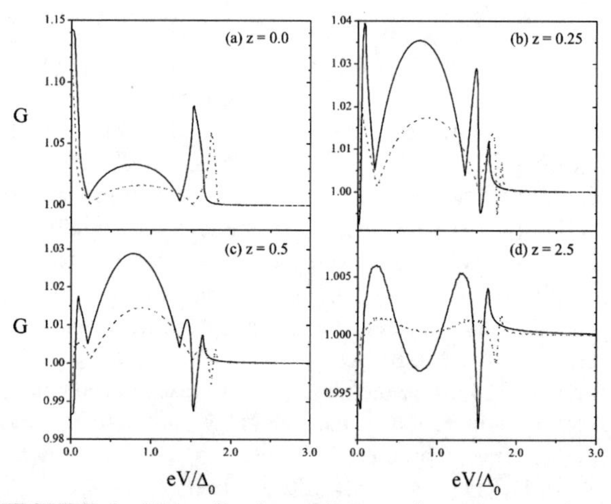

FIGURE 3. Normalized tunneling conductance vs voltage for (normal metal)-(d-wave FF superconductor) junctions with $\mathbf{q}$ along the antinodal direction. Solid, $H = 0.78$, $Q = 0.808$; dash, $H = 0.88$, $Q = 0.9$.

2. A.I. Larkin and Y.N. Ovchinnikov, *Sov. Phys. JETP* **20**, 762 (1965).
3. H. Radovan, *et al.*, *Nature* **425**, 51 (2003).
4. A. Bianchi, *et al.*, *Phys. Rev. Lett.* **91**, 187004 (2003).
5. K. Yang and D.F. Agterberg, *Phys. Rev. Lett.* **84**, 4970 (2000).
6. G.E. Blonder, M. Tinkham, and T.M. Klapwijk, *Phys. Rev. B* **25**, 4515 (1982).
7. D. Zhang, C.S. Ting, and C.-R. Hu, *Phys. Rev. B* **70**, 172508 (2004)

Fulde-Ferrell-Larkin-Ovchinnikov State in Thin Films with Rough Surfaces

S. Matsuo, S. Higashitani and K. Nagai

Faculty of Integrated Arts and Sciences, Hiroshima University, Higashi-Hiroshima 739-8521, Japan

Abstract. We report the critical magnetic field of the second order transition from the normal state to the Fulde-Ferrell-Larkin-Ovchinnikov (FFLO) state in thin films with rough surfaces. The rough surface effects are treated using the random S-matrix theory. The film thickness dependence of the critical field and the characteristic wave number are reported.

Keywords: superconducting film, FFLO state, quasi-classical Green's function, rough surface
PACS: 74.20.Fg, 74.25.Dw, 74.25.Ha, 74.78.-w

INTRODUCTION

Fulde-Ferrell-Larkin-Ovchinnikov (FFLO) state is a nonuniform superconducting state due to the spin Zeeman splitting. FFLO state is expected to exist in clean Type-II superconductors whose magnetic properties are dominated by spin magnetism. Its possibility in organic materials [1] and heavy Fermion systems [2] has been recently reported. On the other hand, no observation of FFLO state has been reported in metallic films in which orbital magnetic effects are suppressed by finite thickness of the film (but see Ref. [3]).

One of the reasons of suppression of FFLO state in films with sufficiently long mean free path has been thought to be the electron scattering by the rough surfaces of the film. However, no systematic study of surface scattering effects on FFLO state has been reported.

We study the boundary scattering effect on FFLO state in superconducting films with rough surfaces using the quasi-classical theory with random S-matrix model. [4] We report the film thickness dependence of the critical field and the characteristic wave number.

FORMULATION

We consider a plane metallic film along the (x,y) plane with rough surfaces at $z=0$ and $z=L$ and assume that the FFLO order parameter takes a form

$$\Delta(z)e^{iqx}, \tag{1}$$

where q is the characteristic wave number of FFLO state. When the surfaces are specular, we have a solution with $\Delta(z)=$ const. for an s-wave superconductor, therefore, the critical field agrees with that of the bulk system. When the surfaces are diffusive, however, $\Delta(z)$ depends on z in general. At the diffusive surfaces, electrons are reflected isotropically without any memory of the incident momentum. Such effects can be incorporated into quasi-classical Green's function in a form of surface self energy using the random S-matrix model. [4]

In this report, we use (K,α) to specify a Fermi momentum: K is the component of the Fermi momentum parallel to the surfaces and $\alpha=\pm$ denotes the direction of the z-component. According to the random S-matrix theory[4], the linearized gap equation takes a form

$$\Delta(z)=gT\sum_n\sum_{K,\alpha}\frac{1}{2v_{Fz}}(\delta\hat{g}_\alpha(K,z))_{\rm OD}, \tag{2}$$

where g is the s-wave pairing interaction, $v_{Fz}=v_F\cos\theta$ is the z component of the Fermi velocity and $(\delta\hat{g}_\alpha)_{\rm OD}$ is the off diagonal element of the linearized quasi-classical Green's function for the Fermi momentum (K,α). The quasi-classical Green's functions are given by

$$\begin{aligned}\sum_\alpha(\delta\hat{g}_\alpha(K,z))_{\rm OD} &= \int_0^L dz'\frac{2}{v_{Fz}}e^{-2\kappa_n|z-z'|}\Delta(z') \\ &+ \delta\sigma(e^{-2\kappa_n z}+e^{-2\kappa_n(L-z)}),\end{aligned} \tag{3}$$

where $\delta\sigma$ is the surface self energy given by

$$\delta\sigma=\frac{\left\langle\int_0^L dz'\frac{2e^{-2\kappa_n z'}}{v_{Fz}}\Delta(z')\right\rangle}{1-\langle e^{-2\kappa_n L}\rangle}, \tag{4}$$

where $<\cdots>$ denotes the angle average

$$<\cdots>=\frac{\sum_K\cdots}{\sum_K 1}=\int_0^{\pi/2}\frac{d\theta}{\pi}\sin\theta\cos\theta\int_0^{2\pi}d\phi\cdots. \tag{5}$$

In the above equations,

$$\kappa_n=\frac{|\omega_n|-\frac{i{\rm sgn}(\omega_n)}{2}(\omega_L-qv_{Fx})}{v_{Fz}}, \tag{6}$$

CP850, *Low Temperature Physics: 24th International Conference on Low Temperature Physics;*
edited by Y. Takano, S. P. Hershfield, S. O. Hill, P. J. Hirschfeld, and A. M. Goldman

where ω_n the Matsubara frequency, ω_L the electron Larmor frequency and q the wave number of FFLO state.

To obtain the critical field, one should solve the integral equation (2). In this report, we simplify the problem by assuming that $\Delta(z) = \overline{\Delta}$ is a constant. In that case, the surface self energy is given by

$$\delta\sigma = \frac{\left\langle \frac{1-e^{-2\kappa_n L}}{v_{Fz}\kappa_n} \right\rangle}{1-\left\langle e^{-2\kappa_n L}\right\rangle}\overline{\Delta} \quad (7)$$

and the gap equation (2) is reduced to

$$1 = gT\sum_n\sum_K \frac{1}{v_{Fz}}\left(\frac{1}{v_{Fz}\kappa_n} + \frac{1-e^{-2\kappa_n}}{2\kappa_n L}\left(\frac{\delta\sigma}{\overline{\Delta}} - \frac{1}{v_{Fz}\kappa_n}\right)\right) \quad (8)$$

In the $L \to 0$ limit, this equation becomes

$$1 = gT\sum_n\sum_K \frac{1}{V_{Fz}}\frac{1}{|\omega_n| - \frac{i\mathrm{sgn}(\omega_n)}{2}\omega_L}, \quad (9)$$

which coincides with the gap equation for a uniform state under magnetic field and the critical field is given by the bottom curve in Fig. 1.

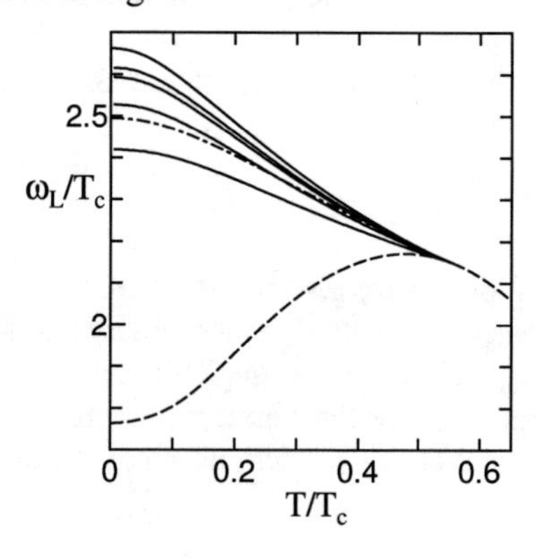

FIGURE 1. Temperature dependence of critical field. Solid curves show critical field of the second order transiton from normal state to FFLO state in films with thickness (from top to bottom) $L = \infty, 6, 4, 2, 1\, v_F/T_c$, dot dashed curve shows critical field of the first order transiton from normal state to uniform superconducting state and dashed curve shows critical field of the second order transiton from normal state to uniform superconducting state.

PHASE DIAGRAM

The numerical results of the critical field are given in Fig.1. For a given temperature, the critical field is given by a highest magnetic field that satisfies the gap equation (8) with appropriate value of q. We find that FFLO state is certainly suppressed by the diffuse scattering by surfaces and the tri-critical point occurs at lower temperature than the bulk value $0.561T_c$. Even when the film thickness L is as small as $2v_F/T_c$, however, the critical field to FFLO state is still higher than the field at which first order transition occurs from the normal state to uniform superconducting state (Pauli limit). In Fig. 2, we show the characteristic wave number q at the critical fields.

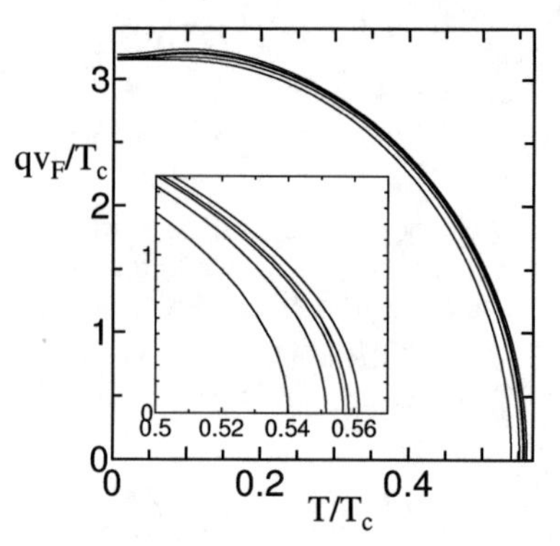

FIGURE 2. Temperature dependence of characteristic wave number q at critical field of FFLO state in films. Curves show characteristic wave number q of films with thickness (from right to left) $L = \infty, 6, 4, 2, 1\, v_F/T_c$.

DISCUSSIONS

We have studied rough surface effects on FFLO state in superconducting films using the random S-matrix theory. We found that the FFLO state is suppressed by the surface roughness but FFLO state can exist in thin films whose thickness L is of order the coherence length v_F/T_c. We have assumed a constant $\Delta(z)$ in the calculations. When one allows for the z dependence of $\Delta(z)$, one will have a higher critical field. On the other hand, FFLO state has not so far been observed in film samples. Further study of simultaneous effects by the surface scattering and the impurity scattering is necessary.

This work is supported in part by a Grant-in-Aid for Scientific Research (No. 16540322) and a Grant-in-Aid for COE Research (No. 13CE2002) from the Ministry of Education, Culture, Sports, Science and Technology of Japan.

REFERENCES

1. M-S Nam et al., *J. Phys.: Condens. Matter*, **11**, L477 (1999).
   S. Manalo and U. Klein, *J. Phys.: Condens. Matter*, **12**, L471 (2000).
   H. Shimahara, *J. Phys. Soc. Jpn.* **71**, 1644 (2002); *ibid.* **66** 541 (1997), **67**, 736 (1998).
2. T. P. Murphy et al., *Phys. Rev. B*, **65**, 100514(R) (2002).
3. P. W. Adams, *Phys. Rev. Lett.* **92**, 067003 (2004).
4. Y. Nagato et al., *J. Low Temp. Phys.* **103**, 1 (1996).

Specific Heat Measurements of Mesoscopic Loops

O. Bourgeois*[1], F. Ong*, S.E. Skipetrov† and J. Chaussy*

*Centre de Recherches sur les Très Basses Températures, CNRS, laboratoire associé à l'UJF et à l'INPG, BP 166, 25 avenue des Martyrs, 38042 Grenoble Cedex 9, France
† Laboratoire de Physique et de Modélisation des Milieux Condensés, CNRS et UJF, BP 166, 25 avenue des Martyrs, 38042 Grenoble Cedex 9, France

Abstract. We report highly sensitive specific heat measurements on mesoscopic superconducting loops at low temperature. These mesoscopic systems exhibit thermal properties significantly different from that of the bulk materials. The measurement is performed on a silicon membrane sensor where 450 000 superconducting aluminium loops are deposited through electron beam lithography, under an applied magnetic field. Each entry of a vortex is associated to a jump in the specific heat of few thousands of Boltzmann constant k_B indicating the existence of phase transitions. The periodicity of this sequential phase transitions is a nontrivial behaviour and varies strongly as the temperature is decreased. The successive phase transitions are well described by the Ginzburg-Landau theory of superconductivity. The presence of metastable states is responsible for the $n\Phi_0$ (n=1, 2, 3...) periodicity of the discontinuities of the measured specific heat.

Keywords: vortex, specific heat, mesoscopy, thermodynamics, nanostructures.
PACS: 74.78.Na, 74.25.Bt, 75.75.+a

INTRODUCTION

Since recent years there has been an increased interest towards the understanding of thermal behavior in nanostructures. For more than two decades, mesoscopic systems have been intensively studied by electronic transport experiments [1]. Although many quantum effects in such systems are now well understood (oscillations of transition temperature in thin superconducting cylinders [2], magnetic flux quantization [3], magnetoresistance oscillations [4], persistent currents [5], etc.), very few experiments shine light on their thermal behaviour [6-7]. Meanwhile, this behaviour is expected to differ significantly from the bulk one, nanometric sample size leading to quantum limitations of thermal transport, new phase transitions and other phenomena specific to mesoscale thermodynamics. Here we report an experimental evidence of nontrivial thermal behaviour of the simplest mesoscopic system – a superconducting loop [7]. By measuring the specific heat of an array of 450 thousands aluminium loops in external magnetic field, we show that they go through a periodic sequence of phase transitions (up to 30 successive transitions) as the magnetic flux Φ threading each loop is increased. The transitions are separated by $\Delta\Phi = n\Phi_0$ (with n integer and Φ_0 the magnetic flux quantum) and each transition is accompanied by a discontinuity of the specific heat of only several thousands of Boltzmann constants k_B. These highly sensitive measurements are made possible by our unique experimental setup that allows us to measure the specific heat with unprecedented accuracy of ~20 femtoJoules per Kelvin, corresponding to energy exchanges as small as few attoJoules (1 aJ = 10^{-18} J) at 0.6 K.

EXPERIMENTAL SET-UP

Our sample is composed of 450 thousands identical aluminium square loops (2 μm in size, w = 230 nm arm width, d = 40 nm thickness, separation of neighbouring loops = 2 μm), patterned by electron beam lithography on a suspended sensor composed of a very thin (4 μm thick) silicon membrane and two integrated transducers: a copper heater and a niobium nitride thermometer [8,7]. The setup is cooled below the critical temperature T_c of the superconducting

[1] olivier.bourgeois@grenoble.cnrs.fr

CP850, *Low Temperature Physics: 24th International Conference on Low Temperature Physics;*
edited by Y. Takano, S. P. Hershfield, S. O. Hill, P. J. Hirschfeld, and A. M. Goldman

transition by a ^{3}He cryostat and then its specific heat is measured by ac calorimetry. The technique of ac calorimetry consists in supplying ac power to the heater, thus inducing temperature oscillations of the thermally isolated membrane and thermometer (see Ref. 8 for more details). For a carefully chosen operating frequency (in our case, the frequency of the temperature modulation is $f \sim 250$ Hz), the temperature of the system (sensor + sample) follows variations of the supplied power in a quasi-adiabatic way, allowing measurements of the specific heat with a resolution of $\Delta C/C = 5\times10^{-5}$ for signal integration times of the order of one minute. It is then possible to measure variations of the specific heat as small as 1-10 fJ/K (which corresponds to 1-10 thousands of k_B per loop), provided that the specific heat of the sensor (silicon membrane, heater, and thermometer) is reduced to about 10-100 pJ/K.

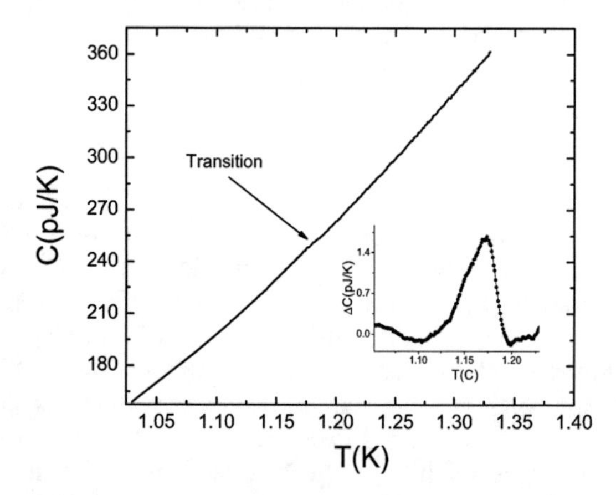

FIGURE 1. Heat capacity of the total system (sensor and aluminium loops). This illustrates the small amplitude of the specific heat jump at the superconducting phase transition. In the inset the background was subtracted to give an order of magnitude of the jump at the transition temperature of 1.2K. The temperature modulation used for the measurement ranges between 1 and 10mK.

We emphasize that working with lithographed thin film systems implies that the mass of the sample is very small, usually few tens of nanograms, leading to major difficulties when one wants to measure thermal signatures of quantum phenomena in such systems. Hence, the above extreme sensitivity is absolutely necessary. Since the temperature oscillates with typical amplitudes of few mK, our experimental apparatus can detect energy exchanges of only few aJ at the lowest temperature of 0.6 K used in our experiments.

In the absence of magnetic field, the transition of the sample into the superconducting state is observed around T_c = 1.2 K with the total specific heat discontinuity of 2 pJ/K as it can be seen in Fig. 1. In terms of jump in the specific heat at the superconducting transition, this corresponds to much less than what is expected for 80 ng of aluminium (4pJ/K) deposited on the sensor. It can be explained either by the fact that some part of the aluminium is oxidized but also by a certain degree of uncertainty when implementing the background subtraction. It gives anyhow an idea of the difficulties of measuring the specific heat signatures arising at the superconducting second order phase transition of a low-mass mesoscopic system.

RESULTS

The mesoscopic thermodynamic behaviour of the superconducting loops is studied through the measurement of the specific heat under an applied magnetic field. The specific heat can be defined as the second derivative of the Ginzburg-Landau free energy F_{GL}:

$$C(H,T) = -T\frac{\partial^2 F_{GL}(H,T)}{\partial T^2} \qquad (1)$$

The free energy is given by:

$$F_{GL} = \alpha|\psi|^2 + \frac{\beta}{2}|\psi|^4 + \frac{1}{4m}\left|\left(\frac{\hbar}{i}\nabla - \frac{2e}{c}\mathbf{A}\right)\psi\right|^2 \qquad (2)$$

Then the only contribution for the flux dependent specific heat comes from the third term; the temperature dependence coming from the order parameter $|\Psi|^2$ and the magnetic field contribution is taken into account in the vector potential **A**.

If we first focus our attention on the temperatures very close to the transition temperature, we do not expect major variations of $C(H)$. Indeed, the coefficient α is linear in temperature and β can be considered as a constant, and because $\Psi = -\alpha/\beta$ then no signal versus magnetic field is expected in the specific heat. This is related to the fact that the critical magnetic field is linear near T_c. One has to go far from T_c and take into account the temperature dependence of β in the expression of the free energy to get nontrivial variation of the specific heat signal versus the magnetic field threading the loops. Since the fluxoid is quantized in such a geometry, we expect that the total specific heat signal to show Φ_0 periodic oscillations versus the magnetic flux in the thermodynamic equilibrium. At a given magnetic flux

Φ threading the loop, we assume that the superconducting order parameter takes the form $\Psi = f(\rho)e^{in\phi}$, where the vorticity n is a number of "giant" magnetic vortices in the loop and we use cylindrical coordinates with the z axis perpendicular to the plane of the loop.

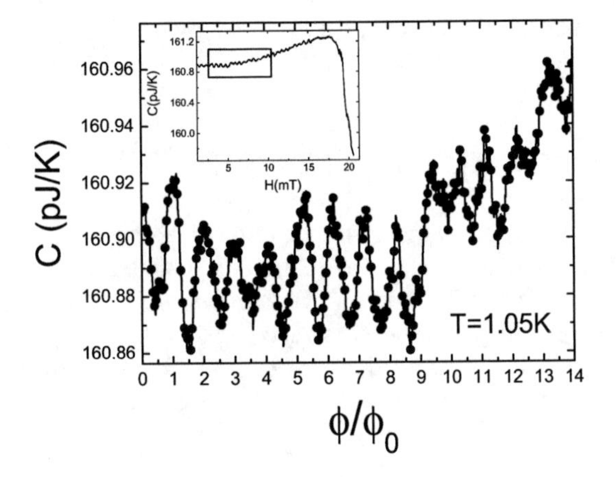

FIGURE 2. Heat capacity of mesoscopic super-conducting loops versus normalized magnetic flux (Φ_0 corresponds to a magnetic field of 0.58 mT in a loop of 1.8 µm in diameter) at 1.05 K. The specific heat exhibits Φ_0 periodic oscillations. In the inset we present the complete variation of specific heat under magnetic field where the critical magnetic field is observed around 19 mT. The rectangle represents the part of the curve enlarged in the main plot.

In Fig. 2 we present the specific heat measurement under magnetic field at 1.05 K. The major jump at 20 mT represents the superconducting-to-normal phase transition at the critical magnetic field of the aluminium loops (see the inset). Meanwhile, significant oscillations appear at low field with a clear periodicity of 0.58 mT which corresponds to the magnetic field required to create one giant vortex (one quantum of flux) in a loop of diameter 1.8 µm. These oscillations, as expected, are the signature of the penetration of vortices inside the loops. Each giant vortex entry is associated to a mesoscopic phase transition and hence a specific heat jump corresponding to 5×10^{-14} J/K, i.e. 9×10^{-20} J/K per loop ($7500k_B$) is observed. The existence of such kind of phase transitions was predicted for holes in a superconductor or for disks in the framework of the Ginzburg-Landau (GL) theory of superconductivity, which can also be applied to superconducting loops [9]. These measurements are in a perfect agreement with the previous measurements by susceptibility [10], micro-Hall probe magnetometry [11,12] or more recent electrical measurements [13] on mesoscopic superconducting disks or rings.

At lower temperatures, we also observed multiple phase transitions, but now between states differing in vorticity by more than unity. Around 0.8 K, the specific heat is $2\Phi_0$ periodic and at 0.6 K the signal becomes $3\Phi_0$ periodic (see Fig. 3). These mesoscopic phase transitions involving a high number of vortices are characterized by a larger specific heat jump as well as a larger magnetization change [14].

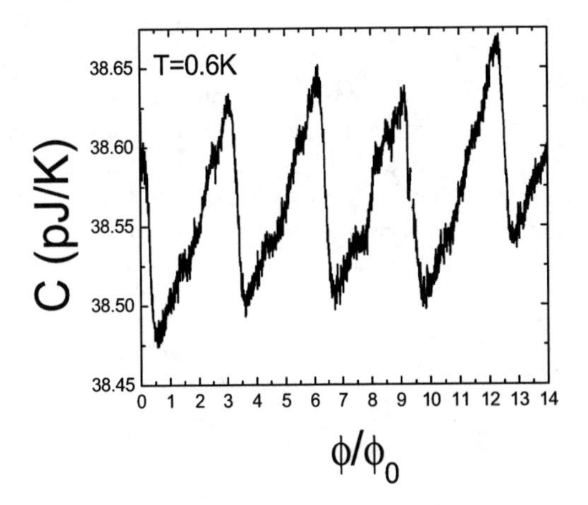

FIGURE 3. Heat capacity of the superconducting loops measured at 0.6 K (the same normalization for the x axis as in Fig. 2). At low temperatures, $3\Phi_0$ periodic specific heat jumps are observed with amplitude of 15×10^{-14} J/K.

These multiple flux jumps or flux avalanches, signatures of simultaneous entrance of several giant vortices in the loops, cannot be explained by equilibrium thermodynamics. If we calculate the position of the specific heat jump in the equilibrium case, the periodicity corresponding to the crossing of two free energy parabola is never greater than Φ_0. The measurement shown in Fig. 3 indicates that the system actually explores metastable states, staying in a local minimum of the free energy, but not in the stable state. The environmental noise (thermal or electromagnetic) is not enough to overcome the energy barrier between the metastable and stable states. We estimate the characteristic time of such an event to be much larger than the total duration of our experiments. This means that the system stays out of thermal equilibrium when we increase the magnetic field, until it reaches an unstable state.

This behaviour has been perfectly described in the framework of the GL theory through numerical calculations. By minimizing the free energy and finding the GL wave function we were able to calculate the specific heat of the system. We assume a circular loop of the same average perimeter as the actual square loop and the value of the GL coherence length at zero temperature $\xi(0)$ = 150 nm, evaluated from transport measurements. The free energy has been minimized numerically and then the specific heat has been calculated using Eq. (1). As in the experiment, several values of vorticity are possible at a given value of Φ. In the thermodynamic equilibrium, the transitions between the states with vorticities n and $n + 1$ (so-called "giant vortex states") occur at $\Phi = (n + 1/2)\ \Phi_0$, which minimizes the free energy and makes the specific heat to follow the lowest of curves corresponding to different n, as it can be seen in the Fig. 4. This results in oscillations of $C(\Phi)$ with a period of Φ_0, in agreement with the experiment. The amplitude of oscillations decreases when T approaches T_c, and hence the oscillations are likely to be masked by noise at T close to T_c, which explains why we did not succeed to observe experimentally any fine structure in C at $T > 0.93\ T_c$.

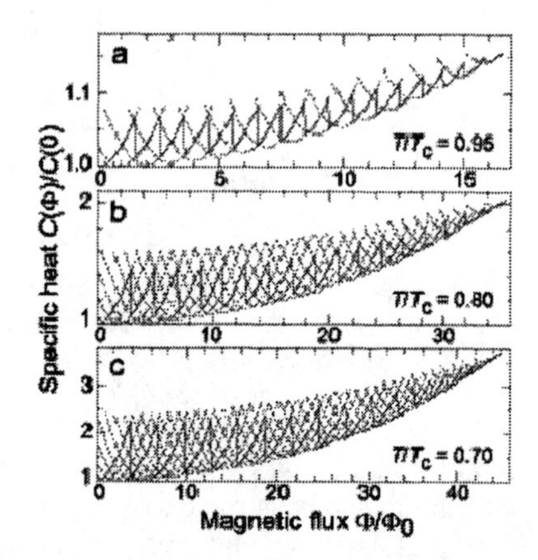

FIGURE 4. Normalized specific heat versus the magnetic flux of a superconducting loop computed from the numerical solution of the GL equation for three different temperatures. Dashed lines correspond to different states with constant vorticity (0, 1, 2 and 3 etc…).

The jump in the specific is found to appear when the current state becomes unstable. This takes place when the circulating current reaches the critical current of the superconducting loop. The numerical calculation gives a rather good qualitative description of what was experimentally observed. When the magnetic field increases, a change of periodicity is observed in both. In Fig. 3, the signal changes from a $3\Phi_0$ periodic signal to a Φ_0 periodic signal. The same kind of change is also observed in Fig. 4.

Finally, the experimental demonstration of the oscillatory behaviour and jumps of the specific heat at vortex entrance in a superconducting loop reveals the great interest of studying thermodynamic behaviour of mesoscopic systems (mesoscale thermodynamics). In fact, the possibility of measuring specific heat of nanoscale objects with very high accuracy opens quite interesting prospects. For instance, at low temperatures great opportunities exist for studying phase transitions between states with the same number but different configurations of vortices in mesoscopic superconductors, the superconducting phase transition in nanometer sized grains [15], thermal signatures of phase coherence in normal metals or quantum heat transfer in single nanocrystals or nanowires [16].

ACKNOWLEDGMENTS

The authors want to thank the Groupe de Biothermique et de Nanocalorimétrie (E. André, P. Lachkar, and J-L. Garden), and P. Gandit, P. Brosse-Maron, J-L. Bret for useful help and assistance in cryogenics and electronics. Fruitful discussions with R. Maynard, G. Deutscher and H. Pothier are acknowledged.

REFERENCES

1. *New Directions in Mesoscopic Physics (Towards Nanoscience)* edited by Fazio, R., Gantmakher, V. F. & Imry, Y., Dordrecht: Kluwer Academic Publishers 2003 and references therein.
2. Little W. A. and Parks R. D. *Phys. Rev. Lett.* **9**, 9-12 (1962).
3. Sharvin D. Yu. and Sharvin Yu. V. *JETP Lett.* **34**, 272-275 (1981).
4. Webb R. A., Washburn S., Umbach C. P. and Laibowitz, R. B. *Phys. Rev. Lett.* **54**, 2696-2699 (1985).
5. Lévy, L. P., Dolan, G., Dunsmuir, J. and Bouchiat, H. *Phys. Rev. Lett.* **64**, 2074-2079 (1990).
6. Schwab K., Henriksen E. A., Worlock, J.M and Roukes, *Nature* **404**, 974-976 (2000).
7. Bourgeois O., Skipetrov S.E., Ong F. and Chaussy J., *Phys. Rev. Lett.* **94**, 057007 (2005).
8. Fominaya F., Fournier T., Gandit P. and Chaussy J. *Rev. Sci. Instrum.* **68**, 4191 (1997).
9. Bezryadin, A., Buzdin, A. & Pannetier, B. *Phys. Rev. Lett.* **51**, 3718 (1995).
10. Zhang, X. and Price, J. C. *Phys. Rev. B* **55**, 3128 (1997).
11. Geim, A. K., Grigorieva, I. V., Dubonos, S. V., Lok, J. G. S., Maan, J. C., Filippov, A. E. & Peeters, F. M. *Nature (London)* **390**, 259 (1997).
12. Pedersen, S., Kofold, G. R., Hollingbery, J. C., Sorensen, C. B. & Lindelof, P. E. *Phys. Rev. B* **64**, 104522 (2001).
13. Kanda, A., Baelus,B.J., Peeters, F.M., Kadowaki, K., and Ootuka, Y. *Phys. Rev. Lett.* **93**, 257002 (2004).
14. Webb R. A., Washburn S., Umbach C. P. and Laibowitz, R. B. *Phys. Rev. Lett.* **54**, 2696-2699 (1985).
15. Neeleshwar S., Chen Y.Y., Wang C.R., Ou M.N. and Huang P.H. *Physica C* **408-410**, 209 (2004).
16. Bourgeois O., Fournier T. and J. Chaussy, *Thermal Conductivity of Silicon Nanowires*, same proceedings (LT24 ref abstract LT1800).

Experimental Distinction Between Giant Vortex and Multivortex States in Mesoscopic Superconductors

Akinobu Kanda*, Ben J. Baelus†, François M. Peeters†, Kazuo Kadowaki** and Youiti Ootuka*

*Institute of Physics and Tsukuba Research Center for Interdisciplinary Materials Science (TIMS), University of Tsukuba, Tsukuba 305-8571, Japan
†Departement Fysica, Universiteit Antwerpen, Groenenborgerlaan 171, B-2020 Antwerpen, Belgium
**Institute of Materials Science, University of Tsukuba, Tsukuba 305-8573, Japan

Abstract. We describe an experimental distinction between giant vortex and multivortex states in mesoscopic superconducting disks by using two methods: the multiple-small-tunnel-junction method and the temperature dependence of vortex expulsion fields. The experimental results are in good agreement with the theoretical simulations based on the non-linear Ginzburg-Landau theory.

Keywords: Mesoscopic superconductor, giant vortex states, multivortex states, Ginzburg-Landau theory
PACS: 74.78.Na, 74.25.Dw, 74.25.Op

INTRODUCTION

Mesoscopic superconductors have sizes comparable to the superconducting coherence length and/or the magnetic penetration depth. The vortex configuration in them is strongly affected not only by the vortex-vortex interaction but also by the boundary, since the shielding supercurrent along the boundary repels vortices. The competition between these interactions leads to corruption of the Abrikosov triangular lattice, which is the lowest energy configuration in bulk type-II superconductors. Theoretical studies have revealed that for mesoscopic disks the novel vortex configurations are divided into two categories: (i) multivortex states (MVSs) with a spatial arrangement of singly quantized vortices, and (ii) giant vortex states (GVSs) with a single multiply quantized vortex in the center [1, 2, 3, 4]. Because of the disk geometry, the spatial distribution of the Cooper-pair density, supercurrent and magnetic field for a GVS is axially symmetric with respect to the disk center. For superconductors without axial symmetry, it is known that the vortex configuration can take a combination of a MVS and a GVS [5], and sometimes include anti-vortices [6].

A variety of experimental techniques have been developed for the observation of theoretically predicted novel vortex states in mesoscopic superconductors. The most preferable one is the direct visualization of the vortices. Such techniques include the scanning SQUID (superconducting quantum interference device) microscopy and the scanning tunnel microscopy. However, the smallest SQUID loop available now has a diameter of several microns, which limits the spatial resolution. This size is not sufficient even for the Al samples having relatively large coherence length ($\xi \approx 0.2$ μm). The STM observation requires mesoscopic samples with atomically flat and clean surfaces, which are not prepared up to now. Alternative indirect methods have been developed; Cusps in the superconducting phase boundary $T_C(H)$ obtained in resistance measurements have been attributed to transitions between different vortex states [7, 8]. Also, ballistic Hall magnetometry for an isolated sample that is placed on a μm-scale Hall bar shows multiple magnetization curves, each of which corresponds to different vortex states [9, 10]. It is noted that these measurements do not provide any information to identify the vortex states (even to determine whether the state is a GVS or a MVS), so that for a long time there was no experimental evidence for the existence of the GVSs.

In the present paper, we describe two methods for the experimental distinction between GVSs and MVSs; The multiple-small-tunnel-junction (MSTJ) method [11] is a direct way for the distinction, taking advantage of the geometrical symmetry of a sample, while the temperature dependence of the vortex expulsion fields is an indirect one, but might be applicable to samples with various shapes [12].

MSTJ METHOD

The MSTJ method is based on a classical technique for estimating the superconducting density of states (DOS) from the current-voltage (*I-V*) characteristics of a superconductor-insulator-normal metal (SIN) tunnel junction. Since the superconducting DOS depends on the magnetic field and the supercurrent density, spatial dis-

CP850, *Low Temperature Physics: 24th International Conference on Low Temperature Physics;* edited by Y. Takano, S. P. Hershfield, S. O. Hill, P. J. Hirschfeld, and A. M. Goldman

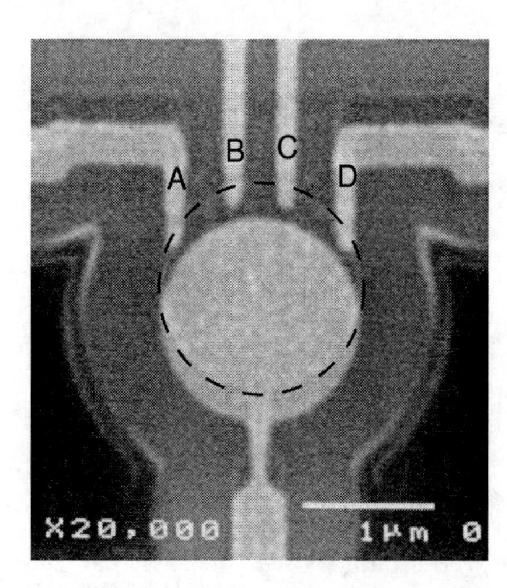

FIGURE 1. Scanning electron micrograph of a disk sample. The superconducting coherence length ξ is estimated to be 0.15 to 0.19 μm from the residual resistance of the Al films prepared in the same way. The superconducting transition temperature was 1.3 - 1.4 K.

tribution of the DOS provides information on the vortex configuration.

Instead of measuring the DOS everywhere over the sample surface, we just measure the DOS at the symmetrical positions in the sample, e.g., at the periphery for a disk sample. When the DOS at the symmetrical positions takes the same value, the state is likely to be a GVS, and otherwise, it is a MVS.

Figure 1 shows a scanning electron micrograph of a disk sample for the MSTJ-measurement. Four normal-metal (Cu) leads are connected to the periphery of a superconducting Al disk (indicated by the dashed circle) through highly resistive SIN tunnel junctions A, B, C and D. Note that the sample structure (including the junction positions) is symmetric with respect to the central axis, so that the junctions A and D (B and C) are at the symmetrical positions. The radius of the disk is 0.75 μm and the disk thickness is 33 nm. The disk is directly connected to an Al drain lead. This structure was fabricated using *e*-beam lithography followed by double-angle evaporation of Al and Cu. Details of the process are described elsewhere [11].

In the measurement, we fixed the current flowing through each junction to a small value, typically 100 pA, and measured simultaneously the voltages between each of the four Cu leads and the drain lead, while sweeping the perpendicular magnetic field at a typical rate of 20 mT/min.

Figure 2 (a) shows the change of the voltages at $I =$ 100 pA in decreasing magnetic fields. V_a and V_d denote the voltages in symmetric junctions A and D, respectively. To make the voltage comparison easier, dV/dB is also shown in Fig. 2 (b). The voltage variation as a function of magnetic field results from two origins: (i) smearing of the energy gap due to pair-breaking by the magnetic field, and (ii) a decrease of the energy gap because of the supercurrent [13]. The former leads to a moderate monotonic decrease in voltage as the strength of the magnetic field increases, so the rapid change in voltage comes from the latter. Especially, each voltage jump corresponds to a transition between different vortex states with a vorticity change of -1 [2, 3, 4]. This allows us to identify the vorticity L as shown in the figure. (Here, $2\pi L$ is the change in the phase of the superconducting order parameter along a closed path around the superconductor.) In Figs. 2 (a) and (b), the difference in V and dV/dB is remarkable for $L = 2$, 4 - 11, indicating that the state is a MVS for these vorticities. For increasing magnetic fields, the difference in V and dV/dB is relatively large between $L = 4$ and 6 (not shown), which is also due to the MVS formation.

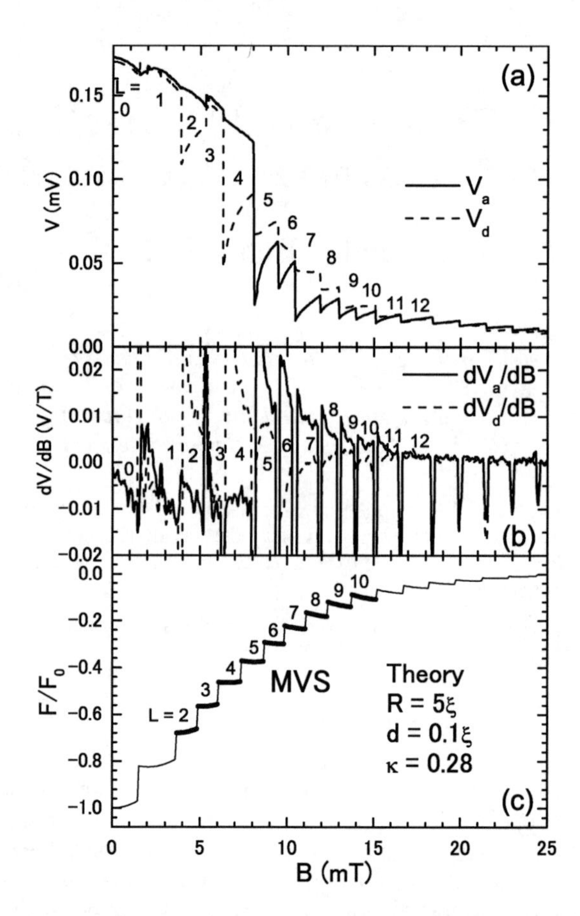

FIGURE 2. (a) Variation of voltages for junction pairs at symmetrical positions A and D in a decreasing magnetic field. The current through each junction is 100 pA. The temperature is 0.03 K. (b) Differential voltage dV/dB for junctions, A and D. (c) Calculated free energy F for a disk with $R = 5.0\xi$, $d = 0.1\xi$, and $\kappa = 0.28$, normalized by the $B = 0$ value, F_0, for decreasing magnetic fields.

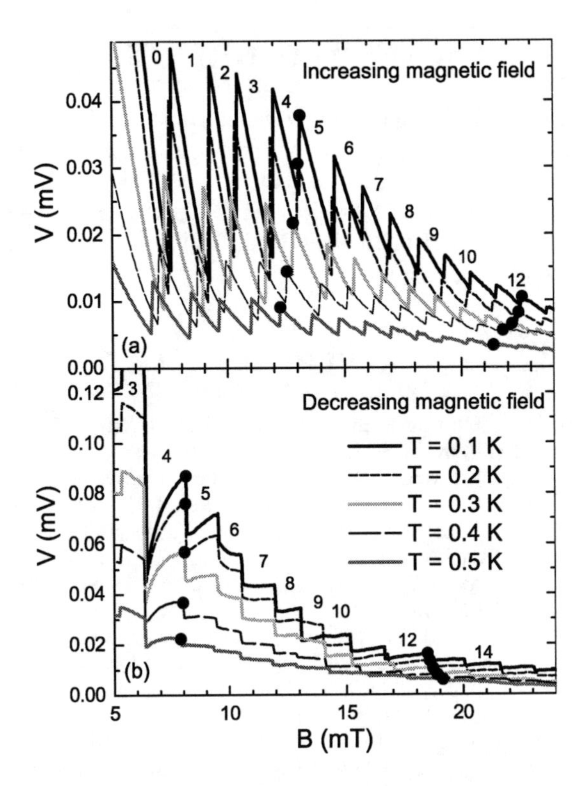

FIGURE 3. Voltage variation of junction D in (a) increasing and (b) decreasing magnetic field. The temperature is 0.1 K (highest curve), 0.2 K, 0.3 K, 0.4 K, and 0.5 K (lowest curve). The filled circles indicate the transition fields between the states with $L = 4$ and 5, and $L = 12$ and 13.

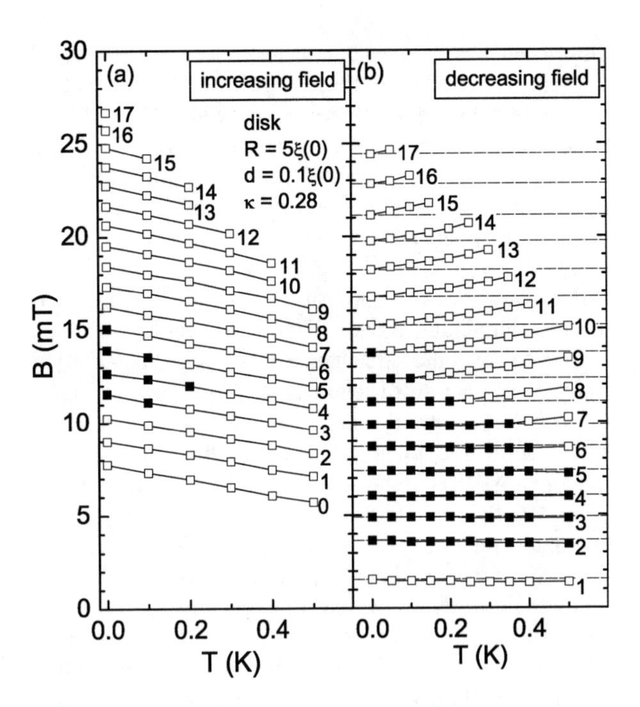

FIGURE 4. (a) Theoretical penetration fields and (b) theoretical expulsion fields of vortices as a function of temperature. The numbers indicate the vorticity before the transitions. The closed symbols correspond to a MVS and open symbols to a GVS before the transition.

This experimental distinction between GVSs and MVSs is supported by a numerical simulation. Figure 2 (c) shows the free energy for a disk with $R = 5.0\xi$, $d = 0.1\xi$, and the Ginzburg-Landau (G-L) parameter $\kappa = 0.28$ as calculated within the framework of the nonlinear G-L theory [1]. These parameters correspond to the experimental sample. Theoretically, MVSs nucleate for vorticity $L = 2$ to 10 for decreasing magnetic fields and $L = 3$ to 6 for increasing fields. Thus, the theoretical calculations confirm the identification of the GVS and MVS by the MSTJ method except for $L = 3$ ($L = 11$), where theoretically the state is predicted to be an MVS (GVS) while experimentally a GVS (MVS) was inferred.

The disagreement for $L = 3$ originates from the junction configuration. The $L = 3$ state has trigonal symmetry, which agrees with the angle $\angle AOD = 120$ degree. (O is the center of the disk.) Thus, the voltage difference for the $L = 3$ MVS is significantly decreased for the A-D junction pair, concealing the $L = 3$ MVS. Although this kind of symmetry induced effect is also expected to appear in $L = 6$ and 9 MVSs, the difference in V and dV/dB is significant for $L = 6$ and 9 in decreasing magnetic field, as shown in Figs. 2 (a) and (b). The difference between the experiment and theory could be attributed to the effects of defects, i.e., the stabilization of a different vortex configuration or distortions of the vortex configurations caused by defects. The experimentally observed MVS for $L = 11$ could be attributed to the effect of defects. Detailed analysis is described in ref. [11].

TEMPERATURE DEPENDENCE OF VORTEX EXPULSION FIELDS

In the MSTJ-measurement described above, we found a curious temperature dependence of vortex penetration/expulsion fields. Theoretical study demonstrated that this is closely related to the types of the vortex states [12]. Figure 3 shows the voltage of junction D of the same sample as in Fig. 2 as a function of the applied magnetic field for several values of the temperature. In increasing magnetic fields [Fig. 3 (a)], the penetration fields, given by the peaks in the voltage, always decrease with increasing temperature. On the other hand, in decreasing magnetic fields [Fig. 3 (b)], the temperature dependence of the expulsion fields shows two regimes: For some vortex states with small L, the expulsion field is almost independent of temperature, while for the other states with larger L, the expulsion field increases with temperature. These behaviors can be clearly seen from the closed symbols that indicate the $L = 5 \rightarrow 4$ and $L = 13 \rightarrow 12$ transition fields. The boundary of the two behaviors at low temperatures is between $L = 11$ and 12, which corresponds to the boundary between MVSs and GVSs obtained from the MSTJ-measurement. This indicates that the temperature dependence of the vortex expulsion fields depends on whether the last state is a MVS or a GVS.

This experimental implication is confirmed by a numerical simulation based on the nonlinear G-L theory. Figure 4 shows the calculated expulsion and penetration fields as a function of temperature. As shown in Fig. 4 (a), all the penetration fields decrease with increasing temperature, i.e., independent of whether the state is a GVS or a MVS. On the other hand, for decreasing fields [Fig. 4 (b)], the temperature dependence of the vortex expulsion fields shows two regimes; at high magnetic fields, the expulsion fields increase with temperature, and at lower fields, they are almost temperature independent. (Actually, they slightly decrease.) The theoretical simulation reveals that the former corresponds to the transition from a GVS and the latter to the transition from a MVS, as indicated by open and filled symbols in the figure. Note that the simulation does not take into account the thermal activation of the transition, but only the temperature dependence of the coherence length, expressed as $\xi(T) = \xi(0)/\sqrt{1-T/T_C}$. From the theoretical simulation [12], it is clear that the shielding current around the sample, which is responsible for the surface barrier for vortex expulsion, has a different temperature dependence for MVSs and GVSs, i.e., in MVSs the magnitude of the shielding current is almost temperature independent while in GVSs it decreases with increasing temperature. But the detailed origin of these behaviors is not fully understood.

CONCLUSION

Two methods for the experimental distinction between GVSs and MVSs are described.

The MSTJ method works almost perfectly for disks, but it crucially depends on the geometrical symmetry of the sample; when the sample geometry is less symmetrical, the distinction becomes more difficult [14].

On the other hand, two types of temperature dependence of the vortex expulsion fields have been observed in samples with various shapes [15]. Although further intensive study on shape and size dependence is needed, this result indicates that the temperature dependence of the transition fields might provide a very powerful tool for obtaining information about MVSs and GVSs by conventional techniques such as ballistic Hall magnetometry and SQUID magnetometry, which can determine the transition fields precisely.

ACKNOWLEDGMENTS

This work was supported by the University of Tsukuba Nanoscience Special Project, Grant-in-Aid for Scientific Research (B) (17340101) and CTC-NES Program of JSPS, the Flemish Science Foundation (FWO-Vl) and the Belgian Science policy. B. J. Baelus acknowledges support from FWO-Vl.

REFERENCES

1. V. A. Schweigert and F. M. Peeters, *Phys. Rev. B*, **57**, 13817–13832 (1998).
2. V. A. Schweigert, F. M. Peeters, and P. S. Deo, *Phys. Rev. Lett.*, **81**, 2783–2786 (1998).
3. J. J. Palacios, *Phys. Rev. B*, **58**, R5948–5951 (1998).
4. J. J. Palacios, *Phys. Rev. Lett.*, **84**, 1796–1799 (2000).
5. B. J. Baelus, L. R. E. Cabral, and F. M. Peeters, *Phys. Rev. B*, **69**, 064506 (2004).
6. V. R. Misko, V. M. Fomin, J. T. Devreese, and V. V. Moshchalkov, *Phys. Rev. Lett.*, **90**, 147003 (2003).
7. V. V. Moshchalkov, L. Gielen, C. Strunk, R. Jonckheere, X. Qiu, C. Van Haesendonck and Y. Bruynseraede, *Nature (London)*, **373**, 319–322 (1995).
8. V. Bruyndoncx, L. Van Look, M. Verschuere, and V. V. Moshchalkov, *Phys. Rev. B*, **60**, 10468–10476 (1999).
9. A. K. Geim, I. V. Grigorieva, S. V. Dubonos, J. G. S. Lok, J. C. Maan, A. E. Filippov, and F. M. Peeters, *Nature (London)*, **390**, 259–262 (1997).
10. A. K. Geim, S. V. Dubonos, I. V. Grigorieva, K. S. Novoselov, F. M. Peeters, and V. A. Schweigert, *Nature (London)*, **407**, 55–57 (2000).
11. A. Kanda, B. J. Baelus, F. M. Peeters, K. Kadowaki, and Y. Ootuka, *Phys. Rev. Lett.*, **93**, 257002 (2004).
12. B. J. Baelus, A. Kanda, F. M. Peeters, Y. Ootuka, and K. Kadowaki, *Phys. Rev. B*, **71**, 140502(R) (2005).
13. M. Tinkham, *Introduction to Superconductivity*, 2nd ed. (McGraw-Hill, New York, 1996), chap. 10.
14. B. J. Baelus, A. Kanda, N. Shimizu, K. Tadano, K. Kadowaki, Y. Ootuka, and F. M. Peeters, in preparation.
15. A. Kanda, M. C. Geisler, K. Ishibashi, Y. Aoyagi, and T. Sugano, unpublished.

Different Temperature Dependence of the Phase Boundary for Multivortex and Giant Vortex States in Mesoscopic Superconductors

B. J. Baelus[1], A. Kanda[2], F. M. Peeters[1], Y. Ootuka[2], and K. Kadowaki[3]

[1]*Departement Fysica, Universiteit Antwerpen, Groenenborgerlaan 171, B-2020 Antwerpen, Belgium*
[2]*Institute of Physics and Tsukuba Research Center for Interdisciplinary Material Science (TIMS), University of Tsukuba, Tsukuba 305-8571, Japan*
[3]*Institute of Materials Science, University of Tsukuba, Tsukuba 305-8573, Japan*

Abstract. Within the framework of the nonlinear Ginzburg-Landau theory, we calculated the full phase diagram for a superconducting disk with radius $R = 4\xi(T = 0)$ and we studied the behavior of the penetration and expulsion fields as a function of temperature for multivortex and giant vortex states.

Keywords: Vortex structure, Ginzburg-Landau theory, mesoscopic
PACS: 74.20.De; 74.25.Dw; 74.78.Na

INTRODUCTION

Mesoscopic superconducting disks have sizes comparable to the coherence length ξ and/or the penetration depth λ. Vortex states in such mesoscopic disks attracted a lot of attention, both theoretically (see e.g. Ref. [1]) and experimentally (see e.g. Ref. [2]). Theoretically it was predicted that two types of vortex states can nucleate in mesoscopic disks: (i) a giant vortex state (GVS) where the order parameter has a single zero with a winding number (or vorticity) L and (ii) a multivortex state (MVS), which is the finite-size version of the Abrikosov vortex lattice. However, for a long time there was no *direct* experimental proof for the existence of these two types of vortex states in mesoscopic disks.

Recently, we were able to experimentally distinguish between MVSs and GVSs in mesoscopic superconducting disks, by using the multiple-small-tunnel-junctions method [3]. Alternatively, we showed later that it is also possible to obtain information about the type of the vortex state by studying the temperature dependence of the transition fields [4]. We found that the expulsion field in a mesoscopic disk with radius $R = 5\xi(T = 0)$ is almost temperature independent when the vortex state is a MVS and that it increases with temperature in case of a GVS.

In the present paper, we present the *full* phase diagram for a slightly smaller superconducting disk with radius $R = 4\xi(T = 0)$ and we study the behavior of the penetration and expulsion fields as a function of temperature for MVSs and GVSs.

To construct the phase diagram, we calculate the free energy of the vortex states in a disk with $R = 4\xi(T = 0)$ for increasing and decreasing magnetic field at various temperature values. Therefore, we solve the two Ginzburg-Landau equations self-consistently, as is described in e.g. Ref. [1]. We assume the following temperature dependences: $\xi(T) = \xi(0)|1-T/T_{c0}|^{-1/2}$, $\lambda(T) = \lambda(0)|1-T/T_{c0}|^{-1/2}$ and $H_{c2}(T) = H_{c2}(0)|1-T/T_{c0}|$, with T_{c0} the zero field critical temperature.

RESULTS

We calculated the H-T phase diagram for a disk with radius $R = 4\xi(T = 0)$, thickness $d = 0.1\xi(T = 0)$ and Ginzburg-Landau parameter $\kappa = 0.28$. The penetration and expulsion fields as a function of temperature are shown by the symbols in Fig. 1. When the last state before the transition is a MVS the transition fields are given by closed symbols, when it is a GVS by open symbols.

CP850, *Low Temperature Physics: 24th International Conference on Low Temperature Physics;*
edited by Y. Takano, S. P. Hershfield, S. O. Hill, P. J. Hirschfeld, and A. M. Goldman

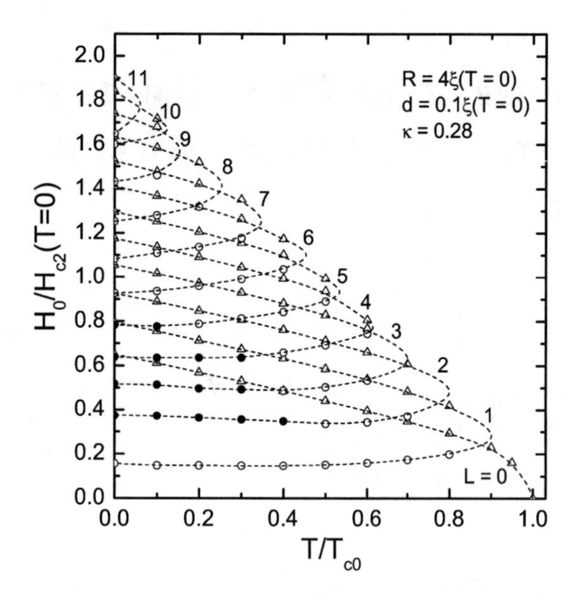

FIGURE 1. *H-T* phase diagram. The penetration fields (triangles) and expulsion fields (dots) as a function of temperature. In case of a GVS (MVS) we use open (closed) symbols. The curves are guidelines for the eye.

We find that the MVSs only stabilize at low temperatures and in this case the expulsion fields are almost temperature independent, while in the case of GVSs they increase with temperature. Notice further that for fixed vorticity the MVSs always transit into a GVS, which means that the state is always a GVS at the penetration field. This was not the case for a disk with radius $R = 5\xi(T = 0)$ [4].

From Fig. 1, it is clear that the stability of the MVSs decreases with increasing temperature. The reason is that the effective size $R/\xi(T)$ of the disk decreases with increasing temperature, since $\xi(T)$ increases with temperature. This results in the fact that the vortices are pushed more towards the center and the GVS becomes more favorable. To show this more clearly we give in Fig. 2 contour plots of the Cooper-pair density for the $L = 3$ state, just before the transition to the $L = 2$ state for different values of temperature. When temperature increases, the three vortices move more towards the center and for $T/T_{c0} = 0.4$ they combine into one giant vortex.

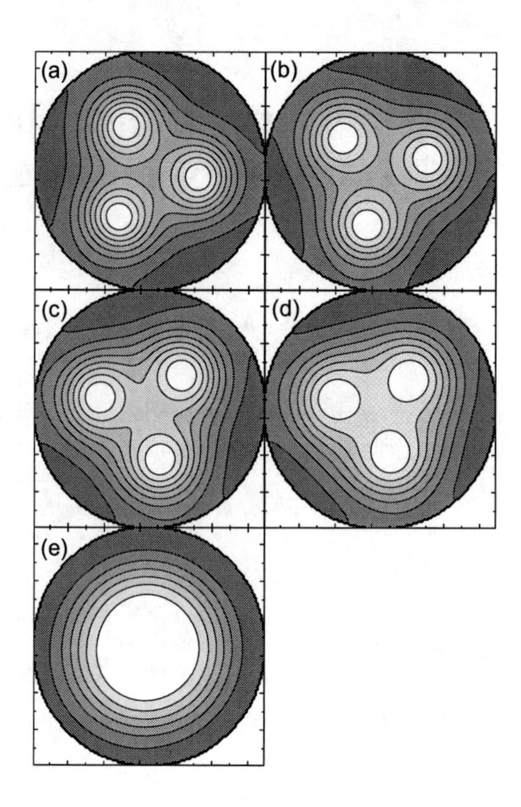

FIGURE 2. Contour plots of the Cooper-pair density of the $L = 3$ state just above the expulsion field, at $T/T_{c0} = 0$ (a), 0.1 (b), 0.2 (c), 0.3 (d) and 0.4 (e). Dark (light) regions correspond to high (low) Cooper-pair density.

CONCLUSIONS

We constructed the phase diagram for a disk with radius $R = 4\xi(T = 0)$. We found that the MVSs stabilize at low temperature and become less stable with increasing temperature. In the case of MVSs we found that the expulsion fields are almost temperature independent.

ACKNOWLEDGMENTS

This work was supported by the University of Tsukuba Nanoscience Special Project, the 21st Century COE Program of MEXT, the Flemish Science Foundation (FWO-Vl), and the Belgian Science Policy. B. J. Baelus acknowledges support from FWO-Vl.

REFERENCES

1. Schweigert, V. A., Peeters, F. M., and Deo, P. S., *Phys. Rev. Lett.* **81**, 2783 (1998).
2. Moshchalkov, V. V., Gielen, L., Strunk, C., Jonckheere, R., Qiu, X., Van Haesendonck, C., and Bruynseraede, Y., *Nature (London)* **373**, 319 (1995).
3. Kanda, A., Baelus, B. J., Peeters, F. M., Kadowaki, K., and Ootuka, Y., *Phys. Rev. Lett.* **93**, 257002 (2004).
4. Baelus, B. J., Kanda, A., Peeters, F. M., Ootuka, Y., and Kadowaki, K., *Phys. Rev. B* **71**, 140502(R) (2005).

Influence of Surface Defects on the Vortex Transitions in Mesoscopic Superconductors

B. J. Baelus[1], K. Kadowaki[2], and F. M. Peeters[1]

[1]Departement Fysica, Universiteit Antwerpen, Groenenborgerlaan 171, B-2020 Antwerpen, Belgium
[2]Institute of Materials Science, University of Tsukuba, Tsukuba 305-8573, Japan

Abstract. Solving the nonlinear Ginzburg-Landau equations self-consistently, we investigate the influence of a triangular surface defect (i.e. a *pacman* shaped sample) on the vortex transitions in mesoscopic superconducting disks. Depending on the size of the defect, vortices may enter/leave one by one or in pairs.

Keywords: vortex entry and exit; meta-stable states; Ginzburg-Landau theory
PACS: 74.20.De; 74.25.Dw; 74.78.Na

INTRODUCTION

The influence of surface defects on the vortex penetration and expulsion in mesoscopic superconductors is not negligible [1]. Due to the interplay between the vortex-vortex repulsion and the vortex-defect interaction, we found e.g. that the vortex does not always enter or leave the sample through the surface defect [2]. Also the penetration and expulsion fields depend on the size, the shape and the relative position of the surface defects.

In the present paper we investigate a superconducting disk with radius $R = 6\xi$ and thickness $d << \xi$ with a triangular surface defect, which penetrates over a distance $R - \Delta$ toward the center of the disk and has an opening of 1.5ξ at the surface (see the upper inset of Fig. 1). Because of the small thickness, we can neglect the demagnetization effects (i.e. the bending of the magnetic field around the superconductor) and we only have to solve the first GL equation. A detailed description of the theoretical model can be found in Ref. [3].

RESULTS

We calculated the free energy as a function of the applied magnetic field for the sample shown in the upper inset of Fig. 1(a), for 5 different values of Δ, i.e. $\Delta/R = 0$ (the surface defect penetrates until the center of the disk), 0.25, 0.5, 0.75 and 1 (no surface defect). The results are shown in Figs. 1(a,b) for increasing and decreasing field, respectively.

Depending on the value of Δ, the vortices may enter or leave the sample one by one or in pairs. With increasing field we find that in a perfect disk ($\Delta = R$), the first few vortices enter by two. So, the vorticity L changes as follows: $L = 0 \rightarrow 2 \rightarrow 4$. When a surface defect is present, the vortices enter one by one, regardless of the value of Δ, i.e. $L = 0 \rightarrow 1 \rightarrow 2 \rightarrow 3$. Also the penetration fields depend strongly on the size of the surface defect, as is shown in the right inset of Fig. 1(a).

With decreasing field the vortices leave the perfect disk ($\Delta = R$) one by one at some transition fields or in pairs at other transition fields, i.e. $L = 6 \rightarrow 5 \rightarrow 4 \rightarrow 2 \rightarrow 1 \rightarrow 0$. For $0 < \Delta < R$, the vortices leave one by one, i.e. $L = 5 \rightarrow 4 \rightarrow 3 \rightarrow 2 \rightarrow 1 \rightarrow 0$. This is in good agreement with the results of Ref. [2], where we studied various sizes and shapes of surface defects. For $\Delta = 0$ we find the following transitions $L = 5 \rightarrow 3 \rightarrow 2 \rightarrow 0$. Notice further that also the expulsion fields are drastically influenced by the size of the triangular surface defect. This is shown in the right inset of Fig. 1(b) and will be discussed below for a specific case.

Notice further that in Fig. 1 the free energy decreases when the defect size increases, because the effective superconducting area decreases.

CP850, *Low Temperature Physics: 24th International Conference on Low Temperature Physics;*
edited by Y. Takano, S. P. Hershfield, S. O. Hill, P. J. Hirschfeld, and A. M. Goldman

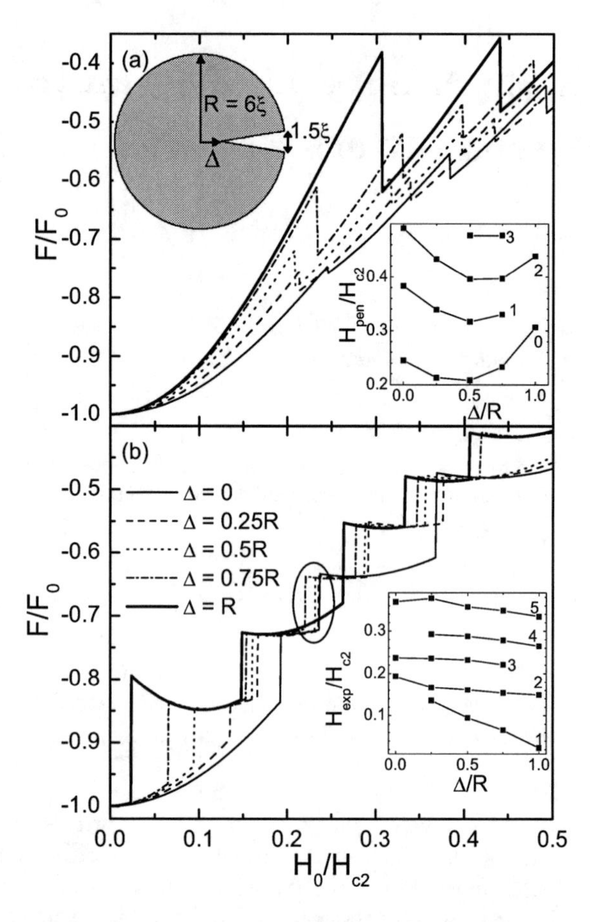

FIGURE 1. Free energy as a function of (a) the increasing and (b) the decreasing magnetic field for a superconducting disk with radius $R = 6\xi$ with a triangular surface defect penetrating over a distance $R - \Delta$ toward the center of the disk with an opening of 1.5ξ at the surface. Δ is varied from 0 to R. The upper inset of (a) shows the sample. The encircled region in (b) indicates the $L = 3 \rightarrow 2$ expulsion. The right insets in (a) and (b) show the penetration fields H_{pen} and the expulsion fields H_{exp} as a function of Δ. The indices denote the vorticity of the vortex state just before the transition.

Surface defects do not only influence the transition fields but also the arrangement of the vortices. In order to show this, we plot in Fig. 2 the (meta-)stable state with 3 vortices for various defect sizes at fixed magnetic field. In a perfect disk, we know that the 3 vortices are on a perfect equilateral triangle (see e.g. Ref. [4]). With increasing defect size, the equilateral triangle of vortices becomes distorted. Moreover, the vortices move more towards the boundary of the sample. This also explains why the $L = 3 \rightarrow 2$ expulsion (see the encircled region in Fig. 1(b)) occurs at higher fields H_{exp} with decreasing Δ. This is also clear from the right inset of Fig. 1(b). For $\Delta/R = 0$, 0.25, 0.5 and 0.75 we find respectively $H_{exp}/H_{c2} =$ 0.237, 0.236, 0.232 and 0.221.

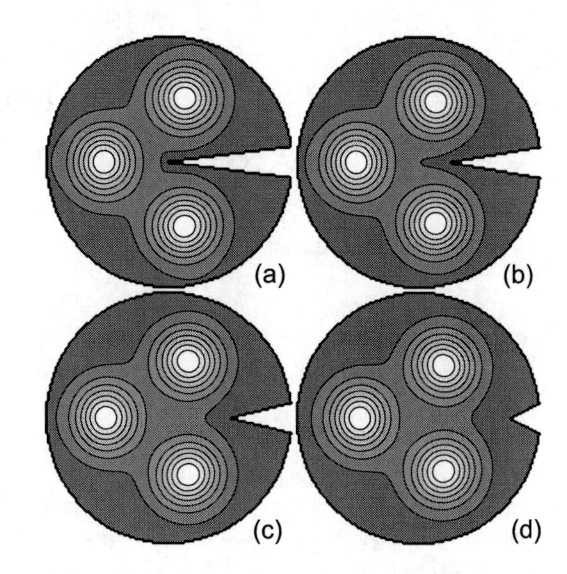

FIGURE 2. Cooper-pair density for the $L = 3$ state at H_0/H_{c2} = 2.5 for Δ = 0 (a), 0.25R (b), 0.5R (c), 0.75R (d). Dark (light) regions indicate high (low) Cooper-pair density.

CONCLUSIONS

We studied the effect of a triangular surface defect on the transition between different vortex states and on the vortex arrangement. Depending on the size of the surface defect, we found that vortices may enter or leave the sample one by one or in pairs. Also the transition fields are influenced by the defect size. We also found that the arrangement of the vortices depends on the size of the surface defect.

ACKNOWLEDGMENTS

This work was supported by the Flemish Science Foundation (FWO-Vl), the Belgian Science Policy, and the 21st Century COE Program of MEXT. B. J. Baelus acknowledges support from FWO-Vl.

REFERENCES

1. Geim, A. K, Dubonos, S. V., Grigorieva, I. V., Novoselov, K. S., Peeters, F. M., and Schweigert, V. A., *Nature (London)* **407**, 55 (2000).
2. Baelus, B. J., Kadowaki, K., and Peeters, F. M., *Phys. Rev. B* **71**, 024524 (2005).
3. Schweigert, V. A., and Peeters, F. M., *Phys. Rev. B* **57**, 13817 (1998).
4. Baelus, B. J., and Peeters, F. M., *Phys. Rev. B* **65**, 104515 (2002).

Superconducting Wire Network under Spatially Modulated Magnetic Field

Hirotaka Sano, Akira Endo, Shingo Katsumoto and Yasuhiro Iye

Institute for Solid State Physics, University of Tokyo, 5-1-5 Kashiwanoha Kashiwa, Chiba 277-8581 Japan

Abstract. A two-dimensional (2D) superconducting square network under spatially modulated magnetic field is studied. The super/normal phase boundaries were measured with field modulation varied. The dependence on the strength of field modulation exhibited the behavior reproducing the calculation we had done before. In addition, *I-V* characteristics measurements were also conducted.

Keywords: superconducting network, Hofstadter butterfly, modulated magnetic field, Little-Parks oscillation
PACS: 74.78.-w, 74.81.Fa, 75.75.+a, 73.21.-b

The super/normal phase boundary of a two-dimensional (2D) superconducting wire network exhibits characteristic Little-Parks oscillation[1] against the magnetic frustration α (perpendicular magnetic flux through the unit cell). This $T_c(H)$ curve reflects the edge of the tight binding energy spectrum for Bloch electrons on a corresponding 2D lattice[2] which is known as Hofstadter butterfly.[3] Reflecting the fine structures of the edge of the Hofstadter butterfly, the $T_c(H)$ curve has local maxima at simple fractional values of α, which correspond to stable vortex configurations[4]. In this work, we address the case in which superconducting network is subjected to both spatially modulated magnetic field and uniform magnetic field. We measured the dependence of the $T_c(H)$ curve on the field modulation amplitude β (expressed in terms of flux per plaquette) and compared the results with the calculation we had done before[5].

In a 2D superconducting network, two types of phase transitions are expected. One is Kosterlitz-Thouless (KT) transition[6] and the other is vortex glass (VG) transition[7]. The hallmark of the KT transition is a "universal jump" of the exponent of the current-voltage (*I-V*) characteristics at $T=T_{KT}$. On the other hand, the VG transition is characterized by scaling behavior of the *I-V* curves above and below the glass transition temperature T_g. Although *I-V* characteristics of superconducting network have been studied for decades, there still remains inconsistency between these experimental results. In order to shed light on the issue, we conducted *I-V* characteristics measurement in the present system with spatially varying magnetic field which is different from the usually studied case of uniform applied field.

The sample used in this study consisted of a square network of Al wire decorated with a regular array of ferromagnetic Co dots. The network pattern consisted of 120 × 120 unit cells, made of Al wire 70 nm wide and 35 nm thick. The lattice period was 500 nm. The ferromagnetic Co dots were oval and the size was 130 nm × 200 nm. This Co dots were 80 nm thick. The Co dots were arranged in such a way that the fringing field from them imposes a sign-alternating (checkerboard pattern) flux pattern to the superconducting network. Figure 1(a) shows the arrangement of the Co dots. Between these two components, Ge layer was deposited 35 nm thick. The role of this Ge layer was to prevent oxidation of Al network and to keep it from direct contact with Co dots.

Measurements were conducted using a cross-coil superconducting magnet system, which consisted of a 7T Helmholtz coil (horizontal field) and a 1T solenoid (vertical field). The vertical field defined the uniform field α for the network. The horizontal field was used to control the magnetization of the Co dots. The spatially modulated field (parameter β) was induced from the stray field from Co dots and its amplitude was tuned by the azimuthal angle φ of the horizontal field, the details of which are given in ref. 5.

The experiment consisted of two steps. The first step was the determination of the super/normal phase boundary $T_c(\alpha)$ for different values of β. The magnetoresistance was measured by a standard ac method for different settings of the azimuthal angle, and the data was converted to $T_c(H)$. The second step

CP850, *Low Temperature Physics: 24th International Conference on Low Temperature Physics;*
edited by Y. Takano, S. P. Hershfield, S. O. Hill, P. J. Hirschfeld, and A. M. Goldman

was the measurement of the I-V characteristics with a programmable dc current source. Temperature stability throughout the experiment was better than 1 mK.

Figure 1(b) shows the Little-Parks oscillation of T_c as a function of the uniform perpendicular field. Each trace is vertically shifted for clarity, and the azimuthal angle φ was rotated by 5° for each trace. Horizontal shift of the traces is a spurious effect due to a small angular misalignment of the sample relative to the horizontal plane. Figure 1(c) is the comparison between the experiment (solid curves) and the calculation (dotted curves) over a single period, which corresponds to the region between the dotted lines in Fig. 1(b). The traces cover the range from $\beta = 0$ (bottom) to $\beta = 1/2$ (top). It is seen that as β is changed from 0 to 1/2, the peak at $\alpha = 1/2$ become pronounced relative to those at integer α[5]. They crossover at $\beta = 1/4$. The discrepancy between the experimental data and the calculated curves is presumably attributable to lithographical irregularity, in particular imperfect registration between the Co dot array and the Al network.

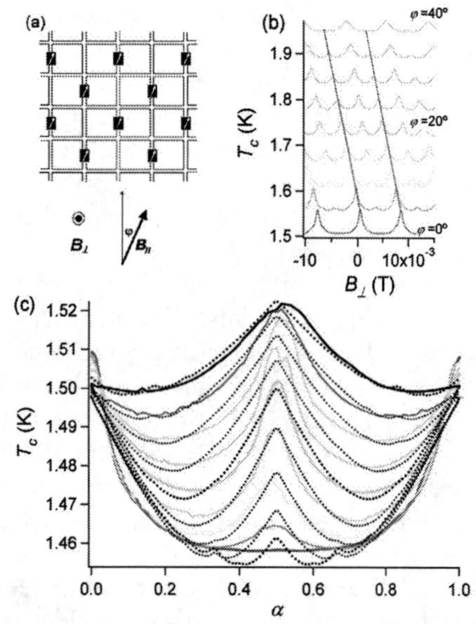

FIGURE 1. (a) Schematic picture of the sample. (b) Result of $T_c(H)$ measurements. Each trace is vertically offset by 0.05K. (c) Comparison between the experiment (solid curves) and the calculation (dotted curves). The value of the parameter β is changed from 0 (bottom) to 1/2 (top).

Figure 2(a) shows the I-V curves at different temperatures for α=0.61 and β=0. The inset of Fig. 2(b) shows a scaling plot of the data by the standard procedure of vortex glass model, which results in significant discrepancy for $T>T_g$. The origin of the discrepancy may be traced to the fact that T_g in this system is rather close to the mean field T_c. In order to take this into account, the correlation length ξ is multiplied by an extra factor $|T_c-T|^{-1/2}$ assuming it is scaled with the diverging length scale near T_c. The main panel of Fig.2(b) is the resulting scaling plot. Thus, the superconducting transition at these values of parameters α and β is consistent with the VG transition, although it by no means imply the only interpretation. Measurements at other values of α and β generally yielded similar results, (even for the nominally α=β=0 case, where a KT transition is expected). The requirement for the lithographical perfection is more stringent in the study of dynamics because ever increasing length scale is involved as the transition is approached.

In conclusion, the superconducting wire network under a spatially modulated magnetic field exhibits behavior reflecting the corresponding Hofstadter spectra. and the I-V curves are consistent with the VG scaling provided that a correction factor for diverging length scale near T_c is taken into account.

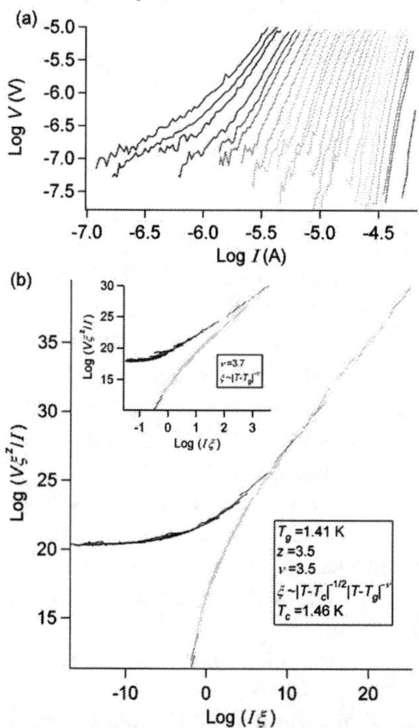

FIGURE 2. (a) I-V curves for α=0.61 and β=0 over the range 1.32 K < T < 1.45 K. (b) VG scaling plot with a correction factor to ξ explained in the text. The inset shows a scaling plot without the correction factor.

The work is supported in part by Grant-in-Aid for Scientific Research from the Ministry of Education, Culture, Sport, Science and Technology (MEXT), Japan.

REFERENCES

1. W.A. Little and R.D. Parks, *Phys. Rev. Lett.* **9**, 9 (1962).
2. S. Alexander, *Phys. Rev. B* **27**, 1541 (1983).
3. D.R. Hofstadter, *Phys. Rev. B* **14**, 2239 (1976).
4. B. Pannetier *et al., Phys. Rev. Lett.* **53**, 1845 (1985).
5. Y. Iye *et al., Phys. Rev. B* **70**, 144524 (2004).
6. H.S.J. van der Zant *et al, Phys. Rev. B* **50**, 340 (1994).
7. D.S. Fisher *et al, Phys. Rev. B* **43**, 130 (1991).

I-V Characteristics in the Superconducting State of a Mesoscopic Al Square

K. Enomoto, T. Yamaguchi, T. Yakabe, T. Terashima, T. Konoike, M. Nishimura, and S. Uji

National Institute for Materials Science, 3-13, Sakura, Tsukuba, Ibaraki 305-0003, Japan

Abstract. We have measured the I-V characteristics in a mesoscopic Al square at low temperatures. The resistance at sufficiently low current is zero in magnetic field $H < 250$ G for $T = 900$ mK. At $H < 250$ G, the current I_c above which a finite voltage appears periodically has minima with increasing magnetic field. These minima are ascribed to the transitions between different vortex states in the sample. The resistance V/I exceeds the normal state value R_N at $|I| > I_c$ for $H < 250$ G. The experimental results suggest that the field dependence of I_c as well as the resistance larger than R_N is attributed to the dynamics of the vortices in the superconducting Al square.

Keywords: superconducting transition, excess resistance, vortex states
PACS: 73.23.-b, 74.25.Qt

INTRODUCTION

In mesoscopic aluminum (Al) samples, the superconducting state is reported to strongly depend on the boundary conditions when their size becomes comparable to the coherence length ξ or the London penetration depth λ [1-5]. The size effect plays a crucial role for the configuration of the vortices inside these samples because the boundary condition depending on the sample size and shape determines the confinement geometry for the superconducting condensate [1-5]. In such systems (square, triangle, and disk etc.), some possible vortex configurations have been proposed: vortex-antivortex states [3], giant vortex states (GVSs) with a single core in the center [4,5], and multivortex states (MVSs) with a spatial arrangement of singly quantized vortices [4,5]. These configurations are reported to show characteristic magnetic field dependence of T_c, which causes oscillatory behaviors in the resistive transitions with field. On the other hand, it has been recently found that the resistance exceeds the normal state value (R_N) under magnetic field below T_c in a mesoscopic Al disk [6]. The origin of the resistance larger than R_N may be attributed to some anomalous energy dissipation caused by the dynamics of the vortices in the confined geometries [6]. In order to further investigate the vortex states in the superconducting Al samples, we have measured the I-V characteristics in a mesoscopic Al square at low temperatures in detail.

EXPERIMENT

The sample (20 nm thick) was deposited by thermal evaporation of high purity Al (99.999%) at room temperature followed by a lift-off process. The structure of the square sample is depicted in the inset of Fig. 1. The length of the square side is 1.0 μm and the width of the leads is 0.1 μm. All the measurements were performed using *ac* (~20 Hz) or *dc* techniques. The resistance ratio R(300 K)/R(4.2 K) is about 2.1 and the coherence length ξ_{GL} is estimated to be about 0.2 μm from the residual resistance of the sample.

RESULTS AND DISCUSSION

Figure 1 shows the typical I-V characteristics under different magnetic fields at $T = 900$ mK ($T_c = 1.27$ K). The field is applied perpendicular to the sample plane. As the current increases from zero, first the transition from the zero- to finite-voltage states is observed around 1.4 μA at H = 170.8 G, then the voltage exceeds $R_N I$ at higher currents (the normal state value R_N = 0.41 Ω). For comparison, we plot the linear dependence $R_N I$ by the thin solid line in Fig. 1. The voltage larger than $R_N I$ suggests that some anomalous energy dissipation takes place in the square part of the sample. Note that an enhancement of the voltage larger than $R_N I$ has been observed irrespective of the location

CP850, *Low Temperature Physics: 24th International Conference on Low Temperature Physics;*
edited by Y. Takano, S. P. Hershfield, S. O. Hill, P. J. Hirschfeld, and A. M. Goldman

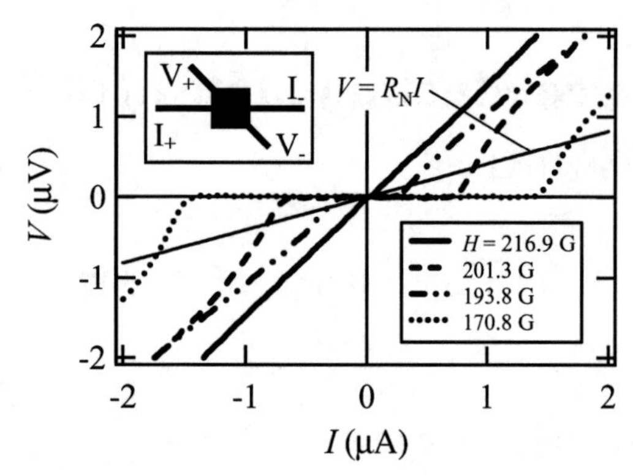

FIGURE 1. Typical I-V characteristics under different magnetic fields at T = 900 mK. The thin solid line indicates the linear dependence $R_N I$. Inset : Schematic view of the Al sample.

of voltage leads [cf. Ref. 7]. Further increase in current gives $V = R_N I$. This characteristic behavior is observed in the low magnetic field regime (~160 G < H < 250 G).

In Fig. 2 (a), we represent the resistance variations with different *ac* currents (from I_{ac} = 50 nA to 500 nA) at T = 900 mK. The I-V curves indicate that the *ac* resistance R_{ac} at sufficiently low current (I_{ac} < 20 nA typically) is zero in magnetic field H < 250 G. At H > 250 G, R_{ac} increases rapidly and exceeds R_N. The value of R_{ac} shows a maximum (~5R_N) at H ~ 260 G, and then decreases to R_N for H > 320 G. As the *ac* current increases, the resistance peaks become evident in the low magnetic field regime (H < 250 G). The positions of the resistance peaks are almost independent of the temperature. The magnetic field dependence of the critical current I_c, which is determined from a 50 nV criterion, is shown in Fig. 2 (b). At H < 250 G, the current I_c periodically has minima with increasing magnetic field. The magnetic fields of these minima correspond to those of the resistance peaks at higher *ac* currents (I_{ac} > ~20 nA).

The resistance peaks have been explained in terms of the transitions between the different vortex states in the superconducting Al sample [6]. The observation of the finite resistance means that some energy dissipation occurs in the square part of the sample. Since the energy dissipation does not occur as far as the vortices are pinned ($|I| < I_c$), the finite resistance could be attributed to the dynamics of the vortices (i.e., the vortices are depinned at $|I| > I_c$) in the superconducting Al square. At the transition-fields between different vortex states, the vortex arrangements are strongly fluctuated because the free energy of the different vortex states is degenerated. It means that the vortices are depinned by a smaller amount of current at the transitions. This picture well explains the field dependence of the current I_c in Fig. 2 (b). At $|I| > I_c$, the resistance V/I exceeds R_N even in the

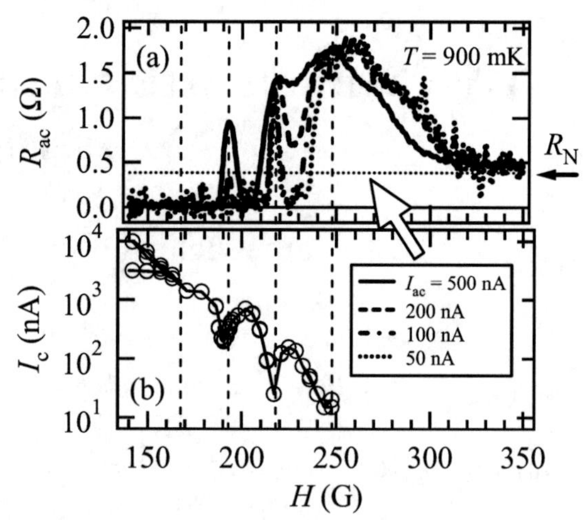

FIGURE 2. (a) Resistance variations with different *ac* currents at T = 900 mK. (b) Magnetic field dependence of the current I_c. The dashed lines indicate the magnetic fields where the transitions between different vortex states occur.

low magnetic field regime (H < 250 G for T = 900 mK). In general, the resistance never exceeds the normal state value R_N for bulk systems [8], but a few mechanisms of the resistances larger than R_N (~1.3R_N) in mesoscopic Al samples have been proposed by several authors [9,10]. These phenomena have been explained by the charge imbalance models based on the non-equilibrium superconductivity near phase slip centers [9] or S/N boundaries [10]. In both cases, however, the resistance anomalies should be observed only in the vicinity of the phase boundary (T_c or H_c). Therefore the large enhancement of the resistance in a wide field range (cf. Figs. 1 and 2 (a)) could not be simply explained by the proposed models. Our experimental results show that the dynamics of the vortices causes anomalously large energy dissipation. Further experimental investigations on the dynamics of the vortices are now in progress.

This work was supported by a Grant-in-Aid for Scientific Research from the Ministry of Education, Culture, Sports, Science and Technology (No.17740201).

REFERENCES

1. V. V. Moshchalkov *et al*, *Nature* **373**, 319 (1995).
2. A. K. Geim *et al*, *Nature* **390**, 259 (1997).
3. L. F. Chibotaru *et al*, *Nature* **408**, 833 (2000).
4. A. Kanda *et al*, *Phys. Rev. Lett.* **93**, 257002 (2004).
5. B. J. Baelus *et al*, *Phys. Rev. B* **69**, 064506 (2004) and references there in.
6. K. Enomoto *et al*, *Physica E*, (2005) in press.
7. Y. Terai *et al*, *J. Phys. Chem. Solids* **63**, 1311 (2002).
8. J. Bardeen *et al*, *Phys. Rev.* **140**, A1197 (1965).
9. C. Strunk *et al*, *Phys. Rev. B* **57**, 10854 (1998).
10. K. Yu. Arutyunov *et al*, *Phys. Rev. B* **59**, 6487 (1999).

Spatial Decay of Nonequilibrium Quasiparticles in Narrow Superconducting Wires with Injection

Ryuta Yagi

Graduate School of Advanced Sciences of Matter, Hiroshima University, Higashi-Hiroshima, Hiroshima 739-8530, Japan.

Abstract. We measured the relaxation length of the quasiparticles in a multi-terminal tunnel junction device fabricated using electron beam lithography, by measuring the spatial distribution of quasiparticle density in a steady injection. Relaxation length estimated assuming exponential decay was found to be much shorter than the early experiment with much larger device size. We argued the possible relation of the relaxation length to the density of quasiparticles.

Keywords: Quasiparticle, Recombination, Squid junction
PACS: 73.23.-b,74.90.+n

Quasi-particles in superconductors are excitations from the superconducting ground state. Excess quasiparticles produced by injection recombine as the system approaches equilibrium. In the early stage of the research on nonequilibrium superconductors, recombination was observed by a transport measurement of a diffusive conduction of quasiparticles.[1][2] A recombination time was compared with a prediction for an equilibrium state at the measured temperature. Since two quasiparticles are involved in a recombination process, a recombination rate should be proportional to the square of a density of the quasiparticles.[3] Recent optical experiments [4] [5] [6]clarified time dependent properties of the quasiparticle recombination and diffusion. In these optical measurements, however, the recombination process is complicated since, as pointed out by Ref.[3], high energy quasiparticles created by light or micro waves produces high energy phonons in the relaxation process which generate bunches of lower energy quasiparticles. In this work, we studied the recombination rate of quasiparticles using transport experiment using tunnel junctions. We estimate a recombination coefficient proportional to the square of the density of quasiparticles. In transport experiment, we expect that the recombination process is much simplified since the quasiparticle energy can be as low as the superconducting energy gap.

Figure 1 is a device structure, which consists of a narrow aluminum wire and SQUID-shaped Josephson junctions serving as injectors and detectors of quasiparticles. The reason why SQUID-junctions were employed was because, by applying magnetic field of a half flux quantum, they can suppress a current of Cooper pair tunneling appearing in sub-gap voltages due to resonating modes of an electromagnetic environment [7] which interfere the detection of a quasiparticle current. Superconducting electrodes composing the junction were designed to have

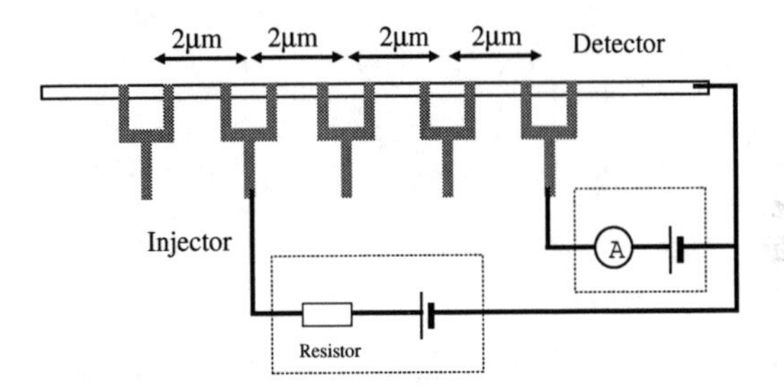

FIGURE 1. A schematic diagram of a device structure. The device consists of a narrow aluminum wire and dc-SQUID junctions.

different superconducting transition temperatures T_c's in order to estimate a precise quasiparticle current with current peaks appearing in the V-I characteristics. The Narrow aluminum wire elongating horizontally in Fig. 1 was evaporated normally and had T_c of 1.5K. T_c of the counter electrode was lifted by evaporating aluminum in the ambient atmosphere of low-pressure oxygen and T_c of 2.1K was attained.

Low temperature measurements were performed using a dilution refrigerator. The device was enclosed in a copper box connected thermally to a mixing chamber to reduce external noise.

A typical voltage-current characteristic of a detector junction is shown in Fig. 2. Quasiparticles are injected to the narrow wire with a junction 6μm away from the detector. In data with quasiparticle injection, a shoulder-shaped structure is discernible at around $V \sim \pm 120\mu$V. This originates from a quasiparticle current peak that occurs when a singularity of BCS density of states matches at $|V| = |\Delta_1 - \Delta_2|/e$ where Δ_1 and Δ_2 are the superconducting energy gaps of each electrode. In a limited range of voltages larger than $|V| = |\Delta_1 - \Delta_2|/e$, the rate of vari-

CP850, *Low Temperature Physics: 24th International Conference on Low Temperature Physics;*
edited by Y. Takano, S. P. Hershfield, S. O. Hill, P. J. Hirschfeld, and A. M. Goldman

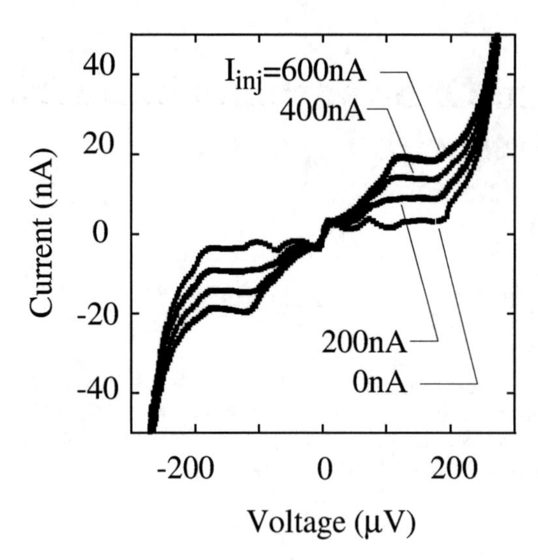

FIGURE 2. Voltage current characteristics of a SQUID junction. Quasiparticles are injected with another junction located 6μm away from this junction serving as a detector. Measured temperature was T =80mK. Magnetic field of a half flux quantum was applied to suppress Josephson current. The current peak appearing at zero-bias is due to the supercurrent arising from the asymmetric properties of junctions of the SQUID.

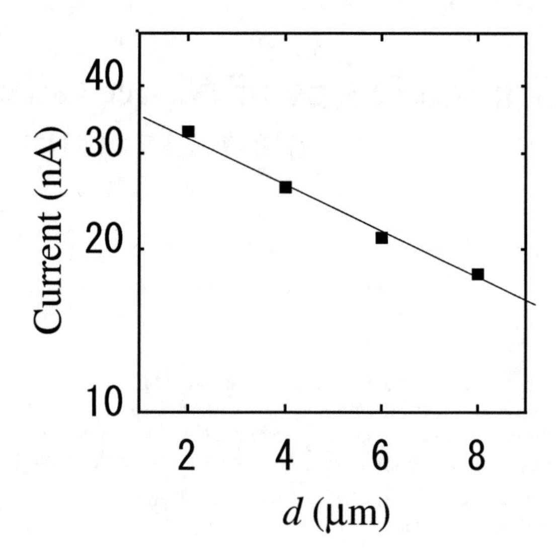

FIGURE 3. Dependence of the current on distance between injectors and the detector. Injection current for this figure is I_{inj} =400nA. Solid squares are measured data. A solid line is a fitting to an exponential decay. T =80mK.

ation of the current against the bias voltages is considerably small. A simulation based on the standard tunneling theory shows that the current in this regime is approximately proportional to a quasiparticle density.

We measured a spatial attenuation of the quasiparticle current using different injectors locating at different distances, d. In Fig. 3, we show a dependence of the quasiparticle current measured at V =160μV. The relaxation length λ was estimated to be about 9.8 μm assuming an exponential decay. The relaxation length is given by $\lambda = \sqrt{D\tau}$ for a transport governed by a diffusion equation with a recombination process approximated by a constant relaxation time. τ was estimated to be about 18nsec using D=5.3$\times$10^{-3}m^2/s estimated from resistivity of the aluminum wire at T =4.2K.

If the recombination rate is proportional to Γn^2, τ in a linear approximation is given by a relation, $1/\tau = \Gamma n$, where Γ is a parameter describing the recombination and n is the quasiparticle density. By solving a diffusion equation, we estimated typical value of n to be about $n = 0.93 \times 10^{22}$m^{-3} for I_{inj}=400nA and $d = 0\mu$m and, hence, Γ was estimated to be Γ =60μm^3/s. This agrees fairly well with Γ =48μm^3/s estimated from τ of the BCS theory for the temperature of experiment that is the averaged recombination time over the energy. [8]

Another evidence is given by comparing our λ with that of Ref. [2] performed in much larger device structures. Their λ (75μm) was much larger than ours. The difference stems partly from the difference in D. However major contribution is due to the recombination time. τ of their experiment was of the order of 2μsec, which is about factor of 100 longer than ours. This difference would be consistently explained by the recombination time dependent on the quasi-particle density since different quasi-particle current density results in different quasi particle density itself.

ACKNOWLEDGMENTS

This work was supported by grant-in-aid from the Ministry of Education, Science, Culture, Sports and Technology of Japan.

REFERENCES

1. S. Y. Hsieh, and J. L. Levine, *Phys. Rev. Lett.*, **24**, 1052-1054 (1968).
2. B. I. Miller, and A. H. Dayen, *Phys. Rev. Lett.*, **18**, 1000-1004 (1967).
3. A. Rothwarf and B. N. Taylor, *Phys. Rev. Lett.*, **19**, 27-30 (1967).
4. M. Johnson, *Phys. Rev. Lett.*, **67**, 374-377 (1991).
5. G. L. Carr, R. P. S. M. Lobo, J. LaVeigne, D. H.Reitze and D. B. Tanner, *Phys. Rev. Lett.*, **85**, 3001-3004 (2000).
6. N. Bulzer, *Phys. Rev. B*, **46**, 1033-1042 (1992).
7. T. Holst, D. Esteve, C. Urbina and M. Devoret, *Phys. Rev. Lett.*, **73**, 3455-3458 (1994).
8. S. B. Kaplan,C. C. Chi, D. N. Lagenberg, J. J. Chang S. Jafarey and D. J. Scalapino, *Phys. Rev. B*, **14**, 4854-4873 (1976).

Magnetic Response of a Mesoscopic Superconducting Disk Surrounded by a Normal Metal

Hidenori Goto, Kazuhito Tsukagoshi, and Kimitoshi Kono

Institute of Physical and Chemical Research (RIKEN), 2-1, Hirosawa, Wako, Saitama, 351-0198, Japan

Abstract. Magnetic response of a superconductor/normal-metal (S/N) concentric disk is studied by use of a ballistic Hall magnetometer, which enables us to investigate the mutual proximity effects in a single and a micrometer-sized sample. The core of the sample is a type-I superconductor whose diameter is comparable to the coherence length and the magnetic penetration depth. The core and the surround are prepared by an improved double-angle evaporation method to realize their metallic contacts. At $T = 1.3$ K, the pair breaking effect on the S has been predominant. We have observed the smaller diamagnetic susceptibility, the larger critical field, and the less stable vortex states in the S/N sample than in only S sample. These results are attributed to the suppression of the surface superconductivity, which effectively decreases the diameter of the superconductor.

Keywords: mesoscopic superconductor, proximity effect, Hall magnetometer
PACS: 74.78.Na, 74.45.+c, 74.25.Ha

INTRODUCTION

Vortex states in a mesoscopic superconductor have attracted much attention in view of configurations of vortices [1] and phase transitions between the different states [2,3]. The arrangement of vortices is stabilized by surface potential that leads to a hysteresis in flux penetration and expulsion. We can decrease this potential barrier by surrounding the superconductor (S) with a normal metal (N). Such a S/N hybrid system opens new possibilities to control the vortex motion in an artificial way. In this paper we report on the stability of the vortex states influenced by the surface suppression of superconductivity. In addition to the above-mentioned problem, magnetic properties of S/N concentric systems have received a revived interest because of paramagnetic Meissner effect (PME) at ultralow temperatures [4]. Whereas it has been suggested that the PME is due to the orbital motion of conduction electrons in the N [5,6], quantitative agreement between the theories and experiments has not been achieved. The purpose of our experiment is to clarify the origin of the PME by studying a single and a smaller S/N concentric disk. The use of such a sample has advantages over the previous studies for the following reasons. Measuring a single sample allows us to study quantitative magnetic properties of individual disks. In addition, decreasing the sample size enables us to raise the characteristic temperature where the proximity effects are observable.

EXPERIMENT

We fabricated a ballistic Hall magnetometer [2] and deposited a mesoscopic S/N disk on its center, as shown in Fig. 1. The Hall sensor was fabricated from a GaAs/AlGaAs heterostructure containing a two-dimensional electron gas (2DEG) 600 nm below the surface. The 2DEG has an electron density of $n_e \sim 3.7\times10^{15}$ /m^2 and a mobility $\mu_e \sim 46$ m^2/Vs at low temperatures, which corresponds to a mean free path of 4.6 μm. By electron-beam lithography (EBL) and shallow wet etch, we prepared the micro Hall probe 1~2 μm in width. The current for measurements was 10~20 μA with a frequency 118 Hz. At this condition our device showed a sensitivity of 100 mG. To provide a good metallic contact for the S/N concentric disk, we applied a double-angle evaporation technique. A single hole 800 nm in diameter was patterned on a 950PMMA/495PMMA/MMA triple-layer resist by using EBL and was used as a mask. We chose gold (Au) for a surrounding normal metal and tin (Sn) for a superconducting core. The Au layer of 100 nm was first evaporated at an angle of 34 ° with respect to the normal of the substrate. Keeping this angle, we rotated the substrate to make the ring structure. As a next step, without breaking the vacuum, the Sn layer of 300 nm was evaporated at the normal to the substrate. The Sn disk had a coherence length of $\xi(0)$ = 160 nm and a

CP850, *Low Temperature Physics: 24th International Conference on Low Temperature Physics;*
edited by Y. Takano, S. P. Hershfield, S. O. Hill, P. J. Hirschfeld, and A. M. Goldman

magnetic penetration depth of $\lambda(0)$ = 50 nm. These parameters were obtained by a resistivity measurement of a macroscopic Sn layer of the same thickness. A reference Sn disk without the Au layer was also prepared. The samples were immersed in a He^4 bath and were cooled down to 1.3 K. Magnetic fields were applied perpendicularly to the disks.

FIGURE 1. Scanning electron micrograph image of our device viewed at an angle of 50 ° with respect to the normal of the substrate. It shows the Hall probe 2.0 µm in width and the concentric disk consisting of the superconducting core and the surrounding normal metal. The inside (outside) radius of the disk is r_{in} = 0.4 µm (r_{out} = 1.2 µm).

RESULTS AND DISCUSSION

Magnetization processes of the S/N and S disk are shown in Fig. 2. Figure 2 shows that the diamagnetic susceptibility is 0.32 times larger and the critical field is 1.2 times larger in the S/N disk than in the S disk. These results are ascribed to the pair breaking effect on the S. This effect is predominant over the superconducting proximity effect on the N, because the measured temperatures are higher than the Thouless temperature, $T_{\text{Th}} \equiv hD/2\pi k_{\text{B}}(r_{\text{out}}\text{-}r_{\text{in}})^2 = 0.46$ K, where D is the diffusion constant in the Au layer. The pair breaking effect reduces the superconducting order parameter at the interface, and hence, changes the effective radius of the S from r to $r\text{-}\xi$. Because of this size effect, the diamagnetic susceptibility decreases by a factor of $(1\text{-}\xi/r)^2 \sim 0.36$, and the critical field increases by a factor of $(1\text{-}\lambda/r)/[1\text{-}\lambda/(r\text{-}\xi)] \sim 1.1$. This thermodynamic calculation is in good agreement with the observed results. Figure 2 also shows that the number of stable vortex states is smaller in the S/N disk than in the S disk. This result is explained by the size effect because the vortex state becomes energetically unstable with decreasing the disk size to the order of ξ [2]. We expect that the vortex states in the S/N disk will be more stable with decreasing temperatures. This is because the size effect is not crucial at lower temperatures where the superconducting order parameter extends to the N layer. To confirm this presumption and to elucidate the unsolved orbital paramagnetism, we plan to measure the samples at lower temperatures.

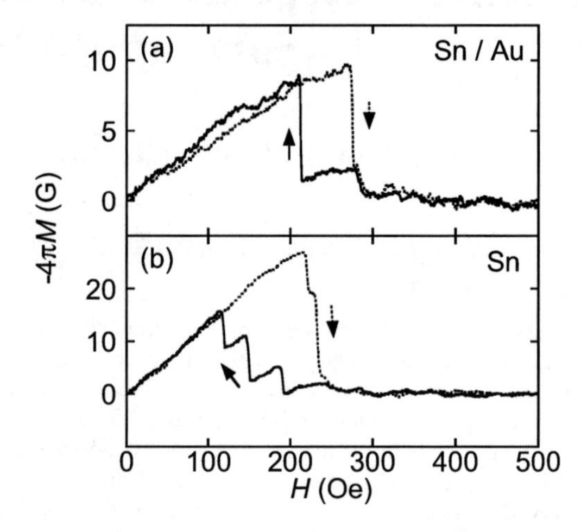

FIGURE 2. Magnetization processes of a Sn/Au concentric disk (a) and a Sn disk (b) at 1.3 K. The radius of the Sn in each disk is 0.4 µm. Arrows show the direction of the sweep. The magnetization is obtained by the Hall voltage at the superconducting state subtracted that at the normal state. A jump in the magnetization corresponds to the flux quantum.

CONCLUSIONS

In summary, by use of the ballistic Hall magnetometer, we have measured the magnetic response of S/N concentric disks at relatively high temperatures. The suppression of the surface superconductivity causes the crucial size effect on the clean mesoscopic superconductor, and thereby, makes the vortex states unstable.

ACKNOWLEDGMENTS

This work was supported by Grant-in-Aid for Scientific Research by the Ministry of Education, Culture, Sports, Science and Technology of Japan.

REFERENCES

1. Chibotaru, L. F., *et al.*, *Nature* **408**, 833 (2000).
2. Geim, A. K., *et al.*, *Nature* **390**, 259 (1997).
3. Kanda, A., *et al.*, *Phys. Rev. Lett.* **93**, 257002 (2004).
4. Visani, P., *et al.*, *Phys. Rev. Lett.* **65**, 1514 (1990). Müller-Allinger, F. B., and Mota, A. C., *Phys. Rev. Lett.* **84**, 3161 (2000).
5. Bruder, C., and Imry, Y., *Phys. Rev. Lett.* **80**, 5782 (1998).
6. Fauchère, A. L., *et al.*, *Phys. Rev. Lett.* **82**, 3336 (1999).

Critical Voltage Of A Mesoscopic Superconductor Between Normal Electrodes

R. S. Keizer*, M. G. Flokstra†, J. Aarts† and T. M. Klapwijk*

*Kavli Institute of NanoScience, Delft University of Technology, 2628 CJ, Delft, The Netherlands
†Kamerlingh Onnes Laboratory, Universiteit Leiden, 2300 RA, Leiden, The Netherlands

Abstract. We investigate the steady state transport properties of a mesoscopic superconducting wire between two normal metallic reservoirs by numerically solving the Usadel equation, going beyond the linear approach. From the calculation of the current-voltage characteristics we find a breakdown of the superconducting state which is characterized by a voltage rather than by a current; in other words, the system cannot be trivially treated as two resistors modelling the normal to supercurrent conversion, with a superconducting element characterized by its depairing current in between. Rather, the change in the longitudinal mode of the quasiparticle distribution function f (f_L) triggers the breakdown, which can be considered as a first order phase transition.

Keywords: Mesoscopic physics, Superconductivity, Non-equilibrium
PACS: 74.78.Na, 74.20.Fg, 74.25.Sv, 74.25.Bt

Under equilibrium conditions, the energy distribution function of quasiparticles in a normal metal is given by the standard Fermi-Dirac distribution. In recent years, it has been demonstrated [1][2] that in a voltage-biased mesoscopic wire an observable non-equilibrium distribution develops which is a two-step function with additional rounding by quasiparticle scattering due to spin-flip and/or Coulomb interactions (Fig. 1a). The question we address here is how the distribution function is modified when the normal metal is replaced by a superconductor (Fig. 1b) and how this affects observable properties such as the current-voltage characteristics of the system and the breakdown of the superconducting state. This question has intrinsic value, but is also important for understanding the current-voltage relation of so called hot-electron bolometer mixers [3][4]. We will calculate this relation and the breakdown of the superconducting state assuming that the superconducting wire is short enough to work only with a steady-state solution.

The transport and spectral properties of dirty superconducting systems ($\ell_e \ll \xi_0$, with ℓ_e the elastic mean free path and ξ_0 the superconducting phase coherence length) are described by the quasiclassical Green functions obeying the Usadel equation [5]. For out of equilibrium systems we use the Keldysh technique in Nambu (particle-hole) space. As usual it will be convenient to discern a symmetric part f_L (energy mode) (Fig. 1d) and an odd part f_T (charge mode) (Fig. 1c) of this quasiparticle distribution function. We ignore inelastic scattering in the wire, and use the time-independent formalism. We assume this to be an acceptable simplification at low temperatures ($T \ll T_c$), and for large enough wire cross section. In this way the role of thermally activated and re-

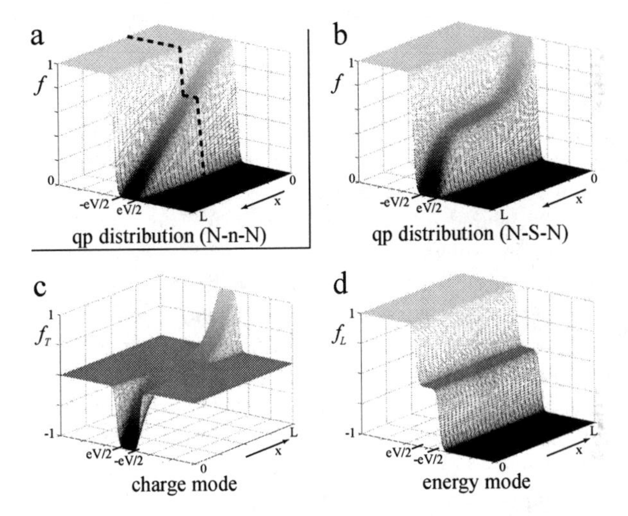

FIGURE 1. Quasiparticle distribution function $f(x,\varepsilon)$ as function of energy ε and position x for a normal wire (a) and a superconducting wire (b) between normal metallic reservoirs for $k_BT \ll eV < \Delta_0$, with (c) and (d) the decomposition of (b) into the charge mode f_T and energy mode f_L

spectively quantum phase slips is ruled out [6][7]. The full set of equations used in the calculation can be found in [8].

To calculate spectral and transport properties, one needs to know the self-consistent solution of the superconducting pair potential Δ. In most practical cases, this has to be done numerically. The solution scheme used here is to first find the Green functions of the system by solving the retarded equations for a certain Δ, next to determine the quasiparticle distribution functions by

CP850, *Low Temperature Physics: 24th International Conference on Low Temperature Physics;* edited by Y. Takano, S. P. Hershfield, S. O. Hill, P. J. Hirschfeld, and A. M. Goldman

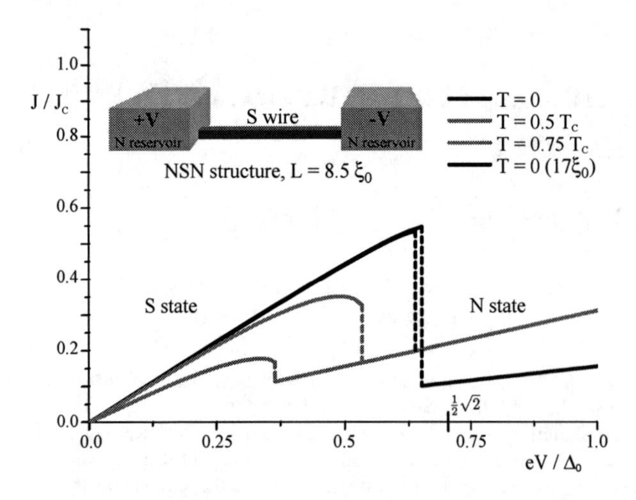

FIGURE 2. The calculated current(J)-voltage(V) relation of a superconducting wire of length $L = 8.5\xi_0$ between normal metallic reservoirs (see inset) at several temperatures, and for a wire of length $17\xi_0$ at $T = 0$. J_c is the critical current density, and Δ_0 the bulk gap energy.

solving the kinetic equations and then calculate a new Δ using the self-consistency relation. This process is repeated until Δ has converged. As a starting value for Δ we use the BCS form at zero temperature. A typical solution employs a grid of in the order of 10^4 energies, 10^2 spatial coordinates, and 10^3 iterations of Δ. The stability of the solution scheme was tested extensively by inserting different initial values. At all the applied voltages self-consistent steady state solutions are found.

In a previous analysis a finite differential conductance was found at zero bias employing a linear response calculation [9]. With the approach introduced here, the full current-voltage relation can be obtained. The result at several temperatures is displayed in Fig. 2, with the voltage normalized to $\Delta_0 (= \Delta_{\mathrm{bulk},T=0})$ and the current density normalized to the critical current density $J_c \approx 0.75 \frac{\Delta_0}{\xi_0 \rho e}$ [10], with $\xi_0 = \sqrt{\frac{\hbar D}{\Delta_0}}$. At T = 0 we observe a linear (universal) resistance at low voltages caused by the decay of f_T [11], and a critical point (voltage) above which the resistance is equal to the normal state resistance.

The current density at which the superconductor switches to the normal state (for T = 0) is much smaller than the critical current density in an infinitely long wire ($J/J_c = 1$). This means that it is acceptable to ignore the occurrence of phase slip centers [12], which are time-dependent solutions that are energetically favorable when $J \approx J_c$. In equilibrium at $T = 0$, a switch to the normal state can only be caused by reaching a critical phase gradient. However, in our case the normal state is reached before this phase gradient starts to play a role. In the presence of quasiparticles, Δ (and thus potentially the state of the system), is also influenced by the distribution functions.

We show explicitly that in our case the breakdown of the superconducting state is triggered by the longitudinal mode of the quasiparticle distribution function f_L. This is done numerically by making a small modification to the system, and analytically by studying the total internal energy of the electron system in the wire in the superconducting state respectively normal state as a function of voltage. This analysis shows that the breakdown can be considered as a first order phase transition.

The authors would like to thank Yuli Nazarov, Andrei Zaikin and Wim van Saarloos for critical and helpful discussions. This work is part of the research programme of the 'Stichting voor Fundamenteel Onderzoek der Materie (FOM)', which is financially supported by the 'Nederlandse Organisatie voor Wetenschappelijk Onderzoek (NWO)'.

REFERENCES

1. H. Pothier, S. Guéron, N. O. Birge, D. Esteve, and M. H. Devoret, *Phys. Rev. Lett.*, **79**, 3490 (1997).
2. F. Pierre, A. B. Gougam, A. Anthore, H. Pothier, D. Esteve, and N. O. Birge, *Phys. Rev. B*, **68**, 085413 (2003).
3. I. Siddiqi, A. Verevkin, D. E. Prober, A. Skalare, W. R. McGrath, P. M. Echternach, and H. G. LeDuc, *J. Appl. Phys.*, **91**, 4646 (2002).
4. J. J. A. Baselmans, A. Baryshev, S. F. Reker, M. Hajenius, J. R. Gao, T. M. Klapwijk, Y. Vachtomin, S. Maslennikov, S. Antipov, B. Voronov, and G. Gol'tsman, *Appl. Phys. Lett.*, **86**, 163503 (2005).
5. For a review, see Nikolai Kopnin, *Theory of nonequilibrium superconductivity*, vol. 110 of *International series of monographs on physics*, Oxford University Press, 2001, and references therein.
6. C. Lau, N. Markovic, M. Bockrath, A. Bezryadin, and M. Tinkham, *Phys. Rev. Lett.*, **87**, 217003 (2001).
7. A. D. Zaikin, D. S. Golubev, A. van Otterlo, and G. T. Zimányi, *Phys. Rev. Lett.*, **78**, 1552 (1997).
8. R. S. Keizer, M. G. Flokstra, J. Aarts, and T. M. Klapwijk, *cond-mat/0502435* (2004), submitted.
9. G. R. Boogaard, A. H. Verbruggen, W. Belzig, and T. M. Klapwijk, *Phys. Rev. B*, **69**, 220503(R) (2004).
10. A. Anthore, H. Pothier, and D. Esteve, *Phys. Rev. Lett.*, **90**, 127001 (2003).
11. S. Sachdev, P. Werner, and M. Troyer, *Phys. Rev. Lett.*, **92**, 237003 (2004).
12. J. S. Langer, and V. Ambegaokar, *Phys. Rev.*, **167**, 498 (1967).

Quantum Oscillations of the Critical Current of Asymmetric Aluminum Loops in Magnetic Field

V.L. Gurtovoi, S.V Dubonos, A.V. Nikulov, N.N. Osipov, and V.A. Tulin

Institute of Microelectronics Technology and High Purity Materials, Russia Academy of Sciences, 142432 Chernogolovka, Moscow Region, Russia. E-mail: nikulov@ipmt-hpm.ac.ru

Abstract. The periodical dependencies in magnetic field of the asymmetry of the current-voltage curves of asymmetric aluminum loop are investigated experimentally at different temperatures below the transition into the superconducting state $T < T_c$. The obtained periodical dependency of the critical current on magnetic field allows to explain the quantum oscillations of the dc voltage as a consequence of the rectification of the external ac current and to calculate the persistent current at different values of magnetic flux inside the loop and temperatures.

Keywords: Mesoscopic superconductor loop, persistent current, quantum oscillations.
PACS: 74.78.Na, 74.78.-w

INTRODUCTION

The persistent current $I_p = sj_p = s2en_sv_s$ should flow along circumference of a superconductor loop l with thin section $s < \lambda_L^2$ at $\Phi \neq n\Phi_0$ since the state with zero velocity of superconducting pairs $v_s = 0$ is forbidden when the magnetic flux Φ inside l is not divisible by the flux quantum $\Phi_0 = h/2e$ [1]. Its equilibrium value and sign $\langle I_p \rangle = s2en_s \langle v_s \rangle \propto \langle n \rangle - \Phi/\Phi_0$ vary periodically with Φ [1]. The Little-Parks resistance oscillations $R_l(\Phi/\Phi_0)$ [2] observed in the loop [3] is experimental evidence of $I_p \neq 0$ at non-zero resistance $R_l > 0$. According to an analogy with the conventional circular current a dc potential difference $V(\Phi/\Phi_0) \propto I_p(\Phi/\Phi_0) \propto \langle n \rangle - \Phi/\Phi_0$ may be expected to be observed on segments of asymmetric superconductor loop at $R_l > 0$. Such quantum oscillations of the dc voltage were observed on segment of asymmetric aluminum loops [4,5] and much before on a double Josephson point contact [6]. The dc voltage $V(\Phi/\Phi_0)$ observed near the critical temperature T_c [4,6] can be induced by switching of the loop between superconducting states with different connectivity [7,8]. The quantum oscillations $V(\Phi/\Phi_0)$ induced at lower temperatures by an external ac current [5] may by interpreted as a result of the rectification because of asymmetry of the current-voltage curves sign and value of which are periodical function of Φ. The results of measurements of this periodical change of the asymmetry of the current-voltage curves with value of magnetic flux Φ inside asymmetric aluminum loop at different temperatures are presented in this work.

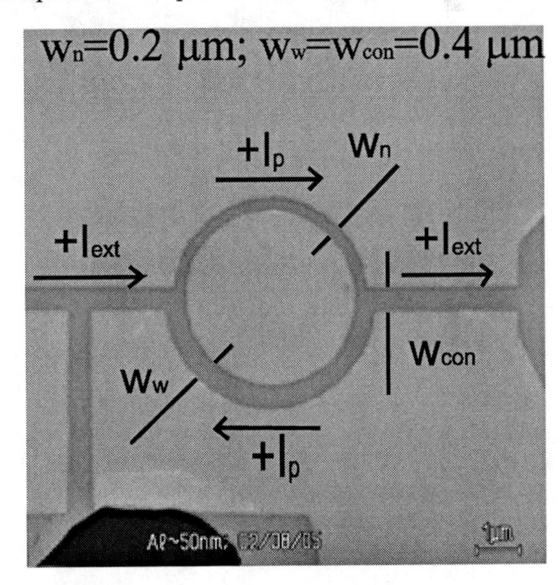

FIGURE 1. SEM image of a typical asymmetric aluminium rings.

EXPERIMENTAL DETAILS AND RESULTS

Microstructures consisting of asymmetric Al rings with semi-ring width w_n = 200 nm and w_w = 400 nm for the narrow and wide parts, respectively, see Fig.1,

CP850, *Low Temperature Physics: 24th International Conference on Low Temperature Physics;*
edited by Y. Takano, S. P. Hershfield, S. O. Hill, P. J. Hirschfeld, and A. M. Goldman

were investigated. 4 μm diameter single asymmetric superconductor ring (ASR) and 20 ASR structures, see Fig.1, were fabricated by e-beam lithography and lift-off process of film d = 45-50 nm in thickness, thermally evaporated on oxidized Si substrates. For these structures, the sheet resistance was 0.23 Ω/□ at 4.2 K, the resistance ratio R(300K)/R(4.2K)=2.7, and the critical temperature was T_c = 1.24-1.27 K.

Current -Voltage Curves

The structure as a whole jumps into the resistive state R > 0 (at a low temperature $T < 0.985T_c$) when the current density exceeds the critical value j_c in any of its segment and the irreversibility of the current-voltage curves is observed at $T < 0.99T_c$. The value of the external current I_{ext} corresponding to this jump to the state with R > 0 is measured as the critical current $|I_{ext}|_c$ of the structure.

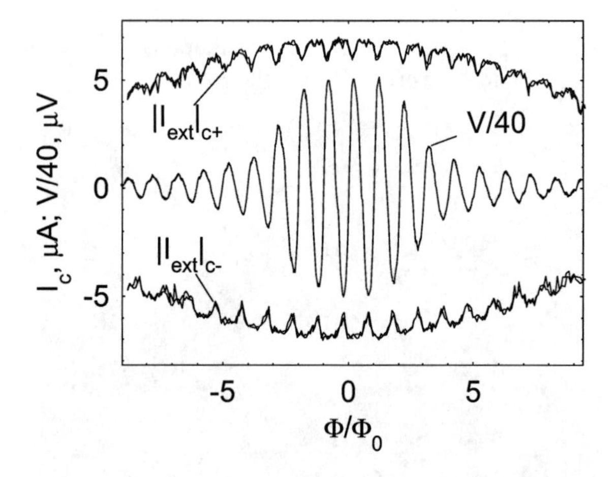

FIGURE 2. Quantum oscillations of the critical current $|I_{ext}|_c(\Phi/\Phi_0)$ of system of 18 asymmetric Al loops with w_{con} = 400 nm < $w_n + w_w$ measured in opposite directions $|I_{ext}|_{c+}$, $|I_{ext}|_{c-}$ at $T \approx 0.978T_c$. The quantum oscillations of the dc voltage $V(\Phi/\Phi_0)$ induced by the external ac current with the frequency f = 40 kHz and the amplitude 7 μA are shown also.

Periodical Dependence of $|I_{ext}|_c(\Phi/\Phi_0)$

Our measurements have revealed the periodical magnetic dependencies $|I_{ext}|_c(\Phi/\Phi_0)$ of both single rings and systems of rings with both $w_{con} < w_n + w_w$, see Fig.2, and $w_{con} \geq w_n + w_w$. These periodical dependencies may be explained as a consequence of superposition of the external I_{ext} and persistent $I_p(\Phi/\Phi_0)$ currents. The current density in the narrow j_n and wide j_w semi-rings is determined by both the I_{ext} and I_p currents: $j_n = I_{ext}/d(w_n + w_w) \pm I_p/dw_n$ and $j_w = I_{ext}/d(w_n + w_w) \pm I_p/dw_w$ [5]. The summation "+" takes place when the direct I_{ext} and circular I_p currents have the same direction in the semi-rings, see Fig.1. The current density mounts the critical value j_c first of all in the narrow semi-rings at $|I_{ext}|_c = d(w_n + w_w)(j_c - I_p/w_n)$, in the wide ones at $|I_{ext}|_c = d(w_n + w_w)(j_c - I_p/w_w)$ or in the stripes connecting the rings at $|I_{ext}|_c = dw_{con} j_c$ depending on the I_{ext} and I_p directions and the j_p/j_c and $w_{con}/(w_n + w_w)$ values.

Our results of the measurement of $|I_{ext}|_c(\Phi/\Phi_0)$ are evidence of the periodical dependence not only value but also sign of $I_p(\Phi/\Phi_0)$ since the $|I_{ext}|_c(\Phi/\Phi_0)$ value depends on the I_{ext} direction, $|I_{ext}|_{c+} \neq |I_{ext}|_{c-}$, i.e. the current-voltage curves are asymmetric, at some Φ/Φ_0 values, see Fig.2. For example $|I_{ext}|_{c+}$ (at the right I_{ext} direction) has a minimum value at $\Phi/\Phi_0 \approx 0.2 \pm n$ whereas $|I_{ext}|_{c-}$ (at the left I_{ext} direction) the minimum is observed at $\Phi/\Phi_0 \approx 0.8 \pm n$, see Fig.2. This means that the persistent current has the clockwise direction at $\Phi/\Phi_0 \approx 0.2 \pm n$ and the counter-clockwise one at $\Phi/\Phi_0 \approx 0.8 \pm n$ since the minimum of $|I_{ext}|_c$ is observed when the I_{ext} and I_p have the same direction in the narrow semi-rings, see Fig.1. The experimental dependencies $|I_{ext}|_{c+}(\Phi/\Phi_0)$, $|I_{ext}|_{c-}(\Phi/\Phi_0)$ obtained in our work allow to explain the quantum oscillations of the dc voltage $V(\Phi/\Phi_0)$ as consequence of the rectification and to calculate the $I_p(\Phi/\Phi_0;T)$ dependencies at $T < T_c$.

ACKNOWLEDGMENTS

This work has been supported by a grant of the Program "Low-Dimensional Quantum Structures", the Russian Foundation of Basic Research (Grant 04-02-17068) and a grant of the program "Technology Basis of New Computing Methods".

REFERENCES

1. M. Tinkham, *Introduction to Superconductivity,* New York: McGraw-Hill Book Company, 1975.
2. W.A. Little and R.D. Parks, *Phys. Rev. Lett.* **9**, 9 (1962).
3. H. Vloeberghs et al., *Phys. Rev. Lett.* **69**,1268 (1992).
4. S.V. Dubonos, V.I. Kuznetsov, and A.V. Nikulov, in *Proceedings of 10th International Symposium "NANOSTRUCTURES: Physics and Technology"*, St Petersburg: Ioffe Institute, 2002, pp. 350-354; e-print arXiv: cond-mat/0305337.
5. S.V. Dubonos et al., *Pisma Zh. Eksp. Teor. Fiz.* **77**, 439 (2003) *(JETP Lett.* **77**, 371 (2003)).
6. A. Th. A. M. de Waele et al., *Physica* **37**, 114 (1967).
7. A.V. Nikulov and I.N. Zhilyaev, *J. Low Temp. Phys.* **112**, 227-236 (1998).
8. A.V. Nikulov, *Phys. Rev. B* **64**, 012505 (2001).

Transport Properties of a Quantum Wire - Superconductor Hybrid Junction

Nobuhiko Yokoshi and Susumu Kurihara

Department of Physics, Waseda University, 3-4-1 Ohkubo, Shinjuku, Tokyo 169-8555, Japan

Abstract. We theoretically investigate transport properties of a one-dimensional quantum wire (QW) connected with superconducting reservoirs, where the carriers in the QW are assumed to behave as a Tomonaga Luttinger liquid. With use of functional bosonization procedure, the system can be regarded as usual normal metal-superconductor junction in fluctuating 'environment' induced by electron-electron interactions. We show that this 'environment' suppresses the Andreev reflection processes at interfaces when the QW is adiabatically connected with the superconductors.

Keywords: Josephson effect, Andreev reflection, Tomonaga Luttinger liquid
PACS: 74.50.+r, 71.10.P

INTRODUCTION

Recently, proximity-induced superconductivity in one-dimensional conductors such as multi walled carbon nanotubes [1.2] and quantum wires (QWs) on semiconductors is investigated experimentally. It is expected that those systems can be theoretically examined by studying Josephson current through them, since coherent multiple Andreev reflection (MRA) processes simulate the dynamics of Cooper pairs. On the other hand, in one-dimensional configuration, the carriers are believed to behave as a Tomonaga Luttinger liquid where the Coulomb interactions play crucial roles in transport properties. Here, we study the effects of electron-electron interactions on the MRAs and the transport properties of hybrid structure composed of a QW and superconductors.

MODEL

The system under consideration is a superconductor-quantum wire-superconductor (S-QW-S) junction where the QW and superconductors are adiabatically connected. Order parameters of the superconductors have a constant amplitude with conventional s-wave symmetry. As for the QW region, we assume that Hamiltonian can be written with use of usual g-ology

$$H_{\rm QW} = H_0 + H_{\rm int}\,, \tag{1}$$

where $H_0=\Sigma_{a,s}\, a\hbar v_{\rm F}(-i\partial_x)$ is the linearized Hamiltonian for free electrons. The subscripts $a,s=\pm1$ represent directions of propagation and spin indices, and $v_{\rm F}$ is Fermi velocity. For the sake of simplicity, back scattering and umklapp scattering processes are neglected. By applying a so-called Hubbard-Stratonovich transformation to the forward scattering terms, the action is given by

$$S[\varphi] = \int dx\, dt\, [L_0(\psi,\psi^\dagger) + L_1(\varphi) + \sum_{a,s} \varphi_{a,s}\, \rho_{a,s}], \tag{2}$$

where $L_0(\psi,\psi^\dagger)$, $L_1(\varphi)$ are the actions for the free electrons $\psi_{a,s}$ and the auxiliary fields $\varphi_{a,s}$, and $\rho_{a,s}$ are the electron densities. Then, one can regard excitations in the QW as the free electrons propagating in a kind of fluctuating 'environment'. We treat $\varphi(x,t)$ as local scholar potential for a time, and take average of it with respect to the fluctuations $S_1=\int dx\, L_1$ [3].

The Eilenberger equation in this system can be written in (4×4) matrix form; it is

$$i\hbar v_{\rm F}\frac{\partial}{\partial x}\hat{g}^{\rm j} + \left[\left((i\hbar\frac{\partial}{\partial t}\hat{\sigma}_z + \varphi(x,t))\hat{\tau}_z + \hat{\Delta}(x)\right), \hat{g}^{\rm j}\right]_- = 0\,, \tag{3}$$

$$\hat{\sigma}_z = \tilde{1}\otimes\tilde{\sigma}_z\,,\ \hat{\tau}_z = \tilde{\sigma}_z\otimes\tilde{1}\,,\ \hat{\Delta}(x) = \tilde{\sigma}_z\otimes\tilde{\Delta}(x)\,, \tag{4}$$

$$\tilde{\Delta}(x) = \begin{pmatrix} 0 & \Delta \\ -\Delta^* & 0 \end{pmatrix}, \tag{5}$$

where [,]- stands for taking commutation as well as internal time integration. The first and the third row refer to right and left moving electrons with spin up, and the second and the forth rows to left and right moving holes with spin down. Also, j={R,A,K} specifies the retarded, advanced and Keldysh Green's function. The quantities like ô denote (4×4) matrices

CP850, *Low Temperature Physics: 24th International Conference on Low Temperature Physics;*
edited by Y. Takano, S. P. Hershfield, S. O. Hill, P. J. Hirschfeld, and A. M. Goldman

and those like õ (2×2) matrices. Here, $\tilde{\sigma}_i$'s are usual Pauli matrices. We assume that superconducting gap energy Δ(x) changes abruptly, and neglect the proximity effect into the QW. Therefore, the total Green's function is determined by matching the formal solutions of Eq. (3) at interfaces between the QW and the superconductors.

RESULTS AND DISCUSSION

In stationary limit, it is straightforward to calculate the Green's function. If the QW's length L is larger than coherence length of the superconductors, the Keldysh Green's function has poles, which correspond to Andreev bound states in the QW, at

$$E_{a,s} = \frac{\hbar v_F}{2L}\left((2n+1)\pi + a\chi + \Phi_{a,s}\right), \quad (6)$$

$$\Phi_{a,s} = \theta_{a,s}(\frac{L}{2},t) + \theta_{-a,-s}(\frac{L}{2},t) - \theta_{a,s}(-\frac{L}{2},t) - \theta_{-a,-s}(-\frac{L}{2},t), \quad (7)$$

with χ being the phase difference between two superconductors. In Eq.(7), θ's are gauge fields with respect to the auxiliary fields, which satisfies the relation $\hbar(\partial_t + v_F\partial_x)\theta_{a,s} = \varphi_{a,s}$ [3]. Equation (6) implies that the Andreev bound states become ill-defined by the 'environment' characterized at the boundaries, when the Coulomb interactions are relevant. Indeed the Green's functions for the process with m Cooper pairs tunneling, are renormalized as

$$\hat{g}_m^i = \Lambda^{m^2} \hat{g}_{0,m}^i, \quad (8)$$

where $\hat{g}_0$ is the Green's function with no interactions. At zero temperature, one can see that $\Lambda = (\hbar v_F/DL)^{\lambda}$, where $\lambda = K_\rho^{-1} - 1$, and D is ultraviolet cut-off [4,5]. Here, $K_\rho = (1 + U/\pi\hbar v_F)^{-1/2}$ is usual Luttinger parameter, where $U > 0$ represents the strength of the Coulomb interactions. Thus, the 'environment' counteracts the phase coherence and suppresses the critical currents.

When finite bias eV is applied to the system, the Green's function is a combination of harmonics with the period of the Josephson oscillation $T_J = \pi\hbar/eV$ [6]. We can set $\chi = 0$ without loss of generality. Because of the phase shift $\Phi_{a,s}$ in Green's functions, one can find that the MRA processes are inhibited by the interactions as in stationary case. For example, taking the average with respect to the fluctuations, the retarded Green's function of m-th harmonics at the left interface becomes

$$\hat{g}_m^R(t=t') \propto 2\Lambda^{m^2} e^{i2meVt/\hbar} \int d\varepsilon d\varepsilon' \prod_{l=1}^{m} e^{i2(\varepsilon+(2l-1)eV)L/\hbar v_F} \times \prod_{l=1}^{2m} \gamma(\varepsilon + leV)\, \delta(\varepsilon - \varepsilon' + 2meV), \quad (9)$$

where $\gamma(\varepsilon)$ is bare Andreev reflection amplitude of the electrons with energy ε. On the other hand, $\Phi_{a,s}$ does not appear in the term proportional to $\delta(\varepsilon-\varepsilon')$. It leads to the well-known result that the process with no Andreev reflection is not influenced by the electron-electron interactions [7]. As a result, under the influence of the strong electron-electron interactions, current-voltage characteristics approach the linear one even at small current regime.

CONCLUSION

In summary, we investigated effects of Coulomb interactions on Andreev reflections in S-QW-S junctions. It was found that the repulsive interactions make the Andreev reflections irrelevant in both DC and AC Josephson effect. Here, we have considered the case umklapp scattering is irrelevant. If the umklapp scattering process becomes relevant, the charge is carried by the kink or anti-kink. The transport properties in such a case may be related with physics in Mott type insulator. We will address this situation in a future publication.

ACKNOWLEDGMENTS

We thank Kenji Kamide for significant comments and discussions. This work was partially supported by The 21st Century COE Program (Holistic Research and Education Center for Physics of Self-Organization System) at Waseda University from the Ministry of Education, Sports, Culture, Science and Technology of Japan.

REFERENCES

1. A. Y. Kasumov *et al.*, *Science*, **294**, 1508 (1999).
2. J. Haruyama, A. Tokita, N. Kobayashi, M. Nomura, and S. Miyadai, *App. Phys. Lett.* **84**, 4714 (2004).
3. D. Lee and Y. Chen, *J. Phys. A*, **21**, 4155 (1988).
4. R. Fazio, F. W. J. Hekking, and A. A. Odintsov, *Phys. Rev. Lett.*, **74**, 263 (1995).
5. N. Yokoshi and S. Kurihara, *Phys. Rev. B*, **71**, 104512 (2005).
6. U. Gunsenheimer and A. D. Zaikin, *Phys. Rev. B*, **50**, 6317 (1994).
7. S. Tarucha, T. Honda, and T. Saku, *Solid State Commun.*, **94**, 413 (1995).

Current-Phase Relation of Fully Characterized Atomic-Size Contacts

Zhenting Dai and Alexei Marchenkov

School of Physics, Georgia Institute of Technology, Atlanta, GA 30332, U.S.A.

Abstract. We demonstrate a novel device geometry for experiments in superconducting atomic-size contacts (ASC) that allows to measure both their current-voltage ($I-V$) and current-phase ($I(\phi)$) characteristics. A mechanically controlled break junction (MCBJ) has been incorporated into a composite superconducting loop. By adjusting the temperature, the device can be switched between two states: a resistively shunted superconducting ASC and an ac SQUID. In the first configuration, we can measure transport properties of the ASC, including individual transmission coefficients of the contact's conductance channels [1, 2]. In the second configuration, the current-phase relation of the junction can be measured using the phase-bias technique [3].

Keywords: superconducting point contacts, current-phase relation
PACS: 73.63.Rt, 74.78.Na, 85.35.Ds

The dependence of the supercurrent I of a superconducting weak link on the phase difference ϕ of the order parameter is a basic characteristic of the weak link and usually called the current-phase relationship. In the tunnel junction, the behavior $I \propto \sin(\phi)$ was first predicted by Josepshon in 1962 [4]. In the limit of low temperatures, for a single conduction channel point contact with the normal transmission τ several authors using different techniques predicted [5]:

$$I(\phi) = \frac{e\tau\Delta}{2\hbar} \frac{\sin(\phi)}{\sqrt{1-\tau\sin^2(\frac{\phi}{2})}} \quad (1)$$

where Δ is the superconducting gap. This formula interpolates between the tunnel limit and the perfect transmission limit ($\tau = 1$), where $I \propto \sin(\phi/2)$ behavior is expected. In the pioneering experiment, Koops *et al.* [6] measured $I(\phi)$ of niobium ASC; however, they could not determine both the number of conductance channels and their individual transmission coefficients. Due to a number of experimental challenges, verification of Eq. (1) remains an outstanding experimental problem.

In this paper, we describe a thin-film device in which both the full characterization of transport properties of ASC and measurements of the current-phase relation are possible. The device's geometry and operation principle are illustrated in Fig. 1. Most of the composite loop is made of niobium whose superconducting transition temperature is higher than of tantalum. In thin films, critical temperatures are reduced compared to those of the bulk materials. In our device, we achieved $T_c(Nb) = 7.6$ K and $T_c(Ta) = T_c(shunt) = 4.2$ K (Fig. 2). The configuration and the atomic structure of the niobium MCBJ can be altered mechanically between completely opened ($R_{MCBJ} > 1$ GΩ) and completely closed

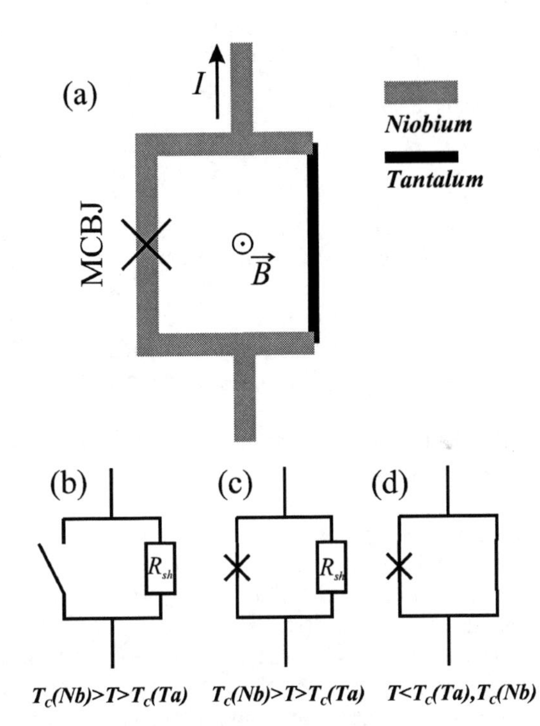

FIGURE 1. Schematics and operation principles of the device. (a) The device is made of two superconducting materials with different critical temperatures. (b) MCBJ open, shunt in the normal state. (c) ASC transport measurement configuration. (d) The SQUID configuration.

($R_{MCBJ} \sim 10\ \Omega$). Initially, the contact can be opened and the resistance (R_{sh}) and the transition temperature of the shunt can be measured (Fig. 1b and Fig. 2). By setting the temperature between $T_c(Nb)$ and $T_c(Ta)$ and closing the junction, we obtain a resistively shunted ASC.

CP850, *Low Temperature Physics: 24th International Conference on Low Temperature Physics;*
edited by Y. Takano, S. P. Hershfield, S. O. Hill, P. J. Hirschfeld, and A. M. Goldman

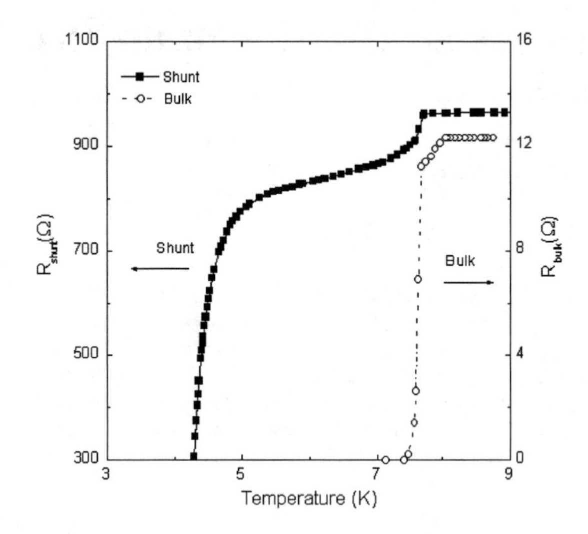

FIGURE 2. Superconducting transitions in the device. With the junction closed, we first observe the "bulk" niobium transition (configuration shown in Fig. 1c). The shunt properties can be measured with the open junction (configuration shown in Fig. 1b).

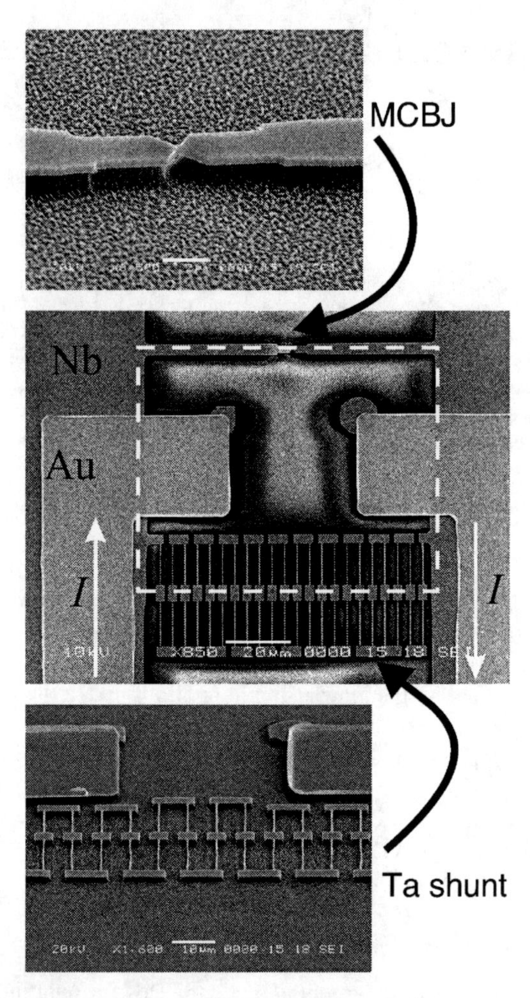

FIGURE 3. SEM images of the device. Metallic structures are deposited on top of a thin polymide film covering a flexible bronze substrate. Details of the fabrication procedure will be published elsewhere. (top) The break junction area. The white horizontal scale bar is 2 μm. (middle) The effective SQUID loop is outlined by the dashed line ; its loop's total area is $\sim$ 100 μm$\times$100 μm. The bottom part of the loop is the tantalum shunt. Its room temperature resistance is $\sim$ 2700 Ω; just above the superconducting transition the resistance is $\sim$ 850 Ω (Fig. 2). Current can be injected into the device through two gold leads; another pair of leads (not visible in the picture) is used for voltage measurements. (bottom) Detailed view of the tantalum shunt. In order to keep the shunt from breaking on a flexible substrate, stretches of thin tantalum wire are interrupted by rectangular pads.

Knowing (R_{sh}), the MCBJ can be adjusted to the desired resistance, usually corresponding to one or two conductance quanta in order to minimize the number of conductance channels. In a separate paper, we describe the details of transport measurements in this configuration [2]. At temperatures below $T_c(Ta)$, the device becomes an ac SQUID.

In the device described in this paper, the resistance of the shunt is too low to reliably determine the composition and the set of transmission coefficients of the ASC [2]. One of the problems may be associated with niobium residuals in the shunt, which reduces its resistance. We plan to enhance the manufacturing procedure and increase the shunt resistance to several kΩ. This will be achieved by adding the energy dispersive x-ray spectroscopy (EDS) microanalysis of the junction and the shunt areas to our manufacturing procedure.

This research is supported by the Georgia Institute of Technology through the Nanoscience/Nanoengineering Research Program (NNRP) and the National Science Foundation CAREER Grant No. DMR - 0349110.

REFERENCES

1. E. Scheer, P. Joyez, D. Esteve, C. Urbina, and M. H. Devoret, *Phys. Rev. Lett.* **78**, 3535–3538 (1997).
2. Z. Dai, B. Donehoo, J. Howard, and A. Marchenkov, *these proceedings*.
3. E. de Wolff, H. C. Roobeek, M. C. Koops, and R. de Bruyn Ouboter, *Physica B* **179**, 295–308 (1992).
4. B. Josephson, *Phys. Lett.* **1**, 251–253 (1962).
5. W. Haberkorn, H. Knauer, and J. Richter, *Phys. Stat. Sol.* **47**, K161-K164 (1978). G. B. Arnold, *J. Low Temp. Phys.* **59**, 143-183 (1985). C. W. J. Beenakker and H. van Houten, *Phys. Rev. Lett.* **66**, 3056–3059 (1992). A. Martín-Rodero, F. j. García-Vidal, and A. Levy Yeyati, *Phys. Rev. Lett.* **72**, 554-557 (1994).
6. M. C. Koops, G. V. van Duyneveldt, and R. de Bruyn Ouboter, *Phys. Rev. Lett.* **77**, 2542–2545 (1996).

Properties of Superconducting Atomic-Size Contacts at Finite Temperatures

Zhenting Dai, Brandon Donehoo, John Howard and Alexei Marchenkov

School of Physics, Georgia Institute of Technology, Atlanta, GA 30332, U.S.A.

Abstract. We investigate properties of microfabricated superconducting atomic-size contacts (ASC) made of transition metals. Typically, the smallest transition metal ASC have the conductance of the order of two conductance quanta, $G_0(G_0^{-1} = 12906\ \Omega)$, and accommodate up to five conduction channels [1], whose transmission coefficients, $\tau_1, ..., \tau_5$, can be determined from the analysis of the subgap structure due to the multiple Andreev reflections (MAR) in the current-voltage $(I-V)$ characteristics [2]. We determined the transmission coefficients of a niobium ASC in a wide temperature range and found them temperature-independent. In addition, we observed the signature features of MAR in the current-voltage characteristics of a *resistively shunted* superconducting ASC.

Keywords: superconducting point contacts, multiple Andreev reflections
PACS: 73.63.Rt, 74.78.Na

Physics of atomic-size contacts has received a lot of attention in the last decade both for the fundamental and practical reasons (see comprehensive review by Agraït, Levy Yeyati, and van Ruitenbeek [3]). Particular attention was paid to ASC made of superconducting materials. Presently, the accepted point of view is that in the superconducting state and at small voltages the transport through ASC occurs through multiple Andreev reflection (MAR) processes. Due to the presence of MAR, the highly non-linear current-voltage curves contain detailed information on the different properties of superconducting ASC. In particular, Scheer *et al.* [2] demonstrated that the number and individual transmission coefficients of conductance channels can be extracted experimentally. Following their procedure, we assume that the total current through the junction is the sum of currents flowing through individual channels:

$$I(V) = \sum_{j=1}^{N} i(V, \tau_j, \Delta)) \qquad (1)$$

where N is the number of conductance channels in the ASC, Δ is the superconducting gap of the metallic sample, and $i(V, \tau_j, \Delta)$ is the current-voltage characteristic of a single channel with the transmission coefficient τ_j. For a given temperature, we generate these model $i(V, \tau_j, \Delta)$ curves numerically using the code developed by Ake Ingerman and Jonn Lantz, Chalmers University, based on the theoretical model by Bratus', Shumeiko, and Wendin [4]. Consequently, our fitting procedure, in essence, selects a set of five model curves with different transmission coefficients, $\tau_1, ..., \tau_5$, which minimizes the χ^2 deviation between the experimental data and the curve obtained using Eq. (1).

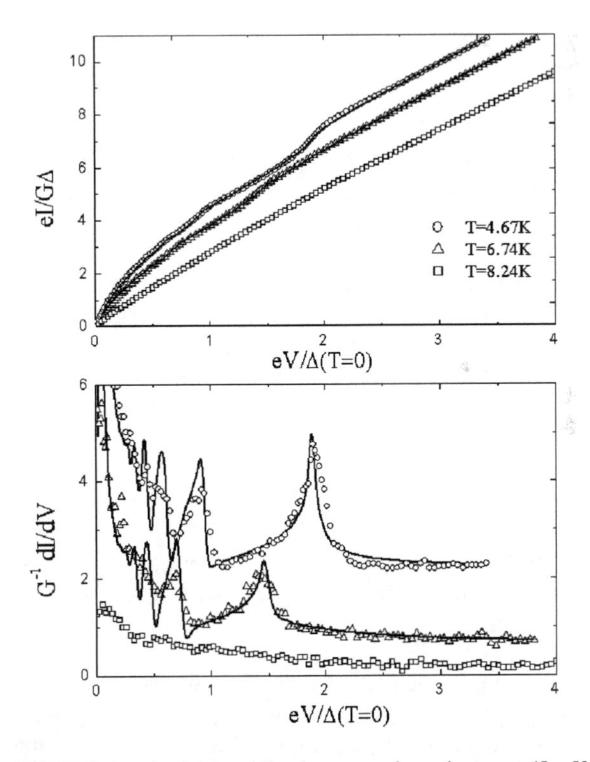

FIGURE 1. (top) Normalized measured conductance ($I-V$ curves) of a niobium contact with the total conductance of $\sim 2.2\ G_0$. Solid lines represent best fits using Eq. (1) with five conductance channels. The gap, Δ, was estimated using the superconducting transition temperature of the niobium thin film. (bottom) Normalized differential conductance (dI/dV) calculated from the same data. For clarity, $T = 6.74\ K$ results are shifted vertically by -1.5, and $T = 8.24\ K$ results are shifted by -2.0. Near T_c, the MAR peaks are not clearly distinguishable.

CP850, *Low Temperature Physics: 24th International Conference on Low Temperature Physics;*
edited by Y. Takano, S. P. Hershfield, S. O. Hill, P. J. Hirschfeld, and A. M. Goldman

Examples of the superconducting $I-V$ characteristics are shown in Figure 1. Atomic-size contacts were produced by the mechanically controlled break junction (MCBJ) technique. The measurements were performed on niobium ASC in the temperature range between 2.4 K and 8.4 K ($\sim T_c$ of the Nb film). For temperatures up to $0.8T_c$, we find a satisfactory agreement of experimental data and theoretical fits (Fig. 1). In this range, our calculations show that the transmission coefficients of the junction's five conductance channels are temperature-independent within our fitting precision: $\tau_1 = 0.97, \tau_2 = 0.77, \tau_3 = 0.33, \tau_4 = 0.10, \tau_5 = 0.04$. We estimate the precision in definition of each of these coefficients to be ± 0.02. When the temperature is increased, the subgap structure is smeared out due to the thermal activation of quasiparticles and the positions of MAR resonances are shifted due to the reduction of the superconducting gap. Above $0.8T_c$, our fitting procedure is not sufficiently sensitive.

In a separate experiment, we measured transport properties of atomic-sized MCBJ shunted by a normal resistance of the order of 850 Ω. The composite niobium-tantalum device where these measurements were performed is described in the paper by Dai and Marchenkov [5]. In Figure 2, we show experimental results for three junctions with different atomic configurations. The device with the resistance of 8.2 $k\Omega$ should correspond to the tunnel junction, since transition metals ASC usually have the lowest conductance above $2.0G_0$ (corresponds to ~ 6 $k\Omega$) before breaking into the tunnel regime [1].

All the measurement were performed at the temperature of 5.4 K. If we associate the reproducible MAR peak in the differential conductance (Fig. 2 (bottom)) with the first-order process, its position, ~ 1.3 mV, is consistent with the superconducting gap of tantalum. Due to the multilayer manufacturing procedure, in this particular device we anticipated that ASC could be made of either niobium or tantalum. In the future, we plan to enhance the sample manufacturing procedure and eliminate this ambiguity.

In summary, we have measured and analyzed multiple Andreev reflection processes in straight and resistively shunted superconducting atomic-size contacts in a wide temperature range. We have verified that the normal transmission properties remain temperature-independent. In future, we plan to use this circumstance in the measurement of the current-phase relation of fully characterized superconducting atomic-size junctions [5].

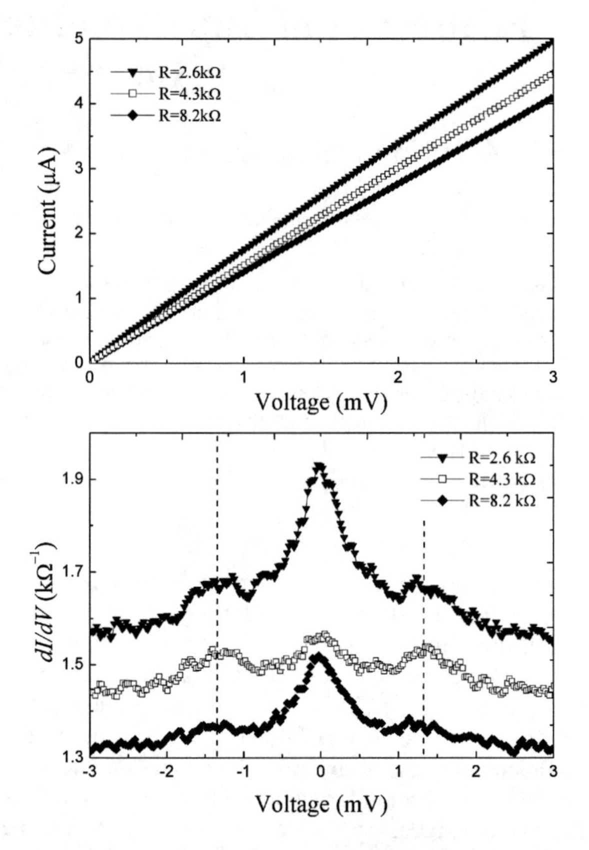

FIGURE 2. Conductance (top) and differential conductance (bottom) measurements performed on three different resistively shunted ASC. We associate the clearly reproducible feature in the differential conductance (vertical dashed lines) with the first-order MAR process. Higher-resistance shunt will result in more sensitive measurement which will allow to clearly distinguish higher-order MAR features and determine the transmission coefficients of resistively shunted ASC.

We thank Dr. Michael Pustilnik for valuable discussions. We are indebted to Dr. Vitaliy Shumeiko and Dr. Ake Ingerman of Chalmers University for providing us the code for calculations of transport properties of mesoscopic contacts. This research is supported by the Georgia Institute of Technology through the Nanoscience/Nanoengineering Research Program (NNRP) and the National Science Foundation CAREER Grant No. DMR - 0349110.

REFERENCES

1. B. Ludoph, N. van der Post, E. N. Bratus, E. V. Bezuglyi, V. S. Shumeiko, G. Wendin, and J. M. van Ruitenbeek, *Phys. Rev. B* **61**, 8561–8569 (2000).
2. E. Scheer, P. Joyez, D. Esteve, C. Urbina, and M. H. Devoret, *Phys. Rev. Lett.* **78**, 3535–3538 (1997).
3. N. Agraït, A. Levy Yeyati, and J. M. van Ruitenbeek, *Physics Reports* **377**, 81–279 (2003).
4. E. N. Bratus', V. S. Shumeiko, and G. Wendin, *Phys. Rev. Lett.* **74**, 2110–2113 (1995). V. S. Shumeiko, E. N. Bratus', and G. Wendin, *J. Low. Temp. Phys.* **23**, 181-195 (1997).
5. Zhenting Dai and Alexei Marchenkov, *these proceedings*.

Current-Phase Relation of a Well-Characterized Superconducting Atomic Point Contact

Hisao Miyazaki*, Yamaguchi Takahide†, Akinobu Kanda* and Youiti Ootuka*

*Institute of Physics and TIMS, University of Tsukuba, Tsukuba 305-8571, Japan
†National Institute for Materials Science, Tsukuba 305-0047, Japan

Abstract. According to theories of a superconducting quantum point contact (QPC), Cooper pairs are transferred via a few channels which are fully characterized by a set of transmission coefficients $\{\tau_i\}$. Indeed, the *I-V* characteristics and the current-phase relation $I(\varphi,\tau)$ of a single-channel QPC have been calculated as a function of the coefficient τ. However, the experimental quantitative verification has not been done yet. In this paper, we propose an experiment in which both the transmission coefficients $\{\tau_i\}$ and $I(\varphi,\{\tau_i\})$ relation can be determined independently.

Keywords: superconducting atomic point contact, current-phase relation
PACS: 73.63.-b, 74.50.+r

INTRODUCTION

A superconducting quantum point contact (QPC) is a test bed to investigate Josephson effect with arbitrary transmission coefficient τ, and its theoretical descriptions are expected to give unified representation of short Josephson junctions. An atomic point contact (APC) is thought to be a realization of the QPC. Indeed, Scheer et al. [1] demonstrated that a voltage branch of *I-V* characteristics of superconducting APC is explained well by expanding a theoretical prediction about a single channel QPC to a multi channels one, where it was assumed that the total current via the multi channels was given by the simple summation of the current via the channels. Namely, one can determine a set $\{\tau_i\} = \{\tau_1, \tau_2, \cdots \tau_N\}$ by taking it as a fitting parameter, where τ_i is the transmission probability of *i*-th channel.

The current-phase relation $I(\varphi)$ is another fundamental property of a Josephson junction. Koops et al. [2] determined the $I(\varphi)$ by measuring the self-induced flux of a superconducting loop closed by an APC as a function of external magnetic field. They made APCs using mechanical break junction (MBJ) technique. This $I(\varphi)$ was distinctly different from the sinusoidal one for tunnel junctions. Theoretically, taking τ as a parameter, $I(\varphi)$ of a single-channel QPC in thermal equilibrium at temperature T is predicted as

$$I(\varphi,\tau) = \frac{e\Delta^2\tau}{2\hbar}\frac{\sin\varphi}{\varepsilon(\varphi,\tau)}\tanh\frac{\varepsilon(\varphi,\tau)}{k_B T}, \quad (1)$$

$$\varepsilon(\varphi,\tau) = \Delta\sqrt{1-\tau\sin^2(\varphi/2)}, \quad (2)$$

where Δ is the superconducting energy gap [3, 4]. The amplitude of $I(\varphi)$ measured by Koops et al. was larger than that of Eq. (1) for $\tau = 1$, a fact which favors the existence of multiple channels to explain the experimental results. Thus, the existence of multiple channels is necessary to explain the experimental results. However, $I(\varphi)$ do not have marked structure to know $\{\tau_i\}$. In fact, Koops et al. estimated only some kind of averaged transmission probability. If the *I-V* curve of the APC could be measured, they could determine $\{\tau_i\}$ following to the method by Scheer et al. [1]: The *I-V* curve was not measurable because the APC was shorted with a superconductor. As stated above, the quantitative evaluation of Eq. (1) was yet to be done. In this paper, we report an experimental study to determine $I(\varphi)$ of well-characterized superconducting APC.

FIGURE 1. SEM image of the dc-SQUID consisting of an MBJ and a tunnel junction.

EXPERIMENTAL

We fabricated a dc-SQUID type device shown in Fig. 1. It consists of a tunnel Josephson junction and a superconducting MBJ which can become an APC. The char-

CP850, *Low Temperature Physics: 24th International Conference on Low Temperature Physics;*
edited by Y. Takano, S. P. Hershfield, S. O. Hill, P. J. Hirschfeld, and A. M. Goldman

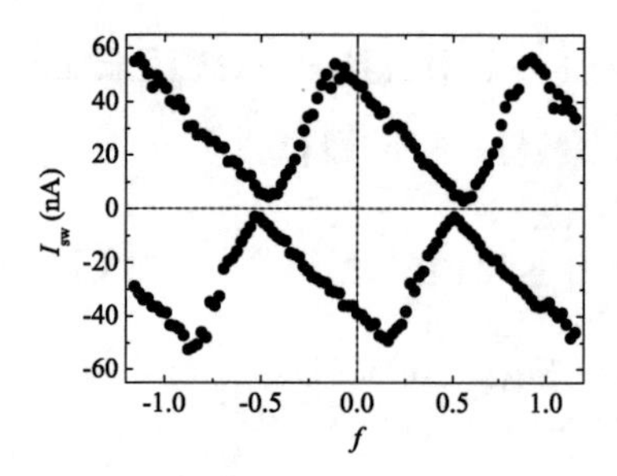

FIGURE 2. Switching current as a function of the magnetic flux $f = \Phi/\Phi_0$ for the APC with $\{\tau_i\} = \{0.97, 0.48, 0.10, 0.09\}$ at 95 mK.

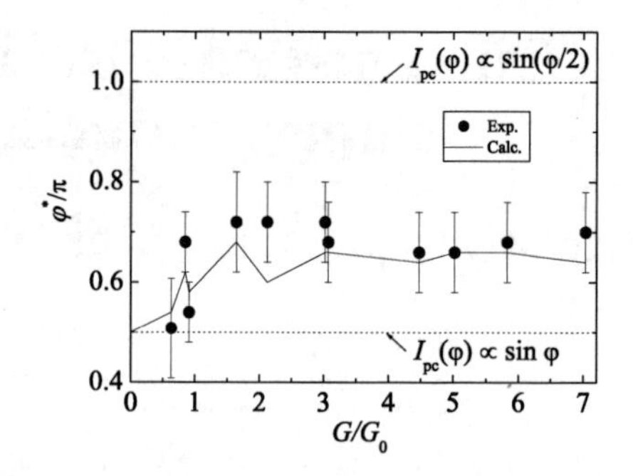

FIGURE 3. Phase φ^* that maximizes $I_{pc}(\varphi)$ as a function of the conductance of the APC. Closed circles are obtained from I_{sw}-f relation, and the solid line is the calculated one using Eq. (1).

acteristics of the tunnel junction can be determined by breaking the MBJ completely. If the loop inductance of the SQUID is small enough, the SQUID can be regarded as one Josephson junction with the phase difference $\varphi = (\varphi_{pc} + \varphi_{tj})/2$, where φ_{pc} and φ_{tj} are the phase difference across the APC and the tunnel junction respectively. When the magnetic flux Φ is applied in the SQUID loop, the supercurrent through the SQUID is given by $I_{sq}(\varphi, f) = I_{pc}(\varphi + \pi f) + I_{tj}(\varphi - \pi f)$, where $I_{pc}(\varphi_{pc})$ and $I_{tj}(\varphi_{tj})$ are the current phase relation of the APC and the tunnel junction respectively, and $f \equiv \Phi/\Phi_0$ ($\Phi_0 \equiv h/2e = 2.1 \times 10^{-15}$ Wb: superconducting magnetic flux quantum). If $I_{pc}(\varphi_{pc})$ is non-sinusoidal, maximum supercurrent $I^{\pm}_{max}(f) \equiv \pm\max\{\pm I_{sq}(\varphi, f) | 0 \le \varphi \le 2\pi\}$ as a function of f should become asymmetric with respect to the magnetic field inversion. Thus, in principle, we can determine $I_{pc}(\varphi)$ from $I_{max}(f)$. On the other hand, taking advantage of the strong suppression of the sub-gap current in the tunnel junction, we can determine τ_i of the APC from the *I*-*V* curve following the method by Scheer et al.[5]. In this way, we can discuss the $I(\varphi)$ of the well-characterized APC quantitatively.

We measured the *I*-*V* characteristics and the switching current I_{sw} as a function of the magnetic field. A typical example of I_{sw} as a function of f is shown in Fig. 2. We see asymmetry with respect to the magnetic field inversion with clarity. The asymmetry is an evidence of the non-sinusoidal $I(\varphi)$ of the APC. If $I_{pc}(\varphi)$ takes the maximum at $\varphi = \varphi^*$, the switching current I_{sw} of the SQUID is largest at $f = n - 1/4 + \varphi^*/2\pi$ (n is an integer). Thus we can obtain φ^* from I_{sw}-f relation. In Fig. 3, φ^* is plotted as a function of the dimensionless conductance $G/G_0 = \sum_i \tau_i$ ($G_0 \equiv 2e^2/h = 7.75 \times 10^{-5}$ S) of the APC. Using $\{\tau_i\}$ determined from the *I*-*V* curve, we calculate φ^* that is shown as the solid line in the figure. Here, we assume that the supercurrent through the APC is a simple sum of the contribution of the each conduction channel expressed by Eq. (1). As G increases, φ^* deviates from $\pi/2$ and saturates around 0.7π, which agrees well with the calculation. If all the open channels are ballistic ($\tau_i = 1$ or 0), $I_{pc}(\varphi)$ is proportional to $\sin(\varphi/2)$ and φ^* becomes π. The saturation at around 0.7π suggests the existence of non-ballistic channels even for $G/G_0 > 1$ as reported by Scheer et al. [1].

In conclusion, we investigated $I(\varphi)$ of the superconducting APC by examining the transport properties of a dc-SQUID which consisted of an APC and a tunnel junction. When the conductance increases, the *I*-φ relation of the APC deviates from the sinusoidal one. The theory based on the Andreev bound states explains the phase that maximizes the supercurrent through the APC. Detailed analysis of the experiment will be published [6].

ACKNOWLEDGMENTS

One of the authors (H.M.) acknowledges the support of the research fellowship from the Japan Society for the Promotion of Science for Young Scientists. This work was supported in part by a Grant-in-Aid for Scientific Research from the Japan Ministry of Education, Culture, Sports, Science and Technology.

REFERENCES

1. E. Scheer, P. Joyez, D. Esteve, C. Urbina, and M. Devoret, *Phys. Rev. Lett.*, **78**, 3535 (1997).
2. M. Koops, G. van Duyneveldt, and R. de Bruyn Ouboter, *Phys. Rev. Lett.*, **77**, 2542 (1996).
3. C. W. J. Beenakker, *Phys. Rev. Lett.*, **67**, 3836 (1991), *ibid.* **68**, 1442 (1992) (Errata).
4. P. Bagwell, *Phys. Rev. B*, **46**, 12573 (1992).
5. E. Scheer, J. Cuevas, A. L. Yeyati, A. Marín-Rodero, P. Joyez, M. Devoret, D. Esteve, and C. Urbina, *Physica B*, **280**, 425 (2000).
6. H. Miyazaki, T. Yamaguchi, A. Kanda, and Y. Ootuka, in preparation.

Conductance Properties of Superconducting Aluminum Point Contacts

H. Tsujii and K. Kono

RIKEN, Wako, Saitama 351-0198, Japan

Abstract. We have investigated the electrical transport properties of superconducting aluminum point contacts, using a mechanically controllable break junction technique in a dilution refrigerator. The current-voltage characteristics of the point contacts show a behavior of either a weak link or a tunnel junction, depending on the contact strength. The weak-link regime exhibits a true zero-voltage supercurrent, which decreases with decreasing contact strength. In the tunnel-junction regime, we find a supercurrent-like feature and a non-linear behavior due to multiple Andreev reflections. A crossover from the point-contact regime to the tunnel-junction regime is signalled by a rapid decrease of the supercurrent as a function of the normal resistance.

Keywords: point contacts, quantum wires, multiple Andreev reflection
PACS: 73.40.Jn, 74.50.+r, 73.21.Hb

Current-voltage (I-V) characteristics provide useful information on superconducting junctions. A Josephson junction [1] shows a strong nonlinear characteristic near the critical supercurrent at zero voltage. In contrast, in a tunnel junction, single particle transport causes a sudden increase of current at $V = 2\Delta/e$, where Δ is the superconducting energy gap. Additionally, current steps at fractional voltages $V = 2\Delta/ne$ appear with integers $n \geq 2$, owing to a transport of n electrons due to multiple Andreev reflections of a quasiparicle [2].

A particularly interesting junction is a small tunable contact of superconductors obtained by a scanning tunneling microscope or the mechanically controllable break junction (MCBJ) technique [3]. The electrical conductance of a small contact in the normal, non-superconducting state is given by $G = (2e^2/h)\Sigma T_i$, where T_i is the transmission probability of channel i. For an adiabatic constriction placed in the path of a free-electron gas, the conductance shows steps of the quantum unit $G_0 = 2e^2/h \simeq (12.9\ \text{k}\Omega)^{-1}$, as the constriction is narrowed. In a superconducting aluminum contact, at least two conduction channels contribute to the current even in the last plateau of the conductance steps before the wire breaks [4–6]. This behavior has been found by comparing experimental data with calculated I-V characteristics of a single channel superconducting contact.

To investigate the properties of superconducting aluminum contacts over a wide range of constriction, we have performed transport measurements using the MCBJ technique in a dilution refrigerator.

Aluminum wire of a 0.1 mm diameter is fixed on a phosphor-bronze bending substrate with two drops of epoxy very close to each side of the center of the substrate. The substrate is mounted in a three-point bending configuration consisting of two counter supports and a drive at the center. The wire, which has been notched in the midpoint of the glued section in advance, is broken by pushing the drive with a precision screw combined with a piezo element, at low temperature in high vacuum to prevent contamination. The electrodes thus formed are then brought back to make a contact, whose strength is adjusted by controlling the electrode separation with the piezo. The normal-state conductance G_N, both above the superconducting transition temperature $T_c = 1.18$ K and below T_c but for an applied voltage $V \geq 5\Delta/e$, shows quantized steps in units of G_0, as the contact strength is varied, as shown in the left inset of Fig. 1.

The I-V characteristics in the superconducting state at about 15 mK, well below T_c, are shown in the main frame of Fig. 1. In the weak-link regime, true zero-voltage supercurrent is observed, with the critical current I_c decreasing with increasing G_N. In the tunnel-junction regime, nonlinear transport due to multiple Andreev reflections is observed at $V = 2\Delta/ne$ with a superconducting gap of $\Delta/e = 185\ \mu$V, and the I-V characteristics around $V = 0$ show a "skew supercurrent", a supercurrent-like feature with a conductance peak. The right inset of Fig. 1 shows the skew supercurrent features observed in both a current-biased and a voltage-biased I-V curve of a $G_N = 0.7G_0$ tunnel junction. The two curves have identical slopes at $V = 0$. The transition from a zero-voltage supercurrent to a skew supercurrent takes place at approximately $4G_0$, which we consider to be still in the weak-link regime.

The critical supercurrent I_c as a function of G_N is shown in Fig. 2. In the skew supercurrent region of $G_N < 4G_0$, I_c is defined as the peak height in the voltage-biased I-V curve. The critical supercurrent for $G_N >$

CP850, *Low Temperature Physics: 24th International Conference on Low Temperature Physics;*
edited by Y. Takano, S. P. Hershfield, S. O. Hill, P. J. Hirschfeld, and A. M. Goldman

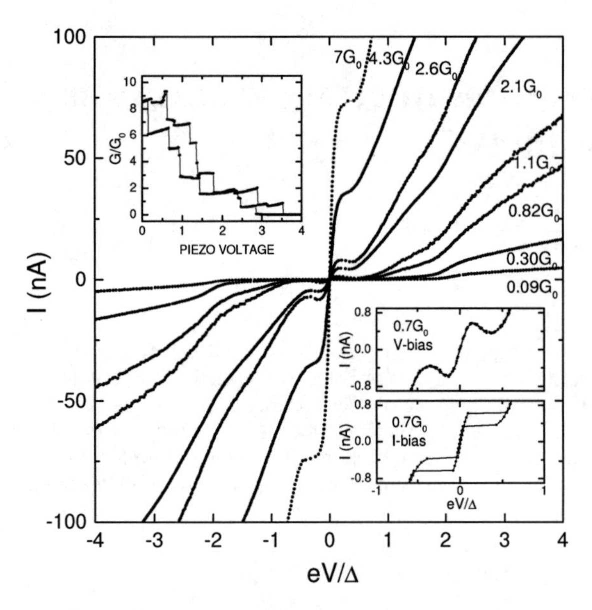

FIGURE 1. *I-V* characteristics of Al junctions of various normal conductances obtained by changing the electrode separation. Left inset: typical conductance traces in the normal state, as a function of the piezo voltage. Right inset: voltage-biased and current-biased *I-V* curves of a $0.7G_0$ tunnel junction.

$10G_0$ agrees with the theoretical Ambegaokar-Baratoff value $I_c = \pi\Delta G_N/2e$ [7]. Below $10G_0$, I_c is reduced by a factor of less than 2 from the theoretical value. At around $2G_0$, it starts to decrease rapidly with decreasing G_N, reaching a value more than two orders of magnitude smaller than the theoretical value. The rapid decrease between $2G_0$ and $1G_0$ indicates a crossover from the weak-link regime to the tunnel-junction regime. These behaviors are quite similar to those of niobium junctions [8–10], except that niobium junctions exhibit an additional change in the tunnel-junction regime [8].

In addition, we have observed that, even within a given conductance step in the unit of G_0, the *I-V* characteristic depends on the electrode separation. This behavior indicates rearrangement of atoms in the point contact.

According to Ingold *et al.* [11], zero-voltage supercurrent is completely suppressed when the charging energy $E_C = 2e^2/C$ exceeds the Josephson coupling energy $E_J = (\hbar/2e)I_c$, where C is the capacitance of the junction. However, some Cooper pairs gain enough energy from an external impedance and stochastically tunnel through the Coulomb barrier, leading to a skew supercurrent peak at a small voltage at low temperature. This scenario qualitatively explains our results, since E_C is expected to increase rapidly in the crossover regime, where the contact becomes atomic in size. Further studies using a controlled environment will be required to elucidate the coupling of point contacts to the external impedance.

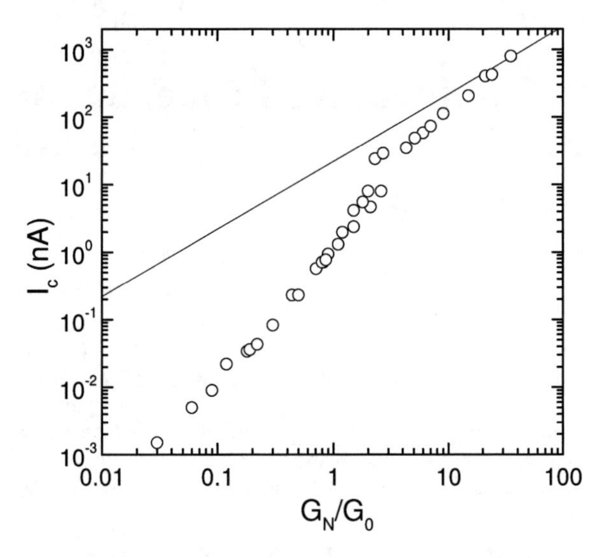

FIGURE 2. Critical supercurrent of the Al junctions as a function of the normal conductance. The solid line shows the theoretical Ambegaokar-Baratoff value.

We thank H. Akimoto, A. Furusaki, M. Kawamura, Y. Takano, and Y. Utsumi for useful discussions. This work was partially supported by a Grant-in-Aid for Scientific Research from the Japanese Ministry of Education, Culture, Sports, Science, and Technology.

REFERENCES

1. B. D. Josephson, *Phys. Lett.* **1**, 251 (1962).
2. T. M. Klapwijk, G. E. Blonder, and M. Tinkham, *Physica B* **109-110**, 1657 (1982).
3. N. Agraït, A. Levy Yeyati, and J. M. van Ruitenbeek, *Phys. Rep.* **377**, 81 (2003) and references therein.
4. E. Scheer, P. Joyez, D. Esteve, C. Urbina, and M. H. Devoret, *Phys. Rev. Lett.* **78**, 3535 (1997).
5. J. C. Cuevas, A. Levy Yeyati, and A. Martín-Rodero, *Phys. Rev. Lett.* **80**, 1066 (1998).
6. E. Scheer, N. Agraït, J. C. Cuevas, A. Levy Yeyati, B. Ludoph, A. Martín-Rodero, G. R. Bollinger, J. M. van Ruitenbeek, and C. Urbina, *Nature* **394**, 154 (1998).
7. V. Ambegaokar and A. Baratoff, *Phys. Rev. Lett.* **10**, 486 (1963); **11**, 104 (1963).
8. C. J. Muller and R. de Bruyn Ouboter, *Physica B* **194-196**, 1043 (1994).
9. C. J. Muller, M. C. Koops, B. J. Vleeming, R. de Bruyn Ouboter, and A. N. Omelyanchouk, *Physica C* **220**, 258 (1994).
10. B. Ludoph, N. van der Post, E. N. Bratus, E. V. Bezuglyi, V. S. Shumeiko, G. Wendin, and J. M. van Ruitenbeek, *Phys. Rev. B* **61**, 8561 (2000).
11. G.-L. Ingold, H. Grabert, and U. Eberhardt, *Phys. Rev. B* **50**, 395 (1994).

Nanostructured I And II Type Superconducting Materials Based On Opal Matrix

R.V. Parfeniev, D.V. Shamshur, M.S. Kononchuk, A.V. Chernyaev, S.G. Romanov and A.V. Fokin

A.F.Ioffe Physico-Technical Inst., RAS, St.Petersburg, Russia

Abstract. Electrical properties in normal and superconducting states of new nanocomposite materials based on opal matrix - In-opal and SnTe:In-opal - were investigated at low temperatures down to 0.4 *K* and magnetic fields up to 6 *T*. Chemical synthesis method and pressure insertion method were used for embedding of superconductor material into voids of opal regular dielectric matrix. Critical magnetic fields H_c of the superconducting nanocomposites sufficiently exceed the critical parameters of bulk materials. Specifity of the size effect for type-I and type-II superconducting materials in spatially modulated opal cavity lattice was discussed.

Keywords: semiconductor, artificial opal, superconductivity, nanocomposite, 3D lattice.
PACS: 81.07.Bc, 74.25.Fy.

INTRODUCTION

Possibility of observation of a size effect, collective effects and heterogeneous superconductivity arouses interest to study electrophysical properties of nanoscale superconductor ensembles in normal and superconducting (SC) states when grain sizes are comparable with the coherence length ξ. The technique [1] of superconductor material insertion into voids of a dielectric opal matrix was used to produce the nanostructured material samples with sizes of conductor grains in a nanometeric scale.

EXPERIMENT AND DISCUSSION

Artificial opal represents face-centered cubic package of amorphous SiO_2 beads with diameter $D \approx$ 230 *nm* has octahedral ($d_O = 0.41D$), tetrahedral ($d_T = 0.23D$) inter-bead voids and bridges ($d_b = 0.15D$). It was used as the template matrix forming a regular void lattice. Indium – type-I superconductor and $Sn_{1-x}In_xTe$ – type-II [3] superconductor were taken as impregnation material. Nanostructured samples for the investigation were prepared by the chemical synthesis method (i) and the pressure insertion method (ii). In the case (i) the opal voids were filled initially with In_2O_3 with subsequent annealing in hydrogen atmosphere at $T = 550^oC$ for 20 hours. Weight measurements of impregnated In showed that the void filling rate k_{fill} varies from 30% to 55%. For $Sn_{0.84}In_{0.16}Te$ – opal samples a matrix and initial components corresponding to the chemical formula $Sn_{0.84}In_{0.16}Te_{1.02}$ were held in a steel ampula at $T = 820°C$ for 40 minutes. In the case (ii) melted In ($T = 180°C$, $P = 10$ *bar*) [3] or the initial solid solution $Sn_{0.95}In_{0.05}Te$ ($T = 750°C$, $P = 3$ *kbar*) was introduced into opal voids under pressure. In this case matter fills all free void spaces in the sample ($k_{fill} = 1$) forming 3D continuous network in the volume amid silica beads. The volume fraction of voids filled by In was varied from 0.26 to 0.08 by the method of preliminary monolayer deposition of TiO_2 on opal inner surface [4].

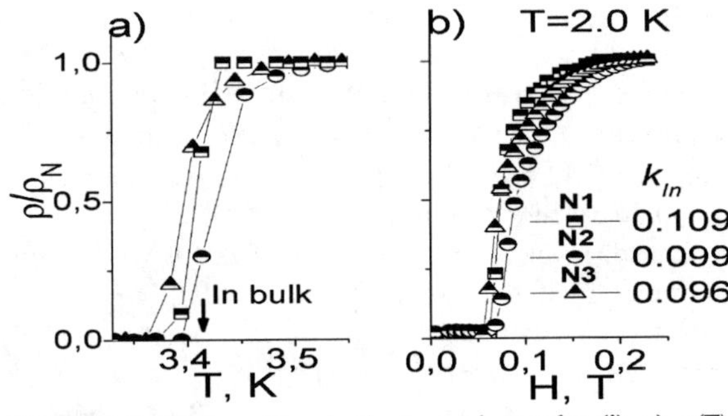

FIGURE 1. SC transition in the In-opal samples (i). a) *ρ(T)* curves; b) *ρ(H)* curves; normal resistivity, *Ohm·cm*: N1 – 0.24, N2 – 0.74, N3– 0.53.

CP850, *Low Temperature Physics: 24th International Conference on Low Temperature Physics;* edited by Y. Takano, S. P. Hershfield, S. O. Hill, P. J. Hirschfeld, and A. M. Goldman

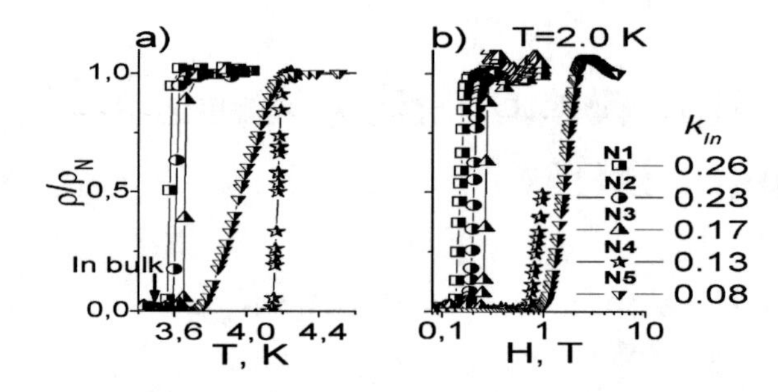

FIGURE 2. SC transition in the In-opal nanocomposites (ii). a) *ρ(T)* curves; b) *ρ(H)* curves. Characteristic size of bridges d_b, *nm* (resistivity, *Ohm·cm*): N1 – 36 (0,0089), N2 – 30 (0,0069), N3 – 21 (0,035), N4 – 19 (0,26), N5 – 10 (8,9).

All In-opal nanocomposite samples show transition to the SC state with the SC parameters dependent upon the volume fraction of In k_{In}. Observed critical temperatures T_c are close to one for bulk material (T_c^{bulk} = 3.41 *K*) or exceed it in the series (ii) (fig 2a). Critical magnetic fields (fig. 1b, 2b) of all samples exceed the corresponding value for bulk In ($H_c^{bulk}(0)$ = 280 *Oe*) in 3-10 times.

TABLE 1. Parameters of the SC transition for investigated SnTe<In>-opal nanocomposite samples and bulk material samples with related composition.

Samples	$\rho_{300K}/\rho_{4.2K}$	T_c,K	$H_{c2}(0)$,T
$Sn_{0.95}In_{0.05}Te$ (bulk)	1,44	1,20	3,8
$Sn_{0.95}In_{0.05}Te$ - opal	1,36	0,92	6,6
$Sn_{0.84}In_{0.16}Te$ (bulk)	1,46	2,14	7,8
$Sn_{0.84}In_{0.16}Te$ - opal	1,82	2,09	13,2

For In-opal samples (ii) the size effect was distinctly traced that reveals in monotonous dependence SC transition parameters on the volume fraction of In in the sample (fig. 2 a,b), i.e. on In nanograin dimensions. For samples (ii) any value of $k_{In} \leq 0.26$ we can confront with definite nanograin dimensions d_0, d_T, d_b. The effective resistivity of samples (ii) was ρ = (50÷100)·10^{-6} *Ohm·cm*. The values of *ρ* for the samples (i) exceeded those for samples (ii) by 1 ÷ 3 orders. Thus In may be allocated in opal voids randomly and may not fill whole voids in the case (i), in contrast to the samples (ii) which have more defined grain structure. According to [5], the curves of dependences *R(T)*, *R(H)* within the SC transition interval change drastically with In granule sizes decreasing especially at $k_{In} \leq 0.13$ when the coherence length becomes comparable with the dimension of superconductor granules.

Thus the method ii allows producing In-opal nanocomposite with stable properties. We can change electrical properties of these nanocomposites including SC transition parameters by a preliminary adjustment of geometrical dimensions of matrix [5].

As seen in the Table 1 and fig.3 observed values of the effective resistivity of the SnTe<In>-opal samples have the same order as ρ for the bulk material as a result of the predominant influence of the resonant scattering of holes into the impurity band in both bulk and nanocomposite SnTe<In>-opal samples. Slight changes of T_c in nanocomposite materials seem to be connected to stoichiometric composition deviation in the $Sn_{1-x}In_xTe$ compound [2].

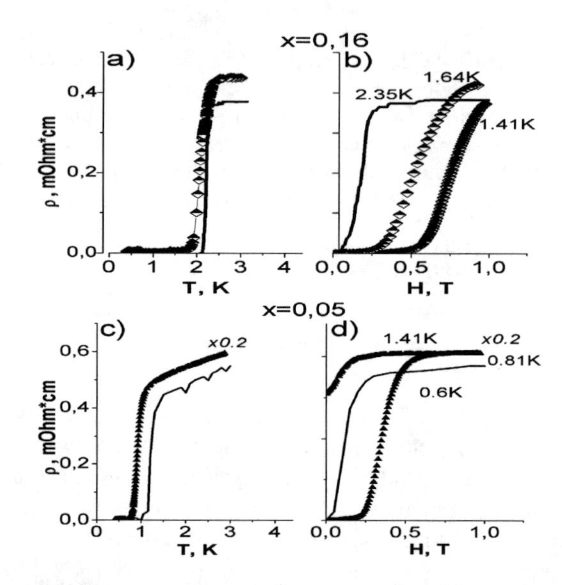

FIGURE 3. SC transition in the sample $Sn_{1-x}In_xTe$-opal, prepared by method i (a, b) and method ii (c, d); a), c) *ρ(T)* and b), d) *ρ(H)* curves. The lines are the bulk material data.

Evident increase of SC parameters for investigated samples connects to both a modulation of superconductor grain dimensions and presence of the collective effect related to complicated geometrical nanodimensional network structure [4].

ACKNOWLEDGMENTS

The authors acknowledge the support of the Presidium of RAS grant, RFBR 04-02-16638a, 05-02-16975 and the LSS grant - 2200. 2003. 2.

REFERENCES

1. V.N.Bogomolov, et al., *JETP Lett.* **36**, 365 (1982).
2. D.V.Shamshur, et al., *Phys. Solid State* **28**, 1094 (1986).
3. S.G.Romanov, et al., *Mikrochim. Acta*, **S15**, 211 (1998).
4. S.G.Romanov, et al., *JETP Lett.* **58**, 824 (1993).
5. A.V.Chernyaev, et al., *Phys. Sol. St.* **47**, in print (2005).

Rectifying Non-Gaussian Noise with Incoherent Cooper Pair Tunneling

P. Virtanen and T. T. Heikkilä

Low Temperature Laboratory, P.O. Box 2200, FIN-02015 HUT, Finland

Abstract. We show how a superconducting Josephson junction in Coulomb blockade can be used to convert voltage fluctuations with a skew distribution (i.e., with a finite third cumulant) to an average current through the junction. Using this scheme, one may access the Fano factor for the third cumulant in the fluctuation source.

Keywords: Josephson junction, full counting statistics, third cumulant, environmental Coulomb blockade
PACS: 74.40.+k,05.40.Ca,72.70.+m,74.50.+r

INTRODUCTION

Apart from the direct supercurrent in superconductors, there are a few different ways of inducing a finite average current through mesoscopic conductors without an applied bias over them. In current pumps, this is done by cycling a few control gate voltages [1] or magnetic fields [2] in a specified sequence to pump electrons through a system typically consisting of several small metallic islands. In this case, it is the asymmetry in these voltages or fields that breaks the symmetry between the currents going opposite directions (say, "left" or "right"), thus resulting in a finite average current. In ratchets [3], the current is induced by a periodic variation of a field in an asymmetric potential. The latter then serves to break the left–right symmetry and create the finite current.

The third alternative for breaking the left-right symmetry is to induce asymmetric fluctuations in the environment [4, 5, 6] and use these to control the current through a mesoscopic junction. In [6], it was shown that a Josephson junction (JJ) in Coulomb blockade can be used as an accurate and wide-band detector of non-Gaussian current fluctuations. For a schematic depiction of this detector, see Fig. 1. The detector output consists of the function $I_m(V_m)$, the current I_m through the junction for a given average voltage V_m through it. This function depends on a characteristic function $I_D(V_m)$ of the detector (current through the junction in the absence of the measured excess fluctuations) and on the excess fluctuations that one wants to measure.

Without the excess fluctuations, the detector is symmetric, and thus $I_D(V_m) = -I_D(-V_m)$. One may characterize the excess noise by the change in $I_m(V_m)$ from $I_D(V_m)$, $I_m(V_m) = I_D(V_m) + \delta I_A(V_m) + I_S(V_m)$, where $\delta I_A(V_m) = -\delta I_A(V_m)$ and $I_S(V_m) = I_S(-V_m)$ are the antisymmetric and symmetric parts of the change in the detector output. Below, we will concentrate on the limit of a vanishing bias in the detector circuit, i.e., $V_m = 0$. Then,

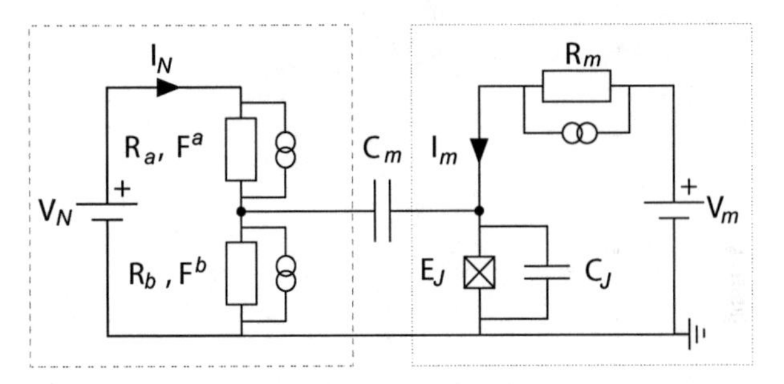

FIGURE 1. Schematics of the non-Gaussian noise detection. The current-voltage characteristics $I_m(V_m)$ of the Josephson junction on the right is measured with and without the current I_N flowing in the noise source on the left. The difference signal can be shown to measure the excess noise in the noise source, as discussed in the text.

in the limit where the excess fluctuations are small, we get [6]

$$I_0 \equiv I_m(V_m = 0) = -2\int_0^\infty \frac{d\omega}{4\pi} I_D\left(\frac{\hbar\omega}{2e}\right) K_\phi(\omega). \quad (1)$$

Here $K_\phi(\omega) = \int d\omega' \mathrm{Im} S_{3\phi}(\omega, \omega', -\omega-\omega')/\pi$, and

$$S_{3\phi}(\omega_1, \omega_2, \omega_3) \equiv \int dt_1 dt_2 dt_3 e^{i(\omega_1 t_1 + \omega_2 t_2 + \omega_3 t_3)} \langle \delta\phi(t_1)\delta\phi(t_2)\delta I\phi(t_3)\rangle \quad (2)$$

describes the third cumulant of phase fluctuations. The latter can be related to the current fluctuations as detailed in Ref. [6]. Assuming that the dispersion of the third cumulant is only due to the circuit and not due to some intrinsic dispersion (which typically takes place within the frequency scales specified by the voltage over the junctions or the inverse time of flight through them), we obtain in the limit $R_S C_m \ll R_m C_J$

$$K_\phi(\omega) = 2\tau\lambda^3 (F_3^b - F_3^a) \frac{I_N}{e/\tau} \frac{\omega\tau}{4 + 5(\omega\tau)^2 + (\omega\tau)^4}, \quad (3)$$

CP850, *Low Temperature Physics: 24th International Conference on Low Temperature Physics;*
edited by Y. Takano, S. P. Hershfield, S. O. Hill, P. J. Hirschfeld, and A. M. Goldman

where $R_S = R_a R_b/(R_a + R_b)$, $\lambda = \pi R_S C_m/(R_K(C + C_m))$, $R_K = h/(4e^2)$, $\tau = R_m(C_m + C_J)$, and the resistances R_i, capacitances C_i, Fano factors F_3^i for the third cumulant and the current I_N are indicated in Fig. 1.

In this way, the Coulomb-blockaded Josephson junction rectifies the skew voltage fluctuations in the source, producing a finite average current.

EXAMPLES OF THE ZERO-BIAS CURRENT

Figure 2 shows the detector current-voltage curve for a few temperatures and for a few resistances R_m of the detector and the source. The function $K_\phi(\omega)$ is plotted as a solid line. The rectified current I_0 is obtained as an integral over these two functions.

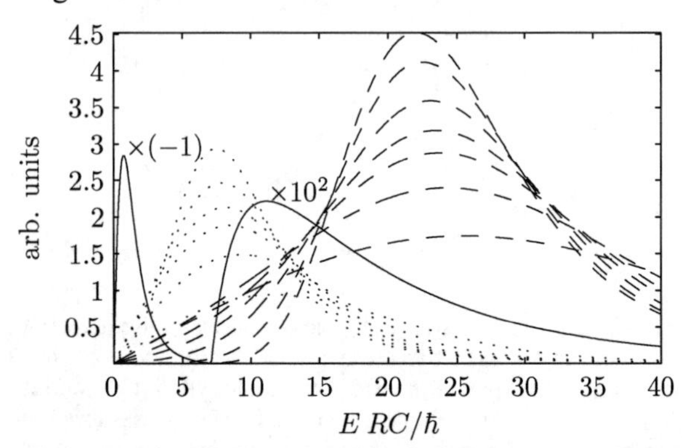

FIGURE 2. Detector current-voltage curve for a few temperatures and for $R_m = 8R_K$ (dashed lines) and for $R_m = 3R_K$ (dotted lines). The curves become lower and wider with an increasing temperature. The parameters in the curves are $C_m = 0.1C_J$, $R_1 = R_2 = R_m$, and $F_3^a = 1$, $F_3^b = 0$. Solid line shows the function K_ϕ, characterizing the third cumulant of voltage fluctuations. The negative part of K_ϕ is turned upside down, and the positive part scaled by a factor 10^2 for clarity as it causes a sign change in the rectified current as a function of temperature.

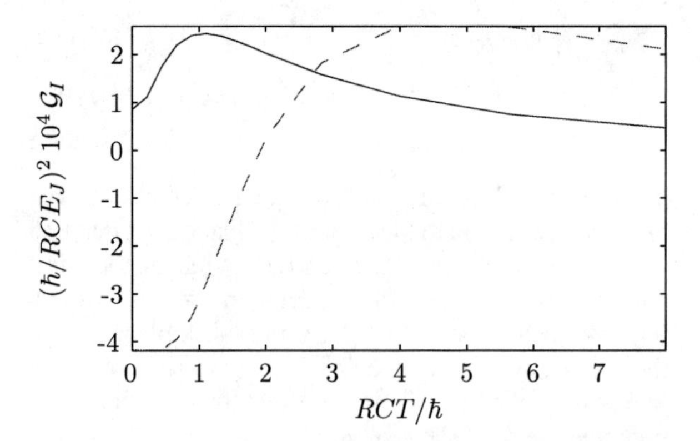

FIGURE 3. Zero-bias current through the detector junction as a function of temperature for $R_m = 3R_K$ (solid line) and $R_m = 8R_K$ (dashed line).

Figure 3 shows the resulting rectified current through the detector as a function of temperature for a few values of R_m, scaled by I_N. In the limit where the excess noise couples weakly, the current I_0 is directly proportional to I_N, and hence the results can be presented in terms of the current gain

$$\mathscr{G}_I \equiv \frac{I_S}{I_N}. \tag{4}$$

We have neglected the (possible) temperature dependence of the intrinsic noise. The negative value of $\mathscr{G}_I$ for $R_m = 8R_K$ is due to the small positive part of $K_\phi(\omega)$ at large energies (see Fig. 2), which arises from the corrections to the weak-coupling result Eq. (3).

This theory is based on expansion in the Josephson coupling energy E_J [7]. For a vanishing temperature, the expansion is valid provided that $E_J R_m C_J/\hbar \ll \sqrt{R_m/R_Q}$. This implies that within the strict validity limits of the theory, the gain $\mathscr{G}_I \ll c10^{-4} R_m/R_Q$, where c is the number plotted in Fig. 3. For example, for $R_m = 8R_Q$, I_0 is up to a fraction of a percent of I_N and should still be observable. Note that this is a pessimistic estimate as the perturbation expansion works better at finite temperatures and we did not try to optimize $\mathscr{G}_I$.

To conclude, we discussed the rectifying effect of a Coulomb-blockaded Josephson junction on a skew distribution of current fluctuations in its environment. We showed that the rectified current may change sign as a function of temperature for some values of the detector resistances R_m.

ACKNOWLEDGMENTS

We thank Pertti Hakonen, Göran Johansson and Frank K. Wilhelm for discussions. TTH is supported by the Academy of Finland.

REFERENCES

1. M. W. Keller, J. M. Martinis, N. M. Zimmerman, and A. H. Steinbach, *Appl. Phys. Lett.* **69**, 1804–1806 (1996).
2. A. O. Niskanen, J. P. Pekola, and H. Seppa, *Phys. Rev. Lett.* **91**, 177003 (2003).
3. P. Reimann, *Phys. Rev. Lett.* **86**, 4992 (2001).
4. R. K. Lindell, J. Delahaye, M. A. Sillanpaa, T. T. Heikkila, E. B. Sonin, and P. J. Hakonen, *Phys. Rev. Lett.* **93**, 197002 (2004).
5. E. B. Sonin, *Phys. Rev. B* **70**, 140506 (2004).
6. T. T. Heikkila, P. Virtanen, G. Johansson, and F. K. Wilhelm, *Phys. Rev. Lett.* **93**, 247005 (2004).
7. G.-L. Ingold, and Y. V. Nazarov, "Charge tunneling rates in ultrasmall junctions," in *Single Charge Tunneling: Coulomb Blockade Phenomena in Nanostructures*, edited by H. Grabert, and M. Devoret, Plenum, New York, 1992.

Fluctuations in a Superconducting Wire

Jorge Berger

Physics Unit, Ort Braude College, 21982 Karmiel, Israel

Abstract. A recent study, that treats weak links as Josephson junctions and treats thermal fluctuations by standard methods, found that an asymmetric superconducting loop can rectify noise. This surprising result is the motivation for the development of a consistent treatment of the Langevin "forces" in the time-dependent Ginzburg-Landau equations, based on the fluctuation-dissipation theorem and appropriate for the integration of these equations using finite difference methods. Previous treatments using Langevin "forces" are not consistent in two respects: the size of these "forces" does not take into account the volume of the elements into which the sample is divided and ignores fluctuations of the electromagnetic potential.

Keywords: fluctuations, noise rectification, fluctuation-dissipation, superconducting wire, paraconductivity.
PACS: 74.40.+k, 02.70.Bf, 05.10.Gg, 05.40.-a

INTRODUCTION

Recent experiments [1] suggest that a nonuniform superconducting ring slightly below its transition temperature that encloses a magnetic flux that is neither an integer nor a half-integer multiple of the quantum of flux supports a nonzero average dc voltage.

As a first attempt to understand these experiments, I modeled the nonuniform ring by a SQUID with unequal junctions [2]. The central assumptions of this model were that the two junctions have different current-phase relationships, that the size of the maximal current in each junction undergoes thermal fluctuations (but the current-phase profiles are fixed), the variance of the normal current across a junction with resistance R during a period of time τ is $2k_BT/R\tau$, and the current is uniform along the circuit. I found that this model predicts, as in the experiments, that for appropriate fluxes and parameters a nonzero dc voltage can be supported across the junctions.

PRESENT METHOD

We would like to describe the superconducting loop by means of a model that is manifestly self-consistent from the point of view of statistical mechanics. A suitable model for this purpose is the time-dependent Ginzburg–Landau model, which in appropriate gauge can be written as a purely dissipative system. The field variables of this model are the complex order parameter ψ and the vector potential **A**. In numeric treatments space is divided into cells and the fields in cell k will be ψ_k and $\mathbf{A}_k$. The Ginzburg–Landau equations, without fluctuations, are then given by

$$\frac{d\psi_k}{dt} = -2\Gamma_{\psi,k}\frac{\partial G}{\partial \psi_k^*};\ \frac{d\mathbf{A}_k}{dt} = -\Gamma_{A,k}\nabla_{\mathbf{A}}G, \tag{1}$$

where G is the thermodynamic potential of the system and the Γ's are constants.

A possible self-consistent method for taking fluctuations into account is the addition of a Langevin term to Eqs. (1). The size of this term is dictated by the fluctuation-dissipation theorem and after a time τ the variance of its contribution has to be $2\Gamma k_BT\tau$. This approach was proposed by Schmid [3] and has been widely used [4]. However, in these references the fluctuations of $\mathbf{A}_k$ are not taken into account and the importance of the size of the cell k is not discussed.

We have focused on the case of a 1D wire of length L divided into N cells of length L/N and cross section w_k. We have considered several discretizations for the thermodynamic potential; one of them (which has the advantage of a symmetric appearance and the disadvantage of decomposing into two sublattices) is

CP850, *Low Temperature Physics: 24th International Conference on Low Temperature Physics;*
edited by Y. Takano, S. P. Hershfield, S. O. Hill, P. J. Hirschfeld, and A. M. Goldman

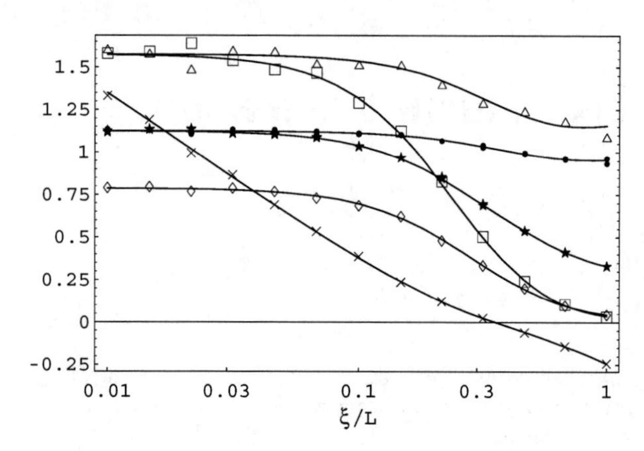

FIGURE 1. Averages of several quantities, as functions of the coherence length. The symbols correspond to values calculated using our method and the lines are ensemble averages. The parameters used in the calculation were N=4, $\alpha\tau/\gamma\hbar = 3\times10^{-5}$, $e^2L^2\gamma/\hbar\sigma = 8000$. The parameter $k_BT/\alpha Lw_0$ factors out and is taken as unit. • $0.25+\langle|\varphi_0|^2\rangle$ and $0.25+\langle|\varphi_2|^2\rangle$;? $0.25+\langle|\varphi_1|^2\rangle$ and $0.25+\langle|\varphi_{-1}|^2\rangle$; λ $\langle|\varphi_0|^4\rangle$; % $\langle|\varphi_1|^4\rangle$; ◊ $\langle|\varphi_0|^2|\varphi_1|^2\rangle$; × $0.5\log_{10}\langle\tilde{A}^2\rangle$.

$$G = \frac{L}{N}\sum w_k\left\{\alpha|\psi_k^2| + \frac{\beta}{2}|\psi_k^4| + |\alpha|\frac{N^2\xi^2}{4L^2}|\psi_{k+1} - \psi_{k-1} + i\tilde{A}_k\psi_k|^2\right\} + I\sum\tilde{A}_k \quad (2)$$

where α and β are Ginzburg-Landau coefficients, ξ is the coherence length, I is the normalized current, and $\tilde{A}_k = 4\pi LA_k/N\Phi_0$ and refers to the direction of the wire. The Γ's for ψ and $\tilde{A}$ are

$$\Gamma_{\psi,k} = \frac{N}{2\gamma\hbar Lw_k};\ \Gamma_{\tilde{A},k} = \frac{16e^2L}{N\hbar^2\sigma w_k}, \quad (3)$$

where e is the electron charge, σ the conductivity, and γ has the same meaning as in [3].

For the trivial case $\beta = I = 0$ and $\tilde{A}_k = \tilde{A}$ and $w_k = w_0$ uniform, we can evaluate exactly the ensemble averages of quantities such as $|\varphi_n|^2$, where

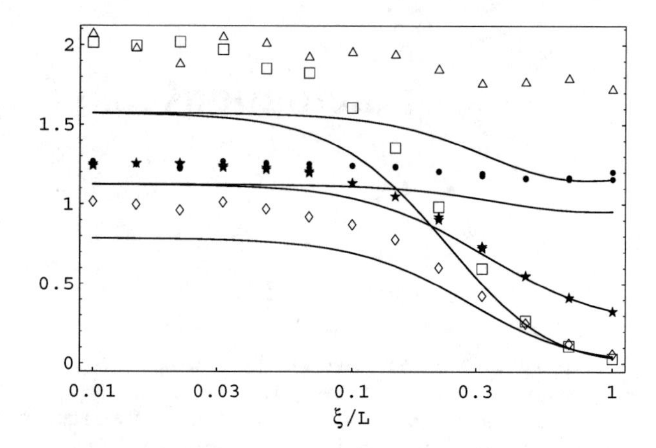

FIGURE 2. Same as Fig. 1, but fluctuations of $\tilde{A}$ were ignored.

φ_n is the space Fourier transform of ψ_k. Figures 1 and 2 compare these exact results with those obtained by the present method. We see that full implementation of this method leads to good agreement, but ignoring the fluctuations of $\tilde{A}$ does not.

Implementation of this method to more challenging situations is underway.

ACKNOWLEDGMENTS

This research was supported by the Israel Science Foundation (grant 4/03-11.7). My participation in LT24 was supported by Ort Braude College.

REFERENCES

1. S. V. Dubonos, V. I. Kusnetsov, I. N. Zhilyaev, A. V. Nikulov, and A. A. Firsov, *Zh. Eksp. Teor. Fiz. Pis'ma Red.* **77**, 439-444 (2003); S. V. Dubonos, V. I. Kuznetsov, A. V. Nikulov, http://arxiv.org/abs/physics/0105059; J. Berger, *Physica E*, in press.
2. J. Berger, *Phys. Rev. B* **70**, 024524 (2004).
3. A. Schmid, *Phys. Rev.* **180**, 527-529 (1969).
4. R. Kato, Y. Enomoto and S. Maekawa, *Phys. Rev. B* **47**, 8016-8024 (1993); C. Bolech, G. C. Buscaglia and A. Lopez, *Phys. Rev. B* **52**, R15719- R15722 (1995).

Phase Transition and Fluctuations in Superconducting Nanostructures

Masahiko Hayashi*,†, Hiromichi Ebisawa*,† and Masaru Kato**,†

*Graduate School of Information Sciences, Tohoku University, 6-3-09 Aramaki, Aoba-ku, Sendai 980-8579, Japan
†JST-CREST, 4-1-8, Honcho, Kawaguchi, Saitama 332-0012, Japan
**Department of Mathematical Sciences, Osaka Prefecture University, 1-1, Gakuencho, Sakai, Osaka 599-8531, Japan

Abstract. Equilibrium and metastable states in nanometer-scale superconducting network systems are studied based on Ginzburg-Landau free energy and the phase transition in a magnetic field is investigated. It is discussed that superconducting fluctuation is affected significantly by the applied field and the strength of the fluctuation varies with respect to the field.

Keywords: Superconductivity, nanostructure, fluctuation
PACS: 74.20.De,74.25.Qt,74.78.Na

With the aide of the developments in nanotechnology, the investigation of the superconductivity in nanometer-scales has been made possible and many interesting features characteristic to the mesoscopic systems have been revealed [1, 2]. In this paper we study the phase transition and the fluctuations in nanometer-scale superconducting (SC) network systems. The SC networks show intriguing behaviors due to multiply-connected nature of the system, especially under a magnetic field. If the size of the network is decreased, the effects of the boundaries become more and more significant and the interplay between multiply-connected effects and the finite-size effects give rise to novel SC states[3]. In this paper we concentrate on the effects of system geometry on the SC fluctuations.

We consider a SC square lattice with a rectangular boundary (two sides are assumed to be parallel to x- and y-direction) under a uniform magnetic field. Ginzburg-Landau (GL) free energy is given by

$$F=\sum_{\langle j,k\rangle}\int dx\left[\frac{\hbar^2}{2m}|(\nabla-i\kappa_{jk})\psi_{jk}|^2-\alpha|\psi_{jk}|^2+\frac{\beta}{2}|\psi_{jk}|^4\right], \quad (1)$$

where j and k number the vertices, $\langle j,k\rangle$ denotes the neighboring vertex, $\alpha\propto T_c-T$ and β are constants, and $\psi_{jk}(x)$ is the order parameter on the link j-k. The vector potential appears in κ_{jk}.

Since we concentrate on the phase transition and the fluctuations, the temperature region where $\psi_{jk}\simeq 0$ is relevant. Therefore the solutions of linearized GL equation, which is called de Gennes-Alexander equation [4], are especially useful. The equation is given by

$$-\frac{\hbar^2}{2m}(\nabla-i\kappa_{jk})^2\psi_{jk}(x)=E\psi_{jk}(x) \quad (2)$$

and the solution is

$$\psi_{jk}(x)=\frac{e^{-i\kappa_{jk}x}}{\sin(l/\xi)}\left[\psi_j\sin\frac{l-x}{\xi}+\psi_k e^{i\kappa_{jk}l}\sin\frac{x}{\xi}\right], \quad (3)$$

where ψ_j is the value of order parameter at vertex j and $\xi=\hbar/\sqrt{2mE}$. The length of the link is denoted by l. ψ_j's should be determined so that the solution satisfy the following conditions at vertices,

$$\sum_{k\in N(j)}\left(i\frac{d}{dx}+\kappa_{jk}\right)\psi_{jk}(x)\bigg|_j=0, \quad (4)$$

where $N(j)$ is the set of vertices connected to the vertex j. Eqs. (4) and (3) give linear simultaneous equations with respect to ψ_j's and the condition for non-vanishing ψ_j's determines the eigenvalues, $E_1,E_2\cdots$.

Since the linearized GL equation is self-adjoint, solutions belonging to all the eigenvalues comprise a complete orthogonal set. Therefore we can expand $\psi_{jk}(x)$ in terms of these functions as $\psi_{jk}(x)=\sum_n\chi_n\psi_{jk}^{(n)}(x)$, where $\psi_{jk}^{(n)}$ is the eigenfunction belonging to E_n. Then the quadratic term in eq. (1) can be written as $\sum_n(E_n-\alpha)|\chi_n|^2$, assuming that $\sum_{\langle jk\rangle}\int dx|\psi_{jk}^{(n)}|^2$ is normalized to unify. The smallest E_n determines the transition temperature since such mode becomes unstable to order formation at the highest temperature.

The SC fluctuation is realized by allowing appearance of all the possible states with probability $\propto e^{-F_n/k_BT}$ where k_B and T are the Boltzmann constant and the temperature, respectively, and F_n is the free energy of the state belonging to E_n. Therefore not only the smallest E_n but also larger E_n's become important. Moreover the

CP850, *Low Temperature Physics: 24th International Conference on Low Temperature Physics;* edited by Y. Takano, S. P. Hershfield, S. O. Hill, P. J. Hirschfeld, and A. M. Goldman

number of E_n's close to the smallest one dominates the strength of the SC fluctuation.

We have calculated the eigenvalues E_n for several different boundary conditions (BC's). The system consists of 10×10 lattice points. Figure 1 is for (a) open BC in both x- and y-direction (b) open BC in y-direction but periodic BC in x-direction. Horizontal axes of the graph is magnetic flux per unit cell, ϕ. (Only $0 < \phi/\phi_0 < 0.5$ is shown, since the graph is symmetric with respect to $\phi = \phi_0/2$. Here ϕ_0 is the magnetic flux quantum.) We can see from the figure that the distribution of the eigenvalues is different between the two. (a) has less eigenvalues close to the lowest one. (b) has a periodic degeneracy of the lowest eigenvalue as indicated by the arrows, which comes from the periodic BC. According to Ref. [5], the SC fluctuation is enhanced when the lowest eigenvalue is degenerate. Therefore the SC fluctuation is significant near the arrows in (b).

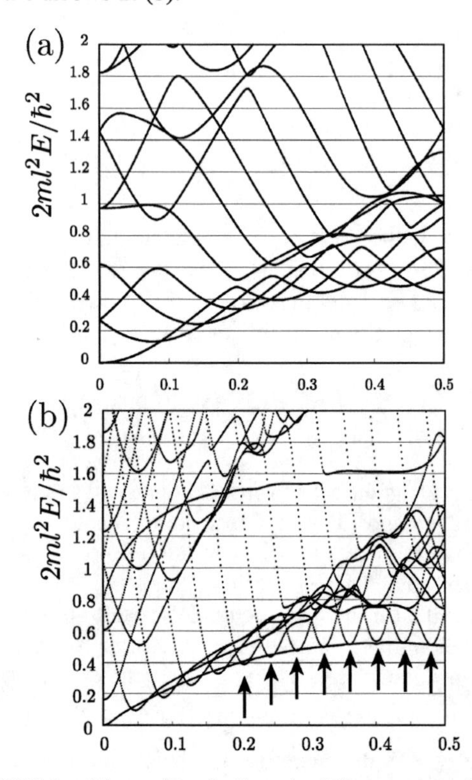

FIGURE 1. Eigenvalues for linearized GL equation with two different boundary conditions, (a) and (b). See text for details.

Figure 2 is shown to study the effects of novel BC. (a) is for open BC in y-direction but periodic BC in x-direction, and (b) is for Möbius boundary condition, namely the BC in x-direction is twisted. In these cases the flux is applied only through the inside of the ring (Aharonov-Bohm flux). As indicated by an arrow, (b) has a characteristic state near $\phi \simeq \phi_0/2$. This state corresponds to the nodal state previously discussed in Ref. [6]. Since this state gives rise to an additional contribution, the SC fluctuation is enhanced near $\phi \simeq \phi_0/2$ in Möbius geometry. This phenomenon may be observed in Möbius crystal created by Tanda *et al.* [7].

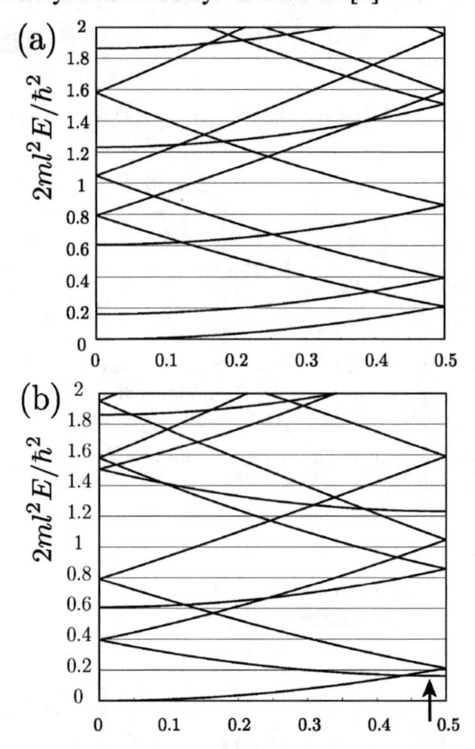

FIGURE 2. Eigenvalues for linearized GL equation for (a) an ordinary ring and (b) a Möbius ring.

In this paper, we argued that the SC fluctuations as well as the transition temperature are largely affected by the system geometry in nanometer scale SC networks. These predictions may be observed by carefully studying resistive transition under a magnetic field.

REFERENCES

1. B. Pannetier, J. Chaussy, R. Rammal, and J. Villegier, *Phys. Rev. Lett.*, **53**, 1845 (1985).
2. T. Ishida, M. Yoshida, S. Okayasu, and K. Hojou, *Supercond. Sci. Technol.*, **14**, 1166 (2001).
3. M. Hayashi, H. Ebisawa, and M. Kato, *Physica C* (2005), to be published.
4. S. Alexander, *Phys. Rev. B*, **27**, 1541 (1983).
5. M. Hayashi, and H. Ebisawa, *Physica C*, **352**, 191 (2001).
6. M. Hayashi, and H. Ebisawa, *J. Phys. Soc. Jpn.*, **70**, 3495 (2001).
7. S. Tanda, T. Tsuneta, Y. Okajima, K. Inagaki, K. Yamaya, and N. Hatakenaka, *Nature*, **417**, 397 (2002).

Quasi-particle Spectrum of Nano-scale Superconductors under External Magnetic Field

Hisataka Suematsu*,†, Masaru Kato*,†, Masahiko Machida**,†, Tomio Koyama‡,† and Takekazu Ishida§,†

*Department of Mathematical Sciences, Osaka Prefecture University, 1-1 Gakuencho, Sakai, Osaka 599-8531, Japan
†JST-CREST, 4-1-8, Honcho, Kawaguchi, Saitama 332-0012, Japan
**CCSE, Japan Atomic Energy Research Institute, Higashi Ueno 6-9-3, Tokyo 110-0015, Japan
‡ Institute for Materials Research, Tohoku University, Sendai 980-8577, Japan
§Department of Physics and Electronics, Osaka Prefecture University, 1-1 Gakuencho, Sakai, Osaka 599-8531, Japan

Abstract. We have solved the Bogoliubov-de Gennes equation of nano scale superconducting square plate under an external magnetic field. The result shows that multiple-vortex states become stable at low temperature. In these states, vortex bound states of these vortices interfere with each other and the vortices form a kind of vortex molecule, and this is reflected in the local density of states.

Keywords: Nano-scale superconductors, Bogoliubov-de Gennes equation, Finite element method
PACS: 74.20.Fg, 74.78.Na

INTRODUCTION

Recently, mesoscopic or nanoscopic superconductors have been studied, especially vortices under the magnetic field. Though a single quantized vortex appears in a bulk superconductor, there are a double quantized vortex or giant vortices in a nanoscopic superconductor [1] [2]. Previous studies based on the phenomenological Ginzburg-Landau (GL) equations. In order to consider quasi-particle spectrum, which is accessible by Scanning Tunneling Microscopy or Scanning Tunneling Spectroscopy, we must solve a microscopic equation of superconductivity, Bogoliubov-de Gennes equation, for a superconductor under the perpendicular magnetic field.

BOGOLIUBOV-DE GENNES EQUATION

For a superconductor that has the s-wave symmetry of the order parameter $\Delta(\mathbf{r})$, the Bogoliubov-de Gennes equation is given as,

$$\left[\frac{1}{2m}\left(\frac{\hbar}{i}\nabla+\frac{e}{c}\mathbf{A}\right)^2-\mu\right]u_n(\mathbf{r})+\Delta(\mathbf{r})v_n(\mathbf{r})=E_n u_n(\mathbf{r}), \tag{1}$$

$$-\left[\frac{1}{2m}\left(\frac{\hbar}{i}\nabla-\frac{e}{c}\mathbf{A}\right)^2-\mu\right]v_n(\mathbf{r})+\Delta^*(\mathbf{r})u_n(\mathbf{r})=E_n v_n(\mathbf{r}), \tag{2}$$

where $u_n(\mathbf{r})$ and $v_n(\mathbf{r})$ are wave functions of nth quasi-particle state, and E_n is the eigenvalue of them and μ is the chemical potential. We assume that $u = v = 0$ at the boundary. The order parameter is given as

$$\Delta(\mathbf{r}) = g\sum_n^{|E_n|\le E_c} u_n(\mathbf{r})v_n^*(\mathbf{r})(1-2f(E_n)), \tag{3}$$

where g is the interaction constant, E_c is the cut-off energy of the attravtive ineration and $f(E)$ is the Fermi-Dirac distribution function. The vector potential is calculated by the Maxwell equation

$$\nabla\times(\nabla\times\mathbf{A}-\mathbf{H}_0)=\frac{4\pi}{c}\mathbf{j}, \tag{4}$$

where $\mathbf{H}_0$ is external magnetic field and $\mathbf{j}$ is the current in the superconductor, which is defined as

$$\mathbf{j}=\frac{\hbar e}{2mc}\sum_n\left[f(E_n)u_n^*\nabla u_n+(1-f(E_n))v_n\nabla v_n^*-h.c.\right] + \frac{e}{mc}\sum_n\left[f(E_n)|u_n|^2-(1-f(E_n))|v_n|^2\right]\mathbf{A}. \tag{5}$$

We impose particle number conservation,

$$N_e=\int\sum_n\left[f(E_n)|u_n(\mathbf{r})|^2+(1-2f(E_n))|v_n(\mathbf{r})|^2\right]d\mathbf{r}, \tag{6}$$

where N_e is the total number of normal electrons. From Eq. (6), the chemical potential μ is determined.

CP850, *Low Temperature Physics: 24th International Conference on Low Temperature Physics;*
edited by Y. Takano, S. P. Hershfield, S. O. Hill, P. J. Hirschfeld, and A. M. Goldman

For solving above equations, we use the finite element method [3].

We consider a square superconductor, of which length is L. We choose the GL parameter is $\kappa = 3.0$, the order parameter at T=0 is $\Delta/E_c = 0.2$, the Fermi wave number is $k_F\xi = 3.0$, the coherence length is $\xi/L = 0.2$ and the temperature is $T = 0.1T_c$. For Nb, $\xi \approx 200$nm.

RESULT AND DISCUSSION

In the following, we show numerical results for multiple-vortex states. In Fig. 1, we show amplitude of the order parameter for four vortices at $H/L^2 = 13\Phi_0$ and five vortices at $H/L^2 = 14\Phi_0$. Increasing magnetic field, the four vortices configuration appears at $H/L^2 = 10\Phi_0$. Calculating the free energy, it is found that four vortices configuration is a stable state between $H/L^2 = 12\Phi_0$ and $H/L^2 = 15\Phi_0$ and that the five vortices configuration is a metastable state.

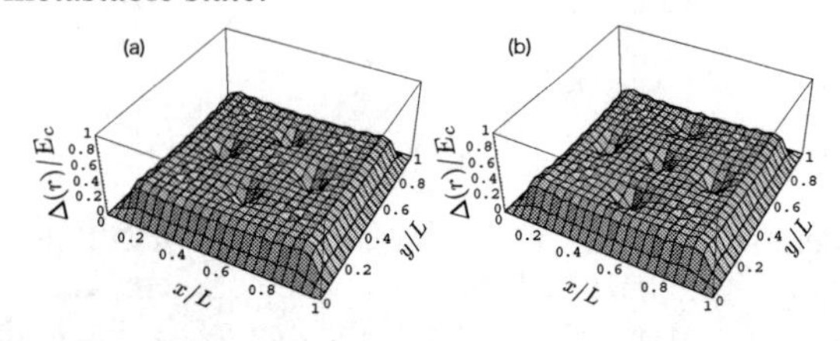

FIGURE 1. Spatial dependence of the order parameter for (a) four vortices and (b) five vortices states.

In Fig, 2, the local density of states (LDOS) of four vortices state is shown. At $E = 0.14E_c$, there are separate

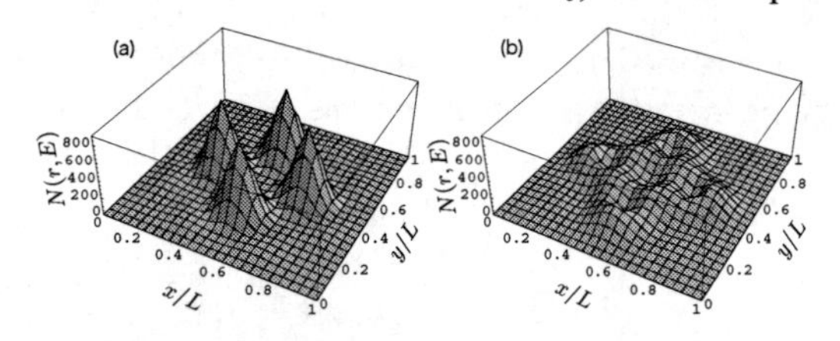

FIGURE 2. Local density of sates of four vortices at (a) $E = 0.14E_c$ and (b) $E = -0.14E_c$.

peaks from bound states of four vortices with angular momentum $m = 0$. At $E = -0.14E_c$, bound states of four vortices have angular momentum $m = 1$, are not separated and vortex bound states interfere mutually.

In Fig. 3, the LDOS of the five vortices state is shown. The peak at the center vortex is different from others. To understand the difference, we investigated five eigen-functions $u_n(\mathbf{r})$ and $v_n(\mathbf{r})$ made from five vortex bound states. Four of five eigenstates show the interference of bound states, but one eigenstate is same as the bound state of a single vortex at center. At $H/L^2 = 14\Phi_0$, the energy eigenvalue of two kinds of eigenstates are $E = 0.13487E_c$, $E = 0.15677E_c$, $E = 0.15840E_c$ and

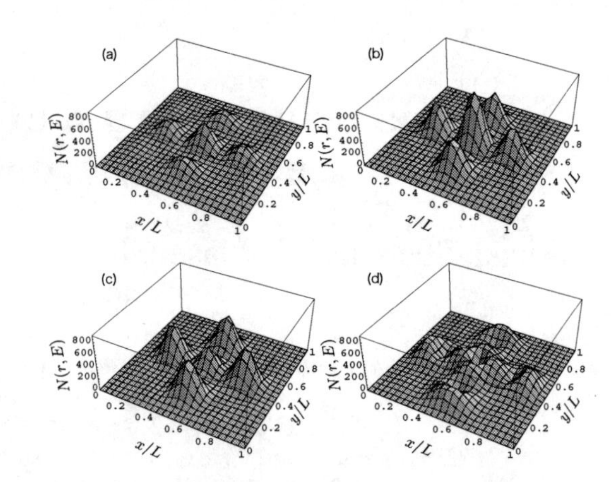

FIGURE 3. Local density of sates of five vortices at (a) $E = 0.20E_c$, (b) $E = 0.17E_c$, (c) $E = 0.13E_c$ and (d) $E = -0.17E_c$.

$E = 0.18368E_c$ for the former states and $E = 0.17601E_c$ for the latter state (and $E = 0.17814E_c$ for the vortex bound state of the single vortex). Therefore the height of peaks of the LDOS at $E \approx 0.17E_c$ becomes different between the center vortex and four vortices around it.

These results show the quasi-particle bound states form a molucular orbital-like states between different vortices. Therefore, we may consider that vortices form a kind of vortex-molecule.

CONCLUSION

We have solved Bogoliubov-de Gennes equation for a nano-scaled superconducting square plate under an external field. We have obtained the order parameter and the local density of states of four vortices and five vortices configurations. They form vortex-molecule-like state.

ACKNOWLEDGMENTS

Part of the numerical computation was carried out on the Computer Center of the Institute for Material Research, Tohoku University.

REFERENCES

1. L. F. Chibotaru, A. Ceulemans, V. Bruyndoncx, and V. Moshchalkov, *Nature*, **408**, 833 (2000).
2. L. F. Chibotaru, A. Ceulemans, V. Bruyndoncx, and V. Moshchalkov, *Phy. Rev. Lett.*, **86**, 1323 (2001).
3. H. Suematsu, M. Machida, T. Koyama, T. Ishida, and M. Kato, *Physica C*, **412-414**, 548 (2004).

Phase Dynamics of a Closed 0-π Josephson Junction

T. Koyama[a,e], M. Machida[b,e], M. Kato[c,e] and T. Ishida[d,e]

[a]*IMR, Tohoku University, Sendai 980-8577, Japan*
[b]*CCSE, Japan Atomic Energy Research Institute, Tokyo 110-0015, Japan*
[c]*Department of Mathematical Science, Osaka Prefecture University, Sakai 599-08531, Japan*
[d]*Department of Physics and Electronics, Osaka Prefecture University, Sakai 599-8531, Japan*
[e]*JST-CREST, Kawaguchi 332-0012, Japan*

Abstract. The phase dynamics of a closed 0-π junction is investigated. Such a system can be realized, for example, in a small $d(x^2\text{-}y^2)$-wave superconductor embedded in a conventional s-wave superconducting matrix. The Hamiltonian of the system is derived. We prove that the system can be mapped on a quantum two-level system when the system size is small enough.

Keywords: pi-junction, two-level system, Josephson effect.
PACS: 74.50+r, 74.20De

INTRODUCTION

In 0-pi Josephson junction systems, which includes both the regions of $j_c > 0$ and $j_c < 0$, j_c being the Josephson critical current density, vortices (a vortex) with fractional flux quantum spontaneously appear(s) [1-3]. Such 0-pi junction systems can be fabricated, using $d(x^2\text{-}y^2)$-wave and s-wave superconductors. In this paper we theoretically investigate the phase dynamics of a closed 0-pi Josephson junction system in which a small square-shape $d(x^2\text{-}y^2)$-wave superconductor is embedded in a s-wave superconducting matrix. We show that the closed 0-pi junction system can be considered as a quantum two-level system if the system size is small enough.

FORMULATION

We study phenomenologically the Josephson phase dynamics in a small square-shape $d(x^2\text{-}y^2)$-wave superconductor (d-dot) embedded in a conventional s-wave superconductor. The crystallographic orientation of the d-dot is shown in Fig.1. Suppose that the interface between the $d(x^2\text{-}y^2)$-wave and s-wave superconductors form the S-I-S Josephson junction. Since in this system the Josephson critical current density j_c changes its sign at the four corners of the square d-dot, this system forms a closed 0-π junction system. It is well known that a vortex with half-flux quantum spontaneously appears in 0-π junctions if the system size is large enough [1- 3]. Let the length of a side of the d-dot be a which is assumed much larger than the in-plane London penetration depths of both d- and s-wave superconductors. In this case the overlap of the magnetic flux on the different sides of the d-dot may be neglected. Hence, the system can be mapped on a one-dimensional Josephson junction model with the boundary condition,

$$\theta(x+4a,t) = \theta(x,t) + \frac{2\pi m}{4a}x. \qquad (1)$$

Where $\theta(x,t)$ is the phase difference at position x ($0<x<4a$) and at time t and m is an integer. Thus, one may assume that the under-damped phase dynamics in this system is described by the generalized sine-Gordon equation as

$$\frac{\varepsilon}{c^2}\partial_t^2\theta(x,t) - \partial_x^2\theta(x,t) = \frac{\eta(x)}{\lambda_J^2}\sin\theta(x,t), \qquad (2)$$

where λ_J is the Josephson penetration depth and $\eta(x)$ is a function taking either of two values, 1 or -1, depending on the sign of j_c, i.e., $\eta(x) = 1$ for the region of $j_c > 0$ and $\eta(x) = -1$ for. $j_c < 0$. Let the thickness of the insulating barrier and the width of the junction along the [001]direction be D and L respectively. In this case the Hamiltonian that leads to Eq.(2) is obtained as,

CP850, *Low Temperature Physics: 24th International Conference on Low Temperature Physics;*
edited by Y. Takano, S. P. Hershfield, S. O. Hill, P. J. Hirschfeld, and A. M. Goldman

$$H = E_J \int_0^{4a} dx\{\frac{c^2}{2\varepsilon E_J^2}\Pi(x)^2 + \frac{1}{2}(\partial_x\theta(x))^2 \tag{3}$$
$$-\frac{1}{\lambda_J^2}\eta(x)\cos\theta(x)\},$$

where $\Pi(x)$ is the canonical momentum of $\theta(x)$ and $E_J=(L/4\pi D)/(\phi_0/2\pi)$ with ϕ_0 being the unit flux. In the following we investigate the case of $\lambda_L< a << \lambda_J$, which may be realized, for example, when $a \sim 1\mu$m In this case the spatial variation of the phase difference is expected to be very weak, since λ_J gives the spatial scale, as seen in Eqs.(2) and (3). In order to describe this situation it is convenient to introduce the Fourier series expansions as

$$\theta(x) = Q + \frac{m\pi}{2a}x + \sqrt{2}\sum_{n=1} q_n \sin\frac{\pi n}{2a}x, \tag{4}$$

$$\Pi(x) = P/4a + (1/2\sqrt{2})\sum_{n=1} p_n \sin\frac{\pi n}{2a}x. \tag{5}$$

In Eqs.(4) and (5) we retained only the sine-components, because the cosines do not appear as a result of the symmetry of the system. In this paper we restrict ourselves to the case of m=0, i.e., the zero external field case. Note that if the canonical commutation relations are imposed for the Fourier coefficients,

$$[Q,P] = i\hbar, \quad [q_n, p_m] = i\hbar\delta_{nm},$$
$$[Q,p_n] = [q_n, P] = 0, \tag{6}$$

one may construct a quantum theory for the phase dynamics of the 0-π junction system. Since the higher harmonics components (q_n,p_n) are small in the present small d-dot system, the Josephson coupling term in the Hamiltonian is approximated as

$$\cos\theta(x) \approx \cos Q$$
$$-\sqrt{2}\sin Q\sum_{n=1} q_n \sin\frac{\pi n}{2a}x + ..., \tag{7}$$

Then, substituting Eqs.(4) and (5) into Eq.(3) and using the approximation (7), one can express the Hamiltonian in terms of the Fourier components. Since the dynamics of the higher harmonics components are linear in this approximation, the variables (q_n, p_n) can be integrated out. Then, after some calculations one can obtain the Hamiltonian in terms of only the uniform component (Q,P) which is valid for the system of $a << \lambda_J$ as

$$H = \frac{P^2}{8Ma_J} - \frac{a_J^4}{6}\sin^2 Q, \tag{8}$$

where $M = (2\pi)^2\varepsilon E_J/h^2c^2$ and $a_J=a/\lambda_J$. Note that the minima of the potential appear at $\sin Q = 1$ and -1, i.e., $Q = \pi/2$ and $-\pi/2$, in the region, $-\pi < Q < \pi$. These minima are the current-carrying states and the direction of the current at $Q = \pi/2$ is opposite to that at $Q = -\pi/2$. Thus, one concludes that the small d-dot system can be mapped on a quantum two-level system. When the quantum mechanical tunneling occurs between these two potential minimum states, the energy difference between the bonding and anti-bonding states is estimated from the Hamiltonian (8) as ΔE = 200 GHz for $L/4\pi D \sim 5$, $a_J \sim 0.1$, $\lambda_J \sim 0.001$cm. The details of the derivation and analysis of Eq.(8) will be published elsewhere.

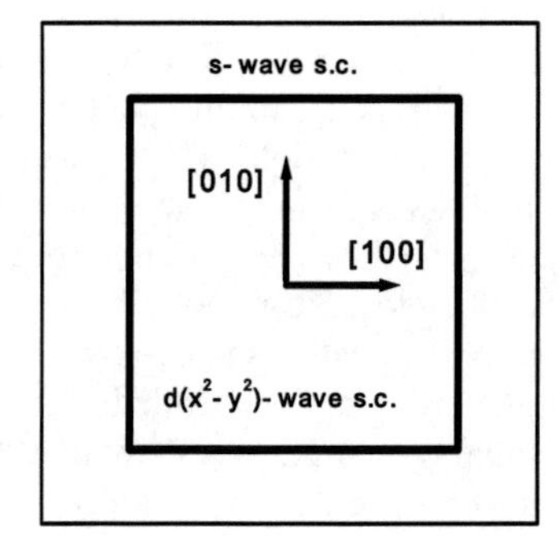

FIGURE 1. Schematic view of the d-dot system. The sign of the Josephson critical current changes at the corners.

REFERENCES

1. C.C. Tsuei and J. Kirtley, *Rev. Mod. Phys.* **72,** 969 (2000).
2. D.J. Harlingen, *Rev. Mod. Phys.* **67**, 515 (1995).
3. H. Hilgenkamp et al., *Nature* **422**, 50 (2003).

Physical Properties of composite structures of *d*- and *s*-wave superconductors (d-dot's)

Masaru Kato[*,†], Masaki Hirayama[*,†], Masahiko Machida[¶,†], Tomio Koyama[§,†], and Takakazu Ishida[**,†]

[*] *Department of Mathematical Sciences, Osaka Prefecture University, 1-1, Gakuencho, Sakai, Osaka 599-8531, Japan*
[¶] *CCSE, Japan Atomic Energy Research Institute, 6-9-3 Higashi-Ueno, Taito-ku, Tokyo 110-0015, Japan*
[§] *Institute for Materials Research, Tohoku University, Sendai 980-8577, Japan*
[**] *Department of Physics and Electronics, Osaka Prefecture University, 1-1 Gakuencho, Sakai, Osaka 599-8531, Japan*
[†] *JST-CREST, 4-1-8, Honcho, Kawaguchi, Saitama 332-0012, Japan*

Abstract. High-T_c superconducting dot embedded in an *s*-wave superconductor, which is called "d-dot", shows unusual properties. Especially, it shows spontaneous half-quantized magnetic fluxes at the corners of the dot because of the anisotropy of *d*-wave superconductivity. The d-dot can be considered as an artificial spin, because most stable states are doubly degenerate. We study stability of these degenerate states, using two-components Ginzburg-Landau equation. We show the energy difference between these degenerate states and excited states is comparable to the condensation energy of *s*-wave superconductor at T=0.

Keywords: High-temperature superconductors, Ginzburg-Landau equation, nano-superconductors, half-quantized vortex.
PACS: 74.78.Na, 74.72.-h, 74.81.Bd, 74.20.De

INTRODUCTION

The composite structure of high-T_c and conventional superconductors show spontaneous half-quantized magnetic fluxes under zero external field [1,2]. This is because of the anisotropy of the d-wave Cooper pairing of the high-T_c superconductors and the corner junction between them is used for determination of the pairing symmetry of high-T_c superconductors [3,4]. Experimentally, such magnetic fluxes were observed by the SQUID microscope in a zigzag junction between YBCO and Nb [5]. Ioffe et al. proposed the qubit that contains the single corner junction [6].

Unlike to the previous studies, we considered the high-T_c superconducting dot embedded in the s-wave superconductor [1,2]. We call it d-dot. Because the state with spontaneous magnetic fluxes breaks the time reversal symmetry, there are always two degenerate states with opposite magnetic fluxes. Therefore such structures also can be used for the two-states devices. It was shown that if the size of the dot is small enough, the transition of two degenerate states becomes quantum [7]. Therefore there is a possibility to use d-dot as a qubit. We also proposed a manipulation method that converts the one of the stable state to another stable state by the external current [8].

In the actual application of the d-dot as the devices, the stability of two degenerate states is important. Especially, stability against the excitation to the meta-stable states must be examined. In this study, we investigate such stability for the square d-dot.

MODEL

In order to investigate the physical properties of the d-dot system, we use phenomenological two-component Ginzburg-Landau (GL) equation. We consider three kinds of regions, which are the high-T_c superconductor, the s-wave superconductor and normal metal layer between them. In the high-T_c (s-wave) superconducting region, we only consider the *d*-wave order parameter $\Delta_d(\boldsymbol{r})$ (*s*-wave order parameter $\Delta_s(\boldsymbol{r})$), respectively. In the normal metal layer, we consider both of two order parameters, which are

CP850, *Low Temperature Physics: 24th International Conference on Low Temperature Physics;*
edited by Y. Takano, S. P. Hershfield, S. O. Hill, P. J. Hirschfeld, and A. M. Goldman

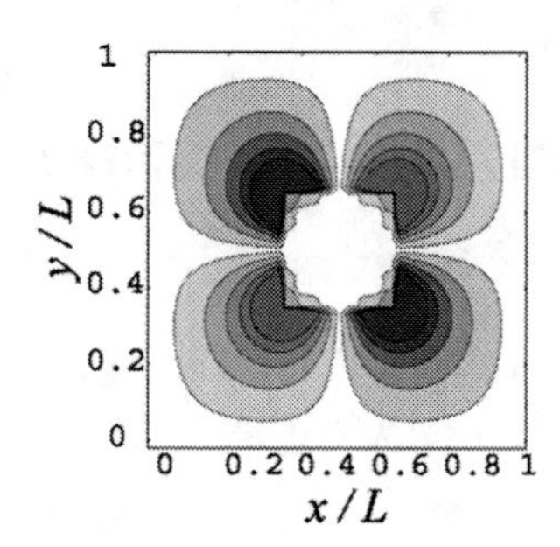

FIGURE 1. Magnetic field distribution of antiferromagnetic like state at T/T_{cs}=0.1 for a square d-dot. Dark (bright) color shows magnetic field is upward (downward).

continuous from the each of superconducting regions. The free energy for each region is given in Ref. [9].

RESULTS

We show numerical results for a square d-dot. We set coherence lengths for *d*- and *s*-wave superconductors as, $\xi_d / L = 0.05$ and $\xi_s / L = 0.5$, respectively, where L is the system size. GL parameters for *d*- and *s*-wave superconductors are set as, $\kappa_d = 15$ and $\kappa_s = 1.3$, respectively, and the ratio of transition temperature of both of superconductors is set as $T_{cs} / T_{cd} = 0.1$. Size of the d-dot is set to 0.3*L*.

The most stable states show the spontaneous magnetic fluxes at the four corners and the fluxes order antiferromagnetically. We show the magnetic field distribution at $T / T_{cs} = 0.1$ in Fig. 1.

In Fig.2, we show the field distribution of an excited state. In this state, the spontaneous magnetic fluxes appear at four corners, but they order as up, up, down and down. We call this state uudd-state.

We compare the free energies of these two states as shown in Fig. 3. In this figure, free energy is normalized by the condensation energy of *s*-wave superconductor in the same region at *T*=0, as a simple

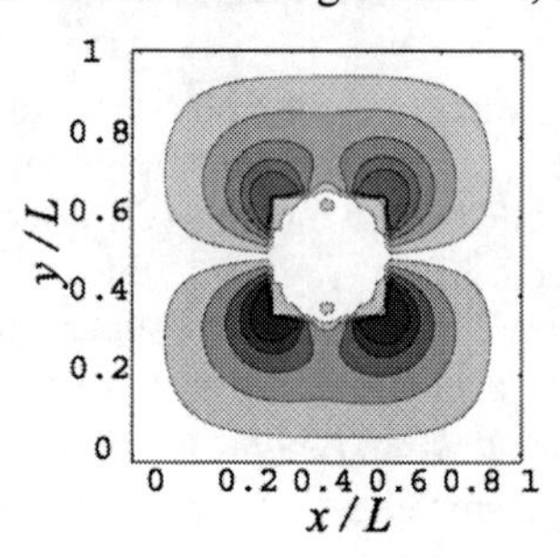

FIGURE 2. Magnetic field distribution of uudd state at T/T_{cs}=0.1 for square d-dot. Dark (bright) color shows magnetic field is upward (downward).

measure. Near the transition temperature of *s*-wave superconductor, difference of the free energies is small.

At low temperature, the excitation energy is comparable to the condensation energy of the *s*-wave superconductor. Therefore two degenerate states are much stable compared with the excited states at low temperature.

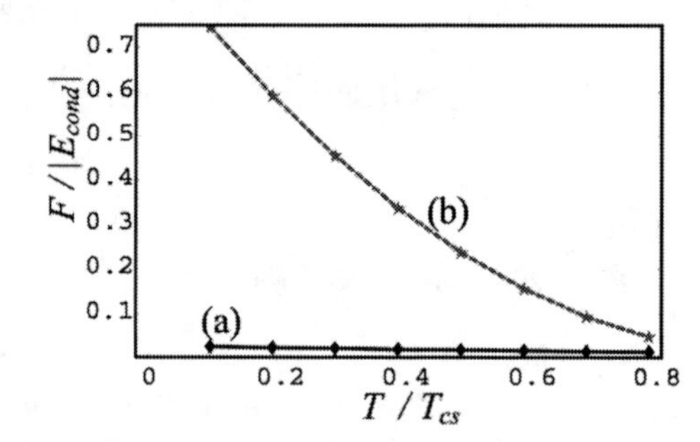

FIGURE 3. Temperature dependences of the free energies *F* for (a) AF-like state (Fig.1) and (b) uudd state (Fig.2). Free energy is normalized by the condensation energy of the condensation energy of the s-wave superconductor at *T*=0.

SUMMARY

We have investigated the stability of two degenerate states of the d-dot, which has spontaneous magnetic flux. The excitation energy at low temperature is comparable to the superconducting condensation energy.

ACKNOWLEDGMENTS

Part of the numerical work has been done on the computer facility of Japan Atomic Energy Research Institute.

REFERENCES

1. M. Kato, M. Ako, M. Machida, T. Koyama and T. Ishida, *Physica C* **412-414**, 352-357 (2004).
2. M. Ako, M. Machida, T. Koyama, T. Ishida and M. Kato, *Physica C* **412-414**, 544-547 (2004).
3. D. J. Van Harlingen, *Rev. Mod. Phys.* **67**, 515 (1995).
4. C. C. Tuei and J. R. Kertley, *Rev. Mod. Phys.* **72**, 969 (2000).
5. H. Hilgenkamp, Ariando, H.-J. H. Smilde, D. H. A. Blank, G. Rijnders, H. Rogalla, J. R. Kirtley and C. C. Tsuei, *Nature (London)* **422**, 50 (2003).
6. L. B. Ioffe, V. B. Geshkenbein, M. V. Fingel'man, A. L. Fauchère and G. Blatter, *Nature(London)* **398**, 679 (1999).
7. T. Koyama, M. Machida, M. Kato and T. Ishida, *Physica C*, **412-414**, .358-361 (2004).
8. M. Ako, M. Machida, T. Koyama, T. Ishida and M. Kato, *Physica C*, in press.
9. M. Hirayama, M. Machida, T. Koyama, T. Ishida and M. Kato, *Physica C*, in press.

Critical Field Of A Type-I Spherical Nanoscale Superconducting Inclusion

O.N. Shevtsova

Institute for Nuclear Research
National Academy of Sciences of Ukraine
Pr. Nauki, 47, Kiev, 03680, Ukraine

Abstract. The method of the critical magnetic field calculation for a type-I superconducting spherical inclusion was developed. The dependence of the critical field as inclusion radius function was calculated for different types of the boundary condition. The proposed method gives the possibility to determine the critical field value with any desirable accuracy.

Keywords: I-type superconductor.
PACS: 74.25.Ha

INTRODUCTION

The finiteness of the inclusion size leads to a change in the properties in comparison with those of bulk superconductors. It is well known that in type-I bulk superconductors the magnetic field suppresses the superconductivity. However, in small samples the threshold field is considerably higher than the bulk critical field [1]. Ginzburg [2] showed that for a type-I superconducting spherical inclusion with a radius less than the coherence length, the critical magnetic field is changed inversely with the radius. Superconducting inclusions in crystals can have artificial origination, and can be caused by different technological processes. Three-dimensional nanoscale structures embedded in material were created by the injection of the melted metal in porous glass [3, 4]. The critical fields of such confined superconductors are essentially higher than the bulk critical fields. The superconducting inclusions imbedded in a crystal lead to change of the host crystal properties at low temperatures. In particular, the conductivity of the crystal increases and a strong dependence of the conductivity on magnetic field takes place. This peculiarity is caused by the superconducting-normal phase transitions of the inclusions upon increase of the magnetic field [5].

CRITICAL FIELD OF A SPHERICAL NANOSIZE INCLUSION

Let us consider a superconducting inclusion embedded in a crystal in the Ginzburg-Landau (GL) approach.

$$\left(-i\nabla-\frac{\mathbf{A}}{\sqrt{2}\kappa}\right)^2\psi-\alpha\psi+|\psi|^2\psi=0 \qquad (1)$$

where ψ is the superconducting order parameter, κ is the GL parameter, $\alpha=1-T/T_{c0}$, T_{c0} is the bulk critical temperature, H is the external magnetic field. We consider the system behavior near the phase transition ($\psi\to 0$), so the deviation of the magnetic field from the external field caused by the current's circulation around the inclusion is small (order of magnitude $|\psi|^2$), and the vector potential of the complete magnetic field can be written as $\mathbf{A}=[\mathbf{H}\times\mathbf{r}]/2$.

Equation (1) shall be supplemented with the general de Gennes boundary condition for the order parameter at the inclusion surface [6]:

CP850, *Low Temperature Physics: 24th International Conference on Low Temperature Physics;*
edited by Y. Takano, S. P. Hershfield, S. O. Hill, P. J. Hirschfeld, and A. M. Goldman

$$\frac{\partial\psi}{\partial\rho}+\frac{1}{b}\psi=0, \quad \text{at } \rho=R, \qquad (2)$$

where 'extrapolation length' b depends on the temperature and type of the material contacting with a superconducting inclusion. The two limit cases $b = 0$ and $b \to \infty$ correspond to the superconductor-normal metal interface and the interface of low-temperature superconductor and vacuum respectively. The critical magnetic field of the superconducting inclusion is defined as the value below which the normal phase is unstable with respect to initiation of the superconducting phase nucleus. So, to determine the critical magnetic field of a spherical form sample it is necessary to solve the linearized equation (1) in spherical coordinates with the origin in the centre of the sphere:

$$-\Delta\psi+\iota\frac{H}{\sqrt{2\kappa}}\frac{\partial\psi}{\partial\varphi}+\frac{H^2\rho^2}{8\kappa^2}\sin^2\theta\psi=\alpha\psi \quad (3)$$

where $\rho = r/\xi_0$ is the dimensionless radius vector with the origin in the centre of the spherical inclusion, the angle θ is counted from the magnetic field direction, the magnetic field is chosen in the units of the thermodynamic magnetic field H_{c0}, ξ_0 is the coherence length.
For an inclusion whose radius is less than the penetration depth the lowest value of α in the case of $b \to \infty$ corresponds to $\psi = const$, i.e. a superconducting nucleus fills up the entire inclusion volume, and the calculated critical field value coincides with the well-known Ginzburg result: $H_c/H_{c0} = \sqrt{20\kappa}/R$. If the inclusion radius is larger than the penetration depth there is occurred magnetic field expelling (Meissner effect) and magnetic field penetrates in a superconductor only in the surface layer. For such inclusions there is realized the flux quanta entrapping (curve 1 in Figure 1). Resulting dependence of H_c is presented in Fig.1. The obtained dependencies indicate that the value of H_c is determined by the values b and R. At each R one should choose the maximal value of H_c among those corresponding to different m. The dependence of H_c at small R coincides with the Ginzburg results and with the increase of R we can observe the entrapping of the magnetic field quanta. Role of "the extrapolation length" value b is very important because at decreasing of b the superconductivity is suppressed (see curves 2 (b =5) and 3 (b =1) in Fig.1). With the increase of the inclusion radius R the amplitude of the oscillations is decreasing and they are practically indistinguishable.

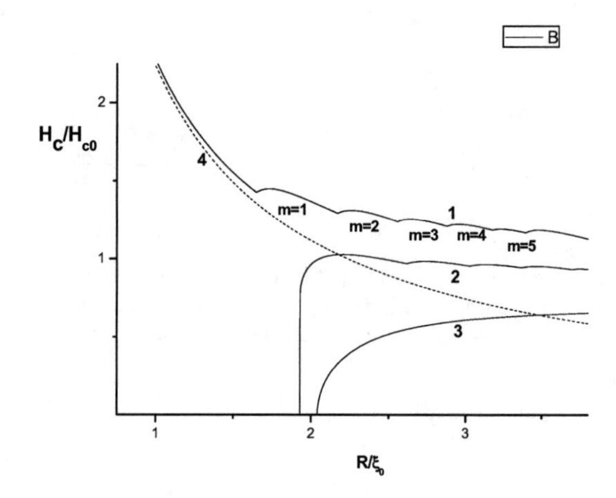

FIGURE 1. Critical field versus the inclusion radius at different values of the de Gennes 'extrapolation length' b: $b \to \infty$ (curve 1), $b = 5$ (curve 2), $b = 1$ (curve 3). Dashed curve (4) describes the Ginzburg solution. The values m below curves denote the number of flux quanta trapped by the inclusion (κ=0.5). H and R are presented in the units of the critical field H_{c0}, and the coherence length ξ_0, respectively.

ACKNOWLEDGMENTS

This research work was partially supported by Ukrainian Ministry of Education and Sciences, Project # 02.07/00147.

REFERENCES

1. R. A. Hein and M. C. Steele, *Phys. Rev.* **105**, 877-882 (1957).
2. V. L. Ginzburg, *Zh. Eksp. Teor. Fiz.* **34**, 113-125 (1958).
3. M. J. Graf, T. E. Huber and C. A. Huber, *Phys. Rev. B* **45**, 3133- 3136 (1992).
4. E. V. Charnaya, C. Tien, K. J. Lin and C. S. Wur, *Phys. Rev. B* **58**, 467-472 (1998).
5. V. I. Sugakov and O. N. Shevtsova, *Supercond. Sci. Technol.* **13**, 1409-1414 (2000).
6. P. G. de Gennes, *Superconductivity of metals and alloys*, New York: W. A. Benjamin, 1966.
7. O. N. Shevtsova, *Supercond. Sci. Technol.* **18**, 726-729 (2005).

Superconducting State of Small Nanoclusters: "Giant" Enhancement of T_c and Other Parameters

Vladimir Z. Kresin* and Yurii N. Ovchinnikov**

*Lawrence Berkeley Laboratory, University of California, Berkeley, CA 94619, USA
** L. Landau Institute of Theoretical Physics, 117334 Moscow, Russia

Abstract. The presence of shell structure and accompanying high level degeneracy leads to strong pairing interaction in some metallic nanoclusters. These clusters form a new family of high temperature superconductors.

Keywords: shell structure, degeneracy, pairing.
PACS: 74.70.-b; 36.40.Cg; 74.78.Na

INTRODUCTION

Small metallic nanoclusters contain delocalized electrons whose states organize in to shells similar to those in nuclei and atoms. As a result, many clusters contain highly degenerate electronic levels, or groups of very close levels (quasi-degenerate case); the picture is very different from equidistant energy levels distribution. High level degeneracy leads to strengthening of the pairing interaction in some "magic" clusters (these clusters contain complete shells), or clusters with slightly incomplete shells.

THEORY

We focus on clusters with $N \approx 10^2$-10^3 (N is a number of delocalized electrons). For example, for the spherical "magic" clusters with N=168 (then the orbital momentum for the highest occupied shell (HOS) is equal to l=7) the degeneracy is $g=2(2l+1)=30$. Such a high degeneracy is a favorable factor for the pairing. In addition, it is essential, that the spacing between the highest occupied shell (HOS) and the lowest unoccupied shell (LUS) is relatively small. If the shell is slightly incomplete, it is realistic that the splitting of the levels is relatively small, and the impact of pairing is still strong.

Pairing is caused by the electron- vibrational interaction. The equation for the pairing order parameter has the form [1]:

$$\phi(x_n) = \sum_m K(x_n, y_m)\phi(y_m) \; ; \text{n, m}>0 \quad (1)$$

Here

$$K(x_n, y_m) = \tilde{\lambda}\left(f^+ + f^- - 4y_m^2 f^+ f^-\right)\times \sum_s \left[y_{n'}^2 + \chi_s^2 + \phi^2(y_m)\right]^{-1}, \quad (2)$$

with

$$f^{\pm} = \left(1 + (x_n \pm y_m)^2\right)^{-1}, \quad (3)$$

$$\tilde{\lambda} = \lambda E_F \left(2\pi N \tilde{\Omega}^2\right)^{-1}, \quad (4)$$

$$x_n = \omega_n \tilde{\Omega}^{-1}, \quad (5)$$

$$\phi(x_n) = \Delta(x_n)\tilde{\Omega}^{-1}, \quad (6)$$

$$\chi_s = \xi_s \tilde{\Omega}^{-1}, \quad (7)$$

$\tilde{\Omega}$ is the characteristic vibrational frequency, $\xi_s = E_s - \mu$ is the electronic energy referred to the chemical potential, index "s" labels different energy levels , V is the cluster volume, $\lambda = \langle I \rangle \nu / M\tilde{\Omega}^2$ is the bulk coupling constant [4], $\omega_n = (2n+1)\pi T$. For "magic" clusters the sum in Eq. 2 is a summation over different complete shells, so that $\sum_s \to \sum_j g_j$, where g_j is the degeneracy of the j_{th} shell. If the shell is incomplete, the label "s" corresponds to the projection of the angular momentum. The position of the chemical potential is determined by the conservation

CP850, *Low Temperature Physics: 24th International Conference on Low Temperature Physics;* edited by Y. Takano, S. P. Hershfield, S. O. Hill, P. J. Hirschfeld, and A. M. Goldman

of the total number of electrons. At $T=T_c$ one should put $\phi=0$ in expression 2. At T=0 K the summation over n in Eq. 1 can be replaced by an integral.

Equations 1 and 2 can be used to calculate the values of T_c and the energy gap parameter. The values of these quantities for specific clusters are determined by the following parameters: $\tilde{\Omega}$, N, λ, E_F, and ξ_s. For example, for the Ga clusters with N=168 we obtain $T_c \approx$ 140K. For similar Zn clusters $T_c \approx 10^2$ K; if N=166 (unoccupied shell) we obtain for $T_c \approx$ 124 K for Zn. For Cd clusters we obtain $T_c \approx$ 73.5 K (N=168) and $T_c \approx$ 90 K (N=166). For Pb clusters with N=380, we obtain $T_c \approx$ 50 K.

It is remarkable that pairing leads to the possibility to observe superconducting state with high value of T_C, much higher than those for the bulk samples. Qualitatively, the high value of T_c is due to the high degeneracy of the HOS which corresponds to a sharp peak in the density of states at the Fermi level (see Ref. [5]).

The value of the gap parameter is also large. For example, for Cd clusters with N=166 we obtain $\Delta \approx$ 20meV. One can study also the region near T_c and derive the Ginzburg-Landau equations.

The impact of fluctuations is not strong, because the large value of the gap parameter leads to relatively small coherence length, which is comparable with the cluster size. A more detailed analysis shows that this factor leads to the broadening of the transition which is of the order of 5-9%.

Pair correlation can manifest itself in odd-even effects for cluster spectra and in their magnetic properties similar to those observed in Ref. [6]. Moreover, the presence of the energy gap at $T<T_c$ leads to a strong temperature dependence of the energy spectrum. A large difference in the energy spectrum near T=0 K and at $T>T_c$ can be observed experimentally. Such observation would demonstrate the presence of pair correlation in small nanoclusters. The phenomenon is promising for the creation of high T_c tunneling contacts. Probably, it would require use of a special matrix similar to that used for Pb nanoparticles in Ref. [7].

ACKNOWLEDGMENTS

The authors are grateful to J.Friedel, A.Goldman, V.V.Kresin and M.Tinkham for fruitful discussions. Research of VZK is supported by DARPA under the Contract 05U716. The research of YNO was supported by the CRDF under Contract No. RP1-2565-MO-03 and by RFBR (Russia).

REFERENCES

1. W. Knight et al., *Phys. Rev. Lett*. **52**, 2141 (1984).
2. W. de Heer, *Rev. Mod. Phys*. **65**, 611 (1993).
3. Y. Ovchinnikov and V. Kresin, *Eur. Phys. J. B* **45**, 5 (2005).
4. W. McMillan, *Phys. Rev.* **167**, 331 (1968).
5. J. Labbe et al., *Phys. Rev. Lett.* **19**, 1039 (1967); J. Friedel, *J. Phys.* **2**, 959 (1992).
6. M. Tinkham et al., *Phys. Rev. B* **51**, 12649 (1995).
7. L. Adams, B. Lang, A. Goldman, *cond-mat*/0502559.

Superconductivity in Multielectron Bubbles in Helium

J. Tempere[1,2], V.N. Gladilin[1], I.F. Silvera[2], and J.T. Devreese[1]

[1]*TFVS, Universiteit Antwerpen, Universiteitsplein 1, B-2610 Antwerpen, Belgium.*
[2]*Lyman Laboratory of Physics, Harvard University, Cambridge MA02318, USA.*

Abstract. Superconductivity on a spherical, two-dimensional surface is investigated in the context of multielectron bubbles in liquid helium, where the electron-ripplon coupling is the underlying mechanism leading to pairing. In this contribution, we apply a BCS-like self-consistent mean-field theory to calculate the superconducting gap. The results obtained here are in agreement with our previous calculations based on Richardson's method.

Keywords: Superconductivity, electron-ripplon coupling, multielectron bubbles, electrons in reduced dimensions.
PACS: 71.10.Pm, 74.20.Fg, 74.78.-w, 74.10.+v

INTRODUCTION

Multielectron bubbles are cavities inside liquid helium, containing from a few up to 10^8 electrons [1]. Such charged bubbles have a radius of about 1 micron for N=10^4 electrons, and this radius scales as $N^{2/3}$ at zero external pressure. The electrons in the bubble do not spread out in the entire volume, but collect in a nanometer thin layer anchored to the surface of the bubble. In the radial direction, only a single mode is occupied so that the system corresponds to a spherical, two-dimensional electron gas [2].

In equilibrium, the surface of the bubble is spherical. Modes of oscillation around this equilibrium surface exist, and can be quantized as 'ripplons'. The dispersion relation for the ripplons on a spherical surface was derived previously [3]. As the bubble surface deforms, the electrons along the surface can move, and redistribute the surface charge density. This leads to a coupling between the electrons and the ripplons, which can result in the formation of a lattice of ripplonic polarons [4]. In a recent work [5], the present authors have investigated how the electron-ripplon coupling can lead to pairing correlations in the electron gas, following the argument first formulated by Cooper for electrons and phonons [6].

In this contribution, we will apply a mean-field self-consistent approach to study the pairing properties of electrons in the coupled electron-ripplon system, and compare the results with those found in Ref. [5] using Richardson's method [7].

BCS HAMILTONIAN ON A SPHERE

In Ref. [5], we show that Cooper's argument for electrons and ripplons leads to the following Hamiltonian for the electronic system:

$$\hat{H} = \sum_L \hat{H}_L \tag{1}$$

with

$$\begin{aligned}\hat{H}_L = & \sum_{m=-L}^{L} \sum_{\sigma=\uparrow,\downarrow} \varepsilon_L \hat{c}^+_{L,m,\sigma} \hat{c}_{L,m,\sigma} \\ & - G \sum_{m,m'=-L}^{L} \hat{c}^+_{L,m',\uparrow} \hat{c}^+_{L,-m',\downarrow} \hat{c}_{L,-m,\downarrow} \hat{c}_{L,m,\uparrow}.\end{aligned} \tag{2}$$

In this expression, $c^+_{L,m,\sigma}$ and $c_{L,m,\sigma}$ are the creation and annihilation operators for an electron in angular momentum state $\{L,m\}$ and with spin σ. The Hamiltonian (2) is of the same type as the Bardeen-Cooper-Schrieffer (BCS) Hamiltonian [8]. Here the angular momentum plays the role of the wave number. The effective BCS interaction strength is given by G and the free-particle energy levels are

$$\varepsilon_L = \frac{\hbar}{m_e R^2} \frac{L(L+1)}{2}, \tag{3}$$

where R is the radius of the bubble, and m_e is the electron mass. As can be seen from (1), the system

CP850, *Low Temperature Physics: 24th International Conference on Low Temperature Physics;*
edited by Y. Takano, S. P. Hershfield, S. O. Hill, P. J. Hirschfeld, and A. M. Goldman

consists of a collection of independent subsystems, one for each subspace of single-particle levels with the same angular momentum L. The underlying reason for this is that the relevant ripplon energies are much smaller than the interlevel spacing $\varepsilon_{L+1}-\varepsilon_L$. Typically, the interaction strength G varies from 1 mK (for large, unpressurized bubbles with radii of 10-100 micron) to 100 mK (for small or compressed bubbles, with radii of 0.1-1 micron). The energy scale $\varepsilon_{L=1}$ is typically an order of magnitude or more smaller than G.

Using Richardson's method [7] the energy spectrum of the Hamiltonians of the subsystems can be derived. The energy is minimized when pairing correlations are present. To destroy these pairing correlations, a pair-breaking energy $\Delta_L=G(2L+1)$ needs to be supplied [5]. In order to complement and check these previous conclusions, we apply in the next section the BCS formalism to obtain a gap equation, and compare the solution of the gap equation to the pair breaking energy obtained in the Richardson formalism.

GAP EQUATION

We introduce a BCS-type wave function, with variational parameters $u_{L,m}$ and $v_{L,m}$:

$$|\Psi\rangle = \prod_L |\Psi_L\rangle \tag{4}$$

with

$$|\Psi_L\rangle = \prod_{m=-L}^{L} \left(u^*_{L,m} + v_{L,m} \hat{c}^+_{L,m,\uparrow} \hat{c}^+_{L,-m,\downarrow} \right) |0\rangle \tag{5}$$

where $|0\rangle$ is the (electron) vacuum. The Cooper pairs consist of electrons with opposite spin and opposite z-component of the angular momentum, but with the same magnitude of the angular momentum. The total energy E is a sum over the energy contributions E_L of each L subsystem. Each energy contribution E_L can be calculated as the expectation value of the Hamiltonian (2) with respect to the wave function (5). Introducing the gap,

$$\Delta_L = -G \sum_{m=-L}^{L} u_{L,m} v_{L,m}, \tag{6}$$

the variational equations for $u_{L,m}$ and $v_{L,m}$ can be solved self-consistently, leading to a gap equation

$$1 = \frac{G(2L+1)}{2\sqrt{(\varepsilon_L - \mu)^2 + \Delta_L^2}}, \tag{7}$$

where μ is the Lagrange multiplier needed to keep the total number of electrons fixed during variation. If we now solve this gap equation at the Fermi level L_F, we obtain:

$$\Delta_{L_F} = G(2L+1)/2. \tag{8}$$

CONCLUSIONS

At first sight, expression (8) seems to be only half of value obtained with the Richardson method. However, the Richardson gap is a spectroscopic gap, which is indeed supposed to be twice the BCS gap. So, we conclude that the straightforward BCS treatment of the Hamiltonian (1) for ripplonic superconductivity leads to the same result for the gap at L_F as the solution obtained in Ref. [5]. Note that the BCS treatment sketched here does not give access to the density of (excited) states, nor does it take into account the redistribution of electrons over different L levels shown in Ref. [5]. Nevertheless, the BCS result constitutes an independent check on the previous work.

J. Tempere gratefully acknowledges support from the FWO-Vlaanderen and from the Special Research Fund of the University of Antwerp, BOF NOI UA 2004. This work is supported financially by FWO-V projects Nos. G.0435.03, G.0306.00, G.0274.01N, the W.O.G. WO.035.04N, the GOA BOF UA 2000, and the Department of Energy, Grant No. DE-FG02-ER45978.

REFERENCES

1. A. P. Volodin, M. S. Khaikin, and V. S. Edelman, *Pis'ma Zh. Eksp. Teor. Fiz.* **26**, 707 (1977) [*JETP Lett.* **26**, 543 (1977)]; U. Albrecht and P. Leiderer, *Europhys. Lett.* **3**, 705 (1987).
2. J. Tempere, I. F. Silvera, and J. T. Devreese, *Phys. Rev. B* **65**, 195418 (2002).
3. J. Tempere, I. F. Silvera and J. T. Devreese, *Phys. Rev. Lett.* **87**, 275301 (2001).
4. J. Tempere, S. N. Klimin, I. F. Silvera and J. T. Devreese, *European Physical Journal B* **32**, 329 (2003).
5. J. Tempere, V. N. Gladilin, I. F. Silvera and J. T. Devreese, submitted to *Phys. Rev. B*, preprint available at cond-mat/0505721.
6. L. N. Cooper, *Phys. Rev.* **104**, 1189 (1956).
7. R.W. Richardson, *Phys. Lett.* **3**, 277 (1963).
8. J. Bardeen, L. N. Cooper and J. R. Schrieffer, *Phys. Rev.* **108**, 1175 (1957).

Size Effect on Flux Creep in $Bi_2Sr_2CaCu_2O_{8+y}$

Z. X. Shi[*,†], N. Kameda[*], M. Tokunaga[*] and T. Tamegai[*]

[*]*Department of Applied Physics, The University of Tokyo, 7-3-1 Hongo, Bunkyo-ku, Tokyo 113-8656, Japan*
[†]*Department of Physics, Southeast University, Nanjing 210096, China*

Abstract. The sample size effect on the flux pinning and vortex dynamics has been studied on micron-sized $Bi_2Sr_2CaCu_2O_{8+y}$ crystals. The second magnetization peak (SMP) and the magnetic relaxation rate change with the sample size. A crossing point is found in field dependence of the normalized magnetic relaxation rate *S* for samples with different sizes. There is a sudden increase of *S* at the SMP field, which possibly originates from the propagation of phase boundary of vortex matter at the SMP field.

Keywords: second magnetization peak, flux creep, size effect
PACS: PACS: 74.25.Qt; 74.25.Sv; 74.78.Na

INTRODUCTION

Vortex lattice (VL) state plays an important role in both static and dynamic properties of type II superconductors. The VL state affects flux creep, flux flow, flux jump as well as the summation of flux pinning force determined by flux bundle size. However, VL state is strongly dependent on crystal defect, superconductivity anisotropy, magnetic field, thermal fluctuation and sample size. VL in high temperature superconductors (HTSC) has a rich phase diagram and there are many vortex phases, such as ordered vortex solid, weakly (elastically) disordered quasi-lattice (called Bragg glass), highly (plastically) disordered glass, vortex liquid, and two-dimensional pancake gas due to the large anisotropy, short coherence lengths, and higher operating temperatures of HTSC. The richness of VL phases, especially in $Bi_2Sr_2CaCu_2O_{8+y}$ (BSCCO) system, determines the shape of magnetization curves. In this paper, we have studied the sample size effect on the flux pinning and the vortex dynamics in BSCCO crystals. A sudden increase of the normalized magnetic creep rate *S* was found at the second magnetization peak (SMP) field, which possibly originates from the propagation of the phase boundary of VL at the SMP field.

EXPERIMENTAL

Single crystals of BSCCO were grown using the floating-zone method. A large crystal was cleaved into thickness of $10-15$ μm and cut into various sizes using a dicing saw. Local magnetization was measured by placing the sample onto a micro-Hall probe (30×30 μm^2). Magnetic hysteresis loops (MHLs) were measured with applied field parallel to the *c*-axis at a fixed temperature and at a constant field sweep rate of 2.8 Oe/s. The relaxations of the trapped magnetic flux were measured at fixed temperature and fields after increasing the field to the maximum and decreasing it to the target value at a sweep rate of 8 Oe/s. The sizes of small, intermediate, and large crystals are $53\times50\times15$ μm^3, $1000\times50\times12$ μm^3, $1000\times150\times12$ μm^3, respectively.

RESULTS AND DISCUSSION

MHLs were measured on three samples with large, intermediate, and small sizes at temperatures from 20 K to 35 K every 2.5 K. The shape and height of SMP are quite different for these samples with different sizes. The height of SMP decreases with increasing temperature and disappears at a higher temperature T_{cr}. Instead of SMP, VL melting appears at temperatures higher than T_{cr}. The height of SMP for the large sample is almost 10 times as large as that for the small sample, indicating that there is a real size effect on the formation of SMP. In other words, SMP only appears below a sample-size-dependent critical temperature.

To clarify the origin of SMP, the sample size effect on dynamic properties of vortices has been investigated further. The magnetization relaxation has been measured on three samples with different sizes. Since the intermediate sample shows intermediate behavior, only the characteristics of the large and the small samples will be compared. The relaxation rate changes with time due to the large thermal fluctuation and collective pinning. For analysis, we calculate and compare the relaxation rate at some short time windows, such as $1\times10^2-2\times10^2$ s and $1\times10^3-2\times10^3$ s. Figure 1 shows the field dependence of the normalized relaxation rate $S=-dlnM/dlnt$ at 27.5 K for two samples with different sizes. It is ob-

CP850, *Low Temperature Physics: 24th International Conference on Low Temperature Physics;*
edited by Y. Takano, S. P. Hershfield, S. O. Hill, P. J. Hirschfeld, and A. M. Goldman

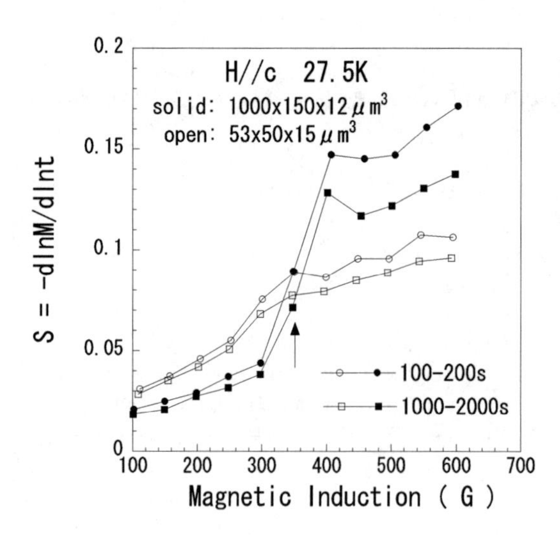

FIGURE 1. Field dependence of the normalized relaxation rate S for crystals with different sizes at 27.5 K.

vious that S increases with the applied field and there is a jump in the field dependence of S at the SMP field. The remarkable feature is a crossing point in Fig. 1. S for the large sample is lower than that for the small sample at smaller fields below the crossing point, and larger than that for the small sample at higher fields beyond the crossing point. This means that the sudden change in S for the large sample is larger than that for the small sample.

To explain the sample size effect on the formation of SMP and the dynamic properties of vortices at fields near SMP, it is necessary to discuss the physical origin of SMP. A lot of experiments have been done to study the mechanism of SMP and many models have been proposed to explain the SMP [1, 2, 3, 4]. All explanations can be divided conceptually into a static and a dynamic picture. Though the physical picture of the SMP formation is not very clear yet, we believe that the real physical origin of SMP is more like a kind of VL phase transition [2, 3] rather than any other model [4]. Due to disorder-induced phase transition of the VL from elastic quasi-lattice to plastic metastable disordered glass, there are two flux pinning regions with lower and higher critical current density separated by the magnetic field B_{OD}, which corresponds to the SMP field H_p. The static and dynamic behavior of SMP as well as the sample size dependence of SMP can be explained by this model.

In this physical picture, the sample size effect on the height of SMP is explained by considering the characteristic time t_0, which is related to the sample size. According to the collective pinning theory, the magnetization relaxes with time non-logarithmically as $M(t) = M(0)\{1+\mu k_B T/U_c ln[1+t/t_0]\}^{-1/\mu}$, where U_c is the collective pinning energy, $t_0 = \pi T d^2/\{2|\partial_j U| c v_0 H\}$, μ is the glass exponent, d is the sample size, v_0 is the vortex vibrating frequency. So the time constant t_0 depends on sample size d as $t_0 \propto d^2$. In the logarithmic approximation, $U = kTln(t/t_0)$ and $U = U_0 ln(J_c/J)$, one obtains: $J = J_c(t_0/t)^{1/n}$, where $n = U_0/kT$. So the large sample has a larger t_0 and a larger J, and it's MHL at the SMP regions is fatter than that of the small sample.

The jump of S at the SMP field is caused by the propagation of the vortex phase boundary between elastic quasi-lattice (below B_{OD}) and plastic metastable disordered glass (above B_{OD}). The motion of the boundary between two different vortex states may accelerate the magnetization relaxation and result in the sudden increase of relaxation rate at the SMP field. For a larger sample, the induction profile is steeper and the phase transition could propagate in a wider space and cause a larger jump of S at fields near SMP. At the same time, the lifetime of the metastable state is shorter for a larger sample, so that the motion of the phase boundary is faster and the relaxation rate is higher for a larger sample. The magnetic relaxation is accelerated by the propagation of vortex phase boundary at the SMP field and the vortex phase transition is limited or suppressed by the sample size so that there is sample-size-effect on the dynamic behavior of SMP and results in the crossing point in Fig. 1. Details of calculation will be published elsewhere.

In conclusion, the sample size effect has been found both on the pinning and dynamics of vortex. The height of SMP and the normalized relaxation rate S depend on the sample size. A sudden change of S happens at the SMP field, which may be a direct evidence that the second magnetization peak is caused by the vortex matter phase transition and the large S at the SMP field originates from the propagation of the phase boundary.

ACKNOWLEDGMENTS

This work is partly supported by Grant-in-Aid for Scientific Research from the Ministry of Education, Culture, Sports, Science, and Technology, Japan; and the Project-sponsored by SRF for ROCS, SEM.

REFERENCES

1. T. Tamegai, Y. Iye, I. Oguro, and K. Kishio, *Physica C* **213**, 33 (1993).
2. B. Kalisky, Y. Bruckental, A. Shaulov, and Y. Yeshurun, *cont-mat*/0311037.
3. B. Kalisky , D. Giller, A. Shaulov , Y. Yeshurun and T. Tamegai, to be published in *Phys. Rev. B*.
4. V. F. Correa, G. Nieva, and F. de la Cruz, *Phys. Rev. Lett.* **87**, 057003 (2001).

Vortex States under Tilted Fields in $Bi_2Sr_2CaCu_2O_{8+y}$ and $YBa_2Cu_3O_{7-\delta}$

T. Tamegai, H. Aoki, M. Matsui, and M. Tokunaga

Department of Applied Physics, The University of Tokyo, 7-3-1 Hongo, Bunkyo-ku, Tokyo 113-8656, Japan

Abstract. We have visualized vortex arrangements under tilted fields in $Bi_2Sr_2CaCu_2O_{8+y}$ (BSCCO) and $YBa_2Cu_3O_{7-\delta}$ (YBCO) with various doping levels by Bitter decorations. In BSCCO, when the in-plane field (H_x) is small, a state with coexisting vortex lattices and vortex chains is observed. At larger H_x (>1 kOe), all pancake vortices are absorbed in the chains and the spacing between them becomes uneven, resulting in the formation of vortex chain bundle. In underdoped YBCO, although the anisotropy parameter is lower than the threshold value for the stability of the crossing-lattices state, we still observe coexisting vortex lattices and vortex chains under tilted fields. We interpret this coexisting state as consisting of normal vortex lattices and tilted vortex chains.

Keywords: crossing lattices, vortex chain, Bitter decoration, magneto-optical method
PACS: 74.25.Qt, 74.72.Hs 74.72.Bk

INTRODUCTION

In anisotropic superconductors, quantized vortices form various arrangements under tilted fields. When the anisotropy parameter, γ, is very large as in $Bi_2Sr_2CaCu_2O_{8+y}$ (BSCCO), vortex lines along the tilted magnetic field are no longer stable and a novel crossing-lattices state with coexisting pancake vortices (PVs) and Josephson vortices (JVs) is realized [1]. In the crossing-lattices state, attractive interactions between PVs and JVs give rise to a formation of one-dimensional arrangements of PVs on JVs, vortex chains (VC), embedded in ordinary vortex lattices (VLs) [2]. On the other hand, in the case of less anisotropic $YBa_2Cu_3O_{7-\delta}$ (YBCO), tilting the field results in the attractive interaction between tilted vortices along the tilting direction. They also form one-dimensional arrangements consisting solely of VCs [3]. VCs in BSCCO has been observed by several methods, like Bitter decoration [2,4], scanning Hall probe microscopy [5], Lorentz microscopy [6], and magneto-optical (MO) method [7,8]. However, these observations are limited in the low in-plane field (H_x) region. We have extended H_x and found a new arrangement of vortices. Recent theoretical studies predict that the crossing-lattices state is unstable when $\lambda/\gamma s>1$ (λ: in-plane magnetic penetration depth, s: layer spacing) [9,10]. However, it has been reported that VCs and VLs coexist under tilted field also in underdoped YBCO with moderate anisotropy [11]. We have revisited the vortex arrangements in underdoped YBCO and discuss the origin of vortex arrangements under tilted field. We have also studied the interaction between VCs and artificially introduced defects using magneto-optical method.

EXPERIMENTAL

BSCCO single crystals are grown by the traveling-solvent floating zone method [12]. Al-doped YBCO single crystals are grown by using a gold crucible [13]. We use slightly underdoped BSCCO crystals with T_c=89 K and the second magnetization peak field H_p=330 Oe at 27.5 K. T_c of Al-doped YBCO is 72 K. Bitter decorations are performed by diffusing small iron particles on *ab*-plane of the crystal after field cooling to 4 K. H_x and the out-of-plane field (H_z) are applied independently above T_c by two orthogonal magnets. Decorated vortex arrangements are observed by a scanning electron microscope (Hitachi S-4300) at room temperature. The decorated image visualizes vortex arrangements at the freezing temperature of vortices, which is estimated as ~60 K, following the procedure of ref. [14]. Magneto-optical imaging of VCs are performed using an in-plane magnetized garnet film. Light intensities produced by the Faraday rotation, which is proportional to the local B, is visualized by a cooled-CCD camera. Grooves with 0.38 μm deep and 20 μm spacings on the surface of a BSCCO crystal are created by Ar-ion milling.

CP850, *Low Temperature Physics: 24th International Conference on Low Temperature Physics;*
edited by Y. Takano, S. P. Hershfield, S. O. Hill, P. J. Hirschfeld, and A. M. Goldman

RESULTS AND DISCUSSION

Figure 1 shows decoration images of BSCCO at (H_x, H_z)=(150 Oe, 10 Oe) and (1250 Oe, 10 Oe). The direction of H_x is parallel to the arrows. At lower H_x, we clearly observe VCs+VLs as shown in Fig. 1(a) [2]. When H_x is increased above 1 kOe, the spacing between VCs becomes uneven and some of VCs form bundles as shown by solid arrows in Fig. 1(b). Let us discuss the formation mechanism of the VC bundles. In the VC state at low H_z, a PV stack feels attractive force from JVs and repulsive one from neighboring PVs in the same VC. This competition causes a minimum of potential energy for PVs at finite distance away from the center of a JV [9]. If the repulsive interaction between JVs is sufficiently small, two VCs spaced at a certain distance have lower energy than two isolated VCs. The same mechanism for attractive interactions between VCs applies to more than three VCs and leads to the formation of VC bundle.

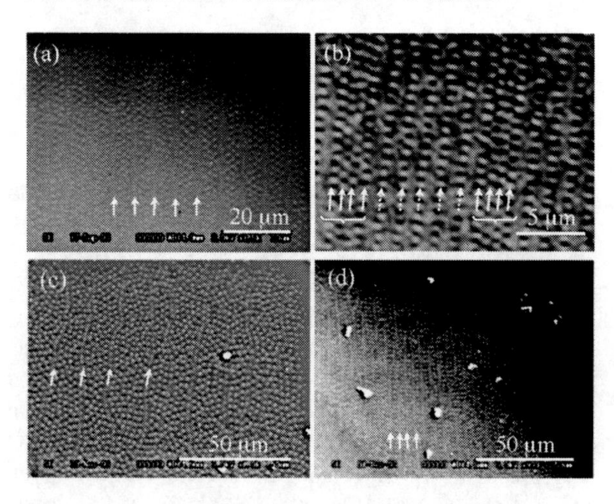

FIGURE 1. Bitter decoration images of BSCCO at (a) (H_x, H_z)=(150 Oe, 10 Oe) and (b) (1250 Oe, 10 Oe), and Al-doped YBCO at (c) (H_x, H_z)=(10 Oe, 5 Oe) and (d) (20 Oe, 5 Oe). Vortices shown in solid arrows in (b) form bundles.

Figures 1(c) and (d) show decoration images of Al-doped YBCO at (H_x, H_z)=(10 Oe, 5 Oe) and (20 Oe, 5 Oe). The direction of H_x is parallel to the arrows. VCs+VLs states are clearly observed as in ref. [11]. The spacing between VCs decreases systematically as H_x is increased. This H_x dependence suggests the presence of underlying JVs. However, according to recent theoretical studies, the crossing-lattices state is stable only for $\gamma>\lambda/s$ [9,10]. In Al-doped YBCO, λ/s is estimated to be 100~140, which is larger than γ in this material. These results may request reconsideration of vortex states in moderately anisotropic superconductors. When the ratio $\lambda/\gamma s$ is larger than 0.7, crossing chains composed of PV stacks located on JVs turn into tilted chains [10]. So, the observed arrangements of vortices in Al-doped YBCO may correspond to a state with coexisting tilted chains and normal lattices.

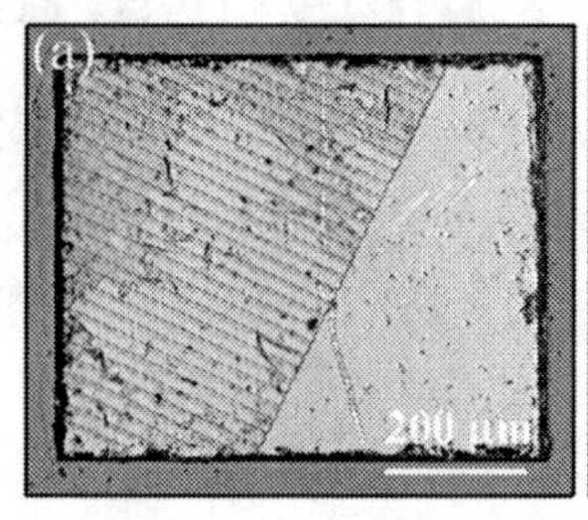

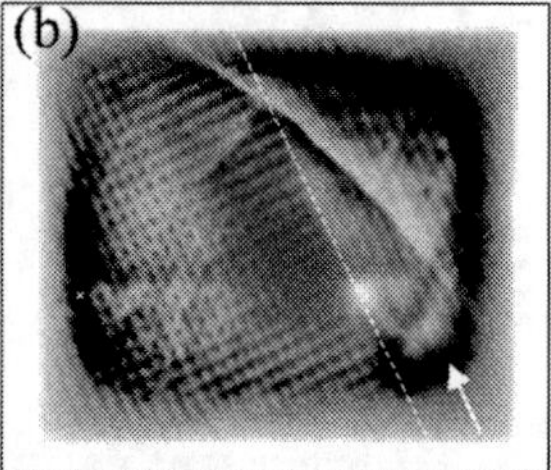

FIGURE 2. (a) BSCCO crystal with grooves, and magneto-optical image of VCs observed in the opposite face of the crystal at T=68 K and (H_x, H_z)=(8 Oe, 32.2 Oe). The direction of H_x is shown by a white arrow.

Figure 2(a) shows a BSCCO crystal with grooves. A magneto-optical image of vortices at T=68 K and (H_x, H_z)=(8 Oe, 32.2 Oe) is shown in Fig. 2(b). Since there are two attractive potentials for PVs from the grooves and JVs along H_x, two kinds of one-dimensional arrangements of PVs are observed.

In summary, we have visualized vortex arrangements under tilted fields in BSCCO and Al-doped YBCO. Under large H_x, a novel state consisting of bundles of vortex chains are found in BSCCO. Although the vortex chain coexisting with vortex lattices in Al-doped YBCO is similar to the case of BSCCO. Correspondence with the theory suggest that it consists of tilted chains and normal lattices.

ACKNOWLEDGMENTS

This work is supported by Grant-in-aid for Scientific Research from the Ministry of Education, Culture, Sports, Science, and Technology.

REFERENCES

1. A. E. Koshelev, *Phys. Rev. Lett.* **83**, 187 (1999).
2. C.A. Bolle *et al.*, *Phys. Rev. Lett.* **66**. 112 (1991).
3. P. L. Gammel *et al.*, *Phys. Rev. Lett.* **68**. 3343 (1992).
4. M. Tokunaga *et al.*, *Phys. Rev. B* **67**, 134501 (2003).
5. A. Grigorenko *et al.*, *Nature* **414**, 728 (2001).
6. T. Matsuda *et al.*, *Science* **294**, 2134 (2001).
7. V.K. Vlasko-Vlasov *et al.*, *Phys. Rev. B* **66**, 014523 (2002).
8. M. Tokunaga *et al.*, *Phys. Rev. B* **66**, 060507 (2002).
9. M.J.W. Dodgson, *Phys. Rev. B* **66**, 014509 (2002).
10. A. E. Koshelev, *Phys. Rev. B* **71**, 174507 (2005).
11. I. V. Grigorieva *et al.*, *Phys. Rev. B* **48**, 16865 (2003).
12. S. Ooi *et al.*, *Physica C* **302**, 339 (1998).
13. F. Holtzberg *et al.*, *J. Solid State Inorg. Chem.* **27**, 107 (1990).
14. M. Marchevsky *et al.*, *Physica C* **282**, 2083 (1997).

'Flux Waves' in $Bi_2Sr_2CaCu_2O_{8+\delta}$

B. Kalisky[1], A. Shaulov[1], B. Ya. Shapiro[1], T. Tamegai[2] and Y. Yeshurun[1]

1. Department of Physics, Institute of Superconductivity, Bar-Ilan University, Ramat-Gan 52900, Israel
2. Department of Applied Physics, The University of Tokyo, Hongo, Bunkyo-ku, Tokyo 113-8656, Japan

Abstract. We observed a new behavior of vortex instabilities in $Bi_2Sr_2CaCu_2O_{8+\delta}$ crystals, namely a quasi-periodic motion of the flux front separating the unstable vortex state from the thermodynamic vortex phase. These "flux waves" were shown to be a direct result of a unique defect pattern that characterizes $Bi_2Sr_2CaCu_2O_{8+\delta}$ crystals grown by the floating zone method.

Keywords: $Bi_2Sr_2CaCu_2O_{8+\delta}$, Vortex phase transition, Magnetic relaxation.
PACS: 74.25.Qt 74.62.Dh 76.60.Es

Local measurements of magnetic relaxation in high temperature superconductors have shown unusual phenomena such as negative relaxation near the edge of the sample and zero relaxation at the 'neutral line'[1,2]. Also, accelerated magnetic relaxation was observed and attributed to the annealing process of metastable disordered states coexisting with the quasi-ordered thermodynamic phase near the vortex order-disorder transition line[3].

Here we report on a remarkable phenomenon of oscillating magnetic relaxation. Time resolved magneto-optical imaging of the spatial distribution of magnetic induction across the sample, revealed a quasi-periodic motion of the flux front separating the unstable vortex state from the thermodynamic vortex phase. In this paper we demonstrate these 'flux waves' and suggest an explanation to this unique phenomenon.

Measurements were performed on a 1.4×8×0.05 mm³ $Bi_2Sr_2CaCu_2O_{8+\delta}$ single crystal (T_C = 92 K). The crystal was grown using the floating zone method[4]. The external magnetic field, H, was raised abruptly to a target value between 140 and 840 Oe with rise-time < 50 ms. Immediately after reaching the target field, magneto optical (MO) snapshots of the induction distribution across the sample surface were recorded at time intervals of 40 ms for 5.5 seconds, using iron-garnet MO indicator with in-plane anisotropy[5] and a high speed Charge Coupled Device (CCD) camera. This procedure was conducted at several temperatures between 20 and 25 K. In a different type of experiment, a sequence of MO images was taken during a sweep of the temperature in the presence of a constant external magnetic field, which was applied for long enough time so that the system is relaxed. The spatial resolution of the presented MO data is 4 μm/pixel.

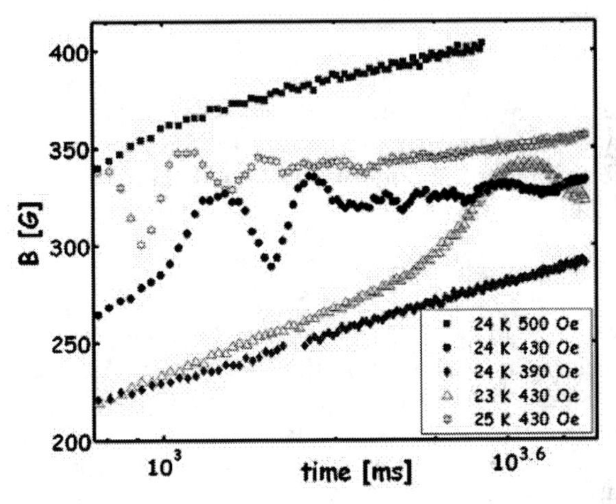

FIGURE 1. Time dependence of local induction B measured at 240 μm from the sample edge, after abrupt application of external magnetic field at constant temperature.

Figure 1 shows the time evolution of the local induction B at a certain location in the sample (240 μm from the left edge) after application of an external magnetic field. At 24 K, the expected behavior of B is observed after application of 500 or 390 Oe (square and diamond symbols, respectively), B increases logarithmically with time. However, when an intermediate field is applied, e.g. 430 Oe, marked by full circles, a unique behavior is observed, namely an oscillatory behavior of B with decaying amplitude. This phenomenon, limited to a certain field range, is also limited by the temperature: Data for 23 and 25 K (triangles and stars, respectively) demonstrates that at

CP850, *Low Temperature Physics: 24th International Conference on Low Temperature Physics;*
edited by Y. Takano, S. P. Hershfield, S. O. Hill, P. J. Hirschfeld, and A. M. Goldman

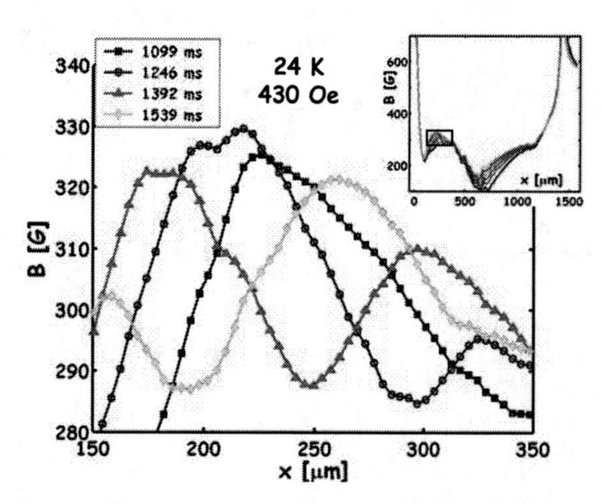

FIGURE 2. Local induction B at the indicated times after application of 430 Oe at 24 K, plotted as a function of the location across the sample (x=0 is the left edge of the sample). In the inset B(x) is plotted across the entire sample width. The small square marks the region which is enlarged in the main figure.

high temperatures the phenomenon is too fast to be measured (we see only its 'tail' at 25 K) and at low temperatures it is very slow (we observe only the first oscillation at 23 K).

The inset to Figure 2 shows the induction profile across the sample at different times after application of 430 Oe at 24 K. The main panel of the figure is focusing on the region marked by a square. The time evolution of induction profiles demonstrates 'flux waves', i.e. induction oscillations in both time and space. We note that the 'waves' move in the opposite direction to the incoming flux from the edge of the sample due to relaxation in the presence of the applied field.

These 'flux waves', observed when the magnetic field was changed at a constant temperature, were also observed when the temperature was changed at a constant applied field.

In the following we suggest an explanation to the observed 'flux waves'. Our explanation is based on the dynamic behavior of transient disordered vortex states (TDVS)[6], which exist below the order disorder transition induction, B_{od}. This dynamics is determined by the difference between the local induction B and B_{od}. We also noticed that all our $Bi_2Sr_2CaCu_2O_{8+\delta}$ samples are characterized by pronounced defect patterns that were already described in the literature[7] and attributed to the inevitable local oxygen disorder[8]. Presumably, such defect patterns cause a spatial distribution of the vortex order-disorder transition induction, B_{od} [9].

When the magnetic field is raised abruptly, the sample is filled with TDVS. After a certain time the ordered phase nucleates at a point where $B-B_{od}$ is minimal (for uniform B_{od} this point is the sample center). The annealing process is revealed by the appearance of a break (change of the local slope) in the profile. This break separates between the ordered phase at the center of the sample and the TDVS, which are characterized by a higher persistent current. With time the annealing process of TDVS continues. The annealing is revealed by the movement of the break towards the sample edge.

In our sample, due to the spatial distribution of B_{od}, TDVS in certain parts of the sample are less stable than in other parts of the sample because they are found at inductions that are further away from B_{od}. This results in nucleation and front propagation of the thermodynamic vortex phase at different parts of the sample at different times. The collective movement of several fronts towards the sample edge creates the impression of flux waves.

In summary, 'flux waves' are observed in a limited field and temperature region near the vortex order-disorder phase transition. Spatial variation of B_{od} across the sample and a relatively flat profile are essential conditions for observation of this phenomenon.

We acknowledge support from the German-Israel Foundation (GIF). Y.Y. acknowledges support from the Wolfson Foundation. This research is supported by the ISF Center of Excellence Program (Grant No. 8003/02) and the Heinrich Hertz Minerva Center for High Temperature Superconductivity.

REFERENCES

1 Y. Yeshurun *et al.*, Rev. Mod. Phys. **68**, 911 (1996).
2 Y. Abulafia *et al.*, J. Appl. Phys. **81**, 4944 (1997).
3 B. Kalisky *et al.*, Physica C **408**, 384 (2004).
4. N. Motohira *et al.*, J. Ceram. Soc. Jpn. **97**, 994 (1989)
5. V. K. Vlasko-Vlasov *et al.*, in Physics and Materials Science of Vortex States, Flux Pinning and Dynamics, edited by R. Kossowsky *et al.*, NATO ASI, Ser. E, Vol. **356** (Kluwer, Kordrecht, 1999), p. 205.
6. D. Giller *et al.*, Phys. Rev. Lett. **84**, 3698 (2000); B. Kalisky, A. Shaulov, and Y. Yeshurun, Phys. Rev. B **68**, 012502 (2003); B. Kalisky *et al.*, Phys. Rev. B **67**, R140508 (2003); B. Kalisky *et al.*, Phys. Rev. B **68**, 24515 (2003).
7. R. Gerbaldo *et al.*, Physica C **354**, 173 (2001).
8. Tsu I-Fei *et al.*, Physica C **349**, 8 (2001).
9. B. Khaykovich *et al.*, Phys. Rev. Lett. **76**, 2555 (1996).

Enhancement Of Critical Current Density Under Low Magnetic Fields Around 70 K In $Bi_2Sr_2CaCu_2O_y$ Superconducting Whiskers

M. Tange*, Y. Arao*, H. Ikeda*† and R. Yoshizaki*†

*Faculty of Frontier Science, Graduate School of Pure and Applied Sciences, University of Tsukuba, Tsukuba, Ibaraki 305-8571, Japan
†Cryogenics Division, Research Facility Center for Science and Technology, University of Tsukuba, Tsukuba, Ibaraki 305-8577, Japan

Abstract. Electrical transport measurements for $Bi_2Sr_2CaCu_2O_y$ whiskers with a single-step superconducting transition were carried out at high temperature ranging from 70 to 77 K below the superconducting transition temperature (= 81.5-82.5 K) to investigate critical current density (J_c) under magnetic fields along the crystal *c* axis. The J_c parallel to the crystal *a* axis, which was estimated from the *V-I* curves, enhances under several ten oersteds, and then decreases with increasing the magnetic field. This peak effect in the low-field regime at high temperature has temperature dependence in contrast to the case of matching effects.

Keywords: Bi-2212, whisker, peak effect, high-T_c superconductor
PACS: 74.25.Fy, 74.25.Qt, 74.25.Sv, 74.72.Hs

INTRODUCTION

$Bi_2Sr_2CaCu_2O_y$ (Bi-2212) superconducting whiskers are striplike crystals with the tiny cross section on micro- and submicro-meter scale. The dimension of the as-grown whisker is related to the crystal axes: the *a* axis, *b* axis, and *c* axis, which are the direction of the length, of the width and of the thickness, respectively. With respect to the vortex dynamics, the vortex penetration is suppressed by Bean-Livingston surface barriers (BLSB), as described in Ref. [1]. Moreover, at high temperature where bulk pinning is weak, geometrical barriers (GB) affect the entry and exit of vortices according to the shape of a sample [2]. In particular for the platelet sample as Bi-2212 whiskers, the effects of GB are pronounced and provide irreversible flux distribution in the sample. The superconducting properties of the whisker, *e.g.*, critical current density (J_c) in the presence of magnetic fields, are complicated with such surface barriers, and further characterizations may be needed. In the present work, electrical transport measurements with driving current in the direction of the *a* axis were carried out under magnetic fields along the *c* axis at high-temperature region below T_c, and consequently the J_c along the *a* axis was estimated as a function of the magnetic field.

EXPERIMENTAL

Bi-2212 whiskers with a single-step superconducting transition, which were slightly overdoped samples, were synthesized by a self-powder compaction method [3] under the growth conditions of T_{gr} (growth temperature) = 875°C and Po_2 (oxygen partial pressure) = 0.2 atm. Voltage-current (*V-I*) characteristics of the whiskers were measured in the presence of magnetic fields (0-200 Oe) along the *c* axis at high temperature ranging from 70 to 77 K. In addition, the electrical resistance was investigated at fixed transport current in the same temperature range with sweeping the magnetic field. The electrical transport measurements, for which a standard four-probe technique was employed, were performed with PPMS (Physical Property Measurement System: Quantum Design) by applying dc current along the *a* axis.

CP850, *Low Temperature Physics: 24th International Conference on Low Temperature Physics;*
edited by Y. Takano, S. P. Hershfield, S. O. Hill, P. J. Hirschfeld, and A. M. Goldman

RESULTS AND DISCUSSION

The presented data were obtained for the sample with $T_{c,zero}$ = 82 K and the dimension of 680 μm in the length, 7.1 μm in the width and 0.8 μm in the thickness. Figure 1 shows *V-I* curves measured in the presence of various magnetic fields. The result was obtained at 70 K in the ZFC (zero-field-cooled) mode and the FC (field-cooled) mode. Over the range of 20 to 50 Oe for FC state, the critical current (I_c) increases with increasing the field. Concerning the depinning of vortices, the behavior of *V-I* curves seems to have changed from the linear slope to the nonlinear slope in the vicinity of 50 Oe. The triangular flux-lattice constant for 50 G approximately equals to 0.7 μm. This suggests that the magnetic fields are rather too low for the enhancement of I_c to be attributed to the crystal structure as the chemical inhomogeneity. There are two possibilities for the origin of the behavior. One is that the state of vortices confined by surface barriers may be translated at the boundary in the magnetic fields, *e.g.*, from a precursory state of the vortex lattice to Bragg glass or from Bragg glass to vortex glass as observed under a lower critical point for untwinned overdoped $YBa_2Cu_3O_7$ single crystals in Ref. [4]. Another is that the channel of vortex motion may be changed as the intensity of surface barriers varies with the external magnetic field.

The result obtained after the preparation of the ZFC state is almost the same as the behavior observed for the FC state instead of showing remarkable discrepancies. This feature implies that the flux profile for the case of ZFC mode is similar to that for the case of FC mode in respect of the whisker at elevated temperature. It suggests that vortices are present with the dome-shaped profile under the magnetic fields below 50 Oe even for ZFC mode because vortices enter through the direction of the *a* axis due to the thickness-to-length ratio of the whisker. BLSB probably should affect the penetration field with respect to the entry of vortices through the direction of the *b* axis as reported on micron-sized Bi-2212 single crystals [5], whereas effects of GB may be dominant to the entry into the center from the direction of the *a* axis because the dimension of the whisker is long along the *a* axis.

The J_c obtained from the transport measurement for the FC mode is plotted as a function of the applied magnetic field in the inset of Fig. 1. The J_c was estimated by using the criterion that the voltage of 0.5 μV appears in the *V-I* data. The value of voltage corresponds to about 260 μV/m. In the measured temperature range of 70-77 K, the J_c has nonmonotonic dependence on the external magnetic field. The J_c enhances in the low-field region below 30-45 Oe, and then decreases with increasing the field after reaching a peak. In contrast to the case of matching effects, the magnetic filed (H_{peak}) at which the peak occurs is dependent on temperature: *i.e.*, being shifted to higher magnetic fields with decreasing temperature. The H_{peak}s on the whisker are in the lower magnetic field region in comparison with those on second peak effects of Bi-2212 bulk single crystals (400-1000 Oe). The behavior of J_c is consistent with the following variation of the electrical resistance measured at constant temperature with sweeping the magnetic field in the presence of fixed transport current. The resistance appears in the presence of low magnetic fields, and then the further increase in the magnetic field induces zero resistance. The results and discussions on the behavior of the resistance under the low fields will be described elsewhere. The peak effects are also observed for other samples.

In conclusion, the J_c estimated from *V-I* curves for Bi-2212 whiskers enhances in the low-field regime.

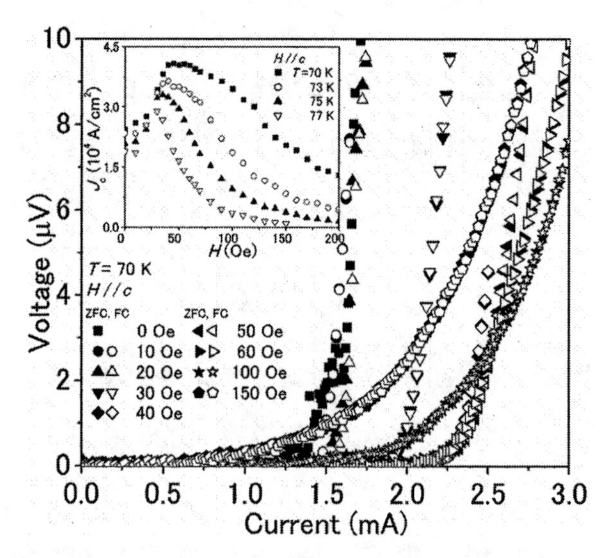

FIGURE 1. *V-I* characteristics measured at transport current (// *a* axis) under magnetic fields (// *c* axis) for the Bi-2212 whisker. The *V-I* curves are obtained at 70 K in the ZFC mode (solid symbols) and the FC mode (open symbols). The inset exhibits the field dependence of J_c (// *a* axis) estimated from the *V-I* curves at high temperature (70-77 K).

REFERENCES

1. J. K. Gregory et al., *Phys. Rev. B* **64**, 134517 (2001).
2. R. A. Doyle et al., *Physica C* **308**, 123 (1998).
3. M. Tange et al., *Physica C* **392-396**, 428 (2003).
4. A. A. Zhukov et al., *Phys. Rev. Lett.* **87**, 017006 (2001).
5. Y. M. Wang et al., *Phys. Rev. B* **65**, 184506 (2002).

Correlation between Vortex Distribution and Electronic Inhomogeneity in $Bi_2Sr_2CaCu_2O_{8+\delta}$ as Probed by STM/STS

Noritaka Fukuo*, Hideki Mashima*, Go Kinoda†, Taro Hitosugi**, Toshihiro Shimada**, Keisuke Shimizu‡, Jun-ichi Shimoyama‡, Kohji Kishio‡, Yuji Matsumoto* and Tetsuya Hasegawa**

*FCRC, Tokyo Institute of Technology, Midori-ku, Yokohama 226-8503, Japan
†Kanagawa Academy of Science and Technology, Takatsu-ku, Kawasaki 213-0012, Japan
**Department of Chemistry, University of Tokyo, Bunkyo-ku, Tokyo 113-0033, Japan
‡Department of Superconductivity, University of Tokyo, Bunkyo-ku, Tokyo 113-8685, Japan

Abstract. In order to investigate the correlation between vortex distribution and inherent electronic inhomogeneity, we have performed scanning tunneling microscopy/spectroscopy (STM/STS) of optimally doped $Bi_2Sr_2CaCu_2O_{8+\delta}$ single crystals at 4.3 K under 8 T. Vortices have been successfully visualized by mapping differential tunneling conductance at 12.5 meV, which was substantially enhanced around the vortex cores. The vortices tended to form an ordinary triangular lattice with a spacing of $a_0 = 1.07\sqrt{\phi/B}$ in a short range of $\sim a_0$, but the triangular correlation was almost missing in a longer scale. By comparing the vortex distribution pattern and gap map, we concluded that pseudogap regions predominantly pin down vortices.

Keywords: electronic inhomogeneity, vortex pinning, Bi2212, STM/STS
PACS: 74.25.Qt; 74.72.Hs; 87.64.Ee

INTRODUCTION

Scanning tunneling microscopy/spectroscopy (STM/STS) has been utilized to visualize vortices trapped in $Bi_2Sr_2CaCu_2O_{8+\delta}$(Bi2212) under intense magnetic field [1–3]. Most of these studies reported that distribution of the vortices is substantially distorted compared with the ideal triangular lattice [1–3]. Recent STM/STS studies on Bi2212 have also revealed nano-scale electronic inhomogeneity composed of superconducting and non-superconducting domains [4–11]. The former shows well-developed conductance peaks at $E = \pm\Delta$, while the latter is characterized by pseudogap-like spectra with larger gap values. As is well known, non-superconducting matrix tends to trap vortices, because they do not lose condensation energy. Therefore, it is anticipated that the irregularly distributed pseudogap regions behave as strong pinning sites, and are responsible for the distortion of vortex lattice as mentioned above.

Here, we have performed cryogenic STM/STS measurements on Bi2212 single crystals under magnetic field of 8 T. We have succeeded in observing both vortex distribution and electronic inhomogeneity at the same time. By comparing vortex images and Δ maps, we provide direct evidence that vortices tend to be located in the pseudogap regions.

EXPERIMENTAL

Bi2212 single crystals with nominal composition of $Bi_{2.1}Sr_{1.9}CaCu_{2.0}O_{8+\delta}$ were grown by the floating-zone method. Superconducting transition temperature T_c was determined to be 90 K (onset) by a SQUID susceptometer. STM/STS measurements were carried out using a home-build UHV-LT STM instrument equipped with a low temperature cleavage stage. The base pressure of the STM chamber was maintained at less than 2.0×10^{-10} Torr during the measurements. All samples examined here were cleaved in situ at 77 K and immediately transferred to the STM head. Mechanically sharpened Pt/Ir wires were used as STM tips.

RESULTS AND DISCUSSION

Figure 1 compares typical tunneling spectra observed inside and outside a vortex core. In the vortex core, coherence peaks are fairly suppressed, whereas tunneling conductance values at ± 12.5 meV, indicated by arrows, are enhanced.

Figure 2(a) is a tunneling conductance map at $E = +12.5$ meV obtained at 4.2 K under 8 T. In Fig. 2(a), dark regions represen vortex cores. At a glance, a triangular lattice is recognizable, although it is fairly distorted. The average interval between vortices is 17.4 nm, which is close to the theoretical value for triangular lat-

CP850, *Low Temperature Physics: 24th International Conference on Low Temperature Physics;*
edited by Y. Takano, S. P. Hershfield, S. O. Hill, P. J. Hirschfeld, and A. M. Goldman

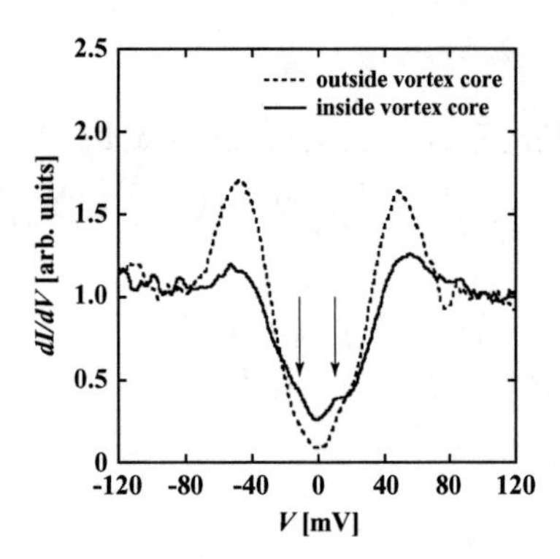

FIGURE 1. Tunneling spectra inside and outside a vortex core at 4.2 K under 8 T. Tunneling conductance is normalized at 100 mV.

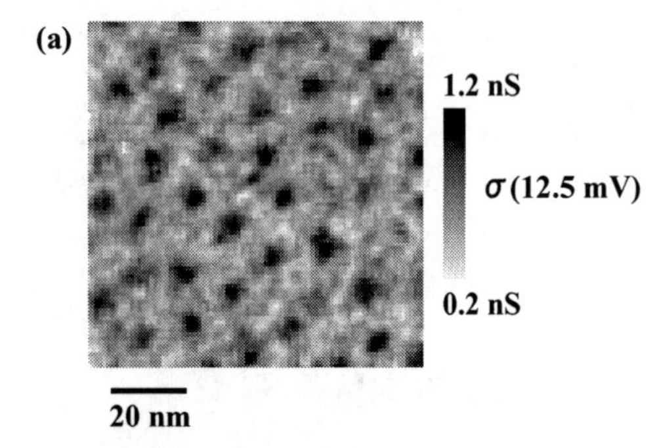

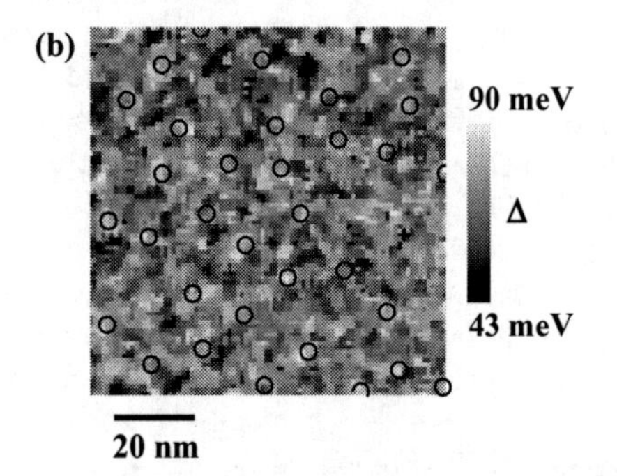

FIGURE 2. (a) Conductance map at +12.5 mV and (b) corresponding Δ map obtained at 4.2 K under 8 T. Image size is 100 nm × 100 nm.

tice, $a_0 = 1.07\sqrt{\phi/B} = 17.3$ nm. The size of vortex cores is estimated to be 3.8 nm from cross sectional line profiles of Fig. 2(a). Autocorrelation analysis of Fig. 2(a) revealed that vortices tend to form a triangular lattice in a short range of $\sim a_0$, while the triangular correlation disappears in a longer range.

Figure 2(b) shows a Δ map, with the imaged region selected to be identical to that in Fig. 2(a). As can be seen, Δ shows spatial variation of a nanometer scale, and ranges from 43 to 90 meV. Open circles indicated in Fig. 2(b) represent locations of vortex cores. Notably, there is a tendency that vortex cores are located in the pseudogap regions with larger Δ.

This behavior can be interpreted in two different ways. One interpretation is that vortices destroy superconductivity and induce pseudogap states around vortex cores. The other assumes that vortices favor to occupy pseudogap regions, because they gain condensation energy. The fact that we occasionally observed vortex cores located in the superconducting region supports the second scenario. In this case, vortex distribution pattern is determined by competition between flux pinning by pseudogap regions and vortex-vortex interactions.

SUMMARY

STM/STS observations of Bi2212 single crystals were performed at 4.3 K under 8 T. We successfully imaged vortices and spatial distributions of Δ. We found that vortices tend to form a triangle lattice within a short scale of $\sim a_0$, although longer scale correlation disappears almost completely. Comparing vortex distribution and Δ map, we tentatively concluded that vortices were dominantly pinned by puseudgap regions.

REFERENCES

1. Ch. Renner, B. Revaz, K. Kadowaki, I. Maggio-Aprile, and Ø. Fischer, *Phys. Rev. Lett.*, **80**, 3606 (1998).
2. S. H. Pan, E. W. Hudson, A. K. Gupta, K. -W. Ng, H. Eisaki, S. Uchida, and J. C. Davis, *Phys. Rev. Rett.*, **85**, 1536 (2000).
3. K. Matsuba, H. Sakata, N. Kosugi, H. Nishimori, and N, Nishida, *J. Phys. Soc. Jpn.*, **72**, 2153 (2003).
4. S. H. Pan, J. P. O'Neal, R. L. Badzey, C. Chamon, H. Ding, J. R. Engelbrecht, Z. Wang, H. Eisaki, S. Uchida, A. K. Gupta, K. -W. Ng, E. W. Hudson, K. M. Lang, and J. C. Davis, *Nature*, **413**, 282 (2001).
5. T. Cren, D. Roditchev, W. Sacks, J. Klein, J. -B. Moussy, C. Deville-Cavelline, and M. Laguës, *Phys. Rev. Lett.*, **84**, 147 (2000).
6. T. Cren, D. Roditchev, W. Sacks, and J. Klein, *Europhys. Lett.*, **54**, 84 (2001).
7. C. Howald, P. Fournier, and A. Kapitulnik, *Phys. Rev. B*, **64**, 100504(R) (2001).
8. K. M. Lang, V. Madhavan, J. E. Hoffman, E. W. Hudson, H. Eisaki, S. Uchida, and J. C. Davis, *Nature*, **415**, 412 (2002).
9. G. Kinoda, T. Hasegawa, S. Nakao, T. Hanaguri, K. Kitazawa, K. Shimizu, J. Shimoyama, and K. Kishio, *Phys. Rev B*, **67**, 224509 (2003).
10. G. Kinoda, T. Hasegawa, S. Nakao, T. Hanaguri, K. Kitazawa, K. Shimizu, J. Shimoyama, and K. Kishio, *Appl. Phys. Lett.*, **83**, 1178 (2003).
11. A. Matsuda, T. Fujii, and T. Watanabe, *Physica C*, **388–389**, 207 (2003).

Peak Effect as Precursor to Lock-in State in $Bi_2Sr_2CaCu_2O_{8+\delta}$ Single Crystal

J. Mirkovic*, K. Murata, A. Nakano, T. Yamamoto, I. Kakeya, and K. Kadowaki

Institute of Materials Science, University of Tsukuba, 1-1-1 Tennodai, 305-8573 Tsukuba, Japan
**Also Faculty of Science, University of Montenegro, Podgorica, Serbia and Montenegro*

Abstract. The vortex phases in high-quality $Bi_2Sr_2CaCu_2O_{8+\delta}$ single crystal have been studied by means of the in-plane resistivity measurements in the Corbino electric contact geometry. At the angle of 89.94° of a magnetic field, titled from the *c*-axis, a peak-effect of the in-plane resistivity was observed in the narrow field and temperature range between 79.4 K and 82 K as a precursor of the lock-in vortex state in the critical angular range near the *ab*-plane. The strong non-Ohmic behavior may suggest the two-stage melting transition in parallel magnetic fields.

Keywords: $Bi_2Sr_2CaCu_2O_{8+\delta}$ single crystal; resistivity; vortex dynamics; peak-effect; lock-in transition.
PACS: 74.25.Qt, 74.72.Hs, 74.25.Op, 74.25.Sv

INTRODUCTION

In tilted magnetic fields, the stacks of vortex pancakes [1] penetrate through the CuO_2 plane perpendicularly, while the Josephson vortex lattice is aligned parallel to the *ab*-plane. These two qualitatively different sublattices interpenetrate each other, weakly interact and form the various tilted and crossing lattice structures including the vortex chains, stripes, etc. [2]. However, despite the developments in the experimental studies of the macroscopic properties of the mixed state in the $Bi_2Sr_2CaCu_2O_{8+\delta}$ superconductors, the nature of the related vortex states and the phase transitions in the nearly parallel magnetic fields has not been understood completely yet.

EXPERIMENTAL

In order to probe the rich physics confined in a narrow window of the experimental conditions, as it is case for highly anisotropic superconductors, it is decisive to have excellent quality of samples. The as-grown $Bi_2Sr_2CaCu_2O_{8+\delta}$ single crystal used has a rocking angle better than 0.01° over the sample. The in-plane resistivity was measured in the Corbino electric contacts geometry on the single crystal with the diameter of 1.9 mm, thickness of 20 μm, and the transition temperature T_c = 84.1 K. The Corbino resistivity measurement is a unique technique to avoid the surface pinning effects since the vortices flow in the concentric circles without crossing the edge of the sample, *i.e.*, avoiding the surface barriers [3]. The resistance was measured by using standard lock-in technique at 37 Hz as a function of magnetic field and temperature at the various field orientations with respect to the *c*-axis. The magnetic field, generated by 60 kOe split s/c coil, was rotated with fine angular resolution of $\delta\theta = 0.01°$.

RESULTS

Figure 1. presents the set of the typical in-plane resistance curves measured as a function of magnetic field on the various orientations with respect to the *c*-axis, at the temperatures of 80.65 K. The first-order vortex lattice melting transition [4] is clearly detected by a distinct resistivity anomaly at low resistance level, which sharply separates the vortex lattice and the vortex liquid across the wide angular range 0° < θ < 89.86° (for the better insight, Figure 1. shows the reduced angular range, after the linear dependence [5] of the first-order vortex melting transition in the H_c-H_{ab} phase diagram, sharply transforms into plateau [6]). With further inclination of the magnetic field, at the angle of 89.94° away from the *c*-axis, a peak of the

CP850, *Low Temperature Physics: 24th International Conference on Low Temperature Physics;* edited by Y. Takano, S. P. Hershfield, S. O. Hill, P. J. Hirschfeld, and A. M. Goldman

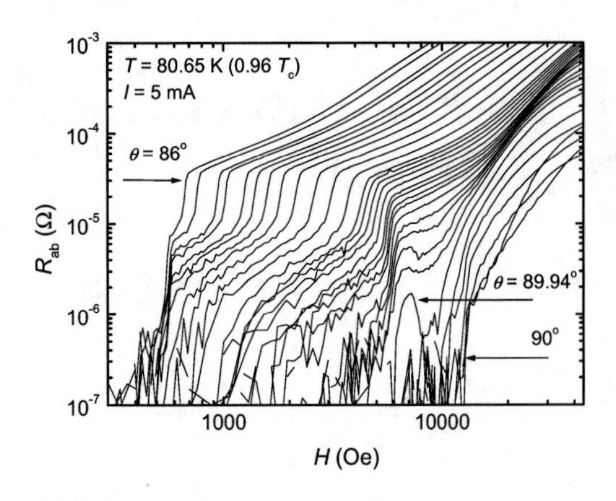

FIGURE 1. Magnetic field dependence of the in- plane resistance measured at the various orientations away from the *c*-axis by current of 5 mA.

resistance was observed. Interestingly, the resistance peak is pronounced in a rather narrow temperature interval, from 79.4 K up to 82 K. Another experimental finding is that the anomaly occurs in a highly non-Ohmic regime, as shown in Figure 2. On the other hand, the current level does not affect much the peak position in magnetic field, at least within the measured range from 2 mA up to 10 mA.

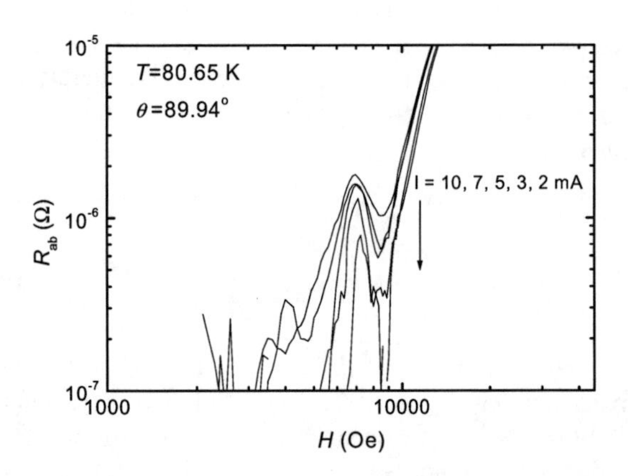

FIGURE 2. The magnetic field dependence of resistance measured by driving currents (from above) of 10, 7, 5, 3, and 2 mA at the orientation of $\Delta\theta = 0.06^{\circ}$ away from the *ab*-plane.

As the melting transition approaches, softening of the shear modulus enables lattice to adjust to a more favorably pinned configuration as it was observed earlier in $YBa_2Cu_3O_{7-\delta}$ single crystal [7]. There, the in-plane resistivity exhibited a peak as magnetic field gets close to the one directional twin boundaries orientation. Instead of twin boundaries, here, the observed anomaly could be attributed to the intrinsic pinning, and may be considered as a precursor of the lock-in state [8] near the *ab*-plane. Approaching the angle of $\theta = 90^{\circ}$, the in-plane resistivity anomaly vanishes and is replaced by a new smooth dependence *R(H)*, and shifted to higher magnetic fields. However, in the parallel magnetic fields, the resistance showed strongly non-Ohmic behavior, and yet there is no clear evidence which phases and transitions may undergo in the vortex lock-in state. One of the possible scenarios is, that near the *ab*-plane, the vortex matter passes through the two-stage vortex lattice - smectic - liquid phase transition [9].

SUMMARY

A peak-effect of the in-plane resistivity was observed at high temperature, $T = 0.96\ T_c$, very close to the *ab*-plane. The anomaly is followed by the continuous magnetic field dependence and the non-Ohmic behavior of the resistivity, suggesting the two-stage melting transition in parallel magnetic field, *i.e*, in the lock-in vortex state.

ACKNOWLEDGMENTS

The authors acknowledge support from the 21st Century COE program under MEXT at University of Tsukuba, Japan.

REFERENCES

1. Buzdin, A. I., and Feinberg, D., *J. Phys.* (*Paris*) **51**, 1971 (1990); Clem, J. R., *Phys. Rev. B* **43**, 7837 (1991).
2. Grigorenko, A. *et al, Nature* (*London*) **414**, 728-731 (2001); Buzdin, A. I., and Baladie, I., *Phys. Rel. Lett.* **88**, 147002-1 (2002); Savel'ev, S. E., Mirkovic, J., and Kadowaki, K., *Phys. Rev. B* **64**, 4521 (2001).
3. Fuchs, D. T., *et al, Nature* (*London*) **391**, 373 (1998).
4. Zeldov, E. *et al., Nature* (*London*) **375**, 373 (1995).
5. Ooi, S., *et al., Phys. Rev. Lett.,* **82**, 4308-4311 (1999); Koshelev, A. E., *Phys. Rev. Lett.* **83**, 187-190 (1999).
6. Mirkovic, J., Sugahara, E., and Kadowaki, K., *Physica B* **284-288**, 733-734 (2000); Mirkovic, J., *et al, Phys. Rev. Lett.* **86**, 886-889 (2001).
7. Kwok, W. K. *et al, Phys. Rev. Lett.* **73**, 2614-2617 (1994).
8. Feinberg, D, and Villard, C., *Phys. Rev. Lett.* **65**, 919-922 (1990).
9. Balents, L. and Nelson, D. R., *Phys. Rev. B* **52**, 12951 (1995).

Phase Transition from Crossing Lattice to Tilted Lattice Near *ab*-plane in $Bi_2Sr_2CaCu_2O_{8+\delta}$ Single Crystal

J. Mirković*, H. Satou, T. Yamamoto, I. Kakeya, and K. Kadowaki

Institute of Materials Science, University of Tsukuba, 1-1-1 Tennodai, 305-8573 Tsukuba, Japan
**Also Faculty of Science, University of Montenegro, 81000 Podgorica, Serbia and Montenegro*

Abstract. The vortex phases in the high-quality $Bi_2Sr_2CaCu_2O_{8+\delta}$ single crystal have been studied by means of the local ac-magnetic permeability measurements by using the miniature coils. The structural transition has been indicated in the vortex solid phase, where the linear dependence of the 1st order vortex melting phase transition $H_c^m(H_{ab})$, a fingerprint of the crossing lattice, sharply changes its character, and separates the strong pinning pancake-vortex phase from the weak pinning phase. The phase transition from the crossing lattice to the tilted vortex lattice near the *ab*-plane is suggested.

Keywords: $Bi_2Sr_2CaCu_2O_{8+\delta}$ single crystal, crossing vortex lattice, tilted vortex lattice, vortex phase diagram
PACS: 74.25.Qt; 74.72.Hs; 74.25.Op; 74.25Sv

INTRODUCTION

While magnetic field can penetrate classical type-II superconductor in flux-quantized vortices and form the triangular Abrikosov vortex lattice, in high-T_c superconductors, such a naive picture becomes more complicated. In highly anisotropic $Bi_2Sr_2CaCu_2O_{8+\delta}$ superconductors, owing to layered structures, the vortex can not be considered as a vortex tube, but as a stack of pancake vortices (PVs) [1]. In tilted magnetic fields, these stacks of PVs penetrate the CuO_2 plane perpendicularly, while the Josephson vortices (JVs) form a lattice parallel to the *ab*-plane. In the first approximation, these two types of vortices (PVs and JVs) interpenetrate each other, and the sub-lattices coexist in the crossing lattice. However, it was found [2] that there is a weak PV-JV mutual pinning energy, leading to the attraction of JVs and PVs, which could be indicated by the linear dependence [3] of $H_c^m(H_{ab})$, where the H_c^m is the *c*-axis component of magnetic field where the first-order vortex-lattice melting transition [4] occurs, and the H_{ab} is the in-plane magnetic field. Moreover, similarly to moderately anisotropic superconductors [5], there is a long range attraction which conglomerates PVs into chains on JV. Nevertheless, despite the observations of the PVs chains by the various visualization techniques [6], and a number of macroscopic measurements [7-8], the nature of the related vortex states and the phase transitions in tilted magnetic fields has not been completely understood yet. Here, we focus on the macroscopic properties and the phase boundaries of the dominant vortex phases in the $Bi_2Sr_2CaCu_2O_{8+\delta}$ single crystal in tilted magnetic fields.

EXPERIMENTAL

To get information about the vortex phases in the oblique magnetic fields, the local ac permeability measurements $\mu = \mu' - i\mu''$ have been performed. Since the transport measurements basically lose the sensitivity in the vortex solid state due to zero resistance, we have developed a new technique to probe the vortex solid by the local ac-mutual inductance measurements by using a simple set of two miniature coils. One is used generating local magnetic fields, to excite the vortex system, while the second coil is glued on the other side of sample, for detection of the transmitted ac response. The size of coils was 0.35 mm, sufficiently smaller than the size of sample (3×3 mm^2, $T_c = 84$ K), which means that the edge and the surface barriers effects [9] could be neglected. Since the coils are fixed for the sample, the sensitivity for all directions is kept constant even in the case of the exact parallel magnetic fields. Magnetic field generated by 60 kOe s/c split magnet was rotated with a fine angular resolution of $\delta\,\theta = 0.01°$.

CP850, *Low Temperature Physics: 24th International Conference on Low Temperature Physics;*
edited by Y. Takano, S. P. Hershfield, S. O. Hill, P. J. Hirschfeld, and A. M. Goldman

RESULTS

Figure 1 shows the magnetic field dependence of the local ac-magnetic permeability measured at the temperature of 65 K, at the various field orientations away from the *c*-axis. The local magnetic field was applied with the intensity of 1.3 G at frequency of 2 kHz. As the magnetic field is swept slowly at rate of 5 G/sec, the local ac magnetic response suddenly jumps up to the maximum value of μ' (the real part) at H^m, where the first-order vortex lattice melting transition occurs, while the imaginary part μ'' sharply drops down to nearly zero. By varying the field angle from the *c*-axis towards the *ab*-plane, the H_m transition is traced as a function of magnetic field and its orientation, and the H_c-H_{ab} phase diagram is constructed (the inset of Fig. 1). The linear dependence of $H_c^m(H_{ab})$ [3], attributed to the crossing lattice [2], abruptly changes the slope [7] at $\theta = 82°$, in the magnetic field marked as H^* ($\approx$ 1.5 kOe) while the melting transition is still clearly seen up to $\theta = 89.8°$, in the magnetic field denoted as H_{cusp} ($\approx$ 20 kOe). The magnetic permeability anomaly, observed at the characteristic field H^*, is recognized deeply inside the vortex solid. By decreasing the *c*-axis magnetic field component, H_c, the H^* phase line gradually bends towards the *c*-axis on the H_c-H_{ab} phase diagram, suggesting the correlation or possible matching of the PVs and JVs lattice parameters. The next finding is that the H^* phase line sharply separates the strong pinning (PVs dominated) phase, in the low in-plane magnetic fields ($H_{ab} < H^*$), from the weak pinning phase, in the higher in-plane magnetic fields. It seems that across the magnetic field region $H > H^*$ ($\theta > 82°$), the PVs lattice is not pinned well enough to prevent the motion of the JVs. Therefore, it could be expected that the probed vortex phase is not the crossing lattice, combined of the JVs and PVs lattices, but possibly the unique tilted vortex lattice near the *ab*-plane, as it was considered in the theory [10] previously.

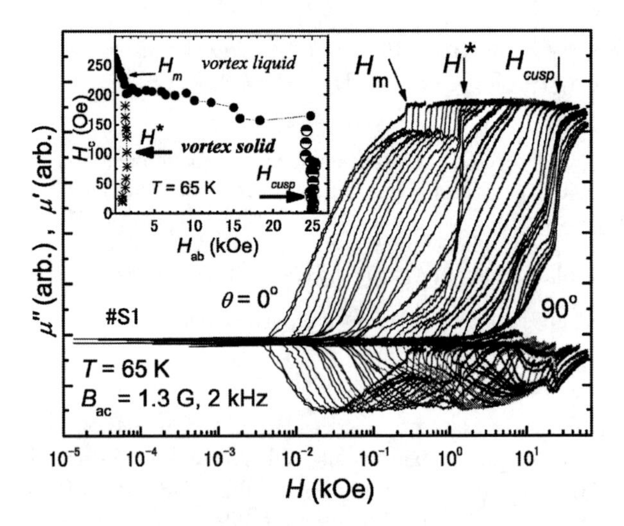

Fig. 1 Real (μ') and imaginary (μ'') parts of the ac-magnetic permeability as a function of H at various magnetic orientations with respect to the *c*-axis. Inset: the H_c-H_{ab} phase diagram at 65 K.

SUMMARY

The phase transition between the crossing lattice of PVs and JVs, near the *c*-axis, and the tilted vortex lattice, near the *ab*-plane ($\theta > 82°$), has been indicated. The crossing vortex lattice dominated by the PVs, exhibited the stronger pinning properties than the tilted vortex lattice.

ACKNOWLEDGMENTS

The authors acknowledge support from the 21st Century COE program under MEXT at University of Tsukuba, Japan.

REFERENCES

1. Buzdin, A. I., and Feinberg, D., *J. Phys. (Paris)* **51**, 1971 (1990); Clem, J. R., *Phys. Rev. B* **43**, 7837 (1991).
2. Koshelev, A. E., *Phys. Rev. Lett.* **83**, 187-190 (1999).
3. Ooi, S., *et al., Phys. Rev. Lett.*, **82**, 4308-4311 (1999).
4. Zeldov, E. *et al., Nature (London)*, **375**, 373 (1995).
5. Buzdin, A. I., and Simonov, A. Yu., *Sov. Phys. JETP* **71**, 1165 (1990).
6. Bolle, C. A. *et al., Phys. Rev. Lett.* **66**, 112 (1991); Grigorenko, A. N. *et al, Nature (London)*, **414**, 728-731 (2001); Tonomura, A. *et al, Phys. Rev. Lett.* **88**, 237001 (2002); Vlasko-Vlasov V. K, *et al, Phys. Rev. B* **66**, 014523 (2002).
7. Mirković, J., Sugahara, E., and Kadowaki, K., *Physica B* **284-288**, 733-734 (2000); Mirković, J., *et al, Phys. Rev. Lett.* **86**, 886-889 (2001).
8. Tokunaga, M. *et al, Phys. Rev B* **66**, 220501 (2002).
9. Fuchs, D. T., *et al, Nature (London)*, **391**, 373 (1998).
10. Savel'ev, S. E., Mirković, J., and Kadowaki, K., *Phys. Rev. B* **64**, 4521 (2001); Buzdin, A. I., and Baladie, I., *Phys. Rev. Lett.* **88**, 1470021 (2002).

Suppression of Magnetic Relaxation Processes by a Transverse ac Magnetic Field

L. M. Fisher,[a] A. V. Kalinov,[a] I. F. Voloshin,[a] and V. A. Yampol'skii[b]

[a] *All-Russian Electrical Engineering Institute, 12 Krasnokazarmennaya Str, Moscow, Russia*
[b] *Institute for Radiophysics and Electronics NASU, 12 Proskura Street, 61085 Kharkov, Ukraine*

Abstract. The effect of the transverse ac magnetic field on relaxation process in $YBa_2Cu_3O_x$ melt-textured superconductor was studied. A factor of 50 suppression of the relaxation rate could be achieved at the expense of some reduction in the maximum trapped field, with the magnetic-induction gradient being unchanged. This phenomenon is interpreted as a result of an increase of the pinning force after the action of the transverse ac magnetic field.

Keywords: Relaxation; YBCO; Collapse
PACS: 74.25.Nf, 74.25.Qt

The suppression of static magnetization under the action of a transverse ac magnetic field was first observed and interpreted by Yamafuji and coauthors [1]. The other origin of the suppression of the static magnetization by the transverse ac magnetic field (the collapse phenomenon) was considered in Refs. [2, 3]. It is the flux-line cutting that provides the homogenization of the static magnetic flux in the wide areas of the sample bulk where the ac field has penetrated [2]. The inhomogeneous magnetic flux distribution is known to be metastable [4]. Within the existing concept of the collapse, one could expect a noticeable decrease of the relaxation rate of the magnetic moment after the action of the ac field. Indeed, the "collapsed" region of the sample should be filled by the vortices and the critical gradient should be reestablished before the vortices start to leave the sample. In the present paper we have checked this assumption and observed that the relaxation rate of the dc magnetic moment is actually decreased significantly by the action of the transverse ac magnetic field. Surprisingly, not only does the magnetic moment not change,, but the spatial distribution of the static magnetic flux holds the shape without significant relaxation for a long time after the action of ac field.

Melt-textured YBCO sample of $9.3\times7.4\times1.5$ mm^3 in sizes was cut out from a bulk textured cylinder grown by the top seeding technique. The **c**-axis is perpendicular to the largest sample face. The homogeneity was checked by a scanning Hall-probe. The critical current density J_c in the **ab**-plane is about 13 kA/cm^2 (T = 77 K). The zero-field cooled sample was exposed to the magnetic field of 12 kOe (parallel to the **c**-axis) which was further reduced to 5 kOe; after that the relaxation measurement was started. A commercial Hall-probe was fixed on the sample to measure the temporal evolution of the magnetic induction locally. The other Hall-probe with the sensitive zone of 0.3×0.3 mm^2 was used to scan the magnetic induction distribution on the sample surface. The Hall measurements were performed in the zero field after exposition of the sample to H = 12 kOe. The ac magnetic field **h** || **ab** was a computer-generated triangle-wave with the frequency F = 140 Hz.. The measurement was performed at T = 77 K.

Figure 1 shows the influence of the orthogonal ac magnetic field on the relaxation of M and B_{tr}. The conventional (without the ac field) relaxation is shown to follow the logarithmic law, that implies an exponential current-voltage characteristics (CVC) (see inset). For the other run, at $t \approx 20$ s the ac field of the amplitude h was applied, followed by the sharp drop in the magnetization due to the collapse-effect [3]. When the ac field action has finished, we can see almost no relaxation for the first 100 seconds and essentially reduced logarithmic relaxation rate $S = dM/d(\ln t)$ for the rest of the observation period.

To clarify the influence of the ac field on the relaxation process the temporal evolution of the dc magnetic field distribution by the scanning Hall-probe was investigated. The spatial distribution $B(x)$ (across

CP850, *Low Temperature Physics: 24th International Conference on Low Temperature Physics;*
edited by Y. Takano, S. P. Hershfield, S. O. Hill, P. J. Hirschfeld, and A. M. Goldman

the ac field direction) of the trapped dc magnetic field is shown in Fig. 2. The unperturbed curves (closed symbols) demonstrate Bean-like profiles of the magnetic flux distribution which relaxes obviously in one hour. After the action of the ac field (open symbols) the relaxation is suppressed significantly in accordance with the data of Fig. 1(b).

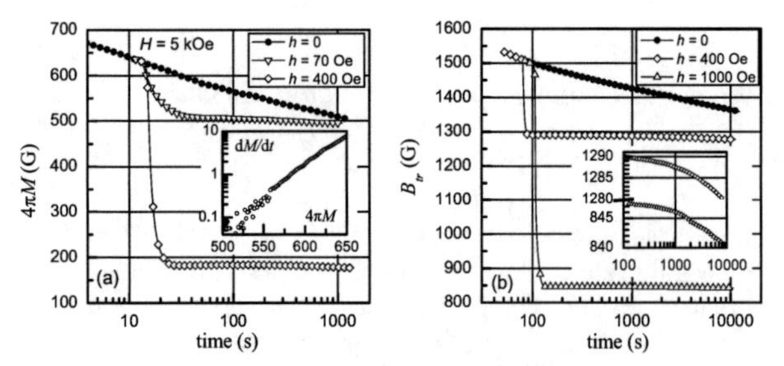

FIGURE 1. Relaxation of the magnetization M (a) and trapped magnetic induction B_{tr} (b) without (closed symbols) and after the action of the ac field (open symbols), the left inset shows the dependence of $dM/dt \sim E$ on $M \sim J$.

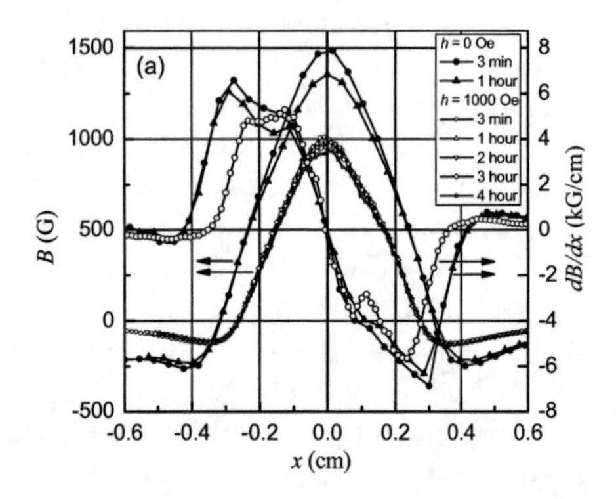

FIGURE 2. Relaxation of the magnetic-induction distribution across the ac field direction on the sample surface without (closed symbols) and after the action of the ac field (open symbols). The right scale shows derivatives $\partial B / \partial x$.

The change in the profile width is due to the collapse of the dc shielding currents in the regions of the sample where the ac field has penetrated. The magnetic field distribution in the central part of the sample before and after this action differs by the vertical displacement, with the gradient being the same. Numerically calculated dB/dx curves shown in Fig. 2 confirm this observation to be true everywhere except in the most outer regions. This means that the current densities remain at their initial values in all sample regions where the ac field has not penetrated. The decrease of the local induction could affect the relaxation rate S due to an increase of the J_c or a change in the CVC exponent. However, the change of the CVC exponent with the magnetic field [4] is too low to explain the huge change in S. The influence of local B on S through the J_c change could be estimated comparing the data of Fig. 1 (a) and (b). The external magnetic field in Fig. 1 (a) is higher than the penetration field and the ac field. In this case, the change of the local B is essentially lower than for zero-field measurement Fig. 1(b), but the effect of the relaxation suppression does exist and is qualitatively the same.

We suppose the penetration of the ac field from the largest (orthogonal to the dc field) faces of the sample to be of great importance. The first possible reason is an increase in the vortex length accompanied by the pinning energy gain, tending to the increase of J_c, because the vortices turn out to be "anchored" by their tails. The elongation of the vortices occurs in such a way that there is no additional Lorentz force directed outside the sample. Thus, the shielding currents become subcritical resulting in the exponential decrease of the relaxation rate. However, to explain the significant decrease of S by the vortex-length increase along, the relative vortex elongation must be of the order of the logarithmic relaxation-rate decrease (up to 50 times), which is hardly possible. Moreover, in the collective pinning approach the vortex elongation gives only sublinear term in the free energy. The other explanation could be based on the anisotropy of the pinning force in the YBCO superconductors. After the action of the ac field, parts (segments) of the vortices can be aligned and locked-in in the **ab**-plane leading to an increase of the pinning force. The action of the ac field could also result in the bend or entangling of the vortices at the scale of 100 – 1000 vortex-lattice periods. As a result, a significant increase of the vortex-bundle size can take place. This, in turn, can lead to the increase of the collective pinning potential and to the observed suppression of the relaxation rate.

ACKNOWLEDGMENTS

This work is supported by NATO (grant PST.CLG.980307) and RFBR (grant 03-02-17169).

REFERENCES

1. K. Funaki and K. Yamafuji, *Jpn. J. Appl. Phys.* **21**, 299 (1982); **21**, 158 (1982).
2. L. M. Fisher, A. V. Kalinov, S. E. Savel'ev, et al., *Physica C* **278**, 169 (1997).
3. L. M. Fisher, K. V. Il'enko, A. V. Kalinov, et al., *Phys. Rev B* **61**, 15382 (2000).
4. H. Kupfer, S. N. Gordeev, W. Jahn, et al., *Phys. Rev. B* **50**, 7016 (1994).

Instability of the Vortex Matter in YBCO Single Crystals

L. M. Fisher[a], T. H. Johansen[b], A. Bobyl[b], A. L. Rakhmanov[c], M. L. Nesterov[d], and V. A. Yampol'skii[d]

[a] *All-Russian Electrical Engineering Institute, 12 Krasnokazarmennaya Str, Moscow, Russia*
[b] *Department of Physics, University of Oslo, P.O. Box 1048, Blindern, 0316 Oslo 3, Norway*
[c] *Institute for Theoretical and Applied Electrodynamics RAS, 13/19 Izhorskaya Street, 125412 Moscow, Russia*
[d] *Institute for Radiophysics and Electronics NASU, 12 Proskura Street, 61085 Kharkov, Ukraine*

Abstract. A theory of the macroturbulent instability in the system containing vortices of opposite directions (vortices and antivortices) in hard superconductors is proposed. The origin of the instablity is connected with the anisotropy of the current capability in the sample plane. The anisotropy results in the appearance of tangential discontinuity of the hydrodynamic velocity of vortex and antivortex motion near the front of magnetization reversal. The examination is performed on the basis of the anisotropic power-law current-voltage characteristics. The instability is shown to be observed even at relatively weak anisotropy. The physical nature of the macroturbulent instability in the vortex matter in YBCO superconductors is verified by means of magnetooptic study of the instability in a single crystal prepared specially for this purpose. The instability develops near those sample edges where the oppositely directed flow of vortices and antivortices, guided by twin boundaries, is characterized by the discontinuity of the tangential component of the hydrodynamic velocity.

Keywords: Macroturbulence; YBCO; Current anisotropy
PACS: 74.25.Op, 74.25.Qt, 74.40.+k

Magnetic flux dynamics in type-II superconductors has been extensively studied since the end of the 50-s, starting with the pioneering work by A.A. Abrikosov. An avalanche of new research activity in this field was triggered by the discovery of high-T_c superconductivity (HTS), which brought into play thermal fluctuations and the strong anisotropy of superconductors. The use of high-resolution magneto-optical (MO) technique enabled an in-depth study of the dynamics of magnetic flux in superconductors. Among the most important features revealed by means of this method are the flux structures behaving strongly irregular both in time and space. Such structures arise usually due to the development of characteristic instabilities such as macroturbulence in 1-2-3 systems [1-2]. Surprisingly, these dramatic instabilities in the flux line lattice (FLL) have so far been investigated to a relatively small extent. Perhaps, the macroturbulence is the most interesting phenomenon observed in the dynamics of the magnetic flux in HTS. It appears like a turbulization of the FLL motion near the front of magnetization reversal that separates regions of vortices of opposite directions (vortices and antivortices). When magnetic flux is trapped in a superconductor and a moderate field of the reverse direction is subsequently applied, a boundary of zero flux density will separate the regions containing vortices and antivortices. At some range of magnetic fields and temperatures, this flux-antiflux distribution becomes unstable. A disordered motion of magnetic flux arises at the front of magnetization reversal, which resembles a turbulent fluid flow. This process rapidly develops in time and is accompanied by the formation of fingers via which the antivortices penetrate into the region occupied by the vortices. In other words, the front of magnetization reversal takes on an irregular shape. The annihilation of vortices and antivortices occurs at the front, and the process of macroturbulence is soon terminated after a complete disappearance of the vortices. This pattern of penetration of the magnetic flux differs qualitatively from the steady-state slow motion of the front of magnetization upon initial turning on of the magnetic field, when vortices of only one direction are present in the sample.

An explanation of the macroturbulence should focus on the experimental fact that the instability was reported for YBCO and other 1-2-3 single crystals

CP850, *Low Temperature Physics: 24th International Conference on Low Temperature Physics;*
edited by Y. Takano, S. P. Hershfield, S. O. Hill, P. J. Hirschfeld, and A. M. Goldman

only, which are characterized by the anisotropy in the basal **ab** plane. This anisotropy can be related to a specific crystallographic structure of these HTS and, in particular, to twin boundaries. Our approach to understanding the mechanism of the macroturbulence was elaborated in [3, 4] taking into account of the specific features of the flux motion in the anisotropic superconductors. The anisotropy gives rise to the motion of the flux lines at some angle with respect to the Lorentz force direction. In the presence of twin boundaries, vortices and antivortices move preferably along these guiding boundaries. As a result, the flux lines move at some angle with respect to the magnetization reversal front. It is exactly this circumstance that is of paramount importance to ascertain the nature of macroturbulence. The vortices and antivortices are forced to move towards each other in such a way that the tangential component of their velocity becomes discontinuous at the flux-antiflux interface. According to a classical paper of Helmholtz, a stationary hydrodynamic flow can be unstable and turbulent under such conditions.

A theoretical model is based on the hydrodynamic approach and operates with the vortex and antivortex densities $N_{1,2}(x,t)$ and the vortex velocities $V_{1,2}(x,t)$. The set of equations includes the continuity equation and the anisotropic current-voltage characteristics (CVC) of a sample. It is assumed that the CVC is a power-law function. The boundary conditions include the law for annihilation of vortices and antivortices at the magnetization front reversal. This relation was derived from the microscopic consideration. The analysis of the problem has shown that the quasistatic evolution of the vortex-antivortex system becomes unstable under definite conditions. The dispersion equation for the instability increment at different values of the anisotropy parameter is obtained and solved numerically. The necessary condition for the instability is the anisotropic flow in the vortex system.

The theoretical results are supported by the experiment carried out on a sample of a special shape and twin structure. In a single crystal of YBCO having the form of right triangle, the twin boundaries were directed along the hypotenuse [5]. The size of the sample along the hypotenuse is about 1.1 mm. The crystallographic **ab** plane coincides with the sample plane.

In order to search for instability, the sample was first cooled in a transverse magnetic field H=680 Oe. Then the field was abruptly reversed and MO images were recorded and analyzed. A set of MO images in Figure obtained at temperature 27 K demonstrate the consecutive stages of the development of the instability. The time intervals between frames are 40 ms. The brighter regions of the image correspond to higher values of the magnetic induction.

FIGURE 1. .Evolution of the magnetic flux distribution under the condition of the development of instability

The key point here is to observe the significant difference in the interface behavior near the hypotenuse and the legs. The interface near hypotenuse remains essentially static whereas substantial motion takes place elsewhere, e.g., for the interface along the upper cathetus, where the velocity is estimated to 5 mm/s at the initial stage of the development of the instability. This result is in accordance with our model. Indeed, owing to the guiding effect, the tangential discontinuity of the flux flow velocity arises only for the flux-antiflux interfaces near the triangle legs. Therefore, the macroturbulence is observed near the legs and does not arise near the sample hypotenuse.

ACKNOWLEDGMENTS

This work was supported by NATO (grant PST.CLG.980307) and RFBR (grant 03-02-17169).

REFERENCES

1. V.K. Vlasko-Vlasov, V.I. Nikitenko, A.A. Polyanskii, et al., *Physica* (Amsterdam) **222C**, 361 (1994).
2. M.V. Indenbom, Th. Schuster, M.R. Koblischka, et al., *Physica* (Amsterdam) **209C**, 259 (1993).
3. L.M. Fisher, P.E. Goa, M. Baziljevich, et al., *Phys. Rev. Lett.* **87**, 247005 (2001).
4. A.L. Rakhmanov, L.M. Fisher, A.A. Levchenko, et al., *JETP Lett.* **76**, 291 (2002).
5. L.M. Fisher, A. Bobyl, T.H. Johansen, et al., *Phys. Rev. Lett.* **92**, 037002 (2004).

Frequency Dependence of Vortex Lattice Elastic Moduli in the Hollow of Superconducting YBaCuO Cylinder

Robert A. Vardanyan, Mkrtich T. Ayvazyan, and Armen A. Kteyan

Solid State Division, Institute of Radiophysics and Electronics, Ashtarak-2, 378410, Armenia

Abstract. Vortex lattice compression moduli in ceramic YBaCuO are studied by measurement of ac response U in the hollow of a cylinder existing in the mixed state. Investigations were performed at stationary magnetic fields H up to 1200 Oe, when the elastic moduli have a non-local character resulting in saturation of the $U(H)$ dependence at large values of H. The analysis of the frequency dependence of the studied response permits to conclude that the wave-vector of vortex lattice deformation decreases with increase of excitation frequency. The penetration of the signal into the hollow increases at reduction of the frequency down to values about 5 kHz; at the further frequency reduction abrupt strengthening of the signal screening is observed.

Keywords: vortex, oscillation, compression.
PACS: 74.25.Nf, 74.25.Qt

We carry study of compression moduli of vortex lattice (VL) in superconductors, which allows to reveal their dependence on the frequency of a distorting perturbation. With this purpose, the response of the hollow superconducting cylinder to an external ac field has been measured in the cavity of the sample.

Magnetic field h^{cav} in the cavity of a superconducting cylinder is composed of the field related to the trapped flux, and of the field created by vortices. The last one as a function of a vortex distance from the cavity decreases exponentially [1]; therefore one can assume that only the vortices placed in the immediate vicinity of the cavity (on the distance r_o) contribute to the magnetic field in the cavity. If a coil is attached to the surface of the cavity, then a small displacement u^{cav} of vortices near the cavity will generate in the coil a voltage which is proportional to the field alteration

$$U \propto \frac{\partial h_o}{\partial t} \propto -\frac{\Phi_o}{\pi\lambda}\cdot\frac{e^{-r_o/\lambda}}{R}\cdot\frac{\partial u^{cav}}{\partial t}. \qquad (1)$$

The hollow cylinders molded of the ceramic YBaCuO were used for the experiment. On the central part of the external surface of the cylinder the excitation coil was wound creating the ac field h on this surface. At the middle part of the cavity the receiving coil was attached to the surface, registering the magnetic field alteration in the hollow (details of experimental setup are described elsewhere [2]). The sample was converted into the vortex state by the dc magnetic field H directed along the cylinder axis. The measurements were performed in the range $H = 0 \div 1200$ Oe at liquid nitrogen temperature. The alternating voltage (with frequency range $\omega = 1 \div 600$ kHz) was supplied to the excitation coil and created the ac magnetic field h up to 1 Oe.

When the dc field is turned off, the ac field $h(\omega)$ created by the excitation coil is screened completely by the supercurrents. By increasing of the field H, the cylinder is driven into vortex state, and the receiving coil registers an ac signal with frequency ω of the exciting field. According to the Eq. (1), the induction of the signal U in the cavity denotes that vortices near the cavity are displaced reversibly: $u^{cav}(t) = u_o^{cav} e^{i\omega t}$. Let us analyze now the amplitude of these oscillations.

The excitation $h(\omega)$ induces ac currents on the external surface of the cylinder and causes oscillations of near-surface vortices $u^{ext}(t) = u_o^{ext} e^{i\omega t}$. As the thickness d of the cylinder wall is small compared to the radius of curvature, and the compression modulus c_{11} much excesses the shear modulus $c_{11} >> c_{66}$, the oscillation amplitude of a vortex near cavity is

$$u_0^{cav} \propto u_0^{ext} c_{11} / d . \qquad (2)$$

Figure 1 shows the dependence of the signal in the cavity on the dc magnetic field at different values of

CP850, *Low Temperature Physics: 24th International Conference on Low Temperature Physics;*
edited by Y. Takano, S. P. Hershfield, S. O. Hill, P. J. Hirschfeld, and A. M. Goldman

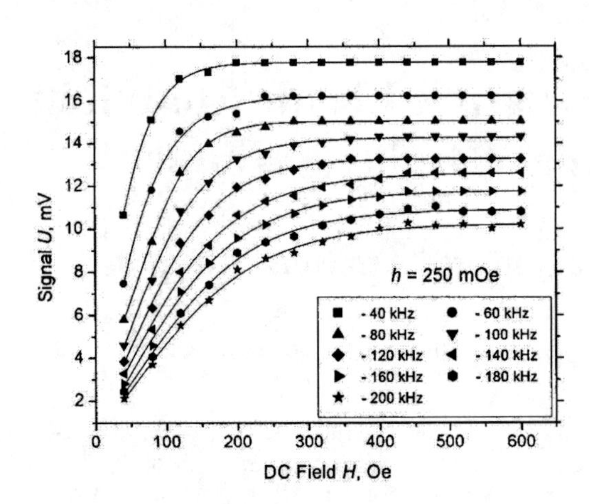

FIGURE 1. Signal in the cavity versus dc magnetic field, plotted with account of the frequency dependence of voltage generation in the receiving coil.

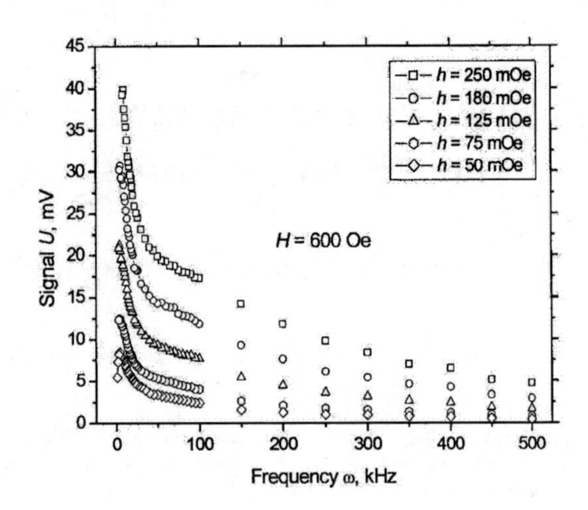

FIGURE 2. Amplitude – frequency characteristics of the signal in the cavity.

the exciting frequency. The signal increase at small values of the field is followed by the saturation at some value H^{sat}, which is approximately 200 Oe at 40 kHz and increases with the excitation frequency. Note that at such fields ($H_{c1} << H_{dc} << H_{c2}$) the inter-vortex distance becomes less than the London penetration depth: $a << \lambda$, and the compression modulus c_{11} has a non-local character, i.e. it depends on the wavelength of the lattice deformation $\boldsymbol{q}^{-1}$ [3]:

$$c_{11} = \frac{B^2}{8\pi} \frac{1}{1+q^2\lambda^2}, \tag{3}$$

where B is the magnetic field in the superconductor. Apparently, the magnitude of the vector $\boldsymbol{q}$ must depend on VL period a, and hence on H, as well as on the amplitude and frequency of the exciting field: $q = q(H, h, \omega)$. Therefore, when the dc field achieves the value H^{sat}, the deformation vector increases until the inequality $q\lambda >> 1$ is satisfied. Then as it follows from (3) the growth of the modulus c_{11} with the increase of dc field can cease (i.e. the VL softening occurs) and according to (2) the oscillation amplitude u_0^{cav} of a vortex near the cavity do not more increase. Thus the growth of H^{sat} at higher frequencies denotes the decrease of the lattice deformation vector.

The obtained frequency dependence of the signal in the cavity can be interpreted by the vortices radial oscillations in the cylinder (neglecting their bending). The oscillation amplitude of a vortex subjected to the external force $f = f_o e^{i\omega t}$ is equal to

$$u_o = \frac{f_o}{m\sqrt{\left(\omega_0^2 - \omega^2\right)^2 + 4\omega^2\eta^2}}, \tag{4}$$

where m is the vortex mass, η is the damping constant, and ω_0 is the proper frequency. If we assume that ω_0 is in the range of microwave frequencies, then it follows that the frequency dependence of the signal $U_0(\omega)$ is conditioned by the increase of the damping force at higher frequencies, provided that $2\omega\eta >> \omega_0^2$. As a result, the frequency increase reduces the signal in the cavity, while the higher dc fields are required to discover the dispersion of the compression modulus.

The frequency dependence of the signal in the cavity is shown on the Fig.2. The behavior of the curve fits well the law $U \propto 1/\omega$ which follows from Eq.(4) when $\omega << \omega_o$. It is notable that the response increases while the frequency decreases down to about 5 kHz, and then it drops abruptly. It is obvious that the signal in the hollow must disappear at zero frequency; however the nature of certain level of the applied frequency below which the screening properties of the cylinder strengthen is not yet clear.

REFERENCES

1. V. V. Shmidt and G. C. Mkrtchian, *Usp. fiz. nauk* **112**, 459-490 (1974).
2. M. T. Ayvazyan, A. A. Kteyan, and R. A. Vardanyan, *Physica C* **405**, 163-172 (2004).
3. E.H. Brandt, *J. Low Temp. Physics* **26**, 735-753 (1977).

Specific Heat of the Vortex Melting Transition in $YBa_2Cu_4O_8$

O.J. Taylor*, A. Carrington* and S. Adachi†

*H. H. Wills Physics Laboratory, University of Bristol, Tyndall Avenue, BS8 1TL, United Kingdom.
†Superconductivity Research Laboratory, ISTEC 1-10-13 Shinonome, Koto-ku, Tokyo 135-0062, Japan.

Abstract. We report high resolution field dependent specific heat measurements of single-crystal $YBa_2Cu_4O_8$ close to the vortex melting temperature T_m. Measurements were performed in fields up to 14 T (applied parallel to either the b or c axes). For both field directions close to T_m a broadened step is seen which is consistent with a second order phase transition.

Keywords: superconductivity, vortex lattice melting
PACS: 74.72.Bk,74.25.Qt,74.25.Bt

In $YBa_2Cu_3O_7$ (Y-123) and $Bi_2Sr_2CaCu_2O_{8+\delta}$ (Bi-2212) it is well established that in the absence of disorder the vortex lattice melts by a first order phase transition [1]. These two compounds span a wide range of anisotropies ranging from $\gamma \simeq 7.5$ (Y-123) to $\gamma \simeq 250$ (Bi-2212). $YBa_2Cu_4O_8$ (Y-124) presents an interesting intermediate case ($\gamma \sim 12-15$) and further it is close to being stoichiometric with very few oxygen vacancies, and is underdoped.

The vortex phase diagram of Y-124 has been investigated by several techniques and resistivity measurements have shown evidence for a first order melting transition [2, 3, 4]. However, the most unambiguous evidence for a first order phase transition is usually found by measuring the specific heat C. The very small size of the Y-124 crystals makes such studies difficult. Indeed, the latent heat at the vortex-lattice melting has only clearly been seen in Y-123 to date [5, 6].

Here we report a study of the specific heat of Y-124 close to the vortex melting temperature T_m. Measurements of a 42 μg single crystal of Y-124 were carried out at temperatures between 60 K and 90 K in fixed magnetic fields (set at $T > T_c$) between 0 T and 14 T for $H||b$ and between 0 T and 2 T for $H||c$. The sample was placed on a flattened 12μm, chromel-constantan thermocouple and heated either with a modulated light source or a small resistive heater. Although this technique is very sensitive ($\sim$ 1 part in 10^5) the absolute values cannot be determined accurately and so our data are scaled at $T > T_c$ and $H = 0$ on to published data for polycrystalline Y-124 [7]. Measurements were performed on another sample with very similar results.

The raw specific heat data are shown in the inset of Fig. 1. We have attempted to isolate superconducting anomaly by subtracting a third order polynomial fitted to the 0 T data in the temperature intervals 55 K< T <65 K and 82.5 K< T <90 K (i.e., well away from T_c). The data with this background subtracted are shown in Fig. 1.

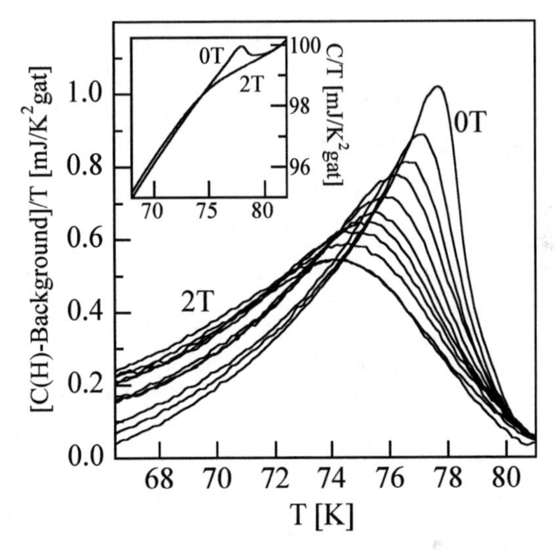

FIGURE 1. Field dependence of the specific heat of Y-124 after subtraction of the background for fields from 0 T to 2.0 T on 0.2 T increments for $H||c$. The inset shows the total specific heat 0 T and 2 T data.

The superconducting anomaly is somewhat asymmetric, with the peak in C broadening and shifting to lower temperature with increasing field. This behavior is closer to that found for Y-123 than that found for Bi-2212 [8].

To be able to observe the signature of vortex-lattice melting we subtract the zero field data from that in finite field, $\Delta C(H,T)/T = [C(H,T) - C(0,T)]/T$. Well below T_c we find that ΔC is linear in T with a gradient which increases monotonically with H. To clarify the behavior near T_m this linear term has been subtracted off the $\Delta C(H,T)/T$ data and the result is shown in Fig. 2. Approaching T_c from below it can be seen that ΔC first increases slightly (by $\sim$ 0.02 % of the total) then decreases rapidly. The behavior closely resembles that found for slightly disordered Y-123 samples, where the initial rise was attributed to the melting transition [11]. We find then

CP850, *Low Temperature Physics: 24th International Conference on Low Temperature Physics;*
edited by Y. Takano, S. P. Hershfield, S. O. Hill, P. J. Hirschfeld, and A. M. Goldman

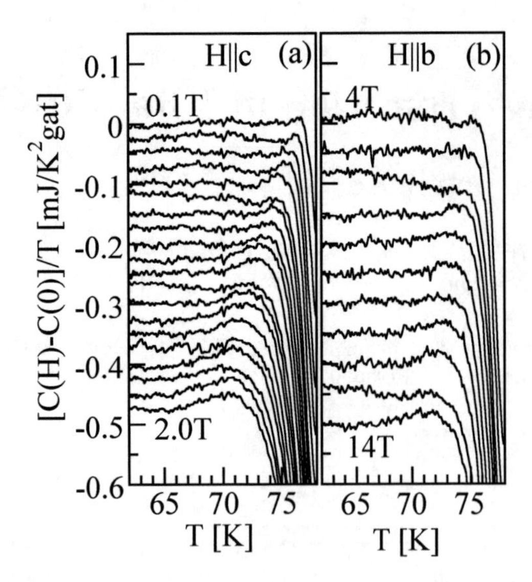

FIGURE 2. Specific heat differences $[C(H) - C(0)]/T$ vs T with a field dependent linear term subtracted. (a) $H||c = 0.1$ to 2.0T in 0.1T increments. (b) $H||b = 4$ to 14T in 1T increments. The curves have been shifted arbitrarily for clarity.

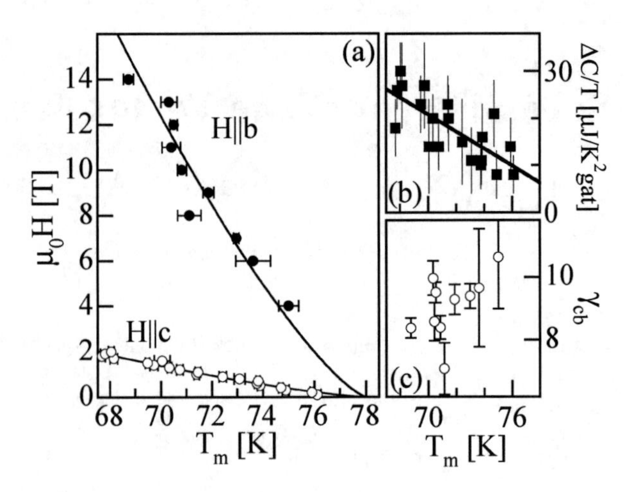

FIGURE 3. (a) $T_m(T)$ for Y-124 for $H||c$ (open circles) and $H||b$ (closed circles). (b) The step height at T_m for $H||c$. (c) The anisotropy of the melting line γ_{cb}.

that the anomaly in C at T_m is a broadened step rather than a (broadened) peak. This suggests that for both $H||b$ and $H||c$ the transition is of second order.

In Fig. 3(a) we show the position of the melting transition temperature (taken as the midpoint of the step) as a function of H for both field directions. $H_m(T)$ obeys the empirical relation $H_m = H_0(1 - T/T_c)^n$. For $H||c$, $\mu_0 H_0 = 28 \pm 5$ T and $n = 1.32 \pm 0.08$ with $T_c = 78.0 \pm 0.1$ K. The superconducting transition temperature T_c was determined by fitting the 0 T data, with the background subtracted, to the critical fluctuation 3D-XY model. The $H||b$ data was fitted to the same function with the same value of n and $\mu_0 H_0 = 252 \pm 50$ T. The exponent found is close to that found in previous studies [3, 4].

The anisotropy of the melting line is shown in Fig. 3(c); the average value of $\gamma_{cb} = H_m^{||b}/H_m^{||c}$ is 9.0±0.3. The same value of γ was found by analyzing the field dependence of the main superconducting anomaly in C for the two field directions. Resistivity measurements [3] found that $\gamma_{cb} \sim 15$, with an in-plane anisotropy γ_{ab} between 1.05 and 1.2. These latter measurements compare favorably with values found from torque magnetometery [9, 10]. The value found here is smaller than these other reported values, which is surprising given that our field alignment is estimated to be with ±2° of the c-axis. Further angle dependant measurements are needed to verify this.

The step height ΔC_m for $H||c$ increases approximately linearly with increasing field; $\Delta C_m = 6(3) - 1.8(4)(T - T_c)$ μJ/(K^{2}gat) [see Fig. 3(b)]. This is ~ 6 times smaller than that observed in slightly disordered Y-123 [11].

The absence of a first order melting transition in these samples is surprising given the sharp anomalies seen in resistivity data in similar crystals [3]. We recall however that Y-124 is underdoped and first-order peaks in C were only seen in the highest quality Y-123 samples with high oxygen doping [6].

In summary, we have measured the specific heat of single crystal Y-124 and observed the anomaly due to the vortex-solid melting transition. To our knowledge this is the first time the specific heat of this transition has been observed in a compound other than Re-123. The transition observed was of second-order, suggesting that the vortex-solid is a glass.

This work was supported by EPSRC (UK) and NEDO (Japan).

REFERENCES

1. G. Blatter *et al.*, *Rev. Mod. Phys.*, **66**, 1125–1388 (1994).
2. X. G. Qiu *et al.*, *Phys. Rev. B*, **58**, 8826–8829 (1998).
3. N. E. Hussey *et al.*, *Phys. Rev. B*, **59**, R11668–R11671 (1999).
4. X. G. Qiu *et al.*, *Phys. Rev. B*, **62**, 4119–4123 (2000).
5. A. Schilling *et al.*, *Phys. Rev. Lett.*, **78**, 4833–4836 (1997).
6. M. Roulin *et al.*, *Phys. Rev. Lett.*, **80**, 1722–1725 (1998).
7. A. Junod *et al.*, *Physica C*, **168**, 47–56 (1990).
8. A. Junod *et al.*, *Physica B*, **280**, 214–219 (2000).
9. D. Zech *et al.*, *Phys. Rev. B*, **54**, 12535–12542 (1996).
10. N. Kagawa *et al.*, *Physica C*, **357**, 302–304 (2001).
11. M. Roulin *et al.*, *Science*, **273**, 1210–1212 (1996).

Dimensional Crossover in the Vortex System of Untwinned YBa_2Cu_3Oy Single Crystals with Highly Oxygen Deficiency

Hiromi Fujita, Terukazu Nishizaki, Kuniaki Kasuga and Norio Kobayashi

Institute for Materials Research, Tohoku University, Sendai 980-8577, Japan

Abstract. We have investigated the vortex phase diagram of high-quality untwinned $YBa_2Cu_3O_y$ (YBCO) single crystals in the underdoped regime. The in-plane resistivity near the transition temperature T_g follows a power law $\rho \propto (T - T_g)^s$, indicating a second-order vortex glass to liquid transition. The temperature dependence of the glass transition field is found to follow $H_g(T) \propto (1-T/T_c)^n$ with $n \sim 4/3$ below a certain field H^* and change to $H_g(T) \propto \exp(-aT/T_c)$ above H^*. Moreover, the critical exponent s also changes from 6-7 below H^* to a smaller value ~3.6 above H^*. These changes in the temperature dependence of the resistivity are attributed to the dimensional crossover in the vortex system.

Keywords: High-T_c superconductivity, $YBa_2Cu_3O_y$, Vortex state, Dimensional crossover.
PACS: 74.72.Bk , 74.25.Fy, 74.25.Qt

INTRODUCTION

In high-temperature superconductors (HTSC), the nature of vortex state has attracted a considerable amount of attention and numerous experimental and theoretical studies have been performed [1]. Especially, the vortex phase diagram of $YBa_2Cu_3O_y$ in the vicinity of optimal doping has been extensively studied. These studies have revealed that the vortex state exhibits various types of vortex phase, such as the vortex lattice, the vortex glass, the vortex liquid states and so on, and is strongly affected by crystalline disorder caused by oxygen deficiencies [2]. However, there are only a few studies of the vortex state of underdoped YBCO [3,4].

In this study, we prepare high-quality untwined YBCO single crystals in the underdoped regime and investigate the vortex phase diagram on the basis of the scaling behavior of the electrical resistivity.

EXPERIMENTAL

YBCO single crystals were prepared by self-flux method using Y_2O_3 crucibles. Samples used in this study were detwinned and then the underdoped samples with T_c = 61 (8t#4 and 77Yh3), 57 (8t#4-2) and 34 K (8t#4-3) were prepared by annealing at 680 °C in 1 bar flowing oxygen gas, and at 455°C and 500 °C in 1 bar flowing nitrogen gas, respectively. The in-plane resistivities ρ_a, ρ_{ab} and ρ_b were measured for the current directions of [100], [110] and [010], respectively, by an eight-probe technique [5] in magnetic field H//c-axis up to 15 T. In this technique, the currents J were oriented in both [100] and [010] directions using two independent current sources and the voltage drops along both directions are simultaneously measured. The resistivity for any direction was obtained by varying the ratio of flowing currents for both [100] and [010] directions.

RESULTS AND DISCUSSION

Figure 1 shows the temperature dependence of the electrical resistivity ρ_b for the current direction of [010] as a typical example. The resistivity curve is broader than that of optimally and slightly overdoped YBCO with the first-order vortex lattice melting transition. Furthermore, in the transition region, the ρ_b values at high magnetic fields above 11 T are slightly larger than the ρ_a and ρ_{ab} values, and have an upward curvature. The resistivity below 11 T, however, is almost isotropic for all current directions studied.

In the lower temperature region of the resistivity transition, the resistivity decreases continuously toward $\rho = 0$ in the presence of a magnetic field. This continuous behavior in the $\rho(T)$ curve indicates that a second-order vortex liquid-to-glass transition occurs. Both vortex glass and Bose glass theories[6,7] predicts that the resistively is descried as $\rho \propto (T - T_g)^s$ near a

CP850, *Low Temperature Physics: 24th International Conference on Low Temperature Physics;*
edited by Y. Takano, S. P. Hershfield, S. O. Hill, P. J. Hirschfeld, and A. M. Goldman

vortex glass transition temperature T_g, where s is the critical exponent. The values of T_g and s are derived from a linear extrapolation of $[d(\ln\rho)]/dT]^{-1}$ to 0 and an inverse of the slope in $[d(\ln\rho)]/dT]^{-1}$ vs. T, respectively.

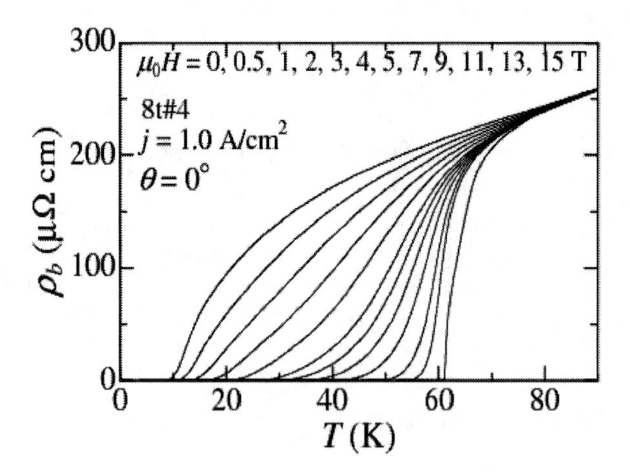

FIGURE 1. Temperature dependence of the resistivity for *J*//b-axis.

Figure 2 shows the temperature dependence of H_g determined from the $T_g(H)$. At the high temperature region near T_c, $H_g(t)$ follows a power law behavior of $H_g \propto (1 - t)^n$ with $n \sim 1.3$, as shown by a solid line, where $t = T/T_c$. This behavior is in good agreement with that of the vortex glass-to-liquid transition line in the 3-dimensional (3D) vortex system predicted by the vortex glass theory [6], and also that of experimental results for YBCO with strong disorder near the optimally doped region. However, at lower temperature, $H_g(T)$ deviates from the power law behavior and exhibits an exponential dependence $H_g(t) \propto \exp(-at)$ (dashed line). This result indicates that a crossover in the vortex system occurs at the crossover field $H^* \sim 2$ T for the sample with $T_c = 61$ K. Similar crossover behavior is also observed at ~ 1.5 and ~ 0.6 T for the samples with 57 and 34 K, respectively, though such a behavior never been observed in YBCO in the vicinity of optimal doping so far [1,2].

Inset of Fig. 2 shows the critical exponent s of the sample with $T_c = 61$ K as a function of magnetic filed for all current directions studied. At low magnetic fields below 1 T, the s value is approximately 6, which is in agreement with reported values of $s = 6 - 8$ in the optimally doped YBCO [1,2]. In the presence of a higher magnetic field, the s value is ~3.6 and is independent of magnetic field. This result also indicates an occurrence of a crossover in the vortex system at ~ 2 T. This is consistent with the crossover behavior in the temperature dependence of $H_g(t)$.

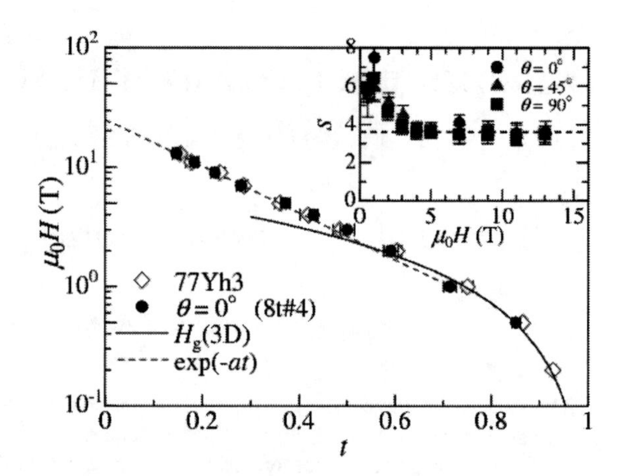

FIGURE 2. Temperature dependence of the glass transition field H_g in 60-K YBCO (8t#4 and 77Yh3). Inset shows the field dependence of the critical exponent s.

As for the strongly anisotropic layered superconductors such as HTSC, the collective pinning theories have predicted that the magnetically induced dimensional crossover from 3D to quasi-2D vortex system occurs at the dimensional crossover field $H_{2D} = \Phi_0/(\gamma^2 l^2)$ [8], where Φ_0 is the magnetic flux quantum, γ^2 the effective mass ratio m_c/m_{ab}, and l the interlayer distance. Using this relation with $\gamma \sim 30$ [4] and $l \sim 12$ Å, the dimensional crossover in the underdoped YBCO with $T_c \sim 60$ K is expected to occur at around 2.4 T. This value is consistent with our experimental results.

In conclusion, We have investigated the vortex phase diagram of high-quality untwinned YBCO single crystals in the underdoped regime and the first time observed the dimensional crossover behavior in the vortex system of underdoped YBCO.

This work was partly supported by a Grant-in-Aid for Scientific Research (15340107) from the Ministry of Education, Science, Sports, and Culture of Japan.

REFERENCES

1. T. Nishizaki and N. Kobayashi, *Supercond. Sci. Technol.* **13**, 1 (2000), and references there in.
2. K. Shibata, et al., *Phys. Rev.* B **66**, 214518 (2002).
3. B. Lundqvist, et al., *Phys. Rev.* B **57**, 14064 (1998).
4. L. Hou, et al., *Phys. Rev.* B **55**, 11806 (1997).
5. G. D'Anna, et al., *Phys. Rev.* B **61**, 4215 (2000).
6. D. S. Fisher, et al., *Phys. Rev.* B **43**, 130 (1991).
7. D. R. Nelson and V. M. Vinokur, *Phys. Rev. Lett.* **68**, 2398 (1992); *Phys. Rev.* B **48**, 13050 (1993).
8. L. I Glazman and A. E. Koshelev, *Phys. Rev.* B **43**, 2835 (1991).

In-Plane Anisotropy of the Vortex Motion in $YBa_2Cu_3O_y$ Single Crystals

Terukazu Nishizaki, Hiromi Fujita, Kuniaki Kasuga, and Norio Kobayashi

Institute for Materials Research, Tohoku University, Sendai 980-8577, Japan

Abstract. In-plane anisotropy of the vortex motion has been studied by a technique of rotating the transport current in the *ab* plane ("a vector current method") in detwinned $YBa_2Cu_3O_y$ single crystals. In the vortex liquid phase ($H//c$), the angular dependence of the in-plane resistivity $\rho(\theta)$ agrees with the flux flow model which is characterized by the anisotropic viscosity η. With decreasing temperature toward the vortex lattice melting line $T_m(H)$, $\rho(\theta)$ deviates from the flux flow model and the anisotropy of $\rho(\theta)$ is enhanced. These results indicate that an additional anisotropy appears in the Bragg glass phase due to the anisotropic pinning potential with the two-fold symmetry.

Keywords: Vortex State, Anisotropy, $YBa_2Cu_3O_y$
PACS: 74.25.Op, 74.25.Qt, 74.72.Bk

INTRODUCTION

High temperature superconductor $YBa_2Cu_3O_y$ shows a pronounced anisotropy in the *ab* plane due to the anisotropic crystal structure and electronic state. In twinned $YBa_2Cu_3O_y$ single crystals, the anisotropy of the vortex motion has been studied and a guided motion along the twin planes has been observed [1, 2]. Since untwined $YBa_2Cu_3O_y$ shows an anisotropy parameter of $\gamma_{ab} \sim 1.5$ in the *ab* plane which results from the electronic state in the CuO chain, it is expected that the vortex lattice structure, dynamics, and pinning show the anisotropic nature. In this paper, we focus our attention on the vortex dynamics as a function of the driving force direction that can be controlled by the rotation of the transport current.

EXPERIMENTAL

High quality $YBa_2Cu_3O_y$ single crystals were grown by self-flux method using Y_2O_3 crucible [3-5]. As-grown single crystals were detwinned under uniaxial stress [3] and the oxygen content y ($\simeq$6.99) was controlled by the annealing under 1 bar oxygen at ~300 °C for 30 days [4]. In order to study the in-plane anisotropy, a technique of rotating the transport current ("a vector current method") [1, 2] has been employed. In this method, four electric contacts along the *x*-direction (//*b*-axis), which are used for usual four-terminal transport measurements, are oriented 90° with respect to the other four contacts along the *y*-direction (//*a*-axis). Therefore, the direction of both the transport current density J ($\simeq$ 0.5 - 12 A/cm^2) and the vortex motion can be controlled using the two current sources. The angular dependence of the resistivity $\rho(\theta)$ can be obtained by simultaneous measurements of the voltages in *x*- and *y*- directions. Here, θ is an angle between the *b*-axis and the current direction.

RESULTS AND DISCUSSION

Figure 1 shows the temperature dependence of the resistivity $\rho_a(T)$ at 11 T ($H//c$) in $YBa_2Cu_3O_{6.99}$. In the lower J (= 0.5 A/cm^2) region, the resistivity shows a sharp drop at the vortex lattice melting $T_m(H)$ between the vortex liquid and the Bragg glass, independently of θ. In the high-J region, however, the transition width becomes broader and the resistivity depends on both the direction and the value of J as shown in the inset of Fig. 1. Since the anisotropy of $\rho(T)$ becomes remarkable below $T_m(H)$, the anisotropy mainly results from the anisotropic vortex dynamics.

Figure 2 shows the angular dependence of the resistivity $\rho(\theta)$ at 11 T under high-J (= 12 A/cm^2). In the vortex liquid phase above 75 K, $\rho(\theta)$ shows weak angular dependence with a maximum at θ = 90° and 270° (i.e., J //*a*-axis) and a minimum at θ = 0°, 180° and 360° (i.e., J//*b*-axis); the behavior is very similar to $\rho(\theta)$ in the normal state. The anisotropic linear resistivity in the vortex liquid phase is closely

CP850, *Low Temperature Physics: 24th International Conference on Low Temperature Physics;*
edited by Y. Takano, S. P. Hershfield, S. O. Hill, P. J. Hirschfeld, and A. M. Goldman

connected with the anisotropy of the flux flow (FF) resistivity. Since the FF resistivity is expressed as $\rho_{FF} = B\Phi_0/\eta = \rho_N B/B_{c2}$ and the magnetic flux density B is always parallel to the c-axis in this configuration, the anisotoropy of $\rho_{FF}(\theta)$ is characterized by the anisotropy of the vortex viscosity $\eta(\theta)$ and the normal state resistivity $\rho_N(\theta)$. Thus, the angular dependence of $\rho(\theta)$ can be described by the equation, $\rho(\theta) = \rho_b(\sin^2\theta + \gamma_{ab}^2 \cos^2\theta)$, with the effective mass ratio $\gamma_{ab}^2 = m_a/m_b = \rho(90°)/\rho(0°)$. The experimental data in high-T region (T > 75 K) in Fig. 2 well agree with the FF model using a reasonable parameter of $\gamma_{ab}^2 \sim 1.5$ for $YBa_2Cu_3O_{6.99}$.

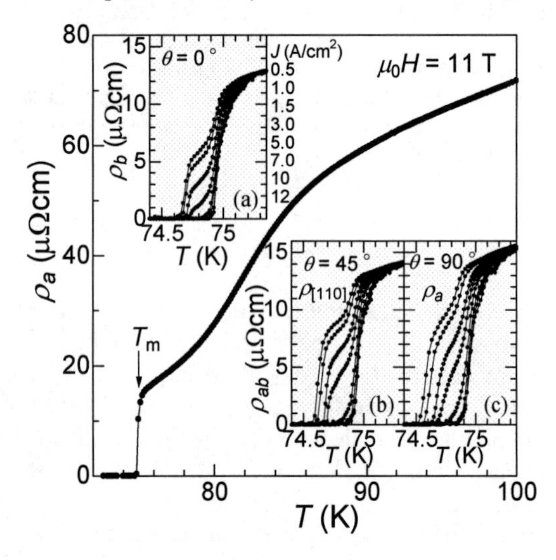

FIGURE 1. Temperature dependence of the resistivity at 11T. Inset: Transport current density J dependence of the resistivity in (a) θ = 0°, (b) θ = 45°, (c) θ = 90°.

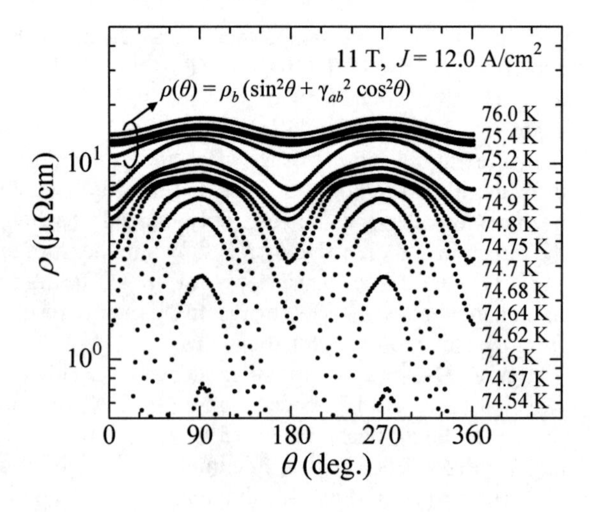

FIGURE 2. Angular dependence of the resistivity ρ(θ).

With decreasing T toward the vortex lattice melting line $T_m(H)$, however, $\rho(\theta)$ deviates from the FF model and the anisotropy of $\rho(\theta)$ is enhanced. Below 74.7 K, $\rho(\theta)$ shows sharp dip structure around $J//b$. The dip grows in low temperatures and finally the resistivity disappears. The strong anisotropy in the Bragg glass phase indicate the existence of the anisotropic pinning potential. The resistivity $\rho(\theta)$ with partly zero value below 74.64 K is due to that the constant driving force $F = JB$ is larger than the pinning force F_p in b-direction, but smaller than that in a-direction. Therefore, the vortices move only along the b-axis in spite of the rotation of current direction under constant-J near the solidification temperature to the Bragg glass phase.

In the Bragg glass phase, the additional anisotropy has a two-fold symmetry with the same phase as compared with $\rho(\theta)$ in the vortex liquid phase. In $YBa_2Cu_3O_y$, the CuO chain is a main origin of the in-plane anisotropy, so the vortex motion is affected by the one dimensionality of the CuO chain and the anisotropy of critical current density J_c [6]. Our findings indicate that the CuO chain shows an anisotropic superconductivity with the modulated order parameter and works as a washboard potential for the vortex motion in the ab-plane, similar to the case of the intrinsic pinning [7] for $H//ab$ which originates from the layered structure of the CuO_2 plane.

SUMMARY

We have measured the in-plane anisotropy of $\rho(\theta)$ in $YBa_2Cu_3O_{6.99}$ single crystals using "the vector current method" in $H//c$. We have found that the anisotropy of the vortex motion increases with decreasing temperature. The results indicate that the Bragg glass phase has the anisotropic pinning potential due to the modulation of the superconducting order parameter by one dimensionality of the CuO chain.

ACKNOWLEDGMENTS

This work was partly supported by a Grant-in-Aid for Scientific Research from the Ministry of Education, Science, Sports, and Culture of Japan.

REFERENCES

1. H. Pastoriza et al., *Phys. Rev. Lett.* **83**, 1026-1029 (1999).
2. G. D'Anna et al., *Phys. Rev.* B **61**, 4215-4221 (2000).
3. T. Naito et al., *Advances in Superconductivity IX*, Tokyo: Springer-Verlag, 1997, pp. 601-604.
4. T. Nishizaki and N. Kobayashi, *Supercond. Sci. Technol.* **13**, 1-11 (2000), and references there in.
5. T. Nishizaki et al., *J. Low Temp. Phys.* **117**, 1375-1379 (1999).
6. T. Tamegai et al., *Critical Currents in Superconductors*, Singapore: World Scientific, 1996, pp. 125-128.
7. M. Tachiki and S. Takahashi, *Solid State Commun.* **70**, 291-294 (1989).

Scaling of Conductivity through the Critical Temperature in $Y_{0.54}Pr_{0.46}Ba_2Cu_3O_7$

V. Sandu,[1,2] T. Katuwal,[1] B. J. Taylor,[3] M. B. Maple,[3] and C. C. Almasan[1]

[1]*Department of Physics, Kent State University, Kent, OH 44242, USA*
[2]*National Institute of Materials Physics, 077125 Bucharest-Magurele, Romania*
[3]*Department of Physics, University of California at San Diego, La Jolla, CA, USA*

Abstract. In-plane conductivity σ_{ab} curves of $Y_{0.54}Pr_{0.46}Ba_2Cu_3O_7$ single crystals, measured at constant temperature T in magnetic fields H tilted with an angle θ relative to the c-axis, were found to map onto a single curve in a σ_{ab} vs $H\cos\theta$ plot. This scaling occurs both below and above the critical temperature T_c, maintaining the same convexity, and is consistent with the dissipation caused by the motion of two-dimensional (2D) vortices. This indicates the presence of 2D vortices above T_c. A second scaling $\sigma_{ab}(H\cos\theta)$ takes place at higher T, with an opposite convexity. This scaling is consistent with the dissipation of quasiparticles. There is a T range between these two regimes where no scaling is found. We attribute the absence of scaling in this intermediate regime to a 2D to 3D crossover in the vortex matter.

Keywords: angular magnetoresistivity, $Y_{1-x}Pr_xBa_2Cu_3O_7$, phase fluctuations, anisotropy, scaling.
PACS: 74.25.Fy, 74.40.+k, 74.72.Bk, 72.15.Gd

Underdoped cuprate superconductors display an extended regime of fluctuations of the order parameter $\Delta=|\Delta|\exp(i\varphi)$ due to the reduced dimensionality and charge carrier density. Experimental findings [1-3] and simulations using the XY-model [4] show that phase φ fluctuations prevail over amplitude $|\Delta|$ fluctuations above the superconducting transition temperature T_c, over an extended temperature range up to $T_\varphi = \zeta T_c$, with $\zeta = 1.5$ for 3D and $\zeta = 2$ for 2D systems. As a result, up to T_φ, the correlator of the order parameter mirrors only the decay of phase correlations. As long as the phase correlation length does not vanish, i.e., up to T_φ, vortices can exist even above T_c. If so, their motion under a driven current produces an additional dissipation, besides that of quasiparticles.

In angle dependent charge transport experiments, flux-flow and quasiparticle resistivities follow different scaling laws as a function of magnetic field and angle. Therefore, it is expected that each one imposes its scaling in the temperature range where it dominates. Here, we present the evolution of the scaling function of the conductivity with increasing T in an attempt to find the limits of the regime of phase fluctuations of the superconducting order parameter.

Resistivity measurements in magnetic field were performed on $Y_{0.54}Pr_{0.47}Ba_2Cu_3O_{7-\delta}$ (T_c = 38 K) single crystals. The experimental procedure is presented elsewhere [5].

Figures 1(a) and 1(b) are plots of the conductivity as a function of $x = H|\cos\theta|$. All data points fall, at constant T, on the same curve with positive convexity, $d^2\sigma/dx^2 > 0$ for T ≤ 44 K. For 45 < T < 60 K, the conductivity still shows a positive convexity over this large temperature range, but the above scaling fails [(see data of Fig. 1(b) at 45 K]. The failure of the scaling most likely arises from the crossover to the 3D vortex regime as the c-axis coherence length ξ_c exceeds the interlayer spacing. It is noteworthy that this effect occurs at a higher temperature than expected from the Ginzburg-Landau theory, in which ξ_c diverges at the critical temperature. At even higher temperatures, the conductivity scales with $H^2\cos^2\theta$ (see 60 and 70 K data of Fig. 2).

To understand the behavior of conductivity over the above three regimes, we have to include in dissipation both quasiparticle and flux flow dissipation channels; namely, the conductivity is given by [5]:

$$\sigma = \sigma_0 - ax^2 + \frac{b}{x}, \qquad (1)$$

with $x = H|\cos\theta|$, and a ($ax^2 << \sigma_0$) and b temperature dependent coefficients. At low temperatures, b, which is of the order of H_{c2}, is much larger than H and the

CP850, *Low Temperature Physics: 24th International Conference on Low Temperature Physics;*
edited by Y. Takano, S. P. Hershfield, S. O. Hill, P. J. Hirschfeld, and A. M. Goldman

last contribution in Eq. (1) is dominant, hence $\sigma \approx \sigma_{FF} = b/x$. Figure 1(a) shows the fits of the data with a $1/x$ function for 20 and 30 K (solid lines). This functional form of conductivity is reasonable. In the intermediate temperature range, the conductivity is given by the whole expression of Eq. (1). For $T \geq 60$ K, the conductivity behaves like $\sigma = \sigma_0 - ax^2$. Figure 2 shows the plot of conductivity data as a function of x^2 measured at 60 and 70 K in a magnetic field of 12 T.

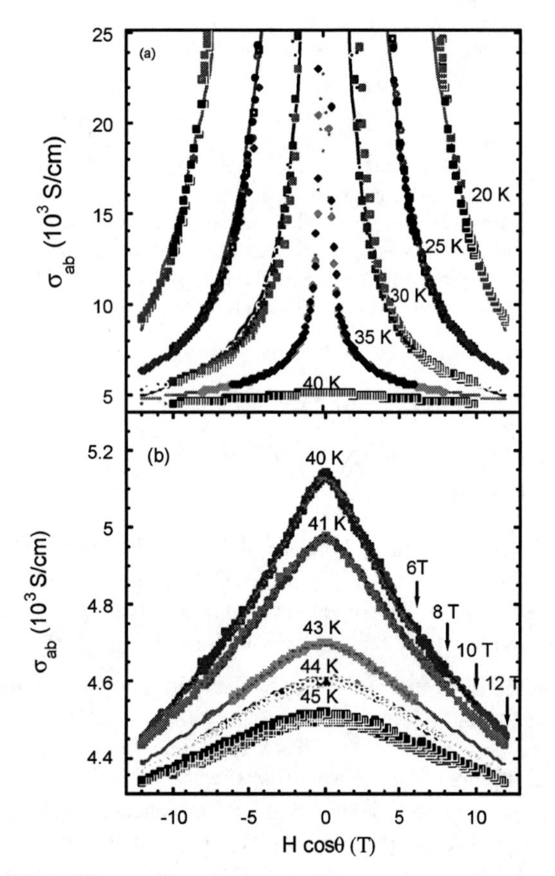

FIGURE 1. In-plane conductivity σ_{ab} as a function of $H\cos\theta$ for applied magnetic fields of 6, 8, 10 and 12 T and a temperature range (a) $20 \leq T \leq 40$ K and (b) $40 \leq T \leq 45$ K. The solid lines in panel (a) are fits of the data with the function $F(x) = const + b/x$, with $x = H\cos\theta$.

The main result of these experiments is that phase fluctuations survive up to a temperature of $1.15 \times T_c = 44$ K as $2D$ vortex like excitations and, further as 3D vortex lines up to $1.5 \times T_c = 57$ K. This is in agreement with the assumption that, in underdoped superconducting cuprates, there is an extended regime above T_c in which the phase fluctuation of the superconducting order parameter is an important ingredient in the dissipation process. It is noteworthy that some experimental data [6] suggest that the amplitude of the order parameter keeps fluctuating up to a much higher temperature, identified with the pseudogap temperature.

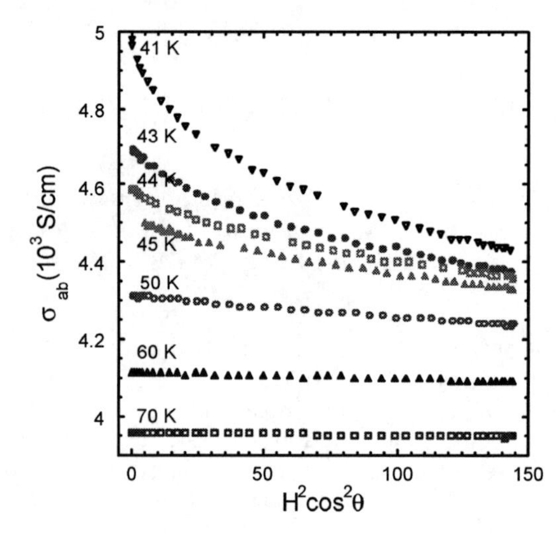

FIGURE 2. In-plane conductivity σ_{ab} measured in a field of 12 T as a function of $H^2 \cos^2\theta$.

In summary, we have found that vorticity is manifested as a continuous process through the critical temperature up to temperatures as high as $T_\varphi = 57$ K which we attribute to the vanishing of the phase correlation length. A two dimensional to three dimensional crossover in the vortex matter was found in the angular dependence of the conductivity at a temperature much higher than the one predicted by the Ginzburg -Landau theory.

ACKNOWLEDGMENTS

This research was supported by the National Science Foundation under Grant No. DMR-0406471 at KSU, the US Department of Energy under Grant No. DE-FG02-04ER-46105 at UCSD, and by the Romanian ME under Grant MATNANTECH No. 260/2004.

REFERENCES

1. J. Corson, *et al.*. *Nature* **398**, 221-223, (1998).
2. Z. A. Xu, *et al.*, *Nature* **406**, 486-488, (1998).
3. Y. Wang, *et al.*, *Phys. Rev. Lett.* **88**, 257003 (2002).
4. C. Meingast, *et al.*, *Phys. Rev. Lett.* **89**, 229704 (2002); R. Markiewicz, *Phys. Rev. Lett.* **89**, 229703 (2002).
5. V. Sandu, *et al.*, *Phys. Rev. Lett.* **93**, 177005 (2004).
6. T. Shibauchi, *et al.*, *Phys. Rev. Lett.* **86**, 5763-5766 (2001).

Critical State Simulation of the Superconducting Layered Structures Based on Numerical Solution of the Ginzburg-Landau Equations.

A.N.Lykov and A.Yu.Tsvetkov

Lebedev Physical Institute of RAS, Leninsky prosp., 53, 119991 Moscow, Russia.

Abstract. The critical state of the superconducting layered structures consisting of isolated superconducting plates is investigated. The critical current density, pinning force, the distribution of the local magnetic field, and the current in the structures are found via numerical solution of the Ginzburg-Landau equations. The external magnetic field and the transport current are directed along plates perpendicularly to each other. The vortices are believed to be absent in the plates. A method for analyzing the critical state of superconducting layered structures in a parallel magnetic field based on simple transformation of the current distribution in a zero external magnetic field is found.

Keywords: superconducting layered structure, critical state, numerical solution, Ginzburg-Landau equations.
PACS: 74.25.Sv, 74.78.-w, 74.78.Fk

The usual operational approach to analyze the critical state in superconductors is based on the study of vortex interaction with structural defects. A variety of properties of a vortex system, which is a quantum elastic medium with nonlinear electrodynamics, is responsible for the complexity of the problem [1]. A new approach to analyze the critical state in layered superconductors, which is based on a numerical solution of the Ginzburg-Landau equations, is offered in this study.

Consider a set of long, wide superconducting plates with a thickness D in a magnetic field H. The field is assumed to be parallel to the plate surfaces. A transport current I flowing through the plates is also parallel to the plate surfaces and perpendicular to the external field. As the transport current, we used the product of the current density and the plate thickness, i.e., current per unit of plate wideness. The problem of finding the critical current of the structure was divided into two parts. First, using a self-consistent solution to the Ginzburg-Landau equations, the dependence of the critical current of a superconductiong plate as a function of the magnetic field was calculated. The superconducting plates are assumed to be in the vortex-free state [2].

Consider a Cartesian co-ordinate system (x, y, z), and let y- and z-axes lie in the plane of the plate surface (z is directed along the external magnetic field H). Then, the vector potential of the magnetic field has only a y component, $\mathbf{A} = \mathbf{e}_y A(x)$. A transport current flowing through the plate is directed along the y-axis. In this case, the system of Ginzburg-Landau equations can be written in the dimensionless form

$$\frac{d^2U}{dx_\lambda^2} - \psi^2 U = 0, \qquad (1)$$

$$\frac{d^2\psi}{dx_\lambda^2} + \kappa^2(\psi - \psi^3) - U^2\psi = 0, \qquad (2)$$

where ψ is the module of the order parameter and κ is the Ginzburg-Landau parameter. Here, we introduce dimensionless parameters $U(x_\lambda)$, $b(x_\lambda)$, and $j(x_\lambda)$:

$$A = \frac{\phi_0}{2\pi\lambda}U,\quad B = \frac{\phi_0}{2\pi\lambda^2}b,\quad x_\lambda = \frac{x}{\lambda}, \text{ and}$$

$$j(x_\lambda) = j_s\left(\frac{c\phi_0}{8\pi^2\lambda^3}\right)^{-1} = -\psi^2 U, \qquad (3)$$

where c is the speed of light in a vacuum, ϕ_0 is the flux quantum, λ is London penetration depth, and B is magnetic induction (**B**=rot**A**). On the plate surfaces we take the conventional boundary conditions. The iteration procedure was used for the numeric solution of the problem.

CP850, *Low Temperature Physics: 24th International Conference on Low Temperature Physics;* edited by Y. Takano, S. P. Hershfield, S. O. Hill, P. J. Hirschfeld, and A. M. Goldman

At the second stage, the critical current of the layered structure was found. We believe that the length and width of the plates are significantly greater than the total thickness of the structure. In this case we can consider every plates are infinite to find the magnetic field induced by the transport current I_t in the plate. Thus, the current in a plate induces the magnetic field: $H_I = \frac{2\pi}{c} I_t$. At the same time, we sought such a distribution of the transport current throughout the layers when all the layers turn into the normal state simultaneously. If h_i is the magnetic field, in which the i-th layer is located, then, in the critical state, this layer passes a current per unit of plate wideness equal to the critical current of the plate $I_c(h_i)$, which was found using the self-consistent solution to the Ginzburg-Landau equations in the first stage. A current flowing through the i-th plate produces a magnetic field. In accordance with the superposition principle, it is necessary to sum up the contributions from each layer to find the magnetic field, in which the i-th plate is situated. The expected distribution of the magnetic field was also calculated by using an iteration method [2].

Figure 1 shows an example of the average critical current dependences as functions of the magnetic field, which were obtained by the method. The magnetic field induced by the plates increases with increasing of their amount. Thus, their influence also increases. This is revealed in decreasing critical current density in the range of small magnetic fields, which corresponds to the Meissner behavior for a layer; on the contrary, for large magnetic fields, the transport properties of the layers are caused by surface superconductivuty.

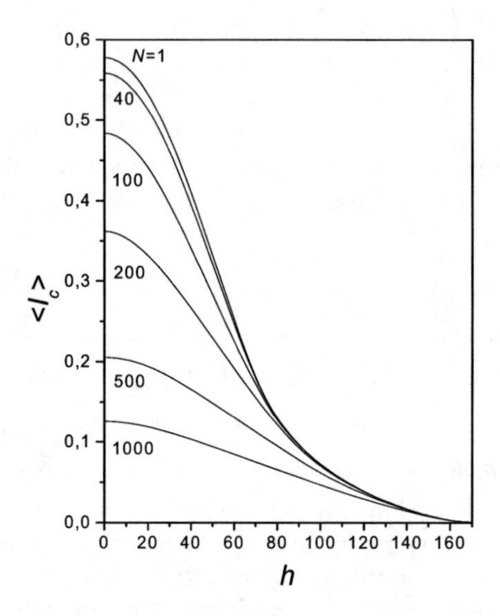

FIGURE 1. Average critical current versus magnetic field for layered structures with different numbers of layers (their number (N) is indicated in the vicinity of the corresponding curve). In this case, the layers are assumed to be identical and their thicknesses are 0.3λ and κ=10.

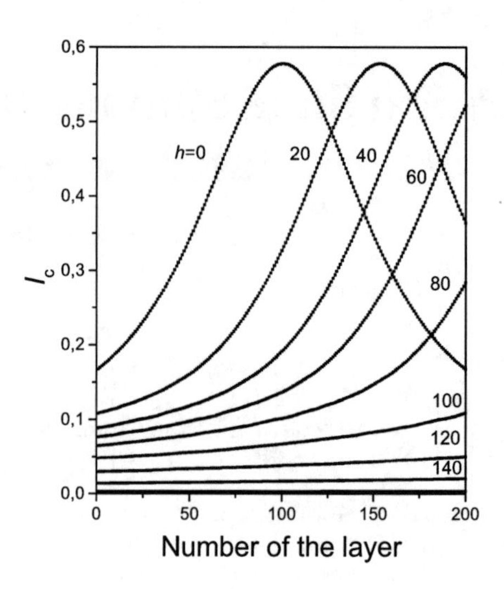

FIGURE 2. Critical current density versus the number of the layer for the layered structure composed of 200 layers in different magnetic fields (h), we believe $D = 0.3\lambda$ and κ=10.

This method of studying the critical state in layered superconductors gives us the possibility of finding the transport current and magnetic field distributions throughout the layers (Fig. 2). It should be noted that in the magnetic field, the transport current distribution in the critical state can be obtained by shifting this distribution in a zero magnetic field to a certain value. Thus, we can easily obtain the average critical current dependence as a function of the magnetic field and the transport current distribution in the layered superconductor in the critical state if critical current dependence as a function of the magnetic field $I_c(h)$ for a layer was found earlier. The dependence $\langle I_c \rangle(h)$ obtained in the work can be considered the upper limit of the critical current of the layered superconductors in the parallel field.

REFERENCES

1. G. Blatter, M.V. Feigel'man, V.B. Geshkenbein, A.I. Larkin, and V.M. Vinokur. *Rev. Mod. Phys.*, **66**, 1125-1388 (1994).
2. A.N. Lykov, A.Yu. Tsvetkov, and G.F. Zharkov. *JETP* to be published (2005).

Angular Effects of the Critical Current in Nb/Pd Multilayer Structures.

S.Yu.Gavrilkin[a], A.N.Lykov[a], A.Yu.Tsvetkov[a], Yu.V.Vishniakov[a], C.Attanasio[b], C.Cirillo[b], and S.L.Prischepa[c]

[a] *Lebedev Physical Institute of RAS, Leninsky prosp., 53, 119991 Moscow, Russia.*
[b] *Departimento di Fisica "E.R. Caianiello" and Laboratorio Regionale SuperMat, INFM Salerno, Università degli Studi di Salerno, Baronissi (Sa) I-84081, Italy.*
[c] *Belarus State University of Information and Radioelectronics, P.Brovka Street 6, Minsk 220013, Belarus.*

Abstract. The critical current density dependencies on the angular orientation of the applied magnetic field ($J_c(\theta)$) in superconducting multilayers based on Nb/Pd were investigated in this work. We have demonstrated that the procedure of measurements is an important factor for the $J_c(\theta)$. In rotating of the sample at a fixed magnetic field, it was found that the direction of the external field at which the maximum of $J_c(\theta)$ was obtained formed an angle with the planar direction. The results can be explained in terms of magnetic field suppression of percolation currents in a weak link network. Flux trapped within grains by flux pinning processes is relatively stable at low fields and its contribution to transport current suppression is added vectorially to the applied field.

Keywords: superconducting layered structure, critical current, magnetic field, angular measurements
PACS: 74.25.Sv, 74.78.-w, 74.78.Fk

INTRODUCTION

The recent developments of Nb-based layered structures, where the role of effective pinning centres was played by insulating NbO layers, demonstrated anomalous angular dependences of critical current [1]. Later, the similar effect was observed on niobium films [2] and high-temperature superconducting tapes [3]. This phenomenon manifests itself in the fact that, under small intensities of the external magnetic field, the critical current attains its maximum in a slightly tilted magnetic field. This results from the difficulties of the transition of the flux vortex lattice into the ground state with minimum energy. Bringing of the lattice into the ground state before every measurement at a new angle results in a symmetrical angular dependence of the critical current To understand the nature of the phenomena observed in [1-3] in greater detail, we carried out precision angular measurements of the critical current in Nb/Pd multilayers. The choice of Pd is related to its large spin susceptibility.

METHODS OF MEASUREMENTS

The Nb/Pd with 10 number of bilayers were grown on Si(100) substrates at room temperature by using a dual source magnetically enhanced *dc* triode sputtering system as described earlier [4]. The samples had equal Nb thickness (d_{Nb} = 20 nm), while Pd thickness was in the range from 1.7 nm to 20 nm. Using the photolithography technique, we fabricated narrow strips with bonding pads from these films. The width of the strips (w) was 10 μm. The critical current I_c is determined from the current–voltage characteristic as the current at which the voltage drop on a sample reaches 1μV. The $I_c(H)$ dependencies were measured at different temperatures. The temperature stabilization was better than 10^{-3} K. For angular measurements the sample was placed into the centre of the solenoid onto the gear-wheel, and could be rotated in such a way that H could make various angles with the film plane, with the transport current being always perpendicular to H. The angular resolution for the sample orientation is 0.04^0. The whole equipment is placed in a helium bath at temperature of 4.2 K.

CP850, *Low Temperature Physics: 24th International Conference on Low Temperature Physics;*
edited by Y. Takano, S. P. Hershfield, S. O. Hill, P. J. Hirschfeld, and A. M. Goldman

RESULTS

Figure 1 shows an example of the angular dependencies of the critical current (I_c) for our multilayers in constant magnetic fields (sample with d_{Pd}=18nm). Here θ is the angle between the direction of the external magnetic field and the surface of the structure. There are several noteworthy features in these dependencies. First, the peak in $I_c(\theta)$ is observed in the external magnetic field tilted to the layers. In The curves marked by squares correspond to the clockwise rotation, while the curves marked by crosses correspond to the rotation in the counter clockwise. Secondly, the positions of the clockwise and counter clockwise maxima do not coincide; moreover, the positions are different for different values of the external magnetic field. The angular shift of the peak position from θ=0^0 increases with decreasing external magnetic field and with increasing the thickness of the Pd layers. Figure 2 shows the value of the shift as a function of H^1 for the three samples. As it can be seen from the figure, the value of the shift is proportional to H^1. Thirdly, the positions of the maxima depend on the initial conditions (memory effect) of a sample, as well as on the way in which it was brought to the conditions. These results are qualitatively analogous to the results obtained on Nb/NbO multilayer structures and high-temperature superconducting tapes [1-3]. The results show that the conventional method for determining a parallel orientation of a sample in a magnetic field, which is based on finding the maximum of the angular dependence of critical current, may give erroneous results.

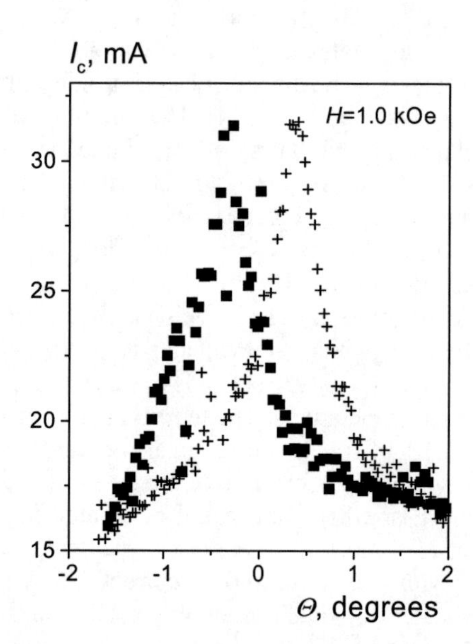

FIGURE 1. Critical current through a sample as a function of the angle between the surface of the structure and the direction of the external magnetic. The squares correspond to the angular dependence recorded when the sample is rotated in the clockwise direction. The crosses represent the experimental results recorded when the sample is rotated in the counter clockwise direction.

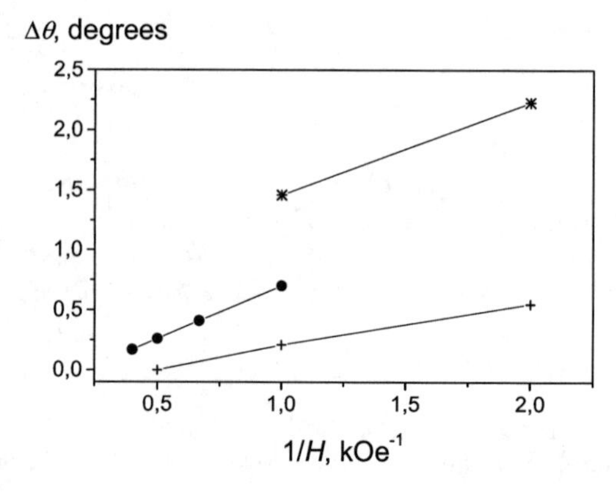

FIGURE 2. Angular difference between the clockwise and counter clockwise peaks as a function of the inverse of the magnetic field intensity for samples with *d* Pd layers 20nm (•), 18 nm (*), and 1.7 nm (+).

The $I_c(H)$ curves is found to present a clear peak [4]. This is so-called peak effect. It is interesting to point out that the peak increases with increasing the Pd thickness. In this work a correlation was found: the more the Pd thickness, the more peak effect on the $I_c(H)$ and the more the angular shift in these structures. The peak effect on the $I_c(H)$ and the angular shift in these structures are explained in terms of magnetic field suppression of percolation currents in a weak link network controlled by trapped flux which is retained by strong pinning regions [5]. We believe that Nb layers is highly granular, and the normal or insulating tunnel barriers are formed by grain boundaries. In this case the critical current is defined by a system of Josephson junctions which are formed by superconducting Nb granules. Co-existence of the superconductivity and magnetism reveals itself in enhancement of the observed effects.

REFERENCES

1. A.N. Lykov and Yu.V. Vishnyakov, *Europhys. Lett.* **36**, 625 (1996).
2. Yu.V. Vishnyakov, A.N. Lykov, and A.Yu.Tsvetkov, *Sov. Phys. JETP* **98**, 1404 (2004).
3. F. Warmont, H. Jones, Supercond. *Sci. Technol.* **14**, 145 (2001).
4. C. Coccorese *et al., Phys.Rev.* B, **57,** 7922 (1998).
5. J.E. Evetts and B.A. Glowacki, *Cryogenics* **28**, 641 (1998).

Vortex Dynamics and Superconducting Fluctuation Effects on Magneto - Oscillations in Extremely Type-II Layered Superconductors

Tsofar Maniv* and Vladimir Zhuravlev*

*Department of Chemistry, Technion, Haifa 32000, Israel

Abstract. We study magneto-oscillations in the vortex state of an extremely type-II quasi 2D superconductor. It is argued that under the driving force generated by the sweeping magnetic field the superconductor tends to form 1D highly correlated channels of moving vortices, which reflect an underlying structure of easily sliding chains along the main principal axis in the Abrikosov vortex lattice. We present a model for calculating the friction coefficient of moving vortices in the presence of a dilute ensemble of strong pinning centers by exploiting the condition of minimal power dissipation. The model seems to account for the remarkable phenomenon of weak hysteresis loops, observed recently in magneto-oscillations measurements far above the major irreversibility field of a nearly 2D organic superconductor.

Keywords: layered superconductors, magnetic oscillations, vortex dynamics
PACS: 74.25.Qt, 74.25.Bt, 74.40.+k

It has been shown recently [1] that the vortex lattice in an extremely type-II 2D superconductor at high magnetic field is strongly anisotropic with respect to shear deformation. This feature characterizes also a typical atomic lattice. However, while the energy required for distorting an atomic lattice on a macroscopic scale is always of the order of (or larger than) the crystal dissociation energy, irrespective of the selected direction, the shear distortion energy along the major principal axis in a vortex lattice is only about 2% of the lattice 'dissociation' energy. This peculiar feature is associated with a fundamental property of the vortex-vortex interaction at small distances, namely its relatively small repulsive core with respect to the effective core of a typical interatomic potential [3] (see Fig. 1). One therefore expects high fluidity of a vortex crystal along certain directions under the driving force generated by the sweeping external magnetic field or in the presence of thermal excitations. It has been therefore argued [2] that just above the melting point the corresponding vortex matter is not an isotropic liquid but a smectic-like phase containing crystallites of easily sliding (but otherwise strongly correlated) vortex chains, which can be stabilized by small number of pinning centers.

Experimental support for this peculiar picture seems to be found in the irreversible magnetization oscillations observed very recently in the quasi 2D organic superconductor β''-$(ET)_2SF_5CH_2CF_2SO_3$ [4]. The unusual feature of the observed hysteresis loops is associated with their appearance deep in the vortex-liquid phase, where one usually expects unrestricted motion of flux lines through the entire superconducting (SC) sample, due to the establishment of thermodynamic equilibrium. Furthermore, a simple analysis of the observed jump, $\Delta M_{\uparrow\downarrow}$, of the measured magnetization at the points where the field sweep is reversed, shows that the work (per unit volume) done by the driving Lorentz force is $W_L \approx \left(\frac{3}{2}\right) B\Delta M_{\uparrow\downarrow}$, which in units of the maximal SC condensation-energy density, $E_{cond} = \frac{H_{c2}^2}{16\pi\kappa^2}$, at $B \approx H_{c2}$, equal to $\widetilde{\varepsilon}_L \equiv W_L/E_{cond} \approx 24\kappa^2\Delta M_{\uparrow\downarrow}/H_{c2}$. Using the experimental value of the magnetization jump at $B = 2$ T, that is $\Delta M_{\uparrow\downarrow} \approx 4\times10^{-8}$ T, it is found that for $\kappa \approx 46$ [5] the resulting value $\widetilde{\varepsilon}_L \approx 4\times10^{-3}$ is of the same order of magnitude as the characteristic energy scale of sliding chains along the main principal axis of the vortex lattice in a 2D superconductor.

In the light of these findings the following model seems plausible: Imagine that under the influence of an external slowly sweeping magnetic field the vortex matter responds, in a way that minimizes its power dissipation, by forming a series of long, parallel vortex crystallites, with common principal lattice vector aligned along the external driving force direction, so that easily sliding vortex chains are injected into the SC region (see Fig. 1). We assume that pinning centers exist only at the crystallites boundaries, where the pinning force is much larger than the driving force. Focusing on a single crystallite all the vortex chains within its interior move under the action of the driving force, subject only to the intrinsic vortex chain-chain interactions, and the boundary conditions imposed by the strong pinning, fixing chains to the laboratory frame. In this model the frictional force, which opposes the action of the driving force, arises solely from the vortex-vortex interactions activated in shear distortions along the principal chain axis, as suggested by the

CP850, *Low Temperature Physics: 24th International Conference on Low Temperature Physics;*
edited by Y. Takano, S. P. Hershfield, S. O. Hill, P. J. Hirschfeld, and A. M. Goldman

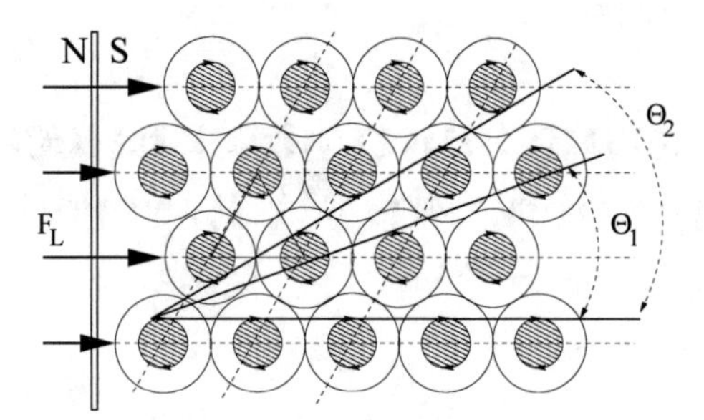

FIGURE 1. Schematic picture illustrating the extreme fluidity of vortices along the major principal axis $\Theta = 0$ in a vortex crystal under a driving Lorentz force, as arising from the relatively small effective size of the vortex cores (filled circles), and the spatial uniformity imposed on the quantized magnetic flux by the external magnetic field. For comparison atomic cores (empty circles) in a closely packed equivalent atomic lattice are also shown. Note that the small vortex core leads to a relatively large vortex fluidity also along the minor principal axis, $\Theta = \Theta_1$, in contrast to all other crystallographic directions such as $\Theta = \Theta_2$

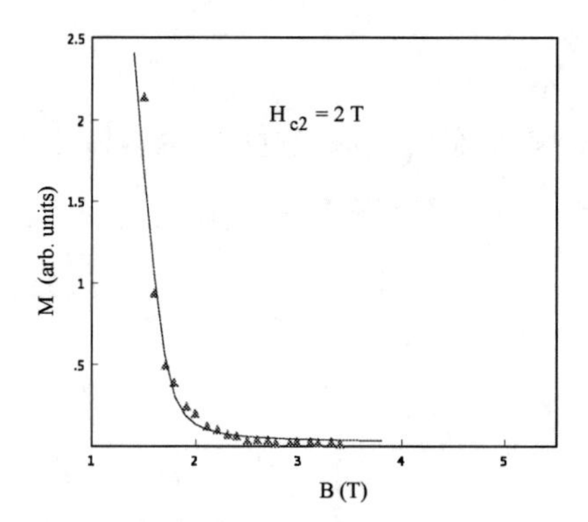

FIGURE 2. Field dependence of the experimental magnetization jumps (triangles) and the theoretically calculated curve (solid line).

above analysis.

Now consider a critical state situation when the macroscopic driving Lorentz force, F_L, supplies to a vortex chain just the minimal kinetic energy $m_v v_{\phi 0}^2/2$ required to overcome the energy barrier for the motion along the main principal axis (Fig. 1). Here m_v is the dynamic mass per vortex, and $v_{\phi 0}$ its initial velocity at the bottom of a potential well. The energy barrier is of the order of the phase-dependent minimal coupling, $\varepsilon_{sl} \approx 4\lambda^2 \varepsilon_0$, which is of the order of the characteristic vortex-lattice melting energy, with ε_0 being the SC condensation energy per vortex and $\lambda \approx 0.066$ [2].

This immediately implies that the work, $\tilde{\varepsilon}_L \equiv W_L/E_{cond}$, done by the driving Lorentz force (normalized with respect to the SC condensation energy) is equal to $\tilde{\varepsilon}_{sl} = \varepsilon_{sl}/\varepsilon_0 \approx 4\lambda^2 \sim 10^{-2}$, as indeed found from the comparison with the experiment mentioned above. It also enables one to derive a simple analytic expression for the friction coefficient, η, of a moving vortex chain (per unit vortex). Suppose that the entire energy supplied to the vortex chain in a jump over a single barrier is dissipated through a frictional force with the other chains so that the energy dissipated in a single jump is $\varepsilon_{dis} \sim \eta v_{\phi 0} \Delta x$, where $\Delta x \sim a_H = \sqrt{c\hbar/eH}$ is the distance between neighboring vortex sites along the chain. Using the condition $\varepsilon_{dis} = \varepsilon_{sl}$, we thus find that $\eta \approx \sqrt{m_v \varepsilon_{sl}}/a_H$, which, up to a numerical factor of the order unity, is identical to the result obtained from the Caldeira-Leggett theory [3].

The magnetic-field dependence of the jump $\Delta M_{\uparrow\downarrow}$ well above the major irreversibility field H_{irr} [6] can also be accounted for in the framework of the proposed model. Since the generators of the flux-density gradient in this model, i.e. the pinning forces and the vortex-vortex interaction, all originate in the existence of local SC order at the vortex position, it should be proportional to the mean-square SC order parameter, $\langle |\Delta(B)|^2 \rangle$, and so the magnetization jump can be written as $\Delta M_{\uparrow\downarrow} \propto \frac{1}{B} \langle |\Delta(B)|^2 \rangle$. The latter quantity involves the effect of SC fluctuations in a broad region above the melting point, which can be calculated quite accurately within the discrete chain model. The accuracy has been tested recently versus the optimized perturbation theory, developed in Ref.[7], yielding good quantitative agreement in both the low and high temperature regions. Employing the material parameters, $T_c = 4.4$ K, $\frac{m_c}{m_e} = 2.0$, $E_F/k_B = 133$ K, $T = 40$ mK, where m_e, m_c are the free electron and cyclotron mass respectively, and treating H_{c2} as adjustable parameters, the best fit is obtained for $H_{c2} = 2$ T (Fig. 2). The resulting curve reflects the crossover between the vortex state below H_{c2}, which is well described by mean-field theory, and the normal state far above H_{c2}, with the long tail of the field-dependent magnetization hysteresis corresponding to the enhanced influence of 2D SC fluctuations.

ACKNOWLEDGMENTS

This research was supported by the fund from the promotion of research at the Technion.

REFERENCES

1. T. Maniv, V. Zhuravlev, I. D. Vagner, and P. Wyder, *Rev. Mod. Phys.* **73**, 867 (2001).
2. V. Zhuravlev and T. Maniv, *Phys. Rev. B* **60** 4277 (1999) ,*ibid.* **66**, 014529 (2002).
3. V. Zhuravlev and T. Maniv, unpublished.
4. J. Wosnitza *et al*, *Phys. Rev. B* **67**, 060504(R) (2003).
5. S. Wanka *et al*, *Phys. Rev. B* **57**, 3084 (1998).
6. T. Maniv *et al*, *J. Phys.: Condens. Matter* **16**, L429 (2004).
7. D. Li and B. Rosenstein, *Phys. Rev. Lett.* **86**, 3618 (2001).

Electronic Polarizability of Superconductors and Inertial Mass of a Moving Vortex

Armen A. Kteyan and Robert A. Vardanyan

Solid State Division, Institute of Radiophysics and Electronics, Ashtarak-2, 378410, Armenia

Abstract. The problem of a vortex electromagnetic mass in a superconductor is considered accounting for the self-interaction effect conditioned by the coupling of the moving vortex to the excited fluctuations of the superfluid density. In the framework of the phenomenological model used, the self-interaction is defined as an interaction of the singular phase with the induced polarization of the charged superfluid. The obtained polaron-type mass exceeds the earlier obtained electromagnetic mass in view of the large value of the light speed relation to the Fermi velocity c/v_F, and can dominate over the vortex core mass.

Keywords: Superuid, vortex, compressibility, dielectric function
PACS: 74.20.De, 74.25.Qt

Inertial mass of a vortex in a superfluid characterizes the increase of the internal energy of the superfluid when the vortex moves with a velocity **v**, owing to origination of the kinetic energy $F_{kin} = Mv^2/2$. We discuss here the electromagnetic mass of a vortex in superconductors (first studied by Suhl [1]), which is conditioned by the polarization of the superfluid around the vortex core, and take account of the vortex self-interaction, representing its interaction with dynamical polarization of the background. The problem is discussed at absolute zero temperature, when the concentration of normal electrons outside the vortex core vanish.

A vortex in a superfluid is a topological object that can be described by the order parameter phase χ, which is identified with the azimuthal angle $\theta = arctg(y/x)$. When the vortex moves uniformly with a small velocity **v**, the phase is determined as $\chi(\boldsymbol{r}, t) = \theta(\boldsymbol{r} - \mathbf{v}t)$. This phase space and time derivatives $\nabla\chi = \hat{e}_\varphi/r$ and $\dot{\chi} = -\mathbf{v}\nabla\chi$ are the sources of the fields and currents around the moving vortex.

In the region far from the core, where the order parameter modulus is constant, the vortex energy is described by the non-homogeneous part of the energy of the superconductor. The energy functional used in the papers [1,2] can be transformed to the form

$$F - F_0 = \int d\boldsymbol{r}\left(-\frac{1}{2c}\boldsymbol{j}^t\boldsymbol{Q} + \frac{\boldsymbol{B}^2}{8\pi}\right) + \left(-\frac{1}{2}\rho\Phi + \frac{\boldsymbol{E}^2}{8\pi}\right). \quad (1)$$

Here F_0 is the energy of the homogeneous superconductor; the integration goes over the two-dimensional radius-vector $\boldsymbol{r}$. The gauge-invariant potentials $\boldsymbol{Q} = \boldsymbol{A}^{ind} - (\phi_0/2\pi)\nabla\chi$ and $\Phi = \varphi^{ind} + \mathbf{v}\boldsymbol{A}^{ind}/c + (\phi_0/2\pi c)\dot{\chi}$ (here $\phi_0 = hc/2e$ is the flux quantum) determine the transversal current and the charge density

$$\boldsymbol{j}^t = -\frac{c}{4\pi}\lambda^{-2}\boldsymbol{Q}, \quad \rho = -\frac{c^2\lambda^{-2}}{4\pi s^2}\Phi, \quad (2)$$

where λ is the London penetration depth and s is the characteristic velocity (of order of Fermi velocity). The magnetic and electric fields are determined by the induced potentials: $\boldsymbol{B} = \nabla\times\boldsymbol{A}^{ind}$ and $\boldsymbol{E} = -\partial_t\boldsymbol{A}^{ind}/c - \nabla\varphi^{ind}$.

The total electric field around the moving vortex represents the sum of gradient of the induced scalar potential and of the electric field generated by the moving magnetic field $\boldsymbol{B}$:

$$\boldsymbol{E}^{total} = -\frac{1}{c}\frac{\partial\boldsymbol{A}^{ind}}{\partial t} - \nabla\left(\varphi^{ind} + \frac{1}{c}\mathbf{v}\boldsymbol{A}^{ind}\right). \quad (3)$$

Here the second term on the right-hand side is the longitudinal electric field which arises due to the transformation of the scalar potential in the laboratory frame $\varphi = \varphi^{ind} + \mathbf{v}\boldsymbol{A}^{ind}/c$.

The first term in the energy functional (1) is the static vortex energy. The second term describes the

CP850, *Low Temperature Physics: 24th International Conference on Low Temperature Physics;*
edited by Y. Takano, S. P. Hershfield, S. O. Hill, P. J. Hirschfeld, and A. M. Goldman

kinetic energy due to the superfluid polarization. The kinetic energy must be completed with account of the energy of the longitudinal field $[\mathbf{v} \times \boldsymbol{B}]^l/c = \nabla(\mathbf{v}\boldsymbol{A}^{ind})/c$. Besides, our aim is to include the effect of the vortex coupling to the superfluid polarization. Recalling that the total energy of a medium influenced by external electric field contains the interaction energy of this field with the polarization $\boldsymbol{P}$ of the medium, we introduce now the vector $\boldsymbol{P}$ for the charged superfluid surrounding the moving vortex, which is defined by the polarization charge $\rho^p = -(c^2\lambda^{-2}/4\pi s^2)(\varphi^{ind} + \mathbf{v}\boldsymbol{A}^{ind}/c)$ by means of the equation $\rho^p = -\nabla\boldsymbol{P}$. Considering the vortex as the source of the external field $\varphi^{ext} = (\phi_0/2\pi c)\,\dot{\chi}$, the coupling of the vortex to the excitations of the superfluid can be clearly determined: the interaction energy of the polarized medium with the external field is given by the expression [3]

$$\widetilde{F}_p = \frac{1}{2}\int \boldsymbol{P}\widetilde{\boldsymbol{E}}\, d\boldsymbol{r}\,. \tag{4}$$

Here $\widetilde{\boldsymbol{E}}$ is the field which will remain in the superconductor if the polarization vanishes, i.e. the longitudinal component of the "external" electric field $\widetilde{\boldsymbol{E}} = -\nabla\varphi^{ext}$.

The field φ^{ext} determined by the singular phase time derivative is generated by the vortex motion, just as φ^{ind}. Therefore, the energy $\widetilde{F}_p$ (4) is the measure of the vortex inertia along with the second term in (1), and the complete kinetic energy is the sum of these two terms. This sum can be transformed to a more cogitable form by means of the electric field induction $\boldsymbol{D} = \boldsymbol{E} + 4\pi\boldsymbol{P}$, which allows rewriting the Poisson's equation as $div\boldsymbol{D} = 4\pi\rho^{ext}$, where the "external" charge density is $\rho^{ext} = -(c^2\lambda^{-2}/4\pi s^2)\varphi^{ext}$. According to the theory of a flowing conducting medium in a magnetic field [3], the longitudinal component of the induction vector is equal to

$$\boldsymbol{D}^l = \varepsilon\boldsymbol{E}^l + \frac{\varepsilon - 1}{c}[\mathbf{v}\times\boldsymbol{B}]^l\,, \tag{5}$$

where ε is the longitudinal dielectric response function of the fluid. After some manipulations the complete electromagnetic kinetic energy is converted to the following form

$$F_{kin} = \int d\boldsymbol{r}\left\{\frac{\boldsymbol{D}^l\boldsymbol{E}^l}{8\pi} - \frac{1}{2}\rho^{ext}\Phi\right\}. \tag{6}$$

The last expression does not include the energy of the transverse electric field which is negligible due to the large value of the light speed c.

The computation of the kinetic energy is straightforward with the use of the Fourier transformations, such as $\nabla\chi = \int d\boldsymbol{k}\; e^{ikr}(\nabla\chi)_k$, so that $(\nabla\chi)_k = \hat{e}_\varphi\; i/2\pi k$ and $\dot{\chi}_k = -\mathbf{v}(\nabla\chi)_k$. To obtain the induced potentials, it is convenient to introduce the dielectric response function

$$\varepsilon_k = 1 + \frac{c^2\lambda^{-2}}{s^2k^2}\,. \tag{7}$$

This function describes the condensate response to a longitudinal perturbation [4]. The model used is valid for the wave vectors $k < \xi^{-1}$ (ξ is the correlation length). We obtain

$$F_{kin} = \frac{1}{8\pi}\left(\frac{\phi_0}{2\pi\lambda s}\right)^2 \int d\boldsymbol{k}\frac{k^2\dot{\chi}^2}{k^2+\lambda^{-2}} = \left(\frac{\phi_0}{4\pi\lambda}\right)^2 \frac{\mathrm{v}^2}{2s^2}\ln\frac{\lambda}{\xi} \tag{8}$$

and the vortex mass is

$$M = \left(\frac{\phi_0}{4\pi\lambda}\right)^2 \frac{1}{s^2}\ln\frac{\lambda}{\xi}\,. \tag{9}$$

This mass is generated by the vortex interaction with the condensate polarization (caused by the vortex motion), and can be regarded as the polaron-like mass.

In contrast to [1,2], the mass of the electromagnetic fields coupled to the condensate is determined by the velocity $s \sim v_F$, but not by the light speed c, that leads to the negligible mass. The velocity v_F enters also the core mass obtained in [1], and the condensate mass (9) exceeds the core mass due to the logarithmic factor $\ln(\lambda/\xi) > 1$.

REFERENCES

1. H. Suhl, *Phys. Rev. Letters* **14**, 226-229 (1965).
2. J. M. Duan, *Phys. Rev. B* **48**, 333-341 (1993).
3. L. D. Landau, E. M. Lifshitz, and L. P. Pitaevskii, *The Electrodynamics of Continuous Media,* Oxford: Pergamon, 1984.
4. A. V. Bezuglyi, E. N. Bratus, and V. P. Galaiko, *J. Low Temp. Physics* **47**, 511-544 (1982).

Thin Superconductors of Complex Shape with and without Vortices

Ernst Helmut Brandt

Max-Planck-Institut für Metallforschung, D-70506 Stuttgart, Germany

Abstract. A method is presented that allows to compute from Maxwell-London theory the static and dynamic electromagnetic properties of thin flat superconducting films of any shape, also for SQUIDs, with and without vortices.

Keywords: superconductors, SQUIDs, Meissner state, vortices, geometry effects
PACS: 74.78.-w, 74.25.Ha, 74.25.Op

Typical SQUIDs (Superconducting Quantum Interference Devices) have the shape of a thin-film disk or a rectangular film, with a radial slit leading to a central hole. This slit is bridged at some point by one or two "weak links" that can carry a limited supercurrent. A perpendicular applied magnetic field $H_a(\mathbf{r})$ generates complicated screening currents that will pass or not pass the weak link and thereby modify the voltage response to a small test current [1]. The penetration and motion of two dimensional (2D) Pearl vortices [2] cause noise in the SQUID. For SQUIDs without and with vortices, the 2D sheet current $\mathbf{J}(x,y) = (J_x, J_y)$ and the related complicated magnetic field $H_z(x,y)$ in the film plane can be calculated for any value of the 2D magnetic penetration depth $\Lambda = \lambda^2/d$ (λ = London depth, d = film thickness, $d < \lambda$) as follows [3]. See also the different methods [4, 5] and similar work on slitted rings [6] and double-strip SQUIDs [7].

Since $\nabla \cdot \mathbf{J} = 0$ in the film except at small contacts, one can express $\mathbf{J}$ in terms of a scalar potential or stream function $g(x,y)$ as $\mathbf{J} = -\hat{\mathbf{z}} \times \nabla g = \nabla \times (\hat{\mathbf{z}} g) = (\partial g/\partial y, -\partial g/\partial x)$. This function $g(x,y)$ has interesting properties:
(a) $g(x,y)$ is the local magnetization or density of tiny current loops; (b) the contour lines of $g(x,y)$ are the current stream lines; (c) on the boundary of the film one may put $g(x,y) = 0$ since the boundary coincides with a stream line; (d) the integral of $g(x,y)$ over the film area equals the magnetic moment of the film if $g = 0$ on its edge. (e) the difference $g(\mathbf{r}_1) - g(\mathbf{r}_2)$ is the current that crosses any line connecting the two points $\mathbf{r}_1$ and $\mathbf{r}_2$; (f) if the film contains an isolated hole or slot such that magnetic flux can be trapped in it or a current I can circulate around it, then in this hole one has $g(x,y) = \text{const} = I$ if $g(x,y) = 0$ is chosen outside the film; (g) in a multiply connected film with n holes, n independent constants $g_1 \ldots g_n$ can be chosen for the values of $g(x,y)$ in each of these holes. The current flowing between hole 1 and hole 2 is then $g_1 - g_2$; (h) a vortex with flux Φ_0 in the film moves in the potential $V = -\Phi_0 g(x,y)$, since the Lorentz force on a vortex is $-\mathbf{J} \times \hat{\mathbf{z}}\Phi_0 = -\Phi_0 \hat{\mathbf{z}} \times (\hat{\mathbf{z}} \times \nabla g) = \Phi_0 \nabla g(x,y) = -\nabla V$; (i) a vortex moving from the edge of the film into a hole connected to the outside by a slit, at each position (x,y) couples a fluxoid $\Phi_0 g(x,y)/I$ into this hole, where $g(x,y)$ is the solution that has $g(x,y) = I$ in this hole (with closed slit) and $g = 0$ outside the film.

For numerics I introduce a 2D grid spanning the film area with (preferrably non-equidistant) points $\mathbf{r}_i = (x_i, y_i)$ and weights w_i such that any integral is approximated by a sum: $\int d^2r\, f(\mathbf{r}) \approx \sum_{i=1}^{N} w_i f(\mathbf{r}_i)$. From Maxwell and London equations one then obtains the static relations

$$H_z(\mathbf{r}_i) = H_a(\mathbf{r}_i) + \sum_j Q_{ij} w_j g(\mathbf{r}_j) \quad (1)$$

$$g(\mathbf{r}_i) = -\sum_j K_{ij}^{\Lambda} H_a(\mathbf{r}_j) \quad (2)$$

with the inverse matrix $K_{ij}^{\Lambda} = (Q_{ij} w_j - \Lambda \nabla_{ij}^2)^{-1}$ and

$$Q_{ij} = (\delta_{ij} - 1) q_{ij} + \delta_{ij} \Big(\sum_{l \neq i} q_{il} w_l + C_i \Big) / w_j \quad (3)$$

where $q_{ij} = 1/(4\pi |\mathbf{r}_i - \mathbf{r}_j|^3)$, ∇_{ij}^2 is the Laplacian such that $\sum_j \nabla_{ij}^2 f(\mathbf{r}_j) \approx \nabla^2 f(\mathbf{r})$ at $\mathbf{r} = \mathbf{r}_i$, and

$$C_i = \frac{1}{4\pi} \sum_{p,q} \Big[(a - p x_i)^{-2} + (b - q y_i)^{-2} \Big]^{1/2}. \quad (4)$$

Here $p, q = \pm 1$ (yielding four terms), and the grid fills the rectangle $|x| \le a$, $|y| \le b$ that should contain the film, i.e., the film may be this rectangle or smaller, containing holes, slits, or rounded corners.

The current stream lines in the Meissner state for various geometries are shown in Fig. 1, while Fig. 2 shows magnetic field profiles in a square film ($|x| \le a$, $|y| \le a$) for various Λ/a. The current stream lines of a vortex pair are depicted in Fig. 3 as the contour lines of

CP850, *Low Temperature Physics: 24th International Conference on Low Temperature Physics;*
edited by Y. Takano, S. P. Hershfield, S. O. Hill, P. J. Hirschfeld, and A. M. Goldman

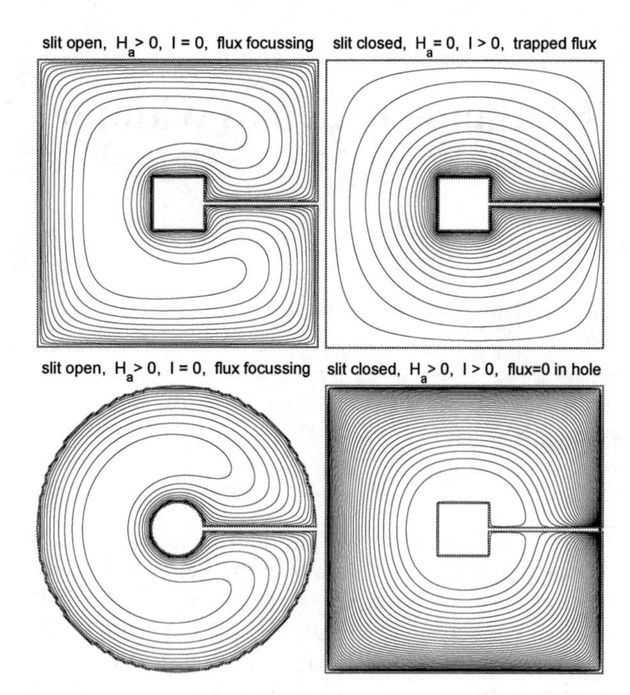

FIGURE 1. The current stream lines in the ideal Meissner state $\Lambda = 0$ for a thin film square with square hole and radial slit, and in a circular disk with circular hole and slit. Top left and bottom left: Slit open, applied field $H_a > 0$, magnetic flux enters the slit and is focussed into the hole where $H(x,y) > H_a$. Top right: Slit bridged at the edge, circulating current $I > 0$ flows due to flux trapped in the hole and slit, no applied field $H_a = 0$. Bottom right: Closed slit, applied field $H_a > 0$, some current $I > 0$ flows such that the flux in hole and slit is exactly zero (ideal screening); this state is a superposition of the two upper states. The current near the hole circulates in opposite direction, except in the trapped-flux case.

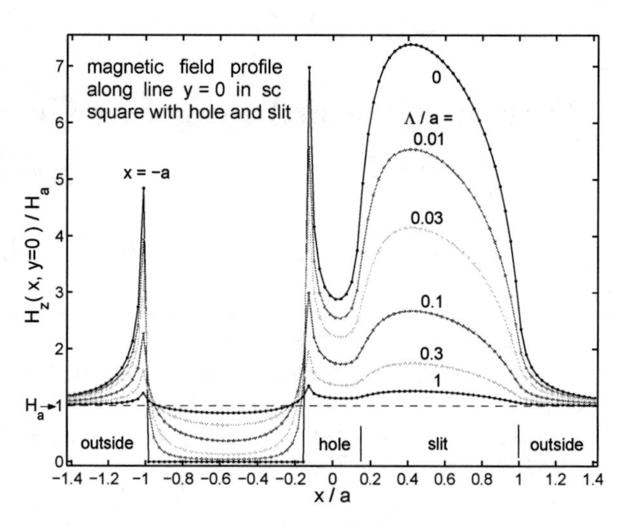

FIGURE 2. Profiles of the magnetic field $H_z(x,0)$ in the thin square of Fig. 1 (top right) taken along the x axis that passes through the hole and slit, in units of the applied field H_a, shown for several values of the 2D magnetic penetration depth $\Lambda = \lambda^2/d = 0, 0.01, 0.03, 0.1, 0.3$, and 1 ($\Lambda$ in units of the half width a of the square, 100×100 grid points). In the center of the square hole the magnetic field is enhanced by a factor of 3 when $\Lambda = 0$, $H_z(0,0) \approx 3H_a$ (flux focussing); in the narrow slit the maximum $H_z(0.4a,0) = 7.4H_a$ is even higher. Finite Λ reduces this enhancement and the spatial variation of H.

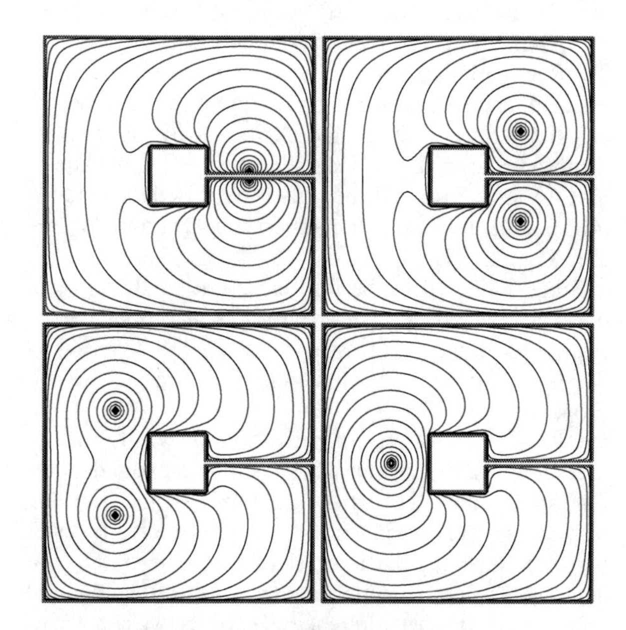

FIGURE 3. Current stream lines of vortex pairs in a square film with slit and hole in the Meissner state at $H_a = 0$. Shown are the contour lines of $\ln|K_{ij}^{\Lambda}|$ at constant j.

K_{ij}^{Λ} for fixed $\mathbf{r}_j$, cf. Eq. (2). Computation of the dynamics $\mathbf{J}(x,y,t)$, $H_z(x,y,t)$ (t = time) for any history $H_a(t)$ and given flux-motion resistivity $\rho(J)$ is described in [3, 8].

ACKNOWLEDGMENTS

This work was supported by the German Israeli Research Grant Agreement (GIF) No G-705-50.14/01.

REFERENCES

1. D. Koelle, R. Kleiner, F. Ludwig, E. Dantsker, and John Clarke, *Rev. Mod. Phys.* **71**, 631-686 (1999).
2. J. Pearl, *Appl. Phys. Lett.* **5**, 65-66 (1964).
3. E. H. Brandt, *Phys. Rev. B*, in print.
4. G. Hildebrandt and H. Uhlmann, *IEEE Trans. Magn.* **32**, 690-693 (1996).
5. M. M. Khapaev, M. Yu. Kuprianov, E. Goldobin, and M. Siegel, *Supercond. Sci. Technol.* **16**, 24-27 (2003).
6. E. H. Brandt and J. R. Clem, *Phys. Rev. B* **69**, 184509, 1-12 (2004).
7. J. R. Clem and E. H. Brandt, unpublished.
8. E. H. Brandt, *Phys. Rev. B* **52**, 15442-15457 (1995); *Phys. Rev. B* **64**, 024505, 1-15 (2001).

Critical State in Type-II Superconductors of Complex Shape

E. H. Brandt* and G. P. Mikitik†,*

*Max-Planck-Institut für Metallforschung, D-70506 Stuttgart, Germany
†B. Verkin Institute for Low Temperature Physics & Engineering, National Ukrainian Academy of Sciences, Kharkov 61103, Ukraine

Abstract. We analyze the critical state problem for superconductors of any shape. A simple example demonstrates that in the general case, a perturbation of the current distribution in the critical state propagates into the sample smoothly in a diffusive way, while the usual Bean critical state with a sharp current front is only a special case of the general critical state.

Keywords: superconductors, vortex pinning, critical state
PACS: 74.25.Sv, 74.25.Qt

The critical state in type-II superconductors is characterized by the component of the current density flowing *perpendicular* to the flux lines, $j_{c\perp}$, since only this component generates a driving force acting on the vortices. Below, to explain the physics with the least mathematical complications, we shall imply the simplest form for the critical current density $j_{c\perp}$: $j_{c\perp} = \text{const}$. For simplicity, we also assume that the magnetic fields $\mathbf{H}$ in the superconductor considerably exceed the lower critical field H_{c1}, and so we put $\mathbf{B} = \mu_0\mathbf{H}$. When the critical state is established in superconductors, in general a current-density component $j_\parallel$ *parallel* to the local magnetic field is also generated. Since the magnitude of this $j_\parallel$ remains undefined (if $j_\parallel$ does not reach the flux-cutting threshold [1]), one cannot in general find the distributions of the magnetic field $\mathbf{H}(\mathbf{r})$ and current density $\mathbf{j}(\mathbf{r})$ in the critical state. Indeed, to solve the Maxwell equations for $\mathbf{H}$,

$$\text{rot}\mathbf{H} = \mathbf{j}, \quad \text{div}\mathbf{H} = 0, \tag{1}$$

it is necessary to know the magnitude and direction of the currents $\mathbf{j}(\mathbf{r})$ in the sample [2]. However, one has only the *two* conditions:

$$j_\perp = j_{c\perp}, \quad \text{div}\mathbf{j} = 0, \tag{2}$$

for *three* quantities: $j_\perp$, $j_\parallel$ and the angle defining the direction of $j_\perp$ in the plane normal to $\mathbf{H}(\mathbf{r})$. Thus, the existing critical-state theory based only on Eqs. (1), (2) is *not complete*. The above Maxwell equations with conditions (2) can provide the description of the critical state only when the shape of the superconductor is sufficiently symmetric and the external magnetic field is applied along a symmetry axis, so that some constraint on the directions of the currents is known in advance. (For example, the direction of the currents is obvious for a slab in an external magnetic field parallel to its surface or for a cylinder in a field parallel to its axis). It is these symmetric situations that are commonly used in analyzing the critical state in superconductors. However, we emphasize that even for *simple* experimental situations equations (1) together with conditions (2) can be insufficient for solving the critical-state problem.

We now show how the critical state problem can be solved for superconductors of arbitrary shape. Let the critical state be known at some moment of time t, i.e., one has $\mathbf{H} = H(\mathbf{r})\boldsymbol{\nu}(\mathbf{r})$ inside the superconductor where the magnitude of the magnetic field, H, and the unit vector $\boldsymbol{\nu}$ are both known functions of the coordinates $\mathbf{r}$ at some external magnetic field $\mathbf{H}_a(t)$. The current density $\mathbf{j}(\mathbf{r})$ in the critical state follows from the Maxwell equation $\mathbf{j} = \text{rot}\,(H(\mathbf{r})\boldsymbol{\nu}(\mathbf{r}))$, while the component of the current density perpendicular to the magnetic field is given by $\mathbf{j}_\perp = \mathbf{j} - \boldsymbol{\nu}(\boldsymbol{\nu}\mathbf{j}) \equiv j_{c\perp}\mathbf{n}_\perp(\mathbf{r})$. Here the last equality defines the unit vector $\mathbf{n}_\perp$. Let the external field infinitesimally change by $\delta\mathbf{H}_a = \dot{\mathbf{H}}_a\delta t$. We now shall find the new critical state at the new external magnetic field $\mathbf{H}_a + \delta\mathbf{H}_a$.

Under the change of $\mathbf{H}_a$, the critical currents locally shift the vortices in the direction of the Lorentz force $[\mathbf{j}\times\boldsymbol{\nu}]$; this shift generates an electric field directed along $[\boldsymbol{\nu}\times[\mathbf{j}\times\boldsymbol{\nu}]] = \mathbf{j}_\perp$, i.e., along the vector $\mathbf{n}_\perp$. Thus, we can represent the electric field $\mathbf{E}(\mathbf{r})$ in the form $\mathbf{E} = \mathbf{n}_\perp e$ where the scalar function $e(\mathbf{r})$ is the modulus of the electric field. Note that the electric field generally is not parallel to the total current density $\mathbf{j}(\mathbf{r})$. Using the Maxwell equation

$$\text{rot}(e\mathbf{n}_\perp) = -\mu_0\dot{\mathbf{H}}, \tag{3}$$

where $\dot{\mathbf{H}} \equiv \partial\mathbf{H}/\partial t$, one can express the change of the magnetic fields (and hence currents) via *one scalar* function $e(\mathbf{r})$. This function can be found from the condition that in the critical state the absolute value of $\mathbf{j}_\perp$ is a given function of $\mathbf{B}$. In our case when $j_{c\perp}$ =const., this condition reads

$$\mathbf{j}_\perp \cdot \frac{\partial\mathbf{j}_\perp}{\partial t} = 0. \tag{4}$$

CP850, *Low Temperature Physics: 24th International Conference on Low Temperature Physics;* edited by Y. Takano, S. P. Hershfield, S. O. Hill, P. J. Hirschfeld, and A. M. Goldman

Taking into account the definition of $\mathbf{j}_\perp$ we arrive at an equation for $e(\mathbf{r})$, [3]

$$\mathbf{n}_\perp \cdot \{\mathrm{rot}\,\mathrm{rot}(e\mathbf{n}_\perp) - (\boldsymbol{\nu}\cdot \mathrm{rot}\boldsymbol{\nu})\,\mathrm{rot}(e\mathbf{n}_\perp)\} = 0. \quad (5)$$

Continuity of $\partial \mathbf{H}/\partial t$ on the surface of the superconductor, S, yields the boundary condition for e. If in the critical state of the superconductor there are also boundaries at which the direction or magnitude of the critical currents changes discontinuously, the function $e(\mathbf{r})$ has to vanish at these boundaries.

After determining the function $e(\mathbf{r})$, one can find the new critical state $\mathbf{H}(\mathbf{r}) + \delta\mathbf{H}(\mathbf{r})$ using the definition $\delta\mathbf{H}(\mathbf{r}) = \dot{\mathbf{H}}\delta t$ and Eq. (3). Note that in agreement with the meaning of the critical state, the new state depends only on the *product* $\dot{\mathbf{H}}_a\delta t = \delta\mathbf{H}_a$, while the electric field e plays an auxiliary role in the above description since it is proportional to $\dot{\mathbf{H}}_a$ rather than to $\delta\mathbf{H}_a$.

To illustrate the obtained results, we now consider an infinite slab of thickness d. Let this slab fill the space $|x|, |y| < \infty$, $|z| \le d/2$, and be in a constant and uniform external magnetic field H_a ($H_a \gg J_c \equiv j_{c\perp}d$) directed along the z axis. Let then a constant field h_{ax} ($J_c/2 \le h_{ax} \ll H_a$) be applied along the x axis, and after that the magnetic field h_{ay} ($h_{ay} \ll H_a$) is switched on in the y direction. The condition $\mathrm{div}\mathbf{j} = 0$ yields $j_z = 0$, i.e., the currents flow in the x-y planes. Then, to describe the critical state, we may use the parametrization:

$$\mathbf{j} = j_c(\varphi,\theta,\psi)(\cos\varphi(z),\ \sin\varphi(z),0),$$
$$\mathbf{H}(z) = \mathbf{H}_a + \mathbf{h}(z),$$
$$\mathbf{h}(z) = (h_x(z), h_y(z), 0), \quad (6)$$

where $j_c(\varphi,\theta,\psi)$ is the magnitude of the critical current density when a flux-line element is given by the angles ψ and θ, $\tan\psi = h_y/h_x$, $\tan\theta = (h_x^2+h_y^2)^{1/2}/H_a$, while the current flows in the direction defined by the angle φ; all these angles generally depend on z. At $H_a \gg h_{ax}, h_{ay}, J_c$, the field $\mathbf{H}$ is practically normal to the x-y planes where the currents flow ($\theta \approx 0$), and we may put $j_c(\varphi,\theta,\psi) = j_{c\perp}$. With this parametrization, equations (1), (3), (5) give *four equations for four functions* $h_{ax}(z)$, $h_{ay}(z)$, $e(z)$, $\varphi(z)$.

The solution of these four equations is shown in Fig. 1. From this solution one can see the evolution of the current direction $\varphi(z)$ in the sequence of the critical states developed in the process of increasing h_{ay}. Interestingly, the change of the angle $\varphi(z, h_{ay})$ has *diffusive character*. This is in stark contrast to the usual Bean critical state, in which any change of the current direction occurs inside a narrow front. This Bean critical state is also obtainable from the above equations if the fields h_{ax} and h_{ay} are switched on simultaneously so that $h_{ax}(t)/h_{ay}(t)$ =const., and it corresponds to a *discontinuous* solution $\varphi(z)$. Hence the Bean state is only a *special case* of the general critical state.

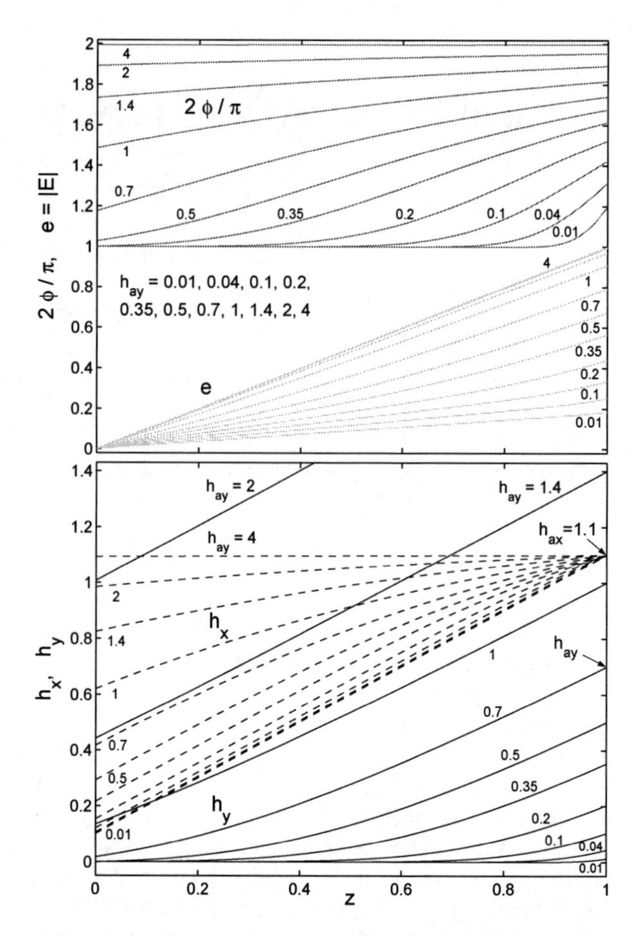

FIGURE 1. Profiles of the angle of currents $\varphi(z)$, magnitude of electric field $e(z)$ (top) and magnetic field components $h_x(z)$ (dashed lines), $h_y(z)$ (solid lines) (bottom) in the critical states of the slab. Here $h_{ax} = 1.1$ and $h_{ay} = 0.01, 0.04, 0.1, 0.2, 0.35, 0.5, 0.7, 1, 1.4, 2, 4$. We start at $h_{ay} = 0$ with $h_x(z) = h_{ax} - 1 + z$, $h_y(z) = 0$, and $\phi(z) = \pi/2$. The coordinate z is in units of $d/2$, h_x and h_y in units of $j_{c\perp}d/2 = J_c/2$, and e in units of $\mu_0(dh_{ay}/dt)d/2$.

ACKNOWLEDGMENTS

This work was supported by the German Israeli Research Grant Agreement (GIF) No G-705-50.14/01.

REFERENCES

1. A. M. Campbell and J. E. Evetts, *Adv. Phys.* **21**, 199-443 (1972), see pp. 261 ff.
2. L. D. Landau and E. M. Lifshits, *Electrodynamics of Continuous Media, Course in Theoretical Physics* Vol. 8 (Pergamon, London, 1959).
3. G. P. Mikitik and E. H. Brandt, *Phys. Rev. B* **71**, 012510, 1-4 (2005).

Determination of the B-Dependent Critical Current Density in Thin Flat Superconductors by Magneto-Optics

G. P. Mikitik*,† and E. H. Brandt†

*B. Verkin Institute for Low Temperature Physics & Engineering, National Ukrainian Academy of Sciences, Kharkov 61103, Ukraine
†Max-Planck-Institut für Metallforschung, D-70506 Stuttgart, Germany

Abstract. Analyzing the critical state problem for a thin superconducting strip, we explain how one can determine the critical current density j_c from the magnetic-field profiles measured by magneto-optical imaging at the upper surface of the strip in the case when this j_c depends on the magnitude and direction of the local magnetic field **H**. We show that measurements of the profiles and their asymmetry in oblique magnetic fields enable one to determine the **H**-dependent j_c.

Keywords: superconductors, critical current density, magneto-optical imaging
PACS: 74.25.Qt, 74.25.Sv

One of the methods for analyzing flux-line pinning in high-T_c superconductor thin platelets is measuring the magnetic-field profiles at the upper surface of these superconductors when they are placed in a perpendicular external magnetic field; see, e.g., the recent review [1] and references therein. It is clear that the investigation of the profiles not only in a perpendicular but also in an oblique magnetic field could, in principle, provide additional information on vortex pinning in thin flat superconductors.

Consider a thin superconducting strip which fills the space $|x| \leq w$, $|y| < \infty$, $|z| \leq d/2$ with $d \ll w$. A constant and homogeneous external magnetic field H_a is applied in the x-z plane at an angle θ_0 to the z axis ($H_{ax} = H_a \sin\theta_0$, $H_{ay} = 0$, $H_{az} = H_a \cos\theta_0$). Let the lower critical field H_{c1} be sufficiently small as compared with $j_c d$, i.e., with self fields of the currents in the critical state, where j_c is the critical current density. Then, we may put for the magnetic induction $B = \mu_0 H$. The thickness of the strip, d, is assumed to exceed the London penetration depth. We shall consider the situation when j_c depends either on the magnitude of the local magnetic field **H** or on the angle θ between the direction of this **H** and the z axis (or on both reasons). The critical current density for various types of flux-line pinning satisfies this assumption (for example, this assumption is valid for the pinning by columnar defects if they are along the z axis, and for pinning by twin planes parallel to the y axis; it is also valid for anisotropic superconducting materials with defects which cannot be considered as point pinning centers). In this case j_c is not constant across the thickness of the strip since **H** and thus j_c change with z in a strip of finite thickness, and the critical value J_c of the sheet current J is not simply $j_c d$ (the sheet current J is the current density integrated over the thickness of the sample). Thus, in experiments the critical current density cannot be found as $j_c = J_c/d$, and the problem arises how to determine $j_c(\mathbf{H})$ from experimental data. In the present paper we address just this problem.

Various procedures were elaborated [1] which enable one to extract the spatial distribution of the sheet current $J(x)$ from the experimental $H_z(x)$ profile measured at the upper surface of a thin superconducting strip placed in a perpendicular magnetic field. Eliminating x from $H_z(x)$ and $J(x)$, one can find the function $J_c(H_z)$ from experimental data. In a similar manner, one can extract the critical sheet current $J_c(H_z, H_{ax})$ from the magnetic field profiles in oblique magnetic field. [In this procedure one has to deal with those regions of x where $H_z(x)$ changes when the external field $\mathbf{H}_a$ is varied.] We now discuss the problem how to reconstruct $j_c(\mathbf{H})$ from a thus obtained $J_c(H_z, H_{ax})$. Our analysis is based on the solution of the critical state problem for anisotropic strips in an oblique magnetic field [2, 3].

In oblique magnetic fields the magnetic field profiles on the upper surface of the strip, $H_z(x)$, can be slightly asymmetric relative to the central line of the strip, $x = 0$. Then, the profiles can be decomposed into their symmetric and antisymmetric parts, $H_z(x) = H_z^s(x) + H_z^a(x)$, with the antisymmetric part $H_z^a(x)$ being relatively small (of the order of d/w). The function $J_c(H_z, H_{ax})$ mentioned above is extracted from the symmetric part $H_z^s(x)$ of the profiles. This function can be expressed in terms of the critical current density $j_c(H_x, H_z) = j_c(\theta, |H|)$ with the following implicit relation [2]:

$$d = \int_{H_x^-}^{H_x^+} \frac{dh}{j_c(h, H_z)}, \tag{1}$$

where $H_x^- = H_{ax} - 0.5J_c(H_z, H_{ax})$, and $H_x^+ = H_{ax} + 0.5J_c(H_z, H_{ax})$ are the x component of the magnetic field

CP850, *Low Temperature Physics: 24th International Conference on Low Temperature Physics;*
edited by Y. Takano, S. P. Hershfield, S. O. Hill, P. J. Hirschfeld, and A. M. Goldman

at the lower and the upper surfaces of the strip. As one might expect, within the simple Bean model when j_c is independent of $\mathbf{H}$, equation (1) yields $J_c = j_c d$ for any H_{ax}. We emphasize that since the function $J_c(H_z, H_{ax})$ extracted from the magnetic-field profiles in oblique magnetic fields generally depends on two variables H_z and H_{ax}, this function, in principle, enables one to determine $j_c(H_x, H_z) = j_c(\theta, |H|)$ depending on two variables, too. In this determination one may first assume some model of the $\mathbf{H}$-dependence of j_c, and then find parameters of this model from Eq. (1), using $J_c(H_z, H_{ax})$ extracted from the experimental data. The result of such determination can be verified by calculation of the antisymmetric part of the magnetic field profiles, $H_z^a(x) \equiv [H_z(x) - H_z(-x)]/2$, which is described by the formula [3]

$$H_z^a(x) = -0.5\frac{d}{dx}\left(\int_{H_x^-}^{H_x^+} \frac{h\,dh}{j_c(h, H_z)}\right), \tag{2}$$

in those regions of x where $J_c(H_z, H_{ax})$ has been extracted from $H_z^s(x)$.

Consider more closely the case when the thickness d of the strip is sufficiently small, $d \ll H_0/j_c$, where H_0 is the characteristic scale of the $|H|$ dependence of j_c. In this case the characteristic field in the critical state of the strip, $j_c d$, is considerably less than H_0, and at H_{az}, $H_{ax} \sim j_c d$ one has $j_c(\theta, |H|) \approx j_c(\theta, 0)$. Note that in general one cannot neglect the angular dependence of j_c even in such thin samples since the characteristic angles θ in the strip are of the order of $j_c d/H_z$ and essentially change at $H_z \sim j_c d$. In the case of thin samples, $j_c d \ll H_0$, one can extract $j_c(\theta, 0)$ from the experimental H_z-dependence of J_c at $H_{ax} = 0$, i.e., from the function $J_c(H_z, 0) \equiv J_c(H_z)$:

$$j_c(\theta)d = J_c(H_z) - H_z\frac{dJ_c(H_z)}{dH_z},$$

$$\tan\theta = \frac{J_c(H_z)}{2H_z}. \tag{3}$$

Note that these formulas are *model-independent*, and they give the dependence $j_c(\theta) \equiv j_c(\theta, 0)$ in parametric form, with H_z being the parameter. Inserting this dependence in Eqs. (1), (2) and using the critical state equations [2], one can calculate $J_c(H_z, H_{ax})$ and $H_z^s(x)$, $H_z^a(x)$ in oblique magnetic fields. Comparing the obtained $H_z^s(x)$, $H_z^a(x)$ with appropriate experimental data, one can verify the fulfilment of the assumption $j_c d \ll H_0$. If this comparison does not reveal a good agreement with the data in oblique magnetic fields for the superconductor under study, the calculated $j_c(\theta)$ may be considered as a first step in determining $j_c(\theta, |H|)$. In this case one should supplement the obtained $j_c(\theta)$ with some $|H|$ dependence and repeat the calculations.

Finally, consider the special case when j_c depends only on the combination $|H|\cos\theta = H_z$ rather than on θ and $|H|$ separately. This case occurs in anisotropic superconductors for weak collective pinning by point defects in the so-called small-bundle pinning regime [4]. In this situation the scaling approach is valid [4], and if the anisotropy parameter ε is small, j_c depends only on H_z, $j_c = j_c(H_z)$, practically for all θ. Then, for thin samples, $j_c d \ll H_0$, one arrives at the Bean model $j_c = j_c(0)$. In this case $J_c(H_z)$ reduces to a constant, one has $j_c(0) = J_c/d$, and measurements of $H_z(x)$ in oblique magnetic fields do not give any new information about flux-line pinning (they all lead to the same value of J_c). For thicker samples, an H_z-dependence of J_c appears, but Eq. (1) still yields $j_c(H_z) = J_c(H_z)/d$. Note that the application of Eqs. (3) to the appropriate experimental data would not give this exact result for j_c. However, this discrepancy will immediately come to light in oblique magnetic fields. Indeed, when $j_c = j_c(H_z)$, the magnetic-field profiles in oblique magnetic fields depend only on H_{az}, and Eq. (2) gives $H_z^a(x) = 0$. On the other hand, if $j_c = j_c(\theta)$, the profiles $H_z(x)$ and the penetration depth of the magnetic field into the sample depend both on H_{az} and on H_{ax}, and $H_z^a(x) \neq 0$ in oblique magnetic fields [2, 3]. Thus, although Eqs. (3) can be always used as a starting point in the determination of $j_c(\mathbf{H})$, they do not provide the best first approximation in the special case $j_c = j_c(H_z)$ when the magnetic field profiles are independent of H_{ax}. In this situation the usual formula $j_c = J_c/d$ is applicable.

ACKNOWLEDGMENTS

This work was supported by the German Israeli Research Grant Agreement (GIF) No G-705-50.14/01.

REFERENCES

1. Ch. Jooss, J. Albrecht, H. Kuhn, S. Leonhardt, and H. Kronmüller, *Rep. Prog. Phys.* **65**, 651-788 (2002).
2. E. H. Brandt and G. P. Mikitik, *Phys. Rev. B* **71**, (2005) (in print).
3. G. P. Mikitik and E. H. Brandt, *Phys. Rev. B* **72**, (2005) (in print).
4. G. Blatter, M. V. Feigel'man, V. B. Geshkenbein, A. I. Larkin, and V. M. Vinokur, *Rev. Mod. Phys.* **66**, 1125-1388 (1994).

How Many Long-Range Orders Are in the Abrikosov State

A.V. Nikulov

Institute of Microelectronics Technology, RAS, 142432 Chernogolovka, Moscow Region, Russia.

Abstract. It is argued that only a single spontaneous long-range order, namely the phase coherence exists in the Abrikosov state, and the prediction of the crystalline long-range order of vortex lattice does not correspond to the facts.

Keywords: Abrikosov vortex state, thermal fluctuations, vortex lattice melting.
PACS: 74.25.Qt, 74.25.Op

INTRODUCTION

Superconductivity is one of the best known macroscopic quantum phenomena. The first experimental evidence of this wonderful fact is the Meissner effect observed in 1933, and the most marvelous one is the Abrikosov vortex state [1]. A superconductor expels the magnetic flux $\oint dlA = \Phi$ when the macroscopic wave function $\Psi = |\Psi| exp(i\varphi)$ does not have singularities, and the integral $\oint dl\nabla\varphi$ of the gradient $\nabla\varphi = (mv + 2eA)2\pi/h$ of its phase φ along any closed path l within superconductor should be equal zero: $\oint dl\nabla\varphi = 2e\Phi 2\pi / h = 0$. When, however, the phase φ of the wave function Ψ has singularities inside l, its circulation equals

$$\oint dl\nabla\varphi = 2\pi n \qquad (1)$$

where n is the number of singularities, and consequently $\Phi = hn/2e = n\Phi_0$ Therefore, numerous direct observations of the proportionality $\Phi = n\Phi_0$ between the magnetic flux Φ and the number n of the Abrikosov vortices prove that the Abrikosov state is a mixed state possessing a long-range phase coherence.

PREDICTION OF THE VORTEX LATTICE AND EXPERIMENTAL DATA

The attention of most people was attracted to the additional spontaneous crystalline long-range order predicted by the famous work of Abrikosov [1] and others [2]. Observations of the ordered vortex lattices created an illusion that theoretical prediction of a vortex lattice with a long-range crystalline order received an experimental corroboration.

Spontaneous or Constrained Order

One should, however, realize the fundamental difference between the spontaneous long-range order predicted by [1,2] in homogeneous infinite symmetric space and an order observed in non-homogeneous asymmetric space of a real superconductor sample of finite sizes.

No observation can corroborate the prediction [1,2] of the spontaneous crystalline long-range order of vortex lattice, since it cannot be realized in any real sample with pinning disorders [3]. Moreover, many experimental results are an evidence of non-spontaneous nature of this order constrained by the asymmetry of the underlying atomic lattice. The question about the possibility of the Abrikosov state in the ideal case of the homogeneous infinite symmetric space is important first of all for numerous theories of the vortex lattice melting, since this phase transition should be connected with spontaneous, unconstrained long-range order.

Two Long-Range Order and a Single Phase Transition

There is a discrepancy between the prediction of two long-range order predicted in [1,2] and only a single second order phase transition assumed to occur at the second critical field H_{c2}. The situation became

CP850, *Low Temperature Physics: 24th International Conference on Low Temperature Physics;*
edited by Y. Takano, S. P. Hershfield, S. O. Hill, P. J. Hirschfeld, and A. M. Goldman

more dramatic in the seventies, when the consideration of the thermal fluctuations showed that this transition, assumed during a long time, can not exist because of the reduction of the effective dimensionality of thermal fluctuations on two near H_{c2} [4]: two long-range orders but no transition. The lost transition into the Abrikosov state was found first in bulk superconductors at $H_{c4} < H_{c2}$ in the early eighties [5]. This result was repeated in ten years on high-Tc superconductors (HTSC) [6] and it was shown that this phase transition is first order [7].

VORTEX LATTICE MELTING OR DISAPPEARANCE OF THE PHASE COHERENCE

Thus, according to all experimental results [5,6] only one phase transition is observed at $H_{c4} < H_{c2}$. The interpretation of this transition as vortex lattice melting became popular, since the Abrikosov state is first of all the vortex lattice for most people. Most naive theorists used the Lindemann criterion for description of this transition [8,9]. But some experts realize that the Abrikosov state is first of all the mixed state with long-range phase coherence [10]. Nevertheless, they do not reject the concept of vortex lattice melting. The reason of such an attitude is the definition of phase coherence used by all theorists.

Definitions of Phase Coherence in the Abrikosov State

According to this definition using a correlation function, i.e. when coherence between two points is considered, the long-range phase coherence can exist in the Abrikosov state described by multi-connected wave function only if this function has a periodical long-range order. In this case the vortex lattice melting is simultaneously the disappearance of long-range phase coherence and the concept of vortex lattice melting should not be rejected. But the definition by the use of a correlation function comes into conflict with the experimental evidence $\Phi = n\Phi_0$ of the phase coherence (1) according to which the existence of the long-range phase coherence can not depend on an arrangement of vortices inside a closed path l. The observation of the vortices at any arrangement is evidence of phase coherence since singularities in the mixed state with phase coherence can not exist without phase coherence.

The Abrikosov State Is Not Vortex Lattice with Spontaneous Long-Range Order

The comparison [11] of the position H_{c4} of the transition into the Abrikosov state experimentally found in bulk [5] and thin film [12] superconductors with weak pinning corroborates the result [13] according to which the mean field approximation [1,2] is not valid just for the infinite homogeneous space. The experimental results [12] show that the prediction [1,2] is not valid at least for two-dimensional superconductor. A mixed state without long-range phase coherence, but no Abrikosov state, is observed down to $H_{c4} \approx 0.005H_{c2}$ in a film with weak pinning [12]. According to [14], the long-range order cannot be realized also in three-dimensional superconductor. The observation of the Abrikosov state below $H_{c4} \approx 0.98H_{c2}$ [5] in bulk superconductor with weak pinning does not refute this result since the real sample has finite sizes. Thus, the experimental results and fluctuation theory are indicative of the non-validity of [1,2] for the ideal case and crystalline long-range order of vortex lattice cannot be realized in real samples [3].

ACKNOWLEDGMENTS

The work was supported by RFBR, Grant 04-02-17068. I thank LT24 Financial Support Committee for the offer of financial assistance to attend the LT24 Conference.

REFERENCES

1. A.A. Abrikosov, *Zh. Eksp. Teor. Fiz.* **32**, 1442 (1957) (*Sov. Phys.- JETP* **5**, 1174 (1957)).
2. W.H. Kleiner, L.M. Roth, and S.H. Autler, *Phys. Rev.* **133**, A1226 (1964).
3. A.I. Larkin, *Zh. Eksp. Teor. Fiz.* **58**, 1466 (1970) (*Sov. Phys. - JETP* **31**, 784 (1970)).
4. D.J. Thouless, *Phys. Rev. Lett.* **34**, 946 (1975); S.P. Farrant and C.E. Gough, *Phys. Rev. Lett.* **34**, 943 (1975).
5. V.A. Marchenko, and A.V. Nikulov, *Pisma Zh. Eksp. Teor. Fiz.* **34**, 19 (1981) (*JETP Lett.* **34**, 17 (1981)).
6. H. Safar et al., *Phys. Rev. Lett.* **69**, 824 (1992).
7. E. Zeldov et al., *Nature* **375**, 373 (1995).
8. G. Blatter et al., *Rev. Mod. Phys.* **66**, 1125 (1994).
9. T. Giamarchi and S. Bhattacharya, cond-mat/0111052.
10. A.I. Larkin (private communication).
11. A.V. Nikulov, e-print arXiv: cond-mat/0312641
12. A.V. Nikulov et al., *Phys. Rev. Lett.* **75**, 2586 (1995); *J. Low Temp. Phys.* **109**, 643 (1997).
13. K. Maki and H. Takayama, *Prog. Theor. Phys.* **46**, 1651 (1971).
14. M. A. Moore, *Phys.Rev. B*, **45**, 7336 (1992).

Dynamics of the Vortex-Glass Transition

Golan Bel[1] and Baruch Rosenstein[1,2]

[1] *Physics Department, Bar-Ilan University, Ramat-Gan 52900, Israel.*
[2] *National Center for Theoretical Sciences and Electrophysics Department, National Chiao Tung University, Hsinchu 30050, Taiwan, R.O.C.*

Abstract. The dynamic of moving vortex matter is considered in the framework of the time dependent Ginzburg - Landau equation beyond linear response. Both disorder and thermal fluctuations are included using the Martin-Siggia-Rose formalism within the lowest Landau level approximation. We determine the critical current as function of magnetic field and temperature. The surface in the *J-B-T* space defined by the function separates between the dissipative moving vortex matter regime (qualitatively appearing as either the vortex creep and flux flow) and dissipation less current state in which vortices are pinned creating an amorphous vortex "glass". Both the thermal depinning and the depinning by a driving force are taken into account. The static irreversibility line is compared to experiments and is consistent with the one obtained in the replica approach. The non-Ohmic *I-V* curve (in the depinned phase) is obtained and resistivity compared with experiments in layered superconductors and thin films.

Keywords: Glass transition, Irreversibility line.
PACS: 74.25.Op, 74.25.Qt, 74.25.Sv

INTRODUCTION AND MODEL

As a result of a delicate interplay between disorder, interactions and thermal fluctuations even the static *B-T* phase diagram of HTSC is very complex and is still far from being reliably determined. Once electric current J is injected into the sample, it makes the analysis far more complicated and the phase diagram should now be drawn in the three dimensional space *T-B-J*. Generally there are two phases, the pinned phase in which the vortices are pinned and thus the resistivity vanish (perfect superconductivity exists), and the unpinned phase in which vortices can move due to Lorentz force and thus a finite resistivity appears. The surface is determined by the critical current as function of magnetic field and temperature. Great efforts have been made both experimentally and theoretically to obtain the surface in *T-B-J* space which separates the two phases [1]. When the critical current vanishes, the intersection of the surface with the *B-T* plane gives the irreversibility line.

Most of the theoretical works consider the vortices as elastic lines; this assumption is valid far from the upper critical field H_{c2}. An alternative simplification is the lowest Landau level (LLL) approximation to the vortex matter near H_{c2} where many vortices are presented and due to overlaps between fields of the vortices the magnetic field is nearly homogeneous.

Dynamics in the presence of thermal fluctuations and disorder is described using the time dependent Ginzburg - Landau (TDGL) equation [1].

$$\frac{\hbar^2\gamma}{2m^*}D_\tau\psi = -\frac{\delta}{\delta\psi^*}F+\zeta, \qquad (1)$$

γ is the inverse diffusion constant and ζ is a thermal white noise. The free energy is

$$F = \int d^3x\left[\frac{\hbar^2}{2m^*}|\vec{D}\psi|^2 - a'(1+U)|\psi|^2+\frac{b'}{2}|\psi|^4\right], \qquad (2)$$

U represents the disorder, with correlation

$$\langle U(x)U(y)\rangle = \delta(x-y)\xi^2 n, \qquad (3)$$

where n is the dimensionless density of pinning centers. The covariant derivatives are given by $D_\tau \equiv \frac{\partial}{\partial\tau}+\frac{ie^*}{\hbar}\Phi, \vec{D}\equiv\vec{\nabla}+\frac{ie^*}{\hbar c}\vec{A}$, where A and Φ are the vector and scalar potentials.

This model in the absence of electric field was considered by Dorsey, Fisher and Huang [2] in the homogeneous (liquid) phase using the dynamic Martin-Siggia-Rose approach [3]. They obtained the irreversibility line and claimed that it is inconsistent with experiments in YBCO.

In this paper we study the glass transition using the dynamic approach within the TDGL model at finite

CP850, *Low Temperature Physics: 24th International Conference on Low Temperature Physics;* edited by Y. Takano, S. P. Hershfield, S. O. Hill, P. J. Hirschfeld, and A. M. Goldman

electric field. This allows us to obtain the *I-V* curve beyond the linear response. The GT line for zero electric coincide with the one obtained using the replica method [4]. Comparison of the irreversibility line and resistivity with experimental results in layered superconductors and thin films is made.

CRITICAL CURRENT

We solved this model using the LLL and Gaussian approximations [5]. The following expression for the critical current as function of temperature and magnetic field was obtained

$$J_c = \frac{\hbar c^2}{e^* \lambda^2 \xi}\left(\pi\sqrt{2Gi}tb/4r\right)^{1/2} \left\{1-t-b+4\left(r\pi tb\sqrt{2Gi}\right)^{1/2}(2-1/r)\right\}^{1/2} \quad (4)$$

where $t=T/T_c$ and $b=B/B_{c2}$. The disorder and thermal fluctuations characterized by the parameters:

$$r = \frac{n(2Gi)^{-1/2}}{2\pi^2}\frac{(1-t)^2}{t}, Gi \equiv \frac{1}{2}\left(\frac{2T_c e^{*2}\lambda^2}{\pi L_Z c^2 \hbar^2}\right)^2,$$

respectively (*Gi* is the *2D* Ginzburg parameter).

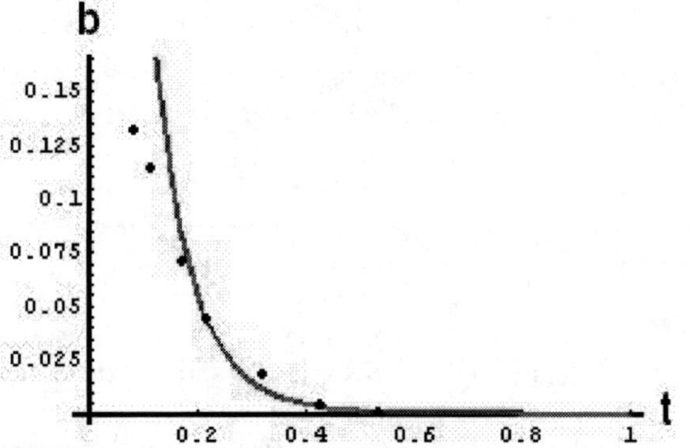

FIGURE 1. Irreversibility line for BSCCO, the dots represent data from [6], while the solid curve corresponds to the theoretical predicted line. The fitting parameters are given in the text.

In the special case of no electric field an equation for the irreversibility line is obtained in the form

$$b = 2\left(\pi rtb\sqrt{2Gi}\right)^{1/2}(2-1/r)+1-t. \quad (5)$$

This line is in agreement with the line obtained using the replica method [4].

I-V AND RESISTIVITY

The *I-V* curve is given by

$$J_y = \frac{\hbar c^2 v}{\xi e^* \lambda^2}\left(\pi tb\sqrt{2Gi}\right)^{1/2} \frac{-a_T(v)+\sqrt{a_T^2(v)+16(1-r)}}{8(1-r)}. \quad (6)$$

The dimensionless velocity is $v = e^*\gamma E\xi^3/(4\hbar b)$, and the dimensionless scaled temperature is $a_T(v) \equiv -\left(1-t-b-v^2\right)/\left(\pi tb\sqrt{2Gi}\right)^{1/2}$.

In order to check our results we compared the irreversibility line and resistivity with experimental results in layered superconductor (BSCCO) [6]. The parameters we used are: H_{c2} =195T, T_c=93K, Gi =4.4x10^{-4} and n=0.005.

In Fig.1 we show that the theoretical and experimental results for the irreversibility line are in a good agreement.

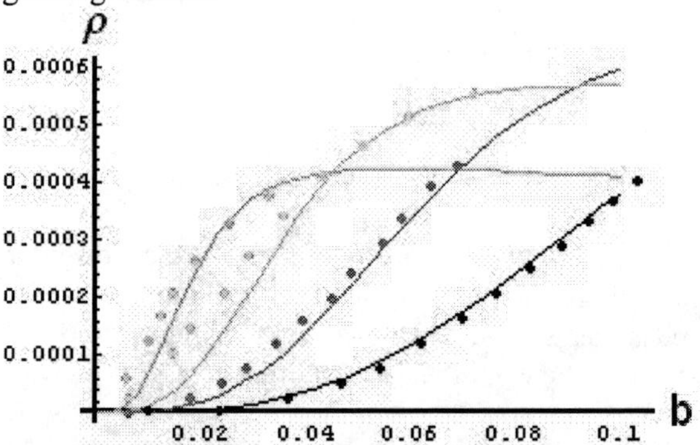

FIGURE 2. Resistivity as function of the magnetic field is plotted for different temperatures. The dots correspond to the experimental results of [6], and the solid lines to the theoretical prediction. The lines from left to right correspond to T=60, 50, 40, 30 K.

In Fig.2 we compare the resistivity as function of the magnetic field for different temperatures with experimental results of [6] for BSCCO. A non trivial temperature dependence of the inverse diffusion constant was used in order to fit the results.

REFERENCES

1. G. Blatter, M. V. Feigelman, V.B. Geshkenbein, A. I. Larkin and V. M. Vinokur, *Rev. Mod. Phys.* **66**, 1125 (1994).
2. A. T. Dorsey, M. Huang and M. P. A. Fisher, *Phys. Rev B* **45**, 523 (1992).
3. P. M. Martin, E. D. Siggia and H. A. Rose, *Phys. Rev. A* **8**, 423 (1973). H. Sompolinsky and A. Zippelius, *Phys. Rev. B* **25**, 6860 (1982).
4. D. Li and B Rosenstein, *cond-mat/0411096.*
5. G. Bel and B. Rosenstein, in preparation.
6. Y. Ando et al. *Phys. Rev. B* **60**, 12475 (1999).

Shock Waves and Avalanches in Type-II Superconductors

B.Ya. Shapiro, B. Rosenstein and I. Shapiro

Institute of Superconductivity, Department of Physics, Bar-Ilan University, Ramat Gan 52900, Israel
National Chiao Tung University, Department of Electrophysics, Hsinchu, Taiwan, R.O.C.

Abstract. Rapid penetration of magnetic flux into a Meissner phase of the type-II superconductor is studied analytically and numerically. A sharp shock wave front of the magnetic induction is formed due to the non-analyticity of the resistivity at the transition from the mixed to the normal state. The shock wave magnetic induction and the temperature profiles move with constant velocity determined by the Joule heat produced by the electric current in the normal domain at the flux front. If the normal state resistance of the sample material is temperature dependent (like in HTSC), then the straight front shows instability. For a sufficiently small thermal diffusion constant a finger shaped avalanches are formed.

Keywords: Vortex matter, flux instabilities, avalanches in superconductors
PACS: 74.20.De 74.25.Ha 75.25.Qt

The dynamics of magnetic flux penetration into type II superconductor and its instabilities has been studied over the years by variety of techniques. Magneto-optics experiments demonstrate that in wide range of situations there exists a well defined interface (front) between the magnetic flux penetrating a sample and the flux-free Meissner state. Magneto-optical technique has been further perfected and revealed a number of instabilities, including generation of avalanches. It is widely excepted now that this instability appears in a critical state of type-II superconductors as a result of heat released by the vortices rolling down over the magnetic induction landscape.

Pinning plays a crucial role in formation of instabilities, hence it is natural to conjecture that the current density at the front is of order of the critical current J_c. It is much smaller than the depairing current J_d at which the superconductivity is destroyed [1]. In recent experiments however [2] the system rapidly forced out of equilibrium created a situation in which fluxons move with huge velocities of order of $10^6 - 10^7$ cm/sec far exceeding those for which the vortices are dragged by the critical currents. Such a dynamic regime was achieved by applying a femtosecond laser pulse to a narrow stripe of the YBCO film subjected to magnetic field perpendicular to the film. It locally destroyed superconductivity in a completely nonadiabatic fashion. The field does not exceed the first critical field H_{c1}, so that initially fluxons cannot penetrate the rest of the sample. Recovery of superconductivity occurs in two stages. After the short pulse has past the stripe is cooled the flux nucleates into a dense system of Abrikosov vortices. On the larger (mesoscopic) time scale the rapidly created vortices are pushed into the superconducting part of the sample, sometimes splitting into the avalanches. A peculiarity of the avalanches is that they exist despite absence of the "mountains", since the initial critical state is not formed yet.

In the present paper we study both numerically and analytically the dynamics of the non-adiabatically created magnetic flux in sufficiently thick (thickness larger than the magnetic penetration length) superconducting films. In particular, effects of dissipation and the heat transport on the motion and stability of the flux front are considered. We found that in its resistive state, the magnetic induction penetrating a flux free superconductor forms a sharp front. It is shown that the Joule heat released at the flux front can produce front propagation at constant velocity. Heating of the front by moving magnetic flux is essential. Strong superconducting currents in the vicinity of the front suppress superconductivity in this area and create a normal domain at the front, see Fig.1. The velocity is determined by the depairing current of

CP850, *Low Temperature Physics: 24th International Conference on Low Temperature Physics;*
edited by Y. Takano, S. P. Hershfield, S. O. Hill, P. J. Hirschfeld, and A. M. Goldman

the material. The straight front line shows an instability with respect to the local temperature fluctuations. The instability is caused by excessive Joule heat released there. Numerical simulation of the set of nonlinear equations allows us to study the evolution of the instability and demonstrates the emergence and development of the corrugated interface.

The set of equations describing dynamics of the (dimensionless) magnetic induction b and temperature θ in the Meissner state of the type-II superconductor is (see [3] for details):

$$\frac{\partial b}{\partial t}=\frac{\partial}{\partial x}\left(\rho\frac{\partial b}{\partial x}\right)+\frac{\partial}{\partial y}\left(\rho\frac{\partial b}{\partial y}\right)$$
$$\frac{\partial \theta}{\partial t}=\kappa\nabla^2\theta+\rho J^2;J=\sqrt{\left(\frac{\partial b}{\partial x}\right)^2+\left(\frac{\partial b}{\partial y}\right)^2} \qquad (1)$$

Here the nonlinear resistivity ρ contains the non-analicity at $J\rightarrow J_d$, separating the behaviour typical to the mixed state $\rho\propto\left[b/(1-\theta)\right]^{\nu}\left[J/(1-\theta)\right]^{\mu}$ from that of the normal state $\rho\propto\rho_0+\rho_1\theta$. In one-dimensional case, a shock wave moving with constant velocity V is a solution. The flux shock wave front moving with velocity $V\simeq\left[J_d/(1-\theta_0)\right]^{1+\mu/\nu}$ has a normal domain of width $w_n\sim\left[(1-\theta_0)/J_d\right]^{1+\mu/\nu}$, where θ_0 is the equilibrium temperature. The structure of the shock wave front is presented in Fig.1. If the normal resistance of the material is temperature dependent then the essential dependence of the front velocity on the Joule heat released near the interface leads to instability of the straight front. Looking for a solution in the form

$$b=b_0(x-Vt)+\eta(x)\exp(iky+\Omega t);$$
$$\theta=\theta_0(x-Vt)+\zeta(x)\exp(iky+\Omega t), \qquad (2)$$

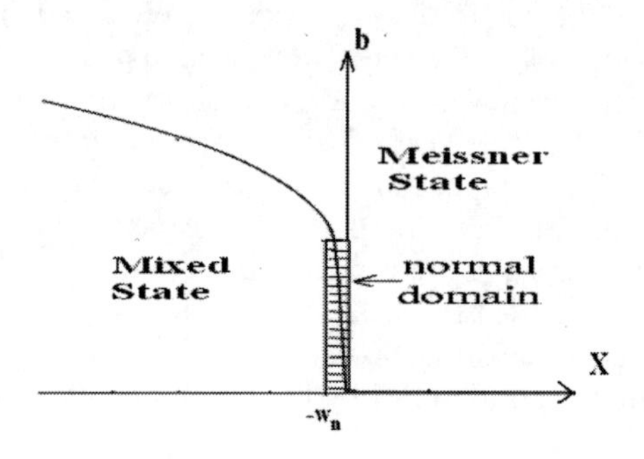

FIGURE 1. Spatial distribution of the magnetic induction for a shock wave.

one obtains, in the special case of zero heat diffusion the rate grow, $\Omega\equiv t_0^{-1}=\rho_0 J_d^2$ exhibiting flux front instability for any wave vectors [3]. Dispersion appears for the case of a non-zero heat diffusion coefficient. Small fluctuations cannot destroy the straight line front. It becomes unstable due to the large amplitude fluctuations. Let us consider the evolution of the instability in this case. Due to diffusion along the flux front interface (y direction in Fig.2), the instability develops under the condition: $ut_0>\sqrt{\kappa t_0}$. This requirement allows us to determine the critical velocity of the fluctuation for which an avalanche is developed: $u>u_c=J_d\sqrt{\kappa\rho_1}$

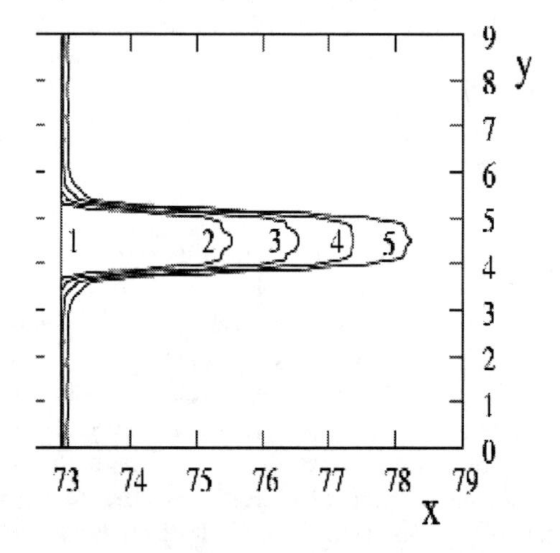

FIGURE 2. Five snapshots of an avalanche, finger-shaped instability in magnetic induction are shown at equal time interval $\Delta t=0.05$ (in dimensionless units of Eq.(1)) for small heat diffusion constant ($\kappa=0.05$)

In order to study the development of instability for arbitrary κ, the set of Eqs.(1) has been solved numerically. The results are presented in Fig.2, where the profile of the magnetic induction and avalanches obtained numerically support our theoretical predictions.

REFERENCES

1. R. G. Mints and A. I.Rahmanov, *Rev. Mod. Physics.* **53**, 551 (1981).
2. U. Bolz et.al., *Physica C* **388-389**, 715 (2003).
3. B.Rosenstein, B.Ya. Shapiro and I. Shapiro, *Europhysics Letters* **70**, 506-512 (2003); *Phys. Rev, B* **71**, 184508 (2005).

Thermal Fluctuations and Disorder Effect on The Square-rhombic Structural Phase Transition of The Vortex Lattice in Tetragonal Superconductors

Fareh PeiJen Lin[1,4], Anton Knigavko[2], DingPing Li[3] and Baruch Rosenstein[2,4]

[1] *Department of Electro-physics, National Chiao Tung University, Hsinchu 30043, Taiwan, R. O. C.*
[2] *Department of Physics and Astronomy, McMaster University, Hamilton, Ontario, Canada L8S 4M1* [2]
[3] *Department of Physics, Peking University, Beijing 100871, China*
[4] *Department of Condense MatterPhysics, Weizmann institute of Science, Rehovot 76100, Israel*

Abstract. Influence of the thermal fluctuations and weak disorder on the square--rhombic transition of the vortex lattice in tetragonal superconductors is investigated using the Ginzburg Landau approach which is valid near the $H_{c2}(T)$ line. In both cases the phase transition line in the T-H plane shifts in a way that makes the region occupied by the more symmetric square phase larger.

Keywords: Superconductivity, phase transitions, fluctuations
PACS: 74.25.Qt, 74.40.+k

INTRODUCTION

It has been known for a long time that in certain anisotropic low T_c[1] and high T_c[2] superconductors the vortex solid phase undergoes structural phase transformations. Theoretically, it was shown that a structural phase transition leads to a peak effect[3] due to softening of certain elastic modulus identified in the square-to-rhomb transition as the "squash" modulus[4]. The square-to-rhomb transition is the simplest possible structural phase transition with the fourfold symmetry Z_4 broken down to Z_2. In the less symmetric phase there are two rhombic lattices differing by a 90° rotation, while in the more symmetric phase, i.e. the square lattice, the vector between two closest vortices may be either parallel to the crystallographic axis **a** of the atomic lattice, or rotated by 45° with respect to it. Physically, the coupling between the crystal lattice and the vortex lattice in a fourfold symmetric superconductor such as LaSCCO or low T_c material originates in two somewhat related anisotropies on a microscopic scale[5]. The first is the Fermi velocity dependence, while the second is the anisotropy of the gap function. The anisotropy was described on the mesoscopic scale either within the Ginzburg - Landau approach[6,7] or the nonlocal London[4,8]

We analyze the GL approach including the thermal fluctuations on the mesoscopic scale. The distinct general feature of our result is that the slope of the transition line is negative. The main result of the symmetry breaking, namely the slope of the SPT line, is in fact independent of the details of the theoretical approach. It rather reflects a more general property of the system: its symmetry. We start therefore the discussion with symmetry and entropy considerations.

In low T_c superconductors the slope of the square – rhombic can be positive. This corresponds to a situation in which for a fixed magnetic field at low temperature the symmetric phase is stable, while upon heating two degenerate asymmetric ground states appear. Assuming that disorder can be neglected, this situation is unphysical. Quite generally, in statistical physics a symmetry breaking second order phase transition proceeds the other way around: from a degenerate asymmetric (symmetry H lower that of Hamiltonian G) vacuum to a symmetric one whose symmetry coincides with that of the Hamiltonian. For example, upon increasing temperature ferromagnet becomes paramagnet, superconductor - normal metal, solid-liquid etc. Although, to our knowledge, a rigorous proof does not exist, the reason for this is that upon heating excitations "across the energy barrier" separating the multiple symmetry broken ground states are generated and eventually the Gibbs state becomes symmetric and degeneracy disappears (the system regains ergodicity).

In vortex physics outside the domain of SPT this general rule holds. For example in very clean materials the melting line has a negative slope. Note that the

CP850, *Low Temperature Physics: 24th International Conference on Low Temperature Physics;*
edited by Y. Takano, S. P. Hershfield, S. O. Hill, P. J. Hirschfeld, and A. M. Goldman

phenomenon of inverse melting was observed. And shown to be caused solely by disorder for which the previous entropy argument does not apply.

In light of this, it is quite surprising to find out that many theoretical papers (which generally do not consider disorder) arrive at a conclusion that the square lattice upon heating becomes a less symmetric rhombic lattice. Therefore, in order to explain the positive slope of numerous SPT one should explore other ideas. As we show disorder can provide such an explanation.

MODELS AND RESULTS

The isotropic GL model is modified by a term describing fourforld symmetric anisotropy

$$F_{anisot}=\frac{\eta}{4}\Psi^*[(D_x^2-D_y^2)^2-(D_xD_y+D_yD_x)^2]\Psi, \quad (1)$$

where the anisotropy strength is η, the covariant derivative $D_i\equiv\nabla_i-i2\pi A_i/\Phi_0, i=x,y$, $A=(0,Hx,0)$. We use perturbation theory in η.

Thermal Fluctuations

Thermal fluctuations on the mesoscopic scale are described by the GL statistical sum $Z=\exp[-F_{GL}/T]$ and treated perturbatively in temperature. The fluctuation correction is:

$$F_{fl}=\frac{\pi tb}{4}\sqrt{Gi(1-t-b)}\int_{BZ}dk\left(\sqrt{\varepsilon_k^A}+\sqrt{\varepsilon_k^O}\right), \quad (2)$$

where $\varepsilon_k^{A(O)}$ is the phonon contribution of vortex spectrum[5] which depends on the two dimensional wave vector k.

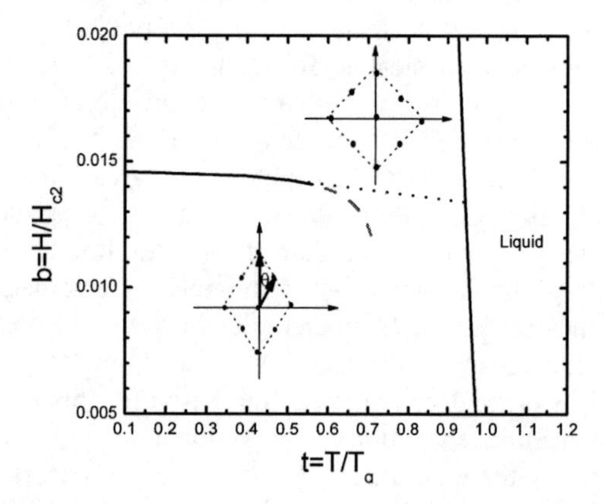

FIGURE 1. Transition line in the presence of thermal fluctuations (Gi =10^{-4}). The dashed line denoted where the perturbative method may fail. The dotted line shows the result of the non-perturbative method.

Quenched Pointlike Disorder

To investigate the weak disorder effect, the white disorder (random potential U(x)) is added in the GL equation. We adapt the method developed in ref.9 to include the anisotropy. By using the bifurcation method and averaging over the disorder in space, the correction to the free energy due to the disorder is given by

$$F_{dis}=-2^{-9/2}\pi^{-1}\gamma^2(1-t-b)^{5/2}b\int_{BZ}dk\left(\frac{B_k-|G_k|}{B_0\sqrt{\varepsilon_k^A}}+\frac{B_k+|G_k|}{B_0\sqrt{\varepsilon_k^O}}\right), \quad (3)$$

where γ is impurity strength, B_k and G_k are functions describing the anisotropic lattice[5].

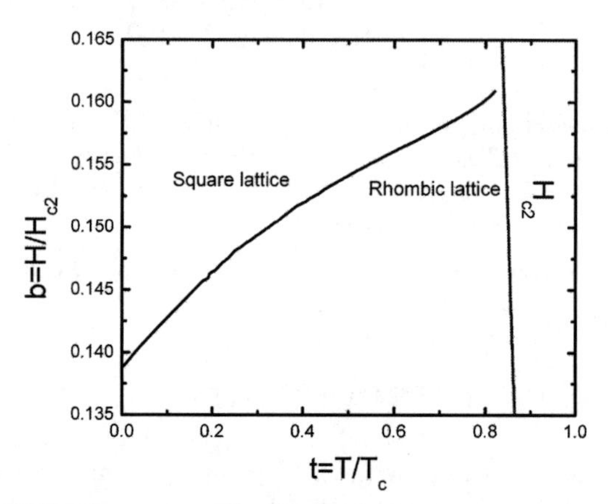

FIGURE 2. Transition line with disorder strength γ=.12.

REFERENCES

1. D. McK. Paul et al, *Phys. Rev. Lett.,* **80**, 1517 (1998); M.R. Eskildsen et al. *Phys. Rev. Lett,.* **86**, 5148 (2001); L.Ya. Vinnikov et al, *Phys. Rev. B* **64**, 220508 (2001).
2. B. Keimer et al, *Phys. Rev. Lett.,* **73**, 3459 (1994); S. T. Johnson et al, *Phys. Rev. Lett.* **82**, 2792 (1999); U. Divakar et al, *Phys. Rev. Lett.*, **92**, 237004 (2004); R. Gilardi et al, *Phys. Rev. Lett.*, **88**, 217003 (2002).
3. B. Rosenstein and A. Knigavko, *Phys. Rev. Lett.*, **83**, 844 (1999).
4. V.G. Kogan, *Phys. Rev. B* **54**, 12386 (1996); P. Miranovich and V.G. Kogan, *Phys. Rev. Lett.*, **87**, 137002 (2001).
5. N. Nakai et al, *Phys. Rev. Lett.*, **89**, 237004 (2002).
6. D. Chang et al, *Phys. Rev. Lett.*, **80**, 145 (1998); *Phys. Rev. B* **57**, 7955 (1998); K. Park and D.A. Huse, *Phys. Rev. B* **58**, 9427 (1998).
7. A.D. Klironomos and A.T. Dorsey, *Phys. Rev. Lett.*, **91**, 097002 (2003).
8. A. Gurevich and V.G. Kogan, *Phys. Rev. Lett.*, **87**, 177009 (2001).
9. D. Li and B. Rosenstein, *Phys. Rev. B* **65**, 024514 (2002).

Triplet Vortex Lattice Solutions of the Bogoliubov-de Gennes Equation in a Square Lattice

Yoshiki Hori, Akira Goto, and Masa-aki Ozaki [a]

Kochi National College of Technology, Nankoku 783-8508, Japan
[a] *Kumakouji 19-21 Uji City, Kyoto 611-0002, Japan*

Abstract. Various self-consistent triplet vortex lattice states are obtained for a two-dimensional extended Hubbard model with nearest-neighbor ferromagnetic exchange interaction in a uniform magnetic field. There are four types of triplet superconducting classes, axial, up-spin, planar, and bipolar state, with maximal magnetic translational symmetry for the magnetic flux $\phi = \phi_0/p^2$ in a square crystal lattice, where $\phi_0 = hc/2e$ is the flux quantum and p is an integer. We diagonalize the mean-field Hamiltonian numerically with self-consistency conditions for each symmetry class, and obtain various meta-stable vortex lattice states. The temperature dependence of the free energy of these meta-stable states is compared.

Keywords: triplet superconductivity, vortex lattice, symmetry
PACS: 71.10.Fd, 74.25.Qt

Recently much attention has been focused on spin-triplet superconductivity in quasi-two dimensional electron systems. In our previous paper [1], we gave a group theoretical classification of spin-triplet vortex lattice states of the two-dimensional extended Hubbard model with ferromagnetic exchange interaction with a perpendicular magnetic field B in a square lattice:

$$H = -\sum_{<i,j>,\alpha} t_{ij} a_{i\alpha}^{\dagger} a_{j\alpha} - \mu \sum_{i,\alpha} a_{i\alpha}^{\dagger} a_{i\alpha} + H_I, \qquad (1)$$

$$H_I = \sum_{<i,j>}\sum_{\lambda=1}^{3} J\Big(\sum_{\alpha,\beta} a_{i\alpha}^{\dagger}\sigma_{\alpha\beta}^{\lambda} a_{i\beta}\Big)\Big(\sum_{\gamma,\delta} a_{j\gamma}^{\dagger}\sigma_{\gamma\delta}^{\lambda} a_{j\delta}\Big) + V\sum_{<i,j>} n_i n_j, \qquad (2)$$

where

$$t_{ij} = 2t \exp\Big[-i\frac{e}{\hbar c}\int_i^j \mathbf{A}(\mathbf{r})\cdot d\mathbf{r}\Big], \qquad (3)$$

$\mathbf{A}(\mathbf{r}) = (-By/2, Bx/2, 0)$, $a_{i\alpha}^{\dagger}$ is the electron creation operator of spin α at site i, σ^{λ} ($\lambda = 1, 2, 3$)'s are the Pauli matrices, and σ^0 is a unit matrix. The symmetry group of this Hamiltonian is given by $G_0 = (e + tC_{2x})\,\mathbf{C}_4\mathbf{T}\mathbf{S}\mathbf{\Phi}$, where t is time reversal, $\mathbf{T}$ is the group of the magnetic translation consisting of the elements $T(Ma\mathbf{e}_x + Na\mathbf{e}_y)$ such that

$$T(Ma\mathbf{e}_x + Na\mathbf{e}_y)\cdot a_{(m,n)s}^{\dagger} = e^{i\pi\phi(Mn-Nm)/2\phi_0} a_{(m+M,n+N)s}^{\dagger}, \qquad (4)$$

(M and N are integers, $\phi = a^2B$, $\phi_0 = hc/2e$), $\mathbf{C}_4$ is four-fold rotation group around the origin (0, 0), $\mathbf{S}$ is the group of the spin rotation, and $\mathbf{\Phi}$ is the global gauge transformation group. We take the mean-field Hamiltonian as follows:

$$H_m = \sum_{\lambda=0}^{3}\sum_{i,\alpha} x_i^{\lambda} a_{i\alpha}^{\dagger} a_{i\alpha} + \sum_{\lambda=0}^{3}\sum_{\substack{<i,j>\\ \alpha,\beta}} x_{ij}^{\lambda} a_{i\alpha}^{\dagger}\sigma_{\alpha\beta}^{\lambda} a_{j\beta} + \sum_{\lambda=0}^{3}\sum_{\substack{<i,j>\\ \alpha,\beta}} y_{ij}^{\lambda} a_{i\alpha}^{\dagger}(i\sigma^{\lambda}\sigma^{y})_{\alpha\beta} a_{j\beta}^{\dagger}. \qquad (5)$$

Minimizing the thermodynamic potential within the mean-field approximation, we obtain the self-consistent field conditions:

CP850, *Low Temperature Physics: 24th International Conference on Low Temperature Physics;*
edited by Y. Takano, S. P. Hershfield, S. O. Hill, P. J. Hirschfeld, and A. M. Goldman

$$x_i^0 = -\frac{\mu}{2} + \frac{J}{2} \sum_{j\,\text{around}\,i} \sum_{\alpha} < a_{j\alpha}^\dagger a_{j\alpha} >, \qquad (6)$$

$$x_i^\lambda = \frac{J}{2} \sum_{j\,\text{around}\,i} \sum_{\alpha,\beta} < a_{j\alpha}^\dagger \sigma_{\alpha\beta}^\lambda a_{j\beta} >, \qquad (7)$$

$$x_{ij}^0 = -t_{ij} - \frac{1}{2}(V+3J) \sum_{\alpha} < a_{i\alpha}^\dagger a_{j\alpha} >^*, \qquad (8)$$

$$x_{ij}^\lambda = -\frac{1}{2}(V-J) \sum_{\alpha,\beta} < a_{i\alpha}^\dagger \sigma_{\alpha\beta}^\lambda a_{j\beta} >^*, \qquad (9)$$

$$y_{ij}^0 = \frac{1}{2}(V-3J) \sum_{\alpha,\beta} < a_{i\alpha}^\dagger i\sigma_{\alpha\beta}^y a_{j\beta}^\dagger >^*, \qquad (10)$$

$$y_{ij}^\lambda = \frac{1}{2}(V+J) \sum_{\alpha,\beta} < a_{i\alpha}^\dagger (i\sigma^\lambda \sigma^y)_{\alpha\beta} a_{j\beta}^\dagger >^*. \qquad (11)$$

Hereafter we restrict our consideration to the case $\phi = \phi_0/p^2$ (p is an integer) for an illustrative purpose. Then we can define an invariance magnetic translation group $\mathbf{L}$, which is a subgroup of $\mathbf{T\Phi}$ consisting of elements $L(Ma\mathbf{e}_x + Na\mathbf{e}_y)$ such that

$$\begin{aligned} L(Mpa\mathbf{e}_x + Npa\mathbf{e}_y) \cdot a_{(m,n)s}^\dagger &= e^{i\frac{\pi}{2}(MN+M+N)} \\ &\times T(Mpa\mathbf{e}_x + Npa\mathbf{e}_y) \cdot a_{(m+M,n+N)s}^\dagger. \end{aligned} \qquad (12)$$

Using similar method with our previous paper [2], we obtain four types of triplet vortex lattice states with maximal translational symmetry $\mathbf{L}$. These states are listed in Table 1.

Diagonalizing the mean-field Hamiltonian, we obtained all of these states as self-consistent solutions. Figure 1. shows the temperature dependence of free energy for each meta-stable solution. In this calculation, we set $t = 1$, $p = 3$, $n_e = 1.3$, $V = -3$, and $J = -1$. For simplicity, only the mean-field of triplet superconductivity is considered here. In this parameter region, the axial state $G^z_{(0)}$ is the most stable. The meta-stable states with higher free energy disappear as the temperature becomes higher. These states seem to have very shallow local minima of the free energy.

TABLE 1. Invariance groups of triplet vortex lattice

state	Invariance group
axial	$G_{(l)}^z = (1+tC_{2x})(e+u_{2x}\tilde{\pi})\mathbf{C}_4^{(l)}\mathbf{A}(\mathbf{e}_z)\mathbf{L}$
up-spin	$G_{(l)}^{\tilde{z}} = (1+tC_{2x}u_{2x})\mathbf{C}_4^{(l)}\tilde{\mathbf{A}}(\mathbf{e}_z)\mathbf{L}$
planar	$G_{II} = (1+tC_{2x}u_{2x})(e+\tilde{\pi}u_{2z})_{II}\mathbf{C}_4\mathbf{L}$
bipolar	$G_{sII}^{\pm} = (1+tC_{2x})(e+\tilde{\pi}u_{2z})\ _{II}\tilde{\mathbf{C}}_4^{\pm}\mathbf{L}$

$\mathbf{C}_4^{(l)} = \{(\tilde{\pi}/2)^{lj} C_{4z}^{lj}; j = -1,0,1,2\}$

$\mathbf{A}(\mathbf{e}_z) = \{u(\mathbf{e}_z,\theta); 0 \le \theta \le 2\pi\}$

$\tilde{\mathbf{A}}(\mathbf{e}_z) = \{u(\mathbf{e}_z,\theta)\tilde{\theta}; 0 \le \theta \le 2\pi\}$

${}_{II}\mathbf{C}_4 = \{C_{4z}^j u(\mathbf{e}_z,(\pi/2)^j); j = -1,0,1,2\}$

${}_{II}\tilde{\mathbf{C}}_4^{\pm} = \{(\tilde{\pi}/2)^{\pm j} C_{4z}^j u_{2a}^j; j = -1,0,1,2\}$

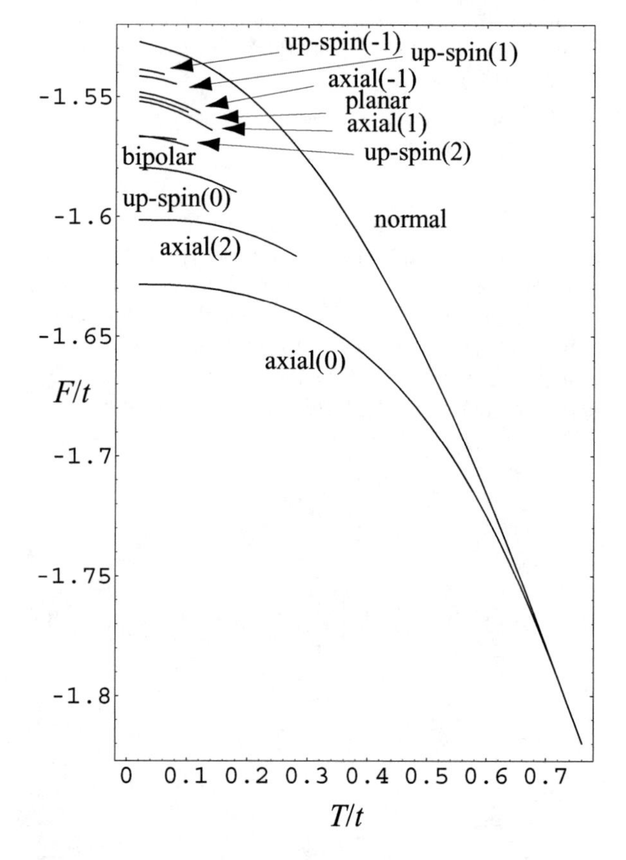

FIGURE 1. Temperature dependence of free energy for each self-consistent solution.

REFERENCES

1. A. Goto, Y. Hori, and M. Ozaki, *Physica* C **388-389**, 663-664 (2003).
2. M. Ozaki, M. Yamazaki, A. Goto, and Y. Hori, *Prog. Theor. Phys.* **100**, 253 (1998).

Double First Order Transitions between Vortex States in Columnar Defects

Shigeru Koikegami* and Ryusuke Ikeda†

*Advance Soft Corporation, Tokyo 160-0017, Japan
†Department of Physics, Kyoto University, Kyoto 606-8502, Japan

Abstract. Through experiments in cuprate superconductors, the vortex state just below the first order transition (FOT1), corresponding to the melting line in clean limit, often has no positional order, and the presence of a *lower* critical point (LCP) of this FOT line is found at least in YBCO. We have performed a Monte Carlo simulation of the layered Ginzburg-Landau model within the lowest Landau level (LLL), by including a *low* density of columnar (line) defects parallel to the applied magnetic field. We have found another first order transition (FOT2) curve lying at lower temperatures, presumably corresponding to the Bragg-Bose glass to the Bose glass melting, and the absence of FOT1, i.e., the presence of a LCP, in lower fields. These findings are consistent with the predictions in a previous theory.

Keywords: Vortex States, Glass Phase, Superconductor
PACS: 74.20.De, 74.25.Dw, 74.25.Qt, 74.72.-h

INTRODUCTION

Through extensive experimental and theoretical works on the vortex states of type II superconductors with point-like impurities, it has been understood that these systems in nonzero magnetic fi elds ($B \neq 0$) seem to have two glass (i.e., superconducting) phases: One is the elastic Bragg-glass (BrG), and the other is the amorphous-like vortex glass (VG). It was argued in [1] that, for a moderately disordered case, the phase just above the discontinuous melting transition of BrG should be the VG so that there may be *two* separate fi rst order transition (FOT) curves. In real systems with only *point* defects, this situation was realized in, e.g., Ni or Zn doped YBCO [2]. Further, at least in the lowest Landau level (LLL) analysis of the Ginzburg-Landau (GL) model, the presence of a *lower* critical point (LCP) of the *higher* FOT line was predicted [1] and has been verifi ed in the overdoped YBCO [3].

The separated two FOTs and a LCP are also expected in the case with *weak* line-like columnar defects parallel to the applied fi eld [1]. Actually, a LCP was clearly seen in YBCO [4], and the phase just below a higher FOT line (FOT1) has no positional (quasi-) long-range order in BSCCO [5]. Here, we report on results of our numerical simulation for the layered GL model in LLL with a low density of line defects parallel to the fi eld and perpendicular to the layers. From the internal energy data, the presence of *both* double FOTs and the LCP of the FOT1 line are found, in agreement with the theoretical prediction [1]. Our results are not the same as those for a different model [6], where two consecutive FOTs have been detected similarly, in that the LCP has been clearly found in the present work.

MODEL AND RESULTS

We performed a Monte Carlo simulation for the partition function $Z = \mathrm{Tr}_{\Psi}\exp(-\mathscr{H}/T)$, where $\mathscr{H}$ is the GL hamiltonian *defined within* LLL

$$\frac{\mathscr{H}}{T} = \sum_j \int d^2r_\perp \left[\alpha|\Psi_j|^2 + \gamma|\Psi_j - \Psi_{j+1}|^2 + \frac{\beta}{2}|\Psi_j|^4 + \delta\alpha(\mathbf{r}_\perp)|\Psi_j|^2\right]. \quad (1)$$

$\Psi_j = \Psi_j(\mathbf{r}_\perp)$ is the superconducting (SC) order parameter belonging to LLL and defi ned on the j-th SC layer, $\beta > 0$, $\gamma > 0$, and the length and fi eld were rescaled appropriately. Equation ((1)) includes the quenched structure disorder due to the columnar defects expressed by the random potential, $\delta\alpha(\mathbf{r}_\perp)$, where the relations $\overline{\delta\alpha(\mathbf{r}_\perp)} = 0$ and $\overline{\delta\alpha(\mathbf{r}_\perp)\delta\alpha(\mathbf{r}'_\perp)} = \xi_\alpha^2 n_\alpha(\mathbf{r}_\perp)\delta(\mathbf{r}_\perp - \mathbf{r}'_\perp)$ are satisfi ed, and the overbar denotes the random average and $n_\alpha(\mathbf{r}_\perp) = \sum_i \delta(\mathbf{r}_\perp - \mathbf{r}_\perp^{(i)})$. Further, $\left\{\mathbf{r}_\perp^{(i)}\right\}_{i=1,2,\ldots}$ denote the position of columnar defects, which distribute randomly with probability p.

We show in Fig. 1 the difference between the computed internal energy measured by Monte Carlo method for cooling process and the one for heating process, $\Delta E = \langle\mathscr{H}\rangle_{\mathrm{cool}} - \langle\mathscr{H}\rangle_{\mathrm{heat}}$, for the case with $p = 0.01$ and with eight SC-layers. We fi nd the presence of a single and weak fi rst order transition at low temperature (a) in the case $N_s = 56$, i.e. in a lower fi eld. On the other hand, when $N_s = 64$ and $N_s = 72$, i.e. in higher fi elds,

CP850, *Low Temperature Physics: 24th International Conference on Low Temperature Physics;*
edited by Y. Takano, S. P. Hershfield, S. O. Hill, P. J. Hirschfeld, and A. M. Goldman

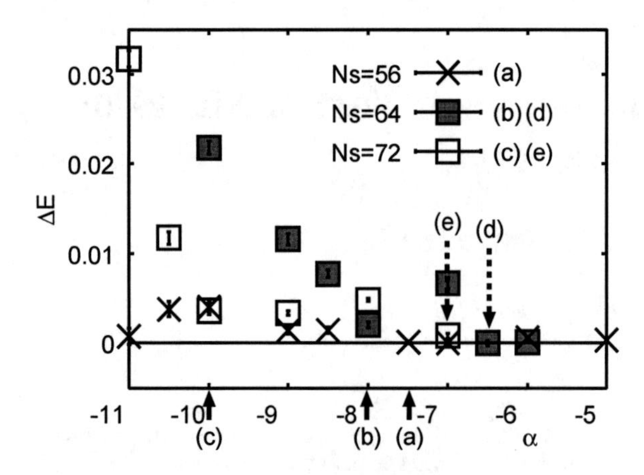

FIGURE 1. ΔE v.s. α (i.e., temperature) data. N_s denotes the number of vortices. In all data for $N_s = 56$ (cross), $N_s = 64$ (open square), and $N_s = 72$ (closed square), the values $\beta = 2\pi$ and $\gamma = 0.05$ are used.

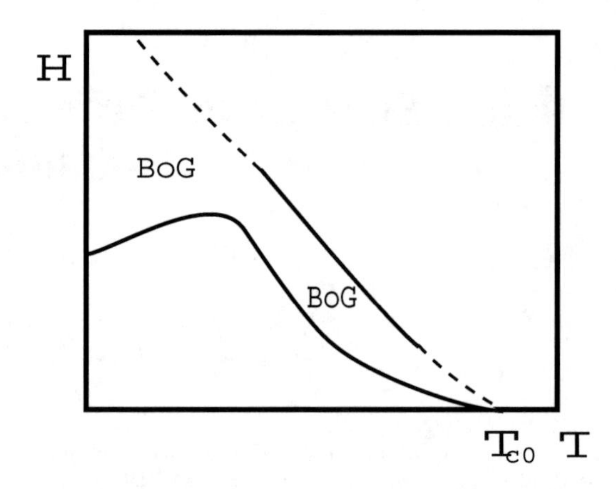

FIGURE 2. Expected H-T phase diagram in the case with *weak* columnar defects parallel to the applied field. The FOT lines are indicated by solid curves, while the dashed curves are second order transition lines into the so-called Bose-glass (BoG).

not only the FOT at (b) and (c), respectively, but also another FOT is found to occur at (d) and (e), respectively, i.e., at higher temperatures. It is natural to expect that the FOTs at (d) and (e) correspond to the FOT1 mentioned in Introduction, while another FOT line, FOT2, consisting of the FOTs at (a), (b), and (c), is present as a separated line in the B-T phase diagram. The absence of the FOT1 in $N_s = 56$ implies the presence of a LCP [1] of the FOT1 line in $56 < N_s < 64$ in the phase diagram, while the FOT2 at lower T seems to correspond to the melting transitions of the Bragg-Bose glass phase [7] in the type II limit, because the glass phases due to line defects should have the transverse Meissner effect [8, 9]. Further, we have preliminarily verifi ed that the FOT1 line lies at a slightly higher temperature than the corresponding one in clean limit with no defects.

The expected phase diagram is sketched in Fig. 2. The topologically same phase diagram was argued in Ref.1 (Fig.4 (a) there).

ACKNOWLEDGMENTS

The authors are grateful to H. Adachi and K. Myojin for thier contributions to the development of the computer program for the computation in this work. The computation in this work has been done using the facilities of the Supercomputer Center, Institute for Solid State Physics, University of Tokyo.

REFERENCES

1. R. Ikeda, *J. Phys. Soc. Jpn.*, **70**, 219 (2001).
2. T. Nishizaki, unpublished (2005).
3. T. Nishizaki, K. Shibata, T. Sasaki, and N. Kobayashi, *Physica C*, **341-348**, 957 (2000).
4. W. K. Kwok, R. J. Olsson, G. Karapetrov, L. M. Paulius, W. G. Moulton, D. J. Hofman, and G. W. Crabtree, *Phys. Rev. Lett.*, **84**, 3706 (2000).
5. M. Menghini, Y. Fasano, F. de la Cruz, S. S. Banerjee, Y. Myasoedov, E. Zeldov, C. J. van der Beek, M. Konczykowski, and T. Tamegai, *Phys. Rev. Lett.*, **90**, 147001 (2003).
6. C. Dasgupta, and O. T. Valls, *Phys. Rev. B*, **72**, 094501 (2005).
7. T. Giamarchi, and P. L. Doussal, *Phys. Rev. B*, **55**, 6577 (1997).
8. R. Ikeda, *J. Phys. Soc. Jpn.*, **69**, 559 (2000).
9. T. Giamarchi, and P. L. Doussal, *Phys. Rev. B*, **53**, 15206 (1996).

Mode Locking of Vortex Matter in $NbSe_2$ Pure Single Crystals

N. Kokubo, K. Kadowaki, and K. Takita

Institute of Materials Science, University of Tsukuba, 1-1-1, Tennoudai, Tsukuba, Ibaraki 305-8573, Japan

Abstract. We present the mode locking (ML) phenomenon of vortex matter observed in a layered superconductor of 2H-$NbSe_2$ pure single crystal. On application of ac current on top of dc current, clear ML resonance appears in differential conductance voltage curves, indicating the presence of elastic modes excited collectively over driven vortices. On increasing magnetic field, the ML voltage shows square root dependence on field, consistent with the ML condition for vortex matter.

Keywords: mode locking, dynamic ordering,
PACS: 74.25Sv, 74.25 Qt

One of the powerful techniques to probe elasticity in periodic media driven over pinning environments is the mode locking (ML) experiment [1-5]. ML is a dynamic resonance between an ac drive on top of a dc drive and lattice (elastic) modes excited collectively over the driven media at an internal frequency given by $f_{int}=qv/a$ with the average velocity v, the lattice spacing a along the flow direction and an integer q. The elastic modes are dynamically locked on the ac drive when the internal frequency and the ac drive frequency are harmonically related, i.e.

$$f_{int} = pf \quad \text{or} \quad v = \frac{p}{q} fa \tag{1}$$

with another integer p, This resonance results in various anomalies like multiple steps in force-velocity characteristics (current–voltage characteristics for vortices). Thus, the observation of such resonance evidences the presence of the elasticity in driven media.

While the ML phenomenon has been extensively studied on charge density waves [1], studies on vortices are quite rare, despite the first observation by Fiory was made three decades ago [2]. In this study we present the ML phenomenon of vortex matter observed in 2H-$NbSe_2$ pure single crystals.

We used platelets of 2H-$NbSe_2$ pure single crystal grown by an iodine vapor transport method. The crystals were cut into the strip shape and cleaved with no significant optical surface damage. Contacts for four-probe method were made by indium solder.

The measurements discussed in this paper were mainly performed on a crystal with a dimension of 0.72mm(l)×0.63mm(w)×0.9μm(t). It shows a sharp resistive transition around 7.2 K. The transition width is typically about 50 mK between 10 % and 90 % of normal state resistance. Residual resistance ratio is 33.

For ML experiment, we measured dc voltage by ramping dc-current up and down with superimposing a constant ac current. The transmission lines for ac current were terminated by a matching circuit near the crystal. To avoid any heating, both the matching circuit and the crystal were immersed in liquid ^{4}He at 4.2 K. Magnetic field was applied parallel to the c axis.

The ML resonance may be better displayed by plotting differential conductance dI/dV vs. voltage V curves, instead of current voltage curves. In Fig, 1 (a), a series of dI/dV curves measured in 1.2 T with superimposing 3 MHz ac current of various amplitudes I_{ac} are shown. Clear conductance ML peaks appear at equidistant voltages denoted by p/q=1/1 and 2/1. First peak corresponds to the fundamental (p/q=1/1), and the other is a higher harmonic (p/q=2/1). Small sub-harmonics are also detected. As expected from the ML velocity condition v=$(p/q)fa$, those resonance voltages depend linearly on frequency as

CP850, *Low Temperature Physics: 24th International Conference on Low Temperature Physics;*
edited by Y. Takano, S. P. Hershfield, S. O. Hill, P. J. Hirschfeld, and A. M. Goldman

$$V_{p/q} = \frac{p}{q} f\Phi_0 \frac{l}{b} \quad (2)$$

where Φ_0 is the flux quantum and b is another lattice parameter related to vortex density B, given by $Bab=\Phi_0$. The results for the fundamental ML are shown in Fig. 1(b).

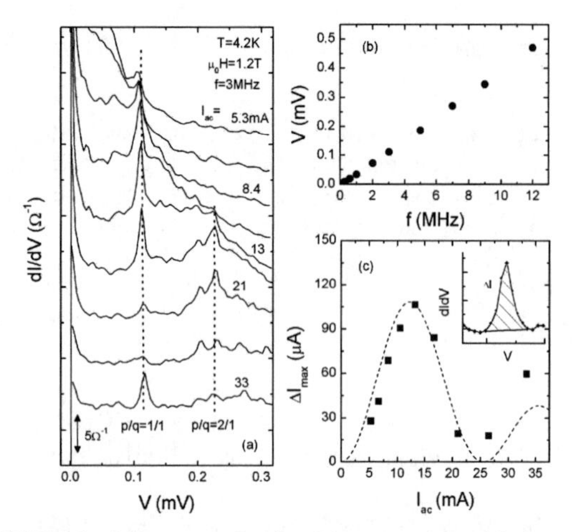

FIGURE 1. ML resonance for driven vortices. (a) A series of differential conductance-voltage curves measured by superimposing 3MHz ac current with various amplitudes. The ML conditions are denoted as *p/q*. In (b) frequency dependence of the fundamental resonance voltage is given. (c) Ac current dependence of the resonance current width determined by integrating the ML peak with respect to the flux flow base line (see the inset).

Next, we show how the ML resonance depends on ac current I_{ac}. Figure 1 (c) shows a plot of the fundamental ML current width vs. I_{ac}. Here, the current width is obtained by integrating the conductance ML peak with respect to the flux flow base line (see inset to Fig. 1 (c)). After showing a large peak, it oscillates with I_{ac}, which agrees qualitatively with a squared Bessel function of the first kind displayed by a broken curve. This behavior is expected when the random pinning due to the disorder quenched in a host material excites elastic modes in driven elastic lattices [3,4].

Finally, we turn to the field dependence of ML phenomenon. In Fig. 2, we show how the fundamental ML voltage evolves with magnetic field. As observed, the fundamental ML voltage measured with superimposing a 7MHz ac current shows an upward increase with magnetic field. This behavior is in good quantitative agreement with the ML voltage condition of Eq. (2) displayed as a solid curve. Here we use the equilibrium lattice parameter of $b=b_0$ where

$$b_0 = \frac{\sqrt{3}}{2} a_0 = \sqrt{\frac{\sqrt{3}\Phi_0}{2B}} \quad (3)$$

with $B=\mu_0 H$. Note that no fitting parameter is used for this comparison. On approaching the second critical field H_{c2} of 2.14T, the ML resonance disappears suddenly at a certain magnetic field (below H_{c2}). Above this field, no ML resonance appears at any amplitude of ac current, indicating the absence of the elasticity in driven vortices. The disappearance of ML marks the dynamic melting transition from a moving solid state into a liquid like incoherent flow state [3]. The details will be discussed elsewhere [5].

N. K used facilities in cryogenic center in university of Tsukuba. This work was partly supported by the grant in Aid for Scientific research (Grant No. 16710063) from MEXT (the Ministry of Education, Culture, Sports, Science and Technology), and also partly by the 21st Century COE (Center of Excellence) Program, *"Promotion of Creative Interdisciplinary Materials Science for Novel Functions"* under MEXT Japan.

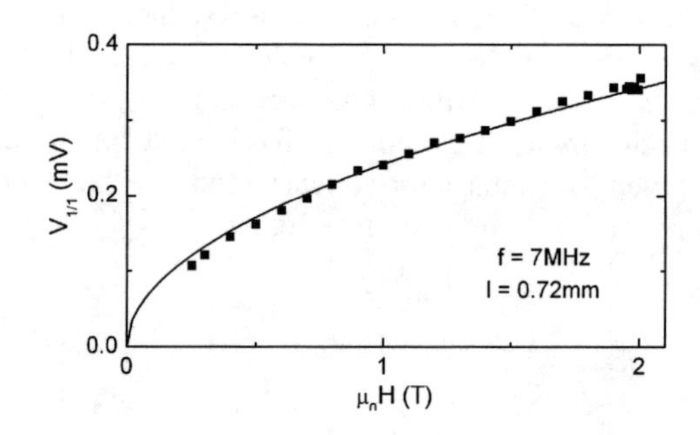

FIGURE 2. Magnetic field H dependence of the fundamental ML voltage $V_{1/1}$. A solid curve represents the fundamental ML voltage condition.

REFERENCES

1. G. Grüner, *Rev. Mod. Phys.* **60**, 1129 (1988).
2. A. T. Fiory, *Phys. Rev. Lett.* **27**, 501 (1971).
3. N. Kokubo, *et al.*, *Phys. Rev. Lett.* **88**, 247004 (2002). R. Besseling, N. Kokubo, and P. H. Kes, *Phys. Rev. Lett.* **91**, 177002 (2003). N. Kokubo, R. Besseling, and P. H. Kes, *Phys. Rev. B* **69**, 064504 (2004).
4. A.Schmid and W. Haüger, *J. Low. Temp. Phys.* **11**, 667 (1973).
5. N. Kokubo, K. Kadowaki, and K. Takita, *Phys. Rev. Lett.* **95**, 177005 (2005).

Plastic Flow and Dynamic Correlation in the Driven Vortex Glass in the Corbino Disk

S. Okuma, S. Morishima, and Y. Watanabe

Research Center for Low Temperature Physics, Tokyo Institute of Technology, 2-12-1, Ohokayama, Meguro-ku, Tokyo 152-8551, Japan

Abstract. We report the dynamic properties of the vortex glass (VG) driven by a radial current in a thick amorphous Mo_xSi_{1-x} film with the Corbino-disk (CD) geometry. In CD vortices rotate around the center of the sample by feeling a frustrating Lorentz force (f_L) inversely proportional to the radius (r) of rotation. We examine the spatial (r) dependence of the vortex motion for different f_L (current) by measuring the voltage and voltage noise generated at the two voltage probes placed radially. An initial dynamic state of the driven VG is the plastic flow. As f_L (current) is increased in the plastic-flow state, the velocities of vortices rotating at the two probes show dynamic correlation in a narrow current regime, and eventually the plastic flow gives way to the liquid-like flow.

Keywords: Vortex dynamics, Plastic flow, Frustration, Corbino disk
PACS: 74.40.+k, 74.25.Dw, 74.78.Db

INTRODUCTION

The dynamic properties of vortices in type-II superconductors have been actively studied in recent years. Of particular interest are studies using the Corbino-disk (CD) geometry, in which an applied radial current I yields a nonuniform Lorentz force and the vortices rotate around the center of the sample [1-3]. Using CD, we have recently studied the dynamics of the vortex glass (VG) driven by the frustrating Lorentz force f_L, which is inversely proportional to the radius r of rotation ($f_L \propto 1/r$) [4]. Our study is motivated by earlier work by López *et al.*, who have studied the vortex velocity profile on a single-crystal $YBa_2Cu_3O_{7-\delta}$ CD [1]. They have shown that in the vortex-solid regime, all vortices move elastically as a rigid disk at low driving currents, while above a certain characteristic current, vortex-vortex shear stresses exceed the elastic limit and the dynamics is characterized by plastic motion. In our amorphous (a-)Mo_xSi_{1-x} films containing relatively strong pinning, however, we have observed the different vortex dynamics.

In this work, to clarify the change in the vortex dynamics of driven VG associated with an increase in f_L (I), we examine the spatial (r) dependence of the vortex motion by measuring the I-V characteristics and voltage-noise spectra $S_V(f)$ (f is a frequency) at the two voltage probes placed radially. We will discuss whether the vortex dynamics (voltage fluctuations) at the two sites are correlated spatially.

EXPERIMENTAL

We prepared a thick (100 nm) a-Mo_xSi_{1-x} film with x=58 at. % by coevaporation of pure Mo and Si [5,6]. The zero-resistivity temperature at zero field (B = 0) is 3.30 K and the VG-transition field is 2 T at 2.16 K. The arrangement of the electrical contacts is shown in the inset of Fig. 1. For the CD geometry the current flows between the contact, +C, of the center and that, -C, of the perimeter of the disk [4]. When measuring in the strip-like geometry, the contacts +S and -S were used. The inner radius of CD is 2.3 mm. The distances from the center of the sample to (the center of) the contacts P_1, P_2, and P_3 (indicated with 1, 2, and 3 in the inset) are 0.9, 1.4, and 1.9 mm, respectively, which serve as two voltage probes (V_{in} and V_{out}) with different r (r_{12} and r_{23}) [1]. The voltage-noise spectrum $S_V(f)$ and the cross-spectrum enhanced with the preamplifier were recorded using a fast-Fourier transform spectrum analyzer [7]. The film was directly immersed in liquid ^{4}He to ensure good thermal contact. The field B was applied perpendicular to the plane of the sample.

CP850, *Low Temperature Physics: 24th International Conference on Low Temperature Physics;*
edited by Y. Takano, S. P. Hershfield, S. O. Hill, P. J. Hirschfeld, and A. M. Goldman

RESULTS AND DISCUSSION

Let us first summarize the results in our earlier work [4] on a thick (100 nm) a-Mo_xSi_{1-x} film, which is similar to the one used in this study. In the vortex-liquid phase the I-V characteristics at low I are almost linear both for CD and strip-like geometries and there is no difference in the shape of I-V curves. By contrast, in the VG phase a significant difference appears between the two geometries. The I-V curves for the strip-like geometry exhibit the usual smooth downward curvature, whereas for CD the anomaly in the I-V curves is visible at a certain current I^*, that is dependent on B. There is the current region just below I^* where the slope of the I-V curve is slightly smaller than that in the strip-like geometry, while just above I^* the slope shows a rapid increase. We define I^* as the current at which $d\log V / d\log I$ takes a local minimum. Based on the simulation work [8] and on our additional experiments, we have proposed that a change in the vortex dynamics from plastic flow to liquid-like(laminar) flow occurs at around I^*.

We now turn to the results of present work. Figure 1 depicts the I-V characteristics measured at r_{12} and r_{23} in the VG phase (at 2.16 K in 1.0 T). Three points are noted. (i) With increasing I, the detectable voltage ($V = 10^{-8}$ V) appears first at r_{12} (in the inner portion of the sample). (ii) In the voltage state ($V > 10^{-8}$ V), V measured at r_{12} is always larger than that at r_{23}, which is more remarkable at smaller I. (iii) I^* described above is again visible in this film at either position. These results are interpreted as follows. As I is increased in the VG phase, the vortices rotating in the inner portion (at r_{12}) are first depinned. This result clearly indicates that an initial dynamic state of driven VG is not the elastic flow (rigid-disk-like rotation), as reported in the vortex lattice system for a clean superconductor [1], but the plastic flow. The larger $V(I)$ at V_{in} (r_{12}) than that at $V_{out}(r_{23})$ further supports the view. In the plastic-flow regime (i.e., at I smaller than I^*), we observe the $1/f$-type noise spectra $S_V(f)$ at either probe position in the f range (1 Hz-1 kHz) measured. These spectra are similar to those which have been observed in the *strip* samples [7].

We attempt to examine the coherence function $h(f)$ extracted from the cross-spectra between r_{12} and r_{23}, where $h(f)$ is defined as the amplitude of the normalized cross-spectral density [9]. When the voltage fluctuations generated at the two voltage probes (at r_{12} and r_{23}) are completely correlated, h is unity. On the other hand, when they are uncorrelated, h is zero. As depicted in Fig. 1, $h(f)$ at low f (<10 Hz) takes substantially large values (0.5-0.9) in a narrow current range (I = 25-30 mA) below I^*. This result

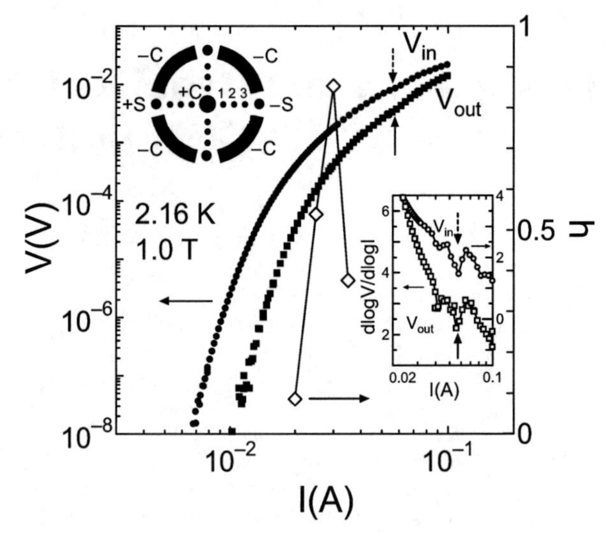

FIGURE 1. I-V characteristics of the thick a-Mo_xSi_{1-x} film with the CD geometry measured at the inner probe $V_{in}(r_{12})$ (filled circles) and the outer probe $V_{out}(r_{23})$ (filled squares) in the VG phase (at 2.16 K in 1.0 T). The values of coherence $h(f)$ at f = 3 Hz extracted from the cross-spectra between r_{12} and r_{23} are also plotted with open diamonds. Inset: (Left) Arrangement of the electrical contacts. (Right) $d\log V / d\log I$ at $V_{in}(r_{12})$ (open circles) and $V_{out}(r_{23})$ (open squares) plotted against I. Arrows indicate the location of I^*.

suggests that in this I range the velocities of vortices rotating on circles with $r = r_{12}$ and r_{23} are correlated with each other. Although an exact picture is not available at present, taking account of the fact that no elastic rotation is realized in our vortex system, it is considered that in this particular I region some domains of vortex solid may flow in the plastic-flow state.

This work was supported by a Grant-in-Aid for Scientific Research from the Ministry of Education, Culture, Sports, Science, and Technology.

REFERENCES

1. D. López *et al., Phys. Rev. Lett.* **82**, 1277 (1999).
2. Y. Paltiel *et al., Phys. Rev. Lett.* **85**, 3712 (2000).
3. G. W. Crabtree, *Nature Mat.* **2**, 435 (2003).
4. M. Kamada and S. Okuma, *J. Phys. Soc. Jpn.* **73**, 526 (2004).
5. S. Okuma, T. Terashima and N. Kokubo, *Solid State Commun.* **106**, 529 (1998):*Phys. Rev. B* **58**, 2816 (1998).
6. S. Okuma, Y. Imamoto and M. Morita, *Phys. Rev. Lett.* **86**, 3136 (2001).
7. S. Okuma and M. Kamada, *Phys. Rev. B* **70**, 014509 (2004).
8. M. –C. Miguel and S. Zapperi, *Nature Mat.* **2**, 477 (2003).
9. A. Maeda *et al., Phys. Rev. B* **65**, 054506 (2002).

Voltage Oscillation due to Vortex-Antivortex Fluctuations in the Corbino-Disk Superconductor

S. Okuma and S. Morishima

Research Center for Low Temperature Physics, Tokyo Institute of Technology, 2-12-1, Ohokayama, Meguro-ku, Tokyo 152-8551, Japan

Abstract. We report on the dynamic properties of vortices and antivortices created and driven by the dc (radial) current in the Meissner phase of thick amorphous Mo_xSi_{1-x} films with the Corbino-disk geometry. We have observed unusual large voltage pulses $V(t)$ that oscillate almost periodically, which originate from large vortex-density fluctuations. We find from the spatial dependence of $V(t)$ that the voltage pulses are generated almost simultaneously within the sample. Seeing more closely, the rise in $V(t)$ detected at the outer radius is slightly delayed compared to that at the inner radius. This means that the increase in number of depaired vortices at the inner portion triggers off unbinding of vortex-antivortex pairs at the outer portion.

Keywords: Vortex-antivortex pairs, Fluctuations, Meissner phase, Corbino-disk
PACS: 74.40.+k, 74.25.Fy, 74.25.Qt

It is important to study fundamentally as well as practically how the dissipation occurs near the superconducting transition temperature T_c. In zero field (B=0) below T_c, the linear resistivity vanishes, instead, the nonlinear resistivity appears. In two dimensions (2D) the nonlinear dissipation is described by the well-known vortex unbinding theory [1], while in 3D it is explained by the vortex-loop model [2], where each grown vortex loop is dissociated into a vortex and an antivortex similarly to the 2D case. Physics related to unbinding of vortex-antivortex pairs has been studied using a variety of superconductors as well as liquid ^{4}He films. However, no experimental work probing in a direct manner the number fluctuations of vortices have been performed yet. Numerous experiments using the usual strip-shaped samples have failed to observe such a phenomenon.

Recently, we have studied a Corbino-disk (CD) superconductor [3,4], where dissociated free vortices with opposite vorticity are driven to the opposite directions by the radial current I and continue to rotate in the sample. Since one vortex can annihilate only by colliding with another vortex with opposite vorticity, we can expect enhanced vortex-number fluctuations δn in CD than in the conventional strip-shaped samples. Moreover, a nonuniform Lorentz force ($f_L \propto J \propto 1/r$, where r and J are a radius of rotation and a radial current density, respectively) in CD may also assist to enhance δn. In fact, in the Meissner phase of a thick amorphous (a-)Mo_xSi_{1-x} film with the CD geometry, we have recently observed unusual large voltage pulses that oscillate almost periodically under the dc I [3,4]. This suggests large δn of vortices and antivortices, whereas there is no available theory to account for the observed oscillation comprehensively. In this work, to explore the origin responsible for the unusual $V(t)$ oscillation, we have measured the spatial (r) dependence of $V(t)$. The detailed data concerning present work have been published elsewhere [3-5].

A thick (100 nm) a-Mo_xSi_{1-x} film with x=58 at. % was prepared by coevaporation of pure Mo and Si. The zero-resistivity temperature T_c at B = 0 is 3.30 K. The arrangement of the silver electrical contacts is shown in the inset of Fig. 1(c). The current flows between the contact, +C, of the center and that, -C, of the perimeter of the disk, which produces J that decays as $1/r$ [6]. The voltage V was measured using the contacts P_1-P_3, which were evaporated at 0.5 mm intervals along a radius r. These contacts serve as the two voltage probes (V_{in} and V_{out}) with different r (r_{12} and r_{23}) [7]. The inner diameter of CD was about 5.5 mm. The time(t)-dependent voltage $V(t)$ enhanced with the preamplifier was recorded using a fast-Fourier transform spectrum analyzer [8]. The film was directly immersed in liquid ^{4}He to ensure good thermal contact.

CP850, *Low Temperature Physics: 24th International Conference on Low Temperature Physics;*
edited by Y. Takano, S. P. Hershfield, S. O. Hill, P. J. Hirschfeld, and A. M. Goldman

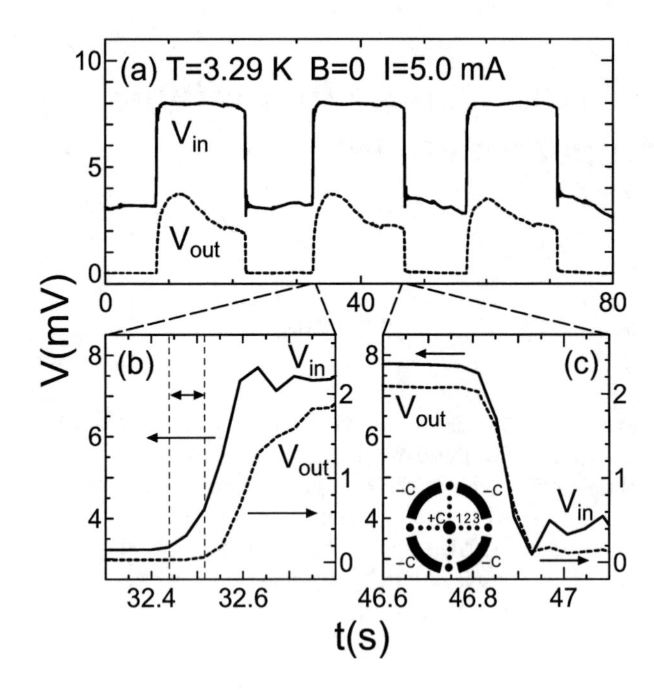

FIGURE 1. (a) $V(t)$ at $V_{in}(r_{12})$ (full line) and at $V_{out}(r_{23})$ (dotted line) at I = 5.0 mA, B = 0, and T = 3.29 K. (b) The rise and (c) fall of the voltage pulses [at $V_{in}(r_{12})$ (full line) and at $V_{out}(r_{23})$ (dotted line)] are enlarged and shown. The vertical dashed lines in (b) represent the delay time. The inset in (c) illustrates the arrangement of the electrical contacts.

Figure 1(a) shows the voltage oscillation $V(t)$ observed at the inner $V_{in}(r_{12})$ (full line) and outer $V_{out}(r_{23})$ (dotted line) voltage probes at fixed I = 5.0 mA in B = 0 at T =3.29 K. We can see that the voltage pulses at r_{12} and r_{23} are synchronized with each other. This implies that the large vortex-number fluctuations δn occur almost simultaneously over the macroscopic distance in the sample. Seeing more closely, we find that the rise in the individual voltage pulses at r_{23} is slightly delayed compared to that at r_{12}, as shown in Fig. 1(b). The delay time is typically 0.1 s, which is smaller than the period or width of the voltage pulse (1-100 s) [3], but much larger than the period (0.1 ms) for individual vortices to rotate (half) on a circle with the radius r_{12} or r_{23}.

We propose based on these results that the increase in number of depaired vortices at the inner portion may trigger off unbinding of vortex-antivortex pairs at the outer portion [5]. Here we assume the finite (nonzero) radial component of the velocity. This assumption is not unreasonable, if we take account of (i) the possible inhomogeneity in the distribution of the radial current density and (ii) the interaction between vortices rotating on circles with different r [9]. In the low-voltage state of the $V(t)$ oscillation a small number of free vortices and antivortices are present. Each free vortex has a chance to cross between a vortex and an antivortex of a bound pair and dissociates it into free vortices, which in turn can dissociate other bound pairs as well. Thus a successive dissociation process (i.e., proliferation of free vortices) takes place [10]. This process continues, until $V(t)$ reaches another "stationary"[9] high-voltage state. Considering that both the number and velocity of free vortices and antivortices rotating in the inner portion are larger than those in the outer portion, the depaired vortices at the inner portion predominantly induce unbinding of vortex-antivortex pairs at the outer portion. Since this picture does not apply to the annihilation process of vortices and antivortices, the fall of the voltage pulses measured at r_{12} and r_{23} is considered to occur simultaneously. This is indeed seen in Fig, 1(c). The result obtained here in Fig. 1(b) also suggests that the nonuniform driving force as well as the confined geometry in CD may play a crucial role in generating the unusual voltage oscillation.

The authors thank M. Hayashi, H. Ebisawa, and Y. Ootuka for useful discussions. This work was supported by a Grant-in-Aid for Scientific Research from the Ministry of Education, Culture, Sports, Science, and Technology.

REFERENCES

1. J. M. Kosterlitz and D. J. Thouless, *J. Phys. C* **6**, 1181 (1973).
2. D. S. Fisher, M. P. A. Fisher and D. A. Huse, *Phys. Rev. B* **43**, 130 (1991).
3. M. Kamada, Y. Watanabe and S. Okuma, in *Proceedings of the 16th International Symposium on Superconductivity, Tsukuba, 2003,* edited by M. Tachiki and M. Nisenoff, *Physica C* **412-414**, 535 (2004).
4. S. Okuma, S. Morishima and M. Kamada, in *Proceedings of the 17th International Symposium on Superconductivity, Niigata, 2004,* edited by K. Kishio, *Physica C*, in press, available on line 23 June 2005.
5. S. Okuma, S. Morishima and M. Kamada, *Nature Mat.*, submitted for publication.
6. Y. Paltiel, E. Zeldov, Y. Myasoedov, M. L. Rappaport, G. Jung, S. Bhattacharya, M. J. Higgins, Z. L. Xiao, E. Y. Andrei, P. L. Gammel and D. J. Bishop, *Phys. Rev. Lett.* **85**, 3712 (2000).
7. D. López, W. K. Kwok, H. Safar, R. J. Olsson, A. M. Petrean, L. Paulius and G. W. Crabtree, *Phys. Rev. Lett.* **82**, 1277 (1999).
8. S. Okuma, M. Kobayashi and M. Kamada, *Phys. Rev. Lett.* **94**, 047003 (2005).
9. M. Hayashi and H. Ebisawa, *J. Phys. Chem. Solid*, to be published.
10. S. Okuma and M. Kamada, *J. Phys. Soc. Jpn.* **73**, 2807 (2004).

Induced In-plane Order in the Vortex Liquid by Periodic Pinning Arrays

S. Ooi, T. Mochiku, S. Yu, E. S. Sadki, and K. Hirata

National Institute for Materials Science, 1-2-1 Sengen, Tsukuba, Ibaraki, Japan

Abstract. The vortex phase diagram of high-T_c superconductors in the field parallel to the c-axis have been already extensively explored. It has been commonly observed that pinning effects are quite weak at high temperatures and high fields in high quality single crystals. Such weak pinning phases, like the vortex liquid and the Bragg glass, are good candidate to study the influence of additional artificial pinning arrays on vortex phases, because the original quenched disorders do not work well as effective pinning centers in these phases. We have introduced several types of hole arrays in $Bi_2Sr_2CaCu_2O_{8+y}$ single-crystal thin film using a focused ion beam. To preserve the quality of the crystals we have made film samples from bulk crystals by the cleaving. In samples with triangular and square pinning arrays of 1μm lattice spacing, the flow resistance of pancake vortices exhibit a matching effect at multiples of the matching fields. This matching effect occurs in the vortex liquid phase. It is difficult to explain these results by the idea of the normal vortex liquid because of the lack of in-plane symmetries. There is a possibility that the in-plane order is induced in the vortex liquid by the introduction of the periodic pinning arrays.

Keywords: Matching effect, Artificial periodic defects, FIB, $Bi_2Sr_2CaCu_2O_{8+y}$
PACS: 74.25.Qt, 74.72.Hs, 74.25.Fy, 74.25.Ha

INTRODUCTION

Recent progress of micro-fabrication techniques has made it possible to understand the vortex matter in mesoscopic superconductors [1, 2]. These techniques are also employed to introduce artificial pinning sites. Studies of the commensurability effect between nano- or micro-scale artificial periodic pinning centers and vortices have importance in both engineering and academic interest. Although in conventional superconductors many studies have been performed using lithography techniques [3], they are hardly found on high-T_c superconductors [4, 5]. A high-T_c superconductor, $Bi_2Sr_2CaCu_2O_{8+y}$ (Bi2212) is a good candidate to study the effects of artificial pinning centers because effects of the quenched disorders, which exist even in high-quality crystals, are very weak in this material at high temperatures. Actually, interactions of vortices and randomly arranged defects, such as columnar defects, have been studied in the past decade. Researchers have succeeded to obtain insights about some new glass phases like Bragg glass or Bose glass. However, these kinds of defects are always randomly located. To our knowledge, there has been no research into Bi2212 single-crystal films with artificially controlled defects.

To study the effect of the regular hole array on vortices in high-T_c superconductors, we have previously reported some results for the local magnetization and the vortex-flow resistance in Bi2212 single-crystal films with triangular arrays of sub-micron holes [6]. As a comparison, we show the vortex-flow resistance in samples with holes arrays of a square lattice in this paper. The matching effect is also observed in this case and dose not depend on the symmetries of defect lattice.

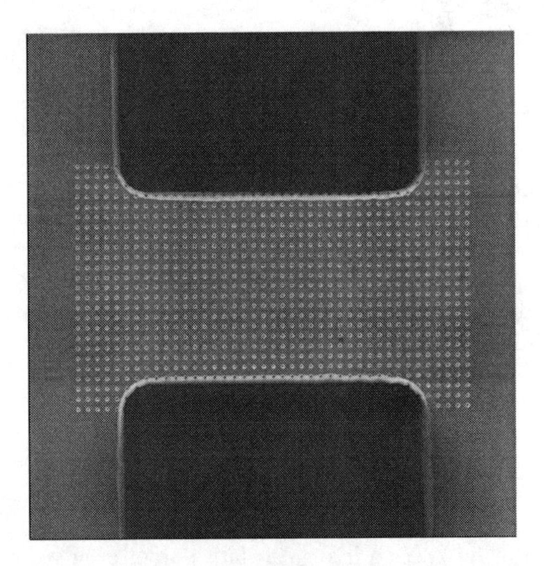

FIGURE 1. Picture of a prepared Bi2212 single-crystal film which has a square lattice of artificial defects. The holes are fabricated by FIB milling. The lattice spacing a_p of the hole array is $1\mu m$.

EXPERIMENTS

High-quality single crystals of Bi2212 were grown by the traveling-solvent floating-zone technique [7]. To measure the transport properties, we need through-holes in

CP850, *Low Temperature Physics: 24th International Conference on Low Temperature Physics;*
edited by Y. Takano, S. P. Hershfield, S. O. Hill, P. J. Hirschfeld, and A. M. Goldman

the crystals to keep a uniform current flow in the bulk. Since it is easy to cleave Bi2212 single crystals like mica, cleaving is a useful technique to make thin single-crystal films. At first, the bulk crystals were fixed on MgO by polyimide. After the cleaving, thin Au layer (200 Å) were evaporated on Bi2212 as electrodes. Patterns of the films for four-terminals measurements were made using a photolithographic and an Ar-ion milling processes. Holes were milled by the focused ion beam (FIB) using Micrion JFIB-2100 in the central of each film between the voltage electrodes. Since the thickness of Bi2212 films was between 100 and 300 nm, the holes perforated the sample thickness perfectly by the FIB milling of efficient dose amounts. Fig. 1 shows an image of a sample prepared for transport measurements. The diameter of each hole in this study was about 300 nm.

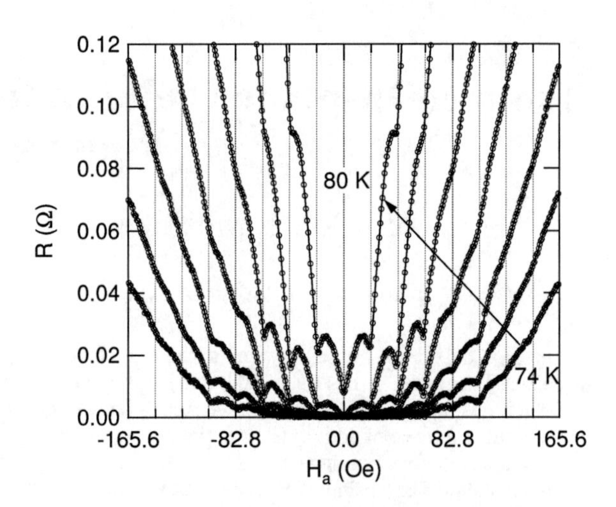

FIGURE 2. Pancake-vortex flow resistance as a function of field between 74 K and 80 K with a 1 K step. The vertical grid lines show the theoretical matching fields $n \cdot H_m$, where n is an integer and H_m is the first matching field.

RESULTS AND DISCUSSION

Figure 2 shows the vortex-flow resistance as a function of magnetic field which is parallel to the *c*-axis in the Bi2212-crystal film with square arrays of holes. Many dip structures are observed in low fields periodically. The matching field of square lattice of 1μm spacing is different from that of triangular lattice, i.e., $B_m = \Phi_0/a_p^2$, here Φ_0 is the magnetic flux quantum, and a_p is lattice spacing of defect arrays. From this equation, H_m is 20.7 Oe in this sample. Experimentally, the interval of magnetic fields of neighboring dips is 20.5 Oe on the average, which is consistent with the estimated value of the matching field. These dip structures become faint with increasing field. The dips in 6th or 7th multiples of the matching field can be identified at higher fields. This is similar to that in the case of triangular arrays.

In the previous study, we have confirmed that this matching effect in samples, which have triangular holes arrays, occurs in a vortex liquid phase, and discussed two possibilities [6]: (1) a vortex liquid phase in Bi2212 intrinsically keeps a triangular lattice order even without artificial pinnings, namely, vortex slush phase [8], (2) a defect-lattice symmetry is induced in the liquid phase by the defect arrays. In present study, the matching effect appears even in holes arrays of square symmetry. Therefore, the second possibility is better to explain the results naturally.

CONCLUSION

To study the influence of the periodic pinning array on the vortex matter in Bi2212 single crystals, we have measured the electrical resistance in the samples containing a square hole array. In the vortex-flow resistance, matching effects have been observed at periodic fields $n \cdot H_m$. It seems that matching effects appear in the vortex liquid phase as shown by transport measurements. Present experimental results combined with previous ones on triangular symmetry of defect array suggest the existence of a kind of vortex liquid which has an in-plane order induced by the defect lattice.

ACKNOWLEDGMENTS

A part of this work was conducted in AIST Nano-Processing Facility, supported by "Nanotechnology Support Project" of the Ministry of Education, Culture, Sports, Science and Technology (MEXT), Japan.

REFERENCES

1. A. K. Geim, I. V. Grigorieva, S. V. Dubonos, J. G. S. Lok, J. C. Maan, A. E. Filippov, and F. M. Peeters, *Nature* **390**, 259 (1997).
2. A. Kanda, B. J. Baelus, F. M. Peeters, K. Kadowaki, and Y. Ootuka, *Phys. Rev. Lett.* **93**, 257002 (2004).
3. A. T. Fiory, A. F. Hebard, and S. Somekh, *Appl. Phys. Lett.* **32**, 73 (1978).
4. A. Castellanos, R. Wordenweber, G. Ockenfuss, A. v.d. Hart, *Appl. Phys. Lett.* **71**, 962 (1997) .
5. R. Wordenweber, P. Dymashevski and V. R. Misko, *Phys. Rev. B* **69**, 184504 (2004) .
6. S. Ooi, T. Mochiku, S. Yu, E. S. Sadki, and K. Hirata, *Physica C* to be published.
7. T. Mochiku, K. Hirata, and K. Kadowaki, *Physica C* **282-287**, 475 (1997).
8. Y. Nonomura, X. Hu, *Phys. Rev. Lett.* **86**, 5140 (2001).

Scaling Law and Irreversibility Fields in Low Temperature Superconductors

N. Sakamoto*, T. Akune* , Y. Matsumoto† , and T. Matsushita¶

*Dept. of Electrical Engineering, Kyushu Sangyo Univ., 2-3-1 Matsukadai, Fukuoka 813-8503, Japan
†Dept. of Electrical Engineering, Fukuoka Univ., 8-19-1 Nanakuma, Fukuoka 814-0180, Japan
¶Dept. of Computer Sci. and Electronics, Kyushu Institute of Technology,680-4 Iizuka, 820-8502, Japan

Abstract. Theoretical analysis of the irreversibility fields B_i was obtained from the flux creep theory based on a depinning mechanism caused by thermally activated flux creep and predicted to vary according to power of (1- $(T/T_c)^2$). The measured B_i, however, increases more rapidly at low temperatures. This enhancement from the power law in low temperatures and high fields has been ascribed to the different pinning mechanism. The irreversibility fields B_i of MgB_2 superconductors are estimated from the onsets of imaginary parts of AC susceptibilities and shown to agree well with the numerical estimation of the original flux creep equation with the upper critical field term. The magnetization characteristics are also successfully obtained from the pinning parameters.

Keywords: scaling law, irreversibility field, magnetization width
PACS: 74.72

INTRODUCTION

Superconductors cannot carry nonresistive transport current outside the irreversibility line. A high irreversibility field B_i is a necessary condition for transporting large current densities in high magnetic fields. Many research works on the irreversibility fields and temperatures are performed in high-T_c superconductors [1,2]. Application of low-T_c superconductors at an elevated liquid H_2 temperature is now supposed. The irreversibility characteristics in low-T_c superconductors should be investigated in more detail.

Matsushita deduced an expression to describe the irreversibility field B_i based on the thermally activated fluxoids (flux creep) model [3] :

$$B_i^{\frac{3-2\gamma}{2}} = \left(\frac{K}{T}\right)^2 \left[1-\left(\frac{T}{T_c}\right)^2\right]^{m-\gamma} \left(1-\frac{B_i}{B_{c2}}\right)^{\delta}, \quad (1)$$

where K is approximately constant determined by the voltage criterion of the irreversibility. Parameters m, γ and δ are introduced by the assumed temperature and magnetic field dependence in the scaling law of creep-free critical current density J_c:

$$J_c = J_\gamma \left[1-\left(\frac{T}{T_c}\right)^2\right]^{m-1} \left(\frac{B}{B_{c2}(T)}\right)^{\gamma-1} \left(1-\frac{B}{B_{c2}(T)}\right)^{\delta}. \quad (2)$$

The empirical temperature dependence of the upper critical field: $B_{c2}(T)$ is given by $B_{c2}(T)= B_{c2}(0)(1-(T/T_c)^2)$.

In high-T_c superconductors B_i is small compared to B_{c2} and Eq. (1) reduces to

$$B_i(T) = B_{i0}\left(\frac{T_c}{T}\right)^p \left[1-\left(\frac{T}{T_c}\right)^2\right]^n, \quad (3)$$

where $B_{i0} = (K/T_c)^p$, and index n and p are

$$n = \frac{2(m-\gamma)}{3-2\gamma}, \quad p = \frac{4}{3-2\gamma}. \quad (4)$$

At high temperatures of $T \sim T_c$, Eq. (3) is further transformed to

$$B_i(T) = B_{i0}\left[1-\left(\frac{T}{T_c}\right)^2\right]^n. \quad (5)$$

This expression is widely used to discuss the irreversibility line and log B_i depends linearly on log $(1-(T/T_c)^2)$, where the gradient of the line equals to n.

RESULTS AND DISCUSSION

The temperature characteristics of B_i is plotted in Fig. 1 as a function of $(1-(T/T_c)^2)$ for three cases of (a) high T ($\sim T_c$) with Eq. (5) as dash lines, (b) low B (<< B_{c2}) with Eq. (3) as dotted lines and (c) general cases

CP850, *Low Temperature Physics: 24th International Conference on Low Temperature Physics;*
edited by Y. Takano, S. P. Hershfield, S. O. Hill, P. J. Hirschfeld, and A. M. Goldman

with Eq. (1) as solid lines, where the pinning parameter γ is varied from 0 to 1. All lines agree well in the range of $(1-(T/T_c)^2) \leq 0.1$ (i.e. $T/T_c \geq 0.95$), deviate significantly below $T/T_c \leq 0.9$. It was reported that temperature dependence of B_i of high-T_c superconductors is well described using Eq. (3) at low temperature range down to $T/T_c = 0.2$ [4]. However, B_i computed with Eq. (3) diverges as T tends to 0.

The irreversibility field B_i of large powder sample MgB_2 was reported [5], where B_i is estimated from the onset of the imaginary part χ'' of the AC susceptibilities and shown as solid circles in Fig. 2 as a function of $(1-(T/T_c)^2)$. The calculated curve of the case (a) is plotted with the dash line, where $T_c = 38.5$ K, $\gamma = 0.56$, $m = 1.7$, $n = 1.21$ and $B_{i0}/B_{c2}(0) = 0.285$. At low temperatures, B_i increases more rapidly as shown in Fig. 2 and has been attributed to the result of different pinning sites. The dotted line shows the computed result by Eq. (3) which diverges when T approaches to 0 and results in an excess increase. The theoretical curve by the original equation (1) depicted by the solid line agrees well in a wide temperature region.

The magnetization width ΔM proportional to the critical current density is calculated using the scaling law of Eq. (2) with the above-estimated values and plotted in Fig. 3. The theoretical results of Eq. (2) agree well with the measured magnetization width.

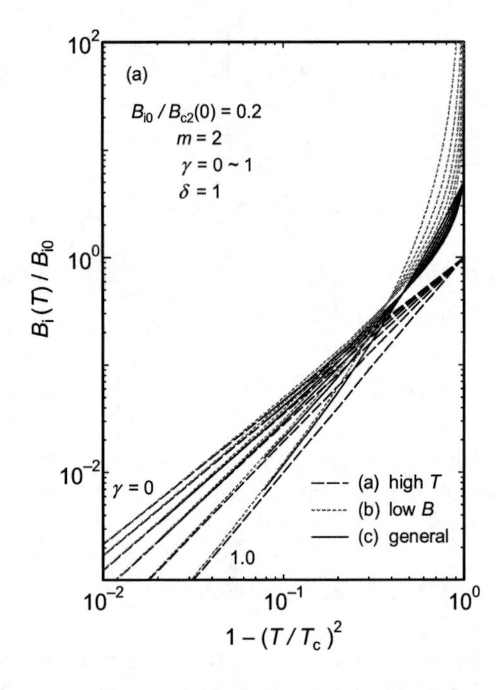

FIGURE 1. Theoretical irreversibility field $B_i(T)$ as a function of $1-(T/T_c)^2$. The dotted lines for the case of (b) show the computed values of Eq. (3) and the solid lines for (c) indicate the numerically estimated results of Eq. (1), which converge to $B_{c2}(0)/B_{i0}$ at $T = 0$.

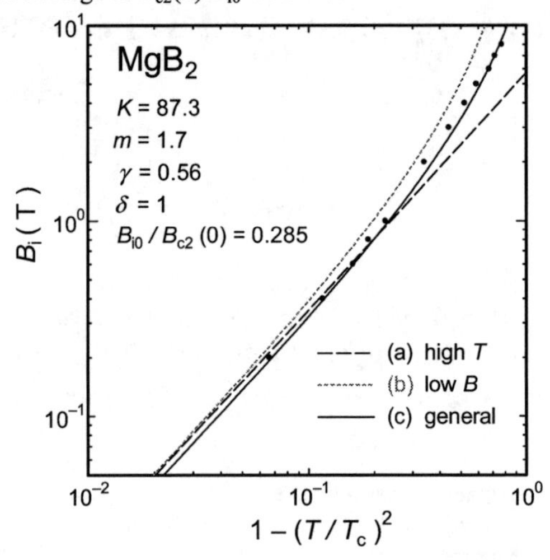

FIGURE 2. Temperature dependence of the irreversibility field B_{irr} in MgB_2 superconductor.

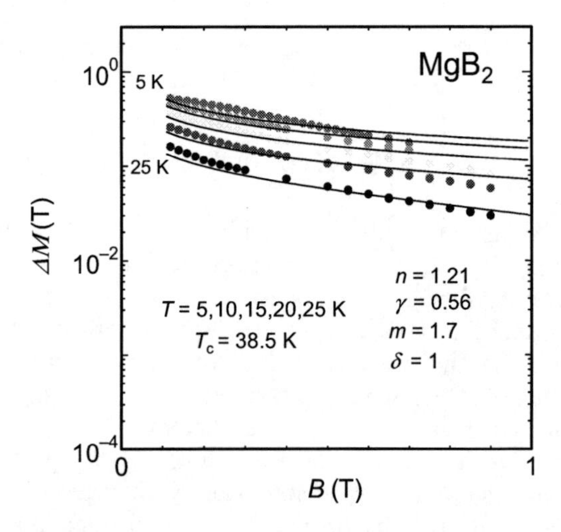

FIGURE 3. Dependence of the magnetization width ΔM of MgB_2 on magnetic field at $T = 5$-25 K.

REFERENCES

1. Y. Yeshurun and A.P. Malozemoff, *Phys. Rev. Letters* **60**, 2202-2205 (1988).
2. Y.C. Kim, J.R. Thompson, D.K. Christen, Y.R. Sun, M. Paranthaman and E.D. Specht, *Phys. Rev. B* **52**, 4438-4445 (1995).
3. T. Matsushita, T. Fujiyoshi, K. Toko and K. Yamafuji, *Appl. Phys. Letters* **56**, 2039-2041 (1990).
4. G. Fuchs, K.A. Nenkov, A. Attenberger, K. Lueders, M. Baenitz, C. Ecker, K. Kajikawa, E.V. Antipov and H.R. Khan, *Physica C* **355**, 299-306 (2001).
5. T. Akune, H. Abe, A. Koga, N. Sakamoto and Y. Matsumoto, *Physica C* **378-381**, 234-238 (2002).

Dynamic Ordering of Vortex Matter in Amorphous MoGe Films

N. Kokubo[a], S. Okayasu[b], and K. Kadowaki[a]

[a]*Institute of Materials Science, University of Tsukuba, 1-1-1, Tennoudai, Tsukuba, Ibaraki 305-8573, Japan*
[b]*JAERI, Tokai-mura, Naka-gun, Ibaraki 319-1195, Japan*

Abstract. We report flow properties of vortex matter near the peak effect regime of amorphous MoGe films detected by mode locking (ML) experiment. The ML features allow us to define clearly the dynamic ordering transition separating incoherent flow state at low velocity from a coherent flow state at high velocity. On raising temperature the ordering velocity increases and seems to show a diverging behavior in the higher part of the peak effect regime.

Keywords: mode locking, dynamic ordering, amorphous films
PACS: 74.25Sv, 74.25 Qt, 74.81Bd

One of the intriguing phenomena in vortex matter in type II superconductors is the dynamic ordering (DO) transition [1-5] separating the coherent solid state where the elastic interaction dominates and positional and temporal order develops well, from an incoherent flow state where the periodicity in flow structure is disrupted by the random pinning due to disorder quenched in a host material and the shear rigidity (elasticity) in driven vortices vanishes.

The experimental indication for this phenomenon was first obtained in a transport measurement in the lower part of the peak effect regime of $NbSe_2$, in which the S-shaped anomaly of current-voltage (*IV*) curve marks the current induced DO. This has triggered a number of theoretical [2] and numerical studies [3] on DO, focusing mainly on the physical meaning of the characteristic force at the inflection point in the *IV* curve. However, the latter studies have pointed out that extrinsic origins like surface barrier, edge contaminations, flow morphology etc. would lead to similar S-shaped *IV* curves. Thus, more microscopic experimental evidence for DO has been sought.

Recently, mode locking (ML) experiment has been focused as a powerful technique to probe the DO phenomenon. ML is a dynamic resonance between rf drive on top of dc drive and lattice (elastic) modes excited collectively over the driven vortices at internal frequency given by $f_{int}=qv/a$ with integer q, the average velocity v and the lattice spacing a. When $f_{int}=pf$ or $v=(p/q)fa$ with integer p, the elastic modes are dynamically locked on the rf drive. This resonance appears as current steps in *IV* characteristics. As the velocity is decreased, the ML step decreases due to the influences of the random pinning and disappears when the flow becomes fully disordered. Thus, the onset of ML marks the DO for driven vortices.

In this study, employing this technique, we investigate DO phenomenon of vortex matter near the peak effect regime of amorphous MoGe films. A minimum frequency (velocity) for ML resonance is clearly observed, marking the onset of crystallization (or DO) of driven vortices. On raising temperature T, this velocity increases and exhibits a diverging behavior, in qualitative agreement with the DO picture proposed by Koshelev and Vinokur (KV).

We sputtered amorphous $Mo_{1-x}Ge_x$ ($x = 0.2$) films on Si substrates held at room temperature. The films were structured lithographically into the Hall bar shape with 0.3 mm width and 1.1mm length between voltage probes by lift off technique. Thickness is 0.330 μm. We deposited silver on top of current and voltage pads to reduce contact resistance. The superconducting transition temperature T_c is 6 K. The slope of the second critical field on T at T_c is -2.7 T/K.

The measurements discussed in this paper were done by varying T at a fixed magnetic field of 7 T which was applied above T_c. No significant temperature history effect appears on *IV* characteristics after application of rf current I_{rf}.

CP850, *Low Temperature Physics: 24th International Conference on Low Temperature Physics;*
edited by Y. Takano, S. P. Hershfield, S. O. Hill, P. J. Hirschfeld, and A. M. Goldman

In the inset to Fig. 1 a series of *IV* curve measured with superimposing 10 MHz rf current with different amplitudes are given. Clear ML current steps appear at equidistant voltages. The step at the lower voltage is the fundamental (p/q=1/1) and the other at the higher voltage is harmonic ML (2/1). As expected from the ML velocity condition $v=(p/q)fa$, these ML voltages increase linearly with f.

Next we show how the ML current width depends on f (or v). To show this, we take the ML current width maximized by I_{rf} at each f since it exhibits a Bessel function like oscillatory behavior on I_{rf} [4]. As observed (see Fig.1), on decreasing f, the maximized width decreases monotonically and seems to disappear around 0.5MHz. We define the vanishing frequency f_c by a linear extrapolation to zero displayed as a broken line. Below f_c, no ML resonance is observed at any amplitude of I_{rf}, indicating the absence of the elasticity of driven vortices. Thus, the onset velocity v_c (=$f_c a$) for ML should mark the dynamical ordering transition [4].

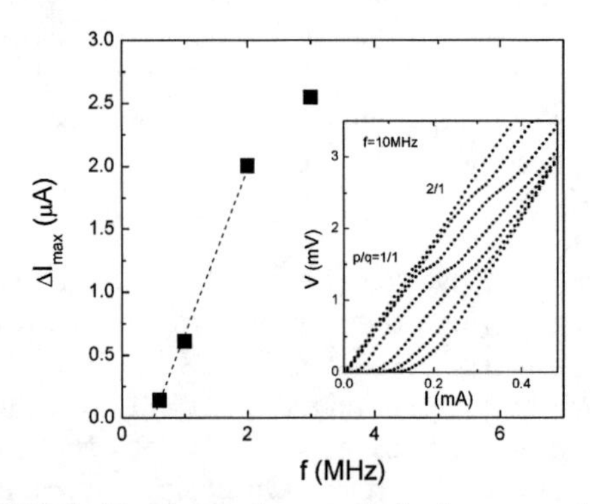

FIGURE 1. Frequency f vs. maximized current width ΔI_{max} measured at 1.8 K. Typical dc current-voltage characteristics measured by superimposing 10 MHz rf current with various amplitudes are shown in the inset.

On elevating T, a remarkable change appears in v_c. In Fig. 2 a plot of v_c (squares) vs T is given. Here, results of v_c determined from the temperature dependence of the ML width are also shown (circles). As observed, it increases monotonically and seems to diverge toward ~2.5 K. Regarding the onset velocity as the crystallization velocity, this behavior is qualitatively similar to the KV picture, in which the combined influences of the quenched disorder and thermal fluctuations on DO near the thermodynamic melting temperature T_m are taken into account. In this model, the crystallization velocity v_c diverges as

$$v_c = v_0/(1-T/T_m) \qquad (1)$$

on approaching T_m from below. As displayed by a solid curve, the v_c data are well approximated by Eq.(1) with T_m=2.46 K (indicated by a broken line) and the onset velocity at T=0 v_0=5.5mm/s over nearly a decade of velocity. Below 2.2 K, v_c decreases much faster than that in the model.

Finally, let us compare the results of v_c to the temperature dependence of a critical current I_c, which is displayed in the inset to Fig.2. I_c is determined from dc *IV* curves by a 1 μV criterion. It shows a broad peak around 2.1 K, above which a sharp drop appears. We find that the diverging behavior of v_c appears in the higher part of the peak effect, while in the lower part the large deviation appears. Similarity has been seen in results of $NbSe_2$ pure crystals [5]. Thus, the KV picture would be applicable only to the dynamic ordering observed in the higher part of the peak effect.

N. K used facilities in cryogenic center in university of Tsukuba. This work was partly supported by the grant in aid for scientific research (Grant No. 16710063) from MEXT (the Ministry of Education, Culture, Sports, Science and Technology), and also by the 21st Century Center of Excellence Program, *"Promotion of Creative Interdisciplinary Materials Science for Novel Functions"* under MEXT Japan.

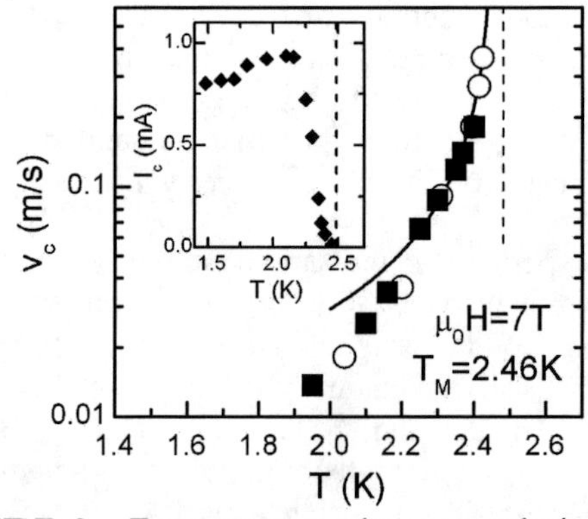

FIGURE 2. Temperature vs the onset velocity. A solid curve represents the crystallization velocity of the dynamic ordering picture. Inset shows the temperature dependence of the critical current.

REFERENCES

1. Bhattacharya and M. J. Higgins, *Phys. Rev. Lett.* **70**, 2617 (1993).
2. A. E. Koshelev and V. M. Vinokur, *Phys. Rev. Lett.* **73**, 3580 (1994).
3. For instance, M. C. Faleski. M. C. Marchetti, and A. A. Middleton, *Phys. Rev. B* **54** 12427 (1996).
4. N. Kokubo, e*t al., Phys. Rev. B.***69**, 064504 (2004). R. Besseling, e*t al., Phys. Rev. Lett.* **91**, 177002 (2003). N. Kokubo, e*t al., Phys. Rev. Lett.* **88**, 247004 (2002).
5. N. Kokubo, K. Kadowaki, and K. Takita, *Phys. Rev. Lett.* **95**, 177005 (2005).

Determination of Current Profiles in Flat Superconductors Using Hall Probe Array

Ryuji Kondo*, Toshihiro Shige*, Takeshi Fukami† and Tsuyoshi Tamegai**

*Faculty of Engineering, Oita University,700 Dannoharu Oita, Japan
†Department of Materials Science and Engineering, Himeji Institute of Technology,Himeji, Japan
**Department of Applied Physics, The University of Tokyo, Tokyo, Japan

Abstract. The current profiles in $YBa_2Cu_3O_7$ films when the magnetic field is perpendicular to the surfaces are studied. The current profiles are calculated from the field profiles, which are measured using a micro Hall-probe array. In this geometry demagnetization effects are important, therefore Bean critical state model in the usual longitudinal geometry cannot be used to interpret the experimental results. In this work we utilize a model, which has a spatial distribution of local current density. According to the model, the specimen is divided in two regions. One region, in which flux penetrates, has a constant current like Bean-model. The other is flux free and has a current distribution. This current distribution is determined by one parameter "J_c". The measurements are performed in increasing and decreasing field. Analyzing the measurement results using this model, the field profiles are reproduced well in both conditions, in particular, at low field regions. This coincidence implies the appropriateness of this model.

Keywords: superconductivity, $YBa_2Cu_3O_7$, Hall probe, local field
PACS: 74.25.Sv

INTRODUCTION

The critical current density in high-T_c superconductors is measured by various ways. In them, the methods using magnetization measurements require interpreting experimental results according to proper models. These are done mainly via Bean's model[1]. This model is originally applied to long superconductors in a parallel field. In this geometry, demagnetization effects are disregarded and the gradient of flux-density are constant. But these assumption can not be applied to flat superconductors in a perpendicular field. Recently, some useful methods has been developed to investigate the spatial distribution of flux-density [2] [3]. In these methods, demagnetization effects are crucial and it is necessary to take acount of a current distribution.

To calculate current profiles, we utilize a critical state model, which has a spatial distribution of local current density [4]. This model, in case a thin superconductor strip has a width $2a$ along the x axis and ∞ length along y axis, allows the sheet current $J(x)$

$$J(x) = \begin{cases} -J_c & (-a \leq x \leq -b) \\ (2J_c/\pi)\arctan\frac{cx}{\sqrt{b^2-x^2}} & (-b < x < b) \\ J_c & (b \leq x \leq a), \end{cases} \quad (1)$$

where J_c is a sheet critical current density and $J(x)$ saturates at $x = b$. b and c are determined by J_c, a and external field. Therefore J_c is the only fitting parameter. In $|x| < b$, there is no flux and this region has a current distribution.

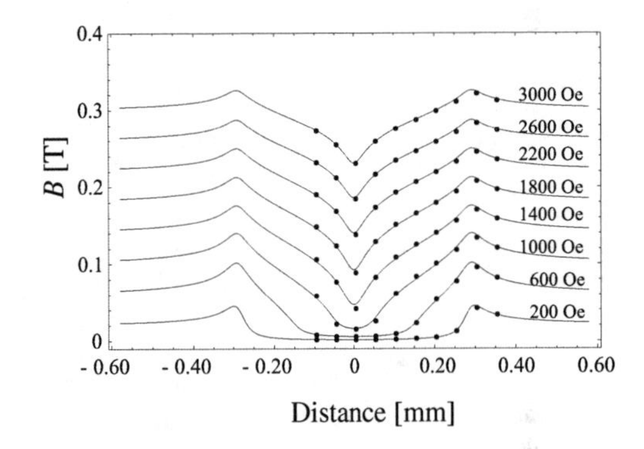

FIGURE 1. Profiles of flux-density B at T=30 K. The field is increased from 0 Oe to 3080 Oe. The solid curves show calculated B at 20 μm distance from the sample surface.

EXPERIMENT

In this experiment, we use a c-axis oriented $YBa_2Cu_3O_7$ epitaxial film having a rectangular shape of size $600\times3800\times0.8$ μm^3. To measure the surface field of the sample, a micro Hall probe array is used. The probe is made of GaAs doped with Si and has 10 elements. Each element has 10×10 μm^2 active area. The sample is placed on the probe directly. Therefore a dc field parallel to the c axis of the crystal is measured by the probe. The sample is cooled to 30K at zero field and applied a dc field H. Figure 1 shows a surface field measurement in increasing external field. The solid curves show calculated B and Fig. 2 shows the

CP850, *Low Temperature Physics: 24th International Conference on Low Temperature Physics;*
edited by Y. Takano, S. P. Hershfield, S. O. Hill, P. J. Hirschfeld, and A. M. Goldman

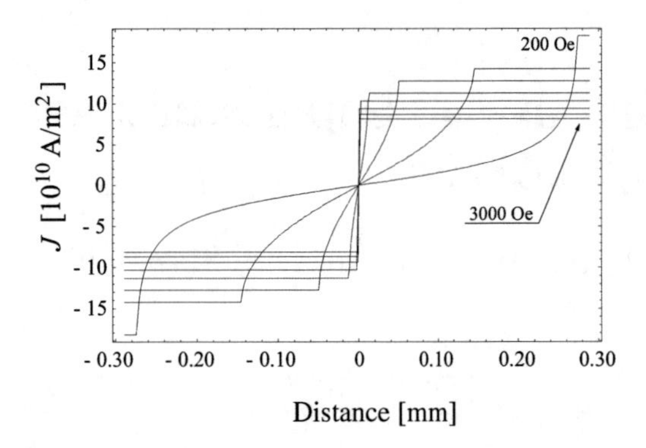

FIGURE 2. Profiles of calculated current density J in increasing the external field.

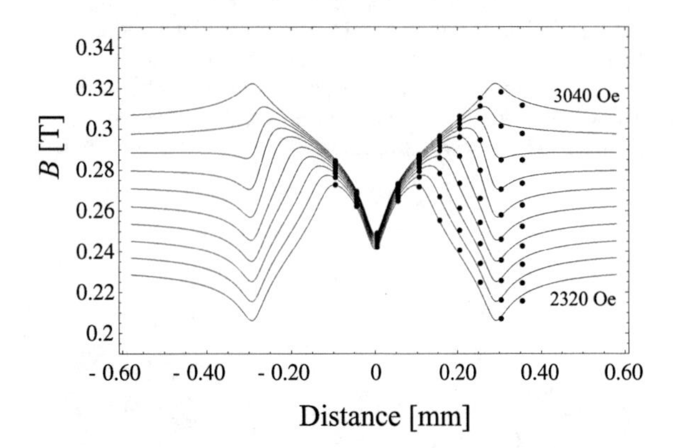

FIGURE 3. Profiles of flux-density B. Closed circles show the experimental data for every 80 Oe. The solid curve show calculated B.

screening-current profiles for calculating B. The current profiles reproduce the measured field quite well. In Fig. 2, with increasing the external field, b in Equation (1) decreases and the profiles approach to a uniform current like Bean's model.

After keeping the external field at the maximum for

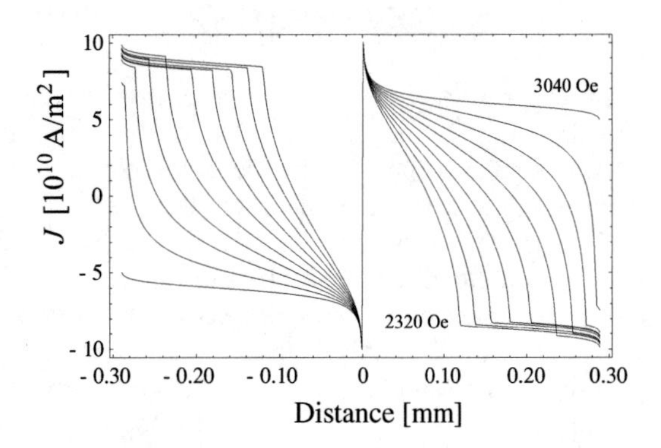

FIGURE 4. Profiles of calculated current density J in decreasing the external field.

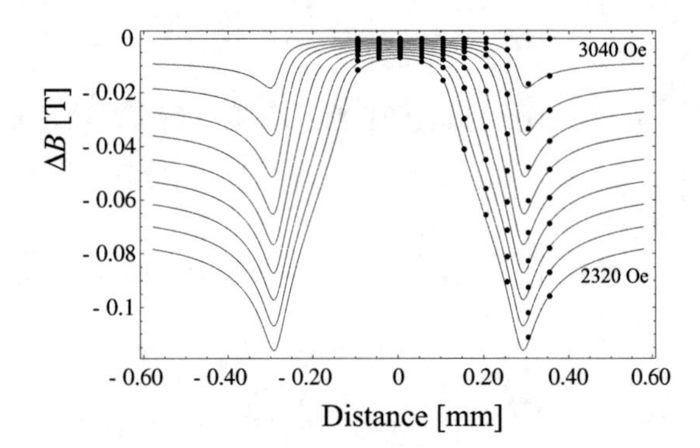

FIGURE 5. Profiles of ΔB. ΔB is a value which is the difference between the initial local field and the present local field. The solid curves show calculated ΔB.

one hour, it is decreased from 3040 Oe to 2320 Oe [Fig. 3]. In this step, the initial field profile, before changing the external field, can not be reproduced by the uniform current profile. Therefore we adopt a model which $J_c(B)$ decays exponentially with B and then calculate the initial current profile. Figure 4 shows estimated current profiles used for the calculation of the field profiles of Fig. 3. The current profiles in Fig. 4 are calculated by adding the initial current profile to the profile which is derived from Equation (1). In order to clarify the meaning of Fig. 3, we replot it after subtracting the initial local field from a present local field [Fig. 5]. These subtracted fields are reproduced quite well by the current profile derived from Equation (1).

In conclusion, the current profile model, proposed by E. H. Brandt *et al.*, explains the experimentally obtained field profiles for both field-increasing and field-decreasing branches. This implicates the appropriateness of the model.

ACKNOWLEDGMENTS

We thank Yasuki Akimoto for connecting lead wires to the Hall probe.

REFERENCES

1. C. P. Bean, *Phys. Rev. Lett.* **8**, 250 (1962).
2. T. Tamegai *et al.*, *Phys. Rev. B* **45**, 8201 (1992).
3. Y. Abulafia *et al.*, *Phys. Rev. Lett.* **75**, 2404 (1995).
4. E. H. Brandt *et al.*, *Phys. Rev. B* **48**, 12893 (1993).

Guiding of Vortices and New Voltages in Ratchet Washboard Pinning Potential

Valerij A. Shklovskij

Institute for Theoretical Physics, National Science Center- Kharkov Institute of Physics and Technology, 61108, Kharkov, Ukraine and Kharkov National University, Physical Department, 61077, Kharkov, Ukraine

Abstract. Two-dimensional vortex dynamics in a ratchet planar pinning potential (PPP) in the presence of thermal fluctuations is considered on the basis of a Fokker-Planck equation. Explicit expressions for two new nonlinear anisotropic voltages (longitudinal and transverse with respect to the current direction) are derived and analyzed. The physical origin of these odd (with respect to magnetic field or transport current direction reversal) voltages is caused by the interplay between the even effect of vortex guiding and the ratchet asymmetry. Both new voltages are going to zero in the linear regimes of the vortex motion (i.e. in the thermoactivated flux flow (TAFF) and ohmic flux flow (FF) regimes) and have a bump-like current or temperature dependence in the vicinity of the highly nonlinear resistive transition from TAFF to FF.

Keywords: Vortex, Guiding, Pinning, Ratchet.
PACS: 74.60.Ge, 75.15.Gd, 74.25.Fy, 74.72.Bk, 05.60.-k, 05.40-a

INTRODUCTION

The common feature of superconducting ratchet systems is their rectifying property: the application of an alternating current to a superconductor patterned with a periodic asymmetric pinning potential can induce vortex motion whose direction is determined only by the asymmetry of the pattern. Although considerable theoretical work exists [1], only few experiments have been realized. Recently a vortex lattice ratchet effect has been investigated in Nb films grown on arrays of nanometric Ni triangles, which induce the periodic asymmetric pinning potential [2]. Similar effects were also discussed for YBCO films with antidots [3]. So far a full theoretical description of the ratchet properties of the superconducting devices proposed in [2, 3] is not available due to the complexity of both, the origin and the two-dimensional structure of the pinning potential.

Due to this reason we propose below to study experimental ratchet properties of superconductors on the basis of a more simple ratchet device for which exists a full theoretical description (at least, in the single-vortex approximation) of its two-dimensional vortex dynamics within the framework of a Fokker-Planck approach. Such a device has already been exploited many years ago by Morrison and Rose in their experiments on controlled asymmetric (as now we say "ratchet") surface pinning in superconducting-alloy films [4]. Recent progress in fabrication of submicrometric structures with a periodic ratchet modulation of their thickness by methods of electron-beam lithography [5] or molecular-beam epitaxy on facetted substrates [6] allows to prepare Nb films with a similar well controlled asymmetric washboard pinning structure.

MAIN RESULTS

I consider the simplest case of nonlinear two-dimensional vortex dynamics in a ratchet PPP in the presence of thermal fluctuations ignoring the Hall effect, vortex viscosity anisotropy, and a competition between planar and additional point-like disorder (see, however, Refs. [7,8]). My objective is to demonstrate the new features which appear in a Fokker-Planck approach for the two-dimensional vortex motion due to presence of a *ratchet* washboard PPP.

The Langevin equation for a vortex moving with velocity $\mathbf{v}$ in a magnetic field $\mathbf{B} = \mathbf{n}B$ ($B \equiv |\mathbf{B}|$, $\mathbf{n} = n\mathbf{z}$, $\mathbf{z}$ is the unit vector in the z direction, and $n = \pm 1$) is

$$\eta \mathbf{v} = \mathbf{F}_L + \mathbf{F}_p + \mathbf{F}_{th}\ , \qquad (1)$$

CP850, *Low Temperature Physics: 24th International Conference on Low Temperature Physics;* edited by Y. Takano, S. P. Hershfield, S. O. Hill, P. J. Hirschfeld, and A. M. Goldman

where $\mathbf{F}_L = n(\Phi_0 / c)\,\mathbf{j} \times \mathbf{z}$ is the Lorentz force (Φ_0 is the magnetic flux quantum, c is the speed of light, and $\mathbf{j}$ is the transport current density), $\mathbf{F}_p = -\nabla U_p$ is the pinning force (U_p is the pinning potential), $\mathbf{F}_{th}$ is the thermal fluctuation force represented by a Gaussian white noise, η is the vortex viscosity. Our PPP is assumed to be one-dimensional periodic, and *ratchet-like*, i.e. $U_p(x) \neq U_p(-x)$. As the pinning force in a PPP is directed perpendicular to the pinning planes, the vortices tend to move along these planes if the driving force has a nonzero component in any in-plane direction. Such a *guided motion of vortices in a PPP* leads to the appearance of a $\rho_\perp^+$ contribution to the transverse (with respect to the current direction) resistivity which is even with respect to the magnetic field or transport current direction reversal. After some calculations the two-dimensional Fokker-Planck equation, consistent with (1) can be solved exactly and current- and temperature-dependent longitudinal $\rho_\|$ and transverse $\rho_\perp$ resistivities (in units of the flux flow resistivity) are

$$\rho_\|^+ = \nu^+ \cos^2\alpha + \sin^2\alpha, \quad (2)$$

$$\rho_\perp^+ = (\nu^+ - 1)\sin 2\alpha / 2, \quad (3)$$

$$\rho_\perp^- = \nu^- \sin 2\alpha / 2, \quad (4)$$

$$\rho_\|^- = \nu^- \cos^2\alpha\,. \quad (5)$$

Here $\nu^\pm(f,t) = [\,\nu(f,t) \pm \nu(-f,t)\,]/2$ are the even (+) and odd (–) parts of the $\nu(f,t)$–function which has *the physical meaning of the probability to overcome the potential barrier of the PPP* [7] under the action of the dimensionless temperature t and effective motive force $f = mnj\cos\alpha$ where $m = \pm 1$ determines the transport current reversal, $j = |\mathbf{j}|$, and α is the angle between the current and PPP channels directions. It is easy to show that at small current and temperature ν^+ has usual (j, t) – steplike behavior [7] whereas ν^- has a bell-shape (j, t)-dependence. The width of the bell as a function of j is defined roughly by the difference of the critical currents along the channels of the ratchet PPP. It is essential that for a symmetric PPP, where $U_p(x) = U_p(-x)$, $\nu^- = 0$, whereas for the ratchet PPP $\nu^- \neq 0$. For an asymmetric sawtooth PPP as was created in Ref. [4], a rather simple expression for $\nu(f,t)$ has been derived in [9].

Even resistivities, as given by Eqs. (2), (3), describe the guiding of vortices also in a symmetric PPP [7], whereas the odd ones, defined by Eqs. (4), (5), appear only due to the ratchet form of the PPP and change their sign with the current or magnetic field reversal. Their origin follows from the emergence of a certain equivalence of the *xy*-direction for the case, that a guiding of vortices along the channels of the washboard PPP is realized at $\alpha \neq 0, \pi/2$. Note also that for $\alpha = 0$ Eq. (5) gives in fact the *ratchet signal* measured in Ref.[2].

In conclusion, an exactly solvable [7–9] two-dimensional model system is proposed for study of the ratchet effect in superconducting film with asymmetric thickness modulation as was studied earlier in Refs. [4, 6, 10]. This opens up the possibility for a variety of experimental studies of directed motion of vortices simply by measuring longitudinal and transverse voltages. Experimental control of amplitude and frequency of the external force, damping, anisotropy parameters, and temperature can be easily provided. In contradistinction with other vortex-based ratchet models, the one presented here allows to separate the Hall and ratchet voltages which are similar in their (j, t) behavior, but have different origin and magnitude. Note also that the new ratchet voltages disappear during the procedure of the "current averaging" frequently used in experiments [10] for the cancellation of parasitic thermoelectric voltages.

ACKNOWLEDGMENTS

This research was partially supported by a DFG grant 436 UKR 17/18/05.

REFERENCES

1. R. Reimann and P. Hanggi, *Appl. Phys. A* **75**, 169-178 (2002).
2. J. E. Villegas, E. M. Gonzales, M. P. Gonzales, J. V. Anguita, and J.L. Vincent, *Phys. Rev. B* **71**, 024519 (2005).
3. R. Wördenweber, P. Dymashevski, and V. R. Misko, *Phys. Rev. B* **69**, 184504 (2004).
4. D. D. Morrison and R. M. Rose, *Phys. Rev. Lett.* **25**, 356-359 (1970).
5. J. I. Martin, Y. Jaccard, A. Hoffmann, J. Nogues, J. M. George, J. L. Vincent, and Ivan K. Shuller, *J. Appl. Phys.* **84**, 411-415 (1998).
6. M. Huth, K. A. Ritley, J. Oster, H. Dosch, and H. Adrian, *Adv. Funct. Mater.* **12**, 333-341 (2002).
7. V. A. Shklovskij, A. A Soroka, A. K Soroka, *JETP* **89**, 1138-1153 (1999).
8. V. A. Shklovskij, *J. Low Temp. Phys.* **131**, 899-905 (2003).
9. V. V. Sosedkin, Diploma thesis, Kharkov National University, 2003.
10. A.K.Soroka, "Vortex Dynamics in Superconductors in the Presence of Anisotropic Pinning" Ph. D. Thesis, J. Gutenberg University, Mainz, 2004.

H-T Phase Diagram of Flux Line Lattice Structure in YNi_2B_2C

N. Sakiyama[*], H. Tsukagoshi[†], F. Yano[†], T. Nagata[†], H. Kawano-Furukawa[†], H. Yoshizawa[*], M. Yethiraj[**], H. Takeya[‡] and J. Suzuki[§]

[*]*Neutron Science Laboratory, I.S.S.P., The University of Tokyo, Ibaraki 319-1106, Japan*
[†]*Department of Physics, Ochanomizu Univ., Tokyo 112-8610, Japan*
[**]*Center for Neutron Scattering, Oak Ridge National Laboratory, TN 37831-6393, USA*
[‡]*National Institute for Materials Science, Ibaraki 305-0047, Japan*
[§]*Advanced Science Research Center, Japan Atomic Energy Research Institute, Ibaraki 319-1195, Japan*

Abstract. The detailed flux line lattice (FLL) structure in YNi_2B_2C was investigated using small angle neutron scattering and the complete H-T phase diagram was determined. The FLL in YNi_2B_2C shows a change of symmetry only in the low magnetic field region between 0.05 to 0.2 T. The observed square lattice is governed by an anisotropic Fermi velocity. Contrary to the theoretical prediction, a square lattice driven by an anisotropic superconducting gap does not appear below 5 T.

Keywords: borocarbide, small angle neutron scattering, FLL structure
PACS: 74.70.Dd, 61.12.Ex, 74.25.Qt

INTRODUCTION

The quaternary borocarbides RNi_2B_2C (R = Y, Rare Earth) have attracted a great deal of interest because of the interplay between superconductivity and magnetism. Another interesting feature of these compounds is the change of symmetry of the flux line lattice (FLL). YNi_2B_2C is a nonmagnetic type II superconductor displaying a relatively high superconducting transition temperature $T_c \sim 15$ K and a Ginzburg-Landau parameter $\kappa \sim 20$[1]. Although this material has a s-wave order parameter, extensive studies revealed that its superconducting gap has a point node along the a- and b- axes[2] in addition to a fourfold anisotropic Fermi velocity distribution [3], reflecting its underlying crystalline symmetry.

Recently, Nakai *et al.* reported a theoretical H-T phase diagram of the FLL structure[4] for a system with fourfold anisotropies in the Fermi velocity and superconducting gap. Each anisotropy stabilizes a square FLL. If the preferred FLL orientation for each anisotropy is different, it may lead to successive FLL transitions from a triangular ($\triangle$) $\rightarrow$ a square ($\square_v$) $\rightarrow$ a triangular ($\triangle$) $\rightarrow$ another square ($\square_g$) with increasing a field. Here $\square_v$ and $\square_g$ denote FLL structures with different anisotropies. These authors propose that this sequence of FLL structures may be realized in YNi_2B_2C and $LuNi_2B_2C$.

It has been shown that $LuNi_2B_2C$ exhibits rhombic to square to rhombic transitions in the low magnetic field region[5]. In the high magnetic field region, however, the symmetry of the FLL has not been explored and there is no proof of the existence of the $\square_g$ phase. In this paper we report the detailed FLL structures in YNi_2B_2C and compare these results with the theoretical predictions. To our knowledge, this is the first FLL study which covered almost the entire H-T plane in YNi_2B_2C.

EXPERIMENT

A single crystal of $YNi_2^{11}B_2C$ was grown by the floating zone method. The ^{11}B isotope was used to reduce the severe neutron absorption of ^{10}B. $T_c(0)$ = 14.2 K and an extrapolated upper critical magnetic field $H_{c2}(0) \sim 6.5$ T were determined by magnetization measurements using a Quantum Design MPMS SQUID magnetometer. Small angle neutron scattering (SANS) measurements were performed at SANS-J and SANS-U installed in the JRR-3 at JAERI, Tokai, Japan. The c-axis of the crystal and an external magnetic field were set along the incident neutron beam. All measurements were performed after cooling in a magnetic field. An oscillating field (≤ 0.15 T) was applied at each measuring temperature to bring a FLL alignment to the equilibrium state[6]. Above 0.5 T, the sample was rotated to satisfy the Bragg condition.

CP850, *Low Temperature Physics: 24th International Conference on Low Temperature Physics;*
edited by Y. Takano, S. P. Hershfield, S. O. Hill, P. J. Hirschfeld, and A. M. Goldman

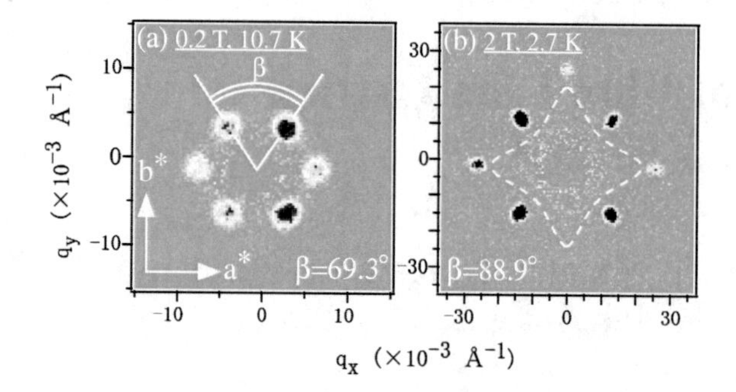

FIGURE 1. FLL diffraction patterns at (a) 0.2 T, 10.7 K and (b) 2 T, 2.7 K. The normal state data was subtracted as a background. The opening angle of the FLL scattering patterns is defined as β (double line in (a)). Broken line in (b) schematically indicates the anisotropic distribution of the Fermi velocity[3, 4].

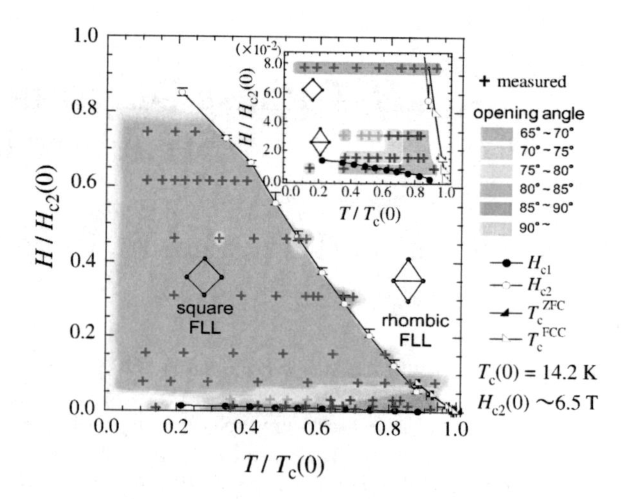

FIGURE 2. *H-T* phase diagram of FLL in YNi_2B_2C. The temperature and magnetic field are normalized by $T_c(0)$ and $H_{c2}(0)$, respectively. The inset shows and enlarged view of the lower magnetic field region. A gradual symmetry transition from rhombic to square is observed in the low magnetic field region (0.05 T $\sim$ 0.2 T) as the magnetic field is increased. At higher magnetic fields $\square_v$ is observed even very close to the H_{c2} line. $\square_v$ is stable up to a high magnetic field of 5 T.

RESULTS AND DISCUSSION

Fig.1 exemplifies the FLL diffraction patterns under magnetic fields of (a) 0.2 T and (b) 2 T, respectively. There is a change of the FLL symmetry from rhombic to square. At 2 T the nearest neighbor FLL is along the [110] direction, where the Fermi velocity has its minimum value, indicating the square lattice is of $\square_v$ type. Although the rhombic FLL is distinct from the square FLL in Fig.1, most of the diffraction patterns show some distortion from an exact triangle or square symmetry. To estimate the distortion quantitatively, we evaluated the opening angle β of the FLL patterns (see definition in Fig.1(a)). We observed that β ranges from 65° to 90°, depending on magnetic field and temperature. Even at our experimentally accessible minimum magnetic field 0.05 T, the exact rhombic symmetry with $\beta = 60°$ was not observed.

Using the defined β we constructed a complete *H-T* phase diagram of the FLL structures (Fig.2). The $\triangle$ lattice occurs only in a narrow region near H_{c1}. Below $H/H_{c2} \sim 0.03$ the $\triangle$ lattice has a tendency to distort to a $\square_v$ lattice with decreasing temperature, while in a high magnetic field region above $H/H_{c2} \sim 0.08$ there is little temperature dependence in β. Our measurements of the temperature dependence under various magnetic fields up to $H/H_{c2} \sim 0.77$ revealed that β is insensitive to temperature even very close to the H_{c2} line, where thermal fluctuations are supposed to suppress the square anisotropy induced by the nonlocality[7]. It seems that the FLL thermal fluctuations in YNi_2B_2C are effective only in the very low magnetic field region. The $\square_v$ lattice also has a striking stability against the magnetic field dependence up to 5 T, and we could not detect any sign of a transition either towards the $\square_g$ lattice or to the $\triangle$ lattice in higher magnetic fields. Note that the latter transition was reported in $LuNi_2B_2C$[5].

CONCLUSION

We elucidated the FLL structure over almost the entire *H-T* phase diagram of $YNi_2^{11}B_2C$ by SANS measurements. The FLL in YNi_2B_2C shows a gradual transition from $\triangle$ to $\square_v$ only in the low magnetic field region from 0.05 to 0.2 T, and $\square_v$ is stabilized otherwise. We observed no evidence of the $\square_g$ FLL phase up to 5 T. The *H-T* phase diagram indicates that the FLL in YNi_2B_2C is different from the theoretical predictions or from the FLL reported in $LuNi_2B_2C$ in a low magnetic field region. This is a rare experimental example showing an intrinsic difference between $LuNi_2B_2C$ and YNi_2B_2C.

REFERENCES

1. K. Ghosh et al., *Physica B*, **223&224**, 109 – 111 (1996).
2. K. Izawa et al., *Phys. Rev. Lett*, **89**, 137006 (2002).
3. S. B. Dugdale et al., *Phys. Rev. Lett*, **83**, 4824 – 4827 (1999).
4. N. Nakai et al., *Phys. Rev. Lett*, **89**, 237004 (2002).
5. M. R. Eskildsen et al., *Phys. Rev. Lett*, **86**, 5148 – 5151 (2001).
6. S. J. Levett et al., *Phys. Rev. B*, **66**, 014515 (2001).
7. A. Gurevich et al., *Phys. Rev. Lett*, **87**, 177009 (2001).

Finite-Sized Square Network of Superconducting Pb

Takekazu Ishida [a,e], Hiroshi Noda [a,e], Osamu Sato [b], Masaru Kato [c,e], Kazuo Satoh [d,e], and Tsutomu Yotsuya [d,e]

[a] *Department of Physics and Electronics, Osaka Prefecture University, 1-1 Gakuen-cho, Sakai, Osaka 599-8531, Japan*
[b] *Osaka Prefectural College of Technology, Neyagawa, Osaka 572-8572, Japan*
[c] *Department of Mathematical Sciences, Osaka Prefecture University, 1-1 Gakuen-cho, Sakai, Osaka 599-8531, Japan*
[d] *Technology Research Institute of Osaka Prefecture, 2-7-1 Ayumino, Izumi, Osaka 594-1157, Japan*
[e] *JST-CREST, 4-1-8, Honcho, Kawaguchi, Saitama 332-0012, Japan*

Abstract. The vortex behavior of superconducting finite-sized Pb networks with 2x2-, 3x3-, 5x5-, and 10x10-hole-arrays have been investigated in a small DC magnetic field. The Pb networks have been fabricated by electron beam lithography of photoresist layer and a lift-off process after depositing Pb film on the resist patterns. The magnetization measurements of the Pb networks by a SQUID magnetometer reveal the matching effect in the magnetization curve. Vortex image observations are also carried out by a SQUID microscope to compare with the theoretical predictions.

Keywords: Superconducting network, matching field, Pb, electron beam lithography, SQUID microscope
PACS: 74.20.De; 74.62.-c; 74.81.Fa

INTRODUCTION

Superconducting networks have been investigated with considerable interests for a long time in connection with the Little-Parks effect [1]. Pannetier et al. [2] found a T_c oscillation of infinite Al superconducting network as a function of the magnetic field. Alexander [3] gave a sound interpretation for the Al network experiment. To our knowledge, most of the existing works are concerned with infinite periodic network systems, for which it is not necessary to consider the effect of sample edges. Vortex configurations in a Pb/Cu microdot with a 2x2 anti-dot cluster were investigated by Puig et al. [4] by means of transport properties. The vortex distribution in finite-sized networks might be quite exotic compared to infinite networks due to the edge effect [5-7].

Studies on superconducting networks have been inspired by developments in nanofabrication techniques. In this paper, we describe the fabrications of the several finite-sized superconducting networks to investigate the vortex behavior by means of a SQUID magnetometer, a scanning SQUID microscope, and theoretical calculations.

EXPERIMENTAL

We fabricated the finite-sized Pb network patterns with 2x2-, 3x3-, 5x5-, and 10x10-hole-arrays. Each network pattern has a lattice constant of 4 μm (a matching field H_φ = 1.29 G) with a line width of 2 μm. After applying electron beam lithography for development of resist layer, we deposited Pb film on the reversed resist patterns of the networks. Then, samples were soaked into organic solvent in an ultrasound bath to dissolve resist and extra Pb films. We coated superconducting networks by a 300-nm resist layer to prevent superconducting networks from being deteriorated during storage and measurements. The thickness of Pb was chosen as 200-250 nm. .

FIGURE 1. SEM photographs of 2x2, 3x3, 5x5, and 10x10 networks of Pb with line width 2 μm.

In Fig. 1, we show the secondary electron microscope (SEM) photographs for four samples. There are (a) the 94x94 arrays of the 2x2 squares (b) the 71x71 arrays of the 3x3 squares, (c) the 48x48

CP850, *Low Temperature Physics: 24th International Conference on Low Temperature Physics;* edited by Y. Takano, S. P. Hershfield, S. O. Hill, P. J. Hirschfeld, and A. M. Goldman

arrays of the 5x5 squares, and (d) the 57x57 arrays of the 10x10 squares on a single Si substrate (4x4 mm).

RESULTS AND DISCUSSION

In Fig. 2, we show the magnetization curve of the finite-sized 10x10 superconducting Pb networks (the lattice constant of 4 μm, the line width of 2 μm, and the thickness of 200 nm) at 7.17 K by using a SQUID magnetometer. The matching effect can be seen in the figure. Vertical lines in Fig. 2 indicate the multiples of the matching field (H_ϕ = 1.29 G). The magnetization curve suggests that vortices at nH_ϕ are expected to have an ordered configuration in the 10x10 holes.

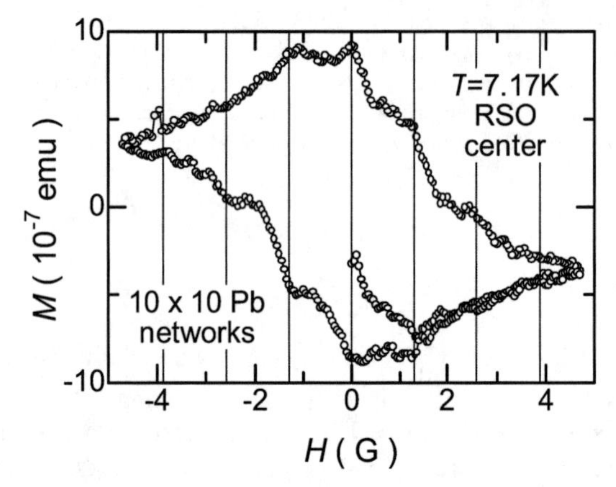

FIGURE 2. Magnetization curve of 10x10 Pb networks.

We measured vortex distribution of the finite-sized superconducting networks by using a scanning SQUID microscope (Seiko Instruments, SQM2000), where a sensor coil has a 10-μm diameter. Typical scanning area is 64x64 μm or 128x128 μm in a 1 μm or 2 μm step. The magnetic field was chosen as a fractional number of the integral multiples of the matching field H_ϕ. However, we could not see a clear vortex patterns because of the lack of spatial resolution.

THEORETICAL CALCULATIONS

We employ the de Gennes–Alexander's network equation to calculate the distribution of quantized vortices. In Fig. 3, we show vortex configurations of the 10x10 square lattices. A vortex first enters from the center. A sort of giant vortex appears at the center by breaking the superconductivity of the four bonding wires. In this case, a giant vortex $2\Phi_0$ appears in the four squares around the center node, breaking superconductivity of four bonding wires. Vortices diffuse toward peripheral regime. An aniti-vortex is surrounded by vortices (see Figs. 3 (b) and (c)). Local structure of vortices-anitvortex appears has a similarity to small superconduqting square plates [8].

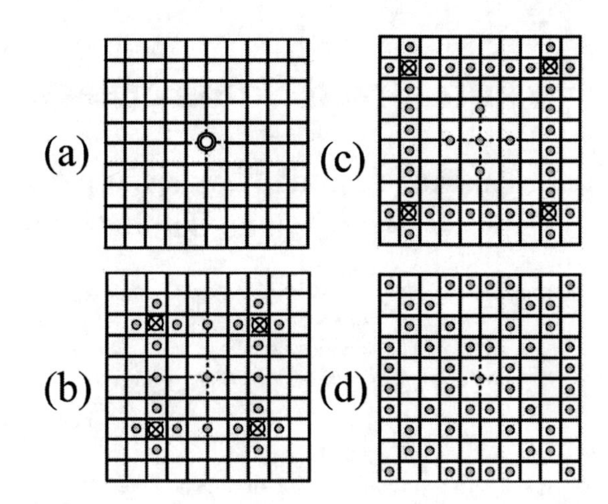

FIGURE 3. Spatial distribution of vortices for the occupied fractions of (a) 0.033, (b) 0.1833, (c) 0.333, and (d) 0.4833. Double circle represent $2\Phi_0$ vortex while the circle with cross is anti-vortex $-\Phi_0$.

CONCLUSIONS

We successfully fabricated the finite-sized square Pb networks by means of electron beam lithography and lift-off technique. The clear matching effect was found in magnetization curve of the finite-sized Pb networks. The SQUID microscope observations are inclusive due to the lack of spatial resolution. Theoretical calculations indicate that finite-sized networks behave quite differently from infinite ones.

ACKNOWLEDGMENTS

This work was partly supported by a Gant-in-Aid for Scientific Research from the Ministry of Education, Culture, Sports, Science and Technology of Japan (Grant No. 16360477).

REFERENCES

1. R. D. Parks and W. A. Little, *Phys. Rev.* **133**, A97 (1964).
2. B. Pannetier, J. Chaussy, R. Rammal, and J. C. Villegier, *Phys. Rev. Lett.* **53**, 1845 (1984).
3. S. Alexander, *Phys. Rev. B* **27**, 1541 (1983).
4. T. Puig, E. Rossel, Van Look, M. J. Van Bael, V. V. Moshchalkov, and Y. Bruynseraede, *Phys. Rev. B* **58**, 5744 (1998).
5. M. Kato and O. Sato, *Physica C.* **392-396**, 396 (2003).
6. O. Sato and M. Kato, *Phys. Rev. B* **68**, 094509 (2003).
7. H. Noda, H. Yoshikawa, O. Sato, M. Kato, K. Satoh, T. Yotsuya, and T. Ishida, *Physica C*, in press.
8. V. B. L. F. Chibotaru, A. Ceulemans, and V. V. Moshchalkov, *Nature* **408,** 833 (2001).

Experimental Vortex Ratchet Effect in Nanostructured Superconductors

J. E. Villegas, E. M. Gonzalez, M. P. Gonzalez, N. O. Nunez, J. V. Anguita* and J. L. Vicent

Departamento Física Materiales, Facultad Ciencias Físicas, Universidad Complutense, 28040 Madrid, Spain.
**Instituto Microelectrónica, Centro Nacional Microelectrónica, Consejo Superior de Investigaciones Científicas, Isaac Newton 8, PTM Tres Cantos, 28760 Madrid, Spain.*

Abstract. Superconducting Nb thin films were grown on different arrays of triangle-shape metallic islands. The vortex lattice dynamics could be strongly modified by these asymmetric vortex traps. These asymmetric pinning potentials lead to a rectification effect on the vortex motion: Injecting an ac supercurrent on the sample yields a net dc vortex flow. This vortex ratchet effect is adiabatic and reversible: The effect is frequency independent and the polarity of the dc voltage output could be tuned by the applied magnetic fields and the input ac currents.

Keywords: Nanostructured superconductors, ratchet effect, vortex pinning.
PACS: 74.78.Na, 74.25.Qt, 85.25.Am

INTRODUCTION

Ratchet mechanisms span inside many realms from Nature to research laboratories, from Applied Mathematics to Molecular Biology. We are alive because ratchet mechanisms are present in our cells, for example RNA polymerase moves by complex Brownian ratchet mechanisms during the crucial transcription step from DNA to RNA messenger (1). The core of ratchet mechanisms could be summarized as directional motion of out-of –equilibrium particles induced by a periodic asymmetric potential, without the need of being driven by non-zero average forces or temperature gradients.
Therefore, two ingredients are needed

a) Periodic structures which lack reflection symmetry.
b) Input signal yielding fluctuation motion of particles with zero-average oscillation.

For example, in proteins motors the room temperature induces the non-equilibrium background, and the asymmetric potential is provided by an on-off mechanism of charged and neutral proteins due to chemical reactions. Usually the biological motors belong to the so-called flashing ratchets. In this paper, we are going to deal with a different type of ratchet. In our case, the temperature is not playing the crucial role that plays in flashing ratchets. We are going to focus in a ratchet system which needs an applied zero-average driving force to be out of equilibrium and besides, in this ratchet device, the asymmetric potential (ratchet potential) is not time dependent. These types belong to the so-called tilted ratchets.

Recently, Villegas *et al.* have reported (2, 3) the fabrication and basic properties of a tilted ratchet system based on the motion of the superconducting vortex lattice in nanostructured superconducting films. They show that a Nb thin film deposited on a nanometric array of mesoscopic Ni triangles behaves as a tilted ratchet. An input ac current yields an output dc voltage and most remarkable the polarity of the dc voltage could be tuned at will with external parameters as the input signal and the applied magnetic field. The system could be easily modeled assuming that the pinned vortices in the Ni triangles and the interstitial vortices out of the Ni triangles are two repulsive particles according to the analysis of Savel'ev *et al* (4).

The temperature and field dependences of this ratchet system were reported in Ref. 2. In this paper we are going to focus in some of the main characteristic of this nanostructured device, mainly in its adiabatic behavior and the interplay between the interstitial and the pin vortices and the effect dependence with the asymmetric pinning center dimensions.

CP850, *Low Temperature Physics: 24th International Conference on Low Temperature Physics;* edited by Y. Takano, S. P. Hershfield, S. O. Hill, P. J. Hirschfeld, and A. M. Goldman

EXPERIMENT

The samples are fabricated by the combination of different techniques, sputtering, electron beam lithography, photolithography and finally ion etching. First, the array (triangles) pattern is written on PMMA resist using an electronic microscope, after the development a Ni film is sputtered. The final lift-off step defines the array of Ni triangles. On top of this array a Nb film is grown by dc sputtering technique. On this film, using conventional photolithography, a micrometric bridge is defined, and finally ion etching on the Nb + Ni triangles sample allows obtaining the bridge for transport measurements. The thicknesses of the Nb film and the Ni nanotriangles are always the same 100 nm and 40 nm respectively. The separation between triangles and the triangle side vary from sample to sample. More experimental details, as array and bridge dimensions are reported in (2).

Magnetotransport measurements allow us to know the number of vortices per unit cell of the array and the position of these vortices. Dissipation (resistivity) sharp minima occur when the applied magnetic field (number of vortices) matches the array unit cell. Therefore, selected values of the magnetic field, which correspond to the minima positions, could be used to control the number of vortices (2, 5). More subtle is to know the position of these vortices, that is, if they are pinned in the Ni triangles or they are interstitial vortices between Ni triangles, but taking into account the so-called filling factor (6) which estimates the saturation number of vortices per defect and the sample (I,V) curves (2), this subject could be addressed. In summary, for chosen magnetic field values we know the number of vortices per array unit cell and where they are located. The only experimental data needed are measurements of resistance versus magnetic field at constant temperature. The equal spaced minima, in these data, are the crucial parameter to know the superconducting mixed state relevant properties. Data from two samples will be analysis in this work. The main characteristics are for sample A: Triangle side 620 nm, array period 770 nm, and magnetic field at first minimum H = 32 Oe (n=1 one vortex per triangle) and sample B: Triangle side 430 nm, array period 602 nm and magnetic field at first minimum H = 54 Oe (n=1 one vortex per triangle).

EXPERIMENTAL RESULTS AND DISCUSSION

In the INTRODUCTION two conditions needed having ratchet mechanisms were presented. In Figure 1 we show the results when only one of the conditions is accomplished. The out of equilibrium condition is reached injecting an ac current on the sample in the Y-

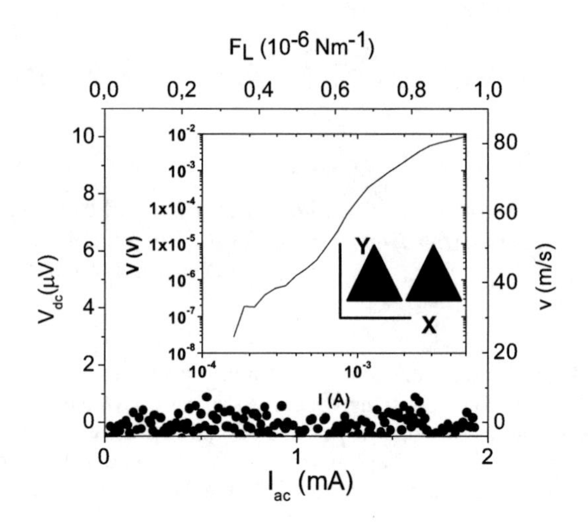

FIGURE 1. Sample A. T/T_c =0.99, H = 32 =Oe, n = 1 (one vortex per triangle). Vortex lattice motion parallel to the X-axis. There is not rectification effect. Y- axis: Left output dc voltage, right vortex lattice velocity extracted from the Josephson expression v = B x E (E, B being the electrical and magnetic fields respectively, see Ref. 2 for details). X-axis: Bottom input ac current (frequency 10 kHz), top Lorentz force on the vortex lattice, extracted from $F_L = J_{ac}$ x Φ_0 (J_{ac} being the applied ac current density and Φ_0 the flux quantum, see Ref. 2 for details). Inset shows (I, V) curve at the same temperature, we observe that the applied current is much larger than the critical current.

-axis direction, therefore the vortex lattice motion is in the X-axis direction. Since there is not broken symmetry in that direction the dc voltage recorded is zero, there is not ratchet effect, even the motion of the vortex lattice is driven by an ac force. However, Figure 2 shows a clear ratchet effect for the same sample A, for applied magnetic field of 64 Oe that corresponds to the second matching field (n=2 vortex per triangle). The ac current is now being applied in the X-axis direction, therefore the vortex lattice is moving in the Y-axis direction, and the second ratchet conditions is fulfilled, the vortices are feeling asymmetric potentials and a rectification is occurring from an ac current source to a dc voltage. Worth a while to note that the ratchet effect is not frequency dependent in the range attainable in our experiment up to 10 kHz. Villegas *et al.* (3) have explored this effect, recently van Vondel *et al* (7) has found the same behavior in superconducting Al films with array of asymmetric antidotes (holes). Both found that the vortex ratchet effect is in the adiabatic limit, but in both cases the analysis is done for n=2, (2) and n=1

(7). But the most remarkable effect happens when the applied magnetic field is increase and interstitial vortices appear. According to Villegas *et al.* (2) the number of vortices tunes the polarity of the ratchet effect, because pinned vortices and interstitial vortices move on opposite ratchet potentials and the interplay between the number of interstitial and pinned vortices as well as the input current strength determine the sign of the output voltage.

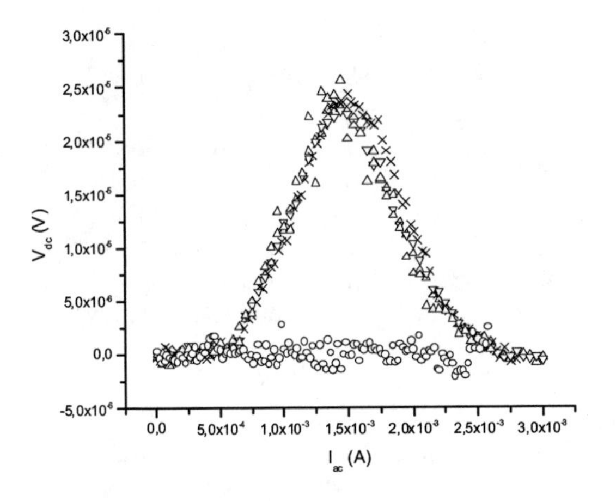

FIGURE 2. Ratchet signal for sample A at $T/T_c = 0.99$ and applied field H = 64 Oe (n=2). X-axis is the input ac current. Y-axis represents the output dc voltage. O experimental data for driving ac (frequency 10 kHz) current parallel to Y-axis (see inset Fig. 1), △, ▽, X, experimental data for driving ac current parallel to X-axis with ac frequencies of 10, kHz, 1 kHz and 1 kHz respectively.

We are going to study the possible adiabatic behavior when both types of vortices develop, that is vortices pinned in the Ni triangles and interstitial vortices.

In sample A the vortex filling factor is 3, see Ref. (2). Therefore n =6 is the situation more interesting, since the sample has 3 pinned vortices and 3 interstitial vortices. Following the analysis of Villegas *et al.* (3), the real ac driving ratchet effect could be mimic using dc (I,V) curves. First, a dc current is applied in the + X direction and the (I,V) data are recorded, after a dc current is applied in –X direction and the corresponding (I,V) curve is measured. Finally, both curves are subtracted and the net voltage is $V = V_p\text{-}V_n$. Fig. 3 shows this plot. The *dc ratchet* extracted mimics the behavior of the real ratchet. The interstitial vortices move first, because they do not feel the potential due to the Ni triangles and they need weaker driving forces. The interstitial vortices are located in *ghost and reversed* triangles among the real Ni triangles, see Ref. 2 and they move in opposite direction to the motion of the pinned vortices. The pinned vortices need stronger depinning forces (higher currents) to move, because they are *placed* in the real Ni triangles, that are pointing in the +Y direction. In conclusion this vortex ratchet is in the adiabatic limit. The frequency of the ac driving force does not play any role in this effect.

Finally, sample B is a good candidate to explore the effect of the pinning center dimensions and the

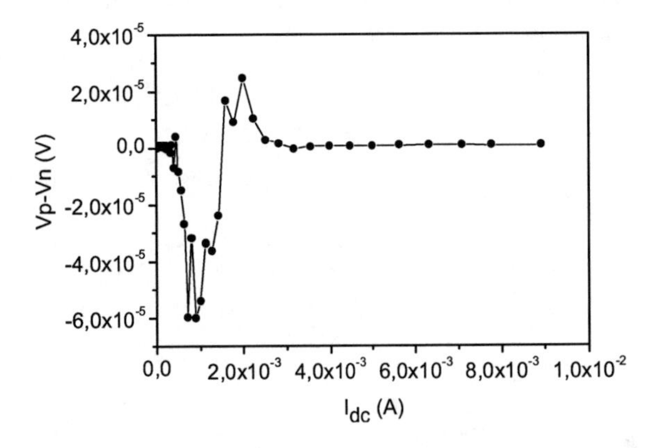

FIGURE 3. Sample A: *dc ratchet effect* extracted from (I,V) curves (see text for explanation). N = 6, applied field correspond to 3 pinned vortices and 3 interstitial vortices (H = 190 Oe). $T/T_c = 0.98$.

interplay between interstitial and pinned vortices is crucial to understand the reversible ratchet effect.

The magnetoresistance data of sample B allows us obtaining the matching magnetic field. At this field the vortex lattice is commensurable with the unit cell of the Ni triangle array. The matching field of sample B is 54 Oe, as well the pinning centers (triangles) are smaller than in sample A. We have larger matching field and lower pinning potential dimension. Sample B has larger separation and smaller pinning centers than sample A. The ratchet effect could be vanishing in this sample.

Taking into account the size of the defects (Ni triangles) and the expression for the filling factor (6) we obtain that in sample B the estimated filling factor is less than 3. Hence the triangles could not accommodate easily three pinned vortices as happens in sample A. The reversible rectification could be expected to occur for lower number of vortices than before. The vortex number three could be a good candidate to be an interstitial vortex. If this happens the situation n = 4 (two interstitial and two pinned vortices) could give the same result as n = 6 in sample A (see Ref. 2). Hence, the contribution of the positive (pinned vortices) and negative (interstitial vortices) could produce similar output dc voltage values, as happens in sample A for n = 6. That means that the positive and negative rectifications have similar amplitude.

Figure 4 shows that sample B is showing exactly the same trend than sample A, but with lower number of vortices per unit cell. Although sample B is not as good candidate as sample A to develop a reversible ratchet effect. The dimension of sample A are almost perfect, the real Ni triangle and the *ghost* triangles between the Ni ones are very similar. The vortex ratchet effect seems to be a very robust phenomena that could be easily observed in samples with array dimension far from ideal.

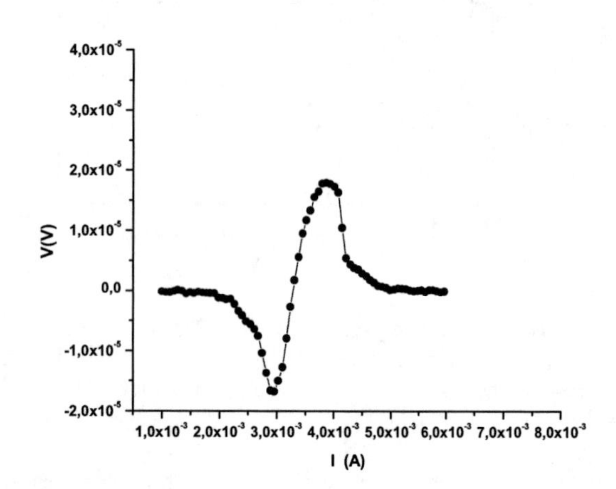

FIGURE 4. Sample B. Reversible rectification. X-axis is the input ac current. Y-axis represents the output dc voltage. N= 4: Two pinned vortices and two interstitial vortices per unit cell (H = 215 Oe). $T/T_c = 0.98$.

CONCLUSIONS

Nanostructured Nb film grown on top of array of Ni triangles rectified the vortex lattice motion. Injecting an ac current in the film an output dc voltage could be observed. This voltage is due to the net motion of the vortex lattice. This happens when the vortex lattice is moving on a landscape of asymmetric pinning potentials. The combination of zero average driving forces and asymmetric pinning centers is the clue to have a tilted vortex ratchet effect. This ratchet effect is adiabatic, that is the frequency of the zero average driving force does not play any role. The interplay between interstitial and pinned vortices governs the effect. The number of vortices per unit cell of the array can tune the polarity of the rectification effect.

ACKNOWLEDGMENTS

We want to thanks support from Spanish grants MAT2002-04543, NAN2004-09087, MAT2002-12385-E, CAM under grant 6R/MAT/0617/2002.Two of us, EMG and NON, want to thank Spanish Ministerio de Educacion y Ciencia for a Ramon y Cajal and Juan de la Cierva contracts respectively.

REFERENCES

1. G. Bar-Nahum, V. Epshtein, A. E. Ruckenstein, R. Rafikov, A. Mustaev, and E. Nudler, *Cell* **120**, 183-193 (2005).
2. J. E. Villegas, S. Savel'ev, F. Nori, E. M. Gonzalez, J. V. Anguita, R. Garcia, and J. L. Vicent, *Science* **302**, 1188-1191 (2003). E. M. Gonzalez, J. E. Villegas, M. P. Gonzalez, J. V. Anguita, and J. L. Vicent, *IEEE Trans. Appl. Supercon.* **15**, 888-891 (2005).
3. J. E. Villegas, E. M. Gonzalez, M. P. Gonzalez, J. V. Anguita, and J. L. Vicent, *Phys. Rev. B* **71**, 024519 (2005).
4. S. Savel'ev, F. Marchesoni, and F. Nori, *Phys. Rev. Lett.* **91**, 010601 (2003).
5. J. I. Martin, M. Velez, A. Hoffmann, I. K. Schuller and J. L. Vicent *Phys. Rev. Lett.* **83**, 1022-1026 (1999).
6. V. V. Moshchalkov, M. Baert, V. V. Metlushko, E. Rossell, M. J. van Bael, K. Temst, R. Jonckheere, and Y. Bruynseraede, *Phys. Rev. B* **54**, 7385-7393 (1996).
7. J. van de Vondel, C. C. de Souza Silva, B. Y. Zhu, M. Morelle, and V. V. Moshchalkov, *Phys. Rev. Lett.* **94**, 057003 (2005).

STM Observation of Vortex Lattice Transitions in Superconducting Single Crystals with Periodic Pinning Arrays

G. Karapetrov, J. Fedor*, M. Iavarone, D. Rosenmann, and W.K. Kwok

Materials Science Division, Argonne National Laboratory, Argonne, IL 60439, USA
**Institute of Electrical Engineering, Slovak Academy of Sciences, Dubravska Cesta 9, 84104 Bratislava, Slovakia*

Abstract. Superconductors containing mesoscopic artificially-engineered defects are orders of magnitude more resistant to the detrimental effect of the magnetic field. Here we present scanning tunnelling microscopy study of the vortex lattice configurations in a mesoscopic single crystal superconductor - normal metal heterostructure. We observe co-existence of a strongly interacting multiquanta vortex lattice and interstitial Abrikosov vortices that form a composite magnetic flux distribution which undergoes a series of transitions between different topological configuration states. The vortex configuration states are strongly dependent on the nanoscale architecture of the superconductor and applied magnetic field. Scanning tunnelling spectroscopy images show the evolution of vortex topological states when the number of flux quanta per unit cell changes.

Keywords: Mesoscopic superconductors, vortices, scanning tunneling microscopy, STM.
PACS: 74.50.+r, 74.25.Jb, 74.60.Ge

INTRODUCTION

Nanoscale superconductors are on the verge of being accepted for widespread use in communications and power distribution. This fact can be attributed to two main reasons: nanoscale superconductors are easier to cool because their mass is negligible and they are orders of magnitude more resistant to the destructive effect of the magnetic field. The latter mesoscopic effect, based on the interaction of magnetic vortices with artificial nano-engineered defects in the superconductor [1,2] have been exploited in model systems elucidating new methods of controlling vortex dynamics [3,4]. Understanding and control of vortex dynamics in mesoscopic superconductors could lead to novel nanoscale superconducting devices [5]. We use scanning tunneling microscopy (STM) to understand the influence of the pinning centers on the vortex distribution and vortex dynamics in the superconductor.

RESULTS

$NbSe_2$ single crystals were grown in evacuated quartz ampoules by an iodine vapour transport method [6]. SQUID magnetization measurements show typical superconducting transitions of 7.2K with transition width of 10mK. As-grown single crystals were cleaved to expose a clean flat surface. Using a focused beam of gallium ions we etched a set of 1 μm deep parallel grooves spaced equidistantly 500 nm apart. A second set of identical grooves, orthogonal to the original one was inscribed in the same area. Two micrometers of gold was electroplated on the patterned surface. The undisturbed backside of the patterned single crystal of $NbSe_2$ was subsequently cleaved several times until the underlying etched/electroplated gold pattern was uncovered. This novel approach in fabricating patterned atomically flat single crystal surfaces free of contaminants that are usually present when using conventional thin film fabrication techniques is a key requirement in order to perform tunnelling spectroscopy scans when imaging vortices [7].

To search for Abrikosov vortices in the superconductor, we performed conductance map of an area 1.25 x 1.25 μm^2. Visualisation of vortices by STS is based on the fact that inside a vortex core, the local density of states is gapless and therefore the conductance spectrum exhibits no gap around the Fermi energy.

The final structure consists of periodic array of elliptical gold islands that has a lower conductance at

CP850, *Low Temperature Physics: 24th International Conference on Low Temperature Physics;*
edited by Y. Takano, S. P. Hershfield, S. O. Hill, P. J. Hirschfeld, and A. M. Goldman

the superconducting peak energy. The gapless tunnelling spectra inside the island indicate that they consist of a normal metal. The square periodicity of the array of submicron Au islands forms an ideal periodic trapping potential for the Abrikosov vortex lattice with matching field of 8.2mT. The elliptical pinning centers on the other hand, start to define superconducting "channels" where the Abrikosov vortices can freely move and adopt the configuration that minimizes the total energy of the combined system of multiquanta and Abrikosov vortices.

When magnetic field is applied beyond 8th matching field, the Abrikosov vortices form a single chain parallel to the major axis of the elliptical pinning centers (Fig.1a). At this point the interaction between the Abrikosov vortices in each channel becomes strong and determines their distribution within the channel. The chains consist of periodically spaced vortices positioned in the middle of the superconducting channels. As the external magnetic field increases, so does the vortex density. This process takes place continuously up to a critical point where a new geometrical configuration becomes more energetically favorable. At this critical vortex density, the linear chain splits into a double chain (Fig.1b). This geometrical phase transition originates from the mesoscopic one-dimensional confinement of the vortex motion. The mesoscopic property of the system is linked to the width of the channel where the vortices can freely move. The confinement is on the order of the magnetic size of the vortex – the penetration length λ. This confinement of the Abrikosov vortex system in one dimension leads to quantized geometrical configurations with distinct symmetries that transition at specific magnetic fields. As the magnetic field is further increased, the vortex separation within each chain decreases while the distance between the chains remains constant. At higher fields, when the inter-vortex separation within a chain reaches a critical value equal to one quarter of the channel width, the double-chain system transitions into a triple chain, (Fig.1c) splitting the superconducting channel into four equal parts. Similar geometrical phase transitions were first theoretically predicted by Brongersma et al. [8] and this is the first direct experimental evidence confirming the chain configurations.

ACKNOWLEDGMENTS

We would like to acknowledge M.Marshall and the Center for Microanalysis of Materials at University of Illinois, Urbana-Champaign for support with the focus ion beam instrumentation. We would also like to thank R.Divan of the CNM, ANL and M. Moldovan (Northwestern, Evanston, IL) for their help with sample preparation and V.Vlasko-Vlasov for useful discussions. Part of this work was carried out in the Center for Microanalysis of Materials, University of Illinois at Urbana-Champaign, which is partially supported by the U.S. Department of Energy under grant DEFG02-91-ER45439. This work has been supported by the U.S. DOE, BES-Material Sciences under contract No. W-31-109-ENG-38.

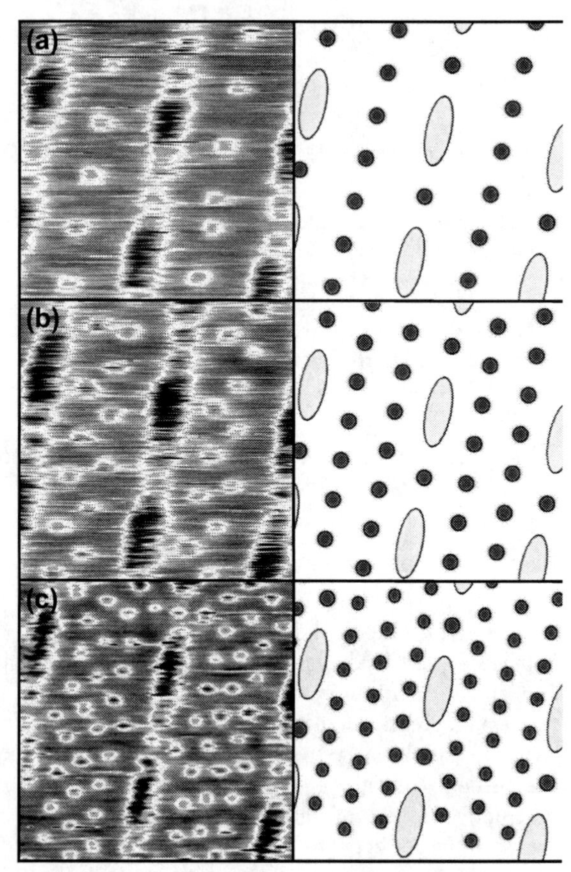

FIGURE 1. Images of vortex geometrical transitions in the periodic array of pinning centres using current image tunnelling spectroscopy (left panels) and corresponding schematic distributions (right panels) on an area of 1.25 x 1.25 μm^2 at 4.2 K. The applied magnetic field varies from 82.5 mT (a), 120 mT (b) to 212.5 mT (c).

REFERENCES

1. G.S. Mkrtchyan and V.V. Shmidt, *Sov. Phys. JETP* **34**, 195 (1972).
2. A.F. Hebard, A.T. Fiory, and S. Somekh, *IEEE Trans. On Magnetics* **1**, 589 (1977).
3. A.K. Geim, et al., *Nature* **390**, 259 (1997).
4. M. Baert, et al., *Europhys. Lett* **29**, 157 (1995).
5. A.S. Mel'nikov and V.M. Vinokur, *Nature* **415**, 60 (2002).
6. C.S. Oglesby, et al., *J. Cryst. Growth* **137**, 289 (1994).
7. G. Karapetrov et al., to be published in *Appl. Phys. Lett.*
8. S.H. Brongersma, et al., *Phys. Rev. Lett.* **71**, 2319 (1993).

Peak Effect and the Rhomb To Square Structural Transition In the Vortex Lattice

Y. Bruckental, B. Rosenstein, B.Ya. Shapiro, I. Shapiro, A. Shaulov and Y.Yeshurun

Institute of Superconductivity, Department of Physics, Bar-Ilan University, Ramat Gan 52900, Israel
National Chiao Tung University, Department of Electrophysics, Hsinchu, Taiwan, R.O.C.

Abstract. The theory of structural transformation of the vortex lattice in a fourfold symmetric type II superconductor in the presence of both thermal fluctuations and quenched disorder is constructed. We show that in the absence of pinning, the slope of the square to rhomb transition in the T-H plane is generally negative: thermal fluctuations favor a more symmetric square lattice. Disorder's influence on the slope is just the opposite- it favors a less symmetric rhombic phase in which the fourfold symmetry is spontaneously broken. The second magnetization peak line in LaSCO in a wide range of doping is well described as a result of the transition.

Keywords: Vortex phase transition, Peak effect, Thermal fluctuations.
PACS: 74.20.De 74.25.Ha 74.25.Qt

It has been known for a long time that both in anisotropic low T_c and in the high T_c type-II superconductors, the vortex solid phase undergoes structural phase transformations (SPT). In particular, in the overdoped LaSCO at high magnetic fields and low temperature the square lattice was observed using SANS by Gilardi et al [1]. When the field is lowered, the rhombic vortex lattice appears. On the other hand, it was shown theoretically that the SPT leads to a peak effect due to softening of certain elastic modulus identified in the case of the square-to-rhomb transition as the "squash" modulus. It is natural to conjecture that the second magnetic peak clearly observed in tetragonal superconductor LaSCO marks SPT in the vortex lattice. The square-to-rhomb transition is by far the simplest possible structural phase transition. In the less symmetric phase there are two rhombic lattices differing by a 90° rotation, while in the more symmetric phase, i.e. the square lattice the vector between two closest vortices may be either parallel to the crystallographic axis a of the atomic lattice, or rotated by 45° with respect to it. Physically, the coupling between the crystal lattice and the vortex lattice in a fourfold symmetric superconductor such as LaSCO originates in two somewhat related anisotropies on a microscopic scale. The first is the Fermi velocity dependence on the angle θ. The second is the anisotropy of the gap function resulting in the anisotropy of the vortex-vortex interaction on the scale of the coherence length ξ. This anisotropy is obviously present and perhaps dominant in the d-wave superconductors due to the nodes in the order parameter. It is this asymmetry, which is effectively taken into account in the Ginzburg - Landau approach to the rhomb-to-square structural phase transition and it is taken into account within the non-local London approach as an asymmetric moment cutoff [2]. At the first glance, the structural phase transition in such a system, even at finite temperature (below the melting temperature of course), is driven by four-fold anisotropy of the inter-vortex interactions on scales smaller that the inter-vortex distances, and consequently have nothing to do with anisotropy of the vortex core. Thermal fluctuations generally prefer a more symmetric square lattice. The interplay between the four fold anisotropy and thermal fluctuations is extremely delicate, so the problem should be considered from a more fundamental stand point. A standard approach to the crystal structure of point-like (or rigid rods) objects at finite temperature requires a sufficiently sophisticated account of the lattice anharmonicity. The simplest version of such a theory takes into account interacting phonon excitations self-consistently (the self consistent harmonic approximation - SCHA).

We adapt in this contribution SCHA to consider structural transformations in tetragonal superconductor and apply the results to $La_{2-x}Sr_xCuO_4$ with different doping concentrations x.

CP850, *Low Temperature Physics: 24th International Conference on Low Temperature Physics;*
edited by Y. Takano, S. P. Hershfield, S. O. Hill, P. J. Hirschfeld, and A. M. Goldman

We obtain a structural phase transition line with a negative slope in the B-T plane unlike other theories of thermal fluctuations. In our theory, unlike the preceding ones no ultraviolet cutoff is required. Using the intervortex interaction for a d-wave superconductor derived microscopically by Yang results compare well with experimental data in LaSCO in a wide range of doping.

The vortex-vortex potential at distances larger than the core size was derived from a microscopic model of the d-wave superconductor by Yang [3]

$$E = \frac{1}{2} W\{\vec{R}_a - \vec{R}_b + \vec{u}_a - \vec{u}_b\}; u_a^\alpha = \int_{BZ} u_q^\alpha \frac{d^2 q}{(2\pi)^2}$$

$$W(q_x, q_y) = \{1 + \eta[2bh/(1+bg)]^2\} V(g), \quad (1)$$

$$V(g) = \frac{L_z \Phi_0^2}{4\pi}\left[\frac{1}{1+bg^2} - \frac{1}{\kappa^2 + bg^2}\right]$$

$$b = \frac{4\pi^2 \lambda^2 B}{\Phi_0}; g = q_x^2 + q_y^2; h = q_x^2 - q_y^2.$$

Here B is the magnetic induction, Φ_0 is the unit flux, $\kappa = \lambda / \xi$, R_a is the vortex coordinate in the vortex lattice, while u_a notes vortex displacement. To develop a SCHA one has to take into account in the statistical sum the interaction between phonons to the third and fourth orders in vortex displacement. Performing gaussian integration over u_q and taking into account the long range order fluctuations of the displacement correlator $\Delta_{\alpha\beta}(q) = L_z T^{-1} \langle u_q^\alpha u_{-q}^\beta \rangle$ for tetragonal lattice in the form $\Phi_{\alpha\beta} = (\Delta_{\alpha\beta})^{-1}$

$$\Phi_{xx} = c_{11} q_x^2 + c_{66} q_y^2; \Phi_{xy} = c q_x q_y;$$
$$\Phi_{yy} = c_{11} q_y^2 + c_{66} q_x^2; c = c_{11} - \frac{c_{sq}}{2} + c_{66}; \quad (2)$$

where c_{11}, c_{66}, c_{sq} are the variation parameters related with elastic moduli in the crystallographic axes $c_{11} = C_{11} + C_{66} - C_{sq}/4; c_{66} = C_{sq}/4; c_{sq} = 4C_{66}$. one obtains the variational free energy for the square lattice. In order to find the line of SPT one has to find the minimum of this energy with respect to the variational parameters. It was found for the SPT line

$$b = (1-t^4) f\left(\frac{at}{1-t^4}\right) \simeq \frac{(1-t^4)^\nu}{t^{\nu-1}}\left[s_1 \frac{(1-t^4)}{t} - s_2\right], \quad (3)$$

here $t = T/T_c$, $a = (4\pi^3 \lambda_0^2 T_c / \Phi_0^2 L_z)$; $s_1, s_2 \sim 1$ while s_1 and s_2 are the fitting parameters.

Three $La_{2-x}Sr_xCuO_4$ single crystals, with different amount of Sr, were grown by the traveling-solvent floating-zone method: underdoped with doping concentration: $x = 0.126, T_c = 32K, \lambda \approx 2 \cdot 10^{-5} cm$ optimally doped $x = 0.154, T_c = 37K, \lambda \approx 10^{-5} cm$, and overdoped $x = 0.194, T_c = 30K, \lambda \approx 10^{-5} cm$. Samples of these crystals, were cut into parallelepiped shape with dimensions (c×a×b) 1.05×1.7×2.3 mm , 1.08×0.7×1.17 mm, 2.08×0.83×0.96 mm, respectively. Measurements were performed using a commercial superconducting quantum interference device (SQUID) magnetometer (Quantum Design MPMS-XL) utilizing the RSO technique with 1-cm scans. Magnetization was measured at constant temperature as a function of the external field applied parallel to c axis and being swept up to 5 T and down to zero in steps of 200 Oe. The structural reconstruction of the vortex lattice leads to softening of the squash elastic modulus and thus to peak effect [4]. The position of the second peak magnetization versus temperature presented in Fig.1 is in a good agreement with theoretical prediction (solid lines).

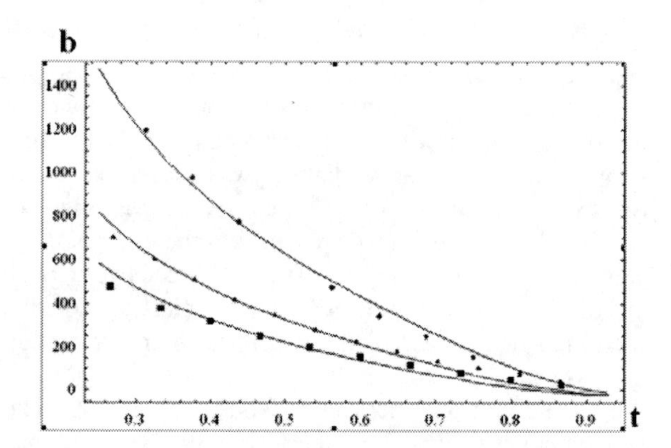

FIGURE 1. Comparison of the experimental second magnetization peak line of $La_{2-x}Sr_xCuO_4$ with the theoretical square-to-rhomb transition line. The rhombs, stars squares, and represent the underdoped, optimally doped and overdoped samples ($\eta = 0.03; 0.02; 0.01$) respectively. The theoretical curves (solid lines) are all for κ=75 .

REFERENCES

1. R. Gilardi et al., *Phys. Rev. Lett.* **88**, 217003 (2002).
2. V.G. Kogan, *Phys. Rev. B* **54**, 12386 (1996); *Phys. Rev B* **55**, R8693 (1997); P. Miranovich and V.G.Kogan, *Phys. Rev. Lett.* **87**, 137002 (2001).
3. M.C. Dai and T.J. Yang, *Physica C* **305**, 301 (1998).
4. B. Rosenstein and A. Knigavko, *Phys. Rev. Lett.* **83**, 8444 (1999).

GLASS TRANSITION IN VORTEX MATTER

Baruch Rosenstein,[1] Dingping Li,[2] and Valery M. Vinokur[3]

[1]*Electrophysics Department, National Chiao Tung University, Hsinchu 30050, Taiwan, R.O.C.*
[2]*School of Physics, Peking University, Beijing 100871, China, P.R.C.*
[3]*Materials Science Division, Argonne National Laboratory, Argonne, Illinois 60439, USA*

Abstract. We develop a unifying approach bridging between the spin and vortex glasses, based, on one hand, on the application of the replica method to Ginzburg-Landau description of the system in terms of the superconductor order parameter near Tc and, on the other hand, on the notion of the extreme value statistics that govern the distribution of the energy barriers controlling the glassy slow dynamics of the vortices. Focusing on the vortex systems not very far from the upper critical field we demonstrate that the glassy behavior is associated with the continuous replica symmetry breaking, which is the inherent feature of the long range spin glasses. We apply our results to the description of vortex phase diagram and find the vortex glass transition line, which appears different and well separated from both the melting line and from the so-called second peak lines.

Keywords: Abrikosov vortex, pinning, vortex glass, replica symmetry.
PACS: 74.20-z, 74.25.Fy, 74.25.Dw, 74.25.Qt

The concept of a glass is a paradigm for the disordered strongly correlated systems. The exemplary model representing glasses is the so-called *spin glass* characterized by a hierarchical structure and the "fragmented" architecture of the energy landscape which consists of the energy valleys separated by the infinite barriers leading to the loss of ergodicity and slow glassy dynamics. In the same glassy domain of the phase diagram, the relevant activation barriers obey the so-called Gumbel statistics.

Not too close to the upper critical field, vortices are well defined and vortex dynamics is described by the dependence of the mean vortex velocity on the applied force. At low temperatures and small driving forces well below the depinning threshold the dynamics is controlled by thermally activated jumps of correlated regions of the vortex elastic manifold over the pinning energy barriers separating different metastable states. It was shown in ref.1 that the creep dynamics originates from the extreme value statistics of barriers controlling thermally activated vortex motion. This points to hierarchical structure of the vortex glass configuration space. Since the waiting time $\tau(E)$ for hops between metastable states separated by energy barriers is thermally activated, the distribution of waiting times scales as a power law at large τ: $\Psi(\tau)\propto\tau^{-1-\alpha}$. The distribution of waiting times is cut off at the waiting time corresponding to the optimal barrier. At larger times the dynamics is no longer thermally activated as string segments slide freely. Thus that the glassy dynamics relies on the divergence of the average waiting times at low temperatures where $\alpha<1$. Now turn to the high-temperature region. Above the depinning temperature the motion of the strings is then governed by the waiting times distribution function with $\alpha= 1$. The typical waiting time now is logarithmically divergent, leading to the power-law I-V characteristics. Therefore, above the single vortex depinning temperature the single vortex dynamics becomes marginally corresponding to the logarithmically divergent barriers.

Within the Ginzburg-Landau (GL) approach the quenched disorder is incorporated by making coefficients of the GL equations random variables. We use the replica method of averaging of the disorder[2] within the lowest Landau level (LLL) approximation[3]. The resulting model is analyzed nonperturbatively using Gaussian approximation. The glass state implies the loss of ergodicity and reversibility with respect to dynamic processes mathematically expressed by the replica symmetry breaking (RSB). The correct solution minimizing the free energy and accounting for RSB is given by the subclass of the matrices which has a hierarchical structure and which can be parameterized by the Parisi function describing the overlap between different "valleys" in the potential landscape[2]. The

CP850, *Low Temperature Physics: 24th International Conference on Low Temperature Physics;*
edited by Y. Takano, S. P. Hershfield, S. O. Hill, P. J. Hirschfeld, and A. M. Goldman

function is non-constant in RSB state. In the disordered phases (domain to the right of the irreversibility line in Fig.1) the replica symmetric solution is stable, while in the glassy phases (the left side of the line) the RSB solution is stable. In dynamics the energy barrier scales translate into the relaxation time scales. The necessary characteristic time scale $\tau_0=\xi^2/D$, follows from the time dependent GL equation containing the diffusion constant D. Therefore Parisi function corresponds to the distribution of the relaxation times which we discussed within the framework of the single vortex approach.

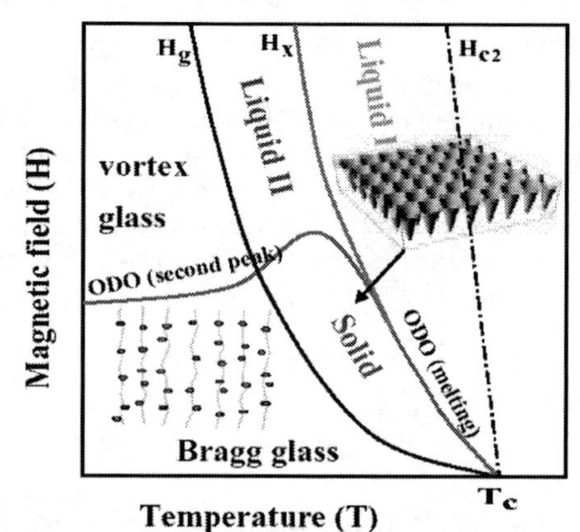

FIGURE 1. Generic phase diagram of the vortex matter. The order-disorder line separates the crystalline phase from the homogenous phase. The glass transition line separates the glass from the weakly pinned phases, while line H_x is a crossover between two homogeneous phases, locally pinned Liquid I and essentially unpinned Liquid II. The generic phase diagram is recently confirmed in ref. 4.

The general picture for the glass state allows to illuminate several delicate features of the vortex matter phase diagram, in particular, to differentiate between the glass (irreversibility) line and the order-disorder transition line (ODO).

1. *The positional order-disorder line.* The ODO transition is defined as the moment of the loss of the translation and rotation symmetry; and the intensity of the first Bragg peak can be used as an order parameter. The unified first order ODO line in Fig.1 comprising the melting and the "second magnetization peak" segments, separates the homogeneous and the crystalline phases. The broken symmetry is not directly related to pinning, however the location of the line is sensitive to the strength of disorder.

2. *The glass transition line.* The glass transition line, shown in Fig.1, is the locus of the RSB continuous phase transition. In the crystalline state the glass line is nearly vertical and is actually the single vortex depinning line mentioned above. The lateral modulation makes a very small difference to the glass line although it is very important for the location of the ODO line. Consequently there are four distinct phases: pinned solid, pinned liquid, weakly pinned solid with marginal glassy dynamics, and unpinned liquid. Shown in Fig. 2 is the comparison of the glass and the order-disorder lines with the experimental phase diagram of a 2D organic superconductor in ref. 5.

3. *Transition to marginal glass.* It has been observed[6] that inside the liquid phase the two qualitatively different dynamical regimes take place (sometimes referred to as liquid I and liquid II). This feature receives a natural explanation within the developed description. The line separating two regimes coincides with the melting line of the pristine material, i.e. clean sample without disorder. In order for the glassy dynamics to set in, the two conditions are to be satisfied: (i) the system has to be in a solid phase and (ii) replica symmetry should be broken. The violation of either of this condition drives the system into the domain of the marginal dynamics. The role of point disorder in this case is to expose the onset of the local crystalline order in the liquid. The irreversibility line cuts the solid part of the phase diagram into two domains, the low temperature /field domain with glassy dynamics due to algebraically divergent barriers and the high-temperature domain of marginal dynamics.

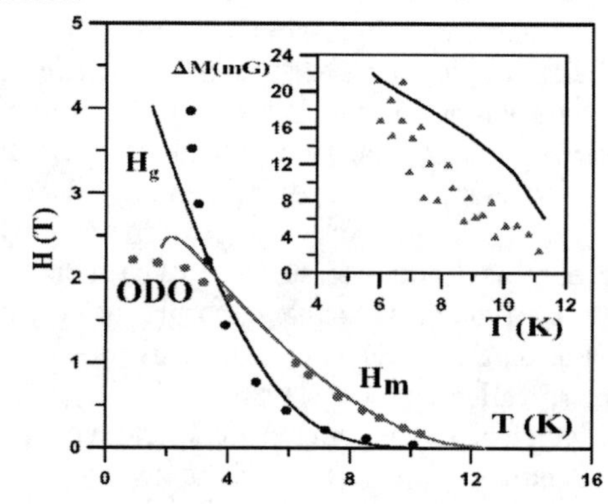

FIGURE 2. Comparison of the melting and the glass line and the magnetization jump in κ-(BEDT-TTF) of ref. 5.

REFERENCES

1. Vinokur, V. M., Marchetti, M. C., and Chen, L.-W., *Phys. Rev. Lett.* **77**, 1845 (1996).
1. Mezard, M., Parisi, G *J. Phys. (France) I* **1**, 809 (1991).
2. Li, D., Rosenstein, B., *Phys. Rev. Lett.* **90**, 167004 (2003); R. Ikeda, J. Phys. Soc. Jap. **69**, 559 (2000).
4. Beidenkopf. H. *et.al.*, Phys. Rev. Lett. **95**, 257004 (2005).
5. Shibauchi T. et al, *Phys. Rev. B* **57**, 5622 (1998).
6. Bouquet F. et al, *Nature* **411**, 448 (2001); Fuchs D.T. *et.al.*, *Phys. Rev. Lett.* **80**, 4971 (1998).

Additional Non Equilibrium Processes in the Dynamic Interaction between Flux Quanta and Defects

S. Pace[a], G. Filatrella[b], G. Grimaldi[a], A. Nigro[a] and M.G. Adesso[a]

[a]*Regional Laboratory SuperMat, INFM Salerno, and Physics Dept., Via S. Allende I-84081Baronissi, Italy*
[b]*Dept. of Biological and Environmental Sciences, Via Port'Arsa 11, I-82100 Benevento, Italy*

Abstract. In this work we discuss the limits of the standard analysis dealing with material defects as potential wells for flux quanta in type II superconductors. Such description, based on the thermodynamic free energy, is appropriate for quasi-static processes only, and its validity should be tested against dynamic conditions for vortices driven by a constant bias above the critical current. We briefly discuss the non equilibrium phenomena that occur during the entry and the exit of vortices from defects, comparing the characteristic times of the vortex motion with the non equilibrium characteristic time scales. Specifically, in BCS superconductors the Cooper pair recombination processes determine phonon emission and corresponding energy losses. For this reason, we underline that non equilibrium processes doubt the validity of the treatment of defects as potential wells, especially well above the critical current when vortices are driven at higher speed.

Keywords: Non equilibrium superconductivity; Vortex dynamics; Pinning; Type-two superconductors.
PACS: 74.20.-z; 74.25.Qt; 74.40.+k

INTRODUCTION

In type II superconductors vortices can move driven by macroscopic currents. In the absence of material defects, vortex dynamics is governed by the well known flux flow motion described by Bardeen-Stephen [1]: dissipations are due to electric fields induced by the time variation of the local magnetic field felt by an observer at rest. In a different approach non equilibrium phenomena have been considered in the past in the description of the flux flow motion in homogeneous superconductors [2,3]. Recently flux flow dissipations have also been ascribed to non equilibrium conditions that establish in the vortex core during vortex motion, owing to the formation of excited states and their subsequent relaxations [4]. Moreover, it has also been verified that vortex motion is accompanied by an entropy flux, producing a local heating and cooling, respectively, in the regions which correspond to the vortex exit and entry into the sample [5]. In this way such processes are clearly related to non equilibrium phenomena.

In the presence of impurities in superconducting materials, the interaction between vortices and defects is a relevant mechanism in the vortex dynamics [6, 7, 8]. Here we focus on the specific case of defects large enough to allow the superconducting order parameter, ψ, to decrease to zero in the normal impurity region. We also limit our analysis to the interaction between a single vortex with a single defect, which is the starting point for the collective pinning description.

NON EQUILIBRIUM DYNAMICS

In the standard analysis of the classical vortex-defect interaction, for the vortex exit from the defect free energy must be provided equal to that necessary to create the normal core. This is the condensation free energy density multiplied by the normal volume, $V_n=\xi^2*L$, where ξ is the coherence length and L is the defect length along the vortex. Therefore the free energy F of the system becomes a function of the vortex center, x, and it results in free energy minima at defects centers. For this reason the defects are considered pinning centers that act as potential wells both in static and dynamic conditions. Nevertheless, in the quasi-static processes only, the driving force is strictly equal to the F derivative [9].

Since in the vortex exit from the defect a destruction of the superconducting state takes place in the volume V_n, in the vortex entry into the defect the opposite process occurs. This corresponds to Cooper pair condensation phenomena in a normal region (that

CP850, *Low Temperature Physics: 24th International Conference on Low Temperature Physics;*
edited by Y. Takano, S. P. Hershfield, S. O. Hill, P. J. Hirschfeld, and A. M. Goldman

was occupied by the vortex core outside the defect) which turns back to the superconducting state when the vortex goes inside the defect, as shown in Figure 1. These condensation processes do not take place in the flux flow vortex motion without defects, since in this case the total Cooper pair number is time constant.

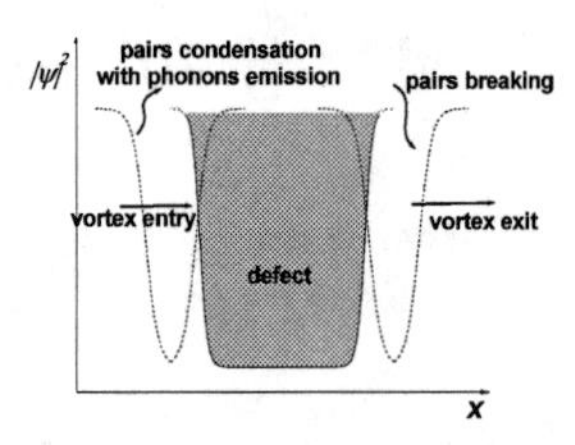

FIGURE 1. Pairs condensation and breaking related to a vortex crossing a defect.

In the vortex entry the dominant non equilibrium processes are: quasiparticles relaxation and Cooper pair condensation [10]. All these processes are usually accompanied by phonon emission and subsequently determine the relaxation of the superconducting gap Δ. Besides, far from equilibrium conditions, the phonon distribution function changes too, with a subsequent local heating. As a consequence, in the last step, the relaxation processes are governed by the thermal diffusion. A similar scenario of the normal to superconducting transition occurring during vortex entry into a defect can also be found in the study of relaxation phenomena following a hot spot [11]. For BCS superconductors, in near equilibrium conditions, quasiparticles relaxation lifetimes and Cooper pair condensation times are of the order of nanoseconds, while the Δ relaxation time is even larger [10, 12]. The whole relaxation process can be estimated to take place over times larger than $\tau_r \approx 10^{-8}$ sec. This time should be compared with the dynamics time of the vortex entry into the defect, τ_{in}. If $\tau_r \ll \tau_{in}$ then far from equilibrium conditions and additional losses due to vortex entry into the defect can be neglected. If $\tau_r > \tau_{in}$ irreversible phenomena occur in the vortex dynamics [9, 13]. A rough estimate of τ_{in} can be done by assuming the flux flow velocity v_f=1-100 m/sec in Nb films [14] and the defect size larger than $2*\xi$, with $\xi \sim$ 100 nm, so that τ_{in} is $\approx 10^{-8}$- 10^{-10} sec, shorter than the relaxation time τ_r. In this case and for all the cases of faster vortex dynamics, the non equilibrium irreversible phenomena of vortex entry into the defects with the subsequent phonon emission have to be considered. As a consequence for a rapid vortex entry into the defects, additional losses are present with respect to the classical flux flow state. In the vortex entry, the free energy difference, between the vortex outside and inside the defect, turns at least partially into heat and its spatial derivative cannot act as a force able to accelerate the vortex motion.

CONCLUSIONS

We remark the presence of additional non equilibrium phenomena associated to the vortex entry into material defects. Although non equilibrium processes have been investigated within the flux flow resistivity state, those related to the vortex motion through normal defects have not yet been considered. In particular if non equilibrium relaxation times are longer than vortex dynamics ones, these relaxation processes play a crucial role determining irreversible phenomena in the vortex motion through defects. We state that it is well known that quasiparticle scattering and Cooper pair condensation occur with phonon emission, and this phonon energy is lost in the thermal bath so it cannot be able to produce any useful work. This establishes non negligible irreversibility conditions during fast vortex motion, so that the standard vortex dynamics should be revised.

REFERENCES

1. Y. B. Kim and M. J. Stephen, "Flux Flow and Irreversible Effects", in *Superconductivity*, edited by R. D. Parks, New York: M. Dekker Inc.,1969, pp.1107-1165.
2. M. Tinkham, *Phys. Rev. Lett.* **13**, 804-807 (1964).
3. J. R. Clem, *Phys. Rev. Lett.* **20**, 735-738 (1968).
4. M. A. Skvortsov, D. A. Ivanov and G. Blatter, *Phys. Rev. B* **67**, 014521-1-11 (2003).
5. T. T. M. Palstra, B. Batlogg, L. F. Schneemeyer and J. V. Waszczak, *Phys. Rev. Lett.* **64**, 3090-3093 (1990).
6. E. H. Brandt, *Rep. Prog. Phys.* **58**, 1465-1594 (1995).
7. G. Blatter, M. V. Feigel'man, V. B. Geshkenbein, A. I. Larkin and V. M. Vinokur, *Rev. Mod. Phys.* **66**, 1125-1388 (1994).
8. A. I. Larkin and Yu. N. Ovchinnikov, "Vortex Motion in Superconductors",in *Non Equilibrium Superconductivity*, edited by D. N. Langenberg and A. I. Larkin, Amterdam: Elsevier Science, 1986, pp. 493-542.
9. H. B. Callen, *Thermodynamics and Introduction to Thermostatics*, New York: John Wiley & Sons, 1985, pp. 96-100, 307-314.
10. S. B. Kaplan, C. C. Chi, D. N. Langenberg, J. J. Chang, S. Jafarey and D. J. Scalapino, *Phys. Rev. B* **14**, 4854-4873 (1976).
11. A. G. Kozorezov, *Nuclear Instr. and Methods in Phys. Res. A* **444**, 3-7 (2000).
12. J. A. Pals, K. Weiss, P. M. T. M. van Attekum, R. E. Horstman and J. Wolter, *Phys. Reports* **89**, 323-390 (1982).
13. S. Pace, G. Filatrella, G. Grimaldi and A. Nigro, *Phys. Lett. A* **329**, 379-384 (2004).
14. M. Velez, D. Jaque, J. I. Martin, F. Guinea and J. L. Vicent, *Phys. Rev. B* **65**, 094509 (2002).

Experimental Evidence for Crossed Andreev Reflection

D. Beckmann* and H. v. Löhneysen¶§

*Forschungszentrum Karlsruhe, Institut für Nanotechnologie, P.O. Box 3640, D-76021 Karlsruhe, Germany
§Forschungszentrum Karlsruhe, Institut für Festkörperphysik, P.O. Box 3640, D-76021 Karlsruhe, Germany
¶Physikalisches Institut, Universität Karlsruhe, D-76128 Karlsruhe, Germany

Abstract. We report on electronic transport properties of mesoscopic superconductor-ferromagnet spin-valve structures. Two ferromagnetic iron leads form planar tunnel contacts to a superconducting aluminum wire, where the distance of the two contacts is of the order of the coherence length of the aluminum. We observe a negative non-local resistance which can be explained by crossed Andreev reflection, a process where an electron incident from one of the leads gets reflected as a hole into the other, thereby creating a pair of spatially separated, entangled particles.

Keywords: superconductivity, ferromagnetism, Andreev reflection, entanglement.
PACS: 74.45.+c, 03.67.Mn, 85.75.-d

INTRODUCTION

We have recently reported on the experimental investigation of electronic transport properties of superconductor-ferromagnet non-local spin-valve structures [1]. On the length scale of the superconductor's coherence length, spin-dependent transport was observed at subgap bias voltages. Our data were explained with a model based on the superimposition of two processes, namely crossed Andreev reflection (CAR) and elastic cotunneling (EC). However, our experimental setup and resolution were not sufficient to delineate the contribution of these two processes. Here, we report on preliminary data of our next generation experiment which overcomes these limitations, and show evidence for dominating CAR at low bias voltages.

EXPERIMENT

Our sample layout consists of two ferromagnetic iron leads A and B which form tunnel contacts to a weakly oxidized superconducting aluminum wire (Fig. 1a). The contact separation is a few 100 nm, comparable to the coherence length of the aluminum wire. Contact A is used to inject a DC current I_A, while contact B measures the voltage U_B with respect to the chemical potential of the superconductor. A spin-up electron incident from contact A on the superconductor in the source-drain voltage window from 0 to U_A can be transmitted to contact B either as spin-up electron at positive energy (EC) or as spin-down hole at negative energy (CAR) (see Fig 1b).

a)

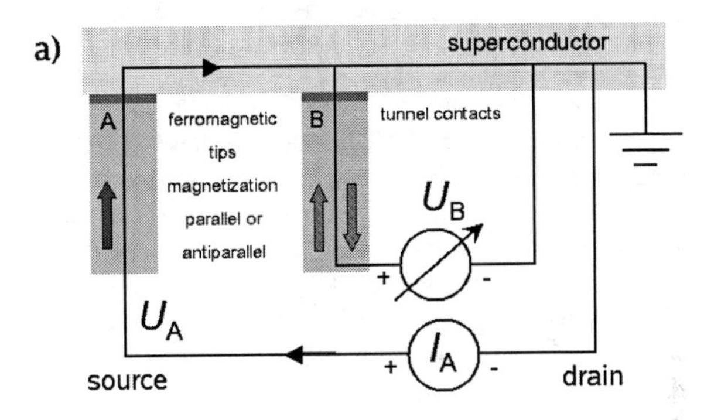

b)

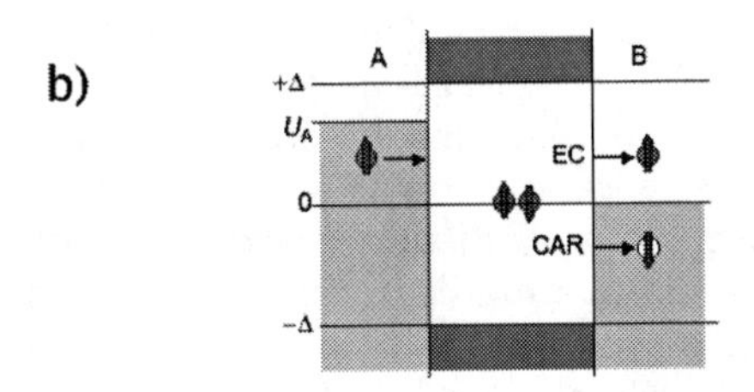

FIGURE 1. Experimental scheme: a) Ferromagnetic leads A and B form tunnel contacts to a superconducting bar. A is used for current injection, B for voltage detection inside the current path. b) Energy scheme for CAR and EC (see text).

For EC (or incoherent electron transmission in the normal state, including the effects of spin accumulation), the voltage U_B is therefore always inside the source-drain window, i.e. for positive U_A

CP850, *Low Temperature Physics: 24th International Conference on Low Temperature Physics;*
edited by Y. Takano, S. P. Hershfield, S. O. Hill, P. J. Hirschfeld, and A. M. Goldman

also U_B will be positive. For CAR, U_B will then be negative, i.e. outside the source-drain window [2]. This issue has been discussed in Ref. 2 in a different setup with only local Andreev reflection, but applies to our situation as well.

RESULTS

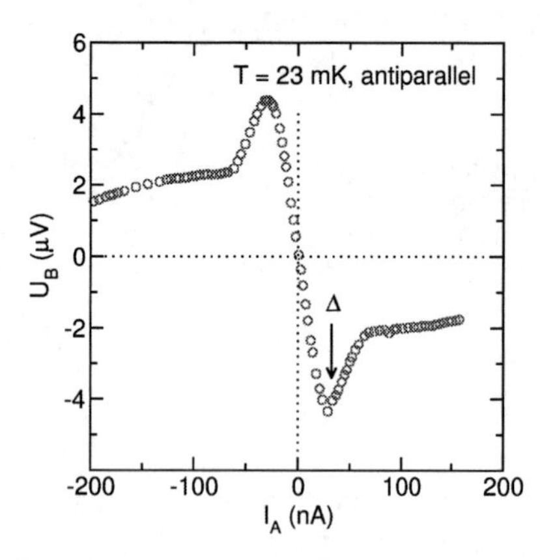

FIGURE 2. Bias dependence of U_B. Below the energy gap of the superconductor, a negative voltage (i.e. outside the source-drain window) is observed.

Figure 2 shows the non-local voltage U_B as a function of injector current I_A for one of our samples at low temperature for antiparallel magnetization alignment, where CAR is favored over EC due to the reversed spin of the hole. At low positive bias currents, a negative voltage is observed, i.e. U_B is outside the source-drain window. At higher bias current, the slope of U_B becomes positive. The turnaround occurs at the current which corresponds to $U_A = 200\mu V$, i.e. at the superconducting energy gap of aluminum, as indicated by the arrow in Fig. 2. Similar behavior was seen for several samples. For one sample, we observed a dominating positive slope at low bias, followed by a negative slope at higher bias (but still below the gap), similar to the observations made by Russo et al. [3] in a different experimental setup using an AC method, as opposed to our DC experiment. The reason for the qualitatively different behavior of some samples is subject to ongoing investigations.

DISCUSSION

Our previous experiment [1] featured a non-local voltage detection (i.e. outside the current path), which has the advantage of being extremely sensitive to both spin accumulation and coherent non-local processes. However, in such a setup the sign of the non-local voltage is not conclusive evidence for CAR, as one compares two different voltages inside the source-drain window. In the presence of spin accumulation, the measured voltage can have either sign even in the normal state without CAR. The most significant change over our previous experiment is therefore the detection of the voltage U_B inside the current path, where the observation of a negative U_B for positive U_A conclusively means that U_B is outside the source-drain window, indicating CAR as the dominating non-local process. The positive slope at bias voltages above the superconducting energy gap can be attributed to the onset of electron transmission through allowed quasiparticle states.

CONCLUSION

We have shown the observation of an unusual negative four-probe resistance occurring in superconductor-ferromagnet spin-valve-like structures. The effect can be explained by crossed Andreev reflection (CAR). Our results show that it may be feasible to create solid-state entanglers with CAR as the dominating transport process. Further systematic investigation is required for a better understanding of CAR compared to competing processes like elastic cotunneling.

ACKNOWLEDGMENTS

We thank D. Feinberg and R. Melin for useful discussions, and especially P. Samuelsson, D. Sanchez, R. Lopez, E. Sukhorukov and M. Büttiker for bringing the source-drain window argument to our attention. This work was partly supported by the Deutsche Forschungsgemeinschaft within the Center for Functional Nanostructures.

REFERENCES

1. D. Beckmann, H. B. Weber and H. v. Löhneysen, *Phys Rev. Letters* **93**, 197003 (2004).
2. F. J. Jedema, B. J. van Wees, B. H. Hoving, A. T. Filip and T. M. Klapwijk, *Phys. Rev. B* **60**, 16549 (1999).
3. S. Russo, M. Kroug, T. M. Klapwijk and A. F. Morpurgo, cond-mat/0501564.

Observation of Non-local Andreev Reflection in Normal Metal/Superconductor Structures

S. Russo, M. Kroug, T. M. Klapwijk and A. F. Morpurgo

Kavli Institute of Nanoscience, Delft University of Technology, Lorentzweg 1, 2628 CJ Delft, The Netherlands

Abstract. We investigate transport through hybrid structures consisting of two normal metal leads connected via tunnel barriers to one common superconducting electrode. We find clear evidence for the occurrence of non-local Andreev reflection and elastic cotunneling through the superconductor when the separation of the tunnel barriers is comparable to the superconducting coherence length. The probability of the two processes is energy dependent, with elastic cotunneling dominating at low energy and non-local Andreev reflection at higher energies. The energy scale of the crossover is found to be the Thouless energy of the superconductor, which indicates the phase coherence of the processes. Our results are relevant for the realization of recently proposed entangler devices.

Keywords: Non-local Andreev Reflection; Superconductivity
PACS: 74.45.+c,74.78.Na,73.23.-b

The process that enables charge transfer across an interface between a normal metal (N) and a superconductor (S) is Andreev Reflection (AR)[1]. The incoming electron in the normal metal pairs with a second electron, 2e charge is transferred into S and a reflected hole is left in N. Recently, it has been theoretically demonstrated that AR can also involve particles at different interfaces: the electron and the hole can be injected in spatially separated normal metal leads. This process is the non-local AR [2, 3, 4, 5], which is equivalent to injecting two spin-entangled electrons forming the singlet state of a Cooper pair into two different normal leads. Thus it enables the realization of solid state entanglers.

In order to investigate the occurrence of this non-local process we use a simple system configuration, as in the inset of Fig. 1. Two normal metal electrodes are connected via two tunnel barriers (junctions J1 and J2) to one common superconducting electrode. The junction J1 is used to inject current into the superconductor and J2 is used as a voltage probe to detect a voltage. When J1 and J2 are closer than the superconducting coherence length ξ, an electron injected at energy $E < \Delta$ from the normal electrode of J1 can propagate as an evanescent wave through the superconductor and pair with an electron in the normal electrode of J2[3]. This process results in a hole "reflected" into the second electrode, i.e. non-local AR. Since holes have the opposite charge of electrons the voltage difference generated across J2 has a sign opposite to that observed when the superconductor is in the normal state ($T > T_c^S$).

However, it is also possible that the electrons injected from J1 are transmitted into J2 without being converted into holes. This other process of conduction is Elastic Cotunneling (EC)[4] and it contributes to generate a voltage across J2 that has the same sign as that observed when the superconductor is in the normal state. Thus, the sign of the voltage measured across J2 depends on whether EC or non-local AR occurs with larger probability. In a recent experiment in which two ferromagnetic leads were used as normal electrodes, only the sign corresponding to EC has been observed[6].

We report clear experimental evidence for both non-local AR and EC using the experimental strategy just outlined. We find that the magnitude and the sign of the measured non-local voltage depend on the bias across the injecting junction. At low bias, the observed sign is the same as when the superconductor is in the normal state, indicating that EC dominates. At higher bias the sign of the voltage is reversed, which indicates the occurrence of non-local AR. The energy scale on which the sign-reversal takes place corresponds to the Thouless energy of the superconducting layer. From this we conclude that the sub-gap microscopic processes of conduction, non-local AR and EC, are phase-coherent.

The structure of the devices used in our experiment is implemented in a Nb/Al multilayer sputtered on an insulating substrate[7]. The multilayer consists of two normal metal layers (N1 and N2, 50nm Al layers) connected via two tunnel barriers to one common superconducor (S). The junctions area is $4 \times 8\ \mu m^2$ and each layer has independent electrical connections. The agreement of a fit of the measured differential conductance based on the BCS density of states is a check of the quality of the tunnel junctions (see inset Fig. 1). In our devices the separation between the two tunnel barriers is determined by the thickness of the S layer. This is crucial, since the separation of the tunnel barriers has to be comparable to the superconducting coherence length in S,

CP850, *Low Temperature Physics: 24th International Conference on Low Temperature Physics;*
edited by Y. Takano, S. P. Hershfield, S. O. Hill, P. J. Hirschfeld, and A. M. Goldman

$\xi \simeq \sqrt{\xi_0 l_e} = 10 - 15$ nm [8].

All the measurements were performed at $T = 1.6$ K or higher, with the aluminum electrodes N1 and N2 in the normal state ($T_c^{Al} \simeq 1.2$ K). In the experiment we send current through one of the junctions (e.g., J1) and measure the non-local voltage V^{nl} across the other junction (J2), while maintaining the superconductor at ground. The current bias has a *dc* component and an *ac* modulation (amplitude 1μA at 19.3Hz), and the *ac* component of the non-local signal is measured with a lock-in technique. This corresponds to measuring the contribution given to the non-local voltage by only those electrons which have an energy $E = eV_{dc}$, where V_{dc} is the *dc* voltage across J1.

Fig. 1 shows the V_{ac}^{nl} measured as a function of V_{dc} at two

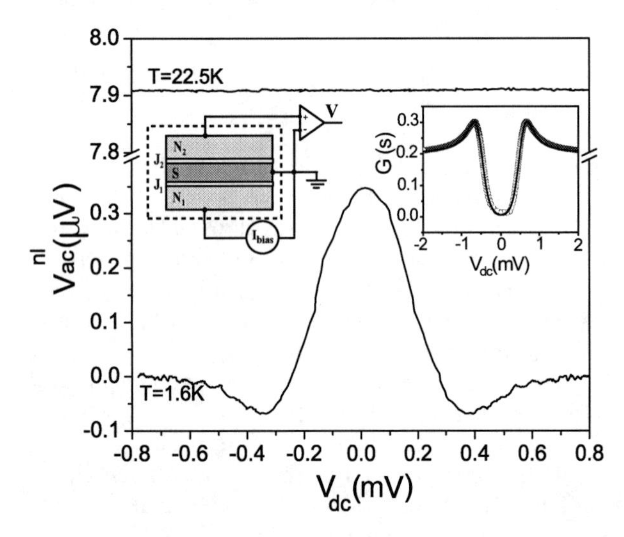

FIGURE 1. Non-local voltage V_{ac}^{nl} is measured across J2, on a device with a $d = 15$ nm thickness of the superconducting layer, for two different temperatures. The upper curve is measured at $T = 22.5$ K -well above T_c^S- and shows a bias-independent non-local voltage due to electrons. At 1.6*K*, below T_c^S), the non-local voltage is much smaller and depends on the bias V_{dc} across J1. In the insets a schema of the non-local measurements and a fit (solid line) based on the BCS density of states of the measured differential conductance (blue dots) for the sample with Nb thickness 15nm[8].

different temperatures (above and below T_c), on a sample in which the superconducting layer is 15 nm thick (approximately equal to ξ). At $T = 22.5$ K, when the Nb is in the normal state, the only charge carriers responsible for the signal are the electrons. At 1.6 K the Nb is superconducting and the Al in the leads is in the normal state. Now the non-local voltage is much smaller and it depends on V_{dc}. Specifically, V_{ac}^{nl} reverses its sign at $V_{dc} = 270\mu$V and eventually vanishes at $V_{dc} \simeq 700\ \mu$V, thus on a bias range much smaller than the superconducting gap (900 μV, see Fig. 2b).

We have measured the non-local voltage in samples with different thickness d of the superconducting layer

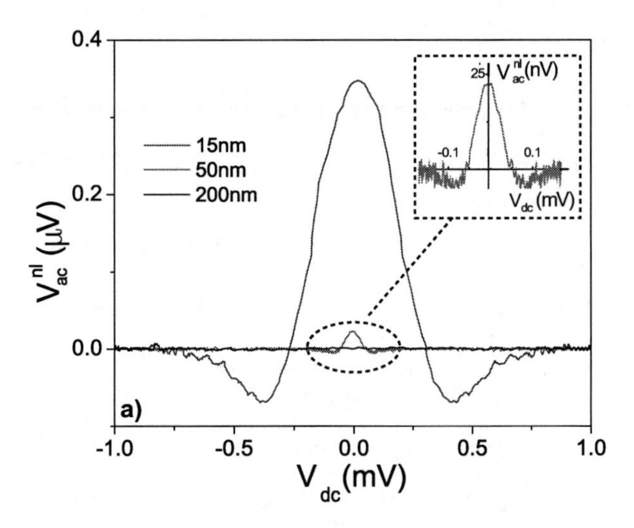

FIGURE 2. (a) Non-local voltage V_{ac}^{nl} measured at $T = 1.6$ K on three samples with different thickness of the superconducting layer ($d = 15$, 50, 200 nm, with a normal state resistance of 4.8, 1.7, and 0.9 Ω respectively).

to demonstrate that this signal originates from evanescent waves propagating below the superconducting gap. Fig. 2a shows the data measured in three samples where $d = 15$, 50, and 200 nm, respectively (for a detailed discussion see Ref.[7].

In conclusion, we have found clear experimental evidence for the occurrence of non-local Andreev reflection and elastic cotunneling through a superconducting layer. Our results show that these processes are phase coherent and strongly depend on the energy of the injected electrons.

This work was financially supported by NWO/FOM and by NOVA. The work of AFM is part of the NWO Vernieuwingsimpuls 2000 program.

REFERENCES

1. A. F. Andreev, *Sov. Phys. JETP* **19**, 1228 (1964).
2. J. M. Byers and M. E. Flatté, *Phys. Rev. Lett.* **74**, 306-309 (1995).
3. G. Deutscher and D. Feinberg, *Appl. Phys. Lett.* **76**, 487-489 (2000).
4. G. Falci, D. Feinberg, and F. W. J. Hekking, *Europhys. Lett.* **54**, 255-261 (2001).
5. P. Recher, E. V. Sukhorukov and D. Loss, *Phys. Rev. B* **63**, 165314 (2001).
6. D. Beckmann, H. B. Weber, and H. v. Löhneysen, *Phys. Rev. Lett.* **93**, 197003 (2004).
7. S. Russo,M. Kroug, T.M. Klapwijk and A.F. Morpurgo, to appear in *Phys. Rev. Lett.*
8. D. Wilms Floet *et al.*, *Appl. Phys. Lett.* **73**, 2826 (1998).

Spin-polarized Tunneling Spectroscopy of YBCO/LSMO Tunnel Junction

Hiromi Kashiwaya[a], Kaori Ikeda[b], Bambang Prijamboedi[a], Satoshi Kashiwaya[a], Akira Sugimoto[a], Itaru Kurosawa[b], and Yukio Tanaka[c]

[a] *Nanoelectronics Research Institute, AIST, Central 2, Tsukuba, Ibaraki, 305-8568, Japan*
[b] *Department of Mathematical and Physical Sciences, Japan Women's University, Bunkyo, Tokyo, 112-8681, Japan*
[c] *Department of Applied Physics, Nagoya University, Nagoya, 464-8603, Japan*

Abstract. The Zeeman magnetic field responses of eptaxially grown $YBa_2Cu_3O_{7-\delta}/La_{0.67}Sr_{0.33}MnO_3$ (YBCO/LSMO) tunnel junction are studied for various doping rates of YBCO. We have fabricated underdoped, optimally doped, and overdoped edge junctions, and the magnetic field is applied along the junction interface (perpendicular to the *c*-axis). The measured conductance spectra at below 1 K showed neither splitting nor broadening up to 14 T with μV resolution for all doping rates. Also the magnetic field response at higher temperature showed no Zeeman splitting. This result apparently contradicts the model based on the weak coupling theory for *d*-wave, and implies the presence of anomalous quasiparticle states in YBCO.

Keywords: superconductivity, tunnel junction, cuprate, magnetic field response
PACS: 74.72.Bk, 74.50.+r, 74.25.Ha

INTRODUCTION

The spin-polarized tunneling spectroscopy has been widely used to detect the spin information on superconducting states. Conductance spectra of a normal metal/insulator/*s*-wave superconductor tunnel junction show Zeeman splitting in the magnetic field applied parallel to the plane of the junction. With using ferromagnet instead of normal metal, the conductance spectra show spin-dependent Zeeman splitting in magnetic field. The amplitude of Zeeman splitting δ is known to be proportional to $N(\chi)/N(\gamma)$, where $N(\chi)$ and $N(\gamma)$ are the densities of states obtained from Pauli susceptibility and normal state electronic specific heat, respectively [1,2]. Detailed information on quasiparticle properties of superconductors, such as the renormalization effect corresponding to the Fermi-liquid parameter G^0, have been successfully determined for s-wave superconductor from the BCS coherent peak.

It is a quite interesting challenge to apply this method to high-T_c superconductor (HTSC) junctions. Although the symmetry of pair potential is established for *d*-wave, other detailed information of electronic states have not been revealed for HTSC. If the Zeeman splitting can be detected in HTSC junctions, we would be able to deduce the Fermi-liquid parameter G^0 in high-T_c superconductors. However, detection of Zeeman splitting in HTSC junctions is not easy because the amplitude of splitting (δ is about 0.8 meV at 14 T if *g*-factor is 2 [1,3]) is quite small and the conductance spectra are seriously smeared for most cases.

In our previous work [3], we have reported the result of optimum doping YBCO/LSMO tunnel junction. In this paper we have fabricated underdoped and overdoped junction and report the results of doping dependence of spin-polarized tunneling spectroscopy.

EXPERIMENTAL

In order to minimize the influence of broadening of conductance spectra, we pay attention to the zero-bias conductance peak (ZBCP) that is known to appear in the conductance spectra in normal metal/insulator/*d*-wave superconductor junction at low temperatures when the junction interface has (110) orientation [4]. Also in order to minimize the effect of the random scattering at the junction interface [5], we adopt the combination of YBCO and LSMO. This is because epitaxial growth of the junction is possible. Actually we have succeeded in getting clear interface between the two layers as presented in our previous work [6].

CP850, *Low Temperature Physics: 24th International Conference on Low Temperature Physics;*
edited by Y. Takano, S. P. Hershfield, S. O. Hill, P. J. Hirschfeld, and A. M. Goldman

All the junctions were fabricated on $SrTiO_3$ substrates with pulsed laser deposition method and photolithography technique [5]. The overdoped YBCO films were obtained by 10% Ca doping and T_c were about 76K. While the underdoped YBCO films were fabricated by Co doping (4%) with T_c of about 50K. The magnetic field is applied parallel to the junction (perpendicular to *c*-axis) to detect the Zeeman splitting effect dominantly. The conductance spectra were measured by conventional four terminal setup with a helium3 refrigerator.

RESULTS AND DISCUSSION

Figure 1 shows the magnetic field response of conductance spectra of (110) underdoped junctions. Consistent with dominant *d*-wave symmetry, the ZBCP is clearly observed. By applying magnetic field up to 14T, the ZBCP did not show splitting. Figure 2 shows the magnetic field response of conductance spectra of (110) edge overdoped junctions. The dominant *d*-wave symmetry is confirmed up to 14T. Based on these results, the symmetry of YBCO is confirmed to be pure $d_{x^2-y^2}$-wave independent of doping rates. Also the lack of magnetic field induced time reversal symmetry breaking is confirmed for all doping rates within our present resolution.

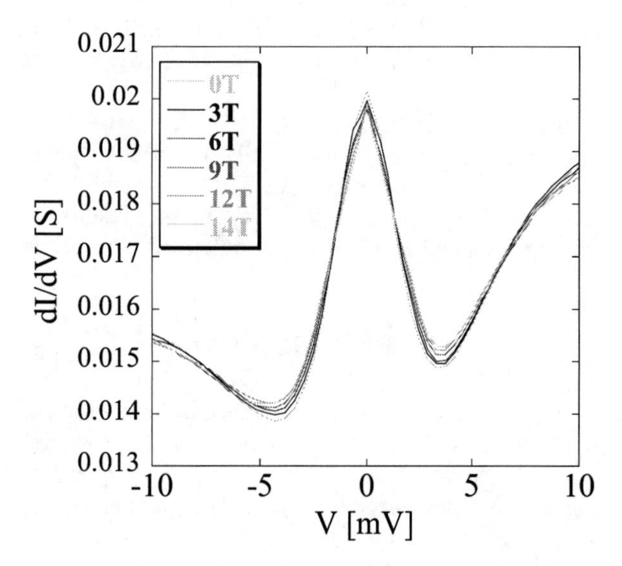

FIGURE 1. Conductance spectra of underdoped YBCO/LSMO tunnel junction measured at 0.5 K. The magnetic field is applied from 0 to 14 T.

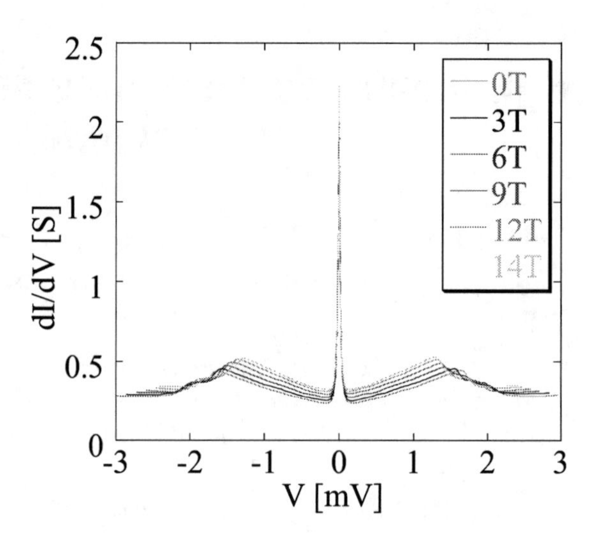

FIGURE 2. Conductance spectra of overdoped YBCO/LSMO tunnel junction measured at 0.5 K. The magnetic field is applied from 0 to 14 T.

Furthermore, it is quite interesting that the Zeeman splitting was not detected for all doping rates. Since the present experimental results contradict the theory based on BCS-type *d*-wave [7], the presence of anomalous quasiparticle states is suggested. A possible scenario is that the spin-gapped metal state lies beneath the superconducting states of HTSC if we can apply the same concept with s-wave analysis that the δ is proportional to the ratio $N(\chi)/N(\gamma)$ [3].

We would like to thank N. Tanuma and A. Asano for fruitful discussion.

REFERENCES

1. P. M. Tedrow, J. T. Kucera, D. Rainer, and T. P. Orlando, *Phys. Rev. Lett.* **52,** 1637 (1984).
2. J. A. X. Alexander, T. P. Orlando, D. Rainer, and P. M. Tedrow, *Phys. Rev. B* **31,** 5811 (1985).
3. H. Kashiwaya, S. Kashiwaya, B. Prijamboedi, A. Sawa, I. Kurosawa, Y. Tanaka, and I. Iguchi, *Phys. Rev. B.* **70,** 094501 (2004).
4. S. Kashiwaya and Y. Tanaka, *Rep. Prog. Phys.* **63,** 1641 (2000).
5. Y. Asano, and Y. Tanaka, *Phys. Rev. B* **65,** 064522 (2002).
6. A. Sawa, S. Kashiwaya, H. Obara, H. Yamasaki, M. Koyanagi, N. Yoshida, and Y. Tanaka, *Physica C* **339,** 287 (2000).
7. Kun Yang and S. L. Sondhi, *Phys. Rev. B* **57,** 8566 (1998).

Half-integer flux quantization in a superconducting loop with a ferromagnetic π-junction

A. Bauer*, J. Bentner*, M. Aprili†, M. Della Rocca†, M. Reinwald*, W. Wegscheider* and C. Strunk*

*Institut für experimentelle und angewandte Physik, Universität Regensburg, D-93025 Regensburg, Germany
†CSNSM-CNRS, Université Paris-Sud, 91405 Orsay Cedex, France

Abstract. Superconducting loops containing a π-junction are predicted to show a spontaneous magnetic moment in zero external magnetic field. In order to confirm this longstanding prediction experimentally we performed magnetization measurements on individual mesoscopic superconducting niobium loops with a ferromagnetic (PdNi) π-junction. The loops are prepared on top of the active area of a micro Hall-sensor based on high mobility GaAs/AlGaAs heterostructures. We observe switching of the loop between different magnetization states at very low magnetic fields, which is asymmetric for positive and negative sweep direction. This is evidence for a spontaneous current induced by the intrinsic phase shift of the π-junction. In addition, the presence of the spontaneous current at zero applied field is directly revealed by an increase of the magnetic moment with decreasing temperature, which results in half integer flux quantization in the loop at low temperatures.

Keywords: Josephson effect, proximity effect, pi-junctions, dilute ferromagnets
PACS: 74.50.+r, 85.25.Cp

The interplay of superconductivity and magnetism in metallic heterostructures constitutes a longstanding problem of condensed matter physics. It has been theoretically predicted that the superconducting pair amplitude in a superconductor-ferromagnet bilayer can penetrate a ferromagnetic metal in an oscillatory way, where the spatial period $\xi_F = \sqrt{\hbar D/I}$ of these oscillations is controlled by the exchange splitting I of the spin up and spin down subbands in the ferromagnet and the diffusion constant D. If a sufficiently thin ferromagnetic layer of thickness d_F is sandwiched between two superconductors, a Josephson junction can be formed. For certain values of d_F/ξ_F the oscillation of the pair amplitude leads to a sign change of $F = |F|e^{i\varphi}$ across the junction [1], implying an intrinsic shift of π of the phase difference $\Delta\varphi$. Such junctions are called π-junctions. In the simplest case of a sinusoidal current-phase relation (CPR) the π-shift leads to a sign change of the supercurrent in the junction:

$$I_S(\Delta\varphi) = I_C \sin(\Delta\varphi + \pi) = -I_C \sin(\Delta\varphi)\,. \qquad (1)$$

In order to experimentally detect this very specific property of ferromagnetic Josephson junctions both the phase difference $\Delta\varphi$ and the resulting supercurrent I_S have to be measured. Already in 1977 Bulaevskii *et al.* predicted that a superconducting loop with an embedded π-junction has to generate a spontaneous supercurrent [2] in order to preserve the uniqueness of the phase along the loop. In the limit of large critical currents such a loop quantizes the enclosed magnetic flux Φ at *half*-integer values $(n+1/2)\Phi_0$ of the superconducting flux quantum $\Phi_0 = h/2e$. In archetypical ferromagnets such as Fe, Ni, Co $\xi_F \approx 1-2$ nm is very short [3]. Even in epitactically grown films the unavoidable surface roughness leads to significant thickness fluctuations, which tend to smear out the effects of π-coupling. A significant advance in the field has been achieved by the use of dilute ferromagnets such as CuNi or PdNi with Curie temperatures in the range of 20-200 K and correspondingly larger values of ξ_F. A pronounced minimum of I_C has been observed as a function of both T [4] and d_F [5], which is a strong but indirect hint towards π-coupling. First phase sensitive experiments have more directly demonstrated the expected $\Phi_0/2$-shift in the magnetic field dependence of the critical current of π-junction networks [6] and dc-SQUIDs [7] made from one 0- and one π-junction. At lower temperatures, experimental evidence for the half-integer flux quantization and the spontaneous supercurrent has been found from direct magnetization measurements on superconducting loops with embedded π-junctions [8, 9].

This work is an extension of Ref. [8] and provides complementary experimental details and a quantitative comparison of the measured spontaneous current with simple models. We measure the supercurrent response of square Nb-loops with a sidelength of 8 μm a to an applied magnetic field. The loops contain an embedded π- or a conventional 0-Josephson junction, respectively. Figure 1 shows a schematic and an scanning electron micrograph of the samples, which are prepared by angle evaporation on top of the active area of a micron sized Hall sensor [10]. The latter is patterned into a high mo-

CP850, *Low Temperature Physics: 24th International Conference on Low Temperature Physics;*
edited by Y. Takano, S. P. Hershfield, S. O. Hill, P. J. Hirschfeld, and A. M. Goldman

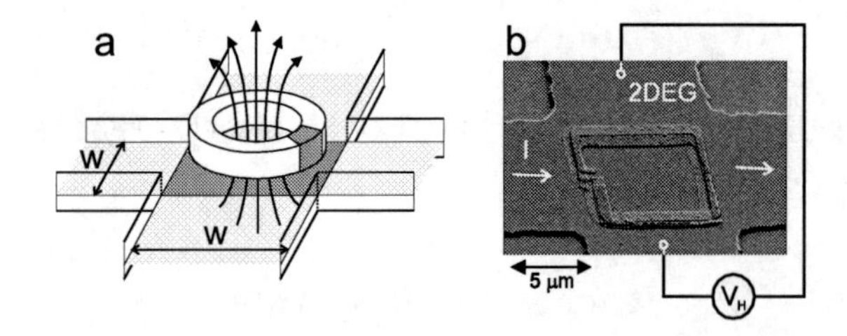

FIGURE 1. a) Schematic of a superconducting loop with an integrated planar Josephson junction on top of a micropatterned Hall-cross of width $w = 8\mu$m. b) Scanning electron micrograph of the sample. The area of the planar SFS-junction is about 180×200 (nm)^2.

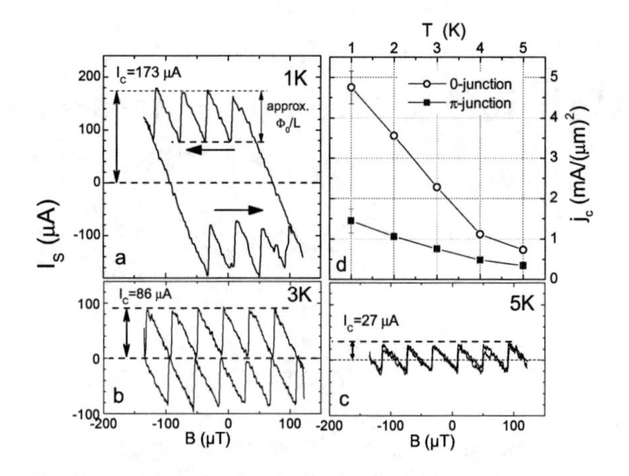

FIGURE 2. Measured supercurrent response of the 0-loops at temperatures a) 1 K, b) 3 and c) 5 K, respectively. The horizontal arrows indicate the sweep direction of the magnetic field. The vertical arrows indicate the extremal values of I_S, i.e. the critical current I_C. d) Temperature dependence of the critical current density $j_c(T)$ for the 0- and the π-loop.

bility GaAs/AlGaAs heterostructure. The classical Hall-voltage of these sensors can be used to detect changes of the magnetic flux in the active area, which are as small as $10^{-3}\,\Phi_0$ or 30 nT. The details of the sample fabrication can be found elsewhere [8]. The sample stage is screened against the earth magnetic field and other external stray fields by a cryoperm shield. Small magnetic fields up to a few mT can be generated by a pair of superconducting Helmholtz coils integrated into the sample stage.

In Figure 2a-c we show the magnetization cycles of the 0-ring at different temperatures. As the magnetic field is ramped up, the diamagnetic screening current in the ring expels the flux from the ring until the critical current $I_C(T)$ at this temperature is reached and the flux in the ring increases step-like. The height of this step is controlled by the dimensionless parameter $\kappa(T) = \Phi_0/(2\pi L I_C(T))$ [11]. In the limit of large critical currents ($\kappa(T) \ll 1$) the external magnetic field is screened completely with negligible phase difference across the junction. Then the jump height corresponds to Φ_0. For smaller values of LI_C the jump height is smaller than Φ_0 because the phase difference $\Delta\varphi$ across the junction contributes significantly to the fluxoid, implying deviations from the full flux quantization. For $\kappa < 1$ (in this loop for $T \lesssim 4$ K) the magnetization curves are hysteretic. From the maximal values of the resulting sawtooth pattern we can determine the critical current of the junction. The self-inductance L can be estimated from the average diameter of the loop, resulting in $L \approx 26$ pH [12].

Like most other magnetometers, the Hall-magnetometer is characterized by a geometric filling factor α. This filling factor results from the fact that the magnetic flux enclosed in the loop only partially threads the active area of the Hall cross: $\Phi_{\text{cross}} = \alpha\,\Phi_{\text{loop}}$. The 2-dimensional electron gas is located 190 nm below the loop. At 1 K we can initially assume $\kappa \ll 1$ and calculate I_C under this condition. Together with the estimated value of L we obtain the next iteration for $\kappa \simeq 0.07$. This procedure leads to a relatively small value of $\alpha \simeq 0.035 \pm 0.005$, which results from the mismatch between the size of the Hall-cross (10 μm) and the outer diameter of the loop (8 μm). In Fig. 2d the critical current $I_C(T)$ of the 0- and the π-loop are presented. The 0-loop was prepared simultaneously with the π-loop but contains no magnetic interlayer. During the deposition of the PdNi-film for the π-loop, a thin layer of adsorbates is gettered on top of the bottom Nb film, which after deposition of the top Nb film forms a weak tunneling barrier. The ratio of the critical current densities of the 0- and the π-loop is about 3.3. The values of the critical currents are important for a later quantitative comparison of the measured spontaneous currents with the theory of Bulaevskii [2].

Now we would like to characterize the ground state of the two loops. In order to avoid complications coming from the hysteresis in Fig. 2 we always warm the sample above T_c and then cool down to 2 K. Each cooling is performed in a certain magnetic field B_{cool}. When the desired temperature of 2 K is reached the applied magnetic field is ramped up, e.g. in positive direction. After ramp down to B_{cool} again, the thermal cycle is repeated to obtain the same initial state again. In Fig. 3 we plot the traces corresponding to different cooling fields $B_{\text{cool}} = 12, 8.6, 5.2, 1.8, -1.6$ and -5.0 μT from top to bottom for the 0-loop (a) and the π-loop (b), respectively. The cooling fields are marked by black dots. We first look at the results for the 0-loop: the switching in the 0-loop occurs nearly symmetric around $B_{\text{applied}} = 0$ as it is expected. In particular, there is no switching in the vicinity of $B_{\text{applied}} = 0$, since switching can occur only, if the screening current is in the vicinity of the critical current. Note that all curves are shifted by about $B_{\text{res}} \simeq 3.5\,\mu$T to the right. This indicates the presence of a small residual field B_{res}, which penetrates the cryoperm shield around

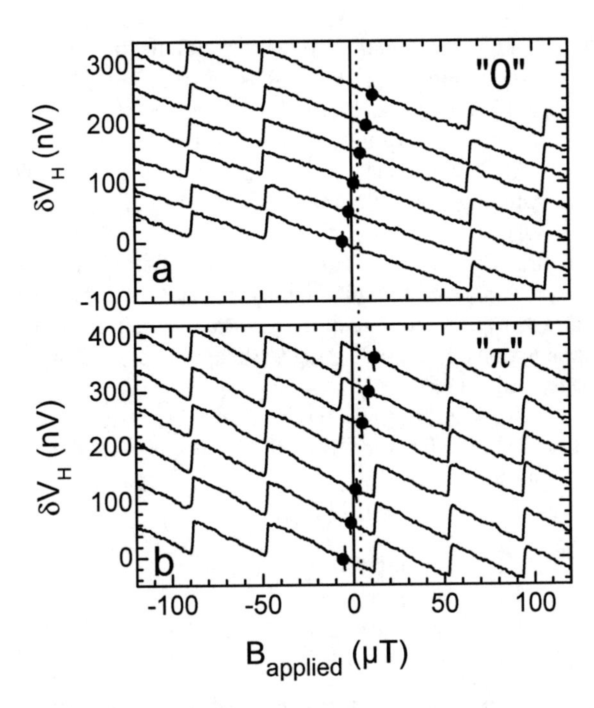

FIGURE 3. Switching characteristics of (a) the 0- and (b) the π-loop at $T = 2$K. The magnetic field has been ramped in positive and negative direction separately after thermal cycling above T_c and cooling down in applied magnetic fields $B_{\text{cool}} = 12, 8.6, 5.2, 1.8, -1.6$ and -5.0 μT from top to bottom. The dashed line indicates the absolute zero of the magnetic field component perpendicular to the loop.

the sample holder. The dashed line indicates the zero of the total external magnetic field.

On the other hand, Fig. 3b reveals that the switching of the π-loop is clearly asymmetric around $B_{\text{applied}} = 0$. Moreover, there is a switching event in the vicinity of $B_{\text{applied}} = 0$, which changes from the negative to the positive side once the applied cooling field B_{cool} is reduced below 5.2 μT. These results can only be understood if there is a significant supercurrent circulating already at these small cooling fields. This *spontaneous* supercurrent must change from clockwise to counterclockwise orientation when the (small) total cooling field is switched from negative to positive values. If a negative cooling field selects a clockwise oriented spontaneous supercurrent, this current has to increase further in order to compensate the rise of the applied field when ramped in positive direction. If the *spontaneous current is close to the critical current, the π-loop has to switch,* even when the compensation of the rising applied field requires an increase of the supercurrent much smaller than $I_C(T)$. Correspondingly, a positive cooling field appears to select a counterclockwise oriented spontaneous current, which forces the π-loop to switch when slightly ramping the applied field in negative direction.

These observations form clear evidence for the correctness of the long-standing prediction of spontaneous currents in π-loops by Bulaevskii *et al.* [2]. Next, we would like to check whether this spontaneous current also has the right magnitude and temperature dependence, when compared with theory. The measurement of the temperature dependence of the magnetic response at fixed field is slightly complicated by a T-dependent contribution of the longitudinal resistance to the measured voltage. This is caused by a small misalignment of the voltage probes and fortunately does not depend significantly on the magnetic field. It can thus be eliminated by subtraction of two temperature sweeps in slightly different magnetic fields, which select opposite orientations of the spontaneous current. In Fig. 4 we plot the result for both loops and (a) an applied magnetic flux close to zero and (b) close to $\Phi_0/2$. The background noise level is enhanced with respect to Fig. 3. This results from a reduction of the measuring current from 20 μA (Fig. 3) to 7 μA required to avoid a saturation of the magnetic response by Joule heating at the lowest temperatures. For zero applied magnetic field the 0-loop shows no magnetic response, while the π-loop builds up a spontaneous magnetic flux below $\simeq 5$ K. With the increase of I_C towards low temperatures the spontaneous flux in the π-loop approaches $\Phi_0/2$. On the other hand, for $\Phi_{\text{applied}} \simeq \Phi_0/2$ the 0-loop has to generate an extra $\Phi_0/2$ in order to maintain the usual *integer* flux quantization, while the π-loop is happy with half a flux quantum and shows no magnetic response. This illustrates the predicted *half-integer* flux quantization induced by the negative Josephson coupling through the ferromagnetic layer.

In order to determine the values of the spontaneous current expected from the measured critical current at a given temperature, we have to assume that the current-phase relation $I_S(\Delta\varphi)$ of both junctions is sinusoidal (Eq. 1). Using $\beta_L(T) = 2\pi L I_C(T)/\Phi_0$ this leads to the following pair of equations for the reduced magnetic flux $\phi = \Phi/\Phi_0$ and the reduced supercurrent $i_s = I_S/I_C(T)$ (see Ref. [11]):

$$i_s = \pm\sin(2\pi\phi) \quad (2)$$

$$\phi = \phi_{\text{applied}} \mp \beta_L(T)\, i_S\,, \quad (3)$$

where the upper sign refers to the 0-loop and the lower sign to the π-loop, respectively.

Using the values of $I_C(T)$ extracted from Fig. 2d, we can calculate the possible values of i_s first for $\phi_{\text{applied}} = 0$. For the 0-junction the only solution is $\phi = 0$, corresponding to the absence of a measured magnetic response for this sample. For the π-junction the number of solutions depends on the value of $\beta_L(T)$. For $\beta_L(T) < 1$, $\phi = 0$ is the only solution. For $\beta_L(T) > 1$, a second solution appears, which in the limit $\beta_L \to \infty$ is $\phi = 1/2$ or $\Phi = \Phi_0/2$.

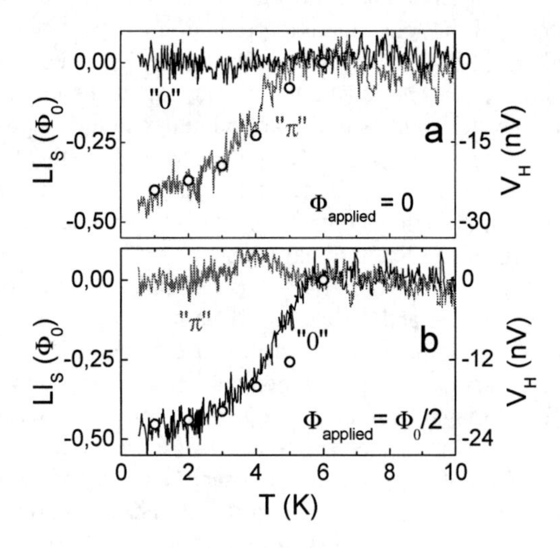

FIGURE 4. Temperature dependence of LI_S for the 0- and the π-loop for applied magnetic fluxes (a) close to zero and (b) close to half a flux quantum. The open circles represent the calculated values of the spontaneous current inferred from the measured critical current (see Fig. 2 and Eqs. 2, 3).

This solution corresponds to a spontaneous current, if its free energy is smaller than that of the solution with $\phi = 0$. The free energy of the junctions reads:

$$\frac{F(T,\phi)}{E_J(T)} = \frac{1}{2\beta_L(T)}(\phi - \phi_{\text{applied}})^2 + 1 \mp \cos(2\pi\phi)\,. \tag{4}$$

The second derivative of F with respect to ϕ corresponds to an effective (reciprocal) inductance $1/L_{\text{eff}} = 1/L + 1/L_J$ of the system, which has two contributions: one from the ring inductance L and the other from the Josephson inductance

$$L_J(T) = \pm\left(\frac{2e}{\hbar} I_C(T)\cos(2\pi\phi)\right)^{-1}. \tag{5}$$

For the 0-loop, L_J is positive, while L_J is *negative* for the π-loop in the vicinity of $\phi_{\text{applied}} = 0$. Thus the condition $\beta_L(T) < 1$ is equivalent to the condition $L > -L_J(T)$ or $L_{\text{eff}}(T) > 0$. The latter condition represents a thermodynamic stability criterion. Its violation implies that the state with $\phi = 0$ becomes unstable for $\beta_L(T) > 1$ and a spontaneous current with $\phi \neq 0$ has to emerge. For very large values of $I_C(T)$ or $\beta_L(T) \to \infty$ the phase difference across the π-junction approaches π and the spontaneous magnetic flux approaches $\Phi_0/2$. At $\Phi_{\text{applied}} = \Phi_0/2$ the roles of the 0- and the π-loop are simply interchanged. In the intermediate cases $1 < \beta_L(T) < \infty$ the numerical solution of Eqs. 2 and 3 for the measured values of $I_C(T)$ (see Fig. 2d) is represented by the open circles in Fig. 4. There is no free parameter involved. The good agreement between the data and the model suggests that the content of higher harmonics in the CPR is not very important in our samples.

In contrast, for values of d_f/ξ_F in the vicinity of the $0-\pi$-crossover, at least for certain ranges of parameters a dominating *second* harmonic in the CPR is expected [13, 14]. The recent observation [15] of half integer Shapiro steps in the IV-characteristics of π-junctions under rf-irradiation in the vicinity of the $0-\pi$-crossover points towards this direction. However, at present other explanations of the half-integer Shapiro steps cannot yet be excluded [16].

We thank Z. Radovic for inspiring discussions. This work was partially funded by the Deutsche Forschungsgemeinschaft.

REFERENCES

1. A. I. Buzdin, L. N. Bulaevskii, and S. V. Panyukov, JETP Lett. 35, **178** (1982); [Pis'ma Zh. Eksp. Teor. Fiz. **35**, 147 (1982)].
2. L. N. Bulaevskii, V. V. Kuzii, and A. A. Sobyanin, JETP Lett. 25, 290 (1977); [Pis'ma Zh. Eksp. Teor. Fiz. 25, 314 (1977)].
3. C. L. Chien and D. H. Reich, J. Magn. Magn. Mater. **200**, 83 (1999).
4. V. V. Ryazanov V. A. Oboznov, A. Yu. Rusanov, A. V. Veretennikov, A. A. Golubov, and J. Aarts, Phys. Rev. Lett. **86**, 2427 (2001).
5. T. Kontos, M. Aprili, J. Lesueur, F. Genêt, B. Stephanidis, and R. Boursier, Phys. Rev. Lett. **89**, 137007 (2002).
6. V. V. Ryazanov, V. A. Oboznov, A. V. Veretennikov, and A. Yu. Rusanov, Phys. Rev. B 65, 020501 (2002).
7. W. Guichard, M. Aprili, O. Bourgeois, T. Kontos, J. Lesueur, and P. Gandit, Phys. Rev. Lett. 90, 167001 (2003).
8. A. Bauer, J. Bentner, M. Aprili, M. L. Della Rocca, M. Reinwald, W. Wegscheider, and C. Strunk, Phys. Rev. Lett. **92**, 217001 (2004).
9. S. M. Frolov, D. J. Van Harlingen, V. A. Oboznov, V. V. Bolginov, and V. V. Ryazanov, Phys. Rev. B **70**, 144505 (2004).
10. A. K. Geim, S. V. Dubonos, J. G. S. Lok, I. V. Grigorieva, J. C. Maan, L. Theil Hansen, and P. E. Lindelof, Appl. Phys. Lett. **71**, 2379 (1997).
11. A. Barone and G. Paterno, Physics and Applications of the Josephson Effect (Wiley, New York, 1982).
12. F. W. Grover, Inductance Calculations: Working Formulas and Tables (Dover Publications, New York, 1962).
13. Z. Radovic, L. Dobrosavljevic-Grujic, and B. Vujiic, Phys. Rev. B **63**, 214512 (2001).
14. A. A. Golubov, M. Yu. Kupriyanov, and E. Il'ichev, Rev. Mod. Phys. **76**, 411 (2004).
15. H. Sellier, C. Baraduc, F. Lefloch, and R. Calemczuk, Phys. Rev. Lett. **92**, 257005 (2004).
16. S. M. Frolov *et al.*, cond-mat/0506003.

Phase Transition from Superconducting to Normal State Induced by Spin Injection in Manganite/Cuprate/Au Double Tunnel Junctions

T. Nojima*, T. Hyodo*, S. Nakamura*, and N. Kobayashi*†

*Center for Low Temperature Science, Tohoku University, Sendai 980-5877, Japan
†Institute for Materials Research, Tohoku University, Sendai 980-8577, Japan

Abstract. We have studied the voltage-current characteristics in $La_{0.7}Ca_{0.3}MnO_3/YBa_2Cu_3O_y$/Au, ferromagnet/superconductor/nonmagnetic metal, double tunnel junctions. Near T_c, the tunnel conductance $G(V)$ reveals a step-like decrease, which can be ascribed to the phenomenon of the phase transition from superconducting state to normal one caused by the spin accumulation effect. With decreasing temperature T, the step in $G(V)$ changes to the sharp peak . Our results indicate that the spin-injection-induced phase transition has a nature of a first order transition at low T.

Keywords: spin injection, tunnel junction, cuprate, manganite.
PACS: 72.25.-b, 74.50.+r, 74.78.Bz, 74.78.Fk

With the rediscovery of lanthanum manganites with half metallicity, the studies of the spin-polarized tunneling [1] have been extended to superconductor /ferromagnet (*S*/*F*) heterostructures consisting of high-T_c cuprates and ferromagnetic manganites. Recently, the strong suppression of critical current by spin injection has been reported in many perovskite *S*/*F* structures [2-4]. It is now important to clarify the nature of phase transition from superconducting to the normal state due to the spin injection. This can be accessible in the $F_{\uparrow}/S/F_{\downarrow}$ or *F*/*S*/*N* (nonmagnetic metal) double tunnel junction [5,6].

In this work, we have prepared *F*/*S*/*N* double tunnel junctions consisting of $La_{0.7}Ca_{0.3}MnO_3$ (LCMO), $YBa_2Cu_3O_y$ (YBCO), and Au, with insulating layers of $LaAlO_3$ (LAO) and amorphous(a)-YBCO. The phase transition phenomena induced by spin-polarized current are examined using the measurements of tunnel voltage-current $V(I)$ characteristics.

We prepared the spin-injection devices as shown in Fig. 1. LCMO/LAO/YBCO heterostructures, with all the layers *c* axis oriented, were grown on (100) $SrTiO_3$ substrates at temperatures of 800, 780, and 750°C, respectively, followed by the deposition of a-YBCO or Au at ambient temperatures, using an Rf sputtering method [7]. Then YBCO/a-YBCO(or /Au) layers were patterned into islands with 200 μm wide and 400 μm long using photolithography and selective wet etching. After sputtering a-YBCO surrounding the island for isolation, Au leads were patterned on the top. In this configuration, the current through the junctions causes excess spin accumulation in the confined *S* (YBCO) layer due to the difference in the density of state for each spin direction between *F* (LCMO) and *N* (Au) layers [5,6]. The critical temperature T_c of YBCO and the Curie temperature of LCMO are typically 86 K and 235 K, respectively.

We compared two kinds of samples, #1 and #2, with YBCO layers of 60 nm and 150 nm thick, respectively. For sample #2, a-YBCO on YBCO layer was replaced by Au, meaning only a natural barrier between YBCO and Au. It is noted the measured $V(I)$ is the sum of the voltage drops at *F*/*S* and *S*/*N* junctions, $V_{FS}(I)$ and $V_{SN}(I)$ for sample #2, while that is only $V_{FS}(I)$ for sample #1.

Figure 2 shows the plots of tunnel conductance $G = dI/dV$ versus V for sample #1 at several temperatures T. At high T, a small step like decrease in $G(V)$ is observed at $\pm V_c$ (denoted by arrows). The large value of V_c more than 50 mV indicates that the anomaly is not the signature of the superconducting gap Δ. Since finite resistance along the *S* layer appears at $I(V_c)$

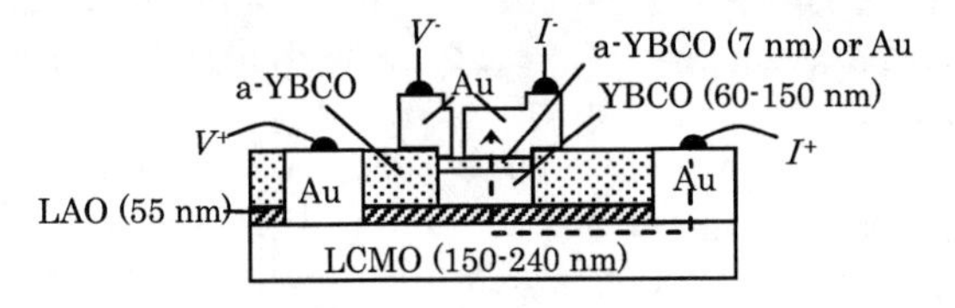

FIGURE 1. Schematic of the device structure. The layer on YBCO is a-YBCO (insulator, sample #1) or Au (sample #2).

CP850, *Low Temperature Physics: 24th International Conference on Low Temperature Physics;*
edited by Y. Takano, S. P. Hershfield, S. O. Hill, P. J. Hirschfeld, and A. M. Goldman

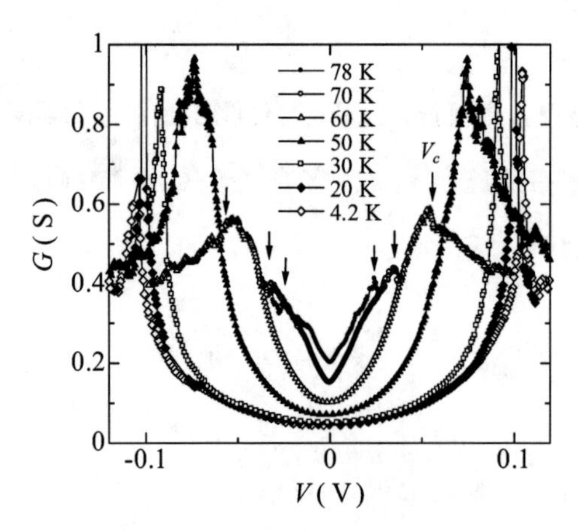

FIGURE 2. $G(V)$ for sample #1. Steps denoted by arrows and peaks are associated with phase transition of the S layer by spin injection.

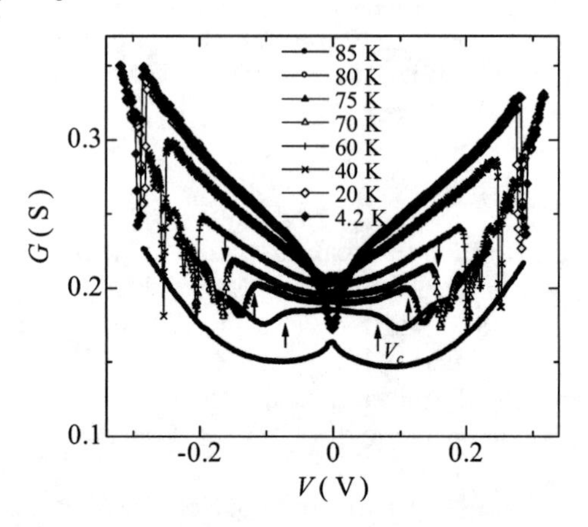

FIGURE 3. $G(V)$ for sample #2. Steps denoted by open triangles and peaks are associated with phase transition of the S layer by spin injection.

in a similar type of device [7], the anomaly at V_c is associated with the transition of the S layer to non-superconducting state. It is noted that the step in $G(V)$ changes to the sharp peak below $T = 0.6T_c$. This implies that nature of the transition alters at low temperatures.

In Fig. 3, $G(V)$ for sample #2 are plotted for several temperatures. As well as sample #1, we observe the step like anomaly in $G(V)$ at $\pm V_c$ (arrows) transfers to the sharp peak. However, we note that the amplitude of V_c is larger and the direction of the peak at V_c is opposite to sample #1. These come from the difference in device structure as discussed below.

The suppression of Δ by the excess spin can happens when the nonequilibrium spin density, $\delta n_s \propto I\tau_s/ed$ with τ_s the spin relaxation time and d the thickness of S, becomes the same order as $N(0)\Delta(T)$ with $N(0)$ the normal density of state of S. In sample #1, the critical injection current I_{inj}^{c} at V_c, is almost constant with 15 mA below 50 K, and decrease above it. This dependence is similar to that of Δ, meaning T dependence of τ_s gives a small effect. On the other hand, in sample #2, I_{inj}^{c} with 70 mA at 4.2 K shows a stronger T dependence. The difference between the two samples may come from the larger d for sample #2, in which the effect of $\tau_s(T)$ may not be negligible.

When the spin-polarized tunnel current suppresses Δ, the $V(I)$ curve approaches the curve of normal state and finally transfers to that. If Δ becomes zero continuously, a slope change in $V(I)$ will occur, resulting in a step like decrease in G [5]. On the other hand, if Δ drops suddenly from a finite value [6], the sudden transfer of $V(I)$ will result in the peak in $G(V)$. Thus, our results indicate the spin-injection-induced phase transition, which is continuous near T_c, becomes a first order transition at low T.

In tunneling limit, it is easily imagined that $G(V)$ show the positive peak at the first order transition as shown in Fig. 2, since $V_{FS}(I)$ and $V_{SN}(I)$ are situated on the higher voltage side than those in the normal state. However, in case that the effect of Andreev reflection is strong, the peak direction becomes opposite due to higher conductivity than that in the normal state. As mentioned above, the sample #2 has no barrier layer between S and N. This may explain the reason why the peak in $G(V)$ with negative direction is observed in Fig. 3, although the degree of $V_{SN}(I)$ contribution to the total $V(I)$ is not clear now.

In conclusion, we observed the superconducting transition controlled by the injection of spin-polarized carriers, which may be caused by spin accumulation effect and be a first or second order transition, depending on temperature.

This work is supported by CREST. We thank to H. Miura, S. Tanno, K. Hosokura, and Y. Watanabe for operation of the low temperature apparatus.

REFERENCES

1. P. M. Tedrow and R. Meservey, *Phys. Rev. Lett.* **26**, 192-195 (1971).
2. V. A. Vas'ko, V. A. Larkin, P. A. Kraus, K. R. Nikolaev, D. E. Grupp, C. A. Nordman, and A. M. Goldman, *Phys. Rev. Lett.* **78**, 1134-1137 (1997).
3. Z. W. Dong, R. Ramesh, T. Venkatesan, M. Johnson, Z. Y. Chen, S. P. Pai, V. Talyansky, R. P. Sharma, R. Shreekala, C. J. Lobb, and R. L. Greene, *Appl. Phys. Lett.* **71**, 1718 -1720 (1997).
4. N. -C. Yeh, R. P. Vasquez, C. C. Fu, A. V. Samoilov, Y. Li, and K. Vakili, *Phys. Rev. B* **60**, 10522 -10526 (1999).
5. S. Takahashi, H. Imamura, and S. Maekawa, *Phys. Rev. Lett.* **82**, 3911-3914 (1999).
6. N. Yoshida, Y. Tanaka, J. Inoue, and S. Kashiwaya, *Phys. Rev. B* **63**, 024509 (2000).
7. T. Nojima, M. Iwata, T. Hyodo, S. Nakamura, and N. Kobayashi, *Physica C* **412-414**, 147-151 (2004).

Transport Properties of Ferromagnetic Semiconductors with Superconducting Electrodes

T. Akazaki[1, 2], H. Takayanagi[1, 2], S. Yanagi[3], H. Munekata[3], and J. Nitta[2, 4]

[1]*NTT Basic Research Labs., NTT Corporation, Atsugi, 243-0198, Japan*
[2]*CREST, Japan Science and Technology Agency, Kawaguchi, 331-0012, Japan*
[3]*Imaging Science and Engineering Lab., Tokyo Institute of Technology, Yokohama, 226-8503, Japan*
[4]*School of Engineering, Tohoku University, Sendai, 980-8579, Japan*

Abstract. We report on the transport properties of ferromagnetic p-$In_{0.96}Mn_{0.04}As$ with Nb superconducting electrodes. We observed conductance reduction within the superconducting gap. This behavior can be qualitatively understood by considering the suppression of Andreev reflection caused by spin polarization in p-$In_{0.96}Mn_{0.04}As$

Keywords: superconductor, ferromagnetic semiconductor, Andreev reflection
PACS: 72.25.Dc, 74.45.+c, 75.50.Pp

INTRODUCTION

Superconductor/ferromagnet (S/F) junctions have attracted considerable interest both theoretically and experimentally [1]. This is because new quantum phenomena can be expected from the interplay between the superconductivity and the spin polarization of the ferromagnet. Ferromagnetic metals such as Ni-Fe and Co have already been used as ferromagnetic materials. However, such ferromagnetic metals have a large number of carriers contributing to the transport and it is difficult to control this number with the electric field effect. In contrast, since p-$In_{0.96}Mn_{0.04}As$ is a kind of ferromagnetic semiconductor [2], it is reasonable to suppose that the number of carriers in p-$In_{0.96}Mn_{0.04}As$ can be changed by using the electric field effect. In this paper, we report on the transport properties of ferromagnetic p-$In_{0.96}Mn_{0.04}As$ with Nb superconducting electrodes.

EXPERIMENT AND DISCUSSION

We fabricated Nb/p-$In_{0.96}Mn_{0.04}As$/Nb junctions with Hall voltage probes as shown in Fig. 1. The p-$In_{0.96}Mn_{0.04}As$ heterostructure was grown by using molecular beam epitaxy (MBE) on a semi-insulating (100) GaAs substrate. The critical temperature T_C of the Nb electrodes was about 8.2 K. The coupling length L between the two Nb electrodes was designed to be 0.8 ~ 10 μm. The superconducting Nb electrodes as well as the normal Ti/Au ones were deposited on p-$In_{0.96}Mn_{0.04}As$ just after cleaning the InMnAs surface. Figure 2 shows the magnetic field dependence of the Hall resistance for p-$In_{0.96}Mn_{0.04}As$ as a function of

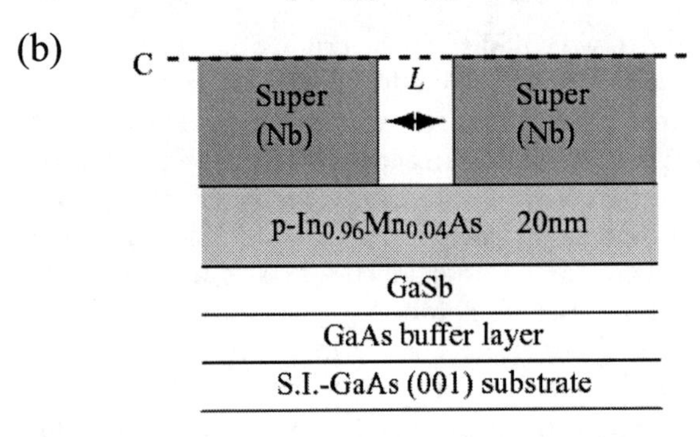

FIGURE 1. Schematic diagram of a superconductor-ferromagnet junction: (a) top view, (b) cross-sectional view.

CP850, *Low Temperature Physics: 24th International Conference on Low Temperature Physics;*
edited by Y. Takano, S. P. Hershfield, S. O. Hill, P. J. Hirschfeld, and A. M. Goldman

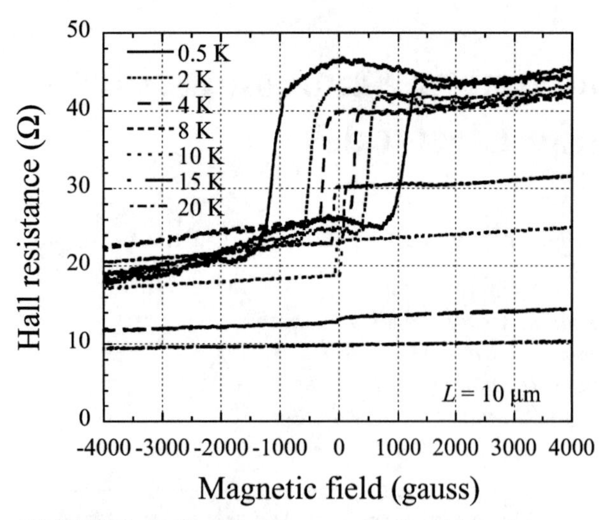

FIGURE 2. Hall resistance curve of p-$In_{0.96}Mn_{0.04}As$ as a function of temperature.

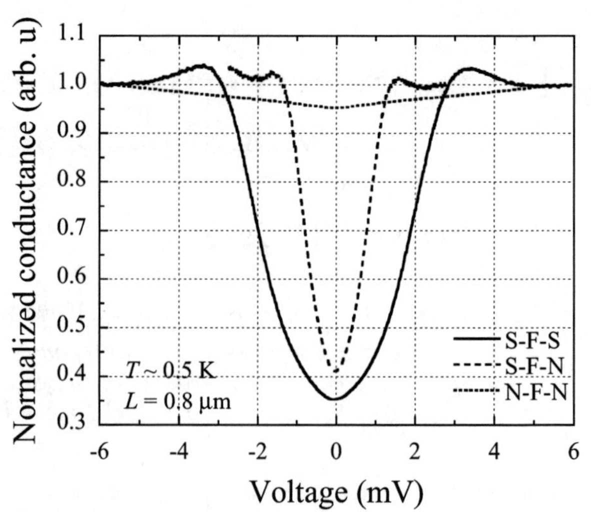

FIGURE 3. Normalized differential conductance as a function of bias voltage for three different measurement setups at 0.5 K.

temperature. We observed the anomalous Hall effect below about 15 K. At 0.5 K, the reverse magnetic field was about 1000 gauss. This result indicates that p-$In_{0.96}Mn_{0.04}As$ becomes ferromagnetic below the T_C of the Nb electrodes (~ 8.2 K). Figure 3 shows normalized differential conductance as a function of bias voltage for three different measurement setups at 0.5 K. The differential conductance of each junction is normalized by the value at the maximum bias voltage. We have obtained nearly linear voltage dependence in the N-F-N junction. This result indicates that we have obtained the ohmic contact without any process after deposition of N and S metals. Thus, we can claim that there is almost no barrier at the interface between metals and p-$In_{0.96}Mn_{0.04}As$. In contrast, we obtained a large conductance reduction within $|V| \leq 1.5$ mV in the S-F-N junction and within $|V| \leq 3$ mV in the S-F-S junction. When we take the Nb superconducting energy gap Δ_S of ~ 1.5 meV into consideration, we can assume that this conductance reduction appears within the superconducting gap.

Next, we will discuss the origin of the conductance reduction. R. J. Soulen Jr. et al. have measured the differential conductance via a superconducting Nb point contact on ferromagnetic CrO_2 [3]. In this case, the differential conductance decreased below the superconducting energy gap of Nb. This behavior is similar to our results and its origin is as follows. In the Andreev reflection process, the incident electron requires the opposite spin electron to be removed from the normal region for conversion to a Cooper pair. Therefore, when there is no opposite spin electron, the conversion to a Cooper pair does not occur. Namely, with S-F junctions, the Andreev reflection is limited by the minority spin population. Therefore, our experimental results can be understood qualitatively by considering the suppression of the Andreev reflection that is caused by the spin polarization in p-$In_{0.96}Mn_{0.04}As$.

CONCLUSIONS

We have investigated the transport properties in Nb/ p-$In_{0.96}Mn_{0.04}As$/Nb junctions. We observed the anomalous Hall effect below ~15 K. The reverse magnetic field was ~ 1000 gauss at 0.5 K. We observed conductance reduction within the superconducting gap. These experimental results arise from the interplay between superconductivity and spin polarization.

ACKNOWLEDGMENTS

We thank Prof. Y. Tanaka, Dr. T. Yokoyama, and Dr. Y. Asano for valuable discussions. We also thank Dr. Y. Hirayama for his encouragement throughout this work.

REFERENCES

1. *Towards the Controllable Quantum States*, edited by H. Takayanagi et al, Singapore: World Scientific, 2003.
2. H. Munekata, *Mater. Sci. Eng.* **B31**, 151-156 (1995).
3. R. J. Soulen Jr., J. M. Byers, M. S. Osofsky, B. Nadgorny, T. Ambrose, S. F. Cheng, P. R. Broussard, C. T. Tanaka, J. Nowak, J. S. Moodera, A. Barry, and J. M. D. Coey, *Science* **282**, 85-88 (1998).

Superconducting Junctions with Ferromagnetic, Antiferromagnetic or Charge-Density-Wave Interlayers

Yuri Barash*, I. V. Bobkova*, Brian M. Andersen†, T. Kopp** and P. J. Hirschfeld†

*Institute of Solid State Physics, Russian Academy of Sciences, Chernogolovka, Moscow reg., 142432 Russia
†Department of Physics, University of Florida, Gainesville, FL 32611 USA
**Center for Electronic Correlations and Magnetism, University of Augsburg, D-86135 Augsburg, Germany

Abstract. Spectra and spin structures of Andreev interface states and the Josephson current are investigated theoretically in junctions between clean superconductors (SC) with ordered interlayers. The Josephson current through the ferromagnet-insulator-ferromagnet interlayer can exhibit a nonmonotonic dependence on the misorientation angle. The characteristic behavior takes place if the π state is the equilibrium state of the junction in the particular case of parallel magnetizations. We find a novel channel of quasiparticle reflection (Q reflection) from the simplest two-sublattice antiferromagnet (AF) on a bipartite lattice. As a combined effect of Andreev and Q reflections, Andreev states arise at the AF/SC interface. When the Q reflection dominates the specular one, Andreev bound states have almost zero energy on AF/ s-wave SC interfaces, whereas they lie near the edge of the continuous spectrum for AF/d-wave SC boundaries. For an s-wave SC/AF/s-wave SC junction, the bound states are found to split and carry the supercurrent. Our analytical results are based on a novel quasiclassical approach, which applies to interfaces involving itinerant antiferromagnets. Similar effects can take place on interfaces of superconductors with charge density wave materials (CDW), including the possible d-density wave state (DDW) of the cuprates.

Keywords: Josephson junctions, Andreev states, competing orders
PACS: 74.50.+r, 74.45.+c, 74.72.-h

Superconducting heterostructures involving ferro- and/or antiferromagnets manifest unusual properties associated with spin effects, and are of both fundamental interest and important for technological applications. Superconductor-ferromagnet-superconductor (SC/F/SC) junctions are known to display $0-\pi$ transitions with varying the temperature or the interlayer width. We have demonstrated theoretically that the $0-\pi$ transition can show up also at fixed temperature and interlayer width in superconducting junctions with a three-layer FIF interface, as a function of the misorientation angle between the magnetizations of two F layers separated by the insulating barrier [1]. The dependence of the Josephson current on the misorientation angle φ becomes especially simple in the tunneling limit, when the critical current takes the form $J_c(T,\varphi) = J_c^{(p)}(T)\cos^2(\varphi/2) + J_c^{(a)}(T)\sin^2(\varphi/2)$. Here $J^{(p)}(T)\&J^{(a)}(T)$ are critical currents in tunnel junctions with parallel and antiparallel orientations of the exchange fields in the three-layer interface. The $0-\pi$ transition can take place with varying φ, if $J_c^{(p)}(T)$ and $J_c^{(a)}(T)$ have opposite signs. This is the case when the junction with parallel magnetizations is in the π-state, since always $J_c^{(a)}(T) > 0$. The transition results in a nonmonotonic dependence of $|J(T,\varphi)|$ on φ. This effect can be used for switching the junction from the zero state to the π state by varying the misorientation angle. The angle is changed in the FIF interlayer with applied magnetic field, if it is larger than the coercive force in one of the F layers and less than in the other layer.

Many fundamental and practical problems involve interfaces with antiferromagnets. In particular, many of the properties of HTSC cuprate materials probably arise from a competition between antiferromagnetic and superconducting order, and many situations involve such natural or fabricated boundaries. We have studied interfaces between itinerant antiferromagnets and normal metals or superconductors and demostrated that a new spin-dependent channel of quasiparticle reflection, the so-called Q reflection, occurs on the interfaces [2]. Parallel to the interface, the momentum component of low-energy normal-metal quasiparticles changes by Q_y in a Q reflection event, where $\mathbf{Q}$ is the wave-vector of the antiferromagnetic pattern and y is the direction parallel to the interface. Assuming comparatively small Fermi velocity mismatches and taking into account the nesting condition $E_F(\mathbf{p}+\mathbf{Q}) = -E_F(\mathbf{p})$ in itinerant antiferromagnets, one can see within the mean-field tight-binding model on the half-filled square lattice that normal metal quasiparticles with energies less than or comparable to the antiferromagnetic gap change their momenta by $\mathbf{Q}$ and reverse the signs of their velocities in a Q reflection event. Consequently, such quasiparticles experience spin-dependent retroreflection at antiferromagnet-normal metal (AF/N) transparent interfaces.

Quasiparticle bound states below the AF and SC gaps at AF/SC interfaces arise as a combined effect of An-

CP850, *Low Temperature Physics: 24th International Conference on Low Temperature Physics;*
edited by Y. Takano, S. P. Hershfield, S. O. Hill, P. J. Hirschfeld, and A. M. Goldman

dreev and Q reflections. Among a variety of subgap states, low-energy states $E_B \ll \min\{m,\Delta\}$ are of special interest since they can result in low-temperature anomalies in the Josephson critical current, as well as low-bias anomalies in the conductance. Here m and Δ are the sublattice electronic magnetization and the superconducting s-wave or d-wave order parameters. In the absence of the interface potential h, the dispersive bound state energies at the (110) and (100) AF - s-wave superconductor (AF/sSC) interfaces can be represented as $E_b(\mathbf{k}_F) = \pm\Delta_s\sqrt{R_{sp}(\mathbf{k}_F)}$. Here $R_{sp}(\mathbf{k}_F)$ is the normal state reflection coefficient for specular reflection from the interface, which occurs even in the absence of any interface potentials due to a mismatch of Fermi velocities in the AF and the sSC. Since normal-metal states are presumably identical in the left and right halfspaces, under the conditions $\Delta \ll m,t$, the mismatch in the model controls the parameter m/t. If the magnetic order parameter m is much less than the hopping matrix element t, Q reflection dominates $R_{sp}(\mathbf{k}_F) \ll 1$, and bound state energies almost coincide with the Fermi level $|E_b| \ll \Delta_s \leq m$. In particular, for the (110) interface on the square lattice we find $R_{sp}(\mathbf{k}_F) = \left[1+\left(\sqrt{2}v_{F,\perp}(\mathbf{k}_F)/am\right)^2\right]^{-1}$, where a is the lattice spacing and $v_{F,\perp}(\mathbf{k}_F)$ the normal-state Fermi velocity component along the boundary normal. In Fig. 1 we plot the quasiparticle spectrum as obtained

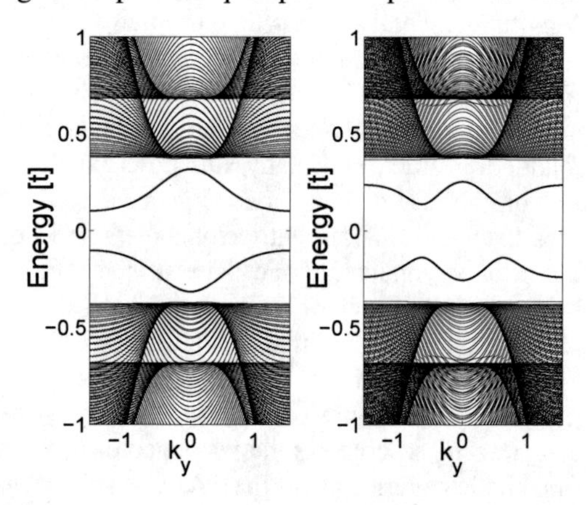

FIGURE 1. Eigenvalues for the (100) AF/sSC interface as a function of $k_y a$ (a is the lattice constant) for (a) $\mu = 0.0$ and $h = 0.0$ and (b) $h = 2.0t$. Order parameter values in the bulk: $\Delta_{s,b} = 0.4t$, $m_b = 0.7t$. Here, one sees explicitly the presence and dispersion of the bound state band inside the gap.

from the selfconsistent eigenvalues of the Bogoliubov-de Gennes equations. Interface bound states show up inside the main gap of the spectrum as a distinct band, which disperses with the momentum component k_y along the interface. The two gap edges seen in Fig. 1 are associated with the superconducting (lesser) and the antiferromagnetic (larger) gaps. Interface potentials h present near the interface tend to suppress the bound states resulting from Q-reflection and move their positions towards the gap edge. In the regime where h is of the order of t, we find that the main effect of the specular reflection channel is to cause a stronger dispersion of the bound state energy. One can identify additional extrema in the wave vector dependence of the bound state energy $E(k_y)$. A typical example is seen in Fig. 1 where $h = 2.0t$. The new stationary points in the dispersion lead to additional LDOS peaks near the interface.

Dispersive bound state energies on an antiferromagnetic - d-wave superconductor interface (AF/dSC) can be represented as $E_b(\mathbf{k}_F) = \pm\Delta_d(\mathbf{k}_F)\sqrt{R_Q(\mathbf{k}_F)}$. They lie near the edges of the continuous spectrum, when Q reflection dominates. This contrasts with the case of a (110) surface of a dSC confined with an impenetrable wall, where zero-energy Andreev states are formed.

For an sSC/AF/sSC junction, due to a finite width l of the AF interlayer, the low-energy bound states are split and carry the supercurrent. If no potential barriers are present on the boundaries, and $l \ll \xi_s$, $\Delta_s \ll m \ll t$, we find the following energies for interface states: $\varepsilon_B = \pm\sqrt{D}|\Delta_s \cos(\chi/2)|$, where χ is the order parameter phase difference, $D(k_y) = 4K(k_y)(K(k_y)+1)^{-2}$ is the transparency of the N/AF/N junction and $K(k_y) = \exp(2ml/|v_{F,x}(k_y)|)$. For large interlayer width, $K,D \ll 1$, there are low-energy states in the junction which result in low-temperature anomalous behavior of the critical current.

Similar effects for CDW/SC interfaces have been studied recently in [3]. Subgap Andreev states arise at CDW/dSC and DDW/sSC interfaces. At the same time there are no subgap states at CDW/sSC interfaces due to the absence of interface-induced pair-breaking processes. The interface states also do not arise at DDW/dSC interfaces since pair-breaking effects from DDW and dSC compensate each other in this case. In dSC/CDW/dSC and sSC/DDW/sSC Josephson junctions, the interface low-energy bound states are split and strongly influence the Josephson current.

This work was supported by grant NSF-INT-0340536 (I.V.B., P.J.H., and Yu.S.B.), and by ONR grant N00014-04-0060 (P.J.H and B.M.A). I.V.B. and Yu.S.B. also acknowledge the support by grant RFBR 05-02-17175. T.K. thanks the support by the DFG through SFB 484, DAAD D/03/36760 and BMBF 13N6918A.

REFERENCES

1. Yu. S. Barash, I. V. Bobkova, and T. Kopp, *Phys. Rev. B* **66**, 140503 (2002).
2. I. V. Bobkova, P. J. Hirschfeld, and Yu. S. Barash, *Phys. Rev. Lett.* **94**, 037005 (2005).
3. I. V. Bobkova, Yu. S. Barash, *Phys. Rev. B* **71**, 144510 (2005).

Spin-dependent Proximity Effects in d-wave Superconductor/Half-metal Heterostructures

Nobukatsu Yoshida* and Mikael Fogelström†

*Department of Physics, University of Tokyo, Hongo,Tokyo 113-0033, Japan
†Applied Quantum Physics, MC2, Chalmers University of Technology, S-412 96 Göteborg, Sweden

Abstract. We report on mutual proximity effects in *d*-wave superconductor/half-metal heterostructures which correspond to systems composed of high-Tc cuprates and manganite materials. In our study, proximity effects are induced by the interplay of two separate interface effects: spin-mixing (or rotation) surface scattering and spin-flip scattering. The surface spin-mixing scattering introduces spin-triplet pairing correlations in superconducting side; as a result, Andreev bound states are formed at energies within the superconducting gap. The spin-flip scattering introduces not only long range equal-spin pairing amplitudes in the half-metal, but also an exotic magnetic proximity effect extending into the superconductor.

Keywords: Superconductor-ferromagnet hybrid structures, Unconventional superconductivity
PACS: 74.72.-h,74.45.+c

The possibility of making heterostructures out of high-T_c cuprate superconductors and strong manganite ferromagnets like LSMO allow us to study the competition between superconducting and magnetic order where both are of equal strength. The magnanites are of special interest since they have the same perovskite structure as the cuprates and e.g. $L_{0.67}Sr_{0.33}MnO_3$ is close to totally spin polarized i.e. a half metal[1]. We report initial results from an on-going study of the electron density of states (DoS) and the induced magnetism and superconductivity in half-metal/*d*-wave superconductor/half-metal trilayers. In particular we wish to identify properties that would allow us to distinguish between a structure having the two half-metals with parallel spin-bands from a one having them anti-parallel.

The study is based on quasiclassical Green's function theory and the key issue is to pose the correct boundary conditions that connects a superconducting half-space with a half-metallic one where only one spin-band is present[2, 3, 4, 5, 6]. In order to have coherent transport between a superconductor and a half-metal the interface needs to be spin-active, i.e. to reflect or transmit quasiparticles (and quasiholes) in a manner that depends on the orientation of their spin relative to the magnetic orientation that defines the spin-active interface. There are two different ways of spin-active scattering the interface may possess that we take into account: **i)** *spin-mixing* or *spin-rotation*, the quasiparticle (quasihole) acquires a phase shift, $\pm\Theta_m/2$, with the sign depending its spin orientation. The presence of spin-mixing introduces Andreev states in the DoS below the gap-energy Δ and an (magnetic) exchange field in the superconductor [3, 4]. **ii)** *spin-flip scattering*, the half-metal allows transmission only of one spin orientation which leads to two channels, with hopping amplitudes $\tau_{\uparrow\uparrow}$ and $\tau_{\downarrow\uparrow}$, of spin-scattering [6]. Through spin-flip scattering equal-spin superconducting correlations can be admitted into the half-metal. These two ways of spin-activeness may be accounted for by the boundary conditions the quasiclassical Greens function must obey at the superconductor/half-metal interface via the set of *phenomenological* parameters $(\Theta_m, \tau_{\uparrow\uparrow}, \tau_{\downarrow\uparrow})$ [6].

The geometry we study is that of two equal length half-metals, length L_{HM}, that sandwich a d-wave superconductor, length L_S. Both types of material are assumed to conduct in two-dimensional planes and the contact between them is made between these planes. Furthermore we consider either that the two half-metals are parallel, both conduct in the same spin band, or they are anti-parallel and conduct in opposite spin bands. The sizes of the systems are taken to be on the order of 10 ξ_0 $(= \hbar v_f/2\pi k_B T_c$, the coherence length of the superconductor) and we assume clean materials so that impurity scattering may be neglected. Finally, since we consider a d-wave superconductor the ab-plane of the superconductor may be oriented with respect to the interface normal. 5A It is well known that the d-wave superconductor admits zero-energy bound states (ZEBS) in the DoS if the ab-plane is not aligned to the interface [7]. These ZEBS will be shifted from zero energy by a finite spin-mixing scattering and the DoS of the 45°-degree rotated (or [110]) d-wave turns out to be very sensitive to weak spin-mixing.

In Figure 1 we display the orderparameter profile in the d-wave superconductor, the triplet pairing correlations (TPC) both in the superconductor and in the half-metal, and the angle averaged (total) DoS at various locations in our structure for a representative set of pa-

CP850, *Low Temperature Physics: 24th International Conference on Low Temperature Physics;*
edited by Y. Takano, S. P. Hershfield, S. O. Hill, P. J. Hirschfeld, and A. M. Goldman

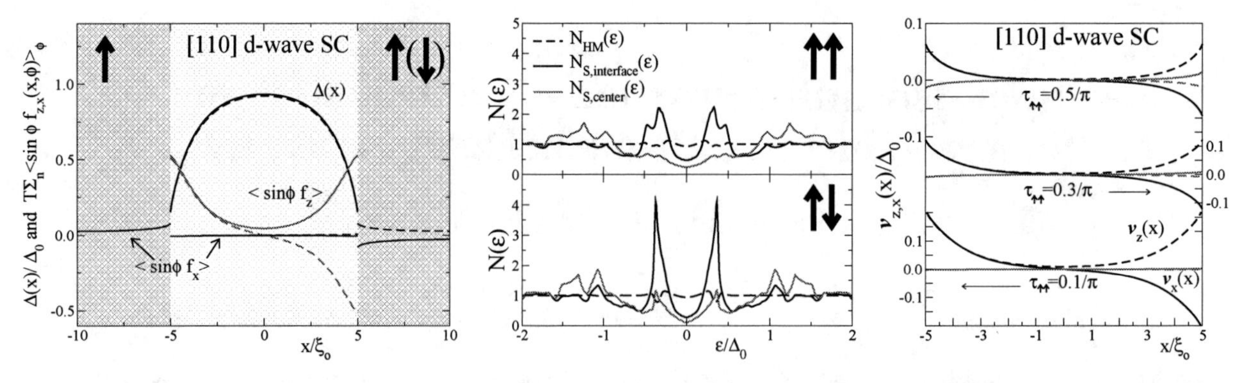

FIGURE 1. To the left we plot the amplitude of the d-wave orderparameter as well as the triplet pairing correlations (TPC). The TPC arise from the (repeated) scattering off the spin-active interfaces and they are also induced as equal-spin TPC ($T\sum_n \langle \sin\phi f_x(x,\phi;\varepsilon_n)\rangle$) in the half-metal. The interface parameters are [$\Theta_m = \frac{\pi}{4}, \tau_{\uparrow\uparrow} = \frac{0.1}{\pi}, \tau_{\downarrow\uparrow} = 0.9\,\tau_{\uparrow\uparrow}$] which corresponds to a low-transmission interface that is strongly spin-active. The full lines correspond to parallel alignment of the conduction bands of the two half-metals and dashed lines to anti-parallel alignment. In the center we show the total density of states (DoS) both in the half-metal and in the superconductor. For the superconductor we show the DoS at the interface and in the center (x=0). Finally to the right we show the induced exchange fi eld in the superconductor for three values of transparency of the interface. Exchange fi elds $\nu_{x,z}$ are calculated assuming a Fermi-liquid parameter $A_0^a = -0.7$. The full (dashed) lines corresponds to (anti-)parallel half-metal alignment.

rameters. The temperature is set to $0.1T_c$. In addition to finite size effects, the orderparameter is strongly suppressed in the vicinity of the interface due to the sign nature of d-wave pairing function. However, its magnitude at the interfaces is slightly enhanced by the spin-mixing scattering compared to that of a d-wave superconductor at a spin-inactive interface. The spin-mixing scattering generates $S_z = 0$ TPC (f_z) in the superconductor while spin-flip scattering induces equal-spin $S_z = \pm 1$ TPC both in half-metal and in the superconductor (f_x). Although parallel or anti-parallel alignment hardly makes a difference on the magnitude of the orderparameter, the TPC in parallel and anti-parallel alignments shows symmetric or antisymmetric behavior with respect to the center, respectively. For more transparent interfaces the equal-spin TPC grow stronger and gives rise to an exchange field which direction-axis rotates in the superconductor. Typical magnitude for the exchange fields is $\sim 10\%$ of Δ_0 at the interface and decaying into the center of the superconductor. Note also that for the [110]-crystal the exchange field has the opposite symmetry with regard to the center-point compared to the TPC. For the [100]-crystal these symmetries are the same and the TPC have a $\cos\phi$ p-wave basis function compared to the $\sin\phi$ for the [110]-crystal.

To investigate alignment effects further, we show the DoS for the two studied cases in the center panel of Figure 1. For both half-metal alignments, the d-wave ZEBS in the DoS split into two peaks in the superconducting side. This due to the spin-mixing scattering at the interface[4]. There is also a change in the DoS in the half-metals via the proximity effect. This is change comes mainly from the spin-flip scattering propreties of the interface. We find the alignment difference most clearly manifested in the DoS in the superconductor. In contrast to the sharp peaks in the DoS the anti-parallel alignment shows, the bound-states peaks in the parallel alignment are broadened or even split by finite size effects. This difference is due to the fact that there is a constructive interference of repeated quasiparticle spin-mixing scattering off the right and the left interface in a parallel alignment that leads to a split bound-state signature in the DoS. For the anti-parallel alignment this possitive interference is canceled due to that the consecutive scattering picks up opposite phase shifts at either interface.

We gratefully acknowledge financial support from "Grant-in-Aid for Scientific Research" from the MEXT of Japan (N.Y) and the Swedish Research Council (M.F).

REFERENCES

1. J.-H. Park. et al, *Nature* **392**, 794 (1998).
2. A. Millis, D. Rainer, and J. A. Sauls, *Phys. Rev. B* **38**, 4504 (1988).
3. T. Tokuyasu, J. A. Sauls, and D. Rainer, *Phys. Rev. B* **38**, 8823 (1988).
4. M. Fogelström, *Phys. Rev. B* **62**, 11812 (2000).
5. J. C. Cuevas and M. Fogelström, *Phys. Rev. B* **64**, 104502 (2001).
6. M. Eschrig, J. Kopu, J. C. Cuevas and G. Schön, *Phys. Rev. Lett.* **90**, 137003 (2003).
7. C.-R. Hu, *Phys. Rev. Lett.* **72**, 1526 (1994).

Nonequilibrium Spin-transfer Torque in SFNFS Junctions

Erhai Zhao and J. A. Sauls

Department of Physics and Astronomy, Northwestern University, Evanston, IL 60208, USA

Abstract. We report theoretical results for the nonequilibrium spin current and spin-transfer torque in voltage-biased SFNFS Josephson structures. The subharmonic gap structures and high voltage asymptotic behaviors of the dc and ac components of the spin current are analyzed and related to the spin-dependent inelastic scattering of quasiparticles at both F layers.

Keywords: spin-transfer torque, magnetic nanopillar, spin filtering, multiple Andreev reflection, ac Josephson effect
PACS: 72.25.Mk, 74.50.+r

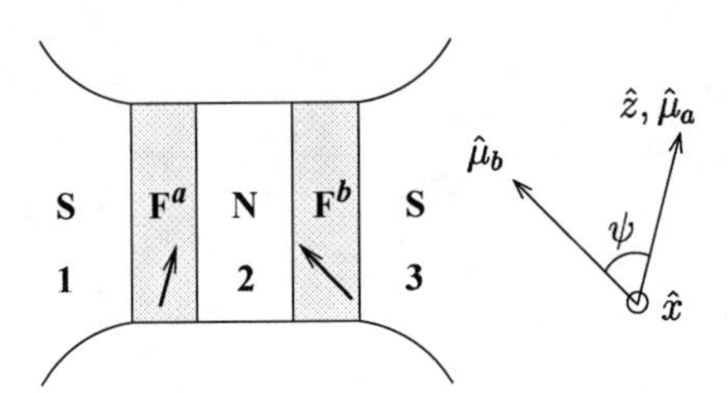

FIGURE 1. Schematic of SFNFS junction. The magnetization $\vec{\mu}_a$ is along the z axis (the quantization axis for spin), while $\vec{\mu}_b$ is at polar angle ψ in the yz plane. The x axis is along $\vec{\mu}_a \times \vec{\mu}_b$. Notice our choice of axis differs from Ref. [6].

Spin-polarized current passing through a ferromagnet can transfer spin angular momentum to the ferromagnet and exert a torque on the magnetic moment [1, 2]. This mechanism offers unique opportunities to manipulate the magnetic state of nanomagnets. Experimentally spin-transfer torque driven magnetization precession and magnetization reversal have been observed in magnetic multilayers [3, 4]. A well studied multilayer system is the magnetic nanopillar which consists of a ferromagnet-normal metal-ferromagnet (FNF) trilayer connected to normal metal electrodes. The typical thickness of each layer is several nm, and the diameter of the pillar is of the order of 50 nm [4]. When sandwiched between superconducting electrodes, the FNF trilayer can mediate finite Josephson coupling to form a SFNFS Josephson junction [5]. In such junctions, scattering of quasiparticle at the magnetic interfaces is sensitive to the phase shift in each F layer as well as the condensate phase difference ϕ across the junction. This indicates the spin momentum transfer between the quasiparticles and ferromagnets can be tuned by varying ϕ. Waintal and Brouwer calculated the phase-sensitive equilibrium torque in SFNFS junctions [6]. They also showed the nonequilibrium torque in NFNFS junctions acquires novel features, e.g. it can favor perpendicular configuration of the two moments [7].

Here we investigate the nonequilibrium spin-transfer torque in SFNFS point contacts under bias voltage V. The setup and the coordinate system are shown in Fig. 1. The two F layers are labelled by indices a and b, respectively. The N spacer is assumed to be transparent and in the clean limit. For simplicity we model each F layer as a delta function barrier with spin-dependent transmission probability $D_\uparrow \neq D_\downarrow$, so the spin mixing angle [8] $\vartheta = \arcsin\sqrt{D_\uparrow} - \arcsin\sqrt{D_\downarrow}$. The scattering matrix of F^a and F^b are related by spin rotation of angle ψ. Since the condensate phase difference evolves at Josephson frequency $\omega_J = 2eV/\hbar$, the spin current in each region, $\mathbf{I}_i$ ($i = 1,2,3$), also oscillates with time. For example,

$$\mathbf{I}_2(t) = \mathbf{I}_0 + \sum_{k=1}^{\infty}[\mathbf{I}_{k,c}\cos(k\omega_J t) + \mathbf{I}_{k,s}\sin(k\omega_J t)]. \quad (1)$$

We use the quasiclassical theory of superconductivity [9] and solve numerically the transport equations for the Green's functions in each region which obey proper boundary conditions at interface F^a and F^b [8]. The spin currents are then determined from the local Keldysh Green's functions. Within this approach the proximity effect and the multiple Andreev reflection (MAR) at both interfaces are fully taken into account.

The spin current in the N layer, measured in unit of $N_f v_f \Delta A\hbar/2$ where A is the contact cross sectional area and Δ is the superconducting gap, is shown in Fig. 2 for $\psi = \pi/2$ and zero temperature. As in the case of nanopillars with normal metal electrodes, the dc spin current flow is a consequence of spin filtering at the F layers. Its voltage dependence, however, is nonlinear and possesses subharmonic gap structures due to the onset and resonances of MAR processes. At $V = 0$, the dc spin current vector is along x direction, which tends to cause the moments to precess around each other [6]. As V is increased, the magnitude of dc currents along y and z direction, which tend to bring the moments towards or away from each other, grow with V. The ac spin current originates from the interference between MAR processes of different order. Its magnitude changes

CP850, *Low Temperature Physics: 24th International Conference on Low Temperature Physics;*
edited by Y. Takano, S. P. Hershfield, S. O. Hill, P. J. Hirschfeld, and A. M. Goldman

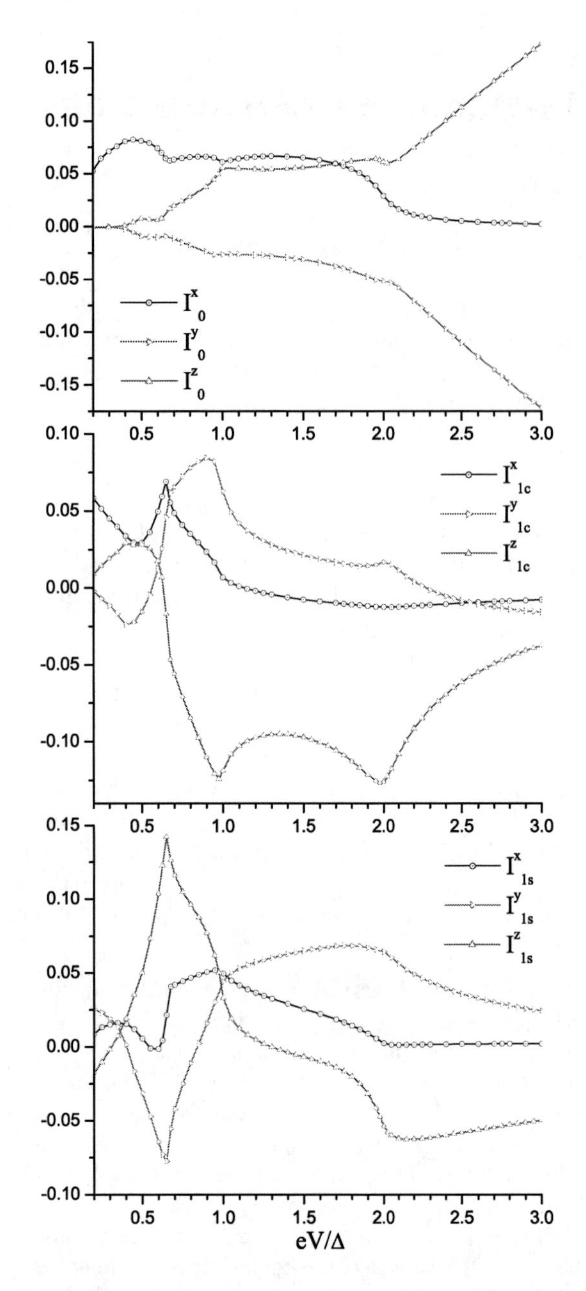

FIGURE 2. Dc (top panel), cosine (middle panel) and sine (bottom panel) part of the first Fourier components of the spin current in the N layer. Each panel features the spin current along x, y, and z direction. $D_\uparrow = 0.81$, $D_\downarrow = 0.64$, $\psi = \pi/2$.

rapidly at voltages below 2Δ and decays to zero at high voltages.

The spin-transfer torque on F^a is given by $\vec{\tau}_a(t) = \mathbf{I}_1(t) - \mathbf{I}_2(t)$, and the torque on F^b is given by $\vec{\tau}_b(t) = \mathbf{I}_2(t) - \mathbf{I}_3(t)$. It is convenient to expand $\vec{\tau}_{a/b}(t)$ in Fourier series similar to Eq. (1). Fig. 3 shows the $k = 0$ and $k = 1$ components of $\vec{\tau}_b$, in unit of $N_f v_f \Delta A \hbar/2$, as functions of the bias voltage at $T = 0$. Notice $\vec{\tau}_b$ is perpendicular to $\vec{\mu}_b$

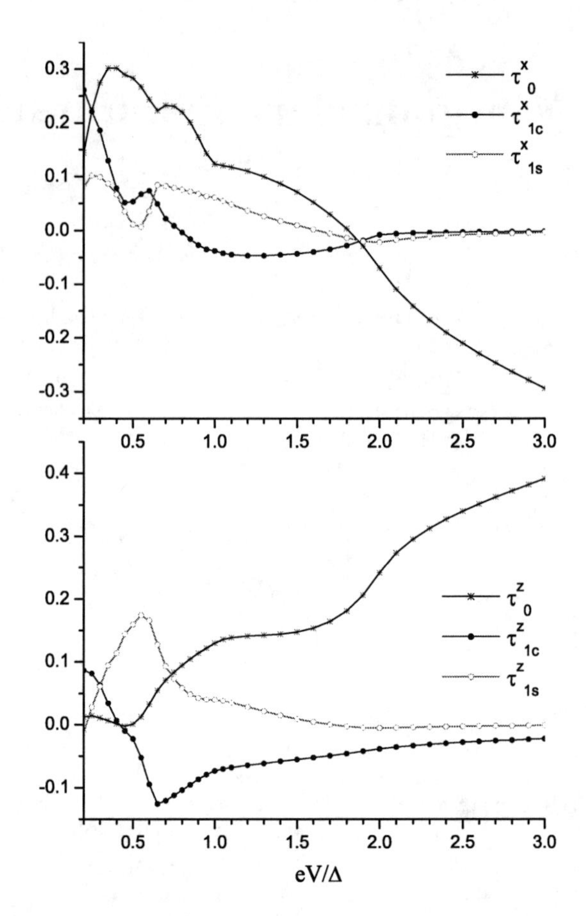

FIGURE 3. Spin-transfer torque on F^b in x (upper panel) and z (lower panel) direction. $D_\uparrow = 0.95$, $D_\downarrow = 0.6$, $\psi = \pi/2$.

and lies in the xz plane ($\psi = \pi/2$). An interesting feature of the dc torque along x direction is that it changes sign around voltage $1.8\Delta/e$. The magnitude of the dc torque becomes linear to V at high voltages. The ac torque becomes vanishingly small compared to the dc torque for $V > 2\Delta$ because the number of subgap Andreev reflections, which constitute the dominant contributions to the ac spin current, is inversely proportional to V.

REFERENCES

1. J. C. Slonczewski, *J. Magn. Magn. Mater.*, **159**, L1 (1996).
2. L. Berger, *Phys. Rev. B*, **54**, 9353(1996).
3. M. Tsoi et al., *Phys. Rev. Lett.*, **80**, 4281 (1998).
4. S. I. Kiselev et al., *Nature*, **425**, 380 (2003).
5. C. Bell et al., *Appl. Phys. Lett.*, **84**, 1153 (2004).
6. X. Waintal and P. W. Brouwer, *Phys. Rev. B*, **65**, 054407 (2002).
7. X. Waintal and P. W. Brouwer, *Phys. Rev. B*, **63**, R220407 (2001).
8. E. Zhao, T. Löfwander, and J. A. Sauls, *Phys. Rev. B*, **70**, 134510 (2004).
9. J. Rammer and H. Smith, *Rev. Mod. Phys.*, **58**, 323 (1986).

Inelastic Scattering Effects in Point Contacts between a Spin Polarized Normal Metal and a Superconductor

Charles W. Smith[a] and Paul J. Dolan, Jr.[b]

a. University of Maine, Orono, ME, USA
b. Northeastern Illinois University, Chicago, IL, USA

Abstract. Conductance curves for normal metal/superconductor point contacts have been modeled in terms of electron spin polarization, for the case in which the normal metal is magnetic [1]. We modify this model to include inelastic scattering, i.e., quasiparticle lifetime effects. Furthermore, a quantitative technique for estimating the inelastic scattering parameter and the spin polarization parameter is presented. Comparison to data for a $Co_2Cr_{0.6}Fe_{0.4}Al$/Sn point contact is discussed.

Keywords: Point Contact, Spin Polarization, Quasiparticle Effects.
PACS: 74.45.+c, 73.23.Ad, 72.25.-b

INTRODUCTION

A recent modification [1] of the BTK model [2] for normal metal/superconductor point contacts has incorporated electron spin polarization, P, and the proximity effect to describe conductance-voltage curves for contacts of the form X/Nb, where X = Ni, Co and Fe. Two energy gaps were used to represent the proximity effect and the contact transparency was modeled using the BTK elastic scattering parameter, Z. However, it turned out that P, which should strictly be a material property of the magnetic metal, had to be varied with Z, which is the contact interface parameter, in order to realize a fit to the data.

We propose a model which uses Z and P, but replaces the two-gap proximity effect representation with quasiparticle finite-lifetime effects, i.e., an inelastic scattering parameter [3].

THE MODEL

Studies [4] have shown that, in addition to elastic scattering, inelastic scattering has a profound effect on charge transport in point contacts. Inelastic scattering is introduced into the BTK model via the inelastic scattering parameter $\Gamma \equiv \hbar/\tau$, where τ is the quasiparticle lifetime. As a consequence [3], the Bogoliubov-deGennes coherence factors become imaginary over the entire energy range, the quasiparticle density of states depends explicitly on τ and Andreev reflection is weakened. For the model, the inelastic scattering, Γ, is parameterized as a fraction of the energy gap.

We write the current in the contact as

$$I = (1 - P)\, I_U + P\, I_P \qquad (1)$$

where the first term is the fully unpolarized part of the current and the second term is the fully polarized part. Here, $0 \leq P \leq 1$. I_U and I_P are calculated using the BTK formalism, with appropriate expressions for the Andreev and ordinary reflection probabilities, i.e., $A_U(Z,\Gamma,P,T)$, $B_U(Z,\Gamma,P,T)$ and $B_P(Z,\Gamma,P,T)$. Note that $A_P(Z,\Gamma,P,T)$ must be zero, since fully polarized Andreev reflection is forbidden.

ANALYSIS

Figure 1 shows the normalized conductance at zero bias, $Y(Z,\Gamma,P,T) = G_{NS}(T)/G_{NN}$, as a function of temperature, for a point contact between the Heusler alloy $Co_2Cr_{0.6}Fe_{0.4}Al$ and Sn [5]. The values of the model parameters ($Z = 0.28$, $\Gamma = 0.10$ and $P = 0.35$) used to fit to the data came from the parameter surface shown in Figure 2. We exploit the fact that there is essentially a one-to-one mapping of Y and dY/dt, here taken from Figure 1 at reduced temperature $t = T/T_c = 0.50$, onto the Γ - P parameter surface. Iteration between the two figures converges to the best fit values of Γ and P that represent the entire data set. We

CP850, *Low Temperature Physics: 24th International Conference on Low Temperature Physics;*
edited by Y. Takano, S. P. Hershfield, S. O. Hill, P. J. Hirschfeld, and A. M. Goldman

point out that an initial estimate for the value of Z must first be made in order to generate Figure 2. For intermediate to high values of Γ and/or P, the model is relatively insensitive to the initial value of Z and therefore one can begin a three parameter iteration with a nominal seed value of Z = 1.00.

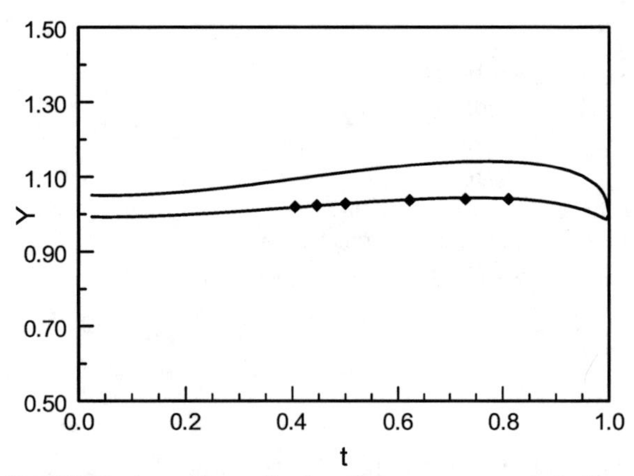

FIGURE 1. Normalized conductance at zero bias, Y(Z,Γ,P,T), versus reduced temperature, t, is plotted for a $Co_2Cr_{0.6}Fe_{0.4}Al/Sn$ point contact. The curve through the data is for values Z = 0.28, Γ = 0.10 and P = 0.35. The curve above the data omits inelastic scattering, i.e., Z = 0.28, Γ = 0.00 and P = 0.35.

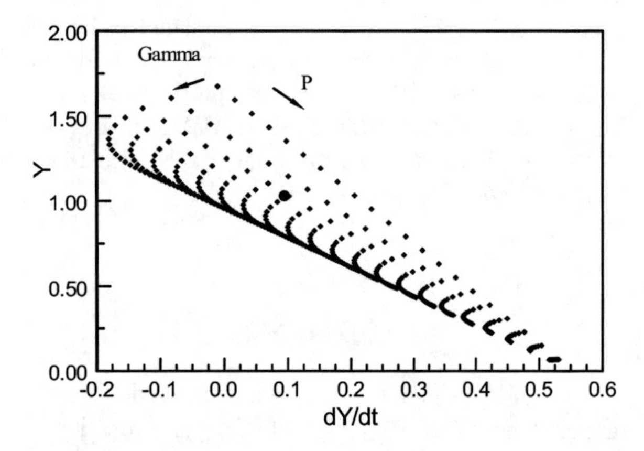

FIGURE 2. The Γ - P parameter surface for Y versus dY/dt at t = 0.50 and Z = 0.28. The values for Γ and P, used to fit the data in Figure 1, are shown as a point, ●, on the surface. P ranges from 0.00 to 1.00 in steps of 0.05 and Γ ranges from 0.00 to 1.00 in steps of 0.05.

CONCLUDING REMARKS

A model for charge transport in a spin polarized normal metal/superconductor point contact is presented, which includes inelastic scattering effects as well as elastic scattering at the contact interface. A general technique is illustrated for quantitatively determining the model parameters from transport data, as a function of temperature. A specific example for a Heusler alloy/tin point contact is discussed.

REFERENCES

1. G.J. Strijkers, Y. Li, F.Y. Yang, C.L. Chien and J.M. Byers, *Phys. Rev. B* **63** 104510 (2001).
2. G.E. Blonder, M. Tinkham and T.M. Klapwijk, *Phys. Rev. B* **25** 4515 (1982).
3. A. Plecenik, M. Grajcar, S. Benacka, P. Seidel and A. Pfuch, *Phys. Rev. B* **49** 10016 (1994).
4. Eugene V. Slobodzian, Charles W. Smith and Paul J. Dolan, Jr., *Physica C* **382** 401 (2002).
5. N. Auth, G. Jakob, T. Block and C. Felser, *Phys. Rev. B* **68** 024403 (2003).

Josephson SFS π-junctions. Potential Applications in Computing

Valeriy Ryazanov, Vladimir Oboznov, Vitalii Bolginov and Alexey Feofanov

Institute of Solid State Physics, Russian Academy of Sciences, Chernogolovka, Moscow reg., 142432 Russia

Abstract. Novel superconducting weak links, 'π-junctions', were realized recently. An origin of the π-state in a Superconductor - Ferromagnet - Superconductor (SFS) junction is an oscillating and sign-reversing superconducting order parameter induced in the ferromagnet close to the SF-interface. The π-behavior in SFS sandwiches was first observed by our group in 2000 [1]. Our recent result was a detection of transitions into π-state and back into 0-state, i.e. a nonmonotonic (with two nodes) behavior of the junction critical current vs. F-layer thickness. π-junctions with critical current density up to 2000 A/cm^2 were achieved that are suitable for applications in future superconducting digital and quantum electronics [2]. Our junctions are based on a niobium thin film technology so they can be incorporated directly into existing architectures of the superconducting electronics.

Keywords: Superconductivity-ferromagnetism coexistence, π-junctions, phase-inverter
PACS: 74.50.+r, 74.80.Dm

INTRODUCTION

At the present time a substantial attention has been paid to a realization of 'π-junctions', i.e. weakly coupled superconducting structures which demonstrate π-shift of a macroscopic phase difference in the ground state. A relation between the superconducting current I_s and a phase difference φ in a Josephson junction is described by a 2π-periodic function; in the simplest case of a tunnel barrier or a barrier made of a dirty normal metal, one finds $I_s = I_c \sin\varphi$. A Josephson π-junction is a weak link with an anomalous current-phase relation $I_s = I_c \sin(\varphi + \pi) = -I_c \sin\varphi$, i.e. it is characterized (nominally) by a negative critical current [3]. An oscillating (sign-reversing) superconducting order parameter in a ferromagnet close to an SF-interface was predicted by Buzdin et al [4]. So the superconducting order parameter does not simply decay in a ferromagnet but also oscillates. A physical origin of oscillations is the exchange splitting of the spin-up and spin-down electron subbands in a ferromagnet. It was discussed in Refs. [1, 5, 6, 7] that in order to observe a manifestation of the transition into π-state one should fabricate an SFS sandwich with the F-layer thickness d_F close to integer numbers of half-periods of the order parameter spatial oscillations. This period equals $\lambda_{ex} = 2\pi\xi_{F2}$, where the "imaginary" length ξ_{F2} can be extracted from the complex coherence length ξ_F in a ferromagnet: $\frac{1}{\xi_F} = \frac{1}{\xi_{F1}} + i\frac{1}{\xi_{F2}}$. In the simplest case the imaginary length ξ_{F2} and the order parameter decay length ξ_{F1} are equal [4]: $\xi_{F1} = \xi_{F2} = \sqrt{\hbar D/E_{ex}}$, where D is the diffusion coefficient for electrons in the ferromagnet and E_{ex} is the exchange energy responsible for the sign-reversal superconductivity in the ferromagnet. Temperature changes of the coherence length related to the thermal energy contribution to pair-breaking processes were introduced in Ref. [1], in which a temperature $0-\pi$-transition was observed for the first time. Expressions for ξ_{F1} and ξ_{F2} are the following:

$$\xi_{F1,2} = \sqrt{\frac{\hbar D}{\sqrt{(\pi kT)^2 + E_{ex}^2} \pm \pi kT}} \simeq \sqrt{\frac{\hbar D}{E_{ex}}}(1 \mp \frac{\pi kT}{2E_{ex}}), \tag{1}$$

The latter approximation corresponds to the case when $E_{ex} \gg kT$, which is valid for experiments discussed below.

In fact a nonmonotonic $I_c(d_F)$ dependence for thicknesses close to the $0-\pi$-transition was observed for the first time in Ref. [1] and has been presented there as a number of $I_c(T)$ curves for different thicknesses d_F. Later Kontos et al [5] for $Nb - Pd_{0.9}Ni_{0.1} - Nb$ and then Sellier et al [6] for $Nb - Cu_{0.52}Ni_{0.48} - Nb$ junctions measured detailed reentrant $I_c(d_F)$ curves for F-interlayer thicknesses also close to the $0-\pi$-transition. In this work we have investigated the thickness dependence of the SFS junction critical current density in a wide thickness range for SFS sandwiches fabricated as described in Ref. [7].

DOUBLE-REVERSAL THICKNESS DEPENDENCE OF CRITICAL CURRENT IN SFS SANDWICHES

Detailed experimental studies of the SFS junction critical current vs. F-thickness dependence were made for $Nb - Cu_{0.47}Ni_{0.53} - Nb$ (SFS) sandwiches. All junctions had

CP850, *Low Temperature Physics: 24th International Conference on Low Temperature Physics;*
edited by Y. Takano, S. P. Hershfield, S. O. Hill, P. J. Hirschfeld, and A. M. Goldman

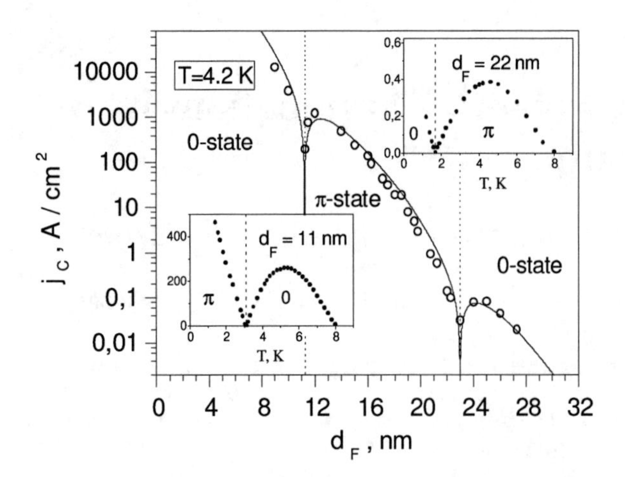

FIGURE 1. The F-layer thickness dependence of the critical current density for $Nb - Cu_{0.47}Ni_{0.53} - Nb$ junctions at the temperature 4.2 K. Open circles presents experimental results and the solid line shows model calculations [8]. Insets show temperature induced '0'-'π'-transitions for F-layer thicknesses 11 nm and 22 nm.

their lateral sizes smaller than the Josephson length and uniform current distribution, thereto junctions with F-layers thicknesses of less than 17 nm were made with the contact area 10×10 μm^2 and all the rest had the area 50×50 μm^2. Weakly-ferromagnetic $Cu_{0.47}Ni_{0.53}$-interlayers had the Curie temperature of about 60 K. In the thickness interval $8 - 28$ nm we had about 6 orders of the critical current density change with vanishing at two d_F values as it is presented in Fig. 1. Undoubtly the curve demonstrates both direct $0 - \pi$-transition and inverse transition from π- into 0-state. In transition points the critical current $I_c(d_F)$ is formally equal to zero and then should change its sign. Since in real experiments we could measure only the magnitude of the critical current, the dependence $I_c(d_F)$ between two sharp cusps is the negative branch of the curve corresponding to π-state which is reflected into the positive region. Due to weak temperature dependence of the order parameter oscillation period in our weak ferromagnet (described by (1)) we could pass through points of the critical current vanishing using samples with critical F-layer thicknesses 11 nm and 22 nm by means of temperature decrease. Temperature $0 - \pi$ and $\pi - 0$ transitions are presented in insets of Fig. 1. Temperature decrease from $9\,K$ down to 1 K is accompanied by decrease of $2 - 3$ nm in the spatial oscillation period and by increase of about 0.2 nm in the decay length. Simple estimations of ξ_{F1} (obtained from the slope of the $I_c(d_F)$ envelope) and ξ_{F2} (estimated from the interval between two minima) show a large difference between these two lengths (1.3 nm and 3.5 nm correspondingly) that can not be explained by the thermal contribution described by (1).

A theoretical analysis of results presented in Ref. [8] takes into account additional depairing processes that increase ξ_{F2} and decrease ξ_{F1}. As the F-layer is an alloy, a role of the magnetic scattering may be quite important [6, 7]. Magnetic inhomogeneity is related above all to Ni-rich clusters arising in $Cu_{1-x}Ni_x$ ferromagnet for x close to 0.5. In the region of these concentrations when the Curie temperature is small, we may expect that the inverse spin-flip scattering time $\hbar\tau_s^{-1}$ could be of the order of averaged E_{ex} or even larger. This circumstance strongly modifies the proximity effect in SF systems. A role of the spin-orbit scattering should be neglected for a $CuNi$ alloy since it is substantial only in ferromagnets with large atomic number Z.

Fig. 1 shows that π-junctions with high critical current density up to 2000 A/cm^2 were achieved that are suitable for possible applications as elements of digital and quantum logic [2]. Using π-junctions makes it possible modifying the existing superconducting rapid single flux quantum (RSFQ) logic and realizing of the superconducting analogy of the well-known complementary metal-oxide-semiconductor (CMOS) digital logic. Moreover, an operation of π-junctions as phase shifters in a passive (non-switching) mode opens a unique possibility to significantly reduce the size of logic cells eliminating their geometrical inductances. Junctions are based on a niobium thin film technology so they can be incorporated directly into existing architectures of the superconducting electronics. Recently a key experiment related to SFS π-junction insertion into a typical cell of the Josephson electronics (dc-SQUID) like a phase inverter was carried out successfully.

REFERENCES

1. A. V. Veretennikov et al, *Physica B* **284-288**, 495 (2000); V. V. Ryazanov et al, *Phys. Rev. Lett.* **86**, 2427 (2001).
2. G. Blatter et al, *Phys. Rev. B* **63**, 174511 (2001); E. Terzioglu, and M. R. Beasley, *IEEE Trans. On Appl. Supercond.* **8** 48 (1998); A. V. Ustinov, V. K. Kaplunenko, *J. Appl. Phys.* **94**, 5405 (2003).
3. L. N. Bulaevskii, V. V. Kuzii, and A. A. Sobyanin, *JETP Lett.* **25**, 290 (1977).
4. A. I. Buzdin, L. N. Bulaevskii, and S. V. Panjukov, *JETP Lett.* **35**, 178 (1982); A. I. Buzdin, B. Vujicic, M. Yu. Kupriyanov, *Sov. Phys. JETP* **74**, 124 (1992).
5. T. Kontos, M. Aprili, J. Lesueur, F. Genet, B. Stephanidis and R. Boursier, *Phys. Rev. Lett.* **89**, 137007 (2002).
6. H. Sellier, C. Baraduc, F. Lefloch, and R. Calemczuk, *Phys. Rev. B* **68**, 054531 (2003).
7. V. V. Ryazanov, V. A. Oboznov, A. S. Prokofiev, V. V. Bolginov, A. K. Feofanov, *Journ. Low Temp. Phys.* **136**, 385 (2004).
8. V. A. Oboznov, V. V. Bolginov, A. K. Feofanov, V. V. Ryazanov, and A. I. Buzdin, 2005, to be published.

2D Inhomogeneous Superconducting States And Umklapp Processes In Ferromagnet/Superconductor Nanostructures

M.G. Khusainov[a,b,c], N.G. Fazleev[b,d], M.M. Khusainov[e], and Yu.N. Proshin[a,b]

[a] *Max-Planck-Institute for the Physics of Complex Systems, Dresden 01871, Germany*
[b] *Kazan State University, Kazan 420008, Russia*
[c] *Kazan State Tupolev Technical University, Chistopol' 422981, Russia*
[d] *University of Texas at Arlington, Arlington, Texas 76019, USA*
[e] *Zavoisky Physical-Technical Institute of RAS, Kazan 420029, Russia*

Abstract. We derive new *three-dimensional* (3D) *boundary-value problem* (BVP) for the proximity effect in ferromagnetic metal/superconductor (FM/S) nanostructures. Our theory takes into account the competition between the one-dimensional (1D) and two-dimensional (2D) realizations of the Larkin-Ovchinnikov-Fulde-Ferrell (LOFF) interface states. It is shown that processes of mutual transformation between BCS and LOFF pairs at the FM/S boundary happen through the *Umklapp processes* during which coherent pair momentum **k** is conserved with exactness up to the reciprocal LOFF lattice vector **g**.

Keywords: proximity effect; superconductivity; ferromagnetism; multilayers; Umklapp processes; boundary conditions.
PACS: 74.78.Fk, 85.25.-j, 74.62.-c, 85.75.-d

In recent years there has been a pronounced interest in unconventional superconductivity, i.e. in superconducting electronic correlations different from the usual singlet BCS pairing with zero total momentum observed in conventional superconductors. One of such examples is the Fulde-Ferrell-Larkin-Ovchinnikov (FFLO) type pairing with a nonzero momentum of pairs [1,2], which may be realized not only in ferromagnetic superconductor but also in the layered ferromagnetic metal/superconductor (FM/S) structures (see review [3]). The existing theories of the proximity effect [3] predict multiple oscillations of $T_c(d_f)$ or even periodically reentrant superconductivity. At the same time, only one local maximum (or minimum) in the $T_c(d_f)$ dependence in the majority of experiments with multilayers FM/S is realized. In our opinion, the reason is that the former theories [3], predicting multiple oscillations of $T_c(d_f)$, are good only for the quasi-one-dimensional FM/S systems, where spatial changes of pair amplitude along the FM/S boundary can be neglected. The real FM/S systems, such as Fe/V, Fe/Pb or Gd/Nb, which were studied in experiments, are three-dimensional (3D). In addition, the recently observed absence of the superconductivity suppression in the short period Gd/La superlattice [4] is a real "physical challenge" to the existent proximity effect theory. Below we will show that full boundary-value problem (BVP) can have solutions which are able to explain this surprising behavior.

Beginning with the pioneer works of Radović et al., Buzdin et al. and in the further papers (see [3]) it was considered that pair amplitude $F(\mathbf{r})$ can change only across the FM and S layers, i.e. along the z axis. This corresponded to the 1D state with $\mathbf{q}_f = 0$, where $\mathbf{q}_f$ is the 2D modulation wave vector in the FM/S boundary plane. The possible realization of only the 1D state with $\mathbf{q}_f = 0$ was reasoned by Demler et al [6] by the necessity of the tangential component of the pairs momentum conservation. We derive a new BVP for the ferromagnetic metal/superconductor (FM/S) nanostructures. This BVP is valid for an arbitrary strength of the exchange field and the nonmagnetic impurity scattering. Our theory takes into account the competition between the one-dimensional (1D) and two-dimensional (2D) realizations of the FFLO interface states. It is shown that processes of mutual transformation between BCS and FFLO pairs at the FM/S boundary are the *Umklapp processes* during which coherent pair momentum **k** becomes confined in the FM layer. This fact is reflected in the crucially new boundary conditions for the Eilenberger equations. On this base we investigate the interplay between the BCS and FFLO states in the pure FM/S bilayers.

CP850, *Low Temperature Physics: 24th International Conference on Low Temperature Physics;*
edited by Y. Takano, S. P. Hershfield, S. O. Hill, P. J. Hirschfeld, and A. M. Goldman

In case of the dirty limit, where the electron mean free path $l_{s(f)}$ becomes much smaller than coherence length $\xi_{s(f)}$ and spin stiffness length a_f, the BVP for Eilenberger function $\Phi(\mathbf{p},\mathbf{q},z,\omega)$ is reduced to the corresponding one for Gor'kov function $F(\mathbf{q},z,\omega)$. This dirty limit BVP (see [3,5]) in the coordinate space includes the Usadel-like differential equations and boundary conditions, which connect the flux of $F(\boldsymbol{\rho},z,\omega)$ with its jump at the FMS interface. Many possible solutions of the dirty limit BVP have been well studied and can be found in Refs. [3,5]. Therefore below we consider the simple examples of the BCS and FFLO pairing mutual accommodation in the clean FM/S nanostructures.

For confinement of paper length we present our results to graphical form and consider here only the case of ideal transparency of metals FM and S with identical electronic parameters, except the fact that interelectron interactions $\lambda_{s(f)}$ and exchange splitting of the Fermi level can be different. The mutual influence of metals FM and S is especially significant in the Cooper limit, where the thicknesses of the F(S) layer $d_{f(s)} << \xi_{f(s)}$, a_f. Then the order parameter and the Eilenberger function $\Phi(\boldsymbol{\rho},z,\omega)$ are practically constant along the z axis inside the FM and S layers, and the spatial variations are possible only in $\boldsymbol{\rho}$ (along the FM/S boundary plane). For simplicity we considered the case of metals FM and S with identical electronic parameters, except the fact that interelectron interactions $\lambda_{s(f)}$ and exchange splitting of the Fermi level can be different. The $T_c(d_f)$ dependence for such an FM/S bilayer is depicted in Fig.1 for case $\lambda_s = \lambda_f$. The concrete value of q_f is found from the T_c maximum condition for each value of d_f (see upper panel of Fig.1). It is easy to see that there are two types of superconducting states in the FM layer. First of them is the BCS state with $q_f = 0$ and constant order parameter Δ_f is realized at small d_f. With d_f increase in the Lifshitz point L this BCS state changes to the FFLO one, in which $q_f \neq 0$ and order parameter $\Delta_f(\boldsymbol{\rho})$ and pair amplitude $\Phi_f(\boldsymbol{\rho},\omega)$ are oscillated in the FM/S boundary plane (x-y) with period of order of the spin stiffness length a_f. Unlike the FM layer, in the S layer only the BCS state takes place. From analysis of Fig. 1 it follows that the clean FM/S bilayers could be used as ideal model systems for direct observation of the FFLO state realization.

In the FM/S/FM trilayers two novel π phase superconducting states, which are realized at the repulsive electron-electron interaction in the FM layers, have been predicted. In the FM/S superlattices there are 0π and $\pi\pi$ states with compensation of the exchange field paramagnetic effect by antiparallel orientation of the neighboring FM layers magnetizations. This fact allows us to explain a surprisingly high $T_c \sim 5$K of the short period Gd/La superlattice and to predict the sign and value of the electron-electron interaction in Gd metal. The 2D modulated FFLO states are possible in such trilayers and superlattices only in presence of external magnetic field and at suitable parameters of the FM and S layers.

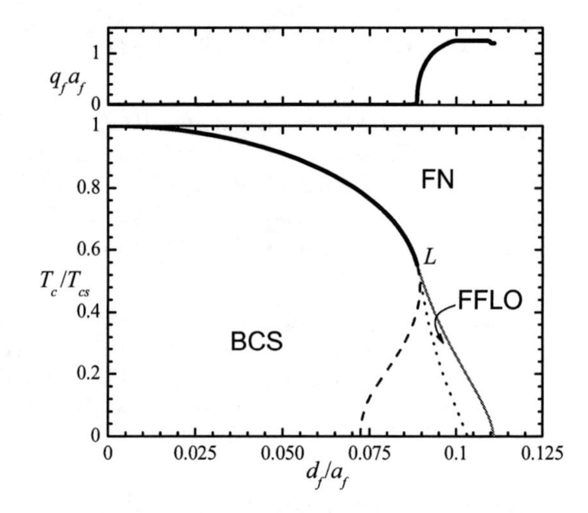

FIGURE 1. Phase diagrams $T_c(d_f)$ (lower panel) and $q_f(d_f)$ (upper panel) of the clean FM/S bilayer. Here T_{cs} is the critical temperature of the S material. The thick (black) solid curve is the second order phase transitions curve, which separates the BCS state ($q_f = 0$) and ferromagnetic normal state (FN). The thin solid (red) line after the Lifshitz point L corresponds to the transition of the second order between the FFLO state with $q_f \neq 0$ and the FN state. The dotted (black) line of the first order transition between the BCS and FFLO states is schematically redrawn from [3]. The thin dashed (blue) curve represents the no realized state with $q_f = 0$. We used the following parameters in numerical calculations: $d_s/\xi_s = 0.25$; $\xi_s/a_f = 10$.

This work was supported in part by RFBR (04-02-16761, 05-02-16369) and CRDF (REC-007).

REFERENCES

1 Fulde, P., and Ferrell, R. A., *Phys. Rev. A* **135**, 550 (1964).
2 Larkin, A. I., and Ovchinnikov, Yu. N., *Sov. Phys. JETP* **20**, 762 (1965).
3. Izyumov, Yu. A., Proshin, Yu. N., and Khusainov, M. G., *Physics-Uspekhi* **45**, 109-148 (2002) (see also references therein).
4. Goff, J. P., Deen, P. P., Ward, R. C. C., Wells, M. R., Langridge, S., Dalgleish, R., Foster, S., and Gordeev, S., *JMMM* **240**, 592-594 (2002).
5 Demler, E. A., Arnold, G. B., and Beasley, M. R., *Phys. Rev. B* **55**, 15174 (1997).
6. Khusainov, M. G., and Proshin, Yu. N., *Physics-Uspekhi* **46,** 1311-1313 (2003).

Decoupled Superconductivity and Hierarchy of Transition Temperatures in the Tetralayer Ferromagnet/Superconductor Nanostructure and Control Devices

Yurii N. Proshin [a,b] and Mansur G. Khusainov [a,b,c]

[a] *Department of Theoretical Physics, Kazan State University, Kazan 420008, Russia*
[b] *Max-Planck-Institute for the Physics of Complex Systems, Dresden, Germany*
[c] *Branch "Vostok", Kazan State Technical University, Chistopol' 422981, Russia*

Abstract. We predict the *decoupled superconductivity* for the four-layered F'/S'/F"/S" nanostructure consisting of dirty superconducting (S) and ferromagnetic (F) metals. The predicted *hierarchy of critical temperatures* is found to manifest itself in its most striking way through arising of different critical temperatures in different superconducting layers S' and S". In common case the phase diagram including four different regions is found. Conceptual sketch of the new control nanodevice based on this tetralayer system are proposed. It is shown that they can have up to ***seven*** various states.

Keywords: proximity effect; superconductivity; ferromagnetism; multilayers; critical temperature; control device.
PACS: 74.78.Fk, 85.25.-j, 74.62.-c, 85.75.-d

In the layered ferromagnetic metal/superconductor (F/S) heterostructures the superconducting order parameter (OP), owing to the proximity effect, can be induced in the F layers; on the other hand, the neighboring pair of the F layers can interact with one another via the S layer (see review [1]). Basically, the F/S structures possess two data-record channels, namely, on the superconducting properties and the mutual ordering of the F layers magnetizations. A conceptual scheme of "spin-valve" devices based on the F/S/F *trilayer* was offered in works [2,3]. This proposed F/S/F device has only ***two*** different states and operates only on transition between the superconducting (S) and normal (N) states controlled by external magnetic field H.

The *multilayered* F/S systems have additional competition between the 0 and π phase types of superconductivity and possesses four different states [1]: two ferromagnetic superconducting (FMS) ones (00, π0), and two antiferromagnetic superconducting (AFMS) ones (0π, ππ). They are distinguished by the phases of the superconducting (the first symbol) and magnetic (the second one) OPs. In the AFMS states the pair-breaking effect of exchange field I of the F layers in the S layers is significantly attenuated, and the transition temperature T_c is higher than in the FMS case. This theoretical prediction of ours has been experimentally confirmed for the Gd/La *superlattice* [4]. For the F/S *superlattices* [1] it is possible to propose the principal scheme of the device (with up to ***five*** different states) that allows us to *separate* the superconducting and magnetic data-record channels in one sample.

Consider the F'/S'/F"/S" tetralayer with alternation of layers along the z axis. The outer F'(S") and inner F"(S') layers occupy regions $-d_f/2 < z < 0$ ($0 < z < d_s$), and $d_s < z < d_s + d_f$ ($d_s + d_f < z < d_s + 3d_f/2$), correspondingly. To find critical temperature T_c we use our theory [1] based on the simultaneous solutions of the Usadel and Gor'kov equations assuming the standard dirty limit conditions. We don't have the opportunity to cite here the full expressions, so let us proceed to the short graphical presentation of the results.

Let the system choose its own state itself according to the second order phase transitions theory. The state with higher critical temperature T_c (lower free energy) is realized. A complete phase diagram for the tetralayer is presented in Fig. 1a. There are four different regions in this diagram: at high temperature both S' and S" layers are in the N state. Then, there are two AFM regions (dark grey). In this *decoupled* state the inner S' layer is superconducting, and the outer S" one is normal. The striped region also corresponds to the *decoupled* state with the with the superconducting

CP850, *Low Temperature Physics: 24th International Conference on Low Temperature Physics;*
edited by Y. Takano, S. P. Hershfield, S. O. Hill, P. J. Hirschfeld, and A. M. Goldman

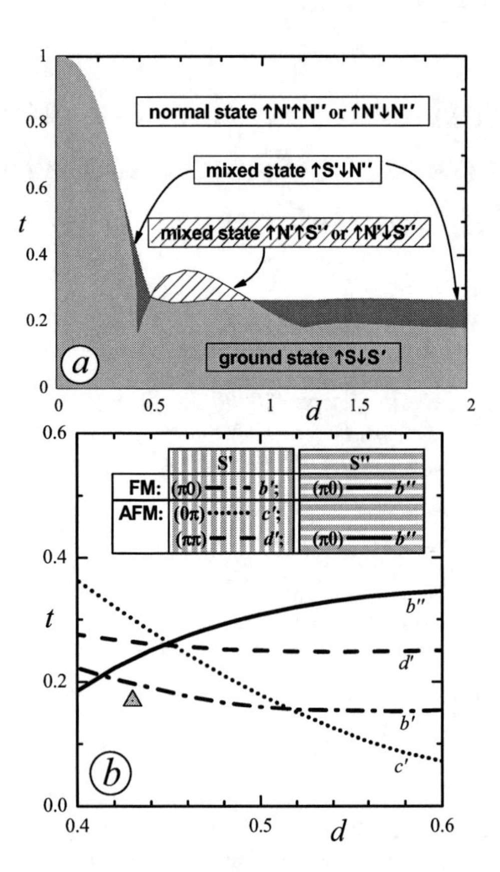

FIGURE 1. The phase diagrams of the F'/S'/F''/S'' system: the reduced critical temperatures $t = T_c/T_{cs}$ for the F'/S'/F''/S'' system versus the reduced F layer thickness $d = d_f/a_f$. Here T_{cs} is the critical temperature of the S material, a_f is the spin stiffness length. The optimal values of parameters are used. (*a*) Vertical arrows indicate the direction of magnetization in the corresponding ferromagnetic layers. For simplicity we assume that the magnetization in the outer F' layer is directed "upwards". (*b*) Control device working point. The *t* curves are indicated by letters with *single (double) prime*, which correspond to the S' (S'') layers.

outer S'' layer and the normal S' layer. There is only the ground AFMS (light grey) state at low temperature and/or at small thicknesses *d*. Note, if the inner S' layer is in the *superconducting* state then the alignment of magnetizations should be *antiferromagnetic*. This is the *inverse* action of *superconductivity on magnetism*. Note, the details of the phase diagram strongly depend on the choice of the system parameters.

Consider a conceptual sketch of the "control device" according to the scheme proposed for the F/S superlattice [1]. For technical convenience [1-3], we add at the left of our system one external layer of a magnetic insulator, whose role is to pin the direction of magnetization in the outer F' layer. The state of the F/S structure can be controlled by small external magnetic field **H**, which slightly changes the phase diagram [1-3]. Thus we can change data recorded on the superconducting property (the first channel) and orientation of magnetizations (the second channel). In the common case there are four specific values of the magnetic field: coercive field H_{coer}; two critical fields H_c' and H_c'' for the S' and S'' layer, correspondingly; and pinning field H_p.

Let the system be in the "triangle" working point presented in Fig. 1b. Suppose that $H_{coer} < H_c' < H_c'' < H_p$. At zero magnetic field the system is in the main AFMS state. The FM curves are absent in this case. When changing external magnetic field **H** at first along the direction of the pinning field, the transition from the ground AFMS state to the FMS one occurs at $H \approx H_{coer}$. Both magnetizations become parallel, and the AFM curves disappear in Fig. 1b. If the orientation of magnetization of the F' layer is pinned upwards (↑), this transition can be presented as ↑S↓S → ↑S↑S. One can say that the data written on the superconducting properties of the S layers is conserved but the information recorded on mutual directions of magnetizations is changed. At $H \approx H_c'$ the second transition ↑S↑S → ↑N↑S occurs, then we have the next transition ↑N↑S → ↑N↑N occurring at $H \approx H_c''$. The data records on the supercurrent are changed. Applying the external magnetic field in the opposite direction one can gain other transitions chain from the AFMS(↑S↓S) to the AFNS(↑N↓S) at $H \approx H_c'$; then, at $H \approx H_c''$, to the AFN(↑N↓N); and at last to another FN one (↓N↓N) at $H \approx H_p$. Thus for this case it is possible to get up to ***seven*** logically various states. For other choice of working points we could get the chains with four, five, or six states. Thus, the tetralayer systems are the most perspective candidates for use in superconducting nanoelectronics.

This work is supported in part by RFBR (04-02-16761, 05-02-16369) and CRDF (REC-007).

REFERENCES

1. Izyumov, Yu. A., Proshin, Yu. N., and Khusainov, M. G., *Physics-Uspekhi* **45**, 109-148 (2002) (see also references therein).
2. Buzdin, A. I., Vedyayev, A. V., and Ryzhanova, N. V., *Europhys. Lett.*, **48**, 686-691 (1999).
3. Tagirov, L. R., *Phys. Rev. Lett.*, **83**, 2058-2061 (1999).
4. Larkin, A. I., and Ovchinnikov, Yu. N., *Soviet Phys. JETP* **47**, 1136 (1964);
   Fulde, P., and Ferrell, R. A., *Phys. Rev.* **A135**, 550 (1964).
5. Goff, J. P., Deen, P. P., Ward, R. C. C., Wells, M. R., Langridge, S., Dalgleish, R., Foster, S., and Gordeev, S., *JMMM* **240**, 592-594 (2002).

Exchange Model of Proximity Effect For F/S Nanostructures

Elena L. Parfenova,[a,b] Damir S. Sattarov,[a,b] Mansur G. Khusainov,[a,b] and Yurii N. Proshin [b]

[a] *Kazan State Tupolev Technical University, Chistopol' 422981, Russia*
[b] *Kazan State University, Kazan 420008, Russia*

Abstract. We propose the exchange model of proximity effect in the ferromagnetic insulator/superconductor/ferromagnetic insulator (F/S/F) trilayers. This model takes into account the long range antiferromagnetic indirect RKKY exchange between localized spins placed at the F/S boundaries through superconducting electrons of the S layers as well as direct ferromagnetic exchange of nearest neighbors in the F layers. We propose new control devices based on the three-layered F/S/F nanostructures and combining the advantages of the superconducting and magnetic data-record channels in a single sample.

Keywords: superconductivity; cryptoferromagnetism; proximity effect; mutual accommodation
PACS:73.21.Cd, 72.80 Sk, 74.78.Fk, 85.25.-j, 74.62.-c.

Crystal structures F/S formed by alternating layers of the ferromagnetic insulator (F) and superconductor (S) constitute a new class of layered materials with unique superconducting and magnetic properties which depend on the characteristics of materials comprising the nanostructures, and the thicknesses of the layers [1]. To explain the existence of a nonuniform internal field, which splits the BCS peak in the aluminum quasiparticles density of states in the EuO/Al and EuS/Al contacts and the unexpectedly weak suppression of superconductivity in the EuO/V multilayers [1,2] a full analysis of the variants of mutual accommodation of superconductivity and ferromagnetism in the F/S systems is required. On the other hand, the three-layered F/S/F nanostructures [1] can be used for creation of nanoelectronics of an essentially new type combining the advantages of the superconducting and magnetic channels of data record in one sample. We emphasize that both channels can be *separately* controlled by a weak external magnetic field.

Let us consider the planar trilayer F/S/F, where the F layers occupy the regions $-d_f < z < 0$ and $d_s < z << d_s + d_f$, and the S layer occupies the region $0 < z < d_s$. We will search for magnetic order of the localized spins (LS) in the F layers in the form $S_r^{\pm} = \langle S \rangle \exp[i(\mathbf{q\rho}+kz)]$. Near the critical temperature T_c an addition to the surface density of the F/S/F trilayer free energy δf, which is connected with superconducting transition in the S layer and transformation of the LS ordering in the F layers, has the form

$$\delta f = 2J\langle S\rangle^2 q^2 \frac{d_f}{a} - \frac{I^2\langle S\rangle^2}{4a} \cdot \left[\delta\chi_S(q,0,0) + \delta\chi_S(q,0,d_S)\cos kd_S\right] - \frac{d_S}{a^3}\left[\alpha_0\frac{\Delta^2}{2} + \beta_0\frac{\Delta^4}{4} + \gamma_0\frac{\Delta^6}{6}\right]. \quad (1)$$

Here the *first term* is the loss in energy of direct exchange due to long-wave ($qa << 1$) modulation of the LS ordering in FI layers. The *second term* can be expressed through the superconducting contribution to the RKKY exchange and spin susceptibility $\delta\chi_s$ of conduction electrons (CE), which has the form [1]

$$\delta\chi_S(q,z,z') = 4\pi N(0)T \cdot \sum_{\omega} \frac{\Delta^2}{\omega^2+\Delta^2} \frac{\cosh\kappa z \cosh\kappa(z'-d_S)}{D\kappa \sinh\kappa d_S}, \quad 0 \le z < z' \le d_S. \quad (2)$$

CP850, *Low Temperature Physics: 24th International Conference on Low Temperature Physics;*
edited by Y. Takano, S. P. Hershfield, S. O. Hill, P. J. Hirschfeld, and A. M. Goldman

Here Δ is the superconducting order parameter; D and $N(0)$ are the CE diffusion coefficient and the CE density of states, correspondingly; $\kappa^2 = q^2 + \xi_\omega^{-2}$; $\xi_\omega^2 = D/2(\omega^2 + \Delta^2)^{1/2}$, $\omega = \pi T(2n + 1)$. The second term in (1) plays double role and describes: i) intralayer and interlayer antiferromagnetic exchange of LS at the F/S interfaces via CE of the S layer and ii) depression of the S layer order parameter Δ due to paramagnetic effect of these LS. The *last term* in the right side of Eq. (1) is the Landau expansion in powers of Δ, which describes the gain in condensation energy due to the Cooper pairing of CE of the S layer. The coefficients α_0, β_0, γ_0 of this expansion are well known from the microscopic theory of superconductivity [1].

Minimization of free energy functional (1) over Δ, q and k leads to *three different ground states*: 1) ferromagnetic normal (FN) state with $A = 0$, $q = 0$, $\Delta = 0$; 2) antiferromagnetic superconducting (AFS) state with $A < 1$ and $q = 0$, $\Delta = \Delta_{AFS}(h)$, $k = \pi/d_s$, and 3) π phase cryptoferromagnetic superconducting (πCFS) state with $A > 1$ and $q \neq 0$, $\Delta = \Delta_{CFS}(h,q)$, $k = \pi/d_s$. The realization of each phase depends on the magnitude of three parameters: the temperature T, the exchange field $h = I\langle S\rangle a/2d_s$ generated by the LS at the F/S boundaries on the CE of the S interlayer and the ratio

$$A = \frac{N(0)h^2 d_S^5}{24J\langle S\rangle^2 d_f a^2 \xi^2}\left(\frac{\Delta}{2T_{CS}}\right)^2 \quad (3)$$

of antiferromagnetic and ferromagnetic molecular fields ratio, which is essentially reduced due to antiparallel orientation of the exchange fields h of the neighboring F layers.

The phase diagrams T_c versus exchange field h of the F/S/F systems (Fig. 1) are found to contain a *tricritical point t* in which the superconducting phase transition changes from the second order to the first one. The conditions for the *Lifshitz point L* to appear on the first-order transition line are established. At this point two commensurate magnetically ordered phases (FN and AFS) and one incommensurate phase (CFS) meet together. Two of three phases are the superconducting phases (AFS and CFS). The origin of the incommensurate magnetic phases in the F/S/F systems lies in the competition between the short-range direct ferromagnetic exchange of localized spins in the F/S boundary and the long-range antiferromagnetic RKKY exchange between these spins via the Cooper pairs. The π phase cryptoferromagnetism induced in the neighbor F layers by superconducting interlayer S is a manifestation of the *inverse proximity effect*, i.e. inverse influence of superconductivity on magnetism. Moreover, we predict the cascade of phase transitions CFS → AFS → FS → FN under the external magnetic field action. Due to this reason the F/S/F nanostructures are the perspective candidates for construction of the superconducting spintronics devices combining the superconducting and magnetic channels of data record in one sample [1]. Analogous chain of transitions explains the increase and subsequent saturation of the exchange splitting of the BCS peak in density of states of aluminum quasiparticles in the EuO/Al and EuS/Al junctions [1,2].

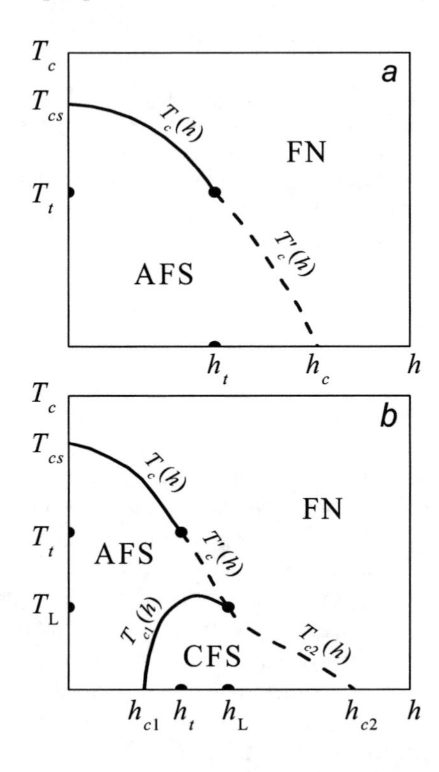

FIGURE 1. Phase diagrams (T_c–h) of F/S/F trilayers of (*a*) the first type with $A < 1$ ($q = 0$), and (*b*) the second type with $A > 1$ ($q \neq 0$). Here indices t and L are labeled the tricritical point and the Lifshitz point, correspondingly.

This work was supported in part by RFBR (04-02-16761, 05-02-16369) and CRDF (REC-007).

REFERENCES

1. Izyumov, Yu. A., Proshin, Yu. N., and Khusainov, M. G., *Physics-Uspekhi* **45**, 109-148 (2002) (see also references therein).
2. Tedrow, M., and Meservey, R., *Phys. Rep.* **238**, 173 (1994).

F/S/F Trilayer: 3D Model Of Proximity Effect

Larisa A. Terentieva,[a] Nikolay M. Ivanov,[a] Damir S. Sattarov,[a] Mansur G. Khusainov,[a,b,c] and Yurii N. Proshin [b,c]

[a] *Kazan State Tupolev Technical University, Chistopol' 422981, Russia*
[b] *Kazan State University, Kazan 420008, Russia*
[c] *Max-Planck-Institute for the Physics of Complex Systems, Dresden 01871, Germany*

Abstract. For the ferromagnetic metal/superconductor/ferromagnetic metal (F/S/F) trilayer we take into account the spatial variations of the pair amplitude not only across the F/S boundaries (the 1D case) but also along these interfaces (the 3D case). Our 3D model of the proximity effect also involves an indirect interaction of the both F layers magnetizations via the S layer. It leads to competition not only between the 1D and 3D LOFF states, but also between the *antiferromagnetic* and *ferromagnetic superconducting* states of the F/S/F trilayer. Possible nanodevices are discussed.

Keywords: superconductivity; ferromagnetism; proximity effect; data recording; multilayers
PACS: 74.78.Fk, 85.25.-j, 74.62.-c, 85.75.-d

In this paper the three dimensional (3D) theory of the proximity effect for the ferromagnet/superconductor/ferromagnet (F/S/F) trilayers is developed. Owing to the proximity effect, the superconducting order parameter can be induced in the F layers; in addition, the neighboring F layers can interact with one another via the S interlayer. Thus, for the F/S/F trilayers the interplay between the *antiferromagnetic superconducting* (AFS) and *ferromagnetic superconducting* (FS) states also takes place along with a competition between the 1D and 3D LOFF states [1,2]. Such systems exhibit rich physics, which can be controlled by varying the thicknesses of the F and S layers or by placing the F/S structure in an external magnetic field.

Consider the three-layered F/S/F system consisting of the S interlayer ($0 < z < d_s$) sandwiched between two the F layers ($-d_f < z < 0$ and $-d_s < z < d_s + d_f$). It is the simplest system allowing complicated competition between the 1D and 3D modifications of the AFS and FS states. We solve the Usadel equations with boundary conditions [1,2] in the dirty limit ($l_s << \xi_s$; l_f, $a_f << \xi_f$) for the F layer. Here $l_{s(f)}$, $\xi_{s(f)}$ and a_f are the mean free path, superconducting coherence and the spin stiffness lengths in the S(F) layer, correspondingly. We take into account the spatial variations of pair amplitude not only across the F/S/F nanostructure (the 1D case) but also along the F/S boundary (the 3D case). It leads to the transformation of the BCS pairs to the LOFF pairs at the passing through the S/F boundary (and LOFF → BCS at the F/S interface). This is immediately connected with the *umklapp processes* at which the LOFF pairs momentum **k** becomes confined in the F layer. We consider two possible configurations different in the alignment of the F layers magnetizations: the ferromagnetic one is the 00 state and the antiferromagnetic one is the 0π state according to known classification of F/S superlattices [1], where first (second) symbol corresponds to the shift in the phase of the superconducting (magnetic) order parameter between adjacent S (F) layers. As result we derive the usual set of equations for the reduced critical temperature $t = T_c/T_{cs}$

$$\ln t^{00} = \Psi\left(\frac{1}{2}\right) - \mathrm{Re}\Psi\left(\frac{1}{2} + \frac{D_s (k_s^{00})^2}{4\pi T_{cs} t}\right). \quad (1)$$

$$D_s k_s^{00} \tan\frac{k_s^{00} d_s}{2} = \frac{\sigma_s v_s}{4 - \dfrac{\sigma_f v_f}{D_f(I) k_f} \cot k_f d_f}, \quad (2)$$

$$k_f^2 + q_f^2 = -\frac{2iI}{D_f(I)}. \quad (3)$$

Here $\Psi(x)$ is digamma function; I is exchange field in the F layer, D_s is diffusion coefficient; $\sigma_{s(f)}$ are the transparencies of the S/F (F/S) boundaries. The true values of transition temperature t and transversal component of the LOFF momentum q_f are found from

CP850, *Low Temperature Physics: 24th International Conference on Low Temperature Physics;*
edited by Y. Takano, S. P. Hershfield, S. O. Hill, P. J. Hirschfeld, and A. M. Goldman

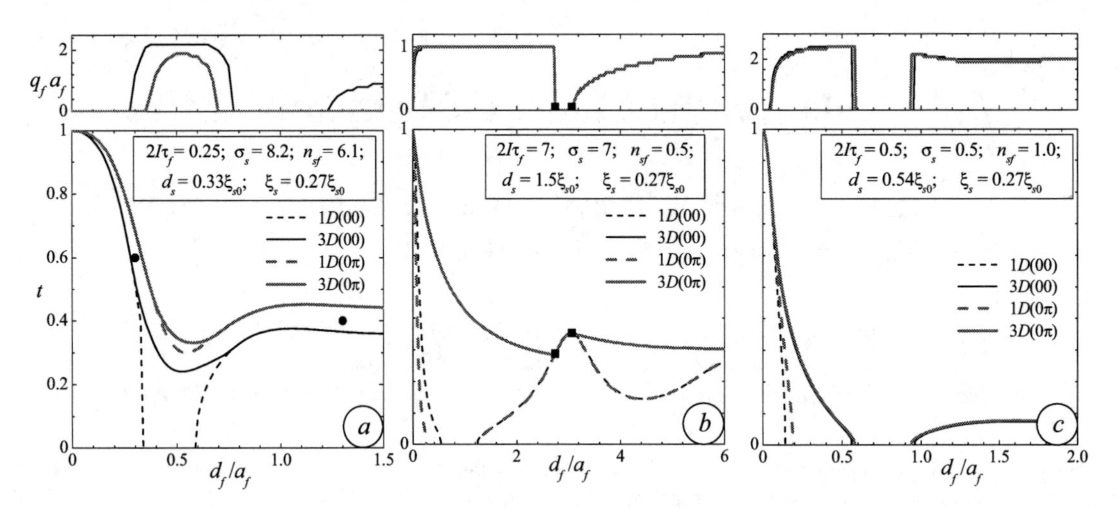

FIGURE 1. Dependences of the reduced critical temperatures t (lower panels) and the q_f value (upper panels) versus the reduced F layer thickness for the F/S/F system. Here ξ_s is the coherence length, parameter $n_{sf} = \sigma_s v_s / \sigma_f v_f$, $v_{s(f)}$ are the Fermi velocities, τ_f is the mean free path time. The thick (red) lines correspond to the AFS states, the thin curves correspond to the FS ones. The solid and dashed lines are responsible for the 3D ($q_f \neq 0$) and 1D ($q_f = 0$) cases, correspondingly. For simplicity the Lifshitz points of transitions 1D ↔ 3D (solid squares) are shown only in the part *b*.

the condition of the free energy minimum (the T_c maximum). We use complex diffusion coefficient $D_f(I)$ according to our review [1] (with small correction) and discussion [2]

$$D_f(I) = \begin{cases} \dfrac{D_f}{1 + i2I\tau_f}, (2I\tau_f << 1); \\ \dfrac{3}{2}\dfrac{D_f}{1 + iI\tau_f}, (2I\tau_f >> 1). \end{cases} \tag{4}$$

The first case corresponds to the extremely dirty limit ($l_f << a_f << \xi_f$), and the last case corresponds to the moderate dirty limit ($a_f << l_f << \xi_f$). The transition temperature for the 0π state ($t^{0\pi}$) can be obtained from Eq. (1) if we replace k_s^{00} (2) by $k_s^{0\pi}$ which is found from the equation

$$\left(k_s^{0\pi}\right)^2 - 2\,\mathrm{Re}\left(k_s^{00}\tan\frac{k_s^{00}d_s}{2}\right)\cdot k_s^{0\pi}\cot\left(k_s^{0\pi}d_s\right) = \left|k_s^{00}\tan\frac{k_s^{00}d_s}{2}\right|^2. \tag{5}$$

In Fig. 1 we see that in the most cases the 3D AFS state has a greater critical temperature T_c in comparison with the 1D AFS state, and also in comparison with the 3D and 1D FS states. Fig. 1a shows *the typical experimental behavior* of the F/S/F trilayers. There is an outcome on the plateau through the deep minimum, which was observed in the Fe/Pb/Fe trilayers [3]. In Fig. 1*b* we see the outcome on the plateau through the single flash due to the 3D-1D-3D phase transitions cascade. Similar behavior was observed, for example, in the Fe/Nb/Fe trilayers [4]. Our Fig. 1c presents the reentrant 3D superconductivity like to observed one in the Fe/V/Fe trilayers [5]. Note, the system in the conditions indicated by solid circles in Fig. 1*a* can work *as control device* with 4 logically various possible states (the superconducting one ↑S↓ and three normal ones ↑N↑, ↑N↓, ↓N↓), and we can change data recorded on the superconducting property (first channel) and mutual orientation of the magnetizations (second channel) by small external magnetic field control.

This work was supported in part by RFBR (04-02-16761, 05-02-16369) and CRDF (REC-007).

REFERENCES

1. Izyumov, Yu. A., Proshin, Yu. N., and Khusainov, M. G., *Physics-Uspekhi* **45**, 109-148 (2002) (see also references therein).
2. Khusainov, M. G., and Proshin, Yu. N., *Physics-Uspekhi* **46**, 1311-1313 (2003).
3. Lazar, L., Westerholt, K., Zabel, Tagirov, L. R., Goryunov, Yu. V., Garif'yanov, N. N., and Garifullin, I. A., *Phys. Rev. B* **61**, 3711-3722 (2000).
4. Mühge, Th., Westerholt, K., Zabel, H., Garif'yanov, N. N., Goryunov, Yu.V., Garifullin, I.A., and Khaliullin, G.G., *Phys. Rev. B* **57** 5071-5079 (1998).
5. Garifullin, I. A., Tikhonov, D. A., Garif'yanov, N. N., Lazar, L., Goryunov, Yu. V., Khlebnikov, S. Ya., Tagirov, L. R., Westerholt, K., and Zabel, H., *Phys. Rev. B* **66**, 020505(R) (2002).

The FM/S/FM Trilayer: Inhomogeneous π-Phase Superconductivity

N. Ivanov,[a] L. Terentieva,[a] D. Sattarov,[a] Yu. Proshin,[b,c] and M.G. Khusainov[a,b,c]

[a] *Kazan State Tupolev Technical University, Chistopol' 422981, Russia*
[b] *Kazan State University, Kazan 420008, Russia*
[c] *Max-Planck-Institute for the Physics of Complex Systems, Dresden 01871, Germany*

Abstract. We proved that the superconductivity in the ferromagnetic metal/superconductor/ferromagnetic metal (FM/S/FM) trilayer is a superposition of the BCS pairing with constant-sign pair amplitude in the S layer and the Larkin-Ovchinnikov-Fulde-Ferrell (LOFF) pairing with the 3D oscillatory pair amplitude in the FM layers. We allow not only the indirect interaction of the FM layers magnetizations via the S layer, but also a possible existence π phase superconducting states in such trilayers. The presence of π magnetic states along with the π superconducting ones allows us to explain the unexpected weak depression of superconductivity which has been found in the short period Gd/La superlattices.

Keywords: superconductivity; ferromagnetism; proximity effect; π phase states; multilayers
PACS: 74.78.Fk, 85.25.-j, 74.62.-c, 85.75.-d, 74.81.-g

According to many experiments [1], coexistence of superconductivity and ferromagnetism in the FM/S multilayers is accompanied by nonmonotonic dependence of the critical temperature T_c, conduction electrons density of states N, and Josephson critical current on the FM layers thickness d_f. In addition, in the FM/S superlattices and S/FM/S trilayers the phenomenon of the π phase superconductivity has been found both theoretically and experimentally. Recently a surprising 3D superconductivity in the short period Gd/La superlattice [2] at zero-field cooling was observed. We will show that boundary-value problem (BVP) for the FM/S/FM trilayers admits the new π superconducting solutions.

Firstly consider the planar contact between the FM layer ($-d_f < z < 0$) and the S layer ($0 < z < d_s$) in the dirty limit ($l_{s(f)} = v_{s(f)}\tau_{s(f)}$, $a_f = v_f/2I << \xi_{s(f)} = v_{s(f)}/2\pi T$). Here $v_{s(f)}$ is the Fermi velocity, $\tau_{s(f)}$ is the mean free path time, and I is the FM layer exchange field. In this case the BVP for the Gor'kov function $F(\boldsymbol{\rho},z,\omega)$ includes the Usadel-like differential equations [1] and boundary conditions [3], which connect the flux of $F(\boldsymbol{\rho},z,\omega)$ with its jump at the FM/S interface. Solutions of BVP in the ideal transparency case [3] are discussed below. The mutual influence of the metals FM and S is especially significant in the Cooper limit, where the thicknesses $d_{f(s)} << \xi_{f(s)}$. In this case the order parameter $\Delta(\boldsymbol{\rho},z)$ and the function $F(\boldsymbol{\rho},z,\omega)$ are practically constant along the z axis and the spatial variations are possible only in $\boldsymbol{\rho}$ (along the FM/S boundary plane). For the reduced critical temperature $t = T_c/T_{cs}$ of the FM/S bilayer we get

$$\ln t = -\frac{c_f(\lambda_s - \lambda_f)}{\lambda_s(c_s\lambda_s + c_f\lambda_f)} + \Psi\left(\frac{1}{2}\right) - \operatorname{Re}\Psi\left(\frac{1}{2} + \frac{c_f\left[2iI + D_f(I)q_f^2\right]}{4\pi T_{cs}t}\right); \tag{1}$$

$$c_{f(s)} = \frac{N_{f(s)}d_{f(s)}}{N_f d_f + N_s d_s}. \tag{2}$$

Here $\Psi(x)$ is the digamma-function, $c_{s(f)}$ is the relative weight of the S(FM) layer, $\lambda_{s(f)}$ is corresponding parameter of electron-electron interaction. In the right side of Eq. (1) the first term describes a decrease of T_c since usually $\lambda_s > \lambda_f$. The second and third terms describe the T_c suppression due to paramagnetic effect of the exchange field I and its partial compensation due to possible creation of the LOFF state. It is easy to check that in the extremely dirty limit with $l_f << a_f$, i.e. $2I\tau_f << 1$, the diffusion coefficient $D_f(I) \approx D_f(1 - i2I\tau_f)$ is almost real [1] and the optimal value of q_f equals to zero. Thus, the LOFF state is not realized in the dirty FM layer. At the same time, in case of the sufficiently

CP850, *Low Temperature Physics: 24th International Conference on Low Temperature Physics;* edited by Y. Takano, S. P. Hershfield, S. O. Hill, P. J. Hirschfeld, and A. M. Goldman

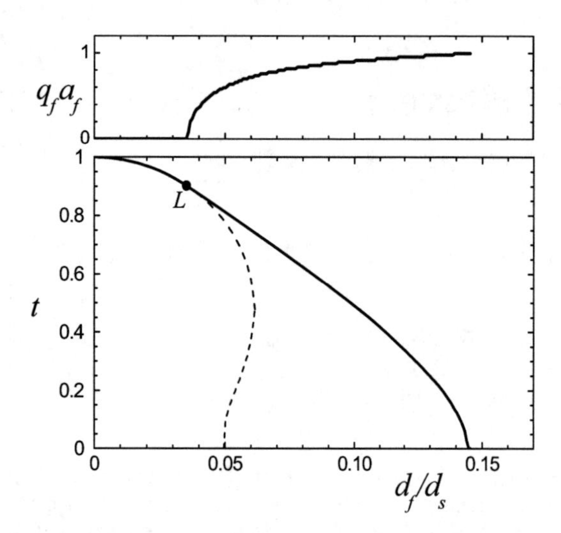

FIGURE 1. Phase diagrams $T_c(d_f)$ (lower panel) and $q_f(d_f)$ (upper panel) of the rather clean FM/S bilayer. The dashed curve represents the unrealized BCS state with $q_f = 0$. The solid curve corresponds to the LOFF state with $q_f \neq 0$. We use here following values $2I\tau_f = 5$; $(I/T_{cs}) = 6\pi$; $N_s = N_f$.

pure FM layer with $2I\tau_f >> 1$ the diffusion coefficient can be approximated [1] as $D_f(I) \approx -iv_f a_f(1+i/I\tau_f)$. This value is almost imaginary and optimal value of q_f can become nonzero due to creation of the LOFF state. The $T_c(d_f)$ dependence for such a FM/S bilayer is depicted in Fig.1 for case $\lambda_s = \lambda_f$. It is easy to see there are two types of superconducting states in the FM layer. First of them is the BCS state with $q_f = 0$ and the constant order parameter is realized at small d_f. With d_f increase in the triple point L the BCS state changes into the LOFF one, in which $q_f \neq 0$ and the order parameter $\Delta_f(\boldsymbol{\rho})$ and pair amplitude $F_f(\boldsymbol{\rho},\omega)$ are oscillated in the FM/S boundary plane $(x - y)$ with period of order of the spin stiffness length a_f.

In the FM/S/FM trilayers case there are ***four*** different states, which are distinguished by phases φ and χ of the superconducting (Δ) and magnetic (I) order parameters in the neighboring FM layers, correspondingly. For the FM/S/FM trilayer with $\lambda_f > 0$ at the antiferromagnetic orientation of the FM layers magnetizations the 0π state is realized with reduced critical temperature $t^{0\pi} = T_c^{0\pi}/T_{cs}$

$$\ln t^{0\pi} = -\frac{c_f\left(\lambda_s - \lambda_f\right)}{\lambda_s\left(c_s\lambda_s + c_f\lambda_f\right)}. \quad (3)$$

For the analogous FM/S/FM trilayer with $\lambda_f < 0$ the different $\pi\pi$ state takes place with its $T_c^{\pi\pi}$, for which

$$\ln t^{\pi\pi} = -\frac{c_f}{c_s\lambda_s}. \quad (4)$$

Possibility of two different in superconductivity 0π and $\pi\pi$ states in the FM/S/FM trilayers is very surprising since it is usually considered that the π phase superconductivity in such systems is impossible in principle. On the one hand, this fact means that in the FM/S multilayers coupling between the neighboring layers is provided by the superconducting correlations, i.e. the role of the true order parameter is played by the pair amplitude $F(\boldsymbol{\rho},z,\omega)$ rather than $\Delta(\boldsymbol{\rho},z)$. On the other hand, the existence of such π magnetic states (3), (4) allows us to explain the unexpected weak depression of superconductivity which has been found in the short period Gd/La superlattices [2]. The measured T_c of the superlattice was 5 K that coincides with the critical temperature of the bulk La sample. This means that the 0π type state is realized rather than the $\pi\pi$ one in the Gd/La superlattice and $\lambda_f \sim \lambda_s$ since $T_c^{0\pi} \approx T_{cs}$ in Eq. (3).

For other possible 00 and π0 states with parallel FM layers magnetizations we get

$$\ln t^{00} = \ln t^{0\pi} + \Psi\left(\frac{1}{2}\right) - \mathrm{Re}\,\Psi\left(\frac{1}{2} + \frac{c_f[2iI + D_f(I)q_f^{\,2}]}{4\pi T_{cs}t}\right), \quad (5)$$

$$\ln t^{\pi 0} = \ln t^{\pi\pi} + \Psi\left(\frac{1}{2}\right) - \mathrm{Re}\,\Psi\left(\frac{1}{2} + \frac{c_f[2iI + D_f(I)q_f^{\,2}]}{4\pi T_{cs}t}\right), \quad (6)$$

correspondingly. These 00 and π0 states possess lower critical temperatures in comparison with 0π and $\pi\pi$ ones, and they could be observed only in the presence of the external magnetic field.

The work was supported in part by RFBR (Grants 04-02-16761; 05-02-16369) and CRDF (REC-007).

REFERENCES

1. Izyumov, Yu. A., Proshin, Yu. N., and Khusainov, M. G., *Physics-Uspekhi* **45,** 109-148 (2002).
2. Goff, J. P., Deen, P. P., Ward, R. C. C., Wells, M. R., Langridge, S., Dalgleish, R., Foster, S., and Gordeev, S., *JMMM* **240,** 592-594 (2002).
3. Khusainov, M. G., and Proshin, Yu. N., *Physics-Uspekhi* **46,** 1311-1313 (2003).

Spin Screening And Inverse Proximity Effect In F/S Nanostructures

M.M. Khusainov,[a] E.L. Parfenova,[b] Yu.N. Proshin,[c] and M.G. Khusainov[b,c]

[a] *Zavoisky Physical-Technical Institute of RAS, Kazan 420029, Russia*
[b] *Kazan State Tupolev Technical University, Chistopol' 422981, Russia*
[c] *Kazan State University, Kazan 420008, Russia*

Abstract. It is shown that the short-range oscillating spin polarization of conduction electrons around magnetic moment embedded in the superconducting film is screened in aggregate by the long-range antiferromagnetic term, which owes its origin to the Cooper pairing. On this base new exchange model and boundary value problem for the proximity effect in ferromagnet/superconductor (F/S) nanostructures are proposed. In the framework of this model we investigate the possible variants of the mutual accommodation of inhomogeneous superconducting and magnetic order parameters in the F/S nanostructures. The F/S systems of the first type allow only homogeneous ferromagnetic ordering in the F layers, which for the weak exchange fields $h < h_c$ coexists with superconductivity in the S layers (FS phase). In the F/S systems of the second type the FS phase exists only for $h < h_{c1}$. For the exchange fields $h_{c1} < h < h_{c2}$ the superconducting layers S induce the nonuniform cryptoferromagnetic modulation (CFS phase) in the spin structure of the F films. This phenomenon can be called as the magnetic (or inverse) proximity effect. The conditions for the coexistence of the inhomogeneous magnetism and superconductivity in the F/S nanostructures EuO/Al, EuO/V, EuS/Al, and LaCaMnO/YBaCuO are investigated and the nontrivial experimental data are interpreted.

Keywords: superconductivity, magnetism, ferromagnet, proximity effect.
PACS: 73.21.Cd, 72.80 Sk, 74.78.Fk, 85.25.-j, 74.62.-c.

For a more complete analysis of the different types of mutual accommodation of superconductivity and ferromagnetism in systems consisting of alternating ferromagnetic and superconducting layers [1], it is necessary to answer the question: by what mechanism do the magnetic layers interact through the superconducting layers. This question, by the way, is very important for the FI/S systems, where FI is a ferromagnetic insulator and S is a superconductor (see review [2]). In particular, the nature of the internal fields causing the splitting of the BCS peak in the density of states of the conduction electrons of aluminum in EuO/Al/Al_2O_3, EuS/Al/Al_2O_3Ag, and Au/EuS/Al tunnel contacts, where EuO and EuS are ferromagnetic insulators, is not clear. This splitting is observed as an extra splitting (in addition to the Zeeman splitting) in the presence of an external magnetic field and saturates with growth of this field. With increase of the magnetic field in the FI/S contacts a first-order phase transition to the normal state occurs, although existing theory [3] predicts a second-order transition. Another interesting variant of coexistence between ferromagnetism and superconductivity in the LaCaMnO/YBaCuO bilayer and superlattice has been found recently [4,5]. In this experiment a surprisingly long-range influence of ferromagnetic manganite LaCaMnO on the critical temperature T_c of unconventional superconductor YbaCuO has been observed.

The origin of spin screening mechanism in dirty superconductors is established. The short-range oscillating spin polarization of conduction electrons around the magnetic moment embedded in the superconducting film is compensated in the aggregate by the long-range antiferromagnetic term, which owes its origin to the Cooper pairing. On this base the exchange model of the proximity effect for the ferromagnetic insulator/superconductor (FI/S) nanostructures is developed. In addition to the direct ferromagnetic exchange of nearest neighbors into the FI layers this model also takes into account the long range antiferromagnetic indirect exchange between localized spins (LS) $\mathbf{S}_\mathbf{r}$ and $\mathbf{S}_{\mathbf{r}'}$ on the FI/S boundary via superconducting electrons of the S layers [1]. This indirect exchange arises due to the effective *s-d(f)* exchange I, which is a result of virtual electron transport from the superconductor to the insulator and vice versa as consequence of the

CP850, *Low Temperature Physics: 24th International Conference on Low Temperature Physics;*
edited by Y. Takano, S. P. Hershfield, S. O. Hill, P. J. Hirschfeld, and A. M. Goldman

corresponding wave functions overlap. In the framework of this model we investigate the ground states of the FI/S nanostructures and also determine possible variants of the mutual accommodation of the superconducting and magnetic order parameters.

It is shown that FI/S bilayers are divided into two different types. This depends on the magnitude of the critical ratio A_c of the antiferromagnetic and ferromagnetic molecular fields acting on each localized spin of the FI/S boundary due to RKKY exchange via CE of the S layer and due to direct exchange in the FI layer, respectively,

$$A_c = A(h = h_c,\ \Delta = \Delta_c); \qquad (1)$$

$$A = \frac{N(0)h^2 d_S^5}{24J\langle S\rangle^2 d_f a^2 \xi^2}\left(\frac{\Delta}{2T_{cs}}\right)^2; \qquad (2)$$

$$h_c = \frac{\Delta_0}{\sqrt{2}}\left(1 - \frac{\pi d_s^2}{12\xi^2}\right); \quad \Delta_c = \Delta_0\left(1 - \frac{\pi d_s^2}{24\xi^2}\right). \qquad (3)$$

The quantity $h = I\langle S\rangle a/2d_s$ is the mean exchange field acting on the CE coming from the LS at FI/S boundary; T_{cs} is the transition temperature of the S material, S is the full spin of the LS, $d_{s(f)}$ are the S(F) layer thicknesses, correspondingly; ξ is the superconducting coherence length; $N(0)$ is the CE density of states; Δ is the superconducting order parameter ($\Delta_0 = \Delta(d_f = 0)$). It is necessary to note that the amplitude of the RKKY exchange is more smaller than one for the direct exchange, $N(0)h^2 << JS^2$. But the region of action of the superconducting part of RKKY exchange is very large in comparison with one for short-range direct exchange, i.e. $\pi\xi^2 d_s >> 4a^2 d_f$. Therefore parameter A can vary over wide limits.

The FI/S bilayers of the first type with $A_c < 1$ allow only the homogeneous ferromagnetic ordering into the FI layer, which coexists with superconductivity in the S layer (FS phase) for exchange fields $h < h_c$ (see Fig. 1). For the corresponding superlattices FI/S with their critical ratio $A_c^* < 1$ the antiferromagnetic superconducting (AFS) phase is realized. This AFS state arises due to the long-range coupling of the adjacent FI layers through the S interlayers. In the FI/S bilayers of the second type with $A_c > 1$ the FS phase exists only for $h < h_{c1}$. For the exchange fields $h_{c1} < h < h_{c2}$ the superconducting substrate S induces the nonuniform cryptoferromagnetic modulation (CFS phase) in the spin structure of the FI film as a result of the long-range indirect antiferromagnetic exchange. This phenomenon can be called the magnetic (or inverse) proximity effect. The wave vector q_f of spin structure $S_{\mathbf{r}} = S\cos(\mathbf{q}_f\mathbf{r})$ in the CFS phase can vary from $q_f = 0$ at $h = h_{c1}$ to $q_f >> 1/\xi$ at $h = h_{c2}$. For the corresponding superlattices with $A_c^* > 1$ the π phase magnetism arises, when the phases of the magnetic order parameters in the neighboring FI layers are shifted by π.

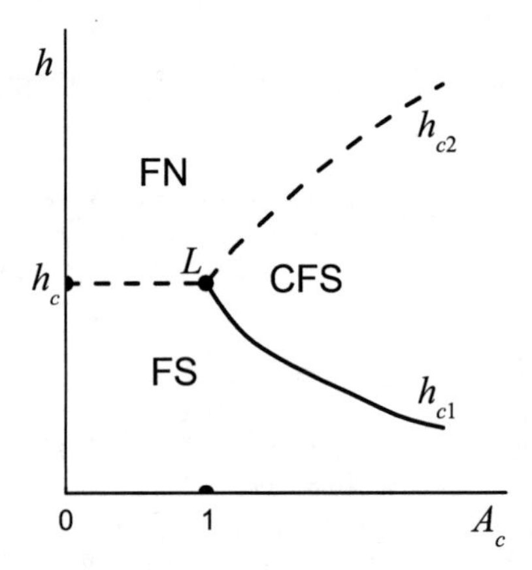

FIGURE 1. Phase diagrams (h–A_c) of FI/S bilayers. The dashed curves represent the lines of the first order phase transitions FS-FN and CFS-FN. The solid curve is the line of the second order phase transition FS-CFS. The three curves meet at the Lifshitz point L.

On the base of our results we can derive the conditions for the simultaneous coexistence of the inhomogeneous magnetism and superconductivity in the FI/S nanostructures: EuO/Al, EuO/V, EuS/Al, and, finally, we give a simple interpretation of the giant proximity effect phenomenon in terms of anomaly large lengths λ_c or ξ in the LaCaMnO/YBaCuO bilayers and superlattices [4,5].

This work was supported in part by RFBR (04-02-16761, 05-02-16369) and CRDF (BRHE REC-007).

REFERENCES

1. Izyumov, Yu. A., Proshin, Yu. N., and Khusainov, M. G., *Physics-Uspekhi* **45,** 109 (2002).
2. Tedrow, P. M., and Meservey, R., *Phys. Rep.,* **238,** 173 (1994).
3. Tokauasu, T., Sauls, J. A., and Rainer, D., *Phys. Rev. B* **38,** 8823 (1988).
4. Soltan, S., Albrecht, J., and Habermeier, H.-U., *Phys. Rev. B* **70,** 14457 (2004).
5. Sefrioui, Z., Arias, D., Pena, V., Villegas, J. E., Varela, M., Prieto, P., Leon, C., Martinez, J. L., and Santamaria, J., *Phys. Rev. B* **67,** 214511 (2003).

Exponentially Suppressed Interlayer Josephson Current in Underdoped $Bi_2Sr_2CaCu_2O_{8+y}$: Bulk Evidence for Inhomogeneous Superconductivity

Takasada Shibauchi*, Satoshi Horiuchi* and Minoru Suzuki*

*Department of Electronic Science and Engineering, Kyoto University, Nishikyo-ku, Kyoto 610-8510, Japan

Abstract. Josephson plasma resonance is used to determine the interlayer Josephson critical current density j_c in $Bi_2(Sr,La)_2CaCu_2O_{8+y}$. We find that j_c shows an exponential doping dependence in a wide doping range. In the underdoped regime, our experimental j_c values are much smaller than expected in the d-wave coherent tunneling model. This discrepancy can be naturally understood if each CuO_2 planes in the bulk have a large portion of non-superconducting regions where Josephson current cannot flow. A simple analysis provides that the fraction of superconducting region gradually decreases at lower dopings, which points to the importance of inhomogeneity on the suppression of T_c in underdoped cuprates.

Keywords: Josephson plasma resonance, microscopic inhomogeneity, intrinsic junctions, critical temperature
PACS: 74.50.+r,74.72.Hs,74.81.-g

INTRODUCTION

Microscopic electronic inhomogeneity in underdoped cuprate superconductors has been recently discussed in a sense that the holes can become concentrated at certain locations. Scanning tunneling spectroscopy data [1] suggest that granular superconductivity (consistent with the above views) occurs at the surface of $Bi_2Sr_2CaCu_2O_{8+y}$ (BSCCO). A serious question remains whether the inhomogeneous superconducting state is realized in the CuO_2 planes deep inside the bulk crystals.

As to a key to this issue, the importance of the interlayer Josephson current was first addressed by Ref. [2]. Here we report on the interlayer Josephson current density j_c in BSCCO, which is a bulk probe for the superconducting coherence. We find that the obtained j_c show an exponential doping dependence, which gives much smaller values in the low doping regime than theoretical predictions based on d-wave coherent tunneling. This implies the existence of a significant fraction of non-superconducting regions which do not contribute to the Josephson interlayer tunneling.

EXPERIMENTAL

To prepare underdoped BSCCO crystals, we substituted La for Sr up to 3 %, and annealed the crystals in vacuum at 750°C for 24 hours. The hole concentration p is estimated from the transition temperature T_c. Josephson plasm resonance (JPR) is used to determine the low-temperature j_c in the zero field [3, 4]. We measured the microwave absorption of BSCCO crystals at 19 GHz which has a peak at the resonance field H_p [5]. Following the procedure in Refs. [3, 4] with the interlayer spacing $s = 1.2$ nm and the dielectric constant $\varepsilon = 6$ [6], we evaluate j_c from the temperature dependence of H_p.

RESULTS AND DISCUSSION

The extracted j_c as a function of p is plotted in Fig. 1. We also plot the j_c values obtained from the same procedure using the literature data in optimally doped and overdoped BSCCO [3, 4]. All the experimental data $j_c^{\rm exp}$ consistently indicate a strong doping dependence; it can be fitted by an exponential dependence $j_c \propto \exp(p)$. The $j_c^{\rm exp}$ values in a range 10^2 to 10^4 A/cm^2 are much smaller than expected in the s-wave Ambegaokar-Baratoff theory $j_c^{\rm AB} = \pi\sigma_{\rm n}\Delta_0/2es$ [7], where $\sigma_{\rm n}$ is the normal-state interlayer conductivity and Δ_0 is the maximum gap. In overdoped BSCCO, it has been pointed out that the small value of $j_c^{\rm exp}$ can be explained if one considers d-wave pairing and a significant contribution of coherent tunneling to the interlayer transport [8]. In this d-wave coherent model, the Josephson current density is expressed by

$$j_c^{\rm coh} \approx \pi\sigma_{\rm q}(0)\Delta_0/2es. \qquad (1)$$

Here $\sigma_{\rm q}(0)$ is the quasiparticle tunneling conductivity in the zero temperature limit, which is much suppressed than $\sigma_{\rm n}$ [8]. To evaluate $\sigma_{\rm q}(0)$ in a wide range of doping, we analyze the high-field interlayer resistivity data in BSCCO [10, 11]. Following Ref. [12], we plot $\sigma_{\rm q}(T)$ as a function of T^2 and extract $\sigma_{\rm q}(0)$. Together with the reported doping dependence of Δ_0 [9], this analysis allows us to estimate $j_c^{\rm coh}$ through Eq. (1) and the results

CP850, *Low Temperature Physics: 24th International Conference on Low Temperature Physics;*
edited by Y. Takano, S. P. Hershfield, S. O. Hill, P. J. Hirschfeld, and A. M. Goldman

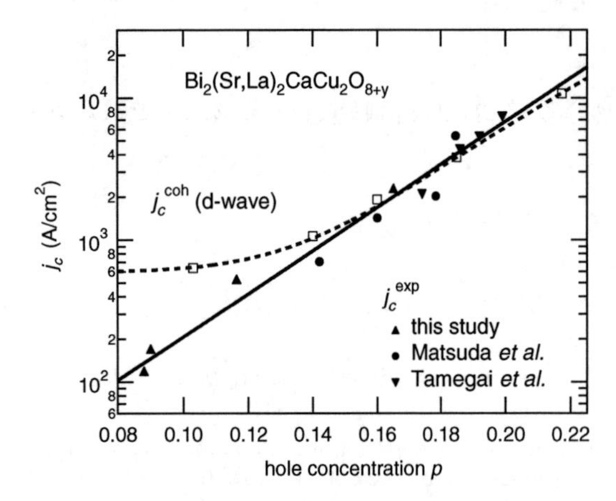

FIGURE 1. Josephson current density j_c as a function of hole concentration p in BSCCO. Closed symbols are the values determined by JPR experiments. Estimates from the reported data [3, 4] are also plotted. Solid line is a fit to the exponential dependence. Open symbols are expected values in the d-wave coherent model [8]. Dashed curve is a guide to the eyes.

are shown in Fig. 1. It is now clear that while in the overdoped regime the theoretical and experimental values are close, j_c^{exp} in the underdoped regime is still an order of magnitude smaller than j_c^{coh}.

Such discrepancy can be naturally understood by simply assuming that each CuO_2 planes in the bulk have a large portion of non-superconducting regions where Josephson current cannot flow. A straightforward argument [2] provides that j_c is reduced by f_s^2, where f_s is the fraction of the superconducting region in the CuO_2 planes. We plot in Fig 2 the doping dependence of $(j_c^{exp}/j_c^{coh})^{1/2}$, which corresponds to f_s in the above analysis. The obtained fraction of superconducting region gradually decreases from almost unity ($p > 0.18$) down to ~ 0.6 ($p \sim 0.1$). Here we point out that j_c^{exp}/j_c^{coh} (which is a measure of the relative coherency between the neiboring CuO_2 planes) scales surprisingly well with the doping dependence of $k_B T_c/\Delta_0$ [9]. This result implies the importance of inhomogeneity on the suppression of T_c in underdoped cuprates.

SUMMARY

In summary, we found that the doping dependence of j_c determined by JPR shows the exponential behavior. j_c in underdoped BSCCO is considerably suppressed than theoretically expected, suggesting that non-superconducting regions exist in CuO_2 planes. From the simple analysis, we argue that the fraction of superconducting region goes down at low dopings, which may be important to understand the doping dependence of T_c.

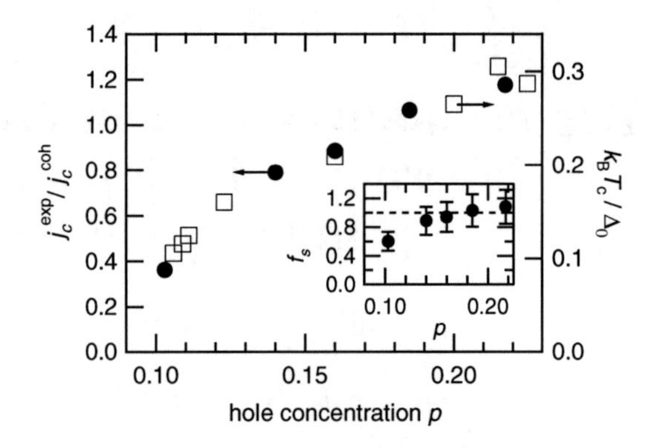

FIGURE 2. $j_c^{exp}/j_c^{coh}(p)$ scales with the the ratio of $k_B T_c$ to the gap energy Δ_0 [9]. Inset: Doping dependence of the superconducting fraction f_s (see text). Dashed line represents the homogeneous superconducting state $f_s = 1$.

ACKNOWLEDGMENTS

We thank L. N. Bulaevskii, M. Graf, I. Vekhter and Y. I. Latyshev for discussion. This work was partly supported by a Grant-In-Aid for Scientific Research from MEXT.

REFERENCES

1. K. M. Lang, V. Madhavan, J. E. Hoffman, E. W. Hudson, H. Eisaki, S. Uchida, J. C. Davis, *Nature* **415**, 412–416 (2002).
2. Y. Yamada, K. Anagawa, T. Shibauchi, T. Fujii, T. Watanabe, A. Matsuda, and M. Suzuki, *Phys. Rev. B* **68**, 054533 (2003).
3. Y. Matsuda, M. B. Gaifullin, K. Kumagai, M. Kosugi, and K. Hitara, *Phys. Rev. Lett.* **78**, 1972–1975 (1997).
4. T. Tamegai, M. Sato, A. Mashio, T. Shibauchi, and S. Ooi, *Advances in Superconductivity* **IX**, 621–624 (1997).
5. T. Shibauchi, T. Nakano, M. Sato, T. Kisu, N. Kameda, N. Okuda, S. Ooi, and T. Tamegai, *Phys. Rev. Lett.* **83**, 1010–1013 (1999).
6. A. Irie, Y. Hirai, and G. Oya, *Appl. Phys. Lett.* **72**, 2159–2161 (1998).
7. V. Ambegaokar and A. Baratoff, *Phys. Rev. Lett.* **10**, 486–489 (1963); **11**, 104 (1963).
8. Y. I. Latyshev, T. Yamashita, L. N. Bulaevskii, M. J. Graf, A. V. Balatsky, and M. P. Maley, *Phys. Rev. Lett.* **82**, 5345–5348 (1999). Here, effects of the pseudogap is not accounted for.
9. N. Miyakawa, J. F. Zasadzinski, L. Ozyuzer, P. Guptasarma, D. G. Hinks, C. Kendziora, and K. E. Gray *Phys. Rev. Lett.* **83**, 1018–1021 (1999).
10. T. Shibauchi, L. Krusin-Elbaum, M. Li, M. P. Maley, and P. H. Kes, *Phys. Rev. Lett.* **86**, 5763–5766 (1999).
11. T. Shibauchi, L. Krusin-Elbaum, G. Blatter, and C. H. Mielke, *Phys. Rev. B* **67**, 064514 (2003).
12. N. Morozov, L. Krusin-Elbaum, T. Shibauchi, L. N. Bulaevskii, M. P. Maley, Y. I. Latyshev, and T. Yamashita, *Phys. Rev. Lett.* **84**, 1784–1787 (2000).

Magnetic Phases of Josephson Vortices in $Bi_2Sr_2CaCu_2O_{8+y}$

K. Hirata, S. Ooi, S. Yu, E. S. Sadki, and T. Mochiku

National Institute for Materials Science, 1-2-1 Sengen, Tsukuba, 305-0047, Japan

Abstract. Josephson vortex (JV) states in $Bi_2Sr_2CaCu_2O_{8+y}$ (Bi-2212) have been studied by measuring the flow resistivity of JV's and *I-V* characteristics with a current along the *c*-axis. In the flow resistivity, periodic oscillations have been found as a function of the parallel magnetic field to the superconducting layers. As this phenomenon is related to the formation of a triangular lattice of JV's as the ground state, the 3D-ordered state (triangular lattice phase) of JV's could be determined. With changing the doping level of Bi-2212, the 3D-ordered phase is shifted towards a higher magnetic field and a higher temperature, which is well explained from the theoretical analyses. Above the 3D-ordered phase, vortex state has been studied with *I-V* characteristic measurements, in which the voltage shows a distinct change in the exponent of the power-law dependence to the current between two phases. This suggests that the 3D-ordered phase changes to the disordered phase along the *c*-axis.

Keywords: Josephson vortex, magnetic phase, $Bi_2Sr_2CaCu_2O_{8+y}$, flow resistance.
PACS: 74.72.Hs, 74.62.Dh, 74.25.Qt, 74.25.Fy

INTRODUCTION

Most of the high T_c superconductors (HTSC's) consist of iterative stacks of the Cu-O superconducting layers and the non-superconducting ones. This causes weak Josephson coupling between the superconducting layers. These Josephson coupling are called as intrinsic Josephson junctions (IJJ's) [1]. The 2D nature of the layered structures shows characteristic features in vortex matter physics. In the perpendicular magnetic fields to the layers, HTSC's show distinguished features of pancake vortices, which are described in Ref.2. In the parallel fields to the layers, magnetic field penetrates into the materials as Josephson vortices (JV's) in superconducting state. Theoretical studies have been made extensively in the past decade [3-8]. However, experimentally, it has been accomplished only on $YBa_2Cu_3O_{7-\delta}$ (YBCO); the melting transition of the JV system [9-10], the oscillatory melting temperature, and the vortex smectic phase [11]. Few experimental results have been reported on the magnetic phase diagram of the JV system in strongly anisotropic HTSC's such as $Bi_2Sr_2CaCu_2O_{8+y}$ (Bi-2212).

Recently we have found a new method to study the magnetic phase diagram of JV system in Bi-2212 by the periodic oscillations in flow resististance of JV's [12], from which we can determine the 3D-ordered state of JV's [13-15]. In this paper, we have studied on the doping effect to the JV phases in a higher magnetic field.

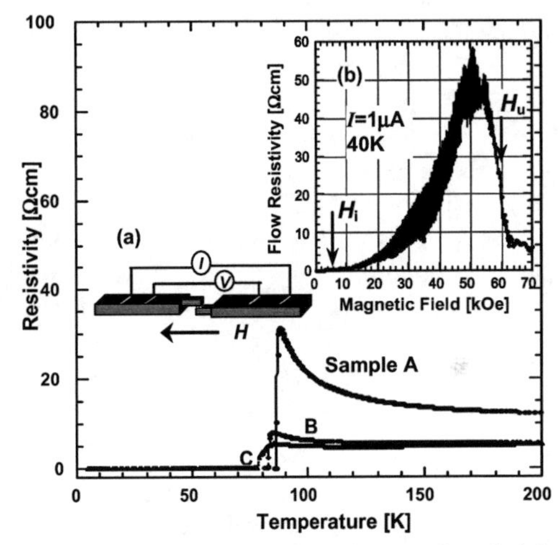

FIGURE 1. Temperature dependence of resistivity in sample A, B and C. Inset (a) shows a schematic drawing of the IJJ structure, and (b) magnetic field dependence of flow resistivity of JV's in sample A at 40K.

EXPERIMENTS

Preparation of the samples is described elsewhere [12]. Three samples for the measurements were prepared by annealing with slightly over-doped sample A (T_c=85.8K) with a size of w=20.8, l=23.0, t=0.9 μm, moderately over-doped sample B (T_c=82.8K) with 11.2, 12.4 and 0.5 μm, and heavily over-doped sample C

CP850, *Low Temperature Physics: 24th International Conference on Low Temperature Physics;* edited by Y. Takano, S. P. Hershfield, S. O. Hill, P. J. Hirschfeld, and A. M. Goldman

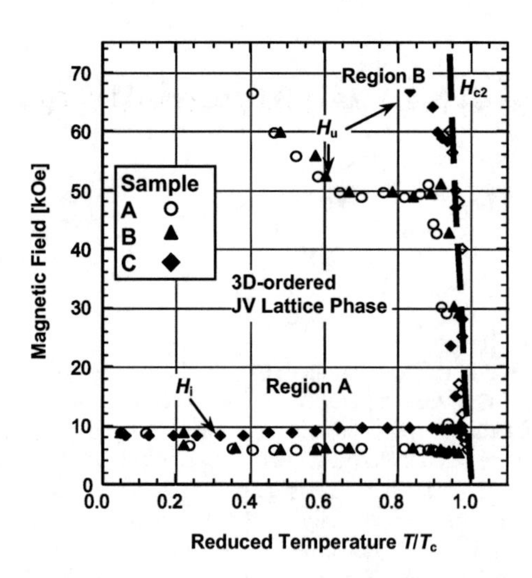

FIGURE 2. Magnetic phase diagram of Josephson vortices in sample A, B and C. Broken line indicates the upper critical field H_{c2}.

(T_c=78.0K) with 14.0, 15.4 and 1.2 μm, respectively, where w is the length perpendicular to the magnetic field, l the length parallel to the magnetic field and t the thickness. Temperature dependence of the c-axis resistivety on these samples without magnetic field is shown in Fig.1. A sharp superconducting transition can be seen in three samples.

Flow resistivity of the IJJ (schematically drawn in the inset (a) of Fig.1) was measured with a four-probe contact configuration in the applied field H. Details of the JV flow measurements are also described in Ref.12.

RESULTS AND DISCUSSION

Typical flow resistivity in sample A is shown in the inset (b) of Fig.1 with the applied current of 1 μA at 40 K. The resistivity begins to oscillate from H_i, and stops at H_u, indicated as arrows in the inset. The region in B-T phase diagram shown in Fig.2, in which the periodic oscillations can be observed, corresponds to the 3D-ordered JV lattice phase. H_i in sample A, B, and C is 4.7, 7.8, and 9.3 kOe, and the anisotropy parameter γ is obtained as 436, 263, and 220 according to the formula $H_i=(1.40\phi_0)/(2\pi\gamma s^2)$ [16], respectively, where ϕ_0 is the flux quantum, s the distance between the superconducting layer. This is the lower boundary of the 3D-ordered phase. At the highest temperature of the boundary H_i may corresponds to the critical point $H_c=\phi_0/(2 3^{1/2}\gamma s^2)$ [7]. In the magnetic field below H_i, several ordered phases were proposed without completely full-filling of JV's to the IJJ [16,17]. However, we could observe neither a large change in the slope of the flow resistivity against magnetic field with no oscillations nor symptoms of the first order transition close to T_c. This may be related to the change in thermodynamic quantity of physical parameters at the transition, which is small enough to be observed in the experiments.

H_u is defined as the upper boundary of the 3D-ordered phase in temperature and magnetic field. In Fig.2, the boundary H_u tends to increase with decreasing the anisotropy γ, which shows a good tendency to relate with weak anisotropic superconductors such as YBCO. To confirm the JV state above/below the upper boundary shown in Fig.2 as the area A/B, respectively, I-V measurements have been made. The power-law dependence of $V=\alpha I^{\beta}$ shows a distinct change in β at H_u from 0.85 to 0.60. In the flow resistivity above H_u, the resistivity always decreases to a finite value after the periodic oscillations in the region B of Fig.2 [15], which may suggest that the JV state is in disordered along the c-axis, as the *in-plane* disorder causes an enhancement of the flow resistivity with less effective potential barrier to the movement of JV's.

REFERENCES

1. R. Kleiner, F. Steinmeyer, G. Kunkel and P. Muller, *Phys. Rev. Lett.* **68**, 2394-2397 (1992).
2. G. Blatter, M.V. Feigel'man, V.B. Geshkenbein, A.I. Larkin and V.M. Vinokur, *Rev. Mod. Phys.* **66**, 1125-1388 (1994).
3. S. Chakravaty, B. Ivlev and Y. Ovchinnikov, *Phys. Rev. Lett.* **64**, 3187-3190 (1990).
4. G. Blatter, B. Ivlev and J. Rhyner, *Phys. Rev. Lett.* **66**, 2392-2395 (1991).
5. B.I. Ivlev, N.B. Kopnin and V.L. Pokrovsky, *Physica B* **169**, 619-620 (1991).
6. L. Balents and D.R. Nelson, *Phys. Rev. B* **52**, 12951-12968 (1995).
7. X. Hu and M. Tachiki, *Phys. Rev. B* **70**, 064506-1-4 (2004).
8. W.K. Kwok, J. Fendrich, U. Welp, S. Fleshler, J. Downey and G.W. Crabtree, *Phys. Rev. Lett.* **72**, 1088-1091 (1994).
9. A. Schilling, R.A. Fischer, N.E. Phillips, U. Welp, W.K. Kwok and G.W. Crabtree, *Phys. Rev. Lett.* **78**, 4833-4836 (1997).
10. T. Ishida, K. Okuda, A.I. Ryukov, S. Tajima and I. Terasaki, *Phys. Rev. B* **58**, 5222-5225 (1998).
11. S.N. Gordeev, A.A. Zhukov, P.A.J. de Groot, A.G.M. Jansen, R. Gagnon and L. Taillefer, *Phys. Rev. Lett.* **85** 4594-4598 (2000).
12. S. Ooi, T. Mochiku and K. Hirata, *Phys. Rev. Lett.* **89**, 247002-1-4 (2002).
13. M. Machida, *Phys. Rev. Lett.*, **90**, 037001-1-4(2003).
14. A. Koshelev, *Phys. Rev. B* **66**, 224514-1-6 (2002).
15. K. Hirata, S. Ooi, S. Yu, E.H. Sadki and T. Mochiku, *Physica C* **412-414**, 449-453 (2004).
16. R. Ikeda, *J. Phys. Soc. Jpn.*, **71**, 587-593 (2002).
17. X. Hu (private communications).

Fiske Resonance-like Behaviors in Intrinsic Junctions of $Bi_2Sr_2CaCu_2O_{8+\delta}$

I. Kakeya, T. Yamazaki, M. Kohri, T. Yamamoto, and K. Kadowaki

Institute of Materials Science, University of Tsukuba, 1-1-1 Ten-nodai, Tsukuba, Ibaraki 305-8573, Japan

Abstract. We have studied the dynamical nature of Josephson vortices in an intrinsic Josephson junction by the current-voltage characteristics along the *c*-axis of $Bi_2Sr_2CaCu_2O_{8+\delta}$. We found periodic current steps with respect to the voltage which resembles the Fiske resonance observed in a single junction. We attributed this phenomenon to the resonance between JV and the Josephson plasma.

Keywords: Intrinsic Josephson junction, Josephson vortex, Fiske resonance
PACS: 74.50.+r, 74.25.Fy, 74.25.Qt

INTRODUCTION

Transport properties of Josephson junctions with Josephson vortices (JV's) have been studied to probe the phase sensitive macroscopic quantum phenomena.[1] One of the remarkable phenomena recently observed is an oscillatory behavior of the flux flow resistance in the intrinsic Josephson junctions of $Bi_2Sr_2CaCu_2O_{8+\delta}$ (Bi2212) single crystals. This phenomenon was interpreted as an oscillatory change of the average of the flow velocity of JV's due to coherent entries of quantized JV's.[2] Furthermore, the dynamical effect of the JV system has more degrees of freedom, which enable us to observe fundamentally interesting phenomena due to collective excitations of the gauge invariant phases associated with the quantized JV's.

In this paper, we report experimental results of current-voltage ($I-V$) characteristics in magnetic fields parallel to the *ab*-plane. The set of $I-V$ curves strongly suggests that the Fiske resonance in Bi2212, which implies that the excitations of Josephson plasma wave inside the junction.

EXPERIMENTAL

A thin and narrow Bi2212 single crystal cleaved from a bulk crystal grown by the TSFZ method was put on a substrate with four electrodes for the standard four-probe current-voltage measurements. Subsequently, the part of the crystal between the voltage electrodes was fabricated with the focused ion beam (FIB) machine in order to make small and rectangular Josephson junctions along the *c* axis. The thickness of the original crystal is 10 μm and dimensions of the junction (measured) part are 4.1, 5.5, and 2.0 μm for the width perpendicular to the field, the length along the field, and the thickness along the *c*-axis, respectively. For $I-V$ measurements, voltage were measured at each current ramped step-by-step. The magnetic field up to 60 kOe was applied by a split-pair superconducting magnet and the angle from the *ab*-plane was set by a precision goniometer. The critical temperature T_c was found to be 84 K with the resistivity measurement at 1 μA. The critical current density J_c at 60 K at $H = 0$ was obtained to be 1.3×10^3 A/cm^2, indicating that this sample is slightly over doped.

RESULTS AND DISCUSSIONS

Figure 1 represents typical $I-V$ characteristics along the *c*-axis at magnetic fields parallel to the *ab*-plane from 27.6 to 34.5 kOe at 60 K. The current was swept up to $0.17J_c$. We found that current steps appear periodically as a function of voltage.

From the $I-V$ characteristics at every 0.2 kOe, we extracted the field dependence of a characteristic current density J_1 which induces 1 mV along the *c*-axis and plotted as a function of applied magnetic field with a solid line in Fig. 2. $J_1(H)$ oscillates with a period of 1.65 kOe below 25 kOe, which is close to a half of $H_0 \equiv \Phi_0/ws = 3.3$ kOe, the field required for a Josephson vortex Φ_0 to fit into an atomic-scale Josephson junction with the thickness s of 15 Åwith the width w of 4.1 μm. The field dependence of J_1 can be considered as that of J_c because all $I-V$ curves are linear below $V = 1$ mV. The half Φ_0 oscillation in J_c has been expected theoretically by assuming a triangular JV lattice slightly deformed by the boundary condition.[3][4]

With increasing H, the peak in $J_1(H)$ at H/H_0 being an integer becomes smaller than the adjacent peaks at H/H_0 being a half odd numbers. Finally, it almost disappears above 45 kOe, resulting that $J_1(H)$ oscillates every Φ_0 similar to the Fraunhofer pattern of J_c in a single

CP850, *Low Temperature Physics: 24th International Conference on Low Temperature Physics;*
edited by Y. Takano, S. P. Hershfield, S. O. Hill, P. J. Hirschfeld, and A. M. Goldman

junction.

In Fig. 1, the $I-V$ curves at characteristic fields approximately corresponding to half integer numbers of Φ_0 par layer are indicated. It was found that current steps appear at around 0.11 and 0.28 V in an $I-V$ curve at H/H_0 being an integer while they appear at 0.18 and 0.35 V at H/H_0 being a half odd number. The voltages at the steps V_p^n and their separations $\Delta V \approx 0.17$ V do not depend on magnetic field, *i.e.*, $V_p^n \simeq n\Delta V/2 + 0.01$, where n corresponds to order of the step. Taking step heights I_s as increases of current by 0.04 V (roughly equal to the full-step-width) around the steps, I_s oscillate in-phase and out-of-phase as $J_1(H)$ for n being even numbers and odd numbers, respectively, as shown in Fig. 2.

These features are quite similar to the Fiske resonance observed in a single junction, which is expected to appear with a period of $\Delta V_{FS} = \Phi_0 c_S/L$ in the $I-V$ curves, where c_S and L are the Swihart velocity and the length of the junction perpendicular to the external field, respectively. Assuming that the period ΔV_{FS} corresponds to the measured voltage separation per layer $\Delta V/N$=0.13 mV, we obtain c_S/c_0 to be 1.6×10^{-3} and the frequency of standing wave resonating with JVs to be $\omega_0/2\pi = c_S/L = 62$ GHz for $N = 1330$ and $L = w = 4.1\mu$m, where c_0 is the light velocity. This value is smaller than the plasma frequency in an over-doped crystal.

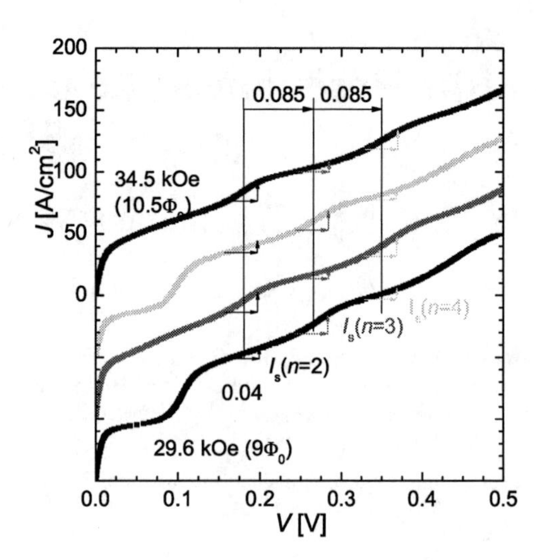

FIGURE 1. I-V curves at fields from 29.6 to 34.5 kOe with a separation of 1.6 kOe. Each curve is shifted 50 A/cm^2 vertically from the adjacent curve. Vertical lines correspond to the position of steps, small horizontal arrows represent the step widths (0.04 V), and vertical arrows indicate the step heights.

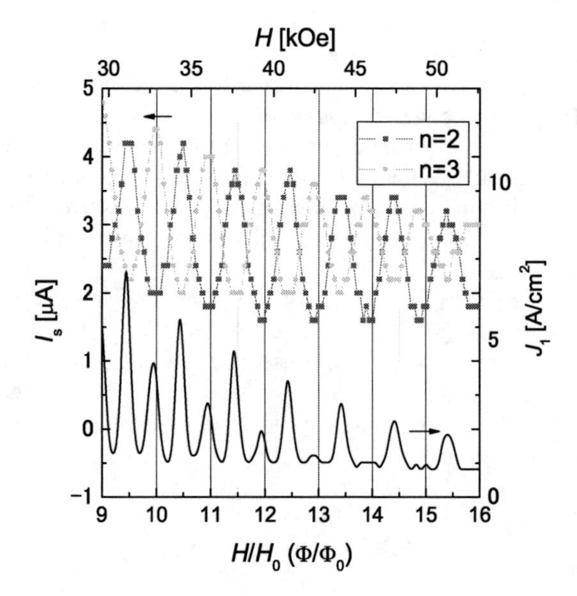

FIGURE 2. Field dependences of J_1 (solid line) and step height in I-V characteristics (solid symbols) for the second and third steps in Fig. 1.

SUMMARY

We found periodic current steps in $I-V$ characteristics of a Bi2212 mesoscopic junction under magnetic fields parallel to the *ab*-plane. The step shows quite similar features to the Fiske step in single junctions. This phenomenon strongly suggests that the Josephson plasma waves may be generated by the resonance with the collective oscillating motion of the JV's in a restricted size of the Josephson junctions. This leads us to the intriguing concept of the new type of millimeter wave generated by high-T_c superconductors.

ACKNOWLEDGMENTS

This work has been supported by the 21st century COE program at University of Tsukuba and Grants-in-Aid for Young Scientists (B) under MEXT, Japan.

REFERENCES

1. Barone, A. and Paterno, G., " *Physics and Applications of the Josephson Effect* ", John Wiley & Sons, 1982.
2. Ooi, S., Mochiku, T., and Hirata, K., *Phys. Rev. Lett.*, **89**, 2427002 (2002).
3. Koshelev, A. E., *Phys. Rev. B*, **66**, 224514 (2002).
4. Machida, M., *Phys. Rev. Lett*, **90**, 037001 (2003).

Termination of Softening of Josephson Plasma Mode in $Bi_2Sr_2CaCu_2O_{8+\delta}$ in the Vicinity of T_c

I. Kakeya*, S. Aman*, T. Yamamoto* and K. Kadowaki*

*Institute of Materials Science, University of Tsukuba, 1-1-1 Ten-nodai, Tsukuba, Ibaraki 305-8573, Japan

Abstract. We measured Josephson plasma resonance in an over-doped single crystal $Bi_2Sr_2CaCu_2O_{8+\delta}$ at low frequencies. The sharp resonance was not observed below a certain finite frequency, while a broad change in impedance attributed to the shielding effect was pronounced. The termination frequency is compared with a theory based on the two-fluid model.

Keywords: Josephson plasma, two-fluid model, quasiparticle scattering
PACS: 74.50.+r, 74.25.Nf, 74.72.Hs

INTRODUCTION

It has been reported that the Josephson plasma frequency in $Bi_2Sr_2CaCu_2O_{8+\delta}$ (Bi2212) at zero magnetic field decreases with increasing temperature and drops in the vicinity of T_c from measurements of Josephson plasma resonance (JPR) by sweeping temperature at fixed frequencies down to 10 GHz[1, 2]. In these papers, this drop is extrapolated to T_c at zero frequency and the whole temperature dependencies including the extrapolation to $\omega = 0$ are explained by using two-fluid models, in which the superfluid density n_s proportional to $\omega_p^2(T)$ decreases with increasing temperature and the quasiparticle scattering rate is larger than the plasma frequency. However, it is not clear experimentally whether $\omega_p(T)$ decreases to zero in the limit of T_c or the decrease terminates at a finite frequency.

This problem is important for the mechanism of the c-axis tunneling of high-T_c superconductors because the weak quasiparticle scattering gives a finite $\omega_p(T_c)$, which may correspond to the normal plasma frequency, according to a microscopic theory[3]. Finite normal plasma frequency($\sim 10^{-1}\omega_p(0)$) is interpreted that the interlayer tunneling is rather coherent in contrast to conventional Josephson junctions, where only superconducting electrons participate in the plasma oscillation.

In this paper, we present experimental data of JPR at low frequencies in an over-doped Bi2212 crystal, which has higher plasma frequency than in an under-doped crystal because of the lower anisotropy. The JPR disappears in the low frequency limit where that in under-doped crystals is still observed.

EXPERIMENTAL

To investigate JPR in the vicinity of T_c, it is indispensable to use crystals with sharp T_c because a broad transition blurs the resonance. Bi2212 single crystals grown by the TSFZ method were annealed at 550 °C in the air for 12 hours then quenched into liquid N_2. A crystal with T_c of 85.8 K and its width of 0.2 K was used for JPR measurements. JPR was detected as changes in impedance of cavity resonators containing the crystal with bridge balanced microwave circuits. Measurement frequencies which were restricted by resonant frequencies of the cavities were 30.54, 34.11, 39.60, and 48.83 GHz. The crystal was placed inside the cavities so as to excite only the longitudinal JPR as reported by the authors[4]. Either magnetic field parallel to the c-axis or temperature was swept to obtain the resonance.

RESULTS AND DISCUSSIONS

Figure 1 represents resonance curves at 48.83 GHz obtained by sweeping the magnetic field at fixed temperatures (a) and by sweeping temperatures at fixed fields (b). Both resonance peaks agree in the plot of resonance field vs. temperature: the resonance field decreases with increasing temperature because the plasma frequency is a decreasing function of both temperature and magnetic field. The disappearance of the resonance just below T_c is considered to be a result that $\omega_p(T)$ becomes lower than the microwave frequency. A step-like feature accompanied by the very sharp and tiny resonance just below T_c in the temperature sweep measurement at $H = 0$ is found in Fig. 1(b). This is consistent with the data obtained by sweeping magnetic field: that background signal level (loss of microwave) at energies (fields or temperatures) higher than the resonance is larger than one at energies lower than the resonance. This feature is pronounced when the resonance is observed at low fields and temperatures close to T_c.

This step-like increase in the cavity impedance indicates an increase in dissipation of microwave accompa-

CP850, *Low Temperature Physics: 24th International Conference on Low Temperature Physics;*
edited by Y. Takano, S. P. Hershfield, S. O. Hill, P. J. Hirschfeld, and A. M. Goldman

nied by the transition from superconducting to normal state. In general, irradiated microwave is expelled below the resonance while it is partly absorbed above the resonance, resulting in such an asymmetric line-shape of the resonance in the field sweep measurements. This feature indicates the c-axis tunneling of superconducting electrons may respond to frequencies higher than the plasma frequency.

At higher frequencies, the small resonance is weaker although the step in impedance does not change as shown in Fig. 1(c). The resonance was not observed below 34.11 GHz. It is not the case that this disappearance is the result that the resonance become sharper than the experimental resolution. Because the resonance peaks has never been observed even at small magnetic fields, where any "sharp" resonance peaks must be broadened by the fields. Therefore we argue that $\omega_p(T)$ of this crystal does not go to zero in the limit of T_c.

According to Ref. [2], $\omega_p(T)$ can be formulated by using two-fluid model with a constant quasiparticle relaxation time τ as:

$$\omega_p(T) = \frac{\omega_p(0)}{\sqrt{2}} \left[1 - \tilde{\gamma}^2 + \sqrt{(1+\tilde{\gamma}^2)^2 - 4\tilde{\gamma}^2 t^4}\right]^{1/2}, \quad (1)$$

where $\tilde{\gamma}$ and t are the reduced quasiparticle scattering rate $1/\omega_p(0)\tau$ and temperature T/T_c, respectively. This indicates that $\omega_p(T)$ decreases with increasing temperature and drops with $T \rightarrow T_c$. The frequency at T_c depends on $\tilde{\gamma}$: $\omega_p(T)$ drops to zero in the case of $\tilde{\gamma} \geq 1$ while it terminates at a finite frequency $\omega_p(0)\sqrt{1-\tilde{\gamma}^2}$ in the case of $0 < \tilde{\gamma} < 1$ and is connected to the temperature-independent constant plasma frequency. Provided $\omega_p\sqrt{1-\tilde{\gamma}^2}/2\pi$ corresponds to 30 GHz, where the JPR disappears, we get $\tilde{\gamma} = 0.99$. Then τ is estimated as 3.3×10^{-12} [sec] from $\omega_p(0)/2\pi \equiv c/2\pi\sqrt{\varepsilon}\lambda_c = 300$GHz which is evaluated from the dielectric constant $\varepsilon = 25$, the c-axis penetration depth λ_c = 20 μm, and the velocity of light c. This value is significantly larger than reported values of Bi2212 $\sim 10^{-13}$ [sec] at low temperatures.[5, 6] We note that it is impossible to detect the normal plasma frequency with our experimental setup, if it is independent of both temperature and magnetic field.

SUMMARY

We found that the decrease of the Josephson plasma frequency with increasing temperature terminates in the vicinity of T_c at a finite frequency $\omega \sim 10^{-1}\omega_p(0)$ in an over-doped Bi2212 single crystal. This result is interpreted that quasiparticles also participate in the plasma oscillation along the c axis of high-T_c superconductors, suggesting that the Josephson tunneling in high-T_c superconductors is not fully incoherent in contrast to one in conventional Josephson junctions.

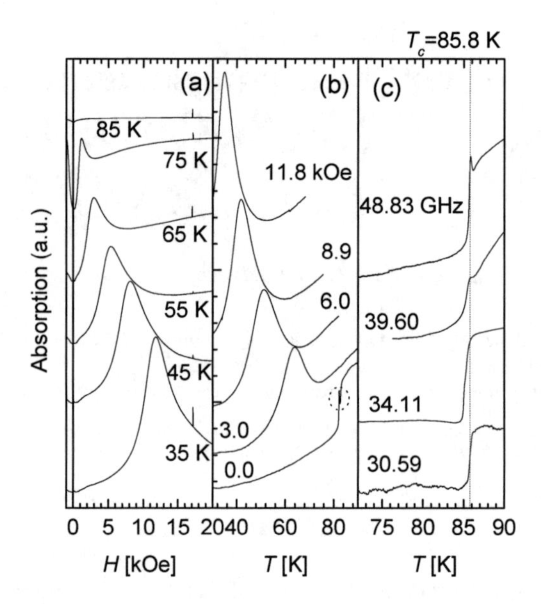

FIGURE 1. JPR data by sweeping magnetic field (a) and temperature (b) at 48.8 GHz and by sweeping temperature at H = 0 at various frequencies(c). It is noted that in the temperature sweep measurements background increases in impedance are included. A dotted circle indicates the very sharp and tiny resonance mentioned in the text and a broken line indicates T_c.

ACKNOWLEDGMENTS

This work has been supported by the 21st century COE program at University of Tsukuba and Grants-in-Aid for Young Scientists (B) under MEXT, Japan.

REFERENCES

1. Gaifullin, M. B., Matsuda, Y., Chikumoto, N., Shimoyama, J., Kishio, K., and Yoshizaki, R., *Phys. Rev. Lett.*, **83**, 3928 (1999).
2. Kadowaki, K., Kakeya, I., Wakabayashi, T., Nakamura, R., and Takahashi, S., *Int. J. Mod. Phys. B*, **14**, 547–554 (2000).
3. Ohashi, Y., and Takada, S., *Phys. Rev. B*, **59**, 4404 (1999).
4. Kakeya, I., Kindo, K., Kadowaki, K., Takahashi, S., and Mochiku, T., *Phys. Rev. B*, **57**, 3108 (1998).
5. Romeo, D. B., et al., *Phys. Rev. Lett.*, **68**, 1590 (1992).
6. Mandrus, D., et al., *Phys. Rev. B*, **46**, 8632 (1992).

Heating-free Interlayer Tunneling Spectroscopy in $Bi_2Sr_2CaCu_2O_{8+x}$ Intrinsic Junctions

Myung-Ho Bae, Jae-Hyun Choi and Hu-Jong Lee

Department of Physics, Pohang University of Science and Technology, Pohang 790-784, Republic of Korea

Abstract. The interlayer tunneling spectroscopy (ITS) in $Bi_2Sr_2CaCu_2O_{8+x}$ (Bi-2212) intrinsic junctions (IJs) reveals the quasiparticle excitation spectrum in the material. In this study, using a proportional-integral-derivative (PID) temperature control scheme incorporated with the *in-situ* temperature measurements of stacks of IJs under study, we eliminate the artifact in the ITS, which is often caused by the self-heating in the high-bias region. Thus-measured spectral weight distribution with the clear peak-dip-hump structure exhibits subtle differences from that obtained without the PID control, which may provide in the long run a crucial guidance to understanding the mechanism of high-T_c superconductivity.

Keywords: Interlayer tunneling spectroscopy, Pseudogap, Density of states, Self-heating effect, PID control scheme
PACS: 74.25.Jb, 74.50.+r, 74.72.Hs

Since the discovery of the tunneling effect in intrinsic junctions (IJs) in a $Bi_2Sr_2CaCu_2O_{8+x}$ (Bi-2212) single crystal [1] the interlayer tunneling spectroscopy (ITS) using a mesa structure fabricated on the surface of a Bi-2212 single crystal [2] has provided a useful means to investigate the density of states (DOS) of quasiparticle excitation. Contrary to the photoemission and scanning tunneling spectroscopy [3] exercised on the surface of single crystals the ITS probes the bulk tunneling properties of IJs embedded inside single crystals, with any surface-sensitive artifact eliminated.

An interesting feature revealed in the DOS of high-T_c superconductors is the existence of the pseudogap (PG), represented by the depletion of the DOS near the Fermi energy, at temperatures even above the superconducting transition temperature, T_c. It has been believed that the PG may offer the key to an understanding of the mechanism of high-T_c superconductivity. To confirm the point the relation between the superconducting gap (SG) and the PG should be clarified. In measurements of the ITS, for instance, the SG and PG were observed simultaneously at a temperature near T_c [2], which led to the suspicion that two gaps may have different origins. The serious problem in the ITS, however, is the local heating in the high bias region [4]. Although the reduction of the junction area and the number of the junctions, and the pulsed-current biasing reduce the heating partially one cannot get rid of the heating problem entirely.

In this study we performed constant-temperature ITS using the combination of *in-situ* monitoring the temperature of IJs under study and employing the computerized proportional-integral-derivative (PID) temperature control [5]. Although the overall feature of the measurements turns out to be similar to the previous ITS results [2], our study reveals delicate differences in the temperature and magnetic-field dependencies of the peak-dip-hump spectral distribution. We believe our results without any artifact due to self-heating will provide a crucial guidance to clarifying the mechanism of high-T_c superconductivity.

Bi-2212 single crystals were grown by the solid-state-reaction method. One gets mostly a slightly overdoped single crystal phases using this method. But, sometimes, one also obtains the locally formed underdoped phases. In this study we used both overdoped and underdoped single crystals, among which we report only the results from an underdoped sample. Instead of a mesa structure, we fabricated a 3×3 μm^2 stack sandwiched between two Au-film electrodes, where the double-side-cleaving of Bi-2212 crystals, micropatterning, and ion-beam etching were employed [6]. The temperature of the sample stack was monitored by placing another stack of IJs (the thermometer stack) in proximity to the sample stack, where the two stacks were in strong thermal coupling through the common bottom Au electrode. The inset of Fig. 1(a) shows the sample configuration; the left (right) part represents the sample (thermometer) stack.

The sample stack contained N=15 IJs, as determined by the number of quasiparticle branches at 4.2 K (not shown). The transition temperature T_c=75.2 K indicates that the sample was a well-underdoped one. The constant-temperature ITS was performed in the following way. We first set up the sample temperature at a certain value using the heater coil wound around the substrate holder, while keeping the bath temperature at 4.2 K. The heat generated by the bias current in the sample stack was directly transferred to the thermometer stack through the bottom Au electrode. The resulting temperature increase of the thermometer stack was monitored by comparing its resistance change in a low-enough bias

CP850, *Low Temperature Physics: 24th International Conference on Low Temperature Physics;*
edited by Y. Takano, S. P. Hershfield, S. O. Hill, P. J. Hirschfeld, and A. M. Goldman

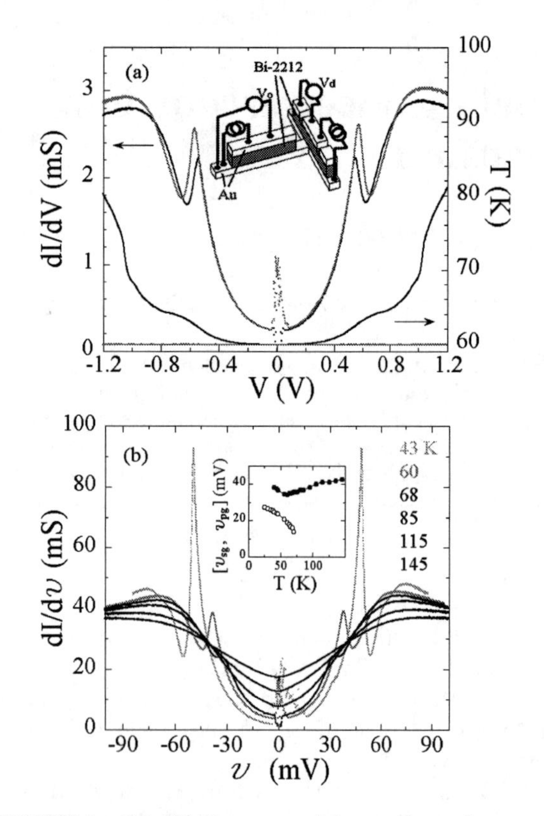

FIGURE 1. (a) dI/dV curves and the sample-stack temperature as a function of the bias voltage, for the setup temperature of 59.9 K, with (gray) and without (black) the PID temperature control. The inset of (a): the sample configuration. (b) The temperature dependence of dI/dv (v; bias across a single junction) curves for a single junction from 43 K to 145 K with the PID control. The inset of (b): the temperature dependence of the SG (open circles) and the PG (closed circles), obtained from one half of the maximum positions of the peak and the hump.

current with the pre-determined resistive transition data of the thermometer stack. The initial sample temperature was then recovered by reducing the heater current with the computerized PID control. Measurements were done along the highest-bias quasiparticle curve, *i.e.*, the last branch.

Fig. 1(a) illustrates the dI/dV and the temperature variation of the sample stack as a function of the bias voltage for a setup temperature of 59.9 K, with (grey curves) and without (black curves) the PID control. The bias current, 5 μA, for the thermometer stack was low enough that it did not cause any self-heating by itself. The distinct peak-dip-hump structure is visible in both cases. The positions of the peak and the hump correspond approximately to the edge of the SG and the PG, respectively. One notices that, without the PID control, the sample stack starts being heated from the bias slightly above 0.2 V, reaching 81 K for the bias of 1.2 V. It implies that the ITS without the PID control can be highly affected by the self-heating, as seen by the reduced spectral weight together with the reduced peak and hump positions in Fig. 1(a).

Fig. 1(b) displays the temperature dependence of dI/dv versus v curves (v=V/N; the bias voltage across a single junction), with the PID temperature control. At temperatures sufficiently below T_c one has a dominant superconducting peak and a clear dip. Closer to T_c, the superconducting peak reduces rapidly and disappears completely above T_c, but the hump retains even above 250 K. The size of the SG also shrinks rapidly as the temperature approaches T_c from below. By contrast, the size of the PG shows a very slow temperature dependence, *with its minimum around* T_c . We observe the coexistence of the SG and the PG near T_c [inset of Fig. 1(b)], which strongly suggests that the two gaps are of different origins. One also notices that the curves at temperatures above T_c converge to a single point inside the PG, the implication of which should be further traced.

The dependence of the spectral distribution on the perpendicular magnetic field up to 6 T was also measured in this study (data not shown). In previous measurements [2] magnetic fields tend to reduce the superconducting peak, with almost no changes in the dip and the hump. In our study, however, the reduced spectral weight in the superconducting peak is transferred to the dip and to the superconducting gap, which poses a major difference from the previous measurements. The spectral weight was found to be conserved in any magnetic fields.

ACKNOWLEDGMENTS

This work was supported by Korea Science and Engineering Foundation through the National Research Laboratory program.

REFERENCES

1. R. Kleiner, F. Steinmeyer, G. Kunkel, and P. Müller, *Phys. Rev. Lett.* **68**, 2394 (1994).
2. M. Suzuki, T. Watanabe, and A. Matsuda, *Phys. Rev. Lett.* **82**, 5361 (1999); V. M. Krasnov, M. Sandberg, and I. Zogaj, *Phys. Rev. Lett.* **94**, 77003 (2005).
3. A. Damascelli, Z. Hussain, and Z. -X. Shen, *Rev. Mod. Phys.* **75**, 473 (2003); Ch. Renner, B. Revaz, J.-Y. Genoud, K. Kadowaki, and Ø. Fischer, *Phys. Rev. Lett.* **80**, 149 (1998).
4. V. N. Zavaritsky, *Phys. Rev. Lett.* **92**, 259701 (2004).
5. M.-H. Bae, J.-H. Choi, and H.-J. Lee, *Appl. Phys. Lett.* **86**, 232502 (2005).
6. H. B. Wang, P. H. Wu, and T. Yamashita, *Appl. Phys. Lett.* **78**, 4010 (2001); M.-H. Bae, H.-J. Lee, J. Kim, and K.-T. Kim, *Appl. Phys. Lett.* **83**, 2187 (2003).

Observation of Collective Transverse Plasma Modes in Stacks of $Bi_2Sr_2CaCu_2O_{8+x}$ Intrinsic Josephson Junctions

Myung-Ho Bae and Hu-Jong Lee

Department of Physics, Pohang University of Science and Technology, Pohang 790-784, Republic of Korea

Abstract. Collective transverse plasma modes in a stack of $Bi_2Sr_2CaCu_2O_{8+x}$ intrinsic Josephson junctions (IJJs) can be examined by observing the resonating Josephson vortex motion in the stack. In this study, we observed distinct multiple branches in the Josephson vortex-flow current-voltage characteristics, which suggested a transformation of Josephson vortex configuration among different collective plasma modes. For a dc bias on one of the multiple branches, we also obtained an evidence for the microwave emission from the transverse Josephson vortex motion in a stacked IJJs. The emission was examined by observing the changes in the quasiparticle branches of another stack of IJJs placed in proximity to the stack under study.

Keywords: Collective transverse plasma modes, Multiple Josephson vortex-flow branches, Microwave emission
PACS: 74.72.Hs, 74.50.+r, 74.78.Fk, 85.25.Cp

The emission of THz-range microwaves using the Josephson vortex motion in intrinsic Josephson junctions (IJJs) of $Bi_2Sr_2CaCu_2O_{8+x}$ (Bi-2212) singe crystals has recently attracted considerable theoretical and experimental research efforts [1]. Local oscillators utilizing conventional $Nb/Al_2O_3/Nb$ junctions have been successfully tested for frequencies up to 700 GHz, which generate power sufficient to pump an SIS mixer [2]. It is, thus, natural to seek a possibility of obtaining THz-range electromagnetic-wave emission from IJJs. Since the transverse plasma oscillations in the periodically and compactly stacked junctions as in Bi-2212 crystals can be in phase the microwave emission in this system would be effectively amplified. To date, however, there still lacks the direct evidence for the emission of THz-range microwaves.

The effective microwave emission has been predicted to occur in a periodically stacked Josephson vortex-flow oscillator when the Josephson vortex flow resonates with the collective transverse plasma modes [1]. This resonating behavior reveals as the multiple collective branches in the Josephson vortex-flow current-voltage *I-V* curves [3]. For the microwave applications, it is thus essential to identify experimentally the collective Josephson vortex-flow branches. In this study, we report the observation of clear collective multiple branches and symptoms of microwave emission induced by Josephson vortex motion.

As-grown slightly overdoped Bi-2212 single crystals were prepared by the conventional solid-state-reaction method. We fabricated, using the double-side cleaving technique, two stacks of IJJs sandwiched between two Au electrodes, respectively, at each stack's top and bottom without the basal part [the inset of Fig. 1(a)]. Here, the bottom electrode serves as a common electrode. This structure is in contrast to the usual mesa structure fabricated on the surface of a single crystal with a large basal part. The detailed fabrication procedure is described in Ref. 5. The inset of Fig. 1(b) illustrates the configuration of the 'oscillator stack' (the left one) and the 'detector stack' (the right one). The lateral size of oscillator stack was $15\times1.4\ \mu m^2$. To generate Josephson vortices in the oscillator stack magnetic fields were applied in parallel with planes of IJJs, facing the longer side of the stack. The width of the detector stack along the direction of the applied field was 0.7 μm.

The inset of Fig. 1(a) shows the zero-field $I-V$ curves. The number of junctions in the oscillator stack, estimated from the number of zero-field quasiparticle branches, was 22. Two important parameters of the Josepshon-vortex dynamics, the Josephson penetration depth λ_J(=0.3 μm), and the plasma frequency f_p(=20 GHz), were also determined from the zero-field *I-V* curves [3]. The Swihart velocity $c_0(=2\pi\lambda_J f_p)$, which is the propagation velocity of the electromagnetic waves in the insulating layer, was $\sim3.7\times10^4$ m/s. Since, in Bi-2212, the London penetration depth along the c axis is much longer than the thickness of the superconducting CuO_2 double layer any junction is strongly coupled to the neighboring junctions. This inductive coupling in a stack consisting of N IJJs leads to the collective transverse plasma modes (eigen-states) with the N-different characteristic collective eigen-frequencies.

Fig. 1(a) shows multiple branches of the tunneling *I-V* characteristics in B=2.5 T. The branches are divided into two regions; the grey and black curve regions. The higher-bias black curves developed from the quasiparticle branches for zero field as the ones shown in the inset of Fig. 1(a). The quasiparticle branches reduce with in-

CP850, *Low Temperature Physics: 24th International Conference on Low Temperature Physics;*
edited by Y. Takano, S. P. Hershfield, S. O. Hill, P. J. Hirschfeld, and A. M. Goldman

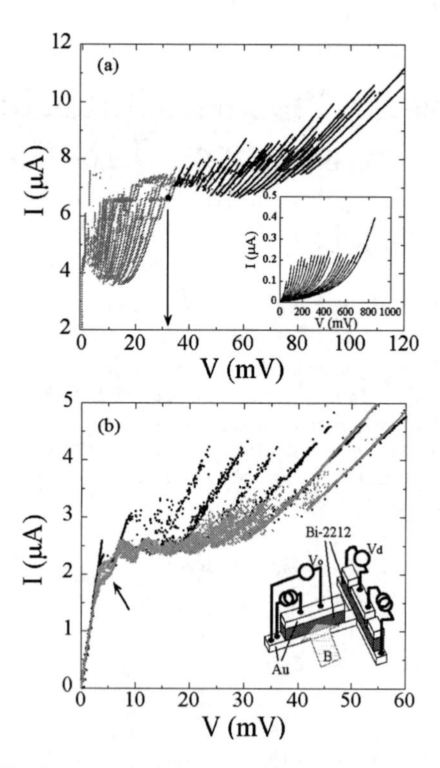

FIGURE 1. (a) Josephson vortex-flow branches (grey curves) and the quasiparticle branches (black curves) of the oscillator stack in B=2.5 T. The inset of (a): the current-voltage curves of the oscillator stack in zero field. (b) The quasiparticle branches of detector stack in B=2.5 T, without a bias (black curves) and with a bias of 32 mV (grey curves) to the oscillator stack. The inset of (b): the sample configuration.

creasing magnetic fields along with the reduction of the Josephson tunneling critical current, disappearing completely for fields higher than ~4 T. On the other hand, the lower-bias grey curves developed only in relatively high field beyond ~2 T and became more prominent for higher magnetic fields. The grey curves are thus believed to be Josephson-vortex-flow branches. The number of Josephson vortex-flow branches in the low-bias region, 22, is almost the same as that of quasiparticle branches or the number of IJJs in the oscillator stack. This feature is consistent with the theoretical prediction that there should exist the same number of collective plasma modes as that of the IJJs, when the Josephson vortex motion is in resonance with the collective plasma oscillation modes. One notices the slight horizontal misalignment in the Josephson-vortex-flow branches at biases a little below and above 6 μA, which may occur when the outermost junctions near the top and the bottom Au electrodes switch to the resistive state at higher biases.

We examined the microwave emission using the detector stack [inset of Fig. 1(b)], with the oscillator stack biased in the Josephson vortex-flow region. The black curves of Fig. 1(b) is the quasiparticle branches of detector stack in B=2.5 T without any bias in the oscillator stack, supposedly corresponds to the absence of microwave emission from the oscillator stack. No noticeable Josephson vortex flow branches are visible in these I-V curves, because the magnetic field was applied facing the narrow side of the detector stack. To confirm the microwave emission the oscillator stack was biased at 32 mV, corresponding to the dot position in Fig. 1(a). The grey curves in Fig. 1(b) is the response of the detector stack for this bias. Two distinct features develop in these quasiparticle branches of the detector stack with a finite bias to the oscillator stack: (i) the reduction of the Josephson critical current of quasiparticle branches and (ii) the appearance of a new wiggle structure denoted by the arrow between the contact-resistance branch and the first quasiparticle branch. In general, the microwave irradiation onto IJJs results in zero-crossing Shapiro steps [5] or a finite voltage along with the reduction of the Josephson critical current [as the response (i) above] due to the motion of Josephson vortices arising from the magnetic-field component of the irradiated microwaves [6]. The observed response (ii) of the detector stack may be a precursor of the Shapiro steps with relatively weak microwave irradiation power. It is, thus, believed that the observed change of the response of the detector stack is a direct evidence for the microwave emission from the collective Josephson vortex flow in the oscillator stack. The detection of this microwave emission can be taken as a direct confirmation of the excited collective plasma modes, which also provides the important key to developing ultra-high-frequency Josephson-vortex-flow oscillators.

ACKNOWLEDGMENTS

This work was supported by KOSEF through the National Research Laboratory program and also by the AOARD of the US Air Force under Contract No. FA5209-04-P-0253.

REFERENCES

1. M. Machida and M. Tachiki, *Current Appl. Phys.* **1**, 341 (2001); G. Hechtfischer *et al.*, *Phys. Rev. Lett.* **79**, 1365 (1997).
2. V. P. Koshelets and S. V. Shitov, *Supercond. Sci. Technol.* **13**, R53 (2000).
3. M.-H. Bae and H.-J. Lee, *Phys. Rev. B* **70**, 52506 (2004).
4. M.-H. Bae *et al.*, *Appl. Phys. Lett.* **83**, 2187 (2003).
5. Y.-J. Doh *et al.*, *Phys. Rev. B* **61**, R3834 (2000); H. B. Wang, P. H. Wu and T. Yamashita, *Phys. Rev. Lett.* **87**, 107002 (2001).
6. Y.-J. Doh *et al.*, *Phys. Rev. B* **63**, 144523 (2001).

Josephson-Vortex Flow in $Bi_2Sr_2Ca_2Cu_3O_{10+y}$ Intrinsic Junctions

S. Yu [a], S. Ooi [a], K. Hirata [a], X. L. Wang [b], C. T. Lin [c], and B. Liang [c]

[a]*National Institute for Materials Science, 1-2-1 Sengen Tsukuba Ibaraki 305-0047, Japan*
[b]*Spintronic and Electronic Materials Group, Institute for Superconducting and Electronic Materials, University of Wollongong, NSW 2522, Australia*
[c]*Max-Planck-Institut für Festkörperforschung, Heisenbergstrasse 1, D-70569, Stuttgart, Germany*

Abstract. We measured the Josephson-vortex flow resistance as a function of in-plane magnetic field in $Bi_2Sr_2Ca_2Cu_3O_{10+y}$ (Bi-2223) intrinsic junctions. The periodic oscillations were observed for the first time in Bi-2223 over a wide range of magnetic fields. The period H_p is almost 5/6 of that for Bi-2212 with same junction width. The smaller H_p is caused by the larger distance between the CuO_2 layers in Bi-2223. We also found that the period doubles when magnetic field exceeds a particular value.

Keywords: Josephson-vortex flow; Periodic oscillations; $Bi_2Sr_2Ca_2Cu_3O_{10+y}$
PACS: 74.60.Ec, 74.72.Hs, 74.25.Dw, 74.25.Ha

INTRODUCTION

In highly anisotropic superconductors, such as $Bi_2Sr_2Ca_{n-1}Cu_nO_{2n+4+y}$ [Bi-22(n-1)n, n=2,3], alternately stacked superconducting CuO_2 layers and insulating layers naturally form atomic-scale Josephson junctions along the c-axis in their crystal structures. When a magnetic field is applied parallel to the CuO_2 layers, Josephson-vortex system can be easily realized in these materials. An applied current along the c-axis exerts a Lorentz force on the Josephson vortices perpendicular to the c-axis direction. Above a critical current, the Josephson vortices start to move, producing a finite voltage. This is the so-called Josephson-vortex flow. Ooi *et al.* have discovered that Josephson-vortex flow resistance oscillates as a function of in-plane magnetic field in Bi-2212 single crystals [1]. The period of the oscillations H_p is inversely proportional to the sample width w perpendicular to the magnetic field and independent of temperature and magnetic field. The value of H_p is expressed as $\phi_0/2ws$ where ϕ_0 ($\approx 2.07\times10^{-7}$ Gcm^2) is the flux quantum and s (15Å for Bi2212, 18Å for Bi2223) is the distance between the CuO_2 layers. Therefore, this phenomenon can be explained by a matching effect between the width of samples and the spacing of the triangular lattice formed by the Josephson vortices. In addition, the periodic oscillations have been theoretically reproduced in both analytical calculations [2] and numerical simulations [3]. Recently, Hirata *et al.* have found that the period of oscillations H_p changes to $2H_p$ above a particular magnetic field which depends on the applied current and sample size [4]. This phenomenon has been theoretically studied by Koshelev [2] and Machida [3].

All the results related to the periodic oscillations were obtained by using Bi-2212 intrinsic junctions. In order to investigate the universality, we measured Josephson-flow resistance as a function of in-plane magnetic field in Bi-2223 intrinsic junctions. The periodic oscillations were observed for the first time in Bi-2223 and were detected over a wide range of magnetic fields.

EXPERIMENTAL

The Bi-2223 crystals used in this study were grown using the traveling solvent floating zone method [5]. Details of the fabrication process and a schematic illustration of the junction are shown in Ref. [1]. The dimensions of the measured sample were w=8.6μm; l=13.6μm; t=1.2μm. The resistance after the fabrication of the junction was ~1000 Ω at 300K, which is three orders of magnitude larger than that before the fabrication. Therefore, the measured resistance is comparable to the c-axis resistance of the

CP850, *Low Temperature Physics: 24th International Conference on Low Temperature Physics;*
edited by Y. Takano, S. P. Hershfield, S. O. Hill, P. J. Hirschfeld, and A. M. Goldman

junction. The superconducting transition temperature T_c is ~100K. The c-axis resistance, the susceptibility, and the Josephson-vortex flow resistance were measured using a customized system (MPMS-5S with EDC option, Quantum Design), which was equipped with a vector magnet. A horizontal magnetic field was used to compensate for the misalignment between the *ab*-plane and the vertical field because it is difficult to align the magnetic field parallel to the *ab*-plane with only a vertical magnetic field [6].

RESULTS AND DISCUSSION

Figure 1 shows the temperature dependence of the *c*-axis resistance for the Bi-2223 sample. Two superconducting transitions are observed at about 100K and 80K, while the susceptibility shows a sharp transition at about 103K. Weak peaks assigned to Bi-2212 are also found in the X-ray diffraction pattern. These results indicate that an intergrowth of Bi-2212 phase coexists in the Bi-2223 sample.

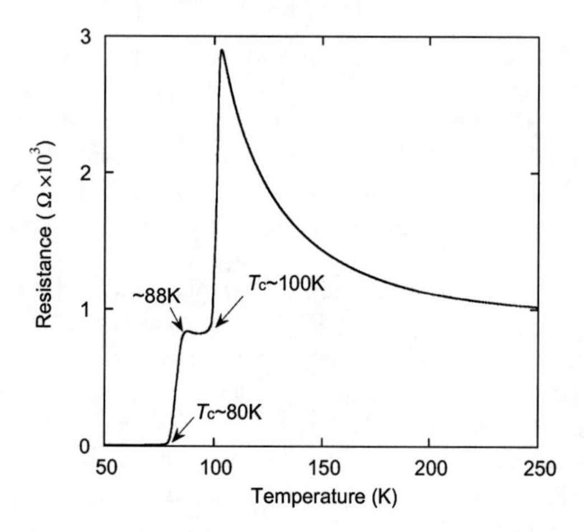

FIGURE 1. Temperature dependence of the *c*-axis resistance for Bi-2223 sample.

Figure 2 shows the in-plane magnetic field dependence of the Josephson-vortex flow resistance with a current of 5μA at 75K for the Bi-2223 sample. As the applied magnetic field rises above a critical point, the flow resistance starts to oscillate periodically. The period H_p is found to be ~630Oe. This value is in agreement with the theoretical value calculated from the expression $\phi_0/2ws$, and is almost 5/6 of that for Bi-2212 with the same junction width. As shown in the inset of Fig. 2, oscillations with the period of ~630Oe were also found at 88K, which is just above T_c of Bi-2212. These results indicate that the periodic oscillations are caused by Bi-2223. On the other hand, the period of oscillations doubles in higher magnetic fields as shown in Fig.2. $2H_p$ corresponds to adding one additional vortex quantum per intrinsic junction. Such a phenomenon has been observed in Bi-2212 [5] and has been theoretically discussed by Koshelev [2] and Machida [3].

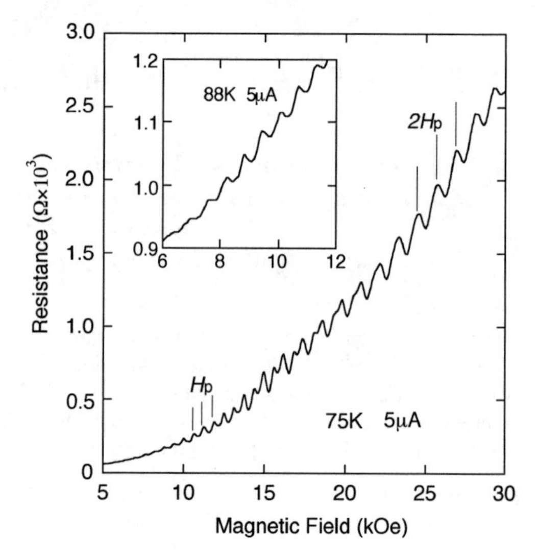

FIGURE 2. Josephson-vortex flow resistance as a function of in-plane magnetic field with a current of 5μA at 75K for the Bi-2223 sample. The inset shows the flow resistance at 88K.

CONCLUSION

To investigate the universality of the periodic oscillations first observed in Bi-2212, we have measured Josephson-vortex flow resistance as a function of in-plane magnetic field in Bi-2223 intrinsic junctions. The periodic oscillations were observed in Bi-2223 for the first time. The period H_p of Bi-2223 is smaller than that of Bi-2212 with same junction width because the intrinsic Josephson junction of Bi-2223 is thicker than that of Bi-2212. We also found that the period of oscillations H_p changes to $2H_p$ above a particular magnetic field.

REFERENCES

1. S. Ooi, T. Mochiku, and K. Hirata, *Phys. Rev. Letters* **89**, 247002 (2002).
2. A. E. Koshelev, *Phys. Rev. B* **66,** 224514 (2002).
3. M. Machida, *Phys. Rev. Letters* **90**, 037001 (2003).
4. K. Hirata, S. Ooi, E. H. Sadki, and T. Mochiku, *Physica B* **329-333**, 1332-1333 (2003).
5. B. Liang, C. T. Lin, P. Shang, and G. Yang, *Physica C* **383**, 75-88 (2002).
6. S. Yu, S. Ooi, T. Mochiku, and K. Hirata, *Physica C* **412-414**, 458-462 (2004).

Reflectivity in the Multi-Josephson Junction Model with Josephson Vortex Lattice

Hideki Matsumoto, Yasuhiro Suzuki and Yoji Ohashi

Institute of Physics, University of Tsukuba, Ibaraki 305-8571, Japan

Abstract. In the intrinsic multi-Josephson model, we calculate reflectivity in order to investigate a coupling between external electromagnetic fields and Josephson plasma oscillations with an influence of a Josephson vortex lattice. We treat electromagnetic fields in the vacuum and the system simultaneously in numerical simulation. With increasing the applied magnetic field, there appear three regions; absorption region in low frequency as the result of Josephson vortices motion, forbidden region in intermediate frequency and plasma edge and above region. The effect of the interlayer coupling induces a sharp change in the boundary of absorption and forbidden regions.

Keywords: Multi-Josephson junctions, Josephson vortex lattice, Josephson plasma, reflectivity
PACS: 74.50.+r, 74.81.Fa, 74.72.-h

INTRODUCTION

The low frequency electromagnetic property of high Tc superconductors is well described by use of the intrinsic multi-Josephson junction model. When the magnetic field is applied along the ab-plane, a Josephson vortex lattice is formed and it affects the electromagnetic excitations in the system. In fact, the observation of two Josephson-plasma modes has been reported in Bi-2212 with the applied magnetic field parallel to the layers[1], and the low frequency mode decreases with increasing the magnetic field. In ref.[2], the presence of two modes was explained from the coupling between the plasma oscillations of the electromagnetic field and triangular Josephson vortex lattice. The numerical simulation also showed that the triangular lattice is a stable for $H \geq \Phi_0/(2\pi\lambda D)$ and the lower oscillation decreases with increasing the applied magnetic field, having a maximum around $H \sim \Phi_0/(2\pi\lambda D)$[3].

In this paper, we calculate reflectivity in order to investigate how the Josephson plasma oscillations and outside electromagnetic field couple. The electromagnetic fields outside and inside of the system are treated simultaneously. The formulation is also useful to investigate properties of emitted waves from junctions and effects on motions of Josephson vortices by an external wave.

FORMULATION

Let us consider a multi-Josephson junction system composed of N-layers stacked along the z-axis (Figure 1). The length of junctions is L along the x-axis, and a magnetic filed H is applied along the y-axis uniformly. We consider the periodic boundary condition along the

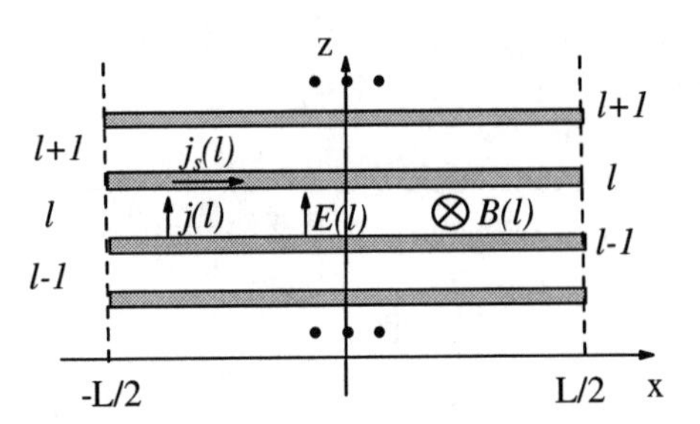

FIGURE 1. Multi-Josephson junction. Shaded layers indicate superconducting ones.

z-axis. Electromagnetism in this multi-junction system is described by the z-component of the electric field at the ℓth insulating layer, $E(x,\ell)$, the y-component of the induction field at the ℓth insulating layer, $B(x,\ell)$, the z-component of the current at the insulating layer, $j(x,\ell)$, and the x-components at the superconducting layer, $j_s(x,\ell)$. The charge on the ℓth superconducting layer is denoted by $\rho_s(\ell)$. We take the currents across the junction in the form of the resistively and capacitive shunted model (RCSJ), $j(\ell) = \sigma E(\ell) + j_c \sin(\phi(\ell))$, where σ is the conductivity of the ℓth insulating layer. The gauge invariant phase difference $\phi(x,\ell)$ satisfies $\partial_t \phi(\ell) = (2eD/\hbar)(E(\ell) - \mu^2 \nabla_z 4\pi\rho_s(\ell))$ and $\nabla_x \phi(\ell) = (2eD/\hbar c)(B(\ell) + \lambda^2 \nabla_z (4\pi/c) j_s(\ell))$, where λ is the in-plane London penetration depth and μ is the shielding length of the charge, which is defined from the charge density expressed by the scalar potential $A_0(\ell)$ and the superconducting phase $\theta(\ell)$, $4\pi\rho_s(\ell) = -\mu^{-2}(A_0(\ell) + (\hbar/2e)\partial_t \theta(\ell))$[4].

We introduce the auxiliary fields, $H_0(x) = (H_{ext}/B_0) + (1/\gamma_0^2)(j_{ext}/j_c)(x/\lambda_0)$ and $\phi_0(x) = (H_{ext}/B_0)(x/\lambda_0) +$

CP850, *Low Temperature Physics: 24th International Conference on Low Temperature Physics;*
edited by Y. Takano, S. P. Hershfield, S. O. Hill, P. J. Hirschfeld, and A. M. Goldman

$(1/2\gamma_0^2)(j_{ext}/j_c)\left(x/\lambda_0\right)^2$, where λ_0 is an arbitrary length scale and $B_0 = (\hbar c/2e)/(D\lambda_0)$ with D being the thickness of the junction. Those fields represent effects of the external current and the applied magnetic field. Define $b(\ell) = B(\ell) - H_0$ and $\varphi(\ell) = \phi(\ell) - \phi_0$. Then the previous equations are rewritten in a non-dimensional form as[5]

$$\begin{aligned}(1/\omega_p)\partial_t\varphi(\ell) &= (1-\alpha\Delta_z^2)(E(\ell)/E_p)\,, \quad (1)\\ (1/\omega_p)\partial_t(E(\ell)/E_p) &= j_{ext}/j_c + \lambda_0\nabla_x\left(\gamma_0^2(b(\ell)/B_0)\right)\\ &\quad -\beta(E(\ell)/E_p) - \sin(\phi_0+\varphi(\ell))\,, \quad (2)\\ b(\ell)/B_0 &= (1-\alpha_s\Delta_z^2)^{-1}\lambda_0\nabla_x\varphi(\ell)\,, \quad (3)\end{aligned}$$

which is a generalization of RCSJ model. Here $\partial_t = \partial/\partial t$, the derivative ∇_z^2 is replaced by the difference $\Delta_z^2 = Ds\nabla_z^2$, which operates on a discretized function $f(\ell)$ as $\Delta_z^2 f(\ell) = f(\ell+1) - 2f(\ell) + f(\ell-1)$. Other parameters are defined as $\omega_p^2 = (8\pi e j_c D)/(\hbar\varepsilon_n)$, $\lambda_p^2 = c^2/(\omega_p^2\varepsilon_n)$, $E_p = \hbar\omega_p/(2eD)$, $\alpha = (\mu^2\varepsilon_n)/(4\pi sD)$, $\beta = (\sigma E_p)/j_c$, $\gamma_0 = \lambda_p/\lambda_0$, and $\alpha_s = \lambda^2/(sD)$. Note that $\alpha \sim 0.5$, $\beta \sim 0.05$, $\alpha_s \sim 10^5$.

The electromagnetic waves of our interest have frequencies around ω_p, and therefore they have a long wave length $\lambda_p(>> D)$ in the vacuum outside of the junctions. We approximately treat the electromagnetic field outside of the system homogeneous in the y- and z- directions, and the Maxwell equation in the medium with a dielectric constant ε_0 becomes $(1/\omega_p)\partial_t(E/E_p) = (\varepsilon/\varepsilon_0)\gamma_0^2\lambda_0\nabla_x(b/B_0)$ and $(1/\omega_p)\partial_t(b/B_0) = \lambda_0\nabla_x(E/E_p)$. Note that $B = b + H_{ext}$ in the medium. At the boundary of the system, we take b/B_0 equal inside and outside the system.

RESULTS

The actual numerical calculation is performed by taking the multi-Josephson junction in the region $-L/2 < x < L/2$ and the vacuum in the regions $x < -L/2$ and $x > L/2$. In this preliminary report, we choose $\lambda_0 = \lambda_p$ (i.e. $\gamma_0 = 1.0$) and $\alpha_s = 10.0$ simply in order to avoid a numerical instability and to save a calculation time. We choose $L/\lambda_p = 50.0$ with $dx/\lambda_p = 0.01$, $\omega_p dt = 0.01$ and time step of 100,000. The reflectivity is obtained from the time average of the poynting vector at the boundaries for an injected plane wave with frequency ω. In Figure 2, we present the reflectivity for a single junction, or for a square vortex lattice with no interlayer coupling. When a magnetic field is applied, the reflectivity shows a broad absorption centered at $\omega/\omega_p \sim 0.5$, which is the result of vortex motion. It recovers near one, indicating the presence of a forbidden region, and then shows the plasma edge. The plasma edge increases with increasing H. In Figure 3, we present the reflectivity for multi-

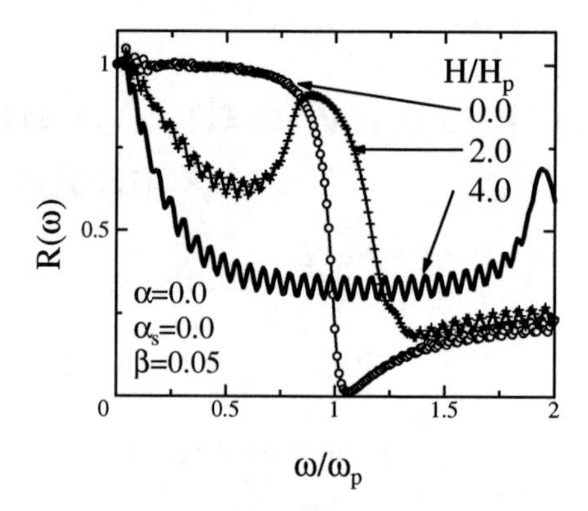

FIGURE 2. Reflectivity for a single junction.

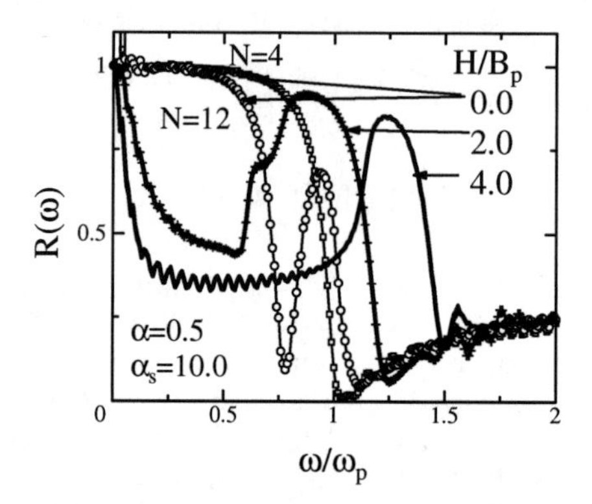

FIGURE 3. Reflectivity for a multi-junction.

junctions (N=4 and N=12) with an interlayer coupling. For $H/B_p = 0.0$, two cases show different results, indicating a coupling with an oscillatory mode in the z-direction. For $H/B_p = 2.0$ and $H/B_p = 4.0$ the reflectivity shows exactly same behavior in $N = 4$ and $N = 12$. We check that the vortex lattice is triangular. In the case of $H/\lambda_p = 2.0$, a sharp change in $R(\omega)$ is observed in the boundary of an absorption region and a forbidden region. It may be closely related with the Josephson plasma oscillation coupled with the triangular lattice. More detailed analysis is in progress.

This work was supported by a Grant-in-Aid for Scientific Research from Japan Society for the Promotion of Science.

REFERENCES

1. K. Kadowaki, et al., *Physica C*, **362**, 71 (2001).
2. T. Koyama, *Phys. Rev. B*, **68**, 224505 (2003).
3. H. Matsumoto, et al., *Physica C*, **412-414**, 444 (2004).
4. T. Koyama, et al., *Phys. Rev. B*, **54**, 16183 (1996).
5. H. Matsumoto, et al., *Physica C*, **388 - 389**, 463 (2003).

Band Engineering in Cooper-Pair Box: Dispersive Measurements of Charge and Phase

Mika Sillanpää[1], Leif Roschier, Teijo Lehtinen and Pertti Hakonen

Low Temperature Laboratory, Helsinki University of Technology, FIN-02015 HUT, Finland

Abstract. Low-frequency susceptibility of the split Cooper-pair box (SCPB) is investigated for use in sensitive measurements of external phase or charge. Depending on the coupling scheme, the box appears as either inductive or capacitive reactance which depends on external phase and charge. While coupling to the source-drain phase, we review how the SCPB looks like a tunable inductance, which property we used to build a novel radio-frequency electrometer. In the dual mode of operation, that is, while observed at the gate input, the SCPB looks like a capacitance. We concentrate on discussing the latter scheme, and we show how to do studies of fast phase fluctuations at a sensitivity of 1 mrad/$\sqrt{\text{Hz}}$ by measuring the input capacitance of the box.

Keywords: quantum measurement, Cooper-pair-box
PACS: 67.57.Fg, 47.32.-y

INTRODUCTION

Josephson junctions (JJ) store energy according to $E = -E_J\cos(\varphi)$, where φ is the phase difference across the junction, and the Josephson energy E_J is related to the junction critical current I_C through $I_C = 2eE_J/\hbar$. Since JJ's also typically exhibit negligible dissipation, they can be used as reactive circuit components. By combining the Josephson equations $I = I_C\sin(\varphi)$ and $\dot{\varphi} = 2eV(t)/\hbar$, where $V(t)$ is the voltage across the junction, we find that a single JJ behaves as a nonlinear inductance,

$$L_J(\varphi) = \frac{\hbar}{2eI_C\cos(\varphi)} = \frac{L_{J0}}{\cos(\varphi)}, \quad (1)$$

where we defined the linear-regime Josephson inductance $L_{J0} = \hbar/(2eI_C)$.

Quantum effects in mesoscopic JJ's [1, 2] may modify Eq. (1) in an important manner. In particular, the Josephson reactance may become capacitive [3, 4]. In this brief communication, we investigate the Josephson reactance in the split Cooper-pair box (SCPB) geometry, with emphasis on detector applications. We first review the inductive susceptibility, and then concentrate on discussing the capacitive susceptibility in the spirit of a novel phase detector. The discussion relies heavily on the energy bands [5] E_k of the SCPB, two lowest of them given in the limit $E_J/E_C \ll 1$ as

$$E_{0,1} = E_C(n_g^2 - 2n_g + 2) \mp \sqrt{(E_J\cos(\varphi/2))^2 + (2E_C(1-n_g))^2} - C_g V_g^2/2 \quad (2)$$

as a function of the classical fields $\varphi = 2\pi\Phi/\Phi_0$ and $n_g = C_g V_g/e$ (see Fig. 1).

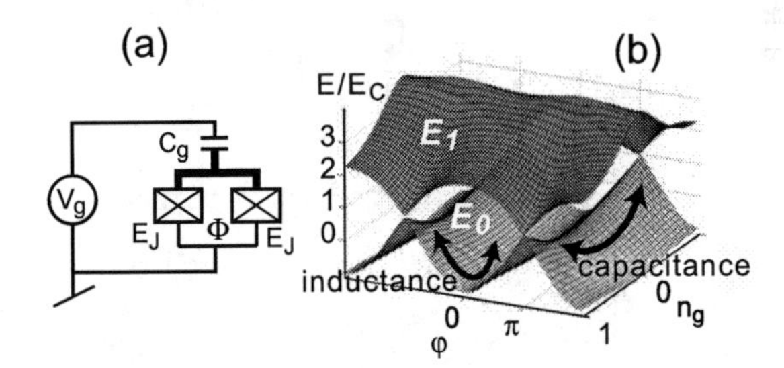

FIGURE 1. (a) Schematics of the SCPB. The mesoscopic island (thick line) has a total capacitance C_Σ and charging energy $E_C = e^2/(2C_\Sigma)$; (b) two lowest energy bands E_k ($k = 0, 1$) of the SCPB , for $E_J/E_C = 1.7$ (without the parabolic background $-(n_g e)^2/(2C_g)$, see Eq. (2)). Inductive and capacitive susceptibilities are illustrated by the arrows parallel to φ and n_g, respectively.

QUANTUM INDUCTANCE

With respect to φ, the SCPT behaves as an inductance (Fig. 2 (b)), dependent, first of all, on the band index k, as well as on n_g and φ:

$$L_{\text{eff}}^k(n_g, \varphi) = \left(\frac{d^2E_k}{d\Phi^2}\right)^{-1} = \left(\frac{\Phi_0}{2\pi}\right)^2 \left(\frac{d^2E_k}{d\varphi^2}\right)^{-1}. \quad (3)$$

The strong n_g dependence of L_{eff}^0 when $E_J/E_C \ll 1$ has been used by the present authors to implement a fast reactive electrometer [6], using the scheme of Fig. 2 (a). The measurements are performed by studying the phase

[1] present address: National Institute of Standards and Technology, 325 Broadway, Boulder, CO 80305, USA

CP850, *Low Temperature Physics: 24th International Conference on Low Temperature Physics;*
edited by Y. Takano, S. P. Hershfield, S. O. Hill, P. J. Hirschfeld, and A. M. Goldman

shift $\Theta = \arg(V_{out}/V_{in})$ of the "carrier" microwave reflected from a resonant circuit containing the SCPB. Denoting by Z the lumped-element impedance seen when looking towards the resonance circuit from the transmission line of impedance $Z_0 = 50\,\Omega$, the reflection coefficient of a voltage wave is

$$\Gamma = \frac{V_{out}}{V_{in}} = \frac{Z - Z_0}{Z + Z_0} = \Gamma_0 e^{i\Theta}. \tag{4}$$

Since the whole setup consists in principle only of reactances, the inductively read scheme should be superior in terms of noise and back-action [7] over the previous fast electrometer, the rf-SET [8], which relies on the control of dissipation.

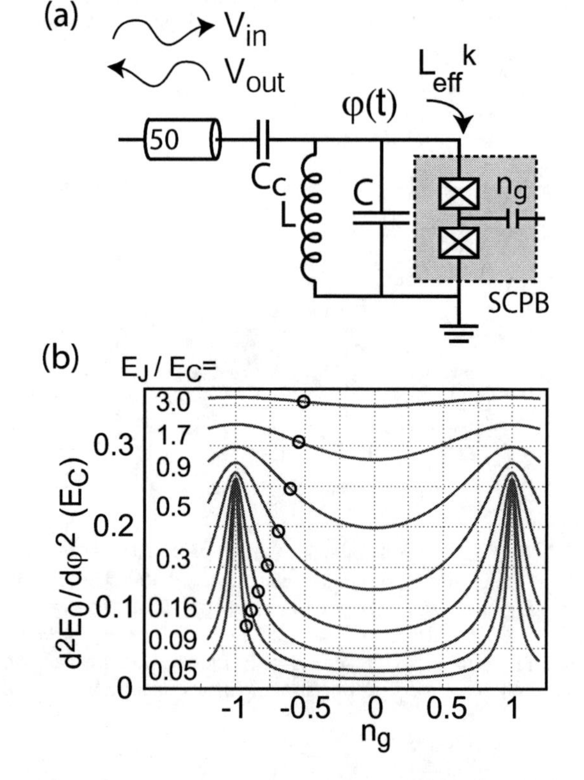

FIGURE 2. (a) Schematics of the capacitively coupled "L-SET" inductive rf-electrometer. The resonance frequency $f_p^{-1} = 2\pi\sqrt{(L \parallel L_{\rm eff}^k)C}$ depends on the SCPB Josephson inductance $L_{\rm eff}^k$; (b) calculated modulation of the second n_g -derivative (inverse $L_{\rm eff}^0$, see Eq. (3)) at the SCPB ground energy band, for different E_J/E_C, and $\varphi = 0$. The circles mark optimal bias points for the electrometer operation.

The crucial number for electrometer operation is the differential modulation of $L_{\rm eff}$ (at the ground band), or dimensionless "gain":

$$g \equiv \frac{\partial}{\partial n_g}\left(\frac{L_{\rm eff}}{L_{\rm eff,0}}\right), \tag{5}$$

which we have presented as normalized by $L_{\rm eff,0}$ which denotes the Josephson inductance at the special point ($n_g = \pm 1, \varphi = 0$). Using Eq. (2), we have $L_{\rm eff,0} = 4L_{J0}$. For the best electrometer performance, n_g should be biased at the points marked by circles in Fig. 2 (b). From Eq. (2) we also find the maximum gain g_m which grows rapidly when $E_J/E_C \ll 1$: $g_m \simeq 2(E_J/E_C)^{-1}$. Another important figure is the value of $L_{\rm eff}$ at the optimal gate bias which yields g_m, denoted here as $L_{\rm eff,m}$ [9]. To some extent, the rapidly growing $L_{\rm eff,m}$ towards lowering E_J/E_C (see Fig. 3) cancels the benefit of growing g_m from the point of view of charge sensitivity.

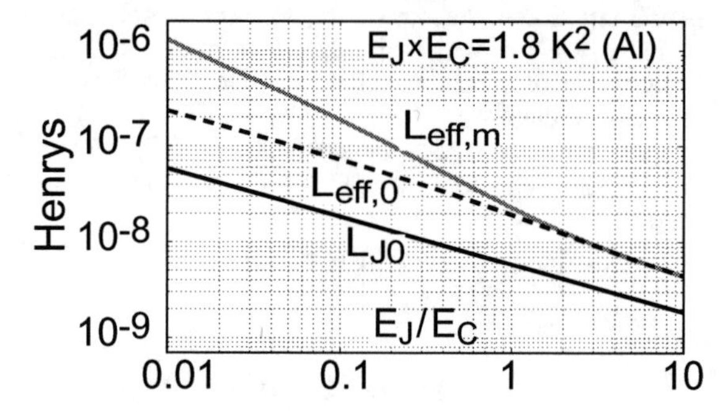

FIGURE 3. Numerical values of the SCPB ground band Josephson inductance $L_{\rm eff,0}$ (that at $n_g = \pm 1, \varphi = 0$), and $L_{\rm eff,m}$ (at maximum gain, at $\varphi = 0$) for a typical aluminium device. Also shown is the "classical" Josephson inductance L_{J0} in Eq. (1).

Without going into details, optimal charge sensitivity limited by *zero-point fluctuations* in the loaded *LC*-oscillator in Fig. 2 (a) is [10]:

$$s_q^{\rm QL} = \frac{16\sqrt{2}e(L_{\rm eff,m})^2\sqrt{2k_B T_N}}{g_m \pi \sqrt{\hbar}\Phi_0 L_{J0}\sqrt{Q_i}}, \tag{6}$$

where T_N is the noise temperature of the rf-amplifier, and Q_i is the *internal* quality factor of the resonator. Evaluating the values in Eq. (6) numerically, we find that $s_q \sim 10^{-7}{\rm e}/\sqrt{\rm Hz}$, order of magnitude better than the shot-noise limit of rf-SET, is intrinsically possible for the L-SET if $Q_i \sim 10^3$ and $T_N \sim 200$ mK. So far, the sensitivity in experiment [10, 11] has been limited by $Q_i \lesssim 20$ down to $s_q \simeq 2\times 10^{-5}{\rm e}/\sqrt{\rm Hz}$. The limit of Eq. (6) is reached when parameter values are chosen so that

$$\omega_p = \frac{\Phi_0^2(L_{\rm eff,m} + L)}{64\hbar L_{\rm eff,m} L}. \tag{7}$$

Equation (7) yields values typically $f_p = \omega_p/(2\pi) \simeq 1-2$ GHz, though dependence of f_p is rather weak.

QUANTUM CAPACITANCE

The band energies of an SCPB depend on the (gate) charge n_g, see Fig. 1 (b), and the SCPB should then behave like a capacitance with respect to changes of n_g

[2, 3], which means that the point of observation is at the gate electrode:

$$C_{\rm eff}^{k} = -\frac{\partial^2 E_k(\varphi, n_g)}{\partial V_g^2} = -\frac{C_g^2}{e^2}\frac{\partial^2 E_k(\varphi, n_g)}{\partial n_g^2}. \quad (8)$$

Phase modulation of the input capacitance $C_{\rm eff}(n_g, \varphi)$ of the SCPB observed in this manner is plotted in Fig. 4 (b). As seen in the figure, $C_{\rm eff}$ has a strong phase dependence in the limit $E_J/E_C \gg 1$ around $\varphi = \pm\pi$. Exactly at $\varphi = \pm\pi$, Cooper-pair tunneling is completely blocked, and $C_{\rm eff}$ reduces to classical series capacitance of the junctions and C_g, that is, $\left[(C_1+C_2)^{-1}+C_g^{-1}\right]^{-1}$.

The input capacitance depends sensitively (quadratically) on the coupling capacitance C_g, and even when C_g is made unusually large such that it practically limits the charging energy, $C_{\rm eff}$ typically remains very small, in the femto-Farad range, see right hand scale of Fig. 4 (b). However, it has been suggested that the extremely strong phase dependence could be used for fast, reactively read phase detection [4]. This "CSET" mode of operation is somewhat dual to the "L-SET" electrometry.

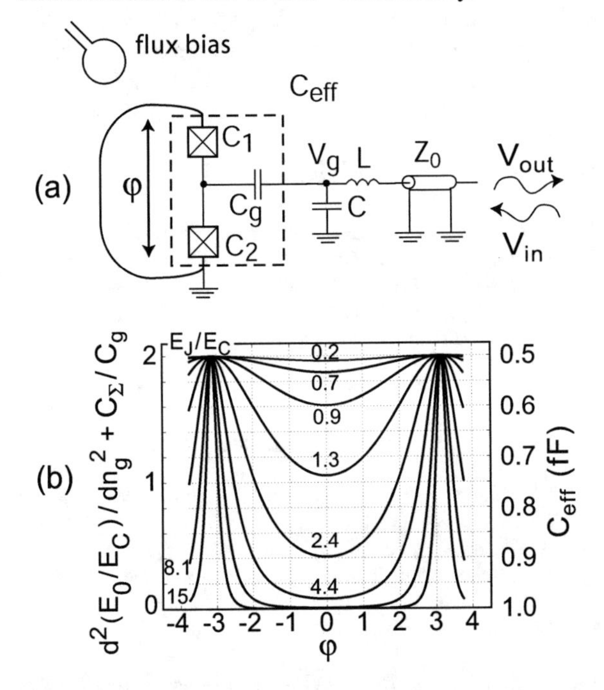

FIGURE 4. (a) Schematics of the experiment used to study how the SCPB appears as a tunable capacitance $C_{\rm eff}$; (b) *left scale* is the calculated second n_g-derivative of the SCPB ground band at $n_g = 0$, and *right scale* is the corresponding effective capacitance if $C_g = 1$ fF and $C_\Sigma = 2$ fF.

An important figure of merit for phase sensitivity is the differential gain, analogous to Eq. (5):

$$f \equiv \frac{\partial}{\partial\varphi}\left(\frac{C_{\rm eff}}{C_g^2/(2C_\Sigma)}\right). \quad (9)$$

The maximum of f w.r.t. φ at $n_g = 0$ is plotted in Fig. 5.

We consider the experimental setup of Fig. 4 (a), where the quantum capacitance $C_{\rm eff}$ is in parallel with a (generally much larger) stray capacitance C, and forms a resonator with an inductance L. In this scheme, it is typical to operate in the limit of vanishing internal dissipation which corresponds to change of phase Θ of the reflected carrier changing by 2π around the resonant frequency f_p.

Similarly as in the inductive readout, there are here no internal noise sources except quantum fluctuations in the resonator. Typically, therefore, sensitivity is again limited by noise of the preamplifier: spectral density of the voltage noise referred to preamplifier input is $s_{V{\rm out}} = \sqrt{2k_BT_NZ_0}$, which can be regarded as a phase noise of the microwave carrier, $s_\Theta = s_{V{\rm out}}/V_{\rm out}$. When the carrier amplitude is optimally large, it can be shown that under the conditions mentioned, $V_{\rm out} = \frac{e}{2C_g}Z_0\sqrt{\frac{C}{L}\frac{1}{2\pi}}$. When referred as an equivalent flux noise at detector input using Eq. (9), the result becomes

$$s_\varphi = \frac{s_\Theta}{\partial\Theta/\partial\varphi} = 2\sqrt{\pi}e\left(\frac{C}{C_g}\right)\frac{\sqrt{k_BT_NZ_0}}{f_mE_C} \simeq \frac{4\sqrt{\pi}C\sqrt{k_BT_NZ_0}}{f_me}, \quad (10)$$

where the last form follows from the assumption that at high E_J/E_C, charging energy is limited by the large gate capacitance. This is the ultimate limit with advanced junction fabrication (very thin oxide). The predicted phase sensitivity is plotted in Fig. 5. Evidently, sensitivity improves with decreasing stray capacitance C, since this results in larger modulation of total capacitance $C + C_{\rm eff}$. We see that $s_\varphi < 10^{-6}{\rm rad}/\sqrt{\rm Hz}$, far beyond an equally fast rf-SQUID, is possible in principle at high $E_J/E_C \sim 10$ and a low stray capacitance $C \sim C_g$.

We investigated the discussed phase detection experimentally in the scheme of Fig. 4 (a), with the parameter values $E_J = 0.30$ K, $E_C = 0.83$ K, $E_J/E_C = 0.36$, $C_g = 0.65$ fF, $C = 250$ fF, $C/C_g = 380$, and $L = 160$ nH. Except C_g, the sample parameters were determined by microwave spectroscopy [12]. To the input bias coil of the phase detector, we applied low-frequency modulation by $0.013\,\Phi_0$ at 80 Hz. Its amplitude was calibrated relying on Φ_0-periodicity of the static response. This way, we obtained a sensitivity of 1.3 mrad/$\sqrt{\rm Hz}$, see the black curve in Fig. 6, limited by the 4 K amplifier noise, which figure is even better than expected (see Fig. 5).

We shall now discuss Fig. 6 in more detail. Both the curves were measured at a flux bias close to $\varphi \sim \pi$ which yields the largest gain f_m. For the black curve, $f(n_g, \varphi)$ was further maximized by tuning n_g close to 1, which also yielded a high level of low-frequency noise as can be seen in the data. Since the low-frequency noise is

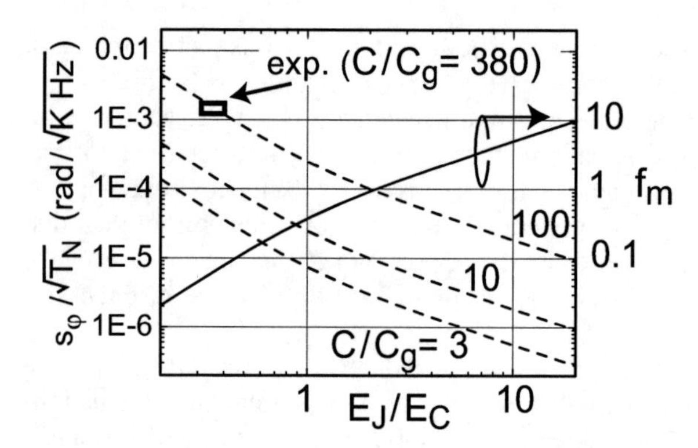

FIGURE 5. Left scale (dashed lines): Phase sensitivity predicted for the CSET, Eq. (10), first form, if $Z_0 = 50\,\Omega$ and $E_C = 1$ K, for different ratios of the gate capacitance to stray capacitance. Right scale: the maximum gain f_m of the phase detector. Experimental point is given by the rectangle (note that it had a larger capacitance ratio of ~ 380).

significantly reduced when we tuned $n_g = 0$ where the response is insensitive to charge fluctuations (the gray curve), we assign the increased noise around $n_g = 1$ to the ubiquitous low-frequency background charge noise.

Since the low-frequency noise at $n_g = 0$ is free from the effect of charge noise, we were able to directly measure in the scheme the apparent flux noise, which we attribute to critical-current fluctuations. The power spectrum of the gray curve shows $1/f^2$ dependence in contrast to typical $1/f$ rule [13] for big junctions. We convert this noise into fluctuations in critical current of either of the junctions, in other words, we ask the question: what would be the I_C fluctuation $\Delta I_C = 2e/\hbar(\Delta E_J)$ in either one of the junctions which would cause a capacitance fluctuation ΔC_{eff}, and hence an apparent phase fluctuation $\Delta\varphi$? Equation (9) implies

$$\Delta C_{\text{eff}}(n_g,\varphi) = f(n_g,\varphi) C_0 \Delta\varphi\,, \tag{11}$$

where we have marked $C_0 = C_g^2/(2C_\Sigma)$. This then converts into E_J fluctuation according to

$$\Delta E_J = \Delta C_{\text{eff}} \left(\frac{\partial C_{\text{eff}}}{\partial E_J}\right)^{-1} \tag{12}$$

We compute the partial derivative in Eq. (12) numerically; the result is $\frac{\partial C_{\text{eff}}}{\partial E_J} \simeq 0.072 \left(\frac{C_g}{e}\right)^2 \frac{E_C}{E_J} \simeq 0.30 \left(\frac{C_g}{e}\right)^2$. We also set $f(n_g,\varphi) \to f_m$ since we had tuned to the maximum gain.

Finally, since the spectral densities of fluctuations are related similarly as the fluctuations itself, we have the amplitude spectrum of I_C noise:

$$s_{IC} = \frac{2e}{\hbar} s_{EJ} = \frac{2e}{\hbar} f_m s_\varphi C_0 \left(\frac{\partial C_{\text{eff}}}{\partial E_J}\right)^{-1}. \tag{13}$$

This yields the gray line in Fig. 6, with the numbers around 10 Hz being comparable to big junctions.

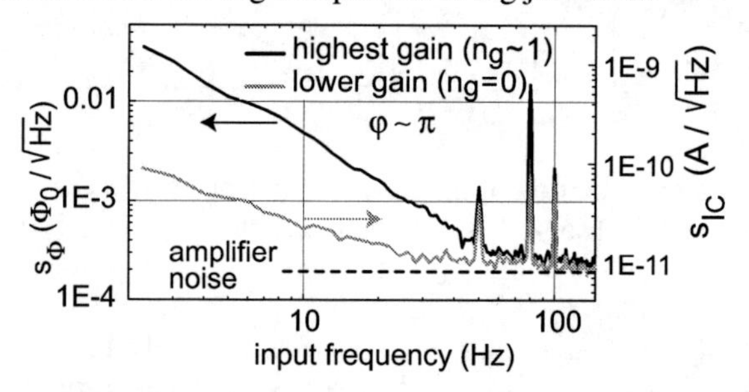

FIGURE 6. Measured equivalent flux noise at CSET input (left scale, black curve) and critical current noise (right scale, gray curve). Low-frequency flux modulalation by $0.013\,\Phi_{0,\text{RMS}}$ at 80 Hz was used as a marker.

ACKNOWLEDGMENTS

We thank T. Heikkilä, F. Hekking, R. Lindell, Yu. Makhlin, M. Paalanen, and R. Schoelkopf for comments and useful criticism. This work was supported by the Academy of Finland and by the Vaisala Foundation of the Finnish Academy of Science and Letters.

REFERENCES

1. A. Widom *et. al.*, *J. Low Temp. Phys.* **57**, 651 (1984).
2. D. V. Averin, A. B. Zorin, and K. Likharev, *Sov. Phys. JETP* **61**, 407 (1985); K. Likharev and A. Zorin, *J. Low. Temp. Phys.* **59**, 347 (1985).
3. D. V. Averin and C. Bruder, *Phys. Rev. Lett.* **91**, 057003 (2003).
4. L. Roschier, M. Sillanpää, and P. Hakonen, *Phys. Rev. B* **71**, 024530 (2005).
5. D. J. Flees, S. Han, and J. E. Lukens, *Phys. Rev. Lett.* **78**, 4817 (1997).
6. M. Sillanpää, L. Roschier, and P. Hakonen, *Phys. Rev. Lett.* **93**, 066805 (2004).
7. A. B. Zorin, *Phys. Rev. Lett.* **86**, 3388 (2001).
8. R. J. Schoelkopf, P. Wahlgren, A. A. Kozhevnikov, P. Delsing, and D. E. Prober, *Science* **280**, 1238 (1998).
9. Equation (2) does not allow for an analytical formula for $L_{\text{eff,m}}$.
10. M. Sillanpää, L. Roschier, and P. Hakonen, *Appl. Phys. Lett.* **87**, 092502 (2005).
11. M.A. Sillanpää, Ph.D. thesis, Helsinki University of Technology (2005); http://lib.tkk.fi/Diss/2005/isbn9512275686/.
12. M. A. Sillanpää, T. Lehtinen, A. Paila, Yu. Makhlin, L. Roschier, and P. J. Hakonen, cond-mat/0504517.
13. F. C. Wellstood, C. Urbina, and J. Clarke, *Appl. Phys. Lett.* **85**, 5296 (2004).

Quasiparticle Poisoning in a Single Cooper-Pair Box

J.F. Schneiderman*, P. Delsing†,‡, G. Johansson†, M.D. Shaw*, H.M. Bozler*, and P.M. Echternach‡

‡*Jet Propulsion Laboratory, California Institute of Technology, Pasadena, CA 91109, USA*
**University of Southern California, Dept of Physics and Astronomy, Los Angeles, CA 90089-0484, USA*
†*Chalmers University of Technology, Microtechnology and Nanoscience, MC2, 412 96 Göteborg, Sweden*

Abstract. We investigate the pheonomenon of quasiparticle poisoning in a single Cooper-pair box (SCB). We have designed, fabricated, and tested an SCB that demonstrates a transition between poisoned and unpoisoned Coulomb staircases, depending on the speed with which the gate charge is swept. Poisoning is shown to be suppressed at moderately high sweep rates. Coulomb staircases were measured for a variety of sweep rates, and quasiparticle tunneling rates were extracted from this data.

Keywords: SCB, quasiparticle, Coulomb blockade.
PACS: 03.67.Lz, 73.23.Hk, 74.50.+r, 85.25.Cp, 85.35.Gv.

The single Cooper-pair box (SCB)[1] consists of a small island of superconducting aluminum coupled to a reservoir of charge through a small-area Josephson junction. The SCB demonstrates the behavior of a two-level quantum system, which is useful both for investigations into fundamental physics and as a building block for quantum computation. Coherent oscillations between charge states corresponding to the number of Cooper-pairs on an SCB island have been demonstrated by several research groups[2,3,4,5].

A major problem associated with using the SCB as a qubit is the presence of odd quasiparticle (QP) states, dubbed "QP poisoning". Attempts to solve this problem have stimulated a number of recent studies[6,7]. QP poisoning occurs when the energy of the odd charge state of the island falls below the energy of the even charge ground state for a particular range of gate voltages. Within this range, quasiparticle tunneling results in transitions between the even and odd charge states, and a "short step" appears in the Coulomb staircase around the degeneracy point, spoiling the 2e-periodidicity. Poisoning prevents manipulation and control of the SCB charge state at the degeneracy point, since the extra quasiparticle state prevents operation as a two-level system. Understanding and eliminating QP poisoning is an essential step towards realizing quantum computation with charge qubits.

In this work, we measure Coulomb staircases from an SCB with an RF-SET[8] readout, and extract quasiparticle tunneling rates from this data. The staircase was measured by sweeping the SCB gate voltage and using the RF-SET to read out the charge transferred to the SCB. By changing the rate at which the gate voltage was swept (across a 7-electron span) from 10.44 Hz to 1044 Hz, and measuring staircases for each sweep rate, we were able to investigate the dynamical behavior of the system.

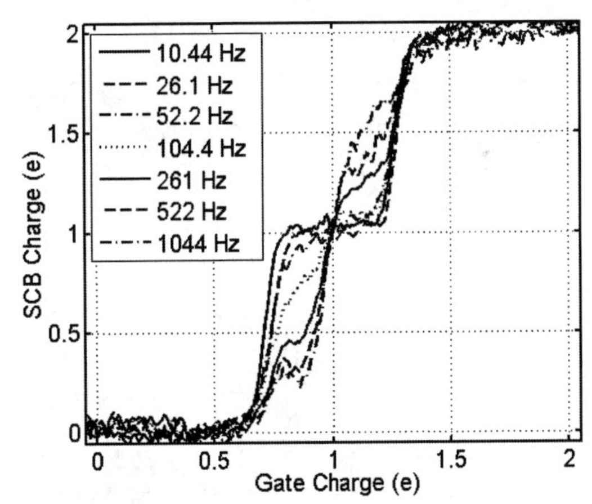

FIGURE 1. Coulomb staircases measured with various gate voltage sweep rates.

For the slower sweep rates, the system settles into the equilibrium state at each bias point, giving rise to a completely poisoned staircase. For faster sweep rates, the system does not have time to reach equilibrium and the staircase is only partially poisoned. This suggests that the problem of QP poisoning in charge qubits can be partially ameliorated by sweeping sufficiently

CP850, *Low Temperature Physics: 24th International Conference on Low Temperature Physics;* edited by Y. Takano, S. P. Hershfield, S. O. Hill, P. J. Hirschfeld, and A. M. Goldman

quickly. A representative set of staircase data is shown in Figure 1.

To extract quasiparticle tunneling rates from the experimental data, we obtain the probability of the system being in the even state (Equation (1)) from the standard solution to the Hamiltonian of the SCB.

$$P_{Even} = \frac{1-<n>}{\left(\frac{4E_C(1-n_G)}{\sqrt{E_J^2+16E_C^2(1-n_G)^2}}\right)} \quad (1)$$

Here, <*n*> is the measured SCB charge at each gate charge value n_G. Figure 2 shows the even-state probability as a function of gate voltage for a number of sweep rates. To obtain the rates, we consider the two-state master equation

$$\frac{dP_{Ev}}{dt} = P_{Odd}(t)\Gamma_{OE}(n_G) - P_{Even}(t)\Gamma_{EO}(n_G) \quad (2)$$

where the time-dependence of the rates arises solely from sweeping the gate charge. The derivative is computed numerically, and the rates Γ_{OE} and Γ_{EO} are extracted by applying a linear regression. Figure 3 shows the rates obtained from a typical data set. Note that Γ_{EO} increases near n_G=1, where the ground state is odd. Meanwhile, Γ_{OE} peaks where the two energy levels cross, which is likely related to the peak in the density-of-states of the reservoir.

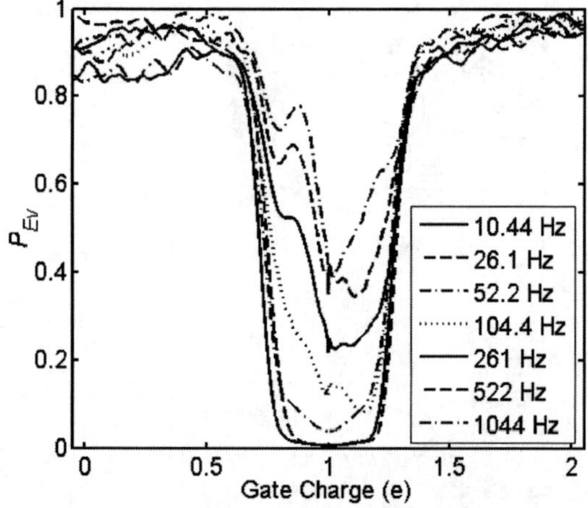

FIGURE 2. Probability of the system being in even state as a function of time for each sweep rate.

In conclusion, we have observed that quasiparticle poisoning can be suppressed in an SCB by quickly sweeping the gate charge. By measuring the staircase at different sweep rates, we extracted the rates of quasiparticle tunneling as a function of gate charge.

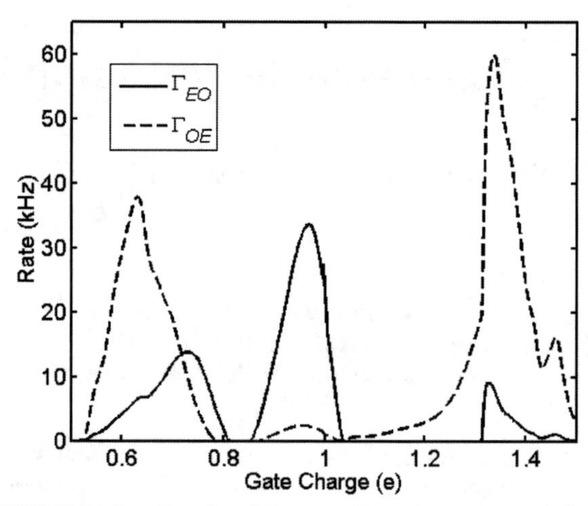

FIGURE 3. Quasiparticle tunneling rates extracted form $P_{Even}(t)$ data.

ACKNOWLEDGMENTS

We thank Richard Muller for the electron beam lithography. This work was performed at the Jet Propulsion Laboratory, California Institute of Technology, under a contract with the National Aeronautics and Space Administration. We acknowledge support from the National Security Agency and the Advanced Research and Development Activity.

REFERENCES

1. D.V. Averin and Yu. V. Nazarov, *Phys. Rev. Lett.* **69**, 1993 (1992).
2. Y. Nakamura, Y.A. Pashkin, and J.S. Tsai, *Nature* **398**, 786 (1999).
3. T. Duty, D. Gunnarsson, K. Bladh, R. J. Schoelkopf, and P. Delsing, *Phys. Rev. B* **69**, 140503 (2004).
4. D. Vion, A. Aassime, A. Cottet, P. Joyez, H. Pothier, C. Urbina, D. Esteve, and M.H. Devoret, *Science* **296**, 886 (2002).
5. A. Guillaume, J.F. Schneiderman, P. Delsing, H.M. Bozler, and P.M Echternach, *Phys. Rev. B* **69**, 132504 (2004).
6. J. Aumentado, M.W. Keller, and J.M. Martinis, *Phys. Rev. Lett.* **92**, 066802 (2004).
7. J. Mannik, and J.E. Lukens, *Phs. Rev. Lett.* **92**, 057004 (2004).
8. R.J. Schoelkopf, P. Wahlgren, A.A. Kozhevnikov, P. Delsing, and D.E. Prober, *Science* **280**, 1238 (1998).

Decoherence Sources of a Superconducting Quantum Bit

G.Ithier[a], N.Boulant[a], E.Collin[a], P.J.Meeson[a,b], P.Joyez[a], D.Vion[a], D.Esteve[a], F.Chiarello[c], A.Shnirman[d], Y.Makhlin[d] and G.Schön[d]

[a]*SPEC-CEA-Saclay, Orme des Merisiers, 91191 Gif sur Yvette, CEDEX, France*
[b]*Department of Physics, Royal Holloway, University of London, Egham, Surrey, TW20 0EX, UK*
[c]*Istituto di Fotonica e Nanotecnologia, CNR, Via Cineto Romano, 42 00156 Roma, Italy*
[d]*Institut für Theoretische Festkörperphysik, Universität Karlsruhe, D-76128 Karlsruhe, Germany*

Abstract. The Quantronium, a superconducting circuit with Josephson Junctions, can be regarded as a solid state qubit prototype with a coupling to the environment over which some degree of control may be exercised. We have demonstrated experimentally that the quantum state of the circuit may be manipulated using pulse sequences derived from NMR and atomic physics and then readout. Measurements of the coherence time obtained in this way are analysed via a general theoretical framework to develop a simple model of the environmental noise sources. A complete picture of decoherence in this quantum electrical circuit can thus be provided.

Keywords: Superconductivity, Josephson Junctions, qubit, decoherence.
PACS: 85.25.Cp, 85.25.Hv

INTRODUCTION

Superconducting Josephson Junction devices have been shown to exhibit quantum coherent behaviour in a number of recent experiments[1-6]. Such circuits are good candidates to become the underlying technology for quantum information processors because they could easily be scaled up to a many qubit design, unlike most other candidate technologies. On the other hand, unlike fundamental particles, these circuits are rather strongly coupled to their external environment and so suffer from quantum decoherence as well as a number of other practical challenges.

EXPERIMENT

In order to address the question of decoherence in superconducting quantum bit circuits we have undertaken a series of experiments designed to measure and control the quantum coherence of a particular circuit, the Quantronium. The Quantronium[7] is a mixed charge-phase qubit composed of a small superconducting island sensitive to charge and a flux loop containing a large Josephson Junction used for determining the state of the quantum circuit. Fig. 1 shows a schematic of the experimental setup, indicating the low frequency charge and flux controls,

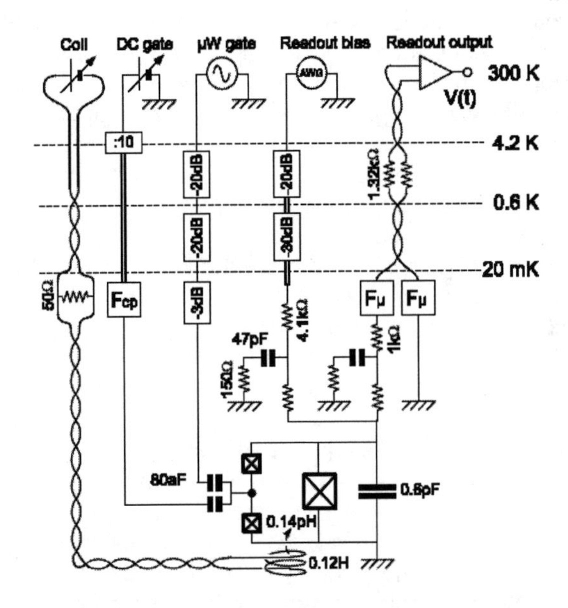

FIGURE 1. Schematic of the experimental setup used to control and measure the "Quantronium", a Josephson Junction qubit circuit. The rectangles labeled in dB are 50 Ω attenuators, :10 is a high impedance voltage divider by 10, F_{CP} is a copper powder filter and the Fμ are microfabricated distributed RC filters. Single lines, double lines and twisted pairs are 50 Ω coax, lossy coax of manganin in a stainless steel capillary and manganin twisted pairs in a stainless steel capillary, respectively.

CP850, *Low Temperature Physics: 24th International Conference on Low Temperature Physics;*
edited by Y. Takano, S. P. Hershfield, S. O. Hill, P. J. Hirschfeld, and A. M. Goldman

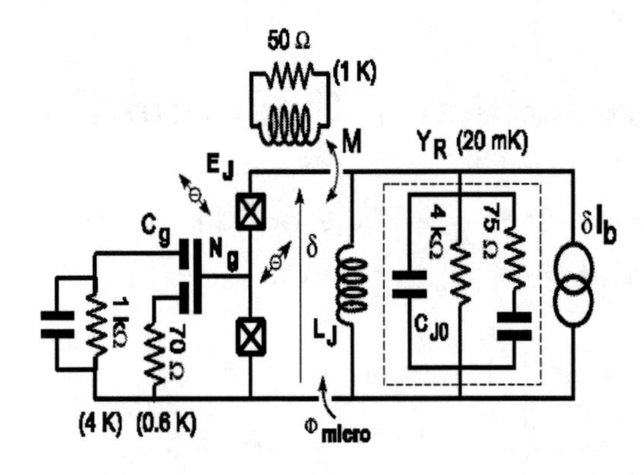

FIGURE 2. Schematic of the noise sources responsible for decoherence in the Quantronium.

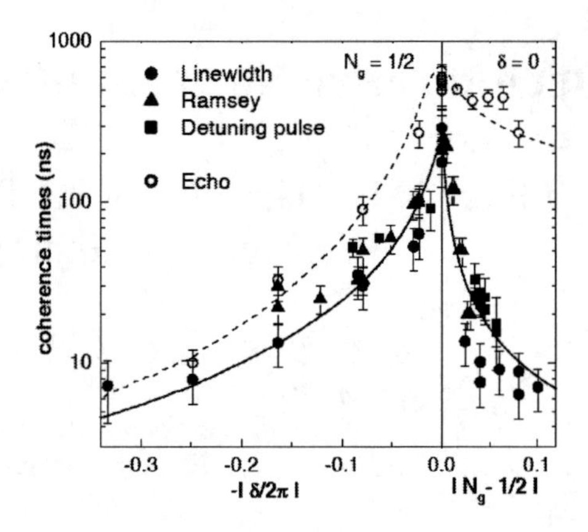

FIGURE 3. Qubit coherence times (1/e decay) measured from a) the spin echo method (open circles), b) the resonance linewidth (solid circles) and c) the decay of the Ramsey signal (triangles) as a function of N_g and δ. Also plotted are data from pulse detuning experiments in which the operating point is changed during a Ramsey sequence (squares). The full and dashed lines are best fits of T_E and T_2 times respectively, using the spectral densities shown in Fig.4.

the (resonant) pulsed microwave line and the readout circuitry. The experiments are based on techniques derived from NMR and atomic physics, such as Rabi, Ramsey and spin echo pulse sequences.

In these experiments the state of the qubit is readout by measuring the bias current required to switch the large Josephson Junction to the voltage state, a technique that provides sensitivity to the qubit current circulating at switching. The qubit state is prepared in the ground state at specified values of the flux and gate charge, then manipulated through the use of resonant or near resonant microwave pulses applied to the gate, and of dc pulses applied to the bias current or dc charge lines for selecting the point where decoherence is measured.

The noise sources that act to decohere the qubit are schematically represented in Fig. 2. They can be parameterized as contributions to the terms in the Hamiltonian that describes the circuit and so appear as noise in E_J, the Josephson energy, N_g, the gate charge or δ, the phase. These sources can be microscopic in origin, represented as two level fluctuators providing charge noise near the island or inducing noise in E_J, or moving vortices in the vicinity of the loop. Macroscopic sources due to the external circuitry are represented as an equivalent circuit across the qubit.

The data from the various experiments is shown in Fig 3. The qubit frequency is stationary w.r.t. N_g and δ at (N_g= 1/2, δ=0) and so the qubit is immune to these noises to first order at this point, leading to the longest coherence times. Theoretical fits to the data indicate $1/f$ noise contributions in both δ and N_g. Noise in δ continues at constant level to high frequency, which we attribute to spurious noise in the flux bias current. Unexpectedly the charge noise N_g appears to have a cut off above about 1/2 MHz. Model noise spectra deduced from and compatible with the data are presented in Fig. 4.

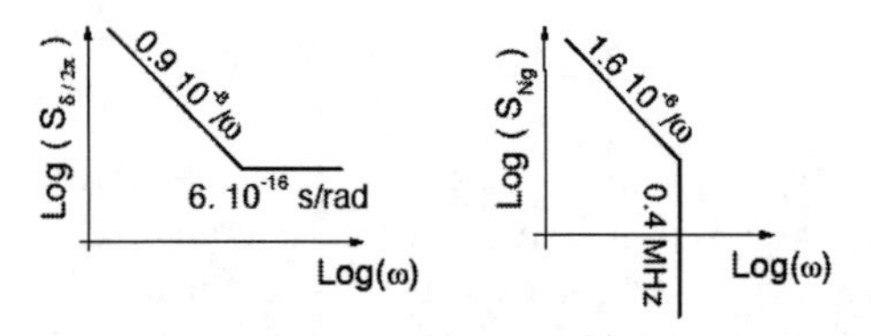

FIGURE 4. Phase and charge noise spectral densities emerging from the fits of Fig.3.

Acknowledgments

We acknowledge discussions with G.Falci, the technical help of P.F.Orfila, P.Sénat, and J.C.Tack, the EU for a Marie Curie fellowship (PJM) and the SQUBIT project, and Yale University (grant DAAD 19-02-1-0044) for financial support.

References

1. Y. Nakamura, Yu.A. Pashkin and J.S. Tsai, *Nature,* **398**, 786 (1999).
2. D. Vion *et.al.*, *Science* **296**, 886 (2002).
3. J.M. Martinis *et.al.*, *Phys. Rev. Lett.* **89**, 117901 (2002).
4. I. Chiorescu *et.al.*, *Science* **299**, 1869 (2003).
5. J. Claudon *et.al., Phys. Rev. Lett.* **93**, 187003 (2004).
6. A. Wallraff *et.al.*, *Nature* **431**, 162 (2004).
7. E. Collin *et.al.*, *Phys. Rev. Lett.* **93**, 157005 (2004).
8. G. Ithier *et.al.*, *Phys. Rev. Lett.* **94**, 057004 (2005).
9. G. Ithier *et.al.*, submitted to *Phys. Rev. B.*

Measurement of Decoherence Time in a Flux Qubit

K. Harrabi*, F. Yoshihara†, Y. Nakamura** and J.S. Tsai**

*CREST-JST, Kawaguchi, Saitama 332-0012, Japan
†The Institute of Physical and Chemical Research (RIKEN), Wako, Saitama 351-0198, Japan
**NEC Fundamental and Environmental Research Laboratories, Tsukuba, Ibaraki 305-8501, Japan

Abstract. We present a measurement of the relaxation and the dephasing times in a flux qubit. In order to improve coherence of the qubit, two external parameters were optimized: the applied flux through the qubit loop and the bias current of the SQUID which serves as a readout device of the qubit state. At the optimal point the dephasing time measured with spin-echo technique was twice longer than the energy relaxation time. By changing one of the two bias parameters while keeping the other at the optimal value, one can separate the contribution of the noise in each parameter to the decoherence of the qubit.

Keywords: Josephson junction; decoherence; flux qubit
PACS: 74.81.Fa; 73.23.Ra

In the last few years, studies of the Josephson junction qubits demonstrated that qubits suffered from different sources of decoherence depending on the types of qubit. The decoherence comes from fluctuations of environmental degrees of freedom. One solution is to set the working point of the qubit at an optimal point giving minimum sensitivity to these fluctuations [1]. In a flux qubit [2] the usefulness of this solution was also confirmed [3]. In order to understand the dominant decoherence mechanism, we investigated coherence of a flux qubit under various bias conditions.

The qubit was fabricated by using e-beam lithography and shadow evaporation technique. It consists of a small superconducting loop (area $\sim 3\ \mu\text{m}^2$) interrupted by four Josephson junctions. Three identical junctions ($Al/AlO_x/Al$) have a size of $0.04\ \mu\text{m}^2$, and the size of the fourth junction is reduced by a factor $\alpha \sim 0.6$. The qubit loop is embedded in a bigger loop of a DC-SQUID which is used as a readout device of the qubit state, and a part of the qubit loop is shared with that of SQUID. The mutual inductance between the SQUID and the qubit was estimated to be ~ 10 pH. The SQUID has two underdamped Josephson junctions ($0.2\ \mu\text{m}^2$ each) and is shunted with a capacitor ($C \sim 10$ pF) fabricated on chip. The critical current of the SQUID and the three larger junctions of the qubit were estimated to be about 7 μA and 0.7 μA, respectively. For the control of the qubit state, microwave current pulses were sent to a control line near the qubit. The resonant microwave flux pulse created in the qubit loop induced Rabi oscillations between the ground and excited states of the qubit. For the qubit readout, we used a switching event of the SQUID from the supercurrent state to the voltage state under application of a bias current pulse [4]. The state-dependent persistent current induced in the qubit loop creates a small magnetic field which can be easily detected by the SQUID inductively coupled to the qubit. To measure the population in each eigenstate of the qubit after a control pulse sequence, we repeated the control and readout procedure 10^4 times and counted the switching probability. The fidelity of the readout characterized by the visibility of Rabi oscillations was about 80% in this particular sample. The measurements were performed at 20 mK using a dilution refrigerator.

The truncated Hamiltonian of the qubit can be written as $H = -\frac{\varepsilon}{2}\sigma_z - \frac{\Delta}{2}\sigma_x$. where $\varepsilon = 2I_p(\Phi - \Phi_0/2)$, and σ_x and σ_z are the Pauli matrix. Φ is a flux through the qubit and Φ_0 is a flux quantum. The energy difference between two eigenstates is $E_{01} = \sqrt{\varepsilon^2 + \Delta^2}$. From the result of spectroscopy measurement, the maximum persistent current and the splitting energy were found to be $I_p \sim 0.34\ \mu$A and $\Delta = 5.08$ GHz. The flux qubit is supposed to be sensitive to the fluctuations of Φ. Especially, if $\Phi \neq \Phi_0/2$, E_{01} linearly fluctuates as Φ does. Thus, it causes strong dephasing of the qubit. However, by adjusting Φ to be $\Phi_0/2$, $\frac{\partial E_{01}}{\partial \Phi}$ becomes zero, and the qubit is decoupled from the noise at least in the first order. This is the optimal condition for the flux bias, i.e., $\frac{\partial E_{01}}{\partial \Phi} = 0$. In the present setup, there were two relevant externally controlled parameters to adjust Φ. The first was the magnetic flux Φ_{ex} applied by an external coil, and the other was the bias current I_b of the SQUID applied during the qubit operation. The bias current is split into two arms of the SQUID, and any imbalance of the current between the two branches produces additional flux Φ_{I_b} in the qubit loop. If the SQUID has a perfectly symmetric structure, small current fluctuations always flow symmetrically, and $\frac{\partial \Phi_{I_b}}{\partial I_b} = 0$ at $I_b = 0$. However, for large I_b and in the presence of finite flux bias of the SQUID loop, the circulating current depends on I_b, and thus, $\frac{\partial \Phi_{I_b}}{\partial I_b} \neq 0$.

CP850, *Low Temperature Physics: 24th International Conference on Low Temperature Physics;*
edited by Y. Takano, S. P. Hershfield, S. O. Hill, P. J. Hirschfeld, and A. M. Goldman

Thus, only at $I_b = 0$, Φ_{I_b} is decoupled from the fluctuation of I_b, and this is the optimal point for the current bias, i.e., $\frac{\partial E_{01}}{\partial I_b} = 0$. In general, if there is a slight asymmetry in the SQUID, the optimal condition is not necessary at $I_b = 0$ but can be shifted.

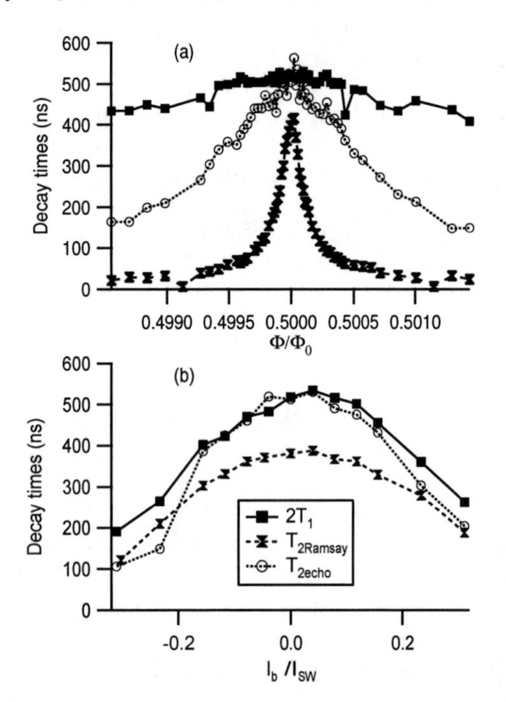

FIGURE 1. Decay times T_1, $T_{2Ramsey}$, and T_{2echo} as a function of (a) the external flux Φ_{ex} under the condition of $\frac{\partial E_{01}}{\partial I_b} = 0$; (b) the SQUID bias current I_b under $\frac{\partial E_{01}}{\partial B} = 0$($I_{SW}$ is the switching current of the SQUID).

We measured the energy relaxation time T_1 by sending a sequence of two π pulses. Population of the qubit ground state was measured as a function of the delay time between the two pulses. The experimental data was nicely fitted with an exponential decay giving T_1. The dephasing time T_2 was determined with Ramsey interference and spin-echo techniques. For Ramsey interference two $\pi/2$ pulses with variable delay time in between were applied to the qubit. The carrier frequency of the pulses was detuned from the qubit frequency E_{01}/h by 10 MHz so that the angle between rotation axes of each $\pi/2$ pulse rotated at the corresponding frequency and the fringes wre visible. The switching probability vs delay time was fitted with a 10 MHz sinusoidal function with an exponential decay envelope which gave $T_{2Ramsey}$. The spin echo technique uses a sequence of two $\pi/2$ pulses with an intermediate π pulse located exactly in the middle of the delay time, all in resonant with the qubit. The echo technique is useful to reduce the effect of low frequency noise on the dephasing of the qubit. The experimental data was fitted with an exponential decay with a time constant of T_{2echo}.

Figure 1 summarizes the results of the decay time measurements. Around the optimal point, $\Phi/\Phi_0 = 0.5$ and $I_b/I_{SW} = 0.018$, T_{2echo} was found to be twice longer than T_1. This indicates that the pure dephasing rate caused by noises at relatively high frequencies is small, allowing the decoherence rate to be domintated by the energy relaxation rate, i.e., $\Gamma_{2echo} = \frac{\Gamma_1}{2}$. On the other hand, $T_{2Ramsey}$ was shorter than T_{2echo} and $2T_1$. Hence, there remained some amount of low frequency noise which dephases the qubit. When the magnetic field was shifted from the degeneracy point while the condition $\frac{\partial E_{01}}{\partial I_b} = 0$ was kept, T_{2echo} and $T_{2Ramsey}$ were reduced considerably (Fig. 1(a)). The strong dependence on the flux bias implies that the relevant noise is a flux noise, and the large difference between T_{2echo} and $T_{2Ramsey}$ suggests that the dephasing is mainly caused by low frequency noise. Indeed, we noticed that for flux bias away from the optimal point, the echo decay curves are not well fitted with a simple exponential decay but with a product of T_1 limited exponential decay and a Gaussian decay. Decrease of decay times was also observed when the system satisfied $\frac{\partial E_{01}}{\partial \Phi}$=0 and I_b was moved from its optimal value (Fig. 1). In this case, however, the difference between T_{2echo} and $T_{2Ramsey}$ became smaller as I_b was shifted further. Also the echo decay curve remained to be exponential even for the largest I_b shift. These facts indicate that the noise coming through the SQUID current bias was not low frequency dominated like $1/f$ noise but has a significant high frequency contribution to the dephasing.

In conclusion, by adjusting the external magnetic field and the bias current of the SQUID we optimized the qubit bias condition. At the optimal point, the observed echo decay time was twice longer than the energy relaxation time. By investigating the decoherence for various operating conditions of the qubit, the different nature of the flux noise and the bias current noise were revealed.

REFERENCES

1. D. Vion, A. Aassime, A. Cottet, P. Joyez, H. Pothier, C. Urbina, D. Esteve, and M. H. Devoret, *Science*, **296**, 886-889 (2002).
2. J. E. Mooij, T. P. Orlando, L. Levitov, L. Tian, C. H. van der Wal, and S. Lloyd, *Science*, **285**, 1036–1039 (1999).
3. P. Bertet, I. Chiorescu, G. Burkard, K. Semba, C. J. P. M. Harmans, D. P. DiVincenzo, and J. E. Mooij, *cond-mat*/0412485.
4. I. Chiorescu, Y. Nakamura, C. J. P. M. Harmans, and J. E. Mooij, *Science*, **299**, 2503–2504 (2003).

Coherent Control of a Flux Qubit with Two-Frequency Microwave Pulses

S. Saito[1, 2], T. Kutsuzawa[3], T. Meno[4], H. Tanaka[1, 2], M. Ueda[1, 2, 5], H. Nakano[1, 2], K. Semba[1, 2], and H. Takayanagi[1, 2, 3]

[1]*NTT BRL, NTT Corporation, 3-1 Morinosato-Wakamiya, Atsugi, Kanagawa 243-0198, Japan*
[2]*CREST, Japan Science and Technology Agency, 4-1-8 Honcho, Kawaguchi, Saitama 332-0012, Japan*
[3]*Department of Physics, Tokyo University of Science, 1-3 Kagurazaka, Shinjuku, Tokyo 162-8601, Japan*
[4]*NTT Advanced Technology, 3-1 Morinosato-Wakamiya, Atsugi, Kanagawa 243-0198, Japan*
[5]*Department of Physics, Tokyo Institute of Technology, 2-12-1 Ookayama, Meguro, Tokyo 152-8551, Japan*

Abstract. Coherent control of a superconducting flux qubit has been achieved by using two-frequency microwave pulses. We have observed Rabi oscillation resulting from parametric transitions between the qubit states when the qubit Larmor frequency matches the sum of the two microwave (MW) frequencies. The Rabi frequency is described by the product of two Bessel functions of the MW amplitudes. The parametric transition observed in our experiment offers useful flexibility for controlling flux qubits.

Keywords: Superconducting flux qubit, nonlinear quantum phenomena, parametric transition
PACS: 74.50.+r, 03.67.Lx, 42.50.Hz, 85.25.Dq

Superconducting qubits have attracted a lot of attention as promising candidates for the implementation of solid-state qubits. Coherent single qubit operation has been demonstrated in several kinds of superconducting qubits [1-5]. In addition to qubit operation, they offer a testing ground for exploring interactions between photons and artificial macroscopic quantum objects. Multi-photon Rabi oscillations [6] and multi-photon absorption [7, 8] have been observed in the strong-driving regime. In this paper, we report two-photon Rabi oscillation stemming from parametric transitions between the ground and first excited states of a superconducting flux qubit.

We fabricated our sample using lithographic techniques that define the structure of the inner aluminum loop with three Josephson junctions forming the qubit and the outer enclosing dc-SQUID loop for the readout. By carefully designing the junction parameters, the inner loop can be made to behave as an effective two-state system [9]. It is described by the Hamiltonian, $H_{qb} = (\hbar/2)\,(\varepsilon\sigma_z + \Delta\sigma_x)$ where $\sigma_{x,z}$ are the Pauli spin matrices. The eigenstates of σ_z correspond to the directions of persistent currents with amplitudes of I_p. The qubit tunneling splitting is described by $\hbar\Delta$, and $\hbar\varepsilon$ is the energy imbalance between the two current states, which is induced by an external dc magnetic field. The energy difference between the ground state $|g\rangle$ and the first excited state $|e\rangle$ of the qubit is $\hbar\omega_{qb} = \hbar\sqrt{\varepsilon^2+\Delta^2}$. We obtained the qubit parameters $\Delta/2\pi$ = 1.73 GHz, I_p = 330 nA from spectroscopy measurements. Assuming that the applied microwaves are classical fields, we can describe a qubit under two-frequency MW irradiation by the Hamiltonian

$$H = H_{qb} + \sum_{k=1}^{2} \frac{\hbar\omega_{MWk}}{2} a_k V_{MWk} \sigma_z \cos\omega_{MWk} t, \quad (1)$$

where ω_{MWk} and V_{MWk} represent the MW frequency and amplitude, respectively and a_k is a scaling parameter determined by the MW attenuation and the coupling between the qubit and the MW. To solve the Schrödinger equation with the Hamiltonian (1), we obtain the time evolution of the probability $P_e(t)$ with which we find the qubit in $|e\rangle$. When the operating point is far from the qubit degeneracy point $\varepsilon >> \Delta$ and the resonant condition $\omega_{qb} = \omega_{MW1} + \omega_{MW2}$ is satisfied, $P_e(t)$ oscillates periodically with the frequency

CP850, *Low Temperature Physics: 24th International Conference on Low Temperature Physics;*
edited by Y. Takano, S. P. Hershfield, S. O. Hill, P. J. Hirschfeld, and A. M. Goldman

$$\Omega_{\mathrm{Rabi}} = \frac{\Delta}{A} J_1(A a_1 V_{\mathrm{MW1}}) J_1(A a_2 V_{\mathrm{MW2}}). \qquad (2)$$

Here J_1 is the first order Bessel function of the first kind and $A = \varepsilon / \omega_{\mathrm{qb}}$.

The measurements were carried out in a dilution refrigerator at a base temperature of 20 mK ($k_B T \ll \hbar\omega_{\mathrm{qb}}$). The resonant MW pulse with a length of t_p prepares the qubit in the superposition state between $|g\rangle$ and $|e\rangle$. After the MW pulse, the qubit state is readout by applying a dc pulse to the detector SQUID. Repeating this measurement 8000 times and averaging the results, we obtain the switching probability of the SQUID P_{sw}, which is directly related to $P_e(t_p)$.

First, we set the qubit Larmor frequency $\omega_{\mathrm{qb}}/2\pi$ at 26.45 GHz by adjusting the external dc magnetic field. Then we observed coherent oscillations of P_{sw} as a function of t_p by using two-frequency MW pulses with $\omega_{\mathrm{MW1}}/2\pi$ = 16.2 GHz and $\omega_{\mathrm{MW2}}/2\pi$ = 10.25 GHz (Fig. 1(a)). The amplitude of MW2 V_{MW2} was 82.5 mV. The open circles represent experimental data and the solid curves are the exponentially damped oscillation fits with single (V_{MW1} = 47.9 mV) and two (V_{MW1} = 95.0 mV) frequency components. Figure 1(b) shows the Rabi frequencies (points) $\Omega_{\mathrm{Rabi}}/2\pi$ obtained from these fittings. The dotted curve represents eq. (2) without any fitting parameters, which reproduces the experimental data qualitatively. Here we used a_1 = 0.0074 [1/mV] and a_2 = 0.013 [1/mV], which was obtained from Rabi oscillations by using single-frequency MW pulse experiments with ω_{MW1} or ω_{MW2}.

The Rabi frequency in Fig. 1(b) exhibits anti-crossing around 280 MHz. To explain this phenomenon, we take account of a two-level resonator coupled to the qubit. Suppose that the resonator with the energy splitting $\hbar\varepsilon_r\sigma'_z$ is coupled to the qubit through the coupling term $\hbar g_r \sigma_z \sigma'_x$. Here, $\hbar g_r$ is the

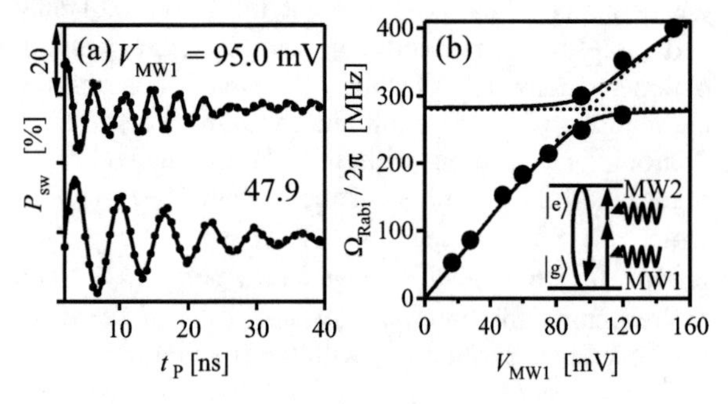

FIGURE 1. (a) Two-photon Rabi oscillations. (b) Two-photon Rabi frequency as a function of the MW amplitude. The inset is a schematic of the parametric transition.

coupling energy between the qubit and the resonator and σ_z and $\sigma'_{x,z}$ are the Pauli matrices for the qubit and the resonator, respectively. We assume that the qubit is excited by a single-frequency microwave with an amplitude of Ω_{Rabi} in eq. (2) and an angular frequency of ω_{qb}. By using the rotating wave approximation, the Hamiltonian is given by $(\hbar/2)[\,\omega_{\mathrm{qb}}\sigma_z + \Omega_{\mathrm{Rabi}}(\sigma_x \cos\omega_{\mathrm{qb}}t + \sigma_y \sin\omega_{\mathrm{qb}}t) + \varepsilon_r\sigma'_z + 2g_r\sigma_z\sigma'_x]$. Solving the Schrödinger equation gives the Rabi oscillations of the qubit with two frequency components, which are given by $\Omega_{\mathrm{Rabi}\pm} = \frac{1}{2}\left(\sqrt{\Omega_+^2 + 4g_r^2} \pm \sqrt{\Omega_-^2 + 4g_r^2}\right)$. Here $\Omega_\pm$ = $\Omega_{\mathrm{Rabi}} \pm \varepsilon_r$. The solid curves in Fig. 1(b) represent $\Omega_{\mathrm{Rabi}\pm}$ when $\varepsilon_r/2\pi$ = 280 MHz and $g_r/2\pi$ = 20 MHz, which well reproduce the measurement data. The coupling $\hbar g_r \sigma_z \sigma'_x$ indicates that the resonator is a truncated two-level system originating in a harmonic oscillator. This resonator is different from those observed in phase qubits, which arise from changes in the junction critical current produced by two-level states in the tunnel barrier [10].

In summary, we demonstrated the parametric controls of a superconducting flux qubit by using two-frequency microwave pulses. This kind of transition widens the frequency range of microwaves for controlling the flux qubit. We also observed evidence of a two-level resonator coupled to the qubit. We can utilize the Rabi oscillations as a quantum spectrum analyzer to investigate the qubit environment.

ACKNOWLEDGMENTS

We thank F. Deppe for useful discussions and for helping with the experimental setup. We also acknowledge useful discussions with J. Johansson, M. Thorwart, Y. Yamamoto, J. E. Mooij, C. J. P. M. Harmans, and Y. Nakamura. This work has been supported by the CREST project of the Japan Science and Technology Agency (JST).

REFERENCES

1. Y. Nakamura *et al.*, *Nature* **398**, 786-788 (1999).
2. D. Vion *et al.*, *Science* **296**, 886-889 (2002).
3. Y. Yu *et al.*, *Science* **296**, 889-892 (2002).
4. J. M. Martinis *et al.*, *Phys. Rev. Lett.* **89**, 117901 (2002).
5. I. Chiorescu et al., *Science* **299**, 1869-1871 (2003).
6. Y. Nakamura *et al.*, *Phys. Rev. Lett.* **87**, 246601 (2001).
7. A. Wallraff *et al.*, *Phys. Rev. Lett.* **90**, 037003 (2003).
8. S. Saito *et al.*, *Phys. Rev. Lett.* **93**, 037001 (2004).
9. J. E. Mooij *et al.*, *Science* 285, 1036-1039 (1999).
10. R. W. Simmonds *et al.*, *Phys. Rev. Lett.* **93**, 077003 (2004).

Coherent Control of Coupled Superconducting Macroscopic Quantum Systems

Kouichi Semba [a], Shiro Saito [a], Takayoshi Meno [b], Jan Johansson [a], and Hideaki Takayanagi [a]

[a]NTT Basic Research Laboratories, NTT Corporation, Atsugi, Kanagawa 243-0198, Japan
CREST, Japan Science and Technology Agency, Kawaguchi, Saitama 332-0012 Japan
[b]NTT Advanced Technology Corporation, Shinjuku Mitsui Bldg., Shinjuku, Tokyo 163-0431, Japan

Abstract. In addition to Rabi, Ramsey, and spin echo experiments for a single qubit[1], we have successfully demonstrated the conditional operations with more complex pulse sequences of a qubit interacting with an LC harmonic oscillator. Rabi oscillations between |e,0> ↔ |g,1> (red sideband) and |g,0> ↔ |e,1> (blue sideband) were observed, where we noted coupled state as |qubit, LC-oscillator>. If we consider an LC-oscillator as a quantum information bus, this is a first step to demonstrate the Cirac-Zoller scheme in a field of macroscopic superconducting circuit.

Keywords: superconducting flux qubit, LC oscillator
PACS: 03.67.Mn, 74.78.Na, 85.25.Cp, 85.25.Dq

INTRODUCTION

The mesoscopic superconducting circuit containing Josephson junctions is one of the promising candidates as a quantum bit (qubit) which is not only an essential building block for quantum computation but also the unique system for exploring physics of macroscopic quantum coherence (MQC) [2].

After the initial breakthrough in the coherent manipulation of a single Josephson qubit [3], the next target has been controlling entanglement in the two-qubit system. Recently, several groups have reported successful entanglement manipulation in the directly coupled two qubit systems[4,5]. The alternative way is using a quantum bus in order to couple arbitrary pair of qubits on demand. This Cirac-Zoller type scheme[6] is commonly used in the trapped ion systems. Very recently, observation of strong coupling in all- solid state implementations of the cavity QED concept using circuit elements were independently reported by the groups from Delft and Yale [7,8].

EXPERIMENT

The sample is fabricated with standard electron beam lithography and aluminium shadow evaporation technique. In Fig. 1, SEM picture of the central part of the sample is shown. The qubit is an inner loop containing three Josephson junctions [9] shown in Fig. 1(b). The qubit is represented by energetically lowest two collective states of macroscopic number ($\sim 10^6$) of Cooper pairs which are linear combination [10] of clockwise and counter-clockwise persistent-current states. The loop just outside the qubit is the readout SQUID. This SQUID is shunted by on-chip capacitors to provide well defined electromagnetic environment and filters to reduce low frequency noise from the lead.

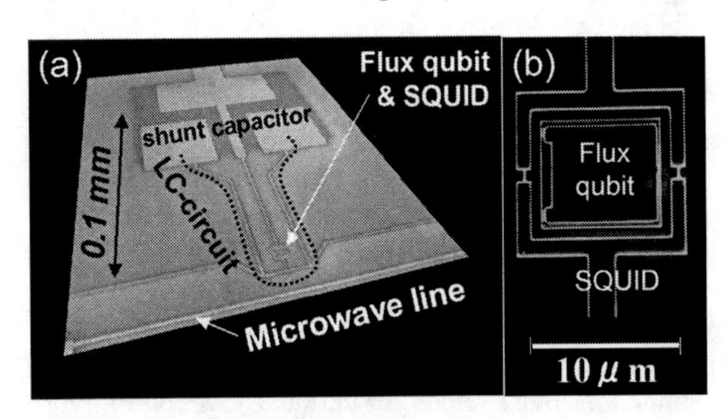

FIGURE 1. Scannning Electron Micrograph of the sample. (a) SEM picture of a flux qubit together with near-by on-chip elements. Note that the size of the LC-circuit made of shunt capacitor and lead inductance is an order of 0.1 mm large. (b) Close-up view of the central part of the device. A flux qubit with a dc-SQUID; a quantum detector of the qubit state.

CP850, *Low Temperature Physics: 24th International Conference on Low Temperature Physics;* edited by Y. Takano, S. P. Hershfield, S. O. Hill, P. J. Hirschfeld, and A. M. Goldman

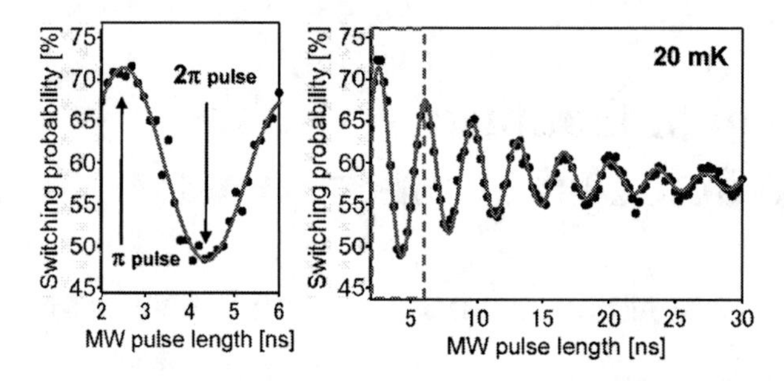

FIGURE 2. Rabi oscillations of the flux-qubit during resonant microwave pulse (ν_E =14 GHz), at temperature of 20 mK. The π- and 2π-pulse for qubit control are determined from Rabi oscillations. The left figure is the close-up of the initial part of the Rabi oscillations framed by the broken line.

The Hamiltonian of the system is composed of three parts; qubit : H_Q, LC-oscillator: H_{LC} and interaction: H_I.

$$\boldsymbol{H = H_Q + H_{LC} + H_I} \quad \textbf{(1)}$$

$$\boldsymbol{= h/2(\varepsilon\sigma_z + \Delta\sigma_x) + \hbar\omega_r(a^{+}a + 1/2) + h\lambda\sigma_z(a^{+} + a).} \quad \textbf{(2)}$$

Here, $\sigma_{z/x}$ are the Pauli spin matrices, $h\varepsilon=2I_P(\Phi_{ex}-\Phi_0/2)$ is the energy bias (I_P is the persistent current in the qubit, Φ_{ex} is the external flux penetrating the qubit loop, and $\Phi_0=h/2e$ is the flux quantum), and Δ is the qubit tunnel splitting. The qubit energy separation, controlled by the external flux, is $h\nu_E=h(\varepsilon^2+\Delta^2)^{1/2}$. The inductances of the leads (L~0.14nH) in series with the capacitors (C~10pF) constitutes an LC-circuit with superconducting plasma oscillation with resonance frequency $\omega_r=2\pi\nu_r=(LC)^{-1/2}$ (dashed line in Fig.1(a)). The qubit is coupled to the LC-oscillator via the mutual inductance M(~5.7pH) which bears the coupling constant $h\lambda$ =M I_P $(h\nu_r/2L)^{1/2}$. The obtained parameters of this sample were Δ=1.7 GHz, λ~0.2GHz, and the LC-resonance frequency ν_r=4.3 GHz.

The measurements were performed in a dilution refrigerator at T=20mK. During the experiment, we biased the qubit resonant frequency ν_E=14GHz in order to avoid unwanted thermal population of the excited state for both qubit and LC-oscillator. Using the π- and 2π-pulse for the qubit which were determined from Rabi oscillations (Fig. 2), we performed conditional spectroscopy as follows. First, a microwave pulse was applied to set the qubit in a well defined initial state, either a 2π-pulse to prepare the state |g,0>, or a π-pulse for the state |e,0>. Where the state of the combined system is noted as |qubit, LC-oscillator>. Immediately after the first pulse, a second microwave pulse, much longer than the qubit coherence time, was applied with variable frequency. Finally, by applying the dc-bias current pulse to the SQUID, the qubit state can be readout. In Fig. 3, the switching probability as a function of applied microwave frequency is plotted for the two different initial pulses. The results can be

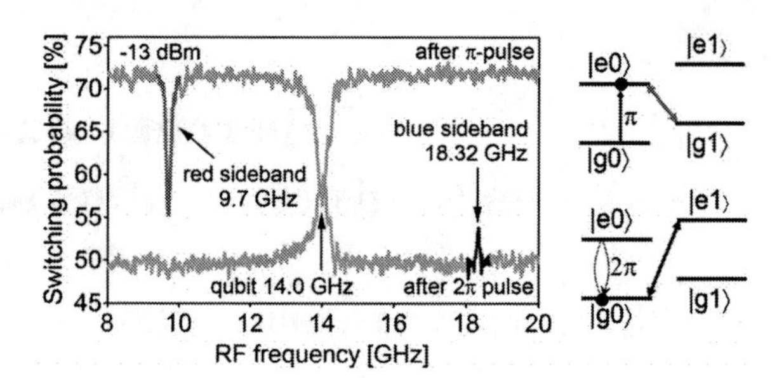

FIGURE 3. Conditional spectroscopy observed in the qubit-LC oscillator coupled system. The qubit resonant frequency is fixed at ν_E =14 GHz. The lower (upper) trace is taken when the qubit is in the ground (excited) state. Only the blue- (red-) sideband is observed except for the qubit transition which can be understood by the level scheme of the coupled system.

interpreted with schematic energy level diagrams shown in Fig. 3. If we prepare the state |e,0>, (|g,0>) by applying π-, (2π-)pulse to the qubit, the red-,(blue-) sideband transition is observed at E-ν_r, (E+ν_r) except for the qubit transition. This result means that the qubit state can be transferred and kept as the quantum state of the LC-oscillator by applying a single sideband pulse.

Further, we studied the dynamics of the coupled system by applying microwave pulses with frequency resonant with the blue or the red sideband transition. We successfully observed dumped sinusoidal Rabi oscillation for both red- and blue-sideband transition (not shown). The characteristic decay time of the sideband Rabi oscillation was faster ~10 ns compared with that of qubit Rabi oscillation ~20 ns (Fig. 2).

CONCLUSION

We have successfully demonstrated conditional operations via composite pulses for a qubit/LC-oscillator coupled system. Obtained coherence time of the coupled system have to be improved, however, this experiment showed that macroscopic superconducting LC-circuit can be a promising candidate for a quantum bus in the multi-qubit system.

REFERENCES

1. T. Kutsuzawa, *et al., Appl. Phys. Lett.* **87**, 073501 (2005).
2. A. J. Leggett, *J. Phys.: Condens. Matter* **14,** R415 (2002).
3. Y. Nakamura, *et al., Nature* **398**, 786 (1999).
4. T. Yamamoto, *et al., Nature* **425**, 941 (2003).
5. R. McDermott, *et al., Science* **307**, 1299 (2005).
6. J. I. Cirac and P. Zoller, *Phys. Rev. Lett.* **74**, 4091 (1995).
7. I. Chiorescu, *et al., Nature* **431**, 159 (2004).
8. A. Wallraff, *et al., Nature* **431**, 162 (2004).
9. J. E. Mooij, *et al., Science* **285**, 1036 (1999).
10. S. Saito, *et al., Phys. Rev. Lett.* **93**, 037001 (2004).

Manipulation And Readout Of A Tunable Flux Qubit With Integrated Readout: Preliminary Results

Carlo Cosmelli[a,b], Pasquale Carelli[c,b], Maria Gabriella Castellano[d,b], Fabio Chiarello[d,b], Lorenzo Gangemi[a], Roberto Leoni[d,b], Stefano Poletto[d], Daniela Simeone[d], and Guido Torrioli[d,b]

[a]*Dipartimento di Fisica, Universita' La Sapienza, p.le Aldo Moro 5, 00185 Roma, Italy*
[b]*Istituto Nazionale di Fisica Nucleare, p.le Aldo Moro 5, 00185 Roma, Italy*
[c]*Dipartimento di Ingegneria Elettrica, Universita' dell'Aquila, Monteluco di Roio, 67040 L'Aquila, Italy*
[d]*Istituto di Fotonica e Nanotecnologie, v. Cineto Romano 42, 00156 Roma, Italy*

Abstract. We show a tunable flux qubit with built-in readout, realized with a double SQUID with a supplementary Josephson junction. State preparation and manipulation of the qubit are achieved by applying pulses of magnetic flux in two externally coupled coils, a feature that suits very well with a future integration with RSFQ integrated logic. We show how the system can be read out and prepared in a definite flux state, moreover we show the results of manipulation (lowering of the potential barrier between states) in incoherent regime at liquid helium temperature.

Keywords: Superconductivity, quantum computing, SQUID.
PACS: 85.25.Cp; 03.67.Lx; 74.50.+r

INTRODUCTION

A superconducting qubit with computational states based on the magnetic flux can be implemented with a double-SQUID [1,2] (Superconducting QUantum Interference Device), i.e. a superconducting loop interrupted by a small dc-SQUID behaving as a tunable Josephson junction (see Fig. 1). Once biased at a magnetic flux of exactly half quantum, the system is represented by a symmetric double well potential, and it has two distinct flux states in correspondence of each potential well.

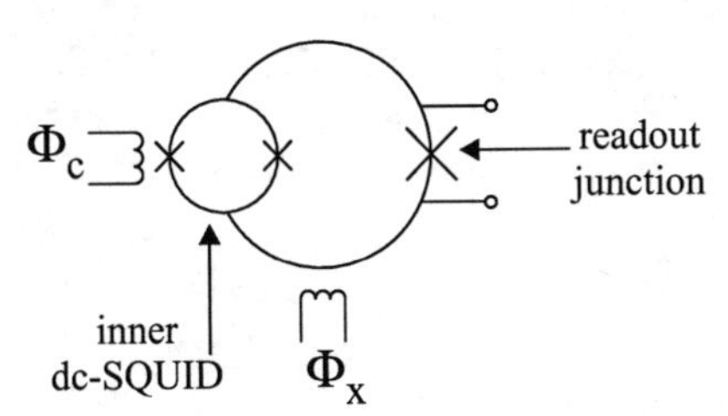

FIGURE 1. Schematic of the double-SQUID with built-in readout. The large readout junction can be biased by an external current through the proper terminals.

The device can be controlled by two magnetic fluxes, applied to the large loop (Φ_x) and to the inner dc-SQUID (Φ_c); the first acts on the tilt of the potential, the second on the barrier height between the two wells. Instead of reading the total flux in the ring by means of another SQUID magnetometer, here we present a built-in readout based on inserting a further Josephson junction in the loop. This readout junction is larger (a square of 10 μm side) than those of the dc-SQUID (3 μm side) and can be externally biased. The current circulating in the qubit ring adds or subtract to this bias current and may cause or not the switching of the junction to the voltage state: this is the signature that allows the discrimination between the two qubit states[3].

The sample presented here has been developed at IFN-CNR, Rome and fabricated by Hypres with a 100 A/cm^2 critical current density process. The inductance of the qubit main ring is 85 pH, while that of the inner dc-SQUID is 6 pH. The whole structure is gradiometric, although not reported in Fig. 1 for clarity.

CP850, *Low Temperature Physics: 24th International Conference on Low Temperature Physics;*
edited by Y. Takano, S. P. Hershfield, S. O. Hill, P. J. Hirschfeld, and A. M. Goldman

READOUT, STATE PREPARATION AND MANIPULATION

The readout is performed by sweeping the large junction with an external current bias and recording the current value I_b at which the junction jumps to the voltage state: this value is a function of the current circulating in the qubit loop. Since the switching process is stochastic, the procedure has to be repeated many times (100-1000) to get the probability distribution. In the following, it is understood that the mean value of the switching current distribution is a measure of the qubit state.

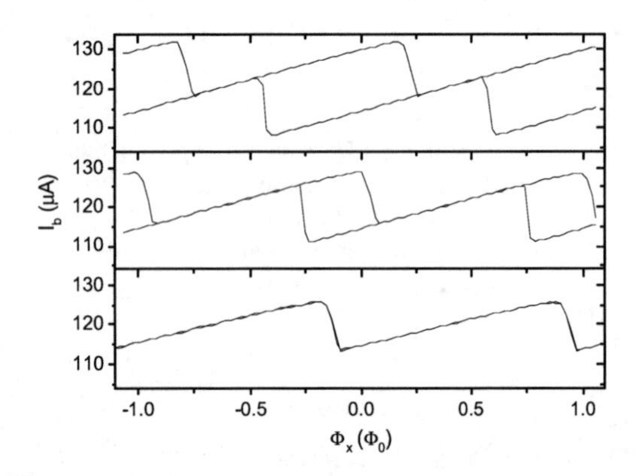

FIGURE 2. Switching current of the built-in junction (mean value of the distribution) as a function of the applied flux, once the system has been prepared in a defined flux state. Graphs from top to bottom correspond to a decreasing barrier height in the potential, controlled by means of Φ_c.

Figure 2. shows I_b versus the applied flux Φ_x for the two flux states and for three different values of Φ_c, that is for different barrier heights. The upper graph corresponds to the highest barrier, where the two flux states are well defined and hysteresis is maximum; by reducing the barrier, hysteresis decreases until it disappears (lowest graph). To prepare the system in a well defined state (say, the left well of the potential), we tilt the potential towards the left applying a pulse to Φ_x until only one well remains and the system relaxes in this state. By switching off the Φ_x pulse, the original symmetry of the potential is restored, with the state now placed in the left well. This procedure is necessary after each readout event, since the measurement procedure destroys the qubit state. After prepararing the system in a particular flux state, we lowered the barrier between the wells during a time *dt* and measured the probability of finding the system in the same or in the opposite state. The results are shown in Fig. 3 and are well fitted by exponential decay curves, as expected for an incoherent regime at 4.2 K. The different curves correspond, top-down, to lower values of the barrier and hence decreasing decay times. Finally the probability relaxes to 50%, as expected for a symmetric double well.

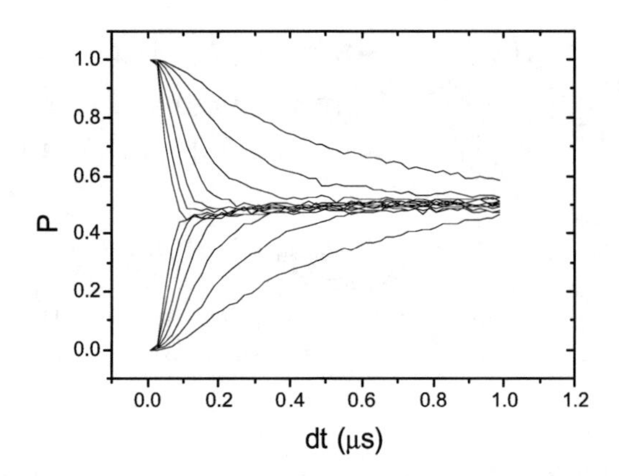

FIGURE 3. Probability of finding the system in the initial or in the opposite flux state, after lowering the barrier for a time *dt*. Different curves correspond to different values of the lowered barrier. The system relaxes incoherently to P=50%.

Work is in progress (in close collaboration with PTB, Braunschweig) to apply the control pulses for both Φ_x and Φ_c by using RSQF logic circuits, which can be built with the same technology as the qubit, provide fast pulses and can be located very close to the qubit. On the other hand, fast RSFQ pulses may induce non-adiabatic escapes of the qubit outside the computational states, while their shunting resistance may add decoherence to the qubit. It is therefore necessary the development of an appropriate interface between RSFQ and qubits, and possible modifications of both in order to achieve this aim [4].

This work was supported in part by the European Commission, contract FP6-502807 (RSFQubit), by INFN under project SQC, and by MIUR-FIRB under project "Nanotechnologies and nanodevices for the information society".

REFERENCES

1. S. H. Han, J. Lapointe and J. E. Lukens, *Phys. Rev. Lett.* **63**, 1712-1715 (1989).
2. J. R. Friedman, V. Patel, W. Chen, S. K. Tolpygo and J. E. Lukens, *Nature* **406**, 43-46 (2000).
3. M. G. Castellano *et al., IEEE Trans. Appl. Supercond.* **15**, 849-851 (2005).
4. F. Chiarello, P. Carelli, M. G. Castellano, C. Cosmelli M. Khabipov, R. Leoni, G. Torrioli, A. B. Zorin, ISEC'05 Extended Abstracts (2005).

Superconducting Cavity Resonator With A Metallic Tip For Realizing Strong Coupling Between Superconducting Qubits And Microwave Photons

H. Kitano[1,2], K. Ota[1], and A. Maeda[1,3]

[1]*Department of Basic Science, University of Tokyo, 3-8-1 Komaba, Meguro-ku, Tokyo, Japan*
[2]*PRESTO, JST, 4-1-8 Honcho Kawaguchi, Saitama, Japan*
[3]*CREST, JST, 4-1-8 Honcho Kawaguchi, Saitama, Japan*

Abstract. We propose that the insertion of a metallic tip into a superconducting cavity resonator can ensure the strong coupling condition between quantum bits (qubits) and microwave photons, which is a requirement for quantum information processing using cavity quantum electrodynamics (QED). By using a 3D electromagnetic numerical simulator, we estimated the strength of the local electric field in a cavity resonator, and found that it can be enhanced up to ~100 mV/m, which is 100 times larger than the value achieved in a Fabry-Perot resonator. As a possible application, we propose a cavity QED experiment using microwave photons and Josephson qubits realized in the intrinsic Josephson junction of cuprate high temperature superconductors.

Keywords: cavity quantum electrodynamics, Josephson phase qubit, high temperature superconductor.
PACS: 42.50.Pq, 03.67.Mn, 74.50.+r, 74.72.Hs

INTRODUCTION

The field of cavity quantum electrodynamics (cavity QED) opens up many intriguing possibilities for realization of entanglement between quantum bits (qubits) and photons, which are crucially important for quantum information processing [1]. One of the most important requirements for cavity QED experiments is the strong coupling between qubits and photons, where the strength of coherent qubit-photon interactions is much greater than the decay rates of qubits and photons. Recently, some cavity QED experiments have been performed both for an atomic system of Rydberg atoms in a Fabry-Perot resonator [2] and for a solid-state system of Josephson qubits embedded in a coplanar waveguide resonator [3]. In each case, the structure of the cavity resonator should be optimized according to the qubits used. In this paper, we present the cavity structure that is appropriate for the phase qubits of the intrinsic Josephson junction (IJJ) in cuprate high temperature superconductors (HTSCs).

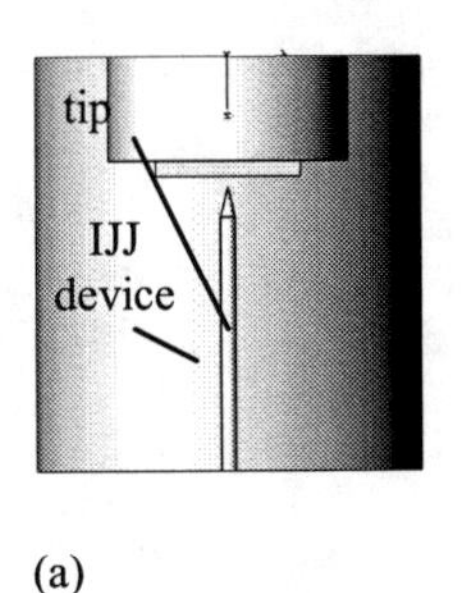

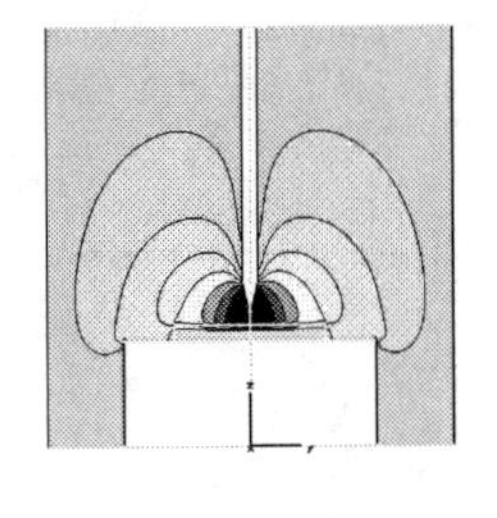

(a) (b)

FIGURE 1. (a) Schematic structure of proposed cavity resonator. (b) Numerical results of spatial distribution of the electric field into the cavity.

BASIC CONCEPTS

In the phase qubit based on the current-biased Josephson junction (JJ), which is one of three basic types of Josephson qubits [4], the transition frequency between |0> and |1> states is mainly determined by the Josephson plasma frequency, ω_p. In the IJJs of HTSCs, ω_p is much larger than that for conventional JJs, suggesting a potential advantage that the phase qubit can be operated at higher temperatures. In fact, a very recent experiment using the IJJs of a $Bi_2Sr_2CaCu_2O_y$ (BSCCO) superconductor suggested that macroscopic quantum tunneling (MQT) was observed up to ~0.8 K [5], which is one order of magnitude larger than the characteristic temperature in Nb JJs [6]. Thus, an IJJ device fabricated using HTSCs is expected to be a promising candidate for phase qubits.

Figure 1(a) shows a schematic view of a proposed reentrant cavity resonator. A metallic tip is mounted

CP850, *Low Temperature Physics: 24th International Conference on Low Temperature Physics;* edited by Y. Takano, S. P. Hershfield, S. O. Hill, P. J. Hirschfeld, and A. M. Goldman

on a center rod along a cylindrical axis, in order to enhance the electric field near the IJJ device. To satisfy the strong coupling condition, it is important to enhance the electric field per photon [3]. In our method, this can be realized by putting the tip very close to the IJJ device and by sharpening the tip.

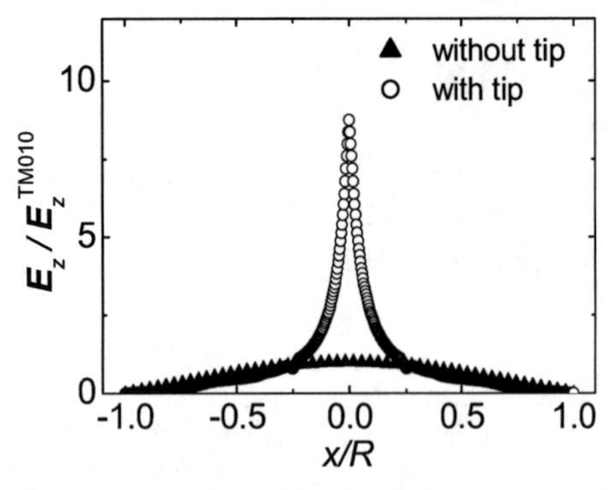

FIGURE 2. Distribution of the normal component of the electric field on the surface of the IJJ device. *R* is a radius of the cavity. *z* is perpendicular to the surface of the IJJ device.

NUMERICAL VERIFICATIONS

By using a 3D numerical simulator (Microwave Studio ver. 5, CST GmbH), we estimated the spatial distribution of the strength of the electric field within the cavity (Fig. 1(b)). We found that the z-component of the electric field on the surface of the IJJ device was strongly enhanced by the tip, as shown in Fig. 2. With the tip 0.1 mm above the surface of the device, the local electric field can be increased up to ~100 mV/m per photon. This value is 100 times larger than that for the Fabry-Perot resonator [2], and is similar to the value achieved for the coplanar waveguide resonator [3]. The quality factor which determines the decay rate of photons in the cavity can also be increased up to at least ~10^5, by using superconducting materials as the cavity wall. Resonant frequencies ranging from 10 GHz to 100GHz can be obtained by changing the cavity size. Moreover, the frequency can be tuned continuously by inserting a dielectric rod into the cavity, implying a great advantage for quantum nondemolition detection of qubits. Thus, it is strongly suggested that the proposed cavity resonator is a good candidate for cavity QED experiments with qubits realized in the IJJs of HTSCs.

EXPERIMENTAL APPROACH

The above numerical results for the distribution of the electric field can also be verified experimentally by using a bolometric technique [7], which is now in progress. Furthermore, toward a future cavity QED

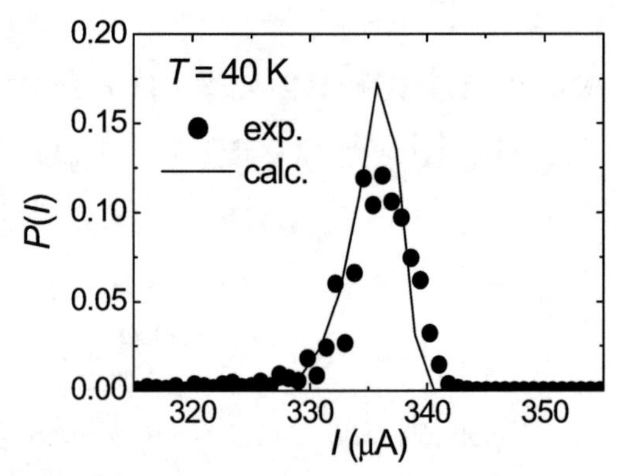

FIGURE 3. Switching current distribution of IJJs at *T*=40 K, which is normalized by the total event counts (2000 counts).

experiment, we have fabricated an IJJ device by etching a small mesa structure (typically 20×20 μm^2) on the top of BSCCO single crystals. Figure 3 shows the switching current distribution of the IJJ device, which is a fundamental measurement to observe the MQT [5,6]. Although we have not yet observed the MQT, we found that the observed switching distribution could successfully be explained by a thermal escape from the zero-voltage state of IJJs at least down to 40 K.

CONCLUSION

The IJJ of HTSCs is expected to a promising candidate for phase qubits. We proposed that the reentrant cavity resonator with a metallic tip is favorable for cavity QED experiments with such phase qubits. The strong coupling condition can be satisfied by putting the tip very close to the IJJ device.

REFERENCES

1. H. Mabuchi and A. C. Doherty, *Science* **298**, 1372-1377 (2002).
2. J. Raimond, M. Brune, and S. Haroche, *Rev. Mod. Phys.* **73**, 565-582 (2001).
3. A. Wallraff *et al.*, *Nature* **431**, 162-167 (2004).
4. M. H. Devoret, A. Wallraff, and J. M. Martinis, *preprint,* cond-mat/0411174.
5. K. Inomata *et al Phys. Rev. Lett.* 95, 107005_1-107005_4 (2005).
6. R. F. Voss and R. A. Webb, *Phys. Rev. Lett.* **47**, 265-268 (1981).
7. P. J. Turner *et al.*, *Rev. Sci. Instrum.* **75**, 124-135 (2004).

Quantum Nondemolition Measurement of a Superconducting Flux Qubit

Kohji Takashima*, Munehiro Nishida†, Shigemasa Matsuo**,† and Noriyuki Hatakenaka**,†

*Ishikawajima-Harima Heavy Industries Co., Ltd., Tokyo 100-8182, Japan.
†AdSM, Hiroshima University, Higashi-Hiroshima, 739-8530, Japan.
**Faculty of Integrated Arts and Science, Hiroshima University, Higashi-Hiroshima, 730-8521, Japan.

Abstract. Quantum nondemolition (QND) measurements allow us to measure an observable of a quantum system without introducing a back-action on this observable due to the measurement itself. Here we propose a method for the QND measurement of a superconducting flux qubit by extending the spin QND measurement. Under an adequate condition, qubit and interaction Hamiltonian satisfy QND conditions and then QND measurement of a superconducting flux qubit is possible.

Keywords: superconductivity, qubit, QND
PACS: 74.50.+r, 03.75.Lm, 03. 67.Lx

INTRODUCTION

When we apply an adequate external magnetic flux through the superconducting loop in an rf-SQUID (Superconducting Quantum Interference Device) at low temperatures, some energy level degenerescences may be lifted. Since the quantum two-level system is utilized as a building block of quantum computers,

superconducting flux quantum bit (qubit) is one promising candidate. In fact, superposition of quantum states associated with different values of the flux was observed [1, 2, 3, 4].

So far the quantum state of superconducting flux qubit is measured by destructive methods. If a measurement of the qubit is carried out in a nondestructive way, it is an advantage, for example, for the confirmation of the initial qubit state before the quantum calculation is started.

We propose a method for the quantum nondemolition (QND) measurement of the superconducting flux qubit by extending the spin QND measurement [5], which is different in the representation from the previous work [6].

QUBIT IN SPIN SPACE

The potential of rf-SQUID that has a Josephson junction in the superconducting loop is described as follows:

$$V = \frac{1}{2L}(\Phi - \Phi_{ext})^2 - \frac{I_c\Phi_0}{2\pi}\cos(\frac{2\pi\Phi}{\Phi_0}), \qquad (1)$$

where L is the inductance of the superconducting loop, Φ is the magnetic flux, Φ_{ext} is the external magnetic flux, I_c is the critical current, and Φ_0 is the flux quantum. The first and second terms stand for magnetic energy and the Josephson coupling energy, respectively. The potential is regarded as a double well structure when an externally applied magnetic flux is near half quantum unit of magnetic flux. At low temperatures, the flux potential involves two energy levels with the bonding state (the ground state) and the anti-bonding state (the excited state) by the tunnel effect.

The effective Hamiltonian of the two-level system is expressed by *two* components of spin in the spin-1/2 representation using Pauli's spin operator σ_x and σ_z:

$$H = E\sigma_z + \Delta\sigma_x, \qquad (2)$$

where $E = I_p(\Phi_{ext} - \frac{1}{2}\Phi_0)$, I_p is the qubit maximum persistent current, and Δ is the tunnel matrix element.

QND MEASUREMENT SCHEME

In spin QND measurement scheme, a QND variable is *one* component of spin, say z component σ_z^s, while the corresponding QND interaction Hamiltonian has the form of $\sigma_z^s\sigma_z^p$ where s and p stand for system and probe, respectively. On the other hand, the qubit is expressed by *two* components of spin as mentioned above.

For making the QND measurement of qubit possible, the spin operators for target qubit in the system Hamiltonian and interaction Hamiltonian must be mapped onto one component of spin by the rotation in spin space. Therefore, an appropriate modification is required for the interaction Hamiltonian. Namely, our superconducting flux qubit is then reformed by an rf-SQUID that contains a 'dc-SQUID-type tunable junction' for spin rotation, while a dc-SQUID magnetometer or a fluxon which

CP850, *Low Temperature Physics: 24th International Conference on Low Temperature Physics;*
edited by Y. Takano, S. P. Hershfield, S. O. Hill, P. J. Hirschfeld, and A. M. Goldman

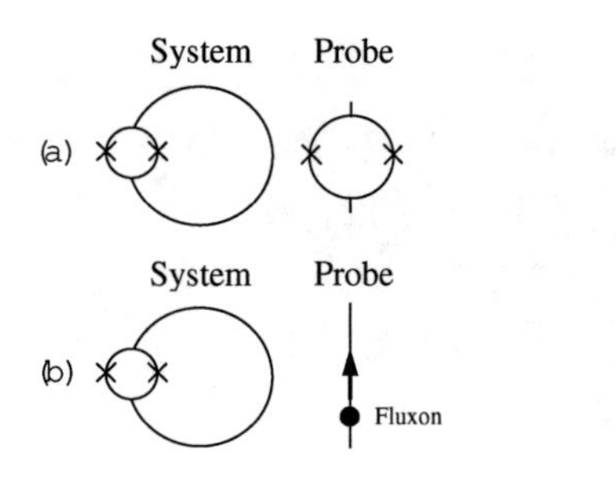

FIGURE 1. Schematic diagrams of a system and a probe

moves near the qubit as shown in Fig. 1 are considered as a probe system.

First, the effective Hamiltonian of the system H_s is expressed as the form

$$H_s = E\sigma_z^s + \Delta\sigma_x^s = \begin{pmatrix} E & \Delta \\ \Delta & -E \end{pmatrix}. \quad (3)$$

By the rotation in spin space, H_s is represented by *one* component of spin along z' axis as follows;

$$H_s = E'\sigma_{z'}^s, \quad (4)$$

where $E' = \sqrt{E^2+\Delta^2}$ and the angle between z' and z axes is $\theta' = \arctan(\Delta/E)$.

Secondly, we assume that the effective interaction Hamiltonian is given as

$$H_{int} = (g_z\sigma_z^s + g_x\sigma_x^s)\cdot\sigma_z^p, \quad (5)$$

where g_z and g_x are coupling constants. By the rotation in spin space as the same way, H_{int} is expressed as

$$H_{int} = g\sigma_{z''}^s\cdot\sigma_z^p, \quad (6)$$

where $g = \sqrt{g_z^2+g_x^2}$ and the angle between z'' and z axes is $\theta'' = \arctan(g_x/g_z)$.

TABLE 1. System, probe and interaction

	Subject to be considered	Effective Hamiltonian
System	Qubit of an rf-SQUID containing a dc-SQUID-type tunable junction	H_s $= E\sigma_z^s + \Delta\sigma_x^s$ $= E'\sigma_{z'}^s$ by rotation
Probe	Flux of dc-SQUID magnetometer or flux of a moving fluxon near qubit	Function of σ_z^p and σ_x^p
Interaction	The flux-flux interaction between system and probe	H_{int} $= (g_z\sigma_z^s + g_x\sigma_x^s)\cdot\sigma_z^p$ $= g\sigma_{z''}^s\cdot\sigma_z^p$ by rotation

In order to carry out QND measurement of spin, the following four conditions are required [5];

1. $[S_z, H_s] = 0$
2. $[S_z, H_{int}] = 0$
3. H_{int} is a function of S_z
4. $[J_x, H_{int}] \neq 0$

where S_z is the QND variable, and J_x is the probe variable which is used for measurement. The dynamics of J_x tells us the information of S_z.

In our system, QND variable S_z and a probe variable J_x correspond to $\sigma_{z'}^s$ and σ_x^p, respectively. Now let us examine the QND conditions;

1. $[\sigma_{z'}^s, H_s] = [\sigma_{z'}^s, E'\sigma_{z'}^s] = 0$
2. $[\sigma_{z'}^s, H_{int}] = [\sigma_{z'}^s, g\sigma_{z''}^s\cdot\sigma_z^p] = 0$ if $\sigma_{z'}^s \propto \sigma_{z''}^s$
3. $H_{int}(= g\sigma_{z''}^s\cdot\sigma_z^p)$ is the function of $\sigma_{z'}^s$ if $\sigma_{z'}^s \propto \sigma_{z''}^s$, i.e., $z' \parallel z''$
4. $[\sigma_x^p, H_{int}] = [\sigma_x^p, g\sigma_{z''}^s\cdot\sigma_z^p] \neq 0$

Therefore, QND measurement is required for the relation $\Delta/E = g_x/g_z$ in our scheme. In other words, the system Hamiltonian and the interaction Hamiltonian are mapped onto one component of spin by the rotation in spin space if $\Delta/E = g_x/g_z$. Even with this condition, the conjugate variable of $\sigma_{z'}^s$ will dephase any superposition of the two qubit's eigenstates. The relation might be established by designing SQUID configurations for tuning an external flux of the dc-SQUID-type tunable junction in the rf-SQUID via the modification of the Josephson coupling energy of the rf-SQUID.

CONCLUSION

Under an adequate condition, qubit and interaction Hamiltonian satisfy QND conditions and then QND measurement of a superconducting flux qubit is possible. Our scheme can be applied to generation of entangled states, i.e., QND measurement-induced entanglement found in quantum optics [7].

ACKNOWLEDGEMENT

This work was supported by a research grant from The Mazda Foundation.

REFERENCES

1. J. E. Mooij et. al., *Science* **285**, 1036 (1999).
2. J. R. Friedman et. al., *Nature* **406**, 43 (2000).
3. C. H.van der Wal et. al., *Science* **290**, 773 (2000).
4. I. Chiorescu et. al., *Science* **299**, 1869 (2003).
5. Y. Takahashi et. al., *Phys. Rev. A* **60**, 4974 (1999).
6. D. V. Averin, *Phys. Rev. Lett.* **88**, 207901 (2002).
7. D. B. Horoshko, et. al., *Phys. Rev. A* **61**, 032304 (2000).

Coherent Tunneling of Cooper Pairs in Asymmetric Single-Cooper-Pair Transistors

Juha Leppäkangas and Erkki Thuneberg

Department of Physical Sciences, P.O.Box 3000, FI-90014 University of Oulu, Finland

Abstract. We have calculated the $I-V$ characteristics of voltage biased asymmetric single-Cooper-pair transistors (SCPT), resulting from coherent Cooper pair tunneling across both the Josephson junctions (JJ), weak dissipation due the electromagnetic environment (EE) satisfying $\mathrm{Re}[Z(\omega)] \ll R_Q$, and quasiparticles. Due to the asymmetry, the smaller JJ is effectively probing the macroscopic quantum states of the island. A resonance occurs whenever the energy released in a tunneling of a single, or several, Cooper pair(s) across the smaller JJ matches to the energy needed to excite the island.

Keywords: Josephson junction, Single Cooper pair transistor, Cooper pair box, Coherent tunneling
PACS: 74.40.+r, 73.23

INTRODUCTION

Coherent tunneling of Cooper pairs across voltage biased SCPTs, disturbed by a dissipative environment, can lead to various phenomena [1]. A lot of research has been focused on charging effects of symmetric SCPTs, but also the $I-V$ characteristics across highly *asymmetric* SCPTs have been measured [2, 3], in order to probe the quantum states of mesoscopic JJs. In this paper we analyze the current across asymmetric SCPTs due to coherent Cooper pair tunneling, perturbed by an electromagnetic environment satisfying $\mathrm{Re}[Z(\omega)] \ll R_Q$, and quasiparticles. Instead to what has been done before, we assume a quantum coherence across both of the JJs and focus on the parameter range $E_{J1} \gg E_{J2} \sim E_c$ and the undergap region $V < 2\Delta_{\mathrm{gap}}/e$.

THE MODEL

We start the analysis from the Hamiltonian of the voltage biased SCPT, which is usually written as [1]

$$H_{\mathrm{SCPT}} = \frac{(Q_1 - Q_2 + Q_0')^2}{2C_\Sigma} - \frac{1}{2}(Q_1 + Q_2)V - E_{J1}\cos(\varphi_1) - E_{J2}\cos(\varphi_2). \quad (1)$$

The first term is the charging energy of the island, characterized by the energy $E_c = e^2/2C_\Sigma$. The second term gives the energy fed by the voltage source and the last two terms describe the Josephson currents across the two JJs. $Q_0' = C_gU + (C_1 - C_2)V/2$ is the polarisation charge due to the applied voltage V and a gate voltage U, $C_\Sigma = C_1 + C_2 + C_g$ where C_i is the capacitance of the i:th JJ, and C_g is the gate capacitance. The phase difference φ_i and the charge (gone through the junction i) Q_i are canonically conjugated variables.

For the case of asymmetric Josephson junctions ($E_{J1} \gg E_{J2}$), it is convenient to do a linear change of variables such that $Q = Q_1 - Q_2$, $Q_\Sigma = Q_2$, $\varphi = \varphi_1$ and $\varphi_\Sigma = \varphi_1 + \varphi_2$. The Hamiltonian (1) is now

$$H_{\mathrm{SCPT}} = \frac{(Q + Q_0)^2}{2C_\Sigma} - E_{J1}\cos(\varphi) - VQ_\Sigma - E_{J2}\cos(\varphi_\Sigma - \varphi). \quad (2)$$

Physically Q is the island charge, Q_Σ the charge tunneled across the small JJ (probe) and $Q_0 = C_gU - (C_2 + C_g/2)V$ is the new polarisation charge. We see that the first two terms describe the Hamiltonian of the Cooper pair box [3] (CPB), whereas the third term describes the charge gone through the probe. The last operator mixes these two subsystems weakly.

An EE satisfying $\mathrm{Re}[Z(\omega)] \ll R_Q$ perturbs the system by small voltage fluctuations V_f across the system, which can be included by a transformation $V \to V + V_f$ in the Hamiltonian (2). Also quasiparticles can tunnel, if an energy $2\Delta_{\mathrm{gap}}$ is released in a proper relaxation process. We calculate the resulting current due to these perturbations by using the golden rule, similarly as in Ref. [1].

RESULTS

At first, we neglect the quasiparticles, and analyze the effect of the dissipative EE. Typical $I-V$ characteristics obtained by numerical calculations are shown in Fig. 1. The nonresonant current, decreasing as a function of V, and the positions and the widths of the resonant peaks can be understood by analysing the eigenstates of the Hamiltonian (2) for $E_{J2} = 0$. These are $|\alpha, Q_0\rangle|n\rangle$, with the eigenenergies $E_{\alpha,Q_0,n} = E_{\alpha,Q_0} - 2eVn$, where $|\alpha, Q_0\rangle$ is the eigenstate of the CPB Hamiltonian with the quasicharge Q_0, eigenenergy E_{α,Q_0} and index $\alpha = 0,1,2\ldots$,

CP850, *Low Temperature Physics: 24th International Conference on Low Temperature Physics;*
edited by Y. Takano, S. P. Hershfield, S. O. Hill, P. J. Hirschfeld, and A. M. Goldman

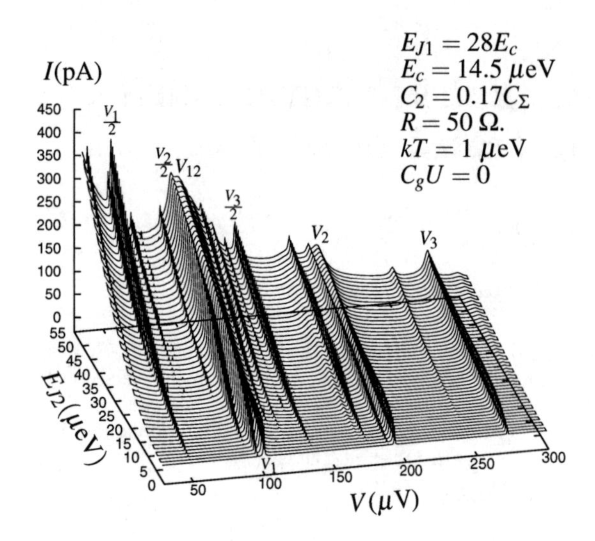

FIGURE 1. $I-V$ curves for different values of E_{J2}. No quasiparticles are taken into account. The first order resonances betwen the CPB ground $|0,Q_0\rangle$ and its excited states $|\alpha,Q_0\rangle$ are seen as current peaks at the voltages $V_1 \approx 100\ \mu$V, $V_2 \approx 190\ \mu$V and $V_3 \approx 270\ \mu$V. The resonance $|1\rangle \leftrightarrow |2\rangle$ (at V_{12}) is also seen nearby the voltage V_1. Higher order resonances (for example at $V_1/2, V_2/2\ V_3/2$ etc.) become stronger for larger values of E_{J2}.

n tells the number of Cooper pairs tunneled across the probe. For small values of E_{J2} these states are very close to the correct eigenstates, expect for the degenerate situations $E_{\alpha,Q_0,n} \approx E_{\beta,Q_0,m}$, when the degenerate states (with the same quasicharge) are mixed srongly.

By taking the last term in the Hamiltonian (2) into account only perturbatively, one obtains for the nonresonant current an approximative result $I \sim 2(eE_{J2}/\hbar)^2\mathrm{Re}[Z(2eV/\hbar)]/V$. A resonant current can flow in a situation, when the CPB ground and its excited state are mixed strongly, i. e. when $E_{0,Q_0,0} \approx E_{\alpha,Q_0,N}$. Resonances between two excited states do not usually produce $I-V$ peaks, since populations of them are small. Exceptions are the situations, when the ground state and two excited states are simultaneously in resonance.

By using a two-state approximation for a resonant situation between the states $|0,Q_0,0\rangle$ and $|\alpha,Q_0,1\rangle$ (a first order resonance) one gets the energy level splitting

$$\Delta E_{0,\alpha,Q_0} = \sqrt{(E_{J2}c_{0,\alpha,Q_0})^2 + (E_{\alpha,Q_0} - 2eV)^2}, \qquad (3)$$

where $c_{0,\alpha,Q_0} = |\langle\alpha,Q_0|\exp(i\varphi)|0,Q_0\rangle|$ and we have set that $E_{0,Q_0} = 0$. The strong mixing of states $|0,Q_0,0\rangle$ and $|\alpha,Q_0,1\rangle$ occurs when the first term inside the square root of eq. (3) is dominant. The corresponding $I-V$ peak has then a width $\Delta V \approx c_{0,\alpha,Q_0}E_{J2}/2e$. Higher order resonant situations ($N > 1$) will also produce $I-V$ peaks but with much smaller widths, since the states are usually connected by terms $\propto E_{J2}^N$.

A physical picture of a resonant situation is the following. Single, or several ($= N$), Cooper pairs tunnel *coherently* back and forth across the probe, and the CPB jumps simultaneously between the states $|0,Q_0\rangle$ and $|\alpha,Q_0\rangle$ cancelling the energy gain. Due to dissipative environment, this coherent process is from time to time interupted by an incoherent tunneling from the state $|\alpha,Q_0\rangle$ to some other state $|\beta,Q_0\rangle$. For $E_{J1} \gg E_c$ the fastest transition is $|\alpha,Q_0\rangle \to |\alpha-1,Q_0\rangle$, which then relaxes to $|\alpha-2,Q_0\rangle$ and so on, until the system is again in the ground state, expect that N Cooper pairs has tunneled across the probe. The slowest is the $|1,Q_0\rangle \to |0,Q_0\rangle$ transition, and therefore the maximum current is of the same order for every same order resonance. Furthermore, $I-V$ areas $A_{0\leftrightarrow\alpha,Q_0}$ of the first order resonances satisfy $A_{0\leftrightarrow\alpha,Q_0}/A_{0\leftrightarrow 1,Q_0} \approx c_{0,\alpha,Q_0}/c_{0,1,Q_0}$, i. e. the ratio of the linewidths. For comparison, if the tunneling across the probe is only incoherent [2], one gets that the areas $\propto (c_{0,\alpha,Q_0}/c_{0,1,Q_0})^2$ drop faster with increasing α, and the linewidths increase $\propto \alpha$.

A new channel for the charge transport can open, whenever the energy $2\Delta_{\mathrm{gap}}$ needed for a quasiparticle to tunnel, is released in a process. A quasiparticle tunneling across the probe (to the positive direction) is associated with an energy release eV and a quasicharge change $Q_0 \to Q_0 - e$, whereas the tunneling across the larger junction changes only Q_0 to $Q_0 + e$. A simultaneous quasiparticle tunneling across the probe and a transition $|\alpha,Q_0\rangle \to |\beta,Q_0 - e\rangle$ is possible if $\delta E = E_{\alpha,Q_0} - E_{\beta,Q_0-e} + eV > 2\Delta_{\mathrm{gap}}$ and occurs with a rate $\Gamma^{qp}_{\alpha\to\beta} \approx \delta E|\langle\beta,Q_0 - e|\exp(-i\varphi/2)|\alpha,Q_0\rangle|^2/e^2R_2$, where R_2 is the normal state resistance of the probe. From this it follows, that the current in the resonant situations above $V > 2\Delta_{\mathrm{gap}}/3e$ can be strongly enhanced due to quasiparticle tunneling.

In conclusion, Cooper pair tunneling across an asymmetric SCPT perturbed weakly by a dissipative EE, leads to resonant current peaks whenever the energy released in the tunneling of a single, or several, Cooper pair(s) across the probe equals the energy needed to excited the equivalent CPB circuit. Compared with the case of only incoherent tunneling across the probe, stronger resonances due to the higher excited states of the CPB are obtained. Quasiparticle tunneling can increase the tunneling rates in resonant situations above $V > 2\Delta_{\mathrm{gap}}/3e$.

REFERENCES

1. A. M. v. d. Brink, A. A. Odintsov, P. A. Bobbert, and G. Schön, *Z. Phys. B* **85**, 459 (1991).
2. R. Lindell, J. Penttilä, M. Sillanpää, and P. Hakonen, *Phys. Rev. B* **68**, 052506 (2003).
3. Y. Nakamura, Yu. A. Pashkin, and J. S. Tsai, *Nature* **398**, 786 (1999).

Tuning the 2D Superconductor-Insulator Transition by Use of the Electric Field Effect

Kevin A. Parendo, K. H. Sarwa B. Tan, A. Bhattacharya, M. Eblen-Zayas, N. Staley, and A. M. Goldman

School of Physics and Astronomy, University of Minnesota, Minneapolis, MN, 55455

Abstract. Some investigations of the superconductor-insulator (SI) transitions in two dimensions have been hindered by aspects of the intrinsic disorder of the studied systems. As a solution to this problem, we have induced superconductivity in insulating, ultrathin films of amorphous bismuth by utilization of the electric field effect. This method of tuning the SI transition does not alter the intrinsic disorder. Analysis of the response to transferred charge density has revealed that screening and the density of states are both involved. This SI transition has been analyzed as a quantum phase transition using a finite size scaling analysis with electron concentration as a tuning parameter, yielding a critical exponent product $\nu z = 0.7 \pm 0.05$. If $z = 1$ as expected, this product is consistent with the universality classes of the (2D+1) XY model and the 2D Boson Hubbard model in the absence of disorder.

Keywords: superconductivity, SIT, electrostatic, FET, 2D
PACS: 74.40.+k, 74.78.Db, 71.30.+h, 72.15.Rn, 74.81-g

INTRODUCTION

Superconductor-insulator (SI) transitions in ultrathin films are believed to be quantum phase transitions. Typical methods of tuning SI transitions involve increasing film thickness or applying perpendicular magnetic fields [1,2] and often yield results that are consistent with percolation. Here, we tune a SI transition by adding charge carriers to an ultrathin film by use of the electric-field effect. Analysis of the data in terms of the transferred electron concentration yields results suggestive that both the density of states as well as screening may be changing as superconductivity is induced and that the transition is not percolative, but is indeed a quantum phase transition.

EXPERIMENTAL

A series of insulating ultrathin films of homogeneously-disordered, amorphous bismuth was deposited in an electric-field effect geometry as described elsewhere [3]. An electric field was applied perpendicular to the film surface across a 45 μm layer of single crystal $SrTiO_3$, using a Pt film as a gate electrode. The thinned substrate served as the gate insulator. As a function of positive gate voltage, an increasing electron concentration was transferred into the film, eventually inducing superconductivity. The transferred electron concentration, n, as a function of gate voltage, was determined using a previous calibration of charge transfer in $SrTiO_3$ performed at T ≈ 1 K [4] together with measurements of the gate insulator thicknesses.

DISCUSSION

In Fig. 1, we show the SI transition of a 10.22 Å thick film as a function of increasing electron concentration. At zero transferred charge, the film at this thickness, as well as thinner films, exhibit Mott variable range hopping conduction, as measured between 60 mK and 10 K. It is believed that this film does not cool below 60 mK, although the dilution refrigerator does. At $n = 3.4 \cdot 10^{13}$ cm^{-2}, superconductivity is induced with $T_c \approx 60$ mK.

Roughly at the concentration when dR/dT first becomes positive at T ≈ 60 mK, the best fit to the conductivity, G, is a logarithmic dependence on temperature. This dependence may be due to the effects of weak localization and electron-electron interactions, despite the high value of normal state resistance, through the relation [5]

CP850, *Low Temperature Physics: 24th International Conference on Low Temperature Physics;*
edited by Y. Takano, S. P. Hershfield, S. O. Hill, P. J. Hirschfeld, and A. M. Goldman

$$G(T) = [\alpha p + (1 - \frac{3F^*}{4})]\frac{e^2}{2\pi^2\hbar}\ln(T). \quad (1)$$

If all changes due to electron doping are assumed to only alter the electron-electron interactions (αp is assumed constant), the screening parameter, F^*, is found to change linearly for *all* electron concentrations. This is illustrated in the inset of Fig. 1. At T > 300 mK, i.e. in the normal state for all superconducting curves, the slope of dG/dn does not change appreciably until n is approximately $1.5 \bullet 10^{13}$ cm^{-2}. dG/dn is proportional to the field effect mobility, which may only depend on the value of the density of states at the Fermi energy. Both screening and changes in the density of states thus appear to affect the development of superconductivity. This is not unreasonable, as the simple BCS formula for the transition temperature involves both the density of states and the electron-electron attractive interaction, which competes with Coulomb repulsion and is thus sensitive to screening.

It is believed that SI transitions are quantum phase transitions (QPTs), i.e. transitions between ground states at zero temperature that are tuned by changing some non-thermal variable in the system's Hamiltonian. Traditionally, analysis of perpendicular magnetic field- and thickness-tuned SI transitions have yielded critical exponents consistent with percolation, a result suggestive of the transitions being strongly influenced by film morphology. However, one example of perpendicular magnetic field tuning in bismuth films yielded results consistent with an XY or Boson Hubbard model in the absence of disorder, suggestive that film morphology was not as important in this case [2].

We have analyzed this electrostatic-tuned SI transition taking the transferred electron concentration as tuning parameter. In the inset to Fig. 2, R vs. n for isotherms between 60 mK and 140 mK show a well defined crossing point, yielding values of critical resistance, R_c, and density, n_c. In Fig. 2, values of R/R_c in the same temperature range are plotted against the scaling function $|(n-n_c)|\ T^{-1/\nu z}$ at the value $\nu z = 0.7$, which collapses the data. The uncertainty in this product is approximately 0.05. This result is consistent with either the (2D + 1) XY model or the Boson Hubbard model in the absence of disorder, but is not consistent with percolation.

Future investigations will involve lowering the high T resistance by increasing film thickness.

This work was supported by the NSF under grant DMR-0455121.

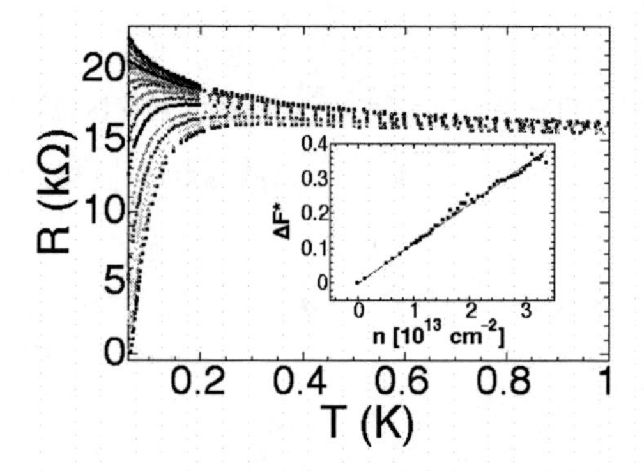

FIGURE 1. Electrostatically induced superconductivity. Top curve corresponds to zero added electron concentration. Bottom curve corresponds to $n = 3.4 \bullet 10^{13}$ cm^{-2}. Inset: Increase in F*, relative to unbiased film, as function of n.

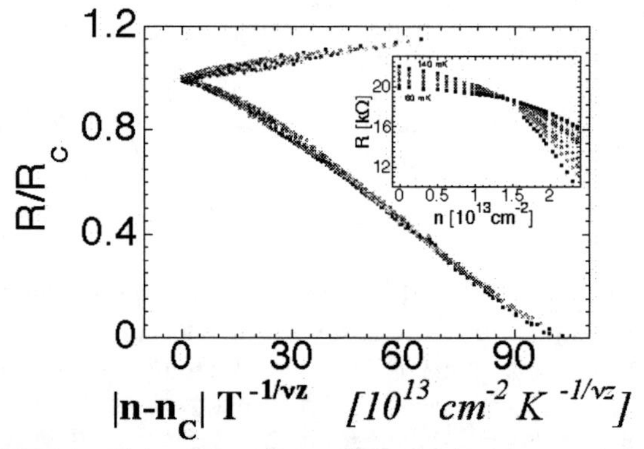

FIGURE 2. Finite size scaling collapse of data for $\nu z = 0.7$ from T = 60 mK to 140 mK, with induced electron density as tuning parameter. Inset: R vs. B for isotherms in same temperature range, yielding a distinct crossing point.

REFERENCES

1. D. B. Haviland, Y. Liu, and A. M. Goldman, *Phys. Rev. Lett.* **62**, 2180-2183 (1989); A. F. Hebard and M. A. Paalanen, *Phys. Rev. Lett.* **65**, 927-930 (1990); A. Yazdani and A. Kapitulnik, *Phys. Rev. Lett.* **74**, 3037-3040 (1995).
2. N. Markovic, et al., *Phys. Rev. B* **60**, 4320-4328 (1999).
3. Kevin A. Parendo, K. H. Sarwa B. Tan, A. Bhattacharya, M. Eblen-Zayas, N. E. Staley, and A. M. Goldman, *Phys. Rev. Lett.* **94**, 197004 (2005).
4. A. Bhattacharya et al., *Appl. Phys. Lett.* **85**, 997-999 (2004).
5. E. Abrahams, P. W. Anderson, D. C. Licciardello, and T. V. Ramakrishnan, *Phys. Rev. Lett.* **42**, 673-676 (1979); B. L. Altschuler and A. G. Aronov, *Solid State Commun.* **46**, 429 (1983).

Anomalous Insulating State Induced By Parallel Magnetic Field In Ultrathin Bismuth Films

K. H. Sarwa B. Tan, Kevin A. Parendo, and A. M. Goldman

School of Physics and Astronomy, University of Minnesota, Minneapolis, MN, 55455, USA

Abstract. Recent investigations of the magnetic-field induced insulator in two-dimensional superconducting systems suggest that unexpected physics occurs. However, direct comparisons with the intrinsic insulator have not been possible, nor have most experiments studied the effects of parallel magnetic fields. We have applied parallel magnetic fields to an insulating, amorphous Bi film, which exhibited Mott variable range hopping in the absence of field. We have also applied fields to an insulating film in which superconductivity was induced by electrostatic electron doping. Comparison of these magnetic-field induced insulators with the intrinsic insulator reveals the appearance of an anomalous insulating phase at low temperatures.

Keywords: superconductor, insulator, parallel magnetic field
PACS: 74.40.+k, 74.78.Db, 71.30.+h, 72.15.Rn, 74.81-g

INTRODUCTION

Recently, various observations have been made of a large magnetoresistance peak in thin superconducting films that have been driven insulating by application of high magnetic fields. Most studies have employed the use of perpendicular magnetic fields [1-5], although one experiment has also been carried out with the use of a parallel magnetic field [6]. The nature of this insulating peak is the subject of some discussion. We present investigations of the effects of high parallel magnetic fields on ultrathin bismuth films that have had superconductivity induced by electrostatic electron doping. The results are compared directly with the undoped, insulating state, revealing two interesting features. First, when superconductivity is destroyed by magnetic field, the resistance becomes higher than in the undoped, insulating state. Second, in high field, R(T) at high temperature has the same hopping form as the undoped state, while at lower temperatures, the resistance becomes unexpectedly higher. In these experiments, the magnetoresistance did not show evidence of a peak at high fields.

EXPERIMENTAL

A series of homogeneously disordered, insulating, ultrathin films of amorphous bismuth was prepared in a geometry in which the substrate served as a gate insulator. Details of the preparation of the field effect device, as well as the superconductivity produced by electric field effect induced electron doping have been previously published [7]. The temperature dependence of the resistance of the undoped insulating films was consistent with Mott variable range hopping (VRH), exhibiting a best fit to $R = R_0 \exp[(T_0/T)^{1/3}]$ from 60 mK to 10 K. Superconductivity was induced in this film by applying a gate voltage of 49 V. Gated films will be referred to here as doped. This voltage was sufficient to induce maximal electron charge transfer into the film, in which the best fit to the resistance in the normal state was that of logarithmic temperature dependence. Magnetic fields of up to 11 T were then applied nominally parallel to the film, both in the doped and undoped states.

DISCUSSION

In low fields, both the doped and undoped films responded to parallel magnetic fields with a regime of negative magnetoresistance, with the same systematics as previously described [8]. In zero magnetic field, the doped film has $T_c \approx 62.5$ mK and a peak in R(T) at about 300 mK, where superconducting fluctuations cease and normal transport begins. In fields from 0.1 to roughly 1.5 T, R(T) in field between 80 mK and 250 mK is smaller than the zero field state. The undoped

CP850, *Low Temperature Physics: 24th International Conference on Low Temperature Physics;*
edited by Y. Takano, S. P. Hershfield, S. O. Hill, P. J. Hirschfeld, and A. M. Goldman

film shows a negative magnetoresistance in the same field range for temperatures between 60 mK and 150 mK. Bismuth is a heavy element and thus electrons in it experience strong spin-orbit scattering. Since each scattering event leads to a spin randomization, it is unlikely that spin polarization develops at such low fields.

In fields larger than approximately 4.5 T, the doped film appears to be insulating (dR/dT < 0 in the low temperature limit). In intermediate fields of 4.5 - 8 T, R(T) is neither a logarithmic or exponential function of temperature. In fields larger than 8T, at temperatures higher than 150 mK, the best description of the behavior is again Mott VRH. Such activated behavior has been observed by Gantmakher *et al.*, slightly below and above their magnetoresistance peak [1]. At lower temperatures, however, the resistance is higher than an extrapolation of the Mott VRH fit. In Figure 1, for a 10.22Å thick film, we illustrate Mott VRH fitting from 60 mK to 1 K in the undoped state in zero magnetic field, and contrast this with the doped state at 9T. The resistance of the doped state in magnetic field deviates from Mott VRH below 150 mK and is about 5% higher than an extrapolation of the Mott VRH at 90 mK. Such excess resistances have been previously observed in "homogenously disordered" films before under the influence of perpendicular magnetic fields [3]. The phenomenon was termed which "quasireentrance" and not studied in detail. We do not believe that we are observing reentrant insulating behavior because the excess resistance at the lowest temperatures is not found as a function of increasing electron density in the absence of field [7]. Nor does it have the $(T_c/T)^4$ dependence required by the creation of normal electrons below the bulk transition temperature [9].

There is no evidence in magnetic fields up to 11 T of the magnetoresistance peak found by various groups in other 2D systems. Resistance versus parallel magnetic field is illustrated in the inset to Fig. 1 at 85 and 120 mK. The resistance appears to saturate at high field. A peak is not found at any temperature, regardless of the presence of excess resistance at high fields or negative magnetoresistance at low fields. Higher magnetic fields than available in our quench condensation apparatus would be required to demonstrate unequivocally that the observed saturation is not the low field side of a peak.

This work was supported by the NSF under grant DMR-0455121.

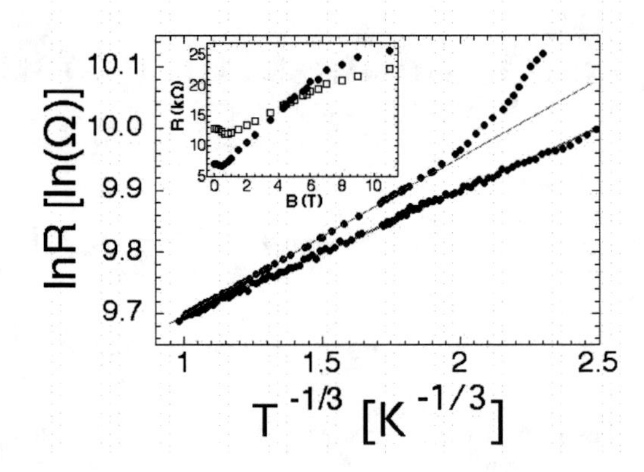

FIGURE 1. Fits to Mott variable range hopping for the doped (top curve) and undoped (bottom) states. Inset: Resistance versus parallel magnetic field for 85 mK (filled circles) and 120 mK (empty squares) isotherms. No peak is observed up to 11 T.

REFERENCES

1. M. A. Paalanen, A. F. Hebard, and R. R. Ruel, *Phys. Rev. Lett.* **69**, 1604-1607 (1992).
2. V. F. Gantmakher, M. V. Golubkov, V. T. Dolgopolov, G. E. Tsydynzhapov, and A. A. Shashkin, *JETP Lett.* **68**, 363-369 (1998).
3. N. Hadacek, M. Sanquer, and J.-C. Villegier, *Phys. Rev. B* **69**, 024505 (2004).
4. G. Sambandamurthy, L.W. Engel, A. Johansson, and D. Shahar, *Phys. Rev. Lett.* **92**, 107005 (2004).
5. Myles Steiner and Aharon Kapitulnik, *Physica C* **422**, 16-26 (2005).
6. V. F. Gantmakher, M. V. Golubkov, V. T. Dolgopolov, A. A. Shashkin, and G. E. Tsydynzhapov, *JETP Lett.* **71**, 473-476 (2000).
7. Kevin A. Parendo, K. H. Sarwa B. Tan, A. Bhattacharya, M. Eblen-Zayas, N. E. Staley, and A. M. Goldman, *Phys. Rev. Lett.* **94**, 197004 (2005).
8. Kevin A. Parendo, L. M. Hernandez, A. Bhattacharya, and A. M. Goldman, *Phys. Rev. B* **70**, 212510 (2004).
9. C. J. Adkins, J. M. D. Thomas, and M. W. Young, *J. Phys. C.* **13**, 3427-3438 (1980).

Quasi-Reentrant Resistance Behavior and Superconducting Transition Temperature of Ultrathin Quench-Condensed Bi Films Overcoated with Au

K. Makise, T. Kawaguti, and B. Shinozaki

Department of Physics, Kyushu University, 4-2-1 Ropponmatsu, chuo-ku, 810-8560 Fukuoka, Japan

Abstract. We have investigated *in situ* the temperature dependence of the sheet resistance $R(T)$ of a series of Bi films overcoated with Au, which is quench-condensed onto a bare glass substrate: The Bi films show the quasi-reentrant resistive behavior characteristic of ultrathin granular superconducting films. The increase in T_c with decreasing normal-state sheet resistance R_n due to Au coating is almost the same as that due to additional condensation of Bi onto a Bi film with similar R_n: T_c is the temperature at which R is a half of R_n. However, the "residual" resistance R_{res} of a tail of the resistive transition, which is less dependent of temperature, does not vanish by Au overcoating, being different from the behavior in the case of additional condensation of Bi. The T_c vs. R_n curve shows no departure to lower T_c with increasing Au thickness, probably because of a small amount of Au up to 2Å

Keywords: granular, S-I transition, quasi-reentrant resistance behavior.
PACS: 74.40.+k; 74.78.Db

INTRODUCTION

The superconductor-insulator (SI) transition has been investigated in various superconducting materials [1]. The SI transition in thin films occurs when film thickness or applied magnetic field is varied through a critical value. The relationship between the SI transition and the film morphology is one of the experimentally important problems. Some metal films quench-condensed (q-c) onto insulating underlayers have been interpreted to be homogeneous. In contrast, a lot of metal films without underlayers have been regarded as a granular film, in which each grain becomes superconducting but the individual phases of the order parameter are weakly connected. Landau *et al.* showed that the increase in transition temperature T_c due to the deposition of Ag onto a q-c Bi film originated in a suppression of the localization of conduction electrons [2]. Their experimental results were obtained for Bi films which showed grobal superconductivity. In this paper, we report the influence of normal-metal Au deposition on the superconducting properties of Bi films with large R_n which show quasi-reentrant behavior. We measured the temperature dependence of the sheet resistance $R(T)$ of a series of Bi films, and Bi films overcoated with Au.

EXPERIMENTAL

Experiments were carried out on Bi films grown on a thin glass substrate held at liquid He temperatures. During the evaporation the pressure was kept better than 5×10^{-8} Torr. The depositing speed of Bi was from 0.1 to 0.2 Å/s. A small amount of Au was deposited on a Bi film which did not exhibit global superconductivity. On the other hand, homogenous Bi films were prepared on SiO of 20Å thickness, which was deposited, prior to the first deposition of Bi, onto glass at liquid He temperatures. The thickness of films was measured with a quartz-crystal thickness monitor.

RESULTS AND DISCUSSION

Figures 1(a) and 1(b) show the evolution of $R(T)$ for a series of Bi films on glass substrate without and with SiO underlayer. Electrical continuity of a q-c Bi film on a SiO underlayer was obtained at a nominal thickness of ~ 10Å, which was thinner than 20Å observed for Bi films without an underlayer. This result has been interpreted as evidence for the formation of homogeneous films disordered on microscopic length scales. $R(T)$ at low temperatures down to 0.9K changes from insulating to

CP850, *Low Temperature Physics: 24th International Conference on Low Temperature Physics;*
edited by Y. Takano, S. P. Hershfield, S. O. Hill, P. J. Hirschfeld, and A. M. Goldman

superconducting behavior at R = 5.2kΩ as shown in Fig.1(b).

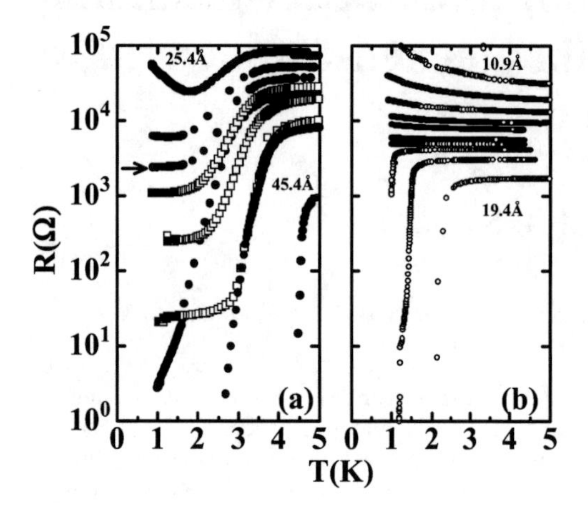

FIGURE 1. Sheet resistance R versus temperature for Bi films without an underlayer (a) and Bi films on SiO (b). Au was deposited on a Bi film with 26.3Å thickness marked by an arrow. Nominal Au thicknesses are 0.1, 0.7, and 1.7Å, from top to bottom (open square).

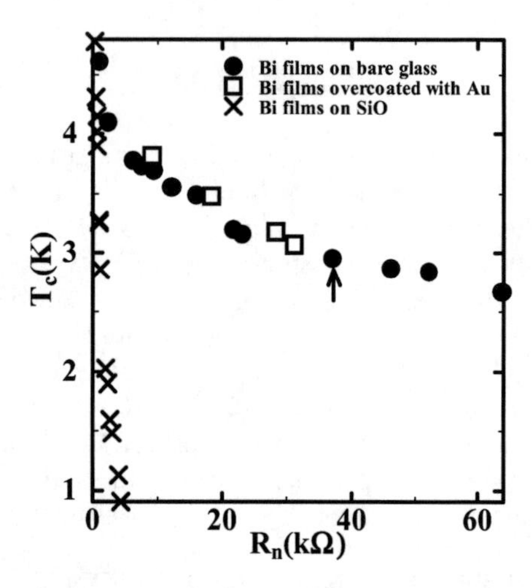

FIGURE 2. Superconducting transition temperature T_c for Bi films on SiO and bare glass, are plotted against the normal-state resistance R_n at 5K.

Bi films quench-condensed on bare glass are significantly different in the $R(T)$ characteristics from those on a SiO underlayer. $R(T)$ of a Bi film on glass with 610kΩ at 5K shows Arrhenius-like insulating behavior, which is not shown in Fig.1(a). With increasing thickness of the Bi film, there appears a resistive transition due to superconductivity, which changes from a quasi-reentrant $R(T)$ to that with a flattening at the lowest temperatures, as shown Fig.1(a). Similar results were reported for In, Ga, and Pb films in previous works [3]. However, it was reported that a flattening of the R-T curve was not observed for granular Pb films [4]. The ultrathin Bi films on bare glass may be not a randomly distributed small superconducting islands on an insulating substrate as in a thin granular Pb film, but the films may have two-dimensional arrays of small crystals whose chinks are partially filled with amorphous Bi [5]. The Bi film of 26.3Å thickness has R_n of 36kΩ and R_{res} of 2.6kΩ. As a small amount of Au equivalent to nominal thickness up to 2Å is deposited onto the Bi film which is marked by an arrow in Fig.1(a), R_n decreases monotonically from 31 to 9.3kΩ and T_c increases from 3.1 to 3.8K. (The open squares refer to the Bi films overcoated with Au.) However, R_{res} does not vanish by Au overcoating, being different from the behavior in the case of additional condensation of Bi. The T_c vs. R_n curve is almost the same as that for Bi films without the Au overlayer, as shown in Fig. 2. The T_c vs. R_n curve shows no departure to lower T_c with increasing Au thickness. A superconducting Bi film overcoated with normal-metal Ag showed an increase in T_c due to weakening of the localization and a decrease in T_c due to the proximity effect [2]. We did not observe a decrease in T_c due to the proximity effect, probably because a small amount of Au is deposited onto the Bi film. The nominal thickness up to 2Å of Au may have little influence on localization and screening effects. It was reported that the results of Bi/Sb/Bi sandwiches did not confirm the Coulomb interpretation and Berezinskii-Kosterlitz-Thouless (BKT) fluctuation might be involved in the reduction of T_c in disordered thin films [6]. If a Bi film on bare glass consists of isolated islands, the deposition of a small amount of Au onto the Bi film would increase the ratio of the Josephson coupling energy and the charging energy, E_J/E_C and consequently the BKT transition temperature [7]. This is consistent with our experimental results. However, the R_{res} remains after Au coating of the Bi film. The results suggest that a flattening of $R(T)$ is relevant to the morphology peculiar to the q-c Bi film on glass.

REFERENCES

1. A. M. Goldman, *Physica E* **18**, 1 (2003) and references cited therein.
2. I. L. Landau, D. L. Shapovalov, and I. A. Parshin, *JETP Lett.* **53**, 263 (1991).
3. H. M. Jaeger, D. B. Haviland, B. G. Orr and A.M. Goldman, *Phys. Rev. B* **40**, 182 (1989).
4. A. Frydman, O. Naaman, and R. C. Dynes, *Phys. Rev. B* **66**, 052509 (2002).
5. B. Kain, and R. P. Barber, Jr, *Phys. Rev. B* **68**, 134502 (2003).
6. E. Yap, and G. Bergmann, *Solid State Commun.* **78**, 245 (1991).
7. O.Bourgeois, A. Frydman, and R. C. Dynes, *Phys. Rev. Lett.* **88**, 186403 (2002).

Magnetoconductance Near The Superconductor-Insulator Transition In Quench-Condensed Be

Wenhao Wu

Department of Physics, Texas A&M University, College Station, TX 77843, USA

Abstract. Near the superconductor-insulator transition, quench-condensed ultrathin Be films show a highly anisotropic magnetoconductance that can drop orders of magnitude in a weak perpendicular field (< 1 T). In the high field regime, 2 ~ 10 T, the magnetoconductance in a perpendicular field is positive and can vary orders of magnitude with increasing field. Such features have been observed in films on both sides of the superconductor-insulator transition. However, they disappears when a small amount of Mn impurities are introduced. Our data show that superconductivity is the origin of the observed magnetoconductance.

Keywords: superconducting film, magnetoconductance, phase transition
PACS: 74.78.-w, 71.30.+h, 75.47.-m

INTRODUCTION

Our measurements of the magnetoconductance (MC) of ultrathin Be films near the superconductor-insulator (S-I) transition have obtained unexpected results [1,2]. These ultrathin Be films were quench-condensed onto bare glass substrates held near 20 K. Such films appear to be amorphous [1]. Therefore, the length scale of disorder in these Be films must be much smaller than the typical grain size in apparently granular films. It should be pointed out that the length scale of the disorder in films considered uniformly disordered is still not understood. Even if microscopy techniques fail to reveal any granular structure, there still can exist clusters, which can support superconductivity and which are connected electrically by relatively narrow insulating or metallic links. For example, the Ge underlayer in Bi/Ge and Pb/Ge films may produce tunneling channels between the superconducting clusters [3]. Kapitulnik *et al.* [4] have also proposed that near the S-I transition in *a*-MoGe there exist superconducting puddles, with transport being dominated by tunneling or hopping between them. Presumably, the size of the superconducting puddles grows approaching the S-I transition and eventually become the longest length scale of the system. The purpose of this report is to present our MC results measured on the Be films in the low-field and high-field regimes. The low-field MC probes long-range phase coherence, and the high-field MC reveals the nature of local superconducting fluctuations. Based on our MC results, we find that superconducting clusters exist on both sides of the S-I transition. Phase coherence is enhanced with increasing film thickness, leading to a robust global superconducting state at the S-I transition.

EXPERIMENTAL RESULTS

In Figure 1, we plot a few sheet resistance-versus-temperature curves measured in a four-probe geometry for films of varying thickness. We chose Be for our studies also because it has a very weak spin-orbit coupling [5]. As a result, a magnetic field applied parallel to the film plane, $H_{\|}$, couples to electron spin only and it does not couple to the orbital motion of the electrons. However, a perpendicular field, $H_{\perp}$, couples to both. Thus the MC can be highly anisotropic in the direction of the applied field. In the left frame in Figure 2, we show the low-field MC data at 100 mK for film (c) in Fig. 1. The MC is negative and can vary orders of magnitude in weak $H_{\perp}$ up to 1 T, but it is insensitive to $H_{\|}$ in the same field regime. The MC in $H_{\perp}$ is an orbital effect. The conductance decreases when the flux by $H_{\perp}$ through a coherent area, $H_{\perp}L_{\phi}^2$, is about a flux quanta, $\Phi_0 = h/2e$, where h is the Plank constant, e is electron charge, and L_{ϕ} is the phase coherence length. This method was applied by Barber and Dynes [6] to their superconducting granular Pb

CP850, *Low Temperature Physics: 24th International Conference on Low Temperature Physics;*
edited by Y. Takano, S. P. Hershfield, S. O. Hill, P. J. Hirschfeld, and A. M. Goldman

films to obtain the L_ϕ, and by us [1] to obtain L_ϕ for Be films across the S-I transition. Therefore, the low-field data in Fig. 2 probes phase coherence.

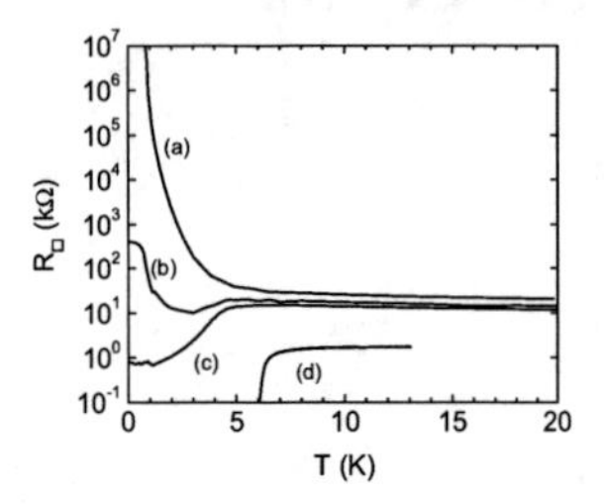

FIGURE 1. Temperature dependence of the sheet resistance for films near the S-I transition. For the top three curves, the films were about 10 Å in thickness. The film for the bottom curve was about 15.5 Å in thickness.

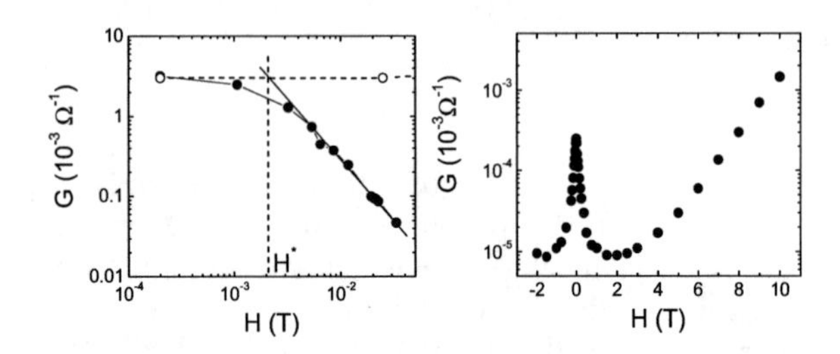

FIGURE 2. Left frame: Low-field MC measured at 100 mK in $H_\perp$ (solid circles) and $H_\parallel$ (open circles) fields. Two curves deviates at a threshold field H* at which the flux through a coherent area L_ϕ^2 is about Φ_0. Right frame: High-field MC measured in $H_\perp$ at 100 mK for a film slightly thinner than film (b) in Fig. 1. In addition to a negative MC at low fields, there is a large positive MC at high fields.

The most unusual part of our MC results is a large and positive MC in high fields, as shown in the right frame in Fig. 2. Other systems, such as thin Be films quench-condensed at liquid nitrogen temperature [7] and *a*:InO films [8], have also shown large and positive MC near the S-I transition. Recently, by dusting a small amount of Mn impurities onto the ultrathin Be films *in situ* during quench-condensation, we have been able to suppress this positive MC, as shown in Fig. 3. Therefore, we are convinced that superconductivity is the origin of this positive MC. This is in agreement with the view suggested by experiments on *a*:InO films [8]. Our results indicate that superconducting clusters exist on both sides of the S-I transition. Phase coherence is enhanced with increasing film thickness, leading to a robust global superconducting state at the S-I transition. With the application of $H_\perp$, phase coherence is suppressed in the low-field regime. As a result, S-I-S tunneling between superconducting clusters dominates transport, leading to a large and negative MC. In the high-field regime, the field reduces the superconducting gap, leading to a large positive MC. However, our maximum field of 10 T is far below the upper critical field of about 18 T [1]. With the addition of Mn impurities, the features in MC disappear as superconductivity is suppressed.

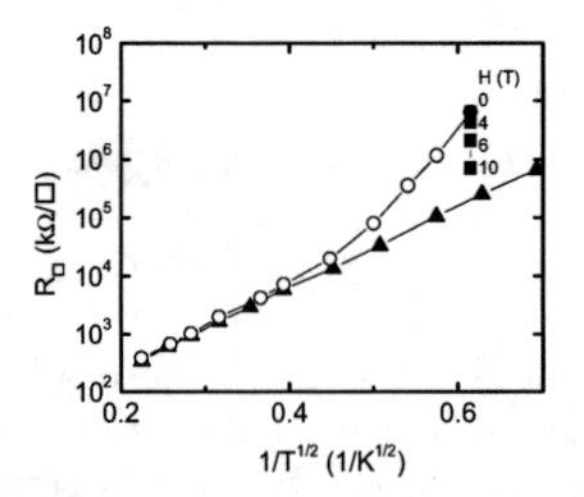

FIGURE 3. Resistance versus $T^{-1/2}$ for a highly insulating film before (open circles) and after (solid triangles) dusting with Mn impurities. MC was measured at 2.5 K for $H_\perp$ values of 4, 6, and 10 T before dusting with Mn impurities, as shown by the solid squares. No significant MC was observed after dusting with Mn impurities.

CONCLUSIONS

Our MC measurements probe phase coherence in the low-field regime and local superconductivity in the high-field regime. We find that superconducting clusters exist on both sides of the S-I transition in the Be films.

ACKNOWLEDGMENTS

This work was supported by NSF DMR-0305428. The experiments described in this report were carried out at the University of Rochester.

REFERENCES

1. E. Bielejec and Wenhao Wu, *Phys. Rev.* B **63**, 100502 (2001).
2. E. Bielejec, *Phys. Rev. Lett.* **88**, 206802 (2002).
3. A. M. Goldman and N. Markovic, *Phys. Today* **51**, 39 (1998).
4. N. Mason and A. Kapitulnik, *Phys. Rev. Lett.* **82**, 5341 (1999).
5. P. M. Tedrow and R. Meservey, *Phys. Lett.* **58** A, 237 (1976).
6. R. P. Barber, Jr. and R. C. Dynes, *Phys. Rev.* B **48**, 10618 (1993).
7. V. Yu. Butko, J. F. DiTusa, and P. W. Adams, *Phys. Rev. Lett.* **85**, 162 (2000).
8. G. Sambandamurthy, L. W. Engel, A. Johansson, and D. Shahar, *Phys. Rev. Lett.* **92**, 107005 (2004).

Current-Induced First-order Superconducting Transitions in Tantalum Thin Films at Zero Magnetic Field

Yongguang Qin, Yongho Seo[1], and Jongsoo Yoon[2]

Department of Physics, University of Virginia, Charlottesville, VA22903, U.S A.

Abstract. We observed hysteretic voltage-current (VI) curves in superconducting tantalum thin films at zero magnetic field (B=0) at temperatures (T) below a characteristic temperature T^*. This is in contrast to the Kosterlitz-Thouless (KT) theory which expects continuous and reversible VI's. We interpret that the hysteresis arises from the critical dynamics of self-generated magnetic vortices, and map the area in B-T plane where the VI's are hysteretic.

Keywords: Superconductivity, thin films, tantalum.
PACS: 74.25.Fy, 74.25.Sv, 74.78.Db.

INTRODUCTION

KT theory has been the framework to understand transport characteristics of superconducting thin films at B=0 [1]. In the KT picture, the superconducting transition corresponds to a thermodynamic instability of vortices, which are geometric disorder in the order parameter induced by thermal fluctuation. The vortices are expected to exist in vortex-antivortex pairs in the superconducting phase, and a KT transition is driven by unbinding of the vortex pairs. Due to the I-induced pair dissociations the VI curves are expected to follow a power law, $V \propto I^{\alpha}$, and a universal jump in α from 1 to 3 at the transition temperature (T_c) is expected [2]. Although it has been shown in simulations [3] that the free vortices induced by finite size effects can suppress the power law and the universal jump, the nonlinear VI's are expected to be continuous and reversible.

We studied VI characteristics of superconducting tantalum films as thin as 5nm at B=0. The films are dc-sputter deposited on quartz substrates, and patterned into bridges (1mm×5mm) for four-point measurements using a shadow mask. In the left column of figure 1 the resistive superconducting transitions are shown for 5nm, 10nm, and 36nm thick films. Even though their T_c's are suppressed to ~1K or below from the bulk value of 4.5K, the transitions are very sharp demonstrating a high homogeneity of the films. The T_c's are found to decrease continuously towards 0K with decreasing film thickness to ~2nm. In contrast to the expectations based on the KT theories, we observed hysteretic voltage jumps in current-biased VI curves at T below a characteristic temperature T^*.

RESULTS AND DISCUSSIONS

Shown in the right column of figure 1 are the current-biased VI curves. The striking feature is a development of a hysteresis with decreasing T below T^* which is 0.568K, 0.808K, and 0.988K for 5nm, 10nm, and 36nm, respectively. There are three known mechanisms for such a hysteresis. The first is the hot spot effects that can cause hysteretic voltage jumps due to a sequential development of Joule-heat driven hot spots arising from inhomogeneities or phase slip centers [4]. The hot spot effects are usually observed in 1D systems such as films in micro-bridge geometry or with filamentary structure, and unlikely in our films because a) the films are highly homogeneous as demonstrated by the sharp transitions, b) the films appear to have no mesoscale clusters as indicated by the continuous decrease of T_c's towards 0K with decreasing film thickness [5], and c) there is no sign of staircase structure in VI curves that are often observed for the hot spot effects. The second mechanism is a uniform Joule heating [6] which can drive the entire system to the normal conducting state via a thermal run-away phenomenon. We can rule out this mechanism because we observe that the critical power

[1] Present address: Department of Physics and Astronomy, Northwestern University, Evanston, IL 60208, U.S.A.
[2] Correspondence to jy2b@virginia.edu.

CP850, *Low Temperature Physics: 24th International Conference on Low Temperature Physics;*
edited by Y. Takano, S. P. Hershfield, S. O. Hill, P. J. Hirschfeld, and A. M. Goldman

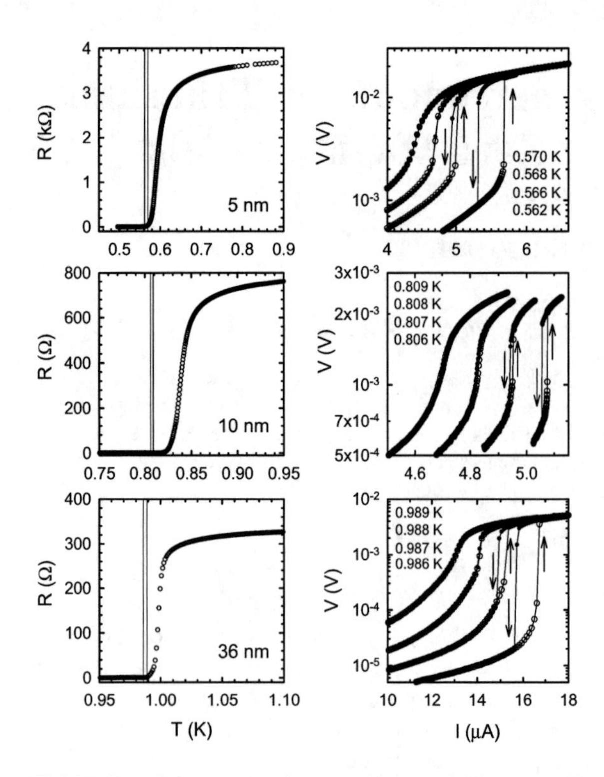

FIGURE 1. Superconducting transitions and VI curves. The vertical lines in the left panels indicate the T range for the VI's shown in the right panels. Open (filled) circles in the right panels are for current increasing (decreasing) branch.

at the onset of the voltage jump, which is expected to be nearly B-independent in the mechanism, changes by more than two orders of magnitude with applying B~10G (36nm thick sample at T=0.970K). The last possibility is associated with critical vortex dynamics which was predicted by Larkin and Ovchinnikov (LO) for the mixed state of a type II superconductor [7], where the Lorentz force on the magnetic vortices is balanced with viscous drag force in a steady state. At large vortex velocities, however, the vortex core size shrinks as quasiparticles escape the core, which in turn causes a reduction in the drag force. Then the vortex motion becomes unstable and runs away to a higher velocity until the system reaches the normal conducting state.

We interpret that the hysteretic voltage jumps at $T<T^*$ arise from the LO-type critical dynamics of magnetic vortices induced by non-uniform current density due to unavoidable inhomogeneities of the samples [8]. Although the appearances of the VI curves at $T>T^*$ are consistent with the KT framework, the systematic evolution of the nonlinear VI's into a hysteretic voltage jump with decreasing T below T^* indicates that the nonlinear transport at $T>T^*$ also arises from the dynamics of the self-generated

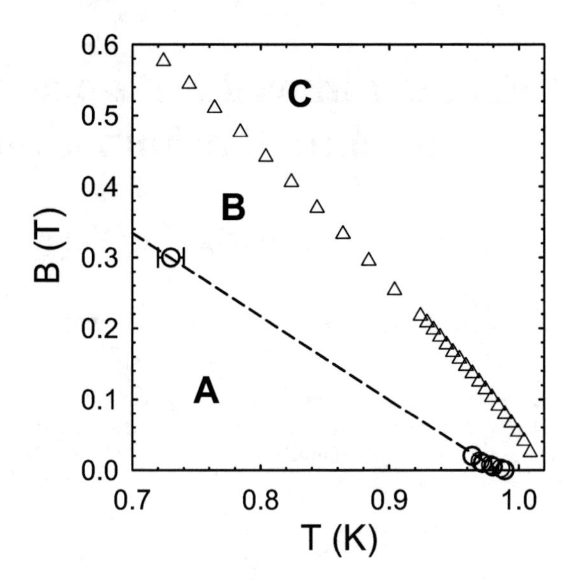

FIGURE 2. B-dependence of T^* (circles) and the upper critical field B_{c2} (triangles). The dashed line is to guide an eye.

vortices. By studying B-dependence of T^* we draw a line in B-T plane as shown in figure 2. The line closely resembles the irreversibility lines in type II superconductors that are related with the vortex dynamics [9]. In region A the VI curves are discontinuous and hysteretic, in region B the VI's are continuous and reversible, and region C corresponds to the normal conducting state. Authors acknowledge informative discussions with E. Kolomeisky and H. Fertig. This work is supported by NSF and Jeffress Memorial Foundation.

REFERENCES

1. J. M. Kosterlitz and D. J. Thouless, *J. Phys. C* **6**, 1181 - 1203 (1973).
2. B. I. Halperin and D. R. Nelson, *J. Low Temp. Phys.* **36**, 599-616 (1979).
3. K. Medvedyeva et al, *Phys. Rev. B* **62**, 14531-14540 (2000).
4. A. V. Gurevich and R. G. Mints, *Rev. Mod. Phys.* **59**, 941-999 (1987).
5. A. M. Goldman and N. Marković, *Phys. Today* **49**, 39-44 (1998).
6. J. Viña et al, *Phys. Rev. B* **68**, 224506 (2004).
7. A. I. Larkin and Yu. N. Ovchinnikov, *Sov. Phys. JETP* **41**, 960-965 (1976).
8. M. A. Topinka et al, *Phys. Today*, **56** 47-52 (2003).
9. Y. Yeshurun et al, *Rev. Mod. Phys.* **68**, 911-949 (1996).

Superconductor to Insulator Transition in Ta Thin Films

Brian Gross, Yongguang Qin, and Jongsoo Yoon*

Department of Physics, University of Virginia, Charlottesville, VA22903, U.S A.

Abstract. We show disorder- and magnetic field-driven superconductor to insulator transitions (SIT) observed in tantalum thin films. At zero magnetic field (B=0), the superconducting transition temperatures (T_c) are found to decrease smoothly towards 0K with decreasing film thickness, and a B=0 disorder-driven SIT is observed at a film thickness of ~2.5nm at which the normal state sheet resistance is close to the value $h/4e^2$ = 6.45kΩ. In the magnetic field driven transition, there appears to be a "metallic" phase intervening the SIT.

Keywords: Superconductor-insulator transition, superconductivity, thin films, tantalum.
PACS: 74.25.Fy, 74.25.Sv, 74.78.Db.

INTRODUCTION

Suppression of superconductivity in thin film geometry by means of disorder or magnetic fields is a well-known phenomenon. The suppression, resulting from the competition between localization and superconductivity, has attracted a heightened attention in recent years [1] because the superconductor-insulator transition (SIT) at a critical disorder or magnetic field is believed to be an example of quantum phase transition (QPT), which is a phase transition occurring at zero temperature between two fundamentally different ground states [2]. Unlike its classical counter part with a finite T_c, the order parameter fluctuation in QPT should be treated quantum mechanically, which leads to an interesting interplay of dynamics and thermodynamics.

In this paper, we present SIT's observed in tantalum thin films. Sputter deposited tantalum films on glass, quartz, silicon, and sapphire substrates without an underlayer are known to exhibit superconductivity at temperatures significantly below the bulk T_c of 4.5K [3,4]. However, transport measurements have not been extended to the insulating phase probing the SIT. We prepared tantalum films in the thickness range from 55nm to 1.6nm. At B=0, we observed a disorder-driven SIT with a critical film thickness of ~2.5nm at which the normal state sheet resistance is close to the value $h/4e^2$ = 6.45kΩ. Our measurements on a 5.75nm thick film at temperatures down to 0.45K with magnetic fields indicate that the film exhibits a "metallic" behavior before it turns insulating at higher fields.

EXPERIMENTS

The tantalum films are dc sputter deposited on Si substrates in argon at 4 mTorr. Before sputtering, the chamber was baked out at ~110°C for at least 2days, giving a base chamber pressure ~10^{-8}Torr. Variation in the deposition rate below ~0.1nA/sec or in the base chamber pressure below ~10^{-7}Torr did not have a noticeable effect on the T_c or the width of the transition. However, films grown in a base chamber pressure above ~10^{-6}Torr showed more than 10 times broader transitions. The films are patterned into bridges (1mm×5mm) for the standard four point measurements using a shadow mask. Electrical leads are attached with indium cold press contacts, and the resistances are measured with a dc current (0.1-10nA) at which the sample response is linear with the current.

Shown in figure 1 are the resistances of 9 samples in the thickness range from 55nm to 1.6nm at B=0. Films with thicknesses of 3nm or above exhibit superconductivity with their T_c's decreasing smoothly towards 0K with decreasing film thickness. This is typical of "amorphous" films such as quench condensed Bi and MoGe films grown with a Ge underlayer, and InOx [1], and in contrast to "granular" films where their T_c's are almost independent of film

* Correspondence to jy2b@virginia.edu

CP850, *Low Temperature Physics: 24th International Conference on Low Temperature Physics;* edited by Y. Takano, S. P. Hershfield, S. O. Hill, P. J. Hirschfeld, and A. M. Goldman

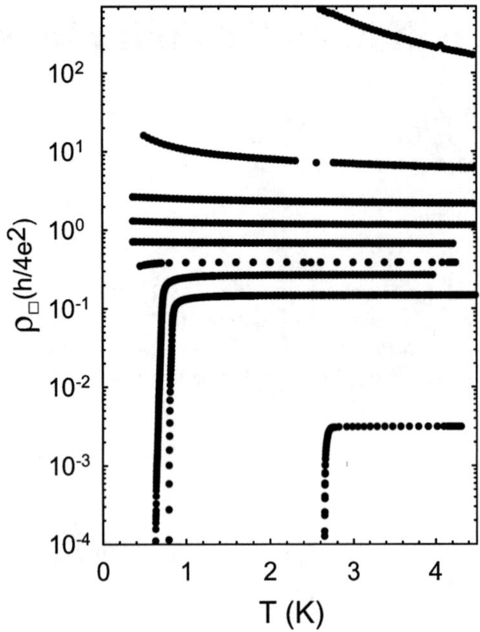

FIGURE 1. Superconductor-insulator transition of tantalum thin films driven by film thickness. The film thicknesses are, from the top, 1.6, 1.75, 2.0, 2.2, 2.6, 3, 4, 6, and 55nm, respectively.

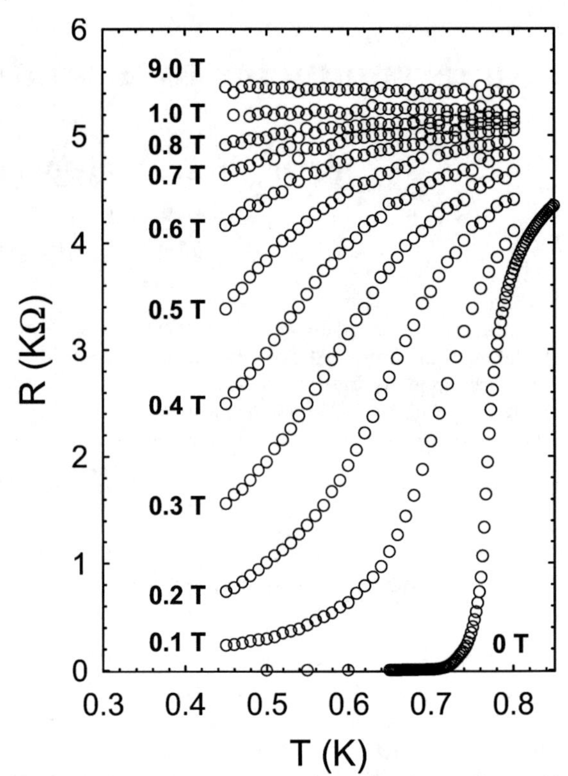

FIGURE 2. T-dependence of resistance in a 5.75nm thick film under magnetic fields as indicated. Note that the magnetic field for the top two traces are 9T and 1T, respectively.

thickness [5]. Many soft metal films such as Al, Sn, Pb, and Ga, made by low-temperature evaporation without an underlayer of amorphous germanium or antimony are examples of granular films, and known to possess mesoscale clusters with their T_c's close to the bulk value. The superconducting transitions in our films are found to be very sharp with no sign of reentrant behavior, demonstrating that the films are highly homogeneous. In terms of $\Delta T_c/T_c$, where T_c is the mid-point of the resistive transition and ΔT_c is the temperature interval for the sample resistance change from 90% to 10% of the normal state resistance, the width of the transition is sharper than 2% for films with $T_c \gtrsim 1$K and ~10% for films with T_c~0.5K. When the film thickness is decreased below ~2.5nm, at which the normal state sheet resistance is close to the value $h/4e^2$ = 6.45kΩ, the temperature coefficient of resistivity (dρ/dT) turns negative manifesting a disorder-, or thickness-driven SIT.

The effect of applied magnetic fields up to 9T on superconductivity in a 5.75nm thick film is shown in figure 2. The striking feature is that, at rather low fields, the film appears to exhibit a "metallic" behavior. For example, at B=0.1T the resistance decreases rather steeply at ~0.7K, but with further decreasing temperature the drop slows down and the resistance appears to saturate to a finite value. As the magnetic field is increased, the resistance drop progressively disappears and eventually the film exhibits an insulating behavior with a negative dρ/dT. Once the resistance drop is suppressed by the magnetic field, the sample resistance becomes almost B-independent. Note that in figure 2, the magnetic fields for the top two traces are 9T and 1T, respectively. Apart from the temperature range, the data shown in figure 2 closely resemble with those obtained in MoGe films (T>0.05K) [6] reporting a metallic behavior intervening B-induced SIT. Authors acknowledge informative discussions with V. Galitski. This work is supported by NSF.

REFERENCES

1. A. M. Goldman and N. Marković, *Phys. Today* **49**, 39-44 (1998).
2. S. Sondi et al., *Rev. Mod. Phys.* **69**, 315-333 (1997).
3. J. J. Hauser and H. C. Theuerer, *Rev. Mod. Phys.* **36**, 80-83 (1964).
4. M. Mohazzab et al, *J. Low Temp. Phys.* **121**, 821-824 (2000).
5. H. M. Jaeger et al, *Phys. Rev. B* **40**, 182-196 (1989).
6. A. Yazdani and A. Kapitulnik, *Phys. Rev. Lett.* **74**, 3037-3040 (1995).

Destructive Regime and Quantum Phase Transition in Doubly Connected Superconducting Cylinders

H. Wang, N.A. Kurz, and Y. Liu

Department of Physics, The Pennsylvania State University, University Park, PA 16802, USA

Abstract. We have studied experimentally the physics of the destructive regime, the loss of superconductivity around half-flux quanta, in ultrathin, doubly connected superconducting cylinders when their diameters become smaller than the superconducting coherence length at zero temperature. We show that the transition from the superconducting to normal behavior at the onset of the destructive regime is very sharp. Near this zero-temperature quantum phase transition, there appears to be a phase separation in which normal regions nucleate in a homogeneous superconducting cylinder.

Keywords: superconductivity, quantum phase transition, 1D.
PACS: 74.78.Na, 74.25.Fy

INTRODUCTION

In addition to strong disorder, a magnetic field, or a Coulomb repulsion, superconductivity at zero temperature can also be destroyed by the rise of kinetic energy in samples with restricted geometry, as demonstrated by the discovery of a destructive regime near half-flux quanta in ultrathin, doubly connected superconductors [1]. The fluxoid quantization in a doubly connected system requires that the equilibrium-state superfluid velocity, v_s, increases continuously as the diameter of the cylinder, d, decreases [2]. When d becomes smaller than the zero-temperature superconducting coherence length, $\xi(0)$, the kinetic energy of the superconducting electrons near half-flux quanta will exceed the superconducting condensation energy, leading to the suppression of superconductivity even at zero temperature [1, 2].

Superconducting cylinders possessing a destructive regime may be considered as an one-dimensional (1D) superconductor, a subject of much renewed interest in recent years [2-8]. Our study focuses on the effects of sample topology (singly *vs.* doubly connected), the quantum phase transition (QPT) [2], and the exotic normal state in the destructive regime [9].

EXPERIMENTAL

Ultrathin, doubly connected cylinders of Al were fabricated by evaporating Al onto rotating quartz filaments. The diameter of the cylinder was calculated from the period in resistance oscillation, using a natural period of $h/2e$ in magnetic flux. Fine Au wires were attached to the cylinder by Ag epoxy, forming normal-metal electrical leads. The length between the two Ag epoxy dots is usually more than 100 μm. An ultrathin cylinder can be made as long as ~mm, allowing the attachment of multiple electrical leads. Unfortunately, it is difficult for contacts with an ultrathin cylinder to survive all the way to low temperatures. Therefore many samples had only three or fewer working leads. In this case, the contact resistance is subtracted. Cylinders were manually aligned to be parallel to the magnetic field. Electric transport measurements were carried out in a dilution refrigerator equipped with a superconducting magnet with a base temperature below 20 mK. All leads entering the measurement enclosure were filtered by RF filters working at room temperatures.

RESULTS AND DISCUSSION

Figure 1 shows the resistance as a function of magnetic field, $R(H)$, of a cylinder showing the destructive regime at two different temperatures. As the upper critical parallel field, $H_{c2}^{\parallel}$ (= 1830 G at 50 mK), is relatively low, fully superconducting state is not recovered even measured at the base temperature. The value of $H_{c2}^{\parallel}$, which is determined by the Zeeman splitting and spin-orbital coupling, is a material property unrelated to destructive-regime physics.

CP850, *Low Temperature Physics: 24th International Conference on Low Temperature Physics;*
edited by Y. Takano, S. P. Hershfield, S. O. Hill, P. J. Hirschfeld, and A. M. Goldman

Our measurements suggest that the zero-temperature quantum phase transition between a superconducting and a normal behavior can be tuned by applied flux. This is rather convenient as the magnetic flux can be controlled very precisely. In addition, since the material properties of the sample, such as the cylinder morphology and the amount of the disorder, remain unchanged across the transition, the analysis of the quantum phase transition is simplified. In Fig. 2, we show the magnetic field dependence of sample resistance near the onset of the destructive regime for Cylinder Al-3 at several temperatures. It can be seen that the transition is very sharp (only the bottom of the transition is shown).

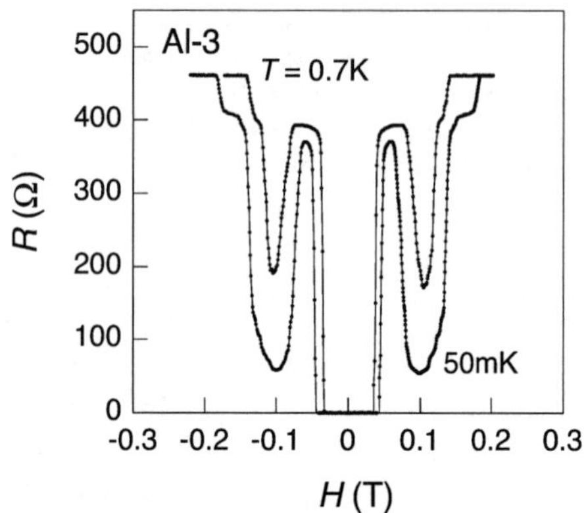

FIGURE 1. a) Resistance as function of magnetic field, *R(H)*, at two different temperatures as indicated for Cylinder Al-3 with $d = 151$ nm, and $\xi(0) = 190$ nm.

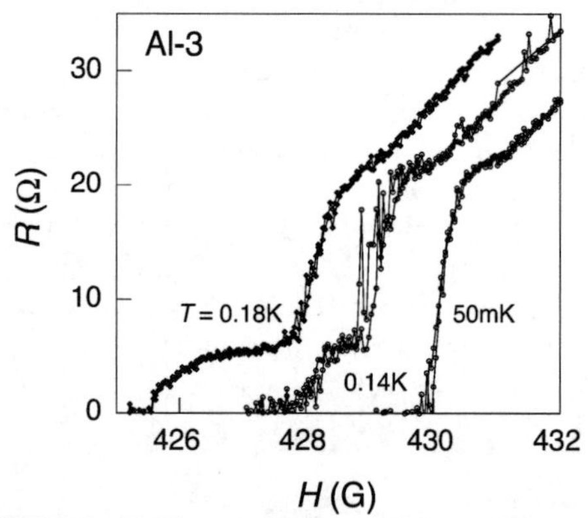

FIGURE 2. *R(H)* near the onset of the resistive state for Cylinder Al-3 measured at different temperatures as indicated. The field was ramped up and down during the measurements. No hysteresis was observed.

Tuning a zero-temperature quantum phase transition between a superconducting and a normal behavior by the applied flux is rather convenient as the magnetic flux can be controlled very precisely. In addition, since the material properties of the sample, such as the cylinder morphology and the amount of the disorder, remain unchanged across the transition, the analysis of the quantum phase transition is simplified.

A distinctive feature seen in Fig. 2 is a resistance step around 6 Ω. This step disappears at lower temperature. Similar step-like features have been seen in resistance *vs.* temperature, *R(T)*, curves in the destructive regime on the superconducting side of the QPT, measured at fixed flux values. These step-like features were attributed to a phase separation in which normal bands nucleate in a homogeneous superconducting background—they did not result from the presence of sample inhomogeneities or phase slip centers [10]. Results shown in Fig. 2 suggest that as the magnetic flux is increased while keeping the temperature fixed, a normal band nucleates in a similar fashion. We speculated that the formation of the normal bands is due to quantum fluctuation near the QPT [10].

Almost all quantum fluctuation in a superconductor can be traced back to the confinement of Cooper pairs, because of the relation, $\Delta N \Delta\phi > 1$, where $\Delta\phi$ is the fluctuation in the phase of the superconducting order parameter, and ΔN is the fluctuation in the number of Cooper pairs. ΔN is not suppressed near the destructive regime in ultrathin superconducting cylinders. It is possible that increasing kinetic energy would result in fluctuation in the amplitude of the superconducting order parameter, which leads to the suppression of superconductivity and the emergence of a phase separation in the destructive regime.

ACKNOWLEDGMENTS

The authors would like to acknowledge useful discussions with Professors M. Beasley, H. Fan, Yuval Oreg, M. Sigrist, M. Tinkham and Drs. Oskar Vafek and Z. Long. The work is supported by NSF through grant DMR-0202534.

REFERENCES

1. P. -G. de Gennes, *C. R. Acad. Sci. Paris* **292**, 279 (1981).
2. Y. Liu *et al.*, *Science* **294**, 2332 (2001).
3. N. Giordano, *Phys. Rev. Lett.* **61**, 2137 (1988).
4. P. Xiong, A. V. Herzog, and R. C. Dynes, *Phys. Rev. Lett.* **78**, 927 (1997).
5. A. Bezryadin, C. N. Lau, and M. Tinkham, *Nature (London)* **404**, 971 (2000).
6. C. N. Lau *et al.*, *Phys. Rev. Lett.* **87**, 217003 (2001).
7. S. Michotte *et al.*, *Phys. Rev. B* **69**, 094512 (2004).
8. M. Tian *et al.*, *Phys. Rev. B* **71**, 104521 (2005).
9. O. Vafek, M. R. Beasley, and S. Kivelson, cond-mat/0505688 (2005).
10. H. Wang *et al.*, to be published.

Observation of Superconductor-insulator Transition Induced by Ge in Ultrathin a-Nb Film

Ryuichi Masutomi, Takashi Ito and Nobuhiko Nishida

Department of Physics, Tokyo Institute of Technology, 2-12-1 O-okayama, Meguro-ku, Tokyo 152-8551, Japan

Abstract. We have observed the superconductor-insulator (S-I) transition induced by over-deposited Ge in the ultrathin amorphous Nb (a-Nb) film. The experiments are performed by depositing Ge onto the *insulating* a-Nb film with a thickness of 1.04 nm. For $d_{Ge} > 0.3$ nm, where d_{Ge} was a thickness for Ge film, the reduction of electrical sheet resistances was seen at low temperature region, suggesting superconductor. Moreover, the normal sheet resistance at 8 K decreased monotonously with increasing d_{Ge}. The simplest explanation of this S-I transition is that the electron localization effect is weakened due to the addition of Ge film.

Keywords: 2D superconductor, Quench-condensed film, Superconductor-insulator transition
PACS: 74.40.+k, 74.78.-W, 74.81.Bd

INTRODUCTION

Superconducting thin films have attracted much attention because of the interplay between superconductivity and electron localization [1, 2, 3, 4]. In addition, similarities with high-T_c superconductors are pointed out [5]. According to the scaling theory by Abrahams *et al.* [6], in two-dimensional systems, electrons are localized for arbitrarily weak disorder as long as electron interactions are neglected. As well known, electron localization due to random potential or disorder enhances Coulomb interaction between electrons. On the other hand, it is essential for superconductivity that electron-electron interaction is attractive and Cooper pair is formed.

Experimentally, uniformly disordered ultrathin films fabricated by quench-condensation are one of the most ideal systems to study the interplay between superconductivity and localization. Haviland *et al.* observed the S-I transition by changing the film thickness of a-Bi and a-Pb deposited onto Ge substrate [7]. Moreover, in order to study a competition between electron localization and proximity effect, we measured the electrical sheet resistances of the *superconducting* a-Nb films with a thickness of 1.5 nm over-deposited by metals or semiconductors and discussed the influences of over-deposited elements [8]. In this paper, we present preliminary results of Ge depositions onto the *insulating* a-Nb film with a thickness of 1.04 nm to investigate S-I transition induced by over-deposited element.

EXPERIMENT

We have prepared the *insulating* a-Nb film with a thickness of 1.04 nm onto a glazed sapphire (Al_2O_3) substrate, and then deposited Ge by 0.05 - 0.1 nm step in vacuum better than 10^{-8} Torr. The mixing of atoms at the interface will not occur as the depositions are preformed below 6 K. An electron-beam gun for Nb and a Kundsen cell source for Ge were used for the evaporation. Deposition rate and film thickness were obtained from a quartz-crystal oscillator. The electrical sheet resistance, $R_{sq}(T)$, was measured by a standard four-terminal method *in situ* down to 0.7 K, using the ^{3}He pumping cryostat precooled by Gifford-McMahon refrigerator [9] combined with an ultra high vacuum evaporator.

RESULTS AND DISCUSSION

Before the Ge/a-Nb system, we discuss the pure a-Nb system. Our films are believed to be homogeneous because an onset of measurable electrical resistance is found to be about 0.7 nm. However, detailed morphological studies have never been carried out on such systems.

CP850, *Low Temperature Physics: 24th International Conference on Low Temperature Physics;*
edited by Y. Takano, S. P. Hershfield, S. O. Hill, P. J. Hirschfeld, and A. M. Goldman

Figure 1 shows the evolution of the temperature dependences of R_{sq} with thickness for a Ge film deposited onto the *insulating* a-Nb film with d_{Nb} = 1.04 nm. For d_{Ge} < 0.3 nm, the R_{sq} increased with decreasing temperature. On the other hand, for d_{Ge} > 0.3 nm, the reduction of R_{sq}, indicating the superconducting transition, was observed at low temperature region as a result of the deposition of Ge. However, we couldn't confirm that the resistance had become zero. This needs experiments at the lower temperatures. The S-I transition occurs when the normal-state sheet resistance is close to a value $h/4e^2 \sim 6.45\ \mathrm{k\Omega}$. In addition, the threshold value seems to be close to that of the pure a-Nb system [10, 11]. The superconducting transitions become sharper with increasing film thickness. Although reentrances shown in granular systems [12] are not observed down to the lowest temperature, the transition exhibits a shoulder in the resistance drop. The inset in Fig. 1 shows the dependence of normal sheet conductance at 8 K (σ_{sq}^{N}) as a function of d_{Ge}. σ_{sq}^{N} keeps up with increasing d_{Ge}, suggesting that the effect of the electron localization is weakened by the deposition of Ge.

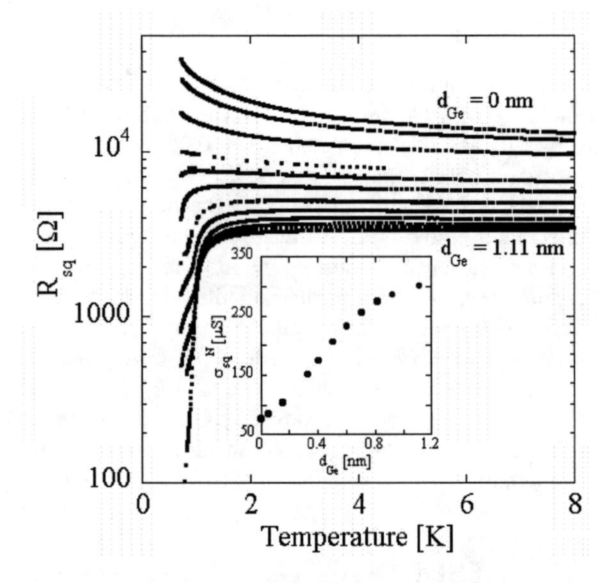

FIGURE 1. R_{sq} vs. temperature for a series of Ge film thicknesses, from top to bottom, of 0, 0.05, 0.15, 0.26, 0.31, 0.33, 0.41, 0.50, 0.60, 0.70, 0.81, 0.92, and 1.11 nm deposited onto the a-Nb film with d_{Nb} = 1.04 nm. Inset: σ_{sq}^{N} vs. d_{Ge} plot.

Finally, superconducting transition temperatures (T_c) were obtained as a temperature at $R_{sq}^{N}/2$, where $R_{sq}^{N} = 1/\sigma_{sq}^{N}$. In this system, T_c was found to increase up to about d_{Ge} = 0.8 nm. This fact also supports that the electron localization is weakened and the electron screening effect is improved. The variation of T_c is very similar to that of our previous report [8].

In summary, we have measured the electrical sheet resistances of ultrathin a-Nb films over-deposited by Ge *in situ* down to 0.7 K. The S-I transition induced by over-deposited Ge in the insulating ultrathin a-Nb film was observed. Judging from the increase of σ_{sq}^{N} and the increase of T_c up to about d_{Ge} = 0.8 nm, the simplest explanation of this S-I transition is that the electron localization effect is weakened due to the addition of Ge film. Further theoretical calculations including both superconducting fluctuation and electron localization are eagerly desired in order to perform quantitative analysis at lower temperatures.

ACKNOWLEDGMENTS

This work is supported by the 21st Century COE Program at Tokyo Institute of Technology "Nanometer-scale Quantum Physics".

REFERENCES

1. S. Maekawa and H. Fukuyama, *J. Phys. Soc. Jpn.* **51**, 1380 (1981).
2. H. Takagi and Y. Kuroda, *Solid State Commun.* **41**, 643 (1982).
3. M. Ma and P. A. Lee, *Phys. Rev. B* **32**, 5658 (1985).
4. A. M. Finkel'stein, *Physica B* **197**, 636-648 (1994).
5. L. Merchant, J. Ostrick, R. P. Barber, Jr., and R. C. Dynes, *Phys. Rev. B* **63**, 134508 (2001).
6. E. Abrahams, P. W. Anderson, D. C. Licciardello, and T. V. Ramakrishnan, *Phys. Rev. Lett.* **42**, 673 (1979).
7. D. B. Haviland, Y. Liu, and A. M. Goldman, *Phys. Rev. Lett.* **62**, 2180 (1989).
8. N. Nishida, T. Fujiki, K. Okada, and H. Ikeda, *Physica B* **284-288**, 1950-1951 (2000).
9. Sumitomo Heavy Industries, Ltd. Japan.
10. N. Nishida, S. Okuma, and A. Asamitsu, *Physica B* **169**, 487-488 (1991).
11. A. Asamitsu, M. Iguchi, A. Ichikawa, and N. Nishida, *Physica B* **194-196**, 1649-1650 (1994); A. Asamitsu, Ph.D. thesis, Tokyo Institute of Technology (1994).
12. For example, B. G. Orr, H. M. Jaeger, and A. M. Goldman, *Phys. Rev. B* **32**, 7586 (1985).

Thickness and Magnetic Field-tuned Superconductor-Insulator Transitions in a-$Nb_{15}Si_{85}$

Claire A. Marrache-Kikuchi*, H. Aubin†, A. Pourret†, K. Behnia†, L. Bergé*, L. Dumoulin* and J. Lesueur†

*CSNSM, CNRS-IN2P3, Bât 108, 91405 Orsay Campus, France
†LPQ, ESPCI (UPR5-CNRS), 10 rue Vauquelin, 75231 Paris, France

Abstract. Results from a study of amorphous superconducting $Nb_{0.15}Si_{0.85}$ thin films are presented. These compounds are subject to a thickness-tuned superconductor-insulator transition at zero-field and, for each given thickness, a magnetic-field tuned superconductor-insulator transition is observed. The field-tuned transition is characterized by an isobestic point (H_c,R_c) that indicates the quantum critical nature of this superconductor-insulator transition and the absence of an intermediate metallic state. Similarly, we find that the thickness-tuned superconductor-insulator transition in zero-field is characterized by a critical thickness d_c. We were thus able to obtain the phase diagram in the (H,d) plane for $Nb_{0.15}Si_{0.85}$.

Keywords: superconductor-insulator transition, amorphous films, quantum phase transition
PACS: 74.25.-q, 74.40.+k, 71.30.+h, 64.60.-i

Superconductor to Insulator Transitions (SIT) in disordered thin films of metal have attracted continuous attention since the pioneering work of Orr et al. [1] for they represent particularly remarkable examples of quantum phase transitions.

The SIT can either be driven by applying a perpendicular magnetic field [2, 3, 4, 5, 6, 7] or by varying the sheet resistance R_{square} of the films, using film thickness [8, 9] or electrostatic field [10]. In this letter, we report on a study of SIT in $Nb_{0.15}Si_{0.85}$ amorphous thin films tuned both by perpendicular magnetic field and by the film thickness in zero-magnetic field.

The amorphous thin films of $Nb_{0.15}Si_{0.85}$ were prepared under ultrahigh vacuum by e-beam co-evaporation of Nb and Si, with special care over the control and homogeneity of concentrations. Such films are known to undergo a transition from insulator to metal with increasing Nb concentration [11, 12, 13].

For this experiment, a series of seven samples with stoichiometry $Nb_{0.15}Si_{0.85}$ and thicknesses ranging from 100 nm down to 2.5 nm, have been deposited onto sapphire substrates with a 50 nm thick SiO underlayer. Resistances were measured by using a standard ac lock-in detection. The three thinnest samples (2.5, 5 and 7.5 nm) were only measured at zero magnetic field, whereas the thicker ones (12.5, 25, 50 and 100 nm) were also measured in a perpendicular magnetic field between 0 kOe and 35 kOe.

Samples with thickness down to 2.5 nm have previously been characterized by atomic force microscopy and showed no sign of granularity or inhomogeneity. Figure 1 shows the temperature dependence of their resistance for a zero magnetic field. For samples with a thickness larger than 5 nm, the superconducting transition temperature decreases for samples of increasing sheet resistance with no sign of reentrant behavior , as usually observed in granular systems, which is another indication of the homogeneity of the films.

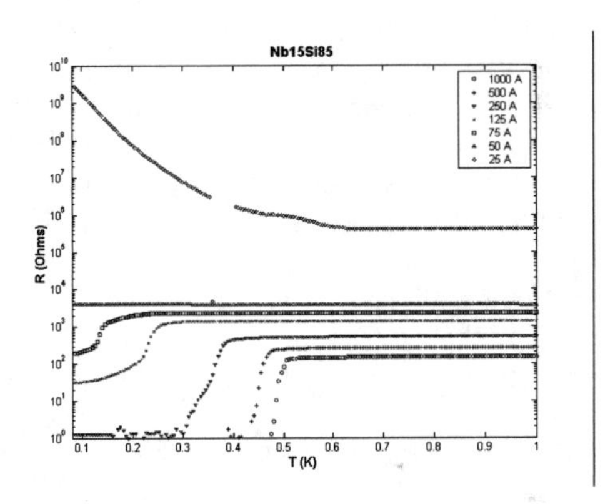

FIGURE 1. R_{square} versus temperature for seven amorphous $Nb_{15}Si_{85}$ samples with thicknesses ranging from 2.5 nm to 100 nm

In contrast, samples with thickness below 5 nm have an insulating behavior, i.e. the resistance increases with decreasing temperature. In between, the sample 5 nm thick displays an almost temperature independent resistance of R = 3980 Ω. This behavior manifest itself in the resistance isotherms - plotted as function of sample thickness - as a crossing point (figure 2). Thus, a critical point can be defined, which critical thickness is d_c = 5 nm and critical resistance 3980 Ω.

Using this value of the critical point (R_c,d_c), we determine the critical exponent product νz - where ν is the

CP850, *Low Temperature Physics: 24th International Conference on Low Temperature Physics;*
edited by Y. Takano, S. P. Hershfield, S. O. Hill, P. J. Hirschfeld, and A. M. Goldman

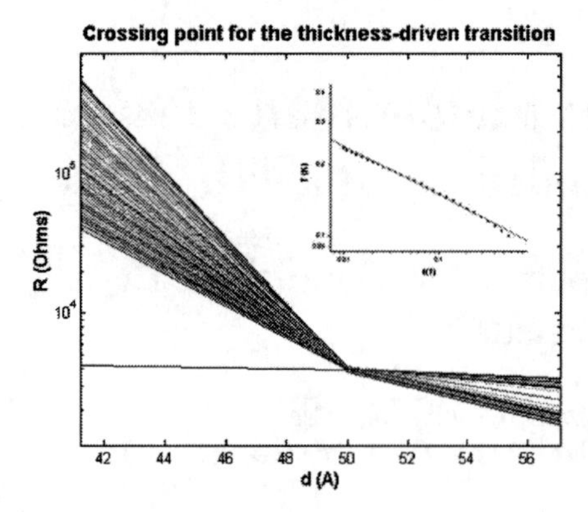

FIGURE 2. R versus thickness displayed for temperatures between 0.01 K and 0.4 K. Inset: log-log plot of the temperature as function of the scaling parameter t(T). Fitting of this data with $t(T) = T^{-\frac{1}{\nu z}}$ gives the product $\nu z = 0.18 \pm 0.05$.

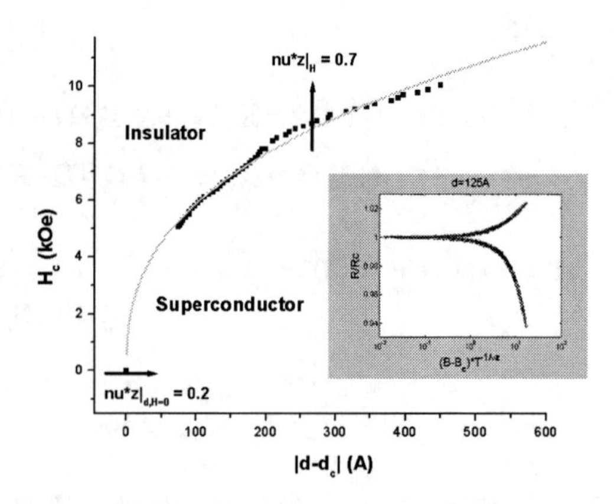

FIGURE 3. Phase diagram in the (d,H) plane. The points were obtained from the magnetic field-tuned transition and the thickness-tuned transition in zero-field. Here, d_c is the critical thickness for the zero-field transition ($d_c = 5$ nm). The solid line is a power-law fit giving $\nu = 0.19$. Inset : Rc versus $\frac{|H-H_c|}{T^{-\frac{1}{\nu z}}}$ for the 12.5 nm sample. All the data at temperatures between 150 mK and 450 mK, magnetic field between 0.5*H_c and 1.5*H_c are shown to collapse.

correlation length exponent and z the dynamical-scaling exponent - using the two different procedures described in [9]. Both procedures give similar results and we obtain the product $\nu z = 0.18 \pm 0.05$ (inset figure 2).

The effect of the magnetic field has been studied for the four thickest samples [14]. As for the thickness-tuned transition, a similar crossing point for the resistance isotherms, measured as function of magnetic field, is observed. Upon increasing the film thickness from 12.5 nm up to 100 nm, the critical magnetic field H_c is observed to change from 5 to 10.0 kOe, and the critical resistance from 1356 to 150 Ω, as found by previous studies [4].

The results were analyzed using the two independent scaling methods previously mentioned, the sample thickness being replaced by the magnetic field [9]. We obtained the product $\nu z = 0.67 \pm 0.05$. When the ratio $\frac{R_{square}}{R_c}$ is plotted against the scaling variable $\frac{|H-H_c|}{T^{-\frac{1}{\nu z}}}$ (inset of figure 3) we find a good collapse of the data for this value of the exponent.

We can then build the phase diagram shown on figure 3 in much the same way as in [9]. We have also evaluated the correlation length exponent ν near the zero-field transition via the relation $H_c \propto |d - d_c|^{2\nu}$. The obtained exponent is $\nu = 0.19 \pm 0.05$.

This result is in agreement with our value of νz found for the thickness-tuned SIT under zero-field and confirms the hypothesis that the dynamical exponent z is equal to 1 [15]. As reported for Bi thin films [9], we find that the critical exponent values characterizing the field-tuned and the thickness-tuned transitions are different. This could be explained by a difference in universality classes due to breaking of time-reversal symmetry induced by the magnetic field.

REFERENCES

1. B. Orr, H. Jaeger, A. Goldman, and C. Kuper, *Phys. Rev. Lett.*, **56**, 378 (1986).
2. A. Hebard, and M. Paalanen, *Phys. Rev. Lett.*, **65**, 927 (1990).
3. M. Paalanen, A. Hebard, and R. Ruel, *Phys. Rev. Lett.*, **69**, 1604 (1992).
4. A. Yazdani, and A. Kapitulnik, *Phys. Rev. Lett.*, **74**, 3037 (1995).
5. N. Markovic, A. Mack, G. Martinez-Arizala, C. Christiansen, and A. Goldman, *Phys. Rev. Lett.*, **81**, 701 (1998).
6. V. Gantmakher, M. Golubkov, V. Dolgopolov, G. Tsydynzhapov, and A. Shashkin, *Physica B*, **284**, 649 (2000).
7. E. Bielejec, and W. Wu, *Phys. Rev. Lett.*, **88**, 206802 (2002).
8. D. Haviland, Y. Liu, and A. Goldman, *Phys. Rev. Lett.*, **62**, 2180 (1989).
9. N. Markovic, C. Christiansen, A. Mack, W. Huber, and A. Goldman, *Phys. Rev. B*, **60**, 4320 (1999).
10. K. A. Parendo, K. H. Sarwa, B. Tan, A. Bhattacharya, M. Eblen-Zayas, N. E. Staley, and A. M. Goldman, *Phys. Rev. Lett.*, **94**, 197004 (2005).
11. L. Dumoulin, L. Bergé, J. Lesueur, H. Bernas, and M. Chapellier, *J. Low. Temp. Phys.*, **93**, 301 (1993).
12. D. Bishop, E. Spencer, and R. Dynes, *Sol. St. Elec.*, **28**, 73 (1985).
13. H. L. Lee, J. P. Carini, D. V. Baxter, W. Henderson, and G. Gruner, *Science*, **287**, 633 (2000).
14. H. Aubin, C. Marrache-Kikuchi, A. Pourret, K. Behnia, L. Bergé, L. Dumoulin, and J. Lesueur, *to be published* (2005).
15. M. Fisher, G. Grinstein, and S. Girvin, *Phys. Rev. Lett.*, **64**, 587 (1990).

Unusual Vortex Motion in the Quantum-Liquid Phase of Amorphous Films

S. Okuma, K. Kainuma, and T. Kishimoto

Research Center for Low Temperature Physics, Tokyo Institute of Technology, 2-12-1, Ohokayama, Meguro-ku, Tokyo 152-8551, Japan

Abstract. The fluctuating component of the flux-flow voltage, $\delta V(t)$, about the average voltage is measured in the low-temperature (T) liquid phase of amorphous Mo_xSi_{1-x} films. For the 100-nm-thick film $\delta V(t)$ originating from the vortex motion is clearly visible in the quantum-vortex-liquid (QVL) phase, where the distribution of $\delta V(t)$ is asymmetric, suggestive of the unusual vortex motion. For the 6- and 4-nm-thick films, in which the QVL phase is not determined from the resistance vs T curves, the similar unusual vortex motion is observed in nearly the same reduced-T regime.

Keywords: Flux flow, Quantum fluctuations, Superconductor-insulator transition, Amorphous films
PACS: 74.40.+k, 74.25.Dw, 74.78.Db

Quantum fluctuations are able to melt the vortex solid into the quantum vortex liquid (QVL) at zero temperature (T = 0). Experimentally, the QVL has been studied in various type-II superconductors with weak pinning [1-6]. For the thick amorphous (a-) Mo_xSi_{1-x} films with moderately strong pinning we have obtained evidence for the QVL phase at low temperatures T [7,8]. For thin amorphous films the existence of the *metallic* QVL phase has been reported by several groups based on the measurements of the resistance(R) vs T at low T [2,3]. For our thin a-Mo_xSi_{1-x} films, however, the T = 0 metallic phase has not been observed from $R(T)$. The QVL in two dimensions (2D) is particularly interesting, because it is related to the field(B)-driven superconductor-insulator transition (SIT).

While a number of studies have been performed to clarify the equilibrium vortex states, as far as we know, there is no paper reporting the vortex dynamics in the low-T and high-B regime, where quantum fluctuations play an important role. We have recently studied the change in vortex dynamics associated with the change in vortex states from the thermal to quantum liquid by measuring, in real time, the fluctuating component of the flux-flow voltage $\delta V(t)$ about the average voltage V_0 [9,10]. For the thick (100 nm) film the contribution of $\delta V(t)$ from the flux-flow motion is clearly visible (only) in the QVL phase, where the distribution of $\delta V(t)$ is asymmetric having a tail along the direction of vortex motion. This means that vortex motion in the QVL phase is not stationary but accompanied by large velocity and/or number fluctuations of moving vortices. For a thin (6 nm) film the similar unusual behavior of $\delta V(t)$ suggestive of the anomalous vortex motion is visible at nearly the same T/T_{c0} ($\approx$ 0.1) (where T_{c0} is the mean-field transition temperature) as that for the thick film. These results suggest that vortex dynamics in the low-T liquid phase of thick and thin films is dominated by common physical mechanisms, related to quantum effects. In this work we perform the same $\delta V(t)$ measurements for an ultrathin (4 nm) a-Mo_xSi_{1-x} film. This film is even closer to the ideal 2D superconductors than the 6-nm-thick film and exhibits clear evidence for the SIT.

The a-Mo_xSi_{1-x} films (x = 0.47-0.70) were prepared by coevaporation of pure Mo and Si. The resistivity and time-dependent voltage $V(t)$ induced by dc current I were measured using a four-terminal method. The voltage $V(t)$ enhanced with a preamplifier was recorded using a fast-Fourier transform spectrum analyzer with a time resolution of 39 or 390 μs [9,10]. The magnetic field was applied perpendicular to the plane of the film using a superconducting magnet in a persistent-current mode. We have measured $\delta V(t)$ and its distribution, $P(\delta V)$, as functions of T and B, where we choose the combination of (T, B) yielding nearly the same flux-flow resistance; $V_0/V_n \approx$ 0.4-0.5 (where V_n is the voltage in the normal state).

CP850, *Low Temperature Physics: 24th International Conference on Low Temperature Physics;*
edited by Y. Takano, S. P. Hershfield, S. O. Hill, P. J. Hirschfeld, and A. M. Goldman

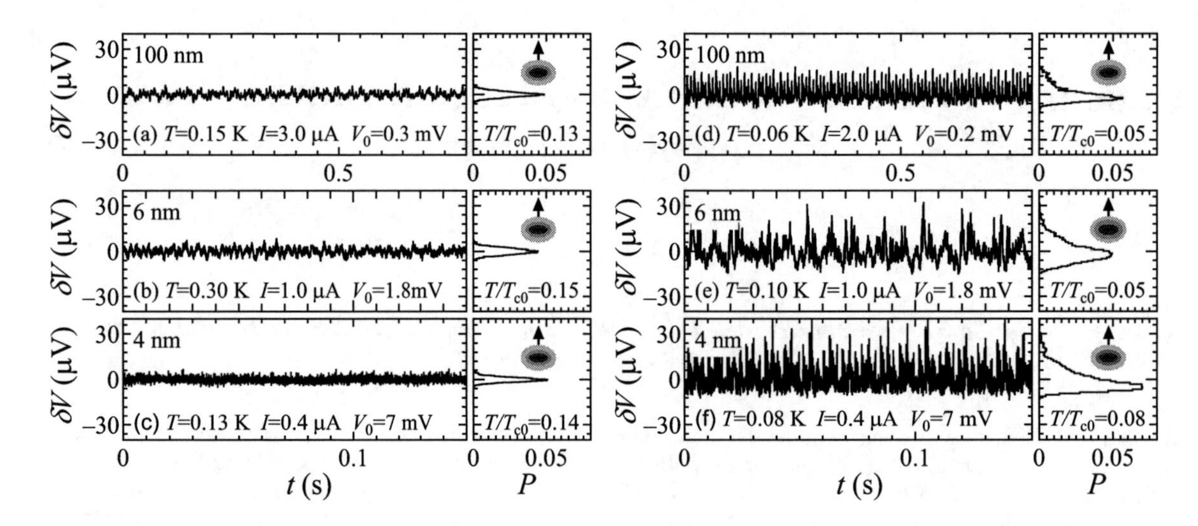

FIGURE 1. $\delta V(t)$ (left) and $P(\delta V)$ (right) (a) for the 100-nm-thick film at 0.15 K (T/T_{c0}=0.13) in 3.12 T, (b) for the 6-nm-thick film at 0.30 K (T/T_{c0}=0.15) in 4.83 T, and (c) for the 4-nm-thick film at 0.13 K (T/T_{c0}=0.14) in 2.22 T. $\delta V(t)$ and $P(\delta V)$ at lower T are also shown (d) for the 100-nm-thick film at 0.06 K (T/T_{c0}=0.05) in 3.24 T, (e) for the 6-nm-thick film at 0.10 K (T/T_{c0}=0.05) in 5.22 T, and (f) for the 4-nm-thick film at 0.08 K (T/T_{c0}=0.08) in 2.35 T. In each figure (right) the direction of the vortex motion is illustrated with an arrow.

Figures 1(a)-(c) depict $\delta V(t)$ and $P(\delta V)$ for the 100-nm-thick film measured at 0.15 K (T/T_{c0} = 0.13) in 3.12 T [i.e., in the thermal-vortex-liquid (TVL) phase or in the crossover regime between the TVL and QVL phases], for the 6-nm-thick film at 0.30 K (T/T_{c0} = 0.15) in 4.83 T, and for the 4-nm-thick film at 0.13 K (T/T_{c0} = 0.14) in 2.22 T. For all the currents I studied, both $\delta V(t)$ and $P(\delta V)$ are nearly identical to the background (I = 0) data and exhibit no anomaly.

In the QVL phase of the thick (100 nm) film, by contrast, the contribution of $\delta V(t)$ from the vortex motion is clearly visible. Figure 1(d) shows $\delta V(t)$ and $P(\delta V)$ for the 100-nm-thick film at 0.06 K (T / T_{c0} = 0.05) in 3.24 T (i.e., in the QVL phase). The amplitude of $\delta V(t)$ measured at nonzero V_0 is remarkably larger than the background data (I = 0, V_0= 0) and the shape of $P(\delta V)$ is highly asymmetric having a tail which extends to the direction of vortex motion. In the T, B, and I regime where $P(\delta V)$ exhibits the unusual asymmetry, large broad-band noise of a Lorentzian type is observed. In the presence of larger I where the film is nearly in the normal state, both $\delta V(t)$ and $P(\delta V)$ are almost identical to the background data. This result implies that the physical origin of large $\delta V(t)$ with asymmetric $P(\delta V)$ is due to the anomalous vortex motion in the liquid phase.

For the thinner (6 and 4 nm) films, in which the QVL phase is not determined from $R(T)$, the similar unusual $\delta V(t)$ is observed in nearly the same reduced-T regime (T / T_{c0} < 0.1) as that for the thick film [9,10]. Figures 1(e) and (f) representatively show $\delta V(t)$ and $P(\delta V)$ for the 6-nm-thick film at 0.10 K (T / T_{c0} = 0.05) in 5.22 T and for the 4-nm-thick film at 0.08 K (T / T_{c0} = 0.08) in 2.35 T, respectively.

All of the results obtained for the three films with different thicknesses (100, 6, and 4 nm) further support our view [9] that, despite the large difference between the static vortex phase diagrams for 3D and 2D, the vortex dynamics in the low-T liquid phase of thick and thin films is dominated by common physical mechanisms related to quantum-fluctuation effects.

REFERENCES

1. G. Blatter *et al.*, *Phys. Rev. B* **50**, 13013 (1994).
2. J. A. Chervenak and J. M. Valles, Jr., *Phys. Rev. B* **54**, R15649 (1996).
3. D. Ephron *et al.*, *Phys. Rev. Lett.* **76**, 1529 (1996).
4. N. Markovic *et al.*, *Phys. Rev. Lett.* **81**, 701 (1998).
5. T. Sasaki *et al.*, *Phys. Rev. B* **57**, 10889 (1998).
6. T. Shibauchi, L. Krusin-Elbaum, G. Blatter and C. H. Mielke, *Phys. Rev. B* **67**, 064514 (2003).
7. S. Okuma, Y. Imamoto and M. Morita, *Phys. Rev. Lett.* **86**, 3136 (2001).
8. S. Okuma, S. Togo and M. Morita, *Phys. Rev. Lett.* **91**, 067001 (2003).
9. S. Okuma, M. Kobayashi and M. Kamada, *Phys. Rev. Lett.*, **94**, 047003 (2005) and references therein.
10. S. Okuma, K. Kainuma and T. Kishimoto, submitted.

Electronic Transport in the Low-Temperature Liquid Phase of a Thin Amorphous Film without the Edge Effects

S. Okuma, T. Kishimoto, K. Kainuma, and M. Morita

Research Center for Low Temperature Physics, Tokyo Institute of Technology, 2-12-1, Ohokayama, Meguro-ku, Tokyo 152-8551, Japan

Abstract. We study the possible edge-pinning effects on the transport properties in the low-temperature (T) liquid phase of a thin amorphous Mo_xSi_{1-x} film with the Corbino-Disk (CD) and strip-like contact geometries. In CD, in the presence of a radial current, field-induced vortices rotate around the center of the sample without crossing the sample edges and, hence, the edge effects are negligible. In both geometries we observe the two-dimensional field-driven superconductor-insulator transition at $T \to 0$. This result indicates that the edge effects are not important and the absence of an intermediate metallic vortex phase at $T = 0$ is intrinsic in our thin amorphous films.

Keywords: Superconductor-insulator transition, Corbino-disk, Edge effects, Amorphous films
PACS: 74.40.+k, 74.25.Dw, 74.78.Db

INTRODUCTION

The electrical resistance R in the low-temperature (T) vortex-liquid phase of two-dimensional (2D) superconductors has been actively studied more than a decade from the viewpoint of the field(B)-driven superconductor-insulator transition (SIT) [1-6]. We have shown so far that thin amorphous (a-)Mo_xSi_{1-x} films exhibit the SIT at $T = 0$ either by increasing a magnetic field or disorder (normal-state resistivity ρ_n) [4-6]. Thus, if the quantum-vortex-liquid (QVL) is present at low temperature ($T \to 0$) in our film, it should lie in an *insulating* phase immediately above the field-driven SIT [5-7]. In the meantime, the existence of a *metallic* QVL phase at $T = 0$ has been reported in several thin amorphous films [1,2]. This result is important, because it challenges the traditional picture of the 2D SIT. Some theories have predicted the $T = 0$ metallic phase in 2D [8,9], while there is a theory which precludes the possibility of an intermediate metallic vortex phase at $T = 0$ [10]. We consider that, independent of the theory, it is important to clarify experimentally what makes the difference between the low-T transport properties in the similar amorphous thin films studied by us and by other groups. In our earlier paper [6] we have pointed out the possibility that the strong pinning effects in our amorphous a-Mo_xSi_{1-x} films may be responsible for the absence of the metallic QVL behavior based on the measurements of the T dependence of the resistance R.

Here we consider the possible edge-pinning (edge-contamination) effects on the transport properties, which have been originally studied in $NbSe_2$ single crystals at relatively high T [11]. If the strong pinning effects are present at the sample edges, they may seriously affect (hinder) the vortex motion and hide the intrinsic properties. In order to examine whether the absence of the metallic QVL behavior in our amorphous films merely originate from the edge-pinning effects instead of the intrinsic properties, we measure $R(T)$ of a thin a-Mo_xSi_{1-x} film in a condition free from the edge effects. This is realized by using a thin a-Mo_xSi_{1-x} film with a Corbino-Disk (CD) contact geometry. In CD, in the presence of a radial current, the field-induced vortices continue to rotate without crossing the sample edges by feeling a Lorentz force and, hence, edge effects are negligible.

EXPERIMENTAL

The sample used in this study was a 6-nm-thick a-Mo_xSi_{1-x} film with $x \approx 0.7$, which was prepared by coevaporation of pure Mo (99.95 %) and Si (99.999 %) onto a glass substrate held at room temperature [4-6]. The structure of the films was confirmed to be amorphous by means of transmission electron microscopy and electron diffraction. The arrangement of the silver electrical contacts is shown in the inset of Fig. 1. The current flows between the contact, +C, of

CP850, *Low Temperature Physics: 24th International Conference on Low Temperature Physics;*
edited by Y. Takano, S. P. Hershfield, S. O. Hill, P. J. Hirschfeld, and A. M. Goldman

the center and that, -C, of the perimeter of the disk, which produces a radial current density that decays as $1/r$, where r is a radius of rotation. For the measurements in the strip-like geometry, contacts +S and -S were used. For both contact geometries, we used the same voltage contacts, +V and -V. The similar contact arrangement was used originally by Paltiel *et al.* to study comparatively the vortex states in the mixed state for the CD and strip-like geometries on the same sample [11]. The distance between the voltage contacts was 0.5 mm and the diameter of the CD was 5.2 mm. Measurements were made by four-terminal ac lock-in techniques at 19 Hz with an applied current of 10 nA, which was well within the ohmic regime. The magnetic field B was applied perpendicular to the plane of the film.

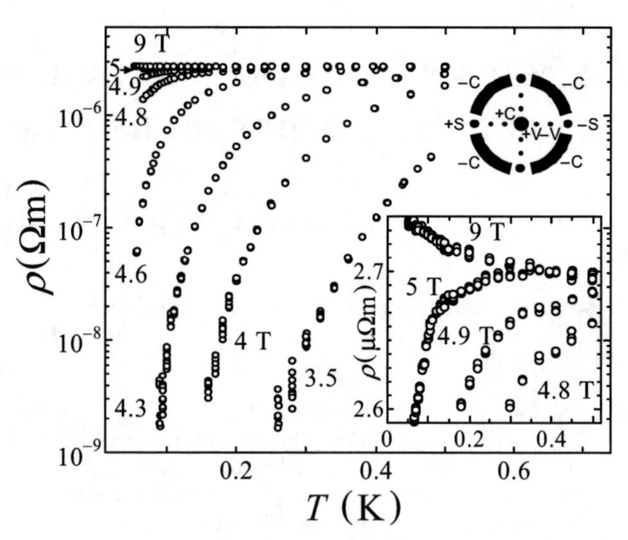

FIGURE 1. Temperature dependence of ρ in different magnetic fields for the 6-nm-thick a- Mo_xSi_{1-x} film with the CD contact geometry. Inset: (Top) The arrangement of the electrical contacts. (Bottom) The high- ρ part is enlarged and shown. These data show the existence of the $T = 0$ field-driven superconductor-insulator transition instead of the metal-insulator transition.

RESULTS AND DISCUSSION

The mean-field transition temperature and the normal-state resistivity ρ_n at 10 K calculated from the resistance (V/I) at the probe position assuming $J \propto 1/r$ were 2.2 K and 2.7 μ Ωm, respectively. Figure 1 depicts the temperature dependence of the resistivity $\rho(T)$ in different fields B for the CD contact geometry. In low fields ($B < 4.5$ T), $\rho(T)$ exhibits a superconducting behavior. With increasing B, the slope of the isomagnetic curves at the lowest T decreases and approaches zero (i.e., ρ becomes T-independent) at around 5 T. In higher B, $\rho(T)$ stays nearly unchanged. The high- ρ part is enlarged and shown in the inset of Fig. 1. We find that $\rho(T)$ in 5 T is superconducting, while that in 9 T is insulating, indicating the field-driven SIT occurs between 5 and 9 T. We notice that the insulating behavior of $\rho(T)$ is not so remarkable. This is because the film with $\rho_n =$ 2.7 $\mu\Omega$m used in this study is not resistive enough (i.e., it is far below the critical resistivity of the disorder-driven SIT) [5] and thus the contribution from the unpaired electrons is important even in the low-T regime (0.06-0.5 K)[4].

It should be noted in Fig. 1 that the weakly T-dependent or nearly T-independent $\rho(T)$ at $T \to 0$, suggestive of the intermediate $T = 0$ metallic phase, is not visible in any B studied. For the strip-like geometry we observe the essentially same behavior of $\rho(T, B)$ as shown in Fig. 1 except for the slight difference in the $\rho(T)$ curves, which probably arises from the different current distribution between the two contact geometries. All of the results obtained in this study indicate that the edge-pining effects are not important in our thin a-Mo_xSi_{1-x} film in the T and B range studied. Accordingly, the picture of the field-driven 2D SIT is not altered.

This work was supported by a Grant-in-Aid for Scientific Research from the Ministry of Education, Culture, Sports, Science, and Technology.

REFERENCES

1. D. Ephron, A. Yazdani, A. Kapitulnik and M. R. Beasley, *Phys. Rev. Lett.* **76**, 1529 (1996).
2. J. A. Chervenak and J. M. Valles, Jr., *Phys. Rev.* B **54**, R15649 (1996); **61**, R9245 (2000).
3. V. F. Gantmakher *et al.*, *JETP Letters*, **71**, 160 (2000).
4. S. Okuma, T. Terashima and N. Kokubo, *Solid State Commun.* **106**, 529 (1998):*Phys. Rev.* B **58**, 2816 (1998).
5. S. Okuma, S. Shinozaki and M. Morita, *Phys. Rev.* B **63**, 054523 (2001).
6. S. Okuma, M. Morita and Y. Imamoto, *Phys. Rev.* B **66**, 104506 (2002).
7. N. Marković, A. M. Mack, G. Martinez-Arizala, C. Christiansen and A. M. Goldman, *Phys. Rev. Lett.* **81**, 701 (1998).
8. P. Phillips and D. Dalidovich, *Phil. Mag.* **81**, 847 (2001).
9. D. Das and S. Doniach, *Phys. Rev.* B **64**, 134511 (2001).
10. H. Ishida and R. Ikeda, *J. Phys. Soc. Jpn.* **71**, 254 (2002).
11. Y. Paltiel *et al.*, *Europhys. Lett.* **58**, 112 (2002).

Superconducting Properties in Granular In/Ge Films

T.Nakamura, B.Shinozaki, and T.Kawaguti

Department of Physics, Kyushu University, Ropponmatsu, Fukuoka 810-8560

Abstract. In order to investigate the crossover from a homogeneous behavior to an inhomogeneous (percolative) one, T and H dependence of the resistance R of granular Indium films has been measured. We prepared films in a wide range of the resistivity ρ by changing the deposition conditions. Except for the films with relatively low values of ρ, H_{C2}-T curves show a downward curvature in a low magnetic field region. To explain this downward curvature, we applied the percolation theory which has the diffusion index θ. The values of θ of a two dimensional (2D) and a three dimensional (3D) systems are given as 0.98 and 2.3, respectively, and that of homogeneous one is zero. θ has been found to correlate well to ρ. θ increases with increase of ρ for the whole range of ρ. For films with ρ between 10 and 1000 $\mu\Omega$ cm, it seems that θ takes values around a certain one unexpected from theory. In the region above $\rho \approx 5\times 10^{-4}\Omega$ cm, the value of θ is bigger than 1.

Keywords: percolation, granular Indium, diffusion index.
PACS: 74.81.-g ,74.81.Bd

INTRODUCTION

Superconductor-insulator mixtures and granular superconductors have been widely investigated in order to clarify the interplay between percolation and superconductivity. Morphologies of the specimens, that is, the granularity, connectedness, and the intergrain spacing are important to discuss the above interplay. A homogeneous-inhomogeneous crossover in granular superconductors has been investigated by using percolation models. Percolative films are characterized by the condition $\xi_p > \xi_S$, where ξ_p and ξ_S are the percolation correlation length and the superconducting coherence length, respectively. According to the model[1-3], the diffusion constant D has a scale dependence $D(L) \propto L^{-\theta}$, where θ is the diffusion index. The ξ_S has the relation $\xi_S^2 = D\tau_{GL}$, where τ_{GL} is the GL relaxation time given as $\tau_{GL} \propto |1-T/T_C|^{-1}$. Therefore, when L is replaced with ξ_S, we obtain the $\xi_S \propto \tau_{GL}^{1/(2+\theta)} \propto |1-T/T_C|^{-1/(2+\theta)}$. As a result, the upper critical field H_{C2} is given by $H_{C2} \propto 1/\xi_S^2 \propto (1-T/T_C)^{2/(2+\theta)}$, although H_{C2} of homogeneous films increases linearly with T near Tc.

In this paper, we report the superconducting property and the surface images by atomic force microscopy (AFM) of In/Ge granular films. We discuss crossovers from a homogeneous to a 2D percolation systems and also from a 2D to a 3D percolation ones.

EXPERIMENT

The In films were made by the vacuum deposition on a glass substrate covered with Ge. The samples were prepared by following four different conditions of changing the In deposition rate and the Ge thickness as (i) fast and thick, (ii) fast and thin, (iii) slow and thick and (iv) slow and thin. During deposition, we kept a substrate temperature of 300K for all conditions. Further, for (ii) and (iv) In was also deposited on substrates kept at 77K. In the R-T curve at a constant H, $T_C(H)$ was defined as a temperature at which half of the normal state resistance was restored. It is found that the T dependence of H_{C2} for films in a range of ρ over 10$\mu\Omega$ cm was not linear in a low field, but had a downward curvature. To explain this dependence, in this paper we applied the percolation theory.

Figure 1 shows the ρ dependence of the index θ. It is found that the θ correlated with ρ in a wide range of ρ, increasing on the whole with increase of ρ. The films with $\theta \approx 0$ below $\rho \approx 30\mu\Omega$ cm correspond to the homogeneous ones. It can be considered that the films with $\theta \approx 1$ in a range of ρ between ~100 and ~200$\mu\Omega$ cm are 2D percolation ones. Although there is no theoretical prediction of the existence of a certain value of θ between 0 and 1, it is interesting that θ seems to take values around 0.6 in a range of ρ between ~10 and ~1000$\mu\Omega$ cm.

CP850, *Low Temperature Physics: 24th International Conference on Low Temperature Physics;*
edited by Y. Takano, S. P. Hershfield, S. O. Hill, P. J. Hirschfeld, and A. M. Goldman

Figure 2 shows the surface of the In films taken by AFM; Figs. 2(a) and 2(b) are the pictures of the surface of the film prepared by (i) and (iii). It is found that the grains of films of faster In deposition rate are larger than those of the slower ones.

DISCUSSION

We will compare the characteristics of surface structures and θ -ρ relation between films by (i) and (iii). Although the extensions of grains parallel to the substrate are about 1.0 μ m and 0.5 μ m for films by (i) and (iii), respectively, the height differences of grains are almost the same, about 0.14μ m for both films. From this result and the total thickness 700Å measured by the quartz crystal oscillator, it is considered that both these films by (i) and (iii) have the structure that the grain is not multiple but single in the direction normal to the substrate, from assuming that each grain does not have the internal construction. Therefore, the current may flow in these films, through interfaces between grains, along a kind of the intermediate conduction path between the homogeneous and the 2D percolative ones. For most films by (i), the values of θ are determined as $0<\theta<1$ in Fig.1. This suggested that these films are in an intermediate state between homogeneous and 2D percolation films. This suggestion is not inconsistent with the observation shown in Fig.2(a). On the other hand, for films by (iii), θ takes values larger than 1 in Fig.1. Taking account of the percolation theory, θ between 1 and 2.3 suggest a possibility of crossover from a 2D to a 3D percolation systems. However, this scenario does not agree with the observation.

For the reason of this discrepancy, followings can be considered. The resistivity of films by (i) is smaller than that of films by (iii) by a factor of about ten, although the Ge thicknesses and the In thicknesses in films by (i) and (iii) are the same and the structures in those films are also almost the same as discussed above. This difference of ρ for films by (i) and (iii) can be understood by assuming that the connectedness between grains changes. Therefore, when θ takes the value suggesting a 3D percolation system, it must be considered that other contributions due to change of the connections than the simple percolation model may occur. Next, we will briefly discuss the crossover from a homogeneous to a 2D percolation systems, where $0<\theta<1$. For Al-Al_2O_3 films prepared by the multilayer method[4], with small grains of diameter less than 40Å , the values of θ changed continuously in the crossover region. On the other hand, for the present In films with $0<\theta<1$, the film is constructed by single layer of large grains in the direction perpendicular to the substrate as discussed above, although films prepared at 77K is not yet confirmed. As a reason which causes the difference of a discontinuous and a continuous changes of θ , we can consider the difference of connectedness due to the grain size of the films prepared by different conditions.

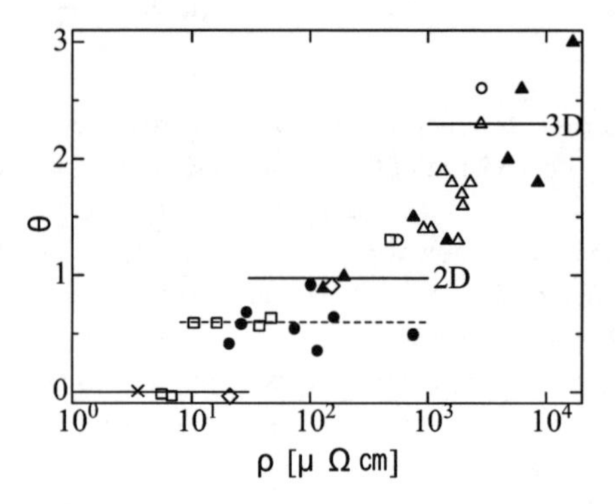

FIGURE 1. Resistivity ρ dependence of index θ : ● (i); ○ (ii); ▲ (iii); and △ (iv), at 300K. □ (ii); and ◇ (iv), at 77K. × represents a film prepared on a substrate covered with SiO at 77K. The solid lines show theoretical values of θ for a homogeneous, a 2D percolation and a 3D percolation systems sequentially from the smaller one about θ . The broken line shows θ =0.6.

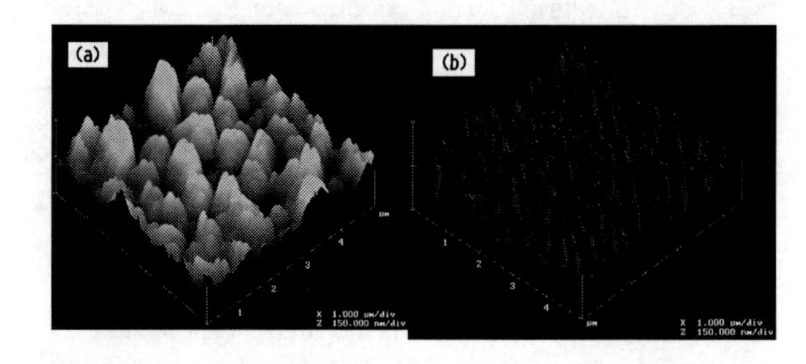

FIGURE 2. Surface images of films taken by AFM. (a) (i); and (b) (iii), at 300K.

REFERENCES

1. A. Gerber and G. Deutscher, *Phys. Rev. B* **35**, 3214 (1987).
2. A. Gerber and G. Deutscher, *Phys. Rev. Lett.* **63**, 1184 (1989).
3. D. Stauffer, *Introduction to Percolation Theory*, London: Taylor and Francis, 1994.
4. K. Yamada, B. Shinozaki and T. Kawaguti, *Phys. Rev. B* **70**, 144503 (2004).

Superconductor -Insulator Transition of In-Sb Granular Films

B. Shinozaki, M. Watanabe, T. Nakamura and T. Kawaguti

Department of Physics, Kyushu University, Ropponmatsu, Fukuoka 810-8560 Japan

Abstract. Measurement of temperature T and magnetic field H dependence of the sheet resistance $R_\square$ of In-Sb thin films have been made up to large magnetic fields. For specimens, we prepared granular films composed of In grains surrounded by normal materials by the method of vacuum depositions. With decreasing the In concentration, we have obtained systematic changes in $R_\square$ -T curves at zero field from global superconductivity to superconductor- insulator, including quasi-reentrant superconductivity between them. For all films showing reentrant or insulating properties, we have observed negative magneto-resistance below T_c. Furthermore, the reentrant film near the insulating region showed double reentrant behaviors in R-H relation at low temperatures.

Keywords: S-I transition, negative magneto resistance, reentrant R-T relation
PACS: 74.78.-w,74.81.-g,74.81.Bd

INTRODUCTION

In the last two decades, experimental and theoretical works in superconductivity in granular films have been intensively investigated from the viewpoint of electron transports, especially, superconductor- insulator (S-I) transition and unusual negative magneto-resistance (NMR). [1-4] The NMR has been observed in not only granular films made by immiscible metallic grains dispersed in insulating continuum but also the quenched films whose structure consists of isolated clusters. At low magnetic fields, this behavior can be explained by the depression of the superconducting energy gap in the superconducting grains.[2] Recently, large NMR signal have been observed near the S-I transition[4] and discussed as the recovery of the depression of state of density due to superconducting fluctuation at large magnetic fields.

In this paper, in order to investigate the interplay between the disorder and the S-I transition and also to clarify the mechanism of anomalous NMR at large magnetic fields in the granular superconducting 2D systems, temperature dependence of $R_\square$ of In-Sb thin films have been measured in a wide range of the magnetic field with change of the In concentration of In-Sb systems.

RESULTS AND DISCCUSSIONS

The present In-Sb films were prepared by following procedure; At first, the In and Sb were deposited alternately on the glass substrate held at liquid N_2 temperature, where we kept the sum of the thicknesses of these materials is 100Å . We repeated this procedure twice, and the total film thickness is about 200Å . Secondly, we increased gradually the substrate temperature to 300K. By the Auger electron spectroscopy method at 300 K, we recognized that the present films show not layer structures but almost the homogeneous distributions of elements In and Sb through the sample thickness. By changing the ratio of these film thicknesses, we prepared films with various concentrations of In. From the Auger-analysis and the following experimental results of the NMR, we suggest that the films are composed of superconducting In grains surrounded by intermetallic compound InSb or pure Sb.

Figure 1 shows the $R_\square$ -T relations at H=0 for films with various In concentrations. With decreasing concentration, the transition from the superconductor to insulator occurs. In the intermediate $R_\square$ region, it can be seen that the reentrant films show abrupt decrease of $R_\square$ due to local superconductor near the T_c of global superconductors. These films showed the strong T dependence of R bellow T_c compared with

CP850, *Low Temperature Physics: 24th International Conference on Low Temperature Physics;*
edited by Y. Takano, S. P. Hershfield, S. O. Hill, P. J. Hirschfeld, and A. M. Goldman

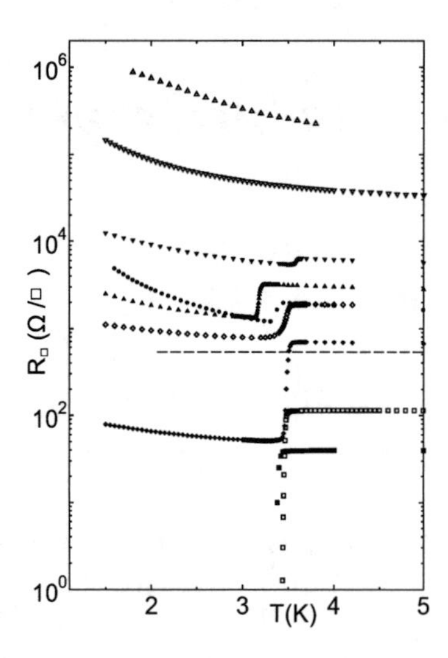

FIGURE 1. *T* dependence of *R* for In-Sb films with various In concentrations. (-----) pure Sb film.

that above T_c. It can be considered that reduction of the density of state in the isolated grains with decreasing *T* inhibits the electron transport, and then *R*□ increases. Under large magnetic fields, the *R*(*T*) in the whole temperature region shows the $R=R_0\exp(\beta/T)^{1/2}$. Such a temperature dependence of *R* can be understood as a hopping conduction in granular systems.

Figure 2 shows the *H* dependence of *R*(*H*)/*R*(0) at various temperatures for the film with *R*□(4.2K)≈1KΩ . The *R* decreases gradually with *H* and takes a minimum at a certain value of H_{min}. With decreasing *T*, magnitudes of the minimum value of the NMR and H_{min} increase. The inset shows the *T* dependence of H_{min}. In the present In-Sb system, we found that the strength of H_{min} is the same order to that of upper critical magnetic field H_{c2} for global superconductors. Therefore, it is reasonable to consider that the energy gap is almost depressed by the H_{min}. On the other hand, as shown in Fg.3, the reentrant film with *R*□(4.2K)≈6KΩ near the insulating region showed the double reentrant behaviors in *R-H* relation at low temperatures. Such a behavior has been observed in granular Al[2] and thin TiN films.[4] The initial lowering of *R*(*H*) is also due to the suppression of the energy gap by the magnetic field as mentioned above. From the fact that the MR becomes positive above T_c, it cannot be considered that the second large NMR at low temperatures is not because of the electron weak localization. For the possibility of this lowering of *R*(*H*), we expect that it is attributed to the suppression of the fluctuation pseudogap in grains by the magnetic field. According to the recent theory[5], the total fluctuation corrections to the normal conductivity are negative at low temperatures, and the magneto-

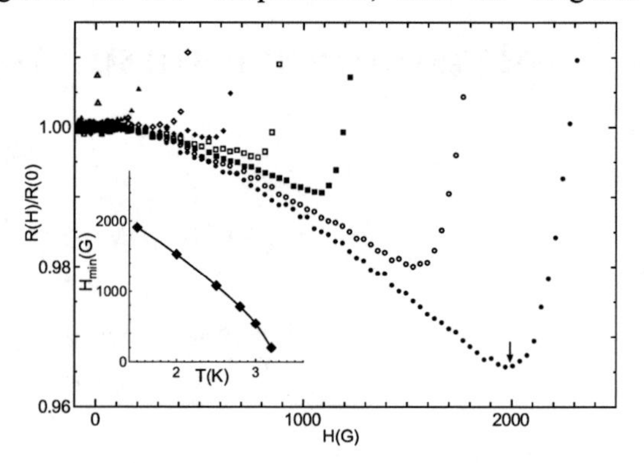

FIGURE 2. *H* dependence of *R* for film with *R*□ (4.20K) ≈1KΩ .

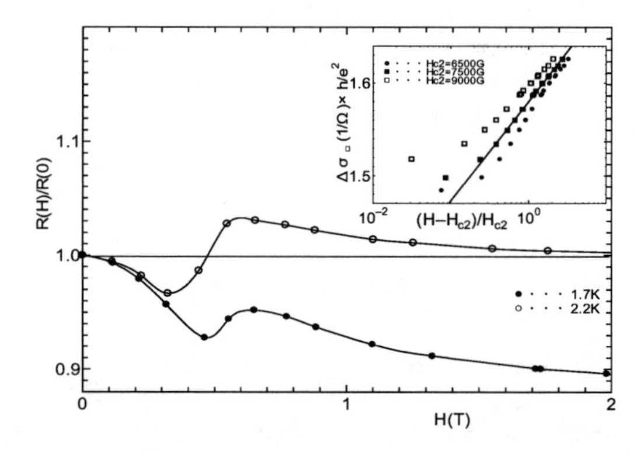

FIGURE 3. *H* dependence of *R* for film with *R*□ (4.20K) ≈6KΩ .

conductance depends logarithmically on *H* at large magnetic fields, $\Delta\sigma\ (h/e^2)=(4/3\pi)[\ln((H-H_{c2})/H)]$. The inset in Fig.3 shows the $\Delta\sigma\ (h/e^2)$ - ln*H* relation, where we try to choose the some values of H_{c2} to obtain the linear relation between $\Delta\sigma$ and ln*H* for the large magnetic field region. Although we obtained linear relation, there is a quantitative discrepancy of an order of magnitude between the theory and experimental results.

REFERENCES

1. B. G. Orr, H. M. Jaeger and A. M. Goldman, *Phys. Rev. B* **32**, 7586 (1985).
2. M. Kunchur, Y. Z. Zhan, P. Lindenfild and W. L. McLean, *Phys. Rev. B* **36**, 4062 (1987).
3. D. C. Kim, J. S. Kim, H. R. Kang, G. T. Kim, A. N. Baranov and Y. W. Park, *Phys. Rev. B* **64**, 64502 (2001).
4. N. Hadacek, M. Sanquer and J.-C. Villegier, *Phys. Rev. B* **69**, 024505 (2004).
5. V. M. Galitski and A. I. Larkin, *Phys. Rev. B* **63**, 174506 (2001).

Scanning Tunneling Spectroscopy on a Disordered Superconductor

C. Chapelier*, W. Escoffier*,†, B. Sacépé*, J.-C. Villégier* and M. Sanquer*

*DRFMC-SPSMS, CEA, Grenoble, F-38054, FRANCE
†Department of Physics and Astronomy, University of Manchester, Manchester, M13 9PL, UK

Abstract. Thin disordered TiN films display a multiple reentrant superconductor-insulator transition. By combining very low temperature scanning tunneling microscopy and spectroscopy on TiN films we have observed a non uniform state made of superconducting and normal areas. These inhomogeneities can be related to fluctuations driven quantum superconductor-metal transition models recently developed. When the disorder of the film is increased or its thickness reduced, electronic interactions begin to take over and the LDOS develops a zero bias anomaly (ZBA). We observed smooth spatial transitions between areas displaying a superconducting LDOS and other ones with a ZBA in the excitation spectrum.

Keywords: tunneling spectroscopy, disordered superconductivity, electronic interactions, proximity effect
PACS: 71.30.+h, 73.23.Hk, 74.45.+c, 74.81.-g

INTRODUCTION

There are two main classes of materials displaying a superconductor-insulator transition (SIT). For homogeneously disordered films, the SIT occurs simultaneously with the monotonous decrease of both the superconducting temperature and gap. The competition between the attraction potential and the disorder-enhanced Coulomb repulsion potential is at the origin of the SIT [1]. On the contrary, when granular films are driven close to the SIT, the superconducting transition broadens but the onset of the critical temperature T_c remains unchanged. In this case, the SIT occurs via the localization of already formed Cooper pairs inside the grains. Superconductivity is fully developed into each grain but as the phase fluctuates from grain to grain, the global superconducting phase coherence is lost [2]. The resistance close to the SIT in most of granular films displays a re-entrant transition as a function of magnetic field or temperature [see for example 3]. We report here tunneling spectroscopy on TiN films for various degrees of disorder. Although granular, these films undergo a disorder dependent sharp superconducting transitions like homogeneous ones down to 500 mK. At lower temperature, films very close to the SIT display however a double reentrant behavior of the thermal dependence of their resisitivity, suggesting a more inhomogeneous superconductivity [4].

TIN FILMS AND STM TECHNIQUE

TiN was prepared by DC reactive magnetron sputtering at 350°C on oxidized Si substrates. By sputtering a Ti target at various nitrogen partial pressures, we obtained TiN_δ compositions with $0.7 \leq \delta \leq 1.2$ and room temperature sheet resistance ranging from $R_{sq} \approx 8\ \Omega$ to $10^4\ \Omega$ for films of different thicknesses.

We combined topography and spectroscopy measurements with a STM cooled down below 100 mK. In order to probe the local density of states (LDOS), the differential conductance was obtained with a lock-in amplifier technique. While scanning the tip, we set the tip-sample DC bias voltage $V = V_g$, so that $e.V_g = \Delta_0$ corresponds to the energy at which the LDOS is maximum when the sample is locally superconducting. Then, by recording simultaneously the topography Z(x,y) and $\frac{dI}{dV}$ (V_g,x,y), we were able to map the spectroscopic properties of the surface and to reveal superconducting spatial inhomogeneities (images can be seen in Ref. [5]).

EXPERIMENTAL RESULTS

Moderately disordered films

For a 100 nm thick film with $R_{sq} = 27\ \Omega$ at room temperature and T_c = 4.7 K, we found a gap Δ = 0.73 mV and a ratio $\frac{\Delta}{kT_c}$ = 1.8 close to the BCS value of 1.76. However, these spectra were not observed everywhere and at other locations a normal-metal-like flat LDOS was measured. We recorded the LDOS evolution along a line across the boundary between a superconducting (S) and a normal region (N) (see Fig.1). Except for the trully gaped spectra on the S side, none of the spectra were BCS-like. We rather observed spectra with peaks pinned at Δ for any position between S and N but with a non zero flat LDOS for energies smaller than Δ which increased as the tip was moved away from S.

CP850, *Low Temperature Physics: 24th International Conference on Low Temperature Physics;*
edited by Y. Takano, S. P. Hershfield, S. O. Hill, P. J. Hirschfeld, and A. M. Goldman

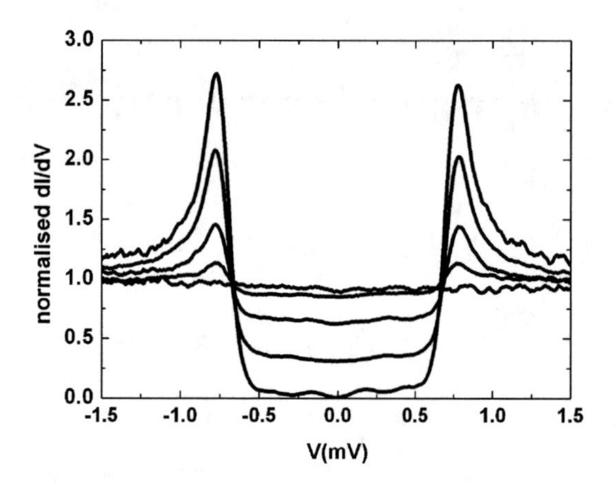

FIGURE 1. Evolution of the LDOS along a 48 nm long line.

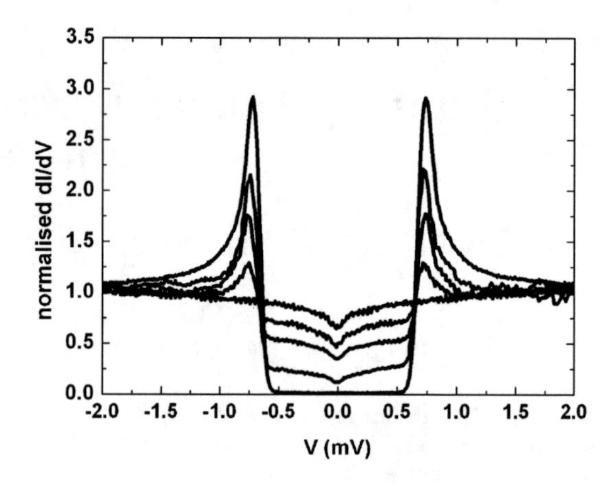

FIGURE 3. Evolution of the LDOS along a 60 nm long line.

Intermediate disordered films

For a 100 nm thick film with $R_{sq} = 135\ \Omega$ at room temperature and $T_c = 4.3$ K we still observed inhomogeneous superconducting properties. While the superconducting spots where characterized by a unique order parameter $\Delta = 0.69$ mV, the normal zones displayed a zero bias anomaly (ZBA) whose strengh could vary drastically from place to place. Two spatial evolutions of the LDOS across the SN interface are shown on Fig.2 and Fig.3.

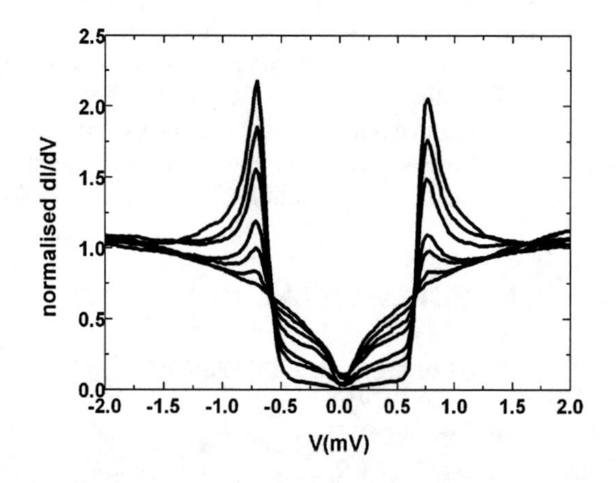

FIGURE 2. Evolution of the LDOS along a 25 nm long line.

Highly disordered films

We also investigated a 10 nm thick film with $R_{sq} = 8400\ \Omega$ at T=1.2 K and $T_c = 0.4$ K. This film was very close to the SIT and displayed a double reentrant behavior in transport measurements. We found almost everywhere a pronounced ZBA (not shown) and tiny superconducting spots with $\Delta = 0.18$ mV.

DISCUSSION

Although, according to transport measurements, our TiN films behaved like homogeneous ones for a not too high disorder, the STM spectroscopy showed that there is always some disorder induced granularity in the superconducting properties as observed for the first time by Kowal *et al* in an amorphous film [6]. Moreover, this inhomogeneous state can't come from variations of the BCS coupling constant due to spatial fluctuations of the disorder (that we obviously saw if we compare Fig.2 and Fig.3). Indeed we would have measured in this case a broad distribution of the order parameter for different positions on a same film. It is therefore more likely that the mesoscopic fluctuations of the disorder drive locally the film into either a normal metal or a fully superconducting cluster following a quantum scenario described by Feigel'man *et al* for superconducting grains embedded in a normal matrix [7].

REFERENCES

1. A. Finkel'stein, *Physica B*, **197**, 636 (1994).
2. A. Goldman, and N. Marković, *Physics Today*, **51**, 39 (1998).
3. H. Jaeger, D. Haviland, B. Orr, and A. Goldman, *Phys. Rev. B*, **40**, 182 (1989).
4. N. Hadacek, M. Sanquer, and J.-C. Villégier, *Phys. Rev. B*, **69**, 024505 (2004).
5. W. Escoffier, C. Chapelier, N. Hadacek, and J.-C. Villégier, *Phys. Rev. Lett.*, **93**, 217005 (2004).
6. D. Kowal, and Z. Ovadyahu, *Solid State Com.*, **90**, 783 (1994).
7. M. Feigel'man, A. Larkin, , and M. Skvortsov, *Phys. Rev. Lett.*, **86**, 1869 (2001).

Superconductor-Insulator Transition Induced by Disorder outside the CuO_2 Plane

Kazuhiro Fujita [a], Shin-ichi Uchida [a, b], and Hiroshi Eisaki [c]

[a] Department Advanced Materials Science, The University of Tokyo Bunkyo, Tokyo 113-0033, Japan
[b] Department Physics, The University of Tokyo Bunkyo, Tokyo 113-0033, Japan
[c] Nanoelectronics Research Institute, National Institute of Advanced Industrial Science and Technology (AIST), Tsukuba, Ibaraki, 305-8568, Japan

Abstract. Novel superconductor-insulator transition has been observed in the in-plane resistivity of "disordered" mono-layer cuprate superconductor $Bi_2Sr_{1.6}Ln_{0.4}CuO_{6+y}$ (Ln: trivalent rare earth element), in which disorder outside the CuO_2 plane is controlled by introducing Ln with different ionic radius with keeping the hole density constant (optimal doping). It is also revealed that this type of disorder produces minimal residual resistivity, while T_c is strongly suppressed. Anomalously divergent resistivity is suggestive of phase competition controlled by disorder.

Keywords: Cuprate, Disorder, Transport.
PACS: 74.25.Fy, 74.72.Hs, 74.72.Dn, 74.25.Jb

Chemical disorder outside the CuO_2 plane always exist in high-T_c cuprates, since doping takes place at this building block, by adding or removing excess oxygen or substituting atoms (e.g. $La^{3+} \rightarrow Sr^{2+}$). The effect of chemical disorder have been overlooked so far. However, recent STM/STS studies show the electronic inhomogeneity in the CuO_2 plane, which is likely to be triggered by the out-of-plane chemical disorder [1]. The effect of disorder has been discussed recently [2,3], but its understanding is still limited.

In this study we have quantitatively investigated the effect of disorder in the building block next to the CuO_2 planes. We have grown a series of $Bi_2Sr_{1.6}Ln_{0.4}CuO_{6+y}$ (Ln-Bi2201) single crystals by the floating zone method. The magnitude of disorder is controlled by changing the ionic radius of the rare-earth (Ln) ions, Ln=La, Nd, Eu, Gd, $La_{0.2}Gd_{0.2}$ and $Gd_{0.3}Dy_{0.1}$. All the samples were prepared to have the same doping level by fixing Ln content (=0.4). The ionic radius of Ln ions shrinks monotonically with increasing atomic number: from 1.216Å (La^{3+}) to 1.103Å (Gd^{3+}) which should be compared with the ionic radius of Sr^{2+}. c-axis length has a trend to be decreased systematically from 24.44Å for La-Bi2201 to 24.35Å for Ga-Bi2201, which were observed by X-ray diffraction measurement, indicating systematic change of disorder. Hence, it is expected that SrO planes are more disordered with increasing Ln atomic number. The degree of disorder is lowest for La-Bi2201, since the difference of ionic radius between Sr^{2+} and La^{3+} is smallest for Ln=La, and actually the maximum T_c value is realized for La-Bi2201(~35K).

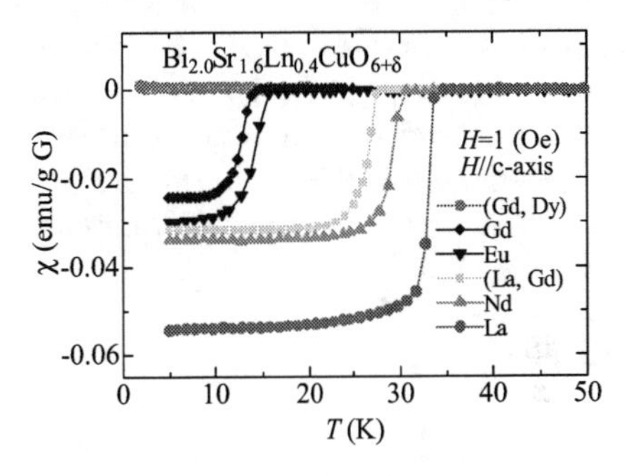

FIGURE 1. Temperature dependence of magnetic susceptibility for Ln-Bi2201.

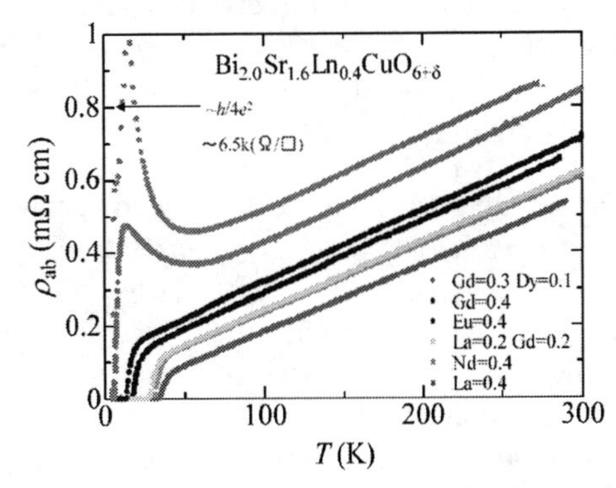

FIGURE 2. Temperature dependence of in-plane resistivity for Ln-Bi2201.

CP850, *Low Temperature Physics: 24th International Conference on Low Temperature Physics;* edited by Y. Takano, S. P. Hershfield, S. O. Hill, P. J. Hirschfeld, and A. M. Goldman

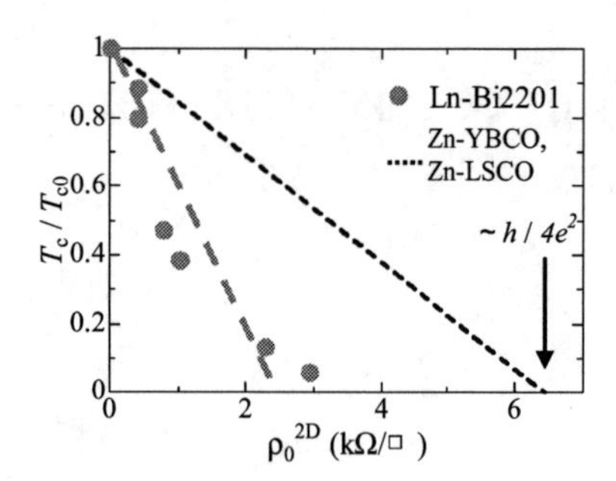

FIGURE 3. Normalized T_c is plotted against two dimensional sheet resistance with Zn doped system.

Temperature dependence of the field-cooled (Meissner) magnetic susceptibility for Ln-Bi2201 is shown in Fig.1. With increasing disorder, T_c is suppressed from 35K for La-Bi2201, to less than 1.8K for (Gd,Dy)-Bi2201. Corresponding to the reduction in T_c, Meissner volume fraction also decreases, suggesting the reduction of the superfluid density, which is consistent with the recent comparative μ SR measurements for La- and Eu-Bi2201[4].

Temperature dependence of the in-plane resistivity are shown in Fig.2. T-linear dependence of resistivity is a characteristic of nearly optimal doping level. With increasing disorder, resistivity curves are parallel shifted adding T-independent term (residual resistivity, ρ_0). The slope of these curves are the same, indicating that the carrier density is almost the same for all of Ln-Bi2201.

Qualitatively, reduction of T_c and increase of ρ_0 is the same effect as in-plane impurity, Zn. But as can be seen in Fig.3 (dotted line is universal line for Zn-doped system, T_c is normalized by maximum T_c of each series), ρ_0 is much small, and superconductor-insulator (SI) transition is observed much below the universal sheet resistance[5]. That is, the mechanism of reduction of T_c due to the out-of-plane disorder is essentially different from the in-plane one.

Normal state transport property is determined by the nodal quasiparticles due to large Fermi velocity and small effective mass[6], while superconducting condensation occurs around the antinodal region, indicating that T_c is determined by the electronic structure of antinodal state[7]. That is, disorder outside the CuO_2 plane (next to apical oxygen site) gives maximum effect on the antinodal state, while minimal effect on nodal state.

Resistivity for (Gd,Dy)-Bi2201 in which T_c observed by magnetic susceptibility is vanished shows divergent resistivity at low temperature with relatively smaller ρ_0 than two dimensional universal sheet resistance (arrow in Fig.2, 3), though somehow residual resistivity depend on sample to sample indicating Bi/Dy intercation mixing (Representative two curves are shown in Fig.2). This divergent behavior is very similar to that of Nd-doped LSCO which shows prototypical competing behavior, stripe order[7]. SI transition with relatively small ρ_0 indicates that some kind of competing order is induced by disorder outside the CuO_2 plane.

T_c suppression in Ln-Bi2201 might arise from local lattice distortion near the substituted Ln atoms. Local tilting of CuO_6 octahedora would lead to local modulation of transfer integrals. Among them, the next nearest neighbor hopping t' most affects the electronic structure in the antinodal region. In the momentum space, it can be easily illustrated that the FS topology is affected near the Brillouin zone edge by a change of t', as the position of the flat band is very sensitive to t'. Thus we speculate that the disorder considered here locally modulates t'. The SI transition would also be related to change of antinodal electronic structure where SC and "Pseudogap" competes. We note that there are some report that suggest a correlation between T_c and t' from the theoretical [9] and experimental [10] point of view. The present results show novel effect of disorder outside the CuO_2 plane. The results indicate that the disorder in the building blocks play a more important role than previously anticipated in high-T_c cuprates.

In-plane resistivity was measured for Ln-Bi2201 with controlled disorder. Superconductivity is completely suppressed at residual resistivity much below the universal 2D resistance. Divergent resistivity is suggestive of disorder-induced phase transformation.

REFERENCES

1. K. McElroy *et al.*, *Nature* (*London*) **411**, 592 (2003).
2. H. Eisaki *et al.*, *Phys. Rev. B* **69**, 064512 (2004).
3. J. P. Attfield *et al.*, *Nature* (*London*) **394**, 157 (1998).
4. Y. J. Uemura *et al.*, unpublished results
5. Y. Fukuzumi *et al.*, *Phys. Rev. Lett.* **76**, 684 (1996).
6. T. Valla *et al.*, *Science* **285**, 2110 (1999).
7. D. L. Feng *et al.*, *Science* **289**, 277 (2000).
8. T. Noda *et al.*, *Science* **286**, 265 (1999).
9. E. Pavarini *et al.*, *Phys. Rev. Lett.* **87**, 047003 (2001).
10. K. Tanaka *et al.*, *Phys. Rev. B* **70**, 092503 (2004).

Magnetostriction Of Charge Density Wave Superconductor

Victor Eremenko*, Peter Gammel+, Gyorgy Remenyi++, Valentyna Sirenko*, Anatolii Panfilov*, Vladimir Desnenko*, Vladimir Ibulaev* and A. Fedorchenko*

*Institute for Low Temperature Physics & Engineering NAS of Ukraine, Kharkov, 61103, Ukraine
+Agere Systems, 1110 American Parkway NE Allentown, PA 18109, USA
++C R T B T and GHMFL, CNRS, Grenoble, 38024, France

Abstract. Single crystals of layered niobium diselenide (2*H*-$NbSe_2$) compound with hexagonally packed layers were studied in a range of temperature 1.5 -300 K by measuring magnetostricrtion and magnetization in field up to 20 T. The data are compared with magnetic susceptibility measurements. Specific features of measured temperature and field dependences were observed near CDW transition temperature (32.5 K) and superconducting transition (7.2 K), sensitive to magnetic field direction and strength.

Keywords: magnetostriction, magnetic susceptibiliry, phase transition.
PACS: 75.80.+q, 33.15.Kr, 64.70.-p, 74.25.-q t

INTRODUCTION

Notwithstanding the broad diversity of measurements performed on the CDW superconductor 2*H*-$NbSe_2$, some important issues are still unclear. Among them there is the behavior of magnetic properties in the vicinity of low temperature phase transitions in a magnetic field. Early studies and the following investigations of magnetic susceptibility [1-4] in the vicinity of CDW transition gave the contradictory results. Here we make attempt to elucidate some aspects of the problem.

RESULTS AND DISCUSSION

The niobium diselenide (2*H*-$NbSe_2$) is well studied superconductor, characterized by the layered crystal structure and moderately anisotropic (uniaxial) superconducting properties (T_{SN}=7.3 K). Its properties near the CDW transition (32.5 K) are less studied. The magnetostriction measurements complemented by magnetization data, appeared to be the effective tool for studying the magnetic behavior of 2*H*-$NbSe_2$ under destruction of superconductivity by magnetic field [5]. They have revealed identical behavior of strain and magnetization change with magnetic field in the singled out regions that is of the peak-effect regime and quantum oscillations. Here the similar effort is focused on the region of transition to the CDW state in 2*H*-$NbSe_2$. The high-quality single crystals of 2*H*-$NbSe_2$ compound were studied here in the temperature range 1.5-300 K by means of the magnetostriction and magnetic susceptibility measuring techniques. Magnetostriction was registered by means of the capacitance bridge in magnetic field up to 20 T applied in plane and in hexagonal axis direction. Parallel (in-plane) and perpendicular components of magnetic susceptibility were measured by Faraday technique in field 0.83 T and the parallel component was measured by SQUID magnetometer in field 300 Gs. Figures 1 and 2 present the low temperature measurements of magnetostriction and magnetic susceptibility. The peculiarity of the magnetic properties measured in high enough magnetic field is clearly seen both for magnetostriction and magnetic susceptibility measurements. The low field measurements in SQUID magnetometer did not reveal such a peculiarity. The quality of the crystals and different geometries of experiment exclude explanation of the observations by the experimental errors. The discrepancy of measurements can be explained by different values and direction of the external magnetic field strengths.

CP850, *Low Temperature Physics: 24th International Conference on Low Temperature Physics;*
edited by Y. Takano, S. P. Hershfield, S. O. Hill, P. J. Hirschfeld, and A. M. Goldman

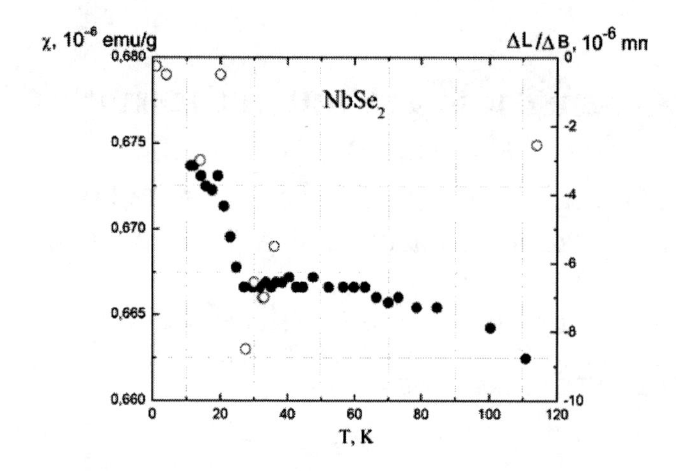

FIGURE 1. Magnetic susceptibility (filled circles) and magnetostriction (empty circles) of 2*H*-$NbSe_2$ single crystal measured in magnetic field parallel to the *ab* plane of the strength 0.83 and 20 T, respectively.

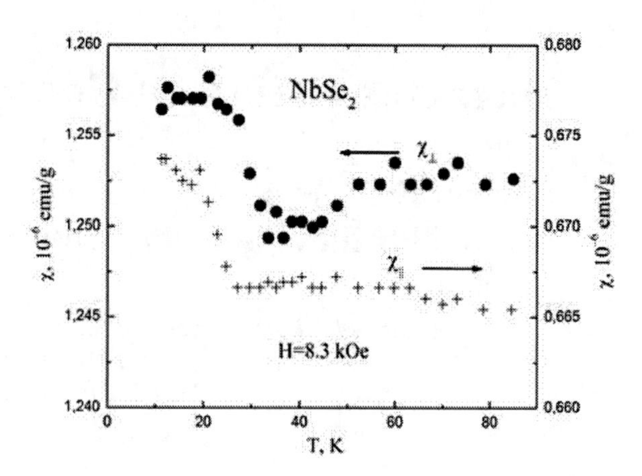

FIGURE 2. Magnetic susceptibility in the *c*-axis (filled circles) and in-plane direction of 2*H*-$NbSe_2$ (The Faraday measurements).

The positions of the features on the measured temperature dependences of both magnetostriction derivative on temperature and of magnetic susceptibility in field parallel to the crystallographic layers coincide and are in the close vicinity to the charge density wave transition in this compound.

ACKNOWLEDGMENTS

This work was supported by grant of the MES of Ukraine #M/257 and of the grants of GHMFL and CRTBT of CNRS in Grenoble.

REFERENCES

1. H. N. Lee, M. Garcia, H. McKinzie and A. Wold *J. Solid State Chemistry* **1**, 190-194 (1970).
2. M. Marezio, P. D. Dernier and G. W. Hull *J. Solid State Chemistry* **4**, 425-429 (1970).
3. S. Shetkin, V. L. Kalihman and Yu. P. Sereda *Low Temperature Physics* **3**, 1272-1279 (1977).
4. S. J. Hillenius and B.V. Coleman *Phys. Rev. B* **20,** 4569-4576 (1979).
5. V. Eremenko, V. Sirenko, R. Schleser and P. Gammel *Low Temperature Physics* **27**, 412-421 (2001).

So, the occurrence of the observed peculiarities can be attributed to the physical processes in the crystals under study arised from the formation of the CDW state.

Time-Resolved Far-Infrared Studies of Superconducting $Nb_{0.5}Ti_{0.5}N$ Film in a Magnetic Field

H. Zhang,[a] H. Tashiro,[a] R.P.S.M. Lobo,[b] D.H. Reitze,[a] C.J. Stanton,[a] D.B. Tanner,[a] and G.L. Carr[c]

[a] *Department of Physics, University of Florida, Gainesville, FL 32611-8440, USA*
[b] *Laboratoire de Physique de Solide, ESPCI, CNRS UPR5, 75231 Paris, France*
[c] *National Synchrotron Light source, Brookhaven National Laboratory, Upton, NY 11973, USA*

Abstract. Time-resolved, optical pump-probe measurements on a thin $Nb_{0.5}Ti_{0.5}N$ film in applied magnetic fields were performed at the National Synchrotron Light Source, Brookhaven National Laboratory. Despite the presence of normal cores from vortices in the films, we find that the relaxation time of photoexcited quasiparticles does not decrease with magnetic field as one might expect. The change in the far-infrared transmittance due to the applied magnetic field is also discussed.

Keywords: time-resolved, superconductor, magnetic field.
PACS: 78. 47. +p

When a superconductor is exposed to an external light source having energy greater than the superconducting energy gap 2Δ, excess quasiparticles are produced. These photoexcited quasiparticles relax primarily through the emission of phonons and eventually recombine into Cooper pairs. The emitted phonons have energies > 2Δ and can break other Cooper pairs unless they decay or escape the superconductor. This cycle leads to a bottleneck in the return to equilibrium superconductivity after photoexcitation. When a magnetic field is applied in a type II superconductor, the quasiparticles may diffuse toward vortices where there is no gap and hence potentially no bottleneck. Thus, vortices would seem to open a new "channel" for the energy relaxation of the photoexcited quasiparticles, and should yield a faster relaxation time.

Our sample was a ~10 to 20 nm thickness film of $Nb_{0.5}Ti_{0.5}N$ on 0.5 mm thick sapphire with T_c approximately 10 K. The far-infrared transmission through the films was measured for both the normal and superconducting states. Figure 1 shows the magnetic field dependence of the superconducting-state transmittance, normalized by the normal state transmission. The superconducting data were measured at 3 K in magnetic fields up to 10 T. The normal state was measured at 12 K at zero magnetic field. Behavior typical of $Nb_{0.5}Ti_{0.5}N$ was observed, e.g. H_{c2} is greater than 10 T at 3 K. The peak transmittance occurs near 28 cm^{-1} (3.47 meV), which agrees with $2\Delta_0 \sim 26.5$ cm^{-1} [1].

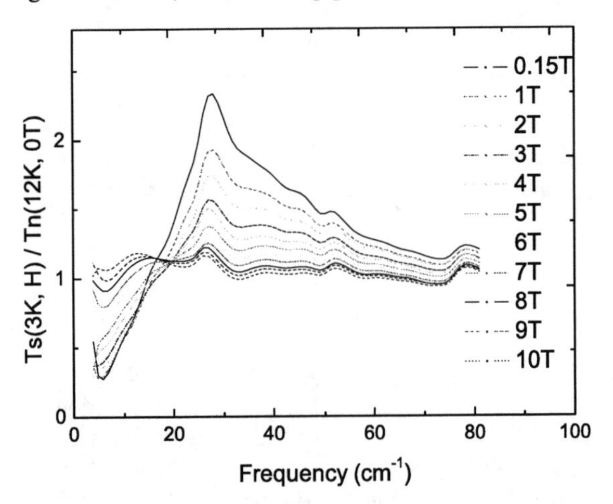

FIGURE 1. Measured ratio of T_S/T_N for a $Nb_{0.5}Ti_{0.5}N$ film at magnetic fields between 0 and 10 T.

Time-resolved pump-probe spectroscopy [2] was performed at the U12IR beamline of the National Synchrotron Light Source (NSLS), Brookhaven National Laboratory. A mode-locked Ti:sapphire laser producing two picosecond duration near-infrared pulses was used as the pump source. Those pulses were synchronized to the far-infrared pulses from

CP850, *Low Temperature Physics: 24th International Conference on Low Temperature Physics;*
edited by Y. Takano, S. P. Hershfield, S. O. Hill, P. J. Hirschfeld, and A. M. Goldman

U12IR that served as the probe source. For high sensitivity to spectroscopic changes, the pump-probe delay time was dithered and a lock-in amplifier then measured the derivative of the photoinduced transmittance signal. This differential was integrated to yield the time-resolved change in transmission. The synchrotron pulse width was about 350 picoseconds and determines the measurement resolution. Figure 2 shows the integrated signal (squares) for a $Nb_{0.5}Ti_{0.5}N$ sample along with a fit.

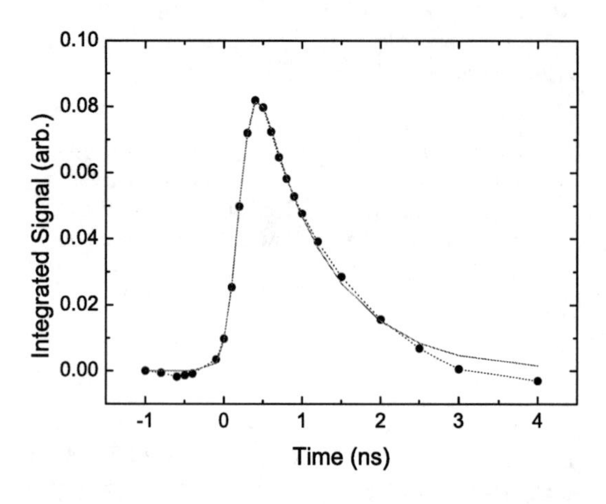

FIGURE 2. Plot of the integrated signal and the fit for the $Nb_{0.5}Ti_{0.5}N$ film, measured at 0 T and 3 K.

The fitting function used is given in Eq. (1) and represents a single exponential, convolved with a Gaussian to account for the width of the probe pulse.

$$S(t) = \frac{A}{T\sqrt{\pi}} \int_0^{+\infty} e^{-\frac{t'}{\tau}} e^{-(\frac{t-t'}{T})^2} dt' \qquad (1)$$

Here, S is the measured temporal response, A is the measured signal magnitude, T is probe pulse duration and τ is the lifetime. For the applied magnetic fields up to 5 T, the fitting parameters are shown in Table 1. It indicates that the lifetime τ does not decrease with increasing magnetic field, in contradiction to our expectation that the vortices would speed up the relaxation dynamics of the photoexcited quasi-particles.

In the dither (derivative) method of measuring the photoinduced signal, one can show that the maximum signal value is approximately inversely proportional to the probe pulse width while the minimum signal is inversely proportional to the quasiparticle relaxation time. By sweeping the magnetic field and collecting maximum and minimum date points for every field, the lifetime can be estimated. The results are shown in Figure 3. The error bars are statistical, from the variations of two trials. At 4.5 K, the lifetime does not show a decrease as the magnetic field increases, in agreement with the results in Table 1.

TABLE 1. Fitting Parameters

T (K)	H (T)	A (arb.)	τ (ps)
3	0	0.114 ± 0.003	880 ± 30
3	1	0.069 ± 0.001	1100 ± 20
3	2	0.063 ± 0.002	1030 ± 50
3	3	0.051 ± 0.001	1050 ± 20
3	4	0.055 ± 0.001	1040 ± 20
3	5	0.048 ± 0.001	1360 ± 50

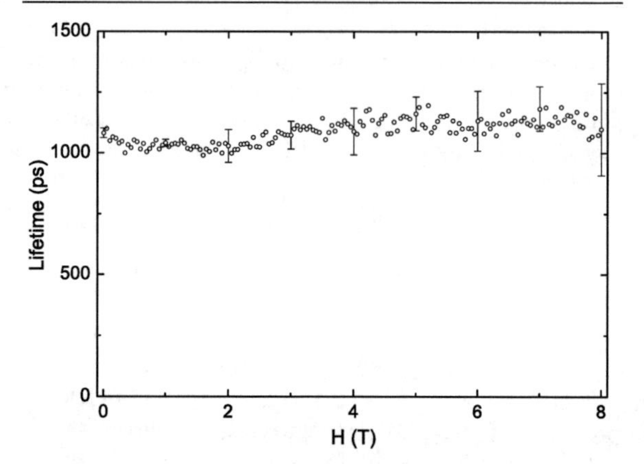

FIGURE 3. Quasiparticle relaxation time from swept measurements for a $Nb_{0.5}Ti_{0.5}N$ sample at 4.5 K.

The data indicate that the vortices do not have a significant effect on the measured lifetime. There are two possible explanations. Either the additional scattering channels made available by the vortex and its core are not an efficient route for relaxing the system back to the true equilibrium state or the excess quasiparticles simply do not interact with the vortices to any appreciable extent.

This work was supported by the U.S. Department of Energy through contract DE-FG02-02ER45984 at UF and DE-AC02-98CH10886 at BNL. Thin films were provided by P. Bosland and E. Jacquesfrom CEA, Saclay, France. We thank Professor L. Mihaly for the use of the superconducting magnet.

REFERENCES

1. R.P.S.M. Lobo et al., *Phys. Rev. B*, in press.
2. G.L. Carr et al., *Phys. Rev. Lett.* **85**, 3001-3004 (2000).

Direct Observation Of The Proximity Effect In The N Layer In A Multi-Terminal SINIS Josephson Junction

I. P. Nevirkovets*, O. Chernyashevskyy*, and J. B. Ketterson¶

*Department of Physics and Astronomy, Northwestern University, Evanston, IL 60208, USA; Institute for Metal Physics NASU, 03680 Kyïv, Ukraine

¶Department of Physics and Astronomy, Department of Electrical and Computer Engineering, and Materials Research Center, Northwestern University, Evanston, IL 60208, USA

Abstract. It is found that the lateral conductivity of a thin N layer inside a multi-terminal SINIS device (at temperatures down to 1.8 K) is dissipative with a small coherent contribution near zero voltage, which can be suppressed with a small parallel magnetic field; here S, I, and N denote a superconductor (Nb), an insulator (AlO_x), and a normal metal (Al) respectively. It is suggested that this contribution is due to the proximity effect through the tunnel barriers, involving the superconducting fluctuations in Al. At the same time, a high Josephson current can flow normal to the structure, which we propose is due to a direct Josephson coupling between the external S electrodes.

Keywords: Superconductivity, tunneling, Josephson effect, proximity effect, fluctuative superconductivity, multilayer.
PACS: 74.45.+c, 74.50.+r, 74.78.Fk, 85.25.Am, 85.25.Cp

INTRODUCTION

In most experiments, the proximity effect has been studied on systems where the current is flowing normal to the N/S interface (here N is a normal metal and S is a superconductor). The electronic spectrum of the normal metal is then inferred indirectly from the characteristics of the entire system [1]. Additional information on the state of the N layer could be obtained if *lateral* electrical transport characteristics were measured.

Here we present results of an experimental study of the lateral conductivity of the N layer in multi-terminal sandwich-type SINIS and NIS junctions.

EXPERIMENT AND DISCUSSION

Our multi-layer Nb/Al/AlO_x/Al/AlO_x/Al/Nb SINIS structure was *in-situ* fabricated using a procedure described elsewhere [3]. A schematic cross-sectional view of the multi-terminal device is shown in Fig. 1*a*. Here we describe devices with a width W =10 μm and lengths L_b=19 μm and L_t=11 μm for the bottom and top junctions respectively; the specific tunneling resistance of the tunnel barriers is of the order of 1×10^{-7} Ω×cm^2. The 150-nm thick Al layer facilitates a non-superconducting electrical contact to the N layer; to suppress the parasitic proximity effect between this Al layer and the topmost Nb layer, 21– 40 nm of Zr was deposited between the Nb and Al. Such contacts were also used in NIS junctions made from the same multilayer by etching away the top Nb layer (Fig. 1*b*).

In Fig. 2, we show the current-voltage characteristics (CVC) for a typical device (referred to as device 1; the thickness of the N layer, d_{Al}, is 13 nm), measured for various configurations (cf. Fig. 1*a*) at 1.8 K. Curve 1 is CVC of the top junction (using 1&8 as current and 2&7 as voltage terminals). Curve 2 is CVC of the bottom junction (using 1&6 as current and as 2&5 voltage terminals). Curve 3 shows the CVC of the N layer, measured between the terminals 1&4 (for current) and 2&3 (for voltage). Curves 4-6 were measured using the terminals 5&7 for the current and terminals 2&8, 6&2, and 6&8 for the voltage (corresponding to the top and bottom junctions, and to the entire SINIS multilayer, respectively). As can be seen from curves 1 and 2, when measured separately, the top junctions have different critical currents, I_{ct} and I_{cb}, such that $I_{ct} > I_{cw} > I_{cb}$, where I_{cw} is the critical current of the whole structure

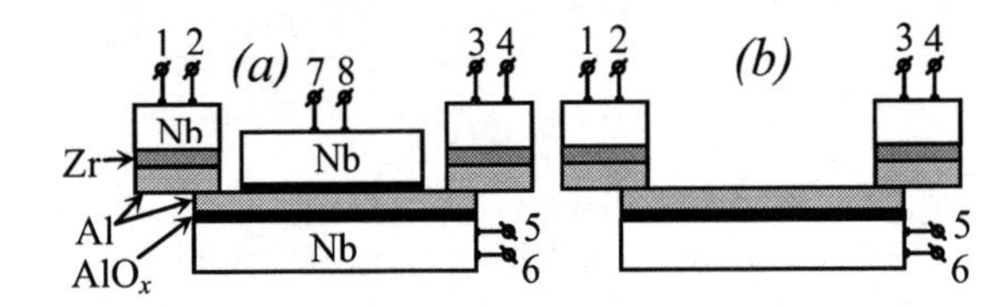

FIGURE 1. Schematic cross-sectional view of the devices.

CP850, *Low Temperature Physics: 24th International Conference on Low Temperature Physics;* edited by Y. Takano, S. P. Hershfield, S. O. Hill, P. J. Hirschfeld, and A. M. Goldman

(cf. curve 6); the respective I_c levels are marked by arrows. However, when current-biased between the terminals 5&7, the top and bottom junctions (curves 4,5) display I_c values identical to I_{cw}. This fact, along with the fact that we were not able to set one junction into the resistive state while keeping the other in the superconducting state, indicates that the Josephson tunneling in our SINIS device takes place directly between the S electrodes.

Note that the critical current vs. magnetic field [$I_c(H)$] dependences for all Josephson junctions contained in the multi-terminal device indicated good junction quality. In the left inset, we show the $I_{ct}(H)$, $I_{cw}(H)$, and $I_{cb}(H)$ dependences (curves 1-3, respectively).

As one can see from curve 3 in Fig. 2, the conductivity of the N layer in the lateral direction a *dissipative*; at the same time, a considerable Josephson current (of about 500 μA) can flow through the film in the vertical direction (at 1.8 K). In the right inset of Fig. 2, an initial portion of curve 3 is shown on a magnified scale for H=0 (curve A) and in a magnetic field of 30 Oe applied parallel to the plane of the structure (curve B). The conductivity of the N film has a small coherent contribution near V=0 that can be completely suppressed in a small magnetic field.

From similar SINIS structures (deposited in a different run), we fabricated the devices shown in Fig. 1*a*,*b*. For the SIN device (cf. Fig. 1*b*), there is a small coherent current in the CVC of the junction itself (measured using the terminals 1&5 for the current and terminals 2&6 for the voltage), and of the N layer (current terminals 1&4 and voltage terminals 2&3); the latter CVC was similar to the CVC of the N layer in the SINIS device, but had lower conductivity. In Fig. 3, we compare the temperature dependence of I_{cw} for a SINIS junction with d_{Al}=18 nm (squares),

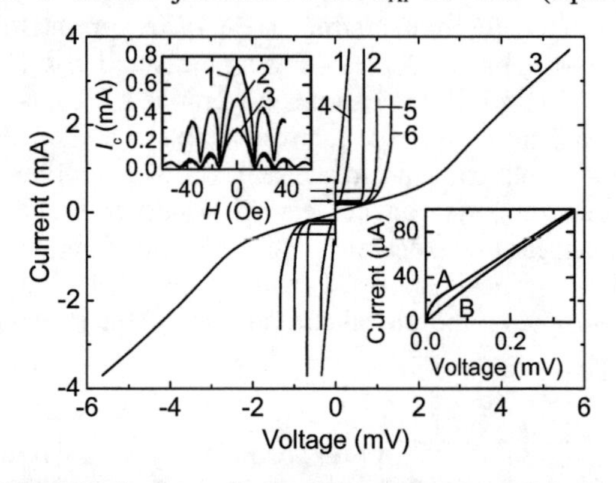

FIGURE 2. CVC of a SINIS device at 1.8 K. Curves 1-6 are recorded using the following terminals for current and voltage, respectively (cf. Fig. 1*a*): 1&8 and 2&7; 1&6 and 2&5; 1&4 and 2&3; 5&7 and 2&8; 5&7 and 6&2; and 5&7 and 6&8. Left inset: I_c vs. H dependences for the top junction, structure as a whole, and bottom junction (curves 1-3, respectively); right inset: initial portion of curve 3 for H=0 (curve A) and for H=30 Oe (curve B).

and the temperature dependence of the maximum coherent current, I_m, through the SIN junction (circles); I_m was determined by subtracting currents at the same voltage in the CVC for H=0 and for H=46 Oe. Note that the ratio I_{cw}/I_m is 3.6 at 1.9 K. The ratio increases for devices with smaller d_{Al}; for devices with d_{Al}=8 nm, I_{cw}=2.2 mA and I_m=80 μA (at 1.9 K), yielding I_{cw}/I_m=27.5. This is in disagreement with a phenomenological model of SINIS junctions [3], according to which one should expect the relation $I_m=I_{cw}/2$; on the other hand, such a behavior is consistent with the idea that Josephson tunneling in the double-barrier device is coherent.

The solid line in Fig. 3 is the theoretical $I_m(T)$ dependence for the SIN junction according to the fluctuation model [4] for an Al critical temperature T_c*=1.4 K and critical magnetic field of 300 Oe (used as fitting parameter). According to [4], the dependence $I_m^{-1}(T)$ is linear near T_c*, crossing the T axis at T_c*. In the inset of Fig. 3 we have plotted both experimental (scattered plot) and theoretical (solid line) $I_m^{-1}(T)$ dependences, which are in agreement with each other at lower temperatures.

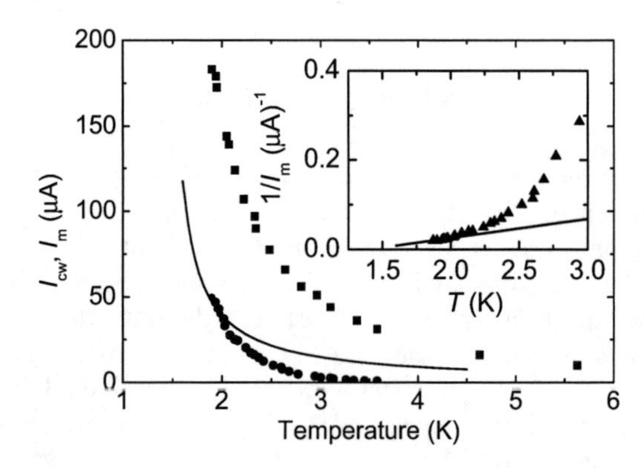

FIGURE 3. $I_c(T)$ dependence of the SINIS device as whole (squares); $I_m(T)$ dependence of the related SIN junction (circles). Solid line is the theoretical $I_m(T)$ dependence for the SIN junction [4]. Inset: experimental (scattered data) and theoretical [4] (solid line) $I_m^{-1}(T)$ dependence.

ACKNOWLEDGMENTS

This work was supported by the National Science Foundation under the grant EIA-0218652 and use was made of facilities operated by the NSF-supported Materials Research Center.

REFERENCES

1. E.L. Wolf and G.B. Arnold, *Phys. Repts.* **91**, 31-102 (1982).
2. I. P. Nevirkovets, O. Chernyashevskyy, J. B. Ketterson, and E. Goldobin, *J. Appl. Phys.* **97**, 123903 (2005).
3. A.M. Kadin, *Supercond. Sci. Technol.* **14**, 276-284 (2001).
4. J. T. Anderson, R. V. Carlson, and A. M. Goldman, *J. Low Temp. Phys.* **8**, 29-46 (1972).

Dynamically Induced Alignment of Magnetic Domains in Type-I Superconductors

V. Jeudy, C. Gourdon, T. Okada

Institut des NanoSciences de Paris, Universités Paris 6 et Paris 7, CNRS UMR 7588 Campus Boucicaut, 140 rue de Lourmel, 75015 Paris, France

Abstract. Normal and superconducting state domains in a 10 μm thick indium slab are observed by the magneto-optical imaging technique. The sample is subjected to a magnetic field H with different sweeping rates $\dot{H}$. The domain patterns are found to depend on the penetration regime of the magnetic flux. For the lowest sweeping rate ($\dot{H}$=1 mT s^{-1}), the domains are randomly oriented. For the highest sweeping rate ($\dot{H}$ =230 mT s^{-1}) and sufficiently high H-values, the domains consist of a nearly straight laminar structure with a well-defined period, oriented perpendicularly to the edge of the sample. These first results open new perspectives for studying the dynamics of formation of magnetic domains.

Keywords: Superconductors, intermediate state, lamellae
PACS: 74.25.Ha ; 05.65.+b ; 75.70.K

INTRODUCTION

A type-I superconducting slab placed in a perpendicular magnetic field presents a spontaneous phase separation in normal and superconducting state domains [1]. A large number of physical systems, as ferrofluids [2] or uniaxial ferromagnetic thin films [3] presents similar self-organized domain structures. Self-organization results from the balance between long-range repulsive magnetic interactions between domains and short-range attractive interactions associated with a positive interface energy. In superconductors, the observed intermediate state (IS) patterns are generally disordered [1]. They are not reproducible and strongly depend on the magnetic history of the sample. To our best knowledge, the only existing method to prepare ordered structures was proposed by Sharvin [4]. It consists in creating a preferential orientation by applying a magnetic field with a small in-plane component. In this letter, we propose an altenative method to prepare ordered laminar domains. We show that lamellae can be aligned dynamically by applying a magnetic field step. The orientation is found to be reproducible and corresponds to the direction of the magnetic flux penetration. The different parameters controlling the alignment are analyzed.

EXPERIMENTAL

The domains are observed by the high resolution magneto-optical imaging technique [5]. A magneto-optic layer (MOL) is placed against the superconducting film and allows one to map the normal component of the magnetic induction at its surface. The sample consists of a 10.0 $\pm$ 0.1 μm thick superconducting indium film. It was grown by Joule evaporation directly onto the MOL [5]. The sample is placed in an immersion-type cryostat in pumped liquid helium. The domains are obtained during the following applied magnetic field cycle: the perpendicular field is raised from zero to H at the sweeping rate $\dot{H}$. An image is acquired when H and the domain structure are stabilized. H is then set back to zero before the next cycle.

RESULTS AND DISCUSSION

Figure 1 shows IS structures observed for H =9 mT and two differents sweeping rates: $\dot{H}$ =1 mT s^{-1} and $\dot{H}$ =230 mT s^{-1}. Normal state domains appear in black.

For the lowest sweeping rate, the patterns is disordered. Two types of domains can be distinguished: nearly circular domains (bubbles) and elongated domains (lamellae) having corrugated shapes. The pattern is not reproducible from one cycle to the next. As it was shown previously, the disorder seen in the orientation of the domains partly results from the pinning of domains walls by defects[7]. However the average lamellar width and the average spacing between lamellae are reproducible.

For the highest sweeping rate, bubble domains are no more observed. The pattern consists in almost straight lamellae with a well-defined period. Lamellae are oriented perpendicularly to the edge of the sample, i.e., along the direction of the penetration of the magnetic flux. This indicates that the domains can be oriented dynamically. Increasing the sweeping rate of the magnetic field reduces the contribution of domain wall pinning to the morphogenesis of domains.

CP850, *Low Temperature Physics: 24th International Conference on Low Temperature Physics;*
edited by Y. Takano, S. P. Hershfield, S. O. Hill, P. J. Hirschfeld, and A. M. Goldman

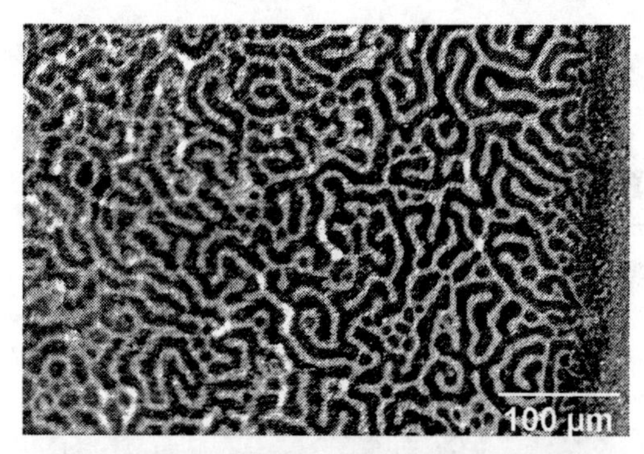

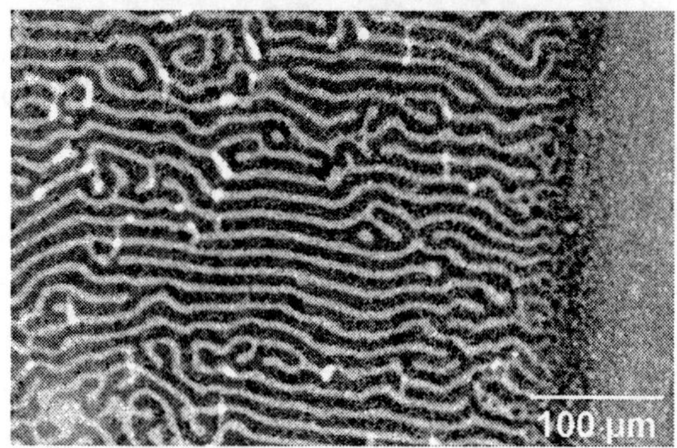

FIGURE 1. IS pattern observed in a 10 μm thick indium slab for two different applied magnetic field sweeping rates $\dot{H}$. The flux-bearing normal phase appears in black. The diamagnetic ($B = 0$) superconducting phase appears in gray. The edge of the film is along the right edge of the image. Top image: $\dot{H} = 1$ mT s^{-1}; the normal state domains are randomly oriented. Bottom image: $\dot{H} = 230$ mT s^{-1}; the domains are nearly straight. Images are taken at $T = 1.85$ K, $H = 9$ mT. The few white spots are due to image processing.

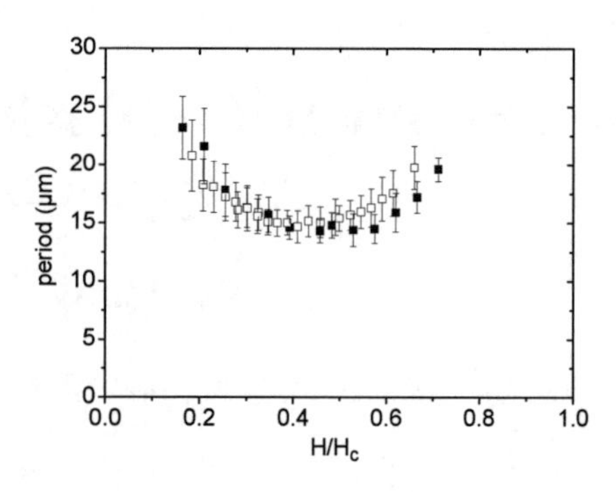

FIGURE 2. Mean period of the lamellar structure as function of the applied magnetic field (in units of the critical field) for $\dot{H}$ =1 mT s^{-1} (empty squares) and $\dot{H}$ =230 mT s^{-1}(black squares).

Let us now examine the parameters that control the alignment of lamellae. The average spacing between domains is only weakly dependent on the sweeping rate $\dot{H}$. The average quasiperiods of lamellae, obtained from the two images of Fig. 1, are identical within the experimental error bars and equal to 15$\pm$1 μm. More generally, as shown in Fig. 2, for ΔH <10 mT, the quasiperiods are found equal for the ordered and disordered cases. For $\Delta H > 10$ mT slightly smaller periods (-10%) are obtained in the ordered case ($\dot{H}$ =230 mT s^{-1}) as compared to the disordered one ($\dot{H}$ =1 mT s^{-1}).

The energy difference between the slow and fast-rising cases is calculated in the framework of the "constrained current-loop" model [7] using the results of Fig. 2. For the reduced applied field H/H_c=0.67 (H_c is the critical field and equals 199 G at T=1.85 K) the energy difference is of the order of 2%. This small difference suggests that structures obtained with slow or fast-rising field both correspond to quasi-equilibrium states.

The degree of alignment was characterized by Fourier analysis of images. It strongly depends on the amplitude of the magnetic field. For $H < 3$ mT, no difference can be seen between the pattern obtained with the low and high sweeping rates of the field. For $H > 3$ mT, the degree of alignement increases with the magnetic field. This threshold may be related to the geometrical energy barrier which governs flux penetration for low applied magnetic field values.

CONCLUSION

We have shown that the normal state domains can be dynamically aligned. This opens new perspectives for studying the dynamics of the formation of magnetic domains in superconductors. Our results indicate that for the fast-rising field, the lamellae are formed simultaneously thus allowing the long-range interactions to order the pattern. On the contrary, for the slow-rising field, the formation of domains is temporally uncorrelated which leads to disordered structures.

REFERENCES

1. R.P. Huebener, *Magnetic Flux Structures in Superconductors*, Springer Verlag, New York, 1979.
2. A. Cebers and M. M. Maiorov, *Magnetohydrodynamics* **16**, 21 (1980)
3. A. Hubert and R. Schäfer, *Magnetic Domains*, Springer, Berlin, 2000.
4. Yu. V. Sharvin, [*Zh. Eksper. Teor. Fiz.* **38**, 298 (1960)]; *Soviet Phys. - JETP* **11**, 216 (1960).
5. C. Gourdon, V. Jeudy, M. Menant, D. Roditchev, Le Anh Tu, E.L. Ivchenko and G. Karczewski, *Appl. Phys. Lett.* **82**, 230 (2003).
6. V. Jeudy, C. Gourdon and T. Okada, *Phys. Rev. Lett.* **92**, 147001 (2004).
7. A. Cebers, C. Gourdon, V. Jeudy and T. Okada, *Phys. Rev. B.* **72**, 014513 (2005).

Observation of a Second Energy Gap in Nb_3Sn Point Contacts

G. Goll*, M. Marz*, R. Lortz†, A. Junod† and W. Goldacker**

*Physikalisches Institut, Universität Karlsruhe, D-76128 Karlsruhe, Germany
†Department of Condensed Matter Physics, University of Geneva, CH-1211 Geneva 4, Switzerland
**Forschungszentrum Karlsruhe, Institut für Technische Physik, D-76021 Karlsruhe, Germany

Abstract. The A15 compound Nb_3Sn is a well-known technically relevant superconductor with critical temperature $T_c \approx 18\,K$. Recently, a low-temperature anomaly in the specific-heat data on a particularly dense and homogeneous polycrystalline sample has been interpreted in terms of the presence of a second superconducting gap. We performed point-contact spectroscopy on samples of the same batch using both the break-junction and the needle-anvil technique with Pt as a normal-metal counterelectrode. The differential conductance as a function of applied voltage at $T \approx 1.5\,K$ shows several characteristic maxima which can be well interpreted under the assumption of two superconducting energy gaps in Nb_3Sn.

Keywords: A15 compounds, superconducting energy gap, point-contact spectroscopy
PACS: 74.50.+r, 74.70.Ad

The discovery of two-gap superconductivity in MgB_2 has reanimated the interest in multi-band superconductivity phenomena and the search for further examples of this extraordinary behaviour. Recent measurements of the specific heat of the A15 superconductor Nb_3Sn in the superconducting state [1] show a deviation from the predictions of the BCS theory [2]. The expected exponential decay of the specific heat at temperatures $T < 0.5T_c$ has not been observed. Instead, a plateau at low T occurs. This extra contribution can be modelled quite satisfactory if two-band superconductivity is assumed. The empirical two-gap model has been used previously to explain similar specific-heat data for MgB_2 [3].

We used point-contact spectroscopy in order to investigate the superconducting gap structure of Nb_3Sn. Point-contact spectroscopy is directly sensitive to the electronic structure, in particular to the superconducting energy gap. Previously, the presence of a second gap in MgB_2 has been verified by the same method [4]. We performed point-contact spectroscopy on samples of the same batch using both the break-junction and the needle-anvil technique with Pt as a normal-metal counterelectrode. In the first case superconductor-superconductor (S-S) contacts are investigated while the latter results in normal metal-superconductor (N-S) contacts. Polycrystalline samples of Nb_3Sn have been prepared at the Forschungszentrum Karlsruhe by hot isostatic pressing (HIP), for details see [1]. The superconducting properties have been verified by resistivity and susceptibility measurements on such a sample. A very sharp step at the transition to superconductivity and the high transition temperature $T_c = 18.0\,K$ confirm the high quality of the sample. The current-voltage characteristics I vs. V and the differential resistance dV/dI vs. voltage V of the N-S and the S-S contacts, respectively, were measured simultaneously using a Lock-In technique. The dI/dV vs. V curves were obtained by numerical inversion of the dV/dI vs. V curves.

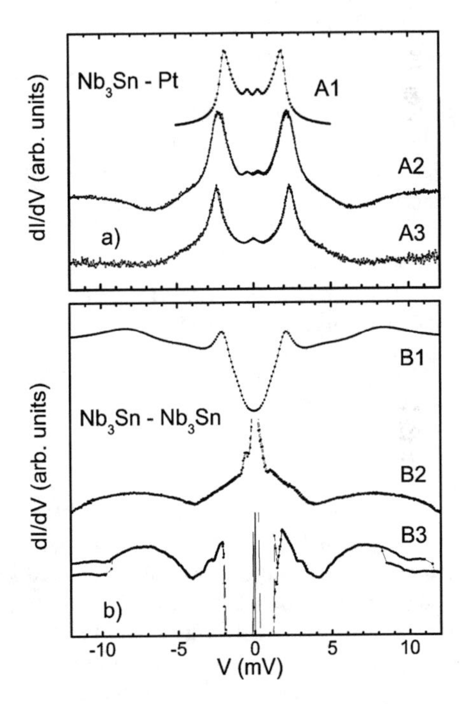

FIGURE 1. dI/dV vs. V spectra at $T < 2.0\,K$ obtained by the needle-anvil technique (panel (a): N-S contacts, curve A2 and A3) and by the break-junction technique (panel (b): S-S contacts, curve B2 and B3). The theoretical curves are calculated under assumption of two energy gaps for a N-S contact (A1) and an S-S contact (B1). For details see text.

CP850, *Low Temperature Physics: 24th International Conference on Low Temperature Physics;*
edited by Y. Takano, S. P. Hershfield, S. O. Hill, P. J. Hirschfeld, and A. M. Goldman

We first focus on the results obtained for the N-S contacts. Representative dI/dV vs. V spectra at $T = 1.5\,\mathrm{K}$ are displayed in Fig. 1a as curve A2 and A3. Clearly visible are two maxima at $V = \pm 2.4\,\mathrm{meV}$ caused by Andreev reflection of charge carriers at the N-S interface [5]. Andreev reflection is the leading mechanism for a current through a N-S contact for $V < \Delta/e$. The presence of a weak interface barrier leads to a reduction of the conductance at low bias, but the characteristic conductance maxima at the gap energy remain [6]. In addition to these maxima weaker structures are observed in our measurements at lower bias. For all data at $T < 2\mathrm{K}$ we read the voltage values of the position of these structures and plot them into a histogram. From the comparison with the calculated curve A1 the cluster points can be identified as two superconducting gaps with $\Delta_l \approx 1.8\,\mathrm{meV}$ and $\Delta_s \approx 0.5\,\mathrm{meV}$. The size of the larger gap is much smaller than the value found in literature [7] and points to a substantial degradation of superconductivity at the surface.

Therefore, we applied the break-junction technique in order to avoid the influence of impurities, oxide layers or a degradation of the superconductivity close to the contact region. Four electrodes have been attached to a tiny bar of Nb_3Sn which was glued on a substrate with stycast epoxy. The bar was broken *in situ* under a ^{4}He atmosphere. Our measuring system allows us to control the contact resistance during the whole measurement, which means that we can open and close the break junction in a controlled fashion even at low T.

Break junctions act as S-S contacts. For S-S contacts with identical material on both sides one characteristic peak in the differential conductance vs. voltage spectra at $V = 2\Delta/e$ is expected for each polarity due to the superconducting energy gap. Whereas, in most of our measurements we find two of these characteristic peaks for each polarity. Typical spectra at $T = 1.5\,\mathrm{K}$ (B2) and 2 K (B3) are shown in Fig. 1b as curve B2 and B3. In addition to the gap structure at finite voltage the dI/dV vs. V curve shows a huge maximum at $V = 0$ caused by Josephson current between the two superconducting electrodes. Again, for all data at $T < 2\,\mathrm{K}$ we read the voltage values of the position of the gap structures and plot them into a histogram. The cluster points can be identified as two energy gaps with:

$$\Delta_l \approx 3.3\,\mathrm{meV} \text{ and } \Delta_s \approx 0.6\,\mathrm{meV}$$

For the break-junction experiment Δ_l is in line with the result of previous tunnelling measurements [7] and the finding of two energy gaps also confirms the interpretation of the specific-heat data [1].

For the theoretical curves A1 and B1 we consider the currents aroused by the two superconducting gaps as independent from each other and neglect interband scattering, so we can write for the total current through the contact as the sum of both contributions:

$$I_{\mathrm{tot}}(V) = \alpha I_{\Delta_l}(V) + (1-\alpha) I_{\Delta_s}(V) \qquad (1)$$

with the weighting parameter α, $0 < \alpha < 1$. $I_\Delta(V)$ through a N-S contact was calculated with the expression derived by Blonder, Tinkham and Klapwijk [6]:

$$I_\Delta(V) \propto \int_{-\infty}^{+\infty} (f(E-eV) - f(E))\,(1 + A(E) - B(E))\,dE \qquad (2)$$

where $A(E)$ denotes the probability for Andreev reflection, $B(E)$ the propability for normal reflection and $f(E)$ the Fermi function at given T. The current through a superconductor-superconductor contact is given by [8]:

$$I_\Delta(V) \propto \int_{-\infty}^{+\infty} \frac{N_s(E) \cdot N_s(E+eV)}{N_n^2(0)} (f(E) - f(E+eV))\,dE \qquad (3)$$

$N_s(E)$ and $N_n(E)$ are the density of states in the superconducting and normal states, respectively. For curve A1 in Fig. 1a we used $\Delta_l = 2.0\,\mathrm{meV}$ and $\Delta_s = 0.4\,\mathrm{meV}$ and a weighting parameter $\alpha = 0.2$. For curve B1 in Fig. 1b we used $\Delta_l = 4.0\,\mathrm{meV}$ and $\Delta_s = 1.0\,\mathrm{meV}$ and a weighting parameter $\alpha = 0.3$. The fit takes also into account an inelastic scattering parametrized by $\Gamma_{e,s} = 0.3 \cdot \Delta_{e,s}$.

In conclusion, our point-contact measurements confirm the existence of two superconducting gaps in Nb_3Sn. Measurements on N-S contacts show evidence for the existence of two superconducting gaps. However, the size of the large gap is significantly smaller than that derived from specific heat and break-junction measurements, which may likely be due to surface degradation or oxide layers. This was prevented by the break-junctions experiment which show evidence for two superconducting energy gaps as well. The results we obtain are in consistence with the specific heat measurement.

We acknowledge stimulating discussions with J. Geerk and H. v. Löhneysen. G.G. acknowledges the support by the Deutsche Forschungsgemeinschaft through GO 651/3-1.

REFERENCES

1. V. Guritanu *et al.*, *Phys. Rev. B*, **70**, 184526 (2004).
2. L. Bardeen, L.N. Cooper, J.R. Schrieffer, *Phys. Rev.*, **108**, 1175–1204 (1957).
3. A. Junod *et al.*, in *Studies of High Temperature Superconductors, Vol. 38*, edited by A.V. Narlikar, Nova Science Publishers, Commack, N.Y., 2002, pp. 179.
4. F. Laube *et al.*, *Europhys. Lett.*, **56**, 296–301 (2001).
5. A. F. Andreev, *Sov. Phys. JETP*, **19**, 1228 (1964).
6. G. E. Blonder, M. Tinkham, T. M. Klapwijk, *Phys. Rev. B*, **25**, 4515–4532 (1982).
7. J. Geerk *et al.*, *Phys. Rev. B*, **33**, 1621–1626 (1986).
8. M. Tinkham, *Introduction to superconductivity*, Dover Publications, Inc., Mineola, New York, 2004, pp. 77.

Onset of Local and Coherent Surface Superconductivity on Niobium-Wires

Lars von Sawilski and Jürgen Kötzler

Institut für Angewandte Physik und Zentrum für Mikrostrukturforschung, Universität Hamburg, Jungiusstrasse 11, D-20355 Hamburg, Germany

Abstract. Recent studies of the ac-conductivity and critical current of pure Niobium, as used for high-quality RF cavities, revealed a percolation-driven transition to a coherent surface superconductivity (c-SSC) at the field $H^c_{c3}(T)=0.81\ H_{c3}(T)$. H_{c3} constitutes the conventional onset field for SSC, which upon surface polishing became increasingly larger than the Saint James-de Gennes value $1.7\ H_{c2}(T)$. Here we extend the investigations of the c-SSC to Niobium wires with larger bulk and surface inhomogeneities and correspondingly higher upper critical fields H_{c2}. After chemical polish, we find now H_{c3} to *decrease* from $1.85\ H_{c2}$ to $1.7\ H_{c2}$, while H^c_{c3} remains near the 'universal' value $0.8\ H_{c3}(T)$. The maximum surface critical current J^s_c turns out to grow upon decreasing H_{c3} reaching the value predicted for a multiply connected surface sheath for $H_{c3}\rightarrow 1.7\ H_{c2}$. This indicates an optimum pinning by surface defects.

Keywords: surface superconductivity, surface critical current, percolation-driven transition.
PACS: 74.25.Op, 74.25.Nf, 74.70.Ad

Since the pioneering prediction of SSC [1] within the framework of the Ginzburg-Landau (GL) theory, the main interest has been devoted to a basic understanding of this phenomenon, e.g. effects of surface properties on the critical field H_{c3} [2] and the surface critical current J^s_c [3]. More recent work [4,5] has been motivated by the fact, that SSC occurs within a correlation length (ξ_{GL}) thick sheath. Therefore, SSC probes the regime to where RF-fields in superconducting cavities penetrate, at least, for the standard material Nb with $\xi_{GL}\approx\lambda$. In exploring the SSC, two novel methods have been applied [4]: (i) from ac-susceptibilities the ac-conductance G'-iG'' of the surface sheath has been determined, which provides clear signatures of a transition to a c-SSC occuring below the standard onset field for SSC, i.e. at $H^c_{cJ} < H_{c3}$, and (ii) by a field gradient technique the surface critical current J^s_c was measured and found to vanish at H^c_{c3}. Since the effects of the bulk and surface inhomogeneities of the samples on H^c_{c3} and J^s_c are not entirely clear, we examine here the SSC of technical, i.e. dirty Nb-wires with larger residual resistance ρ_0 (see Table 1) by the same techniques. For comparison, we present also the results for an electropolished pure Nb-cylinder from Ref.4.

Magnetization isotherms, recorded by a SQUID-magnetometer (QUANTUM DESIGN MPMS2) at 5K, are depicted in Fig.1. Obviously, the upper critical field of the wires H_{c2} is shifted towards higher values, as it is expected for a decreasing coherence length due to impurities. Moreover, the unpolished wire (UW) displays flux jumps, which disappear after a standard buffered chemical polish (BCP) treatment described in Ref.5. As H_{c2} and also the bulk current density J_c, inferred from the irreversible part of the magnetization (see Table 1), remain almost unaffected by the BCP, this treatment seems to influence only the surface. Hence the appearance of the flux-jumps at 7.5K and 3.4 K in the UW and CW samples, respectively [6],

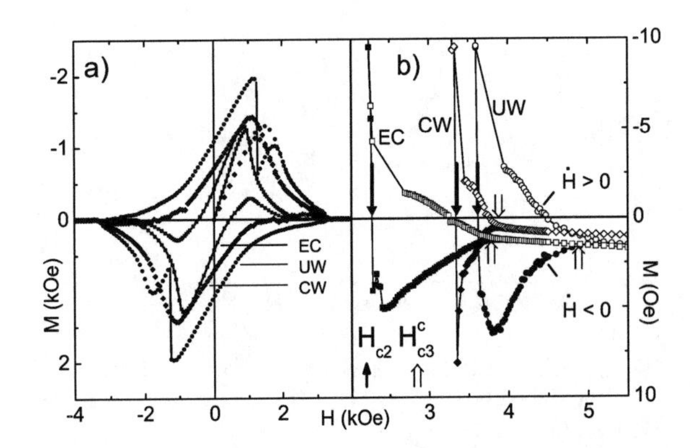

FIGURE 1. Isothermal magnetizations of three Nb-samples (see Table 1): a) full hysteresis loops at 5K, b) results from up- and down scans in a field-gradient [4] above H_{c2} at 5.0 K(EC, CW) and 4.5 K (UW).

CP850, *Low Temperature Physics: 24th International Conference on Low Temperature Physics;* edited by Y. Takano, S. P. Hershfield, S. O. Hill, P. J. Hirschfeld, and A. M. Goldman

TABLE 1. Critical fields and currents of bulk (extrapolated to T=0K) and surface superconductivity.

	ρ_0 ($\mu\Omega$cm)	H_{c2} (kOe)	J_c (A/mm^2)	H_{c3}/H_{c2}	H^c_{c3}/H_{c2}	J^s_c (A/m)
Unpolished wire (UW)	1.14	6.19	1183	1.85	1.48	560
Chem. pol. wire (CW)	1.14	6.10	969	1.69	1.15	320
Electropol. cyl. (EC)	0.05	4.10	176	2.1	1.7	480

may arise from a weaker surface pinning in the unpolished wire. The most interesting effect in Fig.1 is the appearance of additional para- and diamagnetic signals above H_{c2}, depending on the direction of the sample lift [4] through the small built-in gradient of the main magnet of the MPMS2. The signals arise from critical currents J_{sc} induced on the surface, which disappear at the coherent surface critical fields H^c_{c3} introduced in Ref.4. Above H^c_{c3} a reversible paramagnetic background arising from the Pauli susceptibility of Nb and a weak Curie-like contribution from impurities remains. The other characteristic feature of H^c_{c3} is the onset of a strong screening component of the surface conductance, G'-iG'', which is accompanied by a vanishing $1/G'(\omega\rightarrow 0)$, as it was demonstrated in Ref. 4. Both components of G(H) and also J^s_c display power laws in $|H-H^c_{c3}|$, indicating a phase transition. This has been attributed to percolation at the surface [4] and will be discussed elsewhere in more detail for the present samples [6]. Using an inversion routine, the surface conductance G was evaluated from the ac-susceptibility χ'-i χ''.

Here we consider only the dispersion χ' shown in Fig.2. The χ'-data reveal the onset of screening by nucleation of local superconducting regions on the surface at the conventional critical H_{c3} [7]. This field is indicated in Fig.2 (note the logarithmic χ'-scale).

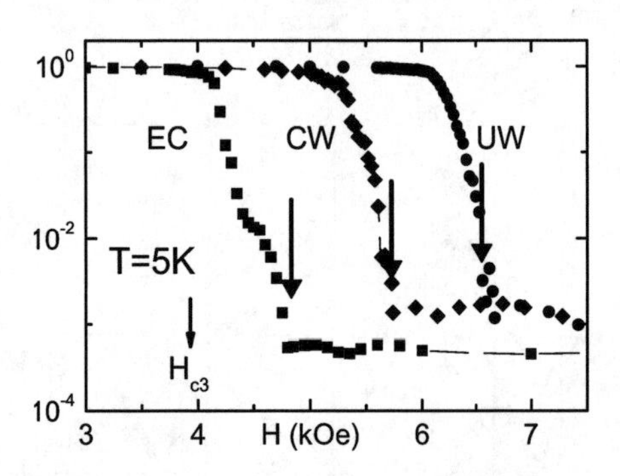

FIGURE 2. Field dependence of the real part of the linear ac-susceptibility at 10Hz defining H_{c3} for the three samples under study at T=5K.

As a summary, the temperature dependences of all surface critical fields are plotted in Fig.3 *vs* H_{c2} (T). Obviously, linear relations, $H_{c3}(T)=r_{32}H_{c2}(T)$ and $H^c_{c3}=r^c_{32}H_{c2}(T)$ are obeyed, the coefficients of which are listed in Table 1.The smallest value, r_{32}=1.69, is obtained for the CW wire which according to the GL

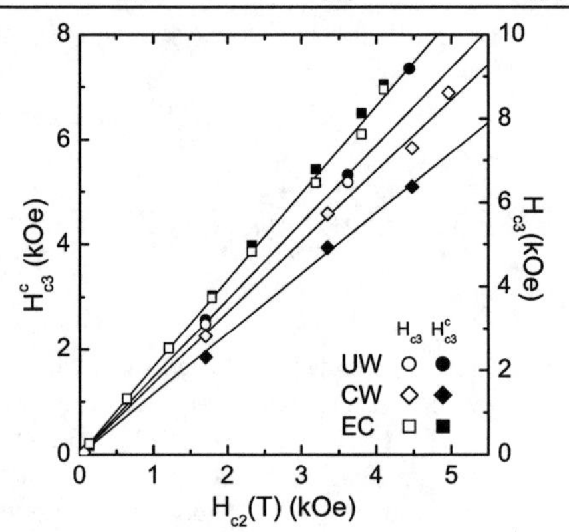

FIGURE 3. Temperature variation of both surface critical fields, here presented in terms of H_{c2}(T).

results of Ref.2 implies, that either the BCP treatment removed all impurities from the surface sheath or leaves them in a very thin layer, $d \ll \xi_{GL}$, at the surface. It is interesting to note, that this effect of polishing on the dirty wires is just inverse to that on the pure cylinders [5], where r_{32} increased towards the maximum GL value 3.8 [2]. Clearly, this observation and also the validity of the relation $H^c_{c3}/H_{c2}=r^c_{32}/r_{32}\approx 0.8$ [4], which for the unpolished wire follows here from the scales chosen for H_{c3} and H^c_{c3} in Fig.3, deserve further attention. Polishing the dirty wire leads to a significant smaller H^c_{c3}/H_{c3}=0.68, see Fig. 3 and Table 1. Here we should note that the existence of surface supercurrents above H_{c2} of Nb, has been observed previously and was rather related to surface inhomogeneities than to the intrinsic SSC [8]. This feature and the effects of surface treatments on J_c will be studied in near future, also by surface imaging.

We are indebted to B. Steffen (DESY) and N. Steinhau-Kuehl (DESY) for polishing the wire.

REFERENCES

1. D. Saint-James and P.G. de Gennes, *Phys. Lett.* **7**, 306 (1963).
2. V.V. Shmidt, *Sov. Phys. JETP* **26**, 566 (1968).
3. H.J. Fink and L.J. Barnes, *Phys. Rev. Lett.* **15**, 792 (1965).
4. J. Kötzler et al., *Phys. Rev. Lett.* **92**, 067005 (2004).
5. S. Casalbuoni et al., *Nucl. Instr. & Methods* **A 538**, 45 (2005).
6. L. von Sawilski and J. Kötzler, *to be published.*
7. R.W. Rollins and J. Silcox, *Phys. Rev.* **155**, 404 (1967).
8. A. DasGupta et al., *J. Appl. Phys.* **47**, 2146 (1976)

Effect of Surface Roughness on Critical Current of Niobium Films

Carsten Mehls, Dale Gill, Christopher Cox, Neel Vora, David Stricker, Eric Berglund, Paritosh Ambekar, Rodney Torii and Suwen Wang

Hansen Experimental Physics Laboratory, Stanford University, Stanford, California, U.S.A.

Abstract. In the development of ultra-high precision superconducting devices, the housing geometry and surface finish that defines the circuit often determines ultimate performance. This can be more critical for non-planar geometries, such as cylindrical surfaces, in which preserving cylindricity can be more important than surface smoothness (our cylindricity requirement is μm along 100 mm cylinder length). Worse than added cost of polishing, there is also the risk that the polishing process will degrade cylindricity. We have investigated the critical current of niobium thin film traces as a function of quartz surface roughness (as defined by ASME standards). Preliminary results show that the critical current decreases with increasing surface roughness, with a more pronounced decrease as the surface roughness becomes comparable to 400 nm film thickness. We will discuss our results based on a simple theoretical model.

Keywords: niobium film, rough surface, critical current.
PACS: 74.25.Sv

INTRODUCTION

In this paper we describe the critical properties of thin superconducting niobium films that are deposited on quartz substrates with different surface roughness. This investigation was done within the scope of the STEP (*S*atellite *T*est of the *E*quivalence *P*rinciple) experiment[1]. The thin film Nb-coating is needed to build high precision superconducting magnetic sensing and suspension coils. The goal was to demonstrate a critical current of 100 mA which is well above the required current of about 10 mA. While it is known that the critical current depends on film thickness it was not clear how it is affected by the substrate roughness. The surface roughness can be characterized by the value R_a[2], the average roughness. Since the normal resistance of thin films is affected by the surface roughness of the substrate[3,4] one might also expect an influence on the critical current of superconductors.

EXPERIMENT

The samples consisted of a meander pattern of 20 parallel Nb traces which had a width of 0.1 mm. Dimensions are chosen to be similar to the magnetic bearing coils used in the STEP experiment. The STEP magnetic bearing suspension force increases with current, 10 mA being the nominal on-orbit value and 100 mA providing a factor of 10 margin. All substrates (with R_a ranging from 0.003 μm to 0.9 μm) were sputtered in a dc magnetron sputtering system, first with a Nb film of either 400 nm or 200 nm thickness, and then with a 50 nm gold layer. The samples were attached to a cryogenic dip stick probe. For critical current measurements at 4.2 K the samples were lowered directly into the LHe bath giving good thermal contact to the liquid to reduce heating effects due to contact resistance. For room temperature resistance and critical current measurements it was sufficient to use a simple 2-point measurement technique. The connection to the traces was made with spring loaded contact pins. The measurements of the critical temperature were done in vacuum using 4-wire technique.

Results and Discussion

The room temperature resistance as a function of roughness for two film thicknesses, 200 nm and 400 nm is shown in FIG. 1. It shows an increase of the resistance with surface roughness. This behavior can be explained by enhanced surface scattering at rough interfaces[3,4]. Beyond Ra ≈ 0.4, the curve changes slope. In this region the roughness becomes comparable to the film thickness. The scattering of electrons is

CP850, *Low Temperature Physics: 24th International Conference on Low Temperature Physics;*
edited by Y. Takano, S. P. Hershfield, S. O. Hill, P. J. Hirschfeld, and A. M. Goldman

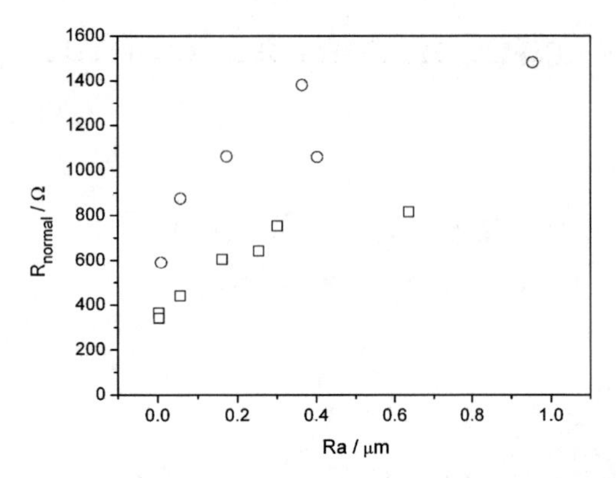

FIGURE 1. Normal resistance as function of surface roughness Ra, for 200 nm (O) and 400 nm ( ) thick samples.

now dominated by defects introduced by the surface roughness. The resistance of the 200 nm film is also approximately 2 times larger than the 400 nm film, which is comparable to the ratio of film thickness. Thus, the higher resistance is due to a smaller cross section.

The measured superconducting transition temperature decreases with film thickness and substrate roughness. The transition temperature decreases from 9 K to 6 K in going from the smoothest to roughest samples. As shown in FIG. 2, the critical current decreases with thinner coatings as expected [5], and also shows a significant decrease with substrate roughness. By using a simple model of substrate geometry in which the film thickness decreases with increasing roughness, we could explain the room temperature resistance dependence on roughness. But the decrease in film thickness in such a model is too small to explain the decrease in critical current. We believe the observed decrease is caused by an increase in number of defects

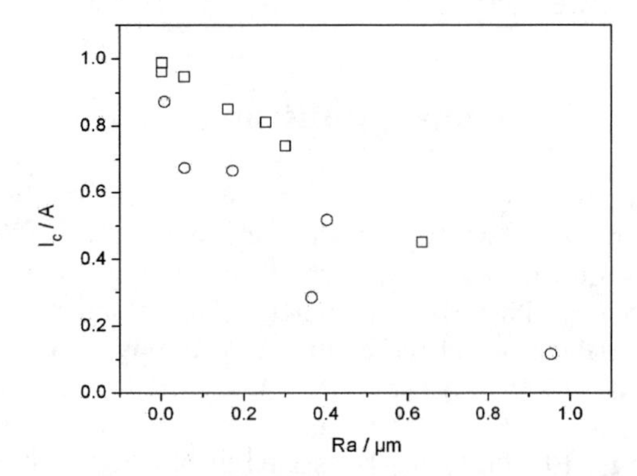

FIGURE 2. Critical current I_c as function of roughness Ra for 200 nm (O) and 400 nm ( ) thick samples at 4.2 K.

in the film as the roughness increases, reducing the effective superconducting film thickness. The primary effect of substrate roughness is a reduction in the quality of the superconducting film. An effective film thickness that is decreasing with substrate roughness is in qualitative agreement with observations that reduced film thickness leads to a decrease in transition temperature and subsequently a reduction in the critical current [6-8].

SUMMARY

We have measured the critical temperature and current of thin superconducting niobium films on quartz substrates with roughness R_a as large as 0.9 μm. We observe a significant decrease in both the critical temperature and current as the surface roughness increases. This decrease is attributed to a degradation in the properties of the film (due to an increase in number of defects). We are planning next to measure samples with 800 nm film thickness. In addition, we are in the process of setting up an experiment to look at flux trapping behavior as a function of surface roughness using a SQUID magnetometer.

ACKNOWLEDGMENTS

This work has been supported by NASA and the Marshall Space Flight Center under the cooperative agreement #NNM04AA18A-01.

REFERENCES

1. J. Mester, R. Torii, P. Worden, N. Lockerbie, S. Vitale and C. W. F. Everitt, *Class. Quantum Grav.* **18,** 2475 (2001).
2. L. Mummery, *Surface Texture Analysis: The Handbook*, VS-Muhlhausen: Hommelwerke GmbH, 1992.
3. G. Choe and M. Steinback, *J. Appl. Phys.*, **85**, 5777 (1999).
4. K. C. Elsom and J. R. Sambles, *J. Phys. F: Metal Phys.*, **11**, 647 (1981).
5. A. C. Rose-Innes and E. H. Rhoderick, *Introduction to Superconductivity*, 2nd ed., Oxford - New York: Pergamon Press, 1978.
6. M. R. Beasley and C. J. Kircher, "Josephson Junction Electronics: Material Issues and Fabrication Techniques," in *Superconductor Material Science,* edited by S. Foner and B. B. Schwartz, New York: Plenum Press, 1981, pp. 605-684
7. R. F. Wang, S. P. Zhao, G. H. Chen, and Q. S. Yang, *Phys. Rev. B*, **62**, 11793 (2000).
8. M. S. M. Minhaj, S. Meepagala, J. T. Chen, and L. E. Wenger, *Phys. Rev. B*, **49**, 15235 (1994).

Low Field ac Susceptibility and High Harmonic Study in Chevrel-Phase and High-T_c Polycrystalline Superconductors

Ioseb. R. Metskhvarishvili*, M.R. Metskhvarishvili*, L. R. Khorbaladze*, S. L. Ginzburg+, É. G Tarovik+ and V. P. Khavronin+

* *Iv. Javakhishvili Tbilisi State University, Physics Faculty, 0128 Tbilisi, Chavchavadze ave.3, Georgia*
+ *St. Petersburg Institute of Nuclear Physics, Russian Academy of Sciences, 188350 Gatchina, Russia*

Abstract. The ac susceptibility and high harmonic response of polycrystalline $SnMo_6S_8$ and $YBa_2Cu_3O_y$ superconductors are measured in the presence of small ac excitation field and dc magnetic field applied on it. Critical state models are used to explain the nonlinear magnetic response in polycrystalline superconductors. For Chevrel-phase superconductors, a field independent Bean's critical state model and for High-T_c superconductor the Anderson-Kim model are observed to work reasonably well.

Keywords: Low field, Josephson junction, critical state models.
PACS: 74.25.Ha, 74.70.Dd, 74.72.Bk.

INTRODUCTION

Granular superconductors have a peculiarity associated to the existence of superconductivity in each grain, parameters of which are characteristic of bulk monocrystals. Besides, between the grains there are weak Josephson junctions determining the superconducting properties of the whole system [1, 2]. For the Josephson junctions in low fields, less than first critical field of grains, a lot of irreversible and nonlinear phenomena are observed. These phenomena are characterized by the presence of high harmonics.

The aim of this work is to study the weak link behavior and make comparison between two systems to understand granular nature of polycrystalline superconductors. The choice of Chevrel-phase superconductor was made because $SnMo_6S_8$ has a high critical field H_{c2} and correspondingly a very short coherence length $\xi \approx 23$ Å [3]. This makes them similar to HTSCs, in which ξ is of order of several angstroms. Therefore one can expect that for such values of ξ any defect even a small one, can act like a Josephson junction. Our experimental results have shown that very weak magnetic fields of order of millioersteds do indeed penetrate into such superconductors.

EXPERIMENTAL DETAILS

The investigated samples SMS and YBCO had form of pellet of ≈ 9.5 mm diameter and ≈ 3.5 mm thickness. The samples were placed in a coil creating ac field h and the same coil was also used for measurements. Constant dc field H was parallel to ac field and was created by external solenoid. The measurements were performed mainly at 20 kHz frequency and in the range of fields $10^{-2} \leq h \leq 1$ Oe, $H \leq 12$ Oe.

For investigation we used fundamental and harmonic ac susceptibilities. For measurement of real χ' and imaginary χ'' parts of ac susceptibility phase method was used. χ' and χ'' measurement errors at higher than 1 kHz frequencies does not exceed 1%, when $4\pi\chi' > 0.1$.

Error of measurements of high harmonics was approximately 2 % when the measured signal was less then 0.2 μV and no more than 0.5 % when the signal was higher.

The Earth's magnetic fielded was shielded to less than 10^{-3} Oe by use of Permalloy screens.

EXPERIMENTAL RESULTS AND DISCUSSION

We measured real and imaginary parts of ac susceptibility in dependence of temperature for h=0.1Oe and H = 0. Our experiments show that the onset temperature of the superconducting transition for SMS is about $T_c \approx 13.5$ K and for YBCO ≈ 93 K

The measurements of temperature dependence of the third harmonics by the method of high harmonics for 0.1, 0.3, 0.6, 0.9 Oe ac field amplitudes and H=0 showed that as for HTSCs as for SMS on $c_3(T)$ curve two peaks are observed, which shows shielding of flux

CP850, *Low Temperature Physics: 24th International Conference on Low Temperature Physics;* edited by Y. Takano, S. P. Hershfield, S. O. Hill, P. J. Hirschfeld, and A. M. Goldman

from and between the grains.

Figure 1a presents the spectrum of high harmonics measured for $SnMo_6S_8$ which was obtained in weak ac and dc magnetic fields. The harmonics decrease weakly as the number of the harmonic increases and the even harmonics are absent for $H \neq 0$. By contrast, for $YBa_2Cu_3O_y$ we observed even harmonics in the dc field (Figure 1b). For $H = 0$ the shape of the odd harmonics remains the same, while the even harmonics actually are vanished.

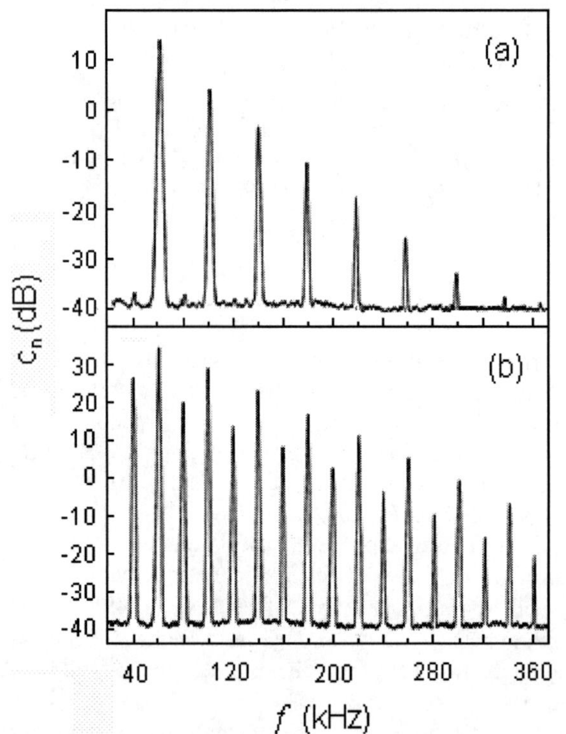

FIGURE 1. High harmonics spectrum: (a) $SnMo_6S_8$, $T \approx 12$ K, $h = 0.6$ Oe, $H = 8$ Oe; (b) $YBa_2Cu_3O_y$, $T \approx 90$ K, $h = 0.6$ Oe, $H = 8$ Oe.

According to Bean's [4] amplitudes of each harmonics are inversely proportional to the critical current density $c_n \approx 1/j_c$. Thus by studying the dependencies of these harmonics on the magnitude of the constant field H at fixed amplitude of h $j_c(B)$ function can be determined.

As one can see from Fig. 2 the amplitude of the third harmonic for SMS does not depend on the magnetic field, at least up to 12 Oe i.e.,

$$j_c(B) = j_0. \qquad (1)$$

The same figure shows that for the YBCO the c_3 is very sensitive to low external dc fields and increases linearly. The observed behavior of the third harmonic can be explained by the corresponding $j_c(B)$ dependence, which has the form [5]:

$$j_c(B) = j_c(0)/(1 + B/B_0). \qquad (2)$$

It must be noted that we carried out such experiments in a large temperature range and $c_3(H)$ dependences do not change for both samples.

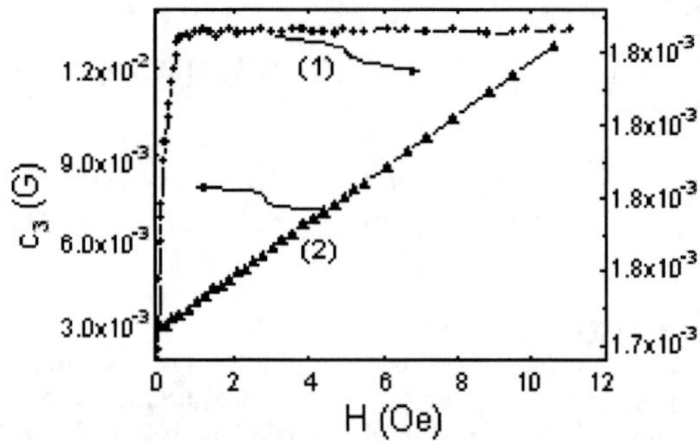

FIGURE 2. Third harmonic c_3 dependence on the dc field. Curve 1 corresponds to $SnMo_6S_8$, $T \approx 12$ K, $h = 0.6$ Oe, and curve 2 - $YBa_2Cu_3O_y$, $T \approx 90$ K, $h = 0.6$ Oe.

CONCLUSION

In summary, for the SMS samples our experimental results can be explained by the Bean's critical state model, where j_c does not depend on magnetic field at least up to 12 Oe, and the even harmonics are absent even for $H \neq 0$. Contrary to SMS, nonlinear magnetic response in $YBa_2Cu_3O_y$ is in a good agreement with the Anderson-Kim model, where even harmonics in constant magnetic field are observed and the dependence of current on the applied dc magnetic field takes place.

ACKNOWLEDGMENTS

One of the authors (I. Metskhvarishvili) acknowledges the Georgian research and Development Foundation, GRDF-№ TGP-06 for supporting of participation on "24th International Conference of Low Temperature Physics". Also, thanks Prof. Mark W. Meisel and all Organizing Committee of LT24 for attention and technical support.

REFERENCE

1. J. R. Clem, *Physica C* **153-155**, 50-51 (1988).
2. E. B. Sonin, *JTETP Letters* **47**, 496-499 (1988).
3. P. Birrer et al , *Phys. Rev. B* **48**, 16589-16599 (1993-II).
4. C. P. Bean, *Rev. Mod. Phys.* **36**, 31-43 (1964).
5. P. W. Anderson and Y. B. Kim, *Rev. Mod. Phys.* **36**, 39-43 (1964).

Theory of Photocurrent in BCS Excitons

Toshiyuki Fujii*, Shigemasa Matsuo†,*, Munehiro Nishida* and Noriyuki Hatakenaka†,*

*AdSM, Hiroshima University, Higashi-Hiroshima 739-8530, Japan
†Faculty of Integrated Arts and Sciences, Hiroshima University, Higashi-Hiroshima 739-8521, Japan

Abstract. We derive a photocurrent formula for a hybrid semiconductor junction with an excitonic Bardeen-Cooper-Schrieffer (BCS) state. The formula provides a new approach to confirm the excitonic BCS state as well as its potential for applications like solar cells.

Keywords: exciton, BCS, photocurrent
PACS: 71.35.Cc, 71.35.Lk

INTRODUCTION

Recently, spontaneous giant emissions from GaAs/AlGaAs superlattices under a current bias have been observed at room temperature [1, 2]. These emissions are considered to be radiative recombinations between Bardeen-Cooper-Schrieffer (BCS)-like condensed states of excitons (an excitonic BCS state) in semiconductors. However, there are no direct evidence in the experiments. Here we propose a new method to observe the excitonic BCS state in a similar system. The unique character of the BCS state appears in its density of states (DOS), which diverges at the energy gap. We extract this feature from photocurrent by photogenerated quasiparticles which reflect the DOS in an excitonic BCS state.

PHOTOABSORPTION OF BCS EXCITONS

We shall derive the photocurrent formula for a hybrid semiconductor junction as shown in Fig. 1. We assume that excitons in the intermediate semiconductor layer with band gap E_g condense into BCS phase written as

$$H_0 = \sum_k E_k \left(\alpha_k^\dagger \alpha_k + \beta_{-k}^\dagger \beta_{-k}\right), \tag{1}$$

where α_k (β_{-k}) is an annihilation operator for an electron-like (a hole-like) quasiparticle with wavenumber $k(-k)$. The quasiparticle energy is defined as $E_k = \mu + \mathrm{sgn}(|k| - k_\mu)E'_k$ where μ is chemical potential and $E'_k = \sqrt{\xi_k^2 + |\Delta|^2}$ with Δ being energy gap under the s-wave assumption and ξ_k being kinetic energies for electrons and holes expressed as $\xi_k = \frac{\hbar^2 k^2}{2m} + \frac{E_g}{2} - \mu$.

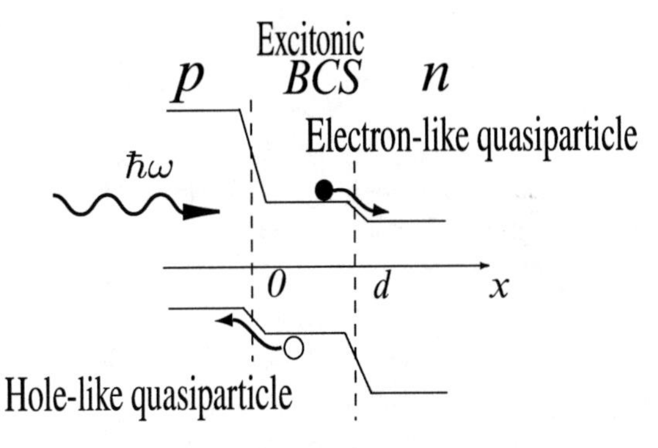

FIGURE 1. Schematic drawing of a *p*-BCS-*n* semiconductor junction

The light-matter interaction Hamiltonian is expressed as

$$\begin{aligned} H_{int} = \sum_k \lambda_k \{ u_k v_k (\alpha_k^\dagger \alpha_k - \beta_{-k}\beta_{-k}^\dagger) \\ + u_k^2 \alpha_k^\dagger \beta_{-k}^\dagger - v_k^2 \beta_{-k}\alpha_k \} e^{-i\omega t} + H.C., \end{aligned} \tag{2}$$

with the coupling constant λ_k in a dipole approximation

$$\lambda_k = \frac{e}{2mc}\mathbf{A}_0(x) \int \phi_{ck}^*(\mathbf{r})\hat{\mathbf{p}}\phi_{vk}(\mathbf{r})d\mathbf{r} \equiv \mathbf{A}_0(x)\cdot \mathbf{M}_\mathbf{k}, \tag{3}$$

where u_k and v_k are the occupation and the unoccupation pair amplitudes with k, respectively. $\mathbf{A}_0(x)$ denotes the amplitude of the vector potential. $\phi_{c(v)}$ is a core-part of the electron wave function in the conduction (valence) band.

Now let us consider photoabsorptions by an interband transition as shown in Fig. 2(b). Using Fermi's golden rule, the transiton rate can be expressed as [3]

$$W \simeq \frac{2\pi}{\hbar}\bar{\lambda}^2 \int_\mu^\infty N(E)u(E)^4 \;\; \delta(2E - \hbar\omega)dE, \tag{4}$$

CP850, *Low Temperature Physics: 24th International Conference on Low Temperature Physics;*
edited by Y. Takano, S. P. Hershfield, S. O. Hill, P. J. Hirschfeld, and A. M. Goldman

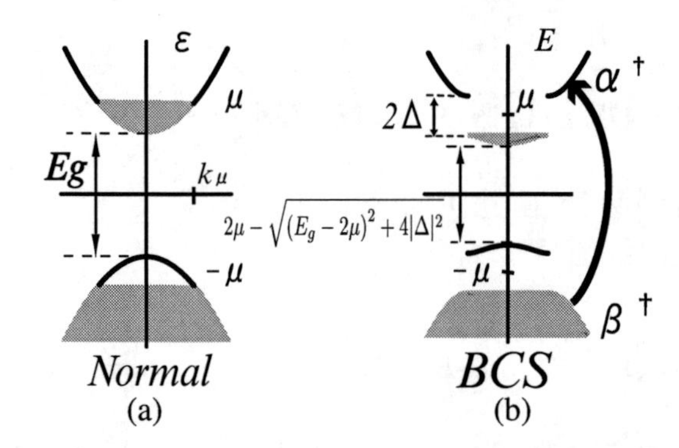

FIGURE 2. Band diagrams for (a) a normal state and (b) an excitonic BCS state

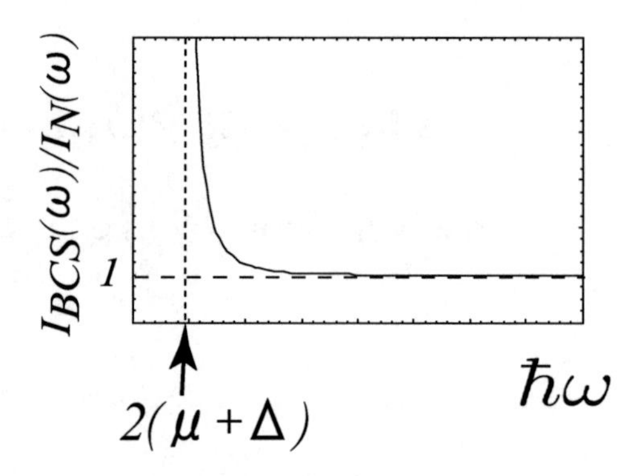

FIGURE 3. $I_{BCS}(\omega)/I_N(\omega)$ v.s. $\hbar\omega$

$\overline{\lambda^2}$ is the mean value of λ_k^2 at the quasi-Fermi surface. Photon flux can be given as

$$\Phi(\omega,x) = \frac{\eta\,\omega}{2\mu_0 c\hbar}|\mathbf{A}_0(x)|^2, \tag{5}$$

by using the vector potential. Here η and μ_0 are the refractive index and the magnetic permeability, respectively. The photon flux $\Phi(\omega)$ attenuates in matter, and is described by

$$\frac{d\Phi}{dx} = \frac{\eta}{c}\frac{d\Phi}{dt} = -\frac{\eta n W}{c} = -\frac{\eta n W}{c\Phi}\Phi, \tag{6}$$

where n denotes density of electron-hole pairs. Thus, the photoabsorption coefficient $\alpha(\omega)$ yields

$$\alpha(\omega) = \frac{\eta n W}{c\Phi} = \frac{4\pi n\mu_0}{\omega}\overline{|\mathbf{M}|^2} N(\hbar\omega/2)u^4(\hbar\omega/2), \tag{7}$$

where N stands for DOS, and $\overline{|\mathbf{M}|^2}$ is the mean value of $|\mathbf{M_k}|^2$ at the quasi-Fermi surface. In two-dimensional excitonic BCS system, DOS is given by

$$N_{BCS}(\hbar\omega/2) \propto \frac{\hbar\omega/2-\mu}{\sqrt{(\hbar\omega/2-\mu)^2-|\Delta|^2}}. \tag{8}$$

The absorption coefficient diverges at the edge of the energy gap. Quasiparticles generated by the incident light in BCS layer diffuse toward p or n layers. Some of these quasiparticles reach to the edge of BCS layer, and contribute to photocurrent as follows;

$$I = e\int_0^\infty \left(\int_0^d e^{-\alpha x}e^{-\frac{x}{L_h}}dx + \int_0^d e^{-\alpha x}e^{-\frac{(d-x)}{L_e}}dx\right) \times \alpha(\omega)\Phi(\omega)\,d\omega \tag{9}$$
$$\equiv \int_0^\infty I(\omega)\,d\omega,$$

where the first term in the parenthesis corresponds to the contribution from the holes, and second the electrons. In the case of $d \ll L_e, L_h, 1/\alpha$, $I(\omega)$ is approximated as

$$I(\omega) \simeq 2ed\alpha(\omega)\Phi(\omega). \tag{10}$$

From this expression, $I(\omega) \propto \alpha(\omega)$, we thus obtain

$$\frac{I_{BCS}(\omega)}{I_N(\omega)} \simeq \frac{\alpha_{BCS}(\omega)}{\alpha_N(\omega)} = u^4(\hbar\omega/2)\frac{N_{BCS}(\hbar\omega/2)}{N_N(\hbar\omega/2)}. \tag{11}$$

This formula is a central result of this work. We can therefore confirm the existence of the excitonic BCS state thourgh the photocurrent reflecting the DOS in p-BCS-n junctions.

SUMMARY

We have derived a photocurrent formula for p-BCS-n semiconductor junctions. The ratio of the photocurrent in BCS-like states and that in normal states is propotinal to the ratio of DOS in two states as shown in Eq. (11). Therefore, it is possible to observe the divergence of DOS in BCS-like state via photocurrent. Conseqently, an excitonic BCS state can be confirmed by measuring photocurrent in p-BCS-n semiconductor junctions. This photocurrent might be applicable to the solar cell devices.

ACKNOWLEDGMENTS

We would like to thank Profs. M. Yamanishi and K. Nagai for valuable discussions. This work was supported in part by a resarch grant from The Mazda Foundation and a Grant-in-Aid (15204029) from the Ministry of Education, Sprots, Science and Technology of Japan.

REFERENCES

1. P. P. Vasil'ev, *et. al*, *JETP* **93**, 1288 (2001).
2. P. P. Vasil'ev, *et. al*, *JETP* **96**, 310 (2003).
3. M. E. Flattè, *et. al*, *Appl. Phys. Lett.* **66**, 1313 (1995).

The Band Structure of Photonic Band-Gap Crystals with Superconducting Elements

Oleg L. Berman[1], Yurii E. Lozovik[2], Sergey L. Eiderman[2] and Rob D. Coalson[3]

[1]*Department of Physics and Astronomy, University of Pittsburgh, Pittsburgh, PA 15260, USA*
[2]*Institute of Spectroscopy, Russian Academy of Sciences, 142190 Troitsk, Moscow Region, Russia*
[3]*Department of Chemistry, University of Pittsburgh, Pittsburgh, PA 15260, USA*

Abstract. The band spectra of a periodic lattice of superconducting particles are studied in the subgap frequency range. We calculate the photonic band structure of the system consisting of an infinite array of identical parallel superconducting cylinders. Dispersion curves, density of states and dependence of the width of the gaps on the filling factor is obtained.

Keywords: Superconductor, photonic crystal.
PACS: 74.25.Gz, 74.78.Fk, 42.70.Qs

INTRODUCTION

Electromagnetic waves in media with periodic dielectric function, i.e., photonic crystals, have frequency spectra that exhibit band structure and coordinate dependence similar to Bloch waves of electrons in natural (atomic) crystals. Photonic crystals are the subject of growing interest due to their unique electromagnetic properties and modern applications in controlling the emission and distribution of light [1-7]. Photonic crystals with different materials as constituent elements, e.g., dielectric, metal, and liquid crystals have been studied. Here we analyze a new type of photonic crystal - with superconducting constituent elements. Superconducting photonic crystals may find interesting applications due to i) possible control of their band structure via magnetic field or temperature and ii) small damping in comparison with normal metal.

THE WAVE EQUATION FOR SUPERCONDUCTING PHOTONIC CRYSTALS

We are interested in the subgap frequency region, i.e. THz or GHz regions for high- and low-temperature superconductors, respectively. At higher, above-gap frequencies, the frequency dependent dielectric function of superconductor elements is close to that of a normal metal and thus the band structure is approximately the same as for normal metal photonic crystals. In the general case, the expression for the dielectric function of a superconductor is extremely complicated [8, 9]. Thus, to obtain a qualitative picture for subgap band structure we consider a London superconductor at $\omega >> 4\pi\sigma/(\varepsilon_\infty + 1)$, $\omega\tau << 2(T - T_c)/T_c$, where σ is the Drude conductivity of a normal metal, τ is the relaxation time, and ε_∞ is the lattice dielectric function. Under these conditions one obtains the following equation for the electric field inside the photonic crystal:

$$-\nabla^2 \vec{E} = \omega^2 / c^2 [\varepsilon_\infty + (1 + c^2 / \delta_L^2 \omega^2 - \varepsilon_\infty) \sum_{\{n_i\}} [\eta(r_0 - |\vec{r} - \vec{a}(\{n_i\})|)]] \vec{E} \quad , \quad (1)$$

where δ_L is the London penetration depth, η is the Heavyside step function, and $\vec{a}(\{n_i\})$ are photonic lattice vectors. There is a direct analogy between this equation and the Schrödinger equation (see, e.g., [10]). The role of the periodic potential which determines the band structure of an electron (moving in accordance with the Schrödinger equation) is played in the case of photonic crystals made from normal metal or superconductor components by the dielectric contrast, i.e., the difference between the dielectric functions of the superconducting elements and the environment medium. One can control the band structure of a superconducting photonic crystal by its geometry, shape and the filling factor of superconducting elements, and also by external magnetic field and temperature.

CP850, *Low Temperature Physics: 24th International Conference on Low Temperature Physics;*
edited by Y. Takano, S. P. Hershfield, S. O. Hill, P. J. Hirschfeld, and A. M. Goldman

BAND STRUCTURE CALCULATION

In this section the band structure of a two-dimensional photonic crystal consisting of an array of infinite superconducting cylinders forming a square lattice is calculated. We consider wave vectors normal to the cylinders analyzing separately E-and H-polarizations.

Expanding the electric field in plane (Bloch) waves, we reduce Maxwell's equations for the superconducting photonic crystal to the following eigenvalue problem:

$$\sum_{\vec{G}'} (\delta_{\vec{G}\vec{G}'}(\vec{k}+\vec{G})^2 - \frac{P^2}{c^2} M_{\vec{G}\vec{G}'}) E_k(\vec{G}') = \frac{\omega^2}{c^2} E_k(\vec{G}) \tag{2}$$

Here $E_k(\vec{G})$ are Fourier components of the electric field $E_k(\vec{x})$, and $P = \delta_L / c$. The Fourier transform of the periodic dielectric function in the superconducting photonic crystal is

$$\varepsilon(\vec{G}-\vec{G}') = \delta_{\vec{G}\vec{G}'} + (P^2/\omega^2) M_{\vec{G}\vec{G}'}, \tag{3}$$

where

$$M_{\vec{G}\vec{G}'} = \begin{cases} f\varepsilon + (f-1), \vec{G} = \vec{G}' \\ 2f(\varepsilon - 1)\dfrac{J_1(|\vec{G}-\vec{G}'|r)}{|\vec{G}-\vec{G}'|r}, \vec{G} \neq \vec{G}' \end{cases}, \tag{4}$$

where $f = S_{\sup ercond} / S = \pi r^2 / a^2$ is the filling factor of superconductor ($S_{\sup ercond}$ is the area occupied by the cylinder in the plane of the electric field distribution, S is the total area occupied by the real-space unit cell, r is the radius of a cylinder, a is the lattice period), and J_1 is a Bessel function.

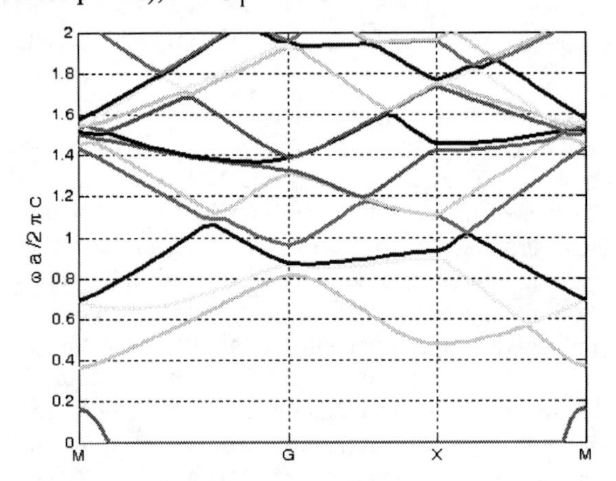

FIGURE 1. Dispersive dependence ω(k) for two-dimensional superconducting photonic crystal with square lattice consisting of infinite cylinders having circular cross section. E-polarization is considered, with f=0.3, P=5. A band gap is clearly apparent.

The band structure for the filling factor f = 0.3 and P =5 in lattice units of frequency $2\pi c/a$ is presented in Fig.1. The density of photonic states for this crystal is presented in Fig.2.

The width of the photonic gap is a nonmonotonic function of the filling factor f. The photonic gap exists only in the range 0.23<f<0.44. The widest gap corresponds to f=0.32. As the filling factor is reduced down to f = 0.23 and increased up to 0.44 the photonic gap disappears.

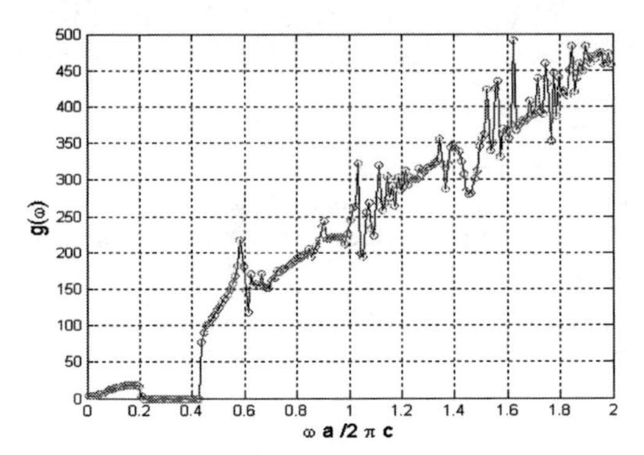

FIGURE 2. Density of states for two-dimensional superconducting photonic crystal with the same geometry as in Figure 1.

ACKNOWLEDGMENTS

Ut Yu.E.L. is grateful to RFBR for support, and R.D.C. thanks the NIRT for support.

REFERENCES

1. E. Yablonovitch, *Phys. Rev. Lett.* 58, 2059 (1987).
2. S. John, *Phys. Rev. Lett.* **58**, 2486 (1987).
3. J.D. Joannopoulos, R.D. Meade, and J.N. Winn, *Photonic Crystals: The Road from Theory to Practice,* Princeton University Press, Princeton, NJ, 1995.
4. A.A. Maradudin and A.R. McGurn, *Phys. Rev.* ***B*** **48**, 17576 (1993).
5. B. Schulkin, L. Sztancsik and J.F. Federici, *Am. J. Phys.* **72**, 1051 (2004).
6. Z. Sun, Y.S. Jung and H.K. Kim, *Appl. Phys.Lett.* **83**, 3021 (2003).
7. Z. Sun and H.K. Kim, *Appl. Phys. Lett.* **85**, 642 (2004).
8. A.A. Abrikosov, *Fundamentals of the Theory of Metals,* Amsterdam, North Holland (1988).
9. A.L. Dobryakov, V.M. Farztdinov and Yu.E. Lozovik, *Physica Scripta* **60**, 474 (1999).
10. Yu.E. Lozovik and A.V. Klyuchnik, in "*The Dielectric Function of Condensed Systems*", eds. L.V.Keldysh et.al., Elsevier Publ., 1987.

Thermal Broadening Effect on the Tunneling Conductance in Diffusive Normal Metal / Superconductor Junctions

Iduru Shigeta*, Takehito Yokoyama†, Yasuhiro Asano** and Yukio Tanaka†,‡

*Department of General Education, Kumamoto National College of Technology, Kumamoto 861-8603, Japan
†Department of Applied Physics, Nagoya University, Nagoya 464-8603, Japan
**Department of Applied Physics, Hokkaido University, Sapporo 060-8628, Japan
‡CREST, Japan Science and Technology Agency (JST), Nagoya 464-8603, Japan

Abstract. Using the circuit theory, we have studied how zero-bias conductance dip (ZBCD) and zero-bias conductance peak (ZBCP) are smeared by increasing temperature in the diffusive normal metal (DN) / s-wave or d-wave superconductor junctions. Tunneling conductance of the junction is calculated by changing the magnitudes of the resistance in DN, the Thouless energy in DN, the transparency of the insulating barrier, and the angle between the normal to the interface and the crystal axis of d-wave superconductors. We present a threshold temperature from a possible observation of the ZBCD and ZBCP in line shapes of tunneling conductance.

Keywords: Andreev reflection, proximity effect, thermal broadening effect, zero-bias conductance peak, zero-bias conductance dip
PACS: 74.20.−z, 74.45.+c, 74.50.+r

INTRODUCTION

The low-energy transport in the mesoscopic system is governed by Andreev reflection [1]. For diffusive normal metal / superconductor (DN/S) junctions, the phase coherence between incoming electrons and outgoing Andreev holes persists in DN at a mesoscopic length scale. This results in strong interference effects on the probability of Andreev reflection through the proximity effect. The circuit theory has recently explained that the interplay between diffusive and interface scattering produces a wide variety of line shapes of the tunneling conductance, not only for d-wave junctions, but also for s-wave junctions: zero-bias conductance dip (ZBCD), zero-bias conductance peak (ZBCP), gap-like, and rounded bottom structures [2, 3, 4, 5, 6, 7].

In the present paper, we will discuss the thermal broadening effect on the tunneling conductance of DN/S junctions, and give a threshold temperature T_{Th}.

THEORY AND RESULTS

We consider a junction consisting of a normal reservoir (N) and a superconducting one connected by a quasi-one-dimensional DN conductor with a resistance R_{d} and a length L, much larger than the mean-free path l. The interface between the DN conductor and the superconductor electrode has a resistance R_{b}, while the N/DN interface has zero resistance. The positions of the N/DN interface and the DN/S interface are denoted as $x=-L$ and $x=0$, respectively. The scattering zone ($x=0$) is modeled as an insulating δ-function barrier with the transparency $T=4\cos^2\phi/(4\cos^2\phi+Z^2)$, where Z is a dimensionless constant, and ϕ is the injection angle measured from the interface normal to the junction. The details of the circuit theory are described elsewhere [6, 7]. Consequently, the tunneling conductance $\sigma_{\mathrm{T}}(eV)$ is given by

$$\sigma_{\mathrm{T}}(eV)=\frac{\displaystyle\int_{-\infty}^{\infty}\sigma_{\mathrm{S}}(E)\,\mathrm{sech}^2\left(\frac{E+eV}{2k_{\mathrm{B}}T}\right)dE}{\displaystyle\int_{-\infty}^{\infty}\sigma_{\mathrm{N}}(E)\,\mathrm{sech}^2\left(\frac{E+eV}{2k_{\mathrm{B}}T}\right)dE}, \quad (1)$$

where $\sigma_{\mathrm{S}}(E)=1/R$ and $\sigma_{\mathrm{N}}(E)=1/(R_{\mathrm{d}}+R_{\mathrm{b}})$. The total resistance R in the superconducting state is written by

$$R=\frac{R_{\mathrm{b}}}{\langle I_{\mathrm{b0}}\rangle}+\frac{R_{\mathrm{d}}}{L}\int_{-L}^{0}\frac{dx}{\cosh^2\theta_{\mathrm{Im}}(x)}. \quad (2)$$

For the equations, we assume $\Delta(T)$ and $\Delta(T)\cos[2(\phi\pm\alpha)]$ as temperature dependent energy gaps of s-wave and d-wave superconductors, where α denotes the angle between the normal to the interface and the crystal axis of d-wave superconductors.

Here, we focus on the thermal broadening effect on the line shapes of the tunneling conductance in the case of no midgap Andreev bound states (MABS). Typical temperature dependence of the ZBCD and ZBCP are presented in Figs. 1(a) and 1(b), respectively. In order that the coherent Andreev reflection (CAR) arises at the DN/S interface [6, 7, 8], we have chosen the parameters as $R_{\mathrm{d}}/R_{\mathrm{b}}=1$ and $E_{\mathrm{Th}}/\Delta_0=0.01$, where E_{Th} is the Thouless energy and Δ_0 is defined as $\Delta(0)$. Figure 1(a) shows the

CP850, *Low Temperature Physics: 24th International Conference on Low Temperature Physics;*
edited by Y. Takano, S. P. Hershfield, S. O. Hill, P. J. Hirschfeld, and A. M. Goldman

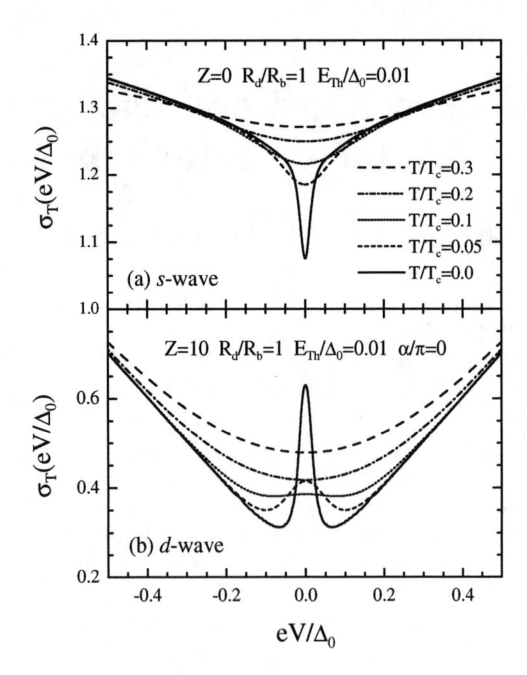

FIGURE 1. (a) Temperature dependence of the ZBCD in tunneling conductance for *s*-wave superconductors. (b) Temperature dependence of the ZBCP in tunneling conductance for *d*-wave superconductors.

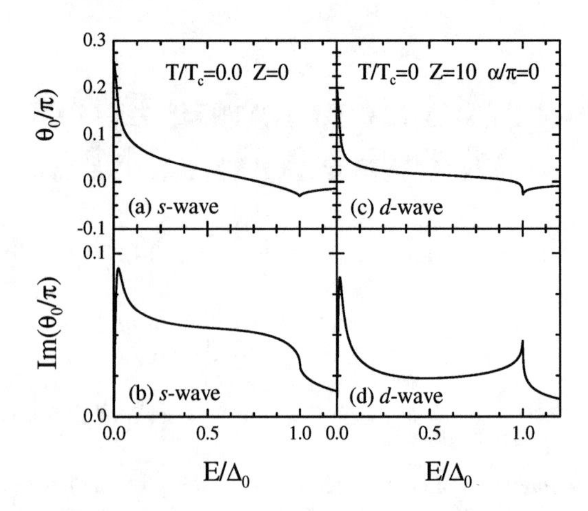

FIGURE 2. Real and imaginary parts of θ_0 are plotted as a function of E with $R_d/R_b = 1$ and $E_{Th}/\Delta_0 = 0.01$ at $T/T_c = 0$. The left panels correspond to θ_0 of *s*-wave superconductors in Fig. 1(a) and the right panels to θ_0 of *d*-wave superconductors in Fig. 1(b), respectively.

ZBCD behavior in the case of *s*-wave superconductors. The ZBCD smears and $\sigma_T(0)$ increases gradually with increasing temperature, and then the ZBCD disappears at $T/T_c \simeq 0.2$. Here, to make actual tunnel junctions with $Z = 0$, the cross section of the DN/S interface should be roughly $(L/l)^2$ times smaller than the cross section of the DN conductor. Figure 1(b) represents the ZBCP behavior in the case of $\alpha = 0$ for *d*-wave superconductors, where the ZBCP originates not due to the MABS, but due to the CAR [7]. With increasing temperature, the ZBCP broadens continuously and the $\sigma_T(0)$ turns to the increase at $T/T_c = 0.1$ after decreasing. Then, the ZBCP disappears at $T/T_c \simeq 0.2$ as well as the ZBCD.

From evaluation of T_{Th}'s in tunneling conductance for various parameters, it is found that T_{Th} depends on R_d/R_b and Z, E_{Th}/Δ_0. The *s*-wave superconductors have the constant and maximum value of $T_{Th}/T_c = 0.2$ less than $E_{Th}/\Delta_0 = 0.1$. On the other hand, the *d*-wave superconductors have the maximum value of $T_{Th}/T_c = 0.2$ around $E_{Th}/\Delta_0 = 0.01$. Therefore, we can estimate maximum T_{Th}'s for actual superconducting materials, for example, $T_{Th} \simeq 8$ K for MgB_2 and $T_{Th} \simeq 18$ K for $YBa_2CuO_{7-\delta}$.

The measure θ_0 of the proximity effect at the DN/S interface is plotted as a function of E with corresponding parameters in Fig. 2 [9]. The Imθ_0's increase sharply from 0 and have a peak at $E \simeq E_{Th}$ in both of *s*-wave and *d*-wave superconductors. The peak does not change for various temperatures. Besides this, both real and imaginary parts suddenly change at $E = \Delta(T)$. With increasing temperature, all the line shapes in Reθ_0 and Imθ_0 immediately shift to low energy near T_c all over the energy range. This is due to θ_0 depending only on an amplitude of the energy gap $\Delta(T)$ in various temperatures, and due to $\Delta(T)$ rapidly decreasing just near T_c.

CONCLUSIONS

Using the circuit theory, we have calculated temperature dependence of the tunneling conductance on the condition that the CAR arises at the DN/S interface. We have presented the maximum threshold temperature as $T_{Th}/T_c = 0.2$ from a possible observation of the thermal broadening behaviors of the ZBCD for *s*-wave superconductors and the ZBCP for *d*-wave superconductors.

REFERENCES

1. A. F. Andreev, *Sov. Phys. JETP*, **19**, 1228 (1964).
2. Yu. V. Nazarov, *Phys. Rev. Lett.*, **73**, 1420 (1994).
3. Yu. V. Nazarov, *Superlattices Microstruct.*, **25**, 1221 (1999).
4. K. D. Usadel, *Phys. Rev. Lett.*, **25**, 507 (1970).
5. Y. Tanaka, Yu. V. Nazarov, and S. Kashiwaya, *Phys. Rev. Lett.*, **90**, 167003 (2003).
6. Y. Tanaka, A. A. Golubov, and S. Kashiwaya, *Phys. Rev. B*, **68**, 054513 (2003).
7. Y. Tanaka, Yu. V. Nazarov, A. A. Golubov, and S. Kashiwaya, *Phys. Rev. B*, **69**, 144519 (2004).
8. A. F. Volkov, A. Z. Zaitsev, and T. M. Klapwijk, *Physica C*, **210**, 21 (1993).
9. T. Yokoyama, Y. Tanaka, A. A. Golubov, J. Inoue, and Y. Asano, *Phys. Rev. B*, **71**, 094506 (2005).

Threshold Resistance in the DC Josephson Effect

Yong-Jihn Kim

Department of Physics, University of Puerto Rico, Mayaguez, PR 00681

Abstract. We show that SIS Josephson junctions have a threshold resistance, above which the Josephson coupling and the supercurrents become extremely small, due to the shrinking of the Cooper pair size during the Josephson tunneling. Accordingly, the threshold resistance is smaller for higher T_c superconductors with small Cooper pair size and for the insulating barrier with higher resistance. This understanding agrees with the observations in SIS junctions of low T_c superconductors, such as Sn, Pb, and Nb. For MgB_2 it explains why the big gap does not show the supercurrents, unlike the small gap. Furthermore, it is consistent with the fact that high T_c cuprates show the Josephson effects only for SNS type junctions, including the intrinsic Josephson effects.

Keywords: Josephson effects, SIS junctions, Threshold resistance.
PACS: 74.50.+r, 74.20Fg, 74.25.Sv

INTRODUCTION

In 1962 Josephson predicted that supercurrents can flow through the insulating barrier between two superconductors due to the Cooper pair tunneling.[1] The supercurrent, j, depends on the relative phase, φ, of two superconductors, i.e.,

$$j = j_1 \sin\varphi. \tag{1}$$

The maximum DC supercurrent, j_1, was given by[2,3]

$$j_1 = \frac{\pi}{2e}\frac{\Delta}{R_n}, \tag{2}$$

where Δ and R_n are the energy gap and the tunneling resistance, respectively.

On the other hand, there are some experiments which show more complicated behavior than Eq. (2) suggests. For instance, the maximum DC supercurrent in low T_c superconductors, such as Pb, Sn, and Nb, decreases much faster than $1/R_n$ above a few ohms.[4,5,6] MgB_2 does not show the supercurent for the big gap of ~7meV.[7,8] High T_c cuprate Josephson junctions are mainly SNS type, including the intrinsic junctions.[9]

We show that SIS Josephson junctions have a threshold resistance, due to the shrinking of the Cooper pair size during the tunneling, which explains the above experiments.

COOPER PAIR WAVEFUNCTION APPROACH TO JOSEPHSON TUNNELING

We present Cooper pair wavefunction approach to the Josephson effects. It is shown that the Josephson coupling energy, E_J, is determined by the overlap of the Cooper pair wavefunctions of two superconductors divided by a thin insulating layer:[10]

$$E_J = V\int F_r^*(x)F_l(x)dx + V\int F_l^*(x)F_r(x)dx, \tag{3}$$

where F_l and F_r are the effective Cooper pair wavefunctions in the left and right sides, and V is the phonon-mediated matrix element. Actually, this equation is closely related to the approximate expression of the supercurrent suggested by Josephson,[1] i.e.,

$$j \cong \frac{1}{2}j_1F_r^*F_l + \frac{1}{2}j_1F_l^*F_r. \tag{4}$$

Note that

CP850, *Low Temperature Physics: 24th International Conference on Low Temperature Physics;*
edited by Y. Takano, S. P. Hershfield, S. O. Hill, P. J. Hirschfeld, and A. M. Goldman

$$j = \frac{2e}{\hbar}\frac{\partial E_J}{\partial \varphi}. \quad (5)$$

THRESHOLD RESISTANCE

It is essential to calculate the tail of the Cooper pair wavefunctions to determine the Josephson coupling energy and the supercurrent. We stress that the Cooper pair size, ξ_0, will shrink during the tunneling for the insulating barriers with high tunneling resistance and for high T_c superconductors, including MgB_2. This is similar to the reduction of the Cooper pair size due to the impurity potential. We can estimate the threshold resistance, R_{th}, above which the shrinking occurs, i.e.,

$$R_{th} \cong Ce^{2\kappa d_{th}}, \quad d_{th} \sim \sqrt{\xi_0 / 2\kappa}, \quad (6)$$

where $\hbar\kappa = \sqrt{2m(U-E)}$, and C is a constant. Accordingly, for the insulator thickness, $d \geq d_{th}$, we find[10]

$$j_1 = \frac{1}{\lambda}\frac{\Delta}{eR_n}\frac{1}{1+2\kappa d} e^{-\frac{1+2\kappa d}{\xi_0}}, \quad (7)$$

where λ is the BCS coupling constant.

COMPARISON WITH EXPERIMENTS

The threshold behavior has been found in experiments.[4-8] Figure 1 shows the comparison of our theoretical calculations with experimental results for Sn-SnO-Pb junction (at 1.4K) by Tinkham's group (Ref. 5), Pb-PbO_x-Pb junction (at 4.2K) by Schwidtal and Finnegan (Ref. 4), and Nb-NbO_x-Pb junction (at 4K) by Octvio's group (Ref. 6). The solid lines are theoretical calculation based on Eq. (7). It is remarkable that above R_n~40Ω the suppercurrent of Sn-SnO-Pb junction becomes larger than those of Pb-PbO_x-Pb and Nb-NbO_x-Pb junctions. It is clear that this is due to the existence of the threshold resistance in the SIS Josephson junctions.

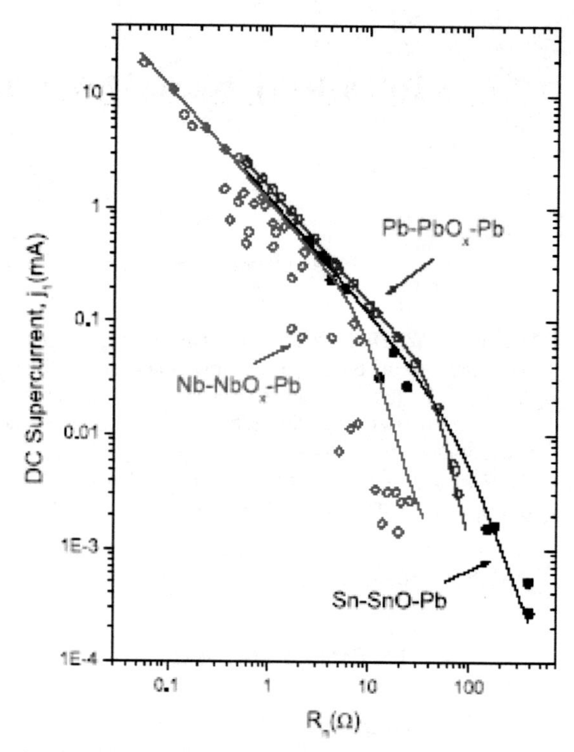

FIGURE 1. Maximum DC supercurrent vs tunneling resistance for Sn-SnO-Pb (Ref. 5), Pb-PbO-Pb (Ref. 4) and Nb-NbO-Pb (Ref. 6) junctions.

ACKNOWLEDGMENTS

Supported by UPRM PREM (Partnership for Research and Education on Materials) program.

REFERENCES

1. B. D. Josephson, *Phys. Lett.* **1**, 251 (1962).
2. P. W. Anderson, *Lectures on the Many-body Problem*, Vol.2, edited by E. R. Caianiello, New York: Academic Press, 1964, p. 115.
3. V. Ambegaokar and A. Baratoff, *Phys. Rev. Lett.* **10**, 486 (1963) and **11**, 104 (erratum).
4. K. Schwidtal and R. D. Finnegan, *Phys. Rev.* B **2**, 148 (1970).
5. W. C. Danchi, J. B. Hansen, M. Octavio, F. Habbal, and M. Tinkham, *Phys. Rev.* B **30**, 2503 (1984).
6. J. Aponte, E. Rivera, A. Sa Neto, and M. Octavio, *J. Appl. Phys.* **62**, 700 (1987).
7. Y. Zhang et al. *Appl. Phys. Lett.* **79**, 3995 (2001).
8. H. Schmidt, J. F. Zasadzinski, K. E. Gray, and D. G. Hinks, *Phys. Rev. Lett.* **88**, 127002 (2001).
9. D. Koelle, R. Kleiner, F. Ludwig, E. Dantsker, and J. Clarke, *Rev. Mod. Phys.* **71**, 631 (1999).
10. Y.-J. Kim, cond-mat/0401374.

NRG Study of the Kondo Effect in a Josephson Junction

Yoshihide Tanaka, Akira Oguri, and A. C. Hewson*

Department of Material Science, Osaka City University, Sumiyoshi-ku, Osaka 558-8585, Japan
**Department of Mathematics, Imperial College, 180 Queen's Gate, London SW7 2BZ, UK*

Abstract. We study the ground-state properties of a quantum dot in a Josephson junction based on an Anderson impurity model connected to two superconducting (SC) leads with the gaps Δ_L and Δ_R. The Kondo effect favors the spin singlet state, while the deficiency of the low-energy excitations in the SC gaps disturbs the screening of the local spin. The phase difference ϕ of the two SC gaps also makes the stability of the singlet state worse. It can drive the quantum phase transition (QPT) to the doublet state, and at the critical point the Josephson current changes the direction discontinuously. In this report, we discuss the ground-state phase diagram obtained with the numerical renormalization group (NRG).

Keywords: Kondo effect, Josephson effect, Quantum dot, NRG
PACS: 72.10.-d, 72.20.-i, 74.50.+r

INTRODUCTION

The Kondo effect in superconducting (SC) materials was studied originally for dilute magnetic alloys. The SC energy gap Δ of the host material disturbs the conduction electron screening of the local moment of the magnetic impurities, and the competition between T_K and Δ determines the low-energy properties, where T_K is the Kondo temperature. The ground state is a singlet for $\Delta << T_K$, while a doublet ground state is stabilized for $\Delta >> T_K$. The Andreev bound state emerging in the energy gap also affects this quantum phase transition (QPT).

In a quantum dot connected to two SC leads as illustrated in Fig. 1, the phase difference ϕ of the two SC order parameters induces the Josephson current, where $\phi = \theta_R - \theta_L$ with $\Delta_R=|\Delta_R|\exp(i\theta_R)$, and $\Delta_L=|\Delta_L|\exp(i\theta_L)$. The phase difference also affects the screening of the local moment [1, 2]. Moreover, when the local moment remains unscreened, the spin-flip tunnellings cause a current flowing in the opposite direction to that in the case of the singlet ground state. Therefore, the phase difference ϕ can derive the QPT, and at the critical point the Josephson current changes discontinuously. Recently efficient numerical approaches such as NRG [3, 4] and quantum Monte Carlo [5] methods, which can capture the low-energy Kondo behavior, have been applied to this problem. The screening of the local moment depends on the environment surrounding the quantum dots. In this report, we discuss how the asymmetries in the Josephson couplings $\Gamma_L \neq \Gamma_R$ and the SC gaps $|\Delta_R| \neq |\Delta_L|$ affect the ground state.

FORMULATION AND RESULTS

We consider an Anderson impurity model as shown in Fig. 1. The impurity site with the on-site energy level ε_d and local interaction U is connected to two SC leads via the hybridization matrix elements v_R and v_L, which cause the level broadning of $\Gamma_L=\pi\rho_L v_L^2$ and $\Gamma_R=\pi\rho_R v_R^2$. Here ρ_L and ρ_R are the local density of states of the leads in the normal limit. The leads are assumed to be s-wave superconductors with the gaps Δ_R and Δ_L. Using the NRG method, we have calculated the low-energy eigenstates of the system [3]. The occurrence of the QPT can be deduced from a level crossing between the singlet and doublet ground states. In the present work, we take the discretization parameter to be $\Lambda=6.0$ in the NRG calculations.

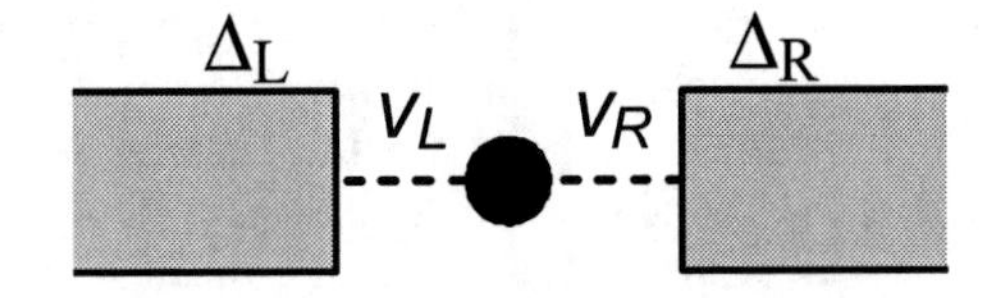

FIGURE 1. Schematic picture of an Anderson impurity connected to two SC leads of the gaps Δ_L and Δ_R, where v_L and v_R are hybridization matrix elements.

CP850, *Low Temperature Physics: 24th International Conference on Low Temperature Physics;*
edited by Y. Takano, S. P. Hershfield, S. O. Hill, P. J. Hirschfeld, and A. M. Goldman

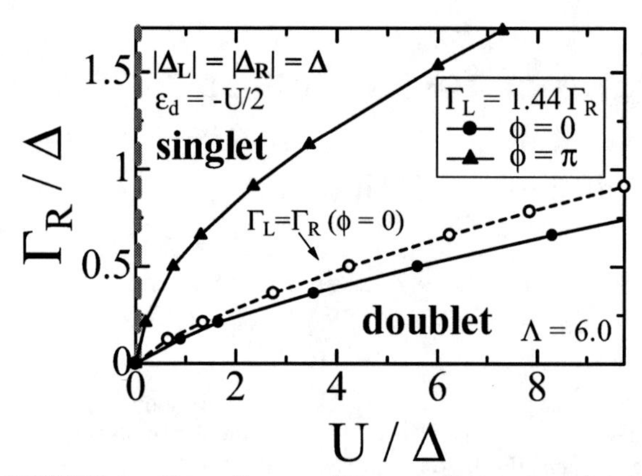

FIGURE 2. Phase diagram of the ground state for (●) ϕ=0 and (▲) ϕ=π, where Γ_L=1.44Γ_R. The phase boundaries for Γ_L=Γ_R are also shown for (○) ϕ=0 and (vertical line) ϕ=π.

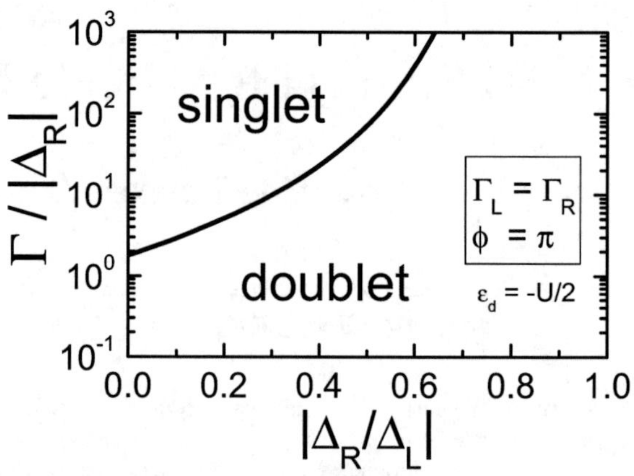

FIGURE 3. Phase diagram of the ground where for ϕ=π, Γ_L=Γ_R (≡Γ), and ε_d=-*U*/2, as a function of $|\Delta_R|/|\Delta_L|$. The 4-fold degeneracy is lifted by an infinitesimal repulsion *U*.

We first of all discuss the asymmetry in Josephson couplings $\Gamma_L \neq \Gamma_R$, taking the amplitude of the two gaps to be symmetric $|\Delta_L|=|\Delta_R|$ (≡Δ). In Fig.2, the phase diagram of the ground state for Γ_L=1.44Γ_R is plotted for (●) ϕ=0 and (▲) ϕ=π at half-liffing ε_d=-*U*/2. The phase boundaries for intermediate values of ϕ, i.e. $0<\phi<\pi$, must be located in between the these two curves for ϕ=0 and π. The Josephson phase ϕ makes the doublet region in the diagram large, and it means that ϕ tends to disturb the screening of the local moment. Furthermore, the Josephson phase ϕ could drive the QPT if other prameters are in the region surrounded by the two curves. For comparison, in Fig.2 the phase boundary for the symmetric coupling Γ_L=Γ_R is plotted for (○) ϕ=0, and the dashed line along the *y*-axis corresponds to the boundary for ϕ=π. Since the odd-parity states do not contribute to the screening in the symmetric-coupling case, the asymmetry in the couplings $\Gamma_L \neq \Gamma_R$ enhances the screening effect to stabilize the singlet ground state.

The dashed vertical line in Fig.2 shows that an infinitesimal repulsion *U* makes the ground state a magnetic doublet. This is because at *U*=0 and at half-filling the in-gap Andreev bound state emerges just on the Fermi level for Γ_L=Γ_R (≡Γ) and ϕ=π. Thus in this case the ground state for non-interacting electrons has a 4-fold degeneracy [6], and the local interaction *U* lifts that. The ground state for infinitesimal repulsion *U* generally depends on Γ and the ratio of the two SC gaps $x=|\Delta_R|/|\Delta_L|$. The details can be described by the degenerate perturbation theory [3], and the result of the phase diagram is plotted in a *x*-Γ plane in Fig. 3. The curve diverges for $|\Delta_L|=|\Delta_R|$, and thus specifically for the symmetric gap the ground state becomes a doublet independent of Γ. In the limit of $|\Delta_L|\rightarrow\infty$, the critical value for the coupling becomes $\Gamma_{cr} \approx 1.77|\Delta_R|$, and the ground state is a singlet for $\Gamma > \Gamma_{cr}$.

In the previous work, we have studied the ground-state properties for the asymmetric gap in the limit of $|\Delta_L|>>|\Delta_R|$ [3]. In this limit the system illustrated in Fig. 1 can be mapped onto a single-cahnnel model, and it gives us a technical advantage for carrying out NRG calculations. Although the limit $|\Delta_L|>>|\Delta_R|$ itself was introduced for a theoretical simplification, the obtained results seem to capture the essential physics of the Kondo-Josephson system. In the present work, we have carried out the calculations for more natural situation $|\Delta_L|=|\Delta_R|$, and have found that the results are similar at least qualitatively. Specifically, the screening of the local moment is sensitive to the asymmetry in the Josephson couplings Γ_L and Γ_R. Furthermore, the Josephson phase ϕ affects the SC proximity effects on the impurity site, and it tends to make the ground state of the quantum dots be a doublet.

AKNOWLEDGMENTS

One of us (ACH) wishes to thank the EPSRC(Grant GR/S18571/01) for financial support. Numerical computation partly carried out using Computer Facility of Yukawa Institute.

REFERENCES

1. A. V. Rozhkov and D. P. Arovas, *Phys. Rev. Lett.* **82**, 2788 (1999).
2. A. A. Clerk and V. Ambegaokar, *Phys. Rev. B* **61**, 9109 (2000).
3. A. Oguri, Y. Tanaka, and A. C. Hewson, *J. Phys. Soc. Jpn.* **73**, 2494 (2004).
4. M.-S Choi, M. Lee, K. Kang, and W. Belzig, *Phys. Rev. B* **70**, R020502 (2004).
5. F. Siano and R. Egger, *Phys. Rev. Lett.* **93**, 047002 (2004).

Author Index

A

B

C

D

E

F

G

H

I

J

K

L

M

N

O

P

Q

R

S

T

U

V

W

X

Y

Z